D0161480

The World of BIOLOGY

FIFTH EDITION

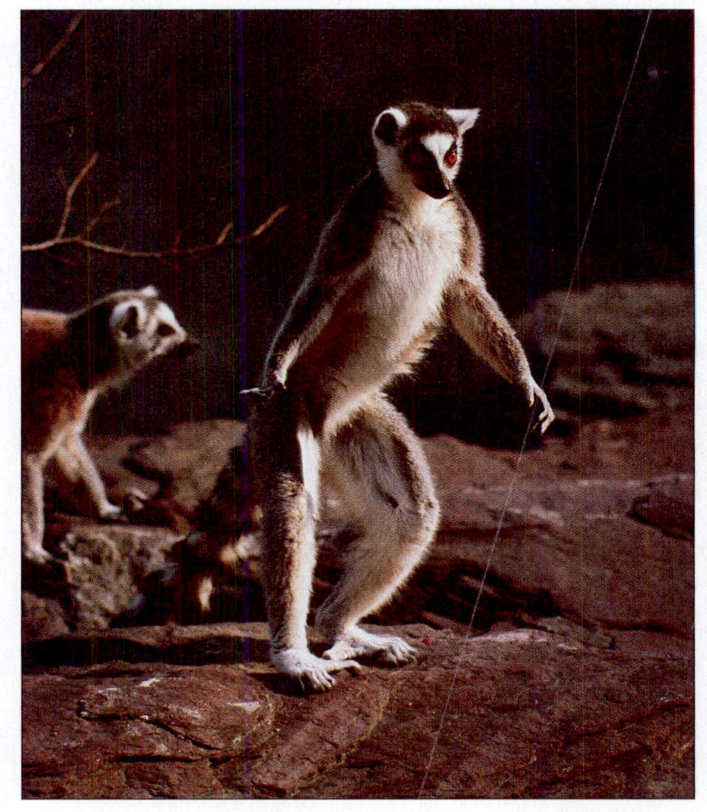

Eldra Pearl Solomon, Ph.D.
University of South Florida

Linda R. Berg, Ph.D.
Pinellas Park, Florida

SAUNDERS COLLEGE PUBLISHING
HARCOURT BRACE COLLEGE PUBLISHERS

Philadelphia • Ft. Worth • Chicago • San Francisco

Montreal • Toronto • London • Sydney • Tokyo

For Amy, Belicia, Mical, Karla, and Jennifer

. . . and for their generation

Text Typeface: Palatino
Compositor: York Graphic Services
Acquisitions Editor: Julie Levin Alexander
Developmental Editor: Gabrielle Goodman
Art Developmental Editor: Ray Tschoepe
Photo Editor: Robin C. Bonner
Managing Editor: Carol Field
Project Editor: Nancy Lubars
Copy Editor: Zanae Rodrigo
Manager of Art and Design: Carol Bleistine
Senior Art Director: Christine Schueler
Art and Design Coordinator: Sue Kinney
Text Designer: Tracy Baldwin/Christine Schueler
Cover Artist: Tom Leonard
Cover Designer: Lawrence R. Didona
Text Artwork: Rolin Graphics
Layout Artist: Rebecca Lemna
Director of EDP: Tim Frelick
Production Manager: Joanne M. Cassetti
Frontmatter photo credits: p. i: Animals Animals © 1995 Henry Avsloos; p. iii: Frans Lanting/Minden Pictures; p. viii: Frans Lanting/Minden Pictures.

Part Opening photo credits: Part 1: Skip Moody/Dembinsky Photo Associates; Part 2: The Stock Market/Thaine Manske, 1995; Part 3: Renee Lynn/Photo Researchers, Inc.; Part 4: Conly Rieder/Biological Photo Service; Part 5: Frans Lanting/Minden Pictures; Part 6: Karl and Jill Wallin/FPG International; Part 7: Skip Moody/Dembinsky Photo Associates; Part 8: Art Wolfe; Part 9: Skip Moody/Dembinsky Photo Associates.

Printed in the United States of America

THE WORLD OF BIOLOGY

0-03-094865-7 (ISBN)

Library of Congress Catalog Card Number: 94:061382

67890123 032 1098765432

About the Cover:

Life forms on Planet Earth are numerous and diverse, but many are increasingly threatened by human activities. The ring-tailed lemur and the rosy periwinkle are both native to Madagascar, a large, lushly rain-forested island nation off the southeastern coast of Africa. The ring-tailed lemur, a primate related to monkeys, apes, and humans, has lost most of its habitat as native people clear the forests for agriculture. Unfortunately, this endangered species does not thrive in captivity because it is a social animal, roaming in large troops over broad areas of land. The rosy periwinkle, which produces chemicals essential in cancer treatment, illustrates the benefits of plants and other living organisms to humans. Many of the Madagascan species, which have not yet been evaluated for their potential usefulness, are found nowhere else in the world.

Preface

As we prepare to enter the 21st century, we are challenged by many global issues. The news media remind us almost daily about overpopulation, world hunger and malnutrition, effects of pesticides, air and water pollution, declining biodiversity and endangered species, AIDS, cancer, and heart disease. All of these issues have biological, as well as cultural and political, aspects. Resolution will require the combined efforts of biologists and other scientists and informed citizens. Whether you are a science major or a nonmajor, an understanding of biological concepts is a vital tool for understanding and helping to meet the pressing challenges that confront us. More than any other discipline, biology is the study of our world.

In this fifth edition of *The World of Biology* we focus on the evolution of life on Earth, the interdependence of the diverse life forms that inhabit Planet Earth, and their interactions with the environment. In particular, we explore the impact of human activity and population expansion on the ecosphere—on evolution, genetics, species diversity, health, and many other aspects of our world. We equip students with the conceptual tools that biologists use to help them understand these issues. We do this with new Chapter Openers that examine real-life problems from a scientific perspective, and demonstrate the larger social relevance of the biological concepts provided in the chapter. We do this by emphasizing connections between biological concepts and human activity throughout the text and in the new Concept Connections boxes, which demonstrate the relationships between various topics in biology.

Resolution of global issues is a scientific process. We encourage students to understand science as a dynamic, creative, and necessary endeavor. A substantially revised chapter on the process of science teaches students how to think critically and logically, and a series of boxes titled Focus on The Process of Science demonstrates how scientists work to solve specific problems.

Conceptual Approach

The World of Biology is known for its interesting, conversational reading style that emphasizes biological concepts rather than specific facts. All of the basic facts of biology are here, and they are integrated to form concepts. In *The World of Biology,* we take the time and space to integrate information so that the student understands how the details fit together to explain cells, organisms, and ecosystems, and how these relate to one another through time and space.

The authors have developed the conceptual approach in the fifth edition by using a variety of pedagogical strategies (see also the section on Learning Aids) that includes: (1) the introduction of basic biological principles in Chapter 1, followed by more detailed discussions of these concepts in later sections of the book; (2) new, issue-oriented Chapter Openers; (3) Key Concepts at the beginning of each chapter; (4) headings and subheadings written in the form of conceptual statements; (5) Concept Connections boxes in each chapter.

Exciting Illustration Program

A completely revised art program features many new photographs and carefully rendered line drawings designed to support the concepts covered in the text. New multipart figures that pair, and sometimes join, line art with photographs are particularly effective in enhancing students' understanding of complex concepts. To provide visual continuity between topics and remind students of the larger context, many micrographs and line drawings have been paired with icons. For example, in Chapter 5 the idealized cell is used as an icon with each micrograph of an organelle, helping students locate structures and relate them to the larger concept—cellular organization.

Learning Aids

The following features are designed to make this book user friendly and to help the student achieve mastery of the concepts presented. An asterisk (*) indicates the features that are new to this fifth edition.

* 1. **Issue-oriented Chapter Openers** explore topical, often controversial issues from today's headlines, helping students relate biology to their own lives. These problem-oriented essays promote understanding of broad biological concepts, and an awareness of current, real-life issues in biology. One or more Thinking Critically questions at the end of each chapter asks students to consider these issues in light of the chapter concepts they've just studied.
2. **Learning objectives** indicate, in behavioral terms, what students must be able to do to demonstrate mastery of material covered in the chapter.
3. **Key Concepts** summarize the main principles presented in the chapter.
4. **Concept-statement heads** introduce each section, previewing for the student the key idea that will be discussed in that section.
5. Numerous **tables,** many of them illustrated, summarize and organize material.

* 6. **Concept Connections** boxes link a concept discussed in the text with one presented in a different section or chapter. Concept Connections demonstrate the interrelationships of biological knowledge. For example, they relate energy flow in cells and energy flow in ecosystems; genetic engineering and ecology; the work of Gregor Mendel and that of Charles Darwin.

* 7. **Four categories of focus boxes** highlight evolution, the environment, the process of science, and health and human affairs. These boxes are written to spark student interest, present applications of concepts discussed, and familiarize students with the directions and methods of research in modern biology.

8. **New terms are boldfaced,** permitting easy identification and providing emphasis.
9. **Chapter Summaries** in outline form at the end of each chapter provide quick reviews of material presented.

*10. **Selected Key Terms** at the end of each chapter list important new words alphabetically with page references to their definitions.

11. **Post-Tests** provide opportunity to evaluate mastery of material in chapters. Answers are provided in an Appendix at the back of the book.
12. **Review Questions** focus on important chapter concepts and applications.

*13. **Thinking Critically** questions can be used for essay assignments or class discussion. They challenge students to apply knowledge learned in the chapter to new situations. At least one Thinking Critically question always addresses the issues examined in the Chapter Opener.

14. **Recommended Readings** provide references for further learning.
15. **Detachable Windows on the Plant and Animal Cells** are overlays located in chapter 4. Rendered with precision, these state of the art drawings present progressively deeper views of the plant and animal cells. The newly annotated Windows will help students visualize the three-dimensional nature of cells, thereby enhancing learning. The Windows are perforated so that students can detach them from the book and add them to study notes for easy reference.
16. In response to student requests, the **Glossary** has been greatly expanded and the definitions updated and extensively revised.

The Organization of the World of Biology, Fifth Edition

Part 1 Introduction: Basic Concepts of Biology

Chapter 1, The World of Biology, introduces several major concepts of biology, including the fundamental similarities of all living things; the organization of life on individual and ecological levels; the evolution of life on our planet; diversity of life and how biologists classify living things; interdependence of producers, consumers, and decomposers; and human impact on the ecosphere. In this first chapter we clearly show how the World of Biology relates to the world of the student.

Thoroughly revised, Chapter 2 lays the foundation for understanding how science works. This chapter examines the evolutionary nature of the process and practice of science, bringing to life both historical and contemporary scientists.

Part 2 The Organization of Life

Chapters 3 and 4 provide the basic tools of chemistry needed to understand biology and introduce the biological molecules that interact in the construction and maintenance of living things. Chapters 5 and 6 focus on the structure and functions of cells and their membranes. Both chapters are enhanced by new illustrations joining photomicrographs and line art, and by the use of icons that relate individual organelles to the context of the larger cell. The Windows on the Animal and Plant Cells are exciting supplementary learning tools for mastering cell structure and understanding the three-dimensional nature of the cell.

Part 3 Energy Flow Through the World of Life

Chapters 7, 8, and 9 focus on the energy transactions involved in life processes. Chapters 8 and 9 introduce complex concepts with simplified overviews, followed by a more detailed explanation for those courses that require more. Much of the art has been revised to clarify the processes of cellular respiration and photosynthesis.

Part 4 The Continuity of Life: Cell Division and Genetics

The genetics unit has been thoroughly updated and revised to clarify complex topics. The processes of mitosis and meiosis are explained and contrasted in Chapter 10. Chapters 11 and 12 present patterns of inheritance and include all new genetics problems at the end of each chapter. Many important principles of inheritance are made relevant for the student by using human applications, particularly in Chapter 12. The molecular basis of inheritance is discussed in Chapters 13–15, and Chapter 16, Recombinant DNA Technology and Genetic Engineering, presents concepts on the cutting edge of genetic research.

Part 5 Evolution

The evolution chapters have been updated and reorganized. Chapter 17 includes a new section on the synthetic theory of evolution. The evolutionary significance of the Hardy-Weinberg law is discussed in Chapter 18, with placement of the mathematical material in a separate box. Chapter 19 includes a simplified discussion of the evolutionary history of life.

Part 6 Evolution and Diversity of Life

Chapter 20 describes how biologists classify organisms and includes a new Focus On the Process of Science describing the cladistic approach. The authors use an evolutionary framework to present the various groups of organisms in Chapters 21 through 25. Primate evolution is included in Chapter 25, Animal Life: Chordates. Icons of evolutionary trees put individual groups in evolutionary perspective.

Part 7 Plant Structure and Life Processes

Context drawings of whole plants, paired line drawings and photomicrographs, and icons are used throughout this section to help students make conceptual connections. Chapter 26 focuses on plant structure, growth, and differentiation. Leaves and their role in photosynthesis are discussed in Chapter 27. The discussion of stem and root structure in Chapter 28 is integrated with the mechanisms of transport in xylem and phloem and mineral nutrition. Chapter 29 discusses reproduction in flowering plants including asexual reproduction, flowers, fruits, and seeds. Chapter 30 focuses on plant hormones and responses.

Part 8 Animal Structure and Life Processes

Chapter 31 describes animal tissues, organs, and organ systems and discusses homeostasis. Chapters 32 through 42 focus on the strategies animals use to carry on life processes. Each chapter begins by comparing how different animal groups carry on digestion, gas exchange, internal transport, or whatever process is being discussed. Then, the human adaptations for carrying on the process are considered. The unit ends with a discussion of development in Chapter 43. The text and art for the immunology chapter have been carefully updated and revised. An icon of the human body is used throughout this Part to facilitate contextual understanding.

Part 9 Behavior and Ecology

The chapters in this section have been extensively revised. Chapter 44 presents the concepts of animal behavior. In Chapters 45 through 49, which have been updated and reorganized, we provide the foundations of ecology. Chapter 45 includes a new section on human population growth. Many new examples have been added to Chapters 46 to 48 to enhance understanding of basic ecological concepts. The final chapter focuses on four major environmental issues: extinction, deforestation, global climate change, and destruction of stratospheric ozone.

Supplements

To further facilitate learning and teaching, a supplement package has been carefully designed for the student and instructor. It includes:

Instructor's Manual contains a chapter overview, list of key terms, lecture outline, class discussion topics, class presentation suggestions, teaching suggestions, audiovisual listings, and suggested readings for each chapter.

Test Bank features approximately 2000 new and challenging multiple-choice and essay questions with references to the text segment from which they are derived.

ExaMaster™ Computerized Test Bank enables instructors to edit, revise, add to, or delete from the printed Test Bank. Available in IBM or Macintosh formats.

Study Guide reinforces concepts from the text in a variety of ways, including chapter summaries and outlines, numerous multiple-choice questions, essay questions, sentence completions, illustrations culled directly from the textbook for labeling, vocabulary-building exercises, an exam at the end of each part, and crossword puzzles.

***Discover* Supplement** is an exclusive collection of feature articles from *Discover* magazine, chosen for their timeliness and for their direct relationship to the concepts presented in *The World of Biology.*

Bio-Art reproduces 100 pieces of art from the text as black-and-white unlabeled line drawings, encouraging students to learn the labeling process and take notes. Bio-Art can serve as a handy study tool or can be used as a test item.

Overhead Transparencies feature 250 pieces of art from the text, using labels with large type for easy classroom viewing.

Saunders General Biology Sequence Overhead Transparencies is a separate set of 50 sequential overhead transparencies, containing topics displayed in a series of stages or layers.

Bio-XL, a unique, computer-assisted tutorial software package, is available in two modes. Test Mode quizzes students about chapter material and assesses students' knowledge; Tutor Mode adds pedagogical support through immediate feedback to responses. Corresponds to specific page references in the text.

SimLife™, an advanced biological simulation, enables students to design animals and plants from a genetic level and manipulate the environment in which they live, testing their ability to survive.

Lecture Outline on Disk, an ASCII version of the lecture outline, is available for IBM and Macintosh computers. Instructors are able to edit, expand the outline, and create study guideline handouts for students.

Electron Micrograph Transparencies include 100 electron micrograph figures from a variety of Saunders biology texts.

The Saunders General Biology Laboratory Manual by Carolyn Eberhard provides a selection of lucid, comprehensive experiments that includes excellent illustrations and pedagogy. An accompanying instructor's manual is available.

Infinite Voyage Videos/Videodiscs from PBS bring the subject of biology to life in the classroom by offering great adventures of scientific exploration and discovery.

Saunders Multimedia Presentation Package includes Saunders General Biology Videodisc Version 3, LectureActive™ Software and User's Guide

- ***The Saunders General Biology Videodisc*** contains approximately 1500 still images and almost one hour of live-action footage and animations.
- ***LectureActive™ Software*** helps instructors customize lectures by providing quick, efficient access to video clips and still frame data on the videodisc. Available for IBM Windows and Macintosh formats.
- ***A User's Guide*** offers suggestions to instructors on how to use the videodisc in lectures, how to get started with LectureActive™, and how to create and print lectures using the software.
- ***The Barcode Manual*** contains complete descriptions, barcode labels, and reference numbers for every still image and video clip.

The Ecology of the World of Biology

The development and production of this new edition of *The World of Biology* was a complex process involving interaction and cooperation among the authors, editors, reviewers, and many individuals in our home and professional environments. We appreciate the valuable input and support from family, friends, editors, colleagues, and students. We thank our families and friends for their understanding, support, and encouragement as we struggled through many revisions and deadlines. We especially thank Dr. Amy Solomon for her contributions to several Chapter Introductions and to some of the Animal Processes chapters. We thank Mical Solomon for his help with the exercise physiology material and for sharing his computer expertise, Belicia Efros for her help in researching, Kathleen M. Heide for her input and support, and Alan Berg for his support and understanding. We thank Jen Berg for her help in expanding the glossary and revising many of the definitions.

The Editorial Environment

Preparing a book of this complexity is challenging and requires a great deal of time and effort. We could not have produced the fifth edition without the help and support of our outstanding editorial and production staff at Saunders. We thank Publisher Elizabeth Widdicombe and Executive Editor for Biology Julie Alexander for their support and help. We are grateful to our Developmental Editor, Gabrielle Goodman, who worked along with us, providing valuable input in every aspect of the project. Our Art Editor, Ray Tschoepe, made a unique contribution by sharing his wonderful artistic talents. Ray reconceptualized and redrew much of the art, providing many creative innovations such as his joined photo-line art figures. We thank Photo Editor Robin Bonner for working hard to find new and exciting photographs that enhance the text.

We greatly appreciate the help of our Project Editor, Nancy Lubars, who guided the project through the complexities of production. We also thank Art Director Christine Schueler for creating the book design, coordinating the art program, and developing and refining the wonderful cover design. All of these dedicated professionals and many others at Saunders contributed importantly to the production of the *The World of Biology, 5th edition.* We thank them for their help and support throughout this project.

Our colleagues and students have provided valuable input by sharing their responses to previous editions of *The World of Biology* with us. We thank them and ask again for their comments and suggestions as they use this new edition. We can be reached through our editors at Saunders College Publishing.

The Professional Environment—Reviewers

We very much appreciate the input of the many professors and researchers who have reviewed the manuscript during various stages of its preparation and provided us with valuable suggestions for improving it. Their work contributed greatly to our final product.

Reviewers for 5th Edition

James K. Adams, *Dalton College*
Michael Bell, *Richland College*
John T. Beneski, *West Chester Univ.*
Latsy Best, *Palm Beach Community College*
Gary Brusca, *Humboldt State Univ.*
Frank Dipino, *College Misericordia*
Herndon Dowling, emeritus *New York Univ.*
Peter Ducey, *SUNY, Cortland*
Sally Frost-Mason, *Univ. of Kansas*
Laszlo Hanzely, *Northern Illinois Univ.*
Wiley Henderson, *Alabama Agricultural and Mechanical Univ.*
Anne Hooke, *Miami Univ.*
Jan Jenner, *Talladega College*
Leonard Kass, *Univ. of Maine-Orono*
Hendrik J. Ketellapper, *Univ. of California, Davis*
Ross Koning, *Eastern Connecticut State Univ.*
William Kroen, *Wesley College*
Norman Leeling, *Grand Valley State Univ.*
Henry McDuffy, *Harold Washington College*
David McMurray, *Texas A&M Univ.*
Debbie Meuler, *Cardinal Stritch College*
Keith Morill, *South Dakota State Univ.*
Jon Morony, *San Antonio College*
Malcolm W. Nason, *North Shore Community College*
Dr. David M. Polcyn, *California State Univ., San Bernadino*
Gerri Seitchik, *LaSalle College*
Salvatore Tavormina, *Austin Community College*
Richard G. Thomas, *Mohawk Valley Community College*
Carol Thorne, *West Virginia Univ./Parkersburg*
Norman Tweed, *Pierce College*
Larry Underwood, *Northern Virginia Community College/Woodbridge*
Paul Van Faasen, *Hope College*
Roland Vieira, *Green River Community College*
Mary Wells-Phillips, *Tulsa Junior College*

Survey Respondents

James K. Adams, *Dalton College*
David F. Blaydes, *West Virginia Univ.*
Michael Bucher, *College of San Mateo*
Dr. Jeffrey C. Burne, *Macon College*
Jim Burnett, *Olney Central College*
Charles T. Collins, *California State Univ., Long Beach*
Dr. Jean DeSaix, *Univ. of North Carolina, Charlotte*
Dr. David W. Inouye, *Univ. of Maryland*
Hendrik J. Ketellapper, *Univ. of California, Davis*
Arthur B. Krupnick, *Quinsigamond Community College*
Paul L. Lago, *Univ. of Mississippi*
Susan J. McDaniel, *East Carolina Univ.*
Debbie Meuler, *Cardinal Stritch College*
Malcolm W. Nason, *North Shore Community College*
Harold Ornes, *Univ. of South Carolina*
Dr. David M. Polcyn, *California State Univ., San Bernadino*
George W. Powell, *Abraham Baldwin Agricultural College*
Dennis Rich, *Mattatuck Community College*
Rosemary Richardson, *Bellevue Community College*
Jay Robinson, *San Antonio College*
Richard G. Rose, *West Valley College*
Michael E. Smith, *Valdosta State College*
Richard G. Thomas, *Mohawk Valley Community College*
Dana L. Wrensch, *Ohio State Univ.*
Paul Wright, *Western Carolina Univ.*

E.P.S
L.R.B
October 1994

Contents Overview

Contents

PART 5
Evolution 299

PART 6
Evolution and the Diversity of Life 359

PART 1

Introduction: Basic Concepts of Biology

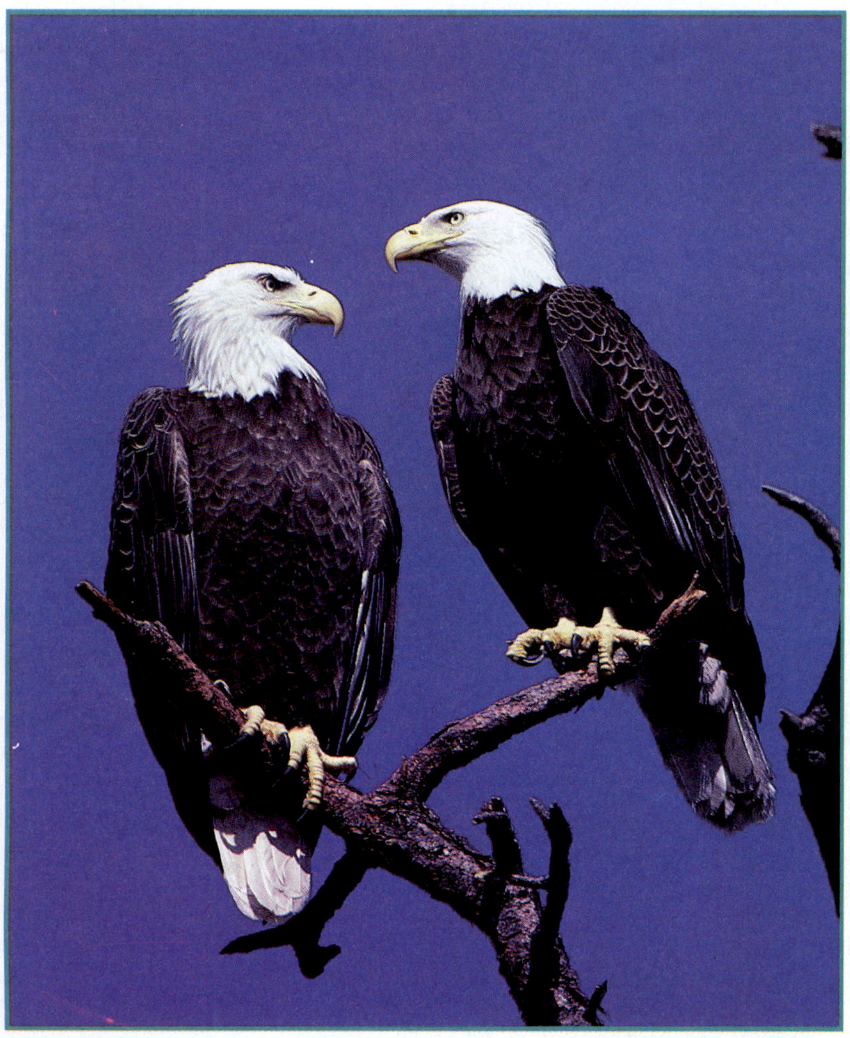

Biologists met the challenge of protecting an endangered species. *A mated pair of bald eagles. (Stan Osolinski/ Dembinsky Photo Assoc.)*

Biological Diversity: The Fall and Rise of the Bald Eagle

The bald eagle—the symbol of the United States—was in danger of extinction in the late 1970s. Fewer than 5000 eagles remained. Decline of this majestic bird was the result of destruction of the eagle's habitat; an expanding human population cleared thousands of acres of forest near lakes and rivers where eagles make their homes. Eagles were also hunted because it was thought they preyed on livestock and on commercially important fish. Further contributing to their decline was the widespread use during the 1940s through the 1960s of pesticides such as DDT, which caused reproductive failure.

Biologists accepted the challenge of protecting the large, stately bald eagle. In the mid 1970s, they began breeding eagles in captivity and then releasing them back into the wild. Such conservation efforts were very successful. Between 1975 and 1990, the number of nesting pairs in the continental United States doubled from 1000 to 2000. The recovery of the bald eagle demonstrates that, if enough people care and take action, we can preserve our biological heritage.

The Earth's biological diversity evolved over millions of years. As a result of human activity, it is decreasing rapidly. Biologists estimate that at least one species becomes extinct each day and that a large part of our planet's biological diversity will disappear within the next few decades. This loss of diversity is entirely irreversible. Once a species becomes extinct, its unique set of characteristics is lost forever.

Why should we, as humans, be concerned about this loss? One reason is that many species represent untapped potential resources for agricultural advances, medical research, and industrial technology. For example, the armadillo is used in research on leprosy because it is the only species other than humans known to be susceptible to the disease. Also, the rosy periwinkle (*Catharanthus roseus*) produces chemicals that are effective against certain cancers. Even if our technology could someday synthesize all of the commercial substances that human society demands, it cannot synthesize a healthy global environment. This can only be accomplished through the continuous, complex interactions that occur between all of the living and nonliving things on this planet. Thus, *every* species contributes in some way to the delicate balance of life on Earth.

Introducing the World of Biology

1

Learning Objectives

After you have studied this chapter you should be able to:

1. Define biology and discuss its applications to human life and society.
2. Distinguish between living and nonliving things by describing the features that characterize living things.
3. Describe the relationship between metabolism and homeostasis, giving specific examples of these life processes.
4. Summarize the basic concepts of the theory of evolution by natural selection.
5. Apply the theory of natural selection to a given adaptation, suggesting a plausible explanation of how the adaptation may have evolved.
6. Construct a hierarchy of biological organization, including individual and ecological levels.
7. Demonstrate use of the binomial system of nomenclature using several specific examples.
8. Classify an organism, such as a human, according to kingdom, phylum or division, class, order, family, genus, and species.
9. Compare the five kingdoms of living organisms and cite examples of each group.
10. Compare the roles of producers, consumers, and decomposers, and cite examples of their interdependence.

Key Concepts

- ☐ Living systems consist of cells; they grow, move, reproduce, carry on self-regulated metabolism, respond to stimuli, and adapt to environmental changes.
- ☐ Evolution by natural selection has become the central unifying theory of biology. It offers the most widely accepted explanation of how organisms are related and how they adapt to changes in their environment.
- ☐ Biologists have developed a hierarchy (ranking) of biological organization on both individual and ecological levels.
- ☐ Each type of organism is classified in one of five kingdoms and is assigned to a genus and species.
- ☐ Humans are part of a delicately balanced environmental system.

BIOLOGY IS THE STUDY OF LIFE

The world of biology encompasses all of the living things that inhabit our planet—from minute bacteria to giant whales and redwood trees. This broad science examines the diversity, structure, and internal processes of living organisms and extends to their origins, relationships with one another, and interaction with the environment. **Biology,** then, is the study of life. The word itself is derived from two Greek word parts, *bio,* meaning "life," and *logos,* meaning "the study of."

This book emphasizes the interdependence of living things and examines our own interactions with other organisms and with the environment. Early humans were a harmonious part of the biological world, for their activities had little impact on the environment. As human society has become increasingly technological, however, our activities have exerted a significant and often damaging effect upon our planet.

The expanding human population, coupled with increasing consumption of natural resources, has transformed the Earth. By clearing millions of acres of land to build homes, roads, and shopping centers, we have destroyed the habitats of our fellow organisms. Chemicals from industries and modern agricultural practices have spread throughout the soil, water, and atmosphere and threaten to disrupt the delicate network of life upon the Earth's surface. As a result, we face many critical problems. Of all the sciences, biology is, perhaps, of the greatest practical interest to us, for a knowledge of the principles of biology may be the key to human survival on planet Earth.

Today, many biologists are working to preserve biological diversity, improve environmental quality, increase the world food supply, identify factors that contribute to health and longevity, and conquer killers like heart disease, cancer, and AIDS (Fig. 1–1). Their work requires the support of informed citizens. This book provides you with tools that will enable you to become a biologically literate member of our society.

The work of biologists is so broad that it affects us at many levels of our lives. Understanding the principles of biology can help us deal more intelligently with a wide range of routine concerns. Health care, nutrition, dieting, smoking, the use of drugs, and the care of domestic plants and animals are a few of the topics that may be of immediate interest. You may be intrigued by specific areas of biological research such as animal communication, biological control of agricultural pests, killer bees, or genetic engineering. The study of biology will also expand your awareness of the millions of diverse life forms that share our planet and your appreciation for the exquisite precision and complexity of living processes and systems.

(*a*)

(*b*)

FIGURE 1–1 Biologists work to improve the quality of life. (***a***) This agricultural biologist is doing genetic research on rice plants. World food supply has been increased by developing high-yield and disease-resistant plants. The nutritional value of plants can also be genetically improved. (***b***) Biologists study the effects of human activities on the environment. Because humans have cut an estimated 90% of the U.S. Northwest's forests and continue to cut down about 70,000 acres of trees each year, this ecosystem is seriously threatened. The biologist shown here is working to protect the spotted owl, a forest inhabitant that lives in the Douglas firs. Only about 2000 northern spotted owl pairs remain. (*a*, Ted Hoffman; *b*, R. M. Collins III)

LIVING THINGS SHARE CERTAIN CHARACTERISTICS

We have defined biology as the study of life, but what is life? What does it mean to be alive? The living things that inhabit our planet are so diverse that it is difficult to lump

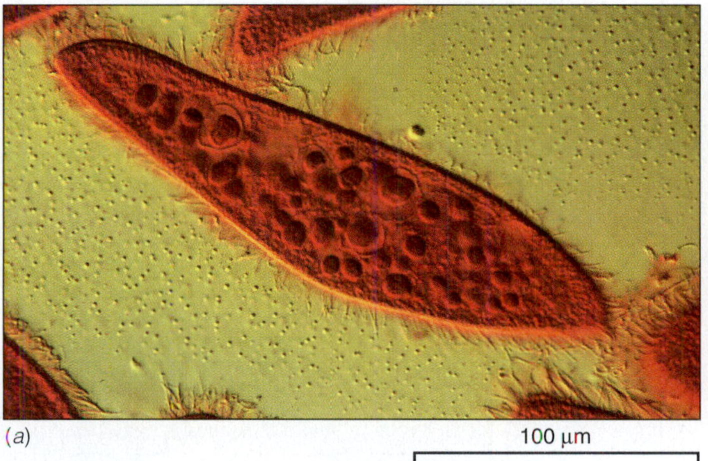

FIGURE 1–2 Single-celled and multicellular organisms. (***a***) Single-celled organisms are generally smaller and less specialized than multicellular organisms. This *Paramecium* is common in pond water and carries on all of its life functions within its one cell. (***b***) The plant (*Passiflora coccinea*) bearing this red passion vine flower is much more complex. Its body parts contain specialized cells that carry on certain tasks. For example, the flower carries on reproduction, the leaves contain cells that carry on photosynthesis, and the roots absorb nutrients. (*a*, Michael Abbey/ Photo Researchers; *b*, James L. Castner)

them together with a simple definition. However, all living things—humans, trees, even single-celled bacteria or amoebas—share certain characteristics and activities. Taken together, these features define life. Living things, more formally referred to as **living systems** or **organisms,** consist of one or more cells that grow, carry on self-regulated metabolism, move, respond to stimuli, reproduce, and adapt to environmental changes. The viruses lack several of these characteristics, and most biologists do not consider them living things. In this section, we explore the characteristics of living systems in some detail.

Living Things Are Composed of Cells

Although living things may vary greatly in size and appearance, all are composed of basic building blocks called cells (see Window on the Plant Cell in Chapter 4 and Window on the Animal Cell in Chapter 4). The **cell** is the simplest unit of living matter that can carry on all the activities necessary for life. Some of the simplest organisms, such as bacteria or certain algae, consist of a single cell. In contrast, the human body or an oak tree is made of trillions of cells (Fig. 1–2). The life processes of such complex organisms depend on the coordinated activities of their component cells.

Living Things Grow and Develop

Some nonliving things appear to grow. A snowball rolling down a hill becomes larger as snow gathers around it, and a stream may swell with accumulating rain water. These inanimate objects increase in size only when preexisting materials are added to them.

In contrast, when a living system grows, it takes in raw materials from the environment and changes them into the specific types of substances that comprise its structure. Biological growth usually proceeds from the inside out. The increase in the amount of living substance can result from an increase in the *size* of the individual cells, in the *number* of cells, or both (Fig. 1–3).

FIGURE 1–3 Biological growth. An organism grows by using raw materials to increase its body size. This process is regulated by its genes. This Laysan albatross chick from the Hawaiian Leeward Islands will grow to adult size by using nutrients from food to increase the size and number of its cells. (Frans Lanting/Minden Pictures)

Some organisms—most trees, for example—continue to grow indefinitely. In contrast, most animals have a growth period that ends when they reach a certain size in adulthood. Each part of the organism continues to function as it grows.

Living things undergo development as well as growth. **Development** includes all of the changes that take place during the life of an organism. Humans and many other organisms begin life as a fertilized egg, which then grows and develops specialized structures and body form.

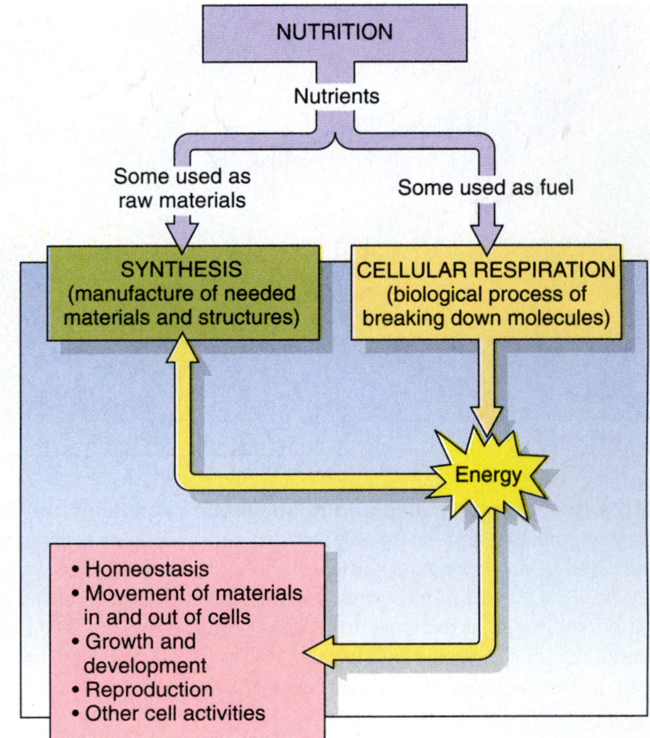

FIGURE 1–4 Metabolic reactions occur continuously in every living cell. Some of the nutrients from food are used to synthesize needed materials and cell parts; other nutrients are used as fuel for cellular respiration, a process that releases energy stored in food. This energy is needed for synthesis and for other forms of cellular work. Cellular respiration also requires oxygen. Wastes from the cells, such as carbon dioxide and water, must be excreted from the body.

Metabolism Includes the Chemical Processes Essential to Growth, Maintenance, and Reproduction

To grow and maintain itself, an organism must be able to convert food materials into living cells. The complex chemical reactions that transform nutrients from food into components needed to build new parts require the expenditure of energy. In the chemical process known as **cellular respiration,** the needed energy is released from nutrients. Many other chemical reactions and energy transformations maintain the routine operations of cells. All of the chemical activities and energy transformations that are essential to growth, maintenance, and reproduction are termed **metabolism** (Fig. 1–4). Metabolic reactions are constantly occurring in every living system. When they cease, the organism dies.

Homeostatic Mechanisms Maintain an Appropriate Internal Environment

Metabolic activities must be carefully regulated so as to maintain a balanced state within the organism. The organism must "know" when to synthesize what and just how much of a particular substance is required. When enough of a product has been made, the synthesizing mechanisms must be turned off. When changes occur in its external environment, an organism's metabolism must adjust appropriately. This automatic tendency to maintain an appropriate internal environment is called **homeostasis.** The mechanisms designed to accomplish this task are known as **homeostatic mechanisms.**

Regulation of body temperature in humans is a good example of a homeostatic mechanism (Fig. 1–5). When body temperature rises above its normal 37° Celsius (C; that is, 98.6° Fahrenheit, F), the increase is sensed by a "thermostat" composed of specialized cells in the brain. This temperature-regulating center sends nerve impulses to sweat glands in the skin. Sweat production then increases. The process of evaporation requires heat, so body heat is lost as sweat evaporates from the body surface. As a result, body temperature is lowered. At the same time, capillaries (tiny blood vessels) in the skin dilate (expand), permitting the blood to carry heat to the body surface more efficiently. The heat radiates from the body surface.

When body temperature falls below normal, messages from the "thermostat" in the brain cause blood vessels in the skin to constrict. This reduces heat loss. Heat may also be generated by the muscular contractions we call shivering. Some animals respond behaviorally (for example, they may orient their bodies to the sun) to increase body heat.

Movement Is a Basic Property of Cells

The living material of cells is itself in continuous motion. Although many living things do not carry on locomotion

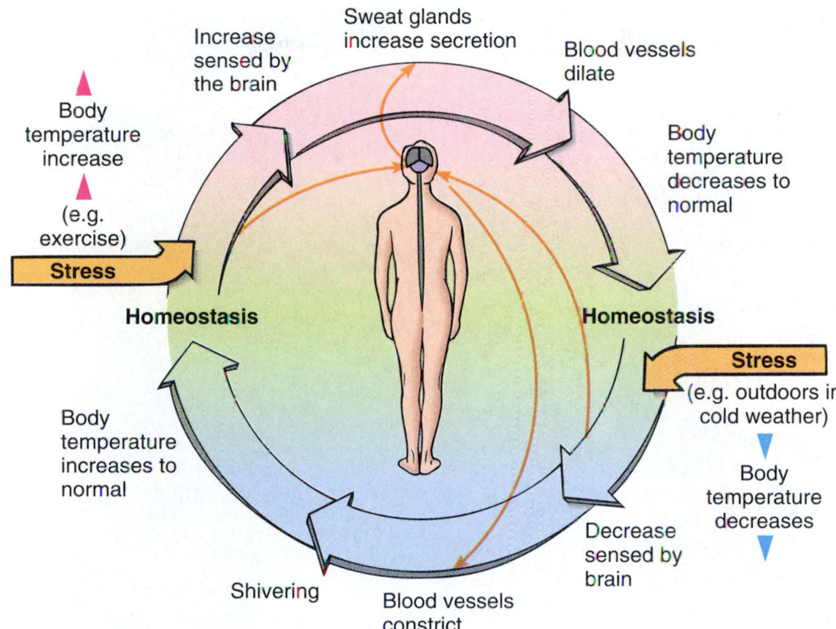

FIGURE 1–5 Regulation of body temperature in the human by homeostatic mechanisms. An increase in body temperature above the normal range stimulates cells in the hypothalamus of the brain to send messages to sweat glands and small blood vessels (capillaries) in the skin. Increased circulation of blood in the skin and increased sweating are mechanisms that help the body rid itself of excess heat. When body temperature falls below the normal range, blood vessels in the skin constrict so that less heat is carried to the body surface. Shivering, in which muscle contractions generate heat, may also occur.

(self-propelled movement from one place to another), internal movement is characteristic of all life. A tree cannot pull up its roots and walk away, but it moves as it grows, its buds open, food is transported, and it responds to changes in its environment. Complex animals, such as insects and reptiles, possess specialized groups of cells called muscles that make complicated, purposeful movements possible (Fig. 1–6).

Living Things Respond to Stimuli

Living things actively respond both to **stimuli** (changes) in the external environment and to changes inside themselves. When in need of nourishment, a one-celled organism, such as an amoeba, reacts positively to food in its watery surroundings by flowing toward it and engulfing it. Most organisms respond to changes in temperature, pressure, sound, variations in light intensity, and the chemical composition of their surroundings.

The responsiveness of plants is often not as obvious as that of animals. However, most of us have, at some time, observed how plants grow toward light, roots tend to grow

FIGURE 1–6 Movement is characteristic of living things. A Mediterranean chameleon (*Chameleo chameleon*) rapidly darts its unusual tongue into a flower to capture an unwary insect. Not all movement in the biological world is as dramatic or as difficult to photograph as what you see here. This photograph was taken at night using a high-speed strobe. (Animals Animals/ Stephen Dalton)

(*a*)

(*b*)

FIGURE 1–7 A few plants can respond to the touch of an insect by trapping it. (***a***) Here a fly lights on a leaf of the Venus flytrap (*Dionaea muscipula*). The leaves of this plant have a scent that attracts insects. Trigger hairs on the leaf surface detect the presence of the insect, and the leaf, hinged along its midrib, folds. (***b***) The edges come together and hairs interlock, preventing the fly's escape. The leaf then secretes enzymes that kill and digest the insect. (David M. Dennis/Tom Stack & Associates)

toward water, and vines wrap around solid objects. A few plants, such as the Venus flytrap of the Carolina swamps, are sensitive to touch and can trap insects (Fig. 1–7).

Living Things Reproduce

Although the life spans of various organisms range from minutes to centuries, they are always limited. However, the death of an individual, or even of a generation of individuals, does not mark the end of that type of organism. Perpetuation of each type of organism is provided for by the process of **reproduction.** Reproduction involves the transmission of information by the remarkable hereditary material known as **deoxyribonucleic acid (DNA).** DNA codes the information responsible for the structure and function of the organism. For example, DNA ensures that cows always give rise to calves—never to cats or rose bushes.

In simple organisms, such as the amoeba, reproduction may be asexual, in which case a single parent divides (or buds), giving rise to offspring (Fig. 1–8). When an amoeba has grown to a certain size, it reproduces by dividing in half to form two new amoebas. Before it divides, an amoeba makes a duplicate copy of its DNA, and one complete set of DNA is distributed to each new cell. Because its DNA is identical, each new amoeba is similar to the parent cell. Unless eaten or destroyed by environmental conditions, such as pollution, an amoeba does not die. It becomes a part of the new generation.

In complex organisms, sexual reproduction is carried out by the production of specialized sperm and egg cells that unite to form a fertilized egg. The fertilized egg develops into the new organism. Generally, though not always, the sexual process involves two individuals, male and female. Each offspring is not a duplicate of a single parent but is, instead, the product of DNA contributed by both the mother and the father (Fig. 1–9). Genetic variation provides raw material for the vital processes of evolution and adaptation.

A Population Has the Potential to Evolve and Adapt to Its Environment

A **species** is a group of organisms with similar structure and function; in nature, they breed mainly with one another, play a similar role, and share a common ancestry. Members of the same species that inhabit a given area make up a **population.** Individuals do not evolve; only populations evolve. The ability of a species to **evolve,** that is, to change over time, enables it to survive in a changing world.

As a population evolves, individuals within that population develop structural, physiological, and behavioral traits that enable them to more effectively grow, reproduce, and maintain homeostasis. Traits that enhance an organism's ability to survive in a particular environment are referred to as **adaptations.** Woodpeckers have structural adaptations—powerful neck muscles, beaks fitted

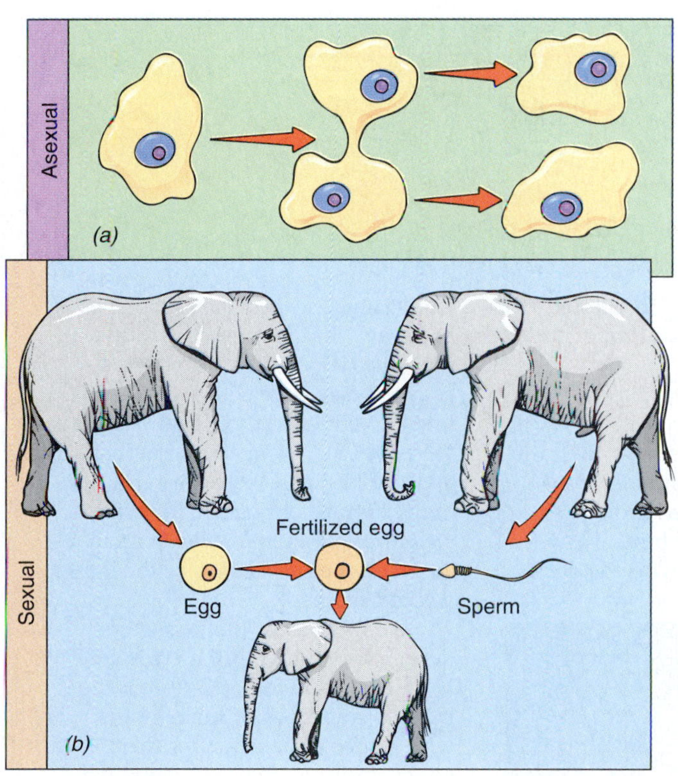

FIGURE 1–8 Approaches to reproduction. (***a***) In asexual reproduction, one individual gives rise to two or more offspring—all identical to the parent. (***b***) In sexual reproduction, two parents each contribute a sex cell; these join, forming a fertilized egg, which gives rise to the offspring. Each new individual is a genetic combination of both parents.

FIGURE 1–9 This calf is white (albino) due to an inherited inability to produce normal pigments. The mother is a black angus cow and the father is also normally colored. The parents of this white calf are both genetic carriers for the albino trait, and the calf is genetically pure for this trait. Sexual reproduction makes it possible for offspring to differ from their parents genetically. Consequently, they can also differ in appearance. (William Munoz/Photo Researchers, Inc.)

for chiseling, and long chisel-like tongues—that enable them to pluck insects from tree trunks. Cactus plants have adaptations that enable them to survive in dry areas (Fig. 1–10). Every biologically successful organism is an impressive collection of coordinated adaptations. Most adaptations evolve over long periods of time and involve many generations; they are the result of evolutionary processes.

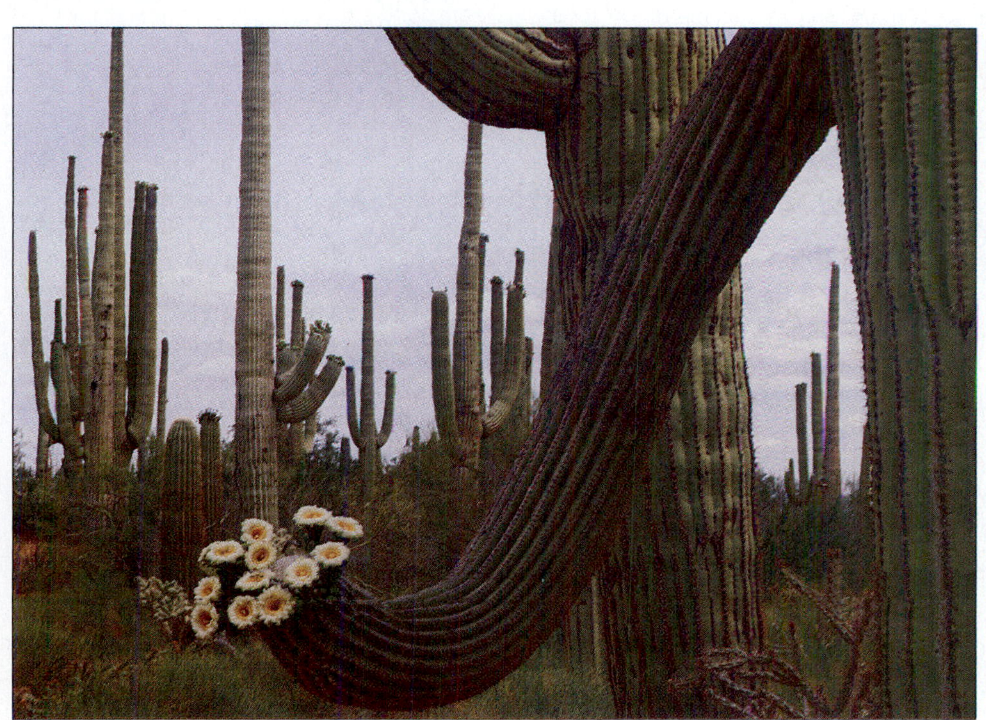

FIGURE 1–10 Adaptations permit organisms to survive. The warmer, moister deserts of North America, like this Arizona desert, are inhabited by large cacti. These Saguaro cactus plants are strikingly adapted to their desert environment. The stem is adapted to carry out photosynthesis and to store water. Leaves are modified into spines that discourage animals from eating them. (David Muench)

FOCUS ON *Evolution*

THE CASE OF THE PEPPERED MOTH

An interesting case of evolution in action has been documented in England since 1850. The tree trunks in a certain region of England were once white because of a type of fungus, a lichen, that grew on them. The common peppered moth was beautifully adapted for landing upon these white tree trunks because its light color blended with the trunks and protected it from predacious birds (see figure). At that time black moths were rare.

Then humans changed the environment. They built industries that polluted the air with soot, killing the lichens and coloring the tree trunks black. The light-colored moths became easy prey to the birds. Now the black moths blended with the dark trunks and escaped the sharp eyes of predators. In these new surroundings, the dark moths were better adapted and were selected for survival. Eventually, more than 90% of the peppered moths in the industrial areas of England were dark. Interestingly, with recent efforts to control air pollution, there has been an increase in the population of the light-colored moths.

Adaptation of the peppered moth was studied in the 1950s by H. B. D. Kettlewell of Oxford, who marked hundreds of male moths with a spot of paint under their wings and then released them in both rural and industrial areas. Observers reported that birds preyed on the moths that were more visible. After a period of time, surviving moths were recaptured by attracting them with light or females. Based on observation and on the percentage of each type of moth recaptured, these studies confirmed that significantly more dark moths survived in industrial areas and more light moths survived in rural areas.

Dark and light peppered moths. Which is most likely to become dinner for the bird? Note that the tree trunk is covered by lichens (a type of fungus). (Visuals Unlimited/John D. Cunningham)

EVOLUTION BY NATURAL SELECTION IS A MAJOR THEORY IN BIOLOGY

A major focus of biology is on how the millions of life forms on our planet are related to one another and how each kind of organism has come to be. Biologists have uncovered a great deal of evidence that the life forms that exist today arose from earlier species. Most biologists accept the theory of **evolution** by natural selection as the best explanation of how this process may have occurred.

Species Evolve in Response to Changes in the Environment

If the environment stayed the same, populations would not change. However, the environment changes continuously and the populations that survive are the ones that change with it. For example, animals that have survived in very cold climates have adaptations such as thick fur and the ability to hibernate.

Natural Selection Is an Important Mechanism of Evolution

Although the concept of evolution has been discussed by philosophers and naturalists through the ages, Charles Darwin and Alfred Wallace first suggested a plausible scientific mechanism to explain it. In his book *The Origin of Species,* published in 1859, Charles Darwin synthesized many new findings in geology and biology and outlined a comprehensive theory that has helped shape the nature of biological science to the present day. Darwin presented a wealth of evidence that the present forms of life on Earth descended with changes from previously existing forms.

Darwin proposed that evolution proceeds by the process of **natural selection.** He based his theory on sev-

FIGURE 1–11 Many organisms produce large numbers of offspring. Strings of eggs of the American toad (*Bufo americanus*) wrapped about Elodea plants. Many more eggs are produced than can possibly develop into adult toads. Random events may be largely responsible for determining which of these developing organisms will hatch, reach adulthood, and reproduce. However, certain traits that each organism has also contribute to its probability for success in its environment. Although not all organisms are as prolific as the toad, the generalization that more organisms are born than survive is true throughout the living world. (Runk/Schoenberger, from Grant Heilman)

eral observations, which are presented here in contemporary form.

1. Members of a population show some variation from one another. For example, there may be differences in size, body structure, and color.
2. Many more organisms are born than survive to reproduce (Fig. 1–11).
3. Resources in any environment are limited, and members of a population compete with one another for necessities like food, sunlight, and space. Individuals who happen to have characteristics that give them some advantage in their competition for resources are more likely to survive.
4. The survivors live to reproduce and pass on their genetic "recipe" for survival to their offspring. Thus, the best-adapted individuals of a population leave (on average) more offspring than do other individuals. Because of this differential reproduction, a greater proportion of the population becomes adapted to the prevailing environmental conditions. Nature *"selects"* the best adapted organisms for survival.

The long neck of the giraffe is an adaptation for reaching leaves on trees (Fig. 1–12). The antelope-like ancestors of modern-day giraffes did not have elongated necks. As with other traits, there was a bell-shaped distribution of neck heights in the giraffe population (i.e., the length of most giraffe necks was about the same, but some giraffes had relatively short necks and others had relatively long necks). Giraffes with the shortest necks could not compete effectively for the leaves on the trees of the African veldt; they were less likely to survive and reproduce. Giraffes with the longest necks were best able to compete for food. These giraffes survived and reproduced, passing on their genes (DNA) for long necks. Through thousands of generations, the giraffe neck became longer and longer.

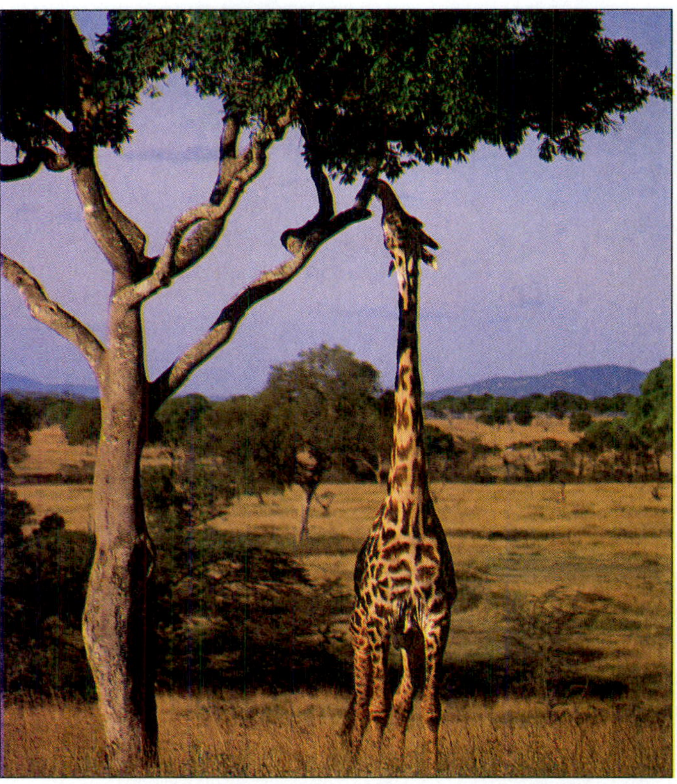

FIGURE 1–12 A successful organism is well adapted to its environment. The long neck of the giraffe is an adaptation for reaching leaves high on trees. (Visuals Unlimited/Walt Anderson)

Evolution Favors Diversity

Although Darwin did not know about DNA or understand the mechanisms of inheritance, we now know that the variations that exist among organisms are genetic. The sources of these variations are random **mutations**, chemical changes in the DNA that persist and can be inherited. Mutations modify the genetic code and provide the raw material for evolution.

If every organism were exactly like every other individual of its population, any change in the environment might be disastrous to all, and the population would become extinct. Differences among individuals—initiated by random mutation, spread by sexual reproduction, and molded by natural selection—enable populations to adapt to an ever-changing environment (see Focus On Evolution: The Case of the Peppered Moth). Those traits

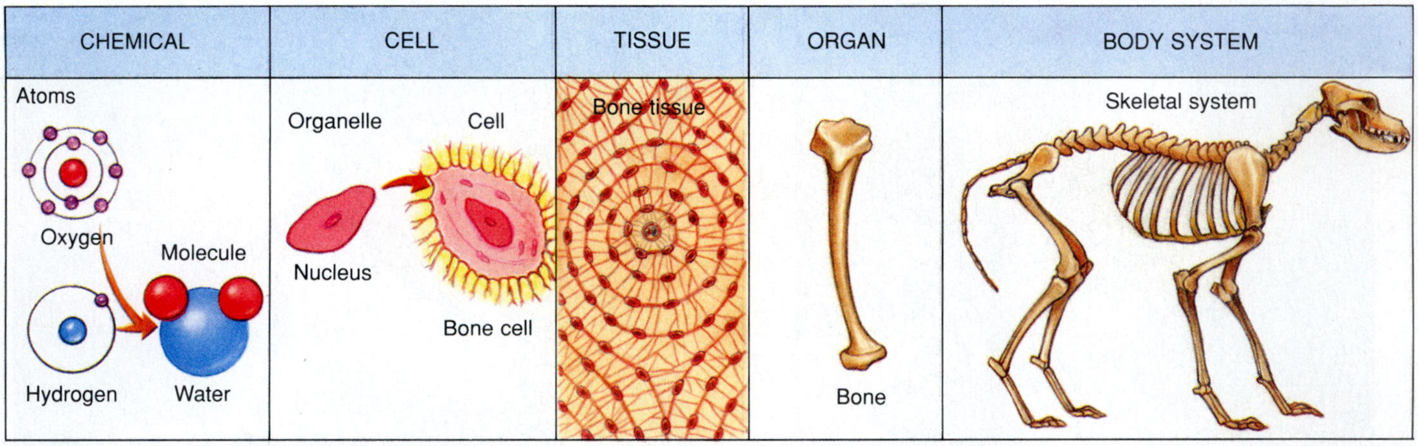

CHEMICAL	CELL	TISSUE	ORGAN	BODY SYSTEM
Atoms Oxygen Molecule Hydrogen Water	Organelle Cell Nucleus Bone cell	Bone tissue	Bone	Skeletal system

FIGURE 1–13 Levels of biological organization. The photograph of Earth taken from space shows Africa. (NASA)

that promote survival become more widely distributed in the population. Over long periods of time, as organisms continue to change in response to changes in the environment, members of the population begin to look increasingly unlike their ancestors.

Concept Connections

Evolution ⟷ *Biology*

An evolutionary perspective is apparent in almost every aspect of biology. Darwin's book raised a storm of controversy in both religion and science, some of which still lingers. It also generated a great wave of scientific research and observation that has provided much additional evidence that evolution is responsible for the tremendous diversity of organisms present on our planet. The details and applications of the theory of evolution will be discussed in Chapters 17 through 25. However, in this first chapter, it is important to emphasize the importance of theory in biology.

Biologists can understand the structure and functions of organisms by considering them in light of the long, continuing process of evolution. These scientists are constantly checking for verification of the evolutionary relationships among different organisms and often reinterpret these in light of new evidence. Biology today is more than a science of describing and naming organisms and life processes. Biologists study, not only the existence of structural similarities, but also what these similarities (and differences) tell us about how organisms might be related to one another.

Biological Organization Reflects Evolutionary Theory

Most biologists think that evolution has proceeded from simple molecules to complex organisms. Whether we study an individual organism or the world of life as a whole, there appears to be a pattern of increasing complexity (Fig. 1–13).

Each Living System Has Several Levels of Organization

We can identify a hierarchy of organization within an organism. The **chemical level** is the simplest level of organization. It includes the basic particles of all matter: atoms and combinations of atoms called molecules. An **atom** is the smallest amount of a chemical element (fundamental substance) that retains the characteristic properties of that element. For example, an atom of iron is the smallest possible unit of iron, and an atom of sodium is the smallest possible unit of sodium. (As we will see in Chapter 3, even atoms consist of subatomic particles and have a very definite organization.) Atoms combine chemically to form **molecules.** For example, two atoms of hydrogen combine with one atom of oxygen to form a molecule of water.

Atoms and molecules associate with one another to form the specialized structures that make up the **cellular level.** Each cell consists of a discrete body of jellylike **cytoplasm** surrounded by a **plasma membrane.** Specialized structures within the cell are called **organelles.** Most types of cells contain a large organelle, the **nucleus,** that stores its DNA. Recall that the *cell* is the basic structural and functional unit of life.

Each type of cell is specialized to perform specific functions. In most multicellular organisms, including humans, similar cells associate to form **tissues,** such as muscle tissue or nervous tissue. In turn, various tissues are

arranged into functional structures called **organs,** such as the heart or brain. In animals, each group of biological functions, such as circulation, is performed by a coordinated group of tissues and organs, known as an **organ system.** The circulatory system and the nervous system are examples of organ systems. Working together with far greater precision and complexity than the most complicated machines designed by humans, the organ systems make up the complex living **organism.**

We Can Also Organize Organisms on the Ecological Level

Biologists can classify the interactions of organisms on the ecological level as well as on the level of the individual organism. The members of a species are not uniformly or even randomly scattered over the Earth. They are grouped together in populations that inhabit particular regions. Various populations of different species occupying the same area interact with one another to form **communities.** A community may be composed of hundreds or even thousands of different types of organisms. All communities of living things on Earth are collectively called the **biosphere.**

A community together with its nonliving environment makes up an **ecosystem.** An ecosystem generally consists of three varieties of organisms—producers, consumers, and decomposers—and has a physical environment appropriate for their survival. An ecosystem can be as small as a pond (or puddle) or as large as a forest. The largest ecosystem of all, and the only truly self-sustaining one known, is planet Earth with all of its inhabitants. This ecosystem is called the **ecosphere.** The ecosphere represents all of the interactions among the biosphere, the Earth's atmosphere, the hydrosphere (water in any form), and the lithosphere (Earth's crust). The study of how organisms interact with one another and relate to their nonliving environment is called **ecology** (derived from the Greek word *oikos,* meaning "house").

Planet Earth and all of its inhabitants

TABLE 1–1
Classification of Domestic Cat, Human and White Oak

Category	*Classification of cat*	*Classification of human*	*Classification of white oak*
Kingdom	Animalia	Animalia	Plantae
Phylum (or division)	Chordata	Chordata	Anthophyta
Subphylum (or subdivision)	Vertebrata	Vertebrata	None
Class	Mammalia	Mammalia	Dicotyledones
Order	Carnivora	Primates	Fagales
Family	Felidae	Hominidae	Fagaceae
Genus and species	*Felis domestica*	*Homo sapiens*	*Quercus alba*

MILLIONS OF KINDS OF ORGANISMS HAVE EVOLVED ON OUR PLANET

The variety of living organism that inhabit the ecosphere challenges the imagination. If we are to study their interrelationships, we need to make some sense of this diversity. To facilitate effective communication with one another, biologists have constructed a formal system of classifying and naming organisms. The science of classifying and naming organisms is known as **taxonomy,** and the biologists who specialize in classification are called **taxonomists.** Taxonomy is a dynamic field, and various methods for classifying specific organisms have been used at different times. We will examine current trends in classification and how taxonomists work in Chapter 20.

Biologists Use a Binomial System of Nomenclature

In the 18th century, Carolus Linnaeus, a Swedish botanist, developed a system of classification that, with some modification, is still used today. The basic unit of classification is the **species.** Closely related species may be grouped together in the next higher unit of classification, the **genus** (plural, genera; adjective, generic).

The Linnaean system is referred to as the **binomial system of nomenclature** because each organism is assigned a two-part name. The first part of the name always designates the genus, and the second part, the **specific epithet,** designates the particular species. The specific epithet is often a descriptive word expressing some quality of the organism and is always used together with the generic name preceding it. Both names are always italicized, or underlined. The generic name is often abbreviated, but always capitalized; the specific epithet is seldom capitalized. For example, the dog, *Canis familiaris* (often abbreviated *C. familiaris*), and the timber wolf, *Canis lupus (C. lupus)*, belong to the same genus. The cat, *Felis domestica,* belongs to a different genus.

Taxonomic Classification Is Hierarchical

Just as species may be grouped together in a common genus, a number of related genera constitute a **family.** In turn, families may be grouped into **orders,** orders into **classes,** and classes into **phyla** (for animals; singular, phylum) or **divisions** (for plants or fungi). For example, the family Canidae includes all doglike carnivores (animals that eat mainly meat). This family includes 12 genera and about 34 living species. Family Canidae, along with family Ursidae (bears), family Felidae (catlike animals), and several other families that eat mainly meat, is placed in order Carnivora (Table 1–1). Order Carnivora, order Primates (the order to which humans belong), order Cetacea (whales), and several other orders belong to class Mammalia (mammals). Class Mammalia, class Aves (birds), class Reptilia (reptiles), and four other classes are grouped together as subphylum Vertebrata. The vertebrates belong to phylum Chordata, which is part of kingdom Animalia.

Most Biologists Recognize Five Kingdoms

From the time of Aristotle until recently, biologists divided the living world into two kingdoms, Plantae and Animalia. After microscopes were developed it became increasingly obvious that many organisms could not easily be assigned to either the plant or the animal kingdom.

According to the system of classification used in this book, every organism is assigned to one of five **kingdoms:** Prokaryotae (formerly known as Monera), Protista, Fungi, Plantae, or Animalia (see Chapter 20). The members of kingdom **Plantae,** the plants, and of kingdom **Animalia,** the animals, are the organisms most familiar to us (Fig. 1–14).

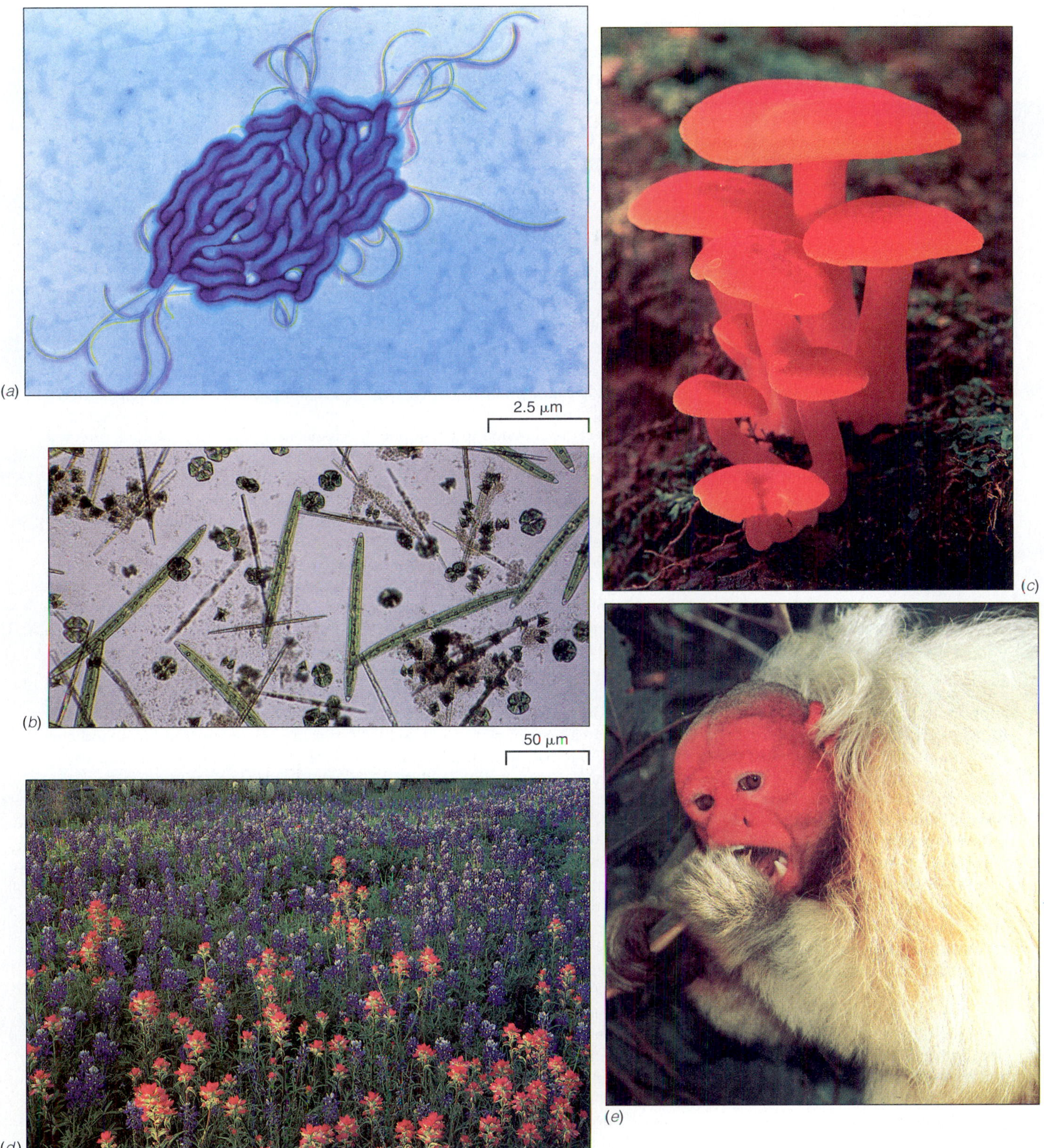

FIGURE 1–14 A survey of life. (***a***) This bacterium, *Campylobacter jejuni,* a member of kingdom Prokaryotae, uses its whiplike tails (flagella) for locomotion. Bacteria of this kind are responsible for food poisoning in humans. This cluster has been magnified several thousand times in this transmission electron micrograph. (***b***) An assortment of algae, organisms belonging to kingdom Protista, inhabit a splash of pond water. (***c***) The scarlet waxy cap mushroom (*Hygrophorus coccineus*) is a member of kingdom Fungi. This edible species is found in deciduous and coniferous forests. (***d***) Members of the plant kingdom in full bloom. This field of paintbrushes and bluebonnets was photographed in Texas. (***e***) The white uakari monkey (*Cacajao calvus calvus*) is an endangered species. A native of Brazil, this monkey's diet consists mainly of seeds of the "matamata" tree, which grows in the Amazon flooded rain forest. Note the distinctive lobes of the head and the baldness of the male monkey shown here. The female has some thin white hair on top of her head and the lobes are not as clearly separated. (*a*, Barry Dowsett/Science Photo Library/ Photo Researchers, Inc.; *b*, Richard H. Gross; *c*, Ed Reschke; *d*, Willard Clay/ Dembinsky Photo Assoc.; *e*, Luiz C. Marigo/Peter Arnold, Inc.)

FOCUS ON *The Environment*

SOCIETY AND THE ECOSPHERE

We are part of a delicately balanced environmental system threatened by human activity. Our technological society exerts tremendous impact on the delicately balanced ecosphere. We are rapidly depleting resources such as oil, coal, minerals, timber, drinking water, and soil. Those of us who live in developed nations make up about 25% of the world population, but we consume nearly 80% of the Earth's resources. With only 5% of the Earth's population, humans in the United States consume about 35% of the world's raw materials.

When we stress one aspect of an ecosystem, we often trigger a chain of events that magnifies the impact. The changes that occur affect the life support systems of our planet. Many forms of air and water pollution destroy plants and algae needed to provide oxygen and food (see figure). These producers serve as the nutritional base for animals that eat and in turn are eaten in the food webs that make up our ecosphere.

When we burn fossil fuels, carbon dioxide is released. The large amount of carbon dioxide released by industry is leading to a greenhouse effect in which more heat is retained in the lower levels of the atmosphere. Some scientists are concerned that the climate of the entire Earth has already been affected and that we are experiencing a gradual increase in temperature. If the global temperature rises by even a few degrees, polar ice caps and glaciers may melt, raising the sea level 1 to 3 meters, and many coastal regions would be flooded.

Another concern is depletion of the ozone layer (in the stratosphere) that surrounds the Earth and absorbs ultraviolet radiation from the sun. A group of commercially important compounds called chlorofluorocarbons (CFCs; e.g., freons) have been shown to cause damage to the ozone layer. These compounds are used as coolants in refrigerators and air conditioners, as propellants in aerosol cans, as foam for insulation and packaging (e.g., Styrofoam), and in many industrial processes (e.g., carbon tetrachloride).

As more ultraviolet radiation reaches the Earth, growth of algae and plants may be affected, and cases of human skin cancer will increase. It has been estimated that for each 1% decrease in ozone concentration, the incidence of skin cancer will increase 2% to 5%. In fact, the incidence of melanoma, a type of skin cancer, has been increasing at the rate of 3.4% a year.

In 1990, after years of debate, more than 90 countries signed an agreement to phase out all use of CFCs by the year 2000. Some nations have taken stricter measures. With some exceptions, the U.S. is scheduled to phase out CFC production by the end of 1995. This action will inconvenience many of us. For example, if you own a pre-1994 car, it will be difficult or expensive for you to have your air conditioner recharged. Do you think the action taken to phase out CFC production is justified? Is it sufficient? Thousands of public policy decisions such as this one are currently being debated. Each of us can also take action on a personal level. For example, we could refrain from using foam plates and cups.

In the early 1970s, there was widespread awareness of environmental problems, and public concern was responsible for the passage of legislation and other actions directed at solving these problems. Unfortunately, just as styles of clothing change, so do intellectual fashions, and the environmental movement became less popular. The need for concern and action, however, is even greater today, for the problem, unlike the fad, shows no signs of disappearing. For example, a significant number of the chemicals we dump into the environment return to haunt us within our own bodies. Each of us now carries strontium-90 and lead in our bones, mercury in our blood and other tissues, DDT in our fatty tissues, asbestos and other particulate matter in our lungs, and an unhealthy concentration of carbon monoxide in our blood.

Any chance of dealing with environmental problems depends upon our degree of commitment, and enlightened commitment demands sound knowledge. However overpopulated the Earth may be, energetic, knowledgeable, and committed people are in short supply. Many aspects of biology are intertwined with the environmental predicament. The principles presented in this book should provide the foundation necessary to understand the critical relationship between society and the ecosphere. The preservation of life itself depends on it.

Biologists study the effects of human activities on the environment. (***a***) Fishing lines and other types of plastic carelessly strewn about the environment prove lethal to thousands of animals each year. This sea turtle has become hopelessly entangled in a fishing net. (***b***) Runoff of fertilizers from nearby farmland has caused a buildup of nutrients in this pond. As a consequence, there has been an explosive growth of protists (*Euglena rubra*). The growth looks like red scum on hot days, then turns green in early evening. Among its many effects, this explosive growth of organisms prevents sunlight from reaching plants and algae below, leading to their death. Their decomposition reduces the concentration of oxygen in the water, resulting in fish kills. (***c***) Massive erosion caused by deforestation in Madagascar. (*a*, Center for Marine Conservation; *b*, Runk/Schoenberger from Grant Heilman; *c*, Frans Lanting/Minden Pictures)

The single-celled bacteria belong to kingdom Prokaryotae (Fig. 1–14a). They differ from all other organisms in that they lack a membrane around their genetic material and lack other cellular organelles. Kingdom **Protista** consists of protozoa, algae, water molds, and slime molds (Fig. 1–14b). These are single-celled or simple multicellular organisms. Kingdom **Fungi** is composed of the mushrooms, molds, and yeasts (Fig. 1–14c). These organisms serve as decomposers, generally absorbing nutrients from dead leaves and other organic matter in the soil.

Plants are complex multicellular organisms adapted to carry out photosynthesis, the process in which light energy is converted to the chemical energy of food molecules (Fig. 1–14d). Plants possess a number of characteristic features, including a waxy covering over aerial parts that reduces water loss (the cuticle), tiny openings in stems and leaves for gas exchange (called stomata), and multicellular sex organs (gametangia) that protect developing reproductive cells. The kingdom Plantae includes mosses, ferns, conifers, and flowering plants.

Animals are multicellular organisms that must eat other organisms for nourishment (Fig. 1–14e). Complex animals have a high degree of tissue specialization and body organization, which have evolved along with the ability to move from place to place. They also have complex sense organs, nervous systems, and muscular systems.

A more detailed presentation of the kingdoms can be found in Chapters 20 through 25. We will refer to these groups repeatedly throughout this book as we consider the many kinds of problems faced by living things and the various adaptations that have evolved in response to these problems.

BIOLOGICAL SYSTEMS ARE INTERDEPENDENT

At every level, the components of biological systems are interdependent. Here, we focus on the interdependence of producers, consumers, and decomposers in an ecosystem and in the ecosphere. (Also see Focus On the Environment: Society and the Ecosphere on pp. 16–17.)

Producers Make Their Own Food

Producers, or **autotrophs** (self-nourishing organisms), can produce their own food from simple raw materials. Algae, plants, and certain bacteria are producers. Most producers carry on **photosynthesis** using sunlight as an energy source.

During photosynthesis, the energy from sunlight is used to make complex food molecules from carbon dioxide and water. The light energy is transformed into chemical energy, which is stored in the food molecules produced. Oxygen, which is required not only by plant cells but also by the cells of most other organisms, is produced as a by-product of photosynthesis.

$$\text{Carbon dioxide} + \text{Water} + \text{Energy (from sunlight)} \longrightarrow \text{Food} + \text{Oxygen}$$

Producers obtain some of the carbon dioxide they need from their own cellular respiration, but they also depend on nonproducers for carbon dioxide.

Consumers Depend on Producers for Food

Animals, including humans, are **consumers.** Consumers, as well as decomposers, are **heterotrophs,** organisms that are dependent either directly or nondirectly on producers for food as well as for energy and oxygen (Fig. 1–15). However, consumers and decomposers also contribute to the balance of the ecosystem. Like all living things (including producers), consumers and decomposers obtain energy by breaking down food molecules originally produced during photosynthesis. When food molecules are broken down by the process of cellular respiration, their stored energy is made available for life processes. Cellular respiration can be represented by the following generalized reaction:

$$\text{Food} + \text{Oxygen} \longrightarrow \text{Carbon dioxide} + \text{Water} + \text{Energy}$$

Gas exchange between producers (autotrophs) and consumers (heterotrophs), by way of the nonliving environment, helps maintain the life-sustaining mixture of gases in the atmosphere (Fig. 1–16).

Decomposers Recycle Nutrients

Bacteria, fungi, and some animals are **decomposers**—organisms that break down wastes and the bodies of dead

FIGURE 1–15 Animals are consumers. This African elephant (*Loxodonta africana*), photographed in Kenya, depends on producers for its nourishment. (Thomas D. Mangelsen/Peter Arnold, Inc.)

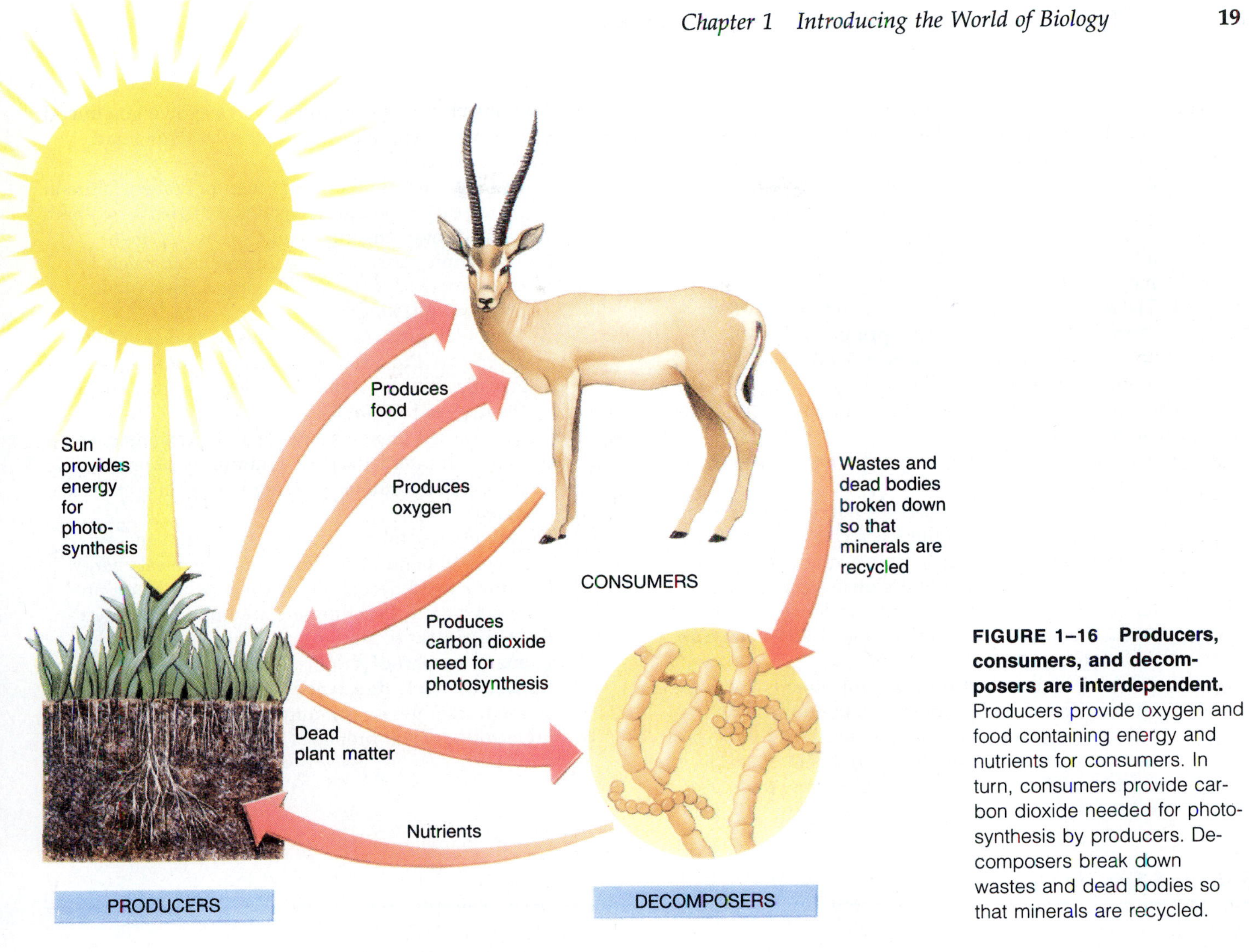

FIGURE 1–16 Producers, consumers, and decomposers are interdependent. Producers provide oxygen and food containing energy and nutrients for consumers. In turn, consumers provide carbon dioxide needed for photosynthesis by producers. Decomposers break down wastes and dead bodies so that minerals are recycled.

organisms. This vital process makes the components of dead organisms available for reuse. If decomposers did not exist, nutrients would remain locked up in dead bodies and wastes. The supply of elements required by living systems would soon be exhausted. Producers and consumers could not survive without the decomposers.

Chapter Summary

I. Life may be defined in terms of the characteristics shared by living things.
   A. All living things are composed of cells.
   B. Living things grow and develop. Growth involves an increase in size and number of cells.
   C. Metabolism includes all of the chemical activities and energy transformations essential to nutrition, growth and repair, and conversion of energy to useful forms.
   D. Metabolic activities are self-regulated so as to maintain an appropriate internal environment; the automatic tendency to maintain a steady state is called homeostasis.
   E. Although not all living things carry on locomotion, all exhibit movement.
   F. Living things respond to stimuli in their internal and external environments.
   G. Reproduction involves the transmission of DNA from one generation to the next.
      1. Reproduction may be asexual, in which the offspring receive a copy of the single parent's DNA and so are identical to the parent.
      2. Reproduction may be sexual, in which the offspring reflect the characteristics of two parents, but are not identical to either parent.
   H. Through the process of evolution, living systems acquire adaptations that enable them to survive in changing environments.

II. The theory of evolution is the most widely accepted explanation of how species of organisms arose from earlier species and how species are related.
   A. Darwin based his theory of natural selection on the following observations:

1. Members of a species vary from one another in size, body structure, color, or other characteristics.
2. More organisms are born than survive to reproduce.
3. Individuals compete for limited resources. Individuals who have characteristics that give them some advantage in their competition for resources are more likely to survive.
4. The survivors live to reproduce and pass on their "survival" genes to their offspring. Thus, the best-adapted individuals of a species leave more offspring than do other individuals.

B. The ultimate source of variation in a population is mutation.

C. Species continuously evolve in response to changes in their environment.

D. Biological organization reflects the course of evolution.
1. In a complex organism there is a chemical level of organization consisting of atoms and molecules, a cellular level consisting of organelles and cells, a tissue level, an organ level, and an organ system level. The organ systems working together make up the organism.
2. The basic unit of ecological organization is the population. The various populations that inhabit an area form a community; a community and its physical environment make up an ecosystem. Planet Earth and all of its inhabitants constitute the ecosphere.

E. Millions of kinds of organisms have evolved on our planet; taxonomy is the science of classifying and naming organisms.
1. Biologists use a binomial system of nomenclature in which each type of organism is assigned a two-part name designating genus and specific epithet.
2. Taxonomic classification is hierarchical: levels include species, genus, family, order, class, phylum or division, and kingdom.
3. Most biologists classify organisms in one of five kingdoms: Prokaryotae (bacteria), Protista (protozoa, algae, water molds and slime molds), Fungi, Plantae, and Animalia.

III. Biological systems are interdependent. For example, an ecosystem is an interdependent community of producers, consumers, and decomposers living within an appropriate environment.

A. During photosynthesis, producers capture energy from sunlight and transform it into chemical energy.

B. Consumers and decomposers are dependent on producers for food. Decomposers are important in recycling nutrients.

C. Humans are part of a delicately balanced environmental system that is threatened by human activity. Increased, enlightened commitment to the task of solving environmental problems is needed.

Selected Key Terms

adaptation, p. 8
atom, p. 12
binomial system of nomenclature, p. 14
biosphere, p. 13
cell, p. 5
cellular level, p. 12
cellular respiration, p. 6
chemical level, p. 12
class, p. 14
community, p. 13
consumer, p. 18
decomposer, p. 18
division, p. 14
ecosphere, p. 13
ecosystem, p. 13
evolution, p. 10
family, p. 14
genus, p. 14
homeostasis, p. 6
kingdom, p. 14
metabolism, p. 6
molecule, p. 12
mutation, p. 11
natural selection, p. 10
order, p. 14
organism, p. 5
photosynthesis, p. 18
phylum, p. 14
population, p. 8
producer, p. 18
reproduction, p. 8
species, p. 8
taxonomy, p. 14

Post-Test

1. The term biology literally means ________________.

2. The chemical and energy transformations that take place within an organism are referred to as its ________________.

3. The automatic tendency to maintain a steady state is called ________________.

4. Reproduction involves the transmission of information by the hereditary material known as ________________.

5. Traits that enhance an organism's ability to survive in its environment are called ________________.

6. Living things are composed of one or more basic units called ________________.

7. Various tissues are organized into functional structures called ________________.

8. Various populations interact with one another and with the environment to form ________________.

9. ________________ is the study of how organisms of a

community relate to one another and to the environment.

10. During ______________, producers use the energy of sunlight to make food from carbon dioxide and water.
11. In cellular respiration, food is broken down in the presence of oxygen, yielding carbon dioxide, ______________, and ______________.
12. Bacteria and fungi function ecologically as ______________.
13. Variation among members of a population can be traced to changes in DNA called ______________.
14. Darwin's theory of evolution emphasizes natural ______________.
15. Protozoa and algae belong to the kingdom ______________.
16. Several closely related species may be assigned to the next highest taxonomic level, a/an (a) genus (b) family (c) phylum (d) kingdom (e) order.
17. Several closely related classes of organisms may be assigned to the next highest taxonomic level, a/an (a) genus (b) family (c) phylum (d) kingdom (e) order.
18. In the scientific name *Canis familiaris*, *familiaris* is the (a) genus (b) family (c) phylum (d) specific epithet (e) order.
19. Which of the following organisms belong to kingdom Prokaryotae? (a) mushrooms (b) mosses (c) bacteria (d) algae (e) two of the preceding answers are correct.
20. Regulation of body temperature is an example of (a) synthesis (b) cellular respiration (c) energy conservation (d) a taxonomic mechanism (e) a homeostatic mechanism.

Review Questions

1. A child might argue that an automobile is alive. After all, it drinks water, guzzles gasoline, moves, and even responds to certain types of stimuli. How would you explain that a car is not alive?
2. How does growth of a living organism differ from "growth" of a snowball as it rolls downhill?
3. Give an example of: (a) homeostasis (b) metabolism (c) adaptation.
4. Compare the environmental roles of producers, consumers, and decomposers.
5. Compare photosynthesis with cellular respiration.
6. In which kingdom would you classify a bacterium? A giraffe? A mushroom?

Thinking Critically

1. What are some arguments for protecting endangered species? Discuss the strategy used by biologists to increase the bald eagle population. How could similar strategies be applied to other endangered species?
2. Discuss some of the ways in which the study of biology could help you to understand and deal with some of the complex issues that confront you in everyday life. How could learning more about biology help you function as a more enlightened citizen?
3. Tigers have sharp claws and teeth. Apply the theory of natural selection to explain how this adaptation may have evolved.
4. What would happen if all of the decomposers on Earth suddenly disappeared?
5. Organisms *A* and *B* are classified in the same species; organisms *X* and *Y* are assigned to the same genus but not to the same species. Which pair of organisms would have the most characteristics in common?

Recommended Readings

Porteous, P. L., "Eagles on the rise," *National Geographic*, Nov 1992. A discussion of the Avian Research Center in Oklahoma, which has a successful captive breeding program for bald eagles.

Raven, P. H., L. R. Berg, and G. B. Johnson, *Environment*. Saunders College Publishing, Philadelphia, 1993. A very readable introduction to environmental issues.

Scientific American, September 1989, Vol. 261, No. 3. This special issue features eleven articles on various environmental issues that focus on the human impact on our planet and strategies for sustaining our world. See especially, "Threats to the Biodiversity," by Edward O. Wilson.

Youth, H., "Flying into trouble," World Watch, Vol. 7, No. 1, Jan–Feb 1994. Bird populations are declining worldwide.

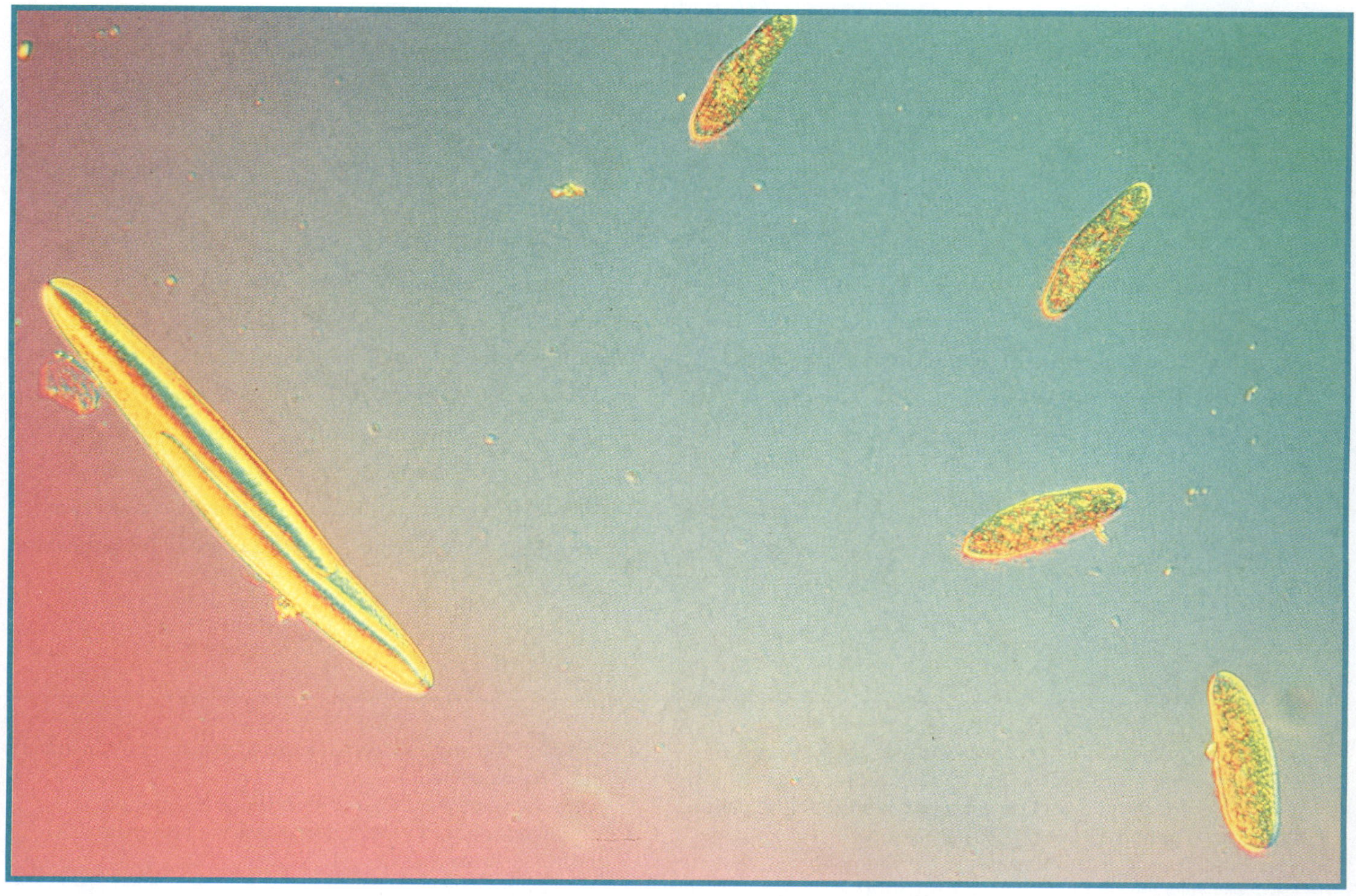

Giant bacterium (*Epulopiscium*). *This bacterium, about a million times larger than a typical bacterium, was photographed with four* paramecia *(members of the protozoa, a group of single-celled protists). (Esther R. Angert and Norman R. Pace, Indiana University)*

Challenging Accepted Ideas: Bacteria Beyond Belief?

Bacteria are microscopic, which means they are too small to be viewed with the naked eye. This generalization has long been accepted by biologists. When investigators discovered a new organism (*Epulopiscium fishelsoni*) in the intestine of surgeonfish caught in the Red Sea and off the Great Barrier Reef of Australia, they thought it was too large to be a bacterium. After all, it was about a million times larger than a typical bacterium. They guessed that it must be a protozoon—a single-celled organism (such as *Paramecium*) belonging to kingdom Protista. Investigators then studied the structure of this new organism with the electron microscope. They found that it lacks a distinct nucleus and has other structural features more typical of bacteria than protists. This new information challenged old beliefs. Could this organism be a giant bacterium?

Esther Angert, a graduate student at Indiana University, and her colleagues carried out a molecular study in which they compared the gene sequences in this organism to those in other microorganisms. Results of these molecular studies by Angert and her coworkers (reported in the journal *Nature* in 1993) confirmed that these organisms are indeed bacteria—the largest bacteria ever imagined by biologists. A million ordinary bacteria could fit inside one of these giants. The newly discovered bacteria live within the intestine of surgeonfish where they are thought to digest the algae upon which the fish graze.

The discovery of the giant bacteria illustrates some important features of the process of science. Scientists must be willing to have their most strongly held beliefs challenged. When new evidence casts doubt on an accepted theory, scientists must reexamine their assumptions and modify the theory. Thus, science is a self-correcting process. Its method encourages continuous observation and questioning and includes recognizing problems; developing hypotheses; making predictions; testing; and, when indicated, modifying accepted theory.

The Process of Science

2

Learning Objectives

After you have studied this chapter you should be able to:

1. Summarize the process of science.
2. Contrast the inductive and deductive modes of reasoning.
3. Distinguish critically between well- and poorly posed hypotheses.
4. Compare and give examples of the concepts of hypothesis, theory, and principle.
5. Design an experiment to test a given hypothesis using the procedures and terminology of the scientific process.
6. Summarize ethical issues related to science.

Key Concepts

- ☐ Science is a process of discovering new facts and developing new hypotheses, theories, and principles.
- ☐ Scientists make careful observations, recognize and state a problem, develop hypotheses (educated guesses), make predictions that can be tested, perform experiments to test their predictions, and use the results to develop theories.
- ☐ The process of science involves the careful use of inductive and deductive logic in the formulation of hypotheses and the testing of predictions by observation and experiment.
- ☐ To be useful, hypotheses must generate predictions that are both testable and falsifiable.
- ☐ Science has ethical dimensions.

Science is a Way of Thinking and a Method of Investigation

You may wonder how biologists came to "know" the many principles discussed in this book. The word *science* comes from the Latin word *scientia,* meaning "to know." To understand not just biology but science itself, it is important to learn how scientists work. The process of science is creative and dynamic, changing over time and influenced by cultural, social, and historical settings, as well as by the personalities of scientists themselves. But science is also a way of thinking. From this manner of thinking, scientists have developed a method for investigating the world around us in a systematic way.

Scientists make careful observations, recognize and state a problem, develop hypotheses (educated guesses), make predictions that can be tested, and perform experiments to test their predictions (Fig. 2–1). If the results do not support the hypothesis, they may still be quite valuable and may lead to new hypotheses. If the results support the hypothesis, the scientist uses them to generate related hypotheses. Consistent results from many experiments can be used to develop a theory or to modify an accepted theory. As new theories emerge, they form the basis of new technology.

Science seeks to understand the complexity of the universe by formulating theories and, eventually, general principles that can be used to solve problems or to suggest new insights. Science is able to uncover ever more about the world we live in and to lead us to an expanded appreciation of our universe.

Some important areas of human life are not scientific. It would be difficult to reduce the arts, for example, to a set of principles, and perhaps we should not try to do so. Yet science can be applied even to the arts. Musical instruments can be improved on or even designed using physics and electronics, for instance, and effective ways of teaching the arts can be investigated by scientific methods.

Observation → Recognize/state problem → Develop hypothesis → Make a prediction that can be tested → Perform experiments to test the prediction → Results → Observation / Develop theory → Principle

FIGURE 2–1 The process of science.

Concept Connections

Science* ⟷ *Technology

For most people, the chief justification for scientific inquiry is the practical, everyday applications that may result. Oddly, though, discoveries that prove to have the greatest practical value often come from purely abstract research. The reason is that very basic discoveries often occur by accident during the course of some general investigation.

The 16th- and 17th-century scientists such as Leeuwenhoek and Hooke could not have seen cells at all without the use of lens-making techniques that were then state of the art (Chapter 5). Our modern knowledge of the cell's structure could not have been obtained without the use of the electron microscope, first developed in the late 1930s with what was then the latest in electronic technology (see figure).

Almost everything we know about photosynthesis and, to a somewhat lesser degree, cellular respiration, has been discovered using radioactively labeled chemical substances. Without nuclear technology, we would probably still be wondering whether the hereditary material of the cell was protein, nucleic acid, or something else. The lensmakers, electron-tube designers, and nuclear physicists who made modern biology possible probably never dreamt of the further scientific advances that would occur because of *their* discoveries. Science nourishes itself, particularly when information from different areas of study is allowed to interact.

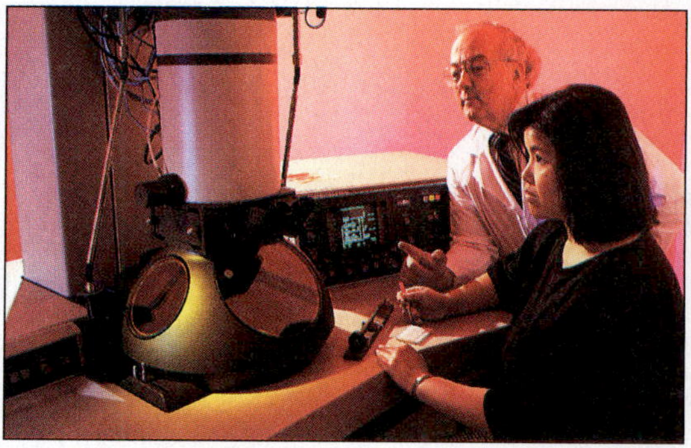

A student and a professor prepare a slide for viewing under the electron microscope. (James Prince/Photo Researchers, Inc.)

Development of the Principle of Biogenesis Illustrates the Process of Science

One of the first important principles of biology to be developed was **biogenesis,** the idea that life always arises from life; living things have living parents. For centuries people believed in **spontaneous generation,** the idea that living things could spring from nonliving matter, especially decaying matter. Some ancient Greeks, for instance, thought that barbarians (by which they meant anyone who was not a Greek) could arise spontaneously from rotting flesh or muck.

Such beliefs persisted throughout the Middle Ages and beyond that time. According to one essay, mice could be produced by placing rags and a handful of grain in a crock. The crock was then to be placed in the thatch of a cottage roof. Before long, a family of mice should appear in the crock. In the unlikely event that someone might actually have wanted mice, and tried the experiment, no doubt mice were obtained. Of course, spontaneous generation was not the only possible explanation.

Scientists must consider alternative explanations. That is one of the things that distinguishes scientific thinking from unscientific thinking. Some individuals may be convinced by the first result that seems to confirm their own views. In fact, they may hope that evidence to the contrary will not be found. Scientists, on the other hand, pride themselves on their attitudes of skepticism not just toward the views of others but also toward their own. The skeptical inquirer must consciously seek alternative, possibly more reasonable, explanations of the data and test them if possible. (The discovery of the giant bacteria discussed in the chapter introduction is an example.)

One way to exclude alternative explanations is to use a comparison group. In doing an **experiment**, a scientist often uses at least two groups. The **experimental group** is tested to determine if and how certain variables change. The **control group** differs from the experimental group in that the experimental variable being tested is either absent or is precisely controlled.

Redi Demonstrated that Flies Do Not Generate Spontaneously

The Italian physician Francesco Redi (1621–1697) used scientific methods to show that flies do not generate spontaneously. He began by observing the natural course of decay of dead animals. Flies visited the corpses, and, in due course, maggots appeared, which eventually changed into flies themselves. To Redi, this suggested that the flies produced had originated as eggs or larvae deposited by other flies (Fig. 2–2). He then performed a controlled experiment with flies. When Redi sealed containers of decaying meat so that flies could not reach the meat, no flies appeared.

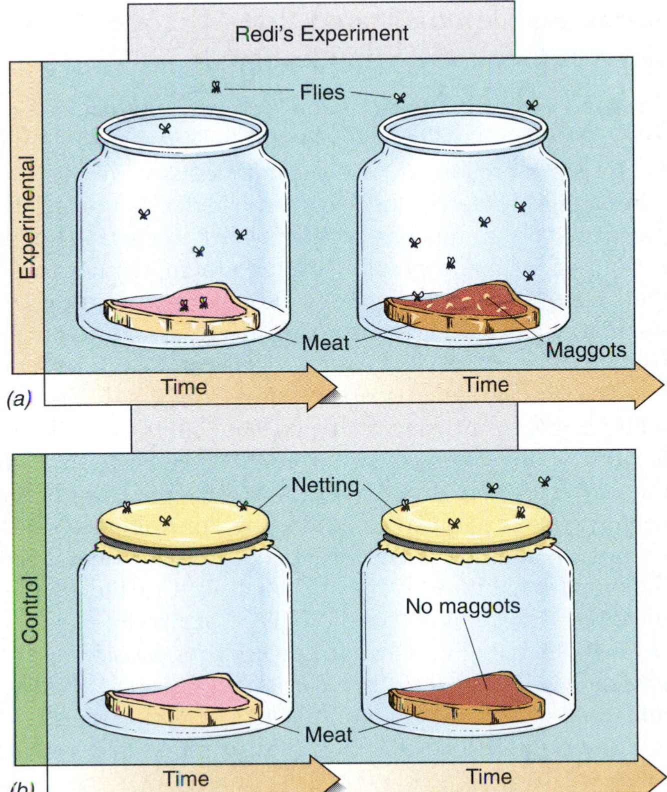

FIGURE 2–2 Can flies appear spontaneously on decaying meat? Redi's experiment supported the hypothesis that living things are produced only by other living things. (***a***) In the experimental group, flies could enter the open jars and lay their eggs. Maggots eventually were seen on the meat and then flies were seen. (***b***) In the control group, the jars were screened. No maggots or flies appeared.

Some observers still felt that spontaneous generation might be true. They claimed that it might require some force or substance from the air, which Redi's seals had excluded along with flies. If the sealed containers are considered controls, they actually differed from the unsealed ones in *two* ways: they had no flies and no air. To make sure that the mere presence of air could not cause spontaneous generation, Redi replaced the seals with fine cotton mesh. Air could move freely through this, but flies could not. The results were the same—maggots appeared only when adult flies had access to the decaying meat.

Notice that without the unsealed containers, Redi could not have been certain that it was the mesh that made the difference. Perhaps some other factor—the season or something that was unusual about the decaying meat—might have been the reason that no maggots appeared.

The appearance of flies on decaying meat had been considered the classic example of spontaneous generation. By showing that flies were not actually produced by spontaneous generation, Redi convinced most informed people of his time that other insects were not produced spontaneously either.

Scientists Demonstrated that Microbes Do Not Generate Spontaneously

Flies and even their eggs can be seen with the naked eye; microbes cannot. Many people still believed that simple microscopic organisms arose spontaneously. Yeasts and other fungi, bacteria, and even protists (single-celled organisms such as amoebas) swiftly swarm in decaying matter. No screen could possibly exclude them; neither could one be sure that they were not initially present in decaying matter. Therefore, an approach like Redi's could not be used to disprove the spontaneous generation of microbes.

The French microscopist Louis Joblot (1645–1723) made a tea, or infusion, of dried hay, boiling the liquid to destroy any organisms that might have been on or in the hay. He then divided the infusion into two parts, keeping one in a sealed container (his control) and one in an open container. Only the open container developed microbial inhabitants. This approach was further developed by the Italian physiologist Lazaro Spellanzani (1729–1799) using very careful procedures for sealing the vessels (Fig. 2–3). When the infusion was properly boiled, microbial growth never took place in sealed vessels.

Some raised the objection that in the sealed vessels there was no oxygen. What if spontaneous generation would take place only if oxygen or some other unknown life-giving substance in the air were available? It was hard to answer this objection, because any treatment one might give air to kill its microbes (bubbling it through lye, for instance, or passing it through a red-hot tube) might also destroy any "vital substance" the air might contain.

FIGURE 2–3 Spellanzani's experiment. (*a*) In the experimental group, flasks are left open and the broth becomes cloudy with microorganisms. (*b*) In the control group, flasks are sterilized and sealed. Microorganisms are not produced in these flasks.

The matter was given its final resolution by the experiments of French investigator Louis Pasteur (1822–1895). Pasteur drew out the neck of a flask containing boiled culture medium (a broth containing nutrients) into a long, curved shape in a flame. Gradual cooling allowed the air to enter slowly; any microorganisms were trapped in the curve of the neck. No life grew in the flask unless the medium inside came in contact with the region of the neck that held the microbes. When that happened, the flask swarmed with life (Fig. 2–4). A more effective control could scarcely be imagined; the control flask was the *same* as the experimental flask. The air in the flask could not have been damaged in any way, yet when microbes were kept out, none developed.

The principle of biogenesis, that all living things had living ancestors, was a major scientific advance, with far-

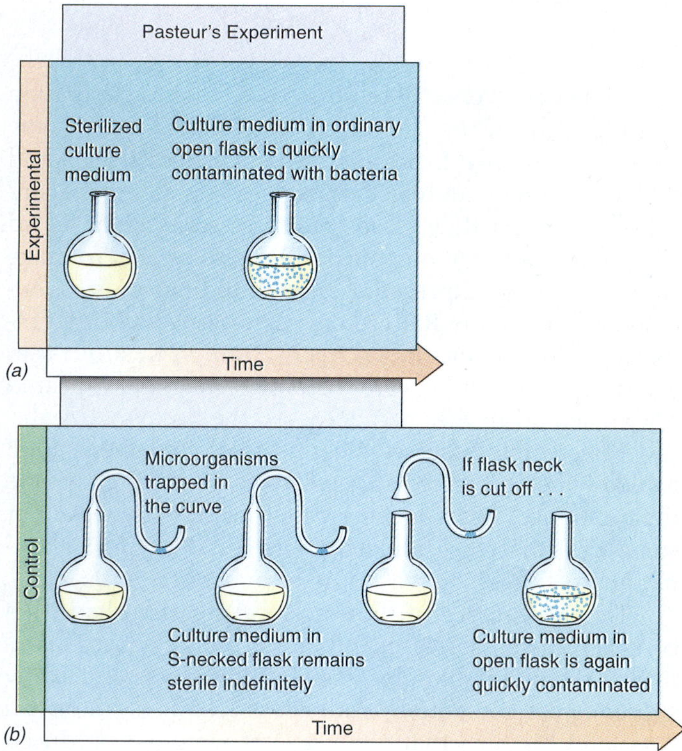

FIGURE 2–4 Pasteur's experiment. (*a*) Experimental group. The flask containing sterilized culture medium was left open. It was quickly contaminated by bacteria. (*b*) In the control group, the sterilized flask was left open, but was designed so that microorganisms could not enter. Pasteur's experiments discredited the spontaneous generation of microorganisms. Even microbes are produced only by other microbes.

reaching practical applications. We take it for granted that canned food will not spoil and that disease microbes will not spontaneously originate in the body, but it took generations of careful work to lay the groundwork for this knowledge. Notice that a new principle was the outcome of this work, that this principle could be used to predict more facts, and that some of those facts were of great practical importance.

Scientists Use Deductive and Inductive Reasoning

Logic is a means of drawing conclusions that are consistent with the evidence used to support them. Deductive and inductive forms of logic may be distinguished. Once a general principle is understood, many of its consequences can be inferred with confidence, using **deduction.** Deduction operates from generalities to specifics. In the scientific process, deduction provides us with predictions about the results of experiments. For example, if we accept the principle of biogenesis, then we can predict that even the giant bacteria discussed in the chapter introduction can be produced only by living ancestors, not by spontaneous generation. This prediction could be tested by experiment.

No matter what undiscovered relationships may exist among the facts and principles that we already know, science's fundamental business is discovering *new* general principles. This is done by using **induction,** a reasoning process that is the opposite of deduction. Inductive reasoning begins with specific examples and seeks to draw a conclusion or to discover a general principle on the basis of those examples. For example, when Redi demonstrated that flies were produced by biogenesis, rather than by spontaneous generation, scientists could induce that other insects were also produced by biogenesis.

Induction begins by organizing raw data into manageable categories and answering the question: What do these facts have in common? It continues by seeking a common explanation for the facts. Then this explanation is brought into some logical relationship with the rest of the known facts and validated explanations of science.

The Scientific Method Is a Series of Ordered Steps

As we have seen in the various experiments described, the *process* of science may be different for every individual scientist. The ideas studied, the design of the experiments, and the range of questions posed are limited only by the creativity of the scientist. In contrast, the **scientific method** involves a set of ordered steps, and is a tool used by all good scientists during the process of their work to check the results of induction against reality. Steps often used in the scientific method are reviewed and applied to the biogenesis question in Table 2–1.

Notice that confirming biogenesis in our example means that its status as a general principle is supported. However, no principle of science is held sacred. If so much as one counterexample could be demonstrated, we would have to modify the biogenesis principle to state: "Life *usually* originates only from life." But isn't that really the case even without a known counterexample? Given long enough—billions of years—and the right set of conditions, most biologists agree that life indeed originated spontaneously on the face of the Earth. So even for

TABLE 2–1
The Scientific Method

Steps in the scientific method	*Illustration*
1. Make careful observations	1. (a) Living things such as birds and humans have living parents, (b) flies seem to spontaneously appear on decaying meat.
2. Recognize and state problem	2. Do living things originate spontaneously or come only from other living things?
3. Develop a hypothesis	3. Living things come only from living things (biogenesis).
4. Make prediction to be tested	4. Flies are produced from pre-existing flies; they will not spontaneously appear in sealed jars.
5. Perform experiments (or make observations) to test prediction	5. Place decaying meat in two sets of jars. Seal one set so flies cannot enter.
6. Results may support or refute the hypothesis; they can be used to generate other hypotheses and may contribute to development of a theory	6. No flies appeared in the sealed jars. Biogenesis is supported.

biogenesis, a mental reservation must be made. In front of even the most secure principle of science there is an unspoken "perhaps."

Scientists Make Careful Observations and Recognize Problems

Significant discoveries are usually made by those who are in the habit of looking critically at nature. Chance and luck are often involved in recognizing a problem, but, as Louis Pasteur said, "Chance favors the prepared mind." On the one hand, the minds of those expert in a field are best prepared by knowledge of that subject area. On the other hand, long immersion in conventional ways of thinking may make it difficult to view a subject area in a new light.

Some of the greatest scientific advances seem obvious in retrospect, and are, in a sense, quite simple. But they must have been very difficult to think of in the first place. The scientific process, in other words, is more than just following a recipe. The personal qualities of the scientist, particularly his or her creativity and originality, determine the effectiveness of the process.

In 1928, the British bacteriologist Alexander Fleming *observed* that one of his bacterial cultures had become invaded by a blue mold. He almost discarded it, but before he did, he *noticed* that the area contaminated by the mold was surrounded by a zone where bacterial colonies did not grow well. His culture looked something like the one shown in Figure 2–5.

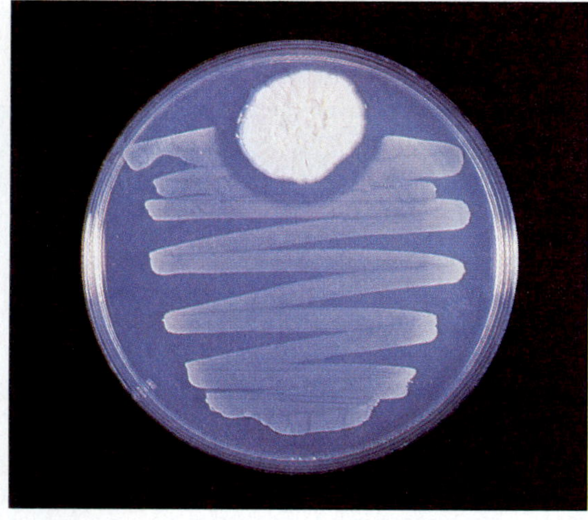

FIGURE 2–5 The effect of penicillin. Bacteria (*S. aureus*) have multiplied to form thick areas of growth along the regions where they were streaked onto this culture dish. However, a zone of inhibited bacterial growth is evident around the colony of *Penicillium* mold on the upper part of the culture dish. (Visuals Unlimited/Christine L. Case)

The bacteria were disease organisms of the genus *Staphylococcus*, which can cause boils and skin infections. Anything that could kill them was interesting! Fleming saved the mold, a variety of *Penicillium* (blue bread mold). What could have been responsible? It was subsequently discovered that the mold produced a substance that prevented successful reproduction of the bacterial population but that was usually harmless to laboratory animals and humans. The substance was penicillin, one of the first antibiotics.

We may wonder how often the same thing happened to other bacteriologists who failed to make the connection and just threw away their contaminated cultures. Fleming benefited from chance, but his mind was prepared to observe it and his pen to publish it. It was left to others, however, to apply it. Although Fleming recognized the potential practical benefit of penicillin, he did not vigorously promote it, and it was more than ten years before the drug was put to significant use.

A Hypothesis Is a Proposed Explanation

A **hypothesis** is a tentative explanation. If a problem is recognized only by a prepared mind, a hypothesis is generated only by a creative one. In the early stages of an investigation, a scientist usually thinks of many possible explanations and hopes that the right one is among them. He or she then decides which, if any of them, could and should be subjected to experimental test. Why not test them all? Time and money are important considerations in conducting research. We must establish priority among the hypotheses so as to decide which to test first. Fortunately, some guidelines do exist. A good hypothesis exhibits the following:

1. It is reasonably consistent with all well-established facts.
2. It is capable of being tested; that is, it should generate definite predictions, whether the results are positive or negative. Test results should also be repeatable by independent observers.
3. It is falsifiable, which means it can be proved false.

A hypothesis cannot really be proved true, but in theory (though not necessarily in practice) a well-stated hypothesis can be proved false. If one believes in an unfalsifiable hypothesis (e.g., the existence of invisible and undetectable angels), it must be on grounds other than scientific ones.

Consider the following hypothesis: *All* female mammals (animals with hair that produce milk for their young) bear live young. Consider further that a dog named Princess is a mammal. Therefore, we can predict that Princess (if female) should bear live young. When Princess has a litter of puppies, this *supports* the hypothesis. Yet it does not really *prove* the hypothesis. Before the Southern Hemisphere was explored, most individuals

FIGURE 2–6 Spiny anteater, *Tachyglossus aculeatus*, a monotreme. These animals are mammals; they have fur and produce milk for their young. However, they lay eggs. (Tom McHugh/Photo Researchers, Inc.)

would probably have believed that hypothesis without question, because all known furry, milk-giving animals did, in fact, bear live young. But it was discovered that some animals (the duck-billed platypus and spiny anteater) that lived in Australia had fur, produced milk for their young, but laid eggs (Fig. 2–6).

The hypothesis, as stated, was false no matter how many times it had previously been supported. As a result, biologists had to either consider the platypus and spiny anteater as nonmammals or had to broaden their definition of mammals to include them (they chose the latter).

A hypothesis is not true just because some of its predictions (the ones we happen to have thought of or have thus far been able to test) have come true. After all, they could be true by coincidence. In particular, failure to observe a prediction does not make a hypothesis false, but it does not show that the hypothesis is true, either.

A Prediction Is a Logical Consequence of a Hypothesis

Since a hypothesis is an abstract idea, there is no way to test it directly. But hypotheses should suggest certain logical consequences, that is, observable things that cannot be false if the hypothesis is true. On the other hand, if the hypothesis is false, *other* definite predictions should disclose that fact. As used here, then, a **prediction** is a deductive logical consequence of a hypothesis. It does not have to be a future event.

Predictions Can Be Tested by Experiment

As an example of an experiment, let us review an investigation of substance abuse treatment described in the *Journal of the American Medical Association* (April 21, 1993). Since its introduction in the 1960s, the use of methadone maintenance in the treatment of heroin addiction has been controversial. Opponents have called methadone treatment "legalized addiction," claiming that methadone is an addictive drug that could also be abused and pointing to failures in many methadone programs. Proponents, on the other hand, claim that methadone treatment reduces intravenous drug use and its attendant risks including transmission of human immunodeficiency virus (HIV) and crime.

A. T. McLellan and his colleagues designed an experiment to test their prediction that the addition of counseling, medical care, and psychosocial services would improve the effectiveness of methadone therapy in the rehabilitation of heroin addicts. The research team randomly assigned 92 male intravenous heroin users to one of three treatment groups for a 6-month clinical trial. Those addicts assigned to group (1) received methadone alone (no other services). This group served as a control group. Addicts assigned to treatment group (2) received the same dose of methadone and also received counseling. Addicts in group (3) received the same dose of methadone, counseling, and received additional on-site medical and psychiatric services including job training/counseling and family therapy.

The results indicated significantly more successful outcomes among the group (2) patients compared to the patients in the control group (1). The patients in group (3) showed significantly more successful outcomes than those in group (2). Conclusions of the experiment were that methadone alone may be effective for only a minority of addicts. The addition of counseling increased the effectiveness of the treatment and the addition of on-site medical and psychiatric services resulted in the most effective treatment. This experiment is an example of a controlled situation that produced results relating to a prediction.

In any experiment, it is important that sufficient numbers of subjects be tested. If the research team testing the treatment had used only one or two subjects per group, the experiment would not be considered statistically significant because the subjects selected may not have been typical of most heroin addicts. Because a total of 92 addicts were included in the study, the results are likely to reflect the outcomes that would occur for most heroin addicts. Some health professionals, however, may be skeptical of the results because only male addicts were included. It is likely that this experiment will be repeated, perhaps many times, using larger numbers of subjects (including females) and with many other variations before health professionals accept the results with complete confidence.

Deciding the significance of data collected in an experiment is sometimes difficult. In one of the minor engagements of the long battle over spontaneous generation, an English clergyman, John T. Needham (1713–1781), boiled hay infusion and sealed it in flasks.

FOCUS ON *The Process of Science*

PILTDOWN MAN

By the early 1900s, many fossil remains of early humans had been found. "Heidelberg man," known from a huge lower jaw found in Germany, plus '"Java man" and "Peking man" found later, were of the species *Homo erectus,* small-brained but clearly human. Numerous large-brained fossils known as "Neandertal man" had been found in Germany and France. Though these humans had apelike features—jutting jaws, heavy brow ridges, low sloping foreheads—we now include them in our own species, *Homo sapiens.* Yet none of these were intermediate between humans and the apelike animals from which humans had supposedly evolved. A "missing link" was in need of discovery.

In 1912, Charles Dawson, an amateur fossil collector, announced that he had found a remarkable fossil, a human skull fragment, in an ancient river bed in a place called Piltdown, England. This discovery, he claimed, at long last filled the evolutionary gap between apes and humans.

Attempts to reconstruct the complete skull were complicated by the fragmentary nature of the remains. Unlike apes and other fossil humans then known, the Piltdown skull had no great brow ridge; the high forehead and the apparently large brain were also characteristic of modern humans. The facial bones were missing, but considering the size of the jaw the face must have sloped sharply forward as did that of Neanderthal man. The jaw was apelike, but the part that would connect with the skull was missing, as were the chin region and teeth. If these missing parts were reconstructed, the total picture was that of a creature with a human skull and an apelike jaw, a mixture of traits that one might well expect to see in a missing link.

Some raised doubts, however. The skull and lower jaw had not been found together and could not be proved to belong to the same individual. The jaw, in fact, could have been that of a young ape. If so, this fact would be evident from the canine teeth, which are much larger and longer in apes than in humans. To prove that the skull and jaw were from the same creature, the missing jaw parts, or at least some of the canine teeth, had to be found.

Not long after this need was voiced, a canine tooth came to light. Though similar to that of a modern chimpanzee, this tooth was smaller and shorter, intermediate, in a way, between those of a human and an ape. Yet since it was found separately, it could not be proved to belong to the jaw.

Another question persisted. Geological evidence suggested that Piltdown man was too young to be an intermediate: The much older Java and Neandertal fossils were already more human-looking than Piltdown man. Two years after Dawson's death in 1916, it was announced that he had discovered a second Piltdown specimen, which closely resembled the first. With this, most skeptics became believers, though the place of Piltdown man in human evolution remained unclear.

Meanwhile, investigators in southern Africa had unearthed the human-like *Australopithecus,* apparently much older than any known human fossils. These primates had chimpanzee-sized brains, long down-slanting faces, and large jaws with essentially human teeth. Seemingly the right age, they were as good a missing link as Piltdown man, though quite different. As more australopithecine bones were found, more and more people doubted Piltdown man.

In 1949, the British Museum tested the Piltdown fossils for their fluorine content, a method of dating fossils developed in the nineteenth century. The results showed that the bones were of various ages and from various sources. Then, in 1952, scientists examined the Piltdown jaw and found that someone had altered the molars by filing them. As for the canine, not only had it been filed, but a cavity had been filled with a rubbery substance. Artificial staining and painting had concealed the tampering. It became clear that the skull

Microorganisms grew in the infusion. Was this evidence for spontaneous generation? These results baffled supporters of biogenesis. Needham and his supporters thought that spontaneous generation had been demonstrated. Eventually, it was shown by experiment that hay contains bacterial spores that can survive ordinary boiling. There was nothing wrong with Needham's data itself; however, it was incomplete and could be accurately interpreted only in light of additional information.

Experimental Samples that Are Not Typical May Be Misleading

One reason for inaccurate conclusions in an experiment is sampling error. Since *all* cases of what is being studied cannot be observed or tested, we must be content with a sample, or subset, of them. Yet how can we know whether that sample is truly representative of whatever we are studying? In the first place, the sample may be too small,

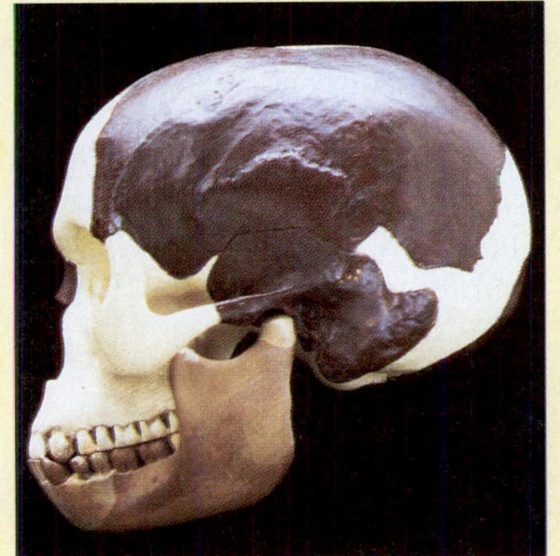

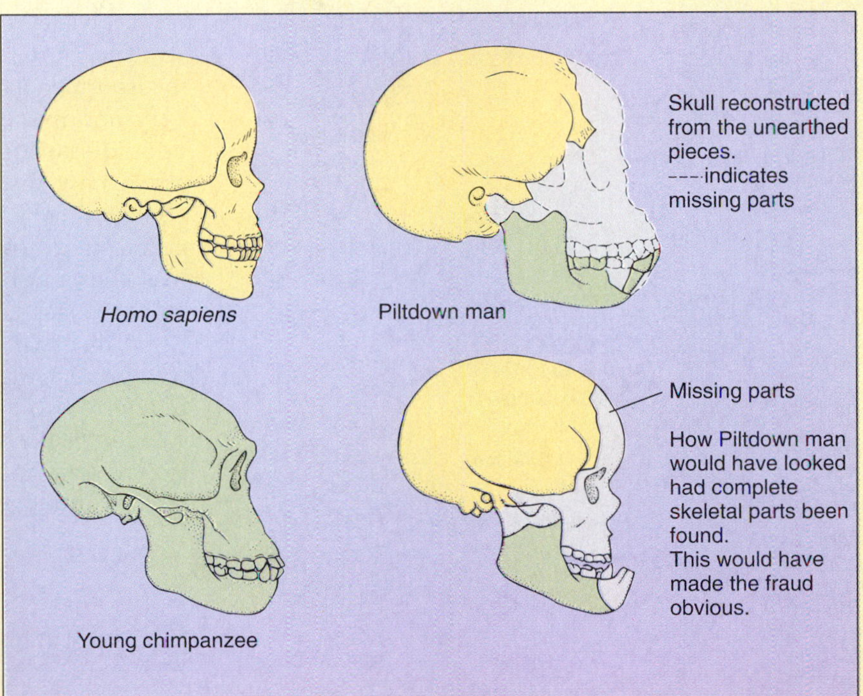

Comparison of the Piltdown skull with both a human skull and the skull of a young chimpanzee. The photograph is of a model reconstruction of the Piltdown skull (lateral view). The dark areas are the original bone fragments. The light-colored areas are the reconstructed parts of the skull. (The National History Museum, London)

was a modern human skull, and the jaw, so apelike, was indeed a chimpanzee's jaw.

It is by no means certain that Dawson was responsible for the clever hoax. Someone else might well have planted the bogus fossils for him to find. Indeed, the whole affair may have been a practical joke that got out of hand. Yet there are lessons we can learn from the story.

Piltdown man appeared to be such a perfect fulfillment of evolutionary prediction that scholars, convinced that such a primate would one day be found, were predisposed to swallow the fraud whole. This postponed for years their recognition of the importance of *Australopithecus*. Decades passed before scientists themselves finally detected the fraud by testing their hypotheses. Piltdown man reminds us that the scientific method depends not only on logical thinking but also on an attitude of open-minded skepticism combined with the strictest honesty.

so that it is likely to be different because of random factors. This problem is usually solvable by applying the mathematics of statistical analysis (Fig. 2–7). In the second place, the sample may not be typical of the group that we intend to study. Again, there are techniques that can be employed to ensure that there is no consistent bias in the way that experimental samples are chosen.

Experimental Bias Can Render Doubtful Conclusions

If we do not take every important factor into account in choosing a sample, our choice may introduce bias into the sample. In one famous case, a telephone poll was taken to determine the results of a forthcoming election.

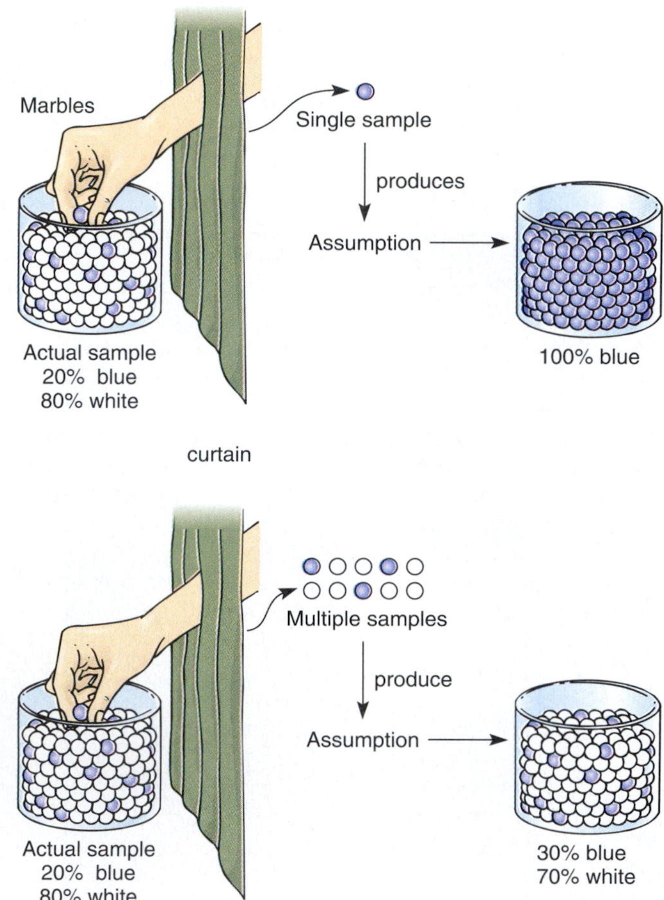

FIGURE 2–7 Sampling error. Taking a single sample can result in sampling error. The greater the number of samples we take of an unknown, the more likely we are to make valid assumptions about it.

At the time, telephones were owned mainly by wealthy people. Because they were not the only ones who voted, the election did not turn out as predicted.

Bias on the part of the scientist can also affect the kinds of conclusions drawn even from excellent, representative data. For example, we could interpret all data in terms of the hypothesis we wish to prove, ignoring or "explaining away" contrary evidence. Needham and his followers had to dismiss a great number of observations that contradicted spontaneous generation, while focusing exclusively on one that seemed to favor it. More skepticism of his own work and a more open mind might have secured for him a far higher place in the history of science. Needham's opponents were skeptical but not really biased. They found an explanation only after repeating his experiments and looking very carefully at the results.

New Information May Be Used to Modify Accepted Theory

A hypothesis supported by a large body of observations and experiments becomes a **theory.** A good theory relates facts that previously appeared to be unrelated and that could not be explained on common ground. A good theory grows as additional facts become known.

The germ theory of infectious disease, for example, incorporated biogenesis, microscopy, mathematics, chemistry, and much else. It does not merely state that infectious disease is caused by microbes but tells us much about how diseases spread and how they can be prevented or cured. Yet at the time of Pasteur and Koch, the germ theory of disease was no more than an elaborate hypothesis. It has been promoted to the status of theory by consensus among scientists.

A theory that, over a long period of time, has yielded true predictions and is thus almost universally accepted is referred to as a scientific **principle.** The term **law** is sometimes used for a principle judged to be of great basic importance as is, for instance, the law of gravity or the law of biogenesis.

SCIENCE HAS ETHICAL DIMENSIONS

It is important for researchers to communicate their results to one another, so that all of their observations can be brought together and compared. In this way, the equivalent of lifetimes of study devoted to a question can be accumulated in a relatively brief time. This is one of the most important reasons for the existence of scientific journals and societies.

Researchers who publish their work in scientific journals describe their experiments in sufficient detail to be independently performed by others. This permits objective observers to detect errors or bias in the original study and helps to guard against the occasional odd result that occurs as a result of random or uncontrolled factors as well as those that may be tainted by unethical dishonesty on the part of the original researcher.

Scientific investigation depends on commitment to such practical ideals as truthfulness and the obligation to communicate results. Honesty is particularly important in science. Consider the great (although hopefully temporary) damage that is done whenever an unprincipled or even desperate researcher (whose career might depend on publication of a research study) knowingly disseminates false data. If the deception is not uncovered, researchers might devote many thousands of dollars or hours of precious professional labor to futile lines of research inspired by untrue reports. (See Focus On the Process of Science:

Piltdown Man pp. 30–31.) Fortunately, the scientific process tends to be self-correcting because it leads to the consistent use of itself. Sooner or later, someone's experimental results are bound to cast doubt on dishonest data.

Scientists face many important ethical issues such as the ethics of human or even animal experimentation, military weapons research, and applications of genetic engineering. Many such ethical issues will be raised throughout this book. As most scientists would agree, no occupation, profession, or field of study is isolated from social concerns, including science.

Chapter Summary

I. Science is a systematic process of investigating the universe and organizing the knowledge gained into a logical structure.

II. We can summarize the method of science:
   A. Make careful observations.
   B. Recognize and state the problem.
   C. Develop a hypothesis.
   D. Make a prediction that can be tested.
   E. Design and perform experiments (or otherwise make observations) to test the prediction.
   F. Use the results to support or falsify the hypothesis, and to develop a theory.

III. Logic is a means of drawing conclusions that are consistent with the evidence used to support them.
   A. Deductive and inductive modes of logic are both employed in the scientific process.
   B. Deduction progresses from a general principle to a specific conclusion. Deduction adds nothing new to knowledge, although it can make relationships among data apparent.
   C. Induction progresses from specific observations to general conclusions; it produces new knowledge and generalizations.
   D. The scientific process tests the proposals of induction by deducing their consequences and testing these by observation and experiment.

IV. Scientists make careful observations and use their creativity and originality to make meaning of what they see. They question what they see and then recognize and state the problem.

V. "Good" hypotheses are reasonably consistent with known facts and generate definite predictions that can be tested.
   A. Total disproof of a hypothesis is rarely possible, and even well-supported hypotheses retain some uncertainty.
   B. Unfalsifiable hypotheses are impossible to disprove.

VI. Theories are usually broadly conceived, logically coherent, and very well supported. Principles are theories that have been almost universally accepted. The term law is sometimes used for principles judged to be of great importance.

VII. Science is subject to ethical constraints.

Selected Key Terms

biogenesis, p. 25
control group, p. 25
deduction, p. 27
experiment, p. 25
experimental group, p. 25
hypothesis, p. 28
induction, p. 27
law, p. 32
principle, p. 32
scientific method, p. 27
spontaneous generation, p. 25
theory, p. 32

Post-Test

1. The principle that life always arises from life is known as ________________.

2. In ________________ reasoning, one proceeds from general principles to specific conclusions.

3. ________________ error occurs because we cannot observe all cases of what we are studying.

4. ________________ reasoning begins with specific examples or facts and seeks to discover a general rule or principle.

5. A/an ________________ is a proposed explanation.

6. Predictions can be tested by observation and ________________.

7. A/an ________________ is a hypothesis supported by many observations and experiments; it relates facts that previously appeared to be unrelated.

8. A/an ________________ is a means of generating results so that a hypothesis can be tested.

9. When Redi sealed containers of decaying meat (a) maggots appeared on the meat (b) no flies appeared on the meat (c) flies eventually appeared on the meat (d) answers (a) and (c) are correct (e) answers (a) and (b) are correct.

10. If flies are not spontaneously generated on decaying meat and are produced only by living things, then other insects must also be produced only by living things. This reasoning is an example of (a) deductive reasoning (b) inductive reasoning (c) predictive reasoning (d) answers (a) and (c) are correct (e) none of the preceding answers is correct.
11. Pasteur's experiments showed that (a) microbes could spontaneously generate (b) even microbes obeyed the principle of biogenesis (c) when microbes were prevented from entering a flask of broth, none developed in the flask (d) answers (a) and (c) are correct (e) answers (b) and (c) are correct.
12. Fleming was able to discover penicillin because (a) the penicillin mold first evolved in his culture dish (b) he observed that bacterial colonies did not grow near the mold (c) he thought that what he observed was meaningful (d) answers (a), (b), and (c) are correct (e) answers (b) and (c) are correct.
13. In the study by A. T. McLellan and his colleagues cited in this chapter (a) the group of addicts that received only methadone served as the control group (b) the group that received methadone and also counseling served as a control group (c) the results allowed the researchers to formulate a new theory (d) answers (a), (b), and (c) are correct (e) answers (b) and (c) are correct.

Review Questions

1. Contrast induction and deduction.
2. Compare hypothesis, theory, and law.
3. What are the advantages of experimentation in science?
4. What is a control group and why is it important?
5. List the sequence of steps used in the scientific method. Use a new example to illustrate the process.

Thinking Critically

1. Describe how Angert's discovery of giant bacteria illustrates the scientific process.
2. Medical and behavioral researchers sometimes deliberately avoid knowing which persons in a study have received a particular treatment. Why might this be desirable?
3. What parts of the scientific method are inductive? Deductive?
4. Critique the following hypothesis: Invisible green imps that cannot be detected by any known means are responsible for hurling objects to the ground when one lets go of them.
5. Does society have the right to govern scientific research? Explain. If so, explain what kinds of research should be so governed and how it should be done.
6. Give four examples of ethical problems in recent scientific research as reported in the media.
7. Many studies have been performed on the relationships among smoking, cancer, and numerous other diseases. All responsible scientists familiar with the issue are convinced that smoking and other tobacco use constitute a very serious health menace. Yet tobacco companies continue to contest these findings. Part of the difficulty has been that some kind of case, however implausible, could be made that "something else," alcohol consumption by smokers, for instance, could have been responsible for the observed facts rather than smoking itself. Propose an experiment that could settle the issue beyond any doubt. Would there be ethical objections to such an experiment?

Recommended Readings

Borg, W. R., and M. D. Gall, *Educational Research: An Introduction,* 5th ed. New York, Longman, 1989. The first chapter contains an extensive discussion of the philosophy of science.

Moore, J. A., *Science as a Way of Knowing: The Foundations of Modern Biology.* Harvard University Press, Cambridge, Mass, 1993. An account of scientific thought as related to the history of modern biology.

Tobias, P. V., "Piltdown unmasked," *The Sciences,* Jan–Feb 1994. An account of how the perpetrators of the scientific hoax of the century were exposed 80 years after they planted their forgery in the ground.

Tudor, A., "Seeing the worst side of science," *Nature,* Vol. 340, August 24, 1989, 589–592. A summary of how horror movies reflect public anxieties about science.

Wilson, D., *In Search of Penicillin,* New York, Knopf, 1976. A thorough and fascinating account of the role played by Fleming and others in the discovery and development of penicillin.

PART 2 The Organization of Life

Acid rain and fog have contributed to widespread forest destruction. *This damaged spruce-fir forest was photographed at an elevation of 6000 feet along the Blue Ridge Parkway in North Carolina. (CubKahn/TERRAPHOTOGRAPHICS)*

Acid Rain: What Goes Up May Rain Back Down on Us

Normal precipitation, such as rain, sleet, and snow, is a slightly acidic solution (pH about 5.6), which means that it is neither highly acidic like lemon juice nor highly basic (alkaline) like oven cleaner. (Acid and basic solutions are discussed later in this chapter.) Acidity and basicity are measured in pH units on a scale of 0 to 14, with 7 being neutral. The lower the pH, the more acidic the solution; the higher the pH, the more basic the solution. Precipitation is slightly acidic because some of the carbon dioxide and water in the atmosphere combine to form a weak acid, carbonic acid.

Acid rain (and other forms of acid precipitation) can result from natural events such as volcanic activity, which adds large amounts of sulfuric acid to the atmosphere. In recent years, industrial activity has become an important contributor to acid rain due to the release of sulfur oxides from power plants and factories. Motor vehicles are also contributors because they release nitrogen oxides. These pollutants combine with water in the atmosphere to form dilute solutions of sulfuric and nitric acids.

During the past few decades, rain has become increasingly acidic. In fact, rain, in parts of the United States and western Europe, is sometimes as acidic as vinegar! Acid rain corrodes monuments, automobiles, and buildings, and poses a serious threat to the environment. It adversely affects soil, forests, lakes, rivers, and the inhabitants of these ecosystems. Living things are very complex chemical systems. The specific organization and precise interaction of their chemical components are essential to maintaining life. Chemical changes in an organism's environment can cause serious damage and even death. On an ecological level, acid rain has affected food chains and upset the delicate balance of entire ecosystems.

Acid rain has been linked to the decline of fish populations in many lakes and rivers and to widespread damage to forests. Due to acid rain, fish no longer swim in many lakes in the northeastern United States and in eastern Canada. About half of the high altitude lakes in the Adirondack Mountains of New York have been acidified and have no fish. Thousands of lakes in Sweden and Norway no longer support fish populations.

Investigators have studied how acid rain changes the chemistry of soils, interfering with the development of plant roots and with their uptake of nutrients and water. Some essential minerals such as calcium and magnesium wash out of acidic soil. Other minerals, including heavy metals such as aluminum, dissolve in acidic soil, accumulate, and may then be absorbed by plant roots in toxic amounts. These metals can be passed along in the food web.

What can we do about the problem of acid rain? Before tough environmental restrictions were imposed in 1985, the United States alone was adding more than 40 million tons of sulfur oxides and nitrogen oxides to the atmosphere each year. A practice that has increased and complicated the problem of acid rain is the construction of tall smokestacks. The intent of industrialists that designed plants with such smokestacks was to release the sulfur-rich smoke high up in the atmosphere where the winds would disperse and dilute it. Unfortunately, the winds only blow the pollutants to some other place, sometimes hundreds of miles away. For example, emissions from the eastern and midwestern United States produce between 50% and 75% of the acid rain that contaminates New England and southeastern Canada.

Export of acid rain to areas far from the source of pollution presents a vexing political problem. Those who export the pollutants are not eager to pay for what they see as some other community's problem. And preventing acid rain is expensive. The pollutants in emissions can be captured and removed, but estimated costs of installing and maintaining the necessary scrubbers in smokestacks in the United States alone are about $4 to $5 billion a year.

Clean air legislative acts passed by the United States in 1985 and 1990 are resulting in a substantial decrease in acid rain. Those concerned about preserving our ecosphere must continue to support legislation that demands industrial responsibility. Yet, they must also be willing to pay for this improved environmental quality.

The Chemistry of Life: Atoms, Molecules, and Reactions

3

Learning Objectives

After you have studied this chapter you should be able to:

1. Diagram the basic structure of the atoms of a given element in accordance with the conventions presented in this chapter, showing the position of protons, neutrons, and electrons.
2. Identify the biologically significant elements by their chemical symbols, and summarize the main functions of each in living organisms.
3. Interpret simple chemical formulas, structural formulas, and equations.
4. Define the term electron orbital and relate orbitals to energy levels; relate the number of valence electrons to the chemical properties of the element.
5. Distinguish between the types of chemical bonds, and give the characteristics of each type.
6. Contrast oxidation and reduction, and explain how these processes are linked.
7. Distinguish between inorganic and organic compounds, and identify several biologically important inorganic compounds.
8. Describe the properties of water molecules and their importance to living things.
9. Contrast acids and bases, and describe how buffers help minimize changes in pH.
10. Describe the composition of a salt, and give reasons why salts are important in living systems.

Key Concepts

- ☐ The atom is the smallest unit of a chemical element that retains the characteristic properties of that element.
- ☐ Valence electrons, the electrons in the outermost energy level of an atom, help determine the chemical properties of the atom.
- ☐ Atoms join by chemical bonds to form molecules and compounds.
- ☐ Chemical bonds are forces of attraction that involve the sharing of electrons by atoms or the donation of electrons from one atom to another. Three basic types of chemical bonds are covalent bonds, ionic bonds, and hydrogen bonds.
- ☐ Life depends on the unique properties of water; many of these properties are related to the polar nature of its molecules and their tendency to form hydrogen bonds with one another.
- ☐ Acids, bases, and salts are common ionic compounds. When dissolved in water, acids dissociate to yield hydrogen ions (protons). Bases accept protons; many bases dissociate in water, producing hydroxide ions.

- ☐ Cells use buffers to maintain appropriate pH by resisting changes in hydrogen ion concentration when acids or bases are added.
- ☐ Salts provide mineral ions essential for fluid balance, nerve and muscle function, and many other processes necessary to life.

MATTER IS MADE UP OF CHEMICAL ELEMENTS

All matter is composed of chemical **elements,** substances that cannot be broken down into simpler substances by chemical reactions. A **chemical reaction** may simply be defined as a chemical change. In a chemical reaction, substances may be joined or broken down to form new substances.

The matter of the universe is composed of 92 naturally occurring elements, ranging from hydrogen, the lightest, to uranium, the heaviest. Despite their great diversity, the chemical composition and metabolic processes of all living things are remarkably similar. About 98% of an organism's mass[1] is composed of only six elements: oxygen, carbon, hydrogen, nitrogen, calcium, and phosphorus. Approximately 14 other elements are consistently present in living things, but in smaller quantities. Some of these, such as iodine and copper, are known as **trace elements** because they are present in such minute amounts. The chemical elements found in living systems cycle between them and the nonliving environment.

Instead of writing out the name of each element, chemists use a system of abbreviations called **chemical symbols,** which are usually comprised of the first one or two letters of the English or Latin name of the element. For example, O is the symbol for oxygen, C for carbon, Cl for chlorine, N for nitrogen, and Na for sodium (its Latin name is *natrium*). Table 3–1 lists the elements that make up the human body, indicates their chemical symbols, and explains the biological importance of each.

ATOMS ARE THE BASIC PARTICLES OF ELEMENTS

Imagine a bit of gold being divided into smaller and smaller pieces. The smallest possible particle of gold that could be obtained would be an atom of gold. The **atom** is the smallest subdivision of an element that retains the characteristic chemical properties of that element. The subdivision of any kind of matter ultimately yields atoms. This is true no matter what physical state matter may assume—solid, liquid, or gas. Atoms are almost unimaginably small, much smaller than the tiniest particle visible under a light microscope. By scanning tunneling microscopy, with magnification as many as five million times, researchers have been able to photograph some of the larger atoms, such as those of uranium.

Atoms Are Composed of Subatomic Particles

An atom is composed of smaller components called subatomic particles. Although physicists have discovered a number of subatomic particles, we need consider only three types: protons, neutrons, and electrons. Each **proton** has one unit of a positive electrical charge; **neutrons** are uncharged particles with about the same mass as protons. Protons and neutrons make up almost all of the mass of an atom and are concentrated in a central area called the **atomic nucleus.** Each **electron** has one unit of a negative electrical charge and an extremely small mass (only about 1/1800 of the mass of a proton). The electrons, as we will see, spin about in the space surrounding the atomic nucleus (Fig. 3–1).

Each kind of element has a fixed number of protons in the atomic nucleus. This number, called the **atomic number,** is written as a subscript to the left of the chemical symbol. Thus, ${}_1H$ indicates that the hydrogen nucleus contains one proton and ${}_8O$ indicates that the oxygen nucleus has eight protons. The atomic number determines the chemical identity of the atom. Any atom with one proton is a hydrogen atom.

The total number of protons plus neutrons in the nucleus is termed the **atomic mass.** It is indicated by a superscript to the left of the chemical symbol. An atom of sodium, ${}^{23}_{11}Na$, has 11 protons; its atomic mass is 23, with 11 protons and 12 neutrons. The common form of an oxygen atom, with eight protons and eight neutrons in its nucleus, has an atomic number of 8 and an atomic mass of 16. It is indicated by the symbol ${}^{16}_{8}O$ (see Focus On the Process of Science: Using Radioisotopes in Biology and Medicine).

[1] For convenience, we will consider weight and mass as equal, although this is not actually true. Mass does not depend on the force of gravity, but weight does. Thus, a person on the moon has the same *mass* as a person on Earth, but, because of the moon's lower gravity, his or her body *weight* is less.

TABLE 3–1
Elements That Make Up the Human Body

Name	*Chemical symbol*	*Approximate composition of human body by mass (%)*	*Importance or function*
Oxygen	O	65	Required for cellular respiration; present in most organic compounds; component of water
Carbon	C	18	Forms backbone of organic molecules; can form four bonds with other atoms
Hydrogen	H	10	Present in most organic compounds; component of water
Nitrogen	N	3	Component of all proteins and nucleic acids
Calcium	Ca	1.5	Structural component of bones and teeth; important in muscle contraction, conduction of nerve impulses, and blood clotting; also in cell walls (structural component) of plants
Phosphorus	P	1	Component of nucleic acids; structural component of bone; important in energy transfer
Potassium	K	0.4	Principal positive ion (cation) within cells; important in nerve function; affects muscle contraction
Sulfur	S	0.3	A component of most proteins
Sodium	Na	0.2	Principal positive ion in interstitial (tissue) fluid; important in fluid balance; essential for conduction of nerve impulses
Magnesium	Mg	0.1	Needed in blood and body tissues; a component of many important enzyme systems; component of chlorophyll in plants
Chlorine	Cl	0.1	Principal negative ion (anion) of interstitial fluid; important in fluid balance
Iron	Fe	Trace amount	Component of hemoglobin, myoglobin, and certain enzymes
Iodine	I	Trace amount	Component of thyroid hormones

Other elements, found in very small amounts in the body (the trace elements), include manganese (Mn), copper (Cu), zinc (Zn), cobalt (Co), fluorine (F), molybdenum (Mo), selenium (Se), and a few others.

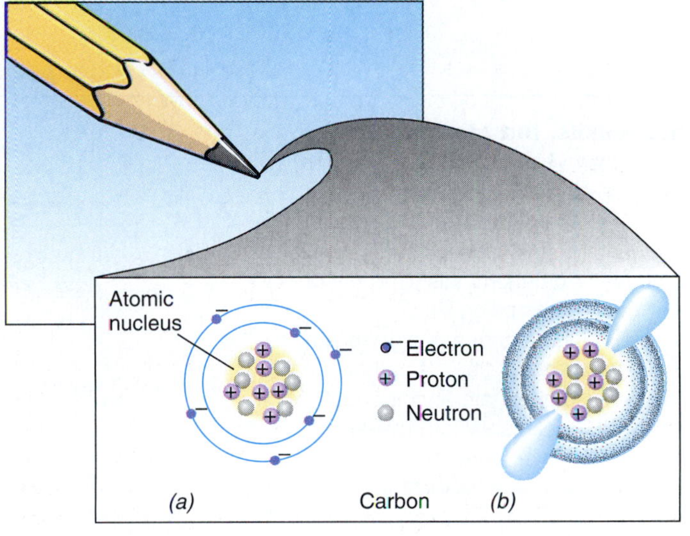

FIGURE 3–1 Two ways of representing an atom. (***a***) Bohr model of a carbon atom. Although the Bohr model is highly simplified and not very accurate, it is a convenient way to visualize the structure of an atom. Each electron is represented by a small sphere. The circles represent the average location of the electrons. (***b***) Electron clouds surrounding a carbon atom. Dots represent the probability of an electron being in that particular location at any given moment.

Electrons Are Located in Energy Levels Outside the Nucleus

An atom contains the same number of electrons and protons so that the negative charges of the electrons equal the positive charges of the protons. Thus, an atom has no net charge. Some kinds of chemical combinations (and certain other conditions) change the number of electrons,

(Text continued on page 42)

FOCUS ON *The Process of Science*

USING RADIOISOTOPES IN BIOLOGY AND MEDICINE

Most elements are made up of atoms with different numbers of neutrons, and thus, different masses. Such atoms are called **isotopes.** The isotopes of the same element have the same number of protons and electrons; only the number of neutrons vary. The three isotopes of hydrogen, ^1_1H, ^2_1H, and ^3_1H, contain zero, one, and two neutrons, respectively. Elements usually occur in nature as a mixture of isotopes.

All isotopes of a given element have essentially the same chemical characteristics. Some isotopes with excess neutrons are unstable and tend to break down, or decay, to a more stable isotope (usually of a different element). Such isotopes are known as **radioisotopes,** since they emit high-energy radiation when they decay. Radioactive decay occurs spontaneously and at a characteristic rate for each type of radioisotope. The time it takes for half the nuclei in any given amount of a radioactive element to decay into another element is its **half-life.** The half-life is not affected by environmental factors such as temperature or pressure. Because the rate of decay is so reliable, scientists use radioisotope content to date fossils (see figure; see also Chapter 17).

Radioisotopes like ^3H (tritium) and ^{14}C have been extremely valuable research tools in biology and are useful in medicine for both diagnosis and treatment. The development of technology for using radioisotopes has been a cooperative effort among scientists from several disciplines including physicists, engineers, chemists, biologists, and anthropologists.

Despite the difference in the number of neutrons, organisms chemically treat all isotopes of a given element the same. The reactions of a chemical—a fat, a hormone, a drug—can be followed in the body by labeling the substance with a radioisotope, such as tritium or carbon-14 (see figure). This is done by replacing one of the nonradioactive atoms in a small sample of molecules of that substance with its radioactive equivalent. Radioactivity can then be traced with an instrument such as a scintillation counter, which detects the location of the radioactive atoms, or molecules containing them. For example, the active component in marijuana (tetrahydrocannabinol) has been labeled and administered intravenously. By measuring the amount of radioactivity in the blood and urine at successive intervals, experimenters determined that this compound remains in the blood, and products of its metabolism remain in the urine for several weeks.

Because radiation from radioisotopes can interfere with cell division, isotopes such as ^{60}cobalt and ^{226}radium are used in the treatment of cancer (a disease characterized by rapid cell division). Radioisotopes are also used to test thyroid gland function. In evaluating thyroid gland function, a small amount of the radioisotope ^{123}iodine can be injected into a patient's blood. The thyroid gland is the only organ in the body that takes up iodine. A radioisotope scan of the thyroid using a scintillation counter can indicate whether the thyroid gland is of normal size or whether it is diseased.

Radioisotopes are used in PET (positron-emission tomography)

Anthropologists use radioisotope content not only to date fossils, but also to study lifestyles. (***a***) The skeleton of this 11th-century inhabitant of a South African village posed an anthropological puzzle. Physically, the man's skeleton was different from those of the other villagers, suggesting that he was not a native of the area. When the skeleton was analyzed for isotopes, its carbon-12 to carbon-14 ratio was found to be similar to that of other skeletons from the same village. Since different kinds of plants incorporate different proportions of isotopes into the food produced from them, this similarity of isotope content indicates that these individuals all ate the same foods. Thus, anthropologists concluded that this man had probably spent most of his life in the village after migrating there from a distant region. (***b***) Two isotopes of carbon: carbon-12 and carbon-14. Carbon-12 accounts for about 99% of carbon atoms. Carbon-14 is relatively rare in nature. (***c***) PET scans, which use radioisotopes, are an important medical and research tool. PET stands for positron-emission tomography. PET scan showing normal, schizophrenic, and depressed brains. (***d***) Patient being positioned in a PET brain scanner.
(*a,* Nikolaas J. van der Merwe, *American Scientist* 70:596–606, 1982; *c,* NIH/SS/PhotoResearchers, Inc.; *d,* Hank Morgan/Photo Researchers, Inc.)

scans to provide diagnostic information about metabolic function. The patient is injected with glucose that has been labeled with a radioisotope of fluorine, ^{18}F. Cells in certain tissues such as brain tissue absorb glucose rapidly. The patient is placed in a PET scanner, which detects the location of radioactive decay. Computer analysis provides images of slices of the body that indicate variations in metabolic activity (see figure).

(a)

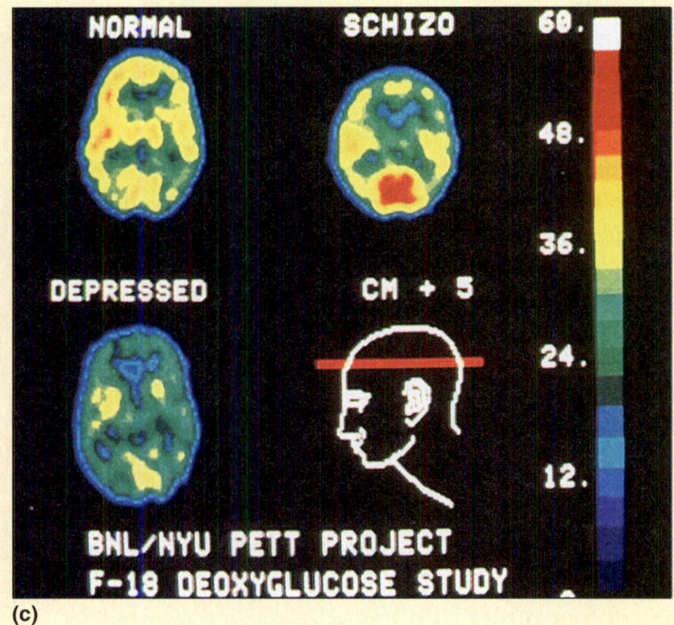

(c)

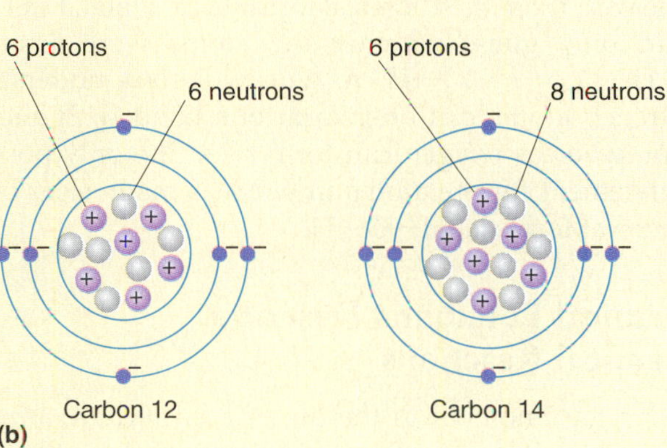

(b)

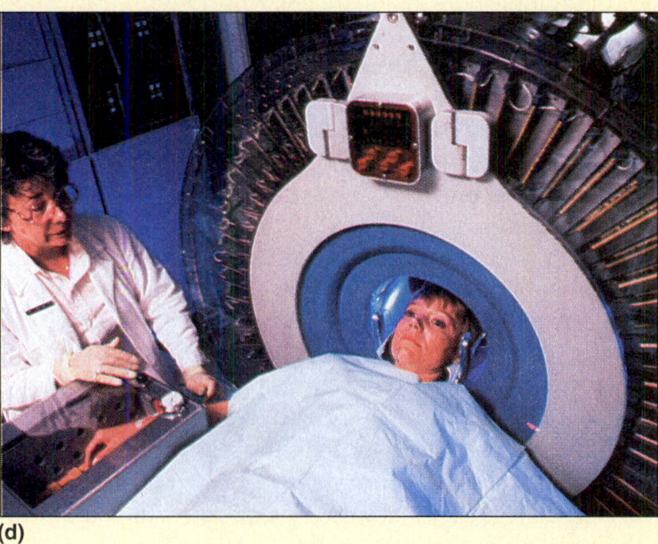
(d)

but *chemical reactions do not affect anything in the atomic nucleus.* (Although nuclear reactions can do so, we may ignore this reservation for the time being, since living things are not nuclear powered!) Because electrons and protons have equal though opposite charges, an uncombined atom is electrically neutral as a whole.

Because electrons are negatively charged, they are attracted to the positively charged protons in the atomic nucleus (opposite charges tend to attract). At the same time, electrons repel one another (like charges tend to repel one another). These considerations help determine the locations of electrons.

The way electrons are arranged around an atomic nucleus is referred to as the atom's **electron configuration.** Knowing the locations of electrons enables chemists to predict how atoms can combine to form different types of chemical compounds.

An atom may have several **electron shells,** or **energy levels,** where electrons are located. An electron shell contains one or more **orbitals,** spaces within the shell where electrons are most likely to be found. Each orbital can hold up to two electrons. Electrons located in the same electron shell have the same energy. The electron shell with the lowest energy level is the one closest to the nucleus. Because this first shell has only one orbital, only two electrons can occupy it. The second shell and outer shells are complete when they have eight electrons, two in each of four orbitals. Atoms with more than one electron shell are most stable when they have a complete outer shell of eight electrons.

The atomic structures of some elements that are important in biological systems are shown in Figure 3–2. A more complete periodic table of the elements may be found in Appendix 3. Although the simple diagrams, called Bohr models, of electron configuration shown in the figure are helpful in understanding atomic structure, they are highly simplified. Electrons do not circle the nucleus in fixed concentric pathways. Instead, they whirl around the nucleus, moving first close to it and then farther away. Orbitals represent the space where electrons are most probably found and they may be represented by spherical, dumbbell-shaped, or more complex three-dimensional coordinates (Fig. 3–3). One way of illustrating an atom is to represent each occupied orbital as an electron cloud as in Figure 3–1. The density of the shaded areas is proportional to the probability that an electron is present there at any given moment.

Moving a negatively charged electron farther away from the positively charged nucleus requires energy. An electron can jump from one level to the next, *but* it cannot stop in the space in between. To move an electron from one level to the next, the atom must absorb a discrete packet of energy known as a **quantum,** which contains just the right amount of energy needed for the transition—no more and no less. The term *quantum leap* is used in everyday language to indicate a sudden move from one level to another. When an electron gives up energy, it sinks back to a lower energy level in an orbital nearer the nucleus.

ATOMS FORM MOLECULES AND COMPOUNDS

Two or more atoms may combine chemically to form a **molecule.** When two atoms of oxygen combine, a molecule of oxygen is formed. Atoms of *different* elements can combine to form chemical compounds. **Chemical compounds** consist of molecules made of two or more *different* elements combined in a fixed ratio. Water is a chemical compound consisting of two atoms of hydrogen combined with one atom of oxygen. A molecule of water is the smallest unit of the compound water.

Chemical Formulas Describe Chemical Compounds

A **chemical formula** is a shorthand method for describing the chemical composition of a molecule or compound. Chemical symbols are used to indicate the types of atoms in the compound, and subscript numbers are used to indicate the number of each type of atom present. The chemical formula for molecular oxygen, O_2, indicates that this molecule consists of two atoms of oxygen. This formula distinguishes it from another form of oxygen, ozone, which has three atoms and is written O_3. The chemical formula for water, H_2O, indicates that each molecule consists of two atoms of hydrogen and one atom of oxygen. (Note that when a single atom of one type is present, it is not necessary to write 1; chemists do *not* write H_2O_1.)

Another type of formula is the **structural formula,** which shows not only the types and numbers of atoms in a molecule but also their arrangement. In any particular chemical compound the atoms are arranged in the same way. From the chemical formula for water, H_2O, you could only guess whether the atoms were arranged H—H—O or H—O—H. As you will soon understand, hydrogen atoms can only form one bond with another atom, whereas oxygen can form two. The only possible arrangement for the atoms in water is made clear by its structural formula, H—O—H.

Chemical Equations Describe Chemical Reactions

During any moment in the life of a bacterium, a maple tree, or any other organism, many complex, highly orga-

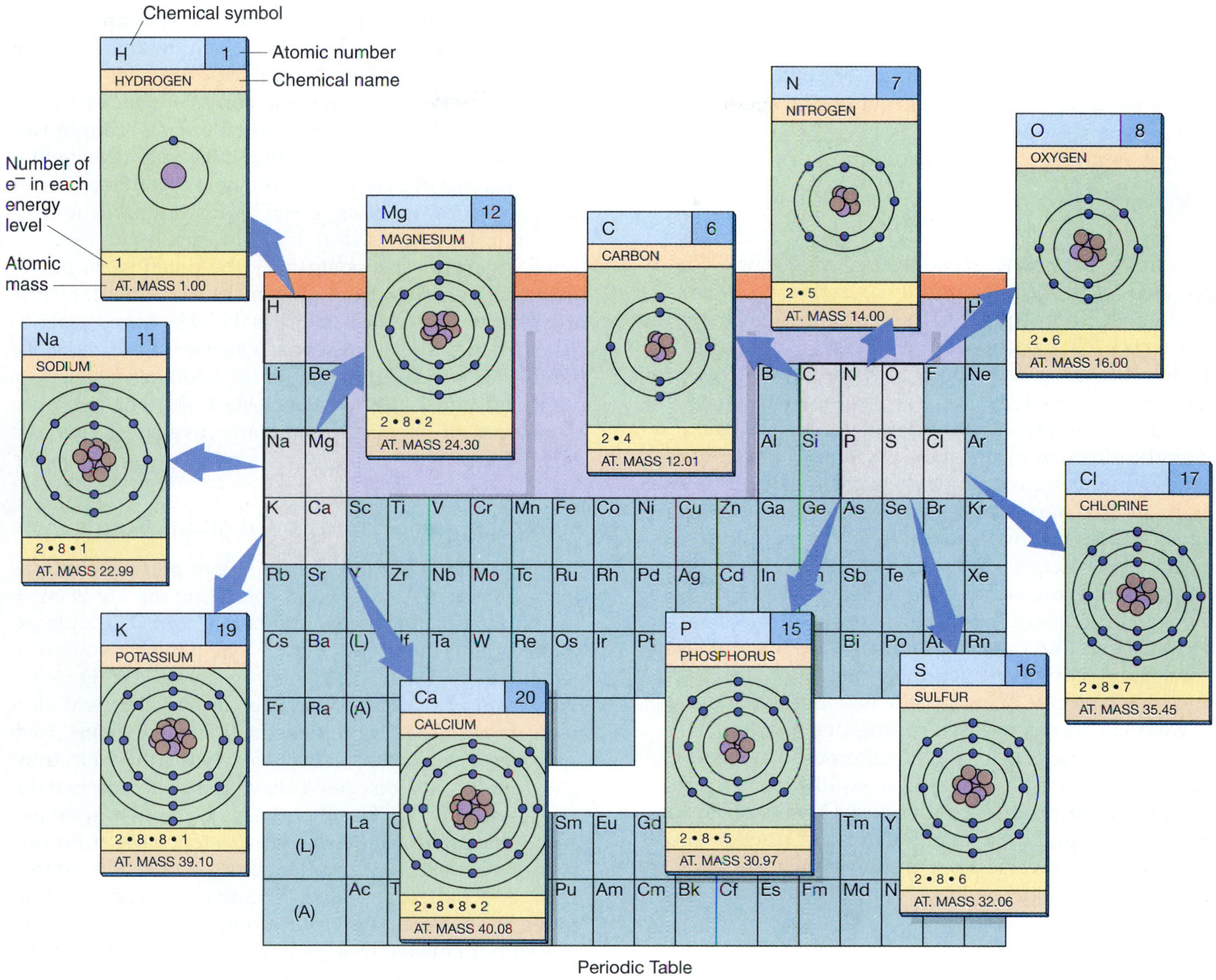

Periodic Table

FIGURE 3–2 Periodic table of the elements showing Bohr models of some biologically important atoms. The periodic table is a chart of atom classification. When atoms are placed in order of atomic number, they fall into groups of similar elements.

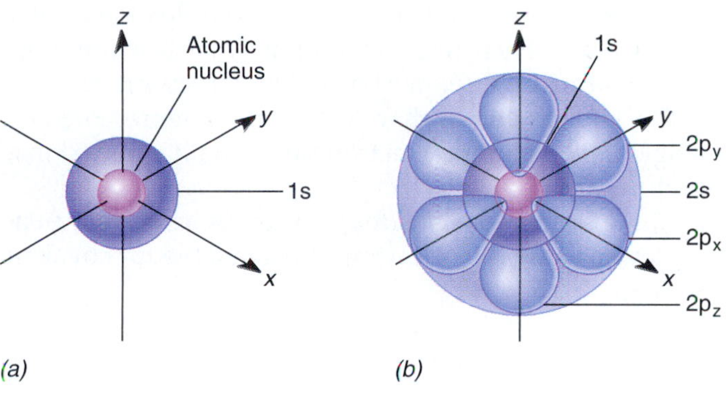

FIGURE 3–3 Orbitals are designated by numbers and letters. They represent the space where electrons are most likely to be located as they whirl around the nucleus in three dimensions. The nucleus (purple) is at the intersection of the three imaginary axes (*x*, *y*, and *z*) of the atom. The first, or lowest, energy level (electron shell) (***a***) is known as the 1*s* orbital; it can contain at most two electrons. (***b***) An atom with two energy levels. The second energy level contains one spherical-shaped orbital, 2*s*, and 3 dumbbell-shaped (2*p*) orbitals at right angles to one another. The 2*p* orbitals have equal energy.

nized chemical reactions are taking place. The chemical reactions that occur between atoms and molecules—for instance, between methane (natural gas) and oxygen—can be described on paper by means of **chemical equations:**

$$\underset{\text{Methane}}{CH_4} + \underset{\text{Oxygen}}{2O_2} \longrightarrow \underset{\text{Carbon dioxide}}{CO_2} + \underset{\text{Water}}{2H_2O} + \text{Energy}$$

As shown, methane, produced during some decay processes, is broken down in this reaction.

In a chemical equation, the **reactants,** which are the substances that participate in the reaction, are written on the left side of the equation. The **products,** which are the substances formed by the reaction, are written on the right side. The arrow means *produces,* or *yields,* and indicates the direction in which the reaction tends to proceed.

The number preceding a chemical symbol or formula indicates the number of atoms or molecules reacting. Thus $2O_2$ means two molecules of oxygen and $2H_2O$ means two molecules of water. The absence of a number indicates that only one atom or molecule is present. Thus, the equation describing the breakdown of methane can be translated into ordinary language as, "One molecule of methane reacts with two molecules of oxygen to produce one molecule of carbon dioxide and two molecules of water. Energy is released during this reaction."

In some cases, the reaction will proceed in the reverse direction as well as forward; at **equilibrium** a certain amount of the product continuously breaks up to form the reactants, and the rate of the forward reaction equals the rate of the reverse reaction. Reversible reactions are indicated by double arrows:

$$\underset{\text{Nitrogen}}{N_2} + \underset{\text{Hydrogen}}{3H_2} \rightleftharpoons \underset{\text{Ammonia}}{2NH_3}$$

In this example, the arrows are drawn different lengths to indicate that, when the reaction is at equilibrium, there is more product than reactant.

ATOMS COMBINE BY FORMING CHEMICAL BONDS

The chemical behavior of an atom is determined mainly by the number and arrangement of electrons in the *outermost* energy level (electron shell). In a few elements, called the *noble gases,* the outermost shell is filled. These elements are chemically inert, meaning that they will not readily combine with other elements. Two such elements are helium, with two electrons (a complete outer shell), and neon, with ten electrons (a complete inner shell of two and a complete outer or second shell of eight).

The electrons in the outer energy level of an atom are referred to as **valence electrons.** When the outer shell of an atom contains fewer than eight electrons, the atom tends to lose, gain, or share electrons to achieve a complete outer shell (of eight for most elements; zero or two for the two lightest elements that have only one shell).

The atoms of a compound are held together by forces of attraction known as **chemical bonds.** A certain amount of chemical energy is stored within each bond. In every living organism, this energy has the potential to power biological processes from movement to the manufacture of new materials. Bond energy is the energy necessary to break a bond. The atoms of each element form a specific number of bonds with the atoms of other elements—a number dictated by the valence electrons. The two principal types of chemical bonds are covalent bonds and ionic bonds.

Electrons Are Shared in Covalent Bonds

Covalent bonds involve the sharing of pairs of electrons between atoms. A compound consisting mainly of covalent bonds is called a **covalent compound.** A simple example of a covalent bond is the one joining two hydrogen atoms in a molecule of hydrogen gas, H_2 (Fig. 3–4*a*). Each atom of hydrogen has one electron, but two electrons are required to complete the first energy level. Each hydrogen atom has the same capacity to attract electrons (electronegativity, discussed shortly), so one does not donate an electron to the other. Instead, the two hydrogen atoms share their single electrons so that each of the two electrons is attracted simultaneously to the two protons in the two hydrogen nuclei. The two electrons are thus under the influence of *both* atomic nuclei: they whirl around both nuclei, thus joining the two atoms together.

A simple way of representing the electrons in the outer shell of an atom is to place dots around the chemical symbol of the element to portray the electrons. In a water molecule, two hydrogen atoms are covalently bonded to an oxygen atom:

$$H\cdot + H\cdot + \cdot\ddot{\underset{\cdot\cdot}{O}}\cdot \longrightarrow H:\ddot{\underset{\cdot\cdot}{O}}:H$$

Oxygen has six valence electrons; by sharing electrons with two hydrogen atoms, it completes its outer level of eight. Each hydrogen atom obtains a complete outer level of two. (Note that in the structural formula H—O—H, each pair of shared electrons is represented by a single line. Unshared electrons are usually omitted in a structural formula.)

The carbon atom has four electrons in its outer energy level. These four electrons are available for covalent bonding:

$$\cdot\dot{\underset{\cdot}{C}}\cdot$$

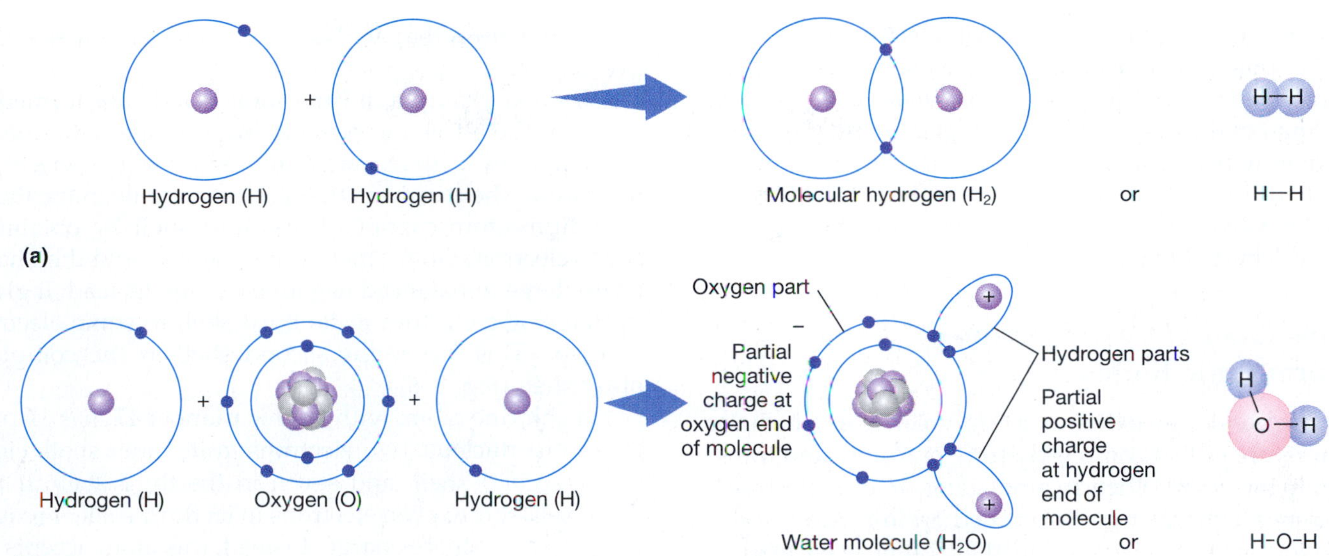

FIGURE 3–4 Formation of covalent bonds. (***a***) Two hydrogen atoms achieve stability by sharing electrons, thereby forming a molecule of hydrogen. The structural formula shown on the right is a simpler way of representing molecular hydrogen. The straight line between the hydrogen atoms represents a single covalent bond. (***b***) When two hydrogen atoms share electrons with an oxygen atom, a molecule of water is formed. Note that the electrons tend to stay closer to the nucleus of the oxygen atom than to the hydrogen nuclei. This results in a partial negative charge on the oxygen portion of the molecule, and a partial positive charge at the hydrogen end. Thus, the water molecule is a polar covalent compound even though as a whole it is electrically neutral.

When one carbon and four hydrogen atoms share electrons, a molecule of methane (CH_4) is formed. Each line in the structural formula represents a single pair of shared electrons.

```
   H                 H
   ..                |
H : C : H    or   H—C—H
   ..                |
   H                 H
```

Each atom shares its outer-level electrons with the others. This sharing completes the first energy level of each hydrogen atom and the second energy level of the carbon atom.

The nitrogen atom has five electrons in its outer shell:

```
 ..
·N·
 ·
```

When a nitrogen atom shares electrons with three hydrogen atoms, a molecule of ammonia, NH_3, is formed:

```
   ..
H : N : H    or   H—N—H
   ..                |
   H                 H
```

When one electron pair is shared between two atoms, the covalent bond is referred to as a **single bond.** Two shared pairs of electrons comprise a **double bond,** shown in diagrams as a pair of lines resembling an equal sign. Double bonds are especially common in compounds of carbon when carbon atoms are linked to one another:

```
   H   H
   |   |
H—C=C—H
```

Two oxygen atoms may become more stable by forming covalent bonds with one another. Each oxygen atom has six electrons in its outer shell. To become stable, the two atoms share two pairs of electrons, forming a molecule of oxygen gas. Note that the covalent bond formed is a double bond. Some atoms, such as carbon, can even form triple bonds with one another, sharing three pairs of electrons.

Electronegativity is a measure of an atom's attraction for the electrons in covalent bonds. When the atoms in a molecule have similar electronegativity, the electrons are shared equally and the covalent bond is referred to as **nonpolar.** The covalent bonds of the hydrogen and oxygen molecules are nonpolar.

In a covalent bond between two different elements, such as oxygen and hydrogen, the electronegativity of the atoms may be different. If so, electrons will be pulled closer to the atomic nucleus of the element with the greater electron affinity (in this case, oxygen). A covalent bond between atoms of different electronegativity is called a **polar covalent bond.** In a water molecule, the electrons tend to be closer to the nucleus of the oxygen atom than to the nuclei of the hydrogen atoms. Although the water molecule as a whole is electrically neutral, it is polar because each hydrogen atom in the molecule has a partial positive charge and the oxygen atom has a partial negative charge (Fig. 3–4*b*).

Atoms Gain or Lose Electrons to Form Ionic Bonds

An **ionic bond** is an extreme case of polarity in which the electrons are pulled completely from one atom and transferred to the other. When an atom gains or loses electrons, it becomes a charged particle called an **ion.** Atoms with one, two, or three electrons in their outer shell tend to lose electrons to other atoms. When these atoms lose electrons they become positively charged because they have more positively charged protons than negatively charged electrons. Positively charged ions are termed **cations.**

Atoms with five, six, or seven valence electrons tend to gain electrons from other atoms and to become negatively charged ions called **anions.** Cations and anions play essential roles in the transmission of nerve impulses, muscle contraction, and many other life processes (Fig. 3–5). An **ionic compound** is a substance consisting of anions and cations that are bonded together by their opposite charges.

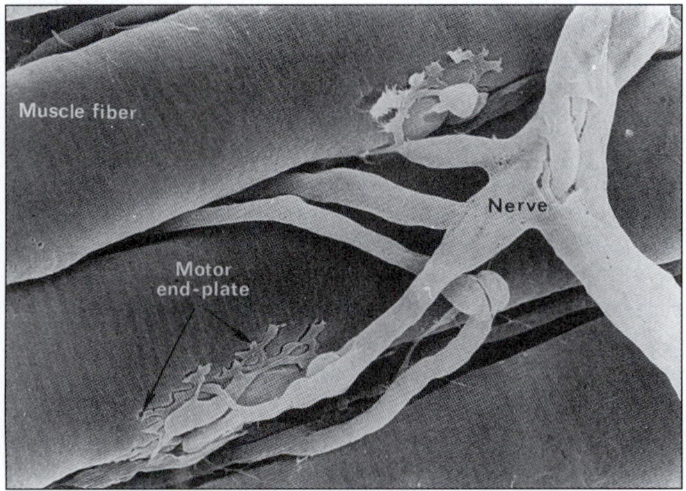

FIGURE 3–5 Sodium, potassium, and chlorine ions are among the ions essential in the conduction of a nerve impulse. This scanning electron micrograph shows a nerve fiber connecting with several muscle cells. The nerve fiber transmits impulses to the muscle cells, stimulating them to contract. The muscle cells are rich in calcium ions, which are essential for muscle contraction. (D. W. Fawcett)

A good example of how ionic bonds are formed is found in the attraction between sodium and chlorine. A sodium atom, with atomic number 11, has two electrons in its inner shell, eight in the second, and one in the third. A sodium atom cannot fill its third shell by obtaining seven electrons from other atoms, for it would then have a very large unbalanced negative charge. Instead, it gives up the single electron in its third shell to some electron acceptor. This leaves the second shell as the complete outer shell (Fig. 3–6).

A chlorine atom, with atomic number 17, has 17 protons in its nucleus, two electrons in its inner shell, eight in the second shell, and seven in the third shell. If the chlorine lost the seven electrons in its third shell, it would have a vast positive charge. Instead, this atom accepts an electron from an electron donor such as sodium to complete its outer third shell.

When sodium reacts with chlorine, its outermost electron is transferred completely to chlorine. The sodium ion now has 11 protons in its nucleus, and 10 electrons circling the nucleus. Its net charge is 1^+. The chloride ion has 17 protons in its nucleus, 18 electrons circling the nucleus, and a net charge of 1^-. These ions attract one another as a result of their opposite charges. They are held together by this electrical attraction, called the ionic bond, to form sodium chloride,[2] common table salt.

When an ionic compound, such as sodium chloride, is in its solid form, considerable energy is required to pull its ions apart. However, ionic compounds have a tendency to **dissociate** (separate) into their individual ions when placed in water.

$$\underset{\text{Sodium chloride}}{NaCl} \xrightarrow{\text{in } H_2O} \underset{\text{Sodium ion}}{Na^+} + \underset{\text{Chloride ion}}{Cl^-}$$

Liquid water is an excellent **solvent,** capable of dissolving many substances. This is because of the polarity of water molecules. The localized partial positive charges (on the hydrogen atom) and the partial negative charges (on the oxygen atom) on each water molecule attract the anions and cations on the surface of an ionic solid. As a result, the solid dissolves. In solution, each cation and anion of the ionic compound is surrounded by oppositely charged ends of the water molecules (Fig. 3–7). This process is known as **hydration.**

[2] In both covalent and ionic binary compounds (*binary* indicates compounds consisting of two elements), the element having the greater attraction for the shared electrons is named second, and an -ide ending is added to the stem name (e.g., sodium chloride, hydrogen fluoride). The -ide ending is also used to indicate an anion, as in chloride (Cl^-) and hydroxide (OH^-).

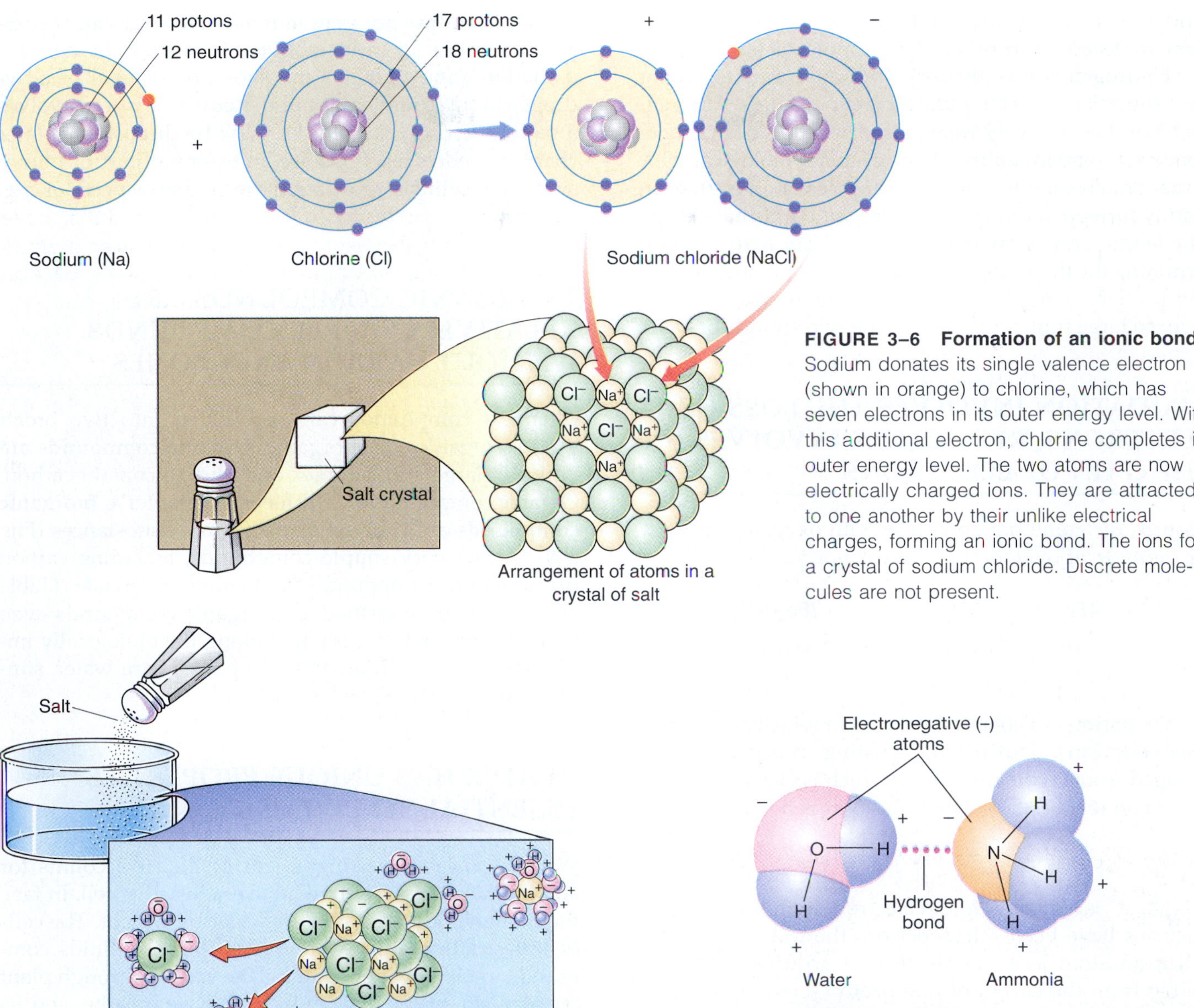

FIGURE 3–6 Formation of an ionic bond. Sodium donates its single valence electron (shown in orange) to chlorine, which has seven electrons in its outer energy level. With this additional electron, chlorine completes its outer energy level. The two atoms are now electrically charged ions. They are attracted to one another by their unlike electrical charges, forming an ionic bond. The ions form a crystal of sodium chloride. Discrete molecules are not present.

FIGURE 3–7 Hydration of an ionic compound. The crystal of NaCl consists of regularly spaced ionic bonds between the Na^+ and Cl^-. When NaCl is added to water, the partial negative ends of the water molecules are attracted to the positive sodium ions and tend to pull them away from the chloride ions. At the same time, the partial positive ends of the water molecules are attracted to the negative chloride ions, separating them from the sodium ions. When the NaCl is dissolved, each of the sodium and chloride ions is surrounded by water molecules electrically attracted to it.

FIGURE 3–8 How hydrogen bonds form. A weakly charged hydrogen atom connected to a negatively charged atom is attracted to the negatively charged part of another polar molecule. In this figure, a hydrogen bond joins a hydrogen atom of a water molecule to the nitrogen atom of an ammonia molecule.

Whether an ionic compound is in solid form or is dissolved in water, its ions do not share electrons. Because of this, the term *molecule* does not adequately explain the properties of ionic compounds such as NaCl. Chemists simply refer to them as compounds.

Hydrogen Bonds Are Weak Attractions

When hydrogen is combined with oxygen (or with another electronegative atom), it has a partial positive charge because its electron is positioned closer to the oxygen atom. **Hydrogen bonds** are weak attractions involving partially charged hydrogen atoms (Fig. 3–8). These bonds tend to form between an electronegative atom and a hydrogen atom that is covalently bonded to oxygen or nitrogen. The atoms involved may be in two parts of the same molecule or in two different molecules.

Hydrogen bonds are very important in determining the properties of water and of large molecules. Although each bond is relatively weak, large numbers of hydrogen bonds act to maintain the shape of many biologically important molecules. Because they are weak bonds, they are readily formed and broken. Hydrogen bonds have a specific length and orientation, an important feature in determining the three-dimensional structure of nucleic acids and proteins (Chapter 4). For example, hydrogen bonds occur in large numbers between the two strands of DNA.

OXIDATION INVOLVES THE LOSS OF ELECTRONS; REDUCTION INVOLVES THE GAIN OF ELECTRONS

Rusting, the combination of iron with oxygen, is a familiar example of oxidation and reduction:

$$\underset{\text{(metal)}}{\underset{\text{Iron}}{4Fe}} + \underset{\text{Oxygen}}{3O_2} \longrightarrow \underset{\text{(rust)}}{\underset{\text{Iron oxide}}{2Fe_2O_3}}$$

Oxidation is a chemical process in which a substance loses electrons. During the rusting process, iron is changed from its metallic state (which is electrically neutral) to an ionic (Fe^{3+}) state. We say it is being oxidized:

$$4Fe \longrightarrow 4Fe^{3+} + 12e^-$$

The e^- is a symbol for an electron. In this reaction, 12 electrons have been released from the four iron atoms. When an atom loses an electron, it acquires a positive charge from the excess of one proton. Loss of two electrons produces an atom with a double positive charge, and so on.

At the same time iron is being oxidized, oxygen is changing from its molecular state to its charged (ionic) state:

$$3O_2 + 12e^- \longrightarrow 6O^{2-}$$

When oxygen accepts the electrons removed from the iron, it is reduced. **Reduction** is a chemical process in which an atom, ion, or molecule gains electrons.

Oxidation cannot take place without reduction because electrons have to go somewhere; free electrons are not usually found in nature. Some substance must accept the electrons that are lost. On the other hand, reduction does not occur without a corresponding oxidation. Oxidation-reduction reactions are sometimes referred to as **redox** reactions.

Electrons are not easy to remove from covalent compounds unless an entire atom is removed. In living cells, oxidation almost always involves the removal of a hydrogen atom from a compound. Reduction often involves a gain in hydrogen atoms. As will be discussed in later chapters, redox reactions are an essential part of photosynthesis, cellular respiration, and other aspects of metabolism.

INORGANIC COMPOUNDS ARE RELATIVELY SIMPLE COMPOUNDS WITHOUT CARBON BACKBONES

Chemical compounds can be divided into two broad groups: organic and inorganic. **Organic compounds** are typically large and complex and always contain carbon. Organic compounds are the focus of Chapter 4. **Inorganic compounds** are relatively small, simple substances (Fig. 3–9). A few very simple compounds, including carbon dioxide and compounds containing carbonate (Table 3–2, p. 53), are classified as inorganic compounds even though they contain carbon. Among the biologically important groups of inorganic compounds are water, simple acids and bases, and salts.

WATER HAS UNIQUE PROPERTIES ESSENTIAL TO LIVING THINGS

Water is an essential ingredient of life. It accounts for about 80% of the weight of an average active cell. In fact, the human body is about 70% water by weight. The cells of terrestrial organisms are bathed in body fluids composed largely of water. Water is the source, through plant metabolism, of the oxygen in the air we breathe, and its hydrogen atoms are incorporated into many organic com-

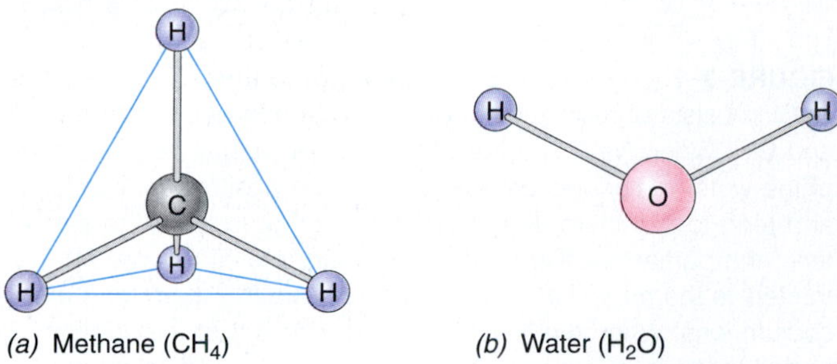

FIGURE 3–9 Molecular shapes. (***a***) Methane. The four covalent bonds of carbon are oriented toward the corners of an imaginary tetrahedron, outlined by the blue lines. Note that methane has a three-dimensional shape. (***b***) Water, an inorganic compound. The two covalent bonds of water have 104.5° angles.

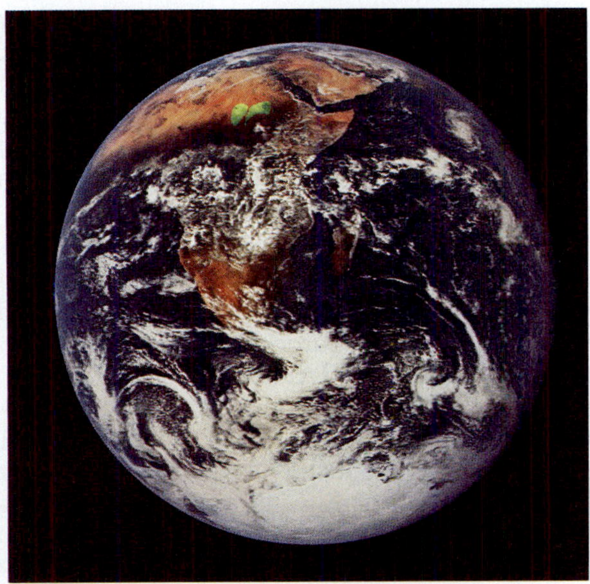

FIGURE 3–10 Planet Earth, the water planet. Most of Earth's surface is covered with water. Here, Earth is seen from Apollo II, about 98,000 miles away. (NASA)

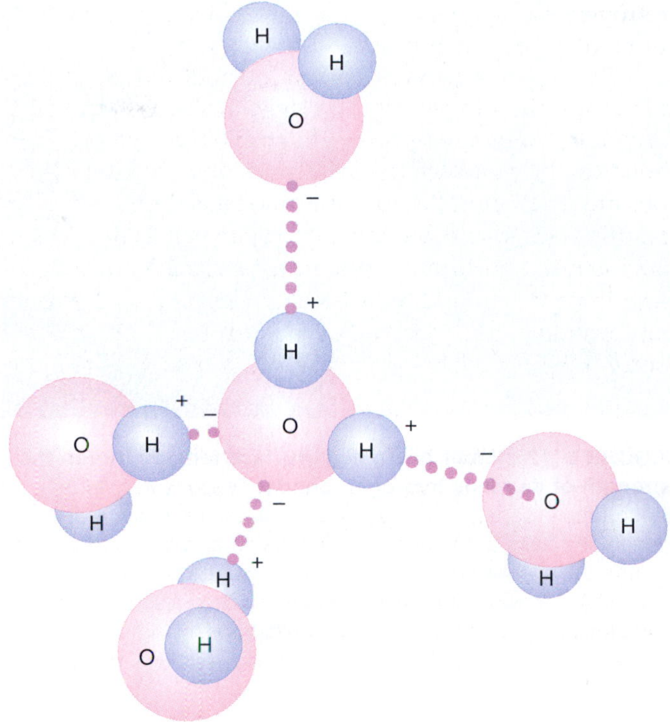

FIGURE 3–11 Hydrogen bonding of water molecules. Each water molecule tends to form hydrogen bonds with four neighboring water molecules. The hydrogen bonds are indicated by dotted lines.

pounds in the bodies of living things. Water is also the solvent for most biological reactions and a reactant or product in many chemical reactions.

Water is not only important inside organisms but is also one of the principal environmental factors affecting them. Many organisms make their homes in the sea or in freshwater rivers, lakes, or puddles. Water's unique combination of physical and chemical properties has been an important factor in the evolution of living things on planet Earth (Fig. 3–10).

Water Molecules Are Polar

The physical and chemical properties of water permit life to exist on our planet. Water molecules are polar; that is, they bear a partial positive and a partial negative charge. The water molecules in liquid water and in ice are held together in part by hydrogen bonds. The hydrogen atom of one water molecule, with its partial positive charge, is attracted to the oxygen atom of a neighboring water molecule, with its partial negative charge, forming a hydrogen bond. Each water molecule can form hydrogen bonds with a maximum of four neighboring water molecules (Fig. 3–11).

Polar compounds are **hydrophilic** (water-loving) and can form hydrogen bonds with water. Nonpolar compounds are **hydrophobic** (water-hating). Water molecules, held together by hydrogen bonds, tend to push nonpolar compounds away. The nonpolar compounds cluster together and are insoluble in water. Hydrophobic interactions explain why oil tends to form globules when it is added to water. Hydrophilic and hydrophobic interactions are very important in biological membranes (Chapter 6).

Water Is an Excellent Solvent

Because its molecules are polar, water is an excellent solvent, particularly for ions and polar compounds. A **solvent** is a fluid capable of dissolving other substances. Consider a salt going into solution. The ions of a salt such as sodium chloride are held together by strong ionic bonds. In fact, they form a three-dimensional structure, called a crystal lattice. Considerable energy is required to pull the positively and negatively charged ions apart. However, when sodium chloride is placed in water, the strong electrical attractions between the polar water molecules and Na^+ and Cl^- ions pull these ions out of their positions. This results in the formation of a solution of dissociated (separated) ions. Because of its solvent properties and its tendency to cause the ionization of compounds in solution, water is important in facilitating chemical reactions.

FIGURE 3–12 Water has a high surface tension due to the strength of its many hydrogen bonds. These water striders, although more dense than water, can walk on the surface of a pond. Fine hairs at the ends of the legs spread the weight over a large area, allowing the body of the insect to be supported by the surface tension of the water. Water striders also use the surface tension of the water for communication, setting up patterns of ripples that other water striders can sense. (Dennis Drenner)

Water Molecules Exhibit Cohesive and Adhesive Forces

Water molecules have a very strong tendency to stick to one another; that is, they are **cohesive.** This is due to the hydrogen bonds among the molecules. Water molecules have a much greater attraction for other water molecules than for molecules in the air. Thus, water molecules at the surface crowd together, producing a strong layer as they are pulled downward by the attraction of other water molecules beneath them. This tendency of water molecules to stick together at the surface is called **surface tension** (Fig. 3–12).

Water molecules also stick to many other kinds of substances (that is, those substances that have charged groups of atoms or molecules on their surfaces). These **adhesive** forces explain how water makes things wet.

Adhesive and cohesive forces account for **capillary action,** the tendency of water to rise in narrow tubes. Water molecules adhere to the tube walls, pulling other water molecules up along with them—against the force of gravity. Capillary action is a mechanism by which water moves through the microscopic spaces between soil particles to the roots of plants.

Water Helps Maintain a Stable Temperature

The temperature of water changes less drastically and more slowly than the temperature of almost all other substances. This ability to minimize temperature changes results from the hydrogen bonds that hold water molecules together.

Raising the temperature of a substance involves adding heat energy to make its molecules move faster, that is, to increase their kinetic energy (energy of motion). **Specific heat** is the amount of heat energy required to raise the temperature of one gram of a substance by 1°C. The hydrogen bonds holding water molecules together tend to restrict their motion, so that it takes more heat energy to raise the temperature of water than it would for a substance that lacks hydrogen bonds. Thus, water has a very high specific heat. It takes a lot of heat gain or loss to raise or lower the temperature of water.

Because so much heat input (or heat loss) is required to raise (or lower) the temperature of water, the oceans and other large bodies of water have relatively constant temperatures. Thus, the aquatic environment provides the organisms that inhabit it with a relatively constant environmental temperature. The water within all organisms, even those that live on land, contributes to a relatively constant internal temperature. This is important because metabolic reactions can take place only within a relatively narrow temperature range.

Because its molecules are held together by hydrogen bonds, water also has a high **heat of vaporization,** another property that helps stabilize temperature. More than 500 calories are required to change 1 gram of liquid water into 1 gram of water vapor (compared to only one calorie to raise the temperature of 1 gram of water 1°C). A **calorie** is a unit of heat energy. Because of the heat of vaporization, we can rid ourselves of excess heat by the evaporation of sweat from our skin, and plants can remain cool in the midday heat by evaporating water from their surfaces.

The Density of Water Is Greatest at 4° Centigrade

Hydrogen bonds contribute another important property of water. Although most substances become more and more dense as the temperature decreases, water reaches its maximum density at 4°C and then begins to expand again as the temperature decreases. Hydrogen bonds become more rigid and ordered, and ice floats on the denser cold water (Fig. 3–13). This important property of water explains why lakes and ponds freeze from the surface down rather than from the bottom up. The sheet of ice that forms at the pond surface insulates the water below from the wintry chill so that it is less likely to freeze. Further, organisms that inhabit northern lakes and ponds are able to carry on their life activities despite the frigid winter and the icy lid at the surface of their home.

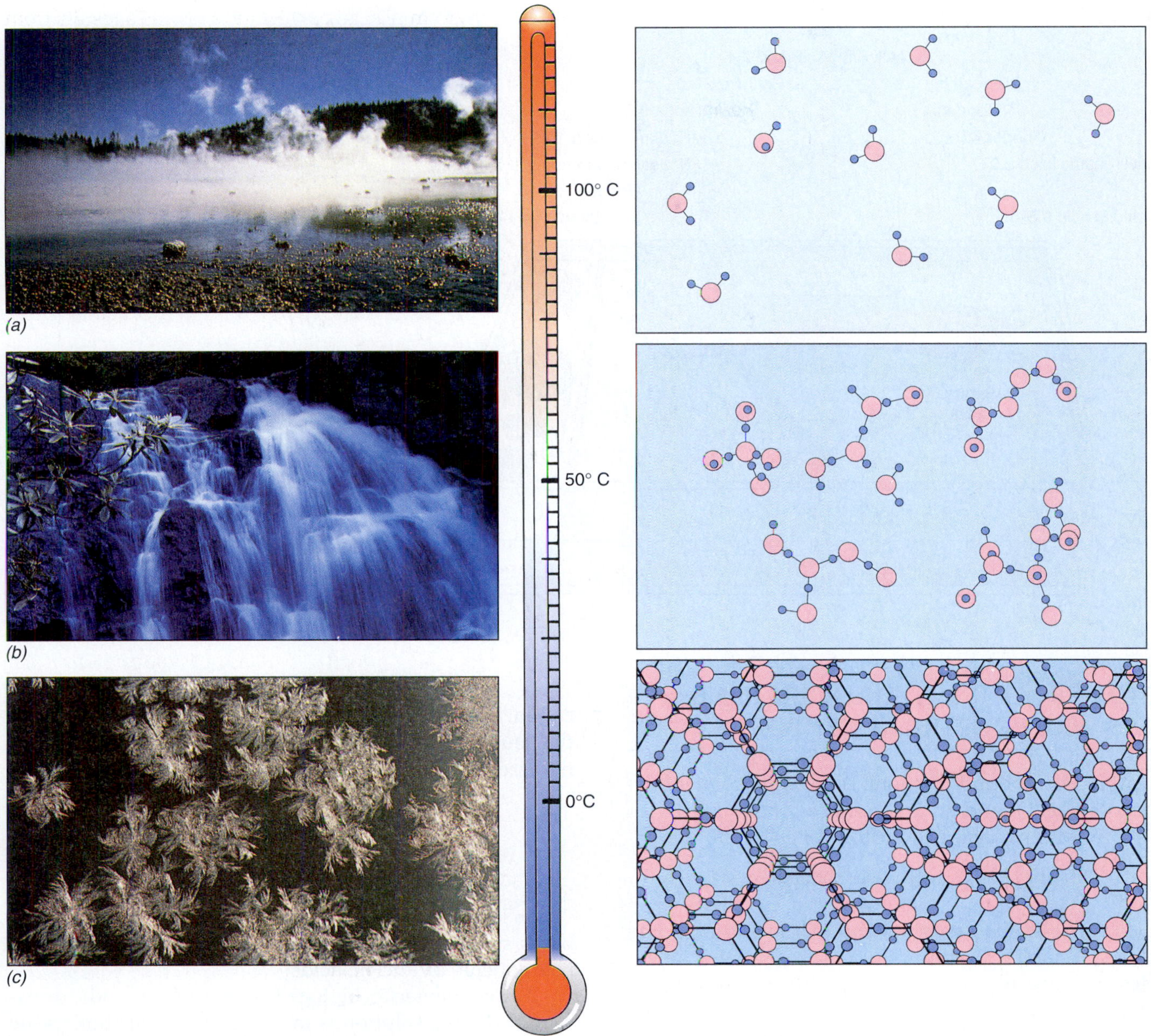

FIGURE 3–13 The hydrogen bonding of three forms of water: steam, liquid water, and ice. (***a***) Hot spring, Yellowstone National Park. When water boils, the liquid becomes steam. In steam, water molecules are far apart and move freely and rapidly. (***b***) Waterfalls, Hickory Run State Park, PA. In liquid water, hydrogen bonds form between molecules packing them close together. (***c***) Ice crystal patterns. Hydrogen bonds are regular and evenly spaced in the superstructure of ice. Water expands as it freezes, leaving empty spaces between the less densely packed ice. Ice is, therefore, one of very few substances that is lighter in its solid than its liquid form. As a result, ice floats on water instead of accumulating on the bottom. When ice melts, the hydrogen bonds occur less consistently and are of unequal length, and the crystal structure collapses. (*a,* Woodbridge Wilson/National Park Service; *b,* Gary R. Bonner; *c,* Stan Osolinski/Dembinsky Photo Associates).

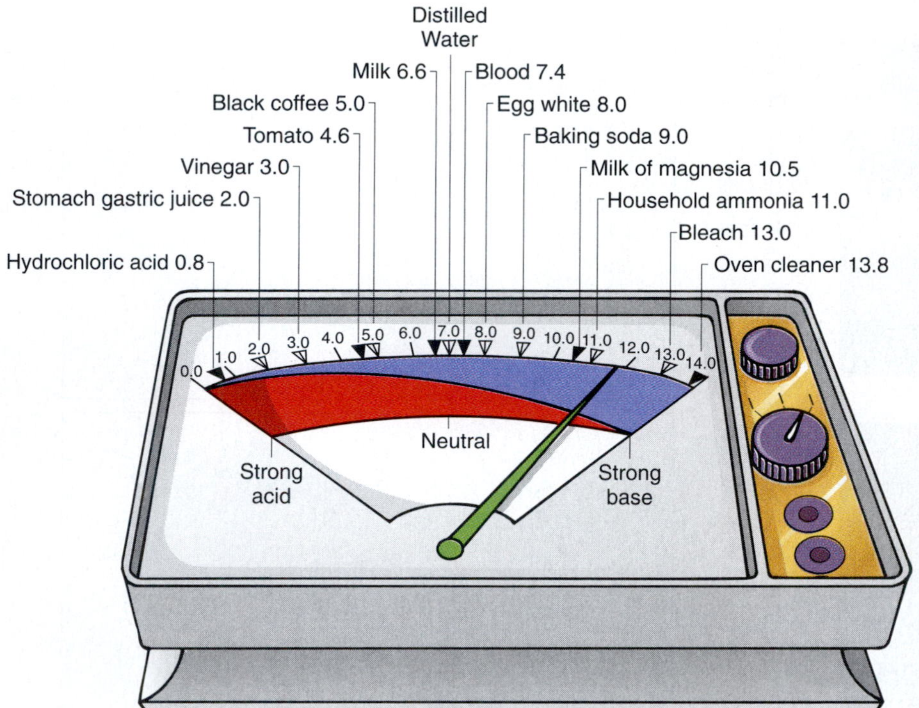

FIGURE 3–14 The pH scale shown on a pH meter. A solution with a pH of 7 is neutral because the concentrations of H^+ and OH^- are equal. The lower the pH below 7, the more H^+ ions are present, and the more acidic the solution. As the pH increases above 7, the concentration of H^+ ions decreases and the concentration of OH^- increases, making the solution more basic (alkaline).

ACIDS PRODUCE HYDROGEN IONS; BASES ARE PROTON ACCEPTORS

An **acid** is a compound that ionizes in water to yield hydrogen ions (H^+)[3] and an anion. Recall that a hydrogen ion is a proton. An acid is a proton *donor.*

$$\text{Acid} \longrightarrow H^+ + \text{Anion}$$

An acid turns litmus paper red and has a sour taste. Hydrochloric acid (HCl) and sulfuric acid (H_2SO_4) are examples of inorganic acids. The strength of an acid depends on the degree to which it ionizes in water, releasing hydrogen ions. Thus, the molecules of weak acids dissociate (separate) into ions just slightly whereas strong acids dissociate to a much greater extent. HCl is a very strong acid because most of its molecules dissociate, producing hydrogen and chloride ions.

$$\underset{\text{Hydrochloric acid}}{HCl} \xrightarrow{\text{in } H_2O} \underset{\text{Hydrogen ion}}{H^+} + \underset{\text{Chloride ion}}{Cl^-}$$

A **base** is defined as a proton *acceptor.* Many bases dissociate in water to yield a hydroxide ion (OH^-) and a cation. Bases turn litmus paper blue and feel slippery to the touch. Sodium hydroxide (NaOH) and ammonium hydroxide (NH_4OH) are common inorganic bases.

$$\underset{\text{Sodium hydroxide}}{NaOH} \xrightarrow{\text{in } H_2O} \underset{\text{Sodium ion}}{Na^+} + \underset{\text{Hydroxide ion}}{OH^-}$$

In later chapters, we will discuss a number of organic bases such as the purine and pyrimidine bases that are components of nucleic acids.

For convenience, biologists express how acidic or basic (alkaline) a solution is in terms of a logarithmic scale called the pH scale. We use the pH scale to measure the relative concentrations of hydrogen and hydroxide ions.

The **pH** of a solution is defined as the negative logarithm (base 10) of the hydrogen ion concentration expressed in moles[4] per liter $pH = -\log[H^+]$[5]. The **pH scale** extends from 0, the pH of a strong acid such as HCl, to 14, the pH of a strong base such as NaOH (Fig. 3–14). The pH of pure water is 7, which means it is neither acidic nor basic, but neutral.

Because the scale is logarithmic (to base 10), a solution with a pH of 6 has a hydrogen ion concentration that

[3] H^+ immediately combines with an electronegative region of a water molecule, forming a hydronium ion (H_3O^+). However, by convention H^+, rather than the more accurate H_3O^+, is used.

[4] A mole is the molecular weight of a substance expressed in grams.

[5] [ ] is a symbol indicating concentration.

is ten times greater than a solution with a pH of 7 and is much more acidic. A pH of 5 represents another tenfold increase, and so a solution with a pH of 4 is 10×10 or 100 times more acidic than a solution with a pH of 6.

Even though water does ionize slightly, the concentrations of H^+ ions and OH^- ions are exactly equal, and each of them has a concentration of 10^{-7} moles/liter. This explains why we can say that water has a pH of 7. Solutions with a pH of *less* than 7 are acidic and contain more H^+ ions than OH^- ions. Solutions with a pH *greater* than 7 are basic, or alkaline, and contain more OH^- ions than H^+ ions.

The contents of most animal and plant cells are neither strongly acidic nor alkaline; they are a neutral or very slightly basic mixture of acidic and basic substances. The pH of living cells ordinarily ranges around the value of about 7.3. Human blood has a normal pH of 7.4. Most living organisms cannot survive if their pH changes very much because the chemical reactions essential to life are very sensitive to the concentrations of hydrogen and hydroxide ions. The pH of the environment must also be appropriate. For example, most fish species die when the pH falls below 4.5. Such a low pH can result from acid rain (see chapter opening).

TABLE 3–2
Some Biologically Important Ions

Name	*Formula*	*Charge*
Sodium	Na^+	1+
Potassium	K^+	1+
Hydrogen	H^+	1+
Magnesium	Mg^{2+}	2+
Calcium	Ca^{2+}	2+
Iron	Fe^{2+} or Fe^{3+}	2+ or 3+
Ammonium	NH_4^+	1+
Chloride	Cl^-	1−
Iodide	I^-	1−
Carbonate	CO_3^{2-}	2−
Bicarbonate	CO_3^-	1−
Phosphate	PO_4^{3-}	3−
Acetate	CH_3COO^-	1−
Sulfate	SO_4^{2-}	2−
Hydroxide	OH^-	1−
Nitrate	NO_3^-	1−
Nitrite	NO_2^-	1−

Salts Form from Acids and Bases

When an acid and a base are mixed together, the H^+ of the acid combines with the OH^- of the base to form a molecule of water. The remainder of the acid (an anion) combines with the remainder of the base (a cation) to form a salt. For example, hydrochloric acid reacts with sodium hydroxide to form water and sodium chloride:

$$\underset{\text{Hydrochloric acid}}{HCl} + \underset{\text{Sodium hydroxide}}{NaOH} \longrightarrow \underset{\text{Water}}{H_2O} + \underset{\text{Sodium chloride}}{NaCl}$$

A **salt** is a compound in which the hydrogen ion of an acid is replaced by some other cation. A salt contains a cation other than H^+ and an anion other than OH^-. Sodium chloride, NaCl, is a compound in which the hydrogen ion of HCl has been replaced by the cation, Na^+.

When a salt, an acid, or a base is dissolved in water, its component ions dissociate. Because these charged particles (ions) can conduct an electrical current, these substances are called **electrolytes.** Sugars, alcohols, and many other substances do not ionize when dissolved in water; they do not conduct an electrical current and are referred to as **nonelectrolytes.**

The cells and extracellular fluids (such as sap or blood) of plants and animals contain a variety of dissolved salts. They are a source of many important mineral ions. Such ions are essential for fluid balance, acid–base balance, and enzyme function in most organisms. In animals they are also important in nerve and muscle function, blood clotting, and many other aspects of body function. Sodium, potassium, calcium, and magnesium are the chief cations present, and chloride, bicarbonate, phosphate, and sulfate are important anions (Table 3–2).

The body fluids of terrestrial animals differ considerably from sea water in their total salt content. However,

Concept Connections

Salt Concentration — ***Evolution***

Most biologists think that living systems arose in the sea. The cells of early organisms became adapted to function optimally in the salt water. As larger animals evolved, their body fluids contained a similar pattern of salts. Later, when organisms migrated into fresh water or onto land, that pattern of salts was retained in their body fluids. In some animals, kidneys and other organs, such as salt glands, have evolved that selectively retain or secrete certain ions (Chapter 40). These organs change the composition of body fluids so that their relative concentration of salts is somewhat different than that of sea water. The concentration of each ion is determined by the relative rates of its uptake and excretion by the organism.

CONCEPT CONNECTIONS

pH ⟷ *Ecology*

Naturally occurring bodies of water, especially sea water, often contain substances that serve as buffers. For instance, streams with limestone bottoms often contain calcium-based minerals that help to remove the hydrogen ions that seep into these streams from decaying vegetation or acid rain. Streams that have granite beds, on the other hand, contain different and less alkaline minerals and are much less able to tolerate acid rain.

One measure that has been attempted to restore acid-damaged waters to health involves **neutralization.** In this process, an inappropriate pH is changed by the addition of a substance that has the opposite effect on pH. In streams with granite beds, a base is required. This is supplied by adding lime, calcium hydroxide, $Ca(OH)_2$. Suppose the offending acid were sulfuric acid, derived from the burning of high-sulfur fuel in electrical generation plants:

$$\underset{\text{Hydrogen}}{2H^+} + \underset{\text{Sulfate}}{SO_4^{2-}} + \underset{\text{Calcium hydroxide}}{Ca(OH)_2} \longrightarrow \underset{\text{Calcium sulfate}}{Ca_2SO_4} + \underset{\text{Water}}{H_2O}$$

This disposes of the hydrogen ions, leaving water in its place. It also removes the sulfate, because calcium sulfate is quite insoluble. The lime is a fairly strong base, however, and must be added carefully so as not to raise the pH of the water too much.

That is the disadvantage of neutralization as opposed to buffering: the alkali or acid that is used to neutralize an inappropriate pH may produce an undesirable pH itself. Generally, body fluid buffering agents are themselves nearly neutral.

---

they resemble sea water in the kinds of salts present and in their relative abundance. The total concentration of salts in the body fluids of most invertebrate (animals without a backbone) marine animals is equivalent to that in sea water, about 3.4%. Vertebrates, whether terrestrial, freshwater, or marine, have less than 1% salt in their body fluids.

Although the concentration of salts in the cells and body fluids of plants and animals is small, these salts are of great importance for normal cell function. Homeostatic mechanisms keep the concentrations of the respective cations and anions remarkably constant under normal conditions. Any marked change results in impaired cellular functions and ultimately in death.

Buffers Minimize pH Change

Many homeostatic mechanisms operate to maintain appropriate pH levels. For example, the pH of human blood is about 7.4 and must be maintained within very narrow limits. Should the blood become too acidic (for instance as a result of respiratory disease), coma and death may result. Excessive alkalinity, on the other hand, can result in nerve cells becoming overexcited and may lead to convulsions.

A **buffer** is a substance or combination of substances that minimizes changes in pH when an acid or base is added. The buffer either accepts or donates hydrogen ions. A buffer consists of a weak acid and a salt of that acid, or a weak base and a salt of that base (Fig. 3–15).

FIGURE 3–15 Buffering is used clinically as a remedy for excess stomach acid. The bubbles are CO_2 from the reaction between an acid (citric acid) and the bicarbonate ion (HCO_3^-) from sodium bicarbonate (Charles D. Winters).

One of the most common buffering systems and one that is important in human blood is carbonic acid and the bicarbonate ion. Bicarbonate ions are formed in the body as follows:

$$CO_2 + H_2O \rightleftharpoons H_2CO_3 \rightleftharpoons H^+ + HCO_3^-$$

Carbon dioxide — Water — Carbonic acid — Bicarbonate ion

As indicated by the arrows, the reactions are reversible.

When excess hydrogen ions are present in blood or other body fluids, bicarbonate ions combine with them to form carbonic acid, a weak acid.

$$H^+ + HCO_3^- \rightleftharpoons H_2CO_3$$

Carbonic acid

In this way, a strong acid can be converted to a weak acid. The carbonic acid is unstable and quickly breaks down into carbon dioxide and water.

Buffers also maintain a relatively constant pH when hydroxide ions are added. A buffer may release hydrogen ions, which combine with the hydroxide ions to form water.

$$OH^- + H_2CO_3 \rightleftharpoons HCO_3^- + H_2O$$

Chapter Summary

I. The atom is the smallest unit of a chemical element that retains the characteristic properties of that element.
   A. An atom consists of subatomic particles, including protons, neutrons, and electrons.
   B. Electrons are found in orbitals located within energy levels. The electrons in the outermost energy level of an atom are valence electrons; they help determine the chemical properties of the atom.

II. A chemical compound consists of two or more elements combined in a fixed ratio.
   A. The composition of a compound may be described by a chemical formula, such as H_2O, or by a structural formula, such as H—O—H.
   B. The chemical reactions that occur between atoms and molecules can be described by means of chemical equations.

III. The atoms of a chemical compound are attached to one another by chemical bonds.
   A. In a covalent bond, atoms share a pair of electrons.
   B. An ionic bond is formed when one atom donates an electron to another. An ionic compound is made up of positively charged ions (cations) and negatively charged ions (anions).
   C. Hydrogen bonds are relatively weak bonds. They form when a hydrogen atom in one molecule is attracted to a highly electronegative element such as oxygen or nitrogen in another molecule or in another part of the same molecule.

IV. When a substance is oxidized, it loses electrons; when a substance is reduced, it gains electrons. In living systems, oxidation-reduction reactions usually involve the loss and gain of hydrogen.

V. Organic compounds are large and complex and they contain carbon; inorganic compounds are relatively small and simple.

VI. Life depends on the unique properties of water: It accounts for a large part of the mass of most organisms, is important in many chemical reactions that occur within living things, and its unique properties help maintain the environment.
   A. Because its molecules are polar, water is an excellent solvent.
   B. Water molecules are cohesive owing to the hydrogen bonding between the molecules; they also adhere to many kinds of substances. Water has a high degree of surface tension because of the cohesiveness of its molecules.
   C. Water has a high specific heat, which helps organisms maintain a relatively constant internal temperature; this property also helps keep the oceans and other large bodies of water at a constant temperature.
   D. Other important properties of water include its high heat of vaporization, its unusual density (ice is less dense than liquid water), its slight tendency to form ions, and its ability to dissolve many different kinds of compounds.

VII. An acid is a substance that dissociates in solution to yield hydrogen ions and an anion. Acids are proton donors. Bases are proton acceptors. Many bases dissociate in solution producing hydroxide ions.
   A. The pH scale extends from 0 to 14, with 7 indicating neutrality. As the pH decreases below 7, the solution is more acidic. As a solution becomes more basic (alkaline), its pH increases from 7 toward 14.
   B. A buffer consists of a weak acid and a salt of that acid, or a weak base and a salt of that base. Buffers resist changes in the pH of a solution when acids or bases are added.
   C. A salt is a compound in which the hydrogen ion of an acid is replaced by some other cation. Salts provide mineral ions essential for fluid balance, nerve and muscle function, and many other body processes.

Selected Key Terms

acid, p. 52
acid rain, p. 36
atom, p. 38
base, p. 52
buffer, p. 54
chemical bond, p. 44
chemical compound, p. 42
chemical reaction, p. 38

electron, p. 38
element, p. 38
hydrogen bond, p. 47
hydrophilic, p. 49
hydrophobic, p. 49
inorganic compound, p. 48
ion, p. 46
ionic bond, p. 46
molecule, p. 42
neutron, p. 38
organic compound, p. 48
pH scale, p. 52
polar covalent bond, p. 46
proton, p. 38
radioisotope, p. 40
redox reaction, p. 48
salt, p. 53
solvent, p. 49

Post-Test

1. A/An ______________ is the smallest amount of an element that retains the chemical properties of the element.
2. Isotopes are atoms of the same element that differ in their number of ______________.
3. Within energy levels, electrons may be found in specific ______________.
4. The chemical symbol for carbon is ______________; for hydrogen, it is ______________; and for oxygen, it is ______________.
5. Atoms with one to three valence electrons generally behave as electron ______________.
6. The type of bond in which atoms share electrons is a ______________ bond.
7. A chemical process in which a substance gains electrons is referred to as ______________.
8. A compound that ionizes in solution to yield hydrogen ions and an anion is a/an ______________.
9. A solution with a pH of 8 is best described as ______________ (acidic, basic, neutral).
10. A substance that resists changes in pH is a/an ______________.
11. The tendency for water to rise in very fine tubes is called ______________ action; it is due to cohesive and ______________ forces.
12. Water molecules in liquid water and ice are held together by ______________ bonds.
13. Which of the following are located in the atomic nucleus? (a) electrons (b) neutrons (c) protons (d) answers a, b, and c; (e) answers b and c only.
14. If you wanted to indicate the types, numbers, and arrangement of atoms in a chemical compound, it would be best to use a (a) chemical formula (b) structural formula (c) chemical equation (d) compound (e) valence diagram.
15. A chemical bond in which an atom gains electrons is called (a) ionic (b) covalent (c) hydrogen (d) oxidation (e) two of the preceding answers are correct.
16. Which of the following bonds do water molecules form with one another? (a) ionic (b) covalent (c) hydrogen (d) reduction (e) none of the preceding.
17. Which of the following is a proton acceptor? (a) salt (b) acid (c) base (d) water (e) nonelectrolyte.
18. Cells help maintain appropriate pH by using (a) nonelectrolytes (b) buffers (c) heat of vaporization (d) trace elements (e) none of the preceding.
19. Large compounds that contain carbon are called (a) buffers (b) salts (c) organic compounds (d) sodium compounds (e) none of the preceding.
20. A solution with a pH of 4 is a/an (a) acid (b) base (c) salt (d) blood (e) two of the preceding answers are correct.

Review Questions

1. Distinguish between (a) an atom and an element, (b) a molecule and a compound, and (c) an atom and an ion.
2. How do isotopes of the same element differ? What is a radioisotope?
3. How do valence electrons help determine the chemical properties of an atom?
4. Compare ionic and covalent bonds and give specific examples of each.
5. Write a chemical equation showing the ionization of (a) sodium hydroxide ($NaOH$) and (b) hydrochloric acid (HCl).
6. How would a solution with a pH of 5 differ from one with a pH of 9? Of 7?
7. Differentiate among acids, bases, and salts. What are the functions of salts in living organisms?
8. Why do oxidation and reduction occur simultaneously.
9. Describe a reversible reaction that is at equilibrium.

Thinking Critically

1. What measures, if any, do you think we should take to reduce the industrial contribution to acid rain?
2. Imagine that rain was extremely alkaline (pH about 11). What, if any, effect do you think that would have on ecosystems?
3. Why are buffers important in living organisms? Give a specific example of how a buffer system works.
4. Imagine that all hydrogen bonds disappeared. What would be some specific consequences?
5. Life on Earth depends on the unique properties of water. Describe three of these properties and explain how they support life as we know it.
6. A person with an upset stomach may take sodium bicarbonate ($NaHCO_3$) to treat the excess acidity. Since stomach acid is hydrochloric acid (HCl), this produces water and carbon dioxide gas (hence the burp that follows!). But before carbon dioxide is emitted, a weak acid, carbonic acid, is produced as an intermediate. Since all that is done is to replace one acid with another, why does sodium bicarbonate reduce excess stomach acidity?

Recommended Readings

Atkins, P. W., *Molecules.* New York, W. H. Freeman and Company (Scientific American Books), 1987. A fascinating and beautifully illustrated collection of different kinds of substances, especially organic substances.

Bettelheim, F. A., and J. March, *Introduction to General, Organic and Biochemistry,* 3rd ed. Philadelphia, Saunders College Publishing, 1991. A very readable reference text for those who would like to know more about the chemistry basic to life.

Kowalok, M.E., "Common Threads," *Environment,* Vol. 35, No. 6, July/August 1993. This article examines research lessons learned from studying acid rain, ozone depletion, and global warming.

Monastersky, R., "Acid Precipitation Drops In United States," *Science News,* Vol. 144, July 10, 1993. The Clean Air Acts are reducing acid rain in the U.S.

Storey, K. B. and J. M. Storey, "Frozen and Alive," *Scientific American,* December 1990. Some animals, including certain lizards, can survive with 60% of their body water in a frozen state.

Each year hundreds of thousands of tons of pesticides are sprayed on farmlands in the United States alone. *Most of these sprays contain organic compounds which are toxic to species other than those targeted. Almost all produce you can buy in the U.S. contains residues of several pesticides. Such residues have also contaminated drinking water supplies in 23 states. (Comstock, Inc./Jack Clark)*

Toxic Mimics

Life and society depend on organic compounds, which are complex chemical compounds that contain carbon atoms. Organic compounds are essential components of living organisms. They also make up the fossil fuels that power our industrial society, many pesticides, and many of the chemicals used in manufacturing.

Although many of the nutrients we require are organic compounds, some organic molecules are toxic to living organisms. For example, 45 environmental pollutants have been identified that affect the reproductive system of humans and other animals. These contaminants include herbicides, insecticides, fungicides, industrial byproducts, and a variety of other chemical compounds used commercially. Studies indicate that these compounds, many of which contain chlorine, mimic steroid hormones, such as the reproductive hormone estrogen. Exposure to some of these compounds has also been linked with an increased risk of cancer.

Many organic compounds persist in the environment for very long periods of time. DDT, the chlorinated organic pesticide that was banned in 1972, is still found in our food supply. A recent study found about 35% higher concentrations of DDE, a breakdown product of the pesticide DDT, in women with breast cancer than in healthy controls. DDE is thought to mimic the reproductive hormone estrogen, which is necessary for growth of many breast cancers.

The chlorinated organic compound dioxin, which also exhibits hormone-like effects, has been linked to suppression of the immune system, increased risk of cancer, and birth defects. Dioxin is an extremely dangerous toxin that forms during the manufacture of certain herbicides. For about 20 years, scientists have monitored the health of 2000 families in Italy who were exposed to dioxin in 1976 when an industrial accident released the toxin into the air. In 1994, researcher Pier Alberto Bertazzi reported that the exposed population has suffered an increased incidence in liver cancer, myeloma, and certain blood cancers.

PCBs (polychlorinated biphenyls), a group of about 70 organic compounds used by industry, may also cause cancer. These compounds have been shown to cause liver and kidney damage and to interfere with reproduction.

Studies suggest pollutants that mimic hormones can affect the development of the male reproductive system during prenatal development as well as interfere with biological processes later in life. Many industrialized countries have recently reported a dramatic rise in cancer of the testes. In Denmark, the rate of testicular cancer has tripled during the past 50 years. In some areas, researchers have reported sharp decreases in sperm counts.

Organic compounds in the environment have been linked with the decline of many animal species, including fish, birds, sea turtles, alligators, and mammals. Florida panthers have almost disappeared. In 1994 less than 50 remained. The reproductive problems that have plagued this species appear to be, at least in part, due to organic compounds in the environment that mimic estrogens.

Important sources of toxic organic compounds include runoff of pesticides and careless disposal of industrial and household wastes. In the United States alone, millions of tons of hazardous organic wastes are stored in landfills, dumps, and underground tanks. Toxic wastes leaking from these sites into surface water and groundwater threaten the health of those who drink the polluted water.

The Chemistry of Life: Organic Compounds

4

Learning Objectives

After you have studied this chapter you should be able to:

1. Compare the major groups of organic compounds—carbohydrates, lipids, proteins, and nucleic acids—with respect to their chemical composition and function.
2. Distinguish among monosaccharides, disaccharides, and polysaccharides, giving examples and functions of each.
3. Distinguish among neutral fats, phospholipids, and steroids, giving the biological functions of each group.
4. Describe the chemical structure and functions of proteins.
5. Describe the chemical structure of nucleotides and nucleic acids and explain the importance of these compounds in living organisms.

Key Concepts

- ☐ Four main groups of organic compounds important in living things are carbohydrates, lipids, proteins, and nucleic acids.
- ☐ Organic compounds are made possible by the ability of carbon atoms to covalently bond to one another, forming long chains that serve as the backbone of the compound.
- ☐ Most carbohydrates are polysaccharides, long polymers consisting of repeating units of a simple sugar. Carbohydrates serve as energy sources and are structural components of cell walls.
- ☐ Lipids are fat-like substances usually composed of fatty acids and glycerol. Animals store energy in the form of neutral fats. Phospholipids make up the bilayer of cellular membranes, and steroids often function as hormones.
- ☐ Proteins are macromolecules consisting of linked amino acids; they are structural components of cells and tissues. Many proteins serve as enzymes, organic catalysts that regulate the chemical life of the organism.
- ☐ The information needed to construct proteins is stored in nucleic acids, usually deoxyribonucleic acid (DNA), a macromolecule composed of nucleotide units.

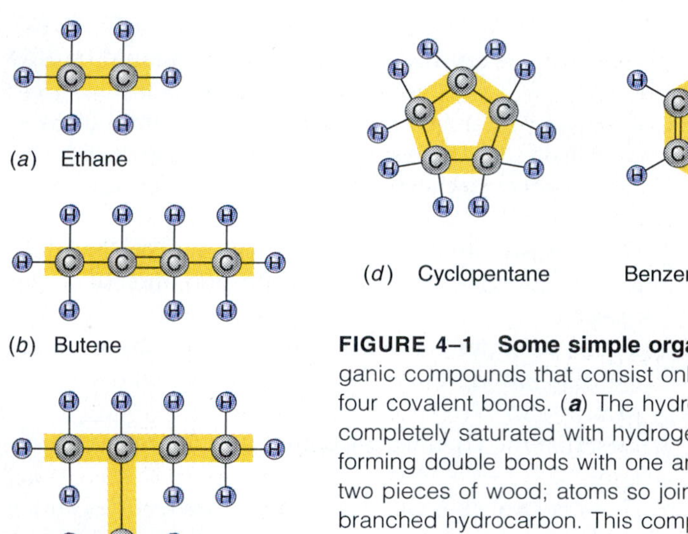

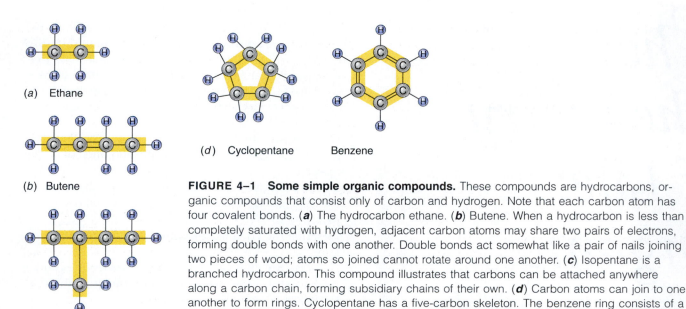

FIGURE 4–1 Some simple organic compounds. These compounds are hydrocarbons, organic compounds that consist only of carbon and hydrogen. Note that each carbon atom has four covalent bonds. (***a***) The hydrocarbon ethane. (***b***) Butene. When a hydrocarbon is less than completely saturated with hydrogen, adjacent carbon atoms may share two pairs of electrons, forming double bonds with one another. Double bonds act somewhat like a pair of nails joining two pieces of wood; atoms so joined cannot rotate around one another. (***c***) Isopentane is a branched hydrocarbon. This compound illustrates that carbons can be attached anywhere along a carbon chain, forming subsidiary chains of their own. (***d***) Carbon atoms can join to one another to form rings. Cyclopentane has a five-carbon skeleton. The benzene ring consists of a six-carbon skeleton with three double bonds. Side chains can attach to the carbons in a ring.

THE CHEMISTRY OF LIVING THINGS IS ORGANIZED AROUND THE CARBON ATOM

Organic compounds are the main structural components of cells and tissues. They participate in and regulate metabolic reactions, transmit information, and provide energy for life processes. In this chapter, we focus on some of the major groups of organic compounds that are important in living organisms, including carbohydrates, lipids, proteins, and nucleic acids (DNA and RNA). Most of these compounds, which are sometimes called the molecules of life, are constructed in the cell from simpler molecular components. For example, protein molecules are built from smaller compounds called amino acids. All of these compounds are organized around carbon atoms.

In a way, the chemistry of the carbon atom is the chemistry of life itself. Perhaps because it can form a greater variety of molecules than any other element, carbon has emerged as the central component of organic compounds. With four electrons in its outer energy level, the carbon atom can share electrons with other atoms, including other carbon atoms, forming four covalent bonds. This ability of carbon atoms yields an immense variety of organic compounds. More than five million have been identified!

Hydrogen, oxygen, and nitrogen are atoms frequently bonded to carbon. Organic compounds that consist only of carbon and hydrogen are known as **hydrocarbons.** Living organisms use hydrocarbon skeletons to build diverse organic compounds. Fossil fuels are hydrocarbons formed from organic compounds originating in organisms that lived and died millions of years ago.

Carbon atoms form covalent bonds with other carbon atoms to produce chains of varying lengths. These chains may be unbranched or branched, or carbon atoms at the ends of short chains may join to form rings. Adjacent carbon atoms may form single bonds, or by sharing additional pairs of electrons, they may form double $(-\overset{|}{C}=\overset{|}{C}-)$ or triple $(-C\equiv C-)$ bonds (Fig. 4–1).

FUNCTIONAL GROUPS FORM BONDS WITH OTHER MOLECULES

The carbon "backbone" of an organic compound is a stable structure that does not interact readily with other compounds. However, one or more of the hydrogen atoms bonded to the carbon skeleton of a hydrocarbon may be replaced by other groups of atoms. These groups of atoms, called **functional groups,** readily take part in chemical reactions. In fact, functional groups help determine the types of chemical reactions in which an organic compound participates.

Each class of organic compounds is characterized by the presence of one or more specific functional groups. For example, as illustrated in Table 4–1, alcohols contain functional groups known as hydroxyl groups. Note that the symbol R is used to represent the *remainder* of the molecule of which the functional group is a part.

An important property of the functional groups found in biological molecules is their solubility in water. A functional group containing only carbon-hydrogen bonds, such as methyl groups (—CH_3), is nonpolar. Water cannot form hydrogen bonds with nonpolar groups. That is why fats such as butter are hydrophobic and do not dissolve in water.

Positively and negatively charged functional groups are polar and hydrophilic; they are water-soluble because they associate strongly with the polar water molecule. Thus, compounds containing polar functional groups tend to dissolve in water. Hydroxyl and amino groups are polar functional groups. Their oxygen-hydrogen and nitrogen-hydrogen bonds are polar; they have a partial positive electrical charge at the hydrogen end of the bond and a partial negative electrical charge at the oxygen or nitrogen end.

Double bonds formed between carbon and oxygen (—C=O, with a single bond drawn up from the C) are also polar; there is a partial positive charge at the carbon end and a partial negative charge at the oxygen end. Consequently, carboxyl and aldehyde groups are polar (Table 4–1).

Most compounds present in cells contain two or more different functional groups. For example, every amino acid (amino acids are molecular subunits of proteins) contains at least two functional groups: an amino group and a carboxyl group. The chemical properties of these functional groups determine the general properties of amino acids. However, many amino acids contain additional functional groups that determine the specific properties of each type of amino acid. When we know what kinds of functional groups are present in an organic compound, we can predict its chemical behavior.

MANY BIOLOGICAL MOLECULES ARE POLYMERS

Many biological molecules such as proteins and nucleic acids are very large, consisting of thousands of atoms. Such giant molecules are known as **macromolecules.** Many macromolecules are **polymers** produced by linking together small organic compounds called **monomers.** Just as all the words in this book have been written by arranging the 26 letters of the alphabet in various combinations, monomers can be grouped together to form an almost infinite variety of larger molecules. And just as we use different words to convey information, cells use different molecules to convey information. The thousands of different complex organic compounds present in living things are constructed from about 40 small, simple monomers. For example, the 20 common types of amino acid monomers can be linked together end to end in countless ways to form the polymers we know as proteins.

The synthetic process by which monomers are covalently linked is called **condensation** (Fig. 4–2). The *equivalent* of a molecule of water is removed during the reactions that combine monomers. Synthetic processes require energy and are regulated by specific cellular enzymes (proteins that regulate chemical reactions).

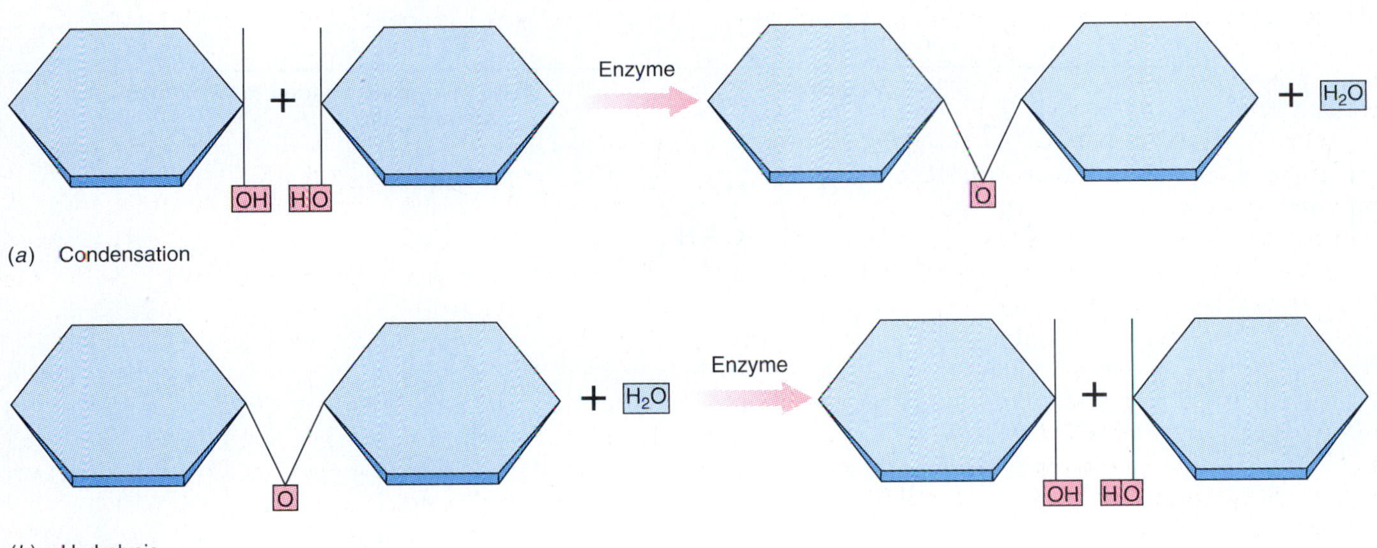

FIGURE 4–2 Making and breaking polymers. (***a***) Two monomers can be joined by condensation. A molecule of water is removed in this process. (***b***) A polymer can be degraded by hydrolysis, a reaction in which water is added. The H^+ and OH^- ions derived from a water molecule are added to the monomer components.

TABLE 4–1
Some Biologically Important Functional Groups

Functional group	*Structural formula*	*Class of compounds characterized by group*	*Description*
Hydroxyl	R—OH	Alcohols $H-CH_2-CH_2-OH$ (H—C(H)(H)—C(H)(H)—OH) Ethanol (the alcohol contained in beverages)	Polar because electronegative oxygen attracts covalent electrons
Amino	$R-NH_2$	Amines R—C(H)(NH_2)—C(=O)—OH Amino acid	Ionic; amino group acts as base
Carbonyl	R—C(=O)—H	Aldehydes H—C(=O)—H Formaldehyde	Carbonyl carbon bonded to at least one H atom; polar
	R—C(=O)—R	Ketones H—C(H)(H)—C(=O)—C(H)(H)—H Acetone	Carbonyl group bonded to two other carbons; polar

Polymers can be degraded to their component monomers by **hydrolysis** (which means "to break with water"). Bonds between monomers are broken by the addition of water. A hydrogen from the water molecule attaches to one monomer and the remaining hydroxyl from the water attaches to the adjacent monomer. Specific examples of condensation and hydrolysis reactions will be presented as we discuss the groups of organic compounds in more detail. The principal groups of biologically important organic compounds are summarized in Table 4–2.

Functional group	Structural formula	Class of compounds characterized by group	Description
Carboxyl	R—C(=O)—OH	Carboxylic acids (organic acids) R—CH(NH_2)—C(=O)—OH Amino acid	Ionic; the H can dissociate as an H^+ ion
Methyl	R—CH_3	Component of many organic compounds H—C(H)(H)—H Methane	Nonpolar
Phosphate	R—O—P(=O)(OH)—OH	Organic phosphates HO—P(=O)(OH)—O—R Phosphate ester (as found in ATP)	Dissociated form of phosphoric acid; the phosphate ion is covalently bonded by one of its oxygen atoms to one of the carbons; ionic
Sulfhydryl	R—SH	Thiols CH_2(SH)—CH(NH_2)—C(=O)—OH Cysteine	Help stabilize internal structure of proteins

Carbohydrates Include the Sugars, Starches, and Cellulose

Familiar to us as sugars and starches, **carbohydrates** serve as fuel molecules and are also important structural components, especially in plant cells. Carbohydrates contain carbon, hydrogen, and oxygen atoms in a ratio of approximately one carbon to two hydrogens to one oxygen (CH_2O). The term carbohydrate, meaning "hydrate (water) of carbon," describes the (approximately) 2 to 1 ratio of hydrogen to oxygen, the same ratio found in water (H_2O). Carbohydrates are classified as monosaccharides, disaccharides, or polysaccharides.

Monosaccharides Are Simple Sugars

Monosaccharides (from two Greek roots, *mono-*, meaning "one" and *sachar-*, meaning "sugar") are simple sugars that usually contain from three to six carbon atoms. **Glucose** and **fructose** are examples of monosaccharides.

TABLE 4–2
Some of the Groups of Biologically Important Organic Compounds

Class of compounds	*Component elements*	*Description*	*How to recognize*	*Principal function in living systems*
Carbohydrates	C, H, O	Contain approximately 1 C : 2 H : 1 O (but make allowance for loss of oxygen and hydrogen when sugar units are linked)	Count the carbons, hydrogens, and oxygens.	Cellular fuel; energy storage; structural component of plant cell walls; component of other compounds such as nucleic acids and glycoproteins
		1. Monosaccharides (simple sugars)—mainly five-carbon (pentose) molecules such as ribose or six-carbon (hexose) molecules such as glucose and fructose	Look for the ring shapes: ⬡ hexose ⬠ or pentose	Cellular fuel; components of other compounds
		2. Disaccharides—two sugar units linked by a glycosidic bond, e.g., maltose, sucrose	Count sugar units.	Components of other compounds
		3. Polysaccharides—many sugar units linked by glycosidic bonds, e.g., glycogen, cellulose	Count sugar units.	Energy storage; structural components of plant cell walls
Lipids	C, H, O	Contain less oxygen relative to carbon and hydrogen than do carbohydrates		Energy storage; cellular fuel, structural components of cells; thermal insulation
		1. Neutral fats. Combination of glycerol with one to three fatty acids. Monoacylglycerol contains one fatty acid; diacylglycerol contains two fatty acids; triacylglycerol contains three fatty acids. If fatty acids contain double carbon-to-carbon linkages (C=C), they are unsaturated; otherwise they are saturated.	Look for glycerol at one end of molecule. H H—C—O— H—C—O— H—C—O— H	Cellular fuel; energy storage

Both are composed of a single six-carbon (hexose) unit with the formula $C_6H_{12}O_6$ (Fig. 4–3). Such compounds as glucose and fructose, which have identical molecular formulas but different arrangements of atoms, are called **isomers.** Because their atoms are arranged differently, the two sugars have different chemical properties.

Glucose, sometimes referred to as blood sugar, is the most abundant hexose in the bodies of humans and other animals. Its concentration is kept at a relatively constant level in the blood, and glucose is used by the cells as a fuel molecule. The **pentoses** (five-carbon sugars) **ribose** and **deoxyribose** are components of nucleotides.

Molecules are not the simple two-dimensional structures that are usually portrayed on a printed page. In fact, the properties of each compound depend in part on its three-dimensional structure. In solution, molecules of glucose and other monosaccharides exist in more than one three-dimensional form. Glucose in solution typically exists as a ring made up of five of its six carbon atoms (Fig. 4–3).

In some compounds, certain atoms can be arranged in more than one position in space around the carbon atom to which they are bonded. Such isomers differ from one another only in some geometric or three-dimensional way so that a right-handed form and a left-handed mirror image form both exist. Only the right-handed form of glucose (sometimes called dextrose from the Latin word, meaning "right") is usable by most organisms. Since it is

Class of compounds	*Component elements*	*Description*	*How to recognize*	*Principal function in living systems*
		2. Phospholipids. Composed of glycerol attached to one or two fatty acids and to an organic base containing phosphorus	Look for glycerol and side chain containing phosphorus and nitrogen.	Components of cell membranes
		3. Steroids. Complex molecules containing carbon atoms arranged in four interlocking rings (three rings contain six carbon atoms each and the fourth ring contains five)	Look for four interlocking rings:	Some are hormones; others include cholesterol, bile salts, vitamin D.
Proteins	C, H, O, N, usually S	One or more polypeptides (chains of amino acids) coiled or folded in characteristic shapes	Look for amino acid units joined by C—N bonds.	Serve as enzymes; structural components; muscle proteins; hemoglobin
Nucleic acids	C, H, O, N, P	Backbone composed of alternating pentose and phosphate groups, from which nitrogenous bases project. DNA contains the sugar deoxyribose and the bases guanine, cytosine, adenine, and thymine. RNA contains the sugar ribose and the bases guanine, cytosine, adenine, and uracil. Each molecular subunit, called a *nucleotide*, consists of a pentose, a phosphate, and a nitrogenous base.	Look for a pentose-phosphate backbone. DNA forms a double helix.	Storage, transmission, and expression of genetic information

sweet to the taste but cannot be utilized by our body cells, the left-handed form of glucose may some day be used as an artificial sweetener in dietetic foods.

Disaccharides Consist of Two Monosaccharide Units

A **disaccharide** (*di-*, meaning "two") consists of two monosaccharide units covalently bonded to one another. The disaccharide **maltose** (malt sugar) consists of two chemically linked glucose units. **Sucrose,** the sugar we use to sweeten our foods, consists of a glucose unit combined with a fructose unit. **Lactose** (the sugar present in milk) is composed of one molecule of glucose and one of galactose, another hexose monosaccharide.

A disaccharide can be hydrolyzed, that is, split with the addition of water into two monosaccharide units. During digestion, maltose is hydrolyzed to form two molecules of glucose:

$$\text{Maltose} + \text{Water} \rightarrow \text{Glucose} + \text{Glucose}$$

Similarly, sucrose is hydrolyzed during digestion to form glucose and fructose:

$$\text{Sucrose} + \text{Water} \rightarrow \text{Glucose} + \text{Fructose}$$

Structural formulas for the compounds in this reaction are shown in Figure 4–4.

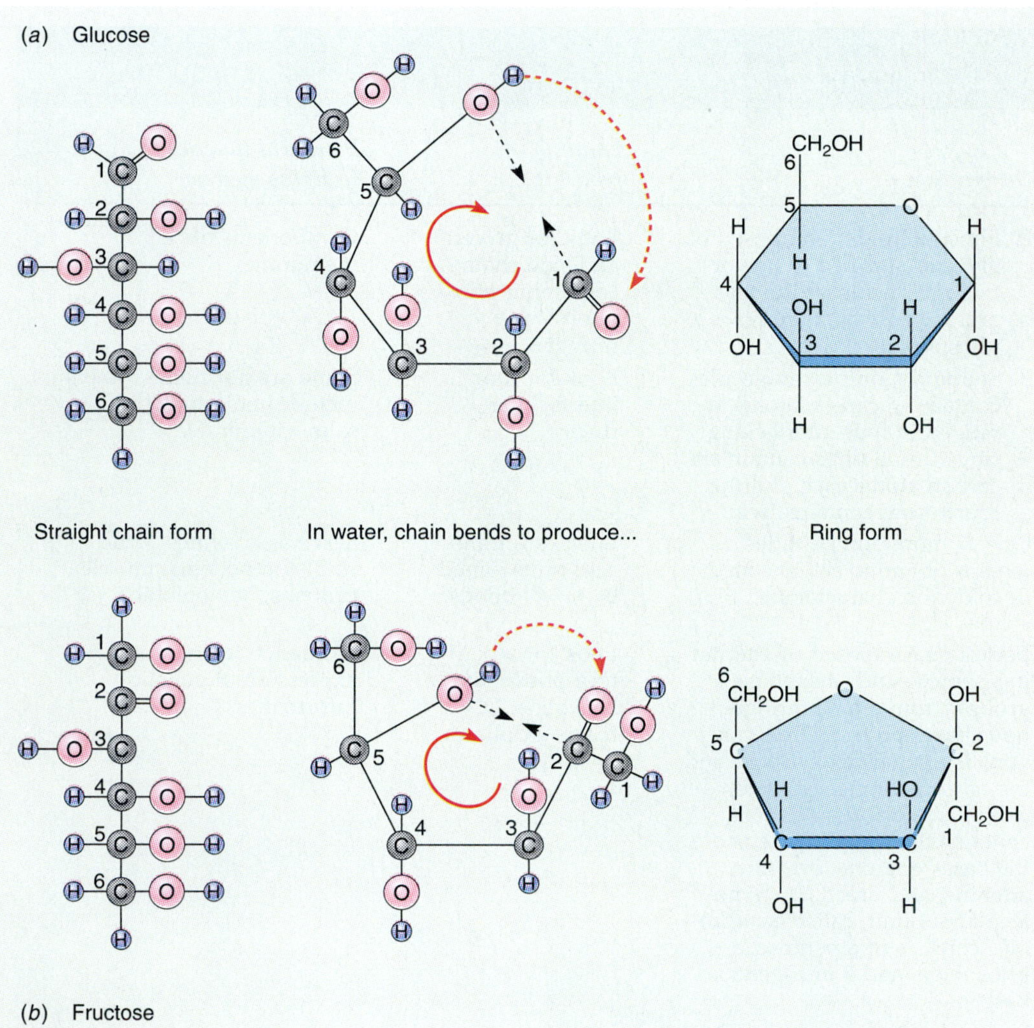

FIGURE 4–3 Straight chain and ring forms of glucose and fructose. The carbon atoms of sugars can be numbered starting with the end of the molecule closest to the aldehyde or ketone group. In the ring form, the thickest part of the bond is interpreted as being the part of the molecule "nearest" the viewer.

FIGURE 4–4 A disaccharide can be hydrolyzed (split with the addition of water) to yield two monosaccharide units. (***a***) These children are enjoying chocolate, sweet due to its sucrose content. (***b***) Hydrolysis of sucrose yields a molecule of glucose and a molecule of fructose. (*a*, R. Bonner)

(*b*)

Sucrose + H_2O ⟶ Glucose + Fructose

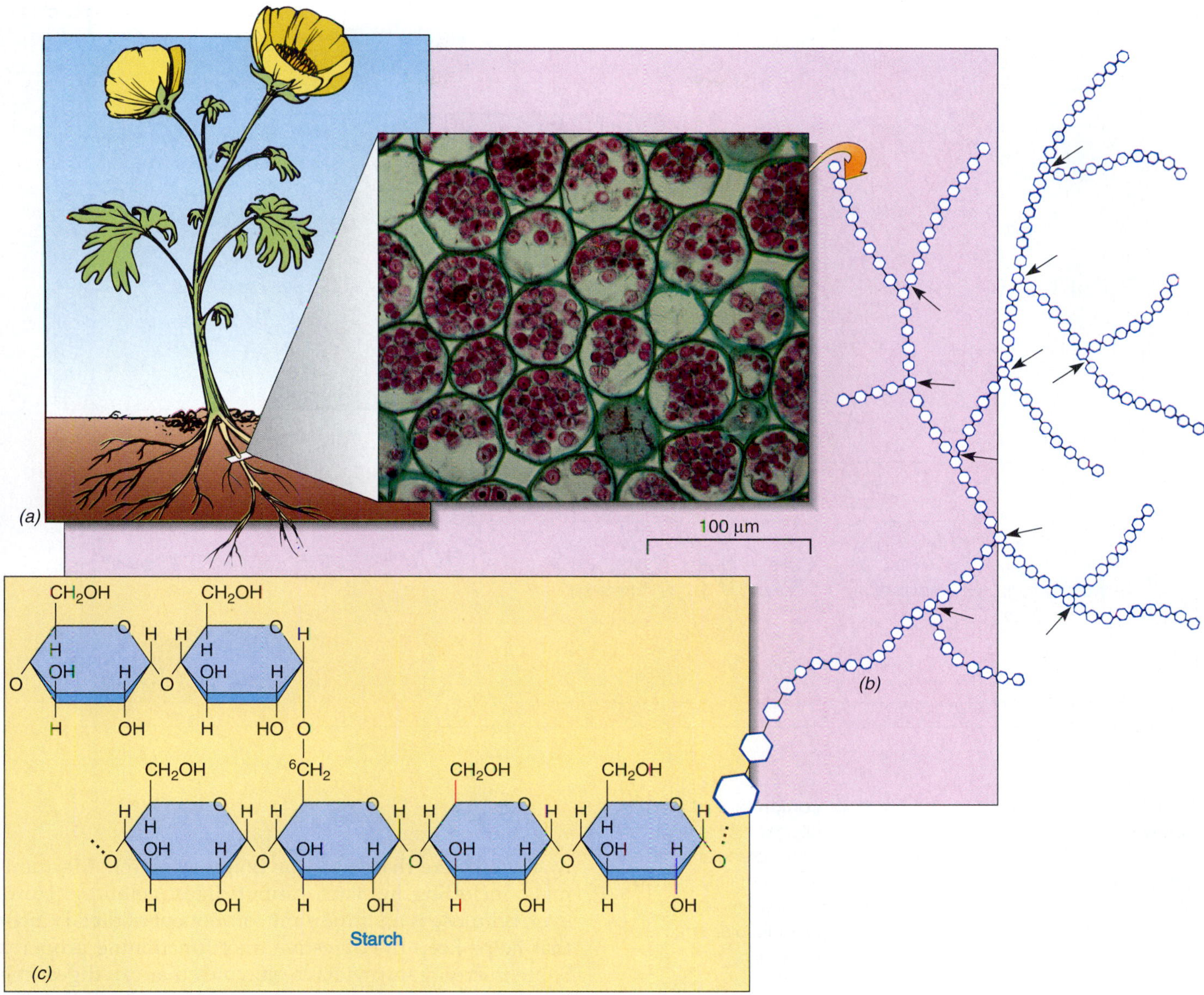

FIGURE 4–5 Starch is a storage form of carbohydrate found in plants. (***a***) Starch (stained purple) stored in specialized organelles in cells of a buttercup root. (***b***) Diagrammatic representation of starch. The arrows represent the branch points. (The structure of glycogen is very similar.) (***c***) Starch molecules are branched polysaccharides composed of glucose molecules joined by covalent bonds known as glycosidic bonds. At the branch points, bonds occur between carbon-6 of a glucose in the straight chain and carbon-1 of the glucose in the branching chain. (Glycogen is more highly branched than starch.) (*a*, Ed Reschke)

Polysaccharides Are Large Polymers

Monosaccharides can be linked to one another to make not only disaccharides, but very large **polysaccharides** (*poly-*, meaning "many"). A polysaccharide is a single long chain, or a branched chain, consisting of repeating units of a simple sugar, usually glucose. Typically, thousands of glucose units may be present in a single molecule of polysaccharide. Because they are composed of different isomers of glucose, or because the glucose units are arranged differently, these polysaccharides have different properties.

The most abundant carbohydrates are polysaccharides, which are the starches, glycogen, cellulose, and the chitin that forms the outer covering (exoskeleton) of insects. **Starch** is the typical storage form of carbohydrate in plants, while **glycogen** (sometimes referred to as animal starch) is the form in which glucose is stored in animal tissues (Fig. 4–5). Glycogen is a highly branched polysaccharide, more water-soluble than plant starch. Be-

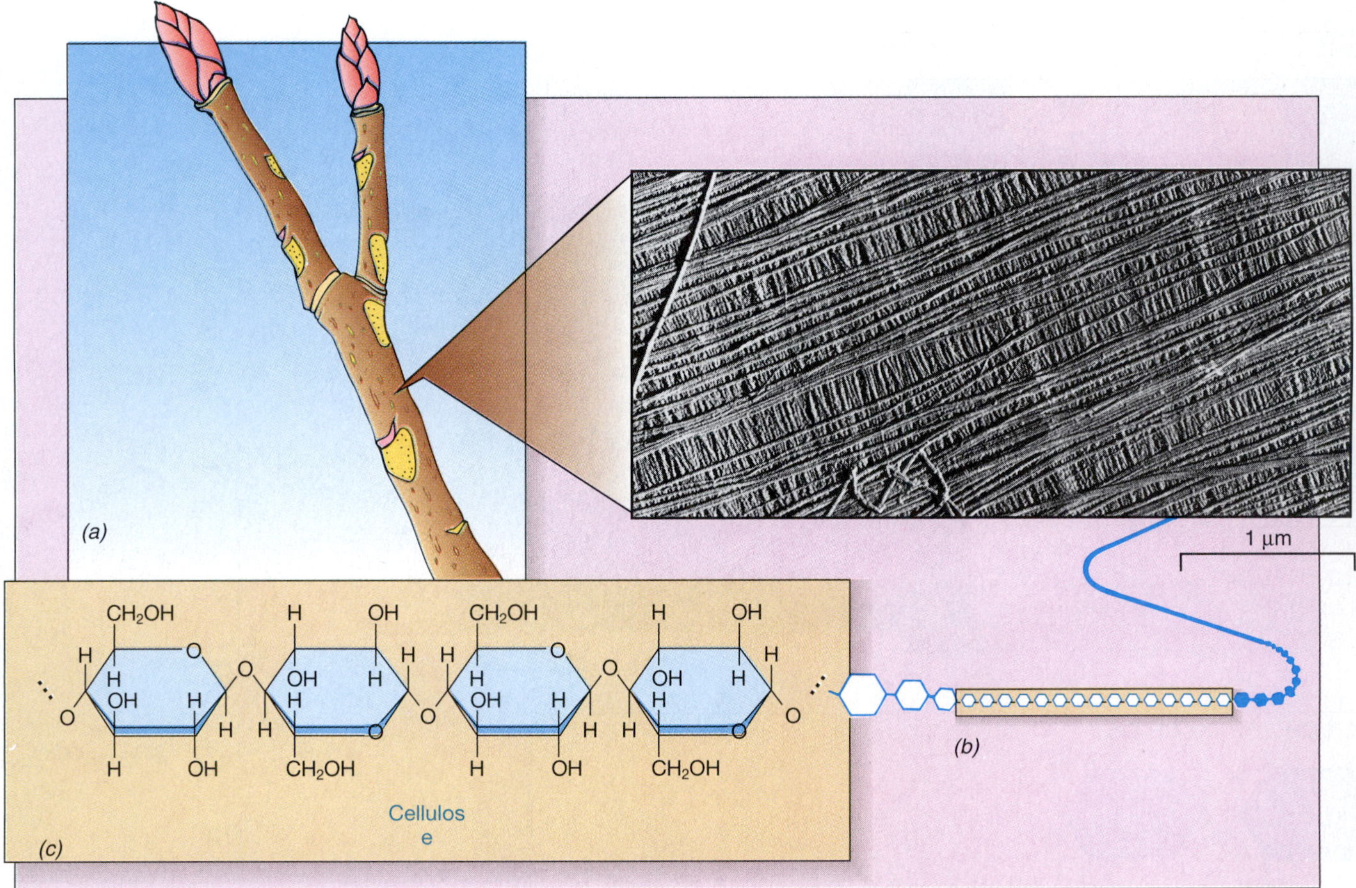

FIGURE 4–6 Cellulose is an important component of plant cell walls. (***a***) An electron micrograph of cellulose fibers from the cell wall of a woody twig. (***b***) The cellulose molecule is an unbranched polysaccharide composed of approximately 10,000 glucose units. Each hexagon represents a glucose molecule bound to the adjacent glucose molecule. (***c***) Molecular structure of cellulose. (*a*, Biophoto Associates/Photo Researchers, Inc.)

cause the glucose molecule is small and readily soluble, it leaks out of cells and cannot be effectively stored. The larger, less soluble starch and glycogen molecules do not readily pass through the plasma membrane. Thus, instead of storing simple sugars, cells store the more complex polysaccharides, such as glycogen, which can be broken down as needed into simple sugars.

Carbohydrates are the most abundant group of organic compounds on Earth, and cellulose is the most abundant carbohydrate, accounting for 50% or more of all the carbon in plants. Wood is about half cellulose, and cotton is at least 90% cellulose. Plant cells are surrounded by a strong supporting cell wall consisting mainly of cellulose. Cellulose is an insoluble polysaccharide composed of many glucose molecules joined together (Fig. 4–6).

The bonds joining the sugar units in cellulose are different from those in starch. These bonds cannot be split by the enzyme that splits the bonds in starch. Most animals, including humans, cannot digest cellulose. However, cellulose is an important component of dietary fiber that helps keep the digestive tract functioning properly. Some microorganisms have enzymes that can digest cellulose to glucose. Animals whose diets are high in cellulose, such as cattle, rabbits, and termites, usually harbor large populations of such microorganisms in their digestive tract.

Chitin, a tough modified polysaccharide, is secreted by the cells of many animals and fungi. It is the main component of the exoskeletons of insects and crustaceans such as crabs, lobsters, and crayfish. The molecular unit that makes up chitin is glucosamine, a sugar in which a hydroxyl group (OH) of a monosaccharide is replaced by an amino group (NH_2; see the section on proteins).

LIPIDS ARE FATS OR FAT-LIKE SUBSTANCES

Lipids are a heterogeneous group of compounds that have a greasy or oily consistency and are relatively insoluble in water. Like carbohydrates, lipids are composed

FIGURE 4–7 Honeybees on a brood comb. The comb is composed of wax secreted by special abdominal glands of the bees. Wax is a compound consisting of fatty acids and alcohols, and although it is classified as a lipid, it can be digested by very few animals. (Charles D. Winters)

of carbon, hydrogen, and oxygen atoms, but they have relatively less oxygen in proportion to the carbon and hydrogen than do carbohydrates. Among the groups of biologically important lipids are the neutral fats, phospholipids, steroids, carotenoids (orange and yellow plant pigments), and waxes (Fig. 4–7). Lipids are important biological fuels; they serve as structural components of cell membranes, and some are essential hormones.

Neutral Fats Are Composed of Glycerol and Fatty Acids

Neutral fats are the most abundant lipids in living things. These compounds are an economical form for fuel reserves storage because they provide more than twice as much energy per gram as do carbohydrates. Fats also can serve as insulation (as in blubber), to produce body contours (as they do in humans), or even as support tissue (they hold the kidneys in place). Carbohydrates and proteins can be transformed by enzymes into fats and stored in the cells of adipose (fat) tissue.

A neutral fat consists of glycerol joined to one, two, or three molecules of a fatty acid. **Glycerol** is a three-carbon alcohol that contains three —OH groups (Fig. 4–8). A **fatty acid** is a long, straight chain of carbon atoms with a carboxyl group (—COOH) at one end. About 30 different fatty acids are commonly found in lipids. They typically have an even number of carbon atoms. For example, butyric acid, present in rancid butter, has four carbon atoms, and oleic acid, the most widely distributed fatty acid in nature, has 18 carbon atoms.

Saturated fatty acids contain the maximum possible number of hydrogen atoms. **Unsaturated fatty acids** contain carbon atoms that are doubly bonded with one another and are not fully saturated with hydrogens. Fatty acids with more than one double bond are called **polyunsaturated fatty acids.** Fats containing unsaturated fatty acids are oils (such as corn oil, olive oil, or peanut oil).

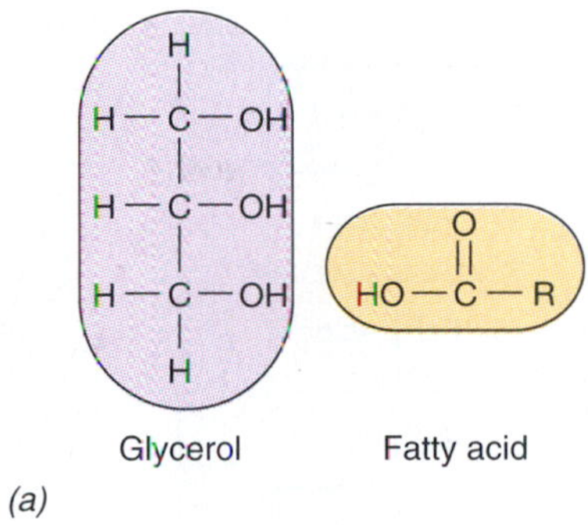

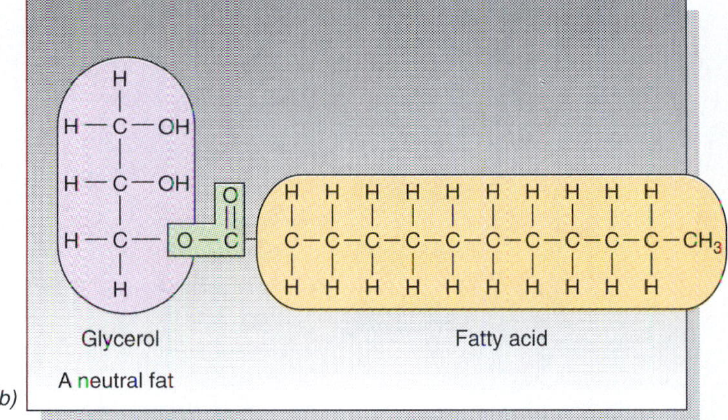

FIGURE 4–8 Neutral fats. (***a***) Structure of glycerol and of a fatty acid. The carboxyl (—COOH) group is present in all fatty acids. The R represents the remainder of the molecule, a carbon chain that varies in length and number of double bonds with each type of fatty acid. (***b***) Structure of a monoacylglycerol. (***c***) The walrus stores large amounts of fat to protect it from the cold. These walruses were photographed in Alaska. (Michio Hoshino/Minden Pictures)

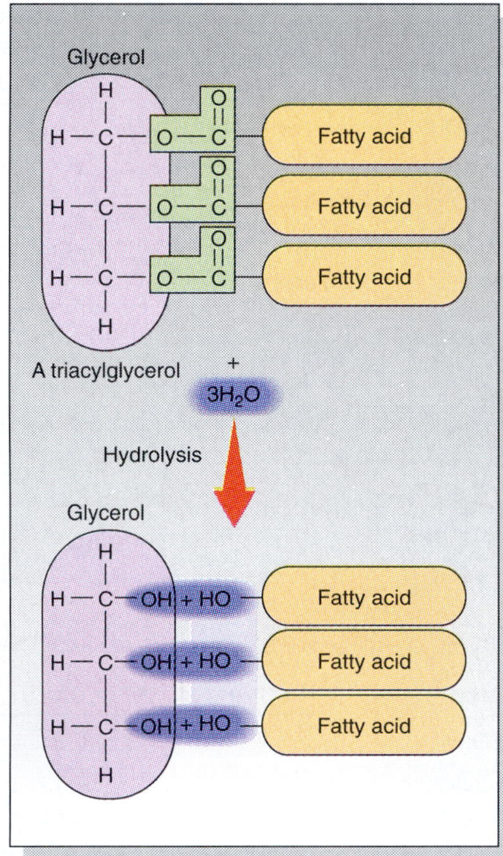

FIGURE 4–9 Hydrolysis of a triacylglycerol. Glycerol plus three fatty acids are produced in this process.

Most oils are liquid at room temperature. In contrast, saturated fats are usually solids; butter and animal fat are examples. At least two fatty acids (linoleic and linolenic) cannot be manufactured by the human body and are therefore essential nutrients, which must be included in the diet.

When a glycerol molecule combines chemically with one fatty acid, a **monoacylglycerol** (sometimes called monoglyceride) is formed. When two fatty acids combine with a glycerol, a **diacylglycerol** (or diglyceride) is formed, and when three fatty acids combine with one glycerol molecule, a **triacylglycerol** (or triglyceride) is formed. Hydrolysis of a triacylglycerol is illustrated in Figure 4–9.

Phospholipids Are Important Components of Cell Membranes

A **phospholipid** consists of a glycerol molecule attached to two fatty acids, and to a phosphate group. The phosphate group can be joined with another small molecular subunit such as choline (shown in Fig. 4–10), which usually contains nitrogen. (Note that phosphorus and nitrogen are characteristic of phospholipids, but are absent in the neutral fats.)

The two ends of the phospholipid molecule differ physically as well as chemically. The fatty acid portion of the molecule is *hydrophobic* (water-hating) and is not soluble in water. However, the portion composed of glycerol and the organic base is ionized and readily water-soluble. This end of the molecule is *hydrophilic* (water-loving). The polarity of these lipid molecules causes them to arrange themselves in the presence of water, with their hydrophilic water-soluble heads facing outward toward the surrounding water. The hydrophobic tails face in the opposite direction. As will be discussed in Chapter 6, the plasma membrane is formed from two layers of phospholipid molecules.

Steroids Contain Four Rings of Carbon Atoms

A **steroid** is made of carbon atoms arranged in four interlocking rings; three of the rings contain six carbon atoms and the fourth contains five (Fig. 4–11). The length and structure of the side chains that extend from these rings distinguish one steroid from another.

Among the steroids of biological importance are cholesterol, bile salts, reproductive hormones, and hormones secreted by the adrenal glands. Cholesterol is a structural component of animal cell membranes. Steroid hormones regulate certain aspects of metabolism in animals, including insects, crabs, and vertebrates.

PROTEINS ARE MACROMOLECULES FORMED FROM AMINO ACIDS

Proteins serve as important structural components of cells and tissues, so growth and repair, as well as maintenance of the organism, depend on an adequate supply of these compounds. Some proteins serve as **enzymes,** catalysts that regulate the thousands of different chemical reactions that take place in living cells.

The protein components of a cell determine its function. Each kind of cell (e.g., muscle, bone, blood, nerve) has characteristic types, distributions, and amounts of protein. The protein composition determines what the cell looks like and how it functions. A muscle cell is different from other cell types because it contains large amounts of the proteins myosin and actin. These proteins contribute greatly to its microscopic structure and are responsible for its ability to contract. The ability of red blood cells to transport oxygen is due to the protein hemoglobin.

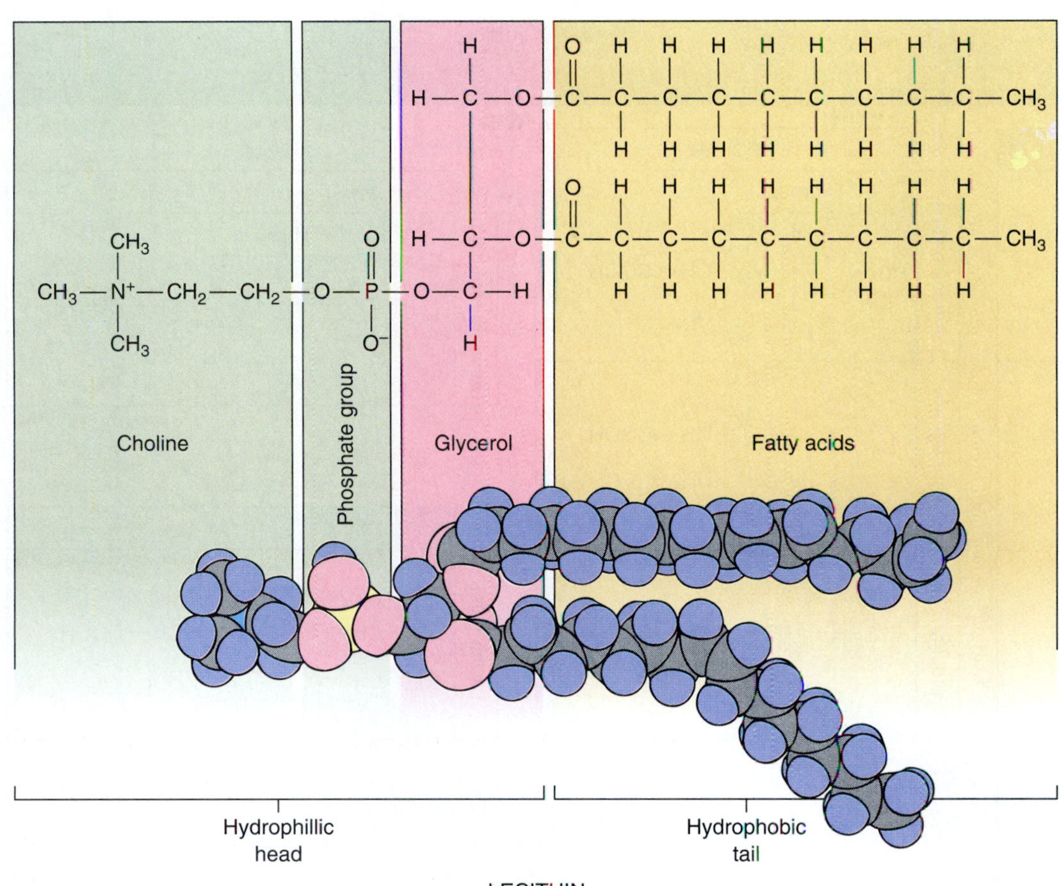

FIGURE 4–10 Lecithin is a phospholipid found in all cell membranes. Like many phospholipids, it is a derivative of phosphatidic acid, a compound consisting of glycerol chemically combined with two fatty acids and a phosphate group. It forms when phosphatidic acid combines with the compound choline. A space-filling model of lecithin is shown below the structural formula.

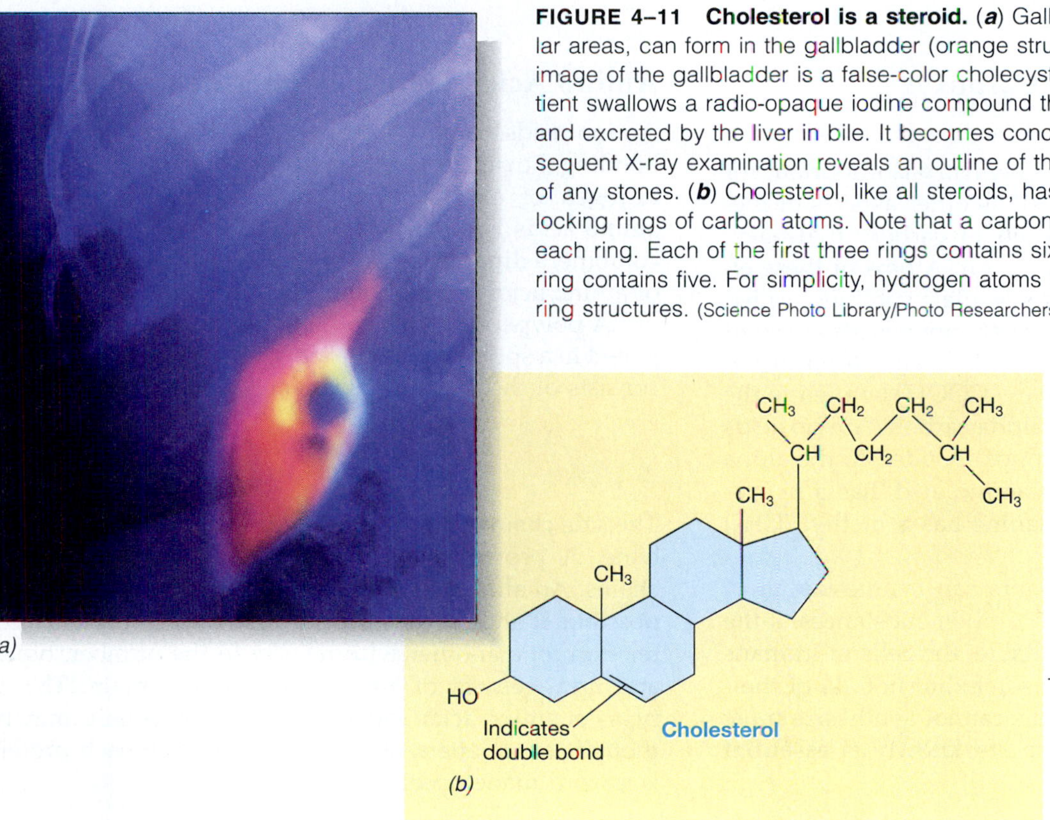

FIGURE 4–11 Cholesterol is a steroid. (***a***) Gallstones, seen here as dark circular areas, can form in the gallbladder (orange structure) from cholesterol. This X-ray image of the gallbladder is a false-color cholecystogram. In this procedure, the patient swallows a radio-opaque iodine compound that is absorbed in the intestine and excreted by the liver in bile. It becomes concentrated in the gallbladder. Subsequent X-ray examination reveals an outline of the gallbladder and the presence of any stones. (***b***) Cholesterol, like all steroids, has the basic skeleton of four interlocking rings of carbon atoms. Note that a carbon atom is present at each point in each ring. Each of the first three rings contains six carbon atoms, and the fourth ring contains five. For simplicity, hydrogen atoms have not been drawn within the ring structures. (Science Photo Library/Photo Researchers, Inc.)

TABLE 4–3
Structure of Some Common Amino Acids

NONPOLAR

	R-groups	
Glycine (Gly)	H	—CH—COOH (with NH_2 on CH)
Alanine (Ala)	CH_3	—CH—COOH (with NH_2 on CH)
Valine (Val)	(H_3C)(H_3C)CH	—CH—COOH (with NH_2 on CH)
Phenylalanine (Phe)	(benzene ring)—CH_2	—CH—COOH (with NH_2 on CH)

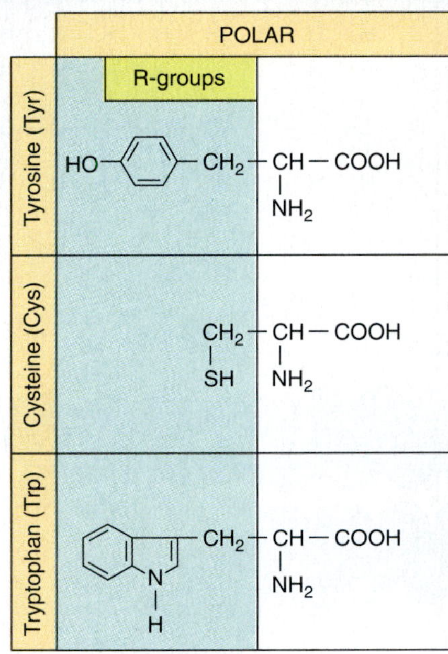

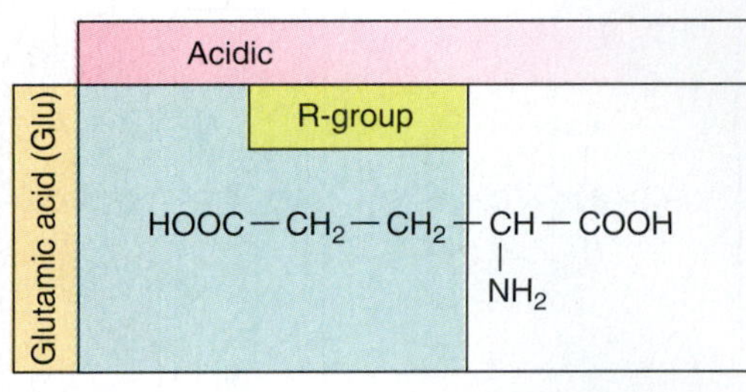

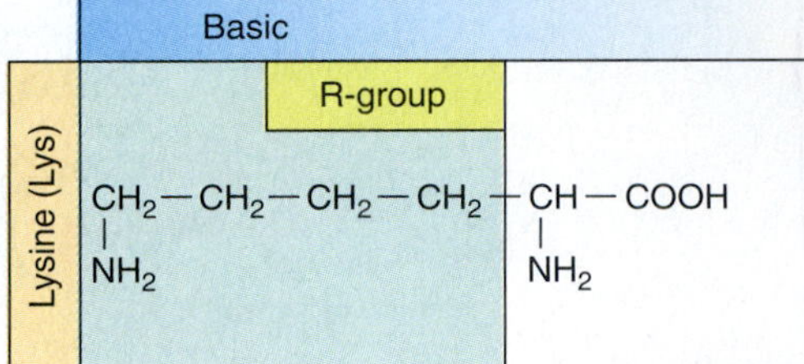

Amino Acids Contain a Carboxyl and an Amino Group

A basic knowledge of protein chemistry is important for understanding nutrition as well as other aspects of metabolism. Proteins are composed of carbon, hydrogen, oxygen, nitrogen, and usually sulfur. Atoms of these elements are arranged into molecular subunits called **amino acids.** More than 20 kinds of amino acids are commonly found in proteins. All contain an amino group (—NH_2) and a carboxyl group (— COOH) bonded to the same carbon atom, called the alpha carbon. Amino acids differ in their *side chains* (R groups) bonded to the alpha carbon. **Glycine,** the simplest amino acid, has a hydrogen atom as its side chain; alanine has a methyl (CH_3) group (Table 4–3).

With some exceptions, plants can synthesize all of their needed amino acids from simpler substances. If the needed raw materials are available, the cells of humans and animals can manufacture some, but not all, of their amino acids. Those that animals cannot synthesize must be obtained in the diet. These are known as **essential amino acids.**

Amino Acids Form Polypeptide Chains

Amino acids combine chemically with one another by bonding the carboxyl carbon of one molecule to the amino nitrogen of another. This covalent bond linking two amino acids is a **peptide bond.** When two amino acids combine, a **dipeptide** is formed (Fig. 4–12); a longer chain of amino acids is a **polypeptide.**

A polypeptide may contain hundreds of amino acids joined in a specific linear order. The backbone of the chain consists of the repeating sequence

—N—C—C—N—C —C—

The side chains of the amino acids extend from this backbone. A protein consists of one or more polypeptide chains. An almost infinite variety of protein molecules is possible. It should be clear that the various proteins differ from one another with respect to the number, types, and arrangement of amino acids they contain. The 20 types of amino acids found in biological proteins may be thought of as letters of a protein alphabet; each protein is a word made up of amino acid letters.

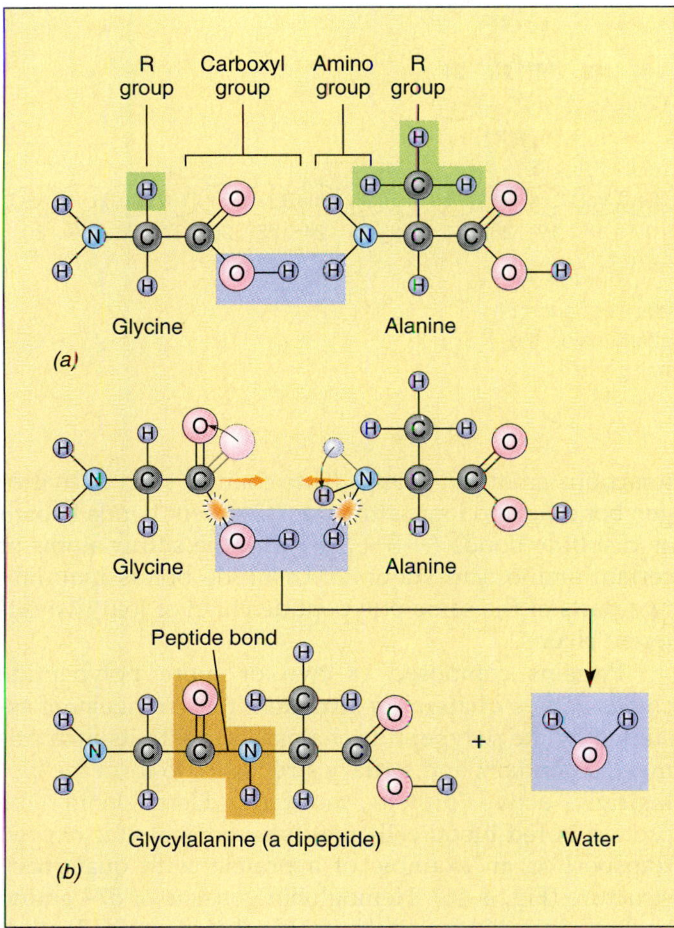

FIGURE 4–12 Formation of a dipeptide. Two amino acids (***a***) combine chemically to form a dipeptide (***b***). Water is produced as a by-product in this condensation reaction. The bond between two amino acids (between the carbon of one and the nitrogen of the second) is a peptide bond.

Concept Connections

Proteins ⟷ *Evolution* ⟷ *Genetics*

All organisms are composed of the same types of biological molecules—carbohydrates, lipids, proteins, and nucleic acids. Yet, each organism is unique due to differences in the nucleotide sequence of its DNA, the polymer that constitutes the genes. Because the DNA, which is different in each organism, determines the protein structure, each organism has different proteins.

Most proteins are species-specific, that is, they vary slightly in each species. As a result, protein makeup (as determined by the instructions encoded in the nucleic acids) is also responsible for differences among species. Thus, the types and distributions of proteins in the cells of a dog vary somewhat from those in the cells of a fox or a coyote. The degree of difference in the proteins of two species is thought to reflect evolutionary relationships. Organisms less closely related by evolution have proteins that differ more markedly than those of closely related forms. As we will discuss in Chapter 20, examining differences in protein structure among species is an important tool used by scientists to classify organisms.

Some proteins differ slightly even among individuals of the same species, so each individual is biochemically unique. In Chapter 13, we will discuss the concept of DNA fingerprints. Only genetically identical organisms, which include identical twins, clones of asexually reproducing organisms, or members of closely inbred strains of organisms, have identical proteins.

Proteins Have Four Levels of Organization

Think about a telephone handset cord (Fig. 4–13). The cord itself is long, thin, and flexible. That is its basic shape or primary structure. To avoid clumsy tangling, it is made in such a way that it coils itself into a long, spring-like shape, which is its secondary structure. Despite that, the cord is usually further twisted or kinked into some tertiary structure.

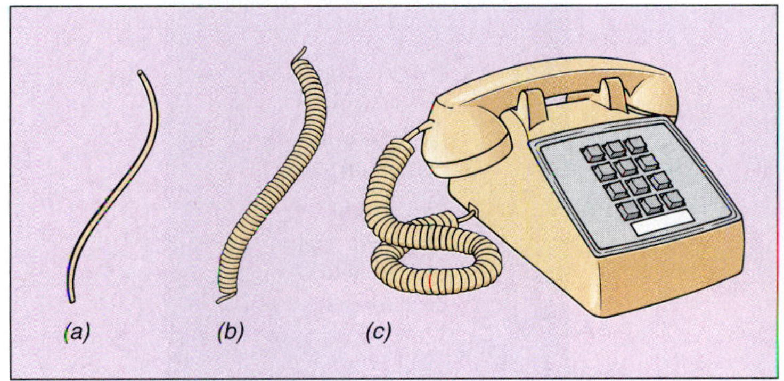

FIGURE 4–13 Levels of organization. A telephone handset cord illustrates levels of structure in proteins and nucleic acids. (***a***) Primary structure, (***b***) secondary structure, (***c***) tertiary structure.

Similar levels of organization can be distinguished in the protein molecule. The sequence of amino acids in a polypeptide chain is its **primary structure** (Fig. 4–14). This sequence is directly specified by the instructions in a gene.

Polypeptide chains do not lay flat or coil randomly; they coil in a predictable way, forming a specific three-dimensional structure. This **secondary structure** of protein molecules involves the coiling of the polypeptide chain into a helix or some other regular shape (Fig. 4–15). The regularity is due to hydrogen bonds between the atoms of the uniform backbone of the polypeptide chain.

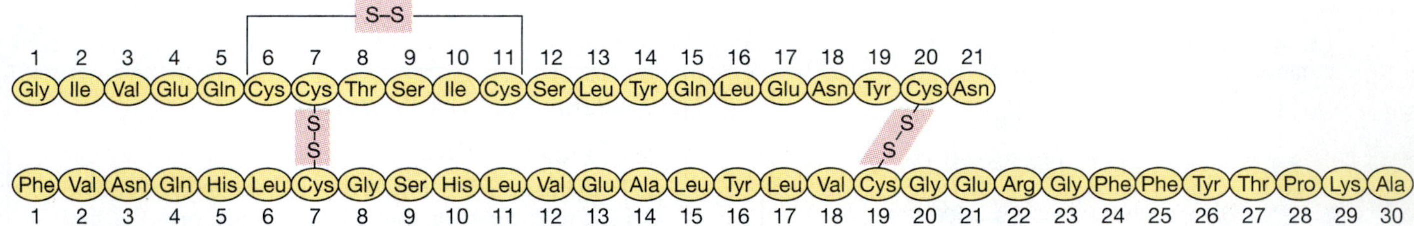

FIGURE 4–14 Primary structure of the protein insulin. The primary structure of the two polypeptide chains that make up insulin. The primary structure is the linear sequence of amino acids. Each oval in the diagram represents an amino acid. The letters inside the ovals are symbols for the names of the amino acids. Insulin is a very small protein.

A common secondary structure in protein molecules is known as the **alpha helix.** The helix is formed by spiral coils of the polypeptide chain (Fig. 4–16). The alpha helix is a very uniform geometric structure with exactly 3.6 amino acids occupying each turn of the helix. The helical structure is determined and maintained by the formation of hydrogen bonds between amino acids in successive turns of the spiral coil.

The alpha helix is the basic structural unit of fibrous proteins such as wool, hair, skin, and nails. The fiber is elastic because the hydrogen bonds can be broken and then reformed. This is why human hairs can be stretched to some extent and will then snap back to their original length.

The **tertiary structure** of a protein molecule is the overall three-dimensional shape assumed by each polypeptide chain (Fig. 4–16). This is determined by interactions among R groups (side chains), such as hydrogen bonding and ionic attraction. Covalent bonds known as disulfide bonds (—S—S—) link the sulfur atoms of certain amino acid subunits. Disulfide bonds may link two parts of the same polypeptide chain or join two different chains.

Proteins composed of two or more polypeptide chains have a **quaternary structure,** the arrangement assumed by the polypeptide chains, each with its own primary, secondary, and tertiary structures, to form the biologically active protein molecule. Hemoglobin, the protein in red blood cells that is responsible for oxygen transport, is an example of a protein with quaternary structure (Fig. 4–16). Hemoglobin consists of 574 amino acids arranged in four polypeptide chains—two identical alpha and two identical beta chains. Its chemical formula is $C_{3032}H_{4816}O_{872}N_{789}S_8Fe_4$.

FIGURE 4–15 The silk of a spider's web is composed of an extremely strong, flexible protein (fibroin) that has a secondary structure that resembles a pleated sheet. The silk fibers harden as they are spun from the glands in the spider's abdomen. To conserve protein, many spiders eat their own webs when they become torn or worn. This banded argiope spider has trapped a grasshopper. (Skip Moody/Dembinsky Photo Associates)

The Biological Activity of a Protein Can Be Disrupted

The biological activity of a protein can be disrupted by changes in the amino acid sequence or in the conformation of a protein. For example, in *sickle cell anemia,* a mutation (a chemical change in a gene) results in the production of hemoglobin with one incorrect amino acid in the chain (Chapters 12 and 14). This substitution makes the hemoglobin less soluble and more likely to form crystal-like structures that change the shape of the red blood cell.

Changes in the three-dimensional structure of a protein also disrupt its biological activity. When a protein is heated or treated with any of a number of chemicals, its tertiary structure becomes disordered and the coiled peptide chains unfold to give a more random conformation. This unfolding is accompanied by a loss of the biological activity of the protein, for example, its ability to act as an enzyme. Such change in shape and loss of biological activity is termed **denaturation** of the protein. Denaturation generally cannot be reversed. A familiar example is the change that takes place in the white of an egg (an albumin protein) when you fry it.

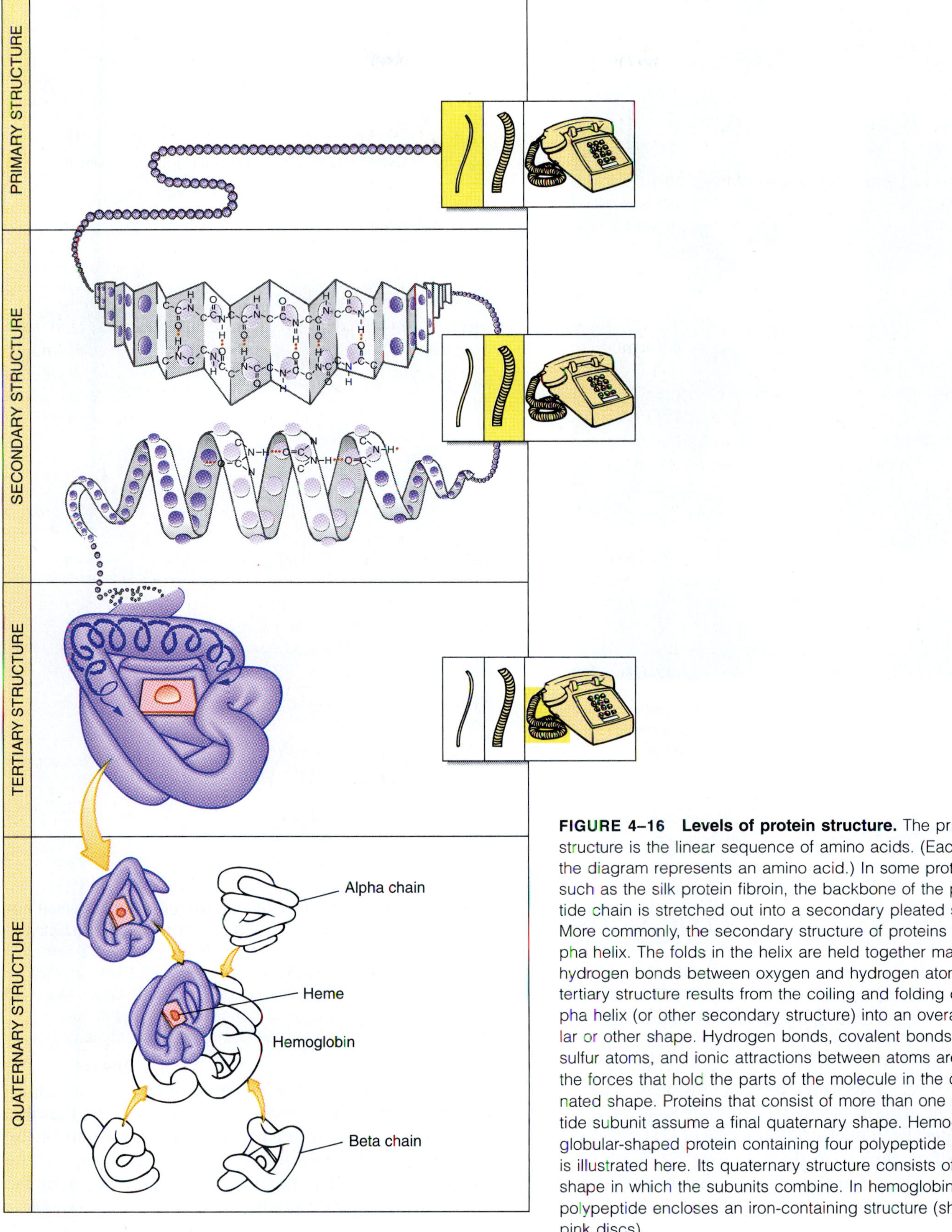

FIGURE 4–16 Levels of protein structure. The primary structure is the linear sequence of amino acids. (Each oval in the diagram represents an amino acid.) In some proteins, such as the silk protein fibroin, the backbone of the polypeptide chain is stretched out into a secondary pleated structure. More commonly, the secondary structure of proteins is an alpha helix. The folds in the helix are held together mainly by hydrogen bonds between oxygen and hydrogen atoms. The tertiary structure results from the coiling and folding of the alpha helix (or other secondary structure) into an overall globular or other shape. Hydrogen bonds, covalent bonds between sulfur atoms, and ionic attractions between atoms are among the forces that hold the parts of the molecule in the designated shape. Proteins that consist of more than one polypeptide subunit assume a final quaternary shape. Hemoglobin, a globular-shaped protein containing four polypeptide subunits, is illustrated here. Its quaternary structure consists of the final shape in which the subunits combine. In hemoglobin each polypeptide encloses an iron-containing structure (shown as pink discs).

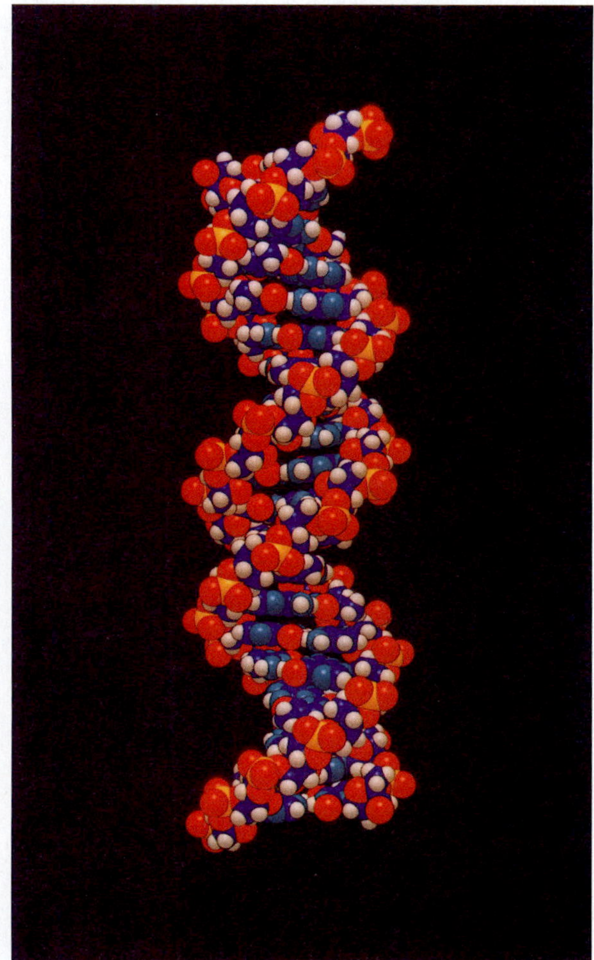

FIGURE 4–17 Structure of DNA. Computer-generated simulation of the colored plastic space-filling molecular models of DNA used by chemists. The colors are red = oxygen, blue = nitrogen, dark blue = carbon, yellow = phosphorus, and white = hydrogen. This photograph shows 20 base pairs of DNA in the crystalline B form first studied by Watson and Crick. (N. L. Max, University of California/Biological Photo Service)

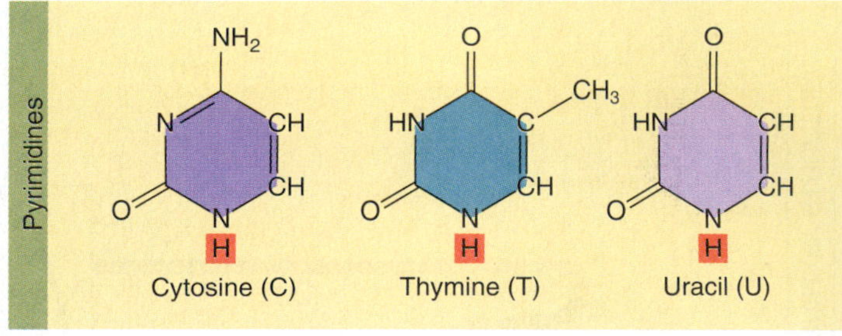

(a)

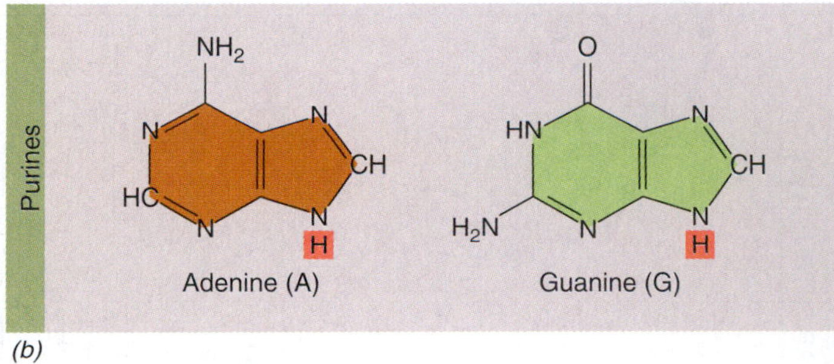

(b)

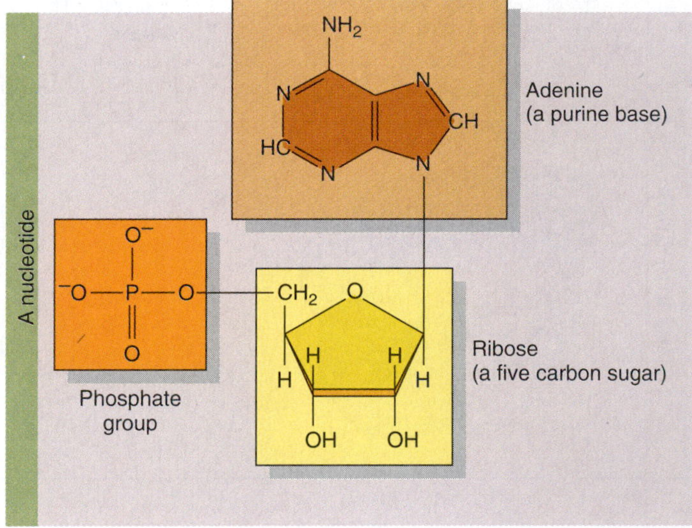

(c)

FIGURE 4–18 A nucleic acid consists of subunits called nucleotides. Each nucleotide consists of (1) a nitrogenous base, which may be either a purine or a pyrimidine; (2) a five-carbon sugar, either ribose (in RNA) or deoxyribose (in DNA); and (3) a phosphate group. (***a***) The three major pyrimidine bases found in nucleotides. (***b***) The two major purine bases found in nucleotides. (***c***) A nucleotide (adenosine monophosphate or AMP).

DNA AND RNA ARE NUCLEIC ACIDS

Nucleic acids, like proteins, are large, complex molecules. Two classes of nucleic acids are **ribonucleic acids (RNA)** and **deoxyribonucleic acids (DNA).** The hereditary material, the genes, is made of DNA (Fig. 4–17). An organism's DNA contains the instructions for making all of its proteins. RNA functions in the process of protein synthesis.

Nucleic acids are composed of molecular units called **nucleotides.** Each nucleotide consists of (1) a five-carbon sugar, either ribose or deoxyribose; (2) a phosphate group; and (3) a nitrogenous base that may be either a double-ringed purine or a single-ringed pyrimidine (Fig. 4–18). DNA contains the purines adenine (A) and guanine (G) and the pyrimidines cytosine (C) and thymine (T), together with the sugar deoxyribose and phosphate. RNA contains the purines adenine and guanine and the pyrimidines cytosine and uracil (U), together with the sugar ribose and phosphate.

WINDOW ON
THE PLANT CELL

(Tear along perforation and add to study notes.)

The molecules of nucleic acids are made of linear chains of nucleotides (Fig. 4–19). The nucleotides are linked by bonds between the sugar molecule of one and the phosphate group of the next. As we will see in our discussion of the genetic code (Chapter 14), the specific information of the nucleic acid is coded in the sequence of the four kinds of nucleotides present in the chain. These sequences specify the order of amino acids in proteins.

A number of other nucleotides serve important functions in living cells. **Adenosine triphosphate (ATP),** composed of adenine, ribose, and three phosphates, serves as the energy currency of all cells (Chapter 7). The two terminal phosphate groups are joined to the nucleotide by unstable bonds, indicated by the ~P symbol. The biologically useful energy of these bonds can be transferred to other molecules.

A nucleotide may be converted by enzymes called cyclases to a cyclic form. ATP, for example, is converted to cyclic adenosine monophosphate (cyclic AMP) by the enzyme adenylate cyclase. Cyclic nucleotides mediate the effects of some hormones and regulate certain aspects of cellular function.

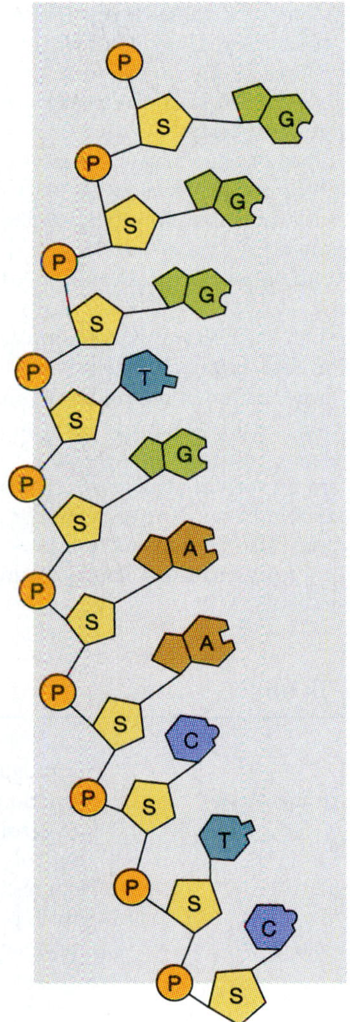

FIGURE 4–19 Schematic diagram of a nucleic acid molecule (DNA or RNA). The four bases of each nucleic acid are arranged in a sequence specific to that particular nucleic acid. P, phosphate; S, sugar; G, guanine; C, cytosine; A, adenine; T, thymine.

Chapter Summary

I. Chains of carbon atoms form the backbone of the organic compounds essential to life. The major groups of organic compounds are carbohydrates, lipids, proteins, and nucleic acids.

II. Organic compounds have specific functional groups with characteristic properties.
   A. Partial charges on atoms at opposite ends of a bond are responsible for the polar property of a functional group.
   B. Polar functional groups interact with other polar groups or with charged ions.

III. Long chains of similar organic subunits linked together are called polymers. For example, proteins are polymers of amino acids and nucleic acids are nucleotide polymers.

IV. Carbohydrates contain carbon, hydrogen, and oxygen in a ratio of approximately 1 carbon to 2 hydrogens to 1 oxygen. Sugars, starches, and cellulose are typical carbohydrates.
   A. Monosaccharides are simple sugars such as glucose, fructose, or ribose. Glucose is an important fuel molecule in living cells.
   B. Most carbohydrates are polysaccharides, which are long chains of repeating units of a simple sugar.
      1. Carbohydrates are typically stored in plants as starch and in animals as glycogen.
      2. The cell walls of plant cells are composed mainly of the polysaccharide cellulose.

V. Lipids are composed of carbon, hydrogen, and oxygen but have relatively less oxygen in proportion to carbon and hydrogen than do carbohydrates. Lipids have a greasy or oily consistency and are relatively insoluble in water.

A. The body stores energy in the form of neutral fats. A fat consists of a molecule of glycerol combined with one to three fatty acids.
   1. Three types of neutral fats are monoacylglycerols, diacylglycerols, and triacylglycerols.
   2. Fatty acids, and therefore fats, can be saturated or unsaturated.

B. Phospholipids are structural components of cell membranes.

C. Steroids, such as cholesterol, contain carbon atoms arranged in four interlocking rings.

VI. Proteins are large, complex polymers of amino acids joined by peptide bonds. They are composed of carbon, hydrogen, oxygen, nitrogen, and sulfur.

A. Proteins are important structural components of cells and tissues. Many serve as enzymes or as hormones, and they may also be used as fuel.

B. Four levels of organization can be distinguished in protein molecules.
   1. Primary structure is the sequence of amino acids in the polypeptide chain.
   2. Secondary structure refers to the coiling of the polypeptide chain in a helix or some other regular conformation.
   3. Tertiary structure is the folding of the chain upon itself.
   4. Quaternary structure is the spatial relationship of the combination of two or more peptide chains.

VII. The nucleic acids DNA and RNA store information that governs the structure and function of the organism. Nucleic acids are composed of carbon, hydrogen, oxygen, nitrogen, and phosphorus.

A. Nucleic acids are composed of long chains of nucleotide units, each composed of (1) a nitrogenous base (a purine or a pyrimidine), (2) a five-carbon sugar (ribose or deoxyribose), and (3) a phosphate group.

B. ATP is a nucleotide of special significance in energy metabolism.

Selected Key Terms

amino acid, p. 72
ATP (adenosine triphosphate), p. 77
carbohydrate, p. 63
condensation, p. 61
denaturation, p. 74
enzyme, p. 70
fatty acid, p. 69
functional group, p. 60
glycerol, p. 69
hydrolysis, p. 61
isomer, p. 64
lipid, p. 68
monoacylglycerol, p. 70
monosaccharide, p. 63
nucleic acid, p. 76
nucleotide, p. 76
peptide bond, p. 72
phospholipid, p. 70
polymer, p. 61
polypeptide, p. 72
polysaccharide, p. 67
protein, p. 70
saturated fatty acid, p. 69
steroid, p. 70
triacylglycerol (triglyceride), p. 70
unsaturated fatty acid, p. 69

Post-Test

1. An example of a monosaccharide is (a) DNA (b) glucose (c) cholesterol (d) cellulose (e) sucrose.

2. Cholesterol is a (a) carbohydrate (b) monosaccharide (c) steroid (d) neutral fat (e) two of the preceding answers are correct.

3. DNA is a (a) carbohydrate (b) monosaccharide (c) steroid (d) nucleic acid (e) two of the preceding answers are correct.

4. Proteins are made up of (a) glucose units (b) monosaccharides (c) amino acids (d) nucleotides (e) two of the preceding answers are correct.

5. Cellulose is a (a) carbohydrate (b) polysaccharide (c) protein (d) nucleic acid (e) two of the preceding answers are correct.

6. Which of the following are important components of cell membranes? (a) phospholipids (b) disaccharides (c) RNA (d) nucleic acids (e) none of the preceding.

7. The energy currency of the cell is (a) maltose (b) ATP (c) steroid (d) DNA (e) none of the preceding.

8. A component of fatty acids: (a) glucose (b) glycerol (c) steroid (d) RNA (e) none of the preceding.

9. Peptide bonds are found linking ________________ ________________.

10. The ________________ structure of a protein is the sequence of amino acids.

11. ________________ is an important polysaccharide component of plant cell walls.

12. Animals store glucose in the form of ________________.

13. A very large molecule is referred to as a/an _______________.

14. _______________ is a modified carbohydrate that is the main component of insect skeletons.

15. The five-carbon sugar found in RNA is _______________.

Review Questions

1. What property of carbon makes it so important in living organisms?
2. Contrast a monosaccharide, such as glucose, with a polysaccharide, such as starch.
3. Why are each of the following biologically important? (a) steroids (b) phospholipids (c) polysaccharides (d) nucleic acids (e) amino acids.
4. Draw a structural formula of a simple amino acid and identify the carboxyl and amino groups.
5. There are thousands of types of proteins. How does one protein differ chemically from another?
6. Compare proteins with nucleic acids.
7. Why are neutral fats important? What are the molecular components of a neutral fat?

Thinking Critically

1. Organic compounds are major components of living systems. Yet, many organic compounds are harmful to living systems. Explain how this apparent paradox could be true.
2. What are some ways that the biological activity of a protein can be disrupted? How might such a disruption affect an organism?
3. Proteins in solution resist changes in acidity and alkalinity and so are important biological buffers. What do you know about the chemical structure of proteins that might account for this property? (*Hint*: Consider side chains.)

Recommended Readings

(Also consult the readings for Chapters 3, 5, and 6.)

Bettelheim, F.A., and J. March, *Introduction to General, Organic and Biochemistry.* Philadelphia, Saunders College Publishing, 1990. A readable reference for those who would like to know more about basic chemistry.

Caplan, A., "Cartilage," *Scientific American,* November 1984, pp. 84–94. This fundamental skeletal tissue has unique properties of strength and resilience that now can be explained in terms of the molecular structure of the chemical constituents of the tissue. An excellent example of the chemical basis of biology.

Raloff, J. "The Gender Benders," *Science News,* Vol. 145, No. 2, Jan. 8, 1994, pp. 24–27, and "That Feminine Touch," *Science News,* Vol. 145, No. 4, Jan. 22, 1994, 56–58. This two-part series is on environmental hormones.

Richards, F.M., "The Protein Folding Problem," *Scientific American,* January 1991, Vol. 264, No. 1, pp. 54–63. A discussion of the mechanisms involved when a protein folds into its biologically active shape.

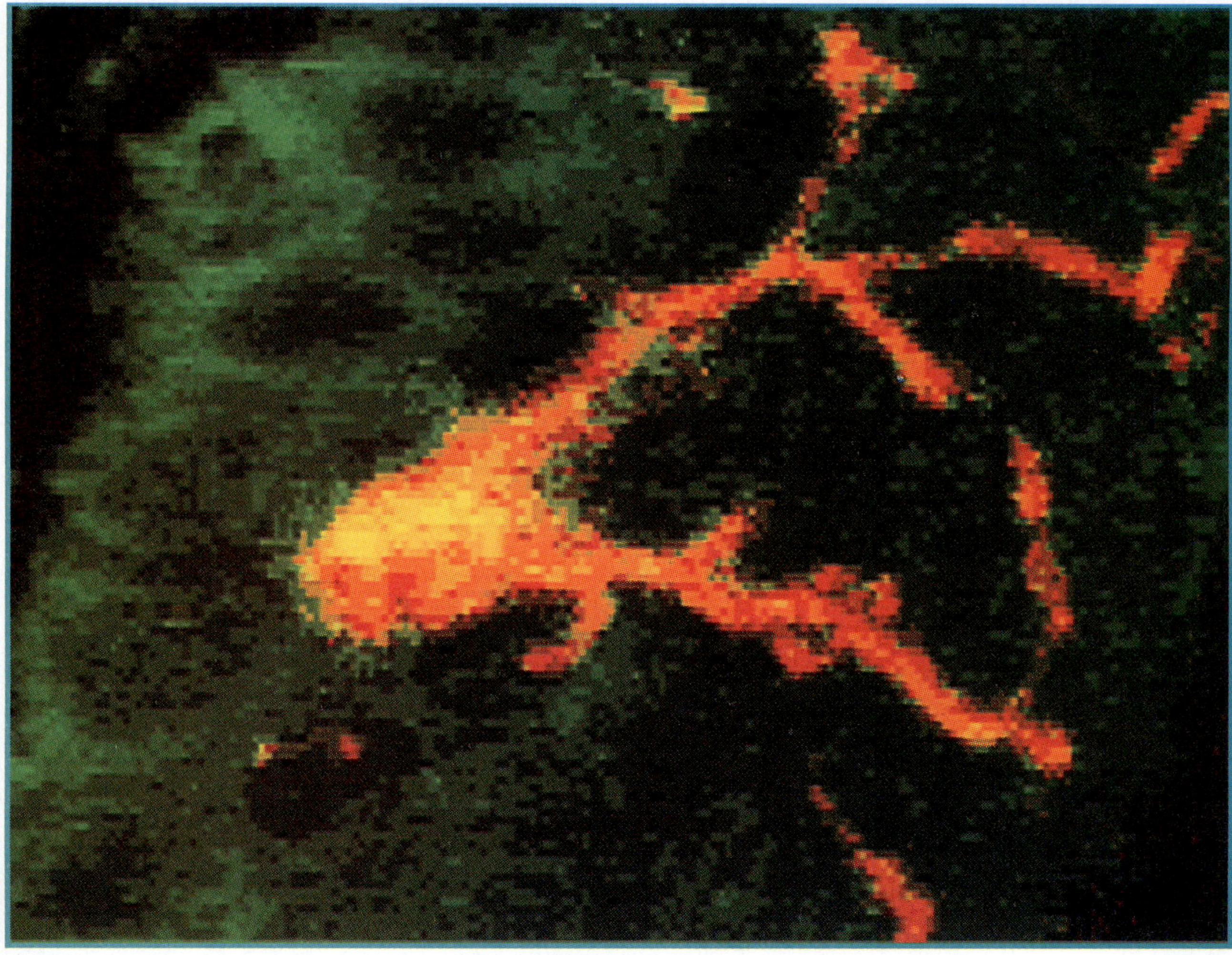

The mind-body connection? *A Langerhans cell (orange), its cytoplasmic extensions stretched toward the skin's surface (green), is poised to attack foreign substances it encounters. Chemical messages from nearby nerve cells may regulate the attack. (Brett Telegan/© 1994 The Walt Disney Co. Reprinted with permission of* Discover Magazine.*)*

The Langerhans Cell: A Mystery for More Than 100 Years

The Langerhans cell (first described by German physician Paul Langerhans in 1868) is a fascinating cell found in the upper layers of human skin. Each Langerhans cell is irregularly shaped; it may be elongated or shaped somewhat like a star. Extensions of the cytoplasm form processes that contact processes from other Langerhans cells, forming a network throughout the upper layer of the skin. These cells contain peculiar, rod-shaped granules surrounded by membranes. Although this cell was identified more than 100 years ago, we are just beginning to understand its function and importance in the body's defense system (immune system).

In May, 1993 Richard Granstein and George Murphy reported that nerve endings in the skin appear to communicate with Langerhans cells by chemical signals. Murphy, a research scientist at the University of Pennsylvania, used a new type of laser microscope to show that nerve endings in the skin make contact with Langerhans cells. Then, Granstein, an investigator at Massachusetts General Hospital, found that a chemical compound called CGRP, which is produced by nerve cells, can suppress the activity of Langerhans cells. These findings suggest that when we experience stress, our brains can send messages by way of nerve cells to the Langerhans cells in the skin. This pathway could prove to be an important mind-body connection.

Langerhans cells are thought to help trap foreign substances or microorganisms that penetrate the skin. For example, when we come in contact with poison ivy, a toxic protein enters the skin. A Langerhans cell might devour the foreign protein, rip it apart, and then display fragments of it on its own cell surface. This action signals other cells of the immune system to launch an attack on the foreign substance.

Langerhans cells may also affect the development of skin allergies. When certain chemical compounds are painted on the skin, they bind to Langerhans cells. These modified Langerhans cells then become the focus of attack for certain sensitized cells of the immune system. It is also of current interest that HIV (the virus that causes AIDS) infects Langerhans cells. Such infection may promote the growth of tumors in the skin.

Cell Structure and Function

5

Learning Objectives

After reading this chapter you should be able to:

1. Justify that the cell is considered the basic unit of life and state the cell theory.
2. Describe the general characteristics of cells, for example, size range and shape.
3. Identify methods by which scientists study cells.
4. Contrast prokaryotic and eukaryotic cells; contrast plant and animal cells.
5. Draw and label a diagram of a prokaryotic cell, a plant cell, and an animal cell. Describe and list the functions of the principal cell organelles.
6. Describe the structure and function of the cell nucleus.
7. Distinguish between smooth and rough endoplasmic reticulum and describe the functional relationship between ribosomes and endoplasmic reticulum.
8. Describe how the Golgi complex packages secretions and manufactures lysosomes.
9. Describe and give the functions of mitochondria and chloroplasts.
10. Describe the cytoskeletal elements, comparing microtubules and microfilaments.
11. Describe the structure and function of the eukaryotic flagellum or cilium.
12. Contrast the plant cell wall with cell walls found in other kingdoms, and give the functions of the cell wall.

Key Concepts

- ☐ The cell, the basic unit of life, consists of a plasma membrane surrounding a microscopic amount of jellylike cytoplasm.
- ☐ Cells can be studied with the microscope or studied by techniques of chemical or physical analysis. The electron microscope makes high magnifications practical by means of its superior resolving power.
- ☐ Two fundamentally different kinds of cells are known: (1) prokaryotic cells, which lack a membrane-bound nucleus and other membrane-bound organelles, and (2) eukaryotic cells, which are equipped with nuclear envelopes and have a variety of membrane-bound organelles.
- ☐ The eukaryotic cell has many membrane-bound organelles including: a distinct nucleus that contains genetic information; rough and smooth endoplasmic reticulum that synthesize proteins and lipids; a Golgi complex that produces lysosomes and that packages and/or stores cell secretions; lysosomes that contain digestive enzymes; mitochondria that package energy for cellular use; and vacuoles that hold fluid.

Key Concepts, continued

- ☐ Some eukaryotic cells (plant cells and algae) contain chloroplasts that function in photosynthesis; prokaryotic and some eukaryotic cells (certain protists, fungi, and plants) have cell walls that maintain cell shape.
- ☐ The eukaryotic cell has a cytoskeleton made of microtubules and microfilaments.

THE CELL IS THE BASIC UNIT OF LIFE

In 1838 and 1839, two German biologists, Matthias Schleiden (a botanist) and Theodor Schwann (a zoologist), proposed that all living things are made up of **cells** and cell products and that the cell is the basic unit of living organisms. This fundamental generalization of biology is known as the **cell theory.**

The cell theory was extended in 1855 by Rudolf Virchow, who stated that new cells come into existence only by the division of previously existing cells. Recall the experiments discussed in Chapter 2; just as flies and microorganisms cannot arise by spontaneous generation from nonliving matter, neither can cells. In 1880, August Weismann pointed out the corollary of this—that all the cells living today can trace their ancestry back to ancient times.

In one-celled organisms, cell division results in the production of two new individuals. In multicellular organisms, the two new cells may remain associated, forming a part of the organism. Like the bricks of a building, cells are the building blocks of a complex organism.

The cell is the smallest unit of living material capable of carrying on all of the activities necessary for life. It has all of the physical and chemical components needed for its own maintenance and growth. When provided with essential nutrients and an appropriate environment, cells can remain alive in laboratory glassware for many years. No cell part is capable of such survival by itself.

CELLS ARE DIVERSE, YET SHARE MANY CHARACTERISTICS

Cells are tiny chemical factories that transfer information through duplication of genetic material, manufacture of proteins, cell division, and communication with other cells. They also transfer energy, regulate movement of materials, and evolve. Some cells produce specialized products that are exported to other cells.

Although thousands of different types of cells have been identified, most cells share certain features. Every cell consists of jellylike **cytoplasm** composed mostly of water (70% to 90%) surrounded by a **plasma membrane.** The plasma membrane is a physical boundary that separates the cell from the outside environment. Most cells have a variety of **organelles,** internal cellular structures that are suspended within the cytoplasm and have specialized functions.

Two Fundamental Cell Types Are Prokaryotic Cells and Eukaryotic Cells

Organisms can be classified into two groups, based on their cell type: prokaryotes and eukaryotes.

Prokaryotic Cells Lack Membrane-Bound Organelles

Only bacteria are **prokaryotes.** They are assigned to their own kingdom, the Prokaryotae. Prokaryotic cells are generally smaller than eukaryotic cells. The term prokaryote, meaning "before the nucleus," indicates that these cells lack a distinct nucleus. Some prokaryotic cells do have one or more nuclear regions, in which DNA is concentrated (Fig. 5–1). However, the DNA is not surrounded by a separate membrane.

Prokaryotic cells also lack most other membrane-bound organelles typical of eukaryotic cells. In some prokaryotic cells, the plasma membrane is folded inward to form a complex of internal membranes. Several functions have been hypothesized for these structures including the reactions of cellular respiration.

Prokaryotic cells typically grow rapidly and divide frequently. Despite the absence of membrane-bound organelles, they are capable of carrying on an amazing number of chemical reactions.

Eukaryotic Cells Have Membrane-Bound Organelles that Divide Them into Compartments

All organisms other than bacteria are classified as **eukaryotes.** The term *eukaryote* means "good nucleus." A eukaryotic cell has a distinct nucleus surrounded by two membranes that make up a nuclear envelope. They also have many types of membrane-bound organelles, structures that are specialized to perform specific functions.

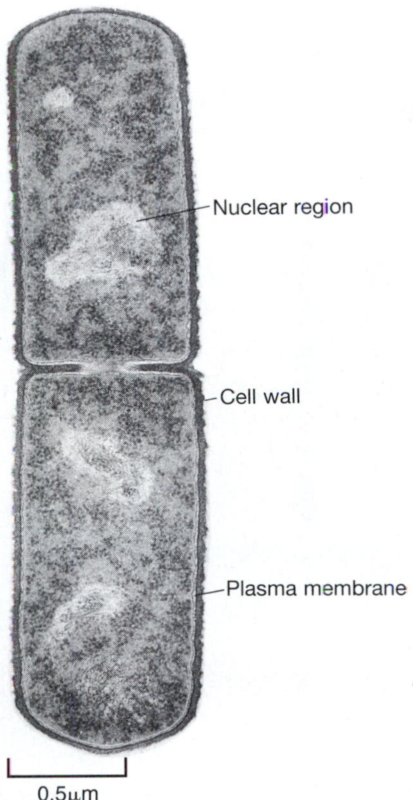

FIGURE 5–1 The structure of a dividing prokaryotic cell, the bacterium, *Bacillus subtilis*. Note that the nuclear region lacks a surrounding membrane and so is not a true nucleus. Prokaryotic cells also lack other membrane-bound organelles. (Courtesy of A. Ryter)

Membrane-bound organelles partition the cytoplasm into compartments.

Dividing the cell into internal compartments separated by membranes has many advantages. Each type of compartment has its own structure and contents and performs specific functions. Certain chemical compounds can be concentrated in one type of organelle, so the compounds that participate in a chemical reaction can "find each other" more easily. This can greatly increase the rate of the reaction. Membrane-bound compartments also separate certain compounds, keeping them away from other parts of the cell that might be damaged by them.

Membranes allow the storage of energy. The membrane provides a barrier that can be compared to a dam on a river. Energy can be stored when there is a difference in the concentration of some substance on the two sides of the membrane. When the molecules move across the membrane from the side of high concentration to the side of low concentration, the energy can be converted to chemical energy in the form of ATP. Cells use this process of energy conversion (discussed in Chapters 8 and 9) to capture and convert energy, which is necessary to sustain life.

Membranes in cells also serve as important work surfaces. For example, many enzymes are located along membranes. Many important chemical reactions are carried out by such membrane-bound enzymes. In fact, groups of enzymes needed to carry out successive steps of a series of reactions are often positioned close together on a membrane surface. This organization permits rapid manufacture of biological molecules required by the cell.

Cell Size Is Limited by Several Factors

Can you imagine amoebas as large as whales (Fig. 5–2)? Or bacteria as large as elephants? Most cells are microscopic (Fig. 5–3). Prokaryotic cells are typically very small (0.2 to 5 micrometers in diameter; micrometer is abbreviated as μm; see inside backcover, The Metric System). Eukaryotic cells are generally about 10 times larger (10 to 100 micrometers in diameter) than prokaryotic cells.

Cells rarely become large because it is inefficient for them to do so. A cell must take in nutrients, gases, and other materials through its plasma membrane. Once inside, materials must be moved to where they are needed. Waste products produced during metabolic reactions must be discharged from the cell. All materials coming into or out of the cell must pass through the plasma membrane. The more surface area the cell has, the faster a given number of molecules can pass through it, so the size of the membrane in comparison with the rest of the cell is critical.

The ratio of the surface area to the volume of the cell is very important in limiting cell size (Fig. 5–2). When the volume of the cytoplasm reaches a certain critical size rel-

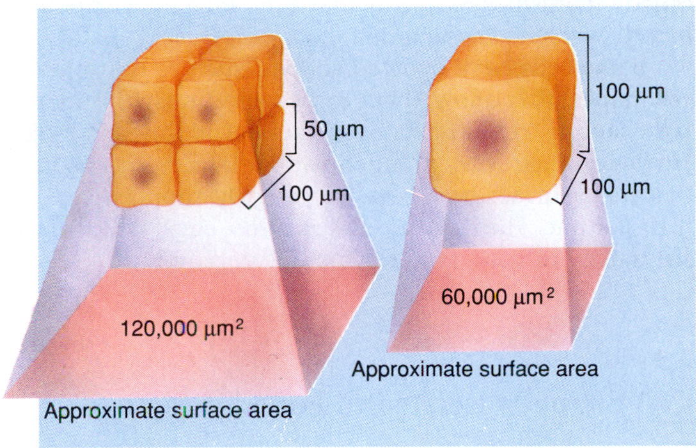

FIGURE 5–2 Surface area-to-volume ratio. Eight small cells have a much greater surface area (that is, the area of the plasma membrane) in relation to their total volume than does one large cell. To make this concept easy to understand, imagine that each of these cells is a potato. You could prepare the same amount of mashed potatoes from eight small potatoes as from one large one, but which would you rather peel?

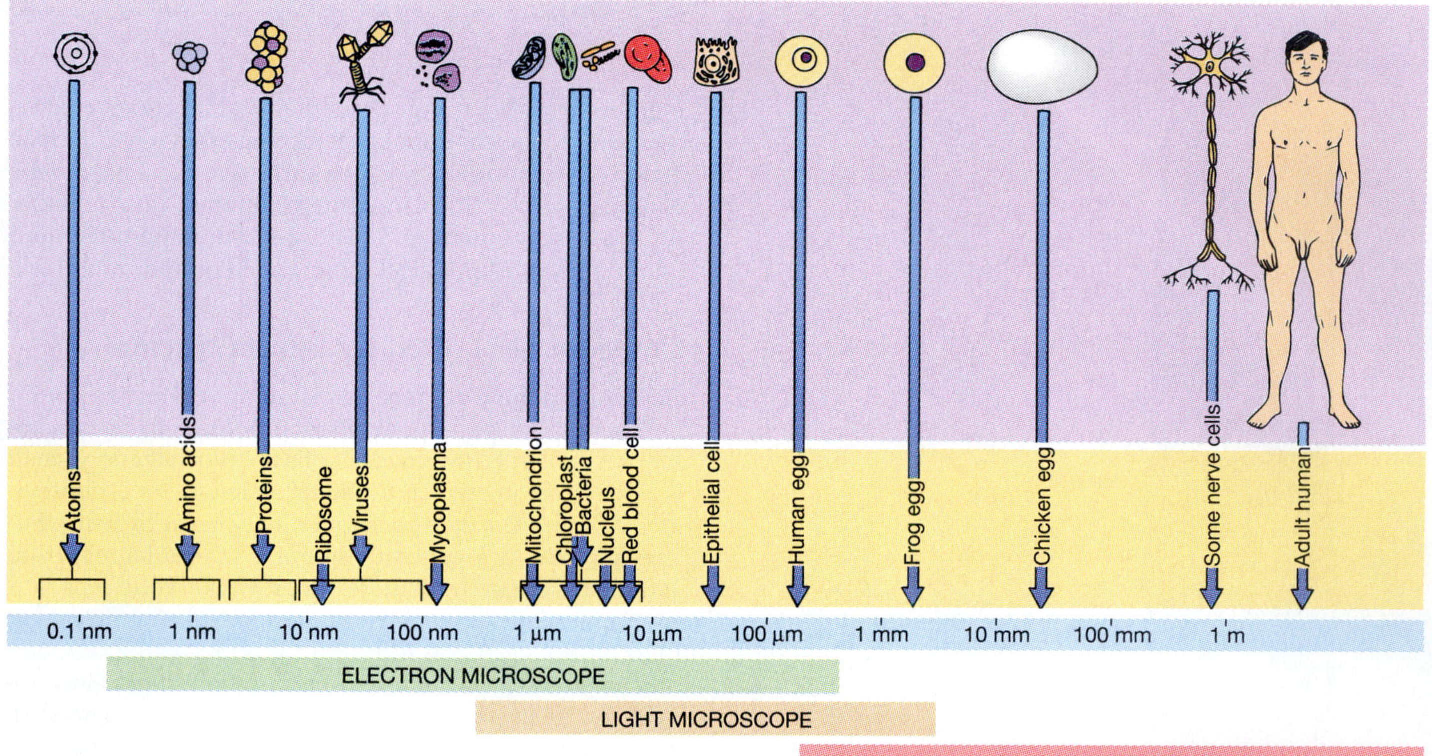

FIGURE 5–3 Biological units of measure. Because the size range is very wide, a logarithmic scale (in which there is a tenfold difference between successive units) is used to make the figure more compact. If a linear scale had been used, the figure would be more than 10,000 kilometers wide! Prokaryotic cells vary in size from less than 1 to 10 micrometers in diameter. Eukaryotic cells generally fall within the range of 10 to 100 micrometers in diameter. Mitochondria are about the size of typical bacteria.

ative to the surrounding plasma membrane, transport of materials is no longer efficient. The cell then divides to form two cells. This explains why we do not see amoebas as large as whales slithering about. In fact, most one-celled organisms are so tiny that even the larger ones are barely visible to the unaided eye.

Some cells have evolved shapes, such as nonspherical shapes, that allow them to become unusually large. For example, certain nerve cells are several meters long. However, their very thin, threadlike structure ensures that their surface area remains roughly proportional to their volume. The largest cells are birds' eggs, which consist mostly of yolk, food for the developing bird.

Cell Shape Is Related to Function

The size and shape of a cell are related to its specific function (Fig. 5–4). The single-celled ameoba has the ability to change its shape as it flows along from one location to another. Similarly, certain white blood cells are able to move about in the body by changing shape. Plant cells have rigid cell walls that maintain their shape.

Epithelial cells may be almost rectangular in shape and may be stacked much like building blocks to form sheetlike structures. The long cytoplasmic extensions of nerve cells are adaptations for transmitting messages long distances through the body. An extension of the sciatic nerve, for example, may extend from the spinal cord to the foot. Although such a nerve cell may be more than a meter long, its diameter is so tiny that no part of it can be seen without the aid of a microscope. The human egg cell, a very large cell, is about 130 µm in diameter, or about the size of the period at the end of this sentence. In contrast, each sperm cell is a very small, elongated cell, equipped with a long tail (flagellum) that it uses to propel itself toward the egg.

Cells are studied by a variety of methods

Cells are so small that you might well wonder how we know so much about what goes on inside them. Biologists use a variety of instruments and techniques to learn what we could not discover with the unaided human eye.

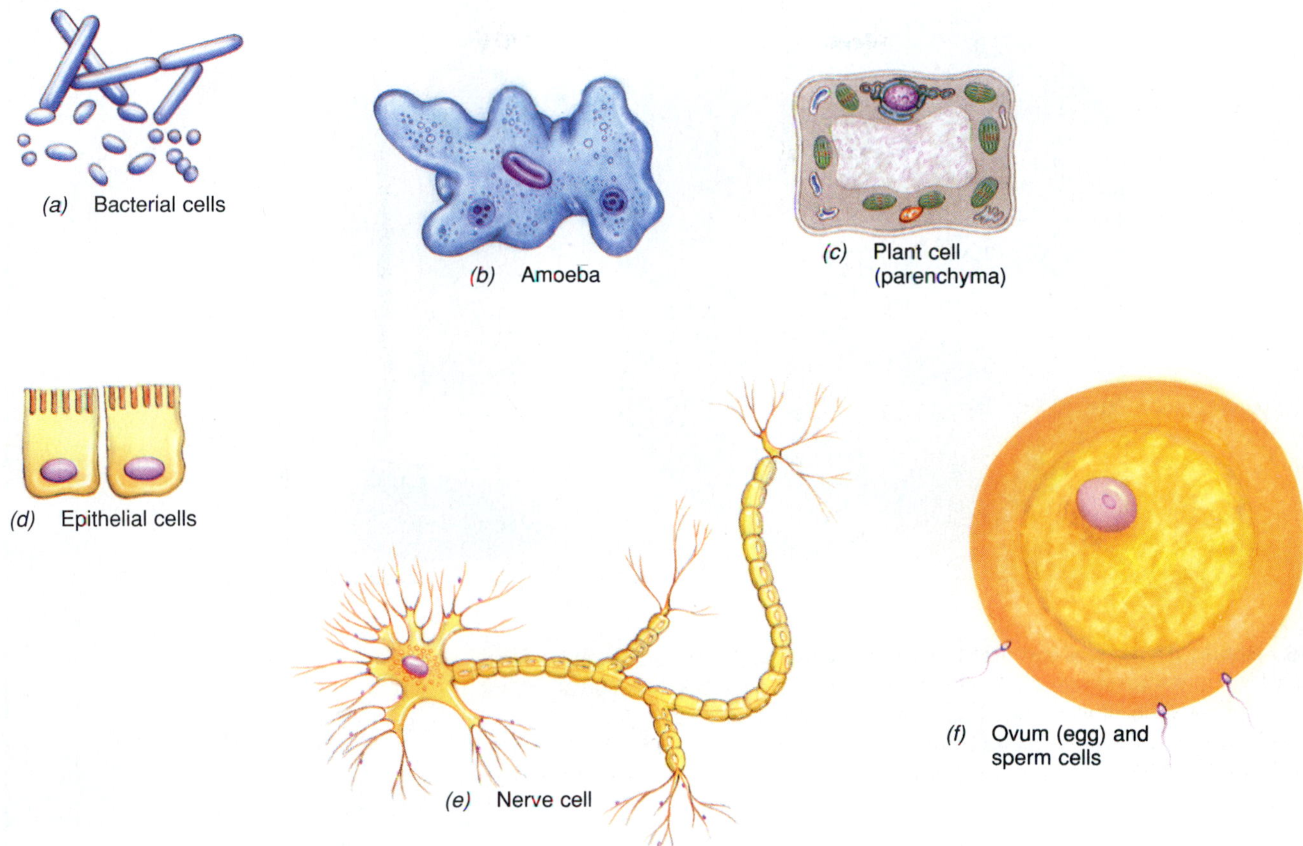

FIGURE 5–4 Cells exhibit a variety of sizes and shapes related to their specific functions. (***a***) Bacterial cells are small, allowing them to grow and divide rapidly. (***b***) The amoeba changes its shape as it moves from place to place. (***c***) The organs of most young plants consist mainly of parenchyma cells. (***d***) Epithelial cells join to form tissues that cover body surfaces and line body cavities. (***e***) Nerve cells are specialized to transmit messages from one part of the body to another. (***f***) The ovum (egg cell) is among the largest cells; sperm cells are comparatively tiny. The sperm propels itself by whipping its flagellum.

The biologist's most important tool for studying the internal structure of cells has been the microscope. Anton van Leeuwenhoek (1632–1723) is credited with developing some of the earliest microscopes and with leaving written records of the structures he studied. In 1665, Robert Hooke (1635–1703) examined a slice of cork with the aid of a crude, homemade microscope. Because the tiny compartments he saw reminded him of the little rooms, or *cells,* of a monastery, he called them cells. What Hooke saw were the cell walls of dead cork cells (see Fig. 5–5). In later observations, he described cell contents. However, it was not until two centuries later that scientists realized that the important part of the cell is its contents and not its outer walls. During the 1800s, scientists studied various cells and observed a variety of intracellular structures. They called these structures organelles (little organs) because they were thought to perform special jobs within the cell, just as our organs perform specific jobs within our bodies.

The ordinary compound **light microscope,** the kind typically used by students in college laboratories, was used to discover most cell structures. The light microscope, gradually improved since Hooke's time, uses visible light as the source of illumination. With the development of the **electron microscope,** which came into wide use in the 1950s, researchers could begin to study

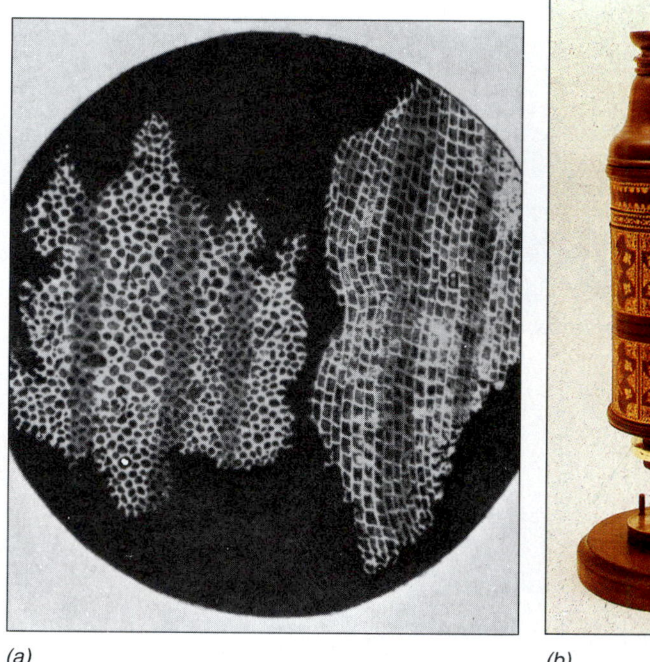

(a)

(b)

FIGURE 5–5 Robert Hooke was the first to describe cells. (***a***) Drawing by Hooke of the microscopic structure of a thin slice of cork from the bark of a tree. Hooke's observations were based on the cell walls of these dead cork cells. (***b***) Microscope belonging to Robert Hooke. (*a*, From the book *Micrographia*, published in 1665, in which Hooke described many of the objects he had viewed using the compound microscope he constructed; *b*, Armed Forces Institute of Pathology)

the fine detail, or **ultrastructure,** of cells. The electron microscope floods the specimen being studied with a beam of electrons rather than with light waves.

Two features of a microscope—magnification and resolving power—determine how clearly you can see a small structure. **Magnification** is the ratio of the size of the image to the size of the specimen. We can use the ordinary light microscope to magnify a structure about 1000 times; the electron microscope can effectively magnify it 250,000 times or more (see Focus On the Process of Science: Viewing the Cell, pages 90–92).

Magnification alone would not be an advantage if we could not distinguish greater detail. This ability to see fine detail is called **resolving power.** It is defined as the minimum distance between two points at which they can be distinguished as separate and distinct points, rather than as one single blurred point. While the best light microscope equipped with very fine lenses can resolve objects about 500 times better than the human eye, the electron microscope increases our resolving power more than 10,000 times.

Microscopes are excellent tools for studying cell structure. To learn about functions of organelles, biologists use physical and chemical methods. **Cell fractionation** procedures are used to purify organelles (Fig. 5–6). Cells can be broken apart and then centrifuged (spun) at high speeds to separate cellular components. Using a variety of techniques, organelles can be separated from one another on the basis of size or density. Once separated from its surrounding cellular components, a given type of organelle can be chemically analyzed to determine what types of proteins and other compounds it contains and what kinds of chemical reactions take place within it. Organelles can also be monitored in test tubes under controlled conditions.

THE EUKARYOTIC CELL HAS A COMPLEX STRUCTURE AND FUNCTION

Early biologists thought that the cell consisted of a homogeneous jelly, which they referred to as protoplasm. Modern research tools like the electron microscope have greatly expanded our understanding of the world within the cell. We no longer view the cell contents as a single

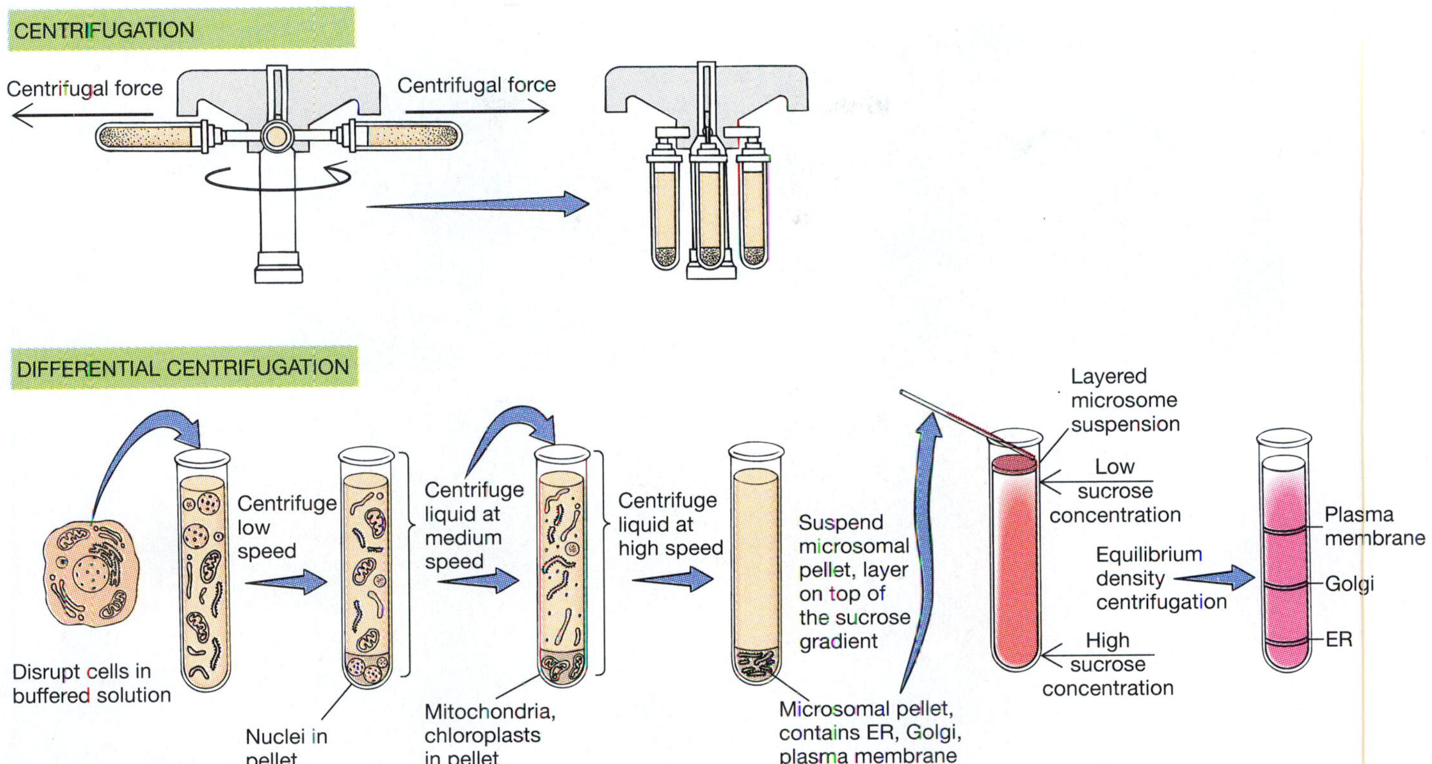

FIGURE 5–6 Cell fractionation. Cell membranes and organelles are usually separated by centrifuges, instruments that can spin test tubes. Spinning the tubes exerts a centrifugal force on the contents. Particles suspended in solution (such as membranes and organelles from disrupted cells) form a pellet at the bottom of the test tube. Different cell parts have different densities. This allows us to separate the cell parts into fractions by centrifuging the suspension at increasing speeds **(differential centrifugation).** Membranes and the organelles from the resuspended pellets can then be further purified by **equilibrium centrifugation,** which involves layering that solution on top of a **density gradient** (a solution in which the concentration of the solute, usually sucrose, increases from top to bottom, forming a range of densities). When the density gradient is centrifuged, organelles and membranes migrate and form bands in the region of the gradient equal to their density. The purified cell fractions are then collected by puncturing the tube bottom and collecting samples of the solution.

substance. We know that the cell is a highly organized, amazingly complex structure (see Windows on the Cell and Figs. 5–7 and 5–8). It has its own control center, internal transportation system, power plants, factories for making needed materials, packaging plants, and even a "self-destruct" system.

As you study this chapter, refer often to Figures 5–7 and 5–8 and to the Windows on the Plant and Animal Cells, which are bound into this book. Note that in Figure 5–7, the animal cell is shown in realistic context surrounded by neighboring cells. Although this is a generalized cell, its shape and context somewhat resemble an epithelial cell that is part of a tissue of closely linked cells. The animal cell illustrated in the Windows on the Animal Cell provides a different perspective, that of a cell that is not joined to other cells.

Today the word protoplasm, if used at all, is used in a very general way. Instead, the jellylike material outside the nucleus is called **cytoplasm,** and the corresponding material within the nucleus is referred to as **nucleoplasm.** Various organelles are suspended within the **cytosol,** which is the fluid component of the cytoplasm. The term cytoplasm includes both the cytosol and all of the organelles except the nucleus.

(Text continues on page 93)

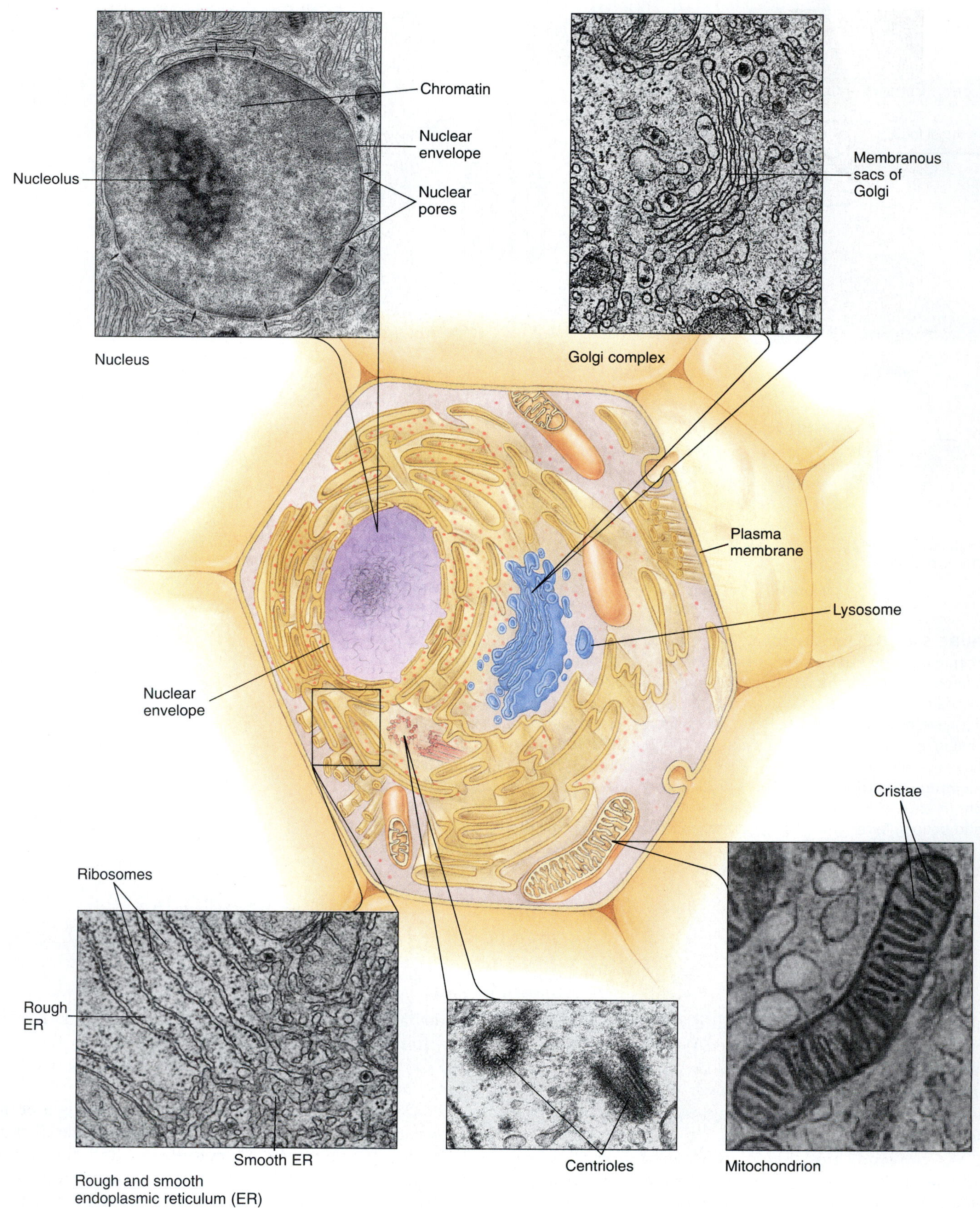

FIGURE 5–7 The structure of an animal cell. This generalized animal cell is shown in realistic context surrounded by adjacent cells that cause it to be slightly compressed. Its four-sided shape somewhat resembles a squamous or cuboidal epithelial cell viewed from the surface. Depending on the cell type, certain organelles may be more or less prominent. (Clockwise from top left, D.W. Fawcett; D.W. Fawcett and R. Bolender-D.W. Fawcett; B.F. King, School of Medicine, U. of California, Davis/BPS; VU/R. Bolender-D.W. Fawcett.)

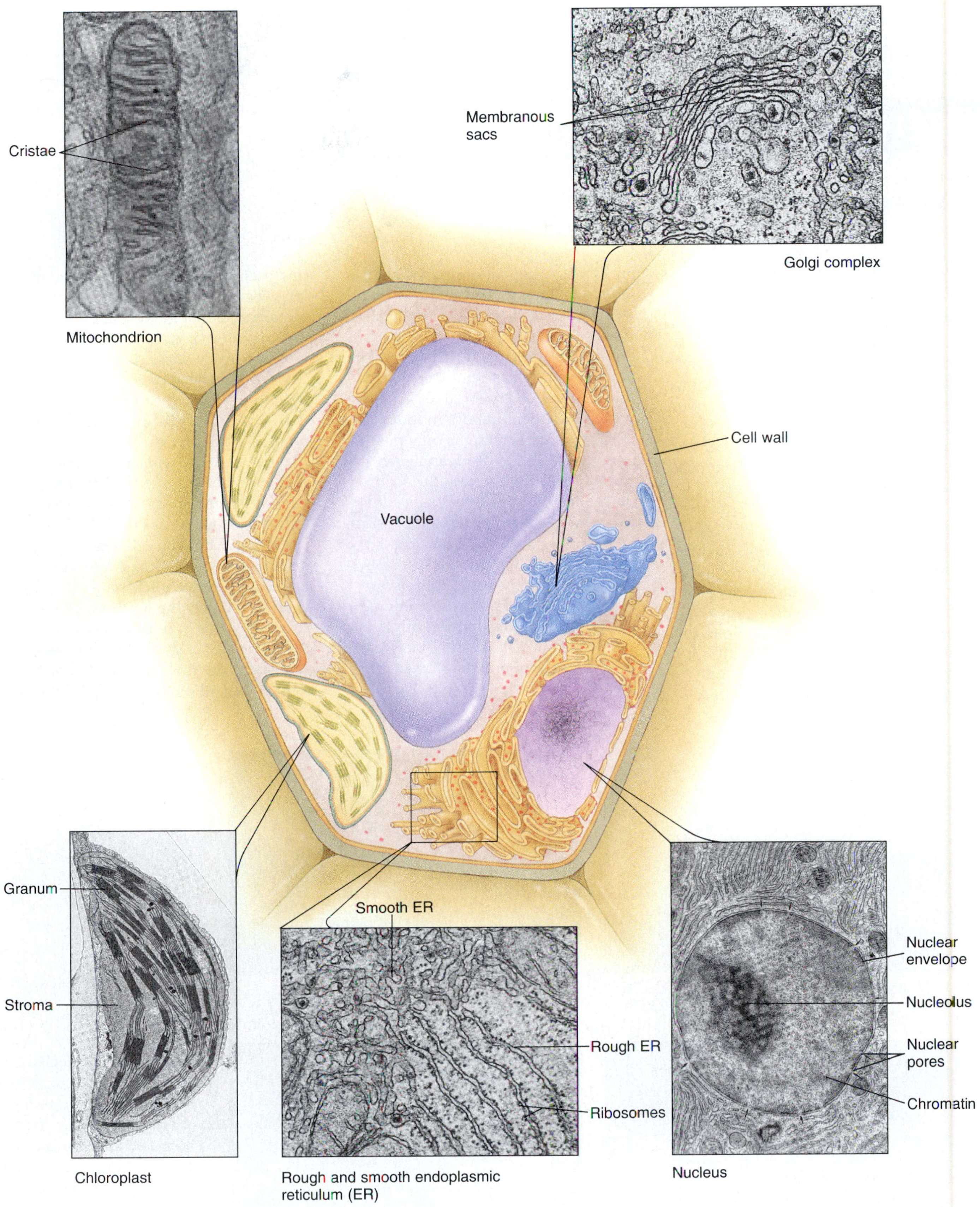

FIGURE 5–8 The structure of a plant cell. A generalized plant cell. Some plant cells do not have all of the organelles shown in this diagram. For example, cells from leaves and stems contain chloroplasts, whereas root cells do not. Chloroplasts, or other plastids, a cell wall, and prominent vacuoles are characteristic of plant cells. (Clockwise from top left, D.W. Fawcett; D.W. Fawcett and R. Bolender; D.W. Fawcett; VU/R. Bolender-D.W. Fawcett; E. H. Newcomb and W.P. Wergin, U. of Wisconsin/BPS)

FOCUS ON *The Process of Science*

VIEWING THE CELL

Viewing Cells with the Light Microscope

The **light microscope** uses visible light as the source of illumination (Fig. A). It has a resolving power of about 200 nanometers (0.2 micrometer). Objects can be magnified about 1000 times with good resolution using the light microscope. Beyond that magnification, structures may appear larger, but they are not clearer.

To be viewed with the light microscope, specimens must be very thin. Single-celled organisms such as amoebas and some cells from multicellular organisms, like sperm cells, can be viewed in the living state. However, tissues generally are too thick and must be *sectioned* (cut) into very thin slices. They are generally preserved and then *stained* with special dyes. Using an ordinary light microscope, we can distinguish certain parts of the cell only when they are stained differently from other parts. Finding ways to produce such differential staining has been a preoccupation of microscopy for at least 150 years and is still important today. Often, however, staining is only possible if the cells are dead, and even when this is not the case the stain may alter cell function. It is best to work with unstained cells, if possible.

In ordinary light microscopy, the absorption of light by the specimen is about the same for all cell parts. But both the **phase-contrast microscope** and the **interference microscope** permit us to readily view parts of living cells that would ordinarily not absorb much light (Fig. A). These microscopes permit us to observe living cells and to watch the constant motion and change in shape of many of their internal structures. The techniques convert small differences in the way specimens refract (bend) light to much greater apparent differences, causing parts of the specimen to appear brighter than others.

In the **dark-field microscope** only the light scattered from edges or particles in the specimen can enter the microscope lenses that produce the image. Thus the cell shows up as bright against a dark background.

Fluorescence microscopy (Fig. B) is used to detect the locations of specific molecules in cells. Like paints that glow under black light, fluorescent stains are molecules that absorb light energy of one wavelength and

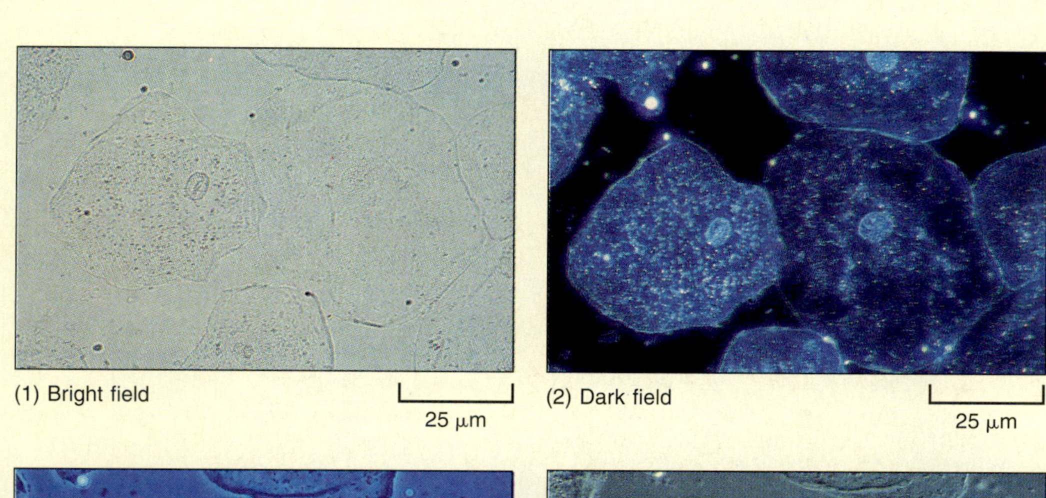

(1) Bright field — 25 μm

(2) Dark field — 25 μm

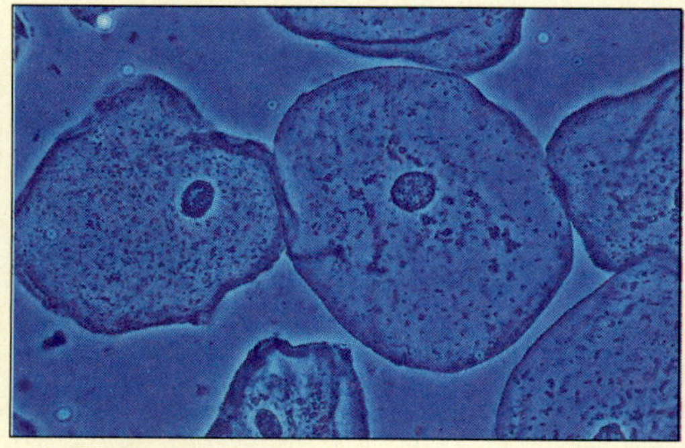

(3) Phase contrast — 25 μm

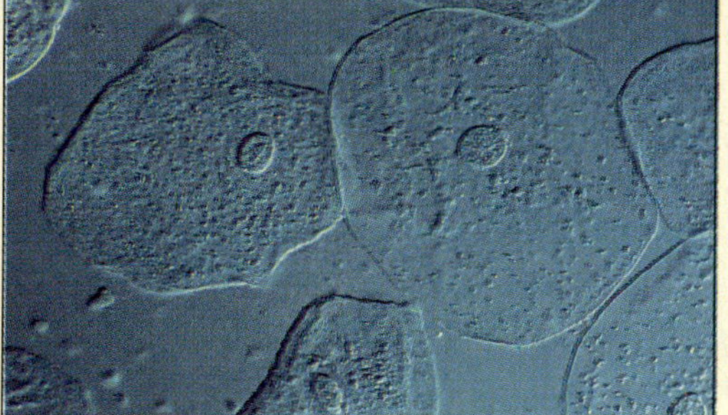

(4) Nomarski differential interference — 25 μm

***A.* Viewing epithelial cells.** Epithelial cells using (1) bright field (transmitted light), (2) dark field, (3) phase contrast, and (4) Nomarski differential interference microscopy. The phase contrast and differential interference microscopes enhance detail by increasing the differences in optical density in different regions of the cells. (Jim Solliday Biological Photo Service)

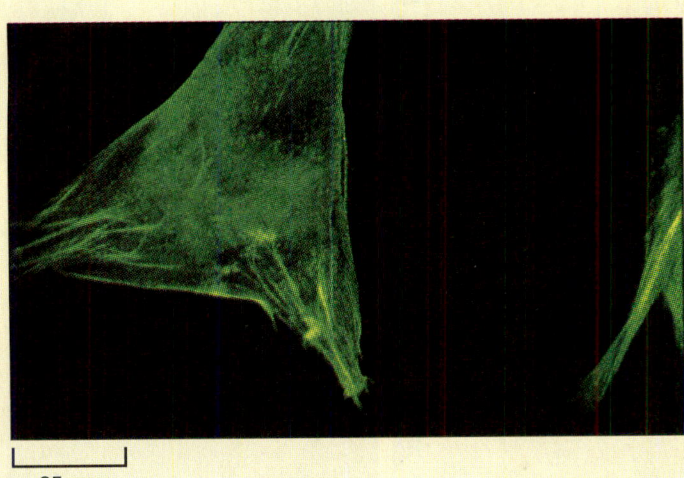

***B.* Epithelial cells in culture stained by a technique that causes them to fluoresce.** (Michael Abby/Photo Researchers, Inc.)

then release some of that energy as light of a longer wavelength. One fluorescent stain binds specifically to DNA molecules and emits green light after absorbing ultraviolet light.

Some fluorescent stains can be chemically bonded to **antibodies,** protein molecules that the body produces in response to foreign proteins. If an animal such as a rabbit is injected with a foreign substance, its immune system will often produce antibodies that will attach to the foreign substance whenever they contact it. If these antibodies are added to a preparation of cells, they will attach to any of the cells that contain the foreign material. Certain dyes are then added that attach to the antibodies, and this preparation is exposed to ultraviolet light. Cells that have the dye then light up brilliantly, or fluoresce. Recently, new computer imaging methods have resulted in development of the **confocal fluorescence microscope,** which improves the resolution of structures labeled by fluorescent dyes.

Viewing Cells with the Electron Microscope

The electron microscope uses a beam of electrons as a source of illumination instead of light (Fig. C). The electron beams have much shorter wavelengths than visible light. Because electrons have electrical charges, a magnetic field can be used to direct them. Two types of electron microscopes in common use are the **transmission electron microscope** and the **scanning electron microscope.** A photograph taken with an electron microscope is called an **electron micrograph (EM).**

Electrons Pass through the Object Being Viewed with the Transmission Electron Microscope

In the transmission electron microscope, a beam of energized electrons is transmitted through the specimen and falls upon a photographic plate or a fluorescent screen that works something like a television screen. Before it can be viewed (Fig. C), the specimen must be embedded in plastic and cut in extremely thin sections so that the beam of electrons can pass through it. This is a disadvantage because live specimens cannot be viewed. However, the transmission electron microscope has been invaluable for studying details of internal cell structure. When you look at transmission electron micrographs (TEMs), remember that they represent only a thin cross section of a cell.

The Scanning Electron Microscope Produces an Illusion of Depth

In scanning electron microscopy (Fig. C) the specimen may be coated with a metal, often gold. The electron beam does not pass through the specimen. Instead, when the electron beam strikes various points on the coated surface of the specimen, secondary electrons are emitted; their intensity varies with the contour of the surface. The recorded emission patterns of the secondary electrons give a three-dimensional picture of the surface of the specimen.

Scanning electron microscopy provides information about the shape and surface of the specimen that could not be gained from transmission electron microscopy. Scanning electrons provide striking views of a specimen's surface but cannot be used to study internal structures without special preparation. As is the case with other forms of electron microscopy, the scanning electron microscope cannot usually be used to view living specimens.

New Kinds of Microscopy May Pave the Way for New Advances in the Understanding of Cell Structure and Function

New techniques involving gamma rays and even sound waves hold promise for examining living cells at magnifications equal to or even greater than those achieved by electron microscopes. This will enable us to observe changes in the cell as they actually occur and can be expected to settle many questions about the development of cellular structures that still are controversial today.

continued

continued

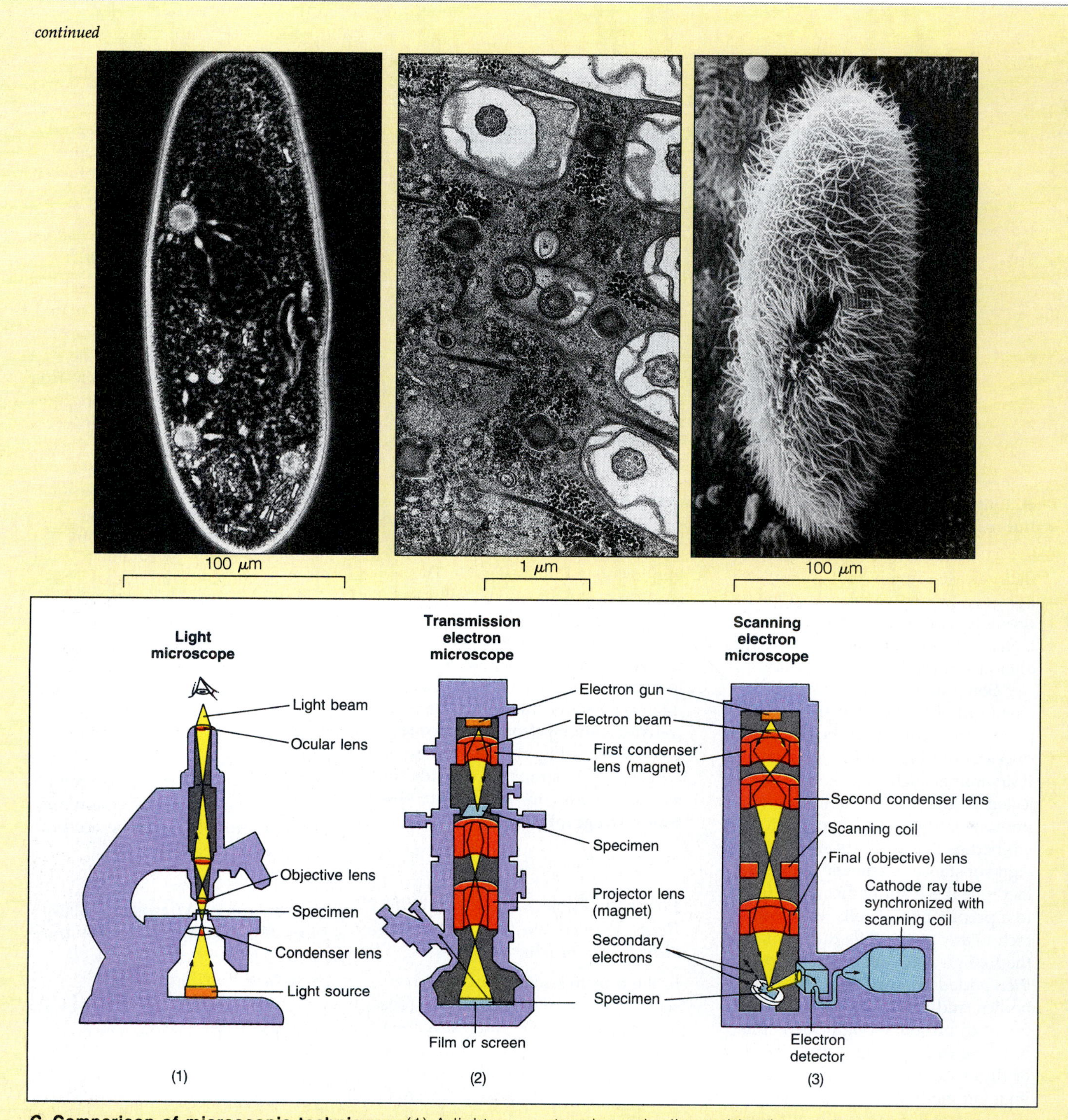

***C*. Comparison of microscopic techniques.** (1) A light microscope can be used to view a living or stained protist such as this *Paramecium.* (2) With a transmission electron microscope the biologist can study the fine structure of cells. This transmission electron micrograph (TEM) shows a very small part of the *Paramecium.* (3) The scanning electron microscope gives a remarkably clear picture, or scanning electron micrograph (SEM), of the surface of the *Paramecium.* All three microscopes are focused by similar principles. A beam of light or an electron beam is directed by the condenser lens onto the specimen and is magnified by the objective lens and the projector lens in the TEM or the objective lens and the eyepiece in the light microscope. The TEM image is focused onto a fluorescent screen, and the SEM image is viewed on a type of "television" screen. Lenses in the electron microscopes are actually magnets that bend the beam of electrons. (*Photos courtesy of T.K. Maugel/University of Maryland*)

Table 5–1 summarizes the types of organelles typical of eukaryotic cells. Remember that some organelles are found only in specific kinds of cells. For example, chloroplasts are found only in cells that carry on photosynthesis, and cilia or flagella usually occur only in cells that swim or where some kind of fluid motion is produced.

The Cell Nucleus Is the Control Center of the Cell

If we were designing a machine that could repair and duplicate itself the way a cell does, we would need to include somewhere within it a complete set of plans, specifications, and directions. In the cell, these functions are carried out by the cell nucleus.

The **nucleus** is usually the most noticeable compartment in the eukaryotic cell (see Windows on the Cell and Fig. 5–9). Perhaps for this reason, early investigators guessed that the nucleus served as the cell's control center even before they had the techniques to test that hypothesis. During recent years, many experiments have confirmed the vital role of the nucleus. In one such experiment, the researcher surgically removed the nucleus from a living amoeba. The enucleated amoeba was unable to eat or grow, and it died after a few weeks. However, if after a day or two an enucleated amoeba was given a new nucleus, it made a complete recovery. These and other experiments show that the nucleus is essential to the well-being of the cell.

A cell nucleus can control only a certain amount of cytoplasm, and this is one of the factors that limits cell size. Some large cells—certain amoebae and algae, for example—have evolved more than one nucleus.

The Nuclear Envelope Separates the Nuclear Contents from the Cytoplasm

The nucleus is separated from the surrounding cytoplasm by the **nuclear envelope,** a double membrane that regulates the flow of materials into and out of the nucleus. The two membranes of the nuclear envelope are fused at intervals, forming **nuclear pores.** These pores are like sieves that allow passage of small molecules between the nucleus and cytoplasm. Passage of larger molecules is selective; only specific molecules are permitted to pass through these openings.

The Nucleolus Assembles Ribosomes

The nucleus may contain one or more prominent **nucleoli** (singular, **nucleolus**). The nucleolus (Fig. 5–9) is not walled off from the rest of the nucleus by any kind of

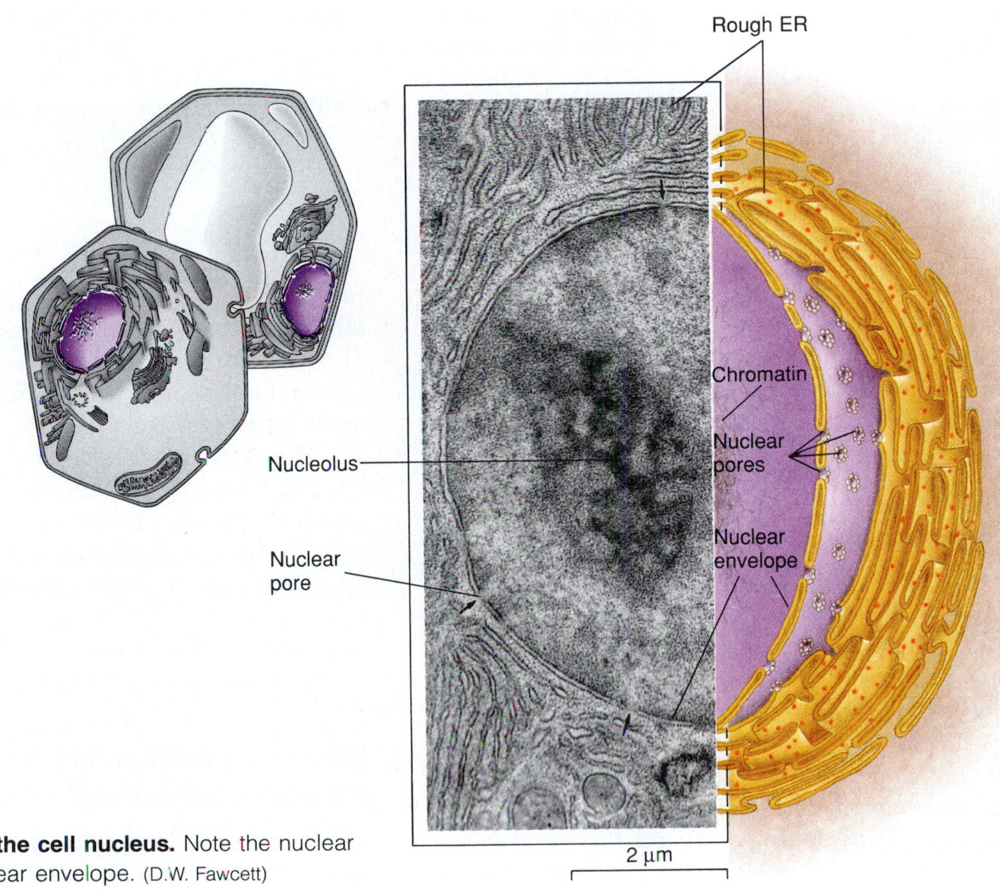

FIGURE 5–9 Structure of the cell nucleus. Note the nuclear pores that perforate the nuclear envelope. (D.W. Fawcett)

TABLE 5–1
Eukaryotic Cell Structures and Their Functions

Structure	*Description*	*Function*
The Cell Nucleus		
Nucleus	Large structure surrounded by double membrane; contains nucleolus and chromosomes	Control center of cell
Nucleolus	Dense region within nucleus	Assembles ribosomes
Chromosomes	Composed of a complex of DNA and protein known as chromatin; visible as rodlike structures when the cell divides	Contain genes (units of hereditary information that govern structure and activity of cell)
The Membrane System of the Cell		
Plasma membrane	Membrane boundary of cell	Encloses cellular contents; regulates movement of materials in and out of cell; helps maintain cell shape; communicates with other cells
Endoplasmic reticulum (ER)	Network of internal membranes extending through cytoplasm	Synthetic site of many lipids and proteins; origin of intracellular transport vesicles carrying proteins to be secreted
Smooth	Lacks ribosomes on outer surface	Lipid synthesis; drug detoxification
Rough	Ribosomes stud outer surface	Manufacture of polypeptides
Ribosomes	Granules composed of RNA and protein; some attached to ER, some free in cytoplasm	Manufacture of polypeptides
Golgi complex	Stacks of flattened membrane sacs	Processes, sorts, packages proteins
Lysosomes	Membrane-bound sacs (in animals)	Contain digestive enzymes
Vacuoles	Membrane-bound sacs (mostly in plants, fungi, algae)	Transport and store materials, wastes, water
Microbodies (e.g., peroxisomes)	Membrane-bound sacs containing a variety of enzymes	Sites of many diverse metabolic reactions

membrane. Rather, it is a dense region where organelles called ribosomes are assembled. Ribosomes leave the nucleus through the nuclear pores and enter the cytoplasm where they function as part of the machinery that manufactures proteins.

Chromosomes Are Packages of Hereditary Information

In a cell that is not dividing, an irregular network of strands and granules, called **chromatin,** can be seen in the nucleus. The chromatin consists of protein and DNA that may be actively synthesizing DNA or RNA. When a cell begins the process of nuclear division (mitosis), the chromatin coils and condenses into discrete, rod-shaped bodies, the **chromosomes** (Fig. 5–10).

Each chromosome contains several thousand genes arranged in a specific linear order; the genes, in turn, are composed of the nucleic acid DNA. Chemically coded within the DNA of the genes are instructions for producing all of the proteins needed by the cell. These proteins determine what the cell will look like and what functions it will perform.

The chromosomes serve as a chemical cookbook for the cell, whereas the genes might be compared to the individual recipes. When condensed, the chromosomes may be compared to a closed cookbook: the recipes are all inside but the pages cannot be read. When the chromosomes elongate, forming chromatin, the book is open and the instructions can be followed. Chromosomes will be discussed in more detail in Chapter 10.

Some Organelles Are Specialized for Manufacturing Products

Cells grow, replace worn-out parts, and repair themselves. They contain efficient chemical factories that produce the proteins and other complex molecules of which

TABLE 5–1
Eukaryotic Cell Structures and Their Functions

Structure	*Description*	*Function*
Energy-converting Organelles		
Mitochondria	Sacs consisting of two membranes; inner membrane is folded to form cristae	Site of most reactions of cellular respiration; converts energy originating from glucose; ATP produced.
Plastids (e.g., chloroplasts)	Double membrane structure enclosing internal thylakoid membranes; chloroplasts contain chlorophyll in thylakoid membranes	Site of photosynthesis
The Cytoskeleton		
Centrosome	Has centrioles in center; microtubules radiate from it	Directs construction of cytoskeleton; important in cell movement and in mitosis
Microtubules	Hollow tubes made of subunits of tubulin protein	Provide structural support; have role in cell and organelle movement and cell division; components of cilia, flagella, centrioles
Microfilaments	Solid, rodlike structures consisting of actin protein	Provide structural support; play role in cell and organelle movement and cell division
Centrioles	Pair of hollow cylinders located within centrosome; each centriole consists of nine microtubule triplets (9 × 3 structure)	Mitotic spindle forms between centrioles during animal cell division; may anchor and organize microtubule formation in animal cells; absent in plant cells
Cilia	Relatively short projections extending from surface of cell; made of two central and nine pairs of peripheral microtubules (9 + 2 structure)	Movement of some single-celled organisms; used to move materials on surface of some tissues
Flagella	Long projections extending from surface of cell made of two central and nine pairs of peripheral microtubules (9 + 2 structure)	Cellular locomotion by sperm cells and some single-celled organisms

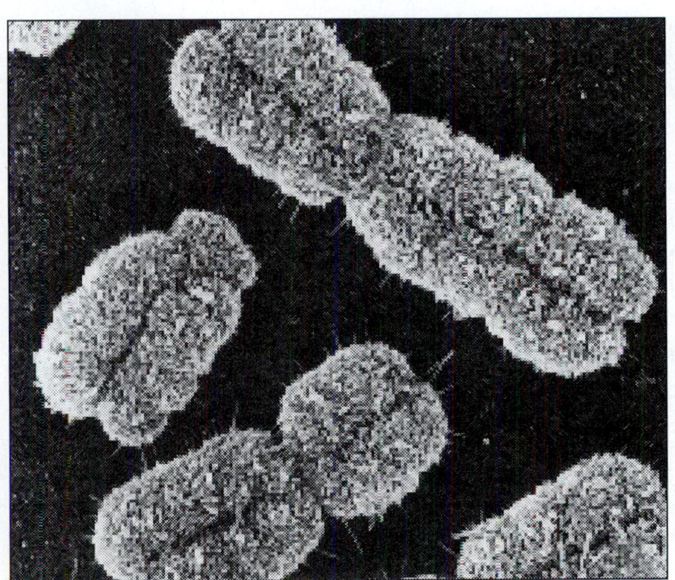

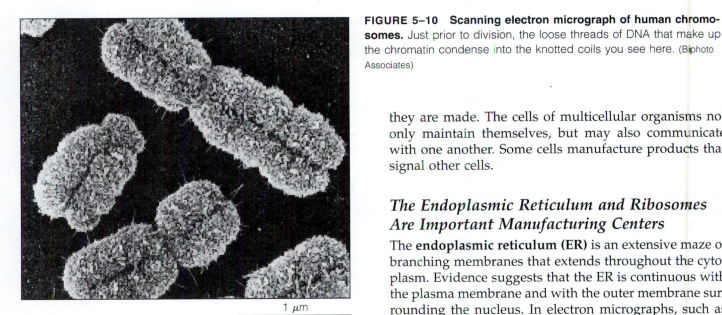

FIGURE 5–10 Scanning electron micrograph of human chromosomes. Just prior to division, the loose threads of DNA that make up the chromatin condense into the knotted coils you see here. (Biphoto Associates)

they are made. The cells of multicellular organisms not only maintain themselves, but may also communicate with one another. Some cells manufacture products that signal other cells.

The Endoplasmic Reticulum and Ribosomes Are Important Manufacturing Centers

The **endoplasmic reticulum (ER)** is an extensive maze of branching membranes that extends throughout the cytoplasm. Evidence suggests that the ER is continuous with the plasma membrane and with the outer membrane surrounding the nucleus. In electron micrographs, such as

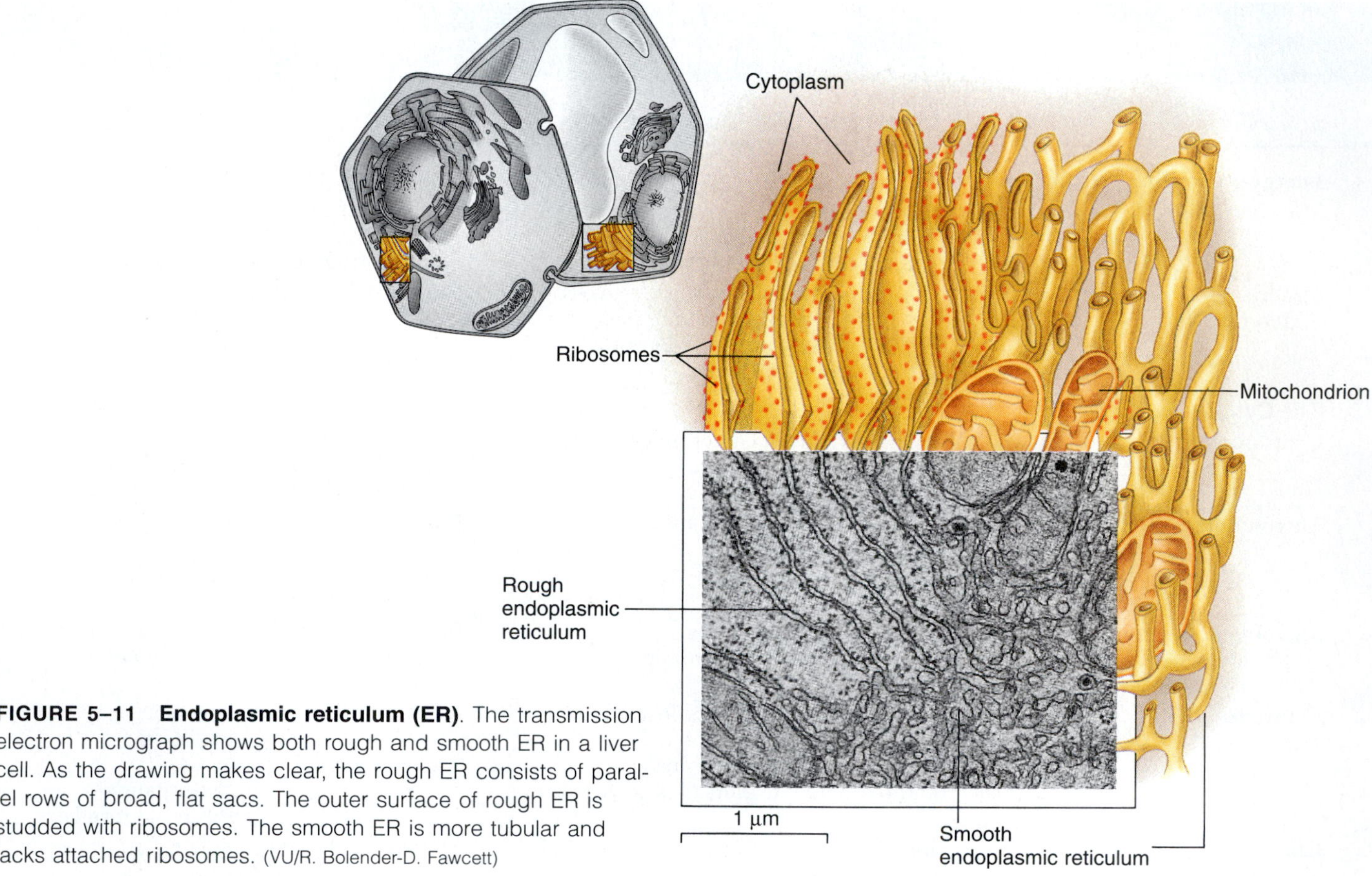

FIGURE 5–11 Endoplasmic reticulum (ER). The transmission electron micrograph shows both rough and smooth ER in a liver cell. As the drawing makes clear, the rough ER consists of parallel rows of broad, flat sacs. The outer surface of rough ER is studded with ribosomes. The smooth ER is more tubular and lacks attached ribosomes. (VU/R. Bolender-D. Fawcett)

Figure 5–11, the ER may appear to be discontinuous. This is because such photographs are taken of a thin slice through the cell, but the ER is continuous in three dimensions.

The membranes of the ER divide the cytoplasm into interconnected compartments in which different types of reactions take place. In fact, the membranes of the ER serve as a framework for groups of enzymes that regulate specific sequences of biochemical reactions. The ER is important in the synthesis of proteins and some lipids. Expanded regions of the ER may serve as temporary storage areas for certain substances. The ER also functions as a system for transporting materials from one part of the cell to another.

Two types of ER can be distinguished: smooth and rough. Although these regions of the ER have different functions, their membranes are connected and their internal spaces are continuous. **Smooth ER** is more tubular and its outer membrane surfaces have a smooth appearance. It is the main site of phospholipid, steroid, and fatty acid metabolism.

Smooth ER also contains enzymes that detoxify harmful chemicals, breaking them down to water-soluble substances that can be excreted from the body. For example, when an experimental animal is injected with the drug phenobarbital, the amount of smooth ER in the liver cells increases over a period of several days. In addition, enzymes known to break down phenobarbital increase in concentration within the smooth ER membranes when this drug is administered over a period of time.

Rough ER has a granular appearance due to the presence of dark particles called **ribosomes** that stud its outer walls. Ribosomes are the site of polypeptide synthesis. Rough ER is especially extensive in cells that synthesize proteins for export from the cell.

Not all ribosomes are attached to the ER. Some float freely in the cytosol or are attached to the cytoskeleton (discussed later in this chapter). Ribosomes are found in all kinds of cells, from bacteria to complex plant and animal cells.

The Golgi Complex Is a Factory for Packaging Proteins

The **Golgi complex** was first described in 1898 by the Italian investigator Camillo Golgi. It consists of stacks of flattened membranous sacs that may be distended in certain regions because they are filled with cell products (Fig.

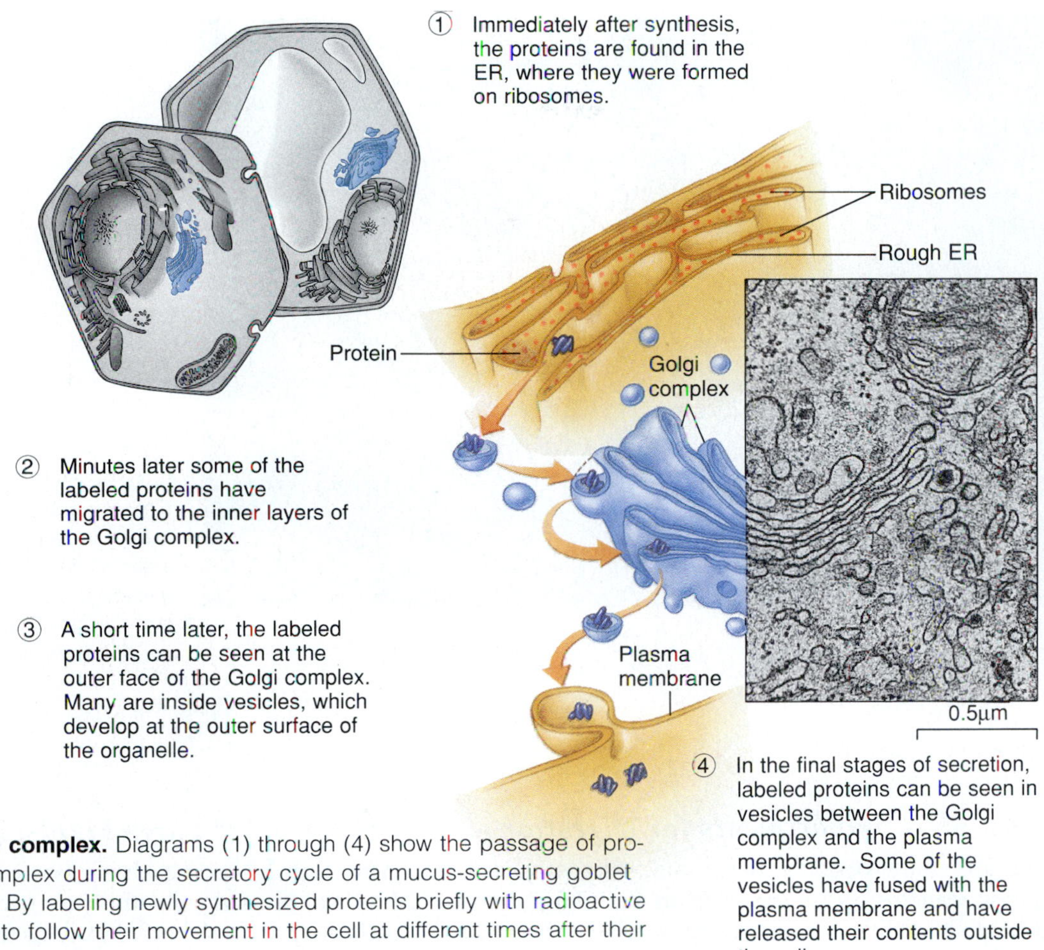

1 Immediately after synthesis, the proteins are found in the ER, where they were formed on ribosomes.
Ribosomes
Rough ER
Protein
Golgi complex
2 Minutes later some of the labeled proteins have migrated to the inner layers of the Golgi complex.
3 A short time later, the labeled proteins can be seen at the outer face of the Golgi complex. Many are inside vesicles, which develop at the outer surface of the organelle.
Plasma membrane
0.5μm
4 In the final stages of secretion, labeled proteins can be seen in vesicles between the Golgi complex and the plasma membrane. Some of the vesicles have fused with the plasma membrane and have released their contents outside the cell.

FIGURE 5–12 The Golgi complex. Diagrams (1) through (4) show the passage of proteins through the Golgi complex during the secretory cycle of a mucus-secreting goblet cell that lines the intestine. By labeling newly synthesized proteins briefly with radioactive amino acids, it is possible to follow their movement in the cell at different times after their synthesis. (D.W. Fawcett and R. Bolender)

5–12). In some animal cells, the Golgi complex is located at one side of the nucleus. Other animal and plant cells may have many Golgi complexes (consisting of separate stacks of membranes) located throughout the cell.

The Golgi complex processes, sorts, modifies, and packages proteins. Most proteins that are secreted from the cell, proteins that are a part of the plasma membrane, and proteins routed to other organelles of the internal membrane system, pass through the Golgi complex. These proteins are synthesized on ribosomes attached to the rough ER. Then, they are transported to the Golgi complex enclosed in small transport vesicles formed from the ER membrane. These vesicles fuse with the membranes of the Golgi complex (Fig. 5–12). The proteins then pass through the separate layers of the Golgi complex (moving by way of membrane transport vesicles).

While moving through the Golgi complex, the proteins are modified in different ways, resulting in the formation of complex biological molecules. Often, carbohydrates are added to the protein or previously added carbohydrates are further modified. (Some sugars are actually added to the protein in the rough ER, but these may be further modified in the Golgi complex.) The resulting **glycoproteins** are proteins with complex branched-chain polysaccharides attached to a number of different amino acids.

Each type of protein is modified in a different way. In some cases, the carbohydrates and other molecules that are added to the protein are used as "sorting signals," allowing the Golgi complex to route the protein to different parts of the cell. The Golgi complex of plant cells also produces some extracellular polysaccharides used as components of the cell wall.

Although it is especially prominent in cells specialized to secrete products, the Golgi complex also performs an important function in nonsecreting cells. It packages intracellular digestive enzymes in the little organelles called lysosomes.

Some Organelles Function in Cell Metabolism

Each cell carries on hundreds of metabolic reactions. These reactions are generally carried out within specialized compartments.

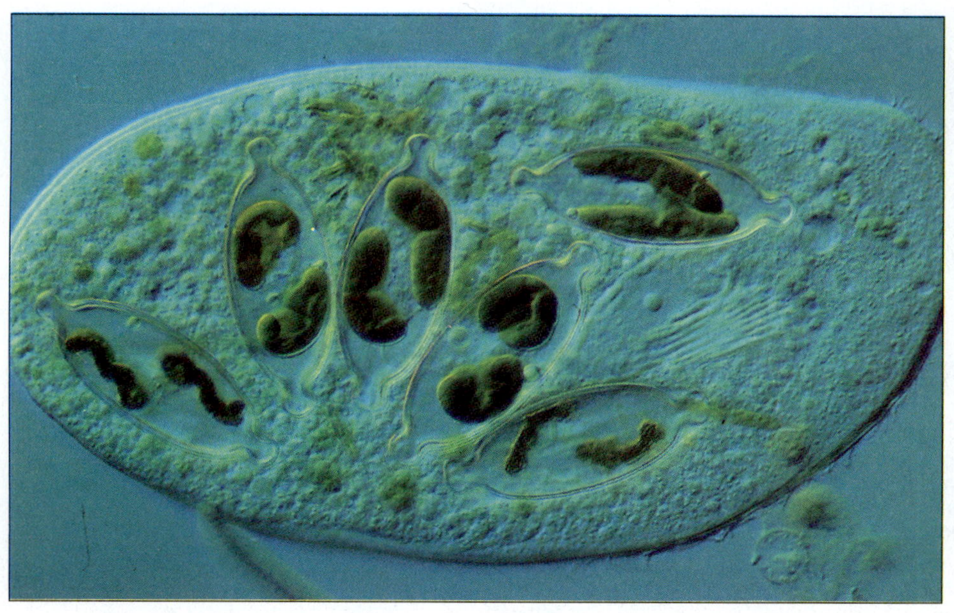

FIGURE 5–13 Vacuoles. The protozoon *Chilodonella* has numerous vacuoles containing ingested diatoms (small, photosynthetic algae). From the number of diatoms scattered about its insides, one might judge that *Chilodonella* has a rather voracious appetite. (M.I. Walker/ Photo Researchers, Inc.)

Lysosomes Are Compartments for Digestion

Lysosomes are small vesicles that contain digestive enzymes. These vesicles are released from the Golgi complex and dispersed throughout the cytoplasm. About 40 different enzymes have been identified in lysosomes. These enzymes are capable of digesting carbohydrates, fats, proteins, and nucleic acids. They destroy bacteria and other foreign matter taken into the cell. In fact, lysosome enzymes can even digest the cell that produces them. In a cell that is low on fuel, lysosomes may break down organelles so that their component molecules can be used as an energy source.

An important function of lysosomes is the digestion and breakdown of old cellular components, which otherwise tend to accumulate and interfere with proper cell function. The lysosome membrane itself is able to resist the digestive action of these powerful enzymes.

When a cell dies, the lysosome membrane breaks down, releasing digestive enzymes into the cytoplasm, where they break down the cell itself. This "self-destruct" system accounts for the rapid deterioration of many cells following death.

Some forms of tissue damage as well as the aging process may be related to leaky lysosomes. Rheumatoid arthritis is thought to result in part from damage done to cartilage cells in the joints by enzymes released from lysosomes. Cortisone-type drugs, which are used as anti-inflammatory agents, stabilize lysosome membranes so that leakage of damaging enzymes is reduced.

Vacuoles Are Large Fluid-Filled Sacs

Although lysosomes have been identified in almost all kinds of animal cells, their occurrence in plant and other cells is open to debate. Many of the functions carried out by lysosomes in animal cells are performed in plant and algal cells by a large, single membrane-bound sac referred to as the **vacuole** (Fig. 5–13). Although the terms vacuole and vesicle are sometimes used interchangeably, vacuoles are usually larger structures, sometimes produced by the merging of many vesicles.

As much as 90% of the volume of a plant cell may be occupied by a large central vacuole containing water, stored food, salts, pigments, and wastes. Plants lack systems for disposing of metabolic waste products that are toxic to the cells. Such waste products often aggregate and form small crystals inside the vacuole, making the vacuole look almost "empty" in the electron micrograph. The vacuole may also serve as a storage compartment for certain chemical compounds in plant cells. Compounds that are noxious to predators may also be stored in some plant vacuoles as a means of defense. In addition to storage, vacuoles in plants are involved in cell enlargement and in maintaining cell rigidity.

Vacuoles can have numerous other functions and are present in many types of animal cells and most commonly in single-celled protists. Most protozoa have food or digestion vacuoles, which fuse with lysosomes so that the food they contain can be digested. Many have contractile vacuoles, which remove excess water from the cell.

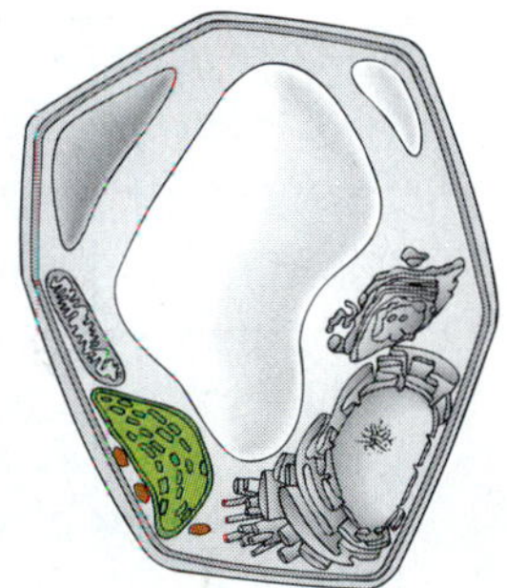

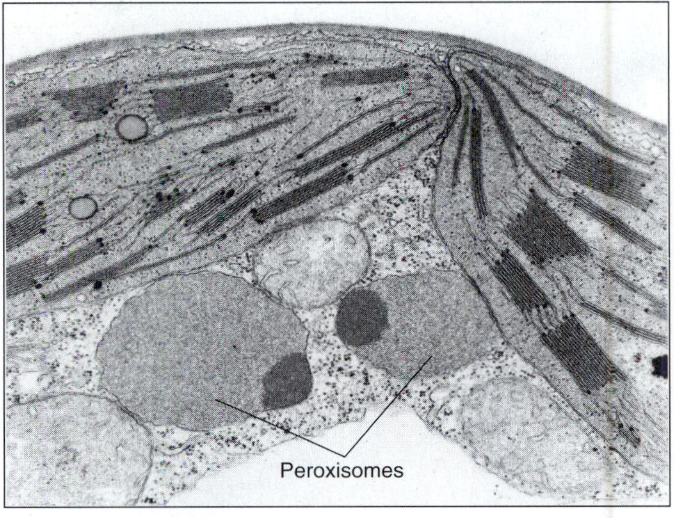

FIGURE 5–14 Peroxisomes. Transmission electron micrograph showing two peroxisomes in a leaf cell of tobacco. Three mitochondria and portions of two chloroplasts are seen adjacent to the peroxisomes. (E.H. Newcomb and S.E. Frederick, U. of Wisconsin, Biological Photo Service)

Microbodies Are Compartments that Contain a Variety of Enzymes

Microbodies are membrane-bound organelles that contain enzymes that regulate many different metabolic reactions. One type of microbody, the **peroxisome,** regulates the conversion of fats to carbohydrates (Fig. 5–14). During the breakdown of fats, hydrogen peroxide, a substance toxic to the cell, is produced. Peroxisomes contain enzymes that split hydrogen peroxide, making it harmless. Peroxisomes in liver and kidney cells may be important in detoxifying certain compounds such as ethanol, the alcohol in alcoholic beverages.

Another type of microbody, the **glyoxysome,** is abundant in the seeds of certain plants. Its enzymes convert stored fats to sugars. These sugars are used as an energy source and as a component for making needed compounds during the germination of seeds.

Some Organelles Convert Energy to Usable Forms

Plant, algal, and some bacterial cells capture energy from sunlight. Consumers, such as animals, obtain energy from the environment in the form of food. These types of energy must be converted into forms that can be used by the cells in metabolic processes. Mitochondria and chloroplasts are specialized to convert energy from one form into another. Generally, energy is ultimately converted to the chemical energy of ATP (Chapter 4).

Mitochondria Are the Power Plants of the Cell

Mitochondria (singular, *mitochondrion*) are the sites of cellular respiration, the chemical process that converts the energy of organic compounds to ATP (discussed in detail in Chapter 8; Fig. 5–15). They are abundant in cells that are very active metabolically. More than 1000 mitochondria have been counted in a single liver cell! Mitochondria are capable of changing size and shape, of fusing with other mitochondria to form larger structures, or of splitting to form smaller ones. They may appear as spheres, rods, sausages, threads, spirals, or even as irregular shapes.

Each mitochondrion is bounded by a double membrane. The area formed between the outer and inner membranes is called the **intermembrane space.** The region enclosed by the inner membrane is the **matrix.** Both the outer and inner membranes consist of lipid bilayers (discussed in the next chapter) in which a variety of protein molecules are embedded. The outer layer of the mitochondrial membrane is smooth, but the inner layer is folded. Its folds, called **cristae,** project into the interior of the mitochondrion. Cristae serve to increase the available surface area, and some of the enzymes needed for cellular respiration are organized along these folds. Other enzymes are found in the semifluid material of the matrix.

Each mitochondrion has a small amount of DNA, enough to code for about 15 proteins, and also contains ribosomes. Some proteins are indeed synthesized within these organelles.

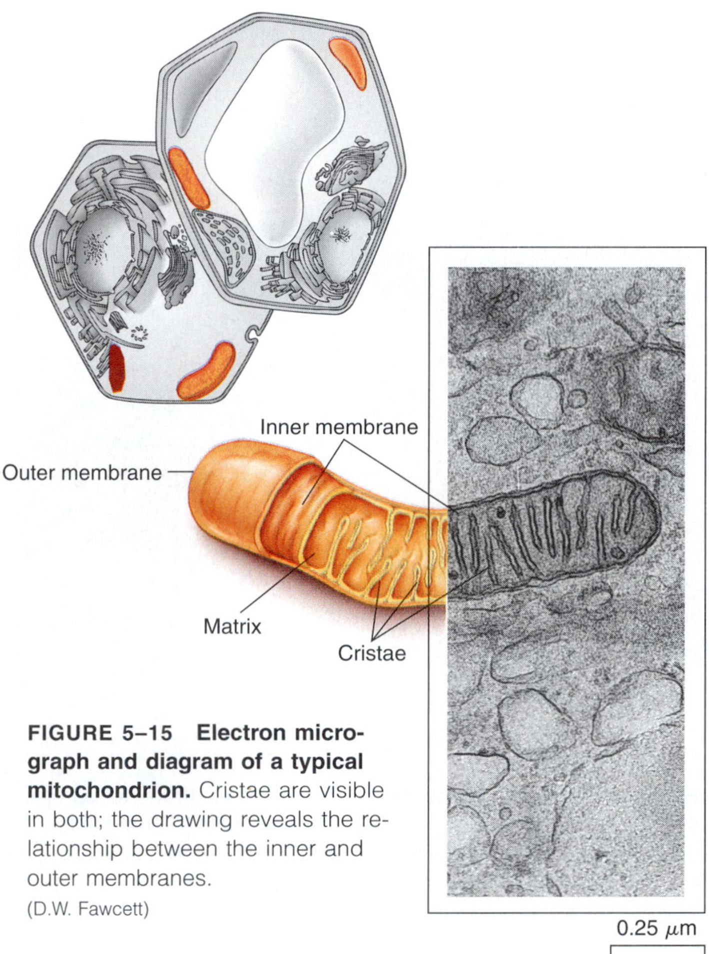

FIGURE 5–15 Electron micrograph and diagram of a typical mitochondrion. Cristae are visible in both; the drawing reveals the relationship between the inner and outer membranes.
(D.W. Fawcett)

Chloroplasts Convert Light Energy to Chemical Energy through the Process of Photosynthesis

Recall from Chapter 1 that plant and algal cells carry out a complex set of energy conversion reactions known as photosynthesis (discussed in detail in Chapter 9). Organelles known as **chloroplasts** contain the pigments, such as **chlorophyll,** that trap light energy for photosynthesis. A unicellular alga may have only a single large chloroplast, whereas a plant leaf cell may have as many as 20 to 100.

Chloroplasts are typically disc-shaped structures bounded by an inner and outer membrane (Fig. 5–16). The space enclosed by the inner membrane, called the **stroma,** contains enzymes needed to manufacture glucose during photosynthesis. The inner chloroplast membrane also encloses a third system of membranes. This system consists of interconnected sets of flat, disc-like sacs called **thylakoids.** The thylakoids contain the chlorophylls and other pigments that trap light energy in photosynthesis. The thylakoids are arranged in stacks called **grana** (singular, *granum*).

Like mitochondria, chloroplasts contain DNA and ribosomes and manufacture some proteins. Chloroplasts are also able to grow and divide to form daughter chloroplasts and sometimes differentiate from plastid precursors that lack the characteristic chloroplast structure.

Chloroplasts are one of several types of organelles known as **plastids,** which produce and store food materials in cells of plants and algae. Plant cells also contain colorless plastids termed **leukoplasts,** which store starch.

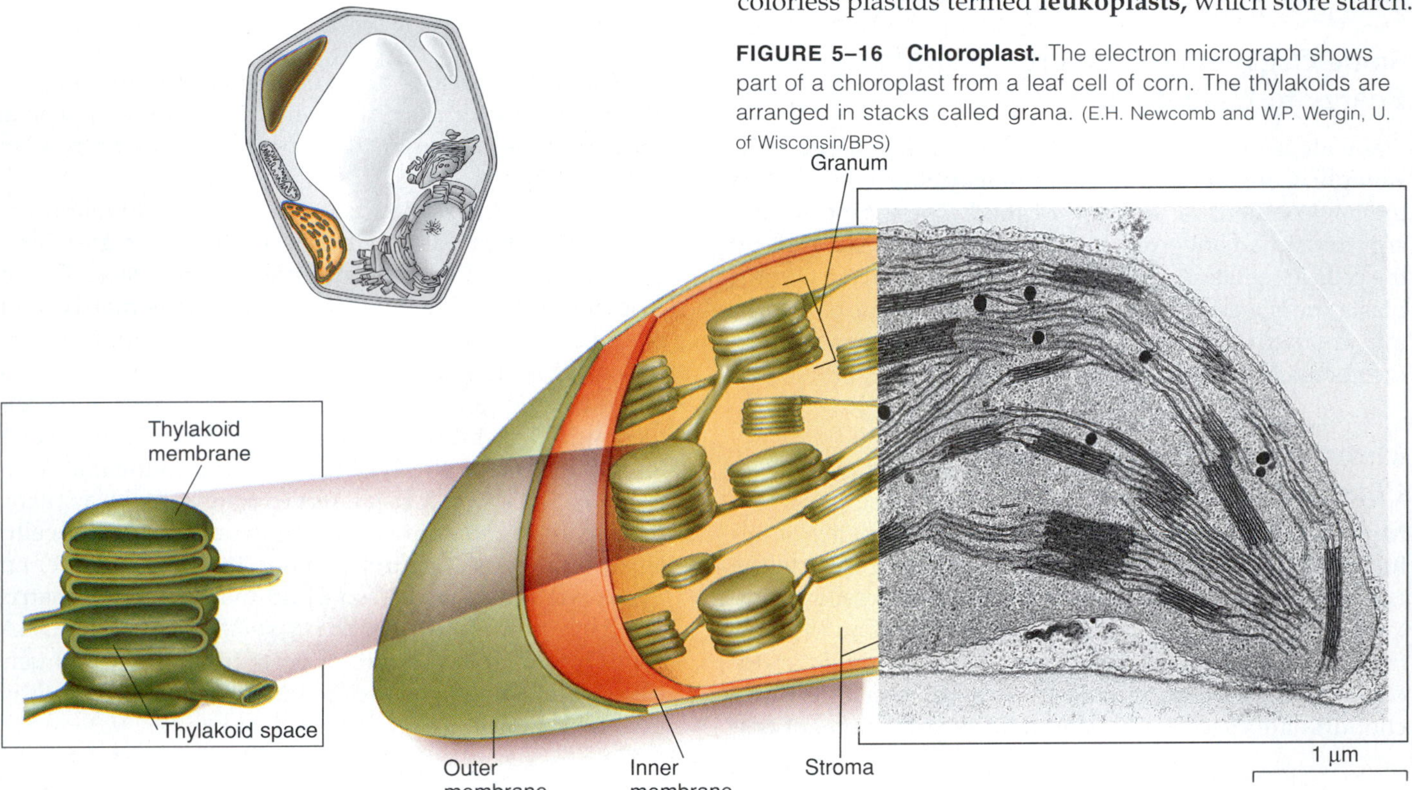

FIGURE 5–16 Chloroplast. The electron micrograph shows part of a chloroplast from a leaf cell of corn. The thylakoids are arranged in stacks called grana. (E.H. Newcomb and W.P. Wergin, U. of Wisconsin/BPS)

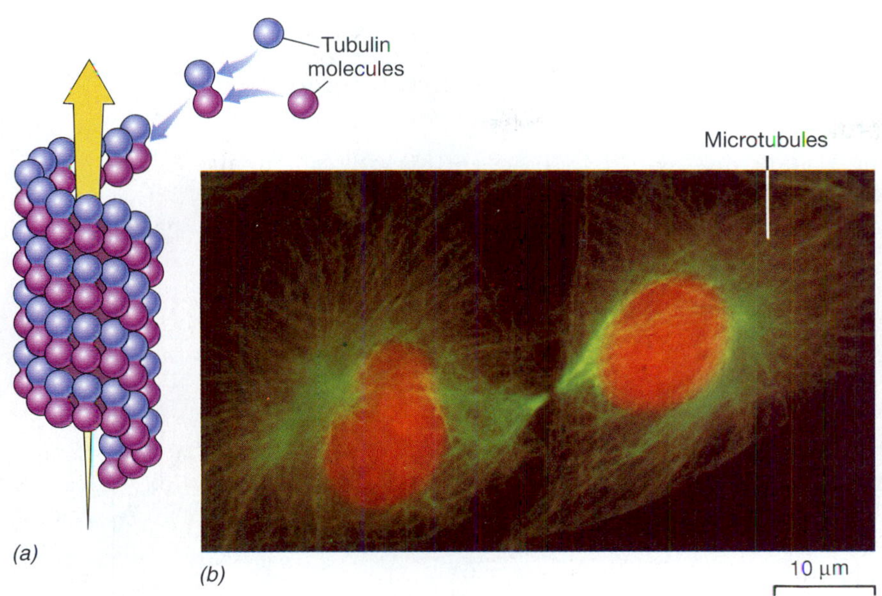

FIGURE 5–17 Microtubules. (***a***) Structure and assembly of a microtubule. Note the spiral arrangement of the tubulin components. Microtubules are constructed by adding tubulin units to an end of the hollow cylinder. Disassembly takes place by removal of subunits from the ends of the filaments. (***b***) Fluorescence micrograph showing the extensive distribution of microtubules throughout these cells. The cells were stained with fluorescent antibodies that bind to the tubulin, permitting the microtubules to be viewed (green). Different fluorescent antibodies were used to stain the DNA (orange). (*b*, Jonathan G. Izant)

A third type of plastid, the **chromoplast,** contains pigments that give color to many flowers, autumn leaves, some ripe fruit, and certain roots.

Concept Connections

Cell Structure ⇄ *Evolution*

Although the origin of cells and organisms will be discussed in Chapter 19, our introduction to cells will be more complete if we begin to think about how eukaryotic cells may have come to be. The oldest fossils are prokaryotic cells, and most biologists agree that they evolved before the more complex eukaryotic cells. In fact, eukaryotic cells are thought to have evolved *from* prokaryotic cells about 1.9 billion years ago. This idea is based on a number of observations about modern cell structure, a few of which are outlined here.

The mitochondria and chloroplasts found in eukaryotic cells look very much like prokaryotic cells. Biologists theorize that chloroplasts evolved from prokaryotic cells. Perhaps they began as bacteria, which could carry on photosynthesis, that were originally ingested by larger cells. These bacteria were not digested by their hosts. Rather, they survived and reproduced along with the host cell. Thus, future generations of the host cell also contained these bacteria. The cells developed a symbiotic relationship, one in which two organisms of two species closely interact. Eventually the bacterial cell lost the ability to live outside its host. Mitochondria may have evolved in the same way from oxygen-requiring bacteria that inhabited other prokaryotes. Evidence for this **endosymbiont theory** comes, in part, from the fact that both chloroplasts and mitochondria contain DNA and are capable of reproducing themselves.

Certain Structures Support the Cell, Connect It with Other Cells, and Help It Move

A cell must be more than a microscopic glob of jelly if it, or the organism of which it may form a part, is to have a complex shape and structure. The flexible, semiliquid membranes we have been discussing are not mechanically suitable for use as cellular struts, girders, or cables. But a complex network of threadlike and tubelike structures form the **cytoskeleton,** which gives shape to cells.

Cytoskeletal Elements Give Cells Their Shape and Allow Movement

The cytoskeleton is constructed from at least three types of components: **microtubules, microfilaments,** and **intermediate filaments.** The **centrosome,** or **microtubule-organizing center (MTOC),** is a somewhat shapeless organelle that is usually located near the nucleus. The centrosome appears to direct the construction of the cytoskeleton, determine the shape of the cell, play an important role in cell movement, and transport materials within the cell. During mitosis (nuclear division) of animal cells, the centrosome sets up the mitotic spindle that separates the chromosomes into the daughter cells.

Microtubules, fibers found in the cytoplasm of all eukaryotic cells, radiate from the centrosome. When the cell is not in the process of dividing, microtubules extend from the centrosome throughout most of the cytoplasm (Fig. 5–17). When an animal cell begins to undergo mitosis, microtubules reorganize to form the mitotic spindle.

Microtubules, the largest of the cytoskeletal components, are very small, hollow cylinders composed mainly of protein subunits called **tubulins** (Fig. 5–17). They can grow by the addition of more tubulin subunits, or they can shorten by disassembly into subunits. Microtubules located just beneath the plasma membrane help determine the shape of many cells. Microtubules are the main structural components of cilia, flagella, and centrioles, structures that are discussed in the following sections.

Microfilaments, the smallest of the cytoskeletal components, can rapidly assemble and disassemble. They are composed mainly of **actin,** a protein important in muscle contraction. Microfilaments are essential in the flowing of cytoplasm that enables a cell to move from one place to another. They are also involved in cytoplasmic streaming, which results in the continuous motion of organelles characteristic of many cells.

Intermediate filaments are stable, tough fibers made of polypeptides. They are thought to help strengthen the cytoskeleton and are abundant in parts of a cell that are subject to mechanical stress.

At the center of the centrosome in animal (and some protist) cells are a pair of **centrioles,** which function in nuclear division (mitosis; Chapter 10). The centrioles are positioned at right angles to one another. Each centriole is a hollow structure composed of nine rods. Each rod consists of three microtubules (Fig. 5–18). In plant cells, centrioles are absent and the centrosome is more diffuse.

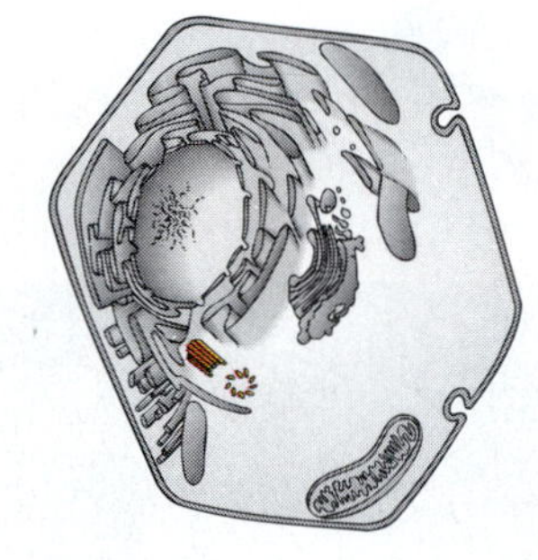

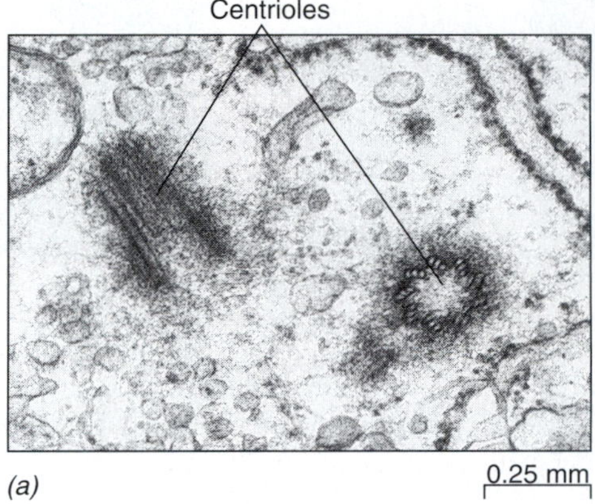

(a)

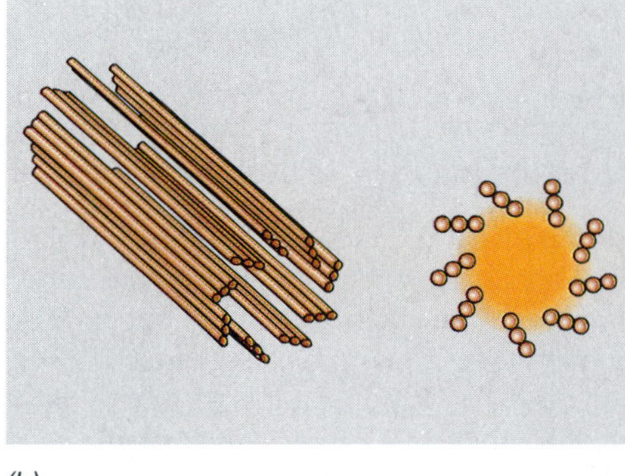

(b)

FIGURE 5–18 Centriole structure. (***a***) Electron micrograph of a pair of centrioles. Note that one centriole has been cut longitudinally and one transversely. (***b***) A drawing of the centrioles. (*a,* B.F. King, School of Medicine, U. of California, Davis/Biological Photo Service)

Cells Swim or Move Materials with Cilia and Flagella

Many cells have movable, whiplike structures projecting from their surface that exhibit a beating motion. If a cell has one, or only a few, of these appendages and they are relatively long in proportion to the size of the cell, they are called **flagella** (singular, *flagellum*). If a cell has many short appendages, they are called **cilia** (singular, *cilium*). Flagella and cilia both function mainly in movement, either to propel the cell through a watery environment or to move liquids and particles across the surface of a layer of cells.

Flagella or cilia are commonly found on one-celled and small multicellular organisms and on the sperm cells of animals and some plants. They are the principal means of locomotion of such cells. Cilia commonly occur on the cells lining the internal ducts, such as respiratory passageways, of animals; their beating assists in moving materials through these passageways. The description that follows is intended to refer only to the flagella and cilia of eukaryotes; prokaryote flagella are fundamentally different.

Each cilium or flagellum consists of a slender, cylindrical stalk covered by an extension of the plasma membrane. The core of the stalk contains a group of microtubules (Fig. 5–19). Nine pairs of microtubules are arranged around the circumference, and two single microtubules are located in the center. This **9+2 arrangement** is characteristic of all eukaryotic cilia and flagella whether they occur on a clam's gill, a gingko tree sperm, or in the human respiratory tract.

Microtubules move by sliding in pairs past each other. The sliding force is generated by special proteins that are attached to the microtubules like small arms. These proteins use the energy stored in ATP in such a way that arms on one pair of tubules are able to "walk" along the adjacent pair of tubules. This causes the entire structure to bend back and forth.

At the base of the stalk of a cilium or flagellum is a **basal body,** which has a 9 × 3 structure similar to that of a centriole. In fact, some cell biologists now consider that the basal body is a centriole. The microtubules in a cilium or flagellum grow directly out of the basal body.

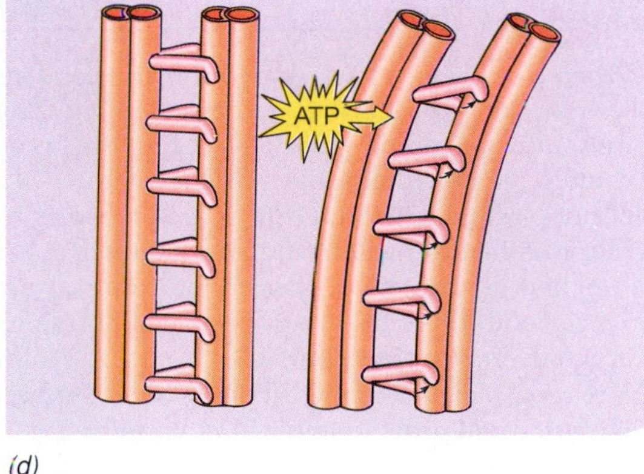

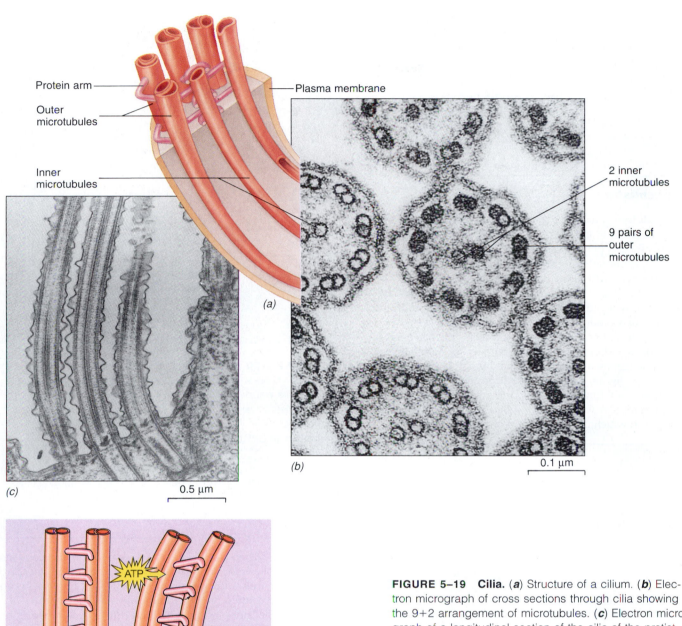

FIGURE 5–19 Cilia. (***a***) Structure of a cilium. (***b***) Electron micrograph of cross sections through cilia showing the 9+2 arrangement of microtubules. (***c***) Electron micrograph of a longitudinal section of the cilia of the protist *Tetrahymena*, an organism often used in biological research. Some of the interior microtubules can be clearly seen. (***d***) The "arms" move the microtubules by forming and breaking cross bridges on the adjacent microtubules so that one tubule "walks" along its neighbor. (*b*, D.W. Fawcett; *c*, W.L. Dentler, U. of Kansas/Biological Photo Service)

Cell Walls Surround Cells in Four of the Five Kingdoms

Cells of prokaryotes, certain protists, fungi, and plants have **cell walls** that lie just outside their plasma membranes (see Window on the Plant Cell). The cell wall protects the cell, gives it shape, and often contributes to the shape and strength of the organism of which the cell may be a part (Fig. 5–20). The cell walls of prokaryotes, protists, and fungi will be discussed in later chapters.

The plant cell wall serves to contain the high internal pressure of plant cells, somewhat as the outer casing of a tire contains the internal pressure of its air. When the internal pressure of plant cells is reduced, the membrane and cell contents no longer press against the cell walls. The plant tissue then becomes limp, and the plant wilts. The rigid plant cell wall is pierced in many places by tiny pores through which water and dissolved materials can pass.

The cell wall is constructed of polysaccharides secreted by the cell. The polysaccharide cellulose is the main component of the plant cell wall and is usually present in the form of long, threadlike fibers.

CONCEPT CONNECTIONS

Chemistry Cell ⇄ *Structure*

Recall from Chapter 4 that the main component of the plant cell wall is the polysaccharide cellulose, a long, unbranched polymer of glucose units. The molecule folds, forming long, thread-like fibers that can be seen under the electron microscope.

The cell wall has been compared to reinforced concrete in which the cellulose fibers are like steel rods and the matrix around them is the concrete. The matrix of the plant cell wall consists of polymers, such as polysaccharides and protein. Among the polymers are pectins (jelling agents used in the preparation of jams and jellies) and lignin, a substance that provides rigidity. Cells whose main function is support contain a great deal of lignin in their walls. In fact, wood consists mainly of cell walls in which cellulose is reinforced by lignin.

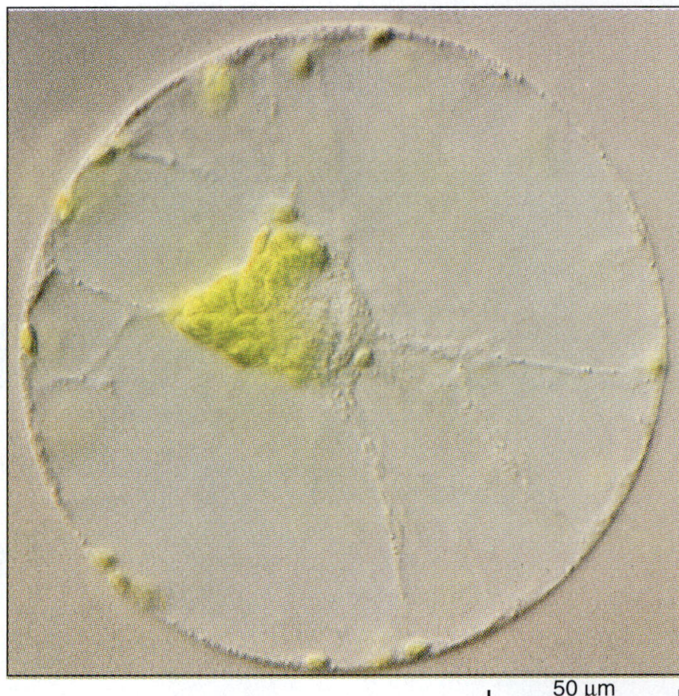

FIGURE 5–20 Cell walls. Many plant cells are polyhedral in shape. However, this *Amaryllis* protoplast was made circular when its walls were dissolved by an enzyme. Enzymes enable biologists to take a plant cell apart without disrupting the organelles. This, in turn, allows the functions of an individual organelle to be studied. (Comstock, Inc./William Marin)

Chapter Summary

I. The cell is considered the basic unit of life because it is the smallest self-sufficient unit of living material and because, as stated in the cell theory, all organisms are composed of cells and their products.

II. Most cells are microscopic, but their size and shape vary according to their function.

III. Biologists have learned about cellular structure by studying cells with light and electron microscopes and by using chemical techniques.

IV. Prokaryotic cells lack a nucleus with a distinct nuclear membrane and other membrane-bound organelles. Eukaryotic cells have distinct membrane-bound nuclei and a variety of other organelles.

V. The cell is bounded by a plasma membrane, and most eukaryotic cells have elaborate organelles specialized to perform specific intracellular functions.

- A. In eukaryotic cells, the nucleus, control center of the cell, is bounded by a double-layered nuclear envelope.
  1. The nucleolus functions in the assembly of ribosomes.
  2. When a cell begins to divide, chromatin condenses, forming long chromosomes. Chromatin and chromosomes are composed of DNA and proteins and contain the hereditary material, the genes.
- B. The endoplasmic reticulum (ER) is a system of internal membranes that transports and stores materials within the cell and divides the cytoplasm into compartments.
  1. The smooth ER is the site of phospholipid, steroid, and fatty acid metabolism.
  2. The rough ER is studded along its outer surface with ribosomes that manufacture proteins.
- C. The Golgi complex modifies substances that are produced in the ER and packages some for export from the cell. It also produces lysosomes.
- D. Lysosomes function in intracellular digestion; peroxisomes contain enzymes that break down hydrogen peroxide.
- E. Mitochondria are the power plants of the cell; the cristae of the inner membrane contain enzymes needed for cellular respiration.

F. Cells of algae and plants contain chloroplasts that trap light during photosynthesis. Some plant cells also contain pigment-filled chromoplasts and colorless leukoplasts.

G. Microtubules, microfilaments, and intermediate filaments help maintain the shape of the cell and play a role in cellular movement. These structures form the cytoskeleton.

H. Centrioles, cilia, and flagella are composed of microtubules. Cilia and flagella move cells through the surrounding medium or move the surrounding medium past the cells.

I. Plant cells secrete cellulose and other polymers to form rigid cell walls. Cell walls protect and give shape to the cells of some organisms.

Selected Key Terms

cell, p. 82
cell theory, p. 82
cell wall, p. 103
chloroplast, p. 100
chromatin, p. 94
chromosome, p. 94
cilia, p. 102
cytoplasm, p. 82
endoplasmic reticulum, p. 95
eukaryotic cell, p. 82
flagellum, p. 102
Golgi complex, p. 96
intermediate filament, p. 101
lysosome, p. 98
microfilament, p. 101
microtubule, p. 101
mitochondrion, p. 99
nuclear envelope, p. 93
nucleolus, p. 93
nucleus, p. 93
organelle, p. 82
plasma membrane, p. 82
plastid, p. 100
prokaryotic cell, p. 82
ribosome, p. 96

Post-Test

1. The ability of a microscope to reveal fine detail is known as ____________ ____________.

2. Proteins are synthesized on the ____________.

3. Fatty acid and phospholipid metabolism take place along the ____________ ____________.

4. The ____________ ____________ modifies proteins and packages cellular secretions.

5. Many of the reactions of cellular respiration take place within the ____________.

6. ____________ are plastids that contain the pigment chlorophyll needed for trapping light energy.

7. In eukaryotic cells, cilia and flagella are composed of ____________ in a 9 + 2 arrangement.

8. ____________ are important in cell movement, such as cytoplasmic streaming.

9. The control center of the cell is the ____________.

10. In eukaryotic cells, the nucleus is bounded by a nuclear ____________.

11. In a cell that is not dividing, the uncoiled DNA and protein are evident as ____________; when the cell begins to divide, this material condenses to form discrete ____________.

12. Each chromosome contains several thousand ____________ that contain the genetic code.

13. In addition to a plasma membrane, plant cells are bounded by a ____________.

14. Many plant cells have a large central ____________.

15. In multicellular organisms, the size and shape of a cell are related to the type of ____________ it performs.

16. Prokaryotic cells lack (a) a nuclear membrane (b) membrane-bound organelles like mitochondria (c) DNA (d) answers a, b, and c are correct (e) only answers a and b are correct.

17. The scientist who gave cells their name based on his observations of dead cork cells was (a) Schleiden (b) Schwann (c) Hooke (d) Leeuwenhoek (e) Weismann.

18. The superior ability of the electron microscope to show cell structure is due to (a) magnification (b) resolving power (c) ultrastructure (d) scanning power (e) none of the preceding.

19. Cellular components that contain digestive enzymes are (a) lysosomes (b) smooth ER (c) nucleolus (d) cell wall (e) microfilaments.

20. The inner folds of the mitochondrion are called (a) grana (b) cristae (c) thylakoids (d) matrix (e) peroxisomes.

Review Questions

1. What limits cell size?
2. Explain the cell theory.
3. Define (a) magnification and (b) resolving power. What are the advantages of the electron microscope over the ordinary light microscope?
4. Draw a nucleus and label its structures. Give the function of each structure.
5. Compare mitochondria with chloroplasts.
6. Why are lysosomes sometimes referred to as the self-destruct system of the cell? Do you think this name is justified? Why?
7. What is the relationship between chromatin and chromosomes? Between chromosomes and genes?
8. Describe the structure of the cytoskeleton.
9. How does a bacterial cell differ from an animal cell?
10. What do centrioles, cilia, and flagella have in common?
11. Label the diagrams of the animal cell and plant cell. What are the fundamental differences between the two cell types?

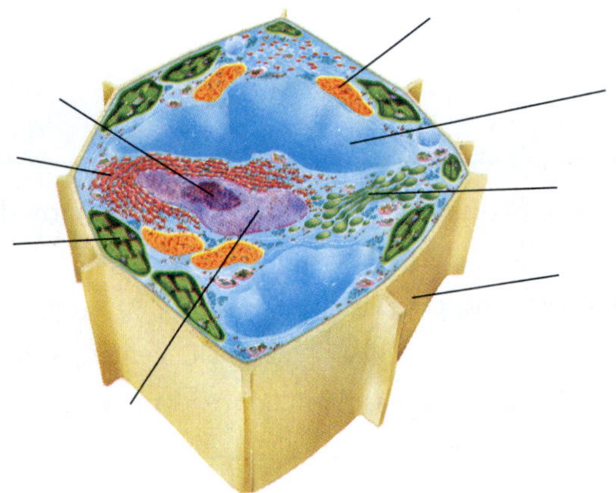

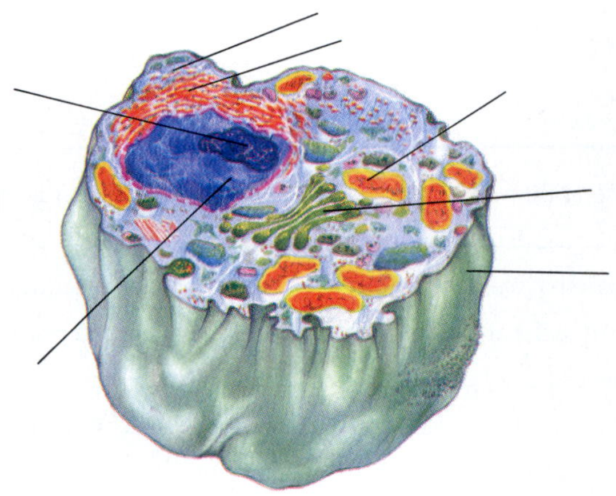

Thinking Critically

1. Design an experiment to learn more about the functions of the Langerhans cell. (*Hint:* Begin by constructing a hypothesis.)
2. Explain the following statement: All living cells can trace their ancestry back to ancient times. Trace a muscle cell in your own body back to a cell of a grandparent.
3. A cell from a gastric (stomach) gland secretes the enzyme pepsin. Trace in sequence its production, transport, packaging, and release from the cell, naming each of the organelles involved.
4. Which do you think are greater: the differences between plant and animal cells, or the differences between prokaryotic and eukaryotic cells? Why?
5. Imagine that a mutant cell was produced that lacked mitochondria. What would be its fate? Why?

Recommended Readings

(Also consult reading lists in Chapters 6 through 10)

de Duve, C., *A Guided Tour of The Living Cell.* San Francisco, W.H. Freeman Company, 1985 (2 vols). The discoverer of the lysosome discusses every organelle.

Fawcett, D.W., *The Cell,* 2nd ed. Philadelphia, W.B. Saunders Company, 1981. A study of cell structure through an exciting collection of electron micrographs.

Glover, D.M., C. Gonzalez, and J.W. Raff, "The centrosome," *Scientific American,* Vol. 268, No. 6, June 1993, pp. 62–68. Centrosomes organize the cellular skeleton and set up the spindle that partitions the chromosomes in cell division.

Loewy, A., P. Siekovitz, J. Menninger, and J. Gallant, *Cell Structure and Function: An Integrated Approach.* Philadelphia, Saunders College Publishing, 1991. A readable introductory cell biology text.

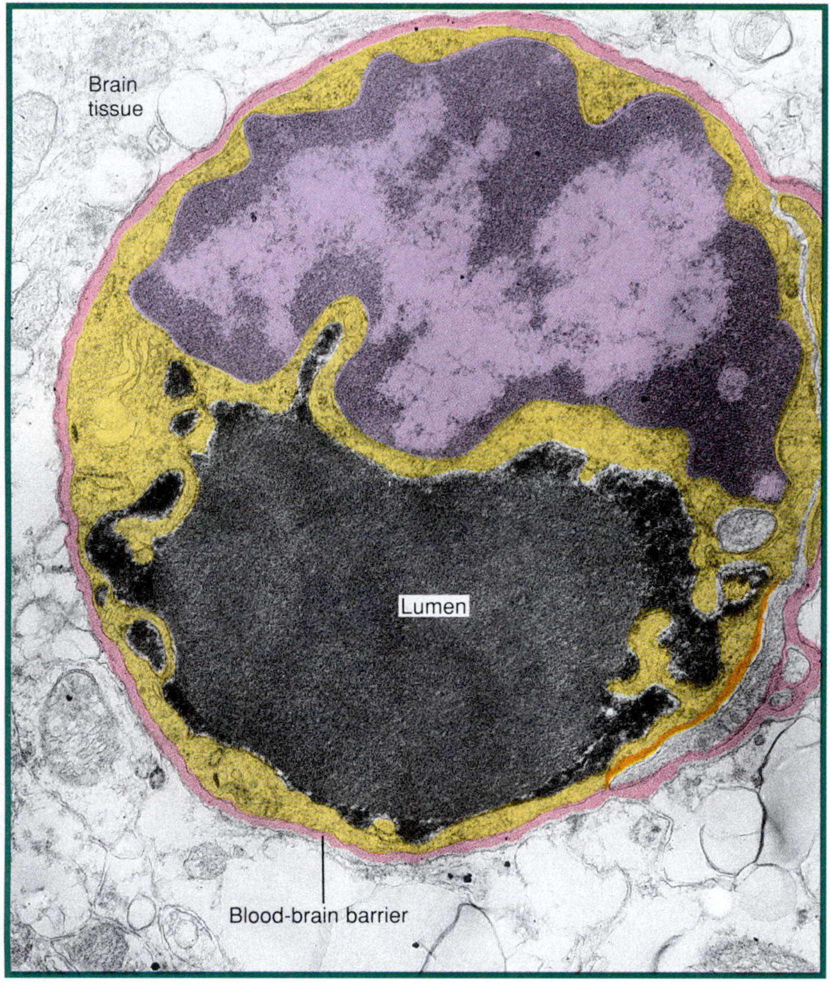

Electron micrograph of cross section of a brain capillary demonstrating an intact blood brain barrier. *The lumen (space) of the capillary contains a dark stain. The stain is prevented from escaping into the adjacent brain tissue by the tight connections, called tight junctions, between the endothelial cells that line the capillary © (Courtesy of Dr. E. Tuomanen, Rockefeller University and Dr. J. Elroy, Tufts University)*

Penetrating the Blood–Brain Barrier

Researchers developing clinical drugs, such as antibiotics, must carefully consider the properties of biological membranes, especially the plasma membrane. Serving as the boundary between the cell and its environment, the plasma membrane regulates the passage of materials into and out of the cell. Of special consideration to investigators is the *blood–brain barrier* that protects the brain from many harmful substances.

The blood–brain barrier is made up of tightly joined cells lining the blood vessels in the brain. The plasma membranes of these cells form connections with one another known as *tight junctions*. Although the blood–brain barrier permits water, oxygen, carbon dioxide, glucose, essential amino acids, and other nutrients to enter the brain, it blocks the passage of many toxic substances and disease organisms from the blood into the brain tissue. Unfortunately, this barrier also blocks the entrance of many healing medications.

Doctor Elaine Tuomanen, head of the laboratory of molecular infectious diseases at the Rockefeller University, and her colleagues are studying how certain bacteria penetrate the blood–brain barrier. Perhaps bacteria will provide clues for developing treatments for brain disorders.

Bacterial meningitis is a deadly disease that attacks about 50,000 people a year. The meningitis bacteria slip through the blood–brain barrier and cause inflammation of the brain, which can lead to coma and death. Survivors of the disease may suffer loss of nerve cells and deficits in hearing or learning ability. Just how the bacteria get through the blood–brain barrier is not known. Dr. Tuomanen has identified a glycopeptide (a sugar linked to a peptide) on the bacterial cell surface that may be a molecular key used by bacteria to unlock the barrier and gain access to the brain. Could similar molecular keys be incorporated into the structure of new drugs?

HIV, the virus that causes AIDS, also infects the brain. In order to reach infected brain cells, any drug developed for treatment of AIDS must be able to cross the blood–brain barrier. Current drugs used to treat AIDS (such as AZT) do traverse the blood–brain barrier, but are thought to do so slowly.

Some drugs that *do* penetrate the blood–brain barrier cause undesirable side effects. For example, antihistamines, such as Benadryl, are used to relieve symptoms of allergies like hay fever. Such symptoms are caused, in part, when histamine, released by activated cells of the immune system, combines with specific receptors on cell surfaces, causing allergic symptoms. Antihistamines combine with the same receptors, blocking histamines from binding. When antihistamines pass through the blood–brain barrier and combine with histamine receptors on brain cells, they cause drowsiness. Some newer antihistamines (such as Seldane) cannot penetrate the blood–brain barrier and so do not cause drowsiness.

As researchers learn more about the blood–brain barrier, they will be able to control by design whether or not drugs penetrate it. Development of drugs that cross the barrier and enter the brain may lead to new treatments for tumors and perhaps even for Alzheimer's disease (discussed in the introduction to Chapter 33).

Biological Membranes

6

Learning Objectives

After you have read this chapter you should be able to:

1. Discuss the importance of the plasma membrane to the cell, describing the various functions it performs.
2. Describe the currently accepted model for the structure of the plasma membrane.
3. Describe the functions of membrane proteins.
4. Contrast desmosomes, tight junctions, and gap junctions.
5. Contrast the physical with the physiological processes by which materials are transported across cell membranes.
6. Solve simple problems involving osmosis. For example, predict whether cells will swell or shrink under various osmotic conditions.
7. Compare exocytosis with endocytosis, and contrast phagocytosis and pinocytosis.

Key Concepts

- ☐ Every cell is surrounded by a plasma membrane that regulates the passage of materials into and out of the cell.
- ☐ Biological membranes consist of phospholipids, which form a bilayer. Proteins are associated with biological membranes; their hydrophobic portions may be embedded in the phospholipid bilayer.
- ☐ The proteins of biological membranes perform many specialized functions; they are vital in maintaining the cell's homeostasis.
- ☐ Materials can pass through membranes nonselectively by diffusion (by osmosis in the case of water) or selectively by facilitated diffusion. These processes, which move materials from regions of high concentration to regions of low concentration ("down" a concentration gradient), do not consume cellular energy.
- ☐ Materials can also move through membranes by active transport, which does consume cellular energy. In active transport, materials can be moved from regions of low concentration to regions of high concentration; that is, active transport concentrates materials.
- ☐ Cells can ingest solid particles by endocytosis, and eject such particles by exocytosis.

BIOLOGICAL MEMBRANES PERFORM VITAL CELLULAR FUNCTIONS

All cells are separated from the outside world by a **plasma membrane** (see Windows on the Plant and Animal Cells). This separation permits the cell to maintain an appropriate internal environment in which its life processes can take place. The plasma membrane is far more than a barrier. It protects the cell and in certain cells it transmits impulses, secretes substances, or is involved in cell movement. Among its many vital functions we can list the following:

1. *The plasma membrane regulates the passage of materials into and out of the cell.* Each cell differs from its surroundings in physical properties and chemical composition. If the plasma membrane were completely permeable, substances would pass freely into and out of the cell. The chemical composition of the cell would be the same as its surroundings and it could no longer function as a cell. This does not happen because the plasma membrane is **selectively permeable;** it can prevent the passage of certain substances while permitting, even facilitating, the passage of others, such as the entrance of nutrients or the elimination of wastes. Its ability to regulate the passage of materials enables the cell to maintain a fairly constant set of internal conditions despite changes in its external environment.
2. *The plasma membrane receives information that permits the cell to sense changes in its environment and to respond to them.* Receptor proteins in the plasma membrane receive chemical messages from other cells. Hormones, growth factors, and neurotransmitters (chemicals released by nerve cells) are among the substances that combine with such receptors. When a compound combines with a receptor, the membrane is stimulated to send a signal into the cell, resulting in some type of behavioral, physiological, or structural response.
3. *The plasma membrane communicates with neighboring cells and with the organism as a whole.* Certain proteins in the plasma membrane permit cells to recognize one another, to adhere to each other when appropriate, and to exchange materials. The plasma membrane is important in internal defense. Cells have peptide receptors on their plasma membranes by which they identify one another. The body recognizes its own receptors as "self" and identifies cells with other molecules as "nonself." Once invaders are identified, the immune system launches a sophisticated chemical and cellular battle. For example, natural killer cells are white blood cells capable of recognizing changes in molecules on the plasma membranes of cells that have been infected by viruses. Once they recognize these cells as "nonself," they are able to attack and destroy them.

The plasma membrane is only one of many types of membranes found in eukaryotic cells. Many of the organelles discussed in the preceding chapter are compartments made of specialized membranes that function together to carry out the life functions of the cell. The wonderfully complex membranes of the mitochondria and chloroplasts hold enzymes that release energy and generate ATP. The membranes of the rough endoplasmic reticulum are essential for efficient synthesis of most proteins. The membranes of the Golgi complex, the lysosomes, and the smooth ER each have their own specialized role to play in the economy of the cell. Membranes make up much of the machinery of cellular life; we are made of membranes. In this chapter we focus on the plasma membrane. However, all biological membranes are basically similar in structure and in the functions outlined above.

BIOLOGICAL MEMBRANES ARE LIPID BILAYERS

The plasma membrane is so thin, less than 10 nanometers (nm), that it can be seen only with the electron microscope (Fig. 6–1; although it appears to be visible with a light microscope, this is actually an optical illusion). In 1972, S. J. Singer and G. L. Nicolson proposed a model of the structure of biological membranes known as the **fluid mosaic model.** According to this model, biological membranes consist of a rather fluid lipid bilayer (a double layer of lipid) in which a variety of proteins are embedded.

The Membrane Bilayer Consists Mainly of Phospholipids

Most of the lipids in biological membranes are **phospholipids.** Recall from Chapter 4 that a phospholipid consists of two fatty acid chains chemically bonded to two of the carbons of a glycerol molecule (see Fig. 4–2). The third carbon of the glycerol is linked to a negatively charged phosphate group that is bonded to a polar organic molecule. The fatty acid chains are nonpolar and **hydrophobic** (water-hating), whereas the polar, organic region of the phospholipid is **hydrophilic** (water-loving). Molecules that have distinct hydrophobic and hydrophilic regions are called **amphipathic molecules.**

The formation of a **lipid bilayer** from phospholipids is a rapid, spontaneous process. The driving force for this process is the hydrophobic interactions of the hydrocarbon chains. When phospholipids are placed in water, the polar heads show an affinity for water, but the hydrocarbon tails are repelled and turn toward one another (Fig. 6–2). Thus, the phospholipid molecules spontaneously become oriented to form a bilayer in a watery environment. No covalent bonds link adjacent lipid mol-

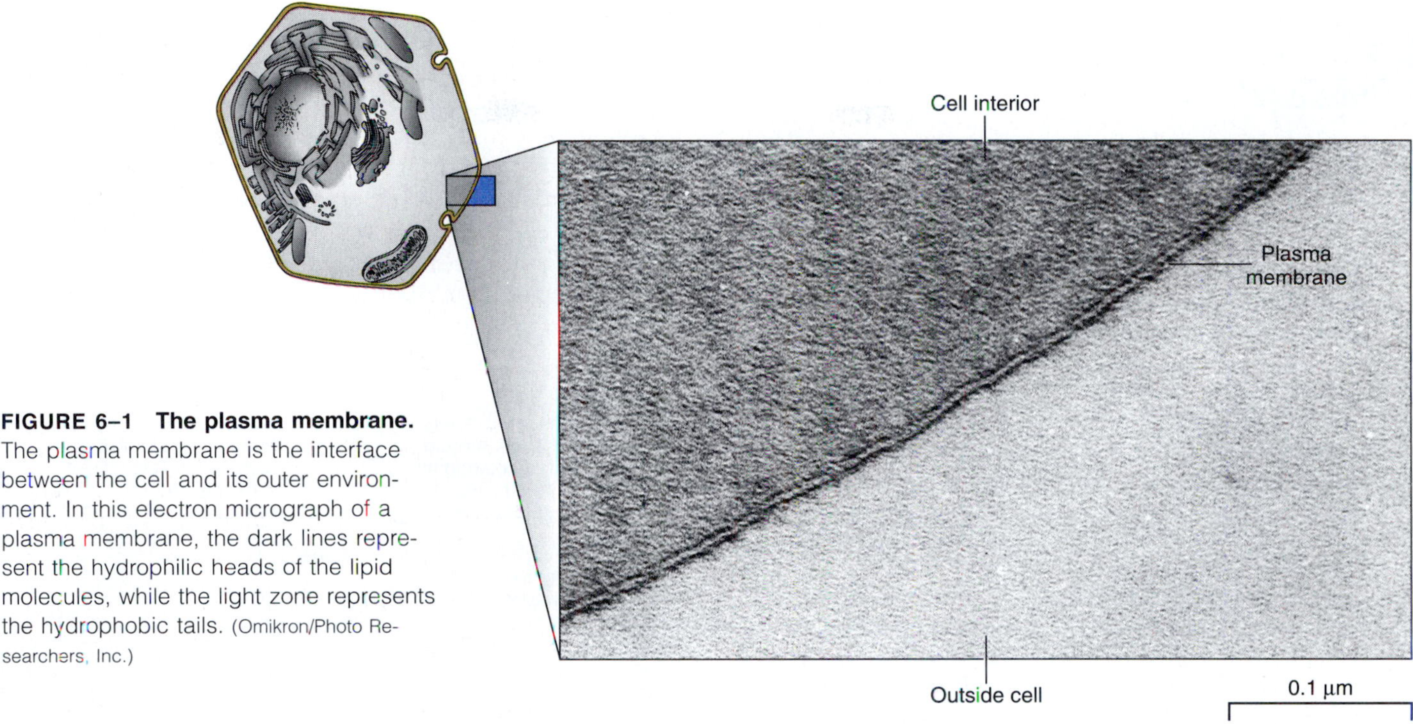

FIGURE 6–1 The plasma membrane. The plasma membrane is the interface between the cell and its outer environment. In this electron micrograph of a plasma membrane, the dark lines represent the hydrophilic heads of the lipid molecules, while the light zone represents the hydrophobic tails. (Omikron/Photo Researchers, Inc.)

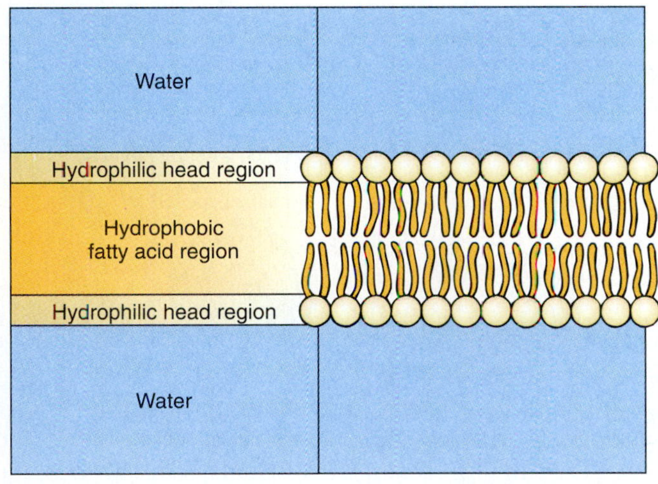

FIGURE 6–2 A phospholipid bilayer. Phospholipids form bilayers in water so that the hydrophobic tails (the fatty acid chains) are not exposed to the water. The hydrophilic heads are in contact with the watery medium.

ecules to one another. Only the hydrophobic forces hold the molecules in place.

Recall that both the inside of a cell (the cytoplasm) and the outside of a cell (an organism's body as a whole or the environment of single-celled organisms) consist mainly of water. In the plasma membrane, the nonpolar, hydrophobic fatty acid chains (the tails) of the phospholipids meet and overlap in the middle, while the polar, hydrophilic heads are directed toward the outside of the membrane (Fig. 6–3). The tails of the phospholipid molecules pack loosely, and the fatty acid chains are in constant motion. This results in a fluid state of the membrane that is essential for its function.

As shown in Figure 6–1, when viewed with the electron microscope, the plasma membrane is seen as two dark lines separated by an intermediate light zone. The dark lines represent the hydrophilic heads of the lipids, and the light zone is produced by the hydrophobic tails. In diagrams, the hydrophilic head groups are represented by circles and the hydrophobic tails by two wavy lines (Fig. 6–3; see also Focus on the Process of Science: Splitting the Lipid Bilayer).

Lipid bilayers are self-sealing as well as self-assembling. If a hole is made in the membrane, as long as the hole is not too large, the tails of the phospholipid molecules in the bilayer will be forced together by repulsion of the surrounding water. In this way, a small hole can be repaired.

The lipid bilayer is very impermeable to ions and polar molecules. Water is an exception and is able to pass in and out of the cell through the lipid bilayer with ease. Additionally, even relatively large nonpolar molecules, such as other lipids (e.g., hormones), pass with relative ease through cellular membranes.

Biological Membranes Contain Proteins and Glycoproteins

The lipids of biological membranes serve as a barrier to polar molecules and ions. The membrane proteins carry

(Text continues on p. 113)

FOCUS ON *The Process of Science*

SPLITTING THE LIPID BILAYER

High-resolution electron micrographs of plasma membranes or other biological membranes typically show a three-layered structure of lines in a dark-light-dark pattern. The two dark lines correspond to the polar heads of the lipid bilayer, and the light area between them corresponds to the hydrophobic region of the fatty acid chains. The plasma membranes of animal, plant, and microbial cells and the membranes of a great many subcellular organelles all appear to have this three-layered structure (Fig. A).

Cells can be rapidly frozen in liquid nitrogen and then fractured with a microtome knife. A small amount of water evaporates from the fracture surface ("etched"). Then a small amount of platinum (or gold) is deposited on that surface, forming a replica of the fractured surface. When the metallic coating is examined with the electron microscope, one can see that the fracture typically splits each plasma membrane into two half-membranes in the middle of the lipid bilayer (see figure). By this process, termed **freeze fracture** electron microscopy, investigators can examine the interior of the split membrane. This method of examining biological membranes has been important in demonstrating that proteins are embedded in the lipid bilayer.

The two faces of the membrane are not identical. The inner half-membrane, the P-face, contains many particles, and the outer half-membrane, the E-face (closer to the outside of the cell), contains many pits (see Fig. B). Studies indicate that the particles are integral proteins (proteins that extend into the lipid bilayer), and the pits are spaces where the proteins had been. When the membrane is treated with enzymes that digest proteins before freeze fracturing, the particles are not seen. The fact that more particles are visible on the P-face does not necessarily mean that there are more proteins on the P-face than on the E-face; it may indicate that proteins are more firmly attached to the inner half-membrane.

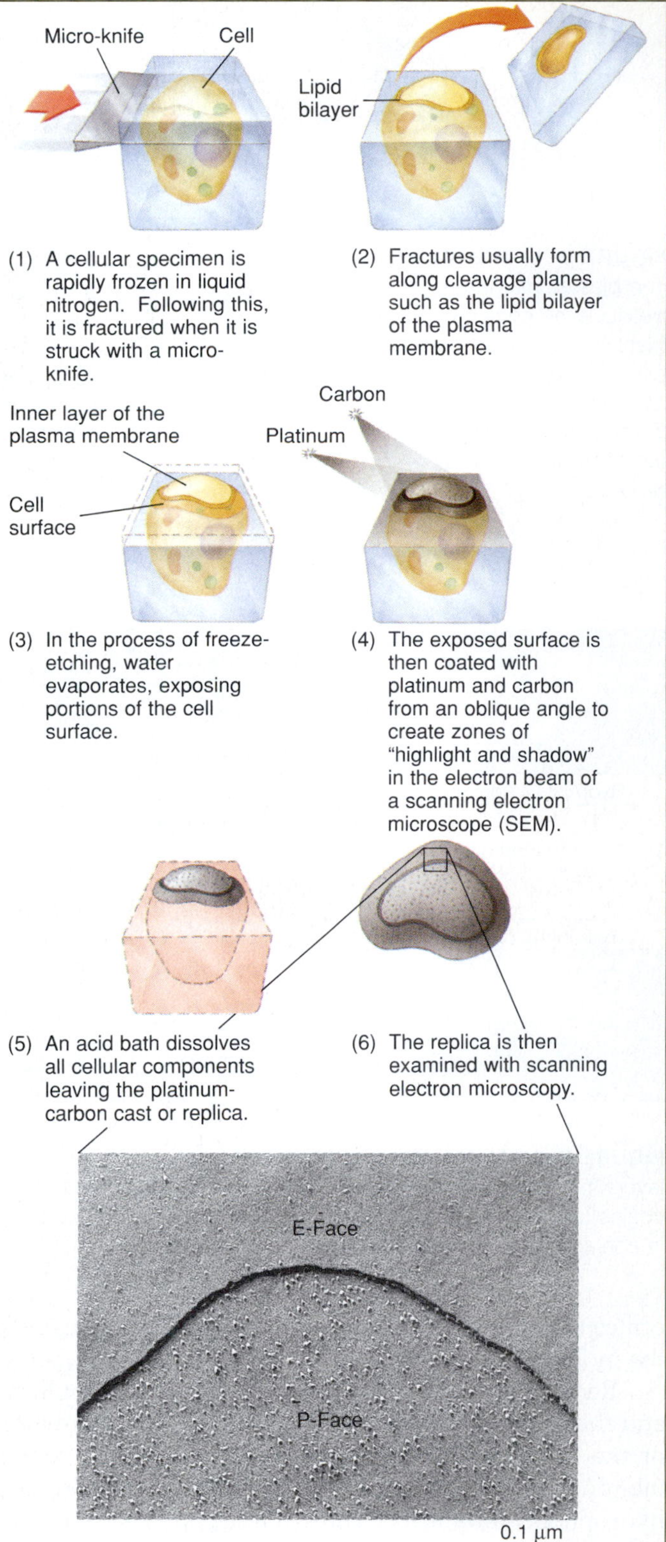

Micro-knife
Cell
Lipid bilayer
(1) A cellular specimen is rapidly frozen in liquid nitrogen. Following this, it is fractured when it is struck with a micro-knife.
(2) Fractures usually form along cleavage planes such as the lipid bilayer of the plasma membrane.
Inner layer of the plasma membrane
Cell surface
Carbon
Platinum
(3) In the process of freeze-etching, water evaporates, exposing portions of the cell surface.
(4) The exposed surface is then coated with platinum and carbon from an oblique angle to create zones of "highlight and shadow" in the electron beam of a scanning electron microscope (SEM).
(5) An acid bath dissolves all cellular components leaving the platinum-carbon cast or replica.
(6) The replica is then examined with scanning electron microscopy.
E-Face
P-Face
0.1 µm

The freeze-fracture method separates the two layers of the plasma membrane. They can be examined with the electron microscope and can be photographed separately. The inner half-membrane, called the P-face, is outwardly directed. Many globular proteins appear to project from it. The outer half-membrane is the inwardly directed E-face. Although it appears relatively smooth, it shows some protein particles. (D.W. Fawcett)

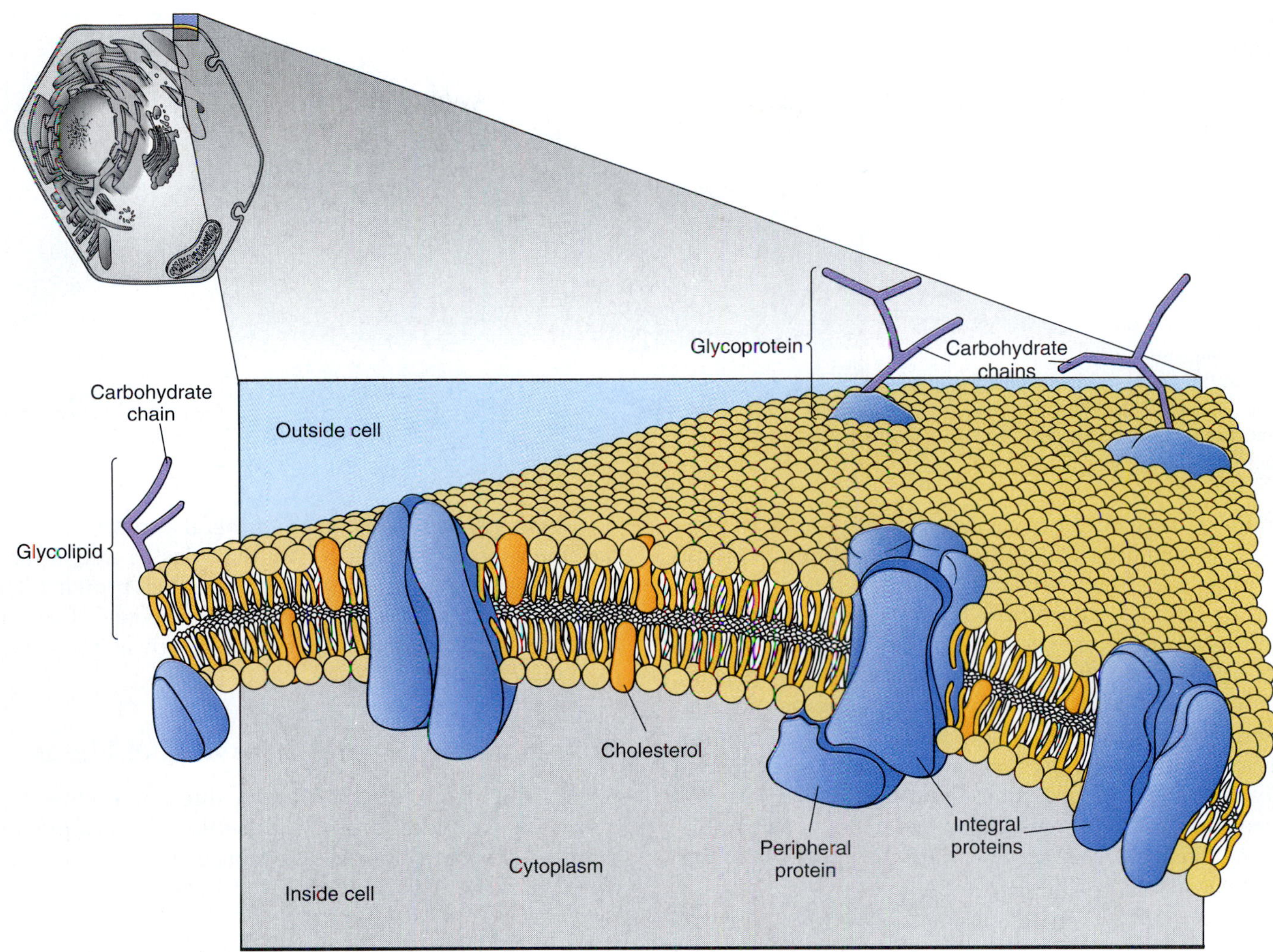

FIGURE 6–3 Structure of the plasma membrane. According to the fluid mosaic model, the plasma membrane consists of a fluid bilayer of lipid molecules in which proteins are embedded.

out specific functions, such as chemical transport, energy transfer, and transmission of messages. All biological membranes are quite similar (although not exactly the same) in their phospholipid makeup, but membranes with different functions contain different proteins.

Proteins must be chemically compatible with the lipids of a membrane in order to be associated with it. Proteins, as you know, are long chains of amino acids. Some amino acids, once they have been incorporated into a protein, are hydrophobic. The parts of a protein that are composed of such amino acids are therefore wholly or partly hydrophobic themselves. Such proteins are attracted to and become embedded in the phospholipid bilayer. The hydrophilic parts of a protein protrude into the watery medium inside or outside of the cell.

If a protein has one hydrophilic end and one hydrophobic end, it usually occurs in just one of the two parts of the bilayer. If only the middle of the protein molecule is hydrophobic, it bridges the entire bilayer, with the two hydrophilic ends protruding on both sides. In some cases, a large and complex protein may thread in and out of the membrane numerous times.

Based on how tightly they are associated with the lipid bilayer, we can classify membrane proteins into two groups: integral proteins and peripheral proteins. **Integral proteins** have regions that are inserted into the hydrophobic regions of the lipid bilayer and are firmly bound to the membrane. Some integral proteins pass all of the way through the membrane, whereas others are located mainly on one side of the membrane. **Peripheral**

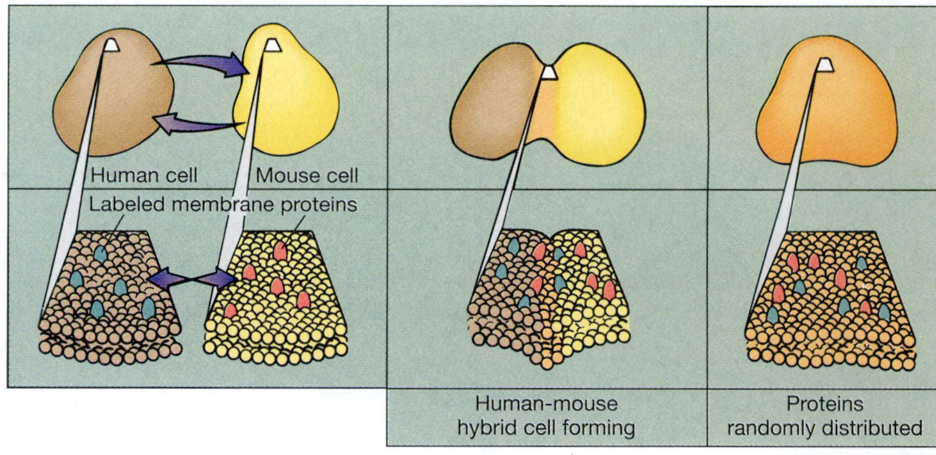

FIGURE 6–4 Movement of proteins in the plasma membrane. In an elegant series of experiments, membrane proteins of mouse cells and human cells were labeled with fluorescent dye markers in two different colors. When the plasma membranes of a mouse cell and a human cell were then fused, mouse proteins migrated to the human side and human proteins to the mouse side. After a short time the proteins from both mouse and human were randomly distributed on the cell surface. This demonstration was convincing evidence that proteins in membranes are not part of a fixed structure like bricks in a wall, but instead are mobile molecules in a two-dimensional fluid.

proteins usually bind to exposed regions of integral proteins. They can be removed from the membrane without disrupting the structure of the bilayer.

The proteins that protrude from the outer surface are mainly **glycoproteins,** proteins to which carbohydrates, such as sugars, are attached. Glycoproteins on the plasma membrane appear to serve as the cell's communication system with its outer environment. Some glycoproteins enable cells to recognize other cells, leading to cell-to-cell adhesion and tissue formation. Certain glycoproteins recognize and bind with messenger molecules such as hormones. Others are very significant in the operation of the immune system.

Some glycoproteins are peripheral proteins and are not themselves embedded in the plasma membrane. Rather, they are attached to other embedded proteins. The glycoprotein coat of a cell can be very extensive, forming a **glycocalyx** (meaning "sugar coating") over the surface of many cells.

Biological Membranes Behave Like Fluids

Are membrane proteins fixed in position or are they free to move from place to place? To settle this question, experimenters fused human and mouse cells in tissue culture (Fig. 6–4). After a brief period of time, the resulting

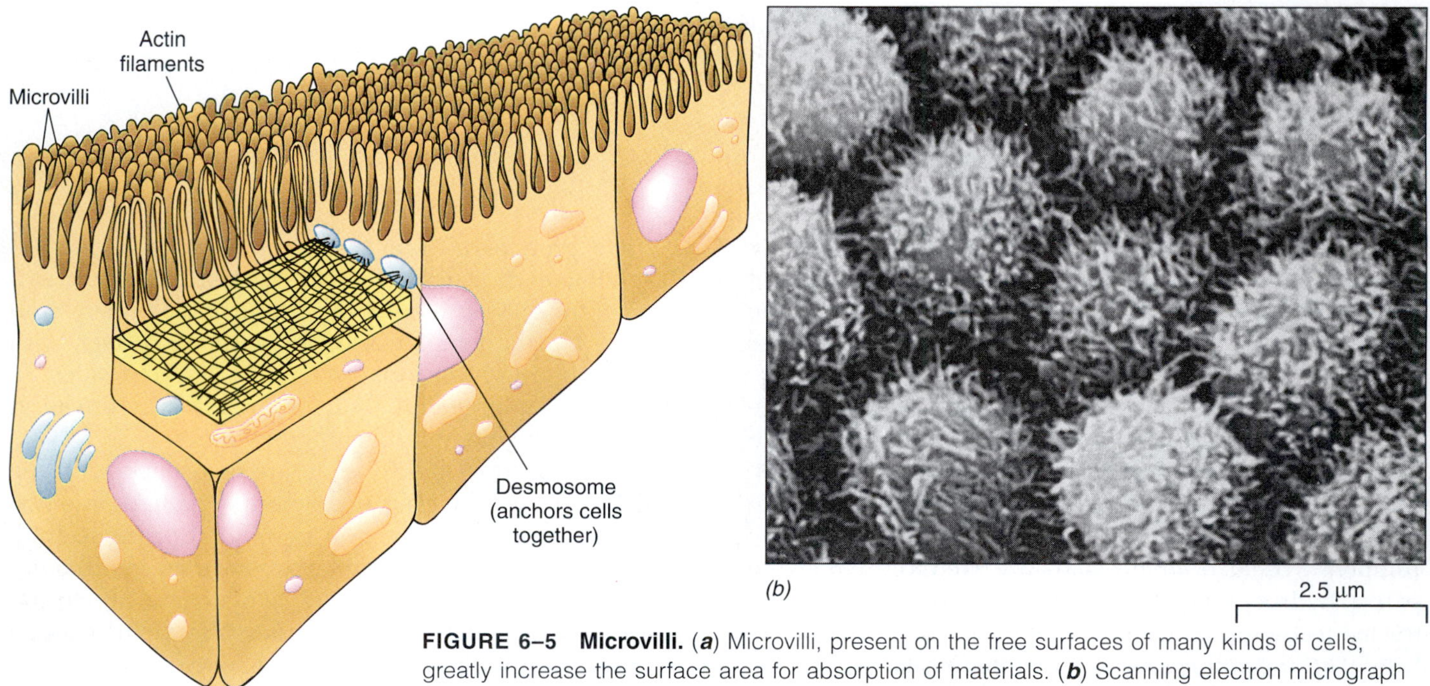

FIGURE 6–5 Microvilli. (***a***) Microvilli, present on the free surfaces of many kinds of cells, greatly increase the surface area for absorption of materials. (***b***) Scanning electron micrograph of ovarian epithelial cells from a mouse. These cells are covered with microvilli. (*b*, Courtesy of Dr. Everett Anderson, Harvard Medical School)

large cell had human and mouse proteins randomly distributed over its surface. This study demonstrated that the proteins in the membrane move about freely. As it turns out, the plasma membrane is quite fluid; it has to be, because many of its proteins function by changing their shape or otherwise moving. Because the plasma membrane is so fluid, the cell depends on its cytoskeleton to give it shape and allow it to attach to other cells.

Microvilli Increase the Surface Area of Some Cells

The plasma membrane on the free surface (a surface not attached to another cell) of many types of cells (e.g., the cells lining the intestine) is marked by numerous tiny extensions known as **microvilli** (Fig. 6–5). Microvilli enormously increase the surface area of the plasma membrane available for the absorption of materials from the cell's environment. Each microvillus has a central cytoskeletal thread composed of the protein **actin.** Microvilli can extend and retract due to polymerization and depolymerization of their actin fibers. Thus, the number of microvilli can rapidly increase or decrease in response to environmental conditions or to changes in the metabolic needs of the cell.

CELLS ARE JOINED BY SPECIALIZED STRUCTURES

Some types of cells communicate directly with one another. Adjacent plant cells may actually be joined by extensions of cytoplasm that pass through openings in the cell walls and plasma membranes. These cytoplasmic extensions, called **plasmodesmata,** provide a pathway for the passage of water, ions, nutrients, and other materials from one cell to another (Fig. 6–6).

Serving a somewhat similar function as plant cell plasmodesmata, **gap junctions** connect the cytoplasmic compartments of animal cells (Fig. 6–7). When a marker substance is injected into one of the cells connected by gap junctions, the marker passes rapidly into the adjacent cell. Gap junctions also provide for electrical communication between certain types of cells. In some species, the gap junction permits the transmission of an electrical impulse from one cell to the next, perhaps in the form of a surge of ions. Such an arrangement is found in the electrical organs of the electric eel and the electric catfish. Cardiac muscle cells are also connected by gap junctions that permit the rapid transmission of neural impulses from one cell to the next so that all of the muscle fibers in the ventricles of the heart can contract one after the other (almost simultaneously) in a rapid, regular fashion.

Adjacent epithelial cells, especially those of the upper layer of the skin, are anchored together by button-like

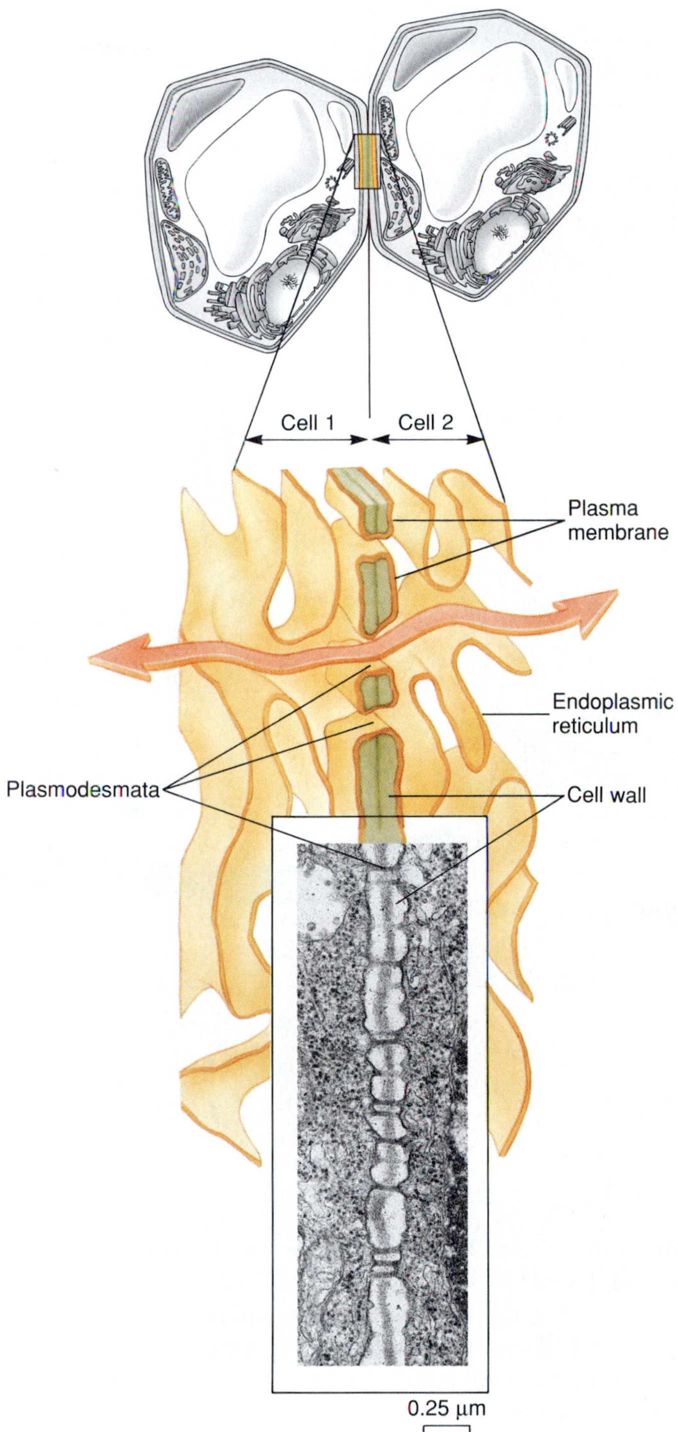

FIGURE 6–6 Plasmodesmata in plant cell walls. Electron micrograph and interpretive drawing of plasmodesmata crossing walls between two cells in the root tip of timothy grass. The fused plasma membranes of the two adjacent cells line channels that form between the cells. Cytoplasmic connections form within these channels, allowing water and small molecules to move between adjacent cells. (E.H. Newcomb, University of Wisconsin/Biological Photo Service)

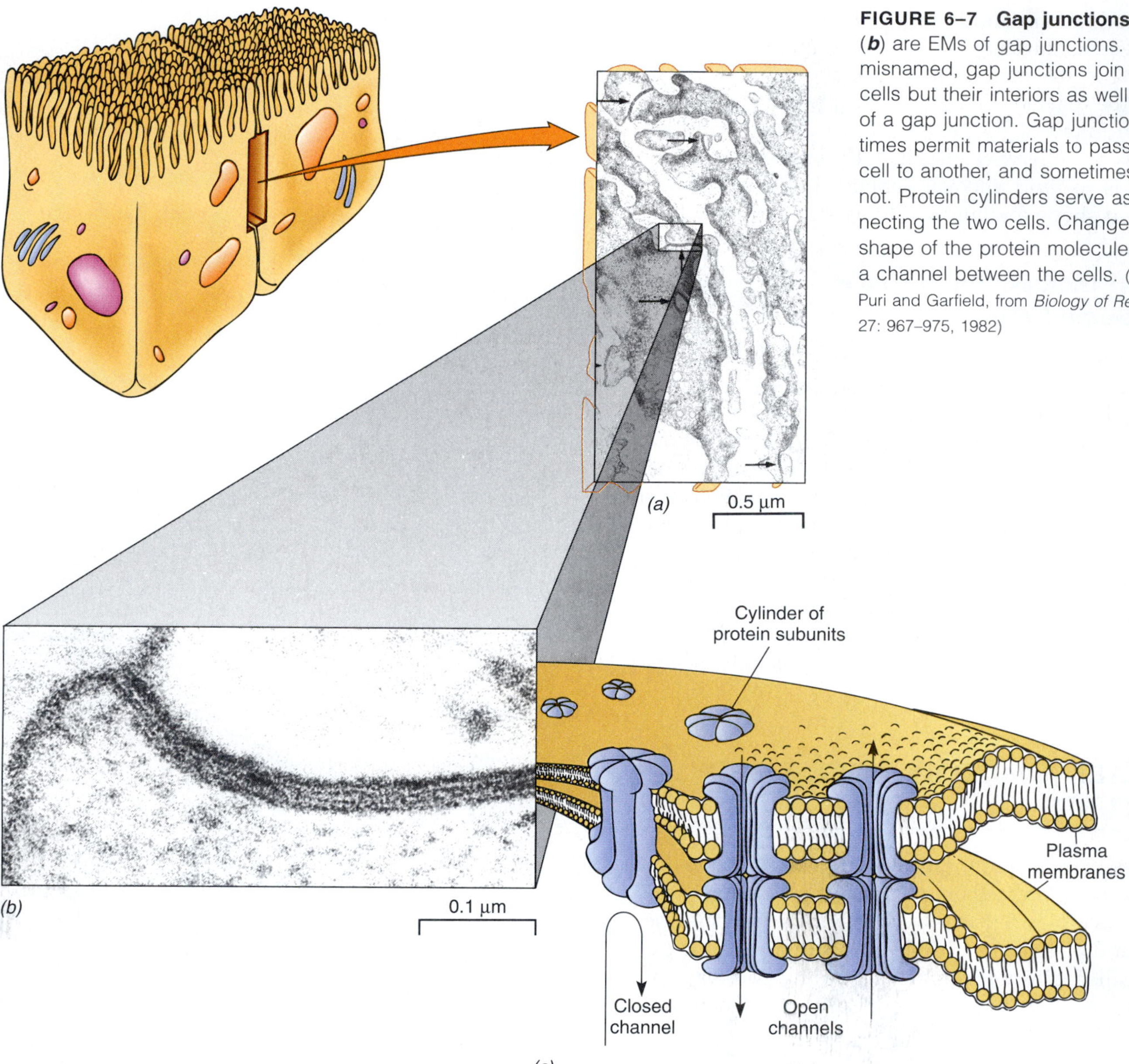

FIGURE 6–7 Gap junctions. (***a***) and (***b***) are EMs of gap junctions. Somewhat misnamed, gap junctions join not only cells but their interiors as well. (***c***) Model of a gap junction. Gap junctions sometimes permit materials to pass from one cell to another, and sometimes they do not. Protein cylinders serve as pipes connecting the two cells. Changes in the shape of the protein molecules may close a channel between the cells. (Courtesy of Puri and Garfield, from *Biology of Reproduction* 27: 967–975, 1982)

structures, called **desmosomes,** on the two opposing cell surfaces (Fig. 6–8). A tiny intercellular space about 24 nanometers wide separates the two opposing cell surfaces. The two cells are held together by protein filaments that cross the intercellular space between the desmosomes. The function of the desmosomes appears to be mechanical—they hold cells together at one point like a spot weld.

In certain tissues, cells are held together by **tight junctions,** connections in which the membranes of adjacent cells are so close that materials cannot pass between the cells. In a tight junction, the two plasma membranes are often fused, and there is no intercellular space (Fig. 6–9). Cells connected by tight junctions form a continuous barrier and are found where a sharp physical separation between two body compartments is essential. The blood–brain barrier, described in the introduction to this chapter, is characterized by tight junctions. Such junctions are also found between cells lining the intestine where they form a barrier that separates the lumen (cavity) of the intestine from the body cavity. Food substances in the intestine must pass through the cells of the intestinal wall rather than leak between the cells.

MATERIALS PASS THROUGH MEMBRANES BY PASSIVE AND BY PHYSIOLOGICAL PROCESSES

Whether a membrane will permit the molecules of any given substance to pass through it depends on the structure of the membrane and the size and charge of the mol-

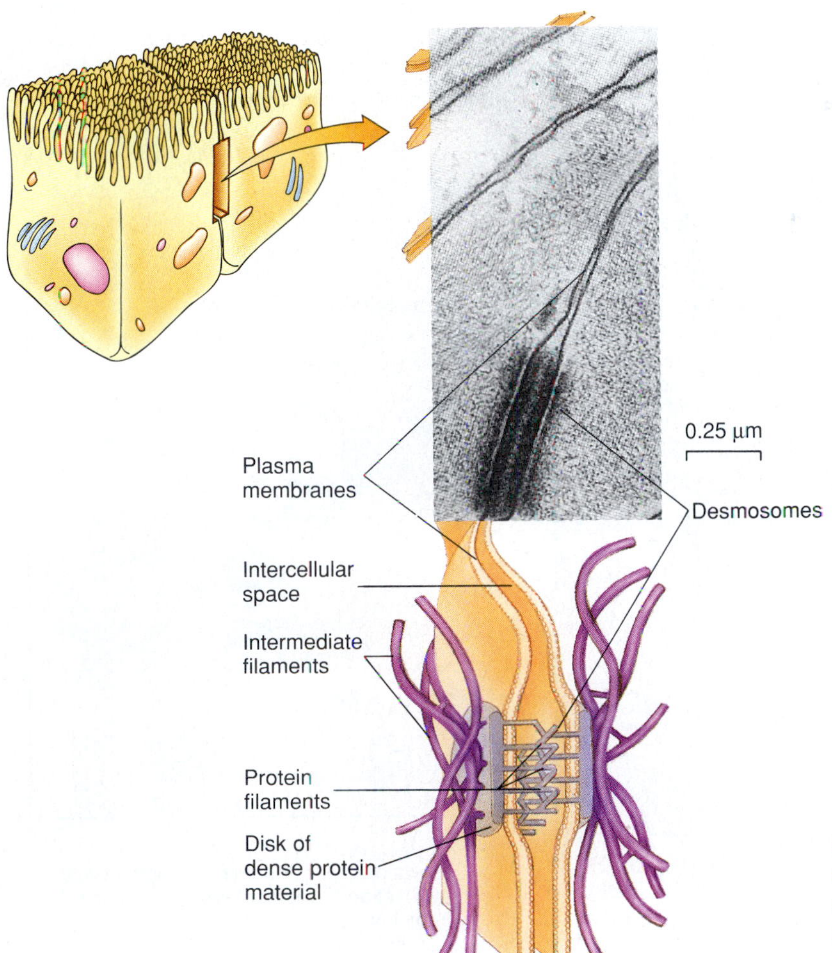

FIGURE 6–8 Structure of a desmosome. Electron micrograph and diagram of a desmosome. Paired discs associated with the plasma membranes of adjacent cells are connected by an interlacing mass of protein filaments. Intermediate filaments inside the cell are attached to the discs and connected to other desmosomes. (D.W. Fawcett)

ecules. A membrane is said to be **permeable** to a given substance if it permits the substance to pass through and **impermeable** if it prevents passage of the substance. A **selectively permeable** membrane allows some, but not other, substances to pass through it. Responding to varying environmental conditions or cellular needs, the plasma membrane may present a barrier to a particular substance at one time and then actively promote its passage at another.

Some materials move passively through biological membranes—as they do through nonliving materials—by physical processes, such as diffusion and osmosis. However, in living cells materials can also be moved actively by physiological processes, such as active transport, or endocytosis (Table 6–1). Such physiological processes require the expenditure of energy by the cell.

Diffusion Is Passive Molecular Movement

Some substances pass into or out of cells by simple diffusion. All molecules in liquids and gases tend to move (diffuse) in all directions until they are spread evenly throughout the available space (Fig. 6–10). **Diffusion** may be defined as the movement of particles (atoms, ions, molecules) from a region of higher concentration to one of lower concentration brought about by the kinetic energy of the particles. Atoms and molecules tend to diffuse down a **concentration gradient,** that is, from where they are more concentrated to where they are less concentrated. Diffusion is a physical process that depends upon the *random* movement of individual particles, propelled by collision with other particles or with the side of the container. As it diffuses, each individual particle moves in a straight line until it bumps into something—another particle or the side of the container. Then it rebounds in another direction. Atoms and molecules continue to move even when they have become uniformly distributed throughout a given space. However, as fast as some particles travel in one direction, others travel in the opposite direction, so that on the whole all of the particles remain uniformly distributed; thus, a dynamic equilibrium exists.

The rate of diffusion is determined by the movement of the molecules, which in turn is determined by their

(Text continues on page 119)

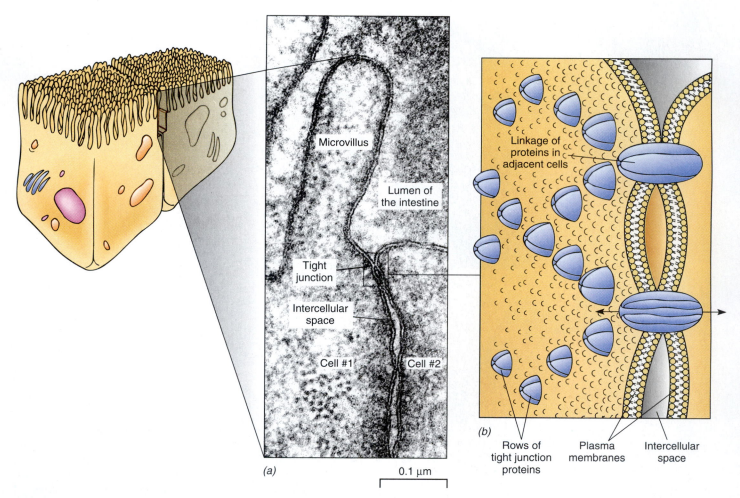

FIGURE 6–9 Structure of a tight junction. (***a***) Thin-section electron micrograph showing points of fusion between the plasma membranes between two cells (designated #1 and #2) lining the intestine. (***b***) A model of the structure of a tight junction between adjacent cells, formed by linkages between rows of proteins. (G.E. Palade)

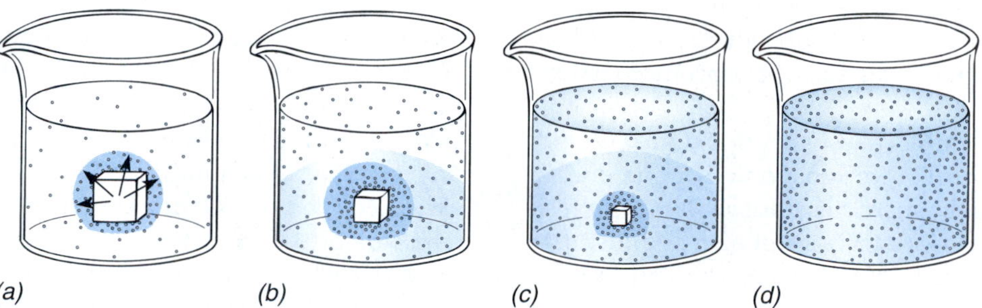

FIGURE 6–10 The process of diffusion. When a small lump of sugar is dropped into a beaker of water, its molecules dissolve (***a***) and begin to diffuse throughout the water in the container (***b*** and ***c***). Over a long period of time, diffusion results in an even distribution of sugar molecules throughout the water (***d***). The arrows in (***a***) indicate net movement of sugar molecules; individual molecules move randomly in all directions.

TABLE 6–1
Mechanisms for Moving Materials through Cell Membranes

Process	*How it works*	*Energy source*	*Example*
Physical process			
Diffusion	Net movement of atoms, molecules, or ions from region of greater concentration to region of lower concentration	Random molecular motion	Movement of oxygen in tissue fluid
Facilitated diffusion	Carrier protein in plasma membrane accelerates movement of relatively large molecules from region of higher to region of lower concentration	Random molecular motion	Movement of glucose into cells
Osmosis	Water molecules diffuse from region of higher to region of lower concentration through differentially permeable membrane	Random molecular motion	Water enters red blood cell placed in distilled water
Physiological process			
Active transport	Protein molecules in membrane transport ions or molecules through membrane; movement is against concentration gradient (i.e., from region of lower to region of higher concentration)	Cellular energy	Pumping of sodium out of cell against concentration gradient
Exocytosis	Plasma membrane ejects materials; vesicle filled with material fuses with cell membrane	Cellular energy	Secretion of mucus
Endocytosis			
Receptor-mediated endocytosis	Certain kinds of molecules combine with receptors in plasma membrane, migrate into coated pits, then form coated vesicles, which are released into the cytoplasm.	Cellular energy	Cholesterol uptake
Phagocytosis	Plasma membrane encircles particle and brings it into cell by forming vacuole around it	Cellular energy	White blood cells ingest bacteria
Pinocytosis	Plasma membrane takes in fluid droplets by forming vesicles around them	Cellular energy	Cell takes in needed solute dissolved in fluid

size and shape, their electric charges, and the temperature. For example, small particles move faster than larger ones at the same temperature. Heat energy causes molecules to move more rapidly, so as temperature rises, the rate of diffusion increases. Atoms and molecules of different substances diffuse independently of one another within the same solution; eventually, all become uniformly distributed.

You might demonstrate the diffusion of gases by opening a bottle of ammonia on a front-row desk of your classroom. Students in the second row would begin to smell ammonia within a few moments because some molecules of ammonia would have left the bottle and begun to diffuse through the air. Some time later the odor would be evident throughout the room. If the room was closed and there were no air currents, the molecules would eventually distribute themselves evenly throughout the room.

Diffusion is important in living systems. A large variety of substances are distributed throughout the cytoplasm by diffusion, and this process is also responsible

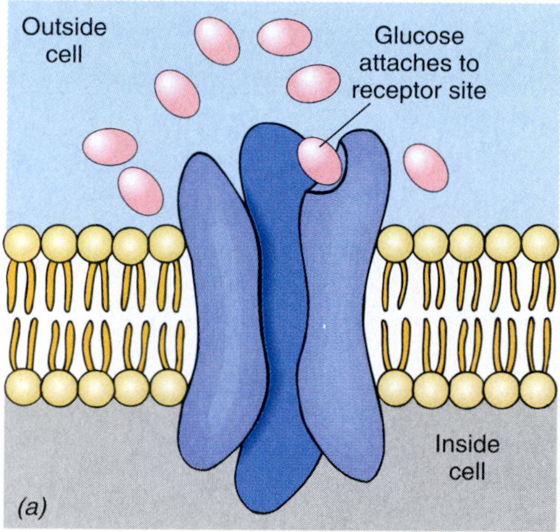

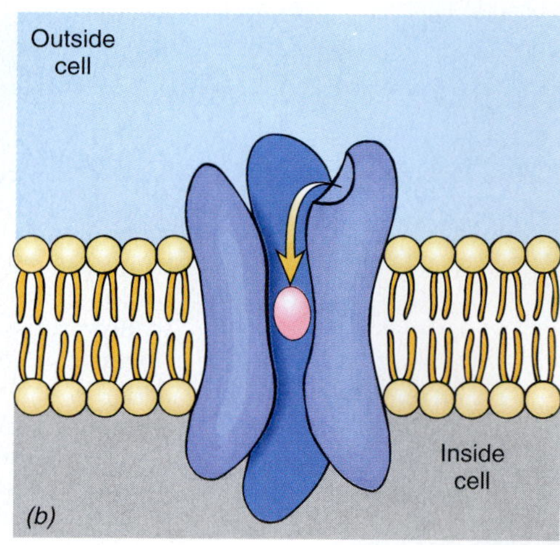

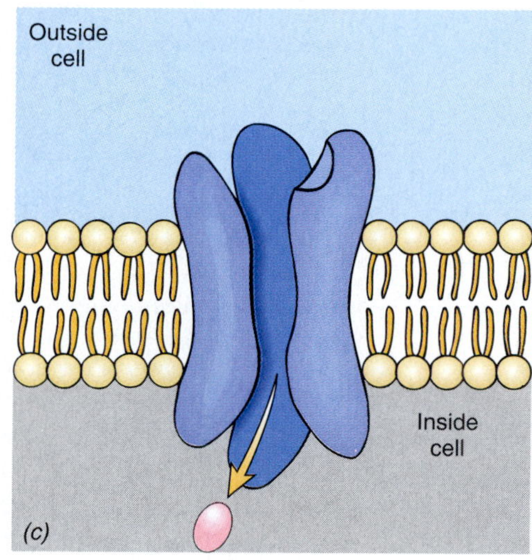

FIGURE 6–11 A model for the facilitated diffusion of glucose. (***a***) Glucose binds to a receptor site on the carrier protein. (***b***) The protein changes shape, opening a channel through the membrane. (***c***) The glucose moves through the channel and is ejected on the other side of the membrane. Net movement is always from a region of higher concentration to a region of lower concentration.

for moving a great many substances in and out of the cell across the plasma membrane. Oxygen, carbon dioxide, water, and numerous other small molecules can diffuse into or out of the cell readily. However, molecules over a certain size cannot diffuse through the membrane, and hydrophilic molecules generally cannot diffuse through it.

Molecules Pass through Special Channels in Facilitated Diffusion

The concentration of a great many materials inside the cell is different from their concentrations outside the cell. In some cases, this results merely from the fact that they are not admitted or not permitted to leave. How hydrophilic compounds, such as glucose and amino acids, can pass through a hydrophobic lipid bilayer had long puzzled biologists. Electron micrographs have revealed proteins prominently attached to the plasma membranes of absorptive cells. Such proteins help transport materials through the membrane.

In **facilitated diffusion,** a membrane protein, called a **carrier protein,** combines temporarily with a solute particle and accelerates its passage through the membrane. The carrier protein serves as a passive conveyor belt that permits the substance to pass in either direction down a concentration gradient. As with simple diffusion, molecules move from a region of greater concentration to a region of lesser concentration (Fig. 6–11).

Carrier proteins probably extend entirely through the membrane. A receptor site on the protein specifically binds to one kind of molecule such as glucose. When the glucose being transported binds with the carrier protein, the protein changes shape. As a result, the glucose is ejected on the other side of the membrane. When the glucose is released into the cytoplasm, the protein assumes its original structure and can transport another glucose molecule.

In facilitated diffusion, molecules are moved because they are more highly concentrated on the outside of the cell than on the inside, much as in simple diffusion. The difference here is that the receptor site on the carrier protein accepts only certain substances whose molecules have the right shape. This kind of diffusion is therefore selective.

Like simple diffusion, facilitated diffusion is a passive process that does not require energy input by the cell. Thus, facilitated diffusion, like simple diffusion, cannot

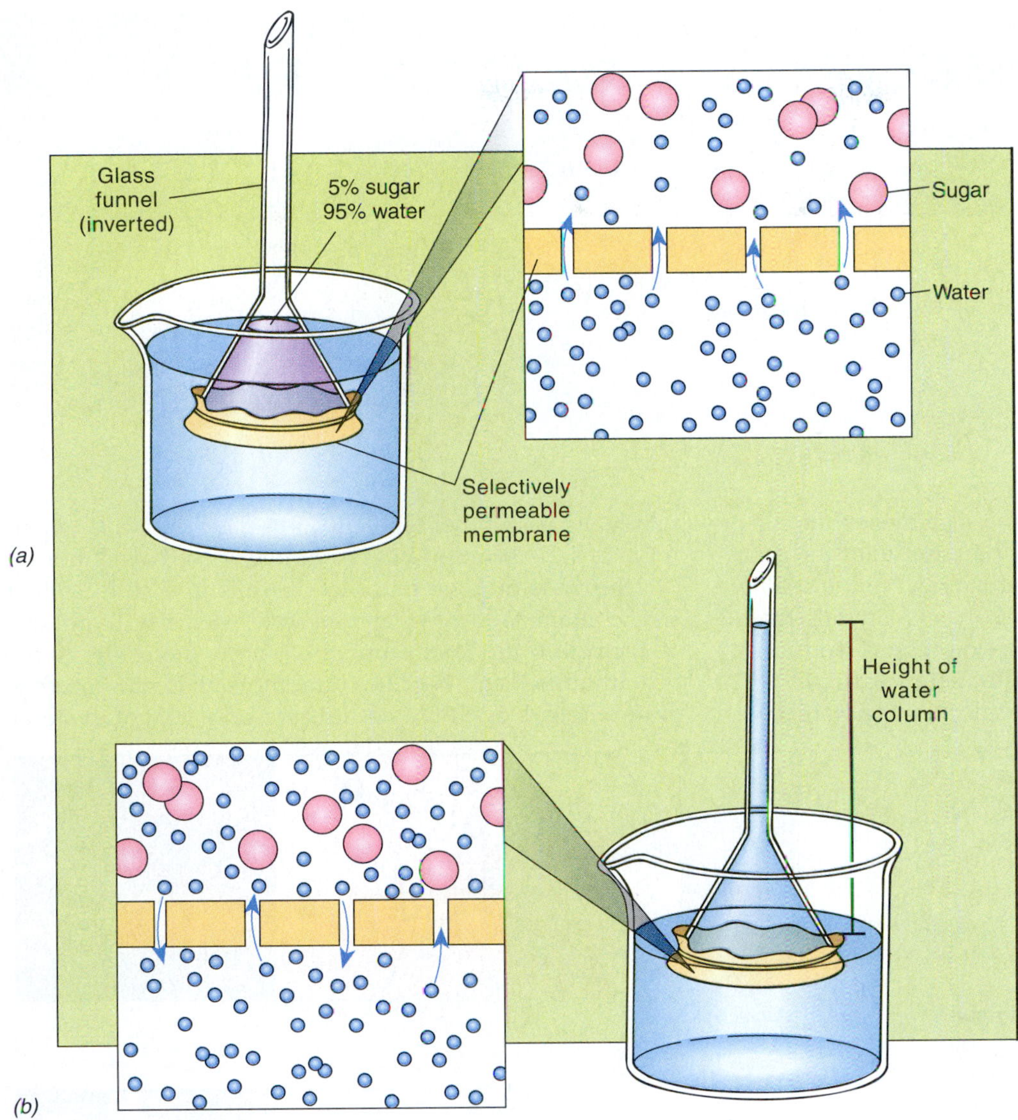

FIGURE 6–12 Osmosis. (***a***) A 5% sugar solution is placed in a sac made of a selectively permeable membrane and suspended in water. The sac is attached to a glass tube. The membrane permits passage of the water molecules but not of the larger sugar molecules. Therefore, the water molecules pass across the membrane into the sac. (We can think of the water as less concentrated inside the sac than outside it.) Movement of water into the sac causes the column of liquid in the glass tube to rise. (***b***) When equilibrium is reached, the pressure of the column of water just equals, and is a measure of, the osmotic pressure of the sugar solution.

concentrate materials. Unlike simple diffusion, however, it is controllable, since only certain molecules can pass through the protein channels. This is a very important property, as we will see when we consider the sodium channels that make the operation of nerve, muscle, and many gland cells possible. The cells lining the blood vessels of the blood–brain barrier (see Chapter introduction) have facilitated diffusion systems for moving molecules such as glucose into the brain.

Osmosis Poses Special Problems for Cells

Osmosis is the diffusion of water molecules across a selectively permeable membrane from a region where water molecules are more concentrated to one where they are less concentrated. Plasma membranes selectively regulate the passage of most **solutes**, dissolved materials such as sugar and salts. However, water is able to move rather freely in and out of the cell by osmosis.

When living cells are placed in a solution that has a solute concentration equal to that in the cells, the water molecule concentration is also equal. Therefore, water molecules move in and out of cells at the same rate, establishing a dynamic equilibrium. The net movement of the water molecules is zero (Fig. 6–12). Such a solution is described as **isotonic** (or isoosmotic) to the cells, that is, of equal solute concentration.

The **osmotic pressure** of a solution is the tendency of water to move into that solution by osmosis. Cells, especially of single-celled organisms, may find themselves in solutions that are of greater or lesser solute concentration relative to the solute concentration within the cytoplasm.

TABLE 6–2
Osmotic Terminology

Solute concentration in solution A	*Solute concentration in solution B*	*Tonicity*	*Solute diffusion*	*Solvent diffusion*
Greater	Less	A hypertonic to B B hypotonic to A	A to B	B to A
Less	Greater	B hypertonic to A A hypotonic to B	B to A	A to B
Equal	Equal	Isotonic	Equal	Equal

If the solution surrounding the cell has a greater solute concentration, it has a higher osmotic pressure than the cell. It is described as **hypertonic** (hyperosmotic) to the cell. If it has a lesser concentration of dissolved materials than that of the cell, it has a lower osmotic pressure and may be described as **hypotonic** (hypoosmotic) compared with the cell. Note that the terms hypertonic and hypotonic are relative to each other (Table 6–2). A 5% solution is hypertonic to a 2% solution but hypotonic to a 10% solution.

Suppose that we place living cells in distilled water that contains 100% water molecules—clearly hypotonic to the cell. If the total number of solute molecules in the cell amounted to 1% of the total molecules present, then water molecules would account for only 99% of the total.

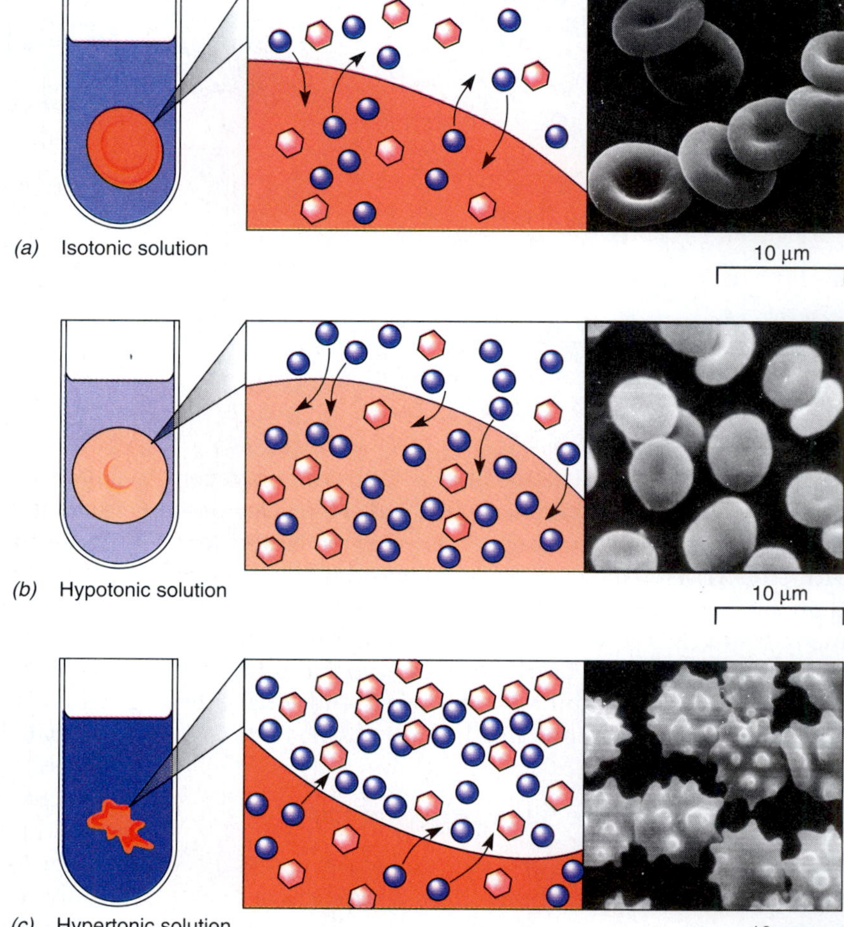

FIGURE 6–13 Osmosis and the living cell. (***a***) A cell is placed in an isotonic solution. Because the concentration of solutes (and thus of water molecules) is the same in the solution as in the cell, the net movement of water molecules is zero. (***b***) A cell is placed in a hypotonic solution. The solution has a lower solute (and thus a greater water) concentration than the cell. The cell contents thus exert an osmotic pressure on the solution, drawing water molecules inward. There is a net diffusion of water molecules into the cell, causing the cell to swell and perhaps even to burst. (***c***) A cell is placed in a hypertonic solution. This solution has a greater solute concentration (thus a lower water concentration) than the cell and therefore exerts an osmotic pressure on the cell. This results in a net movement of water molecules out of the cell, causing the cell to dehydrate, shrink, and perhaps die. Blue-gray spheres represent water molecules. Reddish hexagons represent solute molecules that do not easily pass through the plasma membrane. (Micrographs of human red blood cells courtesy of Dr. R.F. Baker, U. of Southern California Medical School)

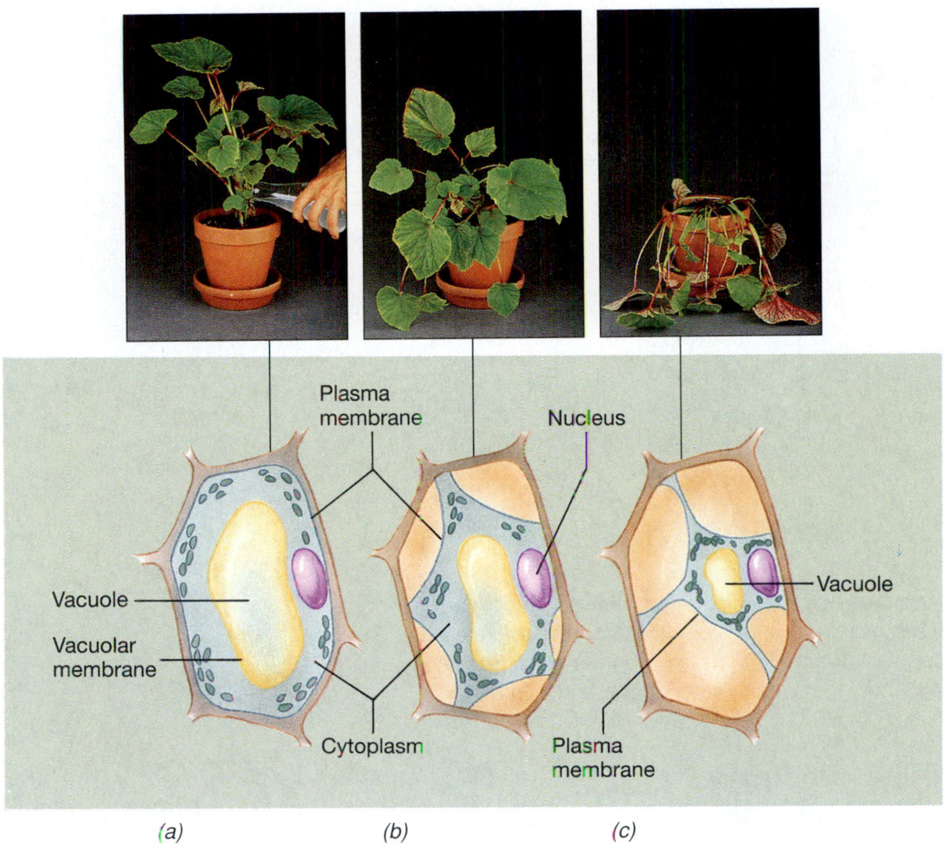

FIGURE 6–14 Effects of solute concentration on plant cells. Plant cells (and other cells with cell walls) are also affected by solute concentrations in their surroundings. (***a***) Begonia and drawing of a single normal cell in hypotonic surroundings. The contents of the cell fill the space within the cell wall. (***b***), (***c***) Effects of exposing begonia to a salt solution. The plant's cells lose water and the cytoplasm shrinks. The begonia wilts and eventually dies due to the hypertonic environment. (Dennis Drenner)

Since water molecules, like solute molecules, tend to move from a region where they are more concentrated to a region where they are less concentrated, they diffuse inward across the plasma membrane (Fig. 6–13). Although the solute molecules have a tendency to diffuse in the opposite direction, the plasma membrane prevents them from "leaking out" to any great extent. Instead, they may be thought of as being trapped within the cell and exerting an osmotic pressure on the less-concentrated solution on the other side of the membrane. Equilibrium can never be reached so water continues to enter the cell. So much water may enter the cell that it swells and bursts.

On the other hand, when cells are placed in a solution that is hypertonic to them, water tends to flow out of them. The cells may become dehydrated, shrink, and die. Can you explain this in terms of the relative concentrations of water molecules in the two solutions? Remember, when a solution contains *more* solute molecules, it has proportionately *fewer* water molecules (the solvent and solute concentrations are reciprocally related). Also remember that water passes freely through the plasma membrane and moves from regions of high water concentration to regions of low water concentration.

The excretory systems of multicellular terrestrial animals like ourselves usually are able to maintain their body fluids in an isotonic state. In contrast, organisms that inhabit lakes and ponds, such as freshwater protists, have a continuous osmotic problem. Their watery surroundings are hypotonic to them, and water tends to pass into their cells. Some protists have an adaptation that solves the problem—a **contractile vacuole** that takes up the excess water and pumps it out of the cell.

Plant cells are adapted to the hypotonic water that often bathes their roots. Their rigid cell walls enable them to withstand, without bursting, pressure exerted by the water that seeps in, filling their central vacuoles. As this internal pressure, called **turgor pressure,** increases, it forces water molecules back out of the cell. The turgor pressure levels off when the outward passage of water molecules equals the rate of inward movement of water molecules. When conditions become dry and the plant cell does not have enough water, the central vacuole decreases in size and the cell shrinks. Loss of turgor pressure explains why plants wilt (Fig. 6–14).

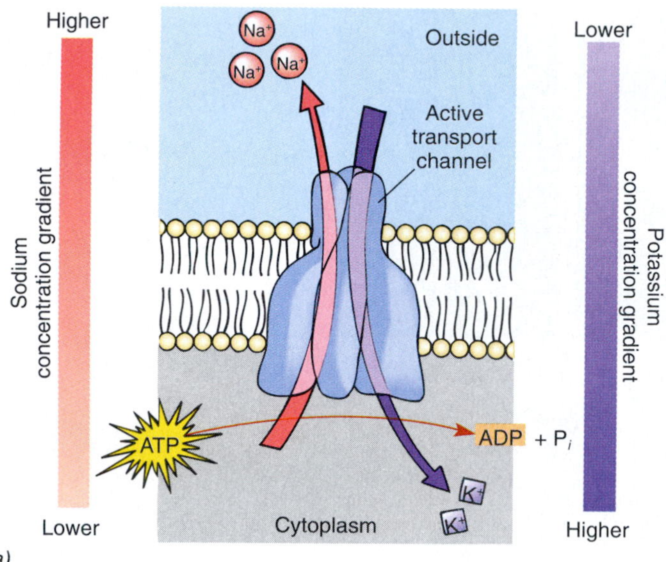

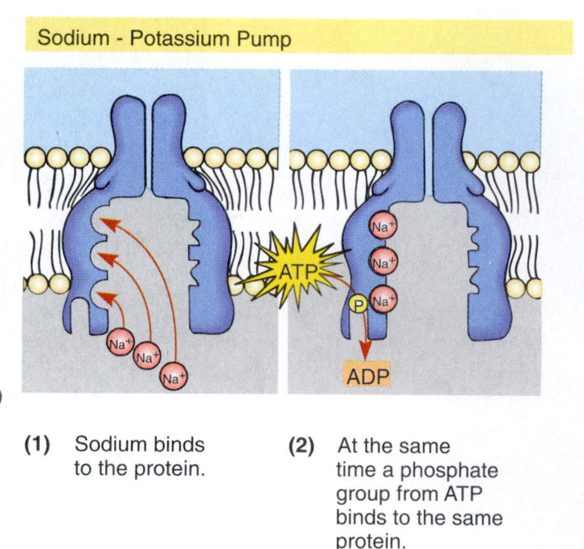

FIGURE 6–15 Active transport by the sodium-potassium pump. (***a***) The sodium-potassium pump is an ATP-driven active transport system. For each molecule of ATP used, three sodium ions are pumped out of the cell and two potassium ions are pumped into the cell. (***b***) A model of how the sodium-potassium pump works.

Active Transport Uses Energy to Concentrate Selected Materials

A cell requires that certain substances, potassium ions, for example, be in greater concentration within the cell than in the cell's surroundings. The potassium concentration is about 35 times greater inside the cell than outside. Other substances, for example, sodium ions, are more concentrated in the environment than could be tolerated inside the cell. Yet sodium and potassium ions are roughly the same size. By itself, diffusion cannot account for the concentration differences of these two ions inside and outside of the cell. Indeed, given the opportunity to do so, diffusion would quickly eliminate such differences in solute concentration, and the cell would die.

In **active transport,** the cell moves materials from a region of lower concentration to a region of higher concentration (the opposite of diffusion or osmosis). Working uphill this way against a concentration gradient requires energy. Expenditure of energy in active transport is an example of how even cells that appear to be resting are actually performing work just to remain alive.

Active transport is carried out by a specific group of integral proteins, sometimes referred to as pumps. These proteins operate somewhat like carrier proteins work in facilitated diffusion, except that they must harness energy to transport molecules against concentration gradients. The needed energy is supplied by **adenosine triphosphate (ATP).**

Sodium-potassium pumps in the plasma membrane produce the different concentrations of sodium and potassium ions required by the cell. As shown in Figure 6–15, sodium ions inside the cell bind to receptor sites on the membrane. A molecule of ATP then transfers its energy to the pump proteins. This causes the protein to change its shape. The change in protein shape results in the transfer of sodium to the outside of the cell. Receptor sites for potassium ions then become accessible to potassium outside the cell. The binding of potassium to the protein causes the phosphate to be released. Then the protein assumes its original shape and the potassium is released inside the cell. Other sodium ions bind to the receptor sites and the cycle repeats.

Cells Can Eject Materials by Exocytosis and Ingest Materials by Endocytosis

Small molecules and ions pass through the plasma membrane by diffusion or by active transport. Larger materials—very large molecules, particles of food, or even whole cells—must sometimes be moved in or out of the cell. Such cellular work requires the cell to expend energy. In **exocytosis** (literally, "outside the cell"), a cell ejects waste products or specific secretion products, such as mucus or hormones (Fig. 6–16). Generally, the material to be ejected is enclosed within a membrane, forming a vesicle. The vesicle fuses with the plasma membrane, then opens at the point of fusion, and the enclosed material is released to the exterior without the loss of other cell contents. The membrane of the secretory vesi-

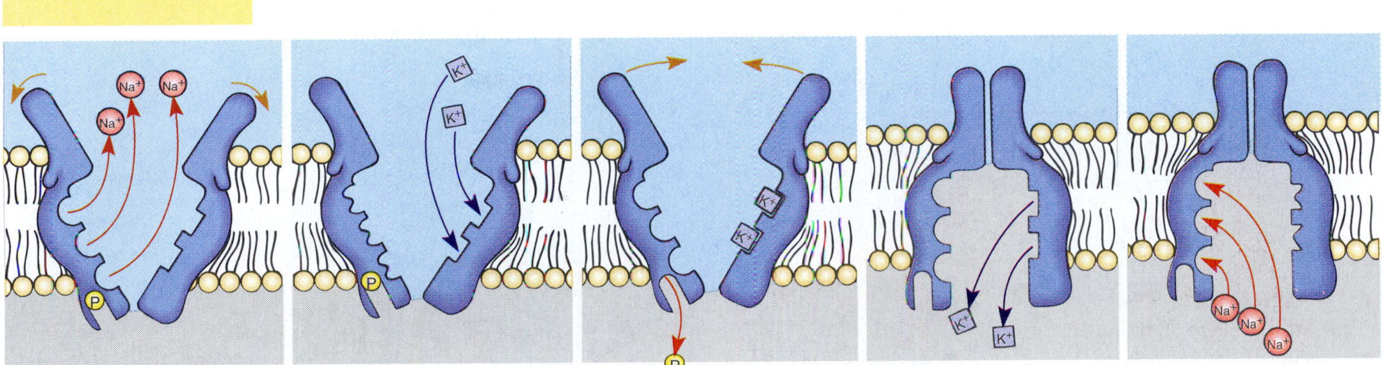

(3) The shape of the protein changes, sodium is transferred to the outside of the cell and ...

(4) ... potassium binding sites are accessible from the outside.

(5) The binding of potassium to the protein causes the phosphate to be released.

(6) The protein then returns to its orginal shape and releases potassium ions to the interior

(7) The process then begins again with the binding of sodium ions.

3

2

1

(a) (b)

0.25 μm

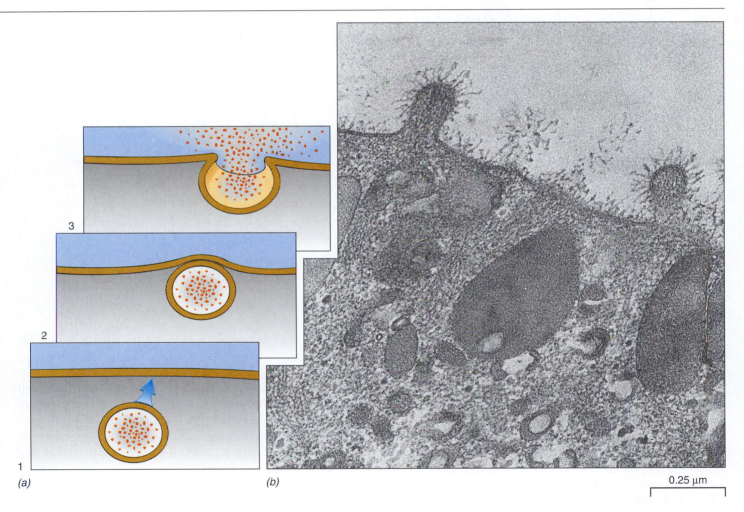

FIGURE 6–16 In exocytosis, materials are exported from the cell. (***a***) Exocytosis (and also movement of molecules between organelles) involves the contact and fusion of a vesicle with a membrane. The contents of the vesicle are released to the outside of the cell (or into another compartment). (***b***) A high-magnification electron micrograph of the upper surface of a secreting cell. Secretion vesicles can be seen in the cytoplasm approaching the plasma membrane. The filaments projecting diffusely from the cell surface are of unknown significance but may be proteins. (J.F. Gennaro/Photo Researchers, Inc.)

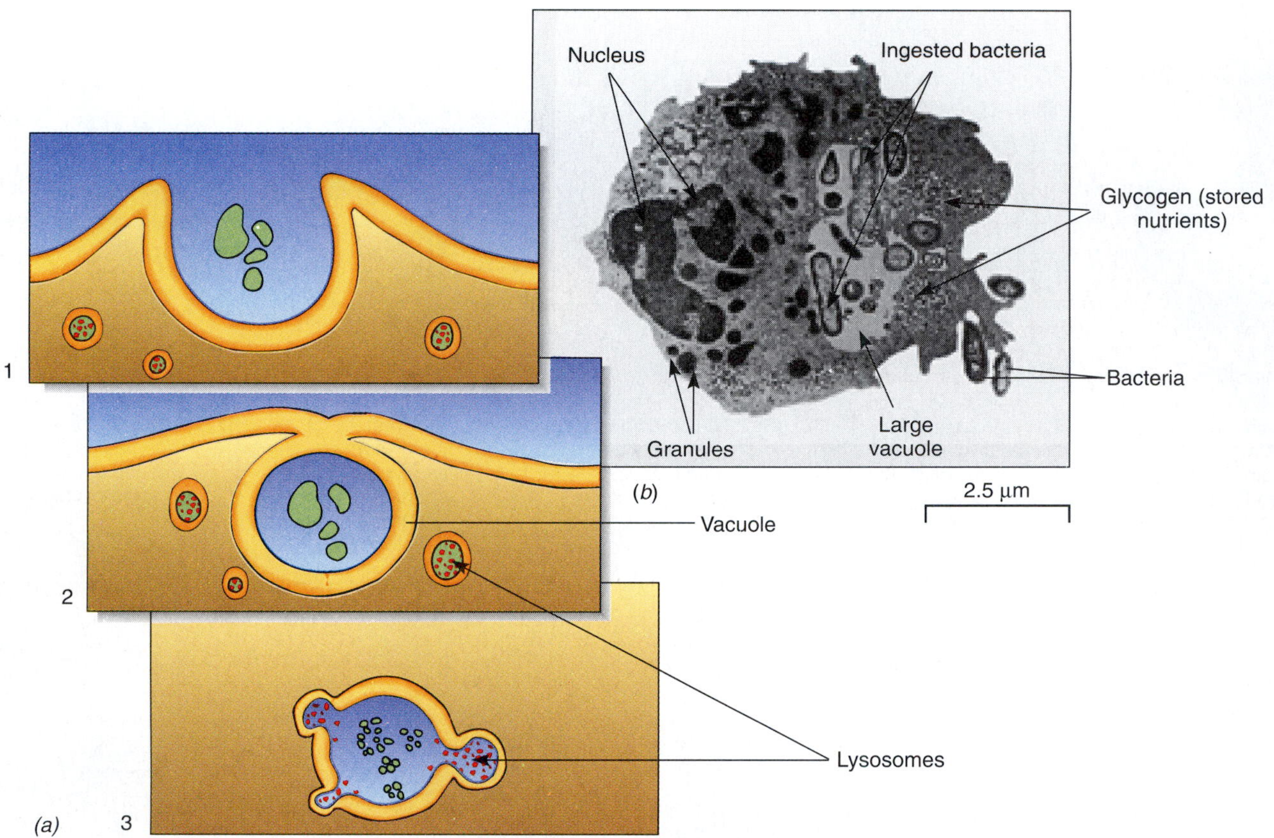

FIGURE 6–17 Phagocytosis. (***a***) In phagocytosis, the cell ingests large solid particles, such as bacteria. (***1***) Folds of the plasma membrane surround the particle to be ingested, forming a small vacuole around it. (***2***) This vacuole then pinches off inside the cell. (***3***) Lysosomes fuse with the vacuole and pour their potent digestive enzymes onto the ingested material. (***b***) This photomicrograph captures a white blood cell (a neutrophil) in the process of phagocytizing bacteria. A bacterium that has already been phagocytized is seen within a large vacuole. Note that the white blood cell's own nucleus (elongated structure) has been partially digested by lysosome enzymes. The granules in the cytoplasm contain enzymes. (D.W. Fawcett)

cle is incorporated into the plasma membrane. This is an important mechanism by which the plasma membrane can grow larger.

In **endocytosis** (literally, "within the cell") materials are taken into the cell. Three types of endocytosis are phagocytosis, pinocytosis, and receptor-mediated endocytosis. In **phagocytosis** (literally, "cell eating"), the cell ingests large solid particles such as food or bacteria (Fig. 6–17). Folds of the plasma membrane extend outward and enclose the particle to be ingested, forming a vacuole around it. The vacuole, still attached to the plasma membrane, bulges into the cell interior. The membrane then tightens like a drawstring purse and fuses together, leaving the vacuole floating freely in the cytoplasm. The vacuole then fuses with lysosomes, which release digestive enzymes that break down the ingested material.

In **pinocytosis** (which means "cell drinking") the cell takes in dissolved materials. Tiny droplets of fluid are trapped by microvilli, which are folds of the plasma membrane. These folds pinch off into the cytoplasm as tiny vesicles of fluid (Fig. 6–18). The contents of these vesicles are slowly transferred into the cytoplasm, and the vesicles themselves may become smaller and smaller until they appear to vanish.

In a third type of endocytosis, called **receptor-mediated endocytosis,** specific proteins or particles combine with **receptor proteins** embedded in the plasma membrane

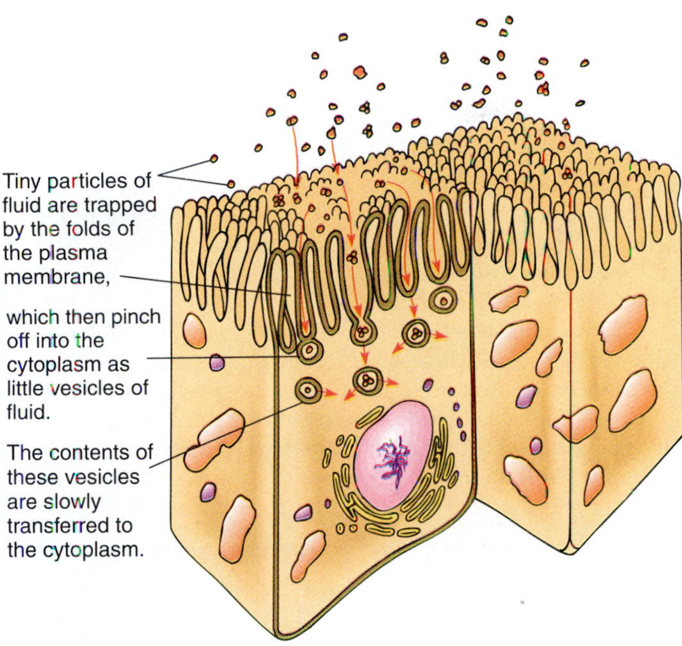

Tiny particles of fluid are trapped by the folds of the plasma membrane,
which then pinch off into the cytoplasm as little vesicles of fluid.
The contents of these vesicles are slowly transferred to the cytoplasm.

FIGURE 6–18 Pinocytosis. In pinocytosis, droplets of fluid are engulfed and absorbed by the cell.

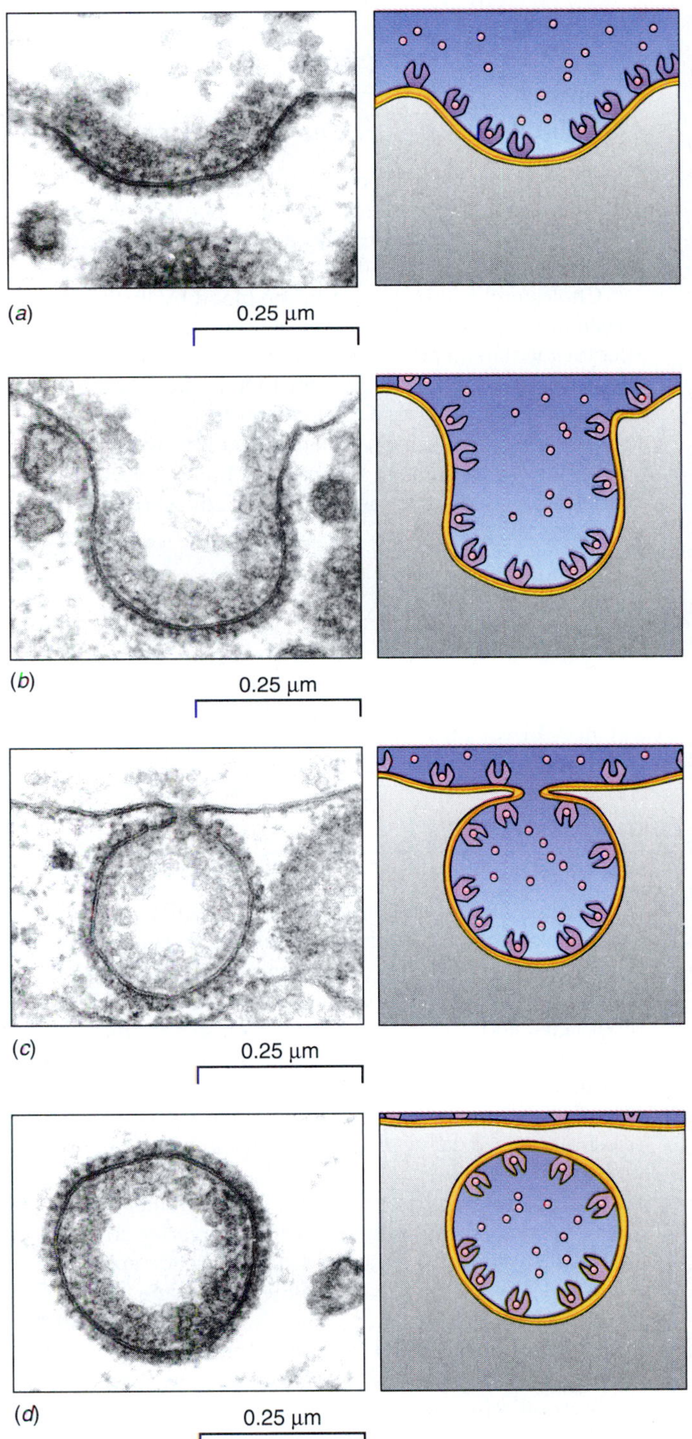

FIGURE 6–19 Receptor-mediated endocytosis. (***a***) Specific proteins combine with receptor proteins in the plasma membrane. (***b***) The receptor-protein complex moves into a coated pit. (***c***) The coated pit forms a coated vesicle and is released into the cytoplasm (***d***) by the process of endocytosis. (From Perry, M.M., and A.B. Gilbert, *J. Cell Sci* 39: 257–272, 1979)

of the cell. The receptor-bound molecules then migrate into *coated pits,* which are regions on the cytoplasmic surface of the membrane coated with whisker-like structures. These coated pits form *coated vesicles* (Fig. 6–19) by endocytosis. The coating on a vesicle consists of proteins, which momentarily form a basket-like structure around it. Seconds after the vesicles are released, however, the coating dissociates from them, leaving the vesicles free in the cytoplasm.

The vesicles then fuse with other similar vesicles to form larger vesicles called **endosomes.** The materials being transported are free inside the endosome, no longer attached to the membrane receptors. An endosome can divide to form two kinds of vesicles. One kind contains the receptors and can be returned to the plasma membrane. The other, which contains the ingested particles, fuses with lysosomes and is then processed by the cell.

Endocytosis, exocytosis, and fusions of organelle membranes within the cell are made possible by the bilayer structure of these membranes. Because hydrophobic interactions force the lipid bilayer together, any hole in the plasma membrane that endocytosis and exocytosis leave is quickly repaired.

CONCEPT CONNECTIONS

Cell Function ⟷ *Cardiovascular Function* ⟷ *Genetics*

Animal cells use receptor-mediated endocytosis to take in cholesterol. Inside the cells, cholesterol is used to manufacture cell membranes and to make certain hormones and other steroids.

Cholesterol is transported in the blood by lipoproteins, particularly by low-density lipoprotein (LDL). Much of the receptor-mediated endocytosis pathway was detailed through studies of the receptor for LDL by investigators M. Brown and J. Goldstein. In 1986, these scientists were awarded the Nobel Prize for their pioneering work.

As we will discuss in Chapter 36, high cholesterol levels in the blood can increase the risk of myocardial infarction (heart attack). When cholesterol remains in the blood instead of entering the cells, it can become deposited in the artery walls. This process can result in atherosclerosis, a cardiovascular disease in which the blood flow to the heart and brain becomes restricted.

In the inherited disorder familial hypercholesterolemia (literally, too much cholesterol in the blood), LDL receptors are missing. Cholesterol cannot enter the cells and instead accumulates in the blood. Atherosclerosis develops and individuals with this disorder have a high risk of early death from myocardial infarction.

Chapter Summary

I. The plasma membrane functions to (1) regulate the passage of materials into and out of the cell, (2) receive information that permits the cell to sense changes in its environment and respond to them, (3) communicate with other cells, and (4) protect the cell. Cell membranes also surround many types of organelles, forming internal compartments.

II. The plasma membrane (and other cell membranes) consist of a fluid lipid bilayer in which a variety of proteins are embedded.
   - A. The nonpolar, hydrophobic fatty acid chains of the membrane phospholipids meet and overlap in the middle, while the polar, hydrophilic heads are directed toward the outside of the membrane.
   - B. Proteins in biological membranes may function in transport of materials through the membrane, in energy transfer, or in the transmission of information. Glycoproteins on the plasma membrane communicate with other cells.
   - C. Microvilli increase the surface area of the cell for absorption of materials from the environment.

III. Plasma membranes of plant and animal cells have specialized structures that permit them to have contact with other cells and to communicate with adjacent cells. These structures are plasmodesmata, gap junctions, desmosomes, and tight junctions.

IV. The selectively permeable plasma membrane allows the passage of some, but not other, substances.
   - A. Some ions and molecules pass through the plasma membrane by simple diffusion; others pass by facilitated diffusion, in which a carrier protein helps move a molecule through the membrane.
   - B. Osmosis is a kind of diffusion in which molecules of water pass through a differentially permeable membrane from a region where water is more concentrated to a region where water is less concentrated.
   - C. In active transport, the cell expends energy to move ions or molecules against a concentration gradient.
   - D. In exocytosis, the cell ejects waste products or secretes substances, such as mucus.
   - E. Three types of endocytosis are phagocytosis, pinocytosis, and receptor-mediated endocytosis.
     1. In phagocytosis, materials, such as food particles or bacteria, can be moved into the cell; a portion of the plasma membrane envelops the material, enclosing it in a vacuole that is released inside the cell.
     2. In pinocytosis, droplets of fluid are engulfed by the plasma membrane, and the contents of the droplets are absorbed into the cell.
     3. In receptor-mediated endocytosis, specific molecules combine with receptor proteins in the plasma membrane, migrate into coated pits, and then form coated vesicles. The vesicles are released in the cytoplasm.

Selected Key Terms

active transport, p. 124
carrier protein, p. 120
concentration gradient, p. 117
desmosome, p. 116
diffusion, p. 117
endocytosis, p. 126
exocytosis, p. 124
facilitated diffusion, p. 120
fluid mosaic model, p. 110
gap junction, p. 115
hypertonic, p. 122
hypotonic, p. 122
isotonic, p. 121
lipid bilayer, p. 110
microvilli, p. 115
osmosis, p. 121
osmotic pressure, p. 121
phagocytosis, p. 126
phospholipid, p. 110
pinocytosis, p. 126
plasma membrane, p. 110
plasmodesmata, p. 115
receptor-mediated endocytosis, p. 126
receptor protein, p. 126
sodium-potassium pump, p. 124
solute, p. 121
tight junction, p. 116
turgor pressure, p. 123

Post-Test

1. A membrane that permits passage of some materials but not of others is described as ______________ permeable.
2. The lipid components of the plasma membrane have a polar, ______________ portion and a nonpolar, ______________ portion.
3. In an electron micrograph, the light zone of the plasma membrane represents the ______________ ______________ of the lipid molecules.
4. ______________ increase the surface area of the cell for absorption of materials.
5. Adjacent plant cells may be joined by ______________, which allow passage of materials from one cell to another.
6. ______________ are buttonlike plaques that hold epithelial cells tightly together.
7. Cardiac muscle cells are connected by ______________ junctions, which permit rapid transmission of impulses from one cell to the next.
8. Red dye poured into a beaker of water spreads throughout the water. This is an example of ______________.
9. The diffusion of water through a differentially permeable membrane from a region of greater concentration to a region of lesser concentration of water molecules is termed ______________.
10. A solution with a greater solute concentration than a tissue is said to be ______________ to the tissue.
11. Cells will neither swell nor shrink if they are placed in ______________ solutions.
12. Waste products may be ejected from a cell through the process of ______________.
13. A white blood cell engulfs a bacterium by the process called ______________.
14. In pinocytosis, a cell takes in ______________.
15. In ______________ transport, the cell moves ions from a region of lower concentration to a region of higher concentration.
16. Protists in a hypotonic medium can rid themselves of excess water by means of their ______________ vacuoles.
17. Coated pits are important in the process of (a) exocytosis (b) facilitated diffusion (c) receptor-mediated endocytosis (d) phagocytosis (e) pinocytosis.
18. A freshwater protist is placed in seawater. Water leaves the cell by the process of (a) exocytosis (b) facilitated diffusion (c) osmosis (d) phagocytosis (e) pinocytosis.
19. A protein that passes all of the way through the plasma membrane is a/an (a) peripheral protein (b) facilitated protein (c) integral protein (d) glycoprotein (e) two of the preceding answers are correct.
20. Active transport (a) requires energy (b) can be used to move materials against a concentration gradient (c) is used by the cell to move sodium ions out of the cell (d) answers (a), (b), and (c) are correct (e) only answers (a) and (b) are correct.

Review Questions

1. Imagine a cell with a cell wall but no plasma membrane. In what ways would it be handicapped? Could it live?
2. The plasma membrane has been described as a "fluid mosaic." Is this a good description? Why?
3. What problems would a cell face if its plasma membrane were permeable rather than differentially permeable?
4. A 0.9% sodium chloride solution is isotonic to red blood cells. A laboratory technician accidentally places a sample of red blood cells in a 1.8% sodium chloride solution. What happens? Explain.
5. Why do carrot and celery sticks become limp after a time? How could they be made crisp once more? Explain in terms of turgor pressure.
6. Carrier proteins are utilized in both facilitated diffusion and active transport, yet these processes are basically very different. Contrast them.
7. Consider the diagram here, and describe all of the events of endocytosis, secretion, and exocytosis that are lettered. This will provide a review of both Chapters 5 and 6.

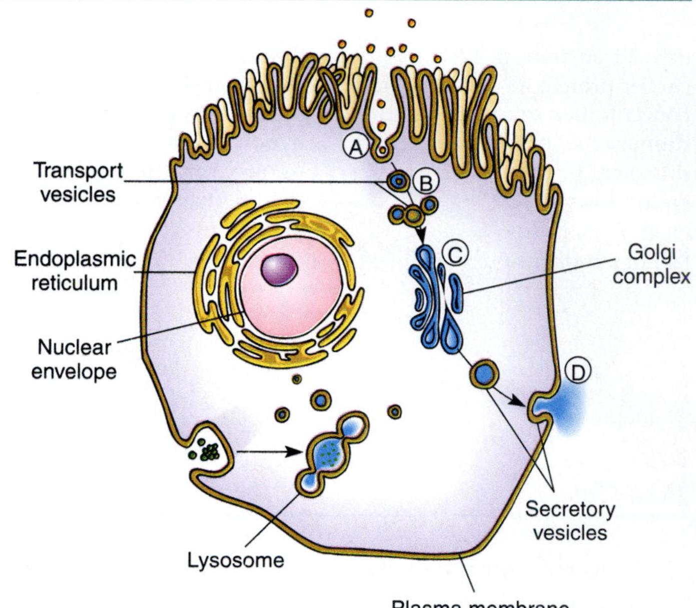

Thinking Critically

1. The blood–brain barrier prevents the passage of certain materials from the blood into the brain tissue. What type of junctions would be found holding the cells of the barrier together? Why?
2. The plasma membrane has been described as having "protein icebergs in a lipid sea." Explain why this is a good description or why it is not.
3. Why is it advantageous for cells lining the digestive tract to be equipped with microvilli?
4. In what way do you think glycoproteins might be important in the function of the immune system?
5. A saltwater amoeba transferred to fresh water forms a contractile vacuole. In what way is this adaptive? Explain. If placed in salt water would a freshwater amoeba form a contractile vacuole?

Recommended Readings

Bretscher, M.S., "The Molecules of the Cell Membrane," *Scientific American,* October 1985. A good discussion of the relationships of membrane molecules.

Loewy, A.G., P. Siekevitz, J.R. Menninger, and J.A.N. Gallant, *Cell Structure & Function: An Integrated Approach,* 3rd ed. Philadelphia, Saunders College Publishing, 1991. A comprehensive reference book for learning more about the cell.

Tuomanen, E., "Breaching the Blood–Brain Barrier," *Scientific American,* Vol. 268, No. 2, February 1993, pp. 80–84. The tight junctions between cells lining blood vessels in the brain block the entrance of bacteria and many substances. Understanding how certain bacteria penetrate this barrier may lead to development of more effective treatment for brain infections.

Unwin, N., and R. Henderson, "The Structure of Proteins in Biological Membranes," *Scientific American,* February 1984, pp. 78–94. A discussion of the configurations of membrane proteins that permit them to be embedded in lipids and yet to function in the watery medium that surrounds the membrane.

PART 3

Energy Flow Through the World of Life

The solar chargeport, a solar energy research project in Southern California. *The solar energy chargeport, which allows electric cars to be recharged in parking lots, addresses the need to recharge electrically-powered cars frequently. It is anticipated that solar chargeports will eventually be located in parking lots of businesses and shopping malls as well as in driveways of private residences. (Courtesy Jean B. Anderson, Southern California Edison.)*

Solar Energy: The Power That Sustains the Planet

Human society depends on energy. We use it to warm our homes in winter and cool them in summer, grow and cook our food, power various forms of transportation, extract and process natural resources, and manufacture items we use everyday. Many of the conveniences of modern living depend on a ready supply of energy.

Energy to power human society comes largely from the sun. Coal, oil, and natural gas, for example, represent radiant energy that was converted to chemical energy millions of years ago. Called fossil fuels, coal, oil, and natural gas are composed of the remnants of ancient organisms. Just as it does today, millions of years ago sunlight provided the energy used by these ancient organisms to form energy-rich chemical compounds. These compounds have been preserved in the Earth's crust, and remain there until they are extracted and burned.

Other energy technologies directly utilize the energy radiating from the sun. Solar energy can be used directly to heat water and buildings and to generate electricity, as shown in the photograph of the solar chargeport. Solar energy is also used indirectly as biomass (plant or animal material used as fuel), wind energy, and hydropower (the energy of flowing water). When you warm yourself by a campfire, the heat energy that you feel came from the sun and was locked into chemical bonds in the wood (a form of biomass) by photosynthesis. The sun also powers the wind (wind energy) and the water cycle that results in precipitation falling on land. As this water flows in rivers and streams toward the ocean, its energy may be harnessed to generate electricity (hydropower).

Energy is clearly important to human society. It has an even more fundamental significance to life because life can be self-sustaining only as long as energy is provided. Living organisms must have a continual input of energy to maintain their ordered structure. Take away an organism's source of energy and the highly organized steady-state that is life quickly degrades into disorder. In other words, the organism dies.

Sunlight is the source of energy that powers almost all life processes. Energy enters the living world as the radiant energy of sunlight. Some of the sun's energy is transformed by plants during the process of photosynthesis. Now in chemical form, this energy is stored in the bonds of organic molecules such as glucose. When living organisms break these molecules apart by cellular respiration, the energy becomes available to do work such as repairing tissues, producing body heat, or reproducing.

The Energy of Life

7

Learning Objectives

After you have studied this chapter you should be able to:

1. Identify the different forms in which energy can exist.
2. Explain the first and second laws of thermodynamics and relate how they apply to living organisms.
3. Describe a chemical reaction that has reached a dynamic equilibrium.
4. Distinguish between endergonic and exergonic reactions and explain how they may be coupled so that the second law of thermodynamics is not violated.
5. Relate the chemical structure of ATP to its role in cellular metabolism.
6. Summarize the role of enzymes as organic catalysts and describe how they work.
7. Explain how factors such as pH and temperature influence enzymatic activity.
8. Compare the actions and effects of the various types of enzyme inhibitors.

Key Concepts

☐ Living things depend on a continuous input of energy to counteract the universal trend toward increasing disorder. They sustain their high level of organization by using energy to maintain a steady state (homeostasis), to synthesize compounds needed for growth and repair, and to reproduce.

☐ Life processes involve energy transformations at the cellular level. In each transaction, some energy dissipates as heat, which is a form of energy that the cell cannot use.

☐ ATP is the energy currency of the cell. All organisms temporarily store energy within the chemical bonds of this compound.

☐ An enzyme is an organic catalyst that increases the rate of a chemical reaction by lowering its activation energy.

BIOLOGICAL WORK REQUIRES ENERGY

Every activity of a living cell or organism requires energy. Movement, transport of materials, synthesis of needed compounds, and manufacture of new cells are a few of the life processes that require energy (Fig. 7–1). Recall from Chapter 1 that all of the chemical processes that occur within a living organism are referred to as its **metabolism.** Energy is involved in all of a living organism's metabolic reactions.

A living organism has no way of creating new energy. It can only extract and use energy from its environment. As the organism performs biological work, some of the energy escapes its body and dissipates back into the environment in the form of heat. Ultimately, this heat energy radiates into space. Thus, once energy has been used by living things, it becomes unavailable for reuse.

For this reason, life depends upon a continuous input of energy, which flows in a one-way direction through the living world. In the one-way energy flow through the cells of living organisms, energy is captured, temporarily stored, and then used to perform biological work. Because the organism cannot reuse this energy, it must continually obtain fresh supplies of energy.

But just how is this accomplished? In this chapter we focus on some principles of energy capture, storage, transfer, and use. In the following two chapters we explore some of the main metabolic pathways—photosynthesis and cellular respiration—used by cells in their continuous quest for energy.

FIGURE 7–1 In order to survive, living things take in and use energy. This grizzly bear is expending energy in an effort to capture the fish. If caught and eaten, the fish will provide nutrients containing energy for future activity. For its part, the fish is expending energy in its effort to escape becoming an energy source for the grizzly. (Michio Hoshino/Minden Pictures)

Energy Is the Capacity To Do Work

We define **matter** as anything that has mass and takes up space, whether it is a subatomic particle or a planet. **Energy** is the ability to produce a change in the state or motion of matter. For example, it takes energy (in the form of heat) to change ice to water, that is, to change the state of water from solid to liquid. It takes energy to produce the increase in the motion of water molecules responsible for this change in state. For our purposes, energy may be simply defined as the capacity to do work.

Energy exists in several different forms. 1. *Chemical energy* is stored in the chemical bonds of molecules. For example, food contains chemical energy. 2. *Nuclear energy* is found within atomic nuclei. 3. *Radiant, or solar, energy* is the transport of energy from the sun as electromagnetic waves. 4. *Heat energy* is the flow of thermal energy from a hotter object to a colder one. 5. *Electrical energy* is the flow of charged particles. 6. *Mechanical energy* is the movement of a body.

Mechanical energy can exist as stored energy (that is, the energy of position, called **potential energy**) or as **kinetic energy**, the energy of motion. A boulder resting at the top of a hill has potential energy because of its position. As the boulder rolls downhill under the influence of gravity, the potential energy is converted to kinetic energy. It would require an input of energy to push the boulder back up the hill and restore the potential energy of its position at the top.

Another example of the conversion of potential energy to kinetic energy is the release of a drawn bow (Fig. 7–2). The tension in the bow and string represents stored energy. When the string is released, this potential energy is converted to kinetic energy as the motion of the bow propels the arrow. It would require the input of additional energy to draw the bow once again and restore the potential energy.

Most of the actions of a complex organism involve a complex series of energy transformations (that is, changing energy from one form to another). For example, to prepare for a running event, athletes eat foods that build up their reserves of glycogen. During the event, the athlete's body continuously converts the chemical energy stored in the glycogen into the mechanical energy used for muscle contraction to run the race.

Heat Is a Form of Energy that Can Be Conveniently Measured

To study energy transformations, scientists must be able to measure energy. How is this done? **Heat** is a form of energy that can be conveniently measured because all other forms of energy can be converted into heat. The study of heat and its relationships with other forms of energy is called **thermodynamics**.

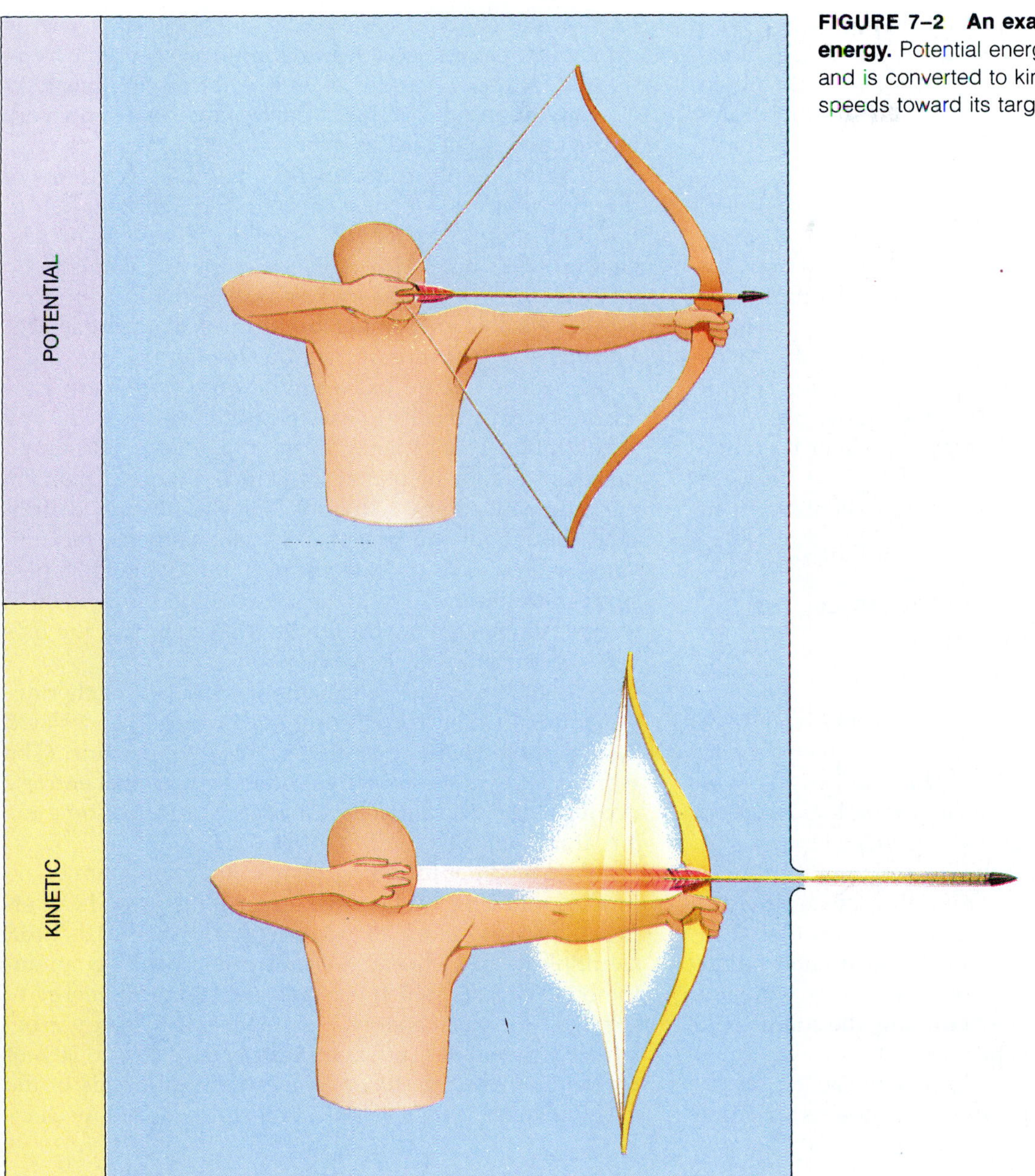

FIGURE 7–2 An example of potential and kinetic energy. Potential energy is stored in the drawn bow and is converted to kinetic energy as the arrow speeds toward its target.

Although several units may be used in measuring energy, the international scientific standard is the **joule (J)**. Until recently, most biologists measured heat in units called calories. A **calorie (cal)** is the amount of heat required to raise the temperature of one gram of water by one degree Celsius.[1] The calorie can be defined in terms of the joule:

1 calorie = 4.184 joules

[1] Nutritionists use the kilocalorie (kcal) to measure the potential energy of foods, and they usually refer to it as a Calorie (with a capital C). A kilocalorie is the amount of heat required to raise the temperature of one kilogram of water by one degree Celsius.

Two Laws of Thermodynamics Govern Energy Transformations

All of the activities of our universe—from the life and death of cells to the life and death of stars—are governed by two laws, the first and second laws of thermodynamics.

The First Law of Thermodynamics Holds that the Quantity of Energy in the Universe Does Not Change

According to the **first law of thermodynamics**, known also as the *law of conservation of energy*, energy cannot be created or destroyed, although it can be transformed from one form to another. As far as we know, the energy present in the universe when it formed some 15 to 20 billion years ago equals the amount of energy present in the universe today. This is all the energy that can ever be present in the universe.

As specified by the first law of thermodynamics, then, living organisms cannot create the energy that they require to live. Instead, they must capture energy from the environment to use for biological work, a process involving the transformation of energy from one form to another. In photosynthesis, for example, plants absorb the radiant energy of the sun and convert it into the chemical energy contained in the bonds of food molecules. Similarly, some of that chemical energy may later be transformed by some animal that eats the plant into the mechanical energy of muscle contraction, enabling it to walk, run, swim, or fly.

The Second Law of Thermodynamics Holds that Entropy Continually Increases in the Universe

As each energy transformation occurs, some of the energy is changed to heat energy that is then given off into the cooler surroundings (Fig. 7–3). This energy can never again be used by the living organism for biological work; it is "lost" from the biological point of view. However, it is not really gone from a thermodynamic point of view because it still exists in the surrounding physical environment. For example, the use of food to enable us to walk or run does not destroy the chemical energy that was once present in the food molecules. After we have performed the task of walking or running, the energy still exists in the surroundings as heat energy.

The **second law of thermodynamics** can be stated most simply as follows: when energy is converted from one form to another, some usable energy (that is, energy available to do work) is degraded into a less-usable form, usually into heat that disperses into the environment. As a result, the amount of usable energy available to do work in the universe decreases over time.

It is important to understand that the second law of thermodynamics is consistent with the first law; that is, the total amount of energy in the universe is *not* decreasing with time. However, the energy available to do work is being degraded to less-usable energy with time.

Less-usable energy is more diffuse, or disorganized. **Entropy** is a measure of disorder or randomness; organized, usable energy has a low entropy, whereas disorganized energy (such as heat) has a high entropy. Entropy is continuously increasing in the universe in all natural processes. It may be that at some time billions of years in the future, all energy will exist as heat uniformly distributed thoughout the universe. If that happens, the universe will cease to operate because no work will be possible; everything will be at the same temperature, so there will be no way to convert the thermal energy of the universe into usable mechanical energy.

Another way to explain the second law of thermodynamics, then, is that entropy, or disorder, in an isolated system spontaneously tends to increase over time. (The word *spontaneous* used in this context means that entropy occurs naturally rather than being caused by some external influence.)

As a result of the second law of thermodynamics, no process requiring an energy conversion is ever 100% efficient, because much of the energy is dispersed as heat, resulting in an increase in entropy. For example, an automobile engine, which converts the chemical energy of gasoline $\longrightarrow$ heat energy $\longrightarrow$ mechanical energy, is between 20% and 30% efficient; that is, only 20% to 30% of the original energy stored in the chemical bonds of the gasoline molecules is actually transformed into mechan-

FIGURE 7–3 During energy transformations, some energy is given into the surroundings. A little owl (*Athene noctua*) in flight, photographed with a strobe light to show the movement of its wings. As a bird flies, chemical energy of food molecules is transformed into the mechanical energy of muscle movement. The chemical energy is not converted entirely to mechanical energy, however; some of its energy is given up as heat. Because the heat energy is gained by the surroundings, the net energy change is zero. (Stephen Dalton/Photo Researchers, Inc.)

FIGURE 7–4 Maintaining order requires energy. Consumers, such as the prairie dog, must eat in order to replace the energy they continuously give up to the surroundings as they carry on life processes. The black-tailed prairie dog, which lives in dry western grasslands, eats mostly seeds and leaves, but also eats grasshoppers and other insects. (Visuals Unlimited/Barbara Gerlach)

ical energy. In our cells, the utilization of energy is about 50% efficient, with the remaining energy given to the surroundings as heat.

Living things have a high degree of organization and, at first glance, they appear to refute the second law of thermodynamics. That is, as living things grow and develop, they maintain a high level of order and do not appear to become more disorganized. However, living things are able to maintain their degree of order over time only with the constant input of energy. That is why plants must photosynthesize and animals must eat (see Fig. 7–4).

METABOLIC REACTIONS INVOLVE ENERGY TRANSFORMATIONS

Metabolic reactions in living things involve changes in the amount and kind of energy. **Free energy** is energy that is available to do work when it is released during a chemical reaction. Free energy is usable energy in biological systems. It can be considered the opposite of entropy, which is a measure of disorganized, less-usable energy.

According to the second law of thermodynamics, the universe is heading toward a condition of maxium entropy—in other words, toward a state of *no free energy.* All physical and chemical processes, therefore, proceed with a decline in free energy until they reach a condition in which the free energy of the system is at a minimum and the entropy is at a maximum.

CONCEPT CONNECTIONS

Energy Flow ⟷ *Ecosystems*

In this chapter we concentrate on the flow of energy as it relates to cellular activities. Energy also flows through higher levels of biological organization, such as the ecosystem, an interacting system that consists of a community of living organisms and its nonliving, physical environment. Energy enters ecosystems as the radiant energy of sunlight. Producers such as plants capture radiant energy during photosynthesis and incorporate some of that energy in the chemical bonds of organic compounds. Then, a portion of that chemical energy is transferred to the consumers, which eat the producers, and to the decomposers, which feed on them all, sooner or later.

In both cells and ecosystems, energy can only be used once to perform biological work. As work is performed, the energy is changed to heat energy and given off into the cooler surroundings (recall the second law of thermodynamics). We sometimes describe the flow of energy, whether in cells or ecosystems, as *linear* because it flows in a line from the environment ⟶ living organisms ⟶ the environment. (Energy flow should never be described as cyclic, because *cyclic* implies that something is recycled, or used over and over again.) The flow of energy through ecosystems is examined in greater depth in the unit on ecology (Chapter 47).

Chemical Reactions May Be Exergonic or Endergonic

A **chemical reaction** is a change involving the molecular structure of one or more substances; the original substances (reactants) are changed into new substances (products) with different properties. During a chemical reaction, energy is released or absorbed. For example, hydrochloric acid (HCl) reacts with the base, sodium hydroxide (NaOH), to yield water (H_2O) and the salt, sodium chloride (NaCl). In the process, energy is released as heat:

$$HCl + NaOH \longrightarrow H_2O + NaCl + Energy$$

The chemical properties of HCl and NaOH are very different from those of H_2O and NaCl. Note that the number of atoms of a given element in the products equals the number of atoms of that element in the reactants. Atoms are neither destroyed nor created in a chemical reaction, but simply change partners, come together, or sep-

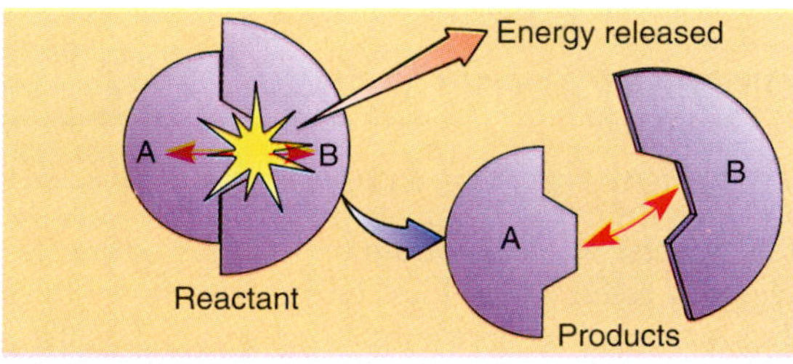

(a) Exergonic reaction

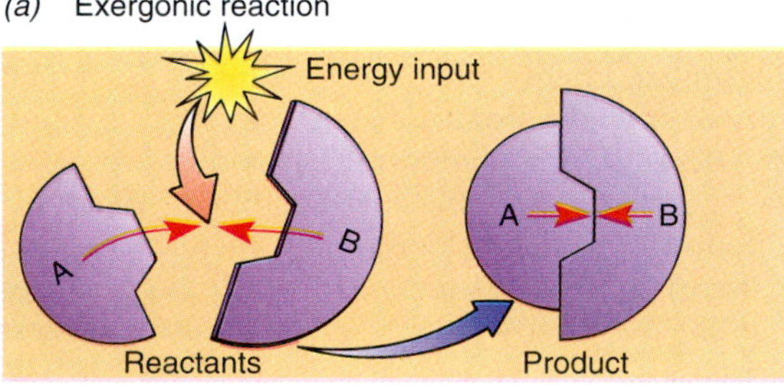

(b) Endergonic reaction

FIGURE 7–5 Energy changes in exergonic and endergonic reactions. (***a***) In exergonic reactions free energy is released, and the products (A and B) have less energy than the reactant (AB). (***b***) In endergonic reactions there is a net input of free energy, so the product (AB) contains more energy than was present in the reactants (A and B).

arate. Likewise, energy is neither created nor destroyed but can be released if the reactants are in a more energetic state than the products, as in this example.

Chemical reactions that release free energy are referred to as **exergonic reactions** (Fig. 7–5). Because energy is released, the products of the reaction contain less energy than the reactants. Reactions that require a net *input* of free energy are called **endergonic reactions**. In these reactions, the products contain more energy than the reactants.

Because exergonic reactions release heat as well as free energy, they are **exothermic** (heat-releasing) reactions (Fig. 7–6). Endergonic reactions, which absorb heat energy from the surroundings, are **endothermic** reactions.

Exergonic reactions proceed spontaneously (remember the second law), but thermodynamics does not predict how *fast* the reactions will occur. Some reactions occur quite rapidly, whereas others take a very long period of time to occur. Sometimes an exergonic reaction proceeds so slowly that it is difficult to measure the release of heat. For example, the rusting of iron occurs over such a long period of time (weeks or months) that the considerable amount of energy which this reaction releases is imperceptible at any point during the reaction.

Chemical Reactions Are Reversible

Many biochemical reactions are reversible (go backward as well as forward) under appropriate conditions. In a reversible reaction, the products are converted back into the reactants. Chemical reactions associated with small changes in heat energy are easier to reverse than reactions involving a large input or release of heat.

Reversibility allows cells to control their release of free energy according to their needs. It also permits cells to resynthesize their large biological molecules for continued use in metabolic processes.

We indicate reversibility by a double arrow, $\rightleftharpoons$. Whether a reaction will occur and whether it will proceed

(a)

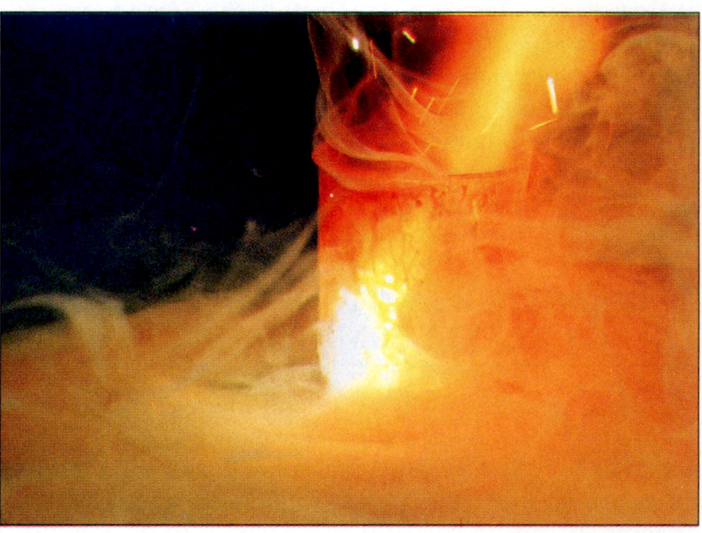

(b)

FIGURE 7–6 An exothermic reaction. (***a***) The element bromine is in the beaker, and the element aluminum is on the table. (***b***) When the aluminum is added to the bromine, the reaction proceeds so vigorously that the aluminum melts and glows white hot. (Dennis Drenner)

from right to left or left to right depends upon factors such as the energy relations of the several chemicals involved, their relative concentrations, and their solubilities.

Reversible Reactions Reach a State of Equilibrium

Imagine that over a ten-year period the population of a city remains the same. Although some new folks have moved into town, others have moved out or perhaps died. However, the net change in the population is zero. We might say that the population in this city is in a state of dynamic equilibrium. In a **dynamic equilibrium**, there is no net change because the rate of change in one direction is exactly equal to the rate of change in the opposite direction.

Consider a chemical reaction in which there is only a small free energy difference between reactants and products. At the beginning of the reaction, only the reactant molecules may be present. These molecules move about and collide with one another with sufficient energy to react and form product molecules. With the production of more and more product molecules, there are fewer and fewer reactant molecules left. As the number of product molecules increases, they collide more frequently and some have sufficient energy to initiate the reverse reaction. The reaction proceeds in both directions simultaneously and eventually reaches a dynamic equilibrium in which the rate of the reverse reaction is about the same as the rate of the forward reaction (Fig. 7–7).

When a chemical reaction is at equilibrium, the free energy difference between the products and reactants is zero.

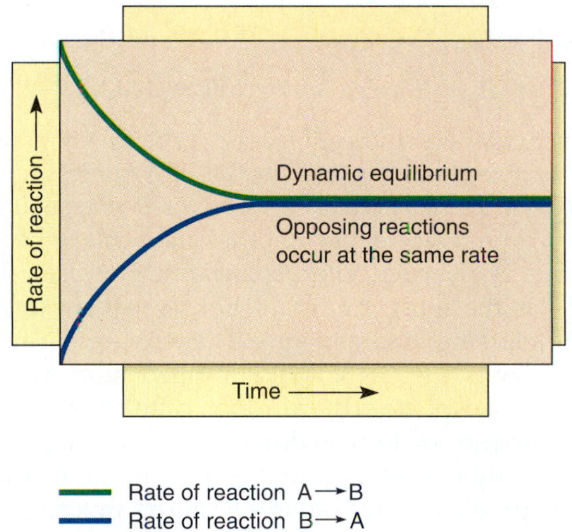

FIGURE 7–7 Dynamic equilibrium. When a chemical reaction reaches a dynamic equilibrium, the rates of the forward (A ⟶ B) and reverse (B ⟶ A) reactions are the same.

Any change, such as a change in temperature or pressure, that affects the reacting system may cause the equilibrium to shift. The reaction may then proceed in one direction or the other until once again the free energy difference is zero, and a new equilibrium has been established.

When there is little free energy difference on the two sides of a chemical equation, the direction of a reaction will be determined mainly by the concentrations of the reactants and products. The reaction tends to proceed in the direction that minimizes the difference in concentration between the reactants and products. If one of them, say the product, is continuously removed as it is formed, the reaction will proceed until all of the reactant is used up. One way that a cell can remove a product (thereby causing the reaction to proceed to completion) is by directing the product down some chemical pathway as soon as it is formed; that is, the product immediately becomes the reactant for another chemical reaction (discussed shortly).

Endergonic and Exergonic Reactions Can Be Coupled in Living Systems

In living organisms, many metabolic reactions such as protein synthesis are endergonic and require an input of energy to proceed. The energy to drive these reactions is often energy released from exergonic reactions. Thus, endergonic and exergonic reactions are often **coupled** to one another. The thermodynamically favorable exergonic reaction provides the energy required to drive the thermodynamically unfavorable endergonic reaction. As we will discuss in the next section, exergonic reactions are often coupled to the *synthesis* of ATP (which is endergonic). Likewise, endergonic reactions are often coupled to the *breakdown* of ATP (which is exergonic).

ATP IS THE UNIVERSAL ENERGY CURRENCY OF THE CELL

In all living cells, energy is temporarily packaged within a remarkable chemical compound called **adenosine triphosphate**, or **ATP** (Chapter 4). This compound stores small amounts of energy for very short periods of time.

When you work you earn money, and so you might say that your "energy" is symbolically stored in that money. The energy of your cells can be compared to the money you earn. ATP is represented by the money in your wallet, which usually remains there temporarily; just as you do not keep the money you carry very long, so too the cell is forever spending its ATP. When you have more money than you need immediately, you might deposit some in the bank. Similarly, your body deposits "excess" energy as lipid in fat cells or as glycogen in liver and muscle cells. Moreover, just as you dare not spend more money than you earn, so too the cell must avoid energy bankruptcy that would mean its death.

ATP is a modified nucleotide compound that consists of three parts: (1) a nitrogen-containing base, **adenine**; (2) a 5-carbon sugar, **ribose**; and (3) three **inorganic phosphate (P_i)** groups identifiable as phosphorus atoms surrounded by oxygen atoms (Fig. 7–8). Notice that the phosphate groups are attached to the end of the molecule in a series, rather like three passenger cars behind a locomotive. The couplings—that is, the chemical bonds attaching the last two phosphates—also resemble those of a train in that they are readily attached and detached.

All three phosphate groups are negatively charged and tend to repel one another. As a result, the phosphate bonds, although rich in chemical energy, are relatively weak and easily broken by **hydrolysis** (the addition of water as the molecule splits). When the phosphate at the very end of the molecule (the third phosphate in the chain) is removed, the remaining molecule is called **adenosine diphosphate**, or simply **ADP**. This is an exergonic (energy-releasing) reaction:

$$ATP + H_2O \longrightarrow ADP + P_i + Energy$$

When the two terminal phosphate groups are removed from ATP, the molecule that remains is **adenosine monophosphate (AMP)**. The hydrolysis of ADP to form AMP is also an exergonic reaction.

When ATP is hydrolyzed, the energy released is used to drive chemical reactions that require the input of energy; in other words, the hydrolysis of ATP is coupled to endergonic processes within the cell. This coupling often involves the transfer of a phosphate group from ATP to some other compound. Addition of a phosphate group to a molecule is referred to as **phosphorylation**.

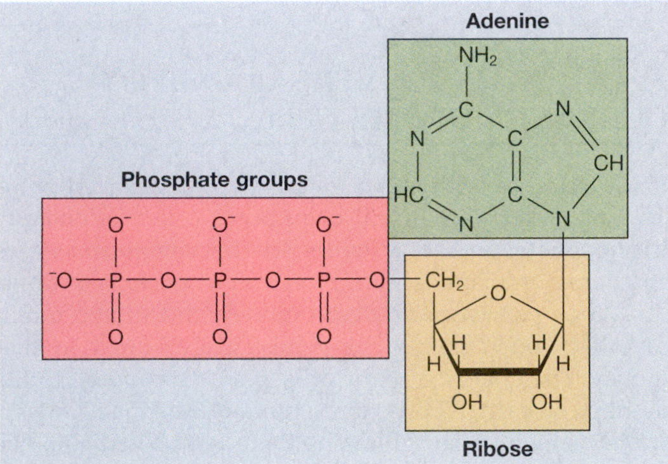

FIGURE 7–8 Chemical structure of ATP. Often described as the energy currency of all living things, ATP provides energy for a wide variety of chemical reactions that take place in cells.

CONCEPT CONNECTIONS

ATP — Condensation and Hydrolysis

In Chapter 4 we learned about condensation and hydrolysis reactions. Recall that condensation reactions join two molecules to form a larger molecule with the elimination of water, whereas hydrolysis reactions break a larger molecule into two smaller ones with the addition of water.

The formation of ATP from ADP and inorganic phosphate (P_i), with the corresponding elimination of water, is an example of a condensation reaction.

$$ADP + P_i \longrightarrow ATP + H_2O$$

Like all condensation reactions, the formation of ATP is endergonic and requires the input of free energy.

The breakdown of ATP into ADP and inorganic phosphate (the reverse of the above reaction) is an example of hydrolysis. Note that water is a reactant in hydrolysis reactions.

$$ATP + H_2O \longrightarrow ADP + P_i$$

The hydrolysis of ATP is exergonic and releases free energy.

When a phosphate is attached to AMP it becomes ADP, and when a phosphate is added to ADP, ATP is produced. Note that these reactions are readily reversible:

$$AMP + P_i + Energy \rightleftharpoons ADP + H_2O$$

$$ADP + P_i + Energy \rightleftharpoons ATP + H_2O$$

As these equations indicate, energy is required to add a phosphate to either the AMP or the ADP molecule; *these reactions are endergonic reactions.* Conversely, the energy is released or transferred to another molecule when the phosphate is detached; *such reactions are exergonic.*

ATP is the most important link between exergonic (energy-releasing) and endergonic (energy-requiring) reactions in cells (Fig. 7–9). ATP is formed from ADP and inorganic phosphate (P_i) during cellular respiration (when nutrients are broken down to release energy). Cellular respiration is exergonic, and the energy it releases is used to produce ATP. Conversely, the hydrolysis of ATP releases energy for use in endergonic reactions, such as those used to produce fats or glycogen, molecules stockpiled for long-term energy storage.

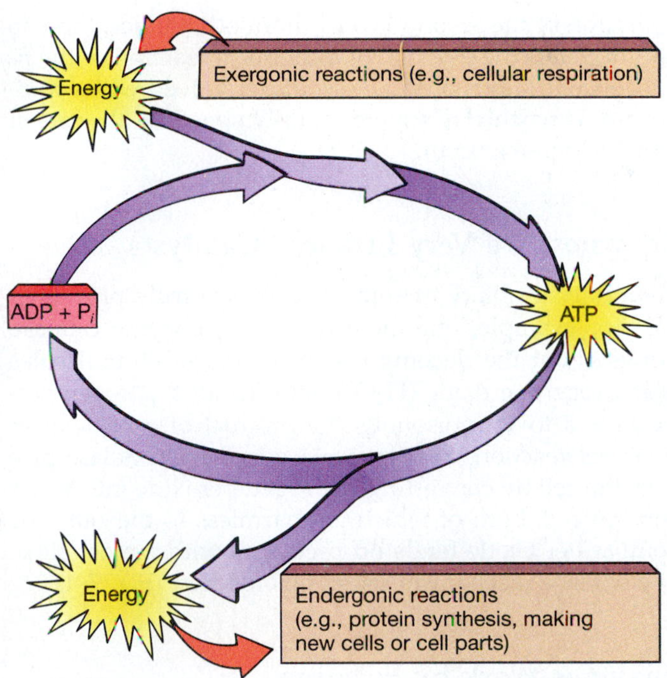

FIGURE 7–9 ATP links many endergonic and exergonic reactions. The energy derived from cellular respiration, an exergonic process, is used to form ATP from ADP and inorganic phosphate. Endergonic reactions, which represent the various kinds of work performed by cells, utilize ATP by removing one of the phosphate groups (and a small amount of energy), thereby forming ADP.

The cell contains a small pool of ADP, ATP, and phosphate, so large quantities of ATP cannot accumulate in the cell. In fact, a bacterial cell may have no more than a one-second supply of ATP. Thus, ATP molecules are used almost as quickly as they are produced. A human at rest uses about 45 kilograms of ATP each day, but the amount present in the body at any given moment is less than 1 gram. Every second in every cell an estimated ten million molecules of ATP are made from ADP and phosphate and an equal number are hydrolyzed, yielding their energy to the life processes that require them.

ENZYMES REGULATE THE RATE OF CHEMICAL REACTIONS

The principles of thermodynamics help us predict whether a reaction can occur but tell us nothing about the *speed* of the reaction. For example, thermodynamics tells us that sucrose should break down into glucose and fructose molecules (because the reaction is exergonic), yet a sucrose solution sealed in a bottle to keep it free of bacteria and molds remains stable indefinitely. In order to break sucrose down we must subject it to high temperature or to strong chemicals such as acids or bases. Living cells can neither wait for years nor use extreme conditions to break down sucrose or any other type of molecule.

In cells, the rates of chemical reactions are controlled by enzymes. An **enzyme** is a biological catalyst, a molecule that increases the rate of a chemical reaction without being consumed in the process. Most enzymes are proteins, and they allow chemical reactions to occur under the conditions that support life, that is, at physiological temperatures and pH. The temperature of cells, for example, is too low to allow chemical reactions to proceed fast enough in the absence of enzymes to meet the cell's needs. When enzymes are present, however, chemical reactions proceed at rates rapid enough to support life.

Enzymes are key elements of the cell's regulation of metabolic processes. For example, fuel molecules, such as glucose, are methodically broken down during the 30 or so chemical reactions of cellular respiration. As a result, energy is released in precise, regulated amounts and homeostatis is maintained. Each of these chemical reactions is controlled by its own enzyme.

We usually name enzymes by adding the suffix *-ase* to the name of the substance acted on. For example, sucrose is split by the enzyme **sucrase** to give glucose and fructose. There are group names for enzymes that catalyze similar reactions: **lipases** catalyze the breakdown of neutral fats into glycerol and fatty acids, **proteinases** break the peptide bonds between amino acids in proteins, and **dehydrogenases** transfer hydrogens from one compound to another.

Enzymes Are Specific

Most enzymes are highly specific and catalyze only a few closely related chemical reactions, or in many cases only one particular reaction. For example, the enzyme sucrase only catalyzes the breakdown of sucrose; it does not act on the similar disaccharides, maltose or lactose.

A few enzymes are specific only in requiring that the substrate have a certain kind of chemical bond. The lipase secreted by the pancreas will split the bonds connecting the glycerol and fatty acids of a wide variety of neutral fats, for example.

An Enzyme Increases the Rate of a Chemical Reaction by Lowering the Activation Energy Necessary to Initiate the Reaction

In order to form new chemical bonds during a chemical reaction, existing bonds must first be broken. The process of breaking these existing bonds represents a barrier to

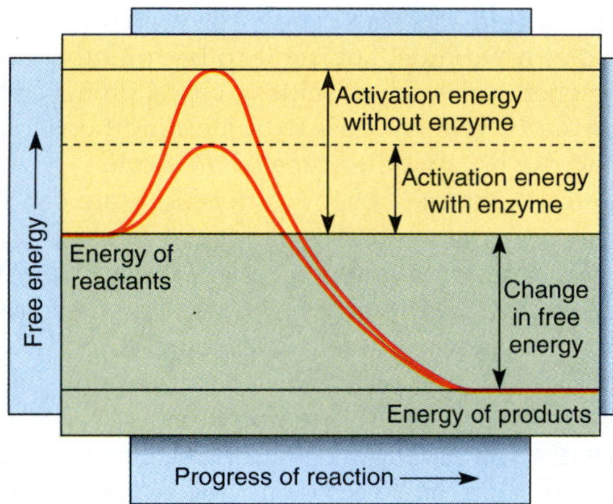

FIGURE 7–10 Activation energy. An enzyme speeds up a chemical reaction by lowering its activation energy (yellow area). A catalyzed reaction proceeds more quickly than an uncatalyzed reaction because it has a lower barrier of activation energy to overcome.

the reaction. The energy required to overcome this barrier and start the reaction is called **activation energy**. An enzyme lowers the activation energy necessary to initiate a chemical reaction.

In a population of molecules of one kind, some have a relatively high energy content, and others have a lower energy content. Only those molecules with a relatively high energy content are likely to overcome the activation energy barrier and form the product. An enzyme lowers the activation energy of the reaction and allows a larger fraction of the population of molecules to react at any one time (Fig. 7–10).

An enzyme can only promote, or speed up, a chemical reaction that could proceed without it (although at a much slower rate). Because enzymes cannot change the operation of the second law of thermodynamics, they do not influence the *direction* of a chemical reaction or the final concentrations of the molecules involved when equilibrium is reached. Enzymes simply speed up the rate at which reactions occur.

Enzymes Are Very Efficient Catalysts

The catalytic ability of some enzymes is truly phenomenal. For example, one molecule of the enzyme **catalase** brings about the decomposition of 40,000,000 molecules of hydrogen peroxide (H_2O_2) per second! Hydrogen peroxide is a toxic (poisonous) by-product of a number of chemical reactions that take place in cells. Catalase protects the cell by converting hydrogen peroxide into water and oxygen, both of which are harmless to the cell. The bombardier beetle fends off predators with a most effective defense mechanism using this reaction (Fig. 7–11).

Enzymes Work by Forming Enzyme-Substrate Complexes

In the first step of an enzyme-catalyzed reaction, the enzyme binds to the **substrate**, the substance upon which it operates, thereby forming an unstable intermediate complex know as an **enzyme-substrate complex** (Fig. 7–12). Then, the enzyme-substrate complex breaks down, forming the product, which is released from the enzyme.

Enzyme + Substrate(s) ⟶
Enzyme-substrate complex ⟶
Enzyme + Product(s)

The enzyme itself is not permanently altered or consumed by the reaction. When the product is released, the enzyme is free to react with a new molecule of substrate.

FIGURE 7–11 The use of catalase as a defense mechanism. An African bombardier beetle uses the catalyzed decomposition of hydrogen peroxide as a defense mechanism. The oxygen gas formed in the reaction forces out water and other chemicals with explosive force. Since the reaction is very exothermic, the water comes out as steam. (Thomas Eisner and Daniel Aneshansley/Cornell University)

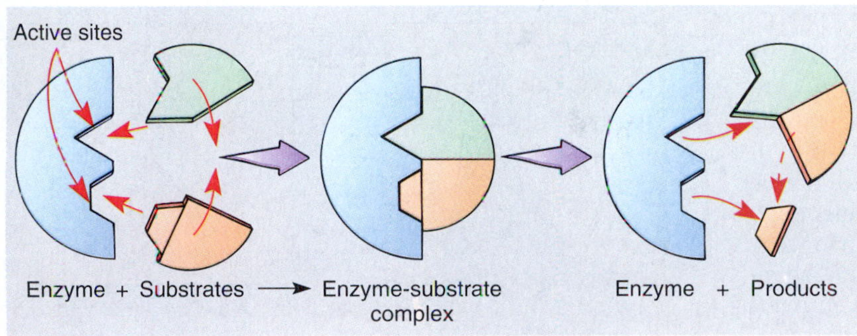

FIGURE 7–12 Mechanism of enzyme action. An enzyme forms a temporary complex with its substrate, which binds to the enzyme's active sites. The enzyme-substrate complex breaks apart to yield enzyme plus product.

Why does the enzyme-substrate complex break up into chemical products that are different than the original substrates? As shown in Figure 7–12, each enzyme contains one or more small regions called **active sites** that have shapes complimentary to the substrates. These active sites are located close to one another on the enzyme's surface. During the course of an enzyme-catalyzed reaction, substrate molecules occupying these sites are temporarily brought close together and so can react with one another.

According to the **induced-fit model** of enzyme action, when the substrate combines with the enzyme, the enzyme molecule is thought to change shape slightly so that it fits the substrate molecule more exactly (Fig. 7–13). This change in shape is possible because the active sites of an enzyme are flexible rather than rigid. The change in shape may produce strain in critical bonds in the substrate molecules so that the bonds break and the products are formed. The products have little affinity for the enzyme and move away from it.

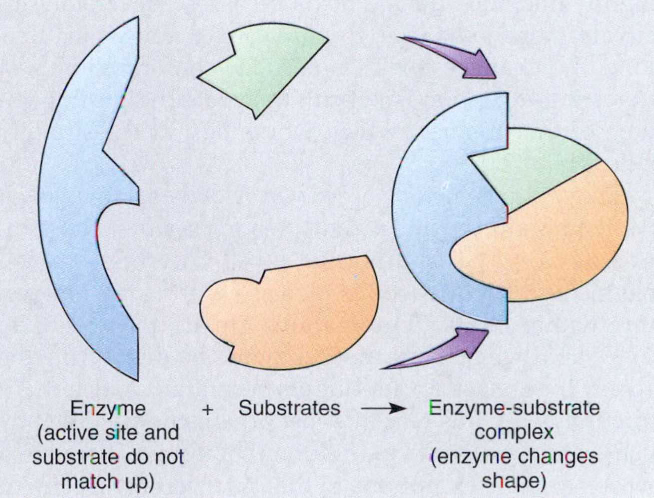

FIGURE 7–13 The binding of an enzyme to its substrate. According to the induced-fit model of enzyme action, the shape of an enzyme's active site becomes complimentary to the substrate only after the substrate has bound to the enzyme.

Many Enzymes Require Cofactors

Some enzymes, for example pepsin secreted by the stomach, consist only of protein. Other enzymes require an additional non-protein component called a **cofactor**, which is essential for the catalytic activity of that enzyme. Like the enzyme, the cofactor is not used up during the reaction and can be used over and over.

The cofactor of some enzymes is a metal ion. Most of the trace elements required in very small amounts in the human diet—elements like copper, zinc, and manganese—function as cofactors. An *organic* nonprotein compound that serves as a cofactor is called a **coenzyme**. Many vitamins are coenzymes or precursors for coenzymes. For example, the vitamin niacin is the precursor of coenzymes involved in oxidation-reduction reactions of photosynthesis and cellular respiration. The vitamin biotin is a coenzyme needed for glucose and fatty acid synthesis.

Enzymes Often Work in Teams

Enzymes usually work in teams to catalyze a series of reactions. The product of one enzyme-controlled reaction may serve as the substrate of the next. We can picture the inside of a cell as a factory with many different assembly lines (and disassembly lines) operating simultaneously. A cellular assembly line is composed of a number of enzymes. Each enzyme carries out one chemical reaction, such as changing molecule A into molecule B, and then passes that molecule along to the next enzyme, which converts molecule B into molecule C and so on.

$$\mathrm{A} \xrightarrow{\text{Enzyme 1}} \mathrm{B} \xrightarrow{\text{Enzyme 2}} \mathrm{C}$$

A short assembly line occurs in germinating barley seeds, where a two-enzyme series converts starch to glucose. The first enzyme, amylase, hydrolyzes starch to maltose (a disaccharide composed of two glucose units), and the second, maltase, splits maltose to glucose.

$$\text{Starch} \xrightarrow{\text{Amylase}} \text{Maltose} \xrightarrow{\text{Maltase}} \text{Glucose}$$

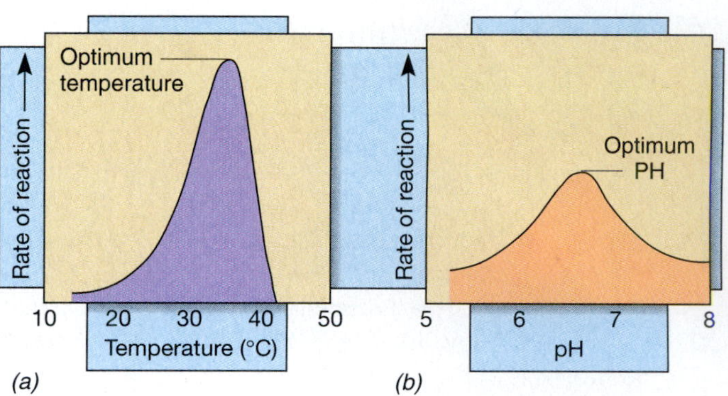

FIGURE 7–14 Enzymes have narrow temperature and pH requirements. The effect of (***a***) temperature and (***b***) pH on the rate of enzyme-catalyzed reactions. Each enzyme has an optimum temperature and pH at which it functions best.

The Cell Regulates Enzymatic Activity in a Variety of Ways

Enzymes affect the rate of chemical reactions in the cell, but the products they form are not usually produced at a constant rate. What controls the activity of enzymes? One mechanism of enzyme control depends upon the *amount* of enzyme produced. The synthesis of each type of enzyme is directed by a specific gene. The gene, in turn, may be switched on or off by a signal from a hormone or some other type of cellular product. When the gene is switched on, the enzyme is synthesized. When more enzyme is present, assuming that pH and temperature are kept constant and an excess of substrate is present, the reaction proceeds faster. Thus, the amount of enzyme present influences the rate of the reaction.

Another mechanism of enzyme control occurs in a linked sequence of enzymatic reactions, in which the product of one enzymatic reaction may alter the activity of another enzyme. For example, in the following series of chemical reactions:

$$\text{A} \xrightarrow{\text{Enzyme 1}} \text{B} \xrightarrow{\text{Enzyme 2}} \text{C} \xrightarrow{\text{Enzyme 3}} \text{D} \xrightarrow{\text{Enzyme 4}} \text{E}$$

each step is catalyzed by a different enzyme. The final product, E, may inhibit the activity of Enzyme 1. When the concentration of E is low, the sequence of reactions proceeds rapidly. However, as the concentration of E increases, it serves as a signal for Enzyme 1 to slow down and eventually to stop functioning. Inhibition of Enzyme 1 stops this entire sequence of reactions, but not permanently. When the concentration of the final product (E) is lowered in the cell, the enzyme resumes its activity. This type of enzyme regulation, in which formation of a product inhibits an earlier reaction in the sequence, is called **feedback inhibition.**

Enzymatic control often depends upon the activation of enzyme molecules that are present, but in an inactive form. Some enzymes, known as **allosteric enzymes**, have two slightly different forms: one form is active and binds to the substrate; the other form is inactive because the active sites of the enzyme are inappropriately shaped so that the substrate does not bind. (The word **allosteric** is derived from the Greek *allo*, meaning "other" and *steric*, meaning "shape.")

Allosteric enzymes possess a site, called an **allosteric site**, located on some region of the enzyme molecule other than the active site. If a substance binds covalently to an allosteric site, it may cause the conformation (shape) of the active site to change, thereby stimulating enzyme action or inhibiting it.

Each Enzyme Works Best at an Optimum Temperature and pH

Enzymes generally work best under certain narrowly defined conditions referred to as optima. These include appropriate temperature, pH, and salt concentration (Fig. 7–14). For example, the starch-digesting enzyme amylase in saliva and pancreatic juice has a pH optimum of 8.5 (slightly alkaline). Strong acids or bases irreversibly inactivate most enzymes by denaturing them, which permanently changes their molecular conformation. An interesting exception is pepsin, the protein-digesting enzyme of the stomach, which works best at the strongly acidic pH of 2.0.

Enzymatic reactions occur very slowly or not at all at low temperatures, but activity resumes when the temperature is raised to a normal level. The rates of most enzyme-regulated reactions increase with rising temperature within limits. Temperatures greater than 50 °C to 60 °C rapidly inactivate most enzymes by denaturing the protein (permanently altering its secondary and tertiary structures). For this reason, most organisms are killed by even a short exposure to high temperatures. There are a few remarkable exceptions to this rule: certain species of bacteria can survive in hot springs, such as the ones in Yellowstone Park, where the temperature exceeds 100 °C.

Enzymes Can Be Inhibited by Certain Chemical Agents

Enzymes can be inhibited or even destroyed by certain chemical agents referred to as **inhibitors** (see Focus On Health and Human Affairs: Designing Drugs to Inhibit Specific Enzymes). Enzyme inhibition by chemical inhibitors may be reversible or irreversible.

Reversible inhibitors are substances that can bind to the enzyme and slow its rate of reaction, but that can be released later, leaving the enzyme in its original state. Reversible inhibitors can be competitive or noncompetitive. In **competitive inhibition**, the inhibitor competes with the normal substrate for the active site of the enzyme (Fig. 7–15). A competitive inhibitor is usually chemically similar to the substrate and so fits the active site and binds with the enzyme. In **noncompetitive inhibition** the inhibitor binds with the enzyme at a site other than the active site. Such an inhibitor renders the enzyme inactive by altering its shape. Many important noncompetitive inhibitors are metabolic substances that help regulate enzyme activity by combining reversibly with the enzyme (recall feedback inhibition).

Many poisons are **irreversible inhibitors** that permanently inactivate the enzyme. A number of insecticides and drugs are irreversible inhibitors. For example, the antibiotic penicillin and its chemical relatives irreversibly inhibit a bacterial enzyme necessary for bacterial cell wall

FOCUS ON *Health and Human Affairs*

Designing Drugs to Inhibit Specific Enzymes

Almost all of the drugs used to combat disease today were discovered in one of two ways: either by chance or as a result of the systematic screening of large numbers of chemicals to identify any with medicinal value. However, a third method of drug development has recently come onto the scene that promises to have a much greater importance in the future—custom-making a drug to fit a specific molecular target in the body. New drugs are currently being designed to treat diseases ranging from cancer to psoriasis, from AIDS to the common cold.

Many of the drugs currently being "designed" are aimed at inhibiting enzymes involved in particular diseases or disorders. That is, it is possible to stop the progress of a disease by inhibiting the activity of an enzyme associated with that disease. Such a designer drug causes few, if any, side effects because the drug devised to inhibit the enzyme is *specific* for that enzyme.

Researchers follow a certain protocol when designing a drug to inhibit an enzyme associated with disease. First, the chemical structure of the enzyme is ascertained, no easy task because enzymes are large, complex molecules. It is not enough to determine the order of amino acids in the enzyme (recall that enzymes are proteins and therefore are composed of amino acids), although this information is helpful. The three-dimensional structure of the enzyme must also be resolved, including the shape of the enzyme's active site. Information about the active site is critical because the drug being developed to inhibit the enzyme must fit into the active site, thereby preventing the enzyme from binding to its substrate.

Computer modeling of the enzyme's three-dimensional structure assists the researchers in the next step—designing a molecule to fit into the active site. The computer simulation usually suggests one or several chemicals that might be possible candidates.

The researchers synthesize these chemical candidates in the laboratory and then test their effects on the activity of the enzme in question. Sometimes a chemical originally recommended by the computer does not inhibit the enzyme as well as hoped. This disappointment occurs because a computer model is only as good as the data and assumptions upon which it is based, and our current models of the three-dimensional structure of molecules such as proteins are far from perfect. When a promising drug fails to live up to its expectations, the researcher returns to the computer, adds new data determined as a result of the failure of the first chemical, and obtains new chemical candidates for the drug.

After promising drugs are demonstrated to inhibit enzyme activity in a test tube, they are tested in cell cultures, then in animals, and finally in clinical trials in humans. How long does this procedure, from identifying the structure of the target enzyme to the initiation of clinical trials, take? Perhaps two to four years, an amount of time that compares very favorably with the ten or more years it takes to develop new drugs by traditional methods.

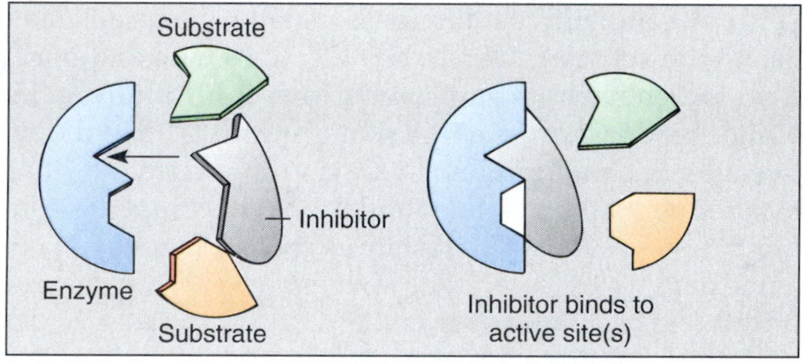

(a)

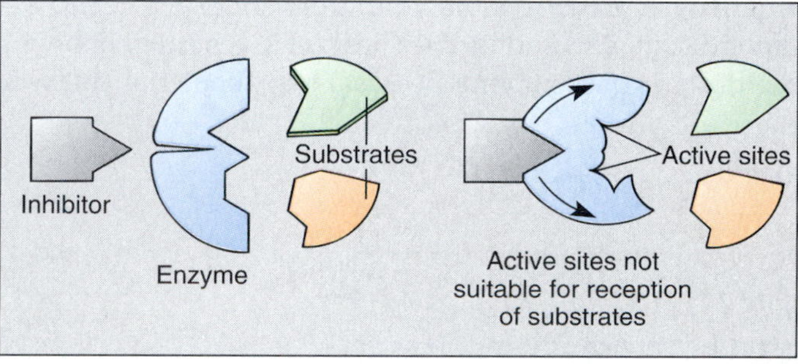

(b)

FIGURE 7–15 Reversible inhibition. (***a***) Competitive inhibition. The inhibitor competes with the normal substrate for the active site of the enzyme. (***b***) Noncompetitive inhibition. The inhibitor binds with the enzyme at a site other than the active site, altering the shape of the enzyme and so inactivating it.

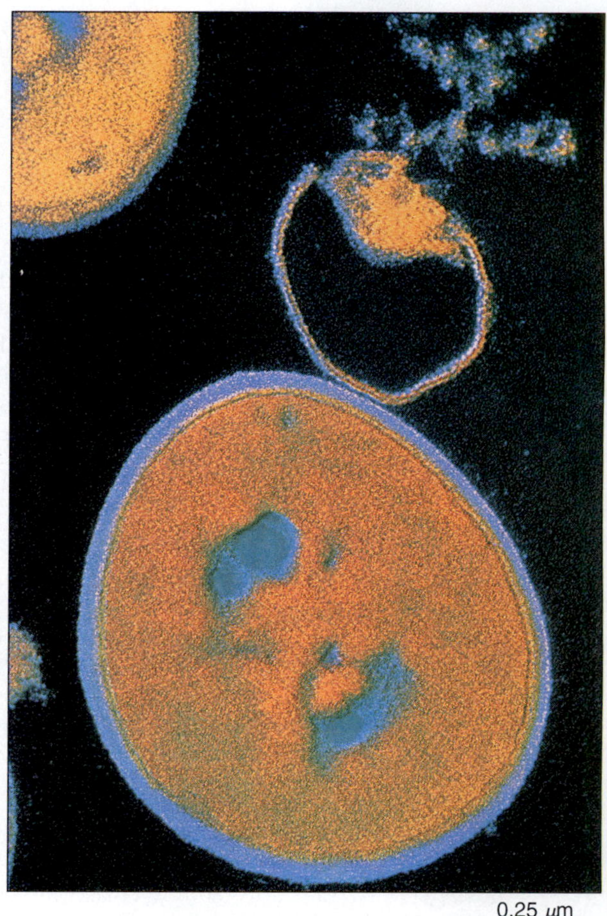

FIGURE 7–16 An example of irreversible inhibition. False-color transmission electron micrograph of antibiotic damage to bacterial cell walls of *Staphylococcus aureus.* The untreated cell (on the bottom) is intact and surrounded by a thick cell wall (blue). The bacterium treated with antibiotic (smaller cell above) lacks the thick cell wall, and without it the cell has lysed, spilling its contents into the surroundings. (CNRI/Science Photo Library/Photo Researchers, Inc.)

construction. Unable to produce new cell walls, susceptible bacteria are unable to multiply effectively (Fig. 7–16). Since human body cells do not possess cell walls (and so do not employ the susceptible enzyme), penicillin is harmless to humans, except for the occasional allergic patient.

Chapter Summary

I. Life depends upon a continuous input of energy.

II. Energy is the capacity to do work.
   A. Energy exists in several forms: chemical, nuclear, radiant, heat, mechanical, and electrical.
   B. Mechanical energy has two forms: potential and kinetic.

III. The first law of thermodynamics states that energy can neither be created nor destroyed but can change from one form to another. The second law of thermodynamics states that entropy, or disorder, in the universe is continuously increasing.
   A. The first law explains why organisms cannot produce energy but must obtain it continuously from the surroundings.
   B. The second law explains why no process requiring energy is ever 100% efficient; in every energy transaction some energy is dissipated into the surroundings as heat.

IV. Metabolic reactions involve energy transformations.
   A. Exergonic reactions release free energy, whereas endergonic reactions require a net input of free energy.
   B. Exergonic reactions are also exothermic; endergonic reactions are also endothermic.
   C. In a chemical reaction that has reached a dynamic

equilibrium, the rate of change in one direction is exactly the same as the rate of change in the opposite direction.

D. In the living cell, endergonic and exergonic reactions are coupled.

V. ATP is the energy currency of the cell; energy is temporarily stored within its chemical bonds.

A. ATP is formed by the phosphorylation of ADP, a process that requires energy.

B. ATP serves as a link between exergonic and endergonic reactions.

VI. An enzyme is an organic catalyst that greatly increases the rate of a chemical reaction without being consumed itself.

A. An enzyme lowers the activation energy necessary to get a chemical reaction going.

B. Enzymes bring substrates into close contact so that they can more easily react with one another.

C. Some enzymes require a cofactor.

1. Certain minerals required in trace amounts in the diet are important cofactors.
2. An organic cofactor is called a coenzyme. Many vitamins are coenzymes.

D. A cell can regulate enzymatic activity by controlling the amount of enzyme produced and by regulating conditions that influence the shape of the enzyme.

E. Each enzyme works best at a specific temperature and pH.

F. Most enzymes can be reversibly or irreversibly inhibited by certain chemical substances. Reversible inhibition may be competitive or noncompetitive.

Selected Key Terms

activation energy, p. 142
active site, p. 143
adenosine triphosphate, p. 139
allosteric enzyme, p. 144
calorie, p. 135
coenzyme, p. 143
cofactor, p. 143
competitive inhibition, p. 145
dynamic equilibrium, p. 139
endergonic reaction, p. 138
endothermic reaction, p. 138
energy, p. 134
entropy, p. 136
enzyme, p. 141
exergonic reaction, p. 138
exothermic reaction, p. 138
feedback inhibition, p. 144
first law of thermodynamics, p. 135
free energy, p. 137
inhibitors, p. 145
joule, p. 135
kinetic energy, p. 134
metabolism, p. 134
noncompetitive inhibition, p. 145
potential energy, p. 134
second law of thermodynamics, p. 136

Post-Test

1. _______________ is the ability to do work.
2. Stored energy is referred to as _______________ energy.
3. The study of energy and its transformations is called _______________.
4. Energy that can be used to do work is referred to as _______________ energy.
5. _______________ reactions require a net input of free energy.
6. In a dynamic _______________, the rate of change in one direction is equal to the rate of change in the opposite direction.
7. ATP contains _______________ (number) phosphate groups.
8. Phosphorylation of ADP, which requires an input of energy, results in the formation of _______________.
9. An organic catalyst that increases the rate of a chemical reaction is a/an _______________.
10. The energy required to break existing bonds and to get a reaction started is known as _______________ _______________.
11. The name of an enzyme generally ends in - _______________.
12. An organic compound that serves as a cofactor is called a/an _______________.
13. In _______________ inhibition, the inhibitor competes with the substrate for the active site.
14. According to the first law of thermodynamics, the energy of the universe is (a) increasing (b) decreasing (c) constant.
15. Each of these statements is true according to the second law of thermodynamics *except*:
(a) During energy transformations, some usable energy is degraded into a less-usable form.
(b) Heat given off to the surroundings is a less-usable form of energy.

(c) Entropy in a system tends to increase over time.
(d) During energy transformations, entropy tends to decrease.

16. In coupled reactions, energy (a) continually increases in a system (b) flows from endergonic to exergonic sources (c) liberated from an exergonic reaction drives an endergonic reaction (d) is unavailable to do work.
17. An enzyme may lower a chemical reaction's activation energy by (a) supplying energy to the reacting molecules (b) putting a strain on the chemical bonds of the substrate (c) placing product molecules in close proximity to each other (d) absorbing extra energy, which activates the enzyme molecule.
18. In the following sequence of reactions, Product D binds to Enzyme 1 at a site other than the active site.

$$A \xrightarrow{\text{Enzyme 1}} B \xrightarrow{\text{Enzyme 2}} C \xrightarrow{\text{Enzyme 3}} D$$

As a result, Enzyme 1 is inhibited. This is an example of (a) feedback inhibition (b) competitive inhibition (c) irreversible inhibition (d) hydrolytic inhibition.

19. Which of the following statements about enzymes is untrue? (a) Enzymes are organic catalysts (b) Most enzymes are highly specific (c) Many enzymes require cofactors (d) Enzymes provide energy for the reactions that they catalyze.
20. The initial barrier to a chemical reaction is the ___?___ energy required to break existing chemical bonds. (a) activation (b) potential (c) kinetic (d) entropy.

Review Questions

1. Give two examples of potential energy. Of kinetic energy.
2. When a consumer eats a producer, it cannot obtain for itself all of the energy captured by the producer during photosynthesis. Explain why in terms of the second law of thermodynamics.
3. The rusting of iron produces heat, but a rusty nail does not feel warm. Explain.
4. Why is it critical that endergonic and exergonic reactions be coupled?
5. What is activation energy? How does an enzyme affect activation energy?
6. Explain the induced-fit model of enzyme action.
7. How are enzymes affected by (a) pH, (b) temperature, (c) inhibitors?
8. In what way does ATP serve as a link between exergonic and endergonic reactions?

Thinking Critically

1. The energy in biomass can be traced to solar energy because biomass is the result of photosynthesis. Given that plants are the organisms that photosynthesize, why are animal wastes such as cow dung also considered biomass?
2. What cannot be seen (although its effects can be seen), cannot be made (although it can be used), and cannot be destroyed (although it can be used inefficiently)? Explain.
3. The laws of thermodynamics predict that our sun will eventually die. Explain.
4. Trace the various forms that energy takes as it flows from sunlight to the heat released during muscle contraction when you exercise.
5. When a lynx eats a rabbit, is the lynx obtaining *all* of the energy in the rabbit's tissue? Why or why not? Suggest how your answer might affect whether food chains (the sequence of who eats whom) are long or short.
6. Jackrabbits have huge ears through which they lose excess body heat. Where does body heat come from? Can you relate body heat to the second law of thermodynamics?
7. How might the high temperature of a fever alter the way a cell functions?

Recommended Readings

Bugg, C. E., W. M. Carson, and J. A. Montgomery, "Drugs by Design," *Scientific American*, December 1993. How new drugs that inhibit enzymes associated with diseases are being designed based on their molecular structure.

Fulkerson, W., R. R. Judkins, and M. K. Sanghvi, "Energy from Fossil Fuels," *Scientific American*, September 1990. Discusses today's technological challenge of extracting as much energy as possible from fossil fuels while harming the environment as little as possible.

Hindrichs, R. A., *Energy*, Philadelphia, Saunders College Publishing, 1992. Chapters 3 and 4 of this very general book consider some of the practical aspects of the first and second laws of thermodynamics.

Joesten, M. D., D. O. Johnston, J. T. Netterville, and J. L. Wood, *World of Chemistry*, Philadelphia, Saunders College Publishing, 1991. Chapter 8, titled "What Every Consumer Should Know about Energy," discusses the fundamental scientific principles of energy before considering energy and human society.

Weinberg, C. J., and R. H. Williams, "Energy from the Sun," *Scientific American*, September 1990. Examines the advantages and disadvantages of various forms of solar energy.

Wheelchair racers. The muscle cells in the arms, shoulders, and backs of these physically challenged athletes are breaking down fuel molecules to obtain energy for muscle contraction associated with wheelchair racing. (David Young-Wolff/PhotoEdit)

The Racer's Thrill: Gasping for ATP

An athlete's body undergoes many physiological changes to permit physical performance during exercise. In a 1.5 kilometer race, for example, the energy production of a runner or wheelchair racer increases nearly eight times normal levels in less than three minutes. In particular, the energy requirements of muscle cells change dramatically during exercise, demanding increased amounts of cellular fuel and oxygen, and releasing increased amounts of carbon dioxide and heat. The blood vessels in actively contracting muscles enlarge so that a greater flow of blood reaches the muscles, supplying needed materials and removing waste products.

Like all other cells, muscle cells obtain energy from the phosphate bonds of adenosine triphosphate molecules (ATP), the energy currency of the cell (Chapter 7). However, the amount of ATP available in muscle cells is only adequate for about five or six seconds of strenuous exercise. Cells must break down cellular fuels such as glucose to regenerate ATP molecules or else the muscle cells will be unable to contract.

Initially, the glucose is converted to pyruvic acid, a three-carbon molecule important in metabolism. Then, if sufficient oxygen is present, the pyruvic acid is completely degraded to carbon dioxide, with enough energy released to form many ATP molecules. During sustained exercise, most of the ATP is regenerated by this pathway.

When exercise is very strenuous, more ATP is needed than can be supplied by the complete breakdown of glucose to carbon dioxide; that is, too little oxygen may be provided by the lungs. In the absence of oxygen, the pyruvic acid is converted to lactic acid (an organic acid produced in actively contracting muscle cells when insufficient oxygen is available), and only a small amount of ATP is produced. Because the lactic acid pathway is less efficient (that is, results in the production of less ATP), it requires many more fuel molecules to supply the muscle cells with an adequate amount of energy.

When muscle cells are producing large amounts of lactic acid, it temporarily accumulates in the muscles, contributing to fatigue and muscle cramps.[1] About 80% of the lactic acid is eventually exported to the liver, where it is used to regenerate more glucose for the muscle cells. The remaining 20% of the lactic acid is metabolized in the muscle cells in the presence of oxygen. This explains why you continue to breathe heavily after you have stopped exercising; the additional oxygen is needed to oxidize lactic acid, thereby restoring the muscle cells to their normal (resting) state.

[1] *Why* exercise causes fatigue and muscle cramps is incompletely understood at this time, but these conditions may be related to the buildup of lactic acid, the presence of insufficient oxygen, and/or the depletion of fuel molecules.

Energy-Releasing Pathways

8

Learning Objectives

After you have studied this chapter you should be able to:

1. Write a general equation illustrating electron transfer from a substrate to an electron acceptor such as NAD^+.
2. Write a summary reaction for aerobic respiration, giving the origin and fate of each substance involved.
3. List and give a brief overview of the four stages of aerobic respiration.
4. Indicate where the reactions of each stage of aerobic respiration take place in the cell.
5. Indicate the number of ATP molecules used and produced in each stage of aerobic respiration during the breakdown of one molecule of glucose.
6. Describe chemiosmosis, explaining (a) how a gradient of protons is established across the inner mitochondrial membrane and (b) the process by which the proton gradient drives ATP synthesis.
7. Summarize how proteins and lipids are broken down to release energy.
8. Compare and contrast aerobic and anaerobic respiration in terms of (a) how much ATP is formed, (b) the final electron acceptor, and (c) the end products.
9. Compare and contrast alcoholic and lactic acid fermentation in terms of (a) the final electron acceptor and (b) the end products.

Key Concepts

- ☐ During cellular respiration, cells break down organic fuel molecules and release their stored energy in a series of steps. The energy is used to make ATP.
- ☐ During aerobic respiration, an organic compound such as glucose is completely degraded to carbon dioxide and water, with a large amount of ATP being produced. Oxygen is required for aerobic respiration.
- ☐ In the absence of oxygen, anaerobic pathways may be used. The fuel molecule is only partially broken down, and only a small amount of energy is released to make ATP.

ENERGY CAN BE RELEASED VIA AEROBIC OR ANAEROBIC PATHWAYS

Every living organism must extract energy from the organic food molecules that it either manufactures (for example, when plants photosynthesize) or captures from the environment (for example, when an animal eats another organism) (Fig. 8–1). In humans and other complex animals, food is first broken down by the digestive system. For example, proteins are split into their component amino acids, carbohydrates are digested to simple sugars such as glucose, and fats are split into glycerol and fatty acids. These nutrients are then absorbed into the blood and transported to all of the cells. Within each cell, these nutrients can be broken down, and some of their chemical energy transferred to ATP for later use in cellular work.

The process of splitting larger molecules to smaller ones is referred to as **catabolism,** and the individual reactions involved are called **catabolic reactions.** Catabolic reactions are exergonic and release free energy (Chapter 7). During the catabolism of glucose, for example, cells break apart glucose one piece at a time in a series of steps that releases the energy in the chemical bonds of glucose in a controlled fashion.

Cells use three different catabolic pathways to extract energy from nutrients: aerobic respiration, anaerobic respiration, and fermentation (another type of anaerobic pathway). The type of environment a cell inhabits determines which catabolic pathway is used to break down nutrients. Cells that live in environments where oxygen is plentiful use aerobic respiration. An **aerobic** pathway requires molecular oxygen (O_2). Cells that inhabit waterlogged soil or polluted water where oxygen is absent must use **anaerobic** pathways that do not require oxygen.

FIGURE 8–1 Animals eat other organisms to obtain energy. Here a male lion consumes a cape buffalo. (Patti Murray)

Most cells extract energy from fuel molecules, such as glucose, fatty acids, and other organic compounds, by using **aerobic respiration.** This exergonic process involves a long sequence of 30 or more chemical reactions, each regulated by a specific enzyme. During aerobic respiration, energy is released efficiently in precise amounts as nutrients are catabolized to carbon dioxide and water. One of the most common pathways of aerobic respiration involves the breakdown of the nutrient glucose. The overall reaction for the aerobic respiration of glucose can be summarized as follows:

$$\underset{\text{Glucose}}{C_6H_{12}O_6} + \underset{\text{Oxygen}}{6O_2} + \underset{\text{Water}}{6H_2O} \longrightarrow \underset{\text{Carbon dioxide}}{6CO_2} + \underset{\text{Water}}{12H_2O} + \text{Energy}$$

ENERGY-RELEASING PATHWAYS USE OXIDATION-REDUCTION REACTIONS

Energy processing by cells involves the transfer of energy through the flow of electrons and protons. Recall from Chapter 3 that **oxidation** is a chemical process in which a substance *loses* electrons, and **reduction** is a chemical process in which a substance *gains* electrons. An oxidation reaction is always accompanied by a reduction reaction because electrons do not exist in the free state in living cells. As quickly as they are released, electrons are accepted by another atom or molecule. *At the same time that it gives up electrons, the oxidized molecule gives up energy; the reduced molecule receives energy when it gains electrons.* **Redox reactions**—reactions in which *red*uction and *ox*idation occur—are characteristic of many cellular processes, including aerobic respiration and photosynthesis. Generally, a sequence of redox reactions takes place as electrons associated with hydrogen pass from one compound to another.

Electrons Associated with Hydrogen Are Transferred to Electron Acceptor Molecules

Electrons are difficult to remove from covalent compounds unless an entire atom is removed. In the cell, oxidation usually involves the removal of a hydrogen atom (and its single electron) from a compound. Reduction usually involves a gain in a hydrogen atom (and thus a gain in an electron).

FIGURE 8–2 The total energy released by a falling object is the same whether the object is released all at once or in a series of steps. Similarly, the energy of an electron liberated from an organic compound is the same whether it is released all at once or whether it is released gradually as it passes through successive electron acceptors. Such electron transport chains permit the controlled extraction of energy from fuel molecules.

When hydrogen atoms are stripped from an organic compound, they take with them some of the energy that had been stored in their electrons. The electrons are transferred to an electron acceptor molecule. One of the most common electron acceptor molecules in cells is **nicotinamide adenine dinucleotide,** more conveniently referred to as **NAD^+.** When NAD^+ temporarily accepts electrons (and protons to make hydrogen), it also accepts free energy. Here is a generalized equation showing the transfer of electrons associated with hydrogen from a given compound (X) to NAD^+.

$$\overbrace{XH_2 + NAD^+ \longrightarrow X}^{\text{Oxidation}} + NADH + H^+ \quad (\text{Reduction: } NAD^+ \rightarrow NADH)$$

NAD^+ is reduced when it combines with hydrogen to form NADH.[2] Some of the energy stored in the chemical bonds holding the hydrogens to molecule X has been transferred as a result of this reaction to the NADH. This energy can now be used to drive some metabolic process, or it can be transferred to ATP.

[2] Although the correct way to write the reduced form of NAD^+ is $NADH + H^+$, for simplicity we present the reduced form as NADH throughout this chapter.

Another important electron acceptor is **flavin adenine dinucleotide (FAD).** Like NAD^+, FAD is capable of being alternately reduced and oxidized as it accepts and then gives up hydrogen.

Electron acceptors are often arranged in a sequence. As electrons are transferred from one acceptor molecule to the next along the chain, a small amount of energy is released. Such **electron transport chains** are the mechanism by which the cell efficiently extracts energy from fuel molecules (Fig. 8–2).

Aerobic Respiration Is a Redox Process

Aerobic respiration is a redox process in which hydrogen is transferred from glucose to oxygen. In a long and complex series of reactions, glucose is oxidized and oxygen is reduced.

$$C_6H_{12}O_6 + 6O_2 + 6H_2O \longrightarrow 6CO_2 + 12H_2O + \underset{(\text{as ATP})}{\text{Energy}}$$

(Oxidation: $C_6H_{12}O_6 \rightarrow 6CO_2$; Reduction: $6O_2 \rightarrow 12H_2O$)

During this process, chemical energy is released and used for ATP synthesis.

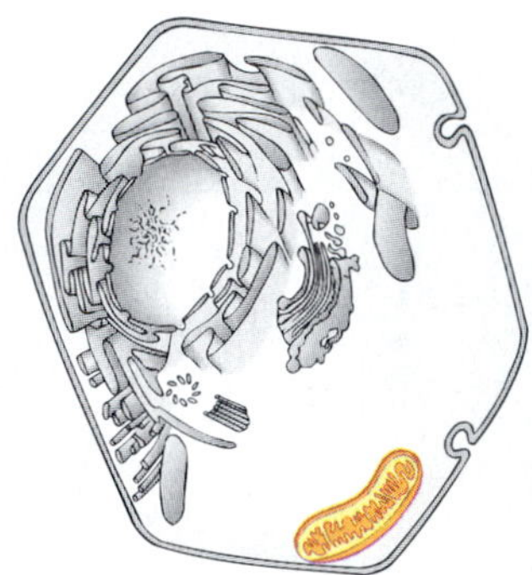

FIGURE 8–3 Stages of aerobic respiration. The four stages of aerobic respiration are glycolysis, formation of acetyl coenzyme A, the citric acid cycle, and the electron transport system with its associated chemiosmotic ATP synthesis. Glycolysis occurs in the cytosol. Pyruvic acid, the product of glycolysis, enters a mitochondrion, where aerobic respiration continues. Most ATP is synthesized during electron transport/chemiosmosis.

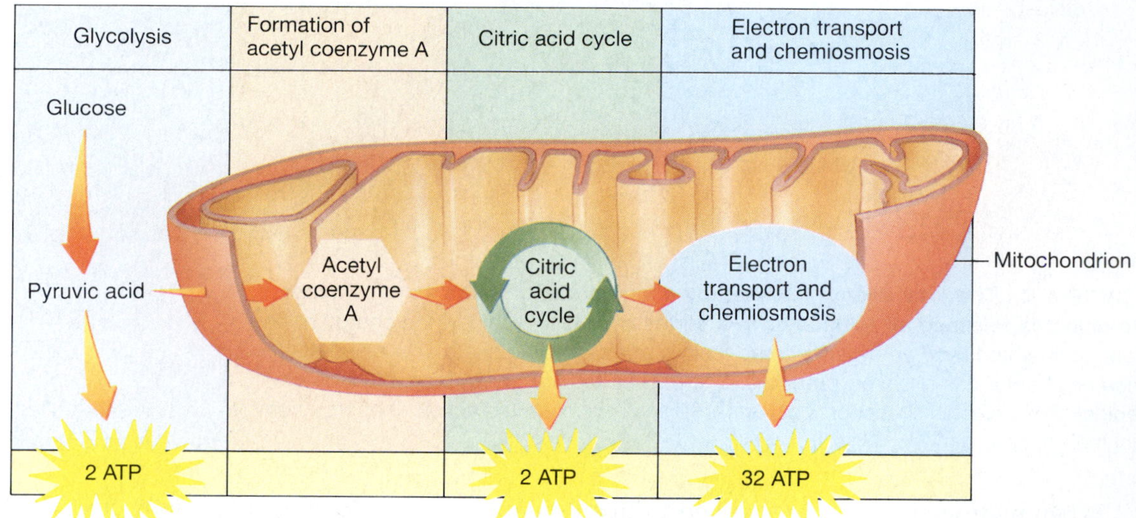

AEROBIC RESPIRATION OCCURS IN FOUR STAGES

The chemical reactions of aerobic respiration are grouped into four stages: (1) glycolysis, (2) formation of acetyl coenzyme A, (3) citric acid cycle, and (4) electron transport and chemiosmosis (Fig. 8–3 and Table 8–1).

1. Glycolysis: The term glycolysis is derived from two Greek words that, taken together, mean "splitting sugar." During glycolysis, the cell splits the six-carbon glucose molecule into 2, three-carbon molecules of **pyruvic acid.**[3] During this sequence of reactions, hydrogens are removed from the fuel molecule and combine with NAD^+ to form NADH. Two molecules of ATP are produced. Glycolysis, which takes place in the cytosol, does not require oxygen, and so it can proceed under aerobic or anaerobic conditions.

2. Formation of acetyl coenzyme A: The second stage of aerobic respiration, the formation of acetyl coenzyme A, links glycolysis to the citric acid cycle. Each molecule of pyruvic acid produced during glycolysis passes into the matrix of the mitochondrion where it is degraded to a two-carbon fuel molecule (an acetyl group) that combines with coenzyme A, forming **acetyl coenzyme A (acetyl CoA).** The carbon removed from pyruvic acid is released as carbon dioxide. Hydrogens are also removed and combine with NAD^+ to produce more NADH. The formation of acetyl CoA is an important intermediate step in the aerobic respiration of energy-rich food molecules because it prepares the fuel molecules to enter the citric acid cycle (as acetyl CoA).

3. Citric acid cycle: Each acetyl CoA enters the citric acid cycle and combines with a four-carbon molecule (oxaloacetic acid) to form a six-carbon molecule (citric acid). Citric acid eventually re-forms oxaloacetic acid, with the release of two molecules of carbon dioxide. A small amount of ATP is produced, but most importantly, the remaining hydrogens are removed from the fuel molecule and picked up by NAD^+ and FAD, forming NADH and $FADH_2$. The entire citric acid cycle occurs in the matrix of the mitochondrion.

[3] Pyruvic acid molecules dissociate to form hydrogen ions (H^+) and pyruvate ions at the pH found in the cell. The same is true for many other compounds in glycolysis. Some textbooks present these compounds in ion form, for example, as *pyruvate* rather than *pyruvic acid*.

TABLE 8–1
Summary of Aerobic Respiration

Stage	*Summary*	*Some starting materials*	*Some end products*
1. Glycolysis (in cytosol)	Series of about ten reactions in which glucose is degraded to pyruvic acid; net gain of 2 ATPs; hydrogens are released; can proceed anaerobically	Glucose, ATP, NAD^+	Pyruvic acid, ATP, NADH
2. Formation of acetyl CoA (in mitochondria)	Pyruvic acid is degraded and combined with coenzyme A to form acetyl CoA; hydrogens are released; CO_2 is released	Pyruvic acid, coenzyme A	Acetyl CoA, CO_2, NADH
3. Citric acid cycle (in mitochondria)	Series of reactions in which the acetyl portion of acetyl CoA is degraded to CO_2; hydrogens are released	Acetyl CoA, H_2O	CO_2, NADH, $FADH_2$, ATP
4. Electron transport and chemiosmosis (in mitochondria)	Chain of several electron transport molecules; H's (or their electrons) are passed along chain; energy released is used to form proton gradient; ATP is synthesized as protons move across the gradient; oxygen is final H-acceptor	NADH, $FADH_2$, oxygen	ATP, H_2O

4. Electron transport and chemiosmosis: The NADH and $FADH_2$ produced in the preceding three stages transfer the electrons that they have accepted to a chain of electron acceptor molecules. These molecules are located within the inner membrane of the mitochondrion. As electrons are passed along the chain of acceptor molecules in a series of redox reactions, energy is released. This energy is used to pump protons (H^+) across the inner mitochondrial membrane, from the matrix of the mitochondrion to the intermembrane space (between the inner and outer mitochondrial membranes). The difference in the concentration of protons between the matrix and the intermembrane space is referred to as a **proton gradient**; this gradient represents potential energy, much like water held behind a dam represents potential energy. In chemiosmosis the energy of the proton gradient (as the protons diffuse across the membrane)—like water spilling over a dam—is used to produce ATP.[4]

AEROBIC RESPIRATION IS A COMPLEX PROCESS

The preceding section gave an overview of aerobic respiration in sufficient detail for some introductory biology courses. The following elaboration is provided for those courses that go into greater detail than the modest discussion previously provided.

In Glycolysis, Glucose Is Converted to Pyruvic Acid

The reactions of glycolysis can be divided into two phases (Fig. 8–4). The first phase requires the input of energy and phosphate from two molecules of ATP. Two phosphates from two ATPs are added to the fuel molecule. As mentioned in Chapter 7, the addition of one or more phosphate groups to a molecule is termed **phosphorylation**. The phosphorylated sugar molecule is then split, forming two molecules of the three-carbon compound **glyceraldehyde-3-phosphate**, or **PGAL**.

In the second phase of glycolysis, each PGAL molecule is oxidized with the removal of two hydrogens, and certain other atoms are rearranged so that each molecule of PGAL is transformed into a molecule of pyruvic acid. During these reactions, enough chemical energy is released from the original sugar molecule to produce four ATP molecules. Because the first phase of glycolysis required two molecules of ATP, but the second phase produced four molecules of ATP, glycolysis yields a *net* energy gain of two ATPs per molecule of glucose.

The two hydrogens removed from each PGAL immediately combine with the hydrogen carrier molecule, NAD^+, forming NADH. The fate of these hydrogens is discussed in the section on the electron transport system.

[4] The term *chemiosmosis* is somewhat misleading in that chemiosmosis has nothing to do with osmosis.

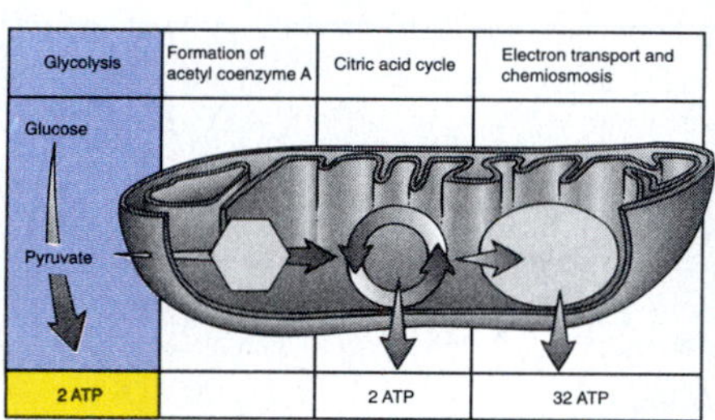
Glycolysis
Formation of acetyl coenzyme A
Citric acid cycle
Electron transport and chemiosmosis
Glucose
Pyruvate
2 ATP
2 ATP
32 ATP

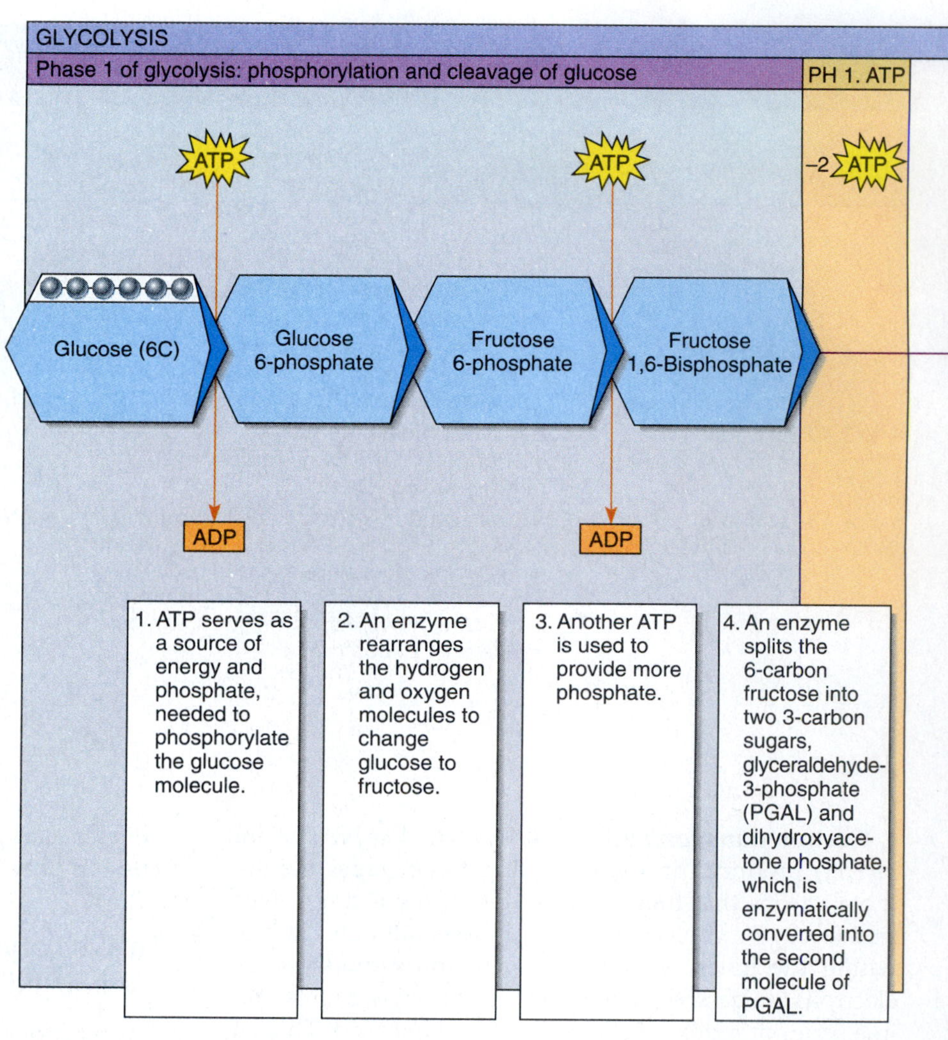
GLYCOLYSIS
Phase 1 of glycolysis: phosphorylation and cleavage of glucose
PH 1. ATP
ATP
ATP
–2 ATP
Glucose (6C)
Glucose 6-phosphate
Fructose 6-phosphate
Fructose 1,6-Bisphosphate
ADP
ADP
1. ATP serves as a source of energy and phosphate, needed to phosphorylate the glucose molecule.
2. An enzyme rearranges the hydrogen and oxygen molecules to change glucose to fructose.
3. Another ATP is used to provide more phosphate.
4. An enzyme splits the 6-carbon fructose into two 3-carbon sugars, glyceraldehyde-3-phosphate (PGAL) and dihydroxyacetone phosphate, which is enzymatically converted into the second molecule of PGAL.

FIGURE 8–4 The two phases of glycolysis. In phase one, glucose is phosphorylated and cleaved to form two molecules of glyceraldehyde-3-phosphate (PGAL), and in phase two each PGAL molecule is converted into a molecule of pyruvic acid. Each reaction is catalyzed by a specific enzyme. Note that we can calculate a net yield of two ATPs per molecule of glucose by subtracting the number of ATPs used in the process (2) from those produced (4).

Pyruvic Acid Is Used to Make Acetyl CoA

Pyruvic acid, the end product of glycolysis, contains most of the energy present in the original glucose molecule. Pyruvic acid molecules move into the matrix of the mitochondrion where all subsequent reactions of aerobic respiration take place.

Once a pyruvic acid molecule enters the mitochondrion, it is converted to the two-carbon compound acetyl coenzyme A, or simply, acetyl CoA. A large, complex enzyme catalyzes the multi-step reaction. First, a carboxyl group is removed and released as carbon dioxide. Then, the two-carbon fragment remaining is oxidized; the hydrogens removed during this oxidation are accepted by NAD^+, forming NADH. Finally, the oxidized fragment, an acetyl group, is attached to coenzyme A.

Note that the original six-carbon glucose molecule has now been oxidized to two acetyl groups and two CO_2 molecules. Hydrogens have been removed and accepted by NAD^+, reducing it to NADH. Two NADH molecules have been formed during glycolysis and two more during the oxidation of pyruvic acid.

The Citric Acid Cycle Oxidizes Acetyl CoA

The citric acid cycle is also known as the **Krebs cycle**, after Sir Hans Krebs, a British biochemist who ascertained the pathway in the 1930s, for which he received a Nobel Prize (Fig. 8–5). This cycle is the common pathway for the final oxidation reactions of the cell's fuel molecules—glucose, fatty acids, and the carbon chains of amino acids—with the carbons being released as CO_2. The citric acid cycle, which takes place in the matrix of the mitochondrion, consists of eight steps, with each reaction being catalyzed by a specific enzyme.

The first reaction of the citric acid cycle occurs when acetyl CoA transfers its two-carbon acetyl group to the four-carbon compound **oxaloacetic acid**, forming **citric acid**, a six-carbon compound. The citric acid then goes

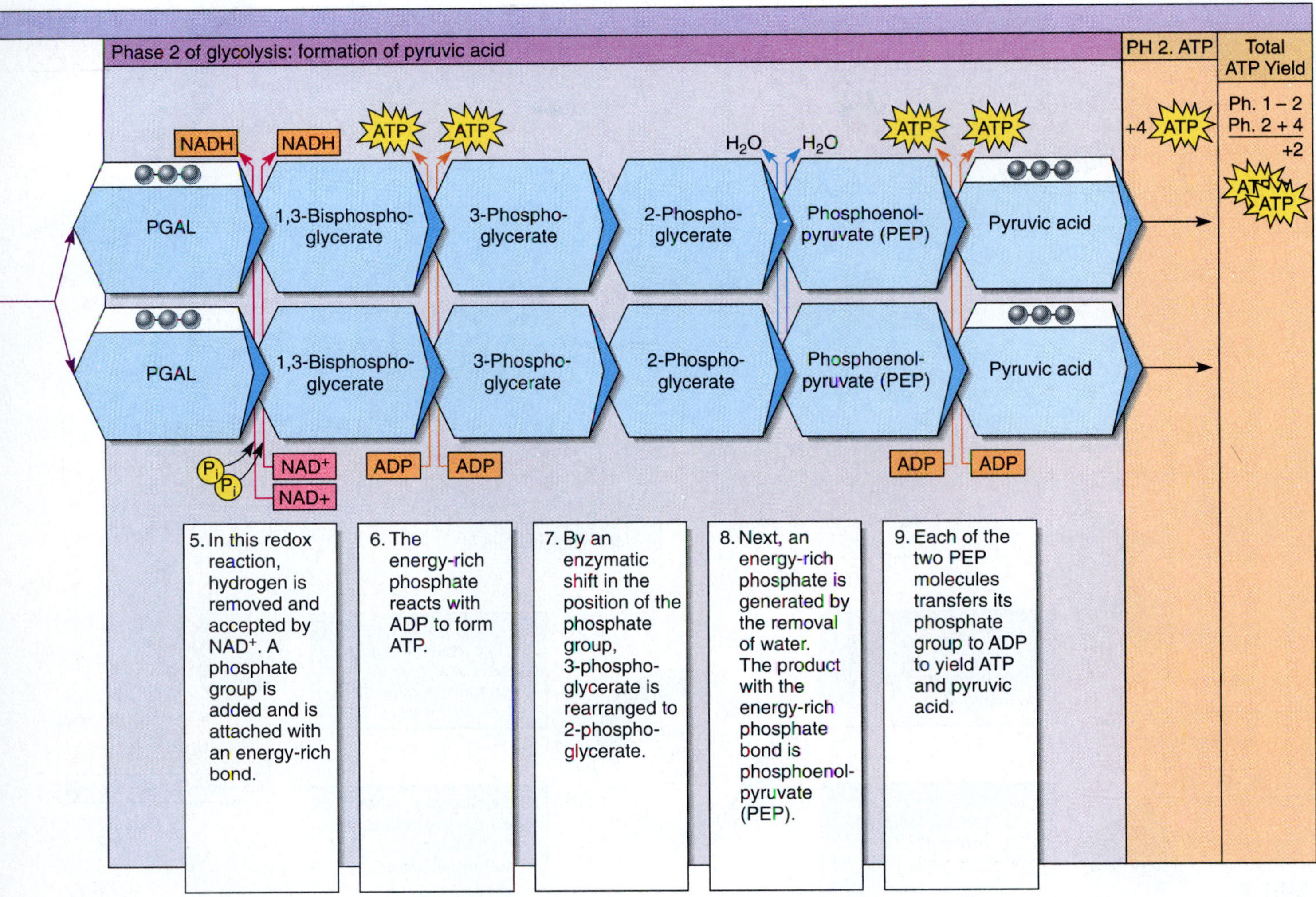

Phase 2 of glycolysis: formation of pyruvic acid
PH 2. ATP
Total ATP Yield
Ph. 1 – 2
Ph. 2 + 4
+2
+4 ATP
NADH
NADH
ATP
ATP
H_2O
H_2O
ATP
ATP
PGAL
1,3-Bisphospho-glycerate
3-Phospho-glycerate
2-Phospho-glycerate
Phosphoenol-pyruvate (PEP)
Pyruvic acid
PGAL
1,3-Bisphospho-glycerate
3-Phospho-glycerate
2-Phospho-glycerate
Phosphoenol-pyruvate (PEP)
Pyruvic acid
P_i
P_i
NAD^+
NAD+
ADP
ADP
ADP
ADP
5. In this redox reaction, hydrogen is removed and accepted by NAD^+. A phosphate group is added and is attached with an energy-rich bond.
6. The energy-rich phosphate reacts with ADP to form ATP.
7. By an enzymatic shift in the position of the phosphate group, 3-phospho-glycerate is rearranged to 2-phospho-glycerate.
8. Next, an energy-rich phosphate is generated by the removal of water. The product with the energy-rich phosphate bond is phosphoenol-pyruvate (PEP).
9. Each of the two PEP molecules transfers its phosphate group to ADP to yield ATP and pyruvic acid.

through a series of chemical transformations losing first one, then another carboxyl (—COOH) group as CO_2. The two CO_2 produced thus account for the two carbon atoms that entered the citric acid cycle as acetyl CoA. Eventually, the original starting molecule, oxaloacetic acid, is reformed; hence the reactions are part of a *cycle*.

Most of the energy made available by the oxidative steps of the cycle is transferred as energy-rich electrons to NAD^+. For each acetyl group that enters the citric acid cycle, three molecules of NAD^+ are reduced to NADH. In addition, electrons are also transferred to the electron acceptor FAD, forming one molecule of $FADH_2$ for each acetyl group entering the cycle.

Because two acetyl CoA molecules are produced from each glucose molecule, the citric acid cycle must turn twice to process each glucose. At the end of a complete cycle, a four-carbon oxaloacetic acid is all that is left, and the cycle is ready for another turn.

Only one molecule of ATP is produced directly with each turn of the citric acid cycle. Thus, at this point in aerobic respiration, the energy of one glucose molecule has resulted in the formation of only four ATPs (two ATPs from glycolysis and two ATPs from two turns of the citric acid cycle). To maintain their highly ordered state, most cells need to expend much more energy than these four ATPs can provide. How, then, is most of the ATP produced?

Most of the ATP Is Produced by the Electron Transport System, which Is Coupled to Chemiosmosis

Now we consider the fate of all the hydrogens removed from the fuel molecule during glycolysis, acetyl CoA formation, and the citric acid cycle. Recall that these hydrogens (and the energy of their electrons) were transferred to the electron acceptor molecules NAD^+ and FAD, forming NADH and $FADH_2$. What becomes of these hydrogens?

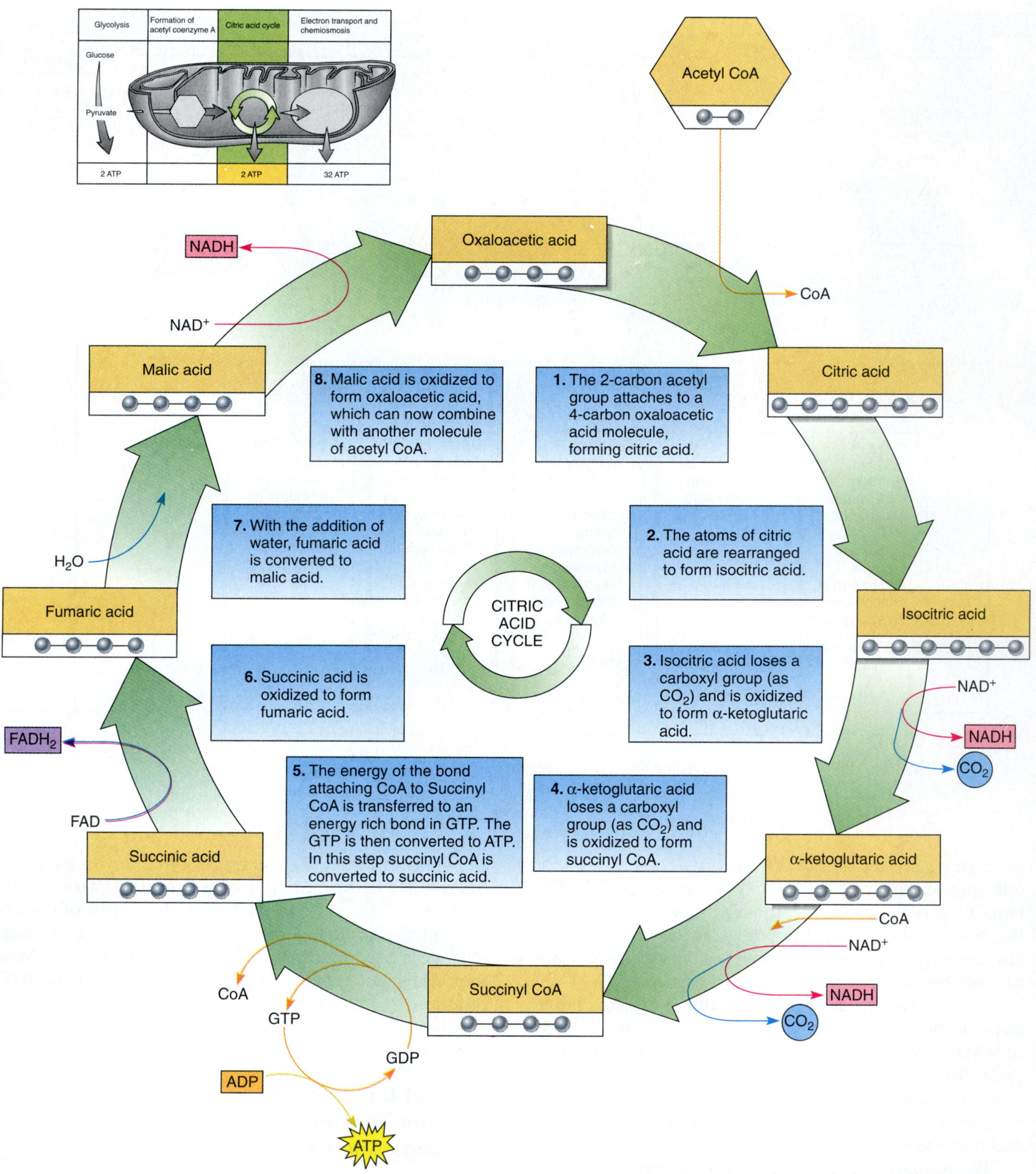
Glycolysis
Formation of acetyl coenzyme A
Citric acid cycle
Electron transport and chemiosmosis
Glucose
Pyruvate
2 ATP
2 ATP
32 ATP
Acetyl CoA
CoA
Oxaloacetic acid
NADH
NAD+
Malic acid
8. Malic acid is oxidized to form oxaloacetic acid, which can now combine with another molecule of acetyl CoA.
1. The 2-carbon acetyl group attaches to a 4-carbon oxaloacetic acid molecule, forming citric acid.
Citric acid
7. With the addition of water, fumaric acid is converted to malic acid.
2. The atoms of citric acid are rearranged to form isocitric acid.
H_2O
Fumaric acid
CITRIC ACID CYCLE
Isocitric acid
6. Succinic acid is oxidized to form fumaric acid.
3. Isocitric acid loses a carboxyl group (as CO_2) and is oxidized to form α-ketoglutaric acid.
NAD+
NADH
CO_2
$FADH_2$
FAD
5. The energy of the bond attaching CoA to Succinyl CoA is transferred to an energy rich bond in GTP. The GTP is then converted to ATP. In this step succinyl CoA is converted to succinic acid.
4. α-ketoglutaric acid loses a carboxyl group (as CO_2) and is oxidized to form succinyl CoA.
Succinic acid
α-ketoglutaric acid
CoA
NAD+
NADH
CO_2
Succinyl CoA
CoA
GTP
GDP
ADP
ATP

FIGURE 8–5 The citric acid cycle. A two-carbon acetyl group combines with four-carbon oxaloacetic acid to form six-carbon citric acid. Two molecules of carbon dioxide are removed, and the four-carbon oxaloacetic acid is ultimately regenerated to begin the cycle again. (The two CO_2 account for the two carbons that entered the cycle as part of one acetyl CoA molecule.) Each turn of the citric acid cycle produces one ATP, three NADH, and one $FADH_2$.

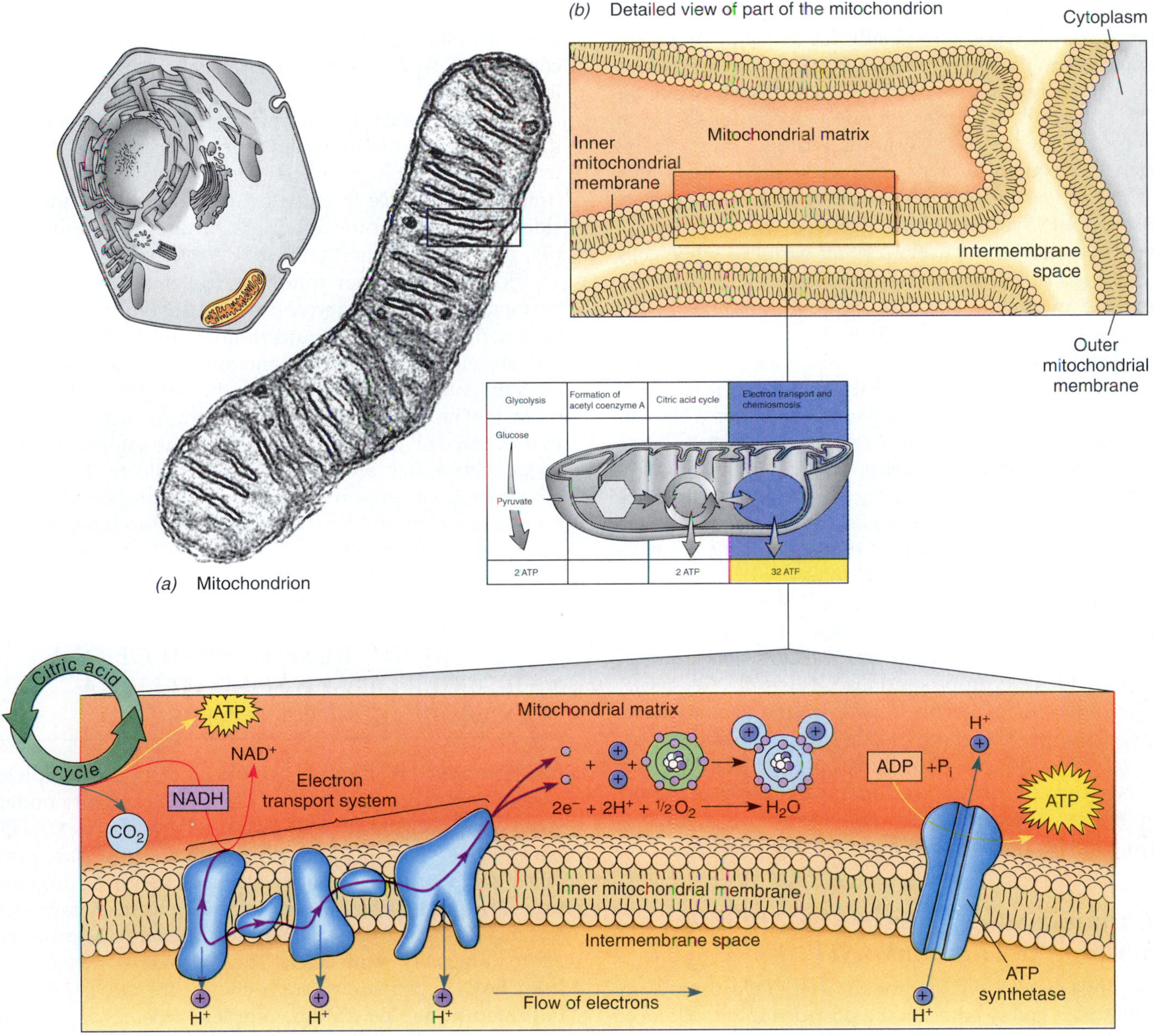

FIGURE 8–6 Electron transport. (***a***) Electron micrograph of a mitochondrion showing its structure. (***b***) Close up of a portion of the mitochondrion, showing the intermembrane space between the inner and outer mitochondrial membranes. (***c***) The electron transport system is located in the inner mitochondrial membrane. As electrons are passed from one electron acceptor to another, protons (H^+) are pumped across the inner mitochondrial membrane from the matrix to the intermembrane space. The inner mitochondrial membrane is impermeable to protons, which can flow back into the matrix only through special channels within the enzyme ATP synthetase. The energy released as the protons move down the energy gradient is used to synthesize ATP. (*a*, D. W. Fawcett)

The Electron Transport Chain Accepts Hydrogens or Their Electrons from the Previous Three Stages

The electron transport system is a chain of electron acceptors embedded in the inner membrane of the mitochondrion (Fig. 8–6). The electrons associated with the hydrogens released from NADH (and $FADH_2$) are passed along the chain of acceptors in a series of redox reactions. Hydrogens are first passed from NADH to **flavin mononucleotide (FMN)**, the first acceptor in the chain. FMN then transfers hydrogens to another acceptor molecule, which passes them on to yet another acceptor.

The electrons entering the electron transport system have a relatively high energy content. As they pass along the chain of electron acceptors, they lose much of their energy. (Some of this energy is used to make ATP, as discussed in the next section.) Finally, the last electron acceptor molecule in the electron transport chain passes the two (now low-energy) electrons on to molecular oxygen. Simultaneously, the electrons reunite with protons (H^+) in the surrounding medium, forming hydrogen, and the hydrogen and oxygen combine chemically, producing water.

Oxygen is thus the final electron acceptor in the electron transport system, which explains why we require oxygen for aerobic respiration. What happens when cells are deprived of oxygen? When no oxygen is available to accept the hydrogen, the last acceptor molecule in the chain is stuck with electrons. As a result, the other acceptor molecules in the chain cannot pass their electrons on. The entire system backs up all the way back to NADH, thus shutting down the Krebs cycle. No further ATPs can be produced by way of the electron transport system. Most cells of complex organisms cannot live long without oxygen because the amount of energy that they can produce in its absence is insufficient to sustain life processes

The passage of each pair of electrons down the electron transport chain from NADH to oxygen is thought to yield enough energy to produce a maximum of three ATPs. The flow of electrons is tightly coupled to ATP synthesis and generally does not occur unless phosphorylation of ADP to yield ATP can proceed also. This prevents waste, for electrons will not usually flow down the electron transport chain unless the energy released can be used to produce ATP. Because the phosphorylation of ADP to form ATP is coupled with the oxidation of electron transport components, ATP synthesis in the electron transport system is referred to as **oxidative phosphorylation.**

In Chemiosmosis, the Energy of a Proton Gradient Is Used to Make ATP

It had long been known that the transfer of electrons from NADH to oxygen resulted in the production of ATP molecules. However, just *how* these ATPs were synthesized remained a mystery until 1961, when Peter Mitchell proposed the **chemiosmotic model,** which states that electron transport and ATP synthesis are coupled by a proton gradient established across the inner mitochondrial membrane. Mitchell was awarded a Nobel Prize for this scientific contribution.

As hydrogens released from NADH and $FADH_2$ are transferred from one to another of the acceptor molecules in the electron transport chain, the protons (H^+) become separated from their electrons. Protons are released into the surrounding medium.

Some of the energy released by the electron transport chain is used to pump the protons across the inner membrane of the mitochondrion from the matrix into the intermembrane space (Fig. 8–6). The proton pumps result in the establishment of a proton gradient (a much greater concentration of protons in the intermembrane space than in the matrix).

The difference in concentration of protons between the matrix and the intermembrane space represents potential energy. This potential energy, which results in part from the difference in electrical charge between the two sides of the membrane and in part from the difference in pH, provides the energy for ATP synthesis.

Because the inner mitochondrial membrane is impermeable to the passage of protons, the protons can flow back to the matrix of the mitochondrion only through special protein channels in the membrane. In this case, the protein channels occur within the enzyme **ATP synthetase**. The protons move down the energy gradient, that is, through ATP synthetase from the intermembrane space where they are highly concentrated to the matrix where they are present in a low concentration. The energy released by the flow of protons is used by ATP synthetase to produce ATP.

THE AEROBIC RESPIRATION OF ONE MOLECULE OF GLUCOSE YIELDS A MAXIMUM OF 36 TO 38 ATPS

Just how much ATP can be produced from the complete breakdown of one molecule of glucose is currently under study. The net gain of ATP from glycolysis is two ATPs per glucose molecule, and two additional ATPs are produced during the citric acid cycle. Each pair of hydrogens from NADH that passes through the electron transport chain is thought to provide sufficient energy to produce up to three ATP molecules. The hydrogens transferred from $FADH_2$ provide enough energy for two ATPs. As indicated in Table 8–2, the complete aerobic respiration of one molecule of glucose is thought to produce a maximum total of 36 to 38 ATPs. Recently, some researchers have challenged this number and have suggested that the actual number of ATPs produced is substantially less.

NUTRIENTS OTHER THAN GLUCOSE CAN BE USED AS ENERGY SOURCES

Many organisms depend on nutrients other than glucose, or in addition to glucose, as sources of energy. Humans and many other animals generally obtain more of their energy by oxidizing fatty acids than by oxidizing glucose. Fatty acids are broken down into two-carbon acetyl

TABLE 8–2
The Energy Yield from Aerobic Respiration of Glucose

1. Net ATP profit from glycolysis			2 ATP*
Also from glycolysis:	2 NADH	⟶	4–6 ATP† (oxidative phosphorylation)
2. 2 pyruvate to 2 acetyl CoA	2 NADH	⟶	6 ATP (oxidative phosphorylation)
3. 2 acetyl CoA through critic acid cycle			2 ATP
	6 NADH	⟶	18 ATP (oxidative phosphorylation)
	2 $FADH_2$	⟶	4 ATP (oxidative phosphorylation)
Total ATP profit			36–38 ATP

* These are the only 2 ATPs that can be generated anaerobically; production of all other ATPs depends on the presence of oxygen.
† Some energy may be expended to transport NADH across the mitochondrial membrane.

groups, which combine with coenzyme A and then enter the citric acid cycle as acetyl CoA (Fig. 8–7).

Amino acids are catabolized by reactions in which the amino group is first removed, in a process called **deamination**. The amino group is converted to urea and excreted in the urine. The carbon chain that remains after deamination is then converted to pyruvic acid, acetyl CoA, or some other metabolic intermediate that can enter the citric acid cycle.

Cells Regulate the Amount of Aerobic Respiration

Aerobic respiration requires a steady input of nutrient fuel molecules and oxygen. Under normal conditions, these materials are adequately provided and do not affect the rate of respiration. Instead, the rate of aerobic respiration is regulated by the amount of ADP and phos-

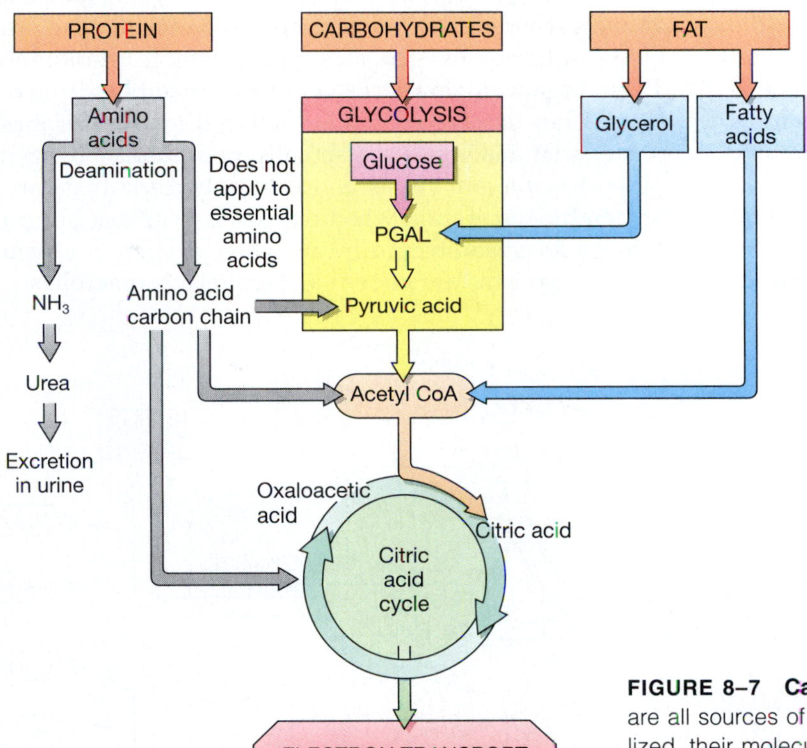

FIGURE 8–7 Catabolism of nutrients. Carbohydrates, proteins, and fats are all sources of energy for the cell. When these compounds are catabolized, their molecular subunits can be converted to metabolic intermediates that enter glycolysis or the citric acid cycle. This diagram is greatly simplified and illustrates only a few of the principal pathways.

phate available. In a resting muscle cell, for example, ATP synthesis continues until all of the ADP has been converted to ATP. Then, when there is no more ADP, oxidative phosphorylation stops. Because electron flow is tightly coupled to phosphorylation, the flow of electrons in the electron transport chain also stops.

As we saw with our wheelchair racer, when an energy-requiring process like muscle contraction occurs, ATP is split to yield ADP and inorganic phosphate plus energy. The ADP just formed can then accept phosphate and energy to become ATP once again. Oxidative phosphorylation resumes and continues until all of the ADP has again been converted to ATP. Because oxidative phosphorylation is tightly coupled to electron flow, the cell possesses a system of control that regulates the rate of ATP production and adjusts it to the rate of energy utilization.

MANY CELLS USE ANAEROBIC PATHWAYS WHEN OXYGEN IS NOT AVAILABLE

Bacteria and some other types of organisms that inhabit waterlogged soil or stagnant ponds where oxygen is absent must engage in anaerobic respiration. In **anaerobic respiration**, energy is released from glucose and other nutrients without oxygen; that is, oxygen is not used as the final electron acceptor. Instead an inorganic compound, such as nitrate (NO_3^-) or sulfate (SO_4^{-2}), serves as the final acceptor of electrons. The end products in anaerobic respiration are carbon dioxide, water, and other inorganic substances.

Fermentation is a kind of anaerobic pathway that also degrades glucose and other organic molecules without oxygen. Unlike anaerobic respiration just described, however, the final acceptor of hydrogen is an *organic* molecule derived from the initial nutrient, rather than an inorganic molecule (Fig. 8–8). Like aerobic respiration, fermentation depends upon the reactions of glycolysis. (Recall that the net gain of two ATPs produced during glycolysis does not require the presence of oxygen.) Two common types of fermentation are alcoholic fermentation and lactic acid fermentation.

Yeasts (unicellular fungi) can carry on **alcoholic fermentation** (Fig. 8–9). First they degrade glucose to pyruvic acid using the process of glycolysis. When deprived of oxygen, yeasts split carbon dioxide off from pyruvic acid, forming a two-carbon compound. Hydrogen from the NADH produced during glycolysis is then transferred to the two-carbon compound, forming **ethyl alcohol**. Alcoholic fermentation is the basis for the production of beer, wine, and other alcoholic beverages. Yeasts are also used in baking to produce the carbon dioxide that causes dough to rise (the alcohol evaporates during baking.)

Certain fungi and bacteria carry on **lactic acid fermentation**. In this pathway, hydrogens removed from glucose as NADH during glycolysis are transferred to pyruvic acid, forming lactic acid. Lactic acid is produced when bacteria sour milk or ferment cabbage to form sauerkraut. Under conditions of insufficient oxygen, human muscle cells can also use lactic acid fermentation (recall the chapter introduction) to get limited amounts of energy.

Most organisms, including the majority of plants and animals, can survive only in an environment that provides sufficient oxygen to support aerobic respiration. These organisms are strict **aerobes**. Anaerobic bacteria that do not require oxygen are referred to as **anaerobes**. Some strict anaerobes are actually poisoned by oxygen. Most versatile are yeasts and certain bacteria that carry on aerobic respiration when oxygen is available but can shift to an anaerobic pathway when oxygen is absent. These organisms are known as **facultative anaerobes**.

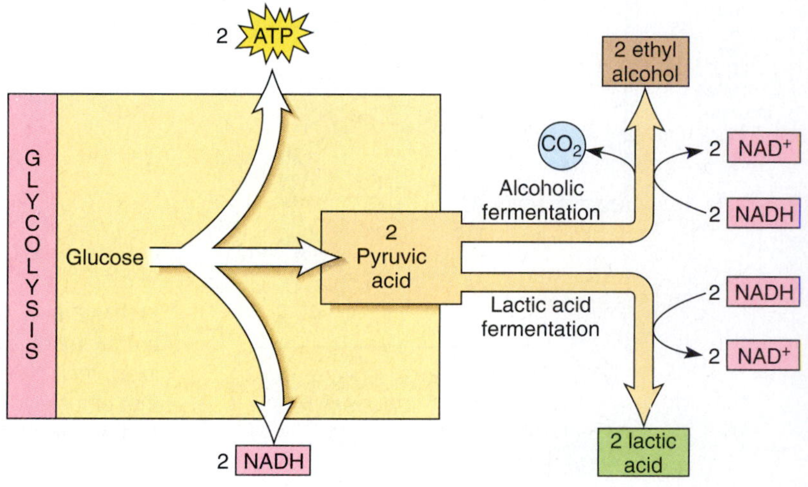

FIGURE 8–8 Fermentation. In fermentation, NADH transfers electrons associated with hydrogen to pyruvic acid, the end product of glycolysis. Thus, pyruvic acid serves as an electron acceptor. In alcoholic fermentation, carbon dioxide is split off, and the two-carbon ethyl alcohol is the end product. In lactic acid fermentation, the final product is the three-carbon compound lactic acid. In both alcoholic and lactic acid fermentation, there is a net gain of only two ATPs; the two NADH molecules produced during glycolysis are used during fermentation.

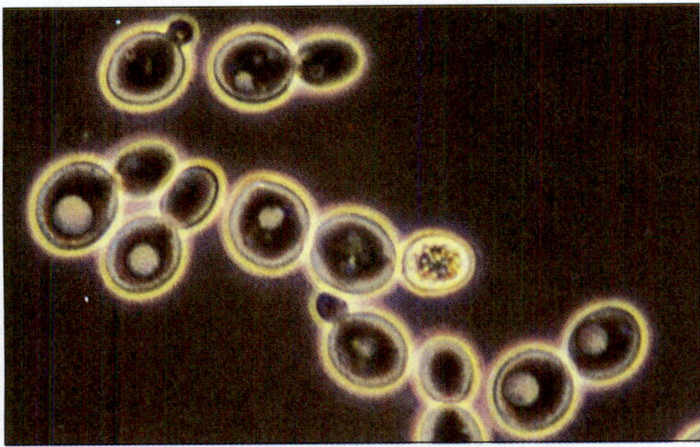

FIGURE 8–9 Baker's yeast. Yeast cells possess mitochondria and carry on aerobic respiration when oxygen is present. In the absence of oxygen, yeasts carry on alcoholic fermentation. (Manfred Kage/Peter Arnold, Inc.)

Fermentation Is Inefficient

Fermentation is very inefficient compared to aerobic respiration because the fuel is only partially oxidized. Alcohol, the end product of fermentation by yeast cells, can be burned and can even be used as automobile fuel. Obviously, it contains a great deal of energy that the yeast cells were unable to extract using anaerobic methods. Lactic acid, a three-carbon compound, contains even more energy than the two-carbon alcohol. In contrast, during aerobic respiration, all available energy is removed because fuel molecules are completely oxidized. A net gain of only two ATPs can be produced from one molecule of glucose by fermentation, compared with a maximum of 36 to 38 ATPs when oxygen is available.

The inefficiency of anaerobic metabolism necessitates a large supply of fuel. By rapidly degrading many fuel molecules, a cell can compensate somewhat for the small amount of energy that can be gained from each. To perform the same amount of work, an anaerobic cell must consume up to 20 times as much glucose or other carbohydrate as a cell metabolizing aerobically. Skeletal muscle cells, which often metabolize anaerobically for short periods, therefore store large quantities of glucose in the form of glycogen.

Concept Connections

Glycolysis ⟷ *Evolution*

Cells are thought to have evolved on planet Earth billions of years ago from assemblages of large biological molecules such as proteins, lipids, and nucleic acids (Chapter 4). These macromolecules remain the constituents of cells today. As the first life forms, ancient cells would have needed energy to survive. Evidence suggests that they obtained the energy they required by fermenting organic compounds such as sugars that were present in the environment. Fermentation is an anaerobic process, and so the first cells were almost certainly anaerobic prokaryotes.

Evidence that glycolysis, which is a major part of fermentation, evolved early in the evolutionary history of cells is found in living organisms today: The glycolytic pathway occurs in *all* eukaryotic cells as well as in many prokaryotic cells. (If glycolysis had not evolved in ancient cells *before* that cell line evolved into the variety of organisms present today, we would not expect to see glycolysis in such a majority of those cells' descendants.)

The enzymes that catalyze the chemical reactions of glycolysis are found universally in cells of many bacteria, all protists, all fungi, all animals, and all plants. The presence of glycolytic enzymes means that all of these organisms have similar genetic information (that is, similar DNA), which instructs their cells to make the enzymes of glycolysis. This also implies that all organisms with similar DNA share a common ancestor (Chapter 17).

Chapter Summary

I. Catabolism is the process of splitting larger molecules into smaller ones.
   A. Cells use three different catabolic pathways to extract energy from nutrients: aerobic respiration, anaerobic respiration, and fermentation.
   B. Aerobic respiration requires molecular oxygen (O_2). Anaerobic respiration and fermentation do not require oxygen.

II. During aerobic respiration, a fuel molecule, such as glucose, is oxidized, forming carbon dioxide and water with the release of energy (a maximum of 36 to 38 ATPs per molecule of glucose).

III. Aerobic respiration is a redox process in which hydrogen is transferred from glucose (which becomes oxidized) to oxygen (which becomes reduced). Oxygen is thus the final electron acceptor molecule in aerobic respiration.

IV. The chemical reactions of aerobic respiration occur in four stages: glycolysis, formation of acetyl CoA, the citric acid cycle, and the electron transport system with its associated chemiosmotic ATP synthesis.
   A. During glycolysis, a molecule of glucose is degraded, forming two molecules of pyruvic acid.
      1. A net gain of two ATP molecules occurs during glycolysis.

2. Four hydrogens are removed from the fuel molecule (as two NADH).

B. The two pyruvic acid molecules each lose a molecule of carbon dioxide, and the remaining acetyl groups combine with coenzyme A, producing acetyl CoA. One NADH is formed as each pyruvic acid is converted to acetyl CoA.

C. Each acetyl CoA enters the citric acid cycle by combining with a four-carbon compound, oxaloacetic acid, to form citric acid, a six-carbon compound.
1. With two turns of the citric acid cycle, the two acetyl CoAs from the original glucose are completely degraded, and two carbon dioxide molecules are released.
2. Hydrogens are transferred to NAD^+ and FAD, forming three NADH and one $FADH_2$ with each turn of the cycle. Only one ATP is produced directly with each turn of the citric acid cycle.

D. The electrons associated with hydrogen released from NADH and $FADH_2$ are transferred from one electron acceptor to another down a chain of acceptor molecules that make up the electron transport system.
1. The final acceptor in the chain is molecular oxygen, which combines with hydrogen to form water.
2. According to the chemiosmotic model, energy liberated in the electron transport chain is used to establish a proton gradient across the inner mitochondrial membrane.
3. Protons flow back through the membrane through a channel within the enzyme ATP synthetase; energy released in this process is used to synthesize ATP.

V. Fatty acids and amino acids can also be used as fuel; they are broken down to metabolic intermediates that can enter the citric acid cycle or some other part of aerobic respiration.

VI. Organisms that inhabit oxygen-poor environments use anaerobic pathways for capturing energy.

A. In anaerobic respiration, fuel molecules are broken down in the absence of oxygen; an inorganic compound serves as the final electron acceptor.

B. Fermentation is an anaerobic pathway in which electrons from NADH are accepted by an organic compound derived from the initial nutrient. There is a net gain of only two ATPs per glucose molecule, as compared to a net gain of 36 to 38 ATPs per glucose molecule by aerobic respiration.
1. Yeasts carry on alcoholic fermentation in which ethyl alcohol and carbon dioxide are the final products.
2. Certain fungi and bacteria, as well as animal muscle cells, carry on lactic acid fermentation, in which hydrogens are added to pyruvic acid, forming lactic acid.

Selected Key Terms

aerobes, p. 162
aerobic respiration, p. 152
alcoholic fermentation, p. 162
anaerobes, p. 162
anaerobic respiration, p. 162
ATP synthetase, p. 160
catabolism, p. 152
chemiosmosis, p. 155
citric acid cycle, p. 154
electron transport, p. 155
facultative anaerobes, p. 162
glycolysis, p. 154
lactic acid fermentation, p. 162
oxidation, p. 152
oxidative phosphorylation, p. 160
phosphorylation, p. 155
redox reactions, p. 152
reduction, p. 152

Post-Test

1. The process of splitting larger molecules into smaller ones is an aspect of metabolism called _______________.

2. An aerobic pathway requires molecular _______________.

3. The reactions of glycolysis take place within the _______________, whereas the oxidation of pyruvic acid, the citric acid cycle, and the electron transport chain take place within the _______________.

4. The hydrogens removed during glycolysis immediately combine with _______________, forming _______________.

5. When oxygen is present, pyruvic acid is converted to the two-carbon compound _______________ _______________.

6. Acetyl CoA reacts with oxaloacetic acid to form _______________ _______________.

7. During the citric acid cycle the fuel molecule is completely oxidized; the products are _______________ _______________, ATP, NADH, and $FADH_2$.

8. As electrons are transferred from one acceptor to another in the electron transport chain, they lose much of their _______________.

9. When protons move down an energy gradient in chemiosmosis, energy is released and used to synthesize _______________.

10. Yeasts and bacteria that can shift to anaerobic respiration or fermentation when oxygen is absent are called facultative _______________.

11. The anaerobic process by which alcohol or lactic acid is produced as a product of glycolysis is referred to as _______________.

12. When deprived of oxygen, yeast cells undergo alcoholic fermentation; the product is _______________ _______________.

13. During strenuous muscle activity, pyruvic acid in muscle cells may accept hydrogen from NADH; this forms ______________ ______________.

14. A net gain of only two ATPs can be produced anaerobically from one molecule of glucose, compared with up to ______________ ATPs that may be produced during aerobic respiration.

15. Anaerobic metabolism is inefficient because the fuel molecule is only partially ______________.

16. In glycolysis, (a) glucose is broken down into pyruvic acid (b) acetyl CoA is degraded into carbon dioxide (c) water is formed when the electrons of hydrogen combine with oxygen (d) pyruvic acid is broken down into acetyl CoA and CO_2.

17. The citric acid cycle (a) consists of the breakdown of glucose into citric acid (b) is the final step in aerobic respiration (c) produces carbon dioxide as acetyl CoA is oxidized (d) requires oxygen, which combines with carbon to form carbon dioxide.

18. When NADH gives up its hydrogens to become NAD^+, the NADH is (a) reduced (b) oxidized.

19. The final electron acceptor in aerobic respiration is (a) O_2 (b) NAD^+ (c) FAD (d) pyruvic acid.

20. According to the chemiosmotic model, (a) energy from the electron transport chain is used to pump electrons across the inner mitochondrial membrane (b) protons accumulate in the intermembrane space between the two mitochondrial membranes (c) ATP is synthesized when electrons flow through the inner mitochondrial membrane (d) ATP synthetase accumulates in the mitochondrial matrix.

Review Questions

1. Justify referring to mitochondria as the "power plants" of the cell. Be specific, using information you have learned in this chapter.
2. Trace the fate of hydrogens removed from the fuel molecule during glycolysis when oxygen is present.
3. What is the specific role of oxygen in the cell? What happens when cells are deprived of oxygen?
4. What aspect of aerobic respiration is explained by the chemiosmotic model?
5. How does a proton gradient contribute to ATP synthesis?
6. Draw a mitochondrion and indicate the locations of (a) enzymes of the electron transport system and (b) the site of the proton gradient that drives ATP production.
7. Calculate how much energy (that is, the number of ATPs) is made available to the cell by the operation of each of the following stages of aerobic respiration: glycolysis, formation of acetyl CoA, the citric acid cycle, and the electron transport system.
8. Trace the fate of hydrogens removed from the fuel molecule when sufficient oxygen is *not* available in muscle cells.
9. Some bacteria and fungi live in oxygen-poor environments. How do they obtain energy?

Thinking Critically

1. Aerobic respiration, which is not 100% efficient, becomes even less efficient in animals during cold winter months. What effect might this decreased efficiency have on runners and wheelchair racers?
2. The poison cyanide works by blocking the electron transport system. Based on what you have learned in this chapter, why is cyanide poisoning fatal?
3. The mechanism of chemiosmosis (generating ATP using the energy of a proton gradient) has been compared to hydroelectric power (generating electricity using the energy of water flowing over a dam). Explain the analogy.
4. Louis Pasteur, an important biologist in the nineteenth century, noted that yeast cells that are added to a closed container of grape juice break down the sugar very slowly as long as oxygen is present in the container. Once the oxygen is used up, however, the yeast cells consume the remainder of the grape sugar very quickly, producing ethyl alcohol in the process. Explain this observation based on what you know about the relative efficiencies of aerobic respiration versus alcoholic fermentation.

Recommended Readings

Solomon, E.P., L.R. Berg, D.W. Martin, and C. Villee, *Biology*, 3rd ed., Philadelphia, Saunders College Publishing, 1993. Chapter 7 presents a slightly more detailed discussion of cellular respiration than is presented in this text.

Solomon, E.P., R.R. Schmidt, and P.J. Adragna, *Human Anatomy & Physiology*, 2nd ed., Philadelphia, Saunders College Publishing, 1990. This text contains a chapter (Chapter 32) on the fundamentals of exercise physiology—discussing what exercise does to the muscles, respiratory system, circulatory system, and other parts of the body.

Tropical rain forest of the upland Amazon region in Brazil. *The trees in the forest are photosynthetic organisms that convert radiant energy to the chemical energy of organic molecules. (Peter Arnold)*

THE HUMAN-RAIN FOREST CONNECTION

The largest tropical rain forest in the world is found in the Amazon basin of Brazil. This lush, densely wooded forest, which stretches for hundreds of kilometers, consists primarily of broadleaf evergreen trees, vines, and small plants such as bromeliads and orchids growing high in the tree branches. The leaves and branches at the tops of the trees form a continuous canopy overhead, and there is little undergrowth at ground level.

Although trees and other plants of the tropical rain forest appear quite passive to the casual observer, a remarkable process is occurring in every green cell of every plant. These trees and other plants are converting the sun's energy into a usable chemical form via **photosynthesis.** Carbohydrates are formed from the simple raw materials of water and carbon dioxide.

The carbohydrates synthesized by photosynthesis are important to the tree for two reasons. First, they are the plant's energy source. That is, carbohydrates can be broken down by cellular respiration to release energy needed for the plant's life processes (Chapter 8). Second, these carbohydrates are modified to form many different types of biologically important molecules. The proteins, nucleic acids, and lipids essential to the plant are formed from carbohydrate building blocks produced by photosynthesis.

The great coastal cities of Brazil, such as São Paulo and Rio de Janeiro, are located far from the steaming Amazon rain forest. These large metropolitan areas are filled with millions of people concerned with daily enterprises far removed from the Amazon basin. Yet the existence of all the humans in these cities is as dependent on photosynthesis as are the trees in the rain forest.

Humans obtain *all* of their food either directly or indirectly from plants. The oxygen we breathe is continually replenished by photosynthesis. And the coal, oil, and natural gas that we use to power our technological society represent the products of photosynthesis that occurred millions of years ago. With few exceptions, all life depends on the energy-transforming abilities of photosynthetic organisms.

Although humans depend on photosynthesis, our activities often harm the very organisms that photosynthesize. For example, the world's tropical rain forests are vanishing at an alarming rate—cut for timber or burned to make pasture or agricultural land. By the early 1990s, more than half of the world's tropical rain forests had been destroyed. What remains, about 6 million square kilometers (2.3 million mi^2), is less than the size of the United States. And each year another 169,000 square kilometers (65,000 mi^2), an area larger than the size of Washington state, is cleared. In the early 1990s we were destroying tropical forests at the rate of 0.54 hectare (1.3 acres) per second!

The importance of the world's tropical rain forests as the repositories of many of the world's species is considered in Chapter 20. Like all forests, tropical rain forests also provide such important environmental services as forming and holding the soil, cleansing the water and air, and providing shelter and food for countless animals and humans (Chapter 49). When a forest is intact, it regulates surface water and thereby controls floods and droughts. In addition, tropical rain forests have a profound effect on the global carbon cycle (Chapter 47), because much of the world's photosynthesis occurs there.

Photosynthesis is clearly an important biological process, whether it occurs in tropical rain forests or wheat fields. In this chapter we examine photosynthesis in some detail.

Capturing Energy: Photosynthesis

9

Learning Objectives

After you have studied this chapter you should be able to:

1. Describe how photosynthesis is important, not only to plants but to the entire web of life on planet Earth.
2. Write a summary reaction for photosynthesis, explaining the origin and fate of each substance involved.
3. Describe the internal structure of a chloroplast.
4. Describe the physical properties of light.
5. Summarize the events of the light-dependent reactions of photosynthesis, including the role of light in the activation of chlorophyll.
6. Distinguish between noncyclic and cyclic photophosphorylation.
7. Describe how proton gradients allow the formation of ATP according to the chemiosmotic model.
8. Summarize the events of the light-independent reactions of photosynthesis.
9. Discuss how the C_4 pathway increases the effectiveness of the Calvin cycle in certain types of plants.
10. Distinguish between C_3, C_4, and CAM pathways for fixing carbon.

Key Concepts

- ☐ Photosynthesis is the most important biological process in the biosphere. Almost all living organisms are directly or indirectly dependent on photosynthesis for energy.
- ☐ Only photosynthetic organisms can convert radiant energy into the chemical energy of organic compounds. In photosynthesis, plants, algae, and cyanobacteria use CO_2 and H_2O to make carbohydrates.
- ☐ Photosynthesis occurs by means of two sets of reactions, the light-dependent and the light-independent reactions. In the light-dependent reactions, radiant energy is used to synthesize ATP and NADPH. During the light-independent reactions, organic compounds are synthesized from carbon dioxide using the energy of ATP and NADPH.

ALMOST ALL LIVING ORGANISMS DEPEND ON PHOTOSYNTHESIS, EITHER DIRECTLY OR INDIRECTLY

Plants, algae, and cyanobacteria are producers (organisms that manufacture complex organic molecules from simple inorganic substances) that obtain their energy directly from the sun (Fig. 9–1). Each year these remarkable organisms collectively produce more than 200 billion tons of organic material from carbon dioxide. The chemical energy stored in this material fuels the chemical reactions (aerobic respiration, for example) that sustain life.

Consumers (organisms that cannot synthesize their own food from inorganic materials) obtain their energy by consuming producers or other consumers that feed on the producers. Consumers therefore obtain their energy from photosynthesis indirectly.

Although almost all life is ultimately dependent upon photosynthesis, there are a few exceptions. A few types of bacteria, for example, metabolize inorganic materials, such as sulfur and iron, and do not depend upon photo-

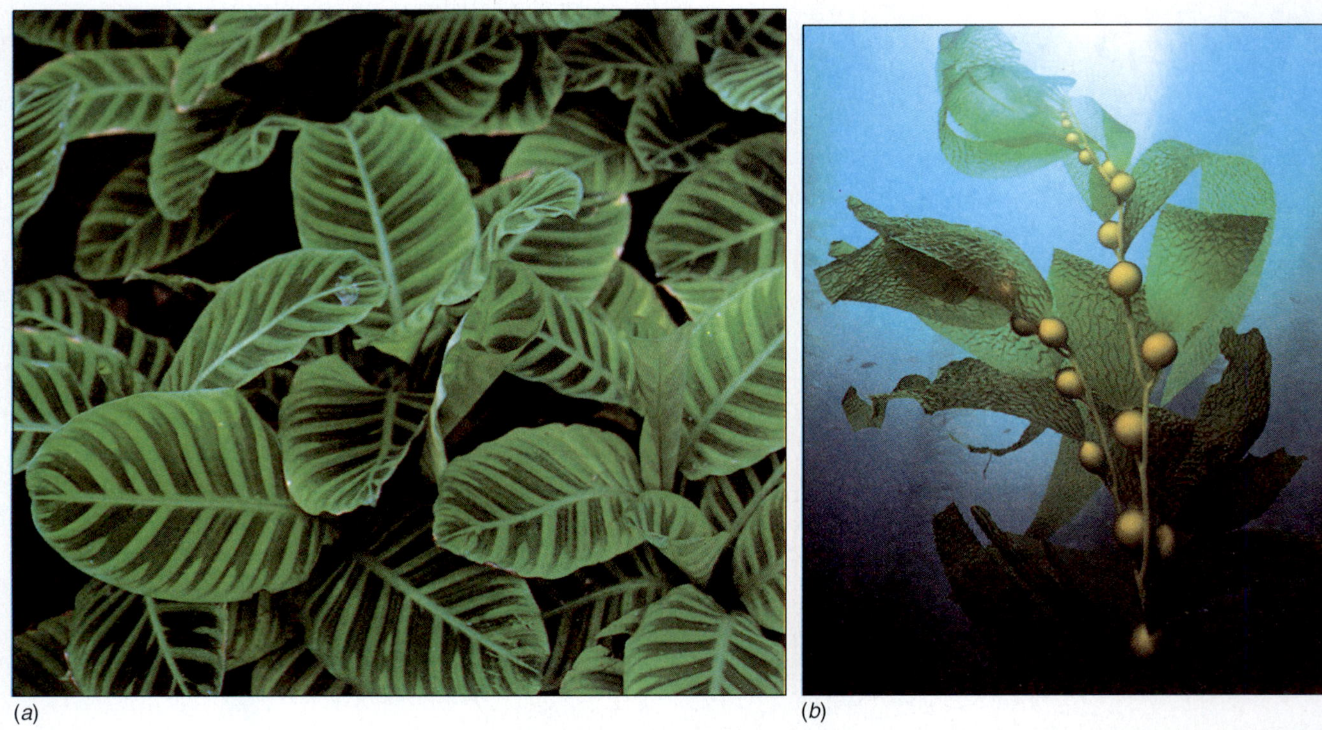

(*a*) (*b*)

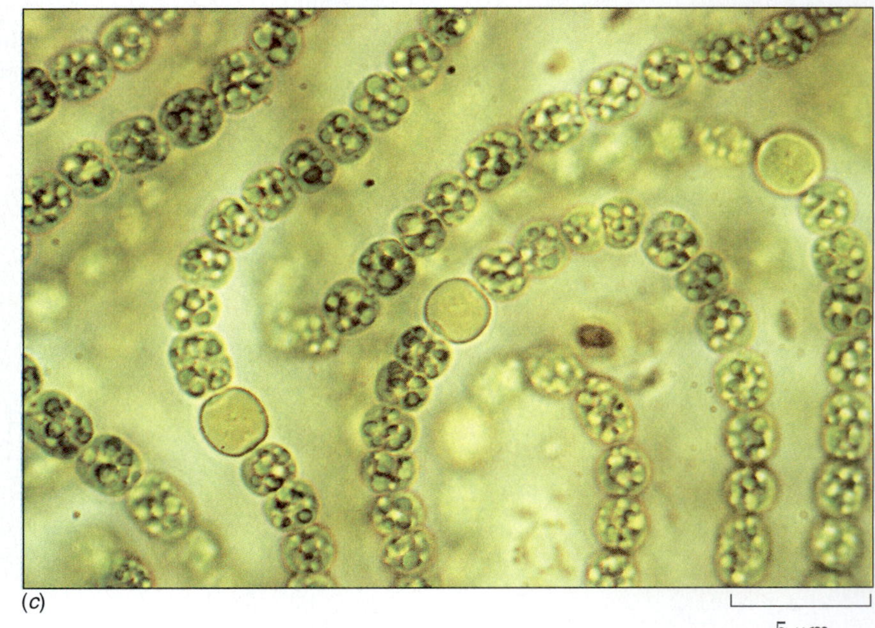

(*c*)

FIGURE 9–1 Photosynthetic organisms. (***a***) Plants, such as the prayer plant (*Maranta* sp), (***b***) algae, such as *Macrocystis integrifolia,* and (***c***) cyanobacteria, such as *Nostoc* sp, are organisms that obtain energy from light by the process of photosynthesis. Plants are primarily terrestrial, whereas algae, which range in size from microscopic single cells to large seaweeds, are primarily aquatic. Cyanobacteria are prokaryotic organisms that photosynthesize like plants and algae. In addition, there are some photosynthetic bacteria that trap the light's energy in a different way than the organisms shown. (*a,* James L. Castner; *b,* Flip Nicklin/Minden Pictures; *c,* Visuals Unlimited/R. Calentine)

Concept Connections

Photosynthesis ⇄ *Aerovic Respiration*

Photosynthesis and aerobic respiration (Chapter 8) are important biological processes that share a number of features. Both are connected with the energy requirements of living things; both are essential for life. Because it is easy to confuse the two processes, let us compare them.

1. **Raw materials.** The starting materials for photosynthesis are CO_2 and H_2O. For aerobic respiration, they are $C_6H_{12}O_6$ and O_2.
2. **End products.** Photosynthesis produces $C_6H_{12}O_6$ and O_2. Aerobic respiration produces CO_2 and H_2O.
3. **Which cells have these processes.** Photosynthesis occurs only in cells that contain chlorophyll. Aerobic respiration (or some other energy-releasing pathway) occurs in *every* actively metabolizing cell of *every* organism.
4. **Organelles involved.** The chloroplast is the site of photosynthesis. Aerobic respiration occurs in the cytosol (glycolysis) and in the mitochondrion.
5. **Pathways of energy.** In photosynthesis, the energy of light passes by way of chlorophyll to NADPH/ATP, and from NADPH/ATP to carbohydrate molecules such as glucose. In respiration, the energy in fuel molecules such as glucose passes to NADH/ATP, and from NADH/ATP to energy for work in the cell.

synthesis for their energy requirements. In addition, certain animals that live in thermal vents in deep sea trenches do not rely on photosynthesis because they obtain their nutrients from the sulfur-metabolizing bacteria present in that specialized environment (Chapter 46).

FIGURE 9–2 The electromagnetic spectrum. Visible light is only a portion of the electromagnetic spectrum and consists of a mixture of wavelengths (from approximately 380 to 760 nm). A prism (shown) sorts light into its component colors by bending light of different wavelengths by different degrees.Photosynthesis uses the energy of visible light to synthesize organic compounds.

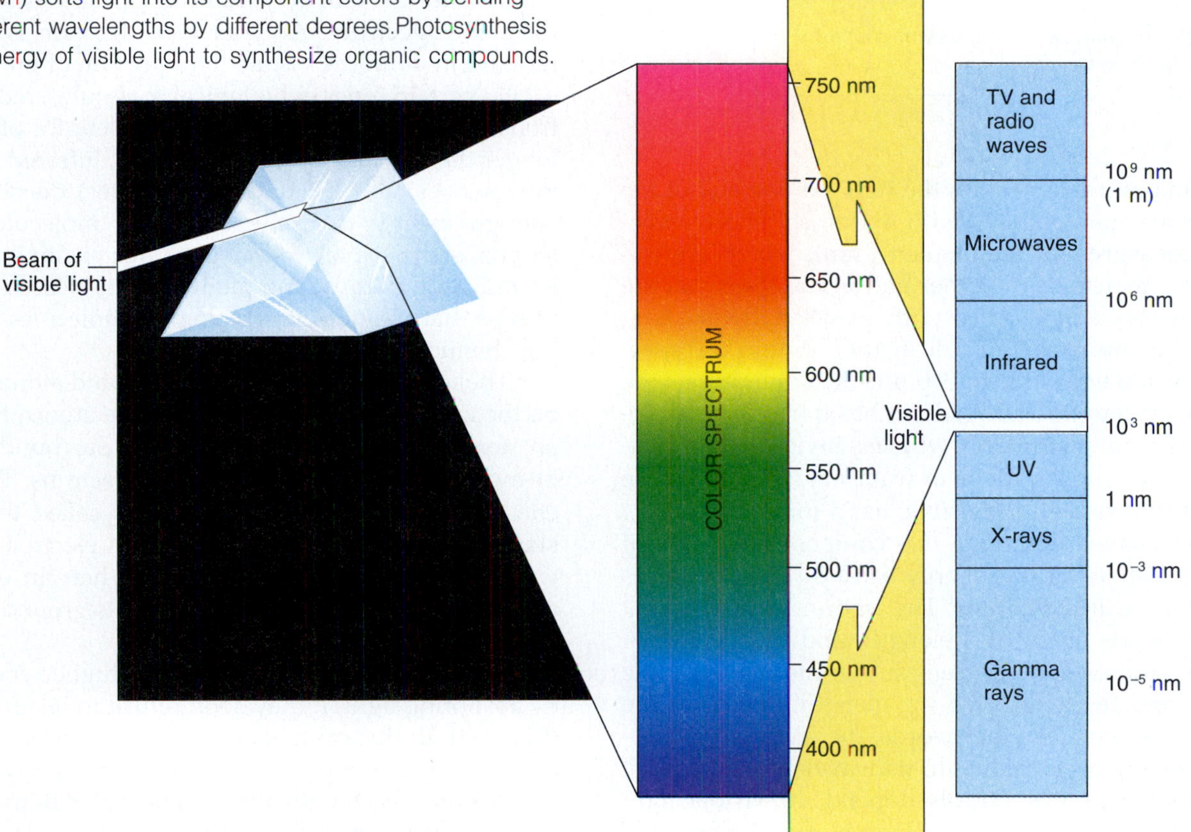

Light exhibits properties of both waves and particles

Light is a very small portion of a vast, continuous spectrum of radiation called the electromagnetic spectrum (Fig. 9–2). All radiations in this spectrum behave as though they travel in waves. A **wavelength** is the dis-

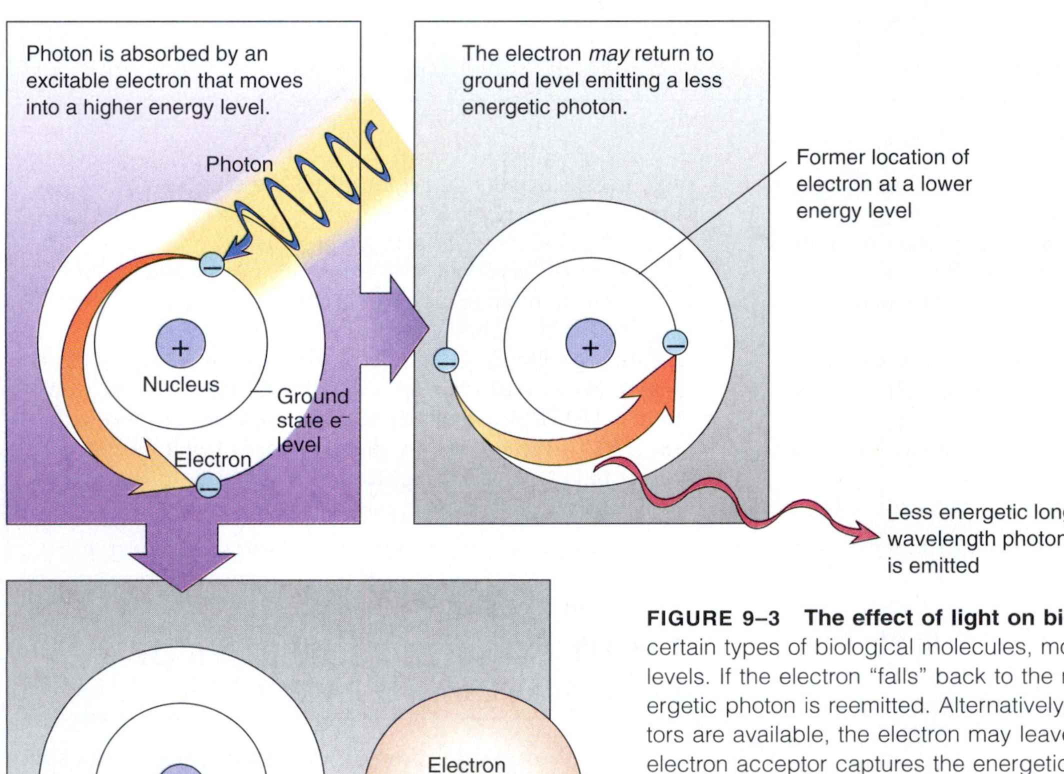
Photon is absorbed by an excitable electron that moves into a higher energy level.
Photon
Nucleus
Ground state e⁻ level
Electron
The electron *may* return to ground level emitting a less energetic photon.
Former location of electron at a lower energy level
Less energetic long wavelength photon is emitted
Electron acceptor molecule
The electron *may* be accepted by an electron acceptor molecule.

FIGURE 9–3 The effect of light on biological molecules. Light excites certain types of biological molecules, moving electrons into higher energy levels. If the electron "falls" back to the next lower energy level, a less energetic photon is reemitted. Alternatively, if the appropriate electron acceptors are available, the electron may leave the atom. In photosynthesis, an electron acceptor captures the energetic electron and passes it along a chain of acceptors.

tance from one wave peak to the next. At one end of the spectrum are gamma rays with extremely short wavelengths, measured in nanometers (nm), which are billionths of a meter. At the other end of the spectrum are low-frequency radio waves, with wavelengths so long that they are measured in full meters. Within the spectrum of visible light (380 to 760 nm), violet light has the shortest wavelength, and red light has the longest wavelength. Ultraviolet radiation, which is invisible to the human eye, has a shorter range of wavelengths than visible light, and infrared, also invisible, has a longer range.

Light behaves as though it is composed not only of waves but also of discrete energy packets. These particles of energy are called **photons**. The amount of energy in a photon depends on the wavelength (and thus, color) of light. The shorter the wavelength, the more energy the light has, and the longer the wavelength, the lower the energy per photon. In other words, the energy of a photon is inversely proportional to its wavelength.

Why does photosynthesis depend on visible light rather than on some other wavelength of radiation? One reason may be that most of the radiation reaching our planet from the sun is within this portion of the electromagnetic spectrum. Another consideration is that only radiation within the visible light portion of the spectrum excites certain types of biological molecules, moving electrons into higher energy levels. Wavelengths of radiation longer than visible light (for example, infrared radiation, microwaves, and TV and radio waves) do not possess enough energy to excite biological molecules. Wavelengths shorter than visible light (for example, ultraviolet radiation, x-rays, and gamma rays) possess so much energy that they disrupt biological molecules by breaking chemical bonds.

The interaction between photons and atoms depends on the arrangement of electrons in the atoms. Recall that an atom consists of an atomic nucleus surrounded by one or more energy levels occupied by electrons. The lowest energy state an electron possesses is called the **ground state**, but energy can be added to an electron so that it will attain a higher energy level. When an electron is raised to a higher energy level than its ground state, the electron is said to be **excited**.

When an electron is raised to a higher energy level by absorbing light, it may soon return to its ground state (Fig. 9–3). In this case, energy is usually released as heat or as light of a longer wavelength (that is, it gives off several photons, each with a lower energy). Alternatively, an excited electron may leave the atom. In this instance, the electron may be accepted by an electron acceptor molecule, which is reduced in the process (Chapter 7). This is what occurs in photosynthesis, resulting in **redox reactions,** which are reactions in which *red*uction and *ox*idation occur.

CHLOROPHYLL IS A PIGMENT THAT ABSORBS LIGHT

Chlorophyll is the green pigment responsible for capturing light, the first step in the conversion of light energy to chemical energy in photosynthesis. A **pigment** may be defined as any substance that absorbs light. Pigments, which are colored, do not absorb different wavelengths (that is, different colors) of light in the same amounts.

Have you ever wondered why most plants are green? The reason is that their leaves reflect most of the green light that strikes them (Fig. 9–4). If green light is reflected, most of it is not being absorbed or used. Chlorophyll absorbs (and uses) light primarily in the violet, blue, and red regions of the visible spectrum rather than in the green region (Fig. 9–5). Hence, chlorophyll and the plant cells that contain chlorophyll appear green in color.

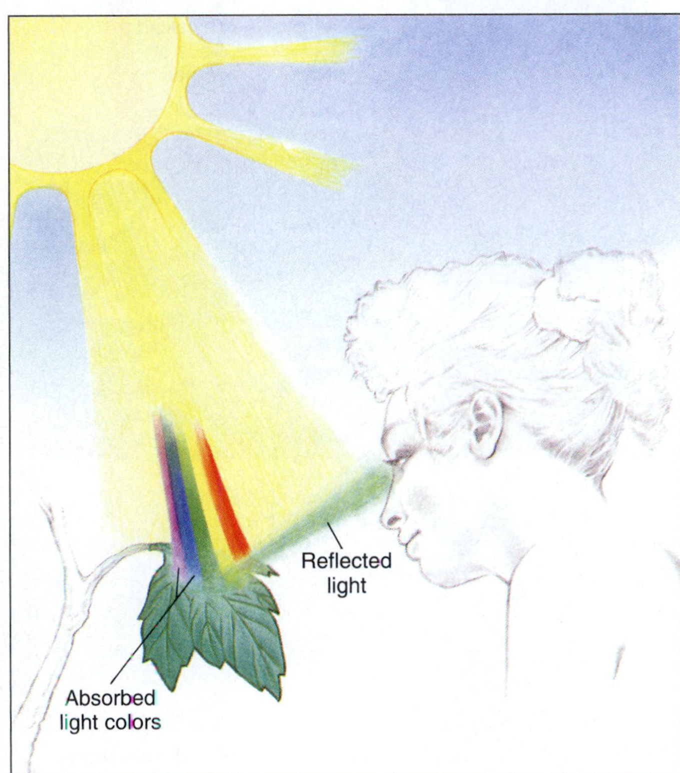

FIGURE 9–4 Why leaves are green. Leaves appear green because when white light strikes a leaf, most of the colors comprising the light are absorbed by chlorophyll. However, chlorophyll does not absorb green light strongly; hence, green light is reflected off the leaf to your eye.

Actually, there are several kinds of chlorophyll. The most common and most important are chlorophyll *a* and chlorophyll *b*. Plants also have accessory photosynthetic pigments, such as the yellow/orange **carotenoids,** that absorb different wavelengths of light besides chlorophyll. (The large quantity of chlorophyll in most leaves usually masks the presence of carotenoids in spring and summer; in autumn, when the chlorophyll breaks down, other pigments, including carotenoids, become visible.)

Chlorophyll and other photosynthetic pigments are located within the membranes of **thylakoids,** tiny membranous sacs located in cells capable of carrying on photosynthesis. In photosynthetic prokaryotes (cyanobacteria), thylakoids often occur as extensions of the plasma membrane and may be arranged around the outer edge of the cell. In photosynthetic eukaryotes (plants and algae), thylakoids are found within **chloroplasts** (Fig. 9–6; also see Chapter 5). The chloroplast is enveloped by a double membrane system and has an interior packed with stacks of thylakoids. These stacks are referred to as **grana** (singular, *granum*). Each granum looks something like a stack of coins, and each "coin" is a thylakoid (Fig. 9–6e). Some thylakoid membranes extend from one granum to another. The interior of the chloroplast surrounding the thylakoids is called the **stroma**.

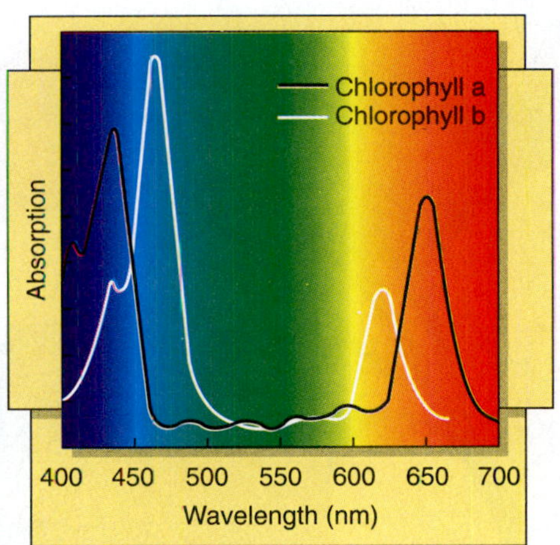

FIGURE 9–5 The absorption spectra of chlorophylls *a* and *b*. These results were obtained by exposing the two types of chlorophyll to different wavelengths of light and measuring how much light was absorbed by chlorophyll at each wavelength. Both types of chlorophyll absorb blue and red light strongly and do not absorb green light very strongly. Note that chlorophyll *a* absorbs blue light more strongly near the violet range, whereas chlorophyll *b* absorbs blue light more strongly near the green region; also, chlorophyll *b* absorbs red light more strongly near the orange region than chlorophyll *a*. Slight differences in the molecular structures of chlorophylls *a* and *b* account for the differences in their absorption spectra.

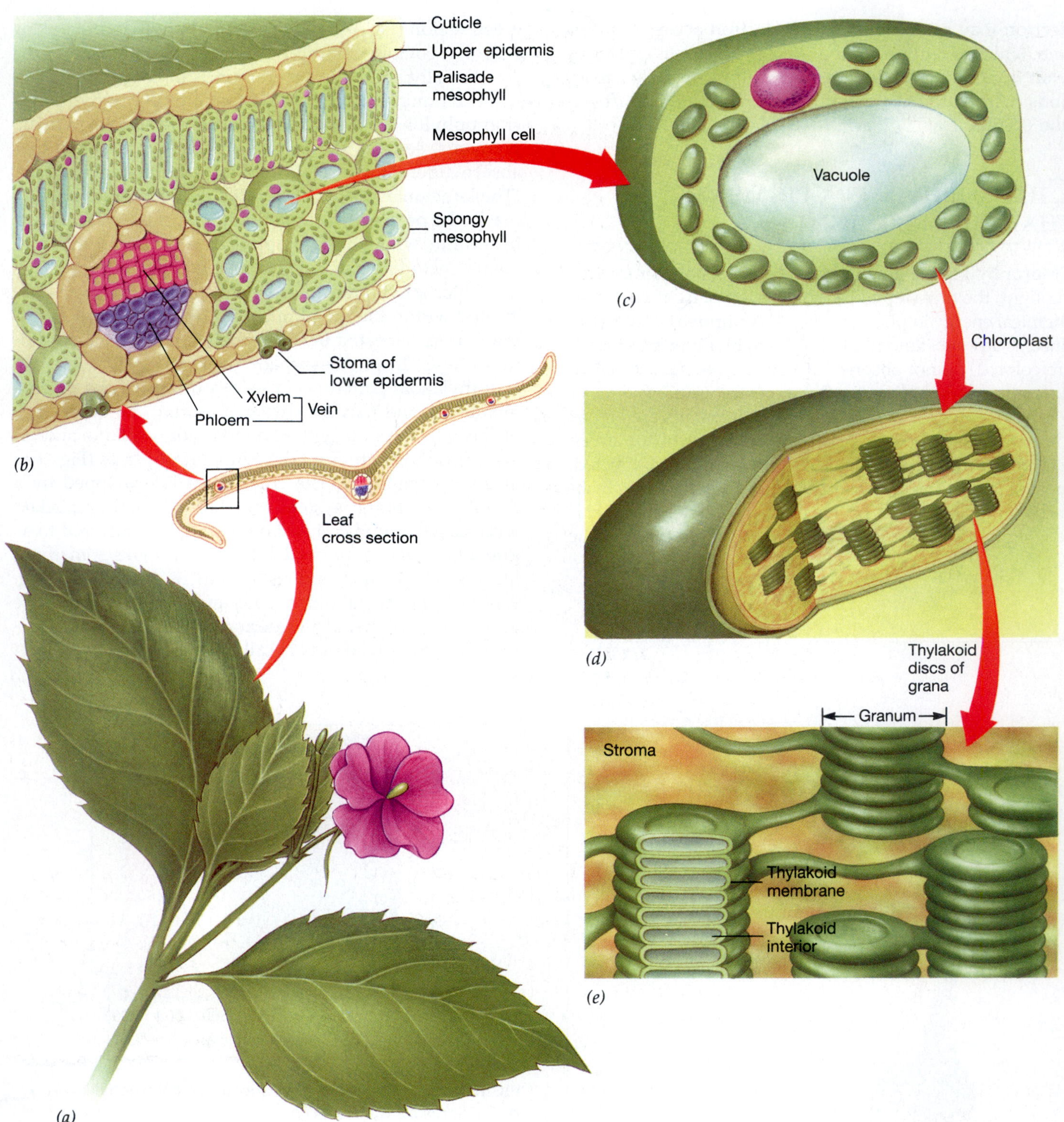

FIGURE 9–6 Photosynthesis from different perspectives. (***a***) Photosynthesis occurs in the green tissues of the plant. (***b***) A cross section of a leaf reveals a structure remarkably adapted for photosynthesis. The middle portion of the leaf, the mesophyll, is the photosynthetic tissue. Carbon dioxide enters the leaf through stomata, and water is carried to the mesophyll in veins. (***c***) A typical mesophyll cell contains numerous chloroplasts. (***d***) Each chloroplast is surrounded by a double membrane. Within the chloroplast, membranous thylakoids are stacked to form grana. The fluid-filled matrix surrounding the grana is the stroma. (***e***) A close-up of the interior of the chloroplast. Chlorophyll is located in the thylakoid membranes. The thylakoids are involved in the light-dependent reactions of photosynthesis, whereas the light-independent reactions of photosynthesis take place in the stroma.

PLANTS USE LIGHT ENERGY TO MAKE CARBOHYDRATES

The principal raw materials for photosynthesis are water and carbon dioxide. The energy that chlorophyll molecules absorb from sunlight is expended in splitting water, which releases the oxygen and hydrogen (Fig. 9–7). The hydrogen joins with carbon dioxide to form the sugar glucose, a six-carbon carbohydrate. Although photosynthesis is a complex process composed of many steps, it may be summarized as follows:

$$\underset{\text{Carbon dioxide}}{6CO_2} + \underset{\text{Water}}{12H_2O} \xrightarrow{\text{Light, chlorophyll}} \underset{\text{Glucose}}{C_6H_{12}O_6} + \underset{\text{Oxygen}}{6O_2} + \underset{\text{Water}}{6H_2O}$$

This equation describes what happens during photosynthesis but not *how* it happens. The reactions of photosynthesis occur in two stages: the light-dependent reactions and the light-independent reactions (Fig. 9–8).

The Light-dependent Reactions Produce ATP and NADPH

The light-dependent reactions can take place only in the presence of light. During this phase of photosynthesis, several important events occur.

1. Chlorophyll absorbs light energy, which triggers a flow of excited, or energized, electrons from the chlorophyll molecule.
2. Some of the energy of the energized electrons is transformed to chemical energy and used to phosphorylate adenosine diphosphate (ADP), forming **adenosine triphosphate (ATP)**.
3. Some of the light energy trapped by the cholorphyll is used to split water, with oxygen from the water being released as molecular oxygen (O_2). Some of the oxygen may be used by the plant for aerobic respiration, but most of it is released into the atmosphere.

FIGURE 9–7 Oxygen production and photosynthesis. On sunny days the oxygen released by aquatic plants is sometimes visible as bubbles in the water. This plant (*Elodea*) is actively carrying on photosynthesis, as evidenced by the oxygen bubbles. (E.R. Degginger)

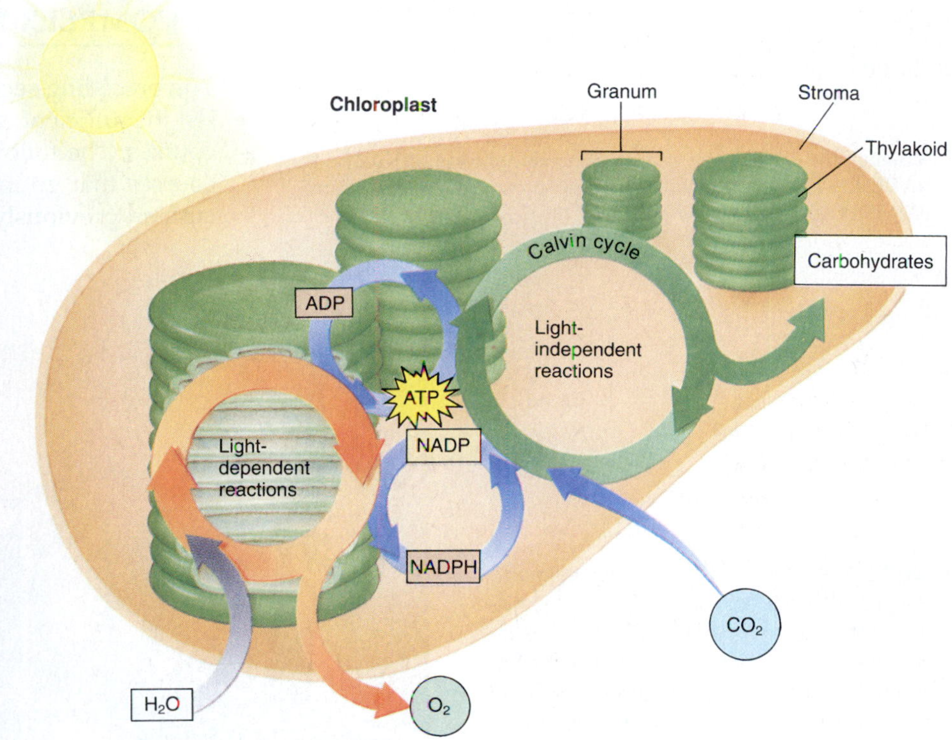

FIGURE 9–8 An overview of photosynthesis. Light energy is used to form ATP and NADPH during the light-dependent reactions of photosynthesis. These reactions, which occur in the thylakoid membranes, also result in the splitting of water and the subsequent release of oxygen. In the light-independent reactions of photosynthesis (that is, the Calvin cycle), carbon dioxide is fixed into glucose and other carbohydrate molecules. The energy and reducing power required for the Calvin cycle are supplied by ATP and NADPH produced in the light-dependent reactions. The enzymes for the Calvin cycle are located in the stroma of the chloroplast.

4. Hydrogen from the water combines with the electron carrier molecule **nicotinamide adenine dinucleotide phosphate ($NADP^+$)**, forming **NADPH** (reduced $NADP^+$).[1] Again, as in step 2, electrical energy is converted to chemical energy.

Thus, in the light-dependent reactions, the energy from sunlight is used to make ATP, and to reduce $NADP^+$, forming NADPH. Some of the captured energy of sunlight is temporarily stored within these two energy-rich compounds. Note that carbon dioxide is not used in the light-dependent reactions of photosynthesis, nor is glucose produced.

The Light-independent Reactions Produce Sugars

During the light-dependent reactions of photosynthesis, light energy was used to produce two high-energy compounds, ATP and NADPH. Although ATP and NADPH provide a quick source of energy and reducing power for various metabolic functions of organisms, neither is useful for the long-term storage of chemical energy. Cells possess very limited amounts of the materials from which ATP and NADPH are made, so neither can accumulate in large quantities.

During the light-independent reactions of photosynthesis, glucose is produced from carbon dioxide. This sugar is a source of energy for the cell and, unlike ATP and NADPH, it can be produced in large quantities and stored for future use.

The light-independent reactions depend on the products of the light-dependent reactions. The energy stored in ATP and NADPH during the light-dependent phase of photosynthesis is used to form carbohydrate molecules from carbon dioxide, a process called **CO_2 fixation**. In other words, during the light-independent reactions of photosynthesis, the energy in ATP and NADPH is transferred to the chemical bonds of sugar molecules.

The reactions of the light-independent phase of photosynthesis proceed by way of a cycle known as the **Calvin cycle**, named after Melvin Calvin, who first described it (Fig. 9–9). Calvin was awarded a Nobel Prize in 1961 for this significant scientific contribution. The enzymes for the Calvin cycle are located in the stroma of the chloroplast.

PHOTOSYNTHESIS IS A COMPLEX PROCESS

The preceding section gave an overview of photosynthesis in sufficient detail for some introductory biology courses. The following elaboration is provided for those courses that go into greater detail than the modest discussion previously provided.

[1] Although the correct way to write the reduced form of $NADP^+$ is $NADPH + H^+$, for simplicity's sake we present the reduced form as NADPH throughout the chapter.

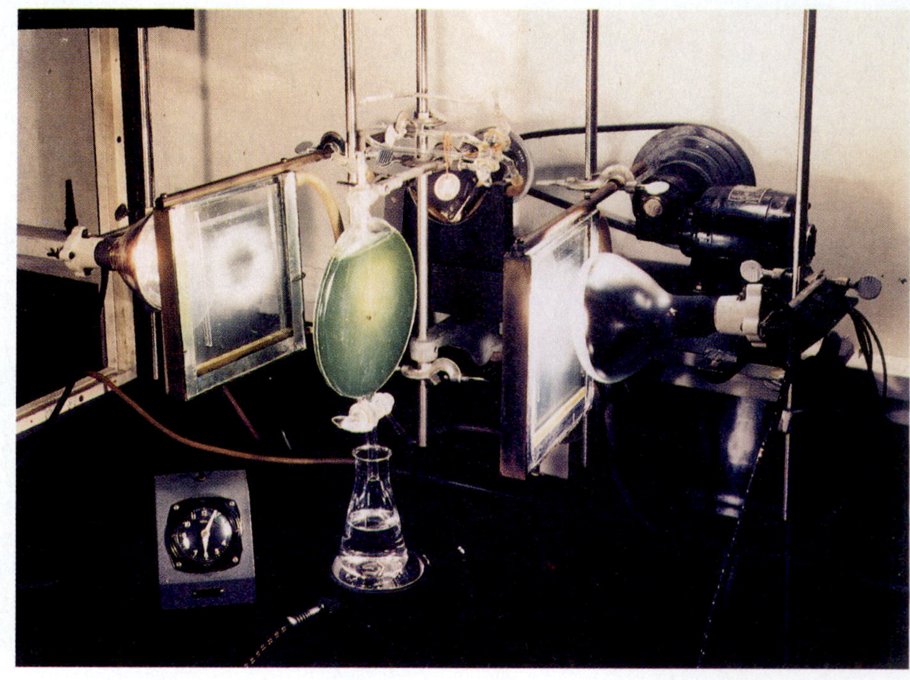

FIGURE 9–9 Calvin's classic experiment. The light-independent reactions of photosynthesis, also known as the Calvin cycle, were elucidated by a series of timed experiments using this apparatus. Calvin and his colleagues grew algae in the green "lollipop." Radioactively labeled $^{14}CO_2$ was bubbled through the algae, which were periodically killed by dumping the "lollipop" contents into a beaker of boiling alcohol. By identifying which compounds contained the radioactive ^{14}C at different times, Calvin was able to determine the steps of carbon dioxide fixation in photosynthesis. (Melvin Calvin, University of California, Berkeley)

Photosystems I and II Are Light-Harvesting Units That Include Chlorophyll Molecules

According to the scientific evidence currently available, chlorophyll molecules and associated electron acceptors are physically organized into units called **photosystems** within the thylakoid membranes. There are two types of photosystems, each containing up to 400 molecules of chlorophyll. Photosystem I contains a reactive pigment (probably a special form of chlorophyll *a*) known as P700 because it absorbs far-red light with a wavelength of 700 nm very strongly. Photosystem II utilizes a reactive pigment, P680 (also probably a special form of chlorophyll *a*) whose absorption maximum is at a wavelength of 680 nm (red light). Photosystems I and II are joined together by an electron transport chain.

All chlorophyll molecules of a photosystem apparently serve as antennae to gather solar energy (Fig. 9–10). Once absorbed, light energy is passed from one chlorophyll molecule to another within the photosystem until it reaches the special P700 or P680 pigment molecule-protein complex, referred to as the **reaction center**. Only the reaction center is able to give up its energized electrons to an electron acceptor compound.

As you read the following details of the light-dependent reactions, use the diagram in Figures. 9–11. The energized electrons that leave photosystem I are transferred to several electron acceptors and finally accepted by $NADP^+$. When $NADP^+$ accepts electrons, the electrons unite with protons present in the chloroplast to form hydrogen, so the reduced form of $NADP^+$ is NADPH. Electrons are restored to photosystem I from photosystem II.

Like photosystem I, photosystem II is activated by photons and gives up electrons to a chain of electron acceptors in a series of redox reactions. The electrons emitted from photosystem II pass from one acceptor to another through an electron transport chain—that is, a chain of alternately oxidized and reduced compounds. The electrons lose some of their energy as they are transferred along this chain. Some of the energy released in this way is used to establish a proton gradient (recall from Chapter 8 that a similar gradient exists in the respiratory electron transport chain), which leads to the synthesis of ATP (see next section). Electrons emitted from photosystem II are eventually donated to photosystem I.

The electrons that leave photosystem II are replaced by electrons from the hydrogen atoms in water. When P680 in the reaction center of photosystem II absorbs light energy, it becomes positively charged and exerts a strong pull on the electrons in water molecules. Water is split

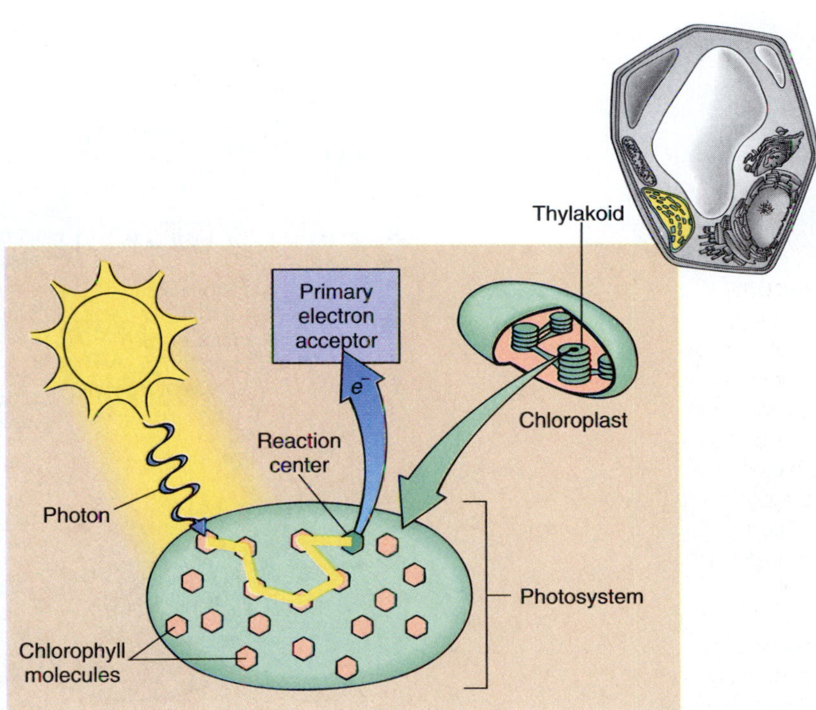

FIGURE 9–10 How a photosystem traps light energy. The many chlorophyll molecules in the photosystem are excited by photons and transfer their excitation energy to the specifically positioned chlorophyll molecule at the reaction center.

into its components: protons (H^+), electrons, and oxygen. The oxygen split from the water is released into the atmosphere as molecular O_2.

The entire process just described, in which NADPH and ATP are produced, is known as **noncyclic photophosphorylation.** It is called photophosphorylation because electrons obtain energy from *photons* of light and then contribute that energy to the *phosphorylation* of adenosine diphosphate (ADP), producing ATP. In other words, in photophosphorylation the energy of light is used to synthesize ATP. It is noncyclic because there is a one-way flow of electrons from water to $NADP^+$:

$$H_2O \longrightarrow \text{photosystem II} \longrightarrow \text{photosystem I} \longrightarrow NADP^+$$

The light-dependent reactions of photosynthesis also include **cyclic photophosphorylation,** a way to make additional ATP by a "shortcut" version of noncyclic photophosphorylation. Only photosystem I is involved in cyclic photophosphorylation. In this pathway, the excited electrons that originate from photosystem I are eventually returned to the same photosystem; that is, they cycle back to photosystem I. In cyclic photophosphorylation, the energy from light is used to synthesize ATP (discussed shortly). Water is not split during cyclic photophospho-

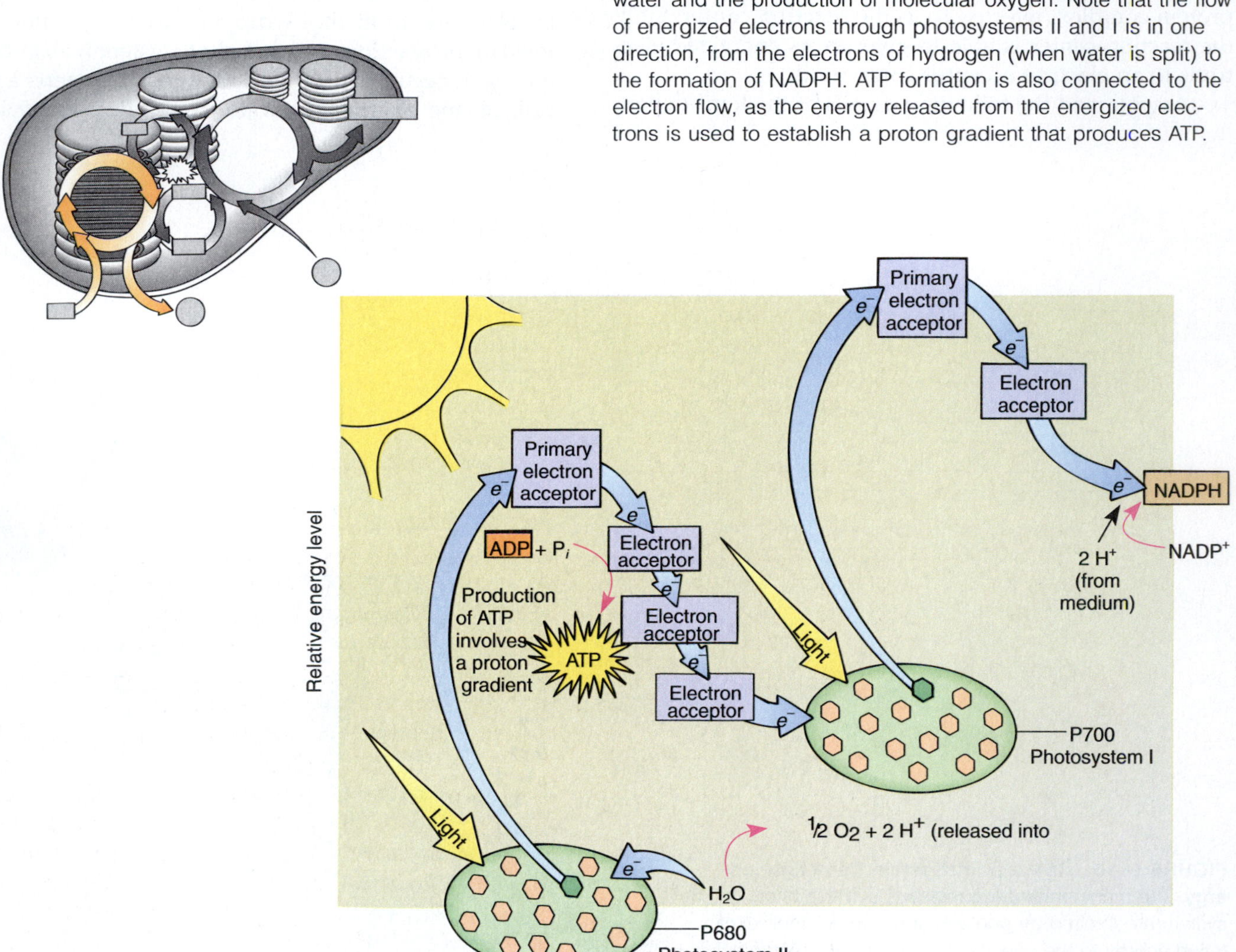

FIGURE 9–11 Light-dependent reactions (noncyclic photophosphorylation). When photosystem I is activated by absorbing photons, electrons are passed along an electron transport chain and eventually donated to $NADP^+$. Photosystem II, also activated by light energy, passes energized electrons along a series of acceptor molecules to replace the ones given up by photosystem I. Photosystem II is also linked to the splitting of water and the production of molecular oxygen. Note that the flow of energized electrons through photosystems II and I is in one direction, from the electrons of hydrogen (when water is split) to the formation of NADPH. ATP formation is also connected to the electron flow, as the energy released from the energized electrons is used to establish a proton gradient that produces ATP.

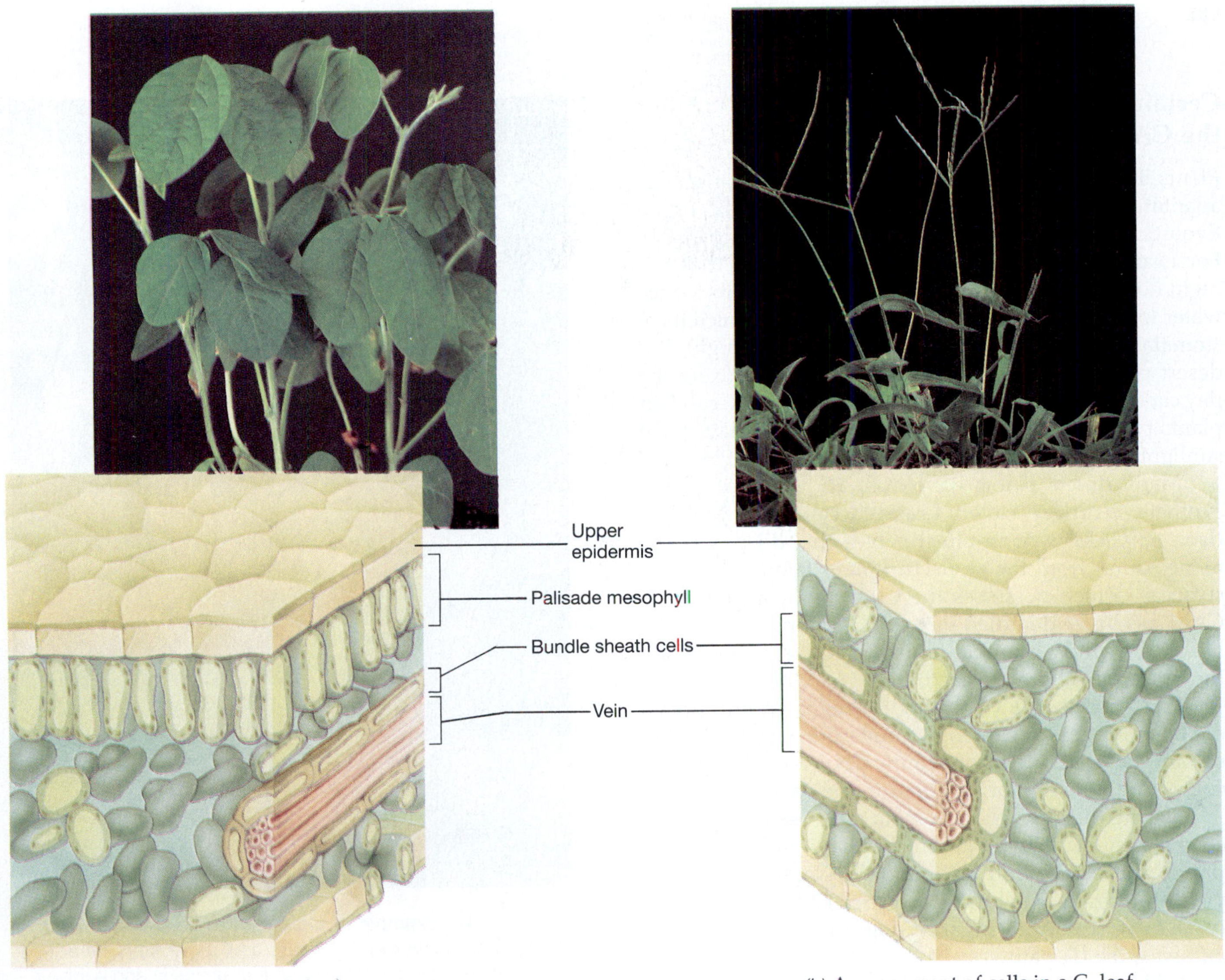

(a) Arrangement of cells in a C_3 leaf

(b) Arrangement of cells in a C_4 leaf

FIGURE 9–15 Comparison of the leaf structures of a C_3 plant and a C_4 plant. (***a***) In C_3 plants, such as soybeans, the Calvin cycle takes place within chloroplasts of the mesophyll cells of the leaf. (***b***) In C_4 plants, such as crabgrass, reactions that fix CO_2 into four-carbon compounds take place within chloroplasts of the mesophyll cells. Then the Calvin cycle occurs within chloroplasts of the bundle sheath cells that surround the veins of the leaf. (Dennis Drenner)

Many Plants with a Tropical Origin Fix Carbon Using the C_4 Pathway

Not all plants have a reduction in photosynthetic efficiency due to photorespiration. Many plants with a tropical origin, including crabgrass, corn, and sugarcane, have evolved mechanisms to bypass photorespiration. One example is the **C_4 pathway**. C_4 plants get their name because the first detectable carbohydrate formed by CO_2 fixation is a four-carbon compound rather than the three-carbon compound produced in the Calvin (C_3) cycle. This four-carbon compound, produced in mesophyll cells (see Fig. 9–6) by the joining of CO_2 to a three-carbon compound, is then rapidly transported to special cells surrounding the veins of the leaf called **bundle sheath cells** (Fig. 9–15). There the CO_2 is removed from the four-carbon molecule and fixed into sugar by the regular C_3 pathway. The C_4 pathway in effect concentrates the amount of CO_2 in the bundle sheath cells, making it higher than it could possibly be as a result of diffusion of CO_2 from the atmosphere.

Photorespiration is negligible in C_4 plants because the concentration of CO_2 in bundle sheath cells (where the enzyme rubisco works) is always high. Some scientists are attempting to transfer the genes coding for the C_4 pathway to crops like soybean and wheat. If this is accomplished, these plants would be able to produce a lot more carbohydrates during hot weather, resulting in increased crop yields.

Certain Desert Plants Utilize the CAM Pathway to Fix Carbon

Plants living in dry conditions have a number of special adaptations that enable them to survive (see Focus on Evolution: Comparative Plant Anatomy in Chapter 27). For example, their stomata may open during the cool night hours and close during the hot day hours to reduce water loss. This is in contrast to most plants, which have stomata that open during the day and close at night. But desert plants that have their stomata closed during the day cannot exchange gases for photosynthesis. Recall that plants typically fix carbon dioxide during the day, when sunlight is available to produce ATP and NADPH.

Many desert plants have evolved a special photosynthetic pathway called **Crassulacean Acid Metabolism**, or **CAM**, that in effect solves this dilemma. The name comes from the stonecrop plant family, the Crassulaceae, which possesses the CAM pathway, although CAM has been identified in over 25 different plant families (Fig. 9–16). A number of unrelated plants, including the pineapple and most cacti, have it.

CAM plants fix CO_2 during the night, when stomata are open for gas exchange, by combining it with a three-carbon compound to form a four-carbon compound. This compound is temporarily stored in the vacuole.

During the day, when stomata are closed and gas exchange cannot occur between the plant and the atmosphere, the four-carbon compound is decarboxylated to yield CO_2 again. Now the CO_2 is available *within the leaf tissue* to be fixed into sugar by the usual photosynthetic pathway, the C_3 pathway.

CAM photosynthesis may sound familiar to you because it is very similar to the C_4 pathway. There are important differences, however. C_4 plants initially fix carbon dioxide into four-carbon compounds in mesophyll cells. The compounds are later decarboxylated to produce CO_2, which is fixed by the C_3 pathway, in the bundle sheath cells of the leaf. In other words, the C_4 and C_3 pathways occur in *different locations* within the leaf of a C_4 plant.

In CAM plants, the initial fixation of CO_2 occurs at night. Decarboxylation of the four-carbon compound and subsequent production of sugar from CO_2 by the C_3 pathway occur during the day. In other words, the C_4 and C_3 pathways occur at *different times* within the same cell of a CAM plant.

FIGURE 9–16 Example of a plant with the CAM pathway. The snake plant (*Sansevieria trifasciata*), a member of the Agavaceae, is a typical CAM plant that grows in arid habitats. (Dennis Drenner)

The CAM pathway is a very successful adaptation to desert conditions. CAM plants are able to have gas exchange for photosynthesis *and* to significantly reduce water loss during hot daylight hours. Plants with CAM photosynthesis can survive in deserts where neither C_3 nor C_4 plants can.

Chapter Summary

I. Light behaves as both a wave and a particle. Particles of light energy, called photons, can be absorbed by and excite pigment molecules such as chlorophyll.

II. Chlorophyll and other photosynthetic pigments are found within the membranes of thylakoids. In photosynthetic eukaryotes, thylakoids are found within chloroplasts and arranged in stacks called grana.

III. During photosynthesis, light energy is captured by chlorophyll and used to chemically combine the hydrogen from water with carbon dioxide to produce carbohydrates. Oxygen is released as a by-product of photosynthesis.

  A. During the light-dependent reactions of photosynthesis, chlorophyll absorbs light and becomes energized. Some of this energy is used to make ATP, and some is used to split water. Hydrogen from the water is transferred to $NADP^+$, forming NADPH.

1. In noncyclic photophosphorylation, ATP and NADPH are formed using the energy of light; water is split.
2. In cyclic photophosphorylation, ATP is formed; water is not split, and NADPH is not formed.
3. The flow of electrons along an electron transport chain in both noncyclic and cyclic photophosphorylation provides energy to establish a proton gradient across the thylakoid membrane. ATP is synthesized chemiosmotically, by using the energy from this proton gradient.

B. During the light-independent reactions, energy stored within ATP and NADPH during the light-dependent reactions is used to chemically fix carbon dioxide.

1. The light-independent reactions proceed in most plants via the Calvin cycle, also known as the C_3 pathway.
2. In the Calvin cycle, carbon dioxide is combined with ribulose bisphosphate, a five-carbon sugar. With each turn of the cycle, one carbon atom enters the cycle. Six turns of the cycle result in the synthesis of two molecules of a three-carbon compound known as PGAL, which combine to produce a molecule of glucose.

IV. In photorespiration, C_3 plants consume oxygen and generate carbon dioxide.

A. This process, which decreases photosynthetic efficiency, occurs on bright, hot days when stomata close to conserve water.

B. The closed stomata prevent the passage of CO_2 into the leaf.

V. Many plants with a tropical origin fix carbon using the C_4 pathway.

A. Carbon dioxide is initially fixed into a four-carbon compound, which moves into bundle sheath cells.

B. In the bundle sheath, CO_2 is removed from the four-carbon compound and enters the C_3 pathway.

C. The C_4 pathway is advantageous in hot climates because it enables plants to form carbohydrates more efficiently than the C_3 pathway. C_4 plants do not photorespire.

VI. Certain desert plants utilize the CAM pathway to fix carbon.

A. CAM plants fix CO_2 at night (when their stomata are open) into a four-carbon compound, which is stored in the vacuole.

B. During the day, when stomata are closed, CO_2 is released from the four-carbon compound and enters the C_3 pathway.

C. The CAM pathway is advantageous in dry conditions because it enables a plant to conserve water (by keeping its stomata closed during the hot daylight hours).

Selected Key Terms

ATP, p. 174
ATP synthetase, p. 178
C_4 pathway, p. 181
Calvin cycle (C_3 pathway), p. 174
CAM pathway, p. 182
chemiosmosis, p. 178
chlorophyll, p. 171
chloroplast, p. 171
CO_2 fixation, p. 174
NADPH, p. 174
photon, p. 170
photophosphorylation, p. 177
photorespiration, p. 180
photosynthesis, p. 166
redox reactions, p. 171
rubisco, p. 178
stroma, p. 171
thylakoid, p. 171
wavelength, p. 169

Post-Test

1. The synthesis of organic compounds using the energy of light is known as ____________.
2. In photosynthesis, light energy is converted to ____________ energy in organic compounds.
3. The welfare of the vast majority of animals, including humans, is dependent upon two photosynthetic products, ____________ and ____________.
4. Light is composed of particles of energy called ____________.
5. In photosynthesis, light is absorbed by the green pigment ____________.
6. Chlorophyll is located within the membranes of ____________.
7. Only the reaction center of a photosystem actually gives up its ____________.
8. In photophosphorylation, light energy is used to add phosphate to ____________, producing ____________.
9. In noncyclic photophosphorylation, both ATP and ____________ are produced.
10. According to the chemiosmotic model, energy released from electrons is used to pump ____________ across

the ______________ membrane.

11. As protons pass through ATP synthetase, ______________ is synthesized.
12. The enzymes for the light-independent reactions of photosynthesis are located in the ______________ of the chloroplast.
13. The light-independent reactions of photosynthesis are also known as the C_3 pathway, or the ______________ cycle.
14. The process of ______________ ______________ involves the chemical combination of carbon dioxide with ribulose bisphosphate.
15. ______________ (number) turns of the Calvin cycle are required to produce one glucose molecule.
16. During photosynthesis, plants produce oxygen by (a) splitting CO_2 (b) digesting $C_6H_{12}O_6$ (c) splitting H_2O (d) reducing $NADP^+$.
17. During noncyclic photophosphorylation, the electrons that pass to $NADP^+$, reducing it to NADPH, are obtained from (a) water (b) sunlight (c) oxygen (d) ATP.
18. Cyclic photophosphorylation (a) fixes carbon dioxide (b) reduces $NADP^+$ (c) regenerates ribulose bisphosphate (d) produces ATP.
19. The part of photosynthesis that actually produces glucose is (a) noncyclic photophosphorylation (b) cyclic photophosphorylation (c) light-dependent reactions (d) the Calvin cycle.

Review Questions

1. Explain the role of light in photosynthesis.
2. Explain the role of chlorophyll in photosynthesis.
3. Why are almost all life forms dependent on the process of photosynthesis?
4. Write the overall equation for photosynthesis.
5. Summarize the two light-dependent reactions of photosynthesis.
6. How is oxygen produced during photosynthesis?
7. Explain how ATP is produced in photosynthesis according to the chemiosmotic model.
8. Summarize the events of the Calvin cycle.
9. Distinguish between C_3 and C_4 pathways.
10. How is CAM photosynthesis different from the C_4 pathway?

Thinking Critically

1. If the world's tropical forests were all destroyed, how would it affect your life?
2. Would placing a plant under green light increase or decrease its rate of photosynthesis (the plant was previously grown in sunlight)? Why?
3. Is photosynthesis exergonic or endergonic? Explain.
4. It has been said that most of the mass of a tree comes from the air. Explain what is meant by this statement.
5. Explain how the term *photosynthesis* provides insight into the process it represents. Then explain why the term *photorespiration* does *not* provide insight into the process it represents.
6. Kentucky bluegrass thrives in eastern lawns during the spring months, but is usually overrun by crabgrass during the hot summer months. Based on what you know about the C_3 and C_4 pathways, which plant would you suppose possesses the C_3 pathway? The C_4 pathway? Explain your reasoning.
7. Rubisco is by far the world's most abundant enzyme. Explain why this is so.

Recommended Readings

Bazzaz, F. A., and E. D. Fajer, "Plant Life in a CO_2-Rich World," *Scientific American*, January 1992. The increase in atmospheric CO_2 caused by human activities will have profound effects on plants.

Emsley, J., "Photochemistry," *New Scientist*, Vol. 137, No. 1856, January 16, 1993. This article discusses the various ways light energy is used by living things, from plant cells (for photosynthesis) to the skin of newborn babies (light helps the baby excrete a waste product called bilirubin).

Govindjee and W. J. Coleman, "How Plants Make Oxygen," *Scientific American*, February 1990. The photosynthetic process of using solar energy to split water molecules into oxygen gas, protons, and electons is probed in this article.

Hendry, G., "Making, Breaking, and Remaking Chlorophyll," *Natural History*, May 1990. The endless process by which plants make chlorophyll in the spring and dispose of it in the fall is examined.

Solomon, E. P., L. R. Berg, D. W. Martin, and C. Villee, *Biology*, 3rd ed., Philadelphia, Saunders College Publishing, 1993. Chapter 8 presents a slightly more detailed discussion of photosynthesis than is presented in this text.

PART 4

The Continuity of Life: Cell Division and Genetics

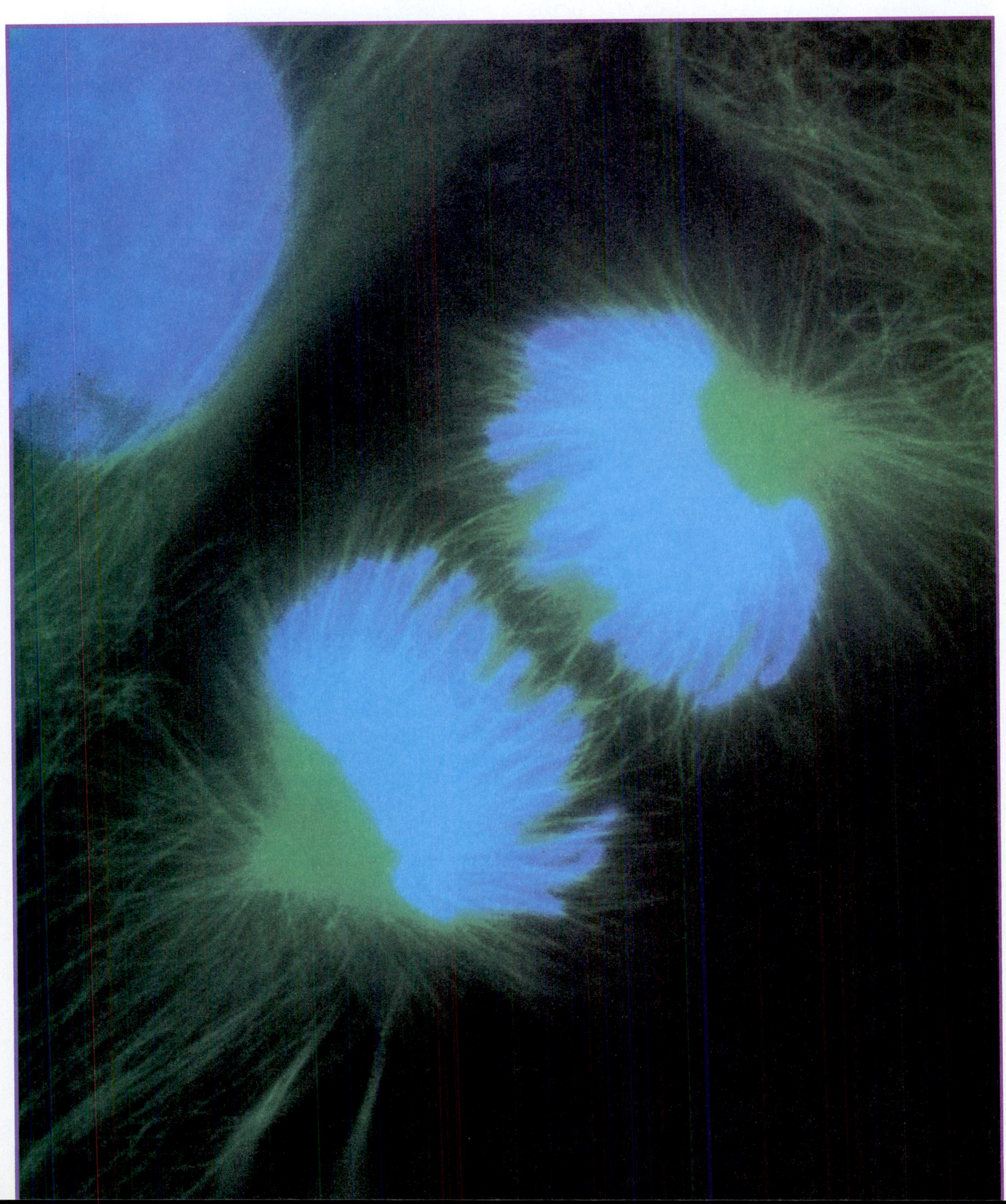

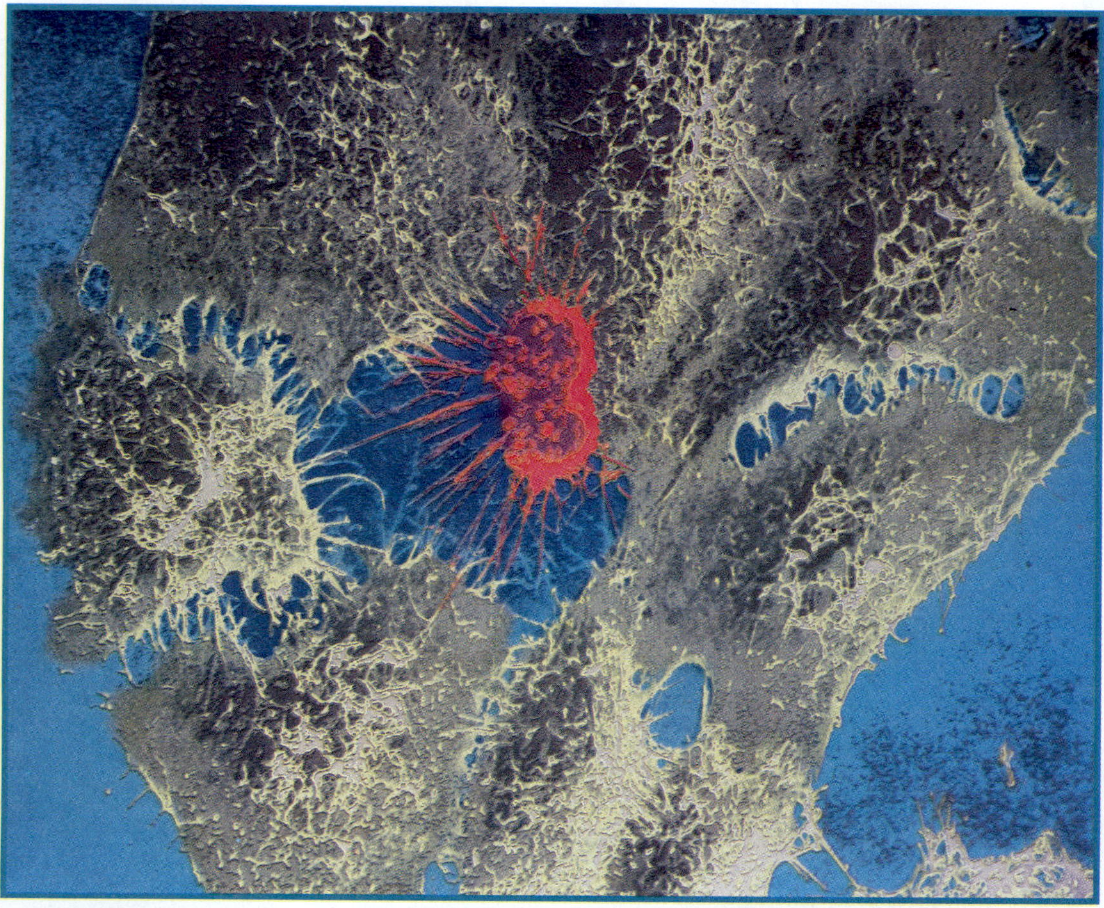

False color SEM of HeLa cells (red) dividing during mitosis (telophase). *(Bill Longcore/ Science Source/ PhotoResearchers, Inc.)*

Cells Out of Control

Cells dividing rapidly, wildly, positioning themselves helter skelter and crowding the normal cells of a tissue. Cells migrating to other parts of the body, invading other tissues, and then dividing again and again. Cells that do not die of old age. These are cancer cells—cells characterized by uncontrolled cell division.

Normal cells also increase in number by dividing. Cells of a multicellular organism divide frequently, enabling the organism to grow and to repair itself. Normal cell division is precisely controlled. Millions of cells are produced every day in an adult human. Each cell divides on schedule, a program dictated by its genes and regulated by its proteins.

Just how normal control mechanisms work and how they fail in cancer are not completely known. Certain genes, known as **oncogenes,** are involved. Oncogenes normally help regulate cellular growth and differentiation. However, certain changes, or mutations, in oncogenes can lead to cancer. Such mutations may be induced by exposure to carcinogens (cancer-causing agents) such as radiation, certain chemicals, and certain viruses.

When DNA is damaged, a **tumor-suppressor** gene becomes active. This gene is thought to code for a protein that inhibits cells from producing new DNA until the damaged DNA is repaired. When the tumor-suppressor gene is not functioning properly, the damaged DNA does duplicate itself, producing more defective copies. Several malfunctioning oncogenes may be necessary to transform a normal cell to a cancer cell that no longer responds appropriately to regulatory signals.

In addition to multiplying rapidly and wildly, cancer cells differ from normal cells in their ability to continue to divide indefinitely. When normal cells from a human or other mammal are grown in laboratory cultures, they generally divide only up to about 50 times before they age and die. In contrast, some lines of cancer cells appear to be "immortal." One of the best-known lines of cancer cells are the HeLa cells used by researchers around the world. This line of cells began as a sample of cervical cancer cells taken from a 30-year-old cancer patient, Henrietta Lacks (thus the name HeLa), in 1951. Lacks died in 1951, but her cells continue to divide in laboratory glassware. Much current research focuses on details of the systems that control cell division and growth and the mechanisms that lead to transformation of normal cells to cancer cells.

Producing a New Generation: Mitosis and Meiosis

10

Learning Objectives

After you have studied this chapter you should be able to:

1. Identify the stages in the cell cycle and describe the main events of each.
2. Distinguish between haploid and diploid cells and define homologous chromosomes.
3. Describe the events occurring in each stage of mitosis, emphasizing the behavior of chromosomes.
4. Summarize the significance of meiosis in sexual reproduction.
5. Contrast the events of mitosis and meiosis.

Key Concepts

- ☐ Each species has a characteristic number of chromosomes. During cell division, chromosomes and the genes they contain are distributed to the resulting cells in an orderly fashion.
- ☐ Mitosis results in daughter cells with the same number of chromosomes as the parent cell.
- ☐ Meiosis reduces the chromosome number by one half, and each new cell has one chromosome from each homologous pair.
- ☐ Genes tend to be inherited as specific portions of chromosomes, but genes can be exchanged between the members of a homologous pair of chromosomes when crossing-over occurs in meiosis.

EACH EUKARYOTIC SPECIES MAINTAINS A CHARACTERISTIC CHROMOSOME NUMBER

Each body cell of a eukaryotic organism of a given species contains a characteristic number of chromosomes. Recall from Chapter 5 that chromosomes are small packages of genetic information that contain the genes. Human cells have 46 chromosomes, for example, and cabbage cells have 20. A certain species of mosquito has only 6 chromosomes per cell, and some ferns have more than 1000. Humans are not unique in having 46 chromosomes. Some other species of animals and some plants also have 46. The number of chromosomes has no simple relationship to the size or complexity of an organism.

Although some organisms consist of only one cell and others of several billion, even the most complex organism begins life as a single cell, the fertilized egg (Fig. 10–1). In most multicellular organisms, including humans, this cell divides to form two cells, and each new cell divides again and again, eventually forming the complex tissues, organs, and systems of the developed organism. Those cells that form a part of the body are called body cells, or **somatic cells**. Examples include muscle cells, blood cells, and nerve cells in animals and leaf and root cells in plants. Through all of these divisions, how is the chromosome number kept constant? Before somatic cells divide, their chromosomes duplicate and are precisely distributed by the process of **mitosis**.

A different process is required to ensure that chromosome number is maintained in offspring produced by sexual reproduction. During sexual reproduction two **gametes** (sex cells; sperm and egg) fuse. If each gamete had the characteristic chromosome number, the **zygote**, or fertilized egg, would have twice the number of chromosomes that it needs. Each generation would have double the number of chromosomes of the preceding generation. Fortunately, gametes are produced by a process called **meiosis**, which reduces the number of chromosomes to half the normal number. Meiosis ensures that the number of chromosomes in the offspring will be the same as in their parents.

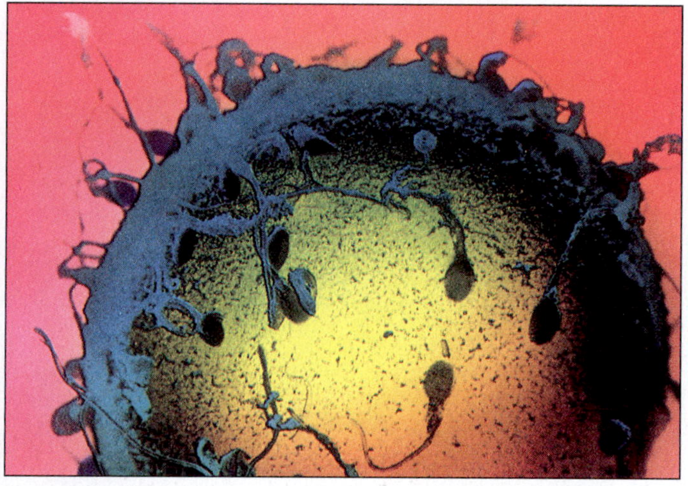

FIGURE 10–1 Fertilization. Color-enhanced SEM of sperm fertilizing an ovum. (David Phillips/Science Source/PhotoResearchers, Inc.)

In both forms of cell division, the new cells inherit genetic information, but not in the same way. In mitosis, each new cell inherits an exact, complete copy of its parent cell's genes. In meiosis, each new reproductive cell inherits only half of its parent cell's genes (Fig. 10–2).

CHROMOSOMES EXIST IN PAIRS

In somatic cells of complex organisms, the chromosomes are paired. Thus, the 46 chromosomes of each human cell are actually 23 pairs. One set of 23 was contributed by the female parent. The chromosomes of this set are referred to as **maternal chromosomes**. The other set of 23 chromosomes, the **paternal chromosomes**, was contributed by the male parent.

Each chromosome pair is different enough in size and shape that biologists can count and identify them. Chromosome shapes and staining patterns are sometimes used by biologists to infer relationships among species. Abnormalities in shape are sometimes associated with inherited defects (Chapter 12).

The members of a given pair of chromosomes, called **homologous chromosomes**, are generally similar in size, shape, and position of their centromeres. Both members carry information governing the same traits, although the information may not be the same. For example, members of a particular pair of homologous chromosomes carry genes that code for the hemoglobin molecule. However, one might have the information that produces normal hemoglobin structure, and the other might specify the abnormal hemoglobin structure that results in the sickle-cell trait.

A cell with chromosomes occurring in pairs is referred to as **diploid**, or **2n**. In humans, the diploid number is 46 chromosomes (23 pairs). Recall that the reproductive cells, or gametes, cannot be diploid, or the zygote (fertilized egg) resulting from their fusion would have twice as many chromosomes as it should. The special type of cell division called meiosis passes along to each cell only one chromosome of each pair, resulting in a **haploid**, or **n**, sperm or egg. When two such haploid gametes join in fertilization, the normal diploid number of chromosomes is restored in the zygote.

Plants also can reproduce sexually (that is, by the union of gametes) but in a somewhat different way. Plants and some other eukaryotic organisms have a life cycle in

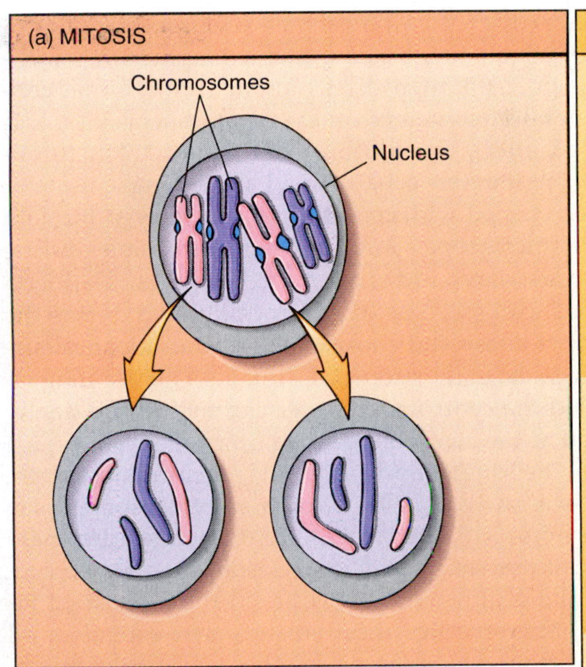

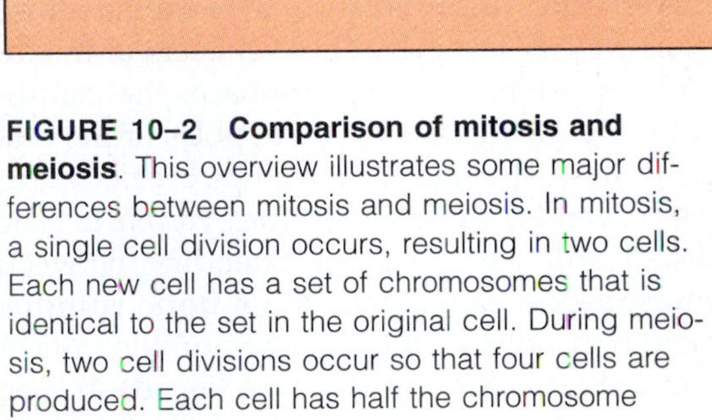

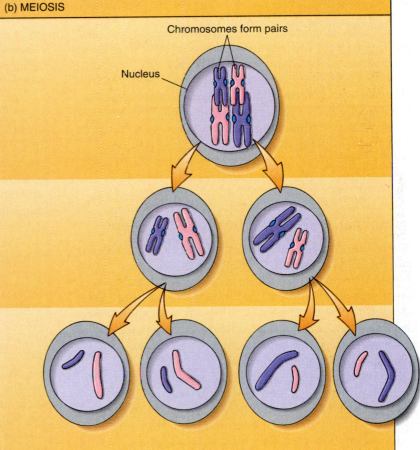

FIGURE 10–2 Comparison of mitosis and meiosis. This overview illustrates some major differences between mitosis and meiosis. In mitosis, a single cell division occurs, resulting in two cells. Each new cell has a set of chromosomes that is identical to the set in the original cell. During meiosis, two cell divisions occur so that four cells are produced. Each cell has half the chromosome complement of the original cell.

which the diploid plant produces **spores** by meiosis. Each spore gives rise to a multicellular organism whose cells *all* have the haploid number of chromosomes. The haploid organism then produces haploid gametes by *mitosis*. When the gametes unite, a new diploid plant is produced. Notice that plant gametes are haploid, just as animal gametes are, but they are not immediate products of meiosis. (The complexities of plant and animal reproduction are more fully discussed in Chapters 23, 29, and 42.)

THE EUKARYOTIC CELL CYCLE IS A SEQUENCE OF CELL GROWTH AND DIVISION

In somatic cells that are capable of dividing, the **cell cycle** is the period from the beginning of one division to the beginning of the next division. The cell cycle may be represented as a circle (Fig. 10–3). The time it takes to complete one cell cycle is the **generation time**. The generation time varies widely. When conditions are favorable, the generation time for bacteria can be as short as 20 min-

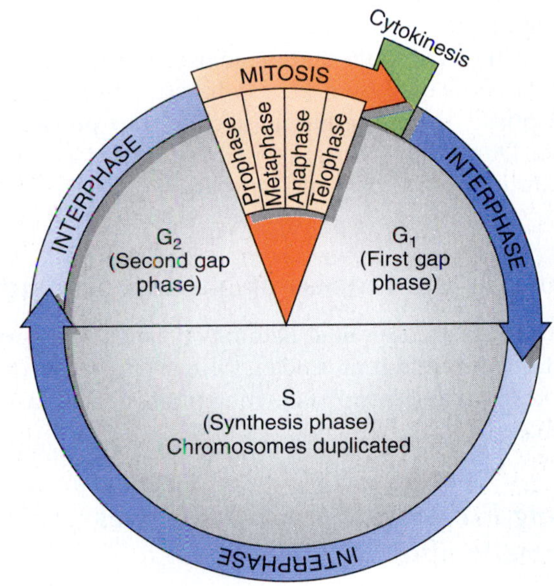

FIGURE 10–3 The cell cycle. The generation time varies widely with the cell type and species. Most cells spend about 90% of their cell cycle in interphase.

utes. For actively growing plant and animal cells, generation time may be several hours or longer.

Cell division involves two main processes, mitosis and cytokinesis. Mitosis ensures that each new cell nucleus contains the identical number and types of chromosomes present in the original nucleus. The division of the cytoplasm to form two cells is **cytokinesis**. If mitosis is not followed by cytokinesis, the cell will have two nuclei. Additional mitoses without cytokinesis can result in a cell with several nuclei. Such a cell is described as *multinucleate*. Some large, specialized cells, for example skeletal muscle cells, are multinucleate.

Interphase Is the Time between Divisions

In a cell that divides, mitosis is followed by cytokinesis, and then the cell enters a growth period, called **interphase**. The cell spends most of its life in interphase, actively synthesizing materials it needs to grow and maintain itself. During the early 1950s, it was shown that chromosomes are duplicated during interphase. They condense into visibly separate structures and are distributed to the daughter nuclei during mitosis.

Interphase consists of three phases: the G_1 phase, the S phase, and the G_2 phase. These phases are not distinct from one another. The time between the end of the previous cell division and the beginning of DNA replication (duplication) (Chapter 13) is termed the **G_1 phase,** or **first gap phase**. During the G_1 phase the cell grows and, late in this phase, synthesizes certain enzymes needed for DNA synthesis.

The **S phase**, or **synthesis phase**, begins with the replication of DNA. During this time, the DNA content doubles. By techniques that make use of radioisotopes, the DNA can now be shown to be half new. After it completes the S phase, the cell enters a **second gap phase**, the **G_2 phase**. During this time, protein synthesis increases in preparation for mitosis and cytokinesis.

Mitosis Is the Process of Nuclear Division

During mitosis, identical sets of chromosomes are distributed into different nuclei. Mitosis may be divided into four stages: prophase, metaphase, anaphase, and telophase.

During Prophase, the Chromosomes Become Visible

The first stage of mitosis, **prophase**, begins when the long threads of chromatin in the nucleus begin to condense and coil into much shorter bundles. The chromatin thus organizes as discrete chromosomes. In this form, the chromosomes can be distributed into the daughter cells without tangling. Although each human chromosome contains several centimeters of DNA, at mitosis, DNA is condensed into a chromosome that is only 5 to 10 micrometers in length—a 10,000-fold shortening!

When stained with certain dyes and viewed through the light microscope, chromosomes are visible during prophase as dark, rod-shaped bodies (Figs. 10–4 and 10–5). Each chromosome has been duplicated during the preceding S phase and consists of a pair of identical **sister chromatids**. Each chromatid contains a nonstaining region called the **centromere**. Sister chromatids are tightly connected at the centromere. Each centromere contains a structure, the **kinetochore**, to which microtubules attach.

Cytoplasmic organelles known as **centrosomes** play a role in mitosis. (Centrosomes were discussed in Chapter 5; recall that animal cell centrosomes have two centrioles.) The centrosomes duplicate in the S phase of interphase. Microtubules form clusters that extend in all directions from the centrosomes. These clusters of microtubules are called **asters**. During prophase, the centrosomes begin to migrate toward opposite poles of the cell.

During prophase, the nuclear envelope breaks apart, and the nuclear contents mingle with the cytoplasm. The nucleolus disappears. Microtubules extending between the two centrosomes organize into the **mitotic spindle**. Toward the end of prophase, spindle microtubules attach to each chromatid at its kinetochore (see Fig. 10–4). The chromatids begin to move toward the equator of the cell, midway between the two poles (opposite ends of the cell) and perpendicular to the axis of the spindle.

During Metaphase, Chromosomes Line Up Along the Equator

The short period during which the chromatids are lined up along the equatorial plane of the cell is called **metaphase**. The mitotic spindle is complete. It is composed of numerous microtubules that extend from each pole to the equator of the cell (see Fig. 10–5). Some microtubules extend from the poles to the kinetochores of the chromatids. Kinetochores of sister chromatids are attached by spindle microtubules to opposite poles of the cell.

During metaphase, each chromosome is completely condensed and appears thick and distinct. Because chromosomes can be seen more clearly at metaphase than at any other time, medical technologists and researchers typically photograph them during this stage. Such photographs can be used clinically to determine possible chromosome abnormalities. Biologists can use them to look for physical similarities and differences in chromosomes among various species.

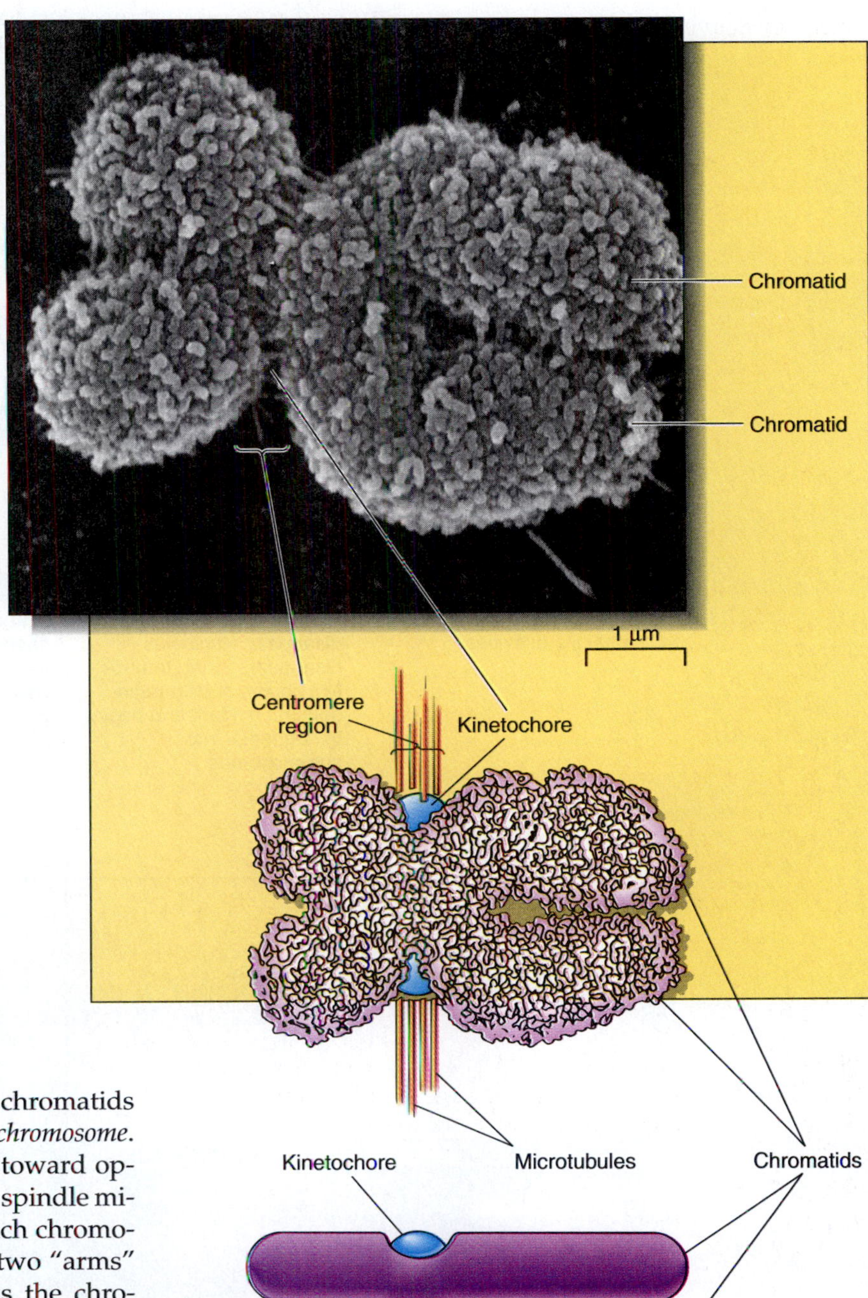

FIGURE 10–4 Chromosome structure. SEM of a human chromosome photographed at the metaphase stage of mitosis. The two members of this pair, referred to as sister chromatids, are held together at their centromeres. The diagrams show the kinetochore, the part of the centromere to which microtubules bind. The microtubules eventually pull the two identical chromosomes apart. The structure of the chromosome is as tightly wound as a clock spring into a compact package that can be distributed to the daughter cells. (Biophoto Associates/PhotoResearchers, Inc.)

During Anaphase, Chromosomes Move to the Poles

Anaphase begins as the centromeres of sister chromatids separate. *Each chromatid is now an independent chromosome.* The separated chromosomes are now pulled toward opposite poles by spindle microtubules. Because spindle microtubules are attached to the kinetochore, each chromosome is pulled with its kinetochore first, the two "arms" trailing behind. The microtubules shorten as the chromosomes are moved toward the poles of the cell. Anaphase ends when a complete set of chromosomes reaches each pole.

Two Nuclei Form during Telophase

During **telophase**, the final stage of mitosis, the cell returns to interphase conditions. The chromosomes elongate by uncoiling, becoming chromatin threads. A nuclear envelope forms around each set of chromosomes, produced at least in part from components of the old nuclear envelope. Nucleoli reappear and spindle microtubules disappear.

Cytokinesis Separates Daughter Cells

Cytokinesis, the division of the cytoplasm to produce two daughter cells, usually overlaps mitosis. Cytokinesis generally begins during telophase and ends soon after the completion of mitosis.

The division of an animal cell is acomplished by a **cleavage furrow** that encircles the cell in the region of the

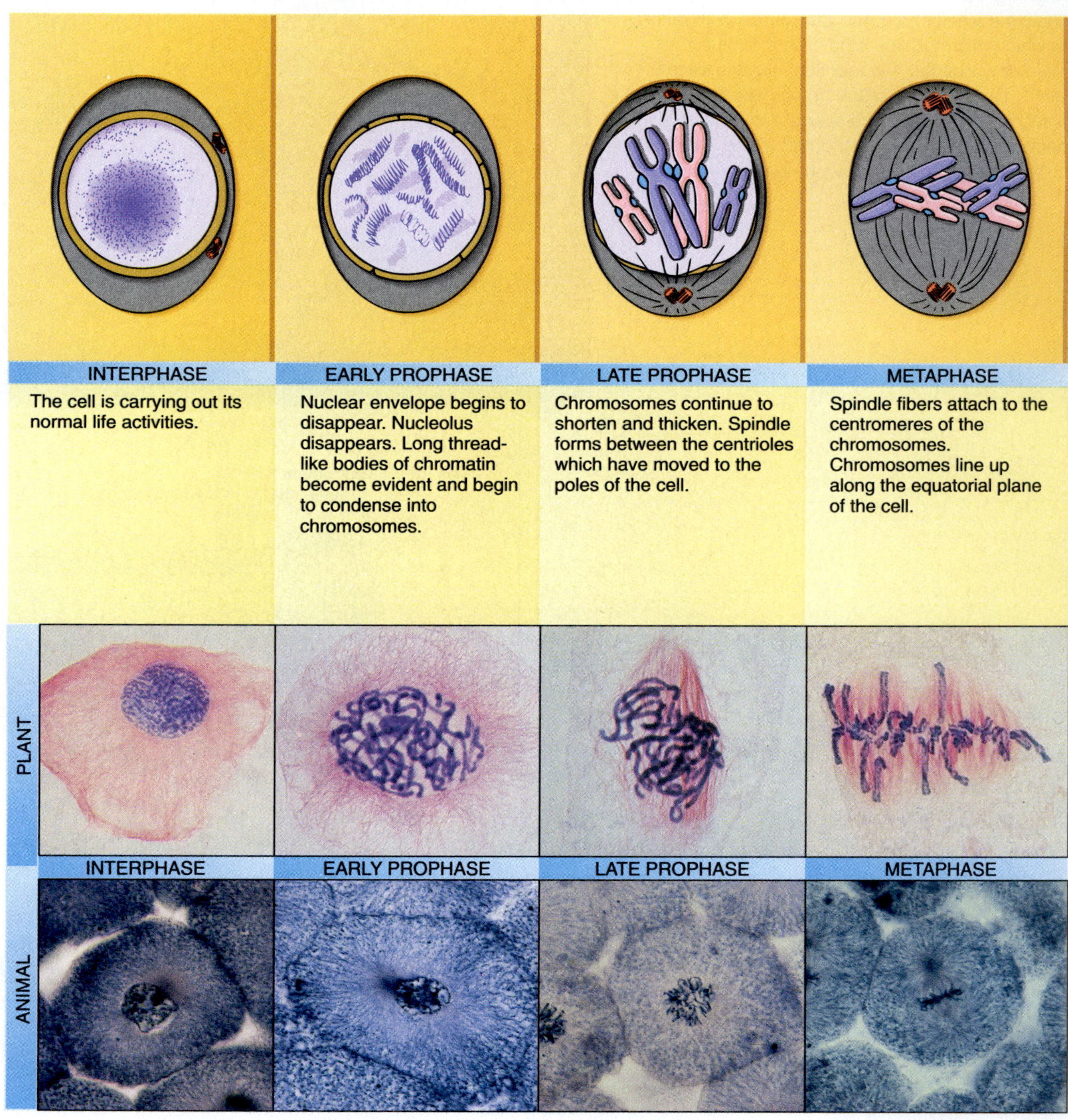

equator. At the furrow, the plasma membrane is attached to a ring of filaments capable of contraction. (These filaments are composed of the contractile proteins, actin and myosin.) As the ring gradually contracts, the furrow deepens, separating the cell into two daughter cells, each with a nucleus.

INTERPHASE
The cell is carrying out its normal life activities.
EARLY PROPHASE
Nuclear envelope begins to disappear. Nucleolus disappears. Long thread-like bodies of chromatin become evident and begin to condense into chromosomes.
LATE PROPHASE
Chromosomes continue to shorten and thicken. Spindle forms between the centrioles which have moved to the poles of the cell.
METAPHASE
Spindle fibers attach to the centromeres of the chromosomes. Chromosomes line up along the equatorial plane of the cell.
PLANT
INTERPHASE
EARLY PROPHASE
LATE PROPHASE
METAPHASE
ANIMAL

In plant cells, cytoplasmic division occurs by the formation of a **cell plate** between the daughter cells. The partition forms in the equatorial region of the spindle and grows laterally to the cell wall. The cell plate forms from vesicles that originate in the Golgi complex (Chapter 5). Each daughter cell forms a plasma membrane and a cell wall on its side of the cell plate.

The Cell Cycle Is Controlled by a Genetic Program

The frequency of mitosis varies greatly not only among cells from different species, but among cells from different tissues within a single organism. Among the more

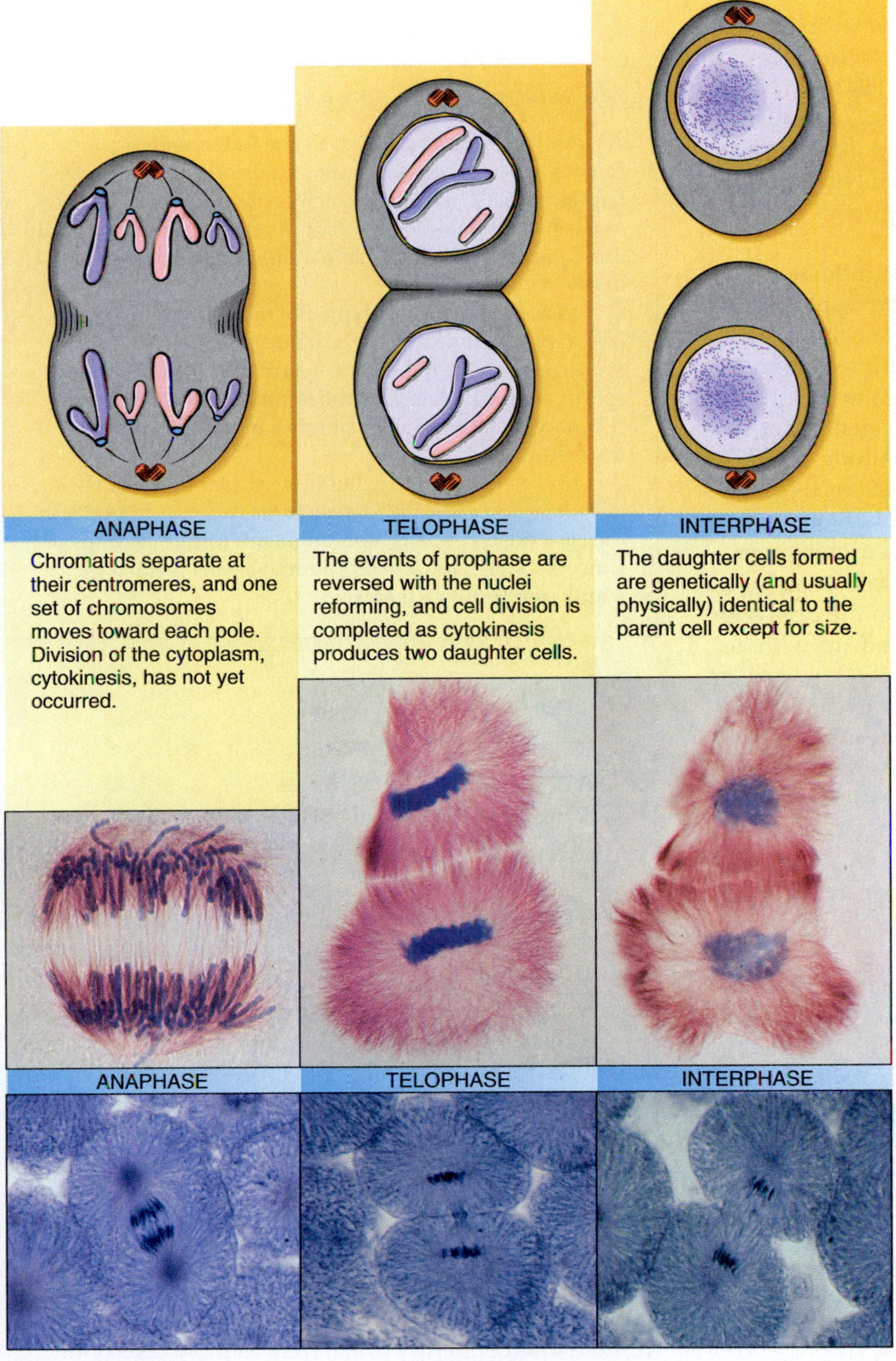

ANAPHASE
Chromatids separate at their centromeres, and one set of chromosomes moves toward each pole. Division of the cytoplasm, cytokinesis, has not yet occurred.
TELOPHASE
The events of prophase are reversed with the nuclei reforming, and cell division is completed as cytokinesis produces two daughter cells.
INTERPHASE
The daughter cells formed are genetically (and usually physically) identical to the parent cell except for size.
ANAPHASE
TELOPHASE
INTERPHASE

FIGURE 10–5 The stages of mitosis within the cell cycle. Individual steps in the cell cycle are explained in labels within the figure. The animal cells drawn here have a diploid chromosome number of four. Photomicrographs of the plant cell cycle illustrate stages in the cell cycle of the blood lily, *Haemanthus*. The photomicrographs of the animal cell cycle show whitefish cells. (Plant cells, Andrew S. Bajer, University of Oregon; animal cells, Ed Reschke)

rapidly dividing cells of humans are the cells that line the digestive tract, cells in the skin, and the stem cells that give rise to blood cells. These cell types divide rapidly and repeatedly throughout life. In contrast, some cells in the central nervous system usually do not divide again after the first few months of life. This explains why we can afford to lose some skin or blood cells, whereas brain damage is often fatal. Under optimal conditions of nutrition, temperature, and pH, the length of the cell cycle is constant for any particular type of cell.

In any multicellular organism the frequency of mitosis must be closely controlled. If it is not, growth abnormalities or even tumors may result. The evidence indicates that a precise program has been built into the cell. The program has several parts, involving DNA synthesis, growth of the cell, and orderly movement from one phase of the cell cycle to the next.

The cell requires a protein called **maturation promoting factor (MPF)** to enter mitosis from G_2. Active MPF consists of two proteins: cdc2 and cyclin. The cdc2 component, which has been found in all eukaryotic cells studied, from yeasts to humans, is an enzyme that by itself is inactive. The concentration of the protein cyclin in the cell varies during the cell cycle. When the cyclin concentration rises, cyclin combines with cdc2, forming a compound known as pre-MPF. This compound is then converted to active MPF. Researchers are only beginning to unravel the details of how the cell regulates mitosis and other events of the cell cycle.

In multicellular organisms, cells are also influenced by signals from other cells and organs. For example, plant cells are stimulated to undergo mitosis by certain hormones, such as **cytokinins**. Cytokinins promote mitosis both in normal growth and in wound healing (Chapter 30).

The cell cycle can also be affected by agents that interfere with the normal function of the mitotic spindle. **Colchicine**, a drug derived from the autumn crocus plant, can be used to block cell division in eukaryotic cells. Colchicine binds with tubulin, the principal protein in microtubules. This causes the spindle to break down and prevents movement of the chromosomes to the opposite poles of the cell. As a result, a cell may end up with an extra set of chromosomes.

In general, plants are relatively tolerant of extra chromosome sets, a condition known as **polyploidy**. In fact, polyploid plants tend to be larger and more vigorous than normal plants. Polyploidy in plants is of commercial importance, especially in connection with the development of new varieties of ornamental and agricultural plants. Although polyploidy can be induced experimentally in certain animals, it rarely occurs naturally. In fact, extra chromosome sets are often lethal in animals.

Recall from the chapter introduction that cancer cells divide rapidly and are not regulated by normal cellular control mechanisms. Some drugs[1] used in cancer chemotherapy block cell division or specifically injure dividing cells. Because cancer cells divide much more rapidly than most normal body cells, they are most affected by these drugs. However, these drugs do affect some types of normal body cells, especially those that multiply rapidly, such as those lining the digestive tract and those that produce the outer layer of the skin and its derivatives such as hair. This is why such drugs produce side effects, like hair loss, nausea, and diarrhea.

[1] Like colchicine, these drugs are plant products. Vincristine and vinblastine are obtained from rosy periwinkle (genus *Vinca*). One argument for conserving plant species is the possibility that some of them may prove to be sources of valuable drugs.

CONCEPT CONNECTIONS

Meiosis* — *Reproduction

As you learn about genetics in the following chapters, remember that the two meiotic divisions produce four haploid cells. Their nuclei each contain one—and only one—of each homologous pair of chromosomes. Each of these haploid cells may have a different combination of genes.

In animals, meiosis typically takes place in specialized reproductive structures called **gonads**. The cells produced mature to become sperm and eggs. When sperm and egg unite in fertilization, forming the single-celled zygote, the diploid number of chromosomes is restored.

In the sporophyte generation of plants, meiosis takes place in specialized structures called **sporangia**. In the case of seed plants, sporangia are found in **cones** or **flowers**. Spores are the reproductive cells formed in plants as a result of meiosis.

DURING MEIOSIS, THE CHROMOSOME NUMBER OF A CELL IS REDUCED BY HALF

Mitosis ensures that each daughter cell recives exactly the same number and kind of chromosomes that the parent cell had. When a diploid cell undergoes mitosis, two diploid cells are produced. When a haploid cell undergoes mitosis, two haploid cells are produced. In contrast, **meiosis** reduces the number of chromosomes by half and produces haploid cells from diploid cells. Meiosis is a special type of cell division that ensures haploid gametes. Through meiosis the chromosome number of a species can remain the same from one generation to the next.

Meiosis also promotes genetic variability. Meiosis *separates* the members of each homologous pair of chromosomes. Any contrasting genetic traits that are held by each homologous pair of chromosomes are separated and distributed independently to different haploid cells (gametes or spores). As a result, no two offspring, even of the same parents, are likely to be exactly alike.

Meiosis Differs from Mitosis

Many of the events of meiosis are somewhat similar to the events of mitosis, but there are several important differences (Fig. 10–6).

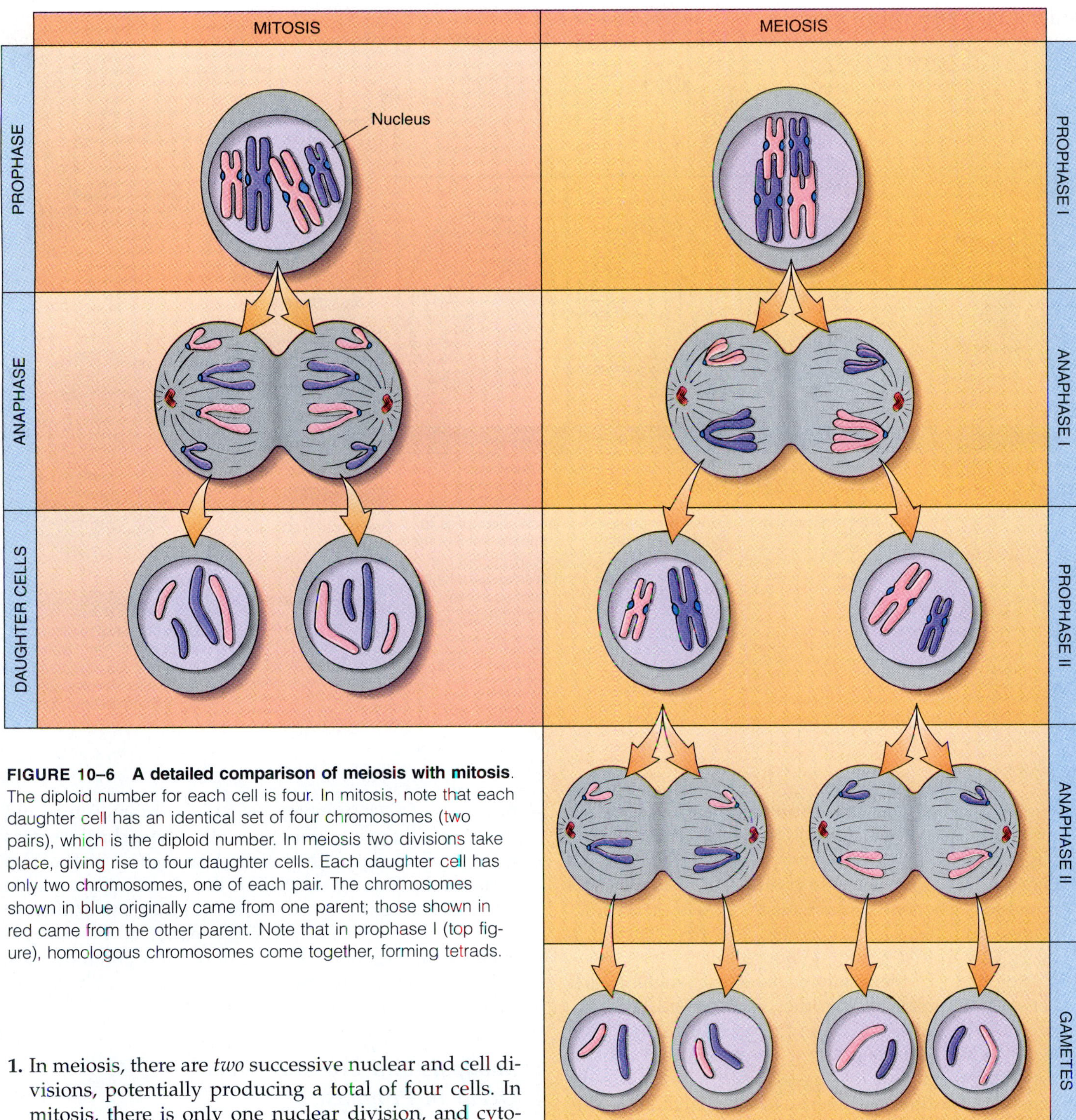

FIGURE 10–6 A detailed comparison of meiosis with mitosis. The diploid number for each cell is four. In mitosis, note that each daughter cell has an identical set of four chromosomes (two pairs), which is the diploid number. In meiosis two divisions take place, giving rise to four daughter cells. Each daughter cell has only two chromosomes, one of each pair. The chromosomes shown in blue originally came from one parent; those shown in red came from the other parent. Note that in prophase I (top figure), homologous chromosomes come together, forming tetrads.

1. In meiosis, there are *two* successive nuclear and cell divisions, potentially producing a total of four cells. In mitosis, there is only one nuclear division, and cytokinesis typically occurs only once, producing two daughter cells.
2. Each of the four cells produced in meiosis contains the haploid number of chromosomes, that is, only one member of each homologous pair. In mitosis, each daughter cell contains the diploid number of chromosomes. (An exception occurs in the plant life cycle, when haploid cells divide by mitosis to form haploid cells.)
3. During meiosis, the homologous chromosomes containing genetic information from each parent are thor-

FIGURE 10–7 The stages of meiosis. The diploid number for the cell shown here is four. Note that the number of chromosomes in each of the four daughter cells is half the number in the original parent cell.

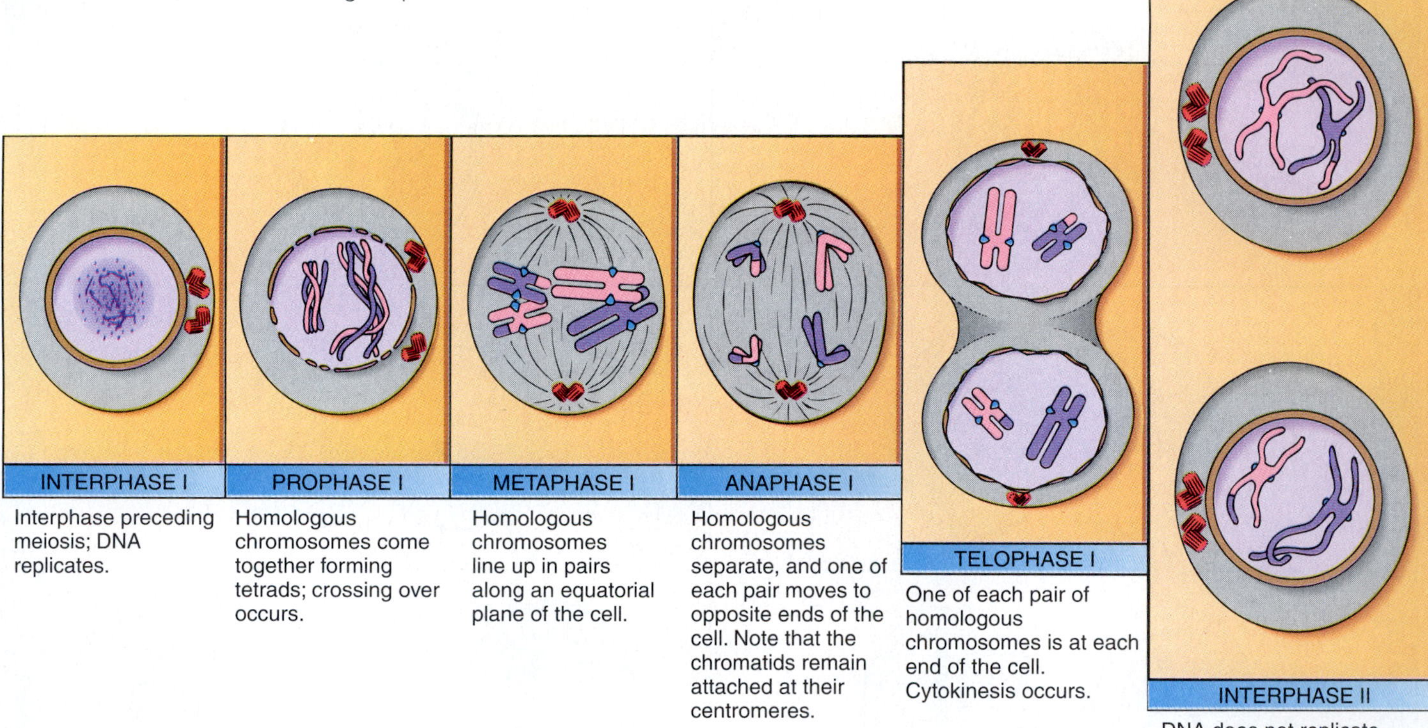
INTERPHASE I
Interphase preceding meiosis; DNA replicates.
PROPHASE I
Homologous chromosomes come together forming tetrads; crossing over occurs.
METAPHASE I
Homologous chromosomes line up in pairs along an equatorial plane of the cell.
ANAPHASE I
Homologous chromosomes separate, and one of each pair moves to opposite ends of the cell. Note that the chromatids remain attached at their centromeres.
TELOPHASE I
One of each pair of homologous chromosomes is at each end of the cell. Cytokinesis occurs.
INTERPHASE II
DNA does not replicate. Each cell is now haploid. Note that the chromatids are still joined. Chromosomes do not completely elongate.

oughly shuffled, and one chromosome of each pair is randomly distributed to each new cell. The resulting haploid cells produced have new combinations of genes and chromosomes. In contrast, mitosis produces daughter cells that contain identical sets of chromosomes, and these chromosome sets are identical in every way to that of the mother cell.

4. During meiosis, there may be some exchange of parts between homologous chromosomes (called crossing-over; see later discussion) so that even the genes originally located together on one chromosome do not always stay together. Crossing-over further increases the genetic reshuffling of meiosis. In mitosis, there is little opportunity for crossing-over.

Meiosis Includes Two Divisions

Meiosis consists of two cell divisions, the first and second meiotic divisions, simply called **meiosis I** and **meiosis II**. Each division may include a prophase, metaphase, anaphase, and telophase stage. Meiosis I separates the members of each homologous pair of chromosomes and distributes them into daughter cells. These chromosomes were duplicated prior to meiosis I, so each consists of two chromatids. During meiosis II, the chromatids separate into individual chromosomes, which then enter different haploid daughter cells. Since it is easier to follow these events in an organism that has only a few chromosomes, we will discus a hypothetical organism with a diploid number of four chromosomes (Fig. 10–7).

Synapsis and Crossing-Over Occur in Prophase I

As in mitosis, the chromosomes are duplicated during the S phase before meiosis actually begins. Recall that when a chromosome is duplicated, for a time it consists of two chromatids joined at their centromeres.

During **prophase I** (prophase of the first meiotic division), while the chromatids are still elongated and thin, the homologous chromosomes come to lie lengthwise side by side. This pairing of homologous chromosomes is called

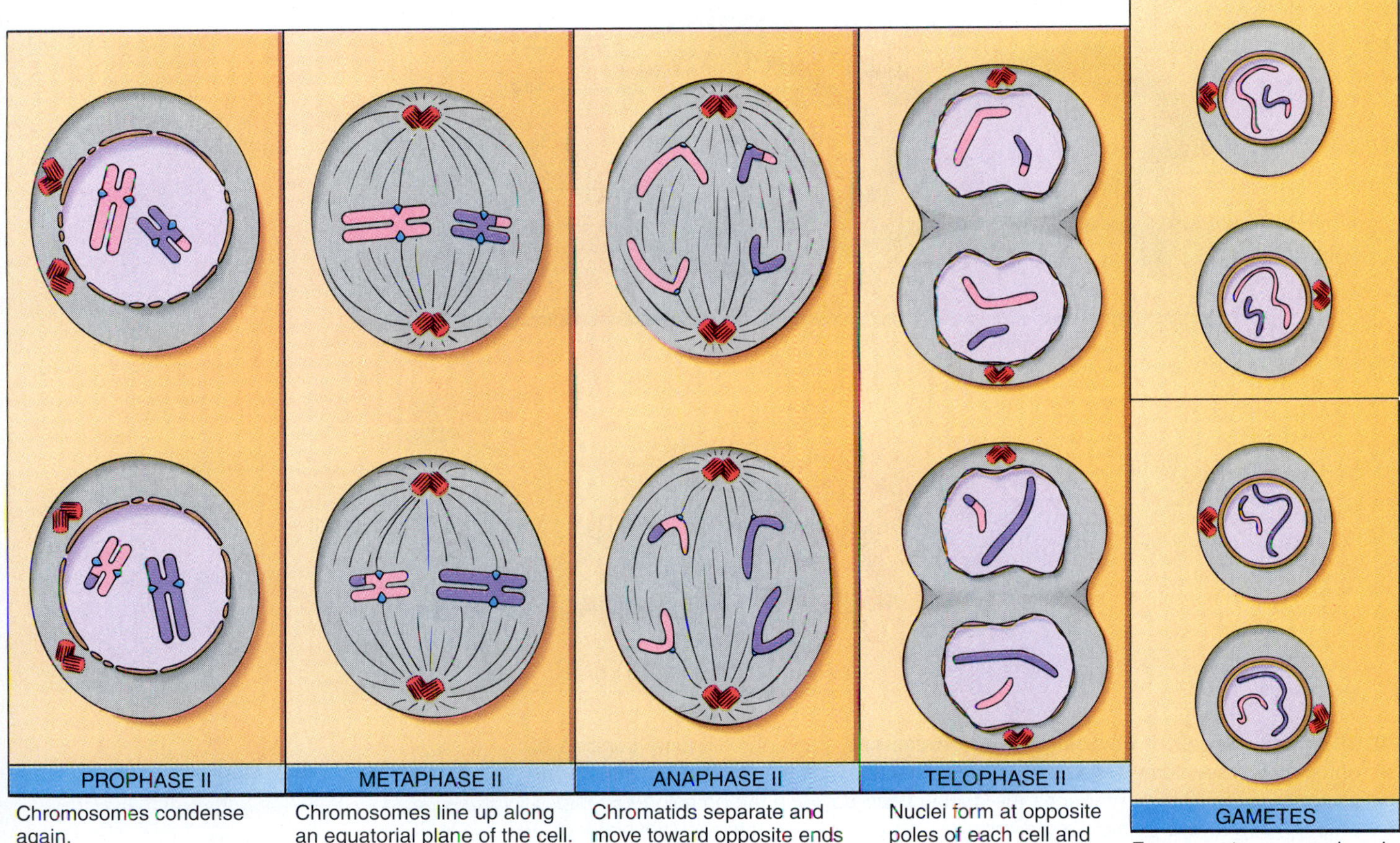
PROPHASE II
Chromosomes condense again.
METAPHASE II
Chromosomes line up along an equatorial plane of the cell.
ANAPHASE II
Chromatids separate and move toward opposite ends of the cell.
TELOPHASE II
Nuclei form at opposite poles of each cell and cytokinesis occurs.
GAMETES
Four gametes are produced. Note that each cell is haploid.

synapsis, which means "fastening together." The diploid number (2n) in our example is four, so at synapsis there are two homologous pairs. One of each pair, the maternal chromosome, was originally inherited from the individual's mother, whereas the other member of each pair, the paternal chromosome, was contributed by the father.

Because each chromosome actually consists of *two* chromatids at this time, synapsis results in the coming together of *four* chromatids. The complex of four chromatids is known as a *tetrad*. The number of tetrads equals the haploid number of chromosomes. In human cells there are 23 tetrads (and a total of 92 chromatids) at this stage.

All the genes located on a particular chromosome are said to be **linked** together and therefore will tend to be inherited together in **linkage groups**. However, this tendency for linked genes to stay together is not absolute. During synapsis, genetic material may be exchanged between homologous chromatids by the process of **crossing-over**. In this process, homologous parts are exchanged between the two chromatids, producing new combinations of genes (Fig. 10–8). The exchange of genetic material between *homologous* chromosomes results in **genetic recombination**. The new combinations of genes greatly enhance the prospects for variety among offspring of sexual partners (Chapter 11).

While the events characteristic of prophase I (synapsis and crossing-over) of meiosis are occurring, other events similar to those of mitotic prophase are taking place. In these events, the centrosomes move to opposite poles, a spindle forms, and the nuclear envelope dissolves.

Homologous Chromosomes Separate during Meiosis I

The tetrads (paired homologous chromosomes) line up along the equator of the spindle during **metaphase I.** Both chromatids of one chromosome are oriented toward the same pole. Their sister chromatids of the homologous chromosome are oriented toward the opposite pole (see Fig. 10–7).

During **anaphase I** the homologous chromosomes of each pair separate, moving toward opposite poles. The

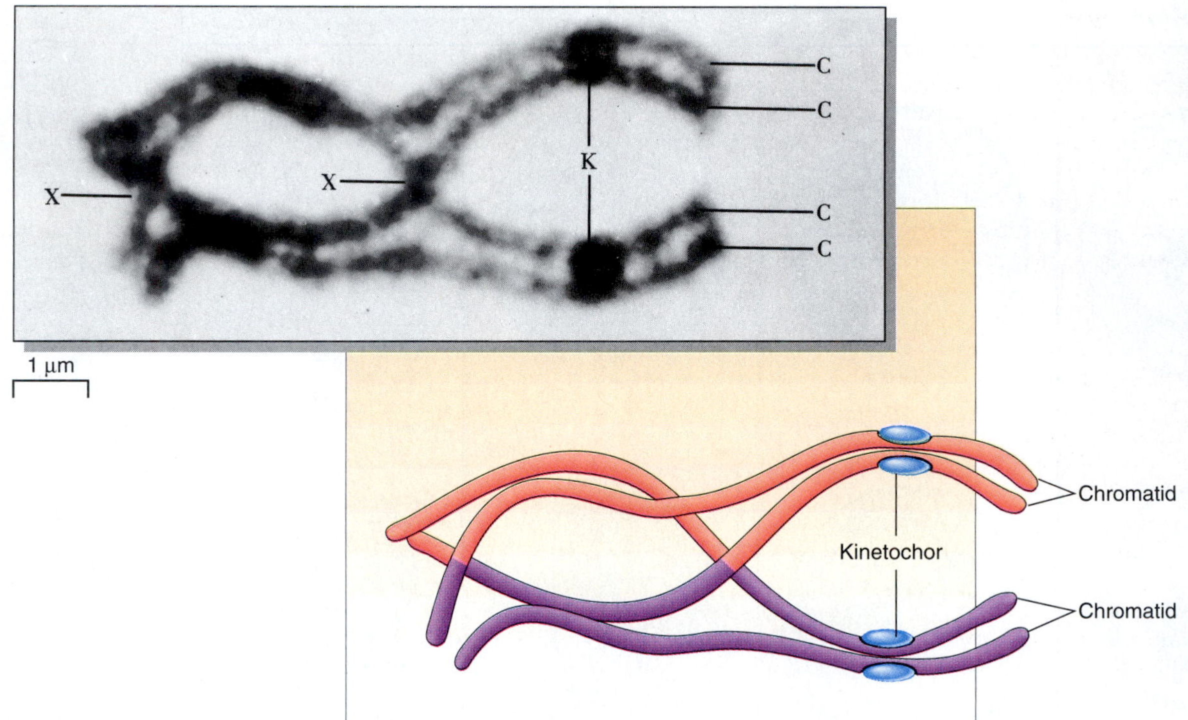

FIGURE 10–8 A pair of homologous chromosomes. Each chromosome consists of two chromatids; they come together during prophase of the first meiotic division, forming a tetrad. Crossing-over produces the connections between the homologous chromosomes shown at two points. Tetrad from a salamander spermatocyte (developing sperm cell) is shown in the photomicrograph. The four chromatids that make up the tetrad are visible. Also note the kinetochores. The interpretive drawing shows the structure of the tetrad. (Courtesy of J. Kezer)

chromatids of each chromosome are still joined at their centromere regions. In **telophase I** in our example, there would be two doubled chromosomes (four chromatids) at each pole. During telophase I, the nuclei reorganize, the chromatids begin to elongate, and cytokinesis (cell division) generally takes place. Notice that the haploid number of chromosomes has been established, although each chromosome still consists of two chromatids.

Chromatids Separate in Meiosis II

During the interphase that follows meiosis I, no further DNA or chromosome replication takes place. In most organisms, meiotic interphase is very brief; in some organisms, it is absent. Since the chromatids do not completely elongate between meiotic divisions, **prophase II** is also brief. In prophase II, only one of each pair of chromosomes is present, so there is no synapsis or crossing-over.

During **metaphase II** the chromatids again line up on the equator. Only one of each homologous chromatid pair is present. This contrasts with metaphase I in which the homologous chromatids are visible as tetrads.

During **anaphase II,** the centromeres split and the sister chromatids, now complete chromosomes, separate and move to opposite poles. Thus, in **telophase II,** one of each kind of chromosome is located at each end of the cell. The haploid condition that existed at telophase I has been maintained, and each chromosome is now in an unduplicated state. Nuclear envelopes then form, the chromosomes gradually elongate into chromatin, and cytokinesis occurs.

Chapter Summary

I. Each somatic cell of every organism of a given species has a characteristic number of chromosomes.

II. In the somatic cells of complex diploid organisms, chromosomes are present in pairs, called homologous chromosomes. One of each pair was contributed by the father and the other was contributed by the mother.
   A. A cell with chromosomes occurring in pairs is diploid, or 2n.
   B. A cell with only one member of each pair of chromosomes is haploid, or n.

III. The cell cycle is the period from the beginning of one cell division to the beginning of the next division.
   A. During interphase, the cell grows and prepares for the next division. The cell also replicates its DNA during interphase (the S phase).
   B. During mitosis, a complete, identical set of chromosomes is distributed to each daughter nucleus.
      1. During prophase, chromatin condenses into chromosomes, the nucleolus and the nuclear envelope break down, and the mitotic spindle begins to form. At the end of prophase, each chromosome is composed of two (sister) chromatids, joined at their centromeres.
      2. During metaphase, the chromatids line up along the equator of the cell. Microtubules of the spindle attach to the kinetochores of the chromatids.
      3. During anaphase, the sister chromatids separate and are moved toward opposite poles of the cell; each chromatid is now a complete chromosome.
      4. During telophase, a nuclear envelope forms around each set of chromosomes, nucleoli reappear, the chromosomes lengthen and become chromatin, the spindle disappears, and cytokinesis generally takes place.

IV. Meiosis reduces the chromosome number by one half. It is a required process in the life of all sexually reproducing organisms because it results in haploid gametes.
   A. During interphase prior to meiosis, DNA replicates and the cell prepares for meiosis.
   B. During prophase I, the homologous chromosomes undergo synapsis and crossing-over, resulting in new combinations of genes (genetic recombination).
   C. The members of each pair of homologous chromosomes separate during anaphase I and are distributed to different daughter cells.
   D. During the second meiotic division, the two chromatids of each chromosome separate, and each is distributed to a daughter cell.
   E. As a result of meiosis, four haploid cells are formed.

V. In sexual reproduction, a haploid set of chromosomes from the sperm and a haploid set of chromosomes from the egg unite, forming a diploid zygote.

Selected Key Terms

anaphase, p. 191
anaphase I, p. 197
anaphase II, p. 198
aster, p. 190
cell cycle, p. 189
cell plate, p. 192
centromere, p. 190
centrosome, p. 190
chromatid, p. 190
crossing-over, p. 197
cytokinesis, p. 190
diploid, p. 188
gamete, p. 188
haploid, p. 188
homologous chromosome, p. 188
interphase, p. 190
kinetochore, p. 190
maternal chromosome, p. 188
maturation promoting factor (MPF), p. 193
meiosis, p. 188
metaphase, p. 190
metaphase I, p. 197
metaphase II, p. 198
mitosis, p. 188
mitotic spindle, p. 190
paternal chromosomes, p. 188
prophase, p. 190
prophase I, p. 196
prophase II, p. 198
sister chromatid, p. 190
somatic cells, p. 188
synapsis, p. 197
telophase, p. 191
telophase I, p. 198
telophase II, p. 198
zygote, p. 188

Post-Test

1. Except during mitosis, eukaryotic chromosomes appear as long, thin, dark-staining threads called ______________.
2. A human somatic cell with 46 chromosomes would be described as ______________ or ______________.
3. A sperm cell with only one of each pair of chromosomes is described as ______________ or ______________.
4. The members of a given pair of chromosomes are referred to as ______________ chromosomes.

5. The period from the beginning of one cell division to the beginning of the next is known as the ______________ ______________.
6. In ______________, a complete diploid set of chromosomes is distributed to each new daughter nucleus.
7. The actual division of the cytoplasm to form two cells is called ______________.
8. During the ______________ stage of its life cycle, a cell grows and carries out its normal functions.
9. DNA is replicated during the ______________ phase of interphase.
10. Sister chromatids are attached at their ______________.
11. During ______________ of mitosis, the chromatids are lined up along the equatorial plane of the cell.
12. During ______________ of mitosis, the chromatids separate and begin to move toward opposite poles.
13. In meiosis, there is/are ______________ division(s), producing a total of ______________ cells.
14. Synapsis results in the formation of complexes of chromosomes known as ______________.
15. During ______________-______________, segments of DNA of homologous chromatids are broken and exchanged.
16. The exchange of DNA segments between homologous chromatids may result in genetic ______________.
17. Crossing-over occurs during (a) prophase of mitosis (b) anaphase of mitosis (c) prophase I of meiosis (d) prophase II of meiosis (e) two of the preceding answers.
18. Homologous chromosomes move toward opposite poles of the cell during (a) metaphase of mitosis (b) anaphase of mitosis (c) anaphase I of meiosis (d) anaphase II of meiosis (e) telophase II of meiosis.
19. The nuclear envelopes reorganize during (a) anaphase of mitosis (b) telophase of mitosis (c) prophase II of meiosis (d) telophase II of meiosis (e) two of the preceding answers.
20. Chromosomes are duplicated during (a) G_1 phase (b) S phase (c) anaphase of mitosis (d) anaphase II of meiosis (e) prophase I of meiosis
21. Maturation promoting factor (MPF) (a) helps regulate the cell cycle (b) consists of two polysaccharides (c) has an action similar to that of colchicine (d) is produced by the asters (e) two of the preceding answers.
22. If the diploid number of a particular species is 40, the number of chromatids in one of its cells that is in prophase II is (a) 40 (b) 80 (c) 20 (d) 10 (e) none of the preceding answers is correct.

Review Questions

1. Describe three ways in which meiosis differs from mitosis.
2. Define the following terms: (a) diploid, (b) haploid, and (c) homologous chromosomes.
3. Draw the stages of mitosis of a cell with a diploid chromosome number of six.
4. Draw the stages of meiosis of a cell with a diploid chromosome number of four. Indicate when and how synapsis and separation of homologous chromosomes occur.
5. Describe the structure of a chromosome. Distinguish between chromatin, chromosome, and chromatid.
6. What is the relationship between genes and chromosomes?

Thinking Critically

1. Propose a hypothesis regarding the relationship between oncogenes and maturation promoting factor (MPF).
2. Two very different species may have the same diploid chromosome number. How can this be explained?
3. What is polyploidy? What is its significance? Give an example.
4. In what way might crossing-over during meiosis be important as an evolutionary mechanism?
5. Imagine that a sperm with six chromosomes fertilizes an ovum with seven chromosomes. How might this affect the zygote and development of the embryo?

Recommended Readings

Glover, D.M., C. Gonzalez, and J.W. Raff, "The Centrosome," *Scientific American*, Vol. 268, No.6, June 1993, 62–68. Centrosomes organize the cellular skeleton and set up the spindle that partitions the chromosomes in cell division.

McIntosh, J.R., and K.L. McDonald, "The Mitotic Spindle," *Scientific American*, October 1989. A review of current understanding of the mechanisms involved in mitotic chromosome separation.

Murray, A.W., and M.W. Kirschner, "What Controls the Cell Cycle," *Scientific American*, March 1991. A review of regulators involved in controlling the cell cycle.

Which is more important in cancer, heredity or environment? *Although colon cancer tends to occur in families, indicating a genetic factor, a lower incidence of colon cancer occurs in those who choose to eat a healthy, sensible diet. (Comstock, Inc./Comstock, Inc.)*

Human Diseases: The Genetic Factor

Beginning in the 1970s, technological breakthroughs have led to thousands of exciting genetic discoveries, many of which have been reports of specific genes linked to human diseases. Consider the following findings, which represent a small fraction of those reported in 1993. A DNA marker for a gene associated with colon cancer was identified on human chromosome 2, and a gene linked with a form of hyperactivity was found on chromosome 3. Also in 1993, the gene associated with Huntington's disease, a progressive hereditary disease discussed in Chapter 12, was located on chromosome 4, while chromosome 7 was determined to be the site of a gene linked with adult-onset diabetes (Type II diabetes). A gene associated with Alzheimer's disease was traced to chromosome 19, whereas a gene involved in ALD, also known as Lorenzo's disease, dramatically portrayed in the movie *Lorenzo's Oil*, was discovered on the X chromosome.

As more and more genes are linked to disease, it is tempting to ascribe *all* human illness to faulty genes. Such a reductionist viewpoint is inappropriate because it fails to take into account the importance of the environment in causing disease. Even when a disease is linked to a faulty gene, that gene usually interacts with several other genes and with environmental factors to determine whether or not a person carrying the gene eventually develops the disease. Sixty-five percent of people carrying the colon cancer gene eventually develop colon cancer, for example, but 35% do not; if the gene was the exclusive cause of colon cancer, all carriers of the gene would eventually develop cancer. It is not known why some carriers of the colon cancer gene get colon cancer while others do not, but environmental factors such as diet may help determine the outcome. For example, the incidence of colon cancer appears to be lower in those who choose a diet low in fat and high in natural fiber (from fruits, vegetables, and whole grains). Thus, although certain genes increase one's risk for disease, in most cases genes are not the single causative agent of human illness.

The basis of the exciting work of identifying genes associated with specific human disorders is found in the 19th-century experiments of Gregor Mendel, an Austrian monk who deduced how inheritable traits are passed from one generation to the next. And so we begin our study of genetics with Mendel, the father of genetics.

Patterns of Inheritance

11

Learning Objectives

After you have studied this chapter you should be able to:

1. Define the following terms relating to genetic inheritance: genotype and phenotype, dominant and recessive, homozygous and heterozygous.
2. Distinguish between genes and chromosomes, and between genes and alleles.
3. Relate the inheritance of genetic traits to the behavior of chromosomes during meiosis.
4. Explain how one generation is genetically connected to the generation that preceded it and to the generation that follows it.
5. Solve simple problems in genetics involving monohybrid and dihybrid crosses and explain your results.
6. Describe an example of linkage.
7. Compare and contrast dominance, incomplete dominance, and codominance.
8. Distinguish among multiple alleles, polygenes, pleiotropy, and epistasis, comparing their effects on gene expression.

Key Concepts

- ☐ The gene is the basic unit of heredity.
- ☐ Genes transmit information from one cell to another, and since genes are located on chromosomes, they are generally inherited in the same manner as are the chromosomes to which they belong.
- ☐ Alternate forms of a gene, called alleles, are responsible for genetic variation. Although alleles occur in pairs, one on each homologous chromosome in a diploid cell, only one allele of each pair normally passes to a gamete (as a result of meiosis).
- ☐ In many cases, an allele is expressed only when both alleles are identical; in other cases, an allele is expressed even when the other allele in the pair is different. Sometimes alleles are *both* partially or fully expressed, even when they are different from one another.
- ☐ Genetic variation may also be affected by interactions among different genes, by a single gene that affects more than one trait, and by factors in the environment.

GREGOR MENDEL FOUNDED THE SCIENCE OF GENETICS

The foundation of our modern knowledge of genetics was laid in the 19th century by Gregor Mendel (1822-1884), an Austrian monk who lived in the town of Brunn, Austria, which is now Brno in the Czech Republic (Fig. 11–1).

The general view that prevailed in Mendel's time was that inheritance was the result of a fusing of two parents' traits. There was no clear-cut understanding of the special role of the reproductive cells in inheritance; instead, the entire body was thought to participate. This idea is still reflected in common expressions linking inheritance to blood, which of course plays no part in it. It was thought that two parents somehow blended or fused their "blood" to produce an offspring that had a mixture of both "bloods" and therefore had traits intermediate between those of its parents.

Mendel changed all of that with a series of elegant experiments involving the garden pea. A number of distinct varieties, differing in such characteristics as height, flower color, seed coat color, and seed shape, exist in garden peas. In each case, the characteristic exhibits two contrasting forms, as for example, *tall* and *dwarf.* Mendel designed his experiments with garden peas carefully, breeding his peas for many generations and carefully counting the number of plants with each trait in each generation. In each experiment, he focused on one or a few pairs of contrasting characteristics.[1] He then analyzed the data and based his conclusions on his analysis. In short, Mendel was a good experimental biologist.

Mendel's conclusions laid the foundation for understanding inheritance in all sexually reproducing organisms, from peas to humans, and revolutionized the field of biology. So what did he find? Mendel's work indicated that inheritance of traits was not due to a blending but rather to the transmission of specific units of inheritance, now called **genes**. Mendel not only predicted the existence of genes, but also that **alleles**, the alternate forms of a gene, normally occur in pairs. His results indicated that, during the formation of gametes (haploid reproductive cells), the two alleles of a gene separate from one another so that each gamete contains only one allele of each pair. During sexual reproduction, the allelic pairs are restored when an egg cell containing one allele combines with a sperm cell containing another allele. Thus, Mendel's research also predicted meiosis and the existence of haploid and diploid cells.

FIGURE 11–1 Gregor Mendel, the father of genetics. (The Bettmann Archive).

[1] Mendel was not the first scientist to study inheritance. He succeeded where others failed in part because he ignored the prevailing view that nature could only be understood by studying all aspects of an organism rather than individual traits. Other breeders obtained results similar to Mendel's, but because they focused on *all* of the traits of the entire organism, they did not comprehend the significance of their own work.

CONCEPT CONNECTIONS

Mendelian Genetics *Evolutionary Theory*

Charles Darwin, the English naturalist credited with originating modern evolutionary theory, lived and worked at the same time as Gregor Mendel, although neither scientist interacted with the other. Darwin proposed that living organisms change over time by a process called natural selection. One of the premises of Darwin's theory of evolution by natural selection is that individuals pass traits on to the next generation. However, Darwin could not explain *how* these traits were inherited, nor could he explain *why* individuals vary within a population. Had Darwin been aware of Mendel's work on peas, he could have answered these questions about the genetic basis of inherited variation, at least partially.

During the 1920s, biologists combined Mendelian genetics with Darwin's theory to produce a comprehensive theory of evolution known as the synthetic theory of evolution. The synthetic theory of evolution is addressed in greater detail in Chapter 17.

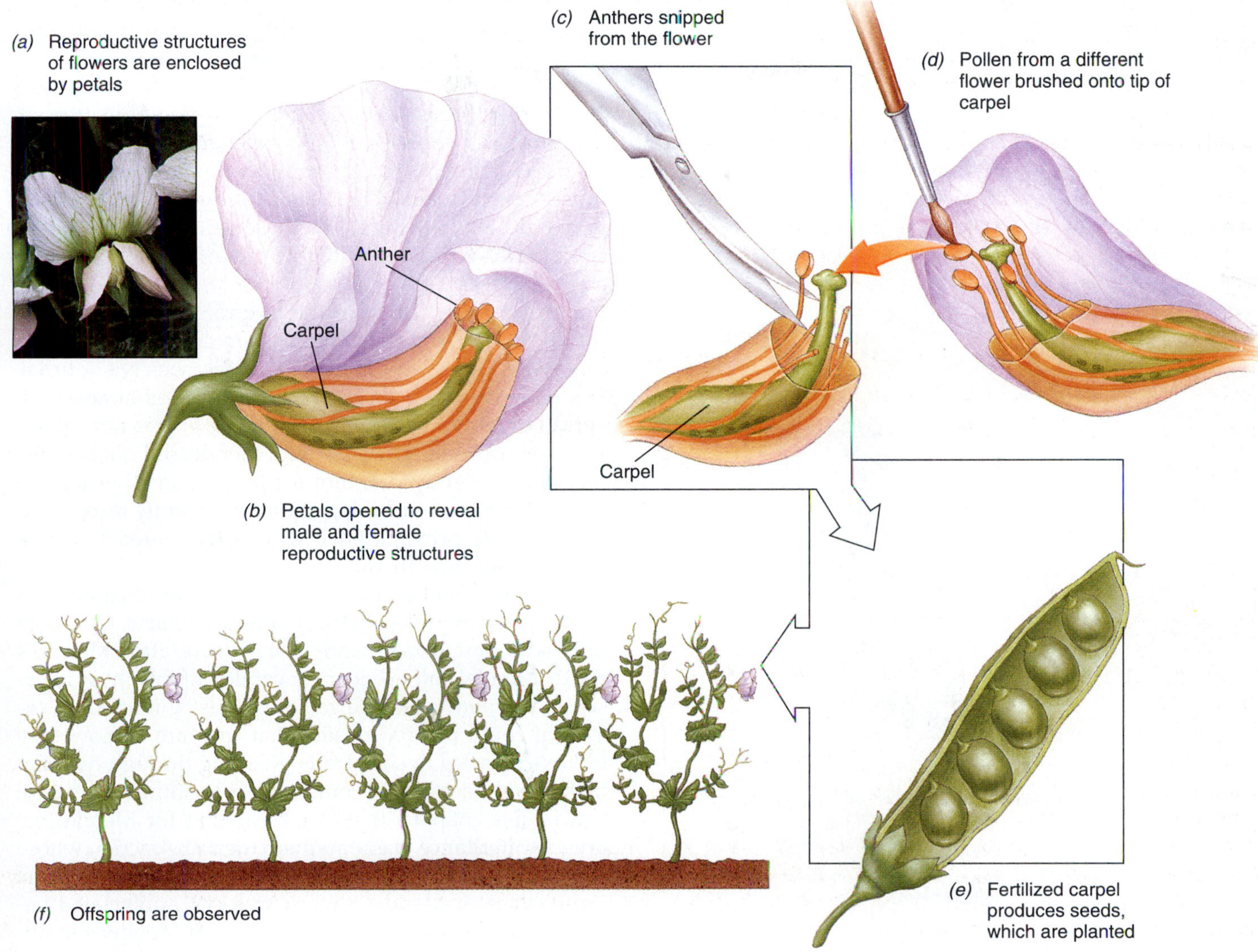
(a) Reproductive structures of flowers are enclosed by petals
(c) Anthers snipped from the flower
(d) Pollen from a different flower brushed onto tip of carpel
Anther
Carpel
Carpel
(b) Petals opened to reveal male and female reproductive structures
(f) Offspring are observed
(e) Fertilized carpel produces seeds, which are planted

FIGURE 11–2 Mendel's research involved the garden pea. (***a***) The open flower (***b***) reveals the pollen-producing anthers and the tip of the carpel, the female part that receives the pollen. Since the petals completely enclose these parts, there is little chance of natural cross-pollination between separate flowers. (***c, d***) After the anthers are removed from the flower, pollen from another plant is dusted onto the tip of the carpel. (***e, f***) The seeds produced from this procedure are planted and the offspring observed. (Dr. Jeremy Burgess/ Science Photo Library/ Photo Researchers, Inc.)

Mendel Experimented with Garden Peas

Like other flowering plants, peas produce pollen grains, each of which contains haploid male gametes (nonflagellated sperm; Chapter 23). When a pollen grain is deposited on the appropriate female flower parts, it grows a long tube into them, which eventually reaches the haploid egg. A sperm passes down the pollen tube to the egg, the two gametes (egg and sperm) fuse, and an embryonic plant develops from the zygote (fertilized egg). As development continues, a seed forms, containing the embryonic plant along with the nutrients that it needs to get a start in life.

Garden peas are normally self-fertilized—that is, the male and female gametes that fuse during sexual reproduction are from the same flower. Therefore, the simple surgical removal of the male flower parts makes the plant incapable of being fertilized except by artificial means (Fig. 11–2). In this way, the crossing of different varieties can be closely controlled.

In a typical experiment, Mendel crossed a tall-growing variety of pea with a short, or bush, variety (Fig. 11–3). The offspring were not of some intermediate height, but instead resembled their tall parent. If two of these offspring were crossed (or one offspring allowed to fertilize itself), then throwbacks to the short variety would occur in a 3:1 ratio, that is, about three tall plants would be seen for every short one in the second generation. No matter what trait Mendel studied in peas, he obtained the same results. The first generation always resembled one of the parents, and the second generation possessed individuals with both traits, but always in a 3:1 ratio.

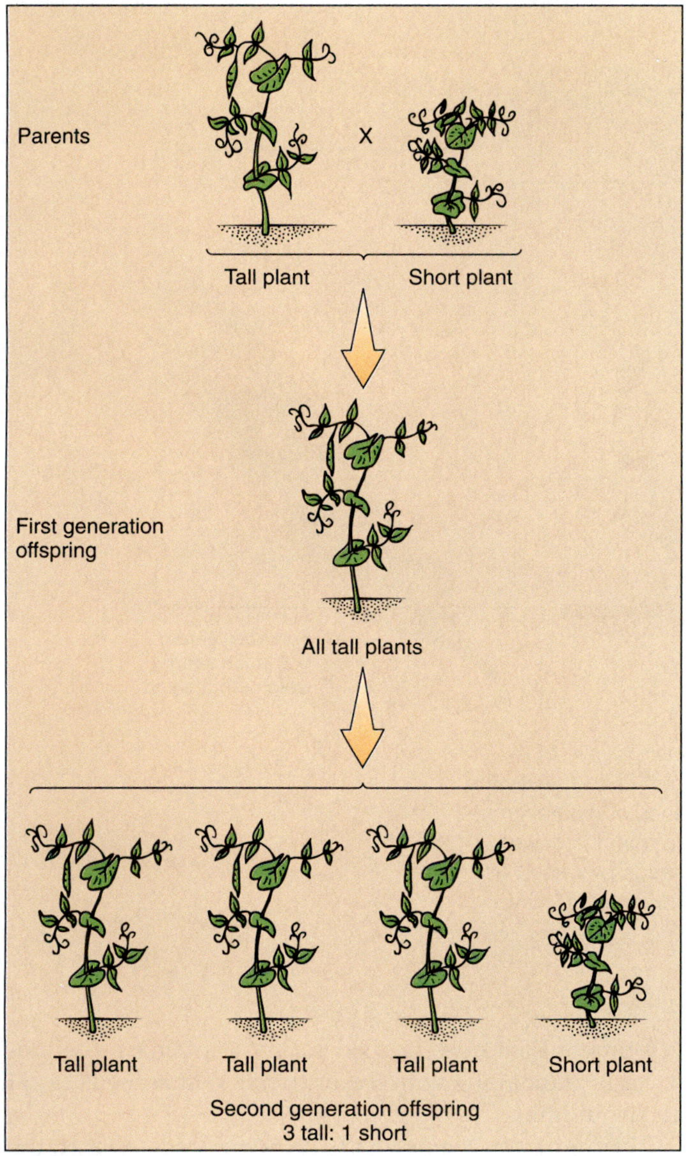

These results led Mendel to conclude that some traits can silently persist, even though they are not expressed. Therefore, an individual must contain two sets of instructions for a particular trait, that is, the instructions must occur in pairs. The 3:1 ratio in the second generation indicated to Mendel that these pairs combine and recombine in accordance with the rules of probability[2] (covered shortly).

Mendel was active in the natural history society of Brunn and presented his findings in a series of research reports, which were published in the society's journal. His research was revolutionary, but it was almost universally ignored because its significance was not understood or appreciated. As a result, Mendel did not gain the recognition rightly due him for more than 30 years. After his death, several biologists independently recognized most of these principles and then rediscovered Mendel's papers describing them.

Because neither the behavior of chromosomes in meiosis nor the role of deoxyribonucleic acid (DNA) in cells was determined until some years after Mendel's death, Mendel had no clear conception of the physical basis of inheritance. He inferred that units of inheritance exist, but could not have said what they are or how they influence an organism's traits. Reading through his original papers (see the Recommended Readings list at the end of this chapter) makes it clear that for Mendel the unit of inheritance was a mathematical abstraction whose behavior could be described by equations. Perhaps the significance of Mendel's work was not appreciated because most biologists of his time lacked Mendel's mathematical background and were not used to thinking in quantitative terms.

Parents

X

Tall plant

Short plant

First generation offspring

All tall plants

Tall plant

Tall plant

Tall plant

Short plant

Second generation offspring
3 tall: 1 short

FIGURE 11–3 One of several kinds of crosses carried out by Gregor Mendel using different varieties of garden pea. If a tall and a short plant were crossed, the offspring were all tall plants. If a member of that generation were self-fertilized, its offspring (the second generation) would yield three tall plants for every short one.

Modern Principles of Inheritance Are Based on Mendel's Work

Mendel's breeding experiments led him to certain conclusions about the mechanisms of heredity, some of which later scholars restated as **Mendel's principles of inheritance**:

1. Inherited traits are transmitted by genes, which occur in pairs, called alleles.

[2] Probability is the likelihood that a given event will occur. For example, when one tosses a coin twice, the probability of getting heads on the first toss is 1/2 (one chance in two), and the probability of getting tails on the first toss is also 1/2. If an event is certain to occur, its probability is 1 (that is, 1/1); if it is certain not to occur, its probability is 0.

2. The **principle of dominance**: When two alternate forms of the same gene (that is, two different alleles) are present in an individual, often only one—the dominant allele—is expressed.
3. The **principle of segregation**: When gametes are formed in meiosis, the two alleles of each gene segregate (separate) from one another, and each gamete receives only one allele.
4. The **principle of independent assortment**: When two or more traits are examined in a single cross, each trait is inherited without relation to the other traits. This occurs because the alleles for each trait assort into the gametes by chance, that is, independently of one another. All possible combinations of genes will thus occur in the gametes.

GENES OCCUR IN PAIRS CALLED ALLELES AND ARE INHERITED AS PARTS OF CHROMOSOMES

A typical gene is now viewed as a portion of DNA that contains the information necessary to manufacture a specific polypeptide or protein.[3] This concept can be illustrated with a common genetic abnormality, albinism. An **albino** is a person or animal that lacks the ability to produce the dark brown or black pigment **melanin**. This pigment is responsible for most of the color of hair, skin, and eyes. Without melanin, an organism appears completely unpigmented, with extremely light skin, white or pale yellow hair, and pink eyes (caused by underlying blood vessels) (Fig. 11–4).

Most albinos are unable to make a functional form of an enzyme needed to form melanin from the amino acid tyrosine. Without that enzyme, no pigment can be formed. Thus, the allele for albinism does not produce an albino directly, but instead results in the production of a defective enzyme. The normal allele of this gene codes for a functional version of the enzyme.

If an individual had both a defective albino allele and a normal allele, the skin, hair, and eyes would be normally colored due to the presence of melanin produced by the functional enzyme. Although the **genotype**, that is, the genetic makeup, of such a genetically mixed individual would contain an albino allele, one would not know this from the individual's appearance. The individual's **phenotype**, the observable features (that is, the portion of the genotype that is actually expressed), would be normal.

[3] Although some DNA (particularly that which is involved in the regulation of genes) may not code for protein directly, it usually does affect the function of protein-producing genes.

FIGURE 11–4 An individual with albinism. Humans with albinism, an inherited lack of pigmentation, have very light skin, pale yellow to white hair, and pink eyes. (John Watney/ Science Source/ Photo Researchers, Inc.)

A Dominant Allele Masks the Expression of a Recessive Allele

Recall that the two members of a pair of chromosomes are said to be **homologous** (Fig. 11–5). Homologous chromosomes contain genes for similar traits arranged in similar order. The gene for each trait occurs at a particular site in the chromosome called a **locus** (plural, loci). In guinea pigs, for example, if one homologous chromosome contains a gene for coat color, so will the other chromosome of that pair.

Now that we understand that alleles of a gene occur on homologous chromosomes, we can define *allele* more precisely. The alternate forms of a gene that govern the same trait and that occupy corresponding loci on homologous chromosomes are known as alleles (Fig. 11–5).

The definition of allele implies that there are at least two alternative forms of the gene that can occupy a specific locus in homologous chromosomes. For bookkeeping purposes, each of these forms is assigned a letter as

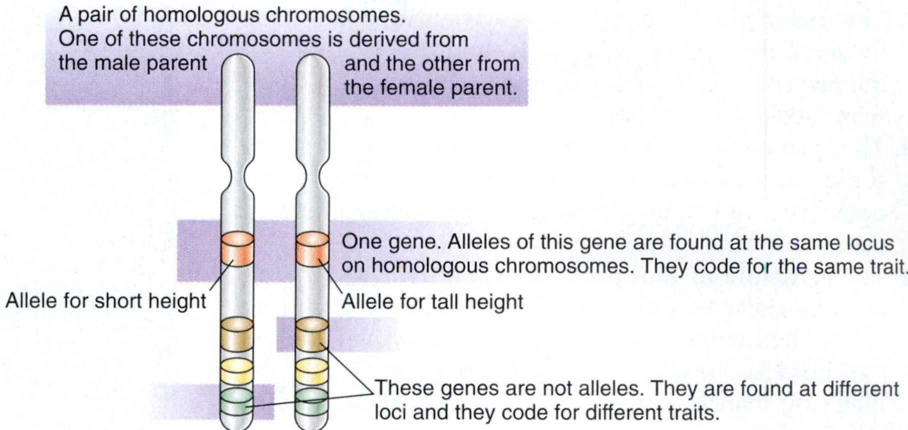
A pair of homologous chromosomes.
One of these chromosomes is derived from
the male parent
and the other from
the female parent.
One gene. Alleles of this gene are found at the same locus on homologous chromosomes. They code for the same trait.
Allele for short height
Allele for tall height
These genes are not alleles. They are found at different loci and they code for different traits.

FIGURE 11–5 Homologous chromosomes, genes, and alleles. Chromosomes occur in pairs in diploid cells. The members of a given pair correspond in shape, size, and type of genetic information, and are referred to as homologous chromosomes. For purposes of illustration, each chromosome is shown in the unduplicated state.

its symbol. It is customary to designate the **dominant** allele, the one that always manifests itself, with a capital letter, and the **recessive** allele, the allele that does not express itself in the presence of a dominant allele, with a lowercase letter. Thus, the letter *B* could be used to represent the allele for black coat color and the letter *b* the allele for brown when specifying the gene that determines coat color in guinea pigs. The black allele is dominant, whereas the brown allele is recessive.

A Monohybrid Cross Involves a Single Pair of Alleles

The usage of genetic terms and some of the basic principles of genetics can be illustrated by considering a simple **monohybrid cross**, that is, a cross between two individuals in which only one trait is being studied. The mating of a genetically pure brown male guinea pig (designated *bb*) with a genetically pure black female guinea pig (designated *BB*) is illustrated at the top of Figure 11–6. The two *bb* alleles separate during meiosis in the male, so each sperm cell has only one *b* allele. In the formation of egg cells in the female, the *BB* alleles separate, so each egg cell has only one *B* allele. The fertilization of this egg by a *b*-bearing sperm results in offspring with the genotype *Bb*, that is, with one allele for brown coat and one allele for black coat. What color would you expect these offspring to be?

Suppose that two black guinea pigs, each having alleles for both brown and black coat color (each *Bb*), are crossed. After meiosis occurs, half of the gametes produced by each guinea pig would have a single allele for black coat color (*B*), and the other half would have a single allele for brown coat color (*b*).

The probable combinations of eggs and sperm may be represented in a checkerboard, or **Punnett square**, as illustrated at the bottom of Figure 11–6. The Punnett square shows all possible combinations of gametes to form offspring. The types of eggs can be represented across the top of the grid, and the types of sperm indicated along the left side. The squares, which are filled in with the resulting combinations of gametes, indicate the genotypes of all possible offspring from this cross.

The generation with which a particular genetic experiment is begun is called the **parental generation**, or **P**. Offspring of this generation are referred to as the **first filial generation**, or $\mathbf{F_1}$. Those resulting when two F_1 individuals are bred constitute the **second filial generation**, or $\mathbf{F_2}$.

A Genotype May Be Either Homozygous or Heterozygous

If both alleles specify brown coat (*bb*), then the individual is said to be **homozygous** for coat color—that is, it has identical alleles for the same gene. In another guinea pig, both alleles may specify black coat (*BB*); this second animal is also homozygous for coat color. But it often happens that one of the alleles carries instructions for black coat and the other carries instructions for brown coat (*Bb*). In that case, the individual is said to be **heterozygous** for coat color—it contains two different alleles for the same gene. In the heterozygous condition, only the dominant allele is expressed; thus, a guinea pig with a genotype designated *Bb* will have a black coat.

To summarize:

1. When an organism is homozygous for the dominant allele, the phenotype (appearance) will reflect the dominant allele.
2. When the organism is homozygous for the recessive allele, the phenotype will reflect the recessive allele.
3. When the organism is heterozygous, the dominant allele will be expressed in the phenotype just as it would

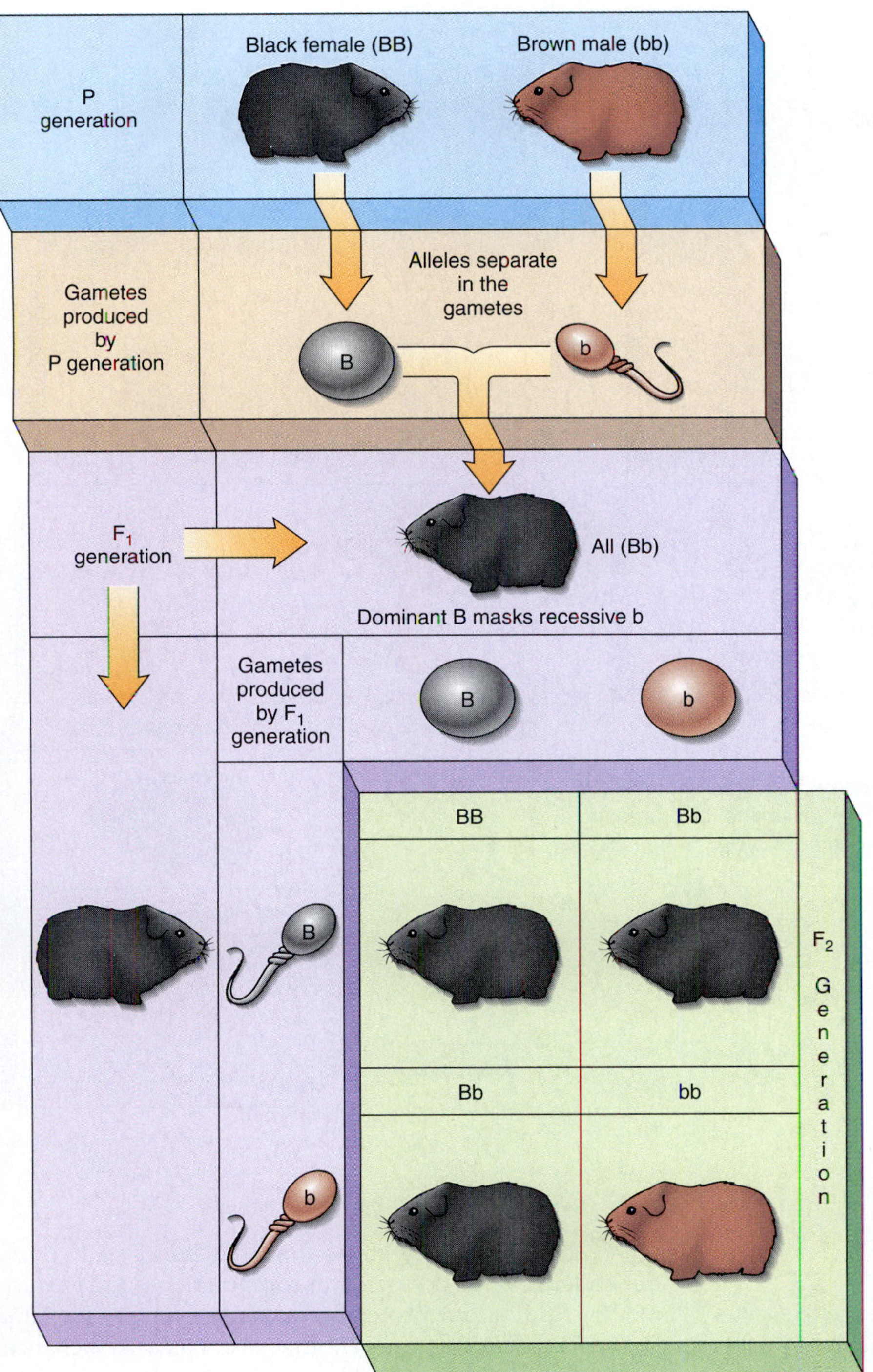

FIGURE 11–6 Monohybrid cross. When a genetically pure black guinea pig is mated with a genetically pure brown guinea pig, all of the offspring are black. When two members of the F_1 generation are crossed, the F_2 generation is produced. A Punnett square is used to determine all possible genotypes in the F_2. The phenotypic ratio of the F_2 is 3:1.

be if the organism were homozygous for the dominant allele. In such cases, one cannot tell from appearances alone whether an individual is heterozygous or homozygous for the dominant allele.

How can we determine the genotype of an organism that displays a dominant phenotype? It might be either homozygous for the dominant allele (*BB*) or heterozygous (*Bb*). One way to discover the answer would be with a **test cross**, an experimental cross with an individual that is homozygous recessive. Suppose, for instance, that a black guinea pig really is heterozygous. If one were to cross it with a brown one (which could *only* be homozygous for the recessive brown trait), at least some of the

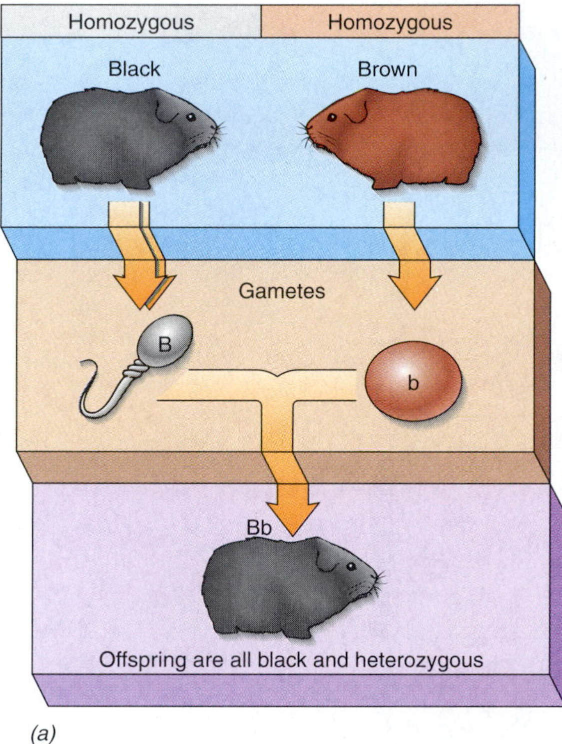

(a)

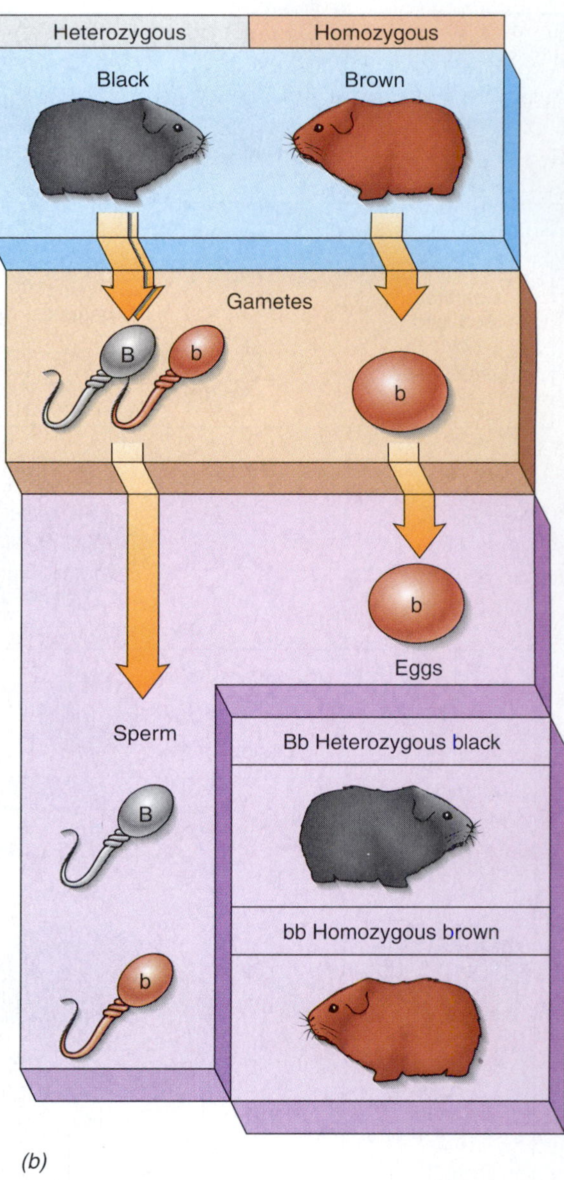

(b)

FIGURE 11–7 Test cross to determine the genotype of a black guinea pig. (***a***) If a homozygous black guinea pig is mated with a brown one, all of the offspring are certain to be black. (***b***) If any of the offspring are brown, the black guinea pig must be heterozygous.

offspring ought to display the recessive trait, that is, be brown (Fig. 11–7). On the other hand, if the unknown animal were homozygous for black coat color, all of the offspring would be black.

THE RULES OF PROBABILITY CAN BE USED TO PREDICT THE OUTCOME OF A GENETIC CROSS

Much of the foregoing is the logical consequence of a simple principle mentioned in the last chapter: During meiosis, each gamete normally receives just one member of a pair of homologous chromosomes and thus just one allele of each gene. This means that a single sperm cell produced by a male guinea pig with a *Bb* genotype, being haploid, will carry either a *B* allele *or* a *b* allele (as chance may have it); normally, no individual sperm cell will carry *both B* and *b*, and no sperm will carry neither. Each sperm has a 50% chance of having a *B* allele and a 50% chance of having a *b* allele. Similar logic indicates that a *Bb* female can produce eggs, each of which has a 50% chance of possessing a *B* allele and a 50% chance of possessing a *b* allele.

Probability theory states that the chance of two independent events occurring in combination is the *product* of their individual probabilities. If an event has a 50% chance of occurring, and another also has a 50% chance of occurring, then the likelihood of them taking place together is 50% × 50%, that is, 50% of 50%, or 25%. Thus, the probability of a *b* sperm finding a *b* egg cell is also about 25%.

A *B* sperm can be expected to find a *b* egg 25% of the time, and similarly, a *b* sperm will find a *B* egg 25% of the time. But genetically, the two *Bb* individuals are the same, so the likelihood of this combination is the *sum* of the probabilities of the two ways in which it can occur, that is, 25% + 25%, or 50%.

Does this mean that out of every four F_2 guinea pigs one will always be brown (*bb*)? Not at all. It is quite possible that a *b* sperm and a *b* egg just won't get together. All the litter might be black, even if both parents are heterozygous. (Less likely, but still possible, the entire litter might be brown, even if both parents are heterozygous.) Given a sufficiently large *number* of instances, though, we should observe very nearly 25% brown offspring in such a cross.

GENES LOCATED ON DIFFERENT CHROMOSOMES ARE INHERITED INDEPENDENTLY

Consider two different human traits, one involving the tongue and the other the ears (Fig. 11–8*a*). Each trait is controlled by a single gene. The ability to curl, or roll, the tongue is determined by a dominant allele; the absence of this talent is recessive. The allele that results in free earlobes is dominant; attached or absent earlobes is recessive.

A mating that involves individuals differing in *two* genes is referred to as a **dihybrid cross**. If a man homozygous for tongue curling (*CC*) marries a woman without this trait (*cc*), all their children (*Cc*) may be expected to have this ability, since tongue curling is dominant and the offspring are heterozygous (*Cc*). Similarly (but quite unrelated), if the father has free earlobes (*EE*) and the mother is homozygous for attached earlobes (*ee*), the children will have free earlobes (*Ee*). The children's genotype for both sets of traits is designated *CcEe* because the children are heterozygous for both traits. Since the two traits are the result of genes found on separate chromosome pairs, they will be inherited (or assorted) independently of one another.

Let us suppose that some of these children were to marry others who have genotypes identical to their own (*CcEe*). As you see in Figure 11–8*b*, four kinds of gametes are possible. Since each gamete will have one allele of each gene, a total of four gametic genotypes is possible, given the genetic makeup of the F_1 generation. If by chance the chromosome bearing *E* is sorted during meiosis into the same gamete as the chromosome bearing *c*, the resulting sperm or egg will have the genotype *Ec*. Similarly, if the chromosomes bearing *e* and *C* are assorted

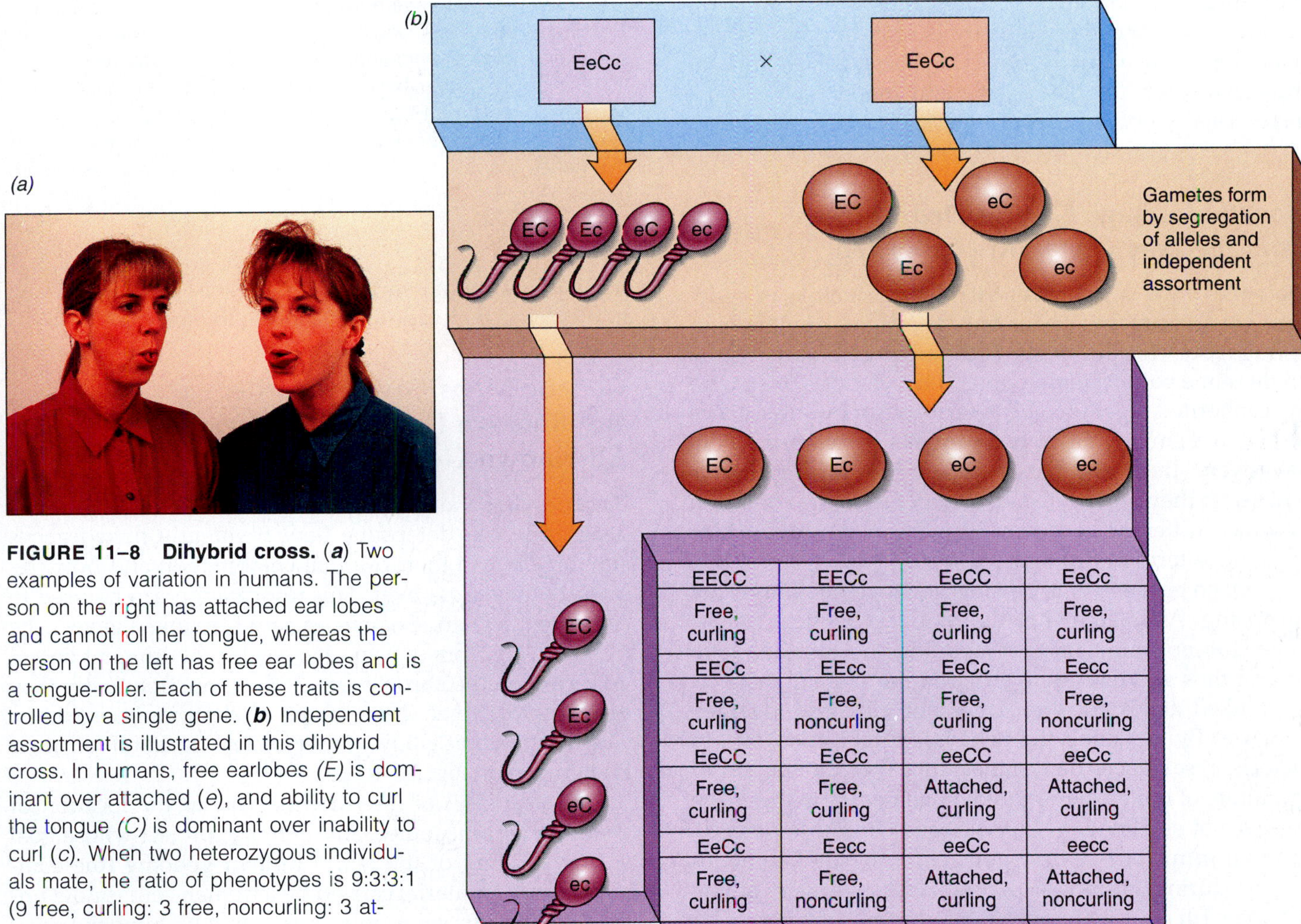

EECC	EECc	EeCC	EeCc
Free, curling	Free, curling	Free, curling	Free, curling
EECc	EEcc	EeCc	Eecc
Free, curling	Free, noncurling	Free, curling	Free, noncurling
EeCC	EeCc	eeCC	eeCc
Free, curling	Free, curling	Attached, curling	Attached, curling
EeCc	Eecc	eeCc	eecc
Free, curling	Free, noncurling	Attached, curling	Attached, noncurling

FIGURE 11–8 Dihybrid cross. (***a***) Two examples of variation in humans. The person on the right has attached ear lobes and cannot roll her tongue, whereas the person on the left has free ear lobes and is a tongue-roller. Each of these traits is controlled by a single gene. (***b***) Independent assortment is illustrated in this dihybrid cross. In humans, free earlobes *(E)* is dominant over attached (*e*), and ability to curl the tongue *(C)* is dominant over inability to curl (*c*). When two heterozygous individuals mate, the ratio of phenotypes is 9:3:3:1 (9 free, curling: 3 free, noncurling: 3 attached, curling: 1 attached, noncurling). (Kenneth Knott/ Fine Light)

together, the gamete will have the genotype *eC*. Other gametes would contain *EC*, and still others *ec*. There are no possible combinations other than these four. Since alleles are always borne on homologous chromosomes that are *separated* from one another in meiosis, a gamete can ordinarily have no more than *one* of each gene (for example, *E* or *e*, but not both; *C* or *c*, but not both).

Since both parents produce four possible kinds of gametes, working out a cross of that kind will require a Punnett square of 16 cells. (Sixteen different combinations of gametes are possible when these gametes come together randomly.) Count the number of individuals exhibiting each of the four possible phenotypes. The phenotypic ratio among the F_2 offspring will be 9:3:3:1. The genotypic ratio is even more complex because there are nine different genotypes.

It is obvious in comparing a monohybrid cross with a dihybrid cross that, as the number of different traits considered increases, the number of possible genotypes in the offspring multiplies rapidly. Imagine, then, how many different genotypes are possible among the offspring of two parents who differ in hundreds of ways! Looking at the human population, one understands why so much diversity exists, and this diversity is only of genes whose phenotypes are visually obvious. Many genes, such as those that code for tissue type and blood type, do not have visually obvious phenotypes.

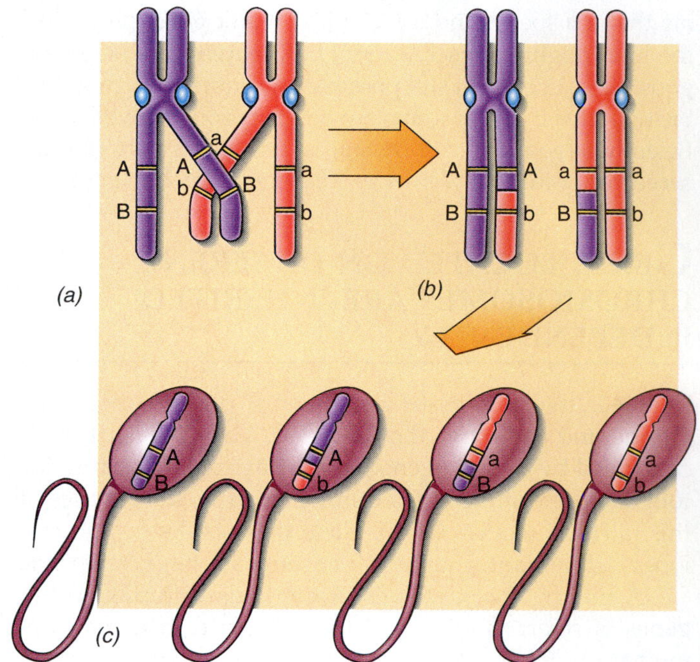

FIGURE 11–9 Linkage and crossing-over. Initially, genes *A* and *B* are linked on a single chromosome, whereas *a* and *b* are linked on a separate chromosome. (***a***) Crossing-over results in the exchange of segments between chromatids of homologous chromosomes (***b***). (***c***) Crossing-over permits the formation of new combinations of genes in the gametes. For example, note that genes *A* and *b* are now linked in one of the gametes, as are genes *a* and *B*.

Genetic Linkage Can Be Deduced from the Way Genes Are Inherited

Biologists now know that Mendel's principle of independent assortment applies only to traits carried on nonhomologous chromosomes. If different genes are carried on the same pair of homologous chromosomes, they tend to be inherited together. The inheritance of multiple genes borne on a single pair of homologous chromosomes behaves very much like that of a *single* gene (at least, as we will see, if they are located close together). The condition in which different genes are located on the same chromosome is termed **linkage**.

Linked genes tend to be inherited together. This is not surprising. After all, genes that happen to occur on the same chromosome tend to remain together during meiosis and thus will end up together in the same gamete.

Linked genes are sometimes not inherited together, however. By watching the behavior of chromosomes in meiosis, researchers determined that the explanation of the failure of linked genes to stay together lay in the crossing-over of segments of homologous chromosomes. Recall that in meiosis homologous chromosomes come together during prophase I, and crossing-over occurs (Chapter 10). As that happens, the chromosomes may exchange parts, as shown in Figure 11–9. Crossing-over, which causes new gene combinations on chromosomes, leads to additional variation in gametes and, as a result, in offspring that arise from those gametes.

Genetic Maps Give the Order and Relative Distances between All Known Genes on Them

By observing a large number of crosses involving linked genes, one can determine how frequently linked genes stay together in their original combination and how frequently they cross over. This information can be used to estimate the relative distances between linked genes: The farther away from one another the loci of different linked genes are on the chromosomes, the more frequently they tend to cross over. Thus, maps of chromosomes can be constructed that show the physical location of their known genes (Fig. 11–10).

A great deal of attention is paid to the location of genes on chromosomes. An immense project whose goal is the mapping of the entire **human genome** (the complete genetic material carried by a human cell) is now under way. This project, which hopes to determine the chromosome location and structure of the 50,000 to 100,000 genes that humans possess, is costing hundreds of mil-

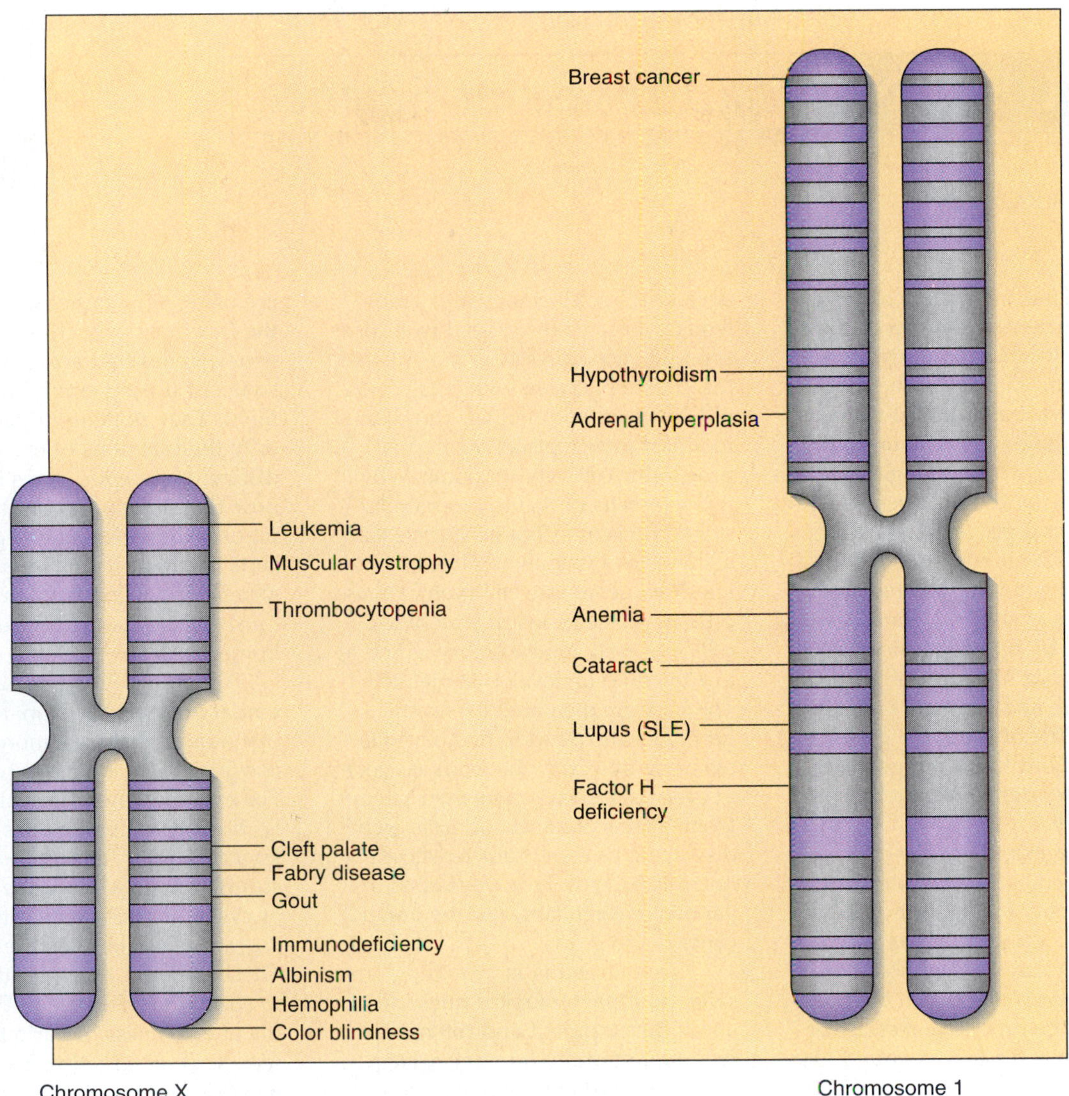

FIGURE 11–10 Genetic mapping. These maps show some of the genetic defects associated with specific genes on the X chromosome and chromosome 1 in humans.

lions of dollars and involves the combined efforts of scientists from many nations. The goal of the project is to know the location of most of the genes by the end of 1995, although the function of the majority of them will still be unknown (See Focus On the Process of Science: Gene Targeting). By 2006 researchers hope to have determined all three billion base pairs that are estimated to make up the human genome.

Why is genetic mapping so important? For one thing, it can help locate the genes responsible for genetic disorders (recall the Chapter Opener), which would assist physicians in their diagnosis and treatment. Genetic mapping is also important in genetic engineering. Precise genetic maps will not only facilitate the redesign of defective human genes but also the development of new varieties of agricultural plants and animals (Chapter 16).

INHERITANCE IS COMPLICATED BY OTHER FORMS OF GENETIC VARIATION

Many traits are not inherited by simple Mendelian principles. Geneticists know of many examples in which an organism's appearance (its phenotype) is the result of complex interactions between genes. These interactions include incomplete dominance, codominance, multiple alleles, epistasis, pleiotropy, and polygenes. (Other genetic interactions exist that are beyond the scope of this text.)

Dominance Can Be Incomplete

Not every gene consists of a dominant allele and a recessive allele. For example, when red-flowered and white-flowered four-o'clocks are crossed, the offspring do

FOCUS ON *The Process of Science*

GENE TARGETING

Although the human genome project is currently under way to determine the location of all the genes on human chromosomes, the functions of the vast majority of these genes are unknown. **Gene targeting**, a procedure in which a single gene is chosen and knocked out (inactivated) in a mouse, is a powerful new research tool. The mouse bearing the knocked out gene can be observed, and the roles of the inactivated gene determined, by figuring out what the mouse does not have or cannot do. Because mice and humans share more than 90% of the same genes, information about knockout genes in mice provides details about human genes as well.

Gene targeting, pioneered by Mario Capecchi, a molecular geneticist at the University of Utah School of Medicine, is a rather complex and lengthy procedure; it takes about a year to develop a new strain of knockout mouse. The process makes use of results that geneticists have found to be fairly typical whenever a modified gene is inserted into a cell. That is, once inside a cell, a modified gene tends to become recognized by a corresponding gene already in the cell, and then the two tend to exchange pieces of DNA with one another. Thus, the modified gene replaces the normal gene in the cell's chromosome. Why this replacement occurs is not understood at this time, but geneticists take advantage of it anyway. They place knockout genes into embryonic mouse cells, inject the genetically modified cells into early mouse embryos, and allow the mice to develop to maturity. The mice are then bred for several generations to develop homozygous offspring that carry the modified gene in every cell. Today more than 100 different strains of knockout mice, each displaying its own characteristic abnormalities, have been developed in various research labs, and the number continues to grow.

Gene targeting is providing answers to basic biological questions about development and the normal functioning of the immune system and other parts of the body. For example, Capecchi and his colleagues are using gene targeting to determine the functions of each of the **homeotic genes**, genes that act as master switches during embryonic development. They hope to understand not only the functions of each of the 38 different homeotic genes in mice, but how they interact to produce a whole mouse from a fertilized egg. The results of much of this research could also be applied to humans, since mice and humans have similar homeotic genes.

Gene targeting also has great potential for learning more about various human diseases, more than 5000 of which have a genetic component. For example, the gene linked to cystic fibrosis in humans has been knocked out in mice, producing a strain of mice with all the symptoms of cystic fibrosis—congested lungs, digestive problems, and early death. These mice are being used as models to study the disease and determine the effectiveness of various treatments. Gene targeting is also being used to study cancer, heart disease, and other health problems.

not have red or white flowers. Instead, all of the F_1 offspring have pink flowers. When two of these pink-flowered plants are crossed, the F_2 offspring appear in the ratio of one red-flowered to two pink-flowered to one white-flowered plant.

In this instance, as in other aspects of science, finding results that differ from those predicted simply prompts scientists to reexamine and modify their assumptions to account for the new experimental results. In Figure 11–11, the pink-flowered plants (R^1R^2) are clearly the heterozygous individuals, and neither the red allele (R^1) nor the white allele (R^2) is completely dominant. When the heterozygote has a phenotype that is intermediate between those of its two parents, the responsible alleles are said to show **incomplete dominance**. In crosses involving incomplete dominance, the genotypic and phenotypic ratios are identical.

Incomplete dominance is not unique to four-o'clocks. Red- and white-flowered snapdragons also produce pink-flowered plants when crossed. The reason is that the single "red" allele in these plants is unable to code for the production of enough red pigment to make the petals look red.

Codominant Alleles Are *Both* Expressed in the Heterozygous Condition

If *both* alleles are *fully* expressed, they are said to be **codominant**. The **ABO blood group**, a classification of human blood into four types (A, B, AB, and O) provides an example of codominance. These blood types vary in the type of certain proteins located on the surface of their red blood cells. The ABO blood group is controlled by a single gene with three alleles, designated I^A, I^B, and *i*. Al-

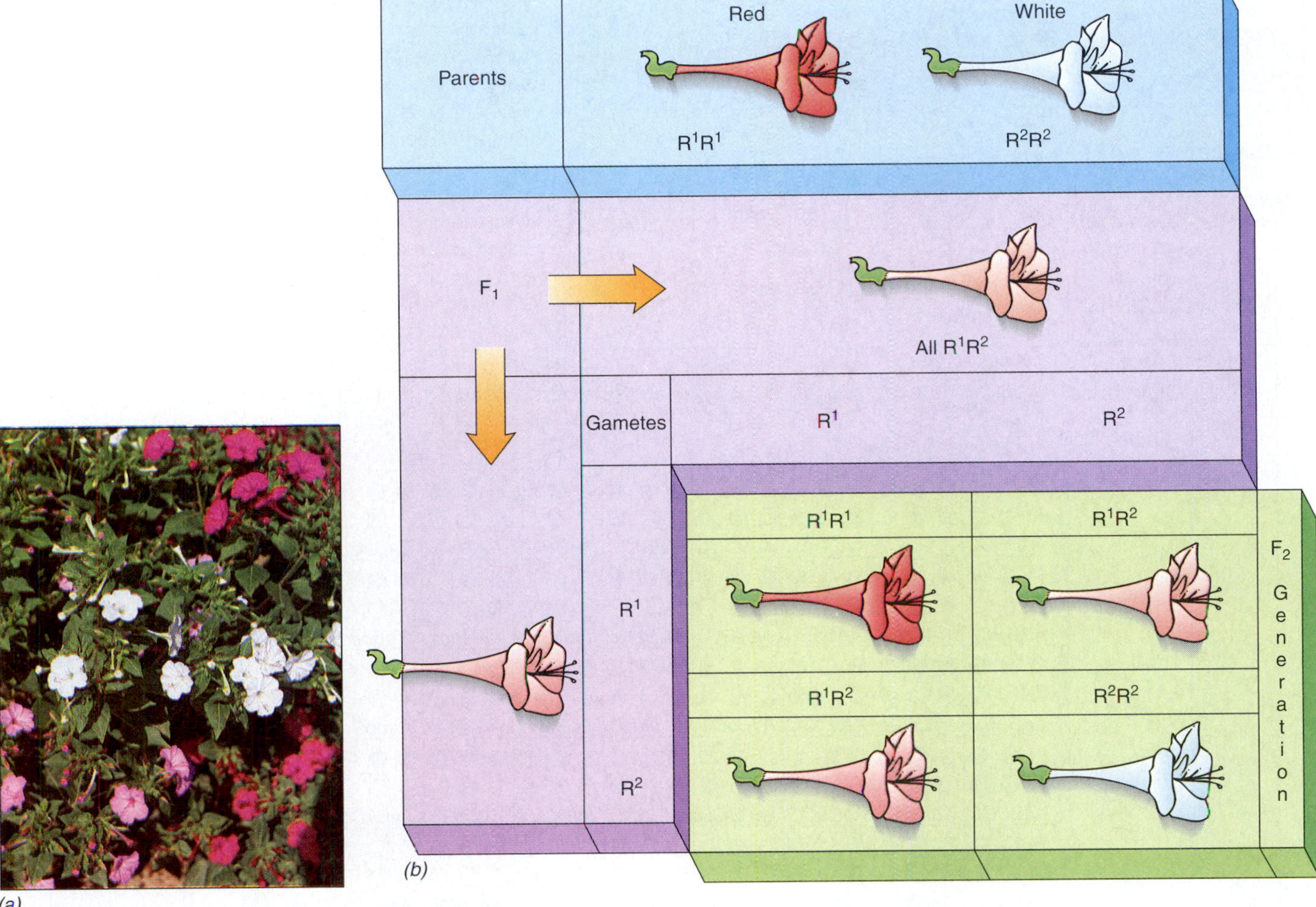

FIGURE 11–11 Incomplete dominance in four-o'clocks. (***a***) Red, pink, and white four-o'clocks. The plant gets its common name from the fact that the flowers open in late afternoon. (***b***) Flower color in four-o'clocks is controlled by a single gene with two alleles, one for red (R^1) and one for white (R^2) flower color. Red is incompletely dominant to white. A plant with the genotype R^1R^2 has pink flowers. Note that there is no allele for pink flowers, but only alleles for red and white flowers. (Eric L. Heyer, from Grant Heilman)

lele I^A in the ABO blood type is codominant to allele I^B, whereas allele i is recessive to both. An individual who possesses both codominant alleles (I^AI^B) produces both the A and B surface proteins. The ABO blood group is considered in greater detail in the following section.

More than Two Alleles of a Gene Sometimes Occur in a Population

Multiple alleles exist when a particular gene has more than two forms. Keep in mind, however, that a diploid organism can have no more than two of a given set of multiple alleles. For example, the ABO blood group has three alleles present in the human population, which by their interaction can produce four blood types. The blood type depends on the exact combination of two alleles possessed by an individual. As mentioned previously, the i allele is recessive to the other alleles, and I^A and I^B exhibit codominance with respect to one another. Thus, two genotypes (I^AI^A and I^Ai) give rise to type A blood and two (I^BI^B and I^Bi) to type B blood. Type AB blood and type O blood each have only one possible genotype (I^AI^B and ii, respectively).

Sometimes One Gene Masks the Phenotypic Effect of a Different Gene

Quite often a single trait is the result of the interactions of many genes. One such interaction occurs between two different genes in which one gene suppresses the expression of the other gene, a phenomenon referred to as **epistasis**. For example, fruit color in summer squash is determined by two genes. One gene codes for yellow (Y) or green (y) fruit, with yellow fruit color being the dom-

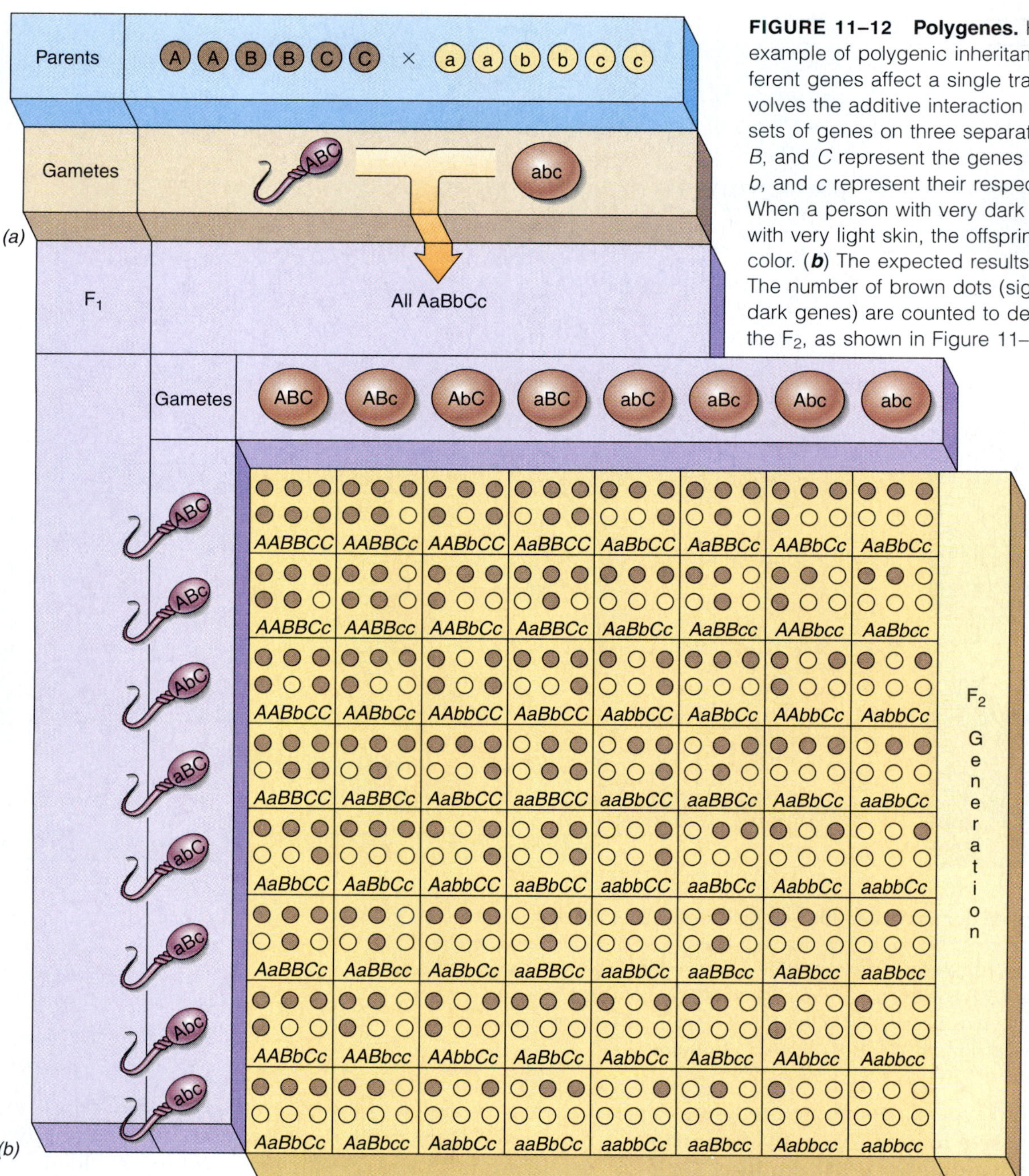

FIGURE 11–12 Polygenes. Human skin color is an example of polygenic inheritance, in which several different genes affect a single trait. Human skin color involves the additive interaction of at least three separate sets of genes on three separate chromosomes. Let *A*, *B*, and *C* represent the genes for dark skin color and *a*, *b*, and *c* represent their respective light alleles. (***a***) When a person with very dark skin mates with a person with very light skin, the offspring are all intermediate in color. (***b***) The expected results in the F_2 generation. The number of brown dots (signifying the number of dark genes) are counted to determine the skin colors in the F_2, as shown in Figure 11–13.

inant allele. The second gene codes for white (*W*) (that is, an absence of color) or colored fruit (*w*), with white fruit color being the dominant allele. If a *W* allele is present in the squash's genotype, its fruit must be white, regardless of what the second gene codes. Thus, the *W* allele blocks the expression of the *Y* and *y* alleles. If a summer squash produces yellow fruit, its genotype must be *wwYY* or *wwYy*. Similarly, green fruit color is exhibited only when a squash has the genotype *wwyy*. Any other possible genotype results in white fruit color (*WWYY*, *WWYy*, *WWyy*, *WwYY*, *WwYy*, or *Wwyy*).

A Single Gene Can Influence More than One Trait

Sometimes a single gene affects more than one trait rather than having a single effect, a phenomenon referred to as **pleiotropy**. Albinism (discussed earlier) is an example of pleiotropy. In addition to a lack of pigment in the skin, albinism also produces defects of vision. The gene for albinism affects vision in two ways—by preventing normal eye pigmentation and by producing disorders in the visual pathways of the brain.

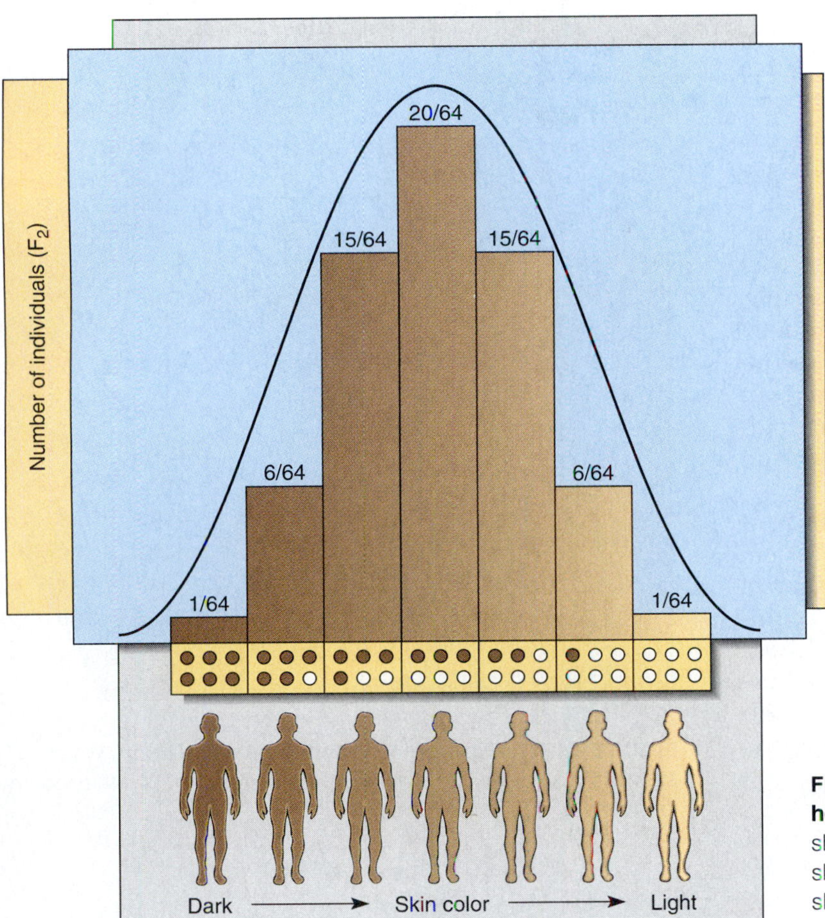

FIGURE 11–13 Continuous variation in polygenic inheritance. The data from Figure 11–12b produce a bell-shaped curve, which demonstrates continuous variation in skin color, from very dark to very light, in the F_2. The bell-shaped curve is typical of the distribution of phenotypes in a trait governed by polygenes.

Some Traits Are Governed by More than One Gene

Some traits are governed in their expression by many different genes, known as **polygenes**. A trait affected by polygenes exhibits a wide range of variability. Each individual gene in polygenic inheritance makes a small contribution to the organism's phenotype. Human skin color is an example of polygenic inheritance, as are height, eye color, and several other traits.

It is well known that if a person of extremely light complexion marries a person of extremely dark complexion, their offspring will be intermediate in color. If that were all there was to it, it would be an obvious case of incomplete dominance. To test whether or not skin color is caused by incomplete dominance, consider the offspring (that is, the F_2 generation) of two such intermediate persons. The assumption of incomplete dominance predicts a 1:2:1 ratio among their offspring. However, what we observe instead (if their family is sufficiently large, or if we observe offspring in several such families) is a complex distribution of skin colors. Although the majority of offspring are intermediate in color, the full range is from very light to very dark and is not easily expressed by any ratio.

The explanation for this is that human skin color is governed by at least three independent genes, each located on a different chromosome. Thus, mating between an extremely dark person and and extremely light person may be represented as *AABBCC* × *aabbcc* (Fig. 11–12). The F_1 offspring can be only *AaBbCc*. What about the F_2? We have worked out the genotypes of the F_2 for you in the illustration. It us up to you to determine their phenotypes, which is not difficult to do. An individual with the genotype *AABBCC* would be very dark, whereas an individual with the genotype *aabbcc* would be extremely light.[4] Intermediate skin colors can be predicted by counting the number of capital letters in each genotype. When the number of F_2-generation individuals with each skin color is plotted against skin color and the points are connected, the result is a bell-shaped curve (Fig. 11–13).

[4] We are not now discussing albinism. Remember that albinism, a complete lack of skin pigment, is an abnormal trait that can occur in any racial group. Albinism behaves as a trait governed by a simple recessive gene.

THE ENVIRONMENT ALSO AFFECTS AN ORGANISM'S PHENOTYPE

We have seen some of the complex ways that genes can interact to form an organism's phenotype. However, the environment in which an individual exists also exerts a strong influence over phenotype (recall the Chapter Opener about the importance of the environment as well as genes in causing certain diseases). As another example, consider human height, which is genetically determined by many genes (that is, by polygenic inheritance). An individual may possess a genotype for very tall height, but if he or she does not eat nutritious foods during the childhood years of growth, that individual will not be able to grow as tall as the genotype would permit.

Certain aquatic plants provide another example of how the external environment affects gene expression. The water starwort (*Callitriche*), for example, produces two kinds of leaves, depending on whether they are under water or above it (Fig. 11–14). The submerged leaves are narrow and finely dissected, whereas the leaves surrounded by air are more "normal" in appearance. Thus, the expression of the genes that determine leaf shape is affected by different environmental conditions.

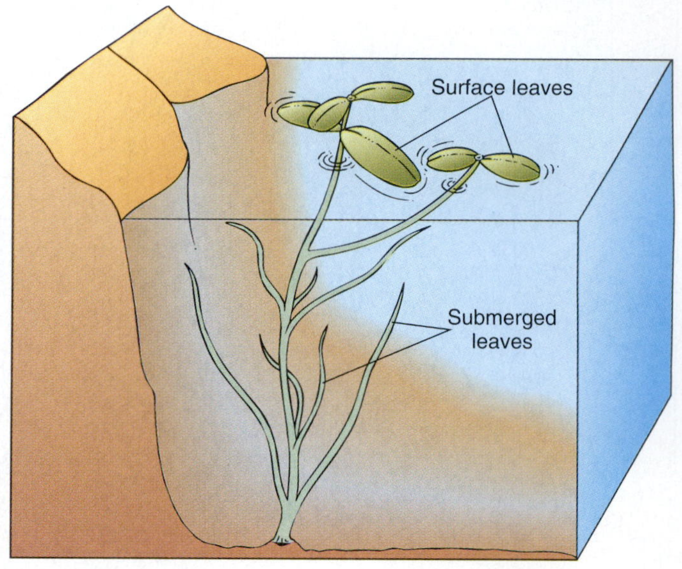

FIGURE 11–14 Environment and gene expression. Water starwort leaves vary as a result of the environment's influence on its genes.

Chapter Summary

I. The science of genetics was founded by the 19th-century research of Gregor Mendel, who used the garden pea to demonstrate important principles of inheritance.
   A. Most inherited traits are governed by a pair of alleles present in each diploid cell. Alleles may be alike (homozygous) or unlike (heterozygous).
   B. An allele expressed in the heterozygous condition is said to be dominant. An allele that is expressed only when homozygous is said to be recessive.
   C. When gametes form during meiosis, the alleles of each gene separate; only one allele of each gene goes to each gamete.
      1. Alleles of a gene are carried on homologous chromosomes.
      2. The behavior of alleles in meiosis is similar to the behavior of homologous chromosomes.
      3. Each offspring receives one allele of each gene from each parent.
   D. The homologous chromosome (and its genes) that a gamete receives is governed entirely by chance. Thus, each trait is inherited without relation to other traits on nonhomologous chromosomes.

II. A monohybrid cross is a cross between two individuals in which only one trait is being studied.
   A. A cross between two individuals in which one is homozygous dominant and the other is homozygous recessive yields F_1 offspring that are all heterozygous.
   B. When two heterozygous individuals are crossed, a 3:1 phenotypic ratio occurs in the F_2 offspring.

III. A dihybrid cross is a cross between two individuals in which two traits are being studied.
   A. A cross between two individuals in which one is homozygous dominant for both traits and the other is homozygous recessive for both traits yields F_1 offspring that are heterozygous for both traits.
   B. When two heterozygous individuals are crossed, a 9:3:3:1 phenotypic ratio occurs in the F_2 offspring.

IV. Many traits are not inherited by simple Mendelian principles.
   A. Two genes that are located on the same chromosome are said to be linked. Linked genes tend to be inherited together.
   B. In incomplete dominance, neither allele is completely dominant. A heterozygote exhibits a phenotype intermediate between those of its two parents.
   C. In codominance, both alleles are fully expressed.
   D. Multiple alleles are three or more forms of the same gene and give rise to more than two phenotypes in a

population. Only two alleles are ever present in an individual, however.

E. Epistasis occurs when two genes affect the same trait, with one of the genes suppressing the expression of the other.

F. Pleiotropy occurs when a single gene affects more than one trait.

G. A trait affected by many genes (that is, by polygenes), in which each individual gene makes a small contribution to the organism's phenotype, is known as polygenic inheritance.

H. The environment interacts with an organism's genotype to produce its phenotype.

Selected Key Terms

Post-Test

1. The modern science of genetics was founded in the 19th century by ______________, whose work was unrecognized until its rediscovery after his death.

2. A specific unit of hereditary information is called a/an ______________.

3. Alternate forms of a gene are called ______________.

4. The appearance of an individual is its ______________, whereas the genetic makeup of an individual is its ______________.

5. An organism with both alleles alike is said to be ______________.

6. An allele that can be expressed even when heterozygous is said to be ______________.

7. Only one allele of each gene occurs in each ______________.

8. Whether a particular sperm finds and fertilizes a given egg is an event determined completely by ______________.

9. A ______________ cross helps determine whether an individual exhibiting a dominant phenotype is homozygous or heterozygous.

10. Genes are inherited independently if they occur on non-______________ chromosomes.

11. Genes that occur on the same chromosome are said to be ______________.

12. In the human blood group, the allele I^A is ______________ to the allele I^B.

13. The inheritance of fruit color in summer squash is an example of ______________.

14. The phenomenon in which a single gene affects more than one trait is called ______________.

15. The ______________ interacts with an organism's genotype to determine its phenotype.

16. Chromosomes occur in pairs, and the two members of a pair of chromosomes are said to be (a) homozygous (b) homologous (c) heterozygous (d) haploid.

17. If two heterozygous organisms are crossed, and the trait being studied displays complete dominance, one may expect the following type of phenotypic ratio in the offspring: (a) 3:1 (b) 1:2:1 (c) 9:3:3:1 (d) no ratio—the offspring are all alike.

18. When red-flowered and white-flowered four-o'clocks are crossed, the offspring exhibit (a) pleiotropy (b) multiple alleles (c) incomplete dominance (d) codominance.

19. The phenomenon that occurs when a particular gene has more than two forms is known as (a) polygenes (b) epistasis (c) linkage (d) multiple alleles.

20. Skin color in humans is determined by (a) epistasis (b) polygenes (c) multiple alleles (d) incomplete dominance.

Review Questions

1. Distinguish between genes and alleles.
2. Describe Mendel's principle of dominance.
3. Why can an individual receive only one member of a homologous pair of chromosomes from each parent? What would happen if you tried to solve genetic problems without realizing this?
4. Describe Mendel's principle of segregation.
5. Distinguish between homozygous and heterozygous individuals. Between dominance and recessiveness. Between genotype and phenotype.
6. Describe Mendel's principle of independent assortment.
7. How are epistasis and polygenes alike? How are they different?
8. Distinguish between incomplete dominance and polygenes.

Thinking Critically

1. The isolation of the colon cancer gene described in the chapter introduction should lead to a screening test sometime in the future. Suppose that such a screening test already exists and that you decide to be tested; further, suppose that you test positive for the colon cancer gene. Using what you have learned in this chapter, what, if anything, would you do? Explain.
2. Would the development of genetics in the 20th century have been any different if Mendel had never lived? Explain.
3. Why do you suppose that most known genetic defects are recessive rather than dominant?
4. Mendel originated the principle of independent assortment on the basis of several crosses involving genes that today are known to be linked. How could the linked genes that he studied appear to be inherited independently?
5. If you wanted to determine whether an organism that showed a dominant trait was homozygous or heterozygous, you would be best advised to cross the unknown organism with one that is homozygous for the recessive allele rather than to cross it with one known to be heterozygous. Why?
6. An albinism gene exists that suppresses the development of all color in guinea pigs. It is located on a different chromosome from the gene (discussed in the text) that determines whether coat color is black or brown. Suppose a black and a brown guinea pig are crossed, and one-fourth of the offspring are white. How can this be explained, assuming that mutation is not the reason?

Genetics Problems

1. Most of the individuals of a certain wildflower population have white flowers, although a few are purple-flowered. Crosses have demonstrated that the allele for white flower color (*W*) is dominant over the allele for purple (*w*).
   a. Give the genotype of a purple-flower plant and show the gametes that it would produce as a result of meiosis.
   b. If two heterozygous white-flowered plants are crossed, what fraction of the offspring would you expect to be purple?
   c. If two purple-flowered plants are crossed, what fraction of their offspring would you expect to be white?
2. In humans, the presence of a widow's peak (frontal hairline that points downward in the middle of the forehead) or straight hairline is determined by a single gene. The allele for widow's peak (designated *W*) is dominant over the allele for straight hairline (*w*). If a man who is heterozygous for the widow's peak hairline marries a woman who has a straight hairline, what fraction of their children would be expected to have a widow's peak?
3. In both cattle and horses, reddish coat color is codominant to white coat color. The heterozygous individuals have roan-colored coats (reddish hairs interspersed with white hairs). If you saw a white mare nursing a roan colt, what would you say was the coat color of the colt's father? Is there more than one possible answer?
4. What blood type(s) could occur in the children of two people with the blood genotypes I^Ai and I^Bi?
5. A baby with type AB blood is born into a family whose wife is type B and whose husband is type O. Can the baby be shown to have a different father than the husband? Explain.
6. In one study, Gregor Mendel crossed a yellow-seeded and tall pea plant with one that had green seeds and was short. The F_1 offspring were all yellow-seeded and tall. Assuming independent assortment of these two genes, what phenotypes and proportions did he find among the F_2 offspring when the F_1 plants were allowed to fertilize themselves?
7. Jimsonweeds produce either purple (*P*) or white (*p*) flowers, and their fruits are either spiny (*S*) or smooth (*s*). These two genes are located on nonhomologous chromosomes.
   a. What is/are the phenotype(s) and genotype(s) of the offspring that result from the cross, *PPss* × *PpSs*?

b. If the cross *PpSS* × *ppSs* is made, which of the following would *not* be represented in the offspring?
(1) *PpSS* (2) *PpSs* (3) *ppSS* (4) *ppSs* (5) *PPSs*

8. In peas, the gene for tall (*T*) is dominant over its allele for short (*t*), and round seed (*R*) is dominant over wrinkled seed (*r*). A homozygous tall, wrinkled-seeded plant is crossed with a homozygous short, round-seeded plant.

a. What is/are the phenotype(s) of the F_1? The genotype(s) of the F_1?

b. What is the phenotypic ratio of the F_2 generation?

9. In corn, a single gene controls height, with the allele for normal height (*N*) being dominant over the allele for short (*n*). A normal corn plant was crossed with a short plant, but the genotypes of the two plants were not known. Of the offspring, 150 were normal height and 153 were short. Based on the information given, determine the genotypes of the parent plants.

10. Sesame plants produce seed pods with one chamber (*O*) or three chambers (*o*), and leaves that are normal (*N*) or wrinkled (*n*). The two traits are inherited independently. Determine the genotypes and phenotypes of the two parents that produced the following offspring: 304 one-chamber pod/normal leaf; 100 one-chamber pod/wrinkled leaf; 298 three-chamber pod/normal leaf; 103 three-chamber pod/wrinkled leaf.

Recommended Readings

Capecchi, M.R., "Target Gene Replacement," *Scientific American*, March 1994. Biologists can now produce mice that contain a mutation of any gene desired, thereby enabling them to determine the function of that gene.

Mendel, G., *Experiments in Plant Hybridization*, edited by J.H. Bennett, Edinburgh, Oliver & Boyd, 1965. The English translation of Mendel's classic paper.

Russell, P., *Genetics*, 3rd ed., New York, Harper Collins, 1992. A well-written genetics text that clearly discusses Mendelian inheritance.

Issues in human genetics. A "perfect" family from outward appearances may carry a detrimental gene, and information about that gene raises many ethical questions. Should an individual be informed of a genetic defect even if he or she exhibits a normal phenotype? Should family members have access to an individual's genetic information? Should a child be informed of a genetic risk? (Comstock, Inc./David Lokey)

GENETIC PRIVACY

A few years ago a story circulated about a man who observed an attractive woman driving in rush hour traffic and wrote down her car's license number. Starting with this piece of information, the man, who was skilled in information retrieval using a computer, was able to determine the woman's telephone number, her address, her employer, her years of employment, and her current salary. He gained access to her confidential financial records, and discovered what airline she flew most often, what stores she frequented, and much, much more about her. This invasion of privacy, discovered after he introduced himself to the woman and bragged about how he "met" her, was universally denounced as one of the dangers brought about by advances in computers.

Today humans face a potentially more serious infringement of individual rights made possible as a result of advances in human genetics. The science of biological variation in humans, or **human genetics**, concerns what genes humans possess and where they are located, how genes are expressed within individuals, and how genes are transmitted from one generation to another. Widespread recognition of the importance of inherited disease has made human genetics one of the most rapidly expanding areas of modern medicine, promising better diagnostic capabilities as well as methods of prevention and treatment of genetic disorders. However, a number of important ethical questions exist.

For example, certain genetic diseases can be precisely diagnosed by genetic tests, and the number of such tests—developed for counseling purposes—is increasing each year. But suppose you tested positive for a genetic disorder and this information was disclosed to outsiders such as your insurance company or your employer. You might be discriminated against by having to pay higher insurance premiums or by being turned down for a promotion or even by being fired. Moreover, because a genetic disorder can be inherited, your positive test results would imply that other members of your family carry the genes for the disorder. Suppose this information was used to discriminate against your parents, your siblings, or your children. How could you be assured that such genetic information remains confidential?

The preceding illustration is only one of many ethical concerns brought about by our expanding knowledge of human genetics. As another example, should we test human embryos for genetic disorders if their families do not have a history of genetic disease? If an embryo tests positive for a particular genetic disease, should pregnancy be terminated? Now that we can determine the sex of an embryo should we select the gender of our offspring by aborting it if it is not the desired gender? What if, years from now, we could select traits for our offspring that have nothing to do with genetic disease—traits such as tallness, attractiveness, or even intelligence? Do parents have the right to "design" their offspring?

Clearly the ethical issues and social consequences caused by advances in human genetics need to be carefully considered. Policies overseeing genetic privacy need to be developed and implemented *now*, before widespread genetic screening becomes routine. Scientists must work closely with public policy-makers to examine the ethical ramifications of certain types of research. Such steps will increase the likelihood that genetic information will be used properly.

Human Genetics

12

Learning Objectives

After you have studied this chapter you should be able to:

1. Describe methods used to study human genetic diseases.
2. Discuss the genetic determination of sex in humans, including the role of the Y chromosome in determining male sex.
3. Describe the inheritance of X-linked genes.
4. Distinguish between birth defects that are inherited and those that are environmentally induced.
5. Distinguish between chromosomal abnormalities and single gene defects.
6. Summarize the characteristics of selected genetic diseases.
7. Describe how amniocentesis is used in the prenatal diagnosis of human genetic abnormalities, and state the relative advantages and disadvantages of amniocentesis and chorionic villus sampling.
8. Explain some of the complex ethical issues associated with human genetics.
9. Define *consanguinity* and give its principal genetic implications.

Key Concepts

- ☐ Sex chromosomes are not the same in males and females. Human females possess two X chromosomes, and human males have one X and one Y chromosome. All other human chromosomes, designated autosomes, are similar in males and females.
- ☐ Chromosomal abnormalities are usually lethal or cause serious defects. Down syndrome, caused by an extra copy of chromosome 21, is one of the most common chromosomal abnormalities in humans. Sex chromosomal abnormalities are less severe than autosomal abnormalities.
- ☐ Most genetic diseases are inherited as autosomal recessive traits; these include PKU, sickle cell anemia, and cystic fibrosis. Huntington's disease is one of the few genetic diseases inherited as an autosomal dominant trait.

METHODS TO STUDY INHERITANCE IN HUMANS INCLUDE PEDIGREE ANALYSIS AND POPULATION STUDIES

Human inheritance was traditionally studied by examining how a trait that exists as two distinct attributes (albino or non-albino, for example) occurred in a particular family. Many early genetic studies were of rare diseases common in certain families but not in the human population as a whole. A **pedigree,** which is a chart that shows the ancestral history of an individual, was prepared in which the trait under study was traced through several

FIGURE 12–1 Pedigrees. (***a***) A pedigree of hemophilia, a rare genetic disorder, among the royal families of Europe. Note that only males exhibit the disorder, a characteristic of X-linked transmission (discussed shortly). (***b***) Prince Alexis, son of Czar Nicholas II, suffered from hemophilia. (*b*, The Bettman Archive)

(*b*)

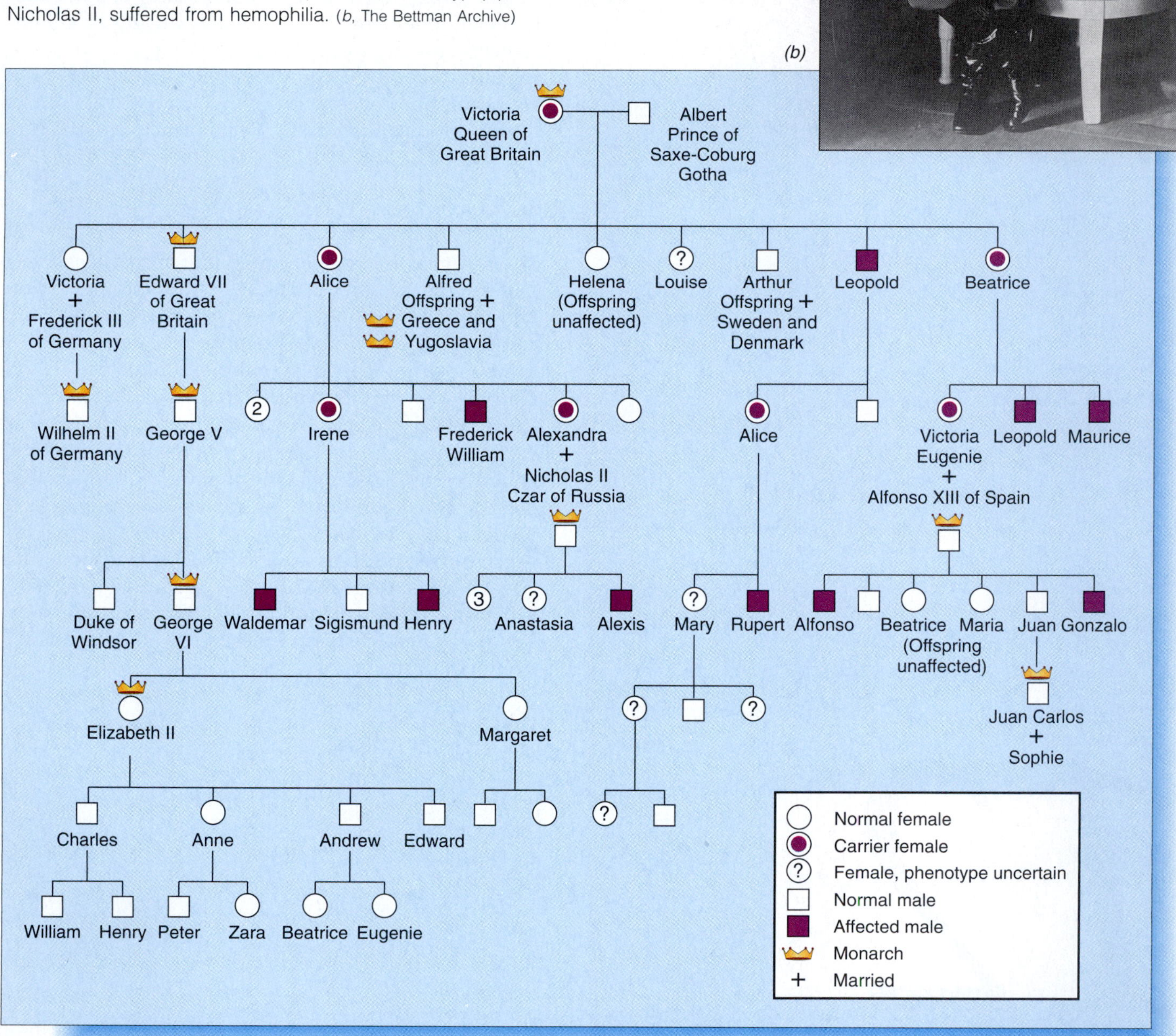

(*a*)

generations of the family (Fig. 12–1). Pedigree analysis is still useful but it has limitations because human families tend to be small and information about certain family members may not be available.

Human geneticists also study the distribution of a particular trait in the entire population. A population geneticist can often determine if a particular trait is inherited in a straightforward Mendelian manner or in a more complex way by using the laws of probability on a large sample of individuals.

More recently, methods of molecular biology and recombinant DNA techniques (Chapter 16), in combination with the use of computers to analyze data, have provided remarkable insights into the structure and function of the 50,000 to 100,000 human genes that each of us possesses. Before this book goes into its next edition, additional information about human genetics will be known, perhaps including how many genes there are, where each is located on the chromosomes, and the exact DNA structure of each (Chapter 11).

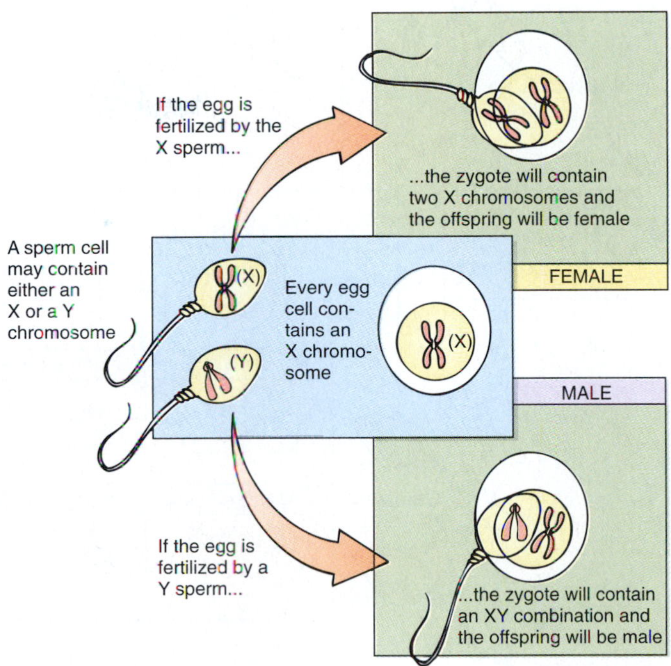

FIGURE 12–2 How sex chromosomes are passed from one generation to the next. A female receives an X chromosome from each of her parents; a male receives an X chromosome from his mother and a Y chromosome from his father.

HUMANS POSSESS AUTOSOMES AND SEX CHROMOSOMES

The diploid chromosome number in humans is 46, or 23 pairs. Of that number, 22 pairs are **autosomes**, chromosomes found in both males and females. The members of a pair of homologous autosomes are identical in size and shape to one another. The autosomes carry most of the genetic information about an individual.

The remaining pair of human chromosomes are the **sex chromosomes**, which are designated as X and Y chromosomes. The sex chromosomes are different in the two sexes. In females, the sex chromosomes consist of two X chromosomes, whereas in males, they consist of an X and a smaller Y chromosome.[1] The sex chromosomes carry genetic instructions that determine whether a human embryo will develop into a male or a female. Specifically, the Y chromosome determines male sex.

As a result of meiosis, each egg produced by a female (XX) carries a single X chromosome (Fig. 12–2). Half the sperm produced by a male (XY) contain the X chromosome, and half carry the Y chromosome. When a sperm carrying a Y chromosome fertilizes an X-bearing egg, the offspring develops into a male; when both sperm and egg carry an X chromosome, the new individual becomes a female.

The Y chromosome is one of the smallest of the human chromosomes and apparently contains only genes related to male sex.[2] At least one gene on the Y chromosome is a male-determining gene that causes testes to develop; when that gene is absent, ovaries develop. Other sexual characteristics develop as a result of hormones secreted by the testes and ovaries.

Although some of the genes on the X chromosome are associated with sexual characteristics, most of its 100 to 200 genes code for nonsexual traits such as eye color and color vision. Since no important Y-linked genes are known in men (other than those producing male sex), the X chromosome, present in both men and women, is what is meant when the terms "sex linkage" and "sex-linked genes" are used. It is more precise to refer to genes located on the X chromosome as **X-linked genes.**

The Expression of X-Linked Recessive Traits Is More Common in Males than in Females

Although the Y chromosome pairs with the X chromosome during meiosis to form sperm, the Y and X chromosome are not truly homologous. Most genes found on the X chromosomes have no known alleles on the Y chromosome. Therefore, in a male (XY) any gene that lies on

[1] An XY mechanism of sex determination is thought to operate in most species of animals, although it is not universal. In some animals, for example, birds and butterflies, the situation is reversed, with males carrying two X chromosomes and females carrying one X and one Y chromosome.

[2] At present no genes other than for male sex determination have been conclusively linked to the Y chromosome, although some evidence exists that a trait known as "hairy ears" occurs on the Y chromosome.

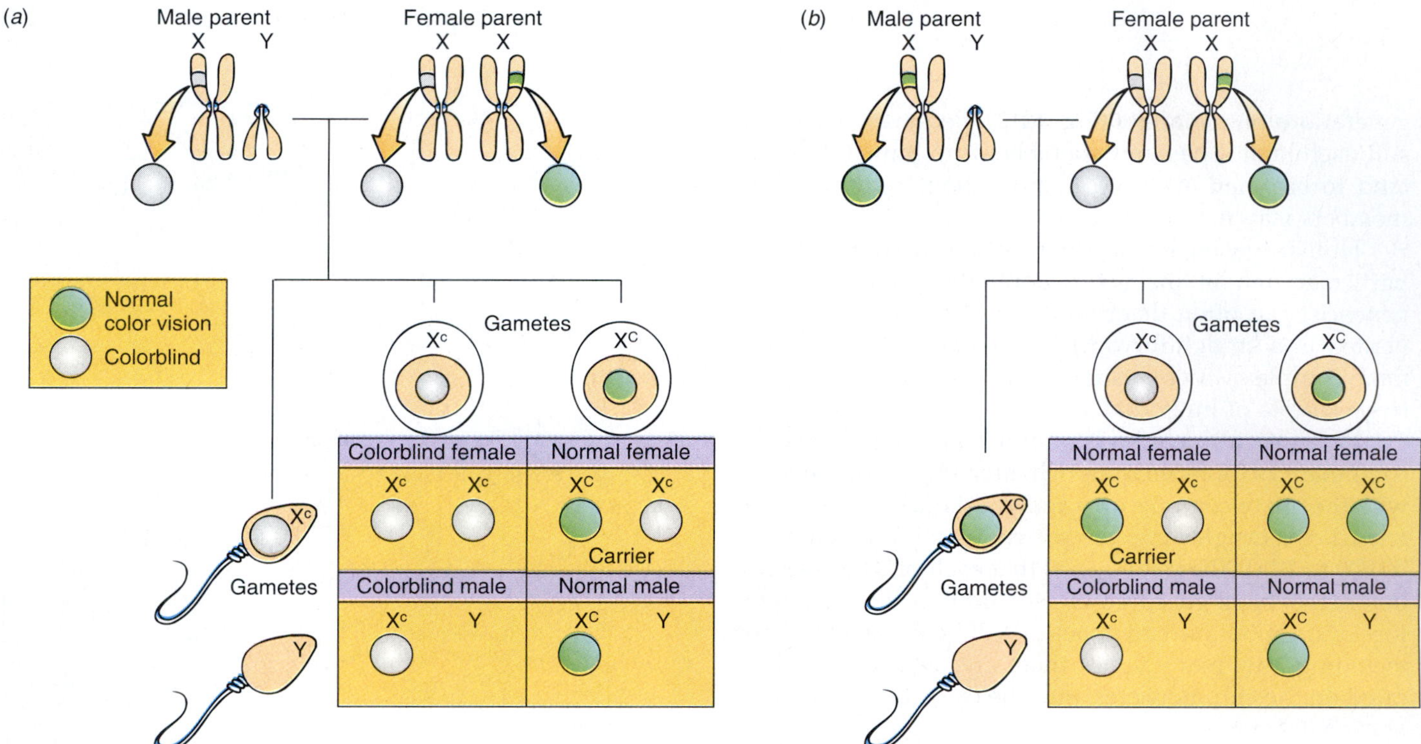

FIGURE 12–3 Two crosses involving color blindness.
X^C represents the gene for normal color vision, and X^c represents the allele for color blindness; the Y chromosome does not carry a gene for color vision. (***a***) The only way a female can be colorblind is if her father is colorblind and her mother is at least heterozygous for the trait. (***b***) More typically, a male is colorblind, having inherited the trait from his mother, who carries a recessive allele on one of her X chromosomes.

the X chromosome will be expressed, regardless of whether it is dominant or recessive. One cannot properly refer to such an X-linked gene in the male as either homozygous or heterozygous. A male is always **hemizygous**, which means that he possesses only one allele of each X-linked gene rather than two (recall that all other alleles occur in pairs).

Examples and problems involving X linkage use various forms of notation. We indicate the X chromosome as *X*, with alleles as superscripts (for example, X^c). The Y chromosome is written as *Y*, without superscripts because it does not carry the allele of interest.

For X-linked genes, the vast majority of abnormal or uncommon alleles are recessive, and the normal or most common alleles are dominant. Thus in a female, two recessive X-linked alleles must be present to be expressed, whereas in the hemizygous male a single recessive allele is expressed. As a result, females may carry recessive X-linked traits (in the sense of being heterozygous for them), but they are usually expressed only in their male offspring.

If a recessive X-linked trait is expressed in a female, the allele must be present on both X chromosomes, which means the alleles must have been inherited from both parents. Consider red-green color blindness, a common genetic abnormality inherited as an X-linked recessive trait. A colorblind girl, as an example, must have a colorblind father and a mother who is at least heterozygous for color blindness (Fig. 12–3). Such a combination of parents is very unusual due to the rarity of the colorblindness allele in the population. Yet a colorblind boy need only have a mother who is heterozygous for color blindness; his father could have normal color vision. This means that X-linked recessive traits are expressed more often in males than in females.

Hemophilia Demonstrates the Importance of Heredity in Human Affairs

Hemophilia is a classic example of X-linked recessive inheritance. This rare disease causes severe bleeding from even a slight injury. It occurred in many of the royal families of Europe. "Our poor family seems persecuted by this awful disease, the worst I know," wrote Queen Victoria, the queen of Great Britain from 1837 to 1901. Hemophilia evidently originated with a mutation (a sudden change in genetic material) in one of Victoria's X chromosomes and was eventually passed on to almost every family of European royalty (Figure 12–1).

Hemophilia was even a contributory factor in the Russian Revolution. Czar Nicholas had married Princess Alexandra of Germany, a granddaughter of Queen Victoria. The couple had a number of daughters but just one son, Prince Alexis. This heir to the throne had hemophilia. The monk Rasputin, a religious mystic, claimed to be able to preserve his life and thereby gained great influence over the czar's family. The resulting public policies are thought to have been among the factors that contributed to the Russian Revolution of 1917 and the family's execution in 1918.

In Prince Alexis's day, neither the cause of hemophilia nor any effective treatment was known. The disease almost invariably led to death in early childhood, caused by a fatal stroke or uncontrollable bleeding, either internally or externally.

A recessive gene carried on the X chromosome produces the disorder. Since males only have one X chromosome, hemophilia is much more common among males than females. (The likelihood that a female would possess the gene for hemophilia on *both* of her X chromosomes, required if she were to exhibit the disorder, is exceedingly unlikely.) No respecter of rank, hemophilia also occurs in those who are not of royal descent.

Hemophilia, produced by the absence of a clotting factor in the blood, affects some 20,000 men and boys in the United States. It is treated today by injecting the blood-clotting factor or by blood transfusions, although there is a risk of contracting blood-borne diseases such as hepatitis and HIV-1, the virus that causes AIDS. (Most of the clotting factors used today come from human blood.) The treatments are prohibitively expensive—as high as $100,000 a year.

Some birth defects are inherited

A birth defect is an abnormality present at birth. Some birth defects are inherited (Table 12–1), whereas others are caused by environmental factors that affect prenatal (before birth) development. For example, when a woman drinks alcohol heavily during pregnancy, the baby may be born with **fetal alcohol syndrome**, that is, deformed and mentally retarded. Low birth weight and certain structural abnormalities have been associated with as few as two drinks a day. Other environmental factors that have been linked with birth defects are discussed in Chapter 43.

Some birth defects are caused by chromosomal abnormalities. For example, persons with Down syndrome usually have three copies of human chromosome 21. Other birth defects are caused by changes in a single gene, as for example, sickle-cell anemia and cystic fibrosis. Relatives are more likely than nonrelatives to carry the same harmful alleles (see Focus on Health & Human Affairs: Consanguineous Matings on page 230).

TABLE 12–1
Examples of Human Genetic Diseases

Disorder or abnormality	*Description*
Change in chromosome number	
Down syndrome	Mental retardation; cardiac defects
Klinefelter syndrome (XXY)	Sterile male; abnormal development of sexual characteristics
Turner syndrome (XO)	Sterile female; abnormal development of sexual characteristics
XYY karyotype	Fertile male; few symptoms
X-linked inheritance	
Hemophilia	Absence of blood-clotting factor
Red-green color blindness	Inability to distinguish between red and green
Autosomal recessive trait	
Albinism*	Absence of melanin (pigmentation)
Cystic fibrosis	Abnormal secretions of body
Phenylketonuria	Inability to metabolize phenylalanine; can cause mental retardation
Sickle-cell anemia	Tissue and organ damage from abnormal form of hemoglobin
Autosomal dominant trait	
Huntington's disease	Progressive deterioration of nervous system; death

*Discussed in Chapter 11.

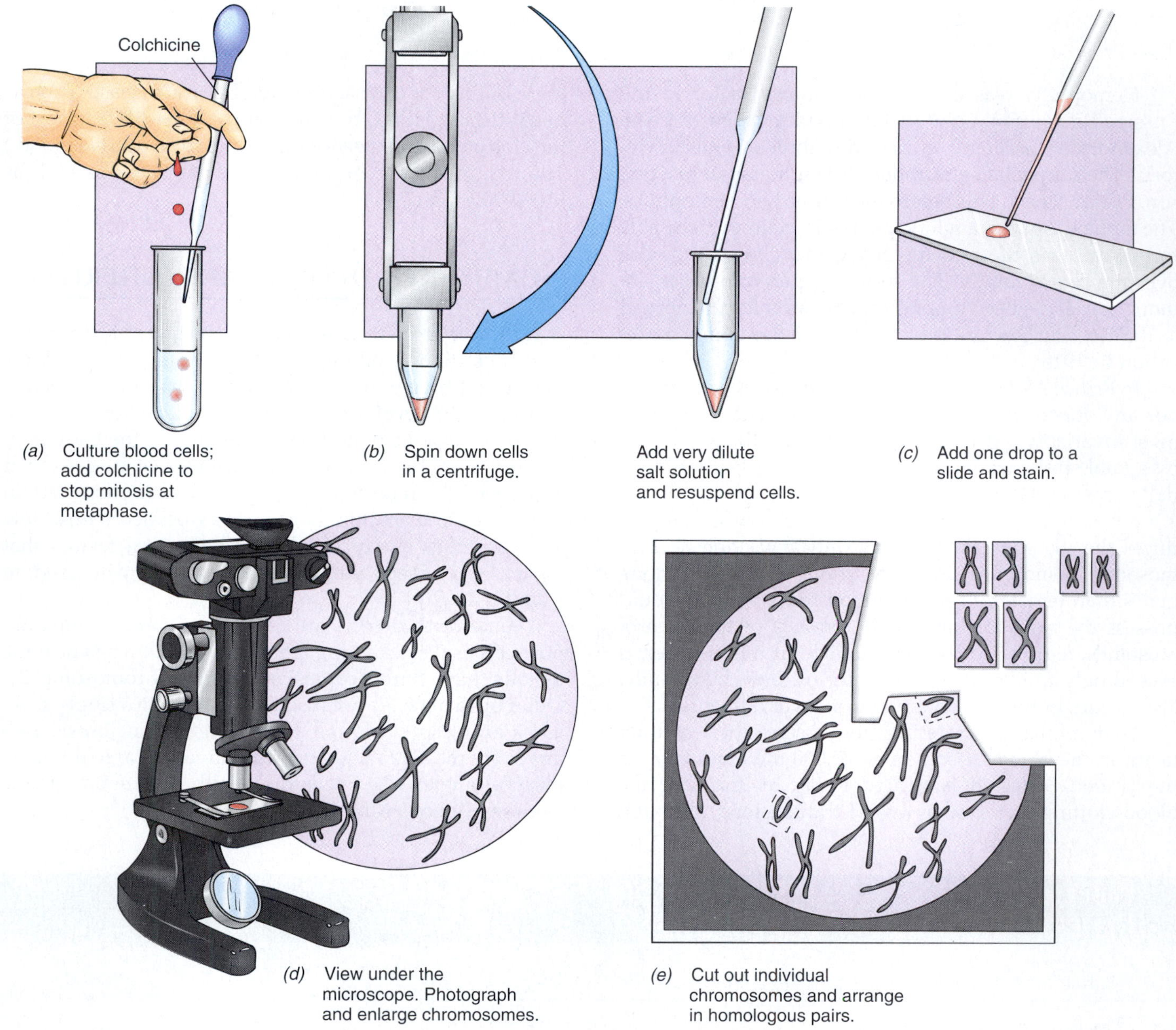

Colchicine
(a) Culture blood cells; add colchicine to stop mitosis at metaphase.
(b) Spin down cells in a centrifuge.
Add very dilute salt solution and resuspend cells.
(c) Add one drop to a slide and stain.
(d) View under the microscope. Photograph and enlarge chromosomes.
(e) Cut out individual chromosomes and arrange in homologous pairs.

FIGURE 12–4 Preparation of a karyotype. An actual karyotype of a normal human male is shown in the lower right. The sex chromosomes (X, Y) are in the lower right corner of the karyotype. (CNRI/Science Photolibrary/Photo Researchers, Inc.)

1
2
3
4
5
6
7
8
9
10
11
12
13
14
15
16
17
18
Y
X
19
20
21
22

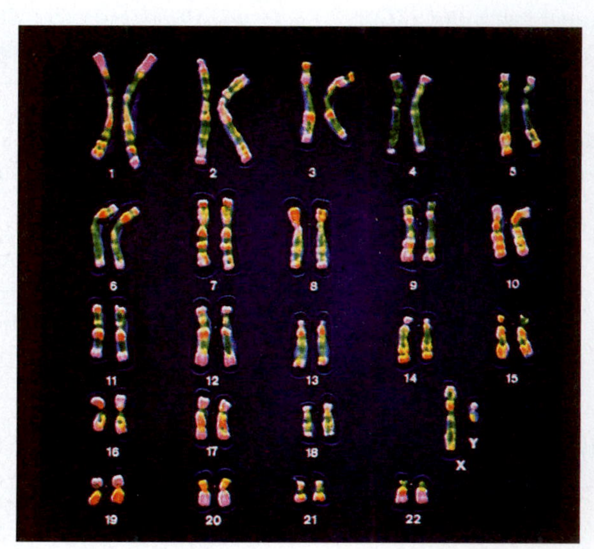

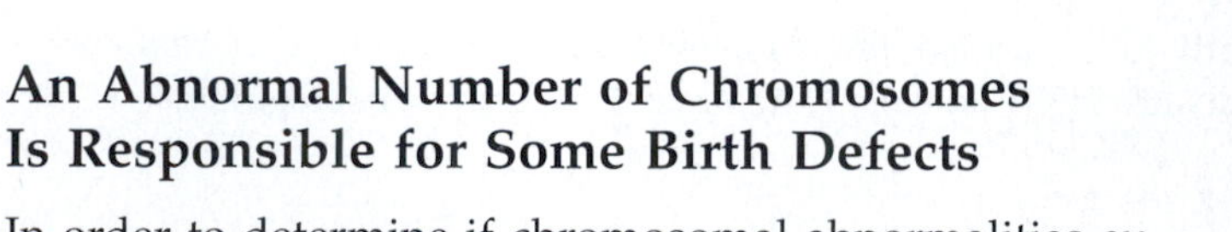

An Abnormal Number of Chromosomes Is Responsible for Some Birth Defects

In order to determine if chromosomal abnormalities exist, a **karyotype**—a photograph that shows the size, number, and characteristic features of the chromosomes—is prepared (Fig. 12–4).

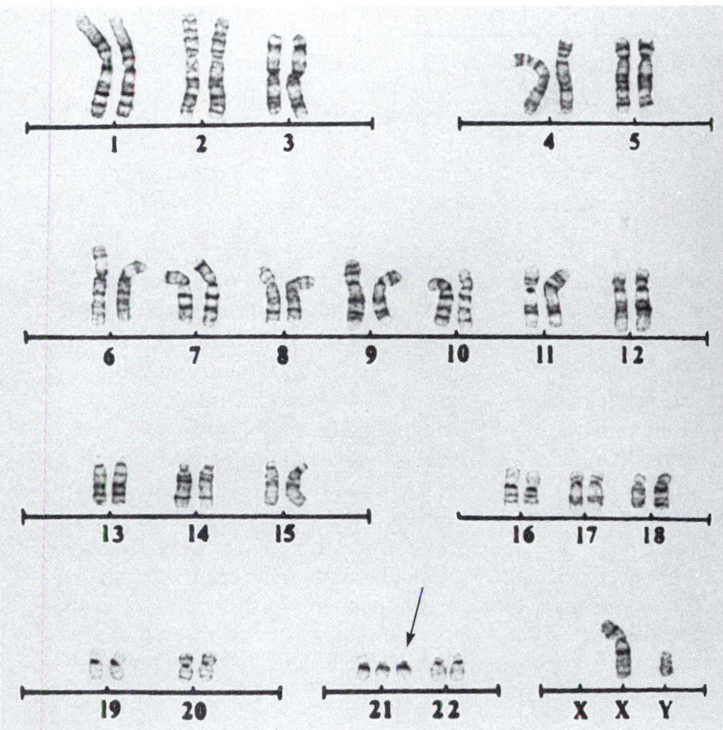

(a)

(b)

FIGURE 12–5 Down syndrome. (***a***) Karyotype of an individual with Down syndrome. Note the presence of the extra chromosome 21. (***b***) An adult with Down syndrome working at McDonald's. People with Down syndrome can be productive members of society. (*a*, Courtesy of Dr. Leonard Sciorra; *b*, Photo Edit)

Down syndrome,[3] one of the most common chromosomal abnormalities in humans, is usually caused by an extra copy of human chromosome 21 (Fig. 12–5). It is characterized by varying degrees of mental retardation, cardiac defects, a fold of skin over the inner corner of the eyelid, short stature, and other deformities. Affected individuals are also very susceptible to certain diseases, such as leukemia and Alzheimer's disease.

The frequency of Down syndrome is about 1 in 700 live births, but the disorder is more common among children of older women; it is 100 times more likely in the offspring of mothers who are 45 years or older than it is in the offspring of mothers who are under 19 years of age. The occurrence of Down syndrome is affected much less by the age of the father. For these reasons, Down syndrome is usually thought to be due to meiotic **nondisjunction** (failure of chromosomes to separate) in the mother.

Abnormalities in the sex chromosomes are usually less severe than autosomal abnormalities. For example, persons with **Klinefelter syndrome** possess 47 chromosomes, including two X chromosomes and one Y chromosome (Fig. 12–6). They are nearly normal males with small testes and enlarged, female-like breasts. Affected persons are usually quite tall, and about half show some

Female
XX
Nondisjunction
Gametes
XX
O
Male
XY
X
Y
Female (Triple X)
Female (Turner syndrome)
XXX
XO
Male (Klinefelter syndrome)
Lethal
XXY
YO

FIGURE 12–6 Nondisjunction. One way that nondisjunction—a failure of chromosomes to separate during meiosis—can occur. This example is of meiotic nondisjunction of the sex chromosomes in a human female, which results in egg cells that each contain two X chromosomes or no X chromosomes. Meiotic nondisjunction of the sex chromosomes also occurs in males. See Table 12-1 (Change in Chromosome Number) for a description of some of the resulting abnormalities caused by nondisjunction.

[3] The term "syndrome" refers to a set of symptoms that usually occur together in a particular disorder.

FOCUS ON *Health and Human Affairs*

CONSANGUINEOUS MATINGS

Each of us is heterozygous for a number of harmful recessive alleles, any of which could cause illness or death in the homozygous state. Why then are genetic diseases not more common? Each of us carries 50,000 to 100,000 different genes, and the chance is remote that you will share any of the same harmful alleles with your mate. This possibility is more likely, however, if you marry a relative, because relatives inherited their genes from a close common ancestor.

Matings between relatives are referred to as **consanguineous matings** (*consanguineous* is derived from two Latin words meaning "related" and "blood"). It can be shown that the risk of bearing a child with a major congenital abnormality of genetic origin is significantly greater if the parents are, for example, first cousins, than if they are unrelated. First cousins share one pair of grandparents, which means that they have a greater chance of passing the same allele for a particular gene to their child. The offspring of consanguineous matings account for a high percentage of those individuals in the population with autosomal recessive disorders.

Because of this perceived social cost, first-cousin marriages and other marriages between close relatives are prohibited in most states in the United States, and less than 1% of U.S. marriages are consanguineous. However, consanguineous marriages are still relatively common in many developing countries, such as parts of Asia and Africa, where economic and social benefits may outweigh the genetic costs. In North Africa, for example, 20% to 50% of all marriages are between biologically related people.

degree of mental retardation. Most live relatively normal lives. About 1 in every 1000 liveborn males has Klinefelter syndrome.

Individuals with **Turner syndrome** have only a single X chromosome and lack a second sex chromosome. They are females, but have undeveloped genitals and are sterile. Affected persons are characterized by a short stature, a webbed neck (extra folds of skin in the neck region), and sometimes slight mental retardation. Turner syndrome occurs at a frequency of about 1 in every 5000 births.

Some males possess one X and two Y chromosomes. Because the effects of this condition are usually mild (tallness, severe facial acne during adolescence) and because they are fertile, men with this condition are described as having an **XYY karyotype** rather than a syndrome. The XYY karyotype occurs with a frequency of about 1 in every 1500 births.

Many Genetic Disorders Are Caused by Single Gene Defects

Hundreds of human disorders involving enzyme or other protein defects are caused by single gene mutations. Most are inherited as autosomal recessive genes and so are only expressed in the homozygous state (Fig. 12–7). Sickle-cell anemia, phenylketonuria (PKU), and cystic fibrosis are well-known diseases transmitted as autosomal recessive genes. Only very rarely is a lethal genetic disease due to an autosomal dominant allele.

Most Genetic Diseases Are Autosomal Recessive Disorders

Although it is not yet possible to cure any genetic disease, some can be successfully treated. Individuals who are homozygous recessive for **phenylketonuria (PKU)** lack an enzyme that converts the amino acid phenylalanine into another amino acid, tyrosine. If untreated, those persons accumulate phenylalanine in their tissues until it reaches toxic levels and damages the central nervous system, resulting in severe mental retardation. A newborn infant with PKU is usually healthy because its mother produces enough enzyme to prevent phenylalanine accumulation. Identifying PKU infants and placing them on low-phenylalanine diets alleviate the symptoms; by adolescence most are able to discontinue special diets because their bodies are not as sensitive to phenylalanine as when they were younger. More than 90% of all infants born in the United States are currently screened for PKU (Fig. 12–8).

In addition to coding for enzymes, some genes code for proteins that have other roles in cells. For example, hemoglobin is a large protein molecule that carries oxy-

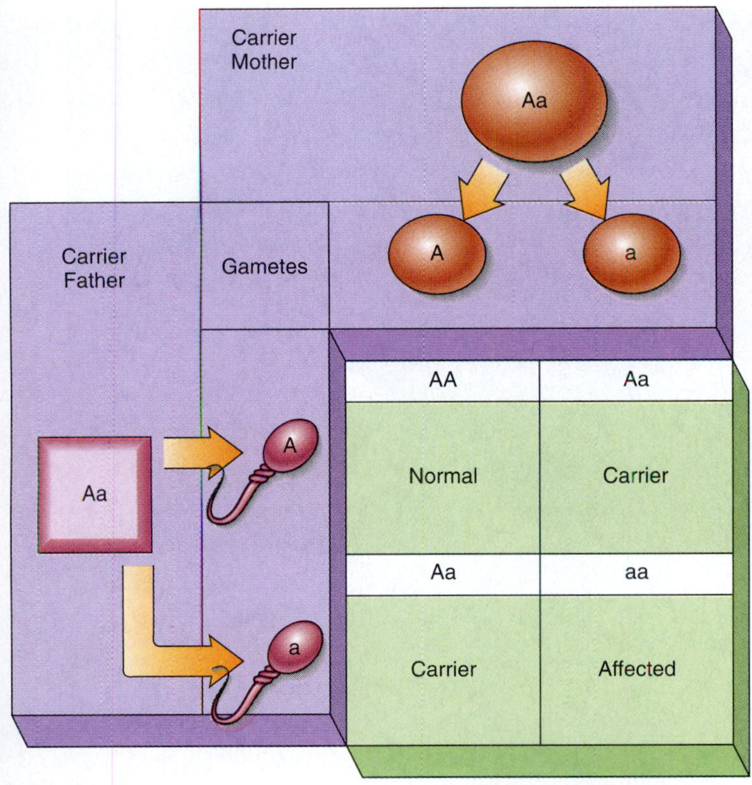

FIGURE 12–7 Autosomal recessive disorders. Because most genetic defects are caused by autosomal recessive genes, an individual must be homozygous recessive in order to be affected. This means that *both* parents must be heterozygous for the trait. There is a 25% likelihood that the child of two carrier parents will be affected by the genetic disorder.

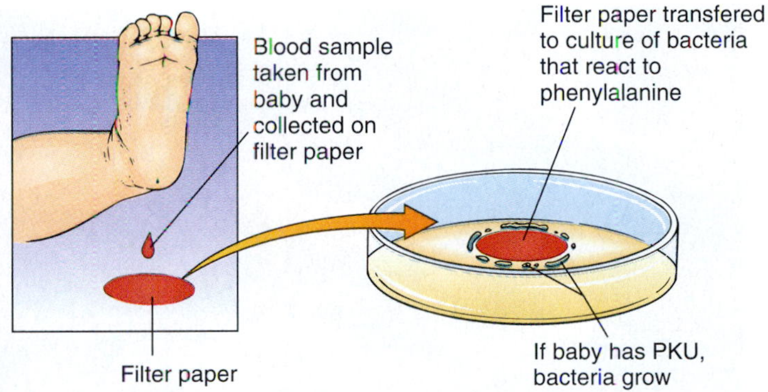

FIGURE 12–8 The PKU test. A small sample of blood is taken from the baby's heel and placed on a plate of agar containing bacteria that only grow in the presence of phenylalanine. Because phenylalanine accumulates in the blood of PKU infants, bacteria grow on the culture, providing a positive test. If the baby is normal, the bacteria do not grow. The PKU test is performed on most newborns before they leave the hospital.

gen in the blood. **Sickle-cell anemia** is an autosomal recessive disorder that causes red blood cells to be sickle-shaped rather than the normal disk-shaped (Fig. 12–9); the change in shape is the result of an abnormal form of hemoglobin. Because of their shape, sickle cells block small blood vessels, causing tissue damage (especially in the lungs, heart, and kidneys) and pain. Sickle-cell anemia can also lead to strokes (when sickle cells block blood vessels leading to the brain). Also, anemia results because the sickle cells do not survive as long as normal red blood cells. Sickle-cell anemia is most common in persons of African descent; some 80,000 African-Americans have the disease and about 1 in 12 African-Americans is a heterozygote carrier for it. (See Chapter 18 for a discussion of the relationship between malaria and being heterozygous for sickle-cell anemia.) Current treatments for sickle-cell anemia include blood transfusions and some forms of drug therapy. However, these procedures are not completely effective, and research efforts are aimed at improving methods of treatment.

Concept Connections

Dominant Lethal Alleles ⟷ *Natural Selection*

Lethal genetic diseases are rarely the result of dominant alleles because natural selection prevents such alleles from becoming established in a population. Natural selection, the mechanism of evolution originally proposed by Charles Darwin in the 19th century, tends to preserve favorable traits and eliminate unfavorable traits from a population. Thus, during the course of time, natural selection helps determine the genetic composition of a population.

How does natural selection eliminate dominant lethal alleles from a population? According to natural selection, individuals possessing the most favorable combination of traits are most likely to reproduce, passing their traits on to the next generation. If a dominant lethal allele is to remain in a population, individuals carrying it must be able to reproduce *before* the allele kills the affected individuals. If the lethal allele kills the affected individuals before they reproduce, the allele will disappear from the population. We say more about dominant lethal alleles later in this chapter and more about natural selection in Chapter 17.

Approximately 1 of every 20 persons in the United States is a heterozygote carrier of **cystic fibrosis**, a disease characterized by abnormal secretions in the body. Individuals who are homozygous recessive for cystic fi-

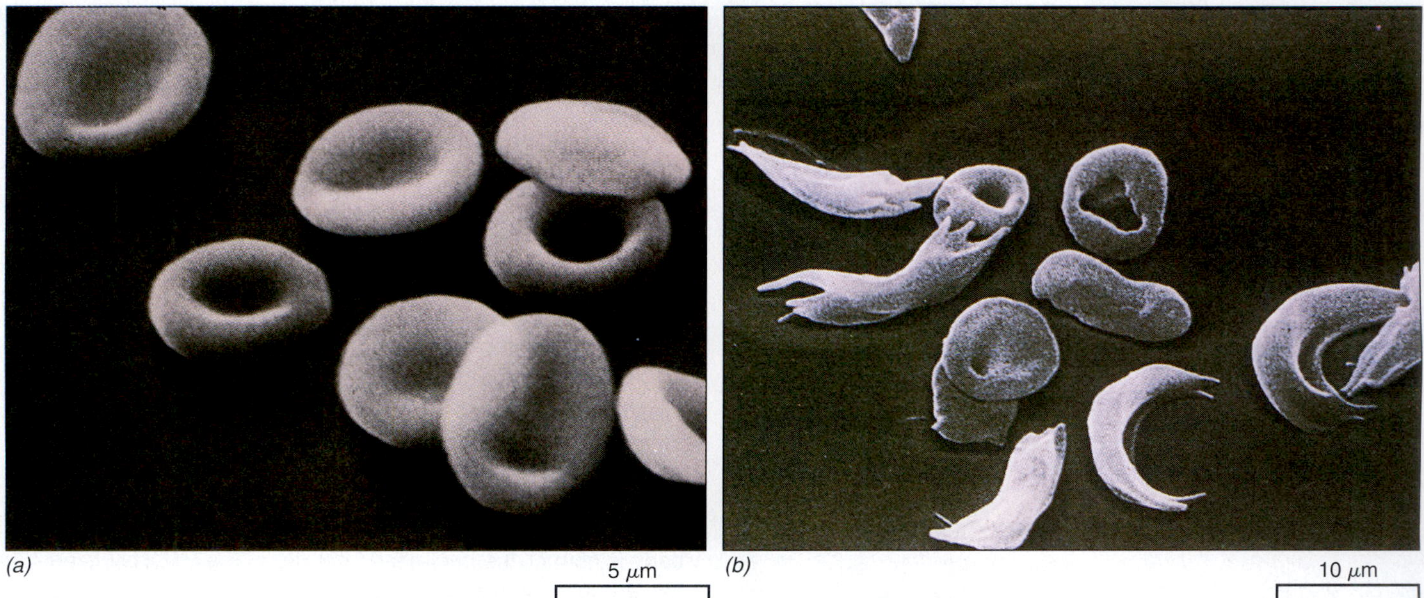

FIGURE 12–9 Sickle-cell anemia. (***a***) Normal red blood cells are shaped as biconcave disks. (***b***) Sickled cells are longer and more fragile than normal red blood cells. Because they clog small capillaries, certain tissues do not receive adequate oxygen. (*a*, Omikron/Photo Researchers, Inc.; *b*, Omikron/Science Source/Photo Researchers, Inc.

brosis lack a functional form of a transmembrane protein that is a chloride ion (Cl^-) channel. Cystic fibrosis affects the mucus-producing glands of the respiratory system, leading to recurring bouts of bronchitis, pneumonia, and other respiratory complications. It also affects the pancreas, which among other things produces enzymes for digesting fats; affected persons must eat low-fat diets and take special medications to aid digestion. Although cystic fibrosis used to be fatal in childhood, modern medical care and ongoing research efforts make the outlook for individuals affected by the disease more promising.

Huntington's Disease Is a Rare Autosomal Dominant Disorder

Although most diseases of genetic origin are recessive, dominant genetic disorders do exist. **Huntington's disease** is a serious hereditary disorder of humans that is both dominant and lethal. Huntington's disease produces mental deterioration, painful paralysis, sensory loss, and ultimately, death. One might think that this disease could only occur as a new mutation and could not be transmitted to future generations (recall the Concept Connections box). However, the symptoms of Huntington's disease do not usually begin until middle age. By that time the sufferer could have produced many children (Fig. 12–10).

The gene that causes Huntington's disease was identified in 1993 after a 10-year search by several research labs. People at risk can be tested to learn whether they carry the gene, which could help them make an informed

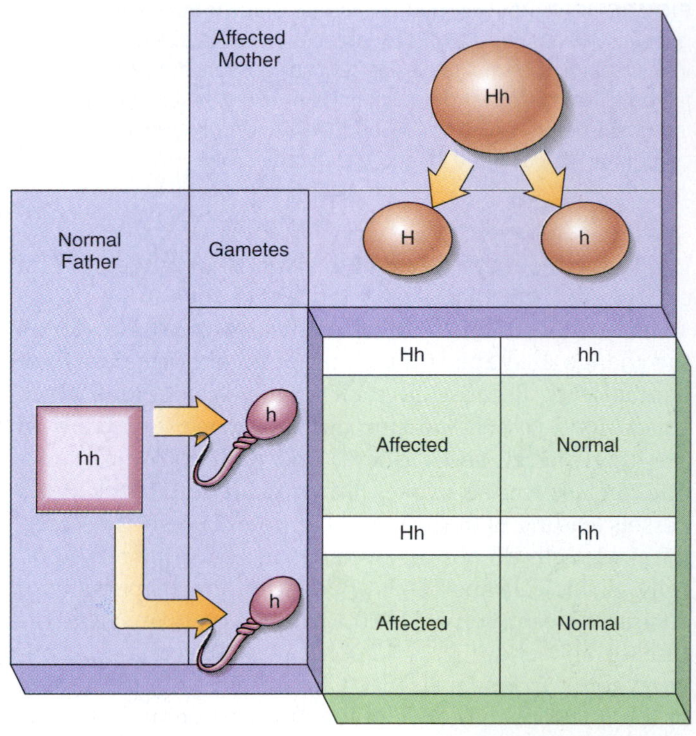

FIGURE 12–10 Inheritance of an autosomal dominant disorder. Half the children of a normal parent and a parent who has the gene for Huntington's disease can be expected to eventually develop this incurable disease.

decision as to whether to have children. However, someone who tests positive for Huntington's disease must then live with the certainty of eventually developing this incurable disease.

Some Genetic Abnormalities Can Be Detected and Treated Before Birth

Early detection increases the possibilities for prevention or alleviation of the effects of genetic abnormalities. Prenatal diagnosis and treatment are now possible for many genetic disorders. However, genetic diseases cannot yet be permanently cured, although genetic engineering (Chapter 16) is bringing cures closer to reality.

Amniocentesis is a technique in which a small amount of amniotic fluid surrounding the fetus is removed, and the fetal cells within the fluid cultured (Fig. 12–11). The cultured cells are then screened for evidence of chromosomal abnormalities or other defects. Amniocentesis is performed mostly on pregnant women over 35 years of age because their offspring have an increased risk of Down syndrome. Amniocentesis is also useful in detecting a noninherited condition known as spina bifida, in which the spinal cord does not close properly during development. Although amniocentesis is a relatively safe procedure, it cannot provide results until well into the second trimester (four to six months) of pregnancy.

Doctors have developed certain prenatal screening tests that yield results earlier in pregnancy. One such test, **chorionic villus sampling (CVS),** involves removing and studying cells from the chorion (the outermost membrane surrounding the fetus) (Fig. 12–12). Although CVS has a slightly higher risk of infection or miscarriage than amniocentesis, it has the advantage of yielding results during the first trimester (first three months) of pregnancy.

Additional tests have been developed to identify a number of other genetic diseases, but these tests are only performed if a particular problem is suspected. The tests for sickle-cell anemia, cystic fibrosis, and Huntington's disease require the use of genetic engineering methods on cells obtained by amniocentesis.

Although genetic testing is continually improving, it is not foolproof, and negative results do not guarantee a normal baby. In addition, many genetic disorders still cannot be diagnosed, although efforts are under way to develop tests for them.

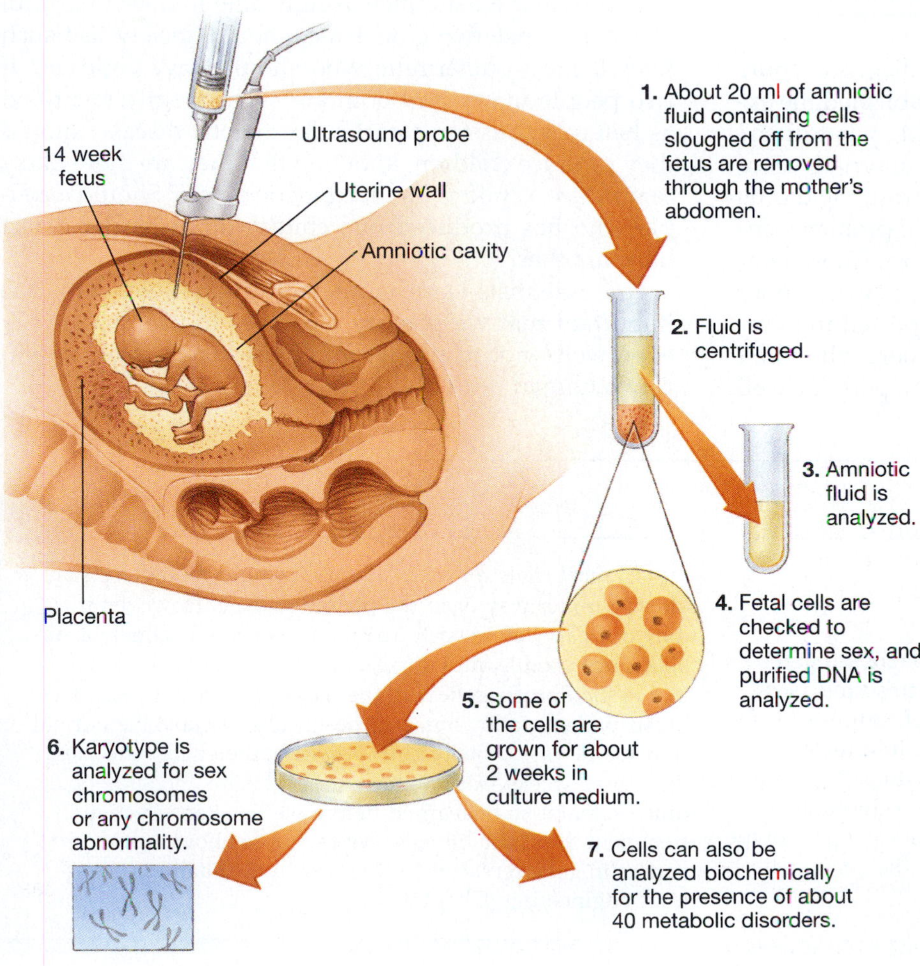
14 week fetus
Ultrasound probe
Uterine wall
Amniotic cavity
Placenta
1. About 20 ml of amniotic fluid containing cells sloughed off from the fetus are removed through the mother's abdomen.
2. Fluid is centrifuged.
3. Amniotic fluid is analyzed.
4. Fetal cells are checked to determine sex, and purified DNA is analyzed.
5. Some of the cells are grown for about 2 weeks in culture medium.
6. Karyotype is analyzed for sex chromosomes or any chromosome abnormality.
7. Cells can also be analyzed biochemically for the presence of about 40 metabolic disorders.

FIGURE 12–11 Amniocentesis. This technique can be used to diagnose genetic disorders before birth.

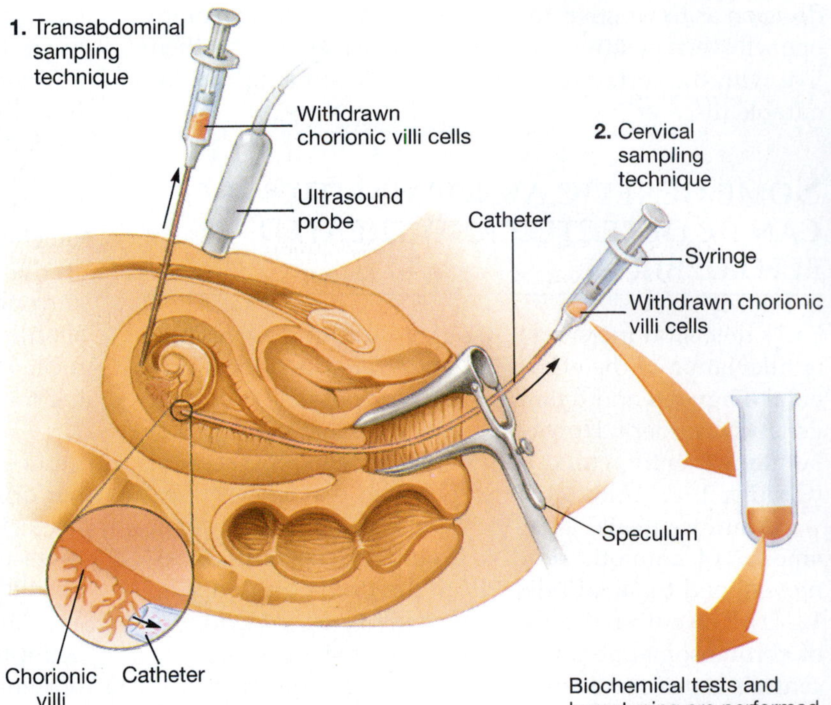

FIGURE 12–12 Chorionic villus sampling (CVS). In the prenatal diagnosis of genetic disorders by chorionic villus sampling, samples may be obtained through the uterine wall (1) or through the cervical opening (2).

HUMAN GENETICS POSES COMPLEX ETHICAL QUESTIONS FOR SOCIETY

It is now possible, as we have seen, to diagnose some hereditary diseases prenatally. Having established the existence of a disease in an unborn child, what then? Prospective parents are faced with the alternatives of abortion (removal of the embryo or fetus from the uterus before it is capable of surviving) versus the problems encountered by allowing the baby to be born. These concerns include crippling medical expenses, worries about the quality of life that the child can be expected to experience, and the strain on the family members who must care perpetually for a chronically ill person, perhaps well into his or her adult years. People in this position are faced with difficult decisions.

Now that we are increasingly able to detect harmful alleles in a heterozygous state, should society use such knowledge to determine who should have children? If two people marry and both are subsequently identified as heterozygous for a particular genetic disease, should they produce children, knowing that they are more likely to produce a child with that genetic defect? Should a couple who has produced one child with a genetic defect have another?

As scientists learn more about human genetics, it is important that we prepare ourselves to use this knowledge well—not to stigmatize or discriminate but to improve human health.

CONCEPT CONNECTIONS

Human Genetics* ⇄ *Gene Replacement Therapy

Although many genetic disorders can be diagnosed and treated, none can be permanently cured at this time. Gene replacement therapy offers the possibility of future cures for single gene disease because it is possible to introduce the normal allele into the cells of the body, in which the allele is expressed. The normal expression of the introduced allele should permit the individual to possess a normal phenotype even though he or she carries the allele for the genetic disorder in most cells.

Clinical trials for many different single gene defects are already under way or in the planning stage. However, it may be years before such treatments become routinely available. For example, initial trials in 1993 introduced the normal cystic fibrosis gene into the nasal passages of cystic fibrosis patients, allowing their nasal cells to make the normal transmembrane protein. Unfortunately, the effect lasted less than three weeks. Although these results are encouraging, much remains to be learned before we have an effective treatment for cystic fibrosis. We say more about recent advances in gene replacement therapy in the chapter on genetic engineering (Chapter 16).

Chapter Summary

I. Human inheritance is studied by pedigree analysis, by population studies, and by using molecular biology and recombinant DNA techniques.

II. Human females have 22 pairs of autosomes plus two X chromosomes; human males have 22 pairs of autosomes plus one X chromosome and one Y chromosome.
   A. The X chromosome is exceptional in that it has no fully homologous member in the male.
   B. An X-linked gene is one that is carried on the X chromosome. A gene located on the X chromosome of a male will be expressed even if it is recessive, whereas a recessive X-linked allele can only be expressed in females in the homozygous condition.
   C. Color blindness and hemophilia are examples of X-linked traits.

III. Some birth defects are caused by an abnormal number of chromosomes.
   A. A karyotype, which is a photograph of chromosomes arranged by size, number, and characteristic features, permits diagnosis of chromosomal abnormalities.
   B. Down syndrome, one of the most common chromosomal abnormalities in humans, is usually caused by an extra copy of human chromosome 21.
   C. Chromosomal abnormalities of sex chromosomes are usually less severe than autosomal abnormalities. Examples include Klinefelter syndrome (XXY males), Turner syndrome (XO females), and the XYY karyotype (XYY males).

IV. Single gene defects, the vast majority of which are recessive, cause many genetic disorders.
   A. Phenylketonuria is caused by a single gene defect in which the body cannot make an enzyme needed in amino acid metabolism.
   B. Sickle-cell anemia is caused by a single gene defect in which the body cannot make normal hemoglobin, the protein needed to carry oxygen in the blood.
   C. Cystic fibrosis is caused by a single gene defect in which the body produces abnormal secretions. The respiratory system and pancreas are particularly affected.

V. Dominant genetic diseases are very rare. Huntington's disease is an autosomal dominant disease that does not have any symptoms until middle age, when it is lethal.

VI. Some genetic abnormalities can be detected before birth.
   A. In amniocentesis, the amniotic fluid surrounding the fetus is removed and the fetal cells in the fluid are cultured and screened for genetic defects. Amniocentesis provides results in the second trimester of pregnancy.
   B. In chorionic villus sampling (CVS), cells from the chorion are removed and studied. CVS provides results in the first trimester of pregnancy.

VII. Genetic diseases cause numerous ethical, social, and medical problems.

Selected Key Terms

amniocentesis, p. 233
autosomes, p. 225
chorionic villus sampling (CVS), p. 233
consanguineous mating, p. 230
hemizygous, p. 226
karyotype, p. 228
nondisjunction, p. 229
pedigree, p. 224
sex chromosomes, p. 225
X-linked genes, p. 225

Post-Test

1. Humans possess 22 pairs of ______________ and one pair of ______________ ______________.

2. The ______________ chromosome is one of the smallest of the human chromosomes and apparently contains only genes related to male sexual characteristics.

3. Humans who carry two copies of the X chromosome are ______________, whereas those who carry XY are ______________.

4. Genes located on the X chromosome are referred to as ______________-______________ genes.

5. A male is considered ______________ with respect to X-linked genes because he possesses only one allele of each gene.

6. If a man is colorblind, he received the gene for color blindness from his ______________ (mother or father).

7. Geneticists prepare a special photograph of the human chromosomes, called a/an ______________, to determine if chromosomal abnormalities exist.

8. Down syndrome is usually thought to be the result of meiotic ______________.

9. Examples of abnormalities in the sex chromosomes include ______________ syndrome, present in XXY males, and ______________ syndrome, present in XO females.

10. Individuals with ______________ lack an enzyme that converts the amino acid phenylalanine into another amino acid.

11. Sickle-cell anemia is caused by a single gene defect in which the body is unable to make a normal form of the oxygen-carrying protein, ______________.
12. In ______________ ______________, individuals produce abnormal secretions in the respiratory system, leading to recurring bouts of pneumonia and other respiratory complications.
13. Analysis of fetal cells within the amniotic fluid is known as ______________.
14. The risk of bearing a child with a major genetic abnormality is greater when two parents are close relatives. Such matings are called ______________ matings.
15. A chart showing the ancestral history of an individual is known as a/an: (a) karyotype (b) DNA marker (c) chorionic villus sampling (d) pedigree.
16. Although a mother can, a father cannot pass an X-linked gene to his: (a) son (b) daughter (c) son or daughter (d) none of these is possible—a father can pass an X-linked gene to any of his offspring.
17. One of the most common chromosomal abnormalities in humans is: (a) phenylketonuria (b) Huntington's disease (c) Down syndrome (d) cystic fibrosis.
18. A serious hereditary disorder that is both dominant and lethal is: (a) cystic fibrosis (b) Huntington's disease (c) sickle-cell anemia (d) Turner syndrome.
19. A prenatal screening test that yields results in the first trimester is: (a) pedigree analysis (b) amniocentesis (c) chorionic villus sampling (d) gene replacement therapy.

Review Questions

1. Are all birth defects hereditary? Explain.
2. How would pedigree analysis aid in determining if a particular birth defect is inherited?
3. What chromosome or chromosomes bear sex-linked traits?
4. Explain why males receive X-linked recessive traits from their mothers but not from their fathers.
5. Why are female hemophiliacs virtually unknown?
6. A hemophilic man has a normal son. He plans never to have any other children, but he worries that he may pass this genetic disease on to his grandchildren through this son. Are these fears justified? Why or why not?
7. Why are most single gene disorders recessive rather than dominant?
8. The father of a 20-year-old man has just been diagnosed as having Huntington's disease (assume that he is heterozygous). Should the young man be concerned that he could get the disease later in life? Explain.
9. Compare amniocentesis with chorionic villus sampling.
10. To be expressed, an autosomal recessive genetic disease must be homozygous. What relationship does this fact have to consanguineous matings?

Thinking Critically

1. Does society have the right to prevent people known to carry a genetic defect from having children? Debate both sides of this issue.
2. How are the methods that geneticists use to study human genetics different from the methods they use to study inheritance in plants and animals? Why is there a difference?
3. Although Tsar Nicholas II, his wife Alexandra, and their daughter Princess Anastasia did not have hemophilia, their son Prince Alexis did. Explain how he inherited the disease. Is it valid to assume that Princess Anastasia was heterozygous for hemophilia?
4. Examine the pedigrees shown to the right and determine whether each disorder is most likely inherited by an autosomal recessive, an autosomal dominant, or an X-linked recessive. Justify your conclusions.
5. Determine the probable genotypes for all persons shown in the pedigrees on the right.

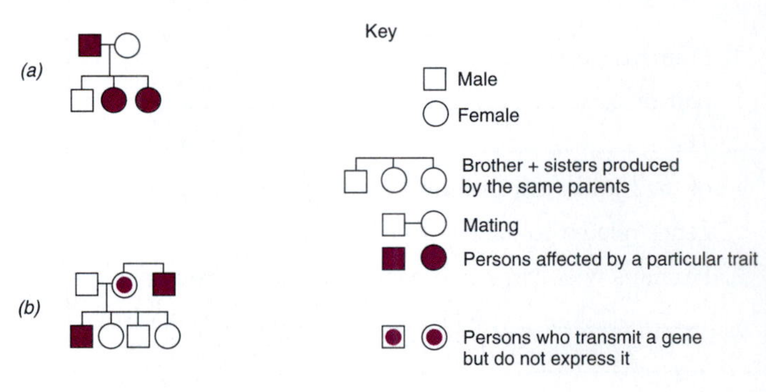

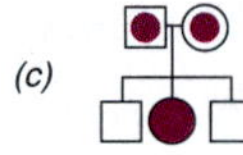

6. It is easier for geneticists to study X-linked recessive traits than autosomal recessive traits. Explain why.

7. Explain why it is not surprising that very few genes occur on the Y chromosome.

Genetics Problems

1. A woman with normal color vision who has a colorblind father marries a man with normal color vision. Calculate the percentage chance that her sons will be colorblind.

2. A colorblind girl has a mother with normal color vision. What was the phenotype of the girl's father? The genotype of both of the girl's parents?

3. A healthy woman has a hemophilic brother. Her husband is healthy and normal. Ultrasound diagnosis determines that the fetus in the next pregnancy is male. What is the probability that this baby is hemophilic?

4. A couple's first child has cystic fibrosis, but neither of the parents is afflicted. Determine the percentage likelihood that their next child will have cystic fibrosis.

5. Both members of a couple are known to be heterozygous for sickle-cell anemia. What is the probability that they will produce a healthy baby who is a carrier of this trait?

Recommended Readings

Annas, G.J., "Privacy Rules for DNA Databanks," *JAMA*, Vol. 270, No. 19, November 17, 1993. Discusses ethical issues involving the privacy of genetic information.

Beckwith, J., "A Historical View of Social Responsibility in Genetics," *BioScience*, Vol. 43, No. 5, May 1993. How genetic information has been abused in the past, from eugenics to Hitler.

Gibbons, A., "The Risks of Inbreeding," *Science*, Vol. 259, February 26, 1993. This article in the News and Comment section of *Science* reviews current worldwide data on the increased risk of genetic disease from consanguineous matings.

Patterson, D., "The Causes of Down Syndrome," *Scientific American*, August 1987. Examines this common genetic defect.

Roberts, L., "To Test or Not to Test," *Science*, Vol. 247, 1990. Considers the social and ethical issues surrounding the question of whether everyone of childbearing age should be screened as a carrier of cystic fibrosis.

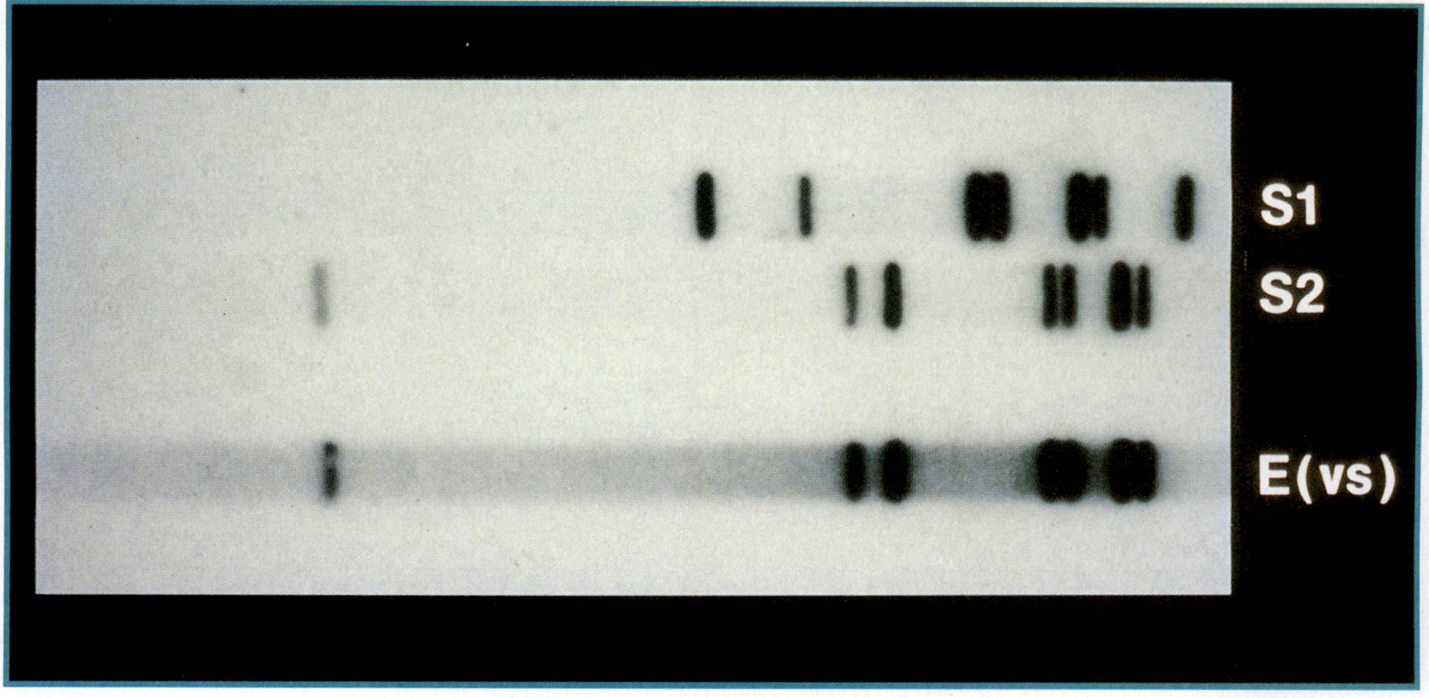

DNA fingerprints from a criminal case involving the rape and murder of a Florida woman. *The DNA fingerprints of two suspects, designated S1 and S2, were compared to the DNA fingerprint of semen obtained from the victim, designated E(vs). Note how the pattern of S2 matches E(vs). This evidence was strong enough to convict the second suspect, who received the death penalty.* (Courtesy of Cellmark Diagnostics)

DNA Fingerprinting and Crime

In many cases involving violent crimes, the suspect is often initially implicated by circumstantial evidence, but conviction by a jury may require additional data linking the suspect to the crime. One of the most powerful tools developed in recent years is **DNA fingerprinting,** the identification of a person based on his or her own unique form of DNA. In DNA fingerprinting, DNA obtained at the scene of the crime from a smear of blood, sample of semen, strand of hair, or other source is compared with the DNA of the suspect.

Generally, medical personnel collect physical evidence from both the victim and the suspect. Forensic scientists isolate nuclei from the cells of each sample, and then extract and purify DNA from the nuclei. The DNA is digested by a restriction enzyme, a special type of enzyme that breaks the DNA into smaller fragments at very specific sites on the DNA molecule (restriction enzymes are discussed further in Chapter 16). A technique called gel electrophoresis separates these DNA fragments by size. Additional techniques help the scientist to visualize and compare the different-sized fragments.

Before DNA fingerprinting, the most accurate blood tests could *implicate* a suspect, but not with the certainty of DNA fingerprinting. For example, if the biological samples from the scene of the crime matched the suspect's blood profile, the match was usually considered strong enough evidence to obtain a jury's conviction. But perhaps one of every thousand people possesses the same blood profile as the suspect, so the evidence was not conclusive.

When the DNA profile of the suspect matches that from biological samples at the crime scene, however, the suspect's involvement in the crime is firmly established. The probability of finding another person with that same DNA profile as the suspect is perhaps one in several hundred million people.[1]

DNA fingerprinting is a reliable way to determine *innocence* of a suspect as well as guilt. For example, a man imprisoned for almost nine years for the rape and murder of a Maryland girl was released in 1993 because new genetic evidence showed that he was not guilty. Had this evidence been available during his trial, it is likely that he would have been found not guilty and would never have been imprisoned.

DNA fingerprinting is a very powerful tool in forensic science. DNA, the molecule of inheritance, contains a precise set of instructions for an individual's traits. Your DNA, for example, not only codes for everything required to make you a member of the human species, but also for everything that makes you distinctive from other humans. With the exception of identical twins, each individual's DNA is unique. In this chapter we consider (1) how DNA was first determined to be the molecule of inheritance and (2) DNA's structure and replication.

[1] The reliability of DNA fingerprinting is sometimes questioned by defense lawyers in some criminal cases because of the improper statistical presentation of the evidence to juries. Scientists and lawyers in the United States, United Kingdom, and other countries are currently debating the best way to present DNA evidence.

DNA: The Molecular Basis of Inheritance

13

Learning Objectives

After you have studied this chapter you should be able to:

1. Trace the history of how DNA was determined to be the genetic material.
2. Describe the experiments performed by Hammerling on *Acetabularia* and interpret the results.
3. Discuss Griffith's experiments with mice and pneumococcus bacteria. Explain how Griffith interpreted his results and what Avery later determined.
4. Describe the Hershey-Chase experiment with bacteriophages and interpret the results.
5. Explain how the structure of DNA was deduced, including the contributions of Wilkins and Franklin, and Watson and Crick.
6. Diagram the basic chemical shape of a DNA molecule.
7. Name the subunits of which the DNA molecule is composed and give the three parts of each subunit.
8. Describe base-pairing in DNA molecules. Given the base sequence of one strand of DNA, predict that of a complementary strand of DNA.
9. Summarize the process of DNA replication, including the role of DNA polymerase. Explain what is meant by "semiconservative replication."
10. Explain how DNA is organized into chromosomes.

Key Concepts

- ☐ Each eukaryotic cell is controlled by its nucleus, which contains a precise plan for the structure and function of the entire organism. The cell's nucleus contains deoxyribonucleic acid (DNA), which stores genetic information.
- ☐ DNA is a double-stranded molecule in which the two strands are twisted about one another to form a double helix. Each strand of DNA is made up of nucleotides that vary only in the sequence of the bases each contains. The two strands are joined together by hydrogen bonds between complementary base pairs; adenine pairs (that is, forms hydrogen bonds) with thymine, and guanine pairs with cytosine.
- ☐ DNA replicates semiconservatively. First the two strands unwind and separate from each other; then each strand serves as a template for the attachment of free nucleotides according to the base-pairing rules.
- ☐ DNA is found in chromosomes where it is organized with proteins into a complex structure (in eukaryotes). Each chromosome contains a single giant molecule of DNA.

THE NUCLEUS CONTAINS HEREDITARY INFORMATION

Each organism contains within it a set of hereditary instructions characteristic of its species. In turn, these instructions are passed on, making offspring of the same species as the parents. This hereditary information must be copied accurately and delivered to each new cell and to each offspring. Once there, it expresses itself by controlling each cell's structure and function and, through the cells, the life of the entire organism.

The hereditary instructions—the blueprint of life—must be stored somewhere in each cell, but in what part of the cell? Biologists had long suspected the cell's nucleus, because almost all eukaryotic cells have a nucleus and cannot survive long without it. But how could this be demonstrated?

The Giant *Acetabularia* Cell Is Ideal for Determining the Role of the Nucleus

Submerged along the rocky shores of warm seas lives a most remarkable organism, *Acetabularia*, commonly known as the mermaid's wine glass (Fig. 13–1*a*). This little organism, a type of alga, consists of a single giant cell, some 2 to 10 cm in length, with three parts: (1) a cup-like cap; (2) a long cylindrical stalk; and (3) a root-like holdfast, which contains the nucleus. There are several species of *Acetabularia*, with caps of different shapes.

If the cap of an *Acetabularia* cell is experimentally removed at the right time in the cell's life cycle, another cap will grow after a few weeks. Such a replacement, common among lower organisms, is called **regeneration.** This fact attracted the attention of J. Hammerling, who in the 1930s became interested in the possible relationship between the nucleus and the physical characteristics of *Acetabularia.*

Because of *Acetabularia's* size and the location of its nucleus, investigators could perform surgery on *Acetabularia* with relative ease. Hammerling and his colleagues performed a brilliant series of experiments that in many ways laid the foundation for much of our modern knowledge of the nucleus. In most of these experiments they employed two species of *Acetabularia: A. mediterranea*, which has a smooth cap, and *A. crenulata*, with a cap divided into a series of finger-like projections.

The kind of cap that is regenerated depends upon the species of *Acetabularia* used in the experiment. As you might expect, *A. crenulata* will ordinarily regenerate a *crenulata* cap when its orginal cap is removed. Likewise, when the cap is removed from *A. mediterranea*, it will regenerate a *mediterranea* cap. Thus, there is evidently something in the lower part of the cell—perhaps the nucleus—that controls cap shape.

The Nucleus Controls the Cell through a Messenger Substance

It is possible to attach the stalk of one species of *Acetabularia* to the holdfast of another species of *Acetabularia.* First the caps are removed from *A. mediterranea* and *A. crenulata*, then the stalks are severed from their holdfasts, and finally, the stalks are exchanged (Fig. 13–1*b*). What happens? The caps that regenerate are characteristic not of the species that donated the holdfasts, but of those that donated the stalks.

However, if the regenerated caps are removed, the second caps to regenerate will be characteristic of the species that donated the holdfasts. This will continue to be the case no matter how many times the regenerated caps are removed!

Hammerling deduced from these and other experiments that the nucleus in the holdfast ultimately controls the cell. But what explains the time lag before a new holdfast gains the upper hand? The simplest explanation is that the holdfast produces a temporary messenger substance, which enters the stalk and controls cap regeneration (Fig. 13–1*c*). Initially, the grafted stalks still contain some of the messenger substance produced by their old nuclei, so they regenerate caps with the old shape. After this brief delay, the nuclei in the new holdfasts begin sending their messenger substances into the stalks and thus begin directing the stalks to grow caps with the new shapes. (The messenger substance, today known to be messenger ribonucleic acid, or mRNA, will be discussed in the next chapter.)

The Nucleus Alone Determines the Shape of the *Acetabularia* Cell

If the nucleus is removed and the cap cut off, a new cap will regenerate. *Acetabularia*, however, is usually able to regenerate only once without a nucleus. If the nucleus of a different species is now inserted and the cap is cut off once again, the new cap that regenerates will be characteristic of the transplanted nucleus (Fig. 13–1*d*). Thus, Hammerling demonstrated that the nucleus controls *Acetabularia's* development.

Experiments with *Acetabularia* helped pave the way for the important findings of other researchers, who discovered the roles of the nucleic acids— deoxyribonucleic acid (DNA) and ribonucleic acid (RNA)—in heredity and cell function.

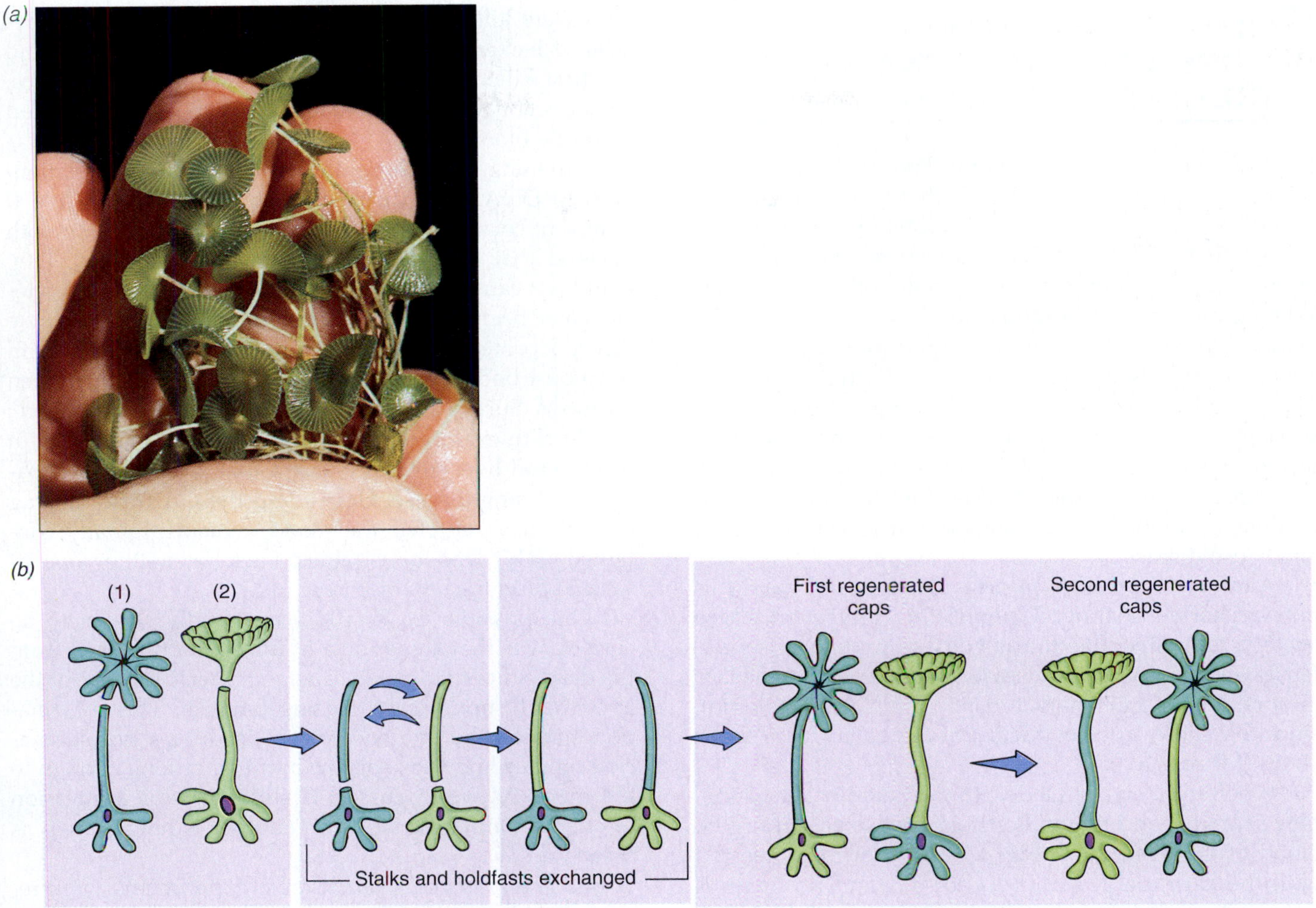

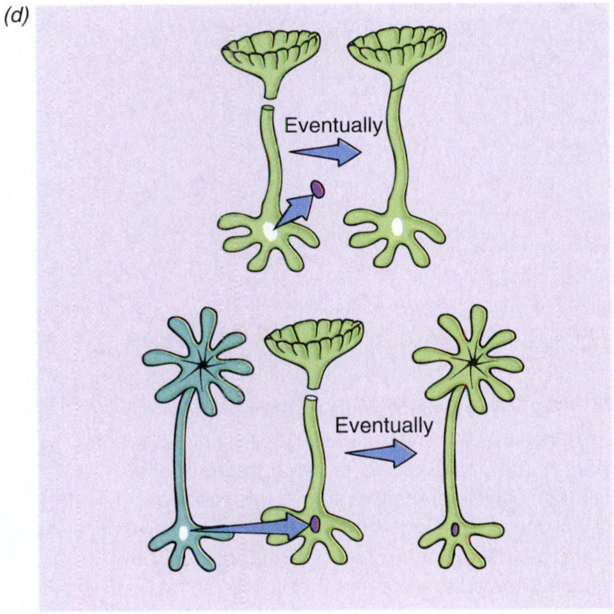

(a)

(b)

(1) (2)

First regenerated caps

Second regenerated caps

Stalks and holdfasts exchanged

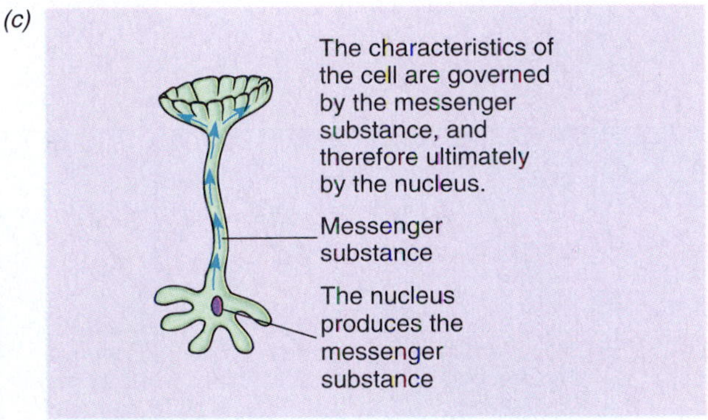

(c)

The characteristics of the cell are governed by the messenger substance, and therefore ultimately by the nucleus.

Messenger substance

The nucleus produces the messenger substance

(d)

Eventually

Eventually

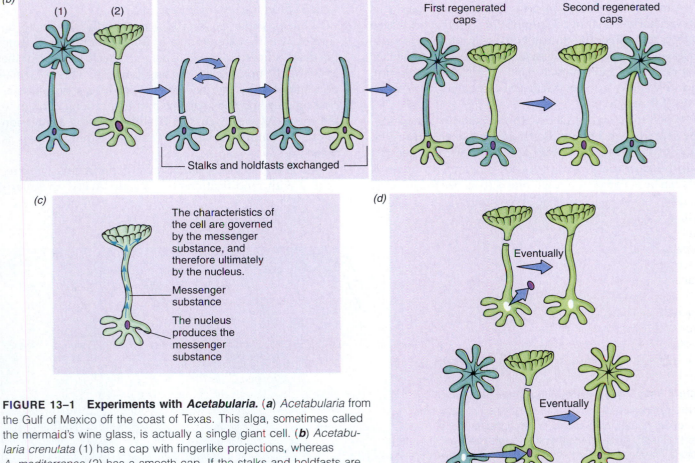

FIGURE 13–1 Experiments with *Acetabularia*. (***a***) *Acetabularia* from the Gulf of Mexico off the coast of Texas. This alga, sometimes called the mermaid's wine glass, is actually a single giant cell. (***b***) *Acetabularia crenulata* (1) has a cap with fingerlike projections, whereas *A. mediterranea* (2) has a smooth cap. If the stalks and holdfasts are exhanged, the cap will eventually take the form dictated by the holdfast, which contains the cell's nucleus. (***c***) How *Acetabularia* is controlled. (***d***) If the nucleus is removed and the nucleus of a different species is inserted, the form of the cap ultimately reflects the species of the nucleus. (Gregory G. Dimijian, M.D./Photo Researchers, Inc.)

GRIFFITH'S EXPERIMENTS DEMONSTRATED THAT DNA IS THE GENETIC MATERIAL

In 1928, the British investigator Frederick Griffith reported the remarkable results of what then seemed to be a very eccentric experiment. Microbiologists had long known that a strain of bacteria—*Streptococcus pneumoniae*—causes pneumonia and certain other infections. When grown in the laboratory on agar, these bacteria develop colonies with a smooth, glistening appearance; hence, Griffith designated them the *S* strain (for *s*mooth). The smooth appearance in culture results from a polysaccharide coating, or **capsule,** that surrounds each bacterium. Inside the body of a human or other mammal, this capsule protects the bacteria from the immunological defenses of the host, allowing the bacteria to stay alive and cause disease.

Another strain of *S. pneumoniae* does not have this polysaccharide coating. Without this protection, these bacteria are vulnerable to the body's immunological defenses and cannot cause disease. When grown in the laboratory, they form characteristic rough, dull colonies, since they lack capsules; Griffith designated them the *R* strain (for *r*ough).

When mice are injected with the S strain of bacteria, they soon die, and living S cells can be isolated from the blood of the dead mice (Fig. 13–2). Other mice injected with the harmless R strain do not die, and no living R cells can be isolated from their blood (because the injected cells were killed by the mice's immune systems). Moreover, if bacterial cells from the disease-causing S strain are first killed by heat and then injected into the mice, they do not cause disease, nor can live cells be isolated from the blood.

Pursuing an investigation with no obvious relationship to DNA, Griffith injected both dead S and live R strains of bacteria together into mice. Because the smooth bacteria that he injected had already been killed, they could not cause disease. Neither should the rough bacteria have been able to produce infection, because they lacked a protective coating and would not survive long in the mouse body. The injection as a whole should have been harmless. Surprisingly, however, many of the mice died.

Griffith examined the blood from the dying mice. In it he found live, smooth bacteria that were clearly capable of causing disease. How could these disease-causing bacteria have entered the blood, which originally contained only dead or disabled bacteria, neither one of which could have harmed the mice?

One possible explanation, eventually shown to be correct, was that some form of information passed from the dead, smooth, disease-causing bacteria to some of the live, rough, nondisease-causing bacteria. This information enabled the live bacteria to produce a polysaccharide capsule and thus survive in the mice long enough to kill them. As we discuss in Chapter 21, this conversion of one type of bacterium into another is now known as *transformation*.

In 1944, Griffith's remarkable experiments were repeated with modifications by Oswald Avery, an Ameri-

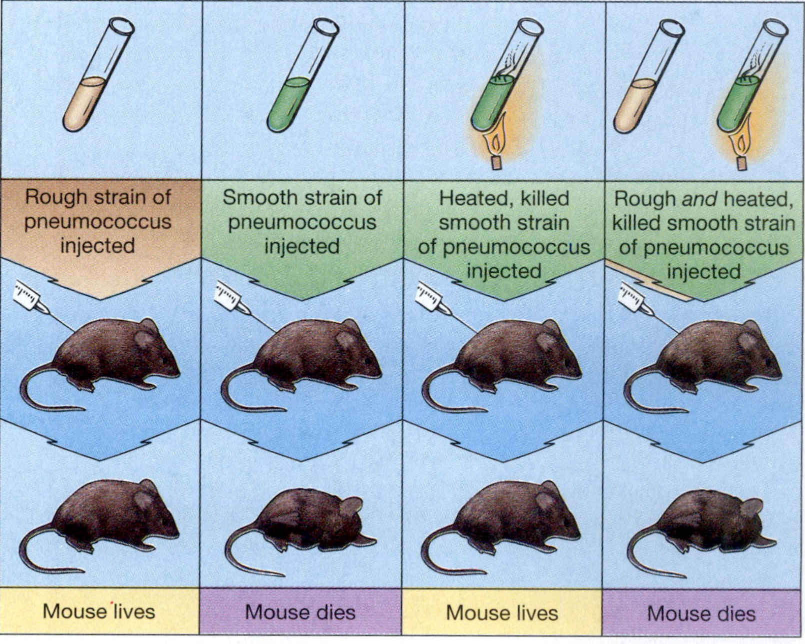

FIGURE 13–2 Bacterial transformation. Frederick Griffith demonstrated the transfer of genetic information from dead, heat-killed bacteria to living bacteria of a different strain. Although neither the rough strain of bacteria nor the heat-killed smooth strain could kill a mouse, a combination of the two did. Autopsy of the dead mouse showed the presence of living, smooth-strain bacteria. Later, Avery demonstrated the restoration of virulence (that is, the ability to cause disease) to rough bacteria by treating them with DNA from smooth bacteria. This established that DNA carries the genetic information necessary for the bacterial transformation.

can physician. Avery reasoned that some "transforming principle" (that is, some chemical substance) had conveyed genetic information from the disease-causing dead bacteria to the harmless living bacteria. He and his research team first suspected that the substance was part of the polysaccharide coat of the bacteria, which was a reasonable idea, but wrong. Further experiments by these scientists revealed that the transforming principle was something from the *interior* of the dead cells. They found that they could produce purified samples of DNA from the disease-causing bacteria. These DNA samples were capable of turning nondisease-causing bacteria into bacteria capable of producing vigorous infections. Avery's revolutionary conclusion was that DNA was the substance that could cause a heritable change in bacteria. Although their evidence that DNA was the genetic material of cells was very convincing, many researchers remained skeptical.

VIRAL NUCLEIC ACID CONTAINS GENETIC INFORMATION TO PRODUCE NEW VIRUSES

Another class of disease-producing agents, the tiny viruses, actually lack most of the traits of living things; they do not metabolize, grow, or move by themselves. They can, however, reproduce with the aid of living cellular hosts.

All cellular life seems to be susceptible to viral infection, including bacteria. Viruses that attack bacteria are called **bacteriophages**, or **phages** for short. Like other viruses, phages consist of a protein coat that surrounds a nucleic acid core (Fig. 13–3). When a bacterium becomes infected, the phage first attaches to its host's cell wall (Chapter 21). Part of it then penetrates the bacterial cell's outer covering and takes over the bacterium's metabolic machinery. The phage causes this machinery to replicate viral genetic material, which then forms new phages like itself. In other words, the phage forces the bacterial host to follow the genetic directions that the phage injects into it. Eventually, what remains of the host cell bursts open, releasing a swarm of phage particles. Some of these may come in contact with other bacterial cells and then repeat the process.

It was not known before 1952 whether the nucleic acid core or the protein coat (or even both) carried the phage's genetic information into the bacterial cell, but Alfred Hershey and Martha Chase settled the question in that year. Hershey and Chase worked with *Escherichia coli*, the common intestinal bacterium, and a phage that attacked it. They knew, as you know from reading Chapter 4, that the phage's DNA core contained phosphorus and that it also contained no sulfur. They also knew that

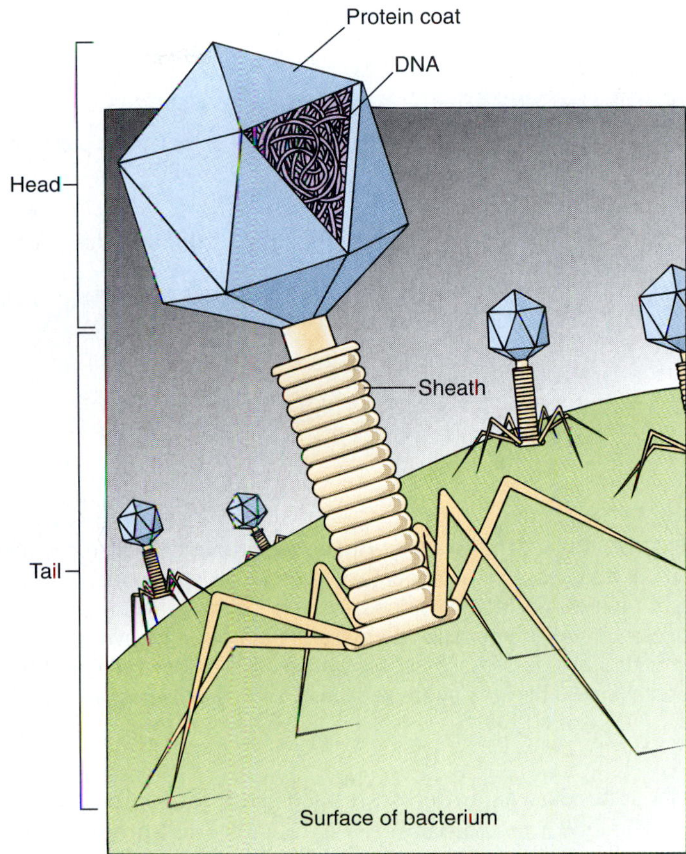

FIGURE 13–3 The structure of a bacteriophage. Each phage consists of a DNA molecule surrounded by a complex protein coat. The protein coat consists of a head region, which encloses the DNA, and a tail region, which is used to land and attach to a bacterial cell wall. The sheath region is hollow.

the phage's protein coat contained sulfur but no phosphorus. They reasoned that if they could find some sulfur from a phage which had gotten inside a bacterial cell, they would know that the phage's protein coat, not its DNA core, had entered the cell. On the other hand, if they found phosphorus inside the bacterium, they would know that the phage's DNA core, not its protein coat, had entered the cell. Whatever substance entered the cell was responsible for replicating the virus and was therefore the genetic material.

Hershey and Chase prepared a batch of phage whose protein coats contained radioactive sulfur (^{35}S). This served as a tracer, allowing them to detect even small quantities of the protein. They also prepared a batch of phage whose DNA core contained radioactive phosphorus (^{32}P). When the phage whose protein was labeled with radioactive sulfur was added to a fresh culture of bacteria, the bacteria did not become radioactive. This showed that the protein did not enter the bacteria. However, when the phage whose DNA was labeled with radioactive

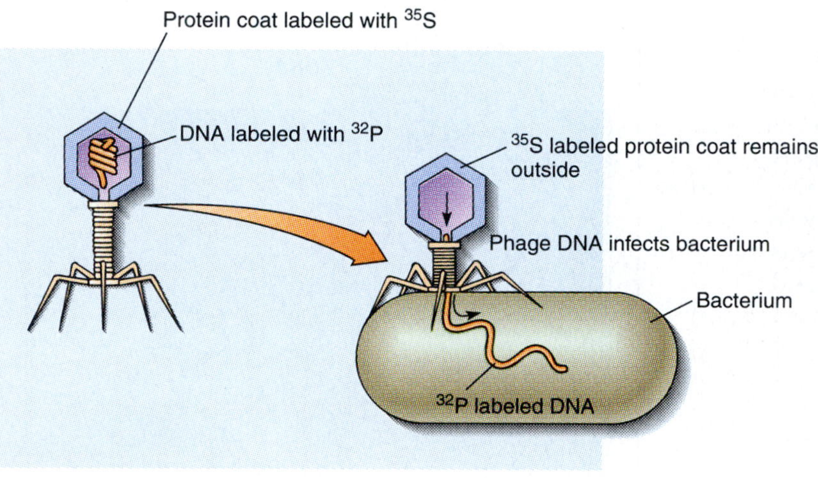

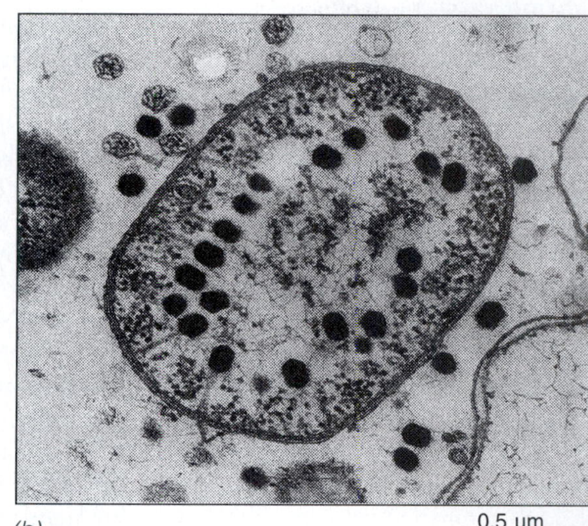

FIGURE 13–4 The Hershey-Chase experiment. (***a***) Hershey and Chase grew bacteriophages on a medium containing radioactive isotopes, thereby labeling the protein and DNA with different isotopes. They found that only the ^{32}P-labeled DNA entered the bacterium. All the genetic information needed for the synthesis of new phages was provided by the DNA. (***b***) Newly synthesized bacteriophages are evident in the host bacterium 30 minutes after infection. (Lee D. Simon/Photo Researchers, Inc.)

phosphorus was added to a bacterial culture, the bacteria did become radioactive (Fig. 13–4). This showed that viral DNA entered the bacteria.

Since phages contain only protein and DNA, all possibilities had now been investigated. It was clear that DNA had to be the material that directed the production of new viruses. DNA, not protein, was the genetic material.[2]

A VARIETY OF METHODS HAVE UNCOVERED THE SHAPE AND STRUCTURE OF THE DNA MOLECULE

Although DNA is a very large molecule, until recently one could not reasonably expect to "see" it in detail. Even the most powerful and sophisticated electron microscopes have limits on how much they can magnify an object; these limits are determined by the wavelengths of radiation by which they operate.[3]

X-Ray Diffraction Provided Clues to the Shape of the DNA Molecule

Much of the progress in analyzing the three-dimensional arrangement of atoms in nucleic acids (and in proteins) has come from the application of a sophisticated form of photography called **x-ray diffraction**. When x-rays of extremely short wavelength are passed through a crystal, they are scattered by the atoms of the crystal much as light rays are scattered by particles of dust or water in a fog. Because of the regularity of the atomic arrangement in a crystal, the scattering of the x-rays is itself regular. The scattered radiation interferes with itself, producing a characteristic pattern that can be photographed—a kind of molecular fingerprint distinctive for any crystalline substance. The photograph, however, is no more an image of its molecules than a fingerprint is an image of a person. Still, the mathematical analysis of such patterns obtained at different angles of exposure enables the researcher to obtain clues to the shape of the molecule.

Two English scientists, Maurice Wilkins and Rosalind Franklin, had already employed x-ray diffraction (mostly in the early 1950s) to investigate the structure of DNA when James Watson and Francis Crick became friends at Cambridge University. Watson, an American molecular biologist who had recently obtained his PhD, wished to gain additional experience in the rapidly developing field of molecular biology. Crick was an English doctoral candidate in the field of protein structure. Viewing Franklin's observations from their own perspective, Watson and Crick concluded that DNA must be a helical (spiral) molecule.

Although inferring the helical structure of DNA was a difficult piece of work, this was not the most challenging part of what they set out to accomplish. The hard part was to demonstrate that the helical shape was consistent with DNA's chemical composition.

[2] In some viruses, RNA stores genetic information rather than DNA, but RNA is, of course, also a nucleic acid.

[3] Recently-developed scanning tunneling microscopes do not have this limitation. With these microscopes, scientists can now generate actual images of DNA. But scanning tunneling microscopes were undreamed of in the 1950s when the basic structure of DNA was deduced.

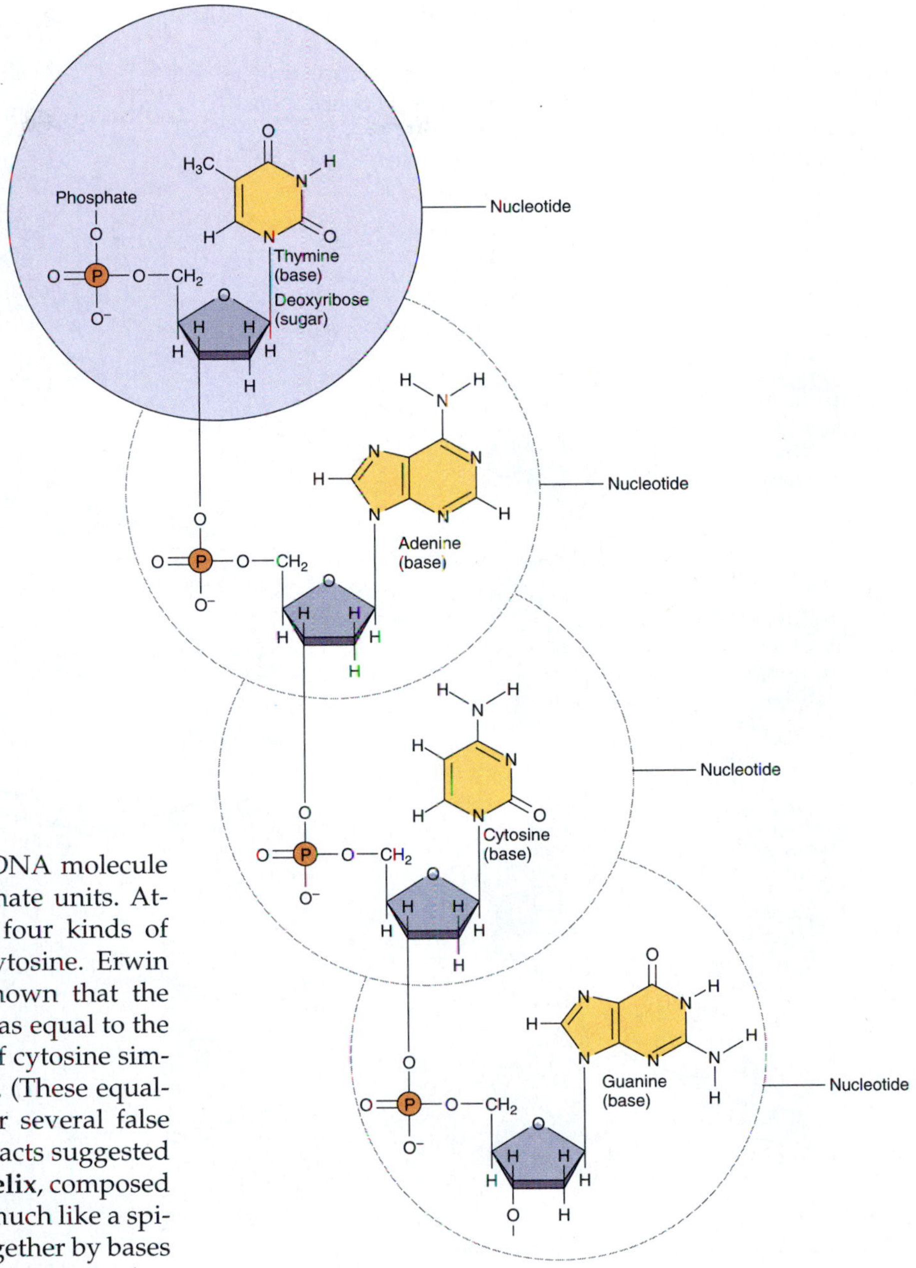

FIGURE 13–5 Nucleotides. DNA is composed of repeating subunits called nucleotides. Each nucleotide consists of deoxyribose (a sugar), phosphate, and a nitrogen-containing organic base. Four nucleotides are shown, one with each of the four bases found in DNA. The nucleotides are joined between the phosphate of one nucleotide and the sugar of another.

Watson and Crick Proposed a Model for DNA Structure

It seemed clear that the backbone of the DNA molecule consisted of alternating sugar and phosphate units. Attached somehow to this backbone were four kinds of bases: adenine, thymine, guanine, and cytosine. Erwin Chargaff of Columbia University had shown that the amount of thymine in a sample of DNA was equal to the amount of adenine, and that the amount of cytosine similarly was equal to the amount of guanine. (These equalities are known as Chargaff's rule.) After several false starts, Watson and Crick found that these facts suggested that DNA was not a single, but a **double helix**, composed of two strands twisted about one another much like a spiral staircase. The two strands were held together by bases that connected with one another, every adenine with a thymine, every guanine with a cytosine, producing the equalities that Chargaff had noted.

Watson and Crick next built a model of the DNA molecule as they understood it. This model was built of accurately scaled models of atoms and groups of atoms, with both sizes and bond angles properly proportioned. If everything could be made to fit, their understanding of the DNA molecule could be shown quite conclusively to be correct. When a huge double helix model stood in their laboratory, they knew that they had been right. It was a great moment of discovery, perhaps one of the greatest in the history of science. Watson and Crick had discovered nothing less than the blueprint of life—a molecule whose structure preprogrammed the composition and functions of every living cell.

DNA Is Composed of Nucleotide Subunits

DNA is composed of molecular subunits called **nucleotides**. Each nucleotide consists of three parts: (1) a five-carbon sugar, **deoxyribose**; (2) a phosphate group; and (3) a nitrogen-containing organic compound called a **base** (Fig. 13–5). DNA actually has four bases: **cytosine (C)** and **thymine (T)** are smaller, single-ring bases (called **pyrimidine** bases); **adenine (A)** and **guanine (G)** are double-ring bases (called **purine** bases). These bases project like the rungs of a ladder more or less at right angles from the sugar-phosphate backbone of the DNA molecule.

Much as the letters forming words and sentences determine the information they carry, the sequence in which

FIGURE 13–6 Physical structure of DNA. (***a***) A space-filling molecular model of a portion of the double helix. Notice that the molecule has two backbones united by their bases to form the double helix. (***b***) The two sugar-phosphate chains run in opposite directions. This orientation permits the complementary bases to pair. (*H* represents hydrogen bonds between complementary bases.)

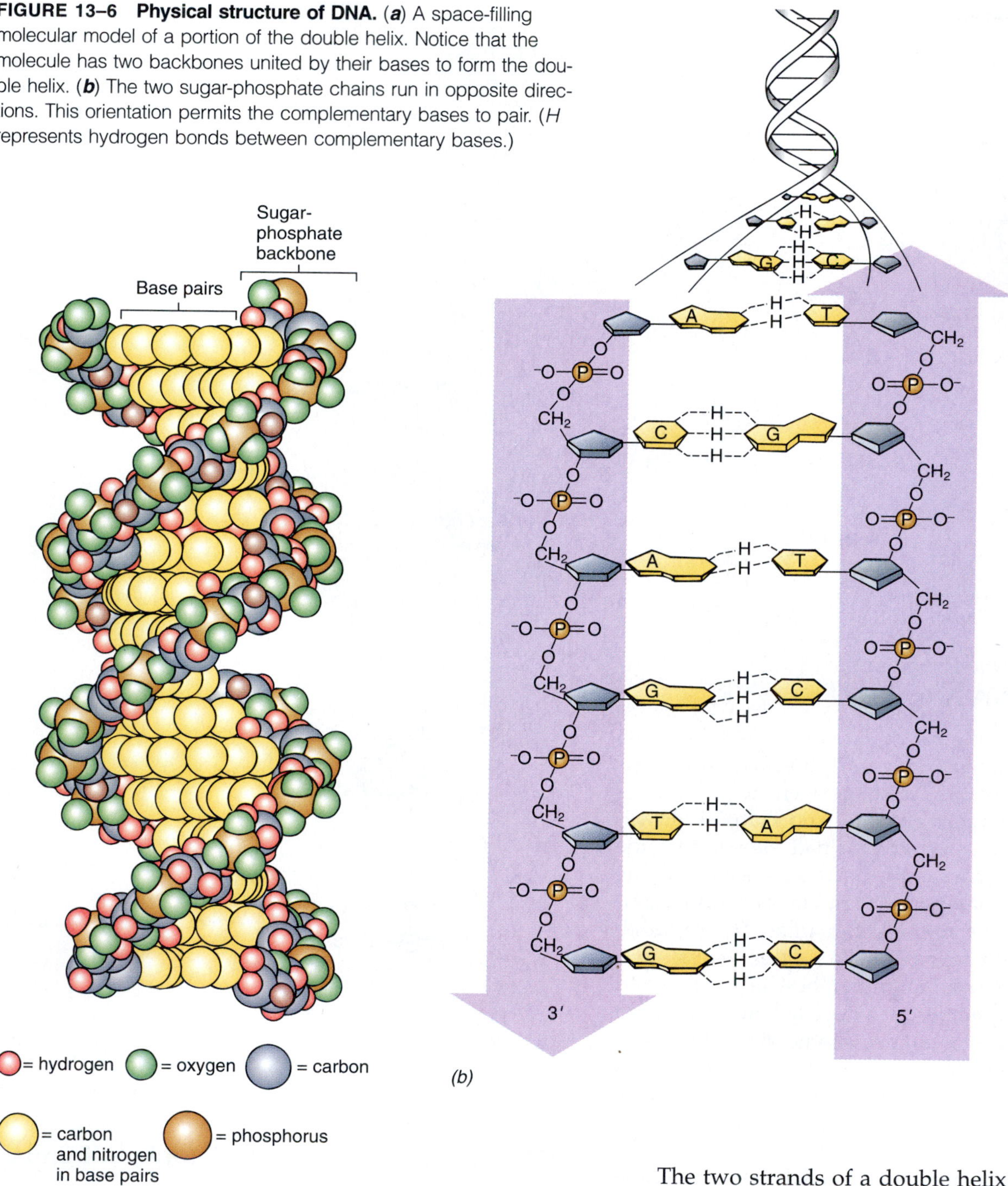

these bases appear determines their genetic message. Thus, the sequence -AGGTAC- bears a different set of information from the sequence -TAGTCC-.

DNA consists of *two* strands, attached to each other by means of their bases as follows: adenine always pairs with thymine, and guanine always pairs with cytosine. The ladder-like double strand of DNA is twisted into a double helix (Fig. 13-6*a*).

The two strands of a double helix are *complementary* but not identical. What that means is that if you know the base sequence of one of the two strands, you can predict the base sequence of the other. This is because adenine on one strand pairs only with thymine on the other strand, and cytosine pairs only with guanine. No other pairing relationship is normally possible because of the nature of the hydrogen bonds that form between the bases. For instance, if a portion of one chain reads -AGGCTA- then the corresponding bases of the other strand must be -TCCGAT-.

The length of the adenine-thymine "rungs" on the DNA ladder is exactly the same as that of the guanine-

Concept Connections

DNA Base Sequences ⟷ Evolution

We have learned that DNA is the molecule of inheritance for all living things (excluding RNA viruses). Further, it is the *order* of bases in a DNA molecule that determines the characteristics of the traits for which DNA codes: Different species owe their individual traits to different sequences of bases in their DNA.

The fact that DNA is the molecule of inheritance in living organisms provides compelling evidence that all life forms are somehow related. Moreover, when two species are closely related, their DNA shares a greater proportion of identical sequences of nucleotide bases. Thus, a pelican and a flamingo have significant stretches of base sequences that are identical because they are both birds and they shared a common ancestor in the not-so-distant past.

On the other hand, a yeast (a single-celled fungus) and a chimpanzee do not share very many identical stretches of bases in their DNA. The fact that they share even a few, however, offers striking evidence that they had a common ancestor in the very distant past. Yeasts and chimpanzees became very different organisms through evolution over hundreds of millions of years along different lines of descent. We say more about the significance of DNA base sequences to evolution in Chapter 17.

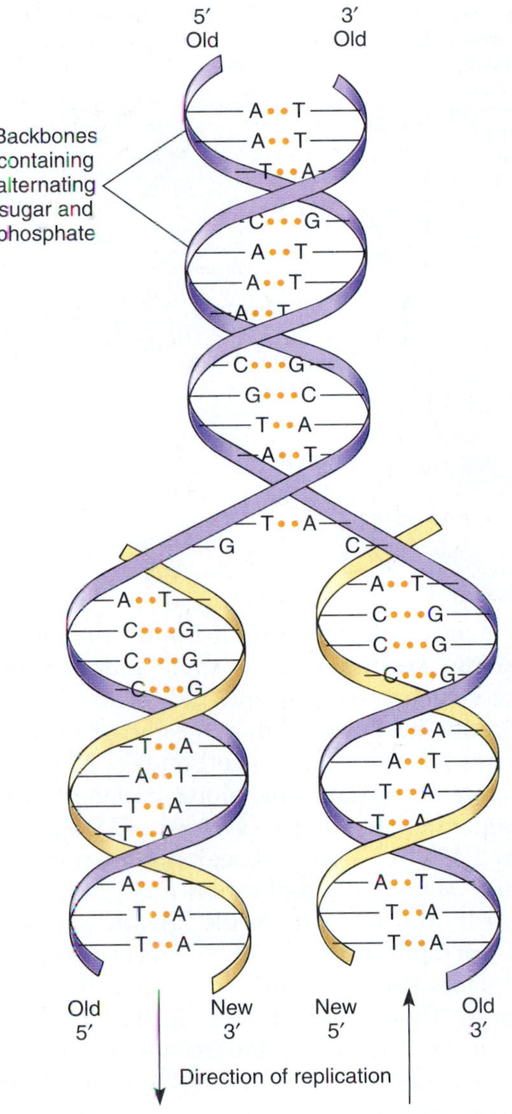

FIGURE 13–7 Mechanism of DNA replication. The two ends of each strand of this helix are labeled 3′ or 5′ to establish the direction of replication. The two strands are shown separating. Both are being copied. The strand whose 5′ end is at the left side of the figure is being copied starting from the right (its 3′ end). This corresponds to the 5′ end of the newly synthesized strand. Replication proceeds in the same direction (3′ to 5′) on both templates. In the new strand, as in the old, adenine pairs with thymine, and cytosine pairs with guanine. The new strands grow in the 5′-to-3′ direction because they are antiparallel to their complementary DNA templates.

cytosine "rungs." Because the distance between complementary base pairs is always the same, a double strand of DNA is the same width from one end to the other, regardless of base sequence.

The two strands extend in *opposite directions*, as you can see from looking carefully at the orientation of the deoxyribose molecules in Figure 13–6*b*. Each strand has a **3′** end and a **5′** end. These are designations that chemists use to distinguish between the ends of a strand. Notice also that the 3′ end of one strand is paired with the 5′ end of the other. Because of this arrangement, the strands are said to be **antiparallel**.

DNA replicates semiconservatively

The most distinctive properties of DNA as the genetic material of life are that it carries genetic information and undergoes precise replication. That is, when replicating, DNA makes an exact duplicate of itself. Before a cell divides, the DNA strands in its double helices "unzip" (separate), and each one is used as a **template** (a mold used to synthesize another molecule) for a new strand of complementary nucleotides. The result is two double-helix molecules, each identical in base sequence to the original double helix. Each contains one of the original strands plus a newly synthesized strand (Fig. 13–7). The two original strands are not simply copied and then left behind but are conserved (that is, kept) to be used as part of the two new double helices. Thus, this process is

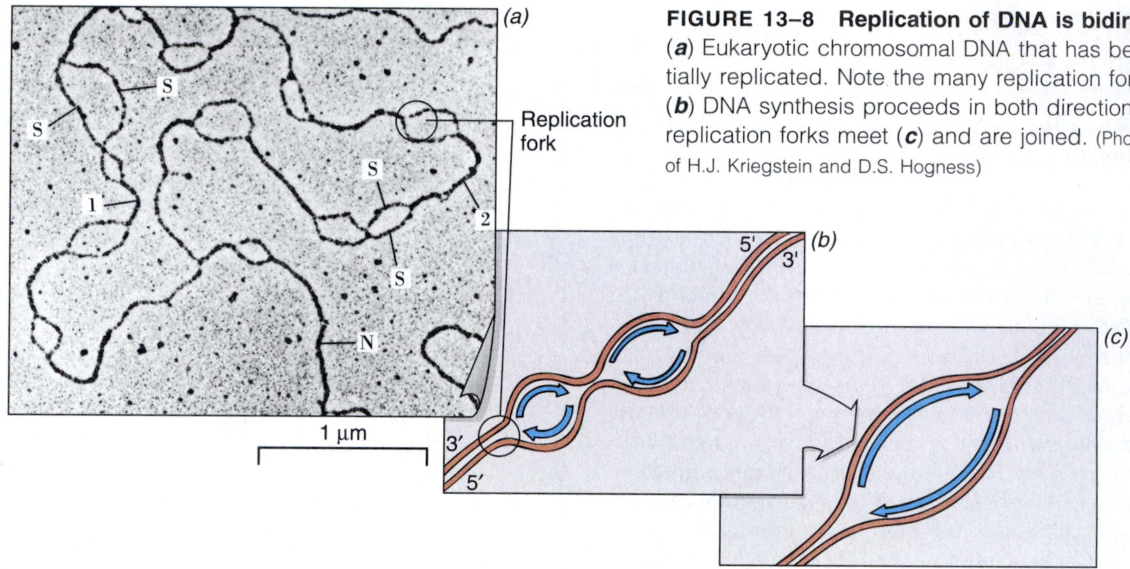

FIGURE 13–8 Replication of DNA is bidirectional. (*a*) Eukaryotic chromosomal DNA that has been partially replicated. Note the many replication forks. (*b*) DNA synthesis proceeds in both directions until two replication forks meet (*c*) and are joined. (Photo courtesy of H.J. Kriegstein and D.S. Hogness)

known as **semiconservative replication** (*semiconservative* here meaning "keeping part the same"). Each new double helix contains one old strand and one new strand.

Replication need not begin at the end of a double helix. Enzymes break the hydrogen bonds linking the two DNA strands at various sites along its length, causing them to separate.[4] As the two strands of DNA separate, they form a kind of Y-shaped region known as a **replication fork**. Many such replication forks may occur simultaneously in a DNA molecule, greatly speeding up the process of replication (Fig. 13–8). Proceeding as it does in many places simultaneously, the replication of all of the enormous fund of information in human DNA can take place in as little as 20 minutes in a very rapidly dividing cell such as some of those in the stomach lining.

After separating, both DNA strands act as templates for the assembly of new complementary strands. This is the task of the enzyme **DNA polymerase**, which adds free nucleotides onto the "unzipped" DNA molecule. DNA polymerase always adds new DNA nucleotides to the 3′ end of the new growing strands. Thus, the new strands are said to grow in the 5′ to 3′ direction. Because the strands are antiparallel, the nucleotides are added in the opposite directions for the two original strands. As the new complementary strands are produced, the original double helix separates further, and the newly synthesized half-helices continue to form hydrogen bonds with the old half-helices until there are two identical double helices of DNA.

The fidelity and regularity of the base pairing are owed to DNA polymerase, which is one of the most important enzymes in living things; DNA replication could not take place without it. DNA polymerase is actually a complex of several enzymes that has a twofold function. Not only does it catalyze the polymerization (chain formation) of nucleotides, but it also does this in a very precise way. Recall from Chapter 7 that enzymes act on specific substances, or substrates. DNA polymerase has *four* different nucleotide substrates that it must add to the growing end of the new strand.

Consider all of the things this remarkable enzyme does. For instance, DNA polymerase recognizes a specific base on a pre-existing strand of DNA and somehow accepts the nucleotide whose base is complementary to that pre-existing one. It then brings that nucleotide to the proper end of the new, elongating strand of DNA. Next it catalyzes the necessary chemical reaction that adds the nucleotide to the new strand, and then it releases the reaction products. The enzyme then moves one base further on the existing strand and repeats the process all over again. And it continues to do this at a rate of about 200 bases per second for hundreds or thousands of bases every time the cell prepares to divide, and occasionally at other times as well!

Mutations Are Random, Permanent Changes in DNA

DNA replication is never 100% accurate, and some errors in replication do occur. What would happen during DNA replication if a base that was not complementary to the template base was substituted in the newly forming strand (as, for example, if C pairs with A instead of G)? What if one to several base pairs were inserted or deleted from a DNA molecule? Such spontaneous changes would be transmitted to all of the DNA strands replicated from the one in which it originally occurred. Thus, all of the descendants of that cell would share its genetic difference.

Such accidental changes in genetic information, called **mutations,** do occur, although infrequently. Muta-

[4] In prokaryotes, double-stranded DNA forms closed circles; so before the strands are separated, other enzymes open up the circle at specific initiation sites.

tions are not necessarily harmful; some are harmless (for example, changes in blood type). Indeed, mutations are the raw material for evolution and sometimes permit new beneficial traits to arise. Beneficial mutations are rare, however, because of the already highly adapted nature of living organisms. It is, after all, unlikely that the random removal or replacement of a part will improve a computer; such a change would more likely damage it. So, too, a random change in the genetic material is unlikely to improve the function of a cell or an organism. Some mutations are quite harmful and produce disease.

Every organism ever investigated has been shown to be subject to mutation. The wonder is that we do not observe more mutations than we do. Fortunately, cells possess a backup system of enzymes whose function is to repair or remove damaged sections of DNA. This backup system snips out erroneous, improperly pairing bases and replaces them with the right ones. As a result, most mutations that occur are not passed on to other cells. (Mutations are discussed further in Chapter 14.)

DNA IS ORGANIZED INTO CHROMOSOMES

Each **chromosome** consists of a single large DNA molecule that codes for hundreds or thousands of different genes (Fig. 13–9). The DNA molecule is so long that it would become hopelessly tangled were it not organized in some fashion. Just like thread is wound around spools to keep it untangled, the DNA of eukaryotes is wound around special protein molecules called **histones**, which help keep the DNA organized. Each histone spool, which consists of eight histone molecules, and its associated DNA (about 140 base pairs) form a structural unit of the eukaryotic chromosome called a **nucleosome**. Additional histones bind to the section of DNA that links two neighboring beads. The nucleosomes make up part of the **chromatin**, the DNA and protein complex that makes up the chromosome. They are further organized into long coiled loops held together by nonhistone proteins called **scaffolding proteins**.

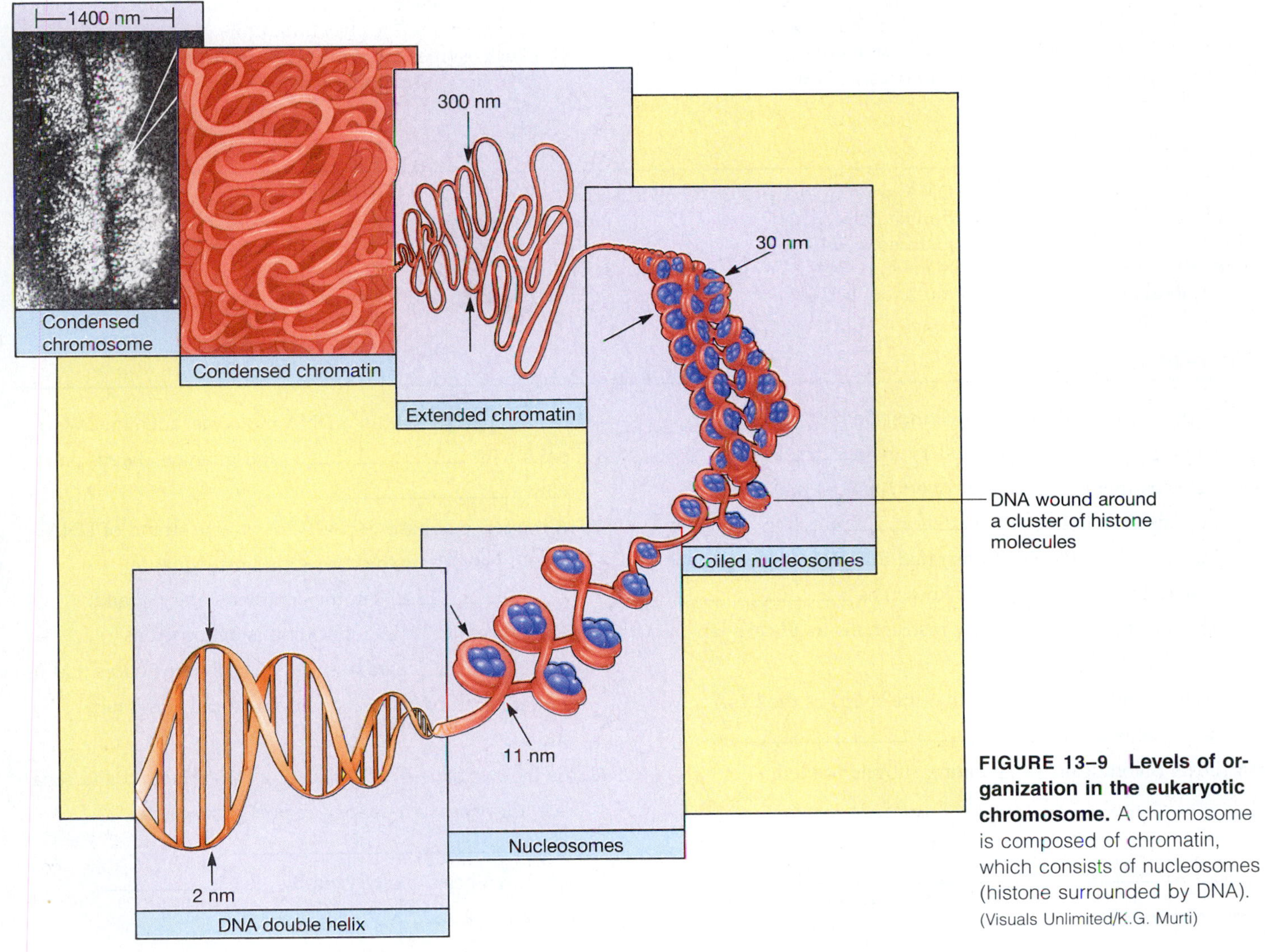

FIGURE 13–9 Levels of organization in the eukaryotic chromosome. A chromosome is composed of chromatin, which consists of nucleosomes (histone surrounded by DNA). (Visuals Unlimited/K.G. Murti)

Chapter Summary

I. The basic plan of all living things and their cells is contained in their DNA, located in the nucleus.
   A. In eukaryotes, the nucleus controls the cell with the aid of a messenger substance that travels to the cytoplasm.
      1. If *Acetabularia* stalks are exchanged, the nucleus in the holdfast determines the form of the regenerated cap, but only after a delay that implies the existence of a temporary messenger substance.
      2. An *Acetabularia* nucleus, when transplanted to a holdfast of a different species, produces a cap characteristic of the nucleus' species.
   B. Transformation experiments involving pneumococcal bacteria and mice showed that DNA is the hereditary material in bacteria.
   C. For a virus such as a bacteriophage, its nucleic acid contains the genetic information, not its protein.

II. The structure of DNA has been elucidated.
   A. X-ray diffraction studies by Wilkins and Franklin demonstrated the helical nature of the DNA molecule.
   B. Watson and Crick worked out the basic structure of DNA in the 1950s.
   C. DNA is an antiparallel double helix; each strand consists of nucleotides joined to the other strand by hydrogen bonds between complementary bases.
      1. Each nucleotide consists of deoxyribose (a sugar), a phosphate group, and an organic base.
      2. There are four bases: adenine, which pairs (that is, forms hydrogen bonds with) with thymine, and cytosine, which pairs with guanine.
      3. Given the base sequence of one strand, that of the other can be predicted.

III. DNA replicates itself by semiconservative replication, in which each double helix contains an old strand and a newly synthesized strand.
   A. DNA strands unwind during replication.
   B. DNA synthesis takes place at replication forks, Y-shaped regions where the two strands of DNA separate and where DNA synthesis occurs on both strands at once.
   C. DNA synthesis always proceeds in the 5′ to 3′ direction; hence, DNA synthesis is bidirectional.
   D. DNA polymerase catalyzes the polymerization of nucleotides, adding the appropriate nucleotide to the strand.
   E. Errors in replication do occur, but infrequently. Such random, permanent changes in DNA are called mutations.

IV. DNA is packaged into chromosomes in a highly organized way.
   A. DNA associates with proteins called histones to form nucleosomes, the structural units of chromosomes.
   B. Nucleosomes are arranged into large coiled loops held together by scaffolding proteins.

Selected Key Terms

adenine, p. 245
chromatin, p. 249
chromosome, p. 249
cytosine, p. 245
deoxyribose, p. 245
DNA polymerase, p. 248
double helix, p. 245
guanine, p. 245
histone, p. 249
mutation, p. 248
nucleosome, p. 249
nucleotide, p. 245
replication fork, p. 248
scaffolding protein, p. 249
semiconservative replication, p. 248
template, p. 247
thymine, p. 245
x-ray diffraction, p. 244

Post-Test

1. Experiments with *Acetabularia* indicate that the ultimate control of the cell is exercised by the ______________.

2. Between them, Griffith and Avery showed that the hereditary material in bacteria is ______________.

3. Hershey and Chase demonstrated that DNA is the genetic material in bacterial viruses known as ______________.

4. The master set of hereditary instructions in all cells is ______________.

5. Watson and Crick showed that the shape of the DNA molecule is a ______________ ______________.

6. DNA is composed of repeating subunits called ______________.

7. Each nucleotide chemically consists of a ______________, a ______________, and a ______________.

8. In base pairing within a DNA molecule, adenine always pairs with ______________, and guanine always pairs with ______________.

9. A sequence of bases -CGGTCA- on one strand of DNA would necessitate the corresponding sequence -______________ - in the complementary strand.

10. The process of DNA replication is described as ______________ because the two double helices that result are each composed of an old strand and a new strand.

11. As the two strands of replicating DNA unwind and separate, they form a Y-shaped region known as a ______________ ______________.

12. DNA synthesis is catalyzed by ______________ ______________.

13. DNA replication is bidirectional, with synthesis in each strand proceeding in the _______________ to _______________ direction.
14. In eukaryotes, DNA is organized with various proteins into tightly coiled structures in the nucleus called _______________.
15. A nucleosome is composed of DNA wound around a protein called _______________.
16. Nucleotides found in DNA contain the five-carbon sugar: (a) ribose (b) phosphate (c) guanine (d) deoxyribose.
17. The backbone of a DNA strand is formed from alternating phosphates and: (a) sugars (b) bases (c) histones (d) nucleosomes.
18. Watson and Crick used the _______________ studies of Franklin and Wilkins to construct their model of DNA. (a) photographic (b) electron microscopic (c) x-ray diffraction (d) biochemical
19. The process of DNA synthesis is called: (a) cell division (b) transformation (c) mitosis (d) replication (e) regeneration.
20. DNA from one species differs from DNA of another species in its: (a) order of base pairs (b) types of sugars in the nucleotides (c) chemical bonds between base pairs (d) all of these.

Review Questions

1. What is *Acetabularia*? What was discovered about the function of the cell's nucleus using this organism, and how was this done?
2. How did experiments by Griffith and Avery help show that the transformation of harmless bacteria into ones that cause disease involves a change in genetic information?
3. What is a bacteriophage? How does it infect a bacterium?
4. Explain how Hershey and Chase used radioactive sulfur and radioactive phosphorus to demonstrate that DNA is the hereditary material in viruses.
5. How did Watson and Crick determine that DNA is a double helix?
6. Draw a sketch of DNA and label the following parts: double helix, nucleotide, sugar-phosphate backbone, and base-pair rungs.
7. Which of these base pairs would not be found in a DNA molecule: A:T, C:G, T:G, G:C, T:A, A:C? Why?
8. How does the molecular structure of DNA allow it to form exact copies of itself during replication?
9. What is meant by semiconservative replication?
10. Would it matter if there was a change in the sequence of bases in a particular portion of DNA? Explain.

Thinking Critically

1. In many states, convicted sex offenders and other felons must now provide a sample of blood from which a DNA fingerprint is made and stored in a database. By matching DNA from a crime scene to a DNA fingerprint in the database, police have sometimes been able to identify a suspect for a crime in which there were no suspects. Some prisoners have filed suit against this program, charging that it is an invasion of privacy. Discuss the ethical issues of such required DNA testing.
2. What characteristics must a molecule have if it is to serve as genetic material? How is the molecular structure of DNA uniquely adapted to its function as hereditary material?
3. Which of the experiments discussed in this chapter (Hammerling, Griffith and Avery, and Hershey and Chase) provided clues about how DNA functions to direct cellular activities? Explain.
4. Which of Griffith's four experiments (see Fig. 13–2) actually represented controls?
5. The DNA in a tiny tree frog is analyzed and it is determined that guanine makes up 20% of the bases in its DNA. What is the percentage of cytosine? Of adenine? Of thymine?

Recommended Readings

Franklin-Barbajosa, C., "DNA Profiling: The New Science of Identity," *National Geographic*, Vol. 181, No. 5, May 1992. Examines some of the varied used of DNA fingerprinting, such as investigating crimes, making breeding decisions for endangered species, reuniting immigrant families separated by bureaucratic red tape, and reconstructing human history.

McElfresh, K.C., D. Vining-Forde, and I. Balazs, "DNA-Based Identity Testing in Forensic Science," *BioScience*, Vol. 43, No. 3, March 1993. Describes how DNA fingerprinting is performed and relates some of the legal challenges it has faced in court.

Wambaugh, J., *The Blooding*, New York, Wm. Morrow and Co., Inc., 1989. A true account of the first use of DNA fingerprinting to determine the murderer of two fifteen-year-old girls in England.

Watson, J.D., *The Double Helix*, New York, Atheneum, 1968. Watson's personal account of the experience of making a Nobel Prize–winning discovery.

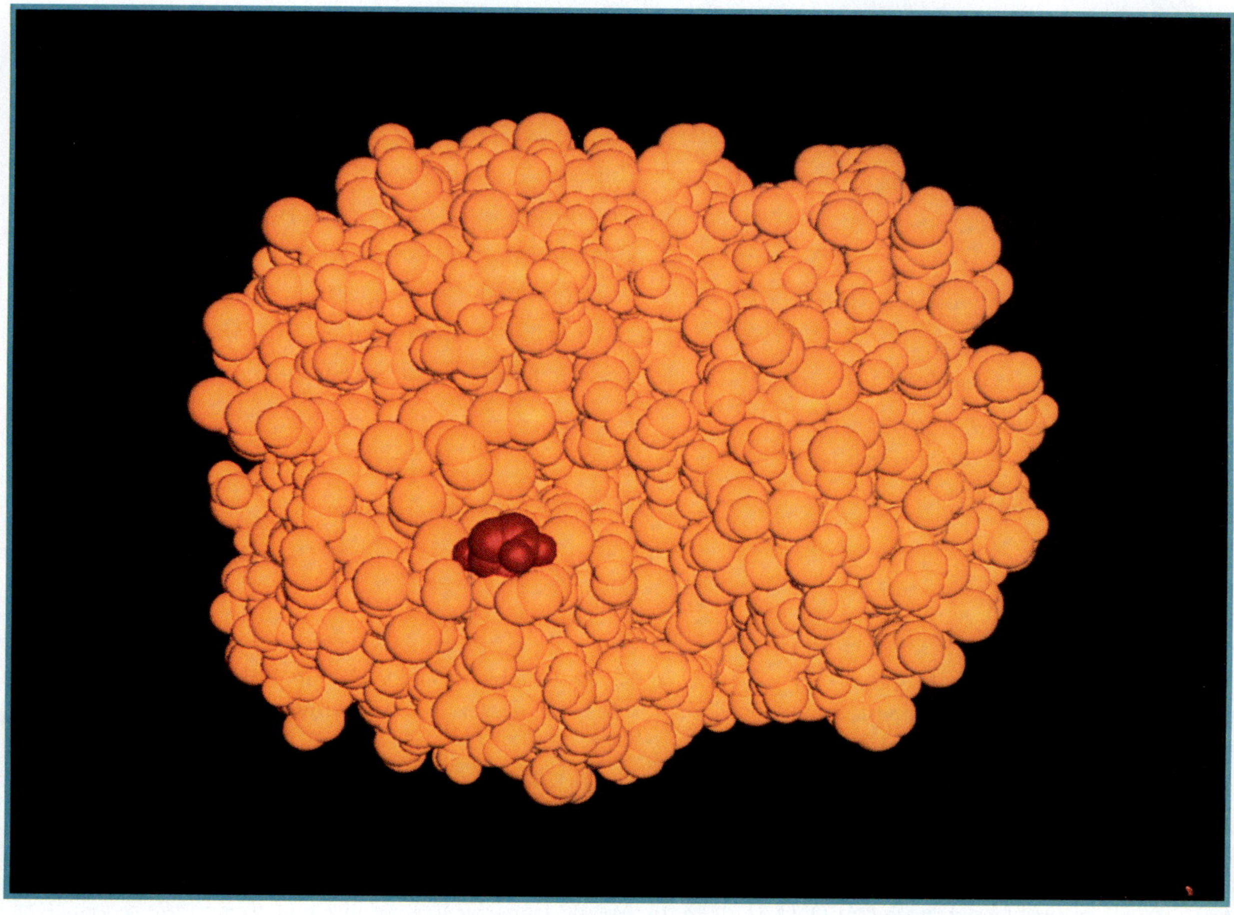

Space-filling model of hemoglobin S from a sickled cell. *The red part indicates the abnormal portion of the molecule. It is interesting that such a tiny part of a large, complex molecule can cause such serious health problems. (Tom Pantages, Courtesy of Molecular Simulations)*

Sickle-Cell Anemia: A Bad Gene Makes a Bad Protein

Hemoglobin, the respiratory pigment of red blood cells, is vital to life because it is able to take up and release oxygen. Hemoglobin is the protein responsible for the color of red blood cells and, more importantly, their ability to transport oxygen from the lungs to all of the cells of the body.

People afflicted with the blood disorder sickle-cell anemia produce an abnormal form of hemoglobin called hemoglobin S. When exposed to low levels of oxygen, hemoglobin S forms crystals that elongate and deform the red blood cell. Normal red blood cells are shaped like biconcave disks (biconcave means thinner in the center than around the edge). Red blood cells in people with sickle-cell anemia are often shaped like sickles (half moons) or some other irregular shape.

The abnormal hemoglobin S crystals often damage the plasma membranes of sickled cells so that the red blood cells rupture easily and are destroyed in great numbers. As a result, the oxygen-carrying capacity of people with sickle-cell anemia is severely reduced.

The sickle shape of the red blood cells also causes serious circulatory problems. Normal red blood cells pass through tiny blood vessels called capillaries in an orderly, single-file fashion. In contrast, sickled cells cannot line up smoothly and often get stuck in capillaries, impeding the flow of blood and sometimes completely blocking the capillaries. This results in tissue injury, damage to major organs, and recurring episodes of intense pain.

Sickle-cell anemia is caused by a change in one of the genes that codes for making hemoglobin. The altered gene contains instructions for placing one incorrect amino acid in the sequence of 146 amino acids in each beta chain of the hemoglobin molecule. (Recall from Chapter 4 that the hemoglobin protein contains four polypeptide chains, two identical alpha chains, and two identical beta chains.) As a result of the amino acid substitution, the three-dimensional structure of hemoglobin is altered. Changes in the three-dimensional structure of a protein also disrupt its biological activity. Thus, a single amino acid change in hemoglobin caused by an altered gene is responsible for this very serious medical condition.

In this chapter, we examine how genes such as the one that codes for hemoglobin are expressed, that is, how their genetic information is decoded and used to make proteins. Gene expression is a complex process that involves the synthesis of RNA molecules complementary to the DNA, as well as the synthesis of proteins.

Gene Function: RNA and Protein Synthesis

14

Learning Objectives

After you have studied this chapter you should be able to:

1. Outline the flow of genetic information in cells from DNA to protein.
2. Explain the universality of the genetic code and its evolutionary significance.
3. Compare the structures of DNA and RNA molecules.
4. Identify the functions of the three types of RNA: messenger RNA, transfer RNA, and ribosomal RNA.
5. Summarize the process of transcription.
6. Describe the structure of a ribosome and explain its role in protein synthesis.
7. Summarize the sequence of events that occur in translation.
8. Distinguish among triplets, codons, and anticodons.
9. Distinguish among the processes of initiation, chain elongation, and termination in protein synthesis.
10. Explain the effect of mutations on protein synthesis.
11. Discuss the effects of transposons and mutagens on genes.

Key Concepts

- ☐ A gene is a sequence of nucleotides in DNA that codes for a product with a specific cellular function. Most genes code for polypeptides (the structural units of proteins). Each specific amino acid in a protein is coded for by a sequence of three nucleotides (that is, a triplet) in DNA.
- ☐ The path of information flow from genes to proteins is as follows: DNA ⟶ RNA ⟶ proteins.
- ☐ Ribonucleic acid (RNA) is synthesized using DNA as a template; the synthesis of RNA molecules complimentary to DNA is called transcription.
- ☐ The three types of RNA are messenger RNA (mRNA), transfer RNA (tRNA), and ribosomal RNA (rRNA). All three are necessary for protein synthesis.
- ☐ Proteins are synthesized at the ribosomes by a process called translation, in which a particular sequence of nucleotides in mRNA is decoded into a specific amino acid sequence of a polypeptide chain.

PROTEINS ARE CENTRAL TO AN ORGANISM'S IDENTITY

Proteins are crucial to the activities of living cells. At different times in its life cycle, each cell uses different proteins in varying amounts. Proteins function directly in elaborate and subtle ways in such important processes as oxygen transport, cell movement, active transport, and facilitated diffusion. Proteins also have an important role as enzymes, the organic catalysts that control the rates of all chemical reactions in the cell (Chapter 7).

Proteins, you will recall, are large, complex molecules made up of amino acids joined by peptide bonds (Chapter 4). A long chain of amino acids is called a **polypeptide**; some proteins consist of a single polypeptide chain, whereas others, like hemoglobin, are composed of more than one polypeptide chain.

Within a typical cell, hundreds of kinds of proteins are constantly in production, with large numbers of protein molecules being produced simultaneously. All this involves not only a multitude of individual protein synthesis machines (the ribosomes), but also a whole associated economy of workers (enzymes), raw materials (amino acids), and blueprints (RNA) of the master plans (DNA).

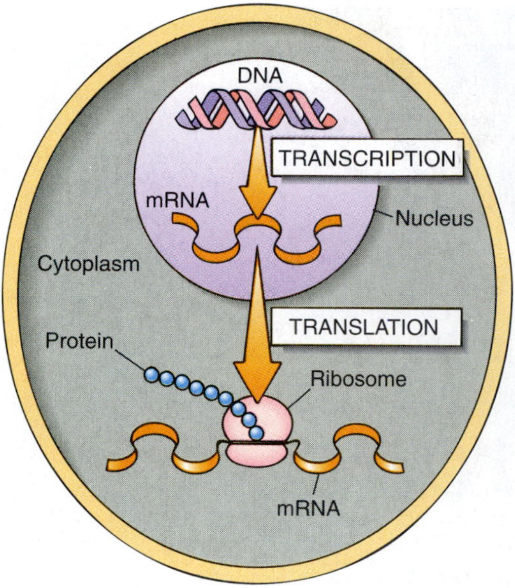

FIGURE 14–1 An overview of gene expression. Most genes code for the production of a polypeptide, or protein, in a two-step process—transcription and translation. Using the analogy developed in the text, the master plans (DNA) are kept in the nucleus. During transcription, blueprints (RNA) of the master plans are made for use by labor. These disposable copies of the master plan are sent to the protein-production area of the cell (the ribosomes) where workers (enzymes) use the blueprint to assemble the correct sequence of amino acids into a protein (translation).

GENETIC INFORMATION FLOWS FROM DNA TO RNA TO PROTEINS

The **gene** is the fundamental unit of heredity; that is, a gene determines a particular hereditary trait. At the molecular level, a gene is a particular segment of a deoxyribonucleic acid (DNA) molecule that codes for a cellular product, usually a polypeptide (Fig. 14–1).[1] The cell's DNA contains the information needed to make all of the proteins that the cell uses.

While we tend to think of obvious traits such as eye color and hair texture as evidence of genes, such traits are visible ultimately because of protein synthesis. At the same time, other aspects of body structure and function, more difficult to observe directly, are also under the control of genes. The messages in an organism's genes control everything about it by governing its production of protein.

Although the sequence of nucleotides in DNA specifies the order of amino acids in a polypeptide chain, DNA does not produce protein directly. Instead a related nucleic acid, ribonucleic acid (RNA), acts as an intermediary between DNA and protein. In the first stage of gene expression, an RNA molecule is made using the information encoded in DNA. RNA synthesis is similar to DNA replication in that the DNA molecule unwinds and unzips; one of its strands serves as a template upon which RNA nucleotides are assembled in the proper order based on complementary base-pairing. The process of making RNA using DNA as a template, called **transcription**, takes place in the nucleus in eukaryotic cells.

The RNA molecule then detaches from its DNA template and leaves the nucleus, while the two strands of the DNA molecule rezip and rewind to form a double helix again. The RNA molecule carries the information about the particular polypeptide to be synthesized from DNA in the nucleus to ribosomes in the cytoplasm.

At the ribosome, the message encoded in the RNA molecule is read and used to assemble amino acids in the proper order to produce a specific polypeptide chain. The formation of a polypeptide chain at the ribosome using the information encoded in RNA is known as **translation**.

To summarize, protein synthesis is a two-step process. In Step 1 (transcription), the genetic message is copied from one nucleotide dialect—that of DNA—to another—that of RNA. In Step 2 (translation), the genetic message is translated from the nucleotide language of RNA into the amino acid language of proteins. Details of these processes follow.

[1] Some genes produce RNA molecules such as rRNA and tRNA instead of coding for polypeptides.

FIGURE 14–2 A gene from a firefly was genetically engineered into this tobacco plant. The gene codes for an enzyme that catalyzes the reaction in fireflies in which light is produced. The tobacco plant was able to express the firefly gene and produce light—a dramatic example that organisms as diverse as fireflies and tobacco share the same genetic code. (From D.W. Ow, et al., *Science* 234:856–859, 1986. ©1986 by the AAAS.)

THE GENETIC CODE IS READ AS A SERIES OF TRIPLETS

Twenty different amino acids are commonly found in proteins. These amino acids can be thought of as the 20 letters of the protein alphabet. How can DNA, with only four nucleotides in its alphabet, serve as a code that specifies all 20 amino acids? Single nucleotides could not be the fundamental unit of such a code, for then only four kinds of amino acids could occur in any protein. Neither could pairs of nucleotides, since a pool of four nucleotides permits only 16 unique pairs,[2] which could specify only 16 different amino acids.

However, trios of nucleotides could—and do—form the **genetic code,** because a pool of four nucleotides permits 64 unique base trios (AAA, AAC, AAG, AAT, etc.), more than enough to specify 20 amino acids. Each such sequence of three nucleotides in DNA—more simply, a sequence of three bases in DNA—is known as a **triplet**. These triplets are the basic units of genetic information. With some exceptions, each triplet codes for one amino acid. Some triplets are equivalent, coding for the same amino acid as another triplet. A few triplets do not code for amino acids but have other roles in protein synthesis.

An important aspect of the genetic code is that it is nearly universal throughout the biological world. The same DNA triplets code for the same amino acids in all living organisms, with only a very few minor variations. For example, a gene removed from a firefly and placed in a different organism, such as a tobacco plant, will still produce firefly protein, indicating that both organisms use the same genetic code (Fig. 14–2). While the types of protein may vary from one species to another, accounting for the specific traits that make each a separate species, the relationship of nucleotides to amino acids is pretty much the same. The universal nature of the genetic code is considered compelling evidence for the evolution of all organisms from the same early life forms.

[2] The sixteen possible combinations of nucleotide pairs are AA, AC, AG, AT, CA, CC, CG, CT, GA, GC, GG, GT, TA, TC, TG, and TT.

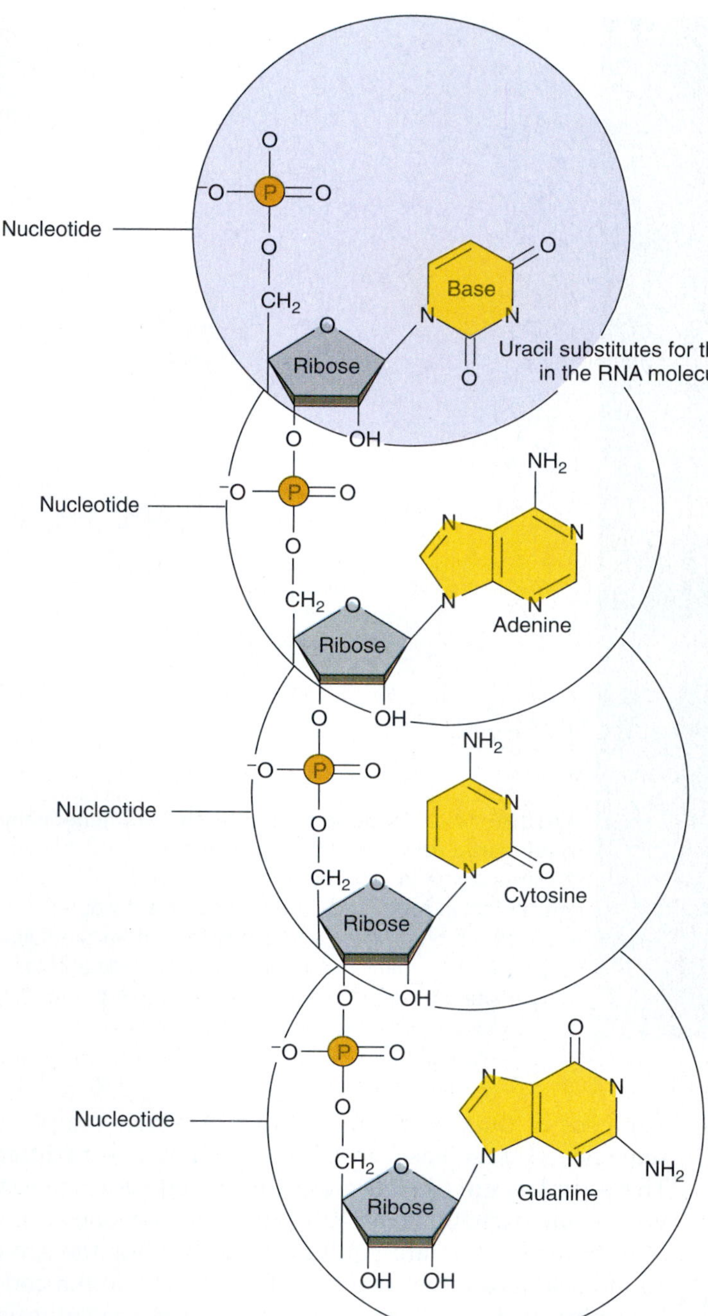

FIGURE 14–3 Part of an RNA molecule, showing the four nucleotides of which it is composed. The sugar in RNA is ribose, rather than deoxyribose. Three of RNA's four bases—adenine, guanine, and cytosine—are also found in DNA, but the fourth RNA base is uracil, rather than thymine. RNA also differs from DNA in that it is composed of a single strand rather than a double helix.

RNA ENCODES AND EXPRESSES THE GENETIC MESSAGE STORED IN DNA

RNA differs from DNA in that each RNA nucleotide contains the sugar ribose rather than deoxyribose (Fig. 14–3). As in DNA, each RNA nucleotide contains one of four bases. However, in RNA the base **uracil** (U) replaces the thymine (T) found in DNA. Uracil behaves chemically like thymine, readily forming base pairs with adenine. Another difference between RNA and DNA is that RNA molecules usually do not form double helices; instead they function as single strands, although some RNA molecules are elaborately folded (Table 14–1).

There are three main types of RNA molecules: messenger RNA (mRNA), ribosomal RNA (rRNA), and transfer RNA (tRNA). All three kinds of RNA molecules are made from DNA by transcription, and each has a role in protein production. **Messenger RNA (mRNA)** carries the coded instruction for protein synthesis from DNA to the ribosome. The ribosome assembles amino acids into a polypeptide guided by the information in the mRNA molecule. Both tRNA and rRNA are also involved in translation. **Transfer RNA (tRNA)** carries specific amino acids to the ribosome during protein assembly. Ribosomes, the sites of protein synthesis, are composed of various proteins and **ribosomal RNA (rRNA)**. Protein synthesis could not occur without rRNA because the ribosome would not be functional. Ribosomal RNA assists in the attachment of the ribosome to the mRNA molecule and in the assembly of amino acids in the proper order to make a polypeptide.

RNA Is Made by Transcription

Recall that, in DNA replication, an existing DNA strand acts as a template, or model, for the synthesis of a new, complementary DNA strand. Something similar to this happens in transcription. A particular sequence of nucleotides on a DNA strand acts as the template for the synthesis of a complementary RNA strand. For example, where a nucleotide containing guanine occurs in the DNA strand, an RNA nucleotide containing cytosine pairs with it, becoming part of the new RNA strand.

There are several important differences between DNA replication and RNA transcription. First, only a single strand of DNA serves as the template, rather than both strands as in DNA replication. Second, RNA con-

TABLE 14–1
A Comparison of DNA and RNA

	DNA	RNA
Subunits	Nucleotides	Nucleotides
Sugar	Deoxyribose	Ribose
Phosphate	Yes	Yes
Bases	Adenine Guanine Cytosine Thymine	Adenine Guanine Cytosine Uracil
Structure	Double helix	Single strand, sometimes folded elaborately
Function	Stores genetic information; used to make RNA	Used to make proteins
Location in cell	Part of chromosomes in nucleus	Made in nucleus; works in cytoplasm (ribosomes)
Structural forms	—	3 forms (Messenger RNA, Transfer RNA, Ribosomal RNA)

tains uracil rather than thymine; thus, an RNA nucleotide containing uracil pairs with the DNA nucleotide containing adenine during transcription. Third, the RNA molecule *detaches* from the DNA template rather than forming a double helix with it.

Let us now look at how a gene produces a molecule of mRNA. Although we focus on the transcription of mRNA, keep in mind that tRNA and rRNA are also synthesized by transcription from special genes in DNA.

Transcription is controlled by a complex of enzymes called **RNA polymerase**. This remarkable enzyme responds to cues built into the base sequence of the DNA molecule. It selects which gene to transcribe, recognizes which of the two paired DNA strands it should copy, and identifies where it should begin and end transcription.

Transcription begins when RNA polymerase binds to a promoter site in the DNA. The **promoter** consists of a special sequence of bases in DNA that indicates where transcription is to begin. RNA polymerase moves along a strand of DNA like a railroad locomotive on a track, constructing a single strand of RNA, or mRNA transcript, as it goes (Fig. 14–4). RNA nucleotides are matched to complementary bases along the DNA template (for example, ATTCGA in DNA = UAAGCU in RNA) and added to the 3′ end of the growing strand as in DNA replication. Thus, as in DNA replication, the new mRNA transcript grows in the 5′-to-3′ direction. Instead of forming a double helix, however, the mRNA strand separates from the DNA template as it forms. Transcription con-

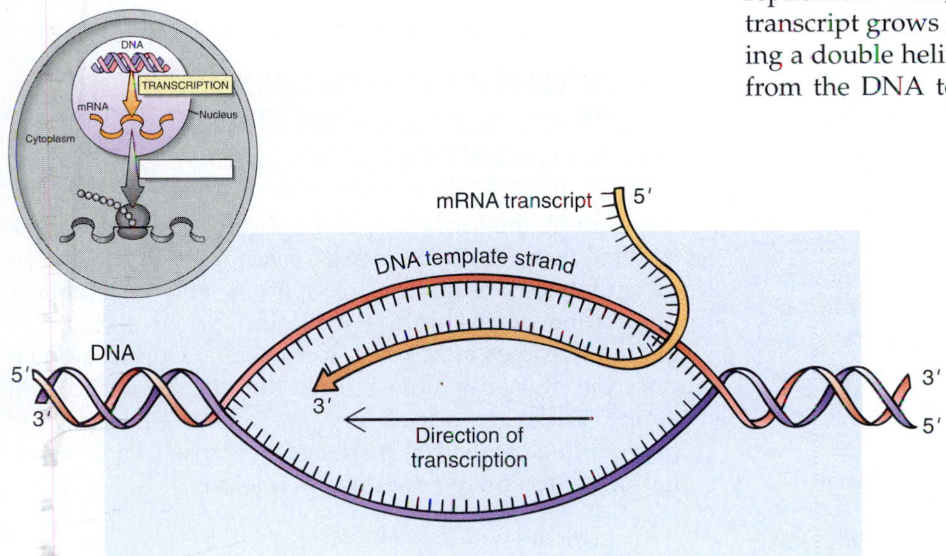

FIGURE 14–4 How RNA is transcribed from a DNA template. A DNA double helix is unwound by RNA polymerase, giving the enzyme access to the nucleotide sequence. The DNA is then rewound behind the moving transcription complex. Specific DNA sequences determine where transcription starts and stops. RNA synthesis depends on base-pairing rules similar to those for DNA synthesis: adenine pairs with uracil, cytosine pairs with guanine.

TABLE 14–2
The Genetic Code: Codons of mRNA (as defined by DNA triplets) that Specify a Given Amino Acid. (For example, the codon UUU designates the amino acid phenylalanine.)

First base	Second base: U	Second base: C	Second base: A	Second base: G	*Third base*
U	phenylalanine	serine	tyrosine	cysteine	U
	phenylalanine	serine	tyrosine	cysteine	C
	leucine	serine	stop	stop	A
	leucine	serine	stop	tryptophan	G
C	leucine	proline	histidine	arginine	U
	leucine	proline	histidine	arginine	C
	leucine	proline	glutamine	arginine	A
	leucine	proline	glutamine	arginine	G
A	isoleucine	threonine	asparagine	serine	U
	isoleucine	threonine	asparagine	serine	C
	isoleucine	threonine	lysine	arginine	A
	(start) methionine	threonine	lysine	arginine	G
G	valine	alanine	aspartic acid	glycine	U
	valine	alanine	aspartic acid	glycine	C
	valine	alanine	glutamine	glycine	A
	valine	alanine	glutamine	glycine	G

tinues until the RNA polymerase comes to a specific nucleotide sequence in the DNA that it recognizes as a stop signal. The RNA molecule then separates completely from the DNA strand.

In any portion of the double helix, only one of the DNA strands (referred to as the template strand) is transcribed. Which of the two strands serves as the template strand varies from region to region. One strand may be transcribed in one region, and the other strand in another region. The presence of a promoter sequence is what determines whether a particular portion of a DNA strand is able to code. Since the DNA strands are complementary, you can see that a promoter sequence on one strand would have a complementary sequence on the other strand that would *not* be a promoter sequence.

Recall that the genetic code is based on triplets of nucleotide bases in DNA. How is this information encoded in mRNA? The sequence of three mRNA bases complementary to a triplet in DNA is known as a **codon**; almost every codon in mRNA codes for a specific amino acid (Table 14–2). During translation, the order of the codons in the mRNA determines the order in which amino acids are added to the polypeptide chain.

In RNA Processing, Introns Are Removed from the New Molecule

Within most eukaryotic genes, some very long DNA base sequences are transcribed into mRNA base sequences that do *not* go on to code for amino acids in the making of a protein. Before the mRNA delivers its protein-making instructions to the ribosome, these sequences are enzymatically snipped out of the newly transcribed mRNA molecules. The remaining portions of the mRNA molecule are then spliced together to form the mature RNA that directs the construction of a polypeptide. The sequences that are discarded (and their DNA complements) are called **introns**; those that are spliced together in the nucleus for use at the ribosome are called **exons** (Fig. 14–5).[3]

IN TRANSLATION, PROTEIN IS MADE USING THE GENETIC MESSAGE ENCODED IN mRNA

Translation, the formation of a polypeptide chain at a ribosome using the sequence of bases in mRNA, consists of three steps: initiation, elongation, and termination. Translation begins during **initiation**, when a strand of mRNA becomes attached to a ribosome. During **elongation**, the ribosome moves down the mRNA strand, and amino acids are added one by one to the growing polypeptide chain. The final step of translation is **termination**, when the polypeptide is released.

[3] Although introns occur in eukaryotic genes, they are not found in prokaryotic genes.

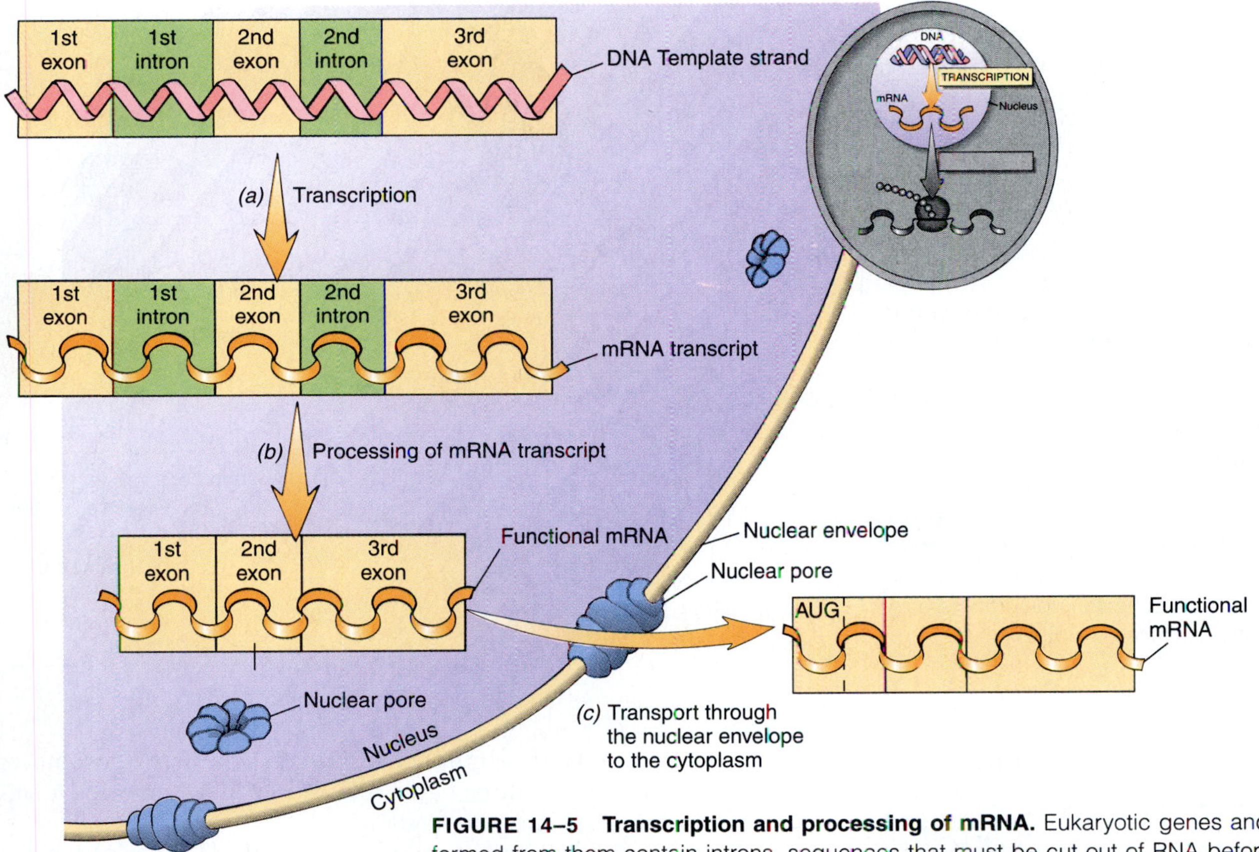

FIGURE 14–5 Transcription and processing of mRNA. Eukaryotic genes and the mRNA formed from them contain introns, sequences that must be cut out of RNA before RNA can be translated into a protein. (***a***) During transcription, a DNA sequence containing both exons and introns is used to make mRNA transcript. (***b***) The introns are removed from the mRNA transcript, and the exons are spliced together to make a fully functional mRNA molecule. (***c***) The mature mRNA is transported through the nuclear envelope into the cytoplasm to be used for protein synthesis.

Translation employs all three types of RNA. While mRNA specifies the order of amino acids in the polypeptide, rRNA and its associated enzymes (as part of the ribosome) play several important roles, and tRNA identifies and transports the amino acids themselves.

Ribosomal RNA Forms Part of the Ribosome

Ribosomes are the cell's protein assembly machines. Each ribosome is structurally composed of a small and a large subunit that do not join together until translation begins; each subunit consists of rRNA and a large number of protein molecules (Fig. 14–6). (The roles of the P-site and A-site shown on the ribosome in Fig. 14–6 will be discussed shortly.)

Ribosomes bring together everything needed for translation—mRNA, tRNAs with their associated amino acids, rRNA, and necessary enzymes. The ribosome in effect serves as an enzymatic matchmaker that ensures not only that peptide bonds form but that they form between the correct amino acids in the order specified by mRNA.

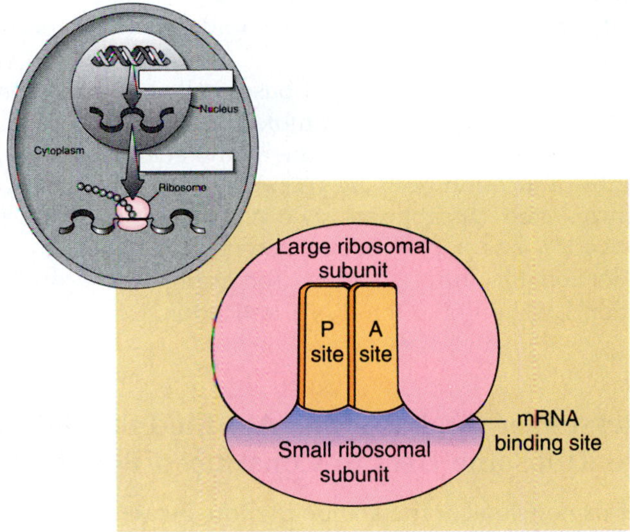

FIGURE 14–6 A ribosome is composed of two subunits. The P-site is where the tRNA fits that is bound to the last amino acid added to the growing polypeptide chain. The A-site is where the next tRNA comes in, carrying an amino acid that will be added to the polypeptide chain.

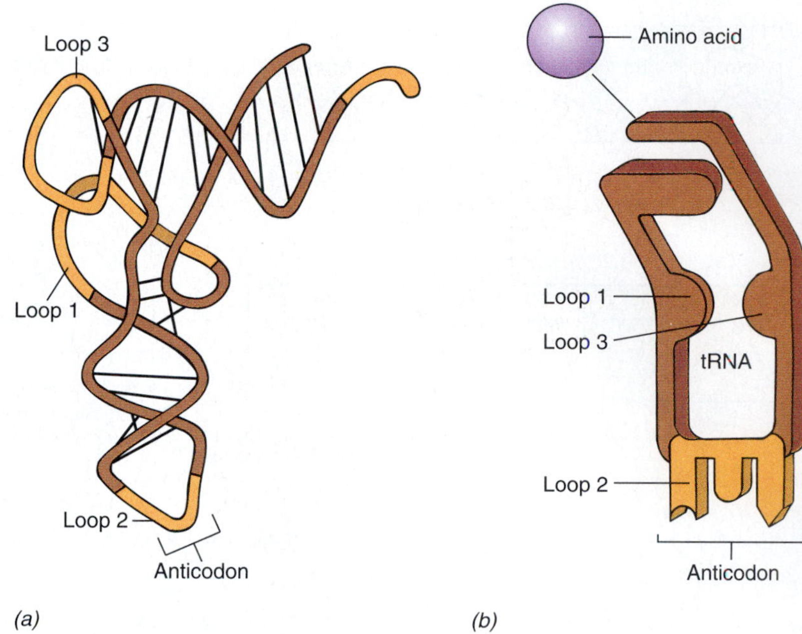

FIGURE 14–7 Two representations of the structure of tRNA, the molecules that "read" the genetic code. (***a***) A diagram of the actual shape of a tRNA molecule. Its three-dimensional shape is determined by intramolecular hydrogen bonds between base-paired regions. One loop contains the triplet anticodon that base pairs with the mRNA codon. The amino acid is attached to the other end of the tRNA molecule. The pattern of folding permits a constant distance between anticodon and amino acid in all tRNAs examined. (***b***) A schematic diagram of how the amino acid is attached to its tRNA.

Transfer RNA Carries Amino Acids to the Ribosome

The individual amino acids that will be assembled into a protein are carried to a ribosome by molecules of tRNA. Each tRNA molecule is elaborately folded and capable of forming a temporary chemical bond with a specific amino acid. Enzymes, one for each kind of amino acid, ensure that the tRNA molecules link up with their specific amino acids. The tRNA molecules thus function as labels or tags. By calling for its tRNA tag, one would obtain the specific amino acid attached to it.

The tRNA tag is a trio of bases called an **anticodon** that projects from the tRNA molecule (Fig. 14–7). The ribosome in effect recognizes the amino acid by recognizing the anticodon of its tRNA carrier. The ribosome recognizes the appropriate tRNA carrier because its anticodon has a base sequence complementary to the codon on the mRNA strand (for example, the mRNA codon UUU pairs to the tRNA anticodon AAA).

Messenger RNA Codons Are the Instructions the Ribosome Uses to Construct a Protein

Let us now look at translation in more detail. During initiation, the mRNA binds to the two ribosomal subunits, which are normally separated when they are not translating mRNA (Fig. 14–8). The mRNA strand attaches to the small subunit, and the large subunit of the ribosome then joins this complex, forming a complete ribosome. A sequence of three bases (the codon *AUG*) on mRNA signals translation to begin.

After the ribosome recognizes the *AUG* codon as the place to start translating the mRNA strand, the elongation of the chain proceeds. As the mRNA molecule passes through the ribosome, each codon on mRNA is read in turn. The tRNA molecule whose anticodon can pair with the next codon on mRNA by complementary base-pairing comes in, carrying its specific amino acid. That amino acid is joined to the growing polypeptide chain by enzymes that are a part of the ribosome. These enzymes catalyze the formation of peptide bonds between two amino acids.

Figure 14–9*a* shows that the amino acid that has been added most recently to the polypeptide chain is located at the **P-site** of the ribosome. Its tRNA molecule is still attached. The **A-site** of the ribosome is where we find the next codon that determines which amino acid will be added to the chain (Fig. 14–9*b*). The ribosome recognizes a complementary anticodon in a tRNA molecule and accepts this tRNA molecule, which together with its attached amino acid temporarily occupy the A-site.

Next, a peptide bond forms between the new amino acid in the A-site and the terminal amino acid in the P-site (Fig. 14–9*c*). The tRNA in the P-site is released (Fig. 14–9*d*), and the newly incorporated amino acid and its tRNA move from the A-site to the P-site (Fig. 14–9*e*). At the same time, the ribosome moves along the mRNA strand so that the next codon occupies the now-vacant A-site, and the process repeats.

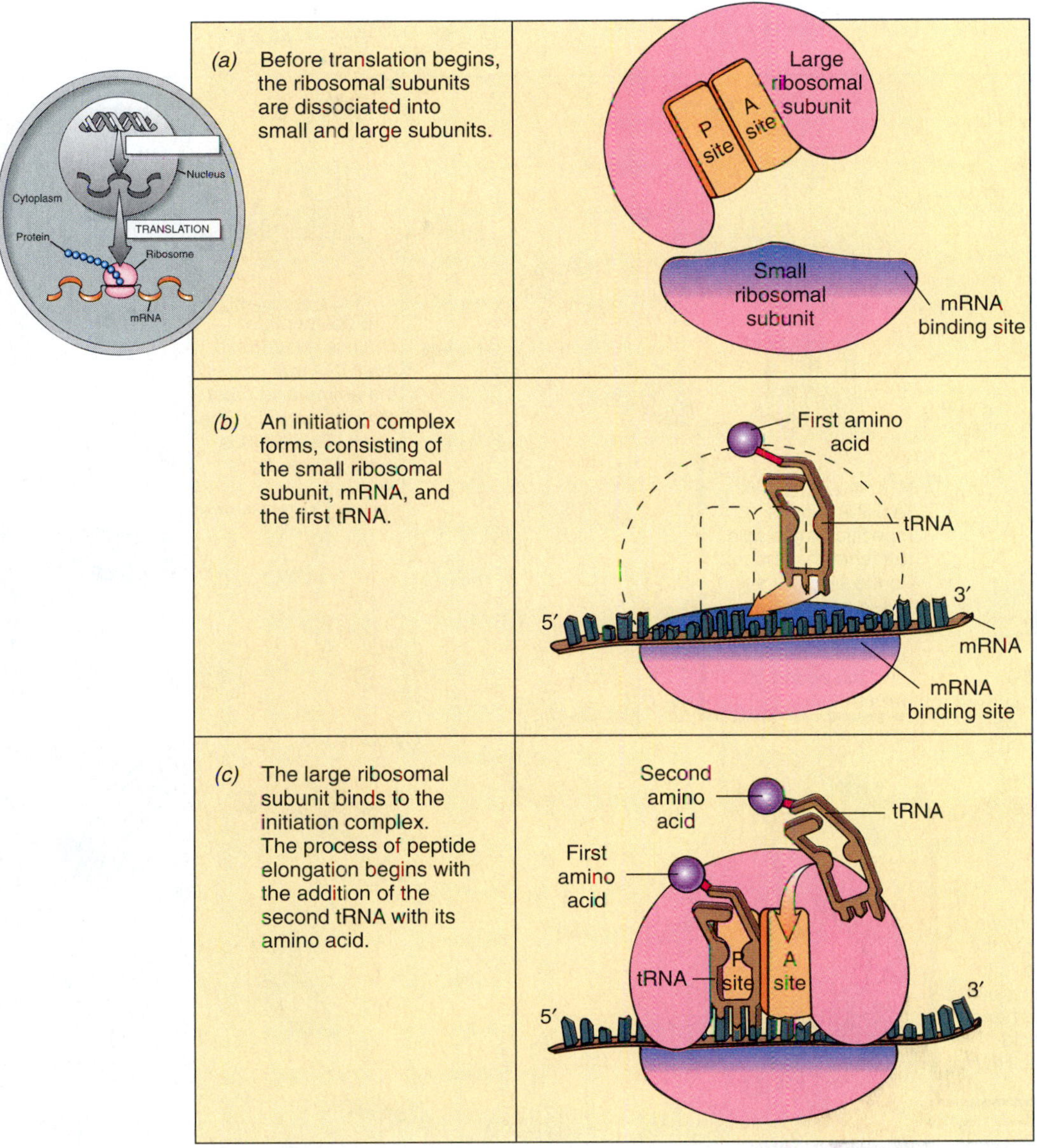
Cytoplasm
Nucleus
Protein
TRANSLATION
Ribosome
mRNA
(a) Before translation begins, the ribosomal subunits are dissociated into small and large subunits.
Large ribosomal subunit
P site
A site
Small ribosomal subunit
mRNA binding site
(b) An initiation complex forms, consisting of the small ribosomal subunit, mRNA, and the first tRNA.
First amino acid
tRNA
5′
3′
mRNA
mRNA binding site
(c) The large ribosomal subunit binds to the initiation complex. The process of peptide elongation begins with the addition of the second tRNA with its amino acid.
Second amino acid
tRNA
First amino acid
tRNA
P site
A site
5′
3′

FIGURE 14–8 Initiation of translation. The overall scheme of events, in which the ribosomal subunits and first tRNA bind to mRNA. Initiation must occur before the message encoded in mRNA is translated into a polypeptide chain.

In summary, elongation involves:

1. Acceptance of a tRNA with its specific amino acid by the A-site of the ribosome.
2. Establishment of a peptide bond between the A-site amino acid and the P-site terminal amino acid of the growing polypeptide chain.
3. Release of the tRNA formerly occupying the P-site.
4. Movement of the ribosome down the mRNA strand by one codon.

Termination of protein synthesis is usually accomplished by special stop signals. These codons do not code for amino acids, but instead trigger special proteins (called release factors), which detach the polypeptide chain and mRNA from the ribosome.

Each strand of mRNA may be used to make multiple copies of a particular protein. A number of ribosomes, as many as 10 or 20, usually bind to a single strand of mRNA. However, each ribosome independently produces a polypeptide. The entire complex, consisting of mRNA and multiple ribosomes, is called a **polysome** (Fig. 14–10).

(Text continues on page 263)

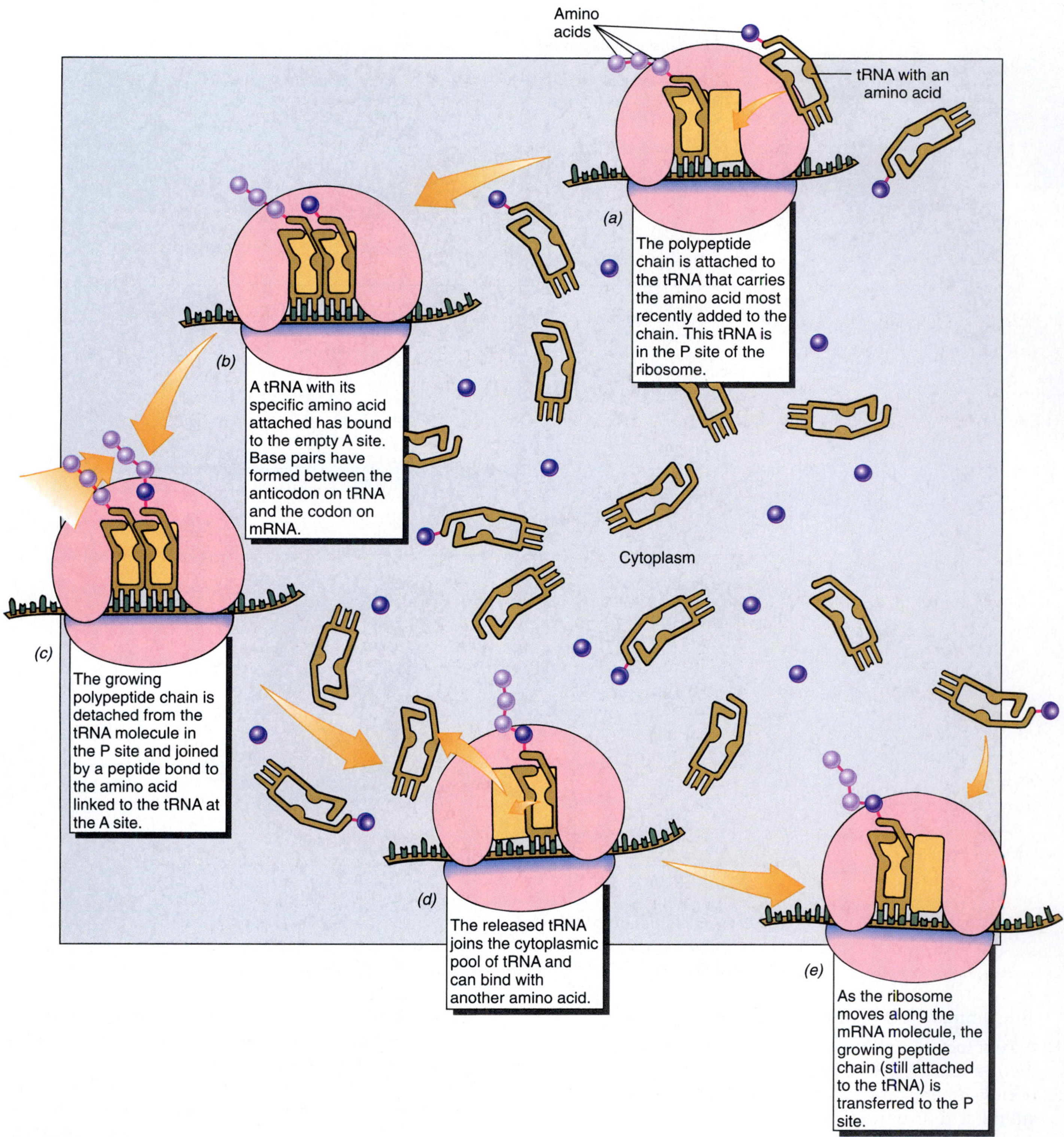

Amino acids
tRNA with an amino acid
(a)
The polypeptide chain is attached to the tRNA that carries the amino acid most recently added to the chain. This tRNA is in the P site of the ribosome.
(b)
A tRNA with its specific amino acid attached has bound to the empty A site. Base pairs have formed between the anticodon on tRNA and the codon on mRNA.
Cytoplasm
(c)
The growing polypeptide chain is detached from the tRNA molecule in the P site and joined by a peptide bond to the amino acid linked to the tRNA at the A site.
(d)
The released tRNA joins the cytoplasmic pool of tRNA and can bind with another amino acid.
(e)
As the ribosome moves along the mRNA molecule, the growing peptide chain (still attached to the tRNA) is transferred to the P site.

FIGURE 14–9 Elongation of a polypeptide chain. The overall scheme of events, which takes place after initiation.

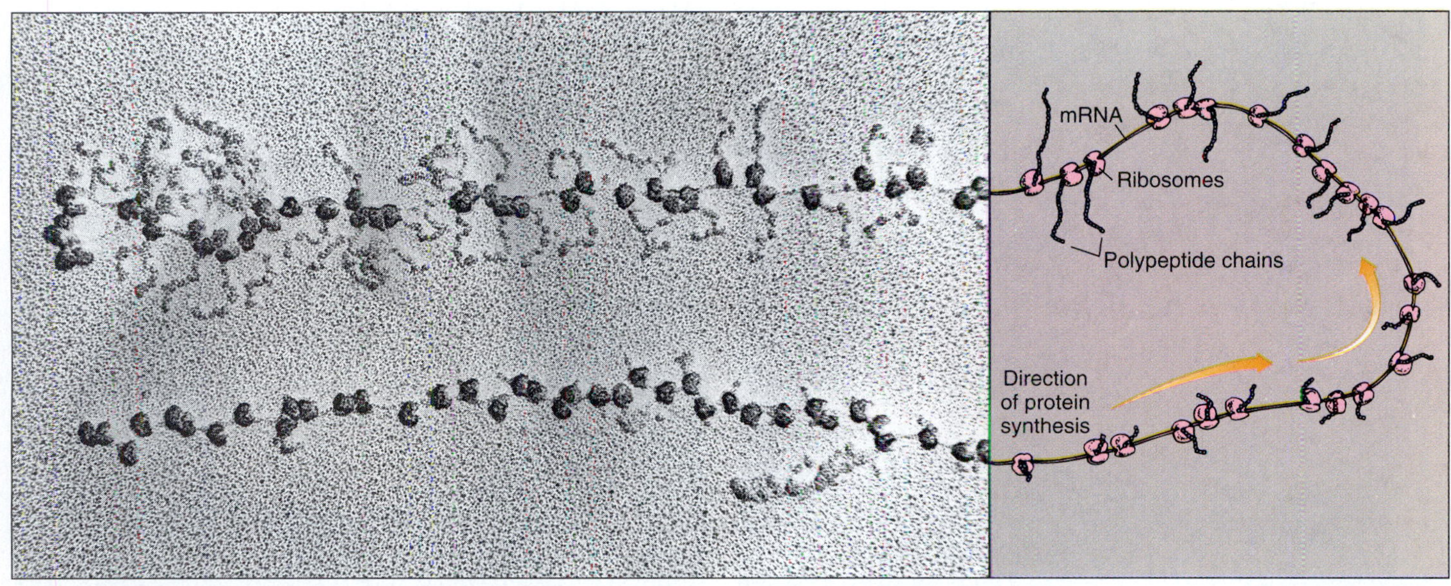

FIGURE 14–10 Polysomes. Multiple ribosomes progress along an mRNA strand, adding amino acids (brought in by tRNA molecules) to growing polypeptide chains as called for by the sequence of codons in the mRNA. (Courtesy of E. Kiseleva)

Changes Can Occur in Some Stages of Genetic Expression

In Chapter 13, you learned that random changes in an organism's genetic information, or DNA, can suddenly appear. Such changes, known as **mutations**, are often harmful. Because the structure of DNA determines the structure of proteins, a change in the order of nucleotides in DNA can alter the structure and function of the protein that DNA codes for.

For example, a mutation in the gene that codes for the protein hemoglobin causes sickle-cell anemia, the genetic disease discussed in the Chapter Opener. The altered gene contains instructions for substituting one amino acid (valine) for another (glutamic acid) in the hemoglobin molecule. As a result of this change, the three-dimensional structure of hemoglobin is altered, and its biological activity is disrupted.

Mutations occur spontaneously and for no apparent reason (as, for example, by errors in replication discussed in Chapter 13). Sometimes mutations result when genes move from one location to another, or they can be caused by factors in the environment.

Genes that Move from One Location to Another Can Cause Mutations

A few genes are mobile and can jump into and out of chromosomes. Such "jumping genes," or **transposons**, are segments of DNA that may move from place to place within a chromosome or even to an entirely different chromosome. When a transposon relocates, it causes a mutation because it often jumps into the middle of a gene, thereby inactivating it. When it leaves the middle of a gene, the gene is reactivated—that is, goes back to its nor-

Concept Connections

Mutations ⇌ *Genetic Disease*

DNA, the genetic material of living organisms, provides a master plan for all of the characteristics of an organism and directs the activities of cells. Because DNA has such crucial roles, mutations are almost always harmful to the organism. When mutations occur in reproductive cells (eggs and sperm), the changes can be passed on to the next generation, where they might result in genetic disease (Chapter 12).

Mutations that occur in the cells of the body are not passed on to the offspring. However, such mutations do have the potential to alter the functioning of the body cells, possibly leading to health problems, including cancer, during an individual's lifetime. For example, changing a protein's structure by changing its amino acid sequence might alter or destroy its performance.

It is likely that each of us contains one or more mutant genes that were not present in either of our parents. Fortunately, most are not noticeable because they are recessive (although they might appear later in our offspring!).

FIGURE 14–11 Transposons. (***a***) A normal, fully-pigmented Indian corn kernel. (***b***) A kernel whose surface layer is colorless (the kernel is yellowish because that is the color of the underlying starch). Note the deeply pigmented specks. Cells in each of these specks grew from single cells in which the transposon moved to another site, restoring the disrupted pigment gene to normal. This transposon was originally identified in corn by Dr. Barbara McClintock. (Courtesy of Nina Fedoroff, Carnegie Institute of Washington)

mal state. Thus, transposons are capable of turning genes "on" and "off."

Transposons were discovered in Indian corn by geneticist Barbara McClintock, who did a complex genetic analysis of the color patterns in corn kernels in the 1940s (Fig. 14–11). McClintock's work, widely considered one of the most significant biological discoveries of the twentieth century, won the Nobel Prize in 1983. Today, transposons are known to occur in many (if not all) organisms, from bacteria to humans.

Factors in the Environment Can Produce Mutations

Mutagens are environmental agents such as ionizing radiation and certain chemicals that induce mutations. For example, a number of common food additives and some natural chemicals, such as the aflatoxins produced by some fungi, increase the rate of mutation. Ionizing radiation in the form of x-rays or gamma rays also produces mutations, as does ultraviolet radiation that reaches the Earth from the sun.

It appears that the *effects* of most mutagens produced by humans are small (although they would not be small in the event of a nuclear war or large-scale nuclear accident). Yet the total human population exposed to human-produced mutagens is large. As a result, *the total number* of mutations caused by human activities may be quite large.

It is sometimes argued that complete removal of artificial mutagens from the environment would be too expensive to be practical and is probably unnecessary. According to this school of thought, low levels of mutagens cause no harm. All we need do is to keep the concentration of each mutagen at a low level, and all will be well.

The biggest objection to this view is that research has failed to determine a minimum threshold level for any known mutagen. (The minimum threshold level is some quantity of a mutagen, above which level mutations occur and below which level mutations do not occur.) What we see instead is called a linear-dose relationship: the smaller the dose, the smaller the effect, with even the smallest dose producing some effect. Therefore, it seems only prudent to reduce our exposure to known mutagens as much as possible.

Chapter Summary

I. DNA contains instructions for inherited characteristics.
   A. A gene is a sequence of nucleotides in a DNA molecule. Most genes code for specific polypeptides.
   B. DNA directs protein synthesis through an intermediary, RNA. Thus, the flow of genetic information is from DNA to RNA to proteins.
   C. Transcription and translation are the two processes involved in using the information encoded in the nucleotide sequence of DNA to specify the sequence of amino acids in protein molecules.

II. In transcription, one DNA strand serves as a template for the synthesis of a complementary RNA molecule. Transcription occurs in the nucleus in eukaryotes.
   A. RNA polymerase catalyzes the synthesis of RNA on a DNA template.
   B. Three kinds of RNA are transcribed and participate in protein synthesis.
      1. Messenger RNA (mRNA) carries the coded genetic message that specifies the amino acid sequence of a polypeptide to a ribosome.
      2. Transfer RNAs (tRNAs) carry specific amino acids to the ribosome for attachment to the growing polypeptide chain.
      3. Ribosomal RNA (rRNA) is a structural part of ribosomes.
   C. The mRNA molecule is modified in eukaryotes before leaving the nucleus.
      1. The entire gene is transcribed, but the introns (noncoding regions) are removed by enzymes.
      2. The coding regions, called exons, are spliced together to make a functional strand of mRNA.

III. Genetic information is stored in sequences of three nucleotide bases called triplets in DNA, codons in mRNA, and anticodons in tRNA.
   A. The base sequence in a DNA triplet is complementary to the base sequence in a mRNA codon.
   B. The mRNA codon specifies a particular amino acid in a polypeptide chain.

IV. In translation, a polypeptide is synthesized using instructions encoded in mRNA. Translation occurs at the ribosome.
   A. Translation consists of three steps: initiation, elongation, and termination.
   B. In initiation, the ribosome attaches to one end of mRNA and begins reading it where instructed to by a start codon.
   C. In elongation, tRNA molecules carry amino acids to the ribosome.
      1. Each tRNA molecule is attached to a specific amino acid, which it carries to the ribosome.
      2. The ribosome recognizes the anticodon of the tRNA-amino acid complex and allows it to base pair with the mRNA codon.
      3. An enzyme forms a peptide bond between the amino acid brought in by its tRNA and the growing polypeptide chain.
      4. The tRNA molecule is then released, and the process is repeated.
   D. In termination, a stop codon prompts the mRNA, ribosome, and polypeptide to separate.

V. Mutations are relatively uncommon.
   A. When they do occur, mutations cause a range of effects, from little-noticeable changes to serious defects and genetic disease.
   B. Mutations can be produced by errors in DNA replication, by transposons, or by mutagens.

Selected Key Terms

Post-Test

1. The synthesis of RNA on a DNA template is called ________________.

2. RNA differs from DNA in that the sugar in RNA nucleotides is ________________ and the base ________________ substitutes for thymine.

3. The three types of RNA are ________________, ________________, and ________________.

4. The information in DNA is carried to the ribosomes by ________________.

5. Ribosomes consist of two subunits, each of which are composed of molecules of ________________ and ________________.

6. Each ________________ molecule carries a specific amino acid to the ribosome.

7. The noncoding sequences of mRNA that are removed before translation are known as ______________.
8. The protein-coding sequences of mRNA, called ______________, are spliced together to form the fully functional mRNA molecule.
9. The process by which the genetic information in mRNA is used to specify the amino acid sequence of a protein is called ______________.
10. A sequence of three nucleotides in DNA, the triplet, corresponds to the ______________ in mRNA and to the ______________ in tRNA.
11. The ______________ ______________, by which the instructions encoded in DNA are eventually translated into proteins, is identical for almost all organisms.
12. Three steps in translation are ______________, ______________, and ______________.
13. In eukaryotes, transcription takes place in the cell's ______________, whereas translation occurs at the ______________.
14. A polysome consists of a single mRNA bound to a group of ______________.
15. Changes in DNA, known as ______________, may arise as mistakes, from transposons, or from exposure to mutagens.
16. Which of the following is a correct summary of the flow of genetic information in living organisms?
(a) DNA ⟶ RNA ⟶ protein
(b) DNA ⟶ protein ⟶ RNA
(c) RNA ⟶ DNA ⟶ protein
(d) protein ⟶ DNA ⟶ RNA
(e) RNA ⟶ protein ⟶ DNA
17. DNA is composed of ______________ different nucleotides, whereas proteins are composed of ______________ different amino acids.
(a) 4, 4 (b) 4, 20 (c) 20, 4 (d) 20, 20
18. A sequence of triplets in DNA -ATT-CGG-TAC- corresponds to what sequence of codons in mRNA?
(a) -UTT-CGG-TUC- (b) -UAA-GCC-AUG-
(c) -TAA-GCC-ATG- (d) -AUU-CGG-UAC-
19. A gene is most often a/an: (a) sequence of DNA nucleotides that codes for a particular polypeptide (b) sequence of RNA nucleotides that codes for a particular nucleic acid (c) enzyme that catalyzes the synthesis of RNA using DNA as a template (d) piece of mRNA after its introns have been removed.
20. A tRNA molecule with the anticodon -AUG- will base pair with which of the following codons on mRNA?
(a) -AUG- (b) -ATG- (c) -GUA- (d) -TAC- (e) -UAC-

Review Questions

1. Outline the sequence of information transfer from DNA to finished protein.
2. How can it be said that DNA determines the nature of the cell's chemical constituents other than protein, given that DNA only directs the synthesis of protein?
3. Contrast DNA replication with transcription.
4. Compare the structures of DNA and RNA.
5. Summarize the roles of mRNA, tRNA, and rRNA in protein synthesis.
6. Outline the steps of initiation, elongation, and termination in protein synthesis.
7. What is the relationship among triplets, codons, and anticodons?
8. What is the logical basis of the fact that the unit of the genetic code is a triplet rather than a single base or pair of bases?
9. Match these terms with the appropriate letter in the diagram: translation, replication, transcription.

(a) ↻DNA $\xrightarrow{(b)}$ RNA $\xrightarrow{(c)}$ protein

10. Explain the evolutionary significance of the fact that almost all living organisms possess the same genetic code.

Thinking Critically

1. A normal individual receives an emergency blood transfusion from a person who has sickle-cell anemia (assume that this is a life-or-death situation, and no other blood is available). Will the recipient develop this disease as a result of the transfusion? Why or why not?
2. Unlike eukaryotic cells, prokaryotic cells do not have to remove introns and splice together exons to make a functional mRNA molecule. Which type of cell has an obvious advantage in terms of energy cost to synthesize proteins? Which type of cell has a greater opportunity to control gene expression? Explain.

3. What would happen to evolution if DNA was transmitted precisely from generation to generation without mutations ever occurring?

4. Is protein synthesis in humans like that in oak trees, field mice, and other organisms? (Do they all have transcription, translation, three kinds of RNA, ribosomes, and so on?) Are there any distinct differences in protein synthesis among different groups of organisms?

5. In a certain fern, a gene composed of 1350 base pairs codes for a polypeptide composed of 336 amino acids. Explain the inconsistency.

6. The following sequence of triplets in DNA codes for what amino acids in a polypeptide chain? (*Hint:* You may use Table 14–2 to help answer the question, but you must do something *before* using the table.)

. . . TAC TGT TTT TCA GGT CTA . . .

Recommended Readings

Barinaga, M., "Introns Pop Up in New Places—What Does it Mean?," *Science*, Vol. 250, 14 December 1990. Introns have now been found in a prokaryote (a cyanobacterium), leading to speculation about the evolutionary significance of introns.

Radman, M., and R. Wagner., "The High Fidelity of DNA Duplication," *Scientific American*, August 1988. How the cell proofreads for genetic mistakes.

Rennie, J., "DNA's New Twists," *Scientific American*, March 1993. Gene expression is not as simple as it has been presented in this chapter. The more we learn about how DNA functions, the more intriguing the DNA molecule becomes.

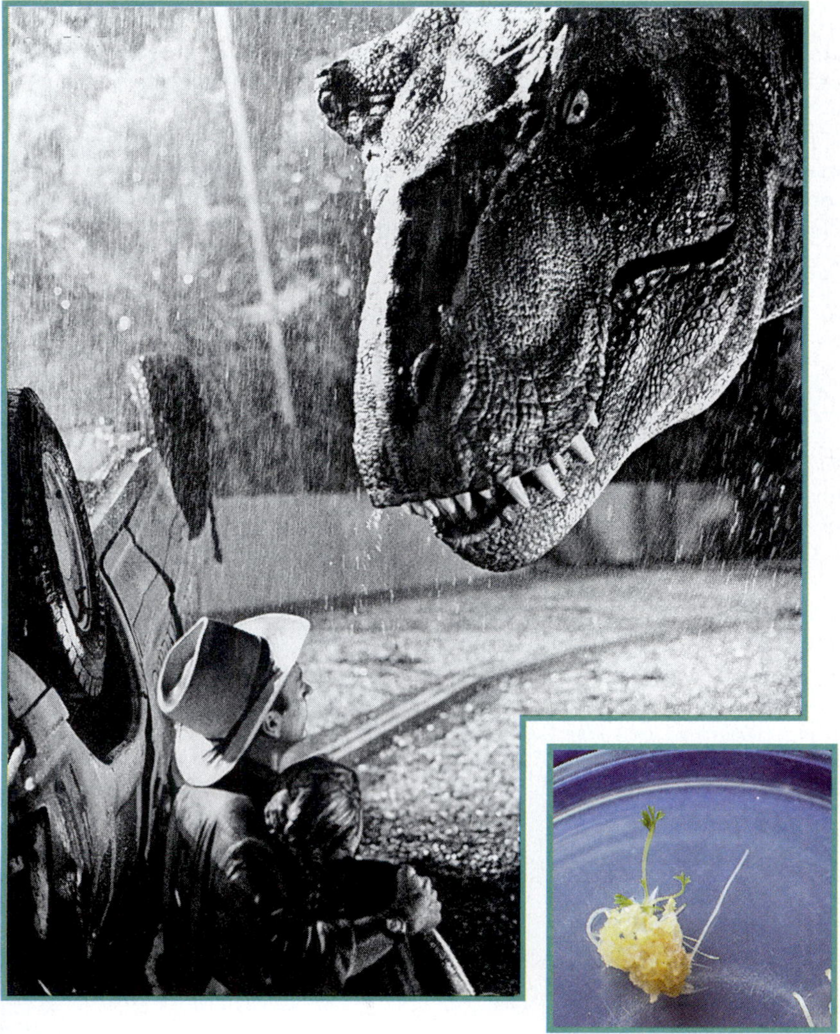

A fearsome Tyrannosaurus rex stares at Dr. Alan Grant and Lex in the film Jurassic Park. *(inset) A clump of undifferentiated carrot cells, when transferred to a different medium, grow tiny roots and shoots. (Universal courtesy Kobal. [inset] Visuals Unlimited/E. Webber)*

Dinosaurs and Carrots

All the cells in a multicellular organism descended from the same cell—a zygote—through cell division (mitosis). Because each daughter cell receives a complete set of genetic information from its parent cell during mitosis, biologists generally accept that all cells in a multicellular organism possess identical genetic blueprints. Thus, even though cells in a multicellular organism may look and function differently, they are thought to be the same genetically.

Is it possible to demonstrate that all cells have a complete genetic blueprint for the entire organism by growing a complete multicellular organism from the DNA contained in a single somatic (body) cell? The 1993 movie *Jurassic Park* portrayed just such a feat—the cloning of dinosaurs from ancient dinosaur DNA that had been preserved in amber for millions of years. However, although amber deposits have provided interesting information about extinct animals and plants, including the sequencing of some genes from ancient DNA, it is not possible to grow dinosaurs from dinosaur DNA. As a matter of fact, *no* animal has been grown from the genetic blueprint stored in single somatic cells.

On the other hand, we *can* demonstrate that plant cells contain a complete genetic blueprint for the entire organism. This is done by growing a complete plant from the DNA in a single cell using **tissue culture techniques**, methods employed to grow large numbers of cells in a sterile, synthetic medium. Using plant tissue culture techniques, it is possible to grow an entire plant that looks and functions like any other from a single mature plant cell, such as a root cell. This property, the ability of a single cell to develop into an entire organism, is called **totipotency**.

Although totipotency in plants had been suggested as early as 1902, it was not actually demonstrated until the late 1950s, when a biologist, F. C. Steward, placed some cells from a carrot root into a medium containing sugar, minerals, vitamins, and coconut milk (known to contain certain growth factors for plants). In this medium, the cells grew and divided but did not **differentiate**, that is, they did not develop into the many different types of cells found in plants. When transferred to a slightly different medium, however, differentiation occurred, and some of the cells formed tiny stems with leaves. Roots later formed on these shoots, and they were transplanted to soil. Thus, Steward demonstrated that a mature cell from a carrot root contains the genetic information needed to grow and develop an entire carrot plant.

Since Steward's pioneering work, plant tissue culture techniques have been successfully used to generate many different kinds of plants, from African violets to California redwoods. These techniques have a lot of potential, from aiding in the improvement of crop plants by genetic engineering (Chapter 16) to preserving biological diversity by cloning a large number of individuals of an endangered species.

Gene Regulation

15

Learning Objectives

After you have studied this chapter, you should be able to:

1. Define totipotency and give evidence that some cells retain the genetic information for the production of an entire multicellular organism.
2. Distinguish between constitutive genes and regulated genes.
3. Describe the process of gene regulation in prokaryotes.
4. Outline the functions of the parts of an operon: structural genes, promoter, operator, and regulator gene.
5. Sketch the lactose operon and describe how it functions.
6. Compare the lactose operon with the tryptophan operon.
7. Distinguish between inducible and repressible systems in prokaryotes.
8. Compare gene regulation in prokaryotes and eukaryotes.
9. Describe the various levels of gene regulation in eukaryotes.
10. Compare the importance of *economy* in prokaryotic gene regulation with that of *variety* in eukaryotic gene regulation.

Key Concepts

- ☐ A bacterium does not produce all of the proteins, including all of the enzymes, that it is capable of producing. Typically, a bacterium produces a specific enzyme only when the substrate for that enzyme is present. Thus, bacterial cells selectively regulate gene activity so that the protein products of genes are only formed at certain times.
- ☐ In prokaryotes, regulated genes are organized into units called operons. Each operon typically codes for several functionally related proteins. Protein synthesis is primarily regulated in prokaryotes by controlling whether or not transcription takes place.
- ☐ A multicellular eukaryotic organism develops by an orderly process from a single cell (the zygote) to an adult composed of many kinds of cells. The cells present in an adult are structurally and functionally distinct from one another because they each produce different kinds of proteins.
- ☐ Gene regulation in eukaryotes is more complex than in prokaryotes. Eukaryotic genes are regulated at the levels of transcription, messenger RNA (mRNA) processing, translation, post-translational processing, and feedback inhibition.

GENE EXPRESSION IS UNDER STRICT CONTROLS

Although plant tissue culture techniques helped demonstrate that a mature cell possesses all of the genetic information necessary to construct an entire plant, there is much we do not know about *how* this information is used. Complex multicellular organisms may be composed of billions of cells, and many of these cells have specialized structures and functions.

Clearly, not all genes are expressed—that is, activated to produce a protein—in all cells all of the time. Just as not all books in a library are in use at any one time, so too not all of the genes possessed by an organism are being "read" and used all at once. Genes are *regulated*, and only certain portions of the total genetic information are expressed in any given cell.

This chapter discusses how gene expression is controlled in prokaryotes and eukaryotes. Cells exert this control by determining which genes are to be switched "on" or "off." Those genes that are switched on will be transcribed and translated into a protein; those genes switched off will not.

Gene regulation is best understood in prokaryotes. Most studies of gene regulation have focused on the common human intestinal bacterium (*Escherichia coli*) as a representative prokaryote. Clearly, other studies must be conducted in order to determine if regulation is comparable in widely divergent organisms. For example, do bacteria living in hot springs or in the soil have the same mechanisms of gene regulation as *E. coli*, which lives inside the body of another organism and obtains nutrients from its host? Are the mechanisms of gene regulation in humans the same as those in simpler eukaryotes, such as yeasts? In sunflowers?

Gene regulation is one of the most intensively studied areas of biological research today, and for good reason. Knowledge of what turns genes on and off will add greatly to the potential of such fields as genetic engineering and medicine. Such information, for example, may eventualy help us cure cancer and prevent certain kinds of birth defects.

CONSTITUTIVE GENES ARE CONSTANTLY TRANSCRIBED

Some genes produce proteins that are continuously used by a cell. For example, the enzymes involved in glycolysis are always needed to provide energy by cellular respiration (Chapter 8). Such genes, which are continually transcribed throughout the life of a cell, are called **constitutive genes**.

Other genes, however, can be turned on and turned off, that is, directed either to produce their protein product or to stop producing it. These regulated genes are the focus of this chapter. Investigation of the ways in which genes are regulated is one of the most exciting and fast-advancing areas of biology, with great practical and theoretical implications.

GENE REGULATION IN PROKARYOTES PRIMARILY INVOLVES CONTROL OF TRANSCRIPTION

The majority of prokaryotes—that is, bacteria—are single-celled organisms that grow rapidly under good environmental conditions. They multiply by splitting in half asexually (Chapter 21). Prokaryotes must be able to utilize whatever food sources they encounter during their short lives. Generally, prokaryotes make whatever proteins they need, including enzymes to digest food. However, little protein is made in anticipation of need. Bacterial cells do not produce all of the enzymes and other proteins that they are capable of making "just in case" they need them; the energy costs and the waste of important raw materials such as amino acids would be too high. Instead, their genes are regulated so that transcription (and ultimately protein synthesis) occurs only when a particular enzyme or other protein is required.

Much of our understanding of gene regulation comes from studies of the intestinal bacterium, *E. coli*. Its availability and rapid growth when cultured in the laboratory have led to its extensive study. The biology of *E. coli* is probably the best understood of any organism.

E. coli live in countless billions in the intestines of mammals. For a short time after the mammal eats, these bacteria are surrounded by the nutrients that the mammal's enzymes have digested. *E. coli* can readily partake of these nutrients without using any digestive enzymes of their own. Their protein production is channeled in other directions.

During times when the mammal's intestine is empty of simple, predigested food molecules, however, the bacteria synthesize enzymes to digest unprocessed food materials. They are able to turn on and off the genes that code for many different digestive enzymes.

Our knowledge of the gene-regulating mechanisms in *E. coli* began with the contributions of the French investigators François Jacob and Jacques Monod. In the course of their research on *E. coli*, Jacob and Monod encountered an oddity in the behavior of *E. coli* bacteria. When fed a mixture of glucose and lactose, a disaccharide found in milk, the bacteria first consumed all of the glucose. Then, after a brief pause, they began to consume the lactose.

Further investigations showed that, in order to consume lactose, the bacteria first had to have the enzyme that breaks down this disaccharide into its two component simple sugars, glucose and galactose. *E. coli* that had not been fed lactose ordinarily did not possess this enzyme. However, once they were fed lactose, they started to make the enzyme.

Operons Are Units of Gene Expression in Prokaryotes

In 1961, Jacob and Monod described how gene regulation in *E. coli* permitted it to produce the lactose-digesting enzyme when it was given lactose as food. Since then, similar mechanisms have been described not only in *E. coli* but in other bacteria as well.

It appears that most, if not all, bacterial genes are organized into clusters called **operons**. Operons are composed of four parts: structural genes, a promoter, an operator, and a regulator gene.

Each operon includes several genes, called **structural genes**, that code for the synthesis of a group of enzymes that are all involved in the same function (for example, the digestion of lactose). Recall from Chapter 14 that transcription of messenger RNA (mRNA) begins when RNA polymerase recognizes the promoter site on a strand of deoxyribonucleic acid (DNA). The **promoter** is that part of the operon where RNA polymerase binds to begin transcription of the structural genes. The structural genes of an operon all share a single promoter site. When RNA polymerase bind to this site, it transcribes all of the structural genes onto one mRNA strand, which may go on to be translated into the individual proteins.

However, other features of the operon may prevent RNA polymerase from binding to the promoter site and thus may prevent transcription from occurring. Whether the structural genes of an operon will be transcribed is under the control of the **operator**. The operator, that part of the operon that acts as a control switch, can switch transcription of the structural genes on or off by allowing or preventing RNA polymerase to bind to the promoter.

The operator, in turn, is regulated by a special part of the operon called the **regulator gene**. This gene codes for the production of a protein that can bind to the operator. When this protein molecule, called the **repressor**, binds to the operator, it prevents RNA polymerase from binding to the promoter region of the operon. Thus, the repressor protein represses the expression of the structural genes in the operon.

The exact mechanisms of gene regulation vary for different operons. We will consider the details of two different operons that have been well characterized in *E. coli*: the lactose operon and the tryptophan operon.

The Lactose Operon Is Activated by the Presence of Lactose

The lactose operon, also designated the *lac* operon, controls the production of enzymes needed to digest lactose in *E. coli*. It consists of three structural genes, a regulator gene, and operator and promoter sites (Fig. 15–1).

The structural genes code for three enzymes used in the digestion of lactose: (1) an enzyme that transports lactose across the plasma membrane into the bacterial cell, (2) an enzyme that breaks down lactose into simpler sugars, and (3) a third enzyme whose role is beyond the scope of our discussion.

Unlike the structural genes (which are regulated), the regulator gene that codes for the repressor protein is constitutive; that is, the regulator gene is always turned on and continually makes small amounts of the repressor protein. Under normal conditions in *E. coli*—that is, when lactose is *absent*—this repressor protein binds to the operator site, blocking the promoter site in the process. (The operator site overlaps the promotor site.) As a result, RNA polymerase cannot bind to the promoter site and therefore cannot begin transcribing the structural genes. Thus, when there is no lactose present, the structural genes that code for the enzymes needed to digest lactose are not even transcribed.

The lactose operon is turned on when lactose is present in *E. coli*'s environment, however. A few molecules of lactose enter the cell and are converted to a derivative of lactose. This derivative attaches to the repressor protein, thereby modifying its shape so that the repressor protein can no longer bind to the operator site. As a result, the promoter site is unblocked, and RNA polymerase binds to the promoter and begins transcribing the structural genes. The resulting mRNA molecule is translated, forming the three enzymes that enable *E. coli* to digest lactose.

When lactose once again becomes scarce in the environment, no lactose derivative is available to bind to the repressor protein. The repressor therefore attaches to the operator site, preventing further transcription of the structural genes for the enzymes.

This sort of control, in which the presence of a substrate induces the synthesis of an enzyme, is known as an **inducible system**. The expression of an inducible system is normally controlled by a repressor that keeps it turned off. The presence of the substrate (here, lactose) inactivates the repressor, allowing the structural genes to be expressed in protein synthesis. Inducible operons usually code for enzymes involved in the breakdown of nutrient molecules that provide the cell with energy.

What is the advantage of *E. coli*'s ability to repress the lactose operon when lactose is not present? Clearly, it takes energy and raw materials to manufacture enzymes.

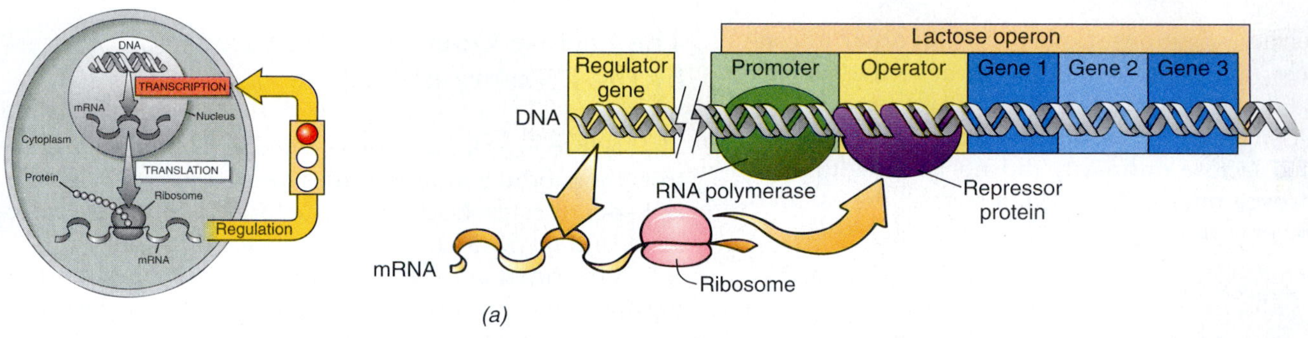

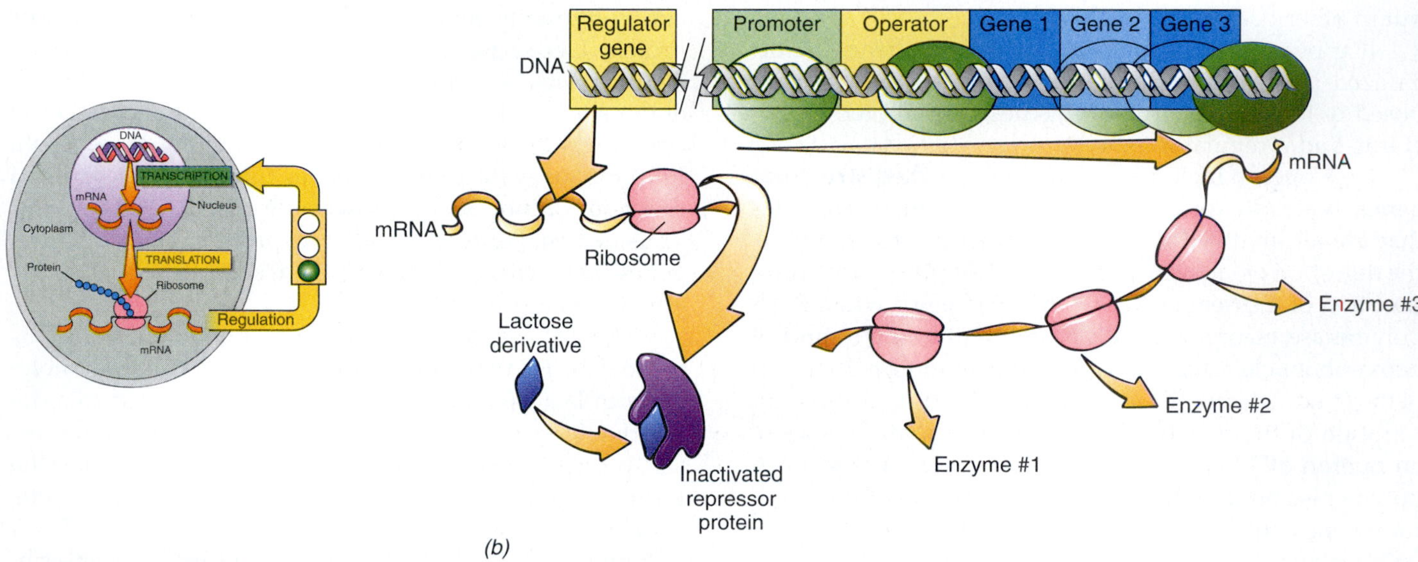

FIGURE 15–1 The lactose operon consists of a group of structural genes, a regulator gene, and operator and promoter sites. (***a***) In the absence of lactose, a repressor protein binds to the operator, blocking RNA polymerase from transcribing the structural genes. (***b***) When lactose is present, a metabolic derivative binds to the repressor protein so that it can no longer bind to the operator. As a result, RNA polymerase binds to the promoter, and transcription of the structural genes takes place. The three enzymes that are ultimately synthesized enable the bacterium to utilize lactose.

If a bacterium were to continue to express those genes when their products would not be used, this wasteful diversion of resources might be the margin between survival and death.

Yet, despite the very small likelihood that any particular *E. coli* will ever encounter lactose, there is an advantage to possessing the genes that code for the enzymes involved in lactose metabolism. An *E. coli* bacterium living in the intestine of a mature cow has probably never been exposed to lactose, because the cow probably has not drunk milk for several years, since it was a calf. Nor have a thousand generations of the *E. coli*'s ancestors fed on lactose (*E. coli* can reproduce several times an hour, so a thousand generations could all have lived in the same cow). Yet at some point, its ancestor used the lactose genes, because at some point the cow was a calf feeding on its mother's milk. As the cow grew and stopped drinking milk, the *E. coli* living inside the cow stopped feeding on lactose and stopped expressing the genes for the lactose-digesting enzymes. However, these genes remained a part of the bacterial DNA and were replicated and passed on through each generation, never once revealing their presence. But when lactose is present in the environment again, they are availible for expression.

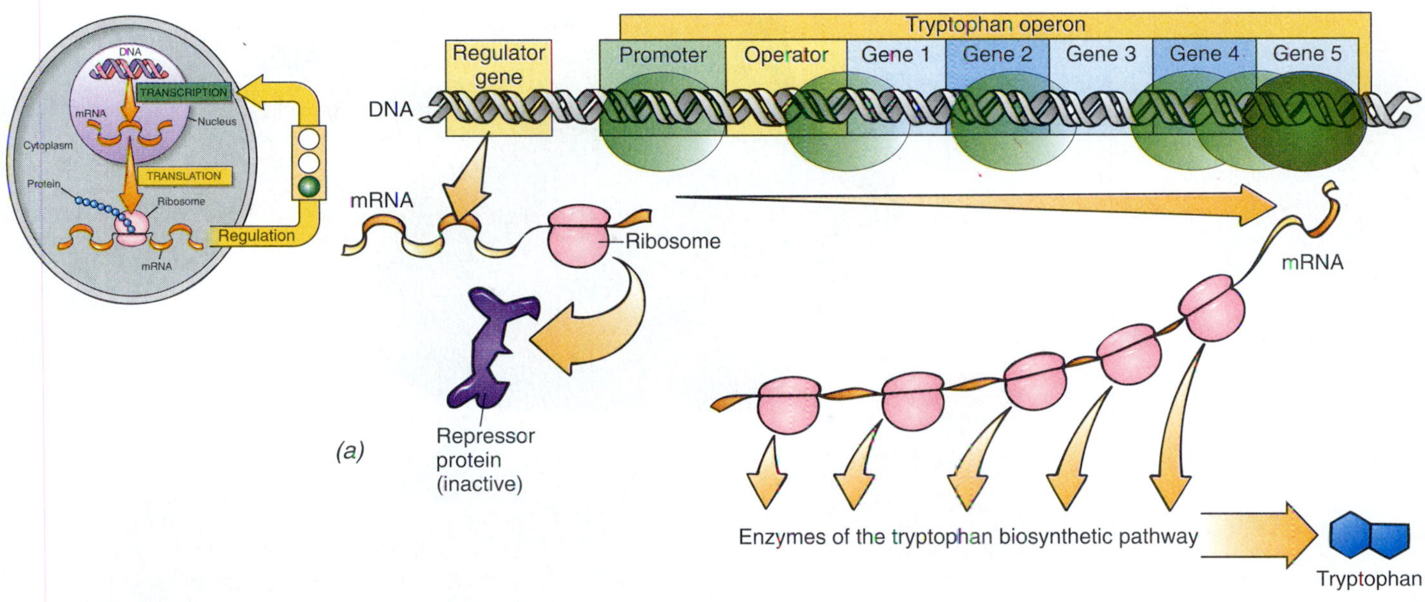

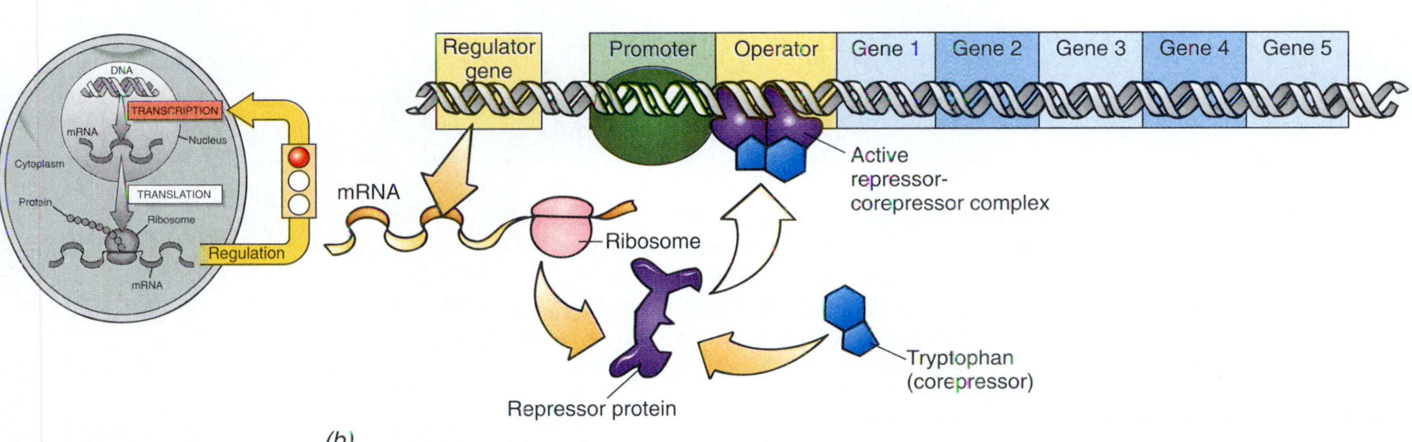

FIGURE 15–2 The tryptophan operon of *E. coli.* (***a***) The regulator gene codes for a repressor that cannot bind to the operator. As a result, the genes for the five enzymes necessary for tryptophan synthesis are transcribed. (***b***) When tryptophan levels are high within the cell, tryptophan attaches to the repressor protein, changing its shape slightly. The resulting active form of the repressor protein binds to the operator site, preventing transcription of the structural genes until tryptophan is again required by the cell.

The Tryptophan Operon Is Activated by the Absence of Tryptophan

In addition to making digestive enzymes, *E. coli* also makes essential molecules such as amino acids. The manufacture of the amino acid tryptophan involve five steps, each catalyzed by a specific enzyme. In order for *E. coli* to make tryptophan, all five enzymes must be present in the cell. In addition, tryptophan itself must be in low supply.

The tryptophan operon, or *trp* operon, in *E. coli* includes five adjacent structural genes that code for the five enzymes needed to produce tryptophan (Fig. 15–2). The tryptophan operon also includes a regulator gene and operator and promoter sites.

As in the lactose operon, the operator site in the tryptophan operon overlaps the promoter region. The regulator gene codes for a repressor protein that can bind to the operator site and block the promoter site. Unlike the repressor in the lactose operon, however, the repressor protein in the tryptophan operon does not by itself bind to the operator site. The repressor attaches to the operator only after tryptophan has attached to the repressor protein, modifying its shape slightly so it has a higher

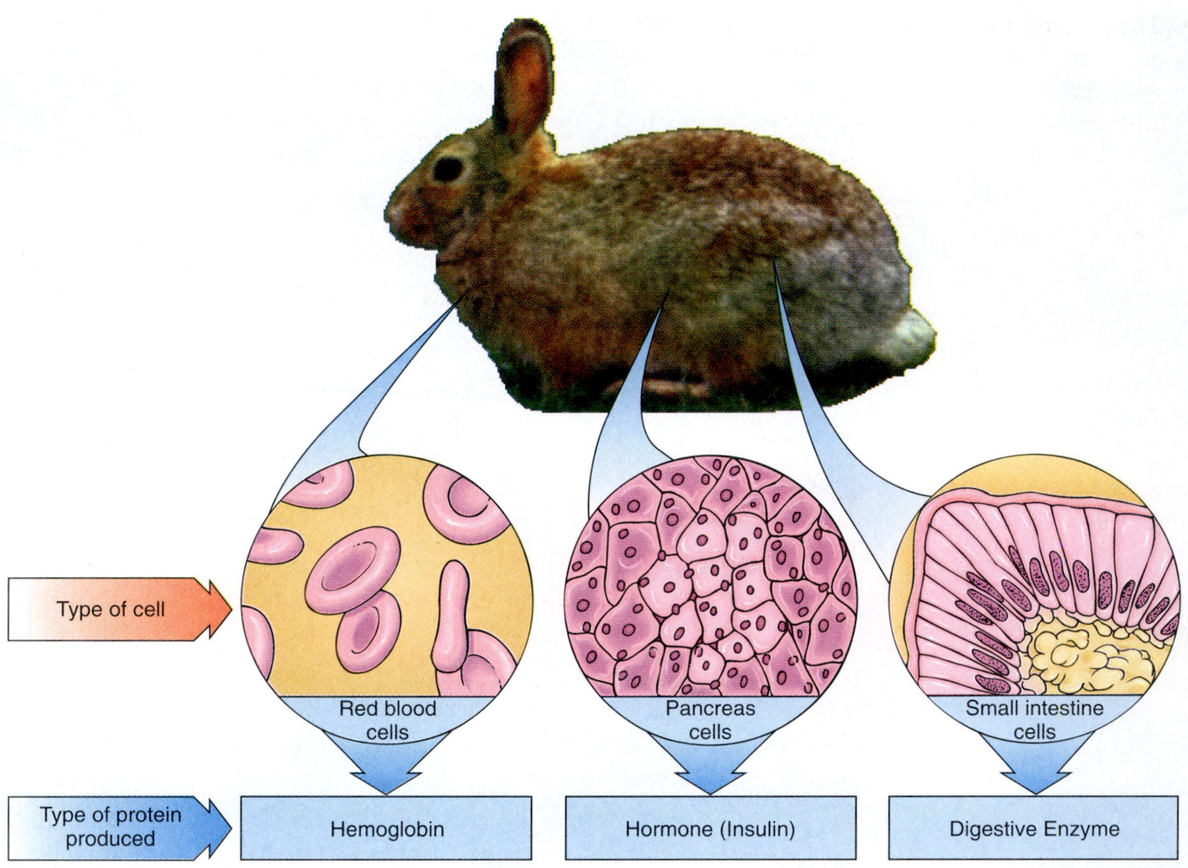

FIGURE 15–3 Different kinds of cells do not produce the same kinds of proteins. This means that different genes are turned on or off in different cells of the body at different stages in development. Each cell of this rabbit contains an identical set of genes. Red blood cells produced in the bone marrow contain the gene for hemoglobin, special cells in the pancreas produce the hormone insulin, and cells lining the small intestine produce certain digestive enzymes. Although the cells in the pancreas and small intestine contain the gene that codes for hemoglobin, neither type of cell produces hemoglobin. Similarly, neither red blood cells nor small intestine cells produce the hormone insulin. Likewise, neither red blood cells nor the hormone-producing cells of the pancreas produce the digestive enzymes manufactured by the small intestine cells. (Don and Esther Phillips/Tom Stack & Associates)

affinity for the operator. Thus, tryptophan enables the repressor protein to function in blocking the synthesis of tryptophan-forming enzymes.

The expression of the tryptophan operon is normally turned on, and as a result, tryptophan is continually made (and used) by the cell. However, when the level of tryptophan is high because the cell has manufactured more tryptophan than it needs, enzyme synthesis is repressed. This occurs because the presence of tryptophan activates the repressor, thereby inhibiting expression of the structural genes that code for the enzyme.

This process, in which the presence of the end product (here, tryptophan) prevents its production at the level of transcription, is called **end-product repression**. An operon involved in end-product repression is called a **repressible system**. Repressible operons usually code for enzymes involved in the synthesis of amino acids, nucleotides, and other essential molecules.

Several important distinctions exist between the tryptophan operon and the lactose operon. The lactose operon is normally turned *off*, whereas the tryptophan operon is normally turned *on*. Also, the lactose operon is activated by the *presence* of the substrate (lactose), whereas the tryptophan operon is activated by the *absence* of the end product (tryptophan).

GENE CONTROL IS IMPORTANT IN CELL GROWTH AND DIFFERENTIATION OF MULTICELLULAR EUKARYOTES

As discussed in the chapter introduction, a multicellular eukaryotic organism grows from a single cell, the fertilized egg, or zygote. In early development, the zygote divides into two, then four, then eight cells, and so on. The cells produced from these early cell divisions are all very similar. Increasingly, however, they begin to differentiate, taking on specific structural characteristics until they become the specialized cells of the mature organism.

The orderly development of multicellular organisms is due to the fact that differentiating cells do not produce the same proteins (Fig. 15–3). The distinctive structures and functions of differentiated cells reflect the fact that each type of cell expresses only part of the genetic information it contains, producing some proteins but not others. That is, selective gene expression occurs in different tissues and at different times during development. As cells differentiate, they gradually suppress certain genes so that these genes cannot later be expressed, or begin expressing genes that were not expressed previously.

EUKARYOTES HAVE MULTIPLE LEVELS OF GENE REGULATION

When molecular biologists first developed the operon concept, they thought that it would be as important in eukaryotes as it is in prokaryotes. However, most eukaryotic genes are not found in operon-like clusters, and gene regulation in eukaryotes has many elements. In view of the greater complexity of eukaryotes, it is perhaps not surprising that eukaryotes do not employ a single, simple method of gene control, but instead use a combination of methods.

One level of gene regulation in eukaryotes occurs at the level of transcription and involves the coiling and uncoiling of DNA. In addition to regulating genes at the level of transcription, eukaryotes have four further levels of control that permit individual cells to become committed to various roles in the multicellular organism: mRNA processing, translation, post-translational processing, and feedback inhibition (Fig. 15–4). Each of these levels of control are considered in the text that follows.

CONCEPT CONNECTIONS

Gene Regulation — *Prokaryotes Versus Eukaryotes*

Prokaryotes and eukaryotes regulate the activity of their genes in different ways that reflect their distinctly different lifestyles. Because prokaryotes are single-celled organisms, each individual cell must be able to perform all of the functions necessary for life. Prokaryotes grow rapidly and have relatively short lifespans. They must be able to exploit whatever food source they encounter in their brief lives. These facts appear to make *economy* a major focus of gene regulation in prokaryotes, and the most cost-effective way to regulate genes is at the level of transcription. Groups of related genes are turned on or off, allowing proteins to be synthesized only when they are required.

In multicellular eukaryotic organisms, the various cells of the body cooperate in a division of labor. It is thus not necessary for every cell in a multicellular organism to perform all of the functions necessary for life. Different cells, for example, may need to regulate a particular gene in different ways. Moreover, because they typically live longer than prokaryotic cells, eukaryotic cells may encounter and have to respond to many different stimuli in their environments.

As a result of the eukaryotic cell's lifestyle, *variety* appears to be more important than economy in eukaryotic gene regulation. Variety is reflected in the multiple levels of gene regulation found in eukaryotes. For example, by possessing inactive forms of certain enzymes, eukaryotic cells can respond very quickly to changes in the environment by simply activating the enzyme rather than having to completely synthesize a new enzyme. We say more about prokaryotic and eukaryotic lifestyles in various chapters throughout the text.

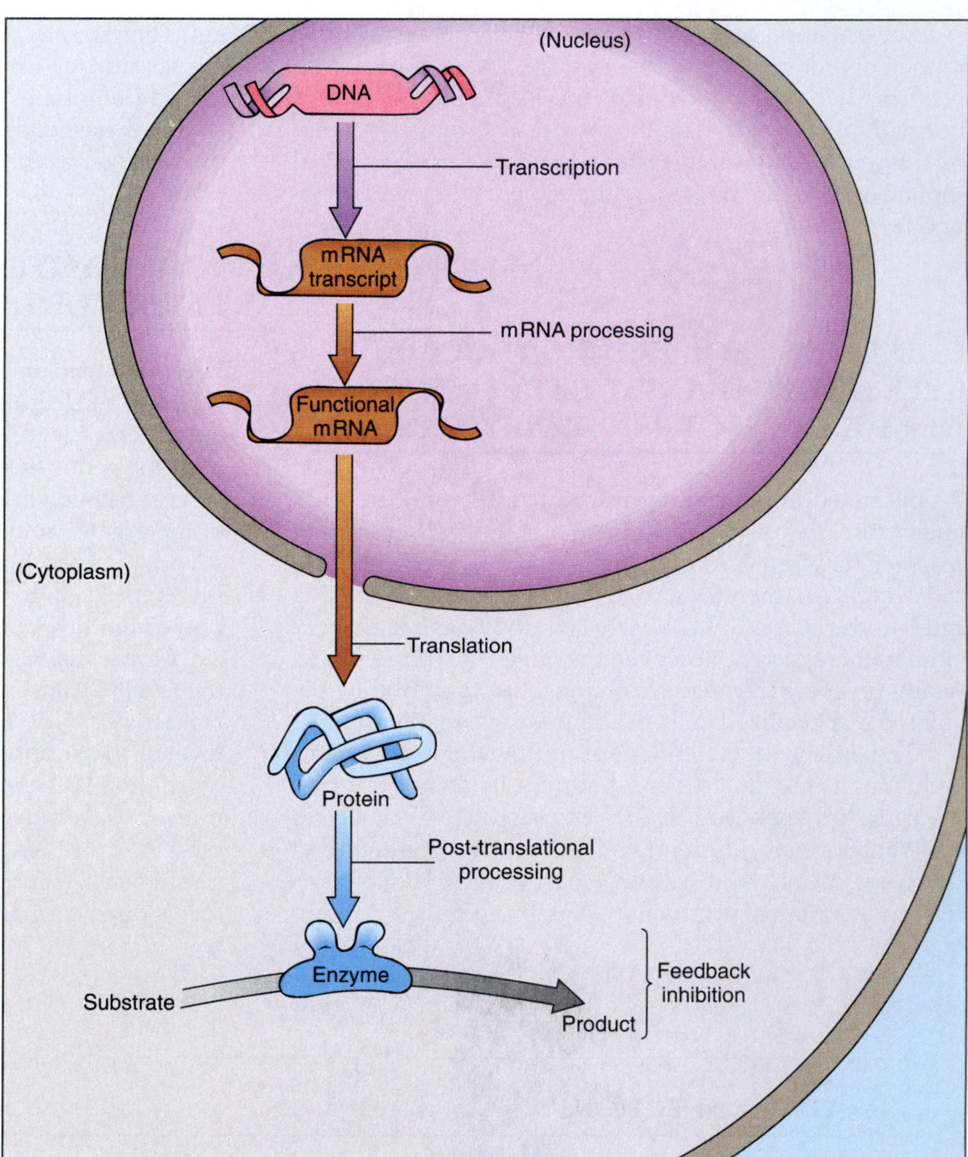

FIGURE 15–4 Gene regulation in eukaryotic cells occurs at the levels of transcription, mRNA processing, translation, post-translational processing, and feedback inhibition. The levels of control are superimposed over a diagram of the cell to show which processes take place in the nucleus and which in the cytoplasm. All are explained in the text.

Eukaryotes Regulate Gene Expression at the Level of Transcription

We have seen that bacteria respond to changes in their environment by turning on and off appropriate sets of genes; prokaryotic gene expression is controlled largely at the level of transcription. Eukaryotic gene control works at the level of transcription as well, by turning genes on and off.

Eukaryotic chromatin structure affects whether a particular gene will be transcribed. It appears that DNA in condensed chromatin—that is, when DNA is coiled around histones to form nucleosomes, as described in Chapter 13—cannot be transcribed into RNA. In other words, genes in condensed regions are inactive and cannot be expressed (Fig. 15–5). Transcription of DNA can only occur if the DNA packaging of the chromosomes is relaxed, or decondensed. Because of the way the DNA molecule is coiled, various portions can be relaxed during transcription without the entire huge molecule being uncoiled.

Once a gene is uncoiled, it can be transcribed. As in prokaryotes, the transcription of a eukaryotic gene requires the presence of a promoter to which RNA polymerase binds in order to initiate transcription. In addition, eukaryotic genes require DNA sequences called **enhancers**, which increase the rate of transcription.

(Text continues on page 278)

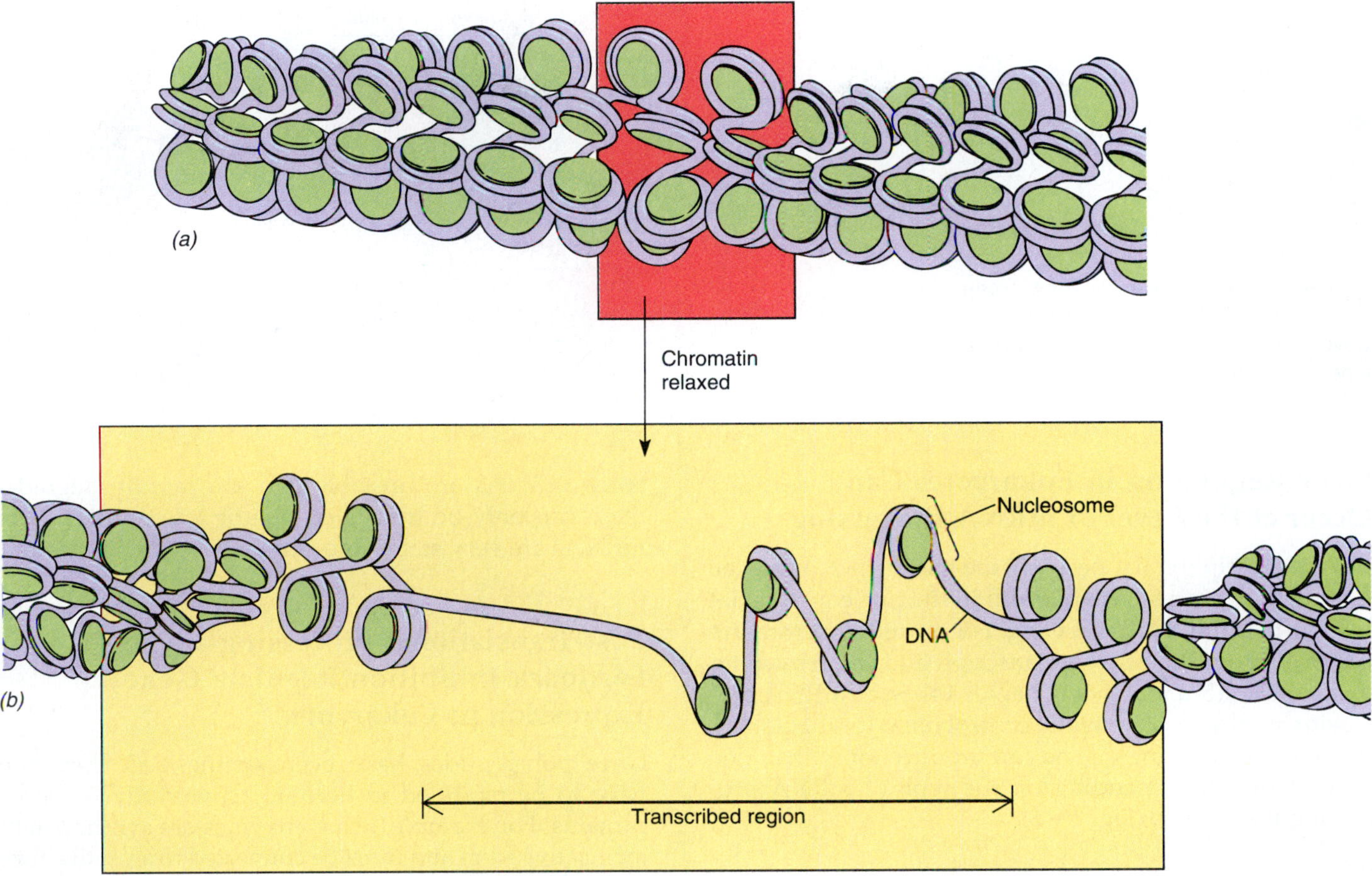

FIGURE 15–5 Chromosome structure affects transcription. (***a***) When eukaryotic DNA is inactive, its nucleosomes are condensed into tight coils. (***b***) Active genes are found in relaxed, or decondensed, chromatin. RNA polymerase can more easily transcribe the information in a section of relaxed chromatin.

Concept Connections

Eukaryotic Gene Regulation — *Chromatin Structure*

A typical eukaryotic cell, such as a human liver cell, contains more than 1000 times the amount of DNA found in an *E. coli* cell, a fact that is not surprising in light of the greater complexity of eukaryotes. However, the vast majority of a eukaryotic cell's genes are turned *off* most, if not all, of the time. In a human liver cell, genes involved with the functions of the liver must be turned on, for example, but all other genes, such as those controlling eye color or sexual traits, are turned off.

The fact that most genes in eukaryotic cells are inactive most of the time is related to the organization of their DNA. Recall from Chapter 10 that DNA in a eukaryotic cell is organized in the nucleus into multiple chromosomes during mitosis or meiosis. At all other times in the cell cycle, DNA exists in a complex with various proteins as fine, threadlike strands called chromatin. DNA, which is negatively charged, is wound around histone proteins, which are positively charged, to form nucleosomes (Chapter 13). When DNA is complexed to histones, it cannot undergo transcription. Thus, histones represent a major part of eukaryotic gene regulation—the off signal that keeps most genes silent.

How, then, do genes ever get turned on? Nonhistone nuclear proteins assist by removing histones, thereby allowing certain genes to be switched on. Then, and only then, do other levels of eukaryotic gene regulation (mRNA processing, translation control, post-translational processing, and feedback inhibition) come into play.

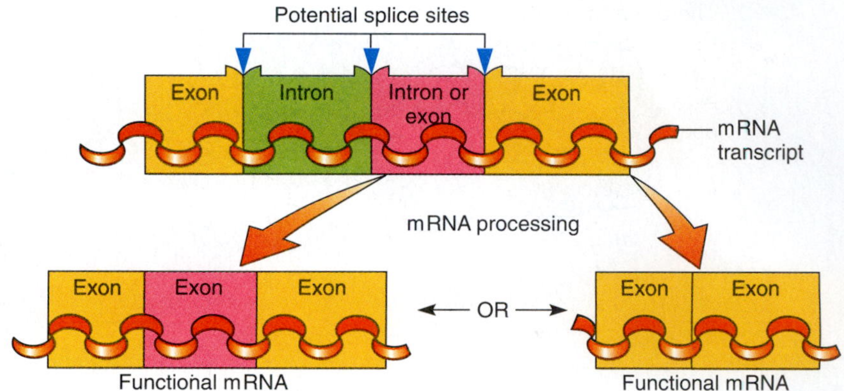

FIGURE 15–6 Messenger RNA processing. Processing mRNA from a single gene in different ways can lead to the production of different types of protein.

Gene Regulation in Eukaryotes Can Occur at the Level of mRNA Processing

Once transcription has occured, the initial mRNA may be modified or processed while still in the nucleus, in what is known as **mRNA processing**. For example, introns are removed from the mRNA transcript, and the remaining exons are spliced together (Chapter 14). Such mRNA processing represents a potential control point for regulation. In some cases, a single gene can produce more than one type of protein depending on the type of mRNA processing that occurs (Fig. 15–6).

Eukaryotic Genes Are Controlled at the Level of Translation

There is also control of translation, in which the message encoded in mRNA is translated into polypeptides by the ribosomes. Gene regulation at the level of translation is related to controlling the *number* of protein molecules translated from mRNA. Gene regulation at the level of translation is related to the stability of mRNA molecules. Some mRNAs are unstable and are rapidly degraded; these can only be translated one or a few times. Other mRNAs are very stable and can be translated repeatedly.

Post-Translational Processing and Feedback Inhibition Regulate Gene Expression in Eukaryotes

Once polypeptides have been synthesized, they often have to be modified to become active, fully functional proteins. For example, some enzymes are synthesized in an inactive form and must be converted to an active form, perhaps by removal of a portion of the polypeptide chain. Such **post-translational processing** represents another level of gene regulation.

Eukaryotic gene activity can also be regulated after a fully-functional protein is synthesized. For example, eukaryotes (as well as prokaryotes) possess many metabolic pathways that are regulated through feedback inhibition, in which the formation of the product inhibits an earlier reaction in the pathway (Chapter 7).

Chapter Summary

I. Constitutive genes are actively transcribed at all times. Regulated genes can be turned on or off.

II. In prokaryotes, genes are regulated primarily by controlling transcription.
   A. The operon is the unit of gene expression in prokaryotes. Each operon consists of:
      1. Structural genes, which code for several related proteins used by the cell.
      2. A promoter, a sequence of bases where RNA polymerase binds to DNA and initiates transcription.
      3. An operator, a sequence of bases that overlaps the promoter and controls whether the operon is turned on or off.
      4. A regulator gene, which codes for a repressor protein that binds to the operator site; when bound, the repressor protein prevents RNA polymerase from binding to the promoter.
   B. An inducible operon, such as the lactose operon, is normally turned off because the repressor is normally bound to the operator.

1. The presence of the substrate (lactose) inactivates the repressor protein (that is, a derivative of lactose binds to the repressor).
2. As a result, the repressor protein cannot bind to the operator, and the structural genes are transcribed.

C. A repressible system, such as the tryptophan operon, is normally turned on.
1. The repressor protein is synthesized in an inactive form that cannot bind to the operator.
2. The presence of the end product of a biosynthetic pathway binds to the inactive repressor, causing it to become active.
3. As a result, RNA polymerase can no longer bind to the promoter, and transcription of the genes ceases.

III. Most eukaryotic genes are not organized into operons. Gene regulation is more complex in eukaryotes than in prokaryotes.

A. Regulation of eukaryotic genes can occur at the level of transcription, by turning genes on and off. Histones play an essential role in both the structure of chromatin and the control of gene expression.
B. Gene regulation can occur as a result of mRNA processing, which includes removing introns and splicing together exons.
C. Eukaryotes also regulate gene expression at the level of translation by regulating the stability of mRNA. When mRNA is more stable, more proteins can be formed per mRNA molecule.
D. Gene regulation can occur after proteins have been synthesized, by post-translational processing (modification of the structure of the protein) and feedback inhibition.

Selected Key Terms

constitutive gene, p. 270
differentiation, p. 268
end-product repression, p. 274
enhancer, p. 276
inducible system, p. 277
mRNA processing, p. 278
operator, p. 278
operon, p. 271
post-translational processing, p. 278
promoter, p. 271
regulator gene, p. 271
repressible system, p. 274
repressor, p. 271
structural genes, p. 271
tissue culture techniques, p. 268
totipotency, p. 268

Post-Test

1. The process of change in unspecialized cells that results in their developing the appearance and functions of a specific type of adult cell is known as ______________.
2. The ability of a cell to develop into an entire organism is called ______________.
3. Genes not under regulatory control are called ______________.
4. Regulation of most prokaryotic genes occurs at the level of ______________.
5. The ______________, a unit of gene expression in bacteria, includes structural genes, a regulator gene, and operator and promoter sites.
6. The portion of DNA in an operon that codes for several protein products used by the cell is known as the ______________ genes.
7. The portion of the operon to which RNA polymerase binds is the ______________.
8. The control region of an operon that overlaps the promoter is called the ______________.
9. The regulator gene codes for a ______________ protein.
10. When the repressor binds to the ______________, transcription is blocked.
11. In a/an ______________ operon, the repressor protein is inactivated by the presence of the substrate, resulting in the synthesis of an enzyme.
12. In a/an ______________ operon, the presence of the end product prevents its further production at the level of transcription.
13. Eukaryotic genes require DNA sequences called ______________, which increase the rate of transcription.
14. Gene regulation in eukaryotes occurs at the levels of transcription, ______________ ______________, translation, post-translational processing, and feedback inhibition.
15. Some metabolic pathways are regulated by ______________ ______________, in which the end product inhibits an earlier reaction in the pathway.

16. Inducible operons are generally turned ______________, and repressible operons are usually turned ______________.

17. The lactose operon is activated by the ______________ of lactose, and the tryptophan operon is activated by the ______________ of tryptophan.

18. If a mutation occurred in the regulator gene of the lactose operon such that the repressor protein that it codes for was rendered permanently inactive, what would happen? (a) The structural genes would be transcribed continuously. (b) The structural genes would be permanently turned off. (c) There would be no effect on the rate of transcription of the structural genes.

19. Under which of the following conditions will the enzymes, whose structural genes make up the lactose operon in *E. coli*, *not* be synthesized? (a) A mutation in the operator site such that the repressor cannot bind to it. (b) Presence of lactose in the environment. (c) A mutation in the promoter that renders it nonfunctional. (d) A mutation in the regulator gene such that the repressor is inactive.

20. After mRNA processing in eukaryotic cells, the final mRNA molecule differs from the mRNA transcript made by RNA polymerase because ______________ have been deleted. (a) exons (b) operons (c) enhancers (d) introns

Review Questions

1. What is an operon? Describe its operation.
2. How are operator and promoter sites similar? How are they different?
3. What binds to the operator? What happens when binding takes place?
4. What binds to the promoter? What happens when binding takes place?
5. What do structural genes code for in the lactose operon? When the structural genes of the lactose operon are transcribed and translated, what can the cell do?
6. What does the regulator gene code for in the operon? Explain why this gene is constitutive.
7. Distinguish among inducible, repressible, and constitutive genes. Give an example of each.
8. Prokaryotic gene regulation occurs primarily at what level or levels?
9. Eukaryotic gene regulation occurs primarily at what level or levels?
10. Explain how feedback inhibition regulates gene expression.

Thinking Critically

1. Based on what you have learned about gene regulation in multicellular eukaryotes, explain why we do not have the ability to clone dinosaurs (assuming the implausible likelihood that a complete set of genetic information for a dinosaur species was discovered).
2. Refer to Figure 15-3. What gene or genes might be turned on in all three kinds of cells?
3. The histidine operon in the *Salmonella* bacterium is responsible for the synthesis of the amino acid histidine. It regulates the expression of nine different structural genes involved in histidine synthesis. Based on what you have learned about the lactose operon and the tryptophan operon, do you think the histidine operon would be activated by the presence or absence of histidine? Why?
4. If bacteria are grown for several hours in a medium containing glucose as the sole source of carbohydrate and then transferred to a medium containing only lactose, what would occur? Why? Would the bacteria still need to metabolize glucose? Why?
5. Suppose that a mutation occurred in certain *E. coli* in which the tryptophan operon is rendered completely inoperable. Under what environmental condition(s) would these bacteria survive? Under what condition(s) would they perish?
6. State whether each of the following bacterial genes would most likely be constitutive, inducible, or repressible, and offer an explanation for your answer.
   a. A gene that codes for an enzyme that breaks down the disaccharide, maltose.
   b. A gene that codes for RNA polymerase.
   c. A gene that codes for an enzyme used to synthesize the amino acid, methionine.
7. Give an example of a situation in which feedback inhibition would be a more efficient control of eukaryotic gene expression than transcription.

Recommended Readings

Grunstein, M., "Histones as Regulators of Genes," *Scientific American*, June 1989. How histones, a protein found in chromosomes, help control gene expression.

Ptashne, M., "How Gene Activators Work," *Scientific American*, January 1989. How regulatory proteins bind to DNA and turn it on.

Ross, J., "The Turnover of Messenger RNA," *Scientific American*, April 1989. How RNA synthesis and degradation are controlled.

Two young girls, Cynthia (left) and Ashanthi (right), were the first humans ever to have gene replacement therapy. (Ted Thai/ Time Magazine)

Out with the Bad Genes, in with the Good

Treating serious human genetic disorders is difficult—in part because there are no cures—and doctors and scientists have long dreamed of actually curing genetic diseases (Chapter 12). Today we are closer to that goal because of **gene replacement therapy,** a variety of techniques that involve introducing normal copies of a gene into some of the cells of the body of a person afflicted with a genetic disorder. Although an individual gene may be present in all cells of the body, it is only expressed in some cells (Chapter 15). The idea behind gene replacement therapy is to introduce normal copies of a defective gene into the cells that express it, in the hope of alleviating the symptoms of the genetic disorder.

In 1990, the first clinical trial for gene replacement therapy in humans was approved. The gene therapy was designed to treat a severe immune deficiency disease caused by the inability of cells in the immune system to produce an enzyme, adenosine deaminase (ADA). Infants born with this rare genetic disease die in childhood because their body lacks a functioning immune system.

Ashanthi Desilva, a little girl with ADA deficiency, was the first human to receive gene replacement therapy. A virus was used to introduce the normal ADA gene into her white blood cells, and the altered cells were then transfused back into her body. The results of this first test were so promising that a second child suffering from the same disorder, Cynthia Cutshall, was also treated by gene replacement therapy.

The results have been encouraging: both Ashanthi and Cynthia now have fully functional immune systems. Both girls attend public school, an activity that would have been impossible without gene replacement therapy because they would have been exposed to (for them) life-threatening diseases such as the common cold. However, current gene replacement therapy is not a permanent cure because white blood cells have limited lifetimes. As a result, the girls must receive gene replacement therapy every two or three months throughout their lives.[1]

Despite its temporary nature, this overall success prompted quick National Institutes of Health (NIH) approval of many other gene replacement therapies now being tested.[2] These include replacing defective genes for other genetic disorders such as hemophilia, sickle-cell anemia, and cystic fibrosis as well as providing new therapeutic abilities for treating diseases such as cancer, Parkinson's disease, and AIDS. For example, one gene replacement therapy trial currently under way involves inserting a gene that causes cancer cells to die into cancer cells of patients in advanced stages of their disease.

Using gene replacement therapy raises both medical and ethical issues. Medical concerns center around the long-term safety of such procedures for patients. Although Ashanthi and Cynthia have responded well to treatment, unforeseen problems and side effects could develop years later.

One of the ethical considerations of using gene replacement therapy centers on whether we are "playing God." This concern does not involve treating genetic defects that cause serious diseases such as ADA deficiency. Rather, this concern is that gene replacement therapy might be used to "enhance" normal individuals. For example, in the future it may be possible to use gene replacement therapy to improve one's characteristics, such as height. (Other ethical issues relating to human genetics were discussed in Chapter 12.)

Although many obstacles remain to be addressed before gene replacement therapy is a routine treatment for genetic diseases, the field is expected to develop at a rapid pace during the next decade. This chapter examines the genetic techniques that have provided a number of practical applications, including gene replacement therapy.

[1] In 1993, a new procedure to introduce the normal ADA gene into stem cells (long-lived cells that produce white blood cells throughout the lifetime of an individual) was tried on a few patients, including Cynthia. This procedure, the success of which is being evaluated as we go to press, may represent the first permanent cure of a genetic disease.

[2] Proposals for gene replacement therapy must currently be reviewed by several groups to ascertain that guidelines established by the Recombinant DNA Advisory Committee (RAC) of NIH are followed. These guidelines consider the safety of the patient, including the potential benefits and risks of using the procedure. In addition, the Food and Drug Administration (FDA) has its own set of guidelines.

Recombinant DNA Technology and Genetic Engineering

16

Learning Objectives

After you have studied this chapter you should be able to:

1. Outline the primary techniques used in recombinant DNA technology.
2. Summarize the problems involved in isolating, identifying, and amplifying a single gene.
3. Explain the actions of restriction enzymes and DNA ligase and their importance in recombinant DNA experiments.
4. Draw a sketch that shows how restriction enzymes cut DNA to produce sticky ends.
5. Identify the role of biological vectors in recombinant DNA technology and give specific examples of such vectors.
6. Give an example of a nonbiological method employed to introduce genes into plant and animal cells.
7. Describe the polymerase chain reaction and list several potential uses of this technique.
8. Distinguish between a gene and cDNA.
9. List at least five potential or realized applications of recombinant DNA technology.
10. Summarize some of the concerns that have been raised about recombinant DNA research and the guidelines that have been formulated to address these concerns.

Key Concepts

- ☐ The deliberate manipulation of genes of different organisms to accomplish human goals is known as genetic engineering. Genetic engineers employ special techniques called recombinant DNA technology.
- ☐ Genetic engineering allows a single gene to be isolated and moved from one organism to another. It results in combinations of genes that are impossible or highly unlikely in nature, thereby providing new opportunities for breeders of plants and animals.
- ☐ Genetic engineering holds great promise for decreasing the amount of time required to produce altered organisms with desirable characteristics, as compared to the traditional method of selective breeding.
- ☐ Recombinant DNA technology promises many practical benefits in the fields of medicine and agriculture.

HUMANS HAVE BEEN MODIFYING LIVING ORGANISMS FOR THOUSANDS OF YEARS

Selective breeding, the breeding by humans of animals and plants with desired traits, has been used for thousands of years to develop a better cow, honeybee, or corn plant. The first farmer to decide which seeds to save for the following year's crop was practicing selective breeding. However, selective breeding requires a long time to obtain results and does not always achieve what one might like to accomplish—cow's milk low in cholesterol, for example, or corn with the protein content of soybeans is beyond the capability of selective breeding.

Using living organisms to produce products that benefit humanity is known as **biotechnology.** Selective breeding of plants and animals is an example of biotechnology, as is using yeast to cause bread to rise during baking. Recent advances in genetics promise an unprecedented control of heredity. When people talk about biotechnology today, they are often speaking about it in a narrower sense, referring to the genetic manipulation of cells so that new strains of organisms with new characteristics can be constructed (Fig. 16–1). Such modification of the DNA of an organism to produce new characteristics is also known as **genetic engineering.**

Genetic engineering differs from selective breeding in several respects. For one thing, genetic engineering allows us to tap a much larger pool of genes. It is now possible to transfer genes from humans to bacteria, or from animals to plants. In contrast, selective breeding allows genetic transfers only between closely related individuals, usually of the same species. Also, genetic engineering allows us to develop new strains of organisms in a fraction of the time required for selective breeding.

FIGURE 16–1 In biotechnology, new combinations of genes produce novel traits in organisms, such as crop resistance to insects. A gene from a common soil bacterium, for example, was engineered into cotton; this gene codes for the production of a protein that acts as a natural insecticide. In a field test, cotton plants were subjected to an insect infestation. The genetically engineered plant (left) received little damage, and as a result of being healthier, was able to produce larger bolls of cotton. The nonaltered plant (right) was weakened by insects and produced much smaller bolls. (Courtesy of Monsanto Company)

RECOMBINANT DNA TECHNOLOGY INVOLVES MANIPULATING GENES IN THE CELLS OF LIVING ORGANISMS

The methods employed in genetic engineering are known as **recombinant DNA technology.** Recombinant DNA technology, a new field of biology that started in the mid-1970s, permits the formation of new combinations of genes by isolating genes from one organism and introducing them into either a similar or an unrelated organism. By these methods, foreign DNA may be inserted not only into the simple single cells of bacteria but also into cells from the bodies of complex organisms. Once inside the cell, this foreign DNA may be expressed. The genetic composition in bacteria may be altered, for example, so that they produce novel proteins such as insulin or human growth hormone.

Recombinant DNA technology is not practiced exclusively by humans. Living organisms have mechanisms for exchanging genes, and such gene transfers have always occurred in nature. For example, bacteria sometimes transfer DNA through a special tube that extends from one cell to another (Chapter 21). The transfer of genetic material from one bacterium to another by means of a bacterial virus is another example of natural genetic exchange. However, these natural "experiments" have occurred randomly and certainly not with human goals and desires in view.

Having a foreign gene expressed in a cell requires several steps. The gene that is being introduced into a different cell must be (1) isolated from its original cell, (2) used to construct a recombinant DNA molecule, (3) transferred into a new (that is, foreign) cell, (4) identified as having been taken up by the cell, and (5) expressed in that cell. We consider each of these steps in detail.

First a Specific Region of DNA Is Isolated from the Rest of the Cell's DNA

To begin the process of genetic engineering, the investigator must isolate the gene of interest. This is done by first breaking up the cell's DNA into more manageable fragments by the use of bacterial enzymes known as **restriction enzymes.** These enzymes cut DNA molecules at

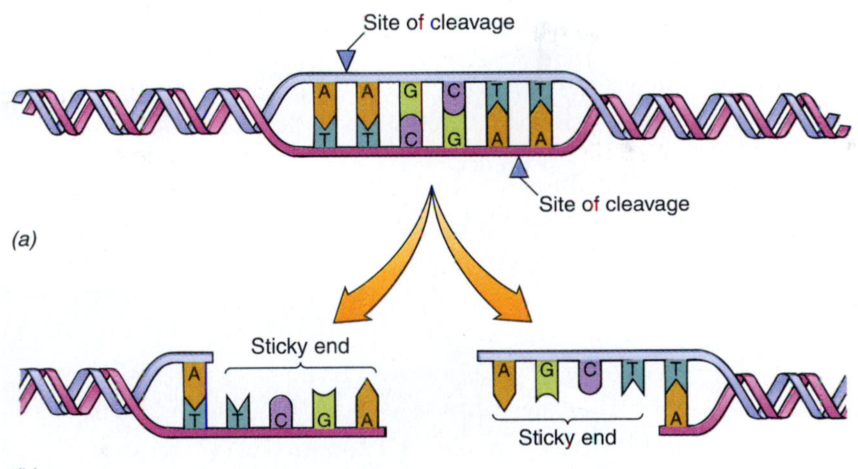

FIGURE 16–2 In the presence of restriction enzymes, DNA is cleaved at specific sites. (***a***) A portion of a DNA molecule (unwound for simplification) that contains the recognition site for a restriction enzyme that recognizes the DNA base sequence -AAGCTT- and its complementary base pairs. Note that -AAGCTT- and its complement -TTCGAA- are palindromic: each strand has the same base sequence, but in the opposite direction. (***b***) The cleavage of DNA by this restriction enzyme results in two DNA fragments that have short single-stranded stubs—that is, sticky ends.

specific base sequences. Restriction enzymes generally recognize a specific DNA sequence four to six base pairs in length. For example, one restriction enzyme recognizes and cuts DNA whenever it encounters the base sequence 5′-AAGCTT-3′, and another cuts DNA at the sequence 5′-GAATTC-3′.

The different restriction enzymes are obtained from various bacteria, which possess these enzymes as a defense against bacterial viruses. When such viruses enter bacterial cells, the bacteria's restriction enzymes destroy the virus by chopping the viral DNA into small pieces, thereby inactivating it. (The bacterial DNA is modified so that it is immune to its own restriction enzymes.) The use of various restriction enzymes gives biologists a way to cut the giant DNA molecule of a chromosome into shorter, gene-sized fragments.

Many of the base sequences recognized by restriction enzymes are **palindromic,** which means that the base sequence of one strand reads the same as its complement strand in the opposite direction. For example, note how the restriction enzyme that recognizes the DNA base-pair sequence 5′-AAGCTT-3′ and its complementary sequence 3′-TTCGAA-5′ are palindromic (Fig. 16–2). When this particular restriction enzyme encounters this sequence, it cleaves the double-stranded DNA between A and A on *each* strand.

This cleavage leaves one strand longer than the other on each fragment of DNA; each segment is said to have **sticky ends** because it can pair with the single-stranded ends of other DNA molecules that were digested by the same restriction enzyme. Thus, DNA segments from entirely different organisms are potentially able to recombine with one another if they've been digested by the same restriction enzyme. The base sequences cut are always the same for each restriction enzyme, no matter what kinds of DNA are cleaved. For this reason, their sticky ends are complementary and, in the presence of a splicing enzyme known as **DNA ligase,** they can attach to one another.

Recombinant DNA Molecules Are Constructed when DNA Is Spliced into a Vector

Restriction enzymes split the potential donor DNA into many fragments. The next step is to construct recombinant DNA molecules by combining our isolated DNA fragments with DNA from another source, which will act as a **vector,** or DNA carrier. We discuss three different vectors: plasmids, viruses, and gene guns.

Some bacteria have small circular DNA molecules known as **plasmids** (Fig. 16–3). When one or more plasmids are present within a bacterial cell, they are replicated along with the bacteria chromosome, and the copies are distributed to daughter cells. Plasmids serve as genetic vectors by carrying desired genes into cells (without a vector, you would have no way to insert the foreign DNA into the target cell). When desired genes are inserted into plasmids and these altered plasmids are incorporated into bacterial cells, the plasmid and its genes are replicated each time the bacterial cells divide. As a result, the entire bacterial population that descended from the original bacterial cells contains the novel genes. Plasmids have been used to introduce foreign genes into bacteria and plant cells.

Viruses can also be used to introduce recombinant DNA into cells. For example, viruses are commonly used as vectors in mammalian cells, including human cells. These viruses have been disabled so that they do not cause disease in the cells to which they are introduced. Instead, their DNA and the foreign DNA that they are carrying are incorporated into a chromosome of the host cell.

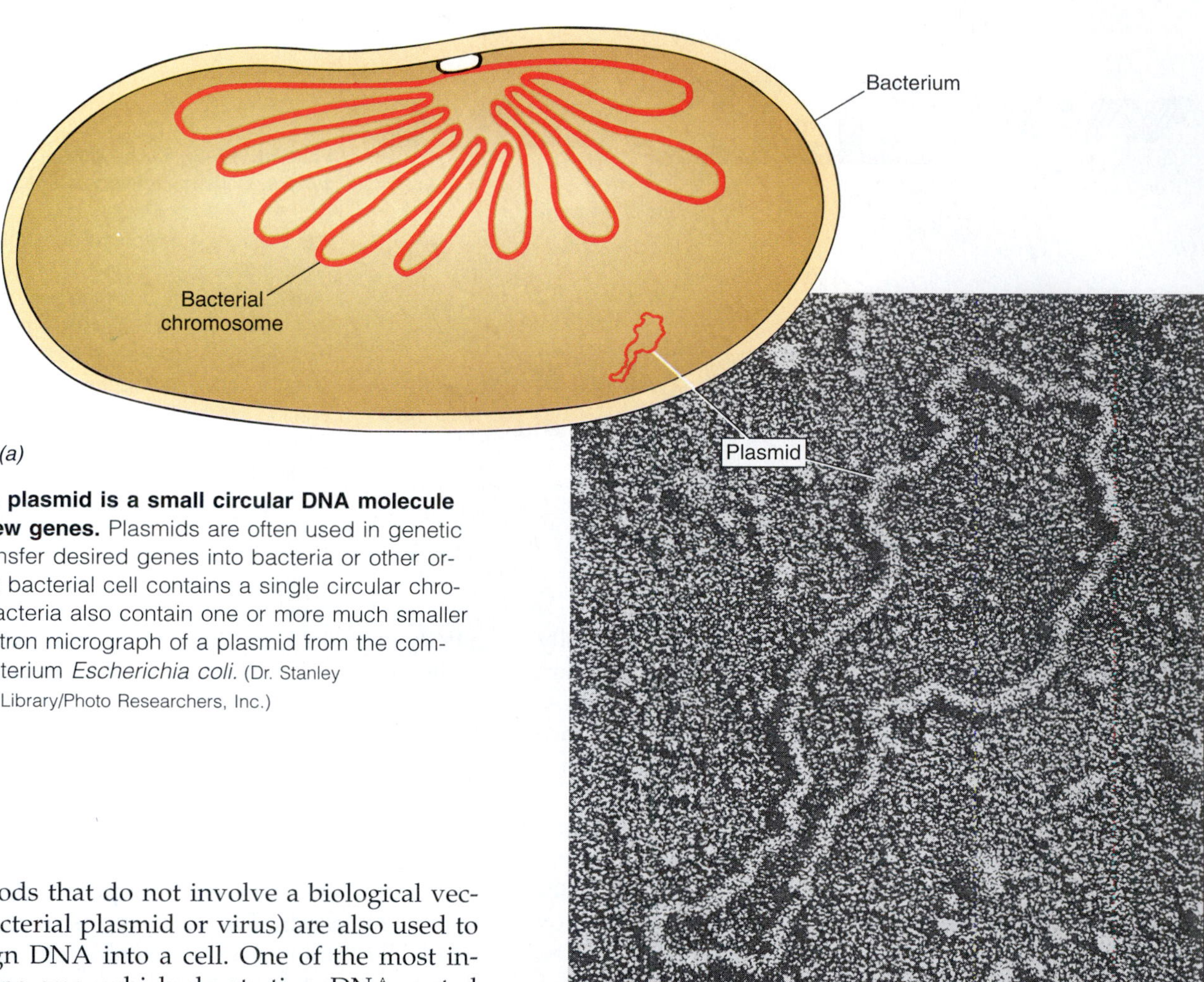

FIGURE 16–3 A plasmid is a small circular DNA molecule that contains a few genes. Plasmids are often used in genetic engineering to transfer desired genes into bacteria or other organisms. (***a***) Each bacterial cell contains a single circular chromosome. Some bacteria also contain one or more much smaller plasmids. (***b***) Electron micrograph of a plasmid from the common intestinal bacterium *Escherichia coli.* (Dr. Stanley Cohen/Science Photo Library/Photo Researchers, Inc.)

Other methods that do not involve a biological vector (that is, a bacterial plasmid or virus) are also used to introduce foreign DNA into a cell. One of the most intriguing is a **gene gun,** which shoots tiny, DNA-coated metallic "bullets" into living cells (Fig. 16–4). Although the gene gun was first used successfully to transform algal and yeast cells, it is now routinely used to introduce genes into plant cells. Recently, the gene gun has been used successfully in animal cells.

Having examined three types of vectors, we now continue our discussion of recombinant DNA technology by focusing on plasmids. To construct recombinant DNA molecules in which a plasmid is the vector, the investigator treats plasmids isolated from bacterial cells with a restriction enzyme (Fig. 16–5). Because the restriction enzyme cuts the DNA, the formerly circular plasmids now become linear molecules of DNA with sticky ends. These strands are then mixed with sticky DNA fragments from the donor cell. By the use of DNA ligase, the ends of the donor and plasmid DNA are joined.

The result is a mixture of recombinant plasmids, with each containing a different fragment of donor DNA. Although the vast majority of plasmids do *not* contain the desired gene, a very few can be expected to possess exactly what is needed. In order to identify the plasmid that contains the gene of interest, all the plasmids must be *amplified,* which will result in millions of copies to work with. Amplification takes place inside bacterial cells (discussed next).

The Recombinant DNA Molecules Are Transferred into a New Host Organism's Cell

The recombinant DNA molecules (in plasmids) are next taken into bacterial cells. Under normal conditions bacteria do not take in plasmids; however, the host bacteria will receive such plasmids under certain conditions, as when they are treated with heat and calcium. Bacteria that take in the plasmid are said to be **transformed.**

Since only a small proportion of the bacteria will take up the plasmids, even when treated with the special con-

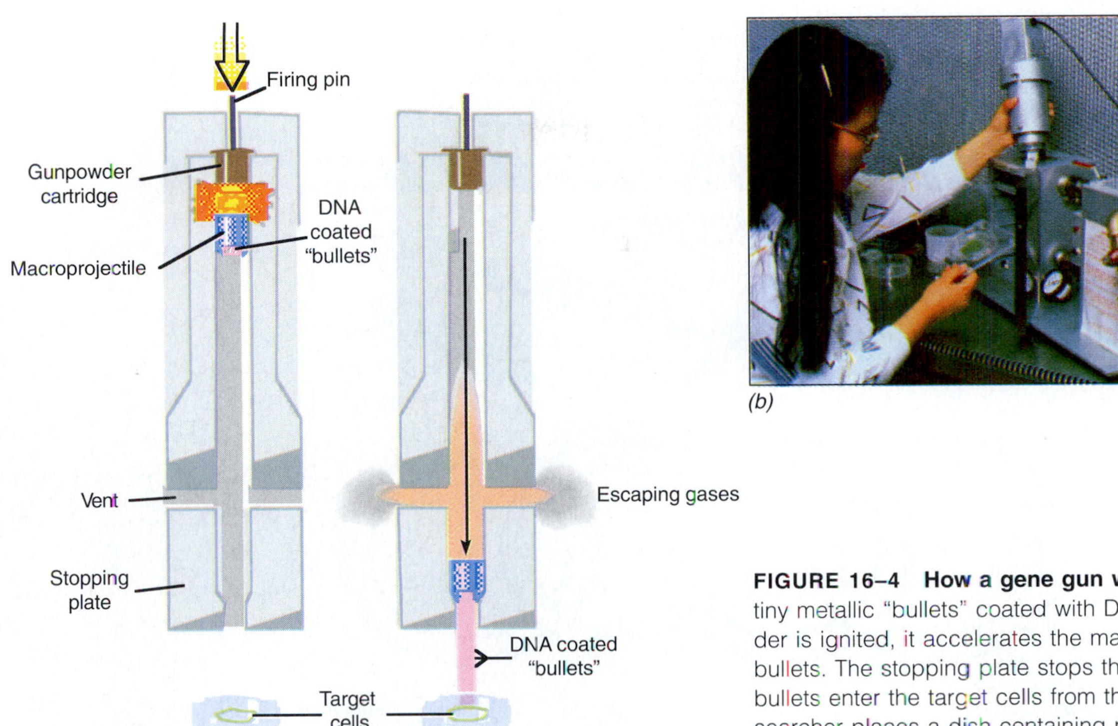

FIGURE 16–4 How a gene gun works. (***a***) A gene gun fires tiny metallic "bullets" coated with DNA into cells. When gunpowder is ignited, it accelerates the macroprojectile that holds the bullets. The stopping plate stops the macroprojectile, but the bullets enter the target cells from the momentum. (***b***) A researcher places a dish containing plant cells into the gene gun. (Courtesy of Dr. John Sanford, Cornell University)

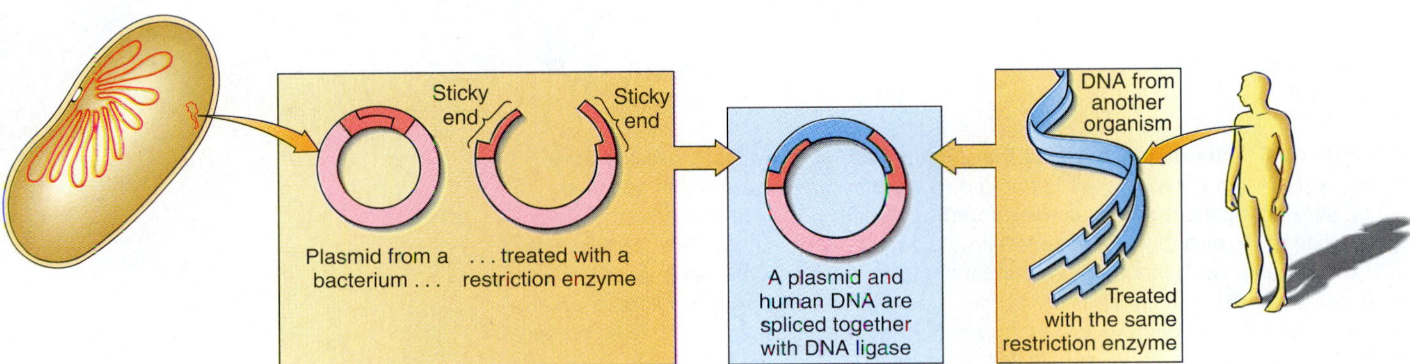

FIGURE 16–5 Constructing a recombinant DNA molecule. DNA molecules from two different sources are cut with the same restriction enzyme, which results in complementary sticky ends. In this example, one of the DNA molecules is a circular plasmid from a bacterium and the other is a segment of DNA from a human cell. When mixed, their sticky ends form base pairs, and DNA ligase splices the two together.

ditions that make them competent to do so, there has to be a way to determine which bacteria contain the genetically altered plasmids. Fortunately, plasmids often carry genetic information that makes the bacterium resistant to certain antibiotics (a natural substance that kills or inhibits microorganisms). If the plasmid that was used also carries a gene for resistance to a particular antibiotic, then you can grow, or culture, the bacterial population in the presence of that antibiotic. Only bacteria that have taken up genetically altered plasmids will resist the antibiotic

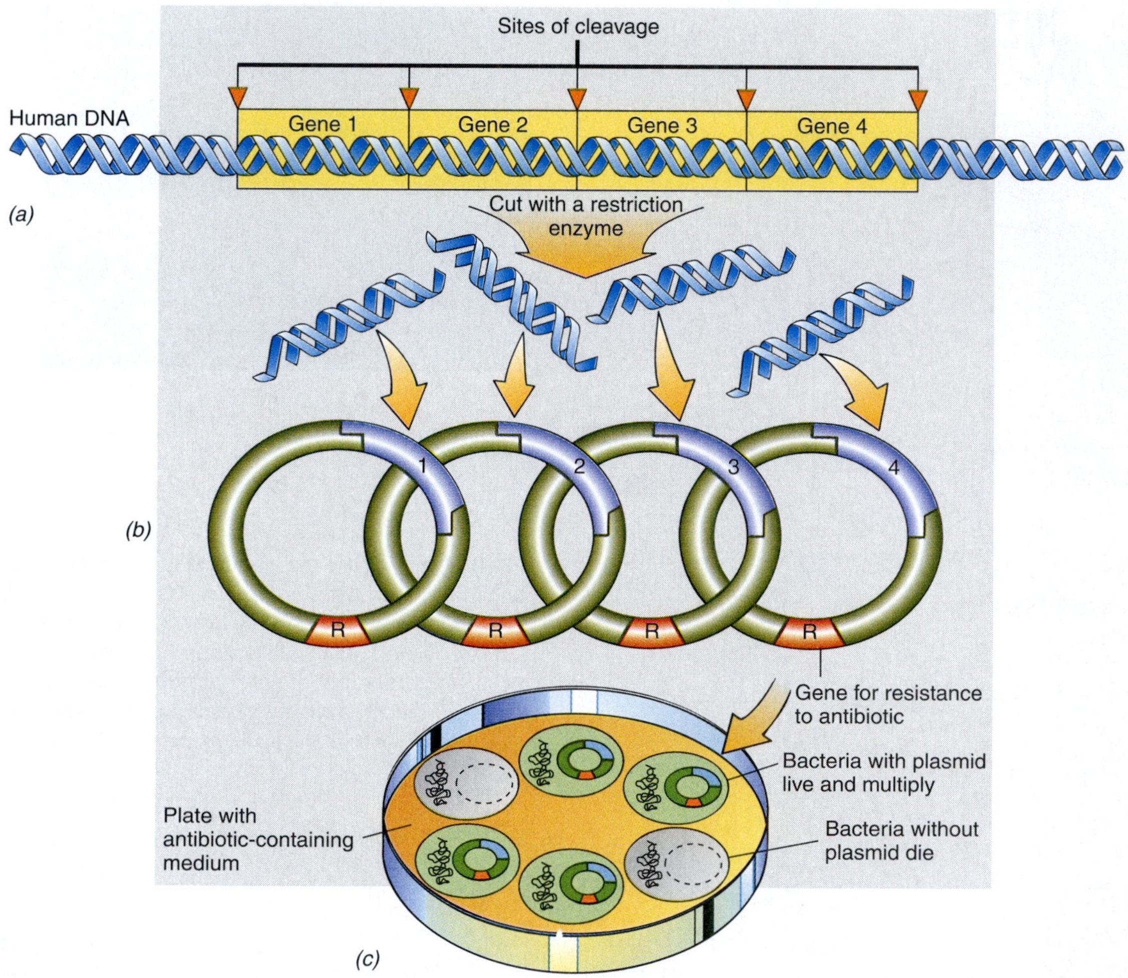

FIGURE 16–6 Identifying bacteria that have taken up genetically altered plasmids. (***a***) DNA from human cells is cut into multiple fragments with a restriction enzyme. (***b***) Recombinant plasmids are formed by cutting plasmids with the same restriction enzyme, mixing the plasmids with the segments of human DNA, and treating with DNA ligase. (***c***) Because the recombinant plasmids contain a gene for resistance to an antibiotic (R), cells of *Escherichia coli* that take up the plasmids will be resistant to that antibiotic. The bacteria are then grown on an antibiotic-containing solid nutrient medium, and only those that receive the recombinant plasmid survive.

and survive. The bacteria that have not taken up plasmids will conveniently die (because they are not resistant to the antibiotic) (Fig. 16–6).

To select the cells with the desired gene from all other genetically altered cells, the researcher may spread a sample of the bacterial culture on plates of nutrient agar, a solid growth medium. If the suspension is dilute enough, each bacterial cell will be separated from other cells on the plate. As each cell reproduces, it gives rise to a **colony,** a cluster of millions of genetically identical cells;[3] each bacterium in the colony contains the same recombinant plasmid. Because there may be as many as several thousand colonies, each containing a different portion of foreign DNA, it is now necessary to identify which colony contains the gene of interest.

[3] The cells in the colony are genetically identical because they arose from a single cell by asexual reproduction.

The Cells that Have Taken Up the Gene of Interest Are Identified

A needle could be found in a haystack of any size if one had a magnet or, better, a metal detector. Fortunately, the genetic equivalent of the metal detector can be used to find just the DNA base sequence desired among the vast

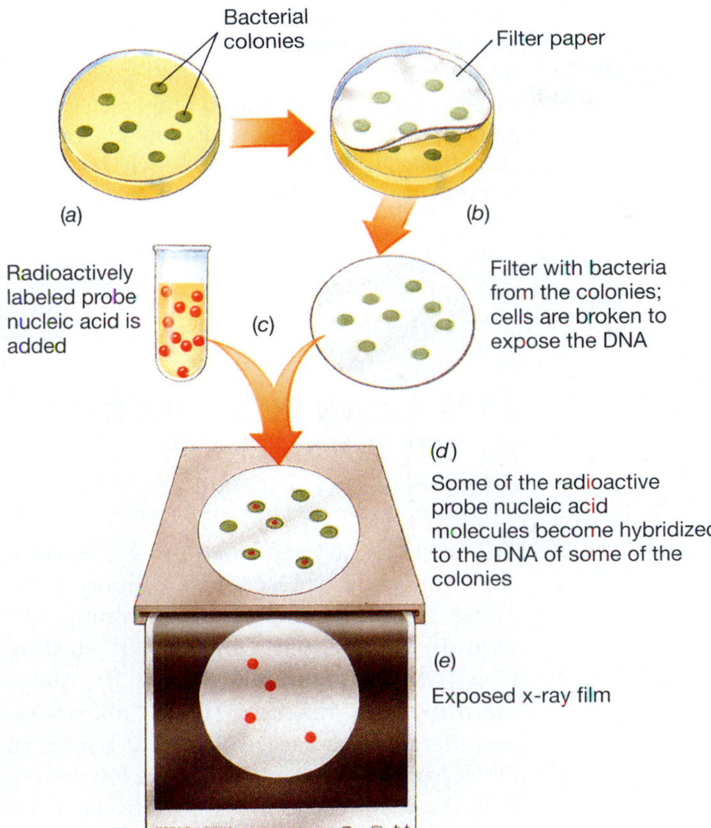
Bacterial colonies
Filter paper
(a)
(b)
Radioactively labeled probe nucleic acid is added
(c)
Filter with bacteria from the colonies; cells are broken to expose the DNA
(d)
Some of the radioactive probe nucleic acid molecules become hybridized to the DNA of some of the colonies
(e)
Exposed x-ray film

FIGURE 16–7 Genetic probes can be used to determine which cells have taken up the gene of interest. (***a***) *Escherichia coli* cells are spread on solid nutrient medium so that only one cell is found in each location. Each cell mutiplies to give rise to genetically identical descendants that form a colony on the medium. (***b***) A few cells from each colony are transferred to special filters. (***c***) Radioactively labeled probe nucleic acid (either mRNA or single-stranded DNA) is added to the filter; the probe nucleic acid contains a sequence of nucleotides complementary to the gene of interest. (***d***) Some of the probe nucleic acid hybridizes (forms base pairs) with the DNA of some of the colonies. (***e***) DNA from cells that contain the sequence complementary to the probe become radioactive and can be detected by x-ray film. The pattern of spots on the film allows the researcher to identify which colonies from the original plate contain the correct plasmid.

tangle of genes in the cell's chromosomes. This device, called a **genetic probe,** is a radioactively labeled segment of RNA or single-stranded DNA that is complementary to the target gene.

For example, suppose we wanted to identify the genes that code for the protein insulin. Because we know the amino acid sequence of insulin, we could synthesize the mRNA that produces it. This mRNA will attach itself to, or **hybridize** (attach by complementary base-pairing), with the DNA that contains the base sequence needed for production of insulin. Because the synthesized mRNA is radioactively labeled, it can be detected by x-ray film. If after such treatment any radioactive DNA can be detected in that of a particular colony of bacterial cells, this is the colony that may be able to produce the desired protein (Fig. 16–7).

The Gene Is Expressed Inside the New Cell

Even though a gene has now been isolated and identified in our example, large-scale propagation of the bacterial strain containing the gene of interest would not necessarily produce the desired protein. The gene of interest will not be transcribed unless it is associated with a set of regulatory and promoter genes. When the lactose operon control region is added to plasmids, the bacterial cells transcribe and translate the gene of interest whenever the cells are grown in the presence of lactose (Chapter 15).

Another complication relating to gene expression is that a bacterial cell may not be able to take a eukaryotic gene and use it to make a fully functional protein. Recall that many eukaryotic genes contain introns (Chapter 14); bacteria lack the enzymes that remove introns from RNA. To avoid having those parts of a gene that do not code for protein, **complimentary DNA (cDNA),** DNA that is complementary to mRNA and therefore does not contain introns, is constructed. Such cDNA molecules are made by isolating mRNA and using the enzyme **reverse transcriptase** to make DNA copies of the mRNA (Fig. 16–8).[4] Inserting cDNA into a bacterium (by the procedures discussed previously) allows fully functional protein molecules to be synthesized.

[4] Reverse transcriptase is a special type of DNA polymerase that catalyzes the synthesis of DNA using RNA as a template.

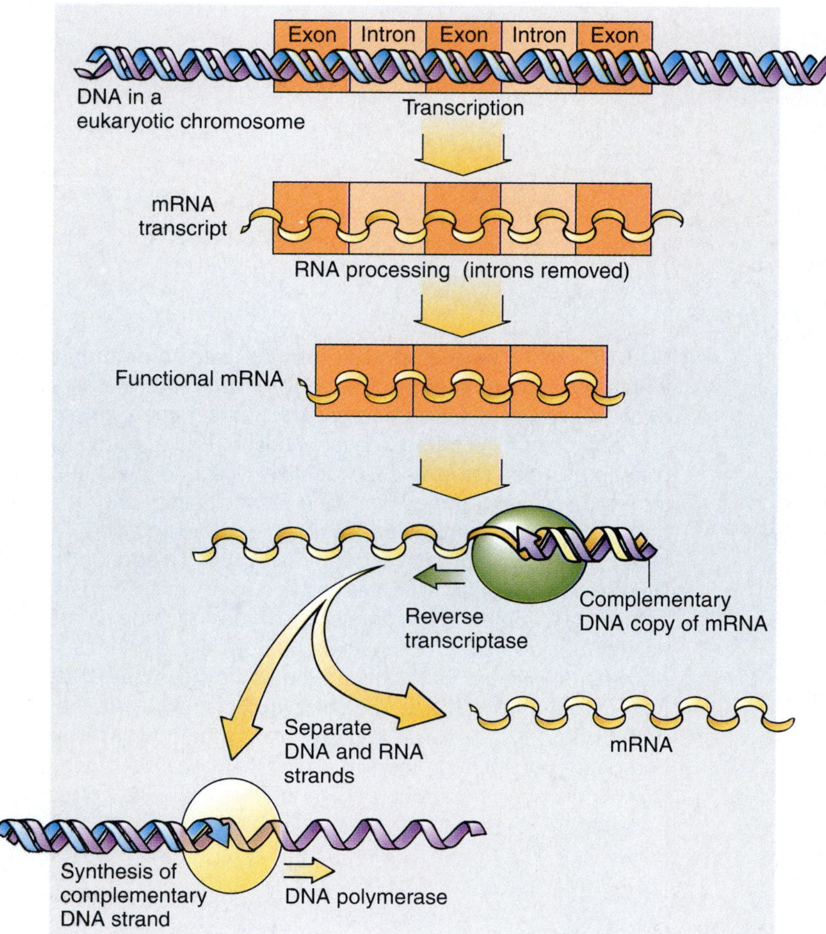

FIGURE 16–8 Construction of complementary DNA (cDNA) from eukaryotic mRNA. First, the mature mRNA is isolated from cells. Then, reverse transcriptase is used to make a copy of DNA by complementary base pairing with the mRNA. The second strand of cDNA's double helix is then assembled by DNA polymerase.

DNA CAN BE AMPLIFIED BY THE POLYMERASE CHAIN REACTION

The methods just described involve making large amounts of DNA inside living cells. These techniques are time-consuming and require an adequate amount of starting DNA. Another method, called the **polymerase chain reaction (PCR),** allows researchers to produce millions of copies of DNA from a tiny sample in just a few hours. PCR has hundreds of applications, from amplifying small amounts of DNA from crime scenes (Chapter 13), to analyzing DNA from fossil leaves and archaeological remains, to helping diagnose disease.

CONCEPT CONNECTIONS

Polymerase Chain Reaction ⟷ *Ecology*

The polymerase chain reaction (PCR) has the potential for many applications in basic biological research in addition to genetic engineering. For example, ecologists can use PCR to help survey the individual partners in symbiotic relationships (intimate associations between members of different species). One such association, known as mycorrhizae, occurs between the roots of forest trees and fungi. This relationship is widespread in nature, and the roots of most, if not all, trees are colonized by such fungi. This association is beneficial for both organisms. The plant provides the fungus with food manufactured by photosynthesis, whereas the fungus supplies the plant with essential minerals that it absorbs from the soil.

Knowing exactly which species or strain of fungus forms mycorrhizae with which tree species has a practical value. Because mycorrhizal associations enhance tree growth, deliberately inoculating tree roots of young seedlings with the specific fungus that colonizes them would cause forest trees to grow faster. The use of mycorrhizae would also help establish vegetation on disturbed or damaged lands, such as abandoned mining lands or clearcut forests.

Identifying specific species and strains of fungi found in mycorrhizal associations is often difficult (for reasons too technical to explain here). However, biologists successfully used PCR techniques to amplify DNA from a single fungus colonizing jack pine roots. They were able to demonstrate that the fungus was a different species from the one with which the jack pine had been thought to form an association. We say more about mycorrhizae and other symbiotic associations in Chapter 46.

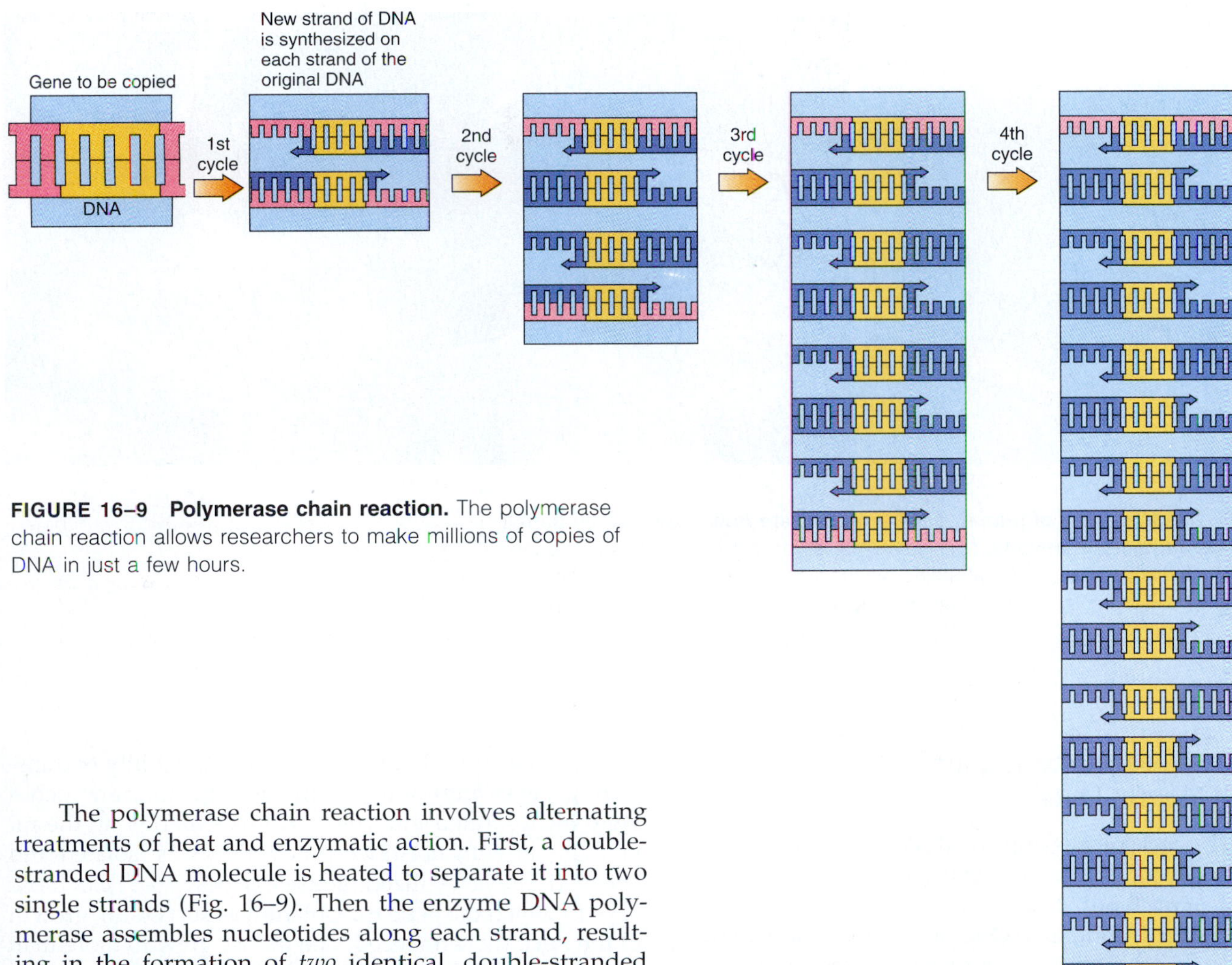

FIGURE 16–9 Polymerase chain reaction. The polymerase chain reaction allows researchers to make millions of copies of DNA in just a few hours.

The polymerase chain reaction involves alternating treatments of heat and enzymatic action. First, a double-stranded DNA molecule is heated to separate it into two single strands (Fig. 16–9). Then the enzyme DNA polymerase assembles nucleotides along each strand, resulting in the formation of *two* identical, double-stranded DNA molecules. The two double-stranded molecules are then heated, which causes them to separate into four single strands, and a second cycle of replication with DNA polymerase results in *four* identical, double-stranded DNA molecules. After the next cycle of heating and replication, there are *eight* identical DNA molecules, and so on, with the number of DNA molecules doubling with each cycle.

GENETIC ENGINEERING HOLDS GREAT PROMISE FOR THE FUTURE

Recombinant DNA technology has provided new approaches to practical problems in many fields. The production of genetically engineered proteins and organisms has already begun to have an impact on our lives, particularly in the fields of pharmaceutics and medicine. Genetic engineering offers future improvements in human health, as well as in livestock and crops.

One of the first genetically engineered proteins to be produced commercially was human insulin produced by *Escherichia coli* (Fig. 16–10). Before human insulin was available from genetically engineered bacteria, it was obtained exclusively from other animals such as cattle. Many diabetics developed allergies to insulin from animal sources because the amino acid sequence differs slightly from that in human insulin. The ability to produce human insulin has resulted in significant benefits to diabetics. Similarly, human growth hormone (a hormone needed for normal growth) and Factor VIII (a blood clotting factor needed to treat hemophilia) have been genetically engineered, and the list of such products continues to grow.

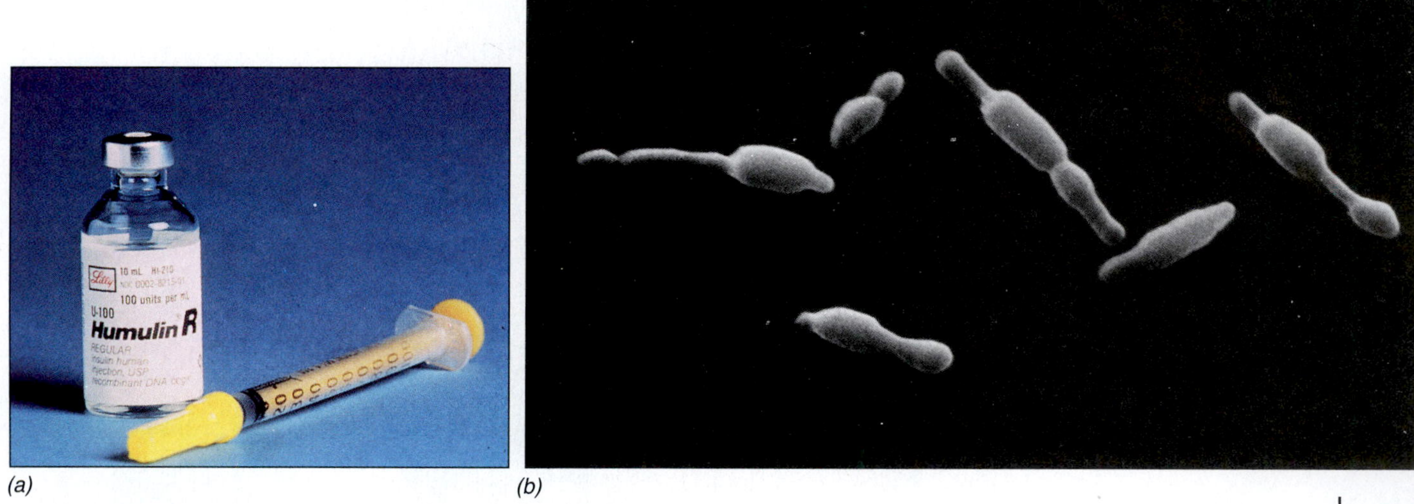

(a) (b)

FIGURE 16–10 A number of human gene products are now produced by genetically engineered microorganisms. (***a***) Human insulin for diabetics is now produced by genetically altered bacteria. (***b***) Scanning electron micrograph of *E. coli* cells "bulging" with human insulin. (*a*, Visuals Unlimited/SIU; b, Dr. Daniel C. Williams and the Lilly Microscope Laboratory)

It May Be Possible to Repair Human Genetic Defects

Human cells each possess up to 100,000 different genes. When errors occur in a particular gene, it can lead to disease. There are about 3500 different genetic disorders known to occur in humans. One of the goals of genetic engineering is gene replacement therapy, the implantation of normal genes into human cells, in the hope of correcting human genetic disease. The single gene defect that causes ADA deficiency (discussed in the chapter introduction) is an example of such a genetic disorder that can be treated by recombinant DNA technology.

Genetically Engineered Animals May Boost Agricultural Production

Two main types of genetic engineering benefit agricultural animals (livestock). One involves using recombinant DNA technology to have microorganisms produce hormones, vaccines, and other drugs for animals. For example, recombinant vaccines to protect cattle against a viral disease known as rinderpest have been developed. Rinderpest is a deadly disease that has reached epidemic proportions in parts of Asia and Africa.

A second type of genetic engineering research involves altering the genetic makeup of agricultural animals themselves. A number of researchers are attempting to develop genetically superior strains of chickens, pigs, and cattle, for example.

In an early study to determine the feasibility of transferring genes into animals, rat growth hormone genes were incorporated into mice. In this experiment, the rat gene for growth hormone production was isolated and transferred to a fertilized mouse egg (Fig. 16–11). In those eggs in which the gene transplant was successful, growth of the mice was enhanced; one mouse grew to more than double the normal size. As might be expected, such mice are also able to transmit their enhanced growth capability genetically to their offspring.

One can imagine all sorts of potentially practical applications of similar experiments in such areas as the breeding of livestock. However, sometimes such genetically engineered animals can develop side effects in addition to the expected improvements. Pigs that had been engineered to produce higher levels of bovine growth hormone gained weight faster and produced more protein and less fat. However, several unanticipated problems offset these desirable effects; for unknown reasons, many of the genetically engineered pigs developed ulcers, arthritis, and other serious health problems.

Scientists Are Using Recombinant DNA Techniques to Improve Plants

Plants have been selectively bred for thousands of years. The success of such efforts depends on the presence of pre-existing genes, either in the variety of plant being selected or in closely related wild or domesticated plants, whose traits can be transferred by crossbreeding.

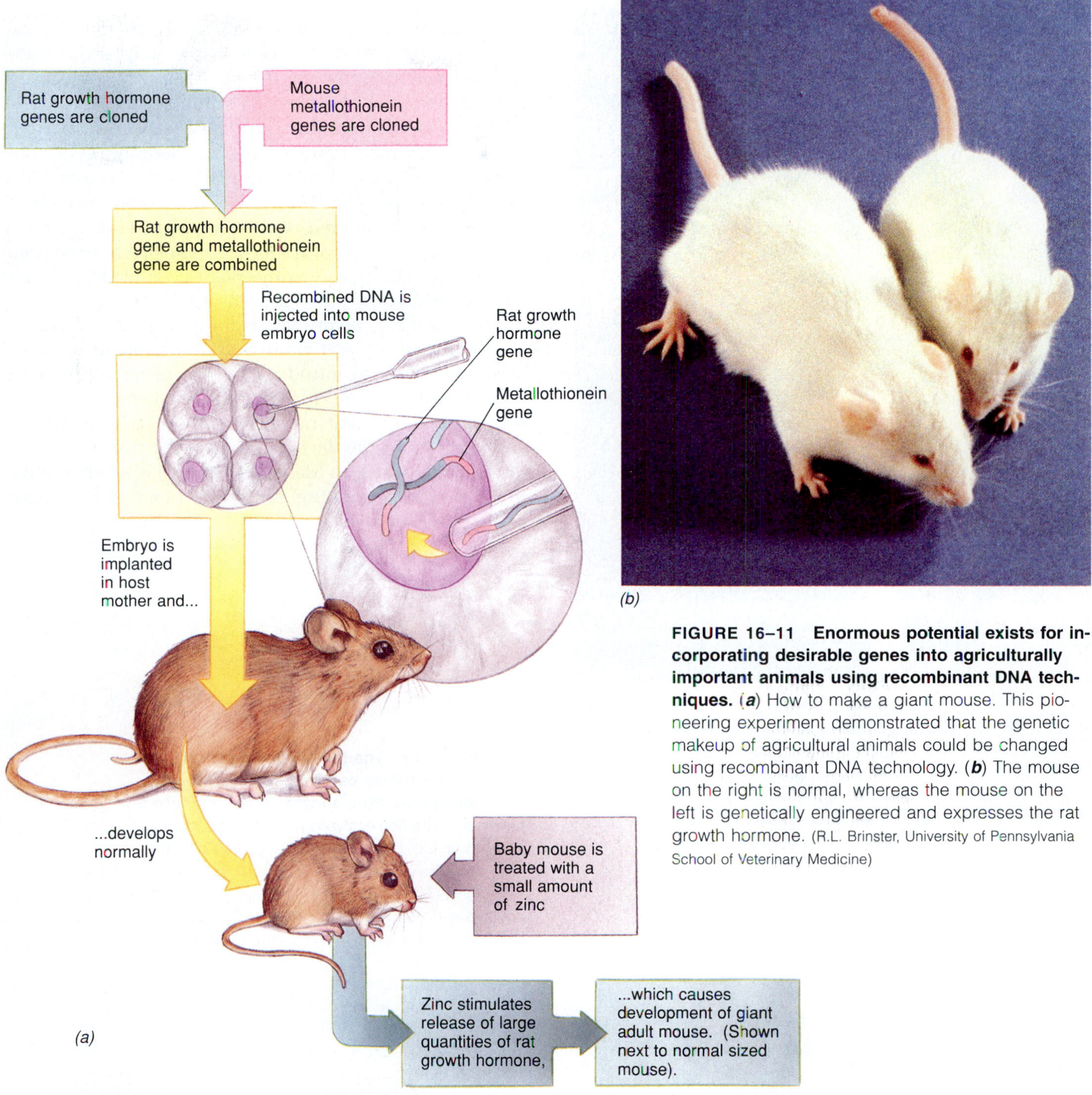
Rat growth hormone genes are cloned
Mouse metallothionein genes are cloned
Rat growth hormone gene and metallothionein gene are combined
Recombined DNA is injected into mouse embryo cells
Rat growth hormone gene
Metallothionein gene
Embryo is implanted in host mother and...
...develops normally
Baby mouse is treated with a small amount of zinc
Zinc stimulates release of large quantities of rat growth hormone,
...which causes development of giant adult mouse. (Shown next to normal sized mouse).
(a)
(b)

FIGURE 16–11 Enormous potential exists for incorporating desirable genes into agriculturally important animals using recombinant DNA techniques. (***a***) How to make a giant mouse. This pioneering experiment demonstrated that the genetic makeup of agricultural animals could be changed using recombinant DNA technology. (***b***) The mouse on the right is normal, whereas the mouse on the left is genetically engineered and expresses the rat growth hormone. (R.L. Brinster, University of Pennsylvania School of Veterinary Medicine)

Because recombinant DNA technology allows genes to be introduced into plants from strains or species with which they do not ordinarily interbreed, the possibilities for improvement are greatly increased. Research funding has been made available to plant geneticists because of the economic potential of increased plant yields, disease resistance, and more nutritious crops.

Geneticists working with plants are at greater liberty to experiment with new techniques than those working with animals, because manipulation of plant genes does not demand the same kinds of ethical considerations.

Although several vectors for the introduction of genes into plant cells have been used, the most widely used vector is the crown gall bacterium, *Agrobacterium tumefaciens,*

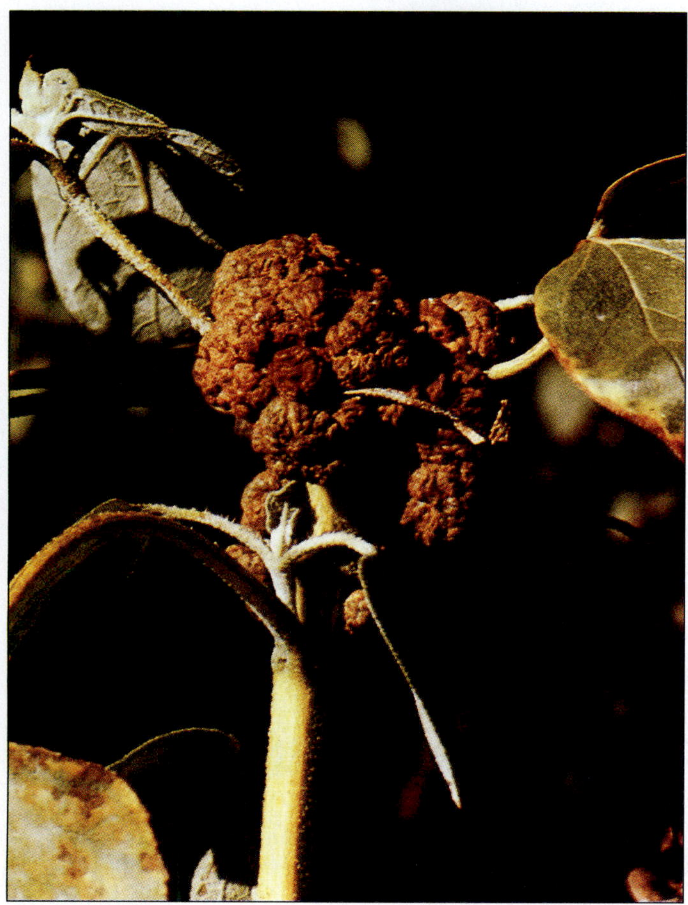

FIGURE 16–12 **Crown gall tumors on a sunflower.** The growth of these tumors is induced by a plasmid carried by *Agrobacterium tumefaciens.* The strain of *Agrobacterium* used in genetic engineering has been modified so that it contains a plasmid that does not cause tumors. (Visuals Unlimited/John D. Cunningham)

which produces plant tumors (Fig. 16–12). This bacterium causes disease by introducing a special plasmid into the chromosomes of host plant cells. The plasmid induces abnormal growth by forcing the plant cells to produce abnormal quantities of a plant growth hormone.

It is possible to disarm the crown gall plasmid so that it does not induce tumor formation. Desirable genes can then be incorporated into the plasmid for introduction into a plant host (Fig. 16–13). The plant cells into which such genetically altered plasmids are introduced are essentially normal except for the new genes that have been inserted. Genes placed in the plant cells in this fashion are transmitted sexually, via seeds, to the next generation, but they can also be propagated asexually if desired.

There are several limitations with using *Agrobacterium* plasmids to introduce foreign genes into plants. First, only dicot plants (Chapter 23) are susceptible to this bacterium; important monocot crops like corn, wheat, and rice are outside the host range of the crown gall bacterium. The gene gun (described earlier) is an alternative method of incorporating genes into plant cells without using *Agrobacterium.*

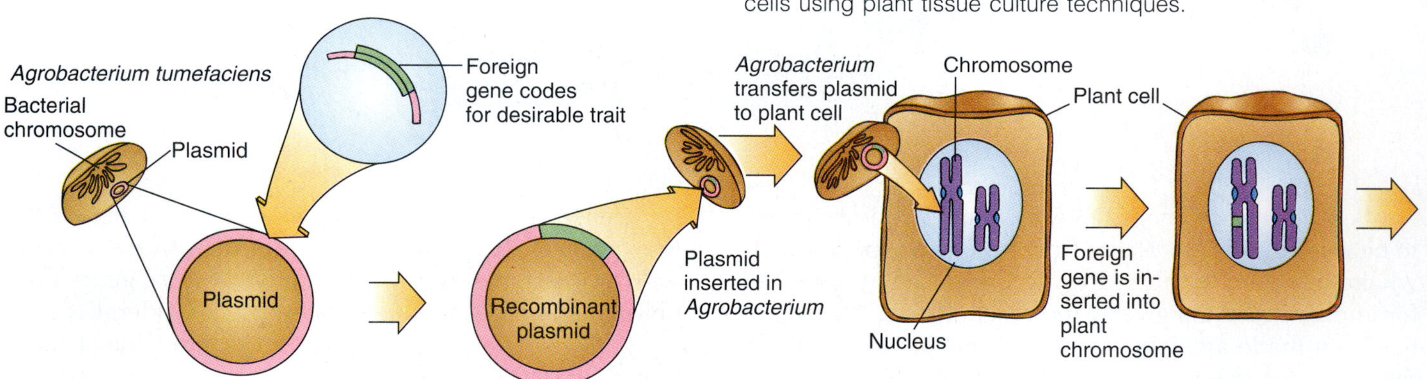

FIGURE 16–13 **The plasmid of *Agrobacterium tumefaciens* can be used to introduce desirable genes from another organism into a plant.** A restriction enzyme and DNA ligase are used to splice the foreign DNA into the crown gall plasmid. The recombinant plasmid is inserted into *Agrobacterium*, which is then used to infect plant cells in culture. The foreign gene is inserted in the plant's chromosome. Genetically engineered plants—also known as transgenic plants—are produced from the cultured plant cells using plant tissue culture techniques.

A second limitation of this method is that plant tissue culture (Chapter 30) must be used to grow multicellular plants after individual cells have been genetically transformed. Tissue culture techniques have not been developed for all plants. Therefore, recombinant DNA techniques cannot be used for plants in which tissue culture is currently not available.

An additional complication of plant genetic engineering is that it allows us to manipulate DNA in the nucleus but not DNA in other cellular organelles. A number of important plant genes are located in the DNA of the chloroplasts. Chloroplasts are essential in photosynthesis, which is the basis of plant productivity (Chapter 5). Obviously, it would be useful to develop ways to improve the genetic information that resides within the chloroplast itself. Methods of chloroplast genetic engineering are currently the focus of intense research interest.

SAFETY GUIDELINES HAVE BEEN DEVELOPED FOR RECOMBINANT DNA TECHNOLOGY

While acknowledging the potential uses of recombinant DNA techniques as important and beneficial, many scientists were initially concerned about potential misuses or unpredictable side effects. An organism might be engineered that would have undesirable ecological or other effects, not by design but by accident. Totally new strains of bacteria or other organisms, with which the world of life has no previous experience, might be difficult to control. This possibility was recognized by those who developed the recombinant DNA methods and led them to propose stringent guidelines for making the new technology safe.

Recent history has failed to bear out these worries. Millions of experiments over the past years have demonstrated that recombinant DNA experiments can be carried out safely, and no reports of such experiments causing human illness or damage to the environment have yet been described.

Scientists recognize the practical importance of recombinant DNA technology and generally agree that the perceived threat to humans and the environment was overestimated. For example, the engineering of crops resistant to pests is widely perceived as posing a minimal threat to the environment. In fact, such genetically engineered crops will *benefit* the environment because they will reduce the use of pesticides.

However, although many regulations have been relaxed, strict guidelines still exist in certain areas of recombinant DNA research where there are unanswered questions about possible effects on the environment. Much research is currently being conducted to assess the effects of introducing genetically engineered organisms into natural environments. For example, could any of the new gene sequences (new in the sense that they are novel to the organism into which they've been engineered) possibly be transferred by natural reproductive processes to closely related species living in the environment? If so, could this cause a previously harmless species to become dangerous? What if genetically altered organisms escaped human control and became established in the natural environment? Could they become "weedy" and outcompete native organisms? In the next few years, as more data accumulate on the environmental effects resulting from the release of genetically engineered organisms, we will know more clearly whether such dangers exist.

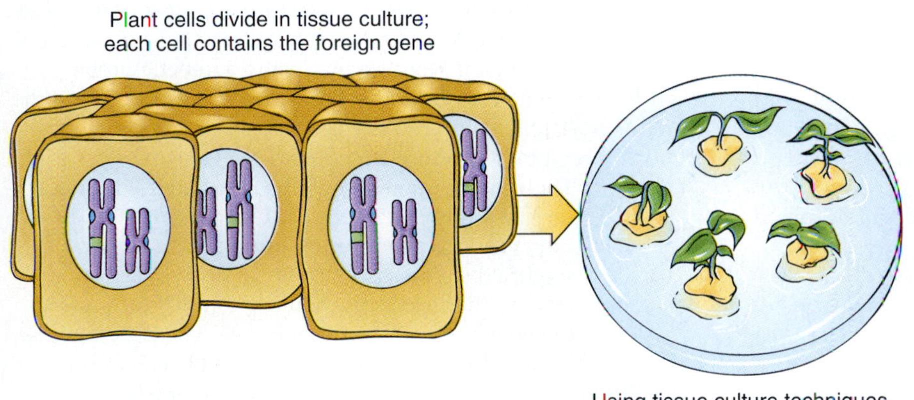
Plant cells divide in tissue culture;
each cell contains the foreign gene
Using tissue culture techniques,
cells are regenerated into plants

Transgenic plant

CONCEPT CONNECTIONS

Genetically Engineered Plants ⟷ *Ecology*

An increasing number of field tests are being conducted on **transgenic** (genetically engineered) crops; as of mid-1993 the U.S. Department of Agriculture (USDA) had issued almost 400 permits for field tests of plants as diverse as alfalfa, cantaloupe, rice, strawberry, and walnut. These and other crops have been engineered to resist insects or diseases or to be more nutritious. Many of the plants in these trials performed so well during field tests that they have demonstrable commercial value. The next step in marketing transgenic plants is deregulation, which means that the USDA will allow them to be grown as any traditional crop. Deregulation will only be approved if it is established that a particular transgenic crop poses no greater risk than the traditional crop from which it was developed.

One of the main concerns with the development of transgenic plants is the possible risk that they could become invasive (that is, become established in places where we do not want them to grow). But, while studies ascertaining whether a particular transgenic plant could become an environmental disaster were widely recognized as being needed, none were performed until recently. In June 1993, a landmark paper appeared in the journal *Nature* that reported on a rigorous field analysis of the invasiveness of a genetically engineered form of oilseed rape.* Populations of transgenic oilseed rape and normal oilseed rape (as a control) were grown for a period of three years in twelve very different environments. These habitats included the presence or absence of the following variables: a moist habitat, direct sunlight, grazing animals, plant-eating insects, other vegetation, and disease-causing fungi. This experiment, executed in the United Kingdom, is considered one of the most comprehensive studies ever performed in ecology; it required the collaboration of ecologists with many different specialties.

The results in this particular experiment demonstrated conclusively that the transgenic oilseed rape poses no risk of invasiveness; it did not grow differently than the control plants under any of the experimental conditions. Although this is good news for the people who wish to market the genetically engineered oilseed rape, the study accomplished much more than providing the green light for the deregulation of transgenic oilseed rape.

This study is significant because it shows how such tests should be designed and conducted in order to determine if the deregulation of a particular transgenic plant is desirable. In addition, it shows how ecology can be used to answer an environmental question about the safety of transgenic plants. (Ecology is presented in Chapters 45 to 49.)

* Oilseed rape is a plant in the mustard family whose seeds yield an oil.

Chapter Summary

I. Genetic engineering refers to the genetic manipulation of cells so that desirable new characteristics can be developed. Recombinant DNA technology encompasses the methods employed in genetic engineering.

II. Recombinant DNA technology enables investigators to isolate, identify, and manipulate genes.
   - A. Restriction enzymes cleave DNA at specific sites, breaking DNA into more manageable fragments. Each fragment cleaved by a restriction enzyme has sticky ends.
   - B Segments of DNA from different sources can be joined by DNA ligase, a splicing enzyme.
   - C. Many bacteria contain small, accessory, circular DNA molecules called plasmids. These plasmids may be used as vectors (DNA carriers) to transfer genes from one organism to another.
      1. Viruses are other common vectors that are particularly useful in introducing genes into mammalian cells.
      2. Another method used to introduce foreign DNA into cells is the gene gun, which shoots DNA-coated “bullets” into cells.
   - D. The recombinant DNA molecules are taken into a host cell, and the cells that have taken up the specific gene of interest are identified by using a genetic probe.
   - E. The gene may be transcribed and translated within the host cell, leading to the production of a protein not previously produced by the host organism.
      1. The gene of interest must be associated with regulatory and promoter genes.
      2. Eukaryotic genes inserted into bacteria must be modified to remove introns. To do this, cDNA is formed that is complementary to mRNA by using the enzyme reverse transcriptase. The cDNA is then introduced into a foreign cell as the gene of interest.

III. The polymerase chain reaction (PCR) provides a way to quickly and easily amplify small amounts of DNA. It has hundreds of applications.

IV. Genetic engineering has great potential.
   A. Genetic engineering has already had an impact on pharmaceutics. Genetically engineered products such as insulin, human growth hormone, and Factor VIII are commercially available.
   B. Gene replacement therapy trials to treat a number of human genetic diseases are under way.
   C. Using genetic engineering to improve agricultural animals, as well as produce hormones, vaccines, and other drugs to enhance animal growth, promises to increase the productivity of livestock.
   D. Recombinant DNA technology is being used to improve crops in a number of ways—by increasing their productivity, making them more nutritious, and developing disease resistance.

V. Concerns of scientists who developed recombinant DNA technology led to safety guidelines to make the new technology safe.
   A. Millions of recombinant DNA experiments have been performed without any adverse effects on human health or the environment.
   B. Strict guidelines still exist for the introduction of genetically engineered organisms into the environment because many unanswered questions remain.

Selected Key Terms

biotechnology, p. 284
colony, p. 288
complementary DNA (cDNA), p. 289
DNA ligase, p. 285
gene gun, p. 286
gene replacement therapy, p. 282
genetic engineering, p. 284
genetic probe, p. 289
hybridize, p. 289
palindromic, p. 285
plasmid, p. 285
polymerase chain reaction, p. 290
recombinant DNA technology, p. 284
restriction enzymes, p. 284
reverse transcriptase, p. 289
transgenic, p. 296
vector, p. 285

Post-Test

1. The methods employed in genetic engineering are termed ________________ DNA technology.
2. The enzymes used to cleave DNA at specific places are called ________________ enzymes.
3. Transferring genes from one organism to another involves a DNA carrier, or ________________.
4. Small accessory loops of DNA, called ________________, that occur in many bacteria can be used to transfer foreign genes into cells.
5. A foreign gene can be incorporated into a plasmid that has first been treated with ________________ enzymes to fragment it and produce sticky ends.
6. The plasmid (from Question 5) can be joined to other DNA fragments by means of a splicing enzyme, DNA ________________.
7. Combining a DNA fragment from one organism with a vector DNA molecule such as a plasmid results in a ________________ DNA molecule.
8. Plasmids typically contain genes for resistance to ________________.
9. A ________________ is a population of genetically identical bacteria grown from a single cell in culture.
10. A ________________ ________________ is a radioactively labeled segment of RNA or single-stranded DNA that is complementary to a specific gene.
11. A genetic probe ________________, or attaches itself by complementary base-pairing, to a target gene.
12. DNA that is complementary to mRNA and therefore does not contain introns is known as ________________.
13. The ________________ ________________ ________________ allows researchers to produce millions of copies of DNA from a tiny sample in just a few hours.
14. ________________ ________________ therapy is the implantation of normal genes into human cells in the hope of correcting human genetic disease.
15. The ________________ ________________ disease has provided biologists with a vector capable of introducing recombinant genes into plants.
16. Which of the following DNA sequences would be a palindromic with its complementary strand?
(a) GACTCA (b) CTTAAG (c) TTAGCA (d) ACCTCC
17. The common intestinal bacterium that is so important in genetic engineering is: (a) *Escherichia coli* (b) *Homo sapiens* (c) *Agrobacterium tumefaciens* (d) plasmid.

18. In order for two DNA molecules to form a recombinant DNA molecule, they must be: (a) obtained from the same organism (b) the same number of nucleotides in length (c) treated with the same restriction enzyme so they have complementary sticky ends.
19. Recombinant DNA technology uses ______________ and ______________ as biological vectors. (a) bacteria, fungi (b) eukaryotic cells, prokaryotic cells (c) PCR, cDNA (d) plasmids, viruses
20. The polymerase chain reaction: (a) cleaves DNA at specific nucleotide sequences (b) splices DNA fragments together (c) makes a cDNA molecule complementary to mRNA (d) uses a tiny amount of DNA to make millions of copies in a test tube.

Review Questions

1. What is genetic engineering? Give some specific examples of improvements biologists can produce in living things.
2. What are restriction enzymes? How are they employed in recombinant DNA research?
3. Explain how recombinant DNA molecules are usually constructed.
4. Describe how a human gene is implanted in a bacterial cell. What might be gained by this procedure?
5. Explain how a foreign gene might be inserted into a plant cell.
6. Distinguish between a eukaryotic gene and cDNA.
7. What is a genetic probe? How is it used to identify genetically modified cells that contain a specific gene?
8. Briefly describe the PCR method to amplify DNA.
9. Why does research involving recombinant DNA technology have strict safety guidelines?

Thinking Critically

1. One of the ethical concerns about manipulating the genes of unborn humans, which may become possible as a result of advances in gene replacement therapy, is over human rights: Does an unborn human have the right to inherit an unaltered set of genes? Think about both sides to this issue, take a stand, and present arguments to support your view.
2. Would genetic engineering be possible if restriction enzymes did not exist? Explain.
3. Think of at least two potential benefits of genetic engineering *not* discussed in this chapter.
4. Think of an environmental problem *not* discussed in this chapter that might be associated with introducing genetically altered organisms into the natural environment. How would you design an experiment to assess the risk involved?
5. What safeguards could be employed in recombinant DNA technology to guard against potential harm and yet not excessively hamper research?

Recommended Readings

Arnheim, N., T. White, and W. Rainey: "Application of PCR: Organismal and Population Biology," *BioScience* Vol. 40, No. 3, March 1990. Some of the biological applications of the polymerase chain reaction.

Barton, J.: "Patenting Life," *Scientific American,* March 1991. Explores the major issues surrounding biotechnology patents.

Beardsley, T.: "From Mice to Men: The Burgeoning Business of Gene Therapy," *Scientific American,* December 1993. Provides an overview of the many clinical trials currently under way that use gene replacement therapy.

Franklin-Barbajosa, C.: "DNA Profiling: The New Science of Identity," *National Geographic* Vol. 181, No. 5, May 1992. Examines some of the varied uses of DNA fingerprinting.

Gasser, C. and R. Fraley: "Transgenic Crops," *Scientific American,* June 1992. Methods used to genetically engineer crops to resist pests and spoilage.

Hoffman, C.: "Ecological Risks of Genetic Engineering of Crop Plants," *BioScience* Vol. 40, No. 6, June 1990. This article and others in this issue of *BioScience* are concerned with possible environmental risks associated with the development of genetically engineered crop plants.

Wrubel, R., S. Krimsky, and R. Wetzler: "Field Testing Transgenic Plants," *BioScience* Vol. 42, No. 4, April 1992. Examines the U.S. Department of Agriculture's supervision of small-scale field tests of transgenic plants.

PART 5

Evolution: Concepts and Processes

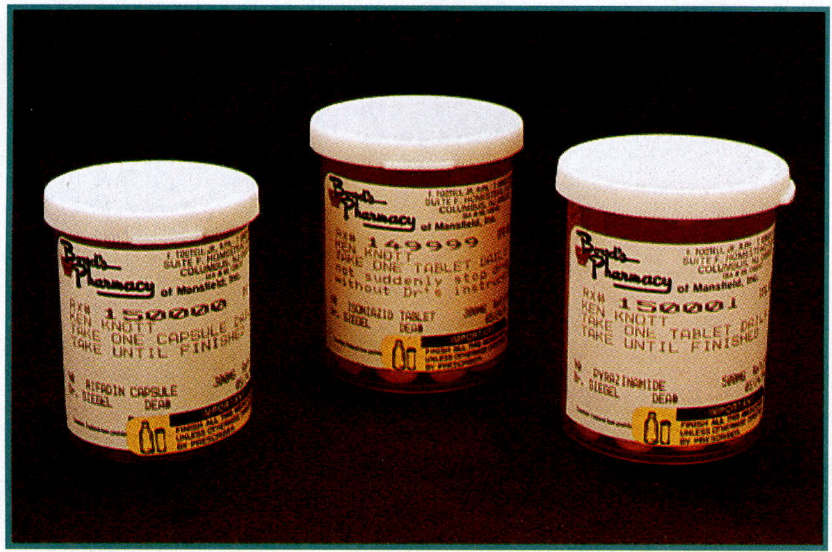

Three different medicines must be taken faithfully for at least six months to successfully treat tuberculosis. *(Kenneth Knott/Fine Light Photography)*

THE NEW TUBERCULOSIS: EVOLUTION OF A KILLER

Beginning in the late 1980s, an alarming increase in tuberculosis (TB) has been documented by the U.S. Centers for Disease Control in Atlanta. The number of cases of TB had declined in the United States during the past thirty or so years, largely as a result of successful treatments with antibiotics (drugs intended to harm or kill bacteria and other microorganisms).

Although many people are exposed to the bacteria that cause TB, only people who are very young, very old, or weakened from some other disease usually exhibit symptoms. One of the reasons for the current outbreak of TB is the HIV virus (the virus that causes AIDS), which incapacitates the immune system. Patients who are infected with HIV and who formerly had symptom-free TB now get deadly infections.

TB is also disturbing because drug-resistant strains of the bacteria that cause TB have recently developed. These strains are unaffected by one or more of the antibiotics traditionally used to treat TB. Drug-resistant TB is deadly: 80% of the people infected with multidrug-resistant TB (MDR-TB) die within two months of diagnosis, even with medical care.

How has bacterial resistance to antibiotics come about? When antibiotics began to be used to treat human and animal infections, it was thought that these drugs would eliminate bacterial diseases. This has not taken place, however. Bacteria are constantly evolving, even inside the lungs of human hosts. Each time an antibiotic is used to treat a bacterial infection, for example, some bacteria survive. The survivors, because of certain genes they have acquired,[1] are genetically resistant to the antibiotic, and they pass this trait on to future generations. As a result, the bacterial population contains a larger percentage of drug-resistant bacteria than before.

Drug resistance is usually found in patients who were previously treated for TB, and quite often human behavior is a factor in the development of drug resistance. A person with TB must take three to ten pills each day for at least six months. After the first week or two of treatment, the person usually feels better; many patients decide to quit taking their pills at this point. When this happens, the TB bacteria still lurking in their bodies—those with a resistance to the antibiotic formerly used—rally. The evolution of a strain of bacteria resistant to *several* drugs is a worst-case scenario. MDR-TB is extremely difficult to treat effectively and, as mentioned previously, is often fatal.

Evolution—any cumulative genetic change in a population of organisms from generation to generation—is exemplified by the changes in the TB bacteria that have caused our current health problem. Evolution is the central theme of biology and links all fields of the life sciences. Biologists attempt to understand all living organisms, from TB bacteria to daisies to sperm whales, in the context of evolution. As you will discover in this chapter, evolutionary theory helps us to make sense of the enormous variety in the living world. Although biologists no longer question that evolution has occurred and is occurring, they continue to study and actively debate the actual mechanisms that control it.

[1] Bacteria are very proficient at acquiring new genes from other organisms. Under certain conditions, genes can be passed from one bacterium to another, or even from a virus to a bacterium. The new genes may give them characteristics that shield them from the defensive cells of the immune system.

Darwin and Natural Selection

17

Learning Objectives

After you have studied this chapter you should be able to:

1. Discuss the historical development of the theory of evolution.
2. Explain the four premises of evolution by natural selection as proposed by Darwin.
3. Explain how the synthetic theory of evolution differs from Darwin's original theory of evolution.
4. Summarize the evidence for evolution from the fossil record.
5. Summarize some of the evidence supporting evolution obtained from the field of comparative anatomy.
6. Define and give examples of vestigial, homologous, and analogous structures.
7. Explain how developmental biology provides insights into the evolutionary process.
8. Define biogeography and summarize the types of evidence it provides for evolution.
9. Describe some of the evidence for evolution obtained from the fields of biochemistry and molecular biology.
10. Relate how scientists make inferences about evolutionary relationships from the sequence of amino acids in specific proteins or the sequence of nucleotides in particular genes in organisms.

Key Concepts

- ☐ Evolution is a genetic change in a population of organisms that occurs over time.
- ☐ Charles Darwin proposed that evolution occurs gradually by natural selection, in which better-adapted organisms (those with a combination of traits better suited to prevailing environmental conditions) are more likely to survive and reproduce, thereby increasing their proportion in the population. Over time the population changes, as favorable traits increase in frequency in successive generations and less favorable traits decrease or disappear.
- ☐ A vast assemblage of scientific evidence supports evolution, including observations from the fossil record, comparative anatomy, biogeography, developmental biology, and biochemistry and molecular biology.

EVOLUTION EXPLAINS HOW PRESENT-DAY ORGANISMS ORIGINATED

All of the life forms on our planet, from microscopic bacteria to giant blue whales, from tropical tree frogs to desert cacti, descended from one or perhaps only a few simple kinds of organisms. This vast diversity of species developed from earlier species by a process Darwin called "descent with modification," or evolution. In biology, **evolution** may be defined as genetic changes in a population of organisms that occur over time. It does not refer to the changes that occur in an individual organism within its lifetime, but to the changes in populations over many generations.

Another way to describe evolution is to say that it involves changes in the frequencies of certain alleles in a population (Chapter 18). All of the alleles of all of the genes present in a population at a given time are known as its **gene pool.** Evolution is a change in the allele frequencies within a gene pool over time.

THE CONCEPT OF EVOLUTION ORIGINATED BEFORE CHARLES DARWIN

Although Charles Darwin is universally associated with evolution, ideas of evolution predate Darwin by centuries. Aristotle (384–322 B.C.) saw much evidence of design and purpose in nature and arranged all of the organisms that he knew in a "Scale of Nature" that extended from the very simple to the most complex. He visualized living organisms as being imperfect but moving toward a more perfect state. Some have interpreted this concept as the forerunner of evolutionary theory, but Aristotle is very vague on the nature of this "movement toward the more perfect state" and certainly did not propose any notion of evolution by natural selection. Furthermore, modern evolutionary theory now recognizes that evolution does *not* move toward more perfect states, nor necessarily toward greater complexity.

Long before Darwin, fossils—odd fragments resembling bones, teeth, shells, and leaves—had been discovered embedded in rocks. Some of these corresponded to parts of familiar living animals and plants, but others were strangely unlike any known form. Fossils of marine invertebrates (animals without backbones) were found in rocks high on mountains. Leonardo da Vinci correctly interpreted these finds in the 15th century as the remains of animals that had existed in previous ages but had become extinct.

The most thoroughly considered view of evolution before Charles Darwin was expressed by Jean Baptiste de Lamarck in 1809. Like most biologists of his time, Lamarck thought that all living things were endowed with a vital force that drove them to change over time toward greater complexity. He also thought that organisms could pass on traits acquired during their lifetimes to their offspring. As an example of this line of reasoning, Lamarck suggested that the long neck of the giraffe developed when a short-necked ancestor took to browsing on the leaves of trees instead of grass (Fig. 17–1). Lamarck speculated that the ancestral giraffe, in reaching up, stretched and elongated its neck. Its offspring, inheriting the longer neck, stretched still further. This process, repeated over many generations, supposedly caused the present-day long necks of modern giraffes.

The mechanism for Lamarckian evolution was an "inner drive" for self-improvement. This concept was discredited when the mechanisms of heredity were discovered. Lamarck's contribution to science is important, however, because he was the first to propose that organisms undergo change over time as a result of some natural phenomenon rather than divine intervention. It remained for Charles Darwin to discover the actual mechanism of evolution—natural selection.

DARWIN WAS INFLUENCED BY MANY DIFFERENT IDEAS WHEN HE MADE HIS VOYAGE

As a young man, Charles Darwin (1809–1882) was appointed naturalist on the ship *H.M.S. Beagle,* which was taking a five-year cruise around the world to prepare navigational charts for the British navy (Fig. 17–2). The *Beagle* left Plymouth, England, in 1831 and cruised slowly down the east coast and up the west coast of South America. While other members of the company mapped the coasts and harbors, Darwin had an opportunity to study the animals, plants, fossils, and geological formations of both coastal and inland regions, areas that had not been extensively explored. He collected and cataloged thousands of specimens of plants and animals and kept notes of his observations. He experienced first-hand the diverse variety of the plants and animals of these regions.

The *Beagle* spent some time at the Galapagos Islands, 965 kilometers (600 miles) west of Ecuador, where Darwin continued his observations and collections (Fig. 17–3). He compared the animals and plants of the Galapagos with those of the South American mainland. He was particularly impressed by their similarities and wondered why, for example, the plants and animals of the

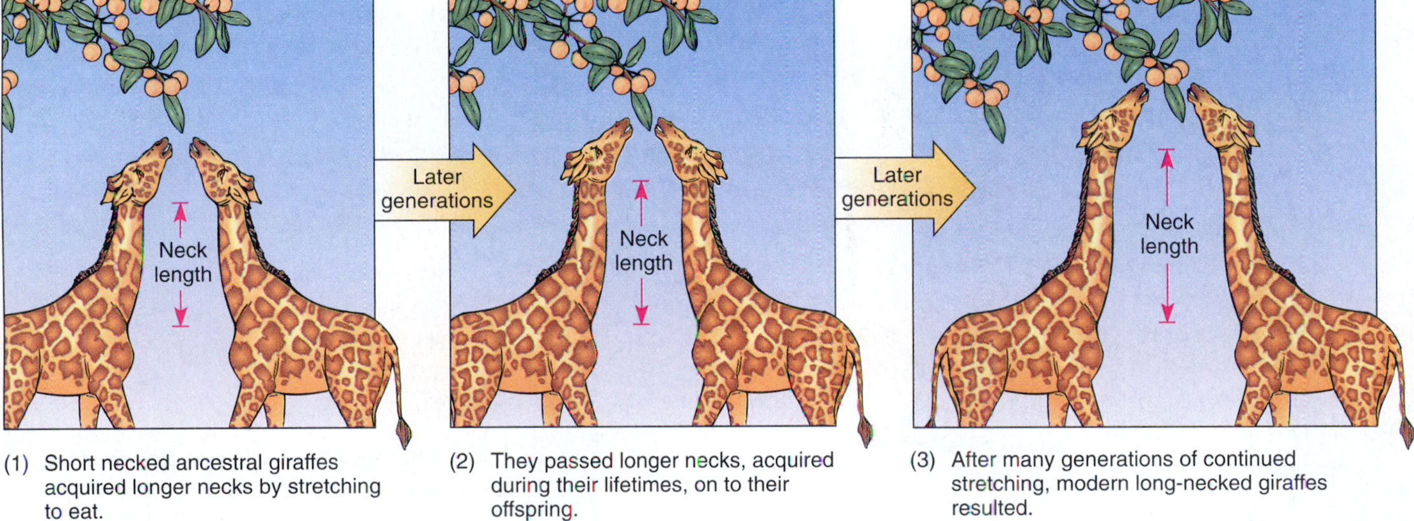

Lamarck's hypothesis
Neck length
Later generations
Neck length
Later generations
Neck length
(1) Short necked ancestral giraffes acquired longer necks by stretching to eat.
(2) They passed longer necks, acquired during their lifetimes, on to their offspring.
(3) After many generations of continued stretching, modern long-necked giraffes resulted.

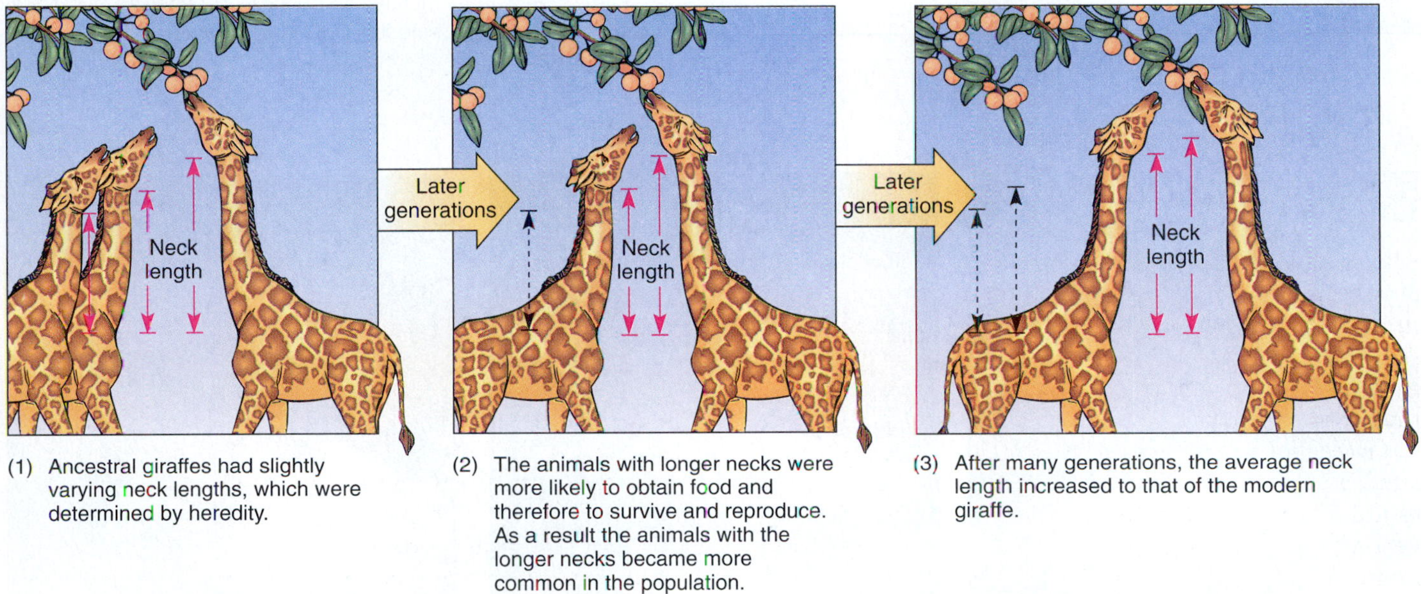

(1) Ancestral giraffes had slightly varying neck lengths, which were determined by heredity.

FIGURE 17–1 How did the giraffe get its long neck? A comparison of Lamarck's (***a***) and Darwin's (***b***) hypotheses on the evolution of giraffes. Scientific evidence supports Darwin's hypothesis. Although Lamarck's mechanism of evolution through the inheritance of acquired characteristics was incorrect, he was the first scientist to propose that organisms undergo evolution by natural means.

Galapagos should resemble those from South America more than those from other islands in different parts of the world. Moreover, although there were similarities between the Galapagos and South American species, there were also distinct differences. There were even differences in the birds and reptiles from one island to the next! Darwin pondered these observations and tried to develop a satisfactory explanation for their distribution.

The general notion in the mid-1800s was that organisms did not change significantly over time, that they

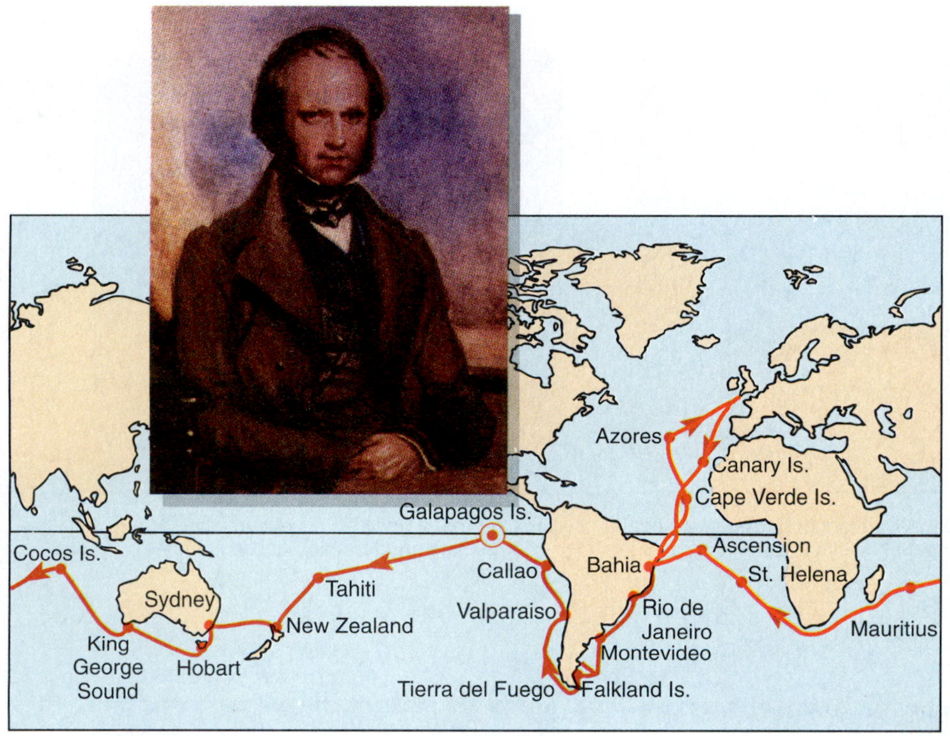

FIGURE 17–2 Charles Darwin as a young man. This portrait was painted shortly after Darwin returned to England from his voyage around the world (see map) on the *H.M.S. Beagle*. Observations made during this voyage helped him formulate the concept of evolution by natural selection. (William E. Ferguson)

(a)

(b)

(c)

FIGURE 17–3 Three species of Galapagos Island finches, birds that apparently evolved from a common ancestral population of seed-eating birds from South America. The 14 known species of Galapagos finches are variously specialized for a variety of lifestyles that are elsewhere filled by birds of different species which never had the opportunity to colonize the Galapagos Islands. The likely evolution of such different birds from a common ancestor suggested to Darwin that species originated by natural selection. (***a***) Cactus finch, *Geospiza scandens*. The cactus finch feeds on the fleshy parts of cacti, including flowers. (***b***) A large ground finch, *Geospiza magnirostra*. This bird has an extremely heavy nutcracker-type bill adapted for eating heavy-walled seeds. (***c***) Woodpecker finch, *Camarhyncus pallidus*. This bird has insectivorous habits similar to those of woodpeckers but lacks the complex beak and tongue adaptations that permit woodpeckers to reach their prey. The adaptations of the woodpecker finch to this lifestyle are almost entirely behavioral: it digs insects out of bark and crevices using cactus spines, twigs, or even dead leaves. (*a* and *b*, Frans Lanting/ Minden Pictures; c, Miguel Castro/ Photo Researchers, Inc.)

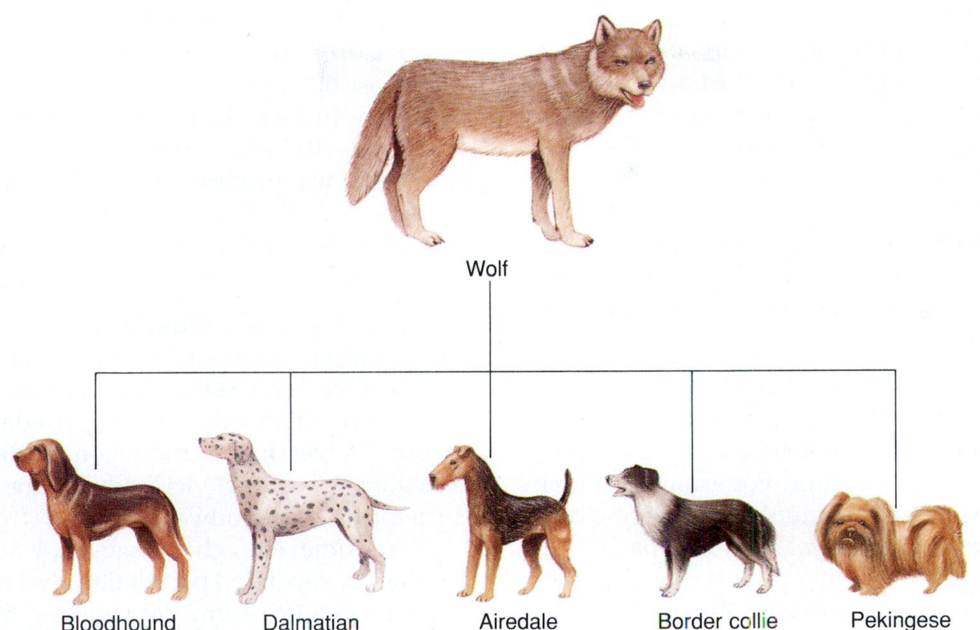

FIGURE 17–4 Variation in dogs as a result of artificial selection. Airedales, bloodhounds, border collies, dalmatians, and Pekingese are some of the numerous dog varieties that have been produced by artificial selection over a period of thousands of years. It is likely that the ancestor of domestic dogs was the timber wolf (*Canus lupus*).

looked the same as the day they were created. There were some obvious exceptions to this idea, however. For example, breeders could induce a great deal of variation in domesticated plants and animals in just a few generations. They did this by selecting desirable traits and breeding only those individuals that possessed the desired traits, in a procedure known as **artificial selection** (Fig. 17–4).

Also, fossil evidence was beginning to contradict the accepted views, as an increasing number of fossil specimens were discovered that did not have living counterparts. Then, too, geological evidence suggested that the Earth was far older than had been previously thought. During the early 19th century, geologists such as Charles Lyell advanced the idea that mountains, valleys, and other physical features of Earth's surface were not created in their present form. Instead, they were formed over long periods of time by the slow geological processes of volcanic activity, uplift, erosion, and glaciation, processes that still occur today. The slow pace of these geological processes indicated that the Earth was very, very old.

The ideas of Thomas Malthus, a clergyman and economist (1766–1834), were another important influence on Darwin. Malthus noted that populations increase in size geometrically ($2 \longrightarrow 4 \longrightarrow 8 \longrightarrow 16$) and thus outstrip the food supply, which increases arithmetically ($1 \longrightarrow 2 \longrightarrow 3 \longrightarrow 4$). In the case of humans, Malthus suggested that famines, wars, and disease served as the inevitable controls on the growth of human populations.

DARWIN DEVELOPED A MECHANISM FOR EVOLUTION: NATURAL SELECTION

Darwin's years of observing the habits of animals and plants had introduced him to the struggle for existence described by Malthus. It occurred to Darwin that, in this struggle, favorable variations would tend to be preserved and unfavorable ones eliminated. As a result, a population would adapt to the environment and, eventually, the accumulation of modifications might cause the origin of a new species. Time was the remaining factor required in order for new species to develop, and the geologists of the era, including Lyell, had supplied evidence that the Earth was indeed old enough to provide an adequate amount of time.

Darwin had at last developed a working theory of evolution, that of evolution by natural selection. He spent the next 20 years accumulating an immense body of evidence to support his theory, and formulating his arguments for natural selection.

As Darwin was pondering his ideas, Alfred Wallace, who was studying the plants and animals of Malaysia and Indonesia, was similarly struck by the diversity of living things and the peculiarities of their distribution. Wallace also arrived at the conclusion that evolution occurred by natural selection. In 1858, he sent a brief essay to Darwin, who was by then a world-renowned biologist, asking his opinion.

Darwin's friends persuaded him to have Wallace's paper presented along with an abstract of his own views, which he had prepared and circulated to a few friends several years earlier. Both papers were presented in London at a meeting of the Linnaean Society in July, 1858. Darwin's monumental book, *Origin of Species by Means of Natural Selection,* was published in November, 1859.

Darwin's mechanism of evolution by **natural selection** consists of four observations about the natural world: overproduction, variation, limits on population growth, and survival to reproduce.

1. **Overproduction:** Each species produces more offspring than will survive to maturity. Natural populations have the reproductive potential to geometrically increase their numbers over time. For example, if each breeding pair of elephants produces six offspring during its 90-year lifespan, in 750 years a single pair of elephants will have given rise to a population of 19 million! Yet elephants have not overrun the planet.
2. **Variation:** The individuals in a population exhibit variation in their traits (Fig. 17–5). Some of these traits improve the chances of an individual's survival and reproductive success, whereas other traits do not. It is important to remember that the variation necessary for evolution by natural selection must be inherited and therefore able to be passed on to offspring. (Although Darwin recognized the importance of variation, he did not know about its genetic basis.)
3. **Limits on population growth:** There is only so much food, water, light, growing space, and so on, available to a population, and organisms compete with one another for the limited resources available to them. Because there are more individuals than the limited resources of the environment can support, not all will survive to reproductive age. Other limits on population growth include predators and disease organisms.
4. **Survival to reproduce,** or **"survival of the fittest":** Those offspring that possess the most favorable combination of characteristics will be most likely to survive and reproduce, passing their traits on to the next generation.

FIGURE 17–5 Variation in offspring. A mother cat with her three kittens, which exhibit variation in coat color that is encoded in their genes. One of the premises upon which natural selection is based is that sexual reproduction results in offspring that are not identical to one another. (Animals Animals © 1995 Fritz Prenzel)

Natural selection thus causes an increase of favorable alleles and a decrease of unfavorable alleles within a population. Over succeeding generations, individual members of a population become better adapted to local conditions. Successful reproduction is the key to natural selection: the "fittest" individuals are those that reproduce most successfully.

Over time, enough changes may accumulate in geographically separated populations (with slightly different environments) to form new species from the ancestral species. Darwin noted that the Galapagos finches evolved in this way. The different islands in the Galapagos kept the finches isolated from one another, thereby allowing them to evolve into separate species. The evolution of new species is considered in greater detail in Chapter 18.

THE SYNTHETIC THEORY OF EVOLUTION COMBINES DARWIN'S THEORY WITH MENDELIAN GENETICS

Darwin's theory of evolution by natural selection was based on the assumption that individuals pass traits on to the next generation. Darwin was a contemporary of Gregor Mendel, who worked out the basic patterns of inheritance (Chapter 11). However, Darwin was apparently not acquainted with Mendel's work, which wasn't recognized by the scientific establishment until the early part of the twentieth century.

In the 1920s, Darwin's theory and Mendelian genetics were joined with other developments in genetics to form a unified explanation of evolution known as **neo-Darwinism** or, more commonly, the **synthetic theory of evolution.** (*Synthesis* in this context refers to the putting together of parts of several previous theories to form a whole.)

The synthetic theory of evolution explains Darwin's observation of variation among offspring in terms of mutation; that is, mutation provides the genetic variability that natural selection acts upon during evolution (Chapter 14). The synthetic theory of evolution, which emphasizes the genetics of *populations* (rather than individuals) as the central focus of evolution, has held up well since it was formulated (Chapter 18). It has dominated the thinking and research of many biologists and has resulted in an enormous accumulation of scientific evidence for evolution.

Most biologists accept the basic principles of the synthetic theory of evolution, but recently have scrutinized some of its features. For example, what is the role of chance in evolution? How rapidly do new species develop? These questions have arisen in part from a reevaluation of the fossil record and in part from discoveries in molecular aspects of inheritance. These debates are an integral part of the scientific process because they stimulate additional observation and experimentation as well as rethinking of older evidence. Science is a continuing process, and information obtained in the future may require us to modify certain parts of the synthetic theory of evolution.

MANY TYPES OF SCIENTIFIC EVIDENCE SUPPORT THE THEORY OF EVOLUTION

Evolution by natural selection is supported by an enormous body of observations and experiments. In this text, we can report only a small fraction of this wealth of evidence that is found in the fossil record and in living organisms. Although biologists still do not agree completely on some aspects by which evolutionary changes occur, the concept that evolution by natural selection has taken place is now well documented. It is consistent with all of the evidence that has been brought to bear upon it.

Fossils Indicate that Evolution Occurred in the Past

Perhaps the most direct evidence for evolution comes from the discovery, identification, and interpretation of fossils. **Fossils** (from the Latin word *fossilis,* meaning "something dug up") are the remains or traces left by previous organisms in rock strata (Fig. 17–6). Although most fossils are preserved in rock strata, some more recent remains have been exceptionally well preserved by being embedded in bogs, tar, amber (ancient tree resin), or ice.

Fossils provide a record of animals and plants that lived earlier, some understanding of where and when they lived, and an idea of the lifestyles that they had. When enough fossils of organisms of different geological ages have been found, we can infer the lines of evolution that gave rise to those organisms; sometimes fossils provide direct evidence of the origin of new species from preexisting species, including extinct intermediate forms. Thus, the fossil record, although incomplete, provides a history of life.

The formation and preservation of a fossil requires that an organism be buried under conditions that will slow the process of decay. This is most likely to occur if an organism's remains are covered quickly by fine particles of soil suspended in water. The soil particles are deposited as sediment and surround the animal or plant. Remains of aquatic organisms may be trapped in bogs, mud flats, sand bars, or deltas. Remains of terrestrial organisms that lived on a flood plain may also be covered by water-borne sediments or, if the organism lived in an arid region, by wind-blown sand. Over time, the sediments harden to form sedimentary rock, and the organism's remains are usually replaced by minerals.

Layers of sedimentary rock occur naturally in the sequence of their deposition, with the more recent layers on top of the older, earlier ones. It is therefore possible to distinguish older fossils from more recent ones, just by observing the layers in which they are found. To determine

(a)

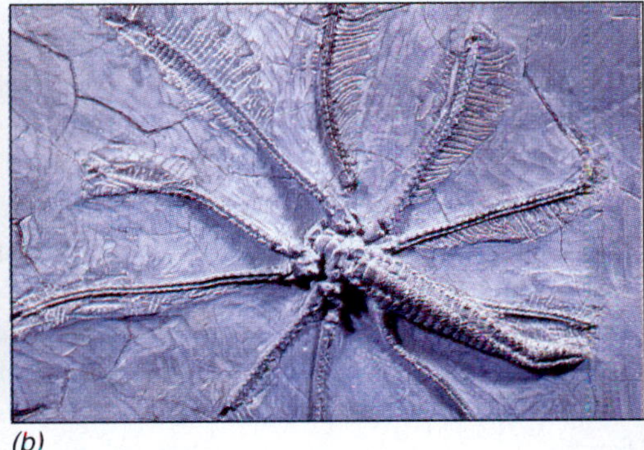

(b)

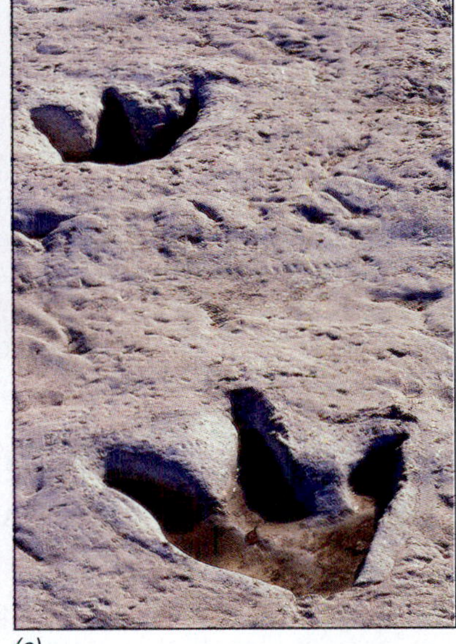

(c)

FIGURE 17–6 Several types of fossils. (***a***) A seed fern leaf. Seed ferns are not related to ferns but are an extinct group of gymnosperms. (***b***) A fossil crinoid. The fossil record indicates that crinoids, echinoderms that spend most of their lives attached in one spot, were far more common in ancient seas than they are today. (***c***) Dinosaur foot prints, each 75 to 90 cm (2.5–3 ft) in length, occur in sedimentary rock in Texas. Dinosaur footprints provide clues about the locomotion, behavior, and ecology of these extinct animals. (*a,* Carolina Biological Supply Company; *b,* Earth Scenes © 1995 Breck P. Kent; *c,* Visuals Unlimited/Scott Berner)

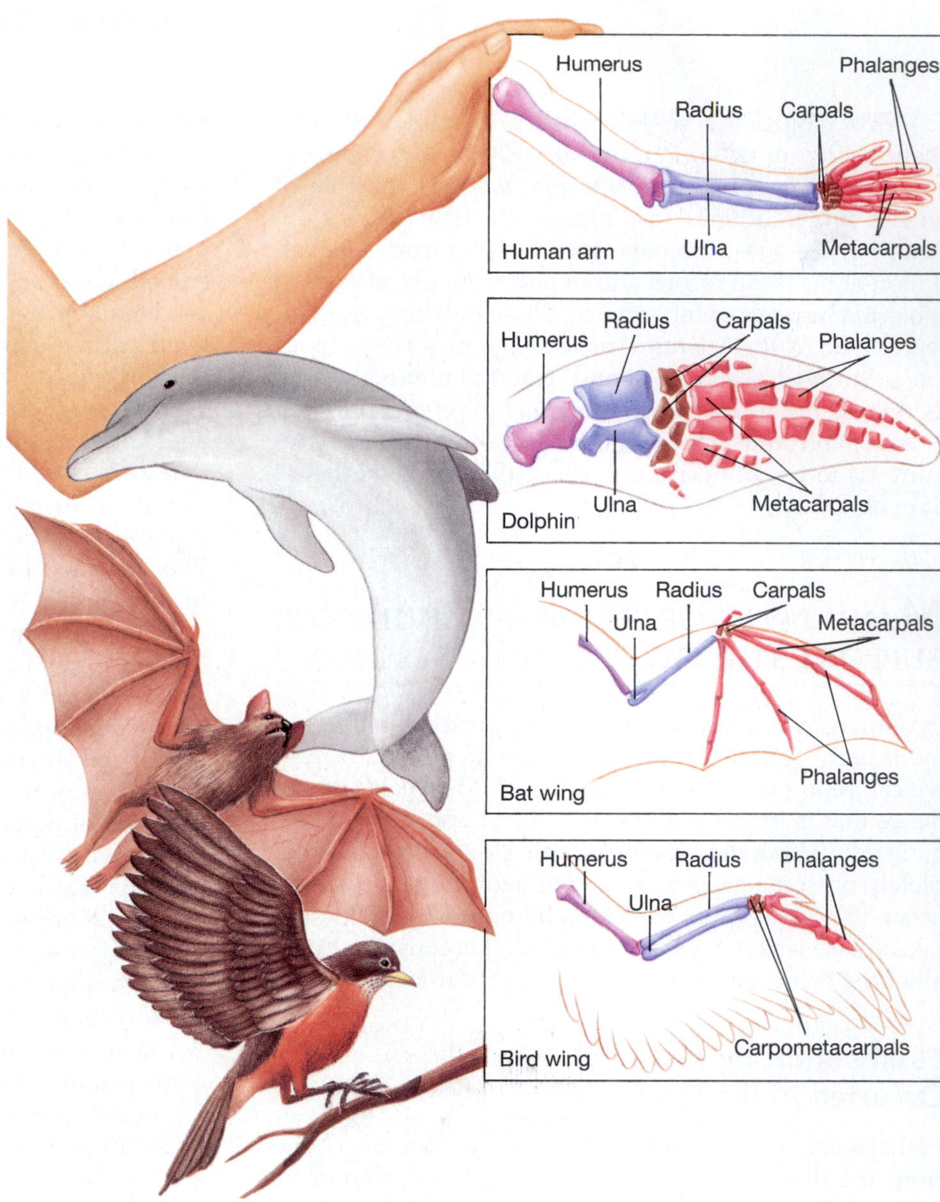

FIGURE 17–7 Homologous organs. The bird wing, bat wing, dolphin flipper, and human arm are homologous because they have a basic, underlying similarity of structure.

the fossils' age, however, requires special laboratory testing (see Focus On The Process of Science: Radioactive Dating).

Because of the conditions required for preservation, the fossil record is not a random sample of past life. There is bias in the record toward aquatic organisms and those living in the few terrestrial habitats conducive to fossil formation. For example, relatively few fossils of tropical forest animals have been found because plant and animal remains decay very rapidly on the forest floor, before fossils can form. Another reason for bias in the fossil record is that those organisms with hard body parts like bones and shells are more likely to form fossils.

Comparative Anatomy of Different Organisms Reveals Evidence of Evolution

Comparing the structural details of any particular organ found in different but related organisms reveals a basic similarity of form that is varied to some extent from one group to another. For example, the forelimbs of a bird, a dolphin, a bat, and a human, although they are quite different in appearance, have a strikingly similar arrangement of bones, muscles, and nerves. Figure 17–7 shows a comparison of their skeletal structures. Each has a single bone (the humerus) in the part of the limb nearest the trunk of the body, followed by the two bones of the forearm (the radius and ulna), then a group of carpals in the wrist, and a variable number of digits. This similarity is particularly remarkable because wings, flippers, and the human arm are used in different ways for different functions, and there is no mechanical reason for them to be so similar. Forelimbs, which are structurally similar in mammals, birds, reptiles, amphibians, and even fishes, also develop in similar ways in the young of these groups.

Darwin pointed out that such basic structural similarities in organs used in different ways are precisely the outcome we would expect if organisms had a common evolutionary origin. The basic structure present in a com-

(a)

(b)

FIGURE 17–8 Analogous organs. The wings of birds (***a***) and insects (***b***), although used for similar functions, have no underlying structural similarity. (*a*, Dennis Drenner; *b*, Skip Moody/Dembinsky Photo Associates)

mon ancestor was modified in different ways as various descendants subsequently evolved. Features similar in underlying structure in different species or groups owing to their descent from a common ancestor are termed **homologous.**

Organs that are not homologous but simply have similar functions in different organisms are termed **analogous.** For example, the lungs of mammals and the tracheae[2] of insects are analogous organs that have developed over time to meet, in quite different ways, the common problem of exchanging gases. Also, the wings of various unrelated flying animals, such as insects and birds, resemble one another superficially, although in more fundamental aspects are quite different (Fig. 17–8). Vertebrate wings are modified forelimbs supported by bones, whereas insect wings are outgrowths of the upper wall of the thorax and are supported by chitinous veins.

Like homologous organs, analogous organs offer crucial evidence of evolution. Analogous organs are of evolutionary interest because they demonstrate that organisms with separate ancestries may adapt in similar ways to similar environmental demands. This is known as **convergent evolution** (Fig. 17–9).

Comparative anatomy also demonstrates the existence of **vestigial organs.** Many organisms contain such organs or parts of organs that are seemingly nonfunctional and degenerate, often undersized or lacking some essential part. Vestigial organs are remnants of more-

[2] Tracheae are tiny tubes that branch throughout the insect's body, conducting air from openings at the surface to the tissues.

(a)

(b)

FIGURE 17–9 Convergent evolution. Two unrelated plant families exhibit structural similarities, such as thick, succulent stems and leaves modified into protective spines, as a result of convergent evolution. These plants evolved in similar desert environments in different parts of the world. (***a***) *Euphorbia ingens*, a member of the spurge family, which is native to Africa. (***b***) A member of the cactus family, which is native to North America. (Dennis Drenner)

FOCUS ON *The Process of Science*

RADIOACTIVE DATING

In order to interpret fossil evidence of Earth's past, the rocks in which they appear must be dated. Radioactive isotopes, also called **radioisotopes,** present in a rock give us an accurate measure of its age. Radioisotopes emit powerful, invisible radiations (see Focus On: Using Radioisotopes in Biology and Medicine in Chapter 3). As a radioisotope emits radiation, its nucleus changes into the nucleus of a different element; this is known as **radioactive decay.** For example, the radioactive nucleus of one isotope of uranium (U-235) decays over time into lead (Pb-207).

Each radioisotope has its own characteristic rate of decay. The period of time required for one-half of a radioisotope to change into a different material is known as its **half-life** (see figure). There is an enormous variation in half-lives of different radioisotopes. For example, the half-life of iodine (I-132) is only 2.4 hours, while the half-life of uranium (U-235) is 704 million years. The half-life of a

Radioactive dating. (***a***) A scientist prepares a fossilized dinosaur skeleton at Dinosaur National Monument in Utah. The age of fossils can be determined by radioactive dating. The decay of radioisotopes and subsequent accumulation of decay products (***b***), (***c***). (***b***) At time zero, the radioactive clock begins ticking. At this point a sample is composed entirely of the radioisotope. After one half-life, only 50 percent of the original radioisotope remains. (***c***) At time zero, the same sample contains no decay product(s). After one half-life, 50 percent of the original radioisotope has decayed into the decay products(s). During each succeeding half-life, one half of the remaining radioisotope is converted to decay product(s). (*a*, The Stock Market/Jeff Gnass, 1995) *(a)*

developed organs that were present in earlier ancestors. In the human body, there are more than 100 structures that have been viewed as vestigial, including reduced tail bones, wisdom teeth, muscles that move the ears, and the appendix. Whales and pythons have vestigial hind leg bones (Fig. 17–10); wingless birds have vestigial wing bones; and many blind, burrowing, or cave-dwelling animals may have vestigial eyes.

Darwin was interested in vestigial organs because they conflicted with the prevailing view of creation. He wondered how organisms that were the product of a "perfect creation" could have seemingly useless parts. Evolution, however, easily explained the existence of vestigial structures. The occasional presence of a vestigial organ is to be expected as a species evolves and adapts to different modes of life. Some organs become much less important for survival and may end up as vestiges. When an organ loses much or all of its function, it no longer has any selective advantage. Since the presence of the vestigial organ is not hurting the organism, however, selective pressure for completely eliminating the vestigial organ is weak, and the organ tends to remain.

The Distribution of Plants and Animals Supports Evolution

The study of the distribution of plants and animals is called **biogeography.** One of its basic tenets is that the evolution of each species occurred only once. The partic-

particular radioisotope is constant and never varies; it is not influenced by temperature or pressure or any other environmental factor.

The age of a fossil is estimated by measuring the proportion of the original radioisotope and its decay product. For example, the half-life of a radioisotope of potassium (K-40) is 1.3 billion years, meaning that in 1.3 billion years half of the radioactive potassium will have decayed into its decay product, argon (Ar-40). If the ratio of potassium (K-40) to argon (Ar-40) in the rock being tested is 1:1, the rock is 1.3 billion years old. (The radioactive clock begins ticking when the rock cools and solidifies. The cooled rock initially contains some potassium, but no argon. Because argon is a gas, it escapes from hot rock as soon as it forms, but when potassium decays in rock that has cooled and solidified, the argon accumulates in the crystalline structure of the rock.)

Several different radioisotopes are used in dating fossils. These include potassium (K-40; half-life 1.3 billion years), uranium (U-235; half-life 704 million years); and carbon (C-14; half-life 5730 years). Because of its relatively short half-life, carbon-14 is useful for dating fossils that are 50,000 years old or less.* In contrast, potassium-40, with its long half-life, can be used to date fossils that are hundreds of millions of years old. Wherever possible, the age of a fossil is independently verified using two or more different radioisotopes.

* Carbon-14 is used to date the carbon remains of anything that was once living, such as wood, bones, and shells. The other isotopes used in radioactive dating are used to date the rock in which fossils may be found.

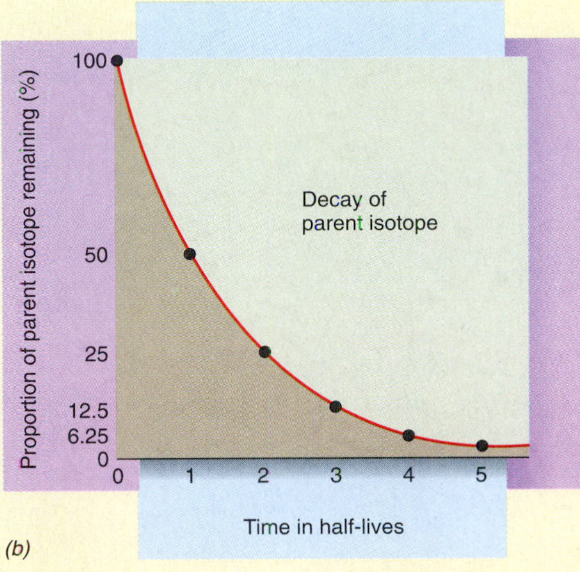

(b)

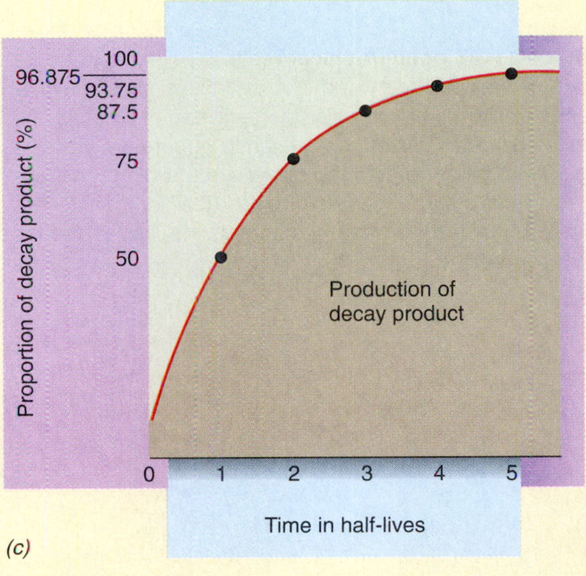

(c)

ular place where this occurred is known as the organism's **center of origin.** The center of origin is not a single point but the range of the population when the new species formed. From its center of origin, each species spreads out until halted by a barrier of some kind: physical, such as an ocean, desert, or mountain; environmental, such as an unfavorable climate; or ecological, such as the presence of organisms that compete with it for food or shelter (Fig. 17–11).

We would expect to find a given species distributed everywhere that it could survive if climate and topography were the only factors determining its distribution and if evolution were *not* a factor. However, species are often not distributed everywhere that they could survive. Central Africa, for example, has elephants, gorillas, chimpanzees, lions, and antelopes, whereas Brazil, with a similar climate and environmental conditions, has none of these. These animals originated in Africa and could not expand their range into South America because the Atlantic Ocean was an impassable barrier. Likewise, South America has prehensile-tailed monkeys, sloths, and tapirs, none of which is found in Africa. Thus, the natural distribution of organisms seems understandable only on the basis of evolution.

Related Species Have Similar Patterns of Development

The resemblance among embryos of different vertebrates is closer than the resemblance among their adults. In fact,

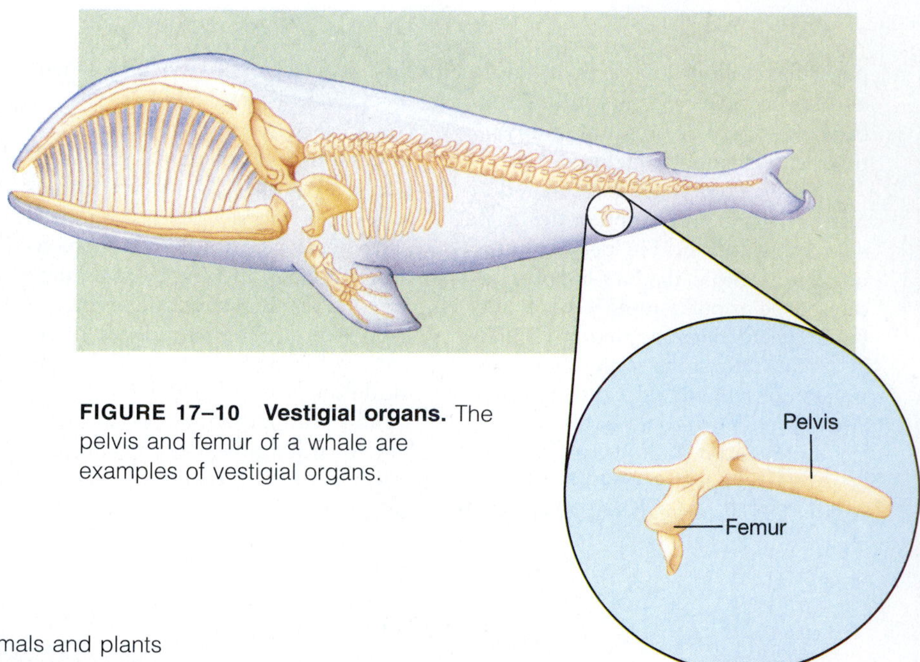

FIGURE 17–10 Vestigial organs. The pelvis and femur of a whale are examples of vestigial organs.

FIGURE 17–11 Biogeographical realms. Animals and plants are distributed around the world in a distinctive pattern, which reveals the existence of six major biogeographical realms. Each of these is characterized by the presence of certain unique species. These biogeographical realms are the direct outcome of the centers of origin of certain species (that is, the location where they originated), of their past migrations, and of the barriers they encountered.

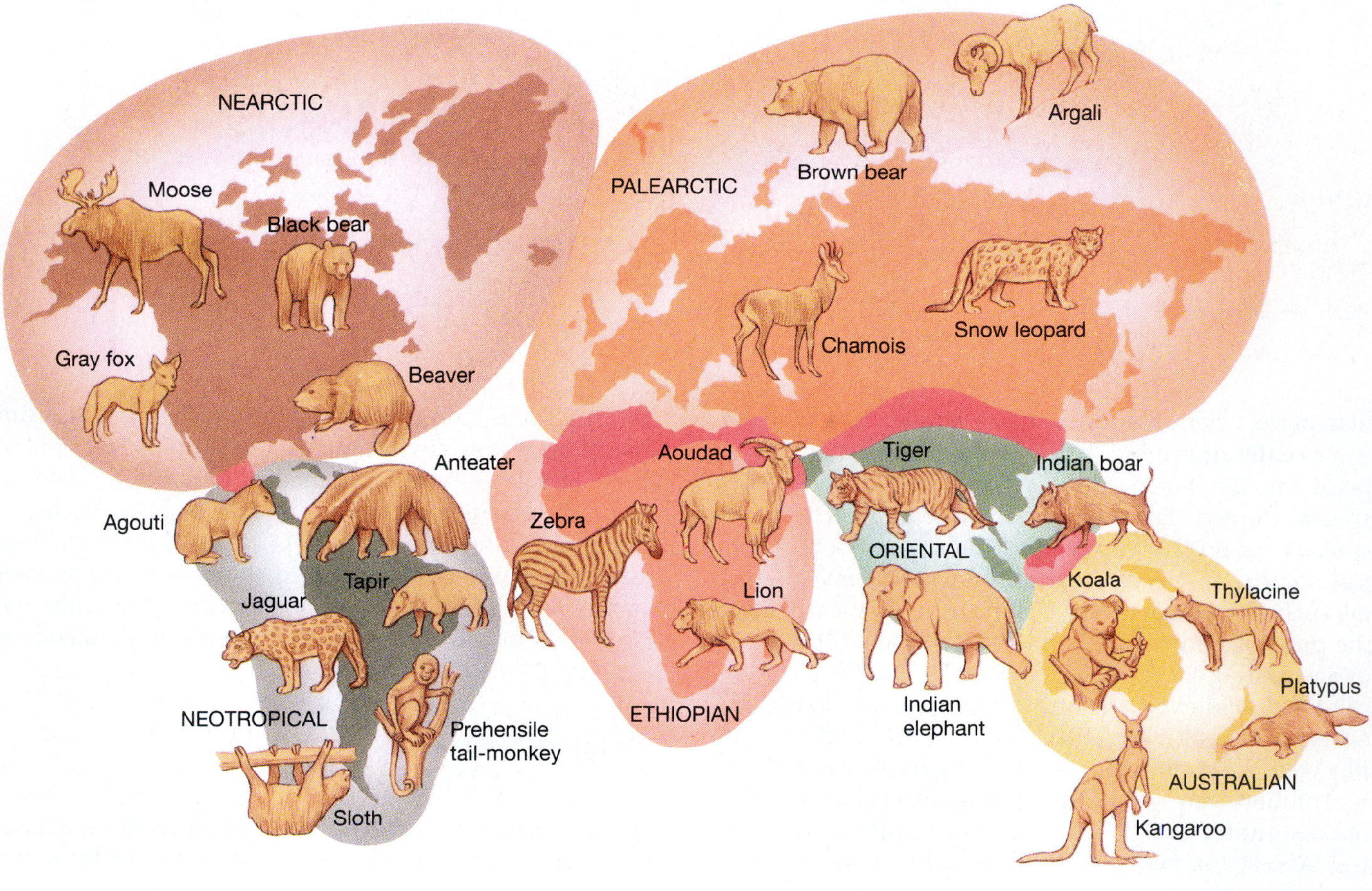

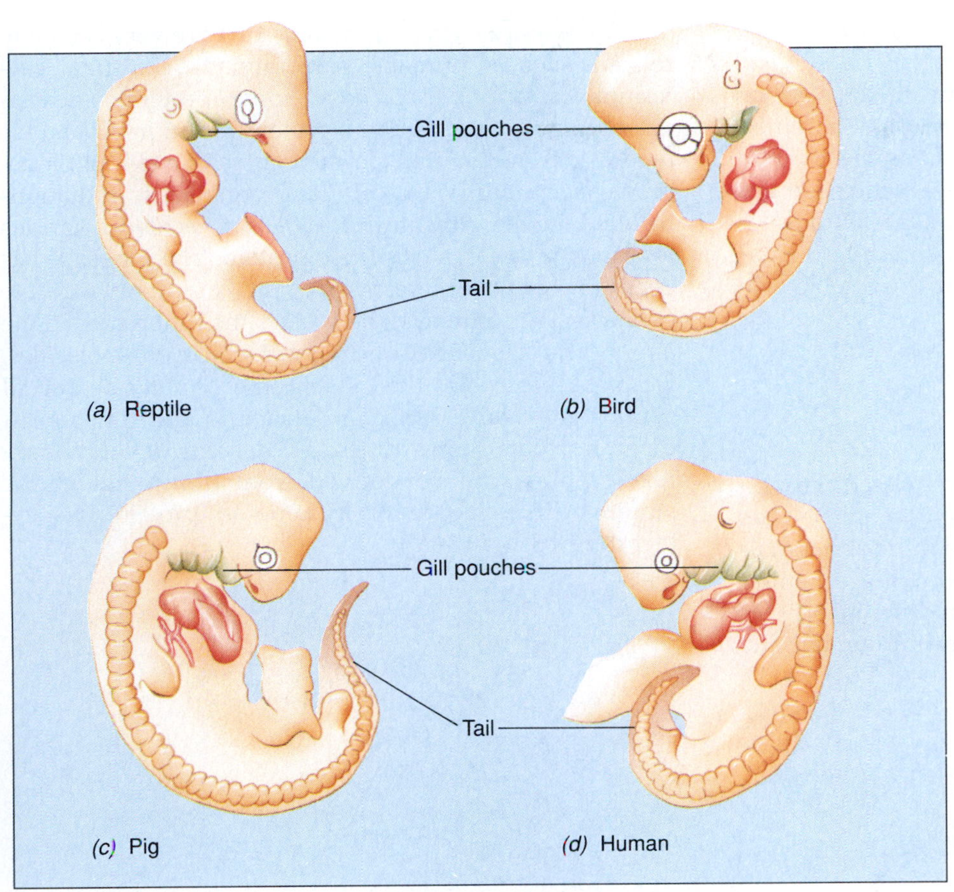

FIGURE 17–12 The early stages of embryonic development in several vertebrates. Numerous structural similarities are shared by the early stages, including the presence of a tail and gill pouches.

it is difficult to distinguish among the early embryos of a reptile, bird, pig, or human (Fig. 17–12). Segmented muscles, gill pouches, a tubular heart without left and right sides, a system of arteries known as aortic arches in the gill region, and many other features are found in the embryos of all vertebrates. All of these structures are necessary and functional in the developing fish. The small segmented muscles of the fish embryo give rise to the segmented muscles used by the adult fish in swimming. The gill pouches break through to the surface as gill slits. The adult fish heart remains undivided because it pumps blood forward to the gills that develop in association with the aortic arches.

However, none of these embryonic features remains in the adults of reptiles, birds, or mammals. Why, then, are these fishlike features present in the embryos of reptiles, birds, and mammals? Evolution is a conservative process, and natural selection builds upon what has come before rather than starting from scratch. Because the higher vertebrates evolved from an ancestral fish, they therefore share some of the fish's basic pattern of development. The genetic changes that have accumulated since the fish line diverged (separated) from the evolutionary line leading to the higher vertebrates modifies the pattern of development of the higher vertebrate embryos.

Biochemical and Molecular Comparisons among Organisms Provide Evidence of Evolution

Compelling evidence for evolutionary relationships is provided by similarities and differences in the biochemistry and molecular biology of various organisms. Indeed, evolutionary lines of descent based solely on biochemical and molecular data closely resemble those based on morphological (structural) and fossil evidence.

The Genetic Code Is Universal

Evidence that all life is related comes from the fact that all organisms use the same genetic code. Recall from Chapter 14 that the genetic code specifies the nucleotide triplets in a segment of DNA that codes for particular codons, (that is, sequences of three nucleotides in mRNA). Codons in turn code for specific amino acids in a polypeptide chain. For example, the codon "UUU" in mRNA codes for the amino acid phenylalanine in organisms as diverse as shrimp, humans, bacteria, and tulips. In fact, "UUU" codes for phenylalanine in *all* organisms examined to date. The universality of the genetic code—no other code has been found in any living organism—is compelling evidence that organisms arose from a common ancestor.

Organisms owe their characteristics to the various proteins that they possess, which in turn are determined by the sequence of nucleotides in mRNA (as specified by the order of nucleotides in DNA). The genetic code has been passed along through all of the branches of the evolutionary tree (another example of the conservative nature of evolution) since its origin in an extremely early (and successful) form of life.

Proteins Contain a Record of Evolutionary Change

Darwin's theory that all forms of life are related through descent with modification from earlier organisms has been further verified as we have learned more about molecular biology. Investigations of the sequence of amino acids in the same protein obtained from different species have revealed both great similarities and certain specific differences.

For example, even organisms that are very remotely related, such as humans, oaks, and the intestinal bacterium *Escherichia coli,* have some proteins such as cytochrome *c* (a respiratory protein found in all aerobic organisms) in common. In the course of the long, independent evolution of different organisms, mutations resulted in the substitution of amino acids at various locations in the cytochrome *c* protein. The longer it has been since two organisms diverged, or took separate evolutionary pathways, the greater the differences in the amino acid sequences of their cytochrome *c* molecules. Despite differences, the cytochrome *c* molecules of all species are clearly similar in structure and function. A diagram that shows lines of descent (evolutionary relationships) can be derived from differences in the amino acid sequence of a common protein like cytochrome *c* (Fig. 17–13).

CONCEPT CONNECTIONS

Evolution ⟷ *Genetics*

Genes that have been highly conserved during evolution are found in most living things. The gene that codes for cytochrome *c* is an example of a highly conserved gene. However, mutations have been occurring in such genes during the course of evolution, and these mutations appear to have occurred at a uniform rate over millions of years. In other words, DNA base-pair changes in a single gene, which cause differences in the DNA sequencing between two different organisms, may occur at a more or less constant rate within a given taxonomic group. Using this knowledge, it is possible to develop a **molecular clock** to complement geological estimates of the divergence of species from a common ancestor. From the number of alterations in the DNA nucleotide sequence of one organism compared with another, we can use a molecular clock to estimate the time of divergence between two closely related species or higher taxonomic groups.

Molecular clocks must be developed and interpreted with care. The rates of mutation appear to vary among different genes and among different taxonomic groups (*why* these variations occur is unknown). Therefore, although some mutations occur at a fairly uniform rate, a single molecular clock for all genes in all organisms has not been demonstrated.

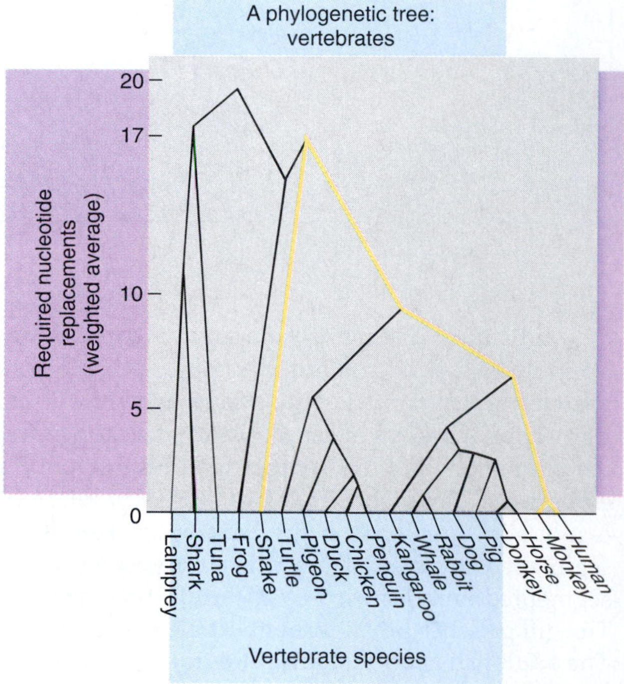

FIGURE 17–13 Lines of descent for selected vertebrates based on differences in the amino acid sequence of cytochrome *c*. Organisms that are closely related, such as monkeys and humans, show fewer differences in their nucleotide sequences (a relative difference of 1 as shown by the point at which monkeys and humans diverge). More distantly related organisms, such as snakes and humans, show greater differences in their nucleotide sequences (a relative difference of 17). There is a fairly close resemblance between this diagram of evolutionary relationships, constructed from molecular evidence, and classical diagrams based on fossil and structural evidence. (Adapted from Fitch and Margoliash, in *Evolutionary Biology,* 4:67–109, Plenum Publishing, 1970).

DNA Contains a Record of Evolutionary Change

Since DNA codes for proteins,[3] the differences in amino acid sequences indirectly demonstrate the nature and number of underlying DNA base-pair changes that must have occurred during evolution. Such molecular information is determined directly by **DNA sequencing,** in which the order of nucleotide bases in a strand of DNA that codes for a gene shared by several organisms is determined. The more closely species are thought to be related on the basis of other evidence, the greater is the percentage of nucleotide sequences that their DNA molecules usually have in common (Table 17–1).

[3] Of course, all genes do not code for proteins (witness regulatory genes). DNA sequencing of nonprotein coding DNA is also useful in determining evolutionary relationships.

TABLE 17–1
Differences in Nucleotide Sequences in DNA as Evidence of Phylogenetic Relationships

Species pairs	*Percentage differences in nucleotide sequences between pairs of species*
Human–chimpanzee	2.5
Human–gibbon	5.1
Human–Old World monkey	9.0
Human–New World monkey	15.8
Human–lemur	42.0

From Stebbins, G. L., *Darwin to DNA, Molecules to Humanity.* San Francisco, W.H. Freeman, 1982.

Chapter Summary

I. Evolution can be defined as a genetic change in a population of organisms that occurs over time.

II. Charles Darwin proposed the theory of evolution by natural selection, which is based on four observations.
   - A. Overproduction: Each species produces more offspring than will survive to maturity.
   - B. Variation: The individuals in a population exhibit heritable variation in their traits.
   - C. Limits on population growth: Organisms compete with one another for the resources needed for life (that is, for available food, space, water, light, and so on).
   - D. Survival to reproduce: The offspring with the most favorable combination of characteristics are most likely to survive and reproduce, passing those genetic characters to the next generation.

III. The synthetic theory of evolution combines Darwin's theory of evolution by natural selection with the genetic mechanisms for explaining evolution.
   - A. Mutation provides the genetic variability that natural selection acts upon during evolution.
   - B. The synthetic theory of evolution emphasizes the genetics of populations rather than individuals.

IV. The concept that evolution has occurred and is occurring is now well documented.
   - A. Perhaps the most direct evidence for evolution comes from fossils, the remains or traces of ancient animals and plants.
   - B. Evidence supporting evolution is derived from comparative anatomy.
     1. Homologous organs have basic structural similarities, even though the organs may be used in different ways. Homologous organs indicate evolutionary ties among the organisms possessing them.
     2. Analogous organs have similar functions in quite different, genetically unrelated organisms. Analogous organs demonstrate that organisms with separate ancestries can adapt in similar ways to comparable environmental demands (convergent evolution).
     3. The occasional presence of a vestigial organ is to be expected as an ancestral species adapts to a different mode of life.
   - C. Biogeography, the distribution of plants and animals, supports evolution.
     1. Each species originated only once, at its center of origin.
     2. From its center of origin, each species spread out until halted by a barrier of some kind.
     3. Areas that have been separated from the rest of the world for a long time have plants and animals specific to those areas.
   - D. Developmental biology provides evidence for evolution.
     1. The embryos of related animals are more similar than the adults.
     2. The accumulation of genetic changes since organisms diverged in evolution modifies the pattern of development in higher vertebrate embryos.
   - E. Biochemistry and molecular biology provide compelling evidence for evolution.
     1. The universality of the genetic code is impressive evidence that all life is related.
     2. The sequence of amino acids in common proteins such as cytochrome *c* reveals greater similarities in closely related species.
     3. A greater proportion of the sequence of nucleotides in DNA is identical in closely related organisms.

Selected Key Terms

analogous organs, p. 309
artificial selection, p. 305
biogeography, p. 310
center of origin, p. 311
convergent evolution, p. 309
DNA sequencing, p. 315
evolution, p. 302
fossils, p. 307
gene pool, p. 302
homologous organs, p. 309
natural selection, p. 306
synthetic theory of evolution (neo-Darwinism), p. 306
vestigial organs, p. 309

Post-Test

1. The fact that all species developed from earlier forms by the accumulation of genetic changes over many successive generations is known as the theory of ______________.
2. The genetic constitution of an entire population of a given organism is called its ______________ ______________.
3. The first scientist to propose that organisms evolve as a result of some natural phenomenon rather than divine intervention was ______________.
4. Thomas Malthus thought that ______________ increase in size geometrically until checked by factors in the environment.
5. Darwin proposed ______________ ______________ as the mechanism by which evolutionary change takes place.
6. Inherent in Darwin's theory of evolution by natural selection is the concept that organisms have the potential to produce more offspring than ______________ to reproductive maturity.
7. In natural selection, the selecting agent is the environment, whereas in artificial selection, the selecting agent is ______________.
8. The synthetic theory of evolution is also called ______________.
9. The remains or traces left by ancient organisms in rock strata are called ______________.
10. The wings of butterflies and bats, which have similar functions but are quite different in underlying structure, are said to be ______________ organs.
11. An organ that appears to have little or no function, and is smaller than a similar, fully functional equivalent in the organ's ancestors or relatives, is known as a/an ______________ organ.
12. The study of the distribution of plants and animals is called ______________.
13. The sequence of ______________ ______________ in common proteins such as cytochrome *c* reveals greater similarities in closely related species.
14. The universality of the ______________ ______________ is evidence for evolution from a common ancestor.
15. Which of the following is *least* likely to have occurred when the first small population of finches reached the Galapagos Islands from the South American mainland? (a) After many generations, the finches became increasingly different from the original population. (b) Over time, the finches adapted to their new environment. (c) After many generations, the finches were unchanged and unmodified in any way. (d) The finches were unable to survive in their new home and died out.
16. Which of the following is *not* part of Darwin's mechanism of evolution? (a) survival to reproduce (b) variation (c) inheritance of acquired (nongenetic) traits (d) limits on population growth.
17. The fossil record: (a) occurs in sedimentary rock (b) sometimes appears fragmentary (c) is relatively complete for tropical forest animals but incomplete for aquatic organisms (d) both a and b (e) a, b, and c are correct.
18. The molecular record inside cells suggests that evolutionary changes are caused by an accumulation of: (a) traits acquired through need (b) alterations in the order of nucleotides in DNA (c) characters acquired during an individual's lifetime.
19. If the same gene from two different species is identical in its order of nucleotides, this suggests that the organisms most likely are closely related. (a) true (b) false

Review Questions

1. Explain briefly the concept of evolution by natural selection.
2. In what ways does Lamarck's concept of evolution not agree with present evidence?
3. Consider the giraffe's long neck. Explain how this came about using Lamarck's explanation of evolution. Then explain the giraffe using Darwin's mechanism of natural selection.
4. Why are only inherited variations important in the evolutionary process?
5. What part of Darwin's theory was he unable to explain? How does the synthetic theory of evolution explain this?
6. Distinguish between artificial selection and natural selection.
7. Discuss at least three different types of evidence used to determine evolutionary relationships among organisms.
8. Discuss the factors that might interfere with our obtaining a complete and unbiased picture of life in the past from a study of the fossil record.
9. List as many vestigial structures on the human body as you can.
10. Explain why marsupials (pouched mammals) are widespread in Australia and almost nonexistent elsewhere.

Thinking Critically

1. One way to help reduce the incidence of drug resistance in bacteria is to avoid prescribing antibiotics indiscriminately. Explain how this measure would help.
2. Sometimes a tuberculosis patient is given two antibiotics, streptomycin and isoniazid, simultaneously. Explain how this procedure might minimize the evolution of drug resistance in TB bacteria.
3. Propose a mechanism whereby mosquitos might develop a genetic resistance to DDT (an insecticide used to kill mosquitos).
4. How can you account for the fact that both Darwin and Wallace independently and almost simultaneously proposed essentially identical theories of evolution by natural selection?
5. Discuss these statements:
   - **a.** Natural selection chooses from among the individuals in a population those that are most suited to *current* environmental conditions. It does not guarantee survival under future conditions.
   - **b.** Evolution is not hierarchical. For example, humans are not "better" or "more highly evolved" than the bacteria living in our intestines.
   - **c.** Individuals do not evolve, but populations do.
   - **d.** Evolution is not purposeful, but is based on chance.
6. Based on what you have learned in this chapter, explain why such genetically different organisms as porpoises and sharks are so similar in body form.

Recommended Readings

Amos, W.H., "Hawaii's Volcanic Cradle of Life," *National Geographic,* July 1990, 70–87. Depicts the colonization of volcanic lava by a few hardy species and portrays some of the endemic (found nowhere else) species that evolved in the Hawaiian Islands.

Darwin, C.R., *On the Origin of Species by Means of Natural Selection or the Preservation of the Favored Races in the Struggle for Life.* New York, Cambridge University Press, 1975. A readily obtainable reprint of one of the most important books of all time. Darwin's long essay is still of great significance to modern readers.

Grant, P.R., "Natural Selection and Darwin's Finches," *Scientific American,* October 1991. A study of the finches of the Galapagos reveals natural selection in action during a recent period of drought.

Swan, L.W., "The Concordance of Ontogeny with Phylogeny," *BioScience* Vol. 40, No. 5, May 1990. Examines embryological evidence for evolution.

Thompson, G.R. and J. Turk, *Modern Physical Geology,* Philadelphia, Saunders College Publishing, 1991. Contains informative material on radioactive dating and fossils.

Wallace, A.R., *The Malay Archipelago,* New York, Oxford University Press, 1987. A reprint of Wallace's classic investigation.

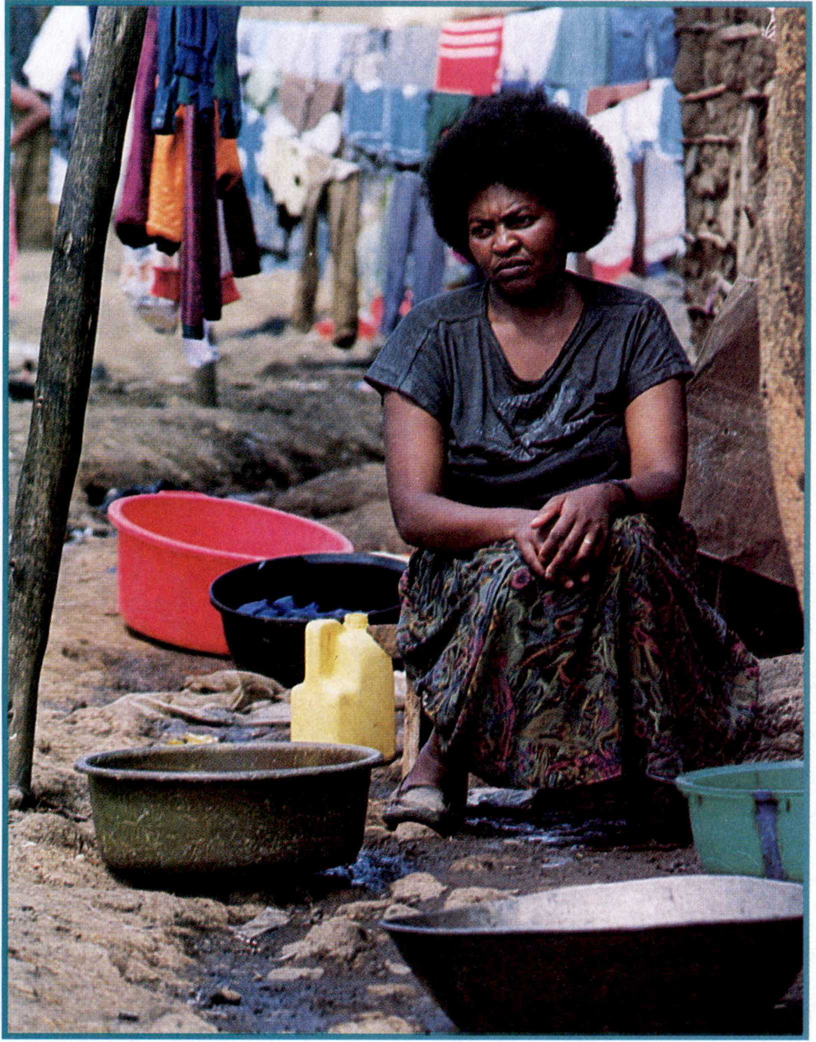

Agnes Munyiva, one of a small group of Kenyan prostitutes who appears to be resistant to the virus that causes AIDS. (Fiona McDougall/Time Magazine)

Evolution in the Headlines: Are Some of Us Naturally Immune to Aids?

Fish to amphibian to reptile to mammal—this popularized image of evolutionary "progression" often leaves the impression that evolution is only ancient history, something that humanity has somehow "outgrown." But genetic variability, the raw material for evolution, is very much present in modern human populations. In fact, this variability may well provide the key to one of our most pressing health concerns.

Acquired immune deficiency syndrome (AIDS) is a deadly immune disease caused by a virus. The virus, called human immunodeficiency virus (HIV-1), infects and destroys cells of the immune system. As the immune system becomes weaker the body becomes susceptible to deadly infections and diseases (Chapter 37).

Medical researchers report that a small group of women in Africa appear to be resistant to HIV-1. These women are regularly exposed to the AIDS virus because they are prostitutes in the slums of Nairobi, Kenya. Each year, they may have as many as 2000 sexual contacts, and 10% to 15% of the men with which they come into contact are HIV positive. The AIDS-resistant prostitutes are tested for HIV infection on a regular basis and repeatedly test negative. (In contrast, most of their peers test positive for HIV infection only a few months after becoming prostitutes. These women subsequently develop AIDS an average of four years after their initial exposures to HIV.)

Why are these women apparently immune to the AIDS virus? Early in their study, researchers ruled out behavior as a possible cause. That is, the AIDS-immune women do not choose their partners more carefully nor do they practice safe sex more regularly than other prostitutes.

One thing that the AIDS-immune women appear to have in common is the presence of up to three specific HLA[1] protein molecules on the surface of their blood cells. Although their exact function is unknown, HLA proteins appear to be important in the recognition of infectious organisms by the body. Once a cell or molecule is recognized as foreign, the body mounts an immune response against the foreign invader. HLA proteins are highly variable in the human population. (This variability is responsible for the rejection of heart, kidney, or other organ transplants when the body's immune system recognizes the foreign HLA proteins on the transplanted tissues.)

Humans possess numerous different genes that code for HLA proteins. Most of these genes have a remarkable number of different possible alleles (recall from Chapter 11 that alleles are alternate forms of the same gene). Because humans are diploid organisms, each individual contains only two alleles for each gene. But the extremely polymorphic nature of the different HLA genes means that the human *population* contains thousands of individual variants. (Genetic polymorphism is defined and discussed shortly.)

If the presence of AIDS immunity in a human population is substantiated, it may ultimately lead to the development of a vaccine to prevent AIDS. However, researchers caution that it will take years to duplicate that immunity in the lab, which is a prerequisite for producing a vaccine.

The discovery of AIDS-immune prostitutes is a timely reminder that humans still possess one of the main requirements for evolution by natural selection—genetic variation in a population. In this chapter we examine genetic variation in populations, including genetic polymorphism, and how the relative abundance of different alleles changes during the course of evolution. We then consider how the mechanisms that lead to evolutionary changes within populations can ultimately lead to the evolution of new species.

[1] HLA stands for human leukocyte antigen (see Chapter 37).

Micro-evolution and Speciation

18

Learning Objectives

After you have studied this chapter you should be able to:

1. Define population, gene pool, allele frequency, and microevolution.
2. Distinguish between the gene pool of a population and the genotype of an individual.
3. Discuss the significance of the Hardy-Weinberg principle as it relates to evolution and list the five conditions required for genetic equilibrium.
4. Explain how each of the following alters allele frequencies in populations: mutation, genetic drift, gene flow, and natural selection.
5. Distinguish among stabilizing selection, directional selection, and disruptive selection, and give an example of each.
6. Explain how genetic variation, the raw material for evolutionary change, is maintained in a population.
7. Define a species and describe some of the limitations of your definition.
8. Explain the evolutionary significance of reproductive-isolating mechanisms.
9. Distinguish among different reproductive-isolating mechanisms and give an example of each.
10. Distinguish between allopatric and sympatric speciation and give an example of each.

Key Concepts

- ☐ Evolution occurs in populations, not individuals. Individuals do not evolve during their lifetimes. Populations evolve over many generations.
- ☐ Individuals in a population exhibit genetic variation for the traits characteristic of the population. Genetic variation—as represented by the number and kinds of alleles in a population—is the raw material for evolutionary change because it provides the diversity upon which natural selection can act.
- ☐ During the course of evolution, some alleles increase in frequency within a population, whereas other alleles decrease or even disappear from the population. Thus, populations evolve by changes in allele frequencies (that is, by changes in the relative proportions of their various alleles) for different genes.
- ☐ A species is a group of more or less distinct organisms that are capable of interbreeding with one another in nature and are reproductively isolated from other species.
- ☐ Two populations that are reproductively isolated for a long period of time may evolve into two separate species. Given enough time, the mechanisms of microevolution (such as the accumulation of changes in allele frequencies) may lead to the origin of new species.

FIGURE 18–1 Genetic variation in a gene pool. These shell patterns and colors occur in a single species of snail (background). A single genetic locus (A), with three different alleles possible at that locus (A_1, A_2, A_3), is depicted. Because each individual (represented by the small shells) is diploid, it possesses only two alleles for each genetic locus. (C. Clark)

INDIVIDUALS IN A POPULATION EXHIBIT GENETIC VARIATION

A **population** consists of all of the individuals of the same species that live in a particular place at a specific time. Individual organisms in a population exhibit genetic variation for the traits characteristic of the population. A population of snails may all possess the same trait—a spiral shell, for example—but they vary from one individual to another in shell patterns and colors (Fig. 18–1).

Each population possesses a **gene pool,** which is the total genetic material of all of the individuals that make up that population at a given time. The gene pool includes all possible **alleles** of all genes present in the population (Fig. 18–1). An individual organism possesses only a small fraction of the genes present in a population's gene pool. In diploid (2n) organisms, each body cell contains only two alleles for each gene, one on each of the **homologous chromosomes** (members of a chromosome pair). The genetic variation evident among individuals in a given population indicates that each individual has a different combination of the alleles that exist in the gene pool.

If a population is not undergoing evolutionary change, frequencies of each allele in the gene pool remain constant from generation to generation. However, changes in allele frequencies over a few successive generations indicate that evolution has occurred. This type of evolution, which involves small, gradual changes within a population, is sometimes referred to as **microevolution.** Microevolution accounts for the differences among the various populations of a species.

Although microevolutionary changes are relatively small or minor, many biologists think that the accumulation of microevolutionary changes over time is substantial. Thus, microevolution may be sufficient to account for the origin of new species and possibly the origin of major groups of living organisms.[2]

THE HARDY-WEINBERG PRINCIPLE DEMONSTRATES THAT ALLELE FREQUENCIES DO NOT CHANGE IN A POPULATION THAT IS NOT EVOLVING

Suppose we are studying a population of well-adapted plants growing in an environment that is relatively constant from year to year. Such a population is said to be in **genetic equilibrium,** which means that it is not undergoing evolutionary change. A population in genetic equilibrium maintains the same allele frequencies over successive generations.

If we were to count the number of red-flowered plants versus white-flowered plants in the population, we might find 1820 plants with red petals and 180 with white petals, which is a ratio of 9 to 1. If we came back to the same location the following summer, we would find a population that is essentially the same as the previous one, with roughly 9 red-flowered plants to every 1 white-flowered plant. If we continued the study for several generations, we should always get the same result.

[2] Biologists distinguish between microevolution, which involves minor evolutionary changes within a species, and macroevolution, which involves major evolutionary changes that lead to new groups of organisms. Macroevolution is discussed in Chapter 19.

The explanation for this stability of successive generations of populations in genetic equilibrium was provided independently by Godfrey Hardy, an English mathematician, and Wilhelm Weinberg, a German physician, in 1908. They pointed out that the frequencies of various **genotypes** (genetic makeups) in a population can be described mathematically. The **Hardy-Weinberg principle** represents an ideal situation that probably never occurs in the natural world. However, it is useful because it provides us with a model to help understand the real world.

How does this principle relate to Mendel's principles of inheritance discussed in Chapter 11? The principles of inheritance worked out by Mendel describe the frequencies of genotypes among offspring of a single mating pair. In contrast, the Hardy-Weinberg principle describes the frequencies of alleles in the genotypes of an *entire breeding population.* The Hardy-Weinberg principle shows that the process of inheritance by itself will not cause changes in allele frequencies. The proportion of alleles in successive generations will always be the same, provided the following criteria are met:

1. **Random mating.** The individuals represented by *AA* (homozygous dominant), *Aa* ((heterozygous), and *aa* (homozygous recessive) must mate with one another at random and must not select their mates on the basis of genotype. That is, matings between genotypes must occur in proportion to the frequencies of the genotypes.
2. **No mutations.** There must be no mutations of *A* or *a*.
3. **Large population size.** Because the Hardy-Weinberg principle is a statistical tool, the population of individuals must be large enough so that the effect of chance is small. Allele frequencies in a small population are more likely to be affected by chance events (genetic drift, discussed shortly) than allele frequencies in a large population.
4. **No migration.** There can be no exchange of genes with other populations that might have different allele frequencies. In other words, there can be no migration of individuals into or out of a population.
5. **No natural selection.** If natural selection is occurring, certain genotypes will be favored over others, and the allele frequencies will change.

The preceding paragraphs gave an overview of the Hardy-Weinberg principle in sufficient detail for some introductory biology courses. An elaboration of the mathematical basis behind the Hardy-Weinberg equilibrium is given for those courses that go into greater detail (see Focus On the Process of Science: The Hardy-Weinberg Principle).

FOCUS ON *The Process of Science*

THE HARDY-WEINBERG PRINCIPLE

A population in genetic equilibrium can be described mathematically. For example, the chemical phenylthiocarbamide (PTC) tastes very bitter to most people, but is tasteless to others. The ability or inability to taste PTC is controlled by a single gene, with the inability to taste being recessive. Thus, this trait has two alleles, *T* being the dominant allele and *t* its recessive counterpart. Individuals possessing the dominant phenotype (tasters) are either homozygous dominant (*TT*) or heterozygous (*Tt*), whereas individuals with the recessive phenotype (nontasters) are homozygous recessive (*tt*).

In a population, these three genotypes (*TT*, *Tt*, and *tt*) may be represented mathematically by the following expression

$$p^2 + 2pq + q^2 = 1$$

where

p = the frequency of the dominant allele (*T*),

q = the frequency of the recessive allele (*t*),

p^2 = the frequency of the homozygous dominant genotype (*TT*),

$2pq$ = the frequency of the heterozygote (*Tt*),

q^2 = the frequency of the homozygous recessive genotype (*tt*).

In mathematical terms, because there are only two possible alleles, *T* and *t*, the sum of their frequencies must equal 100%, or 1, so we may write $p + q = 1$. By simple algebraic manipulation we can determine that

$$p = 1 - q \quad \text{and} \quad q = 1 - p$$

These two equations determine the frequencies of the taster and nontaster alleles in the gene pool.

If we wish to calculate the frequencies of the genotypes in the population, consider the following. A gamete-bearing allele *T* may randomly combine with a similar gamete to form a *TT* zygote. Similarly, *T* combines with *t* to form *Tt*, *t* combines with *T* to form *Tt*, and *t* combines with *t* to form *tt*. The frequency of each of the possible genotypes (*TT*, *Tt*, *tt*) in the offspring is calculated by multiplying the frequencies of the alleles *T* and *t* in eggs and sperm, as follows:

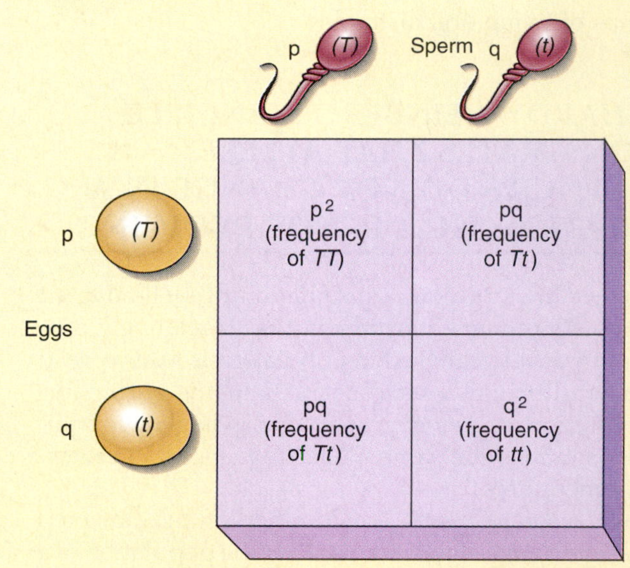

EVOLUTION OCCURS WHEN THERE ARE CHANGES IN ALLELE FREQUENCIES IN A GENE POOL

In studies of populations, the Hardy-Weinberg principle provides a mathematical model of a situation in which no evolution occurs, that is, allele frequencies do not change over successive generations. However, allele frequencies in nature are often significantly different from those that the Hardy-Weinberg principle would predict. *Evolution represents a departure from the Hardy-Weinberg principle of genetic equilibrium.* Changes in the gene pool of a population result from four microevolutionary processes: mutation, genetic drift, migration, and natural selection. When one or more of these processes is acting on the population, allele frequencies will change from one generation to the next.

Again, the frequencies of the 3 possible genotypes must sum to 1, so we may write

$$p^2 + 2pq + q^2 = 1$$

where

p^2 = frequency of *TT* genotype in the population,

$2pq$ = frequency of *Tt* genotype in the population,

q^2 = frequency of *tt* genotype in the population.

For the next generation, current *Tt* people will be able to produce *either* *T* or *t* gametes; that is, one-half of their gametes will contain the *T* allele and one-half the *t* allele. The *TT* people will, of course, produce only *T* gametes, and the *tt* people will produce only *t* gametes. Thus, the total frequency of all *T* and *t* alleles may be expressed by the following equations, where p' stands for the new frequency of *T* and q' for the new frequency of *t* (in the next generation):

$$\begin{aligned} p' &= p^2 + \tfrac{1}{2}(2pq) \\ &= p^2 + pq \\ &= p^2 + p(1 - p) \\ &= p^2 + p - p^2 \\ &= p \end{aligned}$$

and

$$\begin{aligned} q' &= q^2 + \tfrac{1}{2}(2pq) \\ &= q^2 + pq \\ &= q^2 + q(1 - q) \\ &= q^2 + q - q2 \\ &= q \end{aligned}$$

Thus, $p' = p$ and $q' = q$. This demonstrates that, if left undisturbed, allele frequencies in a randomly mating population do not change from generation to generation, regardless of dominance or recessiveness. The process of inheritance does not by itself cause changes in allele frequencies, which remain constant from generation to generation.

Any population in which the distribution of alleles *T* and *t* conforms to the relationship $p^2 + 2pq + q^2 = 1$, whatever the absolute values for p and q may be, and in which those frequencies do not change from generation to generation, is in genetic equilibrium. No evolution is taking place within that population.

How to Solve a Problem Using the Hardy-Weinberg Principle

Try to solve the following problem by yourself before looking at the solution: Approximately 70% of Americans are PTC tasters and 30% are nontasters. Estimate the frequencies of the taster (*T*) and nontaster (*t*) alleles.

Solution: This problem should be solved by first determining the value of q. You do this by starting with what you have been given: $q^2 = 0.30$. The square root of 0.30 is 0.55, so the frequency of the recessive allele is $q = 0.55$. The frequency of the dominant allele p can now be calculated: $p = 1 - q = 1 - 0.55 = 0.45$.

Answer: The frequency of the taster (*T*) allele is 0.55, and the frequency of the nontaster (*t*) allele is 0.45.

Mutation Increases Variation in the Gene Pool

Genetic variation is introduced into a gene pool through **mutation,** which is a random, permanent change in DNA (Chapter 13). Mutations are the ultimate source of all new alleles. However, mutations occurring in body cells are not heritable. When an individual with such a mutation dies, the mutation dies with it. Some mutations, however, alter the DNA in reproductive cells. These mutations may or may not affect the offspring because most of the DNA in a eukaryotic cell is "silent" and does not code for specific polypeptides or proteins (Chapter 14). Even if a mutation occurs in the DNA that codes for a protein, it may still have little effect in altering the structure or function of that protein. However, when the protein is altered enough to change how it functions, the mutation is usually harmful.

Mutations do not determine the *direction* of evolutionary change. Consider a population that is undergoing evolutionary change to adapt to an increasingly dry environment. A mutation that produces a new allele that helps an individual adapt to dry conditions is no more likely to occur than one for adapting to wet conditions or one that has no relationship to the changing environment. The production of new mutations simply increases the genetic variability within a population.

Genetic Drift Causes Changes in Allele Frequencies in Small Populations by Random, or Chance, Events

The size of a population has important effects on allele frequencies because random events, or chance, will tend to cause changes in a small population. If a population consists of only a few individuals, an allele present at a low frequency in the population could be lost purely by chance. Such an event would be most unlikely in a large population.

For example, consider two populations, one with 10,000 individuals and one with 10 individuals. If an uncommon allele occurs at a frequency of 10%, or 0.10, then *2000 individuals* in the larger population possess the allele (10,000 individuals × 2 copies per diploid individual × 0.10 allele frequency = 2000). That same allele frequency, 10%, in the smaller population means that only *2 individuals* possess the allele (10 individuals × 2 copies per diploid individual × 0.1 allele frequency = 2). From this exercise, it is easy to see that there is a greater likelihood that the uncommon allele will be lost from the smaller population than from the larger population. Predators, for example, might kill the two individuals possessing the uncommon allele in the smaller population purely by chance.

The production of random evolutionary changes in small breeding populations is known as **genetic drift.** Genetic drift results in changes in the gene pool of a population from one generation to another. One allele may be eliminated by chance from the population, regardless of whether that allele is beneficial, harmful, or of no particular advantage or disadvantage. Thus, genetic drift can decrease genetic variation *within* a population, although it tends to increase the genetic differences *among* different populations.

The Founder Effect Occurs when a Few "Founders" Establish a New Colony

When one or a few individuals from a large population begin, or found, a new colony—as when a few birds separate from the rest of the flock and fly to a new area—they bring with them only a small fraction of the genetic variation present in the original population. As a result, the only alleles that will be represented among their descendants will be those few that the colonizers chanced to possess. This often results in very different allele frequencies than are possessed by other populations of the same species. The genetic drift that results from a small number of individuals from a large population colonizing a new area is termed the **founder effect** (Fig. 18–2).

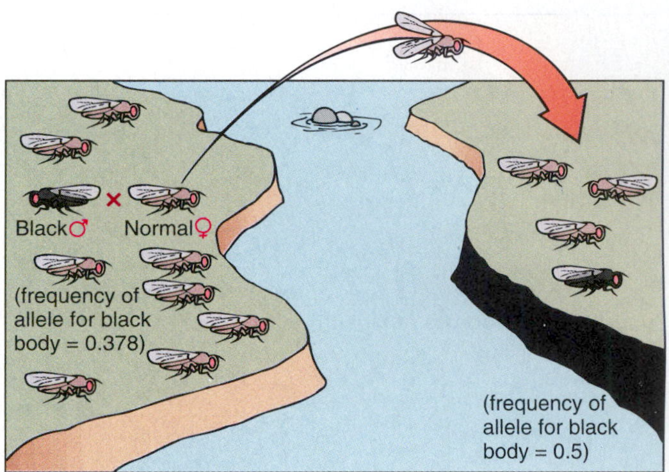

FIGURE 18–2 The founder effect. In this example, the allele for gray body is dominant over the allele for black body. A single inseminated female fruit fly is blown to an island. The alleles of her offspring, which occur in different frequencies than those in the original population, will serve as the foundation for the gene pool of all fruit flies living on that island. (The allele frequencies do not match the phenotypic ratios depicted here because heterozygous individuals exhibit the dominant phenotype but contribute recessive alleles to the allele frequencies.)

Genetic Bottlenecks Cause Genetic Drift

Because of fluctuations in the environment, such as a depletion in food supply or an outbreak of disease, a population may periodically experience a rapid and marked decrease in size. The population is said to go through a **genetic bottleneck** in which genetic drift can occur in the few remaining survivors. As the population again increases in number, the frequencies of many alleles may be quite different from those in the population preceding the decline (Fig. 18–3). A genetic bottleneck that took place in the cheetah population about 10,000 years ago is responsible for its current low genetic variability (Focus On Evolution: The Endangered Cheetah).

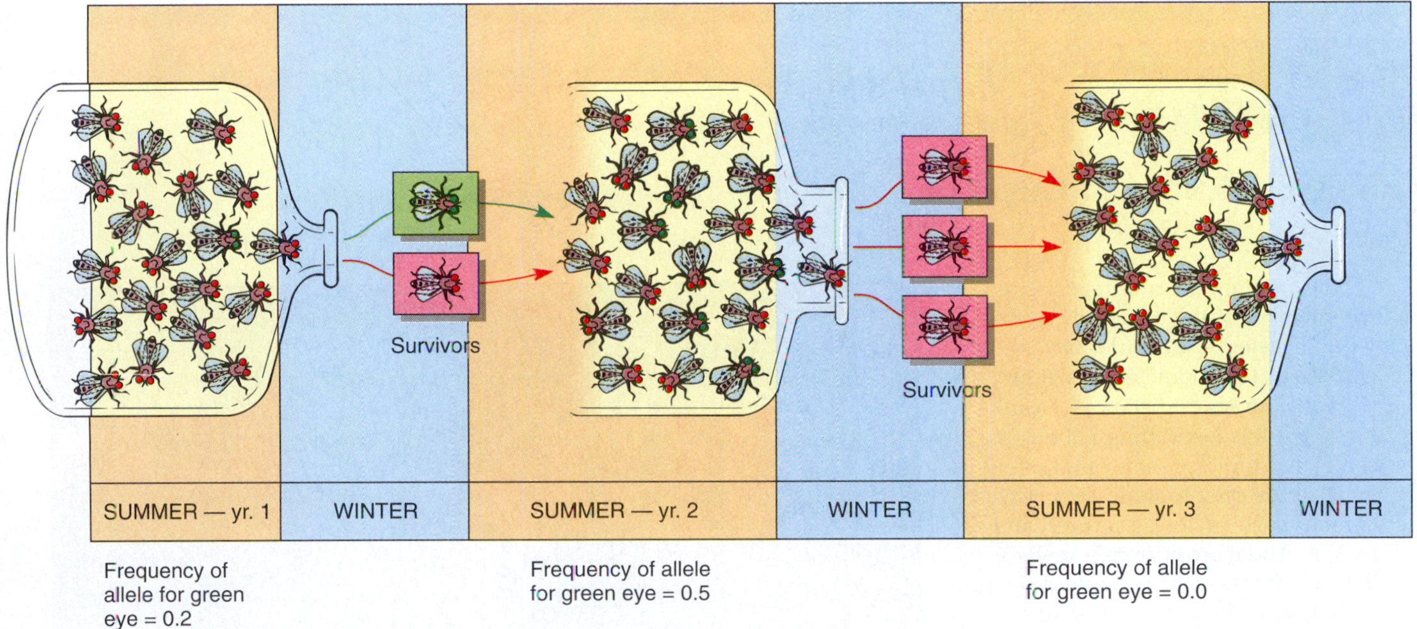

FIGURE 18–3 Genetic bottleneck. Because only a small population of flies survives the winter, its genotypes, not necessarily resulting from natural selection, determine the allele frequencies of the following summer population. (The allele frequencies do not match the phenotypic ratios depicted here because heterozygous individuals exhibit the dominant phenotype but contribute recessive alleles to the allele frequencies.)

Gene Flow, Caused by the Migration of Organisms, Changes the Amount of Variation in a Gene Pool

Members of a species tend to be distributed in local populations that are more or less isolated genetically from other populations. The migration of individuals causes a corresponding movement of alleles, or **gene flow,** that can have significant evolutionary consequences. As alleles "flow" from one population to another, they usually increase the amount of variability within the population that receives them. If the gene flow between two populations is great enough, these populations become more similar genetically. Because gene flow has a tendency to reduce the amount of genetic variation between two populations, it tends to counteract the effects of natural selection and genetic drift, both of which cause individual populations to become increasingly distinct.

Natural Selection Results in Allele Frequency Changes that Increase Adaptation to the Environment

Natural selection is the mechanism of evolution first proposed by Darwin in which members of a population that possess better adaptations to the environment are more likely to survive and reproduce (Chapter 17). Over successive generations, the proportion of more favorable alleles increases in the population. Natural selection checks the random effects of mutation, genetic drift, and gene flow. As a result, a population becomes better adapted to the environment in which it lives.

Natural selection results in the differential reproduction of individuals with different observable traits, or **phenotypes,** (and therefore the underlying genotypes that produce them) in response to the environment. Natural selection functions to preserve individuals with favorable genotypes and eliminate individuals with unfavorable genotypes. Individuals have a selective advantage if they are able to survive and produce fertile offspring. Natural environmental pressures, such as competition for food or water or living space, select the individuals that survive to reproduce.

Natural selection not only explains why organisms are well adapted to the environments in which they live, but also accounts for the astounding diversity of life. Natural selection enables populations to change in order to adapt to different environments and different ways of life.

(text continued on page 327)

FOCUS ON *Evolution*

THE ENDANGERED CHEETAH

The cheetah, the most specialized member of the cat family, is the world's fastest animal and has been clocked at 110 kilometers per hour (70 mph) for short distances. Despite its speed, this fascinating animal is a somewhat timid predator that often gives up its prey to more aggressive animals such as lions. vultures, and hyenas. There are currently fewer than 12,000 cheetahs worldwide, with approximately 650 of them housed in zoos.

The problems faced by cheetahs, both in zoos and in nature, are partly the result of something that occurred at the end of the last ice age, approximately 10,000 years ago. At that time, cheetahs almost became extinct due, perhaps, to overhunting by humans. Only a few cheetahs survived, which greatly reduced their genetic variability as a group. All of the cheetahs alive today descended from the few survivors of the Ice Age. As a result, cheetahs exhibit extreme genetic uniformity.

Why is genetic uniformity undesirable? There are a number of reasons. Such populations cannot adapt well to changing environmental conditions. In contrast, populations with genetic variability maintain evolutionary adaptability and can respond effectively to changing environmental conditions. If all individuals are alike and if the environment changes, it may change in a way that is harmful to *all* individuals of the population. In a genetically diverse population, at least some of the individuals are likely to possess traits that enable them to tolerate or even thrive in the new environment.

Populations that are genetically uniform exhibit low reproductive success. Cheetah males have a very low sperm count compared to other large cats, and many of the sperm have abnormalities. Cheetah females do not bear as many offspring as other feline species. Also, many cheetah offspring have health problems and are more susceptible to disease. If one member of a genetically uniform population is susceptible to a particular disease, they *all* are susceptible, whereas in a genetically diverse population, at least *some* of the individuals may be resistant to that disease.

A number of conservation biologists are developing a comprehensive strategy to help save the cheetah. This strategy includes use of reproductive techniques such as in vitro fertilization, in which an egg is fertilized by sperm in a test tube before being implanted in a female cheetah. Although these techniques may help scientists to maintain the small amount of genetic diversity that remains in cheetahs today, they cannot *increase* the genetic diversity of the sleek cats. The outlook for the cheetah's long-term survival is far from certain.

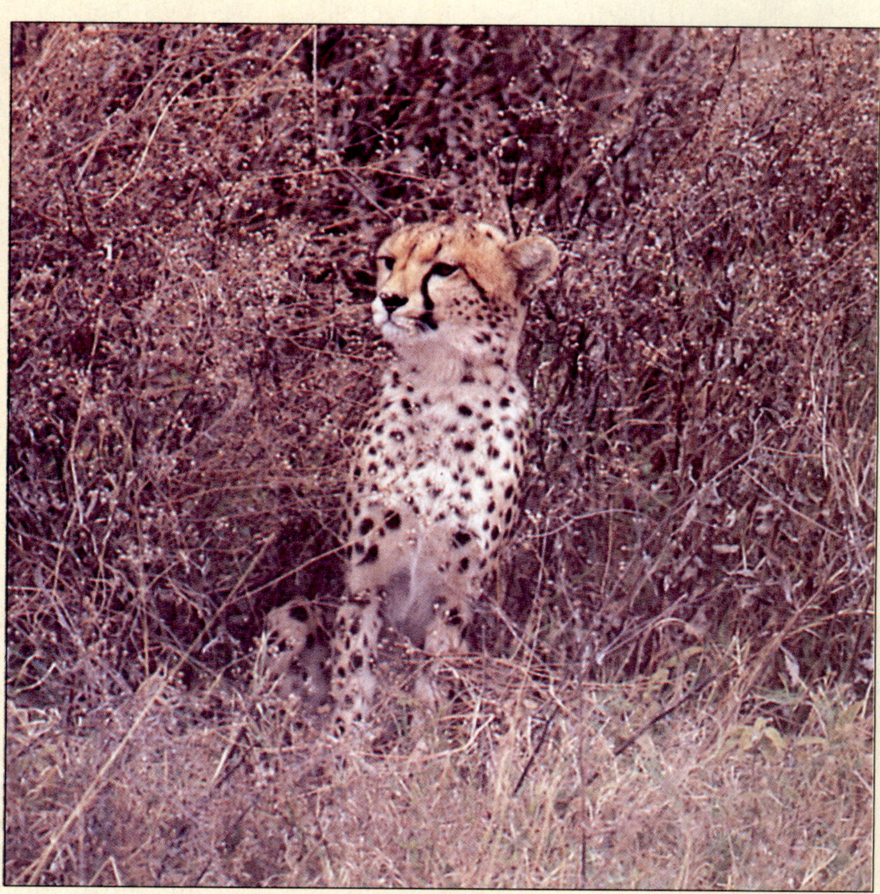

A cheetah concealed in the grass. (Carolina Biological Supply Company)

The mechanism of natural selection does not cause the development of a "perfect" organism. Rather, natural selection weeds out those phenotypes that are less adapted to environmental challenges so that those that are better adapted survive and pass their alleles on to their offspring.

Natural selection is the only process known that brings genetic variation into harmony with the environment and leads to adaptation. By reducing or eliminating alleles that result in the expression of less favorable traits, natural selection changes the composition of the gene pool in a favorable direction. This increases the probability that the favorable alleles responsible for an adaptation will come together in the offspring.

To summarize, using the Hardy-Weinberg principle as our base of comparison, we can see that evolution is occurring in most populations because allele frequencies are always changing. In most instances, microevolutionary changes in allele frequencies are inevitable in natural populations of organisms.

SELECTION CAN OPERATE ON A POPULATION IN SEVERAL WAYS

Three kinds of selection occur that cause changes in the normal distribution of phenotypes in a population (Fig. 18–4*a*)—stabilizing selection, directional selection, and disruptive selection. Although we consider each process separately, their influences generally overlap in nature.

Stabilizing Selection Favors Intermediate Phenotypes and Selects Against Extreme Phenotypes

The process of natural selection associated with a population that is well-adapted to its environment is known as **stabilizing selection.** Most populations are probably under the influence of stabilizing selection most of the time. In stabilizing selection, the frequency of phenotype extremes is selected against. In other words, individuals with a phenotype near the mean are favored.

One of the most widely studied cases of stabilizing selection involves human birth weight. Based on extensive data from hospitals, it has been determined that infants born with intermediate weights are more likely to survive. Infants at either extreme (that is, too small or too large) have higher death rates. When infants are too small, their body systems are immature, whereas infants that are too large at birth have difficult deliveries because they cannot pass as easily through the birth canal. Stabilizing selection operates to reduce the variability in birth weight so that it is close to the weight with the minimum mortality rate.

Because stabilizing selection tends to decrease genetic variation by favoring those individuals near the mean of the normal distribution at the expense of those at either extreme, the bell curve narrows (Fig. 18–4*b*). Although stabilizing selection decreases the amount of genetic variation in a population, variation is rarely eliminated by this process because other forces act against a decrease in

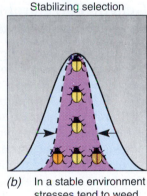

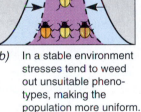

(*a*) Phenotype variation

Stabilizing selection

(*b*) In a stable environment stresses tend to weed out unsuitable phenotypes, making the population more uniform.

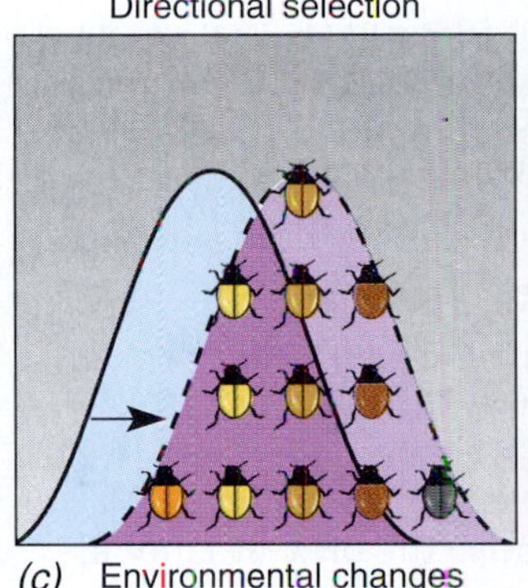

(*c*) Environmental changes favor the selection of more suitable phenotypes, causing the normal distribution to shift.

(*d*) Environmental changes favor the selection of more suitable phenotypes at both extremes of the normal distribution, causing a split.

FIGURE 18–4 Different types of natural selection. (***a***) A trait such as color exhibits a normal distribution of phenotypes. (***b***) As a result of stabilizing selection, the curve is narrower. (***c***) Directional selection moves the curve in one direction. (***d***) Disruptive selection results in two or more peaks.

variation. Mutation, for example, is slowly but continually adding to the genetic variation of a population.

Directional Selection Favors One Phenotype Over Another

If an environment changes over time, **directional selection** may favor phenotypes at one of the extremes of the normal distribution (Fig. 18–4*c*). Over successive generations, one phenotype gradually replaces another. So, for example, if greater size is advantageous in the new environment, larger individuals will become increasingly more common in the population. Directional selection can only occur, however, if the appropriate alleles (those favored under the new circumstances) are already present in the population.

A classic example of directional selection concerns wing color in peppered moths in England (see Focus On Evolution: The Case of the Peppered Moth, page 10). Recall that most of the peppered moths in rural England have a light (black and white peppered) wing color and only a few are melanic, or all black. In industrial regions the situation is reversed: most of the moths are black and only a few are light. In other words, in some places directional selection operates toward the light phenotype, whereas in other places directional selection occurs in the opposite direction toward the black form.

Disruptive Selection Selects for Phenotypic Extremes

Sometimes extreme changes in the environment may favor two or more different phenotypes at the expense of the mean. That is, more than one phenotype may be favored in the new environment, whereas the average, or intermediate, phenotype is selected against. **Disruptive selection** is a special type of directional selection in which there is a trend in several directions rather than one (see Fig. 18–4*d*). It results in a divergence, or splitting apart, of distinct groups of individuals within a population.

Limited food supply during a severe drought caused a population of Galapagos finches to undergo disruptive selection. The finch population initially exhibited a variety of beak sizes and shapes. Because the only food available for the finches during the drought was wood-boring insects and seeds from cactus fruits, selection favored birds with beaks suitable for obtaining these types of foods. Finches with longer beaks survived because they could open cactus fruits, whereas those birds with wider beaks were favored because they could strip off tree bark to expose insects. Thus, finches with beaks at two extremes of the normal distribution were favored over birds with average beaks.

GENETIC VARIATION IS NECESSARY IF EVOLUTION IS TO OCCUR

Change in the types and frequencies of alleles in gene pools, whether by natural selection or other means, is possible only if there is a source of inherited variation. Genetic variation is the raw material for evolutionary change. It provides the diversity upon which natural selection can act. Without genetic variation there can be no heritable differences in the ability to reproduce and, therefore, no natural selection.

The gene pools of populations contain large reservoirs of genetic variation that have been introduced by mutation. Sexual reproduction, with its associated crossing over and independent assortment of chromosomes during meiosis and genetic recombination during fertilization, also contributes to genetic variation (Chapter 10). The sexual process allows the variability introduced by mutation to combine in new ways, which may be expressed as new phenotypes.

The effect of recombination can be surprisingly great. Nine different genotypes are generated in a dihybrid cross (*AaBb* × *AaBb*) involving only 2 genes on different loci, each with only 2 alleles.[3] If we were dealing with 5 different genes, each with 6 alleles, the number of different genotypes possible would be 4,084,101![4] Since most organisms possess thousands of different genes, the number of different possible combinations of alleles is staggering! Some of the combinations that are generated may be adaptively superior, and natural selection could favor them.

Genetic Polymorphism Is an Example of Variation

One way of evaluating genetic variation in a population is to examine **genetic polymorphism,** which is the presence in a population of two or more alleles for a given gene. Gene pools contain a tremendous reservoir of genetic polymorphism, much of which is present at low frequencies and much of which is hidden. Until recently, biologists could not estimate the total amount of genetic polymorphism in populations because they could recognize only those genes that have different alleles conspicuous enough to cause differences in phenotypes. Biologists now take a random sample of proteins from an

[3]The 9 genotypes that result from *AaBb*×*AaBb* are *AABB, AABb, AAbb, AaBB, AaBb, Aabb, aaBB, aaBb,* and *aabb.*

[4]There are 21 different ways that 6 alleles can combine in diploid individuals. Because there are 5 different genes, each with 6 alleles, the total number of different genotypes possible is $21^5 = 4{,}084{,}101$.

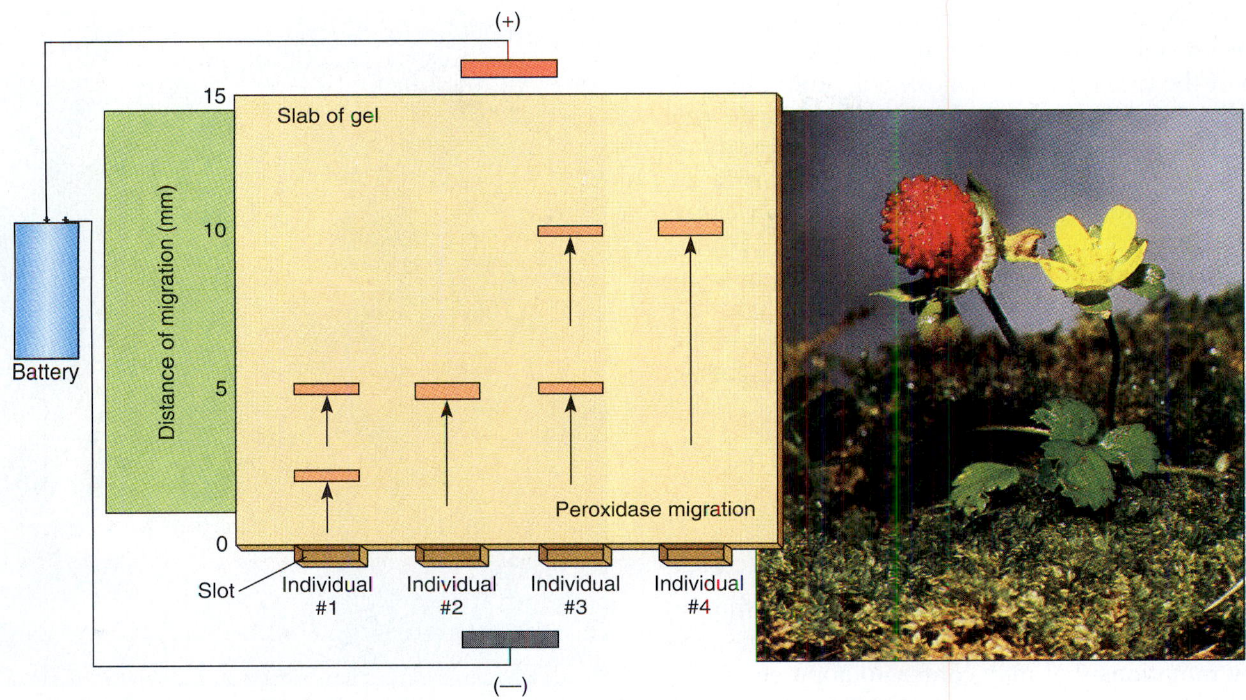

FIGURE 18–5 Genetic polymorphism. Gel electrophoresis demonstrates genetic polymorphism in a wild strawberry population. Tissue extracts containing the enzyme peroxidase from four different individuals were placed in slots in a slab of gel (bottom of drawing). An electric current was applied to the gel, with the positive side at the top of the slab and the negative side at the bottom. Because peroxidase has a net negative charge, it migrated toward the positive side. Slight variations in amino acid sequence in the peroxidase molecules caused them to have slightly different negative charges and, therefore, to migrate at different rates. Three different forms of peroxidase were found in the four individuals studied. Of course, each individual can only possess a maximum of two different forms because strawberries are diploid and therefore carry two alleles for this protein. Individuals with two different forms of peroxidase are heterozygous, whereas those with only one form are homozygous. (Richard H. Gross)

organism and, by biochemical techniques such as gel electrophoresis, measure how many proteins exist in two or more forms as determined by different amino acid sequences (Fig. 18–5). Each variety of a particular protein is coded by a different allele. Using this type of data, it is estimated that 25% of the genes in vertebrate populations are genetically polymorphic.

Genetic Variation Can Be Maintained by Heterozygous Advantage

Often the number of individuals heterozygous for a specific gene (*Aa*) are more common in the population than would be predicted by the Hardy-Weinberg principle. It thus appears that the heterozygous condition, *Aa*, sometimes has a higher degree of fitness than either homozygous state, *AA* or *aa*. This phenomenon, called **heterozygous advantage,** is demonstrated in humans by the selective advantage bestowed on heterozygous carriers of the sickle-cell allele.

Recall that the mutant allele for sickle-cell anemia codes for an altered hemoglobin that deforms or sickles the red blood cell so it is more likely to be destroyed in the liver, spleen, or bone marrow (Chapter 12). Individuals who are homozygous for the sickle-cell allele usually die at an early age.

Heterozygous individuals carry alleles for both normal and sickle-cell hemoglobin. The heterozygous condition causes an individual to be more resistant to a particular type of malaria than are those individuals who are homozygous normal (both alleles are normal) for hemoglobin. In a heterozygous individual, each allele produces its own specific kind of hemoglobin, and the red blood cells contain the two kinds in roughly equivalent amounts. Such cells do not ordinarily sickle, and the red blood cells containing the abnormal hemoglobin are more resistant to infection from the malarial parasite than are the red blood cells containing normal hemoglobin.

Each of the two types of homozygous individuals is at a disadvantage. Those homozygous for the sickle-cell allele are likely to die of sickle-cell anemia, whereas those homozygous for the normal allele may suffer or die of malaria. The heterozygote, thus, is more fit than either homozygote. In certain parts of Africa, India, and Southern Asia where malaria is prevalent, heterozygous individuals survive in greater numbers than do individuals of either homozygote. Both alleles are maintained in the population even though one of the homozygotes is lethal.

Neutral mutations give no selective advantage or disadvantage

Some of the genetic variations observed in a population—the variation in human fingerprints, for example—are caused by mutations that may confer no apparent selective advantage or disadvantage in a particular environment. That is, **neutral mutations** do not alter the fitness of an individual to survive and reproduce and are, therefore, not adaptive.

The extent of neutral mutations in organisms is difficult to determine. It is relatively easy to demonstrate that an allele is beneficial or harmful, provided its effect is observable. But the variation in alleles apparent by gel electrophoresis, often involving very slight differences in the structure of a protein, may or may not be neutral. These alleles may be influencing the organism in subtle ways that are difficult to measure or assess. Also, an allele that is neutral in one environment may be beneficial or harmful in another.

Members of a species form a reproductively isolated group

The concept of distinct groups of living organisms, known as **species** (from the Latin, meaning "kind") is not new. However, every definition of exactly what constitutes a species has some sort of limitation. Linnaeus, the 18th century biologist considered the founder of modern taxonomy, classified plants into separate species based on differences in morphology, or physical form. This method is still used to characterize species, but morphology alone is not adequate to explain what constitutes a species. For example, dogs come in a wide variety of sizes and shapes, but all dogs are members of the same species (Fig. 18–6).

FIGURE 18–6 A beagle and a Labrador retriever exemplify variation in dogs. Many different dog varieties exist, but all are members of the same species, *Canis familiaris.* (Kenneth Knott/ Fine Light.)

The study of population genetics did much to clarify the concept of species: A species is a group of organisms with a common gene pool. Often referred to as the **biological species concept,** this definition of species is based upon reproductive isolation. Members of a species freely interbreed with other members of the same species to produce fertile offspring and do not interbreed with (that is, are **reproductively isolated** from) members of different species. In other words, each species has a gene pool that is isolated from that of other species, and each species is restricted by reproductive barriers from genetically mixing with other species.

One of the problems with the biological species concept is that it only applies to sexually reproducing organisms. Organisms that reproduce asexually do not interbreed in the first place, so we cannot think of them in terms of reproductive isolation. For these organisms and for extinct organisms, the species concept is still valid; they are classified into species on the basis of structural and biochemical characteristics. Two populations that are widely separated geographically may be so much alike that they are placed in the same species, but it is impossible to test whether they will interbreed in nature. Also, organisms assigned to different species in nature may interbreed if they are brought into a zoo, a greenhouse, an

FIGURE 18–7 Members of different species sometimes interbreed in artificial surroundings. A liger (right), a hybrid between a lion and a tiger, in a state circus in former East Germany. Compare the liger's features with the tiger to its left. Although the geographical ranges of lions and tigers overlap in parts of Asia, a hybrid has never been found in nature. Lions and tigers have been known to crossbreed when brought together in zoos, but their offspring are sterile. (Courtesy P. Müller, Zoo Leipzig)

aquarium, or the laboratory (Fig. 18–7). Therefore, we usually include in our definition of species that they do not normally interbreed *in nature.*

To summarize, a species is a group of more or less distinct organisms capable of interbreeding with one another in nature but reproductively isolated from other species. This definition is far from perfect, however, and the biological species concept has a number of limitations.

DIFFERENT SPECIES HAVE VARIOUS MECHANISMS TO ACHIEVE REPRODUCTIVE ISOLATION FROM ONE ANOTHER

A number of biological mechanisms prevent interbreeding between different species. **Reproductive-isolating mechanisms** preserve the integrity of the gene pool of each species because gene flow between species is prevented. Some reproductive-isolating mechanisms work before fertilization occurs, whereas others work after mating.

Sometimes genetic exchange is prevented between two groups because they reproduce at different times of the day, season, or year. Many such examples, called **temporal isolation** (also known as **seasonal isolation**), exist. For example, two very similar species of sage inhabit the same area of southern California, but they do not interbreed. Black sage flowers in early spring, whereas white sage blooms in late spring and early summer (Fig. 18–8).

Many animal species exchange a distinctive series of signals between a male and female before mating. Such courtship behaviors are an example of **behavioral isolation** (also known as **sexual isolation**). Bowerbirds, for example, exhibit species-specific courtship patterns. The male satin bowerbird of Australia constructs an elaborate bower of twigs, adding decorative blue parrot feathers and yellow flowers at the entrance (Fig. 18–8). When a female approaches the bower, the male dances about her, holding a particularly treasured decoration in his beak. While dancing, he sings a courtship song that consists of a variety of sounds, including buzzes and laughlike hoots. These specific courtship behaviors keep other bird species reproductively isolated from the satin bowerbird. If a male and female of two different species begin courtship, the courtship stops when one member does not recognize the courtship signals of the other.

Sometimes members of different species will court and attempt copulation, but the structure of their genital organs is incompatible, so successful mating is prevented.

(a) (b)

FIGURE 18–8 Reproductive isolating mechanisms. (***a***) Temporal isolation. The black sage (left) flowers in early spring, whereas the white sage (right), photographed at the same time of year, has unopened flower buds. (***b***) Behavioral isolation. The male satin bowerbird constructs a bower to help attract a female. Note the flowers and blue decorations that he has arranged at the entrance to his bower. Different species exhibit a variety of highly specialized courtship patterns that prevent mating between closely related animals. (a-1 and 2, Courtesy of Robert Thorne, Rancho Santa Ana Botanic Garden; b, Patti Murray)

Morphological or anatomical differences that inhibit mating between species are known as **mechanical isolation.** For example, many flowering plants have physical differences in their flower parts that help them maintain their reproductive isolation from one another. The sage plants previously discussed as an example of temporal isolation also exhibit mechanical isolation. Black sage, which is pollinated by small bees, has a different floral structure than white sage, which is pollinated by large carpenter bees. The differences in floral structures prevent the insects from cross-pollinating the two species.

Sometimes, despite reproductive-isolating mechanisms to prevent mating, fertilization occurs between gametes of two different species. Even though fertilization has taken place, other reproductive-isolating mechanisms ensure reproductive failure. Generally, the embryo of such a union is aborted. Embryo development is a complex process that requires the precise interaction and coordination of many genes. Apparently, the genes from parents belonging to different species do not interact properly to regulate the mechanisms for normal embryonic development. For example, nearly all the hybrids die in the embryonic stage when the eggs of a bullfrog are fertilized artificially with sperm from a leopard frog.

If a hybrid does live, it may still not be able to reproduce. There are several reasons why this is so. Hybrid animals may exhibit courtship behaviors incompatible with those of either parental species. As a result, they will not mate. More often, the gametes of an interspecific (between-species) hybrid are abnormal because of problems during meiosis. This is particularly true if the two species have different chromosome numbers. For example, a mule is the offspring of a female horse ($2n = 64$) and a male donkey ($2n = 62$). This type of union almost always results in sterile offspring ($2n = 63$) because synapsis, the pairing of homologous chromosomes during meiosis, cannot occur properly (Chapter 10).

THE KEY TO SPECIATION IS THE DEVELOPMENT OF REPRODUCTIVE-ISOLATING MECHANISMS

We are now ready to consider the process of **speciation,** which is the formation of a new species. Speciation occurs when a population diverges, or splits apart, from the rest of the species. A required step in speciation is the reproductive isolation of that population from other members of the species. When a population is sufficiently different from its ancestral species so that no genetic exchange can occur between them, even if the two populations meet, we say that speciation has occurred. Such a situation is thought to arise in two ways: through allopatric or sympatric speciation.

Allopatric Speciation Occurs through the Effects of Long Physical Isolation and Different Selective Pressures

Speciation that occurs when one population becomes geographically separated from the rest of the species and subsequently evolves is known as **allopatric speciation** (from the Greek *allo,* meaning "different," and *patri,* meaning "fatherland"). Allopatric speciation is thought to be the most common method of speciation, and the evolution of new species of animals has been almost exclusively by allopatric speciation.

The geographical isolation required for allopatric speciation may occur in several ways. The Earth's surface is in a constant state of change: rivers shift their courses; glaciers migrate; mountain ranges form; land bridges develop, separating previously united aquatic populations; large lakes diminish into several smaller, geographically separated pools.

What might be an imposing geographical barrier to one species may be of no consequence to another. For example, as a lake subsides into smaller pools, fish are usually unable to cross the land barriers between the pools and so become reproductively isolated (Fig. 18–9). Birds, on the other hand, can easily fly from one pool to another. Likewise, plants such as cattails, which disperse their fruits by air currents, would not be isolated by this barrier.

Allopatric speciation also occurs when a small population migrates and colonizes a new area away from the original species. This colony is geographically isolated from its parent species. The Galapagos Islands and the Hawaiian Islands were colonized by individuals of a few species. The unique species found on each island today descended from these original colonizers (Fig. 18–10).

FIGURE 18–9 Allopatric speciation of pupfish in Death Valley. Shown is a desert pupfish, one of many different species. The desert in the Death Valley region of California and Nevada contains a number of small, isolated pools. Many of these pools are inhabited by pupfish, but each pool contains a different pupfish species. The various pupfish species are thought to have evolved from a common ancestral population when large, interconnected glacial lakes dried up about 10,000 years ago, leaving behind small, isolated pools fed by springs. Each pool contained a small population of pupfish that gradually diverged from the common ancestral population by genetic drift and natural selection. (Steinhart Aquarium, Tom McHugh/Photo Researchers, Inc.)

(a)

(b)

FIGURE 18–10 Allopatric speciation of the nene. (***a***) The nene (pronounced "nay-nay") is a goose found in Hawaii. It is thought to have evolved when a small population of geese from North America colonized the geographically isolated Hawaiian Islands. Although the nene is endangered, strict conservation measures have brought it back from the brink of extinction. (***b***) The Canadian goose is thought to be a close relative of the Hawaiian goose. (*a,* M. J. Rauzon/VIREO: *b,* Mary Clay/Tom Stack & Associates)

Sometimes allopatric speciation occurs quite rapidly. Early in the 15th century, a small population of rabbits was released on Porto Santo, a small island off the coast of Portugal. Because there were no rabbits or other competitors and no carnivorous enemies on the island, the rabbits thrived. By the 19th century, these rabbits were markedly different from their ancestors. They were only half as large, had a different color pattern, and their lifestyle was more nocturnal. Most significantly, they could not produce offspring when bred with members of the ancestral European species. Within 400 years, a new species of rabbit had evolved.

Sympatric Speciation Results from the Divergence of Two Populations in the Same Physical Location

Although geographical isolation is an important factor in many cases of evolution, it is not an absolute requirement. When a population forms a new species within the same geographical region as its parent species, **sympatric speciation** (from the Greek *sym,* meaning "together," and *patri,* meaning "fatherland") has occurred. The divergence of two gene pools in the same geographical range is especially common in plants. Although a few cases of sympatric speciation in parasitic insects have been studied, the role of sympatric speciation is unclear in animals.

How does sympatric speciation occur? We have seen that hybrids formed from the union of gametes from two different species rarely produce offspring and, if offspring are produced, they are usually sterile. Before gametes form, meiosis occurs to reduce the chromosome number. In order for the chromosomes to be parceled correctly into the gametes, homologous chromosome pairs must synapse during prophase I. This cannot occur properly in interspecific hybrid offspring because the chromosomes are not homologous. However, *if* the 2n chromosome number doubles *before* meiosis, then the pairing of homologous chromosomes can take place. This spontaneous doubling of chromosomes has been documented in both plants and animals. It is not a common occurrence, but neither is it rare. It produces nuclei with multiple sets of chromosomes.

Polyploidy, the possession of more than two sets of chromosomes, is a major factor in plant evolution. When polyploidy occurs in conjunction with sexual reproduction between two or more different species, it is known as **allopolyploidy,** and it can produce a fertile interspecific hybrid (Fig. 18–11). This is because the polyploid condition provides the homologous chromosome pairs necessary for synapsis during meiosis. As a result, gametes may be viable. An allopolyploid, that is, an interspecific hybrid produced by allopolyploidy, can reproduce with itself (self-fertilization) or with a similar individual. However, allopolyploids are reproductively isolated from both parents because the gametes of the allopolyploid have a different number of chromosomes than do those of either parent.

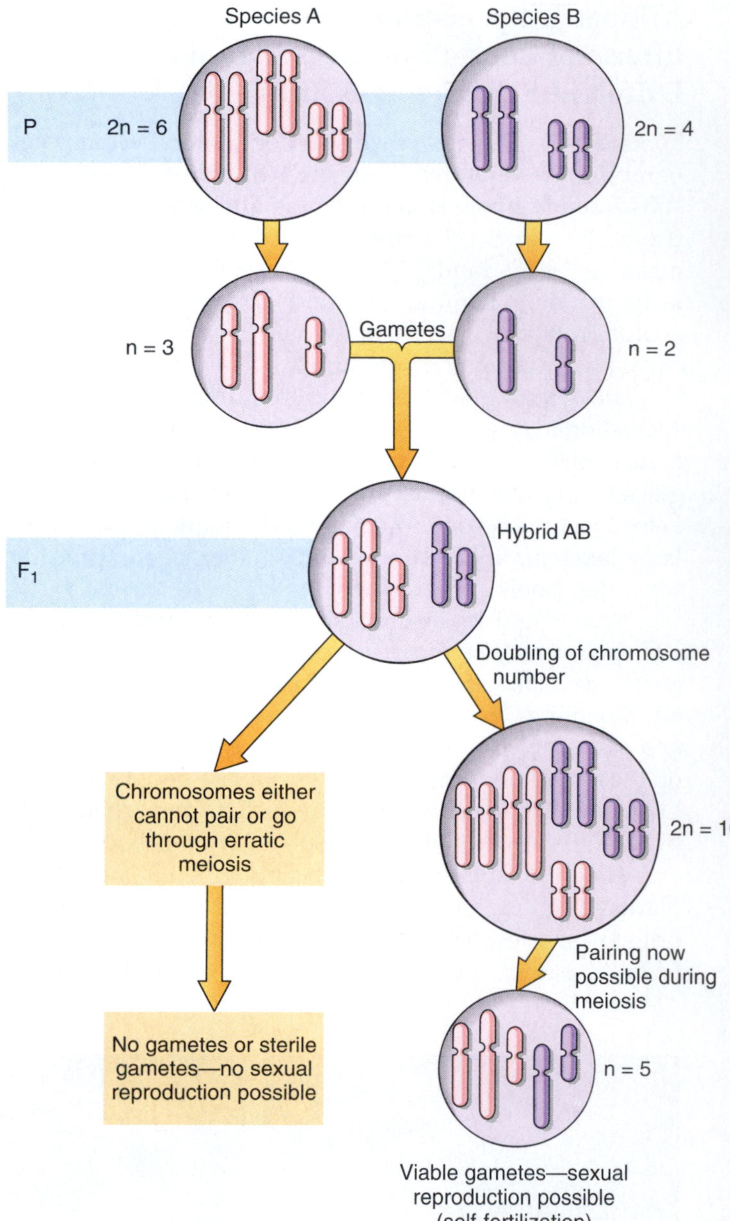

FIGURE 18–11 How a fertile allopolyploid is formed. When two different species (designated the P generation) successfully interbreed, the hybrid offspring (the F_1 generation) are almost always sterile (bottom left). If the chromosomes double, the hybrid is able to undergo meiosis and is fertile (bottom right).

Although allopolyploidy is extremely rare in animals, it has been a significant factor in the evolution of the flowering plants. Almost one-half of all flowering plants are

Concept Connections

Evolution ⟷ *Diversity of Life*

We have seen that mechanisms such as natural selection, mutation, genetic drift, and migration explain microevolution, the evolutionary changes within a population. If a population remains reproductively isolated from its parent species, mutation and natural selection cause it to diverge even more. Given enough changes, the population may become a new species. Thus, new species arise from pre-existing species, which in turn evolved from other species that came before them.

The family tree of life can be traced back in time to the origin of life on planet Earth. All life is connected—that is, all organisms that exist today are related to one another because, if we go back far enough in time, we all share a common ancestor. In Chapter 19 we reconstruct the origin and evolutionary history of life, and in Chapters 20 to 25 we examine life's diversity and continue our examination of how living things today are interrelated.

thought to be polyploids, and most of these are allopolyploids (Fig. 18–12). Morever, allopolyploidy provides a mechanism for extremely rapid speciation. A single generation is all that is needed to form a new, reproductively isolated species. Allopolyploidy is thought to explain the rapid appearance of flowering plants in the fossil record and the remarkable diversity (about 235,000 species) in flowering plants today.

FIGURE 18–12 Evolution of an allopolyploid. An allopolyploid primrose, *Primula kewensis*, arose during the early part of the 20th century. The F_1 hybrid of *P. floribunda* (2n = 18) and *P. verticillata* (2n = 18) was a diploid perennial (2n = 18), which was sterile. Three different times it spontaneously formed a fertile branch, which was *P. kewensis*, a fertile allopolyploid (2n = 36) that produced seeds. (The specific epithet *kewensis* was given in recognition that this species arose at the Royal Botanic Gardens at Kew, England.)

Chapter Summary

I. Evolution is a change in allele frequencies in the gene pool of a population. Each individual within a population contains only a portion of the alleles in the gene pool.
   A. The Hardy-Weinberg principle states that allele frequencies in a population tend to remain constant in successive generations unless certain factors are operating.
   B. Allele frequencies may be changed by mutation, genetic drift, gene flow, and natural selection.
      1. The source of new alleles in a gene pool is mutation.
      2. Genetic drift is the random change in allele frequencies of a small population. The changes are usually not adaptive.
      3. The migration of individuals between local populations causes a corresponding movement of alleles, or gene flow, that can cause changes in allele frequencies.
      4. Changes in allele frequencies that lead to adaptation to the environment are caused by natural selection.

II. Selection can change the composition of a gene pool in a favorable direction for a particular environment.
   A. Stabilizing selection favors the mean at the expense of phenotypic extremes.
   B. Directional selection favors one phenotype over another, causing a shift in the phenotypic mean.
   C. Disruptive selection favors two or more phenotypic extremes.

III. A biological species is defined as a group of similar organisms that have the potential to interbreed with one another in nature but not with members of different species.

IV. Reproductive-isolating mechanisms restrict the gene flow between species.
   A. Temporal isolation occurs when two species reproduce at different times of the day, season, or year.
   B. In behavioral isolation, distinctive courtship behaviors prevent mating between species.
   C. Mechanical isolation is due to anatomical differences in the reproductive structures of species.
   D. Reproductive failure is common even when fertilization has taken place between gametes of two different species.
      1. The hybrid embryo usually dies at an early stage of development.
      2. If a hybrid survives to adulthood, it usually cannot reproduce successfully.

V. Speciation is the evolution of a new species from an ancestral population.
   A. Allopatric speciation occurs when one population becomes geographically isolated from the rest of the species and subsequently evolves.
   B. Sympatric speciation does not require geographical isolation. It is extremely rare in animals; in plants it occurs as a result of allopolyploidy.

Selected Key Terms

allele, p. 321
allopatric speciation, p. 333
allopolyploidy, p. 334
gene flow, p. 325
gene pool, p. 320
genetic drift, p. 324
Hardy-Weinberg principle, p. 321
heterozygous advantage, p. 329
microevolution, p. 321
mutation, p. 323
natural selection, p. 325
population, p. 320
reproductive isolation, p. 330
speciation, p. 332
species, p. 330
sympatric speciation, p. 334

Post-Test

1. A/an ______________ consists of all of the individuals of the same species that live in a particular place at a particular time.
2. The Hardy-Weinberg principle demonstrates that the process of inheritance does not, by itself, cause changes in ______________ ______________.
3. The source of the genetic variability that is the raw material of evolution is ______________.
4. Random genetic events, called ______________ ______________, may have a major effect on allele frequencies.
5. The impoverished gene pool of cheetahs is due to a ______________ ______________.
6. The movement of alleles from one population to another, called gene flow, is caused by the ______________ of breeding individuals.
7. Changes in allele frequencies that lead to adaptation to a particular environment are due to ______________ ______________.
8. In ______________ selection, individuals with a phenotype near the mean are favored over those with phenotype extremes.
9. The ______________ selection of peppered moths is an indirect consequence of air pollution.
10. A human who has alleles for both normal hemoglobin and sickle-cell hemoglobin demonstrates ______________ ______________ in an area where malaria is prevalent.
11. A/an ______________ is a group of similar organisms that interbreed with one another in nature.
12. When two closely related species that live in the same geographical area reproduce at different times of the year, this is known as ______________ isolation.
13. If two different species have reproductive structures that prevent mating, they fail to reproduce because of ______________ isolation.
14. The most important type of speciation in animal evolution is ______________ speciation.
15. An individual that possesses multiple sets of chromosomes, in which one or more of those sets came from a different species, is known as a/an ______________.
16. Which of the following causes changes in allele frequencies? (a) mutation (b) natural selection (c) genetic drift (d) gene flow from migration (e) all of these
17. Mutation: (a) adds to the genetic variation of a population (b) is the result of genetic drift (c) leads to adaptive evolutionary change (d) almost always benefits the organism (e) all of these.
18. Which of the following statements is *not* true about natural selection? (a) Natural selection acts to preserve favorable traits and eliminate unfavorable traits. (b) The offspring of individuals that are better adapted to the

environment will make up a larger proportion of the next generation (c) Natural selection directs the course of evolution by preserving the traits acquired during an individual's lifetime (d) Natural selection depends upon the genetic variability in a population, which in turn arises through mutations.

19. Speciation: (a) only occurs when a population is geographically isolated (b) is a change in allele frequencies in a population (c) is the formation of new species from an ancestral population (d) maintains an intermediate phenotype in a stable environment.

20. Which occurs first during allopatric speciation? (a) temporal isolation (b) geographical isolation (c) mechanical isolation (d) behavioral isolation

Review Questions

1. Define population, species, and gene pool. How did population genetics help clarify the definition of a species?
2. Given that the Hardy-Weinberg equilibrium only occurs under conditions that populations in nature seldom, if ever, experience, why is it important?
3. Explain the effect of each of the following on genetic variation: (a) natural selection (b) mutation (c) sexual reproduction (d) gene flow (e) genetic drift.
4. If a mutation occurs in a body cell, can the new allele become established in a population? Why or why not?
5. Why are mutations almost always neutral or harmful?
6. Draw three graphs representing stabilizing, directional, and disruptive selection.
7. Give an example of each of the following: temporal isolation, behavioral isolation, and mechanical isolation.
8. Describe why reproductive failure usually occurs even when fertilization has taken place between gametes from two different species.
9. Identify at least five geographical barriers that might lead to allopatric speciation.
10. Why is allopatric speciation more likely to occur if the original isolated population is small?
11. Explain how allopolyploidy can cause a new plant species to form in as little time as one generation.

Thinking Critically

1. Discuss the following statement as it relates to AIDS-immune prostitutes: The presence of genetic variability within a population gives it the potential to meet new environmental challenges.
2. Explain the evolution of the giraffe's long neck by directional selection. Explain how stabilizing selection accounts for the giraffe's long neck today.
3. Insect populations that have been exposed to an insecticide such as DDT develop resistance to the insecticide over time. Would this be an example of stabilizing selection, directional selection., or disruptive selection? Explain.
4. Can a species evolve in such a way that it is no longer affected by natural selection? Is human evolution still under the influence of natural selection?
5. Explain why the biological species concept is not valid for certain plants.
6. As a continuation of the problem involving PTC tasters and nontasters in the Hardy-Weinberg focus box, calculate the frequencies of the genotypes *TT* amd *Tt*. (*Hint:* You have already been given the frequency of the *tt* genotype = 0.30. You can determine the frequency of *TT* and *Tt* by solving for p^2 and $2pq$.)

Recommended Readings

Arnold, M. L., "Natural Hybridization and Louisiana Irises: Defining a Major Factor in Plant Evolution," *BioScience* Vol. 44, No. 3, March 1994. Interspecific hybridization appears to be a significant factor in the evolution of Louisiana irises.

Cohn, J. P., "Genetics for Wildlife Conservation," *BioScience* Vol. 40, No. 3, March 1990. Genetic analysis of endangered species offers hope for their survival.

Lipske, M., "Fast Cat in a Marathon," *International Wildlife*, September/October 1993. An update on the status of the cheetah.

Rennie, J. "Are Species Specious?," *Scientific American*, November 1991. A brief essay on the problems biologists still have with the definition of species.

Thompson, J. D. "The Biology of an Invasive Plant," *BioScience* Vol. 41, No. 6, June 1991. A new species of cordgrass that arose approximately 100 years ago has become a common plant in salt marshes of Great Britain.

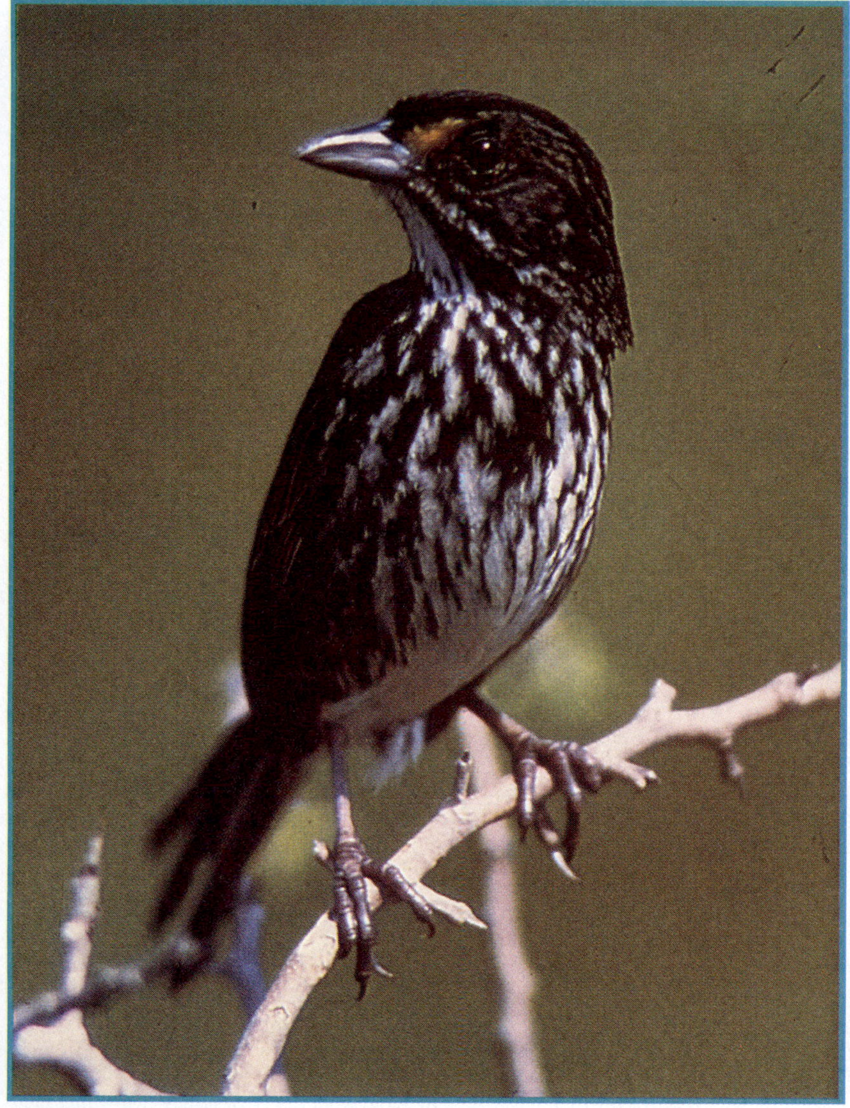

The dusky seaside sparrow became extinct in 1987. (*U.S. Fish and Wildlife Service*)

Death of a Species

The most serious threat to the survival of many species today is habitat disruption caused by human activities. We alter habitats when we build roads, parking lots, and buildings. Forest habitats are altered or destroyed when we log them for timber or clear them to grow crops or graze domestic animals. We drain marshes to build on aquatic habitats, thus making them terrestrial, and we flood terrestrial habitats when we build dikes and dams, making them aquatic. Because most organisms are dependent on a particular type of environment, habitat destruction reduces their biological ability to survive.

The dusky seaside sparrow became extinct in 1987, largely due to human destruction of its habitat. This dark-colored ("dusky") bird originally existed as two populations in Florida, one along the St. Johns River and the other on Merritt Island. Both locations offered an ideal habitat for the dusky seaside sparrow—marshes of cordgrass with widely scattered cabbage palms. Approximately 2000 mating pairs lived on Merritt Island as recently as the 1940s, and a survey in 1968 along the St. Johns River revealed a population almost as large as that on Merritt Island.

The demise of the sparrow population on Merritt Island began with the construction of the Kennedy Space Center on the island during the 1950s. In order to control the mosquitoes that bred on Merritt Island, workers built dikes and flooded the marshes, which destroyed the cordgrass. As the marsh habitat was obliterated, the population of dusky seaside sparrows declined until only a few mating pairs were observed in the 1970s. In the following years, these birds also disappeared.

During the 1950s and 1960s as the marshes along the St. Johns River were drained for residential developments and pastures, the dusky seaside sparrow showed a gradual but steady decline. The birds became so scarce that in 1979 and 1980 biologists searched Merritt Island and St. Johns River for remaining dusky seaside sparrows, which they planned to capture and breed to prevent extinction of this species. However, the biologists were able to find only seven birds, and all of them were males. In a final attempt to preserve some of the dusky seaside sparrow's gene pool, the last six survivors were successfully crossed with a related seaside sparrow subspecies in the mid-1980s. The last pure dusky seaside sparrow died in 1987.

The dusky seaside sparrow is not the only species to become extinct in recent years. Currently, Earth's biological diversity is disappearing at an alarming rate. Biologists estimate that at least one species becomes extinct every day and that it is likely that a substantial portion of Earth's biological diversity will be eliminated within the next few decades.

The current wave of extinction is not the first to occur, however. Later in this chapter we consider major extinction episodes that took place at several times during Earth's history, along with their effects on evolution. Unlike those earlier periods of extinction, however, the current extinction episode is occurring more swiftly in time and is due primarily to human activities. We return to the current problem of declining biological diversity in Chapter 49.

Macroevolution and the History of Life

19

Learning Objectives

After you have studied this chapter you should be able to:

1. Define macroevolution and distinguish between microevolution and macroevolution.
2. Take either side in a debate on the pace of evolution, by representing the opposing views of gradualism and punctuated equilibrium.
3. Define extinction and discuss its biological ramifications.
4. Discuss the evolutionary significance of extinction and adaptive radiation.
5. Describe the conditions that are thought to have existed on early Earth and outline the major steps that are thought to have occurred in the origin of living cells.
6. Explain how the evolution of photosynthetic organisms affected Earth's atmosphere and other life forms.
7. Describe the endosymbiont theory and summarize the supporting evidence.
8. List the four geological eras in chronological order and give approximate dates for each.
9. Briefly describe the geological features and distinguishing plant and animal life for the Precambrian, Paleozoic, Mesozoic, and Cenozoic eras.
10. Explain how the course of evolution was affected by continental drift.

Key Concepts

- ☐ Macroevolution encompasses major evolutionary events that occur in groups of species over geological time. Macroevolutionary changes are so great that the organisms that possess them are assigned to a new genus or higher taxonomic category.
- ☐ Extinction is the eventual fate of all species. Several times during the history of life, mass extinction (a global occurrence in which many species perished abruptly) was followed by adaptive radiation (an "explosion" of speciation in which a single ancestral line produced many new species).
- ☐ The origin of life is linked to the physical and chemical conditions present on early Earth. The sequence of events that led to life started with the synthesis of simple organic molecules, which reacted with one another to form complex organic molecules. These macromolecules in turn reacted to form increasingly complex structures until, eventually, one was complex enough to be considered "alive." All forms of life present today evolved over time from this common ancestor.
- ☐ Geological evidence—in particular, the fossil record—provides us with much of what we know about the history of life: what kinds of organisms existed, and where and when they lived.

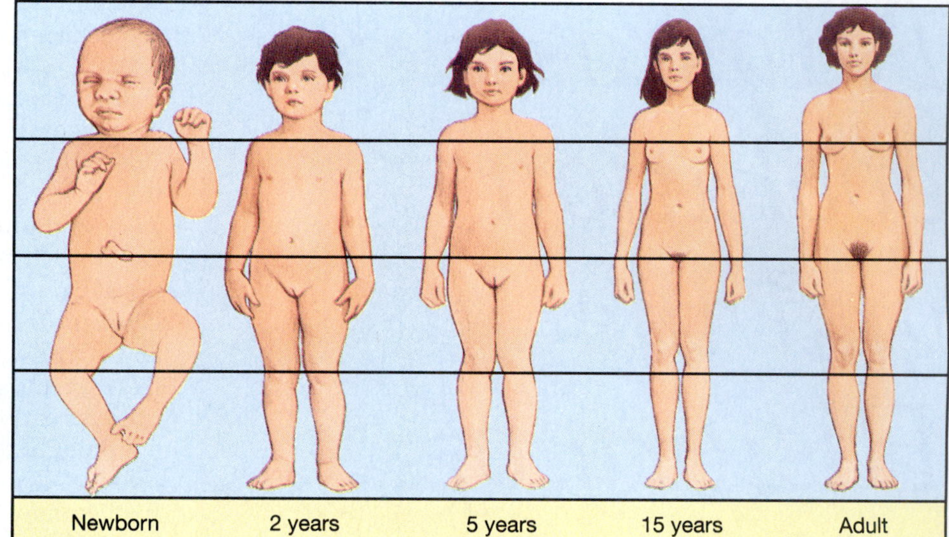

FIGURE 19–1 Differential growth of body parts in humans. Different stages in the growth of a human are drawn the same size to demonstrate varied rates of growth for different parts of the body. As humans develop, for example, their legs grow more rapidly than their heads.

MACROEVOLUTION INVOLVES CHANGES IN THE *KINDS* OF SPECIES OVER EVOLUTIONARY TIME

Macroevolution refers to major evolutionary events that occur in groups of species over geological time. These "major evolutionary events" are large phenotypic changes, such as the appearance of wings with feathers that evolved in birds. The phenotypic changes are so great that the organisms (and their descendants) that possess them are assigned to a new genus or higher taxonomic category.[1] Macroevolution is thus concerned with the origin of taxonomic categories above the species level. The origin of new designs, adaptive radiation, and mass extinction are important aspects of macroevolution.

New Designs Sometimes Arise from Structures Already in Existence

A change in the basic design of an organism can produce something unique. Examples of "new" and unusual features include wings on insects, flowers on flowering plants, and feathers on birds. Usually these "new" structures are variations of some structure that originally fulfilled one role, but that structure changed in such a way that it was adaptive for a different role. Bird feathers, which evolved from reptilian scales, are a good example of an evolutionary novelty. In like manner, lungs evolved from the swim bladders of fish.

How do such novel changes occur? Many are probably due to regulatory changes in development, which is the orderly sequence of changes that take place as a young organism grows and matures. Because regulatory genes responsible for development may exert control over hundreds of other genes, very slight changes in regulatory genes could cause major structural changes in the organism. Thus, evolutionary novelties may originate from mutations that alter the pathway of development.

For example, during development, most organisms have varied rates of growth for different parts of the body. The size of the head in human newborns is large in proportion to the rest of the body. As a human grows and matures, its torso, hands, and legs grow more rapidly than the head (Fig. 19–1). Varied rates of growth is a phenomenon found in many organisms, including the male fiddler crab with its single, oversized claw, and the ocean sunfish with its enlarged tail (Fig. 19–2). If the rates of growth are altered even slightly, drastic changes in the shape of the organisim may occur.

Sometimes novel changes are the result of changes in the *timing* of development. Consider, for example, the changes that would occur should a juvenile characteristic be retained in an adult. Adults of some salamander species have gills, a feature found only in the larval (immature) stages of other salamanders. Retention of gills throughout life obviously alters the adult salamander's behavioral and ecological characteristics (Fig. 19–3). Perhaps such salamanders succeeded because they had a selective advantage over "normal" adult salamanders. The gilled forms could remain aquatic and would not have to compete for food with the adult forms, most of which live on land. The gilled forms would also escape the typical predators of terrestrial salamanders (although they would have other predators in their watery environment).

[1] The taxonomic categories above the level of species are genus, family, order, class, division/phylum, and kingdom.

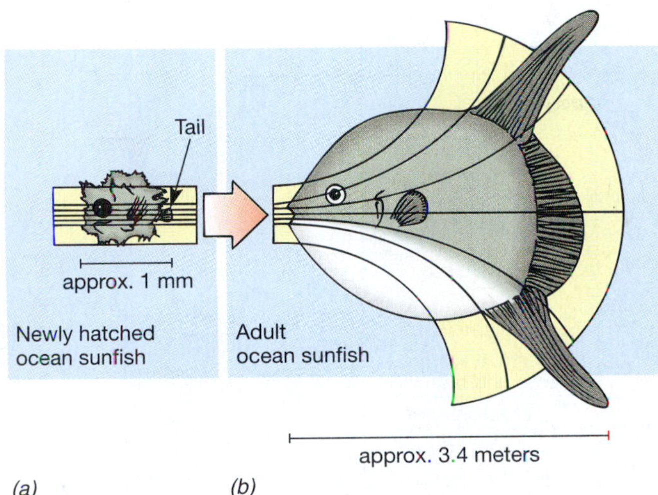

FIGURE 19–2 Differential growth of body parts in the ocean sunfish. The tail end of an ocean sunfish grows faster than its head end, resulting in the unique shape of the adult ocean sunfish. (***a***) A newly hatched ocean sunfish, which is only 1 mm long, has an extremely small tail. (***b***) The transformation to an adult can be visualized by drawing rectangular coordinate lines through the picture of the juvenile fish and then changing the coordinate lines mathematically. The adult ocean sunfish is 3.4 meters (11 feet) long and weighs 1 metric ton (2206 pounds).

Adaptive Radiation Is the Evolutionary Diversification of an Ancestral Species into Many Species

Once a novel feature arises that represents an evolutionary advancement, **adaptive radiation** may occur, in which an ancestral organism diversifies to fill a variety of different ecological roles. Adaptive radiation is the evolution of many related species from one or two ancestral species in a relatively short period of time.

The concept of adaptive zones was developed to help explain why adaptive radiation takes place. **Adaptive zones** are new ecological roles or ways of living that were previously not utilized by an ancestral organism. At the species level, the adaptive zone is essentially identical to the ecological niche (an organism's role within a community; see Chapter 46). Some adaptive zones include nocturnal flying to catch insects, grazing on grass while migrating across a savanna, and swimming at the ocean's surface. At higher levels of classification, such as genera and families, the concept of adaptive zones is more theoretical. An adaptive zone at this higher level represents an ecological pathway along which the taxonomic group evolves.

Each ecological niche can only be occupied by one group of organisms. A newly evolved species can take

FIGURE 19–3 Retention of juvenile characters. An adult salamander that has retained external gills. In most salamander species, the external gills found in the larval stage are not present in the adult. (Jane Burton/Bruce Coleman, Inc.)

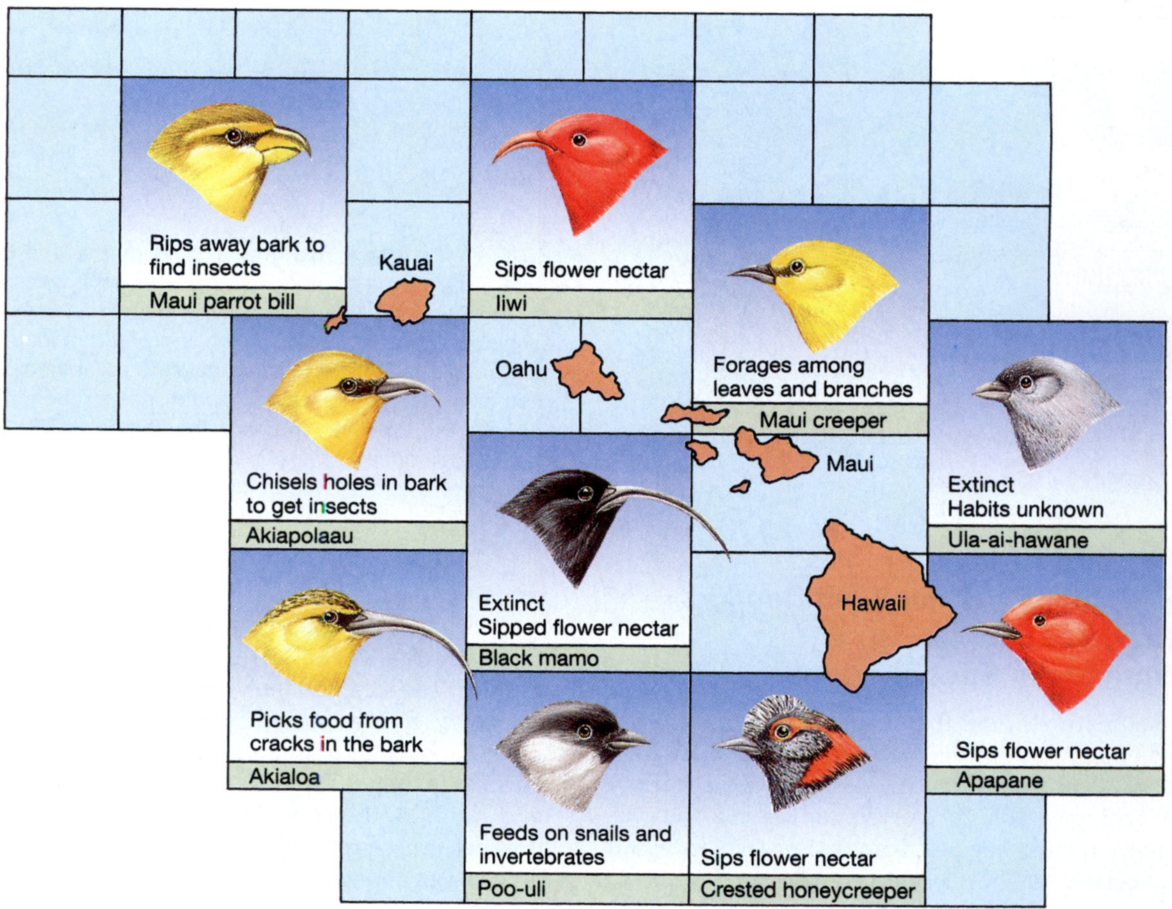

FIGURE 19–4 Adaptive radiation in Hawaiian honeycreepers. Many honeycreepers are now extinct as a result of human activities, including the destruction of habitat and the introduction of predators such as rats, dogs, and pigs.

over an adaptive zone even if it is already occupied, provided the new species has features that make it competitively superior to the original occupants. If a number of ecological niches are empty, they may be exploited by adaptive radiation. For example, consider the honeycreepers, a group of related birds found on the Hawaiian Islands. When the honeycreeper ancestor reached the Hawaiian Islands, there were few birds present. The succeeding generations of honeycreepers quickly diversified to occupy the many available ecological niches. The diversity in their bills is a particularly good illustration of adaptive radiation (Fig. 19–4). For example, some honeycreeper bills are curved to extract nectar out of tubular flowers, whereas others are short and thickened for ripping away bark in search of insects.

Adaptive radiation appears to be more common during periods of major environmental change, but it is difficult to determine if these changes actually trigger adaptive radiation. It is possible that major environmental change has an indirect effect on adaptive radiation by increasing the rate of extinction. Extinction produces empty ecological niches, which are then available for adaptive radiation. Mammals, for example, had existed millions of years before they underwent adaptive radiation, which is thought to have been triggered by the extinction of the dinosaurs. Originally, mammals were small animals that ate insects. In a relatively short period of time after the dinosaurs' demise, mammals diversified, occupying and exploiting a variety of niches. Flying bats, running gazelles, burrowing moles, and swimming whales all evolved from the ancestral mammals as a result of adaptive radiation.

The appearance of novel evolutionary features is associated with each major period of adaptive radiation. For example, shells and skeletons may have been the novel features responsible for a period of adaptive radiation at the beginning of the Paleozoic era (discussed shortly) when most animal phyla, both living and extinct, appeared.

Concept Connections

Natural Selection — *Coevolution*

The Hawaiian honeycreepers were introduced in this chapter as a dramatic example of adaptive radiation because of their diverse bills. At least one honeycreeper species, the iiwi, also provides an example of two principles of evolution considered elsewhere in this book—natural selection and coevolution. Recall that natural selection, first proposed by Darwin, is the differential reproductive success of individuals that are better adapted to the environment in which they live (Chapter 17). Coevolution, which has not been discussed in detail yet, is the interdependent evolution of two different species—such as birds and the flowering plants from which they obtain food. In coevolution, the evolution of each species is affected by the presence of the other species with which the first species interacts (Chapter 29). The result is an increased survivability for both species.

The iiwi is thought to possess its gracefully curved bill in order to sip nectar from flowers of the lobelia. The iiwi's bill fits perfectly into the long, tubular lobelia flowers, and the bird was often observed feeding on lobelias during the 19th century. However, during the 20th century, the number of lobelias in Hawaii declined dramatically. As a result, the iiwi started feeding on the ohia tree, which produces flowers that are short and open (nontubular). By comparing iiwi bills of museum specimens collected before 1902 with live birds (which were captured with nets, measured, and released), it was determined that the iiwi bill has become shorter and straighter during the past 90 to 100 years. This evolutionary change—the result of natural selection—is making the iiwi bill more suited for obtaining nectar from ohia flowers.

Care must be taken in interpreting a cause-and-effect relationship between the appearance of a novel evolutionary feature and adaptive radiation, however. It is tempting to take a simplistic approach and state, for example, that the evolution of the flower triggered the adaptive radiation of thousands of species of flowering plants. Perhaps it is true that flowering plants diversified after the evolution of the flower, which presented a more competitive method of sexual reproduction because it permitted pollination by insects and other animals (Chapter 29). However, adaptive radiation in the flowering plants may be a consequence of their many other advancements, such as efficient conducting tissues, instead of or in addition to flowers (Chapter 23).

Extinction Is an Important Aspect of Macroevolution

Extinction, the end of a lineage, occurs when the last individual of a species dies. It is a permanent loss, for once a species is extinct, it can never reappear. Extinctions have occurred continually since the origin of life. By one estimate, there is only one species living today for every 2000 that have become extinct. Extinction is the eventual fate of all species, in the same way that death is the eventual fate of all individual organisms.

Although extinction has a negative impact on biological diversity, it has a positive evolutionary aspect. As mentioned previously, when species become extinct, the ecological niches that they occupied become vacant. As a result, those organisms still living are presented with new opportunities for radiation and can diverge to fill the unoccupied ecological niches. In other words, the extinct species are replaced by new species.

During the course of life, there appear to have been two types of extinction. The continuous, low-level extinction of species, sometimes called **background extinction,** is one type. The second type has occurred five or six times during Earth's history. At these times, **mass extinctions** of numerous species and higher taxonomic categories have taken place in both terrestrial and aquatic environments (Fig. 19–5). The time period over which a mass extinction occurred may have lasted several million years, but that is a relatively short time period compared with the history of life. Each period of mass extinction, which appears to have been indiscriminate in its choice of which species survived and which became extinct, was followed by a period of "mass speciation," or adaptive radiation.

The causes of past mass extinction episodes are not well understood. Both environmental and biological factors seem to have been involved. Major changes in the climate could have adversely affected plants and animals that were unable to adapt. Marine organisms, in particular, are adapted to a very steady, unchanging climate. If Earth's temperature were to decrease overall by just a few degrees, many marine species would probably perish.

It is also possible that mass extinctions were due to changes in the environment triggered by catastrophes. If Earth was bombarded by a large asteroid or comet, for example, the dust produced by the impact would have filled the atmosphere, blocking much of the sunlight. In addition to killing many plants, the atmospheric dust would have lowered Earth's temperature, leading to the death of many marine organisms.

Biological factors can also trigger extinction. When a new species arises, it may be able to out-compete an older species, leading to its extinction. The human species has had a profound impact on extinction. As a result of the

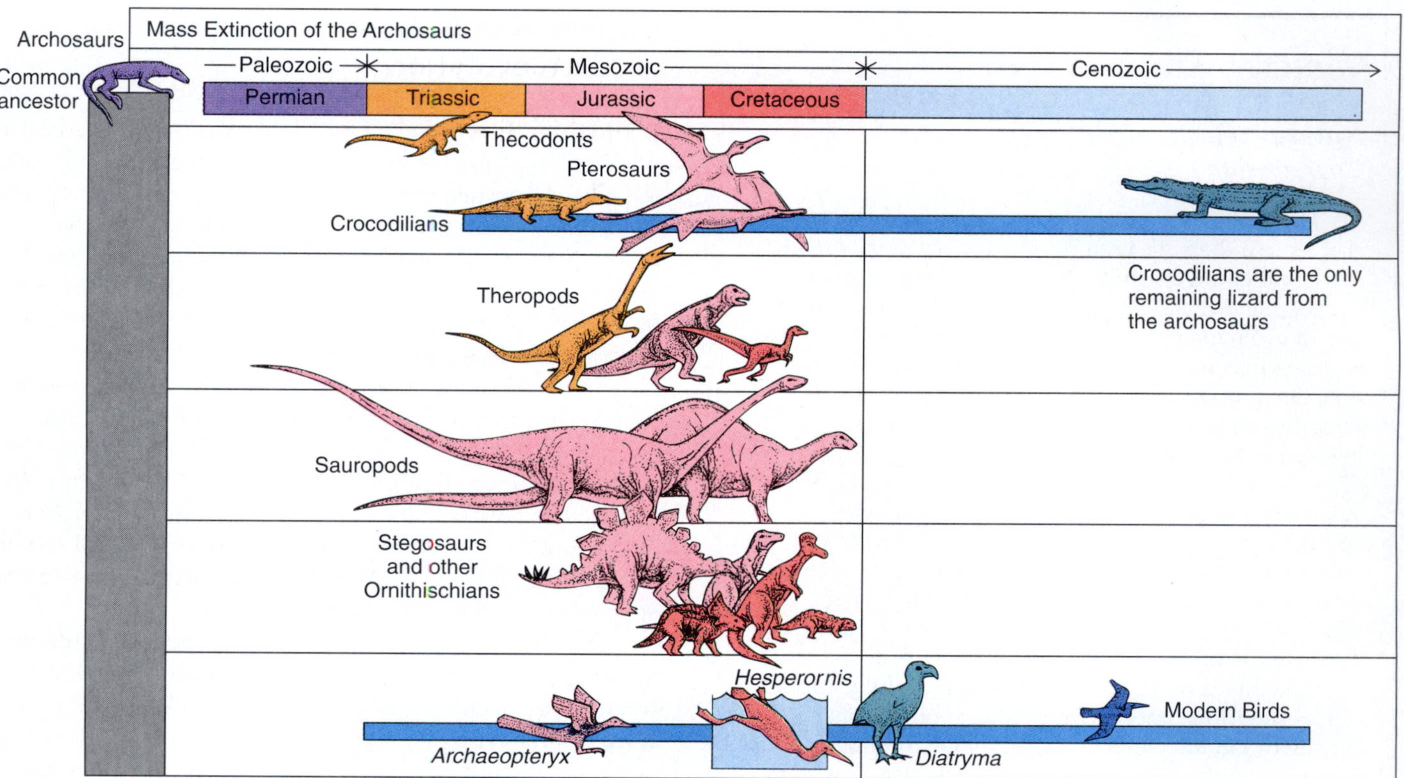

FIGURE 19–5 Episodes of mass extinction. Mass extinctions have taken place several times in Earth's history. For example, at the end of the Cretaceous period, which occurred approximately 65 million years ago, a mass extinction of many organisms, including the dinosaurs, occurred. At that time, the archosaurs (one of five main groups of reptiles) largely became extinct. The only lines to survive were the crocodiles and birds, both of which are archosaur descendants.

enormous increase in the human population, we have spread into areas that were previously not part of our range. The habitats of many animal and plant species have been altered or destroyed by human activities. Habitat destruction can result in an organism's extinction, as in the case of the dusky seaside sparrow discussed in the chapter introduction. Indeed, some biologists think that Earth has entered the largest period of mass extinction in its entire history, and that this has been triggered by human activities.

EVOLUTIONARY CHANGE CAN OCCUR RAPIDLY OR GRADUALLY

Biologists have long recognized that the fossil record often lacks many transitional forms. The starting points and end points are present, but the intermediate stages in the evolution from one species to another are frequently absent. This observation was traditionally blamed on the incompleteness of the fossil record, and biologists have attempted to fill in the missing parts much as a writer might fill in the middle of a novel when the beginning and end are already there.

Two different models have been developed to explain evolutionary change: punctuated equilibrium and gradualism. **Punctuated equilibrium** was proposed by those biologists who have questioned whether the fossil record is as incomplete as it initially appeared. The punctuated equilibrium model maintains that the fossil record accurately reflects evolution as it occurs. This model holds that there are long periods of **stasis** (no evolutionary change) punctuated, or interrupted, by short periods of rapid speciation. Thus, in evolution by punctuated equilibrium, speciation normally proceeds in "spurts." These relatively short periods of active evolution are followed by long periods of stasis; later, when evolution resumes, new species form, and many old ones are out-competed and become extinct.

Concept Connections

Microevolution ⟷ *Macroevolution* ⟷ *Punctuated Equilibrium*

Microevolution, which comprises the various mechanisms that cause changes in allele frequencies in a population over time, is responsible for the differences among different populations of a single species. As Chapter 18 indicates, many biologists think that the accumulation through time of many gradual changes due to natural selection and other microevolutionary events is sufficient to account for the origin of new species. These biologists also think the evolution of most if not all higher taxa (that is, taxonomic categories above the level of species) can also be explained by microevolutionary processes.

However, other biologists ascribe to a different school of thought. They think that evolutionary events quite different from those of microevolution have been responsible for the origin of new species and higher taxa. To these biologists, microevolution is associated with the long periods of stasis in the fossil record, and macroevolution with the rapid bursts of evolution that punctuate stasis.

With punctuated equilibrium, speciation can occur in a relatively short period of time. It is important to realize, however, that a "short" amount of time for rapid speciation may mean thousands of years. Such a period of time is short when compared to the several million years of a species' existence.

Punctuated equilibrium accounts for the abrupt appearance of a new species in the fossil record, with little or no record of intermediate forms. That is, proponents think that there are few transitional forms in the fossil record because there were few transitional forms during speciation.

In contrast to punctuated equilibrium, the traditional, Darwinian view of evolution uses the **gradualism** model. According to the gradualism model, evolution proceeds at a more or less steady rate but is not observed in the fossil record because the record is incomplete. Occasionally, a complete fossil record of transitional forms is discovered and cited as a strong case for gradualism. The gradualism model maintains that populations slowly diverge from one another by the gradual accumulation of adaptive characteristics within each population. These adaptive characteristics accumulate as a result of different selective pressures brought on by the populations living in different environments.

The fact that there is abundant evidence in the fossil record of long periods with no change in a species seems to argue against gradualism. Gradualists, however, think that any periods of stasis evident in the fossil record are the result of stabilizing selection (Chapter 18). They also point out that stasis in fossils is deceptive because fossils do not present all aspects of evolutionary change. Fossils can show changes in external anatomy and skeletal structure, but such characteristics as internal anatomy and changes in physiology, behavior, and ecological roles (which also represent evolution) will not be revealed by fossils. Gradualists recognize rapid evolution as taking place only when strong directional selection occurs.

Thus, while evolutionary biologists generally agree that natural selection is the main mechanism responsible for evolution, there is disagreement as to the timing, or pace, of evolutionary change during a species' existence (Fig. 19–6). Most biologists are of the opinion that both types of evolution may occur, that the pace of evolution may be steady and gradual in certain instances, and abrupt in others.

Early Earth Provided the Conditions for the Origin of Life

We have been concerned with the evolution of organisms, but we have not dealt with a fundamental question involving biological evolution: How did life begin? Scientists generally accept the hypothesis that life developed from nonliving matter. This process involved several stages. First, small organic molecules formed spontaneously and accumulated over time. Then, large organic macromolecules such as proteins and nucleic acids assembled from smaller molecules. The macromolecules interacted with one another, combining into more complicated structures that could eventually metabolize and replicate. As a result of natural selection, these macromolecular assemblages developed into cell-like structures that ultimately became the first true cells. After the first cells originated, they diverged over several billion years into the rich biological diversity that characterizes our planet today. It is thought that life originated only once and that this occurred under environmental conditions quite different from those we experience today.

To understand the origin of life, we must examine the chemical and physical conditions of early Earth. Although we can never be certain of the exact conditions when life arose, scientific evidence from a number of sources provides us with valuable clues.

The atmosphere of early Earth, which contained little or no free oxygen (O_2), included carbon dioxide (CO_2), water vapor (H_2O), carbon monoxide (CO), hydrogen

(Text continues on page 347)

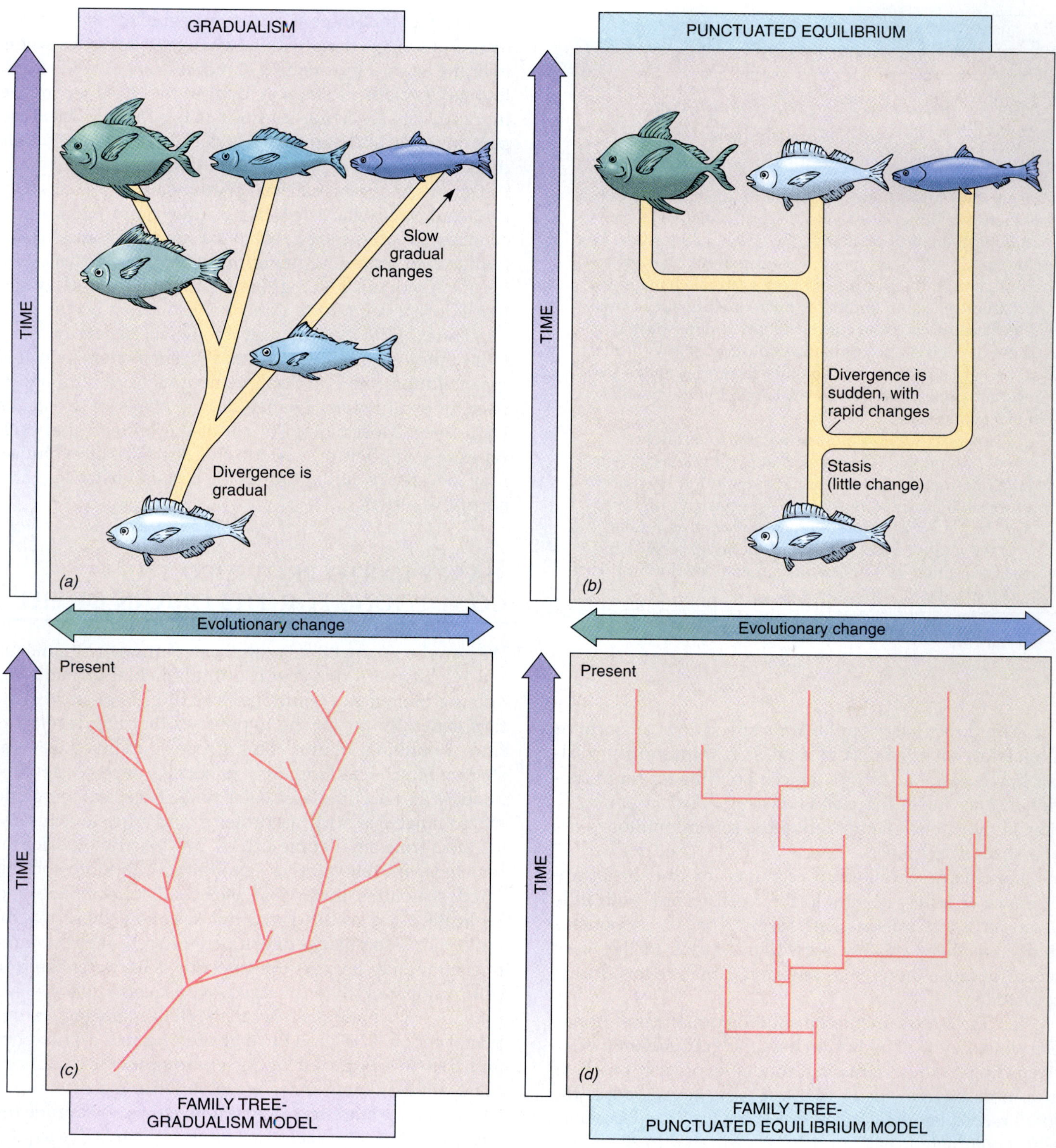

FIGURE 19–6 Gradualism versus punctuated equilibrium. Biologists agree that species evolve from pre-existing species, but there are two different models of the *pace* of evolution. In both models, the evolutionary relationships among species are depicted as branching diagrams, in which each branch is a single line of descent. (***a***) In gradualism, there is a slow, steady change in species over time. (***b***) In punctuated equilibrium, there are long periods of little evolutionary change (stasis) followed by short periods of rapid speciation. (***c***) A "family tree," which is the result of evolution over geological time. The branches that end before the present time indicate species that are extinct. This tree depicts the gradualism model. (***d***) The punctuated equilibrium model of a "family tree."

(H_2), and nitrogen (N_2). The early atmosphere may also have contained some ammonia (NH_3), hydrogen sulfide (H_2S), and methane (CH_4), although these reduced molecules may have been rapidly broken down by ultraviolet radiation from the sun. As the temperature of Earth slowly cooled, water vapor condensed and torrential rains fell, forming the oceans. The falling rain eroded Earth's surface, adding minerals to the oceans, making them salty.

There are four requirements for the origin of life from nonliving matter: no free oxygen, a source of energy, the availability of chemical building blocks, and time. First, life could have begun only in the absence of free oxygen. Oxygen is very reactive and would have broken down the organic molecules that are a necessary step in the origin of life. Earth's early atmosphere was strongly reducing, which means that any free oxygen would have reacted with other elements to form oxides. Thus, oxygen would have been tied up in compounds.

A second requirement for the origin of life was energy. Early Earth was a place of high energy, with violent thunderstorms, widespread volcanism, bombardment from meteorites, and intense radiation, including ultraviolet radiation from the sun (Fig. 19–7). The young sun

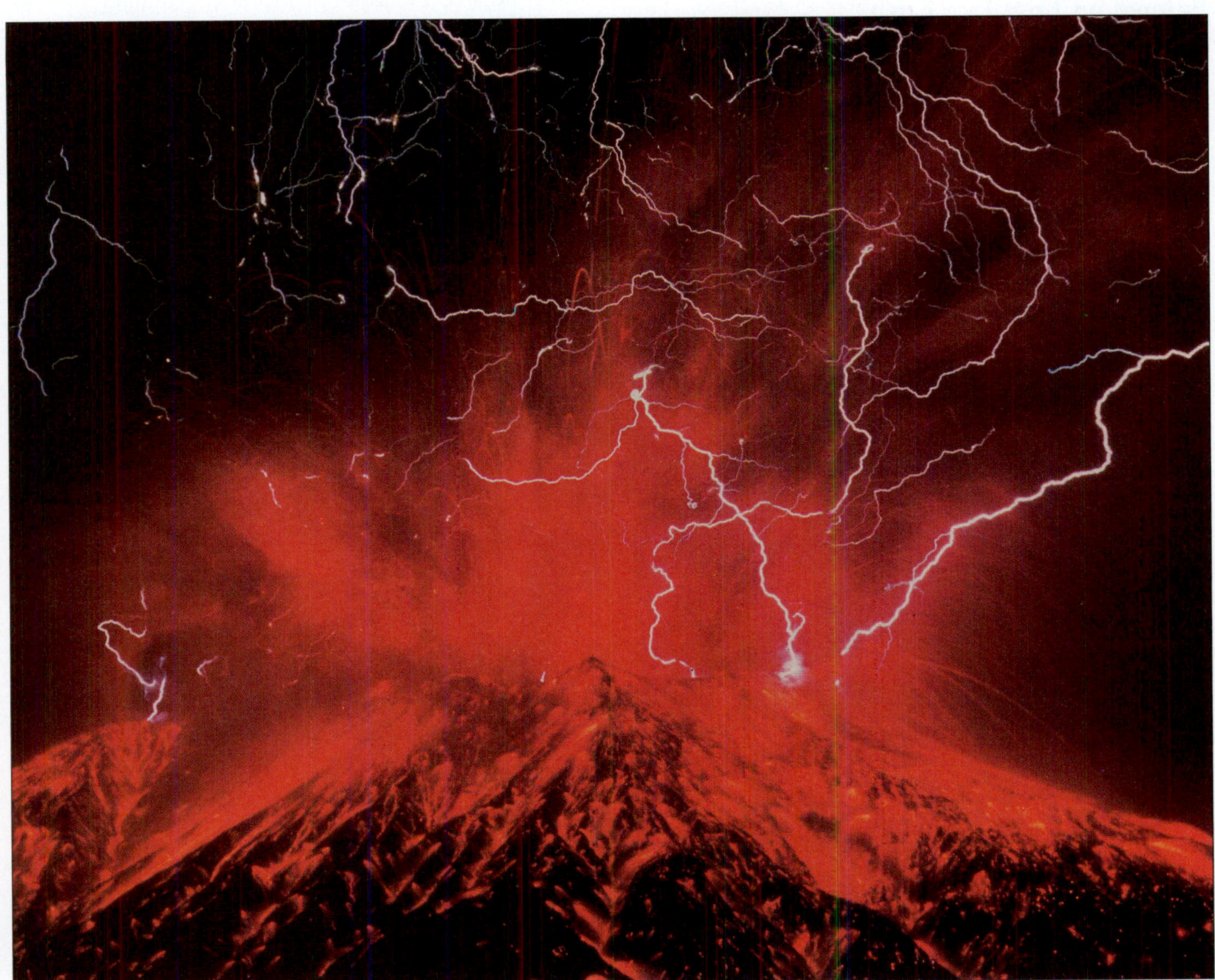

FIGURE 19–7 Conditions on early Earth would have been inhospitable for most of today's life forms. The strongly reducing atmosphere lacked oxygen; volcanoes erupted, spewing gases that contributed to the atmosphere; and violent thunderstorms produced torrential rainfall that eroded the land. Meteorites continually bombarded Earth, causing cataclysmic changes in the crust, oceans, and atmosphere. (Tsuyoshi Nishiinoue and Orion Press)

probably produced more ultraviolet radiation than it does today, and Earth had no protective ozone layer to block much of this radiation.

Third, the chemicals that would supply the building blocks for the origin of life must have been present. These included water, dissolved inorganic minerals (present as ions), and the gases present in the early atmosphere.

A final requirement for the origin of life was time—time for molecules to accumulate and react with one another. The age of Earth, approximately 4.6 billion years, provided adequate time for the origin of life.

Before Cells Existed, Organic Molecules Formed on Primitive Earth

Because organic molecules are the building materials for living organisms, it is reasonable to consider how they might have originated. The concept that simple organic molecules such as sugars, nucleotide bases, and amino acids could form spontaneously from simpler raw materials was first hypothesized in the 1920s by two scientists working independently—A. I. Oparin, a Russian biochemist, and J. B. S. Haldane, a Scottish physiologist and geneticist.

Their hypothesis was tested in the 1950s by Stanley Miller and Harold Urey, who designed an apparatus that simulated conditions then thought to be prevalent on early Earth (Fig. 19–8). They exposed an atmosphere rich in H_2, CH_4, H_2O, and NH_3 to an electrical discharge, which simulated lightning. Their analysis of the chemicals produced in a week revealed that amino acids and other organic molecules had formed. Although more recent data suggest that the early atmosphere was not rich in methane or ammonia, similar experiments using different combinations of gases have produced a wide vari-

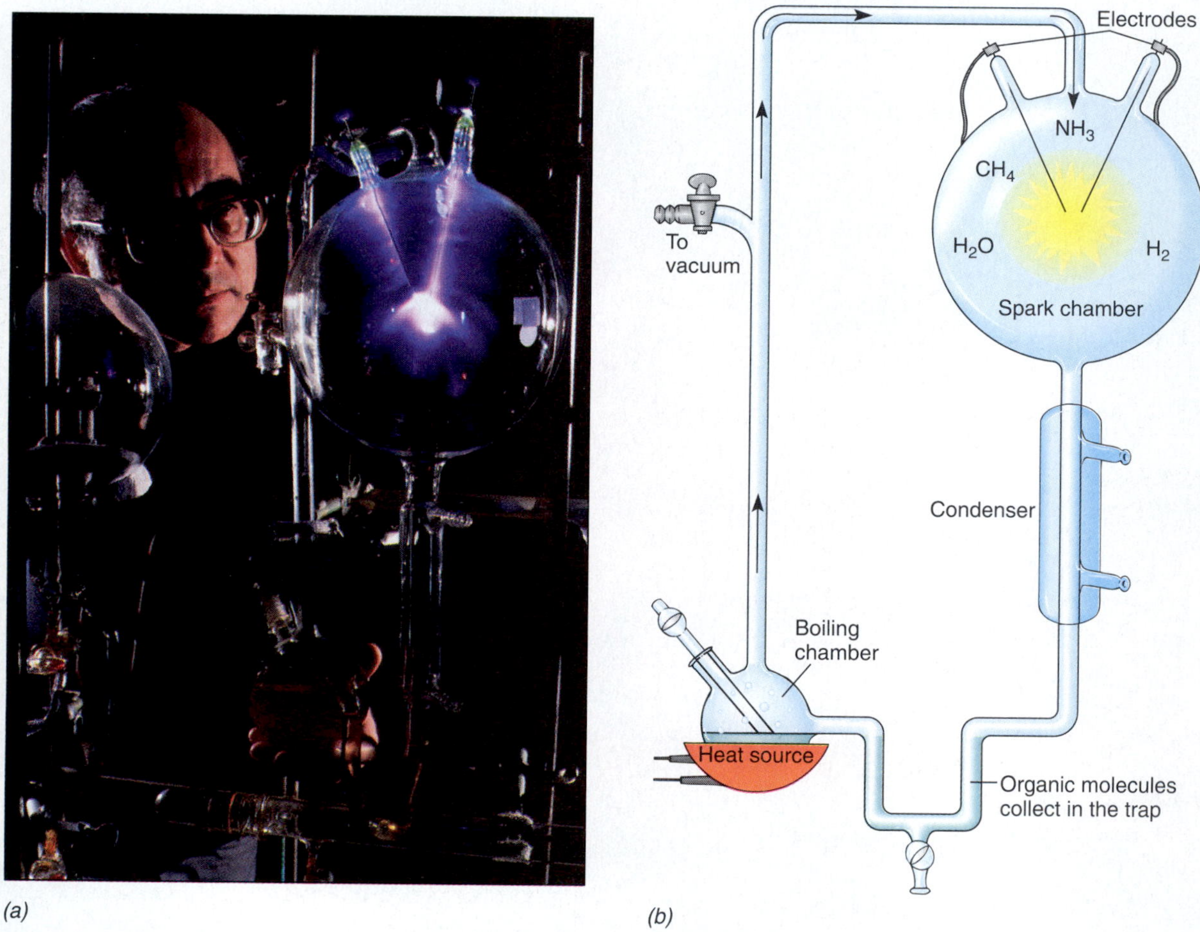

FIGURE 19–8 Demonstrating the synthesis of organic molecules from inorganic substances under the conditions of early Earth. (*a*) American biochemists Stanley Miller (shown) and Harold Urey used this apparatus to simulate the reducing atmosphere of early Earth. (*b*) Diagram of the apparatus. An electrical spark was produced in the upper right flask to simulate lightning. The gases present in the flask reacted together, forming a number of simple organic compounds, which accumulated in the trap at the bottom. (©1988 Roger Ressmeyer/Starlight)

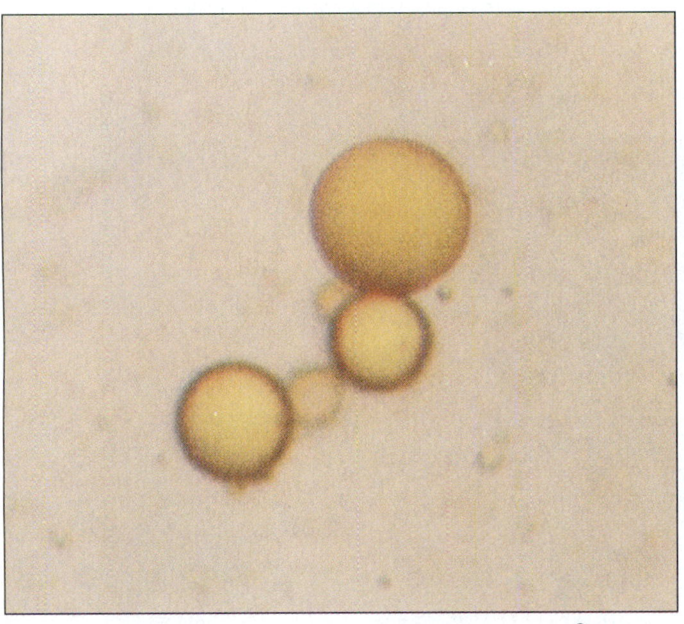

FIGURE 19–9 Protobionts. Protobionts are tiny synthetic spheres (1-2 μm in diameter) composed of organic polymers. They exhibit some of the properties of life. (Steven Brooke and Richard Le Duc)

ety of organic molecules, including the nucleotide bases of RNA and DNA.

Oparin envisioned that the organic molecules would, over vast spans of time, accumulate in the shallow seas, as a "sea of organic soup." Under such conditions, he thought that larger organic molecules (polymers) would form by the union of smaller ones (monomers). Based on scientific evidence gathered since Oparin's time, most scientists think it is more likely that organic polymers formed and accumulated on rock or clay surfaces rather than in the primordial seas. Clay is particularly intriguing as a possible site for early polymerizations because it contains zinc and iron ions that might have served as catalysts. Laboratory experiments have confirmed that organic polymers form spontaneously from single organic molecules on rock or clay surfaces.

After the first polymers formed, could they have assembled into more complex structures? Scientists have synthesized several different **protobionts,** which are assemblages of organic polymers (Fig. 19–9). They have been able to make protobionts that resemble simple life forms in several ways, helping us to envision how complex molecules took that "giant leap" and became living cells. Protobionts often divide in half after they have "grown." Their interior is chemically different from the external environment, and some of them show the beginnings of metabolism. They are highly organized, considering their relatively simple composition.

The First Cells Probably Assembled from Organic Molecules

Studying protobionts can help us appreciate that relatively simple "pre-cells" can exhibit some of the properties of life. However, it is a major step (or steps) to go from simple molecular aggregates such as protobionts to living cells. Although much has been learned about the synthesis of organic molecules on primitive Earth, the problem of how pre-cells developed into cells remains to be solved. Nonetheless, fossil evidence indicates that cells were thriving 3.5 billion years ago.

The earliest cells were prokaryotic. Australian and South African rocks have yielded microscopic fossils of prokaryotic cells 3.1 to almost 3.5 billion years old. **Stromatolites** are another type of fossil evidence of the earliest cells (Fig. 19–10). These rocklike columns are composed of many minute layers of prokaryotic cells. Fossil stromatolite reefs are found in a number of places in the world, including the Canadian Great Slave Lake and the Gunflint Iron Formations along Lake Superior in the United States. Some fossil stromatolites are extremely ancient. Living stromatolite reefs are still found in hot springs and in shallow pools of fresh and salt water.

The earliest cells were probably **heterotrophs** that obtained the organic molecules they needed for energy from the environment, as opposed to synthesizing them. These primitive organisms probably consumed many types of organic molecules that had spontaneously formed: sugars, nucleotides, and amino acids, to name a few. By fermenting these organic compounds, they obtained the energy needed to support life. Fermentation is, of course, an anaerobic process (that is, performed in the absence of oxygen), and the first cells were almost certainly **anaerobes.**

When the supply of spontaneously generated organic molecules decreased, only certain organisms could survive. Mutations had probably already occurred that permitted some cells to obtain energy directly from sunlight (perhaps by using sunlight to make ATP). These cells did not require the energy-rich organic compounds that were now in short supply in the environment, a distinct selective advantage.

Photosynthesis requires not only light energy, but also a source of hydrogen, which is used to reduce CO_2 when organic molecules, such as glucose, are synthesized. Most likely, the first photosynthetic **autotrophs** (organisms that produce their own food from simple raw materials) used the energy of sunlight to split hydrogen-rich molecules like H_2S, releasing sulfur (not oxygen) in the process. Indeed, some types of bacteria still use H_2S as a hydrogen source for photosynthesis.

The first photosynthetic autotrophs to split water in order to obtain hydrogen were the cyanobacteria. Water was quite abundant, and the selective advantage that

(a)

(b)

FIGURE 19–10 Stromatolites. (***a***) These formations at Hamlin Pool in West Australia are composed of mats of cyanobacteria and minerals like calcium carbonate and are several thousand years old. Some fossil stromatolites are 3.1 to 3.4 billion years old. (***b***) Diagram of stromatolites, showing the layers of cyanobacteria and sediments that accumulated over time. (Fred Bavendam/Peter Arnold, Inc.)

splitting water bestowed on them allowed the cyanobacteria to thrive. In the process of splitting water, oxygen was released as a gas, O_2. Initially, the oxygen released from photosynthesis oxidized minerals in the ocean and crust. Over time, however, oxygen began to accumulate in the oceans and atmosphere.

The increase in atmospheric oxygen had a profound effect on life. Some anaerobes were poisoned by the oxygen, and many species undoubtedly perished. Some anaerobes, however, survived in environments where oxygen does not penetrate, whereas others developed ways to neutralize the oxygen so it could not harm them. Some organisms, called **aerobes,** developed a respiratory pathway that *used* the oxygen to extract more energy (as ATP) from food. Aerobic respiration was tacked onto the existing process of glycolysis.

Also, oxygen in the upper atmosphere reacted to form **ozone,** O_3. A layer of ozone eventually blanketed Earth, preventing much of the sun's ultraviolet radiation from penetrating to the surface. The protective ozone layer enabled organisms to live closer to the surface in aquatic environments and eventually to move on land. Because the energy in ultraviolet radiation had been used to form organic molecules, their spontaneous synthesis decreased.

Eukaryotic Cells Descended from Prokaryotic Cells

Eukaryotes appeared in the fossil record 1.9 to 2.1 billion years ago. They arose from prokaryotes. Recall that prokaryotic cells lack nuclear envelopes as well as other membranous organelles such as mitochondria and chloroplasts. How did eukaryotic cells arise from prokaryotes?

The **endosymbiont theory** suggests that organelles such as mitochondria and chloroplasts may have originated from mutually advantageous symbiotic relationships between two prokaryotic organisms (Fig. 19–11). Chloroplasts are thought to have evolved from photosynthetic bacteria that lived inside other cells. Mitochondria are thought to have evolved from aerobic bacteria that lived inside other cells. Thus, the early eukaryotic cell contained a group of formerly free-living prokaryotes.

How did these bacteria come to be **endosymbionts,** which are organisms that live symbiotically inside a host cell? Probably they were originally ingested, but not di-

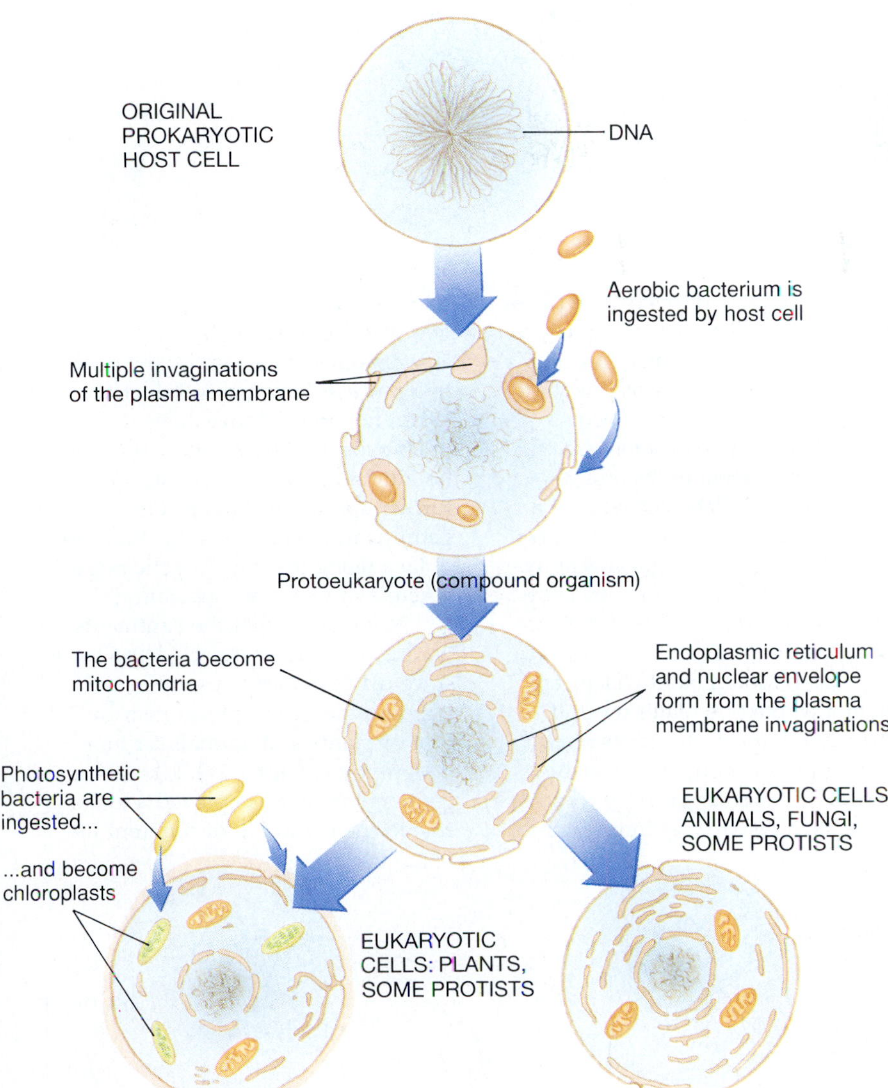
ORIGINAL PROKARYOTIC HOST CELL
DNA
Aerobic bacterium is ingested by host cell
Multiple invaginations of the plasma membrane
Protoeukaryote (compound organism)
The bacteria become mitochondria
Endoplasmic reticulum and nuclear envelope form from the plasma membrane invaginations
Photosynthetic bacteria are ingested...
...and become chloroplasts
EUKARYOTIC CELLS: ANIMALS, FUNGI, SOME PROTISTS
EUKARYOTIC CELLS: PLANTS, SOME PROTISTS

FIGURE 19–11 The endosymbiont theory. Chloroplasts and mitochondria of eukaryotic cells are thought to have originated from various bacteria that lived as endosymbionts inside other cells.

gested, by the host cell. Thus, they survived and reproduced along with the host cell so that future generations of the host also contained endosymbionts. The two organisms developed a mutualistic relationship in which each contributed something to the other, and eventually the endosymbiont lost the ability to exist outside its host.

The principal evidence in favor of the endosymbiont theory is that mitochondria and chloroplasts possess some (although not all) of their own genetic material. They have their own DNA (as a circular molecule much like that of prokaryotes) and their own ribosomes (which resemble prokaryotic ribosomes rather than eukaryotic ribosomes). They have some of the machinery for protein synthesis and are able to conduct protein synthesis on a limited scale independently of the nucleus. They also have double membranes—an outer (host) membrane and an inner (symbiont) membrane.

The endosymbiont theory does not completely explain the evolution of eukaryotic cells from prokaryotes. It does not explain how the genetic material in the nucleus came to be surrounded by a membranous envelope, for example. However, with the advent of eukaryotic cells, the stage was set for further evolutionary developments.

THE FOSSIL RECORD PROVIDES US WITH CLUES TO THE HISTORY OF LIFE

The sediments of the Earth's crust consist of five major rock strata (layers). Each major stratum is subdivided into minor strata, lying one on top of the other. These sheets of rock were formed by the accumulation of mud and sand at the bottoms of oceans, seas, and lakes. Each layer

FOCUS ON *Evolution*

CONTINENTAL DRIFT

In 1912, Alfred Wegener, who had noted a similarity between the geographical shapes of South America and Africa, proposed that all of the land masses had been joined at one time into a single huge supercontinent, which he called Pangaea. He further suggested that Pangaea subsequently broke apart and the various land masses separated, in a process known as **continental drift**. Wegener did not know of any mechanism that could have caused continental drift, so his hypothesis, although debated initially, was largely ignored.

In the 1960s, scientific evidence accumulated that provided the explanation for continental drift. Earth's crust is composed of seven large plates (plus a few smaller ones) that float on the mantle. The land masses are situated on some of these plates. As the plates move about, the continents change their relative positions (see figure). The movement of the crustal plates is termed **plate tectonics.**

Any area where two plates meet is a site of intense geological activity. Earthquakes and volcanoes are common in such a region. Both San Francisco, noted for its earthquakes, and the volcano Mount Saint Helens are situated where two plates meet. If land masses are on the edges of two meeting plates, mountains may be formed. The Himalayas have formed where the plate carrying India rammed into the plate carrying Asia. Where two plates grind together, one of them is sometimes buried under the other, in a process known as **subduction**. When two plates move apart, a ridge of lava forms between them that continually expands as the plates move further apart. The Atlantic Ocean is getting larger because of the buildup of lava along the mid-Atlantic ridge, where two plates are separating.

Knowledge that the continents were at one time connected and have since drifted apart is useful in explaining the geographical distribution of plants and animals, or biogeography (Chapter 17). Likewise, continental drift has played a major role in the evolution of different life

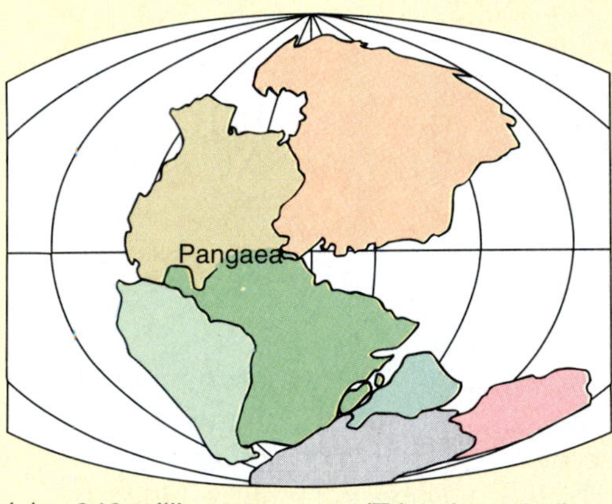

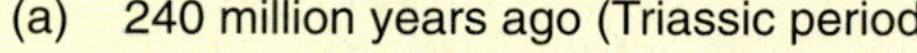

(a) 240 million years ago (Triassic period)

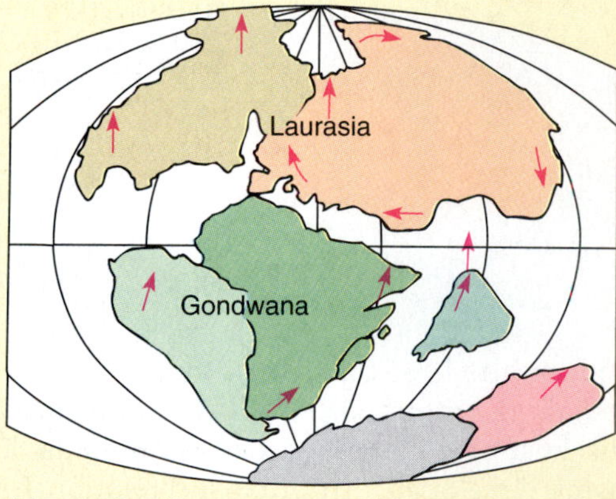

(b) 120 million years ago (Cretaceous period)

contains certain characteristic fossils that serve to identify deposits made at approximately the same time in different parts of the world.

Geological time has been divided into five major intervals, called **eras**. These eras are the Archean, Proterozoic, Paleozoic, Mesozoic, and Cenozoic, and they correspond to the five major rock strata. Each era is subdivided into **periods**, which in turn are composed of **epochs.** Between the major eras, and serving to distinguish them, there occurred widespread geological disturbances that raised or lowered vast regions of Earth's surface and produced or eliminated shallow inland seas. These disturbances altered the distribution of sea and land organisms and may have triggered the mass extinction of many life forms. The raising and lowering of portions of Earth's crust resulted from the slow movements of the enormous plates that compose the crust (see Focus On Evolution: Continental Drift).

Continental drift, as currently envisioned. (***a***) The supercontinent Pangaea of the Triassic period, about 240 million B.P. (before present). (***b***) Breakup of Pangaea into Laurasia (Northern Hemisphere) and Gondwana (Southern Hemisphere) 120 million years B.P., in the Cretaceous period. (***c***) Further separation of land masses, which occurred in the Tertiary period, 60 million years B.P. Note that Europe and North America were still joined, and that India was a separate land mass. (***d***) The continents today. (***e***) Projected positions of the continents in 50 million years.

forms. When Pangaea originally formed, about 250 million years ago, it brought together terrestrial species that had evolved separately from one another, leading to competition and possible extinctions. Marine life was adversely affected, largely because, with the continents joined as one large mass, less coastline existed. (Because coastal areas are shallower, they contain high concentrations of marine organisms.)

Pangaea separated into several land masses approximately 180 million years ago (see figure). As the continents began to drift apart, populations became geographically isolated in different enviornmental conditions, and the evolution of these populations occurred over time.

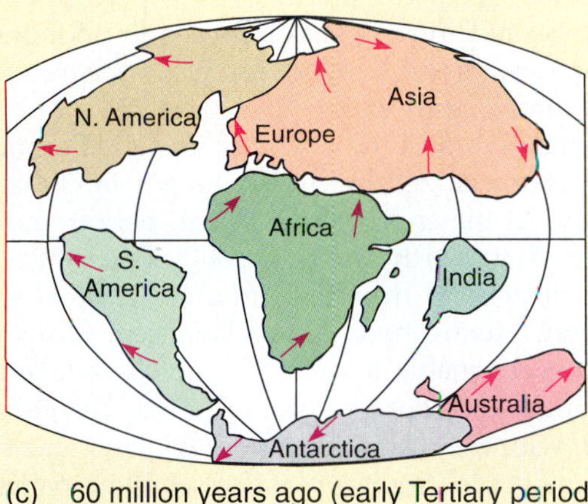

(c) 60 million years ago (early Tertiary period)

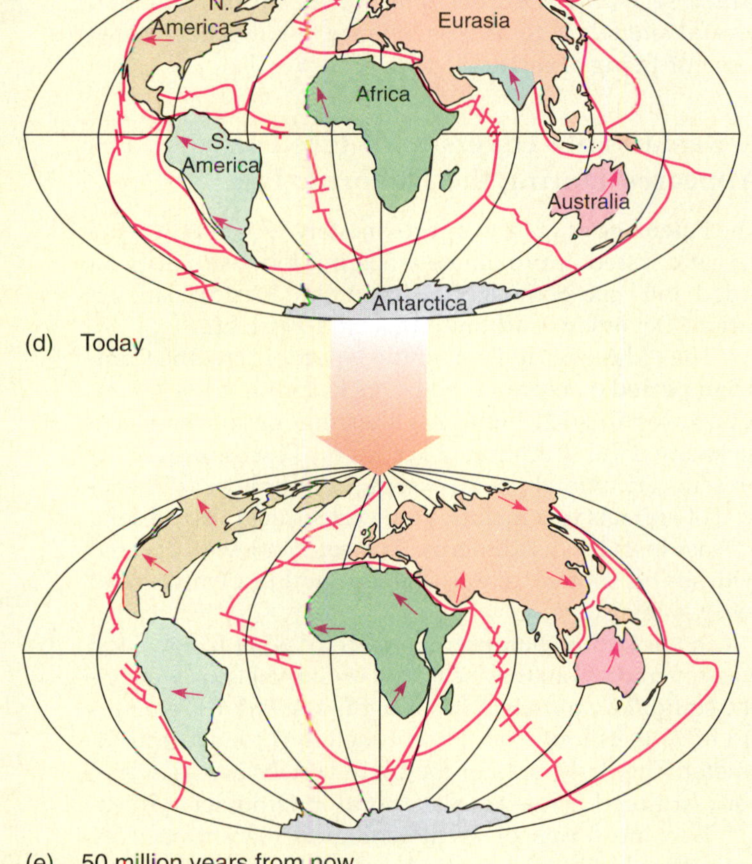

(d) Today

(e) 50 million years from now

Evidence of Living Cells Is Found in the Archean and Proterozoic Eras

Signs of life date back to the **Archean era,** which began about 3.5 billion years ago. The Archean era began after the formation of Earth's crust, when rocks and mountains were already in existence and the processes of erosion and sedimentation had begun. Because the rocks of the Archean era are very deeply buried in most parts of the world, they are considered to be some of the most ancient. However, Archean rocks are exposed at the bottom of the Grand Canyon and along the shores of Lake Superior.

The Archean era lasted 2 billion years and was characterized by widespread volcanic activity and giant upheavals that raised mountains. The heat, pressure, and churning associated with these movements probably destroyed most of whatever fossils may have been formed, but some evidence of life still remains. This evidence consists of traces of graphite or pure carbon, which may be the transformed remains of primitive life. These remains are especially abundant in what were the oceans and seas of that era. Fossils of what appear to be cyanobacteria have been recovered from several Archean formations.

The second era, the **Proterozoic era,** which began approximately 1.5 billion years ago, was almost one billion years in length. It was characterized by the deposition of large quantities of sediment, reflecting massive erosion and perhaps glacial deposits. The fossils found in the later (more recent) Proterozoic rocks show clear-cut examples of some major groups of bacteria, fungi, protists (including multicellular algae), and simple animals. The animals—all invertebrates (lacking backbones)— include jellyfish, segmented worms, soft-bodied arthropods, and several animals with no resemblance to any other known fossil or living animal.

A Remarkable Diversity of Life Forms Appeared During the Paleozoic Era

The **Paleozoic era** began approximately 570 million years ago and lasted approximately 322 million years. It is divided into six periods: the Cambrian, Ordovician, Silurian, Devonian, Carboniferous, and Permian.

The oldest subdivision of the Paleozoic era, the **Cambrian period,** is represented by rocks rich in fossils. Evolution was in such high gear that this period has been nicknamed the "Cambrian Explosion." Fossils of all of the present-day animal phyla, except the chordates (the phylum of animals that includes the vertebrates), are present, at least in marine sediments. The sea floor was covered with simple sponges, corals, snails, clamlike bivalves, and other animals.

In the **Ordovician period,** much of what is now land was covered by shallow seas. Inhabiting the seas were giant squidlike animals with straight shells 5 to 7 meters (16 to 23 feet) long and 30 centimeters (12 inches) in diameter. The first traces of the early vertebrates—jawless, bony-armored fish— are also found in Ordovician rocks.

Two life forms of great biological significance appeared in the **Silurian period**—plants and air-breathing animals. The first known plants resembled ferns in that they possessed vascular (conducting) tissue and reproduced by spores. The evolution of plants allowed animals to begin to colonize the land because the plants provided the first land animals with food and shelter.

A great variety of fishes appeared in the **Devonian period.** In fact, the Devonian is frequently called the "Age of Fishes." More recent Devonian sediments contain fossil remains of large, salamander-like amphibians as well as wingless insects and millipedes. The early vascular plants diversified during the Devonian period, and forests of ferns, club mosses, horsetails, and seed ferns flourished.

FIGURE 19–12 The Carboniferous period. The plants of the Carboniferous period included giant ferns, horsetails, and club mosses. Recent work suggests that Carboniferous swamps were more open than depicted here. (Transp. No. GEO. 85638 C, Field Museum of Natural History)

The **Carboniferous period** is named for the great swamp forests whose remains persist today as major coal deposits (Fig. 19–12). Much of the land during this time was covered with low swamps filled with horsetails, club mosses, ferns, seed ferns, and gymnosperms. The first reptiles appeared in the Carboniferous period, and two important groups of winged insects appeared—cockroaches and dragonflies.

The final period of the Paleozoic era, the **Permian period,** was characterized by great changes in climate and topography. At the end of the Permian period, mountain ranges formed in North America and Europe. An ice sheet, spreading northward from the Antarctic, covered most of the Southern Hemisphere. Many Paleozoic forms of life may have been unable to adapt to the climatic and geological changes and became extinct. The seed plants became dominant, with the diversification of conifers and the appearance of cycads (tropical plants resembling palms with crowns of fernlike leaves and large, seed-containing cones).

The Dinosaurs and Other Reptiles Dominated the Mesozoic Era

The **Mesozoic era** began about 248 million years ago and lasted some 183 million years. It is divided into the Triassic, Jurassic, and Cretaceous periods. The outstanding feature of the Mesozoic era was the origin, differentiation, and finally, the extinction of a large variety of reptiles. For this reason, the Mesozoic era is commonly called the "Age of Reptiles."

FIGURE 19–13 ***Utahraptor*, a formidable predatory dinosaur.** Discovered in 1992 in eastern Utah, *Utahraptor* dates from the early Cretaceous period. (Steven Kirk/ © 1993 The Walt Disney Co. Reprinted with permission of *Discover Magazine*.)

Of all the reptilian branches, the **dinosaurs** are the most famous (Fig. 19–13). Some of these were among the largest animals that ever lived: *Apatosaurus* (formerly known as *Brontosaurus*), with a length of 21 meters (68 feet), and *Diplodocus*, with a length of 29 meters (95 feet). Many traditional ideas about the dinosaurs—that they were cold-blooded, slow-moving monsters that lived in swamps, for example—have been reconsidered. Recent evidence suggests that many dinosaurs may have been warm-blooded and capable of moving very fast. They may have had complex social behaviors, including caring for their young and hunting in packs.

Although the reptiles were the dominant animals of the Mesozoic era, fossils of many other important organisms occur in the same formations. Most of the modern orders of insects appeared during that era. Snails and bivalves increased in number and diversity, and sea urchins reached their peak diversity. The earliest mammals, tiny shrew-like animals, appeared in the **Triassic period,** and birds first appeared in the **Jurassic period**. *Archaeopteryx,* a primitive bird, was about the size of a crow, had rather feeble wings, and a long reptilian tail covered with feathers (Fig. 19–14). During the early Triassic period, the most abundant plants were gymnosperms (seed-bearing plants such as conifers and cycads). By the end of the **Cretaceous period,** flowering plants, many resembling present-day species, were the dominant vegetation.

At the end of the Cretaceous period, a great many animals abruptly became extinct. Most gymnosperms, with the exception of conifers, also perished. Changes in climate may have been a factor in their demise. Other explanations have been proposed, including the catastrophic collision of Earth with a giant asteroid or comet (Chapter 25).

FIGURE 19–14 A fossil of *Archaeopteryx*, a tailed, toothed, primitive bird from the Jurassic period. Despite many reptilian features, *Archaeopteryx* is clearly a bird, as demonstrated by its feathers. (See Fig. 25–12, page 488, for an artist's reconstruction of what *Archaeopteryx* looked like.) (Dennis Drenner)

The Cenozoic Era Is Known as the Age of Mammals

With equal justice, the **Cenozoic era** could be called the "Age of Mammals," the "Age of Birds," the "Age of Insects," or the "Age of Flowering Plants." It is marked by the appearance of all of these life forms in great variety and numbers of species. The Cenozoic era extends from 65 million years ago to the present. It is subdivided into two periods—the Tertiary period, which encompasses some 63 million years, and the Quaternary period, which covers the last 2 million years.

During the **Tertiary period,** grasses, which served as food, and dense forests, which afforded protection from predators, may have been important factors in leading to changes in the mammalian body pattern. Along with the tendency toward increased size, the mammals displayed tendencies toward an increase in the relative size of the brain and toward changes in the teeth and feet.

FIGURE 19–15 Some extinct mammals of the Cenozoic era. (***a***) The wooly mammoth disappeared at the end of the Ice Age. (***b***) The giant ground sloth was nearly the size of a modern elephant. (***c***) A glyptodont, found in what is now the southern United States, weighed about 1 metric ton and resembled a cross between a turtle and an armadillo. (***d***) The saber-toothed cat was found in both North and South America.

The **Quaternary period,** which extends to the present, has had four ice ages. At their greatest extent, ice sheets covered several million square kilometers of North America, extending south as far as the Ohio and Missouri Rivers. During these ice ages, enough water was removed from the sea and locked in the ice to lower the sea level by 65 to 100 meters (213 to 328 feet). This produced land bridges that served as highways for the dispersal of many animals. Examples included a land bridge between Siberia and Alaska at the Bering Strait and one that connected England to the European continent.

A considerable number of mammals, including the saber-toothed cat, the mammoth, and the giant ground sloth, became extinct during the Quaternary period, very possibly as a result of early human hunting (Fig. 19–15). The Quaternary period has also been marked by the extinction of many species of plants, especially woody ones, and by the evolution of numerous herbaceous plants.

Chapter Summary

I. Macroevolution refers to major evolutionary events—large phenotypic changes—that occur in groups of species over geological time. The phenotypic changes are so great that the organisms that possess them are assigned to a new genus or higher taxonomic category.
   A. Macroevolution is concerned with the origin of taxonomic categories above the species level, that is, to new genera or higher-level categories of organisms.
   B. New designs, adaptive radiation, and mass extinction are aspects of macroevolution.
      1. Evolutionary novelties (new designs) may originate from mutations that alter developmental pathways.
      2. Adaptive radiation is the process of evolutionary diversification of an ancestral species into many species.
      3. Extinction is the death of a species. When species become extinct, the adaptive zones that they occupied become vacant, allowing new species to fill those zones.

II. The pace of evolutionary change is currently being debated.
   A. According to the gradualism model, populations slowly diverge from one another by the accumulation of adaptive characteristics within a population.
   B. According to the punctuated equilibrium model, evolution proceeds in spurts. Short periods of active evolution are followed by long periods of stasis.

III. It is thought that life began from nonliving matter. While we may never know exactly how life originated, a number of assumptions about the origin of life are testable.

A. The four requirements for the origin of life include:
   1. Absence of oxygen (because free oxygen would have reacted with and broken down organic molecules).
   2. Energy (to form organic molecules).
   3. Chemical building blocks used to form organic molecules (including water, minerals, and gases present in the atmosphere).
   4. Sufficient time (for molecules to accumulate and react).

B. The four steps hypothesized in the origin of life include:
   1. Small organic molecules formed and accumulated.
   2. Macromolecules assembled from the small organic molecules.
   3. Assemblages of organic polymers (pre-cells) formed from macromolecules.
   4. Cells arose from the assemblages oı organic polymers.

IV. The first cells were prokaryotic anaerobes.
   A. The oldest cells in the fossil record are 3.1 to almost 3.5 billion years old.
   B. The evolution of photosynthesis ultimately changed early life because it generated oxygen, which accumulated in the atmosphere.
   C. Aerobic organisms, which could use oxygen for a more efficient type of cellular respiration, appeared.
   D. Mitochondria and chloroplasts of eukaryotic cells probably evolved from prokaryote endosymbionts.

V. Earth's history is divided into eras, periods, and epochs.
   A. Life began and diverged into different groups of bacteria, protists (including algae), fungi, and simple animals during the Archean and Proterozoic eras.
   B. During the Paleozoic era, all major groups of plants appeared except for flowering plants, and fish and amphibians flourished.
   C. The Mesozoic era was characterized by the evolution of flowering plants and reptiles. Insects flourished, and birds and early mammals appeared.
   D. In the Cenozoic era, which extends to the present time, flowering plants, birds, insects, and mammals diversified greatly.

Selected Key Terms

adaptive radiation, p. 341
adaptive zones, p. 341
aerobes, p. 350
anaerobes, p. 349
Archean era, p. 353
autotrophs, p. 349
background extinction, p. 343
Cenozoic era, p. 355
continental drift, p. 352
endosymbiont theory, p. 350
extinction, p. 343
gradualism, p. 345
heterotrophs, p. 349
macroevolution, p. 340
mass extinction, p. 343
Mesozoic era, p. 354
ozone, p. 350
Paleozoic era, p. 354
plate tectonics, p. 352
Proterozoic era, p. 354
protobionts, p. 349
punctuated equilibrium, p. 344
stasis, p. 344

Post-Test

1. ______________ involves changes in the *kinds* of species over evolutionary time.

2. A novel change in the basic design of an organism is often due to changes in ______________.

3. Several to many species that arose from a single ancestral species is known as ______________ ______________.

4. When the last individual of a species dies, the species is said to be ______________.

5. ______________ extinctions are thought to have occurred during five or six periods of Earth's history.

6. The fact that the fossil record shows few transitional forms during speciation is used to support the ______________ ______________ model.

7. The traditional Darwinian view of evolution is embraced by the ______________ model, in which evolution proceeds at a more or less constant rate.

8. Energy, chemical building blocks, the absence of oxygen, and ______________ were the four main requirements for the origin of life.

9. Although Oparin envisioned life as originating in a "sea of organic soup," it is more likely that it originated on rock or ______________ surfaces.

10. The first photosynthetic autotrophs probably used sunlight to split ______________.

11. The ______________ theory explains the evolution of eukaryotic organelles such as chloroplasts and mitochondria.

12. The most recent era in Earth's history, which includes the present time, is the ______________ era.

13. Earthquakes, subduction, mountain formation, and ______________ may occur where two tectonic plates meet.

14. Adaptive radiation is common following a period of mass extinction, probably because: (a) the survivors of a mass extinction are remarkably well-adapted to the environment (b) the stable environment following a mass extinction drives the evolutionary process (c) many adaptive zones are empty (d) many adaptive zones are filled.

15. This gas was probably least abundant in Earth's early atmosphere (before living organisms appeared): (a) hydrogen (b) oxygen (c) carbon dioxide (d) nitrogen (e) water vapor.
16. Which of the following probable steps in the origin of life from nonliving matter has *not* been duplicated in the laboratory? (a) Organic molecules such as sugars and amino acids spontaneously formed from inorganic chemicals. (b) Complex organic polymers formed from simple organic molecules on rock or clay surfaces. (c) Protobionts assembled from organic polymers. (d) Cells developed from protobionts.
17. Place the following in the correct order of their appearance. (a) autotrophs ⟶ anaerobic heterotrophs ⟶ aerobic organisms (b) aerobic organisms ⟶ anaerobic heterotrophs ⟶ autotrophs (c) anaerobic heterotrophs ⟶ autotrophs ⟶ aerobic organisms (d) anaerobic heterotrophs ⟶ aerobic organisms ⟶ autotrophs
18. Fossil remnants of ______?______ are found in rocks from the Mesozoic era. (a) dinosaurs (b) humans (c) wooly mammoths (d) saber-toothed cats

Review Questions

1. If life originally arose from nonliving matter, why can it not originate now?
2. Distinguish between microevolution and macroevolution.
3. What role does extinction play in evolution?
4. If you were in a debate and were supporting gradualism, what would you say to support your position? What would you say if you were supporting punctuated equilibrium?
5. What are the four requirements for the origin of life, and why is each essential?
6. Provide some evidence used to support the endosymbiont theory.
7. Put the following in chronological order with respect to their evolution: (a) reptiles, mammals, amphibians, fish (b) flowering plants, ferns, gymnosperms.
8. List the five geological eras in chronological order, starting with the earliest.
9. How does continental drift occur?
10. How are honeycreepers an example of adaptive radiation? How does the iiwi, a type of honeycreeper, demonstrate evolution by natural selection? Coevolution?

Thinking Critically

1. Why should humans be concerned about the extinction of species such as the dusky seaside sparrow?
2. If you were experimenting on how protobionts evolved into cells, and you developed a protobiont that was capable of self-replication, would you consider it a living cell? Why or why not?
3. Why would the evolution of complex multicellular organisms such as plants and animals have to have been preceded by the evolution of oxygen-producing photosynthesis?
4. What environmental challenges did the first land-dwelling organisms face?
5. How might the evolution of modern animals be related to the evolution of flowering plants?
6. How might studying outer space help us construct the evolutionary history of life on Earth?

Recommended Readings

Bakker, R. "Jurassic Sea Monsters," *Discover*, September 1993. Extinction occurs differently in the ocean as compared to on the land.

Bean, M. J. "Where Late the Sweet Birds Sang," *Natural History*, February 1993. The tale of the dusky seaside sparrow's demise.

Gore, R. "Dinosaurs," *National Geographic* Vol. 183, No. 1, January 1993. New evidence about these remarkable organisms that roamed the Earth for 165 million years.

Kerr, R. A., "When Climate Twitches, Evolution Takes Great Leaps," *Science* Vol. 257, No. 18, September 1992. How climate changes have driven the evolutionary process.

Levinton, J. S., "The Big Bang of Animal Evolution," *Scientific American*, November 1992. A review of the Cambrian explosion, in which many diverse animal species appeared rather suddenly 600 million years ago.

Rice, J. A., consulting editor, "The Marvelous Mammalian Parade," *Natural History*, April 1994. A special section contains 16 short articles that highlight mammalian evolution over the past 200 million years.

Vickers-Rich, P., and T. H. Rich, "Australia's Polar Dinosaurs," *Scientific American*, July 1993. Dinosaurs living at the South Pole 100 million years ago were superbly adapted to the cold and dark.

PART 6 Evolution and the Diversity of Life

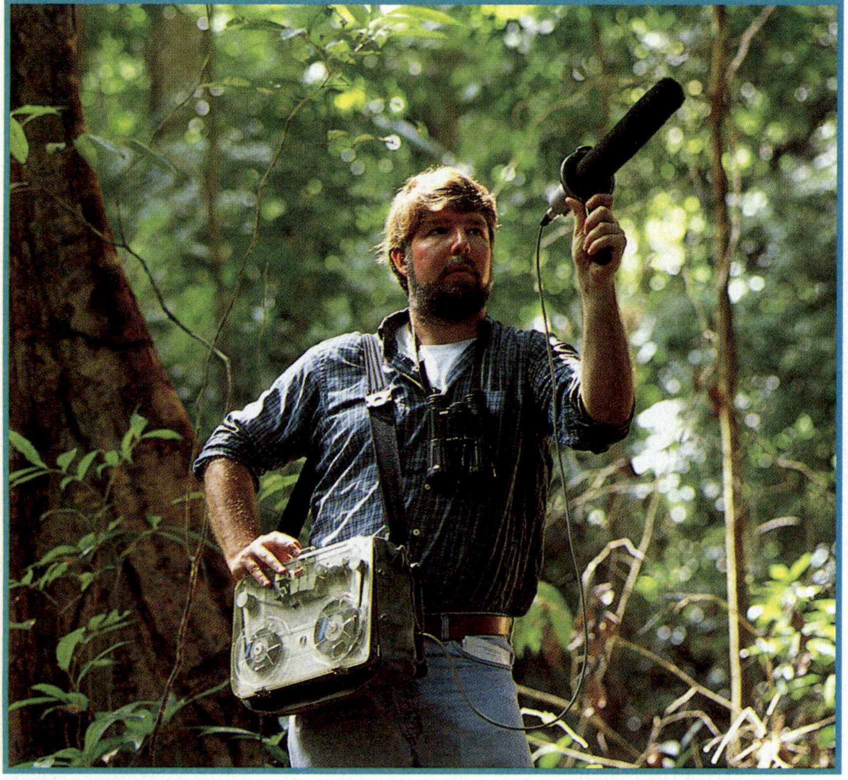

Ted Parker, of the Rapid Assessment Program, recording bird songs in Kanuku Guyana. (Courtesy of Haroldo Castro, Conservation International)

INVENTORY AND ACTION: PRESERVING EARTH'S BIO-DIVERSITY

About one-half of all of the living species on planet Earth inhabit the tropical rain forests (areas popularly referred to as jungles). Few other ecosystems house a comparable number and variety of species. Although about two million species of living organisms have been identified on our planet, biologists speculate that several million additional species remain to be identified. Many of the unidentified species make their homes in the rain forests of South and Central America, Central Africa, and Southeast Asia.

Destruction of habitats by human activities has decreased the extent of the rain forests by about 50% of their original expanse. These forests, which teem with life, now cover only about 7% of the land surface of our planet. Some scientists predict that most tropical rain forests will be destroyed by the end of this century. As the rain forests decrease, many of the species that inhabit them become extinct.

Scientists from many disciplines are working together to preserve as much of Earth's biological diversity (biodiversity) as possible. An important aspect of this enterprise is to identify and quickly classify the living organisms of the tropical rain forests and of other ecosystems of our planet.

Two methods currently employed to accomplish this task are: the Rapid Assessment Program (RAP) and the Neotropical Biological Diversity Program (BIOTROP). RAP utilizes a multidisciplinary, often multinational team that includes local experts to survey the organisms in a specific area as quickly as possible. They generally focus on "hot spots," which are areas in which biodiversity is determined to be at high risk. Although RAP is not a substitute for long-term biological inventory, this program provides important data about habitats that are changing very rapidly. The RAP team shares their findings with local scientists and makes recommendations that can be immediately applied to land-use planning and preservation of species.

The BIOTROP approach surveys broader areas than the RAP approach. The program is designed to study and develop data bases for living organisms found in diverse terrestrial environments that are in immediate danger of destruction in tropical America. The initial focus is on tropical forests. Scientists working in BIOTROP are rapidly gathering data on four groups of organisms—woody plants, vertebrates, ants, and butterflies—at many different sites. After determining the composition and interaction of the species in an area, they will investigate the factors that affect the ecosystem. BIOTROP plans to set up research stations that may eventually become centers for long-term biological research.

Programs such as RAP and BIOTROP are a beginning. Their surveys are providing important information about the organisms that inhabit the tropical rain forests and about the complexities of these ecosystems. We are just beginning to understand how various species depend on one another. For example, more than 90% of the trees of the rain forest depend on fruit-eating mammals to disperse their seeds. When mammal populations are reduced by hunting and by habitat destruction, trees and other plants are also threatened.

How we will respond to our increasing knowledge about tropical rain forests and other ecosystems remains to be seen. International conservation groups are attempting to implement "sustainable development" ("environmentally correct" development that does not destroy ecosystems) in Brazil and other tropical forest areas. These groups are negotiating with government and business interests for assurances that development will proceed in a manner that is ecologically responsible. In exchange, some conservation groups are offering financial incentives. For example, in Brazil, environmentalists are working to return infertile (as a result of slash-and-burn deforestation), abandoned land to productivity through reforestation and other techniques. It is hoped that through such efforts, some of the tropical rain forests will be preserved and even restored.

The Classification of Organisms

20

Learning Objectives

After you have studied this chapter you should be able to:

1. Offer at least two justifications for the use of scientific names and classifications of organisms.
2. Arrange the Linnaean categories in hierarchical fashion, from highest to lowest.
3. List the five kingdoms of organisms recognized by modern biologists, give the rationale for this system of classification, and describe the distinguishing traits of the organisms assigned to each.
4. Determine in which kingdom an organism belongs and summarize the basic characteristics of each kingdom.
5. Critically summarize the difficulties encountered in choosing taxonomic criteria.
6. Apply the concept of shared derived characteristics to the classification of organisms.
7. Describe the methods of molecular biology now used by taxonomists, and summarize their advantages.
8. Contrast the three major approaches to classification—phenetics, cladistics, and classical evolutionary taxonomy.

Key Concepts

- ☐ To make some order out of the millions of kinds of organisms and to communicate knowledge, we need a system of classification. The science of naming and classifying organisms is known as taxonomy.
- ☐ Using the binomial system of nomenclature we assign each species a scientific name composed of two parts—the genus name and the specific epithet.
- ☐ Biologists classify living organisms into five kingdoms—Prokaryotae, Protista, Fungi, Plantae, and Animalia.
- ☐ The hierarchical system of classification currently used includes kingdom, phylum (or division, in plants and fungi), class, order, family, genus, and species.
- ☐ Modern classification is based on evolutionary relationships.

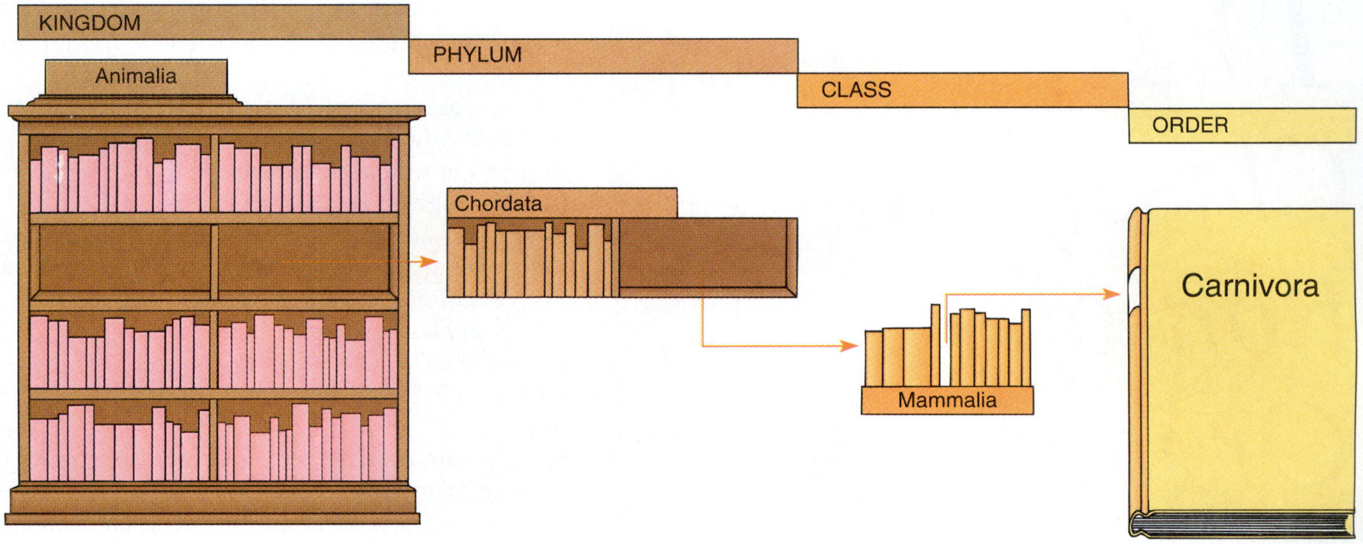

ORGANISMS ARE NAMED USING THE BINOMIAL SYSTEM OF NOMENCLATURE

In the 18th century, Carolus Linnaeus, a Swedish botanist and natural historian, simplified the scientific **classification** of organisms. Before Linnaeus, each organism had a lengthy descriptive name, sometimes composed of ten or more Latinized words! In the Linnaean system, which is also called the **binomial system of nomenclature,** each species is assigned a two-part name (see Chapter 1). The first part of the species name designates the **genus** (plural, genera), and the second part is the **specific epithet** (name). Corn, for example, is assigned to the genus *Zea;* its specific epithet is *mays.* Therefore the proper scientific name for corn is *Zea mays.* The genus name is always capitalized, whereas the specific epithet is not capitalized. Both names must be underlined or italicized. The genus name can be used alone to designate all species in the genus, but the specific epithet can never be used alone; it must always be preceded by the genus name.

Scientific names are generally composed from Greek or Latin roots, or from Latinized versions of the names of persons, places, or characteristics. For example, the genus name for the bacterium *Escherichia coli* is based on the name of the scientist who first described it—Theodor Escherich. The specific epithet *coli* reminds us that *E. coli* live in the colon (large intestine).

Assigning scientific names makes **taxonomy** a truly international study. Although many scientifically important organisms do not have common names, and despite the fact that many common names vary in different locations and languages, the assigning of scientific names allows organisms to be universally identified. As a result, a researcher in Argentina can know exactly which organisms were used in a published study by an Australian, and therefore would be able to repeat or extend the Australian's experiments using the same species.

Each Higher Taxonomic Level Is More General than the One Below

Recall from Chapter 1 that the narrowest category in the Linnaean system is the species, and the most general is the kingdom. The range of categories in between forms a hierarchy (Fig. 20–1).

A **taxon** (plural, taxa) is a taxonomic grouping at any level, such as species, genus, or phylum. For example, the phylum Chordata is a taxon that contains several classes, including Mammalia and Amphibia. Similarly, the class Mammalia is a taxon that includes many different orders.

Recall that a species is a group of similar organisms that can interbreed in their natural environment and are reproductively isolated from other organisms (Chapter 19). Closely related species are assigned to the same genus, and closely related genera are grouped in a single **family.** Families are grouped into **orders,** orders into **classes,** classes into **phyla,** and phyla into **kingdoms.** In classifying plants and fungi, the term **division** is used rather than phylum. These groupings can also be separated into subgroupings, for example, subphylum or subclass.

The **species** is the only one of the taxa that actually exists in nature. Members of a species are defined by their common gene pool and ability to interbreed. All higher taxonomic levels are artificial constructs designed by taxonomists. No operational definition exists for a genus, family, or phylum.

The number and inclusiveness of the principal groups vary according to the basis used for classification

FIGURE 20–1 The principal categories used in classifying an organism. The domestic cat (*Felis catus*) is used to illustrate the hierarchical organization of our taxonomic system.

and the judgment of the taxonomist making the decisions. Some taxonomists ignore minor variations and group organisms into already existing taxa. This practice is referred to as **lumping.** Other taxonomists subdivide taxa on the basis of major differences, establishing separate categories for forms that do not fall naturally into one of the existing classifications. This practice is called **splitting.** Lumpers acknowledge as few as 10 animal phyla and plant divisions, whereas splitters may recognize up to 33 animal phyla and up to 12 plant divisions.

Subspecies May Become Species

The species is the basic unit of classification, but not the smallest taxon in use. Populations inhabiting different geographic areas often display certain consistent characteristics that distinguish them from other populations of the same species. If they can interbreed, however, they are not truly separate species; rather, they are **subspecies.** For some kinds of microorganisms, such as bacteria, the term **strain** is used.

Experts are usually able to distinguish subspecies from one another. Some of these subspecies may be in the process of becoming reproductively isolated and may, in the course of time, become separate species. Thus, they provide an opportunity for field studies of gene pools and of the speciation process.

Many Biologists Recognize Five Kingdoms

For hundreds of years, biologists regarded living things as falling into two broad categories—plants and animals. With the development of microscopes, it became increasingly obvious that many organisms did not fit very well into either the plant or the animal kingdom. For instance, bacteria lack nuclei and other membrane-bound organelles. This difference, which distinguishes bacteria from all other organisms, is far more fundamental than the differences between plants and animals. With our present knowledge, it is difficult to consider bacteria as plants, as

Concept Connections

Classification
Kingdom Protista

Although the five-kingdom system represents a definite improvement over the two-kingdom system, it is not perfect. Many of the problems with the five-kingdom system concern the kingdom Protista. As we will see in Chapter 21, it includes relatively simple eukaryotic organisms such as amebas, algae, and slime molds. Many protists may be more closely related to members of other kingdoms than to certain other protists. For example, the protists called green algae are clearly similar to plants but do not appear to be closely related to other protists such as slime molds or red algae.

Ideally, all members of a kingdom should have a common ancestor. There seems to be no common ancestor for the protist kingdom because it appears that eukaryotic cells evolved several times. Some biologists want to remove this problem by dividing the Protist kingdom into several additional kingdoms along more natural groupings. However, many biologists would rather deal with the limitations of the five-kingdom system than with additional kingdoms.

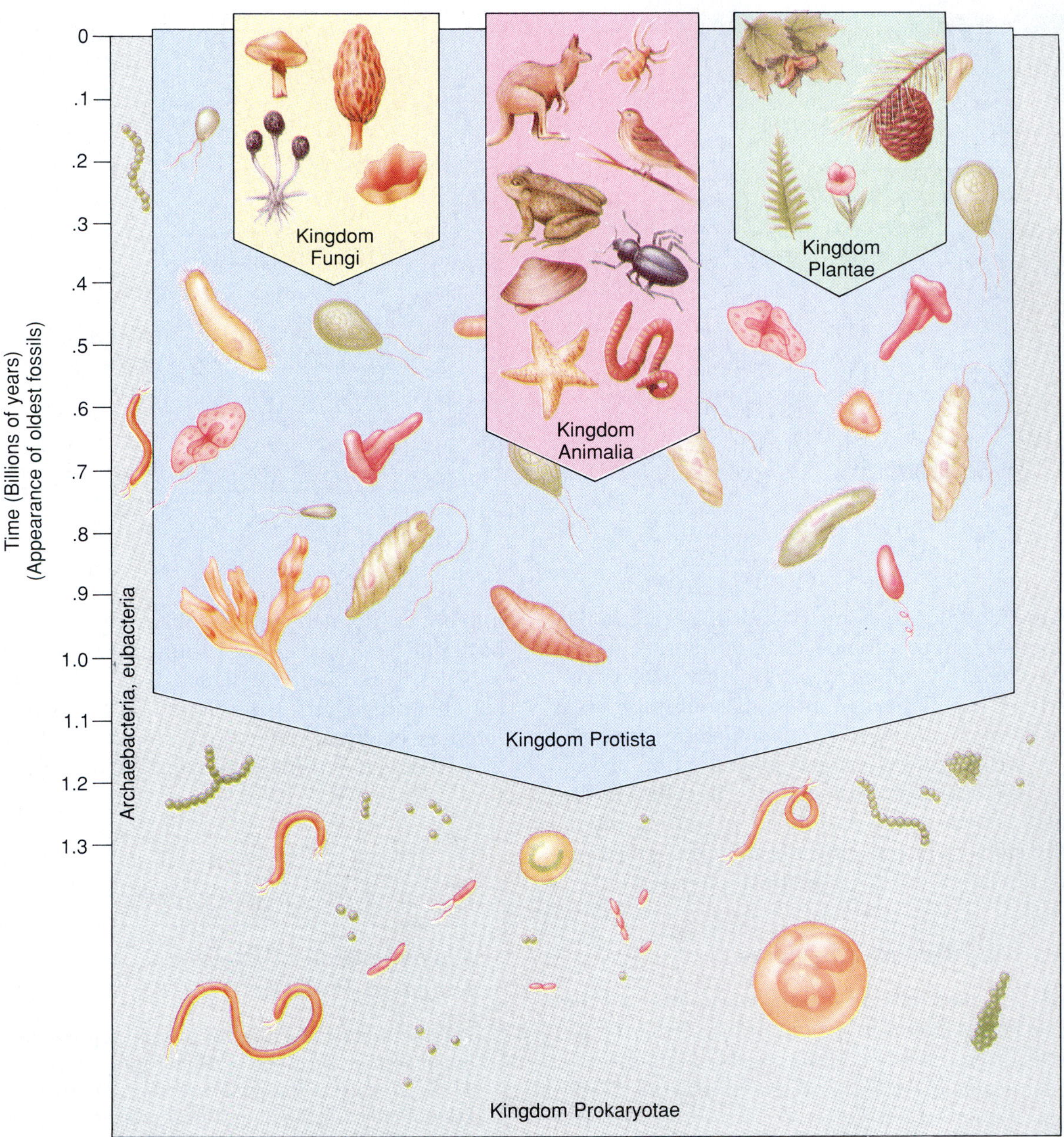

FIGURE 20–2 The five-kingdom system of classification.

was done formerly. Certain organisms—the protist *Euglena,* for example—seem to possess characteristics of both plants and animals. In fact, many single-celled organisms seem to have more in common with one another than with either multicellular plants or multicellular animals.

These and other considerations, including evidence from molecular studies, have led to the five-kingdom system of classification that is used by many biologists today (Fig. 20–2). As outlined in Chapter 1 and in Table 20–1, biologists recognize the kingdoms as Prokaryotae, Protista, Fungi, Plantae, and Animalia.

SYSTEMATICS IS CONCERNED WITH RECONSTRUCTING PHYLOGENY

Modern classification is based on evolutionary relationships. The classification of organisms into groups determined by their evolutionary relationships is called **systematics.** A systematist seeks to reconstruct the evolutionary history, or **phylogeny** (literally, production of phyla), of organisms. Once these relationships are defined, the classification of organisms can be based on common ancestry.

TABLE 20–1
Five Kingdoms: Prokaryotae, Protista, Fungi, Plantae, and Animalia

Kingdom	*Characteristics*	*Ecological role and comments*
Prokaryotae	Prokaryotes (lack distinct nuclei and other membranous organelles); single-celled; microscopic; cell walls composed of peptidoglycan; metabolically varied	Most are decomposers; some parasitic (and pathogenic); some chemosynthetic autotrophs; some photosynthetic; important in recycling nitrogen and other elements; some utilized in industrial processes
Protista	Eukaryotes; mainly unicellular or simple multicellular	
Protozoa	Microscopic; heterotrophic; most move by means of flagella, cilia, or pseudopodia	Important part of zooplankton; near base of many food webs; some are parasitic (and pathogenic)
Algae	Photosynthetic; sometimes hard to differentiate from protozoa; some have brown or red pigments in addition to chlorophyll	Very important producers, especially in marine and freshwater ecosystems
Slime molds	Heterotrophic; reproduce by forming spores	
Fungi	Heterotrophic; absorb nutrients; do not photosynthesize; body composed of threadlike hyphae that form tangled masses that infiltrate fungus's food or habitat	Decomposers; some parasites (and pathogenic); some used as food; yeast used in making bread and alcoholic beverages; some used to make industrial chemicals or antibiotics; responsible for much spoilage and crop loss
Plantae	Multicellular; photosynthetic; plants possess multicellular reproductive organs; pass through alternation of generations; cell walls of cellulose	Terrestrial biosphere depends upon plants in their role as primary producers; important source of oxygen in Earth's atmosphere
Animalia	Multicellular heterotrophs, many of which exhibit advanced tissue differentiation and complex organ systems; most able to move about by muscular contraction; extremely and quickly responsive to stimuli, with specialized nervous tissue to coordinate responses	Almost sole consuming organisms in biosphere; some specialized as herbivores, carnivores, or detritus feeders

Taxa May Be Monophyletic or Polyphyletic

A population of organisms has a dimension in space—its geographical range—and also a dimension in time. Each population extends backward in time, merging with populations of other species much like branches of a tree (Fig. 20–3). Species have various degrees of evolutionary rela-

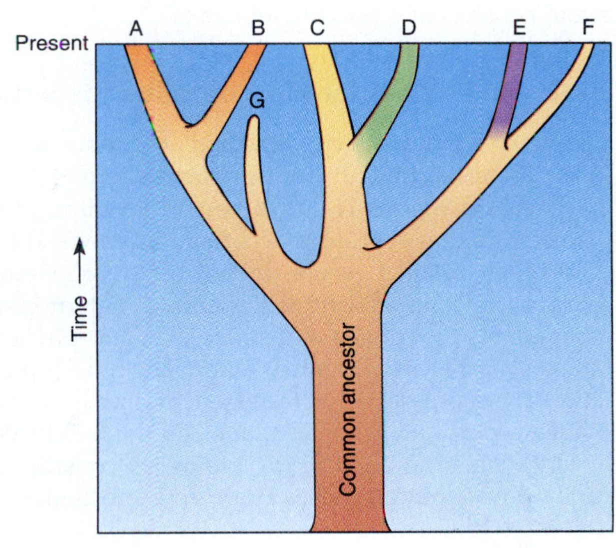

FIGURE 20–3 The evolutionary relationships of several hypothetical monophyletic species. Branches at top of figure represent the species at the present time. Junctions of the branches represent points of common ancestry. Species G is extinct. If you go far enough back in time, all taxa share a common ancestor. The base of the tree represents the common ancestor of species A, B, C, D, E, F, and G. Which groupings of these species might be considered a genus? A family? Many of the groups classified today on the basis of their structural similarities are thought to share a common, although remote, origin and therefore also share genetic similarities.

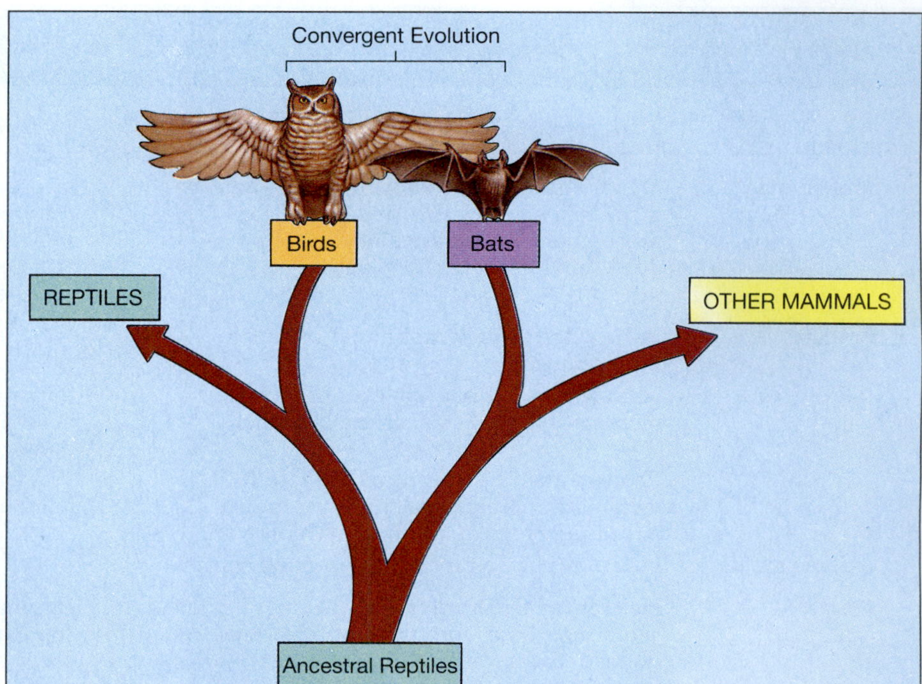

FIGURE 20–4 Convergent evolution. In convergent evolution, distantly related groups such as birds and bats may come to resemble one another in structure and function as they become adapted to fit similar modes of life. Their analogous structures, such as wings, do not indicate common ancestry. In contrast, the presence of homologous structures suggests evolution from a common ancestor.

tionship with one another, depending on the length of time that has elapsed since their populations diverged. Before diverging, they had a common ancestor.

If all of the subgroups within any taxon share the same common ancestor, the grouping is referred to as **monophyletic** (one branch). Monophyletic taxa are, therefore, *natural* groupings as they represent true evolutionary relationships, and they include all close relatives. A taxon containing a common ancestor and all the taxa descended from it is called a **clade.**

Many taxa are **polyphyletic,** consisting of several evolutionary lines and not including a common ancestor. Polyphyletic taxa, therefore, may misrepresent evolutionary relationships. For this reason, taxonomists try to avoid constructing polyphyletic taxa.

Biologists Consider Homologous Structures

Just how to group species into higher taxonomic groups is sometimes a difficult decision. For example, in Figure 20–3, should species A and B be placed within a single genus or do they represent two distinct genera? If species C and E are distinct genera, should D be part of either of these genera? Similar difficulties exist in determining the assignments to families, orders, classes, and phyla. Most biologists base their judgments regarding the degree of relationship of organisms on the extent of similarities among living species, and when available, on the fossil record.

In evaluating similarities, biologists consider structural, physiological, behavioral, and molecular traits. When comparing structural similarities, look for **homologous structures** in different organisms (see Chapter 17). The presence of homologous structures suggests that two species of organisms evolved from a common ancestor. In contrast, similarities among analogous structures result not from shared ancestry but from convergent evolution (Fig. 20–4). This sometimes occurs when unrelated or distantly related organisms become adapted to similar environmental conditions. For example, the shark and the dolphin have similar body forms because they have become adapted to similar environments.

Derived Characters Have Evolved More Recently than Ancestral Characters

Organisms that share many homologous structures are thought to be closely related, and less similar organisms that share few homologous characters are less closely related. However, distinguishing between homologous and analogous structures is not always easy. Therefore, the choice of which similarities should be used to show evolutionary relationships is extremely important.

How does a systematist interpret the significance of these similarities? In making decisions about taxonomic relationships, the biologist first examines the characteristics common to the largest category of organisms and interprets them as indicating the most remote common ancestry. Such characteristics, termed **ancestral characters,** are traits that were present in an ancestral species and that have remained essentially unchanged.

Derived characters are those traits not present in ancestral species because they evolved more recently. A feature viewed as a derived character in a large taxon may also be considered as an ancestral character in a smaller taxon. More recent common ancestry is indicated by classification into smaller and smaller taxonomic groups with more and more specific shared derived characters.

For example, the three small bones in the middle ear are useful in identifying a branching point between mammals and reptiles. The evolution of this derived character was a unique event, and only mammals have these bones. However, if we are comparing mammals with one another, the three ear bones are an ancestral character because all mammals have them. They have no value in distinguishing among mammalian groups. Other derived characters must be used to establish branching points *among* the mammals.

Biologists Carefully Choose Taxonomic Criteria

Although both fishes and porpoises have streamlined body forms, this characteristic is an analogous adaptation and is less important for indicating relationships than homologous structures they share with other organisms. The porpoise shares important derived characteristics with mammals such as humans—the ability to breathe air, nurse young, maintain a constant body temperature, and grow hair. Thus, the porpoise is classified as a mammal and is viewed as descended from a terrestrial mammalian ancestor.

Although the porpoise has more features in common with humans than it does with fishes, some characteristics are shared by all three kinds of animals. Among these are a notochord (skeletal rod) and rudimentary gill slits in the embryo stage, and a dorsal tubular nerve cord. These shared ancestral characteristics indicate a common ancestry and serve as a basis for classification. This ancestry is more remote between the porpoise and the fish than between the porpoise and humans. Therefore, fishes, humans, and porpoises are grouped together in a large taxon, the phylum Chordata, and humans and porpoises are also classified together in class Mammalia, a smaller taxon indicating their closer relationship (Fig. 20–5).

Deciding which traits are most important in illustrating evolutionary relationships is not always simple.

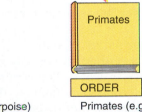

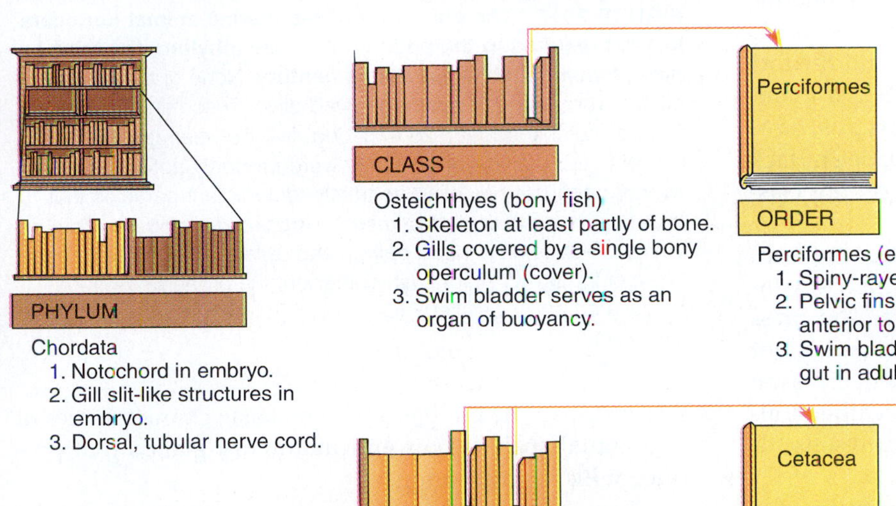
PHYLUM
Chordata
1. Notochord in embryo.
2. Gill slit-like structures in embryo.
3. Dorsal, tubular nerve cord.
CLASS
Osteichthyes (bony fish)
1. Skeleton at least partly of bone.
2. Gills covered by a single bony operculum (cover).
3. Swim bladder serves as an organ of buoyancy.
Perciformes
ORDER
Perciformes (e.g., perch)
1. Spiny-rayed.
2. Pelvic fins either under or anterior to pectoral fins.
3. Swim bladder not connected to gut in adult.
CLASS
Mammalia (mammals)
1. Hair.
2. Mammary glands in the female produce milk for young.
3. Maintain a constant body temperature (endothermic).
4. Muscular diaphragm helps move air in and out of lungs.
Cetacea
ORDER
Cetacea (e.g., porpoise)
1. Adapted to aquatic life.
2. Nostrils modified as blowholes on top of head.
3. Front limbs developed into flippers.
4. Hind limbs absent.
Primates
ORDER
Primates (e.g., human)
1. Highly developed brain and eyes; eyes directed forward.
2. Opposable thumb or great toe.
3. Limbs have ball and socket joints permitting wide range of movement.

FIGURE 20–5 Shared derived characteristics. Members of class Osteichthyes (bony fish) and class Mammalia (the mammals) share many more characteristics with one another and with the members of the other classes of phylum Chordata than they do with members of any other phylum. For example, a perch has more in common with a monkey (a notochord, gill slits, and dorsal nerve cord) than with a sea star or clam. Members of various orders of the same class share more characteristics than members of orders that belong to different classes. Thus, a porpoise has more derived characters in common with a human than with a perch, indicating a more recent common ancestry for the porpoise and the human.

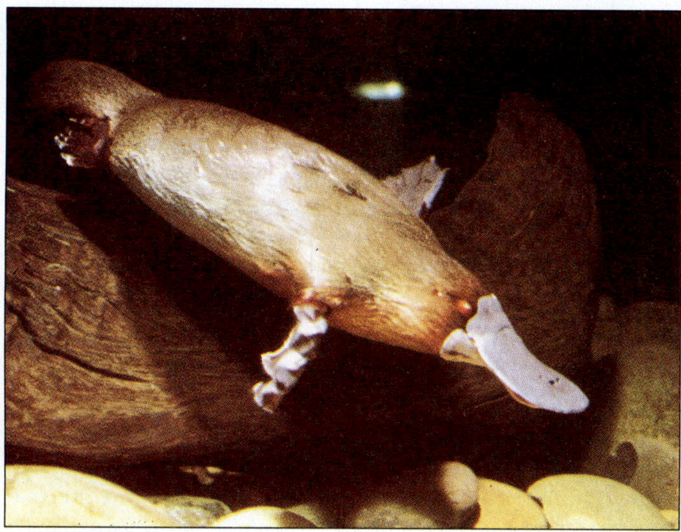

FIGURE 20–6 A few mammals share important characteristics with birds. The duck-billed platypus, a monotreme, lays eggs, has a beak, and lacks teeth. Should we classify it as a bird? (Tom McHugh/PhotoResearchers, Inc.)

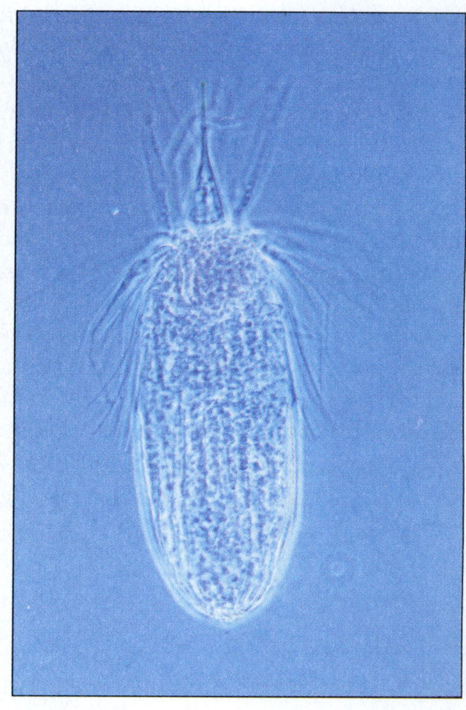

FIGURE 20–7 The discovery of the marine animal Loricifera in 1983 resulted in the addition of a new phylum, the third new phylum established in this century. None of the nine loriciferan species have been observed alive. These tiny animals (adults are only about 0.25 mm long) live between grains of shell gravel in the ocean bottom. They were probably not discovered sooner because they cling so tightly to sediment particles that they cannot be collected by usual extraction techniques. Both larvae and adults have head spines and a flexible, retractable tube-like mouth. Do you think many other unique organisms will be discovered and classified? (Courtesy of Robert P Higgins, Smithsonian Institution)

What, for example, are the most important taxonomic characteristics of a bird? We might list feathers, beak, wings, absence of teeth, the egg-laying trait, and the fact that they are endotherms (warm-blooded). Some mammals (monotremes such as the duck-billed platypus) have many of these same characteristics—they have beaks, lack teeth, lay eggs, and are endotherms. Yet, we do not classify them as birds (Fig. 20–6).

No mammal, however, has feathers. Is this trait absolutely diagnostic of birds? According to the conventional taxonomic wisdom, the answer is yes. The presence of feathers can be used to decide what is and is not a bird. This applies only to modern birds, however. Some extinct reptiles may have been covered in feathers, although they were not "birds" in any meaningful sense of the word.

Usually, organisms are classified on the basis of a combination of traits rather than on any single trait such as the ability to live in water. The significance of these combinations is determined inductively, that is, by an integration and interpretation of data. Such induction is a necessary first step in all science. Taxonomists hypothesize, for example, that birds should all have beaks, feathers, no teeth, and so on. Then they reexamine the living world and observe whether there are organisms that might reasonably be called birds that do not fit the current definition of "birdness." If not, the definition is permitted to stand, at least until too many exceptions emerge. Then the definition may be modified or abandoned. Sometimes, the taxonomist persuades the world that an apparent exception—the bat, for instance—resembles a bird only superficially and should not be considered one. The bat has all of the basic characteristics of a mammal such as hair and mammary glands that produce milk for the young.

Taxonomy is a dynamic science that proceeds by the constant reevaluation of data, hypotheses, and theoretical constructs. As new data are discovered and old data are subjected to reinterpretation, the ideas of taxonomists change. During the 1980s, for example, a type of organism, the Loricifera, was discovered whose combination of traits did not fit those of any existing phylum (Fig. 20–7). A new phylum—Phylum Loricifera—was established to accommodate this animal.

Molecular Biology Provides Taxonomic Tools

When a new species evolves it does not always have obvious phenotypic changes. For example, two distinct species of fruit flies may appear identical. Some of their macromolecules, however are different. Variations in the structure of specific macromolecules among species, just like differences in anatomical structure, result from mutations. Macromolecules that are functionally similar in two different types of organisms are considered homologous if their subunit structure is similar.

Methods that enable biologists to compare the nucleotide sequences of various nucleic acids and the amino acid sequences of proteins have become extremely important taxonomic tools. By comparing nucleotide sequences of DNA, RNA, or amino acid sequences of proteins, we can gain some idea of degree of relatedness of two organisms. The greater the correspondence in amino acid sequences of these organisms the more closely they are thought to be related. The number of differences in nucleotide sequence of DNA or in amino acid sequence of proteins in two groups of organisms, may reflect the time since the groups branched off from a common ancestor. Thus, specific genes and specific proteins can be used as **molecular clocks** (see Chapter 17). Biologists can use such clocks to help date the divergence of two groups from a common ancestor.

Comparison of ribosomal RNA sequences has recently been used to challenge a widely accepted relationship among fungi, plant, and animal kingdoms—that fungi are more closely related to plants than to animals. In a study reported in *Science* (Vol. 260, April 1993) investigators suggested that, based on ribosomal RNA analysis, fungi are more closely related to animals than to plants. These biologists hypothesize that animals and fungi share a more recent common ancestor—perhaps a flagellated protist.

Biologists are using polymerase chain reaction (PCR), a method described in Chapter 16, to amplify small amounts of DNA extracted from fossils. Using this method, researchers can obtain sufficient DNA to compare with DNA from modern organisms, a strategy that may contribute to some important taxonomic decisions.

TAXONOMISTS USE THREE MAIN APPROACHES

In constructing a phylogenetic tree, taxonomists consider branch points that indicate the time at which a particular group of organisms evolved. They also consider the degree of divergence between branches, or how different two groups have become since they originated from a common ancestor and evolved along different pathways. Which of these bits of evolutionary data is utilized more in classifying a group of organisms depends upon one's approach to taxonomy. Three major approaches to the classification of organisms are phenetics, cladistics, and classical evolutionary taxonomy.

Phenetics Is Based on Phenotypic Similarities

Pheneticists argue that we cannot be sure that our view of evolutionary history is correct. These taxonomists base their classification on traits they can measure. The **phenetic (phenotypic) system,** sometimes called numerical taxonomy, is based on similarities of many characteristics. In this system, organisms are grouped according to the number of characteristics they share, without trying to determine whether their similarities arise from a common ancestor or from convergent evolution. Pheneticists suggest that it is not important to try to sort homologous and analogous characteristics, because many more similarities are due to homology rather than analogy. Overall, the number of similarities that two organisms have in common should reflect the degree of homology.

A taxonomist who follows the phenetic system might explain that porpoises are classified along with humans as mammals rather than fishes because they share more similarities with mammals. Pheneticists assign numbers to many arbitrarily chosen traits that are given equal weight. They designate these traits as present (+) or absent (−) in the organisms of a particular taxon. This information is fed into a computer, which indicates which groups have the most traits in common.

Phenetics is not widely used by taxonomists today because the use of analogous similarities can lead to inaccurate conclusions about evolutionary relationships. However, the phenetic emphasis on quantitative comparisons using computers has been an important contribution to biology.

Cladistics Emphasizes Phylogeny

The **cladistic** approach to taxonomy emphasizes phylogeny, focusing on when evolutionary lineages divide into two different branches. Cladists determine branch points using carefully defined objective criteria, and insist that taxa be monophyletic. Thus, common ancestry is the basis for classification rather than data on phenotypic similarity. A cladist would say that porpoises are classified with mammals rather than fishes because porpoises and mammals share a more recent common ancestor than porpoises and fishes.

Cladists develop branching diagrams called **cladograms.** Each branch on a cladogram represents the divergence, or splitting, of two new groups from a common ancestor. Consider the evolutionary grouping of turtles, lizards, snakes, dinosaurs, crocodiles, and birds. Birds, along with dinosaurs, are thought to share a common ancestor with the modern crocodiles and alligators. Crocodiles, dinosaurs, and birds, then, comprise a monophyletic group, and cladists would classify them in the

(Text continues on page 372)

FOCUS ON *The Process of Science*

Building and Interpreting Cladograms

The general goal of cladistics is to reconstruct phylogenies using an analysis of evolutionary changes in specific traits or characters. The kinds of characters used for the analysis can be structural, behavioral, physiological, or molecular. The only requirements are that the characters are homologous (identical by common descent) and that they have evolved independently of each other.

Collecting and preparing the data.

The first step in constructing a cladogram is to select the taxa. The selected taxa may consist of individuals, species, genera, or other taxonomic levels. Here, we use a representative group of eight chordate taxa (Table 20-A).

Next, the homologous characters to be analyzed must be selected. In our example we use seven characters. For each character, we must define all of the different conditions, or states as they exist in our taxa. For simplicity, we will consider our characters to have only two different states (present or absent). Keep in mind that many characters used in cladistics have more than two states. For example, black, brown, yellow, and red may be only a few of the many possible states for the character hair color.

The last, and often the most difficult, step in preparing the data is to organize the character states into their correct evolutionary order. The most common method of accomplishing this task is by outgroup analysis. An **outgroup** is a taxon that is considered to have diverged earlier than any of the other taxa under investigation and thus represents an approximation of the ancestral condition. In our example, *Amphioxus* is the chosen outgroup. Therefore, the

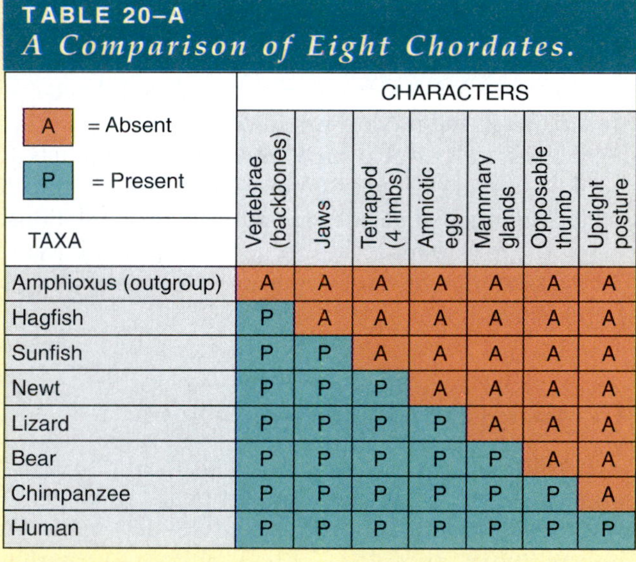

TABLE 20–A
A Comparison of Eight Chordates.

A = Absent; P = Present

TAXA	Vertebrae (backbones)	Jaws	Tetrapod (4 limbs)	Amniotic egg	Mammary glands	Opposable thumb	Upright posture
Amphioxus (outgroup)	A	A	A	A	A	A	A
Hagfish	P	A	A	A	A	A	A
Sunfish	P	P	A	A	A	A	A
Newt	P	P	P	A	A	A	A
Lizard	P	P	P	P	A	A	A
Bear	P	P	P	P	P	A	A
Chimpanzee	P	P	P	P	P	P	A
Human	P	P	P	P	P	P	P

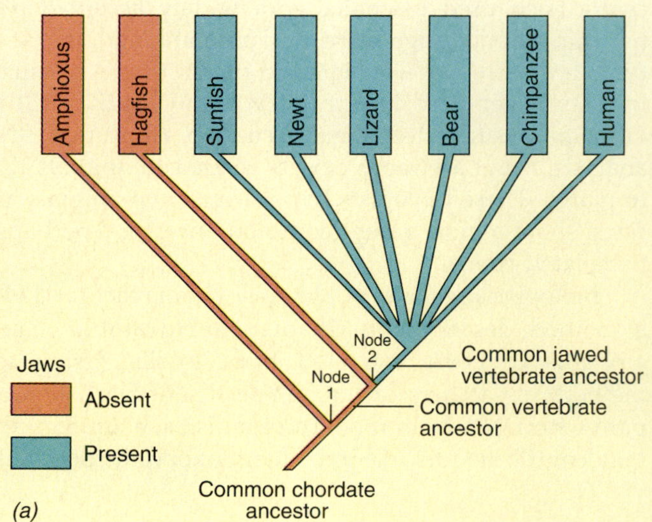

(a)

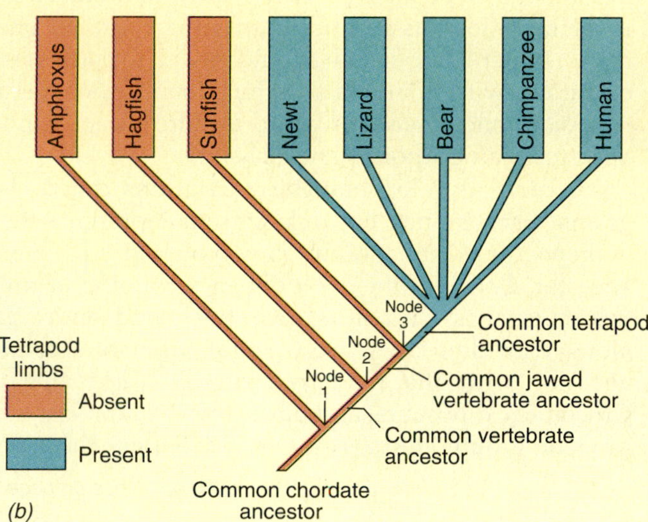

(b)

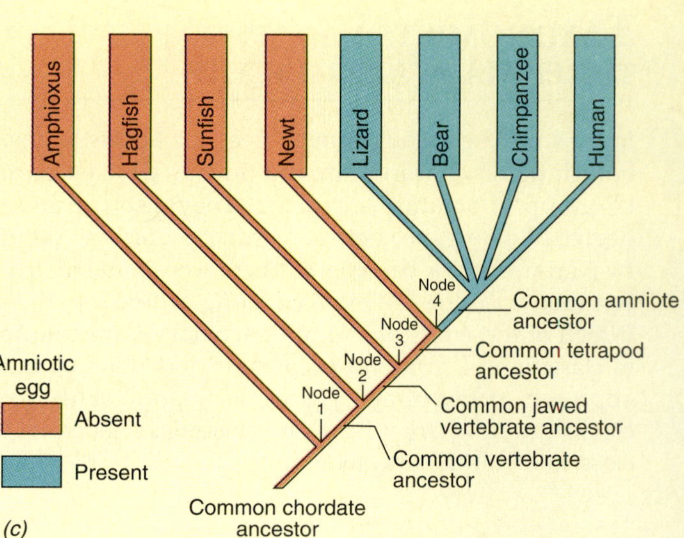

(c)

character state "absent" is the ancestral condition and the character state "present" is the derived condition for all seven characters (Table 20-A).

Constructing the cladogram from the data.

Our objective is to find the cladogram that requires the fewest number of evolutionary changes in the characters. The reason for this criterion is that the cladogram that requires the least evolutionary change is also the most likely cladogram. In cladistics, taxa are grouped by the presence of shared derived character states. In order to form a valid monophyletic group, all members must share at least one derived character state. Membership in a group cannot be established by sharing ancestral character states.

In our example, notice that all taxa, except the outgroup, possess vertebrae. We may therefore conclude that these seven vertebrate taxa form a valid monophyletic group. Next, among the seven vertebrate taxa, notice that jaws are present in all groups except for hagfish. Using these data, we may construct a preliminary cladogram (Figure *a*). The base of the tree represents the common ancestor for all taxa being analyzed and the branch points (referred to as nodes) represent the divergence of the ancestral lineage into two derived lineages, or clades. In Figure *a*, node 1 represents the divergence of the outgroup (*Amphioxus*) and the common ancestor of the seven vertebrate taxa from the common chordate ancestor. Similarly, node 2 represents a subsequent divergence of hagfish and the ancestor of the six jawed vertebrates. Continuing with this procedure, notice that among the six jawed taxa, all but sunfish are tetrapods (Figure *b*). Among the five tetrapods, all but newts have amniotic eggs (Figure *c*). The branching process is continued using Table 20–A data until all clades are established (Figure *d*).

How to interpret the cladogram.

In Figure *d*, notice that humans and chimpanzees are more similar to each other than to any other clade. This relationship is indicated by the presence of a common ancestor at node 7. In the same way, bears are more similar to the human-chimpanzee clade than to any other clade as indicated by the common ancestor at node 6. In comparing the nodes, time of divergence is indicated by distance from the base of the tree. The further a node is located up the tree, the more recent the time of divergence. In our example, node 7 represents the most recent divergence and node 1 represents the least recent (most ancient) divergence. Thus, in our example, humans are closely related to chimpanzees (through node 7) but more distantly related to bears (through node 6). Therefore, humans and chimpanzees would be assigned to a common taxon (Order Primates) whereas humans, chimpanzees, and bears would be assigned to a broader, less exclusive taxon (Class Mammalia). In addition, the cladogram reveals that lizards are more closely related to the clade Mammalia than to newts, sunfish or any other clade. Can you explain why?

In interpreting cladograms, two points must be kept in mind. First, the relationships between taxa can only be determined by tracing along the branches back to the most recent common ancestor (i.e., node) and not by the relative placement of the branches along the horizontal axis. It is possible to represent the same relationships with many different branching diagrams. For example, the cladogram in Figure *e* is equivalent to the one in Figure *d* (you should verify this by comparing the numbered nodes and by checking the relationships described earlier). Second, the cladogram does not establish ancestor-descendant relationships between taxa. In other words, a cladogram does not suggest that a taxon gave rise to any other taxon; rather it tells us which taxa shared a common ancestor and how recently they shared a common ancestor. The ancestor itself remains unspecified.

Essay contributed by Dr. John Beneski, Department of Biology, West Chester University, West Chester, PA

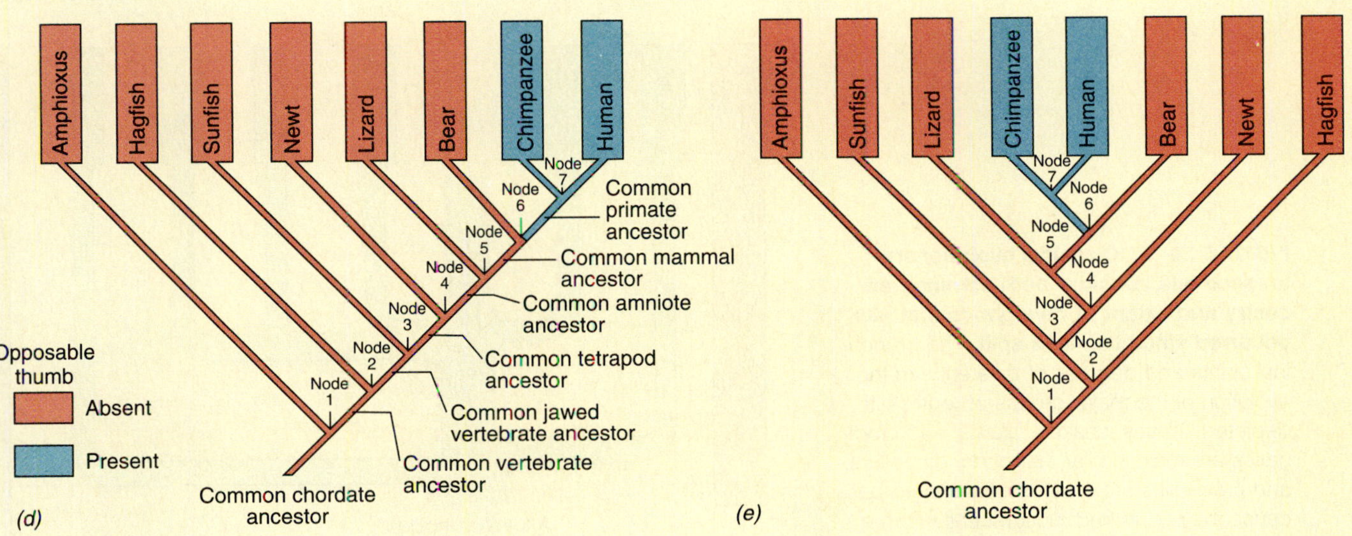

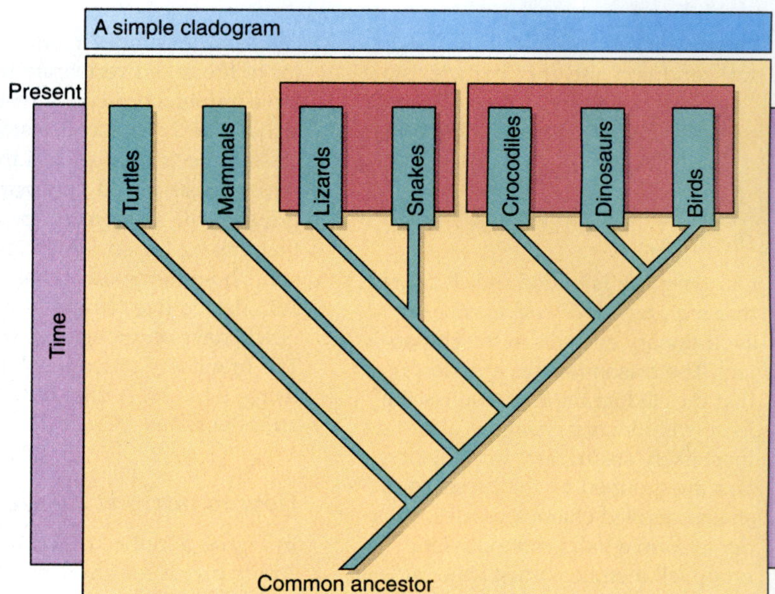

FIGURE 20–8 A simple cladogram. According to the cladistic approach, birds and some reptiles are classified together because they have a common ancestor.

same taxon (Fig 20–8). Similarly, snakes and lizards form a monophyletic group which is a sister group to birds, dinosaurs, and crocodiles. Finally, mammals and turtles form two additional groups.

Notice that in this interpretation, reptiles (snakes, lizards, crocodiles, dinosaurs, and turtles) are not recognized as a natural group since they do not form a monophyletic clade. One explanation for this apparent dilemma is that taxonomists who consider reptiles to be a valid group base their conclusion, in part, on shared ancestral characters such as epidermal scales and ectothermy. In contrast, cladists base their assessment on shared derived characters, such as heart structure, skeletal modifications, and reproductive behavior. The cladistic approach challenges us to reconsider and reevaluate our view of evolutionary relationships.

Classical Evolutionary Taxonomy Uses a Phylogenetic Tree

Classical evolutionary taxonomy uses a system of phylogenetic classification and presents evolutionary relationships in a phylogenetic tree. Classical taxonomists consider both evolutionary branching (like cladists) and the extent of divergence that has occurred in a lineage since it branched from a stem group (Fig. 20–9). They base their

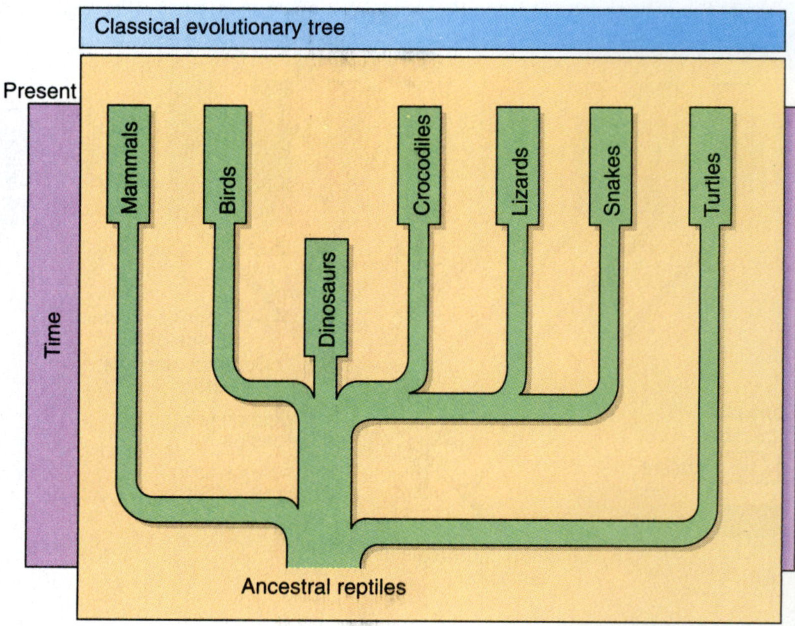

FIGURE 20–9 Classical evolutionary taxonomists consider both common ancestry and extent of divergence that has occurred since two taxa split. The branching points and degrees of difference in the evolution of the major groups of reptiles is depicted. Turtles, snakes, lizards, and crocodiles are most similar, but birds, dinosaurs, and crocodiles are most closely related because they branched most recently from a common ancestor.

classification decisions on shared ancestral characters.

A taxonomist using the classical approach might explain that porpoises are mammals rather than fishes because they share many characteristics with other mammals and because these characteristics can be traced to a common ancestor. Organisms are classified in the same taxon according to their shared characteristics only if those traits are derived from a demonstrable common ancestor. The significance of the adaptations possessed by related organisms is also considered. If, for example, egg-laying mammals could be shown to have a very different ancestry from the other mammals, the classical taxonomist might erect a separate class to accommodate them. On the other hand, common ancestry, although necessary for inclusion in the same category, would not by itself be sufficient grounds for inclusion.

A classical taxonomist would classify birds and crocodiles separately, for example, even though they share a common ancestor. They would place birds in class Aves because they are endotherms (able to maintain a constant body temperature), have feathers, and have other features that indicate extensive divergence since branching from the early reptiles. These taxonomists would consider the shared ancestral characters, such as horny scales and ectothermy (the fluctuation of body temperature with the temperature of the surrounding environment), of turtles, lizards, snakes, and crocodiles and would assign these animals to a separate class, Reptilia.

Chapter Summary

I. The modern system of scientific taxonomy is based on the binomial system first used consistently by Linnaeus.
   A. In this system, the basic unit of classification is the species.
   B. The name of each species has two parts: the genus name and the specific epithet. For example, the scientific name for the human is *Homo sapiens:* that for the domestic cat is *Felis catus*.

II. The hierarchical system of classification currently used includes kingdom, phylum (or division, in plants and fungi), class, order, family, genus, and species.

III. The five-kingdom classification in current use recognizes the kingdoms Prokaryotae, Protista, Fungi, Plantae, and Animalia.

IV. Modern classification is based on evolutionary relationships, or phylogeny.
   A. All of the organisms in a monophyletic taxon have a common ancestor; the organisms in a polyphyletic taxon evolved from different ancestors.
   B. Homologous structures indicate evolution from a common ancestor.
   C. Shared derived characters indicate a more recent common ancestor than shared ancestral characters.
   D. Comparison of DNA and protein structure provides additional evidence about evolutionary relationships.

V. Three main approaches to taxonomy are phenetics, cladistics, and classical evolutionary taxonomy.
   A. The phenetic system is a numerical taxonomy based on similarities of many characters. In this system, organisms are classified according to the number of characteristics they share without trying to determine whether their similarities are homologous or analogous.
   B. The cladistic approach insists that taxa be monophyletic. Each taxon consists of a common ancestor and all its descendants.
   C. Classical evolutionary taxonomy considers both evolutionary branching and the extent of divergence.

Selected Key Terms

ancestral characters, p. 366
binomial system, p. 362
cladistics, p. 369
cladogram, p. 369
class, p. 362
classification, p. 362
derived characters, p. 367
division, p. 362
family, p. 362
genus, p. 362
homologous structure, p. 366
kingdom, p. 362
molecular clock, p. 369
monophyletic group, p. 366
order, p. 362
phenetics, p. 369
phylogeny, p. 364
phylum, p. 362
polyphyletic group, p. 366
species, p. 362
systematics, p. 364
taxon, p. 362
taxonomy, p. 362

Post-Test

1. The science of describing, naming, and classifying organisms is ______________.
2. Using the binomial system of nomenclature, the scientific name of each species consists of two parts, the ______________ and the ______________ epithet.
3. The mold that produces penicillin is *Penicillium notatum*. Its genus is ______________.
4. Closely related genera may be grouped together in a single ______________.
5. The botanical equivalent of a phylum is a/an ______________.
6. The kingdom that includes the algae is ______________.
7. Kingdom ______________ consists of decomposers such as molds and mushrooms.
8. The members of a monophyletic group have a common ______________ that is classified as a member of that group.
9. The presence of ______________ (homologous or analogous) structures in different organisms suggests that they evolved from a common ancestor.
10. The porpoise and the human both have the ability to nurse their young, whereas the less closely related fishes do not. The ability to nurse their young is a shared ______________ character for mammals as compared to fishes.
11. The constancy in DNA and protein evolution permits biologists to use these macromolecules as molecular ______________.
12. The ______________ system is a numerical taxonomy based on phenotypic similarities.
13. Taxonomists who follow the ______________ school of taxonomy might classify crocodiles and birds in the same group based on a common ancestor.
14. A system of classification that attempts to balance data from both phenetics and cladistics is used by ______________ ______________ taxonomists.
15. A species is a group of similar organisms that (a) lacks a common ancestor (b) interbreeds in their natural environment (c) interbreeds with closely related species (d) is reproductively isolated from members of other species (e) (b) and (d) are correct
16. Taxonomists study the classification of organisms based on their: (a) evolutionary history (b) development (c) genetics (d) color
17. All members of kingdom Prokaryotae are (a) plants (b) bacteria (c) animals (d) fungi (e) slime molds

Review Questions

1. Briefly describe the binomial system of nomenclature.
2. Describe (in modern terms) (a) species, (b) class, (c) phylum, (d) division.
3. What are the advantages of a five-kingdom system over a two-kingdom one? What types of organisms are especially difficult to assign a place in the taxonomic hierarchy?
4. In which kingdom would you classify each of the following? (a) an oak tree (b) an ameba (c) *Escherichia coli* (a bacterium) (d) a tapeworm
5. Compare the phenetic, cladistic, and classical evolutionary approaches to taxonomy.
6. Of what use to a taxonomist would be knowledge of the amino acid sequences of the proteins of various organisms?

Thinking Critically

1. Why is it important to survey the organisms that inhabit the tropical rain forests? Why do you think so many of these organisms have never been scientifically described or classified?
2. Imagine that you are a biologist in the 17th century (predating Linnaeus). How might you have classified the organisms of our planet?
3. What taxonomic problems was the five-kingdom scheme intended to solve? Has it created new problems?
4. Why are there some difficulties in attempting to use the concept of a species?
5. If the binomial system of nomenclature had never been developed, what kinds of problems would biologists have?

Recommended Readings

Goulding, M., "Flooded Forests of the Amazon," *Scientific American*, Vol. 268, No. 3, March 1993. Unique adaptations allow organisms to thrive in the aquatic ecosystems of the tropical rain forest.

Holloway, M., "Trends in Environmental Science: Sustaining the Amazon," *Scientific American*, Vol. 269, No. 1, July 1993. Reconciling economic development with the preservation of tropical rain forests.

Margulis, L., and K.V. Schwartz, *Five Kingdoms. An Illustrated Guide to the Phyla of Life on Earth*, 2nd ed., San Francisco, W.H. Freeman, 1988. Presents the great diversity of living things, beautifully illustrated.

May, Robert M., "How Many Species Inhabit the Earth?," *Scientific American*, Vol. 267, No. 4, October 1992, 42–48. An argument for the importance of identifying and classifying the organisms that inhabit our planet; this information impacts on environmental issues.

Sibley, C.G., and J.F. Ahlquist, "Reconstructing Bird Phylogeny by Comparing DNAs," *Scientific American*, February 1986. An interesting account of a modern taxonomic method.

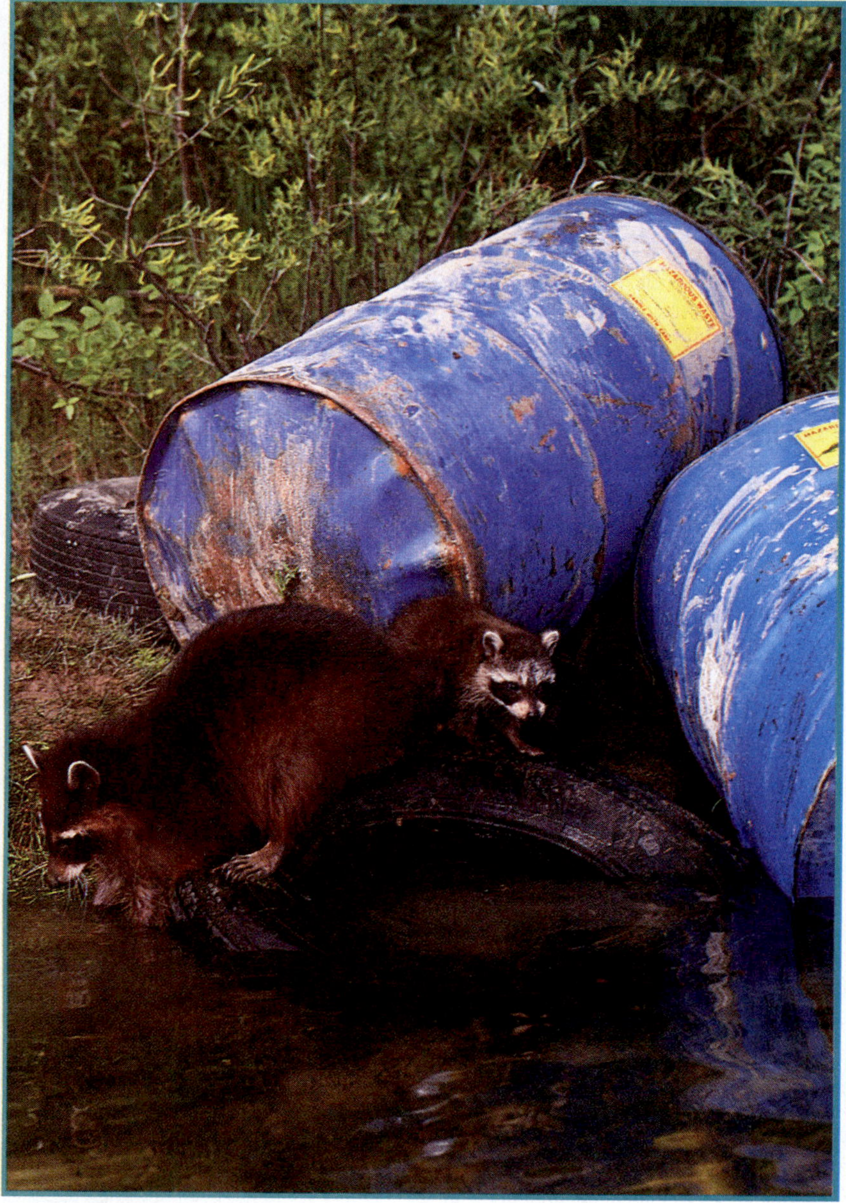

Toxic waste in deteriorating drums. *The drums, located at a site near Washington, D.C., have started to leak into a pond frequented by wildlife (note the raccoons). (Thomas Kitchin/Tom Stack & Associates)*

A Degrading Experience for Toxic Waste

Any discarded chemical that threatens human health or the environment is known as hazardous, or toxic, waste. Hazardous waste includes thousands of chemicals that are dangerously reactive, corrosive, explosive, or poisonous. In 1980, the Comprehensive Environmental Response, Compensation, and Liability Act, commonly known as the Superfund Act, established a program to tackle the huge challenge of cleaning up abandoned and illegal toxic waste sites across the country. At many of these sites, hazardous chemicals have migrated deep into the soil, where they pollute groundwater.

Until recently, companies that were trying to clean up soil contaminated with hazardous wastes had few options. They could dig up the soil and deliver it to a hazardous waste landfill, or they could dig up the soil and incinerate it to break down the poisons with intense heat. Because both processes are prohibitively expensive, innovative approaches for dealing with hazardous wastes are being developed. In **bioremediation,** the contaminated site is exposed to an army of microorganisms that produce special enzymes that break down the poison. This process leaves behind harmless metabolic byproducts such as carbon dioxide and chlorides.

To date, more than 1000 different species of bacteria and fungi have been used to clean up various forms of pollution. For example, bioremediation was used in the early 1990s to clean up an Iowa site that was contaminated by a poisonous wood preservative called pentachlorophenol (or simply, penta). A bacterium called *Flavobacterium* was introduced to clean up the site because it is able to break down the penta molecule.

During bioremediation, conditions at the hazardous waste site are modified so that the bacteria will thrive in large enough numbers to be effective. The site in Iowa was small, so engineers mixed the contaminated soil with sand (to make it more porous) and then injected the soil with *Flavobacterium.* They pumped air through the soil (to increase its oxygen level) and added a few soil nutrients like phosphorus; *Flavobacterium* requires both oxygen and phosphorus in order to flourish. Engineers also built a drainage system at the bottom of the pit to pipe any penta-laden water that leached through the soil back to the surface for another encounter with the bacteria.

Bioremediation takes a little longer to work than traditional hazardous waste disposal methods. One year after *Flavobacterium* was introduced at the Iowa site, two-thirds of the penta was gone. It took the bacteria between two and three years to degrade all of the penta.

How are toxin-degrading microbes like *Flavobacterium* discovered? The penta-degrading bacterium was found by a University of Idaho microbiologist who was working on the environmental effects of penta on stream ecosystems. He found some streams where penta disappeared shortly after it was added. He collected bacteria from these streams and brought them into his lab, where he fed them a steady diet of penta. Eventually, he identified *Flavobacterium* as the champion penta-eater.

This chapter examines the diversity and biological characteristics of microorganisms (viruses, bacteria, and protists). We group the viruses, bacteria, and protists into a single chapter for convenience, but as you will see, these organisms are *not* a natural assemblage of closely related organisms.

Micro-organisms: Viruses, Bacteria, and Protists

21

Learning Objectives

After you have studied this chapter you should be able to:

1. Describe the structure of a virus and compare a virus with a free-living cell.
2. Characterize bacteriophages, animal viruses, and plant viruses.
3. Describe the distinguishing characteristics of members of the kingdom Prokaryotae.
4. Distinguish among each of the following groups of bacteria: archaeobacteria, gram-negative bacteria, and gram-positive bacteria.
5. Discuss the important ecological roles of bacteria.
6. Characterize the common features of members of the kingdom Protista, and discuss in general terms the diversity inherent in this kingdom.
7. Briefly discuss the representative fungus-like protists: plasmodial slime molds, cellular slime molds, and water molds.
8. Briefly characterize the representative groups of algae: dinoflagellates, diatoms, euglenoids, green algae, red algae, and brown algae.
9. Briefly describe the representative protozoan phyla: amebas, foraminiferans, flagellates, ciliates, and sporozoa.
10. Describe how multicellularity may have arisen within the protists, using *Chlamydomonas* as an example.

Key Concepts

- ☐ Viruses are not living organisms in the conventional sense of "living." They are noncellular, infectious agents that can only reproduce inside another living cell.
- ☐ Bacteria possess a prokaryotic cell structure with few or no internal membranes. Although their cellular structure is simple, bacteria as a group exhibit wide metabolic diversity.
- ☐ Protists, which possess a eukaryotic cell structure, are either single-celled organisms or simple multicellular organisms. They are extremely diverse in structure and metabolism. Ancient protists were the ancestors of complex multicellular eukaryotes—the fungi, plants, and animals.

VIRUSES ARE TINY, INFECTIOUS AGENTS THAT ARE NOT ASSIGNED TO ANY OF THE FIVE KINGDOMS

A **virus** is a tiny, infectious particle consisting of a nucleic acid core (its genetic material) surrounded by a protein coat called a **capsid.** Some viruses are also surrounded by an outer membranous envelope that contains proteins, lipids, carbohydrates, and traces of metals.

Viruses lie on the threshold between life and nonlife, and as such, are not true living organisms. Viruses are not cellular, cannot move about on their own, and cannot carry on metabolic activities independently. All living organisms contain both DNA and RNA, but a virus contains *either* DNA *or* RNA, not both. Viruses can reproduce, but only within the complex environment of the living cells they infect. In a sense, viruses come "alive" only when they infect a cell.

Where did viruses come from? The hypothesis currently thought most likely is that viruses are bits of nucleic acid that "escaped" from cellular organisms. According to this view, some viruses may trace their origin to animal cells, others to plant cells, and still others to bacterial cells. Their multiple origins might explain why viruses are species-specific; perhaps viruses infect only those species that are the same or closely related to the organisms from which they originated. This hypothesis is supported by the genetic similarity between a virus and its host cell—a closer similarity, in fact, than exists between one virus and another.

The shape of a virus is determined by the organization of protein subunits that make up the capsid. Viral capsids are generally either helical (rod-shaped) or polyhedral in shape, or a complex combination of both (Fig. 21–1). Helical viruses, such as the tobacco mosaic virus, appear as long rods or threads. The capsid of this virus is a hollow cylinder that encloses its nucleic acid. Polyhedral viruses, such as the influenza virus and adenovirus, appear somewhat spherical in shape. The capsid of a virus that infects bacteria includes both polyhedral and helical shapes.

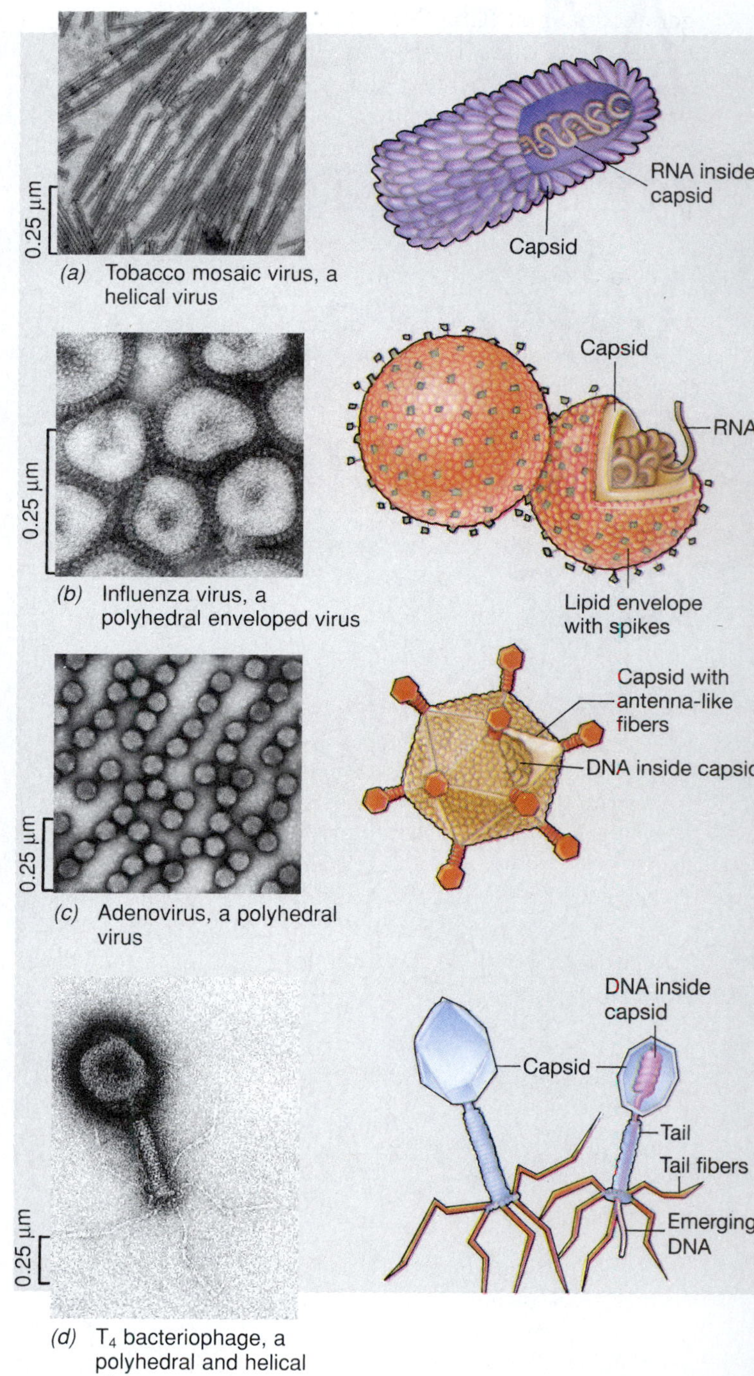

FIGURE 21–1 Viruses are generally either helical or polyhedral in shape, or a complex combination of both. (***a***) Tobacco mosaic virus. (***b***) Influenza virus. (***c***) Adenovirus. (***d***) Bacteriophage. (*a, b,* Visuals Unlimited/K.G. Murti; *c,* Visuals Unlimited/Hans G. Elderblom; *d,* Lee D. Simon/Science Source/Photo Researchers, Inc.)

Bacteriophages Are Viruses that Infect Bacteria

Among the most complex viruses are some of those that infect bacteria. These viruses are known as **bacteriophages** ("bacteria eaters") (Fig. 21–1*d*). The most common bacteriophage structure consists of a single nucleic acid molecule (usually DNA) coiled within a polyhedral head composed of protein. Many bacteriophages have a tail attached to the head. Fibers extending from the tail exist on some bacteriophages and are used to attach the virus to a bacterium.

Because bacteriophages can be easily cultured within living bacteria in the laboratory, much of our knowledge of viruses has come from studying bacteriophages. They were also used to determine that DNA, not protein, is the genetic material (Chapter 13). Currently, bacteriophages are used in genetic engineering research.

Bacteriophages are either lytic or temperate. **Lytic** bacteriophages destroy the host cell. When a lytic virus infects a susceptible host cell, it uses the host cell's metabolic machinery to replicate viral nucleic acid and produce viral proteins.

Several steps in the process of viral infection are common to almost all bacteriophages (Fig. 21–2):

1. **Attachment.** The bacteriophage attaches to the host cell wall.
2. **Penetration.** After the bacteriophage has attached to the cell surface, its tail contracts and pushes a hole through the cell wall. Nucleic acid is then injected through the plasma membrane and into the cytoplasm of the host cell. The capsid of a bacteriophage remains on the outside. (In contrast, most viruses that infect animal cells enter the host cell intact.)
3. **Replication.** Once inside, the bacteriophage DNA takes over the metabolic machinery of the cell. Using the host cell's ribosomes, its energy, and many of its enzymes, the bacteriophage replicates its own molecules. Bacteriophage genes contain all of the information necessary to produce new bacteriophages.
4. **Assembly.** The newly synthesized viral components are assembled into new bacteriophages.
5. **Release.** In a lytic infection, the bacteriophage produces an enzyme that degrades the cell wall of the host cell. The host cell then lyses (ruptures), releasing about 100 bacteriophages. These new viruses infect other bacteria, and the process begins anew.

Unlike lytic viruses that lyse their host cells, **temperate** bacteriophages do not always destroy their hosts. They can integrate their DNA into the DNA of the host cell. When the bacterial DNA replicates, the viral DNA also replicates. Certain external conditions can cause temperate viruses to revert to a lytic cycle and then destroy their host.

An interesting example of a temperate bacteriophage involves the bacterium that causes diphtheria, *Corynebacterium diphtheriae.* Two strains of the bacterium exist, one that produces a toxin (and causes diphtheria) and one that does not. The only difference between these two strains is that the toxin-forming bacteria contain a temperate bacteriophage, designated phage-β. The phage DNA encodes for the powerful toxin that causes the symptoms of diphtheria.

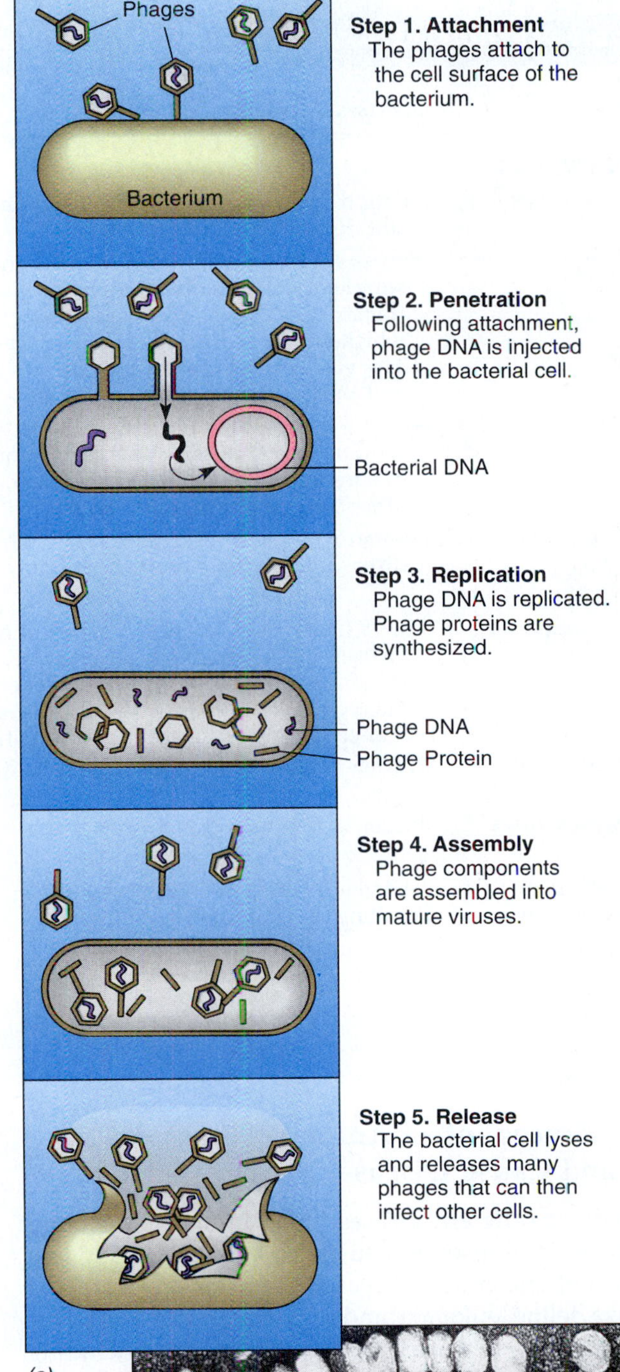
Phages
Bacterium
Step 1. Attachment
The phages attach to the cell surface of the bacterium.
Step 2. Penetration
Following attachment, phage DNA is injected into the bacterial cell.
Bacterial DNA
Step 3. Replication
Phage DNA is replicated. Phage proteins are synthesized.
Phage DNA
Phage Protein
Step 4. Assembly
Phage components are assembled into mature viruses.
Step 5. Release
The bacterial cell lyses and releases many phages that can then infect other cells.
(a)

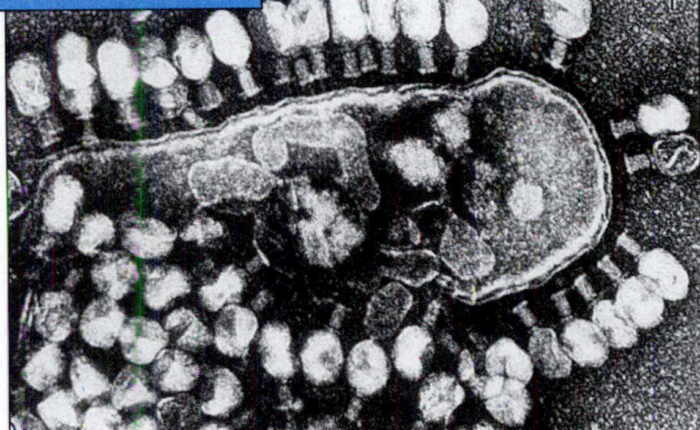
0.25 μm
(b)

FIGURE 21–2 Lytic infection. (***a***) The sequence of events in a lytic infection are attachment, penetration, replication, assembly, and release. (***b***) Phages infecting *Escherichia coli,* a bacterium. (Lee D. Simon/Science Source/Photo Researchers, Inc.)

TABLE 21–1
Some Animal Viruses

Group	*Diseases caused*	*Characteristics*
DNA Viruses		
Poxviruses	Smallpox, cowpox, and economically important diseases of domestic fowl	Large, complex, oval-shaped viruses that replicate in the cytoplasm of the host cell
Herpesviruses	Herpes simplex type 1 (cold sores); herpes simplex type 2 (genital herpes, a sexually transmitted disease); varicella-zoster (chicken-pox and shingles). The Epstein-Barr virus causes infectious mononucleosis and Burkitt's lymphoma.	Medium to large, enveloped viruses; frequently cause latent infections; some cause tumors
Adenoviruses	About 40 types known to infect human respiratory and intestinal tracts; common cause of sore throat, tonsillitis, and conjunctivitis; other varieties infect other animals.	Medium-sized viruses
Papovaviruses	Human warts and some degenerative brain diseases; cervical and other genital cancers	Small viruses
RNA Viruses		
Picornaviruses	About 130 types infect humans including polioviruses; enteroviruses infect intestine; rhinoviruses infect respiratory tract and are main cause of human colds; coxsackievirus and echovirus cause aseptic meningitis.	Diverse group of small viruses
Togaviruses	Rubella, yellow fever, equine encephalitis	Large, diverse group of medium-sized, enveloped viruses; many transmitted by arthropods
Myxoviruses	Influenza in humans and other animals	Medium-sized viruses that often exhibit projecting spikes
Paramyxoviruses	Rubeola, mumps, distemper in dogs	Resemble myxoviruses but somewhat larger
Reoviruses	Vomiting and diarrhea in children	Contain double-stranded RNA
Retroviruses	AIDS, some types of cancer	RNA viruses that contain reverse transcriptase for transcribing the RNA genome into DNA

Some Viruses Infect Animal Cells and Cause Disease

Hundreds of different viruses infect humans and other animals. Animal diseases caused by viruses include hog cholera, foot-and-mouth disease, canine distemper, swine influenza, feline leukemia, and Rous sarcoma in fowl. Humans are prone to a variety of viral diseases, including chicken pox, herpes simplex (one type of which is genital herpes), mumps, rubella (German measles; Fig. 21–3), rubeola (measles), rabies, warts, infectious mononucleosis, influenza, hepatitis, AIDS, and certain types of cancer (Table 21–1).

One way that an animal virus penetrates an animal cell is by **endocytosis,** in which the animal cell engulfs and transports the virus into the cytoplasm (Chapter 6). Like bacteriophages, viruses that infect animal cells replicate and produce new virus particles. Viral nucleic acid is replicated, and viral proteins are synthesized, while host synthesis of DNA, RNA, and proteins is inhibited.

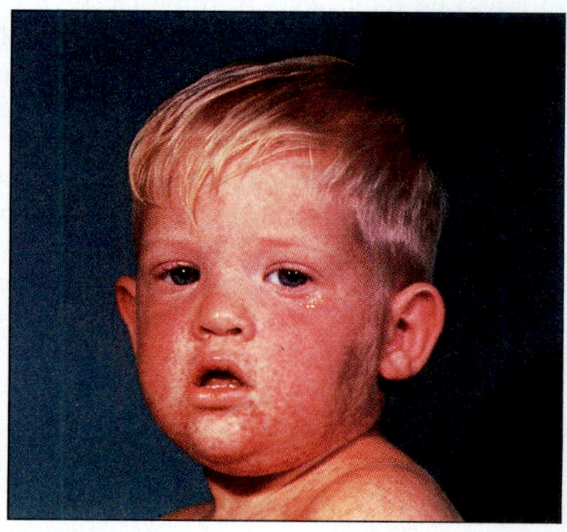

FIGURE 21–3 Rubella (German measles) is caused by an RNA virus spread by close contact. Immunity appears to be lifelong following infection. Vaccination has greatly decreased the incidence of this disease. (Centers for Disease Control and Prevention, U.S. Department of Health and Human Services)

(a)

(b)

FIGURE 21–4 Viral diseases in plants. (***a***) Virus-streaked tulips. The virus that causes this relatively harmless disease affects pigment formation in the petals. (***b***) Tobacco leaves infected with the tobacco mosaic virus are characteristically mottled with light green areas. (Kenneth M. Corbett)

Animal viruses can contain either DNA or RNA. One group of RNA viruses, the **retroviruses,** uses a viral enzyme called **reverse transcriptase** to transcribe the RNA genome into a DNA intermediate. This DNA is then used to synthesize copies of the viral RNA. The human immunodeficiency virus (HIV) that causes AIDS is a retrovirus.

Following replication of the viral nucleic acid, structural proteins needed by the virus are synthesized. New virus particles are assembled and exit the host cell by cell lysis or exocytosis (Chapter 6).

Some Viruses Infect Plant Cells

Viral diseases can be spread among plants by insects such as aphids and leafhoppers as they feed on plant tissues. Plant viruses are also inherited by way of infected seeds or by asexual propagation.

Once a plant is infected, the virus spreads throughout the plant's body by passing through the cytoplasmic connections (plasmodesmata) that penetrate the cell walls between adjacent cells. Most plant viruses contain the nucleic acid RNA. After infecting a host cell, the viral RNA attaches to the host's ribosomes and is translated as though it were mRNA.

Symptoms of viral infection include spots, streaks, or mottled patterns on leaves, flowers, or fruits (Fig. 21–4). Reduced size is another typical symptom, and infected crop plants almost always produce lower yields. Cures are not known for most viral diseases of plants, and so it is common to burn plants that have been infected. Some agricultural scientists are focusing their efforts on prevention of viral disease by developing virus-resistant strains of important crop plants.

Bacteria are Placed in the Kingdom Prokaryotae

Bacteria, most of which are single-celled organisms, are assigned to their own kingdom, the kingdom Prokaryotae.[1] Prokaryotic cells contain ribosomes but lack membrane-bounded organelles typical of eukaryotic cells. Thus, there are no nuclei, no mitochondria, no chloroplasts, no endoplasmic reticulum, no Golgi complex, and no lysosomes (Fig. 21–5).

The genetic material of a bacterium is contained in a single circular DNA molecule that lies in the cytoplasm and is not surrounded by a nuclear envelope. In addition to the bacterial chromosome, a small amount of genetic information may be present as smaller circular DNAs, called **plasmids,** that replicate independently of the chromosome (Chapter 16). Bacterial plasmids often bear genes involved in resistance to antibiotics that can be transmitted from one cell to another.

Most prokaryotic cells have a cell wall surrounding the plasma membrane, but its structure and composition differ from those of eukaryotic cell walls. The cell wall

[1]The kingdom Prokaryotae is the same as the kingdom Monera in previous editions of this text. The name change is to conform with the latest edition of *Bergey's Manual of Systematic Bacteriology,* which has been the definitive work on bacteria since 1923.

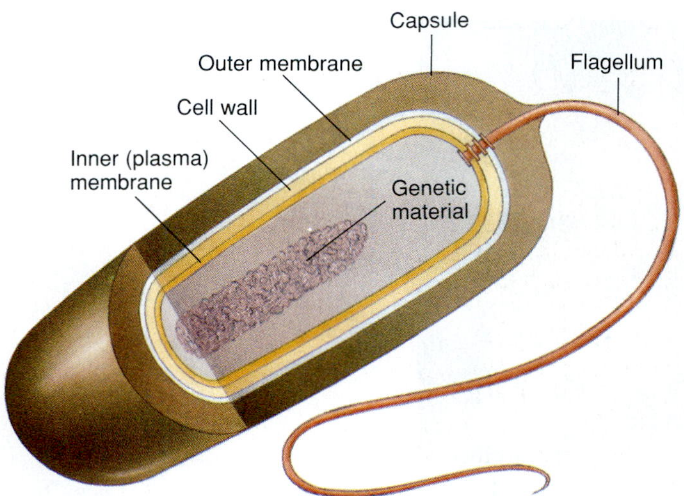

FIGURE 21–5 Structure of a typical bacterium. This bacillus is a gram-negative bacterium (discussed in text). Note the absence of a nuclear envelope surrounding the genetic material.

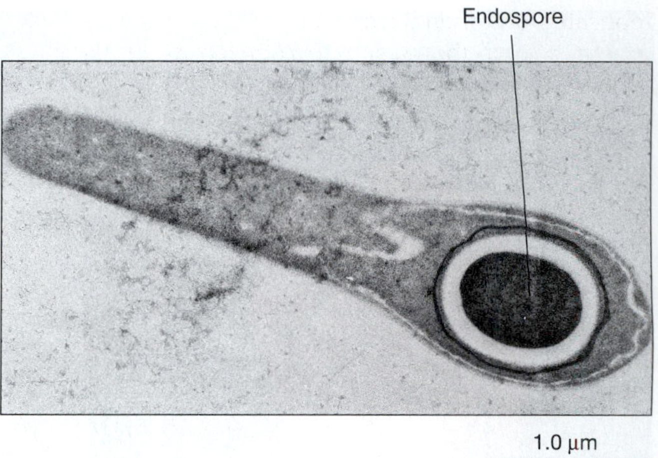

FIGURE 21–6 Endospore within a cell of *Clostridium*. Each bacterial cell contains only one endospore, which is a resistant, dehydrated remnant of the original cell. (T.J. Beveridge, University of Guelph/Biological Photo Service)

provides a rigid framework that supports the cell, maintains its shape, and keeps it from bursting because of osmotic pressure (Chapter 6). The bacterial cell wall is composed of **peptidoglycan,** a complex organic molecule that consists of two chains of unusual types of sugars linked with short peptides.

Some species of bacteria produce a **capsule** or slime layer that surrounds the cell wall. The capsule may provide the cell with added protection against a hostile environment and, in some cases, aids in the disease-causing ability of certain pathogenic bacteria. For example, the pathogenicity of *Streptococcus pneumoniae,* the bacterium that causes bacterial pneumonia, is related to its capsule. A strain of *S. pneumoniae* that lacks a capsule does not cause the disease.

Some prokaryotes move by means of **flagella,** but their structure is quite different from that of eukaryotic flagella (Chapter 5). Bacterial flagella are distinctive in that they consist of a single fibril. At the base of a bacterial flagellum is a complex structure that produces a rotary motion, pushing the cell much as a ship is pushed along by its propeller.

Some bacteria have hundreds of hairlike appendages that are organelles of attachment and help the bacteria adhere to certain surfaces, such as the cells they will infect. Similar filamentous appendages, called **pili** (singular, *pilus*), are involved in the transmission of DNA between bacteria.

When the environment of a bacterium becomes unfavorable, such as when it becomes very dry, many species become dormant. The cell loses water, shrinks slightly, and remains quiescent until water is again available. Other species form dormant, extremely durable structures called **endospores** that are capable of surviving in extremely dry, hot, or frozen environments, or when food is scarce (Fig. 21–6). Some endospores are so resistant that they can survive an hour or more of boiling, or centuries of freezing. When environmental conditions are again suitable for growth, the endospore becomes an active, growing bacterial cell again.

Bacterial Diversity Is Evident in Their Varied Metabolism

Bacteria are either heterotrophic or autotrophic. Most bacteria are **heterotrophs,** which means they must obtain organic compounds from other organisms to survive. The majority of these heterotrophic bacteria are free-living **saprobes,** decomposing organisms that get their nourishment from dead organic matter. Other heterotrophic bacteria obtain their nourishment from living organisms, sometimes harming them by causing diseases, and other times actually providing a beneficial service for their host.

Bacteria that are **autotrophic** are able to manufacture their own organic molecules. Autotrophic bacteria are either photosynthetic or chemosynthetic. Photosynthetic bacteria obtain energy from light, whereas chemosynthetic bacteria obtain energy by oxidizing inorganic chemicals.

Bacteria Reproduce by Binary Fission

Bacteria generally reproduce asexually by **binary fission,** in which one cell divides into two similar cells. First, the DNA that comprises the circular bacterial chromosome replicates, and then a transverse wall is formed by an ingrowth of both the plasma membrane and the cell wall.

Binary fission occurs with remarkable speed. For example, under ideal conditions some bacteria divide every 20 minutes. At this rate, if nothing interfered, one bacterium would give rise to more than *one billion* bacteria within ten hours (see Fig. 45–5 on page 911). However, bacteria cannot reproduce at this rate for very long, because they are soon checked by lack of food or by the accumulation of waste products.

Bacteria Are Ecologically Important

Bacteria are essential for life. Their contributions include **nitrogen fixation,** the ability of some bacteria to convert atmospheric nitrogen to a form that can be utilized by plants (Chapter 47). Nitrogen fixation enables plants and animals (because they eat plants) to manufacture essential nitrogen-containing compounds such as proteins and nucleic acids.

Other bacteria play an essential role in the environment as decomposers, breaking down organic molecules from dead organisms into their simpler components. Bacteria, along with fungi, are nature's recyclers. Without bacteria (and fungi), all available carbon, nitrogen, phosphorus, and sulfur would eventually be tied up in the wastes and dead bodies of plants and animals. Life would soon cease because of the lack of raw materials for the synthesis of new cellular components.

Concept Connections

Nitrogen Fixation — *Mutualism*

Rhizobial bacteria (bacteria in the genus *Rhizobium*) form symbiotic associations with the roots of **legumes,** a large family of herbs, shrubs, and trees. Legumes include such important crops as peas, green beans, kidney beans, lentils, chick peas, and soybeans. Clover and alfalfa, which are grown for livestock feed and to fertilize the soil, are also important legumes.

Rhizobial bacteria are motile rods that live in the soil. After they infect the roots of a legume, nodules are produced on the roots. The nodules consist of plant tissue wherein the bacteria reside and fix nitrogen.

The relationship between rhizobial bacteria and the roots of legumes is **mutualistic,** meaning that both partners benefit (Chapter 46). The bacteria living in nodules supply the plant with all of the nitrogen that it requires, and the plant provides the bacteria with sugar needed for cellular respiration.

Because legumes, like other plants, produce sugar by photosynthesis, a correlation exists between photosynthesis and nitrogen fixation. When a legume is photosynthesizing at a higher rate, its bacterial partners are able to fix larger amounts of nitrogen (because they are receiving more sugar from the plant).

Plants without nodules must obtain the nitrogen they need from the soil (Chapter 28). Many soils are somewhat deficient in nitrogen. Therefore, legumes that have formed a mutualistic association with rhizobial bacteria are able to thrive in nitrogen-deficient soils. This association gives them a decided advantage over other plants.

There Are Two Fundamentally Different Groups of Prokaryotes: The Archaeobacteria and the Eubacteria

Under a microscope, bacteria appear similar in size and form. However, evidence from molecular biology has helped biologists conclude that ancient prokaryotes split into two lineages early in the history of life. The modern descendants of these two ancient lines of bacteria are the archaeobacteria, which include several groups of prokaryotes able to live in extreme environments, and the eubacteria, which comprise all other prokaryotes.

In recognition of the significant differences between archaeobacteria and eubacteria, some biologists have proposed that living organisms be classified into three **domains** (a domain is a level of classification above the kingdom): the Domain Archaea (archaeobacteria), the Domain Bacteria (eubacteria), and the Domain Eucarya (eukaryotes—protists, fungi, plants, and animals). However, in this book we use the more traditional, five-kingdom classification scheme that places both archaeobacteria and their distant relatives, the eubacteria, within the kingdom Prokaryotae.

Biochemically, archaeobacteria are very different from other prokaryotes. One of their most distinguishing features is the absence of peptidoglycan in their cell walls. There are also other important differences in their ribosomal RNA, lipids, and specific enzymes that set the archaeobacteria apart from other bacteria. In addition, the archaeobacteria have certain features in common with eukaryotes (however, these similarities are beyond the scope of this text).

Many of the extreme environments to which the modern archaeobacteria are adapted resemble conditions that were once common on primitive Earth, but are somewhat rare today. These include hot springs whose temperatures may exceed 100 °C and deep sea vents that spew sulfide

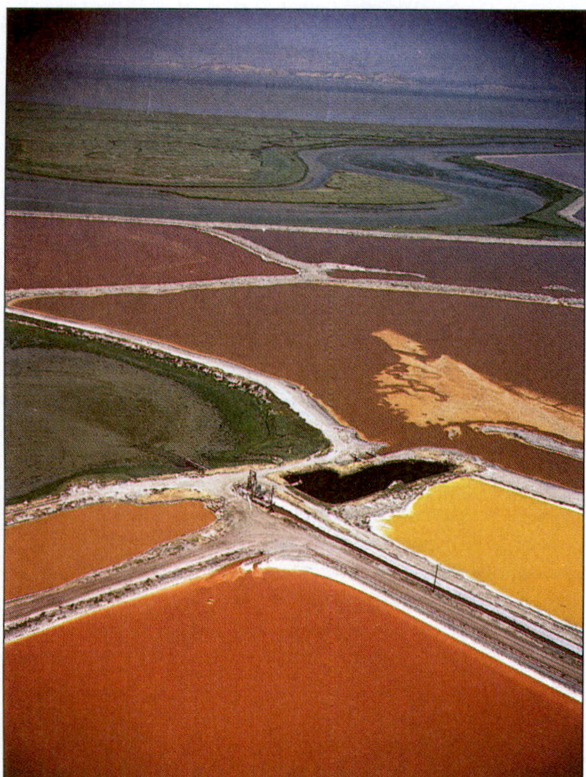

FIGURE 21–7 Salt-loving archaeobacteria. Seawater evaporating ponds near San Francisco Bay are colored pink, orange, and yellow from the large number of extreme halophiles (salt-loving archaeobacteria) growing in them. (Algae also add color to the ponds.) The salt that remains after the water has evaporated has commercial value. (Helen E. Carr/Biological Photo Service.)

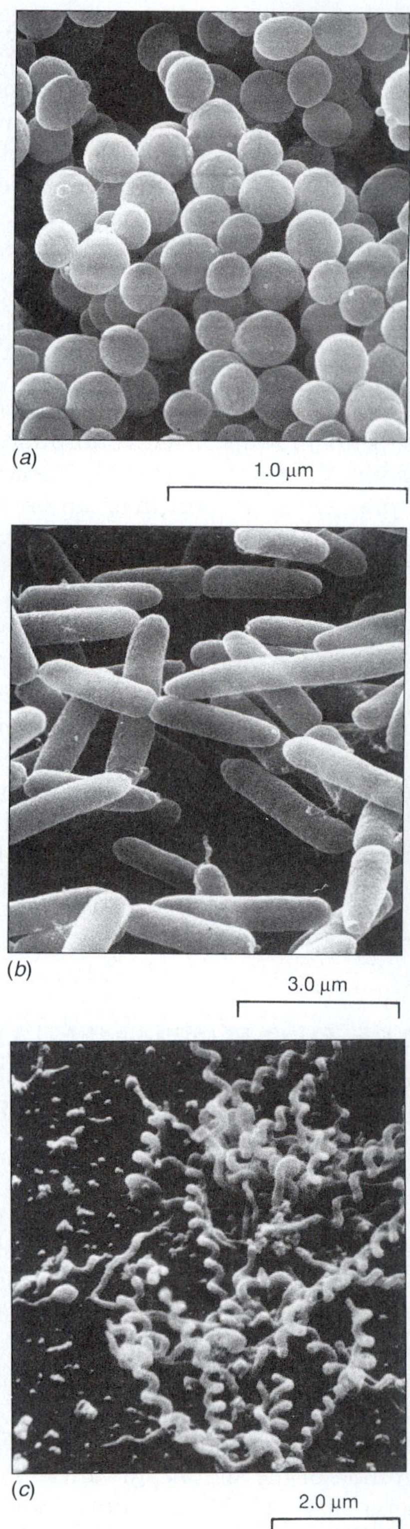

FIGURE 21–8 Three characteristic bacterial shapes. (***a***) Cocci. (***b***) Bacilli. (***c***) Spirilla. (*a, b, c,* Visuals Unlimited/David M. Phillips)

gases. Other archaeobacteria live only in extremely salty environments such as salt ponds (Fig. 21–7).

There are thousands of kinds of eubacteria. Although their metabolisms are quite diverse, most eubacteria have simple morphologies. They have three main shapes: spherical, rod-shaped, and spiral (Fig. 21–8). Spherical bacteria, known as **cocci** (singular, *coccus*), occur singly in some species, in groups of two in others, in long chains, or in irregular clumps that look like bunches of grapes. Rod-shaped bacteria, called **bacilli** (singular, *bacillus*), may occur as single rods or as long chains of rods. Spiral, or helical, bacteria are known as **spirilla** (singular, *spirillum*).

In 1884, the Danish physician Christian Gram developed the gram-staining procedure. Bacteria that absorb and retain crystal violet stain during laboratory staining procedures are referred to as **gram-positive.** Those bacteria that do not retain the stain are **gram-negative.** The cell walls of gram-positive bacteria are very thick and consist primarily of peptidoglycan. The cell walls of gram-negative bacteria consist of two layers, a thin peptidoglycan wall and an outer membrane. The differences in composition of the cell walls of gram-positive and gram-negative bacteria are of great practical importance. For example, the antibiotic penicillin interferes with peptidoglycan synthesis, ultimately resulting in a fragile cell wall that cannot effectively protect the cell. Predictably, penicillin works most effectively against gram-positive bacteria.

There are thousands of kinds of eubacteria. We consider four groups of eubacteria to demonstrate their diversity: enterobacteria, cyanobacteria, lactic acid bacteria, and clostridia. **Enterobacteria** are a group of gram-negative bacilli that include decomposers that live on decaying plant matter, pathogens, and a variety of bacteria that inhabit humans. *Escherichia coli,* a member of the enterobacteria, inhabits the intestinal tracts of humans and other animals as part of the normal microorganism population. However, under certain conditions, such as the consumption of contaminated food or water, *E. coli* can cause moderate to severe diarrhea. For example, in 1993 almost 500 people developed bloody diarrhea in the Pacific Northwest, and three people died. The culprit was a particularly disagreeable strain of *E. coli* that infected hamburger meat sold at a popular fast food restaurant. (The restaurant, which is now testing for the bacterium and has also increased the amount of cooking time for their hamburgers, has had no further problems.)

Cyanobacteria (formerly known as blue-green algae) are gram-negative bacteria that are found in ponds, lakes, swimming pools, and moist soil, as well as on dead logs and the bark of trees. Some also occur in the oceans, and a few species inhabit hot springs. Most cyanobacteria are photosynthetic autotrophs; these contain chlorophyll *a*, which is also found in plants and algae. Many cyanobacteria also fix nitrogen (Fig. 21–9).

Lactic acid bacteria are gram-positive bacteria that produce lactic acid as the main end product of their fermentation of sugars. Lactic acid bacteria may be found in decomposing plant material and in milk and other dairy products. The characteristic taste of yogurt, acidophilus milk, pickles, sauerkraut, and green olives is due to the action of lactic acid bacteria. They are also among the normal inhabitants of the human mouth and vagina.

Clostridia are a notorious group of anaerobic gram-positive bacteria. One species causes tetanus, another causes gas gangrene, and *Clostridium botulinum* can cause botulism, a highly fatal type of food poisoning. *C. botulinum* produces endospores that are quite resistant to heat. Botulism results from consuming foods, such as canned vegetables and smoked meats and fish, that have been inadequately sterilized. Such inadequate sterilization allows the endospores to grow. The bacteria produce poisons that are among the most potent toxins known. Approximately one microgram of toxin is enough to kill a human.

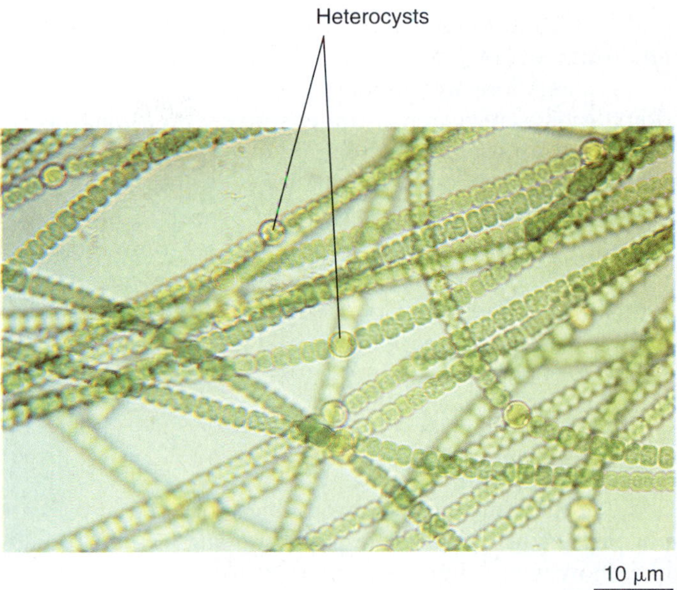

FIGURE 21–9 ***Anabaena*, a filamentous cyanobacterium that fixes nitrogen.** Nitrogen fixation is localized in *Anabaena;* it occurs in the rounded cells called heterocysts. (Dennis Drenner)

Protists are simple eukaryotes

The kingdom Protista consists of a vast assemblage of eukaryotic organisms whose diversity makes them difficult to characterize. The major feature they possess—eukaryotic cell structure—is shared with organisms from three other kingdoms: the fungi, plants, and animals. However, eukaryotic cell structure makes the separation between the protists and the bacteria quite distinct. Eukaryotic cells have true nuclei and other membrane-bounded organelles such as mitochondria and chloroplasts.

Although most protists are microscopic single-celled organisms, some have a colonial organization, some are **coenocytic** (multinucleate but not multicellular), and some are multicellular. Multicellular protists, however, have relatively simple body forms without specialized tissues.

Methods of obtaining nutrients in the kingdom Protista are highly varied. Autotrophic protists (the algae) have chlorophyll and photosynthesize like plants. Some of the heterotrophic protists (the water molds) obtain their food by absorption, like the fungi, while others (protozoa and slime molds) resemble animals and ingest food derived from the bodies of other organisms.

Most protists are aquatic and live in oceans or freshwater ponds, lakes, and streams. They make up part of the **plankton,** the floating, often microscopic organisms that inhabit surface waters. Other aquatic protists attach to rocks and other surfaces in the water. Terrestrial (land-dwelling) protists are restricted to damp places like soil and leaf litter. Even the parasitic protists live in the wet environments of plant and animal body fluids.

The relationships among the organisms in the kingdom Protista are currently under study, with ultrastructure (cell structure studied with the aid of electron microscopy), biochemistry, and molecular biology adding critical information about the various groups of protists. To recognize natural relationships within the protists, some biologists think as many as 50 phyla are needed. Consideration of all protists is beyond the scope of this text, but we discuss several representative groups (Table 21–2).

Slime Molds and Water Molds Are Fungus-Like Protists

Some protists superficially resemble fungi in that they are not photosynthetic and their body form is often thread-like. However, fungus-like protists are not fungi for several reasons. Many fungus-like protists produce flagellated cells, which fungi lack. Many such protists also have centrioles and produce cellulose as a major component of their cell walls, unlike fungi, which lack centrioles and have cell walls of chitin.

Plasmodial Slime Molds and Cellular Slime Molds Are the Two Groups of Slime Molds

The feeding stage of plasmodial slime molds (Phylum Myxomycota) is a **plasmodium,** a multinucleate mass of cytoplasm (Fig. 21–10). The plasmodium, which is slimy in appearance, streams over decaying logs and leaf litter,

(a)

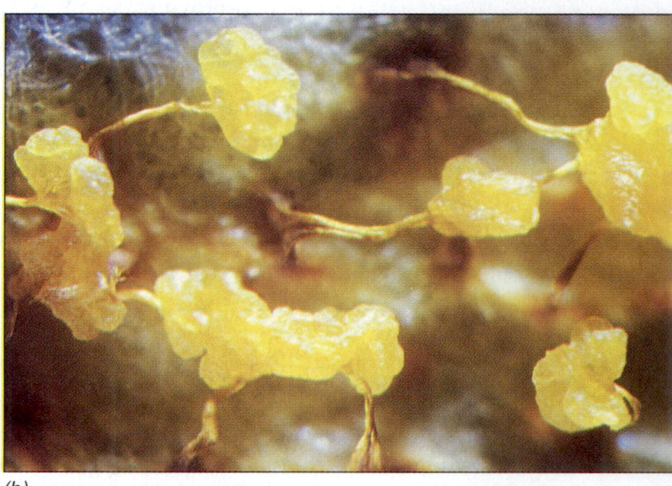

(b)

FIGURE 21–10 ***Physarum*, a plasmodial slime mold.** (***a***) The plasmodium of *Physarum* is colored bright yellow. This naked mass of protoplasm is multinucleate and feeds on bacteria and other microorganisms. (***b***) The reproductive structures of *Physarum* are stalked sporangia. (*a*, P.W. Grace/Photo Researchers, Inc., *b*, Carolina Biological Supply Company)

(Text continues on page 388)

CONCEPT CONNECTIONS

Protists ⟷ *Other Eukaryotes* ⟷ *Evolution*

How are protists related to the other eukaryotic kingdoms? Fungi, plants, and animals (discussed in Chapters 22 to 25) are thought to have their ancestry in the protist kingdom. Although it is generally agreed that the protistan ancestor of plants was a green alga, the origins of animals and fungi are unclear. The fragmented fossil evidence hampers biologists attempting to work out evolutionary relationships. As a result, biologists use other kinds of evidence, such as comparative anatomy and biochemical data, to develop hypotheses about what actually took place during animal and fungal evolution (Chapter 17). However, because clear transitional forms are lacking in the fossil record, definitive answers about the origins of animals and fungi from protists may never be determined.

Many biologists think that one or several flagellates were probably the ancestors of animals. However, biologists are not yet certain whether there was a single protistan ancestor for animals, or several different protistan ancestors for different animal groups. For example, some biologists think that the two different kinds of body symmetry in animals (radial and bilateral) are evidence that these two groups had different protistan ancestors.

The protistan ancestor of the fungi is also uncertain. Both fungi and red algae share a number of similarities, such as a lack of flagellated cells, so it is possible that an ancient red alga was the ancestor of fungi. The molecular analysis of rRNA of fungi and red algae is currently under way and may help resolve this question.

TABLE 21–2
A Comparison of Representative Phyla in the Protist Kingdom

Common name	*Phylum*	*Morphology*	*Locomotion*	*Photosynthetic pigments*	*Special features*
Amebas	Rhizopoda	Single cell, no definite shape	Pseudopodia	—	Some have shells (tests)
Foraminiferans	Foraminifera	Single cell	Cytoplasmic projections		Pore-studded shells (tests)
Flagellates	Zoomastigina	Single cell	One to many flagella; some ameboid	—	Symbiotic forms often highly specialized
Ciliates	Ciliophora	Single cell	Cilia	—	Many possess trichocysts
Sporozoa	Apicomplexa	Single cell	None	—	All parasitic; develop resistant spores
Dinoflagellates	Dinoflagellata	Single cell, some colonial	Two flagella	Chlorophylls *a* and *c*; carotenoids, including fucoxanthin	Many covered with cellulose plates
Diatoms	Bacillariophyta	Single cell, some colonial	Most nonmotile; some move by gliding over secreted slime	Chlorophylls *a* and *c*; carotenoids, including fucoxanthin	Silica in shell
Euglenoids	Euglenophyta	Single cell	Two flagella (one of them very short)	Chlorophylls *a* and *b*; carotenoids	Flexible outer covering
Green algae	Chlorophyta	Single cell, colonial, siphonous, multicellular	Most flagellated at some stage in life; some nonmotile	Chlorophylls *a* and *b*; carotenoids	Share certain features with plants
Red algae	Rhodophyta	Most multicellular, some single cell	None	Chlorophyll *a*; carotenoids; phycocyanin; phycoerythrin	Some reef builders
Brown algae	Phaeophyta	Multicellular	Two flagella on reproductive cells	Chlorophylls *a* and *c*; carotenoids, including fucoxanthin	Differentiation of body into blade, stipe, and holdfast
Plasmodial slime molds	Myxomycota	Multinucleate plasmodium	Streaming cytoplasm, flagellated or ameboid reproductive cells	—	Reproduce by spores formed in sporangia
Cellular slime molds	Acrasiomycota	Vegetative form—single cell; reproductive form—multicellular (slug)	Pseudopods (for single cells); cytoplasmic streaming (for multicellular)	—	Aggregation of cells for reproduction
Water molds	Oomycota	Coenocytic mycelium	Two flagella on reproductive cells	—	Superficially resemble fungi

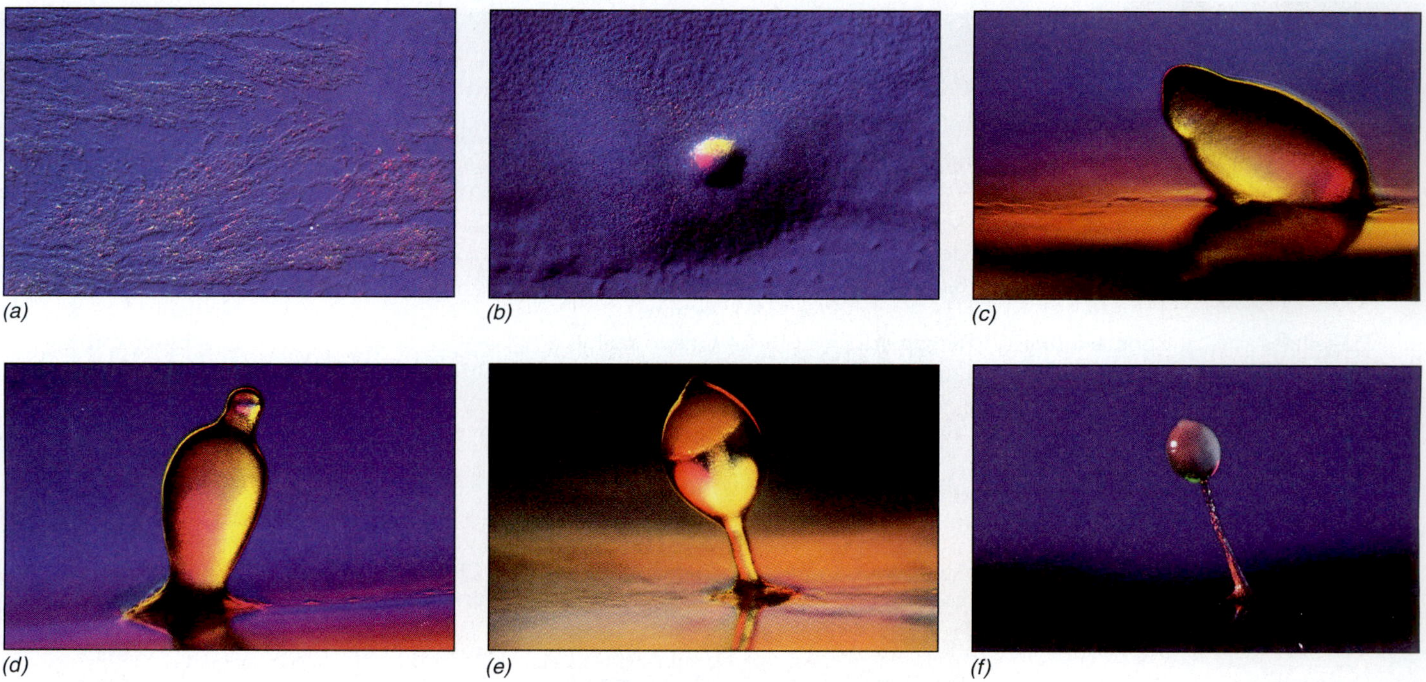

FIGURE 21–11 The life cycle of the cellular slime mold, *Dictyostelium discoideum.* Each ameba-like unicellular organism eats, grows, and reproduces by cell division. (***a***) After their food supply is depleted, the cells stream together. (***b***) An aggregation of cells. (***c***) The aggregation organizes into a slug-shaped, multicellular organism that migrates for a period of time before forming a stalked fruiting body (***d, e,*** and ***f***) that produces spores. Each spore opens to liberate an ameba-like unicellular organism, and the process repeats itself. (Carolina Biological Supply Company)

ingesting bacteria, yeasts, spores, and decaying organic matter that it encounters. When the food supply dwindles or there is insufficient moisture, the plasmodium crawls to an exposed surface and initiates reproduction. Stalked structures usually form from the drying plasmodium. Within these structures, called **sporangia,** meiosis occurs, producing haploid reproductive cells called **spores.**

The resemblance of cellular slime molds (Phylum Acrasiomycota) to the plasmodial slime molds is very superficial. During its feeding stage, each cellular slime mold is an individual ameboid cell that behaves as a separate, solitary organism. Each cell creeps over rotting logs and soil or swims in fresh water, ingesting bacteria and other particles of food. When moisture or food becomes scarce, the individual cells send out a chemical signal that causes them to aggregate by the hundreds or thousands for reproduction (Fig. 21–11). During this stage, the cells creep about for short distances as a single multicellular unit, called a **pseudoplasmodium** or "slug." Eventually, the slug develops into a **fruiting body,** a reproductive structure that bears spores. After being released, each spore opens, a single ameboid cell emerges, and the cycle repeats itself.

Water Molds Produce Flagellated Reproductive Cells

The water molds (Phylum Oomycota) were once classified as fungi because of their superficial resemblance. Both water molds and fungi have a body, termed a **mycelium,** that grows over organic material, digesting it and then absorbing the predigested nutrients (Fig. 21–12). The threadlike **hyphae** that make up the mycelium in water molds are coenocytic; there are no cross walls and the body is like one giant multinucleate cell. However, most biologists classify the water molds as protists rather than fungi. This is because water molds produce flagellated reproductive cells at some point in their life cycles, whereas fungi never produce motile cells.

Some water molds have played infamous roles in human history. For example, the Irish potato famine of the 19th century was caused by the water mold that causes late blight of potatoes. Due to several rainy, cool summers in Ireland in the 1840s, the water mold multiplied unchecked. As a result, potato tubers rotted in the fields. Since potatoes were the staple of the Irish peasant's diet, many people starved. Estimates of the number of deaths

FIGURE 21–12 Water molds are important decomposers in aquatic ecosystems. Some of them are parasitic, as the water mold that has attacked this unhappy goldfish. (Fred M. Rhoades)

resulting from the outbreak of this plant disease range from one quarter million to more than one million. As a consequence, a mass migration followed out of Ireland to such countries as the United States.[2]

[2]Late blight is still a problem today. In 1992, new strains of the water mold that causes late blight of potato were reported in New York and Washington state. These strains are resistant to the fungicide usually used to control the disease.

Algae Are Plantlike Protists

The algae represent a diverse group of organisms that are mostly photosynthetic. They range in size from microscopic, single-celled forms to large, multicellular seaweeds. Like plants, most algae are photosynthetic. In addition to chlorophyll *a* and yellow and orange **carotenoids,** which are photosynthetic pigments that are found in all algae, different groups of algae possess a variety of other pigments that are also important in photosynthesis. Classification into phyla is largely by pigment composition (see Table 21–2) and type of storage product.

Most Dinoflagellates Are Marine Planktonic Forms

One of the most unusual groups of protists are the dinoflagellates (Phylum Dinoflagellata). Most dinoflagellates are unicellular, and their cells are often covered with shells of interlocking cellulose plates (Fig. 21–13*a*). The typical dinoflagellate has two flagella that propel the organism through the water like a spinning top. Indeed, the dinoflagellates' name is derived from the Greek *dinos,* meaning "whirling."

Many dinoflagellates reside in the bodies of marine invertebrates such as jellyfish, corals, and mollusks. These dinoflagellates lack cellulose plates and flagella and are called **zooxanthellae.** Zooxanthellae photosynthesize and

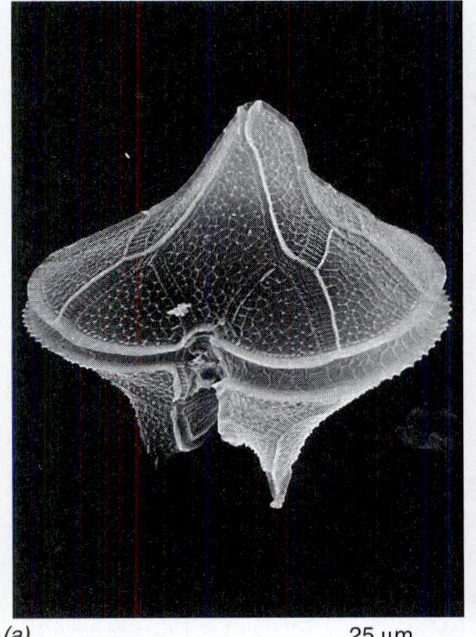

FIGURE 21–13 Dinoflagellates. (***a***) Scanning electron micrographs of *Protoperidinium,* a dinoflagellate. Note the plates that encase the single-celled body. The two flagella are not visible. (***b***) Red tide along the shore of a small bay in central California. The cloudiness in the water is produced by countless billions of dinoflagellates. (*a,* Courtesy of T.K. Maugel, University of Maryland; *b,* Visuals Unlimited/Sanford Berry)

FIGURE 21–14 Diatoms. These algae have strikingly beautiful patterns in their symmetrical shells. (The Stock Market/Phillip Harrington)

provide food for their invertebrate partners. The contribution of zooxanthellae to the productivity of coral reefs is substantial (Chapter 24).

Ecologically, the dinoflagellates are one of the most important groups of producers in marine ecosystems. A few dinoflagellates are known to have occasional population explosions, or **blooms.** These blooms frequently color the water orange, red, or brown and are known as **red tides** (Fig. 21–13*b*). Some of the dinoflagellate species that form red tides produce a toxin that attacks the nervous systems of fishes, leading to massive fish kills. A human condition called paralytic shellfish poisoning is caused by eating oysters, mussels, or clams that fed on certain dinoflagellates. Paralytic shellfish poisoning causes respiratory failure in humans who eat the contaminated shellfish (but does not appear to hurt the shellfish).

Diatoms Have Shells Composed of Two Parts

Most diatoms (Phylum Bacillariophyta) are unicellular, although there are a few colonial forms. The cell wall of each diatom consists of two shells that overlap where they fit together, much like a Petri dish. Silica is deposited in the shell, and this glasslike material is laid down in striking, intricate patterns (Fig. 21–14).

Diatoms are common in both fresh water and ocean water, but they are especially abundant in cooler marine waters. They are major producers in aquatic ecosystems because of their extremely large numbers. When diatoms die, their shells sink and accumulate in layers of what eventually becomes sedimentary rock. After millions of years, some of these deposits were exposed on land by geological upheaval. Called **diatomaceous earth,** these deposits are mined and used as a filtering, insulating, and soundproofing material.

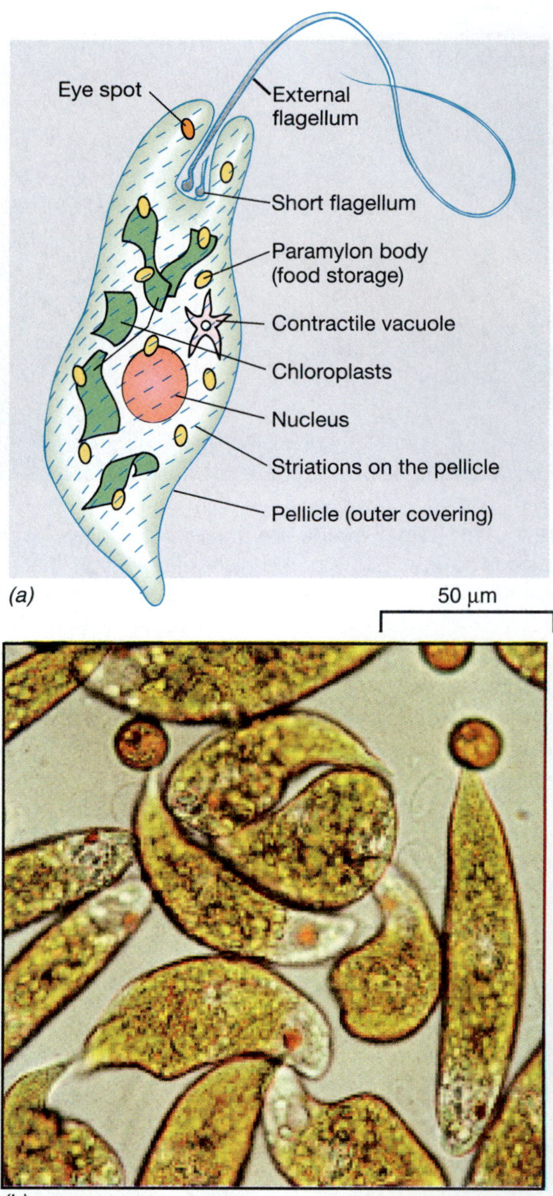

FIGURE 21–15 Euglenoids are motile single-celled algae. (***a***) *Euglena* has a complex cellular structure. The eyespot is a light-sensitive organelle that helps it to react to light. Its outer covering, called a pellicle, is flexible and enables *Euglena* to change shape easily. (***b***) Living euglenoids. (Visuals Unlimited/T.E. Adams)

Euglenoids Are Freshwater Unicellular Flagellates

Most euglenoids (Phylum Euglenophyta) are unicellular flagellates, and about one-third of them are photosynthetic (Fig. 21–15). They generally possess two flagella, one long and whiplike and one so short that it does not protrude outside of the cell. Euglenoids change shape

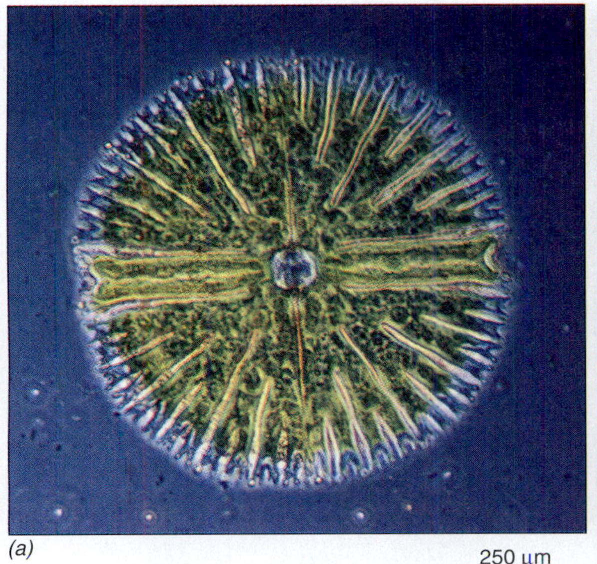

(a)

250 μm

(b)

(c)

FIGURE 21–16 Diversity in the green algae. (***a***) Desmids, such as *Micrasterias*, are microscopic green algae with intricately beautiful cells. (***b***) Underwater view of dead man's fingers (*Codium fragile*), common off the northeastern coast of the United States. Siphonous green algae like *Codium* are coenocytic, which means that its body is composed of one giant cell with multiple nuclei. (***c***) Some multicellular green algae are sheet-like. The thin, leaf-like form has given *Ulva* its common name of "sea lettuce." This sea lettuce lives in the Gulf of Maine. (*a*, Ronald W. Hoham, Colgate University; *b*, Visuals Unlimited/ William C. Jorgensen; *c*, Andrew J. Martinez, Photo Researchers, Inc.)

continually as they move through the water because their outer covering is flexible rather than rigid.

Euglenoids inhabit freshwater ponds and puddles, particularly those with large amounts of organic material. For that reason they are used as an indicator species of organic pollution. If a body of water has unusually large numbers of euglenoids, it is probably polluted. Marine waters and mud flats are also inhabited by some euglenoids.

Green Algae Share Similarities with Plants

If one had to pick a single word to describe the green algae (Phylum Chlorophyta), it would be "variety." These protists exhibit many diverse forms and methods of reproduction. Their body forms range from single cells to colonial forms to coenocytic, **siphonous** (tubular) algae to multicellular filaments and sheets (Fig. 21–16). The multicellular forms do not have cells differentiated into tissues, however. Most green algae are flagellated during at least part of their life history, although a few are totally nonmotile.

There are both aquatic and terrestrial forms of green algae. Aquatic green algae primarily inhabit fresh water, although there are a number of marine species. Green algae that inhabit the land are restricted to damp soil, cracks in tree bark, and other moist places. Regardless of where they live, green algae are ecologically important as the base of the food web, particularly in freshwater habitats where they are quite common.

Green algae share a number of characteristics in common with plants. Their pigmentation, storage products, and cell walls are chemically identical to plants. Because

FIGURE 21–17 A red alga. Most red algae are multicellular, many with complex filamentous bodies, such as *Bonnemaisonia*. (D. P. Wilson/ Eric & David Hosking/ Photo Researchers, Inc.)

of these and other similarities, it is generally accepted that plants descended from green algalike ancestors.

Red Algae Do Not Produce Motile Cells

The vast majority of red algae (Phylum Rhodophyta) are multicellular organisms, although there are a few unicellular species. The multicellular body form of red algae is commonly composed of complex, interwoven filaments that are delicate and feathery (Fig. 21–17). Most multicellular red algae attach to rocks or other materials by a rootlike **holdfast.**

The cell walls of red algae often contain polysaccharides that are of commercial value. For example, **agar** is extracted from certain red algae and used to make a culture medium for growing microorganisms. A second polysaccharide extracted from red algae is **carrageenan,** which is used to stabilize puddings, laxatives, ice creams, and toothpastes. People consume red algae as vegetables, particularly in Japan, Korea, and China.

The red algae are primarily found in warm tropical oceans, although many species occur in cooler ocean waters, in fresh water, and in soil. Some red algae incorporate calcium carbonate into their cell walls from the ocean waters. These coralline red algae are very important in building coral reefs, along with the coral animals themselves.

Brown Algae Are Multicellular Seaweeds

The brown algae (Phylum Phaeophyta) include the giants of the Protist kingdom. All brown algae are multicellular and range in size from several centimeters (an inch or so) to approximately 60 meters (almost 200 ft) in length. The largest brown algae, called kelps, are tough and leathery

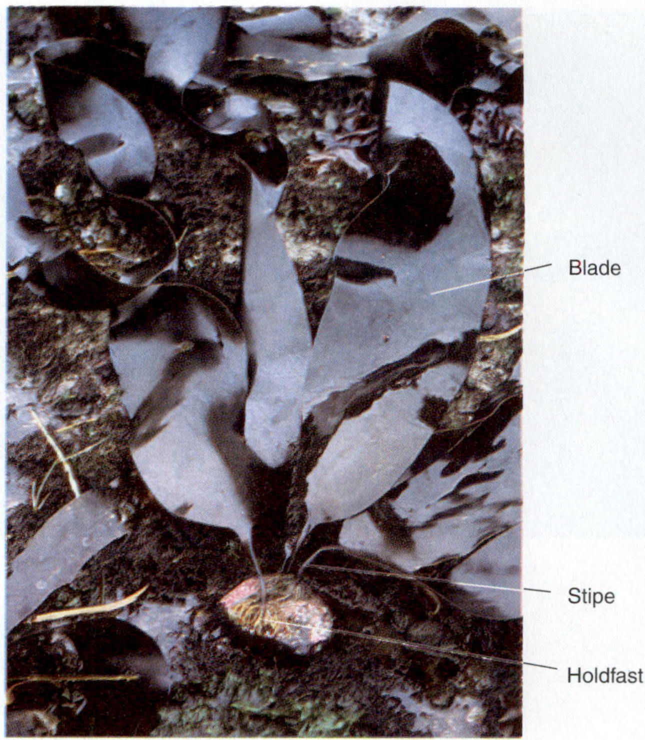

FIGURE 21–18 *Laminaria*, a typical brown alga. Note its blade, stipe, and holdfast, visible in this photograph because *Laminaria* has been removed from the water. Normally, this alga is completely submerged. (J. Robert Waaland, University of Washington/Biological Photo Service)

in appearance; many possess leaflike **blades,** stemlike **stipes,** and rootlike anchoring holdfasts (Fig. 21–18). They often have gas-filled floats that provide buoyancy.

Brown algae are commercially important for several reasons. They have a polysaccharide, called **algin,** in their cell walls, possibly to help cement the cell walls together. It is used as a thickening agent in ice cream, marshmallows, and cosmetics. Brown algae are an important human food, particularly in East Asian countries, and they are rich sources of minerals such as iodine.

Brown algae are common in cooler marine waters, especially along rocky coastlines, where they can be found mainly in the intertidal zone and relatively shallow offshore waters. Kelps form extensive underwater "forests" called kelp beds. They are essential in that ecosystem for two reasons: they are the primary food producer and they also provide habitats for many marine invertebrates, fishes, and mammals.

Protozoa Are Animal-Like Protists

The name **protozoa** (from the Latin, meaning "first animals"; singular *protozoon*) is used today to designate an

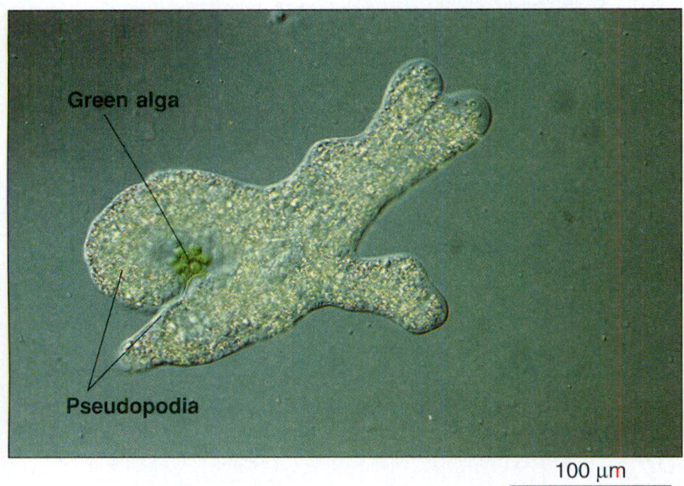

FIGURE 21–19 ***Chaos carolinense*, a giant ameba, ingesting a colonial green alga.** Note the pseudopodia extending to surround the prey. (Michael Abbey, Photo Researchers, Inc.)

informal grouping of protists that ingest their food (like animals). We consider five groups of protozoa: amebas, foraminiferans, flagellates, ciliates, and sporozoa.

Single-Celled Amebas Move by Forming Pseudopodia

Amebas (Phylum Rhizopoda) are single-celled organisms found in soil, fresh water, and oceans. Many members of this group have no definite body shape and change form as they move. An ameba moves by pushing out temporary cytoplasmic projections called **pseudopodia** ("false feet") from the surface of the cell. More cytoplasm flows into the pseudopodia, enlarging them until all the cytoplasm has entered and the organism as a whole has moved. Pseudopodia are also used to engulf and capture food (Fig. 21–19). Parasitic amebas include *Entamoeba histolytica,* which causes serious amebic dysentery in humans.

Foraminiferans Extend Cytoplasmic Projections through Tests

Foraminiferans (Phylum Foraminifera) are almost all marine organisms that produce shells, or **tests.** The oceans contain enormous numbers of foraminiferans, which secrete chalky, many-chambered tests with pores through which cytoplasmic projections can be extended. The cytoplasmic projections form a sticky, interconnected net that entangles its prey (Fig. 21–20).

Dead foraminiferans sink to the bottom of the ocean, where their shells form a grey mud that is gradually transformed into chalk. With geological uplifting, these chalk formations can become part of the land, like the white cliffs of Dover, England.[3]

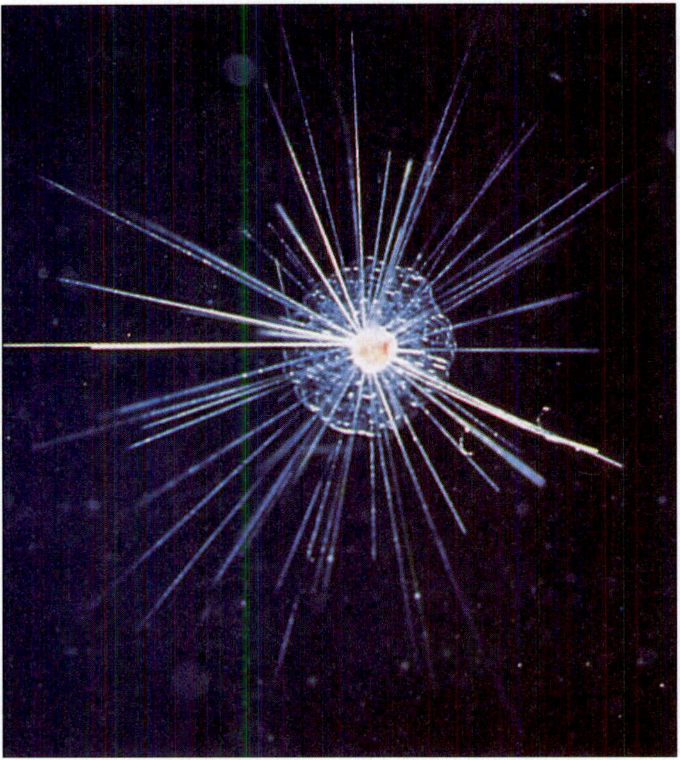

FIGURE 21–20 A foraminiferan. Foraminiferans secrete a shell, or test, that has pores through which cytoplasm extrudes. The cytoplasmic extensions appear as "rays" emanating from the test. (Peter Parks/Peter Arnold, Inc.)

Flagellates Move by Means of Flagella

Flagellates (Phylum Zoomastigina) are single-celled organisms that have spherical or elongate bodies and from one to many whiplike **flagella** that enable them to move. The flagella *pull* the organism through the water, rather than pushing it (unlike the tail of a tadpole, which pushes the tadpole). Some flagellates are also ameboid and engulf food by forming pseudopodia. Others have a definite "mouth," or **oral groove,** and specialized organelles for processing food.

Flagellates are heterotrophic and obtain their food either by ingesting living or dead organisms or by absorbing nutrients from dead or decomposing organic matter.

[3]The white cliffs of Dover are composed of the remains of a variety of calcareous organisms, including foraminiferans.

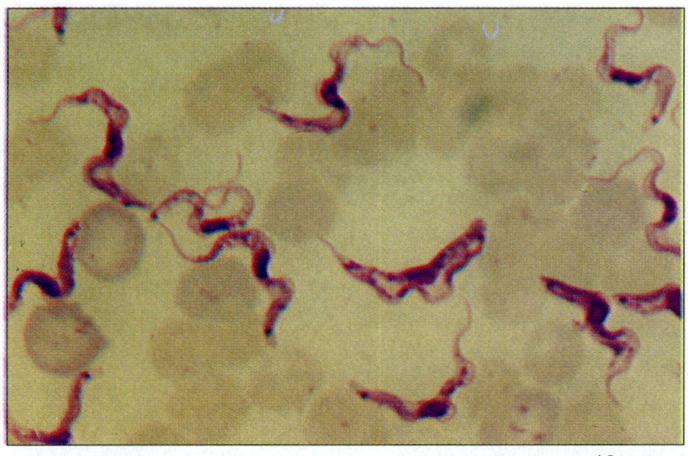

FIGURE 21–21 ***Trypanosoma gambiense*, which causes sleeping sickness, in a blood smear.** The flagellates are visible as dark, wavy bodies among the pale red blood cells. (Ed Reschke)

They may be free-living or symbionts. The flagellate *Trypanosoma* is a human parasite that causes African sleeping sickness (Fig. 21–21).

Ciliates Use Cilia for Locomotion

The ciliates (Phylum Ciliophora) are single-celled organisms with a definite but somewhat changeable shape caused by a flexible outer covering. In *Paramecium,* the surface of the cell is covered with several thousand fine cilia that extend through pores in the outer covering and permit movement (Fig. 21–22). The cilia beat in a coordinated fashion that is so precise that the organism can go forward and also back up and turn around. Near their surface, many ciliates possess numerous small **trichocysts,** which are organelles that discharge filaments thought to aid in trapping and holding prey. Most ciliates ingest bacteria or other tiny organisms.

Not all ciliates are motile. Some forms are stalked and others, while capable of some swimming, are more likely to remain attached to one spot. Their cilia set up water currents that draw food toward them.

Sporozoa Are Spore-Forming Parasites of Animals

The sporozoa (Phylum Apicomplexa) are a large group of parasitic protozoa, some of which cause serious diseases such as malaria in humans. Sporozoa lack structures for locomotion. At some stage in their life many develop a resistant **spore,** which is the infective agent transmitted to the next host. They often spend part of their life in one host species and part in a different host species.

Malaria, which is caused by a sporozoan, is one of the world's most common infectious diseases. According to the World Health Organization, approximately 100 million people currently have malaria, and one to two mil-

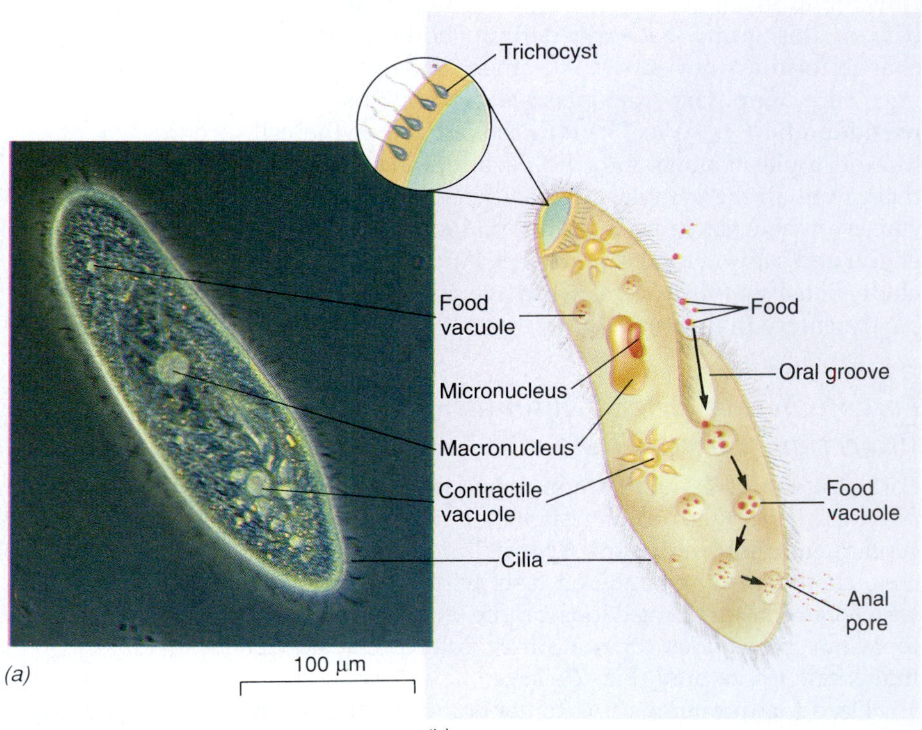

FIGURE 21–22 ***Paramecium* is a freshwater protozoon covered with cilia.** (***a***) Note the complex internal cellular structure of this single-celled protist. Like many ciliates, *Paramecium* has multiple nuclei. (***b***) Food particles, particularly bacteria, are swept into its ciliated oral groove and incorporated into food vacuoles. Lysosomes fuse with the food vacuoles and the food is digested and absorbed; undigested wastes are then eliminated through the anal pore. *Paramecium* has a macronucleus (involved in reproduction) and one or more smaller micronuclei (involved in cell metabolism and growth). *Paramecium* absorbs water by osmosis from its freshwater surroundings, but it doesn't swell up because contractile vacuoles fill up with excess water and then contract to void the contents. (M. Abbey/Photo Researchers, Inc.)

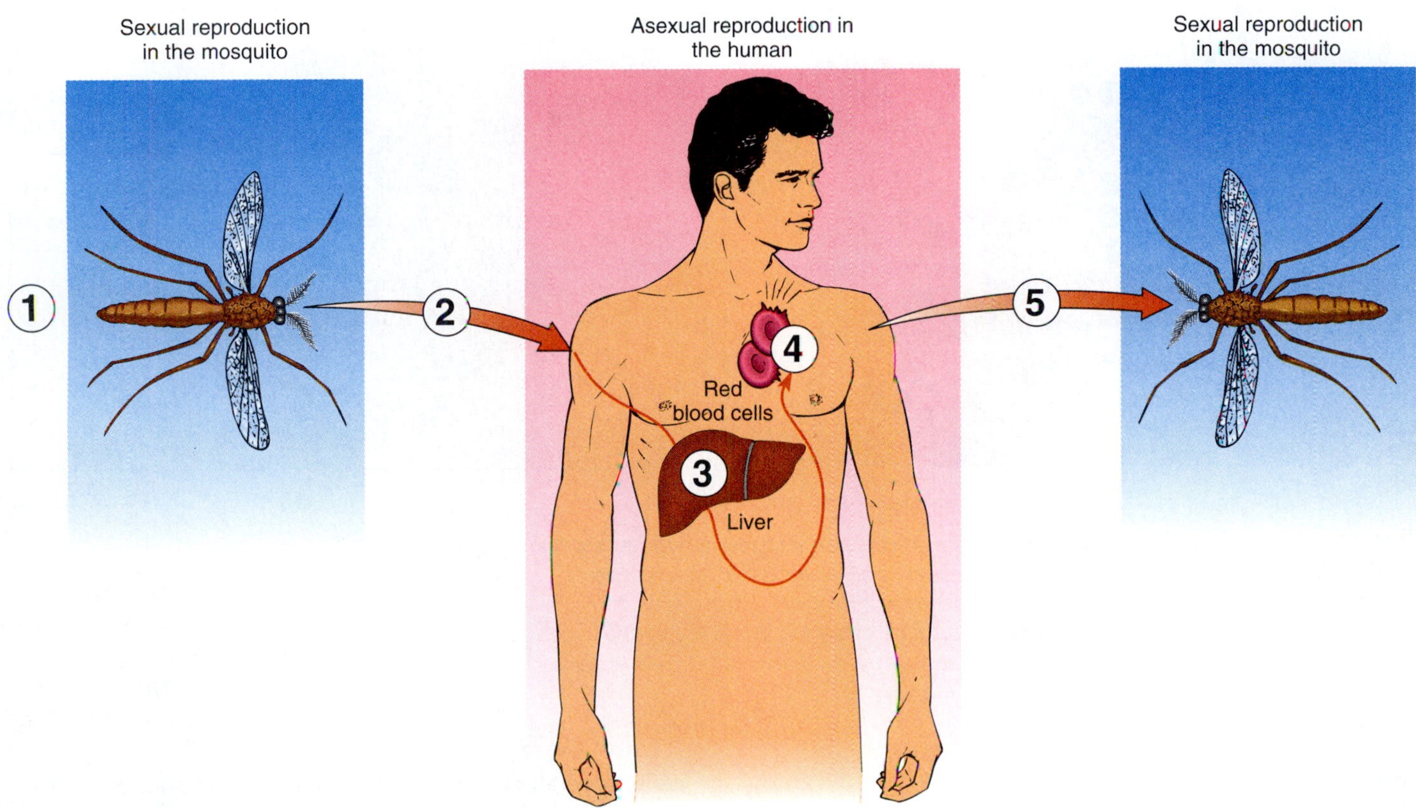

1. A female *Anopheles* mosquito bites an infected person and obtains the parasite, which continues its life cycle in the mosquito by reproducing sexually.
2. Later, the mosquito bites an uninfected human and transmits the parasite to the human's blood.
3. Once inside the human, the parasite continues its life cycle in the liver. From there the parasite re-enters the blood and infects red blood cells, where it reproduces asexually.
4. The red blood cells rupture, causing fever, chills and sweating. The parasites released when the red blood cells rupture infect other red blood cells, causing a repeat of the symptoms.
5. Some red blood cells are filled with the sexual form of the parasite, which can be transmitted to the next mosquito that bites the human.

FIGURE 21–23 Life history of *Plasmodium*, the parasite that causes malaria. *Plasmodium* has a complex life cycle that involves two different hosts.

lion people die each year from it. *Plasmodium,* the sporozoan that causes malaria, enters the human bloodstream when an infected female *Anopheles* mosquito bites a human (Fig. 21–23).

THE EARLIEST EUKARYOTES WERE PROTISTS

Protists are thought to have been the first eukaryotic cells. They may have originated as early as 2.1 billion years ago. There is compelling evidence that two eukaryotic organelles—mitochondria and chloroplasts—arose from endosymbiotic relationships between larger prokaryotes and the smaller prokaryotes that lived within them (Chapter 19).

Multicellularity Arose in the Protist Kingdom

The green algae, red algae, and brown algae are examples of protists that have multicellular species. However, multicellular green algae have more in common with single-celled green algae than with other multicellular protists. Similarly, multicellular forms of both red and brown algae have little in common with one another or with the multicellular green algae. Because these groups are so different, it is likely that multicellularity arose in the Protist kingdom several times. That is, the multicellular green algae, red algae, and brown algae probably had different single-celled protistan ancestors.

Studying the protists living today provides clues about how multicellularity may have evolved. For ex-

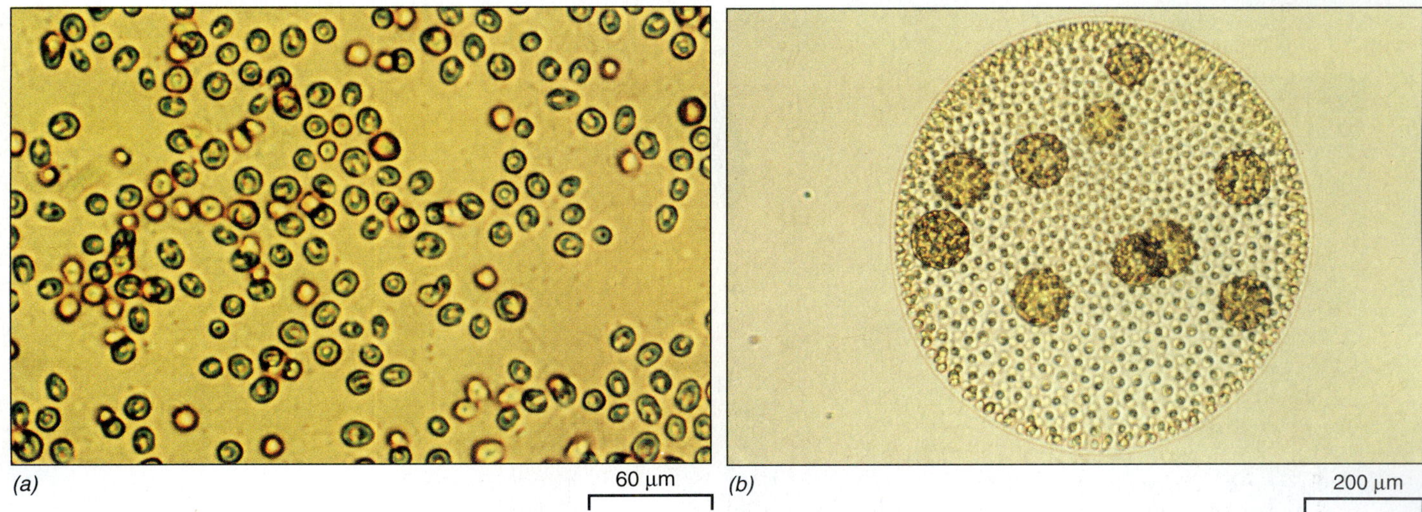

FIGURE 21–24 ***Chlamydomonas* and similar organisms suggest a hypothesis for how multicellularity evolved.** (***a***) *Chlamydomonas* is a biflagellate unicellular green alga. (***b***) *Volvox* is a colonial green alga composed of hundreds to thousands of *Chlamydomonas*-like cells. Daughter colonies can be observed inside the mother colony, which eventually breaks apart to release them. (Visuals Unlimited/James W. Richardson)

ample, *Chlamydomonas* (Fig. 21–24*a*) is a unicellular green alga that uses two flagella for motility. The green algae also include a number of loose aggregations composed of attached *Chlamydomonas*-like cells. These loose aggregations are called **colonies.** For example, there is a colonial green alga that consists of four *Chlamydomonas*-like cells and a different species composed of 16 to 32 *Chlamydomonas*-like cells. The largest colonies—from 1000 to 50,000 *Chlamydomonas*-like cells—are in the genus *Volvox* (Fig. 21–24*b*). As colonies increase in size and number of cells, specialization in cell structure and function occurs, with *Volvox* demonstrating an obvious division of labor among the cells.

The *Chlamydomonas* line is an evolutionary dead end that did not give rise to further organisms with greater complexity. However, the trend in increasing colony size and cell differentiation within the *Chlamydomonas* line indicates one possible way that multicellularity may have originated: single cells ⟶ colonies ⟶ multicellular organisms.

Chapter Summary

I. A virus is a tiny infectious particle consisting of a core of DNA or RNA surrounded by a capsid (protein coat).
  A. Viruses are not cellular and cannot metabolize on their own. They are not considered to be truly living organisms.
  B. Viruses reproduce inside living organisms.
    1. Bacteriophages are viruses that infect bacteria.
    2. Many different viruses infect humans and other animals. Examples of viral diseases in humans include chicken pox, herpes simplex, infectious mononucleosis, mumps, warts, influenza, hepatitis, and AIDS. Viruses also cause certain types of cancer in animals.
    3. Plant viruses cause serious agricultural losses. Viral diseases are spread among plants by insect vectors.

II. Kingdom Prokaryotae contains the prokaryotes (bacteria)
  A. Bacteria have a prokaryotic cell structure and lack membrane-bounded organelles such as nuclei and mitochondria.
    1. The genetic material of a prokaryote is a single circular DNA molecule.
    2. Most bacteria have cell walls composed of peptidoglycan.
    3. Bacterial flagella are structurally different from eukaryotic flagella.

   4. Bacteria are metabolically diverse; some are heterotrophic, whereas others are autotrophic.
   5. Bacteria reproduce asexually by binary fission.
  B. Prokaryotes are divided into two groups—archaeobacteria and eubacteria.
   1. The archaeobacteria have cell walls with an unusual chemical composition, often live in oxygen-deficient environments, and are often adapted to harsh conditions.
   2. The remaining bacteria are collectively known as the eubacteria.
    a. Gram-negative bacteria have thin cell walls of peptidoglycan, but they have an outer membrane surrounding the cell wall.
    b. Gram-positive bacteria have thick-layered cell walls of peptidoglycan.

III. Kingdom Protista is composed of "simple" eukaryotic organisms, most of which are unicellular and live in aquatic environments.
  A. Protists obtain their nutrients autotrophically or heterotrophically.
  B. Protists have various means of locomotion, including pseudopodia, flagella, and cilia. Some are nonmotile.

IV. Fungus-like protists were originally classified with the fungi, but have features that are clearly protistan.
  A. The body of the plasmodial slime molds is a multinucleate plasmodium. Reproduction is by spores.
  B. The cellular slime molds feed as individual ameboid cells. They aggregate into a pseudoplasmodium (slug) for reproduction.
  C. The water molds have a coenocytic mycelium. They produce flagellated reproductive cells at some point in their life cycles.

V. Algae are autotrophic protists.
  A. Dinoflagellates are mostly unicellular, biflagellate, photosynthetic organisms of great ecological importance.
  B. Diatoms are major producers in aquatic ecosystems. These are mostly unicellular, with shells containing silica.
  C. Euglenoids are single-celled, flagellated algae. Many are nonphotosynthetic.
  D. Green algae exhibit a wide diversity in size, complexity, and reproduction.
  E. Red algae, which are mostly multicellular seaweeds, are ecologically important in warm tropical oceans.
  F. All brown algae are multicellular seaweeds; they are ecologically important in cooler ocean waters.

VI. Protozoa are heterotrophic, animal-like protists.
  A. Amebas move and obtain food using cytoplasmic extensions called pseudopodia.
  B. Foraminiferans secrete many-chambered tests with pores through which cytoplasmic projections are ex tended to aid in moving and obtaining food.
  C. Flagellates are heterotrophic protozoa that move by means of flagella.
  D. The ciliates move by cilia.
  E. The sporozoa are parasites that produce spores and are nonmotile. A sporozoan causes malaria.

VII. The protists originated about 2.1 billion years ago and were the first eukaryotes.
  A. Multicellularity arose several times within the kingdom Protista.
  B. Plants, animals, and fungi originated from protistan ancestors.

Selected Key Terms

autotrophic, p.382
bacteriophage, p.378
binary fission, p.382
capsid, p.378
capsule, p.382
coenocytic, p.385
domain, p.383
endospore, p.382
flagellum, p.382
fruiting body, p.388
gram-negative, p.384
gram-positive, p.384
heterotrophic, p.382
mycelium, p.388
nitrogen fixation, p.383
peptidoglycan, p.382
pili, p.382
plasmodium, p.386
pseudoplasmodium, p.388
pseudopodia, p.393
retrovirus, p.381
saprobes, p.382

Post-Test

1. The core of a virus consists of ______________ or ______________, but never both.
2. The protein coat surrounding the nucleic acid core of a virus is called a/an ______________.
3. Bacteriophages are viruses that infect ______________.
4. Some bacteriophages are ______________ and kill the host cell.
5. All of the bacteria are assigned to kingdom ______________.
6. Peptidoglycan is found in the eubacterial ______________ ______________.
7. The majority of heterotrophic bacteria are free-living ______________ that get their nourishment from dead organic matter.

8. Spherical bacteria are referred to as _______________, rod-shaped bacteria as _______________, and helical bacteria as _______________.
9. Bacteria that absorb and retain crystal violet stain (the gram stain) are known as _______________-_______________ bacteria.
10. The feeding stage of the plasmodial slime molds is a multinucleate _______________.
11. The _______________ slime molds behave as single-celled organisms until reproduction, when they aggregate.
12. Agar is obtained from _______________ _______________.
13. The multicellular bodies of _______________ algae are differentiated into blades, stipes, holdfasts, and gas-filled floats.
14. Amebas move and obtain food by using _______________.
15. The _______________ are a group of parasitic protozoa that form spores at some stage in their life.
16. Which of the following is *not* true of the bacteria? (a) Many are ecologically important as nitrogen fixers and decomposers. (b) They are complex multicellular organisms. (c) They lack nuclei and other membrane-bounded organelles. (d) They reproduce by binary fission.
17. Which of the following would *never* be found as part of a bacterial cell? (a) cell wall (b) flagellum (c) nucleus (d) plasma membrane (e) plasmid.
18. In the five-kingdom system, protozoa are classified as: (a) prokaryotes (b) fungi (c) plants (d) protists (e) animals.
19. Red tides are caused by population explosions of: (a) euglenoids (b) foraminiferans (c) dinoflagellates (d) water molds (e) diatoms.
20. Which of the following is *not* characteristic of diatoms? (a) They each possess two flagella. (b) They are photosynthetic. (c) They have cell walls containing silica. (d) They are ecologically important.

Review Questions

1. What characteristics does a virus share with a living cell? What characteristics of life are lacking in a virus?
2. What are the characteristics of bacteria?
3. What are the differences between the archaeobacteria and the eubacteria?
4. Contrast the cell wall of a gram-positive bacterium with that of a gram-negative bacterium.
5. What are the characteristics of a typical protist? Why are protists so difficult to characterize?
6. How are the protists important to humans? How are they important ecologically?
7. Some biologists still classify the water molds as fungi. Why could water molds be considered fungi? Why do most biologists classify them as protists rather than fungi?
8. Some biologists still classify the algae as plants. Why could algae be considered plants? Why do most biologists classify them as protists rather than plants?
9. Why are protozoa not considered animals?

Thinking Critically

1. Why are bacteria more ideal than other organisms as candidates for cleaning up different kinds of toxic waste? Base your answer on what you have learned about evolution as well as the biology of bacteria.
2. Some biologists used to think that viruses, because of their simple structure, evolved before cellular organisms. Based on what you have learned about viruses, present an argument against this hypothesis.
3. Imagine that you discover a new microorganism. After careful study you determine that it should be classified in the kingdom Prokaryotae, with the cyanobacteria. What characteristics might lead you to each classification?
4. How would life on planet Earth be different if bacteria had never evolved?
5. How would life on planet Earth be different if protists had never evolved?
6. Why have algae been referred to as the "grass of the seas"?

Recommended Readings

Angert, E.R., K.D. Clements, and N.R. Pace, "The largest bacterium," *Nature,* Vol. 362, 18 March 1993. Discovery of a novel bacterium so large that it can be seen *without* a microscope.

Canby, T.Y., "Bacteria: Teaching Old Bugs New Tricks, *National Geographic,* August 1993. Considers the industrial potential of bacteria, from cleaning up toxic waste to degrading oil spills.

Carmichael, W.W., "The Toxins of Cyanobacteria," *Scientific American,* January 1994. Many cyanobacteria produce unusual toxins that are deadly, but potentially valuable.

Hively, W., "Life Beyond Boiling," *Discover,* May 1993.Research discoveries about bacteria that thrive at temperatures well above 212 °F. Most of these bacteria live in hot springs at the bottom of the ocean.

Huyghe, P., "Algae," *Discover,* April 1993. A recently discovered dinoflagellate species that causes red tides actually dines on the fish that it kills.

Margulis, L., and K.V. Schwartz, *Five Kingdoms,* 2nd ed., New York, W.H. Freeman & Co., 1988. Discusses the viruses, bacteria, and protists in general terms and introduces each bacterial and protistan phylum.

Rosenthal, E., "Outwitted by Malaria, Desperate Doctors Seek New Remedies," *The New York Times,* February 12, 1991. Excellent overview of the seriousness of malaria, the complexity of the malarial parasite, and hopes for new treatments.

Sharnoff, S.D., "Beauties from a Beast: Woodland Jekyll and Hydes," *Smithsonian,* July 1991. Beautiful photographs of the sporangia of slime molds.

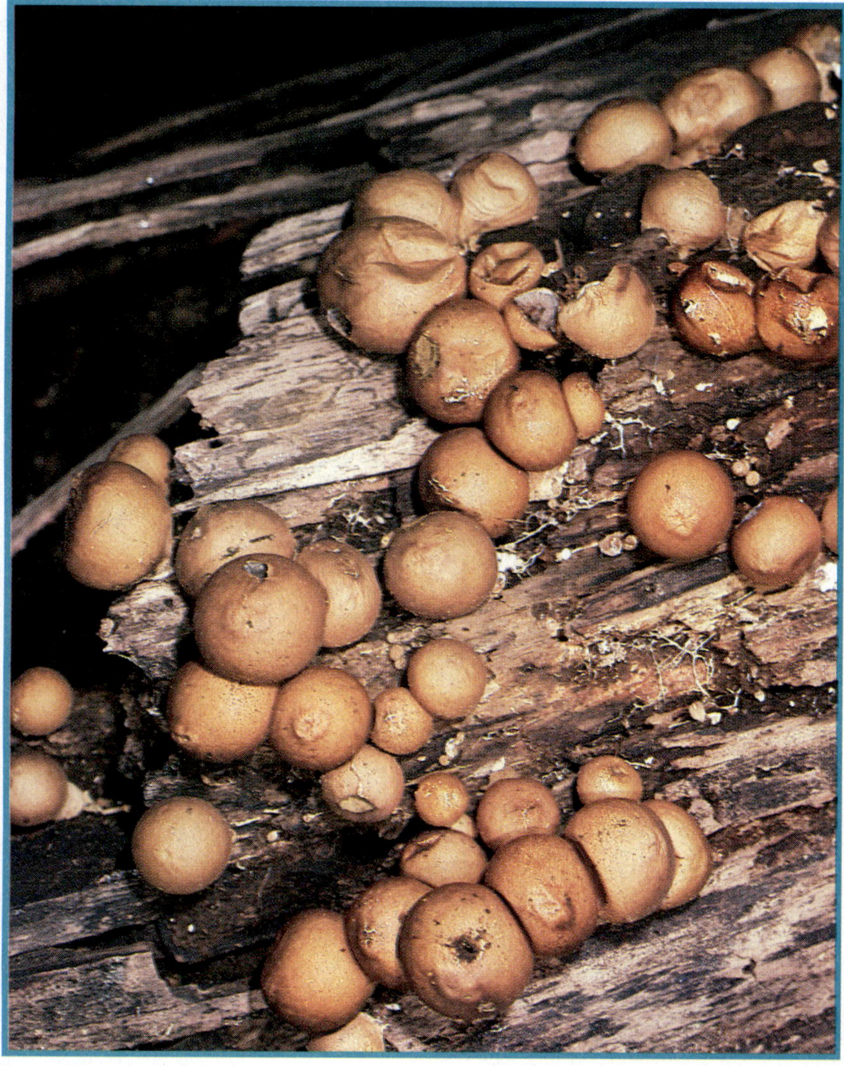

The reproductive structures of pear-shaped puffballs. *The threadlike body of this fungus penetrates the dead wood, decomposing it over time. (Patti Murray)*

Fungi: Slipping Through the Cracks in the Endangered Species Act

The living world functions much like a complex machine. Each ecosystem is composed of many separate parts, the functions of which are organized and integrated to help maintain the ecosystem's overall performance. The activities of all living things are interrelated. We are bound together and dependent upon one another, often in subtle ways. Fungi, the subject of this chapter, are instrumental in a number of environmental processes without which humans and other organisms could not exist. They, along with bacteria, perform the crucial task of decomposition, which allows nutrients to recycle in ecosystems.

In western Europe, fungi are disappearing at an alarming rate. Reports over the past 60 years from mushroom collectors in Austria, Germany, and The Netherlands indicate that between 40% and 50% of the mushroom species are gone from selected sites.

You might think the loss of some species from an ecosystem would not endanger the rest of the living organisms. After all, species have become extinct throughout the history of life on our planet. But, consider the following analogy. Imagine you were asked to drive an automobile in which some of the parts were missing. You might be able to drive it initially, but if more and more parts were removed, over time the automobile would simply stop running. Similarly, the removal of organisms from a community may make an ecosystem function less smoothly. If enough organisms are removed, the entire ecosystem deteriorates and can collapse.

In 1973, the Endangered Species Act was passed in the United States. This act authorizes the U.S. Fish and Wildlife Service to protect from extinction endangered and threatened species in the United States and abroad. The Endangered Species Act makes it illegal to sell or buy any product made from an endangered or threatened species. It is considered one of the strongest pieces of environmental legislation, in part because species are designated as endangered or threatened entirely on biological grounds; economic considerations cannot influence the designation.

And yet the Endangered Species Act is not perfect. It is geared more to saving the few highly visible and popular or unique endangered species such as the giant panda rather than the much greater number of less visible and "unglamorous" species such as fungi that perform valuable environmental services. But while microorganisms, fungi, and insects may not win any public popularity contests, they are the organisms that dominate ecosystems (in that they make up the majority of species) and contribute most to their functioning. Conservationists would like to see the Endangered Species Act strengthened so as to preserve entire ecosystems rather than as it currently is written to save individual endangered species.

Fungal Life

22

Learning Objectives

After you have studied this chapter you should be able to:

1. Describe the distinguishing characteristics of the kingdom Fungi.
2. Contrast the body plan of a yeast with that of a mold.
3. Trace the fate of a fungal spore that lands on an appropriate substrate such as an overripe peach, and describe conditions that permit fungal growth.
4. List distinguishing characteristics and give examples of each: zygomycetes, ascomycetes, basidiomycetes, and imperfect fungi.
5. Characterize the unique nature of a lichen.
6. Explain the ecological significance of fungi as decomposers.
7. Summarize the special ecological roles of mycorrhizae.
8. Summarize the various uses of the fungi.
9. Summarize the destructive nature of certain fungi, especially to food and crops.
10. Identify several different fungal diseases of humans.

Key Concepts

- ☐ Most fungi are multicellular organisms with a body plan consisting of a mass of threadlike filaments called hyphae. Fungi reproduce by forming spores, which may be produced sexually or asexually, in fruiting bodies (reproductive structures) composed of interwoven hyphae.
- ☐ Fungi are heterotrophs whose hyphae obtain nutrients by absorbing predigested food. They do this by secreting digestive enzymes into the substrate on which they are growing.
- ☐ Fungi play a significant role in ecosystems as saprophytes (decomposers), breaking down dead organic material and recycling the nutrients. Other fungi are parasites and obtain nutrients from living organisms. Still others form intimate, mutually beneficial associations with the roots of plants.

MEMBERS OF KINGDOM FUNGI ARE EUKARYOTES WITH UNIQUE MODES OF NUTRITION, BODY PLANS, AND REPRODUCTION

Mushrooms, morels, and truffles—delights of the gourmet—have much in common with the black mold that grows on stale bread and the mildew that collects on damp shower curtains. These life forms belong to the kingdom Fungi, a diverse group of more than 80,000 known species, most of which are terrestrial. Although they vary strikingly in size and shape, all fungi are eukaryotes. This means that their cells contain membrane-bounded nuclei, mitochondria, and other organelles. Fungi were originally classified in the plant kingdom, but biologists today recognize that they are not plants.[1] Fungi are distinct from plants and other eukaryotes in many ways, and thus they are accorded a separate kingdom.

Fungal cells, like plant cells, are enclosed by cell walls. However, fungal cell walls have a different chemical composition than do plant cell walls. In most fungi, the cell wall is composed in part of **chitin,** which is a polymer that consists of subunits of a nitrogen-containing sugar. Chitin is far more resistant to breakdown by microorganisms than is the cellulose that makes up plant cell walls.

Fungi lack chlorophyll and chloroplasts and are not photosynthetic. They are **heterotrophs**, which are organisms unable to synthesize their own organic material, but they do not ingest food as do animals. Instead, fungi secrete digestive enzymes and then *absorb* the predigested food (as small organic molecules) through their cell walls and plasma membranes. They obtain their nutrients from organic matter (as decomposers) or from other living organisms (as parasites).

Fungi grow best in dark, moist habitats, but they are found universally wherever organic material is available. They require moisture to grow, and they can obtain water from the atmosphere as well as from the medium upon which they live. When the environment becomes very dry, fungi survive by going into a resting stage or by producing spores that are resistant to desiccation (drying out). Although the optimum pH for most species is about 5.6, some fungi can tolerate and grow in environments where the pH ranges from 2 to 9. Many fungi are less sensitive to high osmotic pressures than are bacteria. As a result, they can grow in concentrated salt solutions or sugar solutions, such as jelly, that discourage or prevent bacterial growth. Fungi may also thrive over a wide temperature range. Even refrigerated food is not immune to fungal invasion.

[1] Interestingly, recent studies suggest that fungi are more closely related to animals than to plants.

CONCEPT CONNECTIONS

Fungi — Decomposition — Food Webs

Fungi are essential decomposers in the food web.* During decomposition, proteins, lipids, carbohydrates, and other complex organic molecules that make up the remains of dead organisms are broken down. These complex organic molecules are degraded by fungi and other decomposers into simpler organic molecules, which are in turn broken down further. At each step in the process of disassembling the molecules, energy is released and utilized by the decomposers. The end products of decomposition are water, CO_2, and minerals, which are the starting materials used by plants to manufacture complex organic molecules.

When a tree dies in a forest, for example, it undergoes a series of decay steps involving a complex food web. First wood-boring insects and mites open paths through the bark that are later used by other insects, fungi, and plant roots. Fungi and bacteria are the organisms that begin the actual process of decomposition, thus providing nutrients for plant and animal inhabitants. As decay progresses, small mammals burrow into the now-soft wood and eat the fungi, insects, and plants. Ultimately, essential minerals from the tree are returned to the soil. We say more about food webs, decomposition, and the recycling of matter in Chapters 46 and 47.

* A food web is a complex interconnection of all of the food chains in an ecosystem. Each organism eats or decomposes the preceding organism in the series.

Most Fungi Possess a Filamentous Body Plan

The body structures of fungi vary in complexity, ranging from single-celled yeasts to multicellular, filamentous molds (a term used loosely to include mildews, rusts and smuts, mushrooms, and many other fungi). **Yeasts** are unicellular fungi that reproduce asexually mainly by **budding**, in which a small protuberance (bud) grows and eventually separates from the parent cell (see Fig. 8–9, page 163). Each bud can grow into a new yeast. Yeasts can also reproduce asexually by fission and sexually by spore formation. The yeasts are not classified as a single taxonomic group because many different fungi can be induced to form a yeast stage.

Most fungi are filamentous molds (Fig. 22–1). A **mold** consists of long, branched threads (or filaments) of cells called **hyphae** (singular, *hypha*). Hyphae form a tangled mass or tissue-like aggregation known as a **mycelium** (plural, *mycelia*). The cobweb-like mold sometimes seen

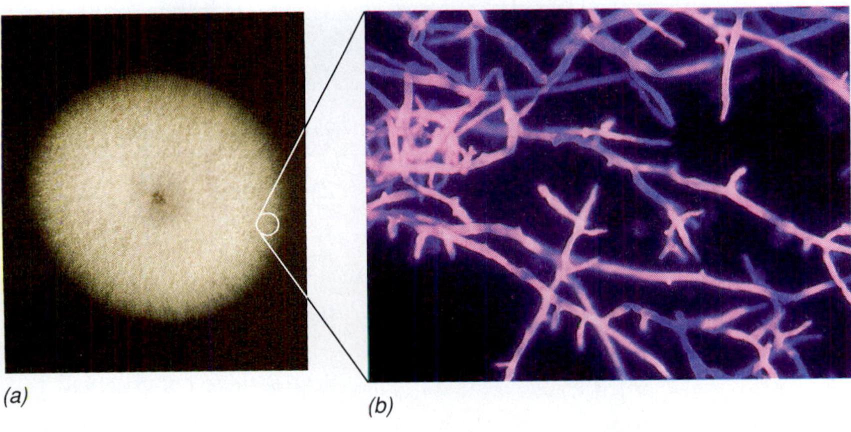

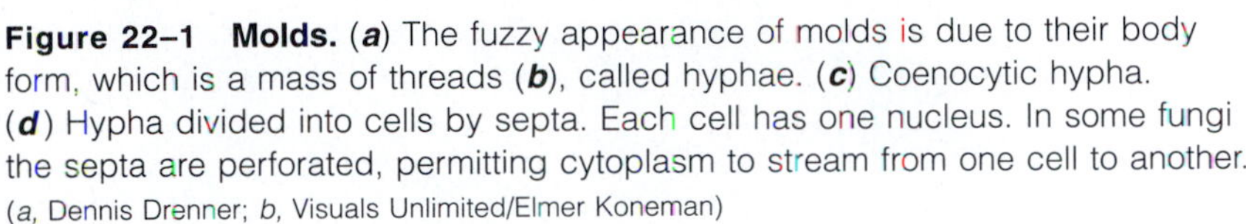

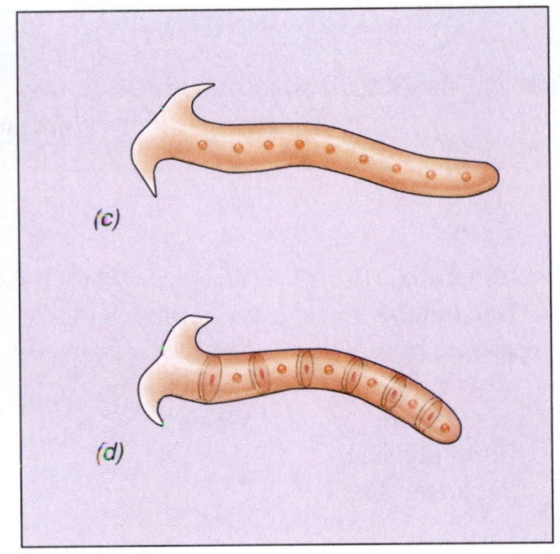

Figure 22–1 Molds. (***a***) The fuzzy appearance of molds is due to their body form, which is a mass of threads (***b***), called hyphae. (***c***) Coenocytic hypha. (***d***) Hypha divided into cells by septa. Each cell has one nucleus. In some fungi the septa are perforated, permitting cytoplasm to stream from one cell to another. (*a*, Dennis Drenner; *b*, Visuals Unlimited/Elmer Koneman)

on bread is the mycelium of a fungus. What is not seen is the extensive mycelium that grows down into the substance of the bread. Some hyphae are **coenocytic**, or undivided into individual cells, and are something like an elongated, multinucleated giant cell. Other hyphal filaments are divided by cross walls, called **septa** (singular, *septum*), into individual cells containing one or more nuclei. The septa of septate fungi often contain large pores that permit organelles to flow from cell to cell. Cytoplasm flows within the hypha, providing a system of internal transport.

Most Fungi Reproduce by Spores

Fungi reproduce by means of microscopic **spores**, which are nonmotile (unflagellated) reproductive cells dispersed by wind or animals. Spores are usually produced on hyphae that project up into the air. This arrangement permits the spores to be blown by air currents and distributed to new areas. In some fungi, these aerial hyphae form large, complex reproductive structures in which spores are produced. These structures are called **fruiting bodies**. The familiar, above-ground part of a mushroom or toadstool is a large fruiting body. We do not normally see the bulk of the fungus, a nearly invisible network of hyphae buried out of sight in the rotting material upon which it grows.

Fungi may produce spores either sexually or asexually. Unlike animal and plant cells, fungal cells usually contain haploid nuclei. In sexual reproduction, the hyphae of two genetically different mating types often come together and their nuclei fuse, forming a diploid zygote. In two fungal groups (the ascomycetes and basidiomycetes), the hyphae fuse but the two different nuclei do not fuse immediately; rather, they remain separate within the fungal cytoplasm. Hyphae that contain two genetically distinct nuclei within each cell are **dikaryotic,** which is described as n+n rather than 2n. Hyphae that contain only one nucleus per cell are said to be **monokaryotic**.

When a fungal spore contacts an appropriate substrate, perhaps an overripe peach that has fallen to the ground, it germinates and begins to grow. A threadlike hypha emerges from the tiny spore and grows, branching frequently. Soon a tangled mat of hyphae infiltrates the peach. Cells of the hyphae secrete digestive enzymes into the peach, degrading its organic compounds to small molecules that the fungus can absorb. Later, other hyphae extend upward into the air to bear spores.

FUNGI ARE CLASSIFIED INTO THREE DIVISIONS

The classification of fungi is based mainly on the characteristics of the sexual spores and fruiting bodies. Biologists do not agree unanimously on how to classify these diverse organisms, but most assign them to three divisions (equivalent to phyla in animal taxonomy): Zygomycota, Ascomycota, and Basidiomycota. In addition, fungi with no known sexual stage are assigned to a form division, Deuteromycota. Members of a **form** division are similar to one another in certain respects but probably do not share a common ancestry; that is, the group is polyphyletic (Chapter 20). They are classified together simply

TABLE 22–1
Divisions of Kingdom Fungi

Division	*Common types*	*Asexual reproduction*	*Sexual reproduction*
Zygomycota	Black bread mold	Nonmotile spores form in sporangium	Zygospores
Ascomycota (sac fungi)	Yeasts, powdery mildews, molds, morels, truffles	Conidia pinch off from conidiophores	Ascospores
Basidiomycota (club fungi)	Mushrooms, bracket fungi, puffballs, rusts, smuts	Uncommon	Basidiospores
Deuteromycota (form division; imperfect fungi)	Molds	Conidia	Sexual stage not observed

as a matter of convenience. Table 22–1 summarizes fungal classification. The slime molds and water molds were traditionally classified as fungi, but are now generally considered protists (Chapter 21).

Zygomycetes Reproduce Sexually by Forming Zygospores

The members of division Zygomycota, which consists of about 800 species, are referred to as **zygomycetes**. They produce sexual spores, called **zygospores**, that remain dormant for a time. Their hyphae are coenocytic; that is, they lack septa. Most zygomycetes are decomposers that live in the soil on decaying plant or animal matter, although some are parasites of plants and animals.

One common zygomycete is the black bread mold *Rhizopus nigricans*, a decomposer that breaks down bread and other foods (Fig. 22–2). Before preservatives were added to food, bread left at room temperature was often covered with a black, fuzzy growth in a few days. Bread becomes moldy when a spore falls on it and then germinates and grows into a tangled mass of threads, the mycelium. Hyphae penetrate the bread and absorb nutrients. Eventually, certain hyphae grow upward and develop **sporangia** (spore sacs) at their tips. Clusters of black asexual spores develop within each sporangium and are released when the delicate sporangium ruptures. The spores give the black bread mold its characteristic color.

Sexual reproduction in the black bread mold occurs when the hyphae of two different mating types (designated as plus and minus) grow into contact with one another. An individual fungal hypha mates only with a hypha of a different mating type. That is, sexual reproduction can occur only between hyphae of a plus (+) strain and hyphae of a minus (−) strain. Because there is no morphological (structural) differentiation between the two mating types, it is not appropriate to refer to them as "male" and "female."

When hyphae of opposite mating types meet, hormones are produced that cause the tips of the hyphae to come together. Plus and minus nuclei then fuse to form a diploid nucleus, the zygote. A zygospore develops, providing a thick protective covering around the zygote. The zygospore may lie dormant for several months and can survive desiccation and extreme temperatures. Meiosis probably occurs at or just before germination of the zygospore. When the zygospore germinates, an aerial hypha develops (by mitosis) with a sporangium at the tip. Mitosis within the sporangium forms haploid spores, which are released and may germinate to form new hyphae. Only the zygote of a black bread mold is diploid; all of the hyphae and asexual spores are haploid.

Ascomycetes (Sac Fungi) Reproduce Sexually by Forming Ascospores

Division Ascomycota, the **ascomycetes**, is a large group of fungi consisting of about 30,000 described species. The ascomycetes are sometimes referred to as **sac fungi** because their sexual spores are produced in little sacs called **asci** (singular, *ascus*). Their hyphae usually have septa, but these cross walls are perforated so that cytoplasm can move from one compartment to another.

Ascomycetes include most yeasts; the powdery mildews; most of the blue, green, pink, and brown molds that cause food to spoil; saprophytic cup fungi; and the edible morels and truffles. Some ascomycetes cause serious plant diseases such as dutch elm disease, chestnut blight, ergot disease (on rye), and powdery mildew (on fruits and ornamental plants).

(Text continues on page 406)

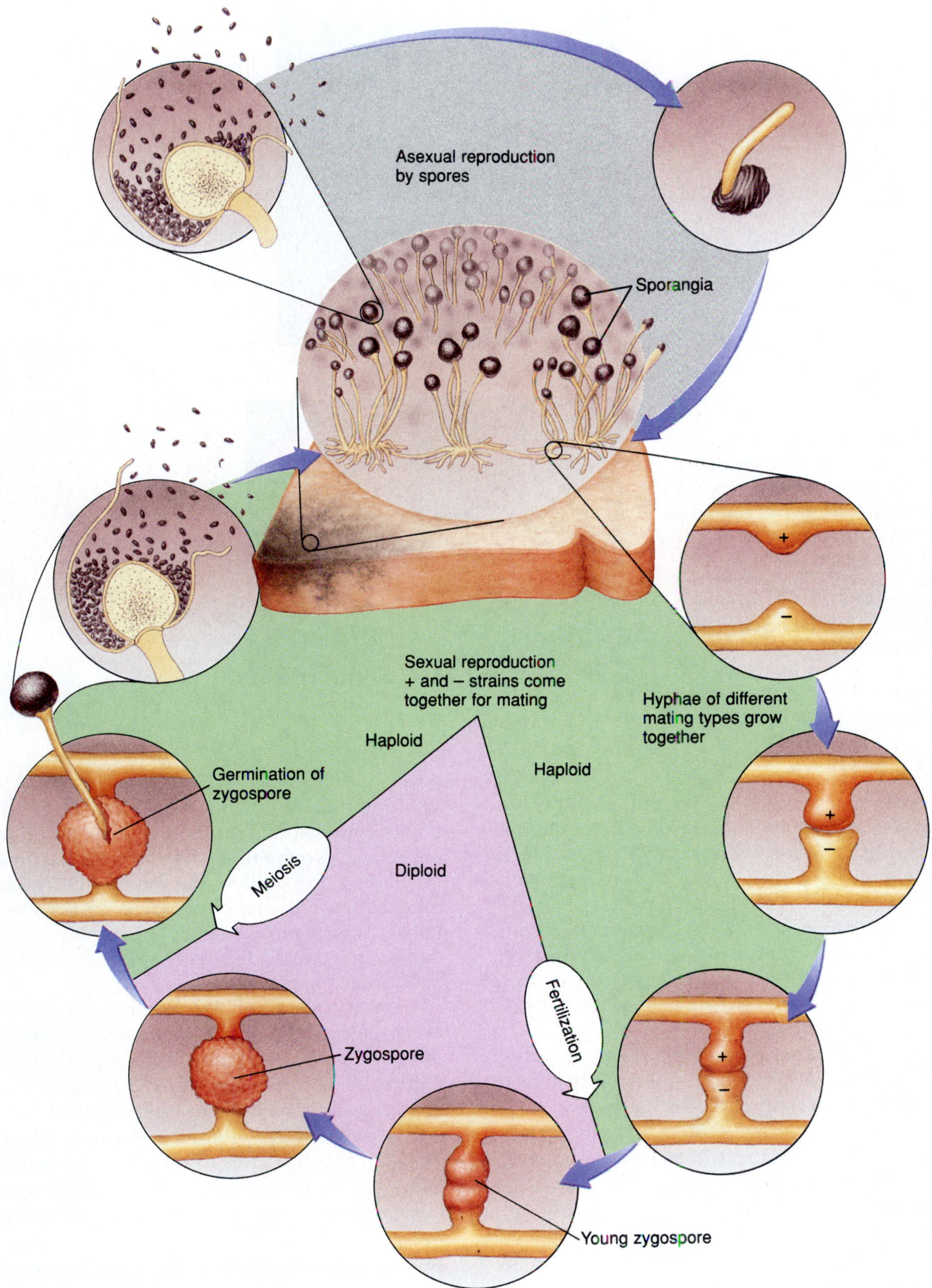

Figure 22–2 Life cycle of the black bread mold, *Rhizopus nigricans*. Sexual reproduction takes place only between different mating types, designated (+) and (−).

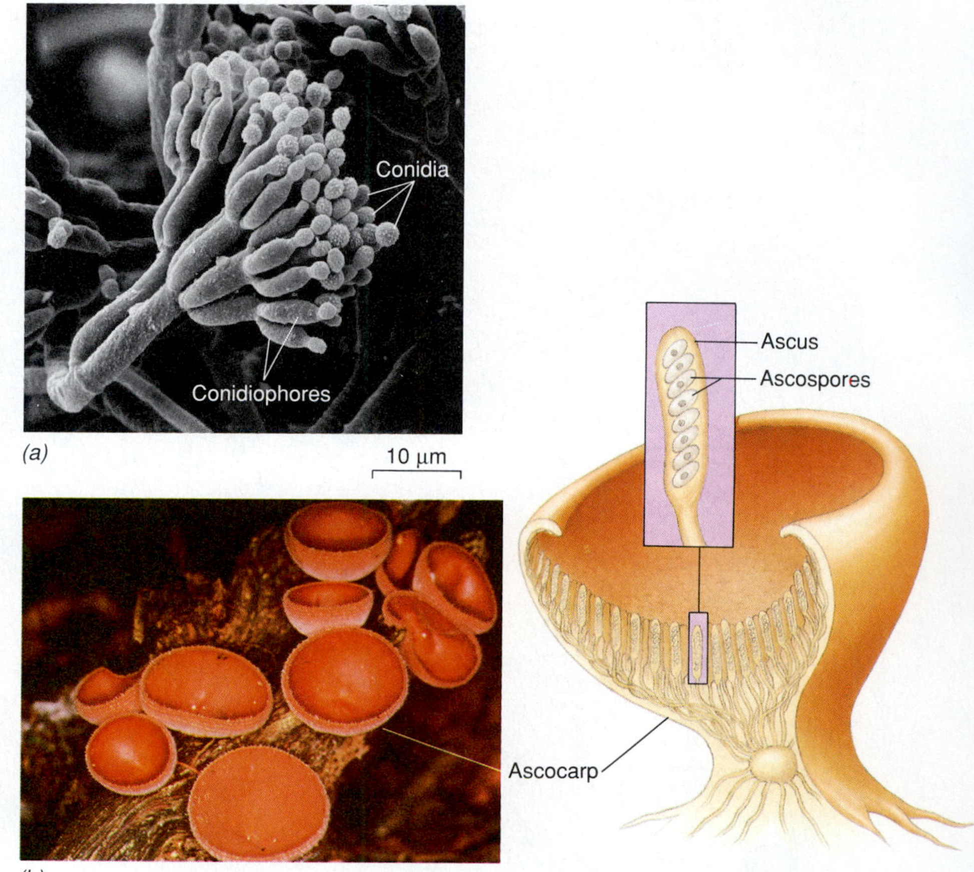

Figure 22–3 The sac fungi. (***a***) Scanning electron micrograph of *Penicillium* conidiophores, which resemble paintbrushes. The arrangement of conidia (asexual spores) on conidiophores varies from species to species and is used to help identify these fungi. (***b***) The ascocarp of a cup fungus. Sexual reproduction in the ascomycetes involves the formation of an asocarp, or fruiting body, in which the sac-like asci develop. (*a*, Biophoto Associates/Photo Researchers, Inc., *b*, James L. Castner)

In most ascomycetes, asexual reproduction involves production of spores called **conidia**, which are pinched off at the tips of certain specialized hyphae known as **conidiophores** (conidia bearers; Fig. 22–3*a*). Sometimes called "summer spores," conidia are a means of rapidly propagating new mycelia when environmental conditions are good. Conidia occur in various shapes, sizes, and colors in different species. The color of the conidia is what gives the characteristic brown, blue, green, pink, or other tint to many of these molds.

Some species of ascomycetes have different mating strains; others are self-fertile and have the ability to mate with themselves. Sexual reproduction takes place after two hyphae grow together and their cytoplasm mingles. Within this fused structure, the two nuclei come together, but they do not fuse. New hyphae develop from the fused structure, and the cells of these hyphae are dikaryotic. The n+n hyphae form a fruiting body, known as an **ascocarp**, that is characteristic of the particular species (Fig. 22–3*b*). The asci develop in the ascocarp.

Within a cell that develops into an ascus, the two nuclei fuse and produce a diploid nucleus, the zygote. Each zygote then undergoes meiosis to form four haploid nuclei. This process is usually followed by one mitotic division of each of the four nuclei, resulting in the formation of eight haploid nuclei. Each haploid nucleus develops into an **ascospore**, so that there are usually eight haploid ascospores within the ascus. The ascospores are released when the tip of the ascus breaks open. Individual ascospores are carried by air currents, often for long distances. If one lands in a suitable location, it germinates and forms a new mycelium.

Yeasts are unicellular ascomycetes that reproduce asexually by budding (already described) and sexually by forming ascospores. During sexual reproduction, two haploid yeasts fuse, forming a diploid zygote. The zygote undergoes meiosis, and the resulting haploid spores remain enclosed for a time within the original diploid cell wall. This sac of spores corresponds to an ascus and ascospores. Yeasts are essential in making bread and fermenting alcoholic beverages (discussed later in this chapter).

Basidiomycetes (Club Fungi) Reproduce Sexually by Forming Basidiospores

The 25,000 or more species that make up the division Basidiomycota include the most familiar of the fungi—mushrooms, bracket fungi, and puff balls (Fig. 22–4).

(a)

(b)

(c)

(d)

Figure 22–4 Representative club fungi. (***a***) Basidia line the gills of the yellow-orange jack o'lantern mushroom, a poisonous species whose gills glow in the dark. (***b***) A giant puffball in White Plains, New York. At maturity a dried-out puffball often has a pore through which the basidiospores are discharged as a puff of dust. (***c***) The stinkhorn is also commonly called the devil's penis. Its foul smell attracts flies that help disperse the slimy mass of basidiospores. (***d***) Bracket fungi grow on both dead and living trees, producing shelf-like fruiting bodies. Spores are produced in pores located underneath each shelf. (*a,* Dennis Drenner; *b,* Ed Kanze/Dembinsky Photo Associates; *c,* Visuals Unlimited/Richard D. Poe; *d,* Richard H. Gross)

Some destructive plant parasites of important crops, such as wheat rust and corn smut, are also basidiomycetes.

Basidiomycetes, or **club fungi**, derive their name from the fact that they develop a **basidium** (plural, *basidia*), which is a structure comparable in function to the ascus of ascomycetes. Each basidium is an enlarged, club-shaped hyphal cell, at the tip of which develop four **ba-**

(Text continues on p. 409)

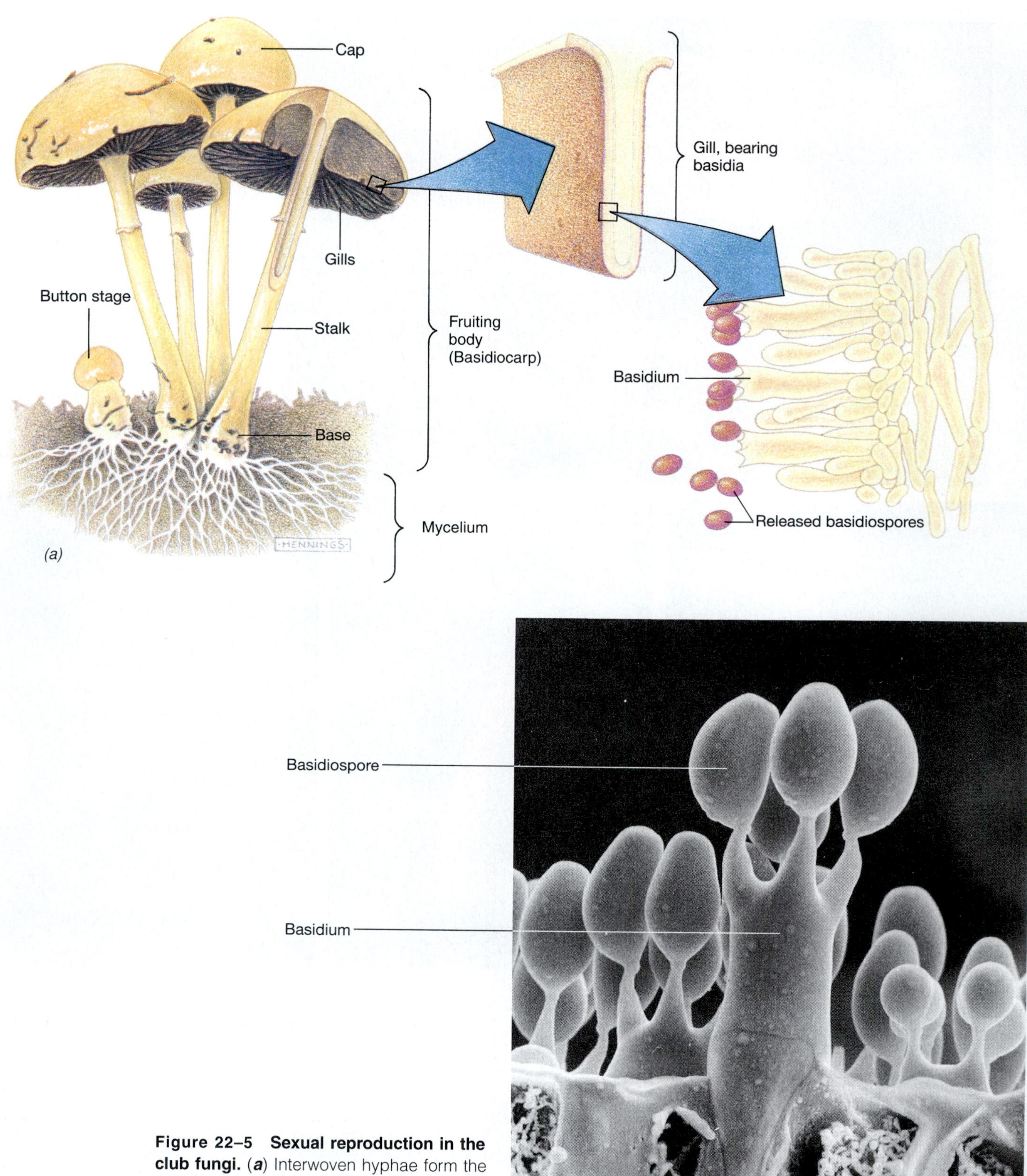

Figure 22–5 Sexual reproduction in the club fungi. (***a***) Interwoven hyphae form the basidiocarp commonly called a mushroom. Numerous basidia are borne along the gills. (***b***) Each basidium produces four basidiospores, which are attached to the basidium. (Biophoto Associates)

FOCUS ON *The Process of Science*

THE GIANT FUNGUS

Giants are known in the plant and animal kingdoms (for example, the giant sequoia and the blue whale), but until recently, the fungi were all thought to be relatively small organisms. In 1992, several biologists reported a saprophytic fungus (*Armillaria bulbosa*) that is one of Earth's largest and oldest organisms, living in a forest in Michigan. Commonly called the honey mushroom, this species of fungus digests and absorbs dead or dying tree roots. The honey mushroom has a vast underground network of hyphae and rhizomorphs (branching cordlike structures). The rhizomorphs, which grow slowly through the upper layers of the soil, are composed of interwoven strands of hyphae, and often resemble shoestrings in thickness.

The biologists used a type of genetic fingerprinting (Chapter 13) to confirm that a single fungus individual extends through at least 15 hectares (37 acres) of the forest floor—an area equivalent to the size of more than 33 football fields! Moreover, they found that the fungus is territorial; no other honey mushroom individual was found in that area. Near the end of each summer, the giant fungus produces thousands of honey-colored mushrooms. But, according to the DNA data, none of the millions of spores produced by these mushrooms appears to have grown into a new individual within the 15 hectares.*

Because rhizomorphs grow at a rate of about 8 inches per year in Northern Michigan, the fungus was calculated to be over 1500 years old. At least one forest fire (in 1928) is known to have destroyed the trees in the area inhabited by this fungal giant, but it survived underground. It probably thrived immediately after the fire because of the large number of dead tree roots that became available for food.

The discovery of this giant fungus has prompted biologists to speculate that even larger fungi may exist elsewhere in the soil. For example, a single fungus of a related species of *Armillaria* in the state of Washington is claimed to occupy over 600 hectares (1500 acres), but so far this claim has not been substantiated with DNA evidence. Clearly, fungi are the giants of the forest, a fact that helps confirm their pivotal role in the forest ecosystem.

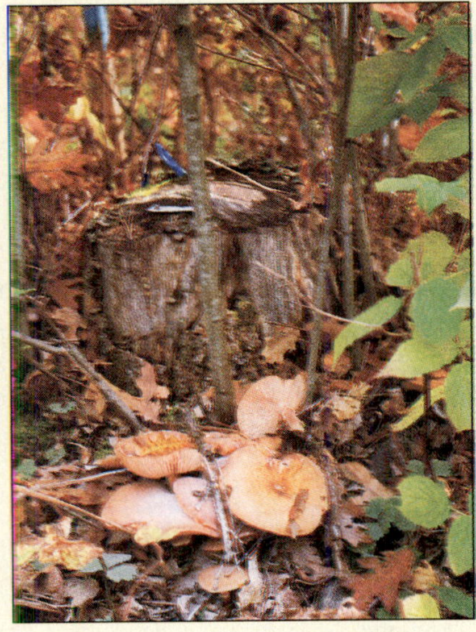

The honey mushroom (*Armillaria bulbosa*) that is among the Earth's largest and longest-lived organisms. (Courtesy of Johann N. Bruhn)

* Of course, many other species of fungi and other organisms cohabit the 15 hectares occupied by the giant honey mushroom.

sidiospores (Fig. 22–5). Note that basidiospores develop on the *outside* of the basidium, whereas ascospores develop *within* the ascus. The mature basidiospores are released, and when they come in contact with the proper environment, each develops into a new mycelium.

The mycelium of a basidiomycete such as the cultivated mushroom consists of a mass of white, branching, threadlike hyphae that occur mostly below ground (see Focus on the Process of Science: The Giant Fungus). The hyphae are divided into cells by septa, but as in ascomycetes, the septa are perforated and allow cytoplasmic streaming between cells.

Compact masses of hyphae, called buttons, develop along the mycelium. Each button grows into the fruiting body that we ordinarily call a mushroom. More formally, a mushroom, which consists of a stalk and cap, is referred to as a **basidiocarp**. The lower surface of the cap usually consists of many thin plates called **gills** that radiate from the stalk to the edge of the cap. The basidia develop on the surfaces of these gills.

Each individual fungus produces millions of basidiospores, and each basidiospore has the potential to give rise to a new **primary mycelium.** Hyphae of a primary mycelium consist of monokaryotic cells. When, in the course of its growth, such a hypha encounters another hypha of a different mating type, the two hyphae fuse. As in the ascomycetes, however, the two haploid nuclei remain separate within each cell. In this way, a **secondary mycelium** with dikaryotic hyphae is produced, in which each cell contains two haploid nuclei.

The n+n hyphae of the secondary mycelium grow extensively and eventually form compact masses, which are the mushrooms or basidiocarps. Each basidiocarp actually consists of intertwined hyphae that are matted together. On the gills of the mushroom, the nuclei in each dikaryotic cell fuse, forming diploid zygotes. These are the only diploid cells present in the entire life history of a basidiomycete. Meiosis then takes place, forming four haploid nuclei, which move to the outer edge of the basidium. Finger-like extensions of the basidium develop, into which the nuclei and some cytoplasm move. Each of these nuclei becomes a basidiospore. A wall forms that separates the basidiospore from the rest of the basidium by a delicate stalk. When the stalk breaks, the basidiospore is released.

Figure 22–6 Lichens vary in their color, shape, and overall appearance. Some lichens grow tightly attached to rocks. It would require a razor blade to scrape this lichen off its substrate. (M.L. Dembinsky, Jr./Dembinsky Photo Associates)

Imperfect Fungi Are Fungi with No Known Sexual Stage

About 25,000 species of fungi have been assigned to a group called the **deuteromycetes** (Division Deuteromycota). They are also known as **imperfect fungi** because many have not been observed to have a sexual stage during their life cycle. Should further study reveal a sexual stage, these species will be reassigned to a different division. Most deuteromycetes reproduce only by means of conidia and so are closely related to the ascomycetes. A few appear to be more closely related to the basidiomycetes.

Lichens Are Dual "Organisms"

Although a **lichen** looks like a single organism (Fig. 22–6), it is actually a symbiotic association between two organisms—a photosynthetic organism and a fungus. (A symbiotic association is an intimate relationship between organisms of different species.) About 20,000 kinds of lichens have been described.

The photosynthetic partner is usually either a green alga or a cyanobacterium. The fungus is most often an ascomycete, although in some lichens from tropical regions, the fungal partner is a basidiomycete. Most of the algae or cyanobacteria found in lichens are also found as free-living species in nature, but the fungal components of lichens are generally found only as a part of lichens.

In the laboratory, the fungal and algal components of a lichen can be isolated and grown separately in appropriate culture media. The alga grows more rapidly when separated, whereas the fungus grows slowly and requires many complex carbohydrates. Neither organism resembles a lichen in appearance when grown separately. The alga and fungus can be reassembled as a lichen, but only if they are placed in a culture medium under conditions that cannot support either of them independently.

What is the nature of this partnership? The lichen was originally considered a definitive example of **mutualism**, which is a symbiotic relationship that benefits both species. The photosynthetic partner carries on photosynthesis, producing food for both members of the lichen, but it is unclear how the photosynthetic partner benefits from the relationship. It has been suggested that the photosynthetic partner obtains water and minerals from the fungus, as well as protection against desiccation. More recently, some biologists have suggested that the lichen partnership is not really an example of mutualism but of controlled parasitism of the alga by the fungus.

Able to tolerate extremes of temperature and moisture, lichens grow everywhere on land that life can be supported at all, except in polluted cities. They exist farther north than any plants of the Arctic region and are equally at home in the steaming equatorial jungle. They grow on tree trunks, mountain peaks, and bare rock. In fact, lichens are often the first organisms to inhabit rocky areas. Lichen growth in these areas plays an important role in the formation of soil from rock because they gradually etch tiny cracks in rock (Chapter 46). This process sets the stage for further disintegration of the rocks by wind and rain.

The reindeer mosses of the arctic region are not mosses but lichens that serve as the main source of food

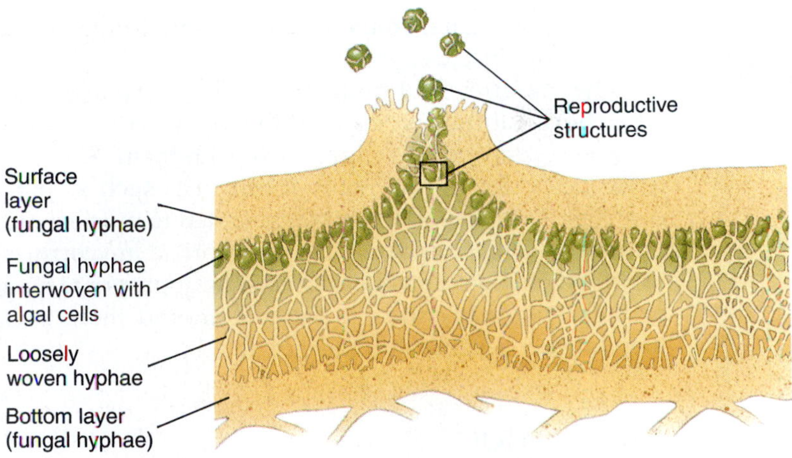

Figure 22–7 Cross section of a typical lichen, showing distinct layers. The asexual reproductive structures are composed of clusters of algal or cyanobacterial cells enclosed by fungal hyphae.

for the caribou of that region. Some lichens produce colored pigments. One of them, orchil, is used to dye woolens, and another, litmus, is widely used in chemistry laboratories as an acid-base (pH) indicator.

Lichens vary greatly in size. Some are almost invisible, whereas others, like the reindeer mosses, may cover miles of land with an ankle-deep growth. Growth proceeds slowly; the radius of a lichen may increase by less than a millimeter each year. Some mature lichens are thought to be thousands of years old.

Lichens absorb minerals mainly from the air and from rainwater, but also directly from the surface on which they grow. They cannot excrete the elements they absorb, and perhaps for this reason, they are very sensitive to toxic compounds in the environment. A reduction in lichen growth has been used as an indicator of air pollution, especially sulfur dioxide. Absorption of such toxic compounds damages the chlorophyll of the photosynthetic partner. The return of lichens to an area indicates an improvement in air quality in that area. (Recall from Chapter 1 that air pollution killed lichens on tree trunks in England, thereby affecting the evolution of peppered moths.)

Lichens reproduce mainly by asexual means, usually by fragmentation. Generally, bits of the lichen break off and, if they land on a suitable surface, establish themselves as new lichens. Some lichens release special dispersal units containing cells of both partners (Fig. 22–7). In others, the alga reproduces asexually by mitosis, while the fungus produces ascospores. The ascospores may be dispersed by wind and find an appropriate algal partner only by chance.

FUNGI ARE ECOLOGICALLY IMPORTANT

Fungi make important contributions to the ecological balance of our world. Like bacteria, most fungi are **saprophytes**, which are decomposers that absorb nutrients from organic wastes and dead organisms. When fungi degrade wastes and dead organisms, carbon (as CO_2) and mineral components of organic compounds are released, and these elements are recycled. Without this continuous decomposition, essential nutrients would soon become trapped in huge mounds of dead animals, feces, branches, logs, and leaves. The nutrients would be unavailable for use by new generations of organisms, and life would cease.

Although most fungi are saprophytes, some species form symbiotic relationships of various kinds. Some fungi are parasites, which are organisms that live in or on other organisms and are harmful to their hosts. Parasitic fungi absorb nutrients from the living bodies of their hosts.

Some types of fungi form mutualistic relationships with other organisms. **Mycorrhizae** (fungus-roots) are mutualistic relationships between fungi and the roots of plants (Fig. 22–8 and Chapter 46). Such relationships oc-

Figure 22–8 An experiment demonstrating that soybeans respond to mycorrhizal fungi. Left (CK), a control plant grows in the absence of the fungus. Right and middle (GM and GE), these soybeans were grown under conditions identical to the control, except that their roots have formed mycorrhizal associations. (Visuals Unlimited/R. Roncadori)

Concept Connections

Fungi · *Flowering Plants* · *Mimicry*

One of the most fascinating phenomena relating to living organisms is **mimicry**, in which, during the course of evolution, one organism comes to resemble another organism or an inanimate object (Chapter 46). For example, a walking stick insect so closely resembles a dead twig that it is frequently overlooked by predators. Able to remain motionless for long periods of time, the walking stick's behavior reinforces its physical resemblance to a dead twig.

In 1993, the journal *Nature* reported an unusual example of mimicry that actually requires *two* different organisms—a fungus and a plant—to produce the deception. The fungus (*Puccinia monoica*) is a plant parasite that causes a rust disease in rock cress (*Arabis holboellii*), a plant widely distributed across the northern part of North America.

When *Puccinia* infects rock cress, it does not kill the plant; rather, it changes the plant's growth pattern. Infected plants grow much taller than normal and produce a cluster of leaves at the top of the plant. The fungus, which is bright yellow in color, covers these leaves, giving them the appearance of buttercup flowers. Even botany students have been fooled by the remarkable resemblance. The fungal mimicry is not only visual. The fungus also secretes a sugary solution and produces a strong scent, imitating the nectar and aroma of flowers (Chapter 29).

What might be the evolutionary advantage of this elaborate mimicry? As you might guess, the answer involves the fungus' reproductive cycle. In the life cycles of flowers and *Puccinia monoica,* insects play the key role of increasing the chances of successful reproduction. The same characteristics that attract insects to real flowers also attract them to the fungal imitation.

In the case of flowers, the insects transfer pollen, the male reproductive part, to the female part of another flower. In the case of *Puccinia,* a basidiomycete, sexual reproduction requires the union of nuclei from two different mating types. Insects, particularly flies, are attracted to the fungal "flowers," where they eat the sugary syrup. As the insects feed, pieces of the fungus cling to their bodies. Flying from one *Puccinia* "flower" to another, the insects distribute complementary mating types from fungus to fungus over a broad range. Thus, for the fungus, flower mimicry leads to better chances for successful reproduction.

cur in more than 90% of all plant families. The mycorrhizal fungus benefits the plant by decomposing organic material in the soil and providing water and minerals such as phosphorus to the plant. At the same time, the roots supply sugars, amino acids, and other organic materials to the fungus.

The importance of mycorrhizae first became evident when horticulturalists observed that orchids do not grow unless an appropriate fungus lives with them. Similarly, it has been shown that many forest trees such as pines die from malnutrition when transplanted to nutrient-rich grassland soils that lack the appropriate mycorrhizal fungi. When forest soil that contains the appropriate fungi or their spores is added to the soil around these trees, they quickly assume normal growth.

Fungi Are Economically Important

The same powerful digestive enzymes that enable fungi to decompose wastes and dead organisms also permit them to reduce wood, fiber, and food to their basic components with great efficiency. Thus, various fungi cause incalculable damage to stored goods and building materials each year. Bracket fungi, for example, cause enormous losses by decaying wood, both in living trees and in stored lumber.

Fungi cause economic gains as well as losses. People eat them and grow them to make various chemicals. At the other extreme, some fungi are harmful from a human perspective because they cause diseases in humans and other animals and are the most destructive disease-causing organisms of plants. Their activities cost billions of dollars in agricultural damage yearly.

Fungi Provide Beverages and Food for Humans

The ability of yeasts to produce ethyl alcohol and carbon dioxide from glucose by fermentation (Chapter 8) is utilized to make wine and beer and to bake bread (Fig. 22–9). Wine is produced when yeasts ferment fruit sugars, and beer is made when yeasts ferment grain, usually barley. During the process of making bread, carbon dioxide produced by the yeast becomes trapped in the dough as bubbles, causing the dough to rise. This process is what gives leavened bread its light texture. Both the carbon dioxide and the alcohol produced by the yeast are driven off during baking.

The unique flavor of cheeses such as Roquefort and Camembert is produced by the action of the ascomycete *Penicillium* (see Fig. 22–9). *Penicillium roquefortii,* for example, is found in caves near the French village of Roquefort, and only cheeses produced in this area can be called Roquefort cheese.

Aspergillus tamarii and other fungi are used in the Orient to produce soy sauce by fermenting soybeans. Soy sauce provides other foods with more than its special flavor. For example, it also adds vital amino acids from both

Figure 22–9 Wine, bread, and distinctive cheeses are produced in part by fungi. Yeasts ferment fruits, producing ethyl alcohol in wine. That same process produces the carbon dioxide bubbles responsible for making bread rise. The bluish splotches in blue cheese are patches of conidia. (Raymond Tschoepe)

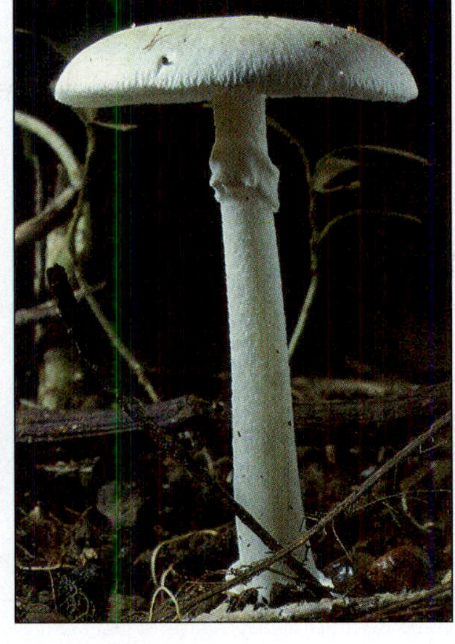

Figure 22–11 *Amanita virosa,* the destroying angel. This extremely poisonous mushroom is recognizable, as are other amanitas, by the ring of tissue around its stalk and the underground cup from which the stalk protrudes. About 50 g (2 ounces) of this mushroom could kill an adult man. Initial symptoms include vomiting, diarrhea, and cramps. If untreated, liver and kidney failure results in death in 5 to 10 days. (Earth Scenes © Dale J. Sarver)

the soybeans and the fungi themselves to supplement the low-protein rice diet.

Among the basidiomycetes are some 200 kinds of edible mushrooms and about 70 species of poisonous ones, sometimes called toadstools. Some edible mushrooms are cultivated commercially. In fact, more than 780 million pounds are produced each year in the United States alone. Morels, which superficially resemble mushrooms, and truffles, which produce underground fruiting bodies, are ascomycetes (Fig. 22–10). These delights of the gourmet are now being cultivated as mycorrhizae on the roots of tree seedlings.

Edible and poisonous mushrooms can look very much alike and may even belong to the same genus. There is no simple way to distinguish edible from poisonous mushrooms; they must be identified by an expert. Some of the most poisonous mushrooms belong to the genus *Amanita*. Toxic species of this genus have been appropriately called such names as "destroying angel" (*Amanita virosa;* Fig. 22–11) and death cap (*Amanita phalloides*). Eating a single mushroom of either species can be fatal.

Ingestion of certain species of mushrooms causes intoxication and hallucinations. The sacred mushrooms of

(a)

(b)

Figure 22–10 Edible ascomycetes. (***a***) *Morchella* sp., commonly called morels, and (***b***) *Tuber melanosporum*, truffles, are expensive gourmet treats. Both are fruiting bodies known as ascocarps that produce ascospores. Truffles are subterranean ascocarps. Here they are shown entire and sectioned. (*a*, James L. Castner; *b*, Visuals Unlimited/John D. Cunningham)

(a)

(b)

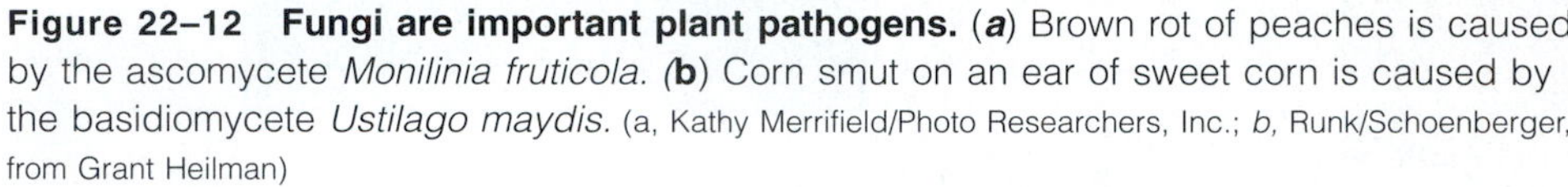

Figure 22–12 Fungi are important plant pathogens. (*a*) Brown rot of peaches is caused by the ascomycete *Monilinia fruticola.* (**b**) Corn smut on an ear of sweet corn is caused by the basidiomycete *Ustilago maydis.* (a, Kathy Merrifield/Photo Researchers, Inc.; *b*, Runk/Schoenberger, from Grant Heilman)

the Aztecs, *Conocybe* and *Psilocybe,* are still used in religious ceremonies by Central American Indians and others for their hallucinogenic properties. The chemical ingredient psilocybin, chemically related to lysergic acid diethylamide (LSD), is responsible for the trancelike state and colorful visions experienced by those who eat these mushrooms.

Fungi Produce Useful Drugs and Chemicals

In 1928, Alexander Fleming noticed that one of his Petri dishes containing bacteria was contaminated by mold (Chapter 2). The bacteria were not growing in the vicinity of the mold, leading Fleming to the conclusion that the mold was releasing some substance harmful to them. Within a decade or so of Fleming's discovery, penicillin produced by the ascomycete *Penicillium notatum* was purified and used in treating bacterial infections. Penicillin is still among the most widely used and most effective antibiotics. Other drugs derived from fungi include the antibiotic griseofulvin, which is used clinically to inhibit the growth of fungi, and cyclosporine, the drug used to suppress immune responses in patients receiving organ transplants.

Fungi Cause Many Important Diseases of Plants

Fungi are responsible for many serious plant diseases, including epidemic diseases that spread rapidly and may result in massive crop failure. Damage may be localized in certain tissues or structures of the plant, or the disease may be systemic and spread throughout the entire plant. Fungus infections may cause stunting of plant parts or of the entire plant, they may cause growths like warts, or they may kill the plant.

Some important plant diseases caused by ascomycetes are powdery mildews, chestnut blight, dutch elm disease, apple scab, and brown rot (Fig. 22–12*a*). Diseases caused by basidiomycetes include smuts and rusts that attack various plants, including the cereals such as corn, wheat, oats, and other grains (Fig. 22–12*b*). Certain imperfect fungi also cause plant diseases.

Fungi Cause Certain Diseases of Animals

Some fungi cause superficial infections in which only the skin, hair, or nails are infected. Ringworm and athlete's foot are examples of superficial fungal infections. Candidiasis, a yeast infection of mucous membranes of the mouth or vagina, is among the most common superficial fungal infections.

Other fungi cause systemic infections, in which they infect internal tissues and organs and may spread through many regions of the body. Histoplasmosis is an infection of the lungs caused by inhaling the spores of a fungus abundant in bird droppings. Most people in the eastern and midwestern parts of the United States have been exposed to this fungus at one time or another. Fortunately, the infection usually stays in the lungs and is of short duration. However, if the infection spreads into the body (carried by the blood), it can be quite serious.

Chapter Summary

I. Fungi are eukaryotes with cell walls composed of chitin.
  A. Fungi lack chlorophyll and are heterotrophic; they absorb predigested food.
  B. A fungus may be unicellular or multicellular.
    1. The body of a multicellular fungus consists of long, branched hyphae, which form a mycelium.
    2. In the zygomycetes, the hyphae are coenocytic (undivided by septa).
    3. In other fungi, perforated septa are present that divide the hyphae into individual cells.
  C. Fungi reproduce by means of spores, which may be produced sexually or asexually. When a fungal spore comes into contact with an appropriate substrate, it germinates and begins to grow.

II. Fungi are classified into divisions based on their mode of sexual reproduction.
  A. Zygomycetes produce sexual spores called zygospores. The black bread mold is a representative of this group.
  B. Ascomycetes produce asexual spores called conidia; sexual spores called ascospores are produced in asci. Ascomycetes include yeasts, cup fungi, morels, truffles, and pink and green molds.
  C. Basidiomycetes produce sexual spores called basidiospores on the outside of a basidium; basidia develop on the surface of gills in mushrooms. Basidiomycetes include mushrooms, puff balls, rusts, and smuts.
  D. The deuteromycetes are the imperfect fungi; a sexual stage has not been observed. Most reproduce asexually by conidia.
  E. A lichen is a symbiotic combination of a fungus and an alga or cyanobacterium in which the fungus benefits from the association.

III. Fungi are ecologically significant.
  A. Many fungi are decomposers that break down organic compounds.
  B. Mycorrhizae are mutualistic relationships between fungi and the roots of higher plants. The fungus supplies minerals to the plant, and the plant secretes organic compounds needed by the fungus.

IV. Fungi are of both positive and negative economic importance.
  A. Mushrooms, morels, and truffles are used as food; yeasts are vital in the production of alcoholic beverages and bread; certain fungi are used to produce cheeses and soy sauce.
  B. Fungi are used to make penicillin and other antibiotics.
  C. Many fungi cause plant or animal diseases.

Selected Key Terms

budding, p. 402
coenocytic, p. 403
dikaryotic, p. 403
fruiting body, p. 403
heterotroph, p. 402
hyphae, p. 402
mold, p. 402
monokaryotic, p. 403
mutualism, p. 410
mycelium, p. 402
mycorrhizae, p. 411
saprophyte, p. 411
septa, p. 403
spore, p.4 03
yeast, p. 402

Post-Test

1. Fungi were originally classified as plants, in part because they possess ______________ ______________.
2. Fungi reproduce both asexually and sexually by forming ______________.
3. Yeasts reproduce asexually mainly by ______________.
4. Some hyphae are divided by cross walls, called ______________, into individual cells.
5. Hyphae that contain two genetically distinct nuclei within each cell are ______________, or n+n.
6. *Rhizopus* and other zygomycetes form sexual spores called ______________.
7. Sexual reproduction in ascomycetes involves production of ______________ within sacs called ______________.
8. The familiar portion of a mushroom is actually a large fruiting body called a/an ______________.
9. The type of sexual spore produced by a mushroom is a/an ______________.
10. In a mushroom, basidia develop on the surface of plates called ______________.
11. The deuteromycetes are known as imperfect fungi because their ______________ stage has not been observed.
12. Ecologically, many fungi serve as ______________ and break down dead, organic material.
13. Fungi that obtain their nutrients from living hosts are known as ______________.
14. Mycorrhizae are fungi that form mutualistic relationships

with the _______________ of plants.

15. _______________ and _______________ are edible (and delicious) fruiting bodies of ascomycetes.

16. Which of the following is *not* true of the fungi? (a) eukaryotic (b) contain chlorophyll (c) ecologically important as decomposers (d) nonphotosynthetic

17. Fungi: (a) are photosynthetic and manufacture organic compounds using the sun's energy (b) ingest their food similarly to animals (c) secrete enzymes that digest food and then absorb the predigested foods (d) can be photosynthetic or ingest food, depending upon the environment they are in.

18. A mass of threadlike hyphae that constitutes the body of a fungus is known as a/an: (a) dikaryon (b) mushroom (c) ascus (d) mycelium.

19. A _______________ is a dual organism that consists of an alga (or cyanobacterium) and a fungus. (a) bracket fungus (b) yeast (c) sac fungus (d) lichen

20. This important fungus is used to make bread, beer, and wine: (a) yeast (b) honey mushroom (c) *Amanita* (death angel) (d) reindeer moss.

21. Label the following:

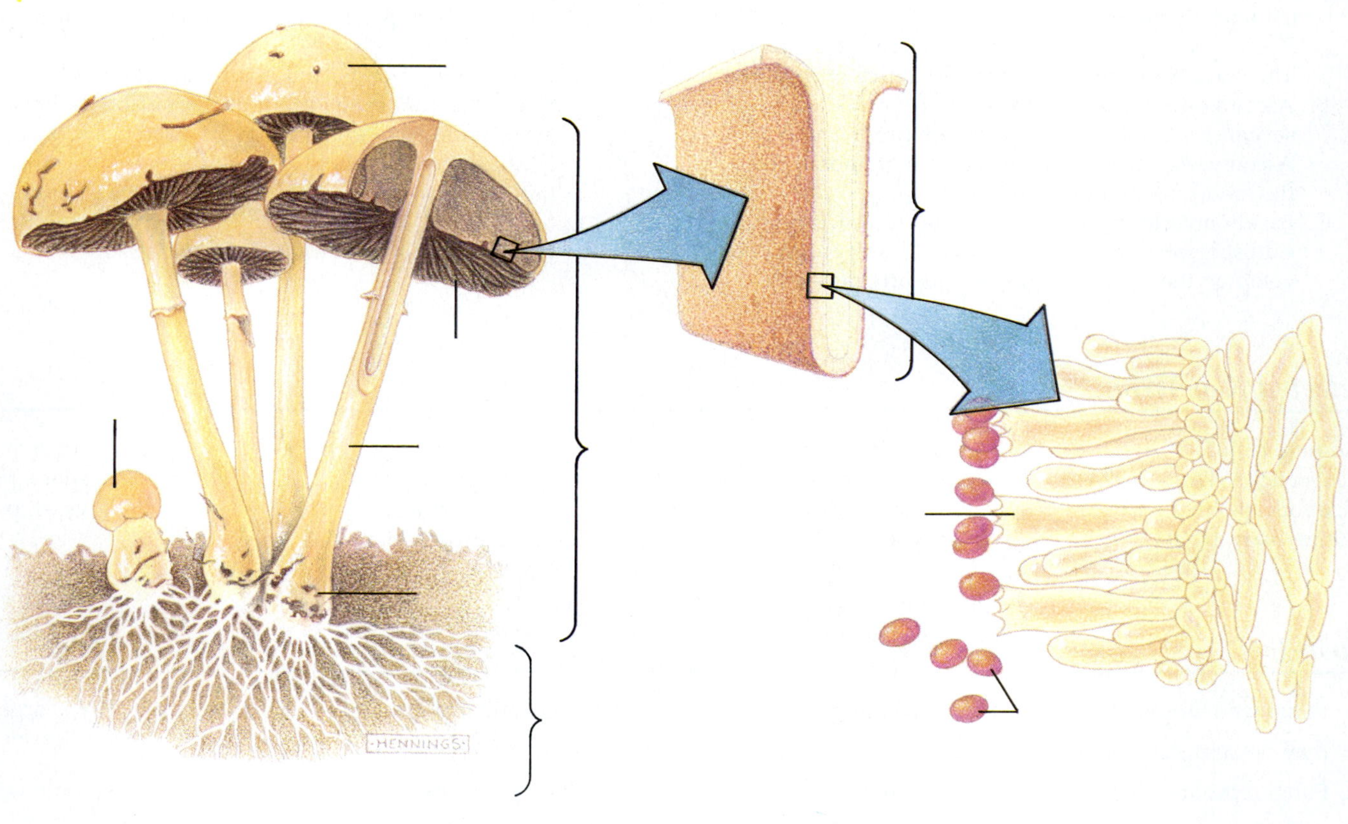

Review Questions

1. What characteristics distinguish fungi from other organisms?
2. How does a fungus differ from a plant?
3. How does the body plan of a yeast differ from that of a mold?
4. What is the difference between a hypha and a mycelium? Between a fruiting body and a mycelium?
5. List the major divisions of fungi and give several characteristic features of each.
6. Describe the life cycle of a typical mushroom.
7. Describe the ecological significance of each of the following: saprophytic fungi, lichens, mycorrhizae.
8. Distinguish between fungi that are saprophytes and those that are parasites.
9. Briefly describe three important fungal diseases of plants and three fungal diseases of humans.

Thinking Critically

1. In the early 1990s, 1116 species were listed as endangered or threatened in the United States: 941 vertebrates (mammals, birds, reptiles, amphibians, and fish); 72 invertebrates (insects, snails, clams, and crustaceans); and 103 plants. Why are there so many more vertebrates than invertebrates? Why are no fungi or bacteria listed as endangered?
2. If you do not see any mushrooms in your lawn, can you conclude that there are no fungi living there? Why or why not?
3. Why are mushrooms and puffballs usually located above ground, whereas the fungal mycelium is below ground?
4. Under what kinds of environmental conditions might it be more advantageous for a fungus to reproduce asexually? Sexually?
5. What measures can you suggest to prevent bread from becoming moldy?
6. Biologists have determined that many mycorrhizal fungi are sensitive to a low pH. What human-caused environmental problem might prove catastrophic for these fungi? How might it affect their plant partners?

Recommended Readings

The Audubon Society Field Guide to North American Mushrooms, New York, Alfred A. Knopf, 1981 (8th printing, 1992). A beautifully illustrated guide to common fungi.

Gould, S.J., "Fungal Forgery," *Natural History,* September 1993. A fascinating account of a fungus that mimics flowers.

Lipske, M., "A New Gold Rush Packs the Woods in Central Oregon," *Smithsonian,* January 1994. A mushroom war is occurring in forests in the Northwest.

Newhouse, J.R., "Chestnut Blight," *Scientific American,* July 1990. An account of the chestnut blight fungus that has almost completely eradicated the American chestnut in the United States.

Radetsky, P., "The Yeast Within," *Discover,* March 1994. The discovery that baker's yeast can be induced to grow like a filamentous mold has far-reaching medical implications.

Scitil, K., "Fungus Among Us," in "Biology 1992," *Discover,* January 1993. Discusses the "monster mushroom" that was discovered in Michigan in 1992.

Coal cars on railroad.
(Louie Psihoyos, © National Geographic Society)

People, Coal, and the Environment

Human society depends on energy. We use it to warm our homes in winter and cool them in summer, to grow and cook our food, to extract and process natural resources, to manufacture items we use daily, and to power various forms of transportation. Many of the conveniences of modern living depend on a ready supply of energy.

Coal, an important source of energy, is composed of the remains of prehistoric plants (see Focus On Health and Human Affairs: Ancient Plants and Coal Formation). When coal is burned, the organic molecules formed hundreds of millions of years ago by photosynthesis are broken down, and heat is released. Coal is used primarily by utility companies to produce electricity and, to a lesser extent, by heavy industries such as steelmaking.

Coal is usually found in underground layers, called seams, of varying thickness. If the coal seam is located fairly close to the surface (within 30 m, or 100 ft), it is usually extracted by a method called strip mining. This process involves using bulldozers, giant power shovels, and wheel excavators to remove the vegetation, soil, and rock covering the coal seam. The coal is then scraped out of the ground and loaded onto railroad cars or trucks. Approximately 60% of the coal mined in the United States is obtained by strip mining.[1]

Strip mining has disastrous effects on the local environment. Existing vegetation and topsoil are completely removed, destroying the habitat for plants and animals and causing soil erosion and water pollution. Before the 1977 Surface Mining Control and Reclamation Act was passed, old strip mines were usually abandoned, leaving large open pits. The removal of topsoil and the leakage of acids and toxic minerals from such mines prevented most plants from recolonizing the land. Some older sites are so severely damaged from topsoil removal and pollution that few plants grow there even though it has been years since mining operations ceased.

Land that has been damaged by strip mining can be restored to prevent further degradation and to make the land productive for other purposes. The 1977 law orders coal companies to restore areas that were strip mined from 1977 to the present. Strip-mined land that was badly damaged prior to 1977 is gradually being restored as well, using money from a tax on currently mined coal.

In this chapter, we examine the various groups of plants, including those related to the ancient plants that formed our present-day coal deposits.

[1] The remaining coal is buried too deeply in the ground to be obtained by strip mining. This coal is acquired from underground mines.

Plant Life

23

Learning Objectives

After you have studied this chapter you should be able to:

1. Discuss some of the environmental challenges of living on land, and relate several adaptations that plants possess to meet these challenges.
2. Name the protistan group from which plants are thought to have descended, and describe some of the evidence that supports this theory.
3. Diagram a generalized plant life cycle, clearly showing alternation of generations.
4. Summarize the features that distinguish bryophytes from green algae and from other plants.
5. Discuss the features that distinguish the ferns and fern allies from other plants.
6. Compare seeds with spores and discuss the advantages of plants that reproduce primarily by seeds rather than spores.
7. Summarize the features that distinguish gymnosperms from other plants.
8. Discuss the evolutionary adaptations of the flowering plants.
9. Contrast the two classes of flowering plants—the monocots and dicots.

Key Concepts

- ☐ Plants are complex multicellular organisms that are photosynthetic autotrophs; they obtain their energy directly from the sun. With few exceptions, virtually all living things depend on plants for energy.
- ☐ Plants possess a number of adaptations that help them survive on land. These adaptations include a waxy covering over aerial parts, stomata for gas exchange, vascular tissues, and specialized structures and methods of reproduction.
- ☐ Bryophytes and seedless vascular plants reproduce by forming spores, whereas gymnosperms and flowering plants form seeds. Each seed contains a well-developed plant embryo and a food supply. Reproduction by seeds is a significant evolutionary adaptation that helps explain why gymnosperms and flowering plants dominate Earth today.

COMPLEX PHOTOSYNTHETIC ORGANISMS ARE PLACED IN KINGDOM PLANTAE

The plant kingdom comprises hundreds of thousands of different species that live in every conceivable environment, from the frozen Arctic tundra to lush tropical rain forests to harsh deserts. Plants range in size from minute, almost microscopic duckweeds to massive giant sequoias. What are some of the features of plants that have permitted them to colonize so many different environments?

Plants Use Photosynthesis to Obtain Energy

One of the unifying characteristics of almost all plants is their mode of nutrition—photosynthesis (Chapter 9). Plants are photosynthetic autotrophs. They absorb radiant energy, which is then converted to the chemical energy found in organic molecules such as carbohydrates. In addition to the photosynthetic pigments chlorophyll *a* and chlorophyll *b*, all plants have accessory photosynthetic pigments, the yellow and orange **carotenoids.** Although the ability to photosynthesize is a key feature of plants, it is not a distinguishing trait because other kinds of organisms, the algae and certain bacteria, also photosynthesize (Chapter 21). Plants, however, possess other traits that help distinguish them from algae and photosynthetic bacteria.

CONCEPT CONNECTIONS

Plants — Vertebrates — Terrestrial Life Style

Life began in the oceans, but many life forms have since adapted to terrestrial life in a sea of air. Every single organism living on land has to meet the same environmental challenges: obtaining enough water; preventing excessive water loss; getting enough energy; and, in temperate and polar regions, tolerating widely varying temperature extremes. How those challenges are met varies from one organism to another, and in large part, it explains the diversity of life encountered on land today.

For example, consider the very different ways that plants and vertebrates (animals with backbones) meet the challenge of obtaining enough water. Terrestrial animals are motile and walk, slither, fly, run, or crawl to water sources. This requires not only the ability to move (skeletal and muscular systems) but also the ability to sense the water's presence (nervous system). Plants adapted in a much different way to this challenge: they have roots that not only anchor the plant in the soil but absorb water and essential dissolved minerals.

Parts VII (Plant Structure and Life Processes) and VIII (Animal Structure and Life Processes) consider the adaptations of these organisms further.

The Ancestors of Plants Were Probably Green Algae

Although plants exhibit a remarkable diversity in size, habit, and form, they are thought to have descended from a common protistan ancestor, an ancient green alga. Because of their common ancestry, the green algae living today share a number of traits with plants. Both contain the same photosynthetic pigments, chlorophylls *a* and *b* and carotenoids, and both store carbohydrates as starch. Cellulose is a major component of the cell walls of both, and certain details of cell division are shared by plants and many green algae but are found in no other photosynthetic organism. All of these shared characteristics point to the likelihood that modern plants descended from an ancient green alga.

Plants Possess Adaptations that Help Conserve Water and Permit the Exchange of Gases

One of the most important adaptations that enable plants to survive on land is a waxy covering, or **cuticle,** over their aerial parts. The cuticle is essential for a terrestrial existence because it helps prevent the drying out of plant tissues by evaporation. Plants are rooted in the ground and, unlike animals, cannot move to wetter areas during dry spells.

Land plants obtain their carbon from the atmosphere as carbon dioxide, which must be accessible to the chloroplasts inside green plant cells. However, gas exchange through the waxy cuticle covering the external surfaces of stems and leaves is negligible. Plants possess tiny openings, or **stomata** (singular, *stoma*), in the surface tissue of stems and leaves that help overcome this difficulty. Stomata permit the gas exchange that is essential for photosynthesis.

The Plant Life Cycle Alternates between Two Different Generations

Plants have a clearly defined **alternation of generations** in which they spend part of their lives in a haploid stage and part in a diploid stage (Fig. 23–1). The haploid portion of the life cycle is called the **gametophyte generation** because it gives rise to gametes (reproductive cells) by mitosis. The diploid portion of the life cycle is called the **sporophyte generation** because it produces spores immediately following meiosis.

The gametophyte plant produces multicellular sex organs, or **gametangia** (singular, *gametangium*), each of which possesses a sterile layer of cells that surrounds and protects the delicate gametes. Each female gametangium

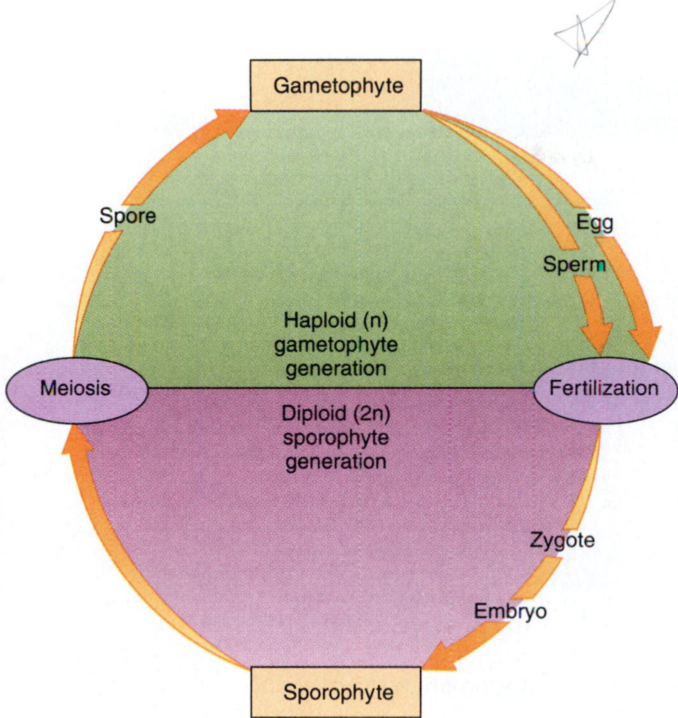

FIGURE 23–1 The basic plant life cycle. All plants have modifications of this cycle. Note that plants alternate generations, spending part of their life in a haploid gametophyte stage and part in a diploid sporophyte stage.

produces a single egg, whereas numerous sperm are produced in the male gametangium.

The sperm reach the female gametangium in a variety of ways, and one sperm fertilizes the egg. This process, known as **fertilization,** results in a fertilized egg, or **zygote.** The diploid zygote is the first cell in the sporophyte generation. The zygote divides by mitosis and develops into a multicellular embryo. Embryo development takes place *within* the female gametangium; thus, during its development the embryo is protected.

Eventually, the embryo matures into the sporophyte plant. The sporophyte plant has special cells that divide by meiosis to form haploid **spores**. The spores represent the first stage in the gametophyte generation. Each spore is capable of growing by mitosis into a multicellular gametophyte plant, and the cycle continues.

Most Plants Have Specialized Tissues for Internal Transport

There are four major groups of plants living today: bryophytes, seedless vascular plants, gymnosperms, and flowering plants (Table 23–1). The mosses and other bryophytes are small plants that lack a vascular, or conducting, system. The other three groups of plants possess vascular tissues, **xylem** for water and mineral conduction, and **phloem** for conduction of dissolved food. Ferns and their allies are seedless vascular plants that reproduce by spores. The gymnosperms and flowering plants are vascular plants that reproduce by forming seeds. Gymnosperms produce seeds borne naked on a stem or in a cone, whereas flowering plants produce seeds enclosed within a fruit.

THE MOSSES AND OTHER BRYOPHYTES ARE NONVASCULAR PLANTS

The **bryophytes,** which comprise over 15,000 species of mosses, liverworts, and hornworts, are the only nonvascular plants (Fig. 23–2).[2] Because they are nonvascular and have no means for extensive internal transport of wa-

[2] Some mosses do have water-conducting cells and food-conducting cells, although they are not as specialized or as effective as are the conducting cells in the vascular plants.

(a) *(b)*

(c)

FIGURE 23–2 Representative bryophytes. (***a***) A closeup of moss gametophytes. Mosses grow in dense clusters. (***b***) The gametophyte of many liverworts is characterized by flattened, ribbonlike lobes. (***c***) A typical hornwort gametophyte. The "horns" projecting out of the flattened lobes are the young sporophytes. (*a*, James Mauseth, University of Texas; *b*, Dennis Drenner; *c*, Ken Davis/Tom Stack & Associates)

TABLE 23–1
The Plant Kingdom

Nonvascular plants

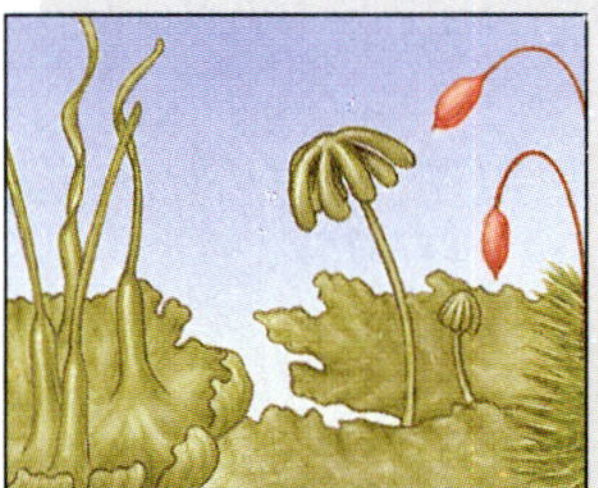

I. Nonvascular plants with a dominant gametophyte generation
- Division Bryophyta (mosses)
- Division Hepatophyta (liverworts)
- Division Anthocerophyta (hornworts)

Vascular plants

II. Vascular plants with a dominant sporophyte generation
- A. Seedless plants
  - Division Pterophyta (ferns)
  - Division Psilotophyta (whisk ferns)
  - Division Sphenophyta (horsetails)
  - Division Lycophyta (club mosses)

- B. Seed plants
  - 1. Plants with naked seeds
    - Division Coniferophyta (conifers)
    - Division Cycadophyta (cycads)
    - Division Ginkgophyta (ginkgo)
    - Division Gnetophyta (gnetophytes)

  - 2. Seeds enclosed within a fruit
    - Division Anthophyta (flowering plants)
      - Class Dicotyledones (dicots)
      - Class Monocotyledones (monocots)

ter, essential minerals, and food, bryophytes are typically quite small. They generally require a moist environment for active growth and reproduction. We confine the following discussion to the mosses.

Mosses (Division Bryophyta) usually live in dense colonies or beds. Each individual plant has tiny rootlike structures, or **rhizoids,** that anchor it to the soil. Each plant also has an upright stemlike structure that bears leaflike blades. (Because mosses lack specialized vascular tissues, they do not possess true roots, stems, or leaves.)

Mosses make up an inconspicuous but significant part of their environment. They play an important role in forming soil (Chapter 46). Because they grow tightly packed together in dense colonies, mosses hold the soil in place and help prevent erosion. They provide food for animals, especially birds and mammals.

Commercially, the most important mosses are the peat mosses in the genus *Sphagnum.* Peat mosses are particularly beneficial as a soil conditioner. For example, when added to sandy soils, they help to hold and retain moisture in the soil. In some countries, layers of dead peat moss that have accumulated for hundreds of years are extracted from bogs, dried, and burned for fuel.

Although all plants apparently descended from green algal ancestors, the mosses are not in a direct evolutionary path to the vascular plants. That is, vascular plants did not have mosslike ancestors. Mosses may represent an evolutionary sideline that evolved from ancestral green algae, or alternatively, mosses may have evolved from vascular plants (by becoming simpler and losing their vascular tissues). The fossil record of ancient mosses does not provide a definitive answer on moss evolution because it can be interpreted in different ways.

Bryophytes Are the Only Plants with a Dominant Gametophyte Generation

An alternation of generations is clear in the life cycle of mosses (Fig. 23–3). The leafy green moss gametophyte bears its gametangia at the top of the plant. Many moss species have separate sexes—male plants that bear male gametangia and female plants that bear female gametangia. Other moss species produce male and female gametangia on the same plant.

In order for fertilization to occur, one of the sperm must fertilize the egg within the female gametangium. The sperm, which are flagellated, are transported from male to female gametangia by flowing water (during rain, and so on). Once in a film of water on the female moss, a sperm swims down into the female gametangium and fuses with the egg.

The diploid zygote formed as a result of fertilization grows into a multicellular embryo by mitosis and matures into a moss sporophyte. This sporophyte plant grows out of the top of the leafy green female gametophyte. The sporophyte remains attached and nutritionally dependent on the gametophyte. Initially green in color and photosynthetic, the sporophyte becomes a golden brown at maturity. It is composed of three main parts: a **foot,** which anchors the sporophyte to the gametophyte; a **seta,** or stalk; and a **capsule,** in which meiosis occurs to form haploid spores.

When the spores are mature, the capsule opens, releasing the spores. These microscopic cells are dispersed by wind or rain. If a spore lands in a suitable spot, it germinates and grows into a filamentous thread that forms buds. Each bud grows into a leafy green gametophyte plant, and the life cycle continues.

The haploid gametophyte generation is considered the dominant generation in mosses because it is capable of living independently of the diploid sporophyte. In contrast, the sporophyte generation in mosses is at all times attached to and dependent on the gametophyte plant.

SEEDLESS VASCULAR PLANTS INCLUDE THE FERNS AND THEIR ALLIES

The **ferns** are an ancient group of plants dating back nearly 400 million years. With over 11,000 living species, they are still successful today. Ferns are especially common in temperate woodlands and tropical rain forests (Fig. 23–4). Three groups of vascular plants—the whisk ferns (several species), club mosses (about 1000 species), and horsetails (15 species)—are considered allies of the ferns because they share similarities in their life cycles.

The main advancement exhibited by the ferns and their allies over mosses and other bryophytes is the presence of specialized vascular tissues—xylem and phloem—for conduction. This system of conduction enables vascular plants to achieve larger sizes than do mosses because water, dissolved minerals, and food can be transported over greater distances to all parts of the plant. Although ferns in temperate environments are relatively small plants, tree ferns in the tropics may grow to heights of 18 meters (60 ft). The ferns and their allies all have true stems with vascular tissues, and most also have true roots and leaves.

Seedless Vascular Plants Have a Dominant Sporophyte Generation and Need Water as a Transport Medium for Fertilization

The life cycle of ferns (Division Pterophyta) involves a clearly defined alternation of generations (Fig. 23–5). A fern (such as the Boston fern or maidenhair fern) that is grown as a house plant represents the sporophyte generation. The fern sporophyte is composed of a horizontal underground stem, or **rhizome,** that bears roots and leaves, called **fronds.** Roots, rhizome, and fronds all contain vascular tissues.

Spore production usually occurs on certain areas on the fronds, which develop **sporangia** (singular, *sporangium*), or spore cases, in which meiosis occurs to form spores. The sporangia are frequently borne in clusters, called **sori** (singular, *sorus*), on the fronds. When the spores are disseminated and land in a suitable place, they may germinate and grow by mitosis into gametophytes.

The mature fern gametophyte is a tiny, leaflike, often heart-shaped structure that grows flat against the ground. This stage has tiny rootlike rhizoids that anchor it, but it lacks vascular tissues. Usually the gametophyte produces both male and female gametangia on its underside. Each female gametangium contains a single egg, whereas numerous sperm are produced in each male gametangium.

(Text continues on page 427)

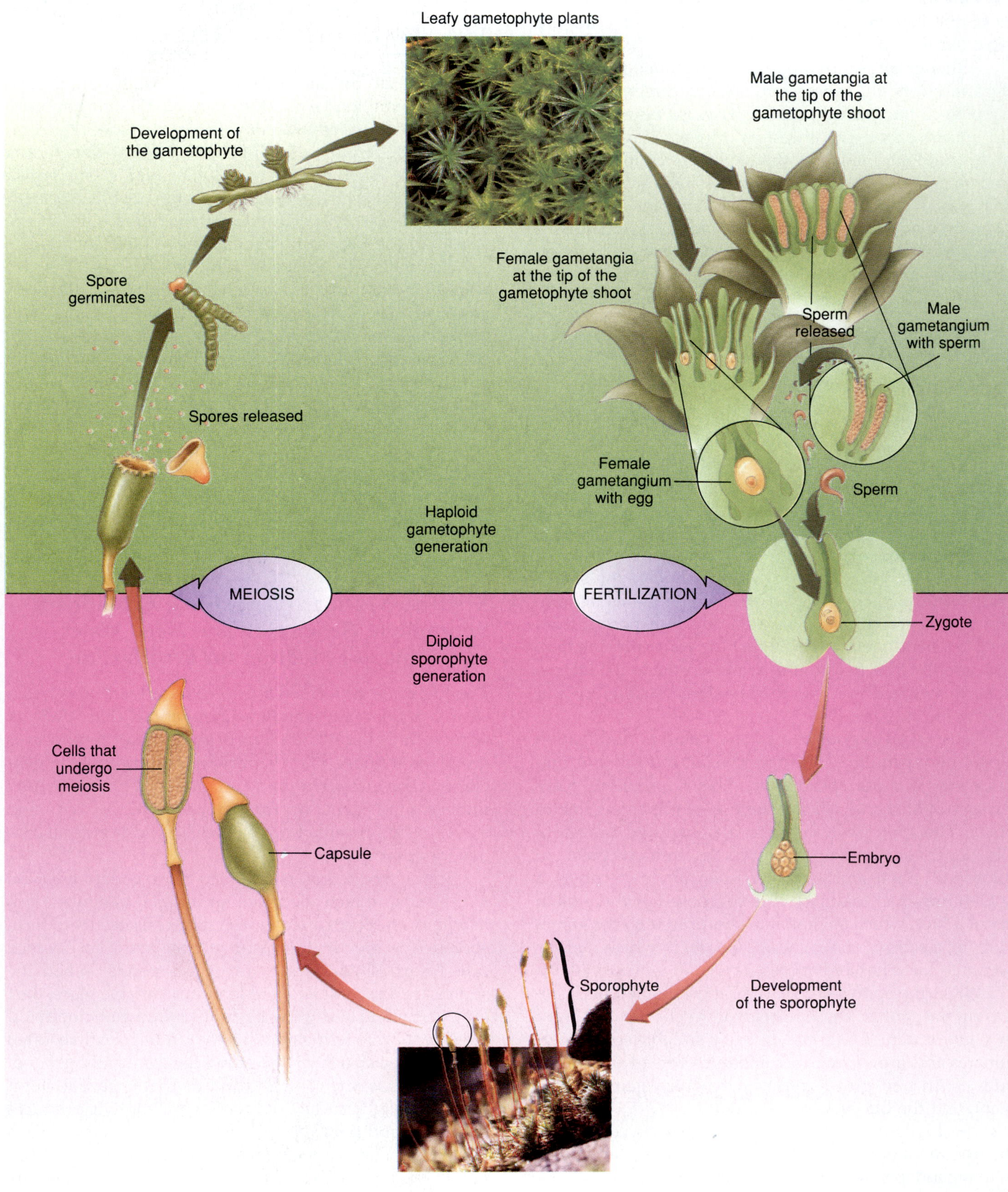
Leafy gametophyte plants
Development of the gametophyte
Male gametangia at the tip of the gametophyte shoot
Female gametangia at the tip of the gametophyte shoot
Spore germinates
Sperm released
Male gametangium with sperm
Spores released
Female gametangium with egg
Sperm
Haploid gametophyte generation
MEIOSIS
FERTILIZATION
Zygote
Diploid sporophyte generation
Cells that undergo meiosis
Capsule
Embryo
Sporophyte
Development of the sporophyte

◀ **FIGURE 23–3 Alternation of generations in the mosses.** The dominant generation is the gametophyte, represented by the leafy green plants. The sporophyte grows out of the top of the gametophyte. Mosses require water as a transport medium during fertilization. (1, Rod Planck/Dembinsky Photo Associates; 2, David Cavagnaro)

(a)

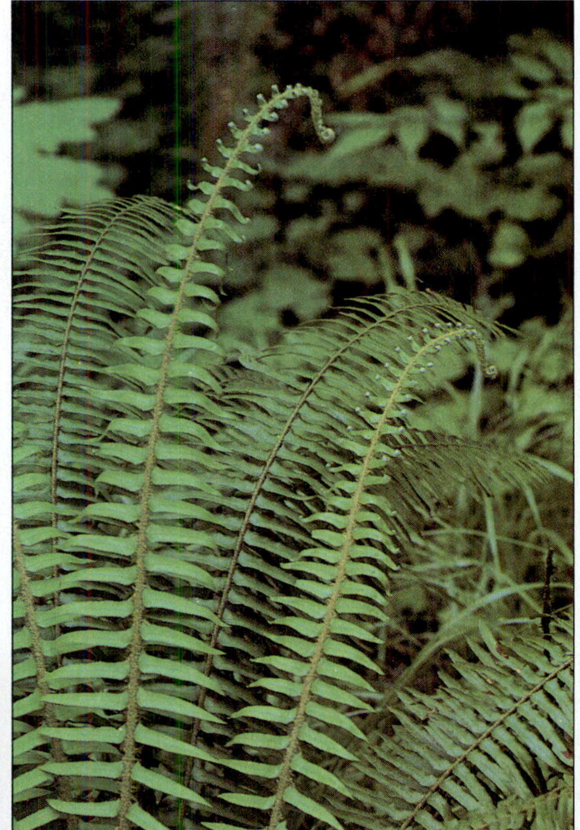

(b)

(c)

FIGURE 23–4 Representative ferns. (***a***) The sword fern *(Polystichum munitum)* is a common fern in Pacific coastal redwood forests. (***b***) The staghorn fern *(Platycerium)* is epiphytic and grows attached to the bark of rainforest trees rather than being rooted in the soil. (***c***) The Tasmanian tree fern, native to tropical rain forests, is found in New Zealand, South Africa, and South America. (*a,* Ed Reschke/Peter Arnold, Inc.; *b,* Richard H. Gross; *c,* Dennis Drenner)

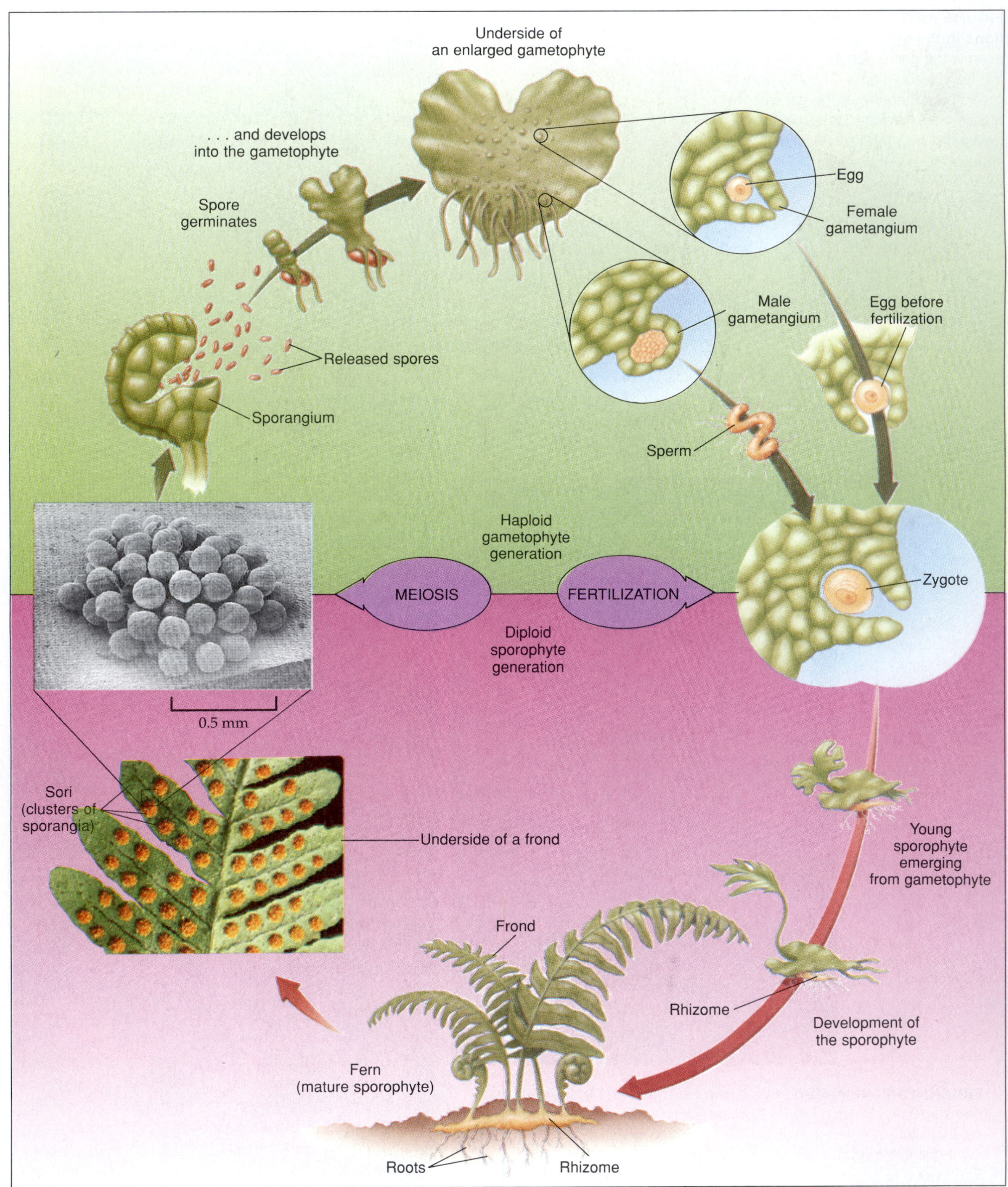

FIGURE 23–5 The fern life cycle. Note the clearly defined alternation of generations between the haploid gametophyte and diploid sporophyte stages. (1, Biophoto Associates/Photo Researchers, Inc., 2, David M. Dennis/Tom Stack and Associates)

(a)

(b)

FIGURE 23–6 Representative fern allies. (***a***) The growth habit of *Psilotum nudum.* The stem is the main organ of photosynthesis; leaves are absent. This ancient-appearing vascular plant bears sporangia directly on the stems. (***b***) *Equisetum telematia,* a horsetail, is widely distributed in Eurasia, Africa, and North America. It has unbranched, nonphotosynthetic fertile shoots bearing conelike structures and separate, highly branched, photosynthetic sterile shoots. Both types of shoots arise from an underground rhizome. (***c***) *Lycopodium,* a club moss. In spite of their name, club mosses are actually fern allies. The sporophyte has reduced, scalelike leaves that are evergreen. Spores are produced in sporangia clustered in a conelike structure. (*a,* James Mauseth, University of Texas; *b,* David Cavagnaro; *c,* Dwight Kuhn)

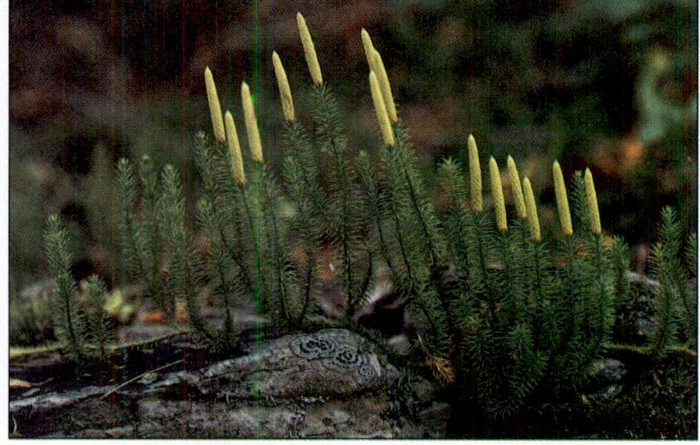

(c)

Although ferns are considered more advanced than mosses because they possess vascular tissues, they have retained a primitive fertilization technique—use of water as transport medium. A thin film of water on the ground underneath the mature gametophyte provides the transport medium in which the flagellated sperm swim to the female gametangium. After one of the sperm fertilizes the egg, a diploid zygote grows by mitosis into a multicellular embryo. At this stage in its life, the sporophyte embryo is attached to and dependent upon the gametophyte, but as the embryo matures into a sporophyte plant, the gametophyte withers and dies.

The fern life cycle has a clearly defined alternation of generations between the sporophyte with its roots, rhizome, and fronds, and the tiny, heart-shaped gametophyte. The sporophyte generation is dominant not only because it is larger than the gametophyte but because it persists for an extended period of time, whereas the gametophyte dies soon after reproducing.

Whisk Ferns, Horsetails, and Club Mosses Are Fern Allies

Most species of **whisk ferns** (Division Psilotophyta) are extinct, and the few species that still exist are found mainly in the tropics and subtropics. *Psilotum,* a representative whisk fern, lacks roots and leaves but does have vascularized stems, both below ground (rhizomes) and above ground (Fig. 23–6*a*). Tiny, round sporangia are borne on the erect, aerial stems, which also are the organs of photosynthesis. Although whisk ferns do not closely resemble the ferns, they are considered fern allies because of similarities in their life cycles.

Millions of years ago, the **horsetails** (Division Sphenophyta) were among the dominant plants and grew to be the size of trees. These ancient horsetails are still significant to us today because they contributed significantly to the Earth's vast coal deposits (see Focus On Health and Human Affairs: Ancient Plants and Coal Formation). The few surviving horsetails, all in the genus *Equisetum,* grow mostly in wet, marshy habitats and are small but very distinctive (Fig. 23–6*b*). Horsetails have roots, stems (both rhizomes and erect, aerial stems), and small leaves. Spores form inside sporangia that are borne on a terminal, conelike structure. The hollow, jointed

FOCUS ON *Health and Human Affairs*

ANCIENT PLANTS AND COAL FORMATION

Much of the coal we use today was formed from the prehistoric remains of primitive plants, particularly those of the Carboniferous Period, approximately 300 million years ago. Five main groups of plants contributed to coal formation. Three of them were seedless vascular plants—club mosses, horsetails, and ferns. The other two important groups of coal formers were seed plants—seed ferns (now extinct) and primitive gymnosperms.

It is hard to imagine that the small, relatively inconspicuous club mosses, ferns, and horsetails of today could have been so significant in forming the vast beds of coal. However, the now-extinct members of these groups that existed during the Carboniferous Period were giants by comparison, and they formed immense forests.

The climate during the Carboniferous Period was warm and mild. Plants in most locations could grow year-round because of the favorable conditions. The forests of these plants often occurred in low-lying, swampy areas that were periodically flooded when the sea level rose. When the sea level receded, these plants would become established again.

When these large plants died or were blown over during storms, they were incompletely decomposed because they were covered by swamp water. (The lack of oxygen in the water prevented wood-rotting fungi from decomposing the plants, and anaerobic bacteria do not decompose rapidly.) Thus, over time the partially decomposed plant material accumulated and consolidated.

Layers of sediment formed over the plant material each time the sea level rose and flooded the low-lying swamps. With time, heat and pressure built up in these accumulated layers and converted the plant material to coal and the sediment layers to sedimentary rock. Much later, geological upheavals raised the layers of coal and sedimentary rock. For example, coal is found in seams (layers) in the Appalachian Mountains.

stems are impregnated with silica, which gives them a gritty texture. In pioneer days, horsetails were called "scouring rushes" and were used to scrub out pots and pans along the stream banks.

Like horsetails, **club mosses** (Division Lycophyta) were important plants millions of years ago when species that are now extinct often attained great size. Like the ancient horsetails, these large plants were major contributors to the coal deposits. The club mosses today, such as *Lycopodium,* are small plants commonly found in woodlands (Fig. 23–6*c*). They possess roots; rhizomes and aerial stems; and small, scalelike leaves. Sporangia are borne in a conelike structure at the tips of stems or scattered along the stems. Club mosses are evergreen and are often fashioned into Christmas wreaths or other decorations. In some areas, they are endangered by overharvesting.

Heterospory Was a Significant Development in Plant Evolution Because It Was the Forerunner of the Evolution of Seeds

In the life cycles examined thus far, plants produce only one type of spore as a result of meiosis. This condition, known as **homospory,** is characteristic of bryophytes, horsetails, whisk ferns, and most ferns and club mosses. However, certain ferns and club mosses are **heterosporous.** As such, they produce two different types of spores—microspores and megaspores. **Microspores** eventually produce male gametophytes, whereas **megaspores** produce female gametophytes. Heterospory is found in the two most successful groups of plants, the gymnosperms and the flowering plants, both of which produce seeds.

THE PRODUCTION OF SEEDS REPRESENTS A MAJOR EVOLUTIONARY ADAPTATION

The primary means of reproduction and dispersal for the most successful plants is seeds, which develop from the female gametophyte and its associated tissues. The seed plants—gymnosperms and flowering plants—show the greatest evolutionary complexity in the plant kingdom and comprise the dominant plants in most terrestrial environments.

Seeds are reproductively superior to spores for three main reasons. First, seeds contain a multicellular, well-developed young plant with embryonic root, stem, and leaves already formed, whereas spores are composed of a single cell. Second, seeds contain a food supply. After

(a)

(b)

(c)

FIGURE 23–7 Conifers. (***a***) Spruce (shown) and other conifers are dominant plants in temperate and colder latitudes. (***b***) This female Torrey pine cone has not yet opened to shed its seeds. (***c***) Male Torrey pine cones produce large quantities of pollen in the spring. These cones have already shed their pollen and will soon abscise, or detach, from the branch. (Dennis Drenner)

germination, the plant embryo is nourished by food stored in the seed until it becomes self-sufficient. Because a spore is a single cell, few food reserves exist for the plant that develops from a spore. Third, seeds are protected by a resistant seed coat.

The two groups of seed plants are the **gymnosperms** and the **angiosperms.** The word *gymnosperm* is adapted from a Greek word meaning "naked seed." These plants produce seeds that are totally exposed or borne on the scales of cones. Pine, spruce, fir, and ginkgo are examples of gymnosperms. The Greek word from which the term *angiosperm* is derived translates as "seed enclosed in a vessel or case." The angiosperms are flowering plants that produce their seeds within a fruit. Angiosperms are very diverse and include such plants as corn, oaks, water lilies, cacti, apples, and buttercups.

Both gymnosperms and flowering plants possess vascular tissues, xylem for conduction of water and dissolved minerals and phloem for conduction of food. Both have a life cycle with an alternation of generations, but the gametophyte generation is significantly reduced in size and is entirely dependent on the sporophyte generation.

THE GYMNOSPERMS ARE THE "NAKED SEED" PLANTS

The gymnosperms include some of the most interesting plants in the plant kingdom. For example, a redwood is probably the world's tallest tree, measuring about 114 meters (380 feet) in height! Another gymnosperm, a giant sequoia known as the General Sherman tree, located in Sequoia National Park in California, is one of the world's most massive organisms (in terms of sheer bulk). Although it is only 82 meters (272 ft) tall, the General Sherman tree has a diameter of 11 meters (36 ft) and an estimated weight of 6167 tons! One of the oldest living trees is a bristlecone pine, in the White Mountains of California, that has been determined by tree ring analysis to be 4900 years old!

Numbering 550 species, the largest division of gymnosperms is the **conifers,** which are woody plants that bear their seeds in cones. Two other divisions of gymnosperms represent evolutionary remnants of groups that were more significant in the past—the cycads and the ginkgoes. The fourth division of gymnosperms, the gnetophytes, is a collection of some very unusual plants that share certain traits not found in other gymnosperms.

Conifers Are Woody Plants that Bear Their Seeds in Cones

The conifers (Division Coniferophyta), which include pines, spruces, hemlocks, and firs, are the most familiar group of gymnosperms (Fig. 23–7). They are woody trees or shrubs, most of which are evergreen. Only a few conifers, such as the larch and the bald cypress, shed their leaves, called **needles,** at the end of the growing season. Most conifers have separate male and female reproductive parts on the same plant. These reproductive parts are generally borne in **cones.**

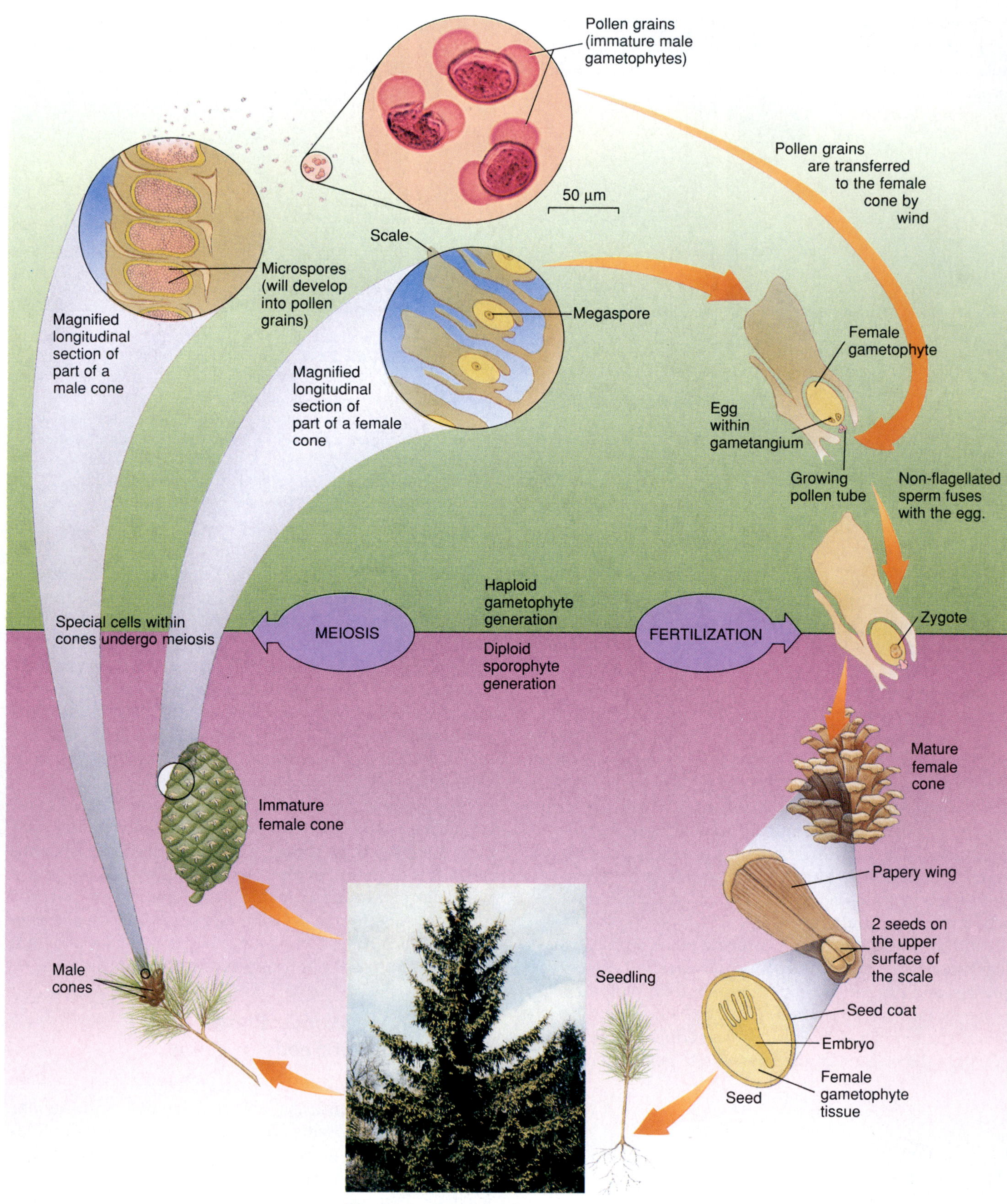

FIGURE 23–8 The life cycle of pine. One major evolutionary advancement of gymnosperms is their wind-borne pollen. Pines and other gymnosperms are not dependent on water as a transport medium for sperm. (1, Dennis Drenner; 2, Manfred Kage/Peter Arnold, Inc.)

Conifers occupy vast areas, ranging from the Arctic to the tropics, and are the dominant vegetation in the forested regions of Canada, Northern Europe, and Siberia. In addition, they are important in the southern hemisphere, particularly in areas of South America, Australia, and Malaysia. Ecologically, conifers contribute food and shelter to animals, and their roots hold the soil in place and help prevent erosion. Humans use conifers for lumber (for building materials as well as paper products), turpentine, and resins. Because of their attractive appearance, conifers are grown for landscape design and for Christmas trees.

A pine tree represents a typical conifer life cycle (Fig. 23–8). The tree is the sporophyte plant and produces microspores and megaspores in separate cones. Male cones are smaller than female cones and are generally produced on the lower branches in the spring. Meiosis occurs within the male cones to produce numerous haploid microspores. Each microspore develops into an extremely reduced male gametophyte; the immature male gametophyte is also called a **pollen grain.** The pollen grains are shed from the male cones in great numbers, and some are carried by wind currents to the immature female cones.

The familiar woody pine cones—the female cones—are usually found on the upper branches of the tree. The woody scales of the female cones bear sacs. Meiosis occurs inside each sac, producing four haploid megaspores. One of these develops into the female gametophyte, which produces an egg within each of several female gametangia. A pollen grain grows a tube that digests its way through the female gametophyte tissue to the egg. Then, a nonflagellated sperm fuses with the egg to form a zygote, which grows into the young pine embryo in the seed. The female gametophyte tissue surrounding the developing embryo becomes the nutritive tissue in the mature seed. The mature seed has a papery wing that enables it to be dispersed by wind currents.

There are two key points to remember about the pine life cycle. One, the sporophyte generation is dominant, and the gametophyte is decreased in size to microscopic structures in the male and female cones. Two, a major advancement in the pine life cycle is elimination of the need for water as a transport medium for the sperm. Instead, the pollen is carried to the female cones by air currents. Once in contact with the female cones, a pollen tube grows, and nonflagellated sperm move through it to the egg. Therefore, gymnosperms are the first plants whose reproduction is totally adapted for life on land.

Cycads, Ginkgoes, and Gnetophytes Are Conifer Allies

The **cycads** (Division Cycadophyta) were a very important plant group in the prehistoric past. Most species are extinct, and the few surviving cycads (only 100 species remain) are considered very primitive seed plants. This is because their seed structure is most like that of the earliest seeds found in the fossil record. Cycads are tropical and subtropical plants. They are slow-growing evergreens with a palmlike or fernlike appearance (Fig.

(a)

(b)

(c)

FIGURE 23–9 Representative conifer allies. (***a***) A cycad growing in South Africa. Cycads are tropical gymnosperms with a palmlike appearance. Note the immense seed cones on this plant. (***b***) *Ginkgo biloba,* the ginkgo or maidenhair tree. The unusual leaves of the ginkgo resemble the maidenhair fern, hence its common name. (***c***) *Gnetum,* a gnetophyte. Several features of *Gnetum* resemble those of flowering plants. (*a,* W.H. Hodge/Peter Arnold, Inc.; *b,* Biophoto Associates/Photo Researchers, Inc.; *c,* Walter Hodge/Peter Arnold, Inc.)

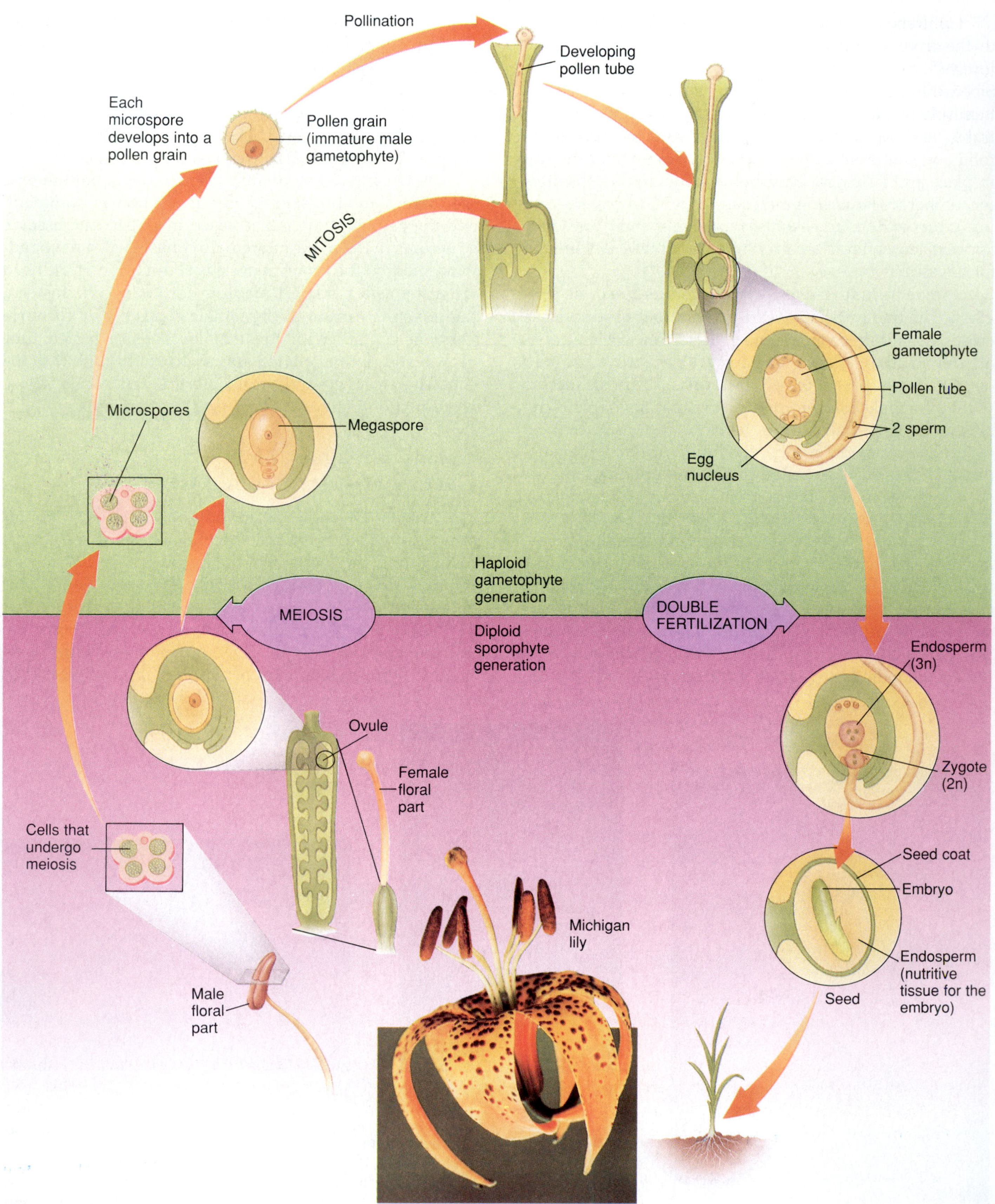

FIGURE 23–10 Generalized life cycle of a typical flowering plant. A significant feature of the flowering plant life cycle is double fertilization. (Visuals Unlimited/John Gerlach)

23–9*a*). Many are endangered species, primarily because they are popular as ornamentals (they are gathered from the wild and sold to collectors).

Ginkgoes (Division Ginkgophyta) are represented by a single species, the **ginkgo,** or maidenhair, tree (Fig. 23–9*b*). It is a native of southeastern China, where it is apparently extinct in the wild but has been under cultivation for centuries. The ginkgo represents the oldest genus of living trees. Fossil ginkgoes 200 million years old have been discovered that are nearly identical to the modern-day ginkgo. The ginkgo is commonly planted in North America today, particularly in cities, because it is hardy and somewhat resistant to air pollution. Male trees are typically planted because the female trees bear seeds that have a foul odor (they smell like rancid butter).

The **gnetophytes** (Division Gnetophyta), with about 70 species, are a remarkably diverse group of gymnosperms that share a number of features that make them clearly more advanced than the rest of the gymnosperms (Fig. 23–9*c*). Gnetophytes have more efficient water-conducting cells, called vessels, in their xylem. (Flowering plants have vessels in their xylem, but gymnosperms, with the exception of the gnetophytes, do not.) Also, the cones produced by some of the gnetophytes resemble some flower clusters.

THE FLOWERING PLANTS PRODUCE FLOWERS AND SEEDS WITHIN FRUITS

The flowering plants (Division Anthophyta), or angiosperms, are the most successful plants today, surpassing even the gymnosperms in importance. They have adapted to almost every habitat, except Antarctica, and, with about 235,000 living species, are Earth's dominant plants. The reproductive structures of angiosperms are flowers. The seeds formed following reproduction are enclosed within a fruit, which protects the developing seeds and often aids in their dispersal.

Like all plants, the life cycle of flowering plants consists of an alternation of generations, but their gametophyte is extremely reduced (Fig. 23–10). The fertilization process in flowering plants, called **double fertilization,** is unique in that *two* nonflagellated sperm are involved. One sperm fertilizes the egg, forming a zygote that develops into the embryo in the seed. The other sperm fuses with two cells in the female gametophyte to form a special nutritive tissue in the seed called **endosperm.** Details on reproduction in the flowering plant are found in Chapter 29.

Flowering plants are extremely important to humans, as our survival as a species literally depends on them. All of our major food crops are flowering plants, including the cereal crops such as rice, wheat, and corn. Flowering plants provide us with fibers like cotton and linen, and medicines like digitalis and codeine. Woody flowering plants, such as oak, cherry, and walnut, provide us with valuable lumber. Plant products as diverse as rubber, tobacco, coffee, and aromatic oils for perfumes come from flowering plants.

(*a*)

(*b*)

FIGURE 23–11 Monocots and dicots are the two classes of flowering plants. (***a***) Most monocots, such as this *Trillium,* have their floral parts in threes. Note the three green sepals, three rose-colored petals, six stamens, and three stigmas (receptive surfaces for pollen located in the center of the flower). (***b***) Most dicots, such as this *Tacitus,* have their floral parts in fours or fives. Note the five petals and ten stamens. (The other floral parts are not visible in this photograph.) (*a,* Don and Esther Phillips/Tom Stack & Associates; *b,* Richard H. Gross)

The flowering plants are divided into two classes—the monocots (class Monocotyledones) and dicots (class Dicotyledones) (Table 23–2). Monocots include palms, grasses, orchids, and lilies. Dicots include oaks, roses, cacti, and sunflowers.

The **monocots** are mostly herbaceous plants with leaves that are usually long and narrow and have parallel veins (the main leaf veins run parallel to one another). The flower parts of most monocot flowers occur in threes

TABLE 23–2
A Comparison of monocots and dicots

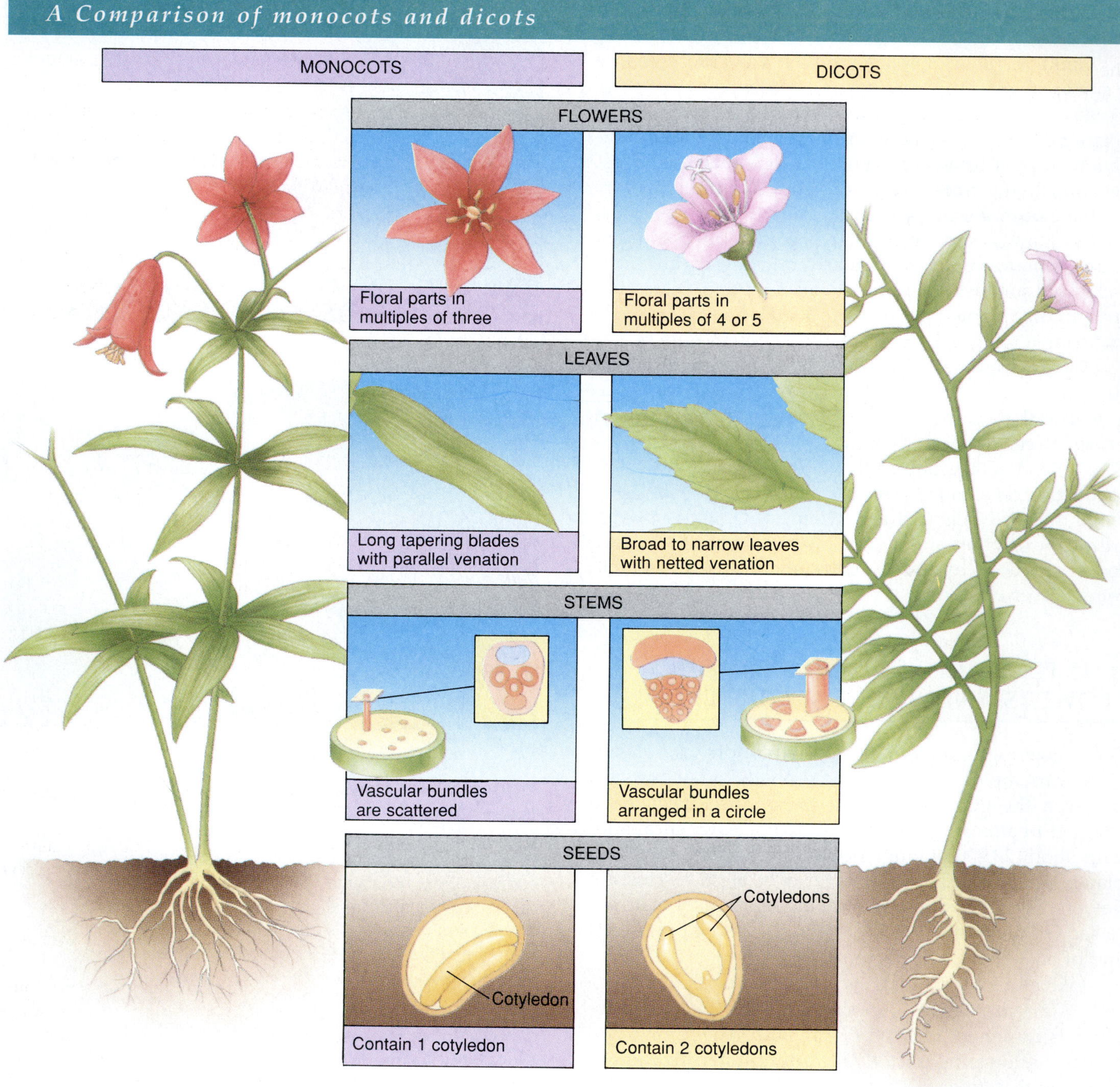

or multiples of three, as for example, three petals (flower parts that are usually conspicuously colored) and six stamens (flower parts that produce pollen; Fig. 23–11*a*). Monocot seeds have a single **cotyledon** (embryonic seed leaf), and endosperm is usually present in the mature seed.

MONOCOTS
DICOTS
FLOWERS
Floral parts in multiples of three
Floral parts in multiples of 4 or 5
LEAVES
Long tapering blades with parallel venation
Broad to narrow leaves with netted venation
STEMS
Vascular bundles are scattered
Vascular bundles arranged in a circle
SEEDS
Cotyledon
Contain 1 cotyledon
Cotyledons
Contain 2 cotyledons

Dicots may be herbaceous (for example, a tomato) or woody (for example, a hickory tree). Dicot leaves are variable in shape, but usually broader than monocot leaves, and have netted veins (branched veins that resemble a net). Flower parts usually occur in fours or fives or multiples of four or five (Fig. 23–11*b*). Two cotyledons are present in seeds of dicots, and endosperm is usually absent in the mature seed, having been absorbed by the two cotyledons prior to germination.

Flowering Plants Are the Most Successful Plant Group

The evolutionary adaptations of the flowering plants account for their success, in terms of both their ecological dominance and their large number of species. The flower, which attracts insects and other animals for pollen dispersal, thereby assuring cross-fertilization, is their main evolutionary advantage (Chapter 29).

Flowering plants possess a number of advanced features in addition to their highly successful reproduction involving flowers, seeds, and fruits. With few exceptions, flowering plants have vessels in the xylem and sieve tubes (food-conducting cells arranged end-on-end to form tubes) in their phloem, making these vascular tissues very efficient at conduction (Chapters 26, 28). The leaves of flowering plants, with their broad expanded blades, are very efficient in absorbing light for photosynthesis (Chapter 27). Shedding of these leaves during cold or dry spells is also an advantage that has enabled some flowering plants to expand into habitats that would otherwise be too harsh for survival. Their roots are often modified for food or water storage (Chapter 28).

Also crucial to the success of flowering plants is the overall adaptability of the sporophyte generation. As a group, flowering plants readily adapt to new habitats and changing environments. This adaptability is evident in the large diversity exhibited by the group. For example, the cactus is remarkably well adapted for desert environments, and the water lily is well adapted for wet environments.

Chapter Summary

I. Plants are complex multicellular organisms that obtain energy by photosynthesis.
   A. Plants probably evolved from green algal ancestors.
   B. Certain adaptations enabled plants to successfully colonize land.
      1. Plants possess a waxy cuticle that protects against water loss and stomata for gas exchange needed for photosynthesis.
      2. Vascular plants possess xylem to conduct water and dissolved minerals and phloem to conduct dissolved food.
   C. Plant life cycles have an alternation of generations.
      1. Plants spend part of their life cycle in the haploid gametophyte stage and part in the diploid sporophyte stage.
      2. An evolutionary trend in plants is toward a larger, more dominant sporophyte and a less dominant gametophyte.

II. Bryophytes are the only plants with a dominant gametophyte generation.
   A. Mosses and other bryophytes have several advancements over green algae, including a cuticle, stomata, and multicellular gametangia.
   B. Mosses are nonvascular plants (they do not possess xylem or phloem) with a dominant gametophyte generation.

III. Ferns and fern allies are seedless vascular plants.
   A. Seedless vascular plants possess vascular tissues and a dominant sporophyte generation.
   B. Whisk ferns, horsetails, and club mosses are fern allies.

IV. Seeds represent an evolutionary advancement over spores.
   A. Heterospory (the production of two kinds of spores, microspores and megaspores) was the evolutionary forerunner of seeds.
   B. Each seed contains a well-developed plant embryo and a food supply.
   C. Gymnosperms and flowering plants reproduce by seeds.

V. The gymnosperms are vascular plants with naked seeds (that is, seeds not enclosed within a fruit).
   A. Gymnosperms produce wind-borne pollen.
   B. There are four divisions of gymnosperms.
      1. The conifers are the largest group of gymnosperms. They are woody plants, usually evergreen, and produce their seeds in cones.
      2. Cycads are palmlike or fernlike in appearance, but they reproduce in a manner similar to pines. There are relatively few surviving members of this once large division.
      3. The ginkgo, a hardy gymnosperm tree, is the only living species in its division.
      4. The gnetophytes share a number of advancements over the rest of the gymnosperms, including vessels in their xylem.

VI. The flowering plants are vascular plants that produce seeds enclosed within a fruit. They are the most diverse and most successful group.
   A. The flower is their organ of sexual reproduction. Double fertilization, which results in the formation of a zygote and endosperm tissue, is characteristic of the flowering plants.
   B. There are two classes of flowering plants.
      1. Most monocots have floral parts in multiples of three, and their seeds contain one cotyledon. The nutritive tissue in their mature seeds is endosperm.
      2. The dicots usually have floral parts in multiples of four or five, and their seeds contain two cotyledons. The nutritive tissue in their mature seeds is usually in the cotyledons, which have absorbed the nutrients in the endosperm.

Selected Key Terms

alternation of generations, p. 420
angiosperm, p. 429
bryophyte, p. 421
cuticle, p. 420
double fertilization, p. 433
endosperm, p. 433
gametangium, p. 420
gametophyte, p. 420
gymnosperm, p. 429
heterospory, p. 428
homospory, p. 428
megaspore, p. 428
microspore, p. 428
phloem, p. 421
pollen grain, p. 431
sorus, p. 423
sporophyte, p. 420
stoma, p. 420
xylem, p. 421
zygote, p. 421

Post-Test

1. The waxy layer that covers the aerial parts of plants is the ______________.
2. Plants probably evolved from an ancient ______________ ______________.
3. The openings in plants that allow gas exchange for photosynthesis are called ______________.
4. Plants have ______________ of ______________ in which they spend part of their life in the haploid stage and part in the diploid stage.
5. ______________ is required as a transport medium in order for fertilization to occur in mosses.
6. The ______________ generation is the dominant generation in mosses.
7. Clusters of sporangia, termed ______________, are often found on fern fronds.
8. ______________ have hollow, jointed stems that are impregnated with silica.
9. Ferns and fern allies possess ______________ for conducting water and dissolved minerals and ______________ for conducting dissolved food.
10. ______________ are better than spores for reproduction because they contain a well-developed embryo and food tissue.
11. Although conifers bear their seeds in cones, they are considered "naked seed" plants because their seeds are not enclosed in a/an ______________.
12. Conifers, ______________, ______________, and ______________ are gymnosperms.
13. The reproductive structures of flowering plants are ______________.
14. This class of flowering plants, the ______________, includes the palms, grasses, and orchids.
15. The ______________ is a nutritive tissue in the seed that formed as a result of double fertilization.
16. What is the main advantage of the waxy cuticle on the outer surface of aerial plant parts? (a) allows for gas exchange (b) conducts water and dissolved minerals (c) houses and protects gametes (d) retards water loss
17. What do bryophytes possess that the algae lack? (a) vascular tissues (b) double fertilization (c) cuticle (d) seeds (e) chlorophyll
18. What do ferns possess that bryophytes lack? (a) vascular tissues (b) double fertilization (c) cuticle (d) seeds (e) chlorophyll
19. What do gymnosperms possess that ferns lack? (a) vascular tissues (b) double fertilization (c) cuticle (d) seeds (e) chlorophyll
20. What do flowering plants possess that gymnosperms lack? (a) vascular tissues (b) double fertilization (c) cuticle (d) seeds (e) chlorophyll

Review Questions

1. What are the most important environmental challenges that plants face when living on land? What adaptations do plants possess to meet these challenges?
2. Compare alternation of generations in the mosses and ferns. Which stage is dominant in each group?
3. Name the three groups of plants known as fern allies. How are these plants similar to ferns?
4. How does heterospory modify the life cycle?

5. Why are seeds such a significant evolutionary development?
6. List several ways that the conifers are advanced over the ferns.
7. Name the four groups of gymnosperms.
8. How are flowering plants different from gymnosperms?
9. What are the two classes of flowering plants, and how can one distinguish between them?
10. Label the diagram to the right.

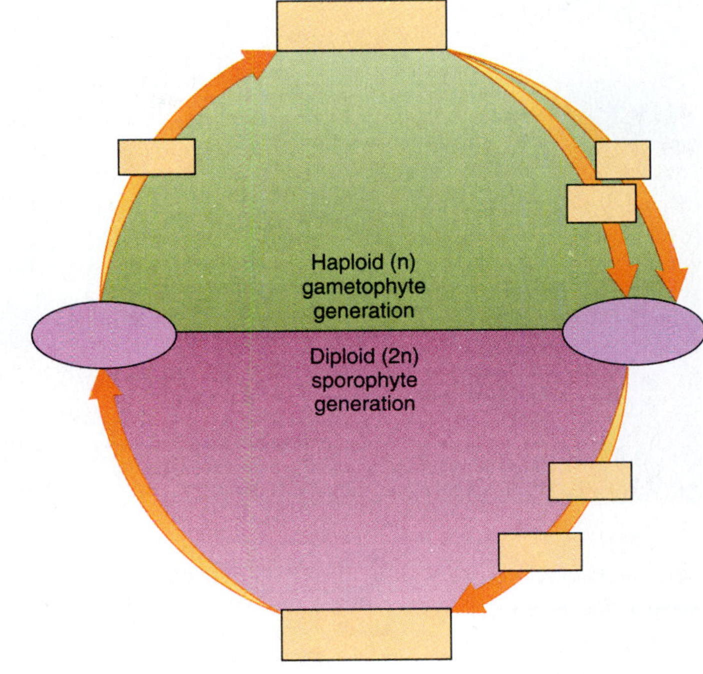

Thinking Critically

1. A company was hired to establish plants on formerly strip-mined land. They planted tree seedlings three times, but each time the young plants died. They decided to hire a fungal biologist as a consultant before replanting the area. Why? (This question ties in with material that you learned in Chapter 22.)
2. Which group would probably have colonized the land first—plants or animals? Explain.
3. How might the following trends in plant evolution be adaptive to living on land?
   a. Dependence on water for fertilization ⟶ no need for water as a transport medium
   b. Dominant gametophyte generation ⟶ dominant sporophyte generation
   c. Homospory ⟶ heterospory

Recommended Readings

Gower, S.T., and J.H. Richards, "Larches: Deciduous Conifers in an Evergreen World," *Bioscience,* Vol. 40, No. 11, December 1990. The biological characteristics and ecological significance of one of the few conifers that is not evergreen.

Heywood, V.H., *Flowering Plants of the World,* New York, Oxford University Press, 1993. This beautifully illustrated guide describes more than 300 families of angiosperms, including their economic uses.

Wolf, T.H., "The Object at Hand," *Smithsonian,* September 1990. This regular feature highlights the dawn redwood, a gymnosperm that was thought to have been extinct for millions of years, but was found living in China.

The coral reef is home for many marine animals. *These sea gold fish* (Anthias squamipinnis) *were photographed along a coral reef in the red sea.* (*Animals Animals © Laurence Gould, Oxford Scientific*

The Decline of Coral Reefs: Coral Bleaching and Other Threats

Strikingly beautiful and among the most productive of all marine ecosystems, coral reefs rival the tropical rain forests in species diversity. A single reef can serve as home for more than 3000 species of fishes and other marine organisms. An estimated one-fourth of all marine species depend on reefs. Coral reefs also form and maintain the foundation of thousands of islands, and by providing a barrier against waves, they protect shorelines against storms and erosion.

But, as important as coral reef ecosystems are, they are also very fragile. Studies indicate that of the 109 countries whose shores are lined with large reef formations, more than 90 have suffered serious reef damage. Some marine biologists predict that as a result of human activity, most coral reefs will disappear within about 50 years. Humans mine coral for building materials, pollute reef waters with industrial chemicals, and smother coral with the silt that washes downstream from clearcut forests. The once beautiful reefs off the island of Oahu, Hawaii, have been largely destroyed by silt produced by beachfront development. Tourists invited to view the reefs find themselves looking at artificial reefs constructed from old tires. The phenomenon of coral bleaching may also be related to human activity. To understand this particular problem, we must look at the process of reef formation.

Coral reefs are made up of colonies of millions of individual corals, which are tiny animals related to jellyfish, and by certain algae (mainly coralline red algae). Coral animals also live in association with photosynthetic algae (called zooxanthellae). Both types of algae contribute to the reef's brilliant colors. The relationship with the zooxanthellae is symbiotic and mutually beneficial. The algae, which live within cells lining the coral's digestive cavity, provide the coral with oxygen and with carbon and nitrogen compounds. In exchange, the coral supplies the algae with waste products such as ammonia, from which the algae make nitrogen compounds for both partners.

Coral bleaching, the whitening of coral, occurs when the corals lose their colorful symbiotic algae. Without their algae, coral become malnourished and die. Although biologists have known about coral bleaching for more than 75 years, bleaching has recently become more widespread.

Although coral bleaching is not well understood, environmental factors that are suspect include pollution, changes in salinity, disease, increased ultraviolet radiation (associated with the destruction of the ozone layer), and unusually high or low temperatures. Many scientists think that one of the most important factors in recent years has been abnormally high water temperature possibly caused by global warming. Healthy coral thrives in a narrow temperature range. An increase of only one or two degrees above the normal summer maximum temperature can cause widespread coral mortality.

Coral bleaching appears to be a response to environmental stress, but some biologists suggest that bleaching may have a positive outcome. Bleaching may be a mechanism that allows a different algal partner to establish residence within the coral. The new partnership may be more resistant to environmental stress. Thus, low-level stress may have adaptive value for the coral and enhance its ability to survive.

Several international monitoring projects are now gathering data that are needed to help us understand coral destruction and take action to protect these beautiful and important ecosystems. Once destroyed, it is difficult to reclaim them due to the long amount of time needed to form new reefs. For example, the major reef builder in Florida and Caribbean waters, a species known as star coral (*Montastrea annularis*), requires about 100 years to form a reef just one meter in height.

Animal Life: Invertebrates

24

Learning Objectives

After you have studied this chapter you should be able to:

1. List the characteristics common to animals; using these characteristics, develop a definition of an animal.
2. Justify classification and proposed relationships of the animal phyla on the basis of (a) symmetry, (b) type of body cavity, and (c) pattern of development (for example, protostomes and deuterostomes).
3. Place a given animal in the appropriate phylum and class.
4. Contrast the animal phyla based on their distinguishing characteristics.
5. Describe the body plan and life-style of one member of each phylum.
6. Trace the life cycle of each parasitic worm described in the chapter, including tapeworm and *Ascaris*. Identify adaptations that contribute to the success of these parasites.
7. Identify factors contributing to the great biological success of the insects.

Key Concepts

- ☐ Animals are eukaryotic multicellular heterotrophs; their cells are specialized to carry on different functions. Most animals are capable of locomotion at some stage of their life cycle, can respond rapidly to stimuli, and can reproduce sexually.
- ☐ Animals can be classified according to type of body symmetry, type of body cavity, and pattern of early development.
- ☐ In sponges and other very simple, small animals, life processes such as gas exchange, circulation of materials, and waste disposal take place by diffusion. In large, complex animals, structures and mechanisms have evolved that are specialized to carry on life processes.
- ☐ Phylum Arthropoda, the largest animal phylum, includes the spiders, crustaceans, and insects. Arthropods have hard exoskeletons and jointed appendages.

ANIMALS ARE MULTICELLULAR HETEROTROPHS

More than one million species of animals have been described, and perhaps several million more remain to be discovered. Most members of the animal kingdom are classified in about 35 phyla. Ten of the animal phyla are described in Table 24–1. The animals most familiar to us—dogs, birds, fishes, frogs, snakes—are *vertebrates.* A vertebrate is an animal with a backbone. Vertebrates, however, account for only about 5% of the known species of the animal kingdom. The majority of animals are the less familiar *invertebrates,* which are animals without backbones. The invertebrates include such diverse forms as corals, earthworms, crustaceans, and spiders (Fig. 24–1).

So many diverse animals are known that exceptions can be found to almost any definition of an animal. Still, there are some characteristics that describe most animals:

1. All animals are multicellular eukaryotes.
2. The cells that make up the animal body are specialized to perform specific functions. In all but the simplest animals, cells are organized to form tissues, and tissues are organized to form organs. In most animal phyla, specialized organ systems carry on specific functions.
3. Animals are heterotrophs. Most are consumers that ingest their food first and then digest it inside the body, usually within a digestive system.
4. Most animals are capable of locomotion at some time during their life cycle. Some animals (the sponges, for example) move about as larvae but are sessile, which means they are firmly attached to a surface, as adults.
5. Most animals have well-developed sense organs and nervous systems and can respond rapidly to stimuli.
6. Most animals reproduce sexually, with large, nonmotile eggs and small, flagellated sperm. Sperm and egg unite to form a fertilized egg, or zygote, that goes through a series of embryonic stages before developing into a larva or immature form.

ANIMALS INHABIT MOST ENVIRONMENTS OF THE ECOSPHERE

Animals occupy virtually every type of environment on the Earth. Of the three principal environments—salt water, fresh water, and land—salt water is the most hospitable. Sea water is isotonic (that is, it has a similar concentration of solute and solvent molecules) to the tissue fluids of most marine animals, so they have little problem maintaining fluid and salt balance. The buoyancy of sea water supports its inhabitants, so they have less need for skeletal support than do terrestrial organisms. The temperature of the sea is relatively constant owing to the large volume of water. Plankton, the protists and tiny animals that are suspended in the water and float with its movement, provide a ready source of food for many marine animals.

Fresh water offers a less constant environment and generally contains less food. Fresh water is hypotonic to animal tissue fluids so water tends to diffuse into the body. Animals that inhabit fresh water must have mechanisms for removing excess water while still retaining salts. These mechanisms require energy expenditure. For these reasons, far fewer animal species make their homes in fresh water than in the sea.

Terrestrial life, or life on dry land, is the most difficult. Dehydration is a serious threat because water is continuously lost by evaporation and is often difficult to replace. Only a few animal groups—most notably, insects, spiders (and some other arthropods), a few mollusks, and higher vertebrates—have adaptations that permit them to make their homes on land.

ANIMALS CAN BE GROUPED ACCORDING TO BODY STRUCTURE OR PATTERN OF DEVELOPMENT

Most biologists agree that animals evolved from protists. Although the relationships among the various animal phyla are a matter of debate, a few of the more widely held hypotheses are presented in this section. The animal kingdom may be divided into two main subkingdoms: the **Parazoa,** which includes the sponges, and the **Eumetazoa,** which includes all other animals. This distinction is made because the sponges are so different that most biologists think they did not give rise to any other animal phylum.

Animals Have Radial or Bilateral Symmetry

Sponges have an interesting variety of shapes, but generally are not symmetrical. Most eumetazoa exhibit radial or bilateral body symmetry. Members of two phyla, the cnidaria (jellyfish and relatives) and the ctenophora (comb jellies), have **radial** (wheellike) **symmetry.** In radial symmetry, similar structures are regularly arranged around a central axis like the spokes of a wheel. Multiple planes can be drawn through the central axis, each dividing the organism into two mirror images. An animal with radial symmetry receives stimuli equally from all directions in the environment.

(Text continues on page 444)

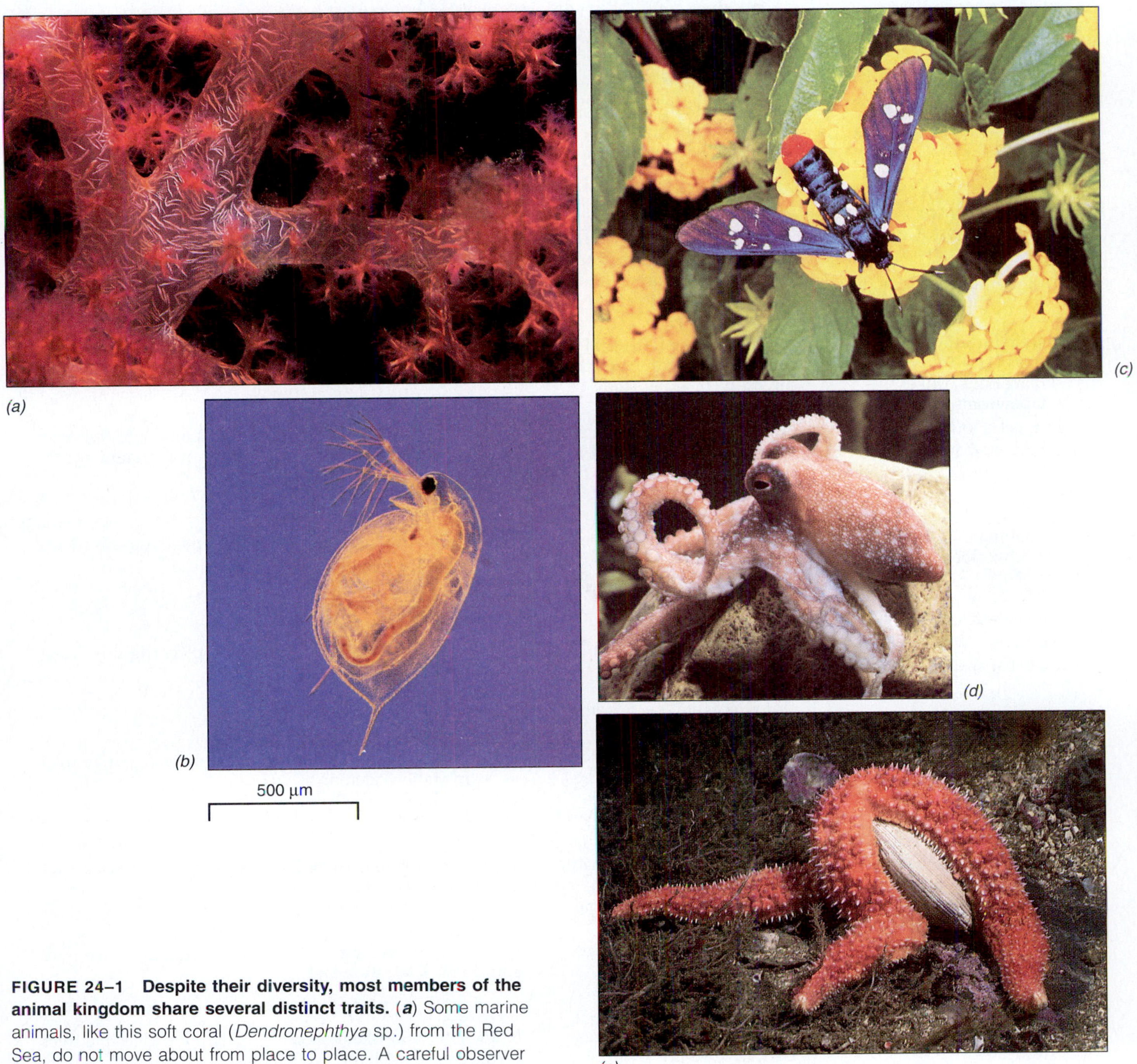

FIGURE 24–1 Despite their diversity, most members of the animal kingdom share several distinct traits. (***a***) Some marine animals, like this soft coral (*Dendronephthya* sp.) from the Red Sea, do not move about from place to place. A careful observer would soon learn that each tiny coral animal extends itself to feed. (***b***) The nearly transparent body of this freshwater crustacean, *Simocephalus vetulus,* or water flea, shows some of its internal organs. (***c***) Polka dot moth *(Syntomeida epilais)*. (***d***) *Octupus macropus*, a cephalopod. The octopod lives in a den among the rocks; it may wait near the entrance of its den to seize a passing crustacean, fish, or snail. (***e***) A painted sea star (*Orthiasterias koehleri*) attacking a clam. (*a*, M. Kazmers, Sharksong, Dembinsky Photo Associates; *b*, Hermann Eiseinbeiss/Photo Researchers, Inc.; *c*, James L. Castner; *d*, Jane Burton/Bruce Coleman, Inc.; *e*, Richard Chesher/Seaphot, Ltd.)

TABLE 24–1
Comparison of the Principal Animal Phyla

Phylum	*Environment and life-style*	*Support and movement*	*Digestion*
Porifera (pore-bearers) (5000) Sponges	Aquatic, mainly marine; ciliated, swimming larvae; adults attach and filter food from water	Support by spicules of calcium carbonate, silica, or spongin; contractile cells can change the size of openings	Intracellular
Cnidaria (10,000) Hydra, jellyfish, coral	Aquatic, mainly marine; some float or swim, others sessile; polyp and medusa forms; some form colonies; capture food with cnidocytes, tentacles	Support by mesoglea, calcareous skeletons (coral), or by fluid in gastrovascular cavity (hydrostatic skeleton); contractile cells in body wall	Gastrovascular cavity with only one opening; intra- and extra-cellular digestion
Platyhelminthes (flatworms)(18,000) Planarians, flukes, tapeworms	Aquatic, some terrestrial in damp areas; many are carnivores; some parasites	Support by its tissues; well developed muscle tissue	Digestive tract with only one opening
Nemertea (900) Ribbon worms	Mainly marine; mainly carnivores; use proboscis for capturing food and defense	Support by its tissues; locomotion by muscle or cilia	Complete digestive tract (with mouth and anus)
Nematoda (roundworms) (12,000) *Ascaris,* hookworms, nematodes	Widely distributed in the soil, sea, and fresh water. Carnivores, scavengers, parasites	Support by tough cuticle; fluid in pseudocoelom serves as hydrostatic skeleton; longitudinal muscle in body wall	Complete digestive tract
Mollusca (100,000) Clams, snails, squids	Mainly marine, some inhabit fresh water or are terrestrial; herbivores, carnivores, scavengers, or filter feeders	Most have hydrostatic skeleton; body usually covered by a shell; most have ventral foot for locomotion	Complete digestive tract
Annelida (segmented worms) (15,000) Earthworms, leeches, marine worms	Marine, fresh water, terrestrial; herbivores, carnivores, scavengers, filter feeders	Fluid-filled coelom serves as hydrostatic skeleton; well developed muscle in body wall	Complete digestive tract
Arthropoda (jointed animals) (1 million) Crustaceans, insects, spiders	Most diverse group in habitat and lifestyle; marine, fresh water, and terrestrial	Tough exoskeleton; jointed appendages (some have wings); well-developed muscles	Complete digestive tract
Echinodermata (spiny-skinned animals) (6000) Sea stars, sea urchins, sand dollars	Marine; mainly carnivores	Endoskeleton bearing spines; muscles; tube-feet	Complete digestive tract
Chordata (47,000) Tunicates, lancelets, vertebrates	Marine, fresh water, terrestrial; diverse life styles; herbivores, carnivores, scavengers, filter feeders	Notochord; endoskeleton of cartilage and/or bone; well developed muscles	Complete digestive tract

Circulation	*Gas exchange*	*Fluid balance waste disposal*	*Nervous system*	*Reproduction*
Diffusion	Diffusion	Diffusion	No nervous system; individual cells respond to stimuli	Asexual by budding; sexual, may be hermaphroditic
Diffusion	Diffusion	Diffusion	Nerve net; no brain	Asexual by budding; sexual, separate sexes
Diffusion	Diffusion	Protonephridia with flame cells	Simple brain; nerve cords; ladder-type system; simple sense organs	Asexual, by fission; sexual, hermaphroditic, but usually cross-fertilize
At least two pulsating blood vessels; no heart; blood cells with hemoglobin	Diffusion	Protonephridia	Simple brain; nerve cords with cross nerves; simple sense organs	Asexual, by fragmentation; sexual, sexes separate
Diffusion	Diffusion	Excretory canals	Simple brain; dorsal and ventral nerve cords; simple sense organs	Sexual, sexes separate
Open system (closed in cephalopods)	Gills and mantle	Kidneys	Three pairs of ganglia; simple sense organs	Sexual; sexes separate
Closed system	Diffusion through moist skin; oxygen circulated by blood	Pair of nephridia in each segment	Simple brain; double ventral nerve cord; simple sense organs	Sexual, hermaphroditic but cross-fertilize
Open system	Trachea in insects; gills in crustaceans; book lungs or trachea in spider group	Malpighian tubules in insects; antennal (green) glands in crustaceans	Complex brain; double ventral nerve cord; well-developed sense organs	Sexual, sexes almost always separate
Open system (reduced)	Skin gills	Diffusion	Nerve rings; no brain	Usually sexual, sexes separate
Closed system; ventral heart	Gills or lungs	Kidneys and other organs	Dorsal nerve cord with complex brain at anterior end	Usually sexual, sexes separate

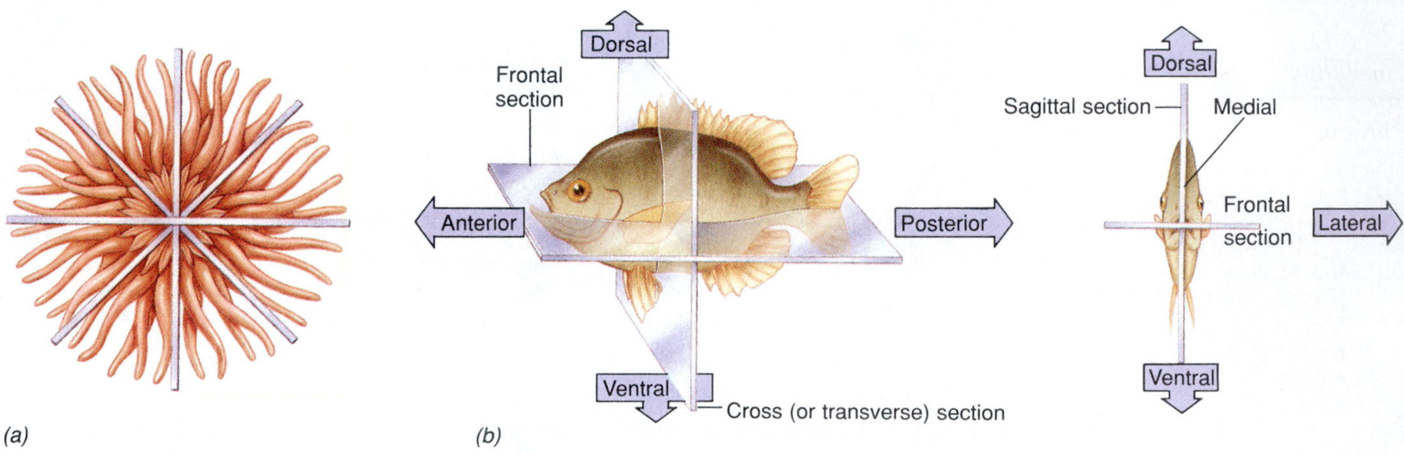

FIGURE 24–2 Types of body symmetry in animals. (***a***) In radial symmetry, multiple planes can be drawn through the central axis; each divides the organism into two mirror images. (***b***) Most animals are bilaterally symmetrical. A sagittal cut (lengthwise vertical cut) divides the animal into right and left parts (see figure to right). The head end of the animal is generally its anterior end, and the opposite end is its posterior end. The back of the animal is its dorsal surface; the belly is its ventral surface. The diagram also illustrates various ways in which the body can be sectioned (cut) in order to study its internal structure. Cross sections and sagittal sections are used in illustrations throughout this book to show relationships among tissues and organs.

Most animals are **bilaterally symmetrical** (at least in their larval stages). A bilaterally symmetrical animal can be divided by only one plane to produce right and left halves that are roughly equivalent mirror images (Fig. 24–2). Bilateral symmetry is considered to be an adaptation to motility. The front end of the animal generally has a head, where sense organs are concentrated; this end receives most environmental stimuli. The rear end of the animal may be equipped with a tail for swimming, or it may just follow along.

To locate body structures in bilaterally symmetrical animals, it is helpful to define some basic terms and directions. The back surface of an animal is its **dorsal** surface; the belly side is its **ventral** surface (Fig. 24-2). **Anterior** means at or toward the front (head end) of the animal. **Posterior** (or caudal) means at or toward the rear (tail end) of the animal. A structure is said to be **medial** if it is located toward the midline of the body and **lateral** if it is toward one side of the body. For example, your ears are lateral to your nose.

A bilaterally symmetrical animal has three planes (flat surfaces that divide the body into specific parts). A **sagittal plane** divides the body into right and left parts; this plane passes from anterior to posterior and from dorsal to ventral. A **frontal plane** divides a bilateral body into dorsal and ventral parts. A **transverse section,** or **cross section,** cuts at right angles to the body axis and separates anterior and posterior parts.

Animals Can Be Grouped According to Type of Body Cavity

A widely held system for grouping animal phyla is based on the presence and type of body cavity. In order to understand the types of body cavities, we must look briefly at the animal's embryonic development.

The structures of most eumetazoan animals develop from three embryonic tissue layers, called **germ layers,** that are present in the embryo. The outer germ layer, called the **ectoderm,** gives rise to the outer covering of the body and to the nervous system. The inner layer, or **endoderm,** forms the lining of the digestive tract. **Mesoderm,** the middle layer, gives rise to most of the other body structures, including the muscles, bones, and circulatory system (when they are present).

The flatworms (and members of a few related phyla) form three germ layers and have a solid body. These animals have no body cavity, and so are referred to as **acoelomates** (the prefix *a* means "without," and the word *coelom* means "cavity"; Fig. 24–3).

Other, more complex animals generally have a **tube-within-a-tube body plan.** The outer tube is the body wall.

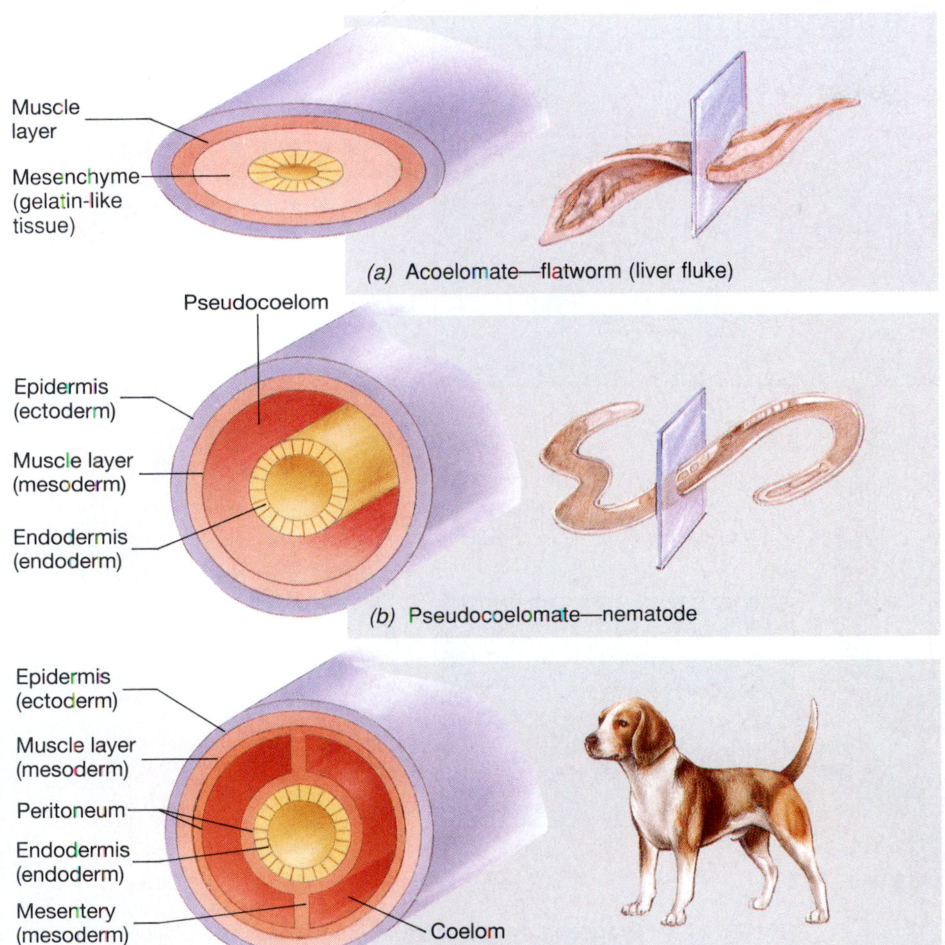

FIGURE 24–3 Three basic animal body plans are illustrated by these cross sections. The term *body cavity* refers to the space between the body wall and the digestive tube. (***a***) An acoelomate animal has no body cavity. (***b***) A pseudocoelomate animal has a "false" body cavity, one that is not completely lined with mesoderm. (***c***) In a coelomate animal the body cavity, or coelom, is completely lined with tissue that develops from mesoderm.

It is covered externally with tissue that develops from ectoderm. Tissue derived from endoderm lines the inner tube—the digestive tract or gut. The digestive tract has an opening at each end—the mouth and the anus. Beneath the ectoderm, the outer tube consists of mesoderm. The space between the two tubes is the body cavity. If the body cavity is not completely lined with mesoderm, it is called a **pseudocoelom** (false coelom). Animals with a pseudocoelom are referred to as *pseudocoelomates.*

In still more complex animals, the body cavity is completely lined with mesoderm. Such a body cavity is a true **coelom.** Animals with true coeloms are referred to as *coelomates.* The tree shown in Figure 24–4 indicates the relationships of the major phyla of animals based on their type of body cavity.

Animals Can Be Grouped Based on Pattern of Embryonic Development

Animals that have a true coelom can be divided into two groups: the *protostomes* and the *deuterostomes.* These groups reflect two main lines of evolution based on their pattern of early development.

Early during embryonic development, a group of cells move inward to form an opening called the *blastopore.* In most of the mollusks, annelids, and arthropods, this opening develops into the mouth. These animals are **protostomes** (meaning "first, the mouth").

In echinoderms (for example, sea stars and sea urchins) and chordates (the phylum that includes the vertebrates), the blastopore does not give rise to the mouth; instead it

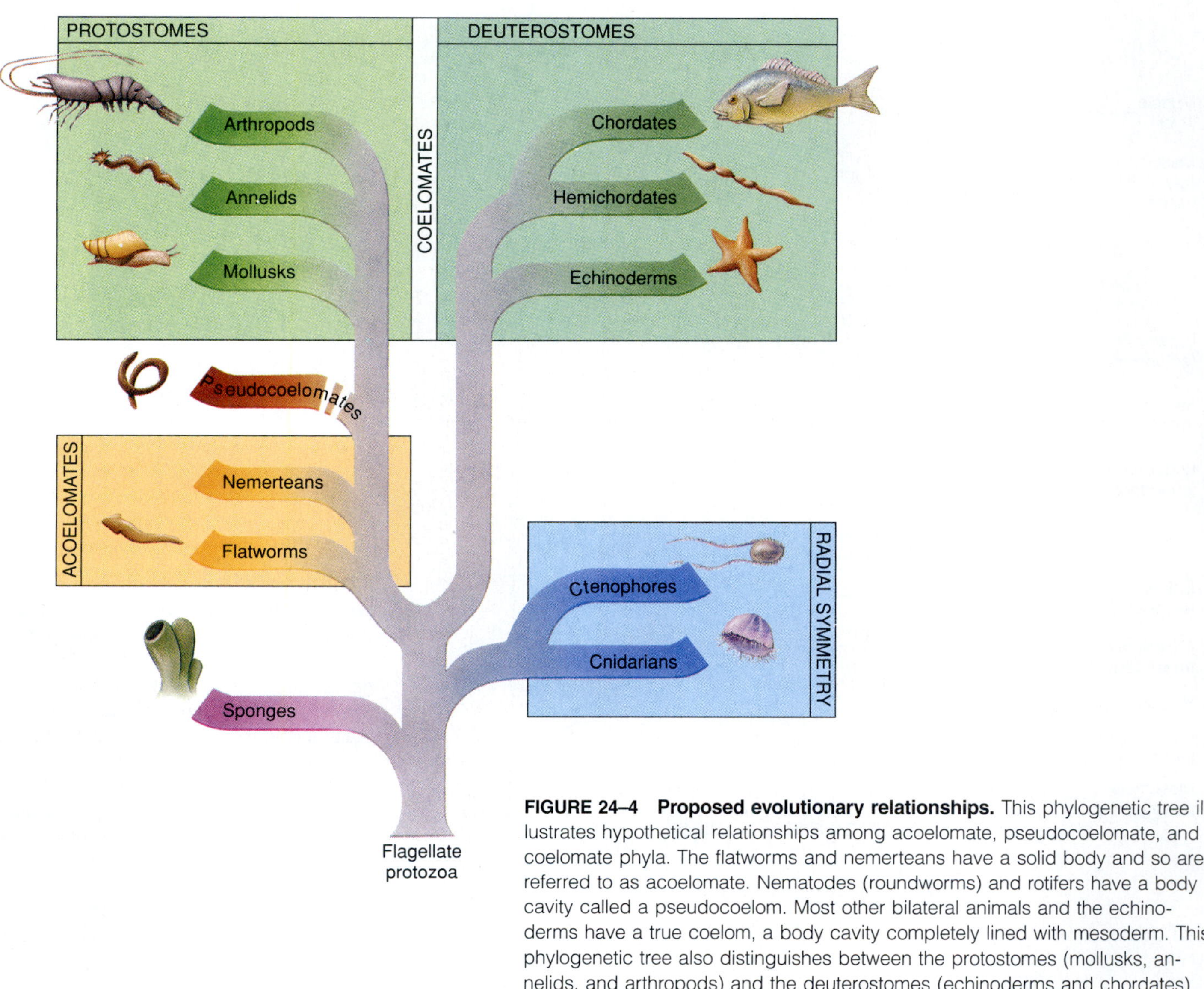

FIGURE 24–4 Proposed evolutionary relationships. This phylogenetic tree illustrates hypothetical relationships among acoelomate, pseudocoelomate, and coelomate phyla. The flatworms and nemerteans have a solid body and so are referred to as acoelomate. Nematodes (roundworms) and rotifers have a body cavity called a pseudocoelom. Most other bilateral animals and the echinoderms have a true coelom, a body cavity completely lined with mesoderm. This phylogenetic tree also distinguishes between the protostomes (mollusks, annelids, and arthropods) and the deuterostomes (echinoderms and chordates).

develops into the anus. The opening that develops into the mouth forms later in development. These animals are the **deuterostomes** (meaning "second, the mouth"). Other differences in the development of protostomes and deuterostomes include the process of mesoderm and coelom formation and the pattern of early cell division.

SPONGES HAVE SPECIALIZED CELLS BUT NO TRUE TISSUES

About 5000 species of sponges make up **phylum Porifera.** The name Porifera, which means "to have pores," aptly describes the sponges, which look like sacs perforated by tiny holes. Sponges occupy aquatic, mainly marine, habitats. Living sponges may be bright red, orange, green, purple, or quite drab. They are generally asymmetrical but may be flat or shaped like fans, balls, or vases (Fig. 24–5).

Sponges are classified on the basis of the type of skeleton they secrete. Members of one class have a chalky skeleton composed of calcium carbonate spikes, called **spicules.** Members of a second class of sponges are glass sponges with skeletons of six-rayed silica spicules. Members of a third class of sponges have skeletons of spongin (a protein material) fibers, and often also have silica spicules. What we recognize as a bath sponge is actually a dried spongin skeleton from which all living cells have been removed.

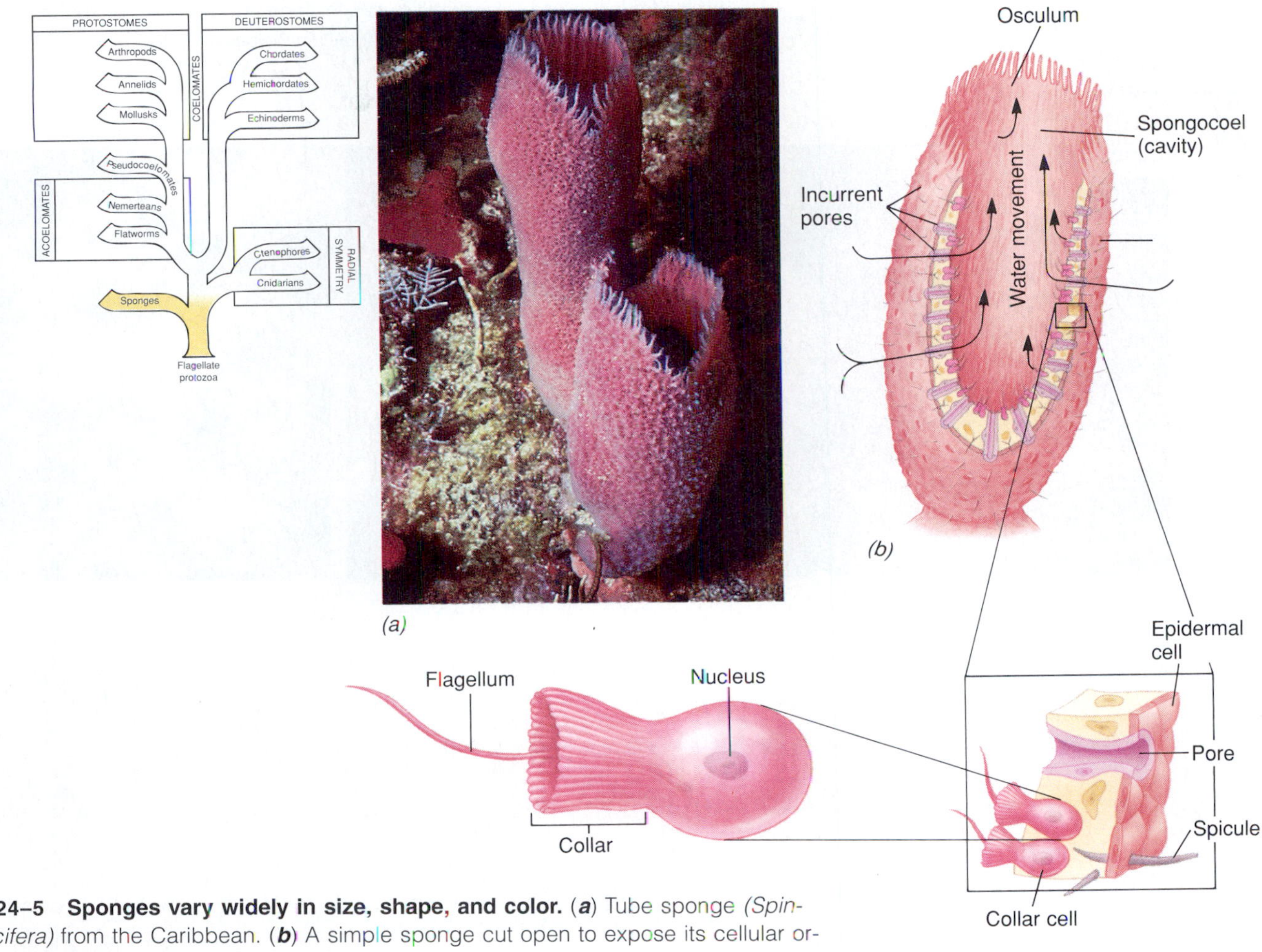

FIGURE 24–5 Sponges vary widely in size, shape, and color. (***a***) Tube sponge *(Spinosella plicifera)* from the Caribbean. (***b***) A simple sponge cut open to expose its cellular organization. (Charles Seaborn/ Odyssey Productions, Chicago)

In a simple sponge, water enters through hundreds of tiny pores; passes into the central cavity, or **spongocoel;** and flows out through the sponge's open end, the **osculum** (Fig. 24–5*b*). In some types of sponges, the body wall is extensively folded and there are complicated systems of water canals.

Although the sponge is a multicellular organism, its cells are so loosely associated that they do not form definite tissues. However, a division of labor exists among the several types of cells that make up the sponge, with certain cells specializing in nutrition, support, or reproduction. Flagellated **collar cells** line the spongocoel and some of the canals. In each collar cell, a tiny collar of microvilli surrounds the base of the flagellum. The flagella create a water current that brings food and oxygen to the cells and carries away carbon dioxide and other wastes. Collar cells also trap and ingest food particles.

The sponge body is composed of two layers of cells separated by a gelatin-like middle layer, which is supported by skeletal spicules or spongin. Ameba-like cells that wander through the middle layer may be specialized to perform specific functions such as storage and distribution of food or secretion of spicules. Some of the cells that surround the opening are capable of contraction (becoming shorter).

Sponge **larvae** are ciliated and can swim about. However, adult sponges remain attached to some solid object on the sea bottom. Sponges are filter feeders, adapted for trapping and eating whatever food the sea brings to them. As water circulates through the body, small particles of food are trapped along the sticky collars of the collar cells. Digestion is intracellular, which means that digestion is the job of each individual cell, rather than of specialized organs. Undigested food is simply eliminated into the water.

Oxygen from the water diffuses throughout the sponge. Gas exchange and waste disposal are carried on by individual cells. Each cell of the sponge body is irritable and can react to stimuli; however, there are no sen-

FIGURE 24–6 Polyp and medusa body forms characteristic of phylum Cnidaria. (***a***) This hydrozoan, *(Gonothyraea loveni)* forms a colony of polyps; (***b***) This jellyfish *(Aurelia aurita)*, a suspension feeder, has a medusa body form; (***c***) Polyps from the coral *Montastrea cavernosa* extended for feeding. (*a*, Robert Brons/Biological Photo Service; *b*, D. J. Wrobel, Monterey Bay Aquarium/Biological Photo Service; *c*, Mike Bacon/Tom Stack & Associates)

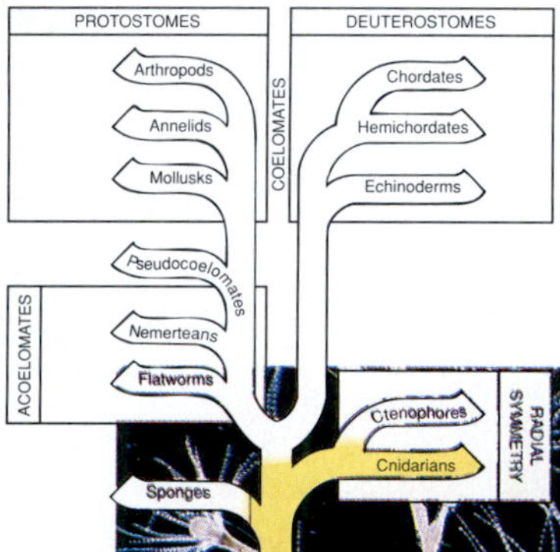

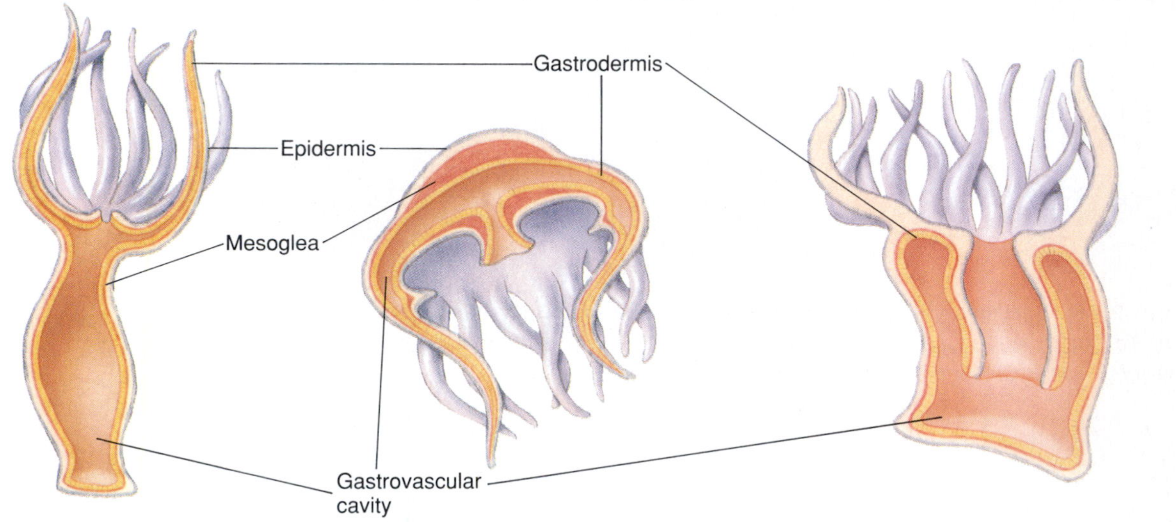

(a) Hydrozoa (polyp form) *(b)* Scyphozoa (jelly fish) *(c)* Anthozoa (sea anemone and coral)

sory or nerve cells that would enable the animal to react as a whole.

Sponges can reproduce asexually. A small fragment or bud may break free from the parent sponge and give rise to a new sponge. Such fragments may remain to form a colony with the parent sponge. Sponges also reproduce sexually. Most sponges are **hermaphroditic,** meaning that the same individual can produce both egg and sperm. Some of the ameba-like cells develop into sperm cells, others into egg cells. Hermaphroditic sponges usually cross-fertilize with other sponges, however. Fertilization and early development take place within the jelly-like middle layer. Embryos become motile larvae that move into the spongocoel and leave the parent along with the stream of outflowing water. After swimming about for a while, the larva finds a solid object, attaches to it, and settles down to a sessile life.

Cnidarians (Hydras, Jellyfish, and Corals) Have Tissue Layers but No True Organs

Most of the 10,000 or so species of **phylum Cnidaria** (pronounced "nie-dare´-e-a") are marine. There are three main groups of cnidarians: the hydras and Portuguese man-of-war (hydrozoans); the "true" jellyfish (scyphozoans); and the sea anemones and true corals (anthozoans; Fig. 24–6 and Chapter Opener). All of the cnidarians have unique stinging cells called **cnidocytes,** from which they get their name (cnidaria is from a Greek word meaning "sea nettles"). These cells contain remarkable organelles, called **nematocysts,** that can be discharged to sting prey or predator (see discussion, which follows). Although many cnidarians live a solitary existence, others group into colonies.

The radially symmetrical cnidarian body is a hollow sac with the mouth and surrounding tentacles located at one end. The mouth leads into a digestive cavity, called the **gastrovascular cavity.** The mouth is the only opening into this cavity and must serve for both ingestion of food and egestion (expulsion) of wastes.

Much more highly organized than the sponge, the cnidarian has two definite tissue layers. The outer **epidermis** is a protective layer. The inner **gastrodermis** functions in digestion. These layers are separated by a gelatin-like **mesoglea,** which is not itself cellular but usually contains a few cells.

Gas exchange and waste disposal occur by diffusion. The cnidaria are the simplest animals to have true nerve cells. These cells are simply arranged, forming irregular **nerve nets** that connect sensory cells in the body wall with gland cells and with cells that contract. A nerve impulse that originates in one part of the body passes in all directions more or less equally, rather than along specific pathways, as occurs in more complex animals.

Cnidarians Have Two Body Plans: A Polyp and a Medusa Form

Cnidarians have two body shapes: the **polyp** and the **medusa,** or jellyfish (Fig. 24–6). The polyp form, represented by *Hydra,* resembles an upside-down, elongated cylindrical jellyfish. Some cnidarian colonies (for example, the Portuguese man-of-war) consist of both polyp and medusa forms.

Some marine cnidarians are remarkable in their ability to alternate between sexual and asexual stages. This alternation of stages differs from the alternation of generations in plants in that both sexual and asexual forms are diploid. Only sperm and egg are haploid. The cnidarian life cycle is illustrated by the colonial marine hydrozoan *Obelia* (Fig. 24–7). In this polyp colony, the

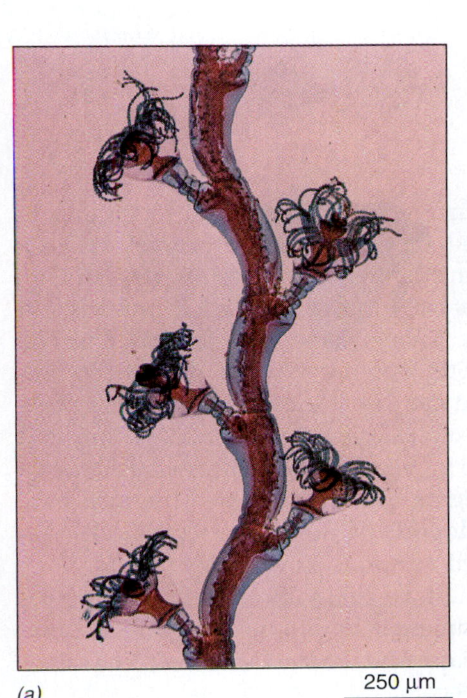

(a)

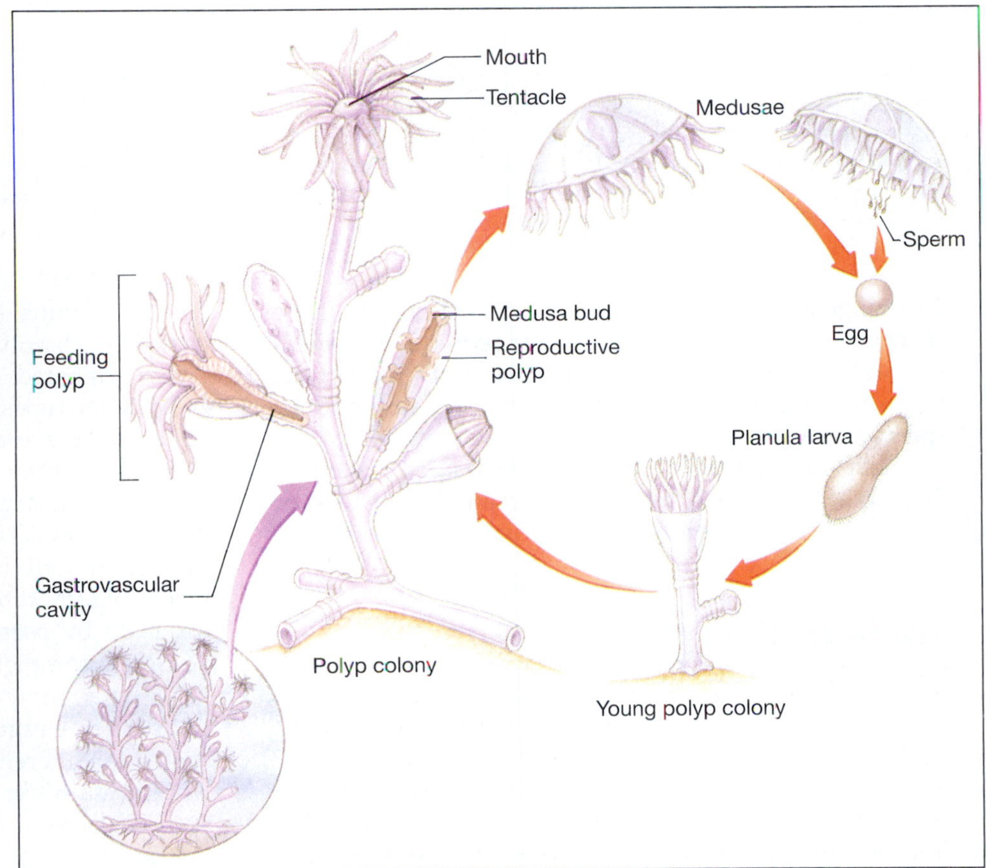

(b)

FIGURE 24–7 Some hydrozoans form colonies. (***a***) Colonial hydrozoan *(Gonothyraea loveni).* (***b***) Life cycle of *Obelia,* a marine hydrozoan. Specialized polyps give rise asexually to medusae. The free-swimming medusae reproduce sexually. The zygote develops into a planula larva which develops into a polyp and gives rise to a new colony. Note the specialization of members of the polyp colony. (Visuals Unlimited/John D. Cunningham)

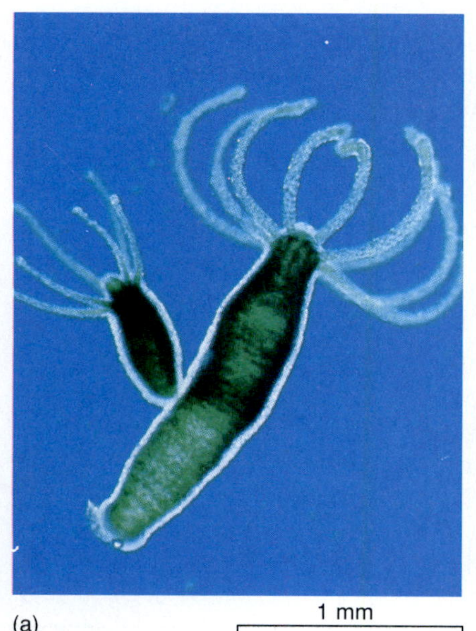

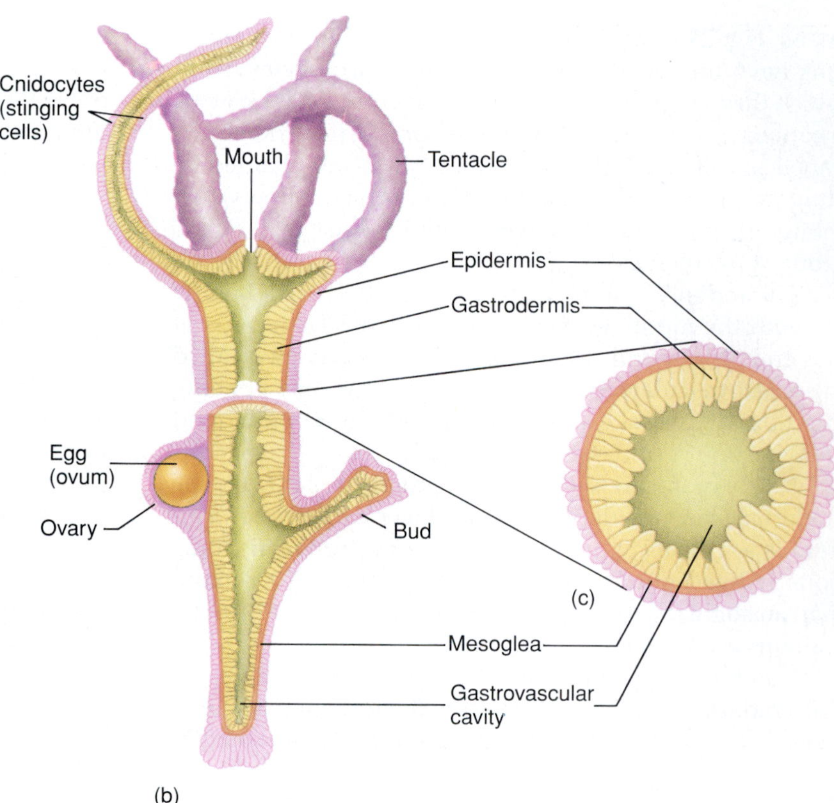

FIGURE 24–8 Hydra. (***a***) A hydra *(Hydra viridis)* in the process of budding. The bud separates from the parent hydra, forming a new animal. (***b***) This hydra is cut longitudinally to reveal its internal structure. Asexual reproduction by budding is evident on the right of the photograph. Sexual reproduction is represented by the ovary on the left of the photograph. Male hydras develop testes, which produce sperm. (***c***) Cross section through the body of a *Hydra.* (Biophoto Associates/Photo Researchers, Inc.)

asexual generation consists of two types of polyps: those specialized for feeding and those for reproduction. Free-swimming male and female medusae bud off from the reproductive polyps. These medusae eventually produce sperm and eggs, and fertilization takes place. The zygote develops into a ciliated swimming larva called a **planula.** The larva attaches to some solid object and begins to form a new generation of polyps by asexual reproduction.

The Hydra Has a Solitary, Carnivorous Life-Style

The freshwater hydra, a common cnidarian, is a solitary polyp seldom more than 1 centimeter long (Fig. 24–8). Although capable of locomotion, an adult hydra generally attaches to a rock, twig, or even a leaf and waits for dinner to come along. When a likely prospect happens to brush by one of its tentacles, the hydra's cnidocytes respond.

Coiled within each cnidocyte is a "thread capsule," or nematocyst. Each cnidocyte has a small projecting trigger on its outer surface that responds to touch and to chemicals dissolved in the water (a form of "taste"). When triggered, the nematocyst fires its thread (Fig. 24–9). Some types of nematocysts have sticky threads that adhere to the prey. Others have long threads that coil around the prey. Still another type of nematocyst has barbs or spines and can inject a protein toxin that paralyzes the prey. The tentacles encircle the snared prey and push it through the mouth into the gastrovascular cavity, where digestion begins. Partially digested fragments are taken up by pseudopods of the gastrodermis cells. Digestion is completed within food vacuoles.

Hydras reproduce asexually by budding during periods when environmental conditions are optimal; however, they differentiate as males and females in the fall or when pond water becomes stagnant. Females develop an *ovary* that produces a single egg, and males form a *testis* that produces sperm. After fertilization, the zygote (fertilized egg) becomes covered with a shell. It leaves the parent and remains within the protective shell throughout the winter.

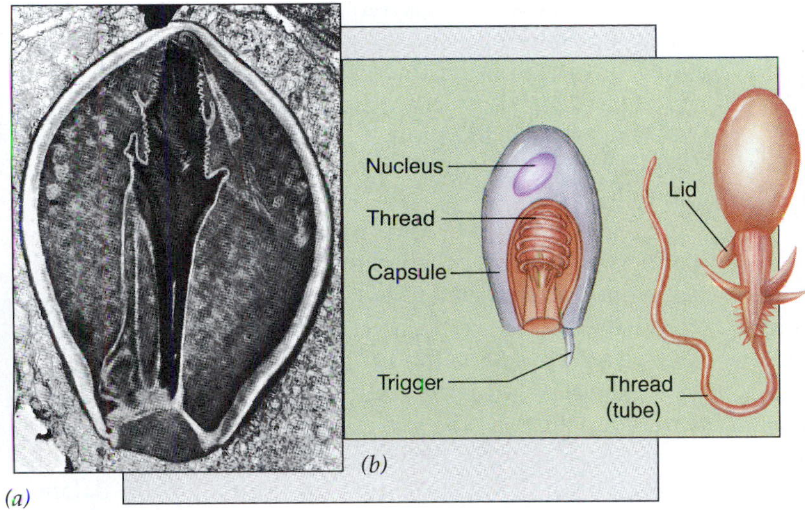

FIGURE 24–9 Nematocysts, the thread capsules within cnidarian stinging cells. (***a***) Electron micrograph of an undischarged nematocyst of *Hydra* (sagittal section). (***b***) Discharge of a nematocyst. When an object comes in contact with the cnidocil, the nematocyst discharges, ejecting a thread that may entangle or penetrate the prey. Some nematocysts secrete a toxic substance that immobilizes the prey. (G. B. Chapman, Cornell University Medical College)

Jellyfish Spend Most of Their Lives as Medusae

Among the jellyfish, the medusa is the dominant body form. The polyp stage is small and inconspicuous. The largest jellyfish, *Cyanea,* may be more than 2 meters (more than 6 feet) in diameter and have tentacles that trail as long as 30 meters (about 97 feet) beneath them. These orange and blue monsters are among the largest of the invertebrate animals. Their nematocysts produce a toxin that makes them a real danger to swimmers in the North Atlantic Ocean.

Corals and Sea Anemones Spend Their Lives as Polyps

Brightly colored corals and anemones inhabit warm shallow seas. The reefs and atolls of the South Pacific are the remains of billions of small, cup-shaped limestone skeletons, secreted during past ages by coral colonies and by coralline algae (see Chapter Opener). Living colonies occur only on the surfaces of such reefs, adding their own skeletons to the forming rock.

Sea anemones and corals have no medusa stage, and the polyps may be either individual or colonial forms. These animals produce a small ciliated larva (planula larva) that may swim to a new location before attaching to develop into a polyp. Sea anemones capture fish and other prey with their tentacles. Although corals can capture prey, many tropical species depend for nutrition mainly on photosynthetic algae that live within their cells (see Chapter Opener).

THE COMB JELLIES MOVE BY MEANS OF CILIA

Phylum Ctenophora consists of about 100 species of comb jellies. These fragile, luminescent marine animals may be as small as a pea or larger than a tomato. Their body plan is somewhat similar to that of a medusa, consisting of two cell layers separated by a thick jelly-like layer.

The outer surface of a ctenophore bears eight rows of cilia, resembling combs (Fig. 24–10). The coordinated beating of the cilia in these combs moves the animal through the water. Ctenophores have only two tentacles, and they lack the cnidocytes characteristic of the cnidarians. However, their tentacles are equipped with adhesive glue cells, which trap their prey.

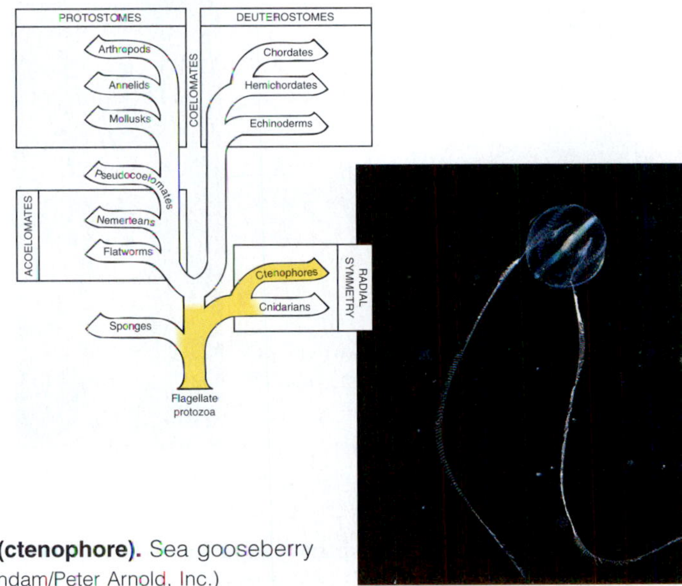

FIGURE 24–10 The comb jelly (ctenophore). Sea gooseberry (*Pleurobranchia pileus*). (Fred Bavendam/Peter Arnold, Inc.)

FLATWORMS HAVE THREE TISSUE LAYERS, A HEAD REGION, AND ORGANS

Flatworms are flat, elongated, legless animals belonging to **phylum Platyhelminthes** (the term *platy* means "flat," and the term *helminthes* means "worms"; Fig. 24–11). About 18,000 species have been identified. This phylum includes the free-living flatworms (turbellarians), for example, the planarian and its relatives; the flukes, which are parasites; and the tapeworms, which are intestinal parasites of vertebrates.

Flatworms have three definite tissue layers. In addition to an outer *epidermis,* which forms from ectoderm, and an inner *endodermis,* derived from endoderm, the flatworms have a middle tissue layer, derived from mesoderm. The mesoderm gives rise to true muscles and reproductive structures. The flatworms are the simplest animals that have well-developed **organs,** which are functional structures of two or more kinds of tissue. They also have very complex reproductive systems. The flatworm body is solid between the outer wall and the gastrovascular cavity. Thus, flatworms are acoelomate.

Along with their bilateral symmetry, flatworms have a definite anterior and posterior end. This is a great advantage to any organism that moves about. With a concentration of sense organs in the part of the body that first meets the environment, the animal is able to detect an enemy quickly enough to escape. The animal is also more likely to detect prey quickly enough to capture it. The beginning of **cephalization,** which is the development of a head, is an important evolutionary adaptation first seen in flatworms.

The simple flatworm "brain" consists of two masses of nervous tissue, called **ganglia,** in the head region. The ganglia are connected to two or more nerve cords that extend the length of the body. A series of nerves connect the nerve cords like the rungs of a ladder.

Flatworms that live in fresh water have an osmotic problem. Because the surrounding water is hypotonic to their tissues, water diffuses into the body more rapidly than it exits. To maintain an appropriate fluid balance, excess water must be discharged. Flatworms have an organ system that regulates the volume and composition of the body fluids. Typically, two tubes extend the length of the body and give off branches called **protonephridia.** Each protonephridium ends in a **flame cell,** which is a collecting cell equipped with cilia that channels fluid (especially excess water) into the system of tubules.

The gastrovascular cavity, when present, is often extensively branched. It has only one opening, the mouth, which is usually located on the middle of the ventral surface.

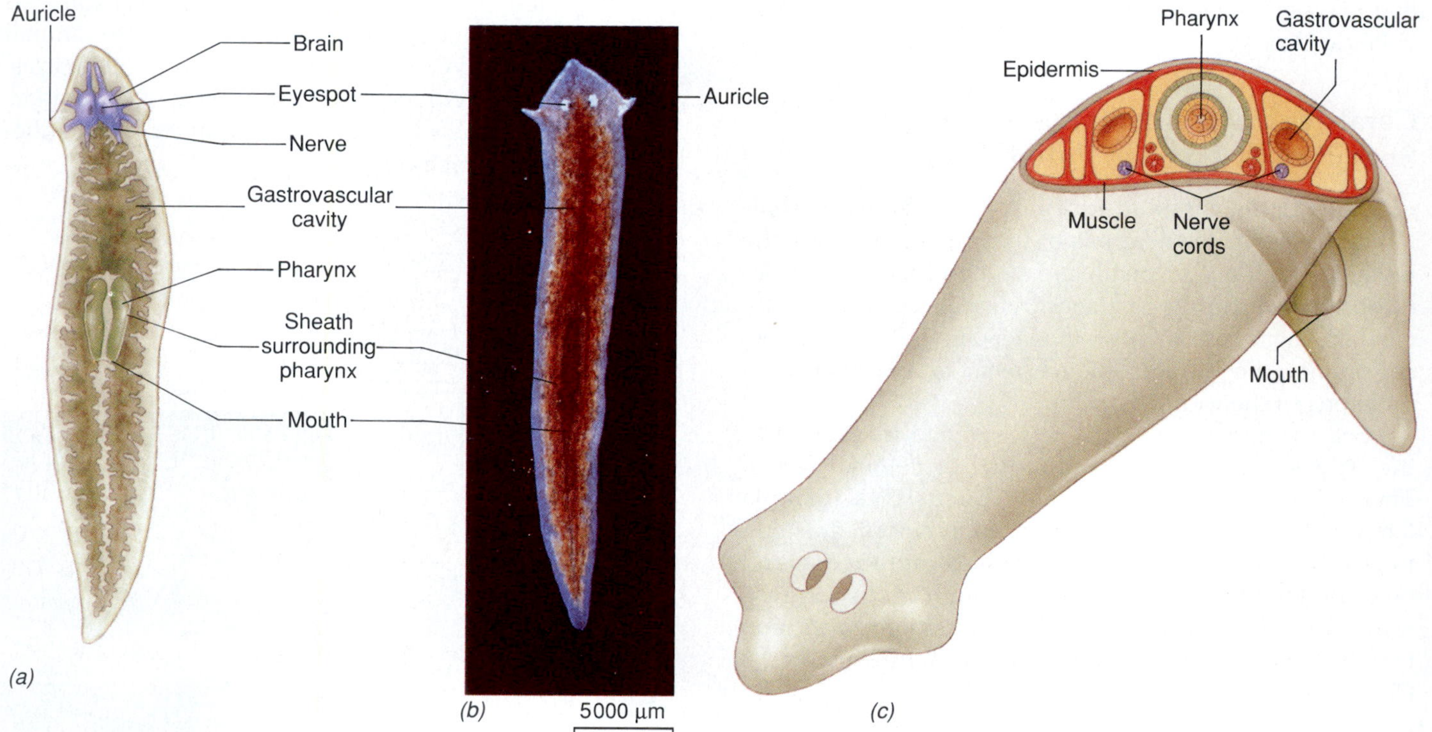

FIGURE 24–11 The common planarian, *Dugesia,* a flatworm. (***a***) Internal structure of *Dugesia (Dugesia dorotocephala).* (***b***) A stained specimen illustrating the internal structure. (***c***) Cross section through a planarian. (T. E. Adams/Peter Arnold, Inc.)

Planarians Are Free-Living Carnivores

Planarians are free-living flatworms, most of which inhabit freshwater or moist land areas. The planarians usually studied in biology laboratories are found burrowing in the mud at the edges of ponds or on the underside of rocks or leaves. The common American planarian, *Dugesia,* is about 15 millimeters (less than 0.1 inch) long, with eyespots that look like crossed eyes and distinct "ears" called **auricles** (Fig. 24–11). The auricles actually serve as organs of smell (chemoreceptors).

Planarians are carnivores that trap small animals in a mucous secretion. The digestive structures include a single opening (the mouth), a pharynx (the first portion of the digestive tube), and a branched gastrovascular cavity. A planarian can project its pharynx outward through its mouth, using it like a vacuum cleaner to suck up its prey. Extracellular digestion takes place in the digestive cavity by enzymes secreted by gland cells. Digestion is completed after the nutrients have been absorbed into individual cells. Undigested food is eliminated through the mouth. The highly branched gastrovascular cavity helps distribute food to all parts of the body, so that each cell is within range of diffusion.

Because the planarian's body is flattened, gases can reach all of the cells by diffusion. No specialized respiratory or circulatory structures are required. Excretion of wastes also occurs mainly by diffusion, although some wastes may be excreted by the flame cells.

Planarians can reproduce either asexually or sexually. In asexual reproduction, an individual constricts in the middle and divides into two individuals. Each regenerates its missing parts. Sexually these animals are hermaphroditic. During the warm months of the year, each is equipped with a complete set of male and female organs. Two planarians come together in copulation and exchange sperm cells. Thus, their eggs are cross-fertilized.

Flukes Have Suckers and Other Adaptations for Their Parasitic Life-Style

Although their body plan resembles that of the free-living flatworms, specialized adaptations have evolved in flukes that contribute to their success as parasites. Both blood flukes and liver flukes have one or more suckers for clinging to their host. Flukes also have extremely complex and efficient reproductive organs.

Flukes have complicated life cycles, involving a number of different forms and the alternation of sexual and asexual stages. During their life cycle, they are parasites on one or more intermediate hosts such as snails and fishes (Fig. 24–12). The aquatic snails that serve as intermediate hosts thrive in ponds and marshy areas, including rice paddies. When dams are built, the marshy areas created often provide habitats for these snails. Blood flukes (genus *Schistosoma*) infect about 200 million people who live in tropical areas.

Tapeworms Are Strikingly Specialized for Their Parasitic Life-Style

Tapeworms are long, flat, ribbon-like animals. As adults, they live as parasites in the intestines of probably every kind of vertebrate, including humans. Sometimes thought of as the most degenerate (changed from a more complex to a simpler form) type of flatworm, the tapeworms are actually strikingly specialized for their parasitic mode of life.

Among their many adaptations are suckers and sometimes hooks on the head (scolex), which attach to the host's intestine (Fig. 24–13). Their reproductive adaptations and abilities are extraordinary. The body of the tapeworm consists of a long chain of segments called **proglottids.** Each proglottid is an entire reproductive unit equipped with both male and female organs and containing as many as 100,000 fertilized eggs, or zygotes. Because an adult tapeworm may have 2000 segments, its reproductive potential is staggering. A single tapeworm may produce 600 million zygotes in 1 year! Segments farthest from the tapeworm's head are the oldest and are shed daily, leaving the host's body along with the feces.

Tapeworms lack digestive organs. They absorb food directly through their body walls from the host's intestine, and they have no mouths and no digestive systems of their own. Some tapeworms have rather complex life cycles, spending the larval stage in the body of an intermediate host and their adult stage in the body of a different, final host.

Let us consider the life cycle of the beef tapeworm. Humans become infected when they eat poorly cooked beef containing the larva (Fig. 24–14). The microscopic tapeworm larva spends part of its life cycle encysted (enclosed in a little sac or cyst) within the muscle tissue of beef. When a human ingests infected meat, the digestive juices break down the cyst, releasing the larva. Soon the larva attaches itself to its host's intestinal lining and within a few weeks matures into an adult tapeworm, eventually growing to a length of more than 8 meters (about 30 feet). The parasite reproduces sexually within the human intestine and sheds proglottids filled with zygotes.

Once established within a human host, the tapeworm may remain there for the rest of its life, as long as 10 years. For the life cycle of the beef tapeworm to continue, its fertilized eggs must be ingested by an intermediate host, a cow. (This requirement explains why the tapeworm must produce millions of eggs to increase the probability that at least a few will survive.) When a cow eats grass or other food contaminated with infected human feces, the

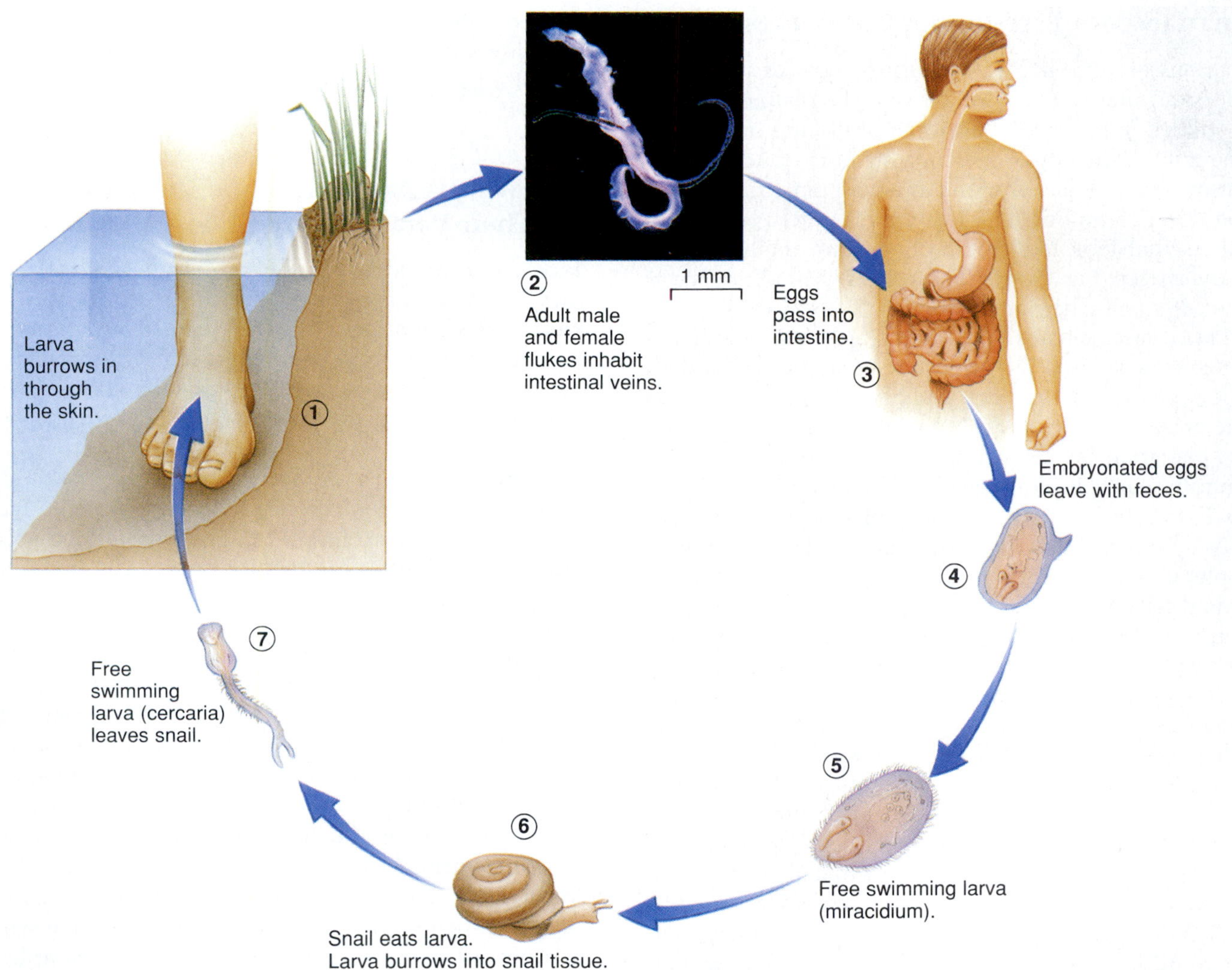
Larva
burrows in
through
the skin.
①
②
1 mm
Adult male
and female
flukes inhabit
intestinal veins.
Eggs
pass into
intestine.
③
Embryonated eggs
leave with feces.
④
⑤
Free swimming larva
(miracidium).
⑥
Snail eats larva.
Larva burrows into snail tissue.
⑦
Free
swimming
larva (cercaria)
leaves snail.

FIGURE 24–12 Life cycle of a blood fluke, a schistosome. Larvae burrow through human skin and make their way to the circulatory system. The male and female are permanently paired throughout life. Sexual reproduction takes place within the human intestine, and the eggs leave the body with the feces. If they find their way to fresh water, the eggs hatch, releasing free-swimming larvae (miracidia). To survive, the larvae must find an intermediate host, an aquatic snail. If successful, they burrow into the tissues of the snail and develop into a form that reproduces asexually. Finally, fork-tailed larvae (cercariae) develop and leave the snail. When they contact a human host they burrow into the skin, completing the cycle. (Photograph of adult male and female, Center for Disease Control, Atlanta, Georgia/Biological Photo Service)

larvae hatch in the cow's intestine. The larvae make their way into muscle, where they encyst. There they await release by a final host, perhaps a human eating a rare steak.

Two other tapeworms that infect humans are the pork tapeworm and the fish tapeworm. The pork tapeworm infects us when we eat poorly cooked infected pork, and the fish tapeworm is contracted when we eat raw or poorly cooked infected fish. Like most parasites, tapeworms tend to be species specific; that is, each can infect only certain species.

RIBBON WORMS HAVE A TUBE-WITHIN-A-TUBE BODY PLAN

The ribbon worms **(phylum Nemertea)** resemble the free-living flatworms, but tend to be more elongated. Their long, narrow bodies may extend to more than 10 meters in length. Some are vivid orange, red, or green, with black or colored stripes (Fig. 24–15). Almost all of the approximately 900 species are marine, although a few inhabit fresh water or damp soil.

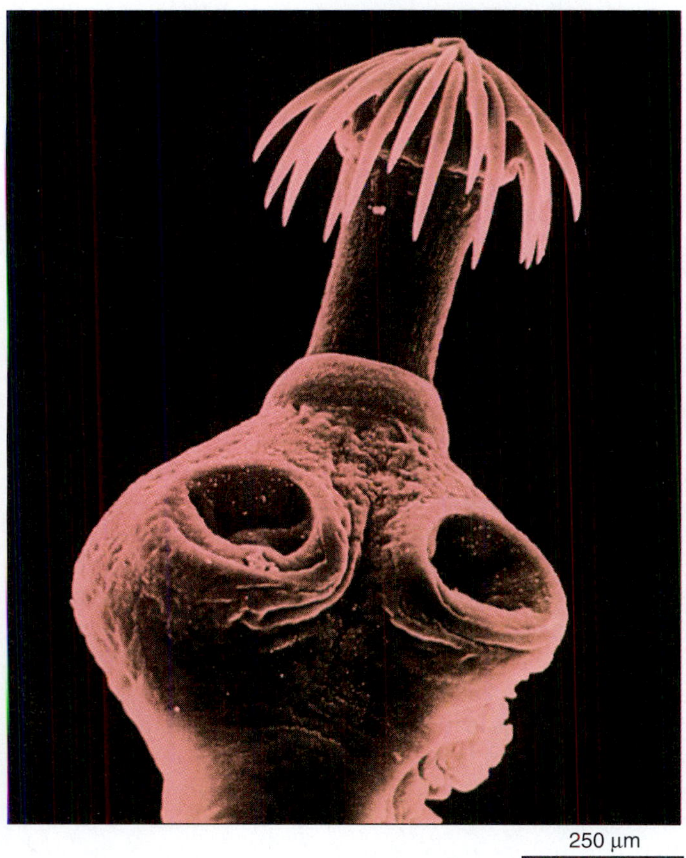

FIGURE 24–13 Some species of tapeworms are armed with powerful hooks. These parasites use their hooks to attach to the host. False-color scanning electron micrograph of the head of the small tapeworm *Acanthrocirrus retrisrostris.* This tapeworm reaches maturity in the intestines of wading birds that eat barnacles. The photograph shows the piston-like rostellum, which can be withdrawn into the head or thrust out and buried in the host's tissue. Beneath the rostellum, two of the four powerful suckers are visible. (Cath Ellis, Dept. of Zoology, University of Hull/Science Photo Library/Photo Researchers, Inc.)

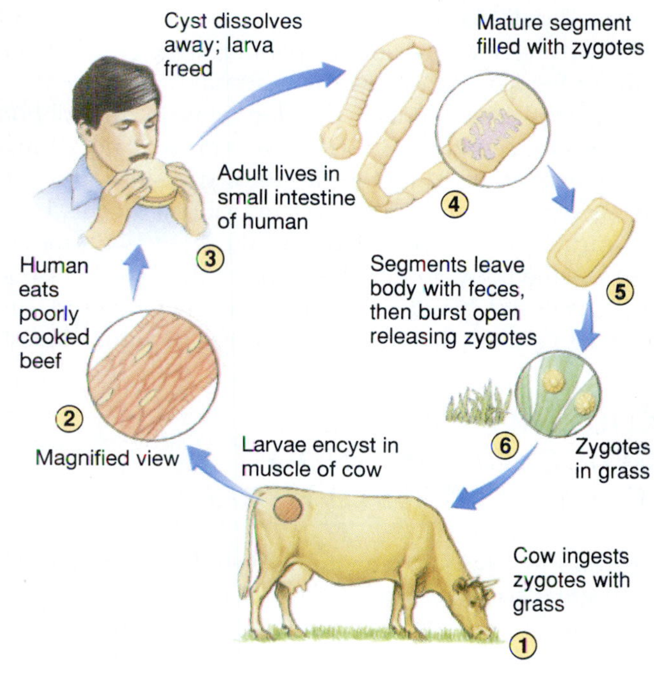

FIGURE 24–14 Life cycle of the beef tapeworm. Humans become infected when they eat poorly cooked beef containing the larvae. Cattle are intermediate hosts for this parasitic flatworm.

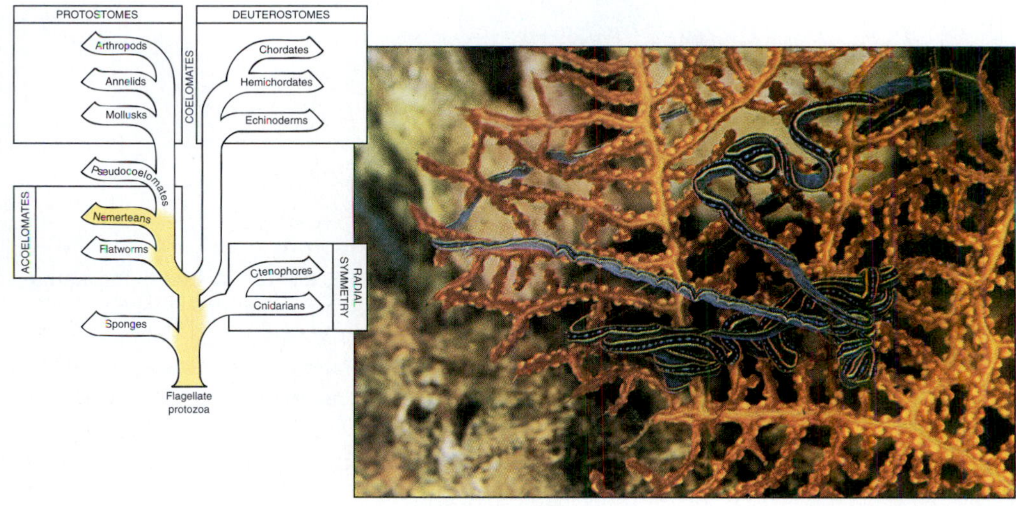

FIGURE 24–15 Ribbon worm or nemertean. Ribbon worm *(Lineus)* from Panama, Pacific. (Kjell B. Sandved)

Ribbon worms have a long, muscular tube called a **proboscis,** which they use to suck in prey or for defense. The proboscis secretes mucus and may be equipped with a barb and with poison-secreting glands.

An important evolutionary development in the ribbon worms is a *tube-within-a-tube* body plan. The digestive tract is a complete tube with a mouth at one end for taking in food and an anus at the other for eliminating wastes. Recall that the digestive tracts of cnidarians and flatworms have only a single opening, which must serve both for food intake and waste disposal.

Another evolutionary development in the ribbon worms is the separation of digestive and circulatory functions. These animals are the earliest organisms known to have a separate circulatory system. Ribbon worms have no heart. Blood is circulated through vessels by movements of the body and contractions of the blood vessels. Some have red blood cells containing hemoglobin, the same red pigment that transports oxygen in human blood.

ROUNDWORMS HAVE A PSEUDOCOELOM AND A COMPLETE DIGESTIVE TRACT

The roundworms, or nematodes **(phylum Nematoda),** are of great ecological importance because they are both numerous and widely distributed in the soil, the sea, and fresh water. More than 12,000 species have been identified. A spadeful of soil may contain more than a million of these mainly microscopic white worms, which thrash around in a coiling and uncoiling manner. Although many are free living (Fig. 24–16), others are important parasites in plants and animals. More than 30 roundworms are human parasites, including hookworms, the intestinal roundworm *Ascaris,* pinworms, trichina worms, and filarial worms.

The cylindrical, threadlike nematode body is pointed at both ends and covered with a tough, flexible **cuticle** (Fig. 24–17). Nematodes are the earliest animals known in which a body cavity, called a pseudocoelom, evolved. It is not a true coelom because it is not completely lined with mesoderm. Like the ribbon worms, the nematodes exhibit bilateral symmetry, a complete digestive tract, three definite tissue layers, and definite organ systems; however, they lack circulatory structures. The sexes are usually separate, and the male is generally smaller than the female.

Ascaris lumbricoides is an example of a parasitic roundworm. A whitish worm about 25 centimeters (10 inches) long, *Ascaris* is a common intestinal parasite of humans. This nematode spends its adult life in the human intestine, where it makes its living by sucking in partly digested food. Like the tapeworm, its reproductive output is enormous. The sexes are separate, and copulation takes place within the host. A mature female may lay as many as 200,000 eggs each day!

Ascaris eggs leave the human body with the feces and, where sanitation is poor (throughout most of the world), they find their way into the soil. In many parts of the world, farmers use human wastes as fertilizer, a practice that encourages the survival of *Ascaris* and many other human parasites. People are infected when they ingest *Ascaris* eggs. The eggs hatch in the intestine, and the larvae then journey through the body. During this migration, the larvae can damage the lungs and other tissues. Eventually, *Ascaris* settles down in the small intestine and may cause symptoms such as abdominal pain, allergic reactions, and malnutrition. Sometimes a tangled mass of these worms blocks the intestine.

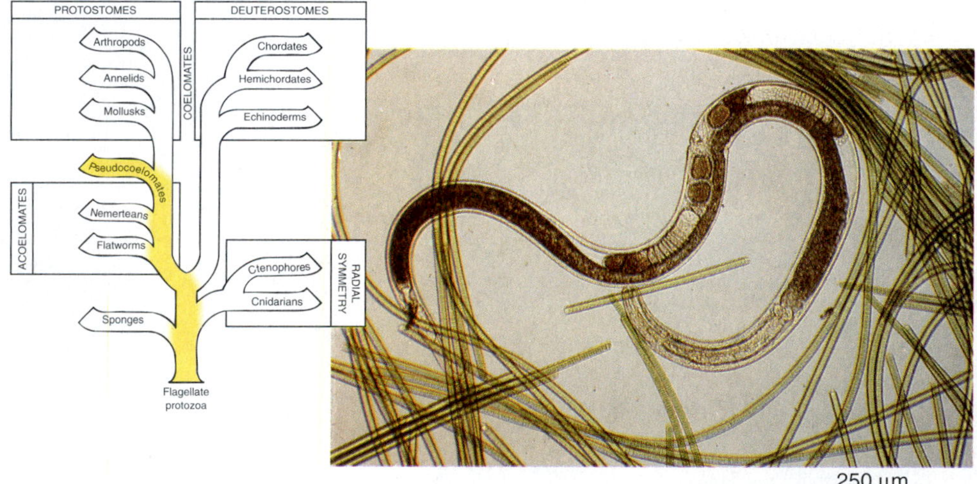

FIGURE 24–16 A free-living nematode. This nematode was photographed among cells of the cyanobacterium *Oscillatoria,* which it eats. (Visuals Unlimited/T. E. Adams)

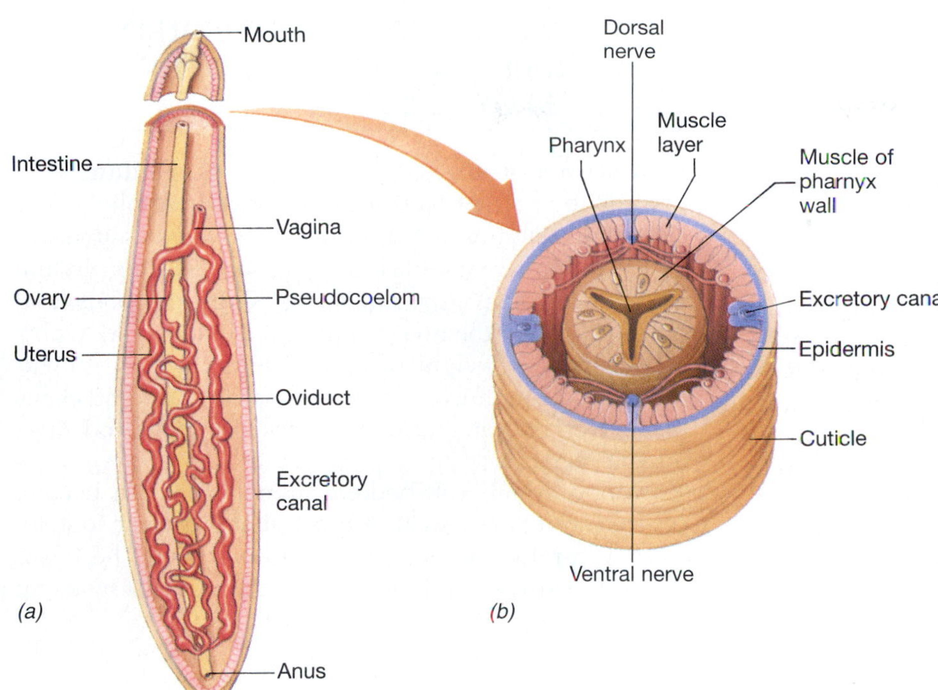

FIGURE 24–17 The structure of the roundworm *Ascaris.* (***a***) Longitudinal section of a female to show internal anatomy. Note the complete digestive tract that extends from mouth to anus. (***b***) Cross section through the body of *Ascaris*.

ROTIFERS ARE KNOWN AS WHEEL ANIMALS

Among the less familiar invertebrates are the 2000 or so species of "wheel animals" of **phylum Rotifera.** Although no larger than many protozoa, these tiny aquatic animals are multicellular. Rotifers have a characteristic crown of cilia on the anterior end. The cilia move rapidly, giving the appearance of a spinning wheel (Fig. 24–18).

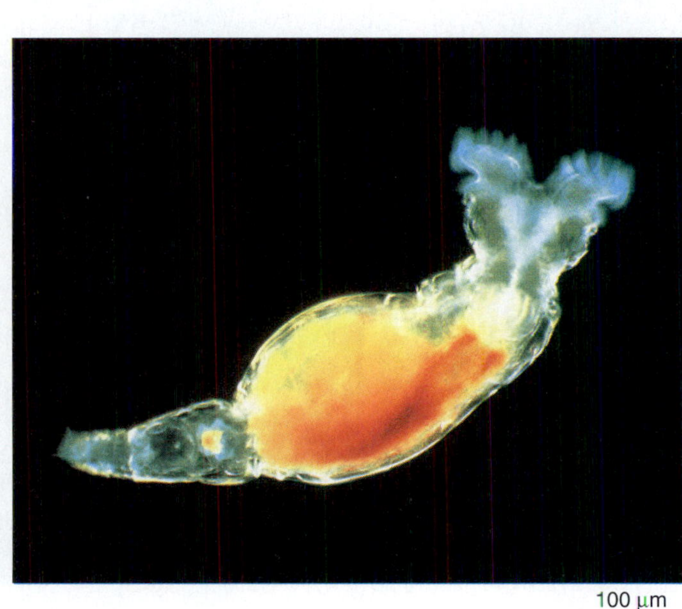

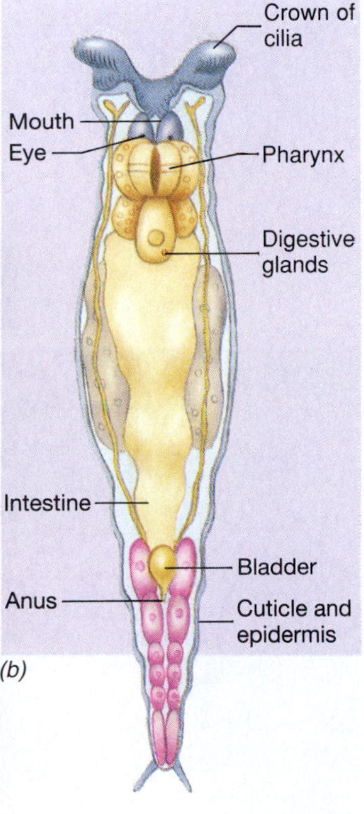

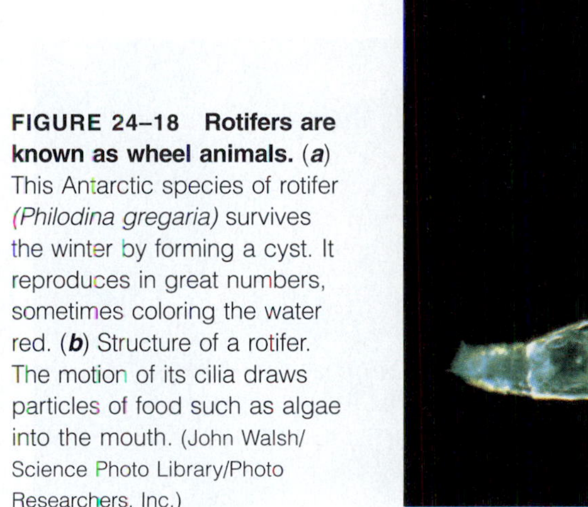

FIGURE 24–18 Rotifers are known as wheel animals. (***a***) This Antarctic species of rotifer *(Philodina gregaria)* survives the winter by forming a cyst. It reproduces in great numbers, sometimes coloring the water red. (***b***) Structure of a rotifer. The motion of its cilia draws particles of food such as algae into the mouth. (John Walsh/ Science Photo Library/Photo Researchers, Inc.)

Rotifers are pseudocoelomate animals that have a complete digestive tract, including a muscular pharynx for grinding food. These animals have a nervous system with a "brain" and sense organs, including eyespots. Protonephridia with flame cells remove excess water from the body of freshwater species, and may also excrete wastes.

Like some other pseudocoelomate animals, rotifers are "cell constant." This means that each member of a given species is composed of exactly the same number of cells. Indeed, each part of the body is made of a precisely fixed number of cells arranged in a characteristic pattern. Cell division does not take place after embryonic development, and mitosis cannot be induced. As a result, growth and repair are not possible. One of the challenging problems of biological research is discovering the difference between such nondividing cells and the dividing cells of other animals. Do rotifers never develop cancer?

MOST MOLLUSKS ARE COVERED BY A SHELL AND HAVE AN OPEN CIRCULATORY SYSTEM

With its more than 60,000 living species, **phylum Mollusca** is the second largest of all the animal phyla. The extensive evolution of the mollusks has led to impressive diversity and remarkable biological success. This phylum includes the clams, oysters, snails, slugs, octopods, and the largest of all the invertebrates, the giant squid, which may achieve a weight of several tons (Fig. 24–19). Although most mollusks are marine, some snails and clams live in fresh water, and many species of snails and slugs inhabit the land.

Mollusks are soft-bodied animals that are usually covered by a dorsal shell. A broad, flat muscular **foot,** located ventrally, can be used for locomotion. The heart, stomach, kidneys, and most of the other organs make up

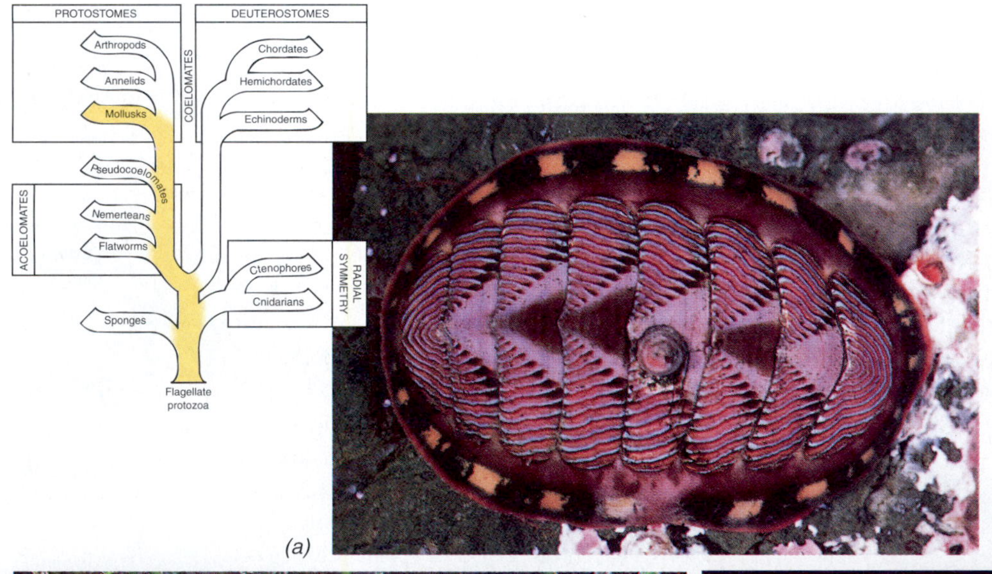

(a)

(b)

(c)

FIGURE 24–19 There are many beautiful forms of mollusks. (***a***) Chitons are sluggish marine animals with shells composed of eight overlapping plates. These sea cradle chitons *(Tonicella lineata)* are from coastal waters off the Pacific Northwest. (***b***) A bay scallop *(Argopecten irradians)* photographed in a sea grass bed in Tampa Bay. (***c***) The squid, a cephalopod, can change its color to blend with the background. This specimen *(Sepioteuthis lessoniana)* is native to Hawaii. (*a*, Visuals Unlimited/Kjell B. Sandved; *b*, Robin Lewis, Coastal Creations; *c*, Mike Severns/Tom Stack & Associates)

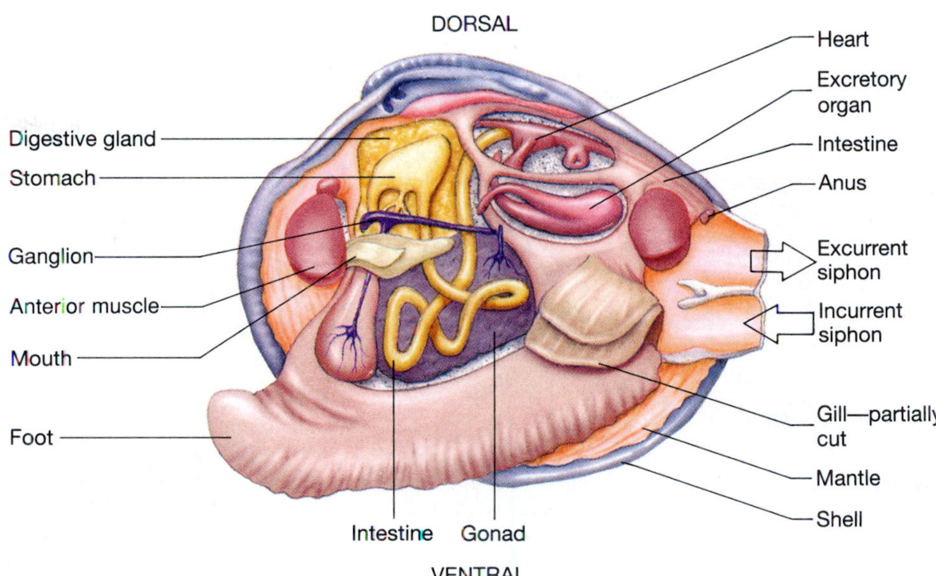

FIGURE 24–20 Internal anatomy of a clam.

a **visceral mass** that is located above the foot. The **mantle,** which is a heavy fold of tissue, covers the visceral mass and, in most species, contains glands that secrete the shell. The mantle generally overhangs the visceral mass, forming a mantle cavity that often contains a pair of **gills.** Most mollusks (except clams) have a **radula,** which is a tonguelike structure with hard teeth that are used to scrape up food.

Mollusks have a true coelom, but it is reduced in size in the adult. Most mollusks have an **open circulatory system,** a system in which blood does not remain within a circuit of blood vessels. Instead, blood leaves the blood vessels and flows into a network of large spaces, or sinuses, where it bathes the tissues directly. This network of sinuses makes up the **hemocoel,** or blood cavity.

Chitons Have Eight Shell Plates

The chitons (class Polyplacophora) are marine mollusks with shells divided into eight dorsal plates (Fig. 24–19*a*). The chiton has a small head and lacks eyes and tentacles. Moving via its large foot, this animal makes its way over rocks, using its radula to graze on algae.

Gastropods Are the Largest Group of Mollusks

The diverse gastropods (class Gastropoda)—snails and slugs—are marine, but some live in fresh water or on land. Gastropods have well-developed heads with eyes and one or two pairs of tentacles. Most gastropods have a single, coiled shell. A few species, such as garden slugs and sea slugs (nudibranchs), lack shells.

During development of the gastropod embryo, one side of the visceral mass grows more rapidly than the other side. This uneven growth results in rotation of the visceral mass. This rotation, called **torsion,** can be so marked that the anus and mantle cavity can end up above the head. The evolutionary advantage of torsion is not known. Some biologists have suggested that it protects the head and others have hypothesized that it helps distribute the weight and bulk so that the animal is better balanced.

Bivalves Typically Burrow in the Mud

Clams, oysters, and other bivalves are mollusks with two-part shells that hinge dorsally and open ventrally. The head is not distinct, but the foot is usually large and used for burrowing in the mud or sand.

The inner pearly layer of some bivalve shells is calcium carbonate secreted in thin sheets by the epithelial cells of the mantle. Known as *mother-of-pearl,* it is valued for making jewelry and buttons. Should a bit of foreign matter become lodged between the shell and the epithelium, the epithelial cells may be stimulated to secrete concentric layers of calcium carbonate around the intruding particle. Many species of oysters form pearls in this way.

Let us look at the clam as an example of a bivalve mollusk. The soft body of the clam is laterally flattened and completely enclosed by its two-part shell (Fig. 24–20). The hatchet-shaped foot protrudes ventrally for burrowing. Large, strong muscles attached to the shell permit the animal to close its shell. Openings are present through which water flows into and out of the mantle cavity. Ex-

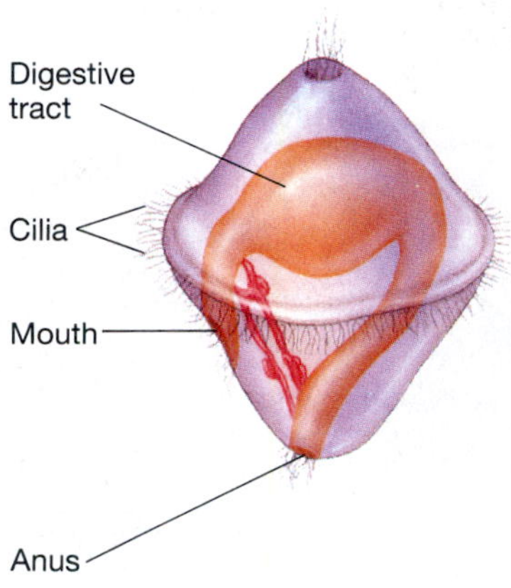

FIGURE 24–21 Trochophore larva. This is the first larval stage of a marine mollusk. The trochophore larva is also characteristic of annelids.

tensions of the mantle, called "siphons," permit the animal to obtain water relatively free of sediment. There is an **incurrent siphon** for water intake and an **excurrent siphon** for water output.

The clam takes its food from the sea water brought in over the gills by the siphon. Food particles are trapped in the mucus secreted by the gills. A stream of water, kept in motion by the beating of cilia on the surface of the gills, carries food particles around to the mouth.

All of the organ systems typical of complex animals are present in the clam. The digestive system is a coiled tube extending from mouth to anus. The open circulatory system consists of a heart that pumps blood into branching blood vessels. Eventually, blood leaves the vessels and flows into a network of large spaces that make up the **hemocoel.** The blood finds its way into special vessels that conduct it to the gills, where gas exchange takes place. It then circulates back to the heart.

Nerve cells connect three pairs of ganglia with one another and with the various organs. Sense receptors include organs of balance and cells sensitive to light and touch.

The sexes are usually separate in clams, and sperm are usually discharged into the water. In some species, fertilization takes place within the mantle cavity of the female. Typically, a free-swimming, ciliated, top-shaped larva, called a **trochophore larva,** develops (Fig. 24–21). The trochophore larva develops further into a **veliger larva** with shell and foot.

Cephalopods Are the Most Complex Mollusks

The cephalopods (class Cephalopoda)—the squids, octopods, nautiluses, and cuttlefish—are active predators. Cephalopods have large heads with conspicuous eyes. They have long tentacles used to capture prey. The octopus has no shell, and the shell of the squid is reduced to a thin "pen" in the mantle. (See Focus On the Environment: Adaptations of Octopods on page 471.)

Nautilus has a coiled shell consisting of many chambers built up over time. Each year the animal lives in the newest and largest chamber of the series. By secreting a gas resembling air into the other chambers, the *Nautilus* is able to regulate its depth in the water.

The Annelids Have Segmented Bodies

Members of **phylum Annelida,** the segmented worms, have bilateral symmetry and a tubular body that may be partitioned into more than 100 ringlike segments. This phylum, comprised of about 15,000 species, includes three major groups: the polychaetes, a group of mainly marine worms; the earthworms; and the leeches (Fig. 24–22).

The body segments of an annelid are separated from one another by transverse partitions, called **septa.** Some structures, such as the digestive tract and certain nerves, run the length of the body, passing through successive segments. Other structures, such as muscles and excretory organs, are repeated in each segment.

One advantage of segmentation is that it facilitates locomotion. Because the coelom is divided into segments, and each segment has its own muscles, the animal can elongate one part of its body while shortening another part.

Segmentation is very important from an evolutionary perspective because it provides the opportunity for specialization of body regions. In many annelids, the individual segments are almost all alike, but in many segmented animals (arthropods and chordates), different segments and groups of segments are specialized to perform different functions. In some groups, the specialization may be so pronounced that the basic segmentation of the body plan may not be apparent.

Bristle-like structures, called **setae,** aid in locomotion. An annelid has a well-developed coelom; a closed circulatory system; and a complete digestive tract, which consists of a tube extending from mouth to anus. Respiration takes place through the skin or by gills. Typically, a pair of excretory structures, called **nephridia,** are found in each segment. The nervous system generally consists of a simple brain composed of a pair of ganglia and a double ventral nerve cord. A pair of ganglia and lateral nerves are repeated in each segment.

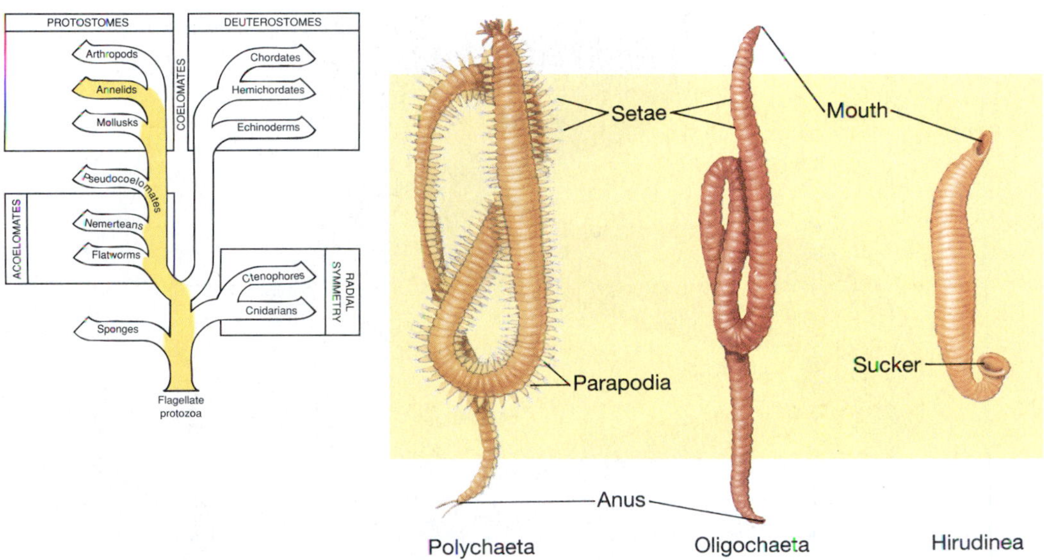

FIGURE 24–22 Comparison of the major classes of annelids. Polychaetes are marine worms with paddle-shaped appendages called parapodia. Oligochaetes, a class that includes the earthworms, inhabit fresh water and moist terrestrial areas. Many leeches, members of class Hirudinea, are bloodsucking parasites equipped with suckers.

Polychaetes Are Marine Worms

Most polychaetes (class Polychaeta; the name means "many hairs") are marine worms such as sandworms and tubeworms (Fig. 24–23). These animals swim freely in the sea or burrow in the mud near the shore. Some live in tubes they construct by cementing bits of shell and sand together with secretions from the body wall.

Most polychaetes have well-developed heads with eyes and antennae. The head may also be equipped with tentacles, bristles, and palps (feelers). Each body segment has a pair of paddle-shaped appendages called **parapo-**

(a)

(b)

FIGURE 24–23 Polychaete annelids. (***a***) The West Indian fireworm *(Hermodice carunculata)* is a crawling polychaete with poisonous setae. (***b***) The Christmas tree worm *(Spirobranchus giganteus)* photographed in a Florida coral reef. (*a*, Charles Seaborn/Odyssey Productions, Chicago; *b*, James H. Carmichael, Coastal Creations)

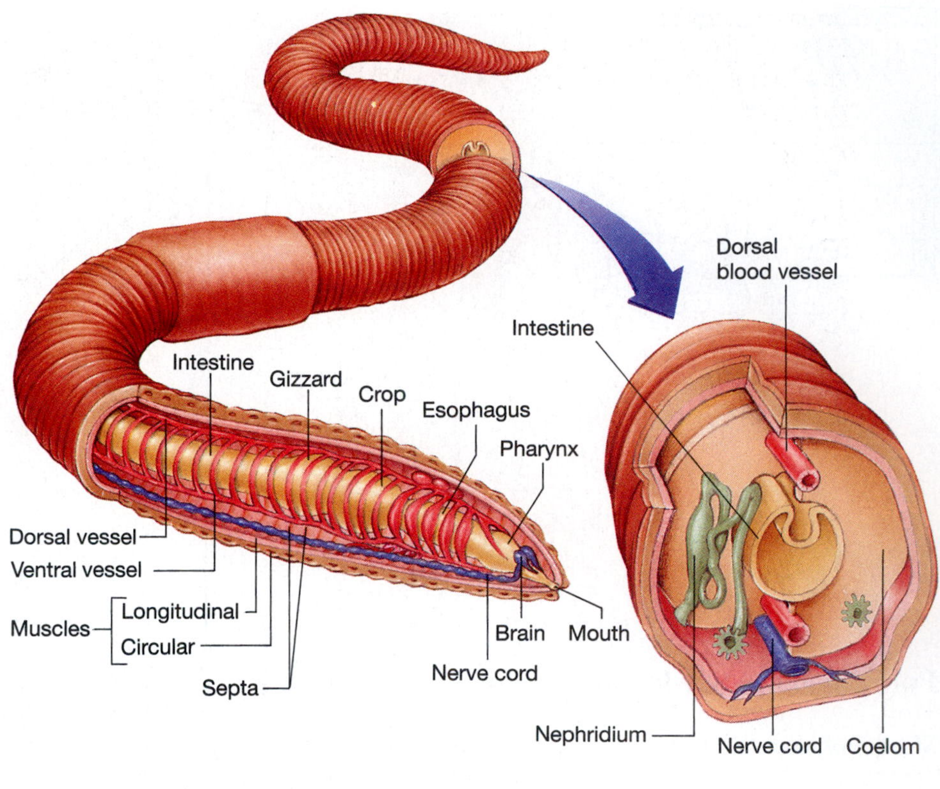

FIGURE 24–24 Structure of the earthworm. (***a***) The internal structure has been exposed at the anterior end of an earthworm. (***b***) Cross section of the earthworm.

dia that bear many stiff setae. These appendages help the worm move about in search of food.

Earthworms Are the Most Familiar Annelids

Earthworms (class Oligochaeta) are found almost exclusively in fresh water and in moist terrestrial habitats. These worms lack parapodia, have few bristles per segment, and lack many sensory structures on the head (Fig. 24–24). All earthworms are hermaphroditic.

Lumbricus terrestris, the common earthworm, is among the most familiar of all invertebrates. Its body is divided into more than 100 segments, and is protected from drying by a thin, transparent cuticle. The body wall has an outer layer of circular muscles and an inner layer of longitudinal muscles.

An earthworm literally eats its way through the soil, ingesting its own weight in soil and decaying vegetation every 24 hours. During this process, the soil is turned, aerated, and enriched by nitrogenous wastes from the earthworm. This is why earthworms are vital to the formation and maintenance of fertile soil. The earthworm's soil meal, containing nutritious decaying vegetation, is processed in its complex digestive system.

Some Leeches Are Blood-Sucking Parasites

Many leeches (class Hirudinea) are blood-sucking parasites that inhabit fresh water, although some species are marine and a few tropical species live on land. Terrestrial blood-sucking leeches drop on their hosts from foliage. Prominent muscular suckers, present at each end of the body, are used for clinging to the host. Some leeches are nonparasitic predators that capture other small invertebrates. Leeches lack both setae and appendages.

Leeches have been used since ancient times for drawing blood from areas swollen by poisonous stings and bites. During the 19th century, leeches were also widely used to remove "bad blood," which was thought to be the cause of many diseases. Recently, the use of leeches has been recognized by modern medicine. Leeches are sometimes used to remove blood that accumulates within body tissues as a result of injury, disease, or surgery (Fig. 24–25).

The leech attaches its sucker near the site of injury, makes an incision, and deposits an anticoagulant called hirudin. The hirudin prevents the blood from clotting and dissolves already-existing clots. In 30 minutes, a leech can suck out as much as ten times its own weight in blood.

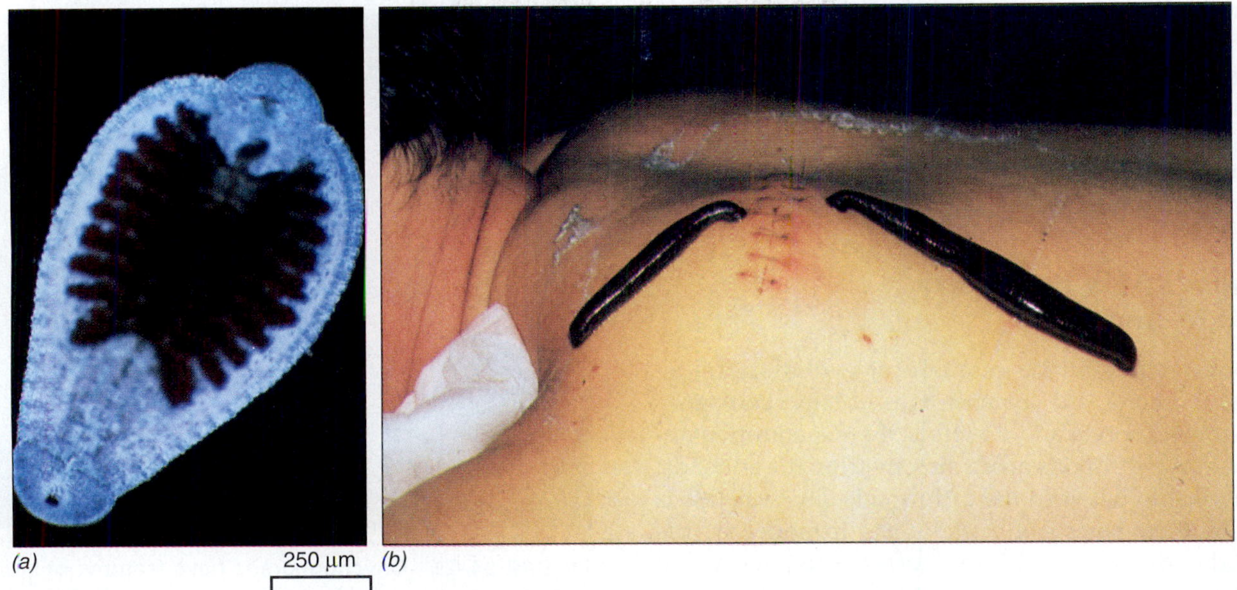

FIGURE 24–25 Most leeches are blood-sucking parasites. (***a***) A blood-sucking leech (*Helobdella stagnalis*) of mammals. The dark area within its swollen body is recently ingested blood. (***b***) The medicinal leech, *Hirudo medicinalis*, is used to treat hematoma, an accumulation of blood within body tissues that results from injury or disease. The leech releases an anticoagulant called hirudin that prevents the blood from clotting and dissolves already-existing clots. (*a*, Visuals Unlimited/T.E. Adams; *b*, St. Bartholomew's Hospital/Science Photo Library/Photo Researchers, Inc.)

ARTHROPODS HAVE JOINTED APPENDAGES AND AN EXOSKELETON OF CHITIN

Most biologists consider arthropods the most diverse and biologically successful group of animals. **Phylum Arthropoda** claims about one million species, and arthropods live in a greater range of habitats than do the members of any other phylum.

Five main adaptations are considered important to their biological success:

1. **Paired, jointed appendages,** from which they get their name (the word *arthropod* means "jointed foot"), may be specialized as swimming paddles, walking legs, mouthparts, sensory structures, or accessory reproductive organs that transfer sperm.
2. A hard, armor-like **exoskeleton,** composed of chitin and protein, covers the arthropod body and appendages. The exoskeleton provides protection against predators and against excessive loss of moisture. It also gives support to the underlying soft tissues. Distinct muscle bundles somewhat comparable to individual vertebrate muscles attach to the inner surface of the exoskeleton. The exoskeleton has an important disadvantage, however. As an arthropod grows, it periodically outgrows this nonliving shell. The process of shedding an old shell and growing another larger one is known as **molting.** The shed exoskeleton represents a net metabolic loss, and molting also leaves the arthropod temporarily vulnerable to predators.
3. The arthropod body, like that of the annelid, is **segmented.** In most arthropods, however, segments become fused together into groups that are specialized to perform certain functions. In many arthropods, segments have organized to form three main body regions: the *head, thorax,* and *abdomen.*
4. Arthropods have specialized respiratory systems. Most of the aquatic arthropods have a system of gills that function in gas exchange. The land forms, in contrast, have a system of fine, branching air tubes, called **tracheae,** or platelike structures, called book lungs.
5. Arthropods have a well-developed nervous system, including a brain, a double ventral nerve cord, and a variety of very effective sense organs. For example, many crustaceans and most insects have compound eyes (discussed in Chapter 35). Many arthropods have antennae sensitive to touch and chemicals.

Arthropods have an open circulatory system. A dorsal, tubular heart pumps blood into a dorsal artery, which may branch into smaller arteries. From the arteries, blood flows into large spaces that collectively make up the hemocoel. Blood in the hemocoel bathes the tissues directly. Eventually, blood finds its way back into the heart through openings, called **ostia,** in its walls.

Arthropods May Be Closely Related to Annelids

Arthropods represent the pinnacle of evolutionary development among the protostomes. Many zoologists think that arthropods and annelids share a common ancestor, or that arthropods arose directly from the annelids. The relationship between the arthropods and annelids is evident in their basic body plans. As in annelids, the arthropod body is divided into a series of segments, and the segments are formed in the same way in both groups. The basic plan of the nervous system is similar in both annelids and arthropods. A double ventral nerve cord extends from a dorsal, anterior brain.

Some zoologists propose that two arthropod subphyla (the chelicerates and crustaceans) descended from a marine ancestor and that the third subphylum (uniramians, the subphylum that includes the insects) descended from a terrestrial ancestor. According to this polyphyletic view, basic arthropod features such as jointed appendages and an exoskeleton of chitin evolved independently at least twice. Zoologists who hold this view suggest that uniramians evolved from members of **phylum Onychophora,** which are wormlike animals that inhabit humid tropical rain forests. Onychophorans have both annelid and arthropod features (Fig. 24–26). Like annelids, they are internally segmented, and many organs are duplicated serially. However, the jaws develop from appendages as in arthropods, and like arthropods, onychophorans have an open circulatory system and a respiratory system, which consists of tracheal tubes.

FIGURE 24–26 Onychophorans have features of both arthropods and annelids. Note the soft, sluglike body and the presence of a series of nonjointed legs (Visuals Unlimited/Thomas C. Boydean)

CONCEPT CONNECTIONS

Diversity* — *Evolution* — *Molecular Taxonomy

Many zoologists have viewed the onychophorans (velvet worms) as the "missing link" between the annelids and arthropods. As described in this chapter, these animals have some annelid and some arthropod characteristics. They are internally segmented like annelids, but have an open circulatory system and tracheal tubes like arthropods. Also, like arthropods they have antennae and grow by molting.

Recent molecular studies bring the evolutionary position of the onychophorans into question. Investigators compared the molecular structure of a mitochondrial gene from a variety of arthropods and other species with six onychophoran species. Their results suggest that, rather than a link between the annelids and arthropods, the onychophorans are arthropods that became adapted to a very specialized life-style. Are these data from molecular taxonomy accurate? Are the conclusions taxonomists have drawn from the fossil record and from comparative anatomy incorrect? A decision on the evolutionary place of the onychophorans must await additional studies.

Many taxonomists disagree about arthropod classification. Here we use a scheme that divides the phylum into three living subphyla: the chelicerates (horseshoe crabs, spiders, scorpions, ticks, and mites); the crustaceans (lobsters, crabs, shrimp); and the insects and their relatives (centipedes and millipedes).

Most Chelicerates Are Arachnids

The chelicerates **(subphylum Chelicerata)** are the horseshoe crabs, spiders, scorpions, ticks, and mites. The chelicerate body is divided into a cephalothorax (fused head and thorax) and an abdomen. These animals have no antennae and no chewing mandibles (which are mouthparts that are characteristic of other arthropod subphyla). Instead, the first pair of appendages, located immediately anterior to the mouth, are the **chelicerae,** which are used to manipulate food and pass it to the mouth. The second pair of appendages, called **pedipalps,** are modified to perform different functions in various groups. Posterior to the pedipalps are usually four pairs of legs.

Two groups in this subphylum are the merostomes (class Merostomata) and the arachnids (class Arachnida; Fig. 24–27). The only living merostomes are the horseshoe crabs.

The arachnids include the spiders, scorpions, mites, ticks, and harvestmen (daddy longlegs). Arachnids are generally carnivorous and prey upon insects and other small arthropods (Fig. 24–27). Arachnids have six pairs of jointed appendages. They use the first pair, the fanglike chelicerae, to penetrate their prey. Some arachnids use the chelicerae to inject poison into the prey. Scorpions use the second pair of appendages, the pedipalps, to capture and hold their prey. In most spiders, the pedipalps are specialized as sensory structures.

The spider has **silk glands** in its abdomen that secrete an elastic protein. Using organs called spinnerets, the spider spins the protein into silk fibers. The silk is liquid as it emerges from the spinnerets but hardens as it is drawn out. Spiders use silk to build nests, to encase their eggs in a cocoon, and in some species, to trap prey. Many spiders lay down a silken dragline as they venture forth. This serves as a safety line and as a means of communication among members of a species. Amazingly, a spider can determine the sex and maturity level of a spinner from its dragline.

Although spiders do have poison glands for capturing prey, only a few have poison that is dangerously toxic to humans. The most widely distributed poisonous spider in the United States is the black widow (*Latrodectus mactans*). Its poison is a neurotoxin that interferes with the transmission of messages from nerves to muscles. The brown recluse (*Loxosceles reclusa*) is smaller than the black widow and has a violin-shaped dorsal stripe on its back. Its venom causes death of the tissues surrounding the bite and can occasionally be fatal. Although painful, spider bites cause fewer than five fatalities per year in the United States (usually in children).

Mites and ticks are among the most serious arthropod nuisances. They eat crops, infect livestock and pets, and inhabit our own bodies. Many live unnoticed, owing to their small size; however, others cause disease. Certain mites cause mange in dogs and other domestic animals. Chiggers (red bugs), the larval form of red mites, attach themselves to the skin and secrete an irritating digestive fluid that may cause itchy red welts. Larger than mites, ticks are external parasites on dogs and other animals. They can transmit diseases such as Rocky Mountain spotted fever, Texas cattle fever, and Lyme disease.

FIGURE 24–27 Subphylum Chelicerata includes the merostomes (horseshoe crabs) and arachnids. (***a***) The only living merostomes are a few closely related species of horseshoe crabs. Seasonally, horseshoe crabs (*Limulus polyphemus*) return to beaches for mating. In this photograph, males are competing for a female. (***b***) This orb spider (*Araneus quadratus*) is shown building a cocoon. (*a*, Visuals Unlimited/Milton H. Tierney, Jr.; *b*, Hans Pfetchinger/Peter Arnold, Inc.)

(*a*)

(*b*)

FIGURE 24–28 Most members of subphylum Crustacea are aquatic. Hermit crab on coral. (The Stock Market/Thomas Dimock, 1995)

Crustaceans Are Vital Members of Marine Food Chains

Crustaceans **(subphylum Crustacea)**—lobsters, crabs, shrimp, and barnacles—are consumers of algae, small animals, and detritus (dead organic debris). Countless billions of crustaceans swarm in the ocean and become food for many fishes and other marine animals such as whales.

Crustaceans have **biramous appendages,** which means that each appendage has two jointed branches. Crustaceans also are the only arthropods to have two pairs of **antennae,** which are sensory organs for touch and taste. Crustaceans are characterized by **mandibles** located on each side of the ventral mouth that are used for biting and grinding food. The most familiar crustaceans have five pairs of walking legs. Other appendages may be specialized for swimming, sperm transmission, carrying eggs and young, or sensation (Fig. 24–28).

Barnacles are sessile animals that differ markedly in their external anatomy from other crustaceans. They are exclusively marine and secrete hard limestone cups within which they live. The larvae of barnacles are free-swimming forms that go through several molts before becoming sessile and developing into the adult form. Barnacles are the bane of marine boaters. They can grow on ship bottoms in such great numbers that their presence may reduce the speed of a ship by more than 30%.

Insects and Their Relatives Have Unbranched Appendages

The insects, centipedes, and millipedes are grouped together in **subphylum Uniramia** because they all possess unbranched **(uniramous)** appendages. They also have only one pair of antennae rather than two pairs, as in crustaceans.

Both centipedes and millipedes have heads and elongated bodies with many segments, each bearing legs (Fig. 24–29). A centipede has one pair of legs on each segment; a millipede has two pairs of legs per adult segment. These animals are all terrestrial and are typically found beneath

FIGURE 24–29 Subphylum Uniramia includes the insects, centipedes, and millipedes. (***a***) A tropical millipede from Rancho Grande, Venezuela. (***b***) A tropical centipede (*Scolopendra* sp.) from the Amazon region of Peru. (***c***) This red dragonfly (*Ambroseli*), a native of Kenya, is a predaceous insect with chewing mouthparts, large eyes, and long narrow wings. (*a*, Patti Murray; *b*, and *c*, James L. Castner)

stones or wood in the soil in temperate and tropical regions. Centipedes are carnivorous and feed on other animals, mostly insects. Larger centipedes have been known to eat snakes, mice, and frogs. The prey is captured and killed with poison claws. Millipedes are generally herbivorous and feed on both living and decaying vegetation.

The insects belong to class Insecta (Fig. 24–29). The earliest fossil insects—primitive, wingless species—date back to the Devonian Period more than 360 million years ago. Insect fossils from the Carboniferous Period include both wingless and primitive winged species. Cockroaches, mayflies, and cicadas are among the insects that have survived relatively unchanged from the Carboniferous Period to the present day.

With more than 750,000 described species, the insects are the most successful group of animals on our planet in terms of number of species and their diversity. Insects have adapted to almost every available environment. Some insect species live in fresh water, some are truly marine, and others inhabit the shore between the tides. Most species, however, are terrestrial. In fact, the insects are one of only a few groups of animals that have adapted to life on land.

More species of insects are known than of all other animals combined. What they lack in size, insects make up for in sheer numbers. It has been calculated that all of the insects in the world would together weigh more than all of the remaining land animals on Earth.

What are the secrets of insect success? One important factor is the adaptability of their body plan, which has been modified and specialized in so many ways that insects have adapted to an incredible number of life-styles. One of their most important adaptations is their ability to fly. Unlike most other invertebrates, which swim or creep slowly along (or under) the ground, many insects fly rapidly through the air. Their wings allow them to migrate into regions that would be inaccessible to flightless animals. Their small size has also facilitated their wide distribution, because they are easily carried by wind or water currents.

The insect body is well protected by its tough exoskeleton, which also helps prevent water loss by evaporation. Other protective mechanisms include mimicry, protective coloration (Chapter 46), and aggressive behavior. Metamorphosis divides the insect life cycle into different stages, a strategy that keeps larval forms from competing with adults for food or habitats.

An insect may be described as an **articulated** (jointed), **tracheated** (having tracheal tubes for gas exchange) **hexapod** (having six "feet"). The insect body consists of three distinct parts: the head, thorax, and abdomen. Three pairs of legs and often two pairs of wings extend from the thorax. A single pair of antennae protrudes from the head (Fig. 24–30). A complex set of mouthparts is present and may be adapted for piercing, chewing, sucking, or lapping.

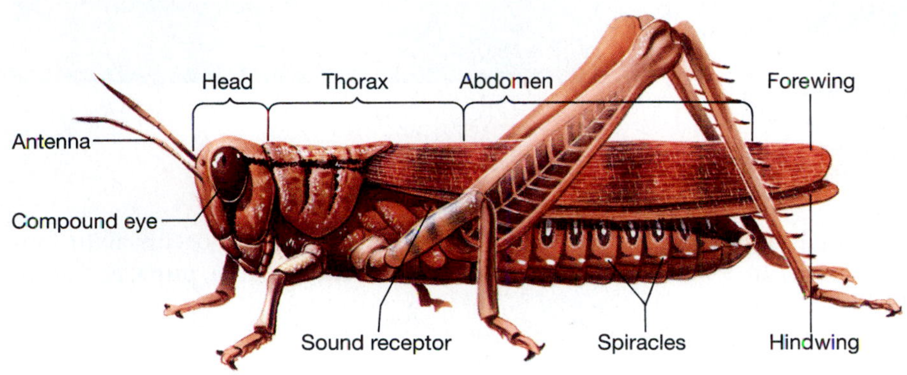

(a)

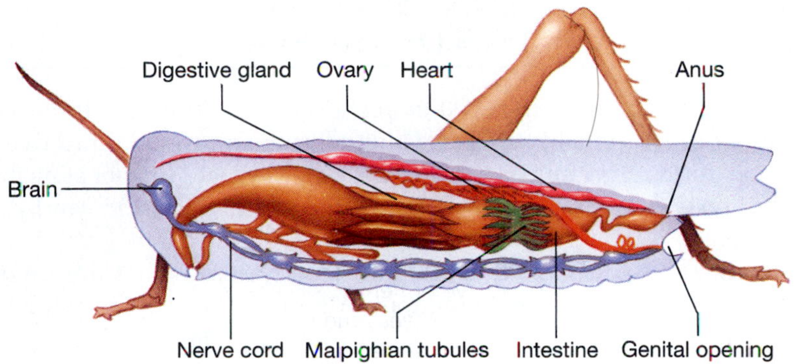

(b)

FIGURE 24–30 Structure of the grasshopper. (***a***) External anatomy. Note the three pairs of articulated legs. (***b***) Internal anatomy of the grasshopper.

FIGURE 24–31 Many insects undergo complete metamorphosis. The life cycle of the morpho butterfly (*Morpho peleides*) includes (***a***) egg, (***b***) larva, (***c***) pupa, and (***d***), (***e***) adult stages. (Patti Murray)

The tracheal system of insects has contributed importantly to their diversity. Air enters the tracheae through tiny openings, called **spiracles,** in the body wall. The trachea permit oxygen to pass directly to the internal organs. This effective oxygen delivery system permits the insects to have the high metabolic rate necessary for activities such as flight.

Excretion is accomplished by two or more slender **Malpighian tubules** that receive metabolic wastes from the blood. After concentrating the wastes, the Malpighian tubules discharge them into the intestine.

The sexes are separate in insects. Fertilization takes place internally, which is an important adaptation for land animals. During development, several molts take place. In some orders, there are several developmental stages, called **nymphal stages,** which result in **gradual metamorphosis** (change in body form) to the adult form. In others, a **complete metamorphosis** occurs with four distinct stages in the life cycle: **egg, larva, pupa,** and **adult** (Fig. 24–31).

THE ECHINODERMS ARE "SPINY-SKINNED" ANIMALS OF THE SEA

All of the 6000 living species of **phylum Echinodermata** inhabit the sea. Echinoderms are divided into five main groups (Fig. 24–32): the sea lilies and feather stars, the sea stars, the brittle stars, the sea urchins and sand dollars, and the sea cucumbers.

Echinoderms are deuterostomes that are unique in the animal kingdom. Although their larvae exhibit bilat-

(Text continues on page 471)

FIGURE 24–32 Some representatives of phylum Echinodermata. (***a***) Feather stars such as this deep sea crinoid, use their slender, jointed appendages to cling to the surface of a rock or coral reef. They are able to crawl and swim, and often swim away to escape predators. (***b***) The leather star (*Dermasterias imbricata*) a seastar, can be found along the Pacific coast from California to Alaska. (***c***) This daisy brittle-star (*Ophiopholis aculeata*) was photographed in Muscongus Bay, Maine. (***d***) With their flattened, circular bodies, sand dollars are well suited for burrowing in the sea bottom. (*Derdraster excentricus*). (***e***) A sea cucumber, *Thelonota sp.*, raising its body to spawn. (*a*, D.J. Wrobel, Monterey Bay Aquarium/Biological Photo Service; *b*, Sharksong/M. Kazmers/Dembinsky Photo Associates; *c*, Robert Dunne/Photo Researchers, Inc.; *d*, D.J. Wrobel, Monterey Bay Aquarium/Biological Photo Service; *e*, Peter Scoones/Seaphot, Ltd.)

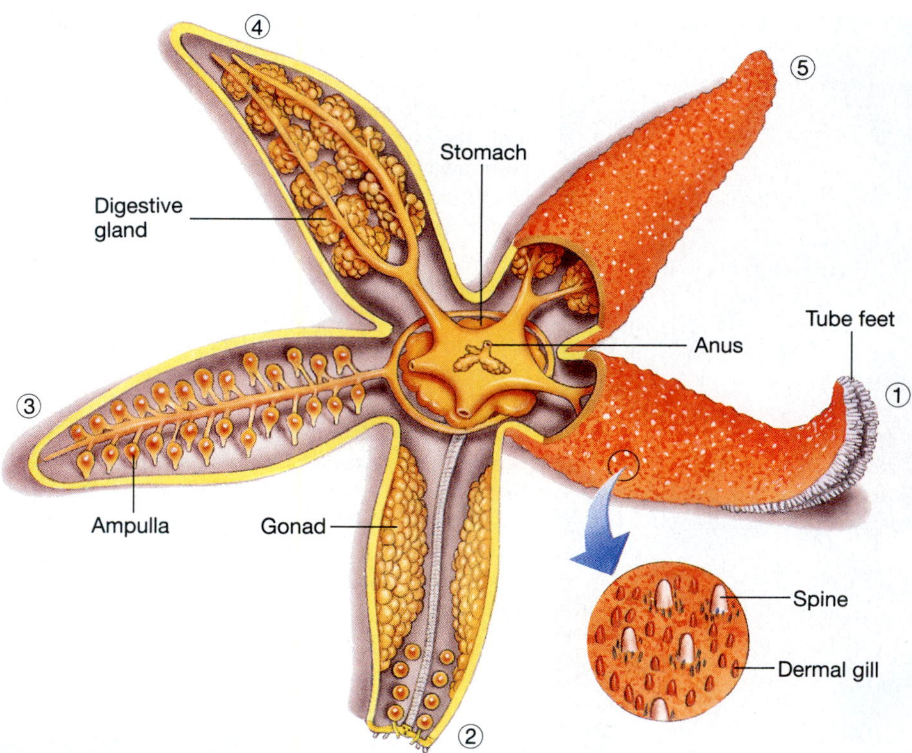

FIGURE 24–33 The sea star. Structure of the sea star *Asterias* viewed from above with the arms in various stages of dissection. (1) Upper surface with a magnified detail showing the features of the surface. The end is turned up to show the tube feet on the lower surface. (2) Arm dissected to show well-developed gonads. Gonads are present in each arm. (3) Upper body and digestive glands have been removed, showing the ampullae of some of the hundreds of tube feet (magnified view). (4) Other organs have been removed to show the digestive glands (actually present in each arm). (5) Upper (dorsal) surface. The two-part stomach is in the central disc with the anus dorsal (uppermost) and the mouth beneath on the ventral surface.

CONCEPT CONNECTIONS

Diffusion ⟷ *Animal Processes*

In Chapter 6, we discussed diffusion as a passive process by which atoms, ions, and small molecules pass into and out of cells. In this chapter, we have seen that small, simple animals depend on diffusion for many of their metabolic needs. Diffusion works well over small distances (micrometers), but if substances need to be moved even a few centimeters, diffusion might require several days. Because diffusion is slow, large, multicellular organisms require specialized structures for transporting materials.

Sponges, cnidarians, flatworms, ribbon worms, and roundworms depend on diffusion for many life processes, including gas exchange. Diffusion alone cannot provide sufficient oxygen for larger, more complex animals because oxygen diffuses slowly through tissues. In an animal more than 1 mm thick, diffusion is insufficient to meet the needs of the body cells for oxygen. Although oxygen can diffuse through the moist surface of the annelid body, in order to reach the cells deep within the body, it must be circulated in the blood (discussed further in Chapters 36 and 38). Specialized respiratory and circulatory structures have evolved in mollusks and arthropods. Terrestrial arthropods have an open circulatory system, as well as a system of air tubes (tracheae) or book lungs.

Disposal of metabolic wastes also takes place by diffusion in small, simple animals. More complex animals require specialized excretory structures. For example, annelids have nephridia in each segment of the body, and insects have Malpighian tubules for collecting and concentrating wastes.

In this chapter, we have traced the evolution of invertebrate adaptations for carrying on many life processes. In Part 8, Animal Structure and Life Processes, we compare in greater detail the structures and mechanisms that have adapted various groups to the challenges of obtaining oxygen; processing food; circulating nutrients, oxygen, and other substances; disposing of metabolic waste; and responding to stimuli.

FOCUS ON *Evolution*

ADAPTATIONS OF OCTOPODS

During the day, the octopus usually hides among the rocks or inside its den, which may be located within a crevice in a reef or rock. At night, it emerges in search of food. The octopus is a master at camouflage. By expanding and contracting pigment cells in its skin, the octopus can display color patterns that allow it to blend into its surroundings where it waits to snare an unwary crab, snail, or clam. It seizes its prey with long, powerful tentacles equipped with suckers. The suckers permit the octopus to hold onto slippery prey while paralyzing it with a nerve poison secreted by the salivary glands. The mouth of the octopod is equipped with a radula, something like a belt of teeth. Its mouth has two strong beaklike jaws that are used to puncture prey.

The octopus has eight tentacles or arms (ten in squids), that surround the central mouth of the large head. The motion of the octopus is incredibly fluid, giving little hint of the considerable strength in its tentacles.

Like other cephalopods, the octopus has large, well-developed eyes that form images. Although they develop differently, the eyes are somewhat like vertebrate eyes.

The octopus has several remarkable adaptations that enable it to avoid or escape predators, which include moray eels, sharks, and dolphins. First, the camouflaging ability that allows it to seize its prey also enables it to avoid detection by its predators. When spotted by a predator, the octopus may confuse its enemy by rapidly changing colors.

When in danger, the octopus uses a kind of jet propulsion to escape. Its mantle is thick and muscular and fitted with a funnel-like structure. By filling the mantle cavity with water and ejecting it through the funnel, the animal can move rapidly in the opposite direction.

Another defense mechanism is the large ink sac, which produces a thick black fluid. The ink sac opens into the rectum near the anus. When the octopus is alarmed, the ink is released through the anus, forming a dark cloud in the water. While its enemy pauses, temporarily confused, the octopus escapes. The ink has been shown to desensitize the sense of smell of some predators.

Octopods have a relatively high degree of intelligence and are capable of learning. For example, in laboratory studies they have learned to distinguish between shapes. Their very adaptable behavior resembles more closely that of the vertebrates than the more stereotyped patterns of behavior seen in most other invertebrates.

eral symmetry, the adults have **pentaradial symmetry** (Fig. 24–33). This means that the body is arranged in five parts around a central axis. A thin, ciliated **epidermis** covers the endoskeleton (internal skeleton), which consists of small plates composed of calcium carbonate. These plates generally bear spines that project outward. The spines are the basis for the name phylum Echinodermata, which means "spiny-skinned."

A characteristic found only in echinoderms is the **water vascular system,** which is a network of canals through which sea water circulates. Branches lead to numerous tiny **tube feet** that extend when filled with fluid and serve in locomotion, obtaining food, and, in some forms, gas exchange. The water vascular system is a hydraulic system. When the echinoderm begins to extend a foot, a rounded muscular sac **(ampulla)** at the upper end contracts, forcing water into the tube of the foot. At the bottom of the foot is a suction-type structure that adheres to whatever surface the animal is perched upon.

Echinoderms have a well-developed coelom. The complete digestive system is the most prominent body system. A variety of respiratory structures are found in the various classes, including dermal (skin) gills in the sea stars. Only a rudimentary circulatory system is present and there are no specialized excretory structures. The nervous system is simple, usually consisting of nerve rings around the mouth with radiating nerves.

Chapter Summary

I. Animals are euykaryotic, multicellular heterotrophs whose cells exhibit a division of labor. Most are capable of locomotion at some time during their life cycle, can reproduce sexually, and respond adaptively to external stimuli.

II. Animals have complex evolutionary relationships.
  A. Most animals exhibit bilateral symmetry; cnidarians and ctenophores have radial symmetry.
  B. Acoelomates are animals that lack a body cavity. Pseudocoelomates have a body cavity that is not completely lined with mesoderm. Coelomates have a true coelom lined with mesoderm.
  C. Coelomate animals can be grouped as protostomes or deuterostomes.

III. Phylum Porifera consists of the sponges, the simplest animals. The sponge body is a sac perforated by tiny holes through which water enters. As water circulates through the sponge, materials are exchanged by diffusion.

IV. Phylum Cnidaria includes the hydras, jellyfish, and corals.
  A. Cnidarians are characterized by radial symmetry, cnidocytes, two definite tissue layers, and a nerve net.
  B. Many cnidarians have both polyp and medusa body forms.

V. Comb jellies (phylum Ctenophora) are fragile, luminescent, biradially symmetrical marine animals.

VI. Flatworms (phylum Platyhelminthes) include the planarians, flukes, and tapeworms.
  A. Flatworms have bilateral symmetry; cephalization; three definite tissue layers; simple brains and nervous systems; and other well-developed organs, including protonephridia with flame cells.
  B. Flukes and tapeworms have suckers, complex life cycles, and other adaptations for their parasitic lifestyles.

VII. Ribbon worms (phylum Nemertea) have a tube-within-a-tube body plan, a complete digestive tract with mouth and anus, and a circulatory system.

VIII. Roundworms (phylum Nematoda) include species of great ecological importance and species parasitic in plants and animals.
  A. Nematodes have three definite tissue layers, a pseudocoelom, and a complete digestive tract.
  B. Nematodes parasitic in humans include *Ascaris*, hookworms, trichina worms, and pinworms.

IX. Members of phylum Rotifera are aquatic, pseudocoelomate, microscopic animals that exhibit cell constancy.

X. Mollusks (phylum Mollusca) are soft-bodied animals usually covered by a shell. They possess a ventral foot for locomotion and a mantle that covers the visceral mass.
  A. The gastropods (snails and slugs) comprise the largest group of mollusks. In gastropods, the body is rotated, and the shell (when present) is usually coiled.
  B. Bivalves include clams and oysters, both of which are enclosed by two shells that are hinged dorsally.
  C. Cephalopods (squids and octopods) are active predatory animals with large heads, long tentacles, and prominent eyes.

XI. Phylum Annelida, the segmented worms, have long bodies that are segmented internally as well as externally. Setae aid in locomotion.
  A. Polychaetes (marine worms) have bristled parapodia that function in locomotion.
  B. Earthworms are segmented worms characterized by a few setae per segment. The body may be divided into more than 100 segments that are separated internally by septa.
  C. Leeches lack setae and appendages, but are equipped with suckers.

XII. Phylum Arthropoda is the largest animal phylum. Arthropods have jointed appendages and armor-like exoskeletons.
  A. Chelicerates include merostomes (horseshoe crabs) and arachnids (spiders, mites, and their relatives).
    1. In the chelicerates, the first pair of appendages are chelicerae, which are used to manipulate food. Chelicerates have no antennae and no mandibles.
    2. The arachnid body consists of a cephalothorax and an abdomen. There are six pairs of jointed appendages, of which four pairs serve as walking legs.
  B. Crustaceans include lobsters, crabs, and barnacles. The body consists of a cephalothorax and an abdomen. Crustaceans have two pairs of antennae and mandibles for chewing.
  C. Insects and their relatives (centipedes and millipedes) have unbranched appendages and a single pair of antennae.
    1. An insect is an articulated, tracheated hexapod. Its body consists of a head, thorax, and abdomen. Insects claim a greater number of species and greater diversity than any other animal group.
    2. The centipedes have one pair of legs per body segment, whereas the millipedes have two pairs of legs per body segment.

XIII. Phylum Echinodermata is made up of the sea stars, sea urchins, sand dollars, and sea cucumbers.
  A. Echinoderms are deuterostomes with a spiny endoskeleton; as adults they exhibit pentaradial symmetry.
  B. Other unique features of the echinoderms are the water vascular system and tube feet.

Selected Key Terms

bilateral symmetry, p. 444
biramous appendages, p. 466
cephalization, p. 452
cnidocytes, p. 449
coelom, p. 445
cuticle, p. 456
deuterostome, p. 446
ectoderm, p. 444
endoderm, p. 444
exoskeleton, p. 463
flame cells, p. 452
ganglion, p. 452
gastrovascular cavity, p. 449
germ layers, p. 444
gill, p. 459
hemocoel, p. 459
hermaphrodite, p. 448
invertebrate, p. 440
larva, p. 447
Malphighian tubules, p. 468
mandibles, p. 466
mantle, p. 459
medusa, p. 449
mesoderm, p. 444
metamorphosis, p. 468
nematocyst, p. 449
nephridia, p. 460
nerve net, p. 449
pedipalps, p. 464
polyp, p. 449
protonephridia, p. 452
protostome, p. 445
pseudocoelom, p. 445
radial symmetry, p. 440
radula, p. 459
setae, p. 460
spiracles, p. 468
tracheae, p. 463
tube feet, p. 471
tube-within-a-tube body plan, p. 444
visceral mass, p. 459
water vascular system, p. 471

Post-Test

1. Animals without backbones are known as ________________.

2. An animal that can be divided by only one plane into roughly right and left halves may be described as ________________ ________________.

3. An acoelomate animal lacks a/an ________________ ________________.

4. Two deuterostome phyla are the ________________ and the ________________.

5. Sponges secrete skeletal structures called ________________.

6. A hermaphroditic animal can produce ________________ and ________________.

7. A distinctive feature of the cnidarians is the presence of stinging cells called ________________.

8. The nerve net is characteristic of (a) sponges (b) cnidarians (c) mollusks (d) echinoderms (e) arthropods.

9. Tube feet are characteristic of (a) sponges (b) cnidarians (c) mollusks (d) echinoderms (e) arthropods.

10. Protonephridia with flame cells are characteristic of (a) sponges (b) cnidarians (c) flatworms (d) echinoderms (e) arthropods.

11. Jointed appendages and an exoskeleton are characteristic of (a) cnidarians (b) flatworms (c) annelids (d) mollusks (e) arthropods.

12. Animals with soft bodies and a ventral foot and mantle are (a) cnidarians (b) flatworms (c) annelids (d) mollusks (e) arthropods.

13. Sponges are members of phylum (a) Porifera (b) Cnidaria (c) Annelida (d) Mollusca (e) Arthropoda.

14. Snails are members of phylum (a) Porifera (b) Cnidaria (c) Annelida (d) Mollusca (e) Arthropoda.

15. Sea stars are members of phylum (a) Porifera (b) Cnidaria (c) Echinodermata (d) Mollusca (e) Arthropoda.

16. Tapeworms are members of phylum (a) Cnidaria (b) Platyhelminthes (c) Echinodermata (d) Annelida (e) Arthropoda.

17. Which of the following animals have Malpighian tubules? (a) coral (b) earthworm (c) clam (d) grasshopper (e) sea star

18. Which of the following animals is an articulated, tracheated hexapod? (a) coral (b) earthworm (c) clam (d) grasshopper (e) sea star

19. The earliest animal in which definite organ systems and a tube-within-a-tube body plan is known to have evolved is the (a) flatworm (b) earthworm (c) ribbon worm (d) grasshopper (e) sea star.

20. Which animal groups include members that are important human parasites? (a) Cnidaria (b) Platyhelminthes (c) Echinodermata (d) Mollusks (e) both (b) and (c)

21. Which of the following animal groups have a true coelom? (a) Cnidaria (b) Platyhelminthes (c) Sponges (d) Annelids (e) two of the preceding
22. Biramous appendages, mandibles, and two pairs of antennae are characteristic of the (a) centipede (b) earthworm (c) crustacean (d) insect (e) sea star.
23. Radial symmetry is characteristic of the (a) hydra (b) rotifer (c) crustacean (d) insect (e) clam.
24. Adaptations for gas exchange include (a) tracheae (b) gills (c) Malpighian tubules (d) (a), (b), and (c) are correct (e) only (a) and (b) are correct.

Review Questions

1. Which animals are considered the most diverse and also claim the greatest number of species? Describe some adaptations that contribute to their success.
2. What are some specific adaptations found in members of phylum Platyhelminthes that are not found in phylum Cnidaria? In what ways are members of these two phyla alike?
3. Which group of animals fits each of the following descriptions: (a) exhibit radial symmetry, (b) have flame cells, (c) have the most primitive brains, (d) were the first known to have complete digestive tubes with mouth and anus?
4. Distinguish between insects and spiders.
5. Which groups of animals are able to depend on diffusion for gas exchange? Why?
6. Draw a diagram that describes the life cycle of the tapeworm.
7. Give two distinguishing characteristics of each of the following: (a) phylum Annelida (b) phylum Arthropoda (c) phylum Mollusca and (d) phylum Echinodermata.
8. Following generally accepted evolutionary principles, draw a hypothetical ancestral tree of the animal kingdom.
9. Which phyla are acoelomate? Which have a pseudocoelom?
10. Relate each of the following to a specific type of animal and give its function: (a) cnidocytes (b) spicules (c) nerve net (d) mantle (e) setae (f) tracheae.
11. What are the distinguishing features of each of the arthropod subphyla?
12. What are the distinguishing features of the principal arthropod groups (classes) within each subphylum? Identify animals that belong to each group of arthropods.
13. Label the diagram.

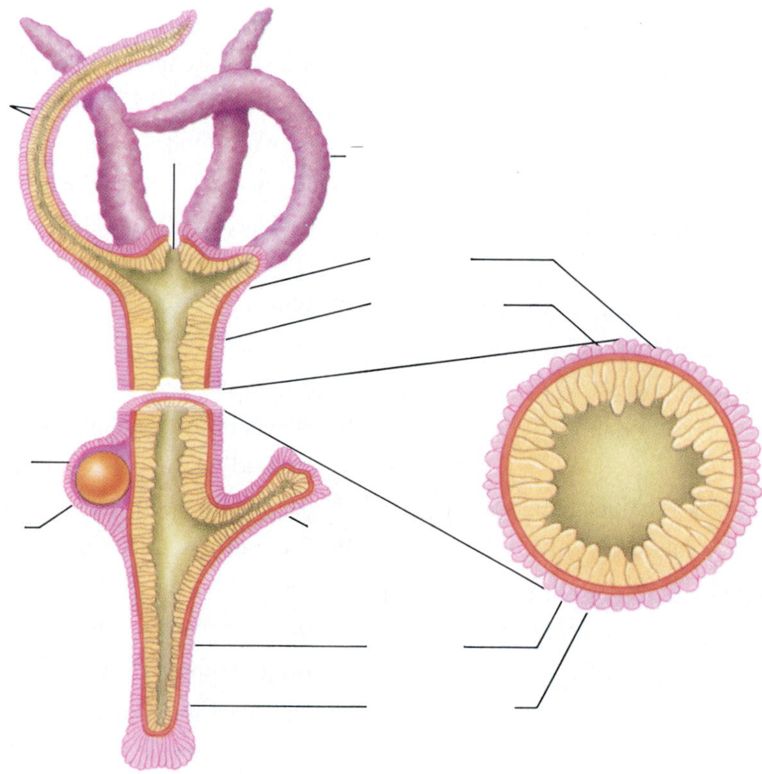

Thinking Critically

1. Design an experiment to determine the cause(s) of coral bleaching. What measures should we take to protect coral reefs? Why?
2. Identify characteristics that mollusks and annelids share. In what ways are these animals different? What do their characteristics suggest about their evolutionary relationship?
3. Discuss the idea that every evolutionary adaptation has both advantages and disadvantages in terms of each of the following characteristics: (a) presence of a coelom (b) the arthropod exoskeleton (c) segmentation (d) complete metamorphosis (e) a complete digestive tract.
4. All animals with radial symmetry are aquatic. Suggest a possible explanation for the lack of radially symmetrical terrestrial animals.
5. The larvae of sponges have external flagella, but flagellated cells in adult sponges are *inside* the central cavity. How is this difference in structure related to differences in lifestyles between larvae and adults?
6. Hypothesize a benefit that might explain why oysters secrete calcium carbonate layers around foreign particles.

Recommended Readings

Brown, B.E., and J.C. Ogden, "Coral Bleaching," *Scientific American,* Vol. 266, No. 1 (Jan 1993). Environmental stressors can cause permanent damage to coral reefs.

Gosline, J.M., and M.E. Demont, "Jet-Propelled Swimming in Squids," *Scientific American,* Vol. 252, No. 1 (Jan 1985). As the squid swims it takes up and expels water by contracting radial and circular muscles in its mantle wall.

Richardson, J.R., "Brachiopods," *Scientific American,* Vol. 255, No. 3 (Sept 1986). One class of these clamlike animals survives by searching out environments suited to an unchanging form; the other class survives by adapting its form or behavior to the local environment.

Evidence for the extraterrestrial explanation? *A coin indicates the relative size of the dark iridium layer. (Lawrence Berkeley Laboratory, University of California)*

WHERE DID ALL THE REPTILES GO? CLUES FROM THE CRETACEOUS

One of the mysteries of modern biology has been the disappearance of the dinosaurs. For almost 200 million years, the reptiles were the dominant land animals on Earth. Dinosaurs, swimming and flying reptiles, and mammal-like reptiles were abundant and highly successful. Then, quite suddenly at the end of the Cretaceous period (about 65 million years ago), many of the reptiles (including all of the dinosaurs) and more than one-half of all animal species disappeared from the fossil record.

Many hypotheses have been proposed to explain this sudden mass extinction. Geologists and biologists working together have found some clues that suggest an extraterrestrial explanation. These investigators hypothesize that about 66 million years ago, one or more extraterrestrial bodies, perhaps an asteroid, or two or more large comets, hit the Earth, drastically altering global ecology. The impact raised a massive cloud of rock particles and dust that blocked out sunlight for several months or even years, plunging the world into cold and perpetual darkness. Photosynthesis came to a halt, causing the extinction of many species of algae and plants. As a result, large animals that eat plants (large herbivores) and also many large carnivores (animals that eat meat) declined. Only animals that ate decaying vegetation or insects and the animals that ate *them* survived. No terrestrial vertebrate larger than about 50 pounds is known to have survived the Cretaceous period.

What are some of the clues that led scientists to the extraterrestrial hypothesis? An important clue has been the discovery of an iridium-rich layer of clay on Earth's surface. The element iridium is extremely rare in the Earth's crust but is comparatively abundant in extraterrestrial objects. Scientists speculate that when the comets crashed into the Earth, their iridium became part of the huge clouds of dust that were thrown into the atmosphere. As the dust from the comet impact gradually settled, iridium was widely deposited. Indeed, the iridium-rich layer has been found at more than 50 locations on Earth. Geologists mark this layer as the boundary between the Cretaceous and Tertiary periods.

Are there other clues that led to the extraterrestrial hypothesis? If an asteroid or large comets had hit the Earth, they would have left evidence of their impact. Several years ago scientists discovered a crater at least 180 kilometers wide near the town of Chicxulub, Mexico in the Yucatan Peninsula. Both the Chicxulub crater and a second crater, the 35-kilometer Manson crater in Iowa, were formed by the impact of a large object. Recently, investigators have used argon–argon dating to determine that these craters are about the same age—an estimated 65.7 million years old. This time coincides with the estimated time of the mass extinction that occurred at the end of the Cretaceous period—dramatic evidence for the extraterrestrial explanation. However, until this hypothesis is supported with further evidence, the cause of this mass extinction will continue to challenge scientists.

Animal Life: Chordates

25

Learning Objectives

After you have studied this chapter you should be able to:

1. Describe the distinguishing characteristics of chordates, and give examples of members of the three subphyla of chordates.
2. Characterize each of the seven classes of vertebrates, and give examples of each group.
3. Describe the course of vertebrate evolution according to contemporary evolutionary theory.
4. Identify adaptations that enabled vertebrates to succeed on land.
5. Compare the three main groups of living mammals and identify specific mammals that belong to each group.
6. Compare the following hominids: australopithecines, *Homo habilis, Homo erectus, Homo sapiens.*
7. Describe cultural evolution and its impact on the ecosphere.

Key Concepts

- ☐ Phylum Chordata consists of three subphyla; two of these are invertebrate subphyla (tunicates and lancelets) and the third is subphylum Vertebrata.
- ☐ At some time in their life cycle, chordates have a notochord; a dorsal, tubular nerve cord, pharyngeal gill grooves, and a postanal tail.
- ☐ Vertebrates include jawless fishes, cartilaginous fishes, bony fishes, amphibians, reptiles, birds, and mammals.
- ☐ Humans are mammals that belong to the order Primates. Primates evolved from small, tree-dwelling, shrewlike mammals.
- ☐ The earliest hominids (humans and their close relatives) belong to the genus *Australopithecus.* They walked on two feet and their brains were larger than their predecessors. Modern *Homo sapiens* made its appearance about 100,000 years ago.

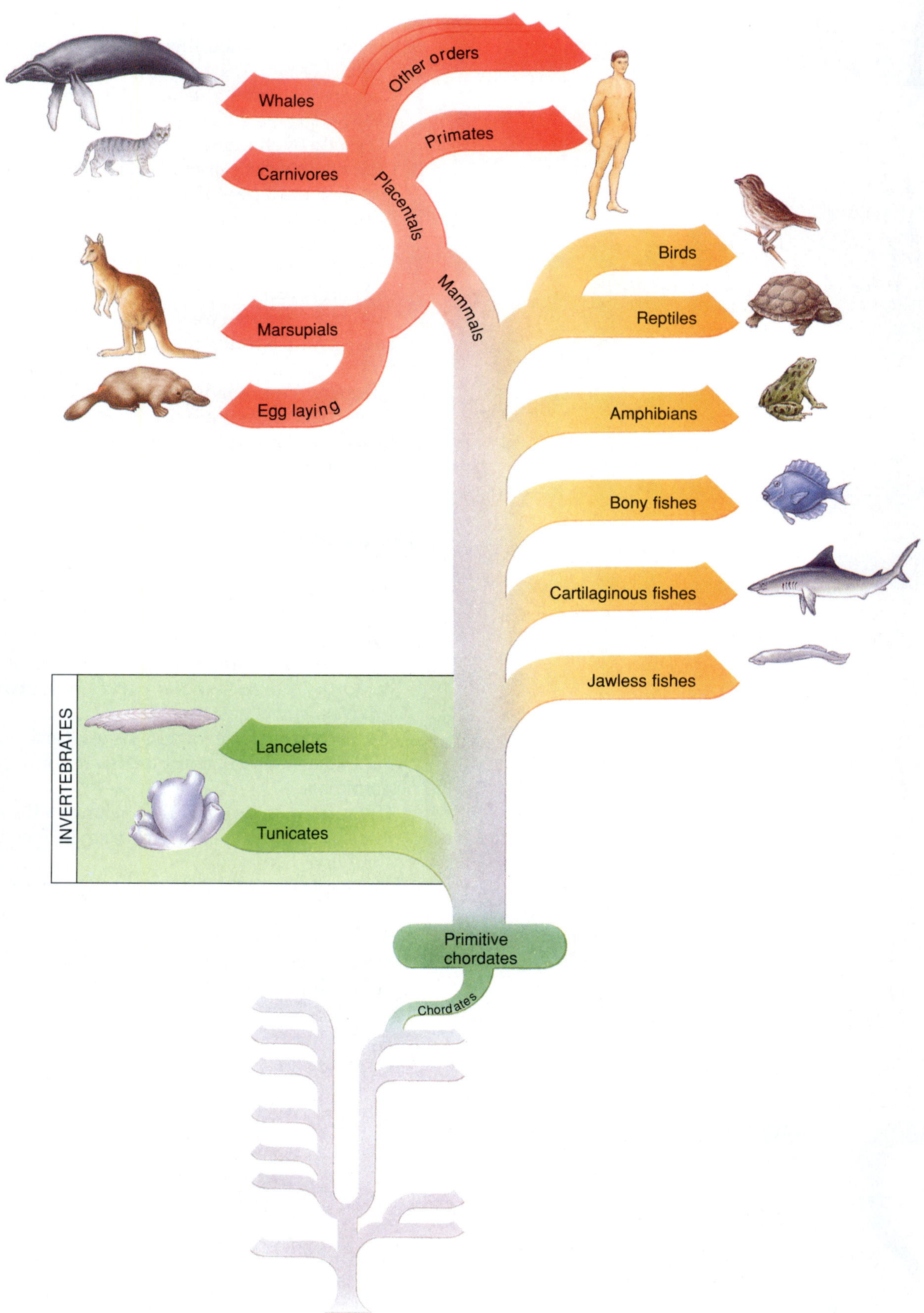

FIGURE 25–1 Proposed chordate relationships.

PHYLUM CHORDATA CONSISTS OF THREE SUBPHYLA

Phylum Chordata, the phylum to which humans belong, is a diverse group of animals that has radiated extensively, filling many types of ecological niches in fresh water, in the sea, and on land (Fig. 25–1). There are three subphyla: (1) **subphylum Urochordata,** which consists of the tunicates (sea squirts, which are filter-feeding marine animals); (2) **subphylum Cephalochordata,** which consists of the lancelets (small, translucent, fishlike animals); and (3) **subphylum Vertebrata,** which consists of the vertebrates (animals with backbones).

No clear fossil record of the ancestors of the chordates exists, but biologists think they were small, soft-bodied animals. A lancelet-like animal, *Pikaia,* was found in the Burgess Shale of British Columbia. These rocks date back to the mid-Cambrian Period (more than 500 million years old).

CHORDATES HAVE FOUR DISTINGUISHING CHARACTERISTICS

Chordates share four characteristics that help distinguish them from other groups:

1. *Chordates have a notochord during some time in their life cycle.* Unique to chordates, the **notochord** is a dorsal longitudinal rod that is firm yet flexible and supports the body.
2. *Chordates have a dorsal, tubular nerve cord.* The nerve cord is different from the nerve cord in most other animals. It is located dorsally rather than ventrally; is hollow, rather than solid; and is single, rather than double.
3. *Chordates have* ***pharyngeal gill grooves*** *during some time in their life cycle.* In many aquatic chordates, the grooves perforate through the wall of the pharynx and become functional gill slits. However, in terrestrial (land) animals, they become modified to form entirely different structures (such as the outer ear canal) more suitable for life on land.
4. *Chordates have a tail that extends beyond the anus (postanal tail) at some time in their life cycle.*

TUNICATES ARE MARINE ANIMALS WITH A PROTECTIVE TUNIC

The **urochordates,** commonly known as **tunicates,** include the sea squirts and their relatives. Members of this subphylum are barrel-shaped marine animals (Fig. 25–2). Adults develop a covering, or **tunic** (hence their name), that protects them. The tunic, made of a carbohydrate much like cellulose, may be soft and transparent or quite leathery.

In their larval stage, tunicates exhibit typical chordate characteristics and somewhat resemble microscopic tadpoles. They have a pharynx with gill slits, a dorsal nerve cord, and a long muscular tail that contains the notochord. In sea squirts, the larva eventually attaches itself to the sea bottom and loses its tail, notochord, and much of its nervous system. In the adult, only the gill slits suggest that the sea squirt is a chordate. Appendicularians, another class of tunicates, retain their chordate features

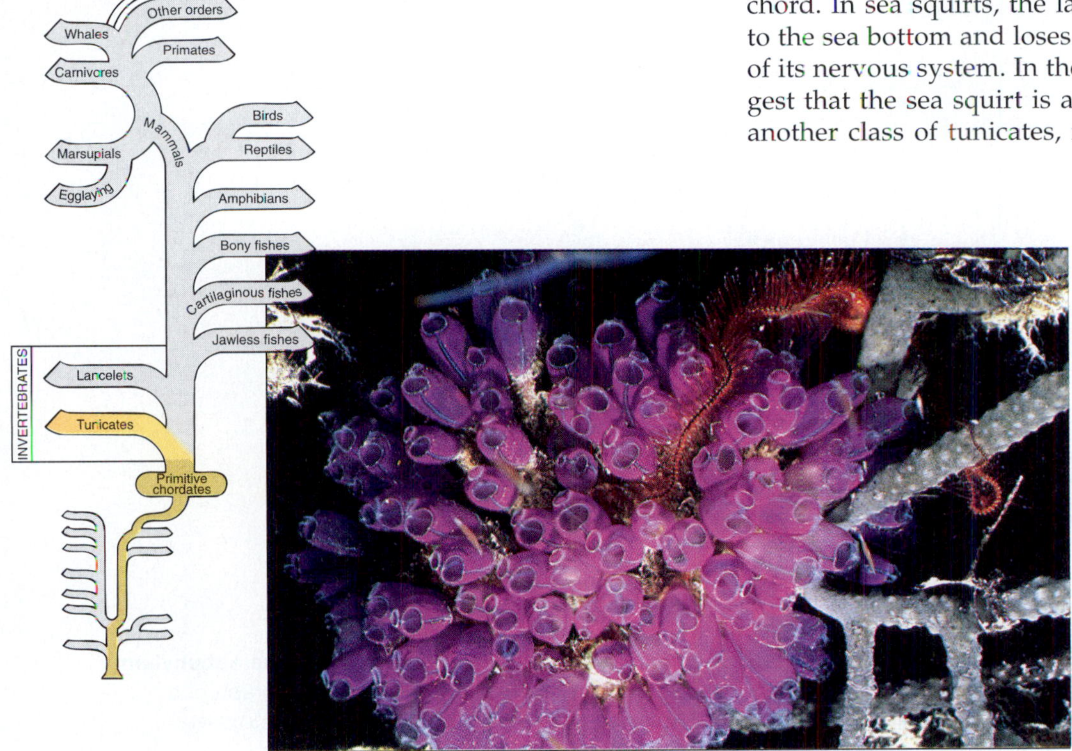

FIGURE 25–2 Tunicates are invertebrate chordates. Tunicates photographed in the British Virgin Islands. (Visuals Unlimited/Hal Beral)

and ability to swim about and are common members of the plankton.

Adult sea squirts are often mistaken for sponges or cnidarians. They are filter feeders that remove plankton from the water as it passes through the pharynx. Some species form large colonies in which members may share a common tunic. Colonial forms often reproduce asexually by budding. Sexual forms are usually hermaphroditic.

THE LANCELETS MAY BE SIMILAR TO PRIMITIVE CHORDATES THAT GAVE RISE TO VERTEBRATES

The chordate characteristics are highly developed in the **lancelets,** which are small, translucent, fish-shaped animals (Fig. 25–3). Members of this subphylum are known as cephalochordates because the notochord extends anteriorly to the tip of the "head." The notochord also extends posteriorly to the end of the tail. The 25 or so species are widely distributed in shallow seas, either swimming freely or burrowing in the sand near the low tide line. The most common genus, *Branchiostoma,* is commonly known as Amphioxus. In some parts of the world, lancelets are an important source of food. One Chinese fishery reported an annual catch of 35 tons (about one billion lancelets).

Although they superficially resemble fishes, lancelets have a far simpler body plan. They lack paired fins, jaws, a well-defined brain, a heart, and sense organs. Biologists speculate that Amphioxus may have a life-style similar to the ancient ancestor from which the vertebrates evolved. Similarities include swimming and obtaining food by filtering seawater.

THE SUCCESS OF THE VERTEBRATES IS LINKED TO THE EVOLUTION OF KEY ADAPTATIONS

Most **vertebrates** are distinguished from other chordates by having a backbone, or **vertebral column,** which forms the skeletal axis of the body. This flexible support develops around the notochord and, in most species, largely replaces the notochord during embryonic development. The vertebral column consists of segments, called **vertebrae,** that are made of cartilage or bone. Dorsal projections of the vertebrae enclose the nerve cord along its length. Located anterior to the vertebral column, a **cranium,** or braincase, encloses and protects the brain, which is the enlarged anterior end of the nerve cord.

The cranium and vertebral column are parts of the **endoskeleton.** In contrast to the nonliving exoskeleton of some invertebrates, the endoskeleton is a living tissue that grows with the animal. Two pairs of appendages are present in most vertebrates. Fish fins stabilize them in the water. As vertebrates moved onto the land, jointed appendages evolved that facilitated locomotion.

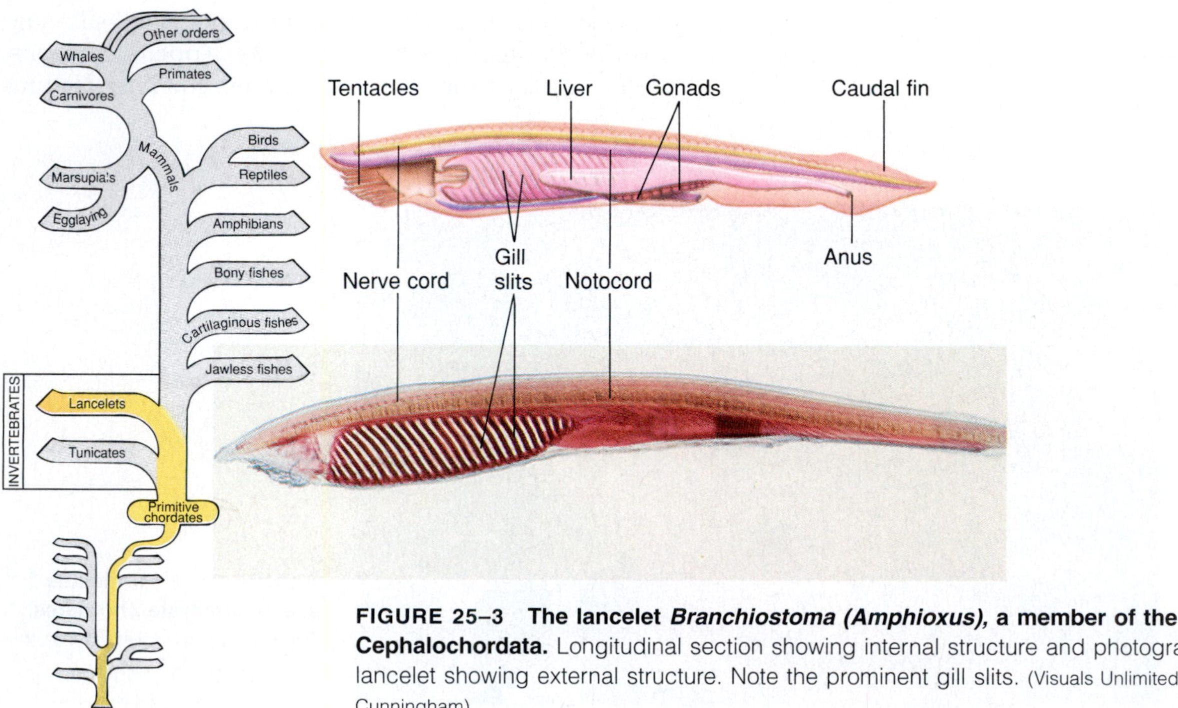

FIGURE 25–3 The lancelet *Branchiostoma (Amphioxus),* a member of the subphylum Cephalochordata. Longitudinal section showing internal structure and photograph of a lancelet showing external structure. Note the prominent gill slits. (Visuals Unlimited/John D. Cunningham)

Recall that in invertebrates there is an evolutionary trend toward **cephalization,** which is the concentration of nerve cells and sense organs in a definite head (Chapter 24). Vertebrate evolution is characterized by *pronounced* cephalization. The brain becomes larger and more elaborate, and its various regions become specialized to perform different functions. Ten or 12 pairs of cranial nerves emerge from the brain and extend to various organs of the body. Vertebrates have well-developed sense organs concentrated in the head. These sense organs include the eyes; ears that serve as organs of balance and, in some vertebrates, for hearing as well; and organs of smell and taste.

Vertebrates have a closed circulatory system with a two-, three-, or four-chambered ventral heart. They have paired kidneys that regulate fluid balance. Vertebrates have a complete digestive tract and large digestive glands (liver and pancreas). The sexes are almost always separate.

The vertebrates, with 43,000 living species, are less diverse and much less numerous than are the insects but they rival the insects in their adaptation to an enormous variety of life-styles. Most vertebrates excel over the insects in their ability to receive and respond appropriately to a wide variety of stimuli.

Living vertebrate species are assigned to seven classes: (1) class Agnatha, comprised of the jawless fishes such as lampreys; (2) class Chondrichthyes, which includes the sharks and rays with skeletons made of cartilage; (3) class Osteichthyes, comprised of the bony fishes; (4) class Amphibia, which includes frogs, toads, and salamanders; (5) class Reptilia, which includes lizards, snakes, turtles, and alligators; (6) class Aves, the birds; and (7) class Mammalia, the mammals. A discussion of each class follows.

THE JAWLESS FISHES ARE THE MOST PRIMITIVE VERTEBRATES

Fossil evidence suggests that vertebrates evolved in the sea 450 to 500 million years ago (Ordovician period). By the Silurian and Devonian periods, primitive fishes had radiated extensively and were mainly small, armored, jawless freshwater fishes. They lived on the bottom and strained their food from the water. Today, the only living members of **class Agnatha** are the lampreys and hagfishes (Fig. 25–4). Several species of adult lampreys are parasites that live on other fishes.

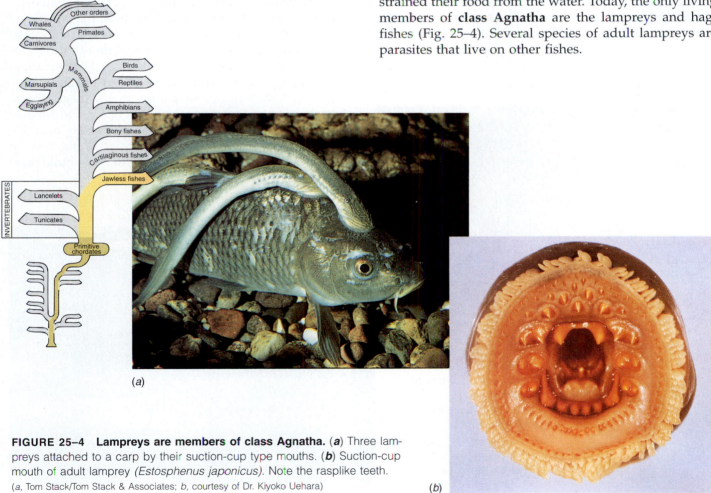

FIGURE 25–4 Lampreys are members of class Agnatha. (***a***) Three lampreys attached to a carp by their suction-cup type mouths. (***b***) Suction-cup mouth of adult lamprey *(Estosphenus japonicus).* Note the rasplike teeth. (*a*, Tom Stack/Tom Stack & Associates; *b*, courtesy of Dr. Kiyoko Uehara)

Fossil evidence suggests that during the Silurian and Devonian periods (more than 400 million years ago), fishes evolved with jaws and finlike, paired appendages. These fishes were able to shift from a bottom-dwelling, filter-feeding life-style to become active predators. The **placoderms,** a group of extinct jawed fishes that lived during the Paleozoic era, were armored with paired fins.

THE CARTILAGINOUS FISHES INCLUDE THE SHARKS, RAYS, AND SKATES

Members of **class Chondrichthyes** have skeletons made of cartilage rather than bone. While early jawed fishes inhabited mainly fresh water, the cartilaginous fishes evolved as successful marine forms in the Devonian period. Most species—the sharks, rays, and skates—still inhabit the oceans (Fig. 25–5). These fishes have paired jaws and paired pectoral (fore) and pelvic (hind) fins, as well as dorsal fins. The fins help stabilize the body and propel the fish through the water. The skin contains toothlike scales, known as placoid scales, that help protect the body.

Most sharks are streamlined, active predators. The largest sharks, like the largest whales, dine on plankton. The whale shark, which may reach a length of 12 meters (more than 39 feet), is the largest fish known. Although some books and films portray sharks as monstrous enemies, most do not go out of their way to attack humans.

Most rays and skates are flattened animals that rest partly burrowed in the sea bottom. Wavelike movements of its large pectoral fins propel the ray or skate along the bottom as it seeks out a meal of mussels and clams. The stingray has a whiplike tail with a barbed spine that can inflict a painful wound. The electric ray has muscles modified as electric organs that can discharge enough electricity (up to 2500 watts) to stun fairly large fishes, as well as human swimmers.

THE BONY FISHES ARE THE MOST NUMEROUS VERTEBRATES

The fishes most familiar to us are the bony fishes of **class Osteichthyes** (Fig. 25–6). With about 24,000 living species, the bony fishes are the most numerous vertebrates. They vary greatly in color and shape. Bony fishes range in size from the Philippine goby, which is only about 10 millimeters (0.4 inch) long, to the ocean sunfish, which may reach 907 kilograms (2000 pounds).

The bodies of most bony fishes are covered with overlapping bony scales (which develop from the inner layer of the skin). Most bony fishes have both median and dorsal paired fins, supported by long rays made of cartilage or bone (Fig. 25–7). A lateral protective flap, the **operculum,** extends posteriorly from the head and covers the gills.

Most bony fishes are **oviparous,** that is, they lay eggs. In fact, many lay impressive numbers of eggs, which they fertilize externally. The ocean sunfish may lay over 300 million eggs! Most eggs and young offspring become food for other animals. Many species of fishes build nests for their eggs and even watch over them.

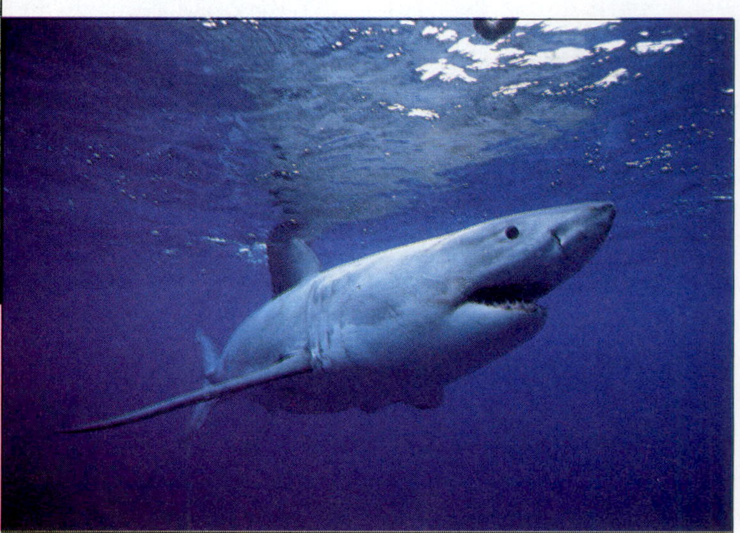

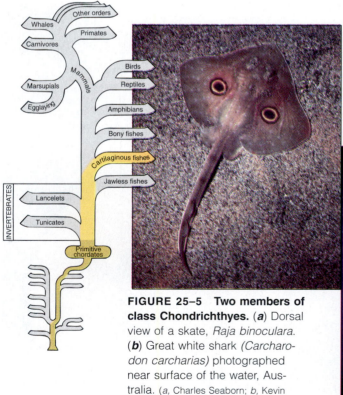

FIGURE 25–5 Two members of class Chondrichthyes. (***a***) Dorsal view of a skate, *Raja binoculara.* (***b***) Great white shark *(Carcharodon carcharias)* photographed near surface of the water, Australia. (*a*, Charles Seaborn; *b*, Kevin Aitken/Peter Arnold, Inc.)

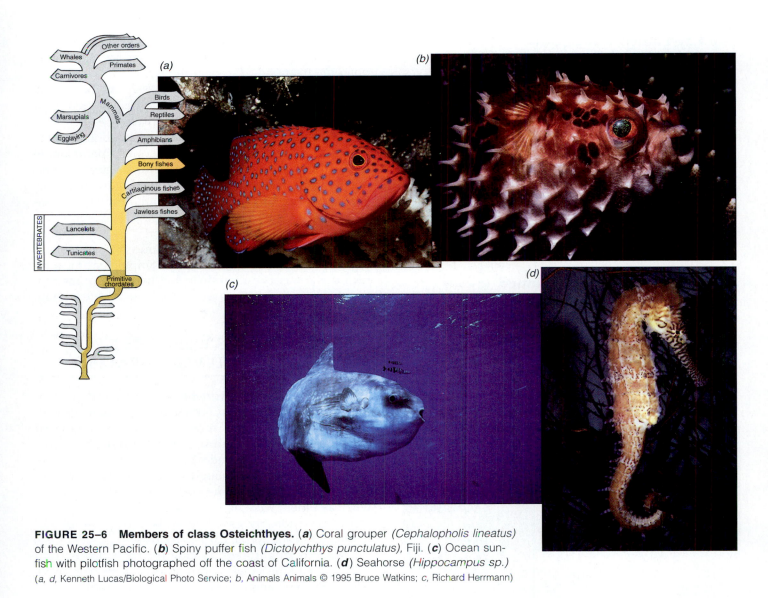

FIGURE 25–6 Members of class Osteichthyes. (***a***) Coral grouper *(Cephalopholis lineatus)* of the Western Pacific. (***b***) Spiny puffer fish *(Dictolychthys punctulatus)*, Fiji. (***c***) Ocean sunfish with pilotfish photographed off the coast of California. (***d***) Seahorse *(Hippocampus sp.)* (*a, d,* Kenneth Lucas/Biological Photo Service; *b,* Animals Animals © 1995 Bruce Watkins; *c,* Richard Herrmann)

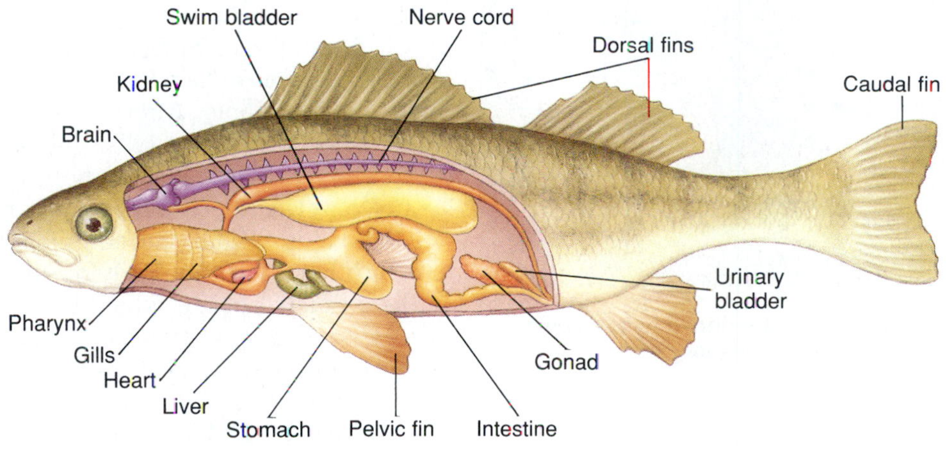

FIGURE 25–7 The perch is a representative bony fish. Anatomy of a perch.

The Ray-Finned Fishes Gave Rise to Modern Bony Fishes

During the Devonian period, the bony fishes diverged into two major groups: the ray-finned fishes and the lobe-finned fishes. The **ray-finned fishes** gave rise to most modern bony fishes. The ancestors of the ray-finned fishes are thought to have had lungs. These lungs later became modified as a **swim bladder,** an air sac that helps the fishes regulate their buoyancy (Fig. 25–7). By secreting gases into the bladder or absorbing gases from them, fishes can change the overall density of their bodies. This ability allows fishes to hover at a given depth of water without muscular effort.

Descendants of the Lobe-Finned Fishes Moved Onto the Land

The **lobe-finned fishes,** which include the lungfishes, are generally thought to have given rise to the land vertebrates (Fig. 25–8). During the Devonian period, there were frequent seasonal droughts, and many ponds dried up. The lobe-finned fishes with lungs adapted for breathing air had a tremendous advantage for survival. Their fleshy lobe fins could support their weight, enabling them to emerge onto dry land and make their way to another pond or stream.

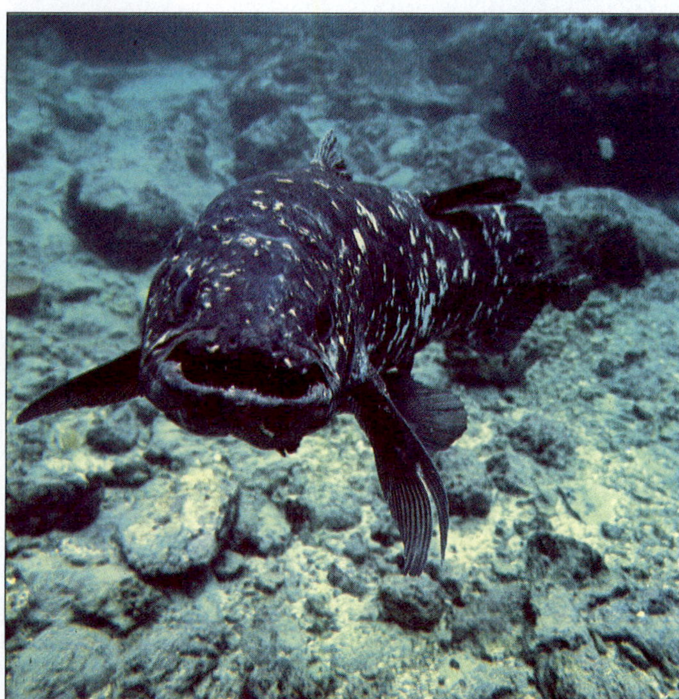

FIGURE 25–8 Ancestors of this lobe-finned fish, a coelacanth, probably gave rise to the amphibians. The paired fins show the basic plan of a jointed series of bones that could evolve into the limbs of a terrestrial vertebrate. Living coelacanths are difficult to observe. Its members (genus *Latimeria*) inhabit deep ocean waters; when brought to the surface, the fish does not survive the change in atmospheric pressure. (Peter Scoones, Planet Earth Pictures)

The ability to move about, however awkwardly, on dry land also gave these animals access to new food sources such as terrestrial plants and insects. A vertebrate that could survive on land had less competition for food. Laying eggs on land away from the many predators in the sea also increased their chances for successful reproduction. Natural selection favored those individuals best adapted for making their way on land, resulting eventually in the evolution of amphibians and, later, reptiles.

Terrestrial vertebrates are referred to as **tetrapods** (meaning, "four-footed"). They include the amphibians, reptiles, birds, and mammals. Most amphibians must return to the water to reproduce, but members of the reptiles, birds, and mammals are fully adapted to life on land.

The Amphibians Were the First Successful Land Vertebrates

The first successful land vertebrates were the **labyrinthodonts,** which were clumsy, salamander-like animals with short necks and heavy, muscular tails. These ancient members of **class Amphibia** somewhat resembled their ancestors, the lobe-finned fishes. However, they had evolved limbs strong enough to support the weight of their bodies on land. Some labyrinthodonts were small, whereas others were as large as 2 meters (more than 6 feet). They probably gave rise to the modern frogs and salamanders and to the earliest reptiles.

Modern **amphibians** include the frogs, toads, salamanders, and the legless, wormlike caecilians (Fig. 25–9). Although some amphibians are quite successful as land animals and can live in dry environments, most return to the water to reproduce. Eggs and sperm are generally released in water.

The embryos of frogs and toads develop into larvae called **tadpoles.** These larvae have tails and gills and feed on aquatic plants. After a time, the tadpole undergoes metamorphosis. The gills, gill slits, and tail disappear, and limbs emerge. When these structural modifications are complete, the amphibian can move onto the land.

Glands within the amphibian skin secrete mucus, which helps keep the body moist. The mucus also makes the animal slippery, which helps it escape from predators. Some amphibians have glands in their skin that secrete poisonous substances harmful to predators.

Adult amphibians do not depend solely on their primitive lungs for the exchange of respiratory gases. Their moist skin, which lacks scales and is richly supplied with blood vessels, also serves as a surface for gas exchange. Some adult salamanders have neither gills nor

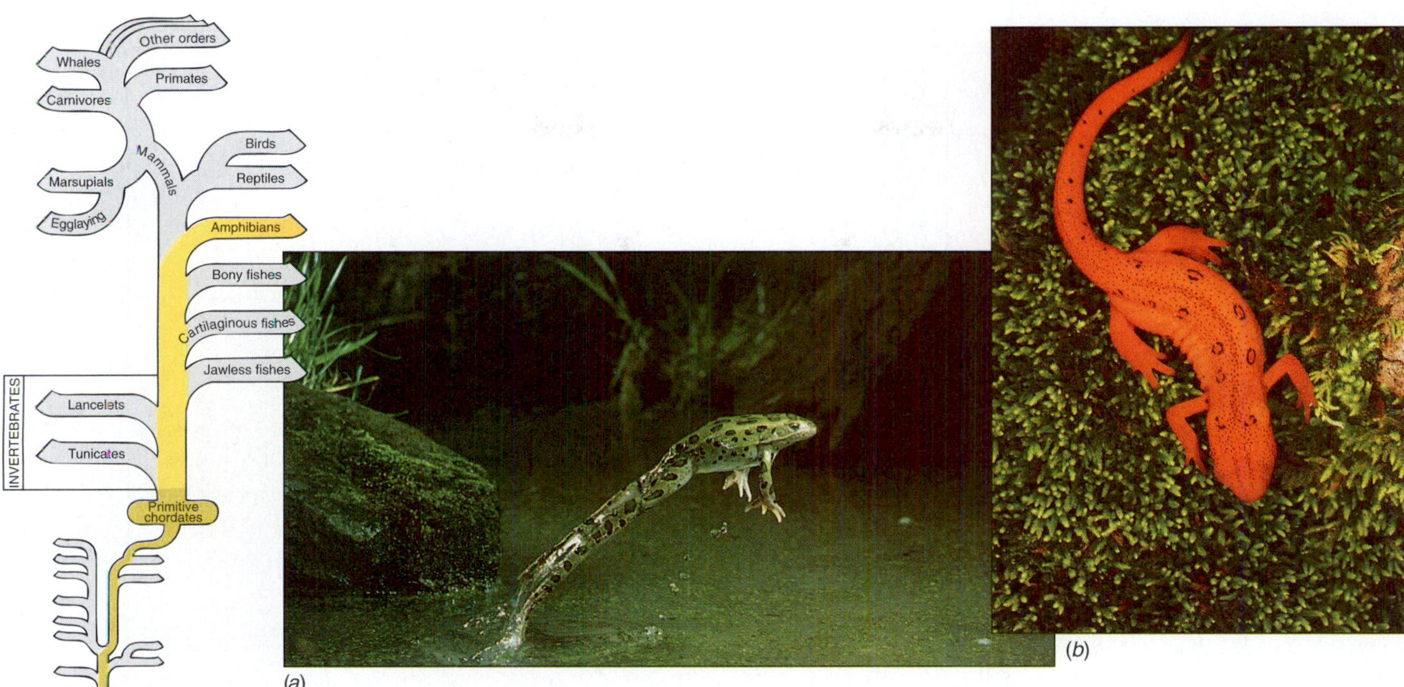

FIGURE 25–9 Modern amphibians. (***a***) This leopard frog *(Rana pipiens)* leaps to evade a predator. (***b***) The red eft form of the red-spotted newt (*Notophthalamus viridescens viridescens*). After an aquatic larval stage, the red-spotted newt enters the terrestrial eft stage, in which it spends one to three years on the forest floor. The eft then returns to the water, where it transforms into an adult. Most newts lack the eft form and remain aquatic throughout their lives. (*a*, Stephen Dalton/Photo Researchers, Inc.; *b*, The Stock Market/Roy Morsch)

lungs. Gas exchange takes place by diffusion through the skin (which is rich in capillaries) and the lining of the mouth. Oxygen enters the blood and is transported by the circulatory system.

The amphibian heart has three chambers—two **atria** that receive blood and a single **ventricle** that pumps blood into the arteries (see Chapter 37). A double circuit of blood vessels keeps oxygen-rich and oxygen-poor blood partly separate. One circuit directs blood to the various tissues and organs of the body, while the other circuit conducts blood to the lungs and skin to be recharged with oxygen.

Reptiles were the dominant land animals for almost 200 million years

Biologists generally agree that the evolutionary lineages leading to the reptiles had a common ancestor—a labyrinthodont amphibian. The ancestral stocks of the reptiles had diverged by the late Carboniferous period (about 290 million years ago). The Mesozoic era, which ended about 65 million years ago, is known as the Age of Reptiles. During that time, reptiles radiated into an impressive variety of ecological niches comparable to those of modern mammals. Many filled terrestrial habitats, some were able to fly, others were marine. Although some reptiles were small, some of the dinosaurs were the largest, most monstrous animals that ever stalked the Earth.

By the end of the Mesozoic era, most species of reptiles had disappeared (see Chapter Opener). Many more extinct than living kinds of reptiles are known. Long before their decline, the reptiles gave rise to both the birds and the mammals (Fig. 25–10).

Reptiles Evolved Adaptations to Life on Land

Reptiles are terrestrial animals that do not have to return to water to reproduce. Many adaptations make this lifestyle possible. The female secretes a protective leathery shell around the egg, which helps prevent the developing embryo from drying out. Because sperm cannot penetrate this shell, fertilization occurs within the body of the female before the shell is added. In this process of **inter-**

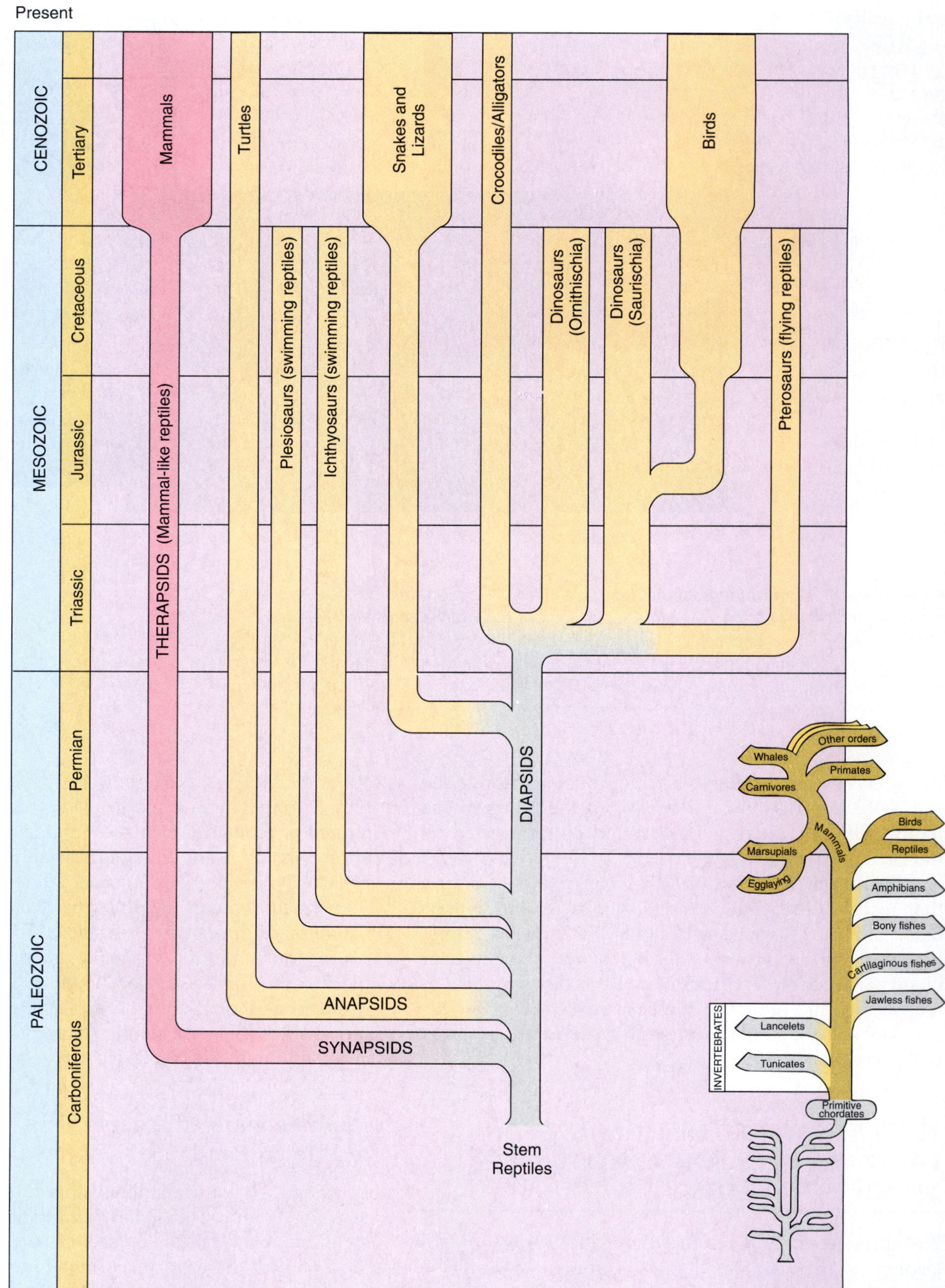

FIGURE 25–10 Proposed evolution of the reptiles. This hypothetical tree illustrates some proposed evolutionary relationships among extinct and present-day animals.

nal fertilization, the male uses a copulatory organ to transfer sperm into the female reproductive tract.

As the embryo develops within the protective shell, a membrane called the **amnion** develops and surrounds the embryo. The amnion secretes amniotic fluid, providing the embryo with its own private pond. This fluid keeps the embryo moist and also serves as a shock absorber should the egg get bumped. The reptile embryo, like that of other terrestrial vertebrates, also has other membranes that protect and support its development.

The hard, dry, horny scales that protect the reptile body from drying are another adaptation to life on land. This scaly protective armor, which helps protect the reptile from predators, must be shed periodically.

Modern Reptiles Are a Diverse Group

Modern reptiles include the turtles, lizards, snakes, alligators, and crocodiles (Fig. 25–11). Lizards and snakes are the most diverse.

Like fishes and amphibians, living reptiles generally lack metabolic mechanisms for regulating body temperature. They are **ectothermic,** which means that their body temperature fluctuates with that of the environment. Some reptiles do have behavioral adaptations that enable them to maintain a body temperature higher than that of the environment. You may have observed a lizard basking in the sun. What you may not have known is that the lizard was waiting for its body temperature to rise so that its metabolic rate would increase. Only then could it actively move about and hunt for food. When the body of a reptile is cold, the metabolic rate is low and the animal tends to be sluggish. Ectothermia probably explains why reptiles are more successful in warm than in cold climates.

The dry skin of reptiles cannot serve as an organ for gas exchange. However, reptile lungs are better developed for gas exchange than the saclike lungs of amphibians. Most reptiles have a three-chambered heart that is more efficient than the amphibian heart (see Chapter 36).

FIGURE 25–11 Modern reptiles. (***a***) The bamboo viper is poisonous. (***b***) Like many turtles, this green turtle, native of Hawaii, is a good swimmer. (***c***) This Nile crocodile is in the process of hatching. (*a*, Sharon Cummings/Dembinsky Photo Associates; *b*, *c*, Frans Lanting/Minden Pictures)

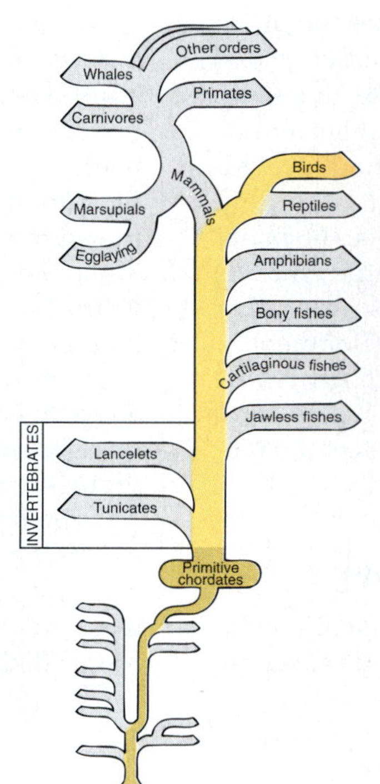

FIGURE 25–12 ***Archaeopteryx*, a very early bird.** This drawing represents the hypothesis that *Archaeopteryx* was at least a climbing animal that had some ability to use its wings and feathers for gliding. Other hypotheses suggest that *Archaeopteryx* was mainly a land animal that used its wings to trap small insects and its feathers for insulation. (Also see Figure 19–14 for the fossil on which this reconstruction is based.) (From a painting by Rudolph Freund, courtesy of Carnegie Museum of Natural History.)

Birds are adapted for flight

Early birds, such as *Archaeopteryx,* resembled small Jurassic dinosaurs that moved about on two legs and may have used their wings for gliding (Fig. 25–12). Modern birds have retained some characteristics in common with the reptiles. For example, their legs have scales like those of reptiles, and they lay eggs. Birds are the only animals with feathers—structures thought to have evolved from reptilian scales.

About 9000 species of birds have been described and classified in 27 orders. Birds live in a wide variety of habitats and can be found on all of the continents, most islands, and even the open sea. The largest living birds are the ostriches of Africa, which may be 2 meters (7 feet) tall and weigh 136 kilograms (300 pounds), and the great condors of the Americas, with wingspreads of up to 3 meters (10 feet). The smallest known bird is Helena's hummingbird of Cuba, which is less than 6 centimeters (2.3 inch) long and weighs less than 4 grams (0.1 ounce). Beautiful and striking colors are found among the birds (Fig. 25–13). Many birds, especially females, are protectively colored by their plumage. During the breeding season, the male usually assumes brighter colors, which help attract a mate.

Birds are beautifully adapted for flight. Their feathers are flexible and very strong for their light weight. Feathers protect the body, decrease water loss through the body surface, decrease the loss of body heat, and aid in flying by presenting a plane surface to the air.

The anterior limbs of birds are usually modified as wings for flight, while the posterior pair are adapted for walking, swimming, or perching. Not all birds fly. Some, such as penguins, have small, flipper-like wings that are used in swimming.

In addition to feathers and wings, birds have many other adaptations for flight. They have compact, streamlined bodies, and the fusion of many bones gives the rigidity needed for flying. Their bones are strong but very light. Many bones are hollow and contain large air spaces. The jaws are light, and instead of teeth, there is a lightweight, horny beak.

The lungs of a bird are very efficient, with thin-walled extensions, called air sacs, that occupy spaces between the internal organs and within certain bones. Birds, like mammals, have four-chambered hearts and a double circuit of blood flow.

The very efficient respiratory and circulatory systems deliver sufficient oxygen to the cells to permit a high metabolic rate. The high metabolic rate provides the energy necessary to support the tremendous muscular activity required for flying. Some of the heat generated by metabolic activities is used to maintain a constant body temperature. This ability permits metabolic processes to

FIGURE 25–13 Modern birds. (***a***) Peacock *(Pavo cristatus)* impressively displaying his feathers. (***b***) This Atlantic puffin *(Fratercula artica)* has just caught a fish. (***c***) The snowy owl *(Nyctea scandiaca)* also has food in its beak. (***d***) Cattle egrets *(Bubulcus ibis)* filmed in Kenya. (***e***) The American goldfinch uses its beak to pluck thistle. (*a,* Frans Lanting/Minden Pictures; *b,* Gary Meszaros/ Dembinsky Photo Associates; *c,* Michio Hoshino/Minden Pictures; *d,* D.W. Fawcett; *e,* George E. Stewart/Dembinsky Photo Associates)

proceed at constant rates and enables birds to remain active in cold climates. Birds and mammals are **endothermic** ("warm-blooded"), meaning that they can maintain a constant body temperature by metabolic mechanisms.

Mammals evolved from reptiles

Mammals, the animals most familiar to us, evolved from a group of reptiles called **therapsids** during the Triassic period some 200 million years ago. The therapsids were doglike carnivores with differentiated teeth (a mam-

FIGURE 25–14 A mammal-like reptile. *Lycaenops*, a therapsid, lived in the late Permian period in South Africa. (Trans. no. 203, Painting by John C. Germann, Department of Library Services, American Museum of Natural History)

malian trait) and legs adapted for running (Fig. 25–14). Some of them may have been endothermic, and some may even have had fur. The fossil record indicates that the earliest mammals were small, about the size of a mouse or shrew.

How did the mammals manage to coexist with the reptiles during the 160 million or so years that the reptiles ruled the world? Many adaptations permitted the mammals to compete for a place on the Earth, but perhaps most important was that the early mammals specialized in being inconspicuous. They were **arboreal** (tree-dwelling) and **nocturnal** (active at night), searching for food (mainly insects and plant material and perhaps reptile eggs) while the reptiles were inactive. This life-style is suggested by the large eye sockets seen in fossil species, indicating that these animals had the large eyes characteristic of present-day nocturnal mammals.

As reptiles died out, the mammals began to move into their abandoned territories. During this time, the flowering plants, including many trees, underwent adaptive radiation, providing new habitats, sources of food, and protection from predators. Larger forms and numerous varieties of mammals evolved. During the early Cenozoic era (perhaps 50 million years ago), the mammals underwent adaptive radiation, becoming widely distributed and adapted to an impressive variety of ecological life-styles.

Three main lines of mammals had evolved by the end of the Cretaceous period. Today, these groups are classified in three subclasses: the egg-laying **monotremes,** like the duck-billed platypus; the **marsupials,** which are pouched mammals such as kangaroos and opossums; and **placental mammals,** animals in which the embryo has a placenta—an organ of exchange between mother and embryo.

Today, mammals inhabit virtually every corner of the Earth—on the land, in fresh and salt water, and in the air. Their sizes range from the tiny pygmy shrew, weighing about 25 grams (less than an ounce), to the blue whale, which may weigh more than 90,000 kilograms (100 tons) and is thought to be the largest animal that ever lived.

The distinguishing characteristics of mammals are the presence of hair; **mammary glands,** which produce milk for the young; and the differentiation of teeth into incisors, canines, premolars, and molars. A muscular diaphragm helps move air in and out of the lungs. Like the birds, mammals are endotherms, which means that they maintain a constant body temperature. This process is supported by the covering of hair, which serves as insulation, and by the four-chambered heart and double circulation. Contributing significantly to the success of the mammals is the complex nervous system, which is more highly developed than in any other group of animals.

Monotremes Are Mammals that Lay Eggs

Monotremes, which inhabit Australia and New Guinea, include two genera, the duck-billed platypus (*Ornithorhynchus*) and the spiny anteater (*Tachyglossus;* Fig. 25–15). The females lay eggs that may be carried in a pouch on the abdomen or kept warm in a nest. When the young hatch, they are nourished with milk secreted by mammary glands. The duck-billed platypus lives in burrows along river banks. It has webbed feet and a flat, beaver-type tail that aids in swimming. The platypus uses its bill to scoop up freshwater invertebrates. As its name suggests, the spiny anteater feeds on ants that it catches with its long sticky tongue.

Marsupials Are Pouched Mammals

Marsupials are pouched mammals such as kangaroos and opossums (Fig. 25–16). Embryos begin their development in the mother's uterus, where they are nourished by yolk and from fluid in the uterus. After a few weeks, still in a very undeveloped stage, the young are born and crawl to the **marsupium** (pouch), where they complete their development. Each of the young attaches itself by its mouth to a mammary gland nipple in the marsupium and is nourished by its mother's milk.

Like the monotremes, marsupials are found mainly in Australia. Only the opossum is common in North and South America. At one time, marsupials may have inhabited much of the world but were largely replaced by the placental mammals. Australia became geographically isolated from the rest of the world before placental mam-

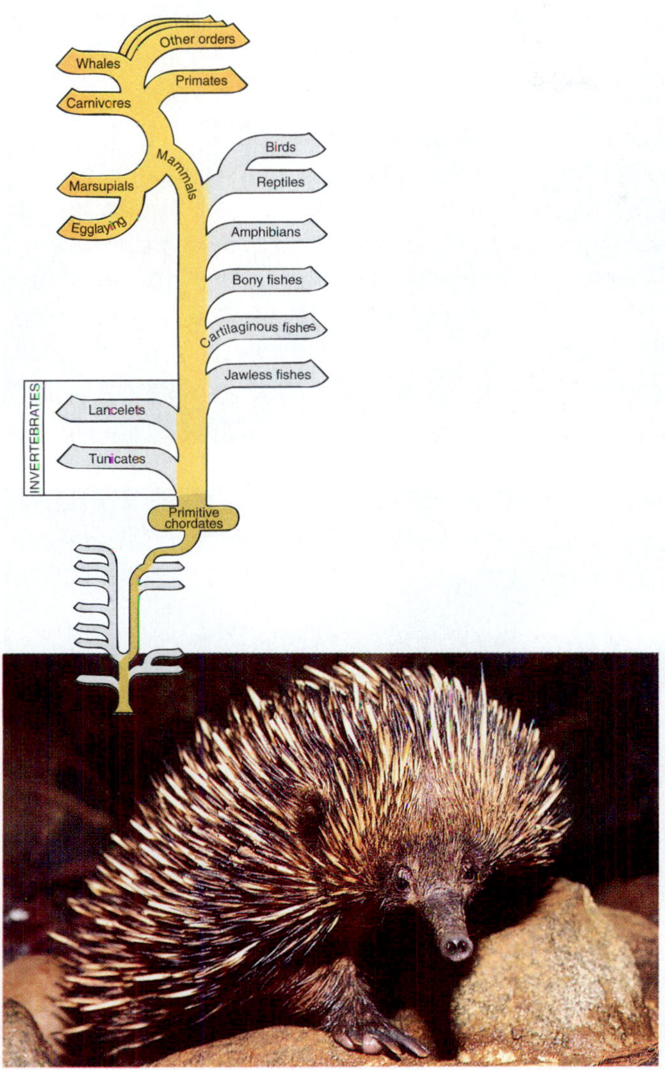

FIGURE 25–15 The spiny anteater, *Tachyglossus*. This egg-laying mammal is a monotreme. (Tom McHugh/Photo Researchers, Inc.)

mals reached it, so the marsupials remained the dominant type of mammal on that continent. Their evolution proceeded in many directions and fitted them for many different life-styles, paralleling the evolution of placental mammals elsewhere. Thus, in Australia and adjacent islands, we find marsupials that correspond to our placental wolves, bears, rats, moles, flying squirrels, and even cats (Fig. 25–17).

When humans migrated to Australia during the Pleistocene and again, more recently, they brought other placental animals with them. Placental mammals competed so effectively with the marsupials that Australian marsupials have declined. Many biologists are concerned that unless they are protected, the marsupials will not survive.

(a)

(b)

FIGURE 25–16 Marsupials are pouched mammals. (***a***) Eastern gray kangaroo *(Macropus giganteus)* with joey. The kangaroo is native to Australia. (***b***) Young marsupials are born in a very immature state. The young continue to develop in the safety of the marsupium (pouch). (*a*, John Cancalosi/Peter Arnold, Inc.; *b*, Robert Anderson, reprinted with permission of Hubbard Scientific Company)

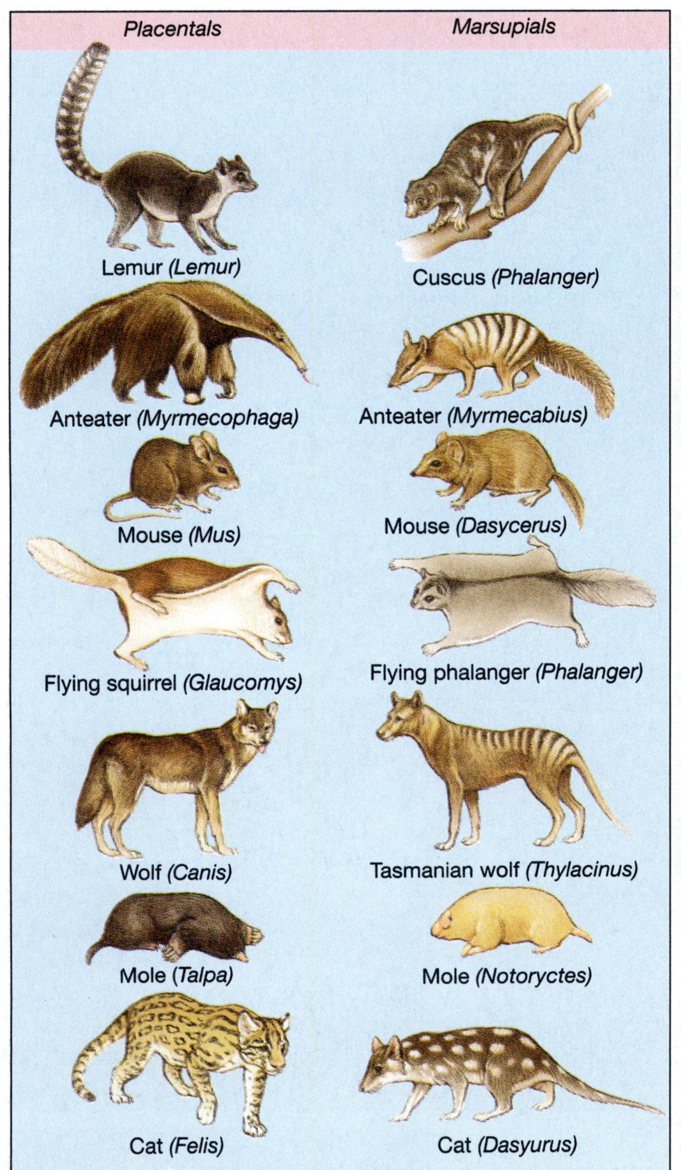

FIGURE 25–17 Comparison of marsupial (right) and placental (left) mammals.

(a)

(b)

(c)

FIGURE 25–18 Placental mammals. (***a***) Bison calf *(Bison bison)* with parent. (***b***) Humpback whale *(Megaptera novaeangliae)*, rare double breach. (***c***) Polar bears *(Ursus maritimus)* are the largest living land carnivores. They feed mainly on seals. (*a*, Mike Barlow/Dembinsky Photo Associates; *b*, Animals Animals © 1995 James D. Watt; *c*, Michio Hoshino/Minden Pictures)

Placental Mammals Complete Embryonic Development Within the Mother

Placental mammals are the animals most familiar to us (Fig. 25–18). In these mammals, an organ of exchange, called the **placenta,** develops from tissues of the embryo and tissues of the mother's uterus. Food and oxygen transported in the mother's blood pass through the placenta into the developing embryo, and wastes from the embryo pass through the placenta into the mother's circulation. This continuous exchange permits the young to remain within the body of the mother until embryonic development is complete. Placental mammals are born at a more mature stage than are marsupials. Indeed, among some species, the young can walk around and begin to interact with other members of the group within a few minutes of birth.

Living placental mammals are classified into about 17 orders. Humans, along with the lemurs, monkeys, and apes, belong to the order Primates.

PRIMATES EVOLVED FROM SHREWLIKE MAMMALS

Human ancestry can be traced back some 65 million years to the earliest primates. These animals probably evolved from the arboreal (tree-dwelling), shrewlike mammals that had appeared during the Age of Reptiles. The living organisms that most resemble these ancient mammals are the tree shrews (Fig. 25–19).

Because of their ancestry, most primates have adaptations for an arboreal existence even if they do not live in trees. One of the most significant features of primates is that they have five digits on their hands and feet—four digits plus an **opposable thumb.** This enables primates to grasp objects such as tree branches.

Another adaptation to life in the trees is long, slender limbs that rotate freely at the hips and shoulders. Highly mobile primates are able climbers, maneuvering easily through the treetops in search of food. The location of the eyes in the front of the head, along with a shortened snout, provides stereoscopic, or three-dimensional, vision. This adaptation is useful for arboreal animals, because an error in depth perception might result in a fatal fall. In addition to sharp sight, hearing is acute in primates. In contrast, the sense of smell is relatively poor compared with that of most other mammals.

FIGURE 25–19 The common tree shrew most resembles the ancient insectivores that gave rise to the primates. (Warren & Genny Garst/Tom Stack & Associates)

Primates share several other characteristics, including complex social behavior. Primate reproduction usually results in only one offspring, which is initially quite helpless and requires a long period of nurturing and protection.

The order **Primates** is divided into two suborders (Fig. 25–20). The **prosimians** (meaning, "before apes") include the lemurs, lorises, and tarsiers. The prosimians were the first primates to evolve (Fig. 25–21). The **anthropoids,** comprising the monkeys, apes, and humans, are primates with larger brains.

The anthropoids descended from a group of prosimians during the Oligocene epoch, approximately 38 million years ago. This divergence took place in Africa or possibly Asia. From there the anthropoids quickly spread throughout Europe, Asia, and Africa. They branched into two main groups: the New World monkeys and the Old World monkeys (Fig. 25–22).

It is not clear how the New World monkeys reached South America because Africa and South America had already split by continental drift (see Chapter 20). Originally, biologists thought that the New World monkeys evolved from a separate prosimian line. Various types of comparative biochemistry and molecular biology, including amino acid sequencing of proteins, indicate that the New World and Old World monkeys share a common ancestor. At any rate, the New World and Old World monkeys have been separated for millions of years, and they evolved along different paths.

Most monkeys are tree dwellers. Different groups eat leaves, fruits, buds, insects, and even small vertebrates. New World monkeys are arboreal and have long, slender limbs that allow easy movement in the trees (Figure 25–22*a*). Many have a **prehensile** tail that is capable of wrapping around branches. Some New World monkeys have a smaller thumb, and in certain cases the thumb is totally absent. Featuring flattened noses with nostrils that open to the side, their facial anatomy is different from that of the Old World monkeys. They live in groups and exhibit social behavior. New World monkeys, which are restricted to Central and South America, include howler monkeys, squirrel monkeys, and spider monkeys.

Many Old World monkeys are arboreal, although some, such as baboons and macaques, are ground dwellers (Fig. 25–22*b*). Biologists think that ground dwellers, which are **quadrupedal** and walk on all fours, evolved from arboreal monkeys. None of the Old World monkeys has a prehensile tail, and some lack tails completely. They have a thumb that is fully opposable, and unlike the New World monkeys, their nostrils are directed downward and are closer together. Old World monkeys are larger than New World monkeys. They are social animals and are distributed in tropical parts of Africa and Asia.

(Text continues on page 495)

Prosimians
Anthropoids
Hominoids
Lemurs
Tarsiers
New World Monkeys
Old World Monkeys
Gibbons
Orangutans
Gorillas
Chimpanzees
Humans

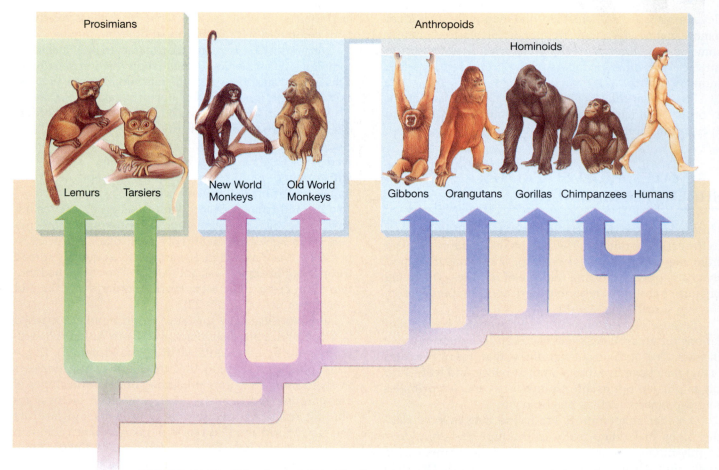

FIGURE 25–20 Primate evolution. The first primates to evolve were the prosimians. The other major group of primates, the anthropoids, evolved from a line of prosimians. They include the monkeys, apes, and humans. (Figures are not drawn to scale.)

(a)

(b)

FIGURE 25–21 Prosimians. (***a***) Mother and baby lemurs eating. Lemurs are native to Madagascar. (***b***) Philippine tarsier *(Tarsius syrichta)*. (*a*, Frans Lanting/Minden Pictures; *b*, Tom McHugh/Photo Researchers, Inc.)

(a) (b)

FIGURE 25–22 New World and Old World monkeys. (***a***) Most New World monkeys have prehensile tails that can function almost as effectively as another limb. Shown here is Geoffrey's spider monkey. (***b***) The lion-tailed monkey *(Macaca silenus)* is an Old World monkey native to India. (*a*, Frans Lanting/Minden Pictures; *b*, Dennis Drenner)

Hominoids Represent a Superfamily of Primates to Which the Apes and Humans Are Assigned

The Old World monkeys were ancestral to the **hominoids,** a group composed of apes and **hominids** (humans and their ancestors). The four genera of apes are classified into two families (Fig. 25–23). The gibbons, sometimes known as lesser apes, are placed in a separate family. The family Pongidae includes the orangutans (*Pongo*), gorillas (*Gorilla*), and chimpanzees (*Pan*).

Gibbons are well adapted for an arboreal existence. They are natural acrobats and can **brachiate,** or swing, with their weight supported by one arm at a time. Orang-

(a) (b) (c) (d)

FIGURE 25–23 Apes have no tails. (***a***) White-handed gibbons *(Hylobates lar).* Gibbons are extremely acrobatic and often move through the trees by brachiation. (***b***) An orangutan in Sumatra. Orangutans are solitary apes that seldom leave the protection of the trees. (***c***) Young lowland gorilla *(Gorilla sp.)* in knuckle-walking stance. (***d***) Chimpanzees live in groups and have a complex social behavior. Female sharing food with infants. (*a*, Visuals Unlimited/Joe McDonald; *b*, Peter Drowne; *c*, David J. Cross/Peter Arnold, Inc.; *d*, Frans Lanting/Minden Pictures)

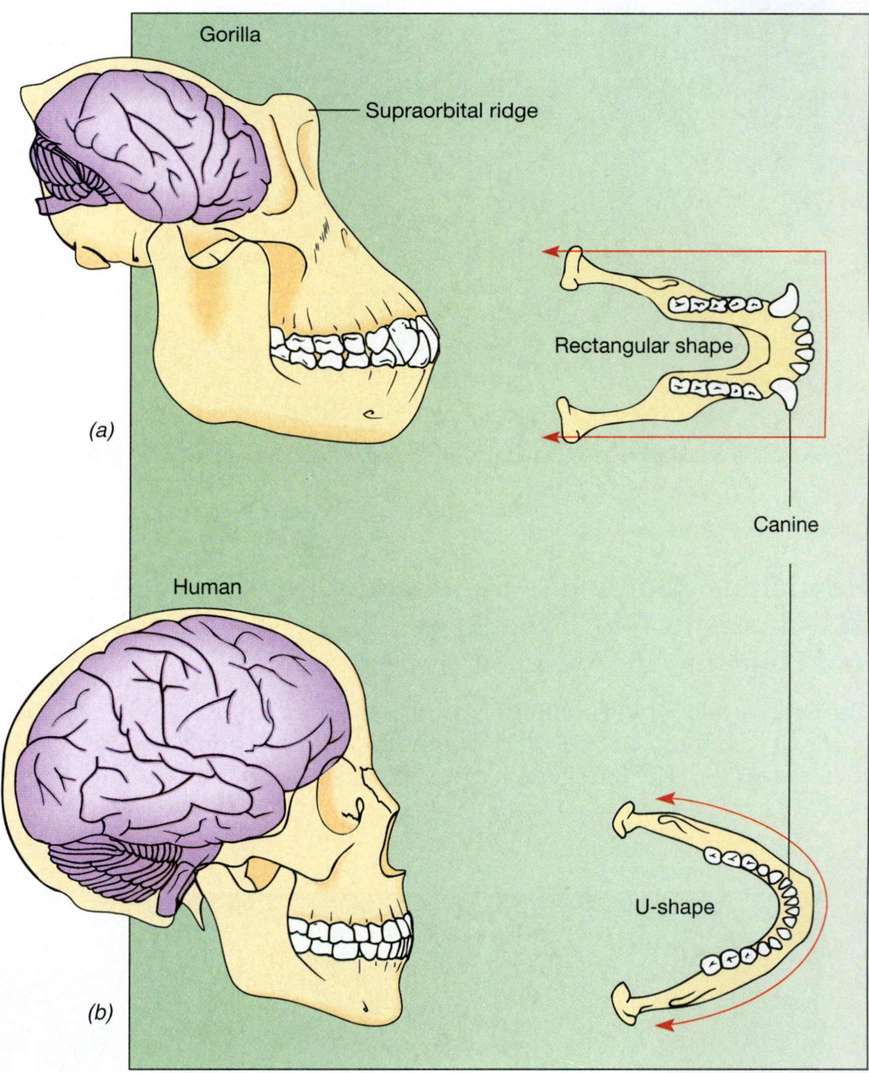

FIGURE 25–24 Comparison of features of the gorilla and human head. (***a***) The gorilla skull has pronounced supraorbital ridges. Note how the human skull is flatter in the front and has a more pronounced chin. (***b***) The human brain, particularly the cerebrum, is larger than that of an ape. The human jaw is structured so that the teeth are arranged in a U-shape. Human canines are smaller than gorilla canines.

utans are also tree dwellers, but both gorillas and chimpanzees have adapted to life on the ground. They have retained elongated forearms typical of tree-dwelling primates but use these to assist in walking. Their locomotion is known as **knuckle-walking** because of the way they fold their digits when moving. Apes, like humans, lack tails. They are generally larger than monkeys, but the gibbons are a notable exception. Probably due in part to their larger brains, ape social organization, particularly among the gorillas and chimpanzees, is highly complex.

Antigen-antibody tests of similarities in serum proteins show that, of all the primates, gorillas and chimpanzees have serum proteins most nearly like those of humans. The amino acid sequence of the chimpanzee's hemoglobin is identical to that of the human; those of the gorilla and rhesus monkey differ from the human's in 2 and 15 amino acids, respectively. Molecular studies of DNA sequences, as well as fossil evidence, suggest that chimpanzees are our nearest living relatives among the apes.

The Fossil Record Suggests General Trends in Hominid Evolution

The hominid line (humans and their ancestors) separated from the ape line approximately 5 to 7 million years ago (see Fig. 25–20). General trends in human evolution are evident from the fossil record, but we do not have enough evidence to make specific conclusions. There are simply too few early hominid fossils, and the ones we have are represented by only a few bones. Moreover, it is difficult to determine many aspects of early hominid biology, appearance, or behavior from fossilized bones. Nevertheless, it is evident that early hominids evolved a **bipedal** (two-footed) posture before their brains enlarged.

Another major trend in human evolution is an increase in the size of the brain relative to the size of the body (Fig. 25–24). In addition, the ape skull possesses prominent bony ridges above the eye sockets. These **supraorbital ridges** are lacking in human skulls. Human faces are flatter than ape faces, and their jaws are different. Compared to humans, apes have larger teeth, and their canine teeth are especially large.

The Earliest Hominids Belong to Genus *Australopithecus*

Human evolution first occurred in Africa. The earliest hominids belong to the genus *Australopithecus,* or "Southern ape." The actual number of species assigned to this genus is a matter of debate. It is very difficult to decide whether differences in the relatively few skeletal fragments that have been discovered indicate individual variation within a species or separate species. Most biologists recognize two to four species of australopithecines.

The most ancient hominids are assigned to the species *A. afarensis.* Several fossils of skeletal remains have been discovered, including a remarkably complete skeleton of a female that was named Lucy by the archaeologists who found her (Fig. 25–25). In addition, fossil footprints of three individuals that walked the African plains over 3.6 million years ago were discovered in 1976 (Fig. 25–26). The footprints, plus pelvis, leg, and foot bones, indicate that the development of an upright posture and bipedalism occurred early in human evolution.

Australopithecus afarensis was a small hominid, approximately 3 feet tall. Its face projected forward, and its apelike skull covered a small brain. The cranial capacity was 450 to 500 cubic centimeters compared to a modern human cranial capacity of 1400 cubic centimeters. Even when differences in body size are taken into account, *A. afarensis* still had a small brain. The number and arrangement of teeth were primitive and included long canines.

Many scientists think *A. afarensis* evolved into the more advanced australopithecine, *A. africanus,* which appeared approximately 3 million years ago. The first *A. africanus* fossil was discovered in South Africa in 1924, and a number of others have since been found. This rather small hominid walked erect and possessed hands and teeth that were distinctly similar to those of a human. Based on characteristics of the teeth, scientists hypothesize that *A. africanus* ate both plants and animals. Like *A. afarensis,* its brain was small, approximately 500 cubic centimeters.

Homo habilis Is the Oldest Member of Genus *Homo*

The first hominid to have enough human features to justify classification in the same genus as modern humans

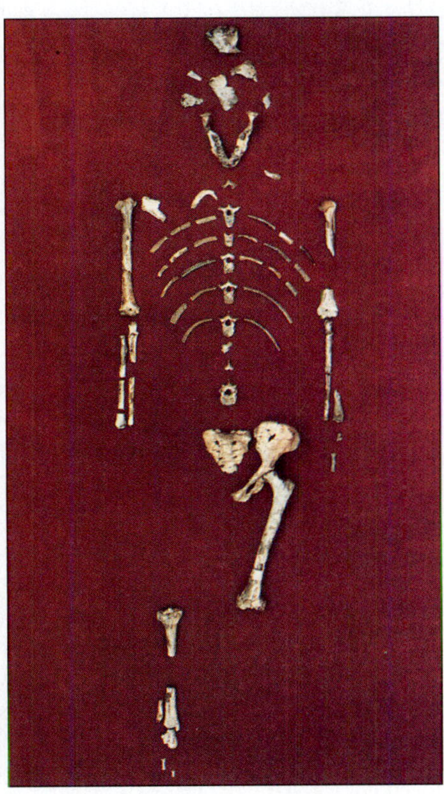

FIGURE 25–25 The skeletal remains of Lucy, a hominid approximately 3.5 million years old. (Institute for Human Origins.)

FIGURE 25–26 Trail of hominid footprints in fossilized volcanic ash. Three hominids *(Australopithecus afarensis)* walked across ash scattered by a volcanic eruption over 3.6 million years ago in Africa. Their footprints were compacted by a rain shower. The footprints, discovered in 1978 by Mary Leakey and her associates, indicate *A. afarensis* had a bipedal gait. (John Reader/Science Photo Library/Photo Reachers, Inc.)

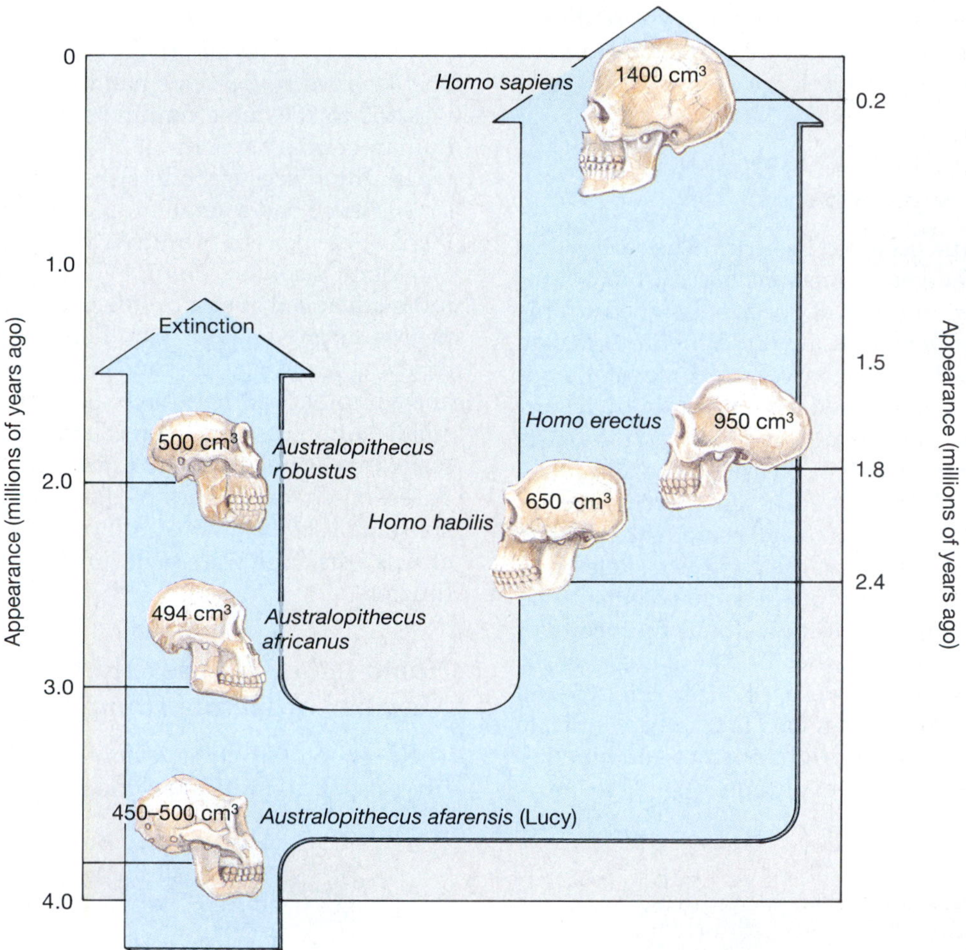

FIGURE 25–27 Skulls of representative hominids. Note the massive bony ridges over the eyes, the receding forehead, and protruding jaw of early hominids.

is *Homo habilis* (Fig. 25–27). *Homo habilis* had a larger cranial capacity (an average of 650 cubic centimeters) than the australopithecines. In 1992, Andrew Hill of Yale University and Steven Ward of Northeastern Ohio University's College of Medicine reported a 2.4 million year old human skull that may have been the earliest *Homo habilis*. Found in Kenya, this skull is half a million years older than any previously found for our genus. Fossils of *H. habilis* have been found in numerous areas in Africa. These sites also contain the first primitive tools, stones that had been chipped to make sharp edges for cutting or scraping. Although other primates occasionally use tools, many consider *H. habilis* the first species to consciously design them.

The relationship between the australopithecines and *H. habilis* is not clear. Using physical characteristics as evidence, some biologists hypothesize that the australopithecines were ancestors of *H. habilis*. Others hold that *H. habilis* and *A. africanus* were contemporaries for much of their existence and that *H. habilis* was in a direct line to humans, but that *A. africanus* was not. Discoveries of additional fossils will help clarify their relationship.

Numerous Fossils of *Homo erectus* Have Been Discovered

Homo erectus evolved in Africa as did the other hominids. For this reason, the oldest fossils of *H. erectus*, which are 1.8 million years old, are found in Africa. Perhaps 200,000 years after it evolved, *H. erectus* migrated from Africa into Europe and Asia. Peking man and Java man discovered in Asia were later examples of *H. erectus*, which existed until approximately 200,000 years ago.

Concept Connections

Cultural Evolution ⟷ *Global Ecology*

Culture is knowledge and experience passed on from one generation to the next, not by genes, but by parents, teachers, books, and microchips. Our cultural evolution has had a profound and far-reaching effect on both the living and nonliving world. In the beginning when we were nomadic hunter/gatherers, children learned what they needed to know to survive by word of mouth from their elders. Culture changed very slowly. With the development of agriculture, permanent settlements developed, changing opportunities for education and culture.

The Industrial Revolution, which began in the 18th century, was a major turning point in cultural evolution. Rapid technological advances have transformed the way we live and the way we learn. One effect of the Industrial Revolution has been a major shift in human population distribution. Displaced from their jobs on farms by new machines, thousands of people flocked to urban areas in search of factory work.

Another effect of the Industrial Revolution has been a greatly increased demand for natural resources to supply the raw materials for industry. New advances in medicine and other technology have greatly enhanced standards of living and increased human life expectancy. As a result, the human population has increased dramatically. It took from the beginning of human evolution until the year 1830 for the population to reach 1 billion. Now our population exceeds 5.6 billion! Many biologists fear the Earth cannot support our numbers. As it is, more than one billion people do not have enough food, and almost all of the fertile land on Earth is already under cultivation.

Cultural evolution has resulted in large-scale disruption and degradation of the environment. The chemical fertilizers, pesticides, and irrigation methods that helped increase agricultural production have also poisoned wildlife and humans and contaminated and depleted our water supplies. In order to provide housing, farmland, and industrial goods for the burgeoning human population, tropical rain forests and other natural environments are being eliminated at a rapid rate. Soil, water, and air pollution now affect most of the ecosphere. Many plant and animal species cannot adapt to the rapid changes humans are causing to the environment and are becoming extinct. The decrease in biological diversity due to extinction is alarming.

On a positive note, we are becoming increasingly aware of the negative impact we have had on the environment, and on the need to choose technologies that are less harmful. Some scientists are predicting a second Industrial Revolution, one in which environmental concerns will be a guiding force. We have the intelligence to modify our behavior so as to sustain our ecosphere. Through education, we can develop environmental sensitivity. If we succeed in educating younger generations, cultural evolution could be our salvation rather than our destruction.

Homo erectus was bipedal, fully erect, and taller than *H. habilis*. Its skull, although larger, did not possess totally modern features. It retained the heavy supraorbital ridge and projecting face that is characteristic of its ape ancestors. During the course of its existence, the *H. erectus* brain got progressively larger.

The increase in mental faculties associated with an increase in brain size enabled these early humans to make more advanced stone tools, including hand axes and other tools that have been interpreted as choppers, borers, and scrapers. Their intelligence enabled them to survive in areas that were cold. *Homo erectus* wore clothing, built fires, and lived in caves or shelters. It is not known for sure whether these early humans were hunters or scavengers. To date, no weapons have been unearthed at their sites.

Homo sapiens Appeared about 200,000 Years Ago

Humans that have modern enough features to be classified within our species appeared approximately 200,000 years ago. Their brains continued to enlarge, developing from 850 cubic centimeters in earliest individuals to the current cranial capacity of 1400 cubic centimeters.

Investigators debate the origin of modern humans. According to one hypothesis, modern humans evolved from an African ancestor and later migrated and replaced humans inhabiting other parts of the world. (Proponents of this hypothesis hold that little interbreeding occurred.) The opposing hypothesis suggests that *Homo sapiens* evolved from *H. erectus* independently in several different regions of Europe and Asia. (According to this hy-

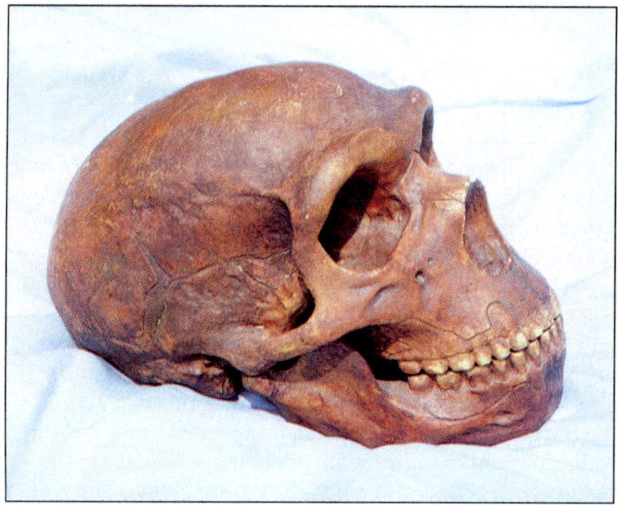

FIGURE 25–28 Neandertal skull. Note the very heavy supraorbital ridge and the protruding face. The size of the Neandertal brain was greater than that of modern human brains.

FIGURE 25–29 The Cro-Magnon people painted animals on cave walls in Europe. These are some of the earliest representations of human art and have been interpreted as having a religious significance, possibly for guaranteeing a successful hunt. (Trans. no. 4722 (2) [Photo by J. Beckett/D. Stipkovitch] courtesy Department of Library Services, American Museum of Natural History)

pothesis, interbreeding occurred so that gene flow took place between various populations.)

The *Neandertals* were among the earliest *H. sapiens.* They were first discovered in the Neander Valley in Germany, but had a widespread distribution in Northern Europe. These early humans had a short, sturdy build. Their brains were slightly larger than modern *H. sapiens,* but their faces still projected slightly, with less pronounced chins and heavy supraorbital ridges (Fig. 25–28).

Neandertal tools, including spear points, were more sophisticated than those of *H. erectus.* Studies of Neandertal sites indicate that their culture included hunting for large animals. The existence of skeletons that were old or had healed fractures demonstrates that they cared for the elderly and sick, an example of advanced social cooperation. They apparently had rituals, possibly of religious significance, and buried their dead. The presence of food, weapons, and flowers in the graves indicates that they possessed the abstract concept of an afterlife.

The disappearance of the Neandertals is a mystery. Other groups of *H. sapiens* with more modern features coexisted with the Neandertals. It is possible that the Neandertals interbred with these humans, or perhaps the other humans out-competed or killed them. It is also possible that the Neandertals could not adapt to the climate changes of the Pleistocene and that their disappearance is unrelated to the presence of other humans.

Homo sapiens with completely modern features existed about 40,000 years ago, and possibly earlier. Some evidence from South Africa suggests that modern *H. sapiens* may have existed 100,000 years ago. Early *H. sapiens* skulls lacked heavy brow ridges and possessed a distinct chin. The **Cro-Magnon** culture in France and Spain exemplifies these humans. Their weapons and tools were complex and often made of materials other than stone, including bone, ivory, and wood. They made stone blades that were very sharp. They developed art, possibly for ritualistic purposes, including cave paintings, engraving, and sculpture (Fig. 25–29). The existence of a variety of complex tools and art is an indication that they may have possessed language abilities, which they used to transmit their culture to younger generations.

Chapter Summary

I. Phylum Chordata consists of three subphyla: Urochordata (tunicates), Cephalochordata (lancelets), and Vertebrata (animals with a backbone).

II. At some time in its life cycle, a chordate has a notochord, a dorsal tubular nerve cord, pharyngeal gill grooves, and a postanal tail.

III. Subphylum Urochordata, the tunicates, are sessile, filter-feeding marine animals covered by protective tunics.

IV. Subphylum Cephalochordata consists of the lancelets, which are small, segmented fishlike animals.

V. Subphylum Vertebrata includes animals with a vertebral column, cranium, pronounced cephalization, differentiated brain, muscles attached to an endoskeleton for movement, and two pairs of paired appendages.

VI. The vertebrate classes include jawless fishes, cartilaginous fishes, bony fishes, amphibians, reptiles, birds, and mammals.
   - A. The jawless fishes (class Agnatha) include the lampreys and hagfishes.
   - B. Class Chondrichthyes, the cartilaginous fishes, consists of the sharks, rays, and skates.
   - C. Class Osteichthyes includes freshwater and saltwater bony fishes. Most modern bony fishes are ray-finned fishes with swim bladders.
   - D. Modern amphibians (class Amphibia) include the salamanders, frogs, toads, and wormlike caecilians.
     1. Most amphibians return to the water to reproduce. Frog embryos develop into tadpoles,

which undergo metamorphosis to become adults.

2. Amphibians use their moist skin as well as lungs for gas exchange.

E. Class Reptilia includes turtles, lizards, snakes, and alligators. Reptiles dominated the Earth during the Mesozoic era. At the end of the Cretaceous period, most of them, including all of the dinosaurs, became extinct.
   1. Some reptiles are completely terrestrial.
   2. Fertilization is internal. Most reptiles secrete a leathery protective shell around the egg. The embryo develops an amnion and other surrounding membranes, which protect it and keep it moist.
   3. A reptile has a dry skin with horny scales, lungs, and a three-chambered heart.

F. Birds (class Aves) have many adaptations for flight, including feathers; wings; and light, hollow bones containing air spaces.
   1. Birds have a four-chambered heart and very efficient lungs.
   2. Birds maintain a constant body temperature and have a high metabolic rate.

G. Mammals (class Mammalia) have hair, mammary glands, and differentiated teeth, and maintain a constant body temperature. They have a highly developed nervous system and a muscular diaphragm.
   1. Monotremes, mammals that lay eggs, include the duck-billed platypus and the spiny anteater.
   2. Marsupials are pouched mammals such as kangaroos and opossums. The young are born in an immature stage and complete their development in the marsupium.
   3. Placental mammals are characterized by an organ of exchange, the placenta, that develops between the embryo and the mother. This organ supplies oxygen and nutrients to the embryo and enables it to complete development within the uterus.

VII. Primates evolved from small, arboreal, shrewlike mammals.

VIII. The hominid line separated from the ape line approximately 5 to 7 million years ago.
   A. The earliest hominids belong to the genus *Australopithecus.* The australopithecines walked on two feet, a human feature.
   B. *Homo habilis* was an early hominid that had some human features the australopithecines lacked, including a slightly larger brain. *Homo habilis* fashioned tools from stone.
   C. *Homo erectus* had a larger brain than *H. habilis,* made more sophisticated tools, and discovered how to use fire.
   D. *Homo sapiens* appeared approximately 200,000 years ago.
      1. The brain continued to enlarge during their evolution.
      2. Modern *H. sapiens* may have evolved from a common African ancestor.

Selected Key Terms

amnion, p. 487
anthropoid, p. 493
cephalochordates, p. 479
cranium, p. 480
endoskeleton, p. 480
hominid, p. 495
hominoid, p. 495
lobe-finned fish, p. 484
marsupial, p. 490
monotreme, p. 490
notochord, p. 479
oviparous, p. 482
pharyngeal gill grooves, p. 479
placenta, p. 492
placoderm, p. 482
primate, p. 493
ray-finned fish, p. 484
swim bladder, p. 484
tetrapod, p. 484
tunicate, p. 479
urochordate, p. 479
vertebrate, p. 480

Post-Test

1. Four distinguishing characteristics of a chordate are a/an ______________, a dorsal tubular ______________ ______________, pharyngeal ______________ grooves, and a postanal ______________.

2. ______________ are sessile, marine chordates often mistaken for sponges.

3. Vertebrates are distinguished from all other animals in having a/an ______________ ______________; anterior to this structure, a/an ______________ encloses and protects the brain.

4. Modern bony fishes are thought to have descended from the ______________ fishes; the lobe-finned fishes are credited with being the ancestors of the ______________.

5. Vertebrates that have a three-chambered heart and moist skin are the ______________.

6. The amnion is an adaptation to ______________ life; its functions include ______________.

7. Reptiles gave rise to ______________ and ______________.

8. Monotremes are mammals that ______________ ______________; marsupials are distinguished by their ______________.

9. Which of the following animals have an amnion? (a) bony fishes (b) amphibians (c) reptiles (d) birds (e) answers (c) and (d) are correct

10. Animals with hair include (a) tunicates (b) reptiles (c) amphibians (d) mammals (e) answers (a) and (d) are correct.

11. Animals that have a four-chambered heart (two atria and two ventricles) include (a) lancelets (b) amphibians (c) birds (d) mammals (e) answers (c) and (d) are correct.
12. Which animals are covered with hard, dry, horny scales? (a) lancelets (b) amphibians (c) reptiles (d) mammals (e) two of the preceding are correct
13. Animals with no teeth and that have bones with air spaces are (a) bony fishes (b) amphibians (c) reptiles (d) birds (e) none of the preceding answers is correct.
14. Animals that have pharyngeal gill grooves at some time in their life cycle are (a) bony fishes (b) amphibians (c) reptiles (d) birds (e) all of the preceding answers are correct.
15. The earliest hominid to be placed in the genus *Homo* was *H.* ______________.

Review Questions

1. What are the four principal distinguishing characteristics of a chordate? How are these evident in a lancelet? In a human?
2. What characteristics distinguish the vertebrates from the rest of the chordates?
3. How do lampreys and hagfishes differ from other fishes?
4. Compare the skins of sharks, frogs, snakes, and mammals.
5. Give the location and function of each of the following: (a) swim bladder (b) placenta (c) amnion.
6. Give the phylum, subphylum, and class for each of the following animals: (a) human (b) turtle (c) lamprey (d) lancelet (e) shark (f) whale (g) frog (h) pelican (i) bat.
7. Which animals are more specialized—birds or mammals? Explain your answer.
8. Cite one anatomical feature and one behavioral feature that distinguishes each of the following from its immediate ancestor: (a) *Australopithecus afarensis* (b) *Homo erectus* (c) *Homo sapiens* (Neandertal) (d) *Homo sapiens* (modern).

Thinking Critically

1. Discuss the extinction of the dinosaurs. How did the disappearance of the large reptiles contribute to the evolution of mammals? Compare the effects on the ecosphere of a natural event like comets hitting the Earth with the effects caused by human events such as the Industrial Revolution and later technological developments.
2. Why are monotremes considered more primitive than other mammals? Some paleontologists consider them to be therapsid reptiles rather than mammals. Give arguments for and against this position.
3. Which vertebrate groups maintain a constant body temperature? Why is endothermy advantageous?
4. Given that reptiles are terrestrial animals that do not have to return to the water to reproduce, how do you account for sea turtles and other aquatic reptiles that live in the oceans but return to land to reproduce?
5. According to current evolutionary theory, give the significance of each of the following: (a) placoderms (b) labyrinthodonts (c) therapsids.
6. Provide at least two reasons that birds inhabit polar regions but reptiles do not.
7. How is cultural evolution related to biological evolution? (Hint: The evolution of what biological characteristic con tributed to cultural evolution?) How has cultural evolution impacted the ecosphere?

Recommended Readings

Blumenschine, R.J., and J.A. Cavallo, "Scavenging and Human Evolution," *Scientific American*, Vol. 267, No. 4 (October 1992), 90–96. Early humans may have been better scavengers than hunters.

del Pino, E.M., "Marsupial Frogs," *Scientific American*, Vol. 260, No. 5 (1989). Marsupial frogs have a long incubation period within the mother's body, resembling pregnancy in mammals. However, the eggs and embryos of these frogs are birdlike.

Eastman, J.T., and A.L. DeVries, "Antarctic Fishes," *Scientific American*, Vol. 255, No. 5 (1986). Most species of fishes died out when the Antarctic Ocean became icy cold, but fishes in the suborder Notothenioidei survive by making biological antifreezes and conserving energy.

Griffiths, M., "The Platypus," *Scientific American*, Vol. 258, No. 5 (1988), 84–91. Everything you might want to know about this interesting monotreme. The platypus has mechanoreceptors and electroreceptors on its beak for detecting prey.

Rismiller, P.D., and R.S. Seymour, "The Echidna," *Scientific American*, Vol. 264, No. 2 (Feb 1991). An interesting account of the natural history and reproductive behavior of the echidna, or spiny anteater.

Welty, J.C., *The Life of Birds*, 4th ed., Philadelphia, Saunders College Publishing (1988). An introduction to the biology of birds.

Wilson, A.C., and R.L. Cann, "The Recent African Genesis of Humans," *Scientific American*, Vol. 266, No. 4 (April 1992), 68–73. One of two articles that appear in this issue debating the origin of humans.

PART 7 Plant Structure and Life Processes

An ethnobotanist consults with a Tirio Indian in Suriname about the uses of a rainforest plant. *(Courtesy of Mark Plotkin, Conservation International).*

The Medicinal Value of Flowering Plants

From extracts of cherry and horehound for cough medicines to chemical compounds in periwinkle and autumn crocus for cancer therapy, derivatives of roots, stems, leaves, and flowers play important roles in the treatment of illness and disease. About 25% of all prescription medicines contain one or more active ingredients extracted from plants.

Many of the plant-produced chemicals with medicinal properties are **alkaloids,** bitter-tasting organic compounds that contain nitrogen. For example, the rosy periwinkle, *Catharanthus roseus,* produces two alkaloids, vinblastine and vincristine. They are used to treat two kinds of cancer—Hodgkin's disease, which generally afflicts young adults, and childhood leukemia. Other important alkaloids with medicinal value include quinine, morphine, and reserpine. Quinine, an antimalarial drug, comes from the bark of yellow cinchona (*Cinchona ledgeriana*). Morphine, used medically to relieve pain, is extracted from the opium poppy (*Papaver somniferum*). The Indian snakeroot (*Rauvolfia serpentina*) is the source of reserpine, an alkaloid used to treat hypertension and some psychiatric disorders.

Various medicinal plants we use today have been used for centuries in folk medicine. Modern medicine has learned to respect reports of medicinal plants from the traditions of indigenous peoples because they provide clues about which plants to test. (Using folklore provides a shortcut in deciding which plants to test. Studies show that plants identified as useful by shamans and traditional plant users are up to 60% more likely to have medicinal value than those that are randomly collected.) **Ethnobotany,** the study of the traditional uses of plants by indigenous people, helps pharmaceutical companies identify medicinal plants. Unfortunately, much of this knowledge is disappearing as tribes are exposed to modern ways and their traditions are quickly forgotten.

Only about 5000 of the 235,000 species of flowering plants have been investigated for their medicinal value. It is likely that the remaining 230,000 species contain other valuable alkaloids with medicinal properties. Unfortunately, we may never have the opportunity to find out because many of the wild plant species that have medicinal potential are themselves threatened as human activities destroy wildlife habitats.

The rosy periwinkle, for example, is native to Madagascar, an extremely poor island nation with a rapidly increasing human population. It is estimated that 80% of the 10,000 plant species on Madagascar are found nowhere else in the world. Yet, during the past 45 years, most of the forests of Madagascar have been destroyed by impoverished people who clear the forests for agriculture.

Many of the world's biological riches are found in economically poor countries. The 13.3 million people living in Madagascar, with a per capita gross national product of $250 (U.S. dollars), have not benefitted from the $180 million earned annually by the sale of vinblastine and vincristine from the rosy periwinkle.

New arrangements seek to distribute proceeds from sales/patents directly to indigenous peoples. The Suriname Biodiversity Prospecting Initiative, announced in 1993 by Conservation International, is the first such agreement of its kind. This five-year enterprise pays money to local people of Suriname, a small country in South America, for new drugs delivered from the plants that they identify in their rain forests. With support from the U.S. government, universities, and various nonprofit organizations, the initiative represents the first partnership between indigenous people and a private pharmaceutical company.

This chapter introduces a unit on the biology of flowering plants, living organisms that share many similarities with Earth's other life forms. Flowering plants, in terms of their great diversity and wide distribution, are the most successful group of plants today. They are crucial to the survival of many other species, including humans. In addition to medicines, humans depend on flowering plants for food, shelter, fibers, fuel, and countless other products. Life as we know it would be impossible without flowering plants.

Plant Structure

26

Learning Objectives

After you have studied this chapter you should be able to:

1. Discuss the plant body, including the basic features of roots, stems, and leaves.
2. List the types of root systems and give an example of each.
3. Describe the ground tissue system (parenchyma tissue, collenchyma tissue, and sclerenchyma tissue) of plants.
4. Outline the structure and function of the vascular tissue system (xylem and phloem) of plants.
5. Describe the dermal tissue system (epidermis and periderm) of plants.
6. Discuss what is meant by "growth" in plants and relate how growth is different in plants and animals.
7. Distinguish between primary and secondary growth.
8. Explain why meristematic tissue is so important in a plant.
9. Differentiate between apical and lateral meristems—where each is located and what tissues originate from each.

Key Concepts

- ☐ The plant body is typically composed of an aboveground shoot system (stem, leaves, flowers, fruits, and seeds) and a below-ground root system. Although flowering plants are complex multicellular organisms that exhibit great diversity, their shoots and roots contain only three tissue systems: ground, vascular, and dermal tissue systems.
- ☐ Functions of the ground tissue system include photosynthesis, secretion, storage, and support. The vascular tissue system conducts various materials throughout the plant body and provides strength and support. The dermal tissue system provides a covering for the plant body that protects the plant from excessive water loss and from invasion by foreign organisms.
- ☐ Plant growth occurs in localized areas of the plant body. Primary growth, which causes an increase in plant length, occurs in all plants and is localized at the tips of stems and roots. Woody dicot plants also have secondary growth, which results in an increase in their girth (thickness). Secondary growth is localized, typically as long cylinders of active growth throughout the length of older stems and roots.

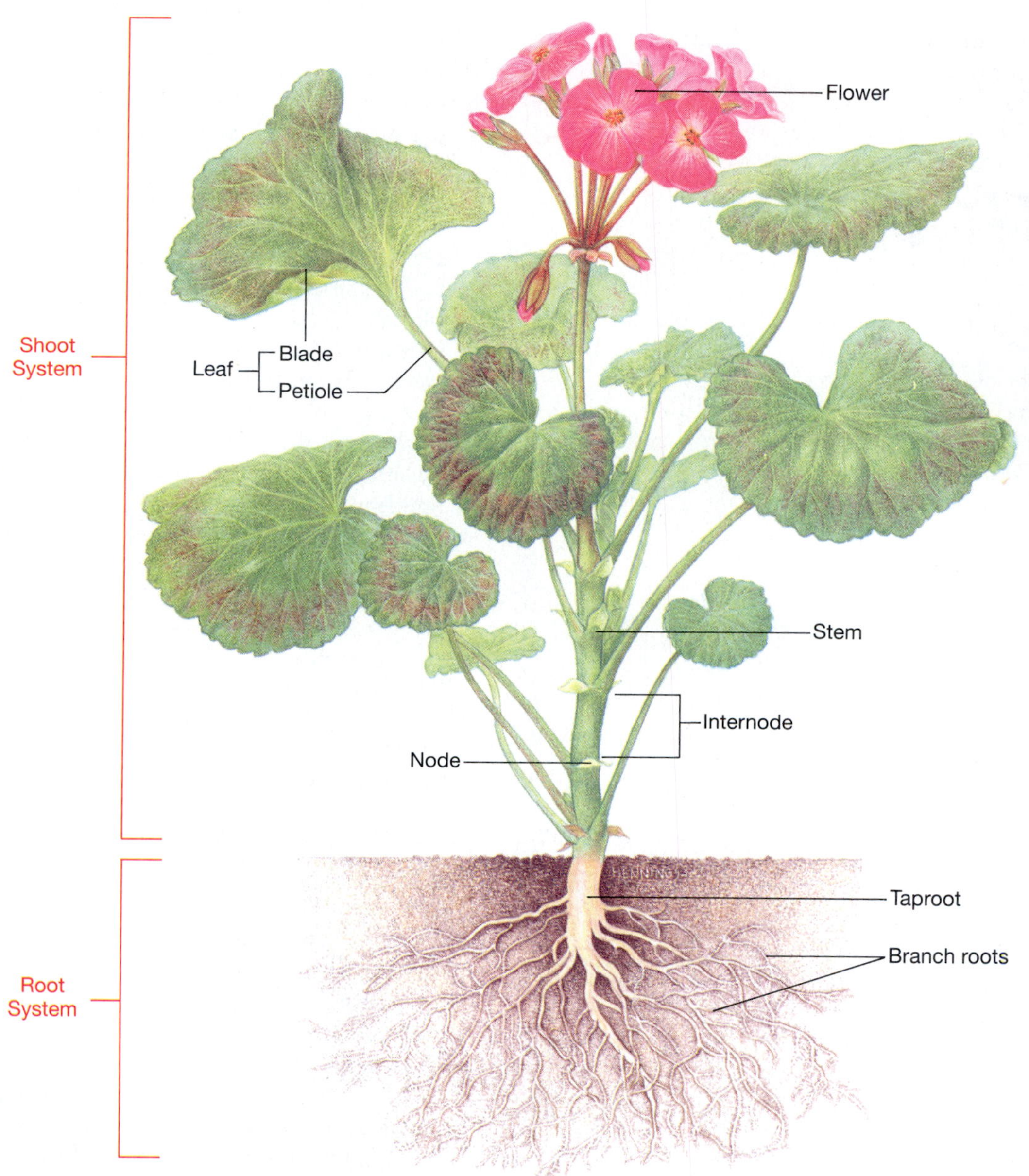

FIGURE 26–1 The plant body of a geranium, a typical herbaceous plant. The plant body, which consists of a root system and a shoot system, possess all of the necessary parts to maintain life.

PLANTS EXHIBIT BOTH SIMILARITY AND DIVERSITY IN STRUCTURE AND LIFESPAN

Remarkable variety is represented in the 235,000 or so species of flowering plants that live in and are adapted to the many environments offered by our planet. Yet all of these plants—from desert cacti with enormously swollen stems to cattails partly submerged in marshes to orchids growing in the uppermost branches of lush tropical forest trees—are recognizable as plants. Regardless of size—from *Wolffia microscopica,* the smallest flowering plant known, to Australian gum trees, some of Earth's tallest plants—all plants have the same basic body plan.

Flowering plants are either herbaceous or woody. **Herbaceous** plants do not develop persistent woody parts above ground, whereas **woody** plants (trees and shrubs) do. Many herbaceous plants (such as corn) are **annuals,** which are plants that grow, reproduce, and die in one year. Some herbaceous plants (such as carrots and beets) are **biennials,** which means they take two years to complete their growth and reproduction before dying. Biennials typically form flowers and reproduce in their second year. Herbaceous plants that are **perennials** (for example, rhubarb and asparagus) live for a number of years. In temperate climates, their above-ground stems die back each winter. During the winter, their underground parts become dormant, sending out new growth each spring. (In dormancy, an organism reduces its metabolic state to a minimum level to survive unfavorable conditions.) Likewise, in certain tropical climates with pronounced wet and dry seasons, the underground parts of herbaceous perennials become dormant during the dry season.

All woody plants are perennials, and some of them live for hundreds or even thousands of years. In temperate climates, the above-ground stems of woody plants become dormant during the winter. Most temperate woody plants shed their leaves before winter and produce new stems with new leaves the following spring.

ROOTS, STEMS, LEAVES, FLOWERS, AND FRUITS MAKE UP THE PLANT BODY

The plant body is organized into a root system and a shoot system (Fig. 26–1). The **root system** is generally the below-ground portion. The above-ground portion, the **shoot system**, usually consists of a vertical stem that bears leaves, flowers, fruits, and seeds. Roots, stems, leaves, flowers, and fruits are considered plant organs because each is composed of several different tissues. (To be more precise, a flower is a *collection* of organs. Each of the parts of a flower—sepals, petals, stamens, and carpels—is an organ.)

Each plant grows in two different environments: the dark, moist soil and the relatively bright, dry air. Plants must have both roots and shoots because they need resources from *both* environments. From the soil, they obtain water and minerals, and from the air they obtain sunlight and atmospheric CO_2. As we discuss the various structures of roots and shoots, try to relate their structures to their various functions and to the different environments in which they live.

The Root System Grows into the Soil

Roots are the underground portion of most plants. They branch extensively through the soil, forming a network that anchors the plant firmly in place and absorbs water and dissolved minerals. Two types of root systems—a taproot system and a fibrous root system—occur in plants (Fig. 26–2). A **taproot** system, which consists of one main root with many smaller lateral roots coming out of it, is characteristic of dicots and gymnosperms.[1] A taproot develops from the embryonic root that was part of the young plant in a seed. A dandelion is a good example of a common plant with a taproot system.

A **fibrous root** system has several to many roots of the same size developing from the end of the stem, with smaller lateral roots branching off these roots. Fibrous root systems form in plants in which the embryonic root is short-lived. Onions, crabgrass, and other monocots have fibrous root systems.

[1] For a discussion of the two groups of flowering plants—dicots and monocots—see Chapter 23. Gymnosperms, also introduced in Chapter 23, are not flowering plants. However, they are such an important group of plants that reference to them will occasionally be made.

(a)

(b)

FIGURE 26–2 Root systems in plants. (***a***) A fibrous root system is characteristic of such plants as daffodils (shown) and grasses. (***b***) The taproot system is common in many flowering plants, including dandelions (shown) and geraniums. Both taproots and fibrous roots may be modified for food storage. (*a*, Dwight R. Kuhn; *b*, Lynwood M. Chace/Photo Researchers, Inc.)

The roots of some plants are modified for food storage. These **storage roots** may be modified taproots (for example, carrots) or modified fibrous roots (for example, sweet potatoes).

The Shoot System Is the Aerial Portion of a Plant

A shoot consists basically of a stem with leaves, the flattened organs of photosynthesis, attached in a more or less regular arrangement. Leaves are positioned on the stem to capture the sun's light. In fact, the presence of leaves is an easy way to distinguish most stems from roots because only stems bear leaves. (Leaves are discussed in detail in Chapter 27.)

The area on a stem where each leaf is attached is called a **node,** and the region of a stem between two successive nodes is an **internode.** Stems have **buds,** which are undeveloped embryonic shoots. A **terminal bud** is the embryonic shoot at the tip of a stem. When a terminal bud is dormant (that is, unopened and not actively growing), it is covered by an outer protective layer of **bud scales,** which are modified leaves. Plants also have **lateral buds** located in the axils of leaves. (An axil is the upper angle between a leaf and the stem to which it is attached.) When terminal and lateral buds grow, they form stems that bear leaves and/or flowers.

A woody twig that is dormant and has shed its leaves can be used to demonstrate stem structure (Fig. 26–3). When the plant resumes growth, the bud scales covering the terminal bud fall off, leaving a bud-scale scar on the stem where the bud scales were attached. Because temperate plants form terminal buds once a year, at the end of the growing season, counting the number of bud-scale scars indicates the age of the twig. A leaf scar shows where each leaf was attached on the stem, and the vascular tissue that runs from the stem out into the leaf forms bundle scars within the leaf scar. Lateral buds may be found above the leaf scars. Also, the bark of a woody twig contains tiny marks. These marks are **lenticels,** sites of loosely arranged cells that allow oxygen to diffuse into the interior of the woody stem.

Most leaves are composed of two parts, a blade and a petiole. The broad, expanded portion of the leaf is the **blade,** and the stalk that attaches the blade to the stem is the **petiole.** Leaves may be **simple** (having an undivided blade) or **compound** (having a blade divided into two or more leaflets; Fig. 26–4). Sometimes it is difficult to tell whether a plant has formed one compound leaf or a small stem bearing several simple leaves. One easy way to determine if a plant has simple or compound leaves is to look for lateral buds. The lateral buds form at the base of the leaf, whether it is simple or compound, never at the base of leaflets. Also, the leaflets of a compound leaf lie in a single plane (you can lay a compound leaf flat on a table), whereas simple leaves are never arranged in one plane on a stem.

Leaves are arranged on a stem in one of three possible ways (Fig. 26–5). Plants such as walnut have an **alternate** leaf arrangement, with one leaf at each node. In an **opposite** leaf arrangement, as occurs in lilac, two leaves grow at each node. In a **whorled** leaf arrangement, as occurs in catalpa, three or more leaves grow at each node.

Leaf blades may possess **parallel** veins (characteristic of monocots) or **netted** veins (characteristic of dicots; Fig. 26–6). Netted veins can be **palmately netted,** in which several major veins radiate out from one point, or **pinnately netted,** in which major veins branch off along the entire length of the main vein.

(Text continues on page 510)

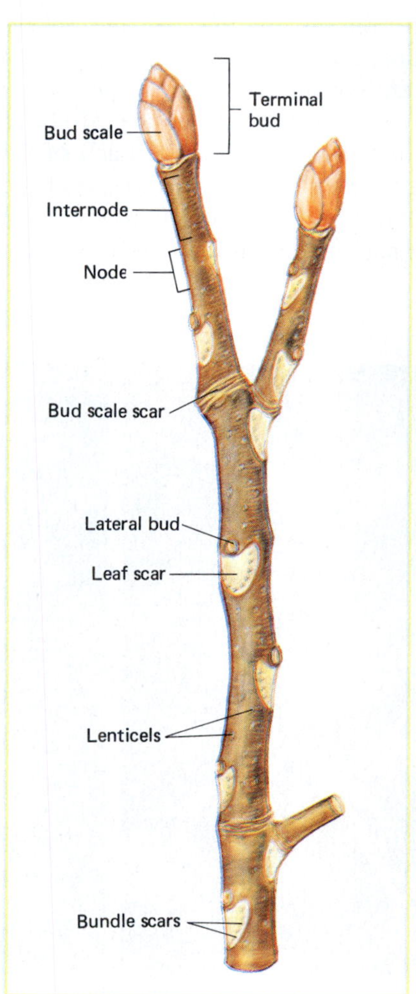

FIGURE 26–3 The external structure of a woody twig. The age of a woody twig can be determined by counting the number of bud-scale scars (do not count side branches). How old is this twig?

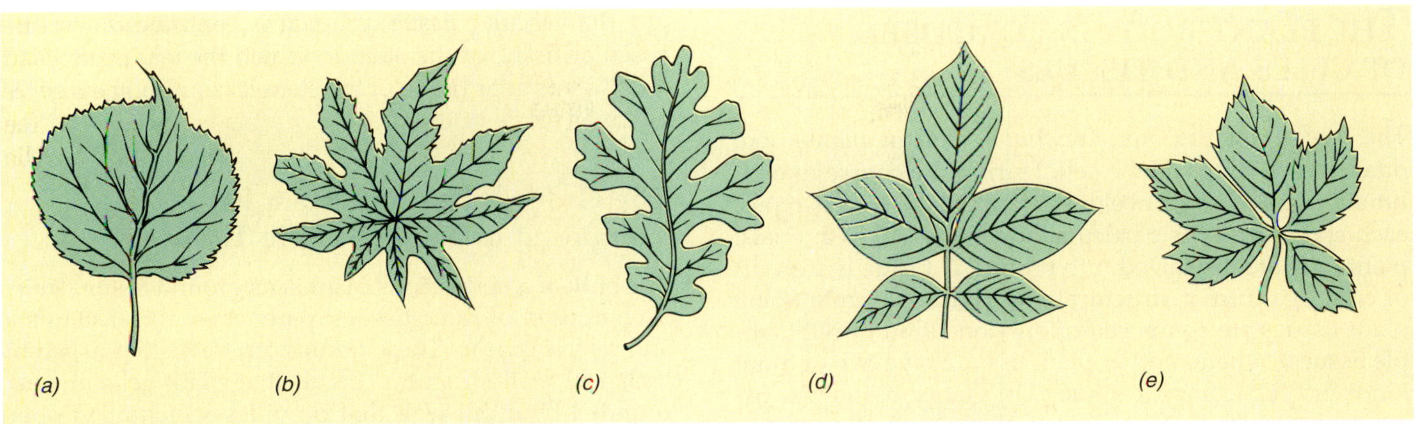

FIGURE 26–4 Simple and compound leaves. Both simple leaves (***a,b,c***) and compound leaves (***d,e***) come in a variety of sizes and shapes.

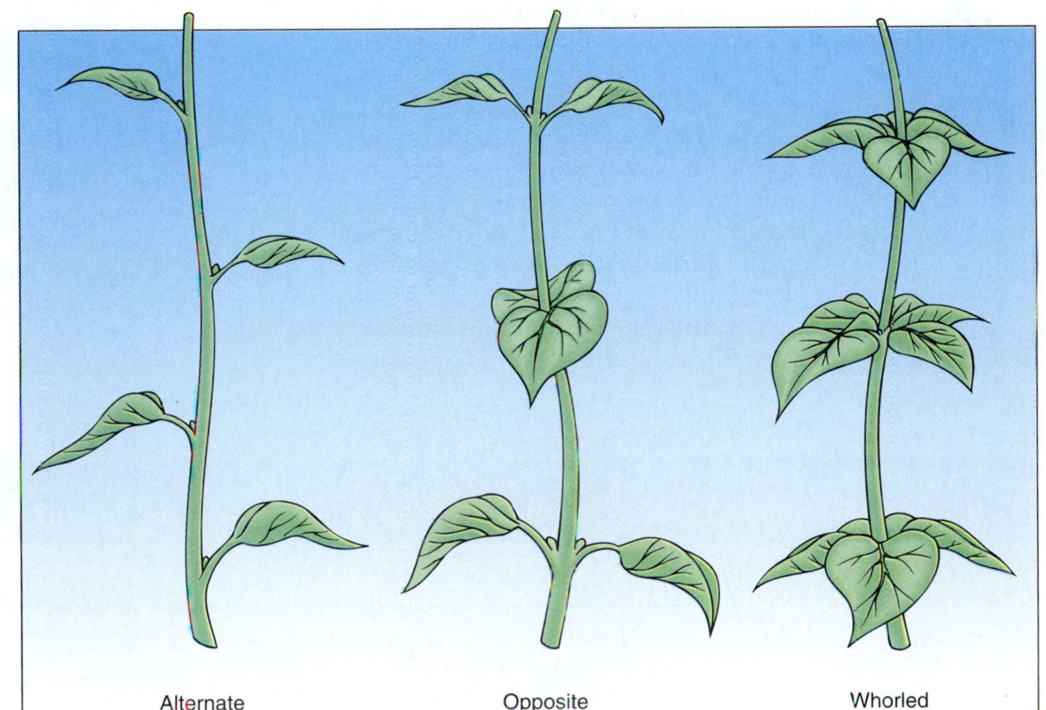

FIGURE 26–5 Leaf arrangement on a stem. Leaves may be alternate, opposite, or whorled, depending on the number of leaves at each node.

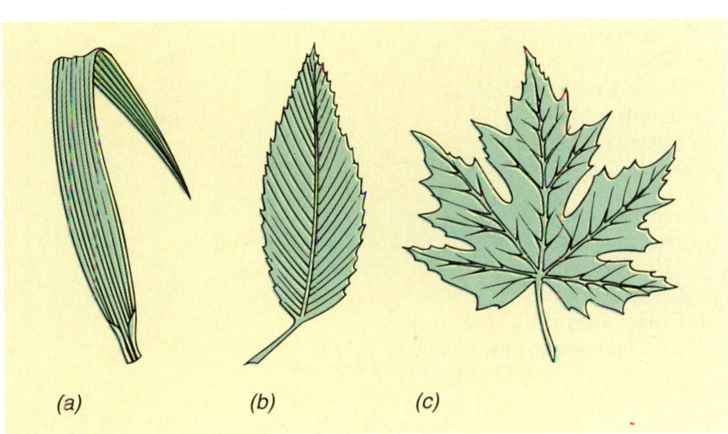

FIGURE 26–6 Vein patterns in leaves. (***a***) Kentucky bluegrass has parallel veins. Siberian elm (***b***) and silver maple (***c***) have netted veins. Siberian elm is pinnately netted and silver maple is palmately netted.

THE PLANT BODY IS COMPOSED OF CELLS AND TISSUES

The basic structural and functional unit of plants, like other living things, is the cell. During the course of evolution, plants have developed a diversity of cell types, each specialized for particular functions. As in animals, plant cells are organized into tissues. A **tissue** is a group of cells that form a structural and functional unit. Some plant tissues are composed of only one kind of cell (simple tissues), whereas other plant tissues have two or more kinds of cells (complex tissues). In plants, tissues are organized into three tissue systems: the ground, vascular, and dermal. All three tissue systems extend throughout the plant body. Each tissue system contains two or more kinds of tissues (Table 26–1).

Most of the plant body is composed of the **ground tissue system,** which is composed of three tissues that exhibit a variety of functions, including photosynthesis, storage, and support. The **vascular tissue system,** an intricate plumbing system that extends throughout the plant body, is responsible for conduction of various substances, including water, dissolved minerals, and food (dissolved sugar). It also functions in strengthening and supporting the plant. The **dermal tissue system** provides a covering for the plant body.

The tissue systems of different plant organs form an interconnected network throughout the plant. For example, the vascular tissue of a leaf is continuous with the vascular tissue of the stem to which the leaf is attached, and the vascular tissue of the stem is continuous with the vascular tissue of the root.

The Ground Tissue System Is Composed of Three Simple Tissues

The bulk of a herbaceous plant is its ground tissue, which is composed of three tissues: parenchyma, collenchyma, and sclerenchyma. These tissues can be distinguished by their cell wall structure. Recall that plant cells are surrounded by a cell wall that provides structural support (Chapter 5). A growing plant cell secretes a thin **primary cell wall,** which stretches and expands as the cell increases in size. After the cell stops growing, it sometimes secretes a thick, strong **secondary cell wall,** which is deposited *inside* the primary cell wall—that is, *between* the primary cell wall and the plasma membrane.

Parenchyma Cells Have Thin Primary Walls

Parenchyma tissue, a simple tissue composed of parenchyma cells, is found throughout the plant body (Fig. 26–7, Table 26–2). The living cells of parenchyma perform a number of important functions for plants, such as photosynthesis, storage, and secretion. Materials

(Text continues on page 513)

TABLE 26–1
Tissue Systems, Tissues, and Cell Types of Flowering Plants

Tissue system	*Tissues*	*Cell types*
Ground tissue system	Parenchyma tissue Collenchyma tissue Sclerenchyma tissue	Parenchyma cells Collenchyma cells Sclerenchyma cells (e.g., fibers)
Vascular tissue system	Xylem	Tracheids Vessel elements Parenchyma cells Fibers
	Phloem	Sieve-tube members Companion cells Parenchyma cells Fibers
Dermal tissue system	Epidermis	Parenchyma cells Guard cells Trichomes
	Periderm	Cork cells Cork cambium cells Cork parenchyma

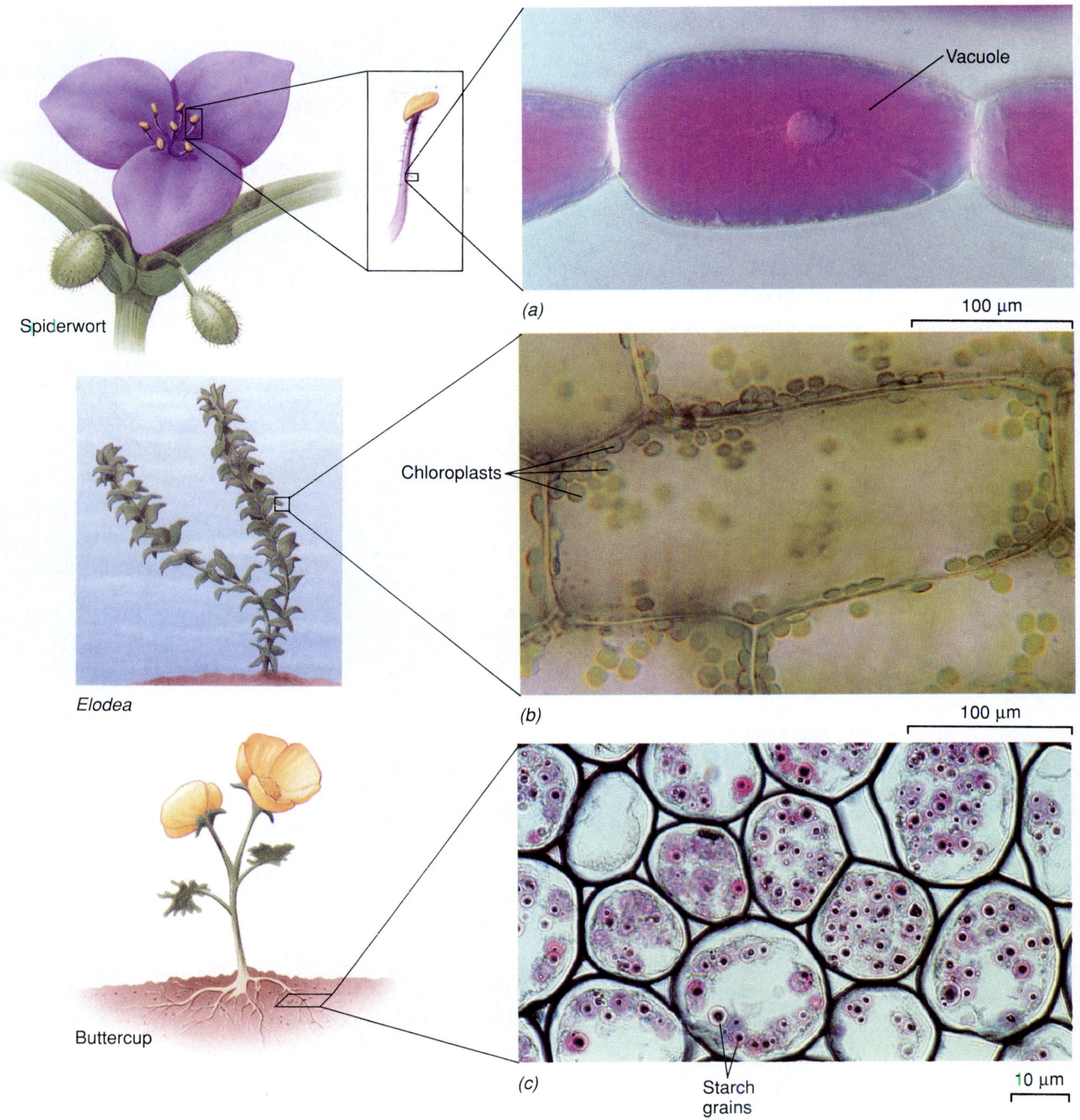

FIGURE 26–7 Parenchyma cells have thin primary cell walls and perform a variety of functions. (***a***) Parenchyma cells from the stamen hairs of the spiderwort (*Tradescantia virginiana*). The large vacuole, which contains a purple pigment, occupies most of the cell. (***b***) Parenchyma cells from the leaves of water weed (*Elodea canadensis*). The primary function of parenchyma cells that contain chloroplasts is photosynthesis. (***c***) Parenchyma cells often function in storage. These parenchyma cells are from the cortex of a buttercup (*Ranunculus*) root. Note the starch grains filling the cells. (*a*, Phil Gates, University of Durham/Biological Photo Service; *b* and *c*, Dennis Drenner)

TABLE 26–2
A Summary of Selected Plant Cell Types

Cell type		*Function*	*Location*
Parenchyma cell	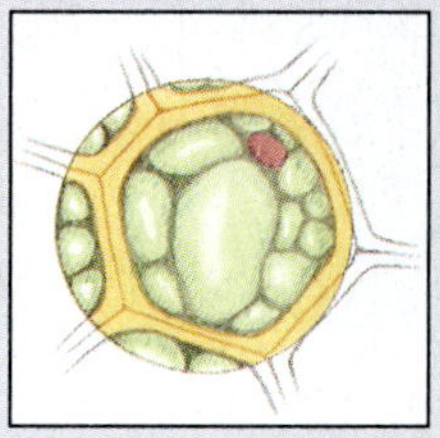	Secretion Storage Photosynthesis	Throughout the plant body
Collenchyma cell	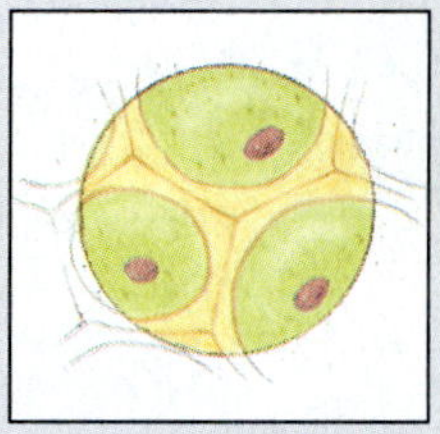	Support	Just under stem epidermis Along leaf veins
Sclerenchyma cell	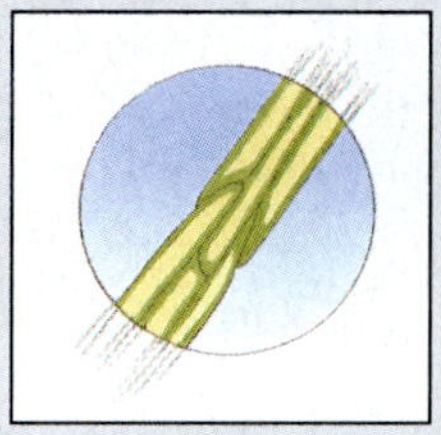	Support	Throughout the plant body Common in stems and certain leaves
Tracheid	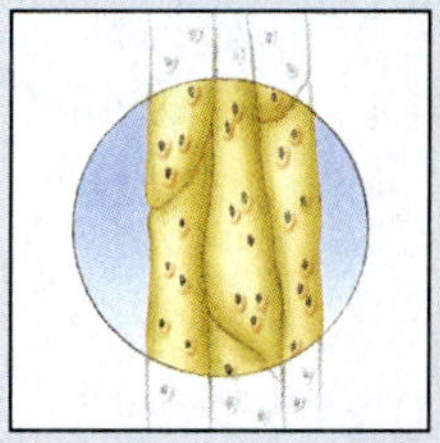	Conduction of water and minerals Also provides support	Xylem of vascular plants
Vessel element	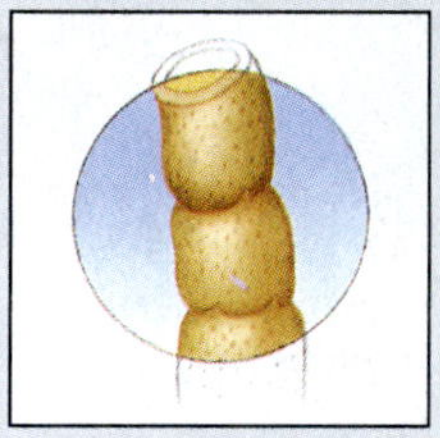	Conduction of water and minerals Also provides support	Xylem of flowering plants
Sieve-tube member Companion cell	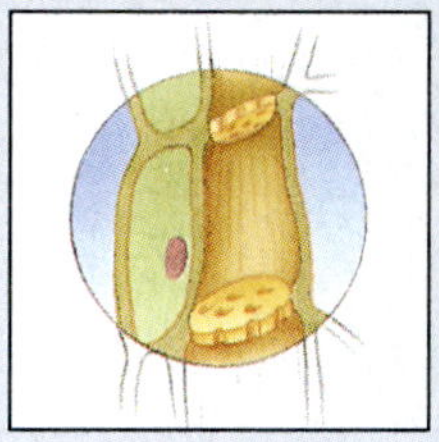	Conduction of food Aids sieve-tube members in food conduction	Phloem of flowering plants Phloem of flowering plants

FIGURE 26–8 Collenchyma cells from the petiole of a water lily leaf. These cells provide support. Note the unevenly thickened cell walls that are especially thick in the corners. (James Mauseth, University of Texas)

stored in parenchyma cells include starch grains, oil droplets, water, and salts (sometimes visible as crystals). The various functions of parenchyma require that they be living, metabolizing cells.

Collenchyma Cells Have Unevenly Thickened Primary Walls

Collenchyma tissue, a simple tissue composed of collenchyma cells, provides structural support (Fig. 26–8). Support is a crucial function in plants because it allows plants to grow upward, enabling them to compete with other plants for available sunlight on a plant-crowded land. Because plants lack a skeletal system (typical of many animals), support of the plant body is provided by individual cells, including collenchyma cells.

Collenchyma cells, which are usually elongated, are alive at maturity. Their primary walls are unevenly thickened, being especially thick in the corners. Collenchyma tissue is not uniformly located throughout the plant; long strands of collenchyma tissue are often found near stem surfaces and along leaf veins. Collenchyma is an extremely flexible support tissue that provides much of the support in soft, nonwoody plant organs.

Sclerenchyma Cells Have Both a Primary Wall and a Thick Secondary Wall

A second simple tissue specialized for structural support is **sclerenchyma** tissue, whose cells have both primary and secondary cell walls. The word *sclerenchyma* is derived from a Greek term meaning "hard." Their secondary cell walls become strong and hard due to extreme thickening. At functional maturity, when sclerenchyma tissue is providing support for the plant body, its cells are often dead.

Sclerenchyma tissue may be located in several areas of the plant body. One type of sclerenchyma cell is a **fiber,** a long, tapered cell (Fig. 26–9). Fibers, which often occur as patches or clumps in various tissues of the plant body, are particularly abundant in the wood and bark of flowering plants.

The Vascular Tissue System Consists of Two Complex Tissues: Xylem and Phloem

The vascular tissue system, which is embedded in the ground tissue, transports needed materials throughout the plant. It does so via two complex tissues: xylem and phloem. **Xylem** conducts water and dissolved minerals from the roots to the stems and leaves and also provides structural support. Conduction of dissolved sugar throughout the plant is accomplished by **phloem,** which also provides structural support. Both xylem and phloem are continuous throughout the plant body. (Chapter 28 discusses the mechanisms of transport in xylem and phloem.)

Xylem Has Two Kinds of Conducting Cells: Tracheids and Vessel Elements

Xylem is a complex tissue composed of four different cell types in flowering plants: tracheids, vessel elements, parenchyma cells, and fibers. Two of the four cell types found in xylem—the **tracheids** and **vessel elements**—ac-

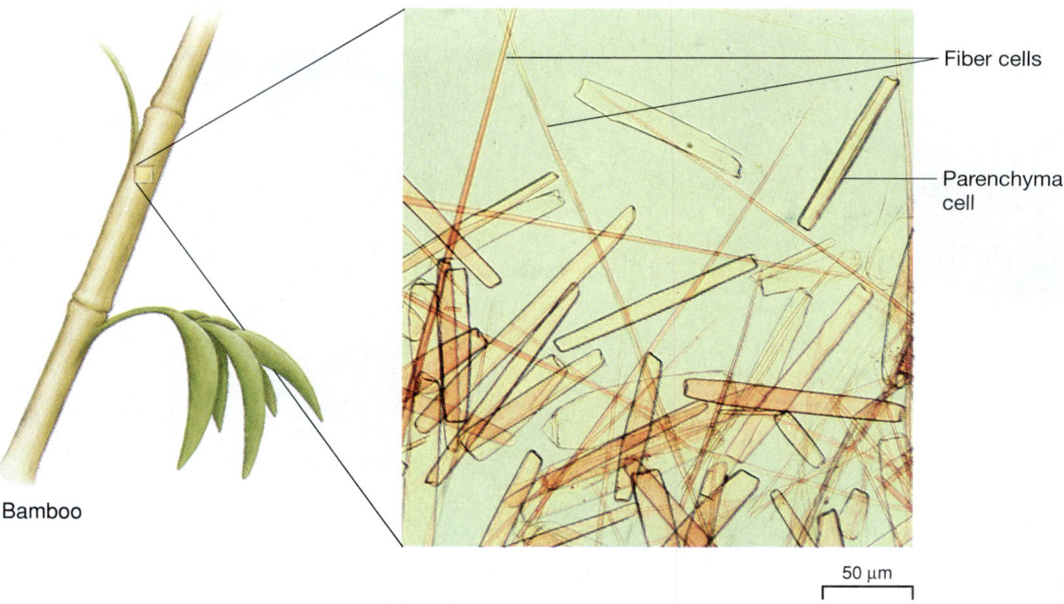

FIGURE 26–9 Long, tapering fibers (a type of sclerenchyma cell) and parenchyma cells from a bamboo stem. The stem was treated with acid to separate the cells. (James Mauseth, University of Texas)

tually conduct water and dissolved minerals (Fig. 26–10). In addition to these cells, xylem also contains parenchyma cells that perform storage functions and fibers that provide support.

Tracheids and vessel elements are highly specialized for conduction. Both cell types are dead and therefore hollow at maturity; only their cell walls remain. Tracheids, the chief water-conducting cells in gymnosperms and seedless vascular plants, are long, tapering cells located in patches or clumps. Water is conducted upward, from roots to shoots, passing from one tracheid into another through **pits,** thin areas in their cell walls where a secondary cell wall did not form.

In addition to tracheids, flowering plants have also developed very efficient water-conducting cells called vessel elements. The cell diameter of vessel elements is usually wider than that of tracheids. Vessel elements are hollow, but unlike tracheids, the end walls of vessel elements have holes, or perforations. Vessel elements are stacked end-on-end, and water is conducted readily through the perforations from one vessel element into the next. A stack of vessel elements, called a **vessel,** resembles a miniature water pipe. Like tracheids, vessel elements also have pits, which permit the lateral transport of water from one vessel to another.

Sieve-Tube Members Are the Conducting Cells of Phloem

Phloem is a complex tissue that in flowering plants is composed of four cell types: sieve-tube members, companion cells, fibers, and parenchyma cells. Fibers are frequently quite extensive in phloem, providing additional structural support for the plant body.

Sugar is conducted in solution (that is, dissolved in water) through the **sieve-tube members,** which are among the most specialized cells in nature (Fig. 26–11). Sieve-tube members are stacked end-on-end to form long **sieve tubes.** The cell's end walls, called **sieve plates,** have a series of holes through which cytoplasmic connections extend from one sieve-tube member into the next. Sieve-tube members are living at maturity, but many cellular organelles disintegrate, including the nucleus, vacuole, and ribosomes, during development.

Sieve-tube members are among the few eukaryotic cells that can function without nuclei. Sieve-tube members typically live for less than a year, although there are notable exceptions. Certain palms have sieve-tube members that have remained alive approximately 100 years!

Adjacent to each sieve-tube member is a **companion cell,** which assists in the functioning of the sieve-tube member. The companion cell is a living cell, complete with nucleus and other organelles. Numerous cytoplasmic connections occur between companion cells and sieve-tube members. Although the companion cell does not conduct dissolved sugar, it plays an essential role in moving sugars into the sieve-tube members for transport to other parts of the plant (Chapter 28).

The Dermal Tissue System Consists of Two Complex Tissues: Epidermis and Periderm

The dermal tissue system provides a protective covering over plant parts. In herbaceous plants, the dermal tissue

(Text continues on page 516)

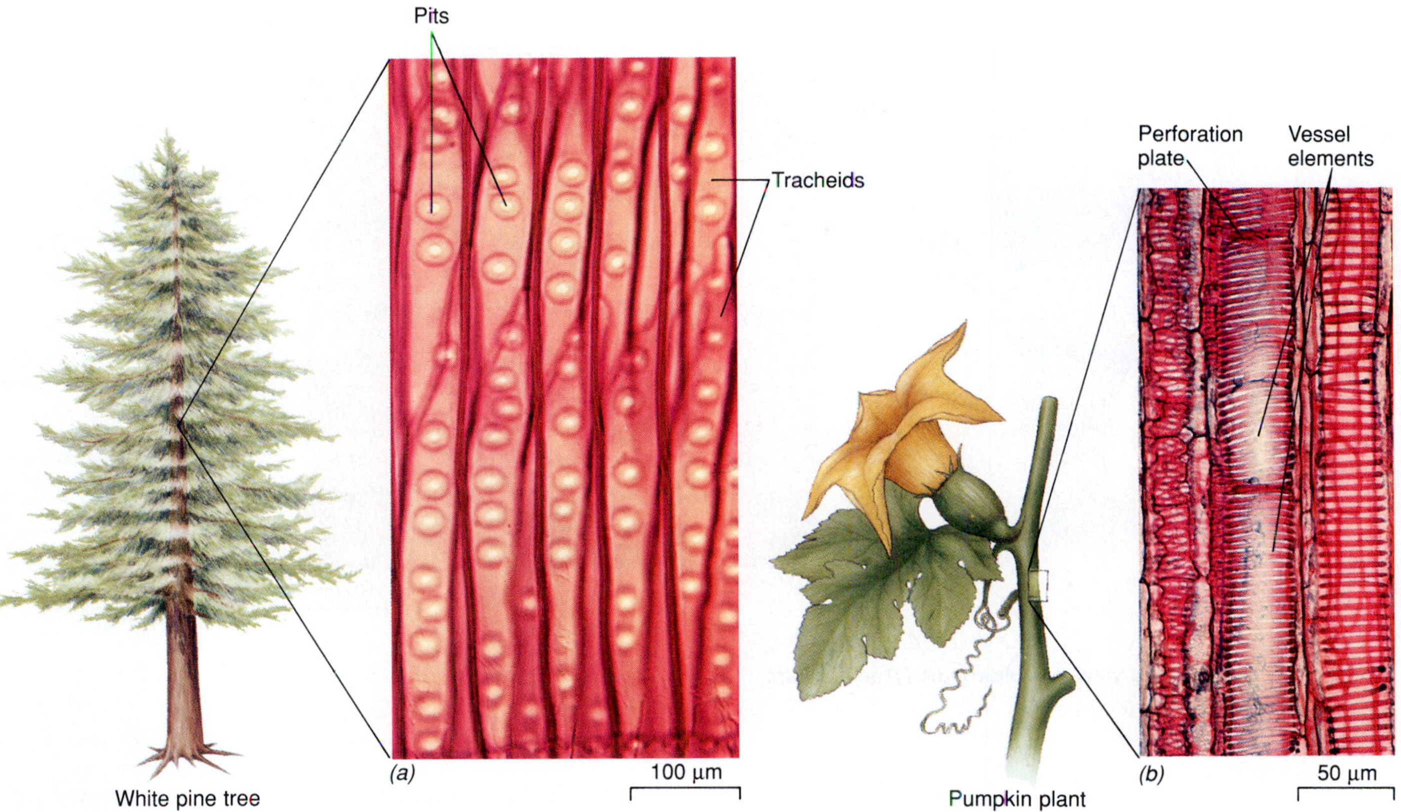

FIGURE 26–10 Conducting cells in xylem. (***a***) Tracheids from a longitudinal section of a pine stem. These cells, which occur in clumps, transport water and dissolved minerals. Water passes readily from tracheid to tracheid through the pits. (***b***) Vessel elements from a longitudinal section of a pumpkin stem. These cells are more efficient than tracheids in conducting water. Note how they are stacked end-on-end. The end walls of vessel elements, called perforation plates, have large holes, and water readily passes through the perforation plate from one vessel element to the next. (*a,* Visuals Unlimited, John D.Cunningham; *b,* J. Robert Waaland, University of Washington/Biological Photo Service)

FIGURE 26–11 Phloem tissue from a longitudinal section of a squash petiole. Dissolved sugar is conducted through sieve-tube members in the phloem. (James Mauseth, University of Texas)

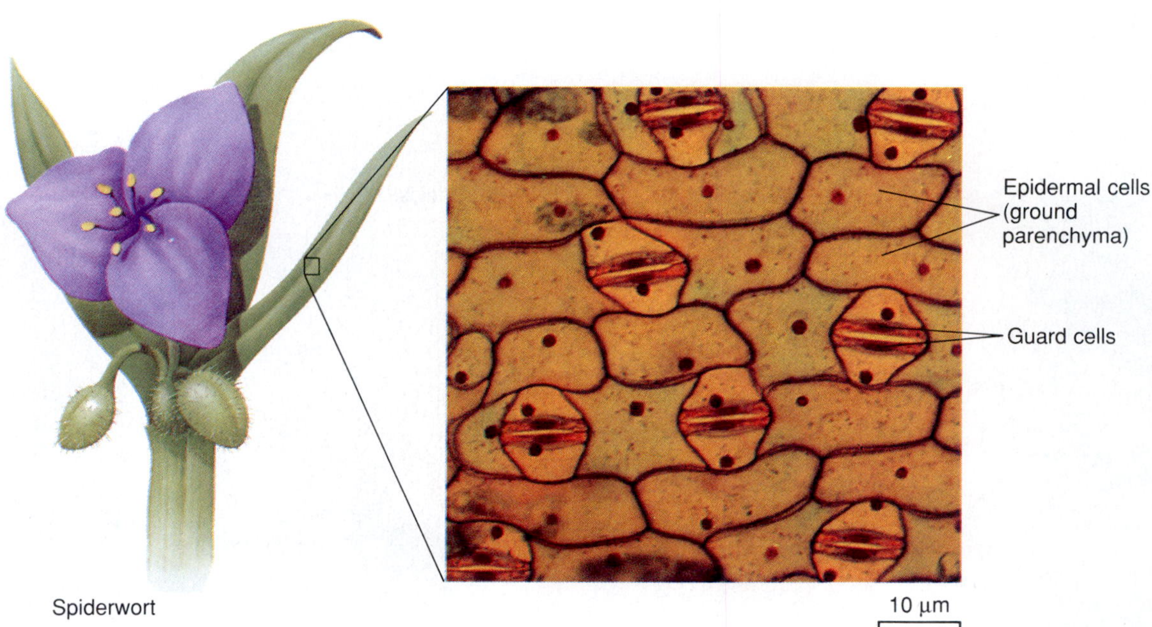

FIGURE 26–12 Surface view of spiderwort (*Tradescantia*) leaf epidermis. Note the pink-colored guard cells that form openings for gas exchange. (James Bell/Photo Researchers, Inc.)

system is usually a single layer of cells called the **epidermis.** Woody plants initially produce an epidermis, but it splits apart as the plant increases in diameter due to the production of additional woody tissues. **Periderm,** a tissue that is several to many cell layers thick, comprises the outer bark. Periderm replaces the epidermis in the stems and roots of older woody plants (Chapter 28).

Epidermis Is the Outermost Layer of Cells on a Herbaceous Plant

The epidermis (Fig. 26–12) is a complex tissue composed mostly of ground parenchyma cells with scattered guard cells and outgrowths called trichomes (discussed below). In most plants, the epidermis consists of a single layer of cells. Epidermal cell walls are somewhat thicker toward the outside of the plant for protection. Epidermal parenchyma cells generally do not contain chloroplasts. Their transparent nature allows light to penetrate into interior tissues of the stem and leaf (discussed further in Chapter 27). (In both stems and leaves, photosynthetic tissues are *inside* the epidermis.)

An important requirement of the above-ground parts of the plant (that is, of stems and leaves) is the ability to control water loss. Epidermal cells of aerial parts secrete a waxy layer called a **cuticle** over the surface of their outer walls. This wax greatly restricts the loss of water from plant surfaces.

Although the cuticle is very efficient at preventing most water loss through epidermal cells, it also prevents the carbon dioxide required for photosynthesis from diffusing into the leaf or stem. The diffusion of carbon dioxide is facilitated by **stomata** (singular, *stoma*). Stomata are tiny pores formed in the epidermis by two rounded cells, called **guard cells** (Fig. 26–12). A number of gases pass through stomata by diffusion, including carbon dioxide, oxygen, and water vapor. Stomata are generally open during the day when photosynthesis is occurring and when the plant must lose water to cool itself. They close during the night, thereby conserving water when photosynthesis is not taking place and cooling is not required. Stomata are discussed in greater detail in Chapter 27.

The epidermis also may contain special outgrowths, or hairs, called **trichomes,** which have a variety of functions. Root hairs are trichomes that increase the surface area of the root epidermis that comes into contact with the soil for more effective water and mineral absorption (Chapter 28). Plants that can tolerate salty environments often have specialized trichomes on their leaves that remove excess salt. The presence of trichomes on the aerial parts of desert plants may increase the reflection of light off the plants and reduce air flow across the surface, thereby keeping the internal tissues cooler and decreasing water loss. Other trichomes have a protective function and discourage herbivorous animals from eating the plant.

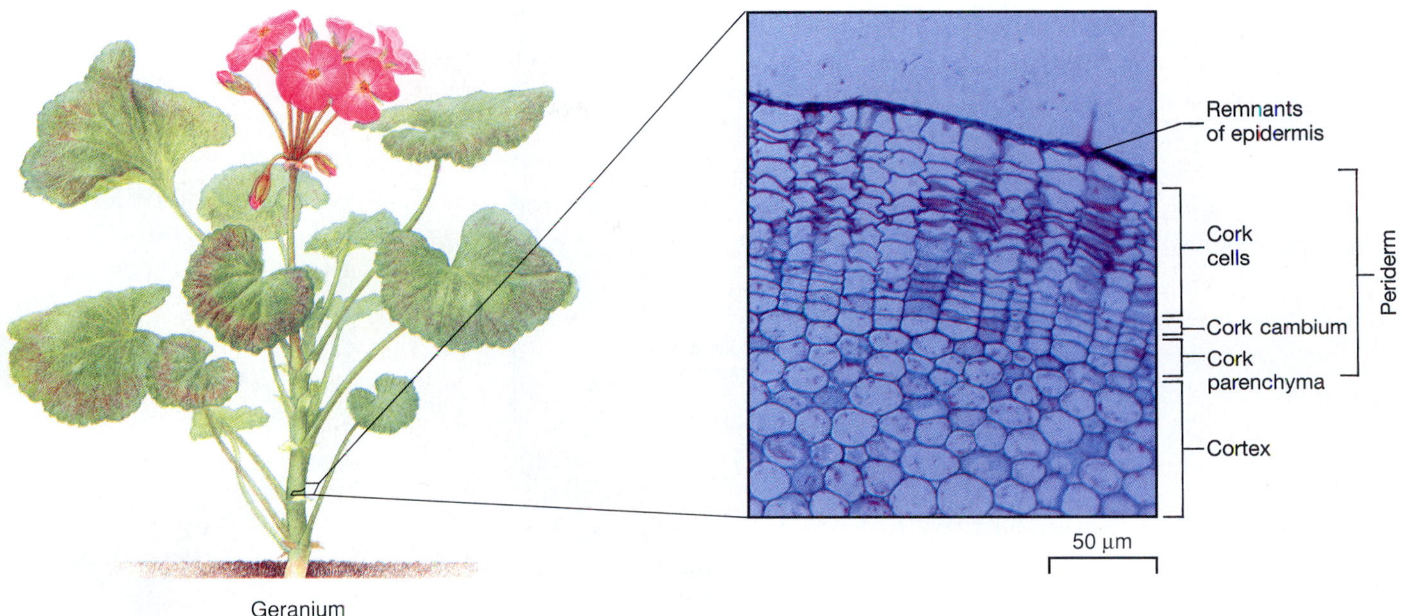

FIGURE 26–13 Periderm is the secondary plant body replacement for epidermis. Formed by the cork cambium, it makes up the outer bark of woody stems and roots. Some herbaceous dicots, such as geranium (shown), form a limited periderm as they age. The cells of periderm are always arranged in stacks in cross section. (Dennis Drenner)

Epidermis Is Replaced by Periderm in Woody Plants

As a woody plant begins to increase in girth, its epidermis is sloughed off and replaced by periderm. Periderm forms the outer bark of older stems and roots (Fig. 26–13). It is a complex tissue composed mostly of cork cells and cork parenchyma cells. **Cork cells** are dead at maturity, and their walls are heavily coated with a waterproof substance to help reduce water loss. **Cork parenchyma** cells function primarily in storage.

PLANTS EXHIBIT LOCALIZED GROWTH AT MERISTEMS

Growth is a complex phenomenon involving three different processes: cell division, cell elongation, and cell differentiation. Cell division is an essential part of growth that results in an increase in the number of cells. However, an increase in cell number without a corresponding increase in cell size would contribute little to overall size increase in a plant. Thus, cell elongation is also an essential part of growth. Plant cells also **differentiate,** or specialize, into the various cell types just discussed. These cell types comprise the mature plant body and perform the various functions required in a complex, multicellular organism.

One difference between plants and animals is the *location* of growth. When a young animal is growing, all parts of its body grow, although all parts do not necessarily grow at the same rate. However, plant growth is localized into specific areas, called **meristems,** that are composed of cells that do not differentiate. Meristematic cells retain the ability to divide by mitosis (Chapter 10). The persistence of meristems means that plants retain the capability for growth throughout their entire lifespans, which can be for hundreds or even thousands of years.

Two kinds of growth may occur in plants. One is **primary growth,** which is an increase in the length of a plant. All plants have primary growth, which comprises the entire plant body in herbaceous plants and the young, soft shoots and roots in woody trees and shrubs. The other kind of growth, **secondary growth,** is an increase in the girth of a plant. For the most part, only gymnosperms and woody dicots have secondary growth. Tissues produced by secondary growth comprise most of the bulk of trees and shrubs.[2]

[2] Monocots do not have secondary growth, although certain monocots (palms) develop thickened trunks by a special form of primary growth.

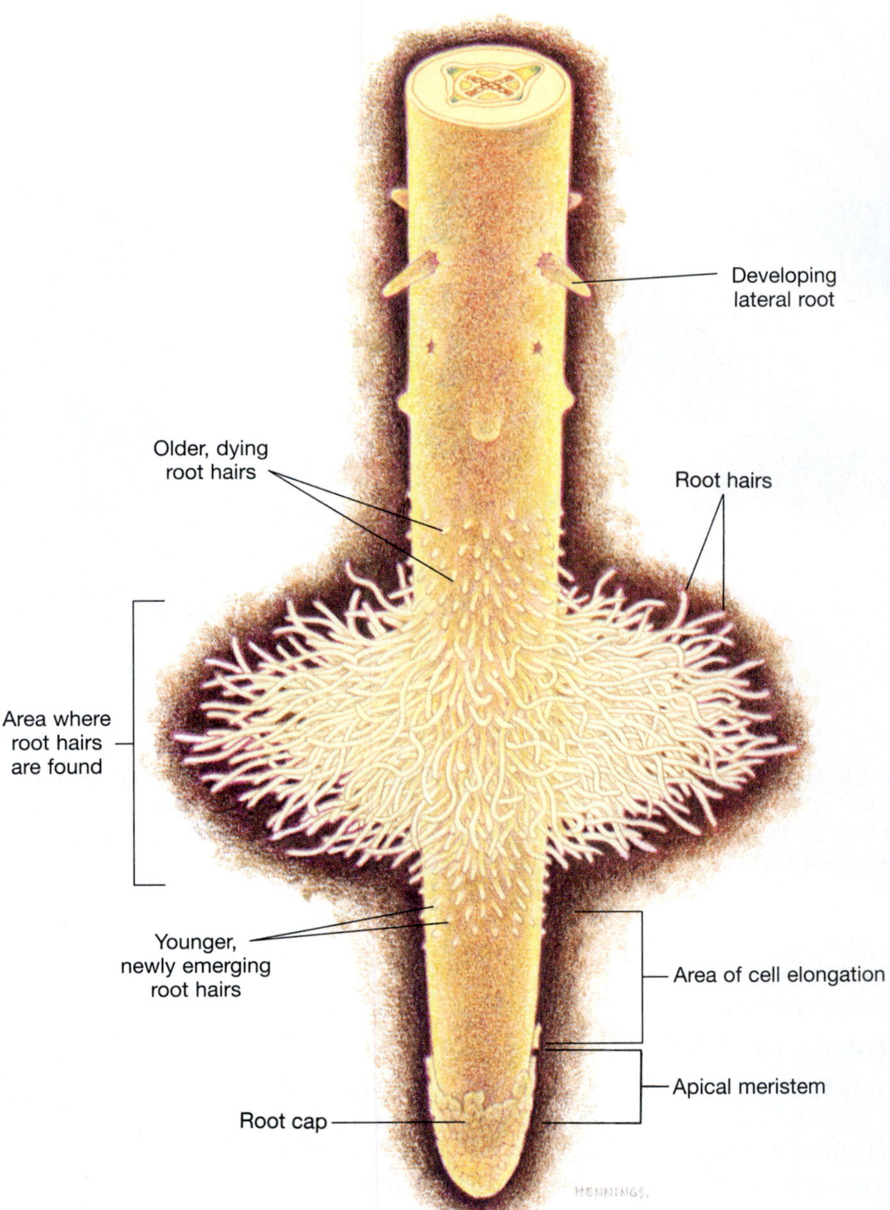

FIGURE 26–14 The root tip. Just behind the root cap is the apical meristem, where cell division occurs. Behind the apical meristem (farther from the tip) is the area of cell elongation, where cells enlarge and begin to differentiate. Behind this region is the area where root hairs are found.

Primary Growth Takes Place at Apical Meristems

Primary growth occurs as a result of the activity of **apical meristems,** which are meristematic areas found at the tips of both roots and stems. Such growth is evident by examining a root tip (Fig. 26–14). The root tip is covered by a protective layer of cells called the root cap. Directly behind the root cap is the root apical meristem, which consists of meristematic cells. Meristematic cells, which are very small and "boxy" in shape, remain small because they are continually dividing. Further back from the tip of the root, just behind the area of cell division, is an **area of cell elongation,** where the cells are no longer dividing but instead are enlarging. Some differentiation also occurs in the area of cell elongation, and immature tissues become evident. The immature tissues continue to develop and differentiate into the mature tissues of the adult plant. Further back from the tip, behind the area of cell elongation, the cells have completely differentiated and are fully mature. Root hairs are evident in this area.

A stem tip is quite different in appearance from a root tip (Fig. 26–15). A dome of tiny meristematic cells—the stem apical meristem—is located in the center at the very tip of the stem. From the stem tip emerges **leaf primordia** (embryonic leaves) and **bud primordia** (embryonic buds). The leaf primordia tend to cover and protect the stem apical meristem. Further back from the tip of the stem, the immature cells enlarge and develop into the three tissue systems.

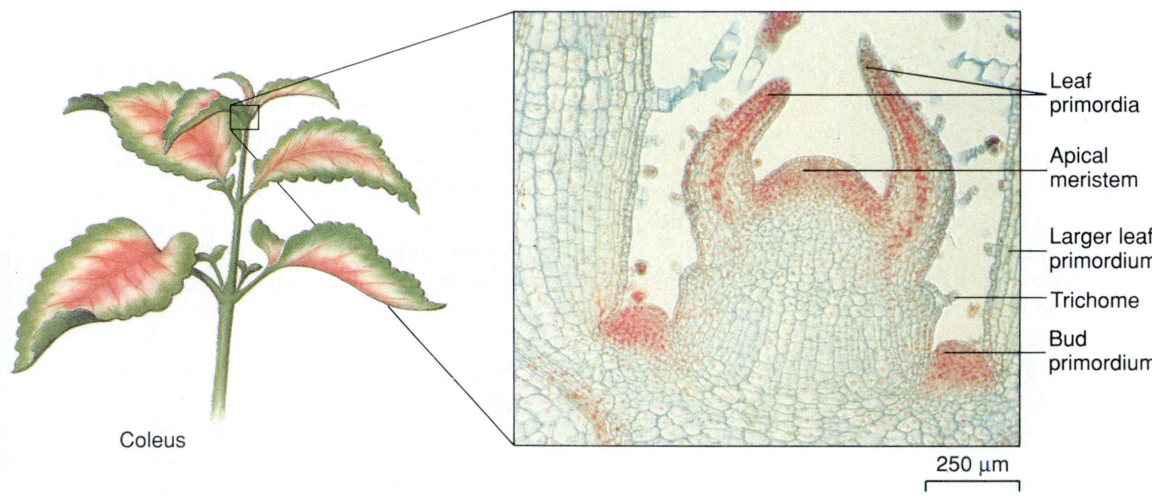

FIGURE 26–15 A longitudinal section through a stem tip. Note the apical meristem, leaf primordia, and bud primordia. (James Mauseth, University of Texas)

Secondary Growth Takes Place at Lateral Meristems

Woody trees and shrubs have secondary growth in addition to primary growth. That is, these plants increase in length by primary growth and increase in girth by secondary growth. The increase in girth, which occurs in areas that are no longer elongating, is due to the activity of **lateral meristems,** which extend the entire length of the stems and roots, except at the tips. Two lateral meristems are responsible for secondary growth, the vascular cambium and the cork cambium (Fig. 26–16 and Chapter 28).

The **vascular cambium** is a layer of meristematic cells that forms a thin cylinder around the stem and root trunk. It is located between the wood and bark of a woody plant. Cells of the vascular cambium divide, adding more cells to the wood and inner bark regions.

The **cork cambium** is composed of a thin cylinder or irregular arrangement of meristematic cells and is located in the outer bark region. Cells of the cork cambium divide to form the cork cells and cork parenchyma cells that make up the periderm (discussed previously).

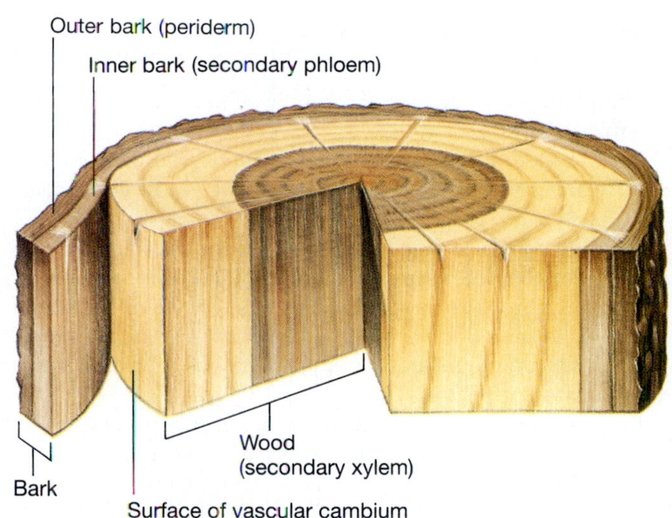

FIGURE 26–16 In secondary growth, plants increase in girth as a result of the activity of two lateral meristems. The vascular cambium produces secondary vascular tissues, the wood and inner bark. The cork cambium produces the outer bark tissues (that is, the periderm) that replace the epidermis in the secondary plant body.

Concept Connections

Annuals & Perennials / Life History Strategies

Woody perennials often live for hundreds of years, whereas some herbaceous annuals may live for only a few weeks or months. Which life history strategy is more advantageous? Biologists have considered the relative advantages of each and concluded that in some environments, a longer life span is advantageous, whereas in others, a shorter lifespan actually increases a plant species' chances for reproductive success.

When an environment is relatively favorable, it is filled with plants competing for available space. Because such an environment is so crowded, it has few open spots in which new plants can become established. When a plant dies, an empty area is produced that is quickly filled by another plant, but not necessarily by the *same* kind of plant as before. Thus, an adult perennial survives well in such a habitat, but young plants (perennials or annuals) do not. A plant with a long lifespan thrives in this type of environment because it can "hold onto" a piece of soil and continue to produce seeds for many years. In a tropical rain forest, for example, competition prevents most young plants from becoming established, and woody perennials predominate.

In an environment that is relatively unfavorable, many possible sites are usually available. This type of environment is not crowded, and young plants usually do not have to compete against large, fully established plants. Here, smaller, short-lived plants have the reproductive advantage. These plants are opportunists—they grow and mature quickly during the brief periods when environmental conditions are most favorable. As a result, all of their resources are put into producing as much seed as possible before dying. For example, annuals are prevalent in deserts where few woody perennials can survive. We return to life history strategies in Chapter 45.

Chapter Summary

I. The plant body consists of a root system and a shoot system.
   A. The root system is generally underground and obtains water and dissolved minerals for the plant. Roots also anchor the plant firmly in place.
      1. A taproot system has one main root with many smaller lateral roots coming out of the taproot.
      2. A fibrous root system has several to many roots of the same size developing from the end of the stem.
      3. Storage roots store food and may be modified taproots or fibrous roots.
   B. The shoot system is generally above ground and obtains sunlight and carbon dioxide for the plant.
      1. The shoot system consists of a vertical stem that bears leaves (the main organs of photosynthesis) and flowers and fruits (reproductive structures).
      2. Terminal and lateral buds (undeveloped embryonic shoots) develop on stems.
      3. Leaves typically consist of a broad, expanded blade and a stalklike petiole. Leaves may be simple or compound; have alternate, opposite, or whorled leaf attachment; and have parallel or netted venation.

II. The plant body is composed of three tissue systems.
   A. The ground tissue system consists of various cell types with a variety of functions.
      1. Parenchyma tissue is composed of living parenchyma cells that possess thin primary cell walls. Functions of parenchyma tissue include photosynthesis, storage, and secretion.
      2. Collenchyma tissue is composed of living collenchyma cells with unevenly thickened cell walls. Collenchyma tissue provides flexible structural support.
      3. Sclerenchyma tissue is composed of sclerenchyma cells, such as fibers, that have both primary and secondary cell walls. Sclerenchyma cells are often dead at maturity, but they provide structural support.
   B. The vascular tissue system conducts materials throughout the plant.
      1. Xylem is a complex tissue that conducts water and

dissolved minerals. The actual conducting cells of xylem are tracheids and vessel elements.
2. Phloem is a complex tissue that conducts sugar in solution. Sieve-tube members are the conducting cells of phloem.

C. The dermal tissue system is the outer protective covering of the plant body.
1. The epidermis is a complex tissue that covers the herbaceous plant body. The epidermis that covers aerial plant parts secretes a layer of wax, called the cuticle, that reduces water loss. Gas exchange between the interior of the shoot system and the surrounding atmosphere occurs through stomata.
2. The periderm is a complex tissue that covers the plant body in woody plants.

D. Although separate organs (roots, stems, leaves, flowers, and fruits) exist in the plant, many tissues are integrated throughout the plant body, providing continuity from organ to organ.

III. Plant growth is localized in regions, or meristems, and involves cell division, cell elongation, and differentiation.
A. Primary growth is an increase in stem and root length and takes place at apical meristems located at the tips of stems and roots.
B. Secondary growth is an increase in the girth of stems and roots and takes place at lateral meristems (the vascular cambium and cork cambium).
C. Herbaceous plants generally have primary growth only, whereas woody plants (gymnosperms and woody dicots) have both primary and secondary growth.

Selected Key Terms

apical meristem, p. 518
collenchyma, p. 513
companion cell, p. 514
cork cambium, p. 519
cuticle, p. 516
epidermis, p. 516
fiber, p. 513
guard cell, p. 516
lateral meristem, p. 519
parenchyma, p. 510
periderm, p. 516
phloem, p. 513
primary growth, p. 517
root system, p. 507
sclerenchyma, p. 513
secondary growth, p. 517
shoot system, p. 507
sieve-tube member, p. 514
tissue, p. 510
tracheid, p. 513
trichome, p. 516
vascular cambium, p. 519
vessel element, p. 513
xylem, p. 513

Post-Test

1. Plants that are ________________ do not have persistent above-ground parts, whereas woody plants do.

2. Grasses have a/an ________________ root system.

3. Dormant terminal buds are covered by ________________ ________________.

4. The area on a stem where each leaf attaches is called a/an ________________.

5. A single leaf composed of several leaflets is said to be ________________.

6. In ________________ leaf arrangement on a stem, two leaves occur at each node.

7. The ________________ tissue system conducts various materials throughout the plant.

8. The two ground tissues specialized for support are ________________ and sclerenchyma.

9. Conduction of water and minerals in xylem occurs in ________________ and vessel elements.

10. Conduction of dissolved sugar in the sieve-tube members of the phloem is aided by ________________ cells.

11. Epidermal cells of the shoot secrete a waxy layer called a/an ________________.

12. Plant growth is localized into specific areas called ________________.

13. Primary growth, the increase in the length of the plant, is a result of the meristematic activity of ________________ meristems.

14. Stem tips differ from root tips in bearing these embryonic structures: ______________ ______________ and ______________ ______________.

15. Plants that complete their life cycle within one year are called ______________, those that complete it in two years are called ______________, and those that live for a number of years are called ______________. (a) perennials, biennials, annuals (b) annuals, perennials, biennials (c) biennials, perennials, annuals (d) annuals, biennials, perennials

16. Herbaceous stems have ______________ for gas exchange, whereas woody stems have ______________. (a) stomata, lenticels (b) stomata, cuticles (c) lenticels, stomata (d) guard cells, trichomes

17. Storage, secretion, and photosynthesis are the functions of: (a) periderm (b) parenchyma tissue (c) sclerenchyma tissue (d) collenchyma tissue.

18. The outer covering of plants with primary growth is ______________, whereas plants with secondary growth are covered by the ______________. (a) cork, cuticle (b) periderm, cuticle (c) epidermis, periderm (d) epidermis, phloem

19. The two lateral meristems responsible for secondary growth are the: (a) vascular cambium and apical meristem (b) cork cambium and periderm (c) wood and bark (d) vascular cambium and cork cambium.

20. Label the following diagram:

Review Questions

1. Distinguish among taproot systems, fibrous root systems, and storage roots.
2. Draw simple diagrams to illustrate the following: (a) simple and compound leaves; (b) alternate, opposite, and whorled leaf arrangement; (c) parallel and netted venation.
3. What are the three ground tissues? Why does a flowering plant need to possess all three?
4. Name the two vascular tissues and describe the main functions of each.
5. What is the function of the cuticle? Stomata?
6. Distinguish between primary growth and secondary growth.
7. What is meristematic tissue, and why is it so important in plants?
8. How is growth in plants different from growth in animals?

Thinking Critically

1. Do you think that tropical countries, which are repositories of much of the world's biological diversity, should charge a fee for the use of their resources? Or do you think that the Earth's biological riches are owned equally by everyone, regardless of the country in which they are found? Present a reasonable argument for your position.
2. How are the functional differences between roots and shoots related to their structural differences?
3. How is the structure of each of the following cell types related to the function(s) that it performs?
   a. Fiber
   b. Vessel element
   c. Parenchyma cell
4. Certain stems grow underground and superficially resemble roots. How would you distinguish between an underground stem and a root?
5. Collenchyma tissue and sclerenchyma tissue are common in shoots but uncommon in roots. Explain why.
6. If a ring of bark is peeled off a tree, it dies. Explain why, based on what you have learned in this chapter.
7. Suppose you built a tree house on a branch 6 feet above the ground. Ten years later, the tree is 15 feet taller. How far above the ground is the tree house now? Explain why.

Recommended Readings

Krajick, K., "Sorcerer's Apprentices," *Newsweek,* January 18, 1993. Pharmaceutical companies hope native healers will help them find new medicines.

Mauseth, J.D., *Botany: An Introduction to Plant Biology,* 2nd ed. Philadelphia, Saunders College Publishing, 1995. A comprehensive introduction to plant biology.

Rensberger, B., "Getting to the Root of Plant Growth," *Washington Post,* Science Section, July 13, 1992. Why roots grow downward and shoots grow upward.

Wilson, E.O., *The Diversity of Life,* New York, W.W. Norton & Company, 1992. Includes discussions of the remarkable diversity of species on Madagascar and of medicinal plants, including the rosy periwinkle.

The effect of air pollution on plants. *The odds are against these small trees in N.Y.C. surviving for many years. (Frank Staub/ProFiles West)*

Air Pollution and The Vulnerable Leaf

The air we breathe is often dirty and contaminated with many pollutants, particularly in urban areas. Air pollution consists of gases, liquids, or solids present in the atmosphere in high enough levels to harm humans, other animals, plants, or materials. Although air pollutants can come from natural sources—as, for example, when lightning causes a forest fire or when a volcano erupts—humans make a major contribution to global air pollution. Thousands of different chemicals pollute the air as a result of human activities. Motor vehicles and industry are the two main human sources of air pollution.

All parts of a plant can be damaged by air pollution, but leaves are particularly susceptible because of their structure and function. The thin blade provides a large surface area that comes into contact with the surrounding air. In addition, carbon dioxide, oxygen, and water vapor pass readily between the atmosphere and the interior of the leaf. The thousands of tiny stomatal pores that dot the epidermis and control gas exchange with the atmosphere also permit pollutants to diffuse into the leaf. Just as the lungs, organs of gas exchange in terrestrial vertebrates, often show the effects of air pollution, so too the leaves of a plant are most affected.

Trees provide a dramatic demonstration of the effect of air pollution on biological longevity. According to the Urban Forest Forum, which is sponsored by the U.S. Forest Service, the average lifespan for a tree living in a rural setting, where air pollution is low, is 150 years. Even in the most ideal urban setting, a tree typically lives only 60 years, whereas in a normal city setting, the average lifespan is 32 years. Trees living in a downtown setting, where air quality is lowest, live only 7 years on average. It should be noted, however, that air pollution is not the sole cause of short lifespan in urban trees. Other factors include inadequate room for growth and polluted runoff from streets and sidewalks.

Many studies have shown that the overall productivity of crop plants is reduced by high levels of most forms of air pollution. The worst pollutant in terms of yield loss is ozone,[1] a toxic gas produced when sunlight catalyzes a reaction between pollutants emitted by automobiles and industries. In plants, ozone inhibits photosynthesis because it damages the photosynthetic cells sandwiched in the middle of the leaf, probably by altering the permeability of their cell membranes. Exposure to low levels of air pollution often causes a decline in photosynthesis without any other symptoms of injury. Lesions on leaves and other obvious symptoms appear at much higher levels of air pollution.

When air pollution is combined with other environmental stresses (such as low winter temperatures; prolonged droughts; insects; and bacterial, fungal, and viral diseases), it can cause plants to decline and die. Many patterns of forest decline, for example, are associated with environmental stresses that interact with one another. In such cases, air pollutants may or may not be the *primary* stress that results in forest decline, but the presence of air pollution lowers plant resistance to the other stress factors.

[1] Ozone (O_3) is a form of oxygen considered a pollutant in one part of the atmosphere but an essential component in another. In the stratosphere, naturally produced ozone prevents much of the ultraviolet radiation from penetrating to the Earth's surface. Unlike stratospheric ozone, ozone produced in the lower atmosphere is a serious air pollutant produced by human activities.

Leaves

27

Learning Objectives

After you have studied this chapter you should be able to:

1. Diagram and label a cross section of a leaf and describe the functions of the various parts.
2. Compare the leaf anatomy of dicot and monocot leaves.
3. Relate leaf structure to its function of photosynthesis.
4. Outline the physiological changes that accompany stomatal opening and closing, and describe the environmental factors that affect these processes.
5. Define transpiration, discuss its effects on plants, and describe the factors that affect the rate of transpiration in plants.
6. Explain how leaves balance the conflicting needs of absorbing light and exchanging gases while at the same time conserving water.
7. Distinguish between transpiration and guttation.
8. Define leaf abscission, explain why it occurs, and discuss the physiological and anatomical changes that precede it.
9. List several modified leaves and in each case explain how the structure of the leaf is related to its function.
10. Outline the relationship among trees, transpiration, and climate.
11. Explain how leaves of some species are modified to tolerate either wet habitats or dry habitats.

Key Concepts

- ☐ Leaf structure is related to its function. The primary function of leaves is photosynthesis, and the structure of a typical leaf is superbly adapted for photosynthesis. Most leaves have a broad, flattened blade that is very efficient in collecting the sun's radiant energy.
- ☐ The cuticle and stomata are two adaptations associated with leaves (and stems). The waxy cuticle coats the epidermis and reduces water loss, enabling the plant to survive the dry conditions of a terrestrial existence. Stomata are small pores in the epidermis that permit gas exchange needed for photosynthesis.
- ☐ Although most leaves are thin and flat, others are modified in various ways. These modifications represent adaptations to specialized environments.

LEAF STRUCTURE CONSISTS OF EPIDERMIS, MESOPHYLL, XYLEM, AND PHLOEM

Leaves are the main photosynthetic organs of most plants. Most leaves are thin and flat, a shape that allows maximum absorption of light energy and efficient internal diffusion of gases. As a result of their ordered arrangement on the stem, leaves efficiently catch the sun's rays. The leaves of plants form an intricate green mosaic, bathed in sunlight and atmospheric gases.

Because the leaf blade has upper and lower sides, the leaf is covered by two epidermal layers, the **upper epidermis** and the **lower epidermis** (Fig. 27–1). Most cells comprising this outer covering of the leaf are living parenchyma cells (Chapter 26). Epidermal cells generally lack chloroplasts and are relatively transparent. One interesting feature of leaf epidermal cells is that the cell wall facing the outside of the leaf is thicker than the cell wall facing inward. This extra thickness may afford the plant additional protection from injury or water loss.

Leaves have such a large surface area exposed to the atmosphere that water loss by evaporation from the surface is unavoidable. The thin expanded blade—the very feature that makes a leaf so efficient for gas exchange and for collecting the sun's rays—is its undoing in conserving water. However, epidermal cells secrete a waxy layer, the **cuticle,** that reduces water loss. Generally, the exposed (and warmer) upper epidermis has a thicker cuticle than the shaded (and cooler) lower epidermis. The cuticle varies in thickness in different plants, in part due to environ-

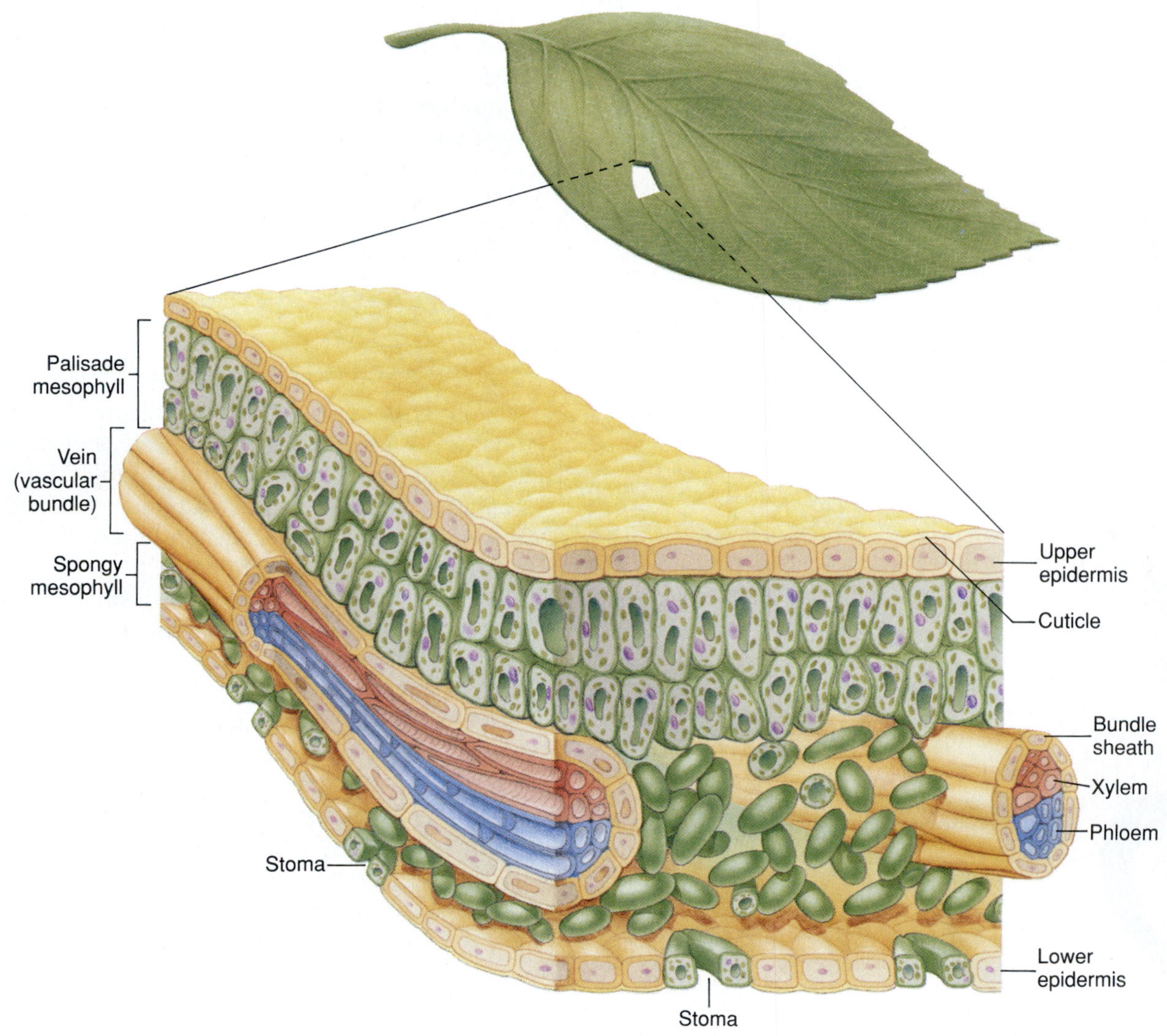

FIGURE 27–1 The internal arrangement of tissues in a typical leaf blade. The blade is covered by an upper and lower epidermis. The photosynthetic tissue, called mesophyll, is often organized into palisade and spongy layers. Veins branch throughout the mesophyll.

FIGURE 27–2 Scanning electron micrograph of a nettle leaf. The leaf epidermis is often covered with trichomes that may limit the transpiration of water, discourage herbivores, sting, or perform other functions. The trichomes on nettles readily break off inside the skin of any animal that brushes against the leaf or attempts to eat it. Irritating substances injected into the skin produce a stinging sensation. (Biophoto Associates)

mental conditions. As one might expect, the leaves of plants living in hot, dry climates have very thick cuticles.

The epidermis of many leaves is covered with various **trichomes** (hairs; Fig. 27–2 and Chapter 26). Some leaves have so many trichomes that they feel quite fuzzy. Trichomes maintain a layer of moist air next to the leaf, thereby reducing evaporation from the leaf's surface. Trichomes also reflect excessive sunlight, thereby protecting the plant. Some trichomes are glands that secrete stinging irritants (for protection from herbivores) and other substances or excrete excess salts absorbed from a salty soil.

The leaf epidermis typically has many tiny openings, or **stomata** (singular, ***stoma***). Each stoma is flanked by two specialized cells in the epidermis called **guard cells** (Fig. 27–3). The variable shape of each pair of guard cells is responsible for opening and closing a tiny pore in the

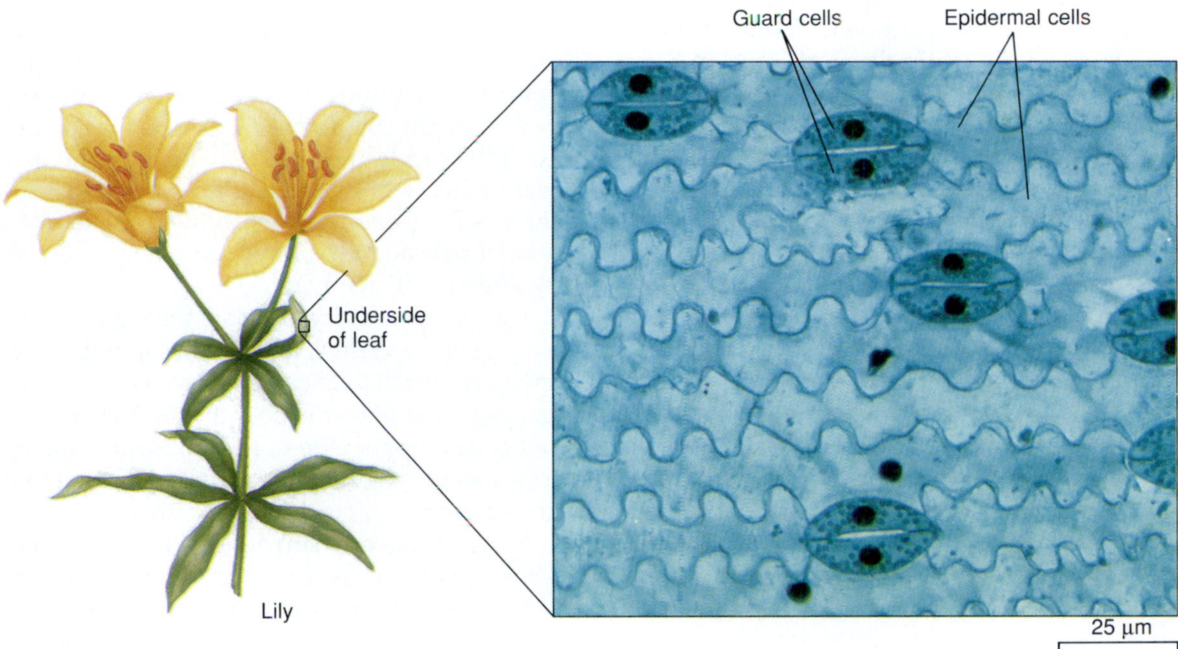

FIGURE 27–3 The lower epidermis of a lily leaf. Note the puzzle-shaped epidermal cells, which are relatively transparent. Each stomatal pore is flanked by two guard cells. (Carolina Biological Supply Company)

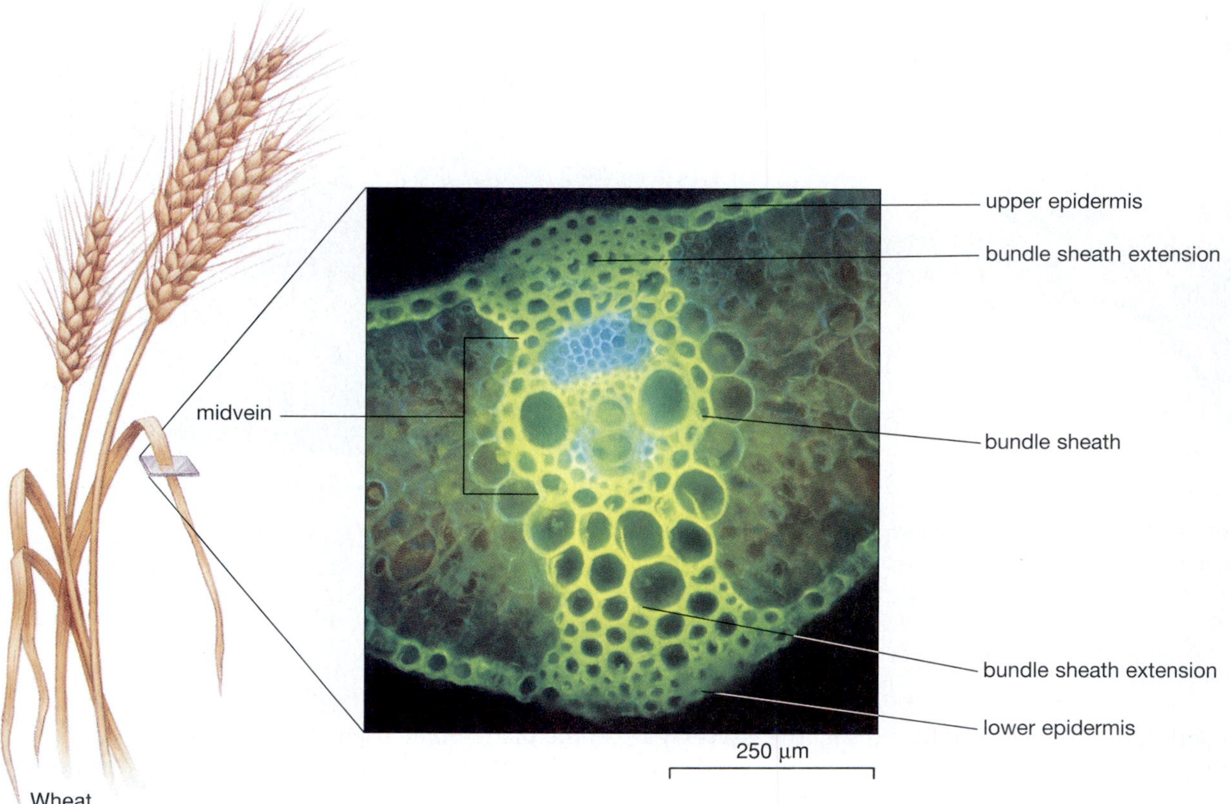

FIGURE 27–4 Bundle sheath extensions. Cross section of a wheat midvein, showing bundle sheath extensions. (Phil Gates, University of Durham/Biological Photo Service)

epidermis. Guard cells are the only cells in the epidermis that have chloroplasts, but the functional significance of this is not clear. Stomata are especially numerous on the lower epidermis (an average of about 100 stomata per mm^2), and in many cases are located *only* on the lower surface. This adaptation reduces water loss because stomata on the lower epidermis are shielded from direct sunlight.

The photosynthetic parenchyma tissue of the leaf, the **mesophyll,** is sandwiched between the upper epidermis and lower epidermis. The word *mesophyll* comes from Greek, meaning "the middle of the leaf." Mesophyll cells, which are packed with chloroplasts, are very loosely arranged with lots of air spaces between them (Fig. 27–1). In many plants, the mesophyll is divided into two specific areas. Toward the upper epidermis the cells are stacked more closely together in a **palisade** layer. In the lower portion, the cells are more loosely and more irregularly arranged in a **spongy** layer. The two layers have different primary functions. Palisade mesophyll is the primary site of photosynthesis in the leaf. Although photosynthesis also occurs in the spongy mesophyll, its primary function is diffusion of gases (particularly CO_2) within the leaf.

Palisade mesophyll may be organized into one, two, three, or even more rows of cells. The presence of additional layers of palisade mesophyll is an adaptation due in part to environmental conditions. Leaves exposed to direct sunlight are thicker because they contain more rows of palisade mesophyll than do shaded leaves on lower branches of the same plant. (In direct sunlight, the light is strong enough to effectively penetrate multiple layers of palisade mesophyll, allowing all layers to photosynthesize efficiently.)

The **veins** of a leaf extend through the mesophyll. Branching is extensive, and no mesophyll cell is very far from a vein. Each vein contains two types of vascular tissue, xylem and phloem (Chapter 26). **Xylem** is usually located on the upper side of a vein, toward the upper epidermis, whereas **phloem** is usually confined to the lower side of a vein.

Veins are usually surrounded by one or more layers of nonvascular cells called a **bundle sheath.** Bundle sheaths are composed of parenchyma or sclerenchyma cells (Chapter 26). Frequently the bundle sheath has support columns, called **bundle sheath extensions,** that extend through the mesophyll from the upper epidermis to

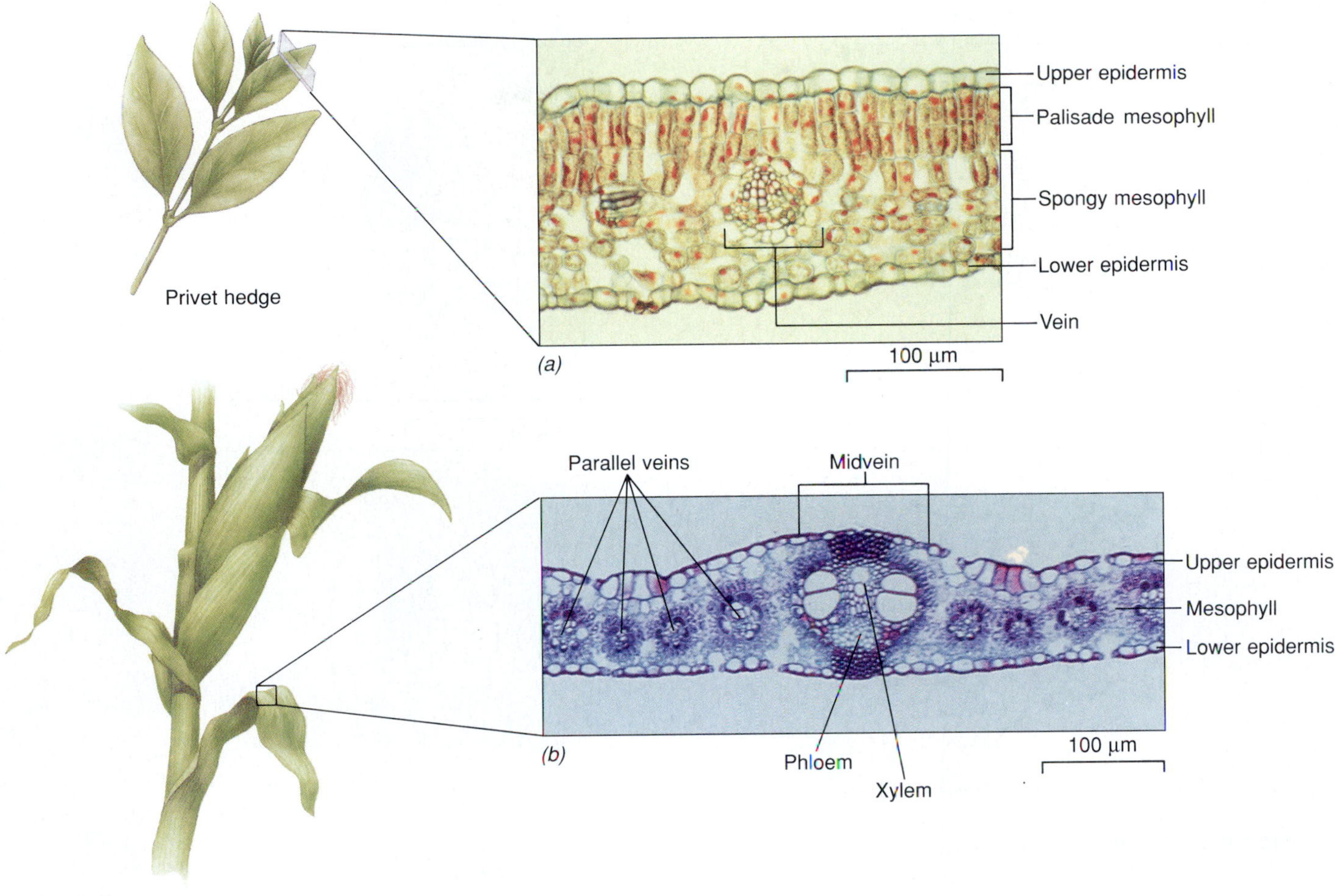

FIGURE 27–5 Leaf cross sections. (***a***) Privet, a dicot, has a mesophyll with distinct palisade and spongy sections. (***b***) Corn, a monocot. Note the absence of distinct regions of palisade and spongy mesophyll. Also evident is the evenly spaced parallel venation characteristic of monocots. (*a*, James Mauseth, University of Texas; *b*, Carolina Biological Supply Company)

the lower epidermis (Fig. 27–4). Bundle sheath extensions may be composed of parenchyma, collenchyma, or sclerenchyma cells.

Leaf Structure Differs in Dicots and Monocots

Flowering plants are divided into two groups, informally called dicots and monocots, based on a variety of structural features (Chapter 23). The leaves of **dicots,** which include such diverse plants as oak, cherry, apple, bean, rose, and snapdragon, are usually composed of a broad, flattened blade and petiole and have netted venation. In contrast, **monocots,** which include plants such as palm, corn, grass, lily, tulip, orchid, and banana, often have narrow leaves that wrap around the stem in a sheath, rather than a petiole. Parallel venation is characteristic of monocot leaves.

The internal anatomy of dicot and monocot leaves is also different (Fig. 27–5). The mesophyll of dicot leaves typically contains both palisade and spongy layers. In contrast, mesophyll in many monocot leaves is not differentiated into palisade and spongy tissue. Because dicots have netted veins, a cross section of a dicot blade often shows veins in cross section as well as lengthwise views. In cross section, the parallel pattern of monocot veins produces evenly spaced veins.

Differences between the guard cells in dicot and certain monocot leaves also occur (Fig. 27–6). The guard cells of dicots and many monocots are shaped like tiny kidney beans. Other monocot leaves (those of grasses, reeds, and sedges) have guard cells shaped like dumbbells. The dif-

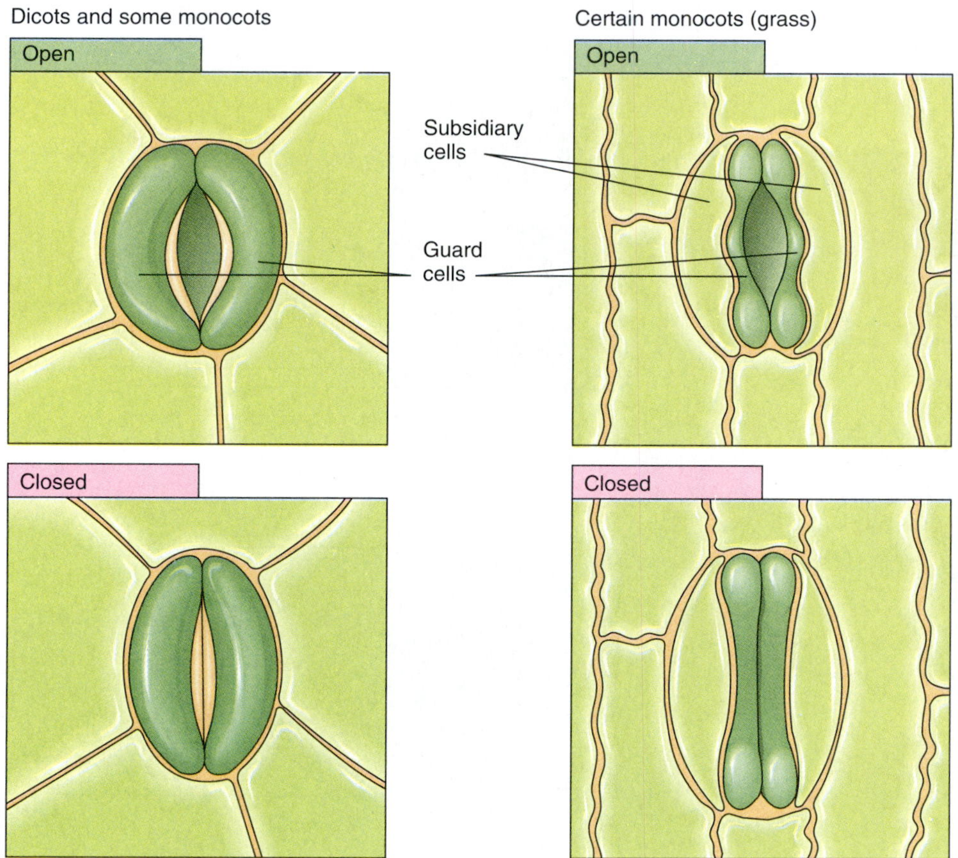

FIGURE 27–6 Variation in guard cells. Guard cells of dicots and many monocots are bean-shaped. Other monocot guard cells (those of grasses, reeds, and sedges) are narrow in the center and thicker at each end. Each narrow guard cell is associated with a special cell in the epidermis called a subsidiary cell.

ference in structure between guard cells in dicots and monocots affects how the cells swell or shrink to open or close the pore. Guard cells of both monocots and dicots may be associated with special epidermal cells called **subsidiary cells** that are structurally distinctive from all other epidermal cells.

STRUCTURE IS RELATED TO FUNCTION IN LEAVES

The primary function of leaves is to collect radiant energy and convert it to the chemical energy stored in the bonds of organic molecules such as glucose. This process, called **photosynthesis,** has been examined in detail in Chapter 9. During photosynthesis, plants take relatively simple inorganic molecules (carbon dioxide and water) and convert them into sugar. Oxygen is given off as a by-product. The sugar formed during photosynthesis is used by the plant in two ways. First, it is broken down by cellular respiration to release the chemical energy stored in its bonds for cellular activities. Second, sugar molecules provide the cell with basic building materials. The cell modifies glucose, converting it into many other important compounds, such as starch and cellulose.

How is leaf structure related to its primary function of photosynthesis? The epidermis of a leaf is relatively transparent and allows light to penetrate to the interior of the leaf where the photosynthetic tissue, the mesophyll, is located. Stomata, which dot the surfaces of the leaf, permit the exchange of gases between the atmosphere and the internal tissues of the leaf. Carbon dioxide, a raw material of photosynthesis, diffuses into the leaf through stomata, and the oxygen produced during photosynthesis diffuses rapidly out of the leaf through stomata. (Stomata, adapted to optimize gas exchange, also permit other gases, including air pollutants, to enter the leaf. Recall the chapter introduction.)

Water required for photosynthesis is obtained from the soil and transported to the leaf in the xylem where it moistens the surfaces of mesophyll cells. The loose arrangement of the mesophyll tissue, with its air spaces between cells, allows for rapid diffusion of carbon dioxide to the surfaces of the mesophyll cells. Carbon dioxide then dissolves in the film of water at the cell surfaces before diffusing into the cells.

FOCUS ON *Evolution*

COMPARATIVE PLANT ANATOMY

Many modifications in leaf anatomy are evolutionary adaptations that enable plants to survive in extreme environments. Figures A and B are cross sections of leaves from plants growing in entirely different environments. One is of a **hydrophyte,** a plant adapted to a very wet habitat. Many hydrophytes live completely or partly submerged in water. The other cross section is of a **xerophyte,** a plant adapted to extremely dry conditions. Let us compare the details of each plant's leaf anatomy to identify which plant is a hydrophyte and which a xerophyte.

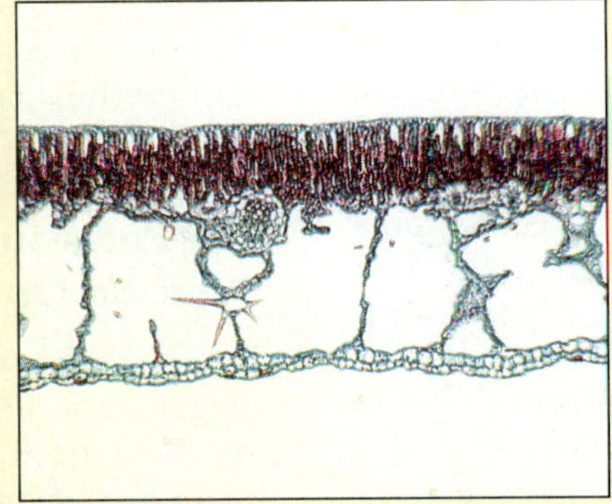

Plant A. (Dennis Drenner)

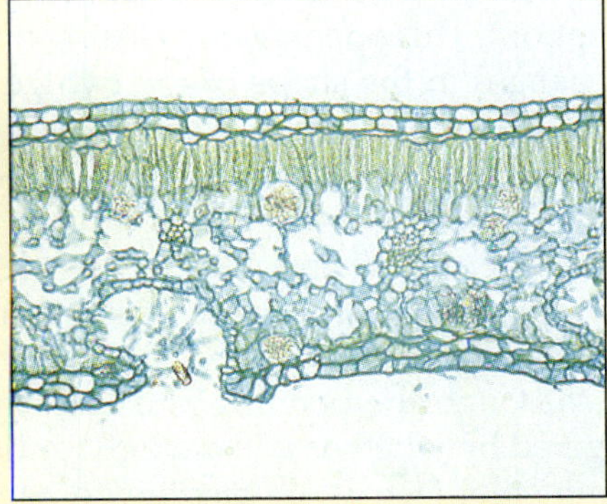

Plant B. (Dennis Drenner)

The epidermis of each plant is quite unusual. You will recall that both the upper epidermis and lower epidermis of "typical" plants are composed of a single layer of cells, with most stomata located on the lower epidermis. Although it is difficult to see at this magnification, plant A has an upper epidermis with *many* stomata, while its lower epidermis lacks stomata entirely! Its cuticle is extremely thin. Plant B is even stranger, with a multiple-layered upper epidermis covered by a thick cuticle. Its stomata are located in indentations of the lower epidermis called **stomatal crypts.** A number of trichomes are also apparent in the stomatal crypts.

The mesophyll of each plant also reveals differences, although perhaps not as pronounced as in the epidermis. Both plants have a palisade layer and a spongy layer, although plant A has larger air spaces between its mesophyll cells. The large pink cells in the A mesophyll are sclerenchyma cells that prevent the leaf's collapse since so much of its interior is air space. Support for the interior of the B leaf is provided by bundle-sheath extensions of the veins (not obvious in this micrograph). The vascular tissue provides a clue on the identity of A and B as well: Plant A has much less vascular tissue than does plant B.

Based on your knowledge of leaf anatomy, can you identify which plant is hydrophytic and which is xerophytic? Turn to page 539 to see if you are a good comparative anatomist!

The veins not only supply water (in the xylem) to the photosynthetic tissue, but also carry the sugar (in the phloem) produced during photosynthesis to other parts of the plant. Bundle sheaths and bundle sheath extensions associated with the veins provide additional support to prevent the leaf, which is structurally weak because of the large amount of air space in the mesophyll, from collapsing under its own mass.

The environment to which a particular plant is adapted also affects its leaf structure. Although aquatic plants and plants adapted to dry conditions both perform photosynthesis and have the same basic leaf anatomy, their leaves are modified to enable them to survive the different environmental conditions (see Focus On Evolution: Comparative Plant Anatomy). The leaves of water lilies, for example, have petioles long enough to allow the blade to float on the water's surface (Fig. 27–7).

FIGURE 27–7 An unusual view of water lily leaves shows their petioles as well as their blades, which float on the water's surface. The length of the petioles depends on the depth of the water. (Frans Lanting/Mendeu Pictures)

Opening and Closing of Stomata Are Due to Changes in Turgidity of the Guard Cells

Stomata are adjustable pores that are usually open during the day when photosynthesis is occurring and closed at night when photosynthesis is shut down (see Chapter 9, the section on CAM photosynthesis, for an interesting exception). This opening and closing of stomata is caused by changes in the shape of the two guard cells that surround each pore. When water moves into bean-shaped guard cells from surrounding cells, they become turgid and the inner cell walls bend outward at the center, producing a pore. When water leaves the guard cells, they become flaccid and collapse against one another, closing the pore.

Although the opening and closing of guard cells is triggered by whether it is daylight or darkness, other environmental factors are also involved, including carbon dioxide concentration. A low concentration of CO_2 in the leaf induces the stomata to open even in the dark. The effects of light and CO_2 concentration on stomatal opening are interrelated. Photosynthesis, which occurs in the presence of light, reduces the internal concentration of CO_2 in the leaf, triggering stomatal opening. Another environmental factor that affects stomatal opening and closing is severe water stress. During a prolonged drought, stomata will remain closed even during the day in order to conserve water. This mechanism is under hormonal control (Chapter 30).

The opening and closing of stomata also appear to be under the control of an internal biological clock that somehow measures time. For example, plants placed in continual darkness persist in opening and closing their stomata at more or less the same times each day. Biological rhythms that follow an approximate 24-hour cycle are known as **circadian rhythms.** Other examples of circadian rhythms in plants are provided in Chapter 30.

The Potassium Ion Mechanism Explains Stomatal Opening and Closing

The data from numerous experiments and observations suggest that stomata open and close by the **potassium ion (K+) mechanism.** Light in some way triggers an influx of potassium ions (K^+) into the guard cells from the surrounding cells of the epidermis (Fig. 27–8). This movement of potassium ions, which occurs through special channels, or gates, in the plasma membrane has been experimentally measured and verified.

The increased concentration of K^+ in the guard cells lowers the relative concentration of water in those cells. Therefore, water passes into the guard cells from surrounding epidermal cells by osmosis. The increased

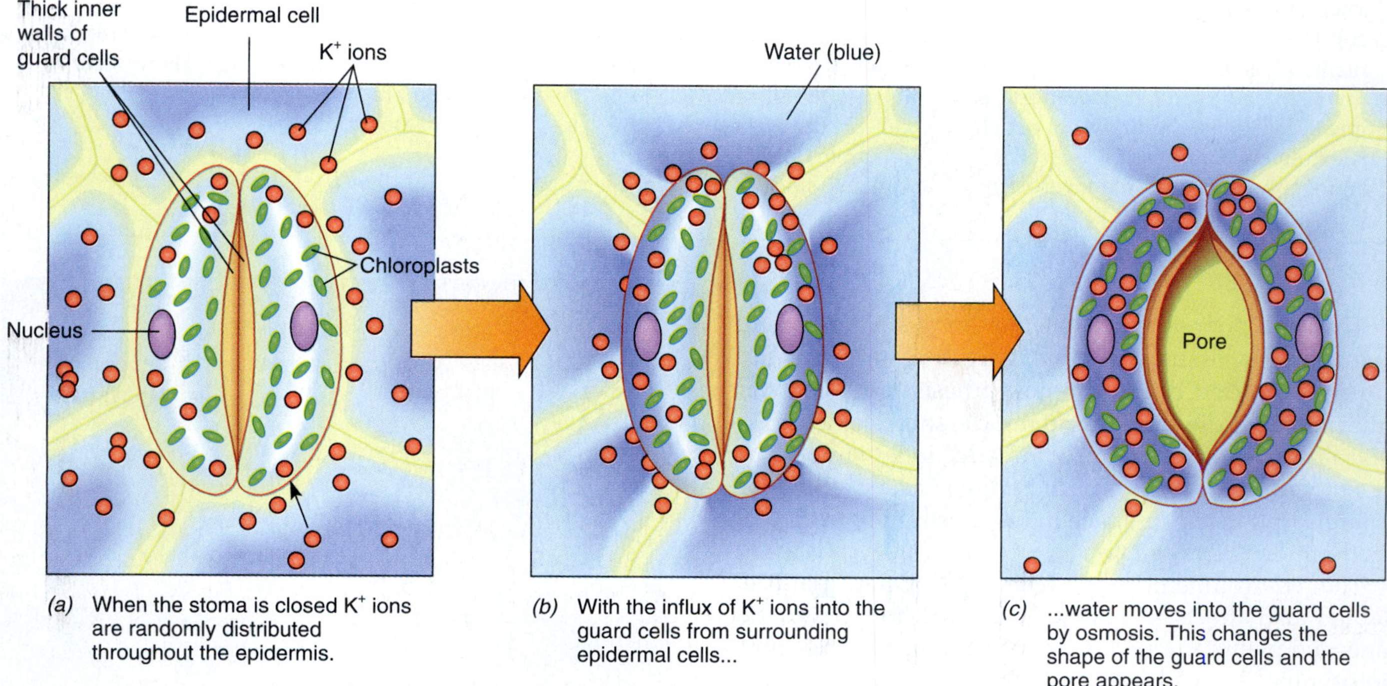
Thick inner walls of guard cells
Epidermal cell
K⁺ ions
Chloroplasts
Nucleus
Water (blue)
Pore
(a) When the stoma is closed K⁺ ions are randomly distributed throughout the epidermis.
(b) With the influx of K⁺ ions into the guard cells from surrounding epidermal cells...
(c) ...water moves into the guard cells by osmosis. This changes the shape of the guard cells and the pore appears.

FIGURE 27–8 Mechanism of stomatal opening and closing. Movement of K^+ ions into and out of the guard cells affects stomatal opening and closing.

turgidity of the guard cells changes their shape, and the pore opens.

In the late afternoon or early evening, the stomata close by a reversal of the process.[2] The potassium ions are pumped out of the guard cells into the surrounding epidermal cells. Water leaves the guard cells by osmosis, thereby causing the cells to lose their turgidity and collapse, and the pore closes.

Like other plant responses to light, stomata vary in their sensitivity to different colors (that is, different wavelengths) of light. Stomatal opening is most pronounced in blue and, to a lesser extent, in red light. Also, dim blue light induces stomatal opening, whereas dim red light does not. Recall from Chapter 9 that any plant response to light must involve a pigment, a molecule that absorbs the light prior to inducing a particular biological response. These and other data suggest that the pigment involved in stomatal opening and closing is yellow (yellow pigments strongly absorb blue light) and is located in the guard cells.

Leaves lose water by transpiration and guttation

Despite leaf adaptations such as the cuticle, approximately 99% of the water that a plant absorbs from the soil is lost by evaporation from the stems and, especially, the leaves. Loss of water vapor from the aerial parts of plants is called **transpiration.**

The cuticle is extremely effective in reducing water loss from transpiration; it is estimated that only 1% to 3% of the water lost from a plant passes directly through the cuticle. Most transpiration occurs through the stomata. The numerous stomatal pores, so effective in gas exchange for photosynthesis, also provide openings through which water vapor escapes. Also, the loose arrangement of the mesophyll cells provides a large surface area within the leaf from which water can evaporate.

A number of environmental factors affect the rate of transpiration. At higher temperatures, more water is lost from plant surfaces. Light increases the transpiration rate, in part because it triggers stomatal opening and in part because sunlight increases the temperature of the leaf. Wind also increases transpiration. A high relative humidity *decreases* transpiration because the air is already saturated, or nearly so, with water vapor.

Although transpiration may seem like a wasteful process, there may be some benefits to the large amount of water that plants lose by transpiration. First, transpiration, like sweating in humans, cools the leaves and stems. When water passes from a liquid state to a vapor, it absorbs a great deal of heat. As the water molecules leave the plant as water vapor, they carry this heat with them. Thus, the cooling effect of transpiration may prevent the plant from overheating, particularly in direct sunlight. On a hot summer day, for example, the internal temperature of leaves is measurably lower than that of the surrounding air.

A second benefit of transpiration is that it ensures that the plant will have sufficient essential minerals. The water a plant transpires is initially absorbed from the soil, where it is not present as pure water, but rather as a very dilute solution of dissolved mineral salts. Water leaves the plant during transpiration, but minerals do not. Many of these minerals are required for the plant's growth. It has been suggested that transpiration enables a plant to take in sufficient water to provide enough essential minerals and that plants cannot satisfy their mineral requirements if the transpiration rate is not high enough.

Concept Connections

Transpiration — Climate — Hydrologic Cycle

Climate comprises the average weather conditions that occur in a place over a period of years. Factors that determine an area's climate include temperature and precipitation, both of which are influenced by transpiration. Transpiration by trees affects the local temperature of forests. If you have ever walked into a forest on a hot summer day, you have probably noticed that the air is cooler and moister there than it is outside the forest. This is true partly because the trees overhead absorb much of the sunlight and partly because transpiration is a biological cooling process. Water transpired from the surfaces of leaves and stems has a cooling effect, not only for the plant but for the local area in which the plant lives.

Transpiration is an important part of the hydrologic cycle, in which water cycles from the oceans and land to the atmosphere, and back to the oceans and land. As a result of transpiration, water evaporates from the surfaces of leaves and stems to form clouds in the atmosphere. Thus, transpiration eventually results in precipitation. As you might expect, forest trees release substantial amounts of moisture into the air by transpiration. When a large forest is burned or cut down, rainfall declines and droughts become more common in that region. We say more about climate and the hydrologic cycle in Chapter 47.

[2] Stomatal closure may not be an *exact* reversal of stomatal opening. There is evidence that Ca^{2+} triggers stomatal closure but inhibits stomatal opening. The actual mechanism whereby Ca^{2+} exerts this effect is under investigation.

(a) (b)

FIGURE 27–9 Temporary wilting. Temporary wilting in pumpkin leaves (***a***) in the late afternoon of a hot day and recovery (***b***) the following morning. Note that wilting helps reduce the surface area from which transpiration occurs. During the night, the plants recovered by absorbing water from the soil while transpiration was negligible. (David Cavagnaro)

There is no doubt, however, that under certain circumstances, transpiration can be harmful to a plant. On hot summer days, plants frequently lose more water by transpiration than is replaced from the soil. Their cells experience a loss of turgor and the plant wilts. If a plant is able to recover overnight, as a result of negligible transpiration (recall that stomata are closed) while water is still being absorbed from the soil, the plant is said to have experienced temporary wilting (Fig. 27–9). Most plants recover from temporary wilting with no ill effects. In cases of prolonged drought, the soil may not contain sufficient moisture to permit recovery from wilting. A plant that cannot recover overnight is said to be permanently wilted and will die unless water is supplied immediately.

Thus, it appears that transpiration is a "mixed blessing" for plants because it has definite benefits and potential hazards. At any rate, transpiration is unavoidable in plants because they possess stomata.

Some Plants Release Water as a Liquid

Many leaves have special structures through which liquid water is literally forced out. This loss of liquid water, known as **guttation,** occurs when transpiration is negligible and available soil moisture is high. Guttation frequently occurs at night because the stomata are closed, but water continues to move into the roots by osmosis. People sometimes incorrectly attribute the early morning droplets of water on leaves to dew (water condensation from the air) rather than guttation (Fig. 27–10).

FIGURE 27–10 Guttation in lady's mantle (*Alchemilla vulgaris*). Many people mistake guttation for early morning dew. (Dennis Drenner)

LEAF ABSCISSION ALLOWS MANY PLANTS IN TEMPERATE CLIMATES TO SURVIVE WINTER

In temperate climates, the leaves of many plants turn color and **abscise,** or fall off, when winter approaches.[3] Most woody plants with broad leaves shed their leaves in order to survive the low temperatures of winter. Dur-

[3] Leaf abscission also occurs during the dry period in tropical climates with pronounced wet and dry seasons.

FIGURE 27–11 A longitudinal section through a maple branch, showing the base of the petiole. Note the abscission zone where the leaf will abscise from the stem. A lateral bud with its protective bud scales is evident above the petiole. (James Mauseth, University of Texas)

ing the winter, water conservation becomes critical for plants. As the ground chills, absorption of water by the roots is inhibited, and when the ground freezes, *no* absorption occurs.

If a plant were to maintain its broad leaves during the winter, it would continue to lose water by transpiration but would be unable to replace the water by absorption from the soil. In the lower temperatures of winter, the plant's metabolism, including its photosynthetic machinery, slows down a great deal. As a result, plants have little need for leaves during the winter.

Leaf abscission is a complex process that involves many physiological changes, all of which are orchestrated by changing levels of plant hormones (Chapter 30). As autumn approaches, the plant reabsorbs sugar (much of it obtained from the breakdown of starch) and many of the essential minerals located in the leaves. Nitrogen, phosphorus, and possibly potassium move into the woody tissues from the leaves. Chlorophyll is broken down, and the orange and yellow accessory pigments in the chloroplast become evident. These pigments were always present in the leaf but were masked by the chlorophyll. In addition, red water-soluble pigments may be synthesized and stored in the vacuoles of leaf cells in some species; their function is unknown. The various combinations of these pigments are responsible for the brilliant colors found in autumn landscapes in temperate climates.

A Deciduous Leaf Possesses an Abscission Zone at the Base of Its Petiole

The area where a leaf petiole detaches from the stem is structurally different from surrounding tissues (Fig. 27–11). This area, called the **abscission zone,** is composed primarily of thin-walled parenchyma cells and is anatomically weak because it contains few fibers. As autumn approaches, a protective layer of cork cells develops on the stem side of the abscission zone. These cells have a waxy, waterproof material impregnated in their walls. Enzymes then dissolve the middle lamella, which is the "cement" holding the cells together, in the abscission zone. By this time, there is nothing holding the leaf to the stem but a few xylem cells. A sudden breeze is enough to make the final break, and the leaf detaches. The protective layer of cork remains, sealing off the area and forming a leaf scar.

FIGURE 27–12 Leaf modifications. (***a***) The leaves of cacti are modified as spines for protection. (***b***) Tendrils, which attach to objects and aid vines in climbing, may be modified leaves or stems. These pea tendrils are modified leaves. (***c***) Overlapping bud scales protect buds. Shown here are a terminal bud and two lateral buds. (***d***) The leaves of bulbs such as the onion are fleshy for storage. (***e***) Some plants have succulent leaves modified for water storage as well as photosynthesis. (*a*, Patti Murray, *e*, James Mauseth, University of Texas; *b*, *c*, *d*, Dennis Drenner)

LEAVES WITH FUNCTIONS OTHER THAN PHOTOSYNTHESIS EXHIBIT MODIFICATIONS IN STRUCTURE

Although photosynthesis is the main function of leaves, certain leaves possess modifications for functions other than photosynthesis. Some plants have leaves specialized for protection. **Spines,** modified leaves that are hard and pointed, may be found on plants like cacti (Fig. 27–12*a*). In the cactus, the main organ of photosynthesis is the stem rather than the leaf. Spines discourage herbivorous animals from eating the succulent stem tissue.

Many vines have **tendrils.** Tendrils, which are usually specialized leaves,[4] are used for grasping and holding onto other structures (Fig. 27–12*b*). Vines are climbing stems that cannot support their own weight, and so they often possess tendrils that help keep the vine attached to the structure on which it is growing.

The winter buds of a dormant woody plant are covered by protective **bud scales,** which are modified leaves (Fig. 27–12*c*). Bud scales protect the delicate meristematic tissue of the shoot from injury and desiccation.

Leaves may also be modified for storage of water or food. For example, a **bulb** is a short underground stem to which large, fleshy leaves are attached (Fig. 27–12*d*). Onions and tulips form bulbs.

[4] Some tendrils are specialized stems.

(a)

(b)

FIGURE 27–13 The pitcher plant has leaves modified to form a pitcher that collects water, drowning its prey. (***a***) Growth habit of a pitcher plant. (***b***) A cutaway view of a pitcher reveals accumulated insect bodies and debris. (*a*, Skip Moody/Dembinsky Photo Associates; *b*, Carolina Biological Supoly Company)

Many plants adapted to arid conditions have fleshy, succulent leaves for water storage (Fig. 27–12*e*). These leaves are usually green and photosynthesize as well.

Some of the Most Remarkable Examples of Modified Leaves Are Those of Insectivorous Plants

Insectivorous plants have leaves adapted to attract, capture, and digest animal prey, usually insects. Most insectivorous plants grow in poor soil that is deficient in certain essential minerals. These plants meet some of their mineral requirements by digesting and absorbing nutrients from insects and other small animals.

Some insectivorous plants have passive traps. For example, the leaves of a pitcher plant are shaped so that rainwater collects within, forming a reservoir that also contains digestive enzymes (Fig. 27–13). Some pitchers are quite large; in the tropics, pitcher plants may hold one liter (more than one quart) or more of liquid.

An insect attracted by the odor or nectar of the pitcher may lean over the edge and fall in. Although it may make repeated attempts to escape, the insect is prevented from crawling out by a row of stiff spines that points downward around the lip of the pitcher. The insect eventually drowns, and part of its body is digested and absorbed.

The Venus's flytrap is an example of an insectivorous plant with active traps. Its leaves resemble tiny bear traps (see Fig. 1–7 on page 8). Each side of the leaf contains three small hairs. If an insect alights and brushes against two of the hairs, the trap snaps shut with amazing rapidity (Chapter 30). After the insect has died and been digested, the trap reopens and the indigestible remains fall off.

Chapter Summary

I. Leaf structure reflects its main function, photosynthesis.
   A. The transparent epidermis allows light to penetrate into the mesophyll, where photosynthesis occurs. Stomata in the epidermis are for gas exchange.
   B. The mesophyll tissue contains air spaces that permit rapid diffusion of carbon dioxide into, and oxygen out of, mesophyll cells.
   C. Leaf veins have xylem to conduct water and essential minerals to the leaf and phloem to conduct sugar produced by photosynthesis to the rest of the plant.

II. Monocot and dicot leaves can be distinguished based on their external structure and their internal anatomy.
   A. Monocot leaves have parallel venation, and mesophyll is not differentiated into palisade and spongy layers.
   B. Dicot leaves have netted venation, and mesophyll is usually differentiated into palisade and spongy layers.

III. Stomata usually open during the day and close at night.

A. The potassium ion mechanism explains the opening and closing of stomata.
   1. Light triggers an influx of potassium ions (K^+) into the guard cells. The resulting osmotic movement of water into the guard cells causes their shape to change, forming a pore.
   2. Stomata close when potassium ions leave the guard cells, followed by the osmotic flow of water out of the guard cells.
B. A number of factors affect stomatal opening, including light or darkness, CO_2 concentration, water stress, and a circadian rhythm within the plant.

IV. Transpiration is the loss of water vapor from plants.
A. It occurs primarily through the stomata.
B. The rate of transpiration is affected by environmental factors like temperature, light, wind, and relative humidity.
C. Transpiration may be both beneficial and harmful to the plant.

V. Guttation is the release of liquid water from leaves that occurs through special structures when transpiration is negligible and available soil moisture is high.

VI. Leaf abscission, the shedding of leaves, is a complex process involving physiological and anatomical changes.

VII. Leaves may be modified for functions other than photosynthesis.
A. Spines are leaves adapted to provide protection.
B. Tendrils are leaves modified for grasping and holding onto other structures (to support weak stems).
C. Bud scales are special leaves that protect delicate meristematic tissue of buds.
D. Bulbs are short underground stems with fleshy leaves specialized for storage.
E. Some leaves are modified to trap insects.

Selected Key Terms

abscission, p. 534
bud scale, p. 536
bulb, p. 536
bundle sheath, p. 528
cuticle, p. 526
epidermis, p. 526
guttation, p. 534
mesophyll, p. 528
potassium ion mechanism, p. 532
spine, p. 536
stoma, p. 527
tendril, p. 536
transpiration, p. 533
trichome, p. 527
vein, p. 528

Post-Test

1. The ______________ is the photosynthetic tissue in the middle of the leaf.
2. Gas exchange occurs through tiny pores surrounded by ______________ ______________.
3. The ______________ of a leaf vein transports water and dissolved minerals.
4. Stomata are usually more numerous on the ______________ epidermis.
5. The leaves of ______________ usually have netted venation, distinct palisade and spongy layers in the mesophyll, and bean-shaped guard cells.
6. The opening and closing of stomata is due to the movement of ______________ ions.
7. Most of the water a plant absorbs from the soil is lost from the leaves by ______________.
8. Biological rhythms that follow an approximate 24-hour cycle, such as the opening and closing of stomata, are called ______________ ______________.
9. The shedding of leaves is known as ______________.
10. ______________ are leaves that are modified for grasping and holding.
11. Onion bulbs have fleshy leaves that are modified for ______________.
12. Chloroplasts are found in: (a) mesophyll cells and guard cells (b) xylem cells and guard cells (c) epidermal cells and phloem cells (d) palisade mesophyll cells and phloem cells.
13. A ring of cells surrounding the vascular bundle in leaves is known as a/an: (a) cuticle (b) bud scale (c) trichome (d) bundle sheath.
14. The cuticle: (a) allows water to exit as a liquid from plants (b) reduces water loss in plants (c) prevents permanent wilting from ever occurring (d) permits gas exchange for photosynthesis.
15. Which of the following statements is *not* true about veins? (a) Veins contain the conducting tissues, xylem and phloem. (b) Veins are part of the network of vascular tissue that extends throughout the plant body. (c) Veins contain the photosynthetic tissue found in the interior of a leaf. (d) Veins are often surrounded by a bundle sheath.
16. Place the following steps in the opening of stomata in the correct order: 1. Increased turgidity of guard cells causes them to change shape. 2. Light triggers an influx of potassium ions into guard cells. 3. Water passes into guard cells by osmosis. (a) 3–1–2 (b) 1–2–3 (c) 2–3–1 (d) 2–1–3 (e) 3–2–1
17. The leaves of a cactus are modified as: (a) tendrils (b) bulbs (c) bud scales (d) spines.

Review Questions

1. Discuss how the leaf is structured to deliver the raw materials and products of photosynthesis.
2. How are stomata related to photosynthesis? To transpiration?
3. Describe the series of physiological changes that occur in guard cells during stomatal opening.
4. Discuss at least two ways that most leaves adapted during the course of evolution to conserve water.
5. How does the environment affect transpiration rate? Stomatal opening and closing?
6. Why do deciduous plants in temperate climates lose their leaves in autumn?
7. Discuss the specialized features of the leaves of insectivorous plants.
8. How are leaves modified to tolerate wet environments? Dry environments?
9. Label the diagram on the right.

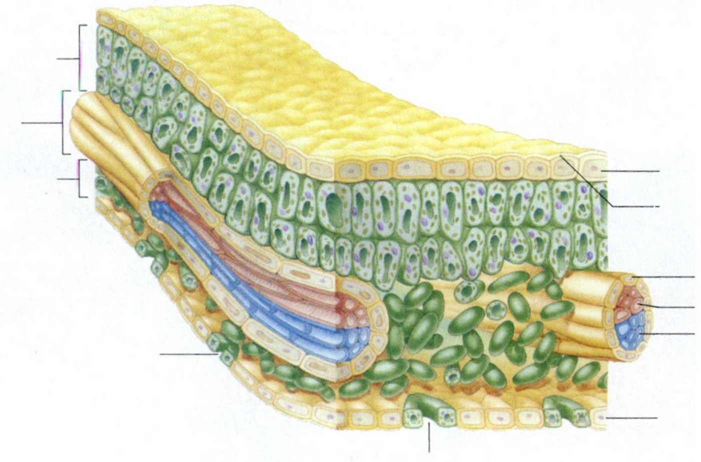

Thinking Critically

1. Why should the impact of air pollution on plants be of concern to you? What, if any, measures would you suggest to protect plants from the effects of air pollution?
2. How is the effect of airborne pollution on leaves analogous to its effect on human lungs?
3. Given that: (a) xylem is located toward the upper epidermis in leaf veins, phloem toward the lower epidermis, and (b) the vascular tissue of a leaf is continuous with that of the stem, suggest one possible arrangement of vascular tissues in the stem that accounts for the arrangement of vascular tissues in the leaf.
4. Suppose you observe a micrograph of a leaf cross section and are asked to determine which side is covered by the upper epidermis and which side by the lower epidermis. How would you make this decision?
5. What might be some advantages of a plant having a few very large leaves? Disadvantages? What might be some advantages of a plant having many very small leaves? Disadvantages?

Recommended Readings

Dyyale, J.E., "How Do Leaves Grow?" *BioScience*, Vol. 42, No. 6, June 1992. Leaf research is branching out in new directions—away from describing leaf variations and toward understanding changes that occur during leaf development.

Lipske, M., "Forget Hollywood: These Bloodthirsty Beauties Are for Real," *Smithsonian*, December 1992. Examines some of the relationships between insectivorous plants and animals.

Mauseth, J.D., *Botany: An Introduction to Plant Biology*, 2nd edition, Philadelphia, Saunders College Publishing, 1995. A comprehensive introduction to plant biology.

Vogel, S., "When Leaves Save the Tree," *Natural History*, September 1993. In high winds, leaves roll or clump together, minimizing drag.

Weiner, J., *The Next One Hundred Years*, New York, Bantam Books, 1990. Addresses the problem of global climate change. Pages 183–185 discuss the relationship among trees, transpiration, and local climate.

Answer to question on page 531: Plant A, a water lily, is a dicot that exhibits hydrophytic characteristics. It has stomata on its upper epidermis only because the leaf blade floats on water. Roots and stems, as well as leaves, have large air spaces because gas exchange is more limiting than water for hydrophytic plants. The air spaces also provide buoyancy. Plant B, oleander, is a xerophytic plant. Many xerophytic plants possess a thick cuticle, multiple layered epidermis, and stomatal crypts to reduce transpiration.

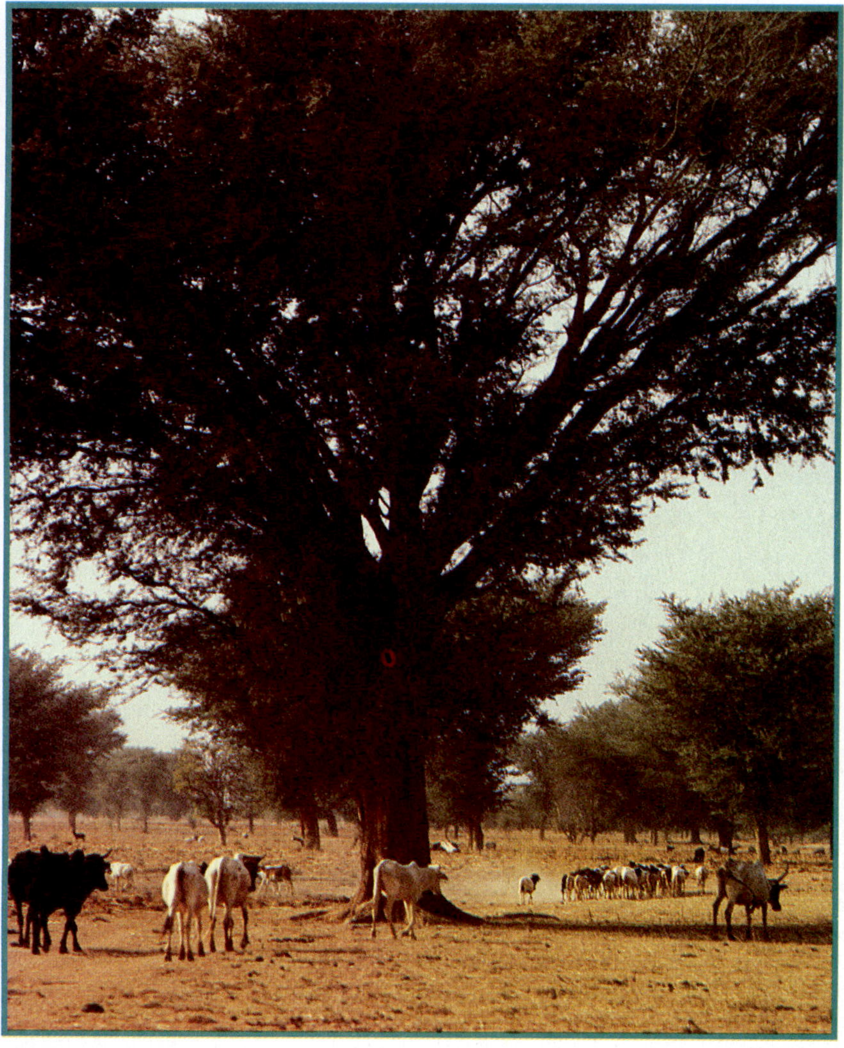

Cattle in Burkina Faso, a country in the African Sahel, have eaten all the ground cover, leaving it exposed and vulnerable to soil erosion. *The trees that remain will probably be stripped of branches to feed the hungry cattle. Over-exploitation of the Sahel is degrading the soil and increasing the amount of unproductive desert area. (Robert E.Ford/Terraphotographics)*

The Erosion of a Precious Resource

Soil, which is composed of mineral particles, organic material, water, and air, is a valuable natural resource upon which humans depend for food. **Soil erosion,** the wearing away, or removal, of soil from the land, is a national and international problem that does not make the headlines very often. To get a feeling for how serious the problem is, however, consider that approximately 2.7 billion metric tons (3.0 billion tons) of topsoil are lost from U.S. farmlands as a result of soil erosion *each year.* The U.S. Department of Agriculture estimates that approximately one-fifth of U.S. cropland is vulnerable to soil erosion damage. Because erosion reduces the amount of soil in an area, it limits the growth of plants. Erosion also causes a loss in soil fertility because essential minerals and organic matter that are part of the soil are also removed. As a result of these losses, the productivity of eroded agricultural soils drops.

Soil erosion and mineral depletion are significant problems worldwide. A global assessment of soil conditions was first completed in 1992 when a three-year study of global soil degradation sponsored by the United Nations Environment Program was released. It reported that 1.96 billion hectares (4.84 billion acres) of soil—an area that equals 17% of the Earth's total vegetated surface area—have been degraded since World War II. Eleven percent of the Earth's vegetated surface area—an area the size of China and India combined—has been degraded so badly that it will be very costly, or in some cases impossible, to reclaim. The main causes of soil degradation are farming practices that are environmentally unsound, overgrazing by livestock, and deforestation (the removal of forest without adequate replanting; discussed in Chapters 9, 20, and 49).

It is possible to reclaim land that is badly degraded. Soil reclamation involves stabilizing the land to prevent further erosion and restoring the soil to its former fertility. In order to stabilize the land, the bare ground is seeded with plants. Once established, these plants cover the soil, and their roots hold it in place. The plants that have been established to stabilize the land start to improve the quality of the soil almost immediately, as dead portions are converted to humus (black or dark brown decomposed organic material). The humus holds mineral nutrients in place and releases them a little at a time. Humus also improves the water-holding capacity of the soil. Restoration of soil fertility to its original level is a slow process, however. During its recovery, the use of the land must be restricted. It cannot be farmed and it cannot be grazed. Disaster is likely if the land is put back to use before the soil has completely recovered.

Asia and Africa have the largest land areas with extensive soil damage, and the problem is compounded by rapid population growth. For example, the Sahel, a semiarid region located south of the Sahara Desert in Africa, is characterized by a rapidly increasing human population. People living in the Sahel must use their land to grow crops and animals for food or they will starve, but the soil is so over-exploited that it is able to support fewer and fewer people. The day is approaching when it will be totally unproductive desert. To reclaim the soil would require restricting its use for many years so that it could recover. However, if these measures are implemented, the people living in the Sahel will have no means of obtaining food.

Restriction of land use for a period of several to many years sometimes poses economic and ethical dilemmas. For example, how can a government tell landowners that they cannot use their own, privately owned land? How can land use be restricted when people's livelihoods and maybe even their lives depend on such use?

Stems and Roots

28

Learning Objectives

After you have studied this chapter you should be able to:

1. Describe three functions of stems and explain how their structure correlates with those functions.
2. Compare the primary structure of dicot and monocot stems, giving at least one function for each tissue.
3. Relate at least three functions of roots and explain how their structure correlates with those functions.
4. Compare the primary structure of dicot and monocot roots, giving at least one function for each tissue.
5. Discuss the structure of stems and roots with secondary growth and describe how secondary tissues form.
6. Trace the pathway of water movement in plants.
7. Discuss root pressure and tension cohesion as mechanisms to explain the rise of water in xylem.
8. Outline the pressure-flow hypothesis of sugar transport in phloem.
9. Distinguish between organic and inorganic fertilizers and relate some of the benefits of each.

Key Concepts

- ☐ Stems support leaves and reproductive structures and transport sugar, water, and minerals. Roots anchor a plant in the soil, absorb water and dissolved minerals, and store food made in leaves and stems by photosynthesis.
- ☐ The arrangement of tissues in stems and roots is related to the different environmental conditions to which they are exposed and to their different functions. Unlike roots, stems bear leaves. Roots differ from stems in that a root possesses root hairs, a root cap, an endodermis, and a pericycle.
- ☐ Although considerable variation exists in stem and root structure, all vascular stems and roots have an outer protective covering (epidermis or periderm), one or more types of ground tissue, and vascular tissues (xylem and phloem).

THE PRIMARY STRUCTURE OF DICOT AND MONOCOT STEMS CAN BE DISTINGUISHED BY THE ARRANGEMENT OF VASCULAR TISSUES

Stems exhibit varied forms ranging from vines to massive tree trunks. They link the roots to the leaves and are usually located above ground (Fig. 28–1). Stems have three main functions. First, stems bear leaves and reproductive structures at regular intervals. The upright position of most stems and the arrangement of leaves on stems allow each leaf to absorb maximum light for use in photosynthesis.

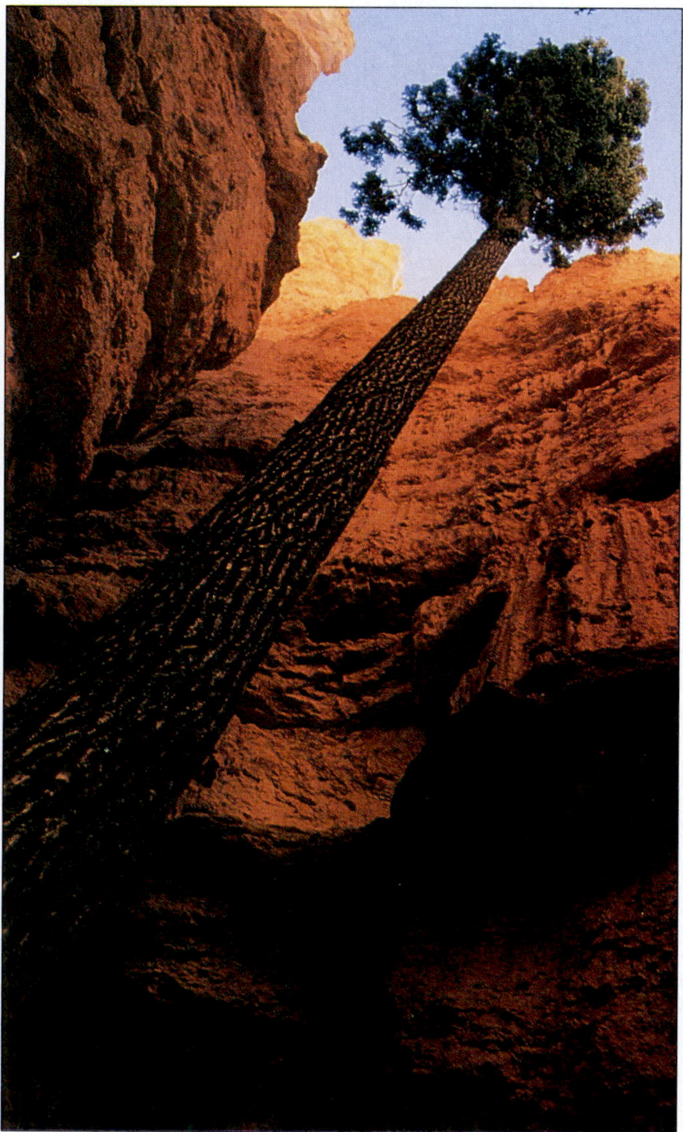

FIGURE 28–1 Stems support the aerial plant body. The stem of a Ponderosa pine grows toward the sunlight at Bryce Canyon, Utah. (Doug Locke/Dembinsky Photo Associates)

Second, stems provide internal transport. Stems conduct water and dissolved minerals from the roots, where they are absorbed from the soil, to the leaves and other plant parts. Stems also conduct the sugar produced in the leaves by photosynthesis to the roots and other parts of the plant. The vascular system is continuous throughout all parts of the plant, and conduction occurs in roots, stems, leaves, and reproductive structures.

Third, stems produce new living tissue. Stems continue to grow throughout a plant's life. In addition to the main functions of support, conduction, and production of new stem tissue, a number of stems are modified for vegetative propagation (asexual reproduction; see Chapter 29). Also, some stems are specialized to store food or, if green, to manufacture sugar by photosynthesis.

In Chapter 26, you learned that plants have two different types of growth. Primary growth is an increase in the length of a plant and occurs at **apical meristems** located at the tips of stems and roots. Secondary growth is an increase in the width of a plant and is due to the activity of **lateral meristems** located along the sides of stems and roots. All plants have primary growth; some plants have both primary growth and secondary growth. Recall that stems with primary growth only are herbaceous, while those with secondary growth are woody (Chapter 26). Although all stems with primary growth only have the same basic tissues, the arrangement of tissues in the stem varies between the two groups of flowering plants, the dicots (for example, sunflower) and the monocots (for example, corn).

The sunflower stem is a representative dicot stem exhibiting primary growth (Fig. 28–2). Its outer covering is the **epidermis.** Inside the epidermis is a layer that is several cells thick, the **cortex.** The vascular tissue is located in patches that are arranged in a circle when viewed in cross section. Each patch, or vascular bundle, contains both **xylem** and **phloem.** The xylem is usually located on the inner side of the vascular bundle and the phloem to the outside. Sandwiched between the xylem and phloem is a single layer of cells, the **vascular cambium.** In sunflowers, there is a cluster of fibers (Chapter 26) called a **phloem fiber cap** directly outside the phloem. The phloem fiber cap is not a part of the primary structure of some dicot stems. Although the vascular bundles are arranged in a circle in cross section, these bundles extend as long strands throughout the length of the stem and are continuous with the vascular tissues of both roots and leaves. The center of the dicot stem is **pith,** a tissue composed of large, thin-walled parenchyma cells (Chapter 26). The areas between the vascular bundles are usually referred to as **pith rays.**

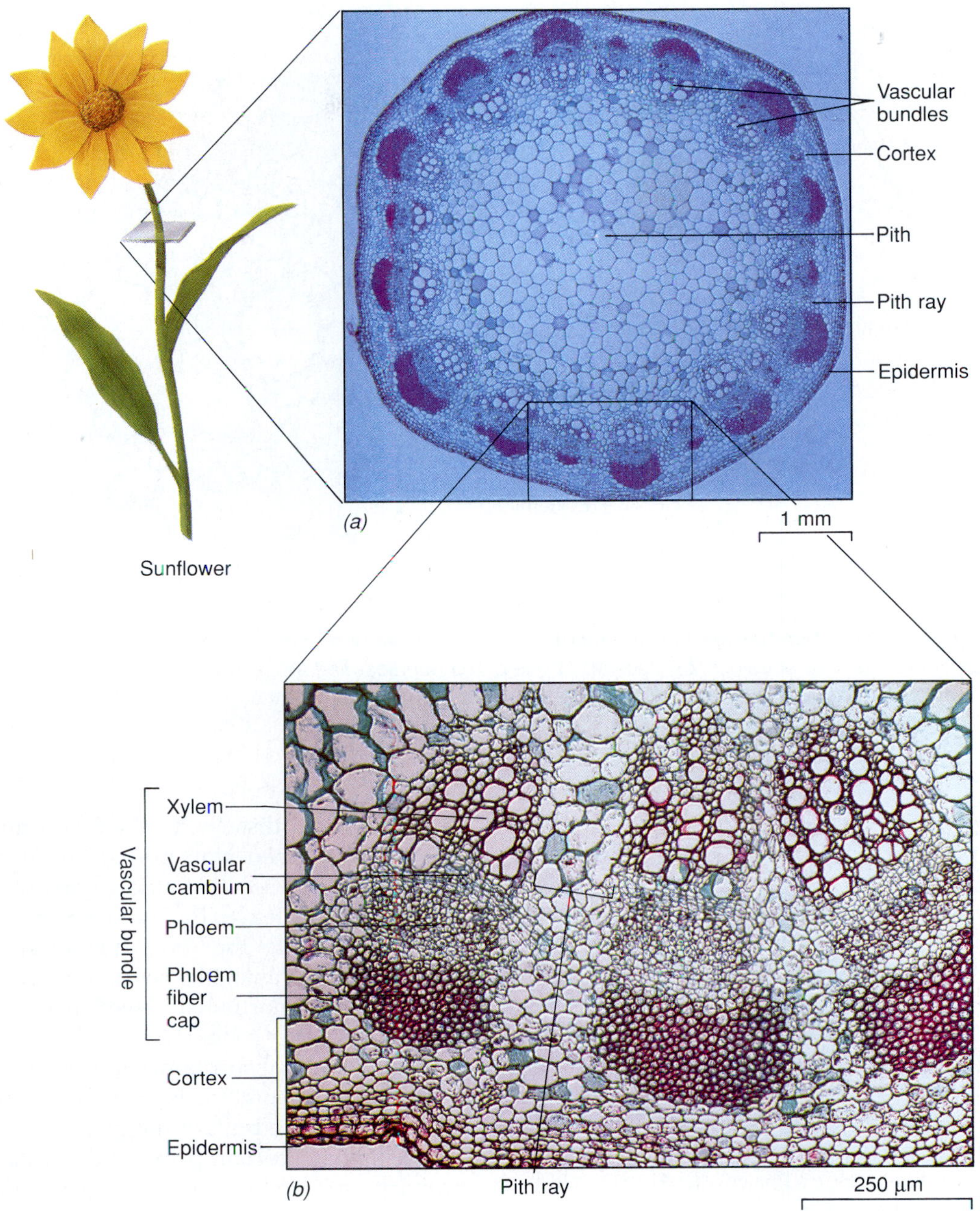

FIGURE 28–2 Primary tissues in a dicot stem. (***a***) Cross section of a sunflower stem showing the organization of tissues. The vascular bundles are arranged in a circle. (***b***) Close-up of vascular bundles. The xylem is located toward the stem's interior and the phloem to the outside. Each vascular bundle is "capped" by a batch of fibers for additional support. (*a*, Dennis Drenner; *b*, Ed Reschke)

Monocot stems, such as a corn stem, are also covered by an epidermis (Fig. 28–3). As in dicot stems, the vascular tissues run in strands throughout the length of the stem. However, in cross section, these vascular bundles are not arranged in a circle but are scattered throughout the stem. Therefore, the monocot stem does not have distinct areas of cortex and pith. The tissue in which the vascular tissues are embedded is called **ground tissue.** Each vascular bundle in a monocot stem contains xylem toward the inside and phloem toward the outside. The bundle is enclosed in a bundle sheath of sclerenchyma cells. Vascular cambium does not occur in monocot stems.

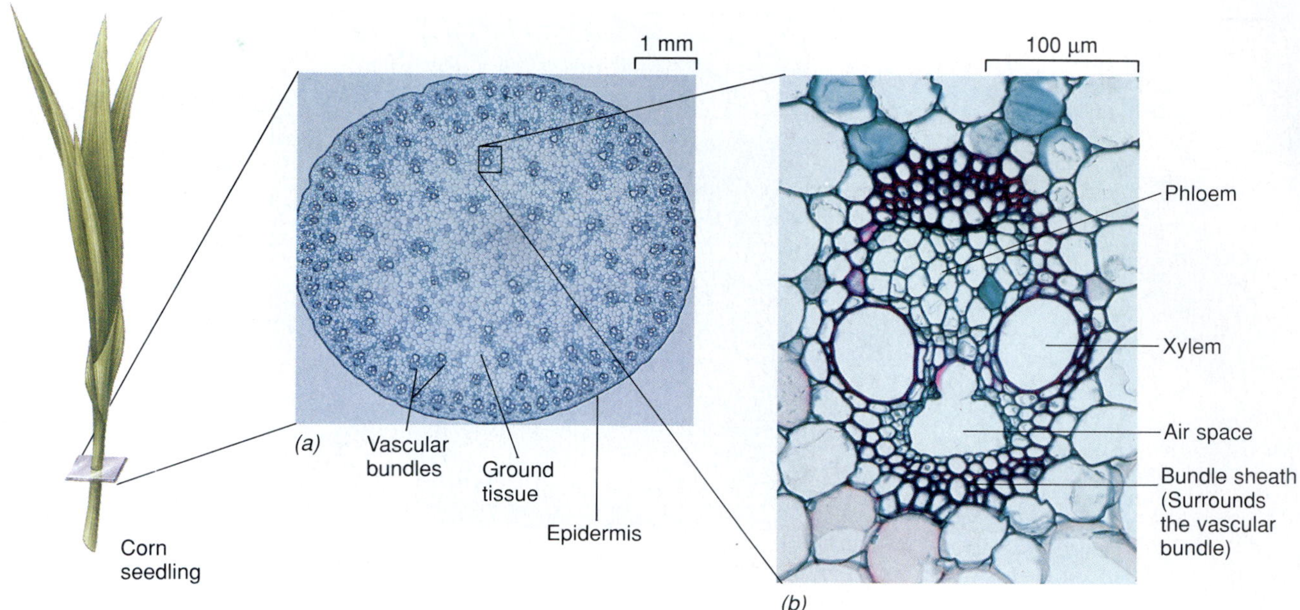

FIGURE 28–3 Arrangement of stem tissues in a monocot. (***a***) Cross section of a corn stem showing the scattered vascular bundles. (***b***) Close-up of one of the bundles. The air space contains remainders of original xylem cells, which were torn apart as elongation of the stem proceeded. The entire bundle is enclosed in a bundle sheath for additional support. (*a*, Carolina Biological Supply Company; *b*, Ed Reschke)

Primary Structure in Stems Is Related to Function

The epidermis is a primary tissue that provides protection in stems. It is covered by the cuticle, a waxy layer that also covers the leaf epidermis. As in leaves, the stem cuticle reduces water loss from the surface of the stem.

In dicots only, the ground tissue in stems is divided into cortex and pith. The cortex in dicot stems and the ground tissue in monocot stems have three functions: photosynthesis, storage, and support. The pith in the center of dicot stems functions primarily for storage.

The vascular tissues function for conduction—xylem transports water and dissolved minerals, and phloem transports dissolved sugar—and support. Because stems grow through the air, they have to be much stronger than roots to support the plant body. Fibers may be found in both xylem and phloem, although they are usually more extensive in phloem. These fibers add considerable strength to the stem body.

ROOTS AND STEMS DIFFER IN THEIR PRIMARY STRUCTURES

Roots are underground and out of sight, so most people do not often think about their importance to plants. Roots are essential plant organs that anchor the plant and absorb water and dissolved minerals from the soil. These materials are then transported throughout the plant. Many roots also serve as storage organs. Surplus sugars produced in the leaves by photosynthesis are transported in the phloem to the roots for storage until they are needed. Some plants, like beets and sweet potatoes, have roots that are enormously swollen with storage tissues (Fig. 28–4). Other roots are modified for functions other than anchorage, absorption, conduction, and storage. For example, certain orchids have aerial photosynthetic roots that comprise the bulk of the plant.

Roots have certain primary tissues also found in the primary structure of stems, such as epidermis, cortex, xylem, phloem, and sometimes pith. Roots also have several tissues and structures not found in stems, including a root cap and root hairs (see Fig. 26–14 on page 518). Each root tip is covered by a **root cap,** a protective layer many cells thick that covers the delicate root apical meristem. As the root grows, pushing its way through the soil, cells of the root cap are sloughed off and replaced by new cells formed by the root apical meristem. **Root hairs** are short-lived extensions of epidermal cells located near the root tip. Root hairs greatly increase the absorptive capacity of the root by increasing the surface area of the root in contact with the moist soil (Fig. 28–5). Table 28–1 summarizes the major differences between dicot stems and roots.

(Text continues on page 546)

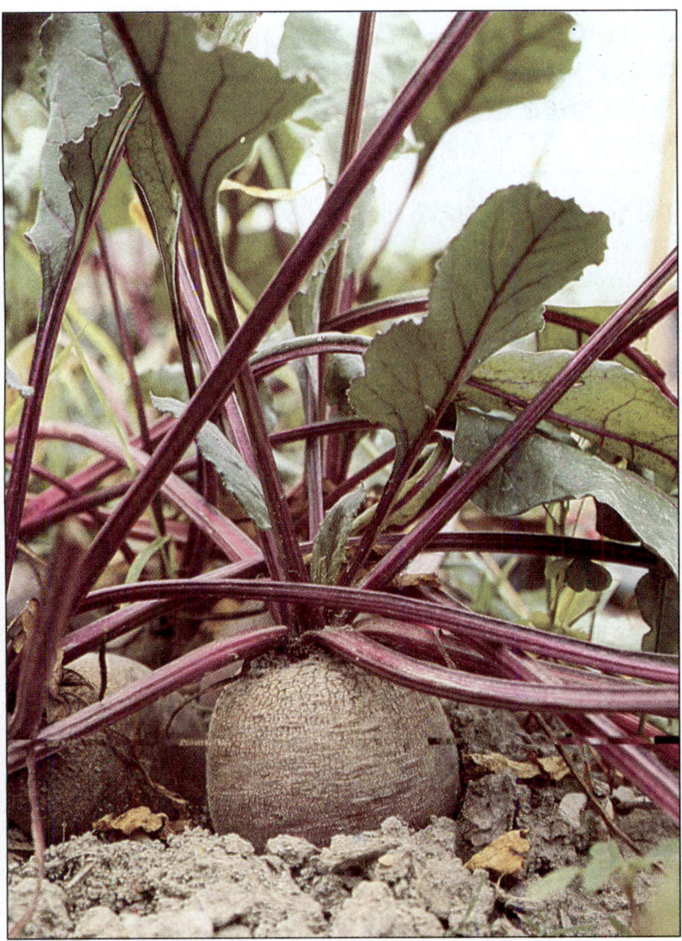

FIGURE 28–4 **Roots as storage organs.** Beets and other root crops are important sources of human food because of the accumulated sugars and starches that they store in their roots. (Dennis Drenner)

TABLE 28–1
General Differences between Dicot Roots and Stems *

Roots	*Stems*
No nodes or internodes	Nodes and internodes
No leaves or buds	Leaves and buds
Nonphotosynthetic	Photosynthetic
No pith	Pith
No cuticle	Cuticle
Root cap	No cap
Root hairs	Trichomes
Pericycle	No pericycle
Endodermis	Endodermis rare
Branches form internally from the pericycle	Branches form externally from lateral buds

* Some exceptions to these general differences exist.

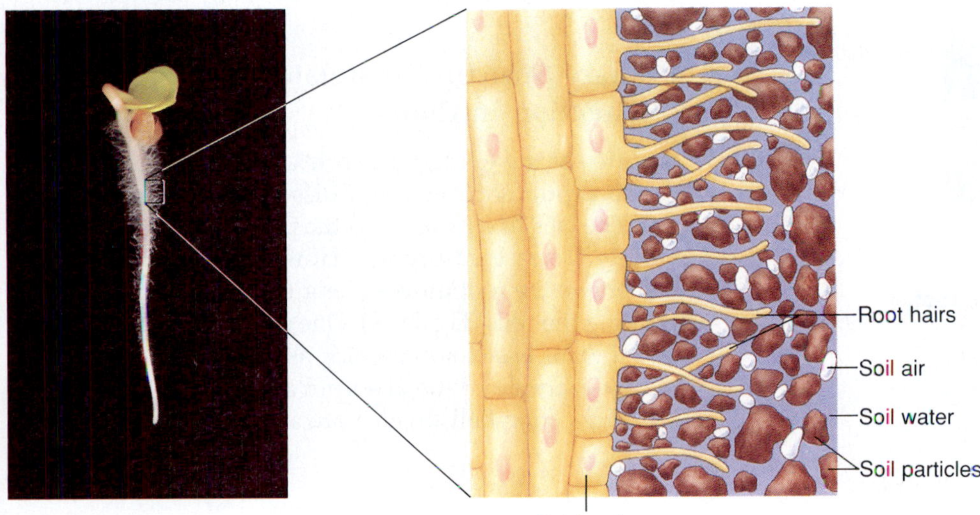

FIGURE 28–5 **Root hairs on a radish seedling.** Each delicate hair is an extension of a single epidermal cell. Root hairs increase the surface area of the root in contact with the soil. (Dennis Drenner)

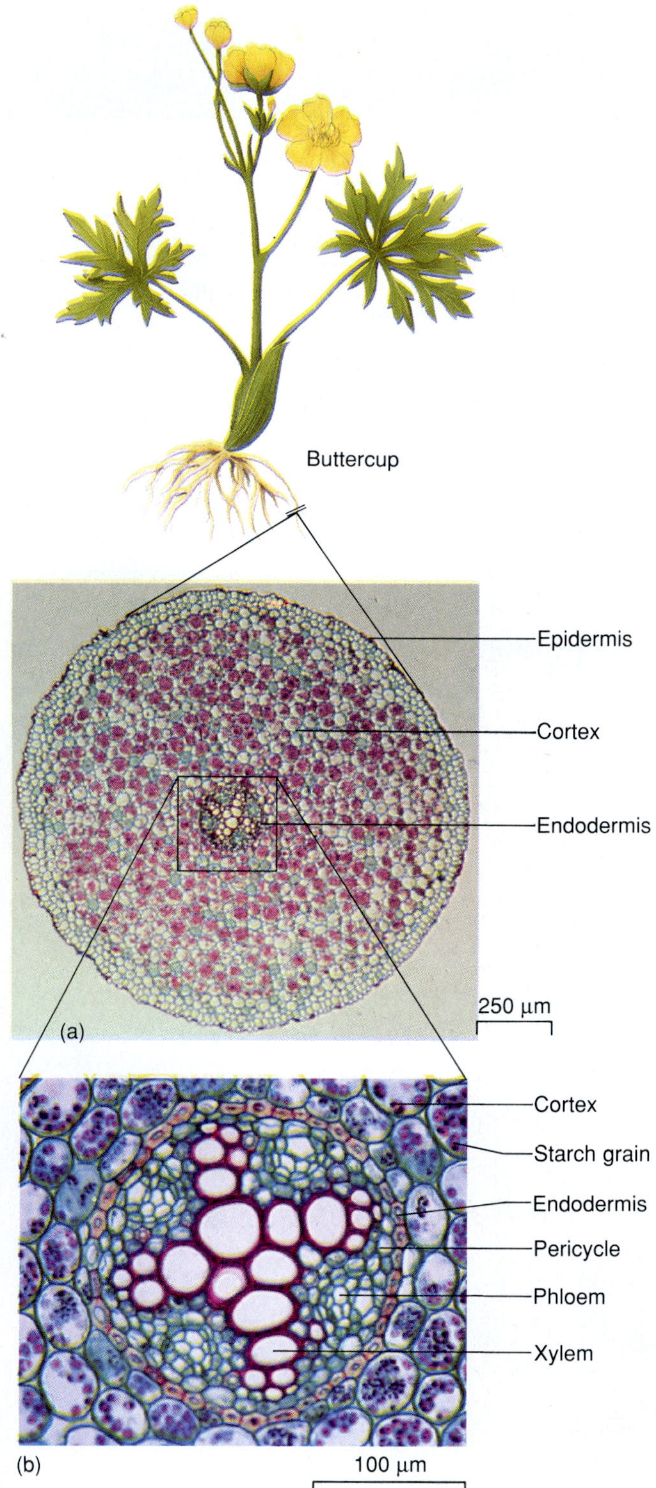

FIGURE 28–6 Primary tissues in a dicot root. (***a***) Cross section of a buttercup root. Note that the bulk of the root is the cortex. (***b***) A close-up of the center of the root. Note the solid core of vascular tissues. (*a, b*; Ed Reschke)

THE PRIMARY STRUCTURE OF DICOT AND MONOCOT ROOTS CAN BE DISTINGUISHED BY THE ARRANGEMENT OF VASCULAR TISSUE

The buttercup root is a representative dicot root with primary growth (Fig. 28–6). Like other parts of the plant, dicot roots are covered by a single layer of protective tissue, the epidermis. The root hairs are a modification of the root epidermis that enables it to absorb more water from the soil. The root epidermis does not secrete a thick waxy cuticle (which would impede the absorption of water from the soil) in the region of root hairs.

The cortex of a dicot root is primarily composed of loosely arranged parenchyma cells. The inner layer of the cortex, the **endodermis,** is different from the rest of the cortex. Endodermal cells fit snugly against one another, and each cell has a special bandlike region on its radial and transverse walls, called a **Casparian strip** (Fig. 28–7). The Casparian strip contains **suberin,** a fatty material that is waterproof. Just inside the endodermis is a single layer of cells called the **pericycle.** The pericycle is composed of cells that remain meristematic. The center region of a dicot root is occupied by vascular tissue. Xylem is found most central and often has two, three, four, or more extensions, or "arms." Phloem is located in patches between the xylem arms. The vascular cambium is sandwiched between the xylem and phloem. Because it possesses an inner core of vascular tissue, the dicot root lacks pith.

The tissues in monocot roots are basically the same as in the dicot root, but they are arranged in a slightly different manner in the center of the monocot root (Fig. 28–8). Starting at the outside of the monocot root, there is epidermis, then cortex, endodermis, and pericycle. However, the vascular tissues inside the pericycle do not form a solid cylinder in the center of the root. Instead, the phloem and xylem are located in separate patches arranged in a circle around the centrally located pith.

Primary Structure Is Related to Function in Roots

Features of the epidermis, cortex, and endodermis aid in the absorption of water and dissolved minerals from the soil. The lack of a cuticle and the presence of root hairs obviously increase absorption. However, most of the water that enters the root moves along the cell walls rather than entering the cells (Fig. 28–9). One of the major components of cell walls is cellulose, which absorbs water like a wick. (As an example of the absorptive properties of cellulose, consider cotton balls, which are almost pure cellulose.)

(Text continues on page 548)

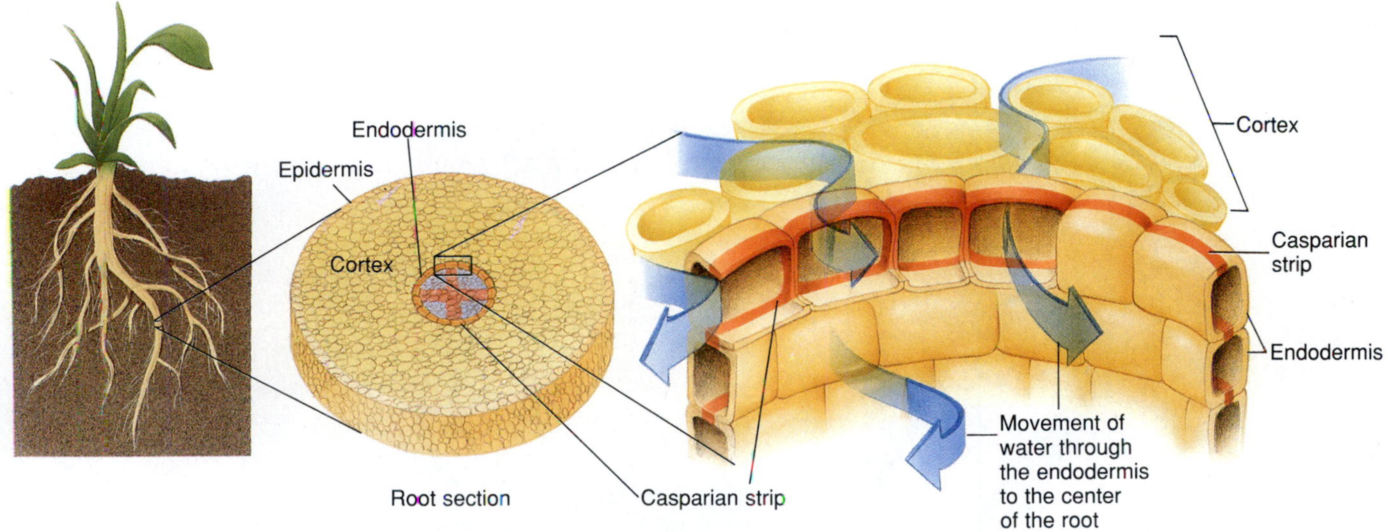

FIGURE 28–7 A few cells of the endodermis. Note the waterproof Casparian strip around the radial and transverse walls. The endodermis controls water uptake by the root.

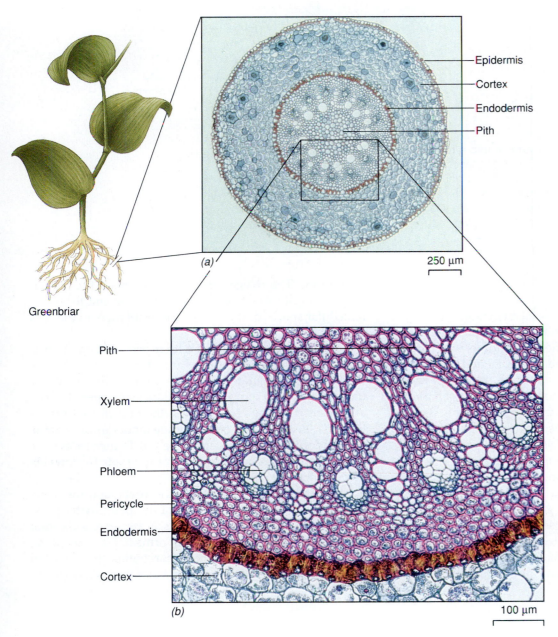

FIGURE 28–8 Arrangement of root tissues in a monocot. (***a***) Cross section of a greenbriar root. (***b***) Close-up of a portion of the center of the root, showing the vascular tissues and the pith. (*a*, Dennis Drenner; *b*, Carolina Biological Supply Company)

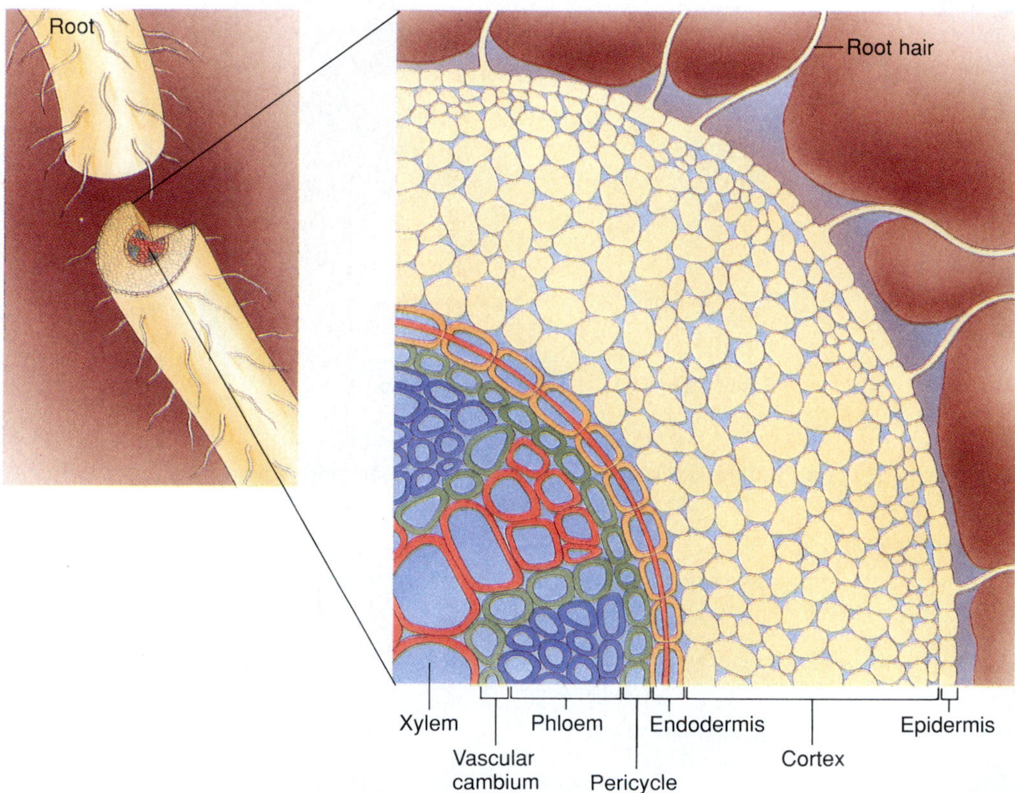

FIGURE 28–9 Pathway of water movement into a root. Most water that enters the root travels along the cell walls and intercellular spaces of the epidermis and cortex. When water reaches the endodermis, it can only continue to move into the root's center if it passes through the plasma membrane and enters an endodermal cell. (The Casparian strip blocks the passage of water along the endodermal cell wall.) Once in the xylem, water moves upward to the rest of the plant.

When water enters the root cortex, it moves along the cell walls and intercellular spaces until it reaches the endodermis. Up to this point, most of the water has never passed through a plasma membrane or entered the cytoplasm of a root cell. The endodermis, with its waterproof Casparian strip running around the radial and transverse walls, blocks water movement along the cell walls. Thus, water can only enter the endodermal cells by passing through their plasma membranes. For this reason, the endodermis controls the movement of water into the root, even though it is an internal tissue and water must pass through other tissues to reach the endodermis.

The principal function of the root cortex is storage. A microscopic examination of the cells that form the cortex often reveals numerous starch grains. Starch, which is an insoluble carbohydrate composed of glucose units, is the most common form of food storage in plants. The large intercellular spaces, another feature of root cortex, provide a pathway for water uptake and allow for aeration of the root. The oxygen that root cells need for cellular respiration diffuses from air spaces in the soil to the intercellular spaces of the cortex, and from there to the cells of the root.

The pericycle, associated with meristematic activities, is the origin of multicellular branch roots (Fig. 28–10). Branch roots originate when a portion of the pericycle, usually at the tip of a xylem arm, becomes meristematic and starts dividing. As it grows, the branch root breaks through several layers of root tissue (endodermis, cortex, and epidermis) before entering the soil. Branch roots have all of the structures and features of the roots from which they branch.

The xylem and phloem of the root have the same functions as they do in the rest of the plant. After passing through the endodermal cells, water enters the root xylem, often at one of the xylem arms. Up to this point, the pathway of water has been horizontal, from the soil into the center of the root. Once water enters the xylem,

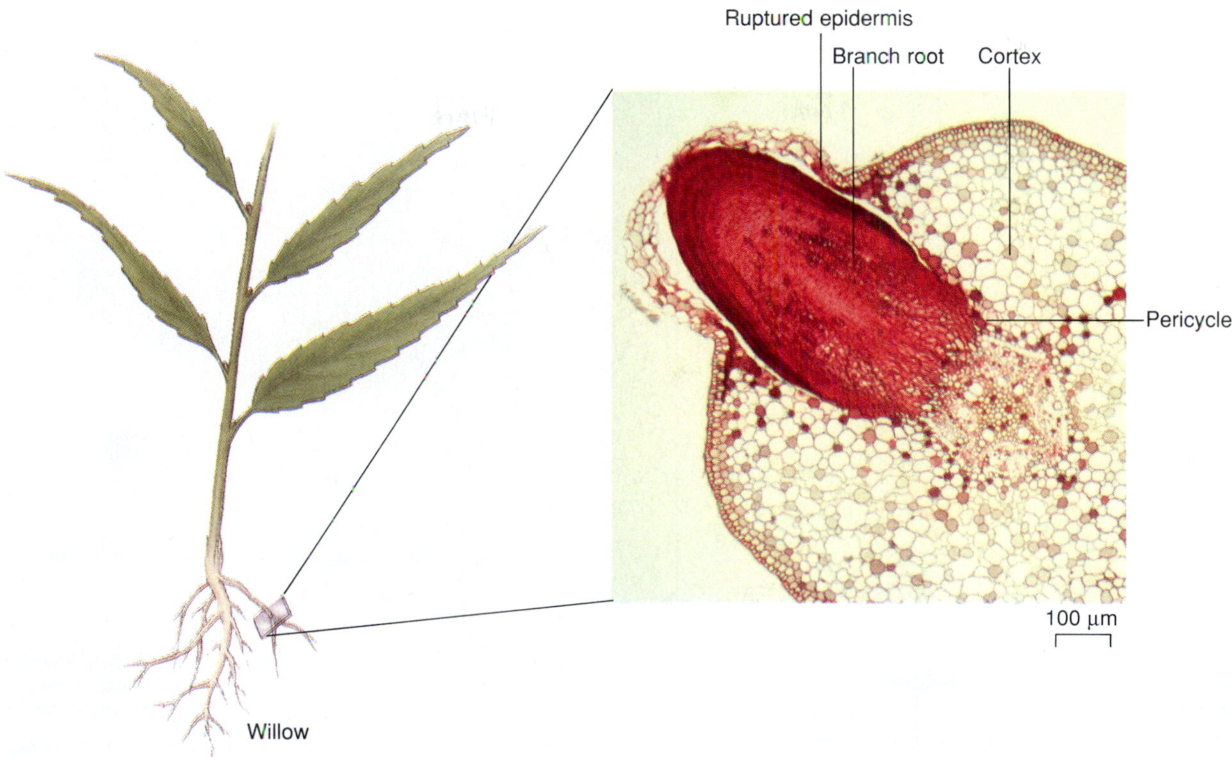

FIGURE 28–10 Branch roots. A multicellular branch root, which originates at the pericycle, emerges from the main root. (James Mauseth, University of Texas)

it is transported upward through root xylem into stem xylem and throughout the rest of the plant. Phloem conducts dissolved sugar (sucrose) to the root, where it is stored (usually as starch), or from the root to other parts of the plant, where it is used.

WOODY PLANTS HAVE STEMS AND ROOTS WITH SECONDARY GROWTH

Secondary growth, an increase in the width of stems and roots, occurs in certain dicots (as well as gymnosperms) as a result of the activity of two lateral meristems, the vascular cambium and cork cambium (Chapter 26). The new tissues formed by the lateral meristems are called secondary tissues.

Cells in the vascular cambium divide and produce secondary xylem (wood) and secondary phloem (inner bark), which become the functional replacements for primary xylem and primary phloem. When a cell in the vascular cambium divides, one of the daughter cells remains meristematic, that is, it remains as a part of the vascular cambium. The other cell may divide again several times, but it eventually develops into mature secondary tissue. Cells in the vascular cambium divide tangentially (that is, perpendicular to the radius) to produce tissues in two directions. The cells formed from the dividing vascular cambium are located either *inside* the ring of vascular cambium (the secondary xylem) or *outside* it (the secondary phloem; Fig. 28–11).

The second lateral meristem, the **cork cambium,** divides to produce cork cells and cork parenchyma. Cork cambium and the tissues it produces are collectively referred to as **periderm** (outer bark), which functions as a replacement for the epidermis.

Among flowering plants, only woody dicots, such as apple, hickory, and maple, have secondary growth. Gymnosperms such as pine, juniper, and spruce also have secondary growth. Plants with secondary growth also produce primary growth. That is, as a woody plant increases in length (by primary growth) at the tips of its stems and roots, the older parts of the plant further back from the tips develop secondary tissues.

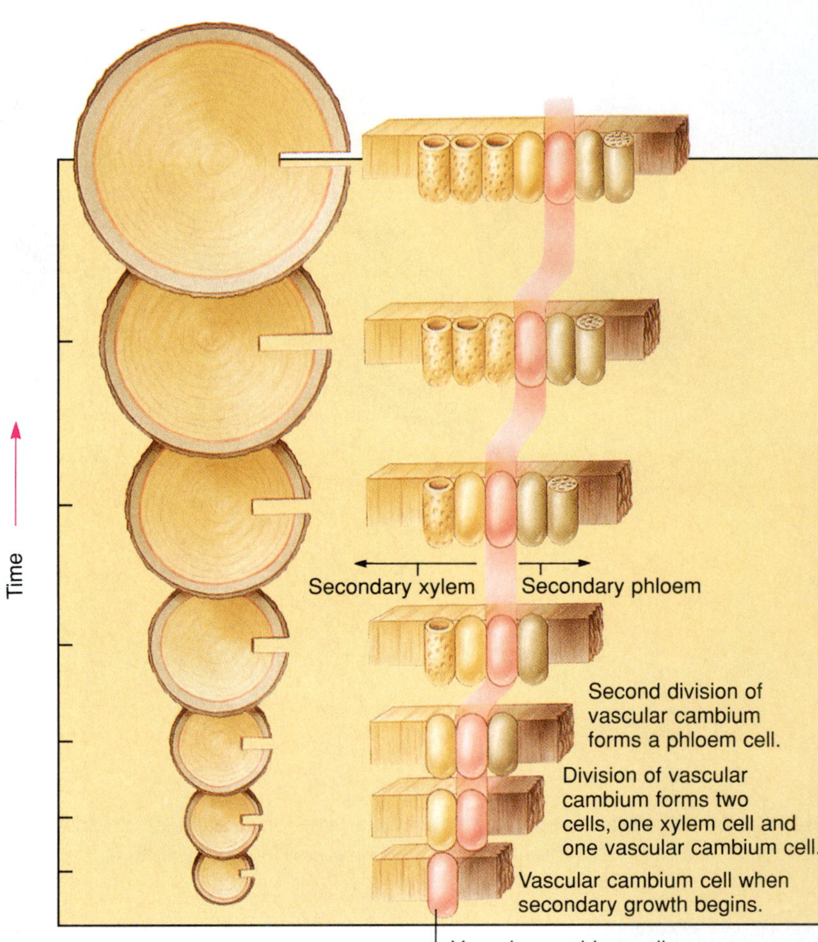
Time
Secondary xylem
Secondary phloem
Second division of vascular cambium forms a phloem cell.
Division of vascular cambium forms two cells, one xylem cell and one vascular cambium cell.
Vascular cambium cell when secondary growth begins.
Vascular cambium cell

FIGURE 28–11 Radial view of a dividing vascular cambium cell. Note that the vascular cambium divides in two directions, forming secondary xylem to the inside and secondary phloem to the outside. These cells differentiate to form the mature cell types associated with xylem and phloem. As secondary xylem accumulates, the vascular cambium moves outward, and the woody stem increases in width. To study the figure, start at the bottom and move up.

TRANSPORT IN PLANTS OCCURS IN XYLEM AND PHLOEM

The movement of sugar, water, and minerals within a vascular plant is called **translocation.** Water and minerals are translocated in xylem, whereas dissolved sugar is translocated in phloem. Translocation in plants does not resemble the transport of materials in animals because nothing *circulates* in a system of vessels in plants. Water and minerals that are translocated in the xylem travel in one direction only (upward), whereas movement of dissolved sugar may occur upward or downward in separate phloem cells. Also, translocation in plants differs from internal transport in animals because plants lack a heart. In plants, translocation in both the xylem and phloem is driven largely by natural physical processes rather than by a pumping organ.

Water and Minerals Are Conducted in Xylem

Water and dissolved minerals form a dilute solution that moves within tracheids and vessel elements, which are hollow, dead cells in the xylem tissue (Chapter 26). The movement of water in the xylem is the most rapid of any transport in plants. On a hot summer day, water has been measured moving upward in xylem at 60 centimeters (2 feet) per minute. How does water move to the tops of plants? It might be either pushed up from the bottom of the plant or pulled up from the top of the plant. Actually, although both mechanisms exist, current evidence indicates that water is transported through the xylem by being *pulled* from the top of the plant.

In the **tension-cohesion mechanism,** water is pulled up the plant as a result of a tension produced at the top of the plant. The tension, which resembles the tension produced by sucking a liquid up a straw, is caused by the evaporative pull of transpiration (Chapter 27), which in turn is driven by solar energy (Fig. 28–12). The tension extends from the leaves down the stem into the roots. It pulls water from root cells into the root xylem and draws water up the xylem to leaf cells that have lost water as a result of transpiration. As water is pulled upward from the roots, additional water from the soil is drawn into the roots.

This upward pulling of water is only possible as long as there is a solid, unbroken column of water in the xylem.

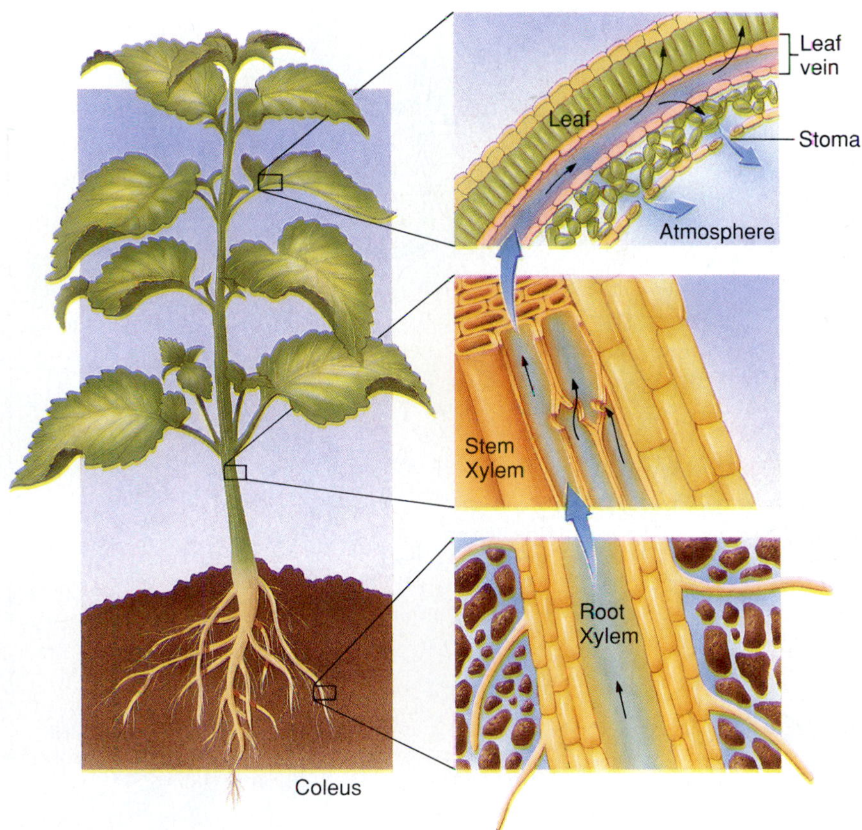

FIGURE 28–12 The tension-cohesion mechanism. The evaporative-pull of transpiration draws water up the plant from the soil to the atmosphere. (*Top*) Water vapor diffuses from the surfaces of leaf mesophyll cells to the drier atmosphere through the stomata. This produces a tension that pulls water out of the leaf xylem toward the mesophyll cells. (*Middle*) The cohesion of water molecules, caused by hydrogen bonding, allows water to be pulled up the narrow vessels and tracheids of the xylem in the stem. (*Bottom*) This in turn pulls water up the xylem of the root, which forms a continuous column of water, from root xylem to stem xylem to leaf xylem. As water moves upward in the root xylem, it produces a pull that causes soil water to diffuse into the root.

Water tends to form an unbroken column because of the cohesiveness of water molecules. (Recall that water molecules are strongly attracted to one another because of hydrogen bonding; see Chapter 3). Also, the adhesion of water to the walls of the xylem cells is an important factor in maintaining an unbroken column of water. Thus, the cohesive and adhesive properties of water enable it to form an unbroken column, which can be pulled.

Is the tension-cohesion mechanism powerful enough to explain the rise of water in the tallest plants? Plant biologists have calculated that the tension produced by transpiration is strong enough to pull water 152 meters (500 feet) in tubes the diameter of xylem vessels. Because the tallest trees are approximately 107 meters (350 feet) high, the tension-cohesion mechanism easily accounts for their water transport.

In the less important mechanism of water transport, known as the **root pressure mechanism,** water that moves into the roots from the soil is *pushed* up the xylem toward the top of the plant. Water moves into the roots by osmosis (Chapter 6), and the accumulation of this water in root cells produces a pressure that forces the water up the xylem. Root pressure is a real phenomenon in plants. For example, guttation (the release of liquid water from leaves, forced out through special openings by root pressure) is a manifestation of this process (Chapter 27). However, plant biologists have measured root pressure and found that it is not a strong enough force to explain the rise of water to the tops of tall trees. Root pressure exerts an influence in smaller plants, particularly in the spring when the soil is very wet, but it clearly does not cause water to rise one hundred meters or more in the tallest plants. Further, root pressure does not occur to any appreciable extent in summer (when water is often not plentiful in the soil), yet water transport is greatest during hot summer days.

Sugar in Solution Is Translocated in Phloem

The sugar produced during photosynthesis is converted into the disaccharide sucrose (common table sugar) before being loaded into the phloem and translocated to the rest of the plant. Translocation in phloem, which is not as swift-moving as xylem transport, has been measured at approximately 2.5 centimeters (1 inch) per minute.

Movement within phloem tissue is both upward and downward. Sucrose is transported in the phloem from the **source,** an area of excess sugar supply (usually a leaf), to the **sink,** an area of storage or metabolism (such as roots, apical meristems, fruits, or seeds).

The **pressure-flow hypothesis** explains how dissolved sugar is translocated in phloem. Phloem transport

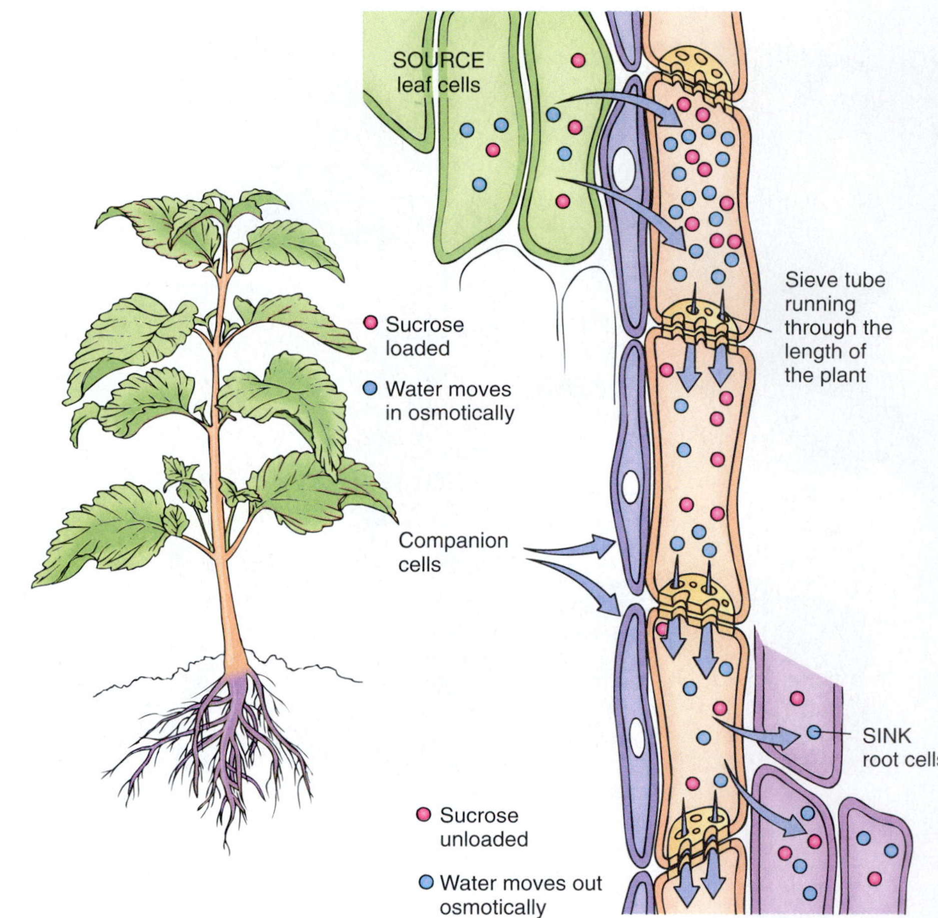

FIGURE 28–13 Pressure-flow hypothesis for phloem transport. Sugar is actively loaded into the sieve-tube member at the source. As a result, water moves osmotically into the sieve-tube member. At the sink the sugar is actively unloaded, and water leaves the sieve-tube member by osmosis. The gradient of sugar from source to sink causes transport through the sieve tube from the area of higher pressure (the source) to the area of lower pressure (the sink).

is due to a pressure gradient (that is, a difference in pressure) between the source, where the sugar is loaded into the phloem, and the sink, where the sugar is removed from the phloem (Fig. 28–13).

At the source, the dissolved sucrose moves from the mesophyll cells where it was manufactured into the companion cells of the phloem by active transport. ATP is required for movement of sugar across the plasma membrane (Chapter 6). Once the sugar is in the companion cell, it readily moves into the sieve-tube member through the many cytoplasmic connections between the two cells. The increase in dissolved sugars in the sieve tubes causes water to move by osmosis into the sieve tubes, increasing the pressure inside the sieve tubes. This pressure pushes the sugar solution through the phloem much as water is forced through a hose. At its destination at the sink, sugar is actively unloaded from the sieve-tube elements, with ATP again being required. With a loss in sugar, water moves out of the sieve tubes by osmosis and into surrounding cells. This decreases the pressure inside the sieve-tube elements in the sink.

Thus, the pressure-flow hypothesis explains the movement of dissolved sugar in phloem by means of a pressure gradient. It is the difference in sugar concentrations between the source and the sink that causes transport in the phloem, as water and dissolved sugar flow down the pressure gradient.

The actual translocation of dissolved sugar in the phloem does not require metabolic energy. However, both loading sugar at the source and unloading sugar at the sink require energy derived from ATP. (The ATP energy is used to transport the sugar across cell membranes.)

Roots selectively absorb minerals

Minerals are available to plants as ions dissolved in water. The concentrations of various minerals are different in the xylem sap (fluid contents) than in soil water, and it appears that plants selectively accumulate the mineral ions that they require. Most minerals are thought to enter the root by passing through the plasma membranes of epidermal cells. Mineral ions then travel through the root tissues from cell to cell, rather than along cell walls as does water.

Dissolved mineral ions pass through plasma membranes by active transport. In active transport, the mineral ions move against the concentration gradient (that is, from

FOCUS ON *Health and Human Affairs*

Commercial Hydroponics

Hydroponics, the practice of growing plants in an aerated solution of chemically defined mineral salts, has been used by scientists to determine which elements are essential for plant growth. Initially, entrepreneurs hailed hydroponics as the scientific way to grow plants in places where soil was poor or unavailable. However, the expenses involved in commercially growing produce for human consumption prevented hydroponics from becoming more than a curiosity. Recent technical improvements have revived an interest in commercial hydroponics (see figure).

Hydroponics has great potential in several places. It is being tried experimentally in desert countries in the Middle East, where the soil is too arid to support cultivation and water is unavailable for irrigation. When plants are grown hydroponically in greenhouses, less water is used compared with traditional agriculture. Hydroponics is also being tried in temperate climates, particularly to produce crops in winter.

Greenhouse with lettuce growing hydroponically. *(Hank Morgan/Photo Researchers, Inc.)*

Hydroponics has several advantages. First, it is possible to grow crops hydroponically under conditions in which disease-causing microorganisms and insect pests are completely absent. This means that the crops are not exposed to chemical pesticides. Also, hydroponics can be used to grow crops near their area of use, saving on transportation costs. (On average, most foods that Americans consume travel 1300 miles.)

The main disadvantage of hydroponics is the expense. Plants grown hydroponically must be supplied with a nutrient solution that is continually monitored and adjusted. Heating and lighting costs are high. Aeration of the roots was a major expense until developments like the nutrient-film technique helped cut costs. In the nutrient-film technique, plants are grown in plastic trenches through which a thin layer (film) of nutrient solution trickles. In this way, the roots get adequate aeration. The nutrient solution is saved and reused, which cuts down on water and mineral costs.

Although hydroponics will probably never replace traditional agriculture, it has been shown to be a viable alternative in certain situations. As new techniques are developed, hydroponics may become even more common.

an area of *low* concentration to an area of *high* concentration of that mineral) through special channels in the membrane. One of many reasons that root cells require sugars for cellular respiration is that active transport of mineral ions across biological membranes requires the expenditure of energy, usually in the form of ATP (Chapter 6).

Sixteen Elements Are Essential for Plant Growth

More than 90 naturally occurring elements exist on Earth, and more than 60 of these, including elements as common as carbon and as rare as gold, have been found in plant tissues. Not all of these elements are considered essential for plant growth, however. How does a biologist determine whether an element is essential? If a biologist suspects that a particular element is essential for plant growth, he or she grows plants in a nutrient solution that contains all known essential elements except the one in question. If plants grown in the absence of that element are unable to develop normally or to complete their life cycle, the element may be essential. Additional criteria are used to confirm whether an element is essential. For example, the element must be shown to have a direct effect on the metabolism of the plant. Also, the element must be demonstrated to be essential in a wide variety of plants.

One of the most useful methods to study plant nutrition is **hydroponics,** which is the growing of plants in aerated water with dissolved mineral salts. (Hydroponics has other applications in addition to its scientific use; see Focus On Health and Human Affairs: Commercial Hydroponics.) It is impossible to conduct mineral nutri-

CONCEPT CONNECTIONS

Nitrogen in Soil ⇌ *Interdependence of Living Organisms*

Because nitrogen is an essential part of biologically important molecules such as proteins and nucleic acids (Chapter 4), all living organisms must have nitrogen in order to survive. Animals, including humans, get their nitrogen from proteins and other nitrogen-containing compounds in the foods they eat. Ultimately, this nitrogen is traced back to plant sources.

Plants obtain their nitrogen in a much simpler form from the soil—as nitrate (NO_3^-) and ammonium (NH_4^+) ions. But where do the nitrate and ammonium ions come from? Nitrogen is converted into those forms from atmospheric nitrogen (N_2) by a few species of prokaryotes, that is, bacteria and cyanobacteria (Chapter 21). Thus, the nitrogen supply of the entire biosphere, including humans, comes from a few, at first glance insignificant, organisms. If these bacteria were to become extinct or substantially decline in number, the supply of nitrogen in a form suitable for plants would be rapidly depleted, and plants would die. With the demise of plants, animals, including humans, would be unable to survive. We say more about the crucial role of microorganisms in the nitrogen cycle in Chapter 47.

tion experiments by growing plants in soil because soil is too complex and contains too many elements. However, one can grow plants in a solution of water and all known required mineral salts.

Sixteen elements have been demonstrated to be essential for plant growth. Nine of these are required in fairly large quantities (greater than 0.05% dry weight) and are therefore known as **macronutrients.** These include carbon, hydrogen, oxygen, nitrogen, phosphorus, potassium, sulfur, calcium, and magnesium. The remaining seven **micronutrients** are needed in very small (trace) amounts for normal plant growth and development. These include iron, boron, manganese, copper, molybdenum, chlorine, and zinc.

Four of the 16 elements—carbon, oxygen, hydrogen, and nitrogen—come from water or gases in the atmosphere. Carbon is obtained from carbon dioxide (CO_2) in the atmosphere by photosynthesis. Oxygen is obtained from oxygen gas (O_2) and water. Water also supplies hydrogen to the plant. Plants get their nitrogen from the soil as ions of nitrogen salts—nitrate (NO_3^-) and ammonium (NH_4^+)—but nitrogen is converted into those forms from atmospheric nitrogen (N_2) by various microorganisms in the soil or in root nodules of plants (Chapter 47). The remaining 12 essential elements are obtained from the soil as dissolved mineral ions. Their ultimate source is the rock from which the soil was formed. Some of the roles of the essential elements are summarized in Table 28–2.

TABLE 28–2
Functions of Essential Elements

Element	*Major functions*
Carbon	In carbohydrate, lipid, protein, and nucleic acid molecules
Hydrogen	In carbohydrate, lipid, protein, and nucleic acid molecules
Oxygen	In carbohydrate, lipid, protein, and nucleic acid molecules
Nitrogen	In proteins, nucleic acids, chlorophyll, certain coenzymes
Phosphorus	In nucleic acids, phospholipids, ATP (energy transfer compound)
Calcium	In cell walls; role in membrane permeability; enzyme activation
Magnesium	In chlorophyll; enzyme activator in carbohydrate metabolism
Sulfur	In certain amino acids and vitamins
Potassium	Osmosis and ionic balance; opening and closing of stomata; enzyme activator (for 40+ enzymes)
Chlorine	Ionic balance; involved in photosynthesis
Iron	Part of enzymes involved in photosynthesis, respiration, and nitrogen fixation
Manganese	Part of enzymes involved in respiration and nitrogen metabolism; required for photosynthesis
Copper	Part of enzymes involved in photosynthesis
Zinc	Part of enzymes involved in respiration and nitrogen metabolism
Molybdenum	Part of enzymes involved in nitrogen metabolism
Boron	Exact role unclear; involved in membrane transport and calcium utilization

Concept Connections

Soil Formation ⟷ Living Organisms

Soil is the complex material in which plants are anchored and from which they receive their water and essential mineral elements. Soils are formed from rock that is gradually broken down into smaller and smaller particles by biological, chemical, and physical weathering processes. Except for nitrogen, the minerals that a plant receives from the soil are obtained from this weathered parent rock. Because different rocks are composed of different minerals, soils vary in their mineral composition.

Two very important factors that sometimes work together in the weathering of rock are climate and living organisms. For example, lichens* growing on exposed rock produce acids that etch tiny cracks, or fissures, in the rock surface. Water seeps into these cracks. During winter, the alternate freezing and thawing of water cause the cracks to enlarge, breaking off small pieces of the rock. Small plants can then become established and send their roots into the larger cracks, fracturing the rock further. The conversion of solid rock into soil takes a very long period of time, usually thousands of years. We say more about the role of living organisms in soil formation in Chapter 46.

*Recall from Chapter 22 that a lichen is a dual organism composed of a fungus and an alga.

Chapter Summary

I. The main functions of stems are support, conduction, and production of new stem tissue.

II. Stems with primary growth have an epidermis, vascular tissue, and cortex and pith, or ground tissue. The epidermis is a protective covering; the xylem conducts water and dissolved minerals; the phloem conducts dissolved sugar; the cortex, pith, and ground tissue function primarily for storage.
   A. Dicot stems have the vascular bundles arranged in a circle (in cross section) and have a distinct cortex and pith.
   B. Monocot stems have scattered vascular bundles that do not divide the ground tissue into a distinct cortex and pith.

III. Anchorage, absorption, conduction, and storage are the main functions of roots.

IV. Roots with primary growth have an epidermis, cortex, endodermis, pericycle, xylem, phloem, and sometimes pith or vascular cambium. The epidermis, cortex, vascular tissues, and pith have the same functions that they do in stems; the endodermis controls the movement of water into the root; the pericycle is the origin of branch roots.
   A. Dicot roots have a solid core of vascular tissue (that is, no pith) and possess a vascular cambium between the xylem and phloem tissues.
   B. Monocot roots often have a pith and do not possess a vascular cambium.

V. Secondary growth occurs in stems and roots of woody dicots and all gymnosperms.
   A. The vascular cambium produces secondary xylem (wood) to the inside and secondary phloem (inner bark) to the outside.
   B. The cork cambium produces cork parenchyma to the inside and cork cells to the outside. These tissues make up the periderm or outer bark of the secondary plant body.

VI. Water and dissolved minerals move upward in the xylem from the root to the stem to the leaves.
   A. The tension-cohesion mechanism causes the rise of water in even the largest plants.
      1. The evaporative pull of transpiration (powered by the sun) causes a tension at the top of the plant.
      2. The column of water that is pulled up through the plant is unbroken as a result of the cohesive and adhesive properties of water.
   B. Root pressure, caused by the movement of water into the root from the soil, helps explain the rise of water in small plants, particularly during the spring when the soil is wet.

VII. Dissolved sugar is translocated upward or downward in the phloem.
   A. Sucrose is the main form of food transported in the phloem.
   B. Translocation in the phloem is explained by the pressure-flow hypothesis.
      1. Sugar is actively loaded (requiring ATP) into the sieve tubes at the source. As a result, water moves into these sieve tubes by osmosis, increasing the pressure inside the sieve tubes at the source.

2. Sugar is actively unloaded (requiring ATP) from the sieve tubes at the sink. As a result, water leaves these sieve tubes by osmosis, decreasing the pressure inside the sieve tubes at the sink.
3. The flow of materials between the source and sink is driven by the pressure gradient produced by water entering the phloem at the source and leaving the phloem at the sink.

VIII. Sixteen elements are essential for normal plant growth.
A. Nine elements are macronutrients: carbon, oxygen, hydrogen, nitrogen, potassium, phosphorus, sulfur, magnesium, and calcium.
B. Seven elements are micronutrients: iron, boron, manganese, copper, zinc, molybdenum, and chlorine.

Selected Key Terms

cortex, p. 542
endodermis, p. 546
epidermis, p. 542
hydroponics, p. 553
pericycle, p. 546
periderm, p. 549
phloem, p. 542
pith, p. 542
pressure-flow hypothesis, p. 551
root cap, p. 544
root hair, p. 544
root pressure mechanism, p. 551
tension-cohesion mechanism, p. 550
translocation, p. 550
xylem, p. 542

Post-Test

1. All plants have ______________ growth, whereas only woody dicots and gymnosperms have ______________ growth.
2. Vascular tissue arranged in a circle (in cross section) is characteristic of the primary structure of stems of ______________.
3. Monocot stems lack a distinct pith and ______________.
4. Roots absorb ______________ and dissolved ______________ from the soil.
5. The ______________ ______________ covers the delicate apical meristem of the root tip.
6. The ______________ is the origin of branch roots.
7. The center of a dicot root is ______________, whereas the center of a monocot root is ______________.
8. The two lateral meristems responsible for secondary growth are the ______________ ______________ and the ______________ ______________.
9. The cork cambium and the tissues it produces are collectively called the ______________.
10. The movement of dissolved sugar, water, and minerals within a plant is called ______________.
11. This mechanism of water movement, ______________ ______________, is not strong enough to explain the rise of water to the tops of the tallest trees.
12. In the tension-cohesion mechanism, the tension is produced at the top of the plant by the evaporative pull of ______________.
13. The pressure produced by water entering the phloem at the source transports, or drives, materials from the source to the sink. This phenomenon is known as the ______________ - ______________ hypothesis.
14. ______________ are essential elements required in fairly large quantities.
15. Which of the following is found in dicot stems but *not* monocot stems? (a) scattered vascular bundles (b) vascular cambium (c) xylem (d) epidermis.
16. What is the correct order of tissues from the inside to the outside of a dicot root? (a) xylem ⟶ phloem ⟶ endodermis ⟶ cortex ⟶ pericycle ⟶ epidermis (b) pith ⟶ xylem ⟶ phloem ⟶ cortex ⟶ epidermis (c) xylem ⟶ phloem ⟶ pericycle ⟶ endodermis ⟶ cortex ⟶ epidermis (d) pith ⟶ xylem ⟶ phloem ⟶ pericycle ⟶ cortex ⟶ epidermis.
17. Which of the following is *not* usually considered meristematic? (a) vascular cambium (b) cork cambium (c) apical meristem (d) epidermis (e) all of these are meristematic.
18. Water enters the roots of plants by the process of: (a) guttation (b) pressure flow (c) active transport (d) osmosis.
19. Growing plants in an aerated, liquid mineral solution: (a) is known as hydroponics (b) has been used by plant biologists to help determine which elements are essential for plants (c) is being done on a limited basis commercially (d) all of these.

Review Questions

1. List three functions of stems and describe the tissue(s) responsible for each function.
2. List three functions of roots and describe the tissue(s) responsible for each function.
3. Trace the pathway of water from the soil through the various root tissues.
4. Compare the root's uptake of water from the soil with its uptake of minerals.
5. Label the various primary tissues of this dicot stem. Give at least one function for each of the tissues.

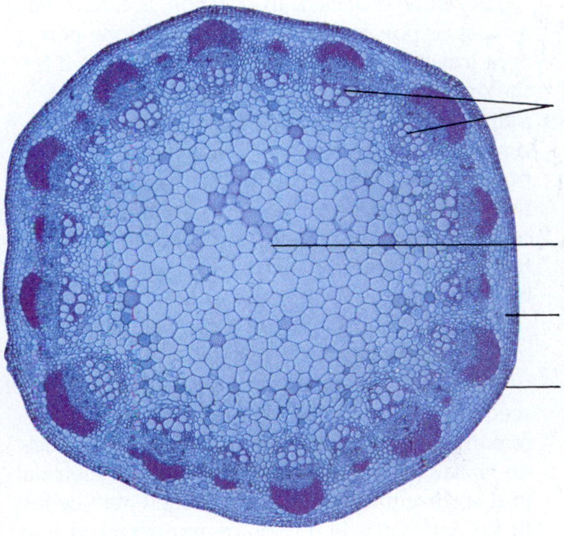

(Dennis Drenner)

6. Explain the tension-cohesion mechanism of water transport. Make sure you consider both the "tension" and "cohesion" aspects of the mechanism.
7. Describe the pressure-flow hypothesis of sugar movement in the phloem, including the activities at the source and sink.
8. What criteria do biologists use to determine which elements are essential for plant growth?
9. Label the various primary tissues of this dicot root. Give at least one function of each of the tissues.

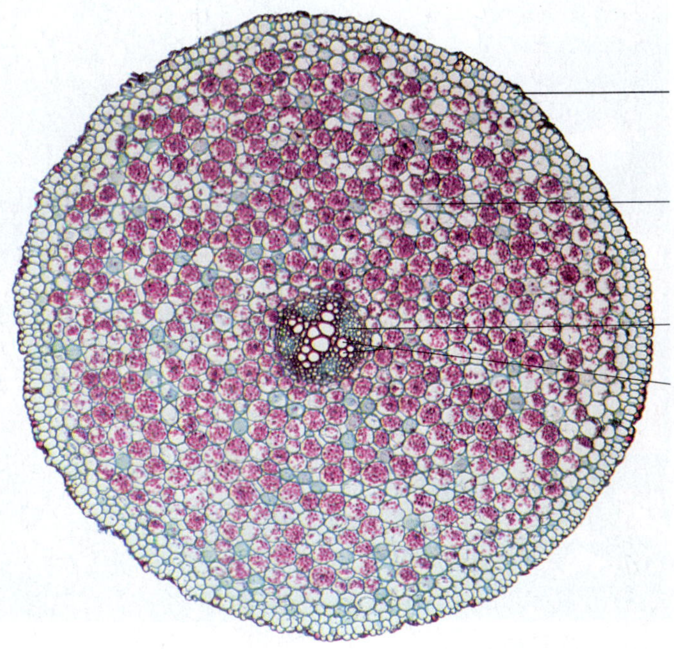

(Ed Reschke)

Thinking Critically

1. Suppose the United Nations appointed you to oversee a 20-year soil reclamation project in the African Sahel. What policies would you implement?
2. Where would you expect to find more fibers, in stems or roots? Why?
3. When secondary growth begins, certain cells dedifferentiate and become meristematic. Could a tracheid ever do this? A sieve-tube member? Why or why not?
4. If you were conducting an experiment to determine whether gold is essential for plant growth, what would you use for an experimental control?
5. Why must hydroponic solutions be aerated?

Recommended Readings

Feldman, L.J., "The Habits of Roots," *BioScience*, Vol. 38, No. 9, October 1988. How roots interact with their soil environment.

Mauseth, J.D., *Botany: An Introduction to Plant Biology*, 2nd ed., Philadelphia, Saunders College Publishing, 1995. A comprehensive introduction to plant biology.

Thybony, S., "Dead Trees Tell Tales," *National Wildlife*, Vol. 25, No. 2, August/September 1987. The potential uses of analyzing the annual rings of tree trunks.

Small vials of seeds from the seed bank in Svalbard, Norway. *(Courtesy of Nordiska Genbanken, Alnarp, Sverige)*

Seed Banks: Treasures in Deep-Freeze

Deep in the frozen soil of an arctic island that is part of Norway lies an international treasury, but this bank does not contain any money, gold, or jewels. Instead, numerous seeds of thousands of different kinds of plants collected from all over the world are stored here, frozen in a giant deep-freeze as a safeguard against future catastrophic plant extinctions.

Most seed collections, called seed banks, help to preserve the genetic variation within different varieties of crops. Farmers typically discontinue planting local varieties when newer, improved varieties become available. The newer varieties have desirable genetic characteristics (a higher yield, for example), but the local, discarded varieties also contain valuable genes that we may want to make use of at some future time. (Although we produce new varieties by crossing plants with desirable traits and therefore desirable genes, we do not have the ability to *make* genes.) Maintaining the genetic diversity present in local crop varieties will help provide genes that we may need in the future.

More than 100 seed banks exist around the world and collectively hold more than 3 million samples. Like the Svalbard International Seedbank in Norway, they offer the advantage of storing a large amount of plant genetic material in a very small space. Seeds stored in seed banks are safe from habitat destruction, climate change, and general neglect. There even have been some instances of seeds from seed banks being used to reintroduce to the wild a plant species that had become extinct.

Some disadvantages to seed banks exist, however. Seeds do not remain alive indefinitely and must be germinated periodically so that new seeds can be collected. In addition, accidents such as fires or power failures can result in the permanent loss of the genetic diversity represented by the seeds in a seed bank. Also, many types of plants—avocados and coconuts, for example—cannot be stored as seeds. The seeds of these plants do not tolerate being dried out, which is a necessary step before the seeds are frozen. (If 20% or more moisture remains in a seed when it is frozen, it probably will die.) Some seeds cannot be stored successfully because they only remain viable for a short period of time such as a few months or even just a few days.

Perhaps the most important disadvantage of seed banks is that plants stored in this manner remain stagnant in an evolutionary sense. They do not adapt in response to changes in their natural environments. As a result, they may be less fit for survival when they are reintroduced into the wild.

Despite their shortcomings, seed banks are increasingly viewed as an important method of safeguarding seeds for future generations. Other international efforts to preserve plant genetic diversity are also being planned and implemented. For example, some farmers may soon be paid to set aside some of their land for cultivating local varieties of crops, thereby preserving the genetic diversity in agriculturally important plants.

Reproduction in Flowering Plants

29

Learning Objectives

After you have studied this chapter you should be able to:

1. Distinguish between sexual and asexual reproduction.
2. Discuss how the following structures are modified for asexual reproduction; rhizomes, tubers, stolons, corms, and bulbs.
3. Label the parts of a flower on a diagram and describe the functions of each part.
4. Relate how eggs and pollen grains are formed within the flower.
5. Distinguish between pollination and fertilization and explain the process of double fertilization.
6. Compare the general features that are characteristic of flowers pollinated in different ways (by insects, birds, and wind).
7. Define coevolution and give two examples of how plants and their animal pollinators have affected one another's evolution.
8. Describe embryonic development in flowering plants.
9. Distinguish among simple, aggregate, multiple, and accessory fruits, and give an example of each.
10. Describe four methods of seed dispersal.

Key Concepts

☐ Flowering plants exhibit remarkable variation in both asexual and sexual reproductive structures.

☐ Sexual reproduction in flowering plants takes place in flowers and results in the formation of seeds enclosed within fruits. Like other plants, flowering plants reproduce by forming spores. Each haploid microspore produced in the male portion of the flower (the anther) develops into a pollen grain that produces two sperm. Haploid megaspores produced in the female portion of the flower (the carpel) develop into female gametophytes that each contain a single egg.

☐ Pollination is the transfer of pollen from the male to the female parts of flowers. Many flowering plants have coevolved with animals that assist in the pollination process.

☐ Fertilization (the fusion of an egg and sperm) follows pollination and results in the formation of a seed. Each seed contains a young, multicellular plant (the embryo) and food that will be used for nourishment during germination (the beginning of growth of a seed).

ONE OF THE REASONS FOR THE SUCCESS OF FLOWERING PLANTS IS THEIR REPRODUCTIVE FLEXIBILITY

Many flowering plants are able to reproduce both sexually and asexually. *Sexual reproduction* involves the formation of flowers and, after fertilization, of seeds and fruits. More specifically, sexual reproduction entails the fusion of haploid gametes—an egg and a sperm. The union of gametes, which is called **fertilization,** occurs within the ovary of the flower.

The offspring formed as a result of sexual reproduction exhibit a great deal of genetic variation. This variation is due in part to the independent assortment of genes that occurs during meiosis. (Recall from Chapter 10 that meiosis must occur in the life cycle if haploid gametes are to be produced.) The union of dissimilar gametes, often from two different plants, is also partly responsible for the variation among offspring, as it results in new combinations of genes not found in either parent. Sexual reproduction, then, offers the advantages of new combinations of genes that might make an individual plant better suited to its environment.

Asexual reproduction in flowering plants does not usually involve the formation of flowers, seeds, and fruits. Instead, the vegetative structures—stems, leaves, and roots—form offspring. In asexual reproduction, part of an existing plant may become separated from the rest of the plant (perhaps by the death of tissues); the separated part subsequently grows to form a complete, independent plant.

Asexual reproduction produces new individuals without the fusion of gametes. The offspring are formed by mitosis, and meiosis is not involved. This means that the offspring of asexual reproduction are genetically identical to the parent plant from which they came.[1]

ASEXUAL REPRODUCTION IN FLOWERING PLANTS MAY INVOLVE MODIFIED STEMS, LEAVES, OR ROOTS

Flowering plants possess various methods of asexual reproduction, most of which involve modified vegetative parts. For example, a number of asexual reproductive structures are modified stems—rhizomes, tubers, bulbs, corms, and stolons. These structures become independent plants when they separate from the parent tissues.

[1] Of course, somatic mutations sometimes occur in the cells of the plant body, but their occurrence is relatively rare. Such mutations, however, would make the offspring of asexual reproduction genetically similar rather than genetically identical.

Figure 29–1 Rhizome. Irises have rhizomes, horizontal, underground stems. New aerial shoots arise from buds that develop along a rhizome.

A **rhizome** is a horizontal, underground stem that may or may not be fleshy for storing food (Fig. 29–1). Although rhizomes resemble roots, they really are stems, as indicated by the presence of scalelike leaves, buds, nodes, and internodes. Rhizomes frequently branch in different directions. Over time, the old portion of the rhizome dies, eventually dividing the two branches into independent plants. Examples of plants with rhizomes include irises, bamboo, and many grasses. Humans propagate plants with rhizomes by dividing or cutting the rhizome into smaller pieces, each with a bud. Each piece is capable of growing into an entire plant.

Another underground stem is the **tuber,** which is greatly enlarged for food storage (Fig. 29–2). When the connection between a tuber and its parent plant breaks (often as a result of the parent plant dying), the tuber grows into an independent plant. White potatoes and *Caladium* (elephant's ear) are examples of plants that produce tubers. The "eyes" of a potato are actually lateral buds, evidence that the tuber is an underground stem rather than a storage root. Potatoes are seldom grown from seed. Instead, a tuber is cut into pieces, each with a lateral bud. When planted, each grows into a plant.

A **bulb** is a shortened underground stem to which fleshy storage leaves are attached (Fig. 29–3). Bulbs are round, and are covered by paper-like scales. They frequently form small daughter bulbs that are initially attached to the parent bulb. Humans separate the daughter bulbs to increase the number of plants, but this process also occurs in nature. If the parent bulb dies and rots away, for example, each daughter bulb can become established as an independent plant. Lilies, tulips, onions, and daffodils form bulbs.

An underground stem that superficially resembles a bulb is a **corm** (Fig. 29–3). The storage organ in a corm is the much-thickened stem, rather than leaves as in a bulb. The entire corm is stem tissue that is covered with papery scales. These scales are modified leaves that are attached to the corm at nodes. Lateral buds that give rise to new corms frequently arise on corms. The death of the parent corm separates these daughter corms. Plants that produce corms include crocus, gladiolus, and cyclamen.

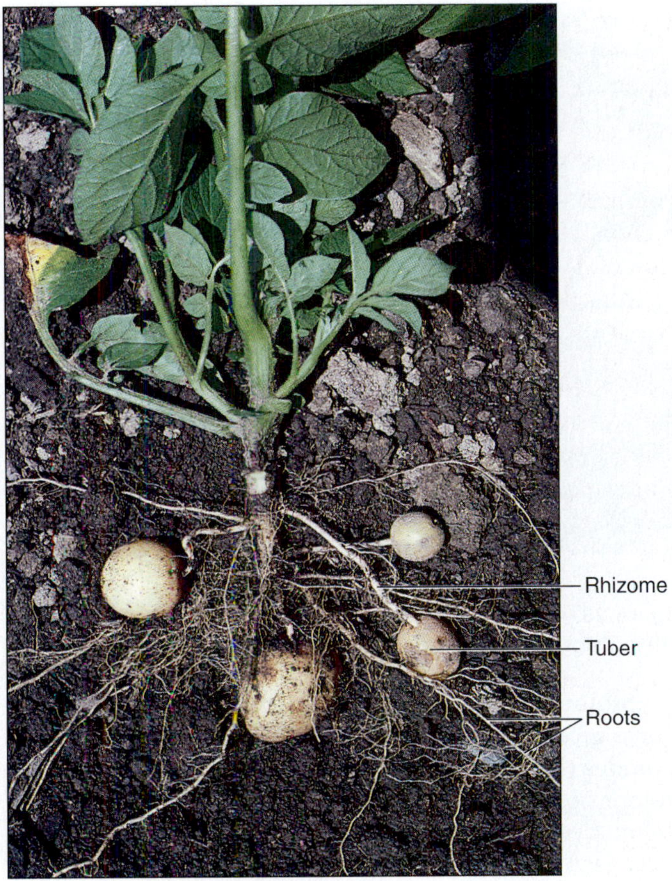

Figure 29–2 Tubers. Potatoes form rhizomes, which enlarge into tubers at the ends. (G. R. Roberts)

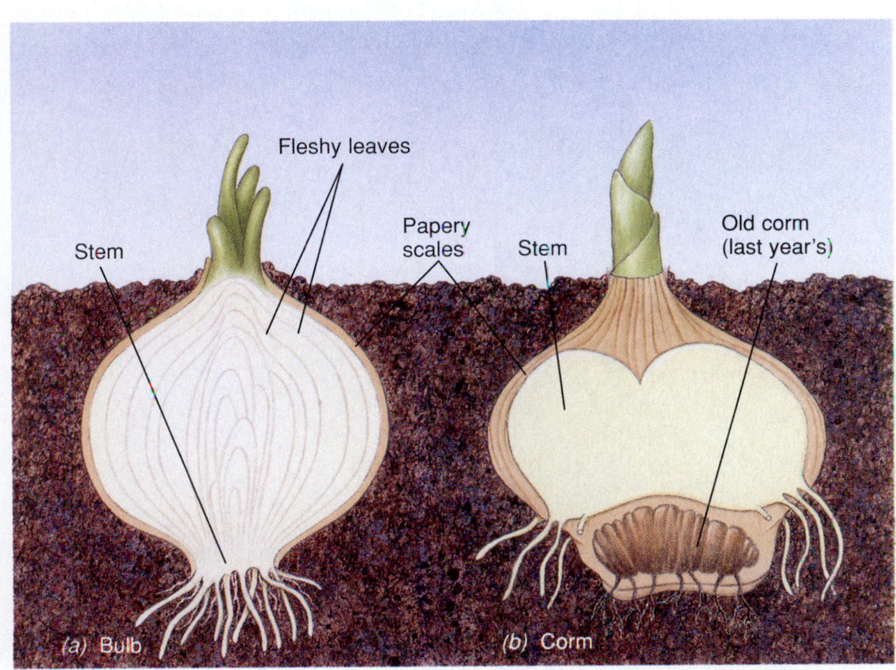

Figure 29–3 Bulb and corm. (***a***) A bulb is an underground stem to which overlapping, fleshy leaves are attached. (***b***) The entire corm is stem tissue, in contrast to the bulb, which is mostly leaf tissue.

Figure 29–4 Stolons. The strawberry reproduces asexually by forming stolons, or runners.

Stolons, or runners, are horizontal stems that run above ground and are characterized by having long internodes (Fig. 29–4). Adventitious buds[2] develop along a stolon, and each bud gives rise to a new plant. When the stolon dies, the daughter plants are separated. The strawberry is an example of a plant that produces stolons.

Some plants are capable of forming young plants (plantlets) along their leaf margins. *Kalanchoe,* commonly called "mother of thousands," has meristematic tissue in

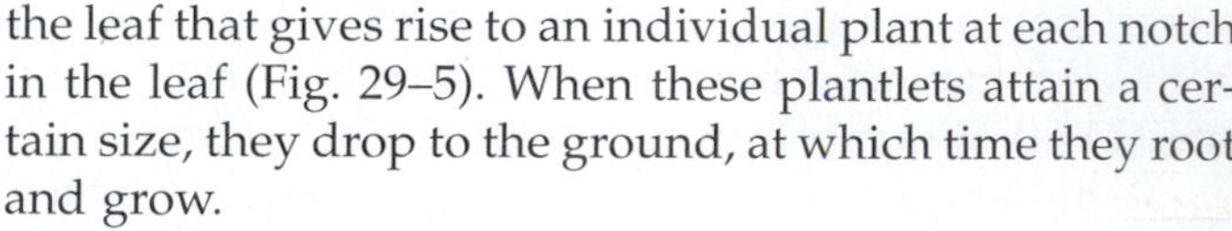
the leaf that gives rise to an individual plant at each notch in the leaf (Fig. 29–5). When these plantlets attain a certain size, they drop to the ground, at which time they root and grow.

Stems can produce roots, but roots cannot normally form stems. Some roots produce **suckers,** however, which are above-ground stems that develop from adventitious buds on the roots. Each sucker grows additional roots and becomes an independent plant when the parent plant dies. Examples of plants that form suckers include black locust, pear, apple, cherry, red raspberry, and blackberry. Some weeds—field bindweed, for example—are able to produce a large number of suckers. These plants are difficult to control, because pulling the plant out of the soil seldom removes all of the roots. The roots of field bindweed can grow as deep as 3 meters (10 feet) in the soil. In fact, in response to wounding, the roots produce additional suckers, which can be a considerable nuisance.

Apomixis Is the Production of Seeds without the Sexual Process

Sometimes plants produce embryos in seeds without meiosis and the fusion of gametes. This phenomenon is known as **apomixis.** The seeds produced by apomixis are produced asexually because there is no fusion of gametes. The embryo is genetically identical to the original parent. However, apomixis has an advantage over other methods of asexual reproduction in that the seeds produced by apomixis can be dispersed by methods associated with sexual reproduction (to be discussed shortly). Examples of plants that reproduce by apomixis include dandelions, citrus trees, blackberries, garlic, and certain grasses.

[2] Recall from Chapter 26 that buds originate in the axils of leaves. An adventitious bud is one that develops at any place on a plant other than a leaf axil.

Figure 29–5 Leaves are reproductive in some flowering plant species. The "mother of thousands" (*Kalanchoe*) produces young plantlets along the margins of its leaves. When the young plants attain a certain size, they drop off and root in the ground. (Dennis Drenner)

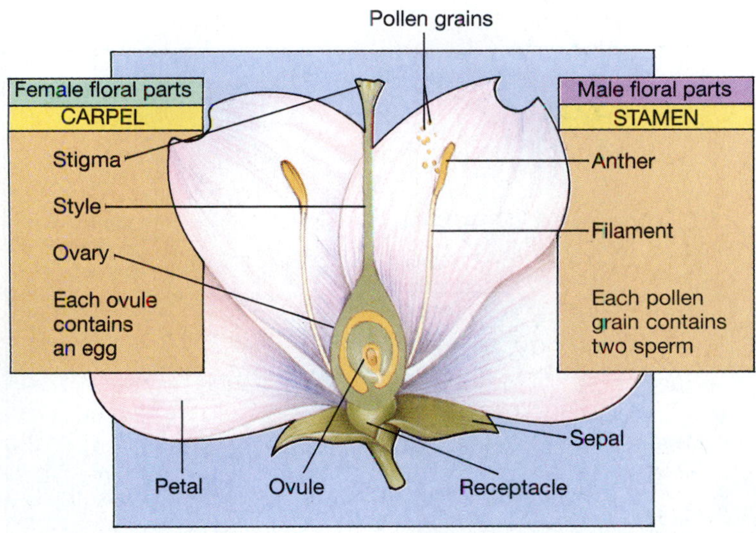

Figure 29–6 Flowers are the reproductive structures of flowering plants. Cutaway view of a "typical" flower.

FLOWERS ARE INVOLVED IN SEXUAL REPRODUCTION

Flowers are reproductive shoots usually composed of four kinds of organs—sepals, petals, stamens, and carpels—arranged in whorls on a shortened stem (Fig. 29–6). Sepals and petals consist of modified leaves that are sterile, and stamens and carpels consist of fertile modified leaves. All four floral parts are important in the reproductive process, but only the stamens (the "male" organs) and carpels (the "female" organs) participate directly in reproduction. A flower that possesses both stamens and carpels is said to be **perfect,** whereas an **imperfect** flower has stamens *or* carpels, but not both.

Sepals, which are the lowermost and outermost whorl on a floral shoot, are leaflike in appearance and often green. Sepals cover and protect the flower parts when the flower is a bud. As the blossom opens from a bud, the sepals fold back to reveal the more conspicuous petals. The collective term for all the sepals of a flower is **calyx.**

Petals are also leaflike in appearance, although they are frequently brightly colored. They play an important role in attracting animal pollinators to the flower. Sometimes petals are fused to form a tube or other floral shape. The petals of a flower are referred to collectively as a **corolla.**

Just inside the petals are the **stamens,** the "male" reproductive organs. Each stamen is composed of a thin stalk, called a **filament,** and a saclike **anther,** where meiosis occurs to form microspores that develop into pollen grains. Each pollen grain produces two male gametes, or **sperm.**

In the center of most flowers are one or more **carpels,** the "female" reproductive organs. Each carpel has three sections: a **stigma,** where the pollen lands; a **style,** or neck; and an **ovary,** which contains one or more **ovules.** Each ovule holds a female gametophyte that produces an **egg.** The carpels of a flower may be separate or fused together into a single structure. The term **pistil** is sometimes used to refer to an individual carpel or to a group of fused carpels.

It is instructive to relate the female and male parts of a flower to the life cycle of flowering plants, which was introduced in Chapter 23. Recall that flowering plants have an alternation of generations in which the diploid sporophyte generation (what you would consider the "plant") is large and conspicuous, and the haploid gametophyte generation is reduced in size to only a few cells within the flower (Fig. 29–7).

Eggs Form within the Ovary

Each ovule within an ovary contains a special cell that undergoes meiosis, producing four haploid cells called **megaspores.** Three of these disintegrate, and one divides by mitosis three times to form a **female gametophyte,** also called an **embryo sac** (Fig. 29–8). The embryo sac consists of eight haploid nuclei, including one egg and two **polar nuclei.** Both the egg and polar nuclei participate directly in fertilization.

Pollen Grains Form within the Anther

The anther contains special cells, each of which undergoes meiosis to form four haploid cells called **microspores** (Fig. 29–9). Each microspore develops into an immature **male gametophyte,** also called a **pollen grain.** A pollen grain consists of two cells surrounded by a tough outer wall. Both cells are involved in fertilization; one cell generates two sperm and the other produces a pollen tube through which the sperm travel to reach the embryo sac.

(Text continues on page 565)

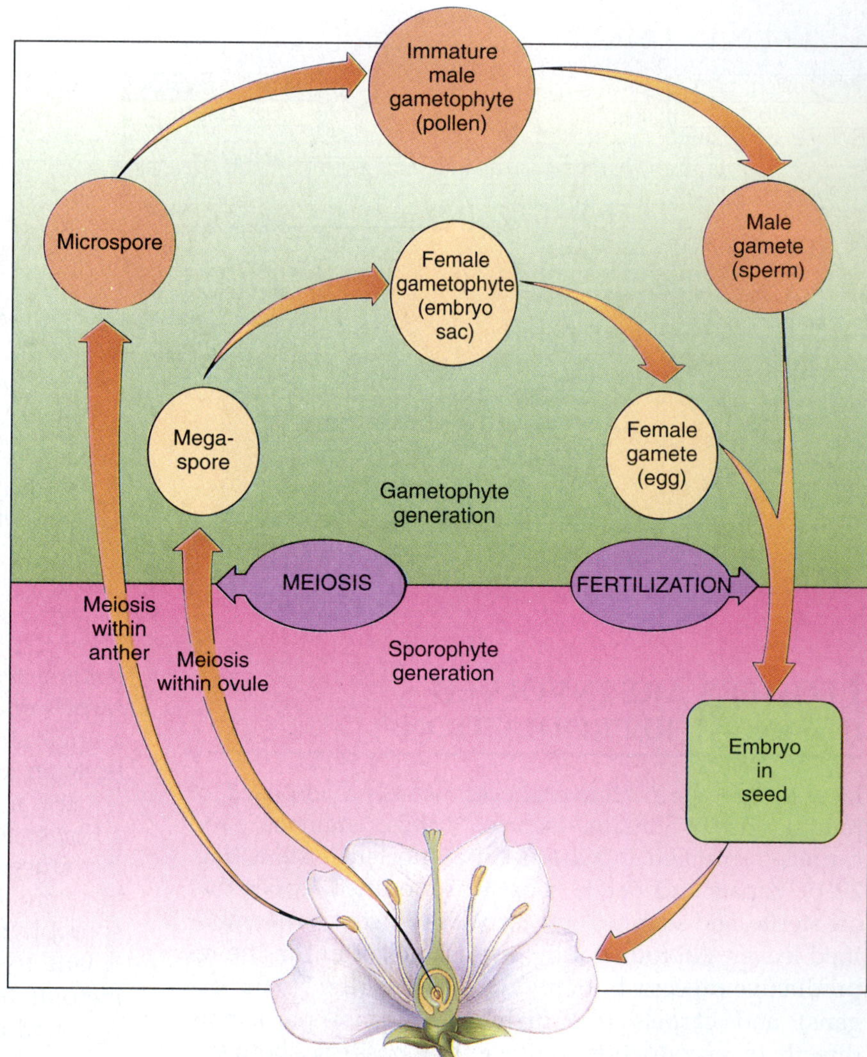

Figure 29–7 A simplified life cycle of flowering plants showing an overview of sexual reproduction. Meiosis within the flower results in haploid spores. Microspores give rise to the male gametophyte, and megaspores produce the female gametophyte. The gametes produced by the male and female gametophytes fuse during fertilization to form a new sporophyte generation.

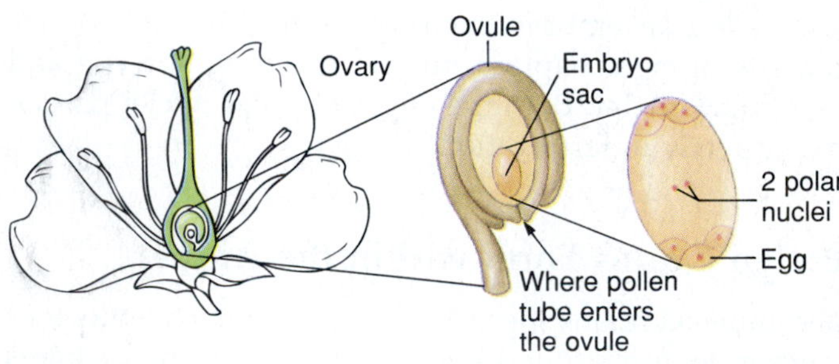

Figure 29–8 The embryo sac, or female gametophyte, of flowering plants. The embryo sac, with its egg and polar nuclei, forms within the ovule. The ovary pictured here contains a single ovule.

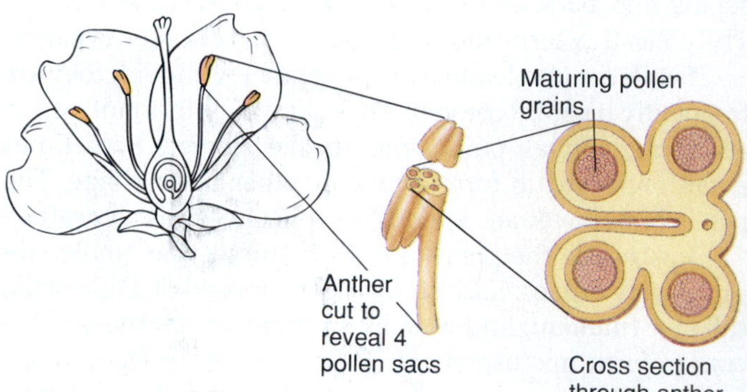

Figure 29–9 Development of pollen. Pollen grains, or immature male gametophytes, develop from microspores within sacs in the anther.

Pollen Is Transferred from the Anther to the Stigma

Before fertilization can occur, the pollen must travel from the anther (where it formed) to the stigma, often on a different flower of the same species. This transfer of pollen is known as **pollination,** and flowering plants possess a variety of mechanisms to accomplish pollination. Some of these involve animals, including insects, birds, and bats, whereas others involve wind and even water.

Flowering Plants and Their Animal Pollinators Have Affected One Another's Evolution

Flowers pollinated by animals have various features to attract them, including showy petals (a visual attractant) and scent (an olfactory attractant; Fig. 29–10). One of the rewards for the animal pollinator is food: nectar (a sugary solution used as an energy-rich food) and pollen (a protein-rich food). As the animal moves from flower to flower searching for rewards, it inadvertently carries pollen, thus facilitating sexual reproduction in plants.

Plants pollinated by insects often have blue or yellow petals. The insect eye does not perceive color in the same manner as does the human eye. Insects see very well in the blue and yellow range of visible light but do not perceive red as a distinct color. Consequently, flowers pollinated by insects are not usually red. Insects can also see in the ultraviolet range, an area that is invisible to the human eye; insects see ultraviolet radiation as a color called *bee's purple.* Many flowers have dramatic ultraviolet markings that are invisible to us but direct the insect to the center of the flower, where the pollen and nectar are located (Fig. 29–11).

Insects have a well-developed sense of smell, and many insect-pollinated flowers have a strong scent that may be pleasant or foul. For example, the carrion plant, which is pollinated by flies, smells like rotting flesh. As flies move from one flower to another looking for a place to deposit their eggs, they transfer pollen.

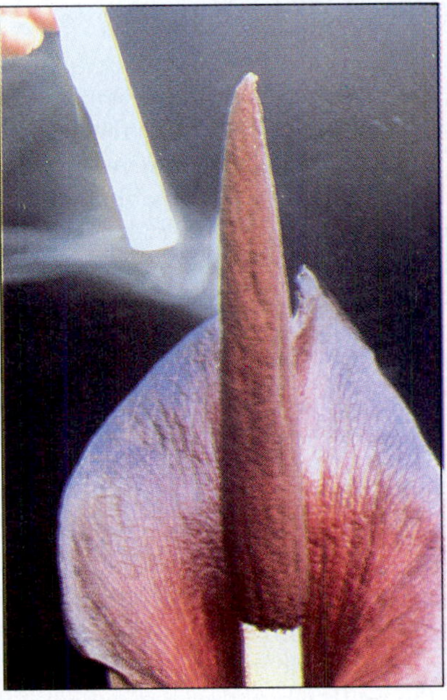

Figure 29–10 Devil's tongue produces a disagreeable floral odor that causes a visible reaction. The chemicals that cause the odor react with HCl on the filter paper to produce a precipitate that is visible. Devil's tongue is pollinated by beetles and flies. (J. M. Patt and B. J. D. Meeuse)

(a)

(b)

Figure 29–11 Many insect-pollinated flowers have ultraviolet markings that are invisible to humans but very conspicuous to insects. (***a***) A flower as seen by the human eye is solid yellow. (***b***) The same flower viewed under ultraviolet radiation provides clues about how the insect eye perceives it. The outer, light-appearing portions of the petals appear purple to the bee's eyes, whereas the inner parts appear yellow. These differences in coloration draw attention to the center of the flower, where the pollen and nectar are located. (*a, b,* Thomas Eisner)

(a) (b)

Figure 29–12 Animal pollinators. (***a***) A ruby-throated hummingbird pollinating a trumpet vine flower. (***b***) Bats are important pollinators in the tropics. The pollen grains on the bat's fur will be carried to the next plant. (*a*, Dan Dempster/Dembinsky Photo Associates; *b*, Merlin D. Tuttle, Bat Conservation International)

Birds such as hummingbirds are important pollinators (Fig. 29–12*a*). Flowers pollinated by birds are usually red, orange, or yellow because birds see well in this region of visible light. Birds do not have a strong sense of smell, however, so bird-pollinated flowers usually lack a scent.

Bats, which feed at night and do not see very well, are important pollinators in the tropics (Fig. 29–12*b*). Bat-pollinated flowers have dull white petals and a strong scent, usually of fermented fruit. Bats are attracted to the flowers by the scent and lap up the nectar. As they move from flower to flower, they transfer pollen. Other animals, including snails and small rodents, sometimes pollinate plants.

Animal pollinators and the plants they pollinate have had such a close interdependent relationship over time that they have affected one another's characteristics. In other words, animal pollinators have been a strong selective force in the evolution of certain features of flowering plants. Flowering plants have likewise affected the evolution of certain features of their animal pollinators. We call such evolution, in which two different organisms interact so closely that they become increasingly adapted to one another, **coevolution.**

One of the most peculiar examples of coevolution involves a group of orchids (*Ophrys*) and their insect pollinators (wasps, bees, and flies). The resemblance between the flowers of one *Ophrys* species and female wasps is so strong that male wasps mount the flowers and mistakenly attempt to copulate with them. During this misdirected activity, a pollen sac is usually attached on the back

CONCEPT CONNECTIONS

Coevolution — *Mutualism*

During the period that specialized features, such as scent and nectar that attract and reward pollinators, were evolving in flowering plants, the animal pollinators were also coevolving. They developed specialized body parts and behaviors that enable them to aid pollination and obtain pollen and nectar as a reward. Coevolution is responsible for the hairy bodies of bumblebees, which catch and hold the sticky pollen for transport from one flower to another. Coevolution is also responsible for the long, curved beaks of certain honeycreepers such as the Hawaiian birds that insert their beaks into tubular flowers to obtain nectar (Chapter 19).

A number of mutualistic (mutually beneficial) relationships have developed between animal pollinators and the flowering plants they pollinate. For example, a yucca plant native to the Southwest can only be pollinated by one species of moth, and the female moth can only lay her eggs in the yucca flower's ovary. Both the yucca and the moth would become extinct if something happened to the other, as neither would be able to reproduce successfully. Other examples of mutualistic relationships between two different species are considered in Chapter 46.

Figure 29–13 Wind pollination. Clusters of male oak flowers, which lack petals, dangle from a tree branch. Wind is an agent of pollination for oaks and many other flowering plants. (Dr. Jeremy Burgess/Science Photo Library/Photo Researchers, Inc.)

of the wasp. When the frustrated wasp departs and attempts to copulate with another orchid flower, the pollen is transferred to that flower.

Some Flowering Plants Depend on Wind Dispersal of Pollen

Flowering plants such as grasses, ragweed, and maples are pollinated by wind and produce many, often inconspicuous, flowers (Fig. 29–13). Wind-pollinated plants do not produce large, colorful petals or scent or nectar. Because wind pollination is a "hit-or-miss" affair, the likelihood of pollen landing on the stigma of the same species of flower is slim. Wind-pollinated plants therefore produce large quantities of pollen.

Concept Connections

Seed Size — *Ecology*

Seed size varies considerably in flowering plants, from the microscopic, dustlike seeds of orchids to the giant seeds of the double coconut, which weigh as much as 27 kilograms (almost 60 pounds). Despite this variation among different species, seed size is a remarkably constant trait within a species.

Assuming that a given plant species invests a fixed amount of its energy in reproduction, is it more advantageous to produce a large number of small seeds or a few large ones? By observing the seed sizes that predominate in different environments, biologists have concluded that in some environments, a smaller seed size appears to be advantageous, whereas in others, a larger seed size may be better.

For example, plants that grow in widely scattered open sites (such as old fields) usually produce smaller seeds, perhaps because smaller seeds can be more easily dispersed over large areas than can larger seeds. On the other hand, wide dispersal is probably less important for plants growing in densely vegetated areas such as forests. These plants generally produce larger seeds, possibly because larger seed size confers a greater likelihood of becoming successfully established in a highly competitive environment.

Other ecological factors are associated with seed size. Larger seeds are typical of plants that live in arid habitats (possibly because the food stored in a large seed allows a young seedling to establish an extensive root system quickly, thereby enabling it to survive the dry climate). Island plants also produce larger seeds than similar species on the nearby mainland. In this case, it is hypothesized that large seeds are less likely to be widely dispersed (and therefore less likely to fall into the ocean). On the other hand, seeds that remain dormant in the soil for a long period of time tend to be smaller. (The reason for this is not known.)

Despite many observations that associate seed size with specific environments, the general principles concerning the adaptive advantages of large seeds versus small seeds remain to be determined.

Fertilization Is Followed by Seed and Fruit Development

Once pollen has been transferred to the stigma, it grows a thin pollen tube down the style and into the embryo sac within an ovule in the ovary. A cell within the pollen grain divides to form two nonflagellated male gametes, the sperm. Both sperm move down the pollen tube and enter the embryo sac.

A Unique Double Fertilization Process Occurs in Flowering Plants

The egg within the embryo sac fuses with one of the sperm, forming a fertilized egg that will develop into an embryonic plant within the seed. The two polar nuclei within the embryo sac are also important in reproduction because they fuse with the second sperm to form **endosperm,** which is the nutritive tissue surrounding the embryonic plant. This process, in which two separate cell fusions occur, is called **double fertilization.** It is, with one exception, unique to flowering plants. (In 1990, double fertilization was reported in *Ephedra,* a gymnosperm.)

Embryonic Development in Seeds Follows an Orderly and Predictable Path

Flowering plants produce a young plant embryo plus stored nutrients in a compact package, the **seed,** that develops after fertilization occurs (Chapter 23). Growth of the embryo is possible because of the constant movement of nutrients into the developing embryo from the parent plant.

Mitotic divisions of the fertilized egg to form a multicellular embryo proceed in a variety of ways in flowering plants. The following description is of dicot embryonic development. (Embryonic development in monocots is identical in the early stages.)

The two cells formed as a result of the first division of the fertilized egg establish polarity, or direction, in the embryo. The bottom cell typically develops into a **suspensor,** which is a multicellular structure that anchors the embryo and aids in nutrient uptake from the endosperm. The top cell develops into the actual embryo.

Initially, the top cell divides to form a short chain of cells, called a **proembryo** (Fig. 29–14). As mitosis contin-

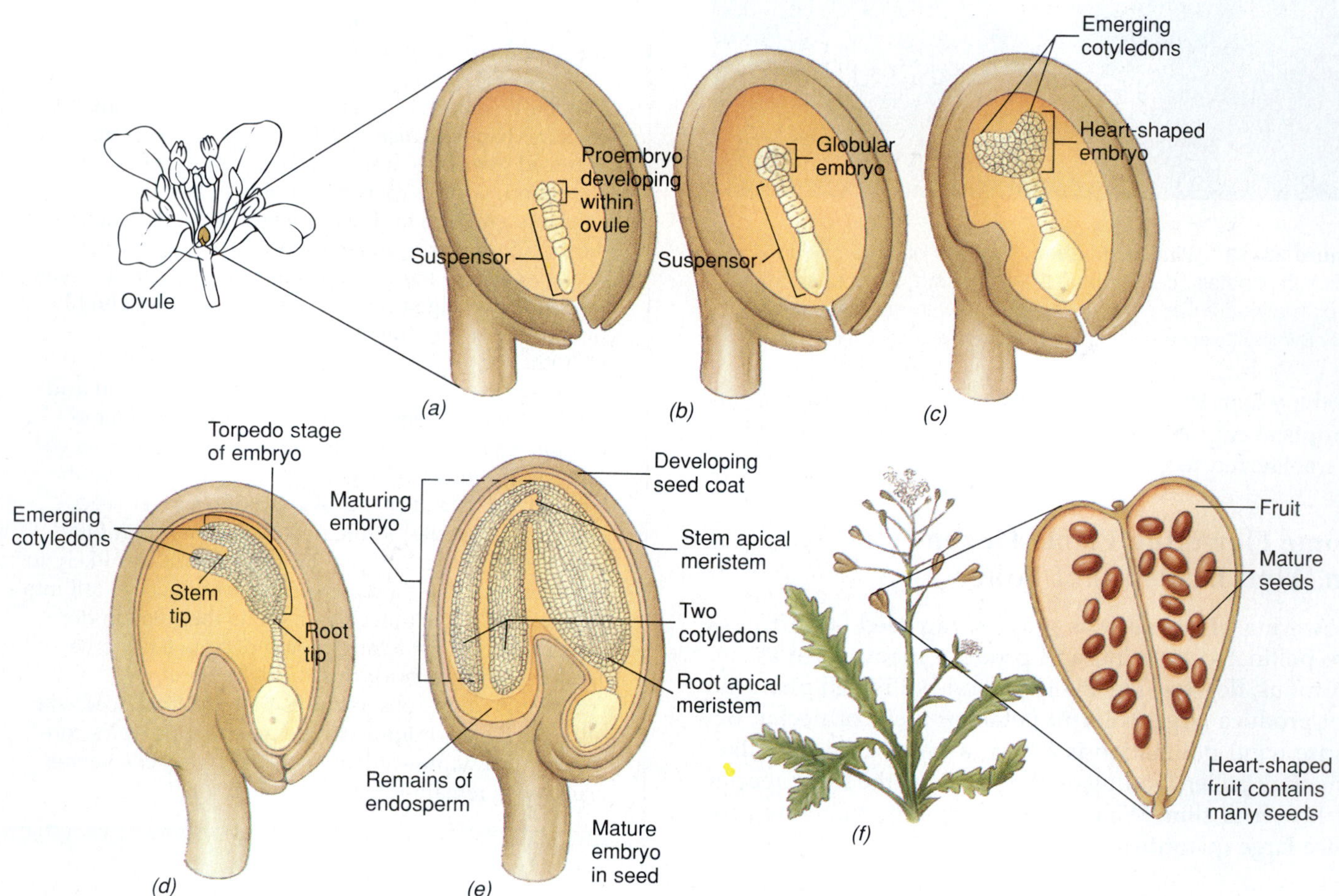

Figure 29–14 Embryonic development in shepherd's purse, a common dicot weed. (***a***) The proembryo is the earliest multicellular stage of the embryo (shown inside the ovule). (***b***) As mitosis continues, the embryo becomes a ball of cells, called the globular stage. (***c***) As the two cotyledons begin to emerge, the embryo is shaped like a heart. (***d***) The cotyledons continue to elongate, forming the torpedo stage. (***e***) A maturing embryo within the seed. The food originally stored as endosperm has been almost completely absorbed by the developing embryo during growth and development. Most of the food for the mature embryo is stored in its two cotyledons. (***f***) A longitudinal section through a heart-shaped fruit of shepherd's purse reveals numerous tiny seeds, each containing a mature embryo.

ues, a small ball of cells, often called a **globular embryo,** develops. Cells begin to develop into specialized tissues during this stage. When the dicot embryo starts to develop its two cotyledons, it has two lobes and resembles a heart; this is often called the **heart stage.** (Because monocot embryos develop a single cotyledon, they are more cylindrical at this stage.) As the cotyledons elongate, the **torpedo stage** develops, which continues to grow into a mature embryo. The embryo continues to enlarge, often curving in the process and crushing the suspensor beyond recognition.

The mature embryo consists of an embryonic root, an embryonic shoot, and two cotyledons (monocots, of course, have a single cotyledon). The cotyledons of many plants function as food storage organs and are large and thick, having absorbed the food initially produced as endosperm. Other plants have thin cotyledons that function primarily to help the young plant digest and absorb food stored as the endosperm.

A seed, when mature, consists of an embryonic plant and food (stored in either the cotyledons or endosperm), surrounded by a tough, protective **seed coat.** Seeds are enclosed within a **fruit.**

Fruits Are Mature, Ripened Ovaries that Enclose Seeds

After double fertilization takes place within the ovule, the ovule develops into a seed (just described), and the ovary surrounding it develops into a fruit. For example, a pea pod is a fruit, and the peas within it are seeds. A fruit may contain one or more seeds. For example, some orchid fruits contain several thousand to a few million seeds. There are several types of fruits, which differ owing to variations in the structure or arrangement of the flowers from which they were formed.

Simple fruits, aggregate fruits, multiple fruits, and accessory fruits are the four basic types of fruits (Table 29–1). Most fruits are simple fruits. A **simple fruit** develops from a single ovary of a single flower. At maturity, simple fruits may be fleshy or dry (Fig. 29–15). Two examples of simple fleshy fruits are berries and drupes. A **berry** is a fleshy fruit that has soft tissues throughout; a tomato is a berry, as are grapes and bananas. A **drupe** is a simple, fleshy fruit that has a hard, stony pit surrounding the seed. Examples of drupes include peaches, plums, and avocados.

Many simple fruits are dry at maturity. At maturity, some dry fruits split open, usually along seams or sutures, to release their seeds. The milkweed pod is an example of a simple, dry fruit that splits open along *one* seam to release its seeds. A pea pod is a simple, dry fruit that splits open along *two* seams (the top and bottom). Other simple, dry fruits—**grains,** for example—do not split open. Each grain contains a single seed. Because the seed coat is fused to the fruit, it appears that a grain is a seed rather than a fruit. Kernels of corn and wheat are fruits of this type.

TABLE 29–1
The Main Types of Fruits

Fruit type	*Definition*	*Example*
Simple fruit	Develops from a flower with a single pistil (and therefore a single ovary)	Peach
Aggregate fruit	Develops from a flower with many separate carpels (and therefore many ovaries)	Raspberry
Multiple fruit	Develops from many flowers borne together on a common floral stalk; the ovaries of these flowers fuse together to form a single fruit	Pineapple
Accessory fruit	Develops from a flower in which the receptacle or floral tube enlarges and becomes part of the mature fruit	Apple

Figure 29–15 Simple fruits. (***a***) The tomato is a berry, a fruit composed of soft tissues throughout. (***b***) The peach is a drupe. The hard, stony pit, which is part of the fruit, encloses a single seed. (***c***) The milkweed fruit, which is dry at maturity, splits open along one seam. (***d***) Corn kernels. Each kernel is a grain, a fruit containing a single seed in which the fruit is fused to the seed coat. (*a,* James Mauseth, University of Texas; *b, d,* Dennis Drenner; *c,* Runk/Schoenberger, from Grant Heilman)

Aggregate fruits are a second main type of fruit. An **aggregate fruit** is formed from a single flower that contains many separate carpels (Fig. 29–16). After fertilization, each ovary from each individual carpel enlarges. As they enlarge, the ovaries fuse to form a single fruit. Raspberries and blackberries are examples of aggregate fruits.

A third type of fruit is the **multiple fruit,** which is formed from the ovaries of many flowers that grow in close proximity on a common floral stalk. The ovary from each flower fuses with nearby ovaries as it develops and enlarges after fertilization. Pineapples are multiple fruits.

Accessory fruits are the fourth type of fruit. They differ from other fruits in that other plant tissues, in addition to ovary tissues, make up the fruit. For example, the edible portion of a strawberry is the red, fleshy **receptacle,** which is the terminal part of the floral stalk. Apples and pears are also accessory fruits; the outer part of each fruit is an enlarged **floral tube**[3] that surrounds the ovary (Fig. 29–17).

Seed Dispersal Is Highly Varied in Flowering Plants

Flowering plants make use of wind, animals, water, and explosive dehiscence to disperse their seeds (Fig. 29–18). Effective methods of seed dispersal have made it possible for some plants to expand their geographical range.

[3] The floral tube consists of receptacle tissue along with portions of the calyx.

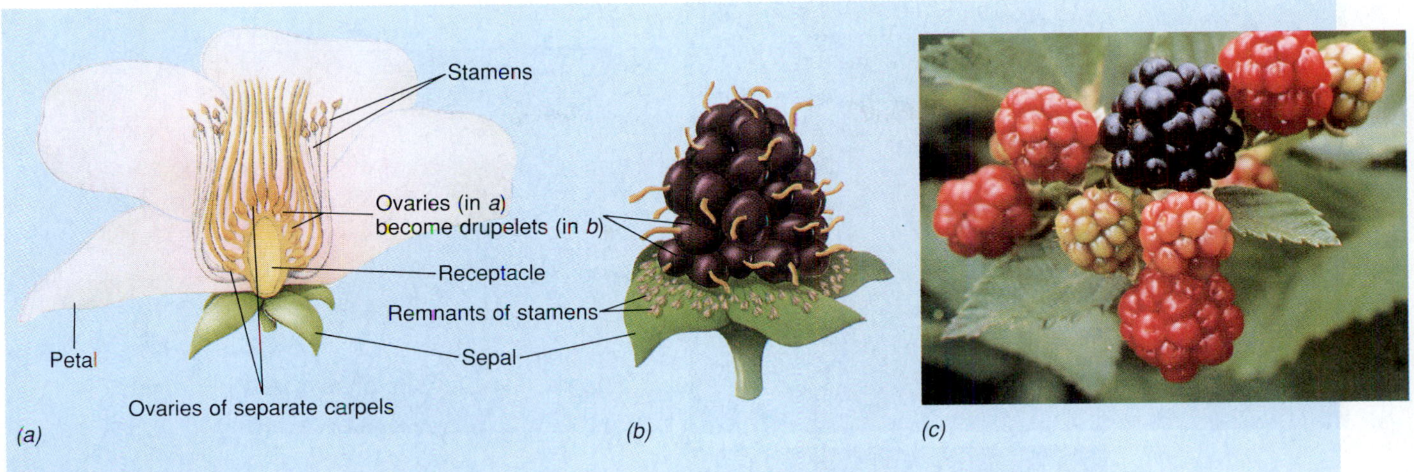

Figure 29–16 Blackberries are aggregate fruits. (***a***) Cutaway view of a blackberry flower, showing the many separate carpels in the center of the flower. (***b***) A developing blackberry fruit is an aggregate of tiny drupelets. (***c***) Developing fruits at various stages of maturity. (Dennis Drenner)

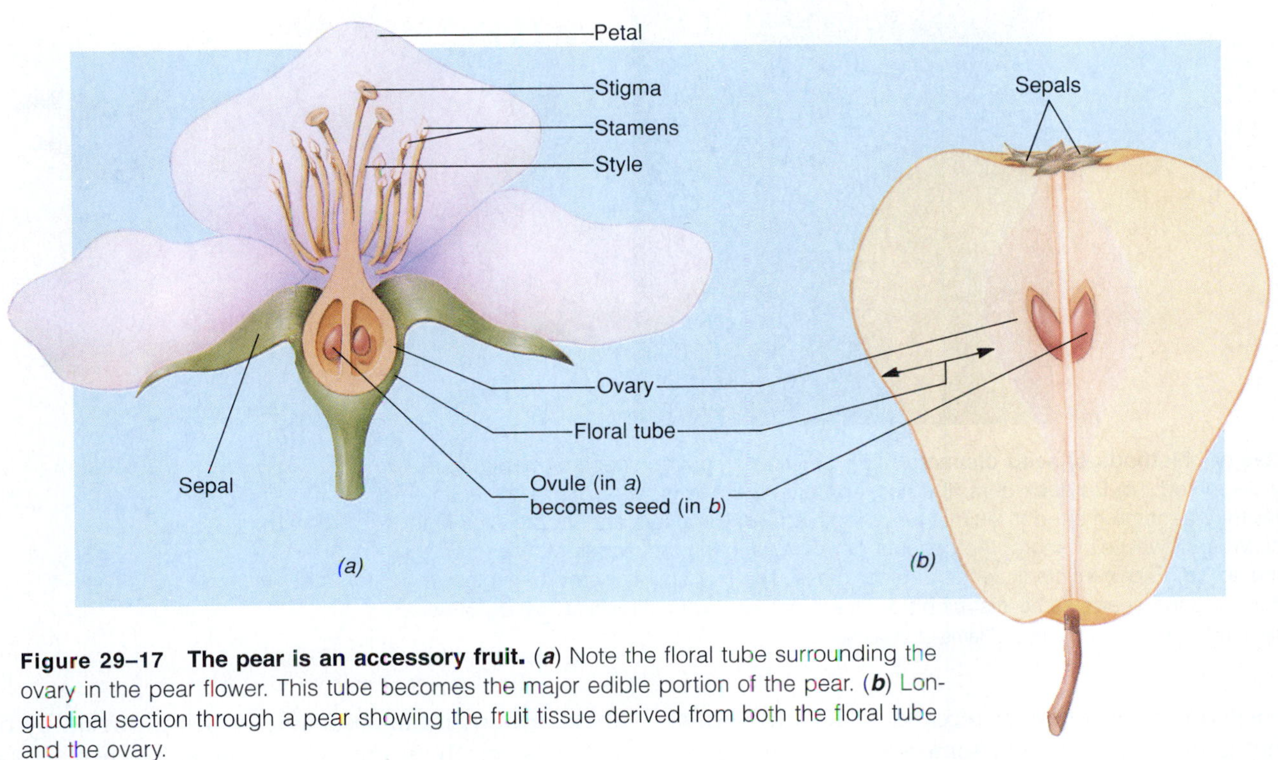

Figure 29–17 The pear is an accessory fruit. (***a***) Note the floral tube surrounding the ovary in the pear flower. This tube becomes the major edible portion of the pear. (***b***) Longitudinal section through a pear showing the fruit tissue derived from both the floral tube and the ovary.

In some cases, the seed is the actual agent of dispersal, whereas in others, it is the fruit.

Wind is responsible for seed dispersal in many plants. Plants such as maple have winged fruits adapted for wind dispersal. Light, feathery plumes are other structures that allow seeds or fruits to be transported by wind, often for considerable distances. Both dandelion fruits and milkweed seeds have this type of adaptation.

Some plants have special structures that aid in dispersal of their seeds by animals. The spines and barbs of cockleburs and similar fruits catch in animal fur and are dispersed as the animal moves about. Fleshy, edible fruits are also adapted for animal dispersal. As these fruits are eaten, the seeds are either discarded or swallowed. Seeds that are swallowed have thick seed coats and are not digested, but instead pass through the digestive tract and

Figure 29–18 Methods of seed dispersal. (***a***) The winged, paired seeds of maple fruits are wind-dispersed. (***b***) Burdock burs (the hooked fruits) are carried away from the parent plant after they become matted in animal fur or clothing. (***c***) Coconuts are adapted for water dispersal. When it washes ashore, the coconut germinates, often thousands of miles from its original home. (***d***) Explosive dehiscence in bitter cress. The fruit splits open with explosive force, propelling the seeds some distance from the plant. (*a*, James Mauseth, University of Texas; *b*, DPA/Dembinsky Photo Associates; *c*, James L. Castner)

are deposited with the animal's feces some distance away from the parent plant. In fact, some seeds will not germinate unless they have passed through an animal's digestive tract.

Animals like squirrels and many bird species help to disperse acorns and other fruits and seeds by burying them for winter use. Many buried seeds are never used by the animal and germinate the following spring (see Focus On Evolution: Seed Dispersal by Ants).

The coconut is an example of a fruit adapted for dispersal by water. It has air spaces and corky floats that make it buoyant and capable of being carried by ocean currents for thousands of kilometers. When it washes ashore, the seed, located within, germinates and grows into a coconut palm tree.

Some seeds are dispersed by neither wind, animals, nor water. These seeds are found in fruits that use explosive dehiscence to forcibly discharge their seeds. Pressures due to differences in turgor or to drying out cause them to burst open suddenly. The fruits of plants like touch-me-not and bitter cress split open so explosively that seeds are scattered a meter or more.

FOCUS ON *Evolution*

Seed Dispersal by Ants

For a plant species to survive, it must be able to disperse its seeds to places where they can successfully germinate and grow. Evolution has resulted in a variety of dispersal methods (such as wind, animals, and water) that increase the chances of seeds landing in suitable locations. Regardless of how seeds are dispersed from the parent plant, most seeds land in places that are unsuitable for growth or are eaten by animals such as mice and squirrels shortly after being dispersed.

Some plant species increase the survivability of their seeds by a dispersal method that buries their seeds underground. Such a dispersal method favors the germination of these seeds and the subsequent growth of seedlings because they are less likely to be eaten by animals.

Yet, how are such seeds buried? For many plant species, the role of burying seeds underground is performed by ants, which collect the seeds and take them underground to their nests. Ants disperse and bury seeds for hundreds of different plant species in almost every terrestrial environment, from northern coniferous forests to tropical rain forests to deserts.

An open fruit pod of Dutchman's breeches (*Dicentra cucullaria*) reveals numerous black/yellow seeds. The black part is the seed proper, whereas the yellow part is the oil body. (Steve Handel, Rutgers University)

Both ants and flowering plants benefit from their association. The ants improve the reproductive success of the plants whose seeds they bury, whereas the plants supply food to the ants. A seed that is collected and taken underground by ants often contains a special structure called an oil body that protrudes from the seed (see figure). Oil bodies are a nutritious food for ants, which carry seeds underground before removing the oil bodies. Once an oil body is removed from a seed, the ants discard the undamaged seed in an underground refuse pile, which happens to be rich in organic material (such as ant droppings and dead ants) and contains the minerals required by young seedlings. Thus, ants not only bury the seeds away from animals that might eat them but they also place the seeds in rich soil that is ideal for seed germination and seedling growth.

Chapter Summary

I. Asexual reproduction involves the formation of offspring without the fusion of gametes. The offspring are genetically identical to the single parent plant.
   A. Rhizomes, tubers, bulbs, corms, and stolons are stems that are specialized for asexual reproduction.
   B. Some leaves have meristematic tissue along their margins and give rise to plantlets.
   C. Roots may develop adventitious buds that develop into suckers. Suckers develop additional roots and may give rise to new plants.
   D. Apomixis is the production of seeds and fruits without sexual reproduction.

II. The offspring produced by sexual reproduction are genetically variable. The flower is the structure in which sexual reproduction occurs.
   A. A flower may contain sepals, petals, stamens, and carpels.
   B. Meiosis of special cells in the anther gives rise to microspores, which in turn develop into immature male gametophytes (pollen grains).
   C. Meiosis of a special cell in the ovule within the ovary gives rise to four megaspores, one of which develops into a female gametophyte (embryo sac).

III. Pollination is the transfer of pollen from anther to stigma.
   A. Some plants rely on animals to transfer pollen.
      1. Coevolution occurs when two different organisms (such as flowering plants and their animal pollinators) form an interdependent relationship and affect the course of one another's evolution.

2. Flowers pollinated by insects are often colored yellow to blue and possess a scent.
3. Bird-pollinated flowers are often colored yellow to red and do not have a strong scent.
4. Bat-pollinated flowers often have dusky white petals and possess a scent.

B. Wind-pollinated plants often have reduced petals or their petals are lacking altogether, and they do not produce a scent or nectar. They make large amounts of pollen.

IV. After pollination, fertilization (fusion of gametes) occurs.
A. A pollen tube grows down the style into the ovary, and two sperm pass through the pollen tube into the embryo sac.
B. The egg fuses with one of the sperm, forming a fertilized egg that eventually develops into a multicellular embryo in the seed.
C. The two polar nuclei fuse with the second sperm, forming a nutritive tissue called endosperm.

V. The seed and fruit develop as a result of successful fertilization.
A. An embryo develops in the seed in an orderly fashion, from proembryo to mature embryo.
B. A mature seed contains a young plant embryo and nutritive tissue (stored in the endosperm or cotyledons). The seed is covered by a protective seed coat.
C. Fruits are mature, ripened ovaries that enclose seeds.
1. Simple fruits develop from a single ovary of a single flower. Some simple fruits are fleshy (grapes, peaches) at maturity, whereas others are dry (milkweed pods, grains).
2. Aggregate fruits such as raspberries develop from many ovaries within a single flower.
3. Multiple fruits such as pineapples develop from many ovaries of many flowers growing in close proximity.
4. In accessory fruits such as strawberries and apples, a major part of the fruit consists of tissue other than ovary tissue.

VI. Seeds and fruits of flowering plants are adapted for various means of dispersal, including wind, animals, water, and explosive dehiscence.

Selected Key Terms

carpel, p. 563
coevolution, p. 566
double fertilization, p. 568
embryo sac, p. 563
endosperm, p. 567
fertilization, p. 560
fruit, p. 569
ovary, p. 563
ovule, p. 563
pistil, p. 563
polar nuclei, p. 563
pollen, p. 563
pollination, p. 565
seed, p. 568
stamen, p. 563
stigma, p. 563
style, p. 563

Post-Test

1. Genetic variability in the offspring is characteristic of ______________ reproduction.

2. A/an ______________ is a horizontal, underground stem that is specialized for asexual reproduction.

3. The white potato is an example of an underground storage stem called a/an ______________.

4. In apomixis, fruits and seeds are produced by ______________ means.

5. The ______________ is composed of a stigma, style, and ovary.

6. Plants with blue petals, nectar, and a strong scent are most likely pollinated by ______________.

7. Plants with reduced or absent petals, no nectar, no scent, and large quantities of pollen are most likely pollinated by ______________.

8. The process of ______________ ______________ in flowering plants involves one sperm fusing with an egg cell and one sperm fusing with two polar nuclei.

9. The ______________ is a multicellular structure that anchors the embryo and aids in nutrient uptake from the endosperm.

10. After fertilization an ovule develops into a/an ______________.

11. A/an ______________ may be defined as a mature, ripened ovary.

12. ______________ fruits form from many ovaries of a single flower, whereas ______________ fruits develop from many ovaries of many separate flowers.

13. Apples, strawberries, and pears are examples of ______________ fruits.

14. Light, feathery plumes on a seed or fruit signify that it is most likely dispersed by ______________.

15. Fleshy, edible fruits are adapted for dispersal by ______________.

16. This part of the flower contains ovules and develops into a fruit: (a) petal (b) anther (c) stigma (d) ovary (e) pollen.

17. The part of the carpel that is receptive to pollen is the

(a) endosperm (b) sepal (c) stigma (d) ovary (e) anther.

18. Pollination involves the (a) fusion of an egg and a sperm (b) development of a pollen tube (c) transfer of pollen from anther to stigma (d) development of pollen in the anther.

19. Which of the following stages in the flowering plant life cycle are in the correct order?
(a) embryo ⟶ meiosis ⟶ embryo sac ⟶ fertilization
(b) meiosis ⟶ embryo sac ⟶ fertilization ⟶ embryo
(c) meiosis ⟶ fertilization ⟶ embryo ⟶ embryo sac
(d) fertilization ⟶ embryo sac ⟶ meiosis ⟶ embryo

20. Place the following events in the correct order, starting with pollination.
(1) pollen tube grows into ovule
(2) insect lands on flower to drink nectar
(3) embryo develops within the seed
(4) fertilization occurs
(5) pollen carried by insect drops onto stigma
(a) 2-1-5-4-3 (b) 4-1-5-2-3 (c) 3-2-5-1-4 (d) 2-5-1-4-3

Review Questions

1. Distinguish between sexual and asexual reproduction.
2. How are stems, leaves, and roots used by certain plants for asexual reproduction?
3. What special advantage does asexual reproduction by apomixis have over other kinds of asexual reproduction (such as corms and bulbs)?
4. Outline the difference between pollination and fertilization. Which process occurs first?
5. Explain the roles of each of the following in the flowering plant life cycle: (a) meiosis (b) pollen tube (c) double fertilization (d) endosperm.
6. Describe the difference between seeds and fruits.
7. Distinguish among simple, aggregate, multiple, and accessory fruits, and give an example of each.
8. Explain some of the features possessed by seeds and fruits.
9. Label the following diagram.

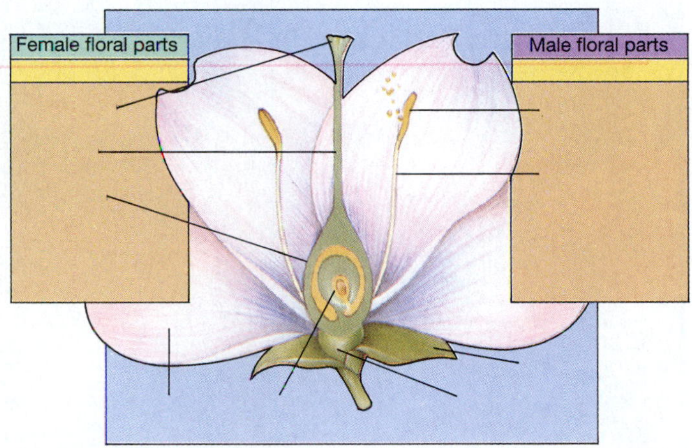

Thinking Critically

1. Many developing nations are reluctant to store seeds in international seed banks such as the one in Norway. They fear that plant breeders and seed manufacturers from developed countries such as the United States may obtain their seeds free of charge from such seed banks. Why would this be a concern to developing nations?
2. Do you think that sexual or asexual reproduction would be more advantageous in each of the following circumstances? Explain.
   a. A constant favorable environment.
   b. A rapidly changing environment.
   c. An extremely crowded environment.
3. That humans have affected the evolution of crop plants such as corn is indisputable (recall artificial selection, discussed in Chapter 17). Yet, has human evolution in turn been affected by these plants? How might you test your idea?
4. Could seed dispersal by ants be considered an example of coevolution? Why or why not?

Recommended Readings

Beattie, A.J., "Ant Plantation," *Natural History,* February 1990. Discusses the role that ants play in dispersing certain seeds.

Cox, P.A., "Water-Pollinated Plants," *Scientific American,* October 1993. Some aquatic plants rely on water currents to transfer pollen to female flowers.

Handel, S.N. and A.J. Beattie, "Seed Dispersal by Ants," *Scientific American,* August 1990. How plants induce ants to disperse seeds without harming them.

Mauseth, J.D., *Botany: An Introduction to Plant Biology,* 2nd edition, Philadelphia, Saunders College Publishing, 1995. A comprehensive introduction to plant biology.

Moore, P.D., "Not Only Gone with the Wind," *Nature,* Vol. 357, 14 May 1992. Some plant species have *two* methods of dispersal instead of one.

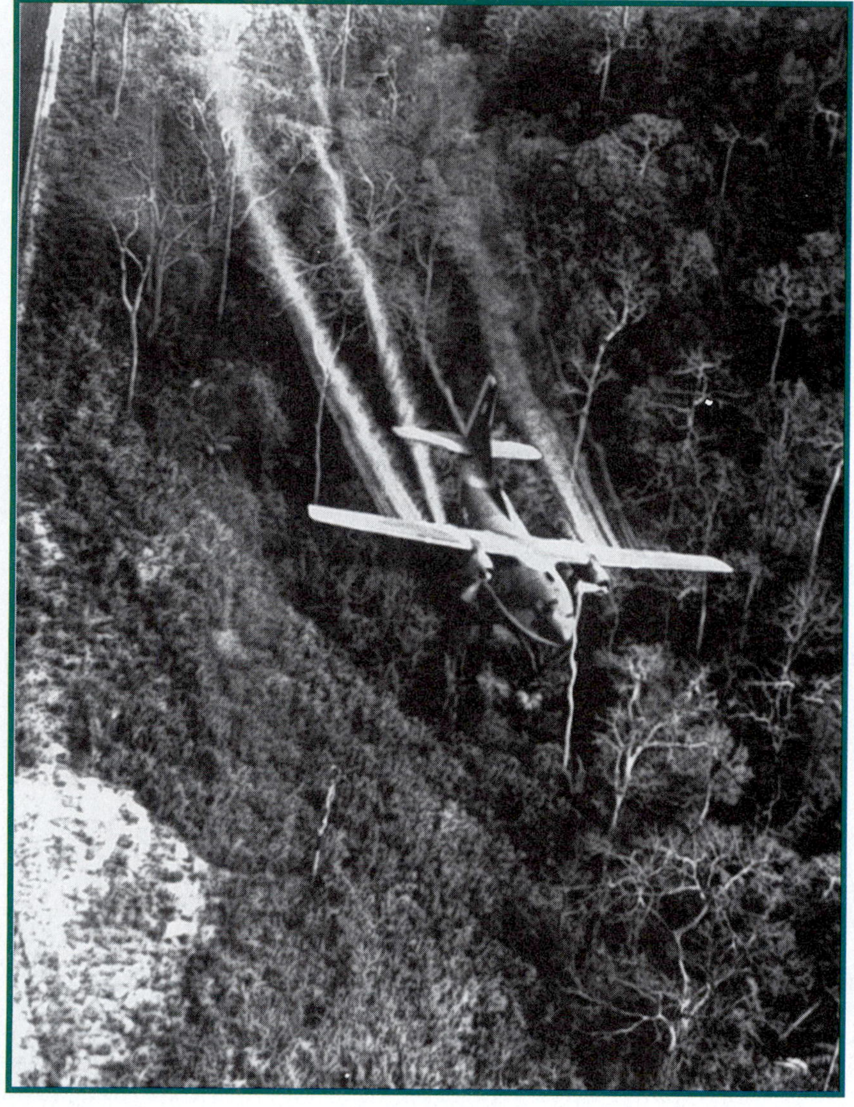

The spraying of herbicides over forested areas during the Vietnam War. (This photo was taken in 1966.) (UPI/Bettman)

Herbicide Applications and Warfare

One of the many controversial aspects of the Vietnam War was the defoliation program carried on by the United States in South Vietnam. From 1961 to 1971, the United States sprayed herbicides (chemicals that kill plants) over large areas of South Vietnam to expose suspected hiding places and destroy crops planted by the Vietcong and North Vietnamese troops. The three mixtures of herbicides used were designated Agent White, Agent Blue, and Agent Orange.

The negative impact of these herbicides on the environment is still being felt today. It is estimated that between 20% and 50% of the ecologically important mangrove forests of South Vietnam were destroyed. Shrubs have replaced these forests, which may take decades to return. Moreover, approximately 30% of the nation's commercially valuable hardwood forests were killed and have been replaced by bamboo and weedy grasses.

In addition to the ecological damage, the herbicide sprays also caused health problems in the native people and members of the U.S. military who were exposed to these chemicals in the Vietnamese jungles. Agent Orange, which contained a mixture of two herbicides (2,4-D and 2,4,5-T), also contained minute traces of dioxin, an extremely dangerous poison that is formed during the manufacture of 2,4,5-T. High doses of dioxin have been shown to cause birth defects in animals. During the period of herbicide spraying, the number of birth defects and stillbirths in Vietnam reportedly increased. It also appears that American veterans who were exposed to high levels of Agent Orange have more health problems than do other veterans.

In 1993, the National Academy of Science's Institute of Medicine released the results of a comprehensive review of 6420 scientific articles on the effects of exposure to herbicides. It concluded that Vietnam veterans with three types of cancer (soft tissue carcinoma, non-Hodgkin's lymphoma, and Hodgkin's disease) and two skin diseases (chloracne and porphyria cutanea tarda) should receive compensation for their exposure to Agent Orange.

The active ingredients in Agent Orange, 2,4-D and 2,4,5-T, kill plants with broad leaves but for reasons not understood at this time, do not kill grasses. Both herbicides are similar in structure to auxin, a natural hormone in plants, and disrupt the plants' normal growth processes. Both cause exaggerated growth in some plant parts and growth inhibition in others. Because many of the world's important crops are grasses (for example, wheat, corn, and rice), both 2,4-D and 2,4,5-T can be used to kill broadleaf weeds that compete with these crops. However, because of its association with dioxin, 2,4,5-T is no longer used in the United States.

In this chapter, we examine the regulation of plant growth and developmental processes by environmental factors and hormones, including auxin, the hormone similar in structure to the herbicides 2,4-D and 2,4,5-T.

Regulation of Plant Growth and Development

30

Learning Objectives

After you have studied this chapter you should be able to:

1. Discuss the genetic and environmental factors that affect plant growth and development.
2. Explain how flowering is affected by varying amounts of light and darkness, and define the role of phytochrome in flowering.
3. Describe how temperature affects flowering.
4. Summarize the influence of internal and external factors on the germination of seeds.
5. Compare turgor movements and tropisms.
6. Distinguish among phototropism, gravitropism, and thigmotropism.
7. Define circadian rhythm, and give an example.
8. Explain what a hormone is, and list several different ways each of the following hormones affect plant growth and development: auxins, gibberellins, cytokinins, ethylene, and abscisic acid.
9. Discuss the physiological responses in plants that may be due to varying ratios of several hormones rather than to one specific hormone.
10. Explain how hormones are involved in regulating each of the following: leaf abscission, seed germination, apical dominance, stem elongation, fruit ripening, and seed dormancy.

Key Concepts

- ☐ Plants, like all living organisms, respond to both internal and external stimuli. Internal stimuli are under genetic control. External stimuli are environmental factors that the plant can detect and respond to (such as the availability of moisture in the soil or the amount of sunlight or shade).
- ☐ A plant's environment affects a number of aspects of its growth and development. Seasonal changes in daylength (length of light period) and temperature, for example, affect the initiation of flowering in many plants. Environmental conditions affect the level and distribution of hormones in the plant. These hormones, in turn, regulate the plant's growth and development.
- ☐ Plant hormones are chemical messengers produced in one part of the plant and transported to another where they cause a physiological response. Auxins, gibberellins, cytokinins, ethylene, and abscisic acid are the hormones known to operate in higher plants.
- ☐ Each plant hormone affects a wide variety of responses in the plant throughout its lifetime. Plant hormones seldom work alone, and it is common for two or more hormones to be involved in a single physiological activity. Often, one hormone stimulates a particular response, and another hormone inhibits the response.

ENVIRONMENTAL CUES MAY INDUCE FLOWERING IN PLANTS

The initiation of sexual reproduction is often under environmental control, particularly in temperate latitudes. Environmental control is important for the plant's survival, because the timing of sexual reproduction is critical to reproductive success. Plants must flower and form seeds before dormancy (a period of arrested growth) is induced by the onset of winter. A number of plants detect changes in the relative lengths of daylight and darkness that accompany the changing seasons, and they flower in response to these changes. Other plants have temperature requirements that induce sexual reproduction.

Flowering May Be Initiated by Changes in Light and Dark Periods

Photoperiodism is any response of a plant to the relative lengths of daylight and darkness. Flowering is one of several physiological activities that are photoperiodic in many plants. Plants are classified into three main groups on the basis of how photoperiodism affects their flowering: short-day, long-day, and day-neutral plants.

Short-day plants were initially defined as plants that flower when exposed to some critical daylength or less. However, the important factor in the initiation of flowering in short-day plants is the long, uninterrupted period of darkness rather than the short period of daylight. In other words, *short-day plants flower when the night length is equal to or greater than some critical length.* Examples of short-day plants are chrysanthemum and poinsettia, which typically flower in late summer or fall (Fig. 30–1).

FIGURE 30–1 The chrysanthemum is a short-day plant. Two identical cuttings from the same plant were planted in separate pots. The plant on the left, which flowered, received 8 hours of daylight and 16 hours of darkness for several weeks. The plant on the right, which remained vegetative, received 16 hours of daylight and 8 hours of darkness during the same time period. (Dennis Drenner).

To summarize, short-day plants are able to detect the shortening days and lengthening nights of late summer or fall, and they flower at that time.

Long-day plants were initially defined as being able to flower when the daylength is equal to or greater than some critical amount. However, a more accurate definition would be that *long-day plants flower when the night length is equal to or less than some critical length.* Plants such as clover, black-eyed Susan, and lettuce flower in late spring or summer and are long-day plants. These plants are able to detect the lengthening days and shortening nights of spring and early summer, and they flower at that time.

Some plants do not initiate flowering in response to seasonal changes in daylength. These **day-neutral plants** flower in response to some other stimulus, either external or internal. Tomato, dandelion, string bean, and pansy are examples of day-neutral plants.

Phytochrome Detects Varying Periods of Daylength and Darkness

For plants, or any living organism, to have a biological response to light, there must be a **photoreceptor** (light-sensitive substance) in that organism to absorb the light. The photoreceptor for photoperiodism and a number of other light-initiated responses of plants is a blue-green pigment called **phytochrome.** Phytochrome, which is present in all vascular plants, has two forms, and it readily converts from one form to the other upon absorption of light of specific wavelengths. One form, designated P_r (for *r*ed-absorbing *p*hytochrome), strongly absorbs red light. In the process, the shape of the molecule changes to the second form of phytochrome, P_{fr}. This form of phytochrome is so designated because it absorbs *f*ar-*r*ed light, which has longer wavelengths than does red light. When P_{fr} absorbs far-red light, it reverts back to the original form, P_r. The P_{fr} form of phytochrome is less stable than the P_r form, so it also reverts spontaneously, albeit slowly, to P_r in the dark. The form of phytochrome that triggers physiological responses such as flowering is P_{fr}.

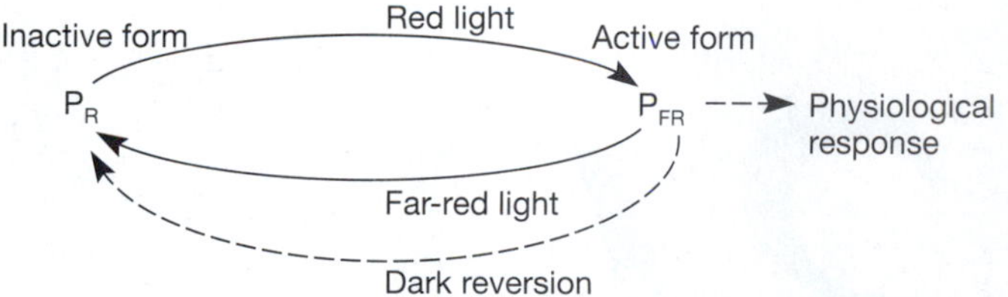

But what does a pigment that absorbs red light and far-red light have to do with daylight and darkness? Sunlight is composed of the entire spectrum of visible light in addition to ultraviolet and infrared (far-red) radiation. However, sunlight contains more red light than far-red

FIGURE 30–2 Photoperiodic responses of a short-day plant and a long-day plant to different periods of light and dark. Note that the short-day plant (top row) does not flower when exposed to 8 hours of daylight and 16 hours of darkness interrupted with a brief flash of light. This same treatment induces the long-day plant (bottom row) to flower.

light. As a result, when a plant is exposed to sunlight, its level of P_{fr} increases, while at night, its level of P_{fr} slowly decreases.

In short-day plants, the P_{fr} form of phytochrome *inhibits* flowering. In order to flower, these plants need long nights. The long period of darkness allows the P_{fr} to revert back to P_r, so the plant has some minimum time during the 24-hour period with *no* P_{fr} present. This initiates flowering. Biologists have experimented with short-day plants by growing them under a short-day/long-night regimen, but interrupting the night with a short burst of red light (Fig. 30–2). Exposure to red light for a few minutes in the middle of the night will prevent flowering in short-day plants. This effect occurs because the brief exposure to red light converts some of the phytochrome from the P_r form to the P_{fr} form. Therefore, the plant does not have a sufficient period of time at night without any P_{fr}.

In long-day plants, the active form of phytochrome, P_{fr}, *induces* flowering. Long-day plants that are exposed to a long-day/short-night regimen flower (Fig. 30–2). The long days cause these plants to produce predominantly P_{fr}. During the short nights, some P_{fr} is changed to P_r, but because the night is short, the plant has little or no time with no P_{fr} present during a 24-hour period; hence, it flowers.

Plant biologists are puzzled by the observation that P_{fr} inhibits flowering in short-day plants and induces flowering in long-day plants. Why different plants respond so differently to P_{fr} is not known at this time. Nevertheless, the importance of phytochrome cannot be overemphasized. Timing of daylength and darkness is the most reliable way for plants to measure the change from one season to the next. This measurement is crucial for survival, particularly in the environments in which the climate goes through a regular, annual pattern of favorable and unfavorable seasons.

Temperature May Also Affect Reproduction

Certain plants have a temperature requirement that must be met if they are to flower. The promotion of flowering by exposure to low temperature for a period of time is known as **vernalization.** The part of the plant that must be exposed to low temperature varies. For some plants, the moist seeds must be exposed to a period of several weeks of low temperature in order for flowering to be induced. For other plants, recently germinated seedlings have a "cold" requirement. Some plants have an absolute requirement for the low-temperature period, meaning that they will not flower unless they have been vernalized. Other plants will flower sooner if exposed to low temperatures, but will still flower at a later time if they are not exposed to low temperatures.

Examples of plants with a low-temperature requirement include annuals like winter wheat and biennials like carrots. Carrots and other biennials grow vegetatively the first year, storing surplus food in their roots. After remaining dormant during the winter, they flower and reproduce during the second year. Carrots left in a warm environment and not exposed to low temperatures continue vegetative growth indefinitely and do not initiate sexual reproduction.

The external stimulus that a plant responds to (in this case, temperature) may be moderated and influenced by internal conditions, such as hormone levels in the plant. For example, it is possible to eliminate the low-temperature requirement for flowering in biennials by treatment with gibberellin (discussed later in this chapter).

CONCEPT CONNECTIONS

Development — *Gene Expression*

Genes exert an important control over the development of all living things. Recall from Chapters 13 and 14 that genes are discrete units of heredity that usually specify proteins. If the genes required for the expression of a particular trait are not present in an organism, that characteristic will not occur. When a particular gene is present, however, it may or may not be expressed in a given cell (Chapter 15).

During development, the descendants of a single cell specialize and organize into a complex multicellular organism. Both plants and animals contain many types of cells specialized both structurally and chemically to carry out specific functions (Chapters 26 and 31).

In most cases, the genetic material of all differentiated cells in an adult organism is genetically identical to the genetic material of the fertilized egg cell from which those cells descended. Differences among the various cell types are apparently the result of differential gene activity, that is, whether or not different genes are expressed in those cells. One factor that has a profound effect on gene expression is the *location* of a cell in the young organism's body. Cell location causes some genes in an immature cell to be turned on and others to be turned off. Thus, cell location helps to determine what each cell ultimately becomes. (Chapter 43 discusses animal development in greater detail.)

A NUMBER OF EXTERNAL AND INTERNAL FACTORS AFFECT SEED GERMINATION AND EARLY GROWTH

When a seed is mature, it is often dormant and may not germinate immediately, even if growing conditions are ideal. A number of factors influence whether a seed germinates. Many of these are environmental cues, including the presence of water and oxygen, proper temperature, and sometimes the presence of light penetrating the soil surface. For example, no seed germinates unless it has absorbed water, because the embryo in the mature seed is dehydrated, and a watery medium in cells is necessary for active metabolism. When a seed germinates, its metabolic machinery is turned on, with numerous materials being synthesized and degraded. Therefore, water is an absolute requirement for germination.

Seed germination and growth also require a great deal of energy. Plants obtain this energy (by converting the energy of food molecules into ATP) through the same aerobic respiratory pathway as do animals, so oxygen is usually needed during germination.[1]

[1] Some plants such as rice are able to respire anaerobically during the early stages of germination and seedling growth. This enables rice to grow and become established in flooded soil, an environment that would suffocate most young plants.

Another environmental factor that affects germination is temperature. In a population of seeds, some germinate at each temperature over a broad range. However, each plant species has an optimal temperature at which the largest number of seeds germinates. For most plants, the optimal germination temperature falls between 25° and 30°C. Some seeds, such as those of apples, require exposure to prolonged periods of cold before their seeds germinate at any temperature.

Also, certain plants, especially those with tiny seeds, require light for germination. Phytochrome is involved in the light requirement, specifically red light, that these seeds have for germination.[2] Exposure to red light converts P_r to P_{fr}, and germination occurs.

Some of the environmental conditions that are needed for seed germination help ensure the survival of the young plant. The requirement of a prolonged cold period ensures that seeds germinate in the spring rather than in the winter. A light requirement ensures that a tiny seed germinates only if it is close to the surface of the soil.

[2] Other physiological functions under the influence of phytochrome include sleep movements in leaves (discussed shortly); shoot dormancy; leaf abscission; and pigment formation in flowers, fruits, and leaves.

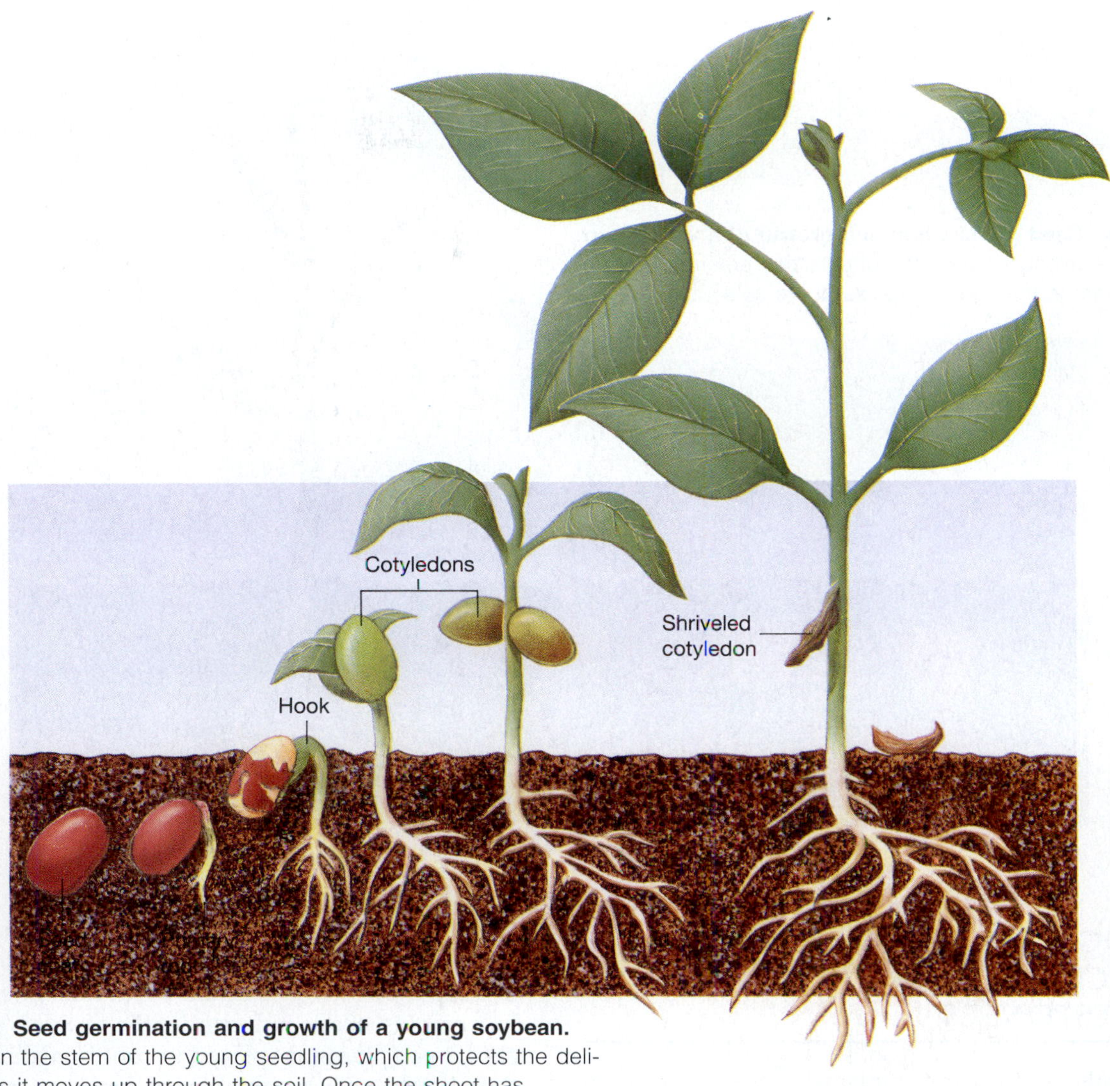

FIGURE 30–3 Seed germination and growth of a young soybean. Note the hook in the stem of the young seedling, which protects the delicate stem tip as it moves up through the soil. Once the shoot has emerged from the soil, the hook straightens.

If such a seed were to germinate several inches below the soil surface, it might not have enough food reserves to grow to the surface. On the other hand, if this seed were to remain dormant until the soil was disturbed and it was brought to the surface, it would have a much greater likelihood of survival.

In certain seeds, internal factors, which are under genetic control, prevent germination even when external conditions are favorable. Many seeds are dormant either because the embryo is immature and must develop further or because chemical inhibitors (such as abscisic acid, discussed later in this chapter) are present. These inhibitors may be washed out of the seed by rain. The presence of chemical inhibitors helps ensure the survival of the plant. For example, the seeds of desert annuals often contain high levels of an inhibitor that is washed out only when rainfall is sufficient to support the plant's growth after germination.

Dicots and Monocots Exhibit Characteristic Patterns of Early Growth

The first part of the plant to emerge from the seed during germination is the embryonic root. As the root grows, it forces its way through the soil, encountering considerable friction. The delicate apical meristem of the root tip is protected by a root cap (Chapters 26, 28). Stem tips are not covered by a protective cap, but plants have different ways to protect the delicate stem tip as it grows up through the soil to the surface. The stem of a bean seedling (a dicot) is curved over to form a hook so that the stem tip is actually *pulled* up through the soil (Fig. 30–3). Corn and other grasses (monocots) have a protective sheath of cells called a **coleoptile** surrounding the young shoot (Fig. 30–4). First, the coleoptile pushes up through the soil, and then the leaves grow up through the middle of the coleoptile.

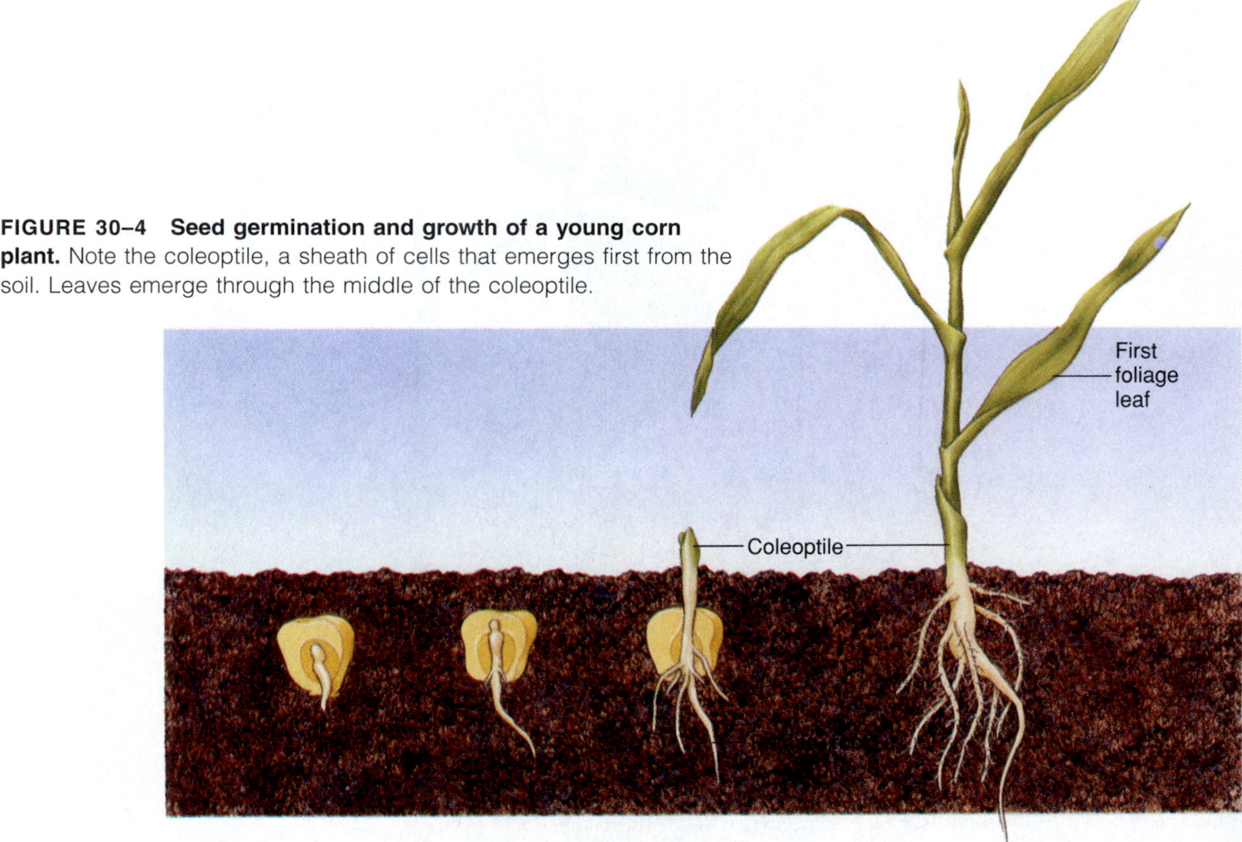

FIGURE 30–4 Seed germination and growth of a young corn plant. Note the coleoptile, a sheath of cells that emerges first from the soil. Leaves emerge through the middle of the coleoptile.

A BIOLOGICAL CLOCK INFLUENCES MANY PLANT RESPONSES

Plants, animals, and microorganisms appear to have an internal timer, or biological clock, that approximates a 24-hour cycle. These internal cycles, known as **circadian rhythms** (from the Latin words *circum,* meaning "around," and *diurn,* meaning "daily," help the organism determine the time of day. (In contrast, photoperiodism enables an organism to detect time of year.)

In the absence of external cues, circadian rhythms repeat every 20 to 30 hours, although in nature the rising and setting of the sun resets the biological clock so that the cycle repeats every 24 hours. (Interestingly, phytochrome has been implicated as the photoreceptor involved in resetting the biological clock for many plants.)

One example of a circadian rhythm in plants is the opening and closing of stomata that occur independently of light and darkness (Chapter 27). Plants placed in continual darkness for extended periods continue to open and close their stomata.

Another example of a circadian rhythm in plants is the **sleep movements** observed in the common bean and other plants (Fig. 30–5). During the day, bean leaves are horizontal for optimal light absorption, but at night the leaves fold down or up, which orients them perpendicular to their daytime position. (The biological significance of sleep movements is unknown at this time.) These sleep movements occur independently of Earth's 24-hour cycle. If bean plants are placed in continual darkness or continual light, sleep movements continue on an approximate 24-hour cycle.

Why do plants and other organisms possess circadian rhythms? A number of predictable environmental changes—sunrise and sunset, for example—occur during the course of each 24-hour period. These predictable changes may be important to an individual organism, causing it to change its behavior (in the case of animals) or its physiological activities. It is thought that circadian rhythms help an organism to synchronize repeatable daily activities so that it does them at an appropriate time each day.

CHANGES IN TURGOR CAN INDUCE TEMPORARY PLANT MOVEMENTS

Mimosa pudica, the sensitive plant, dramatically folds its leaves and droops in response to touch (or to an electrical, chemical, or thermal stimulus; Fig. 30–6). The re-

(Text continues on page 584)

FIGURE 30–5 Sleep movements in the common bean. (***a***) Leaf position at 12:00 noon. (***b***) Leaf position at 12:00 midnight. It is not known why some plants exhibit sleep movements. (Dennis Drenner).

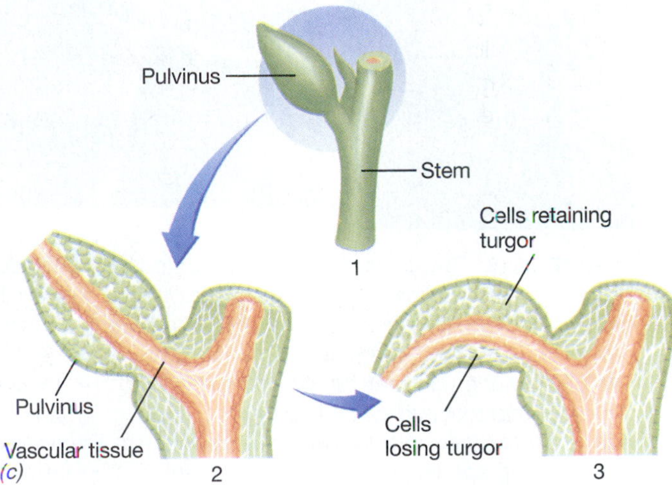

FIGURE 30–6 Turgor movements in the sensitive plant, *Mimosa pudica*. (***a***) *Mimosa* before being disturbed. (***b***) The plant several seconds after being touched. Note how the leaves have folded and drooped. (***c***) The pulvinus is responsible for turgor movements. (1) The base of the petiole, showing the pulvinus. (2) Section through the pulvinus, showing cells when a leaf is undisturbed. (3) Section through the pulvinus, showing loss of turgor that produces the folding of a leaf. (Dennis Drenner).

sponse spreads throughout the plant even if only one leaflet is initially aroused. When a *Mimosa* leaf is touched, an electrical impulse moves down the leaf to special cells located in an organ at the base of the petiole called the **pulvinus** (plural, *pulvini*). The pulvinus is a somewhat swollen joint that acts as a hinge. When the electrical signal reaches cells in the pulvinus, it triggers a loss of turgor in certain pulvinus cells as potassium ions exit, causing water to leave the cells by osmosis. The sudden change in turgor causes the leaf movement. Such **turgor movements** are temporary and reversible; the movement of potassium ions and water back into the pulvinus cells causes the plant part to return to its original position.

Changes in turgor are also responsible for **solar tracking,** the ability of leaves or flowers to follow the sun's movement across the sky (Fig. 30–7). Frequently, the leaves of these plants are arranged perpendicular to the sun's rays, regardless of the time of day or the sun's position in the sky. This allows for maximal light absorption. Many solar trackers have pulvini at the bases of their petioles. Changes in turgor in the cells of the pulvinus help position the leaf in its proper orientation relative to the sun. Sunflower, soybean, and cotton are examples of solar trackers.

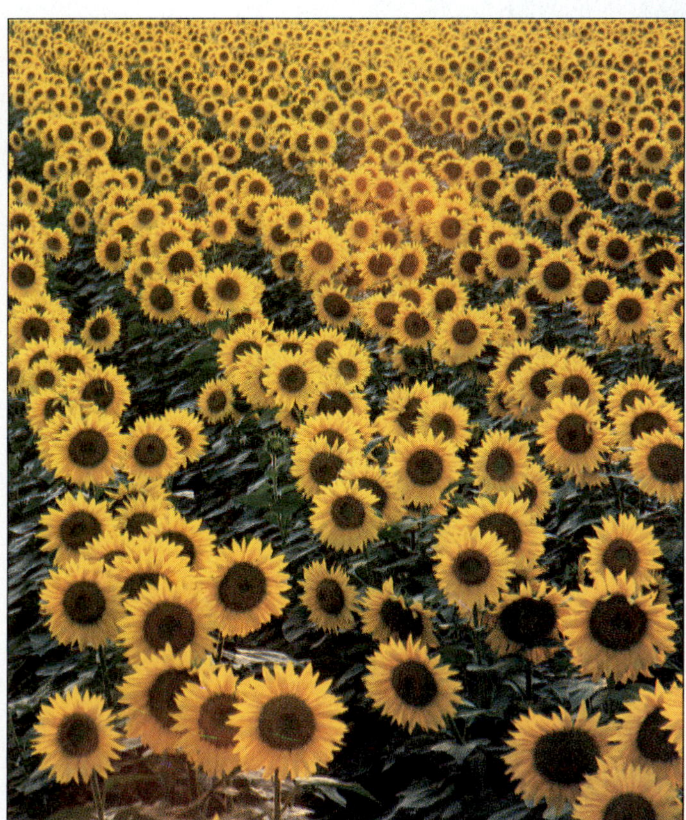

FIGURE 30–7 Solar tracking in sunflowers. Note how all the flower heads are oriented in the same direction. The orientation of the plants toward light in solar tracking is due to changes in turgor. (Grant Heilman, from Grant Heilman Photography).

A TROPISM IS GROWTH IN RESPONSE TO AN EXTERNAL STIMULUS FROM A SPECIFIC DIRECTION

A plant may respond to an external stimulus such as light, gravity, or touch by directional growth. Such directional responses, or **tropisms,** may be positive or negative, depending on whether the plant grows toward (positive) or away from (negative) the stimulus.

Phototropism is the growth of a plant in response to the direction of light. Most growing shoot tips exhibit positive phototropism and bend (grow) toward light (Fig. 30–8*a*). Growth in response to the direction of gravity is

(a)

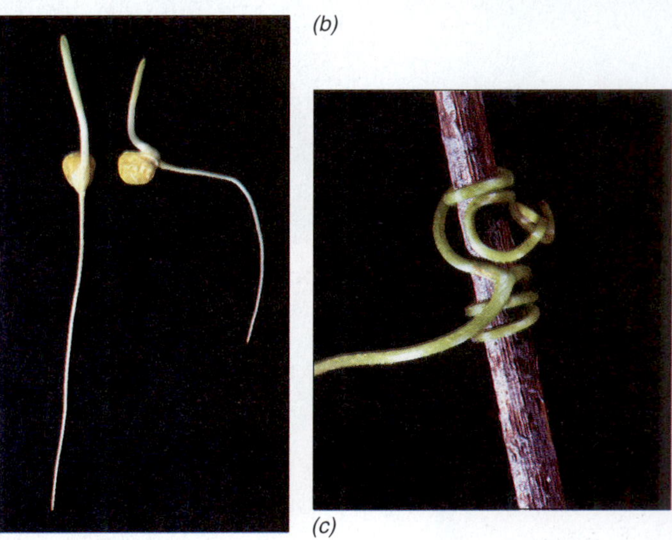

(b)

(c)

FIGURE 30–8 Tropisms. (***a***) Stems exhibit positive phototropism. (***b***) Stems exhibit negative gravitropism, and roots exhibit positive gravitropism. Two corn seeds were germinated at the same time. The 4-day-old plant on the left served as a control. The plant on the right was turned on its side on day 3, making the shoot and root horizontal. In 24 hours, new root growth was downward and new shoot growth was upward. (***c***) The twining motion of a grape tendril is an example of thigmotropism. (*a, b,* Dennis Drenner; *c,* Carolina Biological Supply Company).

TABLE 30–1
Some of the Interactions between Plant Hormones during Various Aspects of Plant Growth

Physiological activity	*Auxin*	*Gibberellin*	*Cytokinin*	*Ethylene*	*Abscisic acid*	*Other factors for some plants*
Seed germination		Promotes	?		Inhibits	Cold requirement, light requirement
Growth of seedling into mature plant	Cell elongation, organogenesis[1]	Cell division and elongation	Cell division and differentiation, organogenesis[1]			
Apical dominance	Inhibits lateral bud development		Promotes lateral bud development	?		
Initiation of reproduction (flowering)		Stimulates flowering in some plants[2]	?			Cold requirement, photoperiod requirement
Fruit development and ripening	Development	Development		Promotes ripening		Light requirement (for pigment formation)
Leaf abscission	Inhibits		Inhibits	Promotes		Light requirement
Winter dormancy of plant		Breaks	?		Promotes	Light requirement
Seed dormancy		Breaks	?		Promotes	

[1] In plant tissue culture.
[2] Gibberellin cannot be considered *the* flowering hormone. There is evidence for a flowering hormone that has not yet been isolated and characterized.

called **gravitropism.** Stems (that is, shoot tips) generally exhibit negative gravitropism (they grow away from the center of the Earth), whereas root tips exhibit positive gravitropism (Fig. 30–8*b*). **Thigmotropism** is growth in response to a mechanical stimulus, such as contact with a solid object. The twining or curling growth of tendrils, which helps attach a plant such as a vine to some type of support, is an example of thigmotropism (Fig. 30–8c). Tropisms in plants may also be caused by other stimuli in the environment such as water, temperature, chemicals, and oxygen. Because tropisms are growth responses, they cause permanent changes in the position of a plant part.

HORMONES ARE CHEMICAL MESSENGERS THAT REGULATE PLANT GROWTH AND DEVELOPMENT

Plant **hormones** are chemical messengers produced in one part of a plant and transported to another part, where they elicit a physiological response. Hormones are effective in extremely minute amounts. For that reason, their study is very challenging. The study of plant hormones is particularly difficult because each hormone elicits *many* different responses. In addition, the effects of different hormones overlap, so that it is difficult to determine which hormone, if any, is the primary cause of a particular response. Also, plant hormones may be stimulatory at one concentration and inhibitory at a different concentration. The five classes of plant hormones that have been identified thus far are auxins, gibberellins, cytokinins, ethylene, and abscisic acid (Table 30–1). Together they control the growth and development of a plant.

Auxins Regulate the Growth of Cells by Promoting Cell Elongation

Although he is known mostly for developing the theory of natural selection to explain evolution, Charles Darwin was a gifted naturalist who experimented on many plants

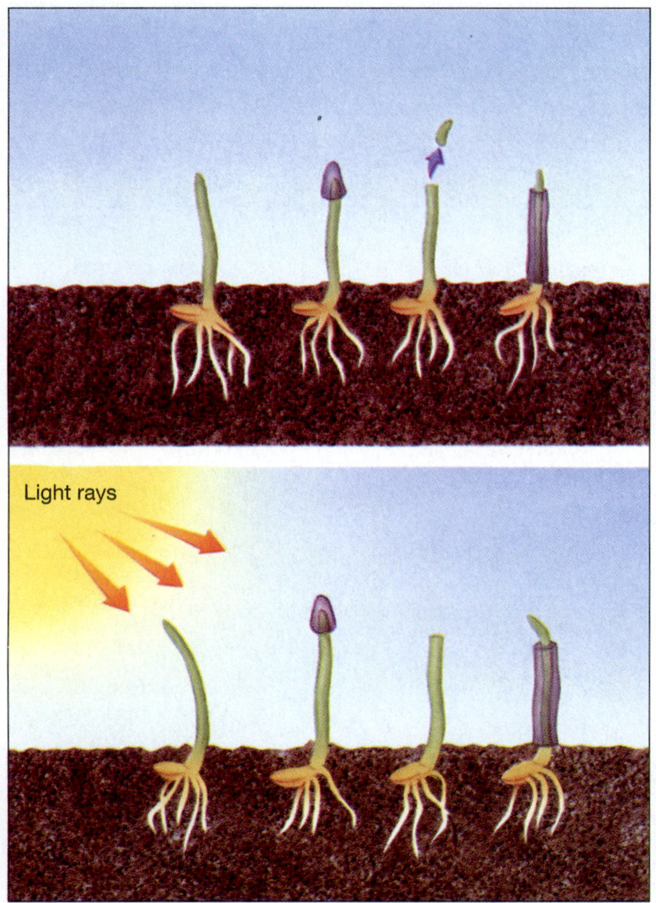

FIGURE 30–9 The Darwins' experiments with coleoptiles of canary grass seedlings. (*Upper row*) Some plants were uncovered, some were covered only at the tip, some had the tip removed, and some were covered everywhere but at the tip. (*Lower row*) After exposure to light coming from one direction, the uncovered plants and the plants with uncovered tips (*far right*) grew toward the light. The plants with covered tips (*center left*) or tips removed (*center right*) did not bend toward light.

and animals. Darwin and his son, Francis, provided the first evidence for the existence of auxins. The experiments that they performed in the 1880s involved positive phototropism, the growth of plants toward light, of newly germinated canary grass seedlings (Fig. 30–9). The first part of the grass seedling to emerge from the soil is the coleoptile, the protective sheath that encircles the stem. When coleoptiles are exposed to unidirectional light, they bend toward the light. The bending occurs close to the tip of the coleoptile.

The Darwins tried to influence this bending in several ways. For example, they covered the tip of the coleoptile as soon as it emerged from the soil. Even though they covered a part of the coleoptile *above* where the bend occurs, the plants treated in this manner did not bend. Likewise, bending did not occur when the coleoptile tip was removed, but when the bottom of the coleoptile was shielded from the light, the coleoptile still bent toward light. From these experiments, the Darwins concluded that "some influence is transmitted from the upper to the lower part, causing it to bend."

It took many years before this substance, named **auxin** (from the Greek word for "enlarge" or "increase"), was successfully extracted and identified. Auxins are any of a group of compounds that stimulate phototropic curvature in coleoptiles and stems. The main auxin found in plants is **indoleacetic acid (IAA).**

Auxin promotes cell growth by triggering cell elongation. Auxin apparently exerts this effect by changing the cell walls so they can expand. Auxin's effect on cell elongation also explains phototropism. When a plant is exposed to a unidirectional source of light, the auxin migrates to the dark side of the stem before moving down the stem (Fig. 30–10). As a result, the cells on the dark

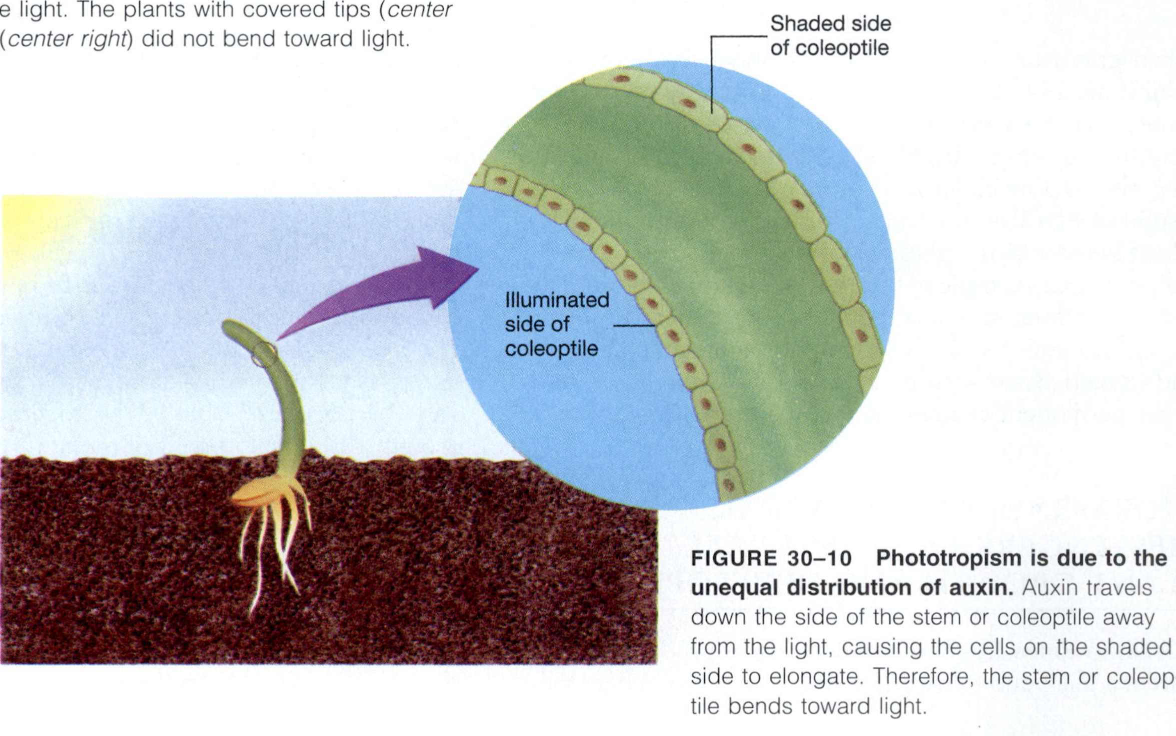

FIGURE 30–10 Phototropism is due to the unequal distribution of auxin. Auxin travels down the side of the stem or coleoptile away from the light, causing the cells on the shaded side to elongate. Therefore, the stem or coleoptile bends toward light.

side of the stem elongate more than do the cells on the light side, and the stem bends toward the light. Auxin is thought to be involved in gravitropism and thigmotropism as well.

Auxin exerts other effects on plants. For example, certain plants tend to branch out very little when they grow. Growth in these plants occurs almost exclusively from the apical meristem, rather than from lateral meristems. Such plants are said to exhibit **apical dominance** (Fig. 30–11). In plants with strong apical dominance, it appears that auxin produced in the apical meristem inhibits the development of lateral buds into actively growing shoots. When the apical meristem is pinched off, the auxin source is removed and lateral buds grow into branches.

Auxin produced by developing seeds stimulates the development of the fruit. When auxin is applied to flowers in which fertilization has not occurred (and in which, therefore, seeds are *not* developing), the ovary enlarges and develops into a seedless fruit. Seedless tomatoes have been produced in this manner. Auxin is not, however, the only hormone involved in fruit development.

A number of manufactured, or synthetic, auxins have been made that have structures similar to indoleacetic acid. One synthetic auxin is used to stimulate root development on stem cuttings for asexual propagation, particularly of woody plants (Fig. 30–12). Another synthetic auxin, 2,4-D (discussed in the chapter introduction) is used as a selective herbicide.

Gibberellins Promote Growth

In the 1920s, a Japanese biologist was studying a disease of rice in which the young rice seedlings grow extremely tall and spindly, fall over, and die. The cause of the disease, called the "foolish seedling" disease, was discovered to be a fungus. The fungus produces a chemical substance, named **gibberellin,** that causes the symptoms. Not until after World War II did scientists in Europe and North America learn of the exciting work done by the Japanese. During the 1960s, gibberellins were isolated from healthy plants and discovered to be involved in many normal plant functions. In the case of the "foolish seedling" disease, the symptoms were caused by an abnormally high gibberellin concentration in the plant tissue.

FIGURE 30–11 Auxin inhibits the development of lateral buds. (***a***) When the tip of a plant (which provides its source of auxin) is intact, the lateral buds do not develop. (***b***) The tip has been removed. Because there is no auxin moving down from the stem tip, lateral buds develop into branches.

FIGURE 30–12 Commercial application of synthetic auxins. Honeysuckle cuttings treated with a synthetic auxin. Many adventitious roots formed on cuttings placed in a higher auxin concentration (*left*), whereas fewer roots formed in a lower auxin concentration (*middle*). Cuttings placed in water (*right*) served as a control and did not form roots in the same time period. (Visual Unlimited/Joe Eakes/Color Advantage).

As in the "foolish seedling" disease, gibberellins promote stem elongation in many plants. When gibberellin is applied to a plant, this elongation may be spectacular, particularly in plants that normally have very short stems. Some dwarf mutants of corn and peas will grow to a normal height when treated with gibberellins (Fig. 30–13). Gibberellins are also involved in **bolting,** the rapid stem elongation that occurs when many plants initiate flowering (Fig. 30–14). In all of these cases, gibberellins cause stem elongation by inducing cells to divide as well as elongate. The actual mechanism of cell elongation appears to be different from that caused by auxin, however.

Gibberellins are involved in several reproductive processes in plants. They stimulate flowering, particularly in long-day plants. In addition, they can substitute for the low-temperature requirement that biennials have before the initiation of flowering. If gibberellins are applied to biennials during their first year of growth, flowering occurs without the period of low temperature. Gibberellins, like auxins, also affect the development of fruits. Commercially, gibberellins are applied to several varieties of grapes to produce larger berries.

Gibberellins are involved in the germination of seeds in many plants. In plants with light or low-temperature

FIGURE 30–13 Effect of the continued application of gibberellin on normal and dwarf corn plants. From left to right: dwarf, untreated; dwarf treated with gibberellin; normal treated with gibberellin; normal, untreated. Note that the dwarf plants respond to gibberellin much more dramatically than the normal plants. In fact, dwarf plants treated with gibberellin resemble normal plants in their growth rate. This dwarf variety is a mutant with a single recessive gene that impairs gibberellin metabolism.) (Courtesy of B.O. Phinney, UCLA)

FIGURE 30–14 Bolting in cabbage. Spectacular stem elongation often accompanied by flowering was caused by treatment with gibberellin. (*left*) Untreated controls. (*right*) Cabbages treated with gibberellin. (Courtesy of Sylvan Wittwer and Michigan State University).

requirements for seed germination, the application of gibberellins can substitute for the specific environmental requirement.

Cytokinins Promote Cell Division

During the 1940s and 1950s, a number of researchers were trying to find substances that might induce plant cells to divide in **tissue culture,** a technique that involves isolating cells from plants and growing them in a nutrient medium. It was discovered that cells would not divide without some substance found in either coconut milk or herring sperm. Finally, the active substance was isolated from herring sperm and called **cytokinin** because it induces cell division, or cytokinesis. In 1963, the first naturally occurring cytokinin in plants was identified from corn, and since that time several similar molecules have been identified from other plants.

Cytokinins promote cell division and differentiation in intact plants. They are also a required ingredient of plant tissue culture media and must be present in order for cells to divide. In tissue culture, cytokinins interact with auxin during the formation of plant organs like roots and stems (Fig. 30–15). For example, in tobacco tissue culture, a high ratio of cytokinin to auxin induces shoots to form, whereas a low ratio of cytokinin to auxin induces roots to form.

Cytokinins and auxin also interact in the control of apical dominance. Here their relationship is antagonistic: auxin inhibits the growth of lateral buds, and cytokinin promotes their growth.

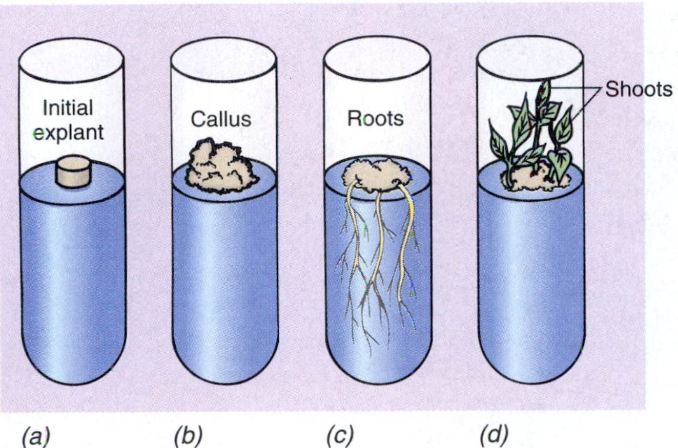

FIGURE 30–15 Growth responses of tobacco tissue culture to auxin and cytokinin. (***a***) The initial explant is a small piece of sterile tissue from the pith of a tobacco stem, which is placed on a nutrient agar medium. After several weeks, the kinds of growth illustrated occur on media supplemented with varying levels of auxin and cytokinin. (***b***) Nutrient agar containing a medium amount of both auxin and cytokinin caused cells to divide and form a clump of undifferentiated cells called a callus. (***c***) When callus is transplanted to media with a higher relative amount of auxin (that is, 2 mg/L auxin, 0.2 mg/L cytokinin), roots differentiate. (***d***) Shoot growth is stimulated by media containing a higher relative amount of cytokinin (that is, 0.02 mg/L auxin, 1 mg/L cytokinin).

One very interesting effect of cytokinins on plant cells is to delay **senescence,** or aging (Fig. 30–16). Plant cells, like all living cells, go through a natural aging process. This process is accelerated in cells of plant parts that are cut such as cut flowers. It is thought that plants must have a continual supply of cytokinins from the roots. Cut flowers, of course, lose their source of cytokinins and therefore age and die rapidly. Commercially, cytokinins are sprayed on cut flowers to prevent their rapid senescence.

FIGURE 30–16 Senescence was delayed in the green leaf by repeated application of cytokinin. Compare this leaf with the rest of the plant, which was not treated with cytokinin. (A.C. Leopold, Cornell University).

Ethylene Stimulates Abscission and the Ripening of Fruits

Ethylene, the only gaseous hormone produced by plants, was first discovered in 1934. Many diverse plant processes are influenced by ethylene. Ethylene inhibits cell elongation, promotes the germination of seeds, and is involved in plant responses to wounding or invasion by disease-causing microorganisms.

Ethylene also has a major role in many aspects of senescence, including the ripening process in fruits. As a fruit ripens, it produces ethylene, which triggers an acceleration of the ripening process. This induces the fruit to produce more ethylene, which further accelerates ripening. The expression "one rotten apple spoils the lot" is true. A rotten apple is one that is overripe. This apple produces a large amount of ethylene, which diffuses and triggers the ripening process in nearby apples. Ethylene is used commercially to promote the uniform ripening of bananas. Bananas are picked while green and shipped to their destination. There they are exposed to ethylene before delivery to grocery stores.

Ethylene has been implicated as the hormone that induces leaf abscission (Chapter 27). However, abscission is affected by two plant hormones that are antagonistic toward one another—ethylene and auxin. As the leaf ages and autumn approaches, the level of auxin in the leaf decreases. Concurrently, cells in the abscission zone begin producing ethylene. To further complicate the process, it is possible that cytokinins may be involved in abscission. Cytokinins, like auxin, decrease in concentration as leaf tissue ages.

Abscisic Acid Promotes Bud and Seed Dormancy

Abscisic acid was discovered simultaneously in 1963 by two independent research teams. Its name was an unfortunate choice because abscisic acid is primarily involved in dormancy and does not induce abscission in most plants.

Abscisic acid is sometimes referred to as the plant "stress hormone" since it promotes change in plant tissues that are stressed, or exposed to unfavorable conditions. (Ethylene also affects plant responses to certain stresses.) The effect of abscisic acid on plants suffering from water stress is best understood. The level of abscisic acid increases dramatically in the leaves of plants that are exposed to severe drought conditions. The high level of abscisic acid in the leaves triggers the closing of stomata, which saves the water normally transpired through the stomata, thereby increasing the plant's likelihood of survival.

The onset of winter could also be considered a type of stress on plants. As winter approaches, woody plants cease growing, and protective coverings of bud scales form over terminal buds. These adaptations are promoted by abscisic acid. Another winter adaptation that involves abscisic acid is dormancy in seeds. Many seeds have high levels of abscisic acid in their tissues and are therefore unable to germinate. In a corn mutant that is unable to synthesize abscisic acid, the seeds germinate as soon as the embryos are mature, while still attached to the ear (Fig. 30–17).

The evidence that abscisic acid is the only hormone involved in both bud and seed dormancy is not conclusive, particularly because the addition of gibberellin reverses the effects of dormancy. In seeds, the level of abscisic acid decreases during the winter, and the level of gibberellin increases. Cytokinins have also been implicated in breaking dormancy. Once again we see that a single physiological activity in plants may be controlled by the interaction of several hormones. The actual response made by a plant may be the result of changing ratios of hormones rather than a result of the effects of each individual hormone.

FIGURE 30–17 Inability to produce abscisic acid can prevent seed dormancy in corn. Some of the white kernels have germinated prematurely, while still on the ear, to produce coleoptiles (see arrows). (Courtesy of M.G. Neuffer).

Chapter Summary

I. Plant development is controlled not only by genetic factors but also by external environmental factors.
   A. Other plant tissues and organs exert a profound influence on plant development, particularly by secreting hormones.
   B. Many factors in the physical environment determine gene expression and affect plant development.

II. Many plants flower in response to specific cues from the environment.
   A. Photoperiodism is the response of plants to the duration and timing of light and dark.
      1. Flowering is a photoperiodic response, with some plants being short-day plants and others being long-day plants. In day-neutral plants, flowering is not affected by photoperiod.
      2. The photoreceptor in photoperiodism is phytochrome, a blue-green pigment with two forms, P_r and P_{fr}.
   B. Vernalization is the promotion of flowering by exposure to low temperatures.

III. Seed germination is affected by both internal and external factors.
   A. External environmental factors that may affect seed germination include requirements for oxygen, water, temperature, and light.
   B. Internal factors affecting whether a seed germinates include the maturity of the embryo and the presence or absence of chemical inhibitors.

IV. Circadian rhythms are regular rhythms in growth or activities of a plant that approximate a 24-hour day and are reset by the rising and setting of the sun.

V. Turgor movements and tropisms are the two kinds of plant movements in response to external stimuli.
   A. Turgor movements are caused by temporary changes in turgor in special cells.
   B. Tropisms are directional growth responses and are permanent.
      1. Phototropism is the growth of a plant in response to the direction of light.

2. Gravitropism is the growth of a plant in response to the influence of gravity.
3. Thigmotropism is the growth of a plant in response to contact with a solid object.

VI. Hormones regulate plant growth and development and are effective in small amounts.
  A. The functions of hormones overlap.
  B. Many effects of hormones may be due to the interactions of several hormones rather than the effect of a single hormone.
  C. There are five classes of plant hormones.
    1. Auxins are involved in cell elongation, tropisms, apical dominance, and fruit development.
    2. Gibberellins are involved in stem elongation, flowering, and seed germination.
    3. Cytokinins promote cell division and differentiation, delay senescence, and interact with auxins in apical dominance.
    4. Ethylene has a role in the ripening of fruits, leaf abscission, and senescence.
    5. Abscisic acid is the stress hormone. It is involved in stomatal closure due to water stress and in bud and seed dormancy.

Selected Key Terms

abscisic acid, p. 589
apical dominance, p. 587
auxin, p. 586
bolting, p. 587
circadian rhythm, p. 582
cytokinin, p. 588
ethylene, p. 589
gibberellin, p. 587
gravitropism, p. 585
hormone, p. 585
photoperiodism, p. 578
phototropism, p. 584
phytochrome, p. 578
pulvinus, p. 584
senescence, p. 589
sleep movements, p. 582
solar tracking, p. 584
thigmotropism, p. 585
turgor movements, p. 584
vernalization, p. 580

Post-Test

1. The response of a plant to the relative amounts of daylight and darkness is known as ____________.
2. The critical factor in the flowering response of short-day and long-day plants is the amount of ____________ (daylight or darkness).
3. The P_{fr} form of phytochrome is produced when P_r absorbs ____________ (red or far-red) light.
4. ____________ is the promotion of flowering by a low-temperature treatment.
5. The ____________ is an organ at the base of the petiole in *Mimosa* that undergoes rapid, reversible changes in turgor, causing dramatic movements of the leaves.
6. ____________ is the growth of a plant in response to the direction of light.
7. The twining of a tendril around a wire is an example of ____________.
8. Sleep movements observed in beans are an example of a/an ____________ ____________.
9. ____________ are chemical messengers produced in one part of the plant and transported to another, where they affect growth and development.
10. A synthetic ____________, 2,4-D, is used as a selective herbicide.
11. Research on a disease of rice provided the first clues about the hormone, ____________.
12. Cytokinins interact with ____________ during the formation of roots and shoots in tissue culture.
13. ____________ delay senescence.
14. The only plant hormone that is a gas is ____________.
15. Abscisic acid promotes the ____________ of woody twigs.
16. The inhibition of lateral buds is known as (a) apical dominance (b) solar tracking (c) photoperiodism (d) thigmotropism (e) phototropism.
17. When a stem tip is removed, one or more lateral buds start to grow because (a) the dormant buds produce phytochrome (b) the removal of the stem tip removed the source of auxin, which inhibits lateral buds (c) cytokinins produced in the roots are transported to the lateral buds after the stem tip is removed (d) ethylene is no longer produced by the stem tip.

18. One commercial use of ethylene is to (a) kill weeds in corn fields (b) cause seedless tomatoes to develop (c) form roots on stem cuttings (d) ripen bananas.

19. Cytokinins (a) inhibit cell division (b) increase membrane permeability (c) promote cell division (d) increase cell elongation.

20. Which hormone has an important role in the regulation of stomatal closure during a drought? (a) auxin (b) cytokinin (c) gibberellin (d) ethylene (e) abscisic acid

Review Questions

1. Why is plant growth and development so sensitive to environmental cues?
2. What is phytochrome? Describe its role in flowering.
3. Define vernalization.
4. What factors influence the germination of seeds? Explain the role that each of these factors plays in influencing germination and discuss why the plant responds the way it does.
5. Distinguish between turgor movements and tropisms.
6. How is auxin involved in phototropism?
7. Define dormancy. Of what value is dormancy in seeds?
8. What is a hormone? Does your definition apply to both plant and animal hormones?
9. Discuss the various hormones that are involved in each of the following physiological processes: (a) germination of seeds (b) stem elongation (c) ripening of fruits (d) abscision of leaves (e) dormancy of seeds.
10. Summarize the roles of auxins, gibberellins, cytokinins, ethylene, and abscisic acid.

Thinking Critically

1. Although difficult to enforce, international law prohibits the use of so-called "germ warfare," that is, weapons designated to infect humans with specific diseases, such as anthrax or viral agents. Should the international community also ban herbicides from wartime use? Why or why not? Consider environmental effects as well as health effects in your discussion.
2. Predict whether flowering would be expected to occur in the following situations. Explain each answer.
   a. A short-day plant is exposed to 15 hours of daylight and 9 hours of darkness.
   b. A short-day plant is exposed to 9 hours of daylight and 15 hours of darkness.
   c. A short-day plant is exposed to 9 hours of daylight and 15 hours of darkness, with a 10-minute flash of red light in the middle of the night.
3. How might solar tracking benefit a plant? How might solar tracking *harm* a plant?
4. Why might some plants have sleep movements?
5. What benefits are conferred on a plant when its stems exhibit positive phototropism and its roots exhibit positive gravitropism?

Recommended Readings

Evans, M.L., R. Moore, and K.H. Hasenstein, "How Roots Respond to Gravity," *Scientific American*, December 1986. A closer look at gravitropism.

"Leaves with Clocks," in "Breakthroughs" section of *Discover*, September 1993. Sleep movements are another example of circadian rhythms in plants.

Mauseth, J.D. *Botany: An Introduction to Plant Biology*, 2nd edition, Philadelphia, Saunders College Publishing: 1995. A comprehensive introduction to plant biology.

Mores, P.B. and N.H. Chua, "Light Switches and Plant Genes," *Scientific American*, April 1988. How an environmental cue is related to gene expression.

Rensberger, B., "Getting to the Root of Plant Growth," in "Science" section of *Washington Post*, July 13, 1992. Why roots grow downward and shoots grow upward.

PART 8

Animal Structure and Life Processes

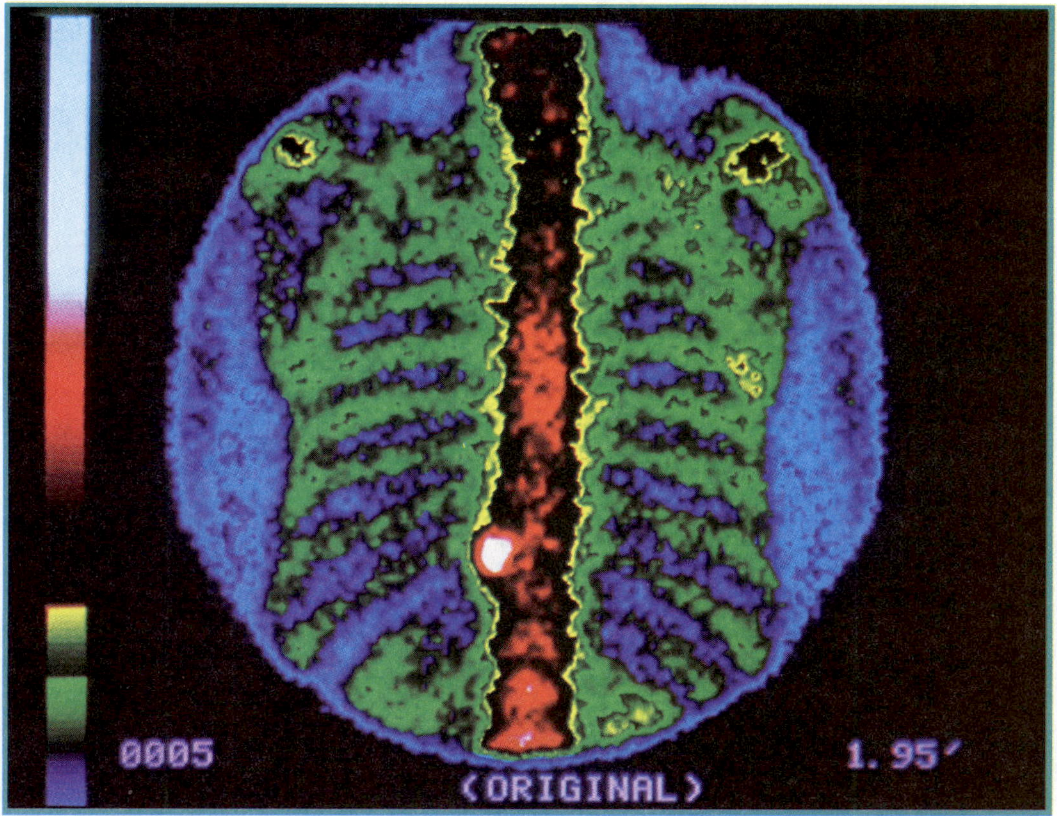

Bone scan of spine and ribs showing a metastatic bone cancer affecting the spine. *The tumor appears as the white area toward the bottom of the image. The bone scan records the distribution and intensity of gamma radiation emitted from a radionuclide that is injected and is concentrated by bone. Cancerous bone concentrates the radionuclide more strongly than normal bone, and appears as brighter "hot spots" on the image. (CNRI/Science Photo Library/Photo Researchers, Inc.)*

UNWELCOME TISSUES: STRATEGIES IN THE WAR AGAINST CANCER

In Chapter 10, we discussed normal mechanisms of cell division and described cancer as a disease that may develop when these mechanisms become abnormal. In this chapter, which focuses on animal tissues, we focus on cancer in the context of the animal body. A **neoplasm** ("new growth"), or **tumor,** is an abnormal mass of cells.

A benign ("kind") tumor tends to grow slowly, and its cells stay together. Because benign tumors form masses with distinct borders, they can usually be surgically removed. A malignant ("wicked") neoplasm, or **cancer,** usually grows much more rapidly and invasively than a benign tumor. Cancers that develop from connective tissue or muscle are referred to as **sarcomas.** Those that originate in epithelial tissues are called **carcinomas.** Unlike the cells of benign tumors, cancer cells do not retain normal structural features.

Death from cancer almost always results from **metastasis,** a migration of cancer cells through blood or lymph channels to distant parts of the body. Once there, they multiply, forming new malignant neoplasms. Such growths interfere with the normal functions of the tissues being invaded. Cancer often spreads so rapidly and extensively that surgeons are unable to locate or remove all the malignant masses.

Studies suggest that many neoplasms grow to a few millimeters in diameter and then enter a dormant stage, which may last for months or even years. At some point, cells of the neoplasm release a chemical substance that stimulates nearby blood vessels to develop new capillaries, which grow into the abnormal mass of cells. The neoplasm, nourished by its new blood supply, may begin to grow rapidly. Newly formed blood vessels have leaky walls. Malignant cells can enter the blood through these walls and are transported by the blood to new sites. Thus, these blood vessels are an important route for metastasis.

Cancer researchers are looking for ways to predict or prevent metastasis and also for treatments for cancer patients whose tumors have already metastasized. Several promising research approaches have been reported. For example, Lance Liotta, Chief of the Laboratory of Pathology at the National Cancer Institute, has studied the process used by cancer cells to invade body tissues. He and his colleagues discovered that increased levels of a class of protein-cleaving enzymes (metalloproteinases) correlated with metastasis of several types of human cancers. These investigators found that the enzymes attack collagen, a protein found in connective tissues (discussed later in this chapter). When they blocked the activity of the enzymes (by treating tumor cells with antibodies to the enzymes), invasion by the cancer cells did not occur. This research could lead to the development of drugs that prevent metastasis.

Animal Tissues, Organs, and Organ Systems

31

Learning Objectives

After you have studied this chapter you should be able to:

1. Discuss the advantages of multicellularity.
2. Define tissue, organ, and organ system.
3. Compare the four principal types of animal tissues (epithelial, connective, muscle, and nervous tissues) with respect to general structure and function.
4. List the functions of epithelial tissue, describe the three main shapes of epithelial cells, and describe how these cells can be arranged into tissues.
5. Compare the main types of connective tissue and their functions.
6. Contrast the three types of muscle tissue and their functions.
7. Identify the cells that make up nervous tissue and give their functions.
8. Briefly describe the organ systems of a complex animal.
9. Define homeostasis and give examples of homeostatic mechanisms.

Key Concepts

- ☐ Animals are multicellular. Their multicellularity makes it possible for animals to achieve large size, and also permits division of labor among cells, with groups of cells specializing to perform specific functions.
- ☐ In most animals, cells are organized into tissues, tissues are organized into organs, and organs are organized into organ systems. Animal tissues may be classified as epithelial, connective, muscle, or nervous tissue. Each type of tissue performs one or more functions.
- ☐ Homeostasis, the body's automatic tendency to maintain a constant internal environment, is maintained by negative feedback mechanisms.

FIGURE 31–1 All animals are multicellular. A large animal, such as the greater kudu *(Tragelaphus strepsiceros)* is composed of more cells than the much smaller yellow-billed oxpeckers (*Buphagus africanus*) riding on its back. (Frans Lanting/Minden Pictures)

Animals Are Multicellular

Animals, like plants and most fungi, are **multicellular** (that is, they are composed of many cells). They are not simply colonies or aggregations of similar cells, but are composed of a number of different types of cells, each with a characteristic size, shape, structure, and function. In most animals, cells are organized into tissues, tissues are organized into organs, and organs are organized into organ systems.

Recall from Chapter 5 that the size of individual cells is limited by the need to efficiently move materials in and out of the cell. When the volume of the cell becomes too large in relation to its plasma membrane, materials cannot be efficiently transported in and out of the cell. The cell divides, forming two cells. In unicellular organisms, cell division produces two new individuals. In multicellular organisms like animals, the two new cells may remain associated, forming a part of the organism.

One result of multicellularity is large size. The number of cells, not their individual dimensions, is responsible for the different sizes of various organisms. The cells of an earthworm and an elephant correspond in magnitude, but the elephant is larger because its genes are programmed to provide for larger numbers of cells (Fig. 31–1).

Consider how different a sea urchin is from a butterfly or tiger. These animals vary in size, body form, and life style. Such marked diversity is another result of multicellularity.

Multicellularity permits specialization. In a one-celled organism, the single cell carries on all life activities. In contrast, the cells of a multicellular organism specialize to perform specific functions. This division of labor permits the organism to become highly proficient at performing a wide variety of activities. Some cells, for example, may become sensory cells of the eyes and specialize in receiving information about the environment. Other cells may become brain cells that interpret incoming sensory information. Still other cells might be specialized to contract, permitting the animal to escape a predator or capture food. How do cells associate and perform such specialized functions as vision or body movement? To answer these questions, we examine the tissues, organs, and organ systems of complex animals.

Tissues of a Complex Animal Are Adapted to Carry Out Specific Functions

Even in the simplest animals, the sponges, there is a division of labor among several cell types. In all other animals, cells are not only specialized but organized, forming tissues. A **tissue** consists of a group of closely associated cells that are adapted to carry out specific functions. Animal tissues may be classified as epithelial, connective, muscle, or nervous tissue. Each type of tissue is composed of cells with a characteristic size, shape, and arrangement.

Epithelial Tissue Covers the Animal Body and Lines Its Cavities

Epithelial tissue (also called **epithelium**) forms the outer layer of the skin; the linings of the digestive, respiratory, and reproductive tracts; and the lining of the kidney tubules. Epithelial tissue consists of cells fitted tightly together, forming a continuous layer, or sheet, of cells. One surface of the sheet typically is free because it lines a cavity such as the intestine or covers the body (the skin). The other surface is attached to the underlying tissue by a

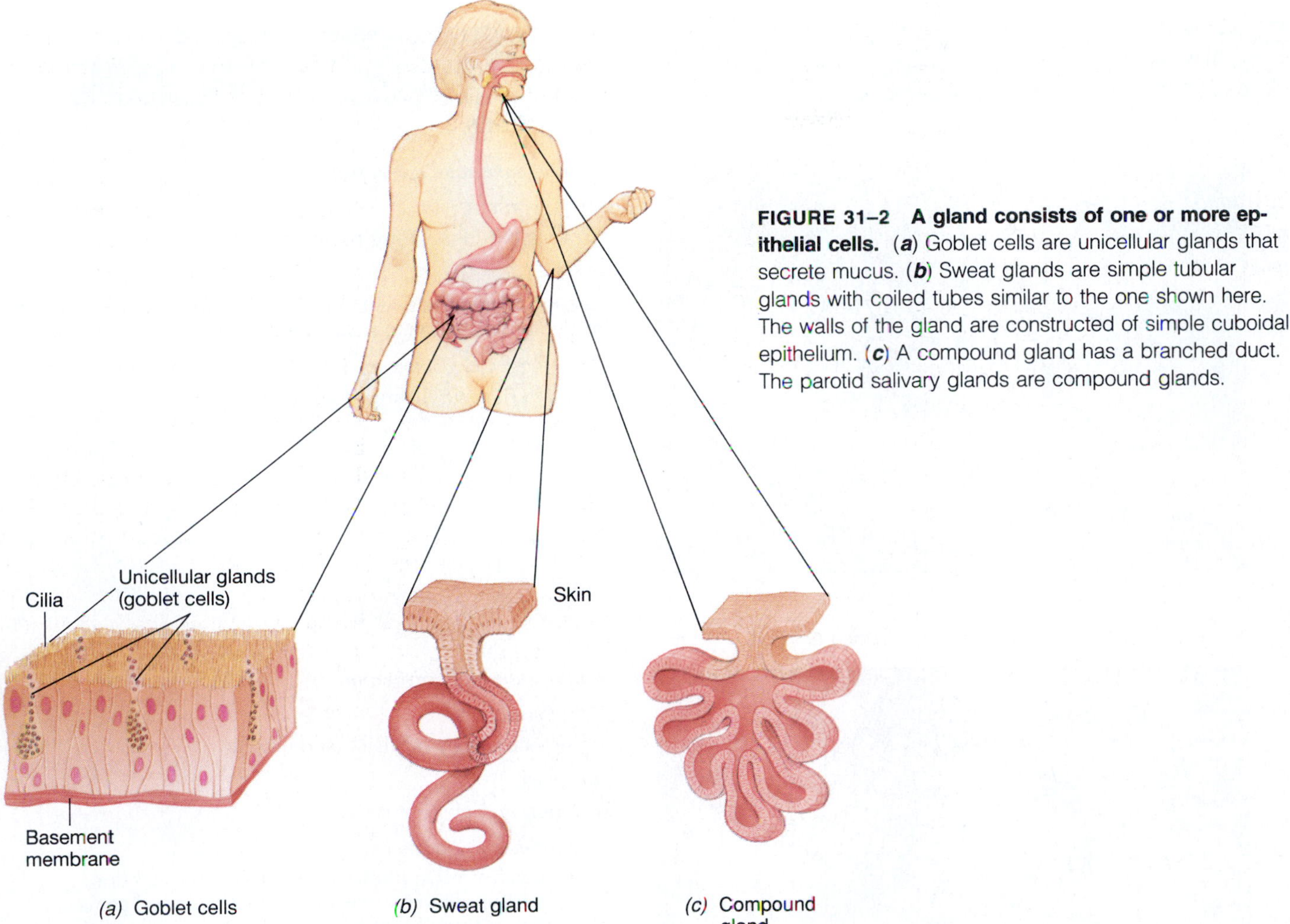

FIGURE 31–2 A gland consists of one or more epithelial cells. (*a*) Goblet cells are unicellular glands that secrete mucus. (*b*) Sweat glands are simple tubular glands with coiled tubes similar to the one shown here. The walls of the gland are constructed of simple cuboidal epithelium. (*c*) A compound gland has a branched duct. The parotid salivary glands are compound glands.

noncellular **basement membrane** that is composed of tiny fibers embedded in polysaccharides. Table 31–1 illustrates the main types of epithelial tissue, indicates their locations in the body, and describes their functions.

Epithelial tissues may function in protection, absorption, secretion, and sensation. As a covering or lining, epithelial tissue protects the body. The epithelial layer of the skin, called the epidermis, covers the entire body and protects it from mechanical injury, invading bacteria, and excessive water loss. The epithelial tissue lining the digestive tract absorbs nutrients and water into the body. Some epithelial cells are organized into glands, which secrete cell products such as hormones, enzymes, or sweat. Other epithelial cells are specialized as sensory receptors and as such receive information from the environment.

Everything that enters or leaves the body crosses one or more layers of epithelium. Even food that is taken into the mouth and swallowed is not really inside the body until it is absorbed through the epithelial lining of the digestive tract. The permeability of various epithelial tissues affects the exchange of substances among the different parts of the body and between the organism and the external environment.

Many epithelial membranes are subjected to continuous wear and tear. As outer cells are sloughed off, they are replaced by new ones from below. Such epithelial tissues generally have a rapid rate of cell division. As a result, new cells are continuously produced, taking the place of those that are lost.

Several types of epithelial tissue may be distinguished by the number of cell layers, the shape of the cells, and their arrangement (Table 31–1). Epithelium may be **simple,** consisting of one cell layer, or **stratified,** consisting of many layers (as in the outer layer of the skin). A third arrangement is **pseudostratified epithelium,** in which the cells (falsely) appear to be layered but actually are not. The shape of the epithelial cells may be **squamous** (flattened), **cuboidal** (cube-shaped), or **columnar** (elongated like columns). The free surface of the outer cells of the tissue may have specialized structures such as cilia or microvilli.

A **gland** consists of one or more epithelial cells that produce and secrete a product, for example, mucus, sweat, milk, saliva, hormones, or enzymes (Fig. 31–2). The epithelial tissue lining the cavities and passageways of the body typically contains **goblet cells,** which are uni-

TABLE 31–1
Epithelial Tissues

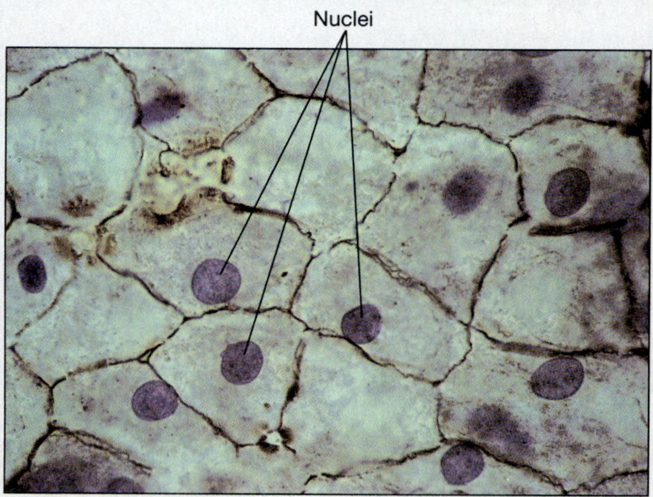

Simple squamous epithelium

Main Locations

Air sacs of lungs, lining of blood vessels

Functions

Passage of materials where little or no protection is needed and where diffusion is major form of transport

Description and Comments

Cells are flat and arranged as single layer

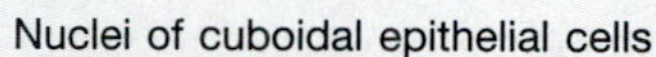

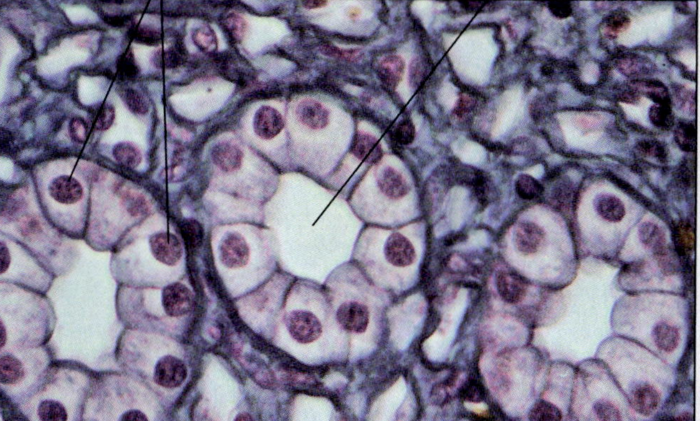

Simple cuboidal epithelium

Main Locations

Linings of kidney tubules, gland ducts

Functions

Secretion and absorption

Description and Comments

Single layer of cells; photograph shows cross section; from the side each cell looks like short cylinder; sometimes have microvilli for absorption

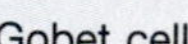

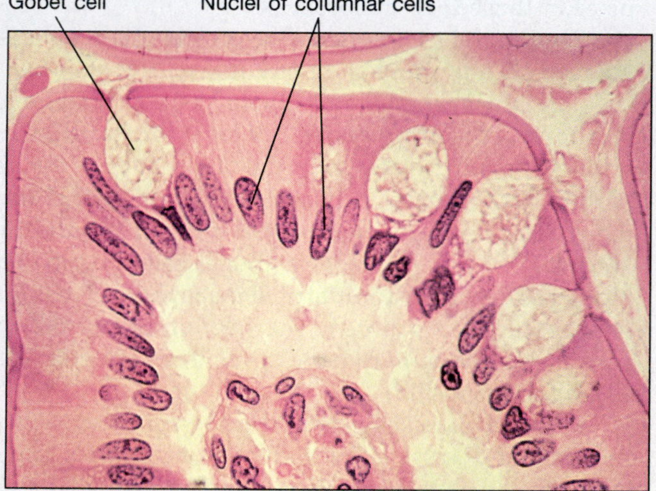

Simple columnar epithelium

Main Locations

Linings of much of digestive tract and upper part of respiratory tract

Functions

Secretion, especially mucus; absorption, protection, movement of mucous layer

Description and Comments

Single layer of columnar cells; sometimes with enclosed secretory vesicles (goblet cells), highly developed Golgi complex; often ciliated

TABLE 31–1
continued

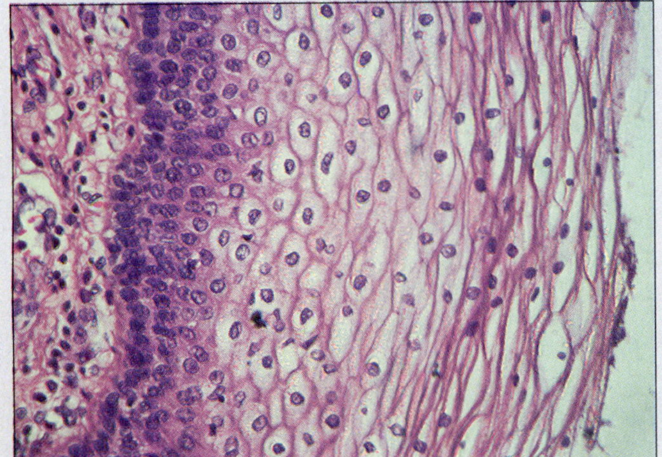

Stratified squamous epithelium

Main Locations

Skin, mouth lining, vaginal lining

Functions

Protection only; little or no absorption or transit of materials; outer layer continuously sloughed off and replaced from below

Description and Comments

Several layers of cells, with only the lower ones columnar and metabolically active; division of lower cells causes older ones to be pushed upward toward surface

25 μm

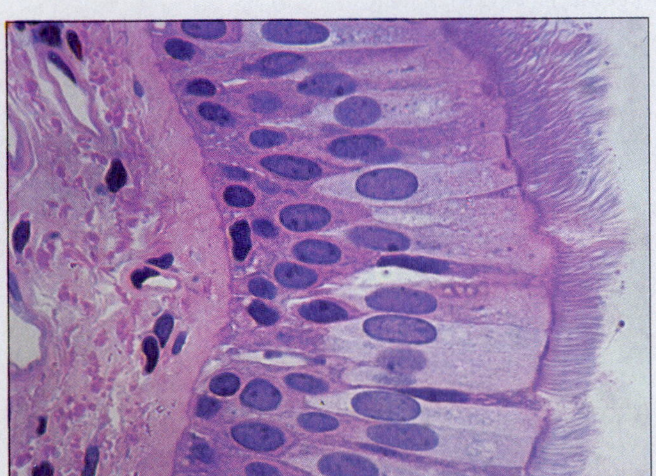

Pseudostratified epithelium

Main Locations

Some respiratory passages, ducts of many glands

Functions

Secretion, protection, movement of mucus

Description and Comments

Comparable in many ways to columnar epithelium except that not all cells are the same height; thus, although all cells contact the same basement membrane, the tissue appears stratified; ciliated, mucous-secreting, or with microvilli

25 μm

(All photos by Ed Reschke)

cellular glands that secrete mucus. The mucus lubricates these surfaces and facilitates the movement of materials.

Glands can be classified as exocrine or endocrine. **Exocrine glands,** like sweat glands, secrete their products onto a free epithelial surface, typically through a duct (tube). **Endocrine glands** lack ducts. As will be described in Chapter 41, endocrine glands release their products (called hormones) into the tissue fluid surrounding them. Hormones are then transported in the blood.

Connective Tissue Joins and Supports Other Body Structures

Almost every organ in the animal body has a framework of connective tissue that supports and cushions its epithelial components. Cartilage and bone are examples of **connective tissues** that support the animal body and protect organs such as the heart and lungs. Blood is a connective tissue that transports materials, and adipose tissue stores fat.

Unlike epithelium, connective tissue consists of relatively few cells separated by large amounts of **intercellular substance.** This intercellular substance consists of threadlike, microscopic **fibers** that are scattered throughout a **matrix.** The matrix is a thin gel composed of polysaccharides. The intercellular substance is secreted by connective tissue cells called **fibroblasts.**

Three types of fibers found in connective tissue are collagen, elastic, and reticular fibers. Most numerous are the **collagen fibers,** which are composed of the protein collagen (familiar in its hydrated form as gelatin). The strength of these flexible fibers has been compared to that of steel.

Elastic fibers stretch easily and then, like a rubber band, snap back to their normal length when the stress

TABLE 31–2
Connective Tissues

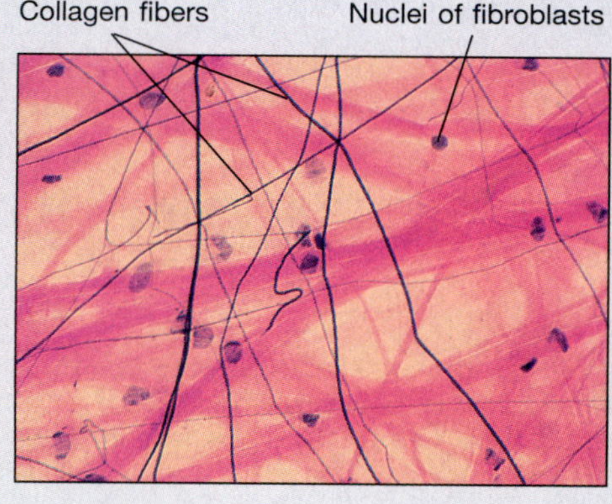

Loose connective tissue

Main Locations

Everywhere support must be combined with elasticity, e.g., subcutaneous layer

Functions

Support; reservoir for fluid and salts

Description and Comments

Fibers embedded in semifluid matrix and mixed with other cells

25 μm

(Photo by Ed Reschke)

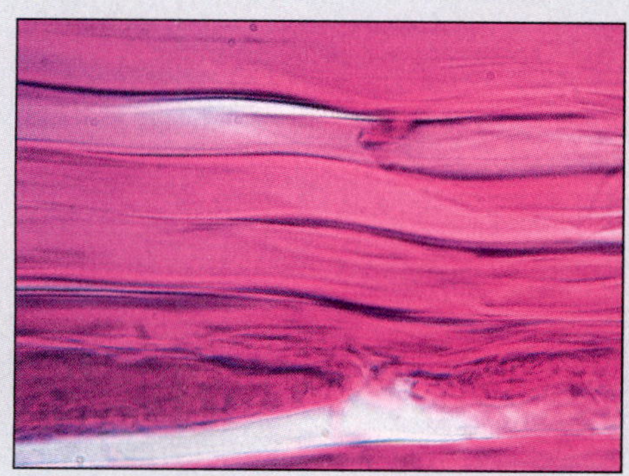

Dense connective tissue

Main Locations

Tendons, many ligaments, dermis of skin

Functions

Support; transmission of mechanical forces

Description and Comments

Collagen fibers may be regularly or irregularly arranged

25 μm

(Photo by Dennis Drenner)

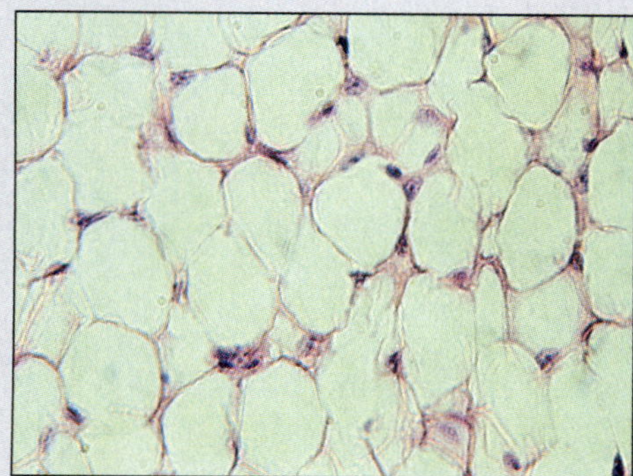

Adipose tissue

Main Locations

Subcutaneous layer; pads around certain internal organs

Functions

Food storage; insulation; support of such organs as mammary glands, kidneys

Description and Comments

Fat cells are star-shaped at first; fat droplets accumulate until typical ring-shaped cells are produced

25 μm

(Photo by Dennis Drenner)

is removed. These fibers are an important component of structures that must stretch. For example, elastic fibers in the walls of a large artery permit the artery to stretch as it fills with blood. As blood moves on through the artery, the expanded region snaps back to its normal size. (When you feel your pulse, you are feeling this alternate expansion and recoiling of an artery.) **Reticular fibers** are very fine, branched fibers that form a supporting network within many tissues and organs.

Some of the main types of connective tissue are (1) loose and dense connective tissues; (2) adipose tissue; (3) cartilage; (4) bone; and (5) blood, lymph, and tissues that produce blood cells (Table 31–2). Each of these tissues is exquisitely adapted to the specific functions it performs.

Loose Connective Tissue Is Found in the Subcutaneous Layer; Dense Connective Tissue Is Located in the Skin and Tendons

Loose connective tissue is the most widely distributed connective tissue in the body. It is found as a thin filling between body parts and serves as a reservoir for fluids and salts. Nerves, blood vessels, and muscles are

TABLE 31–2
continued

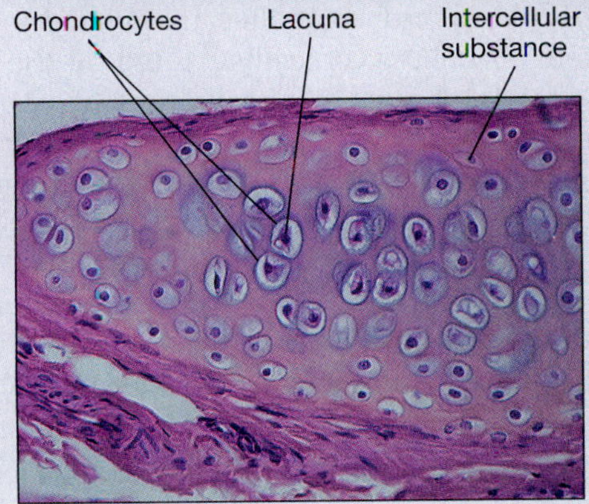

Cartilage

Main Locations

Supporting skeleton in sharks, rays, and some other vertebrates; in other vertebrates, ends of bones; supporting rings in walls of some respiratory tubes; tip of nose; external ear

Functions

Flexible support and reduction of friction in bearing surfaces

Description and Comments

Cells (chondrocytes) separated from one another by intercellular substance; cells occupy lacunae

25 μm

(Photo by Ed Reschke)

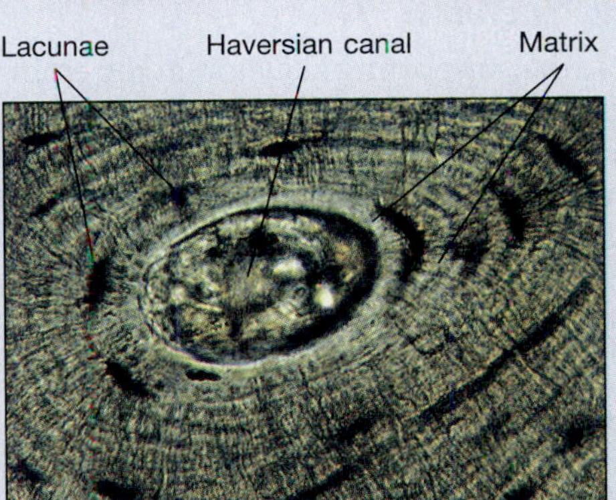

Bone

Main Locations

Forms skeletal structure in most vertebrates

Functions

Support and protection of internal organs; calcium reservoir; skeletal muscles attach to bones

Description and Comments

Osteocytes in lacunae; in compact bone, lacunae arranged in concentric circles about haversian canals

25 μm

(Photo by Dennis Drenner)

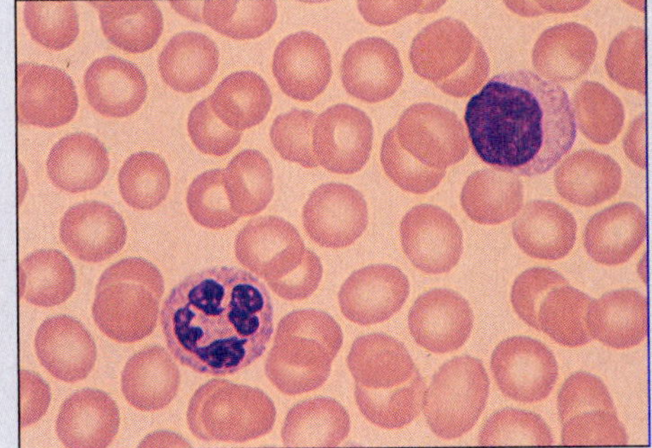

Blood

Main Locations

Within heart and blood vessels of circulatory system

Functions

Transports oxygen, nutrients, wastes, and other materials

Description and Comments

Consists of cells dispersed in fluid intercellular substance

25 μm

(Photo by Ed Reschke)

wrapped in this tissue. Together with adipose tissue, loose connective tissue forms the subcutaneous layer that attaches skin to the underlying structures. Loose connective tissue consists of fibers strewn in all directions through a semifluid matrix. Its flexibility permits the parts it connects to move.

Dense connective tissue is very strong and somewhat less flexible than loose connective tissue. Collagen fibers predominate. In irregular dense connective tissue, the collagen fibers are arranged in bundles that are distributed in all directions through the tissue. This type of tissue is found in the lower layer (dermis) of the skin. In regular dense connective tissue, the collagen bundles are arranged in a definite pattern, making the tissue greatly resistant to stress. Tendons, the cable-like cords that connect muscles to bones, consist of dense connective tissue.

Adipose Tissue Stores Fat

Adipose tissue is rich in cells that store fat and release it when fuel is needed for cellular respiration. Adipose tissue is found in the subcutaneous layer, as well as in tissue that cushions internal organs. An immature fat cell is somewhat star-shaped. As fat droplets accumulate within

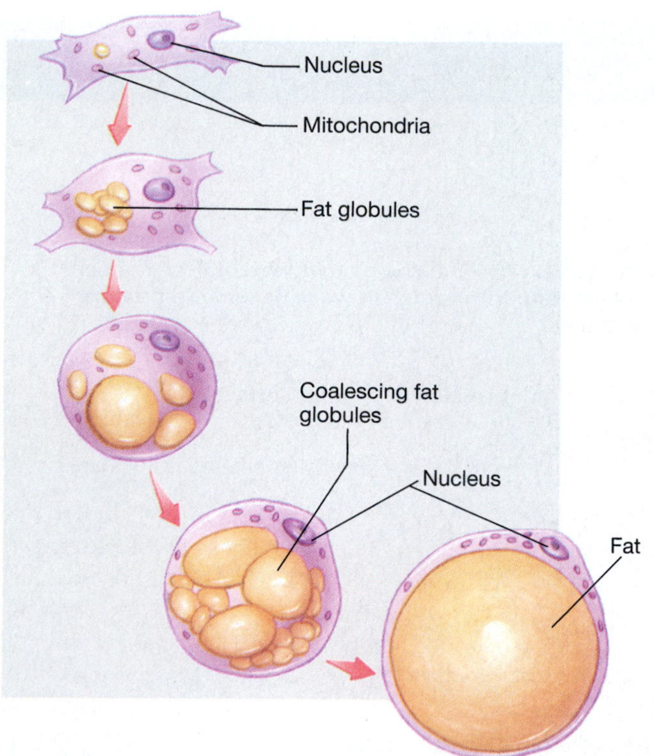

FIGURE 31–3 Storage of fat in a fat cell. As more and more fat droplets accumulate in the cytoplasm, they coalesce to form a very large globule of fat. Such a fat globule may occupy most of the cell, pushing the cytoplasm and the organelles to the periphery. (See Table 31–2 for a photomicrogaph of fat cells filled with fat.)

the cytoplasm, the cell assumes a more rounded appearance (Fig. 31–3). Fat droplets eventually merge with one another until a single large drop of fat occupies most of the cell. The cytoplasm and organelles are pushed to the cell edges, where a bulge is typically created by the nucleus. A cross section of such a fat cell looks like a ring with a single stone.

In a photomicrograph, adipose tissue looks somewhat like chicken wire. The "wire" represents the rings of cytoplasm, and the large spaces indicate where fat drops existed before they were dissolved by chemicals used to prepare the tissue. The empty spaces may cause the cells to collapse, resulting in a wrinkled appearance.

Cartilage and Bone Form a Supporting Skeleton

Cartilage forms the supporting skeleton in the embryo of every vertebrate. During development, it is largely replaced by bone, except in the sharks and rays. In humans, cartilage forms part of the external ear, the supporting rings in the walls of the respiratory passageways, and the tip of the nose. Cartilage also forms a thin layer at the ends of bones where they come together to form joints.

Cartilage is firm yet elastic. Cartilage cells, called **chondrocytes,** secrete this hard, rubbery matrix around themselves and also secrete collagen fibers, which be-

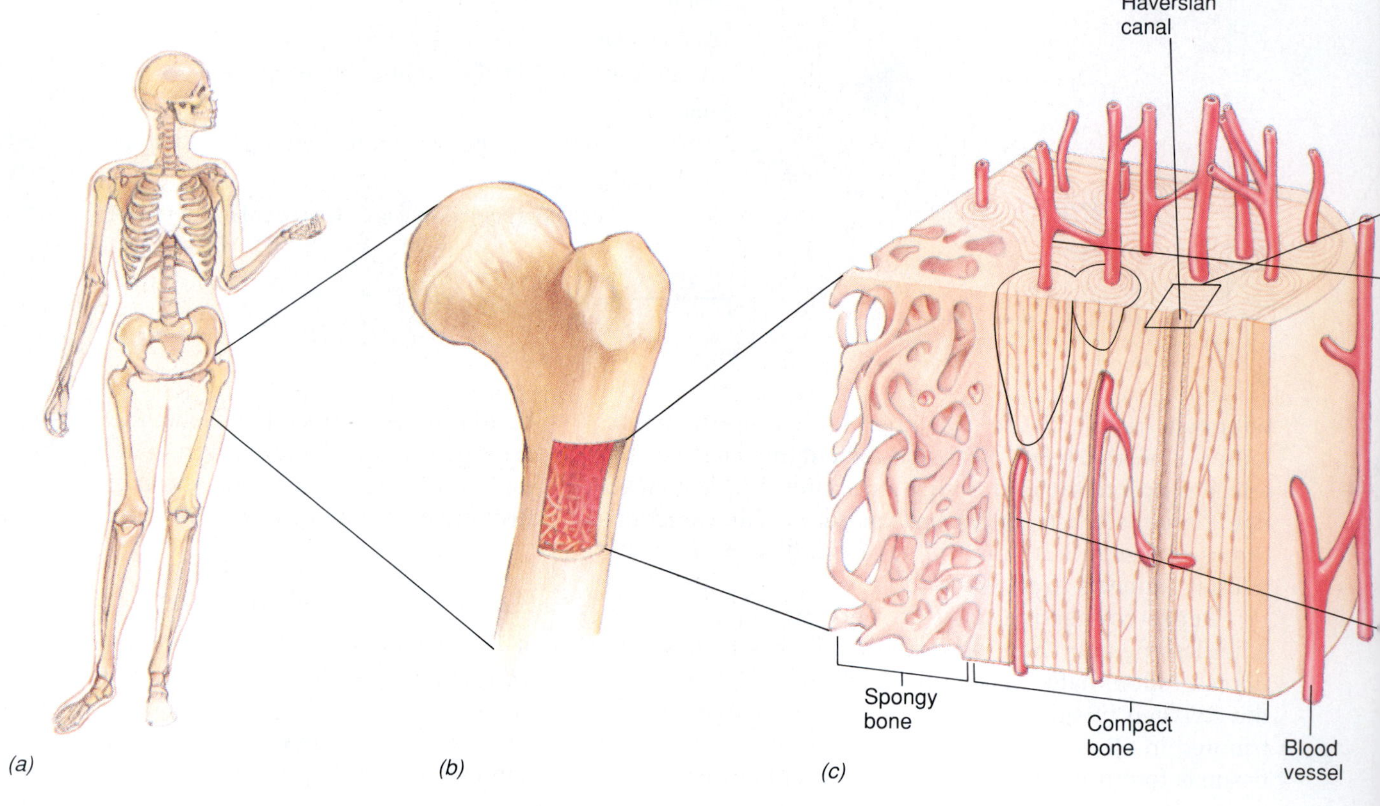

come embedded in the matrix, adding strength. Chondrocytes eventually come to lie singly or in groups of two or four in small cavities, called **lacunae,** in the matrix. The cartilage cells in the matrix remain alive, and are nourished by diffusion of nutrients and oxygen through the matrix. Cartilage tissue lacks nerves, lymph vessels, and blood vessels.

Bone is the principal vertebrate skeletal tissue. It is similar to cartilage in that the **osteocytes** (bone cells) that secrete and maintain the matrix are located in lacunae within the matrix (Fig. 31–4). Unlike cartilage, however, bone has many blood vessels.

In compact bone, the osteocytes are arranged around central capillaries in concentric layers (lamellae), which form spindle-shaped units known as **osteons.** The capillaries, as well as nerves, run through central microscopic channels in the osteons known as **haversian canals.** Bone is discussed further in Chapter 32.

Blood and Lymph Are Circulating Tissues

Like other connective tissues, **blood** and **lymph** consist of specialized cells dispersed in an intercellular substance. Blood and lymph are circulating tissues that enable various parts of the body to communicate. These tissues deliver nutrients and oxygen to the cells and carry away wastes and carbon dioxide.

Vertebrate blood consists of red blood cells, white blood cells, and platelets suspended in a fluid component, called plasma. The functions of the various components of blood will be discussed in Chapter 36.

Muscle Tissue Is Specialized to Contract

In many animals, **muscle** is the most abundant tissue. It accounts for nearly two-thirds of the body weight of a human. Muscle tissue is specialized for contraction and permits a wide range of movement in animals. Because they are long and narrow, muscle cells are referred to as **fibers.** There are three types of muscle tissue: skeletal, cardiac, and smooth (Table 31–3).

Skeletal muscle, which is attached to the bones, can be contracted voluntarily. This muscle tissue permits us to write, walk, and swim. Characterized by a pattern of light and dark stripes (striations), it is also referred to as *striated muscle.* Each skeletal muscle fiber has several nuclei that lie just under the plasma membrane.

Cardiac muscle, the main tissue of the heart, is a kind of striated muscle that is not under voluntary control. The fibers of cardiac muscle are joined end-to-end and branch and rejoin, forming complex networks. One or two nuclei are found within each fiber. A characteristic feature of cardiac muscle tissue is the presence of **intercalated discs**, which are specialized junctions where the fibers join.

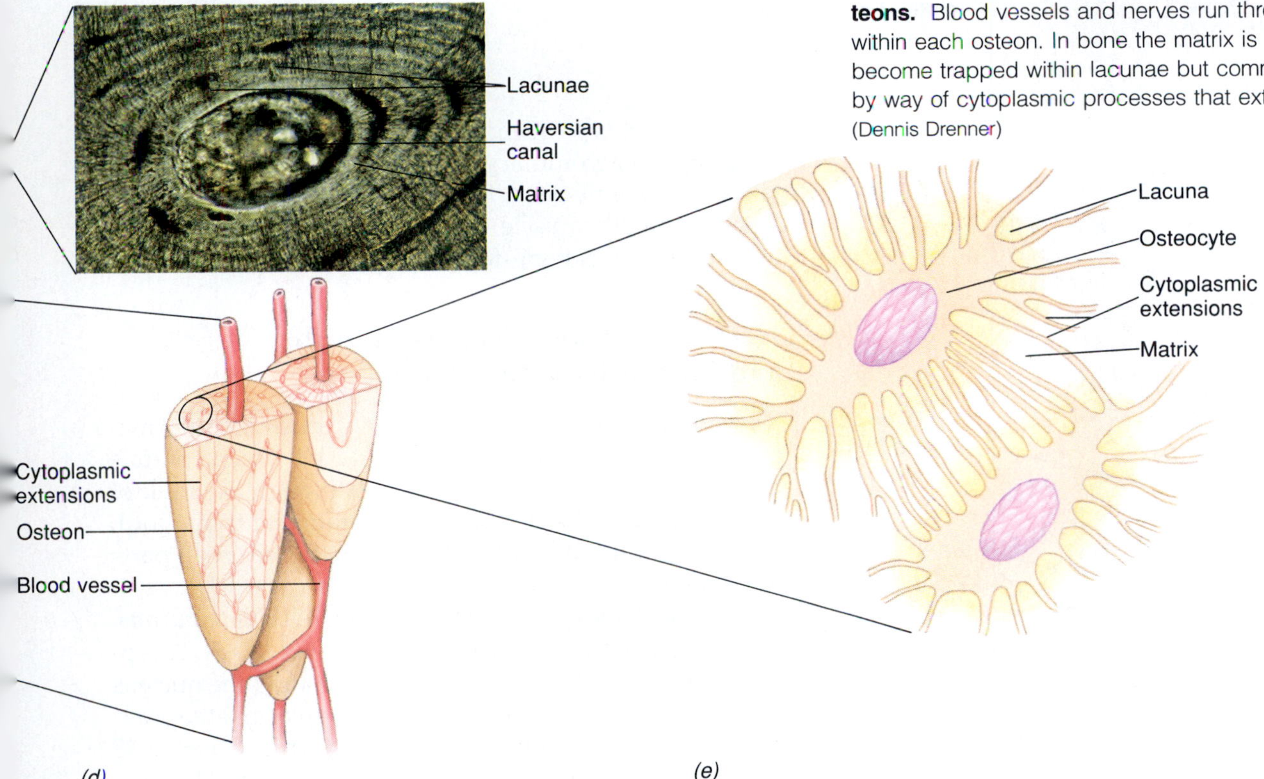

FIGURE 31–4 Compact bone is made up of units called osteons. Blood vessels and nerves run through the haversian canal within each osteon. In bone the matrix is rigid and hard. Bone cells become trapped within lacunae but communicate with one another by way of cytoplasmic processes that extend through tiny canals. (Dennis Drenner)

TABLE 31–3
The Three Types of Muscle Tissues

	Skeletal	*Smooth*	*Cardiac*
Location	Attached to skeleton	Walls of stomach, intestines, etc.	Walls of heart
Type of control	Voluntary	Involuntary	Involuntary
Shape of fibers	Elongated, cylindrical, blunt ends	Elongated, spindle-shaped, pointed ends	Elongated, cylindrical fibers that branch and fuse
Striations	Present	Absent	Present
Number of nuclei per fiber	Many	One	One or two
Position of nuclei	Peripheral	Central	Central
Speed of contraction	Most rapid	Slowest	Intermediate
Ability to remain contracted	Least	Greatest	Intermediate

Nuclei
Striations
Nuclei
Nuclei
Intercalated disks

(a) Skeletal muscle fibers *(b)* Smooth muscle fibers *(c)* Cardiac muscle fibers

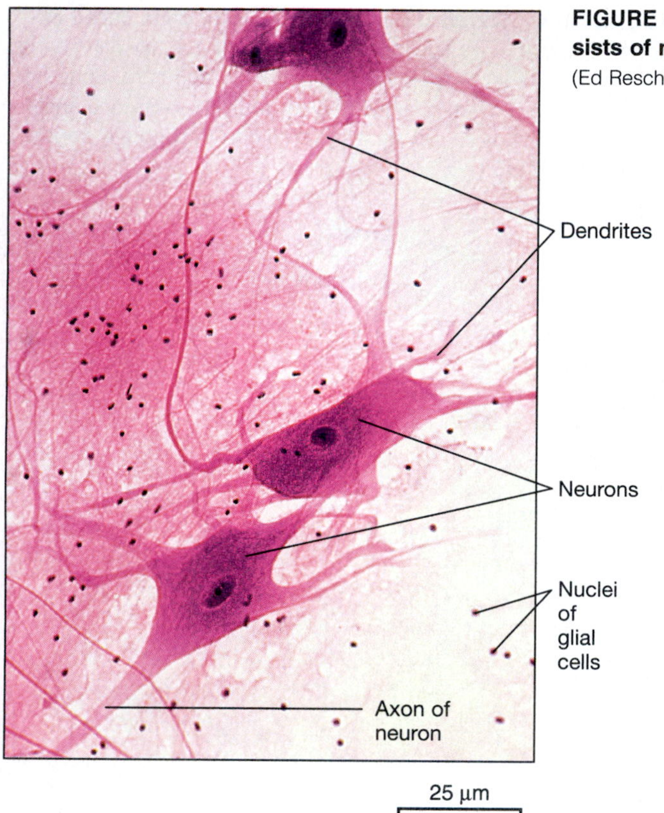

FIGURE 31–5 Nervous tissue consists of neurons and glial cells.
(Ed Reschke)

The third type of muscle, **smooth muscle,** lacks striations and is involuntary. Each spindle-shaped fiber contains only one nucleus. Found within the walls of many organs, smooth muscle is responsible for such internal movements as moving food through the digestive tract.

Nervous Tissue Receives Stimuli and Transmits Information

Nervous tissue receives stimuli and transmits information in the form of nerve impulses. This tissue controls the action of muscles and glands. Although the bulk of nervous tissue is located within the brain and spinal cord, bundles of neurons, called nerves, are found in all parts of the body.

Nervous tissue consists of nerve cells, or **neurons,** and supporting cells called **glial cells** (Fig. 31–5). A typical neuron consists of a cell body containing the nucleus and elongated extensions of the cytoplasm—the dendrites and axon. (Neurons are discussed in Chapter 33 and nervous systems are discussed in Chapter 34.)

An Organ Consists of More Than One Type of Tissue and Performs One or More Functions

Recall from Chapter 1 that structures like the brain, heart, stomach, and eye are **organs,** each composed of various types of tissues that together perform one or more biological functions. Although an organ may consist mainly of one type of tissue, other types of tissues provide support, protection, nourishment (through the blood supply), and communication with other organisms (via nerve impulses). For example, the typical vertebrate intestine is lined with epithelial tissue that secretes digestive enzymes and absorbs nutrients. Layers of muscle make up the bulk of the intestinal wall and contract in waves, moving food through the digestive tube. Nervous tissue places the intestine in communication with other parts of the body, such as the brain. Connective tissue supplies the intestine with blood and holds its tissues together; it also holds the tube in place in the body.

Ten Organ Systems Make Up the Complex Animal Organism

Several tissues and organs may work together, performing a specialized set of functions. Such an organized group of structures is termed an **organ system.** In complex animals, we can identify ten major organ systems, each performing a specific group of activities (Fig. 31–6). Working together, these organ systems make up the complex **organism.**

The ten organ systems of complex animals include the integumentary, skeletal, muscle, nervous, circulatory, digestive, respiratory, urinary, endocrine, and reproductive systems. A summary of their principal organs and functions is found in Table 31–4. As an example of an organ system, consider the digestive system. Organs of the mammalian digestive system include the mouth, esophagus, stomach, small and large intestines, liver, pancreas, and salivary glands. This system digests food, reducing it to simple molecular components. It then absorbs these nutrients, permitting them to enter the blood for transport to all of the body's cells.

The Organ Systems Work Together to Maintain Homeostasis

In a complex animal, trillions of cells are organized to form the tissues, organs, and organ systems. The organism functions effectively, in large part, because very precise control mechanisms maintain a constant, appropriate internal environment. If the organism is to survive and function, the composition of the fluids that bathe its cells must be carefully maintained. An appropriate concentration of nutrients, oxygen and other gases, ions, and compounds needed for metabolism must be available, and internal temperature and pressure must be maintained within narrow limits.

Recall from Chapter 1 that the tendency to maintain a relatively constant internal environment is termed **homeostasis,** and the mechanisms that accomplish the task are **homeostatic mechanisms.** First coined by the physiologist Walter Cannon, the term homeostasis is derived from the Greek words *homoios,* meaning "same," and *stasis,* meaning "standing." Actually, the internal environment never really stays the same; it fluctuates within a normal range. Thus, a dynamic, or constantly changing, equilibrium is maintained.

The internal environment is carefully regulated by the interaction of many homeostatic mechanisms. All of the organ systems participate in these regulatory mechanisms, but most of them are controlled by the nervous and endocrine systems. Homeostasis is a basic concept in physiology. As we study the organ systems, we shall discuss numerous ways in which the systems interact to maintain the organism's steady state.

How do homeostatic mechanisms work? Many homeostatic mechanisms are feedback systems, sometimes called biofeedback systems. Such a system consists of a cycle of events in which information about a change (for example, a change in temperature) is fed back into the system so

(Text continues on page 608)

Concept Connections

Homeostasis* — *Stress

Stressors are changes in the internal or external environment that disturb homeostasis. Examples of external stressors are heat, cold, noise, abnormal pressure, or lack of oxygen. Internal stressors include changes in blood pressure, pH, or salt concentration, and high or low blood sugar level. Many stressors occur routinely and are expertly handled by homeostatic mechanisms, often involving the nervous and endocrine systems. Other stressors are more severe and may cause serious disruption, or stress. When homeostatic mechanisms are unable to restore the steady (normal) state, the stress may lead to a malfunction, which can cause disease or even death. In Chapter 41 we will discuss how the adrenal glands help regulate the homeostatic adjustments necessary for the body to effectively cope with stressors.

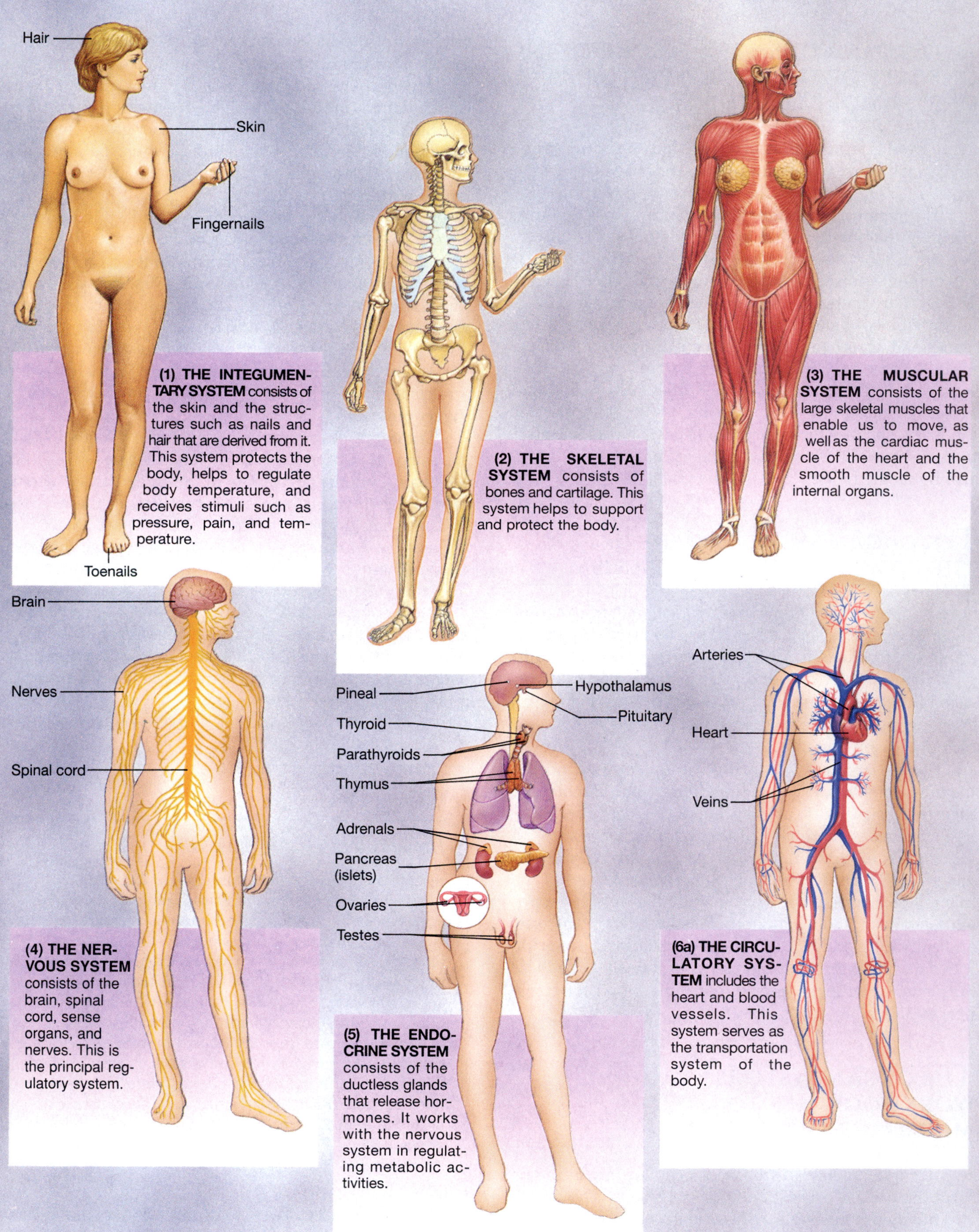

Hair
Skin
Fingernails
Toenails
(1) THE INTEGUMENTARY SYSTEM consists of the skin and the structures such as nails and hair that are derived from it. This system protects the body, helps to regulate body temperature, and receives stimuli such as pressure, pain, and temperature.
(2) THE SKELETAL SYSTEM consists of bones and cartilage. This system helps to support and protect the body.
(3) THE MUSCULAR SYSTEM consists of the large skeletal muscles that enable us to move, as well as the cardiac muscle of the heart and the smooth muscle of the internal organs.
Brain
Nerves
Spinal cord
(4) THE NERVOUS SYSTEM consists of the brain, spinal cord, sense organs, and nerves. This is the principal regulatory system.
Pineal
Hypothalamus
Pituitary
Thyroid
Parathyroids
Thymus
Adrenals
Pancreas (islets)
Ovaries
Testes
(5) THE ENDOCRINE SYSTEM consists of the ductless glands that release hormones. It works with the nervous system in regulating metabolic activities.
Arteries
Heart
Veins
(6a) THE CIRCULATORY SYSTEM includes the heart and blood vessels. This system serves as the transportation system of the body.

FIGURE 31–6 The principal organ systems of the human body.

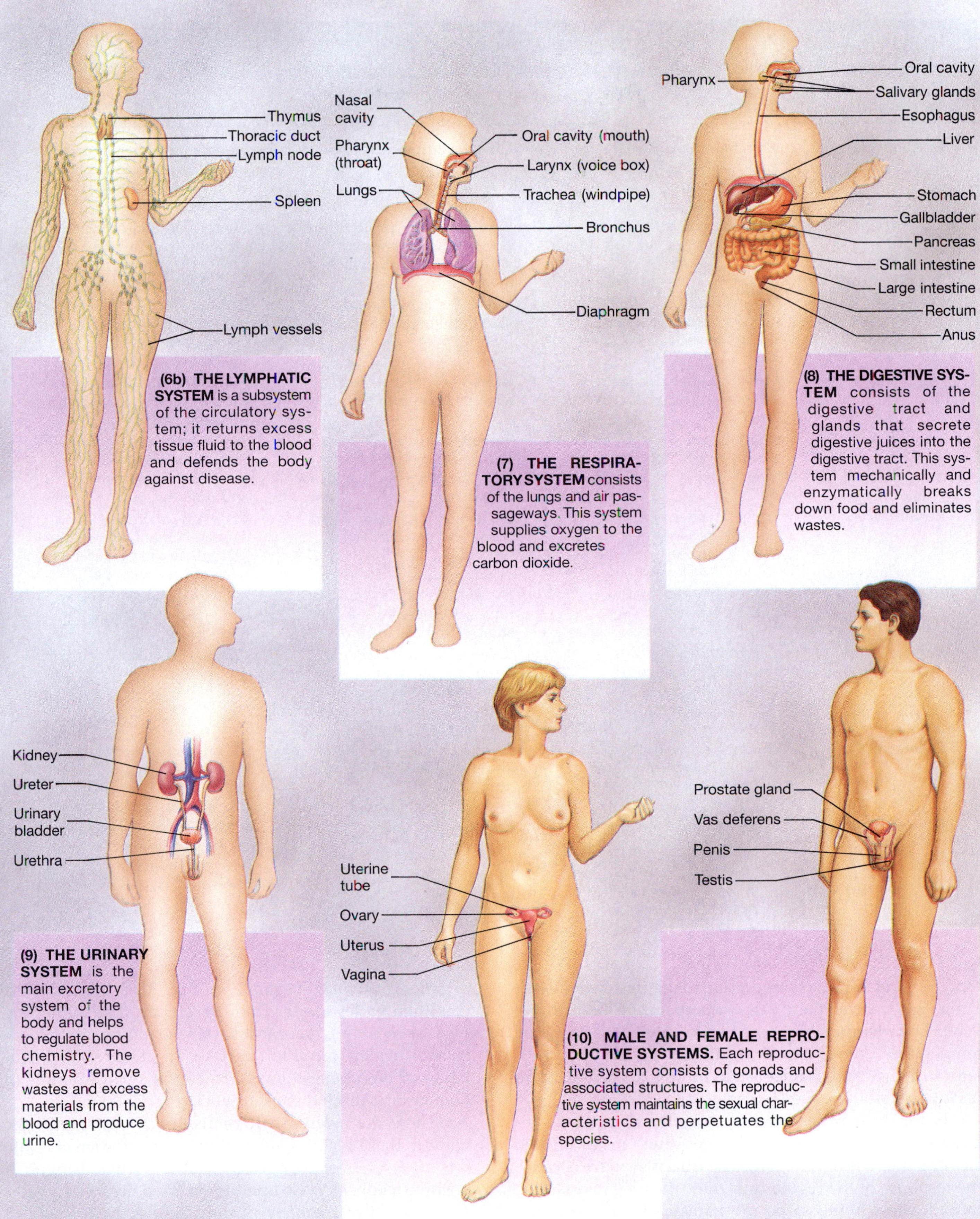

(6b) THE LYMPHATIC SYSTEM is a subsystem of the circulatory system; it returns excess tissue fluid to the blood and defends the body against disease.

(7) THE RESPIRATORY SYSTEM consists of the lungs and air passageways. This system supplies oxygen to the blood and excretes carbon dioxide.

(8) THE DIGESTIVE SYSTEM consists of the digestive tract and glands that secrete digestive juices into the digestive tract. This system mechanically and enzymatically breaks down food and eliminates wastes.

(9) THE URINARY SYSTEM is the main excretory system of the body and helps to regulate blood chemistry. The kidneys remove wastes and excess materials from the blood and produce urine.

(10) MALE AND FEMALE REPRODUCTIVE SYSTEMS. Each reproductive system consists of gonads and associated structures. The reproductive system maintains the sexual characteristics and perpetuates the species.

TABLE 31–4
The Organ Systems of a Mammal and Their Functions

System	*Components*	*Functions*	*Homeostatic ability*
Integumentary	Skin, hair, nails, sweat glands	Covers and protects body	Sweat glands help control body temperature; as barrier, the skin helps maintain steady state
Skeletal	Bones, cartilage, ligaments	Supports body, protects, provides for movement and locomotion, calcium storage	Helps maintain constant calcium level in blood
Muscular	Organs mainly of skeletal muscle; cardiac muscle; smooth muscle	Moves parts of skeleton, locomotion; movement of internal materials	Ensures vital functions requiring movement, e.g., cardiac muscle circulates the blood
Digestive	Mouth, esophagus, stomach, intestines, liver, pancreas	Ingests and digests foods, absorbs them into blood	Maintains adequate supplies of fuel molecules and building materials
Circulatory	Heart, blood vessels, blood; lymph and lymph structures	Transports materials from one part of body to another; defends body against disease	Transports oxygen, nutrients, hormones; removes wastes; maintains water and ionic balance of tissues
Respiratory	Lungs, trachea, and other air passageways	Exchange of gases between blood and external environment	Maintains adequate blood oxygen content and helps regulate blood pH; eliminates carbon dioxide
Urinary	Kidney, bladder, and associated ducts	Excretes metabolic wastes; removes excessive substances from blood	Helps regulate volume and composition of blood and body fluids
Nervous	Nerves and sense organs; brain and spinal cord	Receives stimuli from external and internal environment, conducts impulses, integrates activities of other systems	Principal regulatory system
Endocrine	Pituitary, adrenal, thyroid, and other ductless glands	Regulates blood chemistry and many body functions	In conjunction with nervous system, regulates metabolic activities and blood levels of various substances
Reproductive	Testes, ovaries, and associated structures	Provides for continuation of species	Passes on genetic endowment of individual; maintains secondary sexual characteristics

that the regulator (the temperature regulating center in the brain) can control the process (temperature regulation). When body temperature becomes too high or too low, the temperature change serves as input, triggering the regulator to activate mechanisms that bring body temperature back to normal (see Fig. 1–5, page 7).

In this type of feedback system, the response counteracts the inappropriate change, thus restoring the steady state. This is a **negative feedback system,** because the response of the regulator is opposite (negative) to the output. Most homeostatic mechanisms in the body are negative feedback systems (see Fig. 31–7). When some condition varies too far from the steady state (either too high or too low), a control system using negative feedback brings the condition back to the steady state. Regulation of blood sugar level provides a good example of homeostatic mechanisms at work (Fig. 31–7).

There are a few **positive feedback systems** in the body. In these systems, the variation from the steady state sets off a series of changes that intensify the changes. A positive feedback cycle operates in the delivery of a baby. As the head of the baby pushes against the opening of

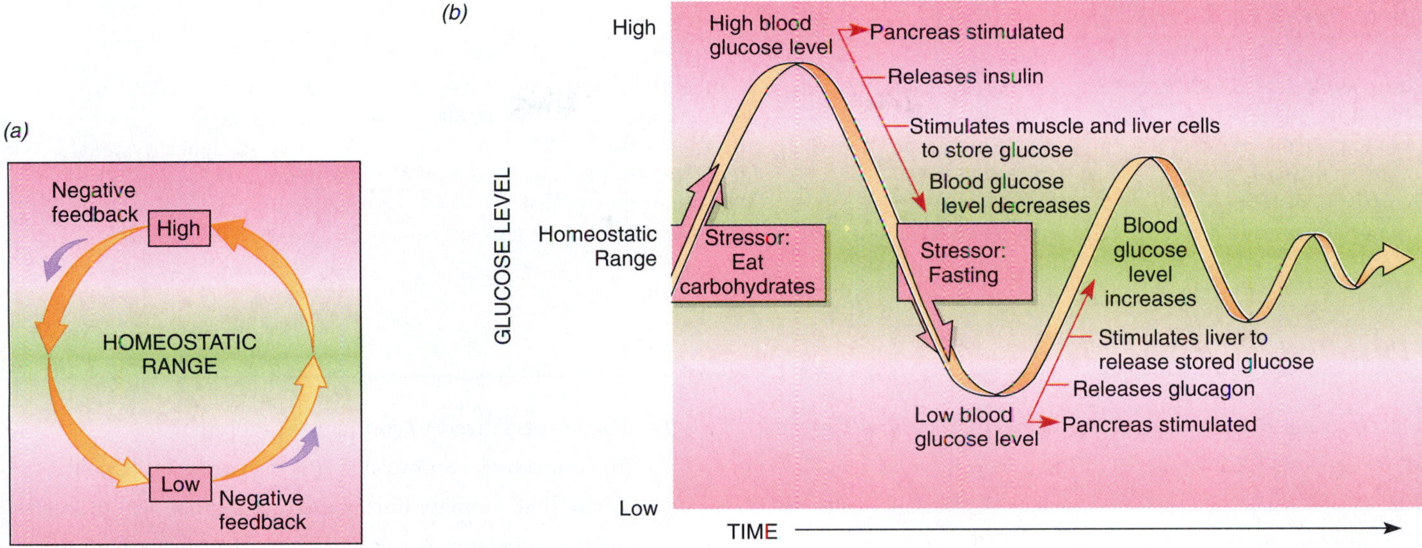

FIGURE 31–7 Maintaining homeostasis by negative feedback mechanisms. (***a***) In negative feedback the response of the regulator is opposite to the output. (***b***) Regulation of blood sugar level. When you eat, the concentration of glucose in your blood increases. This increase stimulates the pancreas to release insulin, a hormone that stimulates cells to take up glucose from the blood and store it. After several hours, when the glucose level of the blood begins to fall below its normal level, the pancreas releases another hormone, glucagon. This hormone stimulates cells to release their stored glucose, thereby increasing glucose concentration.

the uterus (cervix), a reflex action causes the uterus to contract. The contraction forces the baby's head against the cervix again, resulting in another contraction, and the positive feedback cycle is repeated again and again until the baby is delivered. Many positive feedback sequences (such as one that occurs when a person goes into circulatory shock) lead to the disruption of steady states and even to death.

Chapter Summary

I. Multicellular organisms can be much larger and more diverse than unicellular ones, and their cells can specialize, performing specific functions.

II. A tissue consists of a group of closely associated cells that carry out specific functions.
   A. Epithelial tissue consists of cells that are fitted tightly together. It covers the body surface and lines its cavities.
      1. Epithelial tissue protects, absorbs, secretes, or receives stimuli.
      2. Epithelial cells may be squamous, cuboidal, or columnar, and they may be arranged to form simple, stratified, or pseudostratified tissue.
   B. Connective tissue supports and joins structures of the body. It protects underlying structures and may store or transport materials.
      1. Connective tissue consists of relatively few cells separated by intercellular substance, which in turn is composed of fibers scattered through a matrix.
      2. Some main types of connective tissue are loose and dense, adipose, cartilage, bone, blood, lymph, and tissues that produce blood cells.
   C. Muscle tissue contracts, enabling an animal to move. The three types of muscle are skeletal, cardiac, and smooth muscle.
   D. Nervous tissue receives and transmits stimuli; it controls the action of muscles and glands.

III. An organ consists of different types of tissue that work together to perform a particular biological function. The heart, brain, and stomach are examples of organs.

IV. Organs and tissues work together, forming organ systems. In complex animals, ten principal organ systems work together, making up the living organism. Among these are the digestive system, the nervous system, and the skeletal system.

V. Homeostasis is the body's automatic tendency to maintain a constant internal environment, or steady state. It is maintained by negative feedback mechanisms.

Selected Key Terms

adipose tissue, p. 601
blood, p. 603
bone, p. 603
cardiac muscle, p. 603
cartilage, p. 602
dense connective tissue, p. 601
endocrine gland, p. 599
epithelial tissue, p. 596
exocrine gland, p. 599
fibers, p. 599
gland, p. 597
goblet cells, p. 597
homeostasis, p. 605
loose connective tissue, p. 600
muscle tissue, p. 603
negative feedback system, p. 608
nervous tissue, p. 604
organ, p. 605
organ system, p. 605
osteon, p. 603
positive feedback system, p. 608
skeletal muscle, p. 603
smooth muscle, p. 604
tissue, p. 596

Post-Test

1. A group of closely associated cells that carry out specific functions forms a/an ______________.
2. A/An ______________ consists of epithelial cells that produce and secrete a product.
3. Epithelial cells may be flattened, or ______________; cube shaped, or ______________; or elongated like columns, or ______________.
4. In cardiac muscle, intercalated ______________ are the junctions between adjoining fibers.
5. Two types of striated muscle are ______________ and ______________.
6. Supporting cells found in nervous tissue are called ______________ cells.
7. The organ system made up of glands that secrete hormones is the ______________ system.
8. The organ system that covers the body is the ______________ system.
9. In a/an ______________ feedback system, the response counteracts the inappropriate change.
10. Tissue that covers and lines body surfaces is (a) epithelial (b) connective (c) muscle (d) nervous (e) adipose.
11. Tissue that receives stimuli and transmits impulses is (a) epithelial (b) connective (c) muscle (d) nervous (e) smooth.
12. Tissue specialized to contract is (a) epithelial (b) connective (c) muscle (d) nervous (e) smooth.
13. Tissue that contains fibroblasts and a great deal of intercellular substance is (a) epithelial (b) connective (c) muscle (d) nervous (e) smooth.
14. Tissue that forms the subcutaneous layer and is the most widely distributed connective tissue is (a) epithelial (b) loose connective (c) dense connective (d) cartilage (e) smooth muscle.
15. Tissue that has a hard, rubbery matrix and lacks blood vessels is (a) epithelium (b) loose connective (c) cartilage (d) bone.
16. Tissue composed of osteons is (a) epithelium (b) loose connective (c) cartilage (d) bone.
17. Tissue that contains goblet cells is (a) adipose (b) loose connective (c) cartilage (d) epithelial (e) lymph.
18. Exocrine glands (a) lack ducts (b) secrete their product onto a free epithelial surface (c) depend on blood to transport their products (d) are made mainly of connective tissue (e) two of the preceding answers are correct.
19. Homeostatic mechanisms in the body typically (a) depend on negative feedback (b) involve blood sugar levels (c) maintain an appropriate internal environment (d) are often referred to as stressors (e) two of the preceding answers are correct.

Review Questions

1. Distinguish among tissues, organs, and organ systems.
2. What are some advantages to being multicellular?
3. Contrast epithelial tissue with connective tissue.
4. Compare cartilage with bone.
5. What is a gland? What tissue composes a gland?
6. Compare skeletal, cardiac, and smooth muscle with respect to structure and location in the body.
7. How is a neuron adapted for its specific function?

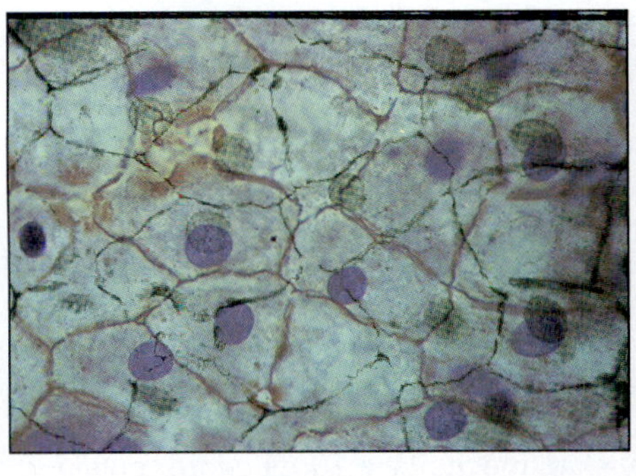
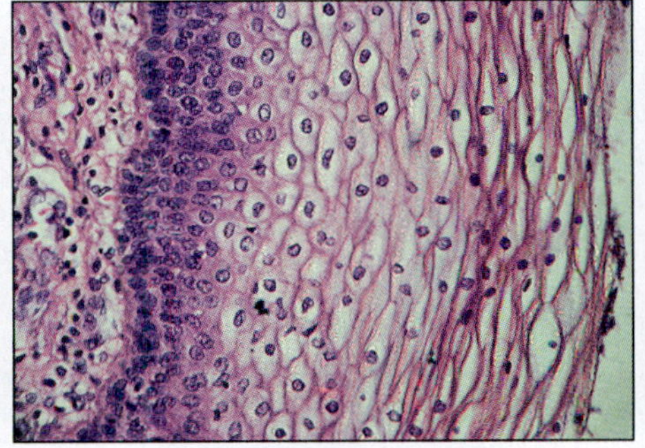
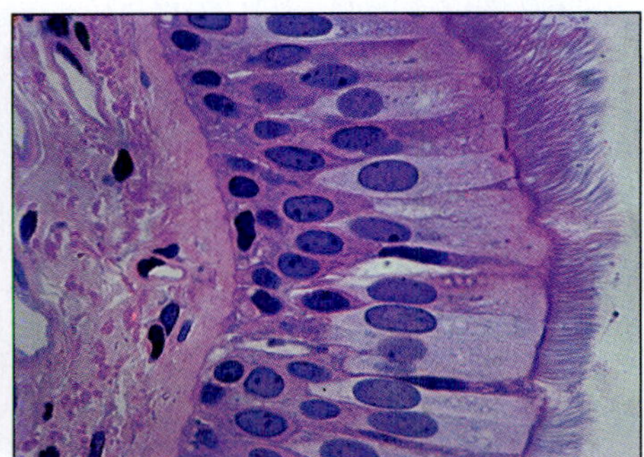
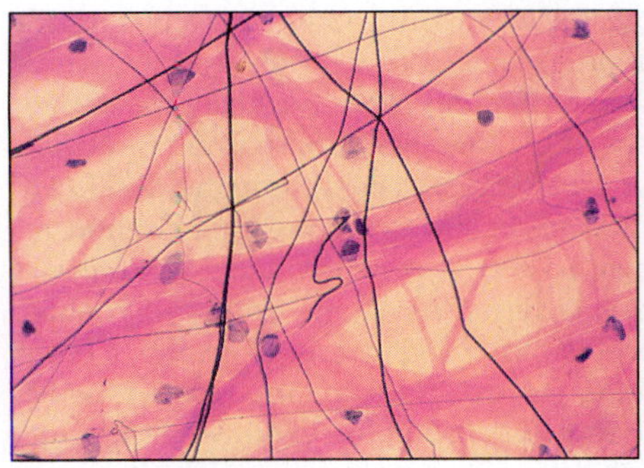
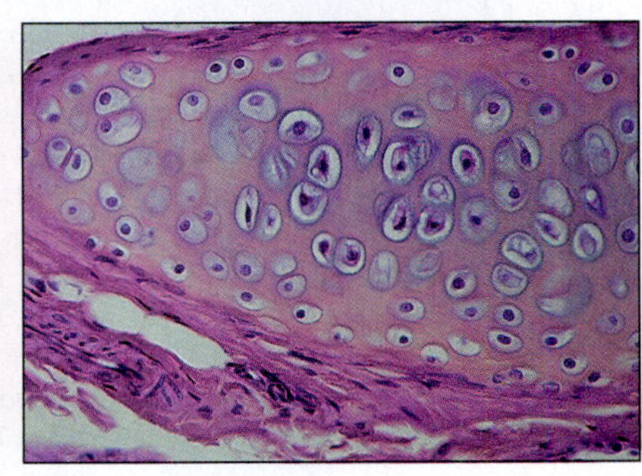

8. Locate each of the following: (a) osteocytes (b) chondrocytes (c) intercalated discs (d) fibroblasts (e) collagen fibers (f) stratified squamous epithelium.
9. Summarize the functions of the following: (a) endocrine system (b) skeletal system (c) circulatory system (d) nervous system.
10. Identify each type of tissue in the figure above.

Thinking Critically

1. How do the cells of a malignant neoplasm differ from the cells of a normal tissue? What is metastasis?
2. Imagine that all of the epithelium in a complex animal, such as a human, suddenly disappeared. What effects would this have on the body and its ability to function?
3. What would connective tissue be like if it had no intercellular substance? What effect would its absence have on the body?

Recommended Readings

Liotta, L.A., "Cancer Cell Invasion and Metastasis," *Scientific American,* Vol. 266, No. 2, February 1992, pp. 54–63. A discussion of current research on the mechanisms and treatment of metastatic cancer.

National Geographic Society Book Service, *The Incredible Machine,* Washington, D.C., National Geographic, 1986. A beautiful and informative introduction to the human body featuring the incredible photographs of Lennart Nilsson.

Solomon, E.P., Schmidt, R., and Adragna, P., *Human Anatomy and Physiology,* 2nd ed., Philadelphia, Saunders College Publishing, 1990. A very readable presentation of human anatomy and physiology that includes a chapter on tissues.

See also the readings for Chapter 32.

Woman wearing nicotine patch (Habitrol), a commonly used transdermal patch. (Courtesy of CIBA-GEIGY)

Transdermal Patches: Delivering Drugs Through the Skin

Have you ever placed a scopolamine patch on the skin behind your ear to avoid seasickness? Do you know a former cigarette smoker who was able to kick the habit with the help of a nicotine patch? Or perhaps you know someone who uses a nitroglycerin patch for angina (pain associated with heart disease). The **transdermal patch** is a method of drug delivery through the skin that appears to be efficient, practical, and user friendly. No injections to bear or pills to gag on. Most important, the user does not need to remember to take daily medication. The transdermal patch is applied easily, much like an adhesive bandage, to the skin, usually on the chest or buttocks. Bypassing the digestive tract, medication delivered through the skin causes fewer side effects and may decrease the risk of liver damage. (The liver is the organ that breaks down drugs.)

Nicotine patches, like Habitrol, are now considered very effective aids in smoking cessation. The patch delivers a steady, low dose of nicotine, reducing the smoker's withdrawal symptoms. When used along with a program of behavior modification, nicotine patches have success rates as high as 86%.

The estrogen patch is another widely used transdermal medication. After menopause, many women are using estrogen patches instead of daily pills for hormone replacement. The clonidine patch is used for hypertension (high blood pressure) and to prevent drug withdrawal symptoms in addicts. The fentanyl patch is used to alleviate chronic pain and to relieve pain after surgery.

The skin, much like a Tupperware container, is famous for its impermeability. It keeps microorganisms and many foreign substances out, while preventing water and other essential substances from escaping. How then do drugs get through this barrier? In order to pass through the skin and to deliver an effective dose of medication, a drug must be able to penetrate the skin's outermost layer (stratum corneum). This layer contains lipids and is composed mainly of keratin, a waterproofing protein. Thus, a transdermal drug must be somewhat soluble in both water and lipid, and have a low molecular weight.

When a transdermal patch is applied, the drug passes through a thin, flexible membrane within the patch itself and into the skin. The membrane within the patch is designed to deliver the drug at effective levels with minimum side effects.

Pharmacology researchers are exploring new methods of enhancing absorption through the skin so that a greater variety of drugs can be administered transdermally. One method, known as iontophoresis, uses charged molecules in the patch to facilitate absorption and may be useful in delivering drugs with larger molecular weight, such as insulin. Another technique is chemical modification of the drug to enhance initial penetration. Then, enzymes in the skin convert the drug back to its original active form.

Skin, Bones, and Muscle: Protection, Support, and Locomotion

32

Learning Objectives

After you have studied this chapter you should be able to:

1. Describe the external epithelium of invertebrates and summarize its functions.
2. Compare the structure and function of vertebrate skin with the external epithelium of invertebrates, and identify the principal derivatives of vertebrate skin.
3. Outline the advantages and disadvantages of different types of skeletal systems, including the hydrostatic skeleton, exoskeleton, and endoskeleton.
4. Identify main divisions of the vertebrate skeleton and the bones that make up each division.
5. Describe the structure of a typical long bone, and summarize the process of bone development.
6. Explain the gross and microscopic structure of skeletal muscle.
7. List in sequence the events that take place in muscle contraction.
8. Compare the roles of glycogen, creatine phosphate, and ATP in providing energy for muscle contraction.
9. Describe the antagonistic action of muscles.
10. Summarize the functional relationship between skeletal and muscular tissues.

Key Concepts

- ☐ The animal body is covered by a protective shield of epithelium that may perform other functions such as secretion, gas exchange, reception of stimuli, or regulation of body temperature.
- ☐ Skeletons protect and support the animal body and may be important in locomotion.
- ☐ The muscular and skeletal systems often work together in movement. Muscles, the body's motors, are anchored to the skeleton, which transmits forces.
- ☐ As muscles contract, they move body parts by pulling on them.

THE ANIMAL BODY IS COVERED WITH A PROTECTIVE EPITHELIAL TISSUE

The epithelial covering of invertebrates and the skin of vertebrates form a protective shield around the body.

The Epithelial Covering of Invertebrates May Be Specialized to Receive Information, for Secretion, or for Gas Exchange

Epithelial cells may be modified as sensory cells that are sensitive to light, temperature, chemical stimuli, or mechanical stimuli such as touch. In many species, the epithelium contains cells that secrete a tough, protective cuticle or that secrete lubricants or adhesives. In some animals, these cells release odorous secretions that allow communication among members of the species. Others produce poisonous secretions that are used in offense or defense. In earthworms, a lubricating, mucous secretion produces a moist slime that aids in the diffusion of gases across the body wall. This lubricating secretion also reduces friction as the earthworm pushes its way through the soil.

In some species, an epithelial secretion may be limited to a particular region of the body surface. In the gastropod mollusk, for example, a mucous secretion is released from the foot, producing a slime track through which the snail glides (Fig. 32–1). Weaver ants and some other insects secrete strong, fine threads that they use to construct nests. Butterflies and moths synthesize silk from amino acids in their silk-forming glands. The spinning glands of spiders also develop from epithelial cells.

Figure 32–1 Some epithelial coverings secrete mucus. The mucus secreted by the foot of the striped land snail *(Helicella candicans)* produces a slime track through which the animal glides. (Y. Momatiuk/PhotoResearchers, Inc.)

The Vertebrate Skin Functions in Protection, Secretion, and Temperature Regulation

In many fish, in some reptiles, and in the African ant-eating pangolin (a mammal), the skin develops into a set of scales that form a protective armor of considerable strength. Human skin is also very strong. It has a variety of structures, including fingernails and toenails, hair, sweat glands, oil (sebaceous) glands, and several types of sensory receptors that permit us to feel pressure, temperature, and pain (Fig. 32–2).

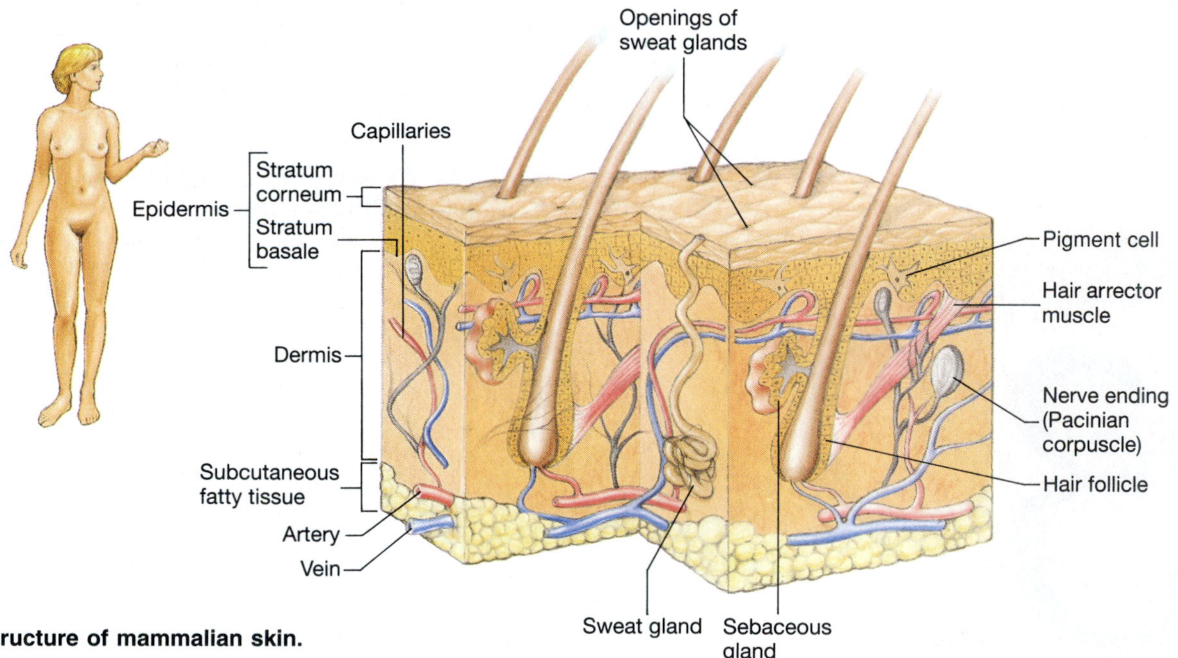

Figure 32–2 The structure of mammalian skin.

Human skin, like the skin of all mammals, contains mammary glands that, in females, secrete milk. Oil glands, which empty into hair follicles, secrete a waxy substance called sebum. In humans, these glands are especially numerous on the face and scalp. The oil secreted keeps the hair moist and pliable and prevents the skin from drying and cracking. (At puberty, increased levels of sex hormones may cause excessive production of sebum, which causes acne.)

In humans, the skin functions as a thermostatically controlled radiator, regulating the elimination of heat from the body (see Chapter 1). About 2.5 million sweat glands secrete sweat, and its evaporation from the surface of the skin lowers the body temperature.

The skin of some other vertebrates varies considerably from ours. For example, instead of hair, birds have feathers, which form in a manner comparable to hairs and provide an even more effective insulation than fur. Fish and reptiles have epidermal scales, and many amphibians have naked skin covered with mucus. The skin of certain tropical frogs is equipped with poison glands. Skin and its derivatives are often brilliantly colored in connection with courtship rituals and territorial displays. The human blush pales alongside the spectacular color changes that occur in the skin of some lizards.

The Epidermis Is a Waterproof Protective Barrier

The outer layer of skin, the **epidermis,** is the interface between the delicate tissues within the body and the hostile universe. The epidermis consists of several strata, or sublayers. The deepest is the **stratum basale,** and the most superficial is the **stratum corneum** (Fig. 32–2). In stratum basale, cells continuously divide, and the new cells are pushed toward the surface as other cells are produced below them. As the epidermal cells move upward in the skin, they mature. Because most vertebrates have no capillaries in the epidermis, the maturing cells receive less and less nourishment. As a result their metabolic activity declines.

As they move toward the body surface, epidermal cells manufacture **keratin,** a protein that gives the skin considerable mechanical strength and flexibility. Keratin is quite insoluble and serves to waterproof the body surface. As epidermal cells move through the stratum corneum, the outermost layer of the skin, they die. When they reach the surface of the skin, the cells of this outer layer wear off and must be continuously replaced.

Concept Connections

Skin — Tissue Culture — Immune Response

Epithelial cells from human skin grow quickly in culture and they retain their ability to organize into the structure of normal skin. Under the appropriate culture conditions, a small bit of skin tissue multiplies itself several thousandfold within a few weeks. Tissue can be generated in this way to treat burn victims. A small piece of undamaged skin can be removed from a burn victim, cultured for three to four weeks, and then grafted to the injured area. Such grafts are most successful when the tissue is taken from the patient's own skin. As will be discussed in Chapter 37, when tissue is taken from another individual, it will most likely cause an immune response. The graft will be rejected by the patient's immune system.

Concept Connections

Skin — Cancer — Environment

Skin cancer is on the rise. Most skin cancer is caused by excessive, chronic exposure to the ultraviolet radiation of sunlight. The ozone layer in the stratosphere, which absorbs incoming ultraviolet radiation from the sun, is being destroyed by pollutants such as chlorofluorocarbons. Environmental scientists are predicting that depletion of the ozone shield around our planet will cause one million extra cases of skin cancer each year, and that about 30,000 of these victims will die. Some forms of this disease (malignant melanoma) spread rapidly through the body and may cause death within a few months after diagnosis. Exposure to the ultraviolet rays causes the epidermis to thicken and stimulates pigment cells in the skin to produce melanin at an increased rate. An increase in melanin causes the skin to become darker. Melanin is an important protective screen against the sun because it absorbs harmful ultraviolet rays. The suntan so prized by sun worshippers is actually a sign that the skin has been exposed to too much ultraviolet radiation. When the melanin is not able to absorb all the ultraviolet rays, the skin becomes inflamed, or sunburned. Because dark-skinned people have more melanin, they suffer less sunburn, wrinkling, and skin cancer.

Most skin cancer can be prevented by protecting the body from sunlight and other forms of ultraviolet radiation. Tanning machines that use artificial light to induce tanning also expose the body to ultraviolet rays and so may not be safer than sunbathing. Those who choose to expose themselves to sunlight can protect their skin by limiting the time they are in the sun and by applying sunscreens containing ingredients such as para-aminobenzoic acid (PABA).

The Dermis Contains Sweat Glands, Hair Follicles, Blood Vessels, and Sense Organs

Beneath the epidermis lies the **dermis,** which consists of a dense, fibrous connective tissue composed mainly of collagen fibers (Fig. 32–2). Collagen imparts strength and flexibility to the skin. The major portion of each sweat gland is embedded in the dermis. The dermis also contains the hair follicles, blood vessels that nourish the skin, and sense organs that are concerned with touch. Mammalian skin rests on a layer of subcutaneous tissue that is composed mainly of adipose tissue. This fatty tissue insulates the body from outside temperature extremes.

SKELETONS ARE IMPORTANT IN LOCOMOTION, PROTECTION, AND SUPPORT

In some of the simplest animals, muscle acts directly on the jelly-like substance of the body itself, or on a fluid-filled body cavity. In more complex animals, however, a skeleton receives, transmits, and transforms the single movement—contraction—of muscular tissues into the variety of motions animals use. In addition to its function in locomotion, the skeleton supports the body and protects the internal organs. In most animals, the skeleton is not a living tissue at all but a lifeless deposit atop the epidermis, called a shell, or **exoskeleton.** In some animals, the skeleton is internal and is composed of plates or shafts of calcium-impregnated tissue.

In Hydrostatic Skeletons, Body Fluids Are Used to Transmit Force

Imagine an elongated balloon full of water. If you were to pull on it, it would lengthen, but it would also lengthen if you squeezed it. Conversely, if you pushed the ends toward the center, it would shorten. Many animals, including cnidarians and annelids, have a **hydrostatic skeleton** that works something like a water-filled balloon. In this type of system, contracting muscles push against a tube of fluid. Because fluids cannot be compressed, the force is transmitted through the fluid and generates movement of the body.

Mechanically, *Hydra* is little more than a simple bag of fluid with contractile cells in its two tissue layers. The fluid acts as a hydrostatic skeleton, transmitting force when the contractile cells push against the fluid. Contractile cells in the outer epidermal layer are arranged longitudinally. Contractile cells of the inner layer (the gastrodermis) are arranged in a circular fashion around the central body axis (Fig. 32–3). These two groups of cells work antagonistically. What one can do, the other can undo. When the epidermal, longitudinal layer contracts,

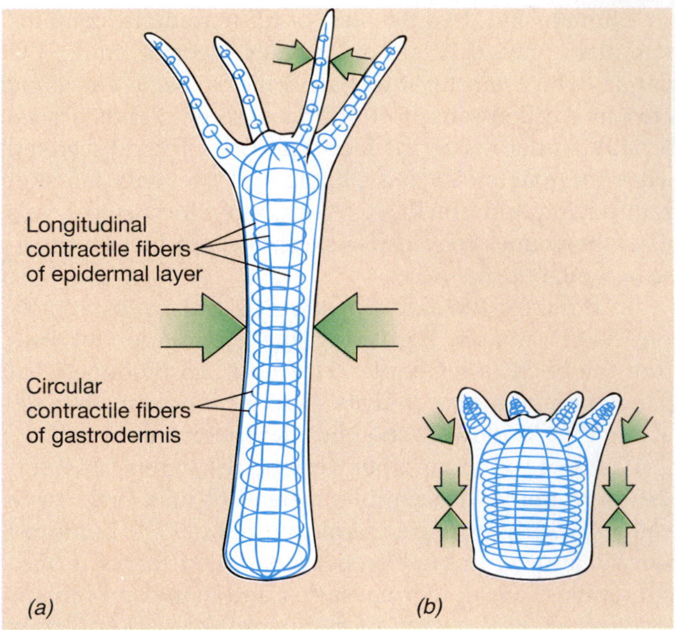

Figure 32–3 Movement in *Hydra*. The longitudinally arranged cells are antagonistic to the cells arranged in circles around the body axis. (***a***) Contraction of the circular muscles elongates the body. (***b***) Contraction of the longitudinal muscles shortens the body.

the hydra shortens, and because of the fluid present in its gastrovascular cavity, force is transmitted so that it thickens as well. On the other hand, when the endodermal circular layer contracts, the hydra thins, and its fluid content forces it to lengthen.

Hydrostatic skeletons permit only crude mass movements of the body or its appendages. Delicate movements are difficult because force tends to be transmitted equally in all directions throughout the entire fluid-filled body of the animal. For example, it is not easy for the hydra to thicken one part of its body while thinning another.

The more sophisticated hydrostatic skeleton of the annelid permits more versatile movement. Recall from Chapter 24 that the earthworm's body consists of a series of transverse partitions, or septa. The septa isolate portions of the body cavity and its contained fluid, permitting the hydrostatic skeletons of each segment to be largely independent of one another. Thus, the contraction of the circular muscle in the elongating anterior end need not interfere with the action of the longitudinal muscle in the segments of the posterior.

Hydrostatic skeletons work fine for many types of small aquatic animals and for animals like the earthworm that do little more than drag themselves along on their bellies. However, some hydrostatic skeletons occur even in complex invertebrates equipped with shells or endoskeletons and in vertebrates that have endoskeletons of cartilage or bone. Among mollusks, for example, the clam extends and anchors its foot by a hydrostatic blood pressure mechanism not too different from that used by the earthworm. Sea stars and sea urchins move their tube feet by an ingenious version of the hydrostatic skeleton. And even the human penis becomes erect and stiff because of the turgidity of pressurized blood in its cavernous spaces.

External Skeletons of Mollusks and Arthropods Are Nonliving Shells

The nonliving exoskeletons of mollusks and arthropods are produced by the cells of the epidermis. In mollusks, the shell is used mainly for protection, serving as an emergency retreat. At other times, the bulk of the naked, tasty body is exposed. Its major muscle attachments serve the skeleton, rather than the reverse. For example, the clam has a pair of muscles that hold the two valves of the shell tightly shut against the attacks of the sea star and chowder maker.

The exoskeleton of arthropods serves not only to protect but also to transmit forces. In this respect, it is comparable to the skeleton of vertebrates. Although the arthropod exoskeleton is a continuous, one-piece sheath covering the entire body, it varies greatly in thickness and flexibility. Large, thick, inflexible plates are separated from one another by thin, flexible joints arranged segmentally. Enough joints are provided to make the arthropod's body as flexible as those of many vertebrates.

A disadvantage of the rigid arthropod exoskeleton is that it interferes with growth. To accommodate growth, the arthropod must **molt,** that is, shed its exoskeleton and replace it with a new one (Fig. 32–4). During molting, the animal is exposed and is vulnerable to predators.

Internal Skeletons Are Living Tissues Capable of Growth

Endoskeletons, or internal skeletons, are extensively developed only in the echinoderms and the chordates. Composed of living tissue, the endoskeleton grows at a similar pace with the growth of the animal as a whole, eliminating the need for molting. The endoskeleton probably also permits a greater variety of possible motions than does an exoskeleton.

The echinoderm endoskeleton consists of plates that are composed of nonliving calcium salts embedded in the

FIGURE 32–4 A greengrocer cicada molting. This insect requires 13 years to mature. (Judy Davidson/Science Photo Library/Photo Researchers)

FIGURE 32–5 The echinoderm endoskeleton provides support and protection. The endoskeleton of this slate pencil sea urchin *(Heterocentrotus mammillatus)* is composed of spicules and plates of nonliving calcium salts embedded in tissues of the body wall. Photographed in Hawaii. (Mike Severns/Tom Stack and Associates)

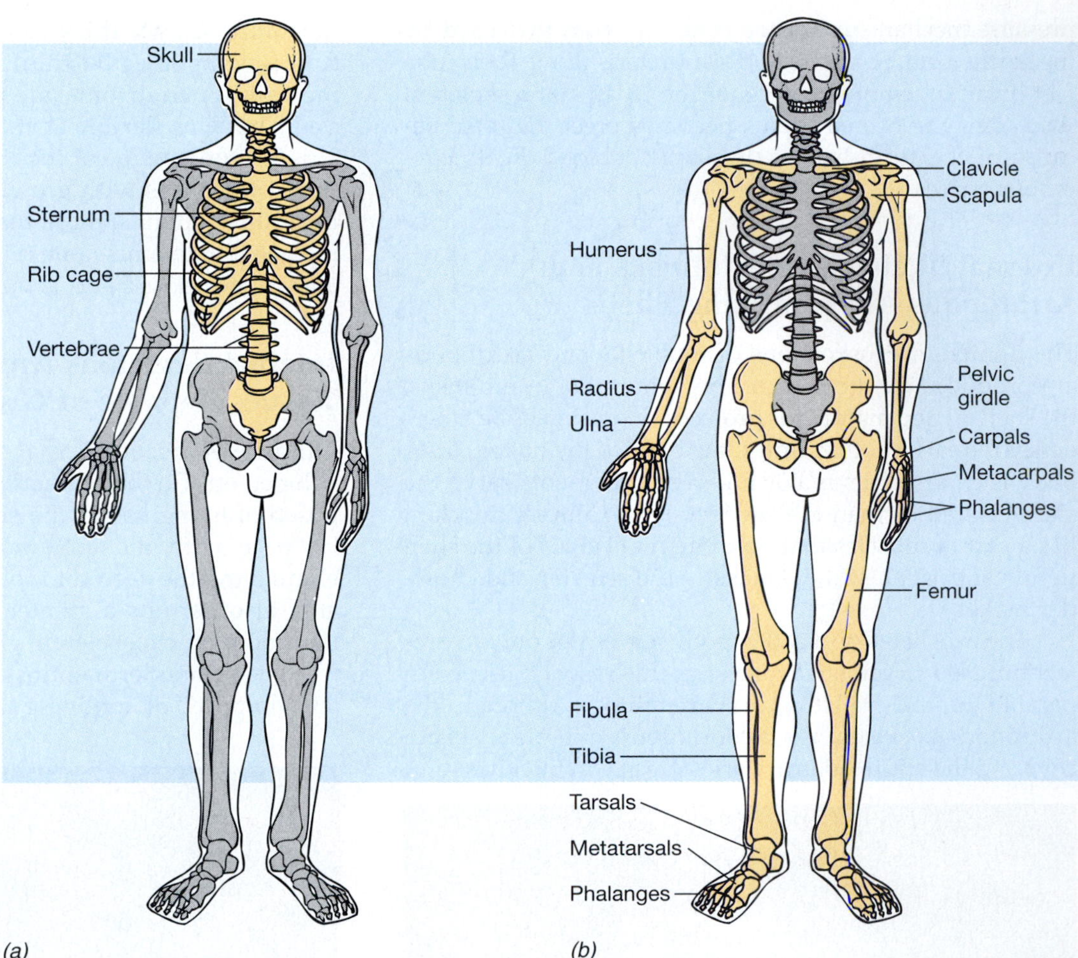

FIGURE 32–6 The human skeleton. (***a***) Bones of the axial skeleton, anterior view. (***b***) Bones of the appendicular skeleton, anterior view.

tissues of the body wall (Fig. 32–5). These form what amounts to an internal shell that provides support and protection. Many echinoderm endoskeletons bear spines that project to the outer surface.

The internal skeleton of vertebrates provides protection and support, and also transmits forces. Members of class Chondrichthyes (sharks and rays) have skeletons composed of cartilage, but in most vertebrates, the skeleton consists mainly of bone.

The vertebrate skeleton has two main divisions: the axial and appendicular skeletons. The **axial skeleton,** located along the central axis of the body, consists of the skull, vertebral column (backbone), ribs, and sternum (breastbone). The **appendicular skeleton** consists of the bones of the appendages (arms and legs), as well as the bones that make up the girdles that connect the appendages to the axial skeleton—the shoulder (or pectoral) girdle and most of the hip (or pelvic) girdle (Fig. 32–6). In the following description of the vertebrate skeleton, we will focus on the human skeleton.

The **skull,** the bony framework of the head, consists of the cranial and facial bones. The vertebrate spine, or **vertebral column,** is the main supporting axis below the skull. In humans, the vertebral column supports the body and bears its weight. It consists of 24 **vertebrae** and the *sacrum* and *coccyx.* The regions of the vertebral column are the *cervical* (neck), which consists of seven vertebrae; the *thoracic* (chest), which consists of 12 vertebrae; the *lumbar* (back), composed of five vertebrae; the *sacral* (pelvic), which consists of five fused vertebrae; and the *coccygeal,* which also consists of fused vertebrae.

Although they differ in size and shape in different regions of the vertebral column, a typical vertebra consists of a bony central portion, called the *centrum,* that bears most of the body weight, and a dorsal ring of bone, called the *neural arch,* which surrounds and protects the delicate spinal cord.

The *rib cage* is a bony basket formed by the *sternum* (breastbone), thoracic vertebrae, and 12 pairs of ribs. It protects the internal organs of the chest, including the heart and lungs, and supports the chest wall, preventing it from collapsing as the diaphragm contracts with each breath. Each pair of ribs is attached dorsally to a separate vertebra. Of the 12 pairs of ribs in the human, the first seven are attached ventrally to the sternum (breastbone); the next three are attached indirectly by cartilages; and the last two, called "floating ribs," have no attachments to the sternum.

The **pectoral girdle** (shoulder girdle) consists of the two collarbones, or *clavicles,* and the two shoulderblades, or *scapulas.* The pectoral girdle is loosely and flexibly attached to the vertebral column by muscles. In contrast, the **pelvic girdle** is securely fused to the vertebral column. The pelvic girdle consists of a pair of large bones, each composed of three fused hipbones.

Each human limb consists of 30 bones, and terminates in five **digits**—the fingers and toes. The more specialized appendages of other vertebrates may be characterized by four digits (as in the pig), three (as in the rhinoceros), two (as in the camel), or one (as in the horse).

Great apes and humans do have a highly specialized feature—the opposable thumb. (Great apes also have an opposable big toe. In humans, although it is not opposable, the big toe is similar enough in structure to the thumb that it can be used as a substitute through surgery.) The opposable thumb can readily be wrapped around objects such as a tree limb in climbing, but it is especially useful in grasping and manipulating objects.

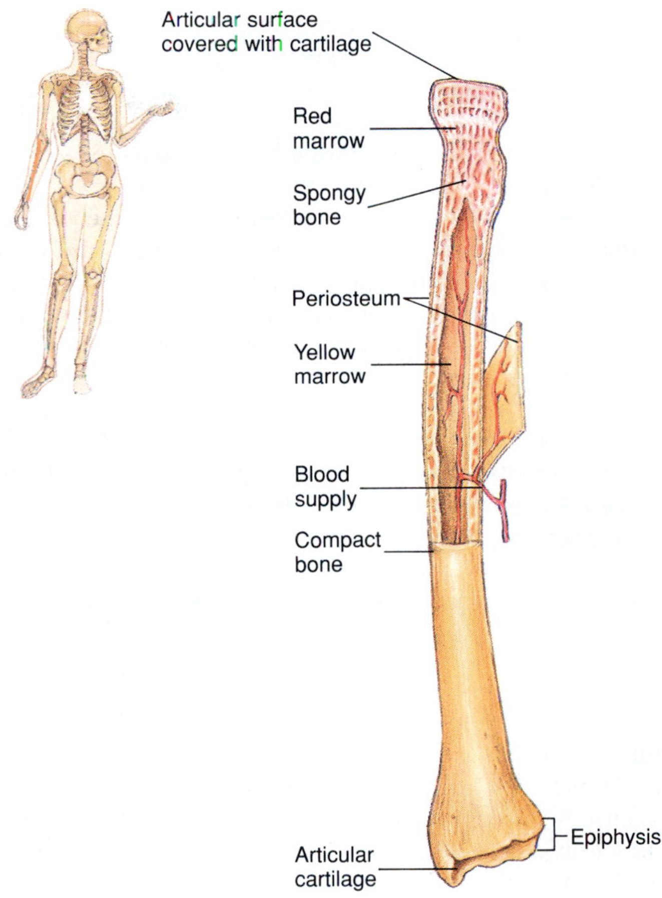

FIGURE 32–7 Anatomy of a bone. Structure of a typical long bone.

A Typical Long Bone Has a Thin Layer of Compact Bone and a Filling of Spongy Bone

The radius, one of the two bones of the forearm, is a typical long bone (Fig. 32–7; see also Fig. 31–4, page 603.) Its numerous muscle attachments are arranged so that the bone operates as a lever that amplifies the motion generated by the muscles. By themselves, muscles cannot shorten enough to produce large movements of the body parts to which they are attached.

Like other bones, the radius is covered by a connective tissue membrane, the **periosteum,** which is capable of laying down fresh layers of bone and thus increasing the diameter of the bone. The main shaft of a long bone is known as its **diaphysis.** The expanded ends of the bone are called **epiphyses.** In children, a disc of cartilage, the growth plate (epiphyseal plate), is found between the epiphyses and the diaphysis. These growth centers disappear at maturity, becoming vague lines. Within a long bone, there is a central cavity filled with **bone marrow.** The **red marrow** found in certain bones produces blood cells. **Yellow marrow** consists mainly of a fatty connective tissue.

The radius has a thin outer shell of **compact bone,** which is very dense and hard. Compact bone is found mainly near the surfaces of a bone, where it provides great strength. Recall from Chapter 31 that compact bone consists of interlocking, spindle-shaped units called **osteons** (or **haversian systems**). Within an osteon, osteocytes (bone cells) are found in small cavities called **lacunae.** The lacunae are arranged in concentric circles around central **haversian canals.** Blood vessels that nourish the bone tissue pass through the haversian canals.

Interior to the thin shell of compact bone is a filling of **spongy bone,** which despite its loose structure provides most of the mechanical strength of the bone. Spongy bone consists of a mesh of thin strands of bone. The spaces within the spongy bone are filled with bone marrow.

Bones Are Remodeled throughout Life

During fetal development, bones form in two ways. Long bones, such as the radius, develop from cartilage replicas, a process called **endochondral** bone development. The flat bones of the skull, the irregular vertebrae, and some other bones develop from a noncartilage connective tissue scaffold. This is known as **membranous** bone development.

Osteoblasts are bone-building cells. They secrete the protein collagen, which forms the strong, elastic fibers of bone. A complex calcium phosphate is present in the tissue fluid. This compound automatically crystallizes around the collagen fibers, forming the hard matrix of bone. The mineral component imparts rigidity to bone, whereas the collagen provides flexibility. As the matrix forms around the osteoblasts, they become isolated within the lacunae. The trapped osteoblasts are subsequently referred to as **osteocytes.**

Bones are modeled during growth and remodeled continuously throughout life in response to physical stresses and other changing demands. As muscles develop in response to physical activity, the bones to which they are attached thicken and become stronger. As bones grow, bone tissue is removed from the interior, especially from the walls of the marrow cavity. This process keeps bones from getting too heavy.

Osteoclasts are very large cells that break down bone in a process referred to as bone resorption. The osteoclasts move about, secreting enzymes that digest bone. Osteoclasts and osteoblasts work side-by-side to shape bones. Most bone is remade as many as ten times during the course of an average lifetime.

Joints Are Junctions between Bones

Joints hold bones together and many of them permit flexibility and movement. One way to classify joints is according to the degree of movement they allow. The sutures found between bones of the skull are **immovable joints.** In a suture, the bones are held together by a thin layer of dense fibrous connective tissue. This tissue may be replaced by bone in the adult. **Slightly movable joints** are found between vertebrae. These joints, which are made of cartilage, help absorb shock.

The most common joints are **freely movable joints** that are enclosed by a joint capsule composed of connective tissue. The joint capsule is lined with a membrane that secretes a lubricant, called **synovial fluid.** Generally, the joint capsule is reinforced by **ligaments,** which are bands of fibrous connective tissue that connect the bones and also limit movement at the joint.

Joints wear down with time and use. In osteoarthritis, a common joint disorder, cartilage repair does not keep up with degeneration and the articular cartilage wears out. In rheumatoid arthritis, the synovial membrane thickens and becomes inflamed. Synovial fluid accumulates, causing pressure and pain, and the joints become stiff.

MUSCLE IS THE CONTRACTILE TISSUE THAT ALLOWS MOVEMENT IN COMPLEX ANIMALS

Some animals run, some jump, and some fly. Others remain rooted to one spot, sweeping their surroundings with tentacles. Many contain internal circulating fluids, pumped by hearts and contained by hollow vessels that maintain their pressure by gentle squeezing. Digestive systems push food along with peristaltic contractions. Each of these actions depends on contractile proteins.

All eukaryotic cells contain the contractile protein **actin.** This protein is the major component of microfilaments and is important in many cell processes such as ameboid movement and attachment of cells to a surface. In most cells, the contractile protein **myosin** is functionally associated with actin. Actin and myosin are most highly organized in **muscle** cells.

We have discussed the contractile cells of the hydrostatic skeleton of *Hydra* and other cnidarians. In flatworms and other animal groups, muscle occurs as a specialized tissue, organized into definite layers. In complex animals, the muscles serve as motors that generate mechanical forces and motion. However varied its effects—locomotion, movement of food through the digestive tract, or the circulation of blood—muscle tissue has but one action: it contracts. The three types of muscle—skeletal, smooth, and cardiac—each specialized for its particular task, were described in Chapter 31. Recall that both skeletal and cardiac muscle are striated (banded).

Bivalve mollusks have both smooth and striated muscle. The smooth muscle, which is capable of slow, sustained contraction, can be used to keep the two shells tightly closed for long periods of time (even days). The striated muscle, which contracts rapidly, shuts the shells quickly when the mollusk is threatened. Arthropod muscles are typically striated.

A Vertebrate Muscle May Consist of Thousands of Muscle Fibers

In vertebrates, each **skeletal muscle** may be considered an organ. Its elongated cells, called fibers, are organized in bundles that are wrapped by connective tissue. The biceps in your arm, for example, consist of thousands of individual muscle fibers and their connective tissue coverings.

Recall that each striated muscle fiber is a long, cylindrical cell with many nuclei (Fig. 32–8). The plasma membrane has multiple inward extensions that form a set of **T tubules** (transverse tubules). The cytoplasm of a muscle fiber is referred to as **sarcoplasm,** and the endoplasmic reticulum as **sarcoplasmic reticulum.**

FIGURE 32–8 Structure of skeletal muscle. (***a***) A muscle such as the biceps in the arm consists of many bundles of muscle cells. (***b***) A bundle of muscle cells wrapped in a connective tissue covering. (***c***) Structure of a muscle cell showing the structure of a myofibril. The Z lines mark the ends of the sarcomeres. (***d***) Light photomicrograph showing striations. (***e***) Electron micrograph of striated muscle. (*d*, D. W. Fawcett; *e*, Ed Reschke) ▶

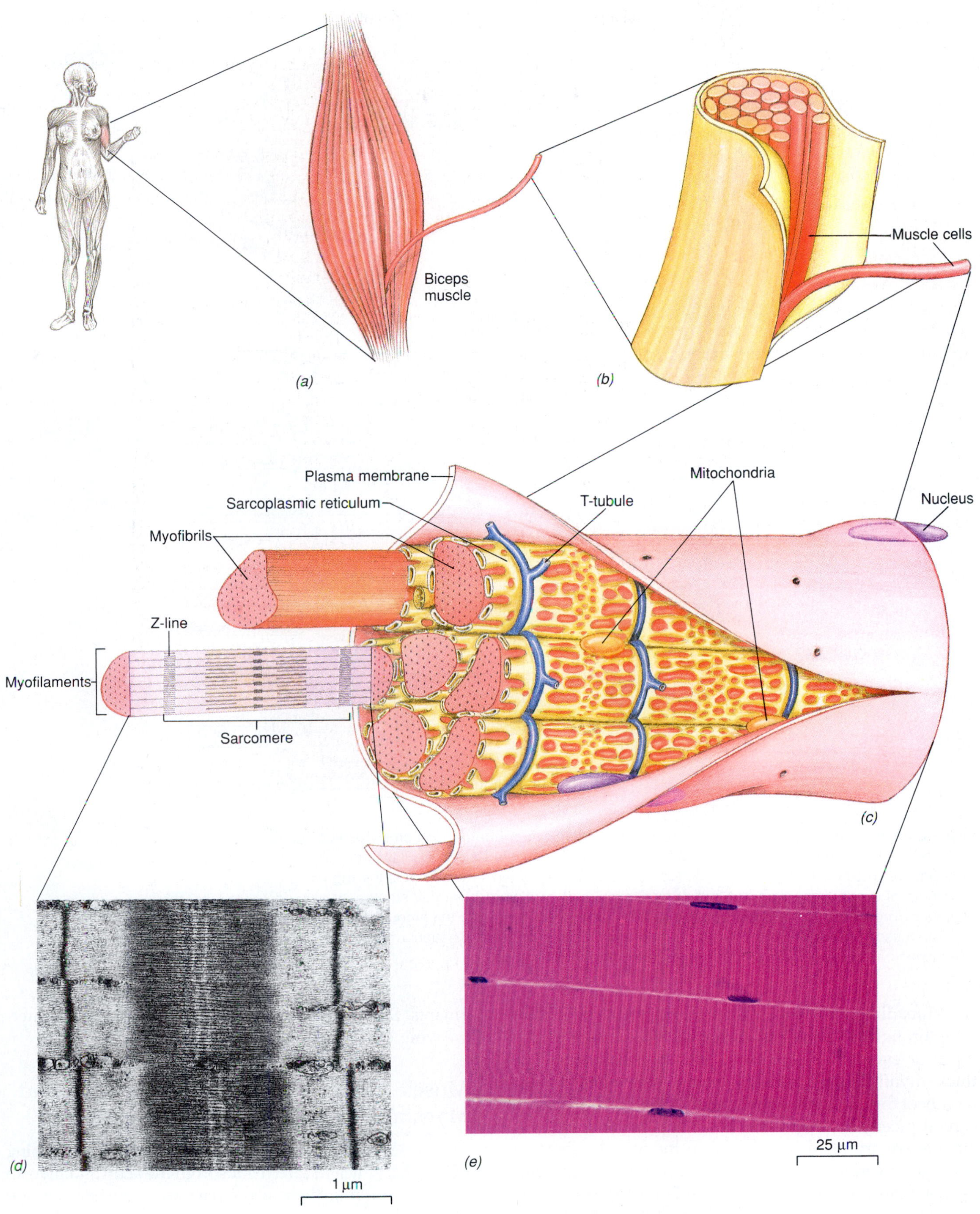

Biceps
muscle
(a)
Muscle cells
(b)
Plasma membrane
Sarcoplasmic reticulum
Mitochondria
T-tubule
Nucleus
Myofibrils
Z-line
Myofilaments
Sarcomere
(c)
(d)
1 μm
(e)
25 μm

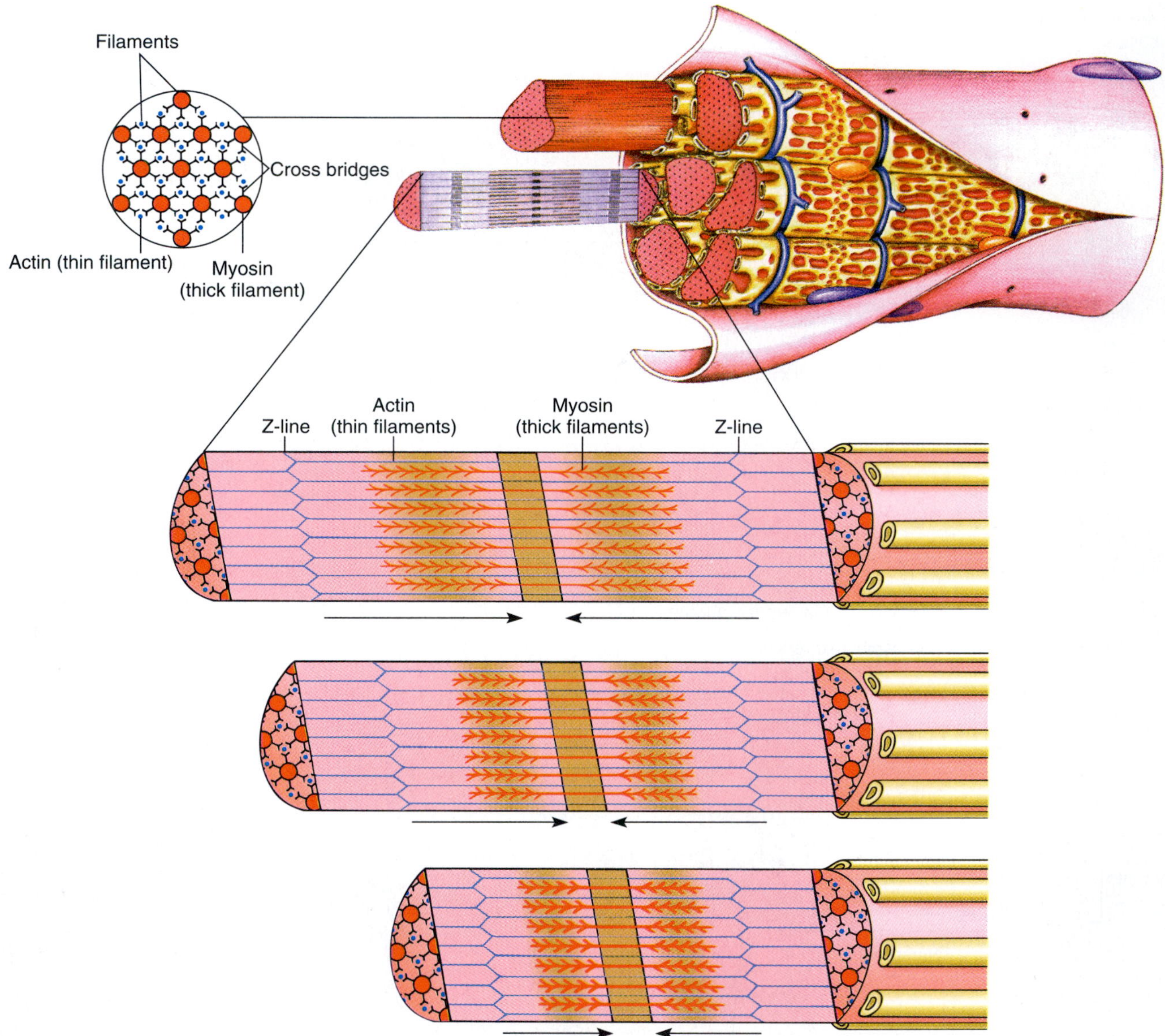

FIGURE 32–9 A myofibril stripped of the accompanying membranes. The cross section at top left of a myofibril shows the arrangement of actin and myosin filaments. Filaments slide past each other during contraction. Notice the way the filaments overlap. It is the regular pattern of overlapping filaments that gives skeletal and cardiac muscle their striated appearance. The filaments are shown sliding toward each other, increasing the amount of overlap. The muscle cell is shortened as the sarcomeres become shorter. At bottom, maximum contraction has occurred; the sarcomere has shortened considerably.

Threadlike structures called **myofibrils** run lengthwise through the muscle fiber. They are composed of two types of even tinier structures, the **myofilaments.** The thick myofilaments, called **myosin filaments,** consist mainly of the protein myosin, and the thin **actin filaments** consist mostly of the protein actin. Myosin and actin filaments are arranged lengthwise in the muscle fibers so that they overlap. Their overlapping produces the pattern of bands, or striations, characteristic of striated muscle (Fig. 32–9). A **sarcomere** is a unit of thick and thin filaments. Sarcomeres are joined at their ends by an interweaving of filaments called the Z line.

Muscle Contraction Occurs When Actin and Myosin Filaments Slide Past One Another

The typical pull of a muscle results from the shortening of its cells, which in turn results from the actin and myosin filaments pulling themselves past and between one another (Fig. 32–9). The rounded heads of the myosin mol-

(1) Myosin head takes up ATP, splits it into ADP and phosphate.

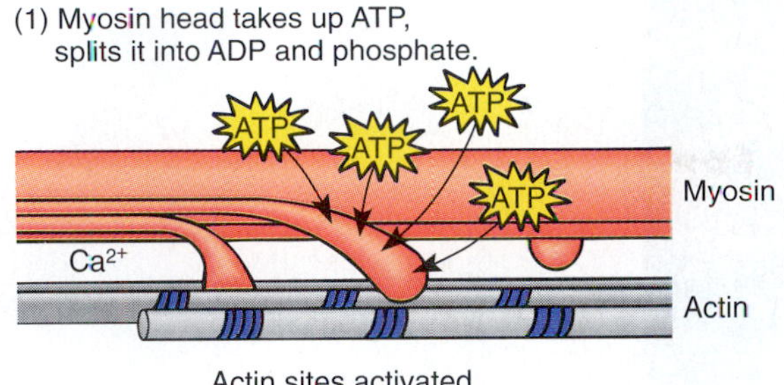

(2) Myosin head forms cross bridge with calcium-activated site on actin.

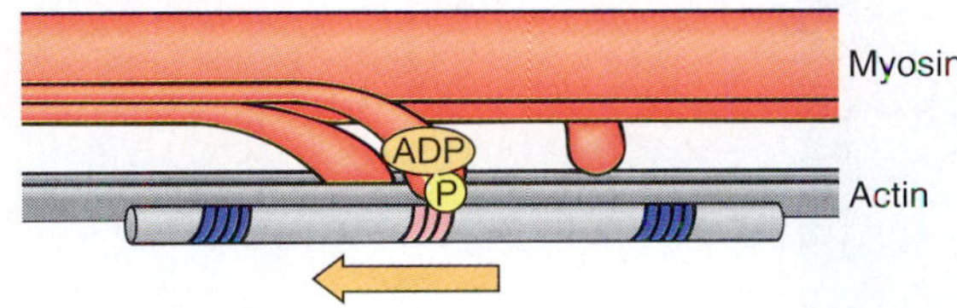

(3) Myosin head pulls actin filament, releases its ADP and phosphate.

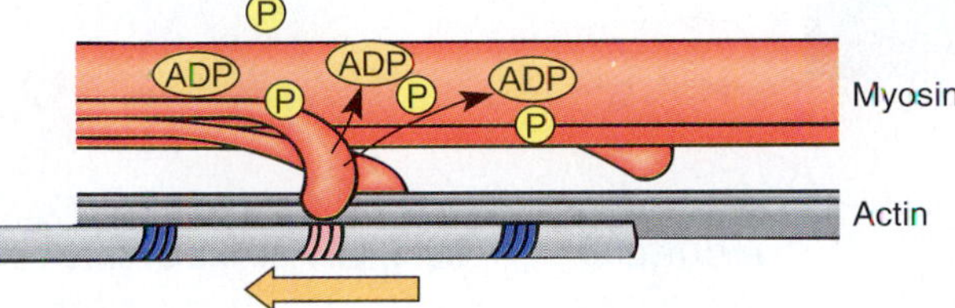

(4) Cross bridge broken.

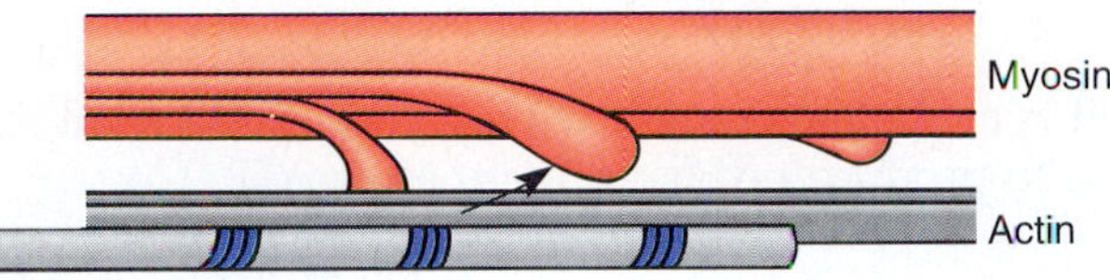

FIGURE 32–10 A model of muscle contraction. The cross bridges move the thin and thick filaments past each other in muscle contraction. Parts of this model are still hypothetical.

ecules extend away from the body of the myosin filament; they are positioned on flexible, armlike appendages (Fig. 32–10). The heads and armlike appendages together form cross bridges with the actin filaments. The head of the myosin molecule bears a binding site that is complementary to binding sites on the actin filament.

Each actin filament is composed of actin and two regulatory proteins. When the muscle cell is resting, molecules of regulatory protein block the active sites, preventing interaction with myosin cross bridges.

The myosin head has the ability to break down ATP in the presence of calcium and use the liberated energy for the contraction process. During contraction, the actin filaments move toward the middle of the myofibril. As this occurs, the muscle shortens. When many sarcomeres contract simultaneously, they produce the contraction of the muscle as a whole. We can summarize the process of muscle contraction as follows (Fig. 32–11):

1. A motor neuron releases the compound **acetylcholine** into the gap between the neuron and muscle fiber (Fig. 32–12).
2. The acetylcholine combines with receptors on the surface of the muscle fiber.
3. This results in depolarization (an electrical change) of the plasma membrane and initiation of an action potential, an electric current that spreads over the plasma membrane.
4. The action potential spreads through the T tubules and stimulates calcium release into the cytoplasm.
5. Calcium initiates a process that uncovers the binding sites of the actin filaments.
6. Myosin splits ATP. Myosin heads attach to the binding sites, forming cross bridges.
7. As the cross bridges flex and reattach to new binding sites, the filaments are pulled past one another and the muscle shortens. A new ATP must bind to the myosin head before the cross bridge can detach itself from the actin and begin a new cycle. The process is repeated with a third set, and so on (see Fig. 32–10). (We can imagine the myosin heads engaging, "hand over hand," in a kind of tug-of-war on the thin filaments.)

Even when we are not moving, our muscles are in a state of partial contraction known as **muscle tone.** At any given moment, some muscle fibers are contracted, stimulated by messages from nerve cells. Muscle tone is an unconscious process that helps keep muscles prepared for action. When the motor nerve to a muscle is cut, the muscle loses its muscle tone and becomes limp.

ATP Powers Muscle Contraction

Muscle cells are often called upon to perform strenuously, and so they must be provided with large amounts of energy. The immediate source of energy necessary for muscle contraction is ATP. ATP is needed both for the pull exerted by the cross bridges and for their release from each active site, as they engage in their tug-of-war on the thin filaments. *Rigor mortis,* the marked muscular rigidity that appears after death, results from ATP depletion, which occurs when cellular respiration ceases.[1]

[1]Rigor mortis does not persist indefinitely, however; the entire contractile apparatus of the muscles degenerates eventually, restoring pliability. The phenomenon is temperature-dependent, so given the prevailing temperature, a police officer can estimate the time of death of a cadaver from its degree of rigor mortis. Rigor mortis is not by itself muscular contraction; it only tends to freeze the corpse in its position at the time of death. Thus, tales of corpses sitting, pointing to their murderers, and otherwise carrying on posthumously may be entertaining but have no factual basis.

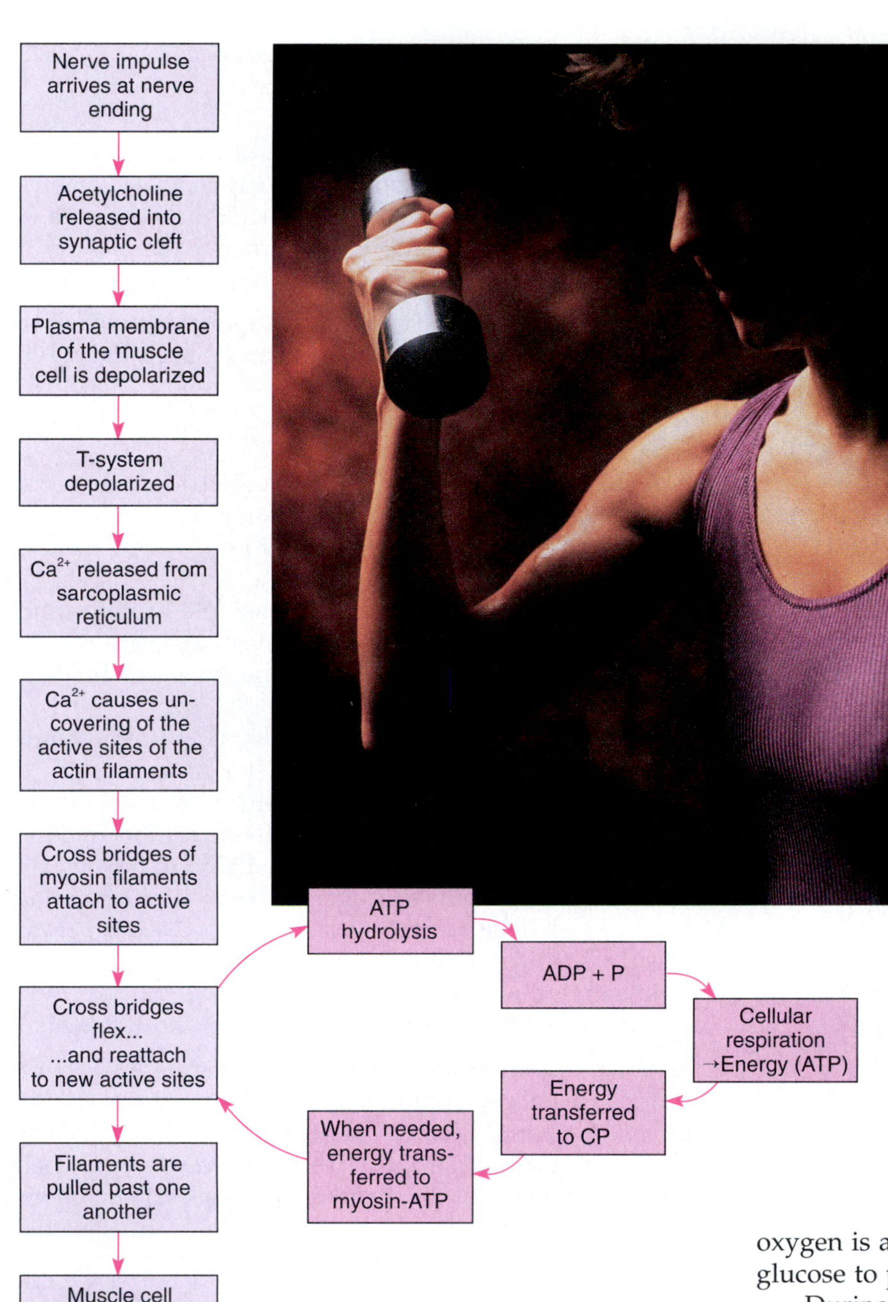
Nerve impulse arrives at nerve ending
Acetylcholine released into synaptic cleft
Plasma membrane of the muscle cell is depolarized
T-system depolarized
Ca2+ released from sarcoplasmic reticulum
Ca2+ causes uncovering of the active sites of the actin filaments
Cross bridges of myosin filaments attach to active sites
Cross bridges flex... ...and reattach to new active sites
Filaments are pulled past one another
Muscle cell shortens
ATP hydrolysis
ADP + P
Cellular respiration →Energy (ATP)
Energy transferred to CP
When needed, energy transferred to myosin-ATP

FIGURE 32–11 Summary of the events of muscular contraction. (Picture Perfect USA/Dembinsky Photo Associates)

Sufficient energy can be stored in ATP molecules for only the first few seconds of strenuous activity. A backup energy storage compound is needed. That compound is **creatine phosphate,** which muscle cells can stockpile. The energy stored in creatine phosphate is transferred to ATP as needed. But during vigorous exercise the supply of creatine phosphate does not last very long, either. As ATP and creatine phosphate stores are depleted, muscle cells must replenish their supplies of these energy-rich compounds.

Fuel is stored in muscle fibers in the form of glycogen, a large polysaccharide formed from hundreds of glucose units. As needed, glucose is released from storage and broken down in cellular respiration. When sufficient oxygen is available, enough energy is captured from the glucose to produce needed quantities of ATP.

During strenuous exercise, oxygen may not be available in sufficient quantity to meet the needs of the rapidly metabolizing muscle cells. Under these conditions, muscle cells are capable of breaking down fuel molecules anaerobically (without oxygen) for short periods of time. Recall from Chapter 8 that lactic acid fermentation is a method of rapidly generating ATP, but not in great quantity. ATP depletion results in weaker contractions and muscle fatigue. Accumulation of the waste product lactic acid also contributes to muscle fatigue. The buildup of lactic acid and depletion of glycogen during muscle exertion make up an **oxygen debt.** The debt is paid back during the period of rapid breathing and aerobic metabolism that typically follows strenuous exercise.

Skeletal Muscle Action Depends on Muscle Pairs that Work Antagonistically

Skeletal muscles produce movements by pulling on **tendons**, which are tough cords of connective tissue that an-

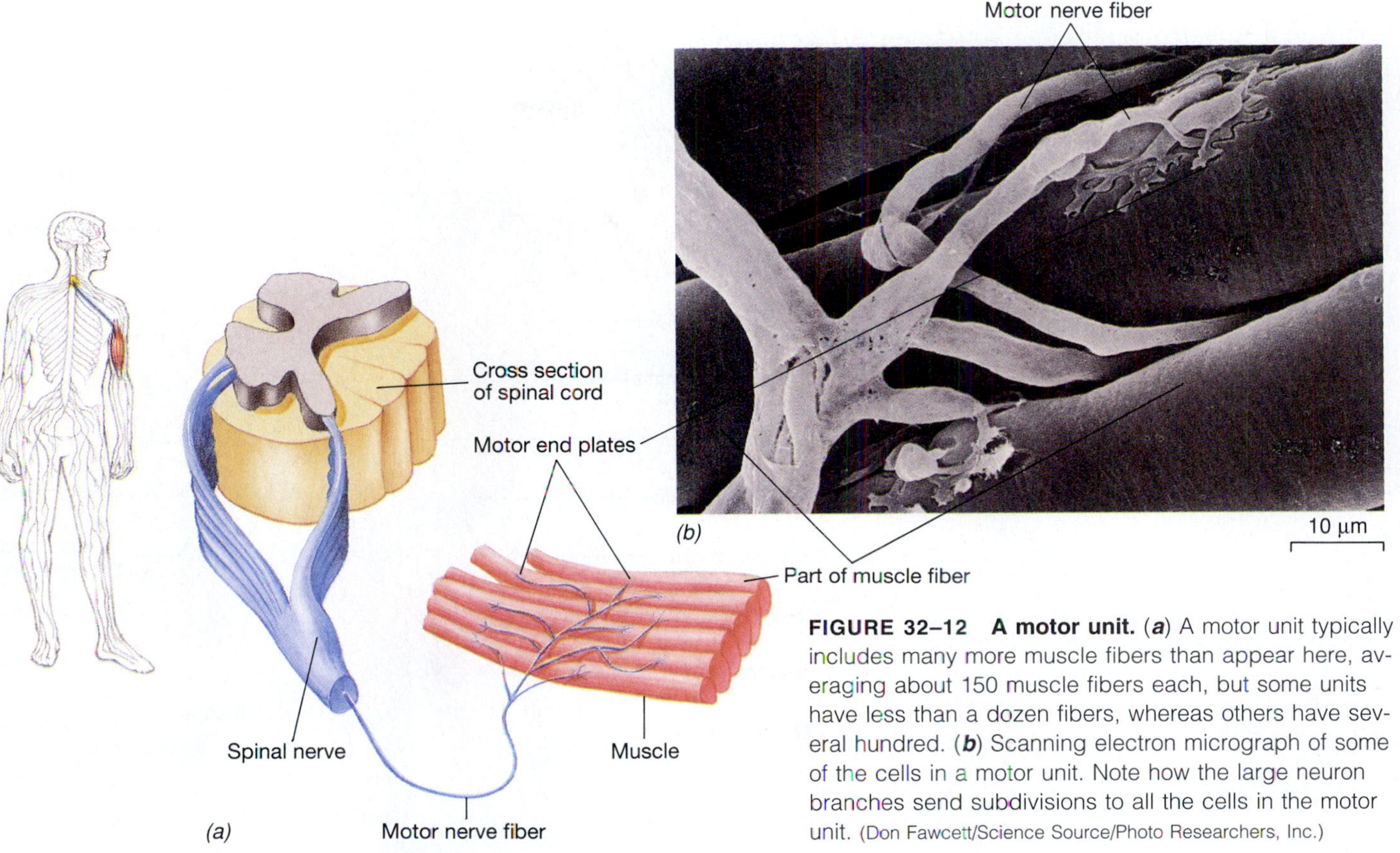

FIGURE 32–12 A motor unit. (*a*) A motor unit typically includes many more muscle fibers than appear here, averaging about 150 muscle fibers each, but some units have less than a dozen fibers, whereas others have several hundred. (*b*) Scanning electron micrograph of some of the cells in a motor unit. Note how the large neuron branches send subdivisions to all the cells in the motor unit. (Don Fawcett/Science Source/Photo Researchers, Inc.)

chor muscles to bone. Tendons then pull on bones. Most skeletal muscles pass across a joint and are attached to the bones that form the joint. When such a muscle contracts, it draws one bone toward or away from the bone with which it articulates.

Muscles can only pull; they cannot push. Muscles act **antagonistically** to one another, which means that the movement produced by one can be reversed by another. The biceps muscle, for example, permits you to flex your arm, whereas the triceps muscle allows you to extend it once again (Fig. 32–13). Thus, the biceps and triceps work antagonistically.

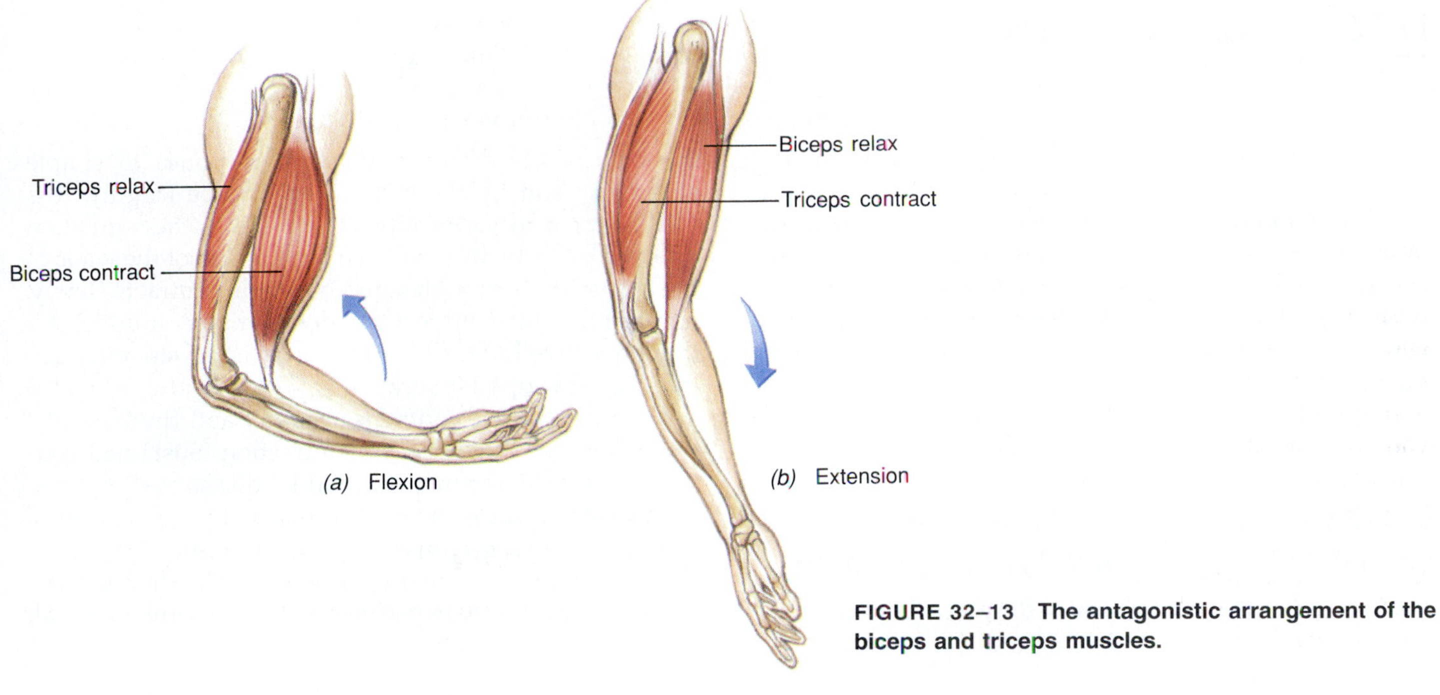

FIGURE 32–13 The antagonistic arrangement of the biceps and triceps muscles.

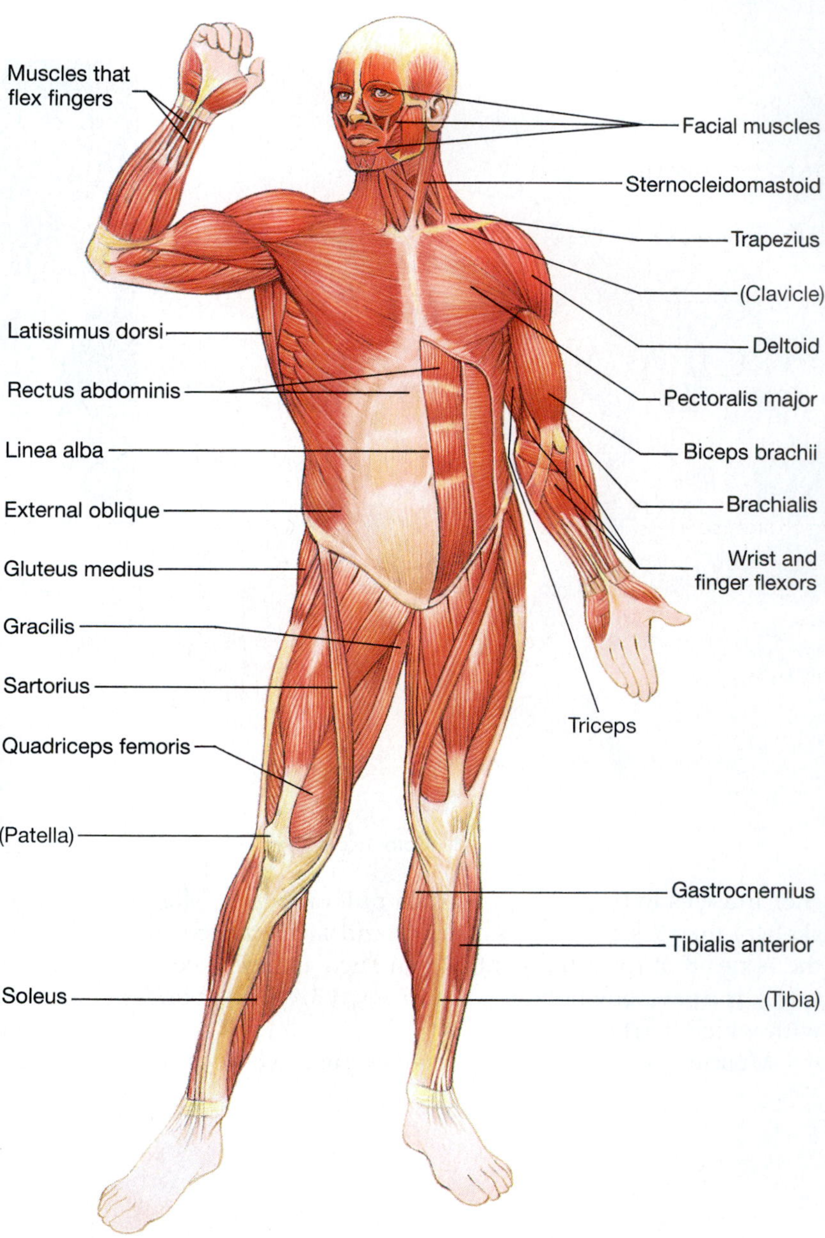

FIGURE 32–14 Superficial muscles of the human body. (*a*) Anterior view. (*b*) Posterior view.

The muscle that contracts to produce a particular action is known as the **agonist.** The muscle that produces the opposite movement is referred to as the **antagonist.** When the agonist is contracting, the antagonist is relaxed. Generally, movements are accomplished by groups of muscles working together, so there may be several agonists and several antagonists in any action. Note that muscles that are agonists in one movement may be antagonists in another. The superficial muscles of the human body are shown in Figure 32–14.

Smooth, Cardiac, and Skeletal Muscle Are Specialized for Particular Types of Response

The three types of muscle differ in the ways they respond. Smooth muscle often contracts in response to simple stretching, and its contraction tends to be lengthy. It is well adapted to performing such tasks as the regulation of blood pressure by sustained contraction of the walls of the arterioles. Although smooth muscle contracts slowly, it shortens much more than does striated muscle. Although not well suited for running or flying, smooth muscle squeezes superlatively.

Cardiac muscle contracts abruptly and rhythmically, propelling blood with each contraction. Sustained contraction of cardiac muscle would be disastrous!

Skeletal muscle, when stimulated by a single brief stimulus, contracts with a single quick contraction, called a **simple twitch.** Simple twitches ordinarily do not occur except in laboratory experiments. In the normal animal,

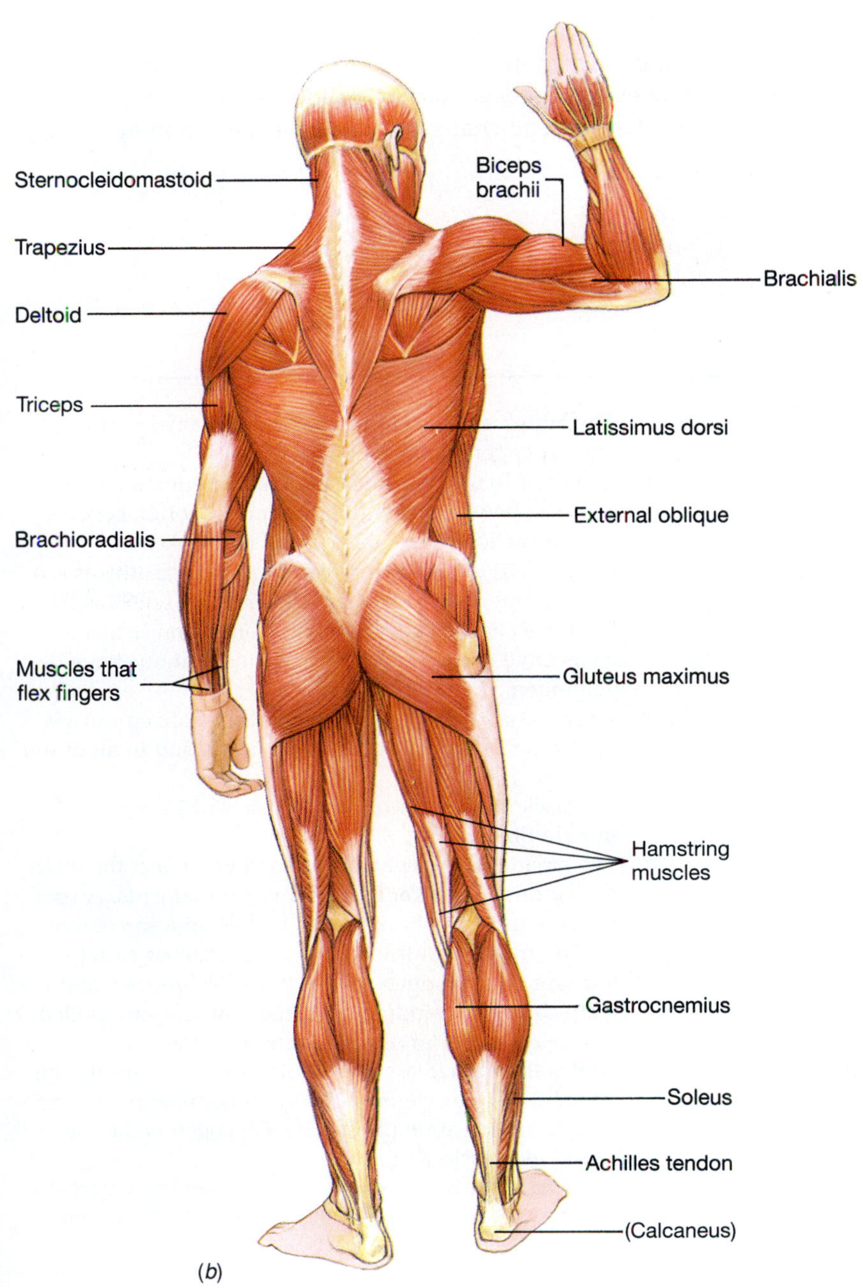

(*b*)

skeletal muscle receives a series of separate stimuli very close together. These produce not a series of simple twitches, however, but a single, smooth, sustained contraction, called **tetanus.** Depending upon the identity and number of our muscle cells tetanically contracting, we might thread a needle, haul a rope, or run a mile.

Not all muscular activities are the same. Dancing and, even more so, typing require quick response rather than the long, sustained effort that might be appropriate in hauling a rope. In many animals, entire muscles are specialized for quick or slow responses. In chickens, for instance, the white breast muscles are efficient for quick responses, adaptive because flight is an escape mechanism for chickens. On the other hand, chickens walk about on the ground all day, and the dark meat of the leg and thigh is composed of muscle specialized for more sustained activity.

Humans do not have light and dark meat. However, like other vertebrates, we do possess **fast-twitch fibers,** which are specialized for fast response, and **slow-twitch fibers,** which are specialized for slow response. Slow-twitch fibers have many mitochondria and a rich supply of capillaries. Slow-twitch fibers also have more **myoglobin,** which is a red pigment similar to hemoglobin. Myoglobin transfers oxygen from hemoglobin to the aerobic metabolic processes of the muscle cell.

The proportions of slow-twitch and fast-twitch fibers vary from person to person and from muscle to muscle in the same person. The relative proportions of the two appear to be genetically determined and influence the kind of athletic activity at which one has the greatest potential proficiency. A person whose leg and thigh muscles contain a high proportion of fast-twitch fibers could, with proper training, become a good sprinter. An athlete with a greater proportion of slow-twitch fibers may be better suited to marathon activities. Recent evidence, however, suggests that the proportions of the two kinds of fibers can be changed by appropriate training.

Chapter Summary

I. In invertebrates, epithelial tissue may be specialized for sensory or respiratory functions, may produce a protective cuticle, secrete lubricants or adhesives, produce odorous or poisonous secretions, or produce threads for nests or webs.

II. Vertebrate skin protects, may prevent dehydration, may be specialized for secretion or reception of stimuli, and may help regulate body temperature.
   - A. In human skin, cells in the stratum basale of the epidermis continuously divide. As they are pushed upward toward the skin surface, these cells mature, produce keratin, and eventually die.
   - B. The dermis, which consists of dense, fibrous connective tissue, rests on a layer of subcutaneous tissue that is composed largely of fat.

III. The skeleton transmits mechanical forces that are generated by muscle, and also supports and protects the body.
   - A. Many invertebrates have a hydrostatic skeleton in which fluid transmits forces generated by contractile cells or muscle.
   - B. Exoskeletons are characteristic of mollusks and arthropods. The arthropod skeleton, composed mainly of chitin, is jointed for flexibility. This nonliving skeleton does not grow. Therefore, arthropods molt periodically, shedding their outgrown skeletons and forming new ones.

IV. Endoskeletons, found in echinoderms and chordates, are composed of living tissue and therefore are capable of growth.
   - A. The vertebrate skeleton consists of an axial portion and an appendicular portion.
   - B. A typical long bone consists of a thin outer shell of compact bone surrounding the inner spongy bone. Within the long bone is a central marrow cavity.
   - C. Long bones develop from cartilage replicas; this is endochondral bone formation. Other bones, such as the flat bones of the skull, develop from a noncartilage connective tissue replica; this is membranous bone development.

V. All animals have the ability to move. Specialized muscle tissue is found in most invertebrate phyla and in all of the vertebrates.
   - A. A muscle such as the biceps consists of hundreds of muscle fibers.
   - B. The striations of skeletal muscle fibers reflect the interdigitations of their actin and myosin filaments. A unit of actin and myosin filaments makes up a sarcomere.
   - C. During muscle contraction, myosin attaches to actin, forming cross bridges. As the cross bridges flex and reattach to new binding sites, the filaments are pulled past one another and the muscle shortens.
   - D. ATP is the immediate source of energy for muscle contraction, but muscle tissue has another energy storage compound, creatine phosphate. Glycogen is the fuel stored in muscle fibers.
   - E. As muscle contracts (shortens), it moves body parts by pulling on them. Muscles act antagonistically to one another.

Selected Key Terms

Post-Test

1. The vertebrate skin consists of two main layers, the outer ________________ and the inner ________________.
2. The cells of the stratum ________________ of the epidermis are dead and almost waterproof.
3. The protein ________________ confers mechanical strength, flexibility, and waterproofing on the skin.
4. ________________ skeletons have the principal or even sole function of transmitting muscular force.
5. Arthropods ________________ from time to time as they grow, replacing their exoskeletons.
6. The internal skeletons of echinoderms and chordates are known as ________________.
7. The radius has a thin outer shell of ________________ bone and a loose filling of ________________ bone.
8. Synovial fluid serves as a/an ________________ in ________________.
9. The two types of myofilaments in muscle tissue are ________________ filaments and ________________ filaments.
10. Unscramble this list of the events of muscle contraction and place them in the correct sequence:
    - a. Calcium release
    - b. T-system depolarization
    - c. Acetylcholine release
    - d. Nerve impulse
    - e. Uncovering of the binding sites of the actin filaments
    - f. Cross bridges flex
    - g. Cross bridges release binding sites
11. The function of creatine phosphate in the muscle cell is ________________________________.
12. Fuel is stored in muscle cells in the form of the polysaccharide ________________; the immediate source of energy for muscle contraction is ________________.
13. Cells that continuously divide are typically found in (a) dermis (b) muscle (c) stratum corneum (d) stratum basale (e) two of the preceding answers are correct.
14. The insoluble protein that gives the skin mechanical strength is (a) keratin (b) sebum (c) oil (d) cuticle (e) actin.
15. The axial skeleton includes the (a) skull (b) vertebral column (c) ribs (d) sternum (e) all of the preceding answers are correct.
16. Large cells that break down bone are called (a) apatites (b) osteoclasts (c) osteoblasts (d) myosins (e) lacunae.
17. Tissue capable of slow, sustained contraction is (a) cardiac muscle (b) skeletal muscle (c) smooth muscle (d) subcutaneous tissue (e) two of the preceding answers are correct.
18. An energy storage compound used by muscle cells to stockpile energy is (a) ATP (b) creatine phosphate (c) lactic acid (d) oxygen (e) apatite.

Review Questions

1. Compare vertebrate skin with the external epithelium of invertebrates.
2. What properties does keratin confer on human skin?
3. What is a hydrostatic skeleton? Which functions does it perform?
4. How do the septa in the annelid worm contribute to the flexibility of its hydroskeleton?
5. What are the disadvantages of an exoskeleton? Advantages?
6. Describe the divisions of the vertebrate skeleton.
7. Draw a typical long bone, such as the radius, and label its parts.
8. Contrast the functions of osteoblasts and osteoclasts. Why is it important that bones be continuously remodeled?
9. Identify three types of joints based on their function.
10. Compare the two types of myofilaments in muscle tissue. What is a sarcomere?

11. Outline the sequence of events that cause a muscle fiber to contract, beginning with the stimulation of its nerve and including cross-bridge action.

12. What is the role of ATP in muscle contraction? What is the function of creatine phosphate? Of glycogen?

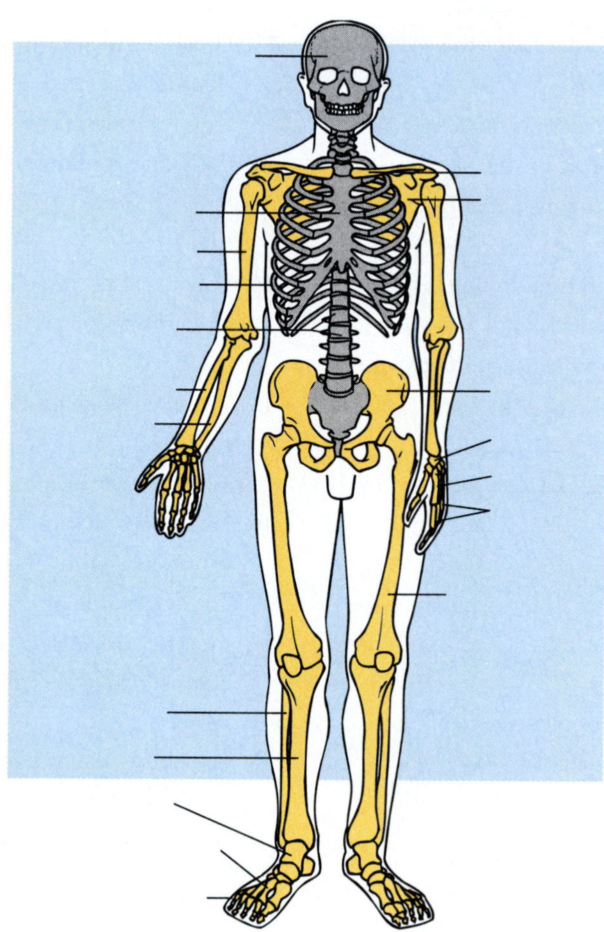

13. Label the bones of the skeleton.

Thinking Critically

1. Many different drugs are being evaluated for use in transdermal patches. How might a person allergic to ragweed pollen benefit from an antihistamine patch? What might be some benefits to a cancer patient of a patch that delivers anticancer drugs? What other transdermal systems might be useful?
2. What, if any, measures are you willing to take to slow the depletion of the ozone layer? What effects do you think depletion of this layer may be having on other organisms?
3. Compare the arthropod exoskeleton to the vertebrate endoskeleton. What are some benefits and some disadvantages of each type of skeleton?
4. What are some examples of hydrostatic support in plants?
5. Why is it important that a muscle be able to switch roles—sometimes acting as an agonist while at other times acting as an antagonist?

Recommended Readings

Hadley, N.F., "The Arthropod Cuticle," *Scientific American,* July 1986, pp. 104–112. The author discusses the properties of the arthropod exoskeleton that have contributed significantly to the adaptive success of the arthropods.

Segal, M., "Patches, Pumps, & Timed Release: New Ways to Deliver Drugs." *FDA Consumer,* October 1991, pp. 13–15. This interesting discussion of drug delivery systems includes a section on transdermal patches.

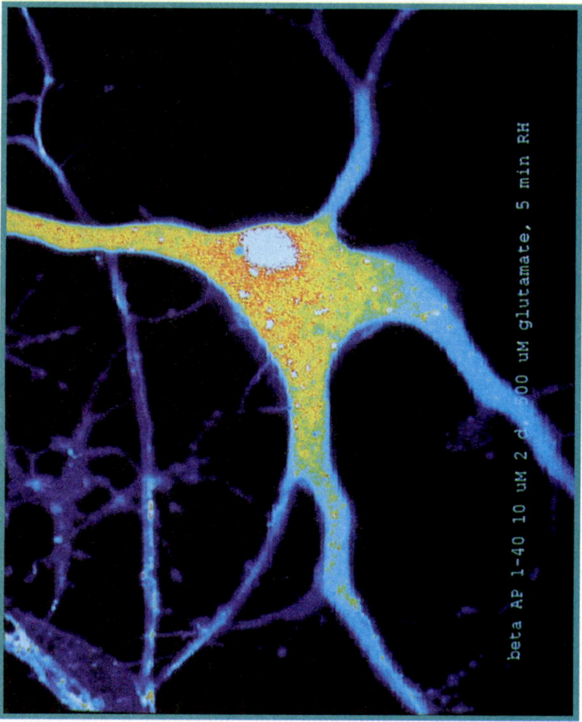

Neuron from rat brain (hippocampus). *The neuron has been loaded with a calcium indicator dye to show the destabilization of calcium homeostasis caused by amyloid protein. The loss of calcium balance makes the neuron vulnerable to toxicity. The cell was imaged by Alzheimer's researchers using a confocal laser scanning microscope.* *(Courtesy of Dr. Mark P. Mattson, University of Kentucky)*

TANGLED NEURONS: CLUES TO UNTANGLING ALZHEIMER'S DISEASE?

When Jane first noticed that she was forgetting the names and phone numbers of her coworkers, she was a successful 61-year-old business executive earning a six-digit income. Within three years, she was unable to walk, talk, or feed herself. She died at age 65. Jane was a victim of Alzheimer's disease, a progressive, degenerative brain disorder that afflicts more than four million people in the United States alone. Although Alzheimer's disease strikes individuals in midlife, it is most common in the over-65 population, affecting an estimated 6% of this population. In fact, it is the leading cause of senile dementia, which is the loss of memory, judgment, and the ability to reason that we associate with aging. For reasons still unknown, women have a higher incidence of Alzheimer's disease.

In Alzheimer's disease, cells are lost in certain areas of the brain, including the cerebral cortex and hippocampus, areas that are important in thinking and remembering. Neurons that secrete the neurotransmitter acetylcholine are especially affected. Two abnormalities that develop in brain tissue as we age are especially characteristic of Alzheimer's disease: *senile plaques* and *neurofibrillar tangles*. These abnormal developments appear to play a role in rendering brain cells nonfunctional, leading to deterioration in function. Researchers are investigating the biochemistry and genetic basis of both plaques and tangles for clues to the causes and cures of Alzheimer's disease.

Senile plaques are clusters of abnormal neurons and glial cells. Investigators have demonstrated that senile plaques have a central core consisting of a protein fragment called beta amyloid peptide. The precursor (beta-APP) of this protein fragment is a large transmembrane protein coded by a gene located on chromosome 21. Normal brain cells make a soluble form of beta amyloid peptide. Alzheimer's disease may develop when an imbalance in brain metabolism results in an insoluble peptide form, which leads to plaque formation. The plaques are thought to disrupt calcium homeostasis, which leads to neural malfunction and brain cell death.

Neurofibrillary tangles consists of abnormal accumulations of certain cytoskeletal proteins in the neuron cytoplasm. One of the proteins involved, called **tau,** normally stimulates the protein tubulin to form microtubules. When too many phosphates attach to tau, it can no longer adhere to microtubules. Instead, tau molecules join with one another to form fibrous deposits that make up the neurofibrillary tangles. Neurons that have such tangles form fewer synapses with other neurons.

In 1993, neurologist Allen Roses at Duke University Medical Center discovered a new clue to the Alzheimer's disease mystery. We all have a gene that codes for a protein called apolipoprotein E, or **Apo-E,** that helps transport cholesterol in the blood. Three different forms of this protein are known. The common form of the protein, Apo-E3, binds to tau and may inhibit phosphate bonding. The Apo-E4 form of the protein (which differs in only one amino acid) does not inhibit phosphate bonding. Apo-E3 may also be important in transporting beta amyloid to cells for processing.

Roses reported that individuals who are homozygous for the Apo-E4 allele (that is, have two copies of the gene that codes for the Apo-E4 form of the protein) are eight times more likely to develop Alzheimer's disease as are individuals with the more common form of the gene, Apo-E3. Thus, the Apo-E4 gene may prove to be an important risk factor. Would you want to know if you had this gene?

Clues to untangling the mysteries of Alzheimer's disease are leading researchers in many different directions. One current focus of study involves certain nerve growth factors that appear to have a protective role. Other investigators are exploring an inherited defect in mitochondria. More than a dozen drugs designed to slow the progress of Alzheimer's disease are in clinical trials. One of the challenges of developing a drug has been getting the medication through the blood–brain barrier (see Chapter 6 opener). Current research on the causes and cures of Alzheimer's disease is providing new insights into the metabolism of neurons and the function of the nervous system.

Responsiveness: Neural Control

33

Learning Objectives

After you have studied this chapter you should be able to:

1. Compare the functions of glial cells and neurons.
2. Draw a neuron. Label each part and give its function.
3. Summarize the process by which an impulse is transmitted along a neuron.
4. Outline the process of synaptic transmission.
5. Describe the factors that determine whether a neuron will transmit an impulse, and summarize their interaction.
6. Draw a reflex pathway consisting of three neurons, label each structure, and indicate the direction of information flow.

Key Concepts

- ☐ In most animals, information flow through the nervous system requires reception, transmission of information to a control center such as the brain or spinal cord, integration of the information, transmission of the "decision," and response by a muscle or gland.
- ☐ Neurons are specialized to receive information and to produce and transmit rapid electrochemical signals, called nerve impulses, or action potentials.
- ☐ Stimulation of a neuron may open sodium channels in the plasma membrane, which may cause depolarization. When the extent of depolarization reaches threshold level, an action potential is generated.
- ☐ Synaptic transmission generally depends on release of a neurotransmitter (such as acetylcholine or norepinephrine) from vesicles in the presynaptic axon. The neurotransmitter can trigger other neurons or stimulate responses in muscle or glands.
- ☐ A simple reflex action such as withdrawal from a painful stimulus involves a sequence of three types of neurons: a sensory neuron, an association neuron, and a motor neuron.

NEURAL RESPONSE INCLUDES RECEPTION, TRANSMISSION, INTEGRATION, AND RESPONSE BY MUSCLES OR GLANDS

An octopus releases a black fluid to confuse a predator. A hungry pelican dives to capture a fish. A college student puts on a jacket before going out in a snow storm and remembers to study for her biology test. The ability of an organism to survive and to maintain its steady state depends on how effectively it can respond to changes in its internal or external environment. Changes within the body or in the outside world that can be detected by an organism are called **stimuli.** In simple animals with simple nervous systems, the range and types of responses to stimuli are limited. In complex animals, varied and sophisticated responses to stimuli are possible because responsiveness is controlled by two highly specialized systems—the nervous and endocrine systems.

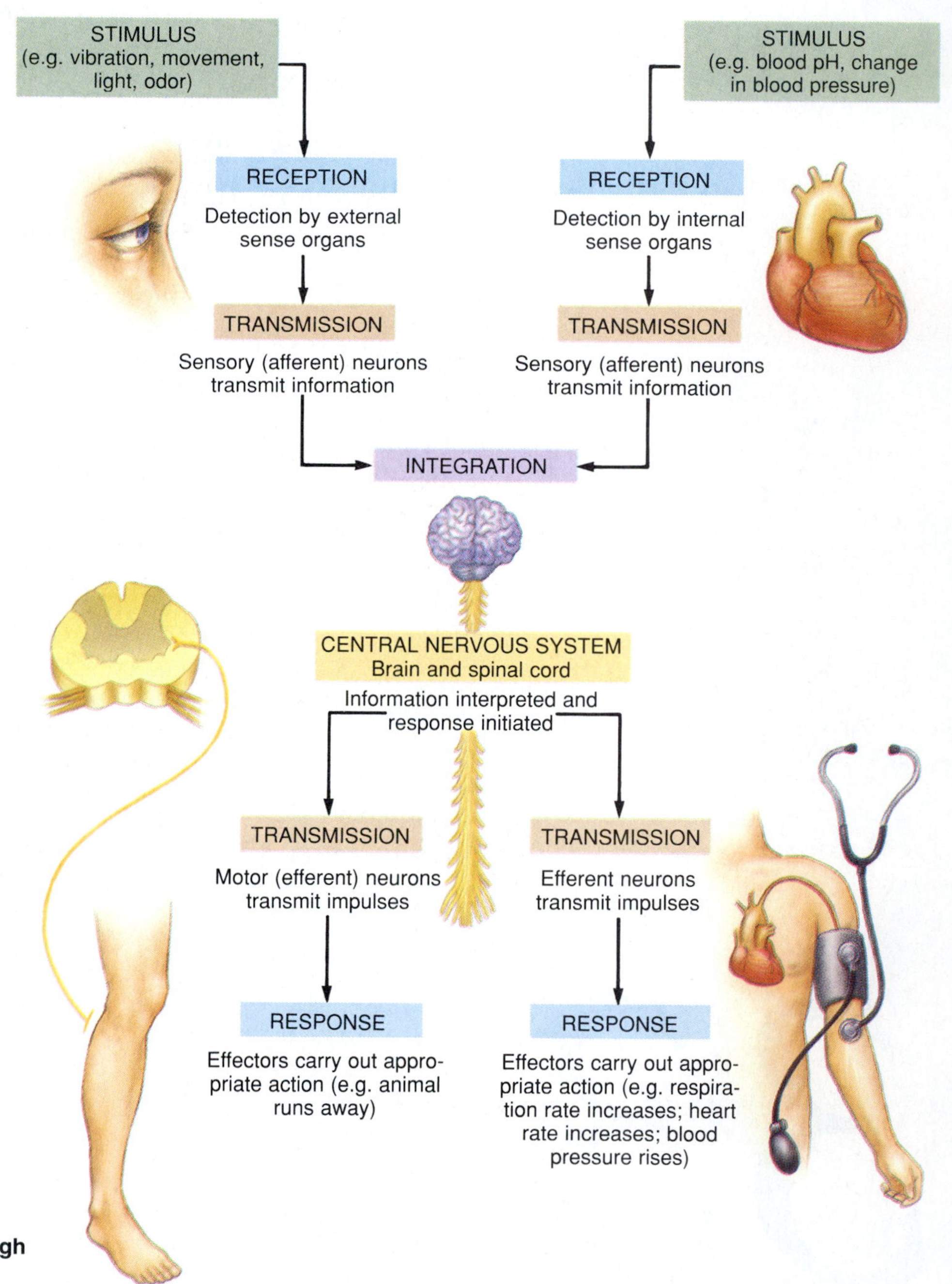

FIGURE 33–1 Flow of information through the nervous system.

The nervous system permits very rapid response, whereas the endocrine system generally responds relatively slowly. As will be discussed in Chapter 41, the endocrine system provides long-lasting chemical regulation. The nervous system receives information, interprets it, and then issues appropriate commands so that responses are coordinated and homeostatic.

Thousands of stimuli bombard an organism each day. Such stimuli include changes in light, temperature, potential food, and appearance of a friend or predator. Whether we consider a lizard darting out its tongue to capture a fly or a human slamming on his automobile brakes to avoid hitting a child, appropriate response to a stimulus involves four processes: reception, transmission, integration, and response by muscles or glands (Fig. 33–1). **Reception** is the process of detecting a stimulus, that is, of receiving information. It is the job of specialized sense organs as well as of the neurons themselves. **Transmission** is the process of sending messages along neurons, from one neuron to another, or from a neuron to a muscle or gland. A neural message is sent from a receptor to the **central nervous system (CNS),** comprised of the brain and spinal cord. Neurons that transmit information to the CNS are called **sensory neurons,** or afferent neurons. **Integration** is the sorting and interpreting of incoming information and the determination of an appropriate mode of response. In vertebrates, integration is primarily the function of the CNS. Neural messages are transmitted from the CNS by efferent neurons, often referred to as **motor neurons.** The actual **response** is carried out by the muscles and glands.

THE CELL TYPES OF THE NERVOUS SYSTEM ARE NEURONS AND GLIAL CELLS

In complex animals, the key unit of the nervous system is the nerve cell, or **neuron.** This cell is specialized to receive and send information. The neuron works by producing and transmitting rapid electrical signals, called **nerve impulses.** A second cell type unique to the nervous system is the **glial cell.** These cells provide support for the neurons.

Glial Cells Support and Protect Neurons

Some glial cells envelop neurons and form insulating sheaths around them. Others are phagocytic and serve to remove debris from the nervous tissue. Another type of glial cell lines the cavities of the brain and spinal cord. **Schwann cells,** which are supporting cells found outside the central nervous system, form sheaths about some neurons. Sometimes glial cells are referred to collectively as the **neuroglia,** which literally means "nerve glue."

A Typical Neuron Consists of a Cell Body, Dendrites, and an Axon

Highly specialized to receive and transmit electrochemical messages, the neuron is distinguished from other cells by its long cytoplasmic extensions. Let us examine the structure of a common type of neuron, the **multipolar** neuron (Fig. 33–2).

The largest portion of the neuron, the **cell body,** contains the bulk of the cytoplasm, the nucleus, and most of the other organelles. Two types of cytoplasmic extensions project from the cell body—the dendrites and a long, single axon.

Dendrites are typically short, highly branched fibers. They are specialized to receive neural information and send it to the cell body. The cell body integrates incoming signals and can also receive information directly.

Although microscopic in diameter, an **axon** may be 3 feet or more in length. The axon conducts nerve impulses from the cell body to another neuron or to a muscle or gland. At its end, the axon branches, forming **axon terminals** that end in tiny structures called **synaptic knobs.** These structures release **neurotransmitters,** chemicals that transmit signals from one neuron to another. Along its course an axon can give off branches.

Axons of many neurons outside the central nervous system have two coverings: an outer **cellular sheath**

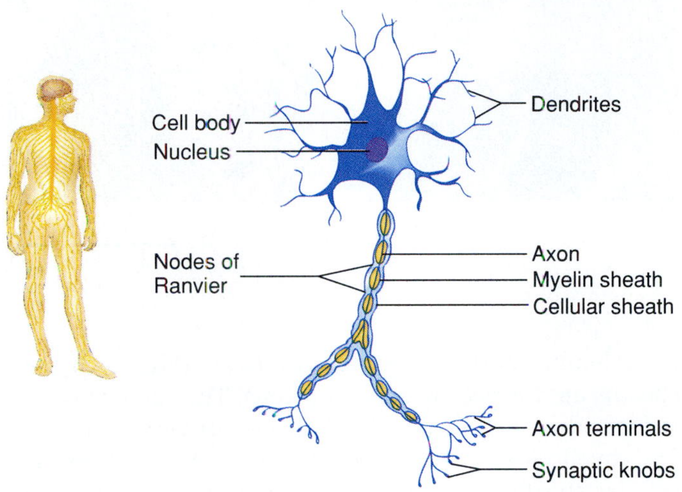

FIGURE 33–2 Structure of a multipolar neuron. The axon of this neuron is myelinated. Both the myelin sheath and the cellular sheath are shown.

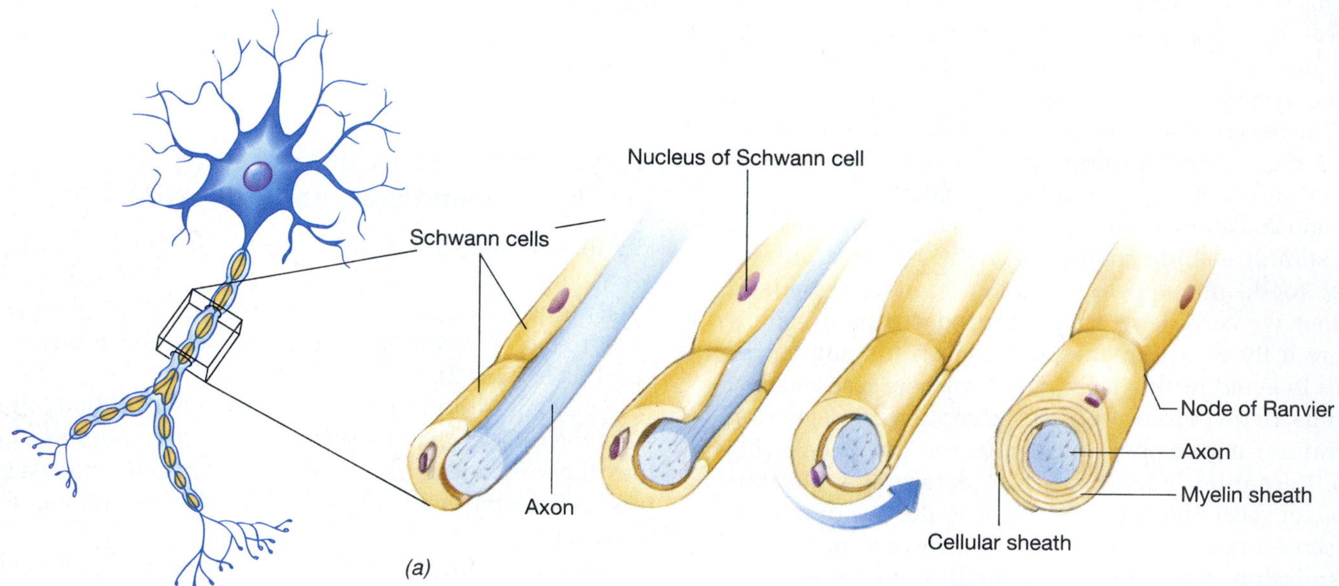

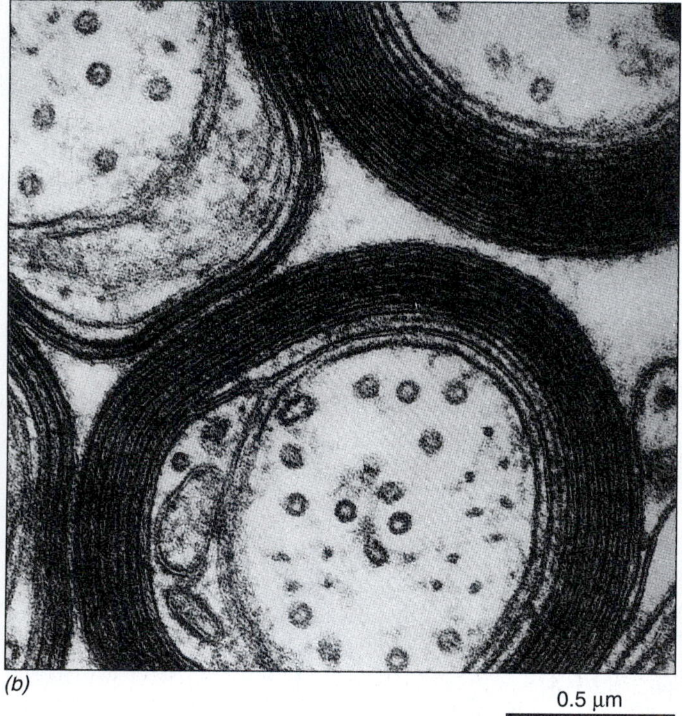

FIGURE 33–3 The myelin sheath. (***a***) Formation of the myelin sheath around an axon. A Schwann cell wraps its plasma membrane around the axon many times to form the insulating myelin sheath. The rest of the Schwann cell remains outside the myelin sheath, forming the cellular sheath. (***b***) Electron micrograph of a section through a myelinated axon showing the myelin sheath. (Visuals Unlimited/C. S. Raine)

(neurilemma) and an inner **myelin sheath** (Fig. 33–3). Both sheaths are formed by Schwann cells. The cellular sheath is formed by Schwann cells that line up along the axon. The myelin sheath, which lies between the axon and the cellular sheath, is formed when the Schwann cells wind their plasma membranes about the axon several times.

The plasma membrane of the Schwann cell contains **myelin,** a white, lipid-rich substance. Myelin is an excellent insulator, and its presence influences the transmission of nerve impulses. Between successive Schwann cells, gaps called **nodes of Ranvier** occur in the myelin sheath. The axon is not insulated with myelin at these points.

Almost all axons more than 2 micrometers in diameter are myelinated, that is, they have myelin sheaths. Those of smaller diameter are generally unmyelinated. In the brain and spinal cord, myelin sheaths are formed by certain glial cells, but cellular sheaths are not present.

In **multiple sclerosis,** a neurological disease that affects about 300,000 people in the United States alone, patches of myelin deteriorate at irregular intervals along the length of the neurons and are replaced by scar tissue. This damage interferes with the conduction of neural impulses, and the victim suffers weakness, numbness, loss of muscle control, and paralysis of parts of the body. The cause of multiple sclerosis has been a medical mystery, but there is some evidence that it is an autoimmune disorder, in which the body attacks its own tissue, in this case myelin (Chapter 37).

The cellular sheath is important in the regeneration of injured neurons. When an axon is cut, the portion separated from the cell body deteriorates and is phagocytized by surrounding cells, but the cellular sheath remains intact. The cut end of the axon grows slowly through the empty cellular sheath, and eventually at least partial neural function may be restored.

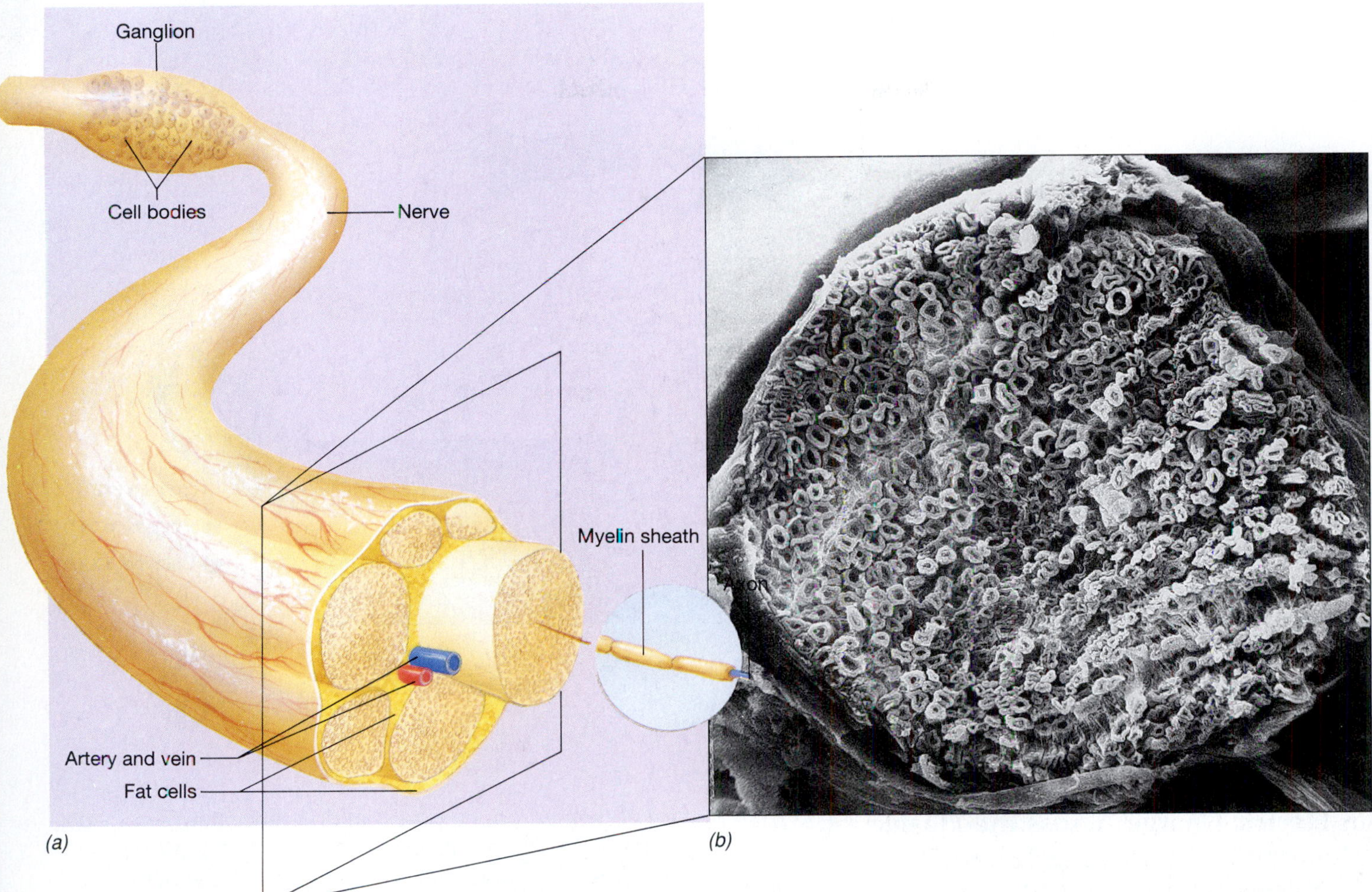

FIGURE 33–4 Structure of a nerve and a ganglion. (***a***) A nerve consists of bundles of axons held together by connective tissue. The cell bodies belonging to these axons are grouped together in a ganglion. (***b***) Electron micrograph showing a cross section through a myelinated afferent nerve of a bullfrog. (E.R. Lewis, University of California/Biological Photo Service)

A **nerve** is a complex cord consisting of hundreds or even thousands of axons wrapped together in connective tissue (Fig. 33–4). We can compare a nerve to a telephone cable. The individual axons correspond to the wires that run through the cable, and the sheaths and connective tissue coverings correspond to the insulation. You might wonder where the cell bodies are that are attached to all of the axons in a bundle. These are often grouped together in a mass known as a **ganglion.**

NEURONS CONVEY INFORMATION BY TRANSMITTING RAPIDLY MOVING ELECTRICAL IMPULSES

When a neuron receives a stimulus that is strong enough, its axon fires a **nerve impulse.** This is an electrical current that travels rapidly down the axon into the synaptic knobs. Once initiated, the electrical impulse is self-propagating. For a nerve impulse to be fired, however, the plasma membrane has to maintain what is called a resting potential.

Concept Connections

Neurons ⟷ ***Evolution***

Paleobiologist Steven Stanley of Johns Hopkins University has suggested that evolution of the neuron was responsible for the tremendous adaptive radiation of animals about 600 million years ago. Stanley suggests that development of a complex cell like the neuron may have taken hundreds of millions of years. Once it evolved, the neuron became the building block of animal nervous systems.

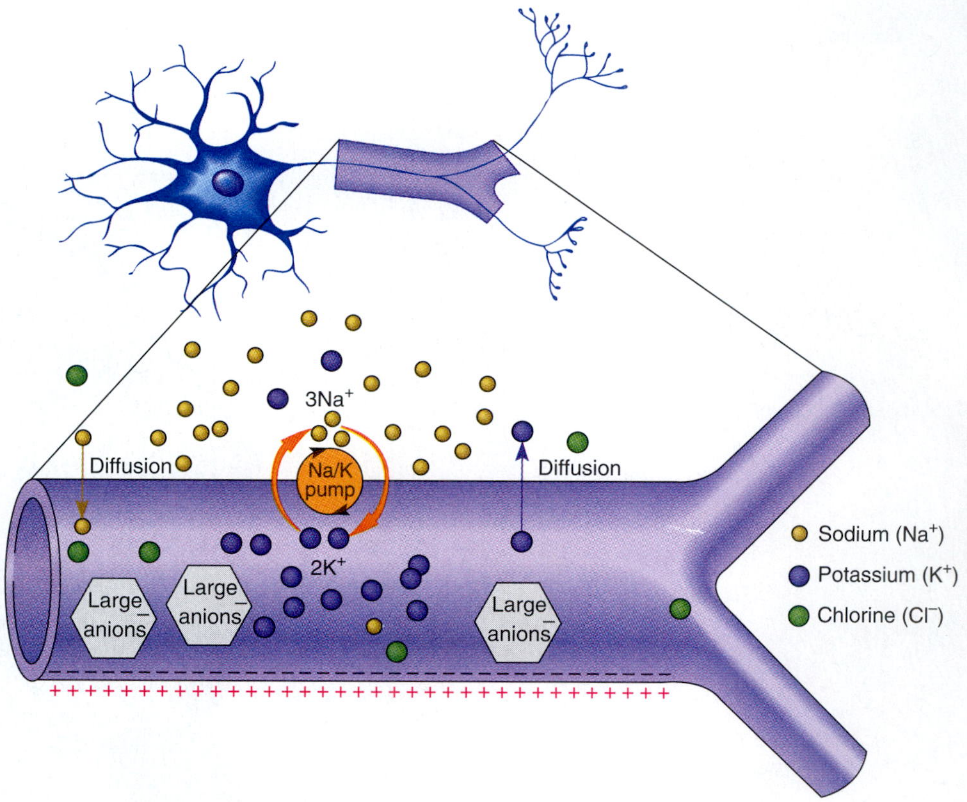

FIGURE 33–5 Segment of an axon of a resting neuron. Sodium-potassium pumps in the plasma membrane actively pump sodium out of the cell and pump potassium in. Sodium is unable to diffuse back to any extent, but potassium does diffuse out along its concentration gradient. Negatively charged proteins and other large anions are present in the cell. Because of the unequal distribution of ions, the inside of the axon is negatively charged compared to the surrounding fluid.

The Resting Potential Is the Difference in Electric Charge Across the Plasma Membrane of a Resting Neuron

In a resting neuron—that is, one that is not transmitting an impulse—the inner surface of the plasma membrane has a negative charge compared with the surrounding tissue fluid (Fig. 33–5). The plasma membrane is said to be **polarized** (i.e., one side, or pole, has a different charge than the other side). When electric charges are separated in this way, there is a **membrane potential,** an electrical potential energy difference across the membrane. Should the charges be permitted to come together, they have the *potential* of doing work. In resting neurons, the membrane potential is called the **resting potential.**

The resting potential can be expressed in units called millivolts (mV). (A millivolt equals one-thousandth of a volt and is a unit for measuring electrical potential.) The resting potential of a neuron amounts to about 70 mV. By convention this is expressed as −70 mV because the inner surface of the plasma membrane is negatively charged relative to the interstitial fluid.

The resting potential can be measured by placing one electrode, insulated except at the tip, inside the cell and a second electrode on the outside surface. The two electrodes are connected with an instrument such as a galvanometer, which measures current by electromagnetic action. If both electrodes are placed on the outside surface of the neuron, no potential difference between them is registered. All points on the outside of the membrane are at the same potential. We can think of the neuron as a biological battery. If its plasma membrane, which is only about one-millionth of a centimeter thick, were 1 cm thick, the resting potential would amount to an impressive 70,000 volts!

How does the resting potential develop? It results from a slight excess of positive ions outside the plasma membrane and a slight excess of negative ions inside the membrane. The K^+ concentration is about 30 times greater *inside* a resting neuron than outside the neuron. In contrast, the Na^+ concentration is about 14 times greater *outside* than inside the neuron.

The ion imbalance is brought about by several factors. The neuron plasma membrane has very efficient **sodium-potassium pumps** that actively transport sodium out of the cell and potassium ions into the cell (Chapter 6). Because the pumps work against a concentration gradient and an electrochemical gradient, ATP is required. For every three sodium ions pumped out of the cell, two potassium ions are pumped in. Thus, more positive ions are pumped out than are pumped in.

Ions also cross the membrane by facilitated diffusion through membrane proteins that form ion-specific channels. Net movement of ions occurs from an area of higher concentration to an area of lower concentration (Fig.

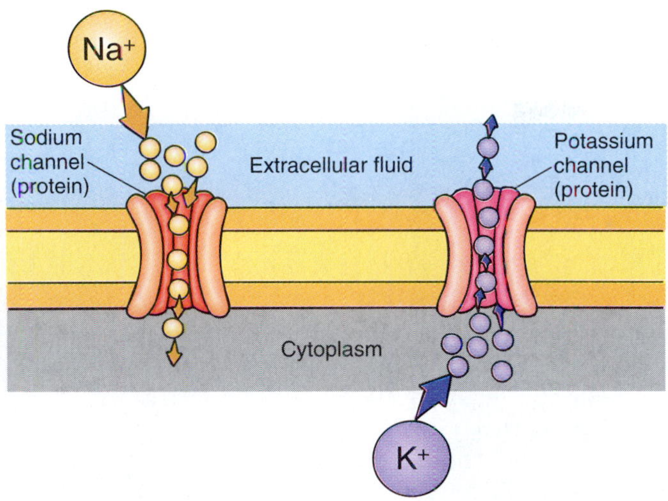

FIGURE 33–6 Proteins in the plasma membrane form ion-specific channels. Ions diffuse through these channels down their concentration gradient, from a region of higher concentration to a region of lower concentration.

33–6). However, the ease of passage through an ion channel varies according to the type of ion. Sodium ions move through sodium channels less easily than potassium ions pass through potassium channels. In the resting neuron, the membrane is up to 100 times more permeable to potassium than to sodium. Consequently, sodium ions pumped out of the neuron cannot easily pass back into the cell, but potassium ions pumped into the neuron can diffuse out.

Potassium ions leak out through the membrane along their concentration gradient until the positive charge outside the membrane reaches a level that repels the outflow of more positively charged potassium ions. A steady state is reached when the potassium outflow equals the inward flow of sodium ions. At this point, a potential difference of about −70 mV has developed across the membrane, establishing the resting potential.

Contributing to the overall ion distribution are large numbers of anions—negatively charged proteins and organic phosphates—within the neuron that are too big to diffuse out. Some of the potassium ions in the cell help to neutralize these negative charges, but there are not enough potassium ions to neutralize them all. The plasma membrane is permeable to negatively charged chloride ions, but because of the positively charged ions that accumulate outside the membrane, chloride ions are attracted to the outside where they tend to accumulate.

In summary, the resting potential is mainly due to the presence of large protein anions inside the cell and to the outward diffusion of potassium ions along their concentration gradient. However, the conditions for this diffusion must first be set by the action of the sodium-potassium pumps. Remember that the active transport of ions by these pumps is a form of cellular work and therefore requires energy.

The Nerve Impulse Is an Action Potential

Neurons are highly excitable cells. They have the ability to convert stimuli into neural impulses. An electrical, chemical, or mechanical stimulus may alter the resting potential by increasing the permeability of the membrane to sodium. If the neuron membrane is only slightly stimulated, only a local disturbance may occur in the membrane. If the stimulus is sufficiently strong, however, it may result in the transmission of a **neural impulse,** or **action potential.**

In addition to the sodium-potassium pumps and passive ion channels already discussed, the plasma membrane of the neuron has specific **voltage-activated ion channels,** which open when they detect a change in the membrane potential (Fig. 33–7). When the voltage reaches a certain critical point, known as the **threshold level,** gates open, allowing the passage of specific ions through these channels.

The membrane of the neuron can **depolarize** up to about 15 mV (that is, to a resting potential of about −55 mV) without actually initiating an impulse. However, when the extent of depolarization is greater than −55 mV, the threshold level is reached. At that point, the voltage-activated sodium ion channels open, and sodium ions pass into the cell. After a certain period of time, a second set of gates, the inactivating gates, closes the channels. The closing is dependent on time rather than on voltage.

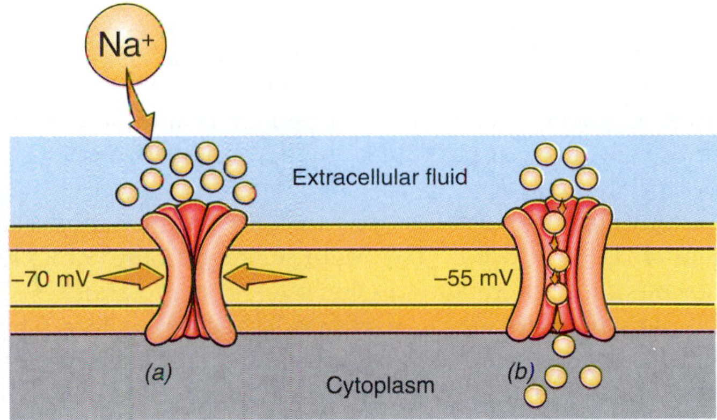

FIGURE 33–7 Voltage-activated ion channels are present in the plasma membrane of the axon and cell body. (***a***) In the resting state, the voltage-activated Na^+ channel is closed. (***b***) When the voltage reaches threshold level, the voltage-activated gate opens, allowing Na^+ to flow into the cell. After a certain amount of time elapses, inactivating gates close the channels.

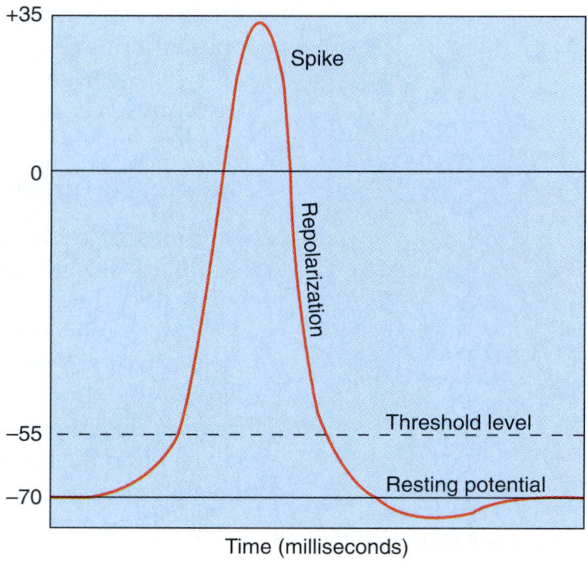

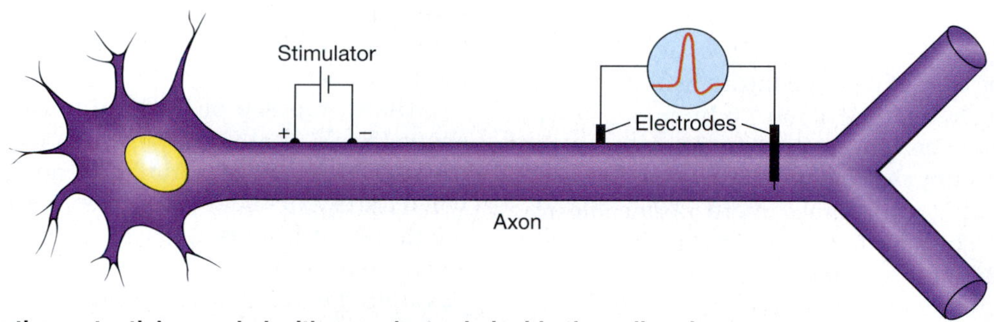

FIGURE 33–8 An action potential recorded with one electrode inside the cell and one just outside the plasma membrane. When the axon depolarizes to about −55 mV, an action potential is generated. (The numerical values are examples. They vary with different nerve cells.)

K^+-sensitive channels also open when the threshold level is reached. They open more slowly and stay open until they sense a particular voltage, that of the resting potential.

An almost explosive action occurs as the action potential is produced. The neuron membrane quickly reaches zero potential and even overshoots to about +35 mV, so a momentary reversal in polarity takes place. The sharp rise and fall of the action potential are referred to as a **spike.** Figure 33–8 illustrates an action potential that has been recorded by placing one electrode inside an axon and one just outside.

The action potential is an electrical current strong enough to cause the resting potential to collapse in the adjacent area of the membrane. The area of depolarization then spreads, like a chain reaction, down the length of the axon (at a constant velocity and amplitude for each type of neuron). Thus, a neural impulse is transmitted as a **wave of depolarization** that travels down the neuron. Conduction of a neural impulse is somewhat like a burning trail of gunpowder. Once the gunpowder is ignited at one end of the trail, the flame moves steadily to the other end by igniting the powder particles ahead of it.

By the time the action potential moves a few millimeters down the axon, the membrane over which it has just passed begins to **repolarize** (Fig. 33–9). The sodium gates close, and the membrane again becomes impermeable to sodium. The potassium gates open (due to the threshold level of voltage), allowing potassium to leak out at the point of stimulation. This leakage of potassium ions returns the interior of the membrane to its relatively negative state, repolarizing the membrane. This entire mechanism—depolarization and then repolarization—can take place in less than 1 millisecond.

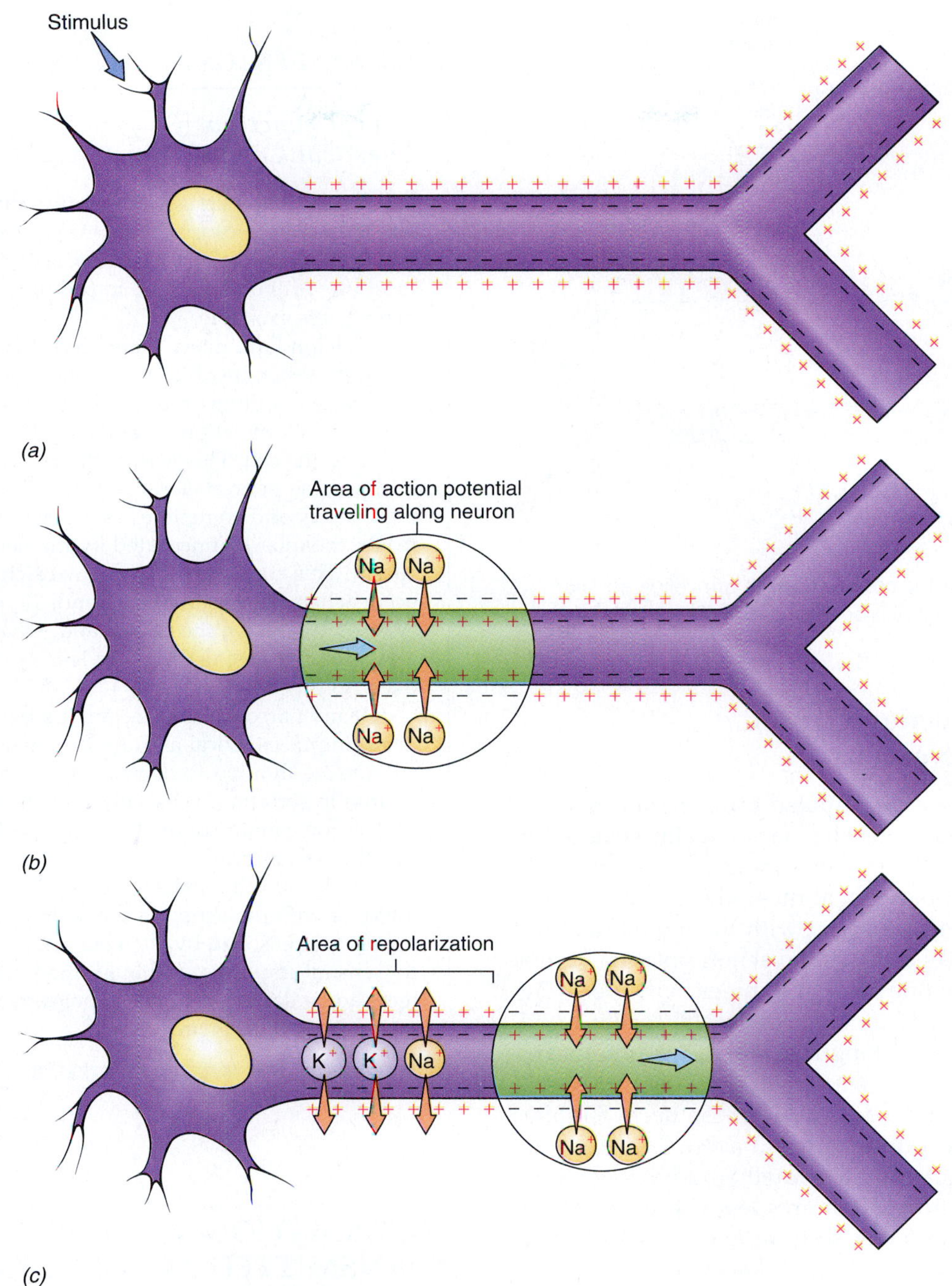

FIGURE 33–9 Transmission of an impulse along an axon. (***a***) The dendrites (or cell body) of a neuron are stimulated sufficiently to depolarize the membrane to firing level. The axon is shown still in the resting state, with a resting potential. (***b***, ***c***) An impulse is transmitted as a wave of depolarization that travels down the axon. At the region of depolarization, sodium ions diffuse into the cell. As the impulse progresses along the axon, repolarization occurs quickly behind it.

During the millisecond or so in which it is depolarized, the axon membrane is in an **absolute refractory period** when it cannot transmit another action potential no matter how great a stimulus is applied. Then, for a few additional milliseconds, while the resting condition is being reestablished, the axon can transmit impulses only when they are more intense than is normally required. This is the **relative refractory period.** Even with the limits imposed by their refractory periods, most neurons can transmit several hundred impulses per second.

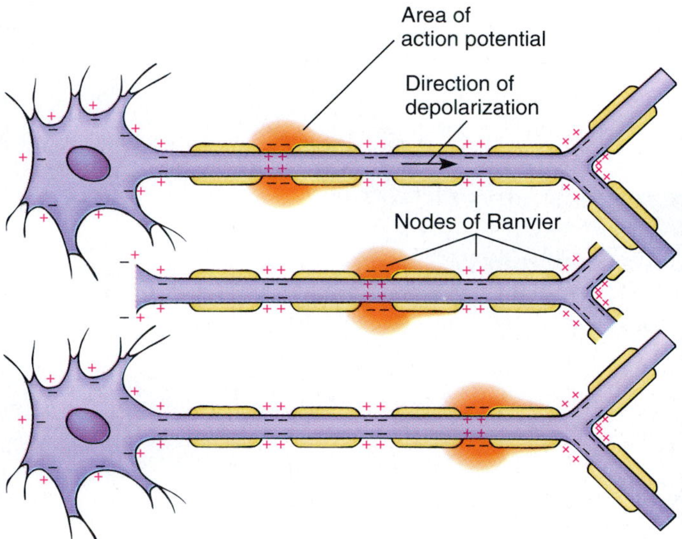

FIGURE 33–10 Saltatory conduction. In a myelinated axon the impulse leaps along from one node of Ranvier to the next.

Saltatory Conduction Occurs in Myelinated Neurons

The smooth, progressive impulse transmission just described occurs in neurons that lack myelin sheaths. In myelinated neurons the myelin acts as an effective insulator around the axon except at the nodes of Ranvier. The plasma membrane is in contact with the interstitial fluid only at the nodes. As a result, the action potential jumps along the axon from one node of Ranvier to the next (Fig. 33–10). The ion activity at the active node serves to depolarize the next node along the axon. This type of impulse transmission is known as **saltatory conduction.** A myelinated axon can conduct an impulse up to 50 times faster than the fastest unmyelinated axon.

Saltatory conduction has another advantage over continuous conduction: it requires less energy. Only the nodes depolarize, so fewer sodium and potassium ions are displaced. As a result, the cell does not have to work so hard to reestablish resting conditions each time an impulse is conducted.

The Neuron Obeys an All-or-None Law

Any stimulus too weak to depolarize the neuron to threshold level cannot fire the neuron. It merely sets up a local response that fades and dies within a few millimeters from the point of stimulus. A stimulus strong enough to depolarize the neuron to its critical threshold level results in the transmission of an impulse along the axon. The neuron either transmits an impulse or it does not. No variation exists in the strength of a single impulse. Thus, the neuron obeys an **all-or-none law.**

CONCEPT CONNECTIONS

Neural Function ***Chemical Compounds***

Substances that increase the permeability of the membrane to sodium cause the neuron to become more excitable than normal. Other substances decrease the permeability of the membrane to sodium, making the neuron less excitable.

Calcium and many other electrolytes affect neural function. When insufficient numbers of calcium ions are present, the sodium gates apparently fail to close completely between action potentials, allowing sodium to leak into the cell. This lowers the resting potential, bringing the neuron closer to firing. The neuron fires more easily and sometimes even spontaneously. As a result, the muscle innervated by the neuron may go into spasm. This condition, called **low-calcium tetany,** can occur when the parathyroid glands do not secrete sufficient hormone. On the other hand, when calcium ions are too numerous, neurons are less excitable and less likely to fire.

Many narcotics and anesthetics block conduction of nerve impulses. Local anesthetics, such as procaine, and cocaine are thought to decrease the permeability of the neuron to sodium. Excitability may be so reduced that the neuron cannot conduct an impulse through the anesthetized region.

DDT and other chlorinated hydrocarbon pesticides interfere with the action of the sodium pump. When nerves are poisoned by such substances, they are unable to transmit impulses. Although the human nervous system can be damaged by these poisons, insects are even more sensitive to them.

INFORMATION MUST BE TRANSMITTED ACROSS SYNAPSES

When an action potential is transmitted down an axon, it eventually reaches the end of the axon. Here, a small gap, the **synapse,** separates the axon from another neuron or from muscle fibers or gland cells. A neuron that transmits an impulse to a synapse is a **presynaptic neuron;** a neuron that transmits an impulse away from the synapse is a **postsynaptic neuron.** The same neuron may be postsynaptic with respect to one synapse and presynaptic relative to another.

There are two types of synapses: electrical synapses and chemical synapses. In an **electrical synapse,** the action potential can be transmitted directly to another cell by means of cell-to-cell connections called **gap junctions**

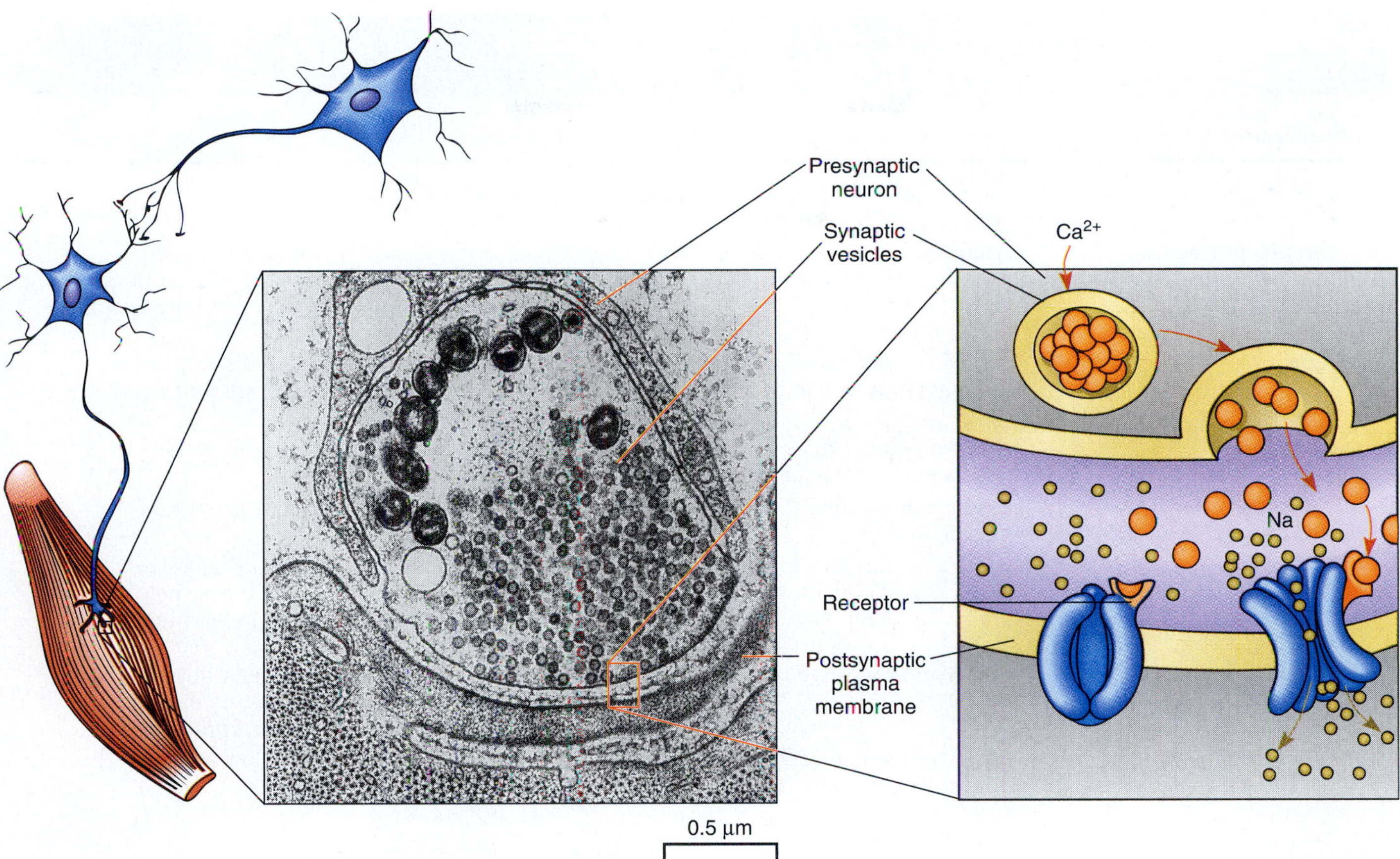

FIGURE 33–11 Transmission of an impulse between neurons or from a neuron to an effector. In most synapses the impulse is unable to jump across the synaptic cleft between two neurons or across the neuromuscular junction between a neuron and muscle fiber. The problem is solved by chemical transmission across synapses. When an impulse reaches the end of a neuron, calcium ions move into the neuron from the interstitial fluid. The calcium ions apparently cause the synaptic vesicles to fuse with the membrane and release neurotransmitter. The neurotransmitter diffuses across the synaptic cleft or neuromuscular junction and may combine with receptors in the membrane of the postsynaptic neuron or muscle fiber. This binding may trigger an impulse in the postsynaptic neuron or muscle fiber by causing depolarization and opening of sodium gates. The electron micrograph shows a motor neuron synapsing with a muscle fiber. The synaptic knob of the motor neuron is filled with synaptic vesicles. (T. Reese and D. W. Fawcett)

(Chapter 6). Electrical synapses allow very rapid transmission of impulses and so are especially important in rapid responses such as escape responses. Because chemical synapses are more common, we will focus on them.

In **chemical synapses,** the synaptic cleft, the gap between the two cells, is more than 20 nm wide (still less than a millionth of an inch). Since depolarization is a property of the neuron membrane, when the impulse reaches the end of the axon, it does not jump the gap. Chemical compounds called neurotransmitters conduct the neural message across the synapse. These chemical messengers diffuse across the synapse from synaptic knobs at the end of the axon. They bind to specific receptors on the plasma membranes of the postsynaptic neurons. Let us look at this process in more detail.

Axons Release Neurotransmitters that Affect Postsynaptic Neurons

Neurotransmitters are continuously synthesized in the synaptic knobs. They are stored in synaptic vesicles (little sacs) in the cytoplasm (Fig. 33–11). Each synaptic vesicle contains only one type of neurotransmitter, but the same neuron can produce two or more different neurotransmitters.

Each time an action potential travels down an axon to the synaptic knob, calcium channels in the membrane open, permitting calcium ions to pass into the axon. The calcium ions induce several hundred vesicles to fuse with the membrane and release their neurotransmitters into the synaptic cleft. The neurotransmitters diffuse across

TABLE 33–1
Some Neurotransmitters

Substance	*Where secreted*	*Comments*
Acetylcholine	Nerve-muscle junctions; autonomic system;* parts of brain	Inactivated by cholinesterase
Norepinephrine	Autonomic system; reticular activating system and other areas of brain and spinal cord	Inactivated slowly by monoamine oxidase (MAO); mainly inactivated by reabsorption by vesicles in the synaptic knob; norepinephrine level in brain affects mood
Dopamine	Limbic system; cerebral cortex; basal ganglia; hypothalamus	Thought to affect motor function; may be involved in schizophrenia;† amount reduced in Parkinson's disease
Serotonin (5-hydroxytryptamine, 5-HT)	Limbic system; hypothalamus; cerebellum; spinal cord	May play role in sleep, schizophrenia, and depression
GABA (gamma-aminobutyric acid)	Spinal cord, cerebral cortex, cerebellum	Acts as inhibitor; may play role in pain perception
Endorphins	CNS and pituitary gland	Neuropeptides that have morphine-like properties and suppress pain; may help regulate cell growth; linked to learning and memory
Enkephalins	Brain and digestive tract	Neuropeptides that inhibit pain impulses by inhibiting release of substance P; bind to same receptors in brain as morphine
Substance P	Brain and spinal cord, sensory nerves, intestine	Transmits pain impulses from pain receptors into CNS

*These and other structures listed in this table will be discussed in Chapter 34.
†Studies suggest that the brains of schizophrenics have more dopamine receptors than do those of nonschizophrenics.

the synaptic cleft. Some may be taken up by specific receptors on the dendrites or cell bodies of postsynaptic neurons. Excess neurotransmitters are reabsorbed into the synaptic vesicles or inactivated by enzymes.

Once they diffuse into the synaptic cleft, neurotransmitters have different effects on other neurons across the synapse. Certain neurotransmitters may cause sodium channels to open in the membrane of another neuron. The resulting influx of sodium ions depolarizes the membrane of this neuron, bringing it closer to firing. As a result, this neuron is more likely to set off an action potential. This type of synapse is called an **excitatory synapse.** A "message" transmitted by one neuron is picked up, in effect, by the next neuron in the sequence.

Other neurotransmitters may cause potassium channels to open. Potassium ions then flow out of the postsynaptic neuron. With additional positive charges outside the membrane, the interior of the neuron becomes even more negative relative to the surrounding fluid. This is called **hyperpolarization.** When neurotransmitters hyperpolarize the membrane of a postsynaptic neuron, an action potential is less likely. This type of synapse is called an **inhibitory synapse.**

Many Neurotransmitters Have Been Identified

More than 60 different substances are now known (or suspected) to be neurotransmitters or chemicals that modify the effects of neurotransmitters. Many types of neurons secrete two or even three different types of neurotransmitters. Furthermore, a postsynaptic neuron can have receptors for more than one type of neurotransmitter. Some of its receptors may be excitatory and some may be inhibitory.

Two neurotransmitters that have been studied extensively are acetylcholine and norepinephrine (Table 33–1). Recall from Chapter 32 that **acetylcholine** is released from motor neurons and diffuses across the neuromuscular junction to trigger muscle contraction. Acetylcholine is also released by some neurons in the brain and by neurons that are part of the autonomic nervous system (Chapter 34). Cells that release this neurotransmitter are referred to as **cholinergic neurons.**

Acetylcholine has an excitatory effect on skeletal muscle, but an inhibitory effect on cardiac muscle, resulting in a decrease in heart rate. Whether a neurotransmitter excites or inhibits depends on the postsynaptic receptors

Concept Connections

Neurotransmitters ***Schizophrenia***

The neurotransmitter dopamine has been linked with the disturbed thought processes and hallucinations characteristic of schizophrenia, a disorder involving decline in general functioning. According to the dopamine hypothesis, individuals suffering from schizophrenia have an abnormally high number of dopamine receptors in certain areas of the brain. More dopamine receptors result in heightened dopamine activity and overactivity of the neurons that respond to dopamine.

When the neurons that respond to dopamine are overactive, they overstimulate areas in the temporal lobes of the brain that can produce hallucinations—perceptual experiences like seeing or hearing things that have no basis in present reality.

The dopamine hypothesis is based on two types of evidence. First, antipsychotic drugs reduce the frequency of hallucinations and delusions by blocking dopamine receptors. Second, amphetamines, cocaine, and other drugs that are biochemically related to dopamine cause symptoms of schizophrenia in normal individuals and increase symptoms in those suffering from schizophrenia.

The neurotransmitter serotonin also plays a role in schizophrenia. Serotonin inhibits activity in some neurons. When serotonin levels are low, dopamine activity is not kept in check. A normal balance of dopamine and serotonin appears to be necessary for normal neural functioning.

with which it combines. After acetylcholine is released by a presynaptic neuron and combines with receptors on the postsynaptic neuron, excess acetylcholine must be removed. The enzyme cholinesterase breaks it down into its chemical components.

Norepinephrine is released by neurons in the brain and spinal cord and by certain neurons of the autonomic system (Chapter 34). Neurons that release norepinephrine are called **adrenergic neurons.** Norepinephrine and the neurotransmitters **epinephrine** and **dopamine** belong to a class of compounds called **catecholamines** (or biogenic amines). After release, most of the excess catecholamines are reabsorbed into the vesicles in the synaptic knobs. Some are degraded by enzymes such as **monoamine oxidase (MAO).** Catecholamines affect mood. For example, low levels of norepinephrine in the brain have been associated with depression. Antidepressant medications, tranquilizers, cocaine, and many other mood-altering drugs work by altering the levels of catecholamines in the brain.

Nerve Fibers May Be Classified in Terms of Speed of Conduction

In the laboratory, we can demonstrate that an impulse can move in both directions within a single axon. In the body, however, an impulse generally stops when it reaches the dendrites, because no neurotransmitter is there to conduct it across the synapse. This limitation makes neural pathways function as one-way streets. The usual direction of transmission is from the axon of the presynaptic neuron across the synapse to the dendrite of the postsynaptic neuron.

Compared with the speed of an electrical current or the speed of light, a nerve impulse travels rather slowly. Its speed varies from about 0.5 meter per second to more than 120 meters (400 feet) per second. What factors affect speed of transmission? In general, the greater the diameter of an axon, the greater its speed of conduction. The largest neurons seem also to be the most heavily myelinated, and the more myelin a neuron has, the faster it transmits impulses. In myelinated neurons, the distance between successive nodes of Ranvier is also important: the farther apart the nodes, the faster the axon conducts.

Neural Impulses Must Be Integrated

Neural integration is the process of sorting and interpreting incoming signals, and determining an appropriate response. Each neuron synapses with hundreds of other neurons. It is the job of the dendrites and cell body of every neuron to integrate the hundreds of messages that continually bombard them.

An **excitatory postsynaptic potential (EPSP)** is a change in potential that brings the neuron closer to firing. In contrast, an **inhibitory postsynaptic potential (IPSP)** brings the neuron farther from the firing level. EPSPs and IPSPs occur continually in postsynaptic neurons; IPSPs cancel the effects of some of the EPSPs. The postsynaptic cell body and dendrites continually tabulate such molecular transactions. When sufficient excitatory neurotransmitter predominates, the neuron is brought to threshold level and an action potential is generated.

Each EPSP or IPSP is not an all-or-none response, and each does not travel down the membrane like an action potential. Rather, each is a local response that may be added to or subtracted from other EPSPs and IPSPs. As a result of these chemical tabulations, the neuron may be inhibited, facilitated, or brought to threshold level. If sufficient EPSPs have been received to bring the neuron to the threshold level, an all-or-none action potential is initiated and travels down the axon. This mechanism provides for integration of hundreds of tiny "messages" (EPSPs and IPSPs) before an impulse is actually transmitted along the axon of a postsynaptic neuron. Such an arrangement permits the neuron and the entire nervous system a far greater range of response than would be the case if every EPSP generated an action potential.

Where does neural integration take place? Every neuron acts as a tiny integrator, sorting through (on a molecular level) the hundreds and thousands of bits of information that continually bombard it. Since more than 90% of the neurons in the body are located in the CNS, most neural integration takes place there—within the brain and spinal cord. These neurons are responsible for making most of the "decisions." In the next chapter, the brain and spinal cord will be examined in some detail.

THE REFLEX ARC IS A SIMPLE NEURAL PATHWAY

Neurons are organized into specific **pathways,** or **neural circuits.** Generally, neurons are arranged so that the axon of one neuron in the circuit forms junctions with the dendrites of the next neuron in the circuit. One of the simplest examples of a neural pathway is the **reflex arc** (Fig. 33–12).

A **reflex action** is a relatively fixed reaction pattern to a simple stimulus. The response is predictable and automatic, which means that it does not require conscious thought. We continue to breathe, for example, even when we are asleep. Breathing and many other body processes are regulated by reflex actions.

Although most reflex actions are much more complex, let us consider a **withdrawal reflex,** in which a neural circuit consisting of only three neurons is needed to carry out a response to a stimulus (Fig. 33–12). Suppose you touch a hot stove. Almost instantly, and before you are consciously aware of the situation, you jerk your hand away from this unpleasant stimulus. But in this brief instant, a message has been carried from pain receptors in the skin to the spinal cord by a **sensory neuron.** Within the spinal cord the message is transmitted from the sensory neuron to an **association neuron.** Then, the message is transmitted to a **motor neuron** that conducts the message to appropriate muscles. These muscles respond by contracting and pulling the hand from the stove. Actually, many neurons located in sensory, association, and motor nerves participate in such a reaction, and complicated switching is involved. We move our hands *up* from a hot stove but *down* from a hot light bulb. Generally, we are not even consciously aware that all of these responding muscles exist.

At the same time that the association neuron sends a message out along a motor neuron, it also sends one up the spinal cord to the conscious areas of the brain. As you withdraw your hand from the hot stove, you become aware of what has happened and feel the pain. This awareness is not part of the reflex response.

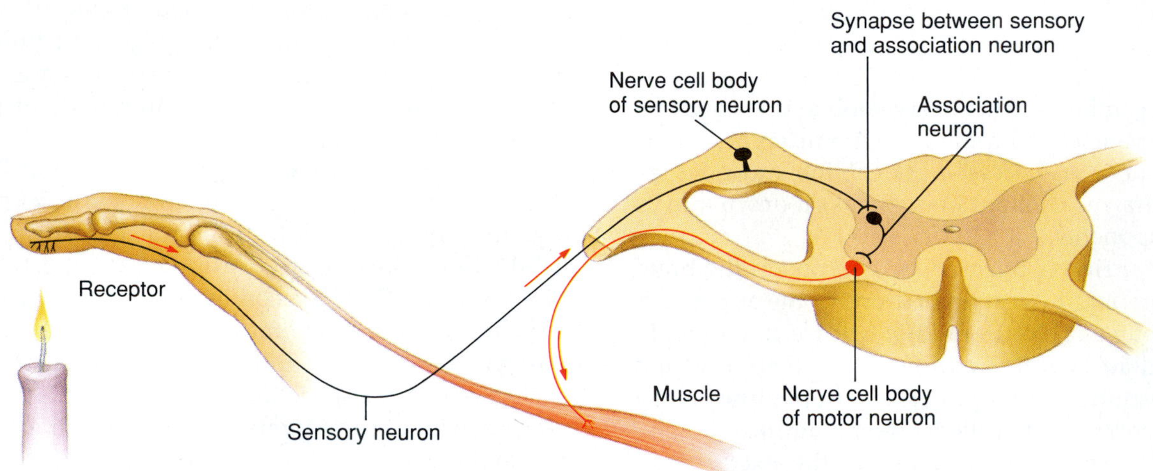

FIGURE 33–12 The withdrawal reflex depends on a chain of three neurons. A sensory neuron transmits the message from the receptor to the central nervous system where it synapses with an association neuron. Then an appropriate motor neuron (shown in red) transmits an impulse to the muscles that move the hand away from the flame (the response).

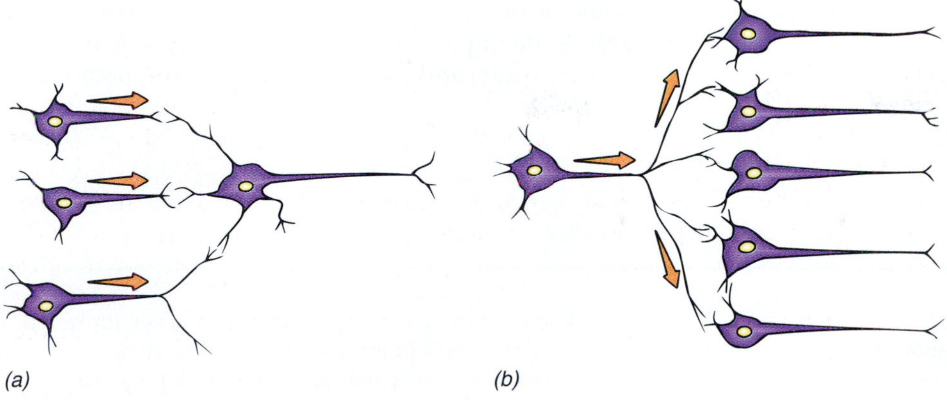

FIGURE 33–13 Organization of neural circuits. (***a***) Convergence of neural input. Several presynaptic neurons synapse with one postsynaptic neuron. This organization in a neural circuit permits one neuron to receive signals from many sources. (***b***) Divergence of neural output. A single presynaptic neuron synapses with several postsynaptic neurons. This organization allows one neuron to communicate with many others.

COMPLEX NEURAL PATHWAYS ARE POSSIBLE BECAUSE NEURONS ASSOCIATE IN A VARIETY OF WAYS

Within a neural pathway, many presynaptic neurons may end at the same synapse and communicate with the same postsynaptic neuron. In **convergence,** the postsynaptic neuron is controlled by signals coming from two or more presynaptic neurons (Fig. 33–13*a*). An association neuron in the spinal cord, for instance, may receive information from sensory neurons entering the cord, from neurons originating at other levels of the spinal cord, and even from neurons bringing information from the brain. Information from all of these converging neurons is integrated before an action potential is generated in the association neuron and an appropriate motor neuron stimulated.

In **divergence,** a single presynaptic neuron stimulates many postsynaptic neurons (Fig. 33–13*b*). Each presynaptic neuron may synapse with 25,000 or more postsynaptic neurons. In **facilitation,** the neuron is brought close to threshold level by stimulation from various presynaptic neurons but is not yet at the threshold level. The neuron can be easily excited by further stimulation. Figure 33–14 illustrates facilitation.

The **reverberating circuit** is a neural pathway arranged so that a neuron branch synapses with an association neuron (Fig. 33–15). The association neuron synapses with a neuron in the sequence that can send

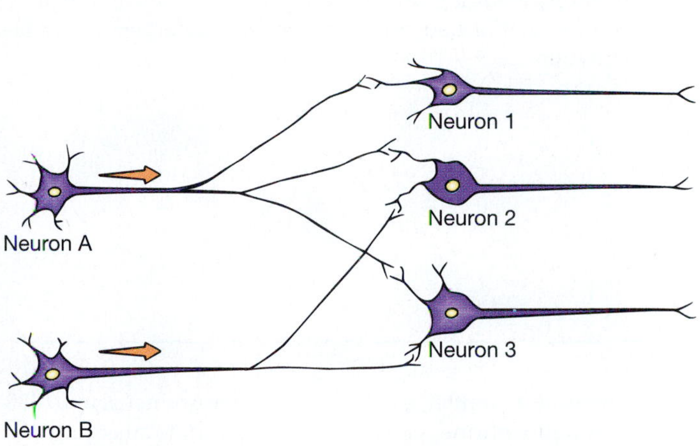

FIGURE 33–14 Facilitation. Neither neuron A nor neuron B by itself can fire neuron 2 or 3. However, stimulation by either A or B does depolarize the neuron toward the threshold level (if the stimulation is excitatory). This facilitates the postsynaptic neuron, so that if another presynaptic neuron stimulates it the threshold level may be reached and an action potential generated.

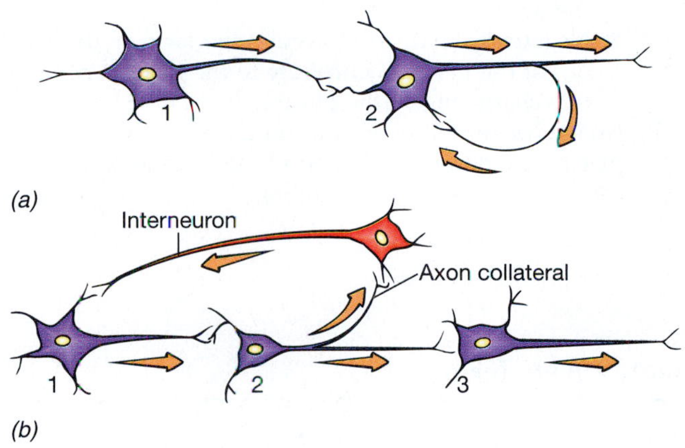

FIGURE 33–15 Reverberating circuits. (***a***) A simple reverberating circuit in which an axon branch of the second neuron turns back upon its own dendrites, so the neuron continues to stimulate itself. (***b***) In this neural circuit an axon branch of the second neuron synapses with an interneuron. The interneuron synapses with the first neuron in the sequence. New impulses are triggered again and again in the first neuron, causing reverberation.

new impulses again through the circuit. New impulses can be generated again and again until the synapses fatigue (from depletion of neurotransmitter) or are stopped by some sort of inhibition. Reverberating circuits are thought to be important in rhythmic breathing, in maintaining alertness, and perhaps in short-term memory.

Chapter Summary

I. Neural function depends on reception, transmission, integration, and response by muscles or glands.

II. Glial cells and neurons are the two types of cells characteristic of neural tissue.
   - A. Glial cells are supporting cells.
   - B. Neurons are specialized to receive stimuli and transmit impulses.
     1. A typical neuron consists of a cell body with many branched dendrites and a single long axon.
     2. Outside the central nervous system axons are surrounded by a cellular sheath, and many axons are also enveloped by a myelin sheath.
     3. A nerve consists of hundreds of axons wrapped in connective tissue; a ganglion is a mass of cell bodies.

III. Neurons transmit rapidly moving electrical impulses.
   - A. A neuron that is not transmitting an impulse has a resting potential.
     1. The inner surface of the plasma membrane is negatively charged compared with the outside; the membrane is polarized.
     2. Sodium-potassium pumps continuously transport sodium out of the neuron and transport potassium into the neuron.
     3. Potassium ions are able to leak out of the neuron more readily than sodium ions are able to leak into the neuron.
     4. Negatively charged molecules too large to diffuse out of the neuron contribute to the relative negative charge inside the plasma membrane.
   - B. Excitatory stimuli open sodium channels in the plasma membrane. This permits sodium to enter the cell and depolarize the membrane.
   - C. When the extent of depolarization reaches threshold level, an action potential may be generated.
     1. The action potential is a wave of depolarization that moves down the axon.
     2. As the action potential moves down the axon, repolarization occurs quickly behind it.
     3. Saltatory conduction takes place in myelinated neurons.
     4. The largest, most heavily myelinated neurons conduct impulses most rapidly.

IV. Synaptic transmission generally depends upon release of a neurotransmitter from vesicles in the presynaptic neuron.
   - A. Neurotransmitter diffuses across the synaptic cleft and combines with receptors on the postsynaptic neuron.
   - B. Neurons that release the neurotransmitter acetylcholine are known as cholinergic neurons. Those that release norepinephrine are adrenergic.

V. Neural integration is the process of adding and subtracting incoming signals and determining whether or not to fire an impulse.

VI. A reflex arc is a simple neural pathway.
   - A. A withdrawal reflex requires a sequence of only three neurons: a sensory neuron, an association neuron, and a motor neuron.
   - B. Reflex action is predictable and automatic; it does not require conscious thought.

VII. Complex neural pathways are possible because of neuron associations such as convergence, divergence, and facilitation.

Selected Key Terms

acetylcholine, p. 644
action potential, p. 639
association neuron, p. 646
axon, p. 635
dendrite, p. 635
excitatory postsynaptic potential (EPSP), p. 645
ganglion, p. 637
glial cells, p. 635
inhibitory postsynaptic potential (IPSP), p. 645
integration, p. 635
motor neuron, p. 635
myelin sheath, p. 636
nerve, p. 637
neuron, p. 635
neurotransmitter, p. 635
norepinephrine, p. 645
reception, p. 635
reflex arc, p. 646
refractory period, p. 641
resting potential, p. 638
saltatory conduction, p. 642
Schwann cell, p. 635
sensory neuron, p. 635
sodium-potassium pumps, p. 638
synapse, p. 642
threshold level, p. 639
transmission, p. 635

Post-Test

1. The process of detecting a stimulus is called ________________.
2. The actual response is carried out by ________________ and ________________.
3. The supporting cells of the nervous system are called ________________ cells; cells specialized to transmit impulses are ________________.
4. The nucleus of a neuron is found within the ________________ ________________.
5. The ________________ transmits impulses from the cell body to the synapse.
6. The ________________ ________________ is important in the regeneration of injured neurons.
7. A/An ________________ consists of a mass of cell bodies outside the central nervous system.
8. Sodium is actively transported out of a resting neuron by ________________-________________ ________________.
9. Any stimulus that increases the neuron's permeability to sodium to threshold level may result in the transmission of a/an ________________ ________________.
10. During the ________________ ________________ period, a neuron cannot transmit another impulse.
11. The junction between two neurons is called a/an ________________.
12. Synaptic knobs release ________________.
13. A/An ________________ action is a predictable, automatic response to a simple stimulus.
14. Impulses are transmitted from sense organs to the central nervous system by ________________ neurons.
15. In ________________, the postsynaptic neuron is controlled by signals from several presynaptic neurons.
16. The plasma membrane of the neuron is polarized when (a) an action potential is being transmitted (b) it has a resting potential (c) one side of it has a different electric charge than the other side (d) answers (a) and (b) are correct (e) answers (b) and (c) are correct.
17. Which of the following contribute to the resting potential of the neuron? (a) presence of anions inside the cell (b) sodium-potassium pumps (c) outward diffusion of potassium ions (d) answers (a), (b), and (c) are correct (e) none of the preceding answers is correct.
18. Voltage-activated ion channels in the plasma membrane of the neuron (a) open when they detect a change in membrane potential (b) are activated by connected sodium-potassium pumps (c) allow facilitated diffusion in the resting neuron (d) answers (a), (b), and (c) are correct (e) none of the preceding answers is correct.
19. Adrenergic neurons release (a) norepinephrine (b) acetylcholine (c) cholinesterase (d) answers (a), (b), and (c) are correct (e) none of the preceding answers is correct.
20. IPSPs (a) excite presynaptic neurons (b) excite postsynaptic neurons (c) cancel the effects of some EPSPs (d) answers (a), (b), and (c) are correct (e) none of the preceding answers is correct.

Review Questions

1. Imagine that you are driving down the street when suddenly a child darts in front of your car. What sequence of events must take place within your nervous system before you can slam on the brakes?
2. Give the functions of (a) myelin, (b) ganglia, (c) glial cells, (d) dendrites, (e) an axon.

3. What is meant by the resting potential of a neuron? How does the sodium-potassium pump contribute to the resting potential?
4. Contrast the resting potential with an action potential. What is responsible for generation of an action potential?
5. How does the all-or-none law affect neural action?
6. Contrast transmission along an axon with transmission across a chemical synapse.
7. Give the functions of (a) acetylcholine, (b) cholinesterase, (c) norepinephrine.
8. Imagine that you have just burned your finger by touching a hot pot. Draw a diagram to illustrate the reflex action that would occur. Label each structure and indicate the direction of information flow.
9. Contrast convergence and divergence. What is their general significance?
10. Label the diagram on the right.

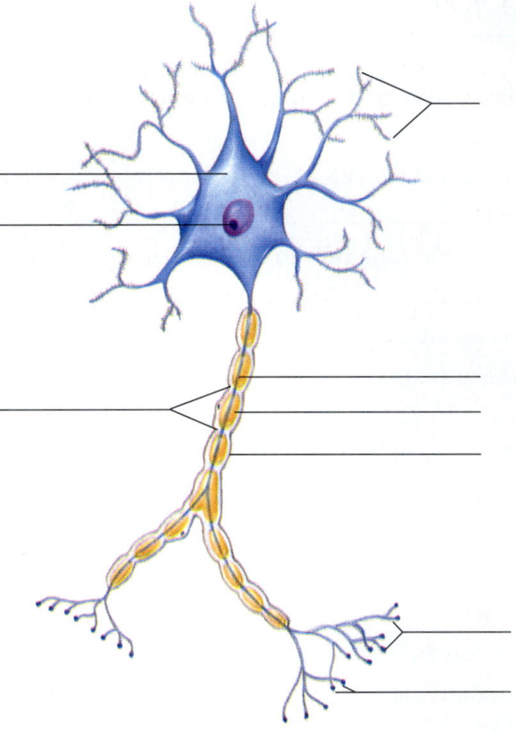

Thinking Critically

1. Researchers have found that Alzheimer's patients have a defective form of the membrane protein that normally removes aluminum from the blood. As a result, aluminum circulates in the blood and enters the brain. Some investigators have suggested that aluminum could cause Alzheimer's disease. Design an experiment to test this hypothesis. Could knowledge about aluminum in Alzheimer's patients be used in diagnosis? In treatment? Support your answers.
2. We have no difficulty distinguishing between the pain of a severe toothache and that of a minor cut on the arm. How can you explain this difference in intensity if the all-or-none law is true?
3. Drugs that are stimulants increase the activity of the nervous system. Propose two mechanisms involving synaptic transmission that could explain the action of stimulants.

Recommended Readings

Gottlieb, D.I., "GABAergic Neurons," *Scientific American*, Vol. 258, No. 2, 1988, pp. 82–89. GABA is an inhibitory neurotransmitter in the brains of all mammals.

Neher, E., and Sakmann, B., "The Patch Clamp Technique," *Scientific American*, Vol. 266, No. 3, March 1992, pp. 44–51. The 1991 Nobel Prize winning authors discuss their technique for isolating ion channels in cell membranes.

Pennisi, E., "A Molecular Whodunit: New Twists in the Alzheimer's Mystery," *Science News*, Vol. 145, No. 1, January 1, 1994, pp. 8–11. An overview of current thinking regarding Alzheimer's disease.

Selkoe, D.J., "Amyloid Protein and Alzheimer's Disease," *Scientific American*, Vol. 265, No. 5, November 1991, pp. 68–78. A discussion of Alzheimer's disease and research that may lead to treatment.

Excessive alcohol intake during pregnancy has been linked with serious birth defects. (Amy C. Etra/PhotoEdit)

ALCOHOL: THE MOST ABUSED OF ALL DRUGS

Our society is saturated with alcohol. Children see more than 100,000 beer commercials on television before they are legally old enough to drink. According to a national study released by the University of Michigan in 1994, we are experiencing an increase in the use of alcohol and other drugs. For example, 28% of high school seniors acknowledged having five or more drinks in a row during the two weeks prior to a survey.

After tobacco, alcohol is the leading cause of premature death in the United States. It is linked to more than 100,000 deaths every year and costs our society more than $100 billion annually. According to the pollster Louis Harris, there are 28 million alcoholics in the United States and about one in three homes includes someone with a serious drinking problem. Alcohol abuse is not limited to adults. About 4.6 million adolescents or nearly one of every three high school students experience negative consequences from alcohol use, including difficulty with parents, poor performance at school, and trouble with the law.

Alcohol abuse results in physiological, psychological, and social impairment for the abuser and has serious negative consequences for family, friends, and society. Alcohol abuse has been linked to the following:

- More than 50% of all traffic fatalities.
- More than 50% of violent crimes.
- More than 50% of suicides.
- More than 60% of cases of child abuse and spouse abuse.
- More than 15,000 babies born each year with serious birth defects because their mothers drank alcohol excessively during pregnancy.
- More breast cancer. Recent studies suggest that as little as three drinks per week increase risk of breast cancer by 50%.
- Greater risk of liver disease and brain impairment for women than men (clinical effects are seen in women consuming about half the alcohol consumed by men).

A single drink—that is, 12 ounces of beer, 5 ounces of wine, or 1.5 ounces of 80 proof liquor—results in a blood alcohol concentration of 20 mg/dL (milligrams per deciliter). This represents about 0.5 ounce of pure alcohol in the blood. Alcohol accumulates in the blood because absorption occurs more rapidly than do oxidation and excretion. Every cell in the body can take in alcohol from the blood. At first the drinker may feel stimulated. But alcohol actually causes depression of the central nervous system, probably by affecting specific neurotransmitters in the brain.

As blood alcohol concentration rises, information processing, judgment, memory, sensory perception, and motor coordination all become progressively impaired. Depression and drowsiness generally occur. Contrary to popular belief, alcohol decreases sexual performance in males. Some individuals become loud, angry, or violent. In most states, a blood alcohol concentration of 100 mg/dL (or 0.10) legally defines driving while intoxicated (DWI). A 150-pound man typically reaches this level by drinking two or three beers in an hour.

In chronic drinkers, cells of the central nervous system adapt to the presence of the drug. This causes tolerance (more and more alcohol is needed to experience the same effect) and physical dependence (discussed in a later section of this chapter). Abrupt withdrawal can result in sleep disturbances, severe anxiety, tremors, seizures, hallucinations, and psychoses.

Congress is currently discussing legislation that would force the alcoholic beverage industry to place warning labels on its products. Such labels would warn that alcohol can cause mental retardation and other birth defects and that it impairs the ability to drive or operate machinery. However, there is strong opposition from the alcohol industry, which spends $2 billion annually on advertising.

Treatment for alcohol problems includes various forms of psychotherapy, including relapse prevention therapy in which individuals are encouraged not to think of lapses from abstinence as failure. The group support offered by Alcoholics Anonymous (AA) has proved effective for many struggling with alcohol abuse.

Responsiveness: Nervous Systems

34

Learning Objectives

After you have studied this chapter you should be able to:

1. Contrast nerve nets and radial nervous systems with bilateral nervous systems.
2. Compare the vertebrate nervous system with a bilateral invertebrate nervous system.
3. Trace the development of the principal regions of the vertebrate brain from the forebrain, midbrain, and hindbrain.
4. Describe the functions and structure of the spinal cord.
5. Locate the following parts of the human brain and give the functions of each: medulla, pons, midbrain, thalamus, hypothalamus, cerebellum, and cerebrum.
6. Compare the reticular activating system with the limbic system.
7. Contrast REM and non-REM sleep.
8. Review current theories of learning and memory.
9. Cite experimental evidence linking environmental stimuli with changes in the brain and with learning ability.
10. Compare the somatic system with the autonomic system.
11. Contrast the sympathetic and parasympathetic divisions of the autonomic system.
12. Discuss the biological actions and effects on mood of the following types of drugs: alcohol, barbiturates, antianxiety drugs, antipsychotic drugs, opiates, stimulants, hallucinogens, and marijuana.

Key Concepts

- ☐ An animal's range of possible responses depends on the complexity of its nervous system. As animal groups evolved, nervous systems became increasingly complex.
- ☐ The vertebrate nervous system consists of the central nervous system (CNS) and the peripheral nervous system (PNS).
- ☐ The CNS (brain and spinal cord) integrates incoming information and determines appropriate responses. The vertebrate brain consists of the cerebrum, cerebellum, and brain stem. The brain stem includes the medulla, pons, midbrain, thalamus, and hypothalamus.
- ☐ The PNS (receptors, afferent nerves, and efferent nerves) transmits information from receptors to the CNS and transmits "decisions" from the CNS to the muscles and glands that must respond. The PNS includes the somatic and autonomic systems.

ALL INVERTEBRATES EXCEPT SPONGES HAVE NERVOUS SYSTEMS

The sponge has no specialized nerve cells and no nervous system. Its activities are limited to basic life support processes like filtering food and reproducing. Whatever responses it makes depend on actions of individual cells. In this chapter, we will see that an animal's life-style is closely linked to the type and complexity of its nervous system.

Some Invertebrates Have Nerve Nets or Radial Nervous Systems

Hydra remains rooted in one spot, waiting for dinner to brush by its tentacles. Yet this animal responds to predator or prey, discharging nematocysts (stinging structures) and coordinating the movements of its tentacles to capture food. Like other cnidarians, *Hydra* has a relatively inefficient **nerve net** (Fig. 34–1). Its neurons are scattered throughout the body, and impulses may be transmitted in more than one direction. If the stimulus is strong, the message will spread to more neurons of the net than if it is weak. Responses involve large parts of the body. An advantage of the nerve net is that the cnidarian can respond to dinner (or a predator) approaching from any direction.

The somewhat more sophisticated nervous system of the sea star and other echinoderms consists of a nerve ring that surrounds the mouth and a large radial nerve that extends into each arm. These nerves coordinate responses such as movement. In sea stars, a nerve net mediates the responses of the dermal gills to touch.

Most Invertebrates Have Bilateral Nervous Systems

In bilaterally symmetrical animals, the nervous system is generally more complex than in radially symmetrical animals. A bilaterally symmetrical animal generally moves forward, head first. With sense organs concentrated at the front of the body, the animal can detect an enemy quickly or sense food in time to capture it. We can identify the following trends in the evolution of bilateral nervous systems:

1. Increased number of nerve cells.
2. Concentration of nerve cells, forming masses of tissue that become ganglia and brain. Thick cords of tissue that become nerve cords and nerves.
3. Specialization of function. For example, one part of the brain may be specialized to receive information from the eyes and another part may direct movement.
4. Increased number of association neurons and more complex synaptic contacts. This permits greater integration of incoming messages, provides a greater range of responses, and allows far more precision in responses.
5. **Cephalization,** or formation of a head with a brain. When an animal has a head, the principal sense organs such as eyes, ears, olfactory (sense of smell) receptors, and taste receptors tend to be concentrated there. Response can be more rapid if these sense organs are linked by short pathways to decision-making nerve cells nearby.

In flatworms, masses of nerve cells in the head region form **cerebral ganglia** (Fig. 34–2). These serve as a prim-

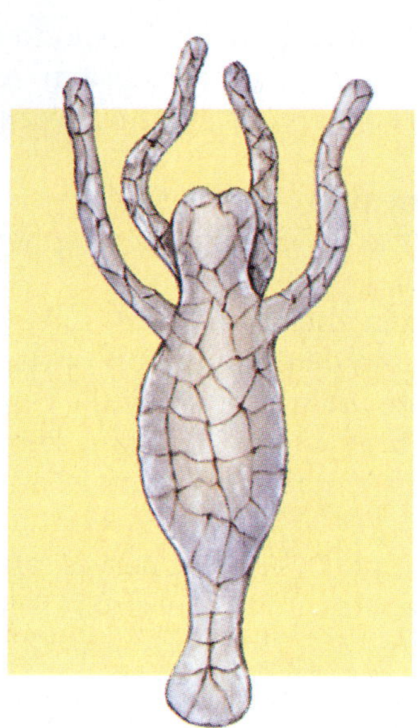

Figure 34–1 The nerve net of *Hydra* and other cnidarians is the simplest organized nervous tissue. No central brain and no definite neural pathways are present.

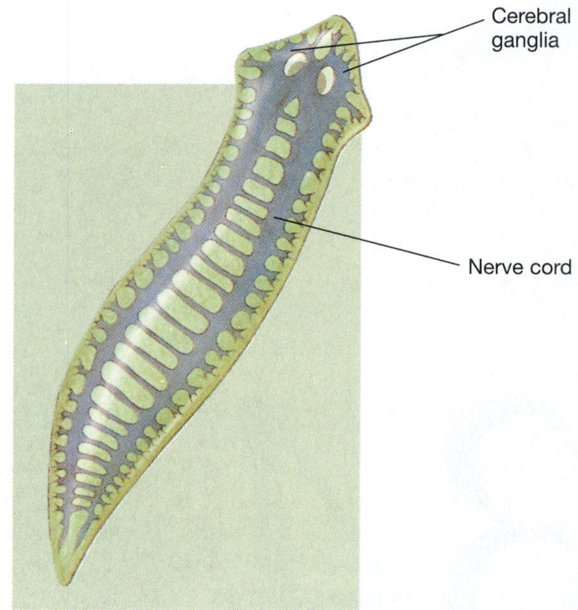

Figure 34–2 Planarian flatworms have a ladder-type nervous system. Cerebral ganglia in the head region serve as a simple brain and exert some control over the rest of the nervous system.

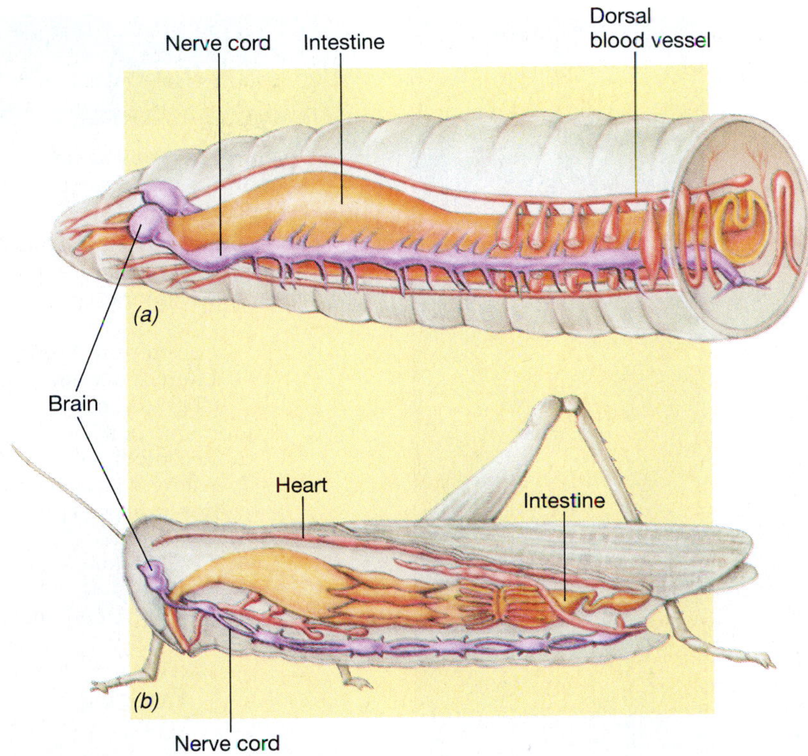

Figure 34–3 Annelid and arthropod nervous systems. (***a***) The nervous system of the earthworm is typical of those found in other annelids. The cell bodies of the neurons are located in ganglia found in each body segment. They are connected by the ventral nerve cord. (***b***) In the insect nervous system the cerebral ganglia (a simple brain) are connected to two ventral nerve cords.

itive "brain" and exert some control over the rest of the nervous system. Two ventral nerve cords extend from the ganglia to the posterior end of the body. The nerve cords are connected by a series of transverse nerves that also connect the brain with the eyespots. This arrangement is referred to as a ladder-type nervous system.

In earthworms and other annelids and in arthropods, the nervous system includes a pair of ventral nerve cords (Fig. 34–3). The cell bodies of the nerve cells are concentrated in pairs of ganglia located in *each* body segment. Lateral nerves link the ganglia with muscles and other body structures. In some arthropods, regions of the cerebral ganglia specialize to perform specific functions.

Clams and other mollusks typically have at least three pairs of ganglia. Each pair has specific functions. In cephalopods, such as the octopus, nerve cells tend to be concentrated in a central region. All of the ganglia are massed in a ring that surrounds the esophagus. The ring contains about 168 million nerve cells. With this complex nervous system, it is no wonder that the octopus is capable of considerable learning and can be taught quite complex tasks. In fact, the octopus is considered to be among the most intelligent invertebrates.

KEY FEATURES OF THE VERTEBRATE NERVOUS SYSTEM ARE THE HOLLOW, DORSAL NERVE CORD AND WELL-DEVELOPED BRAIN

The vertebrate nervous system has two main divisions: the central nervous system (CNS) and the peripheral nervous system (PNS; Fig. 34–4). The **CNS** consists of a complex tubular brain that is continuous with the single, dorsal, tubular spinal cord. Serving as central control, these organs integrate incoming information and determine appropriate responses.

The **PNS** is made up of the sensory receptors (e.g., touch, auditory, and visual receptors) and the nerves, which are the communication lines. Various parts of the body are linked to the brain by cranial nerves and to the spinal cord by spinal nerves. Afferent neurons in these nerves continuously inform the CNS of changing conditions. Then efferent neurons transmit the "decisions" of the CNS to appropriate muscles and glands, which make the adjustments needed to maintain homeostasis.

For convenience we subdivide the PNS into **somatic** and **autonomic** portions. Most of the receptors and nerves

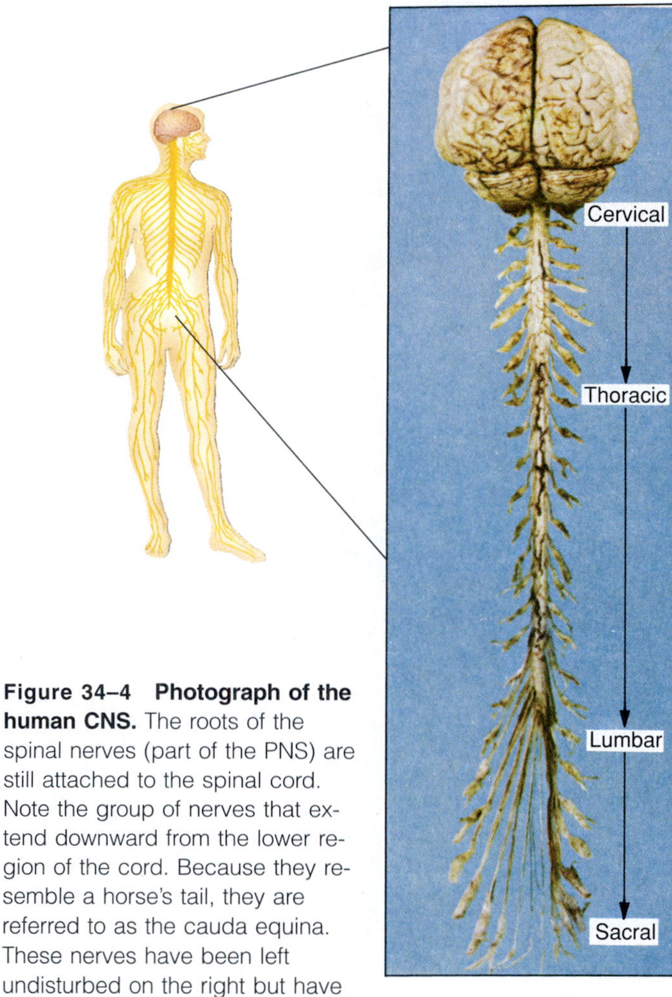

Figure 34–4 Photograph of the human CNS. The roots of the spinal nerves (part of the PNS) are still attached to the spinal cord. Note the group of nerves that extend downward from the lower region of the cord. Because they resemble a horse's tail, they are referred to as the cauda equina. These nerves have been left undisturbed on the right but have been fanned out on the left. (Dissection by Dr. M.C.E. Hutchinson, Department of Anatomy, Guy's Hospital Medical School, London; from Williams and Warwick, eds., *Gray's Anatomy*)

TABLE 34–1
Divisions of the Human Nervous System

Central nervous system (CNS)
- Brain
- Spinal cord

Peripheral nervous system (PNS)
- Somatic portion
  1. Receptors
  2. Afferent (sensory) nerves—transmit information from receptors to CNS
  3. Efferent nerves—transmit information from CNS to skeletal muscles
- Autonomic portion
  1. Receptors
  2. Afferent (sensory) nerves—transmit information from receptors in internal organs to CNS
  3. Efferent nerves—transmit information from CNS to glands and involuntary muscle in organs
     a. Sympathetic nerves—generally stimulate activity that results in mobilization of energy (e.g., speeds heartbeat)
     b. Parasympathetic nerves—action results in energy conservation or restoration (e.g., slows heartbeat)

concerned with changes in the external environment are somatic; those that regulate the internal environment are autonomic. Both systems have sensory (afferent) nerves, which transmit messages from receptors to the CNS, and motor (efferent) nerves, which transmit information back from the CNS to the structures that must respond. The autonomic system has two kinds of efferent pathways—**sympathetic** and **parasympathetic** nerves (see Table 34–1).

THE EVOLUTION OF THE VERTEBRATE BRAIN IS MARKED BY INCREASING COMPLEXITY

All vertebrates, from fish to mammals, have the same basic brain structure. Different parts of the brain are specialized in the various vertebrate classes, and there is an evolutionary trend toward increasing complexity, especially of the cerebrum and cerebellum.

In the early vertebrate embryo, the brain and spinal cord differentiate from a single tube of tissue, called the **neural tube.** Anteriorly, the tube expands and develops into the brain. Posteriorly, the tube becomes the spinal cord. Brain and spinal cord remain continuous and their cavities communicate. As the brain begins to differentiate, three bulges become visible: the hindbrain, midbrain, and forebrain (Fig. 34–5).

The Hindbrain Develops into the Medulla, Pons, and Cerebellum

The **hindbrain,** the posterior portion of the brain in vertebrate embryos, is continuous with the spinal cord. The **medulla** consists largely of nerve tracts that connect the spinal cord with various parts of the brain. In complex vertebrates, the medulla contains **vital centers** that regulate respiration, heart rate, and blood pressure. Other reflex centers in the medulla regulate such activities as swallowing, coughing, and vomiting.

The size and shape of the **cerebellum** vary greatly among the vertebrate classes (Fig. 34–6). In some fish, birds, and mammals, the cerebellum is highly developed, whereas it tends to be small in amphibians and reptiles. The cerebellum coordinates muscle activity and is re-

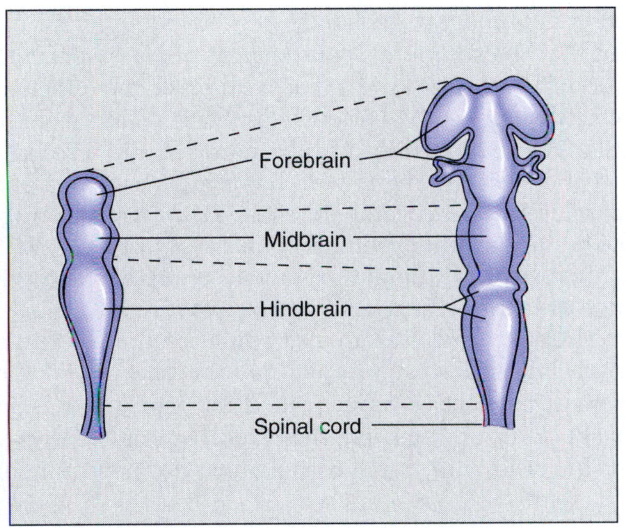

Figure 34–5 Development of the vertebrate brain. Early in development, the anterior end of the neural tube differentiates into the forebrain, midbrain, and hindbrain. These primary divisions subdivide and then give rise to specific structures of the adult brain.

sponsible for muscle tone, posture, and equilibrium. Development of the cerebellum is correlated roughly with the extent and complexity of muscular activity.

Injury or removal of the cerebellum results in impaired muscle coordination. A bird without a cerebellum is unable to fly, and its wings thrash about jerkily. When the human cerebellum is injured by a blow or by disease, muscular movements are uncoordinated. Any activity requiring delicate coordination, such as threading a needle, is very difficult, if not impossible.

In mammals, a large mass of fibers, the **pons,** forms a bulge on the anterior surface of the brain. The pons is a bridge connecting the spinal cord and medulla with upper parts of the brain. It contains nuclei (masses of nerve cell bodies) that relay impulses from the cerebrum to the cerebellum and also contains centers that help regulate respiration.

The Midbrain Is Concerned with Vision and Hearing

In fish and amphibians, the **midbrain** is the most prominent part of the brain, serving as the main association area. It receives incoming sensory information, integrates it, and sends decisions to appropriate motor nerves. The dorsal portion of the midbrain is specialized as the **optic lobes,** which interpret visual information.

In reptiles, birds, and mammals, many of the functions of the optic lobes are assumed by the cerebrum. In mammals, the upper part of the midbrain contains centers for visual reflexes such as pupil constriction, and the inferior portion has centers for certain auditory reflexes. The **red nucleus** is a center in the midbrain that integrates information about muscle tone and posture.

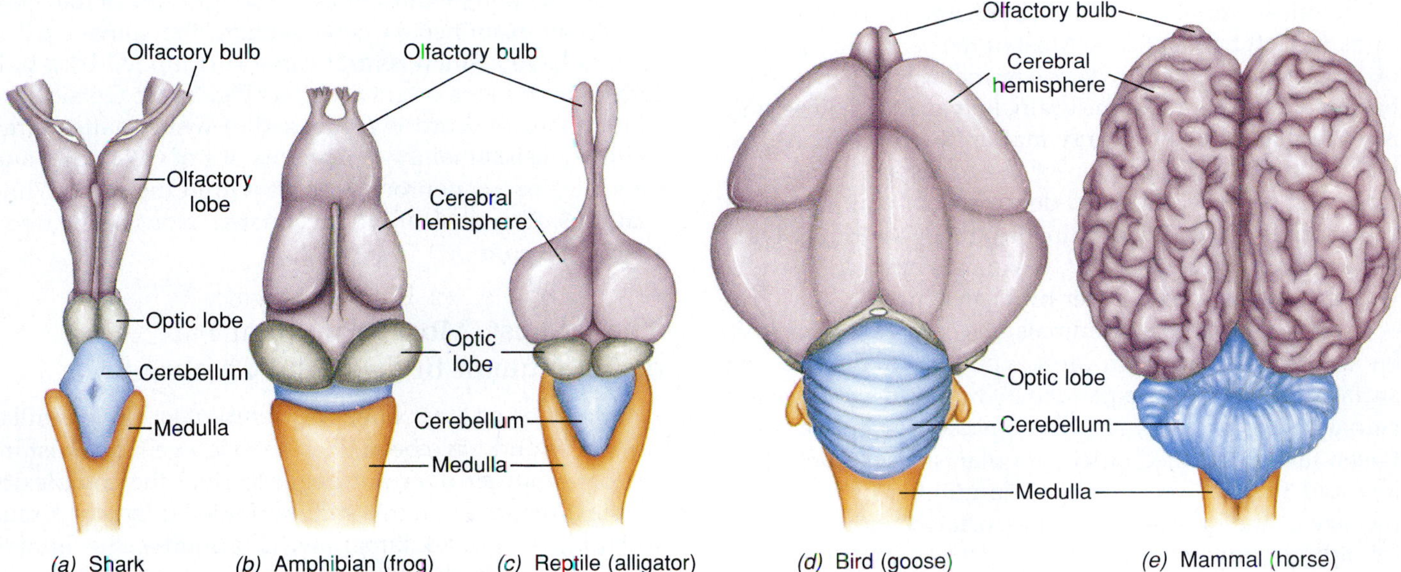

Figure 34–6 Comparison of the brains of several vertebrates. Note the basic similarities and evolutionary trends. Different parts of the brain may be specialized in the various vertebrate classes. For example, the large olfactory lobes in the shark brain (***a***) are essential to this predator's highly developed sense of smell. During the course of evolution, the cerebrum and cerebellum have become larger and more complex (***a-e***). In the mammal (***e***) the cerebrum is the most prominent part of the brain.

The Forebrain Gives Rise to the Thalamus, Hypothalamus, and Cerebrum

The posterior portion of the **forebrain** gives rise to the thalamus and hypothalamus. In all vertebrate classes, the **thalamus** is a relay center for motor and sensory messages. In mammals, all sensory messages (except those from the olfactory receptors) are delivered to the thalamus before being relayed to the sensory areas of the cerebrum.

The **hypothalamus,** which lies below the thalamus, contains olfactory centers and is the principal integration center for the regulation of the internal organs. The hypothalamus links the nervous and endocrine systems. It regulates the pituitary gland (an important endocrine gland) that extends downward from the hypothalamus. In reptiles, birds, and mammals, body temperature is controlled by a "thermostat" in the hypothalamus. The hypothalamus also regulates appetite and water balance and is involved in emotional and sexual responses.

Together the medulla, pons, midbrain, thalamus, and hypothalamus make up the **brain stem,** the elongated portion of the brain that looks like a stalk holding up the large cerebrum.

The anterior region of the forebrain differentiates to form the cerebrum, and, in most vertebrate groups, the **olfactory bulbs.** The olfactory bulbs are important in the chemical sense of smell—the dominant sense in most vertebrates. In fact, much of brain development in vertebrates appears to be focused on integrating information about odors. In fish and amphibians, the cerebrum is almost entirely devoted to the integration of olfactory information.

In most vertebrates, the **cerebrum** is divided into right and left hemispheres. Most of the cerebrum is made of **white matter,** which consists mainly of axons connecting various parts of the brain. In mammals, birds, and most reptiles, a layer of **gray matter,** the **cerebral cortex,** makes up the outer portion of the cerebrum. Gray matter contains cell bodies and dendrites. In mammals, the cerebrum is the most prominent part of the brain. During embryonic development, it expands and grows backwards, covering many other brain structures.

In small or simple mammals, the cerebral cortex may be smooth. However, in large, complex mammals, the surface area is greatly expanded by numerous folds called **convolutions** (or *gyri;* singular, *gyrus*). The furrows between them are called **sulci** (singular, *sulcus*) when shallow and **fissures** when deep. The number of folds (not the size of the brain) has been associated with complexity of brain function.

THE HUMAN CENTRAL NERVOUS SYSTEM IS THE MOST COMPLEX BIOLOGICAL SYSTEM KNOWN

The range of possible responses an animal can make depends in large part on the number of neurons it has and how they are organized in its nervous system. We have seen that the simple nervous system of *Hydra* permits it to do little more than respond with gross movements to prey that brushes by its tentacles. With its more sophisticated nervous system, a frog can hop about in search of food and eject its tongue with lightning speed to capture a passing fly. However, neither the *Hydra* nor the frog is able to solve algebra problems or learn about biology. The highly complex human nervous system permits a wide range of learning and very precise responses.

The soft, fragile brain and spinal cord are well protected. Encased within bone, they are covered by three layers of connective tissue, called the **meninges.** A special shock-absorbing fluid, the **cerebrospinal fluid,** cushions the brain and spinal cord against mechanical injury.

The Spinal Cord Transmits Impulses To and From the Brain and Controls Many Reflex Activities

The **spinal cord** is a hollow cylinder that emerges from the base of the brain and extends downward to about the level of the waist (Fig. 34–4). Its two functions are to (1) control many reflex activities and (2) transmit messages back and forth to the brain through its **ascending** and **descending nerve tracts.** Each tract is a large bundle of axons. As axons pass down through the brain and spinal cord, some of them cross over from one side of the cord (or brain) to the other. For this reason, the right side of the brain mainly controls the left side of the body, and the left side of the brain mainly controls the right side of the body.

When examined in cross section, the spinal cord is seen to have a small central canal surrounded by a butterfly shaped area of gray matter (Fig. 34–7). Outside the gray matter the cord is composed of white matter. Gray matter consists mainly of large masses of cell bodies and dendrites of the neurons present within the cord. White matter is composed of the myelinated axons of the tracts within the cord.

The Largest, Most Prominent Part of the Human Brain Is the Cerebrum

Although computers have been designed along similar principles and have been likened to it, even the most intricate computer does not begin to rival the complexity of the human brain. A soft, wrinkled mass of tissue weighing about 1.4 kilograms (3 pounds), the human brain contains about 100 billion neurons. Each neuron is functionally connected to as many as 100,000 other neurons, and there may be as many as 10^{14} synapses. No wonder scientists have barely begun to unravel some of the tangled neural circuits that govern human physiology and behavior!

Brain cells require a continuous supply of oxygen and glucose. Although the brain accounts for only about 2%

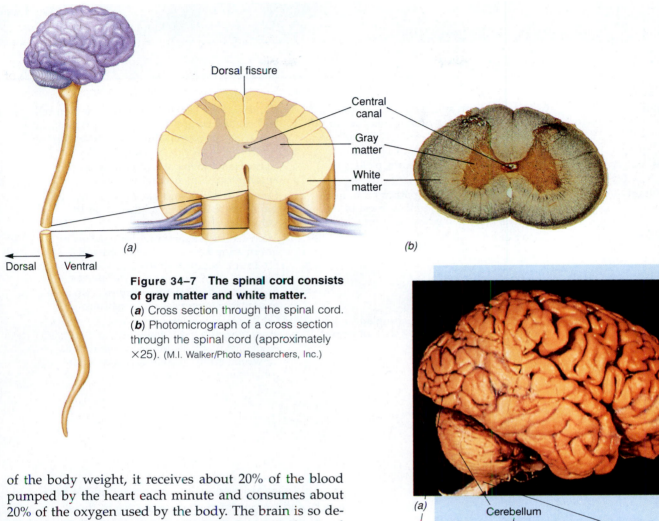

Dorsal fissure

Central canal

Gray matter

White matter

Dorsal Ventral

(a) (b)

Figure 34–7 The spinal cord consists of gray matter and white matter. (***a***) Cross section through the spinal cord. (***b***) Photomicrograph of a cross section through the spinal cord (approximately ×25). (M.I. Walker/Photo Researchers, Inc.)

of the body weight, it receives about 20% of the blood pumped by the heart each minute and consumes about 20% of the oxygen used by the body. The brain is so dependent upon its blood supply that when it is deprived of blood, the person loses consciousness after about 5 seconds, and irreversible damage occurs within a few minutes. In fact, the most common cause of brain damage is stroke (cerebrovascular accident), in which a portion of the brain is deprived of its blood supply (often because a blood vessel has been blocked by a blood clot).

The structure and functions of the main divisions of the brain are summarized in Table 34–2. The brain is illustrated in Figures 34–8 and 34–9.

The cerebrum is divided into right and left cerebral hemispheres, and into regions, called lobes (Fig. 34–8). The occipital lobes contain the visual centers, and the temporal lobes contain centers for hearing. The *central sulcus,* which crosses the top of each hemisphere from medial to lateral edge, separates the frontal lobes from the parietal lobes.

Functionally, the cerebral cortex is divided into three areas: (1) the **sensory areas,** which receive incoming signals from the sense organs; (2) the **motor areas,** which control voluntary movement; and (3) the **association areas,** which link the sensory and motor areas and are responsible for thought, learning, intelligence, language abilities, memory, judgment, and personality. Sensory areas in the

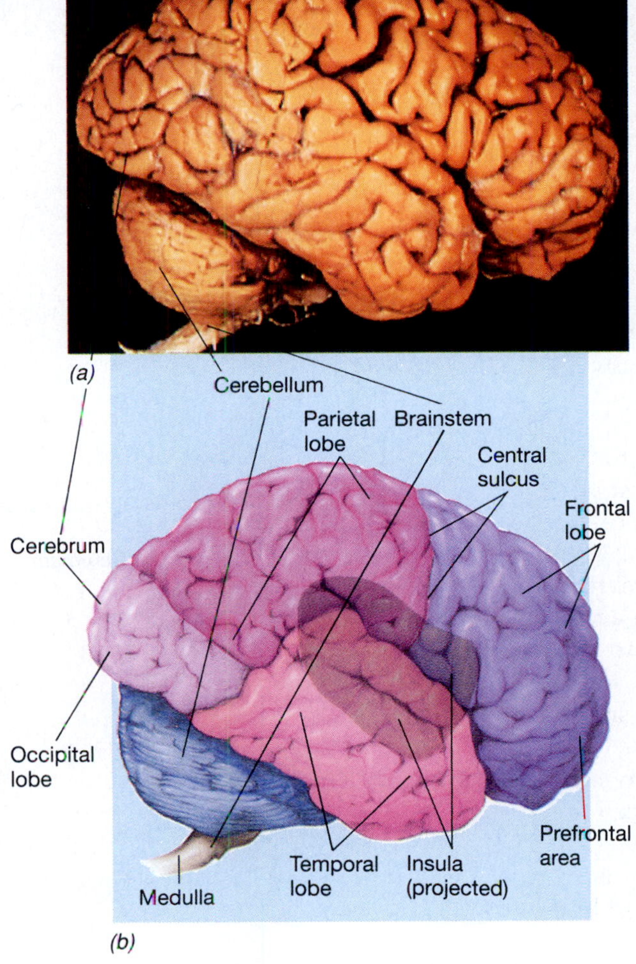

Figure 34–8 Structure of the human brain. (***a***) Photograph of the human brain, lateral view. Note that the cerebrum covers part of the brain stem. (***b***) Lateral view of the human brain showing the lobes of the cerebrum. Part of the brain has been made transparent so that the underlying insular lobe can be located. (Visuals Unlimited/Fred Hossler)

TABLE 34–2
The Brain

Structure	*Description*	*Function*
Brain Stem		
Medulla	Continuous with spinal cord; primarily made up of nerves passing from spinal cord to rest of brain	Contains vital centers (clusters of neuron cell bodies) that control heartbeat, respiration, and blood pressure; contains centers that control swallowing, coughing, vomiting
Pons	Forms bulge on anterior surface of brain stem	Connects various parts of brain with one another; contains respiratory center
Midbrain	Just above pons; largest part of brain in lower vertebrates; in humans, most of its functions are assumed by cerebrum	Center for visual and auditory reflexes (e.g., pupil reflex, blinking, adjusting ear to volume of sound)
Thalamus	At top of brain stem	Main sensory relay center for conducting information between spinal cord and cerebrum. Neurons in thalamus sort and interpret all incoming sensory information (except olfaction) before relaying messages to appropriate neurons in cerebrum.
Hypothalamus	Just below thalamus; pituitary gland is connected to hypothalamus by stalk of neural tissue	Contains centers for control of body temperature, appetite, fat metabolism, and certain emotions; regulates pituitary gland; link between "mind" (cerebrum) and "body" (physiological mechanisms)

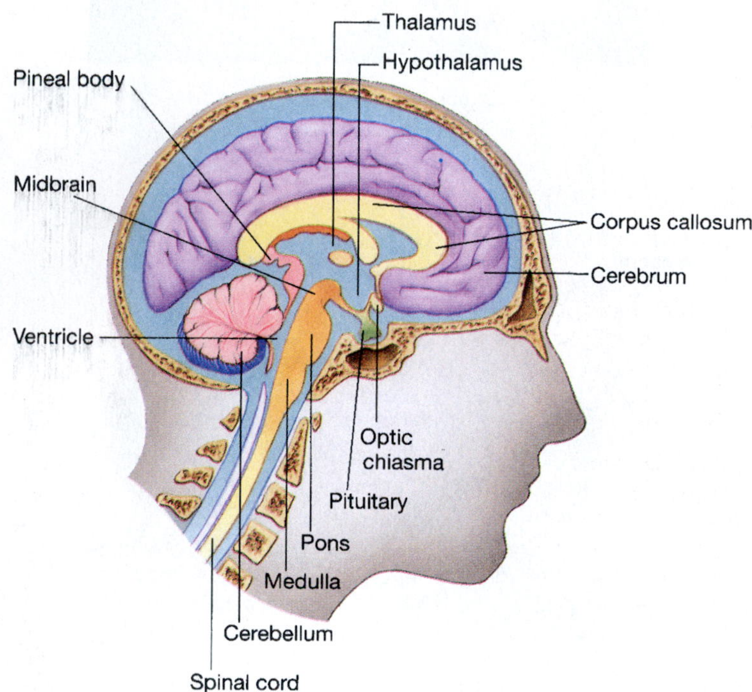

Figure 34–9 Midsagittal view of the brain. Note that in this type of section half of the brain is cut away so that structures normally covered by the cerebrum are exposed.

parietal lobes are responsible for sensations of heat, cold, touch, and pressure from sense organs in the skin.

The primary motor areas, located in the frontal lobes, control skeletal muscles (Fig. 34–10). The size of the motor area in the brain that controls a given part of the body is proportional to the complexity of the movements of that body part. Thus, a relatively large part of the motor area controls the precise movements of the hands and the muscles that give expression to the face. A similar relationship exists between the sensory area in the brain and the region of the skin from which it receives impulses.

The white matter of the cerebrum lies beneath the cerebral cortex. Nerve fibers of the white matter connect the cortical areas with one another and with other parts of the nervous system. A large band of white matter, the **corpus callosum,** connects the right and left hemispheres.

The Reticular Activating System Is an Arousal System

The **reticular activating system (RAS)** is a complex pathway of neurons in the brain stem and thalamus. It receives messages from neurons in the spinal cord and from many other parts of the nervous system and communi-

[T]ABLE 34–2
[co]ntinued

Structure	Description	Function
Cerebellum	Second largest division of brain	Reflex center for muscular coordination and refinement of movements; when it is injured, performance of voluntary movements is uncoordinated and clumsy
Cerebrum	Largest, most prominent part of human brain; more than 70% of brain's cells located here; longitudinal fissure divides cerebrum into right and left hemispheres, each divided by shallow sulci (furrows) into six lobes; frontal, parietal, temporal, insular, occipital, and limbic	Center of intellect, memory, consciousness, and language; also controls sensation and motor functions
Cerebral cortex (outer gray matter)	Arranged into convolutions (folds) that increase surface area; functionally, cerebral cortex is divided into:	
	1. Motor areas	Controls movement of voluntary muscles
	2. Sensory areas	Receives incoming information from eyes, ears, pressure and touch receptors, etc.
	3. Association areas	Site of intellect, memory, language, and emotion; interprets incoming sensory information
White matter	Consists of myelinated axons of neurons that connect various regions of brain; these axons are arranged into bundles (tracts)	Connects: 1. Neurons within same hemisphere 2. Right and left hemispheres 3. Cerebrum with other parts of brain and spinal cord

Concept Connections

Dopamine ⟷ *Motor Function*

The neurotransmitter dopamine plays an important role in motor function. Its function became known through an interesting series of somewhat unrelated events. During the mid-1950s, the drug reserpine became popular as a major tranquilizer used for mental patients. Then, in 1959, investigators noticed that some patients taking reserpine experienced distressing side effects, such as muscle rigidity and persistent tremors (shaking). These symptoms were very similar to those seen in patients with Parkinson's disease, a disorder in which movement is slow, shaky, and difficult. Victims of Parkinson's disease have a shuffling gait and suffer from tremors even when they are not attempting to move. This observation led to studies that showed that reserpine greatly reduces the amount of dopamine within two of the basal ganglia within the white matter of the cerebrum. Investigators then discovered that patients with Parkinson's disease have only about 50% of the normal amount of dopamine in their basal ganglia.

Attempts to administer dopamine to these patients were not successful because dopamine cannot penetrate the blood–brain barrier (Chapter 6 opener). However, L-dopa, a substance from which dopamine is synthesized in the body, does penetrate the blood–brain barrier. Its use has dramatically relieved the symptoms of Parkinson's disease in many patients. Investigators have had some success using animal models of Parkinson's disease to study the effects of transplanting neurons that produce dopamine directly into the brain.

Recent studies suggest that dopamine depletion may occur with aging. As a result, even healthy persons experience changes in motor abilities as they age. Body movements and even reflexes slow, and movement becomes more difficult. Studies suggest that treatment with L-dopa may be helpful.

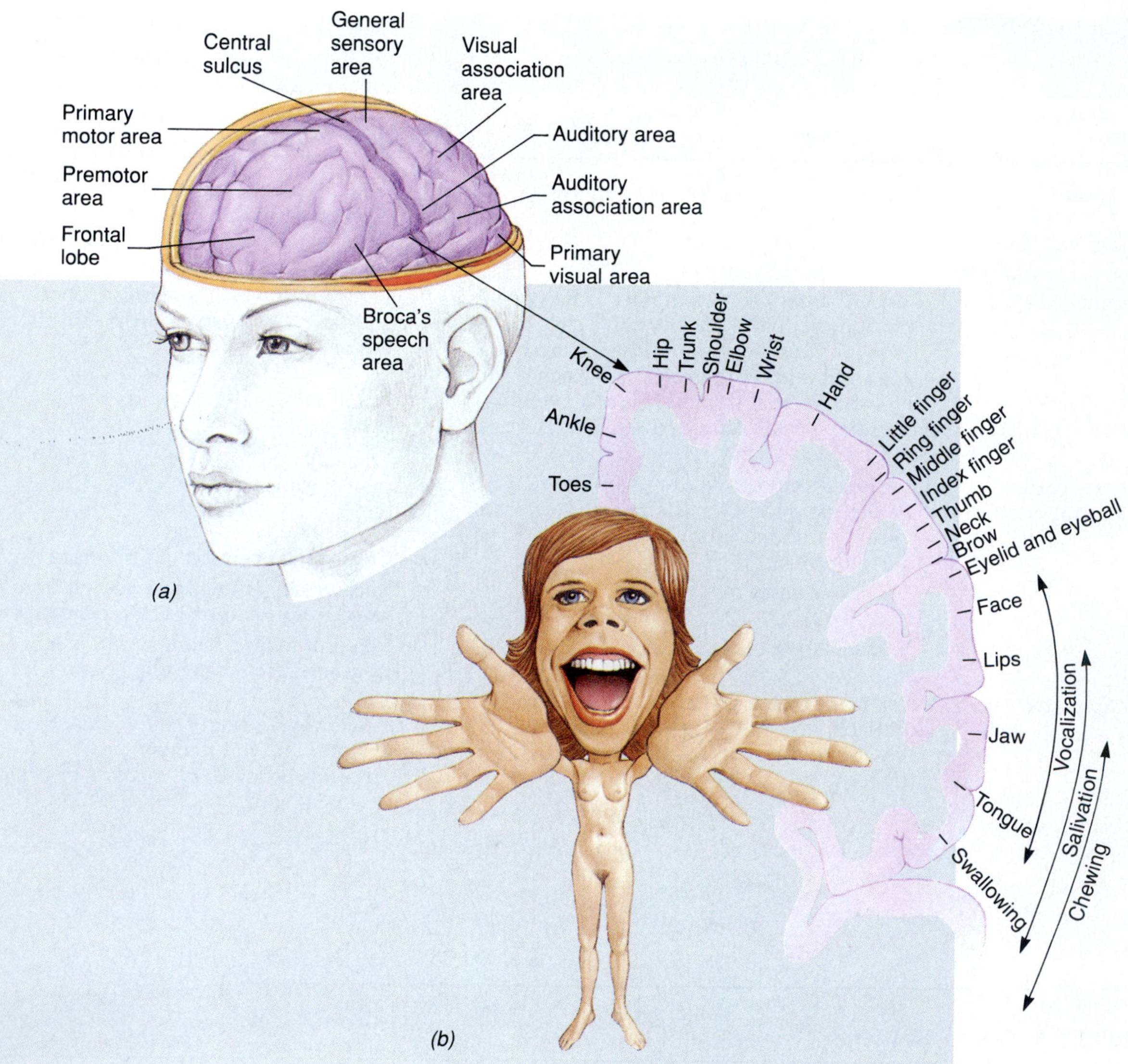

Figure 34–10 Control of movement. (*a*) The cross section through the primary motor area (precentral gyrus) shows which area of cerebral cortex controls each body part. (*b*) The figure (known as a motor homunculus) shown here is proportioned to reflect the amount of cerebral cortex that controls each body part. Note that more cortical tissue is devoted to controlling those body structures capable of skilled, complex movement.

cates with the cerebral cortex by complex circuits. The RAS maintains consciousness (awareness), and the extent of its activity determines the state of alertness. When the RAS bombards the cerebral cortex with stimuli, you feel alert and are able to focus your attention on specific thoughts. When its activity slows, you begin to feel sleepy. If the RAS is severely damaged, the victim may pass into a deep, permanent coma.

The Limbic System Affects Emotional Aspects of Behavior

The **limbic system,** another action system of the brain, consists of certain structures of the cerebrum and diencephalon. It affects the emotional aspects of behavior; sexual behavior; biological rhythms; autonomic responses; and motivation, including feelings of pleasure and punishment. Stimulation of certain areas of the limbic system in an experimental animal results in increased general activity and may cause fighting behavior or extreme rage.

When an electrode is implanted in the so-called reward center of the limbic system, a rat may press a lever that stimulates this area as many as 15,000 times per hour. Stimulation of the reward center apparently feels so good that an animal ignores food and drink and may continue to press the lever until it drops from exhaustion. When an electrode is implanted in the punishment center of the limbic system, an experimental animal quickly learns to press a lever to avoid stimulating that area. The reward and punishment centers appear to be important in influencing motivation and behavior.

The Brain Exhibits Electrical Activity

Brain activity can be studied by measuring and recording the electrical potentials, or "brain waves," given off by various parts of the brain. This electrical activity can be recorded by a device known as an electroencephalograph. To obtain an **electroencephalogram (EEG),** a recording of this electrical activity, electrodes are taped to different parts of the scalp, and the activity of the cerebral cortex is measured. The EEG shows that the brain is continuously active. On the EEG, the most regular indication of activity, **alpha waves,** comes mainly from the visual areas in the occipital lobes when the person being tested is resting quietly with eyes closed. These waves occur rhythmically at the rate of about ten per second (Fig. 34–11).

When the eyes are opened, alpha waves are replaced by more rapid, irregular waves. When some regular stimulus, such as a light blinking at regular intervals, is presented, brain waves with a similar rhythm appear. As you are reading this biology text, your brain is (hopefully) emitting **beta waves.** These have a fast-frequency rhythm characteristic of heightened mental activity such as information processing. During sleep, brain waves become slower and larger as the person falls into deeper unconsciousness. These slow, large waves associated with certain stages of sleep are called **delta waves.** Dreams cause flurries of irregular waves.

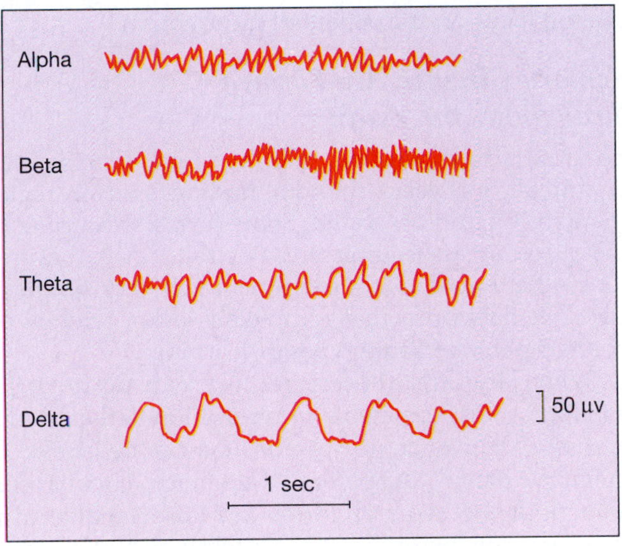

Figure 34–11 Electrical activity of the brain. Electroencephalograms made while the subject was excited, relaxed, and in various stages of sleep. Recordings made during excitement show brain waves that are rapid and of small amplitude, whereas in sleep the waves are much slower and of greater amplitude. The regular waves characteristic of the relaxed state are called alpha waves.

Certain brain diseases change the pattern of brain waves. Individuals with epilepsy, for example, exhibit a readily recognizable abnormal wave pattern. The location of a brain tumor or the site of brain damage caused by a blow to the head can sometimes be determined by noting the part of the brain that shows abnormal waves.

Sleep May Occur when Signals from the RAS Slow

Sleep is a state of unconsciousness during which there is decreased electrical activity of the cerebral cortex, and from which a person can be aroused. When signals from the RAS slow down so that the cerebral cortex is deprived of activating input, a person may lapse into sleep. If there is nothing interesting to occupy the mind, we find it easy to go to sleep even when we are not particularly tired. And, although we tend to be wakeful in the presence of attention-holding stimuli, there is a limit beyond which sleep is inevitable.

Two main stages of sleep are recognized: non-REM and REM. **REM** is an acronym for *rapid eye movements.* During **non-REM** sleep, sometimes called normal sleep, metabolic rate decreases, breathing slows, and blood pressure decreases. Delta waves, thought to be generated spontaneously by the cerebral cortex when it is not driven by impulses from other parts of the brain, are characteristic of non-REM sleep.

Every 90 minutes or so, a sleeping person enters the REM stage for a period of time. During this stage, which accounts for about one-fourth of total sleep time, the eyes move about rapidly beneath the closed but fluttering lids. Brain waves change to a desynchronized pattern of beta waves. Sleep researchers claim that everyone dreams, especially during REM sleep. Release of norepinephrine within the RAS generates stimulating impulses that are fed into the cerebral cortex. This results in dreams.

The hypothalamus and other areas of the brain stem are responsible for the sleep-wake cycle. A nucleus in the hypothalamus is considered the body's biological clock. This nucleus receives input from the retina of the eye regarding light and dark. When stimulated, neurons in the non-REM sleep center release the neurotransmitter serotonin. Serotonin is thought to inhibit signals passing through the RAS, thus inducing sleep.

Why sleep is necessary is not understood. Apparently only higher vertebrates with fairly well-developed cerebral cortices sleep. When a person stays awake for unusually long periods, fatigue and irritability result, and even routine tasks cannot be performed well. Perhaps certain waste products accumulate within the nervous system during waking hours, and sleep gives the nervous system an opportunity to dispose of them.

Not only is non-REM sleep required, but REM sleep is apparently also essential. In sleep deprivation experiments performed with human volunteers, lack of REM sleep makes subjects anxious and irritable. After such experiments, when the subjects are permitted to sleep normally again, they spend more time than usual in the REM stage for a period. Many types of drugs alter sleep patterns and affect the amount of REM sleep. Sleeping pills, for example, may increase the total sleeping time but decrease the time in REM sleep. When a person stops taking such drugs, several weeks may be required before normal sleep patterns are reestablished.

Learning Involves the Storage of Information and Its Retrieval

Learning is a relatively long-lasting adaptive change in behavior that results from experiences. Laboratory experiments have shown that members of every animal phylum can learn. Several types of learning will be discussed in Chapter 44.

Information Processing Involves Three Levels of Memory

According to current theory, there are three levels of memory: sensory, short-term, and long-term. At any moment you are bombarded with thousands of bits of sensory information. At this very moment, your eyes are receiving information not only regarding the words on this page but also about the objects and intensity of the light around you. At the same time you may be hearing a variety of sounds—music, voices, the hum of an air conditioner. Your olfactory epithelium may sense cologne or the smell of dinner. Sensory receptors in your hands may be receiving information regarding the weight and position of your book. The bits of sensory information on which we focus our attention are registered in **sensory memory.** Sensory memory can hold a lot of information, but that information is lost within about 1 second. Attention is an important component of sensory memory. When we attend to information in sensory memory, pattern recognition—the process of identifying the stimuli—begins.

In order for a sensory memory to be identified or recognized, we must relate it to past experience or past knowledge, and this requires further processing. As pattern recognition and encoding occur, we become aware of stimuli. **Short-term memory** is the information that we are aware of at the moment. Short-term memory can hold only about seven chunks of information (a chunk corresponds to some unit such as a word, syllable, or number).

Short-term memory allows us to recall information for a few minutes. Usually when we look up a phone number, for example, we remember it only long enough to dial. Should we need the same number an hour later, we have to look it up again. Keeping information in short-term memory requires rehearsal. If we redirect our attention, the new stimulus interferes with recall of the information already present, and we forget. Short-term memory is necessary for understanding speech.

One theory of short-term memory suggests that it is based on reverberating circuits (Chapter 33). A memory circuit may continue to reverberate for several minutes until it fatigues or until new signals are received that interfere with the old.

Once we have processed information into **long-term memory,** we no longer have to focus attention on it in order to remember. Apparently, when information is selected for long-term storage, the brain rehearses the material and then stores it in association with similar memories. Long-term memory has an unlimited capacity.

Retrieval of information stored in long-term memory is of considerable interest—especially to students. Some researchers think that once information is deposited in long-term memory, it remains within the brain permanently. The trick is to find it when it is needed. When you seem to forget something, the problem may be that you have not effectively searched for the memory. Information retrieval can be improved by careful storage. One method is to form strong associations between items when they are being stored.

Sensory memory, short-term memory, and long-term memory are all part of the information processing system. Sensory memory can be viewed as a component of short-term memory, and short-term memory can be thought of as an activated component of long-term memory (Fig. 34–12). Information can be transferred back and forth from one component to another. For example, information from long-term memory can be retrieved and temporarily transferred to short-term memory, at which time we focus on the activated information.

Memory Circuits Are Formed Throughout the Brain

Researchers have worked with the brains of experimental animals for years without finding specific regions where information is stored. Some forms of learning can take place in association areas within lower brain regions—the thalamus, for example. Even very simple animals like flatworms that completely lack a cerebral cortex are capable of some types of learning.

When large areas of cerebral cortex are destroyed, information is lost somewhat in proportion to the extent of lost tissue. However, no specific area can be labeled the "memory bank." Apparently, memory circuits form throughout the cerebral cortex and also involve many other areas of the brain. Both sensory and motor pathways may be involved in memory.

The association areas of the cerebral cortex, a part of the limbic system known as the **hippocampus,** and the thalamus are involved in learning and remembering. The association areas concerned with the interpretation of vi-

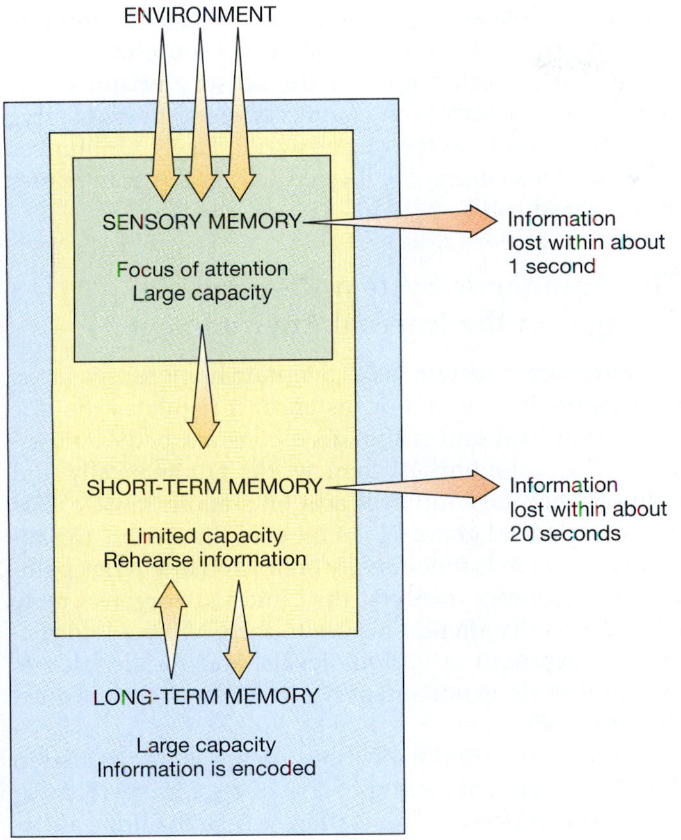

Figure 34–12 Human information processing. Sensory memory is represented here as an activated component of short-term memory. Short-term memory is represented as an activated component of long-term memory.

sual, auditory, and other general sensory information all meet in the **general interpretative area.** The temporal portion of the general interpretative area, called **Wernicke's area,** is an important center for language function because it permits us to recognize and interpret words. When the general interpretative area is damaged, an individual may not understand what he or she hears or reads. He or she may be able to recognize words but may not be able to arrange them into a coherent thought.

Neurons within the association areas form highly complex pathways that permit complicated reverberation. Several minutes are required for a memory to become consolidated within long-term memory. Should a person suffer a brain concussion or undergo electroshock therapy, for example, memory of what happened immediately prior to the incident may be completely lost. This is known as retrograde amnesia.

The limbic system is important in processing stored information. When the hippocampus is damaged, a person can recall information stored in the past but can no longer convert new short-term memories into long-term memories.

Experience Affects the Brain

Studies show that environmental experience may cause physical as well as chemical changes in brain structure. When rats are provided with a stimulating environment and given the opportunity to learn, they exhibit increases in the size of cell bodies and nuclei of brain neurons. A great number of glial cells develop and there is an increased concentration of synaptic contacts. Some investigators have reported that the cerebral cortex becomes thicker and heavier, and biochemical changes take place. Other experiments have indicated that animals reared in a complex environment may be able to process and remember information more quickly than animals not given such advantages. Rats provided with the basic necessities but deprived of stimulation and/or social interaction do not exhibit these changes.

Early environmental stimulation can also enhance the development of motor areas in the brain. Experiments with rats have shown that the brains of animals that are encouraged to exercise become slightly heavier than those of the control animals. Characteristic changes occur within the cerebellum, including the development of larger dendrites. On the basis of this research, investigators have suggested that early experience may help a young child develop his or her physical potential.

Apparently, during early life, certain critical or sensitive periods of nervous system development occur that are influenced by environmental stimuli. For example, when the eyes of young mice first open, neurons in the visual cortex develop large numbers of dendritic spines (structures in which synaptic contact takes place). If the animals are kept in the dark and deprived of visual stimuli, fewer dendritic spines form. If the mice are exposed to light later in life, some new dendritic spines form but never the number that develop in a mouse reared in a normal environment.

Studies linking the development of the brain with environmental experience indicate that early stimulation is important for the sensory, motor, and intellectual development of children. Such studies have led to the rapidly expanding educational toy market and widespread acceptance of early education programs. Researchers have also suggested that continuing environmental stimulation is needed to maintain the status of the cerebral cortex in later life.

THE PERIPHERAL NERVOUS SYSTEM INCLUDES SOMATIC AND AUTONOMIC SYSTEMS

The **peripheral nervous system (PNS)** consists of the sensory receptors, the nerves that link the receptors with the CNS, and the nerves that link the CNS with the muscles and glands (the effectors). The part of the PNS that helps the body respond to changes in the external environment is the somatic system. The nerves and receptors that maintain homeostasis despite internal changes make up the autonomic nervous system.

The Somatic System Helps the Body Adjust to the External Environment

The **somatic nervous system** includes those receptors that react to changes in the external environment, the sensory neurons that keep the CNS informed of those changes, and the motor neurons that adjust the positions of the skeletal muscles. In mammals, 12 pairs of **cranial nerves** emerge from the brain. They transmit information to the brain from the sensory receptors and transmit orders from the brain to the voluntary muscles that control movements of the eyes, face, mouth, tongue, pharynx, and larynx. The largest cranial nerve, the vagus nerve, transmits information to and from many of the internal organs including the heart.

In humans, 31 pairs of **spinal nerves** emerge from the spinal cord (Fig. 34–13). Named for the general region of the vertebral column from which they originate, there are 8 pairs of cervical spinal nerves, 12 pairs of thoracic, 5 pairs of lumbar, 5 pairs of sacral, and 1 pair of coccygeal spinal nerves.

Each spinal nerve has a **dorsal root** and a **ventral root.** The dorsal root consists of sensory (afferent) fibers, which transmit information from the sensory receptors to the spinal cord. Just before the dorsal root joins with the cord, it is marked by a swelling, the **spinal ganglion,** which consists of the cell bodies of the sensory neurons. The ventral root consists of the motor (efferent) fibers leaving the cord en route to the muscles and glands. Cell bodies of the motor neurons are located within the gray matter of the spinal cord.

The Autonomic System Responds to Changes in the Internal Environment

The **autonomic system** helps maintain homeostasis in the internal environment. For instance, it regulates the rate of the heartbeat and maintains a constant body temperature. The autonomic system works automatically and without voluntary input. It acts on smooth muscle, cardiac muscle, and glands. Like the somatic system, the autonomic system is functionally organized into reflex pathways. Receptors within the internal organs relay information via afferent nerves to the CNS, the information is integrated at various levels, and the decision is transmitted along efferent nerves to the appropriate muscles or glands.

Information from the various organs is transmitted to the CNS by afferent neurons. Some of these are part of cranial or spinal nerves. These afferent neurons bring information about blood pressure, respiration, heartbeat, contractions of the digestive tract, and other visceral activities.

The efferent portion of the autonomic system is subdivided into **sympathetic** and **parasympathetic systems.**

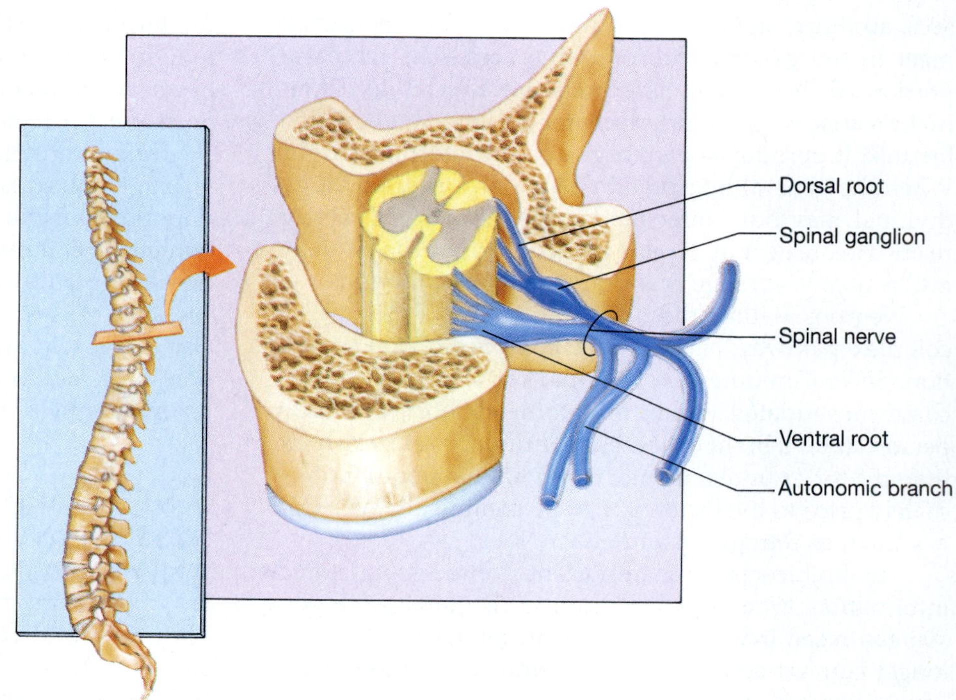

Figure 34–13 A spinal nerve. Dorsal and ventral roots emerge from the spinal cord and join, forming a spinal nerve. The spinal nerve divides into several branches.

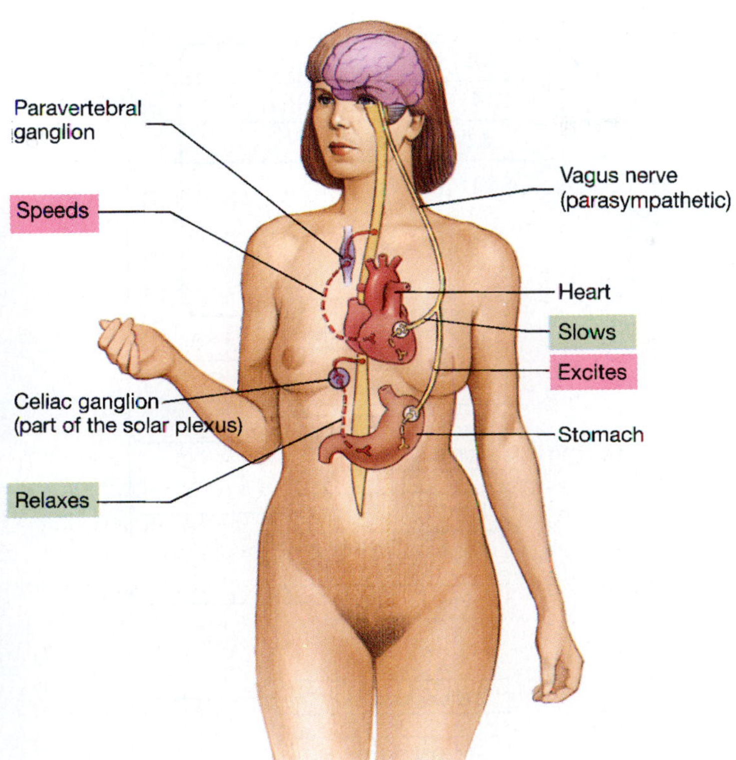

Figure 34–14 Regulation by the autonomic system. Dual innervation of the heart and stomach by sympathetic and parasympathetic nerves. Sympathetic nerves are shown in red.

In general, the sympathetic nerves stimulate organs and mobilize energy, especially in response to stress. In contrast, the parasympathetic nerves influence organs to conserve and restore energy, particularly when one is engaged in quiet, calm activities such as studying biology. Many organs are innervated by both types of nerves, which act on the organ in a complementary way (Figs. 34–14 and 34–15). For example, the heart rate is slowed by impulses from its parasympathetic nerve fibers and speeded up by messages from its sympathetic nerve supply.

MANY MOOD DRUGS CHANGE THE LEVELS OF NEUROTRANSMITTERS IN THE BRAIN

Drugs help us fall asleep; drugs help us stay awake. Drugs help us relax, forget our problems, feel like part of the group. Drugs "expand the mind." Some are purchased freely over the counter; some are prescribed by physicians; others are bought from illegal sources. About 25% of all prescribed drugs are taken to alter psychological conditions, and almost all of the commonly abused drugs affect mood. Many of them act by changing the levels of neurotransmitters within the brain. In particular, levels of norepinephrine, serotonin, and dopamine are thought to influence mood. For example, when excessive amounts of norepinephrine are released in the RAS, we feel stimulated and energetic, whereas low concentrations of this neurotransmitter reduce anxiety. Table 34–3 lists several commonly used and abused drugs and gives their effects. Also see the Chapter Opener and Focus on Health and Human Affairs: Crack Cocaine.

Habitual use of almost all mood drugs may result in **psychological dependence,** in which the user becomes emotionally dependent on the drug. When deprived of it, the user craves the feeling of **euphoria** (well-being) that the drug induces. Some drugs induce **tolerance** when they are taken continually for several weeks. This means that increasingly larger amounts are required to obtain the desired effect. Tolerance occurs when the liver cells are stimulated to produce more of the enzymes that break down the drug. Use of some drugs, such as alcohol, barbiturates, or heroin, results in **addiction,** in which physiological changes take place. When the drug is withheld, the addict suffers physical illness and characteristic withdrawal symptoms. Addiction sometimes occurs because certain drugs, such as morphine, are similar to substances normally manufactured by the body. The continued use of such a drug may depress the body's natural production. Sudden withdrawal from the drug may cause potentially dangerous physiological effects, and it may be some time before homeostasis is reestablished.

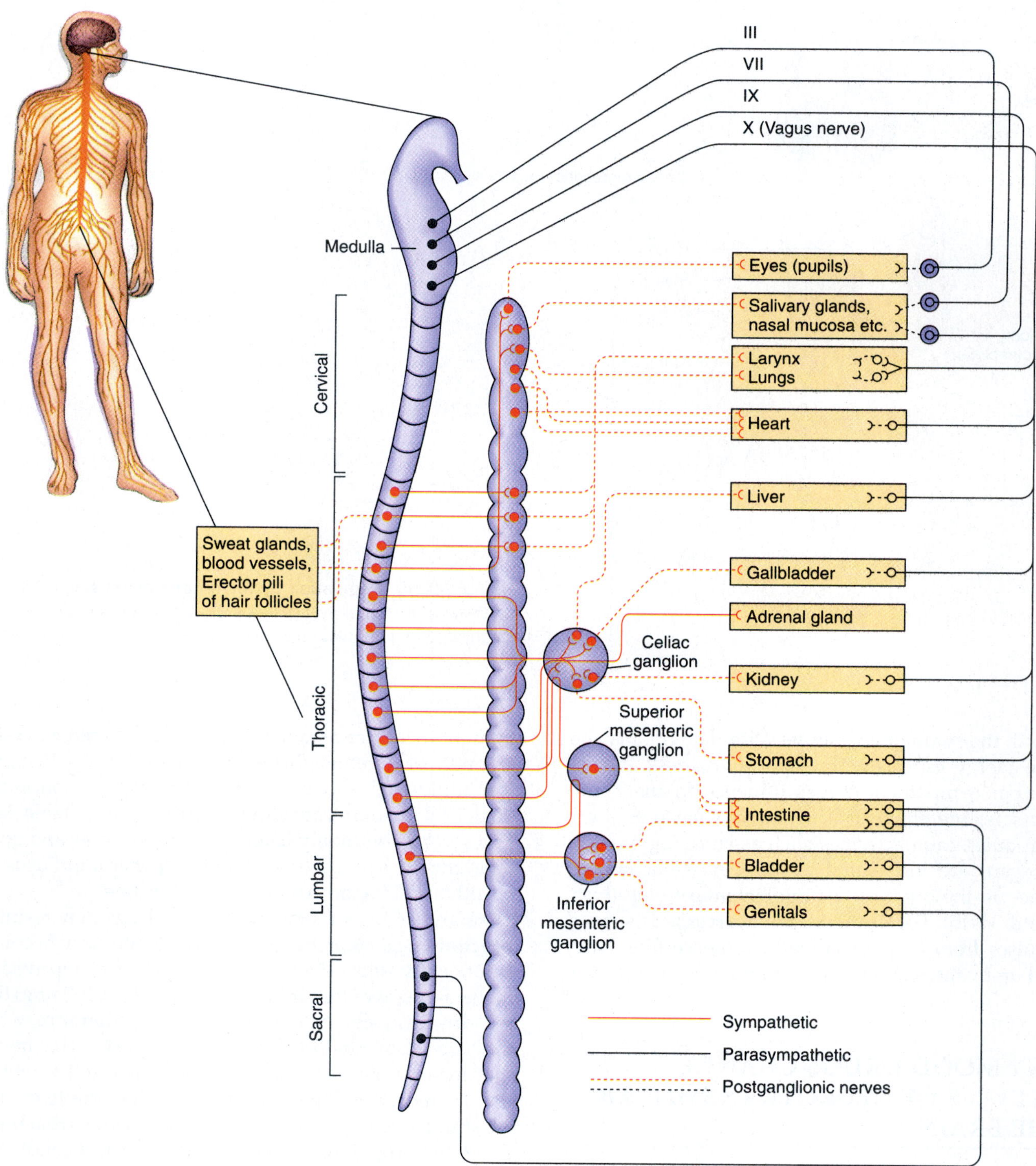

Figure 34–15 Sympathetic and parasympathetic nervous systems. Complex as it appears, this diagram has been greatly simplified. Only some of the nerves supplying certain organs are shown.

FOCUS ON *Health and Human Affairs*

Crack Cocaine

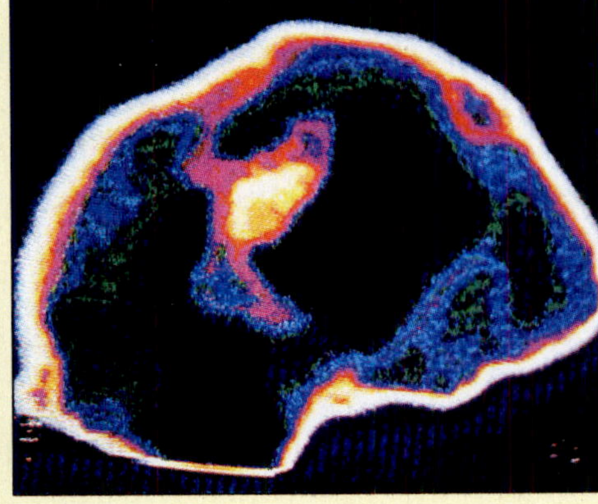

Neurological damage caused by cocaine use. This color-enhanced SPECT (single photon emission computer tomography) scan of the brain of a cocaine abuser shows decreased metabolism (white area) near the frontal region. (Howard Sochurek, Courtesy of Brigham and Women's Hospital, Boston)

With the exception of alcohol abuse, the majority of persons seeking treatment for drug abuse are now crack cocaine addicts. Cocaine use by teenagers alone has increased about 400% during the past 10 years, involving an estimated 2 million youngsters. Crack is a very concentrated and extremely powerful form of cocaine—five to ten times as addictive as other forms of cocaine. This drug is produced in illegal, makeshift labs by converting powdered cocaine into small "rocks," which are up to 80% pure cocaine. Crack is smoked in pipes or added to tobacco or marijuana cigarettes.

Use of crack results in an intense, brief high beginning in 4 to 6 seconds and lasting for 5 to 7 minutes. Physiologically, crack stimulates a massive release of catecholamine neurotransmitters (norepinephrine and dopamine) in the brain and blocks re-uptake. Excitation of the sympathetic nervous system occurs, and users report experiencing feelings of self-confidence, power, and euphoria. As the neurotransmitters are depleted, the high is followed by a "crash," a period of deep depression. The abuser experiences an intense craving for another crack "hit" in order to get more stimulation.

Some abusers spend days smoking crack without stopping to eat or sleep. Although a vial of rocks can be obtained for about $20, many abusers develop habits that cost hundreds of dollars a week. Supporting an expensive drug habit leads many abusers to prostitution, drug dealing, and other forms of crime.

Cocaine addicts report problems with memory, fatigue, depression, insomnia, paranoia, loss of sexual drive, violent behavior, and attempts at suicide. Crack can cause respiratory problems, brain seizures, cardiac arrest, and elevation of blood pressure that leads to stroke. Many users have suffered fatal reactions to impurities in the drug or have died as a result of accidents related to drug use.

Researchers at Columbia University are developing a genetically engineered antibody that can bind to cocaine and destroy it. The antibody can apparently remain in the blood for long periods of time. Clinical trials are planned to test this potential treatment for cocaine addiction.

TABLE 34–3
Effects of Some Commonly Used Drugs

Name of drug	*Effect on mood*	*Actions on body*	*Dangers associated with abuse*
Barbiturates (e.g., Nembutal, Seconal)	Sedative-hypnotic;* "downers"	Inhibit impulse conduction in RAS: depress CNS, skeletal muscle, and heart; depress respiration; lower blood pressure; cause decrease in REM sleep	Tolerance, physical dependence; death from overdose, especially in combination with alcohol
Methaqualone (e.g., Quaalude, Sopor)	Hypnotic	Depresses CNS; depresses certain spinal reflexes	Tolerance, physical dependence, convulsions, death
Meprobamate (e.g., Equanil, Miltown; "minor tranquilizers")	Antianxiety drug;† induces calmness	Causes decrease in REM sleep; relaxes skeletal muscle; depresses CNS	Tolerance, physical dependence; coma and death from overdose
Valium, Librium, Xanax ("mild tranquilizers")	Reduce anxiety	May activate GABA neurons, which inhibit activity in some areas of the brain; relax skeletal muscle	Minor EEG abnormalities with chronic use; very large doses cause physical dependence
Phenothiazines (chlorpromazine; "major tranquilizers")	Antipsychotic; highly effective in controlling symptoms of psychotic patients	Affect levels of catecholamines in brain (block dopamine receptors); depress neurons in RAS and basal ganglia	Prolonged intake may result in Parkinson-like symptoms
Antidepressant drugs (e.g., Paxil, Elavil, Prozac)	Elevate mood; relieve depression	Block re-uptake of serotonin or norepinephrine, so more is available to stimulate nervous system	May cause anxiety, headache, nausea
Alcohol	Euphoria, relaxation, release of inhibitions	Depresses CNS; impairs vision, coordination, judgment, lengthens reaction time	Physical dependence, damage to pancreas, liver cirrhosis, brain damage
Narcotic analgesics (opiates, e.g., morphine, heroin, codeine)	Euphoria, reduction of pain	Depress CNS; depress reflexes; constrict pupils; impair coordination; block release of substance P from pain-transmitting neurons	Tolerance, physical dependence; convulsions, death from overdose

*Sedatives reduce anxiety; hypnotics induce sleep.
†Antianxiety drugs reduce anxiety but are less likely to cause drowsiness than the more potent sedative-hypnotics.

Chapter Summary

I. Among invertebrates, nerve nets and radial nervous systems are typical of radially symmetrical animals. Bilateral nervous systems are characteristic of bilaterally symmetrical animals.
   - A. A nerve net consists of nerve cells scattered throughout the body; no CNS is present. Response in these animals is generally slow and imprecise.
   - B. Echinoderms typically have a nerve ring and nerves that extend into various parts of the body.
   - C. In a bilateral nervous system, there is a concentration of nerve cells to form nerves, nerve cords, ganglia, and (in complex animals) a brain. An increase in numbers of neurons, especially of the association neurons, permits a wide range of responses.

II. The vertebrate nervous system consists of the central nervous system and peripheral nervous system.

III. In the vertebrate embryo, the neural tube gives rise to the brain and spinal cord. The anterior end of the tube differentiates into the forebrain, midbrain, and hindbrain.
   - A. The hindbrain develops into the medulla, cerebellum, and pons.
     1. In complex vertebrates, the medulla contains vital centers regulating respiration, heart rate, and blood pressure.
     2. The cerebellum is highly developed in fish, birds, and mammals—animals that tend to engage in complex movement.
     3. In mammals, the pons serves as a bridge connecting the spinal cord and medulla with the upper parts of the brain.
   - B. In fish and amphibians, the midbrain is the main association area linking sensory input and motor output. In reptiles, birds, and mammals, the midbrain serves as a center for visual and auditory reflexes.

ABLE 34–3
ontinued

Name of drug	Effect on mood	Actions on body	Dangers associated with abuse
Cocaine	Euphoria; excitation followed by depression	CNS stimulation followed by depression; autonomic stimulation; dilation of pupils; local anesthesia; inhibits re-uptake of norepinephrine, dopamine, and serotonin	Mental impairment, convulsions, hallucinations, unconsciousness; death from overdose
Amphetamines (e.g., Dexedrine)	Euphoria, stimulation, hyperactivity ("uppers," "pep pills")	Stimulate release of dopamine and norepinephrine; block re-uptake of norepinephrine and dopamine into neurons; inhibit monoamine oxidase (MAO); enhance flow of impulses in RAS; increase heart rate; raise blood pressure; dilate pupils	Tolerance, possible physical dependence, hallucinations; death from overdose
Caffeine	Increases mental alertness; decreases fatigue and drowsiness	Acts on cerebral cortex; relaxes smooth muscle; stimulates cardiac and skeletal muscle; increases urine volume (diuretic effect)	Very large doses stimulate centers in the medulla (may slow the heart); toxic doses may cause convulsions
Nicotine	Lessens psychological tension	Stimulates sympathetic nervous system; combines with receptors in postsynaptic neurons of autonomic system; has effect similar to that of acetylcholine, but large amounts result in blocking transmission; stimulates synthesis of lipid in arterial wall	Tolerance, physical dependence; stimulates development of atherosclerosis
LSD (lysergic acid diethylamide)	Overexcitation; sensory distortions; hallucinations	Alters levels of transmitters in brain (may inhibit serotonin and increase norepinephrine); potent CNS stimulator; dilates pupils, sometimes unequally; increases heart rate; raises blood pressure	Irrational behavior
Marijuana	Euphoria	Impairs coordination; impairs depth perception and alters sense of timing; impairs short-term memory (probably by decreasing acetylcholine levels in the hippocampus); inflames eyes; causes peripheral vasodilation; exact mode of action unknown	In large doses, sensory distortions, hallucinations; evidence of lowered sperm counts and testosterone (male hormone) levels

C. The forebrain differentiates to form the thalamus and hypothalamus.
   1. The thalamus is a relay center for motor and sensory information.
   2. The hypothalamus controls autonomic functions, links nervous and endocrine systems, and controls temperature, appetite, and fluid balance.
   3. In fish and amphibians, the cerebrum mainly integrates incoming sensory information.

IV. The vertebrate central nervous system consists of the brain and spinal cord.
   A. The spinal cord consists of ascending tracts, which transmit information to the brain, and descending tracts, which transmit information from the brain. The spinal cord also controls many reflex activities.
   B. The brain consists of the cerebrum, cerebellum, and brain stem. The brain stem includes the medulla, pons, midbrain, thalamus, and hypothalamus.
   C. The cerebral cortex contains motor areas, which control voluntary movement; sensory areas, which receive incoming sensory information; and association areas, which link sensory and motor areas and are also responsible for learning, language, thought, and judgment.
   D. The reticular activating system is responsible for maintaining consciousness.
   E. The limbic system affects the emotional aspects of behavior, motivation, sexual behavior, autonomic responses, and biological rhythms.
   F. Metabolic rate slows during non-REM sleep. REM sleep is characterized by rapid eye movements and dreaming.
   G. Learning is a change in behavior that results from experience; memory is the storage of knowledge and the ability to retrieve this knowledge.
      1. Information is transferred from sensory memory to short-term memory and then to long-term memory.
      2. Memories appear to be stored throughout the association areas of the cerebrum.
   H. Experiences can cause physical and chemical changes in the brain.

V. The peripheral nervous system consists of sensory receptors and nerves, including the cranial and spinal nerves and their branches.
   A. The somatic system helps the body to adjust to the external environment.
   B. The autonomic system regulates the internal activities of the body.
      1. The sympathetic system enables the body to respond to stressful situations.
      2. The parasympathetic system influences organs to conserve and restore energy.

VI. Many drugs alter mood by increasing or decreasing the concentrations of specific neurotransmitters within the brain.

Selected Key Terms

addiction, p. 667
association areas, p. 659
autonomic system, p. 666
central nervous system (CNS), p. 655
cerebellum, p. 656
cerebral cortex, p. 658
cerebrospinal fluid, p. 658
cerebrum, p. 658
corpus callosum, p. 660
cranial nerves, p. 666
forebrain, p. 658
gray matter, p. 658
hindbrain, p. 656
hypothalamus, p. 658
learning, p. 664
limbic system, p. 662
medulla, p. 656
meninges, p. 658
midbrain, p. 657
motor areas, p. 659
nerve net, p. 654
parasympathetic system, p. 666
peripheral nervous system (PNS), p. 655
pons, p. 657
reticular activating system (RAS), p. 660
sensory areas, p. 659
somatic system, p. 666
spinal ganglion, p. 666
spinal nerves, p. 666
sympathetic system, p. 666
thalamus, p. 658
tolerance, p. 667
white matter, p. 658

Post-Test

1. The simplest organized nervous tissue is the ______________ ______________, which is found in *Hydra.*
2. Afferent nerves conduct impulses toward the ______________.
3. In a planarian flatworm, the ________ ________ serve as a primitive brain.
4. The vertebrate central nervous system consists of the ______________ and ________ ________.
5. The most prominent part of the amphibian brain is the ______________.
6. In fish and amphibian brains, the well-developed ______________ bulbs receive information about ______________.
7. The function of the cerebrospinal fluid is to ______________________________________.
8. The largest, most prominent part of the mammalian brain is the ______________.
9. Voluntary movement is controlled by ______________ areas in the cerebral cortex.
10. The ______________ ______________ ______________ is responsible for maintaining consciousness.
11. The part of the brain that coordinates and refines muscular movement is the (a) hypothalamus (b) sensory cortex (c) cerebellum (d) medulla (e) pons.
12. The part of the brain that controls the rate of the heartbeat is the (a) thalamus (b) motor cortex (c) cerebellum (d) medulla (e) pons.
13. The part of the brain that regulates the pituitary gland is the (a) thalamus (b) hypothalamus (c) cerebellum (d) medulla (e) pons.
14. The part of the brain that regulates body temperature is the (a) thalamus (b) hypothalamus (c) cerebellum (d) red nucleus (e) pons.
15. The part of the brain that receives incoming information from the sense organs is the (a) sensory area of the cerebral cortex (b) hypothalamus (c) cerebellum (d) red nucleus (e) corpus callosum.
16. The efferent portion of the autonomic system includes (a) sympathetic system (b) parasympathetic system (c) sensory receptors such as those in the eyes and ears (d) answers (a), (b), and (c) are correct (e) only answers (a) and (b) are correct.
17. The need for a drug user to increase the dose of a drug to obtain the desired effect (after several weeks of taking the same drug) is known as (a) euphoria (b) addiction (c) tolerance (d) psychological dependence (e) none of the preceding answers is correct.
18. Drugs that block uptake of serotonin in the CNS are (a) antidepressant drugs (b) barbiturates (c) nicotine (d) the ones that cause psychological dependence (e) none of the preceding answers is correct.
19. Which of the following drugs are CNS depressants? (a) alcohol (b) barbiturates (c) amphetamines (d) nicotine (e) two of the preceding answers are correct.

20. Label the diagram.

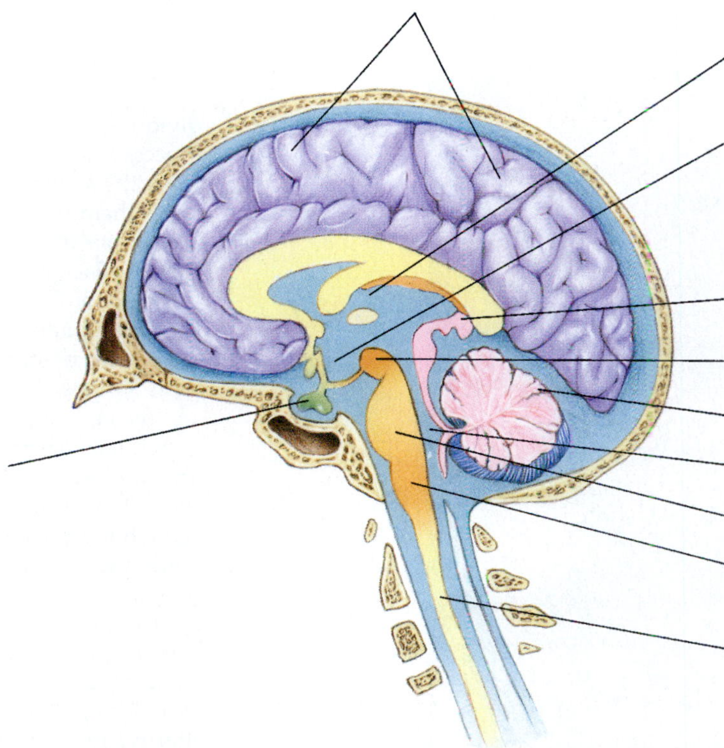

Review Questions

1. Compare the nervous system of a *Hydra* with that of a planarian flatworm.
2. Contrast the flatworm nervous system with that of a vertebrate.
3. Compare the fish brain and the mammalian brain.
4. What structures protect the brain and spinal cord?
5. What are the functions of the spinal cord?
6. Identify the part of the brain most closely associated with each of the following: (a) regulation of body temperature, (b) regulation of heart rate, (c) link between nervous and endocrine systems, (d) interpretation of incoming sensory messages, (e) coordination of movements.
7. What is the RAS? How does it function?
8. What is the function of the limbic system?
9. Contrast short- and long-term memory.
10. Compare the somatic and autonomic systems.
11. Contrast the sympathetic and parasympathetic systems.
12. Describe how these drugs affect the CNS: (a) alcohol, (b) Dexedrine or other amphetamines, (c) barbiturates, (d) antipsychotic drugs, (e) hallucinogens.

Thinking Critically

1. Because about one hour is required to metabolize one alcoholic beverage, you might guess that five drinks could be metabolized in about five hours. However, more than 14 hours after drinking five drinks, a pilot's ability to fly a plane is still impaired. Hypothesize an explanation for this finding.
2. What general trends can you identify in the evolution of the vertebrate brain?
3. In what way does electrical activity of the brain reflect a person's state of consciousness? Of what use might it be to learn to control one's brain wave patterns?
4. Imagine that you have just become a parent. What kind of things could you do to ensure the development of your child's academic abilities?
5. In what sequence is information processed through the levels of memory? In what ways does this system appear to be adaptive?

Recommended Readings

Hinton, G.E., Plaut, D.C., and Shallice, T., "Simulating Brain Damage," *Scientific American,* Vol. 269, No. 4, October 1993, pp. 76–82. When a network of computer simulated neurons is trained to read and is then damaged, errors are made similar to those made by adults with brain damage.

Kalil, R., "Synapse Formation in the Developing Brain," *Scientific American,* Vol. 261, No. 6, December 1989. Developing neurons generate impulses that modify existing synapses and result in the formation of new connections.

Mishkin, M., and Appenzeller, T., "The Anatomy of Memory," *Scientific American* Vol. 256, No. 6, 1987, pp. 80–89. Deep structures in the brain may interact with perceptual pathways in outer layers of the brain to transform sensory stimuli into memories.

Musto, D.F., "Opium, Cocaine and Marijuana in American History," *Scientific American,* Vol. 265, No. 1, July 1991, pp. 40–47. A review of the history of drug use in the United States provides perspective on current reaction to drug use.

A cochlear implant brings sound to this hearing-impaired child. (David Young-Wolff/PhotoEdit)

Helping the Deaf to Hear and the Blind to See

Sensory receptors link all animals with the outside world and enable them to receive information about their external and internal environments. Many people with major hearing and visual impairments have struggled heroically to stay connected with the external world. Investigators have been working to develop electronic devices that would bring sound to the hearing-impaired and sight to the blind.

Hearing is important because, in our culture, much of our social and occupational interaction takes place through spoken language. Hearing is a complex sense discussed in some detail in this chapter. Briefly, when sound waves enter the ear canal, they cause the tympanic membrane (ear drum) to vibrate. Three small bones in the middle ear vibrate, amplifying the vibrations. The vibrations are conducted to the cochlea, a liquid-filled, coiled chamber in the inner ear. When tiny hair cells in the cochlea are stimulated, they produce electric signals that stimulate the cochlear nerve. This nerve transmits impulses to the brain.

One device developed to bring sound to the hearing-impaired is the cochlear implant. An individual with a cochlear implant wears a sound processor around the waist. The processor divides the sound into bands of frequencies and sends them to a receiver implanted in the skull. Signals are transmitted to electrodes in the cochlea. Each electrode receives information in one band of frequencies and then stimulates corresponding nerve endings. The brain interprets the pattern of stimulation so that the individual hears sounds.

By 1993, more than 7000 hearing-impaired individuals worldwide, including 2000 children, had received cochlear implants. However, only about 20% of the recipients could hear well enough to understand most conversations and to use the telephone effectively. Most implant recipients use the sounds they hear to supplement lip-reading and to expand their awareness of what is going on around them. Cochlear implants work best for those who learned to speak *before* losing their hearing. The implants are least effective for those who have been deaf since birth or early childhood. Researchers project that as cochlear implants are improved, the number of hearing-impaired individuals who will benefit will increase dramatically, to perhaps as high as 75%.

For those with tumors of the cochlear nerve, a relay device has been developed that can be implanted in the brain itself. At its current state of development, the implant does not make speech audible, but it does provide the sensation of sound.

For blind individuals, researchers are developing an artificial retina that would allow up to 20% of the blind to see again. The other 80% of blind persons do not have optic nerve function. For them, electrodes are being developed that can be implanted in the visual cortex of the brain. In combination with eyeglasses equipped with tiny cameras, some measure of sight could be restored.

Not everyone is heralding the new technology for artificial senses with enthusiasm. Some health professionals argue that it is risky for individuals to be dependent on electronic devices, especially at the current, rather crude, level of technology. For example, children equipped with cochlear implants may never learn skills such as lip-reading or American Sign Language. And sight-impaired individuals may not learn to read Braille or work with voice-recognizing computers. With only partially restored senses and an inadequate mastery of these other tools, deaf and blind individuals may find themselves at an even greater disadvantage. Until sensory implants are perfected, the debate will continue between those who are eager to embrace the new technology and those who caution against its use.

Sensory Reception

35

Learning Objectives

After you have studied this chapter you should be able to:

1. Distinguish among the five types of receptors classified according to the types of energy to which they respond. Give examples of specific sensory receptors of each type.
2. Summarize how a sensory receptor functions, including definitions of energy transduction, receptor potential, and adaptation in your answer.
3. Describe the following mechanoreceptors: tactile receptors, proprioceptors, lateral line organs, and statocysts.
4. Compare the function of the saccule and utricle with that of the semicircular canals in maintaining equilibrium.
5. Trace the path taken by sound waves through the structures of the ear, and explain how the organ of Corti functions as an auditory receptor.
6. Describe the receptors of taste and smell.
7. Relate the presence of thermoreceptors to the lifestyle of animals that have them.
8. Contrast simple eyes, compound eyes, and the vertebrate eye.
9. Label the structures of the vertebrate eye on a diagram, and give the functions of each of the accessory structures.
10. Compare the two types of photoreceptors in the human retina.

Key Concepts

- ☐ Sensory receptors detect an event in the environment by absorbing energy, converting that energy into electrical energy, and producing a receptor potential that may result in an action potential that transmits information to the central nervous system (CNS).
- ☐ Mechanoreceptors detect touch, pressure, movement, or gravity; chemoreceptors detect specific chemical compounds; thermoreceptors detect heat; electroreceptors detect electrical energy; photoreceptors use pigments to absorb light energy.

SENSORY RECEPTORS DETECT CHANGES IN THE ENVIRONMENT

The kinds of sensory receptors an animal has determine just how it perceives the world. We humans live in a world of rich colors, multiple shapes, and varied sounds. But we cannot hear the high-pitched whistles audible to dogs and cats or the ultrasonic echoes by which bats navigate (Fig. 35–1). And we do not ordinarily recognize our friends by their distinctive odors. Although vision is our dominant and most refined sense, we are blind to the ultraviolet hues that light up the world for insects.

Sensory receptors are structures that respond to information about changes in the internal or external environment. Sensory receptors consist of neuron endings or specialized cells that are in close contact with neurons. Human taste buds, for example, are modified epithelial cells connected to neurons.

Sensory receptors, along with other types of cells, make up the familiar complex sense organs—the eyes, ears, nose, and taste buds. Traditionally, mammals are said to have five senses: sight, hearing, smell, taste, and touch. Balance is now also recognized as a sense, and touch is viewed as a compound sense that allows us to detect pressure, pain, and temperature. In this chapter, we will also consider receptors that enable us to sense muscle tension and joint position.

FIGURE 35–1 Bats navigate by ultrasonic echoes inaudible to the human ear. They emit high-pitched clicking sounds and use the echoes to locate objects. This California leaf-nosed bat (*Macrotus californicus*) uses echoes to locate a flower. (Merlin D. Tuttle/Bat Conservation International/Photo Researchers, Inc.)

Sensory receptors can be classified according to the type of energy to which they respond (Table 35–1). **Mechanoreceptors** respond to mechanical energy—

TABLE 35–1
Classification of Receptors by Stimuli to Which They Respond

Type of receptor	*Examples*	*Effective stimuli*
Mechanoreceptors	Tactile receptors	Touch, pressure
	Pacinian corpuscles	
	Meissner's corpuscles	
	Proprioceptors	Movement, body position
	Muscle spindles	Muscle contraction
	Golgi tendon organs	Stretch of a tendon
	Joint receptors	Movement in ligaments
	Lateral line organs in fish	Waves, currents in water
	Statocysts in invertebrates	Gravity
	Labyrinth of vertebrate ear	
	Saccule and utricle	Gravity, linear acceleration
	Semicircular canals	Angular acceleration
	Hair cells in the cochlea	Pressure waves (sound)
Chemoreceptors	Taste buds, olfactory epithelium	Specific chemical compounds
Thermoreceptors	Temperature receptors in blood-sucking insects and ticks; pit organs in pit vipers; nerve endings and receptors in skins and tongues of many animals	Heat
Electroreceptors	Organs in skins of some fish	Electrical currents in water
Photoreceptors	Eyespots: ommatidia of arthropods; rods and cones in retinas of vertebrates	Light energy

touch, pressure, gravity, stretching, or movement. **Chemoreceptors** respond to certain chemical compounds, while **photoreceptors** detect light energy. **Thermoreceptors** respond to heat or cold. Some fish have **electroreceptors,** which detect electrical energy.

Sensory receptors can also be classified according to the location of the stimuli affecting them. **Exteroceptors** receive stimuli from the outside environment, enabling an animal to know and explore the world, search for food, find and attract a mate, locate shelter, recognize friends, detect enemies, and even learn. **Proprioceptors** are sensory receptors within muscles, tendons, and joints that enable the animal to perceive the position of its arms, legs, head, and other body parts, along with the orientation of its body as a whole. With the help of our proprioceptors, we humans can get dressed or eat in the dark.

Interoceptors are sensory receptors *within* the body that detect changes in pH, osmotic pressure, body temperature, and the chemical composition of the blood. We are usually not conscious of messages sent to the CNS by these receptors. We become aware of their activity when they enable us to perceive such diverse internal conditions as thirst, hunger, nausea, pain, and orgasm. Interoceptors will be described in later chapters with discussions of blood pressure, temperature regulation, respiration, and other specific body functions.

RECEPTOR CELLS WORK BY PRODUCING RECEPTOR POTENTIALS

Receptor cells absorb energy, transduce (convert) that energy into electrical energy, and produce a receptor potential. In its capacity as a sensor, a receptor receives a small amount of energy from the environment. Each kind of receptor is especially sensitive to one particular form of energy. For example, pigment molecules in photoreceptor cells absorb light energy, while temperature receptors respond to infrared thermal energy (heat).

Many kinds of environmental stimuli trigger receptor cells to perform biological work. When unstimulated, the neuron maintains a resting potential; that is, a potential difference exists between the inside and the outside of the neuron. When the receptor is stimulated, the permeability of its plasma membrane is increased. If the potential difference increases, the cell becomes hyperpolarized. If it decreases, the cell becomes depolarized.

In a receptor, the state of depolarization caused by a stimulus is called the **receptor potential.** It is a graded response that spreads relatively slowly down the dendrite, fading as it goes. When the neuron becomes depolarized, the threshold level may be reached and an action potential generated. The action potential travels along the axon to the CNS. In summary,

> Stimulus (e.g., light energy) ⟶ Transduction into electrical energy ⟶ Receptor potential ⟶ Action potential

The receptor performs three important functions: (1) it detects a stimulus in the environment by absorbing energy; (2) it converts the energy of the stimulus into electrical energy; and (3) it produces a receptor potential, which may result in an action potential that transmits the information to the CNS. With minor variations, this is how all receptors operate.

A strong stimulus causes a greater depolarization of the receptor membrane and results in more action potentials than does a weak one. The receptor potential is a *graded* response. In contrast, according to the all-or-none law, the amplitude (size) of each *action potential* has no relation to the stimulus; it is characteristic of the particular neuron.

SENSATION DEPENDS ON TRANSMISSION OF A "CODED" MESSAGE

How do you know whether you are seeing a blue sky, tasting a chocolate cookie, or hearing a note played by a violinist? All action potentials are qualitatively the same. Light of the wavelength 400 nanometers (blue), sugar molecules (sweet), and sound waves of 440 hertz (A above middle C) all cause transmission of similar action potentials. Our ability to differentiate stimuli depends on both the sensory receptors and on the brain. We can discriminate blue from green, a sweet taste from a light breeze, or the sound of a piano from cold because sensory receptors are connected to specific neurons in the brain. Because a receptor normally responds to only one category of stimuli, for example, light or sound, the brain interprets a message arriving from a particular receptor as meaning that a certain type of stimulus occurred (e.g., a flash of color).

Sensation takes place in the brain. Interpretation of the message and the type of sensation depends on which association neurons receive the message. The rods and cones of the eye do not see. When they are stimulated, they send a message to the brain that interprets the signals and translates them into a rainbow, an elephant, or a child. Artificial stimulation of brain centers (e.g., electrical stimulation) results in sensation. Many sensory messages never give rise to sensations at all. For example, certain chemoreceptors sense internal changes in the body but never stir our consciousness.

When stimulated, a sensory receptor initiates what might be considered a "coded" message, composed of action potentials transmitted by nerve fibers. This coded message is later decoded in the brain. Impulses from the sensory receptor may differ in (1) the total number of fibers transmitting, (2) the specific fibers carrying action potentials, (3) the total number of action potentials passing over a given fiber, and (4) the frequency of the action potentials passing over a given fiber. For example, the difference in sound intensity between the gentle rustling of leaves and a clap of thunder depends on the number of neurons transmitting action potentials and by the frequency of action potentials transmitted by each neuron. Just how the sensory receptor initiates different codes and how the brain analyzes and interprets them to produce various sensations are not completely understood.

Receptors Adapt to Stimuli

Many sensory receptors do not continue to respond at their initial rate, even if the stimulation continues at the same intensity. With time, the frequency of action potentials in the sensory neuron decreases. This diminishing response to a continued, constant stimulus, called **sensory adaptation,** occurs when the sensory neuron becomes less responsive to stimulation. Sensory adaptation can also occur because the receptor produces a smaller receptor potential.

Some receptors, such as those for pain or cold, adapt so slowly that they continue to trigger action potentials as long as the stimulus persists. Others, such as olfactory receptors, adapt quickly so that we hardly notice odors that at first smell seemed to assault our senses. Similarly, when you first pull on a pair of tight jeans your pressure receptors let you know that you are being squished and you may feel uncomfortable. Soon, though, these receptors adapt, and you hardly notice the sensation of the tight fit. Sensory adaptation permits us to ignore persistent unpleasant or unimportant stimuli.

MECHANORECEPTORS RESPOND TO TOUCH, PRESSURE, GRAVITY, STRETCH, OR MOVEMENT

Mechanoreceptors are activated when they are mechanically pushed or pulled so that they change shape. **Lateral line organs** found in fish are mechanoreceptors thought to supplement vision. Responding to waves and currents, receptor cells inform the fish of obstacles in its way or of moving objects such as prey or enemies.

Many invertebrates have gravity receptors called **statocysts.** These receptors continually send information regarding the position and movements of the body to the CNS. When displaced, the animal can quickly adjust its body to reassume its normal position.

Touch Receptors Are Located in the Skin

The simplest mechanoreceptors are free nerve endings in the skin. These nerve endings are stimulated by objects that contact the body surface. Thousands of more specialized touch receptors are also located in the skin (Fig. 35–2; see Focus on Health and Human Affairs: Pain Per-

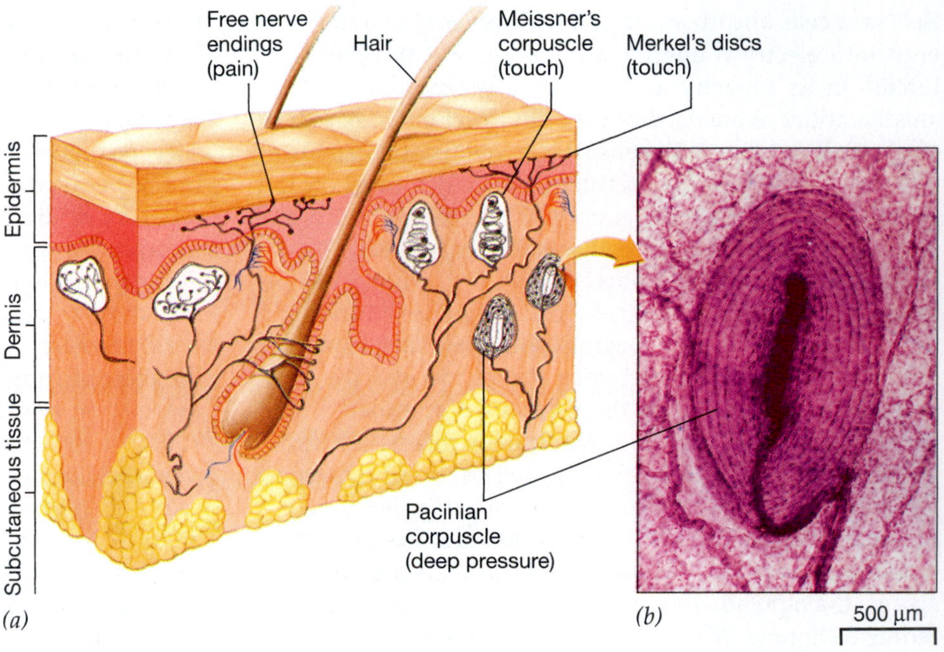

FIGURE 35–2 Sense organs within the human skin. (***a***) Diagrammatic section through the human skin showing the types of sense organs present. The free nerve endings respond to pain; tactile hairs, Merkel's disks, and Meissner's corpuscles respond to touch; pacinian corpuscles respond to deep pressure. (***b***) Pacinian corpuscle, a deep pressure receptor. (Ed Reschke)

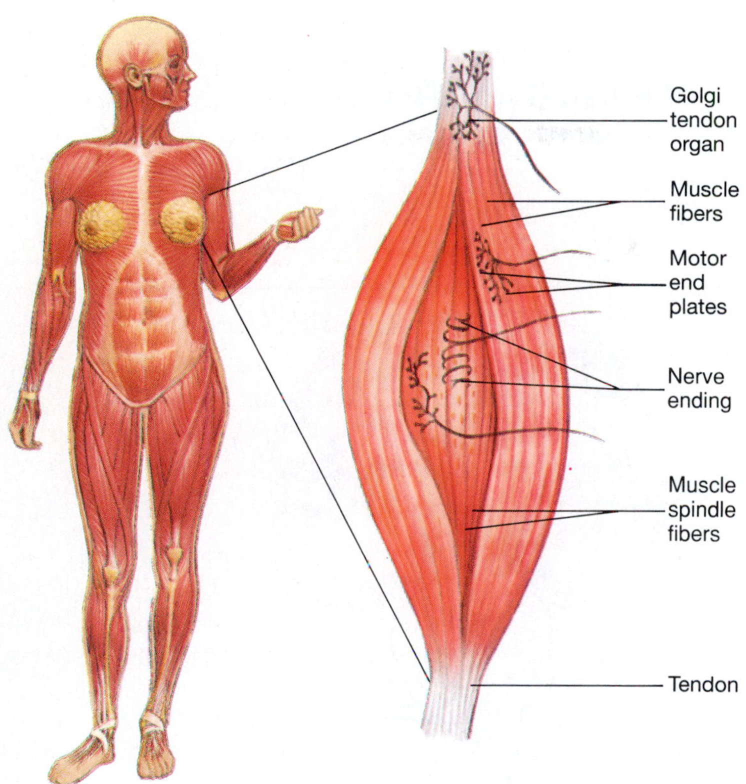

FIGURE 35–3 Proprioceptors. Muscle spindles detect muscle movement. Golgi tendon organs determine stretch in tendons.

ception). Some (Merkel's discs and Meissner's corpuscles) are sensitive to light touch. Others, like the **Pacinian corpuscles,** are especially sensitive to deep pressure applied to the body surface. In the Pacinian corpuscle, a bare nerve ending is surrounded by layers of connective tissue. Pressure on the skin causes displacement of the connective tissue, which in turn deforms the axon.

In many invertebrates as well as vertebrates, tactile (touch) receptors lie at the base of a hair or bristle. They are stimulated indirectly when the hair is bent or displaced. Tactile hairs may be involved in orientation to gravity, in orienting the body in space, and in the reception of vibrations in air and water, as well as in contacts with other objects.

Proprioceptors Help Coordinate Muscle Movement

Proprioceptors are receptors sensitive to changes in movement, tension, and position in muscles and joints. Their continuous reports to the CNS help ensure that muscle movement will be properly coordinated. Vertebrates have three main types of proprioceptors: **muscle spindles,** which detect muscle movement (Fig. 35–3); **Golgi tendon organs,** which are sensitive to stretch in the tendons that attach muscle to bone; and **joint receptors,** which detect movement in ligaments.

By means of these sense organs, we can, even with our eyes closed, perform activities such as dressing or playing the piano. Impulses from the proprioceptors are extremely important in ensuring the smooth contraction of different muscles involved in a single movement. Without such receptors, complicated, skillful acts would be impossible. Impulses from these organs are also important in maintaining balance. Proprioceptors are probably more numerous and more continuously active than are any of the other sensory receptors, although we are less aware of them than of most other sensory receptors.

The Labyrinth of the Vertebrate Ear Is an Organ of Equilibrium

When we think of the ear, we think of hearing. However, in vertebrates, the basic function of the ear is to help maintain equilibrium. Equilibrium is the state of balance that enables an animal to maintain its orientation. Typically, the ear also contains gravity receptors. Although many vertebrates do not have outer or middle ears, all have inner ears.

The **inner ear** consists of a group of interconnected canals and sacs, referred to as the **labyrinth.** In jawed vertebrates, the labyrinth consists of two saclike chambers, the **saccule** and **utricle,** three **semicircular canals,** and a snail-shaped structure known as the **cochlea.** Collectively, the saccule, utricle, and semicircular canals are referred

FOCUS ON *Health and Human Affairs*

PAIN PERCEPTION

Pain is a signal that protects us from harmful stimuli. An excess of any type of stimulus—pressure, heat, or cold—stimulates pain receptors. In the human body, pain receptors are dendrites of certain sensory neurons found in almost every tissue.

When stimulated, pain receptors send a message through sensory neurons to the spinal cord. The message is transmitted to the opposite side of the spinal cord and then sent upward to the thalamus, where pain perception begins (see figure). From the thalamus, impulses are sent into the parietal lobes of the cerebrum. At that time the individual becomes fully aware of the pain and can evaluate the situation. How threatening is the stimulus? How intense is the pain? What can be done about it? From the thalamus, messages are also sent to a region in the limbic system, which is the brain's emotional center. Pain perception is colored by emotional experience.

Pain can be initiated or facilitated at many levels. How intense one's perception of pain is depends on the particular situation and how one has learned to deal with pain. A child with a bruised knee may emotionally heighten the feeling of pain, whereas a professional fighter may virtually ignore a long series of well-delivered blows.

The brain locates pain on the basis of past experience. Generally, pain at the body surface is accurately projected back to the injured area. For example, when you step on a nail, your brain perceives the pain and then projects it back to the injured foot, so that you feel pain at the site of puncture. Artificial stimulation of the leg nerves may produce a sensation of pain in the foot even though the foot is untouched. In fact, for years after an amputation, a patient may feel **phantom pain** in the missing limb. Phantom pain occurs be-

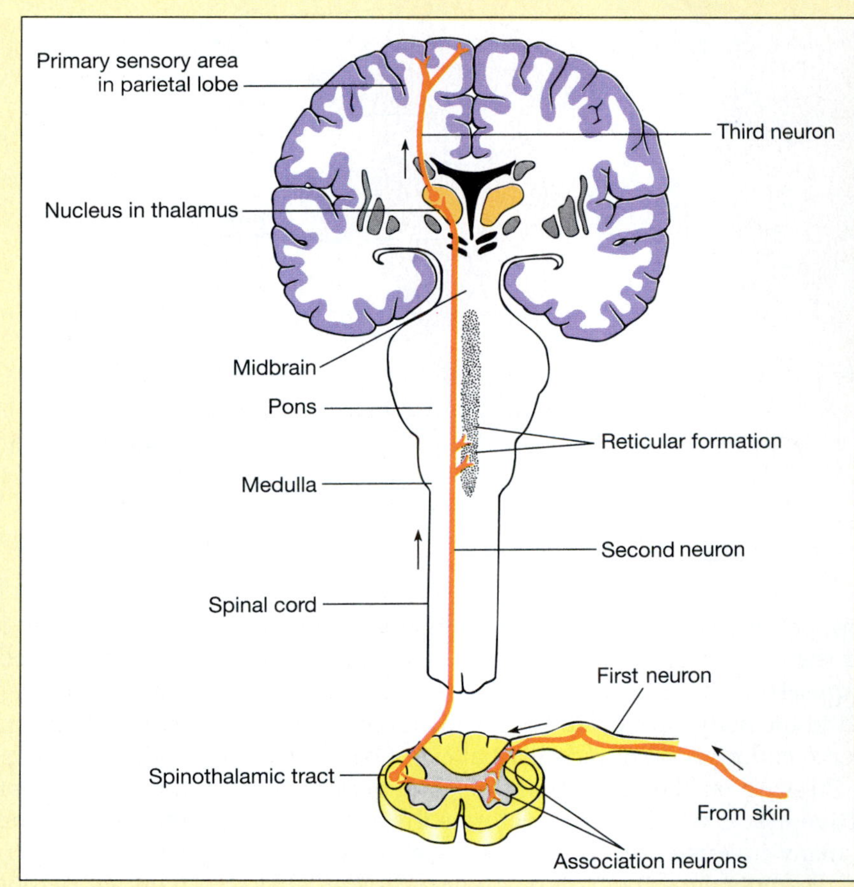

Neural pathway for pain.

to as the **vestibular apparatus** (Fig. 35–4). This structure is responsible for the sense of equilibrium.

The saccule and utricle house gravity detectors, which are small crystals of calcium carbonate called **otoliths** (Fig. 35–5). The receptors consist of groups of **hair cells** surrounded at their tips by a jelly-like mass, the **cupula.** The receptor cells in the saccule and utricle lie in different planes. Normally, the pull of gravity causes the otoliths to press against particular hair cells, stimulating them to initiate impulses. These impulses travel to the brain by way of sensory nerve fibers at their bases. When the head is tilted, or in *linear acceleration* (change in speed when the body is moving in a straight line), the otoliths press upon the hairs of other cells and stimulate them.

cause, when the severed nerve is stimulated and sends a message to the brain, the brain "remembers" the nerve as it originally was—connected to the missing limb.

Most internal organs are poorly supplied with pain receptors. For this reason pain from internal structures is often difficult to locate. In fact, pain is often not projected back to the organ that is stimulated. Instead the pain is *referred* to an area just under the skin that may be some distance from the organ involved. The area to which the pain is referred generally is connected to nerve fibers from the same level of the spinal cord as the organ involved.

A person with angina who feels heart pain in his left arm is experiencing **referred pain.** The pain originates in the heart as a result of ischemia (insufficient blood in the blood vessels of the heart muscle) but is actually felt in the arm. One explanation is that neurons from both the heart and the arm converge upon the same neurons in the central nervous system. The brain interprets the incoming message as coming from the body surface because somatic pain is far more common than pain from internal organs; the brain acts on the basis of its past experience. When pain is felt both at the site of the distress and as a referred pain, it may seem to spread, or *radiate,* from the organ to the superficial area.

The physiology of pain is not completely understood. The sensory neurons that transmit pain impulses to the spinal cord release the neurotransmitter peptide **substance P.** Interneurons that end on the pain neurons release **enkephalin,** which inhibits the release of substance P. In this way, the interneuron can modify the sensation of pain. The interneuron can be activated by neurons from the brain or by sensory neurons associated with touch or pressure. This can be demonstrated by rubbing the skin around an injury. Activation of the surrounding neurons decreases the sensation of pain. The influence of higher brain centers on pain perception is evident when in a crisis situation an individual is not even aware that he or she has been injured until after the emergency is over.

Enkephalin is one of the body's opiate-like peptides. Opiates, such as morphine, are analgesic drugs (drugs that relieve pain). They work by blocking the release of substance P. The body has its own pain control system. In addition to enkephalin released by peripheral pain neurons, the brain and pituitary gland release peptides known as **endorphins** (for "endogenous morphine-like") that are more powerful than the strong opiate morphine. Like the opiate drugs, these peptides are thought to suppress the release of substance P from pain-transmitting neurons. The neurotransmitter GABA (gamma-amino-butyric acid) is also thought to inhibit release of substance P in some areas of the brain. Endorphins are currently being investigated as potential analgesic drugs.

Some neurobiologists think that endorphins may explain how acupuncture works. For thousands of years acupuncture has been used to relieve pain, but how it works has remained a mystery. There is now evidence that acupuncture needles stimulate nerves deep within the muscles, which in turn stimulate the pituitary gland and parts of the brain to release endorphins.

Various clinical methods have been developed for relieving pain. Stimulation of the skin over the painful area with electrodes has been successful in some patients. This procedure is called transcutaneous electrical nerve stimulation. In a few patients, electrodes have been implanted in the appropriate areas of the brain so that the patient can stimulate the brain at will. This procedure is thought to relieve pain by stimulating the release of endorphins.

This enables the animal to perceive its position relative to the ground regardless of what position the head is in at the time.

Information about *angular acceleration* (turning movements) is furnished by the three semicircular canals. Each of these is connected with the utricle and lies in a plane at right angles to the other two. Each canal is a hollow ring filled with fluid, called **endolymph.** At one of the openings of each canal into the utricle is a small, bulblike enlargement, called the **ampulla.** Within each ampulla is a clump of hair cells, called a **crista,** that are similar to those in the utricle and saccule, but lacking otoliths. These

(Text continues on page 683)

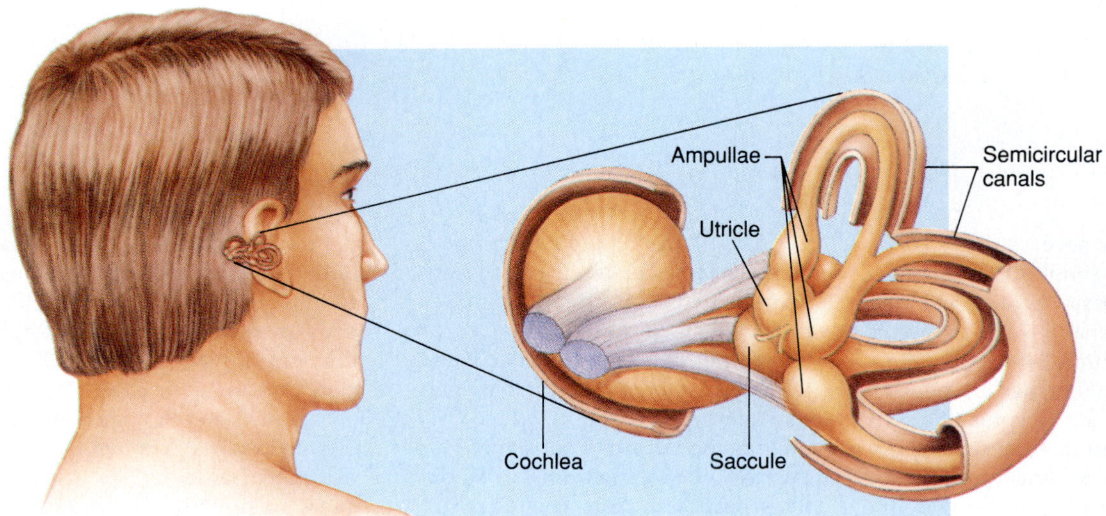

FIGURE 35–4 The human inner ear with the labyrinth exposed. Because this is a posterior view, the utricle and saccule can be seen.

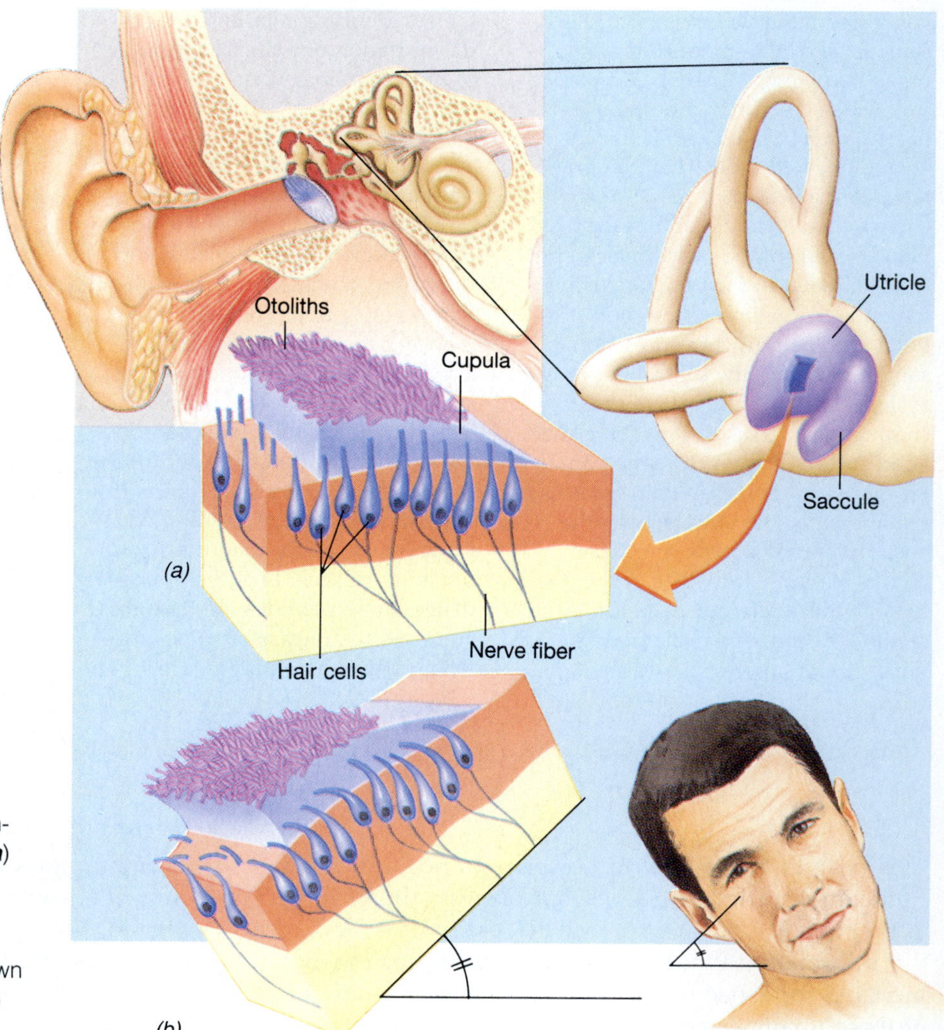

FIGURE 35–5 The saccule and utricle. Compare the positions of the otoliths and hairs in (***a***) with those in (***b***). Changes in head position cause the force of gravity to distort the cupula, which in turn distorts the hairs of the hair cells; the hair cells respond by sending impulses down the vestibular nerve (part of the auditory nerve) to the brain.

receptor cells are stimulated by movements of the endolymph in the canals (Fig. 35–6). Because the three canals are located in three different planes, a movement of the head in any direction stimulates movement of the fluid in at least one of the canals. Different sets of hair cells are stimulated.

We humans are used to movements in the horizontal plane but not to movements such as the motion of an elevator or of a ship pitching in a rough sea. These motions stimulate the semicircular canals in an unusual way and may cause sea sickness or motion sickness. When a person who is motion sick lies down, the movement stimulates the semicircular canals in a more familiar way, and nausea is less likely to occur.

Auditory Receptors Are Located in the Cochlea

Many arthropods and most vertebrates have sound receptors, but for many of them, hearing does not seem to be an important sense. Hearing is important for tetrapods, however, and both birds and mammals have a highly developed sense of hearing. Their auditory receptors, located in the cochlea of the inner ear, are mechanoreceptor hair cells that detect pressure waves.

The cochlea is a spiral tube that resembles a snail's shell (Fig. 35–7). If we uncoiled the cochlea, we would see that it consists of three canals separated from one another by thin membranes. The canals come almost to a point at the apex. Two of these canals, the **vestibular canal** (also known as the scala vestibuli) and the **tympanic canal**

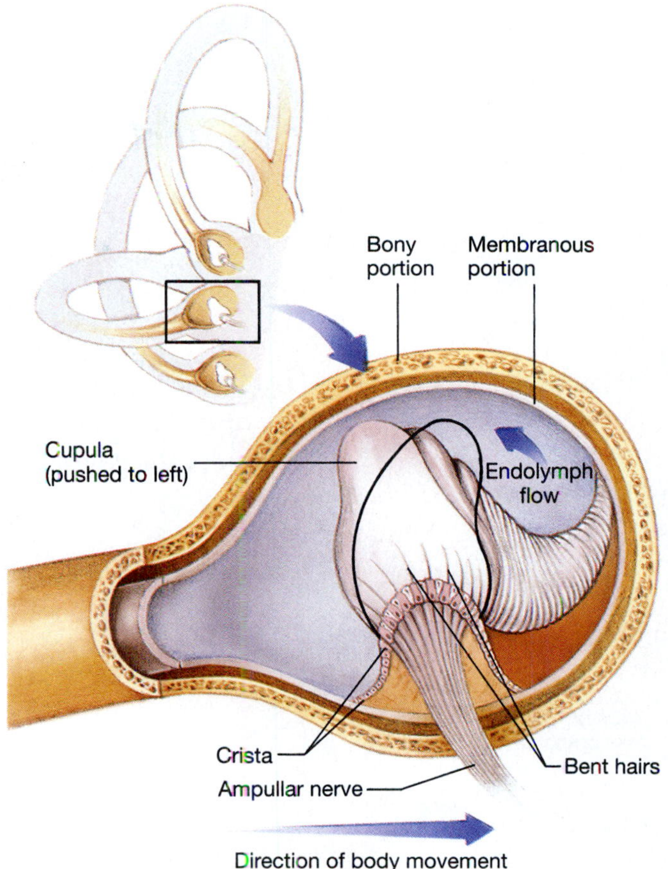

FIGURE 35–6 Movement of the endolymph within the ampulla distorts the cupula. The hair cells of the cupula then are bent. Changes are reported to the brain via the vestibular nerve.

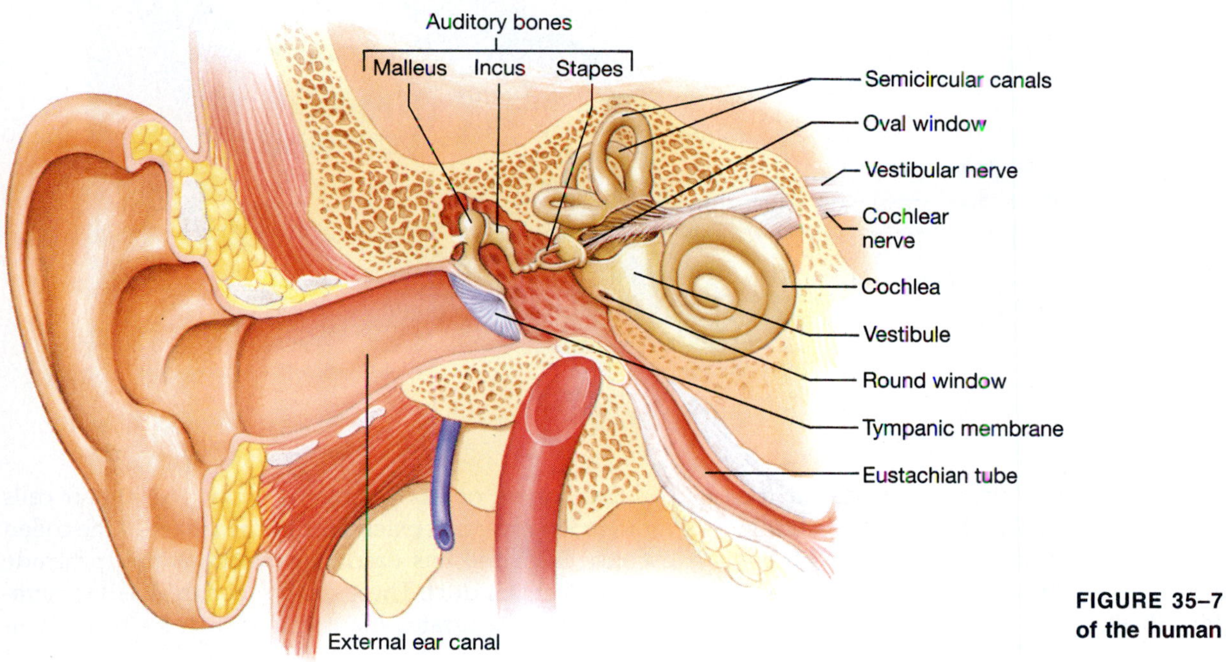

FIGURE 35–7 The anatomy of the human ear.

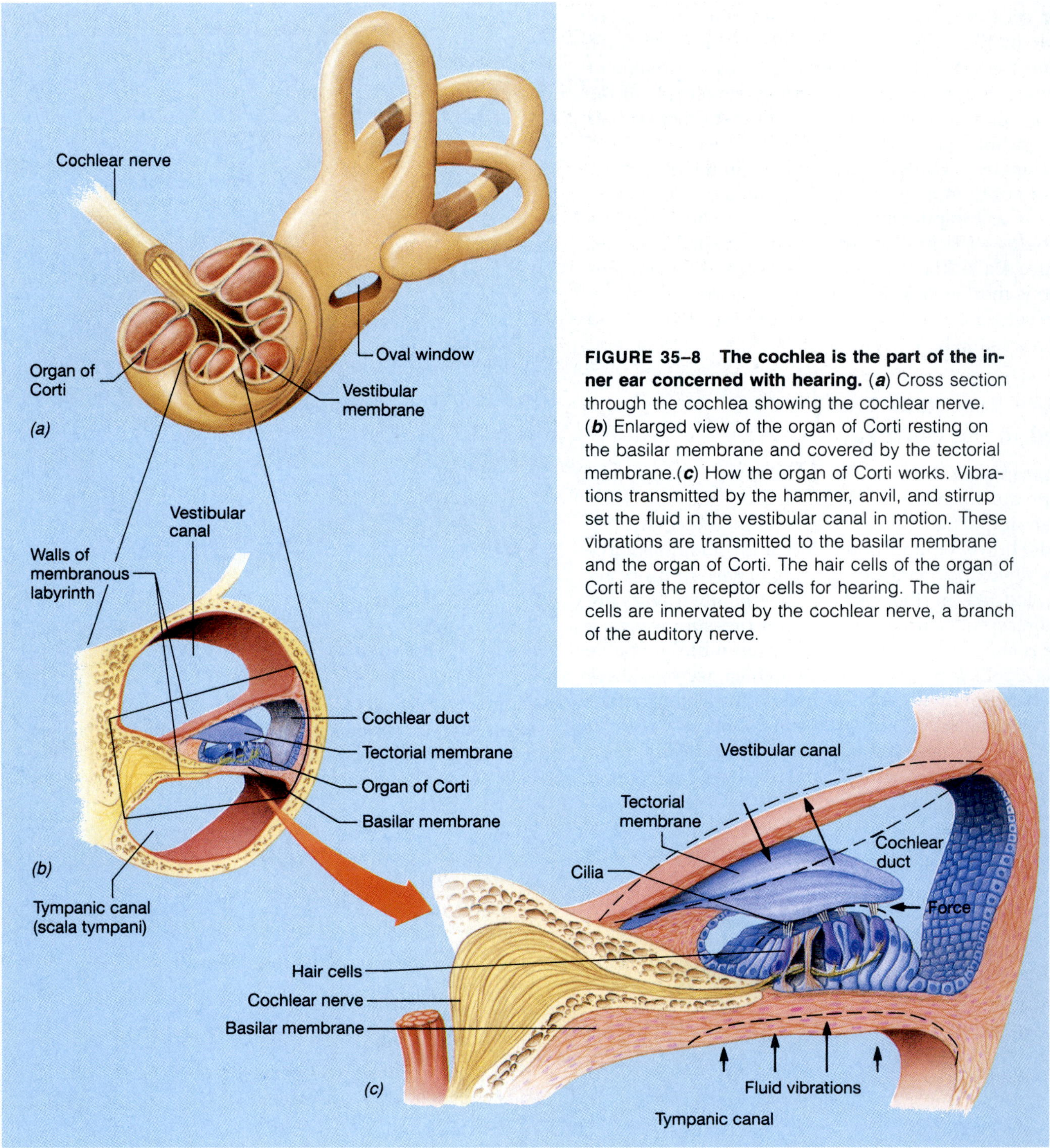

FIGURE 35–8 The cochlea is the part of the inner ear concerned with hearing. (***a***) Cross section through the cochlea showing the cochlear nerve. (***b***) Enlarged view of the organ of Corti resting on the basilar membrane and covered by the tectorial membrane.(***c***) How the organ of Corti works. Vibrations transmitted by the hammer, anvil, and stirrup set the fluid in the vestibular canal in motion. These vibrations are transmitted to the basilar membrane and the organ of Corti. The hair cells of the organ of Corti are the receptor cells for hearing. The hair cells are innervated by the cochlear nerve, a branch of the auditory nerve.

(scala tympani), are connected with one another at the apex of the cochlea and are filled with fluid. The middle canal, the **cochlear duct** (scala media), is filled with endolymph and contains the actual auditory receptor, the **organ of Corti** (Fig. 35–8).

Each organ of Corti contains about 16,000 hair cells arranged in rows that extend the entire length of the coiled cochlea. Each cell has hairlike projections that protrude into the cochlear duct. These cells rest on the **basilar membrane,** which separates the cochlear duct from the tympanic canal. Overhanging the hair cells is the **tectorial membrane,** which is attached along one edge to the membrane on which the hair cells rest, while its other edge remains free. When stimulated by changes in pressure transmitted through the endolymph, the hair cells initiate impulses in the fibers of the cochlear (auditory) nerve.

In terrestrial vertebrates, accessory structures in the outer and middle ear change sound waves in air to pressure waves in the cochlear fluid. In the human ear, for example, sound waves pass through the **external auditory canal** and set the **eardrum** (the tympanic membrane separating outer ear and middle ear) vibrating. These vibrations are transmitted across the middle ear by three tiny bones—the **hammer** (malleus), **anvil** (incus), and **stirrup** (stapes), named for their shapes. The middle ear bones amplify the vibrations, which then pass through the **oval window,** an elastic membrane that covers the passageway from the middle to the inner ear. The vibrations pass to the fluid in the vestibular canal.

Because liquids cannot be compressed, the oval window could not cause movement of the fluid in the vestibular canal unless there were an escape valve for the pressure. This is provided by the **round window** at the end of the tympanic canal. The pressure wave presses on the membranes separating the three ducts and is transmitted to the tympanic canal, where it causes the round window to bulge. The movements of the basilar membrane produced by these pulsations apparently rub the hair cells of the organs of Corti against the overlying tectorial membrane. Stimulation of the hair cells initiates nerve impulses in the cochlear nerve.

We can summarize the sequence of events involved in hearing as follows:

> Sound waves enter external auditory canal ⟶ Tympanic membrane vibrates ⟶ Middle ear bones vibrate ⟶ Intensity of the vibrations is amplified ⟶ Oval window vibrates ⟶ Vibrations are conducted through fluid ⟶ Vibration of basilar membrane ⟶ Hair cells in the organ of Corti in the cochlea are stimulated ⟶ Cochlear nerve transmits impulses to brain

Intense sound can injure the organ of Corti. Members of rock bands and workers subjected to loud, high-pitched noises over a period of years frequently become deaf to high tones because the cells near the base of the organ of Corti become damaged.

CHEMORECEPTORS DETECT TASTE AND SMELL

Two highly sensitive types of chemoreceptors are those for the senses of **taste** and **smell** (olfaction).

Taste Buds Detect Dissolved Food Molecules

The organs of taste in humans and other mammals are the **taste buds,** located in the mouth, mainly on the tongue. In humans, they are found mainly in tiny elevations, or papillae (Fig. 35–9). Each of the 3000 or so taste buds on the human tongue is an oval epithelial capsule,

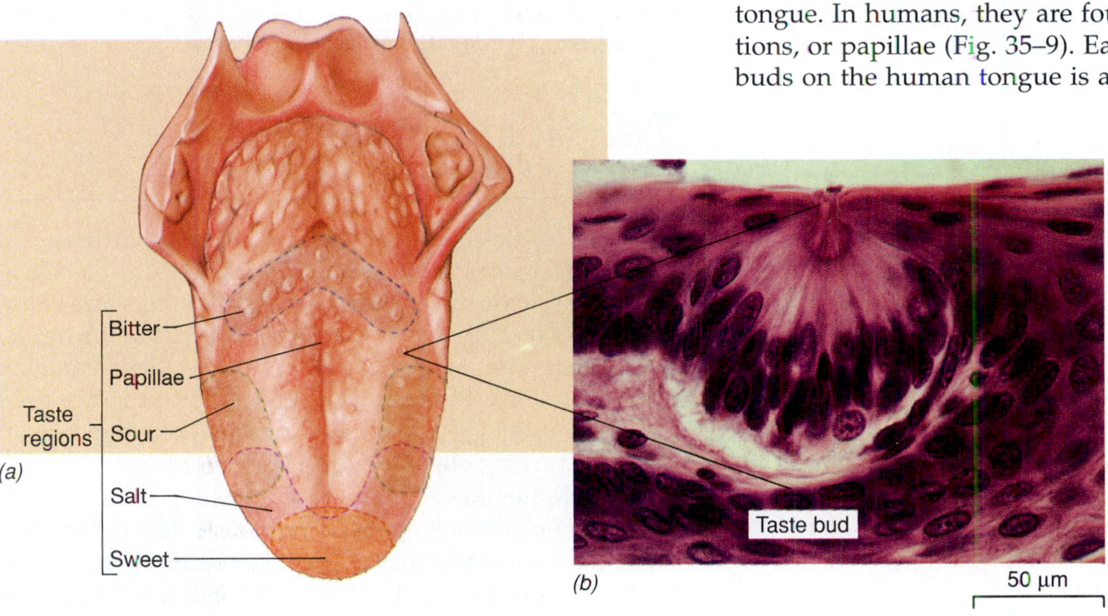

FIGURE 35–9 Taste buds are located mainly on the surface of the tongue. (***a***) The surface of the tongue, showing distribution of taste buds sensitive to sweet, bitter, sour, and salt. A single taste receptor may respond to more than one category of taste. (***b***) A taste bud consists of an epithelial capsule containing several taste receptors. (Ed Reschke)

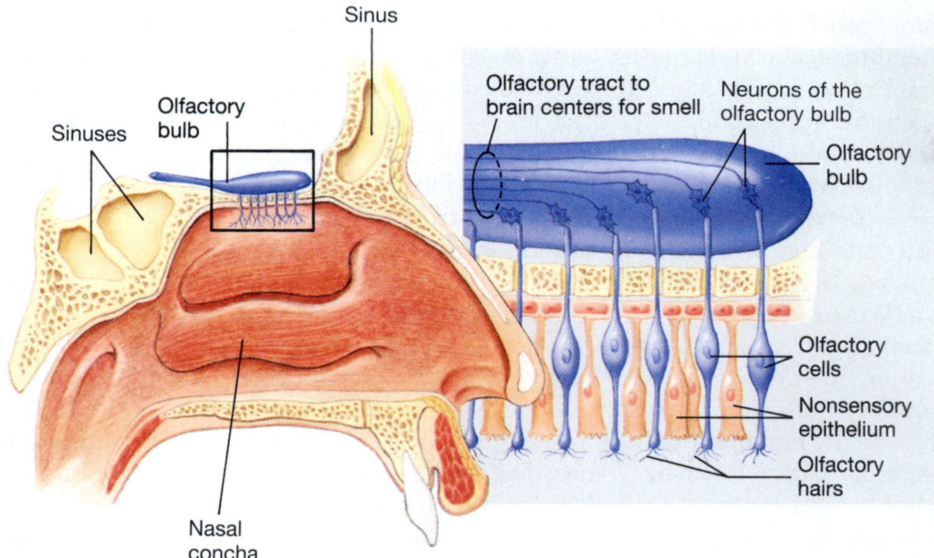

FIGURE 35–10 Location and structure of the olfactory epithelium. Note that the receptor cells are located in the epithelium itself.

containing several taste receptors. The tips of the taste receptor cells extend into a pore on the surface of the tongue. The taste receptors detect food molecules dissolved in saliva, setting up receptor potentials that fire associated neurons.

Traditionally, four basic tastes are recognized: sweet, sour, salty, and bitter. Flavor depends on these four basic tastes in combination with smell, texture, and temperature. When you have a cold and your nose is stopped up, food seems to have little "taste."

The Olfactory Epithelium Is Responsible for the Sense of Smell

Many vertebrates have an acute sense of smell. A polar bear can smell a dead whale 20 miles away. In terrestrial vertebrates, olfaction occurs in the nasal epithelium. In humans, the **olfactory epithelium** is located in the roof of the nasal cavity (Fig. 35–10). This epithelium contains several million smell-sensing neurons with axons that extend upward as the fibers of the olfactory nerves. The end of each olfactory cell bears several olfactory hairs that react to certain molecules (odors) in the air.

CONCEPT CONNECTIONS

Chemoreception ***Animal Behavior***

Many species use chemoreception to communicate information such as danger, ownership of territory, and availability for mating. The chemical messengers are **pheromones,** small volatile molecules that are dispersed into the environment. For example, an ovulating female (one physiologically ready to mate) may release pheromones as part of her vaginal secretion. These chemical odors are smelled by males, and information is transmitted from olfactory neurons to the brain. The effect may be increased sexual interest. Whether or not humans rely on pheromones is not known, but the importance of chemoreception in communication is evident by the use of colognes and perfumes in securing the attention of the opposite sex.

THERMORECEPTORS ARE SENSITIVE TO HEAT

Mosquitoes, ticks, and other blood-sucking arthropods use heat receptors to locate endothermic (warm-blooded) animals. At least two types of snakes, pit vipers and boas, use thermoreceptors to locate their prey. Pits in the heads of pit vipers can detect heat generated by a small animal more than a meter away (Fig. 35–11). Vision is not necessary to locate prey. A blinded rattlesnake will strike at a warm, moving object—even a lightbulb.

Vertebrates have thermoreceptors in their skin and within some internal organs. In mammals, free nerve endings in the skin and tongue detect cold. Certain receptors that consist of encapsulated dendrite ends are thought to be heat receptors. Thermoreceptors in the hypothalamus detect internal changes in temperature and receive and integrate information from thermoreceptors on the body surface. The hypothalamus regulates homeostatic mechanisms that maintain a constant body temperature.

FIGURE 35–11 The pit organ of a yellow eyelash viper. The pit organ is located between each eye and nostril. This sense organ can detect the heat from a warmblooded animal up to a distance of 1 to 2 meters. (G. Dimijian/Photo Researchers, Inc.)

Electroreceptors Detect Electrical Currents in Water

Sharks, rays, and some predatory bony fish have electroreceptors distributed over their bodies (often associated with the lateral line system). These electroreceptors detect small electric currents produced by the heart and other organs of their prey. The electroreceptors of many fish can detect the Earth's magnetic field, an ability that may be important in migration.

Some fish produce electrical fields to locate objects in the water. This is particularly useful in murky water where visibility and olfaction are poor. In some species, electric currents are used for communications among members of a species. The electric eel and electric ray have electrical organs that can deliver a powerful electric discharge that stuns prey or enemies.

Photoreceptors Have Pigments That Absorb Light

Most animals have photoreceptors that use pigments to absorb light energy. **Rhodopsins** are the photosensitive pigments found in the eyes of cephalopod mollusks, arthropods, and vertebrates. Light energy striking a light-sensitive receptor cell containing this pigment triggers chemical changes in the pigment molecules. As a result, the receptor cell may transmit a nerve impulse.

Eyespots, Simple Eyes, and Compound Eyes Are Found Among Invertebrates

All organisms respond to light. The simplest light-sensitive *organs* in animals are found in certain cnidarians and in flatworms (Fig. 35–12). Their **eyespots,** called **ocelli,** detect light but do not see objects. Eyespots are often bowl-shaped clusters of light-sensitive cells within the epidermis. They may detect the direction of the source of light and distinguish light intensity.

Effective image formation requires a more complex **eye,** usually with a lens. A lens is a structure that concentrates light on a group of photoreceptors. Vision also requires a brain that can interpret the action potentials generated by the photoreceptors. The brain must integrate information about movement, brightness, location, position, and shape of the visual stimulus.

Two fundamentally different types of eyes evolved: the camera eye of some mollusks (squids and octopods)

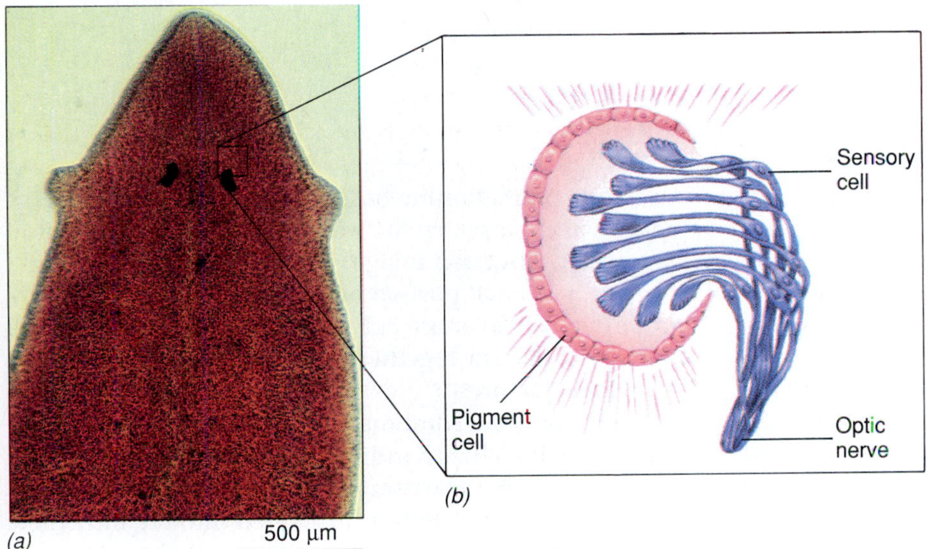

FIGURE 35–12 Simple invertebrate eye. (***a***) Planarian worm *(Planaria agilis)* showing eyespots. (***b***) Structure of the eyespot of a planarian worm. (Terry Ashley/Tom Stack & Associates)

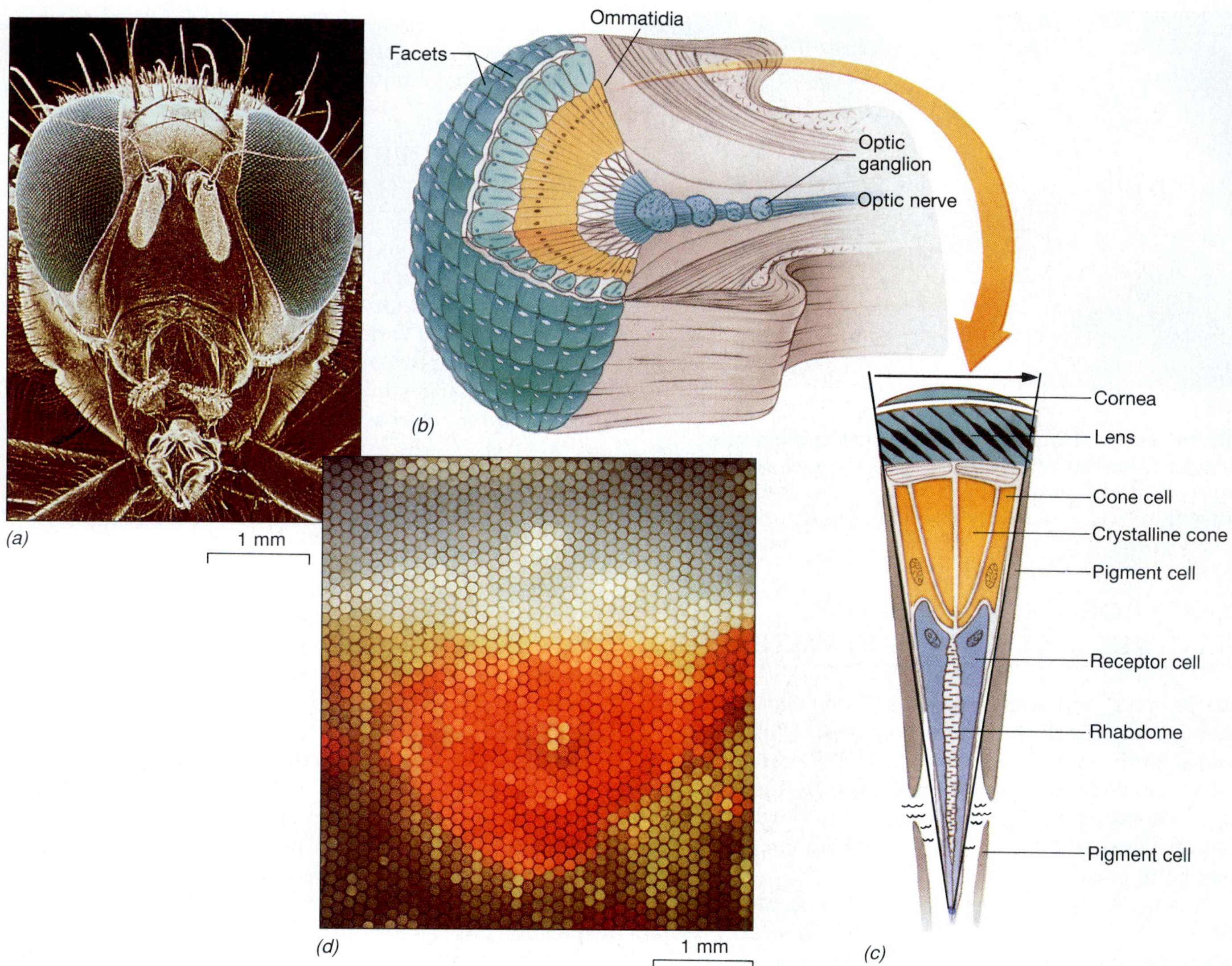

FIGURE 35–13 Compound eye. (***a***) The Mediterranean fruit fly *(Ceratitis capitata).* (***b***) Structure of the compound eye showing several ommatidia. This type of eye registers changes in light and shade so that the animal can detect movement. (***c***) Structure of an ommatidium. The rhabdome is the light-sensitive core of the ommatidium. (***d***) A bee's eye view of poppies. (*a,* David Scharf/Peter Arnold, Inc.; *d,* John Lithgoe/SeaPhot, Ltd.)

and vertebrates, and the compound eye of the arthropods. (Vertebrate and mollusk eyes are analogous structures. They evolved independently of one another.)

Compound eyes are found in crustaceans and insects. Not only do these eyes look different from vertebrate eyes, but they also see differently. The surface of a compound eye appears faceted, which means having many faces, like a diamond (Fig. 35–13). Each facet is the convex cornea of one of its visual units, called an **ommatidium.** The number of ommatidia varies in different species, from just a few in the eye of certain crustaceans to as many as 28,000 in the eye of a dragonfly.

Each ommatidium has a transparent covering, the **cornea,** and a lens that focuses light onto a receptor, forming a small inverted image. Each receptor receives light from only a small portion of the visual field. In this way, it samples the average light intensity from that area. All of the ommatidia together produce a composite image, or **mosaic** picture.

Although the compound eye forms relatively coarse images, a single ommatidium can detect any movement of prey or enemy. The compound eye is sensitive to wavelengths of light ranging from the red into the ultraviolet (UV). Because an insect can see UV radiation, its world

(a)

(b)

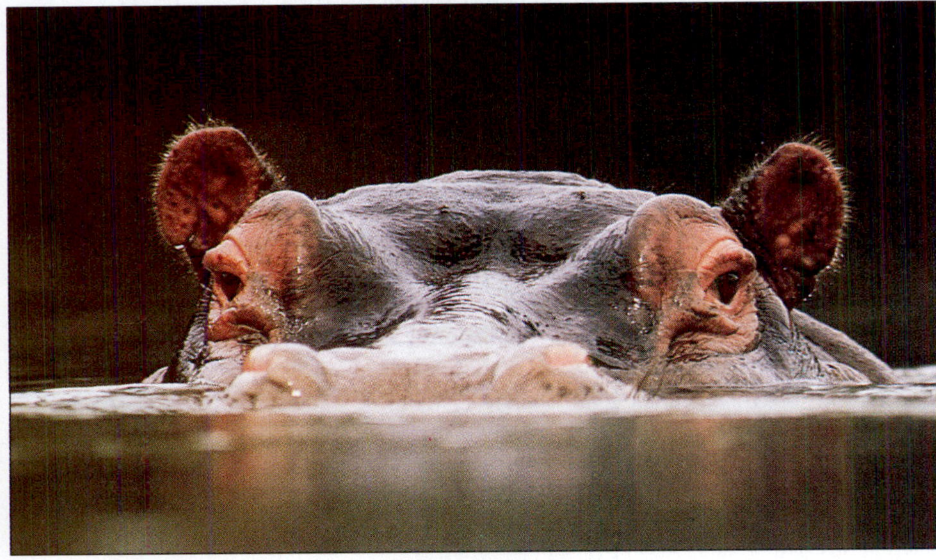

(c)

FIGURE 35–14 The position of the eyes varies in different vertebrates, resulting in differences in vision. (***a***) The eyes of the zebra are positioned laterally, enabling the animal to see on both sides. Even while grazing, it can spot a predator approaching from behind. (***b***) Like many other nocturnal animals, the owl monkey *(Aotus evingatus)* has large eyes. Its eyes are positioned at the front of the head, and it has binocular vision, permitting it to judge distances (***c***) The orbits (bony cavities that contain the eyeballs) of the hippopotamus are elevated, enabling the animal to see even when most of its head is under water. (*a*, Diane Blell/Peter Arnold, Inc.; *b*, Stephen Dalton ©1993 Animals Animals; *c*, Frans Lanting/Minden Pictures)

of color is much different from ours. Since different flowers reflect UV to different degrees, two flowers that appear identically colored to us may appear strikingly different to insects (see Figure 29–11 on p. 565).

Vertebrate Eyes Form Sharp Images

The position of the eyes in the head of humans and certain other higher vertebrates permits both eyes to be focused on the same object (Fig. 35–14). This **binocular vision** is an important factor in judging distance and depth.

The vertebrate eye can be compared to a camera. An adjustable lens can be focused for different distances and a diaphragm, called the **iris,** regulates the size of the light opening, called the **pupil** (Fig. 35–15). The iris is a ring of smooth muscle that appears as blue, green, or brown, depending on the amount and nature of pigment present. The **retina** corresponds to the light-sensitive film used in a camera. Next to the retina is the **choroid layer,** a sheet of cells filled with black pigment that absorbs extra light and prevents internally reflected light from blurring the image. (Cameras are also black on the inside.)

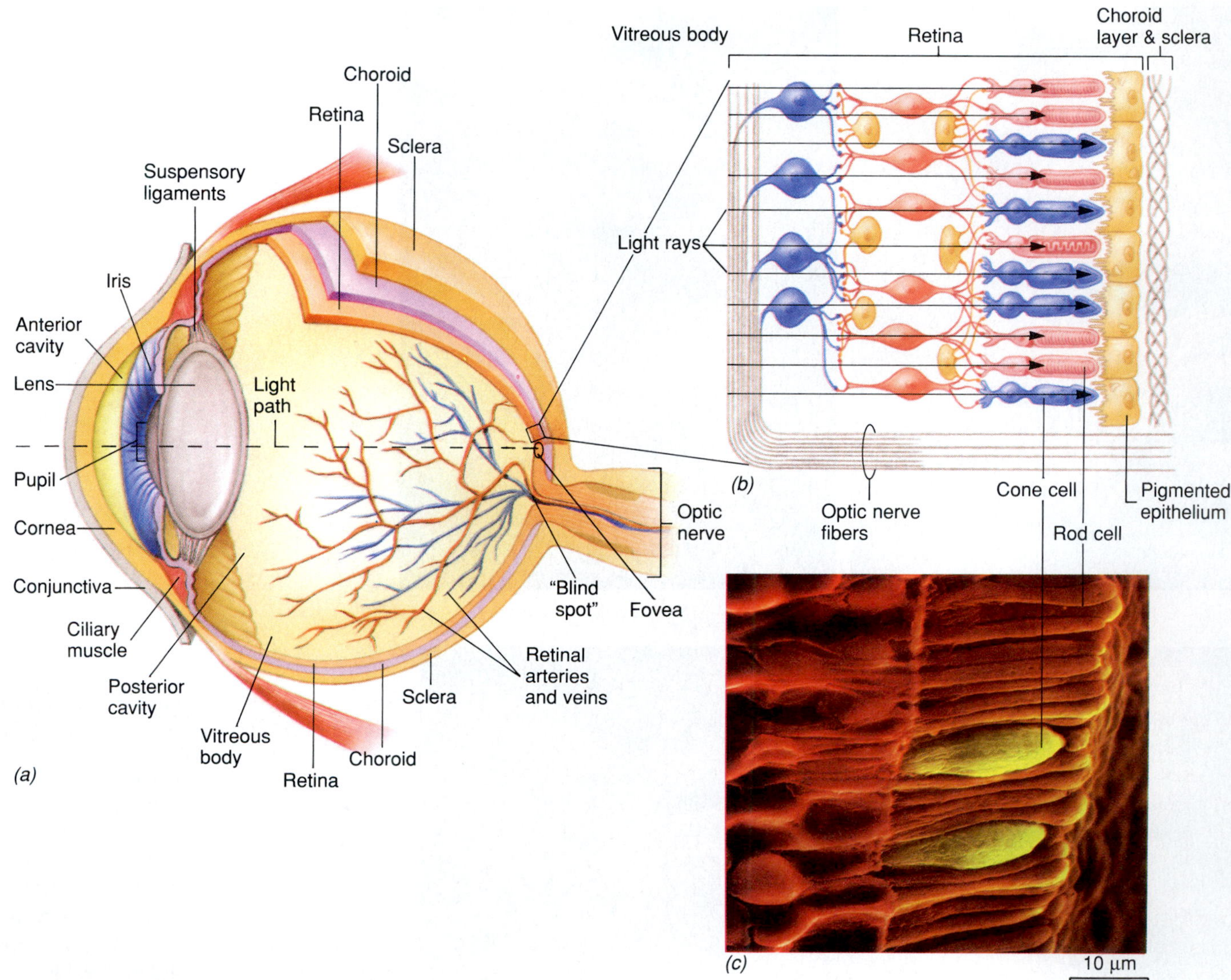

FIGURE 35–15 The human eye. (***a***) Structure of the eye. The retina contains the photoreceptor cells, the rods and cones. (***b***) Neuronal connections in the retina. The elaborate interconnections among the various layers of cells allow them to interact and to influence one another in a number of ways. (***c***) Rods (red elongated structures) and two cones (shorter, thicker, yellow structures) are seen on the right side of this micrograph (magnification × 10,000). The elongated rods permit us to see shape and movement, whereas the shorter cones allow us to view our world in color. (Lennart Nilsson, from *The Incredible Machine,* p. 279)

The outer coat of the eyeball, called the **sclera,** is a tough, opaque, curved sheet of connective tissue that protects the inner structures and helps to maintain the rigidity of the eyeball. On the front surface of the eye this sheet becomes the thinner, transparent **cornea,** through which light enters. The cornea serves as a fixed lens.

The **lens** of the eye is a transparent, elastic ball located just behind the iris. It bends the light rays coming in and brings them to a focus on the retina. The lens is aided by the curved surface of the cornea and by the refractive capabilities (ability to bend light rays) of the liquids inside the eyeball. The gradual loss of transparency of the lens is the condition known as a cataract. The cavity between the cornea and the lens is filled with a watery substance, the **aqueous fluid.** The larger chamber between the lens and the retina is filled with a more viscous

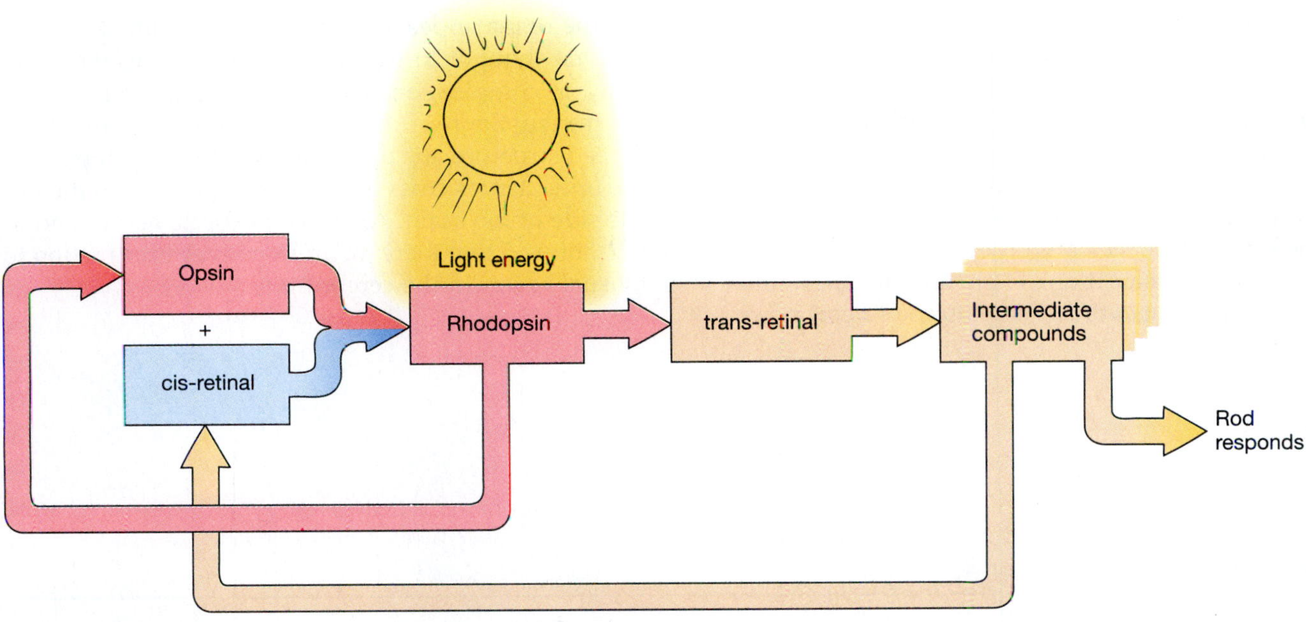

FIGURE 35–16 The visual cycle. When light strikes rhodopsin, it breaks down, depolarizing the rod cell which contains it. This produces an impulse. See text for further explanation.

fluid, the **vitreous body.** Both fluids are important in maintaining the shape of the eyeball by providing an internal, fluid pressure.

The eye has the power of **accommodation,** meaning it can change focus for near or far vision by changing the curvature of the lens. This is made possible by the stretching and relaxing of the lens by the **ciliary muscle.**

In nearsightedness (myopia), the eyeball is elongated. The light rays converge at a point in front of the retina and are diverging again when they reach it. This results in a blurred image. Concave lenses correct for the nearsighted condition by bringing the light rays to a focus at a point farther back. In a farsighted eye, the eyeball is too short and the retina too close to the lens. Light rays strike the retina before they have converged, again resulting in a blurred image. Convex lenses correct for the farsighted condition by causing the light rays to converge farther forward.

The Retina Is the Light-Sensitive Part of the Eye

The light-sensitive retina lines the posterior two-thirds of the eyeball covering the choroid. The retina contains photoreceptor cells, the **rods** and **cones** (Fig. 35–15). The 125 million or so rods function in dim light, allowing us to detect shape and movement. Rods are not sensitive to colors. About 6.5 million cones are responsible for bright-light vision, for the perception of fine detail, and for color vision. The cones permit color vision by being sensitive to different wavelengths (colors) of light. The **fovea,** which is a small, depressed area in the center of the retina, contains a large number of cones and is the region of sharpest vision.

Light must pass through several layers of connecting neurons in the retina to reach the rods and cones (Fig. 35–15). The axons of the sensory neurons extend across the surface of the retina and unite to form the **optic nerve,** which then passes out of the eyeball. This area is called the "blind spot" because it has no rods and cones, and images falling on it cannot be perceived.

In summary, vision involves the following sequence of events:

Light passes through cornea ⟶ Through aqueous humor ⟶ Lens ⟶ Through vitreous body ⟶ Image formed on retina ⟶ Optic nerve transmits nerve impulses to brain

Visual Pigments Absorb Light

Rhodopsin, found in the rod cells, and some very closely related pigments in the cone cells are responsible for the ability to see. Rhodopsin consists of opsin, a polypeptide, that is chemically joined with **retinal,** which is a pigment made from vitamin A. Two isomers of retinal exist: the *cis* form and the *trans* form.

When light strikes rhodopsin, it transforms ***cis*-retinal** to ***trans*-retinal** (Fig. 35–16). This change in shape causes rhodopsin to break down into its components, opsin and retinal. During this process, rhodopsin is converted into a series of intermediate compounds. These reactions result in a change in voltage across the plasma membrane of the rod or cone cell. The photoreceptors release neurotransmitters that can trigger neighboring cells. Neural signals are transmitted through various types of neurons in the retina, eventually resulting in transmission of impulses to the brain by the optic nerves.

Just how visual images are processed is not certain. The size, intensity, and location of light stimuli determine initial processing in the retina. The pattern of neuron firing in the retina appears to be very important. The optic nerves are thought to transmit information to the brain by way of complex, encoded signals. The optic nerves cross in the floor of the hypothalamus, forming an X-shaped structure, the optic chiasma. Many of the axons of the optic nerves cross over and then extend to the opposite side of the brain. Axons of the optic nerves end in the thalamus. From there, neurons convey information to the visual cortex in the cerebrum.

Chapter Summary

I. Sensory receptors are specialized to respond to specific energy stimuli in the environment.
   A. Sensory receptors may be neuron endings or specialized cells in close contact with neurons.
   B. Sense organs are composed of receptor cells and accessory cells.

II. Sensory receptors can be classified as mechanoreceptors, chemoreceptors, photoreceptors, thermoreceptors, and electroreceptors.

III. Exteroceptors are sense organs that receive information from the outside world. Proprioceptors are sense organs within muscles, tendons, and joints that enable the animal to perceive orientation of the body and position of its parts. Interoceptors are sense organs within internal body organs.

IV. Receptor cells absorb energy, transduce that energy into electrical energy, and produce receptor potentials.

V. Impulses from a sense organ may differ in the number of neurons transmitting, which particular neurons are firing, the total number of action potentials transmitted by a given neuron, and the frequency of the action potentials transmitted by a given neuron.

VI. Adaptation of a receptor to a continuous stimulus results in diminished perception.

VII. Mechanoreceptors respond to touch, pressure, gravity, stretch, or movement.
   A. Tactile receptors in the skin are mechanoreceptors that respond to mechanical displacement of hairs or of the receptor cells themselves.
   B. Muscle spindles, Golgi tendon organs, and joint receptors are proprioceptors that respond continuously to tension and movement in muscles and joints.
   C. The saccule and utricle of the vertebrate ear contain otoliths that change position when the head is tilted or when the body is moving forward. Hair cells stimulated by the otoliths send impulses to the brain, enabling the animal to perceive the direction of gravity.
   D. The semicircular canals of the vertebrate ear inform the brain about turning movements. Their hair cells are stimulated by movements of the endolymph.
   E. The organ of Corti within the cochlea is the auditory receptor in birds and mammals.
      1. Sound waves pass through the external auditory canal, cause the eardrum to vibrate, and are transmitted through the middle ear by the hammer, anvil, and stirrup.
      2. Vibrations pass through the oval window to fluid within the vestibular canal. Pressure waves press on the membranes separating the three ducts of the cochlea.
      3. Movements of the basilar membrane rub the hair cells of the organ of Corti against the overlying tectorial membrane, thus stimulating them.
      4. Nerve impulses are initiated in the auditory neurons lying at the base of each hair cell.

VIII. Chemoreceptors detect specific chemical compounds, permitting taste and smell.
   A. Taste receptors are specialized epithelial cells located in taste buds.
   B. The olfactory epithelium contains specialized olfactory cells with axons that extend upward as fibers of the olfactory nerves.

IX. Thermoreceptors provide endothermic animals with information about body temperature. In some animals they are used to locate endothermic prey.

X. Electroreceptors detect electrical currents in water.

XI. Photoreceptors use pigments to absorb light energy.
   A. Ocelli (eyespots) detect light but do not form images.
   B. The compound eye of arthropods consists of ommatidia, which collectively form a mosaic image.
   C. In the vertebrate eye, light enters through the cornea, is focused by the lens, and is sensed as an image by the retina. The iris regulates the amount of light that can enter.

D. When light strikes rhodopsin in the rod cells, a chemical change in retinal occurs that breaks down the rhodopsin, triggering depolarization of the rod cell.

E. Rods form images in black and white, whereas cones are responsible for color vision.

Selected Key Terms

chemoreceptors, p. 677
cochlea, p. 679
compound eye, p. 688
cones, p. 691
cornea, p. 688
electroreceptor, p. 677
eyespot, p. 687
fovea, p. 691
hair cell, p. 680
interoceptors, p. 677
iris, p. 689
labyrinth, p. 679
lens, p. 690
mechanoreceptor, p. 676
muscle spindle, p. 679
olfactory epithelium, p. 686
organ of Corti, p. 684
photoreceptor, p. 677
proprioceptors, p. 677
retina, p. 689
rhodopsin, p. 687
rods, p. 691
semicircular canals, p. 679
sensory adaptation, p. 678
taste buds, p. 685
thermoreceptor, p. 677
vestibular apparatus, p. 680

Post-Test

1. ______________ ______________ are structures specialized to respond to changes in the environment.
2. ______________ enable an animal to perceive position of the body and orientation.
3. ______________ detect light energy; ______________ respond to touch, gravity, or movement.
4. Receptor cells absorb ______________, transduce it into ______________ energy, and produce a/an ______________ ______________.
5. The diminishing response of a receptor to a continued, constant stimulus is called sensory ______________.
6. Statocysts serve as ______________ receptors.
7. Three main types of vertebrate proprioceptors are muscle ______________, which detect muscle movement; ______________ ______________ organs, which determine stretch in tendons; and ______________ receptors, which detect movement in ligaments.
8. The basic function of the vertebrate ear is to help maintain ______________.
9. The inner ear consists of interconnected canals and sacs called the ______________; in jawed vertebrates this structure consists of two saclike chambers, the ______________ and ______________, and three ______________ canals, as well as the cochlea.
10. The rocks in your head (within the saccule and utricle), called ______________, are actually ______________ detectors.
11. Each semicircular canal is filled with fluid, called ______________; at one of the openings of each canal into the utricle is a small enlargement, the ______________.
12. The cochlea, located in the ______________ ear, contains mechanoreceptors (hair cells) that detect ______________ waves.
13. The actual auditory receptor is the organ of ______________; it is located within the ______________ duct.
14. The senses of taste and smell depend upon ______________.
15. The photosensitive pigments in the eyes of vertebrates and arthropods are ______________.
16. The light-sensitive part of the human eye is the (a) ommatidium (b) cochlea (c) retina (d) iris (e) ampulla.
17. Which structure regulates the size of the pupil? (a) ommatidium (b) cone (c) retina (d) iris (e) ampulla
18. Structures specialized to perceive color are (a) rods (b) lens (c) retinas (d) iris (e) cones.
19. The visual unit in the compound eye is the (a) ommatidium (b) lens (c) retina (d) iris (e) ampulla.

20. A structure responsible for the sense of equilibrium is the (a) vestibular apparatus (b) cochlea (c) organ of Corti (d) Golgi tendon organ (e) basilar membrane.

21. Label the diagrams below. (Refer to Figs. 35–7 and 35–15 as necessary.)

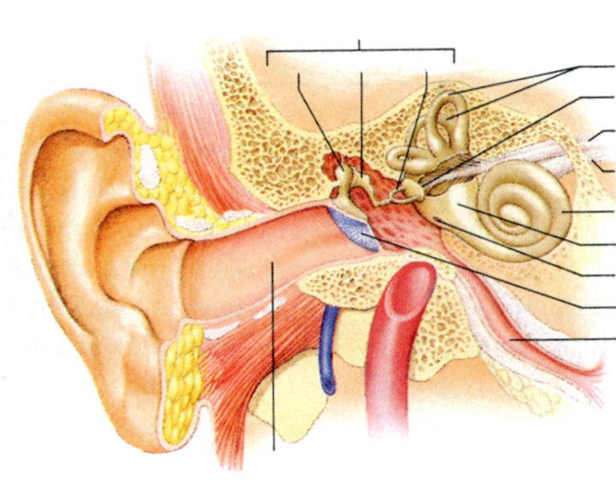

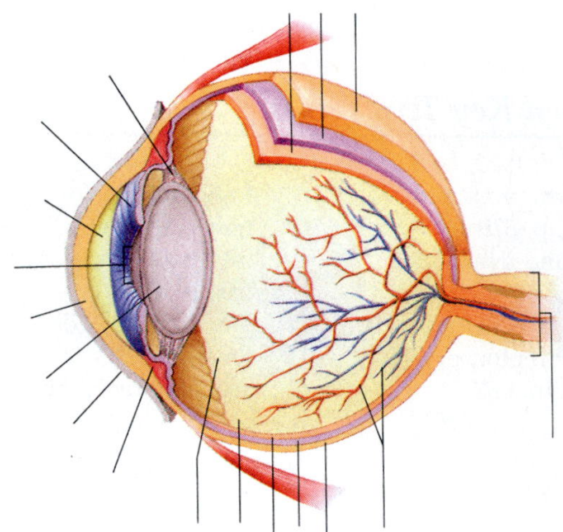

Review Questions

1. Imagine that as you walk into a room you are met by an offensive odor. After a few minutes you hardly notice the smell. Explain.
2. What is a proprioceptor? What is its function in the mammalian body?
3. Discuss the mechanism by which the sensory cells of the ear are stimulated by sound waves.
4. What are the functions of rods and cones? How are they distributed in the retina?
5. Discuss the mechanism by which photoreceptors are stimulated by light. What is the function of rhodopsin?
6. Contrast the function of the insect's compound eye with that of the vertebrate eye.

Thinking Critically

1. What are the benefits of providing hearing-impaired children with devices such as cochlear implants? What are some disadvantages?
2. Artificial retinas do not help individuals with optic nerve damage to see. Explain why. What other types of devices are used to help the visually impaired? What types might be developed in the future?
3. If all neurons transmit the same type of message, how do we know the difference between sound and light? How are we able to distinguish between an intense pain and a mild one?
4. How is it adaptive for organisms to have only four or five basic tastes?
5. Explain the statement: Vision happens mainly in the brain.

Recommended Readings

Abu-Mostafa, Y.S., and D. Psaltis, "Optical Neural Computers," *Scientific American,* Vol. 256, No. 3, 1987, pp. 88–94. The arrangement of neurons in the brain can be used as a model for building a computer that can solve problems, such as recognizing patterns that involve memorizing all possible solutions.

DISCOVER Special Issue, "The Mystery of Sense," Vol. 14, No. 6, June 1993. Several articles discussing vision, hearing, touch, smell, and taste.

Kirchner, W.H., and W.F. Towne, "The Sensory Basis of the Honeybee's Dance Language," *Scientific American,* Vol. 270, No. 6, June 1994, pp. 74–80. Bees can hear and they can detect the sounds associated with the dances they use to communicate.

Long, M.E., "The Sense of Sight," *National Geographic,* Vol. 182, No. 5, November 1992, pp. 3–41. A beautifully illustrated discussion of vision and new technology for restoring sight.

Poggio, T., and C. Koch, "Synapses that Compute Motion," *Scientific American,* Vol. 256, No. 5, 1987, pp. 46–52. Studies of cells in the eye that interpret movement may help clarify mechanisms involved in other neural processes.

Schnapf, J.L., and D.A. Baylor, "How Photoreceptor Cells Respond to Light," *Scientific American,* Vol. 256, No. 4, 1987, pp. 40–47. Discusses how a single photoreceptor cell in the eye registers the absorption of a single photon.

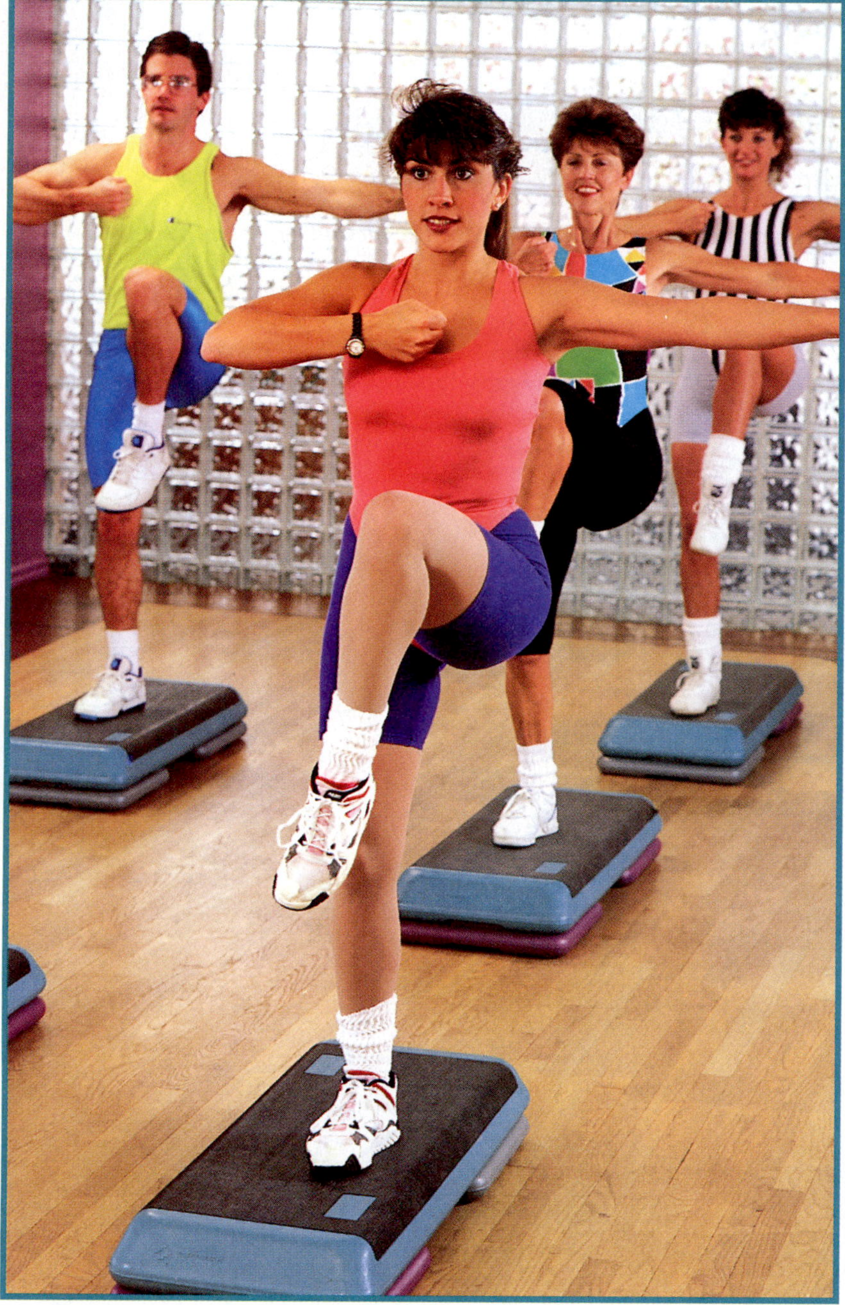

A program of aerobic exercise increases cardiovascular and muscle endurance (The Stock Market/Jon Feingersh 1992)

HOW YOUR WORK-OUT WORKS

Just how often and how intensely should we exercise? What are the benefits to the cardiovascular system and to the rest of the body? Exercise improves fitness. Specifically, exercise improves muscle tone and strength, and enhances muscular and cardiovascular endurance. Recent studies suggest that exercise helps reduce hypertension (high blood pressure) and increases high-density lipoproteins (the good lipoproteins that help reduce cholesterol levels) in the blood. Another benefit is that exercise helps us reduce stress.

Your endurance determines how long you can continue a physical activity, whether it is running, swimming, or dancing. Endurance requires cardiovascular fitness as well as muscle strength. Endurance can be increased by aerobic exercise, which is any activity that increases the heart rate and requires oxygen. Examples of aerobic exercise are rapid walking, in-line skating, and bicycling.

Exercise results in marked physiological changes in the body. For example, it changes the rate and distribution of blood flow to the body's organs. Contracting muscles need more oxygen, so blood flow to the muscles increases dramatically. During aerobic exercise, the rate of blood flow to the muscles can increase more than 20 times. This increase not only helps meet the increased demand for oxygen, but the increased need for removal of CO_2 and other metabolic wastes generated during physical activity. Thus, exercise has a purifying effect on body systems. Although the rate of blood flow to the brain is maintained, blood flow to the organs of the digestive system and to the kidneys is markedly reduced, so more blood is available for the muscles.

Regular aerobic exercise strengthens the heart muscle and decreases the resting heart rate. The heart does not have to work as hard. Athletic training often leads to enlargement (hypertrophy) of the heart, including thickening of the heart walls and an increase in the size of its chambers. This leads to an increase in the amount of blood the heart can pump with each contraction and to an increase in the strength of cardiac muscle contraction. The heart becomes much more efficient in its pumping capacity both at rest and during physical activity.

Another benefit of exercise is that it increases metabolic rate. This makes it easier to maintain desired body weight and body fat. As a result, more food can be eaten without gaining weight.

The American College of Sports Medicine recommends that healthy adults perform aerobic exercise for about 20 to 60 minutes three to five days each week. The aerobic activity should be intense enough to increase the heart rate to 70% to 90% of **maximum heart rate.** You can estimate maximum heart rate by subtracting your age from 220. For example, if you are 20 years old your maximum heart rate would be $220 - 20$, or 200 beats per minute. If you are in good health, and exercise at a level that raises your heart rate to 70% to 90% of maximum, your heart rate would increase to about 160 beats per minute.

If you have not been exercising and are just beginning an exercise program, mild exercise such as walking can raise your heart rate to aerobic levels. As you increase your fitness, you will need to intensify your exercise to reach aerobic range. For example, you might speed up your walking or skating.

Internal Transport

36

Learning Objectives

After you have studied this chapter you should be able to:

1. Compare internal transport in animals that lack a circulatory system, animals with an open circulatory system, and animals with a closed circulatory system.
2. Relate structural adaptations of the vertebrate circulatory system to each function it performs.
3. Compare the structure and functions of red blood cells, white blood cells, and platelets.
4. Describe the structure and function of the different types of blood vessels, including arteries, arterioles, capillaries, and veins.
5. Compare the structure of the heart in each class of vertebrates.
6. Describe the structure and function of the parts of the human heart and label them on a diagram.
7. Summarize how the heart works; include a description of the heartbeat, neural regulation of heart rate, and the sounds produced by the heart.
8. Identify factors that determine blood pressure, and explain the physiological basis of arterial pulse.
9. Compare blood pressure in different types of blood vessels and summarize how arterial blood pressure is regulated.
10. Trace a drop of blood through the pulmonary and systemic circulations, naming in sequence each structure through which it passes.
11. Identify the risk factors of atherosclerosis, trace the progress of the disorder, and summarize its possible complications (including angina pectoris and myocardial infarction).
12. List the functions of the lymphatic system, and describe how the system operates to maintain fluid balance.

Key Concepts

- ☐ Animals that are more than a few cells thick or that have an active life-style have a circulatory system that transports oxygen and nutrients to all of its cells.
- ☐ Some invertebrates (arthropods and mollusks) have an open circulatory system in which blood flows into a hemocoel and bathes the tissues directly. Other invertebrates and vertebrates have a closed circulatory system in which blood is pumped through a continuous circuit of blood vessels.
- ☐ Vertebrate blood consists of a fluid called plasma, which contains red blood cells, white blood cells, and platelets.

Key Concepts continued

- ☐ Arteries carry blood away from the heart; veins carry blood back toward the heart. Capillaries are very small, thin-walled vessels through which materials are exchanged between blood and tissues.
- ☐ The vertebrate heart is a muscular pump. In birds and mammals, the right and left sides of the heart are completely separate and blood flows through a double circuit—the pulmonary circulation and the systemic circulation.

THE CIRCULATORY SYSTEM IS RESPONSIBLE FOR INTERNAL TRANSPORT IN MOST ANIMALS

Each cell of the animal body requires a constant supply of oxygen and nutrients and continuously generates metabolic wastes. In very small organisms, each cell is in close contact with the surrounding environment. Diffusion may be adequate to distribute oxygen, nourish each cell, and dispose of waste products. Large animals need a more effective mechanism for transporting materials to all of their cells.

In complex animals, internal transport of nutrients, oxygen, and wastes is accomplished by a **circulatory system.** Most circulatory systems have three components: (1) blood, a fluid connective tissue consisting of cells and cell fragments dispersed in fluid; (2) a pumping device, generally a heart; and (3) a system of blood vessels or spaces through which the blood circulates. An efficient circulatory system enables large animals to carry on rapid metabolism and thus allows a more active life-style.

SOME INVERTEBRATES HAVE NO CIRCULATORY SYSTEM

In many invertebrates, internal transport depends on diffusion or on other body systems. The success of *Hydras* and other cnidarians does not depend on a high metabolic rate. Diffusion through the thin body wall supplies sufficient oxygen to support their activities and carries wastes from the body.

Cnidarians and many flatworms combine digestive and some internal transport functions in the gastrovascular cavity (Fig. 36–1). Movement of the body, as the animal stretches and contracts, stirs up the contents of the

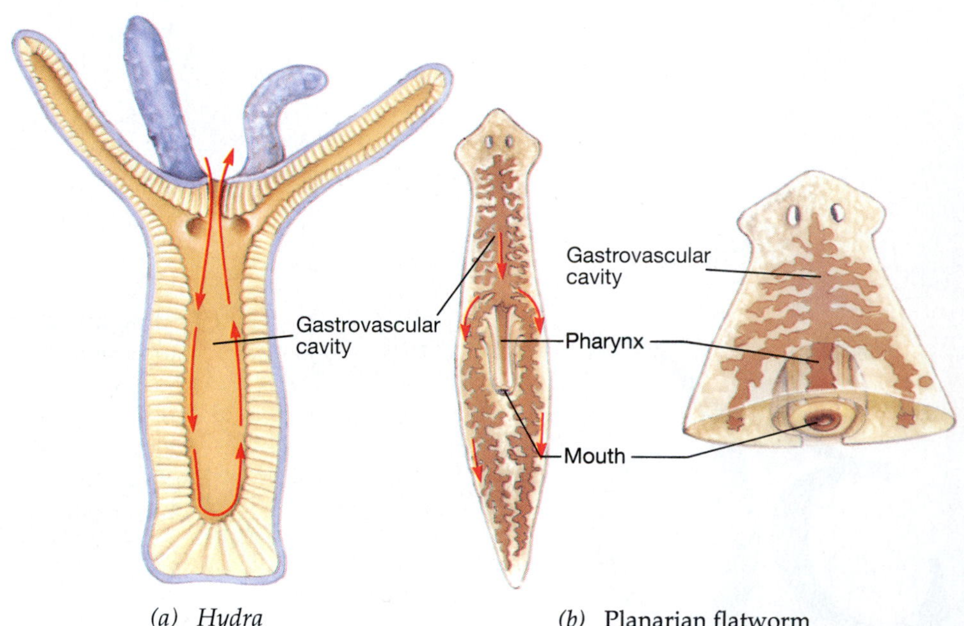

FIGURE 36–1 Two invertebrates with no circulatory system. (***a***) In *Hydra* and other cnidarians, the gastrovascular cavity helps circulate nutrients to all parts of the body. (***b***) In planarian flatworms, the branched intestine circulates food to all regions of the body.

cavity and helps distribute nutrients. The compressed body of the flatworm permits effective gas exchange by diffusion. Wastes are transported by the branching excretory system. Metabolic rate tends to be higher than in cnidarians, allowing a more active life-style.

MANY INVERTEBRATES HAVE AN OPEN CIRCULATORY SYSTEM

Arthropods and most mollusks have an **open circulatory system,** in which the heart pumps blood into vessels that have open ends (Fig. 36–2). Blood spills out of them, filling large spaces that make up the **hemocoel** (blood cavity). The blood bathes the cells of the body. Blood re-enters the circulatory system through openings in the heart (in arthropods) or through open-ended vessels that lead to the gills (in mollusks).

An open circulatory system does not supply oxygen to cells very rapidly even though the blood of some of these invertebrates contains a pigment (hemocyanin) that transports oxygen. For example, the open circulatory system of insects would not provide sufficient oxygen to maintain their active life-style. Instead, oxygen is delivered directly to the cells by a system of air (tracheal) tubes that make up the respiratory system (Chapter 38). Insect blood mainly distributes nutrients and hormones.

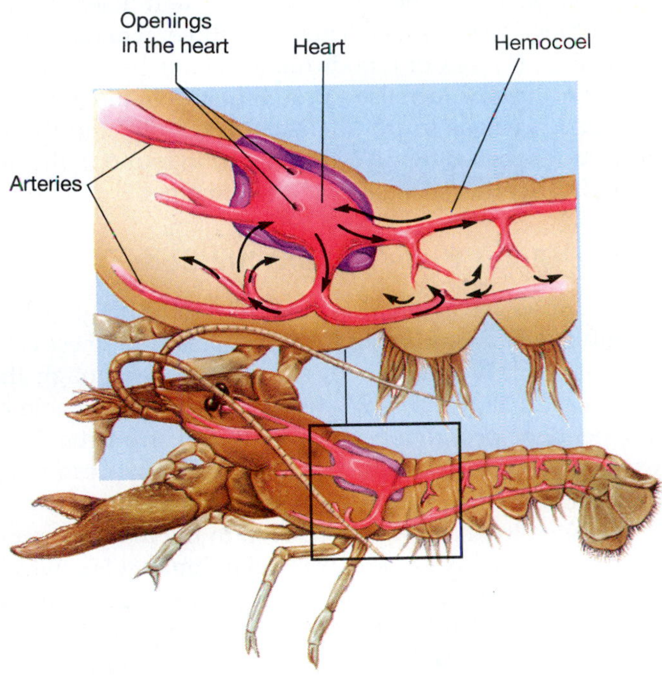

FIGURE 36–2 Open circulatory system of the crayfish. As in other arthropods, the hemolymph bathes the body tissues. Lateral view.

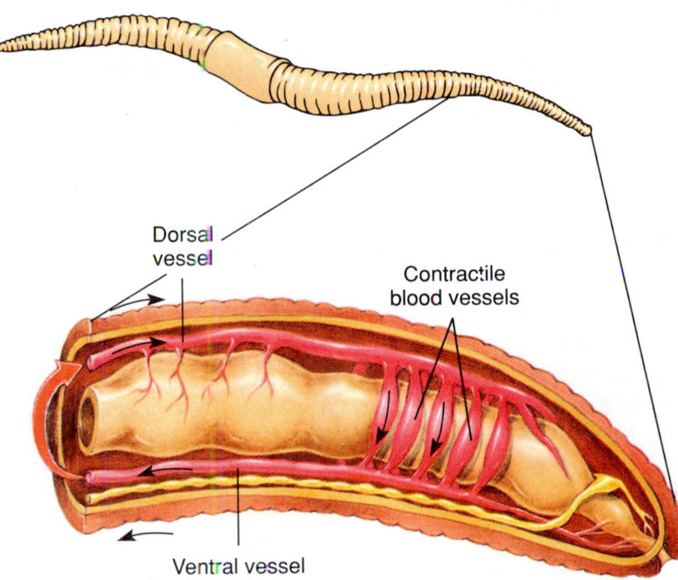

FIGURE 36–3 The earthworm has a closed circulatory system. Five pairs of contractile blood vessels deliver blood from the dorsal vessel to the ventral vessel.

SOME INVERTEBRATES HAVE A CLOSED CIRCULATORY SYSTEM

Annelids, the fast-moving cephalopod mollusks, and echinoderms are among the invertebrates that have a **closed circulatory system** (Fig. 36–3). In them, blood flows through a continuous circuit of blood vessels. The walls of the smallest blood vessels are thin enough to permit the diffusion of gases, nutrients, and wastes between blood in the vessels and the tissue fluid that bathes the cells.

Earthworms and other annelids have no heart. Contractions of certain blood vessels, as well as contractions of the body wall muscles, circulate the blood. Earthworms have hemoglobin, the same red pigment that transports oxygen in vertebrate blood. Earthworm hemoglobin is dissolved in the blood plasma.

VERTEBRATES HAVE A CLOSED CIRCULATORY SYSTEM ADAPTED TO CARRY OUT A VARIETY OF FUNCTIONS

The circulatory system is basically similar in all vertebrates from fishes, frogs, and reptiles to birds and humans. All have a ventral, muscular heart that pumps blood into a closed system of blood vessels. The vertebrate circulatory system consists of the heart; blood vessels; blood; lymph; lymph vessels; and associated organs such as the thymus, spleen, and liver. The tiniest blood vessels, **capillaries,** have very thin walls that permit the exchange of materials between the blood and tissue fluid.

The vertebrate circulatory system functions to:

1. Transport nutrients from the digestive system and from storage depots to each cell of the body.
2. Transport oxygen from respiratory structures (gills, lungs) to the cells of the body.
3. Transport metabolic wastes from each cell to organs that excrete them.
4. Transport hormones from endocrine glands to target tissues.
5. Help maintain fluid balance.
6. Defend the body against invading microorganisms.
7. Help to distribute metabolic heat within the body and to maintain normal body temperature in endothermic (warm-blooded) animals

VERTEBRATE BLOOD CONSISTS OF PLASMA, BLOOD CELLS, AND PLATELETS

In vertebrates, **blood** consists of a pale yellowish fluid, known as **plasma,** in which red blood cells, white blood cells, and platelets are suspended (Fig. 36–4). In humans the total circulating blood volume is about 8% of body weight, which equals about 5.6 liters (6 quarts) in a 70-kg (154-lb) person.

Plasma Is the Fluid Component of Blood

Plasma is mostly water (92%) but also contains plasma proteins, a sprinkling of salts, and a variety of materials being transported such as nutrients, dissolved gases, metabolic wastes, and hormones. **Plasma proteins** may be divided into three groups, or fractions: the albumins, globulins, and fibrinogen. One of the main functions of these plasma proteins is to maintain an appropriate blood volume. As blood flows through the capillaries, some plasma seeps through the capillary walls and passes into the tissues. However, large protein molecules have difficulty passing through the capillary walls, so most of them remain in the blood. There they exert an osmotic pressure, which helps pull plasma back into the blood.

Certain **albumins** transport substances such as specific hormones, keeping them bound in the blood until needed. One group of **globulins,** the gamma globulins, are antibodies, which are substances that provide immunity against invading disease organisms. Alpha and beta globulins transport lipids, hormones, iron, and some other substances. **Fibrinogen** and several other plasma proteins are involved in the clotting process. When the proteins involved in blood clotting have been removed from the plasma, the remaining liquid is called **serum.**

Red Blood Cells Transport Oxygen

Each of us has about 30 trillion (3×10^{13}) **red blood cells,** or **erythrocytes,** suspended in our plasma. These cells are so tiny that about 3000 of them lined up end-to-end would measure only 1 inch. Red blood cells are wonderfully adapted for transporting oxygen.

Red blood cells develop inside certain bones in special tissue called **red bone marrow.** As each cell differentiates, it manufactures great quantities of **hemoglobin,** the pigment that gives vertebrate blood its red color. Red blood cells transport oxygen chemically combined with hemoglobin, and also transport carbon dioxide.

The mature red blood cell is a tiny sac of hemoglobin, shaped with a considerable surface area for gas exchange. In mammals, the mature red blood cell lacks a nucleus and carries on only limited cellular functions. It has a short life span—about 120 days. As blood circulates through the liver and spleen, worn-out red blood cells are removed from circulation and destroyed. Their hemoglobin molecules are taken apart, and some of the components, such as iron, are recycled. In the human body, 2.4 million red blood cells are destroyed every second, so an equal number must be produced in the bone marrow to replace them.

A deficiency of hemoglobin, usually accompanied by a reduced number of red blood cells, is called **anemia.** With less hemoglobin, less oxygen is transported, so the body cells do not receive enough oxygen. An anemic person complains of never having enough energy—the "tired-blood" syndrome. There are three general causes of anemia: (1) loss of blood due to hemorrhage or internal bleeding, (2) decreased production of hemoglobin or red blood cells as in iron deficiency anemia, and (3) increased rate of red blood cell destruction found in the **hemolytic anemias** such as sickle cell anemia.

White Blood Cells Defend the Body

White blood cells, or **leukocytes,** defend the body against harmful bacteria and other foreign substances. These cells are able to leave the blood, squeezing out through the walls of the capillaries. White blood cells are capable of independent locomotion similar to that of an ameba. They wander about through the tissues of the body, destroying invading microorganisms or worn-out cells.

White blood cells are produced in the red bone marrow. Two main types are **agranular leukocytes,** which lack large granules in their cytoplasm, and **granular leukocytes,** which have large, distinctive granules. Two kinds of agranular leukocytes are lymphocytes and monocytes (Fig. 36–4). **Monocytes** leave the circulation and complete their development in the tissues. There, they increase to about five times their original size and become **macrophages,** the giant scavenger cells of the

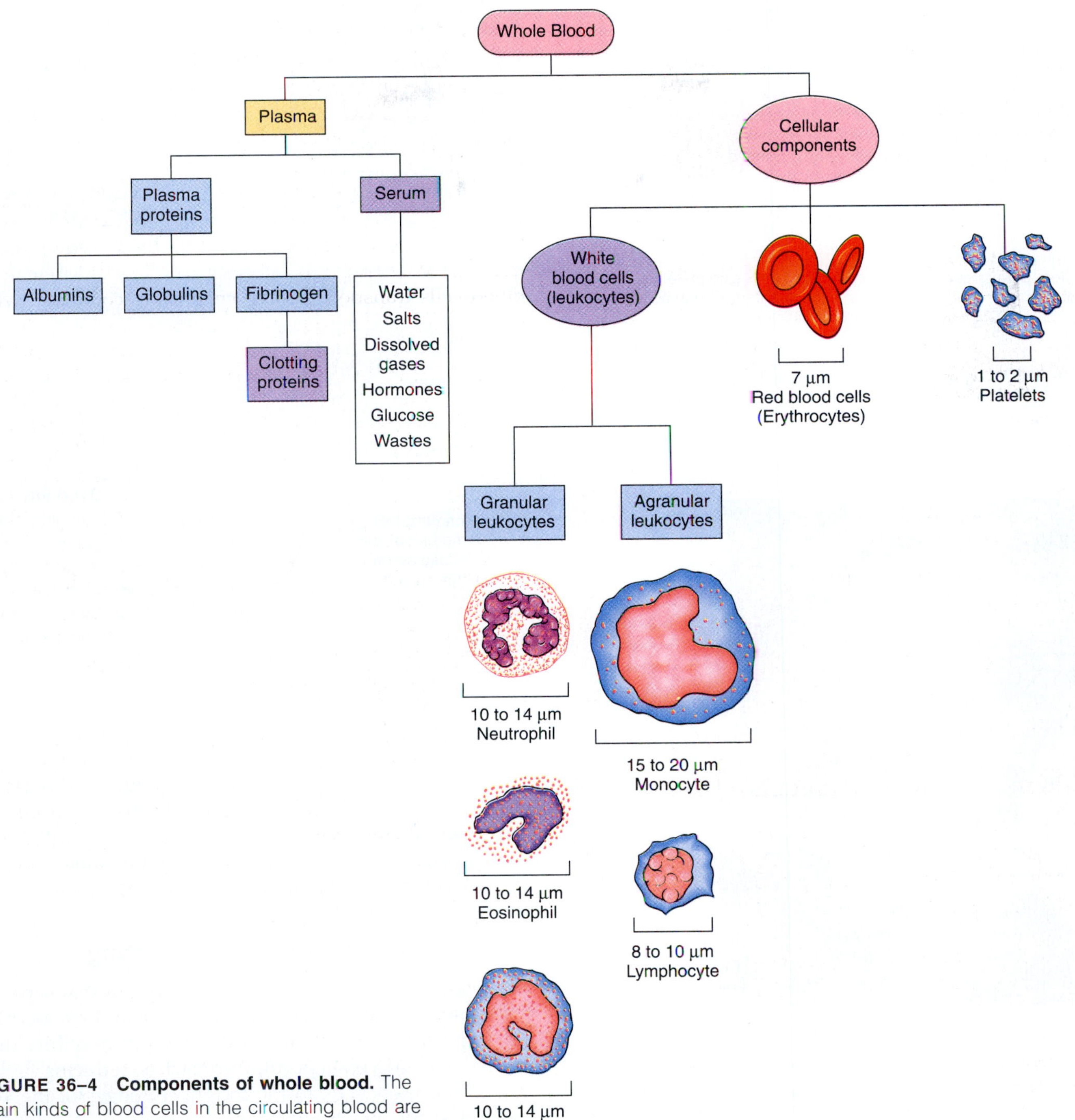

FIGURE 36–4 Components of whole blood. The main kinds of blood cells in the circulating blood are shown.

body. Macrophages have voracious appetites for bacteria, dead cells, and other matter littering the tissues. The role of lymphocytes in immune responses will be discussed in Chapter 37.

Three types of granular leukocytes are neutrophils, eosinophils, and basophils. **Neutrophils** are especially adept at seeking out and ingesting bacteria. **Eosinophils** play a part in allergic reactions. **Basophils** contain large amounts of **histamine,** a chemical released in injured tissues and in allergic reactions. Basophils also contain **heparin,** an anticlotting chemical that may be important in preventing inappropriate clotting within blood vessels.

In human blood, there are normally about 7000 white blood cells per mm^3 of blood (only one for every 700 red blood cells). During bacterial infections, the number may rise sharply, so that a white blood cell count is a useful

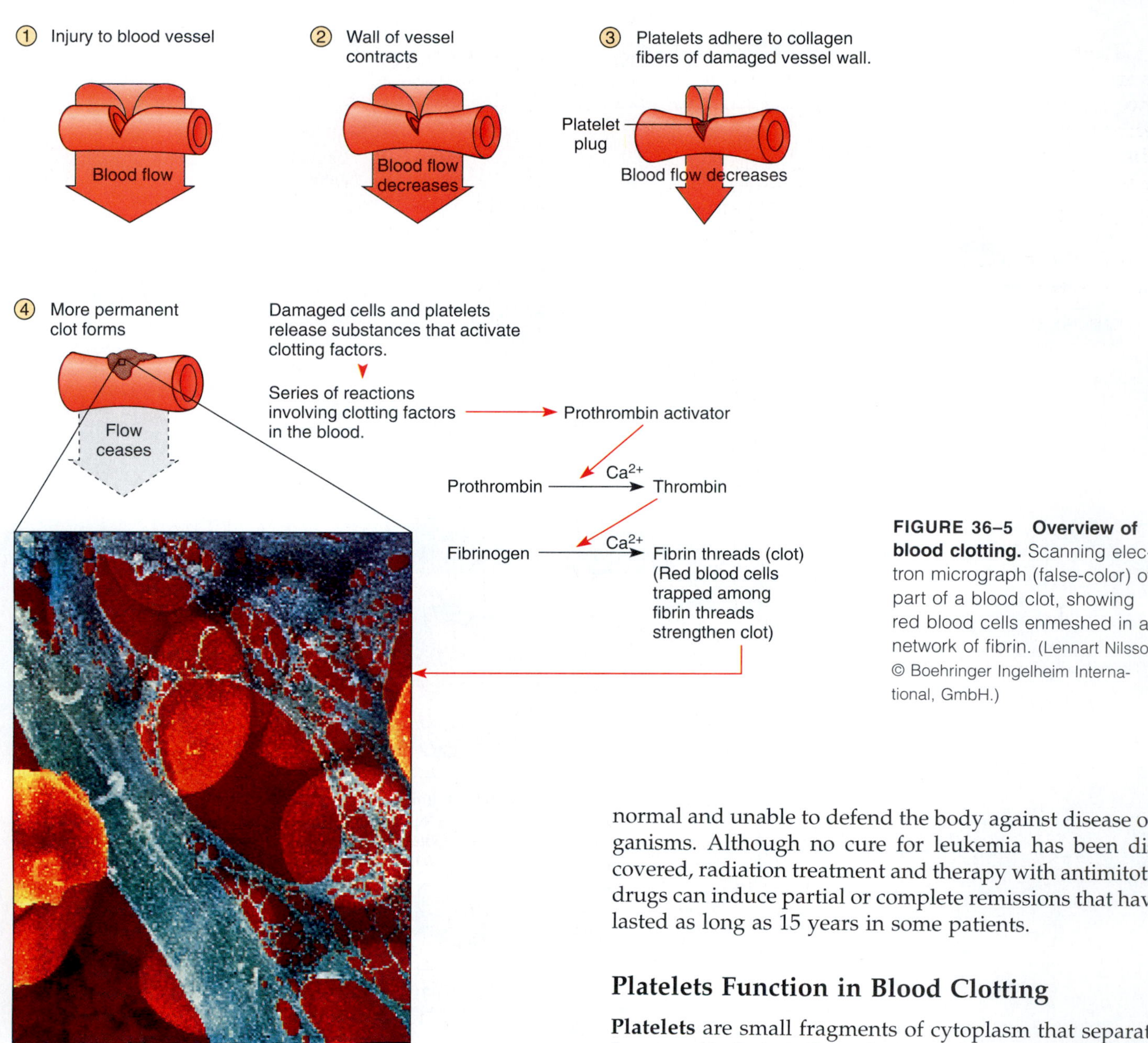

FIGURE 36–5 Overview of blood clotting. Scanning electron micrograph (false-color) of part of a blood clot, showing red blood cells enmeshed in a network of fibrin. (Lennart Nilsson, © Boehringer Ingelheim International, GmbH.)

diagnostic tool. The proportion of each kind of white blood cell is determined by a differential white blood cell count.

Leukemia is a form of cancer in which any one of the kinds of white cells multiply rapidly within the bone marrow. Many of these cells do not mature. Their large numbers crowd out developing red blood cells and platelets, leading to anemia and impaired clotting. A common cause of death from leukemia is internal bleeding, especially in the brain. Another frequent cause of death is infection. Infection occurs because, although there may be a dramatic rise in the white cell count, the cells are immature and abnormal and unable to defend the body against disease organisms. Although no cure for leukemia has been discovered, radiation treatment and therapy with antimitotic drugs can induce partial or complete remissions that have lasted as long as 15 years in some patients.

Platelets Function in Blood Clotting

Platelets are small fragments of cytoplasm that separate from certain large cells in the bone marrow. Platelets play an important role in **hemostasis,** the control of bleeding. When a blood vessel is cut, it constricts, reducing loss of blood. Platelets stick to the rough, cut edges of the vessel, physically patching the break in the wall. As platelets begin to gather, they release substances that attract other platelets. Within about 5 minutes after injury, a complete platelet patch, a temporary clot, has formed.

At the same time that the temporary clot forms, a stronger, more permanent clot begins to develop. More than 30 different chemical substances interact in this process. The series of reactions that leads to clotting is triggered when one of the clotting factors in the blood is activated by contact with the injured tissue. In **hemophiliacs** (persons with "bleeder's disease"), one of the clotting factors is absent as a result of an inherited genetic mutation. The clotting process is summarized in Figure 36–5.

Prothrombin, a plasma protein manufactured in the liver, requires vitamin K for its production. In the presence of clotting factors, calcium ions, and compounds released from platelets, prothrombin is converted to **thrombin.** Then thrombin catalyzes the conversion of the soluble plasma protein **fibrinogen** to an insoluble protein, called **fibrin.** Fibrin produces long threads that stick to the damaged surface of the blood vessel and form the webbing of the clot. These threads trap blood cells and platelets, which help to strengthen the clot.

VERTEBRATES HAVE THREE MAIN TYPES OF BLOOD VESSELS

The vertebrate circulatory system has three main types of blood vessels: arteries, capillaries, and veins (Fig. 36–6). An **artery** carries blood away from the heart, toward other tissues. When an artery enters an organ, it divides into many smaller branches, called **arterioles.** The arterioles deliver blood into the microscopic **capillaries.** After cours-

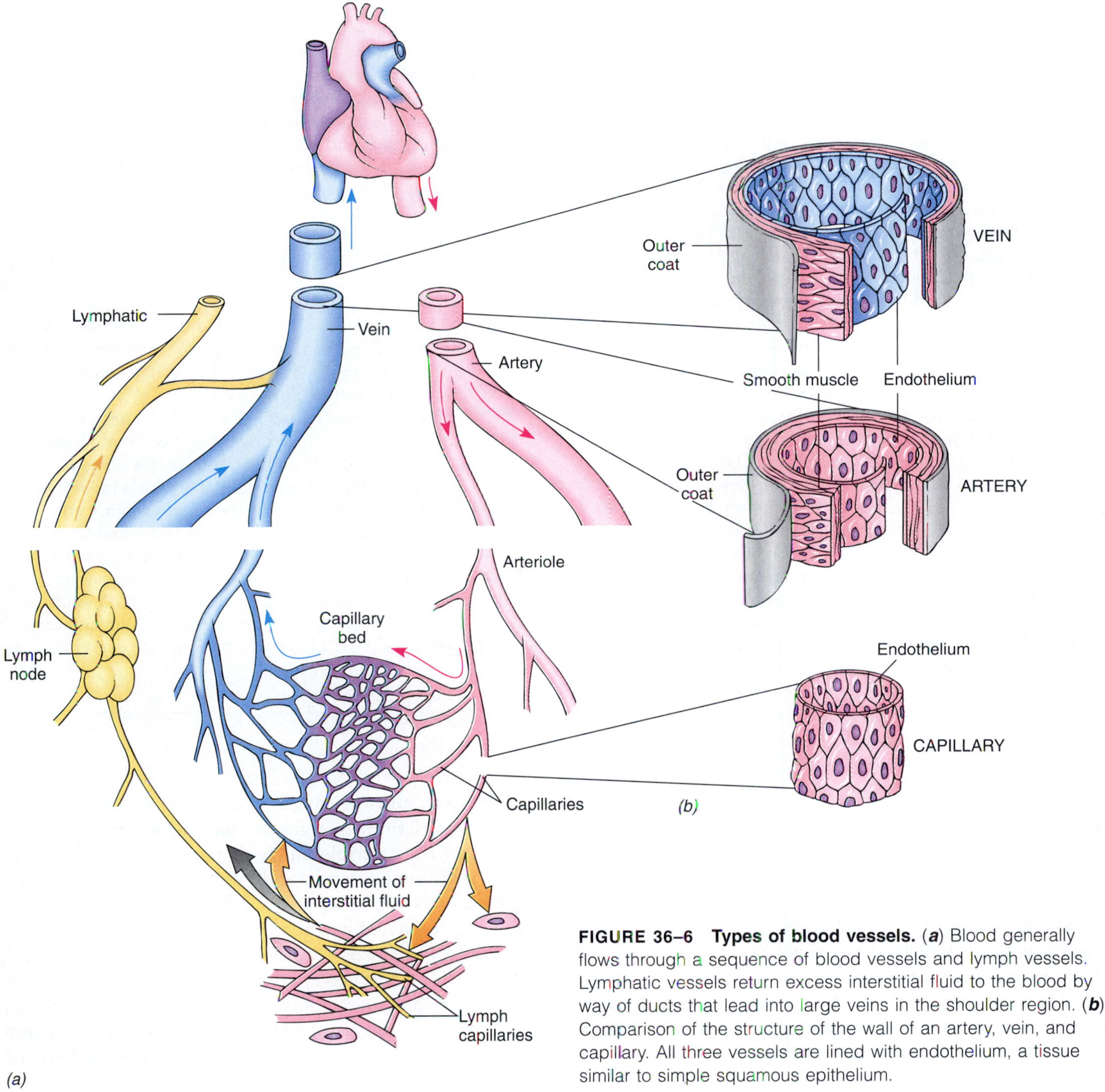

FIGURE 36–6 Types of blood vessels. (***a***) Blood generally flows through a sequence of blood vessels and lymph vessels. Lymphatic vessels return excess interstitial fluid to the blood by way of ducts that lead into large veins in the shoulder region. (***b***) Comparison of the structure of the wall of an artery, vein, and capillary. All three vessels are lined with endothelium, a tissue similar to simple squamous epithelium.

CONCEPT CONNECTIONS

Circulatory System — *Vertebrate Evolution*

The vertebrate circulatory system has changed in the course of evolution as the site of gas exchange changed from gills to lungs and as vertebrates became active, endothermic animals. The vertebrate heart has one or two **atria,** chambers that receive blood returning from the tissues, and a single or divided **ventricle** that pumps blood into the arteries (see figure). Additional chambers are present in some classes.

In fish, blood flows through a single circuit passing through two capillary networks—the first is located in the gills, where it is oxygenated; the second is located in some other organ of the body. The heart pumps blood to the gills. After blood circulates through capillaries in the gills its pressure is low. Blood leaving the gills passes very slowly to the other organs of the body. Circulation is helped by the movements of the fish while it is swimming. This single-circuit, low-pressure circulatory system permits only a low rate of metabolism in fish.

Although it contains only one atrium and one ventricle, the fish heart has four chambers. A thin-walled **sinus venosus** receives blood returning from the tissues and pumps it into the atrium. The atrium then pumps blood into the ventricle. Next, the ventricle pumps blood into an elastic **conus arteriosus,** which does not contract.

In amphibians, blood flows through a double circuit. The **pulmonary circulation** delivers blood to the lungs and skin while the **systemic circulation** transports blood to all of the organs of the body. Oxygen-rich and oxygen-poor blood are kept somewhat separate.

The amphibian heart has two atria and one ventricle. A sinus venosus collects blood returning from the veins and pumps it into the right atrium. Blood returning from the lungs passes directly into the left atrium. Both atria pump into the single ventricle, but oxygen-poor blood is pumped out of the ventricle first. Blood passes into an artery (the conus arteriosus) with a fold that helps keep the blood separate. Much of the oxygen-poor blood is directed to the lungs and skin, where it can be recharged with oxygen. Oxygen-rich blood is delivered into arteries that conduct it to the various tissues of the body.

Reptiles also have a double circuit of blood flow and in the reptilian heart, a wall partly divides the ventricle. Although some mixing of oxygen-rich and oxygen-poor blood occurs, it is minimized by the timing of contractions of the left and right sides of the heart and by pressure differences. (In crocodiles, the wall between the ventricles is complete so that the heart consists of two atria and two ventricles.)

Unlike birds and mammals, amphibians and reptiles do not ventilate their lungs continuously. Therefore, it would be inefficient to pump blood through them continuously. The shunts between the two sides of the heart allow the blood to be distributed to the lungs as needed.

In birds and mammals, the wall between the ventricles is complete, preventing the mixing of oxygen-rich blood in the left side with oxygen-poor blood in the right side. The conus has split and become the base of the aorta and pulmonary artery. No sinus venosus is present as a separate chamber, but a vestige remains as the sinoatrial node (the pacemaker).

Complete separation of right and left sides of the heart makes it necessary for blood to pass through the heart twice each time it makes a tour of the body. As a result, it is possible to maintain higher blood pressures, and materials are delivered to the tissues rapidly and efficiently. Because the blood of birds and mammals contains more oxygen per unit volume and circulates more rapidly than in other vertebrates, the tissues of the body receive more oxygen. As a result, birds and mammals can maintain a higher metabolic rate and can maintain a constant, high body temperature even in cold surroundings.

ing through an organ, capillaries eventually merge to form **veins** that transport the blood back toward the heart.

The walls of arteries and veins are thick, preventing gases and nutrients from passing through them. In contrast, capillary walls are only one cell thick (Fig. 36–7). Materials are readily exchanged between the blood and the tissue fluid (which bathes the cells) through the capillary walls. Capillary networks in the body are so extensive that at least one of these tiny vessels is located close to almost every cell in the body. The total length of all capillaries in the human body has been estimated as more than 60,000 miles!

Smooth muscle in the arteriole wall can constrict **(vasoconstriction)** or relax **(vasodilation),** changing the radius of the arteriole. Such changes help maintain appropriate blood pressure and can help control the volume of blood passing to a particular tissue. Changes in blood flow are regulated by the nervous system in response to the metabolic needs of the tissue, as well as by the demands of the body as a whole. For example, when a tissue is metabolizing rapidly, it needs a greater supply of nutrients and oxygen. During exercise, arterioles within the muscles dilate, increasing by more than tenfold the amount of blood flowing to the muscle cells.

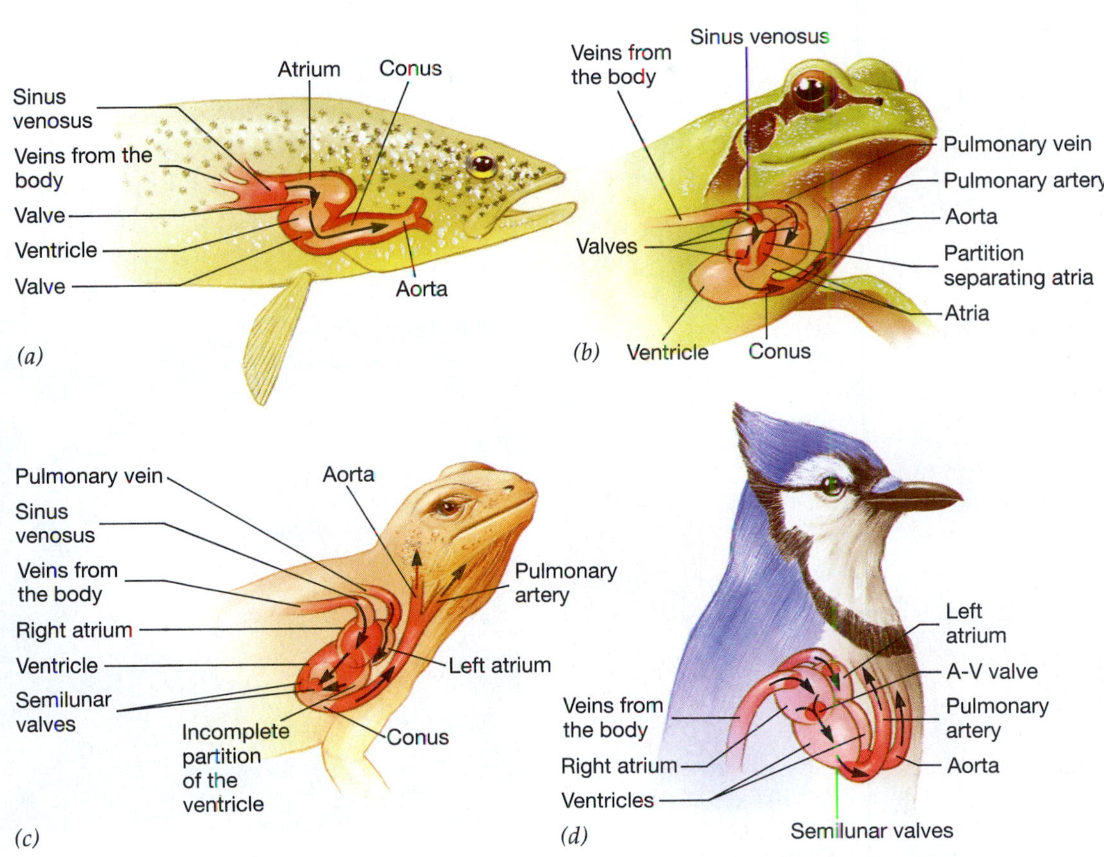

The evolution of the vertebrate heart. (***a***) The fish heart includes one atrium and one ventricle. (***b***) The amphibian heart consists of two atria and one ventricle, (***c***) The reptilian heart has two atria and two ventricles, but the wall separating the ventricles is incomplete so that blood from the right and left chambers mixes to some extent. (***d***) Birds and mammals have two atria and two ventricles. A complete wall separates right and left sides of the heart so that blood rich in oxygen is kept separate from oxygen-poor blood.

THE HUMAN HEART IS WONDERFULLY ADAPTED FOR PUMPING BLOOD

Not much bigger than a fist and weighing less than a pound, the human **heart** (Fig. 36–8) is a remarkable organ that beats about 2.5 billion times in an average lifetime, pumping about 300 million liters (80 million gallons) of blood. The heart is a hollow, muscular organ consisting of four separate chambers (Fig. 36–9). Each side of the heart has an **atrium** (which receives blood from veins) and a **ventricle** (which pumps blood into arteries). The wall of the heart is composed mainly of cardiac muscle covered by a tough connective-tissue membrane. Another membrane, the **pericardium,** surrounds the entire heart but is separated from it by a small space, the **pericardial cavity.** A thin film of lubricating fluid within the pericardial cavity reduces friction, facilitating smooth movement of the heart as it contracts and relaxes.

To prevent blood from flowing backward, the heart is equipped with valves that close automatically (Fig. 36–10). When the ventricles contract, the **atrioventricular (AV) valves** between the atria and the ventricles close. As

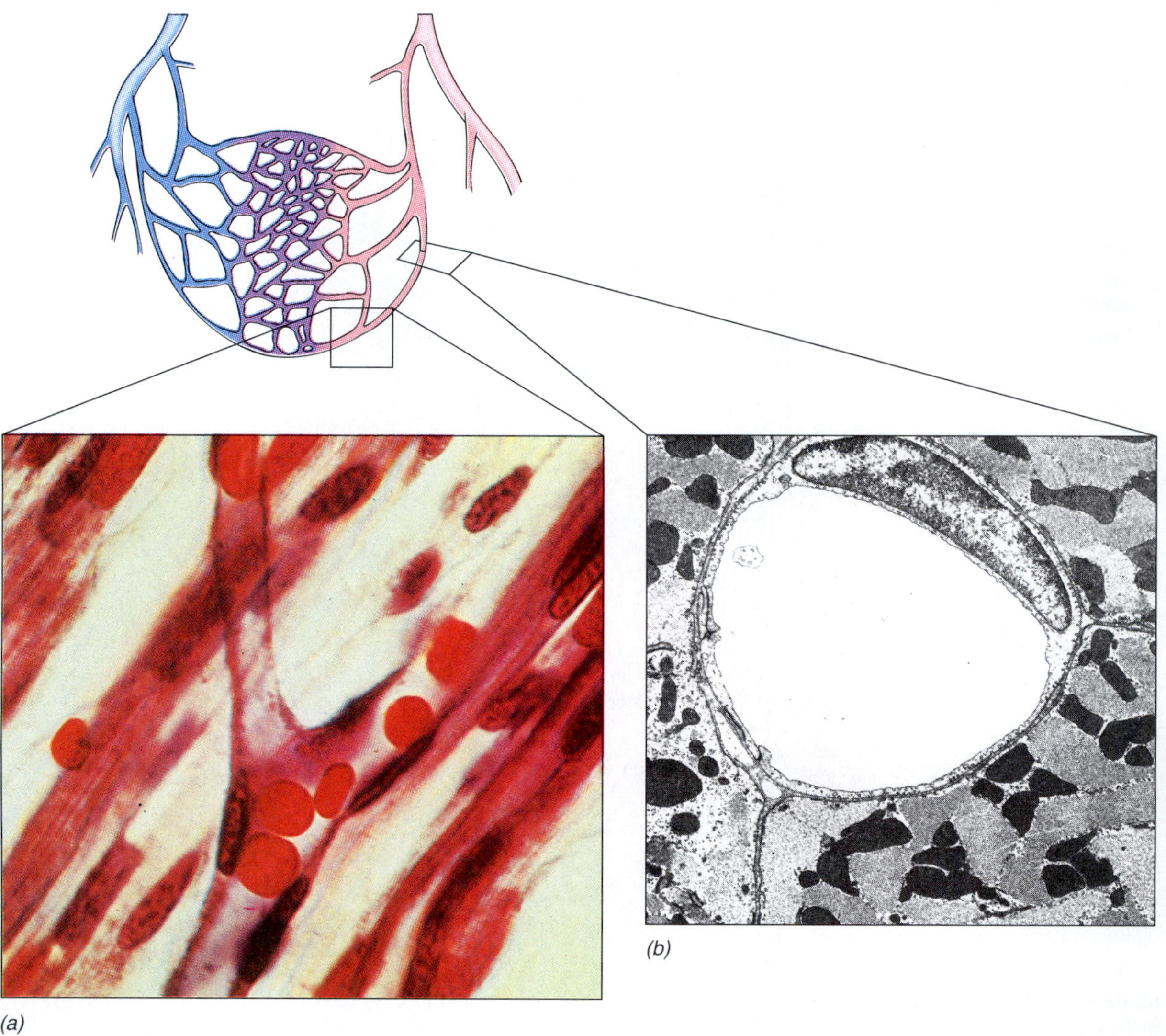

FIGURE 36–7 Capillary structure. (***a***) Red blood cells must pass through capillaries in almost single file. (***b***) Cross section of capillary in the papillary muscle of a rat heart. Notice that the wall of the capillary is composed of a single cell that assumes the shape of a tube. A series of such cells makes up the entire capillary. The light-and-dark pattern surrounding the capillary consists of cross sections of cardiac muscle cells. The heart is too thick to be supplied by the blood it contains, and so requires its own system of capillaries to nourish it and provide it with oxygen. (*a*, Lennart Nilsson, copyright Boehringer Ingelheim International GmbH, *b*, Courtesy of Torsten Mattfeldt and Gerhard Mall, University of Heidelberg, Cardiovascular Research 17, 1983)

a result, blood does not flow backward into the atria. The AV valve between the left atrium and ventricle is known as the **mitral valve.** The AV valve between the right atrium and ventricle is the tricuspid valve. As blood leaves the ventricles and is forced into the great arteries leaving the heart, the **semilunar valves** close, preventing backflow into the ventricles. Both types of valves open and close because of the hydraulic pressure of the blood.

Each Heartbeat Is Initiated by a Pacemaker

You may have watched horror films in which a heart removed from the body of its owner continues to beat spookily. Script writers of such films may have rooted their fantasies in fact, for when removed from the body, the heart will continue to beat for many hours if bathed

(Text continues on page 708)

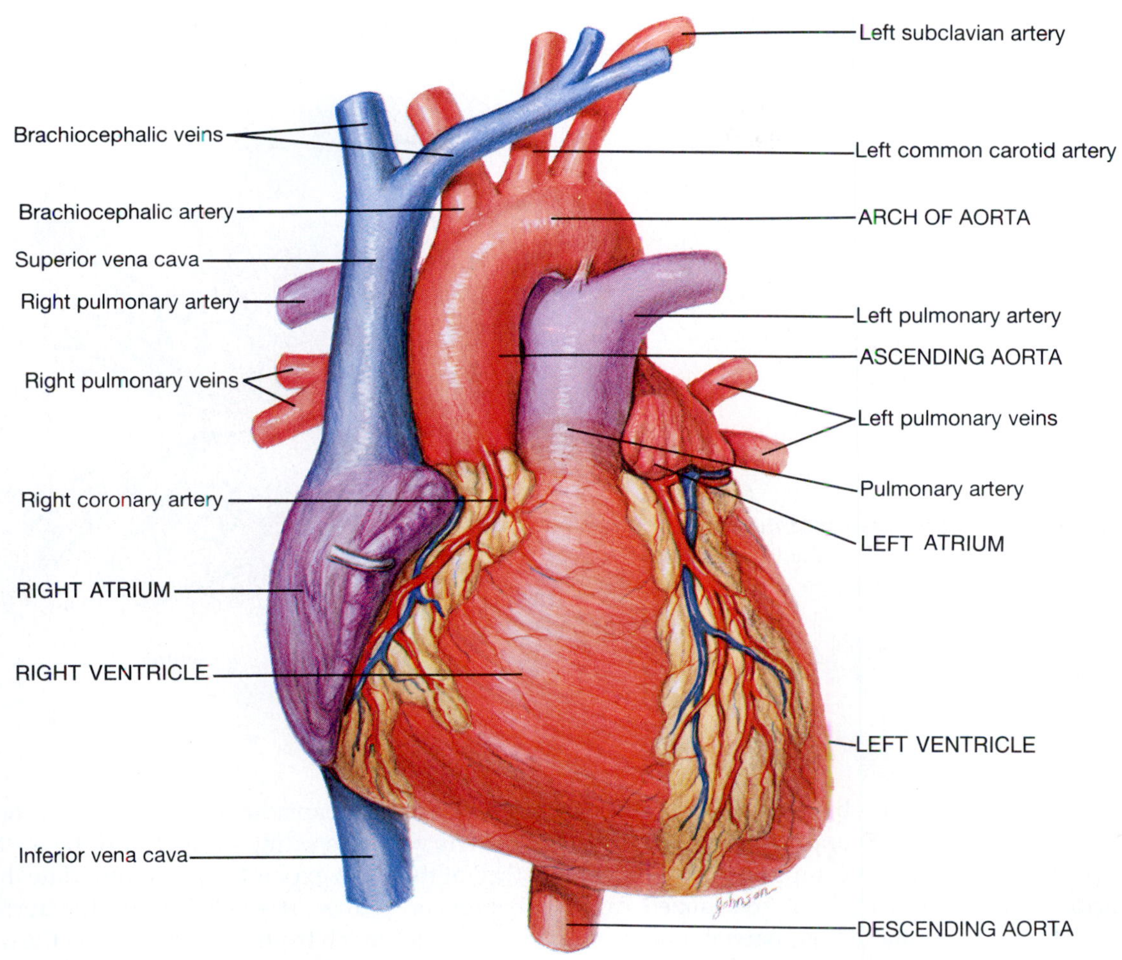

FIGURE 36–8 Structure of the human heart. Anterior view. Note the coronary blood vessels that bring blood to and from the heart muscle itself.

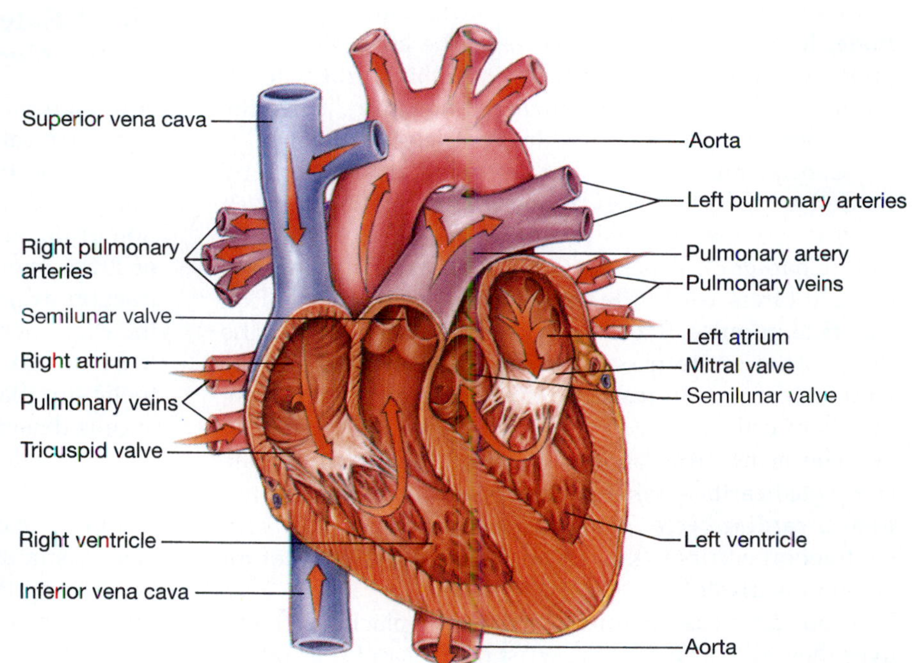

FIGURE 36–9 Internal structure of the heart. Section through the human heart showing chambers, valves, and connecting blood vessels.

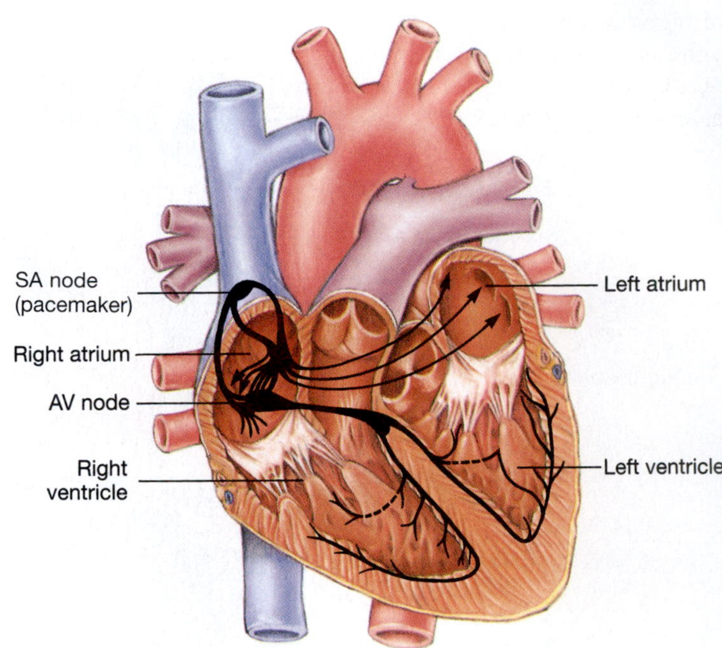

FIGURE 36–10 The conduction system of the heart. The SA node initiates each heart beat. The action potential spreads through the muscle fibers of the atria, producing atrial contraction. Transmission is briefly delayed at the AV node. Then, the action potential spreads through the specialized muscle fibers into the ventricles.

in an appropriate nutritive fluid. This is possible because the heart has its own specialized conduction system and can beat independently of its nerve supply.

Each heartbeat begins in a node of specialized cardiac muscle called the **pacemaker** (**sinoatrial node,** or simply, **SA node**), located in the posterior wall of the right atrium (Fig. 36–10). From the pacemaker, electrical impulses (much like neural impulses) are transmitted through the muscle fibers of both atria, causing them to contract. One group of atrial muscle fibers transmits the muscle impulse directly to the **atrioventricular (AV) node,** located in the wall between the atria. From this node, impulses sweep through specialized fibers to all parts of the ventricles, producing contraction. Thus, the atria contract first, and then the ventricles contract.

Cardiac muscle fibers are separated at their ends by dense bands, called **intercalated discs** (Fig. 36–11). Each intercalated disc is actually a tight junction between two cells (Chapter 6). This junction is of great significance because it offers very little resistance to the passage of an electrical impulse. It allows an impulse to pass across the disc, so the entire mass of atrial or ventricular muscle tends to contract in response to the impulse as if it were one giant cell.

The heart beats about 70 times every minute. One complete heartbeat takes about 0.8 second and is referred to as a **cardiac cycle.** That portion of the cycle in which contraction occurs is known as **systole;** the period of relaxation is **diastole.**

You can measure your heart rate by placing a finger over the radial artery in the wrist or the carotid artery in the neck and counting the pulsations. Arterial **pulse** is the alternate expansion and recoil of an artery. Each time the left ventricle pumps blood into the aorta, the elastic wall of the aorta expands to accommodate the blood. This expansion moves in a wave down the aorta and the arteries that branch from the aorta. When the pressure wave passes, the elastic arterial wall snaps back to its normal size (see Focus on the Process of Science: The Electrical Activity of the Heart).

Heart Rate Is Regulated by the Nervous System

Although the heart is capable of beating rhythmically on its own, it cannot by itself change the strength and rate of contraction to meet the changing needs of the body. Recall from Chapter 34 that this kind of control is the function of the autonomic nervous system. Under conditions of stress, sympathetic nerves can increase strength of contraction as much as 100%. Under more calm conditions, the vagus nerve, a parasympathetic nerve, slows the heart. It is the balance between sympathetic and parasympathetic stimulation that determines heart rate, and this balance is determined by the central nervous system.

The endocrine system also plays a part in regulating heartbeat. When the body is under stress, the hormones epinephrine and norepinephrine, released from the adrenal glands, stimulate the force and rate of the heartbeat.

During the normal heart rate of about 70 beats per minute, the **cardiac output,** which is the volume of blood pumped by one ventricle, is about 5 liters per minute. This amount is approximately equal to the total volume of blood in the body. Cardiac output depends mainly on **venous return,** the volume of blood delivered to the heart by the veins. During aerobic exercise, the heart may beat as many as 200 times per minute and its output may increase to more than 20 liters per minute. As described in the Chapter Opener, in a trained athlete, the heart actually enlarges (in extreme cases up to 50%) and is capable of pumping a greater quantity of blood per beat. An athlete's heart is thus more efficient and does not have to beat as often to distribute the same quantity of blood as does the heart of a person who is not in good physical condition.

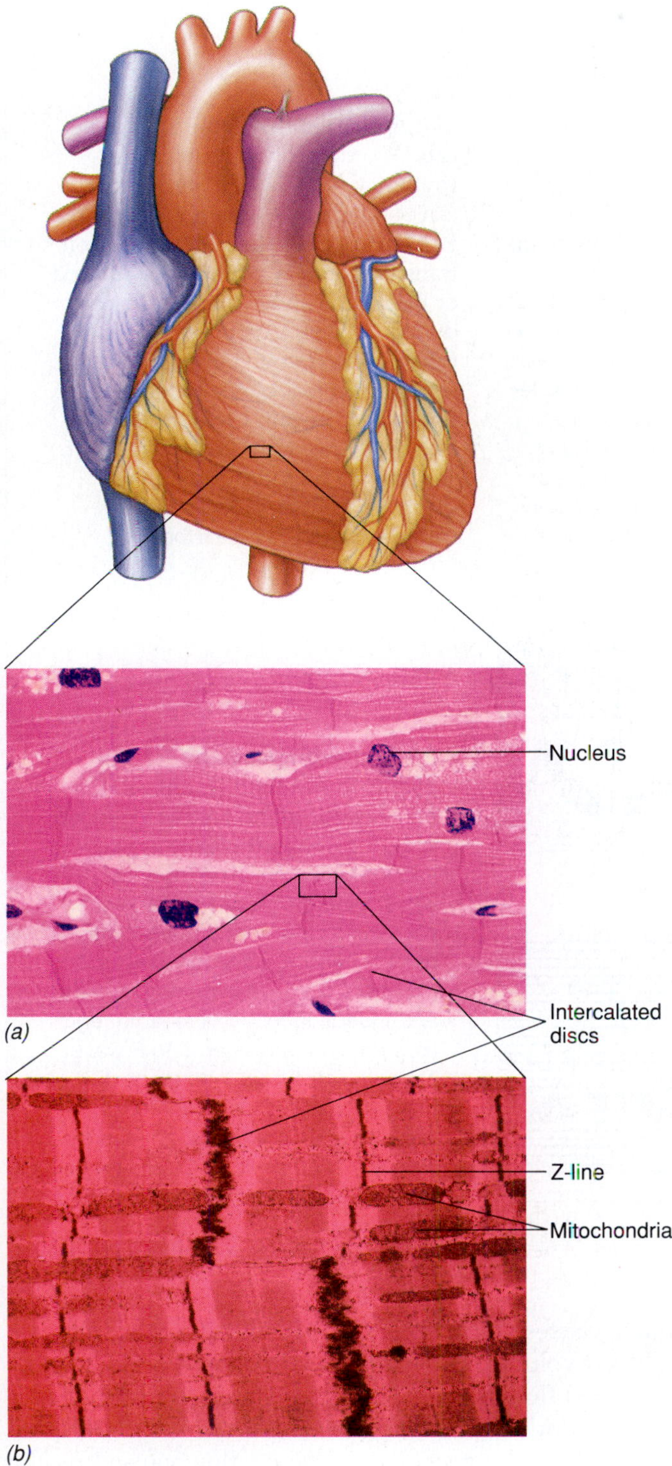

FIGURE 36–11 Cardiac muscle. (***a***) Cardiac muscle as seen with the light microscope (approximately x 250). (***b***) Electron micrograph of cardiac muscle (x 41,000). (*a*, Ed Reschke; *b*, Visuals Unlimited/Don Fawcett)

Two Main Heart Sounds Can Be Distinguished

When you listen to the heartbeat with a stethoscope you can distinguish two distinct sounds, which occur in repeating rhythm. These sounds, usually described as a "lub-dup," are produced each time the heart valves close. The first sound, the "lub," is caused by the closing of the AV valves and marks the beginning of ventricular systole, which is the phase of the heart's cycle when the ventricles contract. The "dup" sound is heard as a quick snap and is caused by the closing of the semilunar valves. This marks the beginning of ventricular diastole, which is the phase of the heart's cycle when the ventricles relax.

A **heart murmur** is a common type of abnormal sound that sometimes indicates a valve disorder. When a valve does not close properly, some blood may flow backwards, creating a hissing sound. Characteristic murmurs may also be heard when a valve is enlarged with scar tissue and is rough, so the passageway is narrowed. Heart murmurs usually do not seriously impair heart function.

BLOOD PRESSURE DEPENDS ON BLOOD FLOW AND RESISTANCE TO BLOOD FLOW

Blood pressure is the force exerted by the blood against the inner walls of the blood vessels. It is determined by the blood flow and the resistance to that flow (Fig. 36–12). Blood flow depends directly on the pumping action of

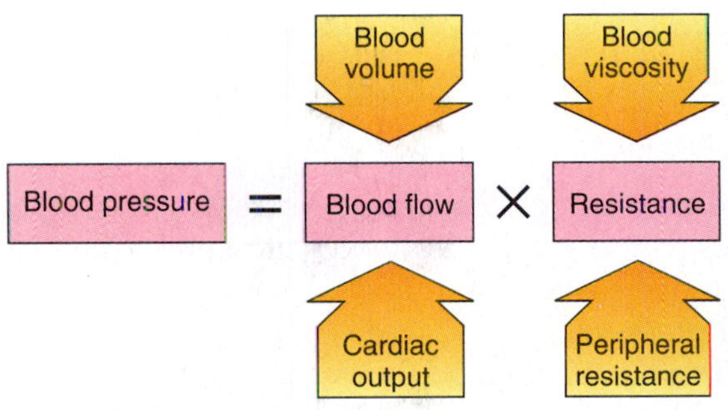

FIGURE 36–12 Some factors that influence blood pressure. Any factor that increases blood flow or resistance raises blood pressure.

FOCUS ON *The Process of Science*

THE ELECTRICAL ACTIVITY OF THE HEART

As each wave of contraction spreads through the heart, electrical currents spread into the tissues surrounding the heart and onto the body surface. This electrical activity can be recorded by placing electrodes on the body surface on opposite sides of the heart. The **electrocardiograph** is the machine used to amplify and record the electrical activity, and the record produced is called an **electrocardiogram (ECG or EKG)**. In intensive-care units and operating rooms, an oscilloscope is often used instead of an electrocardiograph. The oscilloscope continuously monitors the heart, displaying a moving beam of electrons on a screen.

An ECG begins with a **P wave,** which represents the spread of an impulse through the atria just before atrial contraction. Then a **QRS complex** appears, reflecting the spread of an impulse through the ventricles just before they contract. As the ventricles recover, currents generated are reflected upon the graph as a **T wave.** The heart then repeats its pattern of electrical impulses, generating a new P wave, QRS complex, and T wave.

Abnormalities in the ECG indicate disorders in the heart or its rhythm. One class of disorders that can be diagnosed with the help of the ECG is **heart block.** In this condition, transmission of an impulse is delayed or blocked at some point in the conduction system. **Artificial pacemakers** can be implanted in patients with severe heart block. A pacemaker is implanted beneath the skin, and its electrodes are connected to the heart. This device provides continuous rhythmic impulses that avoid the block and drive the heartbeat.

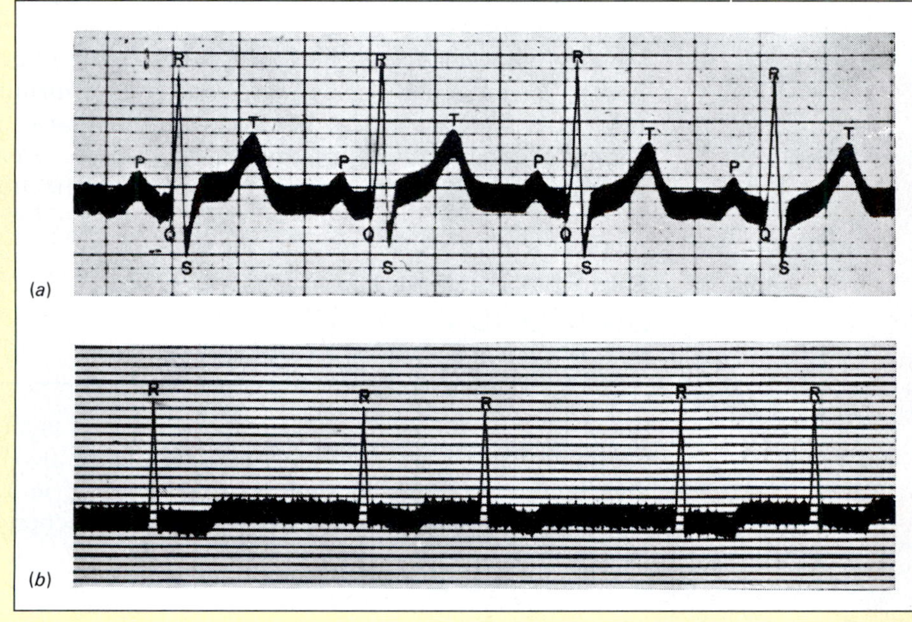

Electrocardiograms. (***a***) Tracing from a normal heart. The P wave corresponds to the contraction of the atria, the QRS complex to the contraction of the ventricle, and the T wave to the relaxation of the ventricle. (***b***) Tracing from a patient with atrial fibrillation. The individual muscle fibers of the atrium twitch rapidly and independently. There is no regular atrial contraction and no P wave. The ventricles beat independently and irregularly, causing the QRS wave to appear at irregular intervals. (Courtesy of Dr. Lewis Dexter and the Peter Bent Brigham Hospital, Boston, Mass.)

the heart. When cardiac output increases, blood flow increases, causing a rise in blood pressure. When cardiac output decreases, blood flow decreases, causing a fall in blood pressure.

The volume of blood flowing through the system also affects blood pressure. When blood volume is reduced by hemorrhage or chronic bleeding, blood pressure drops. On the other hand, an increase in blood volume results in an increase in blood pressure. For example, a high dietary salt intake causes water retention. This results in an increase in blood volume and leads to higher blood pressure.

Blood flow is impeded by resistance; when the resistance to flow increases, blood pressure rises. **Peripheral resistance** is the resistance to blood flow caused by the viscosity of the blood and by friction between the blood and the wall of the blood vessel. The length and diameter of a blood vessel determine the surface area of the vessel in contact with the blood. The length of a blood vessel does not change, but the diameter, especially of an arteriole, does. Even a small change in the diameter of a blood vessel causes a big change in blood pressure. For example, if the radius of a blood vessel is doubled, the resistance is reduced to one-sixteenth of its former value, and the blood flow increases sixteen-fold.

Blood pressure in arteries rises during systole and falls during diastole. A blood pressure reading is expressed as systolic pressure over diastolic pressure. Normal blood pressure of a young adult is about 120/80 (read as "120 over 80" and measured in millimeters of mercury, or mm Hg). When the diastolic pressure consistently reads over 95 mm Hg, the patient may be suffering from **hypertension** (high blood pressure). Hypertension, a common cardiovascular disorder, places a heavy burden on the heart, which must pump harder against greater blood pressure. Heredity, obesity, and possibly high dietary salt intake are thought to be important factors in the development of hypertension.

Blood Pressure Is Highest in Arteries

As you might imagine, blood pressure is greatest in the arteries and lessens as blood flows through the capillaries. By the time blood reaches the veins, its pressure is very low. When you are standing, it is really quite remarkable that blood in the feet manages to move against gravity and make its way back up to the heart. Much of the success of this journey may be attributed to flaplike valves within the veins. These valves prevent the blood from flowing (or falling) backwards. Blood is pushed along through the veins by the pressure of the blood behind it, and by the compression of veins due to muscle contractions as we move about (Fig. 36–13).

In people whose jobs require that they stand for long periods each day, blood accumulates in the veins of the legs. Excessive pooling of the blood stretches the veins, so that the cusps of their valves no longer meet and thus do not close properly. This may lead to **varicose veins,** especially in those who are obese or who have inherited weak vein walls. A varicose vein is dilated and elongated. **Hemorrhoids,** which are varicose veins in the anal region, occur when venous pressure in that region is constantly elevated, as in chronic constipation (because of straining) and during pregnancy (because of pressure of the enlarged uterus).

Blood Pressure Is Carefully Regulated

Several complex homeostatic mechanisms interact to maintain normal blood pressure. Receptors, called **baroreceptors,** located in certain arteries, are sensitive to changes in blood pressure. When stimulated, baroreceptors send messages to centers in the medulla of the brain. Nerves then signal the heart to slow down or speed up, resulting in an adjustment of blood pressure. Other nerves send messages to arterioles and veins, causing them to dilate or constrict. For example, when blood pressure decreases as you get out of bed in the morning, the heart rate increases slightly and blood vessels are constricted, so blood pressure increases. These neural reflexes act continuously to maintain a steady state of blood pressure.

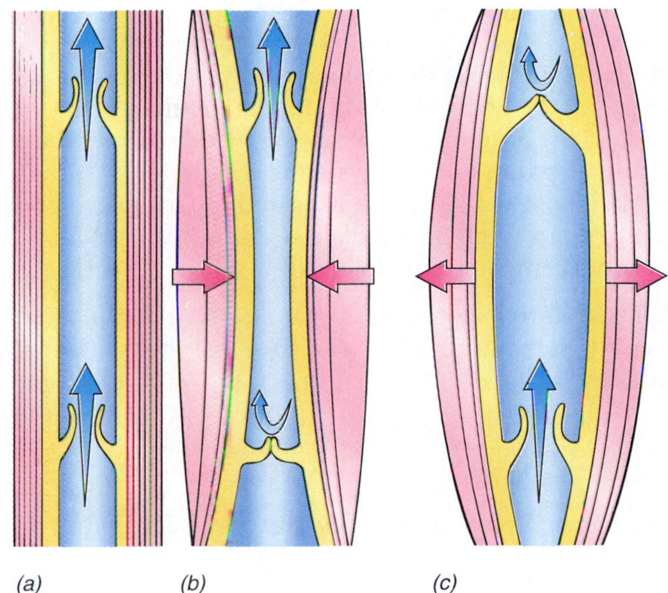

FIGURE 36–13 The action of valves and skeletal muscle in moving blood through veins. (***a***) When blood flows up toward the heart, the valves are pushed open. (***b***) When blood begins to fall downward in the vein, it fills the cup-like flaps of the valve. The flaps are forced together, preventing backflow. When muscles contract and bulge, they compress veins forcing blood toward the heart. The lower valve prevents backflow. (***c***) Muscles relax, and the vein expands and fills with blood from below. The upper valve prevents backflow.

Hormones are also involved in the regulation of blood pressure. The **angiotensins** are a group of hormones that are powerful vasoconstrictors. When blood pressure is low, the kidneys release the hormone **renin,** which stimulates the formation of angiotensins from a plasma protein. Angiotensins cause blood vessels to constrict, which results in an increase in blood pressure. The kidneys also help maintain blood pressure by influencing blood volume. In response to hormones, the kidneys vary the rate of excretion of salts and water, thereby increasing or decreasing the volume of the blood plasma.

IN BIRDS AND MAMMALS, BLOOD IS PUMPED THROUGH A PULMONARY AND A SYSTEMIC CIRCULATION

One of the main jobs of the circulation is to deliver oxygen to all of the cells of the body. In humans, as in other mammals and in birds, blood is charged with oxygen in the lungs. Then it is returned to the heart to be pumped out into the arteries that deliver it to the other tissues and organs of the body. There is a double circuit of blood vessels: (1) the **pulmonary circulation,** which delivers blood to the lungs, and (2) the **systemic circulation,** which delivers blood to all of the tissues and organs of the body. The pattern of blood circulation in birds and mammals may be summarized as follows:

Veins (return blood from organs) ⟶ Right atrium ⟶ Right ventricle ⟶ One of the pulmonary arteries ⟶ Capillaries in the lung ⟶ One of the pulmonary veins ⟶ Left atrium ⟶ Left ventricle ⟶ Aorta ⟶ Arteries (conduct blood to organs) ⟶ Capillaries

This general pattern of circulation may be traced in Figure 36–14.

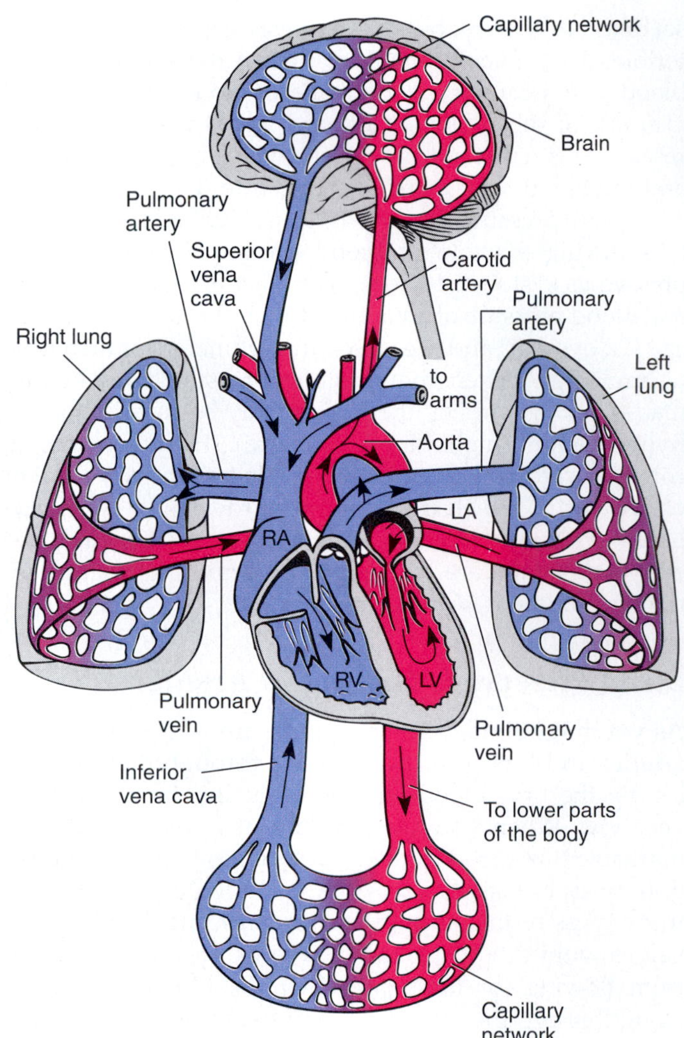

FIGURE 36–14 The pattern of blood flow. This simplified diagram shows the circulation of blood through the systemic and pulmonary circuits. Red represents oxygen-rich blood; blue represents oxygen-poor blood.

The Pulmonary Circulation Oxygenates the Blood

Blood from the tissues returns to the right atrium of the heart partially depleted of its oxygen supply but loaded with carbon dioxide wastes. This oxygen-poor blood is directed to the lungs where oxygen diffuses into the blood and carbon dioxide diffuses out. The right ventricle pumps blood into the pulmonary circulation (Fig. 36–14). The **pulmonary arteries** carry blood to the lungs. These are the only arteries in the human body that carry oxygen-poor blood. In the lungs, the pulmonary arteries branch into smaller and smaller vessels, which finally lead into an extensive network of pulmonary capillaries that bring blood to the air sacs of the lungs. As blood circulates through this capillary network, gases are exchanged. **Pulmonary veins,** the only veins in the body to carry oxygen-rich blood, return blood to the left atrium of the heart.

The Systemic Circulation Delivers Blood to All of the Tissues

Blood returning from the pulmonary circulation enters the left atrium of the heart, then passes into the left ventricle. From there it is pumped into the largest artery of the body, the **aorta.** The aorta divides into arterial branches that carry blood to all regions of the body, including the heart muscle itself. Some of the principal branches include the **carotid arteries** to the brain, the **subclavian arteries** to the shoulder and arm region, the **coronary arteries** to the heart muscle, the **mesenteric arteries** to the intestine, the **renal arteries** to the kidneys, and the **iliac arteries** to the legs (Fig. 36–15). Each of these branches into smaller and smaller arteries that bring blood to the capillary networks within each organ.

Blood returning from the brain is carried back toward the heart by the **jugular veins.** Blood from the shoulders and arms drains into the **subclavian veins.** These veins and others bringing blood from the upper portion of the body merge, forming the **superior vena cava,** a very large vein that empties blood into the right atrium. **Renal veins** from the kidneys, **iliac veins** from the lower limbs, **hepatic veins** from the liver, and other veins returning blood from the lower regions of the body empty blood into the **inferior vena cava,** which returns blood to the right atrium.

The Coronary Circulation Delivers Blood to the Heart

The heart requires a large and continuous supply of nutrients and oxygen. Blood flowing through its chambers cannot serve these needs because the heart wall is too thick to permit effective diffusion. The cardiac muscle is

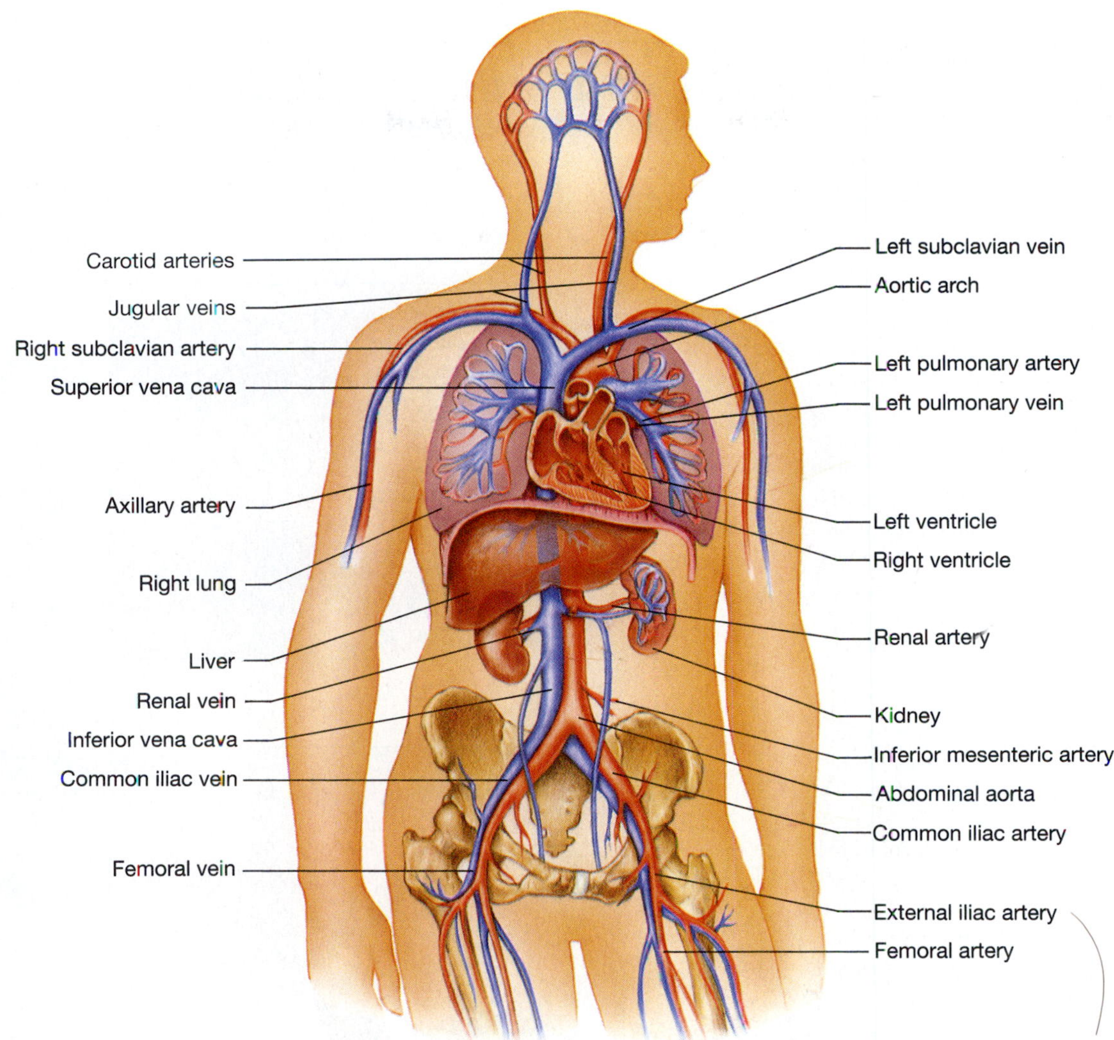

FIGURE 36–15 Circulation of blood through some of the principal arteries and veins of the body. Blood vessels carrying oxygen-rich blood are red; those carrying oxygen-depleted blood are blue.

supplied by its own system of blood vessels, called the **coronary circulation** (see Fig. 36–8).

Two **coronary arteries** branch off from the aorta just as it leaves the heart. These arteries give rise to an extensive system of blood vessels within the heart tissue. Most of the coronary capillaries empty into veins that join, forming a large vein, the **coronary sinus,** which empties into the right atrium. Blockage of the coronary arteries is a principal cause of heart disease (see Focus on Health and Human Affairs: Cardiovascular Disease).

The Hepatic Portal System Delivers Nutrients to the Liver

Blood almost always travels from artery to capillary to vein. An exception to this sequence occurs in the **hepatic portal system,** which delivers blood rich in nutrients to the liver. Blood is conducted to the small intestine by the superior mesenteric artery. Then, as it flows through capillaries within the wall of the intestine, blood picks up glucose, amino acids, and other nutrients. This blood passes into the mesenteric vein and then into the **hepatic portal vein.** Instead of going directly back to the heart (as most veins would), the hepatic portal vein delivers nutrients to the liver.

Within the liver, the hepatic portal vein gives rise to an extensive network of tiny blood vessels. As blood circulates through these vessels, liver cells remove nutrients and store them. These small blood vessels merge to form hepatic veins, which deliver blood to the inferior vena cava.

FOCUS ON *Health and Human Affairs*

CARDIOVASCULAR DISEASE

Cardiovascular disease is the number one cause of death in the United States and in most other industrial societies. Most often death results from some complication of **atherosclerosis*** (hardening of the arteries as a result of lipid and calcium deposition). Although atherosclerosis can affect almost any artery, the disease most often develops in the aorta and in the coronary and cerebral arteries. When it occurs in the cerebral arteries, it can lead to a **cerebrovascular accident (CVA),** commonly referred to as a stroke.

Although there is apparently no single cause of atherosclerosis, several major risk factors have been identified:

1. Elevated levels of cholesterol in the blood, often associated with diets rich in total calories, total fats, saturated fats, and cholesterol.
2. Hypertension. The higher the blood pressure, the greater the risk.
3. Cigarette smoking. The risk of developing atherosclerosis is two to six times greater in smokers than in nonsmokers and is directly proportional to the number of cigarettes smoked daily.
4. Diabetes mellitus, an endocrine disorder in which glucose is not metabolized normally.

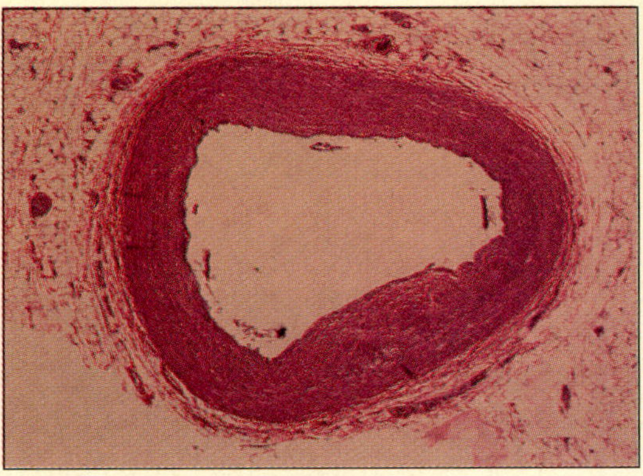

(a)

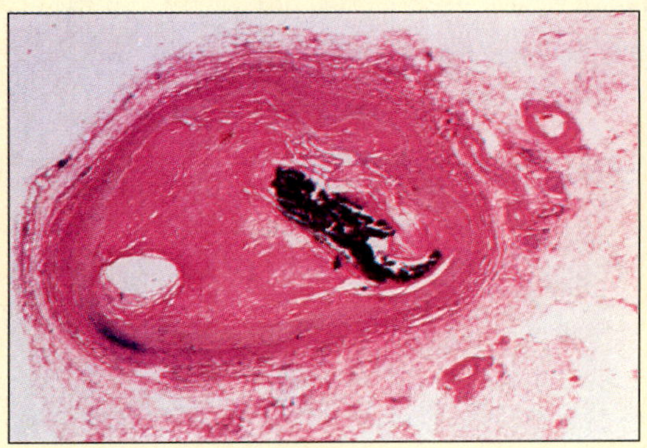

(b)

Progression of atherosclerosis. (***a***) Normal coronary artery. (***b***) Atherosclerosis in coronary artery. The marked thickening of the arterial wall almost completely blocks the passage of blood in this artery. (*a*, Visuals Unlimited/ Cabisco; *b*, Visuals Unlimited/ Sloop-Ober)

* Atherosclerosis is the most common form of arteriosclerosis, any disorder in which arteries lose their elasticity.

THE LYMPHATIC SYSTEM IS AN ACCESSORY CIRCULATORY SYSTEM

The **lymphatic system** is an accessory circulatory system that is connected with blood circulation (Fig. 36–16). Its three principal functions are to (1) collect and return tissue fluid to the blood, (2) defend the body against disease organisms, and (3) absorb lipids from the digestive system. Here we will focus on the first function; the other two are discussed in Chapters 37 and 39, respectively.

The Lymphatic System Consists of Lymphatic Vessels and Lymph Tissue

The lymphatic system has tiny "dead-end" capillaries that extend into almost all tissues of the body (Fig. 36–17).

The risk of developing atherosclerosis also increases with age. Estrogen hormones are thought to offer some protection in women until after menopause, when the concentration of these hormones decreases. Other suggested risk factors that are currently being studied are obesity, hereditary predisposition, lack of exercise, stress and behavior patterns, and dietary factors.

In atherosclerosis, lipids are deposited in the smooth muscle cells of the arterial wall. Cells in the arterial wall proliferate and the inner lining thickens. More lipid, especially cholesterol from low-density lipoproteins, accumulates in the wall. Eventually calcium is deposited there, contributing to the slow formation of hard plaque. As the plaque develops, arteries lose their ability to stretch when they fill with blood, and they become progressively occluded (blocked), as shown in the figure. As the artery narrows, less blood can pass through to reach the tissues served by that vessel and the tissue may become **ischemic** (lacking in blood). Under these conditions, the tissue is deprived of an adequate oxygen supply.

When a coronary artery becomes narrowed, **ischemic heart disease** can occur. Sufficient oxygen may reach the heart tissue during normal activity, but the increased need for oxygen during exercise or emotional stress results in the pain known as **angina pectoris.** Persons with this condition often carry nitroglycerin pills with them for use during an attack. This drug dilates veins so that venous return is reduced. Cardiac output is lowered so that the heart is not working so hard and requires less oxygen. Nitroglycerin also dilates the coronary arteries slightly, allowing more blood to reach the heart muscle.

Myocardial infarction (MI), popularly referred to as heart attack, is a very serious, often fatal, consequence of ischemic heart disease. MI often results from a sudden decrease in coronary blood supply. The portion of cardiac muscle deprived of oxygen dies within a few minutes and is then referred to as an **infarct.** MI is the leading cause of death and disability in the United States. Just what triggers the sudden decrease in blood supply that causes MI is a matter of some debate. It is thought that in some cases an episode of ischemia triggers a fatal arrhythmia such as **ventricular fibrillation,** a condition in which the ventricles contract very rapidly without actually pumping blood. In other cases, a **thrombus** (clot) may form in a diseased coronary artery. Because the arterial wall is roughened, platelets may adhere to it and initiate clotting.

If the thrombus blocks a sizable branch of a coronary artery, blood flow to a portion of heart muscle is impeded or completely halted. This condition is referred to as a coronary occlusion. If the coronary occlusion prevents blood flow to a large region of cardiac muscle, the heart may stop beating—that is, cardiac arrest may occur—and death can follow within moments. If only a small region of the heart is affected, however, the heart may continue to function. Cells in the region deprived of oxygen die and are replaced by scar tissue.

Tissue fluid enters the lymph capillaries and is then referred to as **lymph.** Lymph capillaries conduct the lymph to larger vessels called **lymph veins** (or **lymphatics**).

At strategic locations, lymph veins enter **lymph nodes,** which are small organized masses of lymph tissue. Lymph nodes have two main functions: (1) they filter the lymph as it slowly passes through, and (2) they produce **lymphocytes,** which are white blood cells important in immune responses. Lymph nodes (sometimes called **lymph glands**) are most numerous in the neck region, under the arms, in the groin region, and in the chest and abdomen. Lymph nodes in an infected area enlarge conspicuously and may be felt as hard little knots below the skin.

Lymph veins that leave the lymph nodes conduct lymph toward the shoulder region. Eventually, lymph veins empty their contents into the subclavian veins by way of the thoracic and right **lymphatic ducts.**

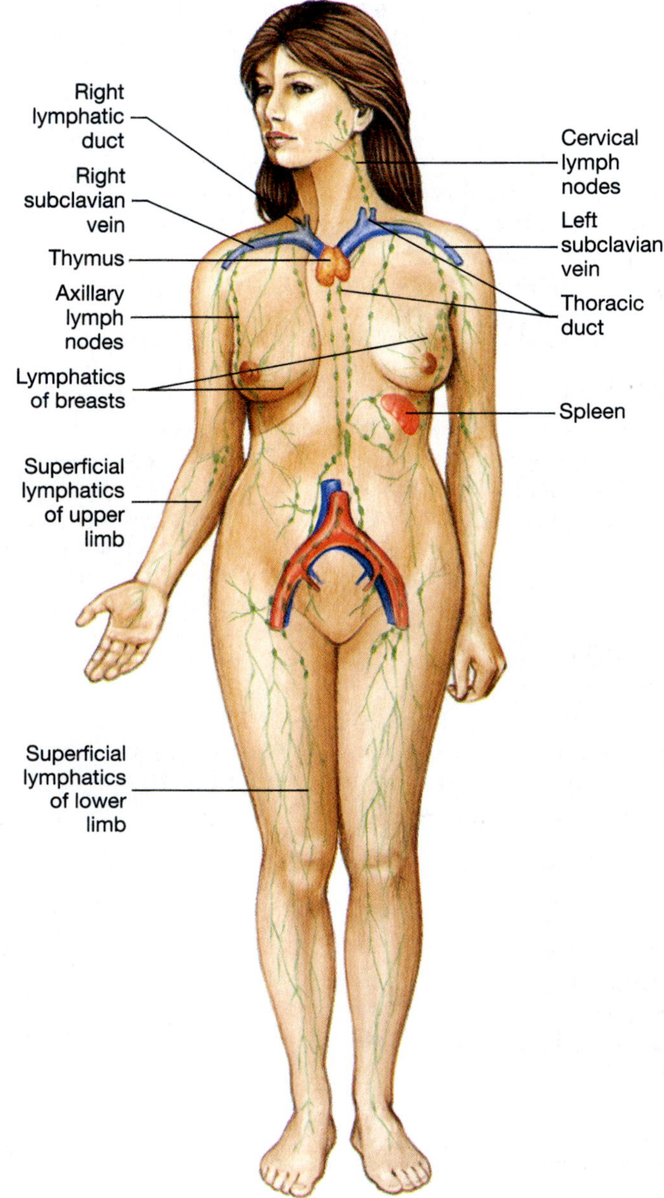

FIGURE 36–16 The lymphatic system. Note that while the lymphatic vessels extend into most tissues of the body, the lymph nodes are clustered in certain regions. The right lymphatic duct drains lymph from the upper right quadrant of the body. The thoracic duct drains lymph from other regions of the body.

Tonsils are masses of lymph tissue under the lining of the oral cavity and throat. (The pharyngeal tonsils in back of the nose are called **adenoids** when they are enlarged.) Tonsils help protect the respiratory system from infection by destroying bacteria and other foreign matter that enter the body through the mouth or nose. Unfortunately, tonsils are sometimes overcome by invading germs, become the site of frequent infection themselves, and then become prime targets for surgical removal.

Some nonmammalian vertebrates, such as frogs, have lymph "hearts" that pulsate and squeeze lymph along. In mammals, however, the walls of the lymph vessels themselves pulsate, pushing the lymph through the vessels. When muscles contract or arteries pulsate, pressure on the lymph vessels enhances lymph flow. Valves within the lymph vessels prevent backflow.

The Lymphatic System Plays an Important Role in Fluid Homeostasis

When blood enters a capillary network, it is under high enough pressure that it forces some plasma out of the capillaries and into the tissues. Once it leaves the blood vessels, this fluid is called **interstitial fluid,** or **tissue fluid.** It is somewhat similar to plasma but contains no red blood cells or platelets and only a few white blood cells. Its protein content is about one-fourth of that found in plasma. This is because proteins are too large to pass easily through capillary walls. Smaller molecules dissolved in the plasma do pass out with the fluid leaving the blood vessels. Thus, interstitial fluid contains glucose, amino acids, other nutrients, and oxygen, as well as a variety of salts. This nourishing fluid bathes all of the cells of the body.

The main force pushing plasma out of the blood is hydrostatic pressure, that is, the pressure exerted by the blood on the capillary wall, which is caused by the beating of the heart (Fig. 36–18). At the venous ends of the capillaries, the hydrostatic pressure is much lower. Here the principal force is the osmotic pressure of the blood, which acts to draw fluid back into the capillary. However, osmotic pressure is not entirely effective. Not as much is returned to the circulation as escapes. Furthermore, protein does not return effectively into the venous capillaries. Instead, it tends to accumulate in the interstitial fluid. These potential problems are so serious that fluid balance in the body would be significantly disturbed within a few hours and death would occur within about 24 hours if it were not for the lymphatic system.

The lymphatic system preserves fluid balance by collecting about 10% of the interstitial fluid and the protein that accumulates in the fluid. The interstitial fluid, called lymph once it enters the lymph capillaries, is returned to the blood.

The walls of the lymph capillaries are composed of endothelial cells that overlap slightly. When interstitial fluid accumulates, it presses against these cells, pushing them inward like tiny swinging doors that can swing in only one direction. As fluid accumulates within the lymph capillary, these cell doors are pushed closed.

Obstruction of the lymphatic vessels can lead to **edema,** a swelling that results from excessive accumulation of interstitial fluid. Lymphatic vessels can be blocked as a result of injury, inflammation, surgery, or parasitic infection. For example, when a breast is removed (mas-

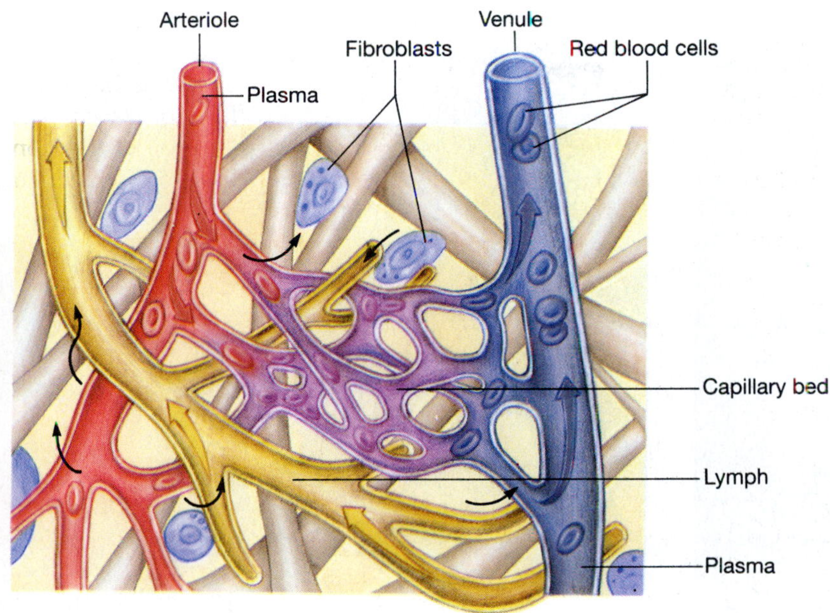

FIGURE 36–17 The relation of lymph capillaries to blood capillaries and tissue cells. Note that blood capillaries are connected to vessels at both ends, whereas lymph capillaries, shown in yellow, are dead-end streets. The arrows indicate direction of flow.

tectomy) because of cancer, lymph nodes in the underarm region often are also removed in an effort to prevent the spread of cancer cells. The disrupted lymph circulation may cause the patient's arm to swell greatly. Fortunately, new lymph vessels develop within a few months and the swelling slowly subsides.

Filariasis, a parasitic infection caused by a larval nematode that is transmitted to humans by mosquitoes, also

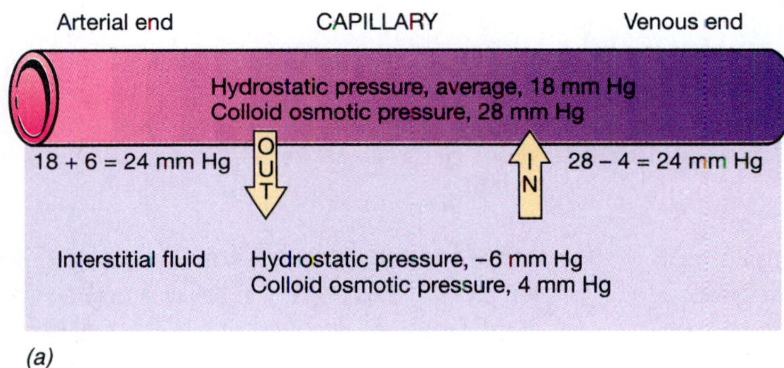

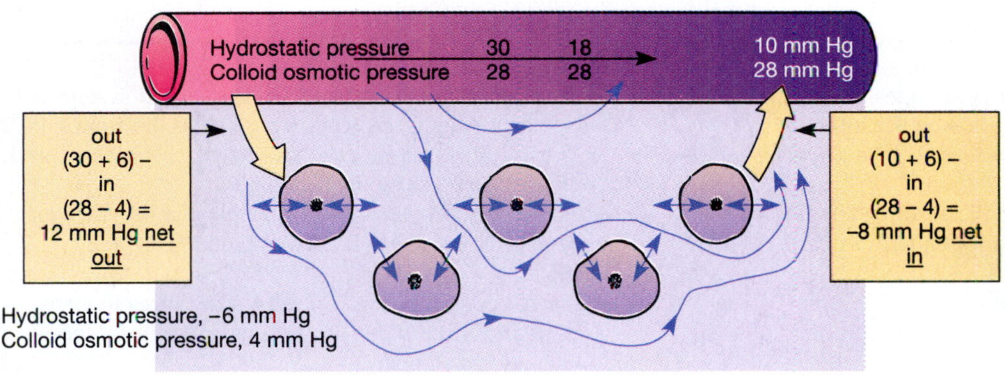

FIGURE 36–18 Movement of fluid. Materials are exchanged between blood and interstitial fluid through the thin walls of the capillaries. (***a***) Hydrostatic and osmotic pressures are responsible for the exchange of materials between blood and tissue fluid. (***b***) The path by which fluid flows between capillaries and tissue fluid.

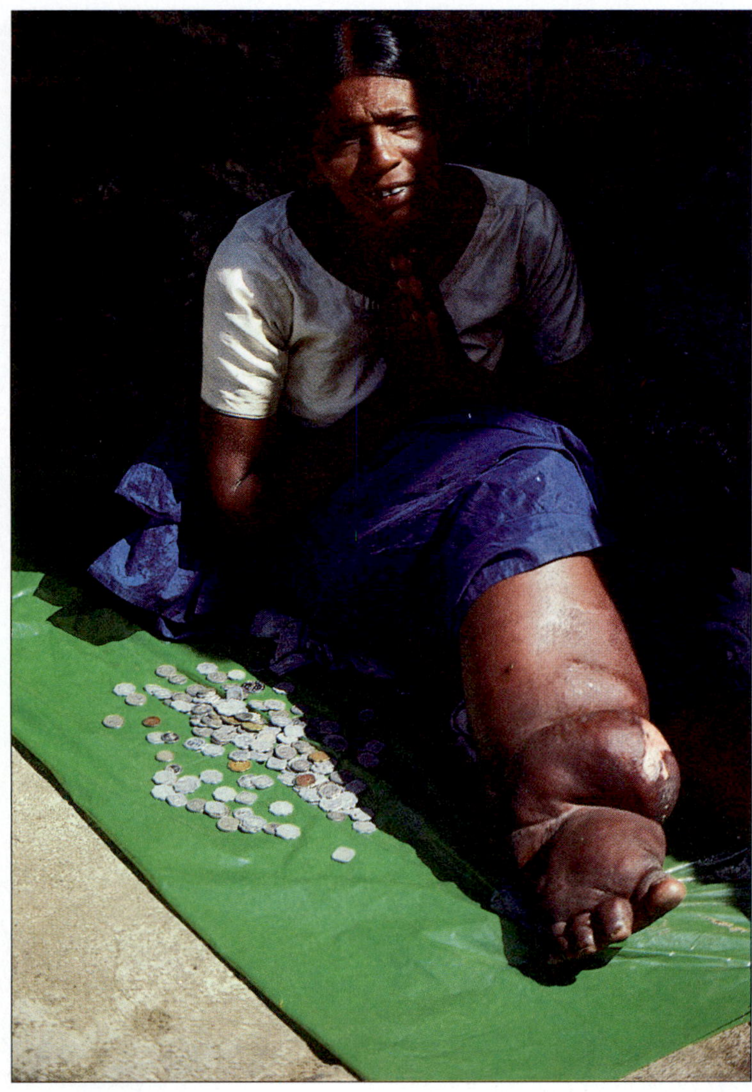

FIGURE 36–19 Lymphatic drainage is blocked in the limbs of this woman because of a parasitic infection known as filariasis. The condition characterized by such swollen limbs is elephantiasis. (Lawrence S. Burr)

disrupts lymph flow. The adult worms live in the lymph veins, blocking lymph flow. Interstitial fluid then accumulates, causing tremendous swelling. The term elephantiasis has been used to describe the swollen legs sometimes caused by this disease, because they resemble the huge limbs of an elephant (Fig. 36–19).

Chapter Summary

I. Small, simple invertebrates, such as sponges, cnidarians, and flatworms, depend on diffusion for internal transport. More complex invertebrates require a specialized circulatory system.

II. Arthropods and most mollusks have an open circulatory system in which blood flows into a hemocoel, bathing the tissues directly.

III. Other invertebrates have a closed circulatory system in which blood flows through a continuous circuit of blood vessels.

IV. The vertebrate circulatory system is a closed system that transports nutrients, oxygen, wastes, and hormones; it helps maintain fluid balance and body temperature, and it defends the body against disease.

V. Vertebrate blood consists of liquid plasma in which are suspended red blood cells, white blood cells, and platelets.

   A. Plasma consists of water, salts, substances in transport, and three types of proteins: albumins, globulins, and fibrinogen.

B. Red blood cells transport oxygen and carbon dioxide.
C. White blood cells defend the body against disease organisms.
D. Platelets patch damaged blood vessels and release substances essential for blood clotting.

VI. Arteries carry blood away from the heart; veins convey blood back toward the heart. Capillaries are tiny, thin-walled vessels through which materials are exchanged between blood and tissues.

VII. The vertebrate heart is a muscular pump.
A. The four-chambered hearts of birds and mammals have completely separate right and left sides that separate oxygen-rich and oxygen-depleted blood.
B. The heart is equipped with valves that prevent back-flow of blood.
C. Although the heart can beat independently of its nerve supply, the heart rate is regulated by sympathetic and parasympathetic nerves.

VIII. Blood pressure is determined by blood flow and resistance to flow.
A. Blood pressure is greatest in arteries and lowest in veins.
B. Neural and hormonal mechanisms act continuously to maintain normal blood pressure.

IX. In birds and mammals, blood flows through a double circuit: the pulmonary circulation and the systemic circulation.
A. The right atrium receives oxygen-depleted blood from the tissues and pumps it into the right ventricle. The right ventricle pumps blood to the lungs through the pulmonary circuit.
B. Blood from the lungs returns to the left atrium via the pulmonary veins and is pumped into the aorta by the left ventricle. The aorta sends arterial branches into all parts of the systemic circulation.
1. In the coronary circulation, blood is transported to the cells of the heart itself.
2. In the hepatic portal system, a vein gives rise to an extensive network of exchange vessels within the tissue of the liver.

X. Atherosclerosis, a form of cardiovascular disease, can lead to ischemic heart disease or cerebrovascular accidents (strokes).

XI. The lymphatic system, a subsystem of the circulatory system, collects tissue fluid and returns it to the blood.

Selected Key Terms

aorta, p. 712
arteriole, p. 703
artery, p. 703
atherosclerosis, p. 714
atrium, p. 705
blood, p. 700
blood pressure, p. 709
capillary, p. 703
cardiac cycle, p. 708
circulatory system, p. 698
closed circulatory system, p. 699
diastole, p. 708
hepatic portal system, p. 713
interstitial fluid, p. 716
lymph, p. 715
lymphatic system, p. 714
lymph node, p. 715
open circulatory system, p. 699
pacemaker, p. 708
plasma protein, p. 700
pulmonary circulation, p. 712
pulse, p. 708
red blood cell, p. 700
semilunar valve, p. 706
systemic circulation, p. 712
systole, p. 708
tissue fluid, p. 716
vasoconstriction, p. 704
vasodilation, p. 704
vein, p. 704
ventricle, p. 705
white blood cell, p. 700

Post-Test

1. In a/an ______________ circulatory system, the heart pumps blood into a hemocoel.

2. Simple invertebrates, such as sponges, depend on ______________ for internal transport.

3. The fluid component of blood is ______________.

4. A deficiency in hemoglobin is referred to as ______________.

5. During clotting, fibrinogen is converted to fibrin by the action of ______________.

6. The force exerted by the blood against the inner walls of the blood vessels is known as ______________ ______________.

7. The pulmonary vein delivers blood to the ______________ ______________.

8. The hepatic portal veins delivers blood to the ______________.

9. Blood pressure is sensed by ______________ within certain arteries.

10. The ______________ valve is located between the left atrium and ventricle.

11. The angiotensins are a group of powerful ______________.

12. When a tissue is ischemic, it lacks sufficient ______________.

13. In atherosclerosis, the _______________ wall thickens and may block the passage of _______________.
14. Lymph is produced when _______________ fluid enters vessels.
15. Red blood cells (a) transport oxygen (b) seek out and ingest bacteria (c) become macrophages (d) initiate clotting (e) two of the preceding answers are correct.
16. Platelets (a) transport oxygen (b) seek out and ingest bacteria (c) become macrophages (d) initiate clotting (e) two of the preceding answers are correct.
17. Neutrophils (a) transport oxygen (b) seek out and ingest bacteria (c) become macrophages (d) initiate clotting (e) two of the preceding answers are correct.
18. Capillaries (a) conduct blood toward the heart (b) have thick walls (c) are the main exchange vessels (d) must always be filled with lymph (e) are located mainly in the liver.
19. An open circulatory system is characteristic of (a) arthropods (b) cnidarians (c) annelids (earthworms) (d) humans (e) two of the preceding answers are correct.
20. Antibodies are (a) albumins (b) globulins (c) lipoproteins (d) fibrinogens (e) two of the preceding answers are correct.
21. Which of the following is true of the amphibian circulatory system? (a) it is an open system (b) it has two atria (c) it has two ventricles (d) all of the preceding answers are correct (e) only two of the preceding answers are correct
22. Label the figure. If you need help, see Figure 36–8.

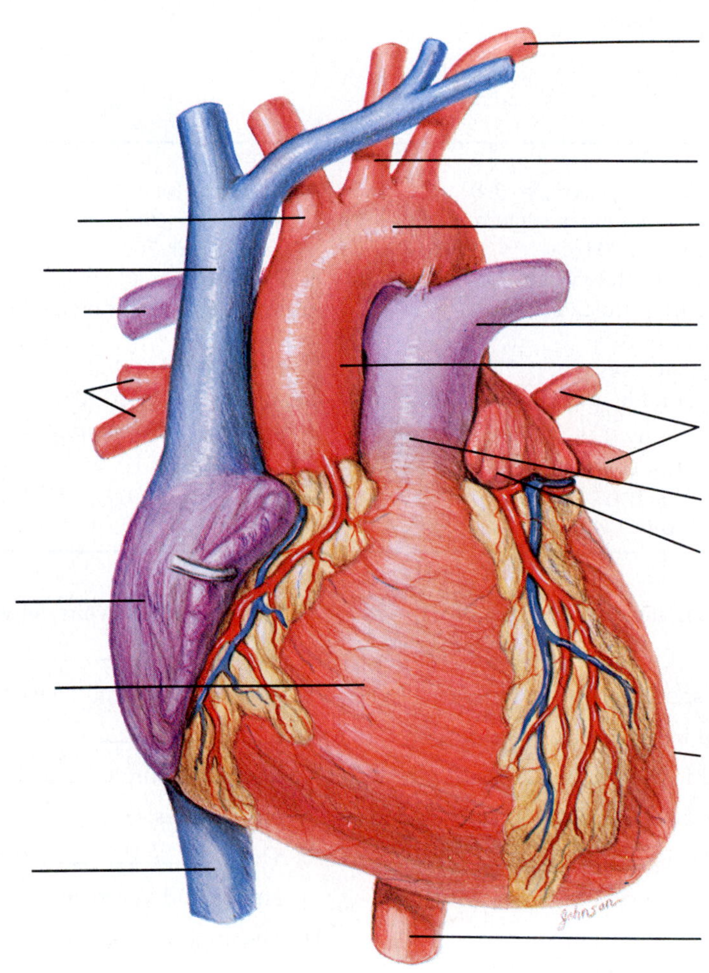

Review Questions

1. Give the functions of (a) red blood cells, (b) plasma proteins, (c) platelets, (d) monocytes, (e) macrophages, (f) neutrophils.
2. Where do red blood cells originate? How are they destroyed?
3. What are three general causes of anemia?
4. How are (a) arterioles and (b) capillaries structurally adapted for carrying out their specific functions?
5. Draw a diagram of the heart and label its chambers, valves, and the principal entering and exiting blood vessels.
6. How is the heartbeat initiated? Regulated?
7. How does blood manage to flow against gravity through veins in the legs on its route back to the heart?
8. Give an example of a normal blood pressure reading and of one from an individual with hypertension.
9. Trace a drop of blood from (a) superior vena cava to aorta, (b) brain to kidney, (c) intestine to lung.
10. How does the lymphatic system help maintain fluid balance?
11. What is the relationship among plasma, interstitial (tissue) fluid, and lymph?

Thinking Critically

1. How do the changes that take place during exercise affect the heart? Compare the heart function of an adult who does not exercise with that of an athlete.
2. Cartilage lacks blood and lymph vessels. How do you imagine its cells are nourished? What effect, if any, do you think a lack of blood vessels would have on healing after an injury?
3. How is a four-chambered heart different from a three-chambered heart? How is the heart of the fish specifically adapted to its life-style?
4. When the nerves to the heart are cut, the heart rate increases to about 100 contractions per minute. What does this indicate about the regulation of the heart rate?
5. Blood pressure is low in capillaries. Explain how this helps to retain fluid in the circulation.

Recommended Readings

Golde, D.W., "The Stem Cells," *Scientific American*, Vol. 265, No. 6, December 1991, pp. 86–93. A discussion of the cells that give rise to the various types of blood cells, with an emphasis on implications for developing new treatments for cancer and immune diseases.

Harken, A.H., "Surgical Treatment of Cardiac Arrhythmias," *Scientific American*, Vol. 269, No. 1, July 1993, pp. 68–74. A surgical procedure is used to correct dangerously rapid heartbeats.

Lawn, R.M., "Lipoprotein *(a)* in Heart Disease," *Scientific American*, Vol. 266, No. 6, June 1992, pp. 54–60. Lipoprotein *(a)* transports cholesterol; this protein can raise the risk of a heart attack.

Zivin, J.A., and D.W. Choi, "Stroke Therapy," *Scientific American*, Vol. 265, No. 1, July 1991, pp. 56–63. New strategies are being evaluated to limit the brain damage that occurs in these cardiovascular accidents.

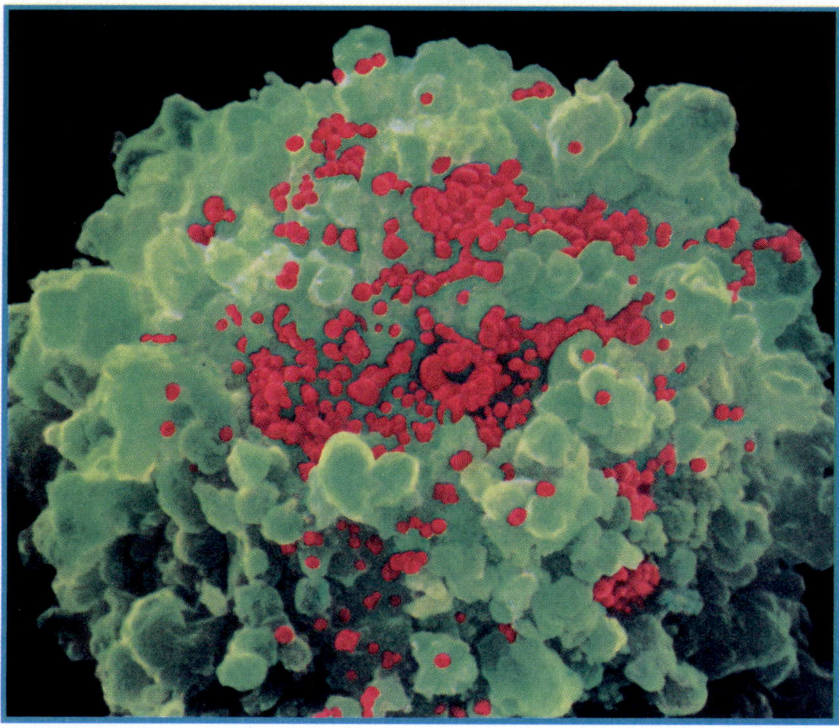

Color scanning EM of T-lymphocyte (green) infected with HIV (human immunodeficiency virus) (red). (NIBSC/Science Photo Library/Photo Researchers, Inc.)

HIV—The World's Most-Studied Virus

Acquired immune deficiency syndrome, or simply **AIDS,** is a deadly disease currently spreading throughout the world's human population at an alarming rate. First recognized in 1981, scientists estimate that more than 20 million people worldwide are now infected with the AIDS virus. Some epidemiologists suggest that these numbers reflect only a small percentage of the actual number of individuals infected. More than three million people have developed the disease and most of them have already died. In the United States alone, more than 200,000 individuals have been diagnosed with AIDS, and more than one million persons are estimated to be infected with the AIDS virus but as yet are symptom-free.

AIDS is caused by infection with a retrovirus called **human immunodeficiency virus (HIV).** (Recall that a retrovirus is an RNA virus that uses its RNA as a template to make DNA with the help of reverse transcriptase.) In a recent issue of *Scientific American,* Warner C. Green, director of the Gladstone Institute of Virology and Immunology and professor at the University of California, San Francisco, wrote that HIV has been studied more than any virus in history.

Several different strains of the virus are known, and some may be more deadly than others. HIV damages the immune system, and AIDS patients die (within several months to about 5 years) from rare forms of cancer, pneumonia, and other opportunistic infections.

Current evidence indicates that AIDS is transmitted mainly by semen during sexual intercourse with an infected person or by direct exposure to infected blood or blood products. Those most at risk are males who engage in homosexual and bisexual behavior and individuals who use intravenous drugs. The fastest rising risk groups are women (who at the time of this writing account for 40% of AIDS cases) and teenagers who contract the virus through heterosexual contact. Heterosexual contact with infected individuals accounts for one-third of HIV infection in women and an increasing number of cases in both men and women. Individuals who require frequent blood transfusions, such as those with hemophilia, and infants born to mothers with AIDS are also at risk. An estimated 10% of AIDS patients are children born to infected mothers.

Effective blood-screening procedures have been developed to safeguard blood bank supplies, markedly reducing the risk of infection from blood transfusion. Use of a latex condom during sexual intercourse provides some protection against the virus. Use of a spermicide containing nonoxynol-9 is thought to provide additional protection.

AIDS is not spread by casual contact. People do not contract the disease by hugging, kissing, sharing a drink, or using the same bathroom facilities. Friends and family members who live with AIDS patients are not more likely to get the disease. AIDS is discussed further in a later section of this chapter.

Internal Defense: Immunity

37

Learning Objectives

After you have studied this chapter you should be able to:

1. Compare in general terms the types of internal defense mechanisms in invertebrates and vertebrates.
2. Distinguish between specific and nonspecific defense mechanisms, and describe nonspecific mechanisms.
3. Describe the physiological changes and clinical symptoms associated with inflammation, and summarize the role of inflammation in the defense of the body.
4. Identify the cells of the immune system; and contrast T and B lymphocytes (T cells and B cells) with respect to life cycle and function.
5. Cite the functions of the thymus in immune mechanisms.
6. Define the terms antigen and antibody, describe how antigens stimulate immune responses, and draw the basic structure of an antibody.
7. Summarize the mechanisms of antibody-mediated immunity, including the effects of antigen-antibody complexes upon pathogens (include a discussion of the complement system).
8. Describe the mechanisms of cell-mediated immunity, including development of memory cells.
9. Contrast a secondary with a primary immune response.
10. Compare active and passive immunity, giving examples of each.
11. Describe how the body destroys cancer cells.
12. Summarize the immunological basis of graft rejection, and explain how the effects of graft rejection can be minimized.
13. Describe the immunological basis of autoimmune diseases, give two examples, and list possible causes.
14. Explain the immunological basis of allergy, and briefly describe the events that occur during (a) a hayfever response and (b) systemic anaphylaxis.
15. Describe the cause of AIDS, the risk factors and the progress of the disease, and summarize the difficulties encountered in developing a vaccine.

Key Concepts

- ☐ Internal defense depends upon the ability of an organism to distinguish between self and nonself.
- ☐ Many invertebrates have only nonspecific defense mechanisms such as phagocytosis; vertebrates have both nonspecific and specific defense mechanisms (immune responses). Specific defense responses

Key Concepts, continued

include antibody-mediated immunity and cell-mediated immunity.

☐ In antibody-mediated immunity, activated B cells multiply, forming a clone of cells. Some cells become plasma cells that secrete specific antibodies. Other cells become memory cells.

☐ In cell-mediated immunity, activated T cells become helper T cells, cytotoxic T cells, suppressor cells, or memory cells. Cytotoxic T cells migrate to the site of infection and destroy cells infected by pathogens.

Internal defense depends on the ability to distinguish between self and nonself

Animals have internal defense mechanisms that protect them against the disease-causing organisms that constantly threaten them. Viruses, bacteria, fungi, and other microorganisms that cause disease are referred to as **pathogens.** Such organisms enter the body with the air we breathe, with the food or water we ingest, or through wounds in the skin. Internal defense depends on the ability of an organism to distinguish between *self* and *nonself.* Such recognition is possible because organisms are biochemically unique. Cells have surface proteins that are different from those on the cells of other species or even other members of the same species. An organism "knows" its own cells and "recognizes" those of other organisms as foreign.

Pathogens have macromolecules on their cell surfaces that the body recognizes as foreign. A single bacterium may have from 10 to more than 1000 distinct macromolecules on its surface. When a bacterium invades an animal, its distinctive surface macromolecules stimulate the animal's defense mechanisms. A substance capable of stimulating an immune response is called an **antigen.** Many macromolecules are antigenic, including proteins, RNA, DNA, and some carbohydrates.

Nonspecific defense mechanisms prevent pathogens from entering the body. Such mechanisms rapidly destroy those pathogens that do penetrate the outer defenses. Phagocytosis of invading bacteria is an example of a nonspecific defense mechanism.

Specific defense mechanisms are highly effective. Their weapons are tailor-made to combat specific antigens associated with each pathogen. Specific defense mechanisms are collectively referred to as **immune responses.** The term *immune* is derived from a Latin word meaning "safe." **Immunology,** the study of specific defense mechanisms, is one of the most exciting fields of medical research today.

Immune responses can be directed to the particular type of pathogen that infects the body. One of the body's most important specific defense mechanisms is the production of **antibodies,** highly specific proteins that help destroy antigens. In complex animals, internal defense includes immunological memory, the capacity to respond more effectively the second time foreign molecules invade the body.

Invertebrates have internal defense mechanisms that are mainly nonspecific

All invertebrate species that have been studied demonstrate the ability to distinguish between self and nonself. However, most invertebrates are able to make only nonspecific immune responses such as phagocytosis and the inflammatory response.

Sponge cells have specific glycoproteins on their surfaces that enable them to distinguish between self and nonself. When cells of two different species of sponges are mixed together, they reaggregate according to species. Cnidarians also have this ability and can reject grafted tissue and destroy foreign tissue.

Invertebrates that have a coelom have ameba-like phagocytes that engulf and destroy bacteria and other foreign matter. Coelomate invertebrates also have nonspecific substances in the hemolymph that kill bacteria, inactivate cells of some pathogens, and cause some foreign cells to clump. In mollusks, these hemolymph substances enhance phagocytosis by the phagocytes.

Certain annelids (e.g., earthworms) and cnidarians (e.g., corals) appear to have specific immune mechanisms and immunological memory. In them, and in some echinoderms and simple chordates, the body appears to remember antigens for a short period of time and can respond to them more effectively when they are encountered again. Echinoderms and tunicates are the simplest animals known to have differentiated white blood cells that perform immune functions.

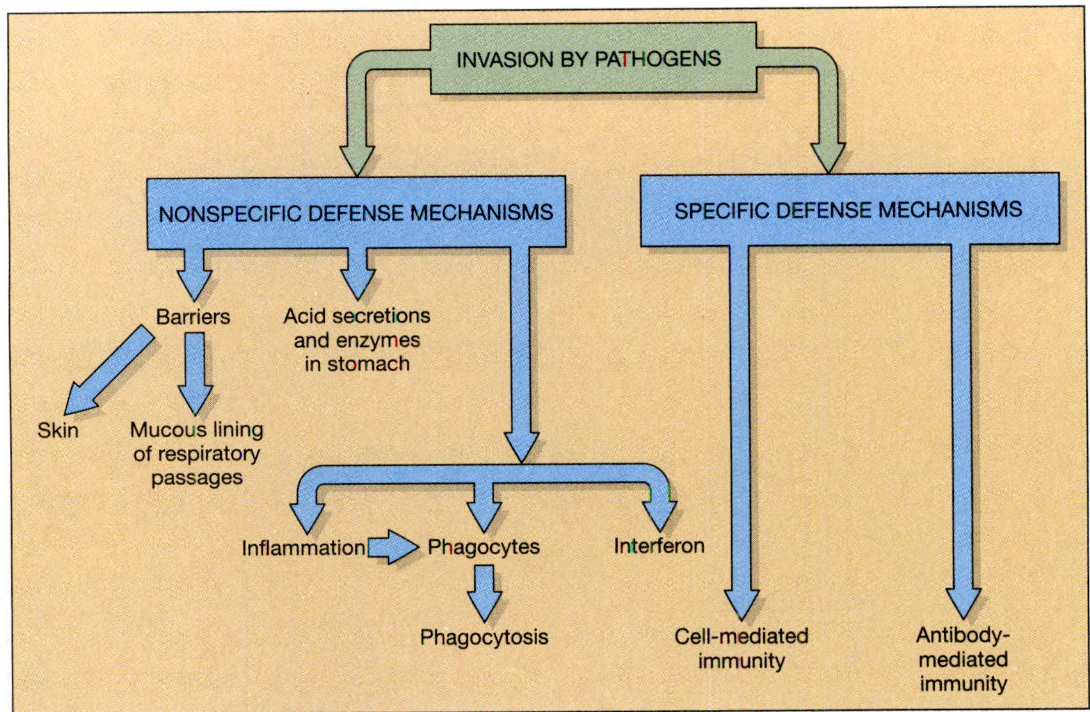

FIGURE 37–1 Nonspecific and specific defense mechanisms. Nonspecific mechanisms prevent entrance of pathogens and rapidly destroy those that manage to traverse the barriers. Specific defense mechanisms take longer to mobilize but are highly effective in destroying invaders.

VERTEBRATE NONSPECIFIC DEFENSE MECHANISMS INCLUDE MECHANICAL AND CHEMICAL BARRIERS AGAINST PATHOGENS

Vertebrates have many of the basic mechanisms present in invertebrates and also have more sophisticated specific defense mechanisms (Fig. 37–1). These are made possible by the development of a specialized lymphatic system. In the discussion that follows, we will focus on the human immune system, with references to those of other vertebrates.

The Skin and Mucous Membranes Are Barriers to Invasion

An animal's first line of defense against pathogens is its outer covering. For example, the *intact* human skin presents both a mechanical and a chemical barrier to microorganisms. Sweat and sebum contain chemicals that destroy certain types of bacteria. **Lysozyme,** an enzyme found in sweat, tears, and saliva, attacks the cell walls of many bacteria.

Microorganisms that enter with food are usually destroyed by the acid secretions and enzymes of the stomach. Pathogens that enter the body with inhaled air may be filtered out by hairs in the nose or trapped in the sticky mucous lining of the respiratory passageways. There, they may be destroyed by phagocytes.

Interferons Are a Nonspecific Defense Against Viral Infection

When infected by viruses or other intracellular parasites (some types of bacteria, fungi, and protozoa), cells respond by secreting proteins called **interferons.** One group of interferons signals neighboring cells to produce proteins that inhibit viral replication. The virus particles produced in cells exposed to interferon are not very effective at infecting cells. Another group of interferons stimulates macrophages to destroy tumor cells and host cells that have been infected by viruses.

Since their discovery in 1957, interferons have been the focus of much research. Recombinant DNA techniques are now used to produce large quantities of some interferons. By mid-1994, the Food and Drug Administration had approved interferons for treating several diseases, including chronic hepatitis, genital warts, a type of leukemia, a type of multiple sclerosis, and Kaposi's sarcoma (a type of cancer that sometimes develops in AIDS patients). Interferons are being tested in clinical trials for treatment of HIV infection and several types of cancer.

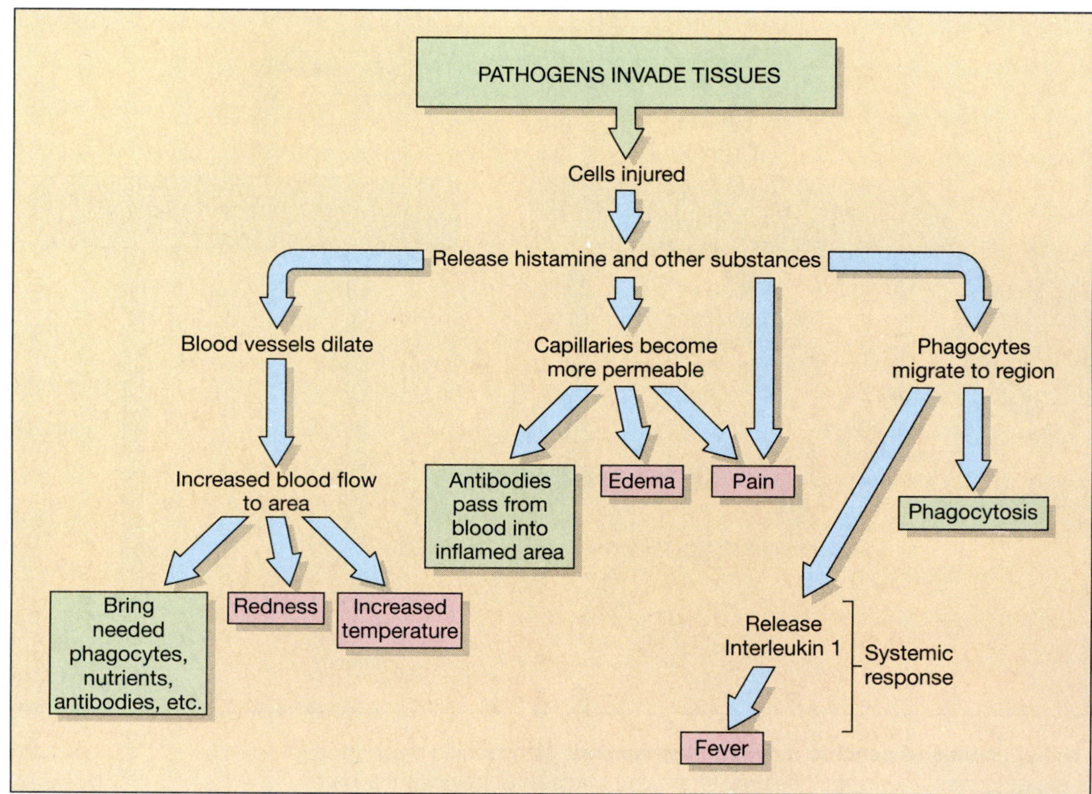

FIGURE 37–2 The inflammatory process. Inflammation is a vital process that localizes immune mechanisms at the area of infection. It permits phagocytic cells, antibodies, and other needed compounds to enter the tissue where microbial invasion is taking place.

Inflammation Is a Protective Mechanism

When pathogens invade tissues, they trigger an **inflammatory response** (Fig. 37–2). Injured cells and certain white blood cells (basophils) release **histamine** and other compounds that dilate blood vessels in the affected area. Blood flow increases to the infected region, bringing in great numbers of phagocytic cells. The increased blood flow makes the skin look red and feel warm.

Capillaries in the inflamed area become more permeable, allowing fluid and antibodies to leave the circulation and enter the tissues. As the volume of tissue fluid increases, **edema** (swelling) occurs. The edema (and also certain substances released by the injured cells) causes the pain that is characteristic of inflammation. Thus, the clinical characteristics of inflammation are *redness, heat, edema,* and *pain.*

Although inflammation is often a local response, sometimes the entire body is involved. **Fever** is a common clinical symptom of widespread inflammatory response. **Macrophages** and certain other cells release compounds, such as the peptide **interleukin-1** (IL-1), that reset the body's thermostat in the hypothalamus, resulting in fever. Prostaglandins (an important group of compounds derived from fatty acids) are also involved in this resetting process. Fever interferes with the growth of some microorganisms, perhaps by decreasing circulating levels of iron that they require. Fever also promotes the production of certain lymphocytes (T cells) and antibodies, and may increase phagocytosis. A short-term low fever may therefore help speed recovery.

Phagocytes Destroy Pathogens

One of the main functions of inflammation appears to be increased phagocytosis (Chapter 6). Two types of phagocytes in vertebrates are neutrophils and macrophages. A neutrophil can phagocytize 20 or so bacteria before it becomes inactivated (perhaps by leaking lysosomal enzymes) and dies. A macrophage can phagocytize about 100 bacteria during its lifespan.

Can bacteria counteract the phagocyte's attack? Certain bacteria release enzymes that destroy the membranes

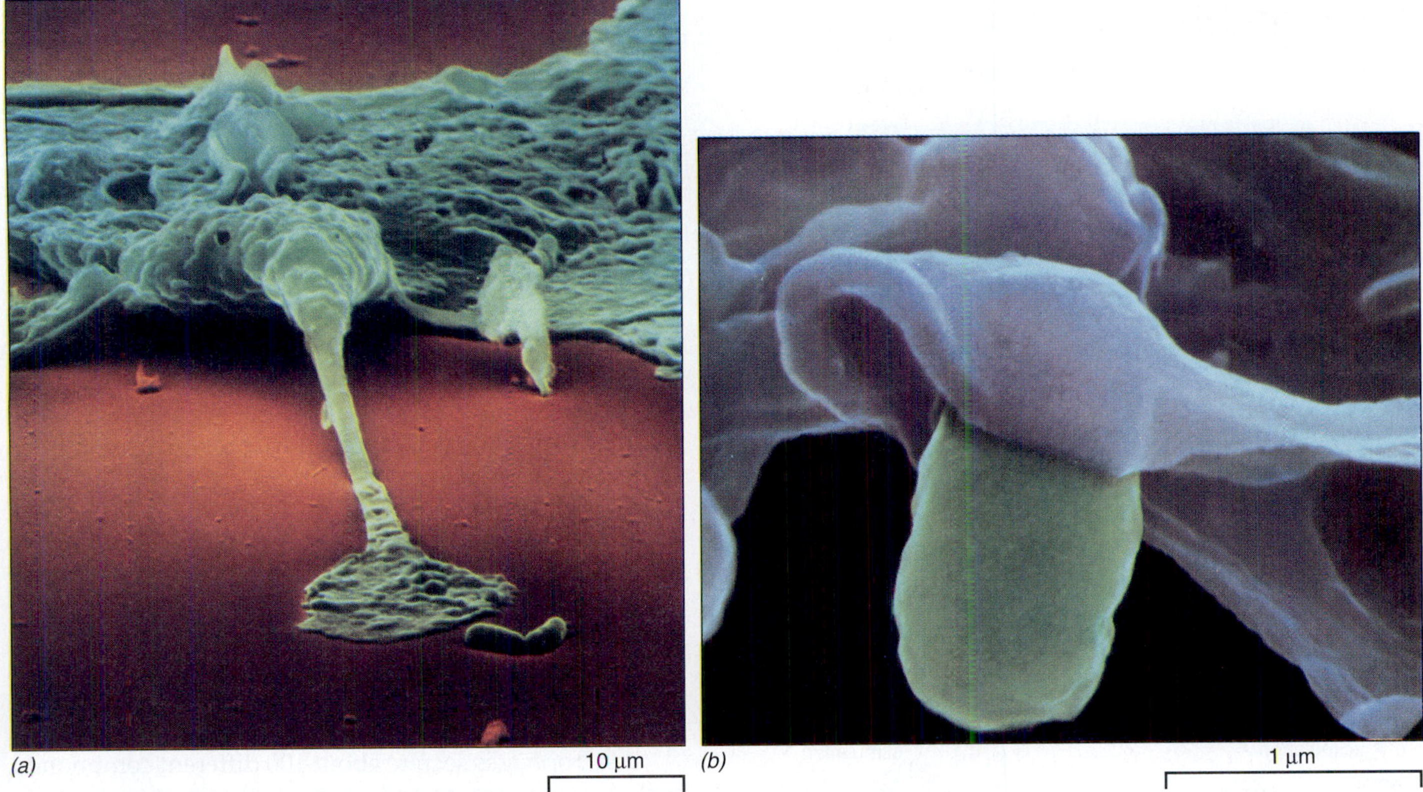

FIGURE 37–3 The macrophage is an efficient warrior. (***a***) A macrophage extends a pseudopod toward an invading *Escherichia coli* bacterium that is already multiplying. (***b***) The bacterium is trapped within the engulfing pseudopod. After the bacterium is taken into the cell, powerful lysosomal enzymes will destroy it. (From Lennart Nillson, © Boehringer Ingelheim International GmbH)

of the lysosomes. The powerful lysosomal enzymes then spill out into the cytoplasm and may destroy the phagocyte. Some bacteria have cell walls or capsules that resist the action of lysosomal enzymes.

Some macrophages wander through the tissue, phagocytizing foreign matter and, when appropriate, releasing antiviral agents (Fig. 37–3). Others stay in one place and destroy bacteria that pass by. For example, air sacs in the lungs contain large numbers of macrophages that destroy foreign matter entering with inhaled air.

SPECIFIC DEFENSE MECHANISMS IN VERTEBRATES INCLUDE ANTIBODY-MEDIATED IMMUNITY AND CELL-MEDIATED IMMUNITY

While nonspecific defense mechanisms destroy pathogens and prevent the spread of infection, specific defense mechanisms are being mobilized. Several days are required to activate specific immune responses, but once in gear, these mechanisms are extremely effective. Two main types of specific immunity are antibody-mediated immunity and cell-mediated immunity. In **antibody-mediated immunity,** lymphocytes produce specific antibodies designed to destroy the pathogen. In **cell-mediated immunity,** lymphocytes attack the cells infected by invading pathogens. Specific immunity depends on several types of cells.

Cells of the Immune System Include Lymphocytes and Phagocytes

Immune responses depend on two main groups of white blood cells: lymphocytes and phagocytes. The millions of lymphocytes are the main warriors in specific immune responses. Lymphocytes are stationed strategically in the lymphatic tissue throughout the body. Two main types of lymphocytes are **T lymphocytes,** or **T cells,** and **B lymphocytes,** or **B cells.** A third type of lymphocyte includes

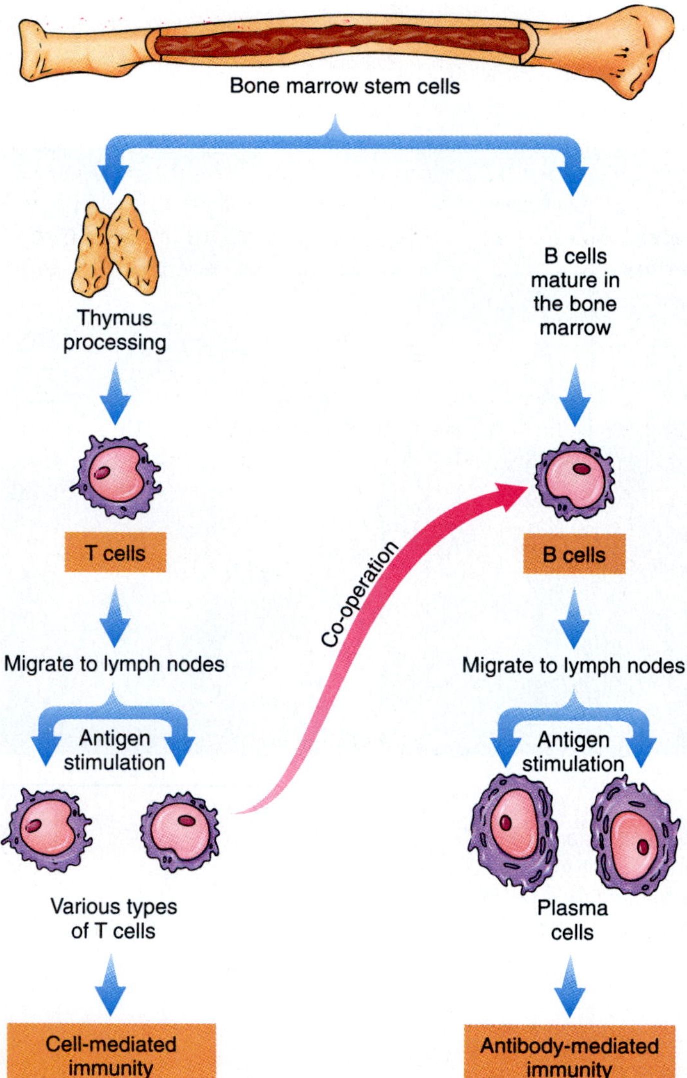

FIGURE 37–4 Origin and functions of T cells and B cells.

the **natural killer (NK) cell**, which kills virally infected cells and tumor cells.

T cells originate from stem cells in the bone marrow (Fig. 37–4). On their way to the lymph tissues, the future T cells stop off in the thymus gland for processing. (The "T" in T cells stands for thymus-derived.) Somehow the thymus gland influences the differentiation of lymphocytes, making them capable of immunological response. T cells are responsible for cellular immunity.

Three main types, or subsets, of T cells have been identified:

1. **Cytotoxic T cells,** also known as killer T cells, recognize and destroy cells with foreign antigens on their surfaces.
2. **Helper T cells** secrete substances that activate or enhance the immune response.
3. **Suppressor T cells** release cytokines (regulatory proteins) that inhibit the activity of other T cells and of B cells.

As with T cells, millions of *B cells* are produced in the bone marrow. B cells mature in the bone marrow which are responsible for antibody-mediated immunity. Each B cell is specialized to bind with a different type of antigen. When a B cell comes into contact with the specific type of antigen to which it is targeted, it divides rapidly, forming a clone of identical cells. These B cells develop into **plasma cells,** the cells that are specialized to secrete antibodies. A plasma cell can produce more than 10 million molecules of antibody per hour!

Macrophages are important in both nonspecific and specific defense responses. When a macrophage ingests a bacterium, it digests most, but not all, of the bacterial antigens. Fragments of the antigens are displayed on the surface of the macrophage. The macrophage can be described as an **antigen-presenting cell (APC)** that displays bacterial antigens as well as its own surface proteins. The displayed antigen activates helper T cells.

Macrophages secrete about 100 different compounds, including interferons and enzymes that destroy bacteria. When macrophages are stimulated by bacteria, they secrete interleukins, which activate B cells and helper T cells. Interleukins also promote a general response to injury, causing fever and activating other mechanisms that defend the body against invasion.

The Thymus "Instructs" T Cells and Produces Hormones

Present in all vertebrates, the **thymus gland** has at least two functions. The first function of the thymus is to confer immunological competence on T cells. Just how this is accomplished is unknown, but somehow when T cells are processed within the thymus they develop the ability to differentiate into cells that can respond to specific antigens. This "instruction" within the thymus is thought to take place just before birth and during the first few months of postnatal life. When the thymus is removed from an animal before this processing takes place, the animal is not able to develop cellular immunity. If the thymus is removed after that time, cellular immunity is not seriously impaired.

The second function of the thymus is that of an endocrine gland that secretes several hormones. One hormone, known as **thymosin,** is thought to stimulate T cells after they leave the thymus, causing them to become immunologically active.

The Major Histocompatibility Complex Permits Recognition of Self

The ability of the vertebrate immune system to distinguish self from nonself depends largely on a group of membrane proteins known as the **major histocompatibility complex (MHC).** In humans, the MHC is called the **HLA** (human leukocyte antigen) **group.** These markers are present on the surface of every cell and are slightly different in each individual. A group of more than 100 genes code for the MHC proteins, and there are perhaps 50 alleles for each gene. With so many possible combinations, no two people, except identical twins, are likely to have the MHC proteins on their cells. Thus, MHC gives each of us a biochemical "fingerprint."

Antibody-Mediated Immunity Is a Chemical Warfare Mechanism

B cells are responsible for antibody-mediated immunity (also called humoral immunity). Each subtype of B cell produces a specific antibody. The antibody molecules serve as receptors that combine with antigens. Only **competent B cells**—the variety of B cell with a matching receptor—can bind with a particular antigen. This binding activates the B cell.

In most cases, activation of B cells is a complex process that involves macrophages and helper T cells (Fig. 37–5). Recall that a macrophage displays fragments of antigen from a pathogen it has engulfed. The foreign antigen forms a complex with MHC antigens of the

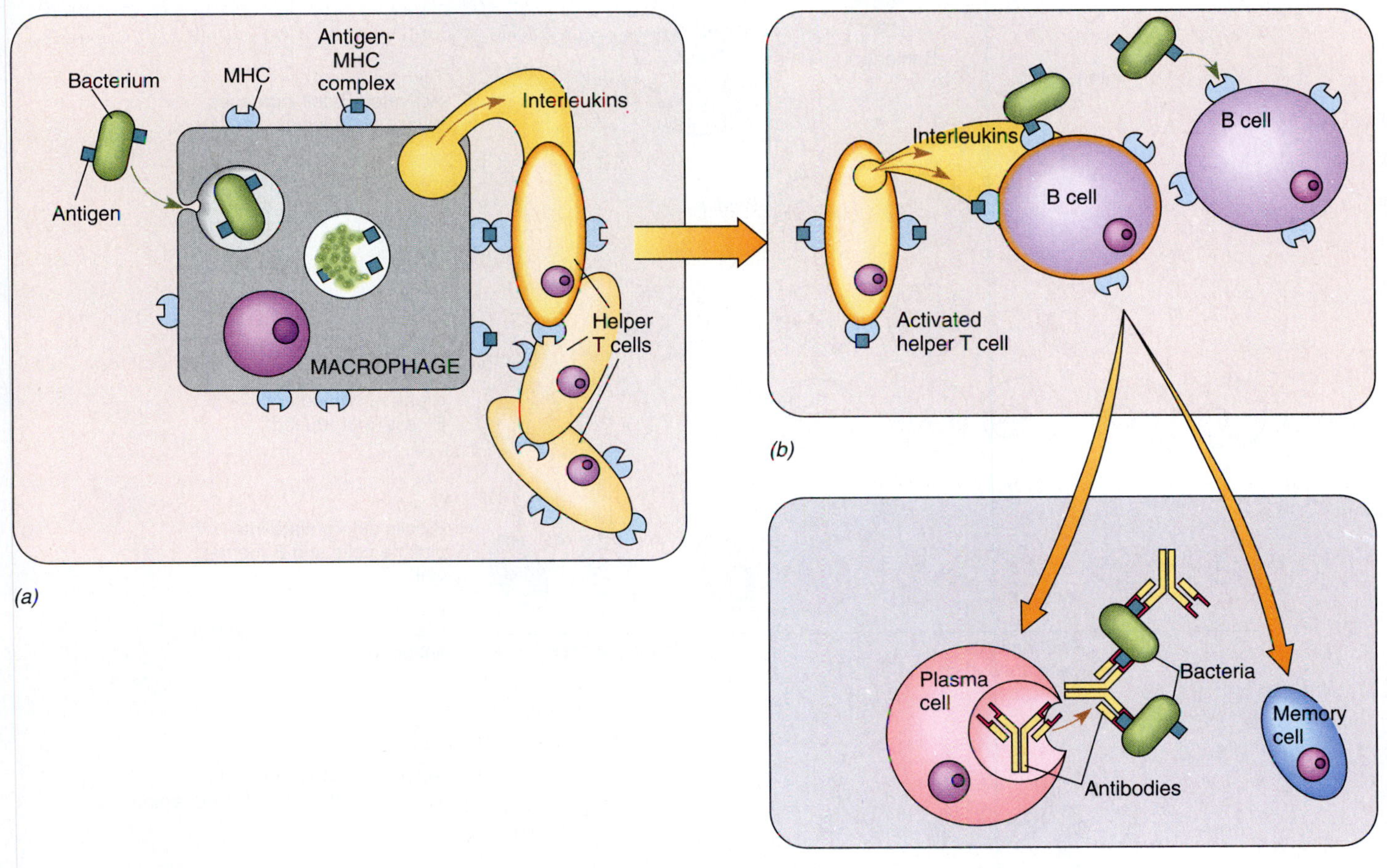

FIGURE 37–5 Activation of B cells. (***a***) The macrophage engulfs a bacterium and presents bacterial antigens on its surface in combination with MHC. Helper T cells are activated when their receptors combine with the foreign antigen-MHC complex and they are stimulated by interleukins secreted by the macrophage. (***b***) Competent B cells can combine with specific antigens. However, B cells are generally activated by interleukins secreted by helper T cells. (***c***) Once activated, the B cell divides, forming a clone of cells. Some of these cells differentiate to form plasma cells that secrete antibodies. Others develop into memory B cells.

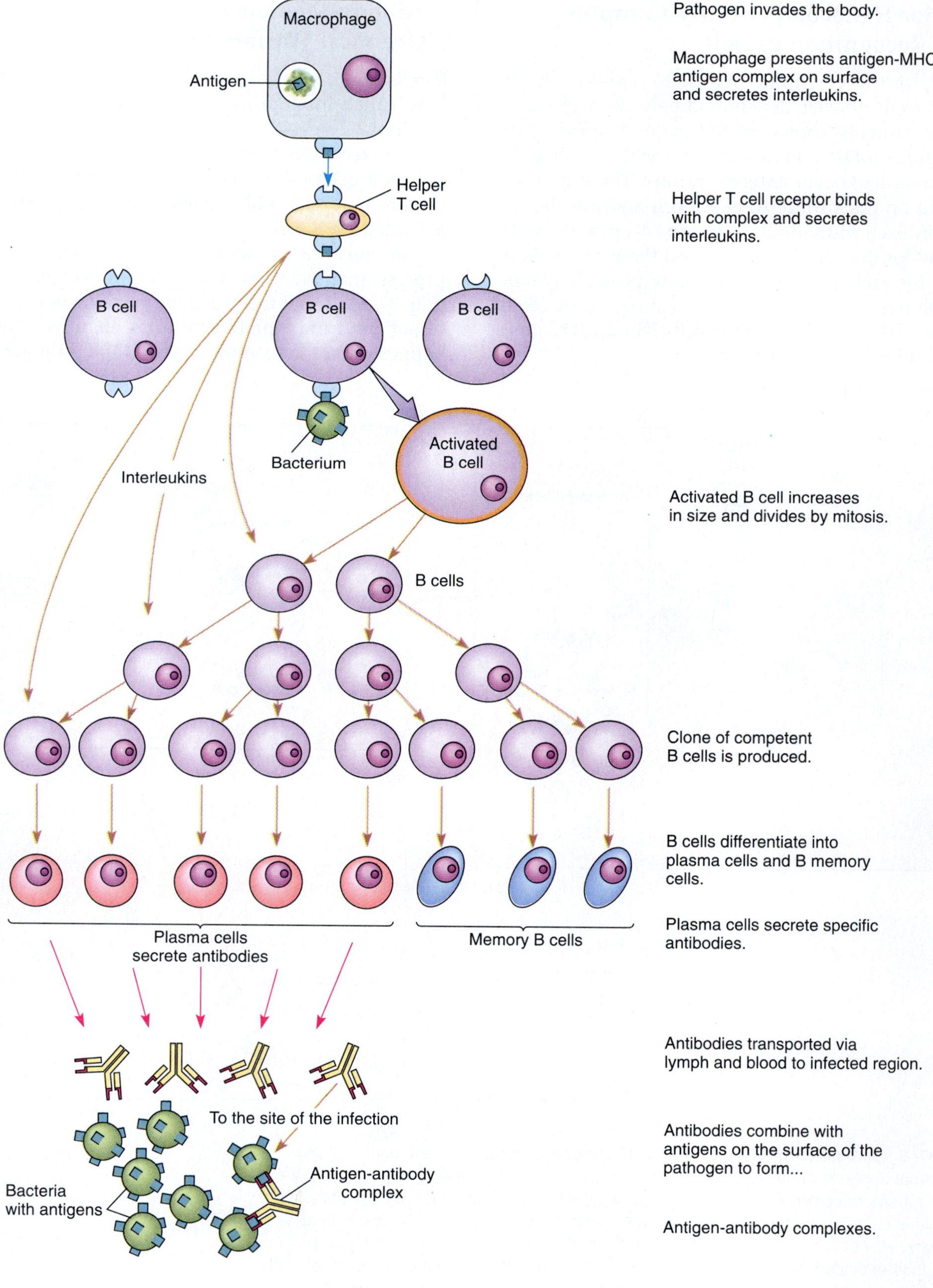

Pathogen invades the body.

Macrophage presents antigen-MHC antigen complex on surface and secretes interleukins.

Helper T cell receptor binds with complex and secretes interleukins.

Activated B cell increases in size and divides by mitosis.

Clone of competent B cells is produced.

B cells differentiate into plasma cells and B memory cells.

Plasma cells secrete specific antibodies.

Antibodies transported via lymph and blood to infected region.

Antibodies combine with antigens on the surface of the pathogen to form...

Antigen-antibody complexes.

FIGURE 37–6 Antibody-mediated immunity. When a B cell binds with a specific antigen and a helper T cell releases interleukins, the B cell becomes activated. Once activated in this way, the competent B lymphocyte multiplies, producing a large clone of cells. Many of these differentiate and become plasma cells, which secrete antibodies. The plasma cells remain in the lymph tissues, but the antibodies are transported to the site of infection by the blood or lymph. The antigen-antibody complexes that form destroy pathogens. Some of the B cells become memory cells that continue to secrete small amounts of antibody for years after the infection is over.

macrophage. It is the foreign antigen-MHC complex that is displayed on the cell surface. Only competent B cells can bind with a particular antigen-MHC complex presented by the macrophage.

When a macrophage displaying antigen-MHC complex contacts a helper T cell with complementary receptors, a complex interaction occurs. One result of this interaction is that the macrophage secretes interleukins. IL-1 activates helper T cells. The activated helper T cells detect B cells that have bound to the antigen-MHC complex on the macrophage, and they bind to the same complex. A T cell does not recognize an antigen that is presented alone. The antigen must be presented to the T cell as part of a foreign antigen-MHC complex on the surface of an antigen-presenting cell. This process activates the competent B cell.

Once activated, B cells increase in size. Then they divide by mitosis, each giving rise to a clone of identical cells (Fig. 37–6). This cell division in response to a specific antigen is known as **clonal selection.** Some of these B cells mature into plasma cells that secrete the type of antibody specific to the antigen. Unlike T cells, most plasma cells do not leave the lymph nodes. Only the antibodies they secrete pass out of the lymph tissues and make their way via the lymph and blood to the infected area. This sequence is summarized as follows:

Pathogen invades body ⟶ Macrophage phagocytizes pathogen ⟶ Foreign antigen-MHC antigen complex displayed on macrophage cell surface ⟶ Helper T cell binds with foreign antigen-MHC complex ⟶ Helper T cell interacts with a B cell that displays the same complex ⟶ Competent B cell activated ⟶ Clone of competent B cells ⟶ Plasma cells ⟶ Antibody

Some activated B cells do not differentiate into plasma cells, but instead become **memory cells.** Memory cells continue to produce small amounts of antibody long after an infection has been overcome. This antibody, part of the gamma globulin fraction of the plasma, becomes part of the body's arsenal of chemical weapons. Should the same pathogen enter the body again, this circulating antibody immediately targets it for destruction. At the same time, memory cells are stimulated to divide, producing new clones of the appropriate plasma cells.

A Typical Antibody Is a Y-Shaped Molecule Consisting of Four Polypeptide Chains

Antibodies, known more formally as **immunoglobulins,** or **Ig,** are highly specific proteins produced in response to specific antigens. The function of an antibody is to bind to an antigen. The antibody does not destroy the antigen directly. Rather, it *labels* the antigen for destruction.

How does an antibody "recognize" a particular antigen? In a protein antigen, specific sequences of amino acids make up an **antigenic determinant** (Fig. 37–7). These amino acids give part of the antigen molecule a specific shape that can be recognized by an antibody or T-cell receptor. Usually, an antigen has 5 to 10 antigenic determinants on its surface. Some have 200 or even more. These antigenic determinants may differ from one another, so several different kinds of antibodies can combine with a single complex antigen.

Certain drugs and some substances found in dust are too small to be antigenic, yet they do stimulate immune responses. These substances, called **haptens,** become antigenic by attaching to the surface of a protein. For example, the antibiotic penicillin is a small molecule that cannot by itself cause an immune response. However, once penicillin is degraded in the body, its components can bind to serum proteins, forming an antigenic determinant. In that form, penicillin is recognized as a foreign substance and can stimulate an immune response. Antibodies produced against it can participate in an allergic response to penicillin.

A typical antibody is a Y-shaped molecule, in which the two arms of the Y are binding sites (Fig. 37–7). This shape permits the antibody to combine with two antigen molecules, and allows formation of **antigen-antibody complexes.** The tail of the Y performs functions such as binding to cells or activating complement (discussed in a later section).

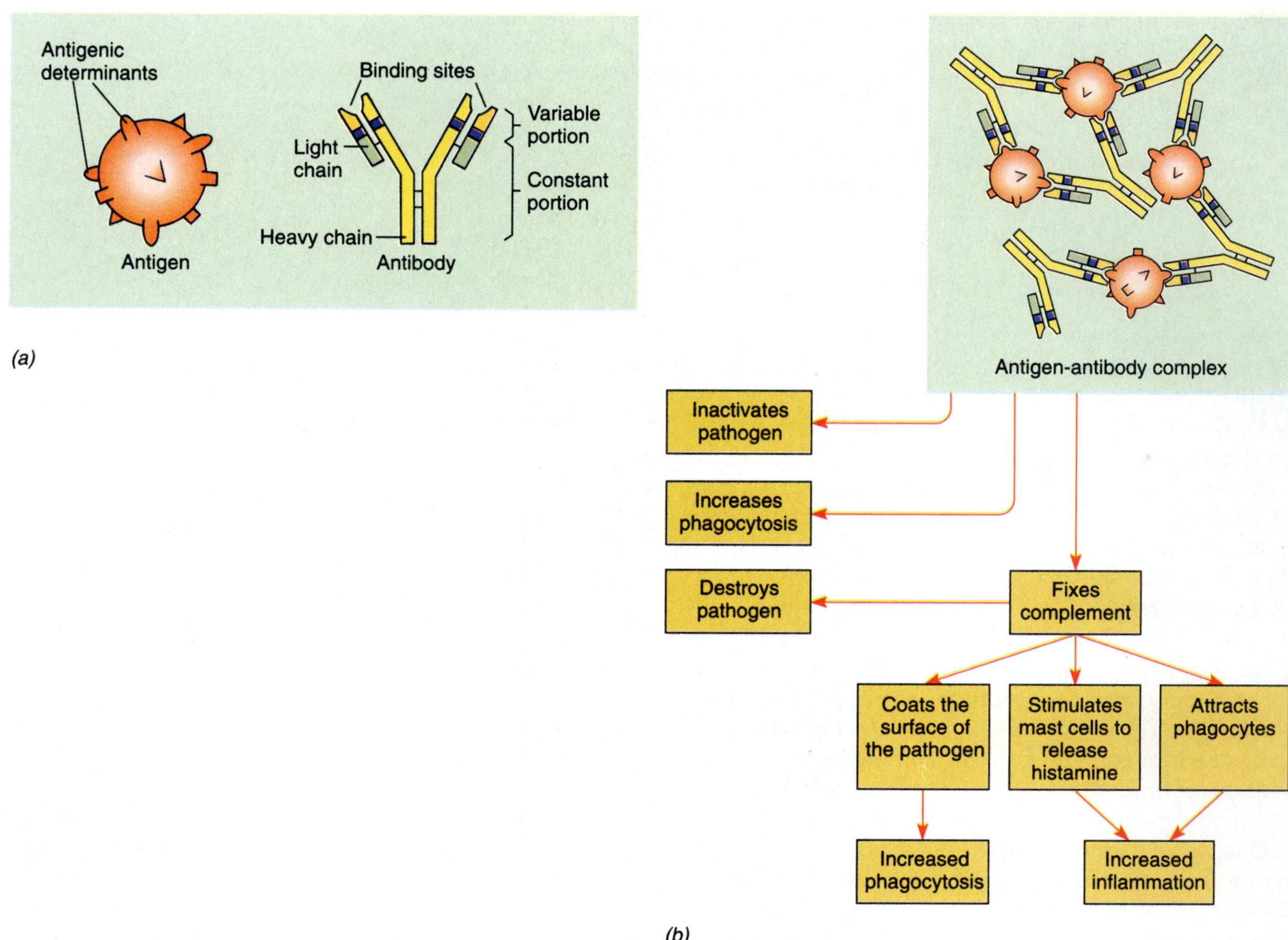

FIGURE 37–7 Antigen, antibody, and antigen-antibody complex. (***a***) The antibody molecule is composed of two light chains and two heavy chains, joined together by disulfide bonds. The constant and variable regions of the chains are indicated. (***b***) Antigen-antibody complexes directly inactivate pathogens and increase phagocytosis. They also activate the complement system.

The antibody molecule consists of four polypeptide chains: two identical long chains called heavy chains and two identical short chains called light chains (Fig. 37–7). Each chain has a constant segment, a junctional segment, and a variable segment. In the **constant segment,** or **C region,** the amino acid sequence is constant within each of the five classes of immunoglobulin. Thus, five types of C regions are known. The C region may be thought of as the handle portion of a door key. Like the elongated part of a key that slides into a lock, the amino acid sequence of the **junctional segment,** or **J region,** is somewhat variable. Finally, like the pattern of bumps and notches at the end of a key, the **variable segment,** or **V region,** has a unique amino acid sequence. In B-cell receptors, the variable region of the immunoglobulin protrudes from the B cell, whereas the constant region anchors the molecule to the cell.

The V region is the part of the key that is unique for a specific antigen (the lock). At its variable regions, the antibody folds three-dimensionally, assuming a shape that enables it to combine with a specific antigen. When they meet, antigen and antibody fit together *somewhat* like a lock and key. They must fit in just the right way for the antibody to be effective (Fig. 37–8). A given antibody can

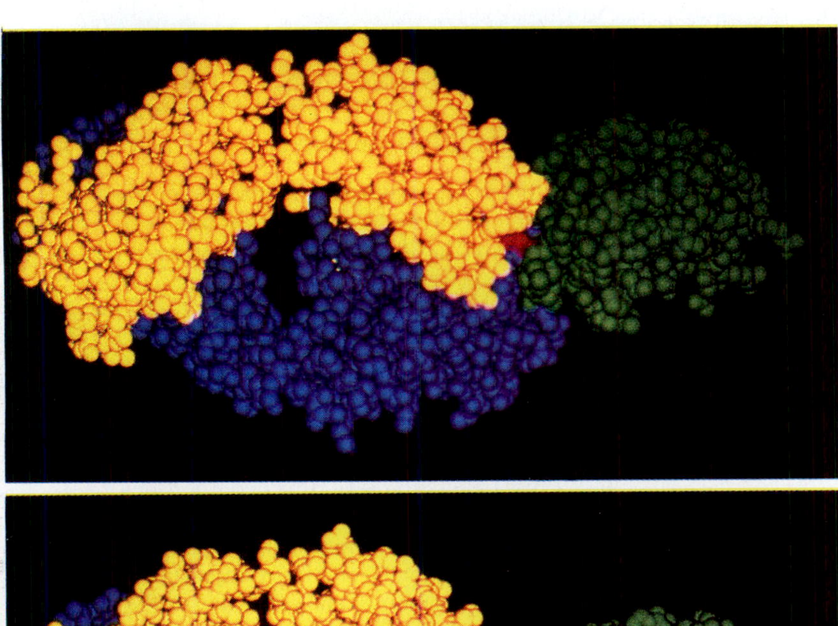

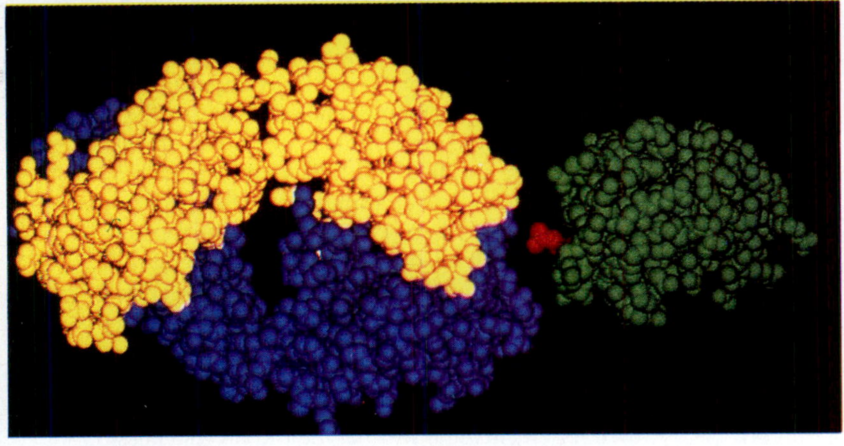

FIGURE 37–8 An antigen-antibody complex. The antigen lysozyme is shown in green. The heavy chain of the antibody is shown in blue, and the light chain in yellow. (***a***) The antigenic determinant, shown in red, fits into a groove in the antibody molecule. (***b***) The antigen-antibody complex has been pulled apart. Note how they fit each other.

bind with different strengths, or **affinities,** to different antigens. In the course of an immune response, stronger (higher affinity) antibodies are generated.

An antibody molecule has two main functions. It combines with antigen and it activates processes that destroy the antigen that binds to it. For example, the antibody may stimulate phagocytosis.

Antibodies Are Grouped in Five Classes

Antibodies are grouped in five classes. Using the abbreviation Ig for immunoglobulin, the classes are designated IgG, IgM, IgA, IgD, and IgE. In humans, about 75% of the antibodies in the blood belong to the **IgG group;** these are part of the gamma globulin fraction of the plasma. IgG and IgM defend against many pathogens carried in the blood, including bacteria, viruses, and some fungi. IgG, along with the **IgM** antibodies, stimulates macrophages and activates the complement system (discussed in the following section).

IgA, present in mucus, tears, and saliva, prevents viruses and bacteria from attaching to epithelial surfaces. IgA is strategically located in the respiratory passageways and digestive tract, and can defend against pathogens that are inhaled or ingested. **IgD,** found on the surfaces of B cells, helps bind antigen to B cells. **IgE** stimulates the release of histamine, which is responsible for many symptoms of allergic responses.

The Binding of Antibody to Antigen Activates Other Defense Mechanisms

Antibodies mark a pathogen as foreign by combining with an antigen on its surface. Generally, several antibodies bind with several antigens, creating a mass of clumped antigen-antibody complexes. The combination of antigen and antibody activates several defense mechanisms:

1. The antigen-antibody complex may inactivate the pathogen or its toxin. For example, when an antibody attaches to the surface of a virus, the virus may lose its ability to attach to a host cell.
2. The antigen-antibody complex stimulates phagocytic cells to ingest it.
3. Antibodies of the IgG and IgM groups work mainly through the **complement system.** This system consists of more than 20 proteins present in plasma and other

body fluids. Normally, complement proteins are inactive. However, an antigen-antibody complex stimulates a series of reactions that activate the system. The antibody is said to "fix" complement. Proteins of the complement system then work to destroy pathogens.

Some complement proteins digest portions of the pathogen cell. Others coat the pathogens. The coating appears to make the pathogens less "slippery" so that macrophages and neutrophils can phagocytize them. Complement proteins also increase inflammation.

Antibodies combine with a specific antigen on a pathogen. Complement proteins *complement* their action by destroying the pathogens. Complement proteins are not specific. They act against any antigen, provided they are activated by antigen-antibody complex.

Cell-Mediated Immunity Provides Cellular Warriors

The T cells and macrophages are responsible for cell-mediated immunity (Fig. 37–9). They destroy cells that are infected with viruses and cells that have been altered in some way, such as cancer cells. How do the T cells know which cells to attack?

T cells do not recognize viruses unless they are presented properly. When a virus infects a cell, some of the viral protein is broken down to peptides and displayed with MHC on the cell surface. T cell receptors can react with such antigen-MHC complexes. Only competent T cells, the variety tailored to react to the specific antigen presented, become activated.

Once stimulated, a T cell increases in size and gives rise to a clone of cytotoxic T cells and memory cells (Fig. 37–9). Cytotoxic T cells make up the cellular infantry. They leave the lymph nodes and make their way to the infected area. These killer cells can destroy a target cell within seconds after contact.

After a cytotoxic T cell combines with antigen on the surface of the target cell, it secretes granules. These granules contain proteins that can destroy the cell. Under some conditions, cytotoxic T cells release proteins called **lymphotoxins,** which are especially toxic for cancer cells. After releasing cytotoxic substances, the T cell disengages itself from its victim cell and seeks out a new target. This sequence is summarized as follows:

Virus invades body cell ⟶ Foreign antigen—MHC complex displayed on cell surface ⟶ Competent T cell activated by this complex (and by helper T cells) ⟶ Clone of competent T cells ⟶ Some become cytotoxic T cells ⟶ Cytotoxic T cells migrate to area of infection ⟶ Release proteins that destroy target cells

(Text continues on page 737)

CONCEPT CONNECTIONS

Immunity ⟷ *Genetics*

How can the immune system recognize every possible antigen, even those produced by newly mutated viruses that have never before been encountered during the evolution of the species? Do our cells contain millions of separate antibody genes, each coding for an antibody with a different specificity? Although each human cell has a large amount of DNA, it is not enough to provide a different gene to code for each specific antibody molecule.

Recombinant DNA technology (Chapter 16) has allowed researchers to make direct comparisons between the DNA of the B cells that produce antibodies and other cells of the body. They have found that separate DNA segments code for different regions of the heavy or light chain of an antibody (see Fig. 37–7). During the development of a B cell, the DNA segments are shuffled and then joined, making one combined gene. By shuffling the gene segments in this way, the number of potential combinations is enormous! Millions of different types of B (and T) cells are produced. By chance, one of those cells may produce just the right antibody to destroy a virus that invades the body. You create diverse combinations of things in your everyday life. A familiar example of creating diversity is when you make your own ice cream sundae. Imagine the diverse possibilities from ten types of ice cream, six types of sauce, and 15 kinds of toppings.

Additional sources of antibody diversity are known. For example, the variable DNA regions of the mature B lymphocytes mutate very readily. These mutations produce genes that code for slightly different antibodies.

We have used antibody diversity here as an example, but similar genetic mechanisms account for the diversity of T cell receptors. Although we may actually use only a relatively few types of antibodies or T cells in a lifetime, the remarkable diversity of the immune system prepares it to attack any antigen that invades the body.

DNA rearrangement in production of an antibody. In undifferentiated cells, gene segments are present for a number of different variable (V) regions, for one or more junction (J) regions, and for one or more different constant (C) regions. During differentiation, the segments are rearranged. A gene segment extending from the end of one of the V segments to the beginning of one of the J segments may be deleted. This produces a gene that can be transcribed. The RNA transcript is processed to remove introns and the mRNA produced is translated. Each of the polypeptide chains of an antibody molecule is produced in this way. This diagram is greatly simplified.

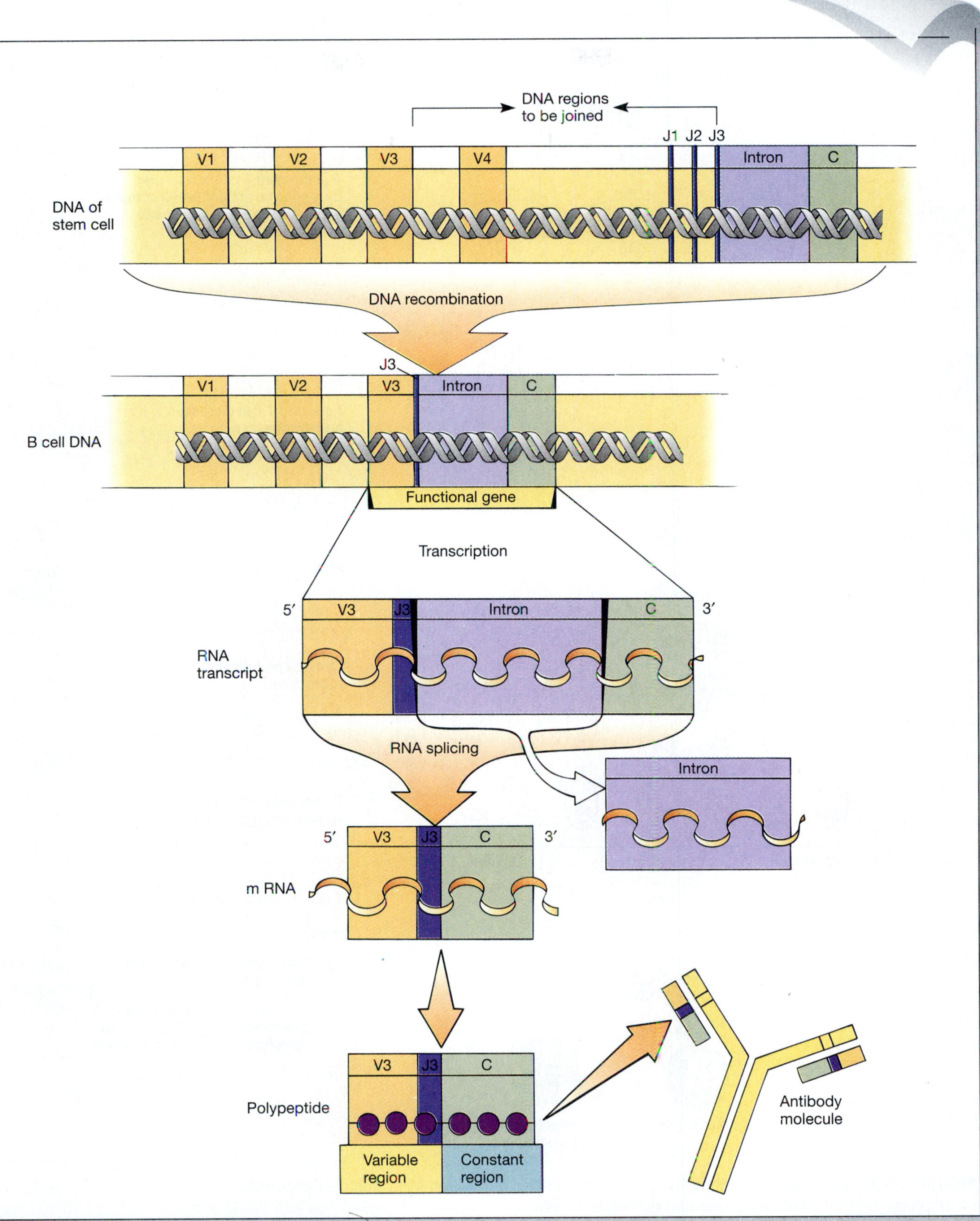
DNA regions
to be joined
J1 J2 J3
V1
V2
V3
V4
Intron
C
DNA of
stem cell
DNA recombination
J3
V1
V2
V3
Intron
C
B cell DNA
Functional gene
Transcription
5′
V3
J3
Intron
C
3′
RNA
transcript
RNA splicing
Intron
5′
V3
J3
C
3′
m RNA
V3
J3
C
Polypeptide
Variable
region
Constant
region
Antibody
molecule

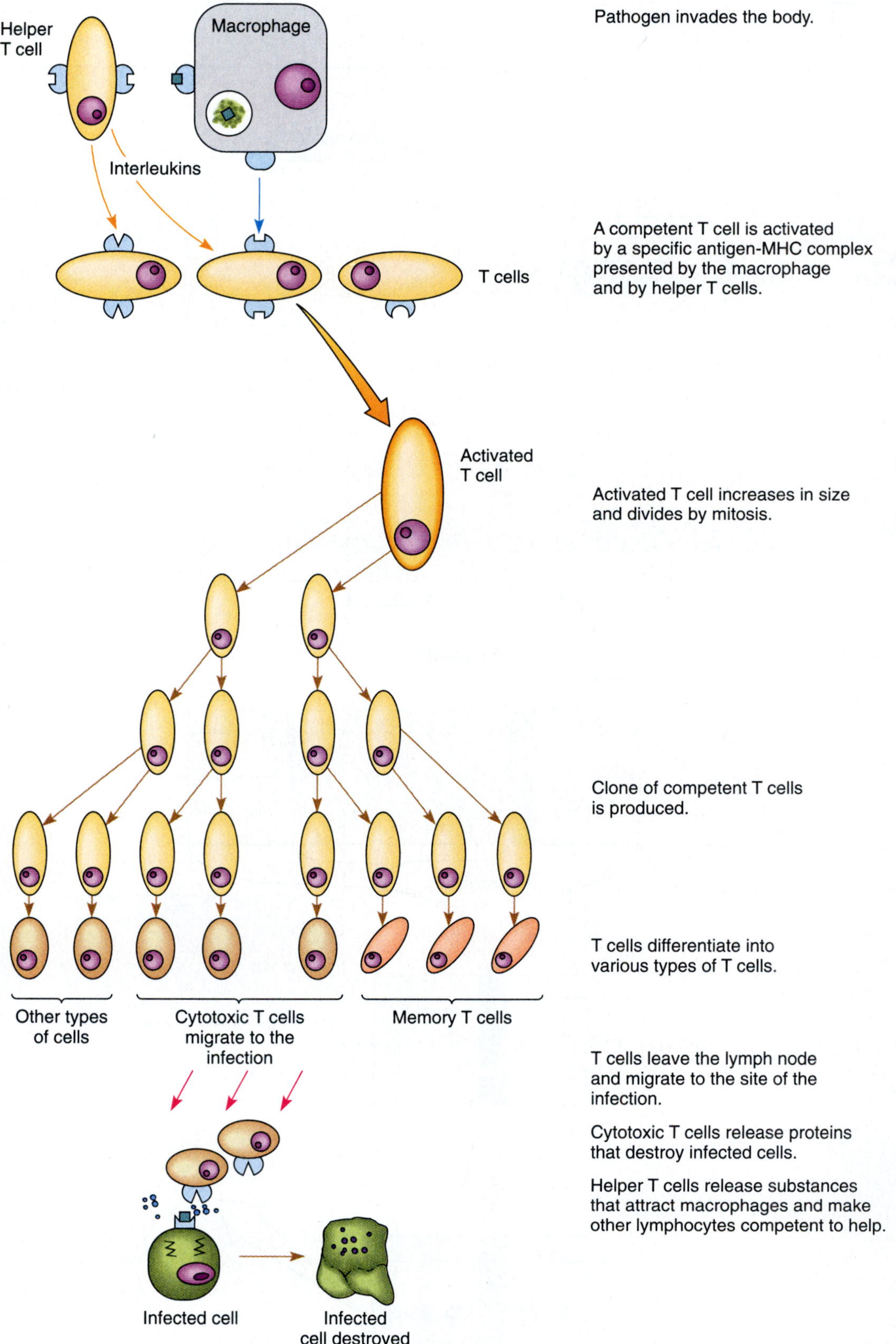
Helper T cell
Macrophage
Pathogen invades the body.
Interleukins
T cells
A competent T cell is activated by a specific antigen-MHC complex presented by the macrophage and by helper T cells.
Activated T cell
Activated T cell increases in size and divides by mitosis.
Clone of competent T cells is produced.
T cells differentiate into various types of T cells.
Other types of cells
Cytotoxic T cells migrate to the infection
Memory T cells
T cells leave the lymph node and migrate to the site of the infection.
Cytotoxic T cells release proteins that destroy infected cells.
Helper T cells release substances that attract macrophages and make other lymphocytes competent to help.
Infected cell
Infected cell destroyed

FIGURE 37–9 Cell-mediated immunity. When activated by the antigen-MHC complex presented by a macrophage and by interleukins secreted by helper T cells, a competent T cell gives rise to a clone of cells. Some of these cells become cytotoxic T cells, which migrate to the site of infection and release proteins that destroy invading pathogens. Other cells become helper T cells, suppressor cells, or memory T cells.

Helper T cells and macrophages at the site of infection secrete substances, like interleukins and interferons, that help regulate immune function. For example, some interleukins stimulate activated lymphocytes to divide. Other interleukins enhance inflammation, attracting great numbers of macrophages to the site of infection.

Suppressor T cells are also stimulated by antigen. These cells inhibit the activity of T cells, B cells, and macrophages. Suppressor T cells multiply more slowly than cytotoxic T cells. In fact, it takes more than a week for them to suppress an immune response. This allows sufficient time for the immune response to effectively defend the body.

A Secondary Immune Response Is More Rapid than a Primary Response

The first exposure to an antigen stimulates a **primary response.** Injection of an antigen into an animal causes specific antibodies to appear in the blood plasma in 3 to 14 days. After injection of the antigen, there is a brief latent period, during which the antigen is recognized and appropriate lymphocytes begin to form clones. Then there is a logarithmic phase, during which the antibody concentration rises rapidly for several days until it reaches a peak (Fig. 37–10). IgM is the principal antibody synthesized. Finally, there is a decline phase, during which the antibody concentration decreases to a very low level.

A second injection of the same antigen, even years later, results in a **secondary response** (Fig. 37–10). Because memory cells bearing antibodies to that antigen may persist throughout an individual's life, the secondary response is generally much more rapid than the primary response and has a shorter latent period. Much less antigen is necessary to stimulate a secondary response than a primary response, and more antibodies are produced. The predominant antibody is IgG.

The body's ability to launch a rapid, effective response during a second encounter with an antigen explains why we do not usually suffer from the same disease several times. Most persons get measles or chicken pox, for example, only once. When exposed a second time, the immune system responds quickly, destroying the pathogens before they have time to multiply and cause symptoms of the disease. Booster shots of vaccine are given in order to elicit a secondary response. Exposure to the vaccine reinforces the immunological memory of the disease-producing antigens.

You may wonder, then, how a person can get influenza (the flu) or a cold more than once. Unfortunately, there are many varieties of these diseases, each caused by a virus with slightly different antigens. For example, more than 100 different viruses cause the common cold, and new varieties of cold and flu virus evolve continuously by mutation (a survival mechanism for them), which may result in changes in their surface antigens. Even a slight change may prevent recognition by memory cells. Because the immune system is so specific, each different antigen is treated by the body as a new immunological challenge.

Active Immunity Follows Exposure to Antigens

We have been considering **active immunity,** immunity developed following exposure to antigens. After you have had chicken pox as a young child, for example, you develop immunity that protects you from contracting

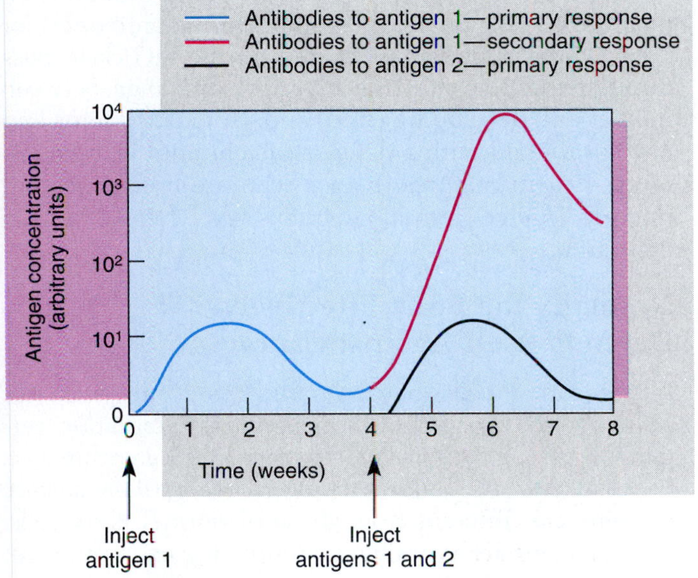

FIGURE 37–10 Antibody production in primary and secondary responses to successive doses of antigens. Antigen 1 was injected at day 0 and immune response was assessed by measuring antibody levels to the antigen. At week 4, the primary response had subsided. Antigen 1 was injected again along with a new protein, antigen 2. Note that the secondary response to antigen 1 was greater and more rapid than the primary response. A primary response was made to the newly encountered antigen 2.

TABLE 37–1
Active and Passive Immunity

Type of immunity	*When developed*	*Development of memory cells*	*Duration of immunity*
Active			
Naturally induced	Pathogens enter the body through natural encounter (e.g., person with measles sneezes on you)	Yes	Many years
Artificially induced	After immunization with a vaccine	Yes	Many years
Passive			
Naturally induced	After transfer of antibodies from mother to developing baby	No	Few months
Artificially induced	After injection with gamma globulin	No	Few months

chicken pox again. Active immunity can be *naturally* or *artificially* induced (Table 37–1). If someone with chicken pox sneezes near you and you contract the disease, you develop active immunity naturally. Active immunity can also be artificially induced by **immunization,** that is, by injection of a **vaccine.** In this case, the body launches an immune response against the antigens contained in the vaccine and develops memory cells so that future encounters with the same pathogen are dealt with swiftly.

Effective vaccines can be prepared in a number of ways. A virus may be weakened by successive passage through cells of nonhuman hosts. In the process, mutations occur that adapt the pathogen to the nonhuman host so that it can no longer cause disease in humans. This is how Sabin polio vaccine, smallpox vaccine, and measles vaccine are produced.

Whooping cough and typhoid fever vaccines are made from killed pathogens that still have the necessary antigens to stimulate an immune response. Tetanus and botulism vaccines are made from toxins secreted by the respective pathogens. The toxin is altered so that it can no longer destroy tissues, but its antigenic determinants are still intact.

Most vaccines consist of the entire pathogen, live or killed, or of a protein from the pathogen. In order to reduce potential side effects, researchers are now designing vaccines that consist of synthetic peptides that are only a small part of the antigen. When any of these vaccines are introduced into the body, the immune system actively develops clones, produces antibodies, and develops more memory cells.

Passive Immunity Is Borrowed Immunity

In **passive immunity,** an individual is given antibodies actively produced by another organism. The serum or gamma globulin that contains these antibodies can be obtained from humans or animals. Animal sera are less desirable because their nonhuman proteins can themselves act as antigens, stimulating an immune response that may result in an illness known as serum sickness.

Passive immunity is borrowed immunity, and its effects are not lasting. It is used to boost the body's defense temporarily against a particular disease. For example, during the Vietnam War, in areas where hepatitis was widespread, soldiers were injected with gamma globulin containing antibodies to the hepatitis pathogen. Such injections of gamma globulin offer protection for only a few months. Because the body has not actively launched an immune response, it has no memory cells and cannot produce antibodies to the pathogen. Once the injected antibodies are broken down, the immunity disappears.

Pregnant women confer natural passive immunity on their developing babies by manufacturing antibodies for them. These maternal antibodies, of the IgG class, pass through the placenta (the organ of exchange between mother and developing child) and provide the fetus and newborn infant with a defense system until its own immune system matures. Babies who are breast-fed continue to receive immunoglobulins, particularly IgA, in their milk.

Normally the Body Effectively Defends Itself Against Cancer

A few normal cells may be transformed into precancer cells every day in each of us in response to radiation, certain viruses, or chemical carcinogens in the environment. Because they are abnormal cells, some of their surface proteins are different from those of normal body cells. Such proteins act as antigens, stimulating an immune response that generally destroys these abnormal cells.

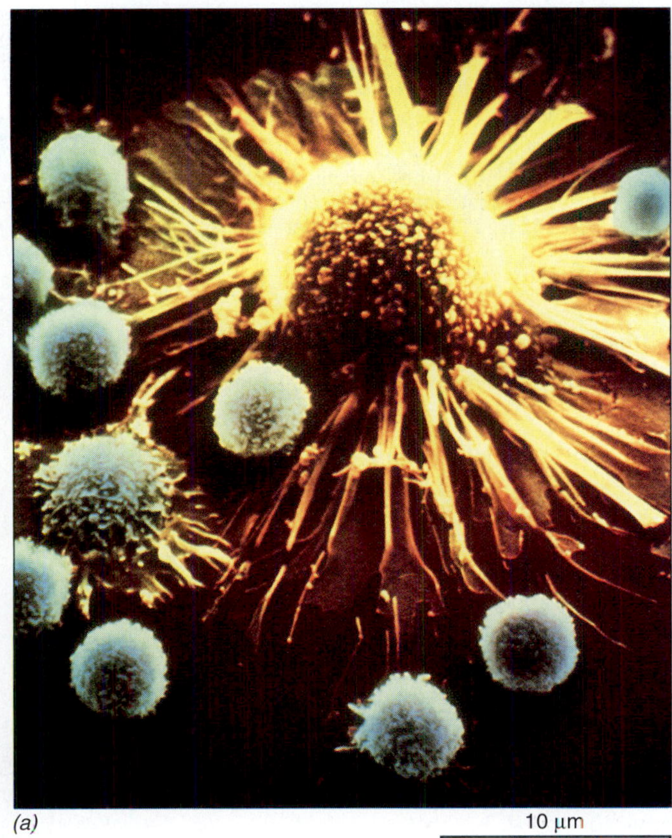

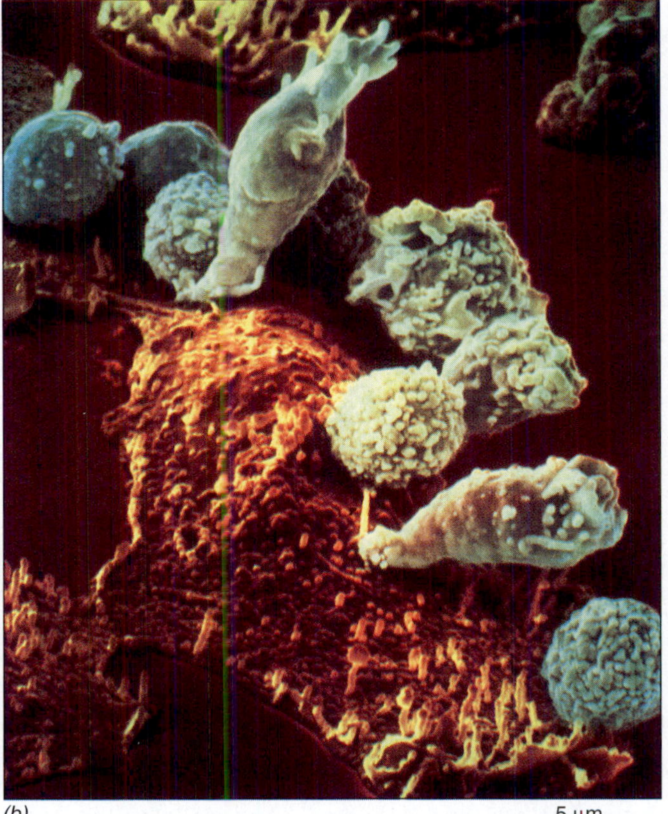

FIGURE 37–11 Cytotoxic T cells attack cancer cells. (***a***) An army of cytotoxic T cells surrounds a large cancer cell. The T cells recognize the cancer cell as nonself because it displays altered antigens on its surface. (***b***) Some of the cytotoxic T cells elongate as they chemically attack the cancer cell, breaking down its plasma membrane. (***c***) The cancer cell has been destroyed. Only a collapsed fibrous cytoskeleton remains. (Lennart Nillson, © Boehringer Ingelheim International GmbH)

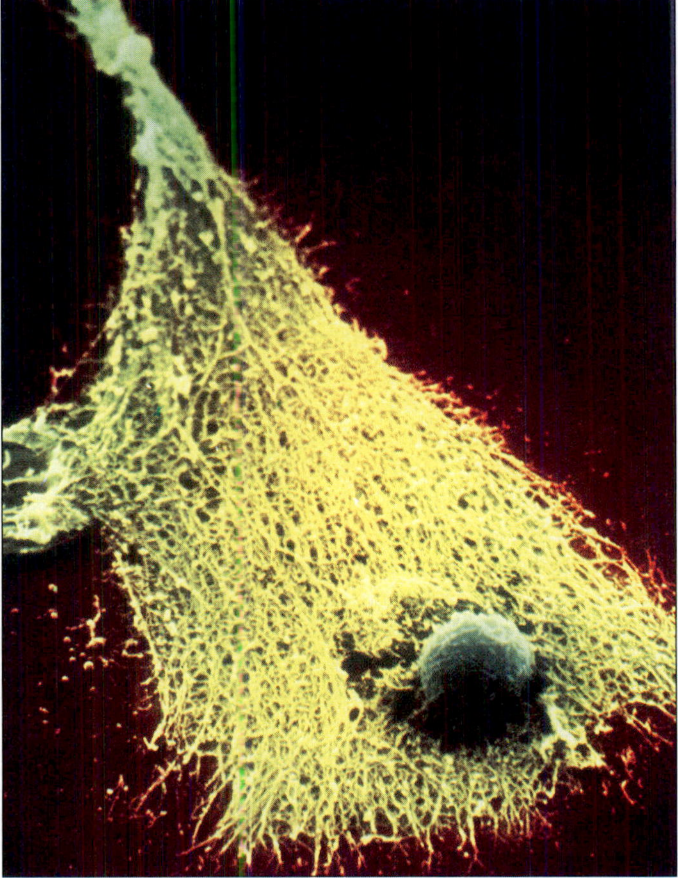

Although many components of the immune system help defend against cancer cells, NK cells and cytotoxic T cells appear to be most critical (Fig. 37–11). T cells produce interleukins, which attract macrophages and NK cells and activate them. The T cells also produce interferons, which have an antitumor effect. The macrophages themselves produce factors, including TNF (tumor necrosis factor), that inhibit tumor growth.

What are the mechanisms by which tumor cells prevent the body from launching an immune response? The cancer cells may remain so similar antigenically to normal cells that the immune system cells may fail to recognize them as foreign. By maintaining a "low profile" in this way cancer cells can multiply until a tumor gets so large that the host is unable to destroy it.

Sometimes, cells of the immune system do recognize cancer cells but are unable to destroy them. Cancer cells can stimulate B cells to produce IgG antibodies that combine with antigens on the surfaces of the cancer cells. These **blocking antibodies** may block the T cells so that they are unable to adhere to the surface of the cancer cells and destroy them. For some unknown reason, the blocking antibodies are not able to activate the complement system that would destroy the cancer cells. Interestingly, in this case the presence of antibodies is harmful.

An exciting approach in cancer research involves the production in the laboratory of **monoclonal antibodies.** In this procedure, mice are injected with antigens from human cancer cells. After the mice have produced antibodies to the cancer cells, B cells containing the antibodies are collected from them. These cells are fused with cancerous B cells from other mice. Unlike normal B cells, these cancer cells can live and divide in tissue culture indefinitely. The hybrid cells produced have properties of the two "parent" cells, and they continue to secrete antibodies.

Researchers select hybrid cells that are manufacturing the specific antibody needed and then clone them in a separate cell culture. Cells of this clone secrete large amounts of the specific antibody needed—thus, the name monoclonal antibodies. When injected into the same patient whose cancer cells were used to stimulate their production, these antibodies are highly specific for destroying the cancer cells. (Monoclonal antibodies specific for a single antigenic determinant can now be produced.) In trial studies, such antibodies are being tagged with toxic drugs that are then delivered specifically to the cancer cells. Such targeted drug therapy would have fewer side effects than current types of chemotherapy, which damage normal cells as well as cancer cells.

Graft Rejection Is an Immune Response Against Transplanted Tissue

Skin can be successfully transplanted from one part of the same body to another or from one identical twin to another. However, when skin is taken from one person and grafted onto the body of a non-twin, the skin graft is rejected and it sloughs off. Why? Recall that tissues from the same individual or from identical twins have identical MHC alleles and thus the same MHC antigens. Such tissues are compatible.

Because there are many alleles for each of the MHC genes, it is difficult to find identical matches among strangers. When a tissue or organ is taken from a donor and transplanted to the body of an unrelated host, several of the MHC antigens are likely to be different. The host's immune system regards the graft as foreign and launches an immune response called **graft rejection.** T cells attack the transplanted tissue and can destroy it within a few days (Fig. 37–12).

Before transplants are performed, tissues from the patient and from potential donors must be typed and matched as closely as possible. Cell typing is somewhat similar to blood typing but is more complex. If all of the MHC antigens are matched, the graft has about a 95% chance of surviving the first year. Unfortunately, not many persons are lucky enough to have an identical twin to supply spare parts, so perfect matches are difficult to find. Furthermore, some parts such as the heart cannot be spared. Most organs to be transplanted, therefore, are removed from unrelated donors, often from patients who have just died.

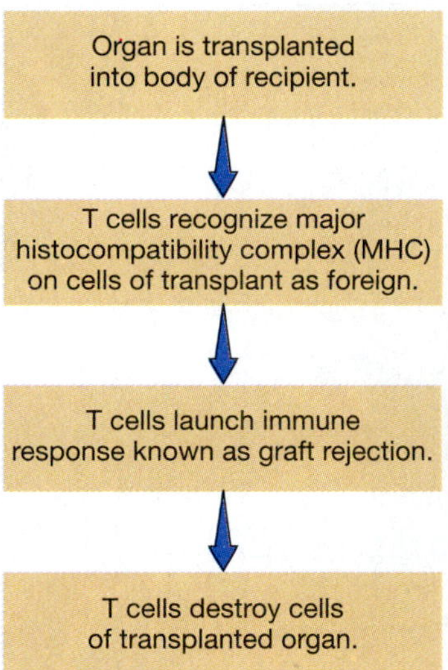

FIGURE 37–12 Graft rejection.

To prevent graft rejection in less compatible matches, physicians use drugs that suppress the immune system. Unfortunately, while these immunosuppressive drugs reduce graft rejection, they also make the transplant patient more vulnerable to pneumonia or other infections, and increase the risk of certain types of tumor growths. If the patient can survive the first few months, dosages of these drugs can be reduced. New, powerful immunosuppressive drugs are increasing the chances for transplants of kidney, heart, bone marrow, liver, pancreas, and intestine.

Certain Sites in the Body Are Immunologically Privileged

A few immunologically privileged locations exist in the body in which foreign tissue is accepted by a host. The brain and cornea are examples. Corneal transplants are highly successful because the cornea has almost no blood or lymphatic vessels associated with it and so is out of reach of most lymphocytes. Furthermore, antigens in the corneal graft probably would not find their way into the circulatory system, and so would not stimulate an immune response.

In an Autoimmune Disease the Body Attacks Its Own Tissues

Sometimes the body reacts immunologically against its own tissues, causing an **autoimmune disease.** Some of the diseases that result from such failures in recognizing

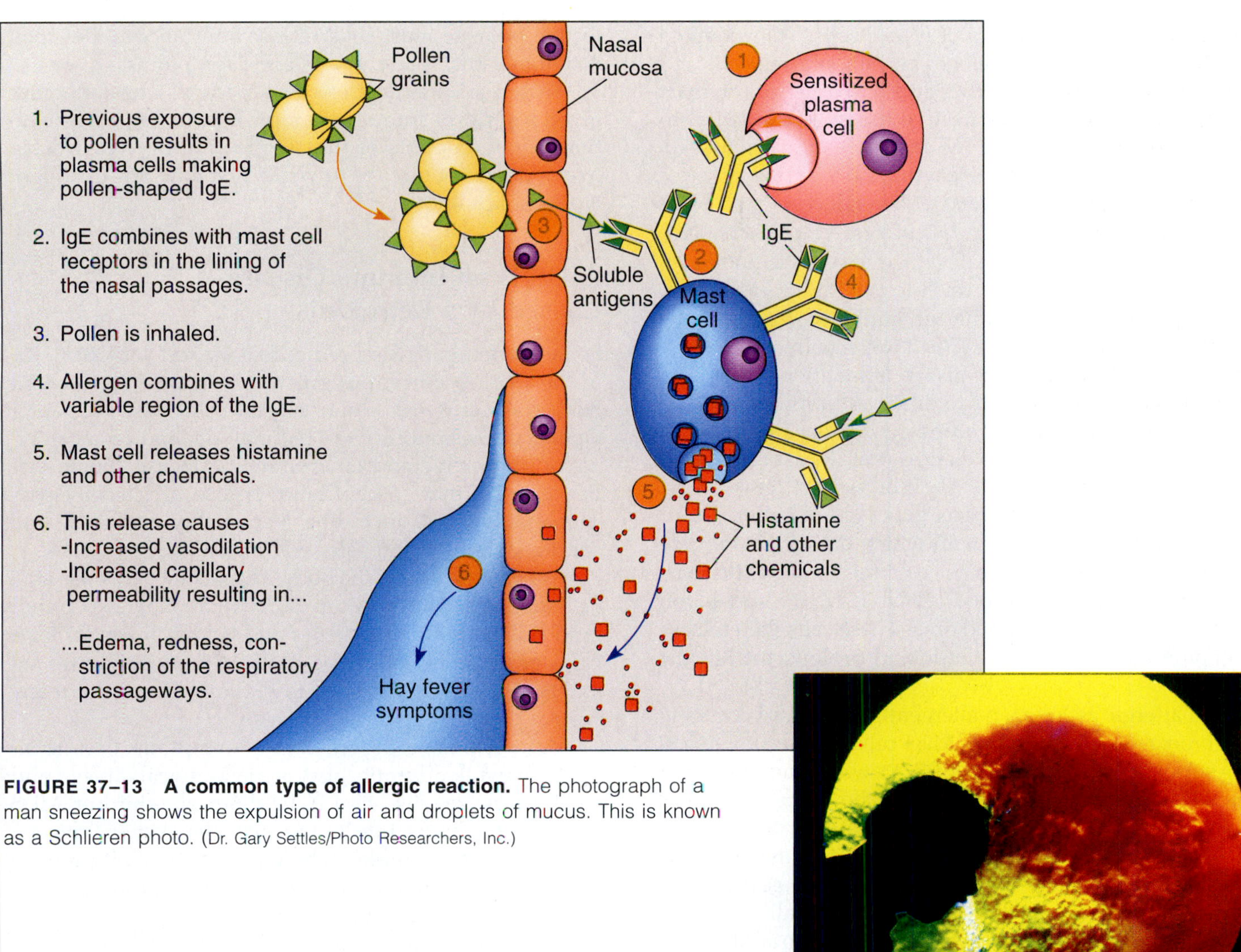

Pollen grains
Nasal mucosa
1. Previous exposure to pollen results in plasma cells making pollen-shaped IgE.
2. IgE combines with mast cell receptors in the lining of the nasal passages.
3. Pollen is inhaled.
4. Allergen combines with variable region of the IgE.
5. Mast cell releases histamine and other chemicals.
6. This release causes
-Increased vasodilation
-Increased capillary permeability resulting in...
...Edema, redness, constriction of the respiratory passageways.
Sensitized plasma cell
IgE
Soluble antigens
Mast cell
Histamine and other chemicals
Hay fever symptoms
1
2
3
4
5
6

FIGURE 37–13 A common type of allergic reaction. The photograph of a man sneezing shows the expulsion of air and droplets of mucus. This is known as a Schlieren photo. (Dr. Gary Settles/Photo Researchers, Inc.)

self (self-tolerance) are rheumatoid arthritis, multiple sclerosis, systemic lupus erythematosus (SLE), insulin-dependent diabetes, psoriasis, and scleroderma.

In multiple sclerosis and other autoimmune diseases the body produces abnormal antibodies. Genetic factors are known to play a role. For example, when one identical twin has multiple sclerosis, the other twin has about a 30% chance of contracting the disease also. In multiple sclerosis, antibodies attack the glial cells that produce the myelin sheath surrounding neurons in the brain and spinal cord. Magnetic resonance scans (MRIs) show the absence of myelin sheaths around axons that are normally myelinated. Patients generally suffer weakness and visual problems, and become progressively more disabled.

Studies also indicate that viral or bacterial infection often precedes the onset of an autoimmune disease. Some pathogens have evolved a tactic known as molecular mimicry. They trick the body by producing molecules that look like self-molecules. For example, an adenovirus that causes respiratory and intestinal illness produces a peptide that mimics myelin protein. When the body launches immune responses to the adenovirus peptide, it may also begin to attack the similar self-molecule, myelin.

Allergic Reactions Are Abnormal Immune Responses

In **allergic reactions,** individuals manufacture antibodies against mild antigens, called **allergens,** that do not stimulate a response in nonallergic individuals. In many kinds of allergic reactions, distinctive IgE immunoglobulins are produced. About 20% of the population of the United States is plagued by an allergic disorder such as allergic asthma or hayfever. A tendency to these disorders appears to be inherited.

Let us examine a common allergic reaction—a hayfever response to ragweed pollen (Fig. 37–13). The first step is *sensitization.* Macrophages degrade the allergen and display fragments of it to T cells. The activated T cells then stimulate B cells to become plasma cells and produce IgE. These antibodies attach to receptors on **mast cells,** which are large connective tissue cells filled with distinctive granules. Each IgE molecule attaches to a receptor by its C region end, leaving the V region end free to combine with the ragweed pollen allergen.

The second step is *activation of mast cells.* When a sensitized, allergic person inhales the microscopic pollen, allergen molecules rapidly attach to the IgE on mast cells. This binding of allergen with IgE antibody stimulates the mast cell to release granules filled with chemicals like histamine and serotonin that cause inflammation. These substances cause blood vessels to dilate and capillaries to become more permeable, leading to edema and redness. Such responses cause the victims' nasal passages to become swollen and irritated. Their noses run, they sneeze, their eyes water, and they feel generally uncomfortable.

A third step may occur in which *the allergic response is prolonged.* Chemical compounds released by the mast cells lure certain white blood cells to leave the circulation and migrate to the inflamed area. These cells then release compounds that damage tissue and prolong the allergic reaction.

In **allergic asthma,** an allergen-IgE response occurs in the bronchioles of the lungs. Mast cells release slow-reacting substance of anaphylaxis (SRS-A), which causes smooth muscle to constrict. The airways in the lungs sometimes constrict for several hours, making breathing difficult.

Certain foods or drugs act as allergens in some persons, causing a reaction in the walls of the digestive tract that leads to discomfort and diarrhea. The allergen may be absorbed and cause mast cells to release granules elsewhere in the body. When the allergen-IgE reaction takes place in the skin, the histamine released by mast cells causes the swollen red welts known as **hives.**

Systemic anaphylaxis is a dangerous allergic reaction that can occur when a person develops an allergy to a specific drug such as penicillin or to compounds in the venom injected by a stinging insect. Within minutes after the substance enters the body, a widespread allergic reaction takes place. Mast cells release large amounts of histamine and other compounds into the circulation. These compounds cause extreme vasodilation and permeability. So much plasma may be lost from the blood that circulatory shock and death can occur within a few minutes.

The symptoms of allergic reactions are often treated with **antihistamines,** drugs that block the effects of histamines. These drugs compete for the same receptor sites on cells targeted by histamine. When the antihistamine combines with the receptor, it prevents the histamine from combining and thus prevents its harmful effects. Antihistamines are useful clinically in relieving the symptoms of hives and hayfever. They are not completely effective because mast cells release substances other than histamines that also cause allergic symptoms.

In serious allergic disorders, patients are sometimes given a form of immunotherapy known as desensitization. Small amounts of the allergen are injected weekly over a period of months or years. Just how this treatment works is not known.

AIDS Is an Immune Disease Caused by a Retrovirus

Some epidemiologists predict that by the year 2000, 2% of the world's current population may be infected with HIV, the retrovirus that causes AIDS (see Chapter Opener). Susceptibility to HIV may depend on a combination of genetic and environmental factors. Some evidence exists that susceptibility is also affected by psychosocial factors, including personality variables and coping styles that intensify environmental stressors.

When HIV infects the body, an immune response may be launched. The infected individual may experience mild flulike symptoms (fever and aching muscles) for a week or so. Some cells infected by HIV are not destroyed and the virus may continue to replicate slowly for many years.

After a time, a progression of symptoms occurs, including swollen lymph glands, night sweats, fever, and weight loss (see Fig. 43–15). In about one-third of AIDS patients, the virus infects the nervous system, causing AIDS dementia complex. These patients exhibit progressive cognitive, motor, and behavioral dysfunction that typically ends in coma and death.

HIV enters a host cell by attaching with a protein on its outer envelope to CD4, a protein present on the surface of several types of immune system cells. Helper T cells appear to be the main target of the virus (Chapter Opening figure and Fig. 37–14). HIV destroys helper T cells and over time causes a dramatic decrease in their numbers. Some evidence suggests that the virus gradually destroys the lymph nodes. When the helper T cell population is depressed, the ability to resist infection is severely impaired, making the AIDS patients more susceptible to cancer and other opportunistic infections (Fig. 37–15).

Researchers throughout the world are searching for drugs that will successfully combat the AIDS virus. Because HIV often infects the central nervous system, an effective drug must cross the blood–brain barrier. AZT (azidothymidine) was the first drug developed to treat HIV infection. It can prolong the period prior to the onset of AIDS symptoms. AZT blocks HIV replication by blocking the action of reverse transcriptase, the enzyme used by the retrovirus to synthesize DNA. This process is essential for the virus to incorporate itself into the host cell's

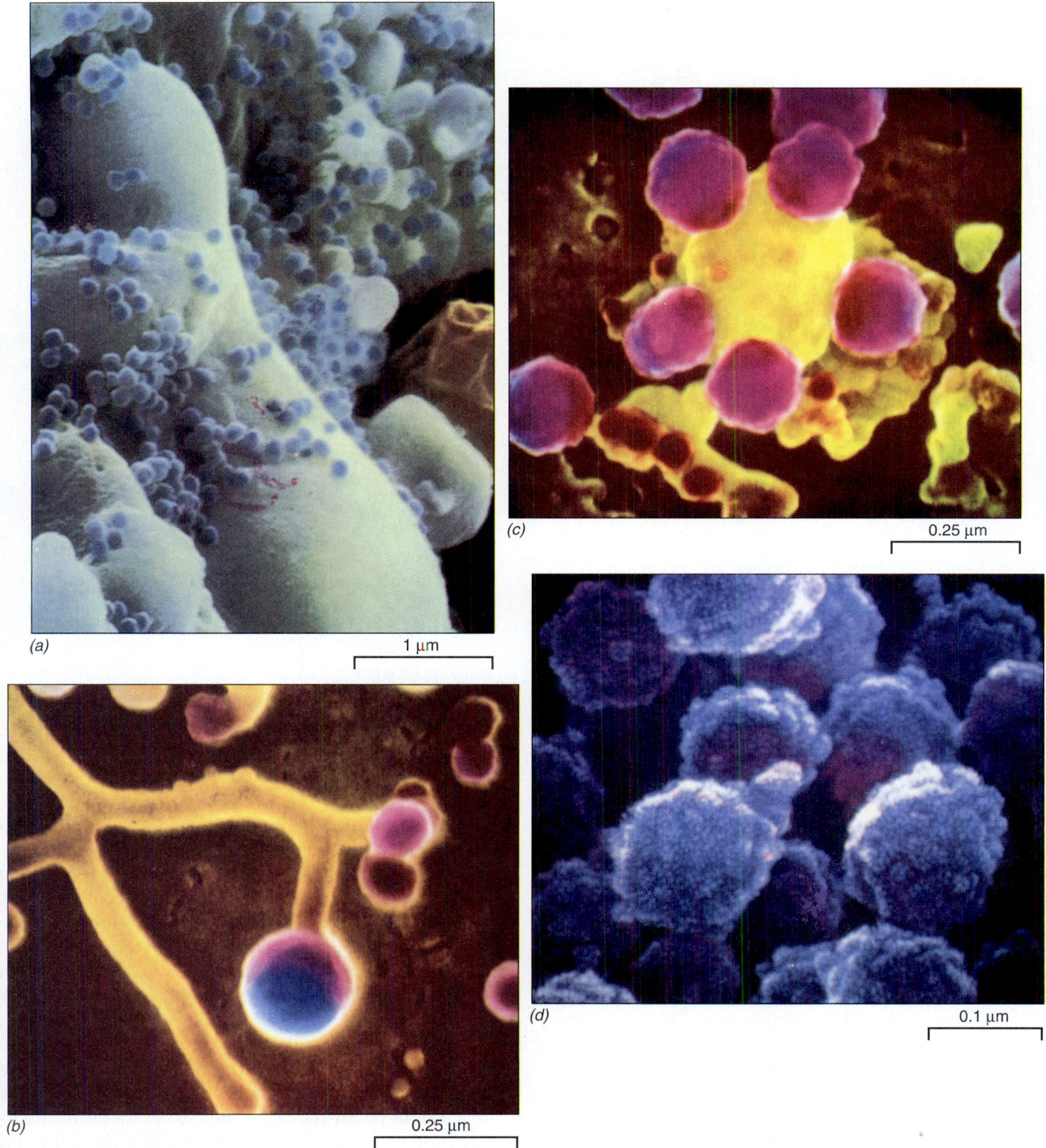

FIGURE 37–14 HIV seriously impairs the immune system by rapidly destroying helper T cells. (***a***) HIV virus particles (blue) attack a helper T cell (light blue). (***b***) HIV-1 virus particles budding from the ends of the branched microvilli of a T cell. (***c***) An even higher magnification of virus particles budding from a "bleb" (a cytoplasmic extension broader than a microvillus). Note the pentagonal symmetry often evident in other biological branching and flowering structures. (***d***) HIV-1 virus particles at extremely high magnification. The surfaces are grainy and the outlines slightly blurred because the preparation was coated with coarse-grained heavy metal salt (palladium). (From Lennart Nillson, © Boehringer Ingelheim International GmbH)

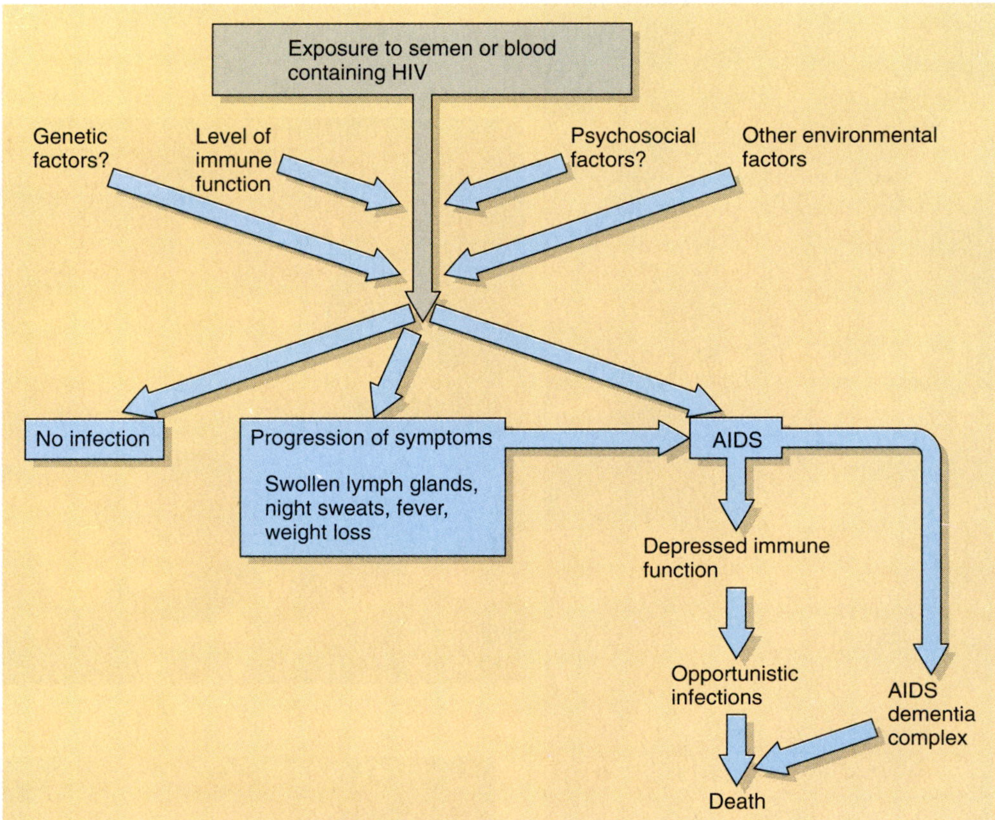

FIGURE 37–15 HIV Infection. Exposure to semen or blood that contains HIV can lead to AIDS. Although some exposed individuals apparently do not become infected, the risk increases with multiple exposures. Many factors apparently determine whether a person exposed to the AIDS virus contracts the disease.

DNA. Unfortunately, the reverse transcriptase used by the virus to synthesize DNA makes a lot of mistakes—about one in every 2000 nucleotides it uses is an incorrect nucleotide. By mutating in this way, the virus has developed strains that are resistant to AZT. As other drugs effective against HIV are developed and approved, a combination of treatments aimed at different parts of the HIV life cycle may prove more effective.

Vaccination is the most effective, simplest way of preventing a disease. However, developing a vaccine against HIV has been a most pressing challenge for virologists. Because of its high rate of mutations, new viral strains with new antigens evolve quickly. A vaccine would not be effective against new antigens, and so would quickly become obsolete. Other barriers to the development of a vaccine include the absence of an effective animal model for AIDS and the ethical and practical difficulties associated with finding human volunteers in whom to test the vaccine.

While immunologists work to develop a successful vaccine and effective drugs to treat infected patients, massive educational programs are being developed to slow the spread of AIDS. Spreading the word that having multiple sexual partners increases the risk of AIDS and teaching sexually active individuals the importance of "safe" sex may help to slow the epidemic. Some have suggested that public health facilities offer free condoms to those who are sexually active and free sterile hypodermic needles to those addicted to drugs. The cost of these measures would be far less than the cost of medical care for increasing numbers of AIDS patients and the toll in human suffering.

Chapter Summary

I. Internal defense depends on the ability of an organism to distinguish between self and nonself

II. Many invertebrates depend on nonspecific responses such as phagocytosis.

III. Vertebrates have nonspecific defense mechanisms such as the skin and mucous lining of the respiratory tract that prevent the entrance of pathogens. Should pathogens break through these first line defenses, other nonspecific defense mechanisms are activated.
   A. When pathogens invade tissues, they trigger an inflammatory response, which brings needed phagocytic cells and antibodies to the infected area.
   B. Neutrophils and macrophages phagocytize and destroy bacteria, and interferons prevent viral replication.

IV. Vertebrates have specific defense mechanisms, called immune responses, that include antibody-mediated immunity and cell-mediated immunity.
   A. In antibody-mediated immunity, competent B cells are activated when they combine with antigen. Activation requires an antigen-presenting cell (such as a macrophage) that presents a foreign antigen-MHC complex. A helper T cell (that secretes interleukins) is also needed. B cells multiply, giving rise to clones of cells. Some differentiate to become plasma cells, which secrete specific antibodies.
   B. Antibodies are highly specific proteins called immunoglobulins. They are produced in response to specific antigens. Antibodies are grouped in five classes according to their structure.
   C. An antibody combines with a specific antigen to form an antigen-antibody complex, which may inactivate the pathogen, stimulate phagocytosis, or activate the complement system. The complement system increases the inflammatory response and phagocyosis; some complement proteins digest portions of the pathogen cell.
   D. In cell-mediated immunity, specific T cells are activated by helper T cells and by a foreign antigen-MHC complex displayed by a macrophage. Activated T cells multiply, giving rise to a clone of cells.
      1. Some T cells differentiate to become cytotoxic T cells, which migrate to the site of infection and chemically destroy cells infected with viruses.
      2. Some activated T cells remain in the lymph nodes as memory cells; others become helper T cells or suppressor T cells.
   E. Second exposure to an antigen evokes a secondary immune response, which is more rapid and more intense than the primary response.
   F. Active immunity develops as a result of exposure to antigens; it may occur naturally after recovery from a disease or may be artificially induced by immunization.
   G. Passive immunity develops when an individual receives injections of antibodies produced from another person or animal, and is temporary.
   H. Normally, the immune system destroys precancer cells when they arise; diseases such as cancer develop when this immune mechanism fails to operate effectively.
   I. Transplanted tissues have MHC antigens that stimulate graft rejection, an immune response that can destroy the transplant.
   J. In autoimmune diseases, the body reacts immunologically against its own tissues.
   K. In an allergic response, an allergen can stimulate production of IgE antibody, which combines with the receptors on mast cells. When allergen combines with the IGE, the mast cells release histamine and other substances that cause symptoms of allergy such as inflammation.

Selected Key Terms

acquired immune deficiency syndrome (AIDS), p. 722
active immunity, p. 737
allergic reaction, p. 741
antibody, p. 724
antibody-mediated immunity, p. 727
antigen, p. 724
antigen-antibody complex, p. 731
autoimmune disease, p. 740
B cell (B lymphocyte), p. 727
complement, p. 733
cytotoxic T cell, p. 728
helper T cell, p. 728
histamine, p. 726
human immunodeficiency virus (HIV), p. 722
immune responses, p. 724
immunization, p. 738
immunoglobulin, p. 731
inflammatory response, p. 726
interferon, p. 725
interleukin, p. 726
macrophage, p. 726
mast cell, p. 742
memory cell, p. 731
MHC (major histocompatibility complex), p. 729
natural killer cell (NK cell), p. 728
passive immunity, p. 738
pathogen, p. 724
plasma cell, p. 728
primary response, p. 737
secondary response, p. 737
suppressor T cell, p. 728
T cell (T lymphocyte), p. 727
thymus, p. 728
vaccine, p. 738

Post-Test

1. An antigen is a substance capable of stimulating a/an ______________ ______________.
2. Specific proteins produced in response to specific antigens are called ______________.
3. When infected by viruses, some cells respond by producing proteins called ______________.
4. The clinical characteristics of inflammation are ______________, ______________, ______________, and ______________.
5. T lymphocytes originate in the ______________ ______________ and they are processed in the ______________.
6. ______________ T cells inhibit immune responses.
7. When the body is invaded by the same pathogen a second time, the immune response can be launched more rapidly due to the presence of ______________ cells.
8. The cells that produce antibodies are ______________ cells.
9. An antigenic determinant gives the antigen molecule a specific configuration that can be "recognized" by a/an ______________.
10. The ______________ confers immunological competence upon T cells.
11. The complement system is activated when a/an ______________-______________ complex is formed.
12. After a macrophage ingests a pathogen, it may display foreign antigen on its surface along with ______________.
13. Although artificially induced, immunization is a form of ______________ immunity.
14. An individual receiving injections of antibodies produced by another organism is receiving ______________ immunity.
15. In graft rejection, the host launches an effective ______________ ______________ against ______________ tissue.
16. Cornea transplants are highly successful because the cornea is an immunologically ______________ site.
17. A/An ______________ is a mild antigen that does not stimulate a response in an individual who is not ______________.
18. In a typical allergic reaction, mast cells secrete ______________ and other compounds that cause ______________.
19. All of the following are examples of nonspecific immune responses *except* (a) interferon release (b) inflammatory response (c) cytotoxic T cell attacks an infected cell (d) macrophage engulfs pathogen (e) enzyme in saliva attacks bacterial cell wall.
20. Cells that inhibit activity of other immune system cells are (a) helper T cells (b) suppressor T cells (c) cytotoxic T cells (d) macrophages (e) B cells.
21. Cells that migrate to the site of infection and then release potent chemicals to destroy an infected cell are (a) helper T cells (b) suppressor T cells (c) cytotoxic T cells (d) memory cells (e) B cells.
22. Cells that are important for a secondary response to occur are (a) helper T cells (b) suppressor T cells (c) cytotoxic T cells (d) memory cells (e) B cells.
23. HIV (a) is a retrovirus (b) attacks mainly suppressor T cells (c) attaches to CD4 (d) is a type of allergen (e) two of the preceding answers are correct.

Review Questions

1. How does the body distinguish between self and nonself? Are invertebrates capable of making this distinction?
2. Contrast specific and nonspecific defense mechanisms. Which type confronts invading pathogens immediately? How do the two systems work together?
3. How does inflammation help to restore homeostasis?
4. Give two specific ways in which cell-mediated and antibody-mediated immune responses are similar and two ways in which they are different.
5. Describe how antibodies destroy pathogens.
6. Why is passive immunity temporary?
7. What is graft rejection? What is the immunological basis for it?
8. List the immunological events that take place in a common type of allergic reaction such as hayfever.
9. What is an autoimmune disease? Give two examples.

Thinking Critically

1. Imagine that you were a researcher developing new treatments for HIV. What approaches might you take? What public policy decisions would you recommend that might help slow the spread of AIDS while new treatments or vaccines are being developed?
2. John is immunized against measles. Jack contracts measles from a playmate in nursery school before his mother gets around to having him immunized. Compare the immune responses in each child. Five years later, John and Jack are playing together when Judy, who is coming down with measles, sneezes on both of them. Compare the immune response in Jack and John.
3. Macrophages can be selectively destroyed in the body by the administration of a certain chemical. What would be the effects of such a loss of macrophages? Which do you think would have a greater effect on the immune system, loss of macrophages or loss of B cells?
4. What are the advantages of having MHC antigens? Disadvantages? What do you think would be the consequences of not having them?
5. Specificity, diversity, and memory are key features of the immune system. Giving specific examples, explain how each of these features is important.

Recommended Readings

Boon, T., "Teaching the Immune System to Fight Cancer," *Scientific American,* Vol. 266, No. 3, March 1993, pp. 82–89. Certain molecules on cancer cells serve as targets for cells of the immune system. These tumor-rejection antigens are the focus of research and may be useful in treating cancer.

Fischetti, V.A., "Streptococcal M Protein," *Scientific American,* Vol. 264, No. 6, June 1991, pp. 58–65. The bacteria that cause strep throat and rheumatic fever depend on a surface protein to evade the body's defenses.

Johnson, H.M., F.W. Bazer, B.E. Szente, and M.A. Jarpe, "How Interferons Fight Disease," *Scientific American,* Vol. 270, No. 5, May 1994, pp. 68–75. An interesting article that discusses how interferons protect against disease and how they are used clinically.

Johnson, H.M., J.K. Russell, and C.H. Pontzer, "Superantigens in Human Disease," *Scientific American,* Vol. 266, No. 4, April 1992, pp. 92–101. Superantigens cause food poisoning and toxic shock. The article discusses how superantigens cause disease.

Scientific American, Vol. 269, No. 3, September 1993. A single-topic issue on life, death, and the immune system.

Smith, K.A., "Interleukin-2," *Scientific American,* Vol. 262, No. 3, March 1990, pp. 50–57. A discussion of the regulatory functions of interleukin-2.

Tizard, I.R., *Immunology: An Introduction,* 3rd ed., Saunders College Publishing, Philadelphia, 1992. A basic introduction to immunology.

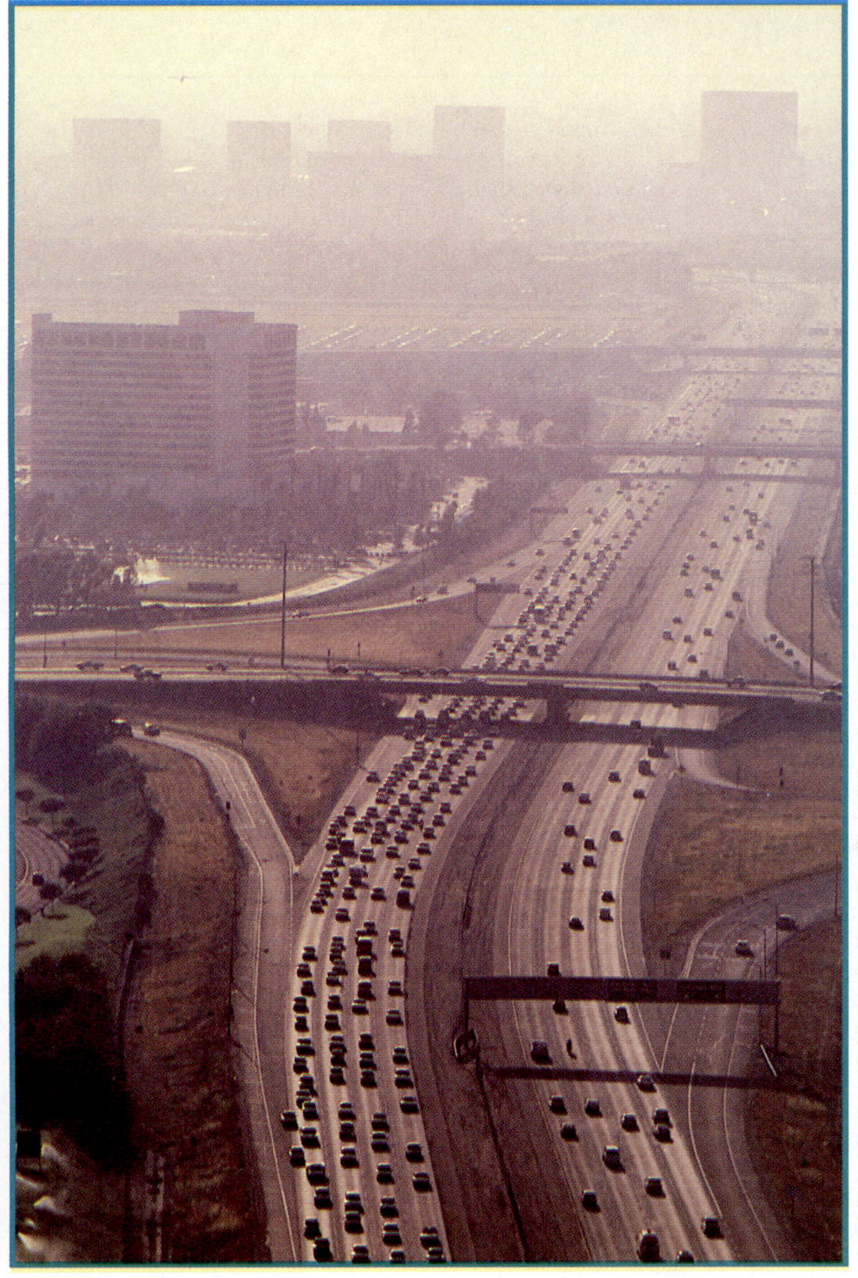

Nitrogen oxides released by automobiles contribute to the brownish orange haze, called photochemical smog, that is common in Los Angeles and some other urban areas. (M. Richards/PhotoEdit)

Cleaning Up The Air We Breathe

We breathe about 20,000 times every day, inhaling about 35 pounds of air—six times more than all the food and drink we consume. Most of us breathe dirty urban air laden with suspended particles, carbon monoxide, ozone, sulfur oxides, nitrogen oxides, and other harmful substances that are damaging to the respiratory system. Inhaling polluted air can lead to illnesses such as chronic bronchitis, emphysema, and lung cancer. Many air pollutants also suppress the immune system, increasing our susceptibility to infections.

The most important sources of air pollution are industrial emissions, motor vehicle emissions, and cigarette smoking. A 1992 study, known as *Urban Air Pollution in the Megacities of the World*, indicates that every large city (more than 10 million people) studied had at least one major air pollutant that exceeded the World Health Organization's health guidelines. For example, Los Angeles, the U.S. city with the worst air pollution, was found to have a serious ozone problem, as well as moderate to heavy pollution by carbon monoxide, nitrous oxides, and suspended particulate matter.

People are becoming more concerned about the quality of the air we breathe. Worldwide, countries are beginning to pass legislation that will help reduce air pollution. In the U.S., the 1990 Clean Air Act provides for a 90% reduction in the atmospheric emissions of 189 toxins by the year 2003. Sulfur dioxide and nitrogen oxide emissions from coal-burning power plants will be greatly reduced by the year 2000.

Legislation is also being passed to regulate pollution caused by tobacco smoke, a pollutant that is responsible for the premature deaths of nearly one-half million people every year in the U.S. alone (see Focus On Health and Human Affairs: Facts about Smoking). Nearly 1000 different chemical components have been identified in cigarette smoke. At least 10 of the components in the tar of cigarette smoke have been shown to cause cancer in animals. Cigarette smoke is a "portable" air pollutant. Smokers move about exhaling tobacco smoke into the very air we all must breathe. This environmental tobacco smoke has been linked to the deaths of about 3000 nonsmokers from lung cancer every year.

Almost every American who takes up smoking is a child. More than one million Americans—3000 every day—take up this habit every year. Former Surgeon General Dr. Antonia C. Novello reported in 1993 in the *Journal of the American Medical Association* that 10% of children who begin smoking start by the fourth grade, and nearly two-thirds begin by the tenth grade. Novello argues that we need to implement the following strategies to control tobacco usage: legislation, increasing taxes on cigarettes, and individual and public health education. Only when we are willing, both individually and as citizens of our respective countries, to take action to reduce the toxins released into the air, will we all have clean air to breathe.

Gas Exchange

38

Learning Objectives

After you have studied this chapter you should be able to:

1. Compare the advantages and disadvantages of gas exchange in air with those in water.
2. Describe the following adaptations for gas exchange: moist body surface, tracheal tubes, gills, and lungs.
3. Trace the route traveled by a breath of air through the human respiratory system from nose to air sacs, and then trace the oxygen from the air sacs to body tissues. Describe the function of each of the human respiratory structures.
4. Summarize the mechanics of breathing and the regulation of breathing.
5. Compare the composition of exhaled air with that of inhaled air, and describe the exchange of oxygen and carbon dioxide in the lungs and tissues.
6. Describe the mechanisms by which oxygen and carbon dioxide are transported in the blood.
7. Summarize the defense mechanisms that protect the lungs and describe the effects of breathing polluted air on the respiratory system.

Key Concepts

- ☐ Gases move in and out of cells by diffusion. Large animals have specialized respiratory surfaces that facilitate gas exchange. Adaptations for gas exchange include the body surface, gills, tracheal tubes, and lungs.
- ☐ In humans and other mammals, the respiratory system consists of a series of air passageways that branch into smaller and smaller tubes, ending in the alveoli within the lungs. Gas exchange takes place across the thin walls of the alveoli by diffusion.
- ☐ In humans and other mammals, oxygen diffuses from the alveoli into the blood in the pulmonary capillaries. Carbon dioxide diffuses from the blood into the alveoli and is expired. In the tissues, each process is reversed.
- ☐ Oxygen is transported to the cells in the form of oxyhemoglobin. Carbon dioxide is transported mainly in the form of bicarbonate ions.

RESPIRATORY STRUCTURES ARE ADAPTED FOR GAS EXCHANGE IN AIR OR WATER

During gas exchange, oxygen from the environment is taken up by the animal and delivered to its individual cells, and carbon dioxide is released into the environment. Gases move in and out of cells by diffusion.

Terrestrial animals have specialized **respiratory structures,** like tracheal tubes and lungs, that are adapted for gas exchange in air. Many aquatic animals have gills that are adapted for gas exchange in water. Whether an animal makes its home on land or water, gas exchange can take place only across a moist surface.

Because gas molecules must be dissolved in water to pass through plasma membranes, an animal that lives in water may seem to have an advantage. However, because water has a much greater density and viscosity (resistance to flow) than does air, a large aquatic animal must expend more energy to move water over its respiratory surface than it would to move air. For example, a fish uses up to 20% of its total energy expenditure to perform the muscular work needed to move water over its gills. An air-breather uses much less energy, only 1% or 2% of its total, to move air in and out of its lungs.

Gas exchange in air has other advantages over gas exchange in water. Compared to water, air contains far more molecular oxygen. Oxygen also diffuses much faster through air than through water. Another advantage is that air is not salty like seawater. Air-breathers do not have to cope with the diffusion of ions into their body fluids along with their oxygen. As a result, they have an easier time maintaining appropriate concentrations of ions in their body fluids.

On the other hand, organisms that respire in air struggle continuously with water loss. Terrestrial animals must have adaptations that minimize drying out, and their respiratory surfaces must be kept moist so that oxygen and carbon dioxide can pass through the plasma membranes. The lungs of air-breathing vertebrates are located deep within the body, not exposed like gills. Air must pass through a long sequence of passageways before reaching the moist respiratory surfaces of the lungs, and expired air must again pass through these passageways before leaving the body. These adaptations help protect the lungs from the drying effects of air.

FOUR MAIN TYPES OF SURFACES FOR GAS EXCHANGE HAVE EVOLVED

The air or water supplying oxygen must be continuously renewed so that as oxygen is used up, more is available. For this reason, animals carry on **ventilation,** that is, they actively move air or water over their respiratory surfaces. Sponges do this by setting up a current of water through the channels of their bodies by means of flagella. Most fish gulp water and then actively pass it over their gills. Terrestrial vertebrates breathe air.

In small, aquatic animals such as sponges, *Hydras*, and flatworms, dissolved oxygen from the surrounding water diffuses through the body surface and into the cells, and carbon dioxide diffuses out of the cells and into the water. Large animals cannot rely solely on the body surface for gas exchange. Specialized respiratory systems have evolved that increase the surface area available for gas exchange. Four main types of respiratory surfaces used by animals are the body surface, tracheal tubes, gills, and lungs (Fig. 38–1).

The Body Surface Can Be Adapted for Gas Exchange

The body surface is specialized for gas exchange in some animals, including some mollusks, most annelids, and a few vertebrates (Fig. 38–2). All of these animals are small, with a high surface-to-volume ratio. They also have a low metabolic rate so smaller quantities of oxygen are used per cell. In aquatic animals, the body surface is kept moist by the surrounding water. In terrestrial animals, the body secretes fluids that keep its surface moist.

How does an animal such as the earthworm exchange gases across its body surface? Gland cells in the epidermis secrete mucus, which keeps the body surface moist while also offering protection. Oxygen, present in air pockets in the loose soil that the earthworm inhabits, dissolves in the mucus. Then, the oxygen diffuses through the body wall. Oxygen diffuses into blood circulating in a network of capillaries just beneath the outer cell layer. Carbon dioxide is transported by the blood to the body surface, from which it diffuses out into the environment.

Tracheal Tubes Are an Adaptation for Gas Exchange in Arthropods

In insects and some other arthropods, the respiratory system consists of a network of **tracheal tubes.** Air enters the tracheal tubes through a series of up to 20 tiny openings, called **spiracles,** located along the body surface (Fig. 38–3). In large or active insects, air moves in and out of the spiracles by movements of the body or by rhythmic movements of the tracheal tubes.

The branching tracheal tubes deliver air to all parts of the body. At their microscopic ends, the tracheal tubes are filled with fluid. Gases are exchanged between this fluid and the body cells.

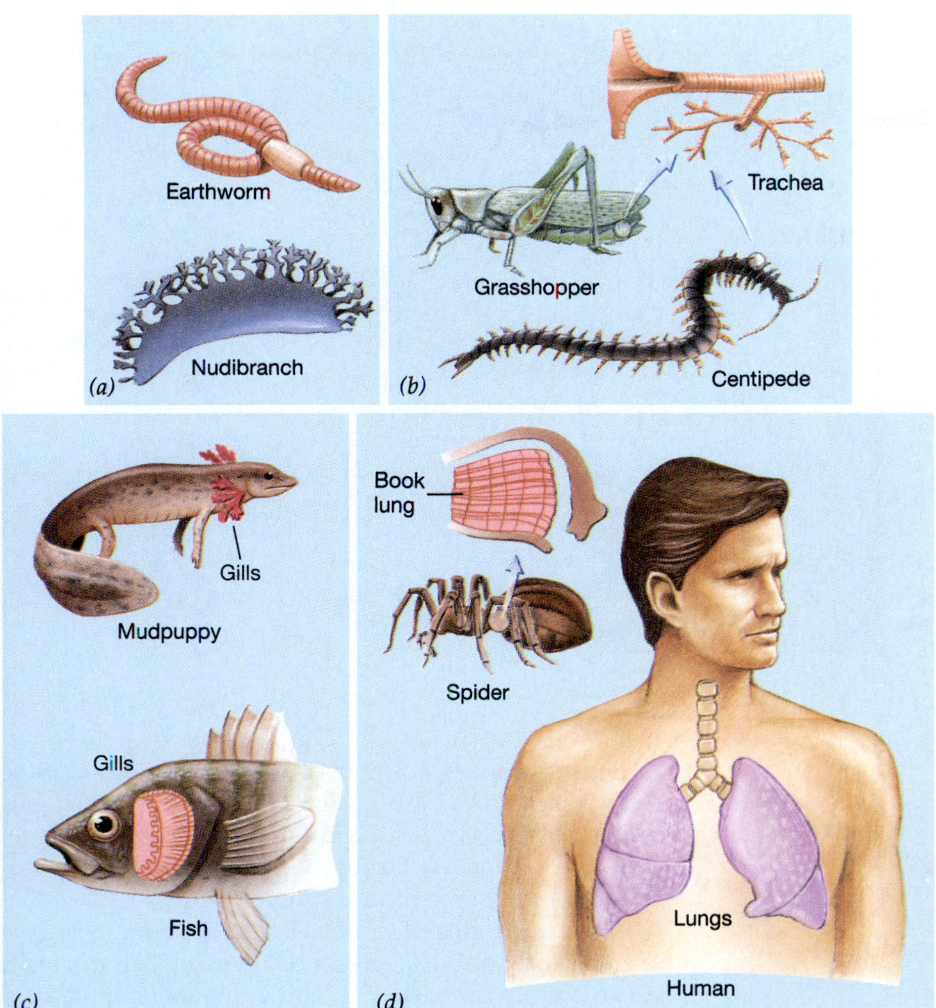

FIGURE 38–1 Types of respiratory structures found in animals. (***a***) Some animals exchange gases through the body surface. (Nudibranchs are mollusks.) (***b***) Insects and some other arthropods exchange gases through a system of tracheal tubes. (***c***) Many aquatic animals have gills for gas exchange. (***d***) Lungs are adaptations for terrestrial gas exchange.

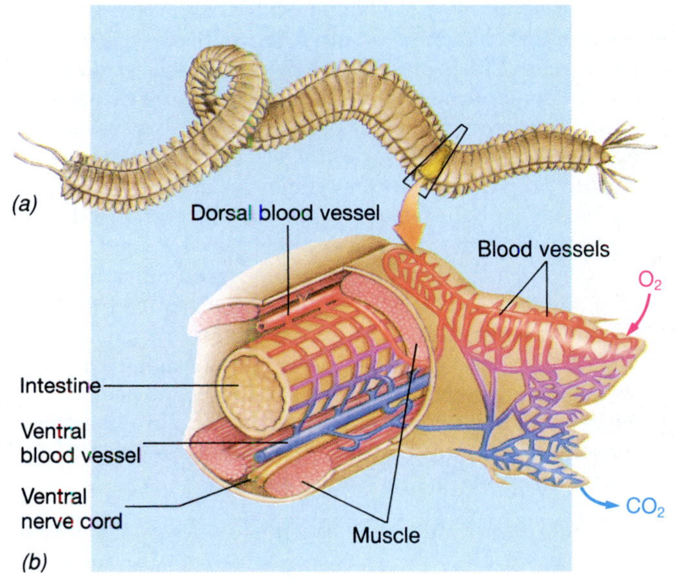

FIGURE 38–2 Gas exchange across the body surface. (***a***) The clamworm *Nereis virens* has parapodia (limblike extensions of each body segment) typical of the polychaete worms. (***b***) The parapodium acts as an extension of the body wall in gas exchange with the surrounding water. Note the rich supply of blood vessels for transport of carbon dioxide and oxygen to and from the body surface.

Many Aquatic Animals Exchange Gases through Gills

Gills are respiratory structures found mainly in aquatic animals. They are moist, thin structures that extend from the body surface. Gills are supported by the buoyancy of water but tend to collapse in air. In many animals, the outer surface of the gills is exposed to water, and the inner side is in close contact with networks of blood vessels.

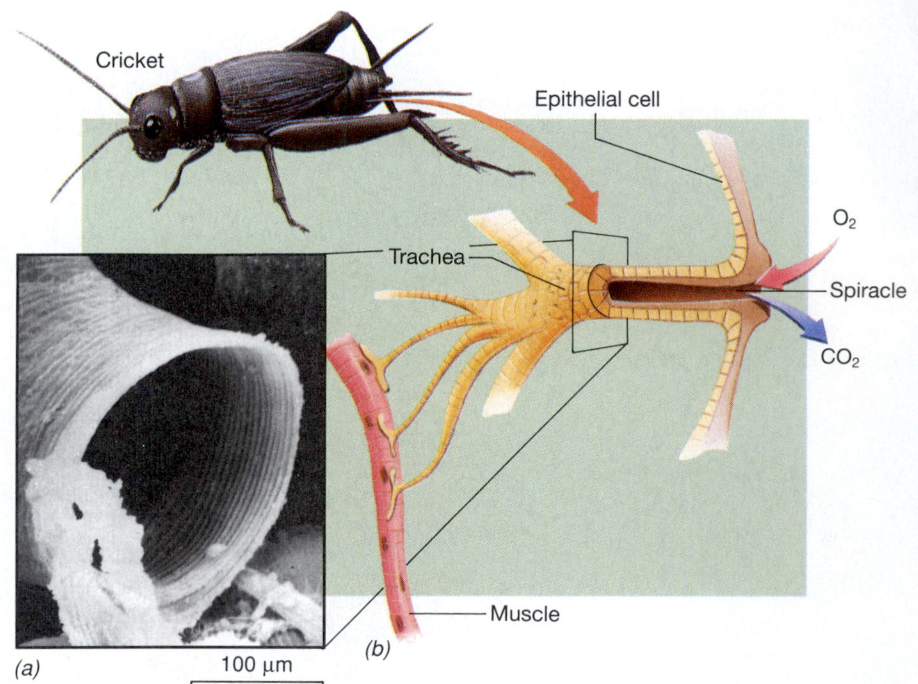

FIGURE 38–3 Tracheal tubes. (***a***) A scanning electron micrograph of a mole cricket trachea (approximately ×1300). The wrinkles are a part of a long spiral around the tube, which may strengthen the tracheal wall somewhat like the spring strengthens the plastic hoses of many vacuum cleaners. The tracheal wall is composed of chitin. (***b***) Each tracheal tube and its branches conduct oxygen to the body cells of the insect. (Courtesy of Dr. James L. Nation, *Stain Technology,* Vol. 58, 1983, No. 6, p. 349)

CONCEPT CONNECTIONS

Gas Exchange ⇔ *Cellular Respiration*

Two important biological concepts discussed in this book are (1) the dependence of most cells on oxygen and (2) the interdependence of producers and consumers, which involves gas exchange. Recall from Chapter 8 that most cells die quickly without oxygen because this element is required for cellular respiration, the complex series of reactions by which cells oxidize organic compounds such as glucose, releasing carbon dioxide and energy. In this process, oxygen serves as the final electron acceptor in the electron transport system. To provide oxygen for cellular respiration and to rid cells of carbon dioxide, gases must be exchanged continuously between cells and their environment.

Gas exchange replenishes the oxygen and carbon dioxide in the atmosphere. Recall from Chapter 1 that most organisms depend on producers for oxygen. Plants, algae, and certain bacteria carry on photosynthesis during which water is split, producing molecular oxygen. On the other hand, carbon dioxide produced by consumers and decomposers during cellular respiration is used by producers for photosynthesis.

Sea stars and sea urchins have simple **dermal gills,** which project from the body wall. Ciliated epidermal cells ventilate the gills by beating a stream of water over them. Gases are exchanged through the gills between the water and the coelomic fluid inside the body.

The folded gills of mollusks provide a large surface for gas exchange. In clams and other bivalve mollusks and in simple chordates, the gills are also adapted for trapping and sorting food. Rhythmic beating of cilia draws water over the gill area, and food is filtered out of the water at the same time that gases are exchanged.

In chordates, the gills are usually located internally. A series of slits perforate the pharynx, and the gills are located along the edges of these gill slits. In bony fish, the fragile gills are protected by an external bony plate, called the **operculum.** Movements of the operculum help to pump water in through the mouth, across the gills, and then out through the gill slits.

Each gill in the bony fish consists of many **filaments,** which provide an extensive surface for gas exchange (Fig. 38–4). The filaments extend out into the water that continuously flows over them. A capillary network delivers blood to the gill filaments, facilitating diffusion of oxygen and carbon dioxide between blood and water. The impressive efficiency of this system is possible because blood flows in a direction opposite to the movement of the water. This arrangement, referred to as a **countercurrent exchange system,** maximizes the difference in oxygen concentration between blood and water throughout the area where the two remain in contact.

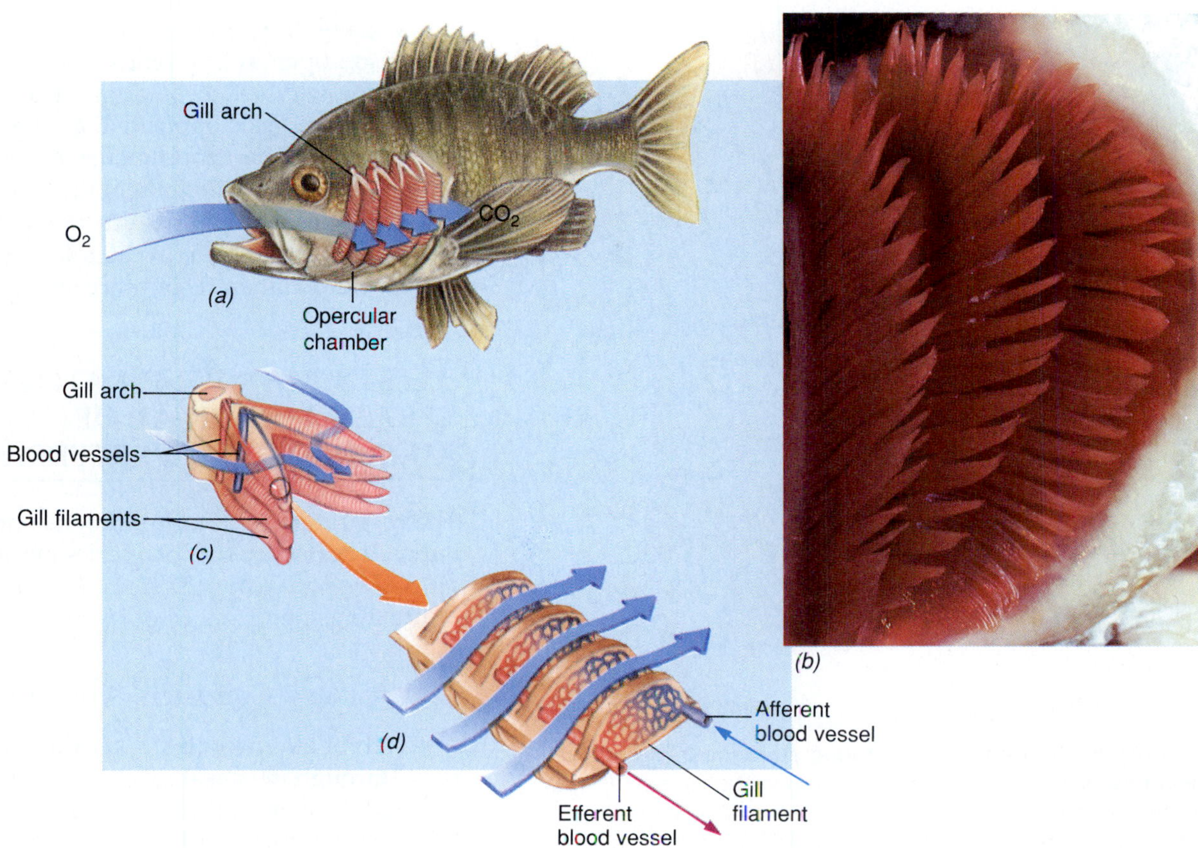

FIGURE 38–4 How the fish gill works. (***a***) The gills are located under a bony plate, the operculum, which has been removed in this side view. The gills occupy the opercular chamber and form the lateral wall of the pharyngeal cavity. The fish respires by movements of its mouth and pharyngeal cavity. (***b***) The gills of the salmon provide an extensive surface for gas exchange. (***c***) Each gill consists of a cartilaginous gill arch to which two rows of leaf-like gill filaments are attached. As water passes among the gill filaments, blood circulates through them. (***d***) Each gill filament has many smaller extensions rich in capillaries. Blood entering the capillaries is deficient in oxygen. The blood flows through the capillaries in a direction *opposite* to that taken by the water. This countercurrent exchange system efficiently charges the blood with oxygen. (G.I. Bernard 1995 Animals Animals)

If blood and water flowed in the *same* direction, the difference between the oxygen concentrations in blood (low) and water (high) would be very large initially and very small at the end. The oxygen concentration in the water would decrease as the oxygen concentration in the blood increased. When the concentrations of oxygen in the two fluids became equal, net diffusion of oxygen would stop. A great deal of oxygen would remain in the water.

In the countercurrent exchange system, however, blood low in oxygen comes in contact with water that is partly oxygen-depleted. Then, as the blood becomes more and more oxygen-rich, it comes in contact with water with a progressively higher concentration of oxygen. In this way, a high rate of diffusion is maintained. As a result, a very high percentage (more than 80%) of the available oxygen in the water diffuses into the blood.

Oxygen and carbon dioxide diffuse in opposite directions at the same time. This is because oxygen is more concentrated outside the gills than within, but carbon dioxide is more concentrated inside the gills than outside. Thus, the same countercurrent exchange mechanism that results in efficient inflow of oxygen also results in equally efficient outflow of carbon dioxide.

Terrestrial Vertebrates Exchange Gases through Lungs

Lungs are respiratory structures that develop as ingrowths of the body surface or from the wall of a body cavity, such as the pharynx. The **book lungs** of some spiders are enclosed in an inpocketing of the abdominal wall. These lungs consist of a series of parallel, thin plates of tissue (like the pages of a book) filled with blood. The plates of tissue are separated by air spaces that receive oxygen from the outside environment.

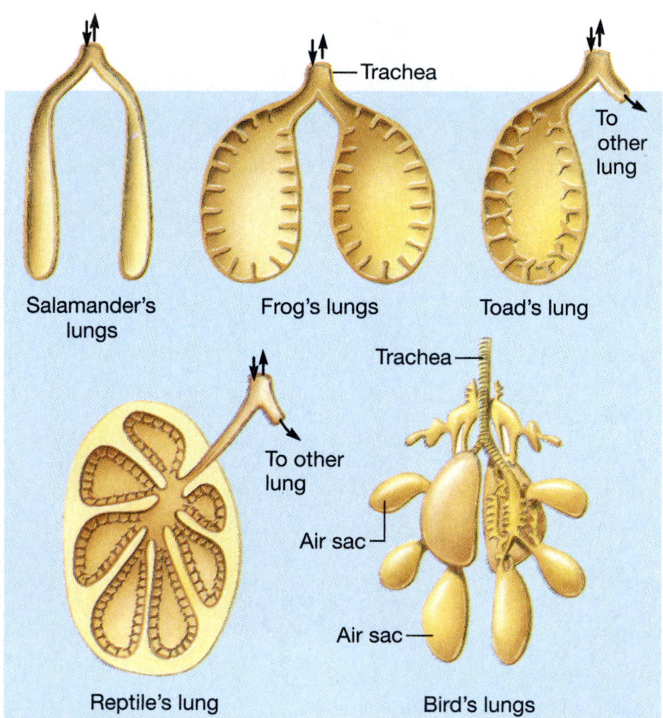

FIGURE 38–5 Comparison of lung structure in various vertebrates. Note the progressive increase in surface area for gas exchange in the different vertebrate classes. Salamander lungs are simple sacs. Frog lungs have small ridges in the lung wall that help increase the surface area, while reptile lungs have a larger surface area. Birds have an elaborate system of lungs and air sacs. The lungs of mammals (Fig. 38–6) have millions of air sacs that increase the surface available for gas exchange.

Not all fish breathe exclusively by gills. African lungfish use lungs as well as gills. Amphibians depend mainly on their body surface for gas exchange, but they do have simple lungs. The lungs of salamanders are two long, simple sacs, covered on the outside by capillaries (Fig. 38–5). The lungs of frogs and toads have ridges that increase the respiratory surface somewhat.

In reptiles, birds, and mammals, lungs are the principal respiratory surface (Fig. 38–5). The lungs of most reptiles are rather simple, with only some folding that increases the surface for gas exchange. In some lizards and in turtles and crocodiles, the lungs have many subdivisions that increase the surface area for gas exchange.

Birds are very active animals with high metabolic rates. They require large amounts of oxygen, and they have highly effective respiratory systems. In birds, the lungs have developed several extensions (usually nine), called **air sacs,** which reach into all parts of the body and even penetrate into some of the bones. The air sacs act as bellows drawing air into the system. Collapse of the air sacs during respiration forces air out. The respiratory system is arranged so that air flows in one direction through the lungs and is renewed during each inspiration. The lungs have tiny, thin-walled ducts, the **parabronchi,** which are open at both ends. Gas exchange takes place across the walls of these ducts. The direction of blood flow in the lungs is opposite that of air flow through the parabronchi. This countercurrent flow increases the amount of oxygen that enters the blood.

The lungs of mammals are very complex and have an enormous surface area. In the following sections we will examine the human respiratory system in some detail.

The Human Respiratory System Is Typical of Air-Breathing Vertebrates

The respiratory system in humans and other air-breathing vertebrates consists of a series of tubes through which air passes on its journey from the nostrils to the air sacs of the lungs and back (Fig. 38–6).

The Airway Conducts Air into the Lungs

A breath of air enters the body through the **nostrils** and flows through the **nasal cavities** of the nose. As air passes through the nose, it is filtered, moistened, and brought to body temperature. The nasal cavities are lined with moist, ciliated epithelium that is rich in blood vessels. Mucous cells within the epithelium produce more than a pint of mucus a day. Inhaled dirt, bacteria, and other foreign particles are trapped in the layer of mucus and pushed along with the stream of mucus toward the throat by the cilia. In this way, foreign particles are delivered to the digestive system, which is far more capable of disposing of such materials than are the delicate lungs. A person normally swallows more than a pint of nasal mucus each day, more if he or she has an allergy or infection.

The back of the nasal cavities is continuous with the throat region, or **pharynx.** Air finds its way into the pharynx whether one breathes through the nose or mouth. An opening in the floor of the pharynx leads into the **larynx,** sometimes called the "Adam's apple." Because the larynx contains the vocal cords, it is also referred to as the voice box. Cartilage embedded in its wall prevents the larynx from collapsing and makes it hard to the touch when felt through the neck.

During swallowing, a flap of tissue, called the **epiglottis,** automatically closes off the larynx so that food and liquid enter the esophagus rather than the lower airway. Should this defense mechanism fail and foreign matter enter the sensitive larynx, a **cough reflex** is initiated, expelling the material from the respiratory system. Despite these mechanisms, choking sometimes occurs.

From the larynx, air passes into the **trachea,** or windpipe. The trachea is kept from collapsing by C-shaped

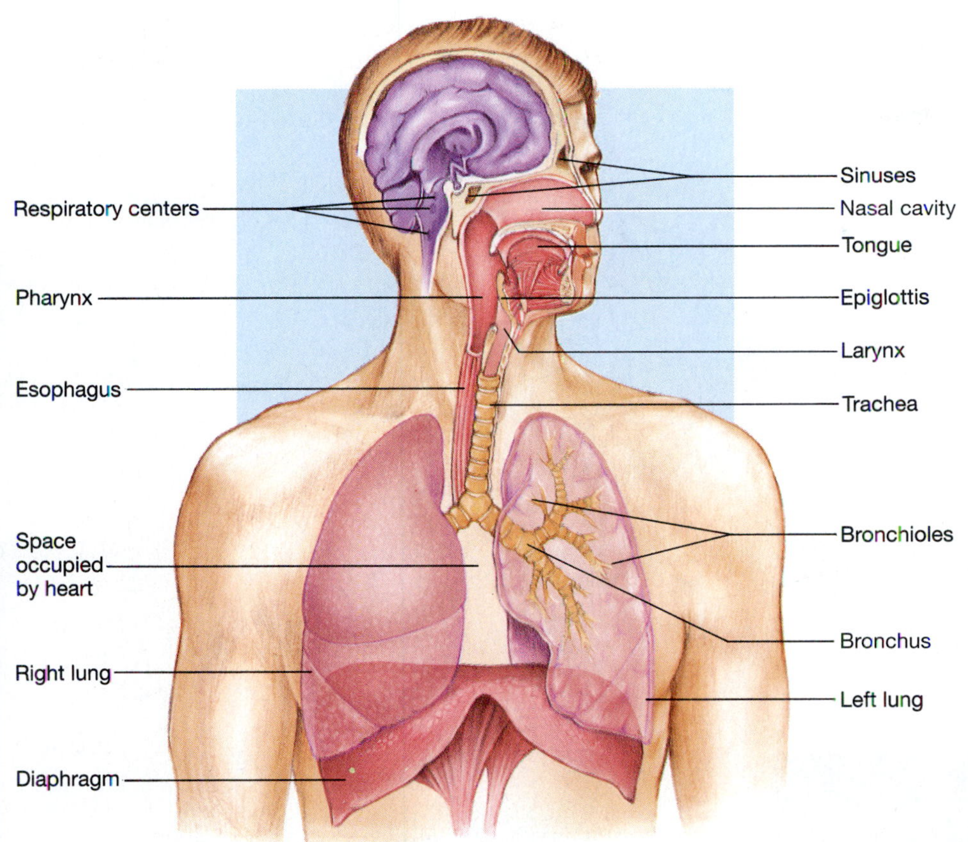

FIGURE 38–6 The human respiratory system. The paired lungs are located in the thoracic cavity. The muscular diaphragm forms the floor of the thoracic cavity, separating it from the abdominal cavity below. An internal view of one lung illustrates its extensive system of air passageways. The microscopic alveoli are shown in Figure 38–7.

rings of cartilage in its wall. The trachea divides into two branches, the **bronchi,** each of which connects to a lung.

Both trachea and bronchi are lined by a mucous membrane containing ciliated cells. Many medium-sized particles that have escaped the cleansing mechanisms of nose and larynx are trapped here. Mucus that contains these particles is constantly beaten upward by the cilia to the pharynx, where it is periodically swallowed. This mechanism, functioning as a cilia-propelled mucous elevator, helps keep foreign material out of the lungs.

Gas Exchange Occurs in the Lungs

The lungs are large, paired, spongy organs occupying the thoracic (chest) cavity. The right lung is divided into three lobes, the left lung into two lobes. Each lung is covered with a membrane, called the **pleural membrane,** which forms a continuous sac enclosing the lung and continuing as the lining of the chest cavity. The space between the pleural membrane covering the lung and the pleural membrane lining the chest cavity is called the **pleural cavity.** A film of fluid in the pleural cavity provides lubrication between the lungs and the chest wall.

Because the lung consists largely of air tubes, air sacs, and elastic tissue, it is a spongy, elastic organ with a very large internal surface area for gas exchange. In normal adults, the surface area of the lung is estimated as approximately that of a tennis court.

Inside the lungs the bronchi branch, becoming smaller and more numerous. These branches give rise to more than one million tiny **bronchioles** in each lung. Each bronchiole ends in a cluster of tiny air sacs, the **alveoli** (singular, *alveolus;* Fig. 38–7). The alveoli are lined by an extremely thin, single layer of epithelial cells. Gases diffuse freely through the wall of the alveolus and into the capillaries that surround each alveolus. Only two thin membranes separate the air in the alveolus from the blood: the epithelium of the alveolar wall and the capillary wall.

In summary, the sequence of structures through which air passes after it enters the body is:

Nostrils → Nasal cavities → Pharynx → Larynx →
Trachea → Bronchus → Bronchiole → Alveolus

Ventilation Is Accomplished by Breathing

Breathing is the mechanical process of moving air from the environment into the lungs and of expelling air from

(Text continues on page 757)

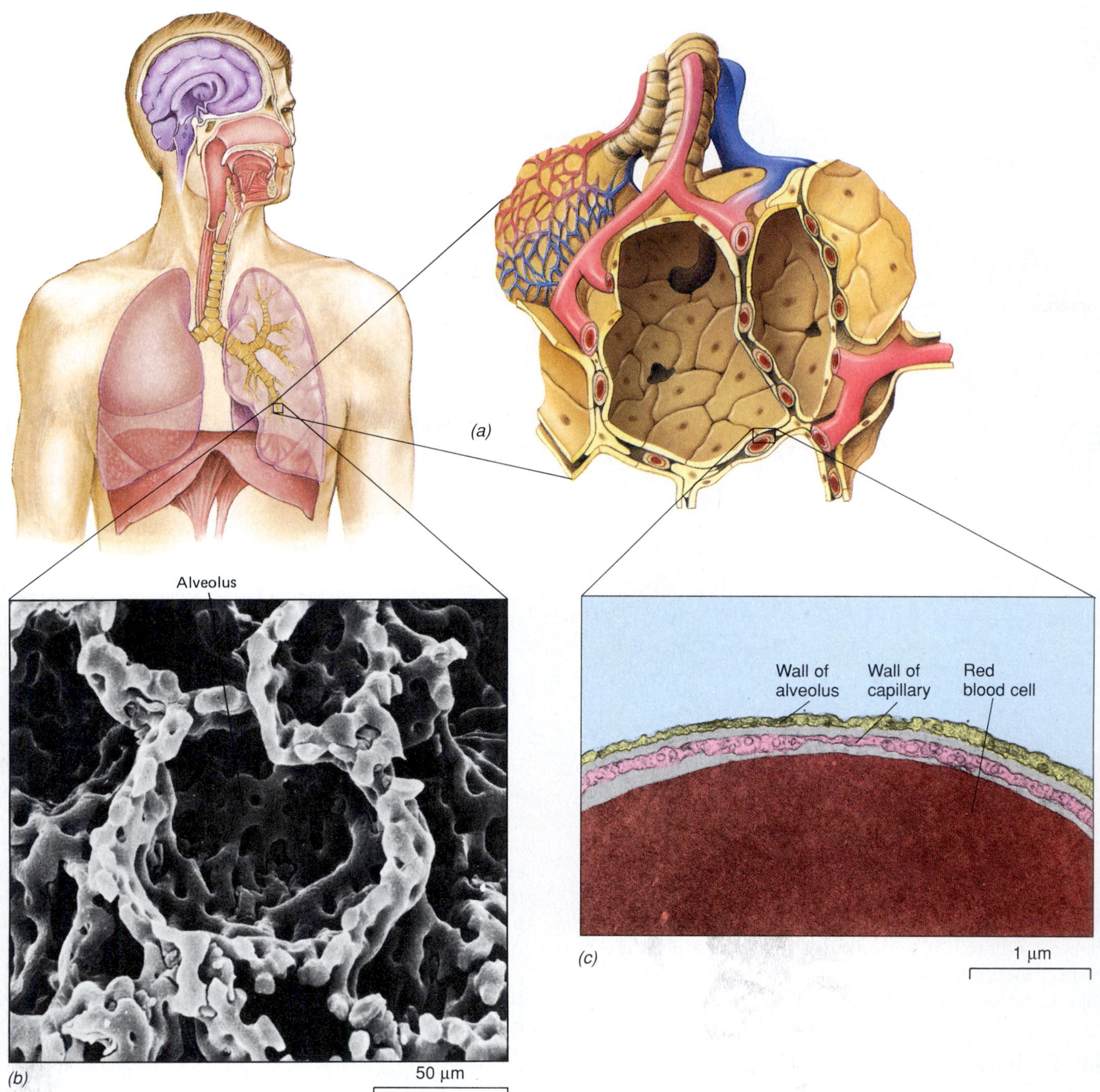

FIGURE 38–7 Gas exchange takes place across the thin wall of the alveolus. (***a***) Structure of the alveolus. The alveolar wall consists of extremely thin squamous epithelium. Between the walls of the alveoli lie extensive capillary networks. (***b***) Scanning electron micrograph showing the capillary network surrounding a portion of several alveoli. (***c***) An enlargement (approximately ×48,100) of a portion of a capillary in the lung. The dark structure extending through the capillary is a part of a red blood cell. The wall of the alveolus is visible just above the wall of the capillary. Notice the very short distance oxygen must travel to get from the air within the alveolus to the red blood cell in which it is transported to the body tissues. (*b*, Keesel, R.G. and Kardon, R.H. *Tissues and Organs, A Text-Atlas of Scanning Electron Microscopy,* San Francisco, W.H. Freeman Co., 1979; *c*, Courtesy of Drs. Peter Gehr, Marianne Bachofen, and Ewald R. Wiebel)

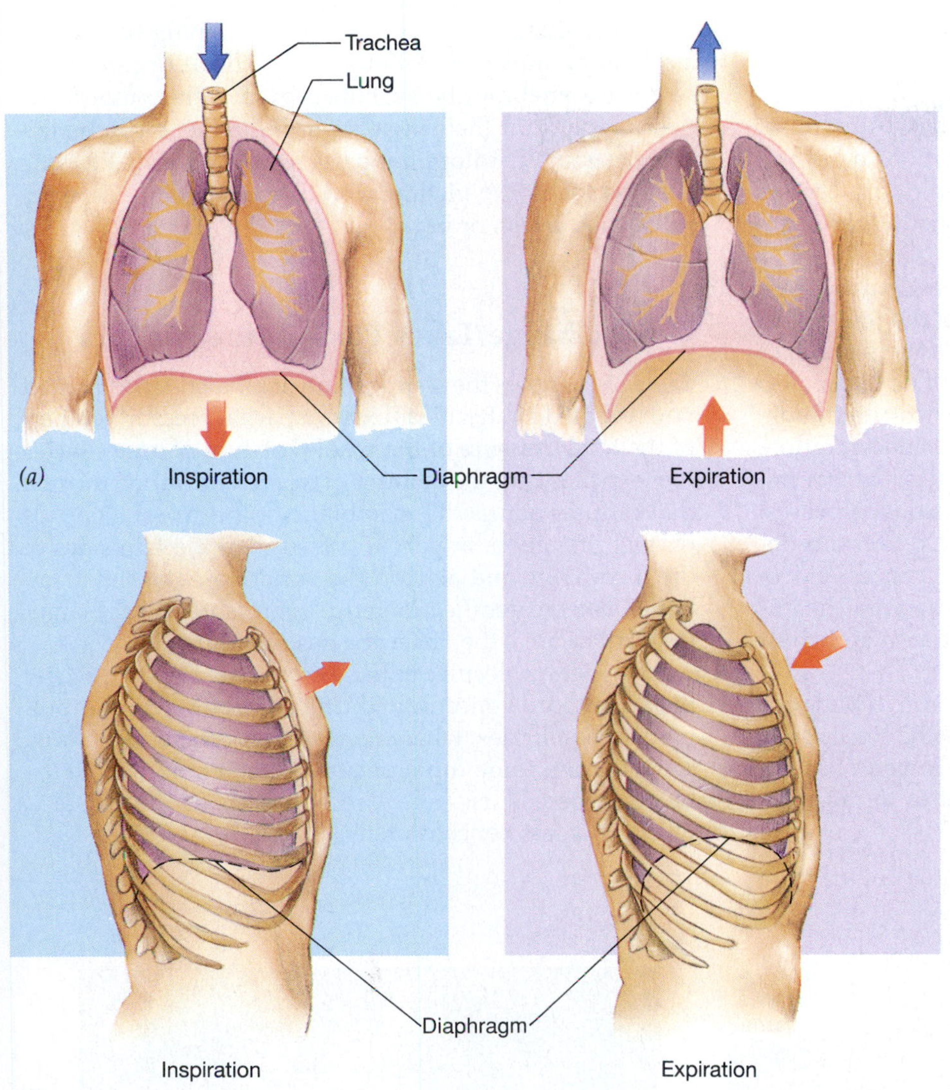

FIGURE 38–8 The mechanics of breathing. (***a***) Changes in the position of the diaphragm in inspiration and expiration result in changes in the volume of the chest cavity. (***b***) Changes in position of the rib cage and diaphragm in expiration and inspiration. The elevation of the front ends of the ribs by the intercostal (chest) muscles causes an increase in the front-to-back dimension of the chest and a corresponding increase in the volume of the chest cavity. When the volume of the chest cavity increases, air moves into the lungs.

the lungs. Inhaling air is referred to as **inspiration;** exhaling air is called **expiration.** A resting adult breathes about 12 times each minute.

The chest cavity is closed so no air can enter except through the trachea. The chest cavity has a muscular floor, the **diaphragm,** and two lateral **pleural cavities,** each of which contains a lung. During inspiration, the chest cavity is expanded by the contraction of the diaphragm, which moves the diaphragm downward, and by the contraction of the rib muscles, which moves the ribs upward. These contractions increase the circumference of the chest cavity (Fig. 38–8).

As the chest expands, the film of fluid on the pleura pulls the membranous walls of the lungs outward along with the chest walls. This increases the space within each lung. The air in the lungs now has more space in which to move about, so the pressure of the air in the lungs decreases. When the pressure in the lungs falls below the pressure of the air outside the body, air from outside rushes in through the respiratory passageways and fills the air sacs in the lungs. This process continues until the two pressures are again equal.

Expiration occurs when the diaphragm and rib muscles relax. The volume of the chest cavity decreases, increasing the pressure in the lungs. The millions of air sacs deflate, expelling the air that was inhaled. The pressure returns to normal and the lung is ready for another change of air.

Breathing Is Regulated by Respiratory Centers in the Brain

The amount of oxygen used by the body varies with different levels of activity. When you are engaged in a strenuous game of tennis, for example, you require more oxygen than when reading quietly. Breathing is controlled by respiratory centers (groups of specialized neurons) in the brain that are indirectly sensitive to increases in the amount of carbon dioxide in the blood (Fig. 38–6). Groups of neurons in the medulla regulate the basic rhythm of respiration: one group regulates inspiration, a second group is concerned with expiration. Respiratory centers in the pons can stimulate or inhibit the medullary respiratory center.

Chemoreceptors in the medulla respond to a decrease in blood pH by stimulating an increase in the rate of respiration. (An increase in carbon dioxide lowers the pH.) Chemoreceptors in the cartoid arteries and aorta are also sensitive to increases in carbon dioxide concentration, and also to large decreases in oxygen concentration. When stimulated, these chemoreceptors send neural impulses to the inspiratory center and respiration increases.

Nerve impulses from the inspiratory center are delivered to the diaphragm by the phrenic nerves and to the intercostal muscles by the intercostal nerves. The impulses stimulate contraction of the diaphragm and chest muscles.

During exercise, large amounts of carbon dioxide are produced. The carbon dioxide stimulates the respiratory centers to produce more rapid and more forceful breathing. In this way, more oxygen is brought in to meet the body's increased need.

Individuals who have stopped breathing because of drowning, smoke inhalation, electric shock, or cardiac arrest can sometimes be sustained by mouth-to-mouth resuscitation until their own breathing reflexes can be initiated again. Cardiopulmonary resuscitation (CPR) is a method for aiding victims who have suffered respiratory or cardiac arrest, or both.

Gas Exchange Takes Place in the Air Sacs

Table 38–1 shows the composition of inhaled air compared with exhaled air. Each gas exerts a partial pressure—a part of the total pressure of the whole mixture of gases. **Fick's law** explains that the amount of oxygen or carbon dioxide that diffuses across the membrane of the air sac depends on the differences in partial pressure on the two sides of the membrane and also on the surface area of the membrane. The greater the difference in pressure and the larger the surface area, the faster the gas will diffuse.

Oxygen molecules diffuse from the air sacs, where they are more concentrated, into the blood in the pulmonary capillaries, where they are less concentrated (Fig. 38–9). At the same time, carbon dioxide moves from the blood, where it is more concentrated, to the air sacs, where it is less concentrated.

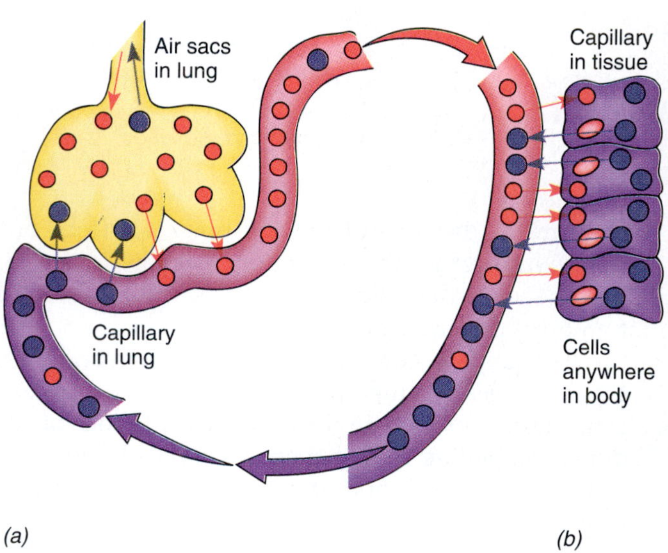

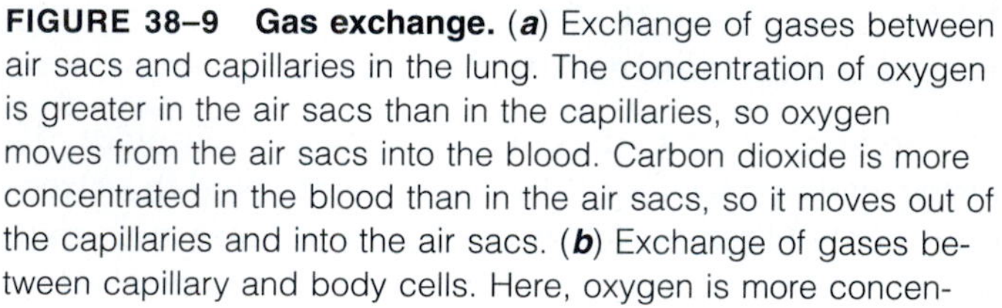
FIGURE 38–9 Gas exchange. (***a***) Exchange of gases between air sacs and capillaries in the lung. The concentration of oxygen is greater in the air sacs than in the capillaries, so oxygen moves from the air sacs into the blood. Carbon dioxide is more concentrated in the blood than in the air sacs, so it moves out of the capillaries and into the air sacs. (***b***) Exchange of gases between capillary and body cells. Here, oxygen is more concentrated in the blood, so it moves out of the capillary and into the cells. Carbon dioxide is more concentrated in the cells, and so it diffuses out of the cells and moves into the blood. (***c***) Like other animals, this whitetail deer buck (*Odocoileus virginianus*) depends on producers for the oxygen it inspires, and in turn, releases carbon dioxide into the environment. (*c*, Visuals Unlimited/Joe McDonald)

TABLE 38–1
Composition of Inhaled Air Compared with That of Exhaled Air

	% Oxygen (O_2)	% Carbon dioxide (CO_2)	% Nitrogen (N_2)
Inhaled air (atmospheric air)	20.9	0.04	79
Exhaled air (alveolar air)	14.0	5.60	79

Note the percentages of oxygen and carbon dioxide present in exhaled air compared with inhaled air as shown in Table 38–1. Because carbon dioxide is produced during cellular respiration, there is more of the gas—100 times as much—entering the alveoli from the blood than there is in air inhaled from the environment.

The movement of gases between air sacs and blood is not completely efficient. Not every molecule of inhaled oxygen actually finds its way into the blood, and not every molecule of carbon dioxide is removed from the blood. Newly inhaled air mixes with air already in the lungs. However, the exchange that does take place is sufficient to support the metabolic well-being of the body. The lungs of some marine mammals that remain underwater for considerable periods of time are more efficient at removing oxygen from inhaled air.

Oxygen Is Transported in Combination with Hemoglobin

When oxygen diffuses into the pulmonary capillaries, it forms a weak chemical bond with hemoglobin in the red blood cells, forming **oxyhemoglobin.** Each hemoglobin molecule can combine with, and transport, four molecules of oxygen. Because the chemical bond formed between the oxygen and the hemoglobin is weak, the reaction is readily reversible. In the body tissues, the reaction proceeds to the left, releasing oxygen.

$$\text{Oxygen} + \text{Hemoglobin} \rightleftharpoons \text{Oxyhemoglobin}$$

Oxyhemoglobin is bright scarlet, giving arterial blood its color. Deoxygenated hemoglobin is purple, giving venous blood a darker hue.

The ability of oxygen to combine with hemoglobin and to be released from oxyhemoglobin is influenced by several factors, including pH, concentrations of carbon dioxide and oxygen, and temperature. The **oxygen-hemoglobin dissociation curves** shown in Figure 38–10 illustrate that, as oxygen concentration increases, there is a progressive increase in the amount of hemoglobin that is combined with oxygen. This is known as the **percent saturation** of the hemoglobin. The percent saturation is highest in the pulmonary capillaries where the concentration of oxygen is greatest. In the capillaries of the tissues where there is less oxygen, the oxyhemoglobin dissociates, releasing oxygen. There, the percent saturation of hemoglobin is correspondingly less.

Carbon dioxide reacts with water in the plasma to form carbonic acid, H_2CO_3. Thus, an increase in the carbon dioxide concentration lowers the pH of the blood. Oxyhemoglobin dissociates more readily in a more acidic

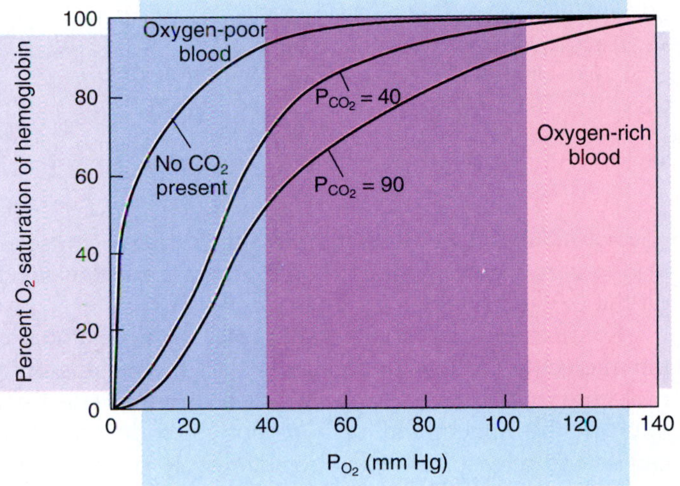

FIGURE 38–10 Oxygen dissociation curves. These curves show that as oxygen concentration increases, a progressive increase occurs in the amount of hemoglobin that is combined with oxygen. The curves also show how carbon dioxide affects the dissociation of oxyhemoglobin. Look at the vertical axis, labeled percentage saturation. If the blood contains 20% O_2 by volume, which is one fourth of the amount it could contain, it is said to be 25% saturated. The left curve shows what happens if no carbon dioxide is present. The middle curve shows the situation when the P_{CO_2} is 40, which is typical of arterial blood. The right curve indicates P_{CO_2} of 90, an unhealthy concentration. Now find the location on the horizontal axis where the partial pressure of oxygen is 40, and follow the line up through the curves. Notice how the saturation of hemoglobin with oxygen differs among the three curves, even though the partial pressure of oxygen is the same.

FOCUS ON *Health and Human Affairs*

FACTS ABOUT SMOKING

- Smoking is the single most preventable cause of death in our society.
- In the U.S. alone, smoking costs us more than a billion dollars every week in health costs and lost productivity.
- The life of a 30-year-old who smokes 15 cigarettes a day is shortened by an average of more than 5 years.
- If you smoke more than one pack per day, you are about 20 times more likely to develop lung cancer than is a nonsmoker. According to the American Cancer Society, cigarette smoking causes more than 75% of all lung cancer deaths.
- If you smoke, you are more likely to develop atherosclerosis, and you double your chances of dying from cardiovascular disease.
- If you smoke, you are 20 times more likely to develop chronic bronchitis and emphysema than is a nonsmoker.
- If you smoke, you are 7 times more likely to develop peptic ulcers (especially malignant ulcers) than is a nonsmoker.
- If you smoke, you have about 5% less oxygen circulating in your blood (because carbon monoxide binds to hemoglobin) than does a nonsmoker.
- If you smoke when you are pregnant, your baby will weigh about 6 ounces less at birth, and there is double the risk of miscarriage, stillbirth, and infant death than if you did not smoke.
- Workers who smoke one or more packs of cigarettes per day are absent from their jobs because of illness 33% more often than are nonsmokers.
- Nonsmokers confined in living rooms, offices, automobiles, or other places with smokers are adversely affected by the smoke. For example, when parents of infants smoke, the infant has double the risk of contracting pneumonia or bronchitis in its first year of life.

environment. Lactic acid released from active muscles also lowers the pH of the blood and has a similar effect on the oxygen-hemoglobin dissociation curve.

Some carbon dioxide is transported by the hemoglobin molecule. Although it attaches to the hemoglobin molecule in a different way and at a different site than oxygen, the attachment of a carbon dioxide molecule causes the release of an oxygen molecule from the hemoglobin. Thus, carbon dioxide concentration affects the oxygen-hemoglobin dissociation curve in two ways. This results in an extremely efficient transport system. In the capillaries of the lungs (or gills in fishes), carbon dioxide concentration is relatively low and oxygen concentration is high, so oxygen saturates a very high percentage of hemoglobin. In the capillaries of other tissues, carbon dioxide concentration is high and oxygen concentration is low, so oxygen is released from the hemoglobin.

Carbon Dioxide Is Transported Mainly as Bicarbonate Ions

As blood flows through the capillary networks of an organ, carbon dioxide moves out of the cells, where it has accumulated, and into the blood, where it is less concentrated. Carbon dioxide is transported in the blood in three ways. About 20% is attached to hemoglobin and 7% is dissolved in the plasma as carbon dioxide itself. However, most of the carbon dioxide combines with water in the plasma to form carbonic acid. This reaction is catalyzed by the enzyme, carbonic anhydrase. The carbonic acid dissociates, forming hydrogen ions and bicarbonate ions (HCO_3^-). Thus, most of the carbon dioxide is transported as bicarbonate ions.

$$\underset{\text{Carbon dioxide}}{CO_2} + \underset{\text{Water}}{H_2O} \longrightarrow \underset{\text{Carbonic acid}}{H_2CO_3} \longrightarrow \underset{\text{Hydrogen ion}}{H^+} + \underset{\text{Bicarbonate ion}}{HCO_3^-}$$

Hemoglobin buffers the hydrogen ions produced so that pH change is minimized.

When it reaches the lungs, carbon dioxide diffuses out of the blood and into the alveoli. Most of the transported carbon dioxide leaves the blood and eventually leaves the respiratory system. However, the human body is adjusted to function at an internal pH of about 7.4, which is maintained by a balance of substances in the blood, including a moderate amount of carbon dioxide. If too much is retained because of respiratory insufficiency, the blood pH becomes abnormally low (acidosis) due to excess carbonic acid formation. If hyperventilation blows off too much carbon dioxide, an equally undesirable state (alkalosis) results from abnormally low concentrations of carbonic acid in the blood.

- When smokers quit smoking, their risk of dying from chronic pulmonary disease, cardiovascular disease, or cancer decreases. (Precise changes in risk figures depend on the number of years the person smoked, the number of cigarettes smoked per day, the age of starting to smoke, and the number of years since quitting.)
- If everyone in the United States stopped smoking, nearly one-half million lives would be saved each year.

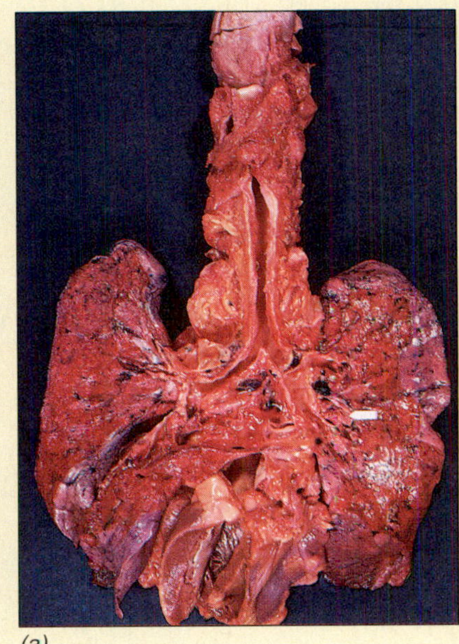
(a)

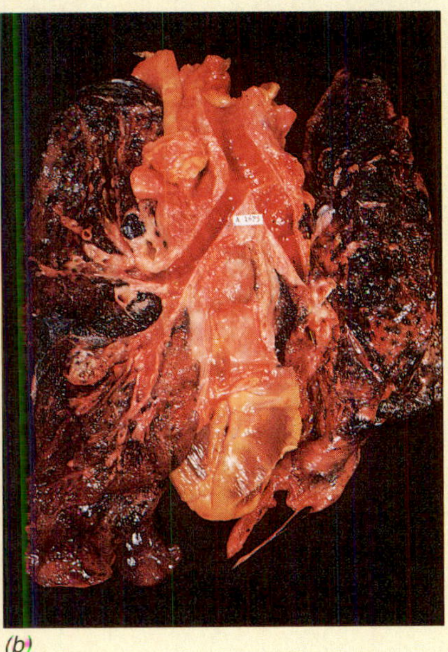
(b)

Comparison of diseased and healthy lungs. (***a***) Normal human lungs and major bronchi. (***b***) Human lungs and heart showing effects of cigarette smoking. (Martin M. Rotker/Taurus Photos, Inc.)

Physiological Adaptation to Changes in Pressure Takes Time

Scuba divers who return to the surface too quickly, or pilots who ascend to over 35,000 feet too rapidly, may suffer from **decompression sickness** (also called the bends). While submerged, a diver breathes gases under high pressure, and because of this pressure, excessive amounts of nitrogen gas dissolve in the blood and tissues. As the diver ascends to a lower pressure, the dissolved nitrogen comes out of solution. If he or she surfaces too rapidly, tiny nitrogen bubbles form in the blood and tissues and may block the flow of blood in capillaries and cause other damage. These bubbles cause the symptoms of decompression sickness: pain, paralysis, even death.

BREATHING POLLUTED AIR DAMAGES THE RESPIRATORY SYSTEM

Several defense mechanisms protect the delicate lungs from the harmful substances we breathe. The hair around the nostrils, the ciliated mucous lining in the nose and pharynx, and the cilia-mucous elevator serve to trap foreign particles in inspired air. One of the body's most rapid defense responses to breathing dirty air is **bronchial constriction.** In this process, the bronchial tubes narrow, increasing the chance that inhaled particles will land on their sticky mucous lining. Unfortunately, bronchial constriction increases airway constriction so that less air can pass through to the lungs. This decreases the amount of oxygen available to body cells. Fifteen puffs on a cigarette during a five-minute period increases airway resistance as much as threefold, and this added resistance to breathing lasts more than 30 minutes. Chain smokers and those who breathe heavily polluted air may remain in a state of chronic bronchial constriction.

Neither the smallest bronchioles nor the alveoli are equipped with mucus or ciliated cells. Foreign particles that get through other respiratory defenses and find their way into the alveoli may be engulfed by macrophages. The macrophages may then accumulate in the lymph tissue of the lungs. Lung tissues of chronic smokers and those who work in dirty fossil fuel–burning industries contain large blackened areas where carbon particles have been deposited (Fig. 38–11).

Continued insult to the respiratory system results in disease. Chronic bronchitis and emphysema are **chronic obstructive pulmonary diseases (COPD)** that have been linked to smoking and breathing polluted air. More than 75% of patients with **chronic bronchitis** have a history of heavy cigarette smoking (see Focus On Health and Human

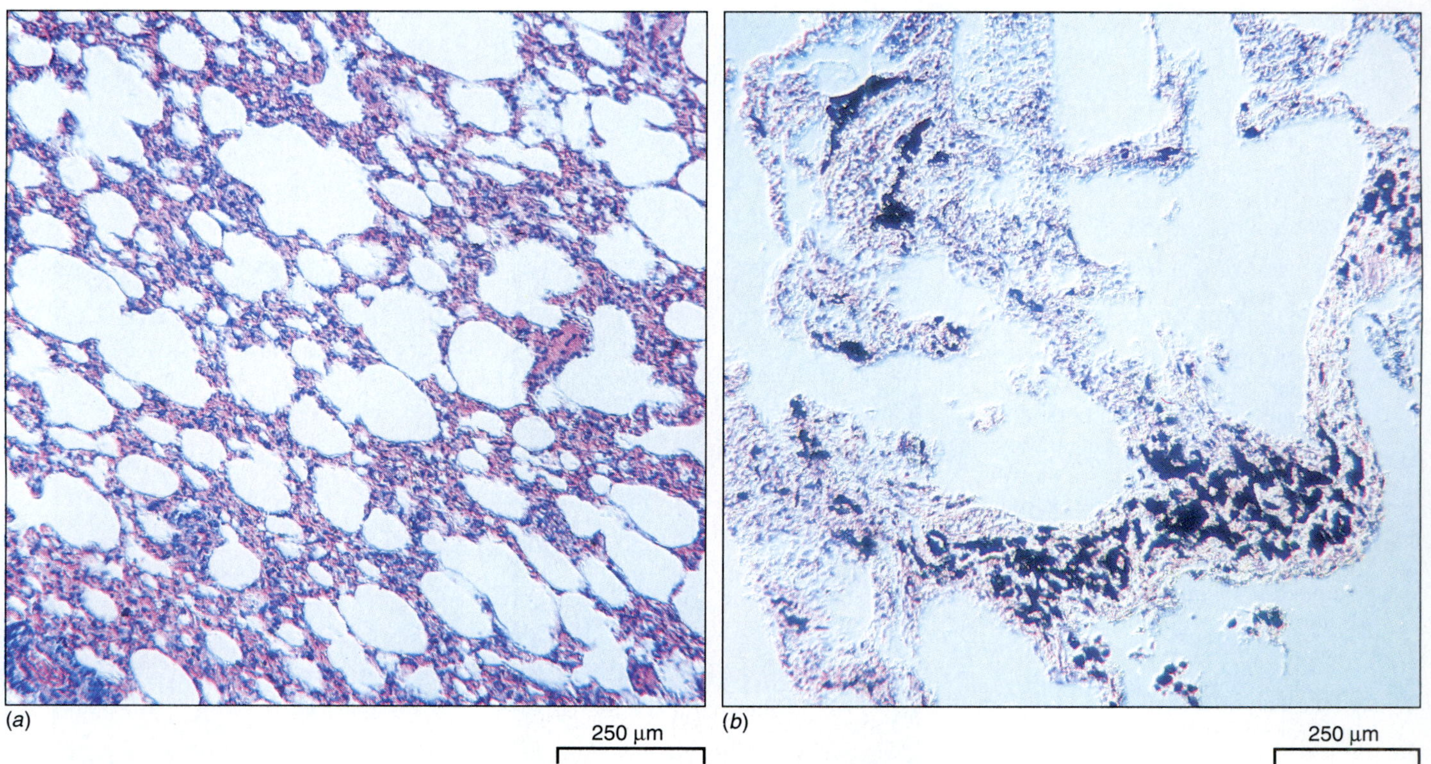

FIGURE 38–11 Healthy and diseased lung tissue. (***a***) Healthy lung tissue. (***b***) Lung tissue with accumulated carbon particles. Despite the body's defenses, when we inhale smoky, polluted air, especially over a long period of time, dirt particles enter the lung tissue and remain lodged there. (Alfred Pasieka/Taurus Photos, Inc.)

Affairs: Facts about Smoking). In chronic bronchitis, irritation from inhaled pollutants causes the bronchial tubes to secrete too much mucus. Ciliated cells, damaged by the pollutants, cannot effectively clear the mucus and trapped particles from the airways. The body resorts to coughing in an attempt to clear the airways. The bronchioles become constricted and inflamed, and the patient is short of breath.

Victims of chronic bronchitis often develop **pulmonary emphysema,** a disease most common in cigarette smokers. In this disorder, alveoli lose their elasticity and walls between adjacent alveoli are destroyed. The surface area of the lung is so reduced that gas exchange is seriously impaired. Air is not expelled effectively and stale air accumulates in the lungs. The emphysema victim struggles for every breath and still the body does not get enough oxygen. To compensate, the right ventricle of the heart pumps harder and becomes enlarged. Emphysema patients frequently die of heart failure. Cigarette smoking is also the main cause of lung cancer.

Chapter Summary

I. Gas exchange in air has advantages and disadvantages over gas exchange in water. Air contains more molecular oxygen than does water, oxygen diffuses more rapidly through air than through water, and less energy is required for ventilation. However, air-dwellers must have adaptations that prevent drying.

II. In very small animals, gas exchange occurs by diffusion across the body surface. Larger animals have specialized respiratory structures, and many animals have a circulatory system that delivers oxygen to each cell of the body.

A. Many invertebrates and some vertebrates exchange gases across the body surface.

B. In insects and some other arthropods, the respiratory system consists of a network of tracheal tubes that extend to all parts of the body.

C. Gills are moist, thin projections of the body surface that occur mainly in aquatic animals.

D. Terrestrial vertebrates have lungs, respiratory structures with moist surfaces for gas exchange.

III. The human respiratory system consists of a system of air passageways that branch into smaller and smaller tubes, ending in the alveoli within the lungs.

A. Air passes through the nostrils, the nasal cavities, and the pharynx, and into the larynx. The larynx helps prevent the entrance of foreign material into the lungs.

B. From the larynx, inhaled air passes into the trachea and then into the right or left bronchus.

C. Within the lungs, the bronchi branch into an extensive system of bronchioles, which eventually terminate in millions of tiny alveoli, through which gas exchange takes place with the blood.

D. Breathing is the mechanical process of moving air back and forth between the environment and the air sacs of the lungs.

1. When the diaphragm and rib muscles contract, expanding the chest, air rushes into the lungs.
2. When these muscles relax, pressure in the lung increases and air is expired.

E. Breathing is normally regulated by respiratory centers in the brain. These centers are sensitive to the pH of the blood, which is directly influenced by the amount of carbon dioxide in the blood.

F. Oxygen diffuses from the air sacs into the blood, while carbon dioxide diffuses from the blood into the air sacs.

G. Oxygen is transported to the body cells in the form of oxyhemoglobin. As oxygen is needed by the cells, the oxyhemoglobin dissociates, releasing oxygen, which diffuses from the blood into the cells.

H. Carbon dioxide is transported mainly as bicarbonate ions in the blood.

I. The respiratory system defends itself against dirt and other foreign matter by trapping particles on mucous surfaces, the cilia-mucous elevator, by bronchial constriction, and phagocytosis by macrophages. Inhaling polluted air or smoking cigarettes can wear down these defense mechanisms and cause serious damage to the respiratory system.

Selected Key Terms

alveolus (plural, *alveoli*), p. 755
breathing, p. 755
bronchus (plural, *bronchi*), p. 755
countercurrent exchange system, p. 752
diaphragm, p. 757
epiglottis, p. 754
expiration, p. 757
Fick's law, p. 758
gill, p. 751
inspiration, p. 757
larynx, p. 754
lung, p. 753
oxyhemoglobin, p. 759
pharynx, p. 754
pleural cavity, p. 755
pleural membrane, p. 755
respiratory structures, p. 750
trachea, p. 754
tracheal tubes, p. 750
ventilation, p. 750

Post-Test

1. Specialized respiratory structures must have thin walls so that ______________ ______________ can easily occur, they must be ______________ so that gases can be dissolved, and they must be richly supplied with ______________ to ensure transport of gases.

2. In insects, air enters a network of ______________ tubes through openings called ______________.

3. Respiratory structures that develop from the wall of a body cavity, such as the pharynx, are called ______________.

4. In birds, the lungs have several extensions referred to as ______________ ______________.

5. In the mammalian respiratory system, inhaled air passing through the larynx would next enter the ______________.

6. In the mammalian respiratory system, gas exchange takes place through the thin walls of the _______________.
7. In mammals, the floor of the thoracic cavity is formed by the _______________.
8. _______________ sickness can result when a diver surfaces too rapidly and nitrogen bubbles form in the blood and tissues.
9. Bronchial constriction is one of the body's most rapid responses to _______________.
10. In _______________, the alveolar walls break down so that several air sacs join to form larger, less elastic alveoli.
11. The main cause of lung cancer is _______________.
12. The epiglottis (a) seals off the larynx during swallowing (b) is a cavity in the bones of the skull (c) is one of the structures through which gas exchange takes place (d) covers the lungs (e) two of the preceding answers are correct.
13. The larynx (a) can initiate a cough reflex (b) is a cavity in the bones of the skull (c) is one of the structures through which gas exchange takes place (d) covers the lungs (e) two of the preceding answers are correct.
14. In humans, oxygen is transported (a) in the red blood cells (b) as oxyhemoglobin (c) as bicarbonate ions (d) answers (a), (b), and (c) are correct (e) only two of the preceding answers are correct.
15. Breathing is regulated by (a) respiratory centers in the medulla (b) respiratory centers in the pons (c) contractions of the larynx (d) answers (a), (b), and (c) are correct (e) only two of the preceding answers are correct.

Review Questions

1. What adaptations for gas exchange are found in fishes? In insects? In terrestrial vertebrates?
2. Why are lungs more suited for an air-breathing vertebrate and gills more effective in a fish?
3. Trace a breath of inhaled air from nose to alveoli, listing each structure through which it passes.
4. Describe the protective mechanisms of the respiratory system, including the cilia-mucous elevator and bronchial constriction.
5. Describe the processes of inspiration and expiration.
6. How is breathing regulated?
7. What role does diffusion play in gas exchange in humans?
8. How is oxygen transported in the blood of humans?
9. In what way does the composition of inhaled air differ from that of exhaled air? Why?
10. Summarize the health effects of smoking.
11. Label the diagram.

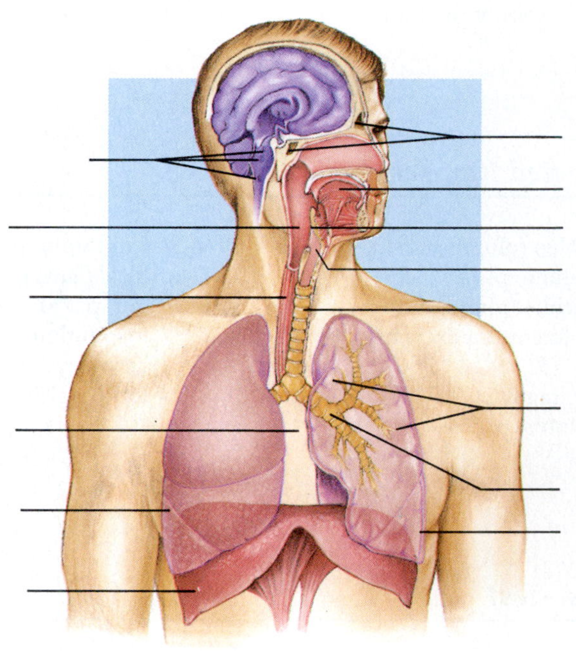

Thinking Critically

1. According to the American Heart Association, the 9 million children under age five who live with a smoker have an increased risk of asthma and respiratory infections. Explain how this is possible. What measures could you recommend to protect children from secondhand smoke?
2. What problems would be faced by a terrestrial animal that had gills instead of lungs?
3. Why do some aquatic mammals such as whales and dolphins use lungs rather than gills for gas exchange?
4. What are the advantages of having millions of alveoli rather than a pair of simple, balloon-like lungs?

Recommended Readings

Feder, M.E., and W.W. Bruggren, "Skin Breathing Vertebrates," *Scientific American,* Vol. 253, No. 2, 1985, pp. 126–142. An interesting account of gas exchange through the body surface in some vertebrates.

Zapol, W.M., "Diving Adaptations of the Weddell Seal," *Scientific American,* Vol. 256, No. 6, 1987, pp. 100–105. Physiological adaptations permit the seal to swim deeper and hold its breath longer than most other mammals.

Scientists work to increase world food supply. *Here, Amnon Jemini examines wheat grown with salt water irrigation in Ashalim, Negev, Israel. (Steve Kaufman/Peter Arnold, Inc.)*

More People, Less Food

Most college students reading these words have not had to face chronic hunger. This is not the case, however, for more than 1 billion people worldwide, nearly one in five, who are malnourished. In fact, about 20 million people (three-fourths of them children) die each year as a result of inadequate nutrition. Even though world food supplies have greatly increased during the past few decades, more people are undernourished today (both total numbers and percentage of the world population) than in the 1950s. The main sources of food—farms, ranches, and oceans—are either approaching their limits of productivity or are already declining in productivity. Can we increase world food supply to meet the needs of a continually expanding population? And, considering the environmental impact of such efforts, *should we try*?

Most land suitable for growing food, from an economically feasible standpoint, is already being cultivated. Crop yields have been improved during the last 20 years or so by new technological advances such as the development of high yield crops and the more intensive use of fertilizers and pesticides. Unfortunately, most countries do not have the economic or natural resources to support such intensive methods of farming. Furthermore, intensive use of fertilizers and pesticides destroys the natural ecology of the soil, causes water pollution, and requires industrial processes (to produce the fertilizer, pesticides, tractors, and other farm equipment) that use energy and contribute to air pollution.

World meat production increased about fourfold between 1950 and 1990. Since 1990, meat production has been declining. Rangelands, which provide most of the feed for cattle, sheep, and goats, have been overused and have become less productive. Many rangelands have disappeared due to development. As a result, meat is becoming more expensive and less available.

Seafood is also becoming more expensive as the annual harvest from the oceans declines. Overfishing and pollution have caused the decline of many marine species. Between 1950 and 1989, the world fish harvest increased from 22 million to 100 million tons. Since that time, the fish catch has declined, dropping more than 7%.

The only long-term solution to the problem of world food supply is population control. Our numbers are currently growing at the rate of 100 million new people each year. Based on current growth rates, the world population of almost 6 billion will double by the year 2037. Because the amount of food that our planet can produce per person is declining, present rates of population growth must be slowed if we are to avoid worldwide famine.

Processing Food and Nutrition

39

Learning Objectives

After you have studied this chapter you should be able to:

1. Compare food processing, including ingestion, digestion, absorption, and elimination in an animal (such as *Hydra*) that has a single-opening digestive system with an animal (such as a vertebrate) that has a complete digestive system (with two openings).
2. Identify on a diagram or model each of the structures of the human digestive system described in this chapter, and give the function of each structure.
3. Trace the pathway traveled by an ingested meal in the human digestive system, describing each of the changes that takes place en route.
4. Summarize the functions of the accessory digestive glands of humans and other terrestrial vertebrates.
5. Trace the step-by-step digestion of carbohydrate, protein, and lipid.
6. Draw and label a diagram of an intestinal villus, and explain how its structure is adapted to its function.
7. Trace the fate of glucose, lipids, and amino acids after their absorption, and discuss their roles in the body.
8. Discuss the roles of vitamins and minerals in the body, and distinguish between water-soluble and fat-soluble vitamins.
9. Contrast basal metabolic rate with total metabolic rate.
10. Write the basic energy equation for maintaining body weight and describe the consequences of altering it in either direction.
11. In general terms, describe the problem of world food supply relative to world population, and describe the effects of malnutrition.
12. Summarize the challenges encountered in obtaining adequate amounts of amino acids in a vegetarian diet, and describe how a nutritionally balanced vegetarian diet could be planned.

Key Concepts

- ☐ Nutrition is so vital that an animal's body plan and life-style are adapted to its particular style of obtaining food.
- ☐ Processing food involves ingestion, digestion, absorption, and elimination of wastes. In animals with a complete digestive tract (a digestive tube with two openings), various regions of the digestive tract are specialized for carrying on food processing functions.

Key Concepts, continued

- ☐ With only slight variation, all animals require the same basic nutrients: carbohydrates, lipids, proteins, vitamins, and minerals.
- ☐ Carbohydrates, lipids, and proteins can all be used as energy sources. Eating too much of any of these nutrients can result in weight gain. Eating too few nutrients or an unbalanced diet can result in malnutrition and death. The basic energy equation for maintaining body weight is:

  Energy (calorie) input = Energy output

ALL ANIMALS MUST PROCESS FOOD

Animals require *nutrients*—substances present in food that are used as an energy source to run the systems of the body, as ingredients to make compounds for metabolic processes, and as building blocks to permit the growth and repair of tissues. Obtaining nutrients is of such vital importance that both individual organisms and ecosystems are designed around the central theme of **nutrition,** the process of taking in and using food. An organism's body plan and its life-style are adapted to its particular mode of obtaining food (Fig. 39–1).

All animals are heterotrophs, organisms that must obtain their energy and nourishment from the organic molecules manufactured by other organisms. Most animals have a digestive system that processes the food they eat. Food processing may be divided into several steps: ingestion, digestion, absorption, and elimination.

After foods are selected and obtained, they are ingested, that is, taken into the digestive cavity. **Ingestion** generally includes taking the food into the mouth and swallowing it. Because animals eat the macromolecules tailor-made by and for other organisms, they must break down these molecules and refashion them for their own needs. We cannot incorporate the proteins in steak directly into our own muscles, for example. The body **digests** the steak, mechanically breaking down the large bites of meat into smaller ones and then enzymatically

(a) (b)

FIGURE 39–1 Adaptations for obtaining food. (***a***) An acorn weevil. The impressively long "snout" of this little beetle is used both for feeding and for making a hole in the acorn through which an egg is deposited. When it has hatched, the larva feeds on the contents of the acorn seed. (***b***) This carnivorous snake *(Dromicus)* is strangling a lava lizard *(Tropidurus)*. (*a*, Darwin Dale/Photo Researchers Inc.; *b*, Frans Lanting/Minden Pictures)

hydrolyzing (breaking down with the addition of water) the proteins into their component amino acids.

The amino acids can then be **absorbed** (passed through the lining of the digestive tract and into the blood) and transported to the muscle cells, which arrange these components into human muscle proteins. Food that is not digested and absorbed is discharged from the body. This process is called **egestion** in simple animals and **elimination** in more complex animals.

SOME INVERTEBRATES HAVE DIGESTIVE SYSTEMS WITH A SINGLE OPENING

Some simple invertebrates have no digestive system at all. Sponges obtain food by filtering microscopic organisms from the surrounding water. Individual cells phagocytize the food particles, and digestion is intracellular within food vacuoles. Wastes are egested into the water that continuously circulates through the sponge body.

Cnidarians (such as hydras and jellyfish) and flatworms have a digestive system with only a single opening. Cnidarians capture small aquatic animals with the help of their stinging cells and tentacles (Fig. 39–2*a*). The mouth opens into a large gastrovascular cavity. Cells lining this digestive cavity secrete enzymes that break down proteins. Digestion continues intracellularly within food vacuoles, and the digested nutrients diffuse into other cells. Undigested food particles are egested through the mouth by contraction of the body.

Free-living flatworms begin to digest their prey even before ingesting it. They extend the pharynx out through their mouth and secrete digestive enzymes onto the prey (Fig. 39–2*b*). When ingested, the food enters the branched intestine where enzymes continue to digest it. Partly digested food fragments are then phagocytized by cells of the intestinal lining, and digestion is completed within food vacuoles. As in cnidarians, the flatworm digestive system has only one opening, so undigested wastes are egested through the mouth.

MOST INVERTEBRATES AND ALL VERTEBRATES HAVE DIGESTIVE SYSTEMS WITH TWO OPENINGS

Most other invertebrates, and all vertebrates, have a digestive tract with two openings, referred to as a complete digestive system (Fig. 39–3). Food enters through the mouth, and undigested food is eliminated through the anus. Waves of muscular contractions push the food in one direction, so that more food can be taken in while previously eaten food is being digested and absorbed farther down the tract.

In a digestive tract with two openings, various regions of the tube are specialized to perform specific functions. In the vertebrate digestive tract, food passes in sequence through the following specialized regions: mouth, pharynx (throat), esophagus, stomach, small intestine, large intestine, and anus. All vertebrates have accessory glands that secrete digestive juices into the digestive tract. These include the liver, the pancreas, and, in terrestrial vertebrates, the salivary glands.

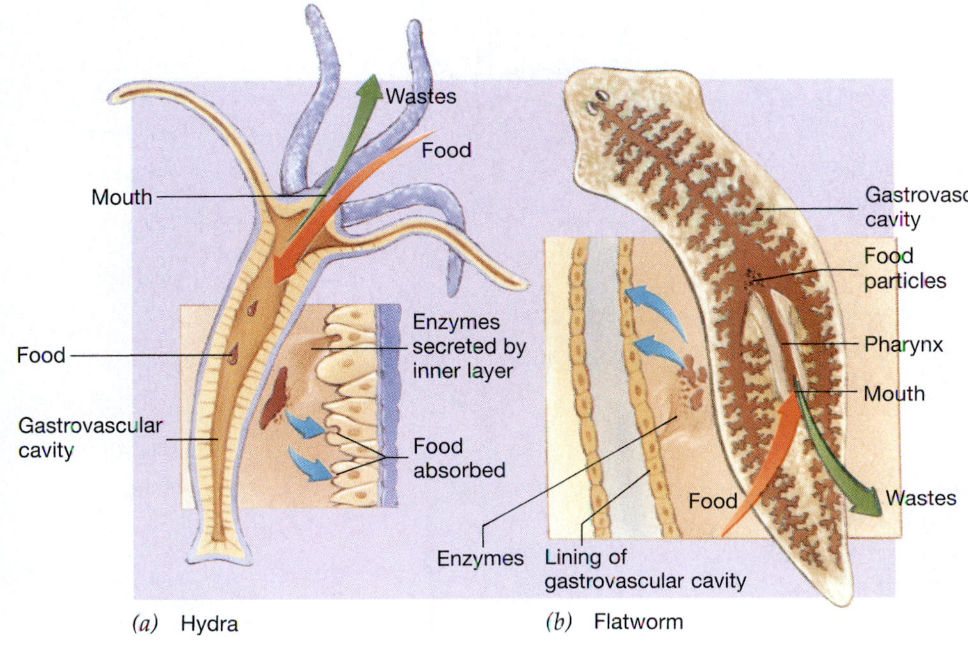

FIGURE 39–2 Processing food with a simple digestive system. The hydra (***a***) and the flatworm (***b***) have a digestive tract with a single opening that serves as both mouth and anus.

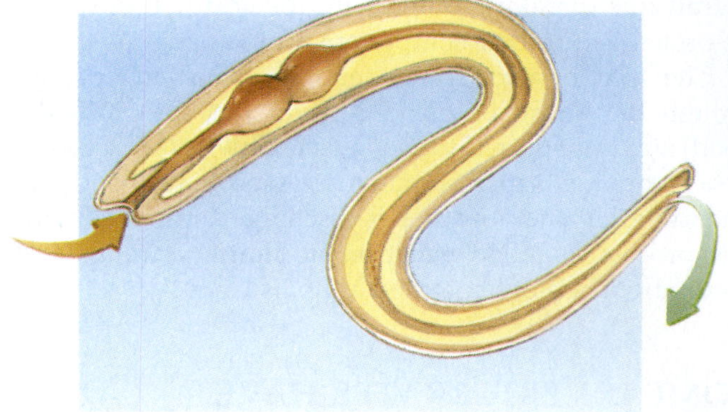

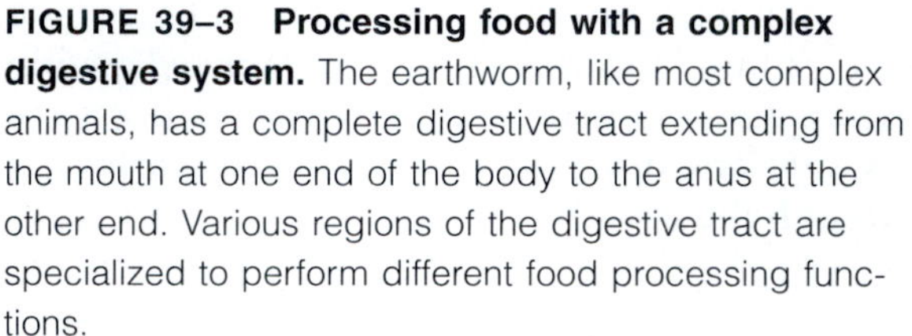

FIGURE 39–3 Processing food with a complex digestive system. The earthworm, like most complex animals, has a complete digestive tract extending from the mouth at one end of the body to the anus at the other end. Various regions of the digestive tract are specialized to perform different food processing functions.

THE HUMAN DIGESTIVE SYSTEM HAS HIGHLY SPECIALIZED STRUCTURES FOR PROCESSING FOOD

In the human digestive system, various regions of the digestive tract have highly specialized structures and functions (Fig. 39–4). The wall of the digestive tract is composed of four layers. Although they vary somewhat in structure in various regions, the layers are basically similar throughout the digestive tract (Fig. 39–5). The **mucosa,** a layer of epithelial tissue and underlying connective tissue, lines the lumen (inner space) of the digestive tract. Surrounding the mucosa is the **submucosa,** a con-

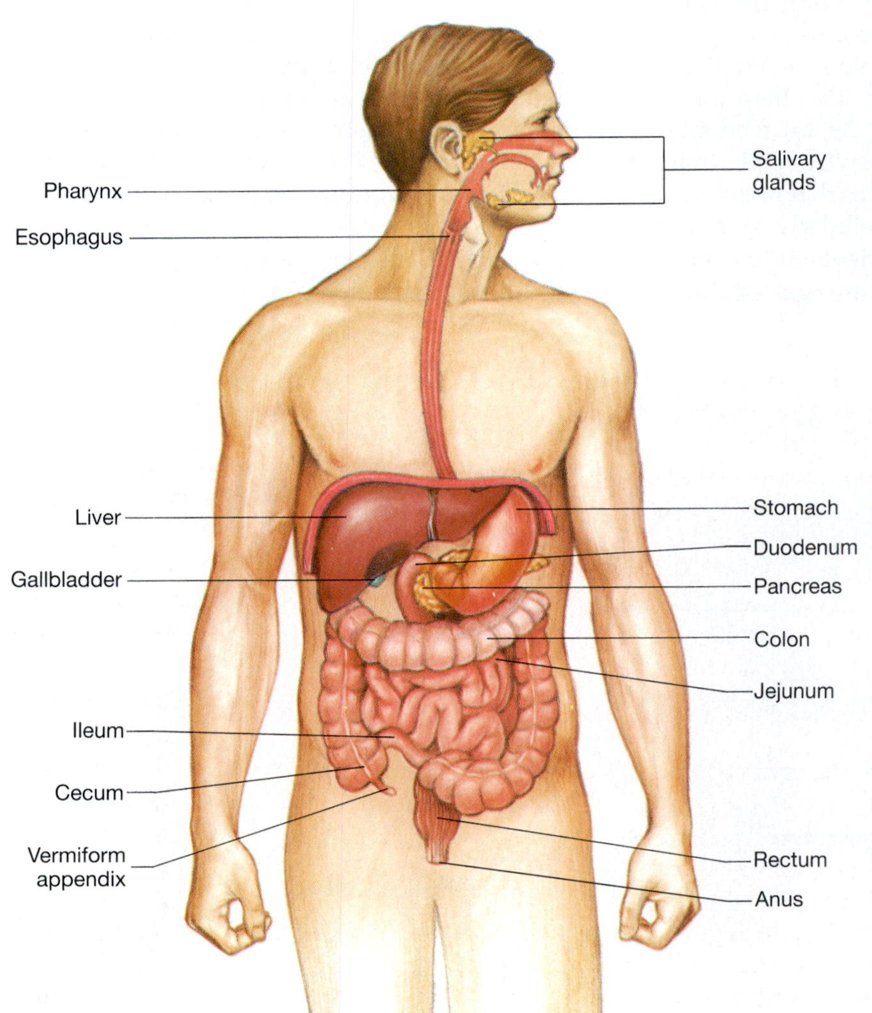

FIGURE 39–4 The human digestive system. Note the complex digestive tract—a long, coiled tube extending from mouth to anus. Locate the three types of accessory glands.

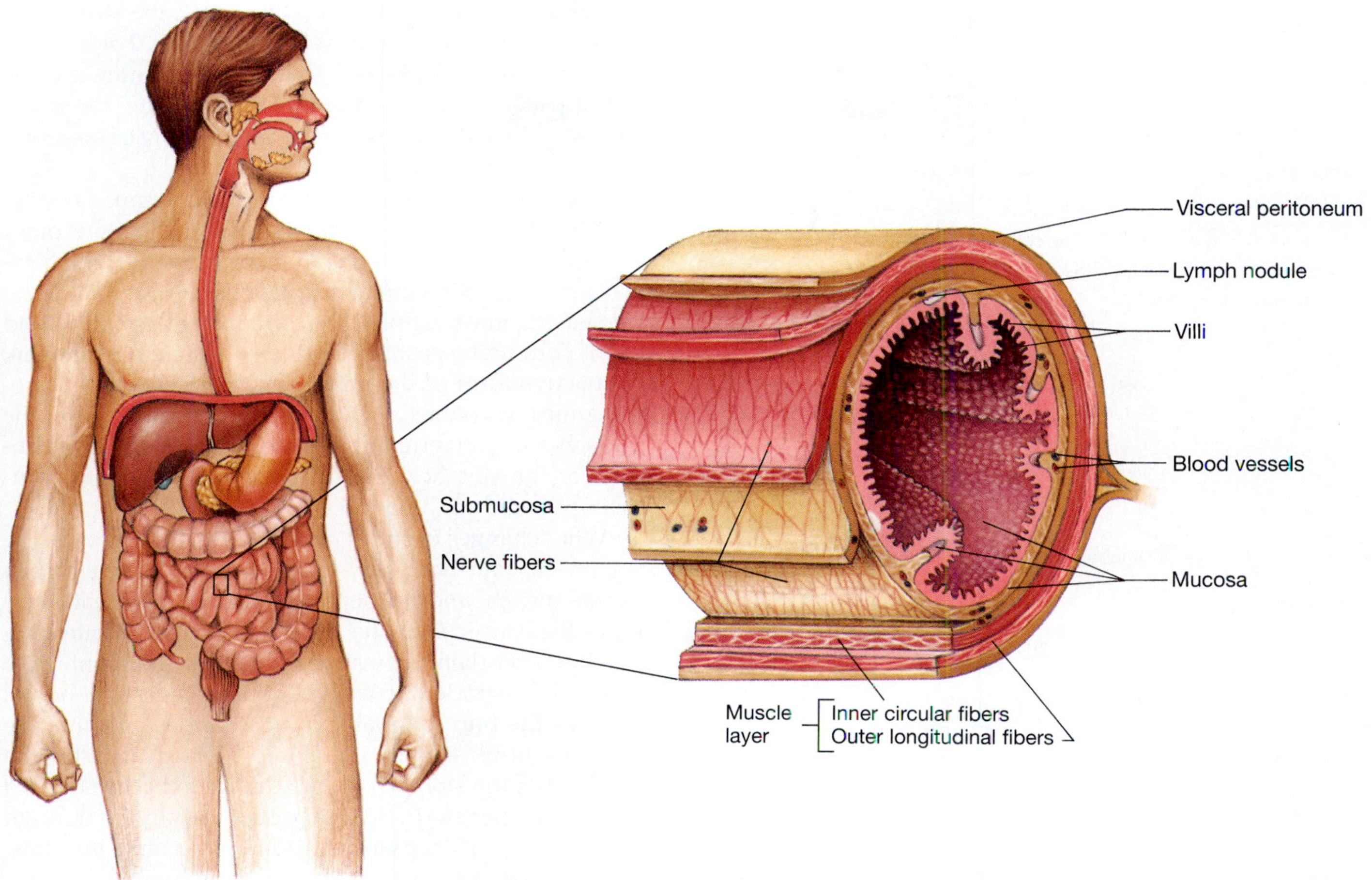

FIGURE 39–5 The wall of the digestive tract consists of four layers. The outer layer is known as the visceral peritoneum in the wall of the stomach and intestine. (Above the stomach it is referred to as the serosa, or adventitia.)

nective tissue layer rich in blood vessels, lymphatic vessels, and nerves. Surrounding the submucosa is a **muscle layer,** consisting of two sublayers of smooth muscle. In the inner sublayer, the muscle fibers are arranged circularly around the digestive tube, whereas in the outer sublayer the muscle fibers are arranged longitudinally. The outer connective tissue coat of the digestive tract is the **adventitia.** Below the level of the diaphragm, the adventitia becomes the **visceral peritoneum.**

Food Begins Its Journey Inside the Mouth

Imagine that you have just taken a big bite of a hamburger. The mouth is specialized for ingestion and for beginning the digestive process. Mechanical digestion begins as you bite, grind, and chew the meat and bun with your teeth. Unlike the simple, pointed teeth of fish, amphibians, and reptiles, the teeth of mammals vary in size and shape and are specialized to perform specific functions. The chisel-shaped **incisors** are used for biting, while the long, pointed **canines** are adapted for tearing food. The flattened surfaces of the **premolars** and **molars** are specialized for crushing and grinding food.

Each tooth is covered by **enamel,** the hardest substance in the body. Most of the tooth consists of **dentin,** which resembles bone in composition and hardness. Beneath the dentin is the **pulp cavity,** a soft connective tissue containing blood and lymph vessels and nerves.

While the food is being mechanically disassembled by the teeth, it is also moistened by saliva. Some of its molecules dissolve, enabling you to taste the food. Recall from Chapter 35 that taste buds are located on the tongue and other surfaces of the mouth. Three pairs of **salivary glands** secrete about a liter of saliva into the mouth cavity each day. Saliva contains an enzyme, **salivary amylase,** which initiates the digestion of starch into sugar.

The Pharynx and Esophagus Conduct Food to the Stomach

After the bite of food has been chewed and fashioned into a lump called a **bolus,** it is swallowed, that is, moved

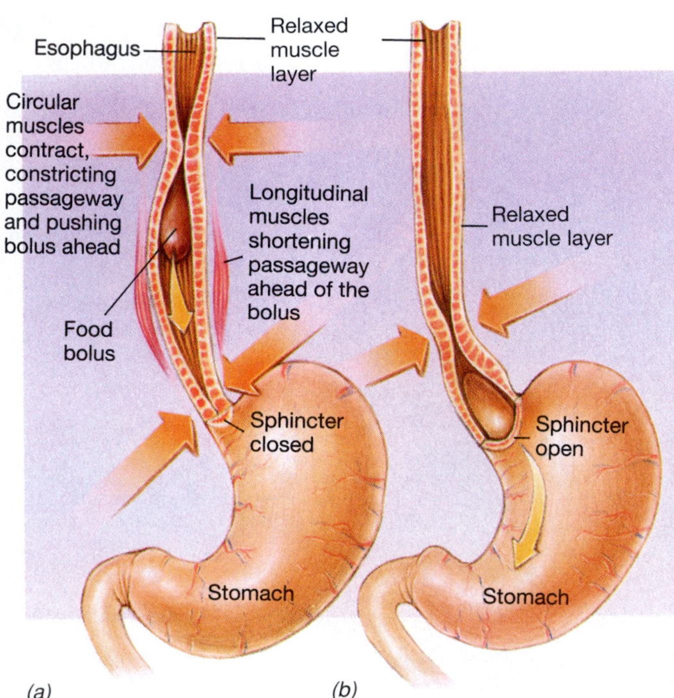

FIGURE 39–6 Peristalsis. (***a***) Food is moved through the digestive tract by waves of muscular contraction known as peristalsis. (***b***) When the sphincter (ring of muscle) at the entrance of the stomach opens, food enters the stomach.

through the **pharynx** and into the **esophagus.** The pharynx, or throat, is a muscular tube that serves as the hallway of the respiratory system as well as the digestive system. During swallowing, the opening to the airway is closed by a small flap of tissue, the **epiglottis.**

Waves of muscular contraction, called **peristalsis,** sweep the bolus through the pharynx and esophagus toward the stomach (Fig. 39–6). Circular muscle fibers in the wall of the esophagus contract around the top of the bolus, pushing it downward. Almost at the same time, longitudinal muscles around the bottom of the bolus and below it contract, shortening the tube.

When the body is in an upright position, gravity helps to move the food through the esophagus, but gravity is not necessary. Astronauts are able to eat in its absence, and even if you are standing on your head, food will reach your stomach.

Food Is Mechanically and Enzymatically Digested in the Stomach

The entrance to the stomach is normally closed by a ring of muscle at the lower end of the esophagus. When a peristaltic wave passes down the esophagus, the muscle relaxes, permitting the bolus to enter the **stomach,** a large muscular organ (Fig. 39–7). When empty, the stomach is collapsed and shaped almost like a hotdog. Folds of the stomach wall, called **rugae,** give the inner lining a wrinkled appearance. As food enters the stomach, the rugae gradually smooth out, stretching the capacity of the stomach to more than a liter.

The stomach is lined with simple columnar epithelium that secretes large amounts of mucus. Tiny pits mark the entrances to the millions of gastric glands, which extend deep into the stomach wall. The gastric glands secrete gastric juice, a fluid that contains hydrochloric acid (HCl). Cells in the gastric glands also secrete **pepsinogen,** the inactive form of the enzyme **pepsin.** When pepsinogen comes in contact with the acidic gastric juice in the stomach, it is converted to pepsin, the main digestive enzyme of the stomach. Pepsin hydrolyzes proteins, converting them to polypeptides.

What changes occur in our bite of hamburger during its three- to four-hour stay in the stomach? The stomach churns and chemically degrades the food so that it assumes the consistency of a thick soup called **chyme.** Protein digestion then begins, and much of the hamburger protein is degraded to polypeptides. Digestion of the starch in the bun to small polysaccharides and maltose continues until salivary amylase is inactivated by the acidic pH of the stomach. When digestion in the stomach is complete, peristaltic waves propel the chyme through the stomach exit, the **pylorus,** and into the small intestine.

Most Enzymatic Digestion Takes Place Inside the Small Intestine

Digestion of food is completed in the **small intestine,** and nutrients are absorbed through its wall. The small intestine has three regions: the **duodenum,** the **jejunum,** and the **ileum.** Most chemical digestion takes place in the duodenum, the first portion of the small intestine, not in the stomach, as is commonly believed. Bile from the liver and enzymes from the pancreas are released into the duodenum and act upon the chyme. Then enzymes produced by the epithelial cells lining the duodenum catalyze the final steps in the digestion of the major types of nutrients (Table 39–1).

The lining of the small intestine appears velvety because of millions of tiny finger-like projections called **villi** (Fig. 39–8). The villi (singular, *villus*) increase the surface area of the small intestine for digestion and absorption of nutrients. The intestinal surface is further expanded by thousands of **microvilli,** folds of cytoplasm on the exposed surface of the epithelial cells of the villi. About 600 microvilli protrude from the surface of each cell, giving the epithelial lining a fuzzy appearance when viewed with the electron microscope.

If the intestinal lining were smooth like the inside of a water pipe, food would zip right through the intestine

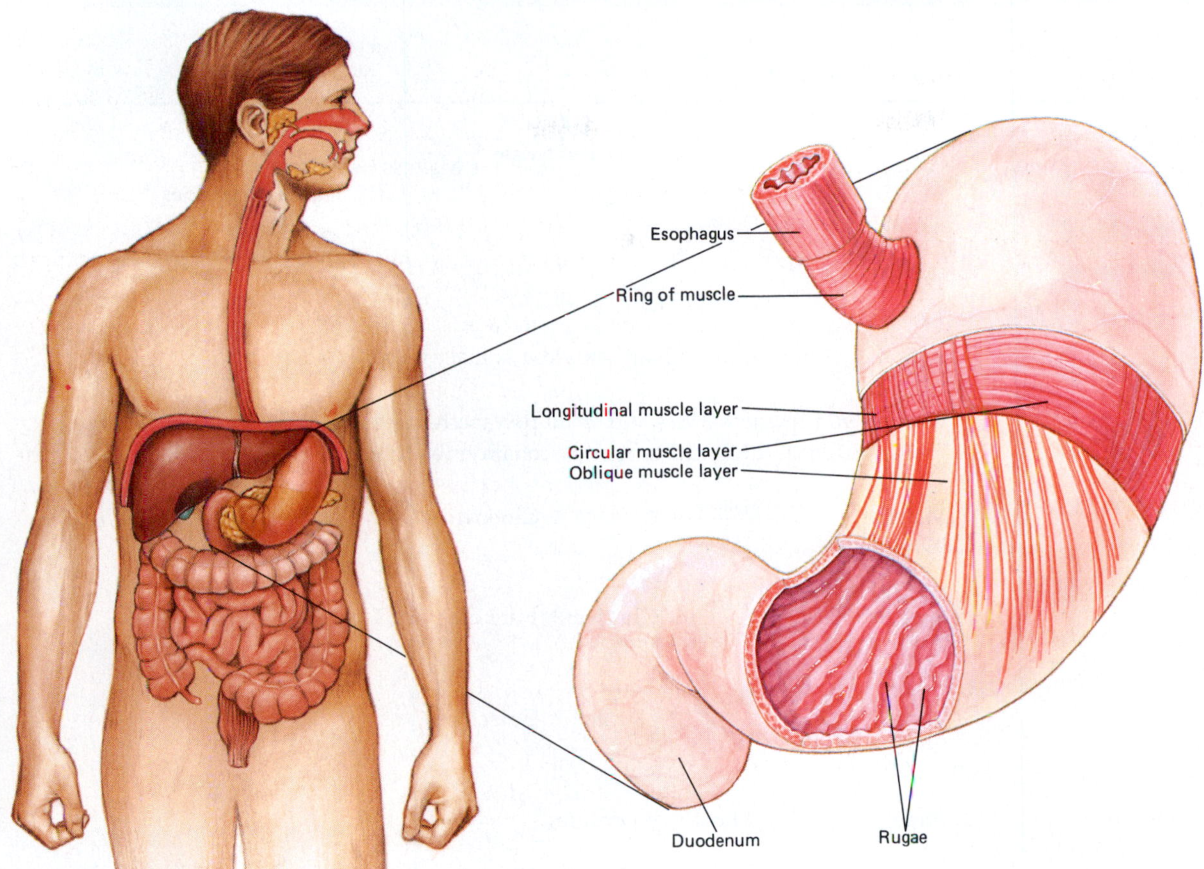

FIGURE 39–7 Structure of the stomach. From the esophagus, food enters the stomach, where it is mechanically and enzymatically digested.

and many valuable nutrients would not be absorbed. Folds in the wall of the intestine, the villi, and microvilli together increase the surface area of the small intestine by about 600 times. If we could unfold and spread out the lining of the small intestine of an adult human, its surface would approximate the size of a tennis court.

The Liver Secretes Bile which Mechanically Digests Fats

Just under the diaphragm lies the **liver,** the largest and also one of the most complex organs in the body (Fig. 39–9). A single liver cell can carry on more than 500 separate metabolic activities. The liver's food processing functions include the following:

1. Secretes **bile,** which is important in the mechanical digestion of fats.
2. Helps maintain homeostasis by removing or adding nutrients to the blood.
3. Converts excess glucose to glycogen and stores it.
4. Converts excess amino acids to fatty acids and urea.
5. Stores iron and certain vitamins.
6. Detoxifies alcohol and many drugs and poisons that enter the body.

Bile consists of water, bile salts, bile pigments, cholesterol, salts, and lecithin (a phospholipid). Bile produced in the liver is stored in the pear-shaped **gallbladder.** The gallbladder concentrates the bile and releases it into the duodenum as needed. Bile mechanically digests fats by a detergent-like action in which it decreases the surface tension of fat particles. This action permits the fat molecules to disperse so they can be attacked by lipases (fat-digesting enzymes). The dispersion of fat globules by bile is called **emulsification.** Bile contains no digestive enzymes and so it does not enzymatically digest food.

The Pancreas Secretes Digestive Enzymes

The **pancreas** is an elongated gland that secretes both digestive enzymes and hormones that help regulate the level of glucose in the blood. Among its enzymes are (1) **trypsin** and **chymotrypsin,** which digest polypeptides to

(Text continues on page 775)

TABLE 39–1
Summary of Digestion

Location	*Source of enzyme*	*Digestive process**
Carbohydrate digestion		
Mouth	Salivary glands	Polysaccharides (e.g., starch) —Salivary amylase→ Maltose + Small polysaccharides 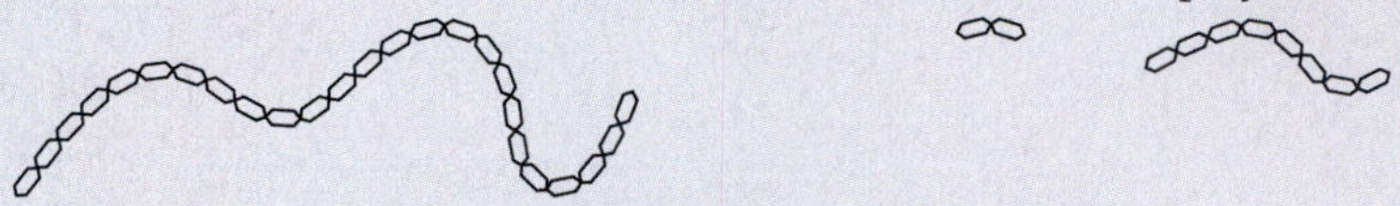
Stomach		Action continues until salivary amylase is inactivated by acidic pH
Small intestine	Pancreas	Undigested polysaccharides and small polysaccharides —Pancreatic amylase→ Maltose
	Intestine	Disaccharides hydrolyzed to monosaccharides as follows:
		Maltose (malt sugar) —Maltase→ Glucose + Glucose
		Sucrose (table sugar) —Sucrase→ Glucose + Fructose
		Lactose (milk sugar) —Lactase→ Glucose + Galactose
Protein digestion		
Stomach	Stomach (gastric glands)	Protein —Pepsin→ Short polypeptides
Small intestine	Pancreas	Polypeptides (A—A—A—A—A / A—A—A—A—A) —Trypsin, chymotrypsin→ Tripeptides (A—A—A) + Dipeptides (A—A)
		Dipeptides (A—A) —Carboxypeptidase, aminopeptidase→ Free amino acids (A A A)
	Small intestine	Tripeptides (A—A—A) + Dipeptides (A—A) —Peptidases→ Free amino acids (A A A A A A)
Lipid digestion		
Small Intestine	Liver	Glob of fat —Bile salts→ Emulsified fat (individual triacylglycerols) 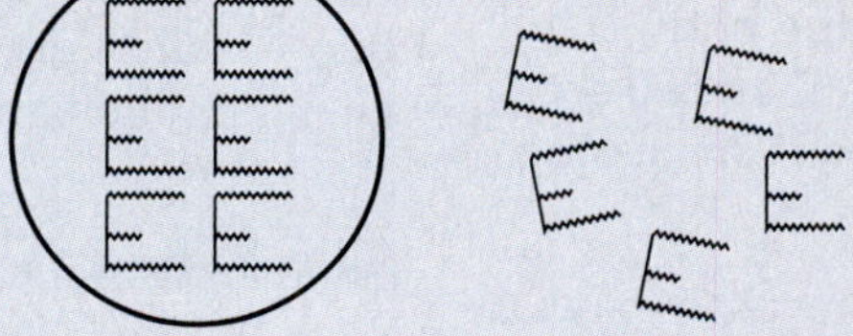
	Pancreas	Triacylglycerol —Lipase→ Fatty acids + Glycerol

* ⬭ = monosaccharide; ⋿ = triacylglycerol; E = glycerol; 〰 = fatty acid; A = amino acid.

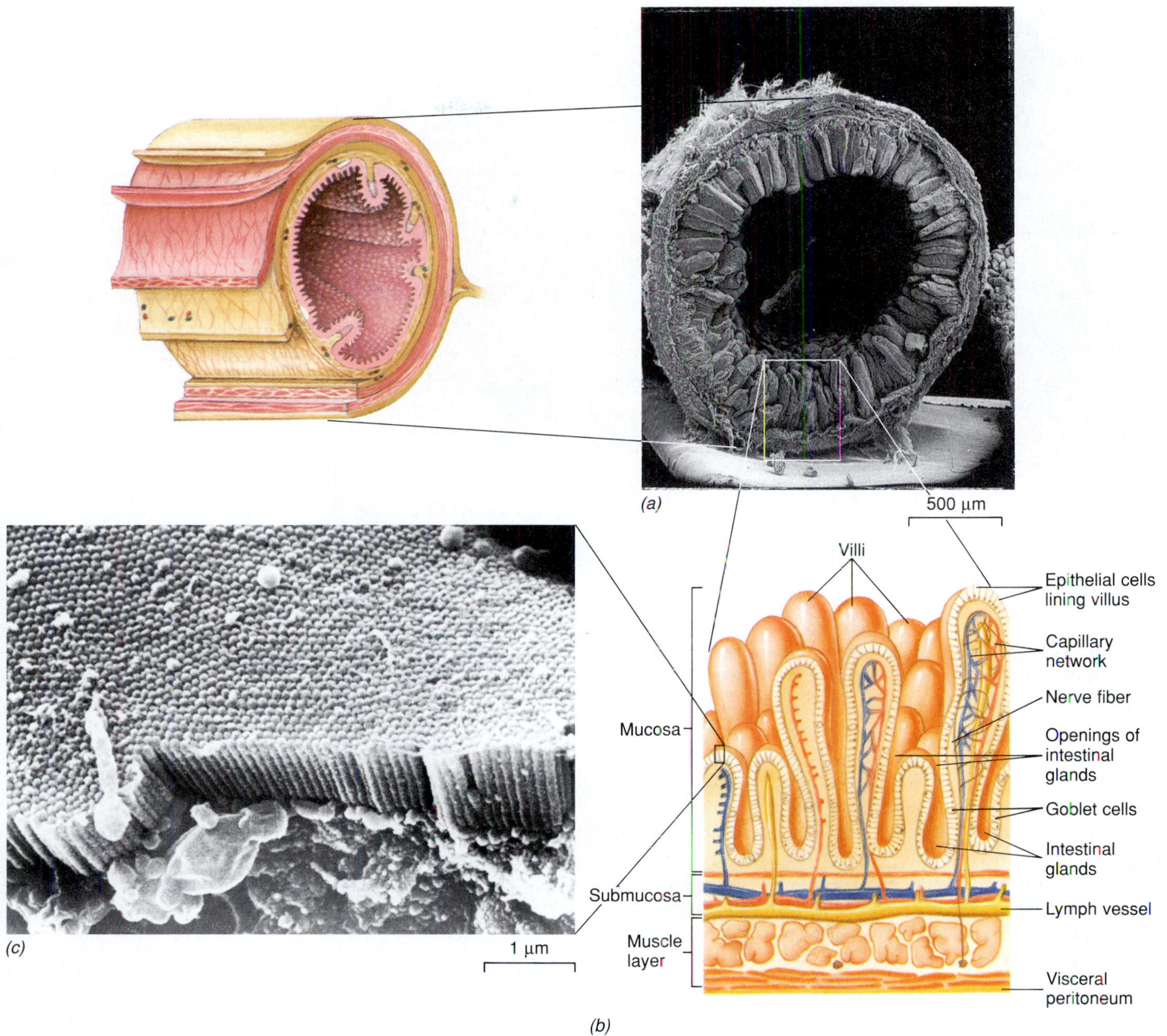

FIGURE 39–8 Structure of the wall of the small intestine. The inner wall of the small intestine is studded with villi. (***a***) Scanning electron micrograph of a cross section of the small intestine. (***b***) Microscopic view of a small portion of the intestinal wall. Some of the villi have been opened to show the blood and lymph vessels within. (***c***) SEM (approximately × 14,000) of the surface of an epithelial cell from the lining of the small intestine showing microvilli. The epithelium has been cut vertically, allowing the microvilli to be viewed from the side as well as from above. (*a*, Visuals Unlimited/G. Shih-R. Kessel; *c*, courtesy of J.D. Hoskins, W.G. Henk, and Y.Z. Abdelbaki, from the *American Journal of Veterinary Research*)

dipeptides; (2) **pancreatic lipase,** which degrades neutral fats; (3) **pancreatic amylase,** which breaks down almost all types of carbohydrates, except cellulose, to disaccharides; and (4) **ribonuclease** and **deoxyribonuclease,** which split the nucleic acids ribonucleic acid (RNA) and deoxyribonucleic acid (DNA) to free nucleotides.

Nerves and Hormones Regulate Digestion

Most digestive enzymes are produced only when food is present in the digestive tract. Salivary gland secretion is controlled entirely by the nervous system, but secretion of other digestive juices is regulated by both nerves and

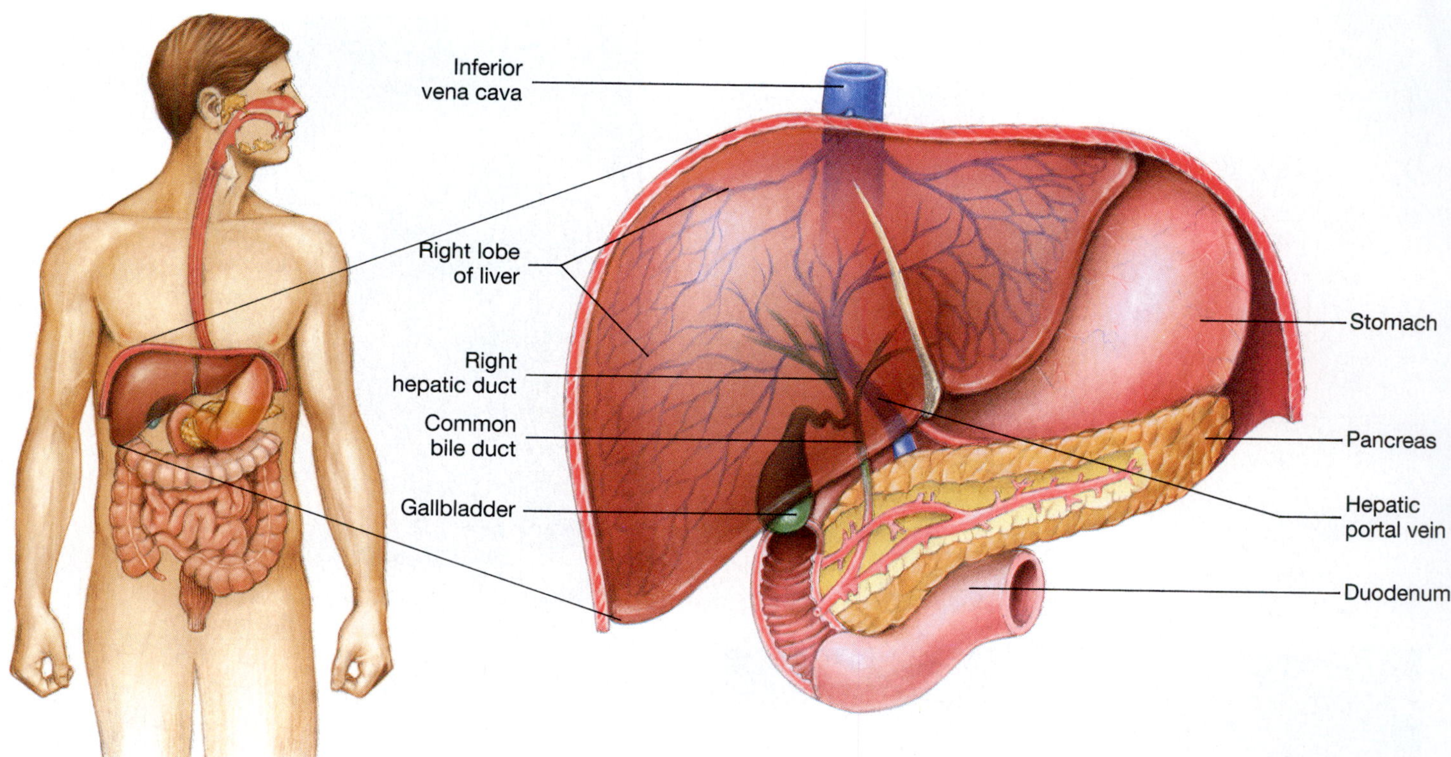

FIGURE 39–9 Structure of the liver and pancreas. Note the gallbladder, which stores bile, and the ducts, which conduct bile to the duodenum. Recall that the hepatic portal vein brings blood rich in nutrients to the liver.

hormones (Table 39–2). As an example, consider the secretion of gastric juice. Seeing, smelling, tasting, or even thinking about food causes the brain to send neural messages to the glands in the stomach, stimulating them to secrete. In addition, when food distends the stomach, it stimulates glands in the stomach wall to release the hormone **gastrin.** Gastrin is absorbed into the blood and transported to the gastric glands, where it stimulates the release of gastric juice.

Absorption Takes Place Mainly through the Villi of the Small Intestine

Only a few substances—water, simple sugars, salts, alcohol, and certain drugs—are composed of molecules small enough to be absorbed through the wall of the stomach. Absorption of nutrients is primarily the job of the intestinal villi. As illustrated in Figure 39–8, the wall of a villus consists of a single layer of epithelial cells. Inside each villus is a network of capillaries and a central lymph vessel.

To reach the blood (or lymph), a nutrient molecule must pass through an epithelial cell of the intestinal lining and through a cell lining the blood or lymph vessel. Glucose and amino acids cannot diffuse through the intestinal lining and are absorbed by active transport. Absorption of these nutrients is coupled with the active transport of sodium (Chapter 6). Lipids enter the lymph system by diffusion and are transported to the upper trunk region, where the lymph fluid and its contents enter the blood.

The Large Intestine Eliminates Waste

Undigested material, such as the cellulose of plant foods, along with unabsorbed chyme, passes into the **large intestine** (Fig. 39–4). Although only about 1.3 meters long (about 4 feet), this organ is referred to as "large" because its diameter is greater than that of the small intestine. The small intestine joins the large intestine about 7 centimeters (2.8 inches) from the end of the large intestine, thereby creating a blind pouch, the **cecum.** The **vermiform appendix** projects from the end of the cecum. (Appendicitis is an inflammation of the appendix.) The functions of the cecum and appendix in humans are not known. They are generally considered vestigial organs, perhaps important in the vegetarian past of the human species. Herbivores such as rabbits have a large, functional cecum that holds food while bacteria digest the cellulose.

From the cecum to the **rectum** (the last portion of the large intestine) the large intestine is known as the **colon.** The regions of the large intestine are the cecum; ascending colon; transverse colon; descending colon; sigmoid

TABLE 39–2
Hormonal Control of Digestion

Hormone	*Source*	*Target tissue*	*Actions*	*Factors that stimulate release*
Gastrin	Stomach (mucosa)	Stomach (gastric glands)	Stimulates gastric glands to secrete pepsinogen	Distention of the stomach by food; certain substances such as partially digested proteins and caffeine
Secretin	Duodenum (mucosa)	Pancreas	Signals secretion of sodium bicarbonate	Acidic chyme acting on mucosa of duodenum
		Liver	Stimulates bile secretion	
Cholecystokinin (CCK)	Duodenum (mucosa)	Pancreas	Stimulates release of digestive enzymes	Presence of fatty acids and partially digested proteins in duodenum
		Gallbladder	Stimulates contraction and emptying of bile	
Gastrin inhibitory peptide	Duodenum (mucosa)	Stomach	Decreases stomach motor activity, thus slowing emptying	Presence of fatty acids or glucose in duodenum

colon; rectum; and anus, which is the opening for the elimination of wastes.

As the chyme passes slowly through the large intestine, water and sodium are absorbed from it, and it gradually assumes the consistency of normal feces. Bacteria inhabiting the large intestine enjoy the last remnants of the meal and return the favor by producing vitamin K and certain B vitamins that can be absorbed and utilized.

A distinction should be made between elimination and excretion. *Elimination* is the process of getting rid of digestive wastes, materials that have not been absorbed from the digestive tract and did not participate in metabolic activities. *Excretion* refers to the process of getting rid of metabolic wastes, and in mammals is mainly the function of the kidneys (Chapter 40). The large intestine, however, does excrete bile pigments.

When chyme passes through the intestine too rapidly, **defecation** (expulsion of feces) becomes more frequent and the feces are watery. This condition, called diarrhea, may be caused by anxiety, certain foods, or by certain disease organisms that irritate the intestinal lining. Prolonged diarrhea results in loss of water and salts, leading to dehydration, a serious condition, especially in infants.

Constipation results when chyme passes through the intestine too slowly. Because more water than usual is removed from the chyme, the feces may be hard and dry. Constipation is often caused by a diet containing insufficient fiber.

In Western countries, cancer of the colon and rectum accounts for more new cancer cases each year than do any other cancers except lung cancer (Fig. 39–10). Studies indicate that a high incidence of this cancer is found in populations whose diets are very low in fiber and high in animal protein, fat, and refined carbohydrate. It has

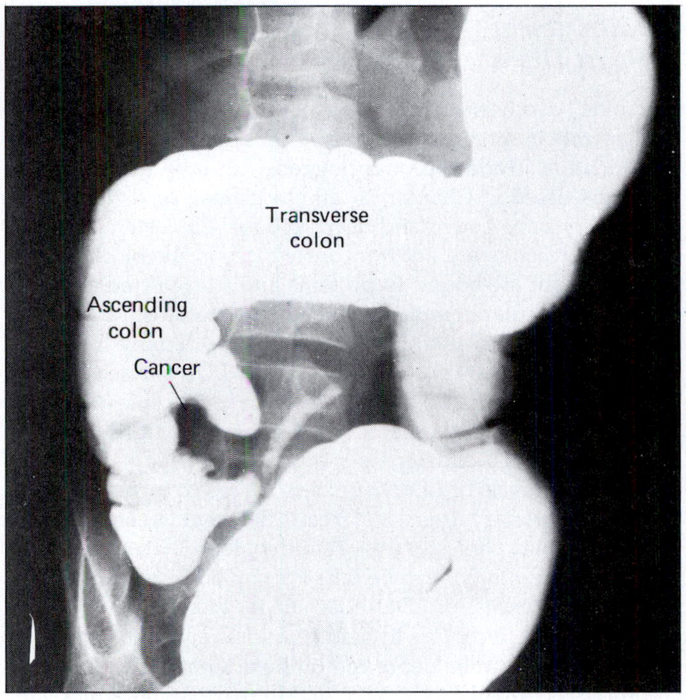

FIGURE 39–10 Radiographic view of the large intestine of a patient with colon cancer. The large intestine has been filled with a suspension of barium sulfate, which makes irregularities in the wall visible. The cancer is evident as a mass that projects from the wall into the lumen.

been suggested that diets low in fiber result in less frequent defecation, allowing prolonged contact between the mucous membrane of the colon and such carcinogens as nitrites (used as preservatives) in foods.

ADEQUATE AMOUNTS OF REQUIRED NUTRIENTS ARE NECESSARY TO SUPPORT METABOLIC PROCESSES

With only slight variation, all animals require the same basic nutrients—carbohydrates, lipids, proteins, vitamins, and minerals. Although not considered a nutrient in a strict sense, water is a necessary dietary component. Sufficient fluid must be ingested to replace fluid lost in urine, sweat, feces, and breath.

Adequate amounts of essential nutrients are necessary for metabolic processes. Recall from Chapter 1 that metabolism refers to all of the chemical processes that take place in the body. Metabolic processes include anabolism and catabolism. Anabolism refers to synthetic processes such as producing proteins. Catabolism includes breakdown processes such as hydrolysis. Nutritionists measure the energy value of food in Calories. (A Calorie, spelled with a capital C, is actually a kilocalorie. It is defined as the amount of heat required to raise the temperature of a kilogram of water from 15° to 16° C.)

Once nutrients are absorbed from the digestive tract, they are transported by the blood or lymph. Surplus nutrients in the blood are taken up by the liver cells, where they are either stored or converted into other materials. Under normal circumstances, blood leaving the liver carries sufficient nutrients to meet the requirements of all of the cells of the body. The blood has been appropriately described as a traveling smorgasbord from which each cell selects whatever nutrients it needs to carry on its metabolic processes.

CONCEPT CONNECTIONS

Nutrition
Heart Disease

Lipids have been the focus of much research because of their role in atherosclerosis, a progressive disease in which the arteries become clogged with fatty material. As discussed in Chapter 36, atherosclerosis leads to circulatory impairment and heart disease. Cholesterol and triacylglycerols are not transported free in the blood plasma but are bound to proteins and transported as large molecular complexes, called **lipoproteins.** Most plasma cholesterol is transported on **low-density lipoproteins (LDLs).** A protein on the LDL surface binds with a protein, an LDL receptor, on the plasma membrane of body cells. Some cholesterol is transported on **high-density lipoproteins (HDLs).**

High levels of LDL have been linked to increased risk for coronary artery and heart disease. LDL receptors help regulate cholesterol level in the blood by combining with LDL so that it can be taken up by body cells. If there is an excess of LDL in the blood, there will not be enough LDL receptors to bind it. Excess LDL cholesterol in the blood can be deposited in the arterial wall. This process is thought to involve highly reactive molecules in the arterial wall that oxidize the LDL cholesterol.

When LDL levels are high, HDL may play a protective role and decrease risk for coronary heart disease. HDL may collect cholesterol and transport it to the liver. A healthy proportion of HDL to LDL can apparently be promoted by a regular exercise program, by diet (reducing intake of animal fats and cholesterol), by maintaining appropriate body weight (obesity has a negative effect on lipoprotein levels), and by not smoking cigarettes. Omega-3 fatty acids, found in fish oils, are thought to decrease LDL levels and play other protective roles in decreasing risk for coronary heart disease.

Genetics plays an important role in risk for cardiovascular disease. Some individuals with high fat diets do not develop high blood cholesterol levels. However, in about one-third of the population, a diet high in saturated fats and cholesterol raises the blood cholesterol level by as much as 25%.

Ingestion of polyunsaturated fats tends to decrease the blood cholesterol level. For these reasons, many people now cook with vegetable oils rather than with butter and lard, drink skim milk rather than whole milk, and eat ice milk instead of ice cream.

Some individuals with high cholesterol levels do not develop coronary heart disease. On the other hand, others who seem to have a safe risk profile suddenly drop dead of coronary heart disease. A recently identified particle, **lipoprotein (a)**—comprised of proteins, cholesterol, and other lipids—may prove a missing piece to this puzzle. Lipoprotein (a) appears to be highly concentrated in the blood of individuals whose coronary heart disease could not be explained. Its concentration does not appear to be affected by changes in diet. Just how this particle affects the cardiovascular system is not known, but researchers are currently studying its actions.

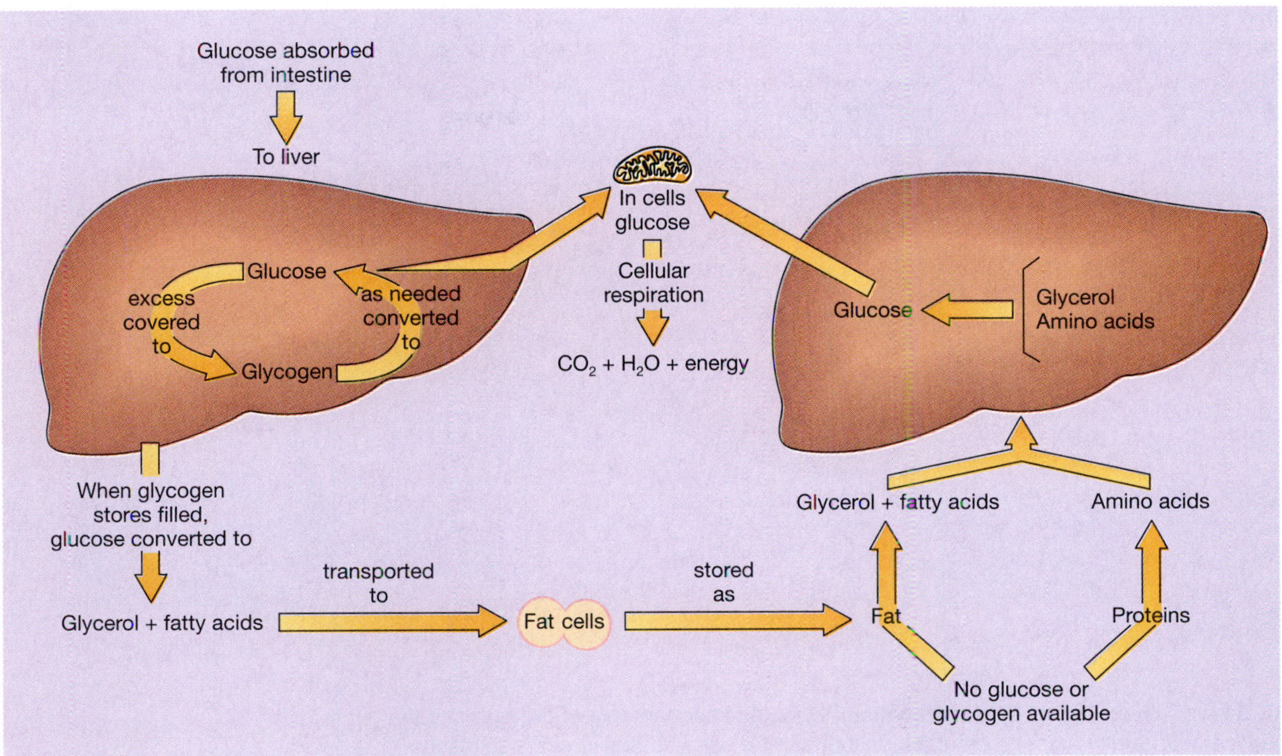

FIGURE 39–11 The fate of glucose in the body. The liver plays a central role in maintaining an appropriate level of glucose in the blood. Note that excess glucose can be converted to fat.

Carbohydrates Are a Major Energy Source in the Human Diet

Sugars and starches are the principal sources of energy in the ordinary human diet. However, they are not considered essential nutrients, because the body can obtain sufficient energy from a mixture of proteins and fats. In the average American diet, carbohydrates provide about 50% of the Calories ingested daily.

Most carbohydrates are ingested in the form of starch and cellulose, both polysaccharides. (You may want to review the discussion of carbohydrates in Chapter 4.) Nutritionists refer to polysaccharides as **complex carbohydrates.** Foods rich in complex carbohydrates include rice, potatoes, corn, and other cereal grains. These are the least expensive foods, and for this reason the proportion of carbohydrate in a family's diet often reflects economic status. Very poor people subsist on diets that are almost exclusively carbohydrate, while the more affluent enjoy the more expensive protein-rich foods, such as meat and dairy products.

Nutritionists suggest that we increase our consumption of complex carbohydrates and fiber by eating more fruits, vegetables, and whole grains. **Fiber** is mainly a complex mixture of cellulose and other indigestible carbohydrates. The U.S. diet is low in fiber due to low intake of fruit and vegetables and use of refined flour. Increasing fiber in the diet may decrease the risk of cancer of the colon. Fiber may also stimulate the feeling of being satisfied with the amount of food intake (satiety), and thus be useful in treating obesity.

Monosaccharides, the product of carbohydrate digestion, are converted to glucose in the liver. Excess glucose is stored as glycogen (Fig. 39–11). Hormones secreted by the pancreas act on the liver to regulate the concentration of glucose in the blood (blood sugar level).

When an excess of carbohydrate-rich food is eaten, the liver cells may become fully packed with glycogen and still have more glucose coming in. Liver cells then convert excess glucose to fatty acids and glycerol. These compounds are synthesized into triglycerides (triacylglycerols) and sent to the fat depots of the body for storage.

Lipids Are Used as an Energy Source and to Make Biological Molecules

Cells use ingested lipids as fuel, as components of cell membranes, and to make lipid compounds such as

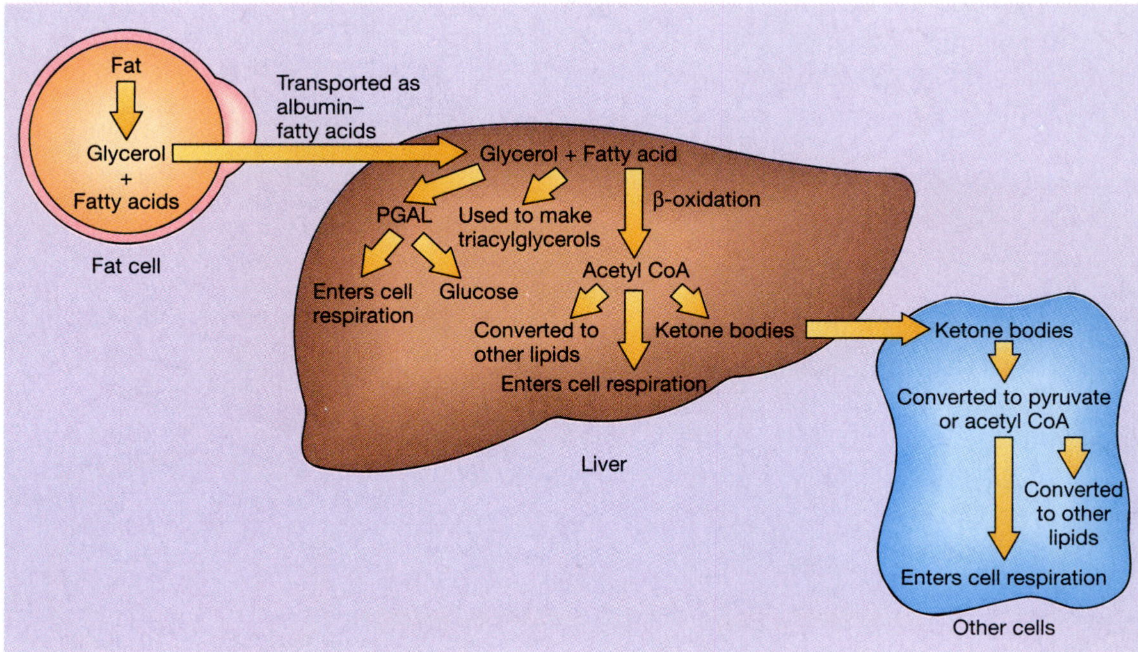

FIGURE 39–12 Overview of lipid metabolism. Note the important role of the liver in converting glycerol and fatty acids to compounds that can be used as fuel in cell respiration.

steroid hormones and bile salts. Lipid accounts for about 40% of the Calories in the average U.S. diet. Three polyunsaturated fatty acids (linoleic, linolenic, and arachidonic acids) are essential fatty acids that must be obtained in the human diet. Given these and sufficient nonlipid nutrients, the body can make all of the lipid compounds (including fats, cholesterol, phospholipids, and prostaglandins) that it needs.

About 98% of lipids in the diet are ingested in the form of triacylglycerols (triglycerides). (Recall from Chapter 4 that a triacylglycerol is a glycerol molecule chemically combined with three fatty acids; see Fig. 4–6.) Triacylglycerols may be saturated, that is, fully loaded with hydrogens. They may be monounsaturated (containing one double bond in the carbon chain of a fatty acid, so two more hydrogen atoms can be added), or polyunsaturated (containing two or more double bonds).

Generally, animal foods are rich in both saturated fats and cholesterol, while plant foods contain unsaturated fats and no cholesterol. Commonly used polyunsaturated vegetable oils are corn, soya, cottonseed, and safflower oils. Olive and peanut oils contain large amounts of monounsaturated fats. Butter contains mainly saturated fats.

The average U.S. diet provides about 700 milligrams of cholesterol each day, whereas only about 300 milligrams is recommended. High cholesterol sources include egg yolks, butter, and meat. The body is not dependent upon dietary sources for cholesterol because it is able to synthesize cholesterol from other nutrients. In fact, dietary intake of saturated fats can increase cholesterol level markedly.

When needed, stored fats are hydrolyzed to fatty acids and released into the blood. Before these fatty acids can be used by cells as fuel, they must be broken down into smaller compounds and combined with coenzyme A to form molecules of acetyl coenzyme A (Fig. 39–12).

Proteins Serve as Enzymes and as Structural Components of Cells

Proteins are essential building blocks of cells, serve as enzymes, and are also used to make many needed substances such as hemoglobin and myosin. Protein consumption is an index of a country's (or an individual's) economic status, because high-quality protein tends to be the most expensive and least available of the nutrients.

The recommended daily intake of protein is about 56 grams—only about an eighth of a pound. In the United States and other developed countries, most people eat far more protein than they require. The average American eats about 300 pounds of meat and dairy products per year, compared to only 2 pounds per person per year in some developing countries. Protein poverty is one of the world's most pressing health problems; millions of humans suffer from poor health, disease, and even death as a consequence of protein malnutrition.

Ingested proteins are degraded in the digestive tract to amino acids. Of the 20 or so amino acids important in

FOCUS ON *Health and Human Affairs*

VEGETARIAN DIETS

Most of the world's population depends almost entirely upon plants, especially cereal grains—usually rice, wheat, or corn—as the staple food. None of these foods contains adequate amounts of all of the essential amino acids. Besides being deficient in some of the essential amino acids, plant foods contain a lower percentage of protein than do animal foods. Meat contains about 25% protein, whereas even the new high-yield grains contain only 5% to 13% protein. What protein is available in plant food is also less digestible than that found in animal foods. Because most of the protein is encased within indigestible cellulose cell walls, much of it passes right through the digestive tract.

Despite these potential nutritional problems, more and more people are turning to vegetarian diets. Meats are becoming increasingly expensive because they are ecologically expensive to produce. About twenty-one kilograms of protein in grain, for example, is required to produce just one kilogram of beef protein. If the human population of our planet continues to expand at a much greater rate than does our food production, more grain will be diverted for human food and less for animal feed. The price of meat will continue to soar and may become unaffordable for many of us.

Can a vegetarian diet be nutritionally balanced? With an awareness of the special nutritional risks associated with a vegetarian diet (especially in growing children), they can be overcome. The most important rule is to select foods that complement one another. This requires knowledge of which amino acids are deficient in each kind of food. Since the body cannot store amino acids, all of the essential amino acids must be ingested at the same meal. For example, if rice is eaten for dinner, and beans for lunch the next day, the body will not have all of the essential amino acids needed at the same time to manufacture proteins. If beans and rice are eaten together, however, all of the needed amino acids are provided, because one food provides what the other lacks. Similarly, if dairy products are not excluded from the vegetarian diet, then macaroni can be paired with cheese, or cereal with milk, and all of the essential amino acids can be obtained.

Balanced vegetarian diets offer many health benefits. Increases in legume, grain, vegetable, and fruit intake increase the fiber content, as well as the amount of some vitamins and minerals. Researchers have studied Seventh Day Adventists, a religious group that excludes animal products in their diet. They report that compared to non–Seventh Day Adventists living in the same area, the incidence of cardiovascular disease is reduced by about 50%.

nutrition, approximately eight (nine in children) cannot by synthesized by humans at all, or at least not in sufficient quantity to meet the body's needs. These, which must be provided in the diet, are referred to as **essential amino acids.**

Complete proteins, those that contain the most appropriate distribution of amino acids for human nutrition, are found in eggs, milk, meat, and fish. Some foods, such as gelatin or soybeans, contain a high proportion of protein but do not contain all of the essential amino acids, or they do not contain them in proper nutritional proportions. Most plant proteins are deficient in one or more essential amino acids (See Focus On Health and Human Affairs: Vegetarian Diets).

Amino acids circulating in the blood can be taken up by cells and used for the synthesis of proteins. Excess amino acids are removed from circulation by the liver. In the liver cells, these are deaminated, that is, the amine group is removed (Fig. 39–13). During deamination, ammonia forms from the amine group. Ammonia, which is toxic at high concentrations, is converted to urea and excreted from the body.

The remaining carbon chain of the amino acid (called a keto acid) may be converted into carbohydrate or lipid and used as fuel or stored. Thus, even people who eat high-protein diets can gain weight if they eat too much.

Vitamins Are Organic Compounds Essential for Normal Metabolism

Vitamins are organic compounds required in the diet in relatively small amounts for normal biochemical functioning. Many are components of coenzymes (see Chapter 7). Vitamins may be divided into two main groups. **Fat-soluble vitamins** are those that can be dissolved in fat and include vitamins A, D, E, and K. **Water-soluble vitamins** are the B and C vitamins. Table 39–3 provides

(Text continues on page 784)

TABLE 39–3
The Vitamins

*Vitamins and U.S. RDA**	*Actions*	*Effect of deficiency*
Fat-soluble		
Vitamin A, retinol 5000 IU	Converted to retinal, a necessary component of retinal pigments, essential for normal vision; essential for normal growth and differentiation of cells; promotes normal growth, reproduction, and immunity	Retards growth; night blindness; 500,000 preschool children are blinded by vitamin A deficiency each year.
Vitamin D, calciferol 400 IU	Promotes calcium absorption from digestive tract; essential to normal growth and maintenance of bone	Bone deformities; rickets in children; osteomalacia in adults
Vitamin E, tocopherols 30 IU	Antioxidant; protects unsaturated fatty acids, cell membranes.	Increased catabolism of unsaturated fatty acids, so that not enough are available for maintenance of cell membranes; prevents normal growth
Vitamin K probably about 1 mg	Essential for blood clotting	Prolonged blood clotting time
Water-soluble		
Vitamin C, ascorbic acid 60 mg	Needed for synthesis of collagen and other intercellular substances; formation of bone matrix and tooth dentin, intercellular cement; needed for metabolism of several amino acids; may help body withstand injury from burns and bacterial toxins	Scurvy (wounds heal very slowly and scars become weak and split open; capillaries become fragile; bone does not grow or heal properly)
B-complex vitamins		
Vitamin B_1, Thiamine 1.5 mg	Active form is a coenzyme in many enzyme systems; important in carbohydrate and amino acid metabolism	Beriberi (weakened heart muscle, enlarged right side of heart, nervous system and digestive tract disorders)
Vitamin B_2, riboflavin 1.7 mg	Used to make coenzymes (e.g., FAD) essential in cellular respiration	Dermatitis, inflammation and cracking at corners of mouth; confusion
Niacin 20 mg	Component of important coenzymes (NAD and NADP) essential to cellular respiration	Pellagra (dermatitis, diarrhea, mental symptoms, muscular weakness, fatigue)
Vitamin B_6, pyridoxine 2 mg	Derivative is coenzyme in many reactions in amino acid metabolism	Dermatitis, digestive tract disturbances, convulsions
Pantothenic acid 10 mg	Constituent of coenzyme A (important in cellular metabolism)	Deficiency extremely rare
Folic acid 0.4 mg	Coenzyme needed for reactions involved in nucleic acid synthesis and for maturation of red blood cells	A type of anemia
Biotin 0.3 mg	Coenzyme important in metabolism	
Vitamin B_{12} 6 mg	Coenzyme important in metabolism	A type of anemia

*RDA is the recommended dietary allowance, established by the Food and Nutrition Board of the National Research Council, to maintain good nutrition for healthy persons.
†International Unit: the amount that produces a specific biological effect and is internationally accepted as a measure of the activity of the substance.

Sources	*Comments*
Liver, fortified milk, yellow and green vegetables such as carrots, broccoli	Can be formed from provitamin carotene (a yellow or red pigment); sometimes called anti-infection vitamin because it helps maintain epithelial membranes; excessive amounts harmful
Fish oils, egg yolk, fortified milk, butter, margarine	Two types: D_2, a synthetic form; D_3, formed by action of ultraviolet rays from sun upon a cholesterol compound in the skin; excessive amounts harmful
Oils made from cereals, nuts, leafy greens	
Normally supplied by intestinal bacteria; leafy greens, legumes	Antibiotics may kill bacteria; then supplements needed in surgical patients
Citrus fruits, strawberries, tomatoes	Possible role in preventing common cold or in the development of acquired immunity(?); harmful in very excessive dose
Liver, yeast, cereals, meat, green leafy vegetables	Deficiency common in alcoholics
Liver, cheese, milk, eggs, green leafy vegetables	
Liver, cheese, milk, eggs, green leafy vegetables	
Liver, meat, cereals, legumes	Many common drugs interfere with vitamin B_6 metabolism
Widespread in foods	
Produced by intestinal bacteria; liver, cereals, dark green leafy vegetables	
Produced by intestinal bacteria; liver, chocolate, egg yolk	
Liver, meat, fish	Contains cobalt; intrinsic factor secreted by gastric mucosa needed for absorption

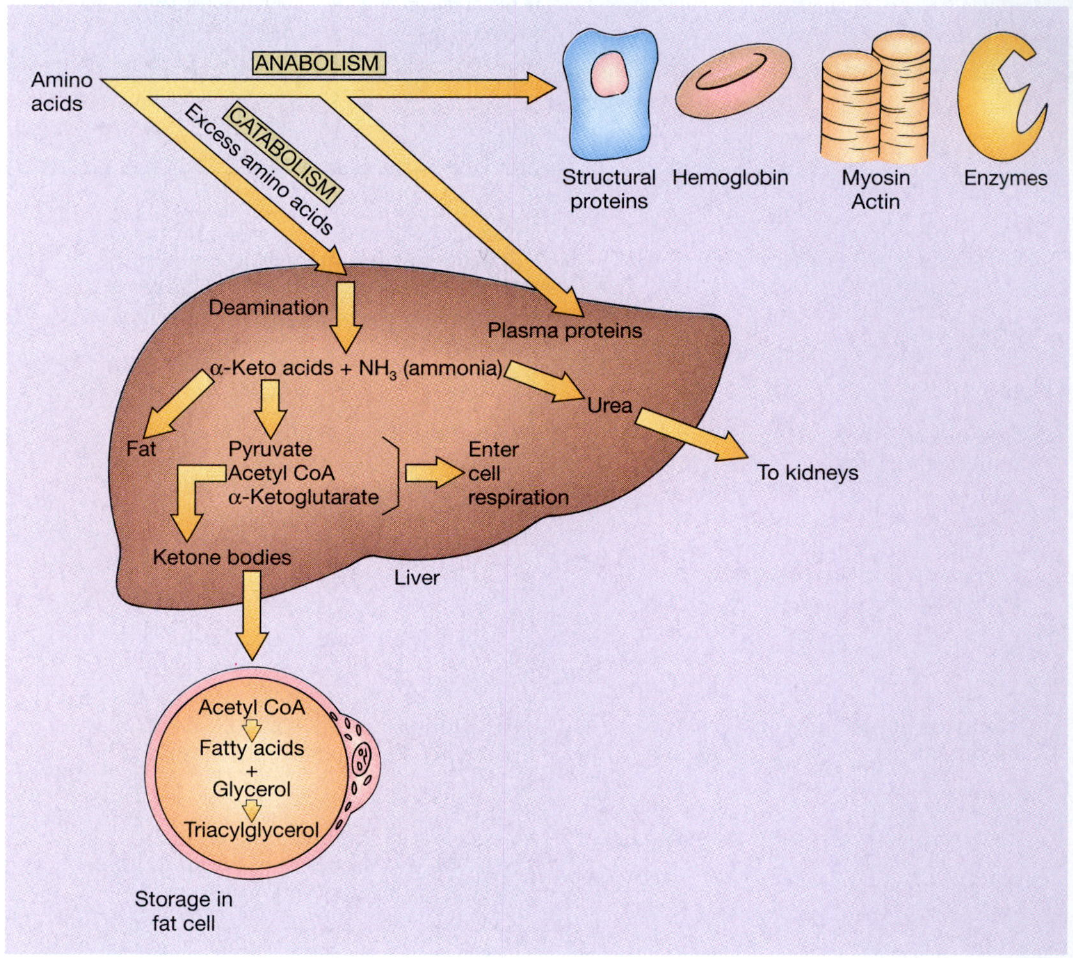

FIGURE 39–13 Overview of protein metabolism. The liver plays a central role in protein metabolism. Deamination of amino acids and conversion of the amino groups to urea takes place there. In addition, many proteins are synthesized in the liver. Note that you can gain weight on a high protein diet because excess amino acids can be converted to fat.

the sources, functions, and consequences of deficiency for most of the vitamins.

Health professionals debate the advisability of taking large amounts of certain specific vitamins, such as vitamin C to prevent colds or vitamin E to protect against vascular disease. Some studies suggest that vitamin A (found in yellow and green vegetables) and vitamin C (found in citrus fruit and tomatoes) may help protect against certain forms of cancer. We do not yet understand all of the biochemical roles played by vitamins or the interactions among various vitamins and other nutrients.

We do know that large overdoses of vitamins, like vitamin deficiency, can be harmful. Moderate overdoses of the B and C vitamins are excreted in the urine, but surpluses of the fat-soluble vitamins are not easily excreted and can accumulate to harmful levels.

Minerals Are Inorganic Nutrients Required by Cells

Minerals are inorganic nutrients ingested in the form of salts dissolved in food and water (Table 39–4). Essential minerals required in amounts of 100 mg or more daily include sodium, chloride, potassium, calcium, phosphorus, magnesium, and sulfur. Several others, such as iron, copper, iodine, fluoride, and selenium, are **trace elements** that are required in an amount of less than 100 mg per day.

Minerals are needed as components of body tissues and fluids. Salt content (about 0.9%) is vital in maintaining the fluid balance of the body, and since salts are lost from the body daily in sweat, urine, and feces, they must be replaced by dietary intake. Sodium chloride (common table salt) is the salt needed in largest quantity in blood and other body fluids. A deficiency results in dehydration.

TABLE 39–4
Some Important Minerals and their Functions

Mineral	*Functions*	*Sources; Comments*
Calcium	Component of bone and teeth; essential for normal blood clotting; needed for normal muscle and nerve function	Milk and other dairy products, fish, green leafy vegetables. Bones serve as calcium reservoir
Phosphorus	Performs more functions than any other mineral. Structural component of bone; component of ATP, DNA, and RNA.	Meat, dairy products, cereals
Sulfur	Component of many proteins and vitamins	High-protein foods such as meat, fish, legumes, nuts
Potassium	Principal positive ion within cells; influences muscle contraction and nerve excitability	Occurs in many foods
Sodium	Principal positive ion in interstitial fluid; important in fluid balance; essential for conduction of nerve impulses	Many foods, table salt; too much ingested in average American diet; in excessive amounts, may contribute to high blood pressure
Chloride	Principal negative ion of interstitial fluid; important in fluid balance and in acid-base balance	Many foods; table salt
Magnesium	Needed for normal muscle and nerve function	Nuts, whole grains, greens
Copper	Component of enzyme needed for melanin synthesis; component of many other enzymes; essential for hemoglobin synthesis	Liver, eggs, fish, whole wheat flour, beans
Iodine	Component of thyroid hormones (hormones that increase metabolic rate). Deficiency results in goiter (abnormal enlargement of thyroid gland)	Seafoods, iodized salt, vegetables grown in iodine-rich soils
Manganese	Necessary to activate arginase, an enzyme essential for urea formation; activates many other enzymes	Whole-grain cereals, egg yolks, green vegetables. Poorly absorbed from intestine
Iron	Component of hemoglobin, myoglobin, important respiratory enzymes (cytochromes), and other enzymes essential to oxygen transport and cellular respiration. Deficiency results in anemia and may impair cognitive function	Mineral most likely to be deficient in diet. Good sources: meat (especially liver), nuts, egg yolk, legumes
Fluoride	Component of bones and teeth; makes teeth resistant to decay; excess causes tooth mottling	In areas where it does not occur naturally, fluoride may be added to municipal water supplies (fluoridation)
Zinc	Cofactor for at least 70 enzymes; helps regulate synthesis of certain proteins; needed for growth and repair of tissues; deficiency may impair cognitive function	Meat, milk, yogurt, some seafood
Selenium	Antioxidant (part of a peroxidase that breaks down peroxides)	Seafood, eggs, liver

Iron is the mineral most likely to be deficient in the diet. In fact, iron deficiency is one of the most widespread nutritional problems in the world. In most developing countries, an estimated two-thirds of children and women of childbearing age suffer from iron deficiency. In the U.S., Europe, and Japan, 10% to 20% of women of childbearing age have this deficiency.

Energy Metabolism Is Balanced when Energy Input Equals Energy Output

The amount of energy liberated by the body per unit time is a measure of the **metabolic rate.** Much of the energy expended by the body is ultimately converted to heat. Metabolic rate may be expressed either in Calories of heat energy expended per day or as a percentage above or below a standard normal level.

The **basal metabolic rate (BMR)** is the rate at which the body releases heat as a result of breaking down fuel molecules. BMR is the body's basic cost of metabolic living, that is, the rate of energy used during resting conditions. An individual's **total metabolic rate** is the sum of his or her BMR and the energy used to carry on all daily activities. For example, a laborer has a greater total metabolic rate than does an executive whose job requirements do not include a substantial amount of movement and who does not exercise regularly.

Concept Connections

Nutrition ⟷ *Cell Function*

Normal cellular processes that require oxygen produce highly reactive molecules such as peroxides and **free radicals.** These compounds snatch electrons from other molecules such as DNA. Free radicals are also generated by tobacco smoke and other forms of air pollution.

Cells have **antioxidants** that destroy free radicals and other reactive molecules. However, when too many free radicals are present, they damage DNA, proteins, and unsaturated fatty acids. Damage to DNA can cause mutations that can lead to cancer. Injury to unsaturated fatty acids can result in damage to cell membranes. Free radicals are thought to contribute to atherosclerosis by causing oxidation of LDL cholesterol. The effects of oxidative damage to the body over the years are thought to contribute to the aging process.

Certain vitamins, including A, C, and E, function as antioxidants. For example, vitamins A and E protect cell membranes from free radicals. Selenium, zinc, copper, manganese, and iron are minerals that are parts of antioxidant enzyme systems. Nutritionists recommend that we increase the antioxidants in our diet by eating fruit, vegetables, and other foods that are high in antioxidant vitamins and minerals.

An average-sized man who does not engage in any exercise program and who sits at a desk all day expends about 2000 Calories daily. If the food he eats each day also contains about 2000 Calories, he will be in a state of energy balance; that is, his energy input will equal his energy output. This is an extremely important concept, because body weight remains constant when

Energy (Calorie) input = Energy output

When energy output is greater than energy input, stored fat is burned and body weight decreases. On the other hand, people gain weight when they take in more energy (Calories) in food than they expend in daily activity—in other words, when

Energy (Calorie) input > Energy output

Obesity Is a Serious Nutritional Problem

Obesity, the excess accumulation of body fat, is a serious form of malnutrition, and in affluent societies it has become a problem of epidemic proportions. An overweight person places an extra burden upon the heart and is susceptible to heart disease and other ailments. Obese persons generally die at a younger age than do people of normal weight. Yet one-third of our working population is 25% or more overweight.

Obesity can result from an increase in the size of fat cells or from an increase in the number of fat cells, or both. The number of fat cells in the adult is apparently determined mainly by the amount of fat stored during infancy and childhood. When babies or small children are overfed, abnormally large numbers of fat cells are formed. Later in life, these fat cells may be fully stocked with excess lipids or may be shrunken, but they are always there. People with such increased numbers of fat cells are thought to be more susceptible to obesity than are those with normal numbers.

Most overweight people overeat due to a combination of poor eating habits and psychological factors. Whatever the underlying causes, overeating is the only way to become obese. For every 9.3 Calories of excess food taken into the body, about 1 gram of fat is stored. (An excess of about 140 Calories per day for a month results in gaining 1 pound).

Because so many people are overweight, dieting has generated a multimillion-dollar industry of diet foods, formulas, pills, books, clubs, slenderizing devices, and even surgical procedures such as gastric stapling. Unfortunately, there is no magic cure for obesity. The only sure (and healthful) way to lose weight is to adjust food intake so that energy intake is less than energy output. Then the body will have to draw on its fat stores for the missing calories. As fat is mobilized and burned, body weight decreases. This can generally be accomplished by a combination of increased exercise and decreased caloric intake (1000 to 1500 Calories daily for the mildly obese). Most nutritionists agree that the best reducing diet is well-balanced and provides the bulk of the calories in the form of complex carbohydrates.

Malnutrition Can Cause Serious Health Problems

While millions of people eat too much, more than a billion humans do not have enough to eat or do not eat a balanced diet (see Chapter Opener). Individuals suffering from malnutrition are weak, easily fatigued, and highly susceptible to infection. Essential amino acids, iron, calcium, and vitamin A are commonly deficient nutrients. An estimated quarter of a million children become permanently blind every year because their diets are deficient in vitamin A.

Of all the required nutrients, essential amino acids are the ones most often deficient in the diet. Millions of people suffer from poor health and a lowered resistance to disease because of protein deficiency. Children's physical and mental development are retarded when the essential building blocks of cells are not provided in the diet. Because their bodies cannot manufacture antibodies (which are proteins) and cells needed to fight infection, common childhood diseases, such as measles, whooping cough, and chicken pox, are often fatal in children suffering from protein malnutrition.

In young children, severe protein malnutrition results in the condition known as **kwashiorkor.** The term, an African word that means "first-second," refers to the situation in which a first child is displaced from its mother's breast when a younger sibling is born. The older child is placed on a diet of starchy cereal or cassava that is deficient in protein. Growth becomes stunted, muscles are wasted, edema develops (as displayed by a swollen belly), the child becomes apathetic and anemic, and metabolism is impaired (Fig. 39–14). Without essential amino acids, the digestive enzymes themselves cannot be manufactured, so what little protein is ingested cannot be digested. Dehydration and diarrhea occur and often lead to death.

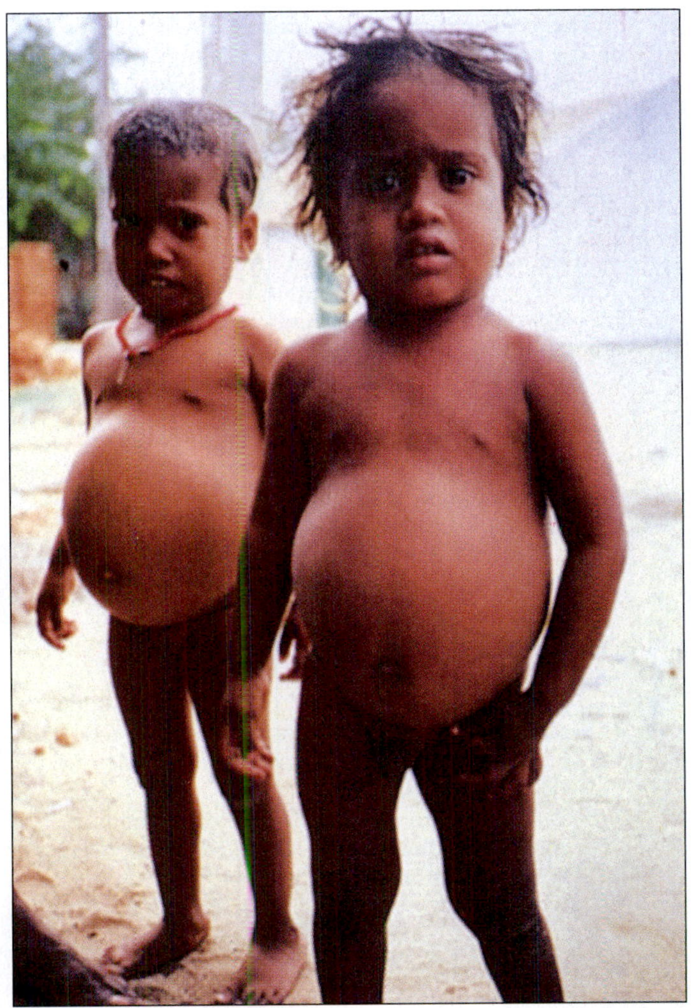

FIGURE 39–14 Children suffering from kwashiorkor. This disease is caused by severe protein deficiency. Note the characteristic swollen belly, which results from fluid imbalance. (*United Nations, Food and Agricultural Organization [FAO], photo by P. Pittet*)

Chapter Summary

I. Processing food includes ingestion, digestion, absorption of nutrients, and elimination of wastes.

II. An organism's body plan and life-style are adapted to its mode of nutrition.
   A. In the simplest invertebrates, the sponges, there is no digestive system; digestion is carried on intracellularly.
   B. Cnidarians and flatworms have digestive systems with only one opening, which serves as both mouth and anus.
   C. In more complex invertebrates and in all vertebrates, the digestive tract is a complete tube with an opening at each end.

III. In vertebrates, various parts of the digestive tract are specialized to perform specific functions.
   A. Mechanical digestion and enzymatic digestion of carbohydrates begin in the mouth.
   B. As food is swallowed, it is propelled through the pharynx and esophagus. A bolus of food is moved along through the digestive tract by peristaltic action.
   C. In the stomach, food is mechanically digested by vigorous churning, and proteins are enzymatically digested by the action of pepsin in the gastric juice.
   D. Most enzymatic digestion takes place in the duodenum, which receives secretions from the liver and pancreas and produces several digestive enzymes of its own.

E. The liver produces bile, which emulsifies fats.
F. The pancreas releases enzymes that digest protein, lipid, and carbohydrate, as well as RNA and DNA.
G. Activities of the digestive system are regulated by both nerves and hormones.
H. Most nutrients are absorbed through the thin walls of the intestinal villi.
I. The large intestine is responsible for the elimination of undigested wastes. It also incubates bacteria that produce vitamin K and certain B vitamins.

IV. For a balanced diet, humans and other animals require carbohydrates, lipids, proteins, vitamins, and minerals.
  A. Most carbohydrates are ingested in the form of polysaccharides—starch and cellulose.
    1. Carbohydrates are used mainly as fuel.
    2. Glucose concentration in the blood is carefully regulated. Excess glucose is stored as glycogen and can also be converted to fat.
  B. Lipids are used as fuel, as components of cell membranes, and to synthesize steroid hormones and other lipid substances.
    1. Most lipids are ingested in the form of triacylglycerols.
    2. Cholesterol is transported on low-density lipoproteins (LDLs) and high-density lipoproteins (HDLs). High levels of LDL are associated with increased risk for heart disease.
    3. Fatty acids are converted to molecules of acetyl coenzyme A and used as fuel. Excess fatty acids are stored as fat.
  C. Proteins serve as enzymes and are essential structural components of cells.
    1. The best distribution of essential amino acids is found in the complete proteins of animal foods.
    2. Excess amino acids are deaminated by liver cells. Amine groups are converted to urea and excreted in urine, and the remaining keto acids are converted to carbohydrate and used as fuel or converted to lipid and stored in fat cells.
  D. Vitamins are organic compounds required in small amounts for many biochemical processes. Many serve as components of coenzymes.
  E. Minerals are inorganic nutrients ingested as salts dissolved in food and water.

V. Basal metabolic rate is the body's cost of metabolic living.
  A. Total metabolic rate is the BMR plus the energy used to carry on daily activities.
  B. When energy (Calorie) input equals energy output, body weight remains constant.

VI. Obesity is a serious nutritional problem in which an excess amount of fat accumulates in the adipose tissues. A person gains weight by taking in more energy, in the form of Calories, than is expended in activity.

VII. Millions of people suffer from malnutrition. Essential amino acids are the nutrients most often deficient in the diet.

Selected Key Terms

absorption, p. 769
bile, p. 773
colon, p. 776
digestion, p. 768
duodenum, p. 772
elimination, p. 769
emulsification,p 773
essential amino acid, p. 781
gallbladder, p. 773
ingestion, p. 768
large intestine, p. 776
liver, p. 773
metabolic rate, p. 785
mineral, p. 784
pancreas, p. 773
peristalsis, p. 772
pharynx, p. 772
salivary gland, p. 771
small intestine, p. 772
stomach, p. 772
villus (plural, *villi*), p. 772
vitamin, p. 781

Post-Test

1. The process of taking food into the body is called ________________.

2. ________________ consists of mechanically and enzymatically breaking down food into molecules small enough to be absorbed.

3. ________________ is the process of getting rid of undigested and unabsorbed food.

4. The most characteristic feature shared by the cnidarian and flatworm digestive system is that it has (one or two?) ________________ opening(s).

5. The vertebrate digestive tract has (one or two?) ______________ opening(s).
6. Salivary ______________ is an enzyme that initiates the digestion of starch to sugar.
7. A mammalian tooth consists mainly of ______________.
8. Protein digestion begins in the (a) mouth (b) pharynx (c) esophagus (d) stomach (e) small intestine.
9. Bacteria that produce vitamin K inhabit the (a) mouth (b) esophagus (c) stomach (d) small intestine (e) large intestine.
10. Bile is secreted by the (a) liver (b) pancreas (c) stomach (d) small intestine (e) large intestine.
11. Gastrin (a) is an enzyme secreted by the stomach (b) is transported by the blood (c) stimulates release of gastric juice (d) answers (a), (b), and (c) are correct (e) only answers (b) and (c) are correct.
12. The surface area of the stomach is increased by the presence of (a) villi (b) microvilli (c) rugae (d) answers (a), (b), and (c) are correct (e) only answers (a) and (b) are correct.
13. Absorption takes place mainly through (a) finger-like villi in the lining of the small intestine (b) gastric glands in the stomach wall (c) rugae (d) lymph vessels in the visceral peritoneum (e) peristalsis.
14. Food leaving the stomach next enters the (a) liver (b) pancreas (c) duodenum (d) colon (e) esophagus.
15. Vitamins (a) function as components of coenzymes (b) are generally ingested in the form of salts (c) stimulate release of gastric juice (d) answers (a), (b), and (c) are correct (e) none of the preceding answers is correct.
16. Amino acids (a) function mainly as components of cell membranes (b) are generally ingested in the form of salts (c) can be used as fuel molecules (d) answers (a), (b), and (c) are correct (e) none of the preceding answers is correct.
17. Lipids (a) function as components of cell membranes (b) are mainly ingested in the form of triacylglycerols (c) are an important source of fiber (d) answers (a), (b), and (c) are correct (e) only answers (a) and (b) are correct.
18. In the digestive tract, proteins are degraded to (a) vitamin D (b) amino acids (c) glucose (d) glycerol and fatty acids (e) cholesterol.
19. Most carbohydrates are degraded to (a) minerals (b) amino acids (c) glucose (d) glycerol and fatty acids (e) cholesterol.
20. The body's rate of energy use during resting conditions is (a) the basal metabolic rate (b) the rate at which the body releases heat as a result of breaking down fuel molecules (c) the sum of all of the energy used to carry on daily activities (d) answers (a), (b), and (c) are correct (e) only answers (a) and (b) are correct.

Review Questions

1. How are digestive structures and methods of processing food in sponges, hydras, and flatworms, adapted to each group's life-style? Give specific examples.
2. Why must food be digested?
3. Trace a bite of food through the digestive tract, listing each structure through which it passes.
4. List the three types of vertebrate accessory glands that secrete digestive juices and give their functions.
5. The inner lining of the digestive tract is not smooth like the inside of a water pipe. Why is this advantageous? What structures increase its surface area?
6. Summarize the step-by-step digestion of (a) carbohydrates, (b) lipids, (c) proteins.
7. What happens to ingested cellulose in humans? Why?
8. Draw and label an intestinal villus.
9. How dow the absorption of fat differ from the absorption of glucose?
10. List the nutrients that must be included in a balanced diet.
11. Draw a diagram to illustrate the fate of carbohydrates in the body.
12. Describe the fate of absorbed amino acids.
13. Write an equation to describe energy balance and tell what happens when the equation is altered in either direction.
14. Summarize the relationship between diet and heart disease.

15. Label the diagram. For correct labeling, see Figure 39–4.

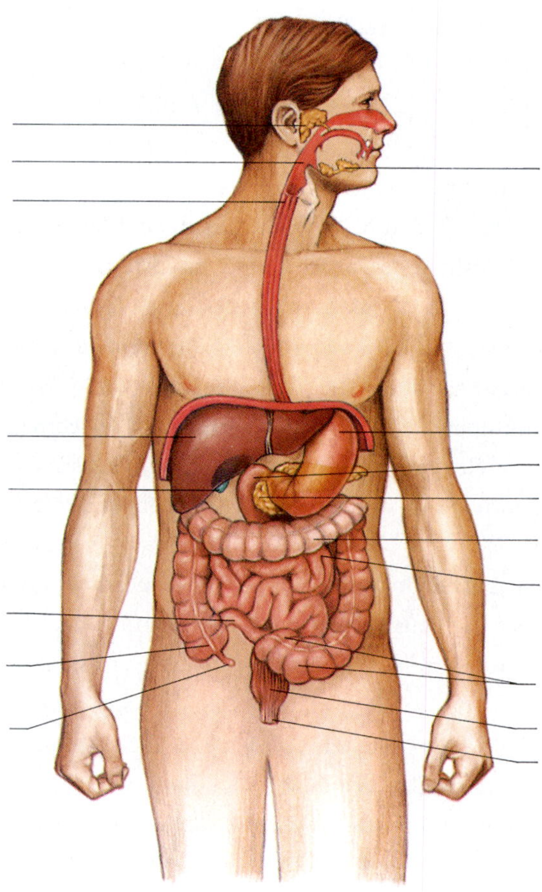

Thinking Critically

1. Some experts predict that more and more people will suffer from malnutrition. Considering recent technological advances that increase food supply, how could this be true? What measures would you suggest to increase the amount of food per person on our planet?
2. What is the adaptive advantage of specialization of different regions in a complete digestive tract?
3. A high percentage of adults suffer from gastrointestinal discomfort, including cramps and diarrhea, when they drink milk. What do you think could cause this condition known as lactose intolerance?
4. Design an experiment to demonstrate the nutritional need for the B vitamin pyridoxine.
5. What are some of the challenges in planning a nutritionally balanced vegetarian diet?

Recommended Readings

Brown, L.R. *et al., State of the World 1994,* W. W. Norton & Company, New York, 1994. A Worldwatch Institute Report on environmental problems and progress. Includes an analysis of our ability to provide adequate food for our increasing population.

Sanderson, S. L., and R. Wassersug, "Suspension-Feeding Vertebrates," *Scientific American,* Vol. 262, No. 3, March 1990, pp. 96–101. Animals that filter food out of the surrounding water can grow to large size.

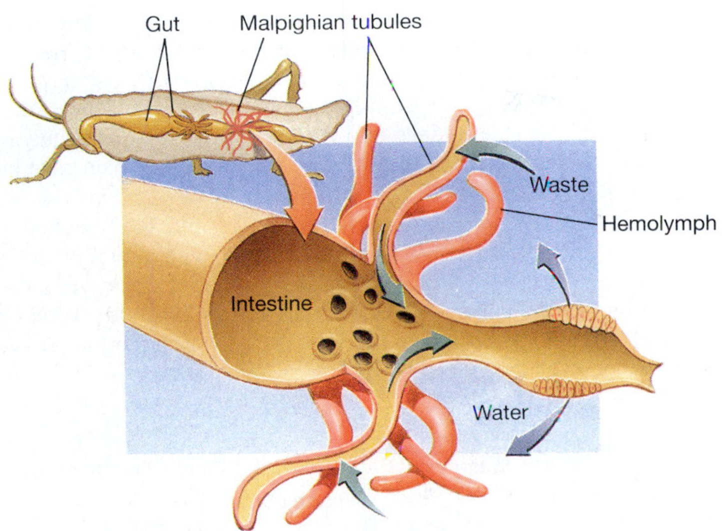

FIGURE 40–1 Malpighian tubules. The slender Malpighian tubules of insects have blind ends that lie in the hemocoel. Their cells transfer wastes from the hemolymph to the cavity of the tubule. Uric acid, the major waste product, is discharged into the gut.

moves into the body, and salt diffuses out. An animal adapted to this environment has excretory structures that actively remove the excess water. Many also have cells in their gills that remove salts from the surrounding water and transport them into the body fluids.

Terrestrial animals have a higher fluid concentration than that found in the surrounding air. They tend to lose water by evaporation from the body surface and from respiratory surfaces. They may also lose water as wastes are excreted. Adaptation to life on land has required the evolution of structures and processes that conserve water.

Nephridial organs, found in many invertebrates, consist of simple or branching tubes that usually open to the outside of the body through pores. The protonephridia of flatworms, described in Chapter 24, are specialized for osmoregulation. Annelids and mollusks have nephridial organs called metanephridia. Fluid from the coelom passes into the tubule, bringing with it whatever it contains—glucose, salts, and wastes. As the fluid moves through the tubule, needed materials, such as water or glucose, are reabsorbed by the capillaries, leaving the wastes behind. In this way, urine is produced that contains concentrated wastes.

The excretory system of insects and spiders consists of **Malpighian tubules** (Fig. 40–1). Two to several hundred tubules may be present, depending on the species. Malpighian tubules have blind ends that lie in the body cavity (hemocoel) where they are bathed in blood. The tubules collect wastes and empty them into the intestine. Water and some salts are reabsorbed into the blood by specialized rectal glands. Uric acid, the major waste product, is excreted as a semidry paste with a minimum of water. By conserving fluid, Malpighian tubules have contributed significantly to the success of the insects in terrestrial environments.

THE KIDNEY IS THE KEY VERTEBRATE ORGAN OF OSMOREGULATION AND EXCRETION

Vertebrates live successfully in a wide range of habitats—in fresh water, the sea, tidal regions, and on land, even in extreme environments such as deserts. In response to the requirements of these environments, vertebrates have evolved adaptations for regulating their salt and water content and for excreting wastes. An extreme example is the desert-dwelling kangaroo rat, which must carefully conserve water. It obtains most of its water from its own metabolism, and its kidneys are so efficient that it loses little fluid as urine.

The main osmoregulatory and excretory organ in vertebrates is the kidney. In most vertebrates, the skin, lungs or gills, and digestive system also help maintain fluid balance and dispose of metabolic wastes. Some reptiles and birds have salt glands in the head that excrete salt that enters the body with the seawater they drink.

Osmoregulation Is a Continuous Challenge for Aquatic Vertebrates

As fishes began to move into freshwater habitats about 400 million years ago, there was strong selection for the evolution of adaptations for effective osmoregulation. Because the body fluids of freshwater animals have higher salt concentrations than (i.e., are hypertonic to) their surroundings, water passes into them osmotically. As a result, they are in constant danger of becoming waterlogged. Freshwater fishes are covered by scales and a mucous secretion that retards the passage of water into the body. However, water enters through the gills. The

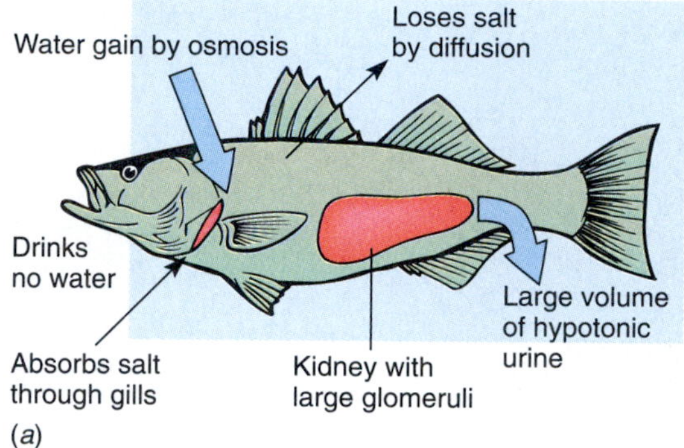

FIGURE 40–2 Osmoregulation in fishes. (***a***) Freshwater fishes live in a hypotonic medium, so water continuously enters the body by osmosis, while salts leave by diffusion. These fishes excrete large quantities of dilute urine, and actively absorb salts through the gills. (***b***) Marine fishes live in a hypertonic medium and so lose water by osmosis. They gain salts in water they drink and by diffusion. To compensate, the fishes drink salt water, excrete the salt, and produce a small volume of urine. (***c***) The shark solves its osmotic problem differently. It accumulates urea in high enough concentration to become hypertonic to the surrounding medium. As a result, some water enters its body by osmosis. A large quantity of hypotonic urine is excreted. Salt, which enters by diffusion, is excreted by a salt excreting gland.

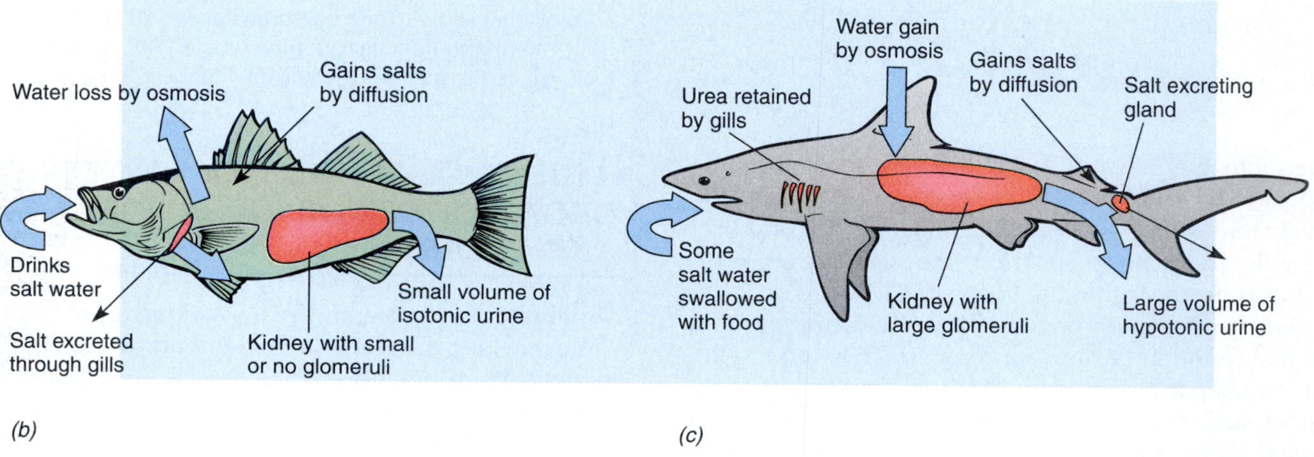

kidneys of these fishes have become adapted to filter out excess water, and they excrete a large amount of dilute urine (Fig. 40–2).

Water entry, though, is only part of the problem of osmoregulation in freshwater fishes. These animals also tend to lose salts to the surrounding fresh water. To compensate, special cells in the gills have evolved that actively transport salt (mainly sodium chloride) from the water into the body.

Most amphibians are at least semiaquatic, and their mechanisms of osmoregulation are similar to those of freshwater fishes. They produce a large amount of dilute urine. For example, through its urine and skin, a frog can lose an amount of water equivalent to one-third of its body weight in one day. Active transport of salt inward by special cells in the skin compensates for salt loss through skin and urine.

An important adaptation of freshwater fishes to their aquatic habitats was the evolution of body fluids more dilute than seawater. Modern vertebrates have a salt concentration in their body fluids about one-third that of seawater. Thus, when some freshwater fishes returned to the sea about 200 million years ago, their blood and body fluids were less salty than (i.e., hypotonic to) their surroundings. They tended to lose water osmotically and to take in salt.

To compensate for fluid loss, many marine bony fishes drink seawater. They retain the water and excrete salt by the action of specialized cells in their gills. Very little urine is excreted by the kidneys, and the nephrons (microscopic units of the kidney) have only small (or no) capillary clusters (glomeruli) for filtering the blood.

Marine chondrichthyes (sharks and rays) have different osmoregulatory adaptations that allow them to tolerate the salt concentrations of their environment. These animals accumulate and tolerate urea. Their tissues are adapted to function at concentrations of urea that would be toxic to most other animals. The high concentration of urea makes the osmotic concentration of body fluids slightly higher than (hypertonic to) seawater. This results in a net inflow of water into their bodies. Their well-developed kidneys excrete a large volume of urine. Excess salt is excreted by the kidneys and, in many species, by a rectal gland.

Whales, dolphins, and other marine mammals ingest seawater along with their food. Their kidneys produce a

FIGURE 40–3 Killer whale (*Orcinus orca*). Marine mammals ingest seawater along with their food. To rid the body of excess salt, their kidneys produce a very concentrated urine, much more salty than seawater. (Ken Lucas/Biological Photo Service)

very concentrated urine, much more salty than seawater. This is an important physiological adaptation, especially for marine carnivores (Fig. 40–3). The high protein diet of these animals results in the production of large amounts of urea, which must be excreted in the urine or, in some cases, by special accessory salt glands.

The Mammalian Kidney Is Important in Maintaining Homeostasis

In mammals, the kidneys are the principal excretory organs and are responsible for the excretion of most nitrogenous wastes and for helping to maintain fluid balance by adjusting the salt and water content of the urine. As in other terrestrial vertebrates, the lungs, skin, and digestive system also are important in mammalian osmoregulation and waste disposal (Fig. 40–4). Most carbon dioxide and a great deal of water are excreted by the lungs. Although primarily concerned with the regulation of body temperature, the sweat glands excrete 5% to 10% of all metabolic wastes.

Most of the bile pigments produced by the breakdown of red blood cells are normally excreted by the liver into the intestine. From the intestine they then pass out of the body with the feces. The liver also produces both urea and uric acid.

THE KIDNEYS, URINARY BLADDER, AND THEIR DUCTS MAKE UP THE URINARY SYSTEM

The mammalian **urinary system** consists of the kidneys, the urinary bladder, and associated ducts. The overall structure of the human urinary system is shown in Figure 40–5. Located just below the diaphragm in the "small" of the back, the kidneys look like a pair of giant, dark-red lima beans, each about the size of a fist. The outer portion of the kidney is called the **renal cortex;** the inner

(Text continued on page 799)

Concept Connections

Waste Disposal — Neural Control

In mammals, the release of metabolic wastes by the urinary system is regulated by the nervous system. **Urination,** or the release of urine from the body, is a reflex action. In humans, two sphincter valves (rings of muscle) keep the opening of the urethra closed. When the volume of urine in the bladder reaches about 300 ml, receptors in the bladder wall are stretched sufficiently to stimulate the urination reflex. The sphincter muscles relax, allowing urination to occur.

The sphincter muscles are not under voluntary nervous control in the sense that our skeletal muscles are voluntary. What we call bladder control depends on the ability to facilitate or inhibit this reflex voluntarily. For example, one can voluntarily empty the bladder at a convenient time even before it is full. On the other hand, even when the volume of urine in the bladder has exceeded 300 ml, one can inhibit urination until a more convenient time. Such voluntary control cannot be exerted by an immature nervous system. Most babies cannot develop urinary control until about age 2, no matter how hard anxious parents try to teach them.

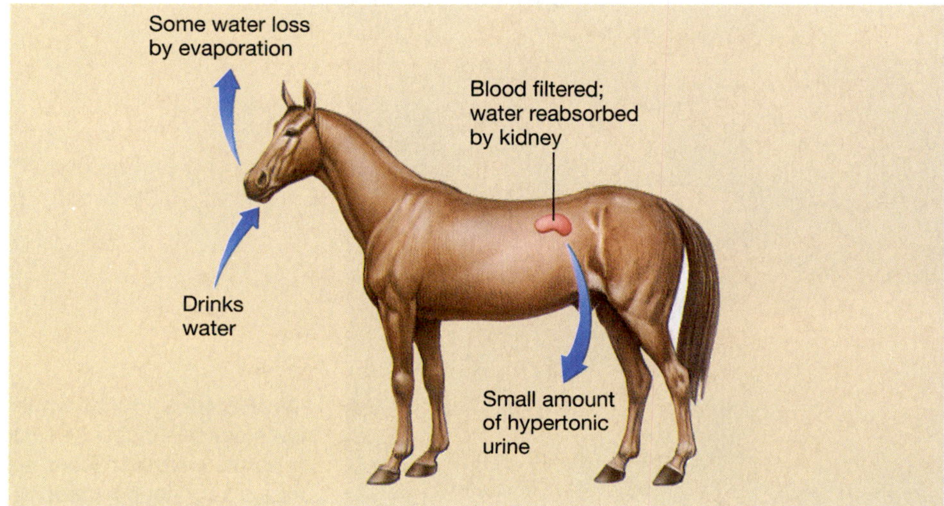

(*a*) Terrestrial mammal

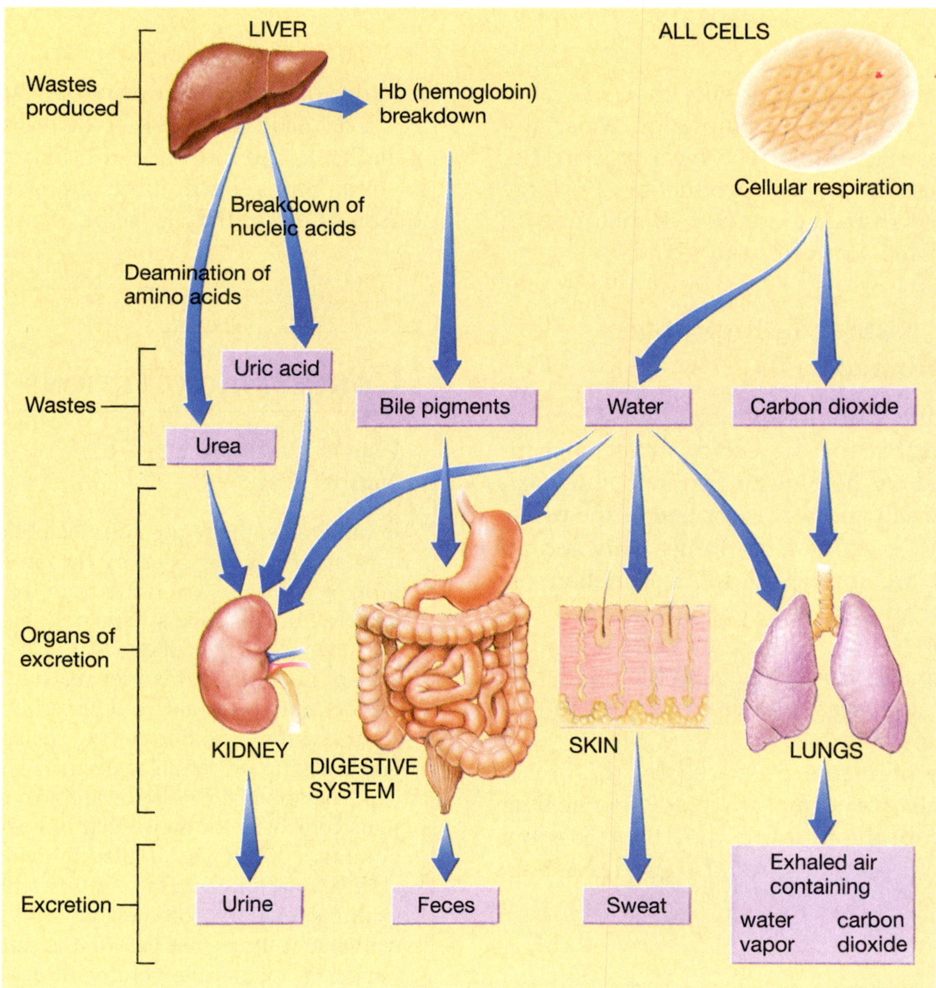

FIGURE 40–4 Disposal of metabolic wastes in vertebrates. In many terrestrial vertebrates, the kidneys, lungs, skin, and digestive system all participate in disposal of metabolic wastes. (***a***) The kidney conserves water by reabsorbing it. (***b***) Disposal of metabolic wastes in humans and other terrestrial mammals. To conserve water, a small amount of hypertonic urine is usually produced. Nitrogenous wastes are produced by the liver and transported to the kidneys. All cells produce carbon dioxide and some water during cellular respiration.

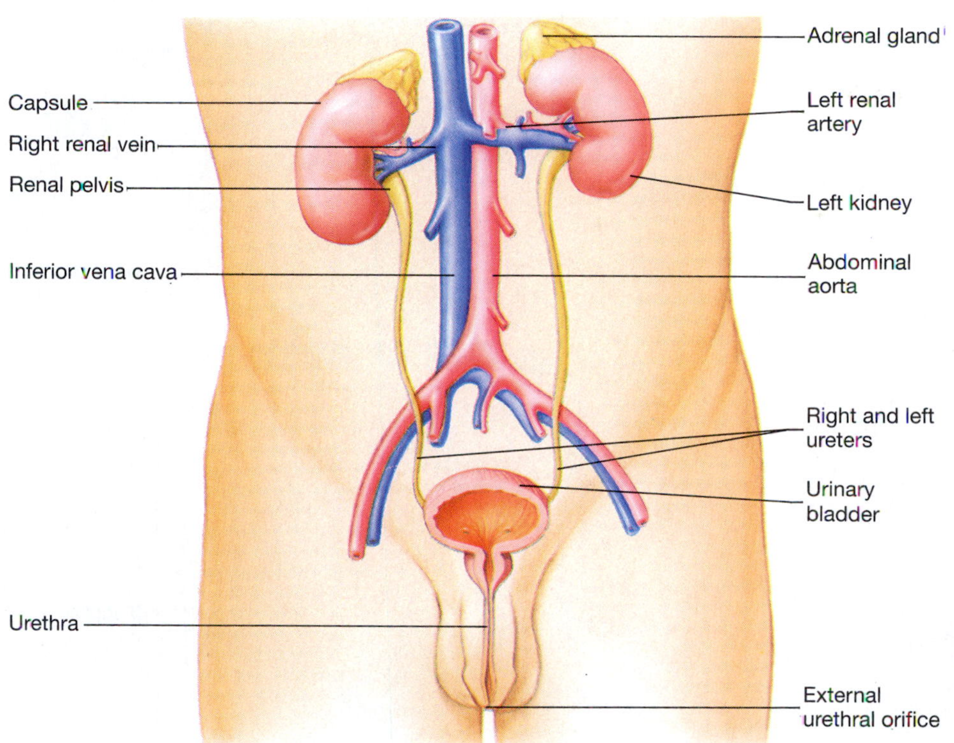

FIGURE 40–5 The human urinary system. Urine is produced in the kidneys, then conveyed by the ureters to the urinary bladder for temporary storage. The urethra conducts urine from the bladder to the outside of the body.

portion is called the **renal medulla** (Fig. 40–6). As urine is produced, it flows into the **renal pelvis,** a funnel-shaped chamber.

From the renal pelvis, urine flows into one of the paired **ureters,** which are ducts that connect the kidney with the **urinary bladder.** The urinary bladder is a remarkable organ capable of holding (with practice) up to 800 ml (about a pint and a half) of urine. Emptying the bladder changes it from the size of a melon to that of a pecan. This feat is made possible by the smooth muscle and special epithelium of the bladder wall, which is capable of great shrinkage and stretching.

During **urination,** urine is released from the bladder and flows through the **urethra,** a duct leading to the outside of the body. In the male, the urethra is lengthy and passes through the penis. Semen, as well as urine, is transported through the male urethra. In the female, the urethra is short and transports only urine. Its opening to the outside is just above the opening of the vagina. The length of the male urethra discourages bacterial invasions of the bladder. This length difference helps explain why bladder infections are more common in females than in males.

The Nephron Is the Functional Unit of the Kidney

Each kidney consists of more than one million functional units, called **nephrons.** A nephron consists of a cuplike **Bowman's capsule** connected to a long, partially coiled **renal tubule** (Fig. 40–7). Positioned within the cup-shaped Bowman's capsule is a cluster of capillaries known as a **glomerulus.** Three main regions of the renal tubule are the **proximal convoluted tubule,** which conducts the filtrate from Bowman's capsule; the **loop of Henle,** an elongated, hairpin-shaped portion; and the **distal convoluted tubule,** which conducts the filtrate to a **collecting duct.** Thus, filtrate passes through the following structures:

Bowman's capsule ⟶ Proximal convoluted tubule ⟶ Loop of Henle ⟶ Distal convoluted tubule ⟶ Collecting duct

Blood is delivered to the kidney by the renal artery. Small branches of the renal artery give rise to afferent arterioles. (Afferent means "to carry toward.") An **afferent arteriole** conducts blood into the capillaries that make up each glomerulus. As blood flows through the glomerulus, some of the plasma is forced into Bowman's capsule.

You may recall that in the usual circulatory pattern capillaries deliver blood into veins. Circulation in the kidneys is an exception in that blood flowing from the glomerular capillaries next passes into an **efferent arteriole,** so-called because it conducts blood *away* from the glomerulus. The efferent arteriole delivers blood to a second capillary network that surrounds the renal tubule. From the first set of capillaries, those of the glomerulus, blood is filtered. In the second set, materials are returned

(Text continued on page 801)

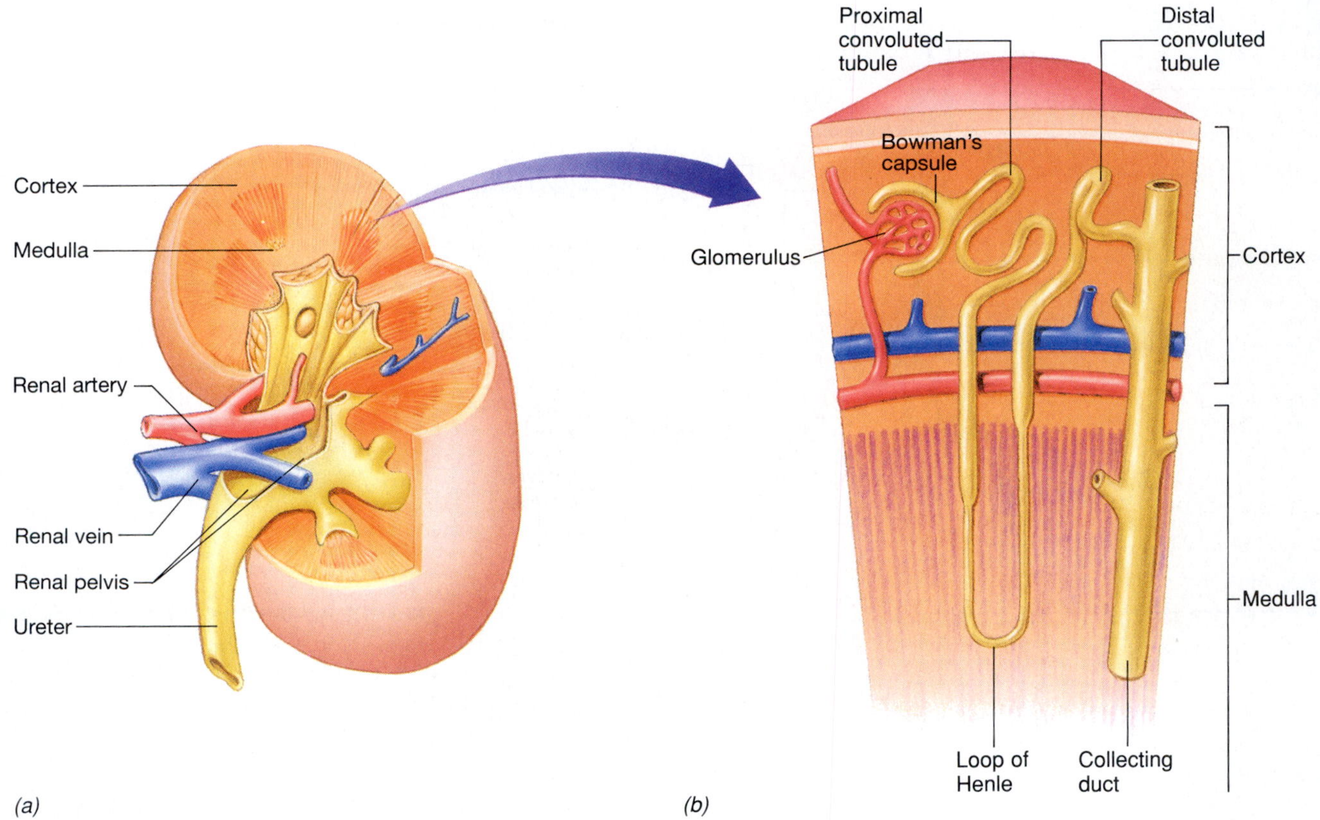

FIGURE 40–6 Structure of the kidney. (***a***) The outer region of the kidney is the cortex; the inner region is the medulla. Urine flows into the renal pelvis and leaves the kidney through the ureter. The renal artery delivers blood to the kidney, while the renal vein drains blood from the kidney. (***b***) Location of a nephron in the kidney.

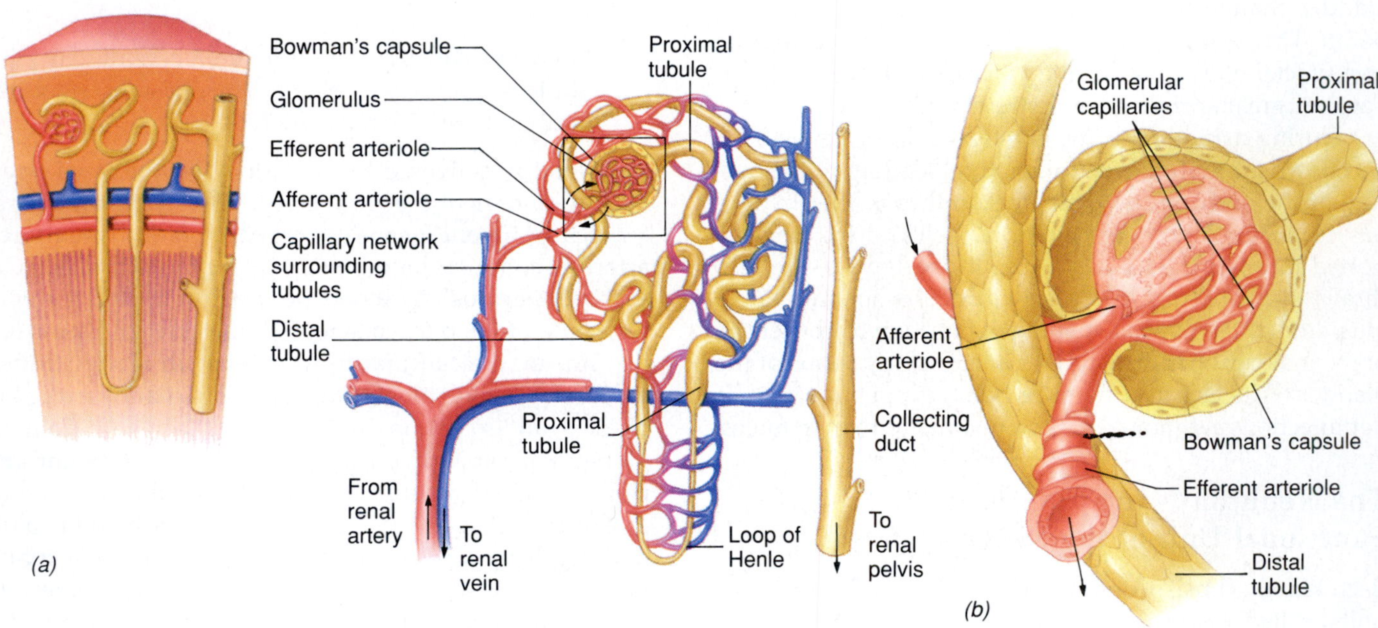

to the blood from the tubule. Blood from the second set of capillaries enters small veins that empty into the renal vein. In summary, blood circulates through the kidney in the following sequence:

Renal artery ⟶ Afferent arteriole ⟶ Capillaries of glomerulus ⟶ Efferent arteriole ⟶ Second set of capillaries ⟶ Veins ⟶ Renal vein

Urine Is Produced by Filtration, Reabsorption, and Secretion

Urine is produced by a combination of three processes: filtration, reabsorption, and tubular secretion.

Filtration Is Not Selective with Regard to Small Molecules

Blood flows through the glomerular capillaries under high pressure, forcing more than 10% of the plasma out of the capillaries and into Bowman's capsule (Fig. 40–8). **Filtration** is somewhat similar to the mechanism whereby tissue fluid is formed as blood flows through other capillary networks in the body. However, much more plasma is filtered in the kidney. Three factors contribute to this process. First, the hydrostatic pressure in the glomerular capillaries is higher compared to other capillaries. The high pressure is due in part to the high resistance to flow presented by the efferent arteriole, which is smaller in diameter than the afferent arteriole. A second factor that contributes to the large amount of filtrate is the large surface area for filtration provided by the highly coiled glomerular capillaries. A third factor is the great permeability of the glomerular capillaries. These vessels are far more porous than typical capillaries.

The filtrate contains plasma and whatever small solutes, such as glucose, salts, and urea, are present in the plasma. However, blood cells, platelets, and proteins do not normally leave the blood.

Almost 25% of the blood pumped by the heart is delivered to the kidneys each minute, and more than 10% of the plasma passing through the glomerulus is filtered out. Every 24 hours about 180 liters (about 45 gallons) of filtrate are produced. Common sense tells us that no one could excrete urine at the rate of 45 gallons per day: within a few minutes, dehydration would become a life-threatening problem.

Reabsorption Is Highly Selective

Fortunately, about 99% of the filtrate is **reabsorbed** into the blood from the renal tubule, leaving only about 1.5 liters to be excreted as urine. Wastes, surplus salts, and water remain in the filtrate. Glucose, amino acids, vitamins, and other useful materials are reabsorbed into the blood. This is accomplished by a combination of active transport, diffusion, and osmosis.

Normally, substances that are useful to the body are completely reabsorbed from the tubules. However, if a large excess of a particular substance is present in the

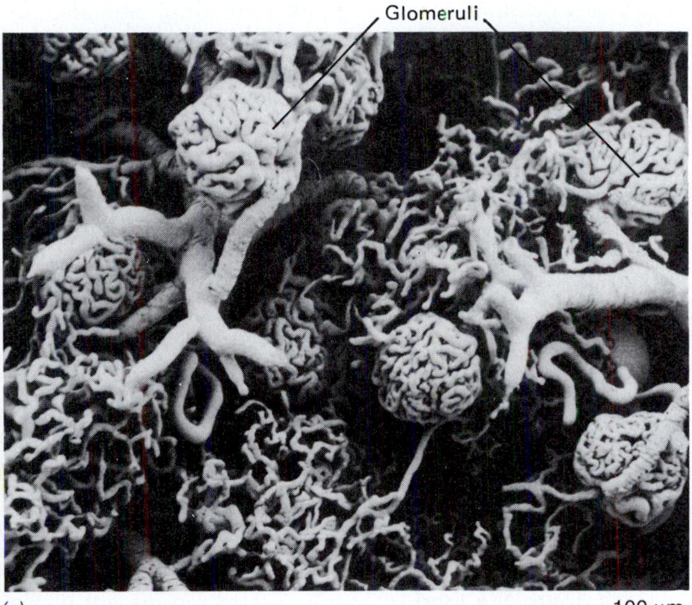

FIGURE 40–7 Structure of a nephron. Each kidney is composed of more than a million microscopic nephrons. (***a***) Diagrammatic view of the basic structure of a nephron. (***b***) Detailed view of Bowman's capsule. (***c***) Low-power scanning electron micrograph of a portion of the kidney cortex, showing glomeruli and associated blood vessels. Urine forms by filtration from the blood in the glomeruli and by adjustment of the filtrate as it passes through the tubules that drain the glomeruli. (CNRI/Science Photo Library/Photo Researchers, Inc.)

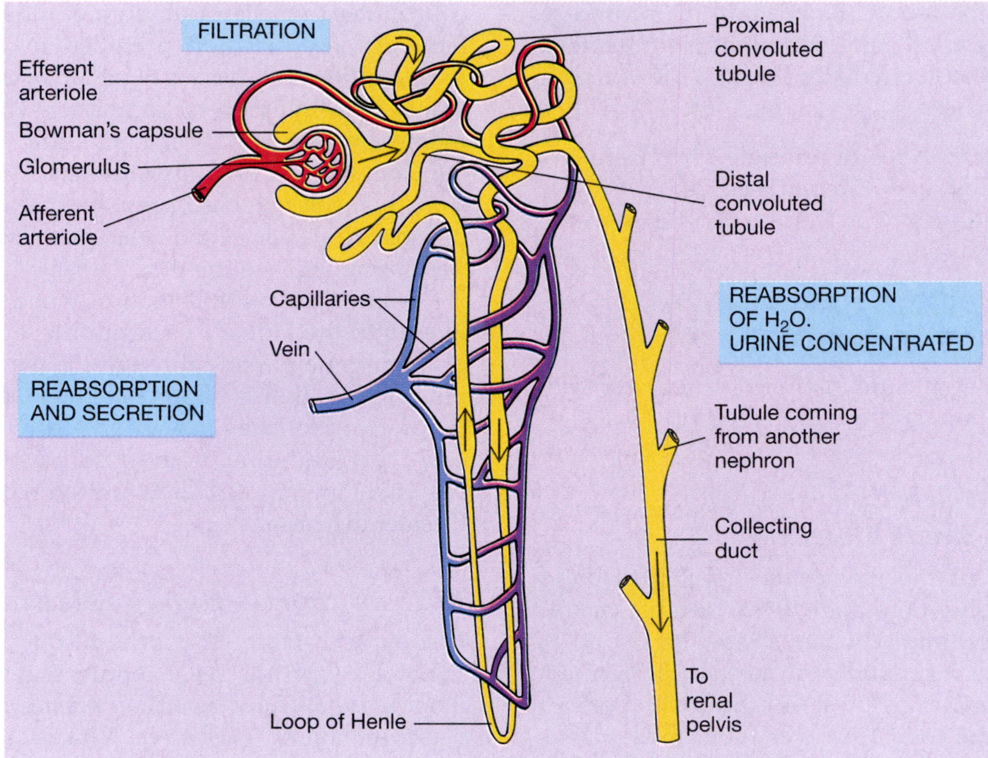

FIGURE 40–8 Function of the nephron. Diagram shows where filtration, reabsorption, and secretion take place.

blood, the tubules may not be able to return it all. The maximum concentration of a specific substance in the blood at which complete reabsorption can take place is termed the **renal threshold** for that substance. When a substance exceeds its renal threshold, the portion not reabsorbed is excreted in the urine.

Some substances, such as urea, have very low thresholds. Even when present in small concentrations, not much of these substances is reabsorbed. Other substances, such as glucose, amino acids, and hormones, have high renal thresholds and are normally completely reabsorbed. Threshold values permit regulation of each component of the internal chemical environment of the body. Every day the tubules reabsorb more than 40 gallons of water, 2.5 pounds of salt, and about 0.5 pound of glucose. Most of this has been reabsorbed many times over.

What happens if a substance in the blood exceeds its threshold value? An important example of this occurs in the condition **diabetes mellitus.** In this disorder, glucose accumulates in the blood instead of being efficiently absorbed and utilized by the cells. The concentration of glucose filtered into the nephron exceeds the renal threshold, so glucose is excreted in the urine. Its presence is evidence of this disorder.

Some Substances Are Actively Secreted from the Blood into the Filtrate

Potassium, hydrogen, and ammonium ions are among the substances actively **secreted** from the blood into the filtrate. Certain drugs, such as penicillin, are also removed from the blood by secretion. Secretion occurs mainly in the region of the distal convoluted tubule.

Urine Is Concentrated as It Passes through the Renal Tubule

The ability of the kidneys to produce a concentrated urine depends on a high salt concentration in the interstitial fluid in the medulla of the kidney. A salt concentration gradient is established, in part, by salt reabsorption from various regions of the renal tubule.

Sodium ions are actively transported out of the proximal tubule, and water follows osmotically. The walls of the descending loop of Henle (the first part of the loop) are relatively permeable to water but relatively impermeable to sodium and urea. There is a high concentration of sodium in the interstitial fluid, so as the filtrate passes down the loop of Henle, water moves out by os-

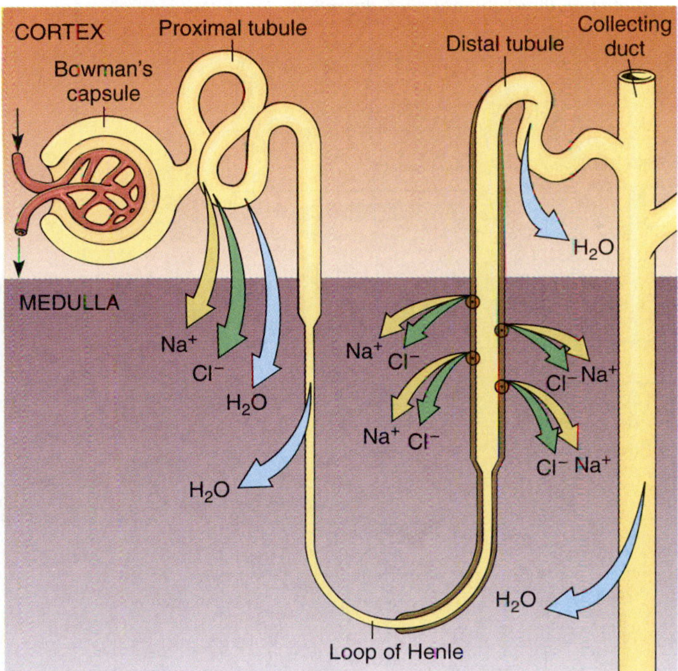

FIGURE 40–9 Urine formation. Water passes out of the descending loop of Henle, leaving a more concentrated filtrate inside. In the ascending loop, salt moves out. Chloride pumps transport chloride out into the interstitial fluid, and sodium follows. The saltier the interstitial fluid becomes, the more water moves out of the descending loop. This leaves a concentrated filtrate inside so more salt passes out. Note that this is a positive feedback system. Urea also moves out into the interstitial fluid through the collecting ducts. Water from the collecting ducts moves out osmotically into this hypertonic tissue fluid.

mosis. This concentrates the filtrate inside the loop of Henle (Fig. 40–9).

At the turn of the loop of Henle, the walls become more permeable to salt and less permeable to water. As the concentrated filtrate moves up the ascending portion of the loop of Henle, salt diffuses out into the interstitial fluid. This contributes to the high salt concentration in the interstitial fluid in the medulla of the kidney surrounding the loop of Henle. Higher in the ascending part of the loop of Henle, sodium is actively transported out of the tubule.

Because water passes out of the descending portion of the loop of Henle, the filtrate at the bottom of the loop has a high salt concentration. However, because salt (but not water) is removed in the ascending portion, by the time the filtrate moves through the distal tubule, it is isotonic (or even hypotonic) to blood.

The filtrate passes from the renal tubule into a larger **collecting duct** that eventually empties into the renal pelvis. The collecting ducts are routed so that they pass through the zone of very salty interstitial fluid. As the filtrate moves down the collecting duct, water passes osmotically into the interstitial fluid where it is collected by blood vessels. So much water may leave the collecting ducts that highly concentrated urine can be produced.

Urine Volume Is Regulated by the Hormone ADH

The amount of urine produced depends on the body's need to retain or rid itself of water. We have seen that salt reabsorption in the loops of Henle establishes a very salty interstitial fluid that draws water osmotically from the collecting ducts. Permeability of the collecting ducts to water is regulated by **antidiuretic hormone (ADH).** When the body needs to conserve water, ADH is released from the posterior pituitary gland (Fig. 40–10). This hormone makes the collecting ducts more permeable to water so that more water is reabsorbed, and a small volume of concentrated urine is produced.

Secretion of ADH is stimulated by special receptors in the hypothalamus. When fluid intake is low, the body begins to dehydrate, causing the blood volume to decrease. As blood volume decreases, the *concentration* of salts dissolved in the blood becomes greater, causing an increase in osmotic pressure. Receptors in the hypothalamus are sensitive to this osmotic change and stimulate the posterior lobe of the pituitary to release ADH. A *thirst center* in the hypothalamus also responds to dehydration, stimulating an increase in fluid intake.

When one drinks a great deal of water, the blood becomes diluted and its osmotic pressure falls. Release of ADH by the pituitary gland decreases, lessening the amount of water reabsorbed from the collecting ducts. A large volume of dilute urine is produced.

Occasionally, the pituitary gland malfunctions and does not produce sufficient ADH. The resulting condition is called **diabetes insipidus** (not to be confused with the more common disorder, diabetes mellitus). This condition can also result from a developed insensitivity of the kidney to ADH. In diabetes insipidus, water is not efficiently reabsorbed from the ducts, so a large volume of urine is produced. A person with severe diabetes insipidus may excrete up to 25 quarts of urine each day, a serious loss of water to the body. The affected individual becomes dehydrated and must drink almost continually to offset fluid loss. Diabetes insipidus can often be controlled by injections of ADH or by use of an ADH nasal spray.

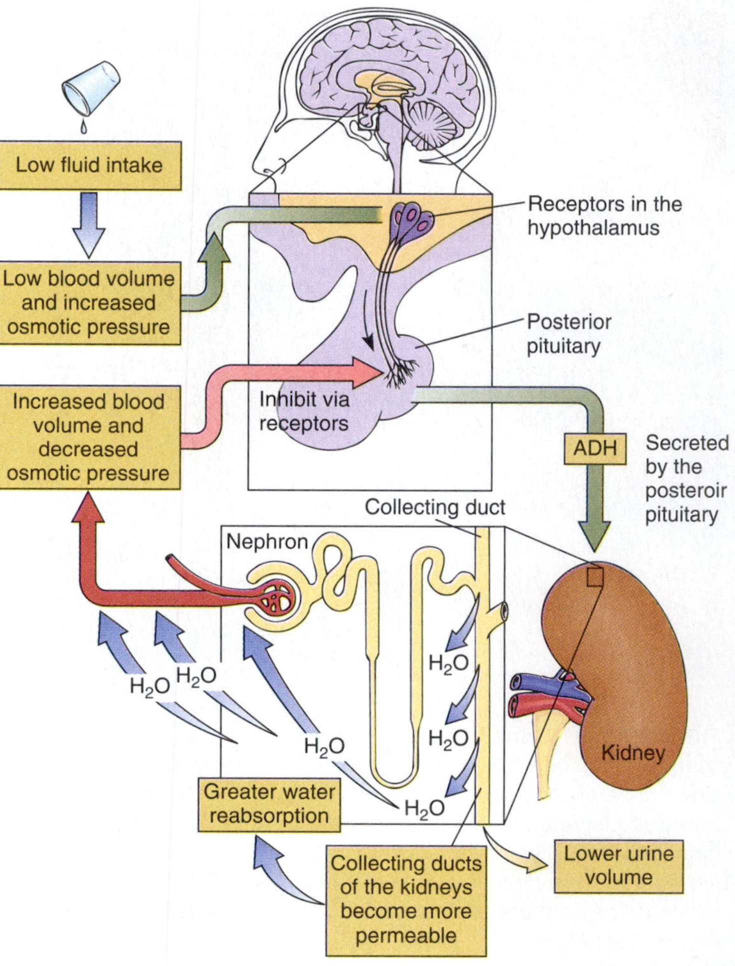

FIGURE 40–10 Regulation of urine volume. When the body is dehydrated, the hormone ADH increases the permeability of the collecting ducts to water. More water is reabsorbed, and only a small volume of concentrated urine is produced.

Sodium Reabsorption Is Regulated by the Hormone Aldosterone

Sodium is the most abundant extracellular ion, accounting for about 90% of all positive ions outside cells. Sodium concentration is carefully regulated by the hormone **aldosterone,** which is secreted by the cortex of the adrenal glands. This hormone stimulates the distal tubules and collecting ducts to increase sodium reabsorption.

Aldosterone secretion can be stimulated by a decrease in blood pressure. When blood pressure falls, certain cells near the glomerulus secrete the enzyme renin, which activates the renin-angiotensin pathway. Renin acts on a plasma protein, converting it to angiotensin, which stimulates aldosterone secretion. Angiotensin also raises blood pressure by constricting blood vessels.

Urine Is Composed of Water, Nitrogenous Wastes, and Salts

By the time the filtrate reaches the renal pelvis, its composition has been precisely adjusted. Useful materials have been returned to the blood by reabsorption. Wastes and excess materials that entered by filtration or secretion have been retained by the tubules. The adjusted filtrate, called **urine,** is composed of about 96% water; 2.5% nitrogenous wastes (mainly urea); 1.5% salts; and traces of other substances, such as bile pigments, which may contribute to the characteristic color and odor.

Healthy urine is sterile and has been used to wash battlefield wounds where clean water is not available. However, urine swiftly decomposes when exposed to bacterial action, forming ammonia and other products. It is the ammonia that produces the diaper rash of infants.

The composition of urine yields many clues to body function and malfunction. **Urinalysis,** the physical, chemical, and microscopic examination of urine, is a very important diagnostic tool and is used to monitor many disorders such as diabetes mellitus. Urinalysis is also extensively used in drug testing because breakdown products of some drugs can be identified in the urine for several weeks.

ANIMALS HAVE OPTIMAL TEMPERATURE RANGES

Snowshoe hares, snowy owls, weasels, and a few other animals inhabit cold arctic regions. Others—like the kangaroo rat— are adapted to desert life. Each animal species has an optimal temperature range and has evolved strategies for regulating body temperature.

Animals produce heat as a byproduct of metabolic activities. Heat can also be gained from the environment or lost to the environment. One way that heat can be transferred to the environment is by **evaporation,** the conversion of a liquid, such as sweat, to a vapor. Evaporation requires energy so that when sweat evaporates from the surface of the body or when a dog pants, body heat is lost to the surroundings.

(a)

(b)

FIGURE 40–11 Animals have behavioral adaptations for thermoregulation. (***a***) This marine iguana (*Amblyrhynochos cristatus*) is an ectotherm that increases its body temperature by sunning itself. (***b***) The body temperature of this sooty tern of Hawaii (an endotherm) is reduced as heat leaves the body through its open mouth. (*a*, Animals Animals, © 1995, Breck P. Kent; *b*, Frans Lanting/Minden Pictures)

Ectotherms Absorb Heat from Their Surroundings

As we discussed in earlier chapters, most animals are **ectotherms,** which means that their body temperature depends to a large extent on heat from the surrounding environment. Metabolic rate tends to change with the weather.

Many ectotherms use structural and behavioral strategies to adjust body temperature. For example, lizards bask in the sun, orienting their body to expose the maximum surface to the sun's rays (Fig. 40–11). Some insects use a combination of behavior and metabolic heat production to regulate body temperature. The "furry" body of the moth conserves body heat. When a moth prepares for flight, it contracts its flight muscles with little movement of its wings. The heat generated enables the moth to sustain the intense metabolic activity needed for flight. Other behavioral strategies for regulating temperature include hibernation and migration.

An advantage to ectothermy is that there is no direct metabolic cost. Most of the heat for thermoregulation comes from the sun. As a result, ectotherms have a much lower daily energy expenditure than do endotherms, and more of the energy in their food can be converted to growth and reproduction. One of the disadvantages of ectothermy is that activity is limited by daily and seasonal temperature conditions.

Endotherms Derive Body Heat from Their Own Metabolism

Birds and mammals are **endotherms**, which means that they have homeostatic mechanisms that maintain a constant body temperature despite changes in the external temperature. Endotherms obtain most of their body heat from their own metabolic processes. They have homeostatic mechanisms for regulating metabolic rate and for regulating heat exchange with the environment.

As described in Chapter 1, mammals have several homeostatic mechanisms for maintaining a constant body temperature (see Fig. 1–5, p. 7 and Fig. 40–12). Mammals can also increase their rate of heat production by contracting muscles or by the action of hormones that increase metabolic rate (e.g., thyroid hormones).

In humans, receptors in the skin, in the hypothalamus, and in certain other areas are sensitive to changes in body temperature. Information about body temperature is sent to the temperature-regulating center in the hypothalamus. Nerves that are part of the somatic sys-

(Text continued on page 807)

FIGURE 40–12 Mechanisms for temperature regulation in the human body. (***a***) How the body restores homeostasis when body temperature increases. (***b***) How the body restores homeostasis when body temperature decreases. (*a*, Merrit Vincent/PhotoEdit; *b*, Jim Brandenburg/Minden Pictures)

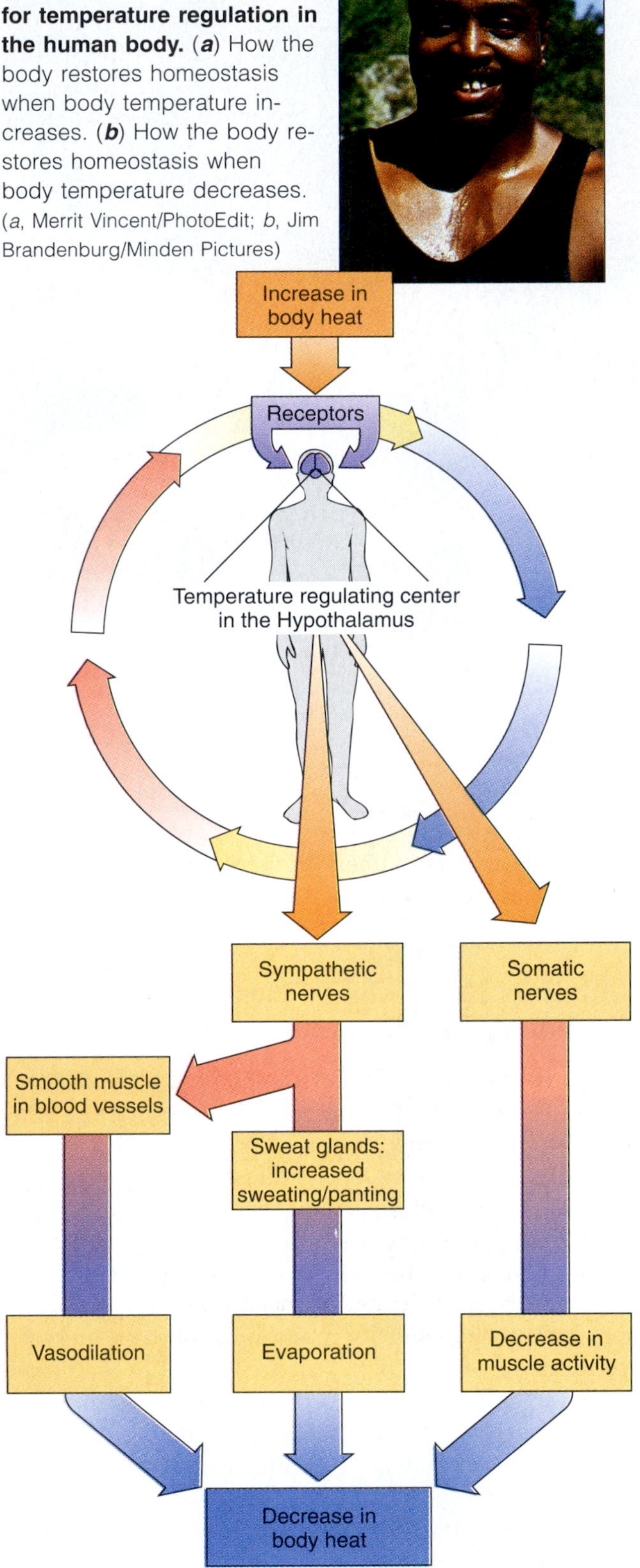

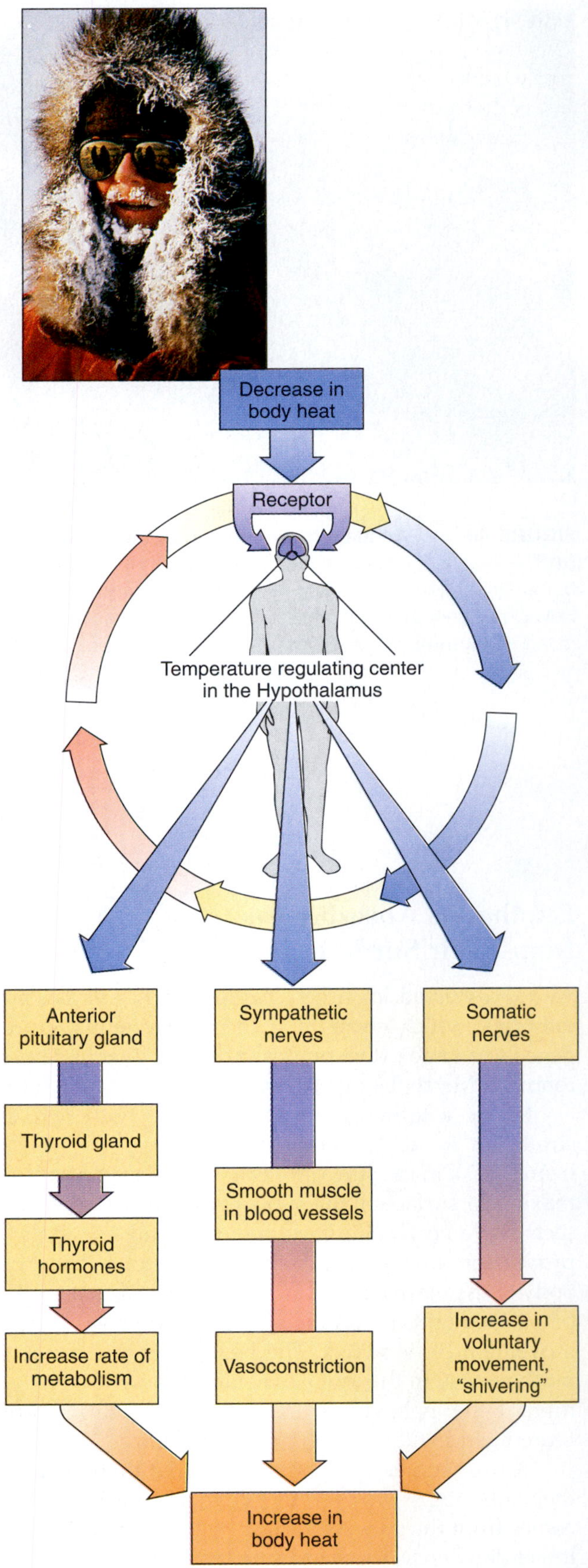

tem signal muscles to shiver or allow us to move muscles voluntarily to increase body temperature. When body temperature increases, sympathetic nerves increase the activity of sweat glands.

The autonomic system also helps regulate body temperature by dilating blood vessels in the skin when we are hot. Increased blood flow to the skin brings body heat to the surface of the body. The skin acts as a heat radiator, allowing heat to leave the body. When we are cold, the autonomic system constricts blood vessels in the skin, reducing heat loss.

Important advantages of endothermy include increased rate of enzyme activity and constant body temperature. Endotherms can carry out their activities even in low winter temperatures. However, endotherms are disadvantaged by the high energy cost of thermoregulation during times when they are inactive. We maintain our body temperature even when we are asleep.

Chapter Summary

I. Excretory systems help regulate the concentration of body fluids by osmoregulation and excretion of metabolic wastes.

II. The principal waste products of animal metabolism are water; carbon dioxide; and nitrogenous wastes, including ammonia, urea, and uric acid.

III. Invertebrate mechanisms of osmoregulation and waste disposal include the nephridial organs of flatworms, metanephridia of annelids, and Malpighian tubules of insects.

IV. The vertebrate kidney maintains homeostasis of body fluids and excretes metabolic wastes.

- A. In mammals, the kidneys produce urine, which passes through the ureters to the urinary bladder for storage. During urination, the urine is released through the urethra to the outside of the body.
- B. Each nephron consists of Bowman's capsule surrounding a cluster of capillaries, called a glomerulus, and a long, coiled renal tubule.
- C. Urine formation is accomplished by the filtration of plasma, reabsorption of needed materials, and secretion of a few substances such as potassium and hydrogen ions into the renal tubule.
    1. Plasma filters out of the glomerular capillaries and into Bowman's capsule. Because filtration is nonselective with regard to small molecules, glucose and other needed materials, as well as metabolic wastes, become part of the filtrate.
    2. About 99% of the filtrate is reabsorbed from the renal tubules into the blood. Reabsorption is a highly selective process that returns usable materials to the blood but leaves wastes and excesses of other substances to be excreted in the urine.
    3. In secretion, certain substances and drugs are actively transported into the renal tubule to become part of the urine.
    4. The ability to produce a concentrated urine depends on a high salt concentration in the interstitial fluid of the kidney medulla. The interstitial fluid in the medulla has a salt concentration that increases progressively, becoming most concentrated around the bottom of the loop of Henle. This concentration gradient is maintained, in part, by salt reabsorption from various parts of the renal tubule.
    5. Water is drawn by osmosis from the filtrate as it passes through the collecting ducts. This permits the concentration of urine in the collecting ducts.
    6. Urine volume is regulated by the hormone ADH, which is released by the posterior lobe of the pituitary gland in response to an increase in osmotic concentration of the blood (caused by dehydration). ADH increases the permeability of the collecting ducts. As a result, more water is reabsorbed and only a small volume of urine is produced.
    7. Aldosterone regulates sodium reabsorption.
    8. Urine consists of water, nitrogenous wastes, salts, and other substances not needed by the body.

V. Body temperature depends on a balance between heat produced by metabolic activities and the exchange of heat with the environment.

- A. In ectotherms, body temperature depends to a large extent on the temperature of the environment. Many ectotherms have structural and behavioral strategies for regulating body temperature.
- B. In endotherms, body temperature depends on the regulation of heat generated by metabolic activities and on the regulation of heat exchange with the environment.

Selected Key Terms

ADH, p. 803
aldosterone, p. 804
Bowman's capsule, p. 799
collecting duct, p. 799
distal convoluted tubule, p. 799
ectotherm, p. 805
endotherm, p. 805
excretion, p. 794
excretory system, p. 794
filtration, p. 801
glomerulus, p. 799
kidney, p. 792
loop of Henle, p. 799
Malpighian tubules, p. 795
nephridial organs, p. 795
nephron, p. 799
osmoregulation, p. 794
proximal convoluted tubule, p. 799
reabsorption, p. 801
secretion, p. 802
urea, p. 794
ureter, p. 799
urethra, p. 799
uric acid, p. 794
urinary bladder, p. 799
urinary system, p. 797
urine, p. 794

Post-Test

1. The process of removing metabolic wastes from the body is called ________________.
2. ________________ is the process by which an animal regulates its fluid content.
3. The principal nitrogenous waste product of insects and birds is ________________ ________________.
4. The principal nitrogenous waste product of amphibians and mammals is ________________.
5. Flatworms have excretory structures called ________________.
6. Earthworms have excretory structures known as ________________.
7. The excretory structures of insects are ________________ ________________.
8. The vertebrate kidney consists of functional units called ________________.
9. The glomerulus consists of a cluster of ________________, which is surrounded by ________________ ________________.
10. Blood is delivered to the glomerulus by the ________________ arteriole and leaves the glomerulus in the ________________ arteriole.
11. Fluid that leaves the glomerular capillaries and enters Bowman's capsule is called ________________.
12. When a substance exceeds its renal threshold, the portion not reabsorbed is ________________.
13. Antidiuretic hormone (ADH) increases the permeability of the ________________ ________________ so that more water is ________________.
14. In ectothermy most of the heat for temperature regulation comes from the ________________.
15. The outer portion of the human kidney is the (a) cortex (b) medulla (c) ureter (d) glomerulus (e) renal pelvis.
16. Urine is delivered to the urinary bladder by the (a) cortex (b) urethra (c) ureter (d) glomerulus (e) renal pelvis.
17. The part of the kidney that receives urine from collecting ducts is the (a) cortex (b) medulla (c) ureter (d) glomerulus (e) renal pelvis.
18. The structure where filtration takes place is the (a) loop of Henle (b) urethra (c) ureter (d) Bowman's capsule (e) proximal tubule.
19. ADH acts on the (a) cortex (b) Bowman's capsule (c) ureter (d) glomerulus (e) collecting ducts.
20. Sweating cools the body by the process of (a) ectothermy (b) endothermy (c) evaporation (d) radiation (e) none of the preceding answers is correct.

Review Questions

1. Compare osmoregulation in flatworms and insects.
2. What type of osmoregulatory problem is faced by marine fish? By freshwater fish? How are these problems met in each case?
3. What are the principal types of nitrogenous wastes?
4. Name the structure in the mammalian body that is associated with each of the following: (a) urea formation, (b) urine formation, (c) temporary storage of urine, and (d) conduction of urine out of the body.
5. Draw a diagram of a nephron and label its parts.
6. Which part of the nephron is associated with the following: (a) filtration, (b) reabsorption, (c) secretion?
7. Contrast reabsorption and secretion. Contrast filtration and secretion.
8. List the sequence of blood vessels through which a drop of blood passes as it is conducted to and from a nephron.
9. How is urine volume regulated? Explain. Why must victims of untreated diabetes insipidus drink great quantities of water?
10. Describe two treatment strategies for kidney failure. What are the advantages of each?
11. What are some benefits of being an endotherm? What are some disadvantages?
12. Label the diagram.

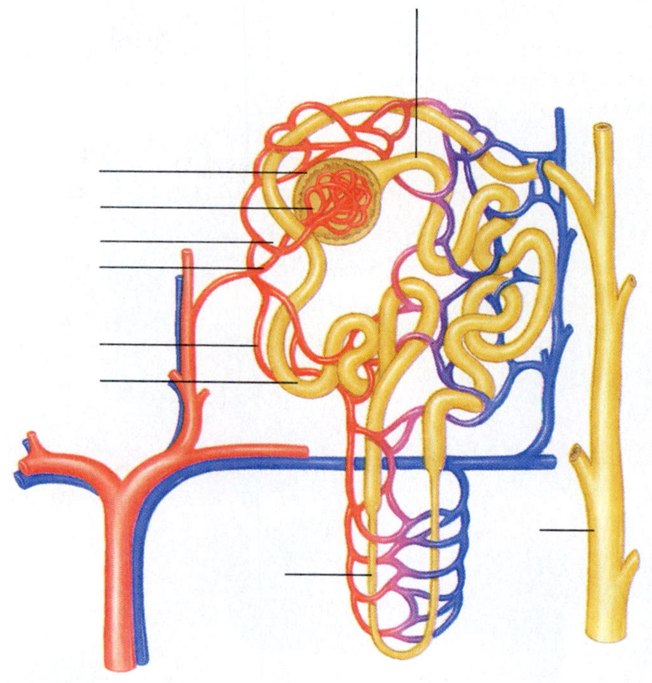

Thinking Critically

1. Glomerulonephritis is thought to be an autoimmune disease. Design an experiment to test that hypothesis.
2. The number of protonephridia in a planarian is related to the salinity of its environment. Planaria inhabiting slightly salty water develop fewer protonephridia, but the number quickly increases when the concentration of salt in the environment is lowered. Explain why.
3. What types of osmoregulatory challenges do humans experience? Explain. What mechanisms do we have to solve these challenges?
4. Why is glucose normally not present in urine? Why is it present in diabetes mellitus? Why do you suppose diabetics experience an increased output of urine?
5. Mammals have a layer of fat tissue beneath the skin. Some mammals, such as the whale, have a large amount of subcutaneous fat. In what way do you think subcutaneous fat is adaptive?
6. The kangaroo rat's diet consists of dry seeds and it drinks no water. Speculate about the adaptations this animal would need to have in order to survive.

Recommended Readings

Dantzler, W.H., "Renal Adaptations of Desert Vertebrates," *Bioscience,* Vol. 32, No. 2, 1982, pp. 108–112. Adaptations in renal physiology are integrated with other physiological and behavioral mechanisms for desert survival.

McClanahan, L.L., R. Ruibal, and V.H. Shoemaker, "Frogs and Toads in Deserts," *Scientific American,* Vol. 270, No. 3, March 1994, pp. 82–88. A discussion of some unusual adaptations for life in the desert.

Solomon, E.P., R. Schmidt, and P. Adragna, *Human Anatomy and Physiology,* Philadelphia, Saunders College Publishing, 1990. Chapters 27 and 28 focus on the human urinary system and fluid balance.

Defensive lineman Lyle Alzado began taking anabolic steroids when he was a college freshman. *By age 33 he was spending $20,000 to $30,000 per year on steroids. In 1992, Alzado died from a rare form of brain cancer, which he believed was caused by his use of steroids.* *(Anthony Neste)*

Pumping up with Anabolic Steroids

Jim desperately wanted to be a star athlete so he could be popular and please his father. In ninth grade, when Jim asked about trying out for the football team, the high school coach explained that his thin, 110-pound body was much too small for high school football competition. The following year, Jim had gained 50 pounds. When he arrived at school, flexing large, well-defined muscles, he soon found himself surrounded by new friends. Noticing Jim's new bulk, the coach approached him to try out for the team.

What had caused such a remarkable change from one year to the next? Jim had discovered **anabolic steroids,** which are synthetic androgens (male reproductive hormones). Such synthetic androgens are used to increase muscle mass, physical strength, endurance, and aggressiveness.

But Jim's new-found popularity came at great cost. His parents had begun to notice his wide mood swings and his unpredictable rage toward his younger brother and sister. Inside Jim's body, the synthetic hormones were stunting his growth by prematurely closing the growth plates in his bones. His testes were shrinking and he was at risk of becoming sterile; his liver was not functioning normally; and his LDL level was increasing, raising his risk for cardiovascular disease.

Jim is one of an estimated one million individuals in the U.S., one-half of them adolescents, who abuse anabolic steroids. The typical anabolic steroid user is a male (95%) athlete (65%), most often a football player, weight lifter, or wrestler. About 10% of male high school students have used anabolic steroids and about one-third of them are not even on a high school team. They use the hormone only to change their physical appearance ("bulk up"). Many anabolic steroid users have difficulty realistically perceiving their body image. They remain unhappy, even after dramatic increases in their muscle mass.

Anabolic steroids were developed in the 1930s to prevent muscle atrophy in patients with diseases that prevented them from moving about. In the 1950s, these hormones became popular with professional athletes, who used them to pump up their muscles and increase endurance. In truth, athletic performance is probably enhanced, at least in part, by drug-induced euphoria and increased enthusiasm for training. When their dangerous side effects became known in the 1960s, steroid use became controversial. In 1973, the Olympic Committee banned the use of anabolic steroids. Their use is now prohibited worldwide by amateur and professional sports organizations.

According to the U.S. Drug Enforcement Administration, a multimillion dollar black market exists for anabolic steroids. They are both injected and taken in pill form. As with other hormones, the amount of steroid hormones circulating in the body is precisely regulated. Use of anabolic steroids interferes with normal physiological processes, and has been linked with a variety of very serious physical and psychological side effects. "Steroid rage" refers to the mood swings, increased aggressiveness, and irrational behavior exhibited by many users. A group of researchers reported in the *Journal of the American Medical Association* (June 2, 1993) that even during short-term, relatively low dose trials, anabolic steroids have a significant effect on mood and behavior. At higher doses, subjects experience radical mood swings, including violent feelings, anger, and hostility. Thought processes are disturbed and users experience forgetfulness and confusion, and often find themselves easily distracted. Anabolic steroids remain in the body for a long time. Their metabolites (breakdown products) can be detected in the urine for up to six months.

In this chapter, we will discuss the hormones normally produced by the body and describe their actions. We will also examine how too little or too much of various hormones interferes with normal functioning.

Endocrine Regulation

41

Learning Objectives

After you have studied this chapter you should be able to:

1. Define the terms *hormone* and *endocrine gland* and distinguish between endocrine and exocrine glands.
2. Compare the mechanisms of action of steroid and protein-type hormones. (Include the role of second messengers, such as cyclic AMP.)
3. Summarize the role of hormones in invertebrates.
4. Identify the principal vertebrate endocrine glands, locate them in the body, and list the hormones secreted by each (consult Fig. 41–4).
5. Summarize the regulation of endocrine glands by negative feedback mechanisms and relate the concept of negative feedback to the specific hormones discussed.
6. Justify the description of the hypothalamus as the link between nervous and endocrine systems and describe the mechanisms by which the hypothalamus exerts its control.
7. Compare the functions of the posterior and anterior lobes of the pituitary; identify their hormones and describe the actions of these hormones.
8. Describe the actions of growth hormone on growth and metabolism and contrast the consequences of hyposecretion and hypersecretion.
9. Define the actions of the thyroid hormones, their regulation, and the thyroid disorders discussed in this chapter.
10. Contrast the actions of insulin and glucagon and describe the disorders associated with the malfunction of the islets of the pancreas.
11. Describe the actions of the mineralocorticoid and glucocorticoid hormones and describe the effects of the malfunction of the adrenal cortex.
12. Summarize the role of the adrenal glands in helping the body to adapt to stress.

Key Concepts

- ☐ Endocrine glands produce and secrete hormones, which are chemical messengers that help regulate various activities of the body.
- ☐ Endocrine activity is regulated by negative feedback mechanisms.
- ☐ Hormones are transported to the cells of the body by the blood; only target cells have appropriate receptors that combine with the hormone and produce a response.
- ☐ In mammals, the nervous and endocrine systems are linked by the hypothalamus, which regulates the activity of the pituitary gland.

Key Concepts, continued

- ☐ Human hormones help regulate growth, metabolism, and reproduction. Pituitary hormones regulate growth and reproduction, and help regulate other endocrine glands. Thyroid hormones stimulate the rate of metabolism. Insulin and glucagon regulate glucose concentration in the blood. The adrenal hormones bring about metabolic changes that help the body cope with stress.

HORMONES ARE CHEMICAL MESSENGERS THAT REGULATE MANY BODY PROCESSES

Several types of chemical messengers are released by cells or glands and stimulate some type of response in other cells. For example, neurotransmitters are chemical messengers released by neurons that relay nerve signals. **Hormones,** another type of chemical messenger, regulate the activities of many tissues and organs of the body. (The word hormone is derived from a Greek word meaning "to excite.") Hormones are secreted by certain tissues and glands of the body that make up the **endocrine system.** The study of endocrine activity, called **endocrinology,** is an exciting field of medical research.

Endocrine glands secrete hormones into the surrounding tissue fluid. Hormones diffuse into capillaries and are transported by the blood. They stimulate responses only in their **target tissues.** The target tissue may be part of another endocrine gland or it may be part of an entirely different type of organ such as a bone or the kidney. Often the target tissue is located far from the endocrine gland that secreted the hormone. Endocrine glands differ from **exocrine glands** (such as sweat glands and gastric glands), which release their secretions into ducts. Chemically, hormones either are lipids (usually steroids) or they belong to the protein family (proteins, peptides, and derivatives of amino acids).

Traditionally, endocrinology focused on the activity of about ten discrete endocrine glands. We now know that specialized cells in the digestive tract, heart, kidneys, and many other organs also release hormones. The scope of endocrinology has been broadened to include the study of chemical messengers produced by these organs and also by cells that are widely distributed in the body, rather than by single, discrete organs.[1]

[1] **Pheromones**, still another type of chemical messenger, are produced by an animal for communication with other animals of the same species. Since pheromones are generally produced by exocrine glands and do not regulate metabolic activities within the animal that produces them, most biologists do not classify them as hormones. Their role in regulating behavior is discussed in Chapter 44.

HORMONE SECRETION IS REGULATED BY NEGATIVE-FEEDBACK MECHANISMS

In vertebrates, most endocrine glands secrete at least small amounts of their hormones continuously. Although present in minute amounts, more than 50 different hormones may be circulating in the blood at all times.

Hormone secretion is regulated internally by **negative feedback** mechanisms. Information about the amount of hormone or of some other substance in the blood or tissue fluid is fed back to the gland, which then responds to restore homeostasis. The parathyroid glands, located in the neck of tetrapod vertebrates, provide a good example of how negative feedback works.

The parathyroid glands regulate the calcium concentration of the blood. When the calcium concentration is not within homeostatic limits, nerves and muscles cannot function properly. For example, when insufficient calcium ions are present, neurons can fire spontaneously, causing muscle spasms. Even a slight decrease in calcium concentration signals the parathyroid glands to release more parathyroid hormone (Fig. 41–1). This hormone increases the concentration of calcium in the blood by stimulating the release of calcium from the bones and by increasing calcium reabsorption by the kidney tubules.

When the calcium concentration rises above normal limits, the parathyroid glands slow their output of hormone. Both responses are *negative* feedback mechanisms, because in both cases the effects are opposite (negative) to the stimulus (that is, more calcium leads to less hormone). Negative feedback is the basis of hormone regulation.

HORMONES COMBINE WITH SPECIFIC RECEPTOR PROTEINS IN TARGET CELLS

A hormone may pass through many tissues seemingly "unnoticed" until it reaches its target tissue. How does the target tissue "recognize" its hormone? Specific receptor proteins in or on the cells of the target tissues bind the hormone. This is a highly specific process. The receptor site is similar to a lock, and the hormones are sim-

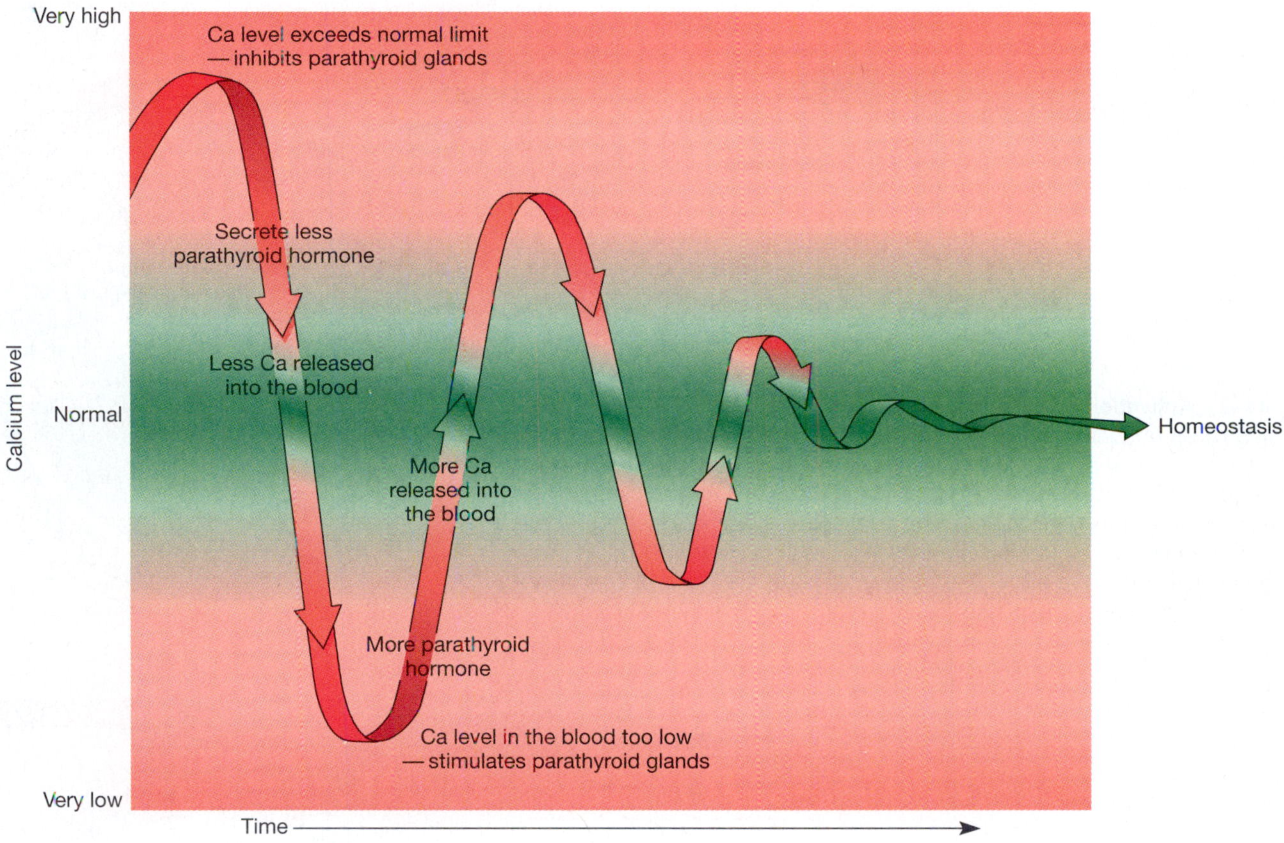

FIGURE 41–1 Regulation of hormone secretion by negative feedback. When the calcium concentration exceeds normal limits, the parathyroid glands are inhibited and slow their release of hormone. When the calcium concentration in the blood falls below normal, the parathyroid glands are stimulated to release more parathyroid hormone. This hormone acts to increase the calcium level in the blood, thus restoring homeostasis. This diagram has been simplified. Calcitonin, a hormone secreted by the thyroid gland, works antagonistically to parathyroid hormone and is important in lowering blood calcium concentration.

ilar to different keys. Only the hormone that fits the lock—the specific receptor—can influence the metabolic machinery of the cell.

Several types of hormones may be involved in regulating the metabolic activities of a particular type of cell. In fact, many hormones produce a synergistic effect in which the presence of one hormone may enhance the effects of another.

Some Hormones Enter the Cell and Activate Genes

Steroid hormones and thyroid hormones (small polypeptides) are relatively small, lipid-soluble molecules that easily pass through the plasma membrane of the target cell. Specific protein receptors in the cytoplasm or nucleus combine with the hormone to form a hormone-receptor complex (Fig. 41–2). This complex then combines with a protein associated with the DNA in the nucleus. This combination activates certain genes and leads to a synthesis of messenger RNAs coding for specific proteins. These proteins then produce the changes in structure or metabolic activity that are the actual effect of the hormone.

Some Hormones Work through Second Messengers

Many hormones do not enter the cell, but combine with receptors on the plasma membrane of the target cell. The hormonal message is then relayed to the appropriate site within the cell by a **second messenger** (Fig. 41–3). In the 1960s, Earl Sutherland identified **cyclic AMP (cAMP)** as

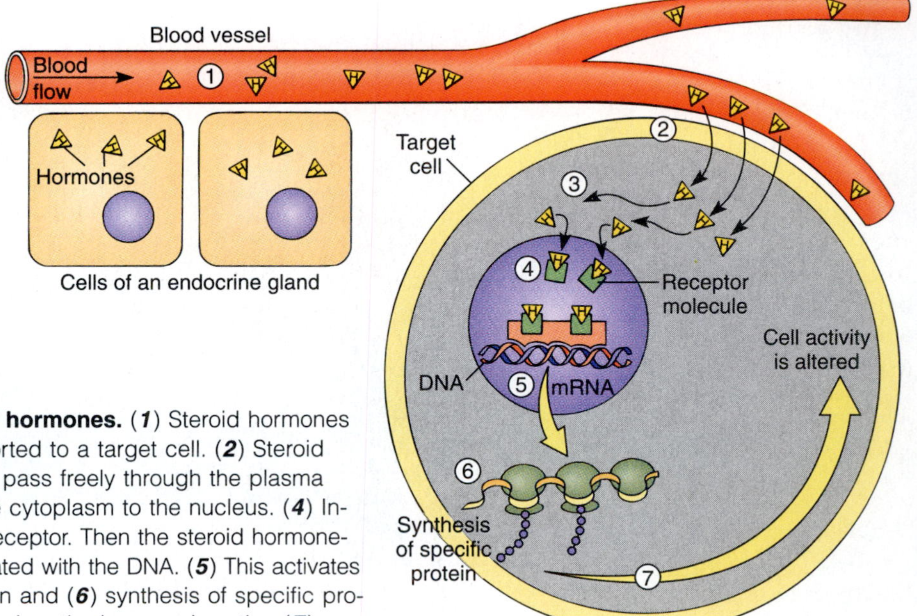

FIGURE 41–2 Activation of genes by steroid hormones. (*1*) Steroid hormones are secreted by an endocrine gland and transported to a target cell. (*2*) Steroid hormones are small, lipid-soluble molecules that pass freely through the plasma membrane. (*3*) The hormone passes through the cytoplasm to the nucleus. (*4*) Inside the nucleus, the hormone combines with a receptor. Then the steroid hormone-receptor complex combines with a protein associated with the DNA. (*5*) This activates specific genes, leading to the mRNA transcription and (*6*) synthesis of specific proteins. The proteins cause the response recognized as the hormone's action (*7*).

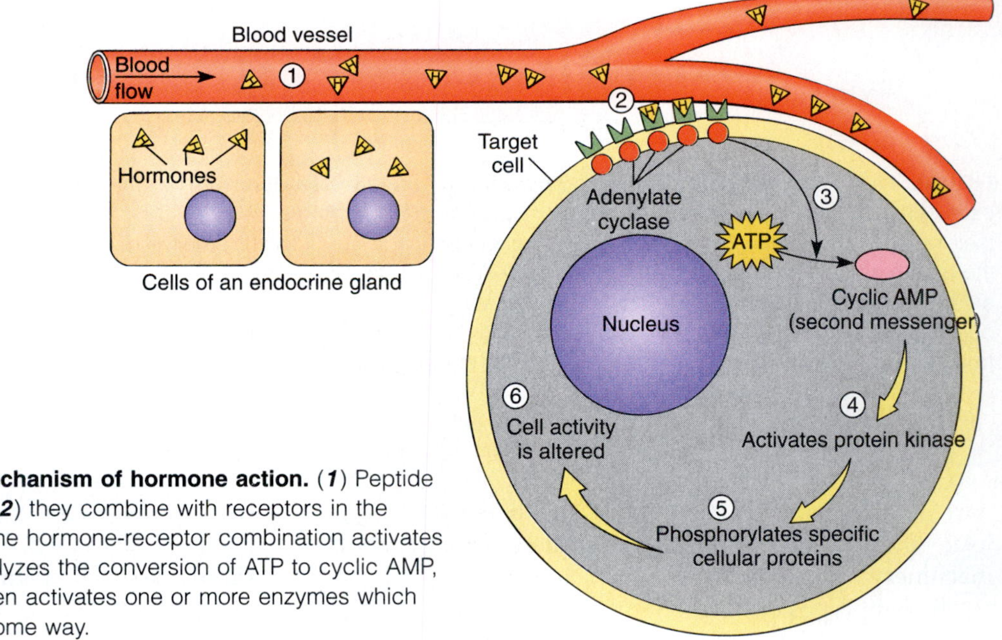

FIGURE 41–3 Second messenger mechanism of hormone action. (*1*) Peptide hormones are transported in the blood. (*2*) they combine with receptors in the plasma membrane of a target cell. (*3*) The hormone-receptor combination activates an enzyme (adenylate cyclase) that catalyzes the conversion of ATP to cyclic AMP, a second messenger. (*4*) Cyclic AMP then activates one or more enzymes which (*5, 6*) change the activity of the cell in some way.

a second messenger, and it is the one that has been most extensively studied.

When the hormone combines with its receptor on the plasma membrane, a membrane-bound enzyme (adenylate cyclase) is activated. This enzyme speeds the conversion of ATP to cyclic AMP. Cyclic AMP activates enzymes that either activate or inhibit the activity of certain proteins in the cell. These proteins affect cell processes. For example, the permeability of the plasma membrane may be affected, or protein synthesis may be stimulated. The particular action initiated by cyclic AMP depends on the specific types of enzyme systems present in the cell. This is why the same hormone can promote different responses in different cell types.

Prostaglandins Are Local Chemical Mediators

Prostaglandins are modified fatty acids released by many different organs, including the prostate gland, lungs, liver, and digestive tract. Although present in very small quantities, prostaglandins affect a wide range of body processes. They are often referred to as **local hormones** because they act on cells in their immediate vicinity.

Prostaglandins, which mimic many of the actions of cyclic AMP, interact with other hormones to regulate various metabolic activities. About 16 prostaglandins are known, and they have different actions on different tissues. Some reduce blood pressure, whereas others raise it. Various prostaglandins dilate the bronchial passageways, inhibit gastric secretion, stimulate contraction of the uterus, affect nerve function, cause inflammation, and affect blood clotting. Those synthesized in the temperature-regulating center of the hypothalamus cause fever. In fact, the ability of aspirin and acetaminophen to reduce fever and decrease pain depends on inhibiting prostaglandin synthesis.

Because prostaglandins are involved in the regulation of so many metabolic processes, they have great potential for a variety of clinical uses. At present, prostaglandins are used to induce labor in pregnant women, to induce abortion, and to promote the healing of ulcers in the stomach and duodenum. Their use as a birth control drug is being investigated. Prostaglandins may someday be used to treat a wide variety of illnesses, including asthma, arthritis, kidney disease, certain cardiovascular disorders, and some forms of cancer.

Invertebrate Hormones Regulate Growth, Development, Metabolism, Reproduction, Molting, and Pigmentation

Among invertebrates, hormones are secreted mainly by neurons rather than by endocrine glands. These **neurohormones** regulate regeneration in hydras, flatworms, and annelids; molting and metamorphosis in insects; color changes in crustaceans; and growth, gamete production, reproductive behavior, and metabolism in other groups.

Concept Connections

Invertebrate Hormones — Insect Metamorphosis

In Chapter 24, we briefly described metamorphosis in insects. Metamorphosis is a complex process regulated by hormones. Although the details vary in different species, development is generally influenced by environmental factors such as temperature change. Such changes affect neurosecretory cells in the brain. Once activated, these cells produce a hormone referred to as **brain hormone (BH)** that is transported down axons and stored in the paired structures called **corpora cardiaca** (see figure). When released from the corpora cardiaca, BH stimulates the endocrine glands in the prothorax (prothoracic glands) to produce **molting hormone** (also called ecdysone). Molting hormone stimulates growth and molting.

In the immature insect, paired endocrine glands called **corpora allata** secrete **juvenile hormone.** This hormone suppresses metamorphosis at each larval molt so that the insect increases in size but remains in its immature state (see figure). After the molt, the insect is still in a larval stage. When the concentration of juvenile hormone decreases, metamorphosis occurs, and the insect is transformed into a pupa. In the absence of juvenile hormone, the pupa molts and becomes an adult. The secretory activity of the corpora allata is regulated by the nervous system.

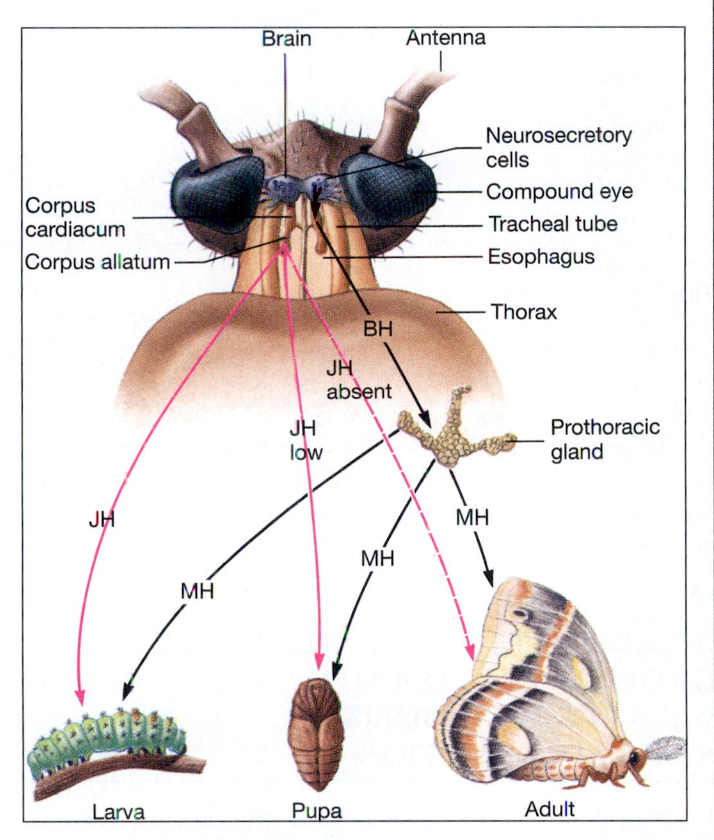

TABLE 41–1
*Some Endocrine Glands and Their Hormones**

Endocrine gland and hormone	*Target tissue*	*Principal actions*
Hypothalamus		
Releasing and release-inhibiting hormones	Anterior lobe of pituitary	Stimulates or inhibits secretion of specific hormones
Hypothalamus (production)		
Posterior lobe of pituitary (storage and release)		
Oxytocin	Uterus	Stimulates contraction
	Mammary glands	Stimulates ejection of milk into ducts
Antidiuretic hormone (ADH)	Kidneys (collecting ducts)	Stimulates reabsorption of water; conserves water
Anterior lobe of pituitary		
Growth hormone (GH)	General	Stimulates growth by promoting protein synthesis
Prolactin	Mammary glands	Stimulates milk production
Thyroid-stimulating hormone (TSH)	Thyroid gland	Stimulates secretion of thyroid hormones; stimulates increase in size of thyroid gland
Adrenocorticotropic hormone (ACTH)	Adrenal cortex	Stimulates secretion of adrenal cortical hormones
Gonadotropic hormones* (follicle-stimulating hormone, FSH; luteinizing hormone, LH)	Gonads	Stimulate gonad function and growth
Thyroid gland		
Thyroxine (T_4) and triiodothyronine (T_3)	General	Stimulate metabolic rate; essential to normal growth and development
Calcitonin	Bone	Lowers blood-calcium level by inhibiting bone breakdown
Parathyroid glands		
Parathyroid hormone	Bone, kidneys, digestive tract	Increases blood-calcium level by stimulating bone breakdown; stimulates calcium reabsorption by kidneys; activates vitamin D

* The gonadotropic hormones (FSH and LH) and the ovaries and testes and their hormones are discussed in Chapter 42. The digestive hormones are described in Chapter 39.

Insects have both endocrine glands and neurons that secrete hormones. Their hormones interact with one another to regulate growth and development, metabolism, and reproduction.

VERTEBRATE HORMONES REGULATE GROWTH, DEVELOPMENT, FLUID BALANCE, METABOLISM, AND REPRODUCTION

Vertebrate hormones regulate such diverse activities as metabolic rate, reproduction, and blood homeostasis, and they help the body adjust to stress. The principal human endocrine glands are illustrated in Figure 41–4. Most vertebrates have similar endocrine glands. Table 41–1 gives the sources, target tissues, and physiological actions of some of the major vertebrate hormones.

Endocrine Disorders May Involve Too Little or Too Much Hormone

When a disorder or disease process affects an endocrine gland, the rate of secretion may become abnormal. If **hyposecretion** (abnormal reduced output) occurs, target cells are deprived of needed stimulation. If **hypersecretion** (abnormal increase in output) occurs, the target cells may be

Endocrine gland and hormone	*Target tissue*	*Principal actions*
Islets of Langerhans of pancreas		
Insulin	General	Lowers glucose concentration in the blood by facilitating glucose uptake and utilization by cells; stimulates glycogen formation; stimulates fat storage and protein synthesis
Glucagon	Liver, adipose tissue	Raises glucose concentration in the blood; mobilizes fat
Adrenal medulla		
Epinephrine and norepinephrine	Muscle, cardiac muscle, blood vessels, liver, adipose tissue	Help body cope with stress; increase heart rate, blood pressure, metabolic rate; reroute blood; mobilize fat; raise blood sugar level
Adrenal cortex		
Mineralocorticoids (aldosterone)	Kidney tubules	Maintain sodium and phosphate balance
Glucocorticoids (cortisol)	General	Help body adapt to long-term stress; raise blood-glucose level; mobilize fat
Pineal gland		
Melatonin	Gonads, pigment cells, other tissues (?)	Influences reproductive processes in hamsters and other animals; pigmentation in some vertebrates; may control biorhythms in some animals; may help control onset of puberty in humans
Ovary†		
Estrogens (Estradiol)	General; uterus	Develop and maintain sex characteristics in female; stimulate growth of uterine lining
Progesterone	Uterus; breast	Stimulates development of uterine lining
Testis‡		
Testosterone	General, reproductive structures	Develops and maintains sex characteristics of males; promotes spermatogenesis; responsible for adolescent growth spurt

† For more detailed description see Table 42–2.
‡ For more detailed description see Table 42–1.

overstimulated. In some endocrine disorders, an appropriate amount of hormone is secreted, but the receptors on target cells do not function properly. As a result, the target cells may not be able to take up the hormone. Any of these abnormalities leads to predictable metabolic malfunctions and clinical symptoms (Table 41–2).

Nervous Regulation and Endocrine Regulation Are Integrated by the Hypothalamus

Most hormone activity is controlled directly or indirectly by the hypothalamus, which links the nervous and endocrine systems. In response to input from other areas of the brain and from hormones in the blood, neurons of the hypothalamus secrete neurohormones that regulate the release of hormones by the pituitary gland.

Because its secretions control the activities of several other endocrine glands, the **pituitary gland** is known as the master gland of the body. Truly a biological marvel, the pituitary gland is only the size of a large pea and weighs only about 0.5 gram (0.02 ounce), yet it secretes at least nine distinct hormones that exert far-reaching influence over body activities. Connected to the hypothalamus by a stalk of nervous tissue, the pituitary gland consists of two main lobes, the anterior and posterior lobes.

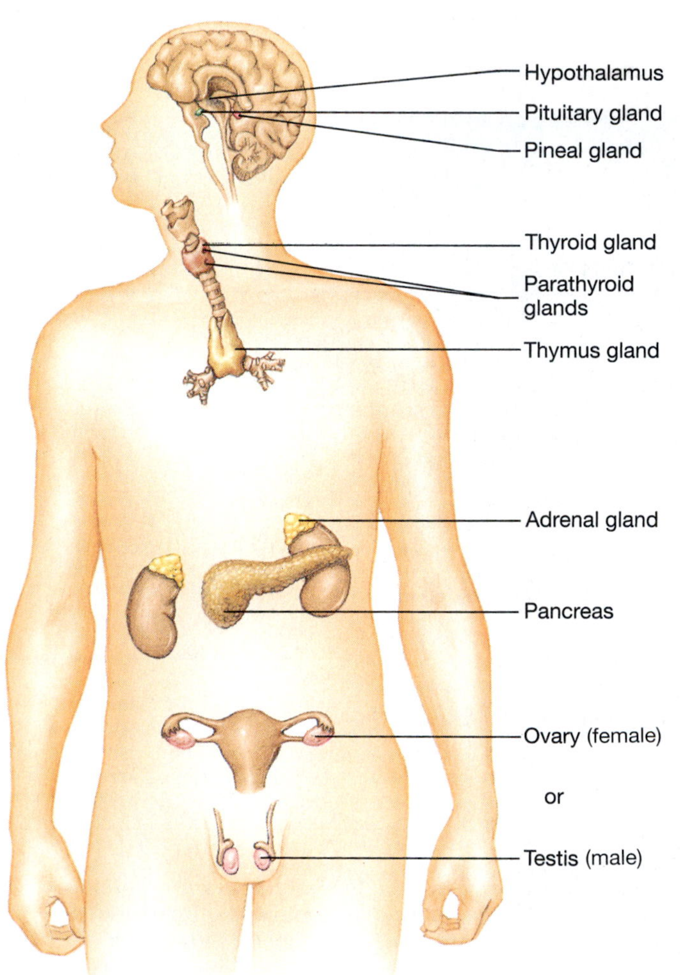

FIGURE 41–4 Location of the principal endocrine glands in the human body.

In some animals, an intermediate lobe secretes hormones that regulate skin color.

The hypothalamus secretes several **releasing** and **release-inhibiting hormones** that regulate the anterior lobe of the pituitary gland. These neurohormones enter capillaries and pass through special portal veins that connect the hypothalamus with the anterior lobe of the pituitary (Fig. 41–5). (These portal veins, like the hepatic portal vein, do not deliver blood to a larger vein directly but connect two sets of capillaries.) Within the anterior lobe of the pituitary, the portal veins divide into a second set of capillaries. The releasing and release-inhibiting hormones pass through the walls of these capillaries into the tissue of the anterior lobe where each regulates the production and secretion of specific pituitary hormones.

The Posterior Lobe of the Pituitary Gland Releases Two Hormones

Two peptide hormones, oxytocin and antidiuretic hormone (ADH; discussed in Chapter 40), are secreted by the **posterior lobe** of the pituitary gland. These hormones are

TABLE 41–2
Consequences of Endocrine Malfunction

Hormone	*Hyposecretion*	*Hypersecretion*
Growth hormone	Pituitary dwarfism	Gigantism if malfunction occurs in childhood; acromegaly in adult
Thyroid hormones	Cretinism (in children); myxedema, a condition of pronounced adult hypothyroidism (metabolic rate is reduced by about 40%; patient feels tired all of the time and may be mentally slow); goiter, enlargement of the thyroid gland (see figure)	Hyperthyroidism; increased metabolic rate, nervousness, irritability
Parathyroid hormone	Spontaneous discharge of nerves; spasms; tetany; death	Weak, brittle bones; kidney stones
Insulin	Diabetes mellitus	Hypoglycemia
Hormones of adrenal cortex	Addison's disease (inability to cope with stress; sodium loss in urine may lead to shock)	Cushing's disease (edema gives face a full-moon appearance; fat is deposited about trunk; blood glucose level rises; immune responses are depressed)

Goiter resulting from iodine deficiency. (John Paul Kay/Peter Arnold, Inc.)

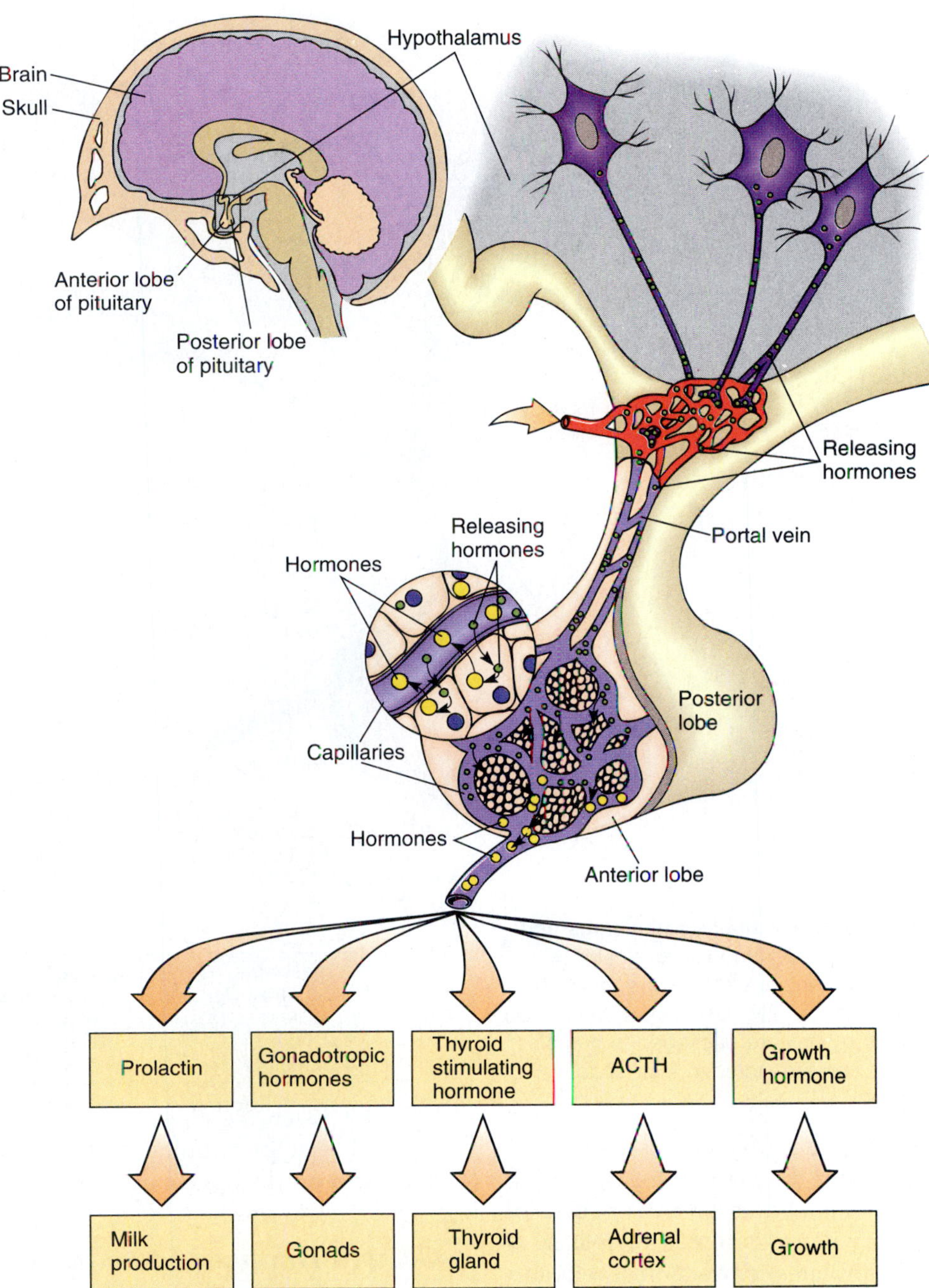

FIGURE 41–5 The hypothalamus regulates the anterior lobe of the pituitary gland. The hypothalamus secretes several specific releasing and release-inhibiting hormones, which reach the anterior lobe of the pituitary gland by way of portal veins. Each releasing hormone stimulates the release of a particular hormone by cells of the anterior lobe.

actually produced by specialized nerve cells in the hypothalamus. They reach the posterior lobe of the pituitary by flowing through axons that connect the hypothalamus with the posterior pituitary (Fig. 41–6). Enclosed within tiny vesicles, the hormones pass slowly down the axons of these nerve cells. The axons extend through the pituitary stalk and into the posterior lobe. The peptide hormone accumulates in the axon endings until the neuron is stimulated; then it is released and diffuses into surrounding capillaries.

In a female, **oxytocin** levels rise toward the end of pregnancy, stimulating the strong contractions of the

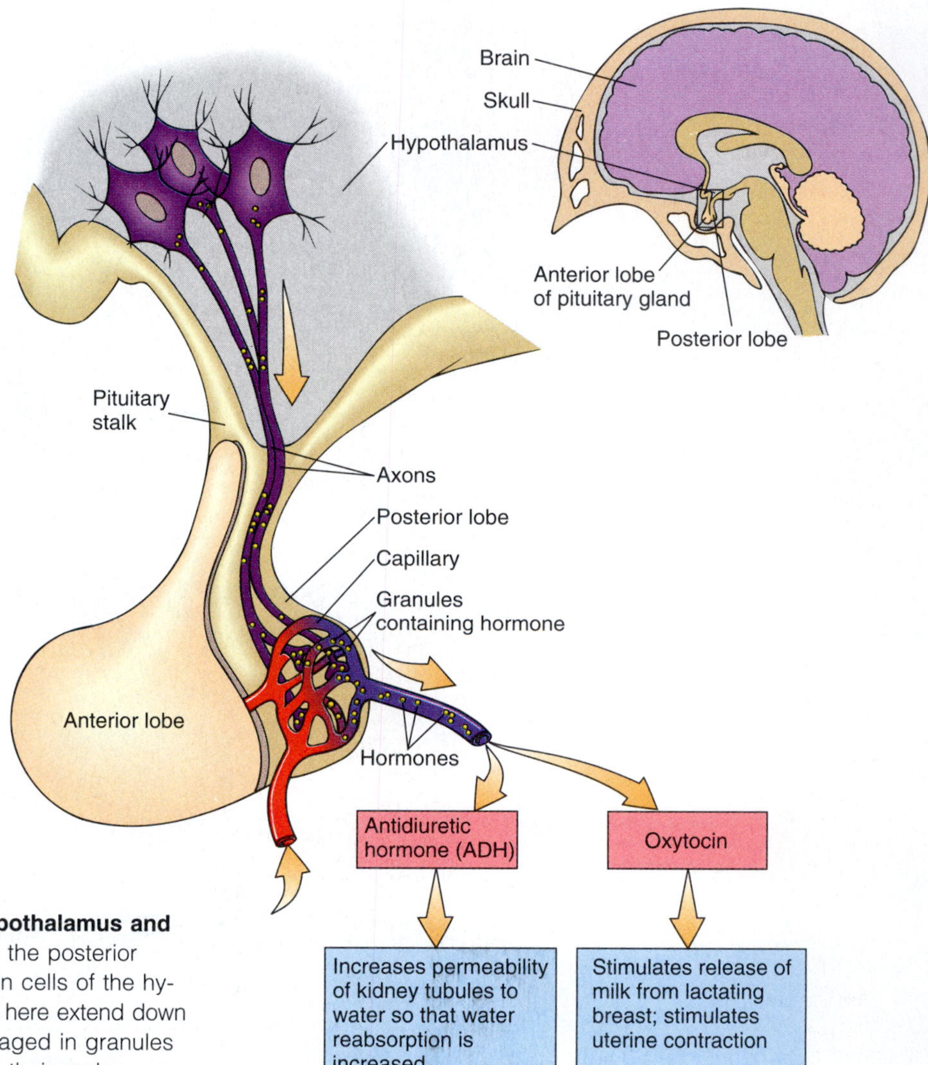

FIGURE 41–6 Relationship between the hypothalamus and the posterior lobe. The hormones secreted by the posterior lobe of the pituitary are actually manufactured in cells of the hypothalamus. The axons of the neurons illustrated here extend down into the posterior lobe. The hormones are packaged in granules that flow through these axons and are stored in their ends.

uterus needed to expel the baby. Oxytocin is sometimes administered clinically (under the name Pitocin) to initiate or speed labor. After birth, when an infant sucks at its mother's breast, sensory neurons signal the release of oxytocin. The hormone stimulates contraction of cells surrounding the milk glands so that milk is let down into the ducts, from which it can be sucked by the infant. Because oxytocin also stimulates the uterus to contract, breast feeding promotes rapid recovery of the uterus to nonpregnant size.

The Anterior Lobe of the Pituitary Gland Regulates Growth and Several Other Endocrine Glands

The **anterior lobe** of the pituitary secretes **prolactin,** growth hormone, and several **tropic hormones,** which are hormones that stimulate other endocrine glands (Fig. 41–7). Prolactin stimulates the cells of the mammary glands to produce milk during lactation.

Growth Hormone Stimulates Protein Synthesis

Small children measure themselves periodically against their parents, eagerly awaiting that time when they, too, will be "big." Whether one will be tall or short depends upon many factors, including genes, diet, and hormonal balance.

Growth hormone (**GH;** also called somatotropin) stimulates body growth by increasing uptake of amino acids by the cells and by stimulating protein synthesis. Growth hormone promotes mobilization of fat from adipose tissues, raising the level of free fatty acids in the blood. Fatty acids become available for cells to use as fuel, an action that conserves proteins. How does this help to promote growth?

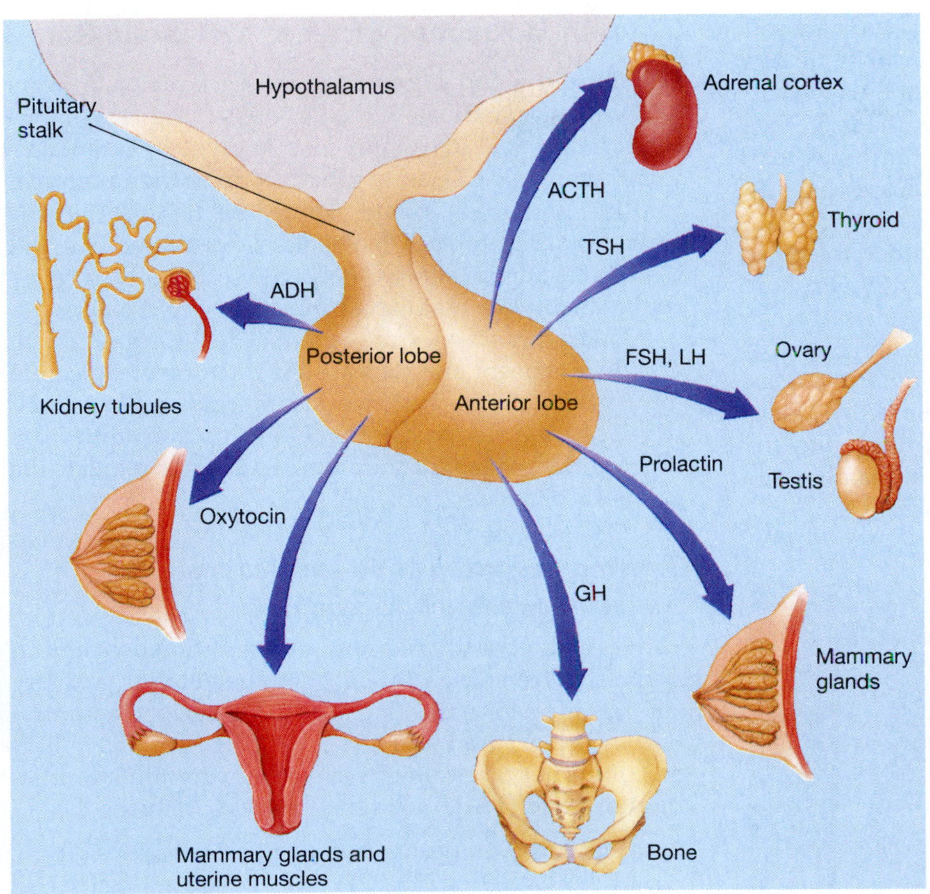

FIGURE 41–7 Hormones released by the pituitary gland. The pituitary gland is suspended from the hypothalamus by a stalk of neural tissue. Shown here are the hormones secreted by the anterior and posterior lobes of the pituitary gland and the target tissues they act on.

Fat mobilization by GH is also important during fasting or when a person is under prolonged stress. In both of these situations, the blood sugar level is low.

Growth Is Affected by Many Factors

Secretion of growth hormone from the anterior pituitary is regulated both by a growth-hormone-releasing hormone and a growth-hormone-release-inhibiting hormone secreted by the hypothalamus. A high level of GH in the blood signals the hypothalamus to secrete the release-inhibiting hormone. As a result, the pituitary releases less GH. A low level of growth hormone in the blood causes the hypothalamus to secrete the releasing hormone. This hormone stimulates the pituitary gland to release more growth hormone. Many other factors stimulate secretion of growth hormone, including low blood sugar, increased amino acid concentration in the blood, and stress.

You may recall your parents telling you to get plenty of sleep and exercise so that you would grow properly. Studies support these age-old notions. Secretion of growth hormone does increase during exercise, probably because rapid metabolism by muscle cells lowers blood sugar level. Growth hormone secretion also increases during non-REM sleep.

Emotional support is also necessary for proper growth. Growth may be retarded in children who are deprived of cuddling, playing, and other forms of nurture, even when their physical needs (food and shelter) are met. In extreme cases, childhood stress can produce actual dwarfism (psychosocial dwarfism). Some emotionally deprived children exhibit abnormal sleep patterns, which may be the basis for decreased secretion of GH.

Other hormones also influence growth. Thyroid hormones appear to be necessary for normal growth hormone secretion and function. Sex hormone must be present for the growth spurt associated with puberty to occur. However, the presence of sex hormone eventually causes the growth centers within the long bones to fuse to the shafts, so further increase in height is impossible even when growth hormone is present.

Inappropriate Amounts of Growth Hormone Secretion Result in Abnormal Growth

Have you ever wondered why some people fail to grow normally? Some may be **pituitary dwarfs**—individuals whose pituitary glands do not produce sufficient growth hormone during childhood. Although miniature, a pituitary dwarf has normal intelligence and is usually well proportioned. If the growth centers in the long bones are still active when this condition is diagnosed, it can be treated clinically by injection with growth hormone, which can now be synthesized commercially through the use of recombinant DNA technology. Growth problems may also result from the malfunction of other mechanisms, such as the regulating hormones from the hypothalamus.

Abnormally tall individuals develop when the anterior pituitary secretes excessive amounts of growth hormone during childhood. This condition is referred to as **gigantism** (Fig. 41–8). If pituitary malfunction leads to hypersecretion of growth hormone during adulthood, the individual cannot grow taller. Instead, connective tissue thickens and bones in the hands, feet, and face may increase in diameter. This condition is known as **acromegaly,** which means "large extremities."

FIGURE 41–8 The world's tallest woman with her family. Sandy Allen is 2.22 meters, or 7 ft, 7 1/4 inches tall. Her gigantism is due to an excess of growth hormone. (Bettina Cirone/Photo Researchers, Inc.)

Thyroid Hormones Increase Metabolic Rate

The **thyroid gland** is located in the neck region, in front of the trachea and below the larynx (see Fig. 41–4). Two of its hormones, **thyroxine,** also known as T_4, and triiodothyronine, or T_3, are synthesized from the amino acid tyrosine and from iodine. Thyroxine has four iodine atoms attached to each molecule; T_3 has three. We will discuss calcitonin, another hormone secreted by the thyroid gland, when we discuss the parathyroid glands.

Thyroid hormones are essential for normal growth and development and increase the rate of metabolism in most body tissues. They are also necessary for cellular differentiation. Tadpoles cannot develop into adult frogs without thyroxine. This hormone appears to regulate the synthesis of needed proteins.

Thyroid Secretion Is Regulated by Negative Feedback Mechanisms

The regulation of thyroid hormone secretion depends on a negative feedback loop between the anterior pituitary and the thyroid gland (Fig. 41–9). When the concentration of thyroid hormones in the blood rises above normal, the anterior pituitary secretes less **thyroid stimulating hormone (TSH),** as in the following sequence:

> Concentration of thyroid hormones increases ---> Inhibits anterior pituitary ---> Secretes less TSH ---> Thyroid gland secretes less hormone ---> Thyroid hormone concentrations decrease

Too much thyroid hormone in the blood also affects the hypothalamus, inhibiting secretion of TSH-releasing hormone. However, the hypothalamus is thought to exert its regulatory effects mainly in certain stressful situations, such as extreme weather change. Exposure to very cold weather may stimulate the hypothalamus to increase secretion of TSH-releasing hormone. This action raises body temperature through increased metabolic heat production.

When the concentration of thyroid hormones decreases, the pituitary secretes more TSH, as follows:

> Low concentration of thyroid hormones ⟶ Stimulates anterior pituitary ⟶ Secretes more TSH ⟶ Stimulates thyroid gland ⟶ Secretes more thyroid hormones ⟶ Thyroid hormone concentration increases

The TSH acts by way of cyclic AMP to promote synthesis and secretion of thyroid hormones and also to promote increased size of the gland itself.

Malfunction of the Thyroid Gland Leads to Specific Disorders

Extreme hypothyroidism during infancy and childhood results in low metabolic rate and can lead to **cretinism,** a condition of retarded mental and physical development.

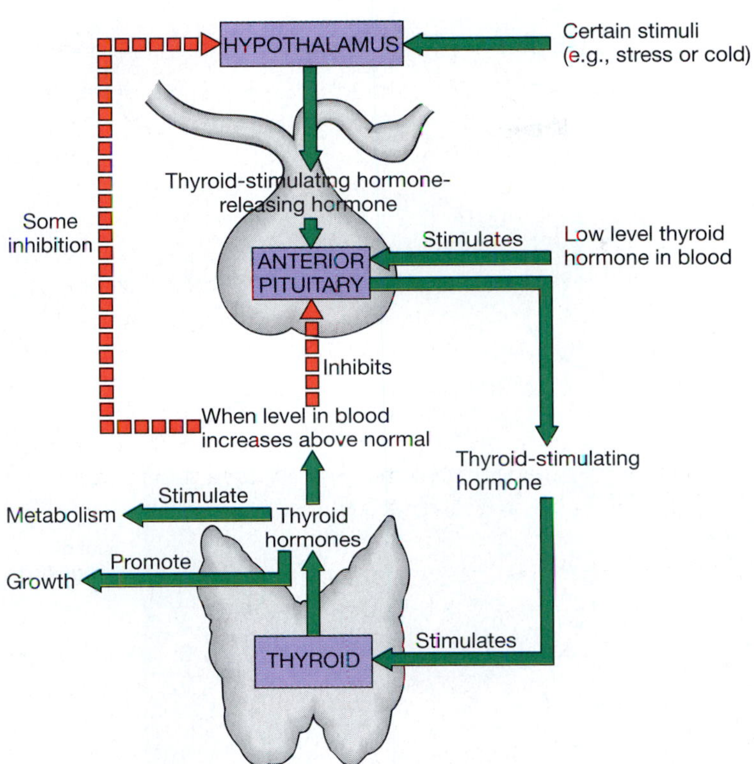

FIGURE 41–9 Regulation of thyroid hormone secretion. Green arrows indicate stimulation; red arrows indicate inhibition.

When diagnosed early enough and treated by the administration of thyroid hormones, the effects of cretinism can be prevented.

An adult who feels like sleeping all of the time, has little energy, and is mentally slow or confused may also be suffering from hypothyroidism. When there is almost no thyroid function, the basal metabolic rate is reduced by about 40% and the patient develops **myxedema,** characterized by a slowing down of physical and mental activity. Hypothyroidism can be treated by oral administration of the missing hormone.

Hyperthyroidism does not cause abnormal growth but does increase metabolic rate by 60% or even more. This increase in metabolism results in the rapid use of nutrients, causing the individual to be hungry and to eat more. But this is not sufficient to meet the demands of the rapidly metabolizing cells, so individuals with this condition often lose weight. They also tend to be nervous, irritable, and emotionally unstable.

Any abnormal enlargement of the thyroid gland is termed a **goiter** and may be associated with either hyposecretion or hypersecretion (see figure in Table 41–2). One cause of hyposecretion is dietary iodine deficiency. Without iodine the gland cannot make thyroid hormones, so their concentration in the blood decreases. In compensation, the anterior pituitary secretes large amounts of TSH. The thyroid gland enlarges, sometimes to gigantic proportions. However, enlargement of the gland cannot increase production of the hormones, because the needed ingredient is still missing. Thanks to iodized salt, goiter is no longer common in the United States and other developed countries. In other parts of the world, however, hundreds of thousands still suffer from this easily preventable disorder.

The Parathyroid Glands Regulate Calcium Concentration

The **parathyroid glands** are embedded in the connective tissue surrounding the thyroid gland. These glands secrete **parathyroid hormone,** which helps regulate the calcium level of the blood and tissue fluid (Fig. 41-1). Parathyroid hormone stimulates the release of calcium from bones and calcium reabsorption from the kidney tubules. It also activates vitamin D, which then increases the amount of calcium absorbed from the intestine.

Calcitonin, secreted by the thyroid gland, works antagonistically to parathyroid hormone. When the concentration of calcium rises above homeostatic levels, calcitonin is released and rapidly inhibits removal of calcium from bone.

The Islets of the Pancreas Regulate Glucose Concentration

In addition to secreting digestive enzymes (Chapter 39), the pancreas is an important endocrine gland. Its hormones, insulin and glucagon, are secreted by cells that

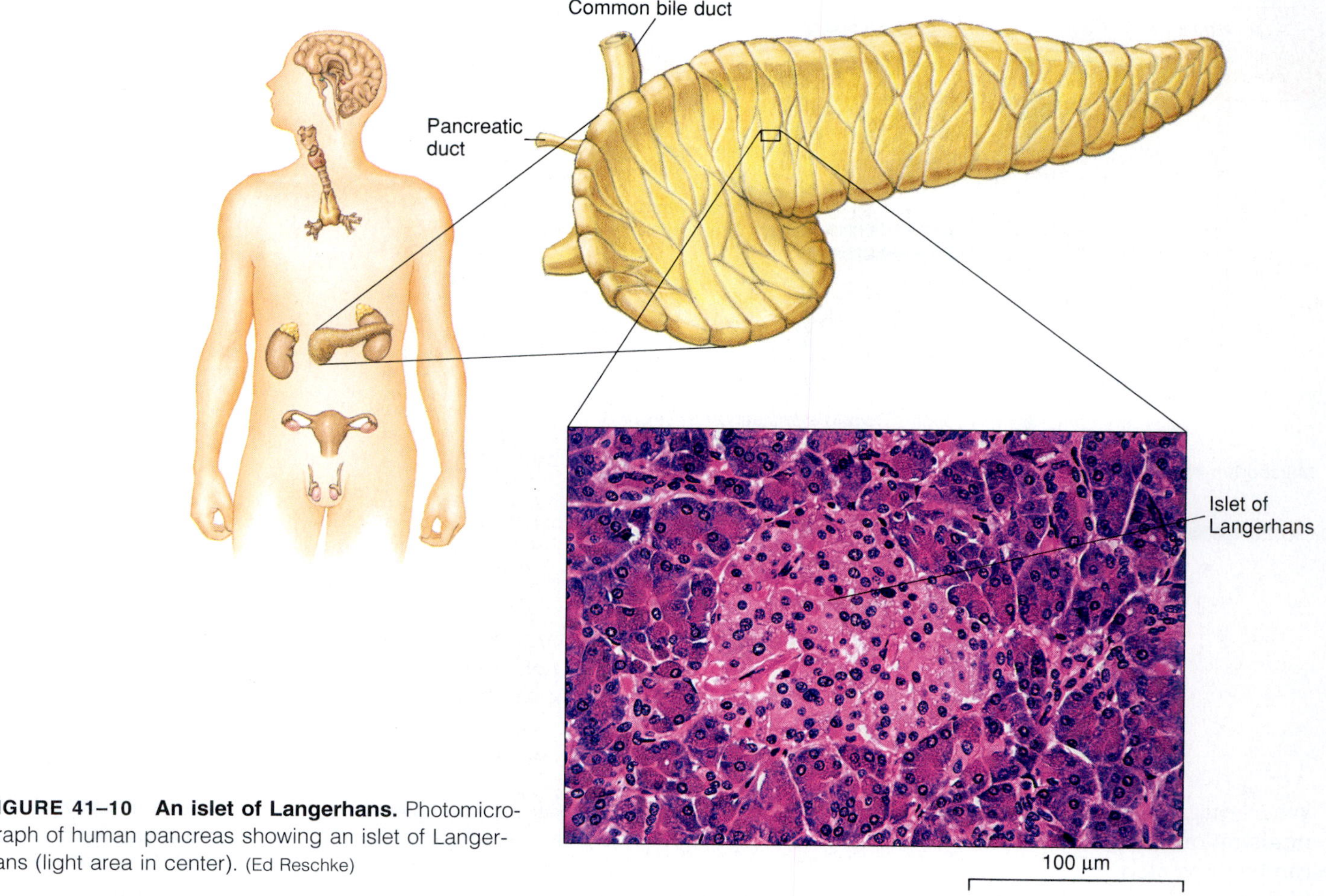

FIGURE 41–10 An islet of Langerhans. Photomicrograph of human pancreas showing an islet of Langerhans (light area in center). (Ed Reschke)

form little clusters, the **islets of Langerhans,** throughout the pancreas (Fig. 41–10). About one million islets are present in the human pancreas. They are composed of **beta cells,** which secrete **insulin,** and **alpha cells,** which secrete **glucagon.**

Insulin Lowers the Concentration of Glucose in the Blood

Insulin stimulates cells of many tissues, including muscle and fat cells, to take up glucose from the blood. Once glucose enters muscle cells, it is either used immediately as fuel or stored as glycogen. Insulin also inhibits liver cells from releasing glucose. Thus, insulin activity results in *lowering* the glucose level in the blood.

Insulin also influences fat and protein metabolism. This hormone reduces the use of fatty acids as fuel and instead stimulates their storage in adipose tissue. It also inhibits the use of amino acids as fuel, thus promoting protein synthesis.

Glucagon Raises the Concentration of Glucose in the Blood

The actions of glucagon are opposite to those of insulin. The main effect of glucagon is to *raise* blood sugar level. It does this by stimulating liver cells to convert glycogen to glucose and by stimulating liver cells to make glucose from other metabolites. Glucagon mobilizes fatty acids and amino acids as well as glucose.

Insulin and Glucagon Secretion Are Regulated by Glucose Concentration

Secretion of insulin and glucagon is directly controlled by the concentration of glucose in the blood (Fig. 41–11). After a meal, when the blood glucose level rises as a result of intestinal absorption, beta cells are stimulated to increase insulin secretion. Then, as the cells remove glucose from the blood, decreasing its concentration, insulin secretion decreases accordingly, as follows:

> Blood glucose level too high ⟶ Stimulates beta cells ⟶ Increased insulin secretion ---→ Blood glucose concentration decreases

When one has not eaten for several hours, the concentration of glucose in the blood begins to fall. When it falls from its normal fasting level of about 90 milligrams of glucose per 100 milliliters of blood to about 70 milligrams of glucose, the alpha cells of the islets increase their secretion of glucagon. Glucose is mobilized from storage

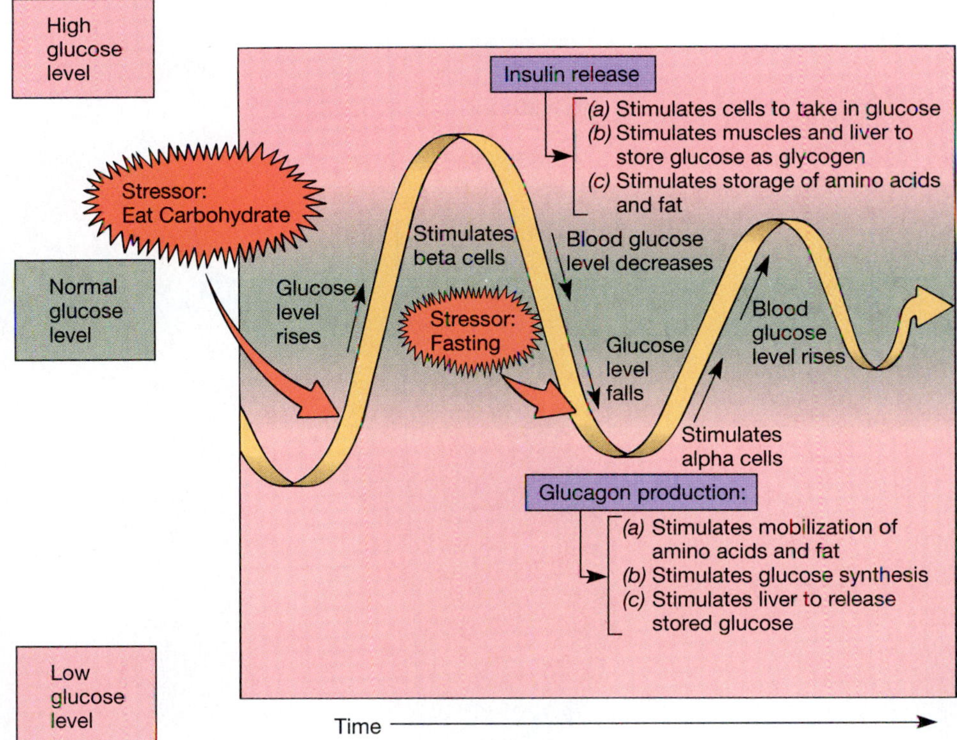

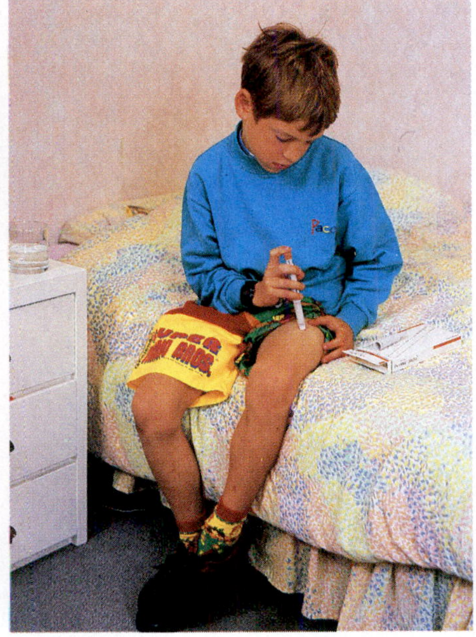

FIGURE 41–11 Regulation of glucose concentration in the blood by insulin and glucagon. The photograph shows a diabetic boy injecting himself with insulin using a novopen. This dispenser contains insulin in portable cartridges and meters the required dose. (Mark Clarke/Science Photo Library/Photo Researchers, Inc.)

in the liver cells, and blood sugar concentration returns to normal:

Glucose concentration too low ⟶ Stimulates alpha cells ⟶ Increased glucagon secretion ⟶ Glucose concentration increases

The alpha cells respond to the glucose concentration within their own cytoplasm, which reflects the blood sugar level. When blood sugar level is high, there is generally a high level of glucose within the alpha cells, and glucagon secretion is inhibited.

It should be clear that insulin and glucagon work oppositely to keep blood sugar concentration within normal limits. When the glucose level rises, insulin release brings it back to normal; when it falls, glucagon acts to raise it again. The insulin-glucagon system is a powerful, fast-acting mechanism for keeping blood sugar level within normal limits.

Diabetes Mellitus Is a Serious Disorder of Carbohydrate Metabolism

Diabetes mellitus, the most common endocrine disorder, is a worldwide health problem. Of the estimated 10 million diabetics in the United States, about 40,000 die each year as a result of this disorder, making it the third lead-

Concept Connections

Hormones ⟷ *Metabolism* ⟷ *Kidney Function*

The blood concentration of glucose is so high in the untreated diabetic that glucose levels exceed the renal threshold. The tubules in the kidneys are unable to return all of the glucose in the filtrate to the blood. As a result, glucose is excreted in the urine. The presence of glucose in the urine is a simple screening test for diabetes mellitus.

Insulin-dependent cells in a diabetic can take in only about 25% of the glucose they require for fuel. The body turns to fat and protein for energy. Increased fat metabolism increases the formation of ketone bodies. These compounds build up in the blood, causing **ketosis,** a condition in which the body fluids and blood become too acidic. If severe, ketosis can lead to coma and death.

When the ketone level in the blood rises, ketones appear in the urine, another clinical indication of diabetes mellitus. Because of osmotic pressure, when ketone bodies and glucose are excreted in the urine, they take water with them, so that urine volume increases. The resulting dehydration causes the diabetic to feel continually thirsty.

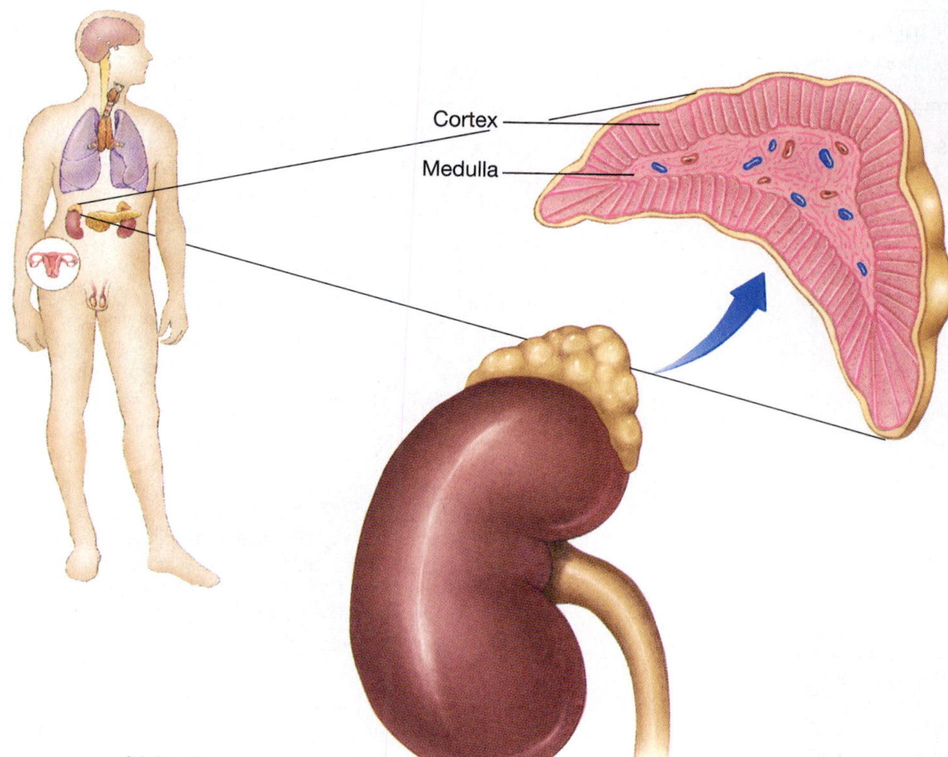

FIGURE 41–12 The adrenal gland.

ing cause of death. Diabetes is a leading cause of blindness, kidney disorders, disease of small blood vessels, gangrene of the limbs, and various other malfunctions.

About 90% of diabetics have type II diabetes. This disorder develops gradually, usually in overweight persons over the age of 40. Some patients with type II diabetes secrete enough insulin, but insulin receptors on target cells cannot bind it. Many patients with type II diabetes can keep their blood sugar level within normal range by managing their diets and exercising, and do not require insulin injection.

Insulin-dependent diabetes, referred to as type I (and formerly known as juvenile-onset diabetes), usually develops before age 30. In type I diabetes there is a marked decrease in the number of beta cells in the pancreas, resulting in insulin deficiency. Daily insulin injections are needed to correct the carbohydrate imbalance that results. Type I diabetes is thought to be an autoimmune disease in which antibodies mark the beta cells for destruction. This disorder may be caused by a combination of genetic predisposition and infection by a virus. Patients with type I diabetes have a shortened life expectancy because atherosclerotic disease develops as a result of impaired lipid metabolism (see Chapter 36).

Similar metabolic disturbances occur in both types of diabetes mellitus:

1. **Decreased use of glucose.** Because the cells of diabetics cannot take up glucose from the blood, glucose accumulates in the blood, causing **hyperglycemia** (an abnormally high concentration of glucose in the blood). Instead of the normal fasting level of 90 milligrams per 100 milliliters, the level may reach from 300 to more than 1000 milligrams.
2. **Increased fat mobilization.** Despite the large quantities of glucose in the blood, most cells cannot use it and must turn to other fuel sources. The absence of insulin promotes the mobilization of fat stores, providing nutrients for cellular respiration. But unfortunately, the blood lipid level may reach five times the normal level, leading to the development of atherosclerosis.
3. **Increased protein use.** Lack of insulin also results in increased protein breakdown relative to protein synthesis, so the untreated diabetic becomes thin and emaciated.

The Adrenal Glands Help the Body Adapt to Stress

The paired **adrenal glands** are small, yellow masses of tissue that lie in contact with the upper ends of the kidneys (Fig. 41–12). Each gland consists of a central portion, the **adrenal medulla,** and a larger outer section, the **adrenal cortex.** Although joined anatomically, the adrenal medulla and cortex develop from different types of tissue in the embryo and function as distinct glands. Both secrete hormones that help to regulate metabolism, and both help the body adjust to stress.

The Adrenal Medulla Initiates an Alarm Reaction

The adrenal medulla develops from neural tissue, and its secretion is controlled by sympathetic nerves. This endocrine gland secretes **epinephrine** (sometimes called adrenaline) and **norepinephrine** (noradrenaline).

Under normal conditions, both epinephrine and norepinephrine are secreted continuously in small amounts. Their secretion is under nervous control. When anxiety is aroused, messages are sent from the brain through sympathetic nerves to the adrenal medulla. Acetylcholine released by these neurons triggers the release of epinephrine and norepinephrine.

During a stressful situation, hormone secretion from this gland initiates an alarm reaction enabling you to think more quickly, fight harder, or run faster than usual. Metabolic rate increases by as much as 100%.

Hormones of the adrenal medulla cause blood to be rerouted to those organs essential for emergency action (Fig. 41–13). Blood vessels going to the brain, muscles,

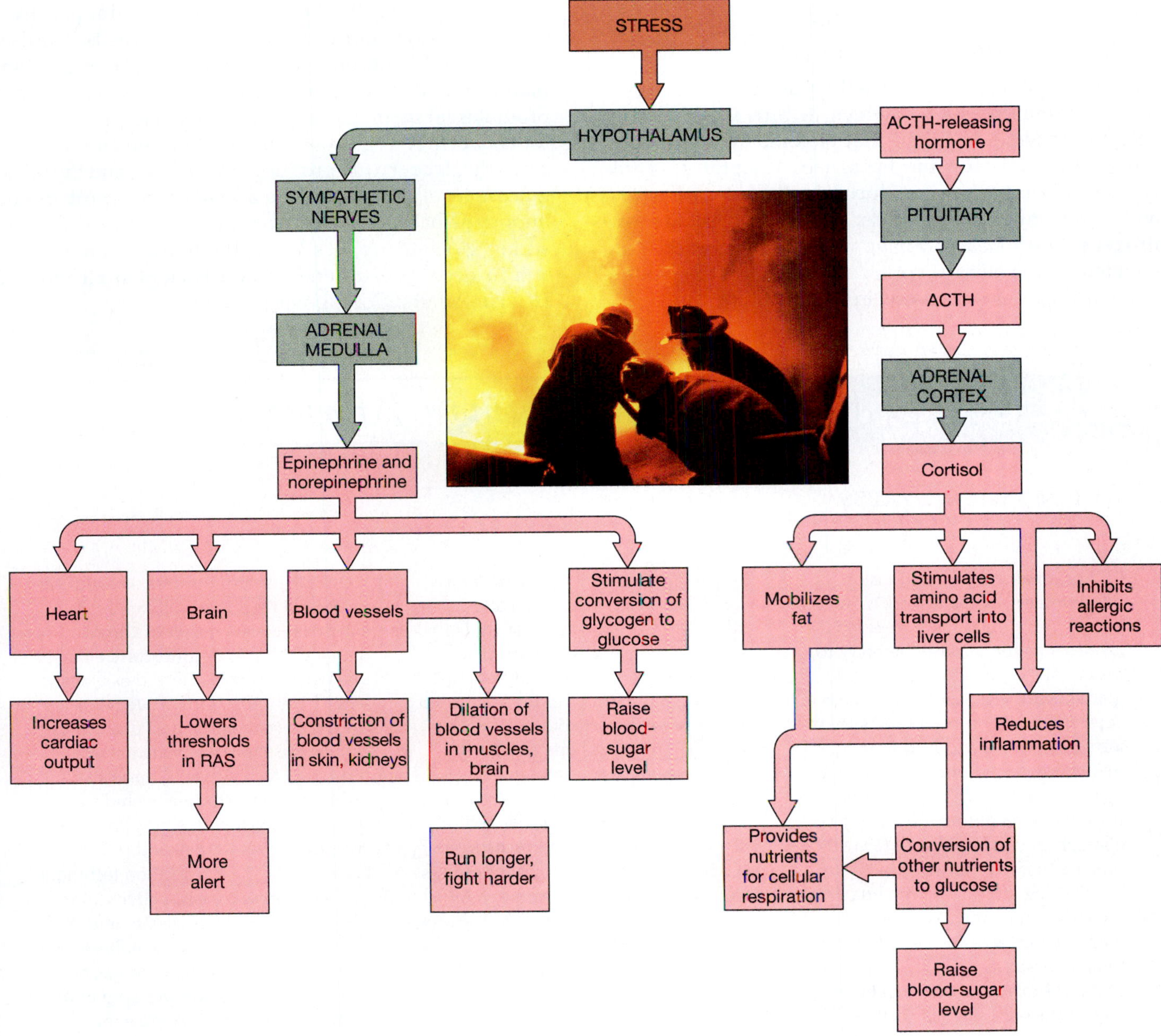

FIGURE 41–13 Some effects of stress. Firefighting is one of many types of careers that are stressful. (Comstock, Inc./Gary Benson)

and heart are dilated, while those to the skin and kidneys are constricted. Constriction of blood vessels serving the skin has the added advantage of decreasing blood loss in case of hemorrhage (and explains the sudden paling that comes with fear or rage). At the same time, the heart beats faster. Thresholds in the reticular activating system of the brain are lowered, so you become more alert. Strength of muscle contraction increases. The adrenal medullary hormones also raise fatty acid and glucose levels in the blood, ensuring needed fuel for extra energy.

The Adrenal Cortex Helps the Body Deal with Chronic Stress

All the hormones of the **adrenal cortex** are steroids synthesized from cholesterol, which in turn is made from acetyl coenzyme A. Recall that steroids are a chemical group classified with the lipids (see Chapter 4). Three types of hormones are produced by the adrenal cortex: androgens, mineralocorticoids, and glucocorticoids. **Androgens,** hormones that have masculinizing effects, are secreted by the adrenal cortex in both sexes. (Reproductive hormones are discussed in Chapter 42.)

Mineralocorticoid hormones help regulate salt balance. The principal mineralocorticoid, **aldosterone,** helps the kidneys reabsorb sodium and excrete potassium. These actions help maintain sodium and potassium balance in the body and an appropriate blood pressure.

When the adrenal glands do not produce enough aldosterone, large amounts of sodium are excreted in the urine. Water leaves the body with the sodium (due to osmotic pressure), and the blood volume may drop so markedly that the patient dies of low blood pressure.

Cortisol, also called hydrocortisone, accounts for about 95% of the **glucocorticoid** activity of the adrenal cortex. The principal action of cortisol is to stimulate liver cells to produce glucose from other nutrients. Cortisol helps provide nutrients for glucose production by stimulating transport of amino acids into liver cells. It also promotes fat mobilization so that fatty acids are available for conversion to glucose. These actions ensure that glucose and glycogen are produced in the liver, and the blood glucose level rises. Thus, the adrenal cortex provides an important backup system for the adrenal medulla, ensuring glucose supplies when the body is under stress and in need of extra energy (see Focus On Health and Human Affairs: Coping with Stress).

FOCUS ON *Health and Human Affairs*

Coping with Stress

The breakup of a relationship, an infection, or the anxiety of taking a test when you are not prepared are all stressors that arouse the body to action. The brain and the adrenal glands work together to help the body cope effectively. Information is transferred by nerves and hormones to many tissues and organs of the body. Neural messages from the brain stimulate the adrenal medulla to release epinephrine and norepinephrine that prepare the body for fight or flight.

During stress, the hypothalamus secretes corticotropin-releasing hormone, which signals the anterior pituitary to secrete ACTH. The release of ACTH increases the secretion of glucocorticoids such as cortisol. These hormones adjust metabolism to meet the increased demands of the stressful situation (see Fig. 41–13).

Some forms of stress are short-lived. We react, quickly resolving the situation. Chronic stress, such as that from an unhappy marriage or toxic work situation, is harmful because of the side effects of long-term elevated levels of glucocorticoids.

Chronic stress has also been shown to damage the brain. Studies indicate that when rodents and monkeys are subjected to prolonged stress, the elevation of glucorticoids leads to the degeneration of neurons, especially in the hippocampus (a part of the brain involved in learning and remembering). Elevated concentration of glucocorticoids also impairs the capacity of neurons in the hippocampus to withstand physiological insult such as reduced blood flow or oxygen to the brain.

Individuals approach stressful situations in their lives differently. A stressor that may result in chronic, damaging levels of glucocorticoids in one person may be viewed as a challenge by another. One strategy for reducing psychological and physiological response to stressors is to learn relaxation techniques. A variety of techniques are effective, including meditation, visual imagery, progressive muscle relaxation, and self-hypnosis. Practicing a relaxation technique can result in decreased activity of the sympathetic nervous system and reduced response to norepinephrine. For example, relaxation training has been shown to lower blood pressure in hypertensive patients, decrease the frequency of migraine headaches, and reduce chronic pain.

Stress stimulates the hypothalamus to secrete **corticotropin-releasing factor (CRF).** In turn, CRF stimulates the anterior pituitary to secrete **adrenocorticotropic hormone (ACTH).** This tropic hormone regulates glucocorticoid secretion (as well as aldosterone secretion). ACTH is so potent that it can result in up to a twentyfold increase in cortisol secretion within minutes. When the body is not under stress, high levels of cortisol in the blood inhibit both the hypothalamus and the pituitary.

Destruction of the adrenal cortex and the resulting decrease in aldosterone and cortisol secretion cause **Addison's disease.** In this condition, a reduction in cortisol prevents the body from regulating the concentration of glucose in the blood because it cannot convert other nutrients to glucose. As a result, the cortisol-deficient patient also has a decreased capacity to cope with stress.

Glucocorticoids are used clinically to reduce inflammation in allergic reactions, infections, arthritis, and certain types of cancer. These hormones help stabilize lysosome membranes so that they do not destroy tissues with their potent enzymes. Because they reduce the effects of histamine, they are used to treat allergic symptoms. When used in large amounts, glucocorticoids can cause serious side effects such as ulcers, hypertension, diabetes mellitus, and atherosclerosis. They also decrease the number of lymphocytes in the body, reducing the patient's ability to fight infections.

Abnormally large amounts of glucocorticoids, whether due to disease or drugs, can result in **Cushing's disease.** In this condition, fat is mobilized from the lower part of the body and deposited about the trunk. Edema gives the patient's face a full-moon appearance.

Many Other Hormones Are Known

Many other tissues of the body secrete hormones. Several hormones secreted by the digestive tract (e.g., gastrin, secretin, and CCK) regulate digestive processes. The **thymus gland** produces a hormone (thymosin) that plays a role in immune responses (Chapter 37). The kidneys release hormones that help regulate blood pressure and stimulate production of red blood cells. Atrial natriuretic factor (ANF), secreted by the heart, helps regulate sodium excretion and so affects blood pressure. ANF acts, in part, by inhibiting aldosterone and renin release. The **pineal gland,** located in the brain, produces a hormone called **melatonin,** which influences the onset of sexual maturity. In Chapter 42, we discuss the principal reproductive hormones.

Chapter Summary

- I. The endocrine system consists of endocrine glands and other tissues that secrete hormones. This system helps regulate many aspects of metabolism, growth, and reproduction. Hormones are transported to their target tissues via the blood.
- II. Hormone secretion is regulated by negative-feedback mechanisms.
- III. Hormones combine with receptor proteins of target tissues.
  - A. Steroid hormones combine with receptor proteins inside the target cell. The hormone-receptor complex may stimulate a particular gene to initiate protein synthesis.
  - B. Some hormones combine with receptors on the plasma membrane of the target cell and trigger formation of a second messenger, such as cyclic AMP.
  - C. Prostaglandins may help regulate hormone action by regulating formation of cyclic AMP.
- IV. Invertebrate hormones influence growth, development, reproduction, and metabolism.
- V. Vertebrate hormones help regulate growth, development, reproduction, salt and fluid balance, and many aspects of metabolism.
- VI. Nervous and endocrine system regulation is integrated by the hypothalamus, which controls the activity of the pituitary gland.
  - A. The hormones oxytocin and ADH are produced by the hypothalamus and released by the posterior lobe of the pituitary.
  - B. Secretion of anterior pituitary hormones is regulated by releasing and release-inhibiting hormones secreted by the hypothalamus.
    1. Prolactin stimulates milk production and secretion by the mammary glands.
    2. Growth hormone stimulates body growth by promoting protein synthesis.
    3. Tropic hormones help regulate other endocrine glands.
- VII. Thyroid hormones increase the rate of metabolism.
  - A. Regulation of thyroid secretion depends on a feedback system between the anterior pituitary and the thyroid gland.
  - B. Hyposecretion of thyroxine during childhood may lead to cretinism; during adulthood it may result in myxedema. Goiter may develop from hyposecretion or hypersecretion.
- VIII. The parathyroid glands help regulate calcium level.
- IX. The islets of the pancreas secrete insulin and glucagon.
  - A. Insulin stimulates cells to take up glucose from the blood and so lowers blood sugar concentration.
  - B. Glucagon raises blood glucose concentration by stimulating conversion of glycogen to glucose and production of glucose from other nutrients.
  - C. Insulin and glucagon secretion are regulated directly by blood glucose levels.
  - D. In diabetes mellitus, insulin deficiency results in decreased utilization of glucose, increased fat mobilization, and increased protein utilization.

X. The adrenal glands secrete hormones that help the body cope with stress.
   A. The adrenal medulla secretes epinephrine and norepinephrine. These hormones increase heart rate, metabolic rate, and strength of muscle contraction, and reroute blood to those organs needed for fight or flight.
   B. The adrenal cortex secretes androgens; mineralocorticoids, such as aldosterone, which increases the rate of sodium reabsorption and potassium excretion by the kidneys; and glucocorticoids, such as cortisol, which promotes synthesis of glucose.
   C. During stress, the adrenal cortex acts as a backup system, ensuring adequate fuel supplies for rapidly metabolizing cells.

Selected Key Terms

adrenal cortex, p. 826
adrenal medulla, p. 826
aldosterone, p. 828
anabolic steroid, p. 810
anterior lobe of pituitary, p. 820
calcitonin, p. 823
cyclic AMP, p. 813
diabetes mellitus, p. 825
endocrine gland, p. 812
endocrine system, p. 812
glucagon, p. 824
growth hormone (GH), p. 820
hormone, p. 812
insulin, p. 824
islets of Langerhans, p. 824
mineralocorticoid, p. 828
norepinephrine, p. 827
oxytocin, p. 819
parathyroid gland, p. 823
parathyroid hormone, p. 823
pineal gland, p. 829
posterior lobe of pituitary, p. 818
release-inhibiting hormone, p. 818
releasing hormone, p. 818
second messenger, p. 813
target tissue, p. 812
thyroid gland, p. 822
thyroid hormones, p. 822
tropic hormone, p. 820

Post Test

1. Endocrine glands lack ______________; they release their secretions into the surrounding tissue fluid and they are transported by the ______________.

2. Endocrine glands produce chemical messengers called ______________.

3. A second messenger important in the action of many hormones is ______________ ______________.

4. The ______________ serves as a link between the nervous and endocrine systems.

5. Hormones bind to specific receptor proteins in their ______________ tissues.

6. Cretinism is caused by ______________-secretion of ______________ hormones during childhood.

7. An abnormal enlargement of the thyroid gland is a/an ______________.

8. The principal action of ______________ is to promote the production of glucose from other nutrients.

9. Calcitonin is secreted by the ______________ gland.

10. In untreated ______________ ______________, glucose utilization is decreased and the body turns to fat and protein for fuel.

11. The parathyroid glands (a) are located in the neck region (b) secrete insulin (c) are sometimes called "emergency glands" (d) regulate other endocrine glands via tropic hormones (e) three of the preceding answers are correct.

12. The posterior pituitary gland (a) is located in the neck region (b) secretes insulin (c) secretes growth hormone (d) releases ADH (e) two of the preceding answers are correct.

13. The hormone(s) that increase(s) metabolism is/are (a) thyroid hormones (b) insulin (c) parathyroid hormone (d) oxytocin (e) none of the preceding answers is correct.

14. The adrenal gland (a) is located in the head (b) secretes insulin (c) secretes growth hormone (d) releases ADH (e) none of the preceding answers is correct.

15. Glucocorticoids (a) are secreted by the adrenal medulla (b) raise blood glucose level (c) are sometimes called emergency hormones (d) are tropic hormones (e) two of the preceding answers are correct.

16. The concentration of calcium in the blood is regulated by (a) parathyroid hormones (b) calcitonin (c) cortisol (d) glucagon (e) two of the preceding answers are correct.

17. Hyposecretion of growth hormone during childhood can result in (a) pituitary dwarfism (b) cretinism (c) diabetes mellitus (d) diabetes insipidus (e) acromegaly.

18. Too much adrenocortical hormone can result in (a) pituitary dwarfism (b) depressed immune system (c) Cushing's disease (d) diabetes insipidus (e) two of the preceding answers are correct.

19. Label the diagram.

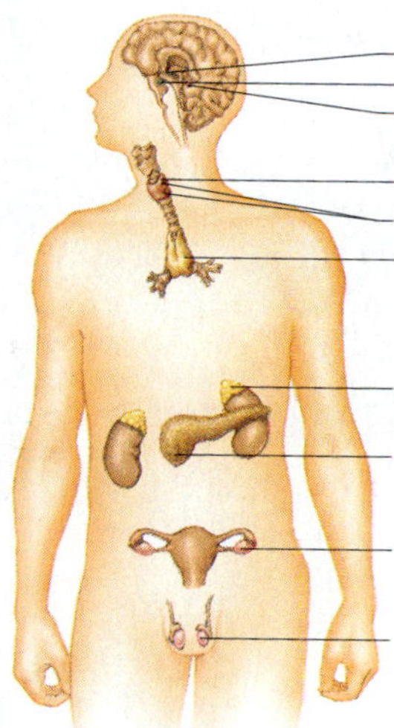

Review Questions

1. What is a hormone? What are some of the important functions of hormones?
2. How are hormones transported? How do they "recognize" their target tissues? What is the role of cyclic AMP in hormone action?
3. What are the functions of prostaglandins?
4. Why is the hypothalamus considered the link between the nervous and endocrine systems? Explain.
5. Describe the actions of (a) prolactin, (b) oxytocin, and (c) thyroid-stimulating hormone.
6. Draw a diagram to illustrate the regulation of thyroid hormone secretion by the anterior pituitary gland.
7. Explain the hormonal basis for (a) acromegaly, (b) pituitary dwarfism, (c) cretinism, (d) hypoglycemia, and (e) Cushing's disease.
8. Explain the antagonistic actions of insulin and glucagon in regulating blood glucose level.
9. Describe several physiological disturbances that result from diabetes mellitus.
10. What are the actions of epinephrine and norepinephrine?
11. How is the adrenal medulla regulated?
12. What types of hormones are released by the adrenal cortex, and what are the actions of each type?
13. Explain how the adrenal glands help the body deal with stress.

Thinking Critically

1. Why do athletes abuse anabolic steroids? In what ways are these hormones harmful? Propose policies to decrease their abuse.
2. Human males have about the same amount of oxytocin circulating in their blood as do females, but its function in males is unknown. Hypothesize its function and design an experiment to test your hypothesis.
3. How do receptors impart specificity within the endocrine system? What might be the advantages of having such complex mechanisms for hormone action as second messengers?
4. Can you think of reasons why it is important to maintain a constant blood sugar level? Several hormones discussed in this chapter affect carbohydrate metabolism. Why do you think it is important to have more than one? How do they interact?
5. Receiving an injection of too much insulin, a diabetic patient may go into insulin shock, a condition in which the patient may appear to be drunk or may become unconscious, suffer convulsions, or even die. From what you know about the actions of insulin, explain the physiological causes of insulin shock.

Recommended Readings

Atkinson, M. and N. Maclaren, "What Causes Diabetes?" *Scientific American,* Vol. 263, No. 1, July 1990, pp. 62–71. An informative discussion of this common disorder.

Lienhard, G.E., J. Slot, D.E. James, and M.M. Mueckler, "How Cells Absorb Glucose," *Scientific American,* Vol. 266, No. 1, 1992, pp. 86–91. A discussion of recent findings on how insulin helps cells transport glucose.

Linder, M.E., and A.G. Gilman, "G Proteins," *Scientific American,* Vol. 267, No. 1, July 1992. G proteins play an important role in the mechanism of hormone action.

Uvnas-Moberg, K. "The Gastrointestinal Tract in Growth and Reproduction," *Scientific American,* Vol. 261, No. 1, July 1989, pp. 78–83. The digestive tract secretes hormones that affect not only digestion but also metabolism of ingested nutrients, growth, and behavior.

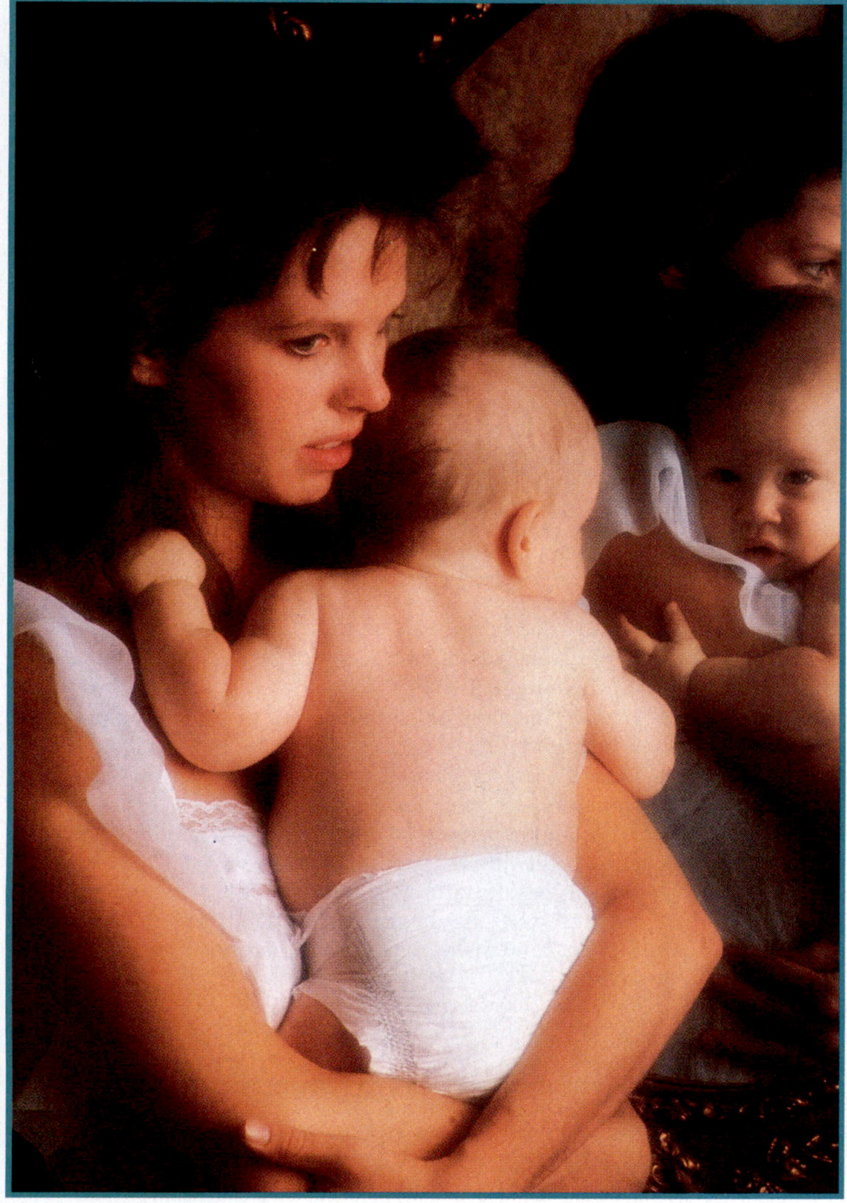

More than 1 million teenagers in the U.S. become pregnant every year. (Leo de Wys, Inc./James K. Hackett)

Babies By Choice: In Search of the Ideal Contraceptive

When a sexually active woman uses no form of birth control, her chances of becoming pregnant during the course of a year are about 90%. Any method for deliberately separating sexual intercourse from reproduction is considered **contraception** (literally, "against conception"). Since ancient times, humans have searched for effective contraceptive methods.

Although many human couples worldwide agree that it is best to have babies by choice, rather than by chance, the majority either are not educated about contraceptive methods or do not have contraceptives available to them. In underdeveloped countries an estimated 88% of women lack the means to limit family size. Studies indicate that many of these women would use modern birth control methods if they were available, affordable, and if someone showed them how.

Young people particularly lack the knowledge and means of protecting themselves from unwanted pregnancy. Every year more than 1 million teenagers in the United States become pregnant, and thousands of girls aged 14 or younger have babies. Yet only one third of sexually active teenagers consistently uses birth control. In addition, the AIDS epidemic poses an increasing risk. One in every 250 Americans is HIV-positive, and 29% with AIDS contracted the virus as teenagers or young adults.

Modern science has developed a variety of contraceptives with a high percentage of reliability, but the ideal contraceptive has not yet been developed. Women in the United States currently have four highly effective, reversible methods of contraception: oral contraceptives, copper IUDs, progestin implants (Norplant), and injectable progestin (DMPA). The progestin implants and injectable progestin became available in the early 1990s. These and other methods are described later in this chapter.

Among the issues that must be considered in developing contraceptives are: effectiveness, cost, convenience, and ease of use. Risks of cancer, birth defects in case the method fails and the woman becomes pregnant, permanent (or long term) infertility, and side effects such as menstrual abnormalities must also be minimized.

Some researchers predict that the contraceptive methods of the 21st century will control regulatory peptides and their genes. The genes that code for the pituitary hormones which stimulate the ovaries to release estrogens could be turned off selectively or other hormone signals could be interrupted. Another approach being studied is molecular interruption of fertilization. Within 25 years, contraception, abortion, and unwanted pregnancy may be replaced by such sophisticated molecular methods.

Reproduction

42

Learning Objectives

After you have studied this chapter you should be able to:

1. Compare asexual and sexual reproduction, and compare external and internal fertilization.
2. Trace the passage of sperm cells through the male reproductive system from their origin in the seminiferous tubules to their expulsion from the body in the semen.
3. Label the structures of the male reproductive system on a diagram, and describe the functions of each.
4. Describe the actions of testosterone and of the gonadotropic hormones in the male.
5. Label the structures of the female reproductive system on a diagram, and describe the functions of each.
6. Trace the development of an ovum (egg) and its passage through the female reproductive system until it is fertilized.
7. Identify the important events of the menstrual cycle such as ovulation and menstruation, explain the actions of each of the hormones involved, and describe the hormonal regulation of the menstrual cycle.
8. Summarize the process of human fertilization.
9. Compare the methods of birth control in Table 42–3 with respect to mode of action, effectiveness, advantages, and disadvantages.
10. Identify common sexually transmitted diseases, and describe their symptoms, effects, and treatment.

Key Concepts

- ☐ In asexual reproduction, a single parent may split, bud, or fragment, giving rise to offspring that are genetically similar to the parent. (Differences can be introduced by mutation.)
- ☐ In animal sexual reproduction, two types of gametes fuse to form a zygote, which develops into an individual that is a unique combination of the genes contributed by each parent.
- ☐ The reproductive role of the male mammal is to produce sperm cells and deliver them into the female reproductive tract.
- ☐ The reproductive role of the female mammal is to produce ova, receive sperm, incubate and nourish the developing embryo, give birth, and nourish the newborn.
- ☐ In both sexes, reproductive hormones maintain sexual characteristics and regulate production of gametes. In females, hormones also regulate the menstrual cycle, maintain pregnancy, and regulate lactation (milk production) for nourishment of the newborn.

ASEXUAL REPRODUCTION IS COMMON AMONG SOME ANIMAL GROUPS

In **asexual reproduction,** a single parent may split, bud, or fragment to give rise to two or more offspring. Except for mutations, the offspring have hereditary traits similar to those of the parent. Sponges and cnidarians are among the animals that can reproduce by **budding,** in which a small part of the parent's body can separate from the rest and develop into a new individual (Fig. 42–1). Sometimes the buds remain attached and become more-or-less independent members of a colony.

Oyster farmers learned long ago that when they tried to kill sea stars by chopping them in half and throwing the pieces back into the sea, the number of sea stars preying on the oyster bed doubled! A sea star can, in fact, regenerate an entire new individual from a single arm. In flatworms, this ability to regenerate a part has become a method of reproduction known as **fragmentation.** The body of the parent may break into several pieces. Each piece then regenerates the missing parts and develops into a whole animal.

CONCEPT CONNECTIONS

Reproduction* ⇌ *Animal Behavior

A special form of parthenogenesis occurs in honeybees. The queen honeybee receives sperm from a male during the "nuptial flight." The queen bee stores the sperm she receives in a little pouch separated from her genital tract by a muscular valve. As she lays eggs, she can either open this valve, permitting the sperm to escape and fertilize the eggs, or keep the valve closed, so that the eggs develop without fertilization. The fertilized eggs become females (queens and workers); the unfertilized eggs become males (drones). Some species of wasps alternately produce a parthenogenetic generation and a generation that develops from fertilized eggs.

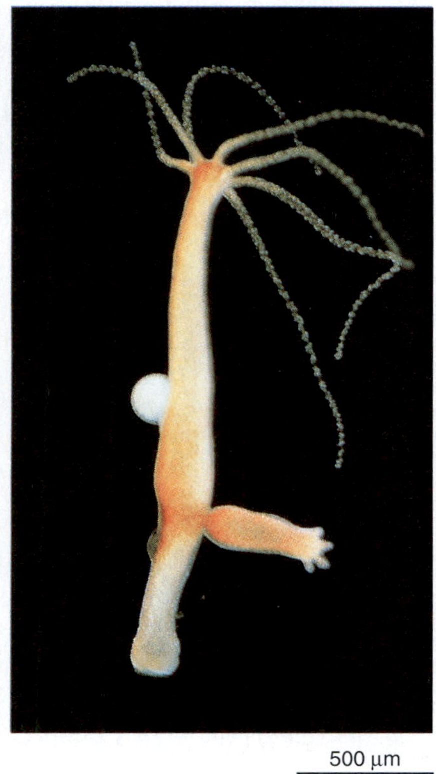

FIGURE 42–1 ***Hydra* reproduces asexually by budding.** A part of the body grows outward and then separates and develops into a new individual. The portion of the parent body that buds is not specialized exclusively for reproduction. The *Hydra* shown here is also reproducing sexually as evidenced by the egg (at left). (Richard Campbell/Biological Photo Service)

Parthenogenesis (virgin development) is a form of reproduction in which an unfertilized egg develops into an adult animal. Parthenogenesis is common among some mollusks, some crustaceans, insects (especially honeybees and wasps) and some reptiles. Parthenogenesis may occur for several generations. At some point males develop, produce sperm, and mate with the females to fertilize their eggs. In some species, parthenogenesis appears to be an adaptation for survival in times of stress or serious population decline.

SEXUAL REPRODUCTION IS THE MOST COMMON TYPE OF ANIMAL REPRODUCTION

Most animals reproduce by sexual reproduction, which occurs through the fusion of gametes (sperm and egg). Sexual reproduction generally involves a male parent that produces sperm and a female parent that produces eggs (ova). The egg is typically large and nonmotile, with a store of nutrients that supports the development of the embryo. The sperm is usually small and motile, adapted to propel itself by beating its long, whiplike flagellum. When a sperm cell fertilizes an egg cell, a fertilized egg, or **zygote,** forms.

In **internal fertilization,** the gametes fuse inside the body. Many aquatic animals practice **external fertilization** in which the gametes meet outside the body (Fig.

(a)

(b)

FIGURE 42–2 External fertilization and internal fertilization. (***a***) External fertilization is illustrated by these spawning frogs (*Rana temporaria*). Most amphibians must return to water for mating. The female lays a mass of eggs, while the male mounts her and simultaneously deposits his sperm in the water. (***b***) Internal fertilization is practiced by mammals such as these lions. (*a*, Zig Leszczynski, copyright 1995 Animals Animals; *b*, Fritz Polking/Dembinsky Photo Associates)

42–2). Mating partners usually release eggs and sperm into the water simultaneously. Many gametes are lost; some are eaten by predators. However, so many gametes are released that sufficient numbers of sperm and egg cells do meet to perpetuate the species.

In internal fertilization, matters are left less to chance. The male generally delivers sperm cells directly into the body of the female. Her moist tissues provide the watery medium required for the movement of sperm. Most terrestrial animals, as well as aquatic reptiles, birds, and mammals, practice internal fertilization.

In **hermaphroditism,** a single individual produces both eggs and sperm. A few hermaphrodites, such as the tapeworm, are capable of self-fertilization. Earthworms are more typical hermaphrodites. Two animals copulate, and each inseminates the other. In some hermaphroditic species, self-fertilization is prevented by the development of testes and ovaries at different times.

HUMAN REPRODUCTION: THE MALE PROVIDES SPERM

The human male, like other male mammals, has the reproductive role of producing sperm cells and delivering them into the female reproductive tract. When a sperm combines with an egg, it contributes its genes and determines the sex of the offspring. The male reproductive system is illustrated in Figure 42–3.

The Testes Produce Sperm

In humans and other vertebrates, **spermatogenesis,** the process of sperm cell production, occurs in the paired male gonads, or **testes.** Spermatogenesis takes place within a vast tangle of hollow tubules, called the **seminiferous tubules,** within each testis (Fig. 42–4). The sem-

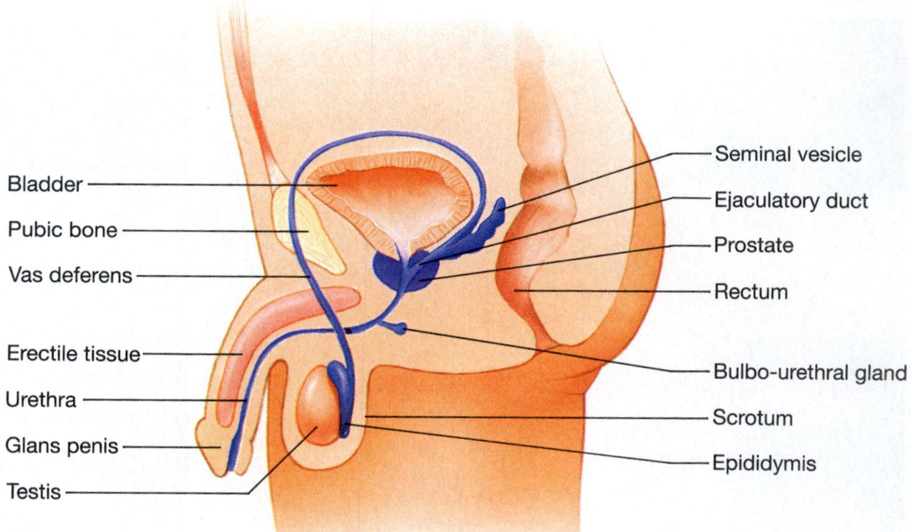

FIGURE 42–3 Anatomy of the human male reproductive system. The scrotum, penis, and pelvic region are shown in sagittal section to illustrate their internal structures.

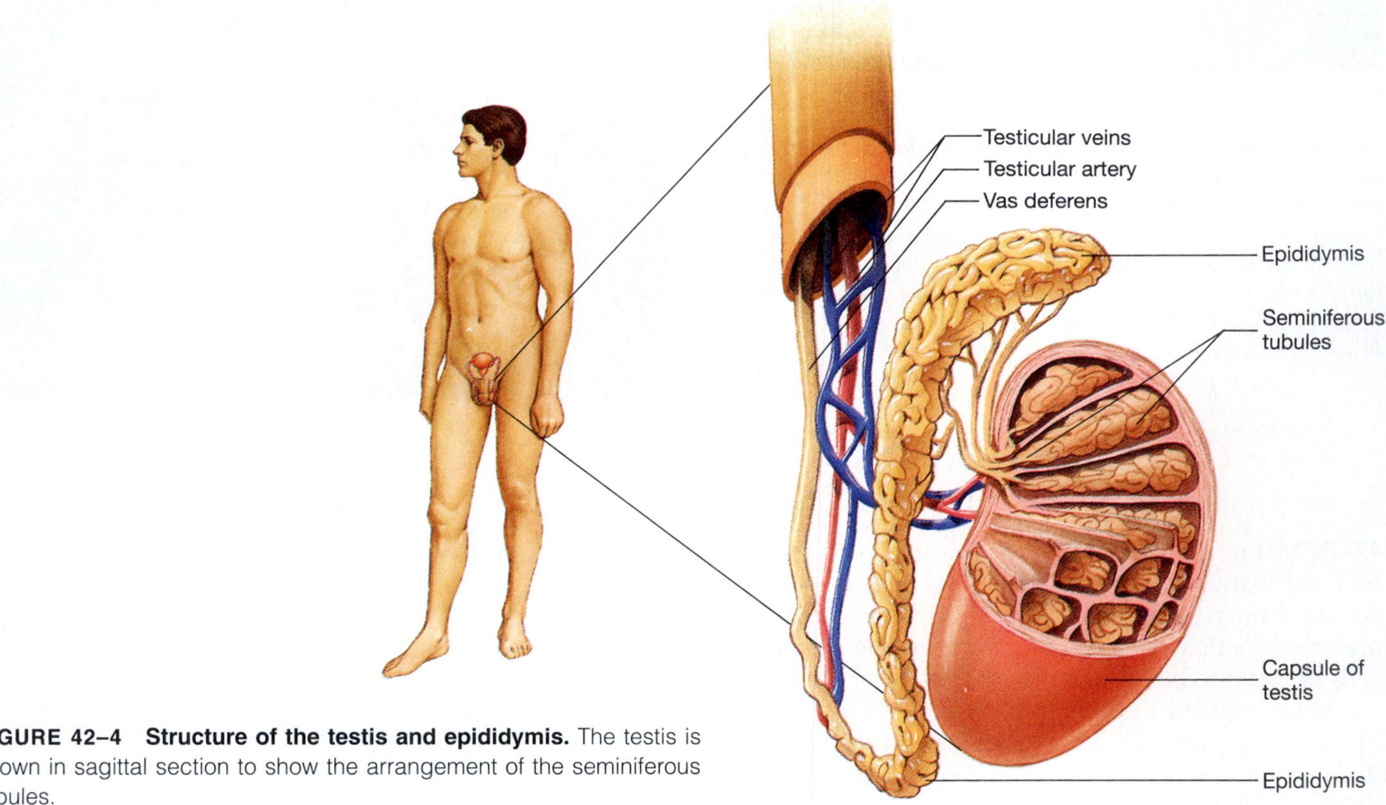

FIGURE 42–4 Structure of the testis and epididymis. The testis is shown in sagittal section to show the arrangement of the seminiferous tubules.

iniferous tubules are partially lined with undifferentiated cells, called **spermatogonia** (Fig. 42–5).

The spermatogonia divide by mitosis, producing more spermatogonia. Some of them enlarge and become **primary spermatocytes,** which undergo meiosis. (You may want to review the discussion of meiosis in Chapter 10.) In many animals, gamete production occurs only in the spring or fall, but humans have no special breeding season. In the human adult male, spermatogenesis proceeds continuously and millions of sperm are produced each day.

Each primary spermatocyte undergoes a first meiotic division producing **secondary spermatocytes** (Fig. 42–6). In the second meiotic division, each secondary spermatocyte gives rise to two **spermatids.** Four spermatids are

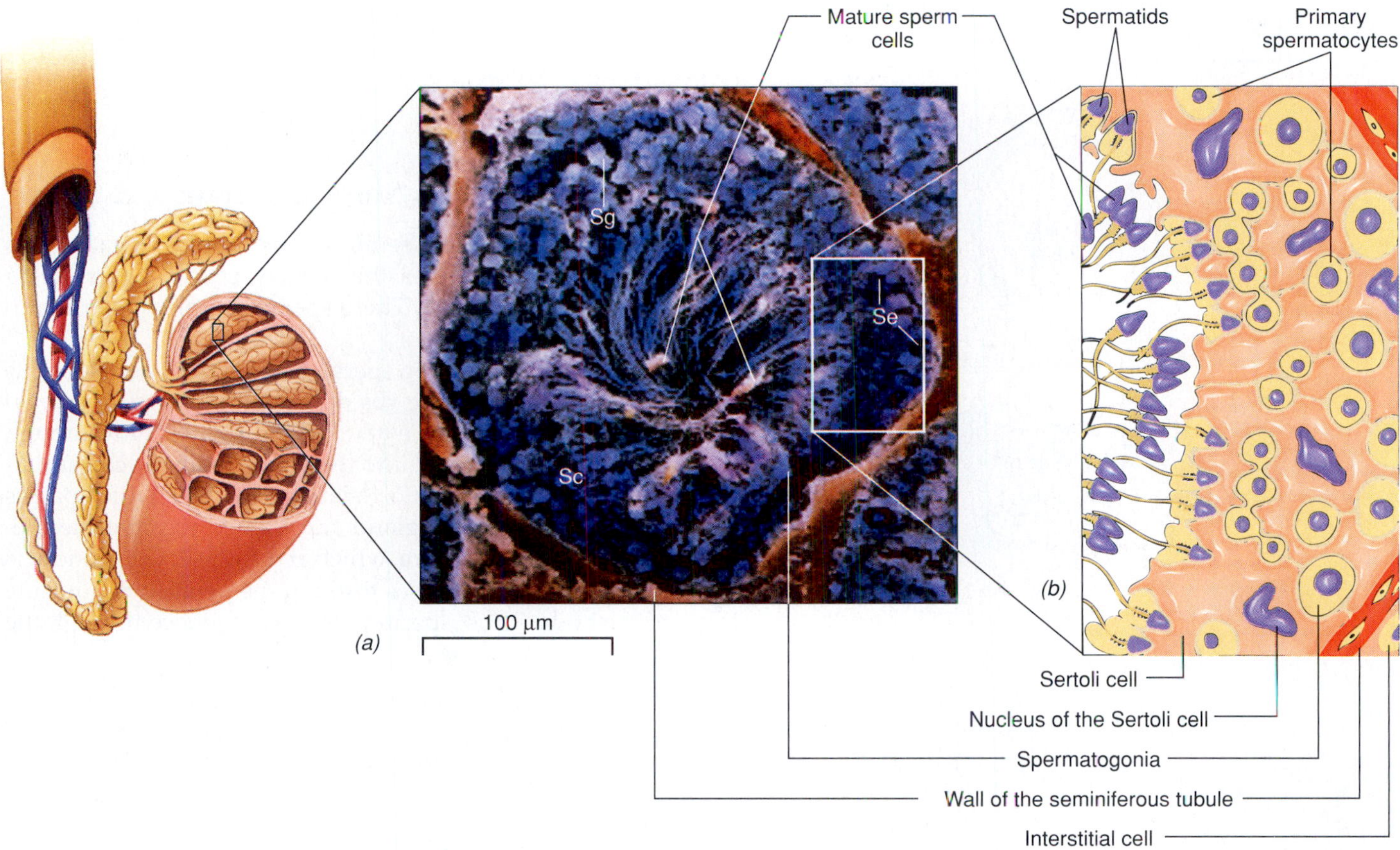

FIGURE 42–5 Structure of a seminiferous tubule. (***a***) Scanning electron micrograph of a cross section through a seminiferous tubule. Se, Sertoli cell; Sc, primary spermatocyte; Sg, spermatogonium. (***b***) Sperm cells can be seen in various stages of development. Note the interstitial cells and the large nutritive Sertoli cells. (*a*, Custom Medical Stock Photo)

produced from the original primary spermatocyte. Each haploid spermatid differentiates into a mature **sperm.** The sequence is as follows:

> Spermatogonium (diploid) ⟶ Primary spermatocyte ⟶ Two secondary spermatocytes (haploid) ⟶ Four spermatids (haploid) ⟶ Four mature sperm (haploid)

Each mature sperm consists of a head, midpiece, and flagellum (Fig. 42–7). The head consists of the nucleus and a cap, called an acrosome, that produces enzymes. The enzymes help the sperm penetrate the egg. Mitochondria, located in the midpiece of the sperm, provide the energy for movement of the flagellum. The sperm flagellum has the typical eukaryotic 9 + 2 arrangement of microtubules. Most of the sperm's cytoplasm is discarded and is phagocytized by the large nutritive Sertoli cells present within the seminiferous tubules.

Human sperm cells cannot develop at body temperature. Although the testes develop within the abdominal cavity of the male embryo, about 2 months before birth they descend into the **scrotum,** a skin-covered sac suspended from the groin. The scrotum serves as a cooling unit, maintaining sperm below body temperature. In rare cases, the testes do not descend. If this condition is not surgically corrected, the seminiferous tubules eventually degenerate and the male becomes **sterile.**

The scrotum is an outpocketing of the pelvic cavity and is connected to it by the inguinal canals. As they descend, the testes pull their blood vessels, nerves, and conducting tubes after them. The inguinal region is a weak place in the abdominal wall. Straining the abdominal muscles by lifting heavy objects sometimes results in tearing the inguinal tissue. A loop of intestine can then bulge into the scrotum through the tear, a condition known as an **inguinal hernia.**

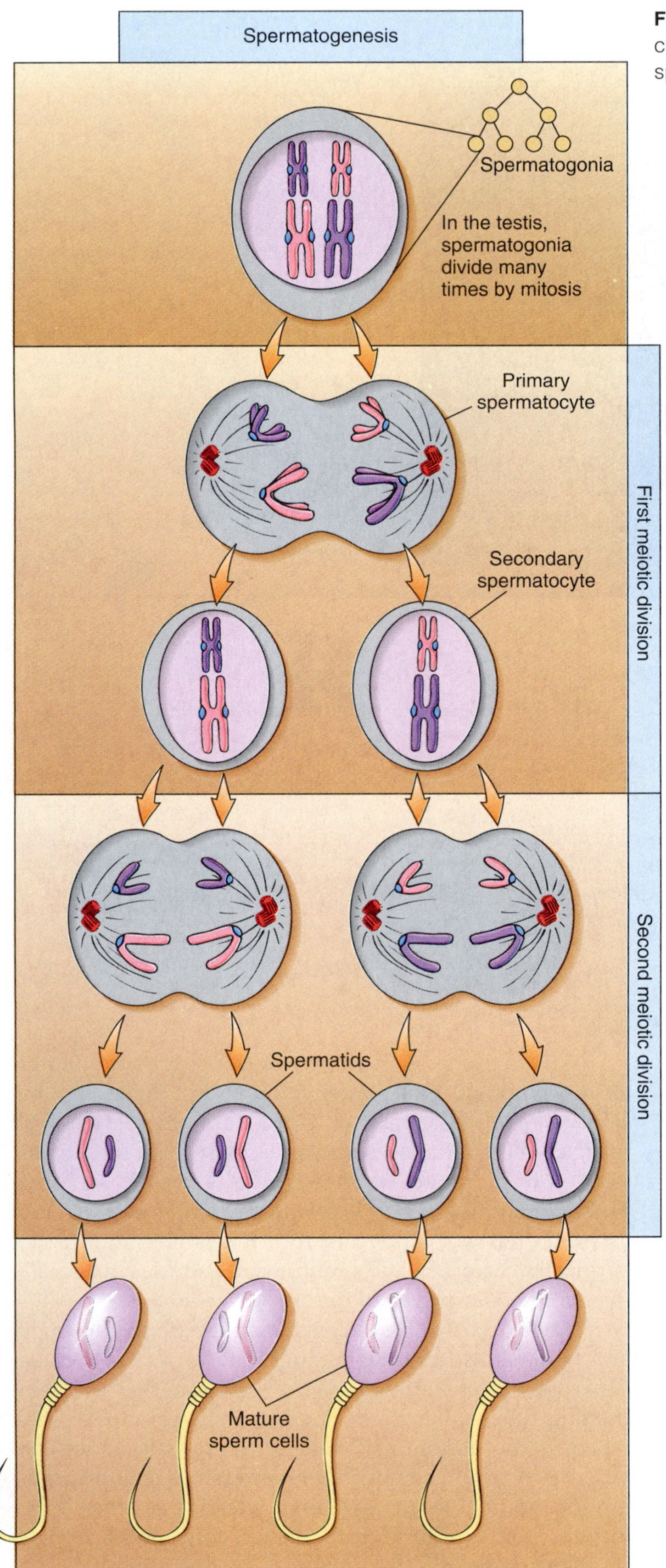

FIGURE 42–6 Spermatogenesis. A primary spermatocyte undergoes meiosis, giving rise to four spermatids. The spermatids differentiate, becoming mature sperm cells.

A Series of Ducts Store and Transport Sperm

Sperm cells leave the seminiferous tubules of each testis through small tubules that empty into a larger coiled tube, the **epididymis.** There sperm complete their maturation and are stored.

During ejaculation, sperm pass from each epididymis into a sperm duct, the **vas deferens** (plural, *vasa deferentia*). The vas deferens extends from the scrotum through the inguinal canal and into the pelvic cavity. Each vas deferens empties into a short **ejaculatory duct,** which passes through the prostate gland and then opens into the urethra. The single **urethra,** which at different times conducts urine and semen, passes through the penis to the outside of the body. Thus, sperm passes in sequence through the following structures:

Seminiferous tubules ⟶ Epididymis ⟶
Vas deferens ⟶ Ejaculatory duct ⟶
Urethra ⟶ Released from body

The Accessory Glands Produce the Fluid Portion of Semen

As sperm are transported through the conducting tubes, they are mixed with secretions from three types of accessory glands. The 3.5 or so ml of **semen** ejaculated during sexual climax consists of about 400 million sperm cells suspended in the secretions of these glands.

The paired **seminal vesicles** secrete a nutritive fluid into the vasa deferentia (Fig. 42–3). The alkaline secretion of the single **prostate gland** is thought to increase the motility of sperm cells. During sexual arousal, the paired **bulbourethral glands** release a mucous secretion that lubricates the penis, facilitating its penetration into the vagina.

The Penis Transfers Sperm to the Female

The **penis** is an erectile copulatory organ that delivers sperm into the female reproductive tract. It consists of a long **shaft** that enlarges to form an expanded tip, the **glans.** Part of the loose-fitting skin of the penis folds down and covers the proximal portion of the glans, forming a cuff called the **prepuce,** or **foreskin.** In the operation termed **circumcision** (commonly performed on male babies for either hygienic or religious reasons), the foreskin is removed.

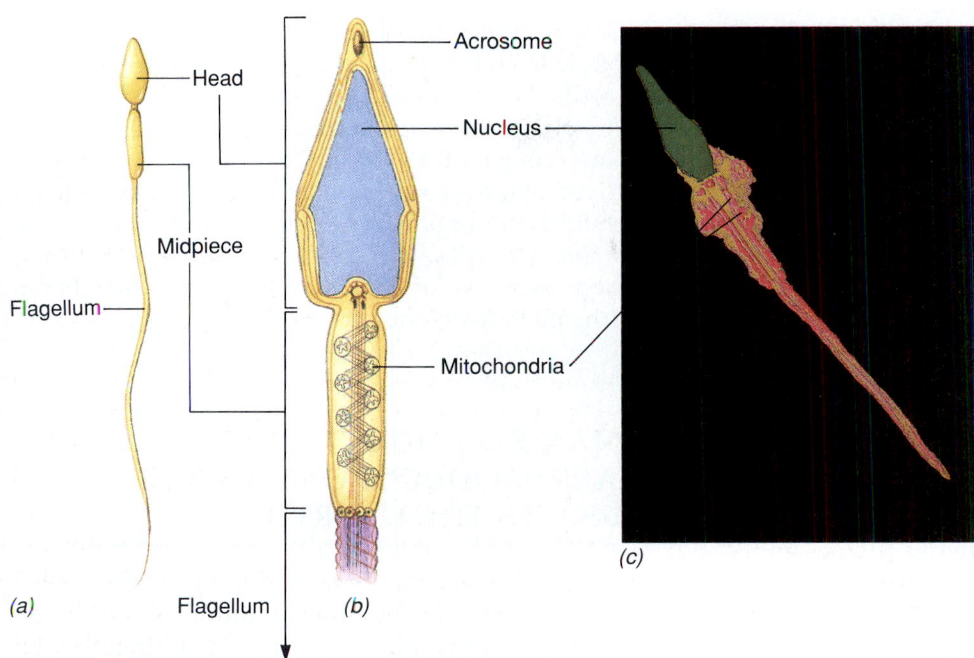

FIGURE 42–7 Sperm cell structure. (***a***, ***b***) Structure of a mature sperm. (***c***) Electron micrograph of a human sperm cell. Mitochondria (colored red) are visible in the midpiece (approximately x 3000). (Dr. Tony Brain/Science Photo Library/Photo Researchers, Inc.)

Under the skin, the penis consists of three parallel columns of **erectile tissue,** sometimes called the **cavernous bodies** (Fig. 42–8). One of these columns surrounds the portion of the urethra that passes through the penis. When the male is sexually stimulated, nerve impulses cause the arteries of the penis to dilate. Blood rushes into the numerous blood vessels of the erectile tissue, causing the tissue to swell. This compresses veins that

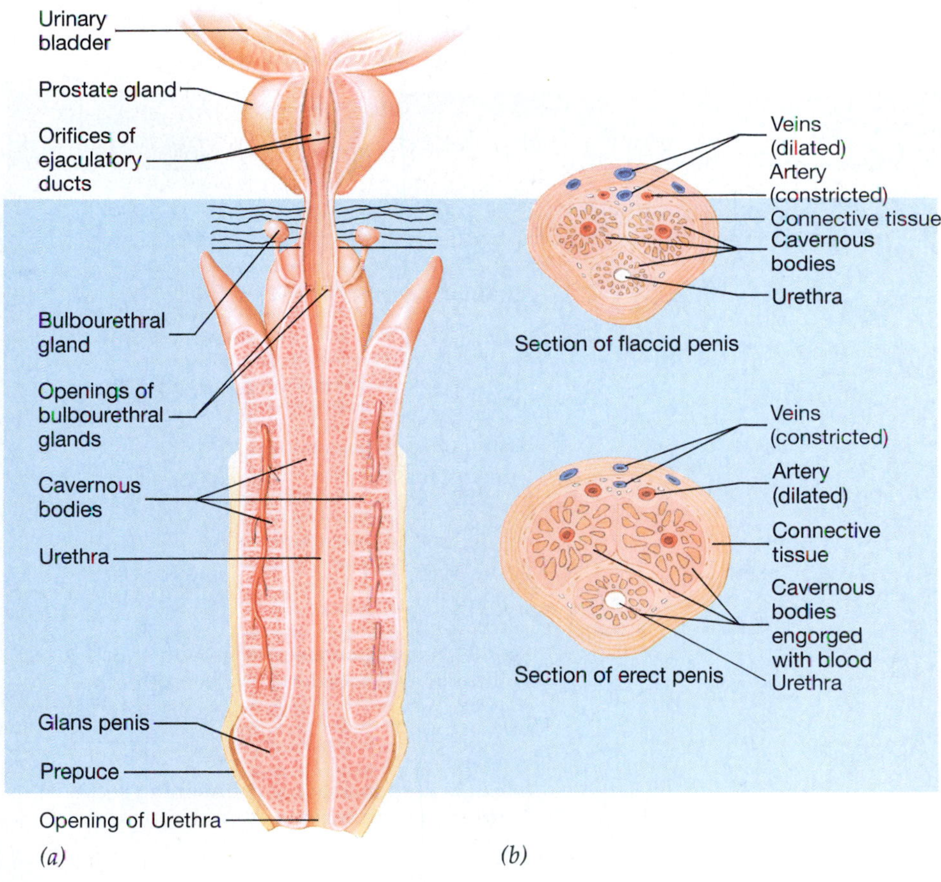

FIGURE 42–8 Internal structure of the penis. (***a***) Longitudinal section through the prostate gland and penis. (***b***) Cross section through flaccid and erect penis. Note that the erectile tissues of the cavernous bodies are engorged with blood in the erect penis.

conduct blood away from the penis, slowing the outflow of blood. Thus, more blood enters the penis than can leave, causing the erectile tissue to become further engorged with blood. The penis becomes erect—that is, longer, larger in circumference, and firm. Although the human penis contains no bone, penis bones do occur in some other mammals, such as bats, rodents, and some primates.

Reproductive Hormones Promote Sperm Production and Maintain Masculinity

At about the age of 10, the hypothalamus begins to secrete **gonadotropin-releasing hormone (GnRH;** Table 42–1). This hormone stimulates the anterior pituitary to secrete the gonadotropic hormones **follicle-stimulating hormone (FSH)** and **luteinizing hormone (LH).** FSH stimulates development of the seminiferous tubules and may promote spermatogenesis. LH stimulates the **interstitial cells,** which lie between the seminiferous tubules in the testes, to secrete the hormone **testosterone.**

Testosterone causes the adolescent growth spurt at about age 13 years. This hormone stimulates growth of the male reproductive organs and so is responsible for the primary male sex characteristics. Testosterone is also responsible for the secondary sexual characteristics that develop at puberty. These include growth of the beard and of pubic and axillary (under the arms) hair, muscle development, and increase in length and thickness of the vocal cords that causes the voice to deepen.

What happens when testosterone is absent? If the testes are removed (castration) before puberty, the male is deprived of testosterone and becomes a eunuch. He retains childlike sex organs and does not develop secondary sexual characteristics. If castration occurs after puberty, increased secretion of male hormone by the adrenal glands helps to maintain masculinity.

HUMAN REPRODUCTION: THE FEMALE PRODUCES OVA AND INCUBATES THE EMBRYO

The female reproductive system produces ova, receives the penis and sperm released from it during sexual intercourse, houses and nourishes the embryo during prenatal development, gives birth, and produces milk for the young (lactation). These processes are regulated and coordinated by the interaction of hormones secreted by the hypothalamus, the anterior lobe of the pituitary gland, and the ovaries. The principal organs of the female reproductive system are illustrated in Figures 42–9 and 42–10.

TABLE 42–1
Principal Male Reproductive Hormones

Endocrine gland and hormones	*Principal target tissue*	*Principal actions*
Hypothalamus		
Gonadotropin-releasing hormone (GnRH)	Anterior pituitary	Stimulates release of FSH and LH
Anterior pituitary		
Follicle-stimulating hormone (FSH)	Testes	Stimulates development of seminiferous tubules; may stimulate spermatogenesis
Luteinizing hormone (LH); also called interstitial cell-stimulating hormone (ICSH)	Testes	Stimulates interstitial cells to secrete testosterone
Testes		
Testosterone	General	*Before birth:* Stimulates development of primary sex organs and descent of testes into scrotum *At puberty:* Responsible for growth spurt; stimulates development of reproductive structures and secondary sex characteristics (male body build, growth of beard, deep voice, etc.) *In adult:* Responsible for maintaining secondary sex characteristics; stimulates spermatogenesis

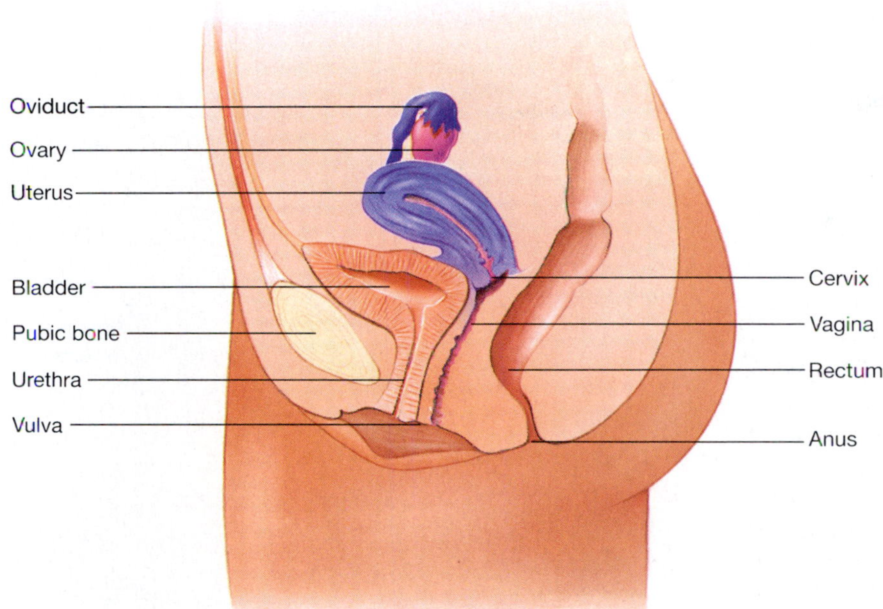

FIGURE 42–9 Midsagittal section of female pelvis showing reproductive organs. Note the position of the uterus relative to the vagina.

The Ovaries Produce Ova and Sex Hormones

Like the male gonads, the female gonads, or **ovaries,** produce both gametes and sex hormones. About the size and shape of large almonds, the ovaries are located close to the lateral walls of the pelvic cavity. The ovaries are held in position by several connective tissue ligaments.

Internally, the ovary consists mainly of connective tissue containing scattered **ova** in various stages of maturation. Each month a secondary oocyte (a developing ovum) is ejected from the ovary, a process known as **ovulation.** The process of ovum formation will be discussed in a later section.

The Oviducts Transport the Secondary Oocyte

Almost immediately after ovulation, the secondary oocyte passes into the funnel-shaped opening of the **oviduct,** or **uterine tube** (also called the fallopian tube). Peristaltic contractions of the muscular wall of the

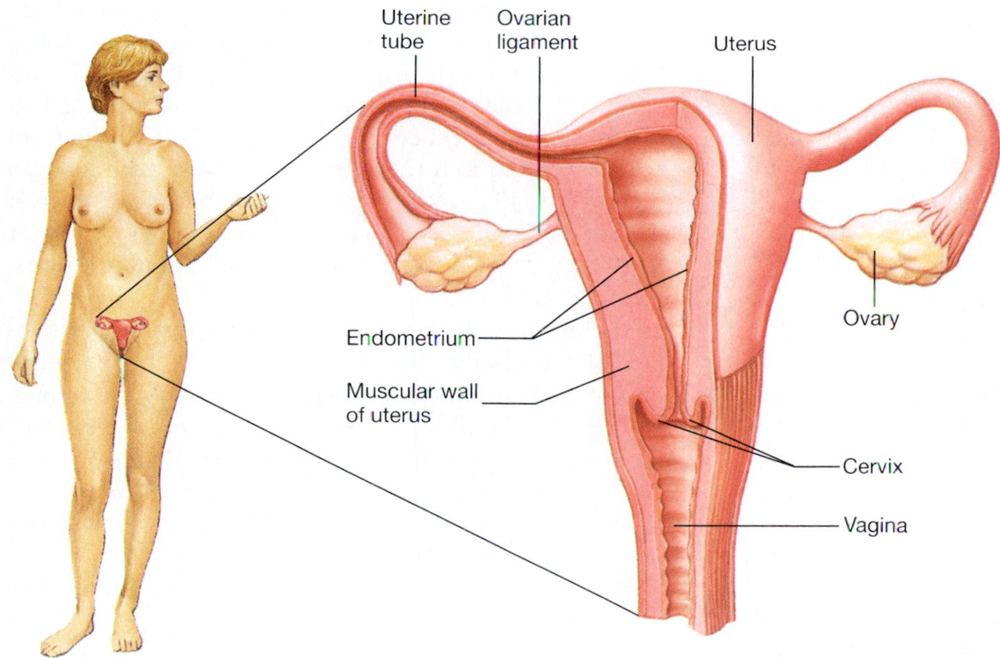

FIGURE 42–10 Anterior view of female reproductive system. Some organs have been cut open to expose the internal structure. The ligaments help to hold the reproductive organs in place.

oviduct and beating of the cilia in its lining help to move the secondary oocyte along toward the uterus. Fertilization takes place within the oviduct. If fertilization does not occur, the secondary oocyte degenerates there.

The Uterus Incubates the Embryo

The oviducts open into the upper corners of the pear-shaped **uterus** (see Fig. 42–10). About the size of a fist, the uterus (or womb) occupies a central position in the pelvic cavity. It has thick walls of smooth muscle and a mucous lining, the **endometrium,** which thickens each month in preparation for possible pregnancy.

If a secondary oocyte is fertilized, the tiny embryo finds its way into the uterus and is implanted in the endometrium. There it grows and develops, sustained by nutrients and oxygen delivered by surrounding maternal blood vessels. If fertilization does not occur during the monthly cycle, the endometrium sloughs off and is discharged in the process known as **menstruation.**

More than five million women in the United States are affected by endometriosis, a painful disorder in which fragments of the tissue lining the uterus migrate to other areas such as the oviducts or ovaries. Endometriosis can cause scarring, which can lead to infertility.

The lower portion of the uterus, called the **cervix,** projects slightly into the vagina. The cervix is a common site of cancer in women. Detection is usually possible by the routine Papanicolaou test (Pap smear) in which a few cells are scraped from the cervix during a regular gynecological examination and studied microscopically. When cervical cancer is detected at very early stages of malignancy, the patient can be cured.

The Vagina Receives Sperm

The **vagina** is an elastic, muscular tube that extends from the uterus to the exterior of the body. The vagina serves as a receptacle for sperm during sexual intercourse and as part of the birth canal when development of the fetus is complete (Fig. 42–10).

The Vulva Are External Genital Structures

The external female sex organs, collectively known as the **vulva,** include liplike folds, the **labia minora,** which surround the vaginal and urethral openings (Fig. 42–11). External to the delicate labia minora are the thicker, hair-covered **labia majora.** Anteriorly, the labia minora merge to form the prepuce of the **clitoris,** a small erectile structure comparable to the male glans penis. Like the penis, the clitoris contains erectile tissue that becomes engorged with blood during sexual excitement. Rich in nerve endings, the clitoris serves as a center of sexual sensation in the female.

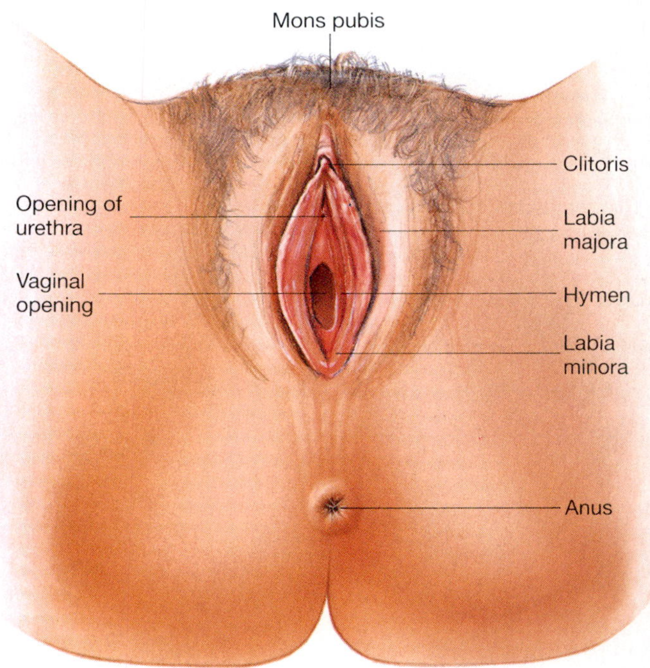

FIGURE 42–11 The vulva. The external genital structures of the female.

The **mons pubis** is the mound of fatty tissue just above the clitoris at the junction of the thighs and torso. At puberty it becomes covered by coarse pubic hair. The **hymen** is a thin ring of tissue that forms a border around the entrance to the vagina.

The Breasts Function in Lactation

Each breast is composed of 15 to 20 lobes of glandular tissue. The amount of adipose tissue around these lobes of glandular tissue determines the size of the breasts and accounts for their softness. Gland cells are arranged in grapelike clusters called alveoli (Fig. 42–12). Ducts from each cluster join to form a single duct from each lobe, so that there are 15 to 20 tiny openings on the surface of each nipple. The breasts are the most common site of cancer in women (see Focus on Health and Human Affairs: Breast Cancer).

Lactation is the production of milk for the nourishment of the young. During pregnancy, high concentrations of female hormones, estrogens and progesterone, stimulate the breasts to increase in size. For the first couple of days after childbirth, the mammary glands produce a fluid called **colostrum,** which contains protein and lactose but little fat. After birth the hormone prolactin stimulates milk production.

Breastfeeding promotes recovery of the uterus, because oxytocin released during breastfeeding stimulates the uterus to contract to nonpregnant size. Breastfeeding offers advantages to the baby as well. It promotes a close

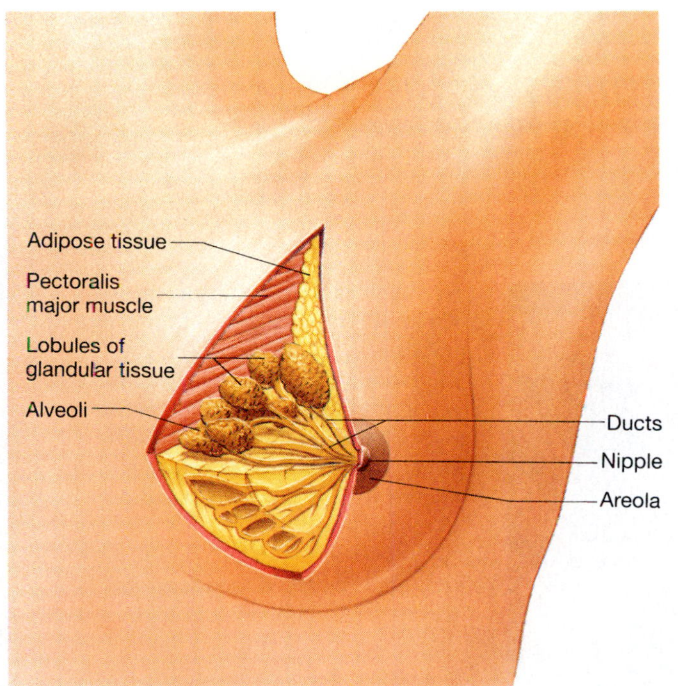

FIGURE 42–12 The mature human female breast.

bond between mother and child, and provides milk tailored to the nutritional needs of the human infant. Breast milk contains antibodies, and breastfed infants have a lower incidence of diarrhea, ear and respiratory infection, and hospital admissions than do botttlefed babies.

Ovum Formation Begins in the Ovaries

The process of ovum production, called **oogenesis,** begins in the ovaries. Before birth, hundreds of thousands of **oogonia** are present in the ovaries. All of a female's gametes originate during embryonic development. No new oogonia are formed after birth. During prenatal development, the oogonia increase in size and become **primary oocytes** (Fig. 42–13). By the time of birth, they are in the prophase of the first meiotic division. At this stage, they enter a resting phase that lasts throughout childhood and into adult life.

A primary oocyte and the cluster of cells surrounding it together make up a **follicle.** With the onset of puberty a few follicles develop each month in response to FSH se-

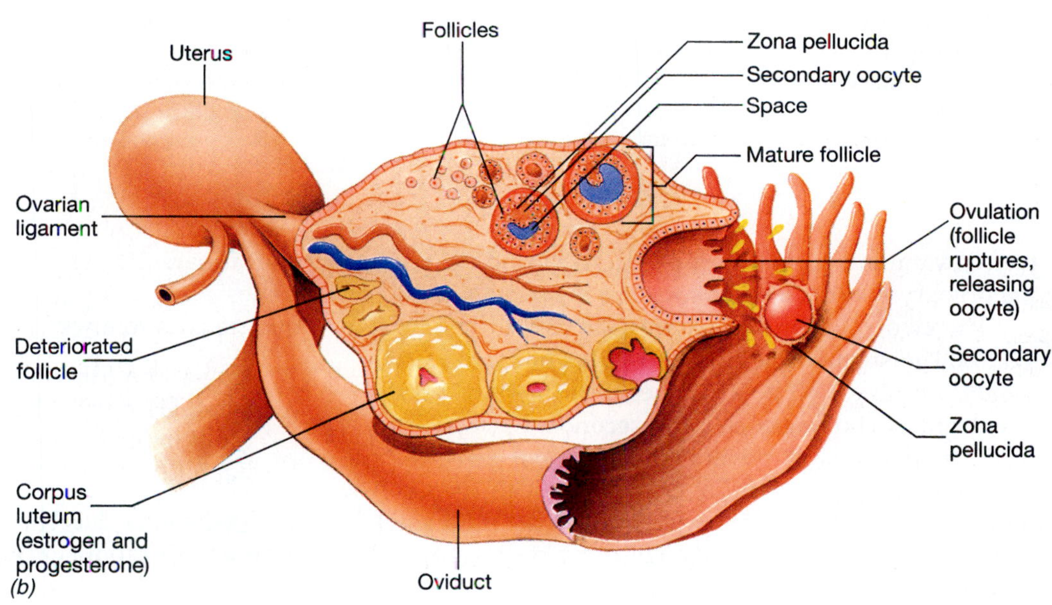

FIGURE 42–13 Microscopic structure of th ovary. (***a***) A stained section through a developing follicle. The ovum is surrounded by a layer of follicle cells that will be released along with it. These follicle cells become the corona radiata, a layer that acts as a barrier to sperm cells. (***b***) Follicles in various stages of development are scattered throughout the ovary. (Biophoto Associates)

FOCUS ON *Health and Human Affairs*

BREAST CANCER

Breast cancer is the most common type of cancer among women. Its incidence has increased in recent years, and now the disease strikes about one in every nine women. Next to lung cancer it is the leading cause of cancer deaths in women. The increased incidence of breast cancer is partly due to increased life expectancy. The causes of breast cancer are not known, but there appears to be a higher risk in women with a family history of the disease. Although no conclusive evidence yet exists, some investigators think that other risk factors include a diet rich in fats, obesity, exposure to radiation, and exposure to certain chemicals. Smoking cigarettes increases a woman's risk of dying from breast cancer by at least 25%. Women who smoke two packs of cigarettes a day or more have a 75% greater risk. Researchers are studying a tumor suppressor gene (called the maspin gene) that may play a role in breast cancer.

About 50% of breast cancers begin in the upper outer quadrant of the breast (see the photograph). As a malignant tumor grows, it may adhere to the deep tissue of the chest wall. Sometimes it extends to the skin, causing dimpling. Eventually the cancer spreads to the lymphatic system. About two-thirds of breast cancers have metastasized (spread) to the lymph nodes by the time they are first diagnosed. When diagnosis and treatment begin early, 80% of patients survive for 5 years and 62% for 10 years or longer. Untreated patients have a 5-year survival rate of only 20%.

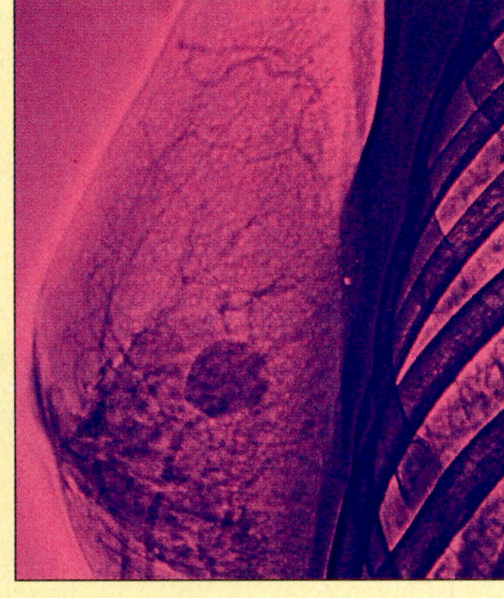

Mammogram showing breast cancer. Note the extensive vascularization. (Visuals Unlimited/SIU)

Mastectomy (surgical removal of the breast) and **radiation treatment** are common methods of treating breast cancer. Lumpectomy (surgical removal of the affected portion of the breast) in conjunction with radiation treatment is thought to be as effective as mastectomy in some cases. **Chemotherapy** is useful in preventing metastasis, especially in premenopausal patients. A recent development in cancer treatment is the use of **biological response modifiers,** which include substances such as interferons, interleukins, and monoclonal antibodies.

About one-third of breast cancers are estrogen-dependent; that is, their growth depends upon circulating estrogens. Removing the ovaries in patients with these tumors relieves the symptoms and may cause remission of the disease for months or even years. The synthetic drug tamoxifen, which blocks estrogen receptors, has been heralded as the most effective drug for preventing new cancers in women who have had breast cancer. However, recent studies suggest that tamoxifen may increase risk of uterine cancer.

Because early detection of these cancers greatly increases the chances of cure and survival, campaigns have been launched to educate women on the importance of self-examination. **Mammography,** a soft-tissue radiological study of the breast, is helpful in detecting very small lesions that might not be identified by palpation. In mammography, lesions show on an x-ray plate as areas of increased density.

creted by the anterior pituitary gland. As the follicle grows, the primary oocyte completes its first meiotic division.

The two cells produced are different in size (Fig. 42–14). The smaller one, the first **polar body,** may later divide, forming two polar bodies, but these eventually disintegrate. The larger cell, the **secondary oocyte,** proceeds to the second meiotic division, but remains in metaphase until it is fertilized. When meiosis does continue, the second meiotic division gives rise to a single ovum and a second polar body. The polar bodies are small and apparently serve to dispose of unneeded chromosomes with a minimal amount of cytoplasm. The sequence is as follows:

Oogonium (diploid) ⟶ Primary oocyte (diploid) ⟶ Secondary oocyte +1 polar body (both haploid) ⟶ Ovum +1 polar body (both haploid)

In the male, large numbers of sperm are necessary to ensure fertilization. Each primary spermatocyte gives rise

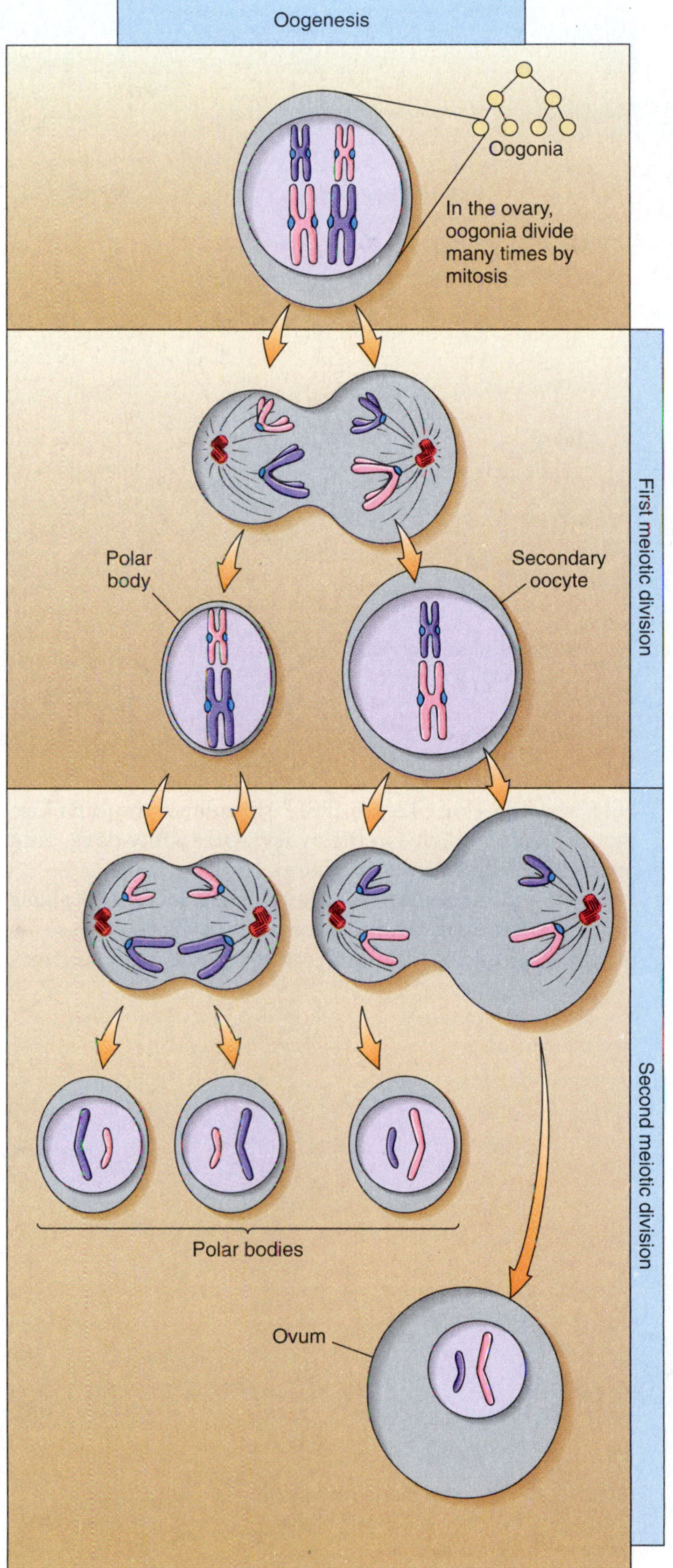

FIGURE 42–14 Oogenesis. Only one functional ovum is produced from each primary oocyte. The other cells produced are polar bodies that degenerate. The second meiotic division takes place after fertilization.

to four sperm. In contrast, each primary oocyte generates only one ovum.

As an oocyte develops, it becomes separated from its surrounding follicle cells by a thick membrane, the **zona pellucida.** As the follicle develops, follicle cells secrete fluid, which collects in the space between them (Fig. 42–13). The follicle secretes **estrogens,** female sex hormones.

As a follicle matures, it moves closer to the surface of the ovary, eventually resembling a fluid-filled bulge on the ovarian surface. Typically, only one follicle fully matures each month. Several others may develop for about a week, and then deteriorate.

At ovulation, the secondary oocyte is ejected through the wall of the ovary and into the pelvic cavity. The portion of the follicle that remains in the ovary develops into the **corpus luteum,** a temporary endocrine gland. The corpus luteum secretes estrogens and **progesterone.**

Reproductive Hormones Maintain Female Characteristics and Regulate the Menstrual Cycle

Table 42–2 lists the actions of the principal female reproductive hormones. Like testosterone in the male, estrogens are responsible for the growth of the sex organs at puberty, for body growth, and for the development of secondary sexual characteristics. In the female, these include the development of the breasts, the broadening of the pelvis, and the characteristic development and distribution of muscle and fat responsible for the shape of the female body.

Hormones of the hypothalamus, anterior pituitary, and ovaries regulate the **menstrual cycle,** the monthly sequence of events that prepares the body for possible pregnancy. The menstrual cycle runs its course every month from puberty until menopause, at about age 50.

Although wide variations exist, a typical menstrual cycle is 28 days long (Fig. 42–15). The first day of the cycle is marked by the onset of **menstruation,** the monthly discharge through the vagina of blood and tissue from the endometrium. Ovulation occurs on about the 14th day of the cycle.

During the **menstrual phase** of the cycle, which lasts about five days, GnRH is released from the hypothalamus. GnRH stimulates the anterior pituitary to release

TABLE 42–2
Principal Female Reproductive Hormones

Endocrine gland and hormones	*Principal target tissue*	*Principal actions*
Hypothalamus		
Gonadotropin-releasing hormone (GnRH)	Anterior pituitary	Stimulates release of FSH and LH
Anterior pituitary		
Follicle-stimulating hormone (FSH)	Ovary	Stimulates development of follicles; with LH, stimulates secretion of estrogen and ovulation
Luteinizing hormone (LH)	Ovary	Stimulates ovulation and development of corpus luteum
Prolactin	Breast	Stimulates milk production (after breast has been prepared by estrogen and progesterone)
Ovary		
Estrogens (estradiol)	General	Growth of sex organs at puberty; development of secondary sex characteristics (breast development, broadening of pelvis, distribution of fat and muscle)
	Reproductive structures	Maturation; monthly preparation of the endometrium for pregnancy; makes cervical mucus thinner and more alkaline
Progesterone (secreted mainly by corpus luteum)	Uterus	Completes preparation of endometrium for pregnancy
	Breast	Stimulates development

FSH and LH (Fig. 42–16). FSH stimulates the development of a few follicles in the ovary. After a few days, only one follicle continues to develop.

During the **preovulatory** (also called follicular) **phase** of the menstrual cycle, the developing follicles secrete estrogens. Estrogens stimulate growth of the endometrium, which thickens and develops new blood vessels and glands. The rise in the concentration of estrogens in the blood signals the anterior pituitary to secrete LH.

FSH and LH together stimulate the final maturation of the follicle. The rise in estrogens secreted by the developing follicle stimulates the anterior pituitary to se-

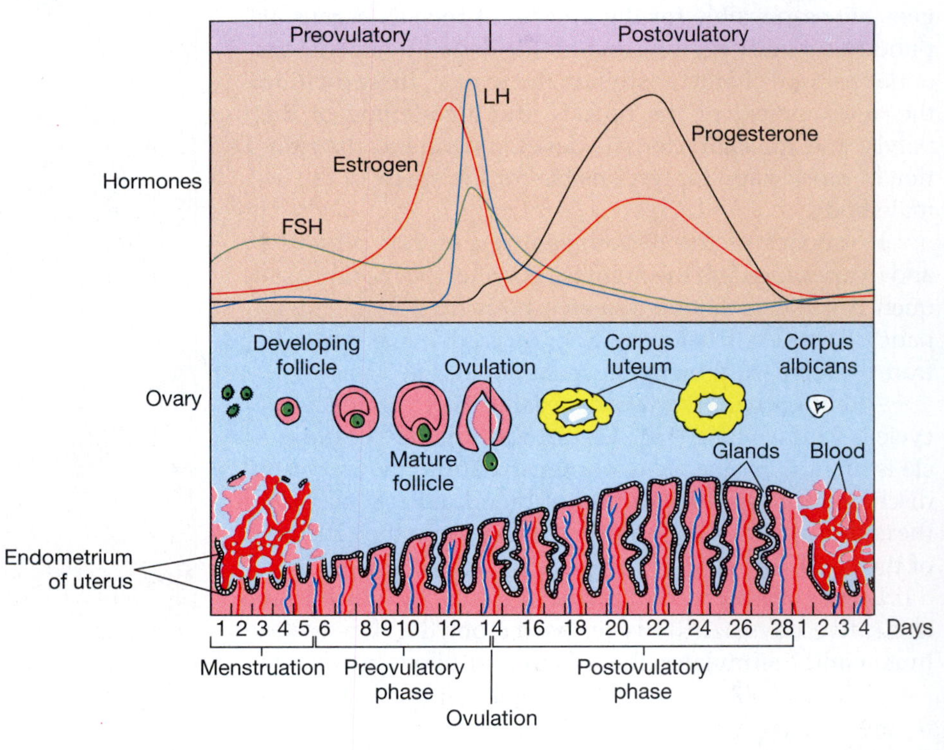

FIGURE 42–15 The menstrual cycle. The events that take place within the ovary and uterus are precisely coordinated by hormones. When fertilization does not occur, the cycle repeats itself about every 28 days. Compare this illustration with Figure 42–17.

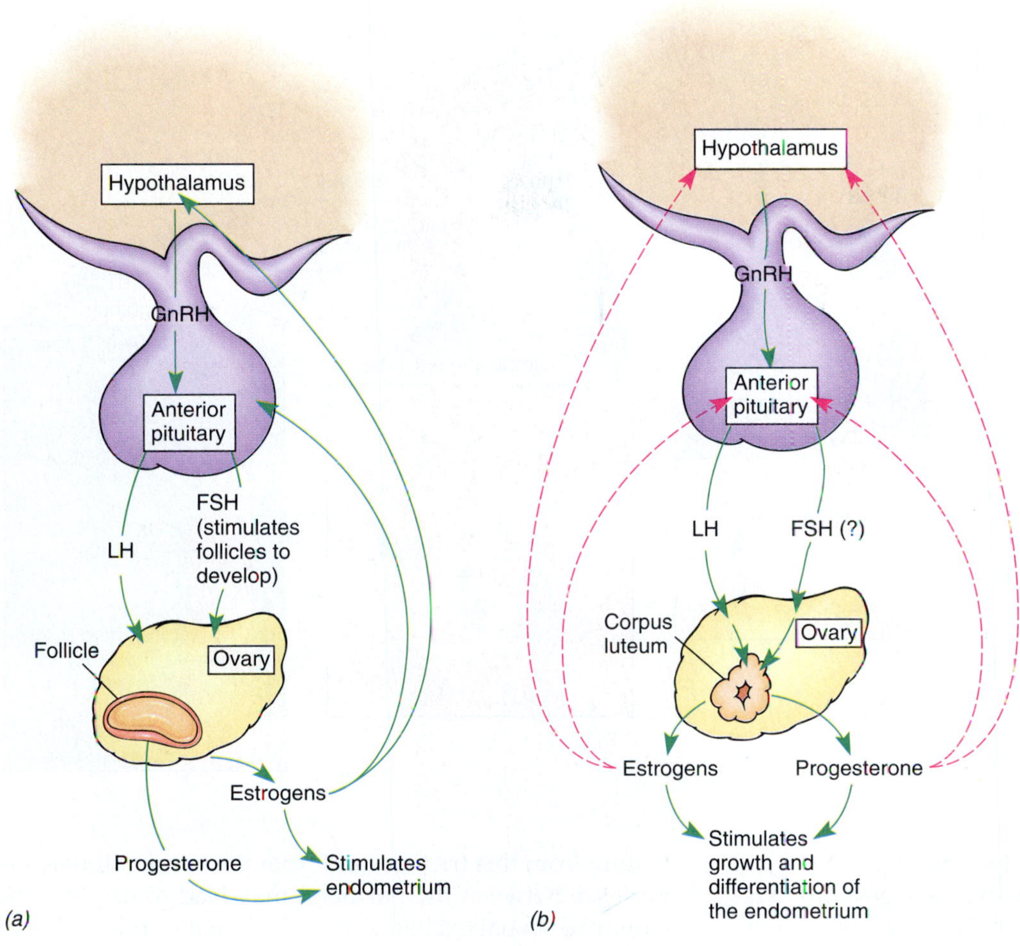

FIGURE 42–16 Hormones that regulate the menstrual cycle. (***a***) Hormonal interactions during the preovulatory phase. (***b***) Hormonal interactions during the postovulatory (luteal) phase. Red arrows indicate inhibition.

crete a surge of LH. This stimulating effect is a positive feedback mechanism. The surge of LH is necessary for the final maturation of the follicle and stimulates ovulation to occur at about day 14 of the cycle.

After the secondary oocyte has been ejected from the ovary, the **postovulatory phase** (also called luteal phase) begins. LH stimulates development of the corpus luteum, which secretes both progesterone and estrogens (Fig. 42–16b). These hormones stimulate the uterus to continue its preparation for pregnancy. Progesterone stimulates tiny glands in the endometrium to secrete a fluid rich in nutrients.

The high constant levels of estrogens and progesterone maintained by the corpus luteum inhibit GnRH, FSH, and LH secretion. If the secondary oocyte is not fertilized, the corpus luteum begins to degenerate and stops secreting progesterone and estrogens. The concentrations of these hormones in the blood fall markedly. As a result, small arteries in the endometrium constrict, reducing the oxygen supply. Menstruation begins again as cells die and damaged arteries rupture and bleed. Low levels of estrogens and progesterone trigger secretion of FSH and LH by the anterior pituitary once again.

If the secondary oocyte is fertilized, it makes its way to the uterus and begins to implant in the thick endometrium on about the seventh day after fertilization (Fig. 42–17). Membranes that develop around the embryo secrete **human chorionic gonadotropin (hCG),** a hormone that signals the mother's corpus luteum to continue to function.

Estrogen-progesterone imbalance has been suggested as a cause of **premenstrual syndrome (PMS),** a condition experienced by some women several hours to 10 days before menstruation and ending a few hours after onset of menstruation. Symptoms include fatigue, anxiety, depression, irritability, headache, edema, and skin eruptions.

SEXUAL RESPONSE INVOLVES PHYSIOLOGICAL CHANGES

During copulation, also called **coitus** or sexual intercourse in humans, the male deposits semen into the upper end of the vagina. The complex structures of the male and female reproductive systems and the physiological, endocrine, and psychological processes associated with sexual activity are adaptations that promote the successful union of sperm and secondary oocyte and the development of the resulting embryo.

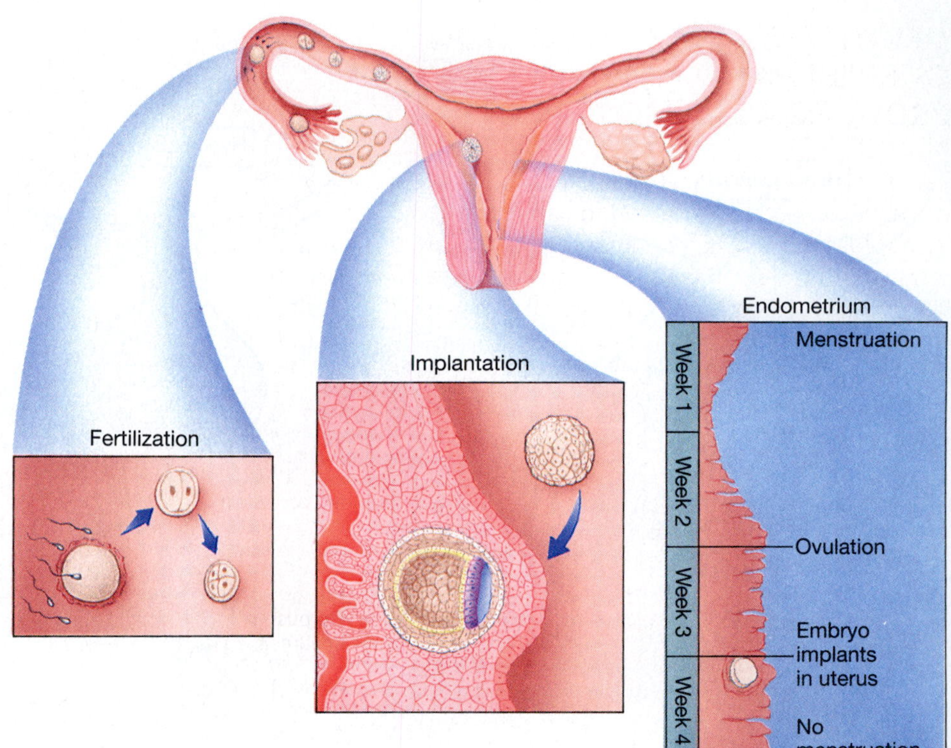

FIGURE 42–17 The menstrual cycle is interrupted when pregnancy occurs. The corpus luteum does not degenerate, and menstruation does not take place. Instead, the wall of the uterus remains thickened so that the embryo can develop within it.

Sexual stimulation results in two basic physiological responses: (1) increased blood flow (vasocongestion) to reproductive structures and certain other tissues such as the skin and (2) increased muscle tension. During vasocongestion, erectile tissues within the penis and clitoris, as well as in other areas of the body, become engorged with blood.

Sexual response includes four phases: sexual desire, excitement, orgasm, and resolution. The *desire* to have sexual activity may be motivated by fantasies or thoughts about sex. This anticipation can lead to (physical) sexual *excitement* and a sense of sexual pleasure. The excitement phase involves vasocongestion and increased muscle tension. Before the penis can enter the vagina and function in coitus, it must be erect. Penile erection is the first male response to sexual excitement. In the female, vaginal lubrication is the first response to effective sexual stimulation. During the excitement phase, the vagina lengthens and expands in preparation for receiving the penis; the clitoris and breasts become vasocongested, and the nipples become erect.

If erotic stimulation continues, sexual excitement heightens. Vasocongestion and muscle tension increase markedly. In both sexes, blood pressure increases and heart rate and breathing accelerate.

Coitus may be initiated during the excitement phase. During coitus the penis is moved inward and outward in the vagina in actions referred to as pelvic thrusts, which create friction. Physical and psychological sensations resulting from this friction (and from the entire intimate experience between the partners) may lead to **orgasm,** the climax of sexual excitement. In the female, stimulation of the clitoris is important in heightening the sexual excitement that leads to orgasm.

Although it lasts only a few seconds, orgasm is the phase of maximum sexual tension and its release. In both sexes, orgasm is marked by rhythmic contractions of the muscles of the pelvic floor and reproductive structures. These muscular contractions continue at about 0.8-second intervals for several seconds. After the first few contractions, their intensity decreases, and they become less regular and less frequent. Heart rate and respiration more than double, and blood pressure rises markedly just before and during orgasm. In the male, orgasm is marked by the **ejaculation** of semen from the penis. No fluid ejaculation accompanies orgasm in the female. Orgasm is followed by the **resolution phase,** a state of well-being during which the body is restored to its normal state.

Sexual dysfunction may be caused by psychological or biological factors. For example, chronic inability to sustain an erection, termed erectile dysfunction (formerly called impotence), is often associated with psychological issues. This disorder prevents effective coitus. Vaginismus is a condition in which, during sexual intercourse, a woman experiences painful involuntary spasms of the outer third of the vaginal muscles. Vaginismus is often associated with a history of sexual abuse.

FERTILIZATION IS THE FUSION OF SPERM AND EGG TO PRODUCE A ZYGOTE

Fertilization serves three functions: (1) the diploid number of chromosomes is restored as the sperm contributes its haploid set of chromosomes to the haploid set in the ovum; (2) in mammals and many other animals, sex of the offspring is determined; and (3) the needed stimulation is provided to initiate the reactions in the egg that permit development to take place. Fertilization and the subsequent establishment of pregnancy together are referred to as **conception.**

When conditions in the vagina and cervix are favorable, sperm begin to arrive at the site of fertilization in the upper oviduct within five minutes after ejaculation. Contractions of the uterus and oviducts help transport the sperm. The sperms' own motility is probably most important in approaching and fertilizing the secondary oocyte.

If only one sperm is needed to fertilize a secondary oocyte, why are millions involved in each act of coitus? For one thing, many sperm lose their way. Others die as a result of unfavorable pH or phagocytosis by leukocytes and macrophages in the female tract. Only a few thousand reach the vicinity of the secondary oocyte. Additionally, large numbers of sperm may be necessary to penetrate the covering of follicle cells (the corona radiata) that surrounds the secondary oocyte. Each sperm is thought to release small amounts of enzymes that break down the cement-like substance holding the follicle cells together (Fig. 42–18).

Fertilization involves several steps. First, the sperm contacts the egg and recognition occurs. If the species are the same, a species-specific protein located on the acrosome attaches to a species-specific receptor on the membrane surrounding the egg.

Next, the sperm enters the egg. As soon as a sperm enters the secondary oocyte there is a rapid electrical change, followed by a slower chemical change in the plasma membrane of the secondary oocyte. These changes prevent the entrance of other sperm. As the fertilizing sperm enters the secondary oocyte, it usually loses its tail. Sperm entry stimulates the secondary oocyte to complete its second meiotic division, resulting in formation of an ovum and another polar body.

The sperm nucleus swells, forming the male pronucleus, and the nucleus of the ovum becomes the female pronucleus. The haploid pronuclei then fuse, forming the diploid nucleus of the zygote (Fig. 42–18*c*).

After ejaculation into the female reproductive tract, sperm retain their ability to fertilize a secondary oocyte

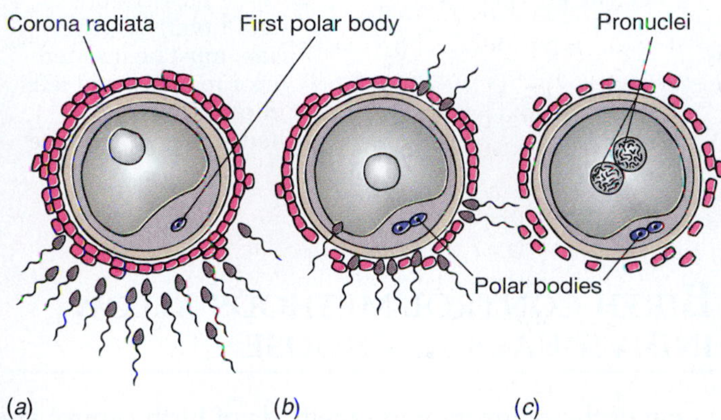

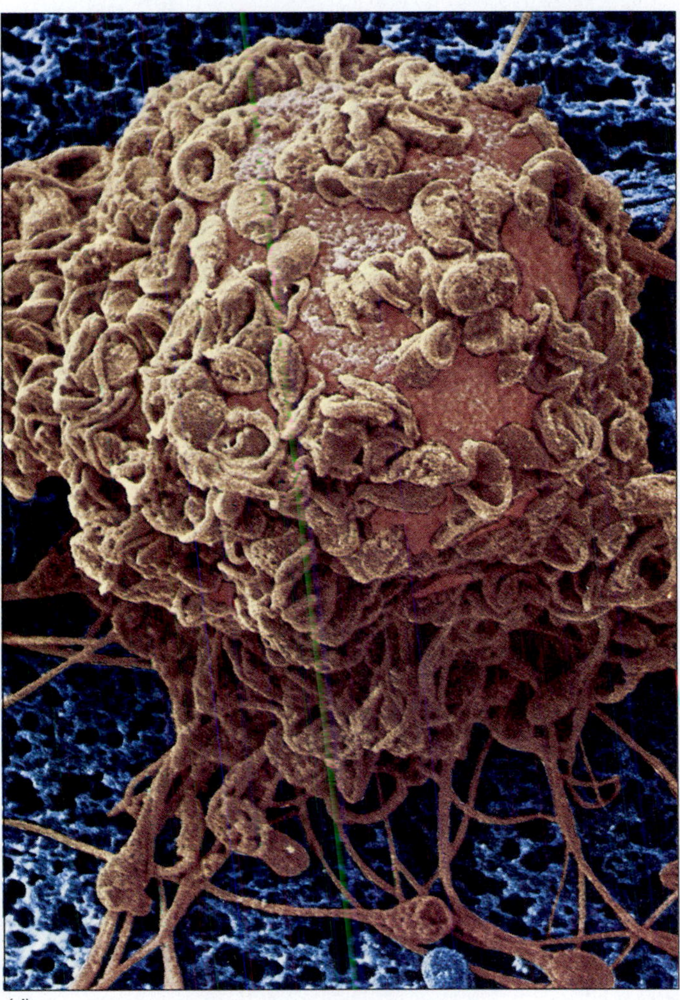

FIGURE 42–18 Fertilization. (***a***) Each sperm is thought to release a small amount of enzyme that helps to disperse the follicle cells surrounding the ovum. (***b***) After a sperm cell enters the ovum, the ovum completes its second meiotic division, producing an ovum and a polar body. (***c***) Pronuclei of sperm and ovum combine, producing a zygote with the diploid number of chromosomes. (***d***) Human sperm on test egg. Sperm are being tested for viability. (David Scharf/Peter Arnold, Inc.)

TABLE 42–3
Birth Control Methods

Method	Failure rate*	Mode of action	Advantages	Disadvantages
Oral contraceptives	0.3; 5	Inhibit ovulation; may also affect endometrium and cervical mucus and prevent implantation	Highly effective; regulate menstrual cycle	Minor discomfort in some women; possible thromboembolism; hypertension, heart disease in some users; possible increased risk of infertility; should not be used by women who smoke
Depo-Provera (medroxyprogesterone acetate)	About 1	Inhibits ovulation	Effective; long-lasting	Fertility may not return for 6–12 months after use is discontinued
Progesterone implantation (Norplant)	About 1	Inhibits ovulation	Effective; long-lasting	Irregular menstrual bleeding in some women
Intrauterine device (IUD)	1; 5	Not known; probably stimulates inflammatory response	Provides continuous protection; highly effective	Cramps; increased menstrual flow; spontaneous expulsion; increased risk of pelvic inflammatory disease and infertility; not recommended for women who have not completed childbearing
Spermicides (sponges, foams, jellies, creams)	3; 20	Chemically kill sperm	No side effects (?); vaginal sponges are effective in vagina for up to 24 hours after insertion; sponges also act as physical barriers to sperm cells	Some evidence linking spermicides to birth defects
Contraceptive diaphragm (with jelly)†	3; 14	Diaphragm mechanically blocks entrance to cervix; jelly is spermicidal	No side effects	Must be prescribed (and fitted) by physician; must be inserted prior to coitus and left in place for several hours after intercourse

for about 48 hours. The secondary oocyte itself remains fertile for only about 24 hours after ovulation. Thus, in a very regular 28-day menstrual cycle, sexual intercourse is most likely to result in fertilization from two days before to one day after ovulation. However, many women do not have regular menstrual cycles. Many factors can result in irregularities, even in women who generally have regular cycles.

In view of the many factors working against fertilization, it may seem remarkable that it ever occurs! Yet the frequency of coitus and the large number of sperm deposited at each ejaculation enable the human species not only to maintain itself but to increase its numbers at an alarming rate.

BIRTH CONTROL METHODS ALLOW INDIVIDUALS TO CHOOSE

Some of the more common methods of birth control are described in the following paragraphs and in Table 42–3 (see also Fig. 42–19). Note that intrauterine devices (IUDs) as well as some types of oral contraceptives may not actually prevent fertilization; they probably destroy the embryo or prevent its implantation in the wall of the uterus.

Hormone Contraceptives Prevent Ovulation

More than 80 million women worldwide (more than 8 million in the United States alone) use **oral contracep-**

Method	*Failure rate**	*Mode of action*	*Advantages*	*Disadvantages*
Condom	2.6; 10	Mechanically prevents sperm from entering vagina	No side effects; some protection against STD, including AIDS	Interruption of foreplay to put it on; slightly decreased sensation for male; could break
Rhythm‡	13; 21	Abstinence during fertile period	No side effects	Not very reliable
Douche	40	Flush semen from vagina	No side effects	Not reliable; sperm are beyond reach of douche in seconds
Withdrawal (coitus interruptus)	9; 22	Male withdraws penis from vagina prior to ejaculation	No side effects	Not reliable; contrary to powerful drives present when an orgasm is approached; sperm in the fluid secreted before ejaculation may be sufficient for conception
Sterilization				
Tubal ligation	0.04	Prevents ovum from leaving uterine tube	Most reliable method	Often not reversible
Vasectomy	0.15	Prevents sperm from leaving scrotum	Most reliable method	Often not reversible
Chance (no contraception)	About 90			

*The lower figure is the failure rate of the method; the higher figure is the rate of method failure plus failure of the user to utilize the method correctly. Based on number of failures per 100 women who use the method per year in the United States.

†The failure rate is lower when the diaphragm is used together with spermicides.

‡There are several variations of the rhythm method. For those who use the calendar method alone, the failure rate is about 35. However, if the body temperature is taken daily and careful records are kept (temperature rises after ovulation), the failure rate can be reduced. Also, if a daily record of the type of vaginal secretion is kept, changes in cervical mucus can be noted and used to determine time of ovulation. This type of rhythm contraception is also slightly more effective. When women use the temperature or mucus method and have intercourse *only* more than 48 hours *after* ovulation, the failure rate can be reduced to about 7.

tives. The most common preparations are combinations of progestin (synthetic progesterone) and synthetic estrogen. (Natural hormones are destroyed by the liver almost immediately, but synthetic ones are chemically modified so that they can be absorbed effectively and metabolized slowly.) A woman takes one pill each day for about 3 weeks. She then takes a sugar pill for 1 week that allows menstruation to occur because of the withdrawal of the hormones. When taken correctly, these pills are about 99.9% effective in preventing pregnancy.

Most oral contraceptives prevent pregnancy by preventing ovulation. When postovulatory levels of ovarian hormones are maintained in the blood, the body is tricked into responding as though conception has occurred. The pituitary gland is inhibited and does not produce the surge of FSH and LH that stimulates ovulation. The chief advantage of oral contraceptives is their high rate of effectiveness.

Studies suggest that women over the age of 35 who smoke or have other risk factors, such as untreated hypertension, should not take oral contraceptives. Women in this category who take oral contraceptives have an increased risk of death from circulatory diseases such as stroke and myocardial infarction. The new low-dose oral contraceptive pills appear to be safe for nonsmokers up to the time of menopause. Oral contraceptives result in

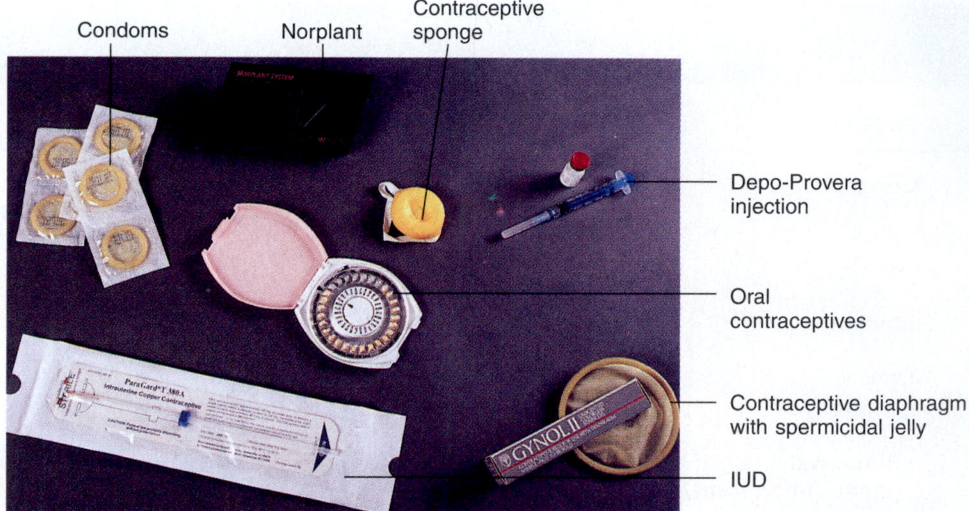

FIGURE 42–19 Some commonly used contraceptives. (Dennis Drenner)

death in about 3 per 100,000 users. This compares favorably with the death rate of about 9 per 100,000 pregnancies (Table 42–4).

Recently, Depo-Provera (a form of progesterone) has been approved for contraceptive use in the United States. This hormone prevents ovulation by suppressing anterior pituitary function. It is generally injected intramuscularly every three months.

Another form of contraception involves the implantation of progesterone (Norplant) in the woman's body. Several soft, flexible capsules are inserted under the skin of the upper arm. There the synthetic hormone is continuously released and is transported by the blood. The hormone inhibits ovulation and stimulates thickening of the cervical mucus (which makes it more difficult for the sperm to reach the egg). The capsules are effective for about five years, after which they must be replaced. Progesterone implants are one of the most effective, reversible contraceptive procedures available. The failure rate is less than 1 in 100. The most common side effect is irregular menstrual bleeding, which may last up to one year.

TABLE 42–4
Deaths in the United States from Pregnancy and Childbirth and from Some Birth Control Methods

	Death rate per 100,000
Pregnancy and childbirth	9
Oral contraception	3
IUD	0.5
Legal abortions—first trimester	1.9
Legal abortions—after first trimester (mainly therapeutic abortions)	12.5
Illegal abortions performed by medically untrained individuals	About 100

Use of the Intrauterine Device (IUD) Has Declined

The **intrauterine device (IUD)** is a small plastic loop or coil that must be inserted into the uterus by a medical professional. Once in place, some types of IUD can be left in the uterus indefinitely or until the woman wishes to conceive. Newer types of IUDs are about 99% effective.

The mode of action of the IUD is not well understood. White blood cells, mobilized in response to the foreign body in the uterus, may produce substances toxic to the fertilized ovum. Some IUDs (copper T and copper 7) contain copper, which dissolves slowly in the uterine secretions and apparently interferes with implantation of the embryo. Disadvantages of the IUD include increased risk of pelvic inflammatory disease, ectopic pregnancy when pregnancy does occur, and abnormal uterine bleeding.

Other Common Contraceptive Methods Include the Diaphragm and Condom

The **contraceptive diaphragm** mechanically blocks the passage of sperm from the vagina into the cervix. It is covered with spermicidal jelly or cream and inserted just prior to sexual intercourse.

The **condom** is also a mechanical method of birth control. The most commonly used male contraceptive device, the condom provides a barrier that contains the semen so that sperm cannot enter the female tract. The condom is

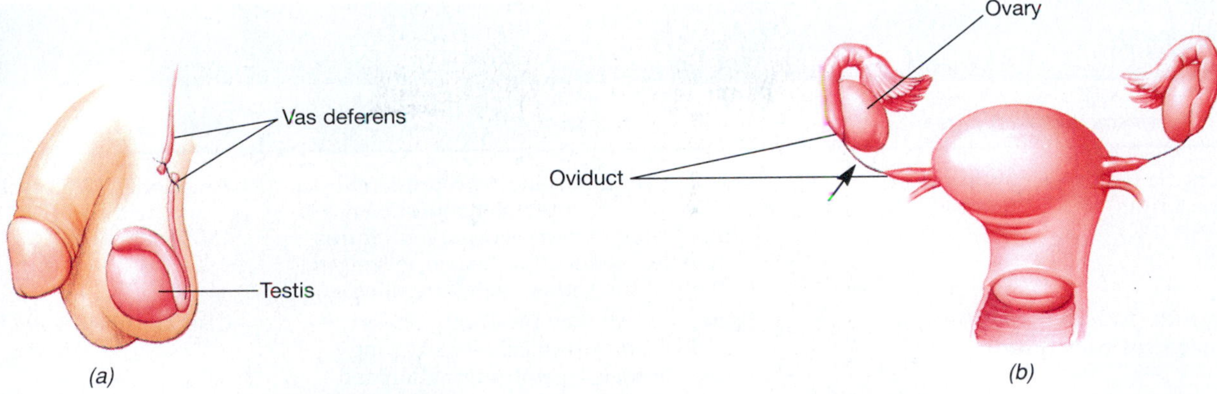

FIGURE 42–20 Sterilization. (***a***) In vasectomy, the vas deferens (sperm duct) on each side is cut and tied. (***b***) In tubal ligation, each oviduct (uterine tube) is cut and tied so that ovum and sperm can no longer meet.

one of the only contraceptives that provides some protection against AIDS and other sexually transmitted diseases.

Sterilization Renders an Individual Incapable of Producing Offspring

Aside from total abstinence, **sterilization** is the only foolproof method of contraception. Sterilization is currently the most popular contraceptive method for couples in which the wife is over the age of 30. About 75% of sterilization operations are currently performed on males.

Male Sterilization Is Performed by Vasectomy

An estimated one million **vasectomies** are performed each year in the United States. Using a local anesthetic, a small incision is made on each side of the scrotum. Then, each vas deferens is cut and its ends are tied or clipped so that they cannot grow back together (Fig. 42–20*a*).

Because testosterone secretion and transport are not affected, vasectomy does not affect masculinity. Sperm continue to be produced, though at a much slower rate, and are destroyed by macrophages in the testes. No change in the amount of semen ejaculated is noticed, because sperm account for very little of the semen volume.

By surgically reuniting the ends of the vasa deferentia, surgeons can successfully reverse sterilization in about 50% of attempts made. Apparently, some sterilized men eventually develop antibodies against their own sperm and remain sterile even when their vasectomies are surgically reversed.

An alternative to reversal of vasectomy is the storage of frozen sperm in sperm banks. If the male should decide to father another child after he has been sterilized, he simply "withdraws" his sperm for use in artificially inseminating his mate. Sperm banks have been established throughout the United States. Not much is known yet about the effects of long-term sperm storage, but there may be an increased risk of genetic defects.

Female Sterilization Is by Tubal Ligation

Several techniques are in current use that prevent the transport of ova. Most of them involve **tubal ligation,** which involves cutting and tying the oviducts (Fig. 42–20*b*). This can be done through the vagina, but it usually is performed through an abdominal incision. As in the male, hormone balance and sexual performance are not affected.

THERE ARE THREE TYPES OF ABORTION

Abortion is termination of pregnancy that results in the death of the embryo or fetus. Worldwide, an estimated 40 million deliberate abortions are performed each year (more than one million in the United States). Three types of abortions may be distinguished: spontaneous abortions, therapeutic abortions, and those undertaken as a means of birth control. **Spontaneous abortions** (popularly called miscarriages) occur without intervention and often are a biological mechanism for destroying an abnormal embryo. **Therapeutic abortions** are performed in order to maintain the health of the mother, or when there is reason to suspect that the embryo is grossly abnormal. The third type of abortion—that performed as a means of birth control—is the most controversial.

Most first-trimester abortions (those performed during the first 3 months of pregnancy) and some later ones are performed using a suction method. After the cervix has

TABLE 42–5
*Some Common Sexually Transmitted Diseases**

Disease and causative organism	*Course of disease*	*Treatment*
Chlamydia (*Chlamydia trachomatis,* a bacterium)	Discharge and burning with urination, or may be asymptomatic; most common cause of nongonococcal urethritis in males; about 10% of male college students in the United States are infected	Antibiotics: Doxycycline and azithromycin (Zithromax)
Gonorrhea (*Neisseria gonorrhoeae,* a gonococcus bacterium)	Bacterial toxin may produce redness and swelling at infection site; symptoms in males: painful urination and discharge of pus from penis; in about 60% of infected women no symptoms occur in initial stages; can spread to epididymis (in males) or uterine tubes and ovaries (in females), causing sterility; can cause widespread pelvic or other infection, plus damage to heart valves, meninges (outer coverings of brain and spinal cord), and joints	Cephalosporin
Syphilis (*Treponema pallidum,* a spirochete bacterium)	Bacteria enter body through defect in skin near site of infection and spread throughout the body by lymphatic and circulatory routes; primary chancre (a small, painless ulcer) forms at site of initial infection and heals in about a month; highly infectious at this stage; secondary stage follows, in which a widespread rash and influenza-like symptoms may occur; scaly lesions may occur that teem with bacteria and are highly infectious; latent stage that follows can last 20 years; eventually, lesions called gummae may occur, consuming parts of the body surface or damaging liver, bone, or spleen; serious brain damage may occur; death results in 5–10% of cases	Penicillin

Clinical symptoms of syphilis. A primary syphilitic chancre. This is usually the first symptom of syphilis. (Custom Medical Stock Photo)

been dilated, a suction aspirator is inserted in the uterus, and the embryo and other products of conception are evacuated. In pregnancies of more than 12 weeks, the method most commonly used is dilation and evacuation ("D & E"). The cervix is dilated, forceps are used to remove the fetus, and suction is used to aspirate the endometrium. Such abortions are mainly therapeutic abortions.

When abortion is performed during the first trimester by skilled medical personnel, the mortality is about 1.9 per 100,000. After the first trimester, the death rate rises to 12.5 per 100,000 (Table 42–4). The U.S. death rate from illegal abortions performed by medically untrained individuals is about 100 per 100,000. These statistics can be contrasted with the death rate from pregnancy and childbirth of about 9 per 100,000.

SEXUALLY TRANSMITTED DISEASES ARE SPREAD BY SEXUAL CONTACT

Sexually transmitted diseases (STD), also called venereal diseases (VD), are, next to the common cold, the most prevalent communicable diseases in the world. The World Health Organization has estimated that more than 250 million people are infected each year with gonorrhea and more than 50 million with syphilis. Currently, the most common STD in the United States is chlamydia. According to Planned Parenthood, one out of four U.S. teenagers becomes infected with an STD by age 21. Some common sexually transmitted diseases are listed and described in Table 42–5 (AIDS was discussed in Chapter 37).

Disease and causative organism	*Course of disease*	*Treatment*
Genital herpes (herpes simplex type 2 virus)	Tiny, painful blisters appear on genitals; may develop into ulcers; influenza-like symptoms may occur; recurs periodically; threat to fetus or newborn infant; may predispose to cervical cancer in females	No effective cure; some drugs may shorten outbreaks or reduce severity of symptoms
Trichomoniasis (a protozoon)	Symptoms include itching, discharge, soreness; can be contracted from dirty toilet seats and towels; may be asymptomatic in males	Flagyl (an antibiotic)
"Yeast" infections (genital candidiasis; *Candida albicans*)	Irritation, soreness, discharge; especially common in females; may be asymptomatic in males	Antifungal drugs
Pelvic inflammatory disease (PID; primarily caused by gonorrhea or chlamydia)	Generalized infection of reproductive organs and pelvic cavity; usually chronic and difficult to treat; may lead to sterility (more than 15% of cases)	Antibiotics, surgical removal of affected organs
Genital warts (certain strains of human papilloma virus [HPV])	Warty growths may be present on the internal or external genitalia; associated with cervical cancer	Difficult to treat; antiviral drugs or injection with agents that destroy infected cells
Acquired immune deficiency syndrome (AIDS)* (caused by a retrovirus known as human immunodeficiency virus, HIV)	Influenza-like symptoms; swollen lymph glands, fever, night sweats, weight loss; decreased immunity leading to pneumonia, rare forms of cancer	AZT and a variety of experimental drugs

*AIDS was discussed in Chapter 37.

Chapter Summary

I. In asexual reproduction, a single parent endows its offspring with a set of genes similar to its own (except for mutations). In animal sexual reproduction, offspring are produced by the fusion of two types of gametes (egg and sperm).

II. The human male reproductive system includes the testes, which produce sperm and testosterone; a series of conducting tubes; accessory glands; and the penis.
   A. The testes, housed in the scrotum, contain the seminiferous tubules, where the sperm are produced, and the interstitial cells, which secrete testosterone.
   B. Sperm complete their maturation and are stored in the epididymis and vas deferens.
   C. During ejaculation, sperm pass from the vas deferens to the ejaculatory duct and then into the urethra, which passes through the penis.
   D. Each ejaculate of semen contains about 400 million sperm suspended in the secretions of the seminal vesicles and prostate gland.
   E. The penis consists of three columns of erectile tissue. When this tissue becomes engorged with blood, the penis becomes erect.

F. The gonadotropic hormones FSH and LH stimulate sperm production and testosterone secretion. Testosterone is responsible for establishing and maintaining primary and secondary sex characteristics in the male.

III. The female reproductive system includes the uterus; ovaries, which produce ova and hormones; the oviducts (uterine tubes); vagina; vulva; and breasts.
   A. After ovulation, the secondary oocyte enters the oviduct where it may be fertilized.
   B. The uterus serves as an incubator for the developing embryo.
   C. The vagina is the lower part of the birth canal, and also receives the penis during sexual intercourse.
   D. The first day of menstrual bleeding marks the first day of the menstrual cycle. Ovulation occurs at about day 14 in a typical 28-day menstrual cycle. Events of the menstrual cycle are coordinated by the gonadotropic and ovarian hormones.
      1. FSH stimulates follicle development; FSH and LH together stimulate ovulation; LH promotes development of the corpus luteum.
      2. The developing follicles release estrogens, which stimulate development of the endometrium and are responsible for the secondary female sex characteristics.
      3. The corpus luteum secretes progesterone and estrogens, which stimulate final preparation of the uterus for possible pregnancy.

IV. Vasocongestion and increased muscle tension are two basic physiological responses to sexual stimulation. The phases of sexual response include sexual desire, excitement, orgasm, and resolution.

V. Fertilization is the fusion of ovum and sperm to form a zygote.

VI. Effective methods of birth control include hormonal contraceptives, intrauterine devices, condoms, contraceptive diaphragms, and sterilization.

VII. Among the important types of sexually transmitted diseases are chlamydia, gonorrhea, syphilis, genital herpes, pelvic inflammatory disease, and AIDS.

Selected Key Terms

clitoris, p. 842
coitus, p. 847
corpus luteum, p. 845
endometrium, p. 842
epididymis, , p. 838
estrogens, p. 845
follicle,, p. 843
FSH (follicle-stimulating hormone), p. 840
GnRH (gonadotropin-releasing hormone), p. 840
hCG (human chorionic gonadotropin), p. 847
interstitial cells (of testis), p. 840
LH (luteinizing hormone), p. 840
oocyte, p. 843
oogonia, p. 843
orgasm, p. 848
ovary, p. 841
oviduct, p. 841
ovulation, p. 841
ovum, (pl. ova) p. 841
polar body, p. 844
progesterone, p. 845
prostate, p. 838
scrotum, p. 837
secondary oocyte, p. 844
semen, p. 838
seminiferous tubules, p. 835
sperm, p. 837
spermatid, p. 836
spermatocyte, p. 836
sterilization, p. 853
testis, (pl. testes) p. 835
testosterone, p. 840
uterus, p. 842
vas deferens, p. 838

Post-Test

1. The type of reproduction in which an animal divides into several pieces and then each piece develops into an entire new animal is called ________________.

2. Parthenogenesis is a type of reproduction in which a/an ________________ (fertilized/unfertilized) egg develops into an adult animal.

3. An individual that can produce both eggs and sperm is described as ________________.

4. A sex cell (either egg or sperm) is called a/an ________________; a fertilized egg is a/an ________________.

5. An adult who is unable to parent offspring is said to be ________________.

6. The testes are contained within the ________________.

7. A/An ________________ abortion is performed in order to maintain the mother's health or when the embryo is thought to be grossly abnormal.

8. Tubal ligation is a common method of ________________.

9. A typical menstrual cycle is about ______________ days long; ovulation takes place about day ______________.
10. Sperm are produced in the (a) prostate gland (b) seminal vesicles (c) seminiferous tubules of the testes (d) vas deferens (e) erectile tissue of the penis.
11. Testosterone is produced mainly by the (a) prostate gland (b) seminal vesicles (c) seminiferous tubules of the testes (d) vas deferens (e) interstitial cells of the testes.
12. Alkaline fluid that becomes part of the semen is secreted into the urethra by the (a) prostate gland (b) seminal vesicles (c) seminiferous tubules of the testes (d) vas deferens (e) erectile tissue of the penis.
13. The correct sequence of structures through which sperm pass is (a) prostate gland ⟶ seminal vesicles ⟶ bulbourethral glands (b) epididymis ⟶ urethra ⟶ seminal vesicle (c) seminiferous tubules of the testes ⟶ epididymis ⟶ vas deferens (d) seminiferous tubules of the testes ⟶ ejaculatory duct ⟶ vas deferens (e) seminiferous tubules of the testes ⟶ urethra ⟶ vas deferens.
14. Estrogens are produced mainly by the (a) ovary (b) cervix (c) testes (d) endometrium (e) oviduct.
15. The structure that thickens each month in preparation for possible pregnancy is the (a) cervix (b) endometrium (c) ovary (d) corpus luteum (e) labia majora.
16. The correct sequence of structures through which an ovum passes is (a) ovary ⟶ oviduct (uterine tube) ⟶ uterus (b) ovary ⟶ urethra ⟶ cervix (c) oviduct ⟶ ovary ⟶ uterus (d) ovary ⟶ oviduct ⟶ cervix (e) ovary ⟶ endometrium ⟶ oviduct.
17. Gametes are produced in the (a) ovary (b) endometrium (c) testes (d) answers (a) and (b) are correct (e) answers (a) and (c) are correct.
18. The structure that produces progesterone is the (a) cervix (b) endometrium (c) corpus luteum (d) clitoris (e) labia majora.
19. The correct sequence of events is (a) FSH secreted ⟶ follicles develop ⟶ corpus luteum secretes estrogen ⟶ ovulation (b) ovary secretes progesterone ⟶ FSH secretion ⟶ corpus luteum development ⟶ estrogen (c) FSH secreted ⟶ follicles develop ⟶ estrogen secreted ⟶ LH secretion ⟶ ovulation (d) ovary secretes FSH ⟶ follicles develop ⟶ estrogen secreted ⟶ menstruation (e) ovary secretes LH ⟶ endometrium thickens ⟶ menstruation.
20. Fertilization generally takes place in the (a) cervix (b) endometrium (c) vagina (d) oviduct (e) ovary.
21. Which of the following is responsible for secondary sexual characteristics in the female? (a) testosterone (b) progesterone (c) estrogens (d) FSH (e) HCG
22. Which of the following blocks passage of sperm from the vagina into the uterus? (a) IUD (b) contraceptive diaphragm (c) Depo-Provera (d) oral contraceptives (e) rhythm method
23. A primary chancre is a sign of (a) early genital herpes (b) early gonorrhea (c) chlamydia infection (d) late syphilis (e) early syphilis.

Review Questions

1. Compare asexual with sexual reproduction, and give two specific examples of asexual reproduction.
2. What are the advantages of internal fertilization?
3. Explain the physiological basis of erection of the penis.
4. Compare the functions of ovaries and testes.
5. Trace the passage of sperm from a seminiferous tubule through the male reproductive system until it leaves the male body during ejaculation. Assuming that ejaculation takes place within the vagina, trace the journey of the sperm until it meets the ovum.
6. What are the actions of testosterone? Estrogens? Progesterone?
7. What is the function of the corpus luteum? Which hormone stimulates its development?
8. What are the actions of FSH and LH in the female?
9. Why are so many sperm produced in the male and so few ova produced in the female?
10. Which methods of birth control are most effective? Least effective?

11. Draw a diagram of the principal events of the menstrual cycle, including ovulation and menstruation. Indicate on which days of the cycle sexual intercourse would most likely result in pregnancy.

12. Label the following diagrams. (Refer to Figs. 42–3 and 42–10 in the text as necessary.)

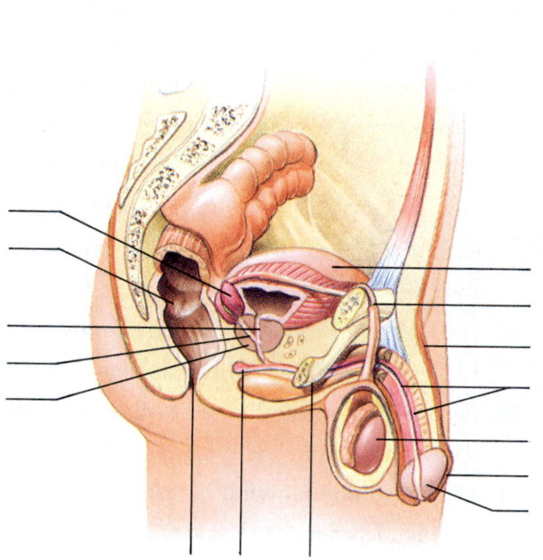

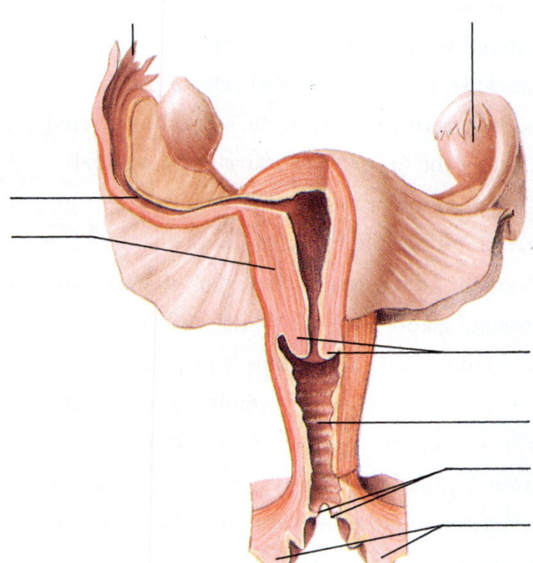

Thinking Critically

1. Imagine that you were designing a program to limit human population expansion. What are some of the measures you would include?
2. What are the biological advantages of hermaphroditism when self-fertilization occurs? When cross fertilization is necessary?
3. What are the relationships among the following conditions: erectile dysfunction, sterility, and castration?
4. What would happen if ovulation occurred, but no corpus luteum developed?
5. Why do you think a variety of effective male methods of contraception have not yet been developed? If you were a researcher, what are some approaches you might take to male contraception?

Recommended Readings

Duellman, W.E., "Reproductive Strategies of Frogs," *Scientific American,* Vol. 267, No. 1, July 1992, pp. 80–87. In addition to the egg-to-tadpole progression, frogs have other reproductive strategies such as egg to froglet, egg brooding, and tadpoles in the mother's stomach.

Frisch, R.E., "Fatness and Fertility," *Scientific American,* Vol. 258, No. 3, March 1988, pp. 88–95. Discusses the possibility that fat tissue exerts a regulatory effect on human female reproduction.

Solomon, E.P., R.R. Schmidt, and P. Adragna, *Human Anatomy and Physiology.* Philadelphia: Saunders College Publishing, 1990. This book has an excellent chapter on human reproduction.

Wassarman, P.M., "Fertilization in Mammals," *Scientific American,* Vol. 259, No. 6, December 1988. A glycoprotein governs many of the events of fertilization, including the process by which the fertilized egg prevents other sperm from entering.

Louise Brown (left) was the first child to be conceived as a result of in vitro fertilization. *Fifteen years old in this photo, she is shown with her sister Natalie, also an in vitro baby. (People Weekly ©, 1994 John Loengard)*

Novel Origins

Human embryo research has been hotly debated by scientists, politicians, and theologians for almost two decades. Federally funded research on **in vitro fertilization,** a technique by which an ovum is fertilized with sperm in laboratory glassware, was halted in 1980 by regulations under effect in the Reagan administration and carried over through the Bush administration. In the mid 1990s the Clinton administration and Congress have moved to support human embryo research. Legitimate ethical concerns need to be addressed, as do misconceptions promoted by the media. Images have been created in the public mind of breeding children as a source of compatible organs for ill parents or older siblings, or of mass-producing armies bred for physical strength and endurance.

Techniques developed through reproduction and human embryo research have been used clinically for several years to help couples who are having difficulty conceiving children. **Infertility,** the inability of a couple to achieve conception, affects about 15% of married couples in the United States.

A major cause of male infertility is **sterility,** which can result from insufficient sperm production. Sometimes semen is found to contain large numbers of abnormal sperm or occasionally none at all. Men with fewer than 20 million sperm per milliliter of semen are usually considered to be sterile.

Permanent sterility occurs in about one-fourth of the cases of mumps in adult males. Low sperm counts have also been linked to chronic marijuana use, alcohol abuse, and cigarette smoking. Studies show that men who smoke tobacco are more likely than nonsmokers to produce abnormal sperm. Exposure to chemicals such as DDT and PCBs may also result in low sperm count. Use of anabolic steroids to accelerate muscle development (see Chapter 41 Opener) can cause sterility in both males and females.

When the male partner of a couple desiring a child is sterile or carries a genetic defect, **artificial insemination** can be used. About 10,000 children born each year are products of this procedure. Although the sperm donor usually remains anonymous to the couple involved, his genetic qualifications are screened by physicians.

Female infertility can result from scarring of the oviducts that blocks the passage of the secondary oocyte to the uterus. One cause of scarring is the inflammation of the oviducts caused by pelvic inflammatory disease (PID). Women with blocked oviducts usually can produce ova and can incubate an embryo normally. However, they need clinical assistance in getting the ovum from the ovary to the uterus.

In vitro fertilization was first used in England in 1978 to help a couple who had tried unsuccessfully for several years to have a child. Since that time, thousands of "test tube babies" have been conceived and have been born to previously infertile women. In this procedure an ovum is removed from a woman's ovary, fertilized with sperm in laboratory glassware, and then reimplanted in the woman's uterus, where it may develop normally.

Oocyte donation is an infertility treatment for women who do not produce ova, including older women who want to become pregnant. By 1993, an estimated 1000 cases of oocyte donation had been performed in the United States, and the number is increasing rapidly.

Another novel procedure is **host mothering.** A tiny embryo is removed from its natural mother and implanted into a female substitute. The foster mother can support the developing embryo either until birth or temporarily until it is implanted again into the original mother or into another host. Although not approved in humans, this technique has already proved useful to animal breeders. For example, embryos from prize sheep can be temporarily implanted into rabbits for easy shipping by air, and then reimplanted into a foster mother sheep, perhaps of inferior quality. Host mothering also has the advantage of allowing an animal of superior quality to produce more offspring than would be naturally possible.

Host mothering may someday be popular with women who can produce embryos but are either unable or unwilling to carry them to term. Today it is possible to freeze the embryos of many species, including humans, and then to successfully transplant them into host mothers. This procedure may someday be used by young women not yet ready to become parents, but who want to preserve embryos for reimplantation at a later time.

Development

43

Learning Objectives

After you have studied this chapter you should be able to:

1. Summarize the roles of cell division, cell growth, morphogenesis, and cell differentiation in the development of an animal.
2. Trace the early development of the embryo from zygote through cleavage, morula, blastula, gastrula, and early organ development.
3. Compare the fate of each of the germ layers in a developing embryo.
4. Explain the functions of the extraembryonic membranes and placenta.
5. Describe the general course of human development from fertilization to birth.
6. Distinguish among the three stages of labor in the birth process.
7. Contrast postnatal with prenatal life, describing several adaptations that the neonate must make in order to live independently.
8. List specific steps that a pregnant woman can take to promote the well-being of her developing child; describe how the embryo can be affected by nutrients, drugs, cigarette smoking, pathogens, and ionizing radiation.
9. Describe the stages of the human life cycle.
10. Identify anatomical and physiological changes that occur with aging, and discuss current hypotheses of aging.

Key Concepts

- ☐ Embryonic development involves cell division, cell growth, morphogenesis, and cell differentiation. All of these processes are involved in the formation of specialized tissues, organs, and organ systems in the new individual.
- ☐ Early development proceeds through the following stages: fertilization of an egg by a sperm, zygote, cleavage, morula, blastula, gastrula, and development of organs.
- ☐ In terrestrial vertebrates, four extraembryonic membranes—chorion, amnion, allantois, and yolk sac—protect the embryo, help supply it with food and oxygen, and eliminate wastes.
- ☐ Development is influenced by genetic factors, by cytoplasmic factors such as distribution of yolk in the egg, and by environmental factors.
- ☐ The human life cycle extends from fertilization to death.

DEVELOPMENT IS A BALANCED COMBINATION OF SEVERAL PROCESSES

Just how does a microscopic zygote give rise to the blood, bones, brain, and all the other structures of a complex animal? As we will see, development is a balanced combination of several processes, including cell division, cell growth, morphogenesis, and cell differentiation.

The single-celled zygote undergoes division, forming two cells. Then each of these cells divides, giving rise to four cells. This process is repeated again and again, producing the trillions of cells of the adult animal. Growth occurs by both an increase in the number of cells and in the size of these cells. An orderly pattern of cell division and growth provides the cellular building blocks of the organism and results in size increase. But these processes alone would produce only a formless heap of cells.

Morphogenesis is the process by which cells organize themselves, shaping the multicellular animal with its intricate pattern of tissues and organs. This process requires precise migrations of cells, and even involves the controlled death of some cells.

Not only must cells organize into specific structures, but they must also become specialized, which is the process known as **cell differentiation.** During early development, cells begin to become different from one another, becoming biochemically and structurally specialized and performing specific tasks. More than 200 distinct types of cells can be found in the adult vertebrate body.

As you read the following sections, notice how cell division, growth, morphogenesis, and cell differentiation are intimately interrelated. Also note that the pattern of early development is basically similar for all animals.

Development actually includes all of the changes that take place during the entire life of an animal from conception to death. In this chapter, we focus on development before birth and we briefly discuss human growth and maturation after birth and during aging.

DURING CLEAVAGE THE ZYGOTE DIVIDES, GIVING RISE TO MANY CELLS

Despite its simple appearance, the zygote has the potential to give rise to all of the cell types of the new individual. Because the ovum is very large compared with the sperm, the bulk of the zygote cytoplasm and organelles come from the ovum. However, the sperm and ovum contribute equal numbers of chromosomes to the zygote.

Shortly after fertilization, the zygote undergoes **cleavage,** which is a series of rapid, mitotic divisions. The zygote divides to form a two-cell embryo (Fig. 43–1). Then, each of these cells undergoes mitosis and divides, bringing the number of cells to four. Repeated divisions increase the number of cells, forming a ball of cells called a **morula.** Eventually, several hundred cells form a hollow ball, the **blastula,** with a fluid-filled cavity (the blastocoel).

Cleavage Provides Building Blocks for Development

Cleavage partitions the zygote into many small cells that serve as basic building blocks. Their small size allows the cells to move about with relative ease, arranging themselves into the patterns necessary for further development. Each cell moves by ameboid motion. Surface proteins are important in helping cells to "recognize" one another and therefore in determining which ones adhere to form tissues.

Genetic and nongenetic factors interact to regulate development. An important nongenetic factor is the distribution of materials in the cytoplasm of the zygote. Because the zygote cytoplasm is not homogeneous, the cytoplasm portioned out to each new cell during cleavage may be different. Such differences help determine the course of development.

In the amphibian egg, fertilization causes rearrangement of some of the superficial cytoplasm that contains dark granules. This shift exposes the underlying lighter-colored cytoplasm, known as the **gray crescent.** The gray crescent lies near the cell equator, on the side opposite of where the sperm penetrated the egg. This region is thought to contain growth factors from both ends of the egg. Normally, the first cell division separates the gray crescent into the first two cells of the embryo. As cleavage continues, the gray crescent becomes partitioned into certain cells. The cells that contain the parts of the gray crescent eventually develop into the dorsal region of the embryo.

Experiments have confirmed the importance of the gray crescent material to development. When the first two cells of the frog embryo are separated experimentally, each cell develops into a complete tadpole (Fig. 43–2). When the plane of the first division is altered so that the gray crescent is completely absent from one of the cells, that cell does not develop normally.

The Amount of Yolk Determines the Pattern of Cleavage

Many animal eggs contain **yolk,** which is a mixture of proteins, phospholipids, and fats that serves as food for the developing embryo. The amount and distribution of yolk vary among different animal groups. Most invertebrates and simple chordates have eggs with relatively small amounts of yolk that is uniformly distributed through the cytoplasm. Many vertebrate eggs have large

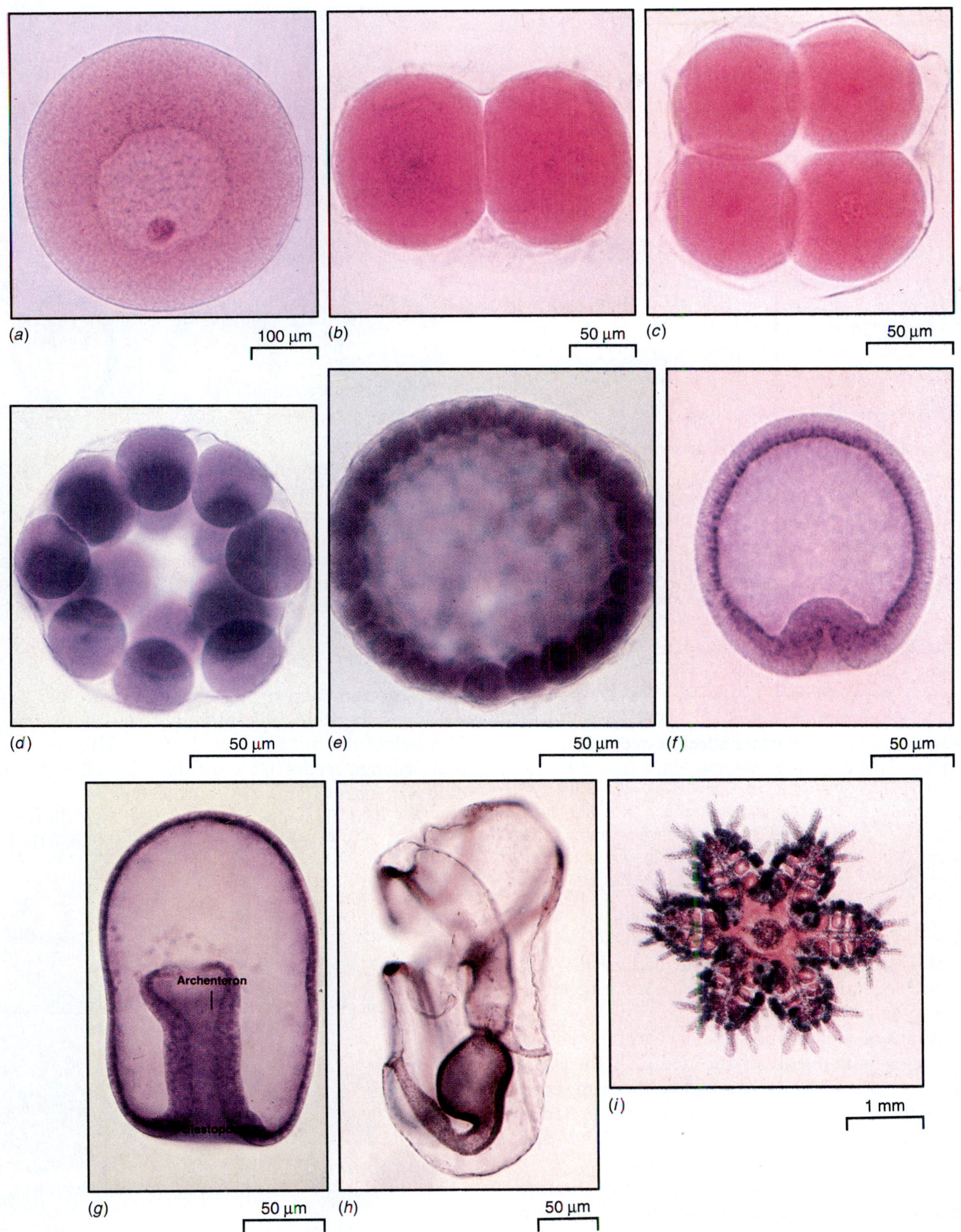

FIGURE 43–1 Development of a sea star. (***a***) Unfertilized sea star egg. (***b***) Two-cell stage. (***c***) Top view of four-cell stage. (***d***) Sixteen-cell stage. (***e***) Cross section through 64-cell blastula. (***f***) Section through early gastrula. (***g***) Section through middle gastrula. (***h***) Sea star larva. (***i***) Young sea star. In this type of cleavage the entire egg becomes partitioned into cells. The blastopore is the opening into the inner cavity, the archenteron. All views are side views with the animal pole at the top except (***c***) and (***i***), which are top views. (Carolina Biological Supply Company)

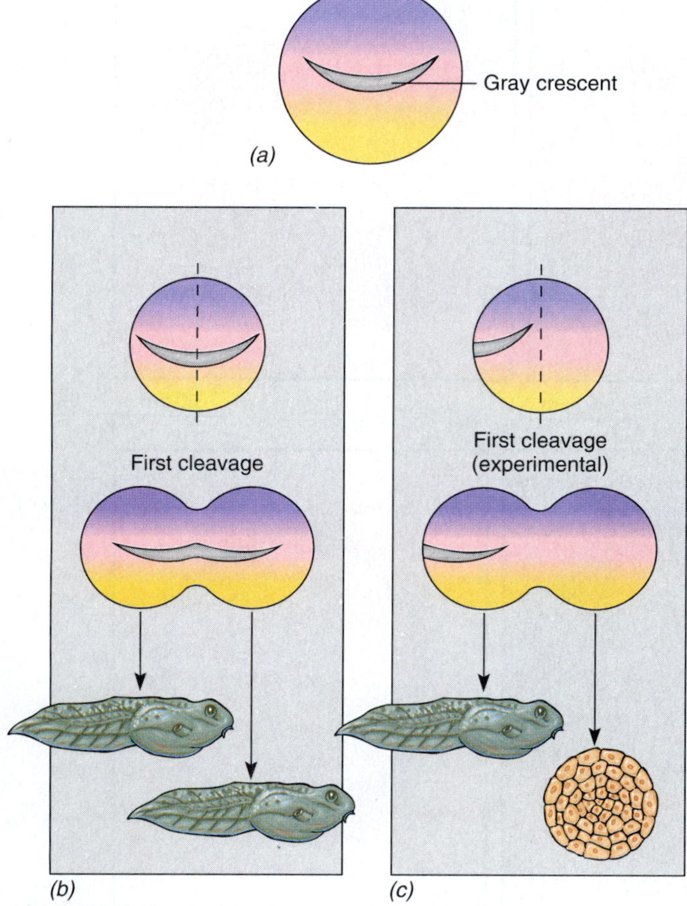

FIGURE 43–2 Cytoplasmic factors affect development. (***a***) The position of the gray crescent in the frog zygote determines the main axes of the body. (***b***) The first division of the zygote partitions the gray crescent into each of the two daughter cells. If these cells are separated, each can develop into a tadpole. (***c***) If the plane of cleavage is changed experimentally, so that only one cell receives the gray crescent, only that cell can develop into a tadpole.

amounts of yolk concentrated at one end of the cell, known as the **vegetal pole.** The opposite, more metabolically active pole is the **animal pole.**

The amount of yolk in the egg affects the pattern of cleavage. In eggs with evenly distributed yolk, the entire egg divides, producing cells of roughly the same size. Cleavage of these eggs can be radial or spiral. **Radial cleavage** is characteristic of deuterostomes, whereas spiral cleavage is characteristic of protostomes.

In radial cleavage, the first division passes through both animal and vegetal poles and splits the egg into two equal cells. The second cleavage division passes through both poles of the egg at right angles to the first and separates the two cells into four equal cells. The third division is horizontal, at right angles to the other two, and separates the four cells into eight cells: four above and four below the third line of cleavage. This pattern of radial cleavage occurs in the sea star and in the simple chordate *Amphioxus* (Figs. 43–1 and 43–3).

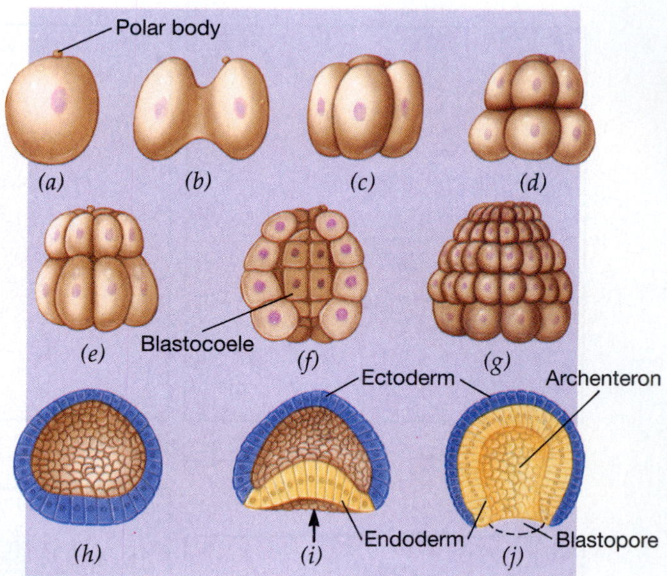

FIGURE 43–3 Early development in *Amphioxus.* Yolk is evenly distributed so the entire egg divides. Cleavage is radial. The embryos are viewed from the side. (***a***) Mature egg with polar body. (***b–e***) Two-, 4-, 8-, and 16-cell stages. (***f***) Embryo at 32-cell stage cut open to show the blastocoel. (***g***) Blastula. (***h***) Blastula cut open. (***i***) Early gastrula showing beginning of invagination at vegetal pole (arrow). (***j***) Late gastrula. Invagination is completed and blastopore has formed.

In **spiral cleavage,** after the first two divisions, the plane of cytokinesis is oblique to the polar axis (Fig. 43–4).

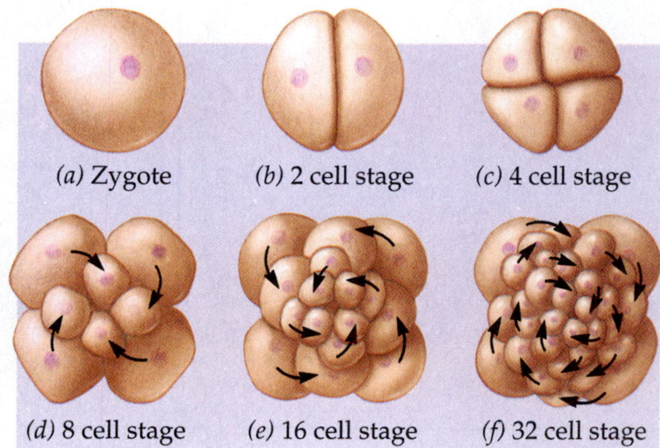

FIGURE 43–4 Spiral cleavage in annelids. (***a***) through (***f***) are top views of the animal pole. The successive cleavage divisions occur in a spiral pattern as indicated.

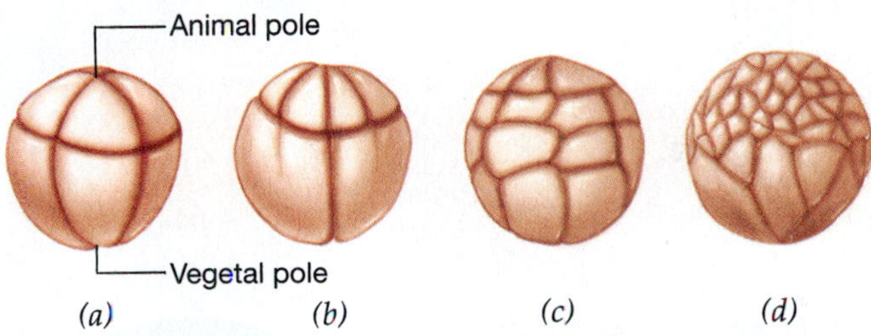

FIGURE 43-5 Cleavage in the frog embryo.

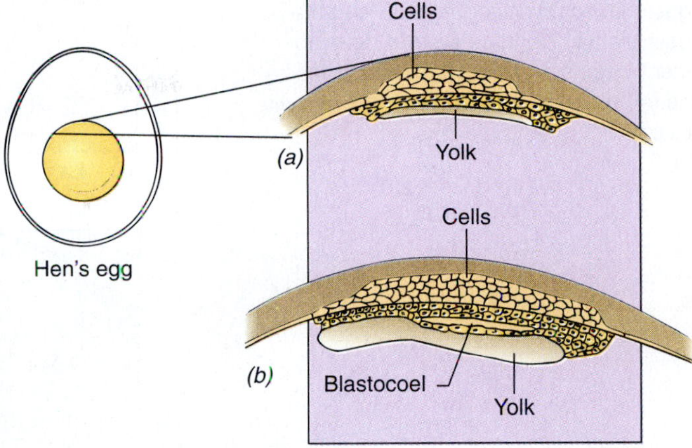

FIGURE 43–6 Cleavage in chick embryo. (***a***) Cleavage is restricted to a small disc of cytoplasm on the upper surface of the egg yolk called the blastodisc. A space appears beneath the blastodisc, separating it from the yolk. (***b***) The blastodisc splits into two tissue layers separated by a space called the blastocoel.

This results in a spiral arrangement of cells, with each cell located above and in between the two underlying cells.

The eggs of bony fish and amphibians contain a large amount of yolk that is concentrated toward the vegetal pole. The cleavage divisions in the vegetal hemisphere are slowed by the presence of the inert yolk. As a result, the blastula consists of many small cells in the animal hemisphere and fewer, but larger, cells in the vegetal hemisphere (Fig. 43–5). The blastocoel is displaced upward.

The eggs of reptiles and birds have so much yolk that they have only a small amount of cytoplasm, which is concentrated at the animal pole. In such eggs, cleavage takes place only in the small disc of cytoplasm at the animal pole (Fig. 43–6).

THE GERM LAYERS FORM DURING GASTRULATION

The process by which the blastula becomes a three-layered embryo, or gastrula, is called **gastrulation.** Thus, early development proceeds through the following sequence of events:

Zygote ⟶ Cleavage ⟶ Morula ⟶ Blastula ⟶ Gastrulation

During gastrulation, the cells arrange themselves into three distinct **germ layers,** or embryonic tissue layers: the ectoderm, mesoderm, and endoderm.

Each Germ Layer Has a Specific Fate

The cells lining the **archenteron** (developing digestive cavity) of the embryo make up the **endoderm.** The endoderm gives rise to tissues that eventually line the digestive tract and organs that develop as outgrowths of the digestive tract (including the liver, pancreas, and lungs). The outer wall of the gastrula consists of **ectoderm.** This germ layer eventually forms the outer layer of the skin and gives rise to the nervous system and sense organs.

A third layer of cells, the **mesoderm,** develops between the ectoderm and endoderm. The mesoderm gives rise to the skeletal tissue, muscle, circulatory system, excretory system, and reproductive system (Table 43–1).

The Pattern of Gastrulation Is Affected by the Amount of Yolk

Gastrulation in the sea star and in *Amphioxus* is illustrated in Figures 43–1 and 43–3, respectively. Gastrulation begins when a group of cells at the vegetal pole flattens and then bends inward. The cells of a section of the vegetal wall of the blastula move inward (invaginate). The invaginated wall eventually meets the opposite wall, obliterating the blastocoel. We can roughly demonstrate this process by pushing inward on the wall of a partly deflated rubber ball until it rests against the opposite wall. In a similar way the embryo is converted into a double-

FIGURE 43–7 Development of a frog embryo. The diagrams show the embryo cut in half so you can view the insides of the embryo. (***a***) Late blastula. (***b***) Early gastrula. (***c***) Middle gastrula. (***d***) Late gastrula. (***e***) Nervous system development begins with the formation of the neural plate. (Carolina Biological Supply Company)

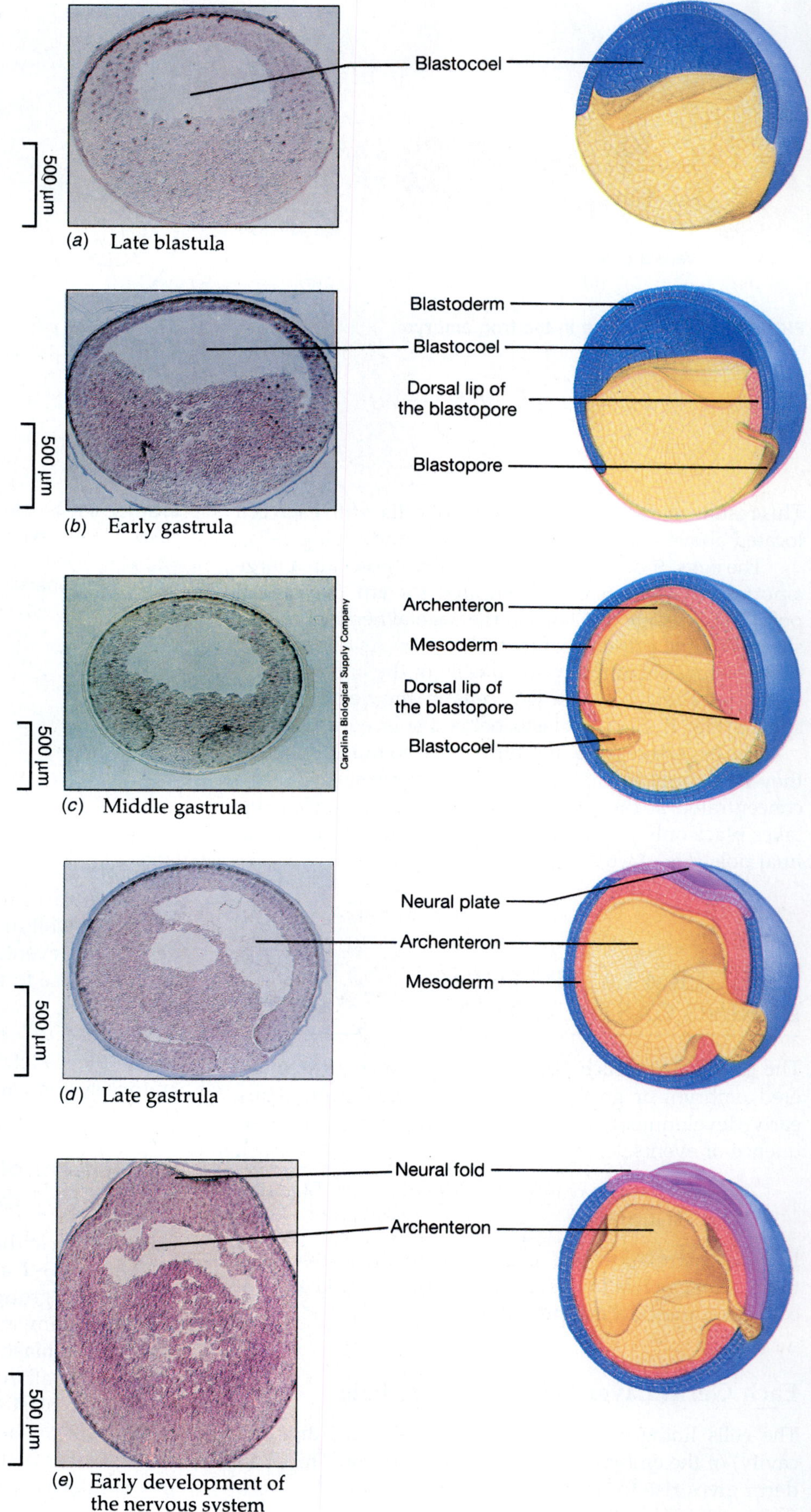

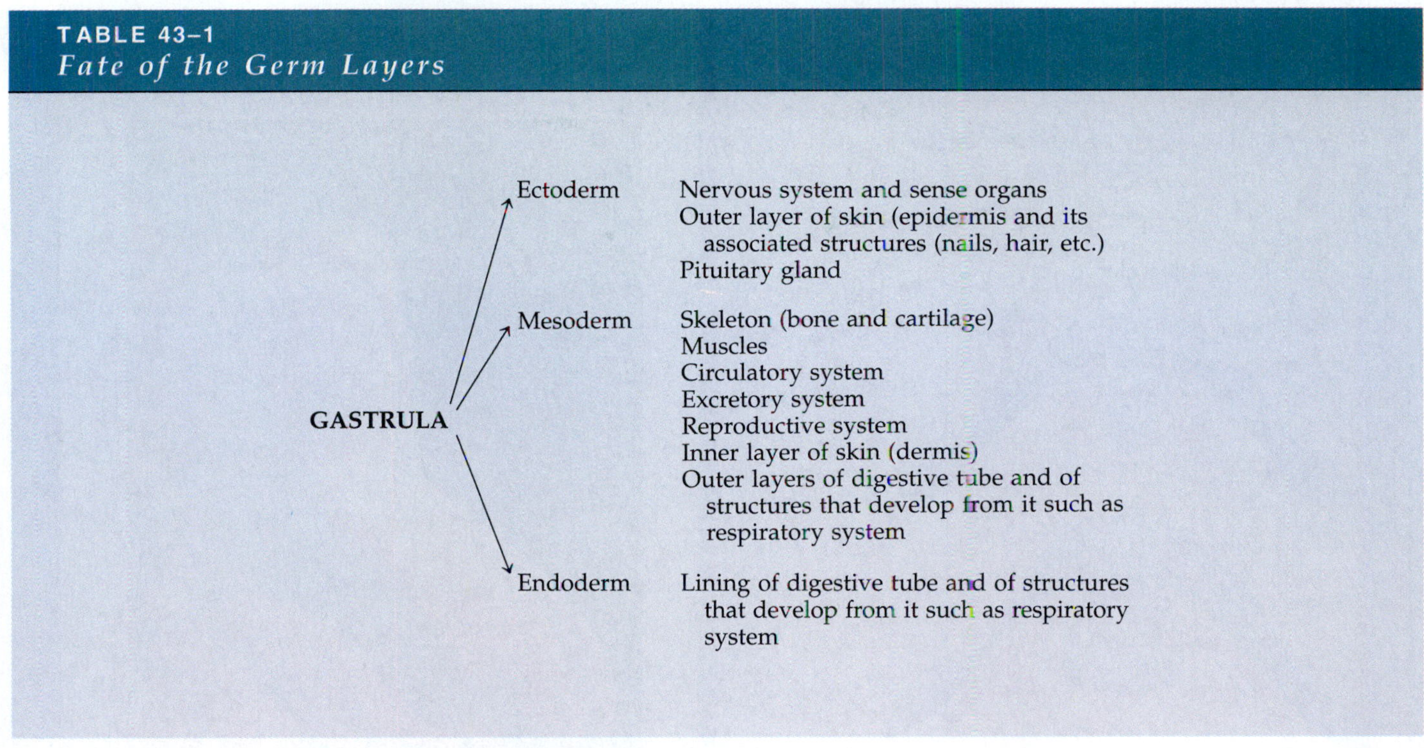
TABLE 43–1
Fate of the Germ Layers

	Germ layer	Fate
GASTRULA	Ectoderm	Nervous system and sense organs Outer layer of skin (epidermis and its associated structures (nails, hair, etc.) Pituitary gland
	Mesoderm	Skeleton (bone and cartilage) Muscles Circulatory system Excretory system Reproductive system Inner layer of skin (dermis) Outer layers of digestive tube and of structures that develop from it such as respiratory system
	Endoderm	Lining of digestive tube and of structures that develop from it such as respiratory system

walled, cup-shaped structure. The cavity of the cup communicates with the exterior on the side that was originally the vegetal pole of the embryo. The internal wall lines the newly formed cavity, the archenteron (cavity of the primitive gut). The opening of the archenteron, the **blastopore,** becomes the anus in deuterostomes.

In the amphibian, the large yolk-laden cells in the vegetal half of the blastula obstruct the inward movement at the vegetal pole. Instead, cells from the animal pole move down toward the yolk-rich cells and then inward and away from the yolk-rich cells, forming the dorsal lip of the blastopore (Fig. 43–7). As the process continues, the blastopore becomes ring-shaped as cells lateral, and then ventral, to the blastopore become involved in the same movements. The yolk-filled cells of the vegetal hemisphere remain as a yolk plug, filling the space enclosed by the lips of the blastopore.

The archenteron forms and is lined on all sides by cells that have moved in from the surface. At first, the archenteron is a narrow slit, but it gradually expands at the anterior end, and eventually obliterates the blastocoel. Although the details differ somewhat, gastrulation in the bird is basically similar to amphibian gastrulation.

Organs of the Nervous System Are Among the First to Develop

In the early vertebrate embryo, the notochord (the flexible skeletal axis in all chordate embryos) grows forward along the length of the embryo as a cylindrical rod of cells. The developing notochord *induces* (stimulates) the overlying ectoderm to thicken, forming the **neural plate** (Fig. 43–8). Central cells of the neural plate move downward, forming a depression, called the **neural groove.** The cells flanking the groove on each side form **neural folds.**

Continued movements of their cells bring the neural folds closer together until they meet and fuse, forming the **neural tube.** In this process, the neural tube comes to lie beneath the surface. The ectoderm overlying it will form the outer layer of skin. The anterior portion of the neural tube grows and differentiates into the brain; posterior to the brain, the tube develops into the spinal cord.

The induction of the neural plate by the notochord is a good example of the importance of the position of a cell in relation to its cellular neighbors. That position is often critical in determining the fate of a cell because it determines its exposure to substances released from other cells

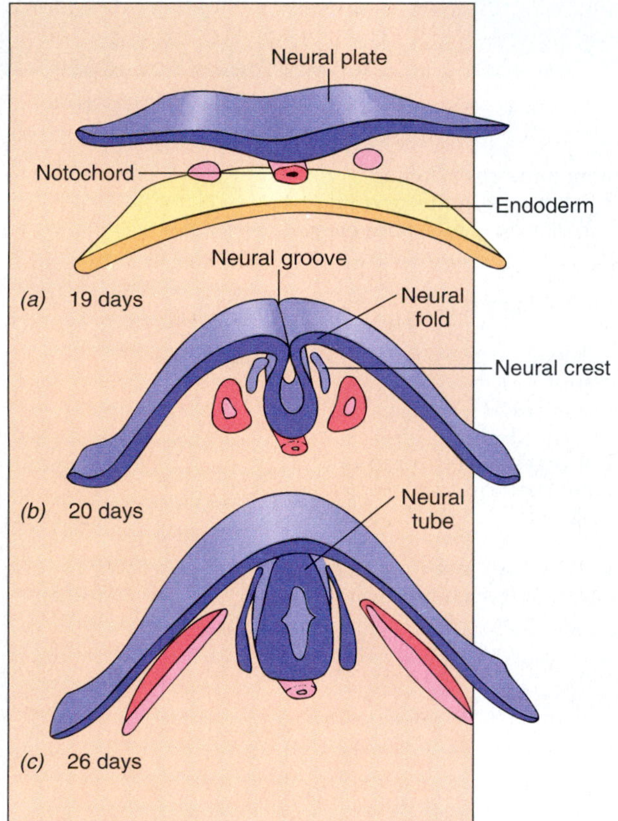

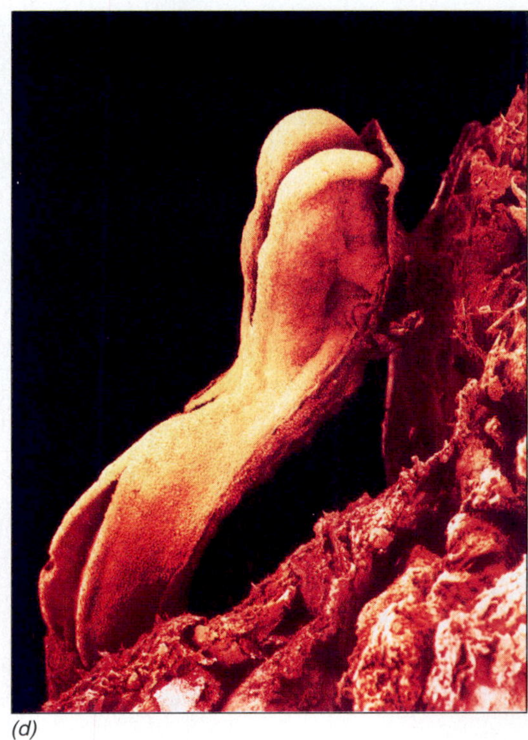

(d)

FIGURE 43–8 Early development of the nervous system. Cross sections of human embryos at successively later stages, illustrating early nervous system development. The neural crest cells give rise to sensory neurons. (***a***) Approximately 19 days. The neural plate has indented to form a shallow groove flanked by neural folds. (***b***) Approximately 20 days. The neural folds approach one another. (***c***) Approximately 26 days. The neural folds have formed the neural tube which will give rise to the brain at the anterior end of the embryo and the spinal cord posteriorly. (***d***) Photograph of a 20-day-old human embryo shows the developing nervous system. (*d*, Lennart Nilsson, copyright, Boehringer Ingelheim International GmbH, from *A Child Is Born*, p. 76)

CONCEPT CONNECTIONS

Development — *Genetics*

Although nongenetic influences are important in determining cell differentiation, the genes exercise ultimate control over development. Consider a specific experimental example. When ectoderm from the mouth region of an early frog embryo is transplanted to the mouth region of a salamander, the surrounding salamander tissue induces the transplanted cells to develop into mouth structures. However, the mouth that forms is not the characteristic mouth of a salamander. Instead it bears the horny rows of teeth and horny jaws of a frog. Why? As you might guess, the frog cells are not *competent* to form structures specific to salamanders.

Cell differentiation is an expression of changes in the activity of specific genes, and genetic activity in turn is influenced by a variety of factors within and outside the cell. Although each cell of an organism has the same set of genes, differential gene expression results in variations in chemistry, behavior, and structure among cells. Through this process, an embryo develops into an organism composed of more than 200 types of cells, each specialized to perform specific functions.

Extraembryonic Membranes Protect and Nourish the Embryo

All terrestrial vertebrates have four **extraembryonic membranes:** the chorion, allantois, yolk sac, and amnion (Fig. 43–9). Although they develop from the germ layers, these membranes are not part of the embryo itself and are discarded at birth. The extraembryonic membranes are adaptations to the challenges of embryonic development on land. They protect the embryo, prevent it from drying out, and help in obtaining food and oxygen and eliminating wastes.

The **chorion** and **amnion** enclose the entire embryo (Fig. 43–9*c*). The amniotic cavity, the space between the embryo and the amnion, becomes filled with amniotic fluid secreted by the membrane. Embryos of terrestrial vertebrates develop within this pool of fluid. The amniotic fluid prevents the embryo from drying out and permits the embryo a certain freedom of motion. It also serves as a protective cushion that absorbs shocks and prevents the amniotic membrane from sticking to the embryo.

The **allantois** is an outgrowth of the developing digestive tract. In reptiles and birds, it stores nitrogenous wastes. In humans, the allantois is small and nonfunctional, except that its blood vessels contribute to the formation of umbilical vessels joining the embryo to the placenta.

In vertebrates with yolk-rich eggs, the **yolk sac** encloses the yolk, slowly digests it, and makes it available to the embryo. Even in vertebrate embryos with little or no yolk, a yolk sac forms. Its walls serve as temporary centers for the formation of blood cells.

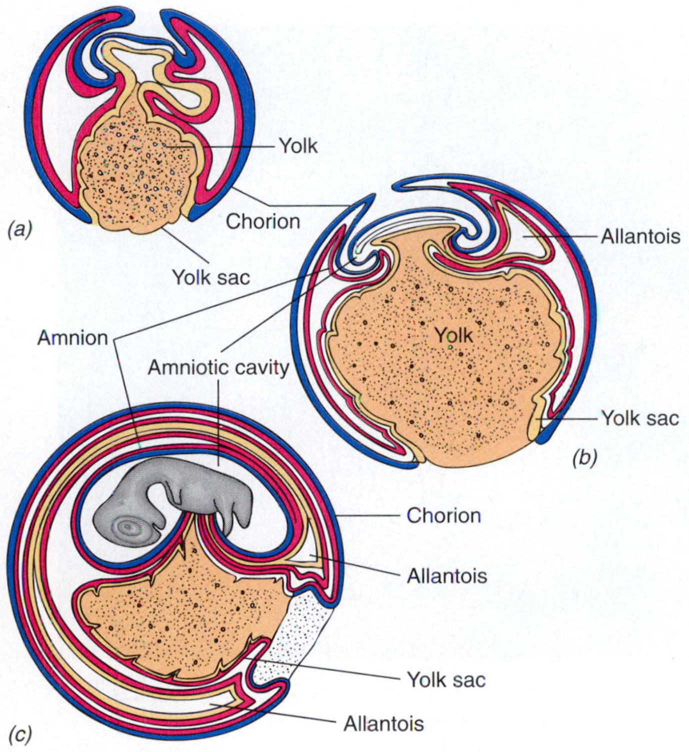

FIGURE 43–9 Development of extraembryonic membranes. (***a, b, c***) Successive stages in the development of the extraembryonic membranes of the chick. Each of the membranes develops from a combination of two germ layers. The chorion and amnion are formed from lateral folds of the ectoderm and mesoderm that extend over the embryo and fuse. The allantois develops from an outpocketing of the gut. The allantois, an elongated sac, and the yolk sac develop from endoderm and mesoderm.

Human Prenatal Development Requires About 266 Days

The human **gestation period,** the duration of pregnancy, averages 280 days (40 weeks) from the time of the mother's last menstrual period to the birth of the baby, or 266 days (about 9 months) from the time of conception (Table 43–2).

Development Begins in the Oviduct

By about 24 hours after fertilization, the human zygote has divided to become a two-cell embryo (Fig. 43–10). Cleavage continues as the embryo is pushed along the oviduct by ciliary action and muscular contraction (Fig. 43–11).

When the embryo enters the uterus on about the fifth day of development, the zona pellucida (its surrounding membrane), is dissolved. During the next few days, the embryo floats free in the uterine cavity nourished by a nutritive fluid secreted by the glands of the uterus. Its cells arrange themselves, forming a blastula, which in mammals is called a **blastocyst** (Fig. 43–12). The outer layer of cells, the **trophoblast,** eventually forms the chorion and amnion that surround the embryo. A little cluster of cells, the **inner cell mass,** projects into the cavity of the blastocyst. The inner cell mass gives rise to the embryo proper.

The Embryo Implants in the Wall of the Uterus

On about the seventh day of development the embryo begins to **implant** in the endometrium of the uterus (Fig. 43–12*a*). The trophoblast cells in contact with the uterine lining secrete enzymes that erode an area just large enough to accommodate the tiny embryo. Slowly the embryo works its way down into the underlying connective and vascular tissues, and the opening in the lining of the uterus repairs itself. All further development of the embryo takes place *within* the endometrium of the uterus.

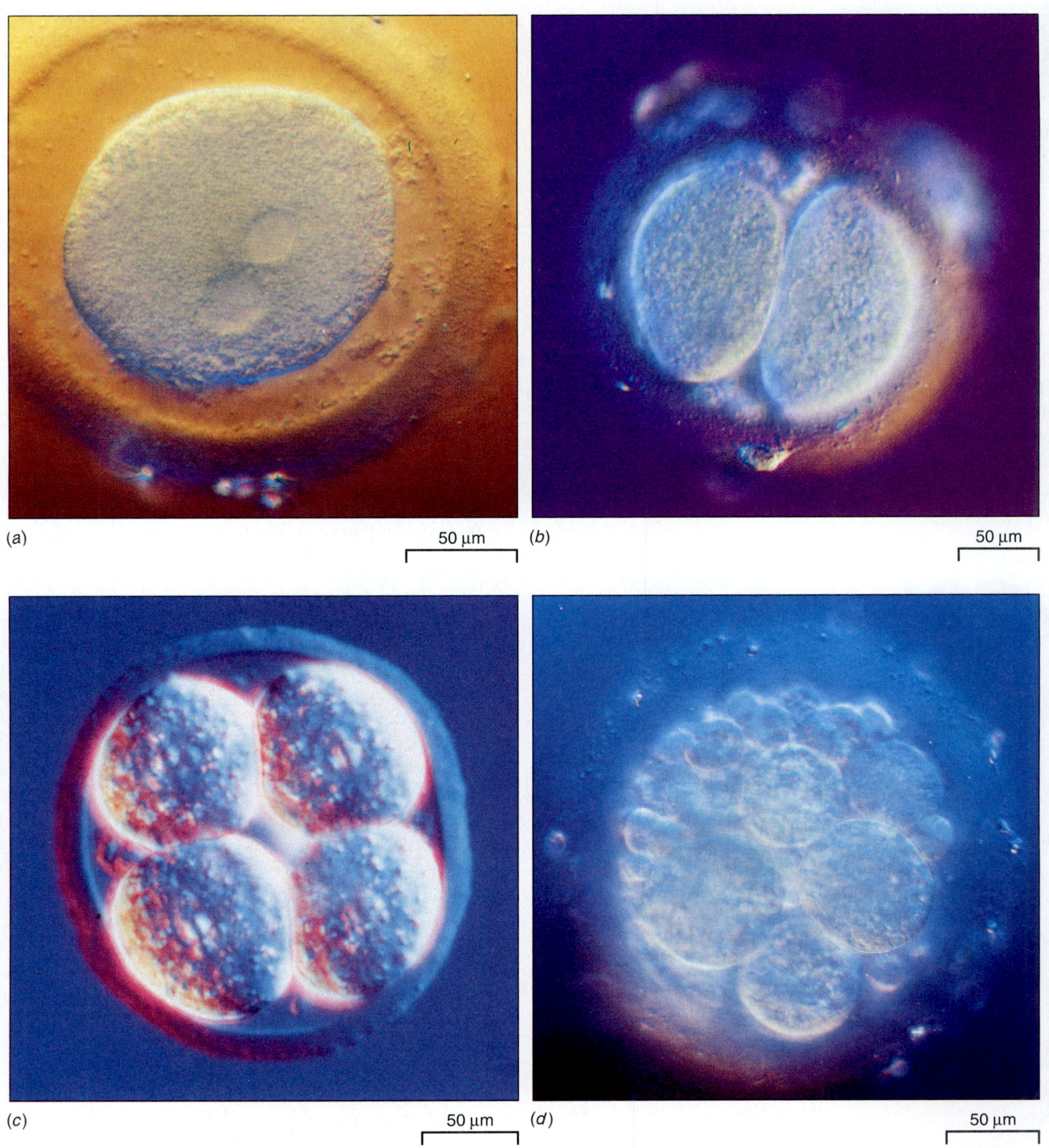

FIGURE 43–10 Early human development. (approximately × 250). (***a***) Human zygote. This single cell contains the genetic instructions for producing a complete human. (***b***) Two-cell stage. (***c***) Eight-cell stage. (***d***) Cleavage continues, giving rise to a cluster of cells. (Lennart Nilsson, from *Being Born*, pp. 14, 15, 17)

TABLE 43–2
Some Important Developmental Events in the Human Embryo

Time from fertilization	*Event*
24 hours	Embryo reaches two-cell stage
3 days	Morula reaches uterus
7 days	Blastocyst begins to implant
2.5 weeks	Notochord and neural plate are formed; tissue that will give rise to heart is differentiating; blood cells are forming in yolk sac and chorion
3.5 weeks	Neural tube forming; primordial eye and ear visible; pharyngeal pouches forming; liver bud differentiating; respiratory system and thyroid gland just beginning to develop; heart tubes fuse, bend, and begin to beat; blood vessels are laid down
4 weeks	Limb buds appear; three primary divisions of brain formed
2 months	Muscles differentiating; embryo capable of movement; gonad distinguishable as testis or ovary; bones begin to ossify; cerebral cortex differentiating; principal blood vessels assume final positions
3 months	Sex can be determined by external inspection; notochord degenerates; lymph glands develop
4 months	Face begins to look human; lobes of cerebrum differentiate; eyes, ears, and nose look more "normal"
Third trimester	A covering of downy hair covers the fetus, then later is shed; neuron myelination begins; tremendous growth of body
266 days (from conception)	Birth

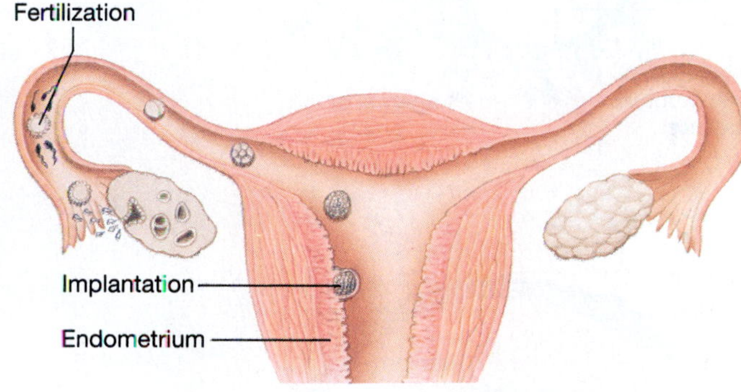

FIGURE 43–11 Cleavage takes place as the embryo is moved along through the oviduct to the uterus.

The Placenta Is an Organ of Exchange

In placental mammals, the **placenta** is the organ of exchange between mother and embryo. The placenta provides nutrients and oxygen for the fetus, and removes wastes, which the mother then excretes (Fig. 43–12). In addition, the placenta is an endocrine organ that secretes hormones that maintain pregnancy.

The placenta develops from both the chorion of the embryo and the uterine tissue of the mother. In early development, the chorion grows rapidly, invading the endometrium (lining of the uterus) and forming finger-like projections called **villi.** The villi become vascularized (infiltrated with blood vessels) as the embryonic circulation develops.

As the human embryo grows, the **umbilical cord** develops and connects the embryo with the placenta (Fig. 43–12). The umbilical cord contains the two umbilical arteries and the umbilical vein. The **umbilical arteries** connect the embryo with a vast network of capillaries developing within the villi. Blood from the villi returns to the embryo through the **umbilical vein.**

The placenta consists of the portion of the chorion that develops villi, together with the uterine tissue underlying the villi, that contains maternal capillaries and small pools of maternal blood. The blood of the fetus in the capillaries of the chorionic villi comes in close contact with the mother's blood in the tissues between the villi. However, they are always separated by a membrane through which substances may diffuse or be actively transported. *Maternal and fetal blood do not normally mix in the placenta or any other place.*

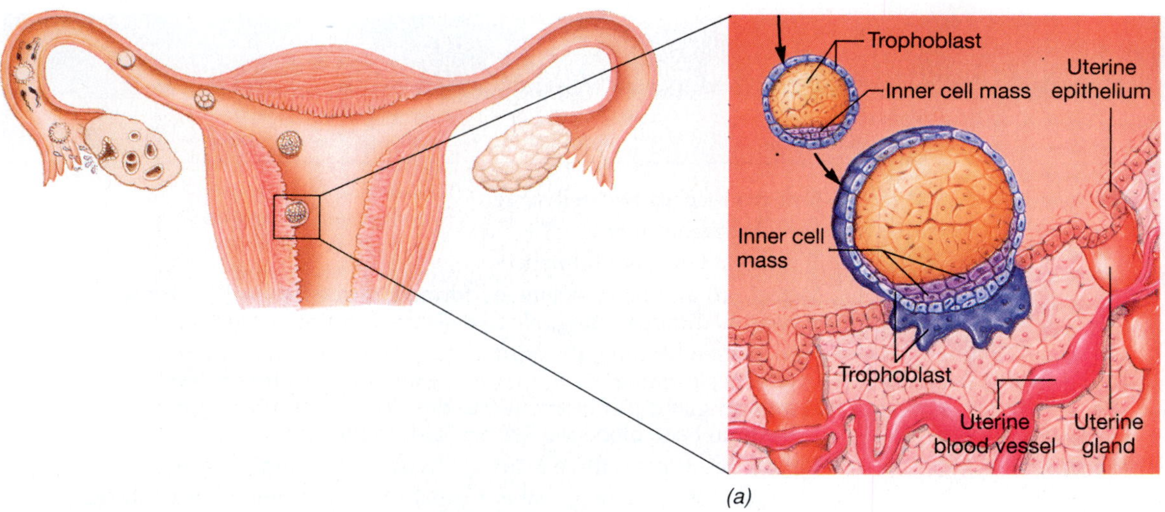

(a)

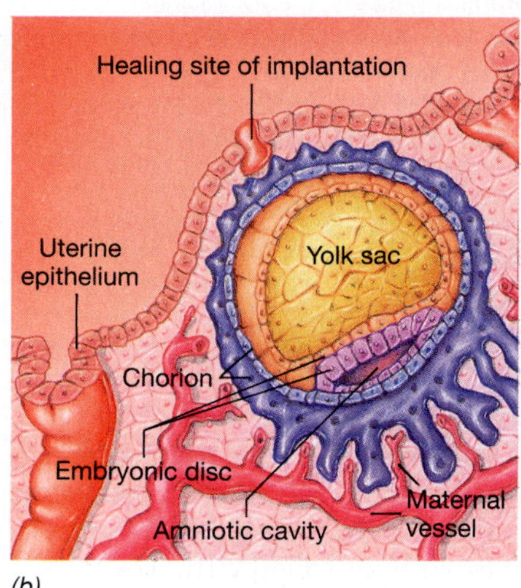

(b)

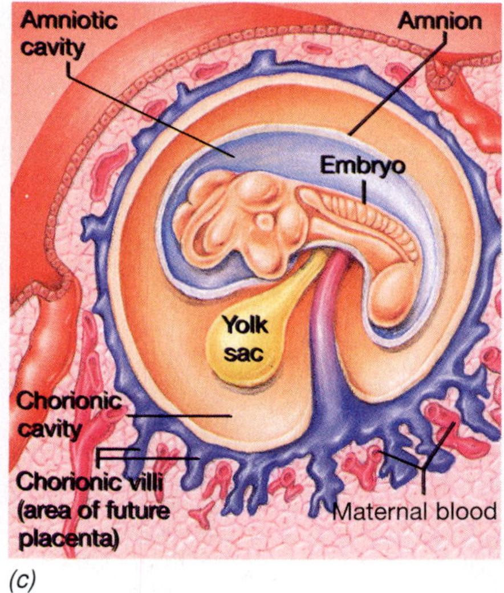

(c)

FIGURE 43–12 Implantation and development of the early human embryo. (***a***) About 7 days after fertilization the blastocyst begins to implant itself in the uterine wall. (***b***) About 10 days after fertilization the chorion has formed from the trophoblast. (***c***) By 25 days intimate relationships have been established between the embryo and the maternal blood vessels. Oxygen and nutrients from the maternal blood are now satisfying the embryo's needs. Note the specialized region of the chorion that will become the placenta. The umbilical cord is beginning to develop. (***d***) At about 45 days the embryo and its membrane together are about the size of a ping-pong ball, and the mother still may be unaware of her pregnancy. The amnion filled with amniotic fluid surrounds and cushions the embryo. The yolk sac has been incorporated into the umbilical cord. Blood circulation has been established through the umbilical cord to the placenta.

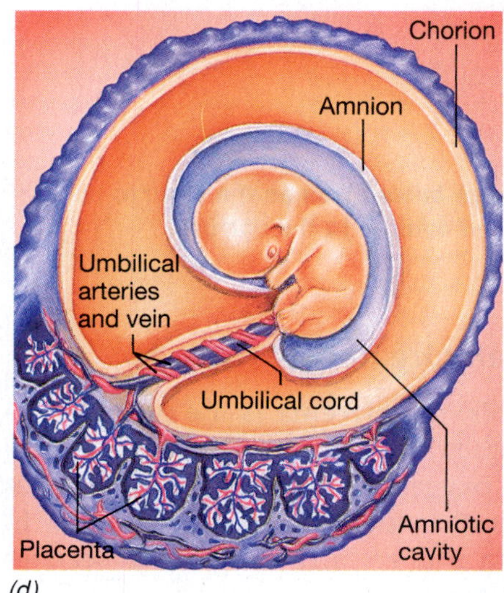

(d)

Several hormones are produced by the placenta. From the time the embryo first begins to implant itself, its trophoblastic cells release **human chorionic gonadotropin (hCG),** which signals the corpus luteum that pregnancy has begun. In response, the corpus luteum increases in size and releases large amounts of progesterone and estrogens. These hormones stimulate continued development of the endometrium and placenta.

Without hCG, the corpus luteum would degenerate and the embryo would be aborted and flushed out with the menstrual flow. The woman would probably not even know that she had been pregnant. When the corpus luteum is removed before about the 11th week of pregnancy, the embryo is spontaneously aborted. After that time, the placenta itself produces enough progesterone and estrogens to maintain pregnancy.

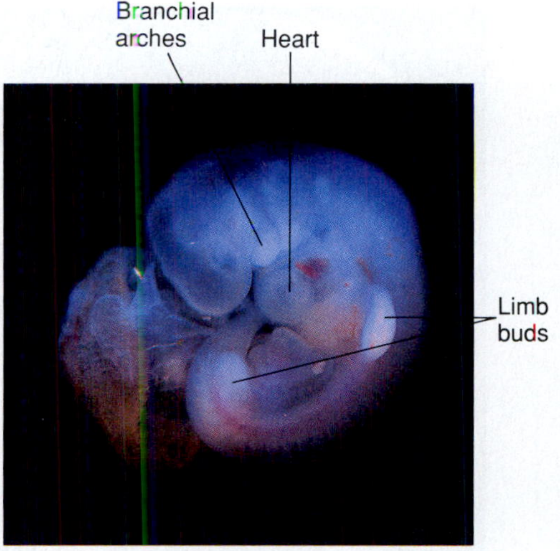

FIGURE 43–13 Photograph of developing human embryo. Human embryo at 29 days, about 7 mm (0.3 inch) long. Note the slender tail and developing limb buds. The tail will regress during later development. The heart can be seen below the head near the mouth of the embryo. Branchial arches appear as "double chins." (Lennart Nilsson, from A Child Is Born, Dell Publishing Co., Inc.)

Organ Development Begins During the First Trimester

Gastrulation occurs during the second and third weeks of development. Then, the notochord begins to form and induces formation of the neural plate. The neural tube develops and the forebrain, midbrain, and hindbrain are evident by the fifth week of development. A week or so later, the forebrain begins to grow outward, forming the rudiments of the cerebral hemispheres.

The heart begins to develop and, at 3.5 weeks of development, begins to beat (Fig. 43–13). In the region of the developing pharynx, pharyngeal pouches, branchial grooves, and branchial arches form. In the floor of the pharynx, a tube of cells grows downward, forming the primordial trachea, which gives rise to the lung buds. The digestive system also gives rise to outgrowths that will develop into the liver, gallbladder, and pancreas. Near the end of the first month, the limb buds begin to differentiate, and eventually give rise to arms and legs.

All of the organs continue to develop during the second month (Fig. 43–14). A thin tail becomes evident during the fifth week but does not grow as rapidly as the rest of the body and so becomes inconspicuous by the end of the second month. Muscles develop, and the embryo becomes capable of movement. The brain begins to send impulses that regulate the functions of some organs, and a few simple reflexes are evident. After the first two months of development, the embryo is referred to as a **fetus.**

By the end of the **first trimester,** the first three months of development, the fetus is recognizably human (Fig. 43–15). The external genital structures have differentiated, indicating the sex of the fetus. Ears and eyes approach their final positions. Some of the skeleton becomes distinct, and the notochord has been replaced by the developing vertebral column. The fetus performs breathing movements, pumping amniotic fluid into and out of its lungs, and even carries on sucking movements. By the end of the third month, the fetus is almost 56 mm (about 2.2 in.) long and weighs about 14 g (0.5 oz).

Development Continues during the Second and Third Trimesters

During the second trimester (months four through six), the fetal heart can be heard with a stethoscope. The fetus moves freely through the amniotic cavity, and during the fifth month, the mother usually becomes aware of fetal movements ("quickening").

The fetus grows rapidly during the final trimester (months seven through nine), and final differentiation of tissues and organs occurs. If born prematurely during this stage, the fetus attempts to breathe and is able to move and cry but almost always dies because its brain is not sufficiently developed to sustain vital functions such as rhythmic breathing and the regulation of body temperature.

During the seventh month the cerebrum grows rapidly and develops convolutions. The grasp and sucking reflexes are evident, and the fetus may suck its thumb. At birth the average full-term baby weighs about 3000 g (7 lb) and measures about 52 cm (20 in.) in total length. Any infant born before 37 weeks' gestation is considered premature.

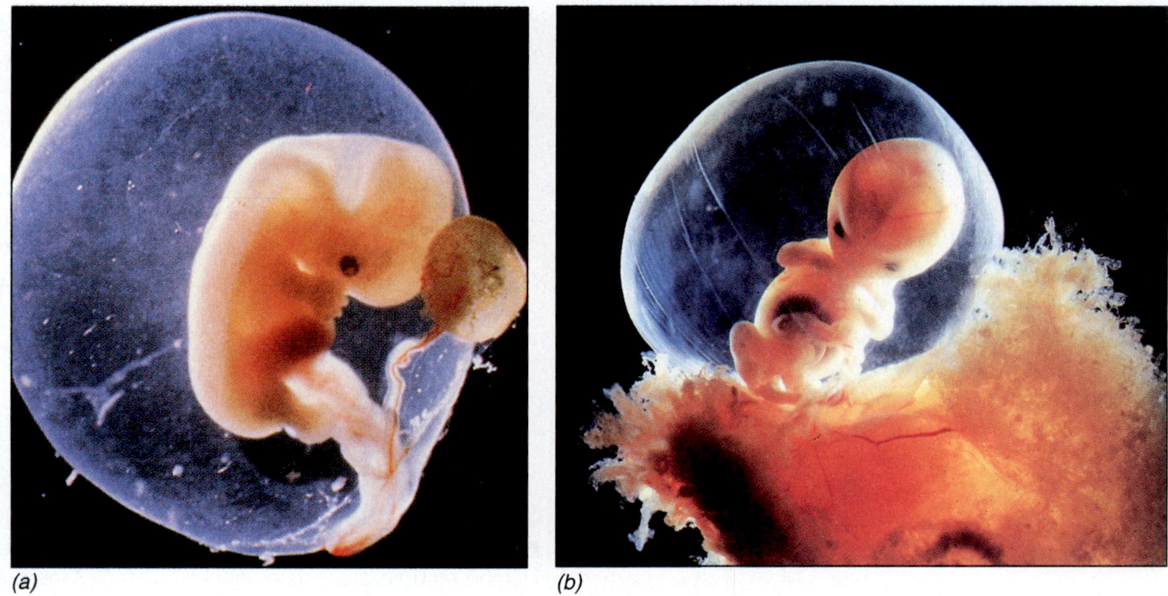

FIGURE 43–14 Development during the second month. (***a***) Human embryo at $5\frac{1}{2}$ weeks, 1 cm (0.4 inch) long. Limb buds have lengthened and the eyes have become prominent. (***b***) In its 7th week of development, the embryo is 2 cm (0.8 inch) long. The dark red object inside the embryo is the liver. (a, Guigoz/Petit Format/Photo Researchers, Inc.; b, Lennart Nilsson, from A Child Is Born, Dell Publishing Co., Inc.)

The Birth Process May Be Divided into Three Stages

The factors that initiate the process of birth, or **parturition,** are not well understood. Muscles in the wall of the uterus become sensitive to the hormone oxytocin and begin to contract. Prostaglandins released by the uterus and nerve reflexes also help regulate the process. A long series of involuntary contractions of the uterus are experienced as the contractions of **labor.**

Labor may be divided into three stages. During the **first stage,** which typically lasts about 12 hours, the con-

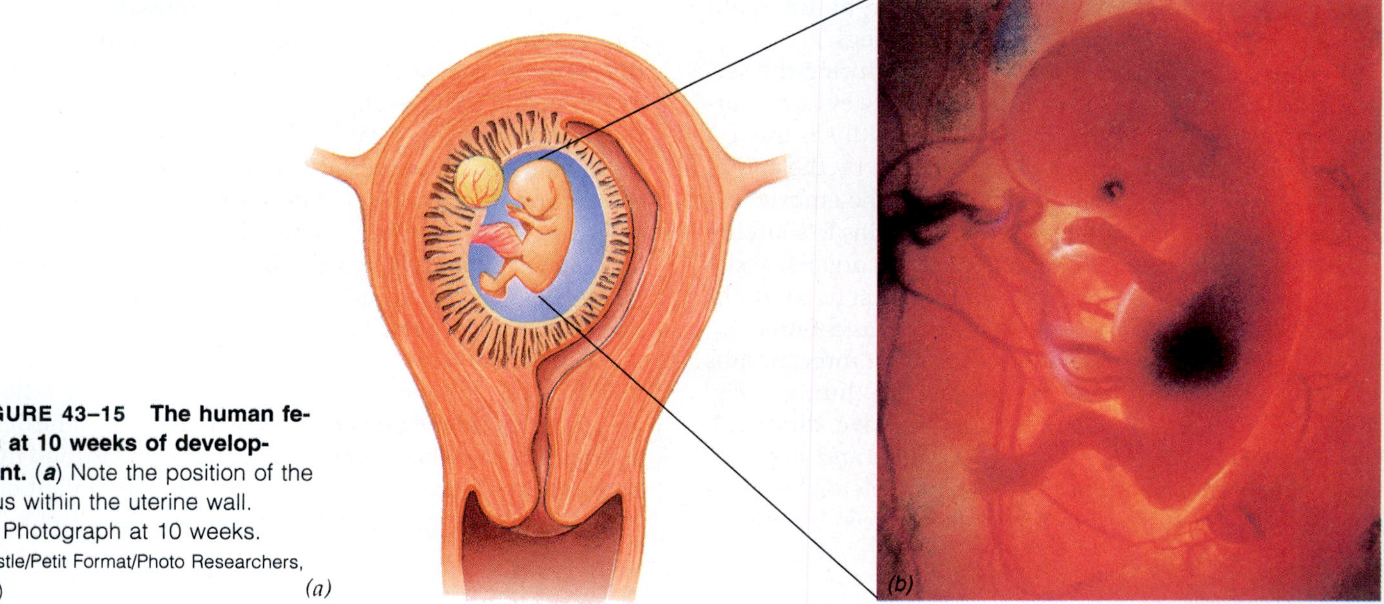

FIGURE 43–15 The human fetus at 10 weeks of development. (***a***) Note the position of the fetus within the uterine wall. (***b***) Photograph at 10 weeks. (Nestle/Petit Format/Photo Researchers, Inc.)

tractions of the uterus move the fetus toward the cervix, causing the cervix to dilate (open). The cervix also becomes effaced, that is, it loses its normal shape and flattens so that the fetal head can pass through. During the first stage of labor, the amnion usually ruptures, releasing about a liter of amniotic fluid, which flows out through the vagina.

During the **second stage,** which normally lasts between 20 minutes and an hour, the fetus passes through the cervix and vagina and is born, or "delivered" (Fig. 43–16). With each uterine contraction the woman holds her breath and bears down so that the fetus is expelled from the uterus by the combined forces of uterine contractions and contractions of abdominal-wall muscles.

After the baby is born, the contractions of the uterus squeeze much of the fetal blood from the placenta back into the infant. The cord is tied and cut, separating the child from the mother. The stump of the cord gradually shrivels until nothing remains but the scar, the **navel.**

During the **third stage** of labor, which lasts 10 or 15 minutes after the birth of the child, the placenta and the fetal membranes are loosened from the lining of the uterus by another series of contractions and expelled. At this stage they are called collectively the **afterbirth.**

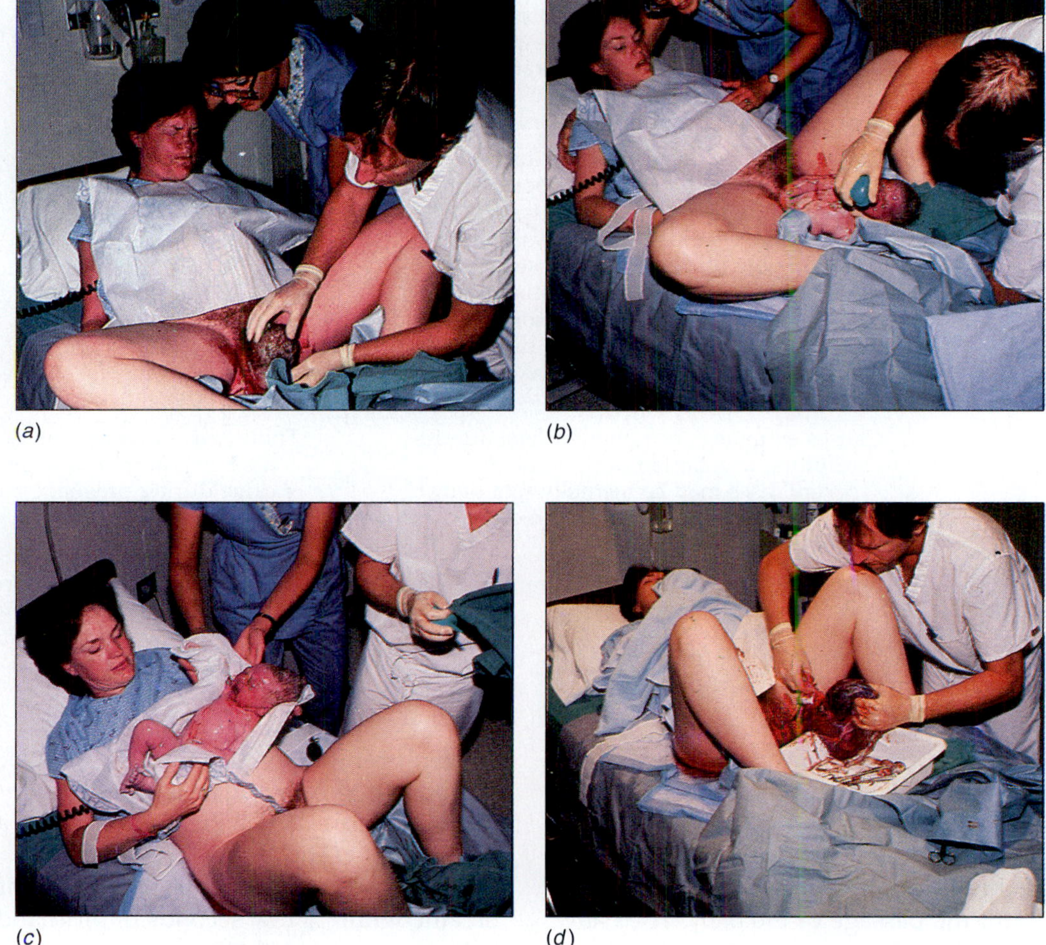

(*a*) (*b*) (*c*) (*d*)

FIGURE 43–16 Birth of a baby. In about 95% of all human births the baby descends through the cervix and vagina in the head-down position. (***a***) The mother bears down hard with her abdominal muscles, helping to push the baby out. When the head fully appears, the physician or midwife can gently grasp it and guide the baby's entrance into the outside world. (***b***) Once the head has emerged, the rest of the body usually follows readily. The physician gently aspirates the mouth and pharynx to clear the upper airway of any amniotic fluid, mucus, or blood. At this time the neonate usually takes its first breath. (***c***) The baby, still attached to the placenta by its umbilical cord, is presented to its mother. (***d***) During the third stage of labor the placenta is delivered. (Courtesy of Dan Atkinson)

TABLE 43–3
Environmental Influences on the Embryo

Factor	*Example and effect*	*Comment*
Nutrition	Severe protein malnutrition doubles number of defects; fewer brain cells are produced, and learning ability may be permanently affected; vitamin deficiencies linked to CNS defects	Growth rate mainly determined by rate of net protein synthesis by embryo's cells; low birth weight
Excessive amounts of vitamins	Vitamin D essential, but excessive amounts may result in form of mental retardation; an excess of vitamins A and K may also be harmful	Vitamin supplements are normally prescribed for pregnant women, but some women mistakenly reason that if one vitamin pill is beneficial, four or five might be even better
Drugs	Many drugs affect development of fetus: even aspirin has been shown to inhibit growth of human fetal cells (especially kidney cells) cultured in laboratory; it may also inhibit prostaglandins, which are concentrated in growing tissue	
Alcohol	When a woman drinks heavily during pregnancy, the baby may be born with fetal alcohol syndrome—that is, deformed and mentally and physically retarded; low birth weight and structural abnormalities have been associated with as little as two drinks a day; some cases of hyperactivity and learning disabilities may be caused by alcohol intake of a pregnant mother	Fetal alcohol syndrome is one of the leading causes of mental retardation in the United States; low birth weight
Cocaine	Causes extreme constriction of fetal arteries, resulting in retarded development; severe cases may be mentally retarded, have heart defects and other medical problems	Thousands of cocaine-addicted babies are being born to mothers who use cocaine during pregnancy; low birth weight
Heroin	High mortality rate and high prematurity rate	Infants that survive are born addicted and must be treated for weeks or months; low birth weight

During labor, an obstetrician may administer drugs such as oxytocin or prostaglandins to increase the contractions of the uterus or may assist with special forceps or other techniques. In some women, the opening between the pelvic bones through which the vagina passes is too small to permit the passage of the baby. The baby must then be delivered by **Cesarean section,** an operation in which an incision is made in the abdominal wall and uterus.

The Neonate Must Adapt to Its New Environment

Important changes take place after birth. During prenatal (before birth) life, the fetus received both food and oxygen from the mother through the placenta. Now the newborn's own digestive and respiratory systems must function. Correlated with these changes are several major changes in the circulatory system.

Normally, the **neonate** (newborn infant) begins to breathe within a few seconds of birth and cries within a half minute. If anesthetics have been given to the mother, however, the fetus may also have been anesthetized, and breathing and other activities may be depressed. Some infants may not begin breathing until several minutes have passed. This is one of the reasons that many women request childbirth methods that minimize the use of medication.

The neonate's first breath is thought to be initiated by the accumulation of carbon dioxide in the blood after the umbilical cord is cut. The carbon dioxide stimulates

Factor	Example and effect	Comment
Thalidomide	Thalidomide, marketed as a mild sedative, was responsible for more than 7000 grossly deformed babies born in the late 1950s in 20 countries; principal defect was phocomelia, a condition in which babies are born with extremely short limbs, often with no fingers or toes	This drug interferes with cellular metabolism; most hazardous when taken during fourth to sixth weeks, when limbs are developing
Cigarette smoking	Cigarette smoking reduces the amount of oxygen available to the fetus because some of the maternal hemoglobin is combined with carbon monoxide; may slow growth and can cause subtle forms of damage; in extreme form, carbon monoxide poisoning causes such gross defects as hydrocephaly	Mothers who smoke deliver babies with lower-than-average birth weights and have a higher incidence of spontaneous abortions, stillbirths, and neonatal deaths; studies also indicate a possible link between maternal smoking and slower intellectual development in offspring
Pathogens	Rubella (German measles) virus crosses placenta and infects embryo; interferes with normal metabolism and cell movements; causes syndrome that involves blinding cataracts, deafness, heart malformations, and mental retardation; risk is greatest (about 50%) when rubella is contracted during first month of pregnancy; risk declines with each succeeding month	Rubella epidemic in the United States in 1963–1965 resulted in about 20,000 fetal deaths and 30,000 infants born with gross defects
	HIV can be transmitted from mother to baby before birth, during birth, or postpartum through breastfeeding	See discussion of AIDS in Chapter 37
	Syphilis is transmitted to fetus in about 40% of infected women; fetus may die or be born with defects and congenital syphilis	Pregnant women are routinely tested for syphilis during prenatal examinations
Ionizing radiation	When mother is subjected to x-rays or other forms of radiation during pregnancy, infant has higher risk of birth defects and leukemia	Radiation was one of the earliest causes of birth defects to be recognized

the respiratory centers in the medulla. The resulting expansion of the lungs enlarges its blood vessels (which previously were partially collapsed). Blood from the right ventricle flows in increasing amounts through the pulmonary vessels. During fetal life, blood bypassed the lungs by flowing through an arterial duct connecting the pulmonary artery and aorta.

Environmental Factors Affect the Embryo

We all know that the growth and development of babies are influenced by the food they eat, the air they breathe, the disease organisms that infect them, and the chemicals or drugs to which they are exposed. *Prenatal* development is also affected by these environmental influences. Life before birth is even more sensitive to environmental changes than it is in the fully formed baby. Anything that circulates in the maternal blood—nutrients, drugs, pathogens, or even gases—may find their way into the blood of the fetus (Fig. 43–17). Table 43–3 describes some of the environmental influences on development. Many of these environmental factors contribute to low birth weight, a condition responsible for a great number of infant deaths.

About 5% of newborns (more than 150,000 babies per year) in the United States have a defect of clinical significance. Such birth defects account for about 15% of deaths among newborns. Birth defects may be caused by genetic or environmental factors or a combination of the two. Genetic factors were discussed in Chapter 12. In this section,

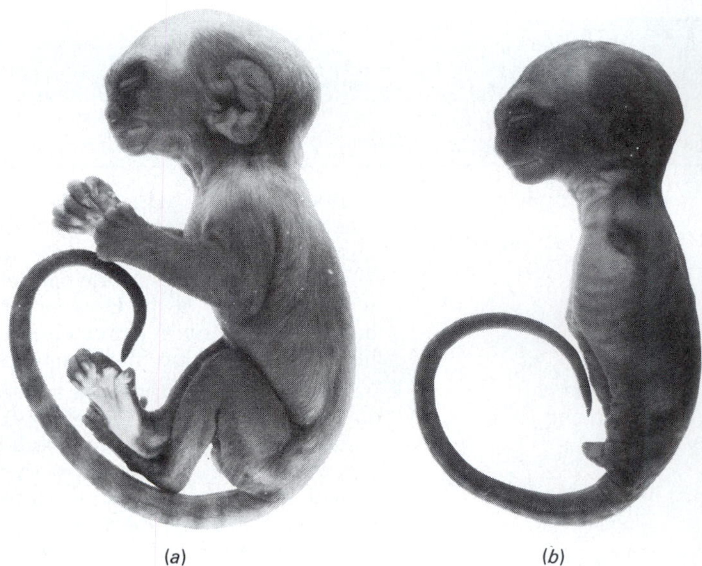

FIGURE 43–17 Effect of drugs on development. Thalidomide administered to the marmoset (Callithrix jacchus) produces a pattern of developmental defects similar to those found in humans. (***a***) Control marmoset fetus obtained from an untreated mother on day 125 of gestation. (***b***) Fetus (same age as control) of marmoset treated with 25 mg/kg thalidomide from days 38 to 52 of gestation. The drug suppresses limb formation, perhaps by interfering with the function of certain nerves. (Courtesy of Dr. W. G. Mcbride and P. H. Vardy, Foundation 41; from *Development, Growth and differentiation,* 25 [4] : 361-373, 1983)

(*a*) (*b*)

we examine some environmental conditions that affect the well-being of the embryo.

Timing is important. Each developing structure has a critical period during which it is most susceptible to unfavorable conditions. Generally, this critical period occurs early in the development of the structure, when interference with cell movements or divisions may prevent formation of normal shape or size, resulting in permanent malformation. Because most structures form during the first three months of embryonic life, the embryo is most susceptible to environmental factors during this early period. During a substantial portion of this time, the mother may not even realize that she is pregnant and so may not take special precautions to minimize potentially dangerous influences.

Physicians are now able to diagnose some defects while the embryo is in the uterus. In some cases, treatment is possible before birth. Amniocentesis and chorionic villus sampling, discussed in Chapter 13, are techniques used to detect certain defects. Figure 43–18 shows a **sonogram,** a photograph taken of the embryo by using ultrasound. Such previews are helpful in diagnosing defects and also in determining the position of the fetus and whether a multiple birth is pending.

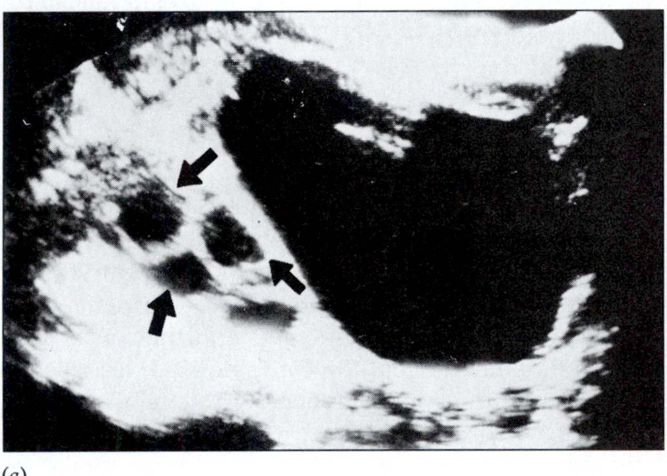

(*a*)

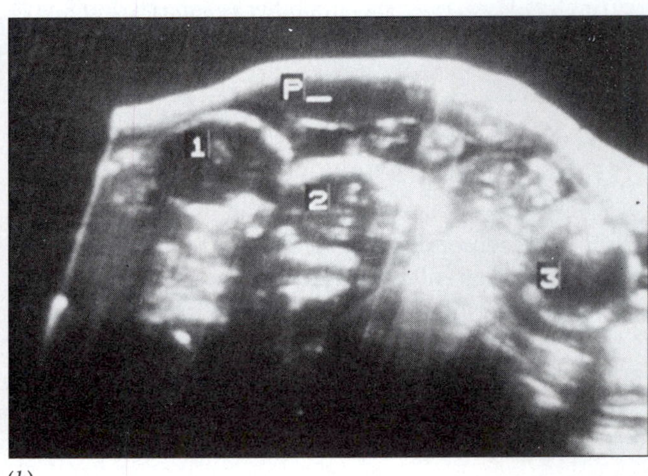

(*b*)

FIGURE 43–18 Sonogram of fetus. Ultrasonic techniques can be used to monitor follicle maturation and ovulation, as well as to give the physician information about the fetus. (***a***) Sonogram taken with ultrasound techniques, showing three follicles of equal maturation in the left ovary of a woman. (***b***) Triplets in the same patient at 16 weeks of pregnancy. P =placenta. Such previews are valuable to the physician in diagnosing defects and predicting multiple births. (Courtesy of Biserka Funduk-Kurjak, from *Acta Obstetrics and Gynecoogy Scan. 61:1982)*

THE HUMAN LIFE CYCLE EXTENDS FROM FERTILIZATION TO DEATH

Development begins at fertilization and continues through the stages of the human life cycle until death (Table 43–4). We have examined briefly the development of the embryo and fetus, the birth process, and the adjustments required of the neonate. The human life cycle then proceeds through the stages of infant, child, adolescent, young adult, middle-aged adult, and old adult.

HOMEOSTATIC RESPONSE TO STRESS DECREASES DURING AGING

Development encompasses any biological change that takes place within an organism with time, including the

TABLE 43–4
Stages in the Human Life Cycle

Stage	*Time period*	*Characteristics*
Embryo	Conception to end of eighth week of prenatal development	Development proceeds from single-celled zygote to embryo that is about 30 mm long, weighs 1 gram, and has rudiments of all its organs
Fetus	Beginning of ninth week of prenatal development to birth	Period of rapid growth, morphogenesis, and cellular differentiation, changing tiny parasite to physiologically independent organism
Neonate	Birth to 4 weeks of age	Neonate must make vital physiological adjustments to independent life: it must now process its own food, excrete its wastes, obtain oxygen, and make appropriate circulatory changes
Infant	End of fourth week to 2 years of age (sometimes, ability to walk is considered end of infancy)	Rapid growth; deciduous teeth begin to erupt; nervous system develops (myelinization), making coordinated activities possible; language skills begin to develop
Child	Two years to puberty	Rapid growth; deciduous teeth erupt, are slowly shed, and replaced by permanent teeth; development of muscular coordination; development of language skills and other intellectual abilities
Adolescent	Puberty (approximately ages 11–14) to adult	Growth spurt; primary and secondary sexual characteristics develop; development of motor skills; development of intellectual abilities; psychological changes as adolescent approaches adulthood
Young adult	End of adolescence (approximately age 20) to about age 40	Peak of physical development reached; individual assumes adult responsibilities that may include marriage, fulfilling reproductive potential, and establishing career; after age 30, physiological changes associated with aging begin
Middle-aged adult	Age 40 to about age 65	Physiological aging continues, leading to menopause in women and physical changes associated with aging in both sexes (e.g., graying hair, decline in athletic abilities, wrinkling skin); this is period of adjustment for many as they begin to face their own mortality
Old adult	Age 65 to death	Period of senescence (growing old); physiological aging continues; maintaining homeostasis more difficult when body is challenged by stress; death often results from failure of cardiovascular or immune system

changes commonly called **aging.** Changes during the aging process result in decreased function in the older organism. The declining capacities of the various systems in the human body, although most apparent in the elderly, may begin much earlier in life, for example, during childhood or even during prenatal life.

The aging process is far from uniform among different individuals or in various parts of the body. The systems of the body generally decline at different times and rates. On the average, compared with his body at age 30, a 75-year-old man has lost 64% of his taste buds, 44% of the glomeruli in his kidneys, and 37% of the axons in his spinal nerves. His nerve impulses are propagated at a rate 10% slower, the blood supply to his brain is 20% less, his glomerular filtration rate has decreased 31%, and the vital capacity of his lungs has declined 44%. The aging process is also marked by a progressive decrease in the body's homeostatic ability to respond to stress.

Although relatively little is known about the aging process itself, this is now an active field of scientific investigation. While marked improvements in medicine and public health have led to survival of a larger fraction of the total human population to an advanced age, there has been no corresponding increase in the maximum life expectancy.

Is the aging process genetically programmed like other aspects of human development? Or is aging the result of wear and tear on the body? Or is aging a *combination* of inheritance and environmental wear and tear? Cells that are genetically programmed to differentiate and stop dividing appear to be more subject to the changes of aging than are those that continue to divide throughout life. For example, nerve and muscle cells, which lose the capacity for cell division at an earlier age, show a decline in their functional capacities at an earlier age than do tissues such as liver and spleen, which retain the capacity to undergo cell division.

Several hypotheses have been advanced regarding the nature of the aging process—that it is affected by hormonal changes; that it involves the development of autoimmune responses; that it involves the accumulation of specific waste products within the cell; that it involves changes in the molecular structure of macromolecules such as collagen; that the elastic properties of connective tissues decrease owing to an accumulation of calcium, resulting in stiffening of the joints and hardening of the arteries; that it results from the oxidation of certain lipids by free radicals; and that cells are destroyed by hydrolases released by the breaking of lysosomes.

Other hypotheses suggest that continued exposure to cosmic radiation and x-radiation leads to the accumulation of somatic mutations that decrease the ability of the cell to carry out its normal functions. The aging process may be part of the program of timed development built into the genome. Like other developmental processes, aging may be accelerated by certain environmental influences and may occur at different rates in different individuals because of inherited differences. Experimental evidence suggests that aging—at least in rats—can be delayed by caloric restriction: thin rats live longer than fat rats. For now, genetic predisposition may be the best guarantee of a long life.

Chapter Summary

I. Development proceeds as a balanced combination of cell division, cell growth, morphogenesis, and cell differentiation.

II. The zygote undergoes cleavage and forms a hollow ball of cells called a blastula.
   - A. The main effect of cleavage is to partition the zygote into many small cells. As cells divide, the distribution of materials in the cytoplasm influences development.
   - B. Yolk is evenly distributed in the eggs of most invertebrates and simple chordates. Cleavage involves division of the entire egg to form cells that are about equal in size.
   - C. In bony fish and amphibians, a concentration of yolk at the vegetal pole slows cleavage so that only a few large cells form there, compared to a large number of smaller cells at the animal pole.

III. During gastrulation, the ectoderm, mesoderm, and endoderm form; each of these embryonic tissues gives rise to specific structures.

IV. The developing notochord induces nervous system development. The brain and spinal cord develop from the neural tube.

V. Development is regulated by the interaction of genes with cytoplasmic factors and environmental factors.

VI. In terrestrial vertebrates, four extraembryonic membranes—chorion, amnion, allantois, and yolk sac—protect the embryo and help in obtaining food and oxygen and in eliminating wastes. The amnion is a fluid-filled sac that surrounds the embryo and keeps it moist; it also acts as a shock absorber.

VII. Early human development follows a fairly typical vertebrate pattern.
   - A. Cleavage takes place as the embryo is moved toward the uterus.
   - B. In the uterus, the embryo develops into a blastocyst and implants itself in the endometrium.
   - C. After the first two months of development, the embryo is referred to as a fetus.

D. In placental mammals, the embryonic chorion and maternal tissue give rise to the placenta, the organ of exchange between mother and developing child.
E. Parturition takes place after about 280 days from the time of the mother's last menstrual period. During the first stage of labor, the cervix becomes dilated and effaced. During the second stage of labor, the baby is delivered. And during the third stage of labor, the placenta is delivered.

VIII. By controlling environmental factors such as diet, smoking, and the intake of alcohol and drugs, a pregnant woman can help ensure the well-being of her unborn child.
IX. The human life cycle includes: embryo, fetus, neonate, infant, child, adolescent, young adult, middle age, and old age.
X. The aging process is marked by a decrease in homeostatic response to stress.

Selected Key Terms

aging, p. 880
amnion, p. 869
animal pole, p. 864
blastula (blastocyst), p. 862
cell differentiation, p. 862
cleavage, p. 862
ectoderm, p. 865
endoderm, p. 865
extraembryonic membranes, p. 869
fetus, p. 873
gastrulation, p. 865
germ layers, p. 865
gestation period, p. 869
host mothering, p. 860
inner cell mass, p. 869
mesoderm, p. 865
morphogenesis, p. 862
neural plate, p. 867
neural tube, p. 867
placenta, p. 871
umbilical cord, p. 871
vegetal pole, p. 864
yolk, p. 862

Post-Test

1. Migration and organization of cells to form a tube such as the neural tube is an example of ______________.
2. Specialization of cells to form neurons or some other cell type is called ______________ ______________.
3. The rapid series of mitoses that converts the zygote to a morula is referred to as ______________.
4. The process by which the blastula becomes a three-layered embryo is called ______________.
5. The germ layer that gives rise to the nervous system is the ______________; the germ layer that gives rise to the lining of the digestive tract is the ______________.
6. The notochord induces the overlying ectoderm to form the ______________ ______________.
7. The neural tube develops into the ______________ and ______________ ______________.
8. The ______________ ______________ prevents the embryo from drying out and acts as a shock absorber.
9. In humans, the ______________ is the organ of exchange between mother and fetus.
10. The cluster of cells that projects into the cavity of the blastocyst is the ______________ ______________ ______________; it gives rise to the ______________.
11. On about the seventh day of development, the human embryo begins to ______________ in the ______________.
12. After the first two months of development, the human embryo is referred to as a/an ______________.
13. A human baby is delivered during the ______________ stage of ______________.
14. Maternal and fetal blood normally (a) do not mix (b) mix only in the placenta (c) are mixed throughout the maternal and fetal circulations (d) flow together only in the umbilical capillaries (e) mix only in the allantois.
15. The human heart begins to beat (a) at birth (b) during the first month of development (c) during the second trimester (d) during the third trimester (e) before the neural tube forms.
16. From birth to four weeks of age, a human is referred to as a/an (a) embryo (b) fetus (c) infant (d) neonate (e) three of the preceding answers are correct.
17. If the corpus luteum is removed before the 11th week of pregnancy, (a) the embryo is spontaneously aborted (b) conjoined twins are produced (c) an excess of estrogen is produced by the embryo (d) an excess of progesterone is produced by the placenta (e) three of the preceding answers are correct.

Review Questions

1. Trace the development of a sea star (or *Amphioxus*) embryo from zygote to gastrula; draw and label diagrams to illustrate your description.
2. Contrast cleavage in the sea star (or *Amphioxus*) and amphibian.
3. Trace some developmental process, such as the formation of the neural tube, and explain how cell division, growth, morphogenesis, and cell differentiation interact in the process.
4. Give examples of adult structures that develop from each of the germ layers.
5. Why do terrestrial vertebrate embryos develop an amnion? What are its functions?
6. Describe human blastocyst formation and implantation.
7. What kinds of adaptations must the neonate make immediately after birth?
8. What are some of the nongenetic factors that influence development?
9. What steps can the pregnant woman take to help ensure the safety and well-being of her developing child?
10. Describe some of the changes that take place during the aging process.

Thinking Critically

1. Construct arguments for and against human embryo research. Consider such techniques as in vitro fertilization or freezing embryos for later implantation in light of concerns about human population expansion.
2. What is the adaptive value of developing a placenta?
3. For almost 200 years, scientists debated the preformation theory that held that the egg cell or the sperm cell contains a completely formed, miniature human. As research techniques improved, the theory of epigenesis was developed. This theory held that the embryo develops from a formless zygote, taking form as development proceeds. Relate these theories to current concepts of development. Can you find any truth in the preformationist view?

Recommended Readings

Browder, L.W., C.A. Erickson, and W.R. Jeffery. *Developmental Biology,* 3rd ed. Philadelphia: Saunders College Publishing, 1991. An excellent introduction to animal development.

Holiday, R., "A Different Kind of Inheritance," *Scientific American,* Vol. 260, No. 6, June 1989. DNA methylation may be an important mechanism by which patterns of gene activity are passed from one generation of cells to another during development.

Wasserman, P.M., "Fertilization in Mammals," *Scientific American,* Vol. 259, No. 6, December 1988. A glycoprotein governs many of the events of fertilization.

PART 9

Behavior and Ecology

A green turtle hatchling **(Chelonia mydas)** *makes its way back to the sea.* (Kelvin Aitken/Peter Arnold, Inc.)

Migration: Costs and Benefits

Birds, butterflies, fishes, and sea turtles are among the many animals that travel long distances and then return home to reproduce. Recent DNA tests confirm that loggerhead sea turtles hatch on Florida beaches along the Atlantic, then swim hundreds of miles across the Atlantic to the Mediterranean Sea. Several years later, those that survive to become adults, navigate back, often to the same beach to lay their eggs.

Green turtles (*Chelonia mydas*) migrate more than 2000 kilometers (1200 miles) across open ocean between their feeding area off the coast of Brazil and their nesting place on Ascension Island, a tiny island less than 10 km (6 miles) wide. There, each turtle will make three to seven trips, about 12 days apart, from the water to an isolated beach, laying about 100 eggs in the sand each time.

Why do animals migrate? Migration is a very precise evolutionary adaptation to seasonal changes in climate, availability of food resources, and safe nesting sites. The dramatic annual migration of millions of monarch butterflies from Eastern Canada and the United States to Mexico—a 2500-kilometer journey (about 1500 miles) for some—appears to be related to the availability of milkweed plants on which females lay their eggs. In cold regions these plants die in late autumn, and grow again with the warmer weather of spring. The wintering destinations of monarchs also seem to be determined by temperature and humidity.

The benefits of migration are not without cost. Many weeks may be spent each year on energy-demanding journeys. Some animals may become lost or die along the way. And migrating individuals are often at greater risk from predators in unfamiliar areas. In recent years, human activities have interfered with migrations of many kinds of animals. For example, after millions of years of biological success, survival of the sea turtles is threatened by the fishing and shrimping industries. As they migrate, as many as 5000 Florida loggerhead turtles drown each year in the nets of fisheries that block their passage. In addition, thousands of sea turtles drown each year in the nets of shrimpers. The National Marine Fisheries Service has asked shrimpers to use turtle excluder devices, but many refuse because they claim that these devices reduce their shrimp harvest.

Just how animals navigate remains somewhat of a mystery. Young sea turtles appear to use the earth's magnetic field to guide them. Studies suggest that they also use wave direction to orient themselves. Many other animals including birds, fishes, insects, and amphibians also migrate with the help of the magnetic field. Adult salmon use the unique odors of different streams to find their way back (sometimes across thousands of miles) to the same stream from which they hatched.

Animal Behavior

44

Learning Objectives

After you have studied this chapter you should be able to:

1. Explain the concept that behavior is (a) adaptive, (b) homeostatic, and (c) flexible.
2. Cite examples of biological rhythms and suggest some of the mechanisms responsible for them.
3. Summarize the contributions of heredity, environment, and maturation to behavior.
4. Classify learned behavior as examples of (a) classical conditioning, (b) operant conditioning, (c) habituation, or (d) insight learning.
5. Discuss the adaptive significance of imprinting.
6. Postulate biological advantages and disadvantages for migration.
7. Discuss the hypothesis that optimal foraging behavior is adaptive.
8. Give a description of an animal society, and identify the adaptive advantages of cooperative behavior.
9. Summarize the modes of communication that animals employ.
10. Present the concept of a dominance hierarchy, giving at least one example, and propose a possible adaptive significance and social function for this hierarchy.
11. Distinguish between home range and territory and give three theories about the adaptive significance of territoriality.
12. Discuss the adaptive value of courtship behavior and describe a pair bond.
13. Compare the society of a social insect with human society.
14. Define kin selection and summarize its proposed role in the maintenance of insect and other animal societies.

Key Concepts

- ☐ Each organism has a set of adaptive behaviors that fit its life-style.
- ☐ The capacity for behavior is inherited. Much inherited behavior can be modified by experience. Learning involves persistent changes in behavior that result from experiences. Types of learning include classical conditioning, operant conditioning, insight learning, and imprinting.
- ☐ Social behavior is the interaction of two or more animals, usually of the same species. Social behavior requires an inhibition of aggression and some system of communication.

Key Concepts, continued

- ☐ Aggression is minimized by the use of dominance hierarchies and by territoriality.
- ☐ Elaborate societies occur among some insects. The behavioral basis of these societies is genetic. More flexible societies occur among vertebrates, especially mammals. Human society, the most elaborate vertebrate society, involves the symbolic transmission of culture from one generation to another.
- ☐ Kin selection may explain the development of altruistic behavior.

BEHAVIOR IS ADAPTIVE

Suppose that your professor were to arm you with a hypodermic syringe full of poison and demand that you find a particular type of insect, one that you have never seen before and that is armed with active defenses. You must then inject the ganglia of its nervous system (about which you have been taught nothing) with just enough poison to paralyze, but not kill, your victim. You would have difficulty accomplishing these tasks, but a solitary wasp no larger than the first joint of your thumb does it all with elegance and surgical precision, and without instruction.

The bee-killer wasp *Philanthus* captures bees, stings them, and places the paralyzed insects in burrows excavated in the sand. She then lays an egg on her victim, which is devoured alive by the larva that hatches from that egg. From time to time, the *Philanthus* returns to her hidden nest to reprovision it until the larva becomes a hibernating pupa in the fall. Her offspring will repeat this, doing it to perfection without ever having seen it done.

Behavior refers to the responses of an organism to signals from its environment, including those from other organisms. Much of what organisms do can be analyzed in terms of specific behavior patterns that occur in response to stimuli (changes) in the environment. A dog may wag its tail, a bird may sing, a butterfly may release a volatile sex attractant. Behavior is just as diverse as biological structure, and is just as characteristic of a given species as its structure and biochemistry. Structure, function, and behavior are all adaptations that help define an organism and equip it for survival.

Biologists study behavior in the laboratory and in natural environments, always keeping in mind that what an animal does cannot be isolated from the way in which it lives. **Ethology** is the study of behavior in natural environments from the point of view of adaptation. Thus, a particular behavior may help an organism obtain food or water, acquire and maintain territory in which to live, protect itself, or reproduce.

Behavior tends to be homeostatic for the individual organism as well as adaptive in the evolutionary sense. Certain behavioral responses, however, may lead to the death of the individual while increasing the chance that copies of its genes will survive through the enhanced production or survival of the offspring. In this chapter, we consider how behavior fits into the total life of the organism so as to enable it to live and pass on its heritage to those that follow.

BIOLOGICAL RHYTHMS ANTICIPATE ENVIRONMENTAL CHANGES

An organism adapts by synchronizing its metabolic processes and behavior with the cyclic changes in the external environment. Its behavior can *anticipate* these regular changes. The little fiddler crabs (Fig. 44–1) of marine beaches often emerge from their burrows at low tide to engage in social activities such as territorial disputes. They must return to their burrows *before* the tide returns

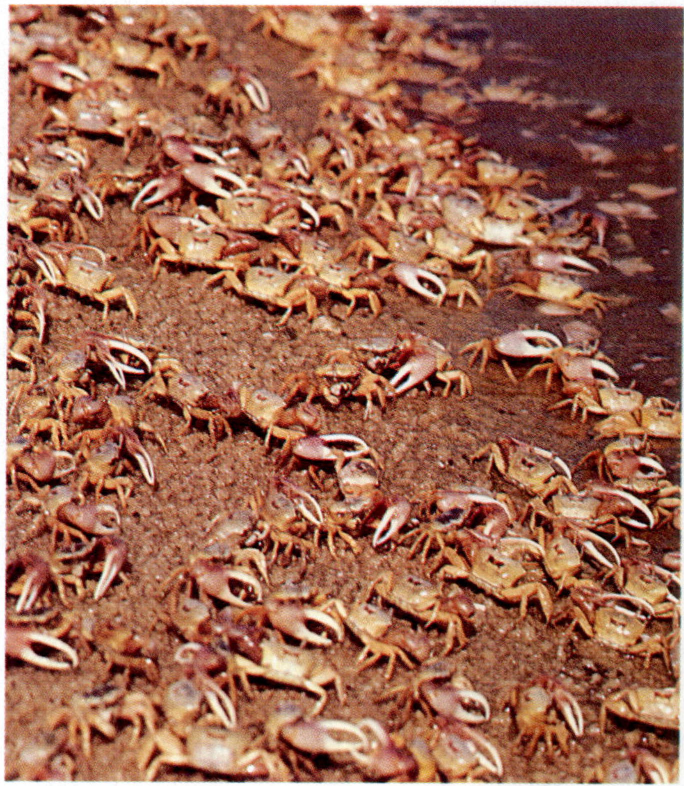

FIGURE 44–1 Behavior synchronized with the tides. Fiddler crabs run about on the surface of the sand at low tide but must return to their burrows before high tide returns. The males have enlarged claws. (Tom Walker, Photo Researchers, Inc.)

or they would probably be washed away. How do the crabs "know" that high tide is about to occur? They cannot consult tide tables!

One might guess that the crabs recognize clues present in the seashore. However, when the crabs are isolated in the laboratory away from any known stimulus that could relate to time and tide, their characteristic behavioral rhythms persist.

A Variety of Behavioral Cycles Occur Among Organisms

In addition to the biological rhythms oriented around tides, other types of biological rhythms occur, including daily, monthly, and annual rhythms. Among many animals, periods of activity and sleep, feeding and drinking, body temperature, and secretion of some hormones have a cycle that is 24 hours long. Human physiological processes also seem to follow an intrinsic rhythm. Human body temperature, for example, follows a typical daily curve. These daily rhythmic activities are **circadian** (meaning "approximately one day"). Such biological rhythms suggest that animals have **biological clocks** that are precisely adjusted or reset by environmental cues (see Chapter 30).

The behavior of many animals appears to be organized around circadian rhythms. **Diurnal** animals, like honeybees and pigeons, are most active during the day. **Nocturnal** animals are most active during the hours of darkness. **Crepuscular** animals, like the fiddler crab, are busiest during the twilight hours, at dawn, or both. As in the case of the fiddler crab, there are ecological reasons for these adaptations. If an animal's food is most plentiful in the early morning, for example, its cycle of activity must be regulated so that it becomes active shortly before dawn.

Some biological rhythms of animals reflect the **lunar** (moon) **cycle.** The most striking rhythms are those in marine organisms that are tuned to the changes in the tides due to the phases of the moon. For instance, a combination of tidal, lunar, and annual rhythms governs the reproductive behavior of the grunion, a small fish of the Pacific coast of the United States.

The grunion swarms from April through June on those three or four nights when the highest tide of the year occurs. At precisely the high point of the tide, the fishes squirm onto the beach, deposit eggs and sperm in the sand, and return to the sea in the next wave. By the time the next tide reaches that portion of the beach 15 days later, the young fishes have hatched and are ready to enter the sea.

Biological Rhythms Are Controlled by an Internal Clock

Biological rhythms are regulated by internal timing mechanisms referred to as biological clocks. Current evidence suggests that most organisms have no single biological clock. Instead, the interaction of a number of biochemical processes (possibly involving cellular membranes) might be responsible for governing physiological and behavioral rhythms. The pineal gland is thought to play a role in the timing systems of birds, rats, humans, and some other vertebrates. Regions of the hypothalamus are also part of the biological clock in mammals. Although some parts of an organism may coordinate or dominate the function of the biological clock, it is likely that every cell has some type of timing mechanism.

BEHAVIOR CAPACITY IS INHERITED AND IS MODIFIED BY ENVIRONMENTAL FACTORS

All behavior has a genetic basis. In other words, the *capacity* for behavior is inherited. Even the capacity to learn is inherited. However, behavior can be modified by the environment so that it is a product of the interaction between genetic capacity and environmental influences. Thus, behavior begins with an inherited framework that experience can modify.

The instructions for the honeybee society are inherited and preprogrammed, and the size and structure of the bee's nervous system permit only a limited range of behavior. Yet, these insects are not automatons. Within those limits, the complex bee society can respond flexibly to food and other stimuli in the environment.

Some behavior of vertebrates is also largely predetermined by genes. Several species of the lovebird (*Agapornis*) differ not only in appearance but in behavior. One species transports nest-building materials in its bill. Another species tucks materials into its feathers. Studies indicate that the method of transporting materials is inherited but somewhat flexible.

Behavior Develops

Behavior involves all body systems, but it is influenced mainly by the nervous and endocrine systems. The capacity for behavior depends on the genetic characteristics that govern the function of these systems. One may think of a continuous scale of behaviors, ranging from the most rigidly programmed, genetically inherited types through those that are somewhat modifiable, to those that, although containing a genetic component, are extensively developed through experience.

Before an organism can exhibit any pattern of behavior, it must be physiologically ready to produce the behavior. For example, breeding behavior does not ordinarily occur among birds or most mammals unless steroid sex hormones are present in their blood at certain concentrations. A human baby cannot walk unless its reflex and muscular development permit it to walk. These states of physiological readiness are themselves produced by a continuous interaction with the environment. The level of sex hormones in a bird's blood may be determined by seasonal variations in day length. The baby's muscles develop in response to exercise. Without the trial-and-error involved in preparing to walk, the ability to walk would be delayed.

A good example of such interaction between readiness and environment is afforded by the white-crowned sparrow, which exhibits considerable regional variation in its song. This bird, even if kept in isolation, eventually sings a very poorly developed but recognizable white-crowned sparrow song. However, if it is allowed to grow up under the care of its parents, it learns the local "dialect" from its parents. When mature, it can sing the more highly developed local song characteristic of the region. If such learning does not take place early in life, it never will, and if the sparrow later consorts with birds of other species, it does not learn their songs. It appears that the white-crowned sparrow is hatched equipped with a rough genetic pattern of its song. The regional details are filled in by learning.

Some Behavior Patterns Have Strong Genetic Components

The term **instinct** refers to *innate* (inborn) behavior, that is, genetically programmed behavior. A classic example of innate behavior in vertebrates is egg-rolling in the European graylag goose. When an egg is removed from the nest of a goose and placed a few inches in front of her, she reaches out with her neck and pulls the egg back into the nest (Fig. 44–2). If, while the goose is rolling the egg back toward the nest, the egg veers off to the side, the goose steers it back toward the nest. Because this type of behavior appears to have a strong genetic component, ethologists refer to it as a **fixed action pattern (FAP).**

A stimulus that elicits an FAP—such as the egg in the egg-rolling behavior or the red-colored "belly" in the attack behavior of the stickleback fish (Fig. 44–3)—is called a **sign stimulus.** A goose will retrieve a wooden egg in the same way as a real egg, so the wooden egg is also by definition a sign stimulus.

Behavior Is Modified by Learning

We can define learning as a change in behavior due to experience. The capacity to learn appropriate responses to

Concept Connections

Learned Behavior — Nervous Systems

The simpler the nervous system, the simpler the range of behavior. Even innate behavior depends on the properties of individual neurons and of their interconnections. For an organism to learn, its neurons must have a large number of potential interactions with one another—not just the few required by stereotyped preprogrammed behavior, such as the sand wasp's ability to sting its victims.

Without the pre-existence of the necessary neural circuitry, learned behavior would be impossible. The kind of learned behavior that an animal typically and most easily develops depends upon the complexity and layout of its neural circuitry. Learning also depends on changes in the readiness of neurons to form circuits with one another.

FIGURE 44–2 A fixed action pattern (FAP). Egg-rolling behavior in the European graylag goose.

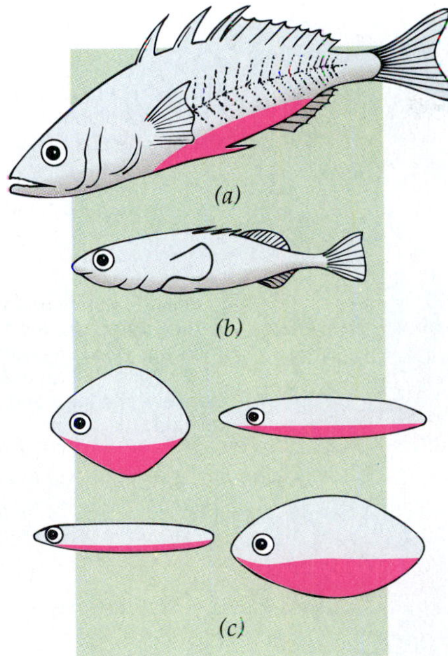

FIGURE 44–3 A sign stimulus triggers a fixed action pattern. A male stickleback fish (***a***) will not attack a realistic model of another male stickleback if it lacks a red belly (***b***), but it will attack another model, however unrealistic, that has a red "belly" (***c***). Therefore, it is the specific red sign stimulus, rather than recognition based on a combination of features, that triggers the aggressive behavior.

new situations is adaptive, because learned responses can be shaped to meet the needs of the changing environment that most animals experience. The wasp *Philanthus* efficiently carries out a complex, although largely genetically programmed, sequence of behaviors. Yet, some of her behavior is learned. When *Philanthus* covers a nest with sand, she takes precise bearings on the location of the burrow before flying off again to hunt. There is no way that her ability to locate the burrow could be genetically programmed. How to dig it, how to cover it, how to kill the bees—these behaviors appear to be genetically programmed. But because a burrow can be dug only in a suitable spot, its location must be learned *after* it is dug.

This distinction between innate behavior and behavior modified by experience in the wasp *Philanthus* was determined by the Dutch investigator Niko Tinbergen. Tinbergen surrounded the wasp's burrow with a circle of pine cones, on which the wasp took her bearings (Fig. 44–4). Before she returned with another bee, Tinbergen moved the circles of pine cones. The wasp could not find her burrow because it was no longer surrounded by pine cones. Only when the experimenter restored the cones to their original location could the wasp find her burrow. In contrast, the existence of complex programmed behavior is hard to demonstrate in humans. We owe the complexity of our behavior to a *generalized* ability to learn. The *Philanthus* wasp's intelligence is as narrowly specialized as is her stinger.

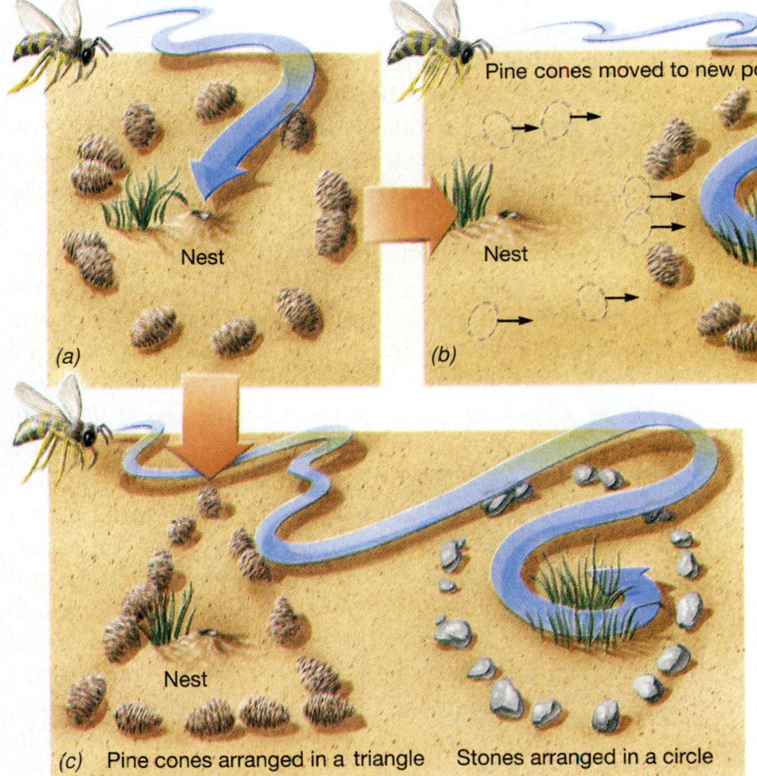

FIGURE 44–4 Tinbergen's sand wasp experiment. When the ring of pine cones is moved from position (***a***) to position (***b***), the *Philanthus* wasp behaves as if her nest were still located at the center. She learned its position in relation to the cones. That it is the arrangement of the cones rather than the cones themselves that the wasp responds to is shown by the substitution of a ring of stones for cones in (***c***). The learning ability of *Philanthus* is quite limited but is adequate for situations that normally arise in nature. (After Tinbergen)

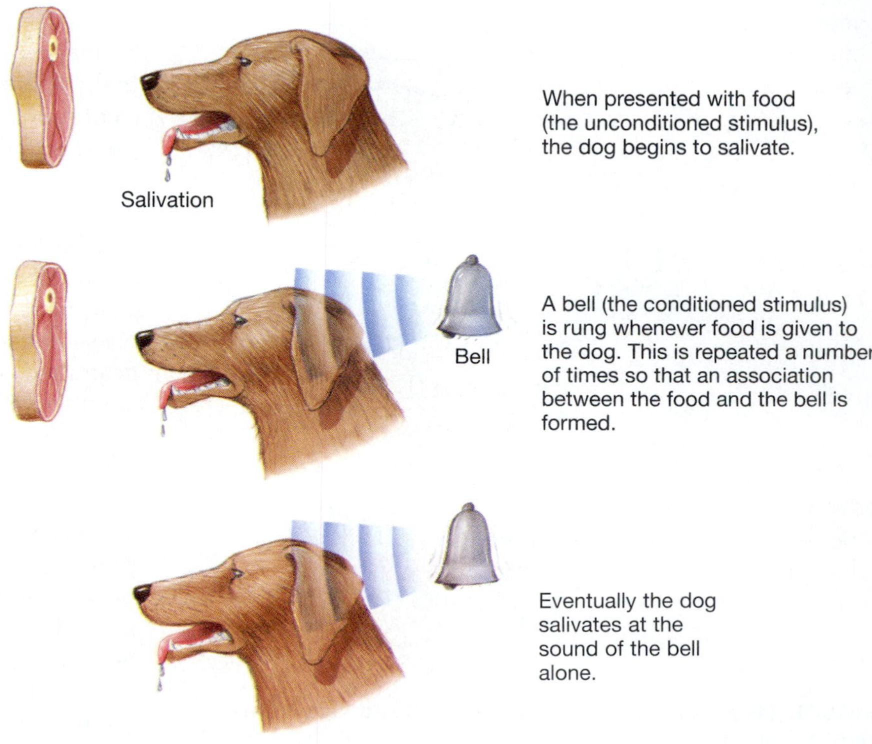

Salivation
When presented with food (the unconditioned stimulus), the dog begins to salivate.
Bell
A bell (the conditioned stimulus) is rung whenever food is given to the dog. This is repeated a number of times so that an association between the food and the bell is formed.
Eventually the dog salivates at the sound of the bell alone.

FIGURE 44–5 Pavlov's experiment. The dog is classically conditioned to salivate at the sound of the bell.

In Classical Conditioning, a Reflex Becomes Associated with a New Stimulus

In a type of learning called **classical conditioning,** an association is formed between some normal body function and a stimulus. Ivan Pavlov, a Russian physiologist, studied dogs in the laboratory early in this century. He discovered that when a bell was rung at the same time a dog was fed, an association formed between the sound of the bell and the secretion of saliva. Eventually (Fig. 44–5), when the bell was rung by itself, the dog salivated. Pavlov called the physiologically meaningful stimulus (food, in this case) the *unconditioned stimulus.* The normally irrelevant stimulus (the bell) that became a substitute for it was the **conditioned stimulus.** Because a dog does not normally salivate at the sound of a bell, the association was clearly a learned one. It could also be forgotten. If it no longer signaled food, the dog would eventually stop responding to the bell. Pavlov called this process **extinction.**

In Operant Conditioning, Spontaneous Behavior Is Reinforced

In **operant conditioning** (also called instrumental conditioning), the animal must do something in order to gain a reward (**positive reinforcement**) or avoid a punishment. In a typical experiment, a rat is placed in a cage containing a movable bar. When random actions of the rat result in the rat pressing the bar, a pellet of food rolls down a chute and is delivered to the rat. Thus, the rat is positively reinforced for pressing the bar. Eventually, the rat learns the association and presses the bar whenever it wants food.

In **negative reinforcement,** removal of a stimulus increases the probability that a behavior will occur. For example, a rat may be subjected to an unpleasant stimulus, such as a low-level electric shock. When the animal presses a bar, this negative reinforcer is removed and the animal experiences relief.

Many variations of these techniques have been developed. A pigeon might be trained to peck at a lighted circle to obtain food, a chimpanzee might learn to perform some task in order to get tokens that can be exchanged for food, or children might learn to stay quietly in their seats at school to obtain some reward. Operant conditioning is probably the way that animals learn to perform complex tasks like walking or perfecting feeding skills.

Operant conditioning plays a role in the development of behaviors that appear to be preprogrammed. An example is the feeding behavior of gull chicks. Herring gull chicks peck the beaks of the parents, which regurgitate partially digested food for them. The chicks are attracted by two stimuli: the general appearance of the parent's beak with its elongated shape and distinctive red spot, and its downward movement as the parent lowers its head. Like the rat's chance pressing of the bar, this behavior is sufficiently functional to get the chicks their first meal, but they waste a lot of energy in pecking. Some pecks are off target and are therefore not rewarded. However, the begging behavior becomes more efficient over time. Thus, a behavior that might appear to be entirely instinctive is perfected by learning (Fig. 44–6).

Imprinting Is a Form of Learning that Occurs during a Critical Period

Anyone who has watched a mother duck with her ducklings must have wondered how she can keep track of such a horde of almost identical little creatures, tumbling about in the grass, let alone tell them from those belonging to another hen (Fig. 44–7). Although she is capable of recognizing her offspring to an extent, basically they have the responsibility of keeping track of her. The survival of the duckling requires that it quickly establish a behavioral bond with its parent. This bond, which is usually formed within a few hours of birth (or hatching), forms by a type of learning known as **imprinting.** An early investigator of imprinting, ethologist Konrad Lorenz, discovered that a newly hatched bird imprints on the first moving object it sees—even a human, or an inanimate object. Although imprinting itself is genetically determined, the bird learns the object.

Among many types of birds, especially ducks and geese, the older embryos are able to exchange calls with their nest mates and parents right through the porous eggshell. When they hatch, at least one parent is normally on hand, emitting the characteristic sounds with which the hatchlings are already familiar. If the parent moves,

FIGURE 44–6 Operant conditioning plays a role in the development of some innate behaviors. The pelican chick learns to be more accurate in begging for food from a parent. (H. Cruickshank/VIREO)

FIGURE 44–7 Formation of parent-offspring bonds. Through imprinting, some young animals follow the first moving object they encounter. Usually, the object is their mother, although it is possible experimentally to imprint many such infants upon unnatural objects. (J. H. Dick/VIREO)

FIGURE 44–8 Habituation. After repeated safe encounters with humans, these pigeons are unperturbed by people. (Janet Goldwater)

the chicks follow. This movement plus the sounds produces imprinting. During a brief critical period after hatching, the chicks learn the appearance of the parent.

Imprinting establishes the bond between mother and offspring among many mammals, as well as among birds. In many species, the mother also establishes a bond with her offspring during a critical period. The mother in some species of hoofed mammals, such as sheep, will accept her offspring for only a few hours after its birth. If they are kept apart past that time, the young are rejected. Normally, this behavior enables the mother to distinguish her own offspring from those of others, evidently by olfactory cues.

Habituation Enables an Animal to Ignore Irrelevant Stimuli

In **habituation,** an animal learns to ignore a repeated, irrelevant stimulus. In Figure 44–8, we see the familiar gathering of pigeons in a city street. These birds have learned by repeated harmless encounters that humans are no more dangerous to them than are cows to crows, and behave accordingly. This is to their advantage. A pigeon intolerant of people might not get enough to eat.

Insight Learning Uses Recalled Events to Solve New Problems

The most complex learning is **insight** learning, which is the ability to adapt past experiences that may involve different stimuli to solve a new problem. A dog can be placed in a blind alley that it must *circumvent* in order to reach a reward. The difficulty of the problem lies in the fact that the animal must move *away* from the reward in order to get *to* it. Typically, the dog flings itself at the barrier nearest the food. Eventually, by trial-and-error, the frustrated dog may find its way around the barrier and reach the reward.

FIGURE 44–9 Insight learning. Confronted with the problem of reaching food hanging from the ceiling, the chimpanzee stacks boxes until it can climb and reach the food.

In contrast to the dog, a chimpanzee placed in a similar situation is likely to see the solution instantly (Fig. 44–9). Primates are skilled at insight learning, but a broad range of animals seem to have this ability to some degree.

Learning Abilities Are Biased

Animals learn some things more easily than others. For instance, one does not really need to teach a baby bird to fly. Learning language comes very naturally to humans. A child will learn the speech of its caretakers, even if no one deliberately instructs it. In general, learning biases reflect an animal's specialized mode of life.

What is most important to survival appears to be most easily learned. The same rat that may have taken a dozen trials to perfect the artificial task of pushing a lever to get a reward learns in a *single* trial to avoid a food that has made it ill. Those who poison rats to get rid of them can readily appreciate the adaptive value of this learning ability. Such quick learning in response to an unpleasant experience forms the basis of warning coloration, which is found in many poisonous insects and brilliantly colored, but distasteful, bird eggs. Once made ill by such an egg, predators quickly learn to avoid them.

BEHAVIORAL ECOLOGY EXAMINES INTERACTION OF ANIMALS WITH THEIR ENVIRONMENTS

Behavioral ecologists focus on how animals interact with their environments and on the survival value of their behavior. Two of the many activities that relate behavior and ecology are migration and foraging for food.

Migration Is Triggered by Environmental Changes

Migration is the periodic long-distance movement from one location to another with a subsequent return to the first location (see Chapter Opener). Some migrations involve astonishing feats of endurance and navigation. Ruby-throated hummingbirds cross the vast reaches of the Gulf of Mexico twice each year, and the sooty tern travels across the entire South Atlantic from Africa to reach its tiny island breeding grounds south of Florida.

The behavioral trigger that sets off migratory behavior varies. Some animals migrate when they mature. In others, certain environmental cues trigger the process. In migratory birds, for example, the pineal gland senses changes in day length. The pineal gland then releases hormones that cause restless behavior. The bird shows an increased readiness to fly and flies for longer periods of time (Fig. 44–10).

The *direction* of travel is obviously important, and this raises the general problem of animal navigation. Honeybees, some birds, sea turtles, and some other animals ap-

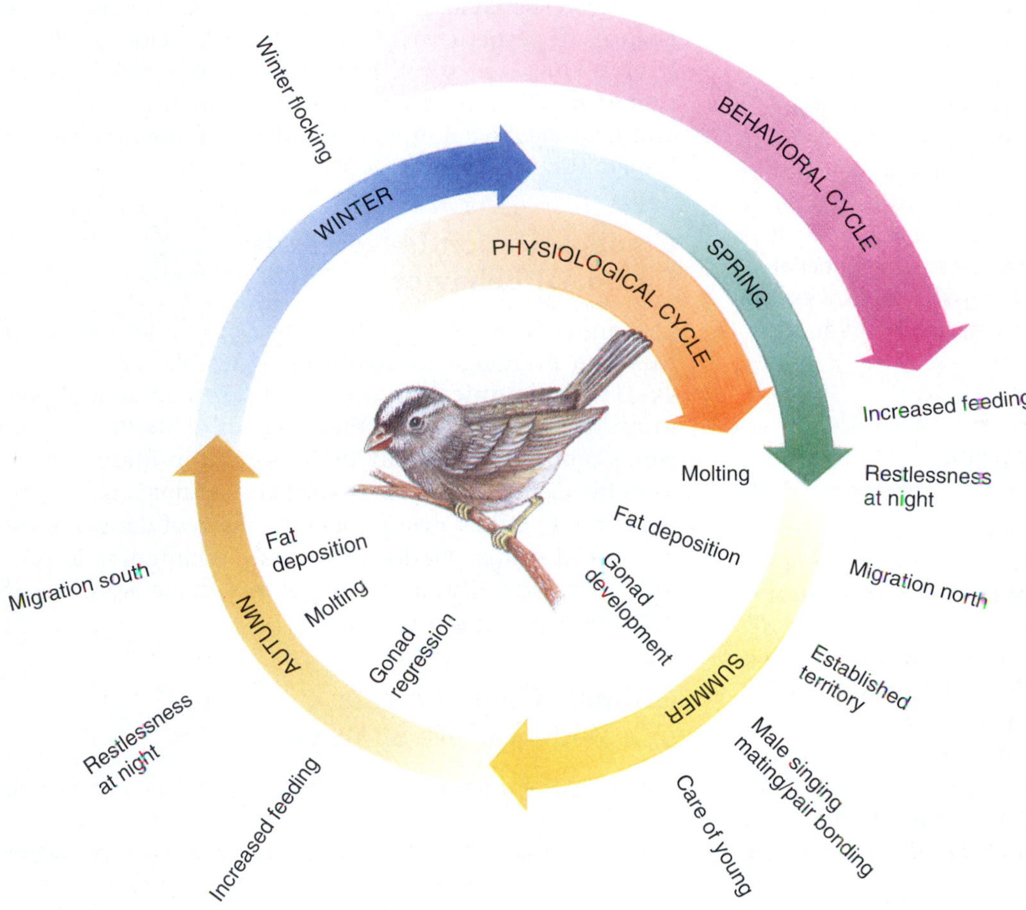

FIGURE 44–10 Seasonal changes in the physiology and behavior of the white-crowned sparrow. Note the increased rate of feeding and then restlessness that precedes each period of migration. (Based on Alcock, J., *Animal Behavior: An Evolutionary Approach,* 4th ed., Sunderland, Mass., Sinauer Associates, 1989)

pear to be sensitive to the Earth's magnetic field. Birds appear to navigate by a combination of celestial (sun- and star-related) and geographical and climatic cues.

Working in the 1950s, Franz and Eleonore Sauer hand reared a number of whitethroats, a species of small European warbler. This ruled out the possibility that the parents had transmitted any information to their offspring. When (and only when) the birds could see the star patterns of the night sky, they attempted to fly in the normal direction of migration for this species, a direction that they had no opportunity to learn. When the birds were brought into a planetarium that simulated the night sky of a different locale, they attempted to fly in a direction that would have taken them to their normal wintering grounds from that locality. The conclusion seemed inescapable: although the direction of migration was not learned, the birds were able to find their way instinctively by means of celestial navigation.

Efficient Foraging Behavior Contributes to Survival

Natural selection produces animals that are efficient at propagating their genes. Such efficiency depends on making the best choices in mate selection, defending territory, and foraging for food. A topic of current interest to many behavioral ecologists is **optimal foraging,** which refers to the most efficient way for an animal to obtain food. According to current hypotheses, when animals maximize energy obtained per unit of foraging time, they maximize reproductive success. Many factors must be considered such as avoiding predators while foraging. Grizzly bears that spend hours digging Arctic ground squirrels out of burrows provide an example of optimal foraging. Why do these bears ignore larger prey such as caribou? It is energetically more efficient to dig for squirrels because the bears' efforts most probably will be rewarded, whereas caribou might escape, leaving the bears hungry.

SOCIAL BEHAVIOR HAS BOTH BENEFITS AND COSTS

The mere presence of more than one individual does not mean that a behavior is social. Many factors of the physical environment bring animals together in **aggregations,** but whatever interaction they experience may be circumstantial. A light shining in the dark is a stimulus that draws large numbers of moths. The high humidity under a log may attract wood lice. Although these aggregations may have adaptive value, they are not truly social, because the organisms are not responding to one another.

We can define **social behavior** as the interaction of two or more animals, usually of the same species. Many animals benefit from living in groups. By cooperation and division of labor, some insects construct elaborate nests and raise young by mass-production methods. Schools of fishes are less vulnerable to predators than are solitary fish. Their large numbers tend to confuse predators, and working together they can often repel attacks. Antelope can more effectively detect or discourage predators when some individuals in the herd are on watch, and drive off predators by collective action.

Social foraging is an adaptive, efficient strategy, and is used routinely by many animal species. A pack of wolves and a pride of lions have greater success in hunting than the individual wolves or lions would have if hunting alone. Among some birds of prey, such as ospreys, cooperative hunting results in locating prey more quickly.

Some species that engage in social behavior form societies. A **society** is an actively cooperating group of individuals belonging to the same species. A hive of bees, a flock of birds, a pack of wolves, and a school of fishes are examples of societies. Some societies are loosely organized, whereas others have a complex structure. Characteristics of a well-organized society include cooperation and division of labor among animals of different sexes, age groups, or castes. A complex system of communication reinforces the organization of the society. The members of a society tend to remain together and to resist attempts by outsiders to enter the group.

Social behavior offers benefits that increase the chances of perpetuating the genes that produce such behavior. However, social behavior also has certain costs. Living together means increased competition for food and habitats. Social interaction also increases the risk of transmitting disease.

Communication Is Necessary for Social Behavior

One animal can influence the behavior of another only if they can exchange mutually recognizable signals (Fig. 44–11). **Communication** is evident when an animal performs an act that changes the behavior of another organism. Communication may be important in finding food, as in the elaborate dances of the bees. Animals may communicate to hold a group together, warn of danger, indicate social status, ask for or indicate willingness to provide care, identify members of the same species, or indicate sexual maturity.

Animals Communicate in a Wide Variety of Ways

Methods of animal communication are extremely varied. The singing of many birds announces the presence of a territorial male. Some animals communicate by scent

(a)

(b)

FIGURE 44–11 Communication. (***a***) Auditory communication. A male hylid frog of Costa Rica calling to locate a mate. (***b***) Communicating with language. Chimpanzee Tatu (top) is signing "food" to Washoe (below). Washoe was the first chimp to learn American Sign Language from a human. She then taught other chimps how to sign. (*a*, L.E. Gilbert, University of Texas at Austin/Biological Photo Service; *b*, April Ottey, Chimpanzee and Human Communication Institute, Central Washington University)

rather than sound. Antelopes rub the secretions of facial glands on conspicuous objects in their vicinity. Dogs mark territory by frequent urination. Certain fishes, the gymnotids, use electric pulses for navigation and communication, including territorial threat, in a fashion similar to bird vocalization. As Edward O. Wilson has said, "The fish, in effect, sing electrical songs."

Pheromones Are Chemical Signals Used in Communication

Pheromones are chemical signals that convey information between members of a species. They are a simple, widespread means of communication. Most pheromones elicit a very specific, immediate, but transitory type of behavior. Others trigger hormonal activities that result in slow, but long-lasting, responses. Some pheromones may act in both ways.

An advantage to pheromone communication is that little energy must be expended to synthesize the simple, but distinctive, organic compounds involved. Members of the same species have receptors that fit the molecular configuration of the pheromone; other species usually ignore it. Pheromones are effective in the dark, they can pass around obstacles, and they last for several hours or longer. Major disadvantages of pheromone communication are slow transmission and limited information content. Some animals compensate for the latter disadvantage by secreting different pheromones with different meanings.

Pheromones are important in attracting the opposite sex and in sex recognition in many species. Many female insects produce pheromones that attract males. We have taken advantage of some sex-attractant pheromones to help control such pests as gypsy moths by luring the males to traps baited with synthetic versions of female pheromones.

Some aspects of the sexual cycle of vertebrates are affected by pheromones. When the odor of a male mouse is introduced among a group of females, the reproductive cycles of the female mice become synchronized. In some species of mice, the odor of a strange male, a sign of high population density, causes a newly impregnated female to abort. Among humans, it appears that some unconsciously perceived body odor is capable of synchronizing the menstrual cycles of women who associate closely (for instance, college roommates or cellmates in prison). As we shall see, pheromones much more strictly govern the reproduction of many social insects.

Animals Often Form Dominance Hierarchies

In the spring, female paper-wasps awake from hibernation and begin to build a nest cooperatively. During the early course of construction, a series of squabbles among the females takes place in which the combatants bite one another's bodies or legs. Finally, one of the wasps emerges as dominant. After that, she is rarely ever challenged. This queen wasp spends more and more time tending the nest, and less and less time out foraging for herself. She takes the food she needs from the others as they return.

The queen then begins to take an interest in raising a family—her family. Because she is almost always at hand, she is able to prevent other wasps from laying eggs in the brood cells by rushing at them, jaws agape.

The queen can bite any other wasp without serious fear of retaliation. The other wasps are not equal in their relationships with one another. Indeed, the wasps organize into a **dominance hierarchy,** an arrangement of status that regulates aggressive behavior within the society:

Queen > Wasp A > Wasp B > . . . Wasp J > Wasp K

Dominance Hierarchies Suppress Aggression

Once a dominance hierarchy is established, little or no time is wasted in fighting (Fig. 44–12). When challenged, subordinate wasps exhibit submissive poses that, in turn, inhibit the queen's aggressive behavior. Consequently, few or no colony members are lost through wounds sustained in fighting one another. This ensures greater reproductive success for the colony.

Many Factors Affect Dominance

In some animals, dominance is a simple function of aggressiveness, which is itself often influenced directly by sex hormones. Among chickens, the rooster is the most dominant; as with most vertebrates, the hormone testosterone increases aggressiveness. If a hen receives testosterone injections, her place in the dominance hierarchy shifts upward. When male rhesus monkeys are dominant, their testosterone levels are much higher than when they have been defeated. Not only can estrogen sometimes reduce dominance and testosterone increase dominance, but dominance may even increase testosterone. It is not always easy to determine cause-and-effect.

In many species, males and females have separate dominance systems. However, in many monogamous animals, especially birds, the female takes on the dominance status of her mate by virtue of their relationship. This is not always the case, however. Like many fishes, some coral reef fishes (labrids) are capable of sex reversal. The most dominant individual is always male, and the remaining fishes within his territory are all female. If the male dies or is removed, the most dominant female will become the new male. Should any harm come to "him," the next-ranking female will take charge. Still other fishes exhibit the reverse behavior and the most dominant fish is always female.

Many Animals Defend Territory

Most animals have a **home range,** a geographical area that they seldom or never leave (Fig. 44–13). Because the animal has the opportunity to become familiar with everything in that range, it has an advantage over both its predators and its prey in negotiating cover and finding food.

Some, but not all, animals exhibit **territoriality.** They defend a **territory,** a portion of the home range, against other individuals of their species (and sometimes against individuals of other species; Fig. 44–14).

FIGURE 44–12 Establishing dominance. Social animals use many signals to convey messages relating to social dominance. This baboon bares his teeth and screams in an unmistakable show of aggression. (Gerald Lacz, Peter Arnold, Inc.)

FIGURE 44–13 Territoriality. A coral reef has many secluded areas in which a territorial animal can establish a home range. Among the most territorial of coral reef fishes is the moray eel, which will attack any animal (including a human diver) that comes too close to its shelter. (IFA-Bilderteam–D. Eichler/Peter Arnold, Inc.)

Territoriality is easily studied in birds. Typically, the male chooses a territory at the beginning of the breeding season. This behavior results from high concentrations of sex hormones in the blood. The males of adjacent territories fight until territorial boundaries become fixed. Generally, the dominance of a male varies directly with his nearness to the center of his territory. Thus, close to "home," he is a lion, but when invading some other bird's territory, he is likely to be a lamb. The interplay of dominance values among territorial males eventually produces a neutral line at which neither is dominant. That line is the territorial boundary.

Bird songs announce the existence of a territory and often serve as a substitute for violence. Furthermore, they

FIGURE 44–14 Nesting sea birds. These Falkland Island shags often defend the territory directly surrounding their nests. This results in a regular spacing of the nests. (Frans Lanting/Minden Pictures)

(a)

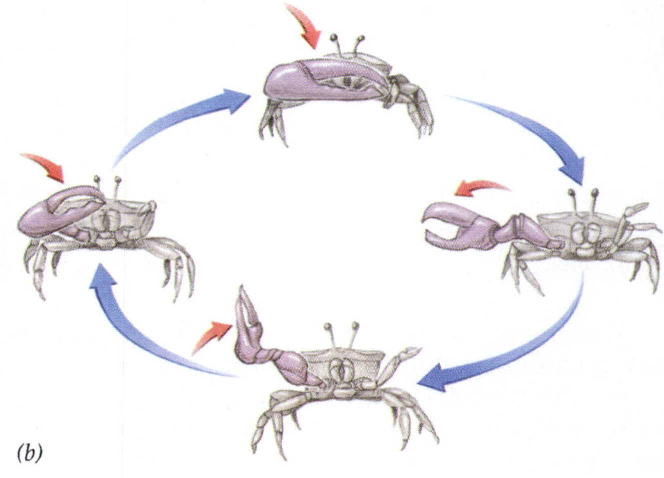
(b)

FIGURE 44–15 Courtship displays. (***a***) The male great frigatebird *(Frigata minor)* displays himself as part of a courtship ritual. (Christmas Island, Pacific.) (***b***) Courtship signals by male fiddler crabs are specific to each species. This particular sequence of the motion of the large right claw is characteristic of the species *Uca lactae.* (a, Sid Bart/Photo Researchers, Inc.)

announce to eligible females that a propertied male resides in the territory. Typically, male birds take up a conspicuous station, sing, and sometimes display striking patterns of coloration to their neighbors, their rivals, and sometimes their mates (Fig. 44–15).

Territoriality among animals appears to be adaptive. It tends to reduce conflict among members of the same species and also helps control population growth. Animals that fail to establish territories fail to reproduce. Territoriality also ensures the most efficient use of environmental resources by encouraging individuals to spread throughout a habitat.

Usually, territorial behavior is related to the specific life-style of the animal and to whatever aspect of its ecology is most critical to its reproductive success. For instance, sea birds may range over hundreds of square miles of open water but exhibit territorial behavior only at crowded nesting sites on an island. The nesting sites are their resource that is in the shortest supply and for which competition is keenest.

Sexual Behavior Is Generally Social

For some animals, mating and perhaps the rearing of young are the only forms of social behavior (Fig. 44–16). Let us consider the sex act as a basic example of social conduct. The sex act is obviously adaptive in that in most animals it is necessary for reproduction. It requires *cooperation,* the *temporary suppression of aggressive behavior,* and a *system of communication.*

In many species, success of a male in dominance encounters with other males indicates his quality to the female. The female allows the victorious male to court her. Courtship ensures that the male is a member of the same species, and also provides the female further opportunity to evaluate the male. Courtship may also be necessary as a signal to trigger nest building or ovulation.

Courtship rituals can last seconds or hours and often involve complex behaviors. The first display of the male releases a counter behavior by the female. This, in turn, releases additional male behavior, and so on until the pair are ready for copulation.

Courtship is often strenuous and dangerous, especially for males. Tremendous energy is often needed for the fights engaged in by some birds and many mammals, including rams and bull seals. Males may expend additional energy in wallowing, roaring, and leaping about. These behaviors appear to be a test of male fitness. The male with the greatest endurance has the opportunity to mate and to perpetuate his genes.

Among some jumping spiders, mating is preceded by a ritual courtship in which the male temporarily paralyzes the female. While she is thus enthralled, the male inseminates her. Should she recover before he makes his escape, he becomes the main course at his own wedding feast. Even so, he would still make the ultimate material contribution to the eggs the female will produce and, therefore, to the perpetuation of his genes. She would otherwise have to bear the metabolic burden of their production by herself.

(a) (b)

FIGURE 44–16 Courtship behavior. Male and female jumping spiders, *Phidippus princeps,* in the courtship behavior that precedes mating (***a***). The male performs an elaborate dance that inhibits the female's natural aggression toward him, allowing him to get close enough to inseminate her (***b***). (Visuals Unlimited/P. Starborn)

Pair Bonds Establish Reproductive Cooperation

A **pair bond** is a stable relationship between animals of the opposite sex that ensures cooperative behavior in mating and the rearing of the young (Fig. 44–17). In an estimated 90% of bird species, males pair bond with a female during the breeding season. The mechanisms involved in establishing and maintaining the pair bond are often remarkably detailed. Specific cues enable courtship rituals to function as reproductive isolating mechanisms among species (see Chapter 18).

FIGURE 44–17 Most birds establish pair bonds. A pair of nesting blue herons *(Ardea herodias).* In many species pair bonds are maintained by grooming or other displays of attraction. (SharkSong/M. Kazmers/Dembinsky PhotoAssociates)

Many Organisms Care for Their Young

Care of the young, an important part of successful reproduction in many species, requires parental investment (Fig. 44–18). The benefit of parental care is the increased likelihood that the offspring will survive. The cost is a reduction in the number of offspring that can be produced. Because of the time spent carrying the developing embryo, the female has more to lose than the male if the young do not develop. Thus, females are more likely than males to brood eggs and young, and usually the females invest more in parental care.

Investing time and effort in care of the young is usually less advantageous to a male (assuming that the female can handle the job by herself), for time spent in parenting is time lost from inseminating other females. In some species, it may not even be certain who fathered the offspring. Raising some other male's offspring is a genetic disadvantage. This is probably why male lions kill cubs of a former pride male whose position they have seized.

In some situations, however, a male can benefit by helping to rear his own young or even those of a genetic relative. Receptive females may be scarce, and gathering sufficient food may require more effort than one parent

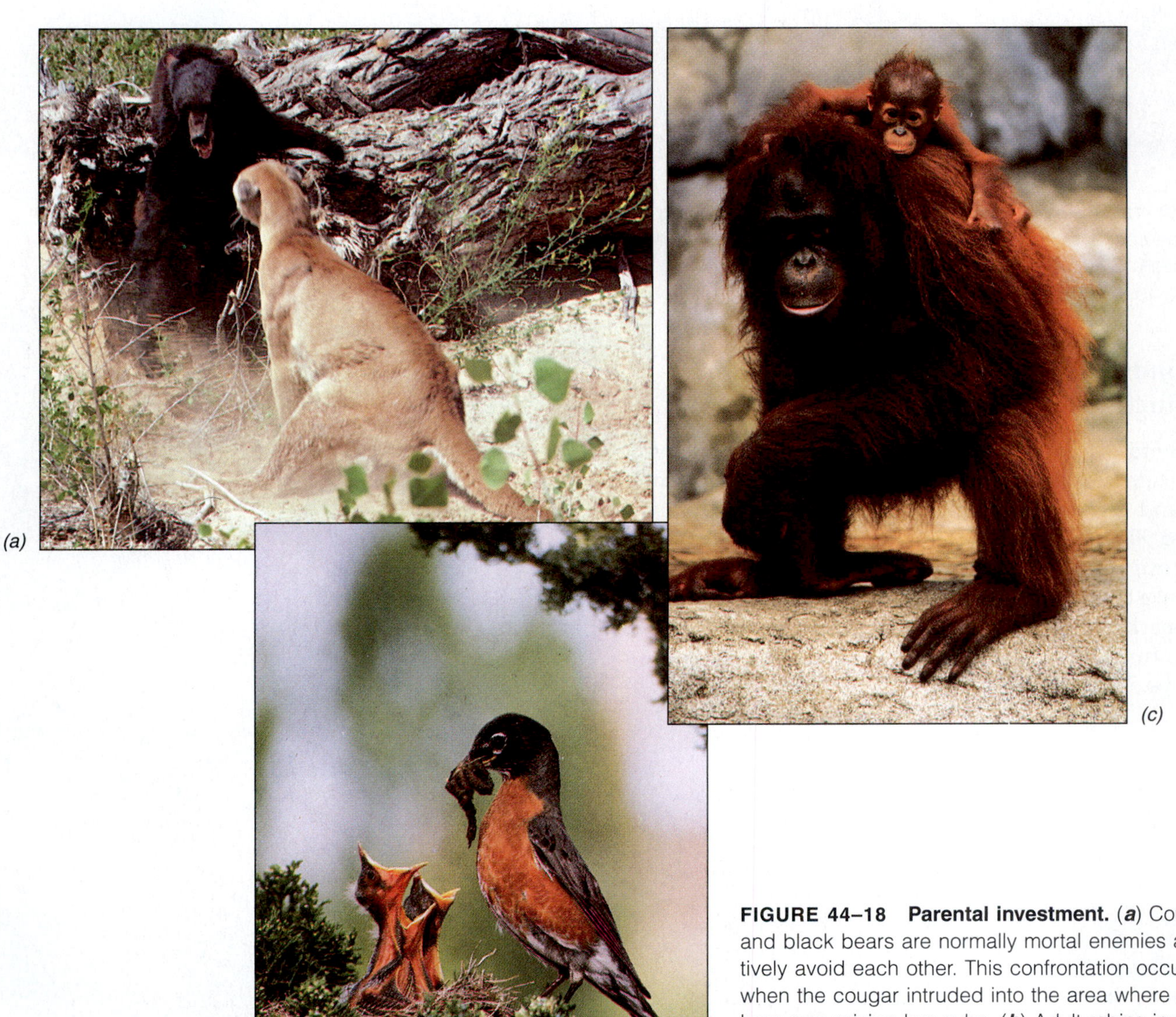

FIGURE 44–18 Parental investment. (***a***) Cougars and black bears are normally mortal enemies and actively avoid each other. This confrontation occurred when the cougar intruded into the area where a female bear was raising her cubs. (***b***) Adult robins invest a great deal of energy in feeding their young. (***c***) A baby orangutan rides on its mother's back. (*a*, E.R. Degginger; *b*, Dominique Braud/Dembinsky Photo Associates; *c*, Schafer and Hill/Peter Arnold, Inc.)

FIGURE 44–19 Arctic pups playing in Alaska. Play often serves as a means of practicing behavior that will be used in earnest in later life, possibly in hunting, fighting for territory, or competing for mates. (Michio Hoshino/Minden Pictures)

can provide. In some habitats, the young may need protection against predators and sometimes against cannibalistic males of the same species.

Play Is Often Practice Behavior

Perhaps you have watched a kitten pounce on a dead leaf. Or witnessed a kitten practicing a carnivore neck bite or using its hind claws to simulate a disemboweling stroke on a littermate without injury. Many animals, especially young birds and mammals, use play to practice adult patterns of behavior (Fig. 44–19). They may perfect means of escape, killing prey, or sexual behavior. In true play, the behavior may not be actually consummated.

Elaborate Societies Are Found Among the Social Insects

Although many insects cooperate socially, such as tent caterpillars, which spin a communal nest, the most elaborate insect societies are found among the bees, ants, wasps, and termites. (The first three of these all belong, not coincidentally, to the order Hymenoptera.) Insect societies are held together by a complex system of sign stimuli that are keyed to social interaction. As a result, they tend to be quite rigid. In addition to other modes of communication, the social insects secrete pheromones that accomplish such tasks as suppressing the ovaries of worker honeybees.

The social organization of honeybees has been studied more extensively than that of any social insect. A honeybee society generally consists of a single adult queen, up to 80,000 worker bees (all female), and, at certain times, a few males called drones that fertilize newly developed queens. The queen's job is reproduction; she deposits about 1000 fertilized eggs per day in the wax cells of a comb.

Division of labor in the bee society is mostly determined by age. The youngest worker bees serve as nurse bees that nourish larval bees. After about a week as nurse bees, workers begin to produce wax and build and maintain the wax cells. Older workers are foragers, bringing home nectar and pollen. Most worker bees die at the ripe old age of 42—days, that is.

The composition of a bee society is controlled by an antiqueen pheromone secreted by the queen. It inhibits the workers from raising a new queen, and prevents development of the ovaries in the workers (Fig. 44–20). If

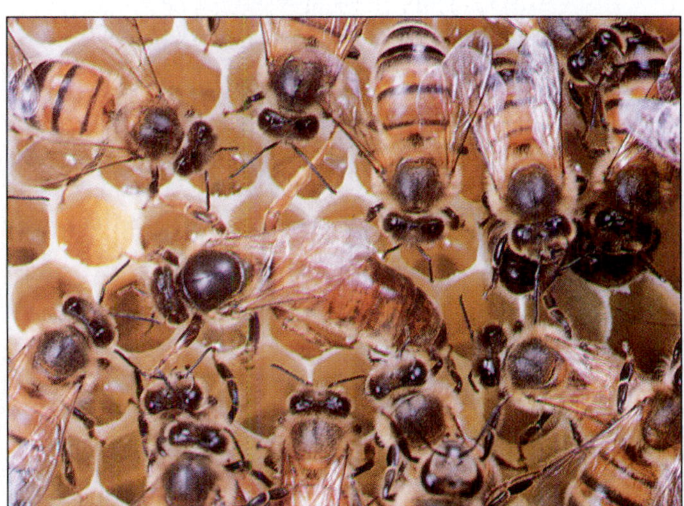

FIGURE 44–20 A queen honeybee inspecting wax cells. Note the numerous workers that surround her. They constantly lick secretions from the queen bee. These secretions are transmitted throughout the hive and act to suppress the activity of the workers' ovaries. (Treat Davidson/Photo Researchers, Inc.)

the queen dies, or if the colony becomes so large that the inhibiting effect of the pheromone is dissipated, the workers begin to feed some larvae special food that promotes their development into new queens.

The most sophisticated known mode of communication among bees is a stereotyped series of body movements known as a **dance.** When a scout locates a rich source of nectar, it communicates the direction and distance of the food source relative to the hive to the other bees. If the food supply is nearby, the scout performs a round dance, which generally excites the other bees and causes them to fly about in all directions (but within a certain distance from the hive) until they have found the nectar. If the source is distant, however, the scout performs a waggle dance.

Karl von Frisch, a pioneer in the study of communication in bees, showed that the orientation of the circular movements indicates the direction of the food source. The frequency of the waggle indicates the distance. Bees use the angle of the Sun and light polarization to orient themselves. The dance locates the food in reference to the position of the Sun.

In social insects such as bees, males develop from unfertilized eggs and are haploid. Reproductive females store sperm cells from previous matings in a seminal receptacle. If they permit a sperm cell to contact the egg as it is laid, the resulting insect is female; otherwise, it is male.

Because a drone is haploid, each one of its sperm cells has *all* of his chromosomes (that is, meiosis does not occur during sperm production). The queen bee stores this sperm throughout her lifetime, and it is used to produce worker bees. Thus, the worker bees of a hive are more closely related than would be sisters born of a diploid father. Indeed, they have three-quarters of their genes in common (they will share half of the queen's chromosomes and all of the drone's). As a consequence, they are more closely related to one another than they would be to their own offspring, if they could have any. Because new queens will also be their sisters, they are actually more likely to be able to pass on copies of their genes to the next generation by raising these individuals than they would by producing their own offspring. A worker bee's offspring would have only one-half of its genes in common with its worker mother.

Vertebrate Societies Tend To Be Relatively Flexible

Vertebrate societies usually contain nothing comparable to the physically and behaviorally specialized castes of termites or ants. What is more, except for humans, individual members of vertebrate societies are not as specialized in their tasks as are the social insects. Although in some ways vertebrate societies seem simpler than insect societies, they are also more flexible.

Vertebrate societies share a great range and plasticity of potential behavior, and can adapt to changing needs. The behavioral plasticity of vertebrates makes possible the symbolic transmission of culture, and this opens up whole new worlds inaccessible to the social insects. Human society is based to a great extent on the symbolic transmission of culture (see Chapter 25, Concept Connections: Cultural Evolution and Global Ecology, page 499).

Kin Selection Could Produce Altruistic Behavior

In **altruistic behavior,** one individual appears to act in such a way as to benefit others rather than itself. This type of behavior is frequently seen in complex social groups (Fig. 44–21). Biologists Watts and Stokes observed a particularly clear case of altruistic behavior in the mating of wild turkeys.

Several groups of male turkeys, each of which has an internal dominance hierarchy, gather in a special mating territory. They go through their displays of tail spreading, wing dragging, and gobbling in front of females who come to the area to copulate. One group attains dominance over other groups as a result of cooperation among the males within the group. The dominant member of the dominant group is the one to copulate most frequently with the females.

Seemingly, the males that have low status within the group they helped to establish gain nothing. Close analysis has shown that members of a group are brothers from the same brood. Because they share many genes with the successful male, they are indirectly perpetuating many of their genes. In this case, altruism is closely related to **kin selection,** the evolutionary effect caused by individuals

FIGURE 44–21 Kin selection in prairie dogs. Low-ranking members of this social rodent group act as sentries, risking their own lives by exposing themselves outside their burrows. However, by protecting their siblings, they ensure that the genes they share in common will be perpetuated in the population. (Tina Waisman)

who favor the survival and reproduction of their relatives (kin). By promoting the breeding success of a close relative, the nonbreeding turkeys perpetuate copies of many of their own genes.

Among some birds (Florida jays and others), nonreproducing individuals aid in the rearing of the young. Nests tended by these additional helpers as well as parents produce more young than nests with the same number of eggs overseen only by parents. The nonreproducing helpers, siblings of the parents, are apparently increasing their own biological success by ensuring the successful perpetuation of their genes via their siblings.

Sociobiology Explains Altruism by Kin Selection

Sociobiology focuses on the evolution of social behavior through natural selection. It represents a synthesis of population genetics, evolution, and ethology. Like many biologists of the past (such as Darwin), E. O. Wilson and other sociobiologists stress the animal roots of human behavior, emphasizing the effect of kin selection on patterns of inheritance. Many of the concepts discussed in this chapter, such as altruism and paternal investment in care of the young, are based on contributions made by sociobiologists.

From the sociobiological perspective, the organism and its adaptations, including its behavior, ensure that genes make more copies of themselves. The cells and tissues of the body support the functions of the reproductive system, and the reproductive system functions to transmit genetic information to succeeding generations.

Most of the controversy that has been triggered by sociobiology seems related to its possible ethical implications. Sociobiology is often viewed as denying that human behavior is flexible enough to permit substantial improvements in the quality of our social lives. Yet sociobiologists do not disagree with their critics that human behavior is flexible. The debate therefore seems to rest on the *degree* to which human behavior is genetic and the *extent* to which it can be modified.

As sociobiologists acknowledge, people can, through culture, change their way of life far more profoundly in a few years than could a hive of bees or a troop of baboons in hundreds of generations of genetic evolution. This capacity to make changes is indeed genetically determined, and that is a great gift. How we use it and what we accomplish with it is not a gift but a responsibility upon which our own well-being and the well-being of other species depend.

Chapter Summary

I. Behavior consists of the responses of an organism to signals from its environment.

II. Ethology is the scientific study of behavior in natural environments from the point of view of adaptation.
   - A. Behavior tends to be adaptive.
   - B. Behavior tends to be homeostatic.

III. It is adaptive for an organism's metabolic processes and behavior to be synchronized with the cyclical changes in the environment.
   - A. Some biological rhythms reflect the lunar cycle or the changes in tides due to phases of the Moon.
   - B. In many species, physiological processes and activity follow circadian rhythms.
   - C. No single biological clock has been found. Biological rhythms are thought to be regulated by both internal and external factors.

IV. The capacity for behavior is inherited, and behavior is modified by environmental stimuli.
   - A. Before an organism can show any pattern of behavior, it must be physiologically ready to produce the behavior.
   - B. Some behavior patterns have strong genetic components. Innate behavior may be triggered by a specific unlearned sign stimulus.
   - C. Learning is a change in behavior resulting from experience.
   - D. Simpler forms of learning are conditioning, both classical and operant, and habituation.
   - E. Imprinting establishes a parent–offspring bond during a critical period early in development.
   - F. Insight learning, characteristic of the more "intelligent" animals, involves the ability to see through a problem.

V. Behavioral ecology focuses on interactions between animals and their environment and on the survival value of the behavior.
   - A. In some birds, the need to migrate and the direction of migration appear to be genetically programmed, but navigation may be learned.
   - B. Optimal foraging, the most efficient strategy for an animal to obtain food, enhances reproductive success.

VI. Social behavior is adaptive interaction usually among members of the same species.
   - A. A society is a group of individuals of the same species that cooperate in an adaptive manner. In a society, there is a means of communication, cooperation, division of labor, and a tendency to stay together.
   - B. Animal communication involves the transmission of signals but does not utilize (as far as is known) symbolic language in the human sense. Pheromones are chemical signals that convey information between members of a species.
   - C. Dominance hierarchies result in the suppression of aggressive behavior.
   - D. Organisms often inhabit a home range, from which they seldom or never depart. This range, or some por-

tion of it, may be defended from members of the same (or occasionally different) species.

1. Defended areas are called territories, and the defensive behavior is territoriality.
2. Territorial defense is often carried out by display behavior rather than by actual fighting.

E. Courtship behavior ensures that the male is a member of the same species, and it permits the female to assess the quality of the male.

F. A pair bond is a stable relationship between a male and a female that ensures cooperative behavior in mating and rearing the young.

G. Parental care increases the probability that the offspring will survive. A high investment in parenting is often less advantageous to the male than to the female.

H. Play gives the young animal a chance to practice adult patterns of behavior.

I. Insect societies tend to be rigid, with the role of the individual narrowly defined.

J. Vertebrate societies are far less rigid than are insect societies. Although innate behavior is important, the role of the individual is generally learned.

K. In altruistic behavior, one individual appears to behave in such a way as to benefit others rather than itself. Altruism may be closely related to kin selection.

L. Sociobiology focuses on the evolution of social behavior through natural selection.

Selected Key Terms

altruistic behavior, p. 886
behavior, p. 902
biological clock, p. 887
circadian rhythm, p. 887
classical conditioning, p. 890
dominance hierarchy, p. 896
fixed action pattern, p. 888
habituation, p. 892
imprinting, p. 891
insight, p. 892
kin selection, p. 902
operant conditioning, p. 890
pair bond, p. 899
pheromones, p. 895
sign stimulus, p. 888
social behavior, p. 894
sociobiology, p. 903
territoriality, p. 896

Post-Test

1. ______________ may be defined as responses of an organism to signals from its environment.

2. ______________ is the study of behavior in natural environments from the point of view of adaptation.

3. A biological rhythm with approximately a 24-hour cycle is a/an ______________ rhythm.

4. Animals that are most active at dawn or twilight are described as ______________.

5. A stimulus that elicits a fixed action pattern is a/an ______________ ______________.

6. ______________ behavior is mainly genetic; ______________ behavior develops as a result of experience.

7. ______________ is a form of learning in which a young animal forms a strong attachment to a moving object (usually its parent) within a few hours of birth.

8. In habituation, an organism learns to ignore a repeated, irrelevant ______________.

9. Secretion of saliva by a student when the noon bell rings is an example of ______________ conditioning.

10. Adaptive interactions among members of a population are referred to as ______________ behavior.

11. A/An ______________ is a group of individuals belonging to the same species that cooperate in an adaptive manner and have a means of communicating with one another.

12. ______________ are chemical signals that convey information between members of a species.

13. An arrangement of members of a population by status is called a/an ______________ ______________.

14. The geographical area that members of a population seldom leave is the ______________ ______________.

15. Territoriality tends to reduce ______________ and control ______________ growth.

16. A/An ______________ ______________ is a stable relationship between animals of the opposite sex that ensures cooperative behavior in mating and rearing the young.

17. In a bee society, the worker bees are all (male, female) ______________.

18. The extensive behavioral repertoire of the bee is almost entirely ______________ (innate or learned).

19. Human society differs from other animal societies in that it depends mainly on the transmission of ______________.

20. In ______________ behavior, one individual appears to act to benefit others rather than itself.

21. ______________ selection favors the indirect perpetuation of an animal's genes by a relative.

22. According to sociobiology, an organism and its adaptations are ways that its genes have of ______________.

23. When a conditioned stimulus no longer signals an unconditioned stimulus, (a) extinction occurs (b) operant conditioning has occurred (c) negative reinforcement takes place (d) insight learning occurs (e) three of the preceding answers are correct.

24. A third grader who has been struggling to understand long division suddenly sees the light and can work problems. This is an example of (a) extinction (b) operant conditioning (c) positive reinforcement (d) insight (e) imprinting.

25. In order for a young child to learn to talk, (a) he or she must be positively reinforced (b) operant conditioning must not occur (c) he or she must be physiologically ready (d) insight learning must occur (e) classical conditioning must take place.

26. Which of the following is/are a/an example(s) of a society? (a) hive of bees (b) pride of lions (c) group of moths attracted to a light (d) answers (a), (b), and (c) are correct (e) answers (a) and (b) only are correct.

Review Questions

1. In what ways are the behaviors of *Philanthus*, the bee-killer wasp, adaptive?
2. Why is it adaptive for some species to be diurnal but others nocturnal or crepuscular?
3. Behavior capacity is inherited and is modified by learning. Give an example.
4. When Konrad Lorenz kept a greylag goose isolated from other geese for the first week of its life, the goose persisted in following humans about in preference to other geese. How could this behavior be explained?
5. How does physiological readiness affect instinctual behavior? How does it affect learned behavior?
6. What distinguishes an organized society from a mere aggregation of organisms? Cite an example of an organized society, and describe characteristics that qualify the society as organized.
7. How does an organism learn its place in a dominance hierarchy? What determines this place? What are the advantages of a dominance hierarchy?
8. What is territoriality? What functions does it seem to serve?
9. What is kin selection? How is kin selection used by sociobiologists to explain the evolution of altruistic behavior?
10. Do animals play just for the fun of it?
11. What are some advantages of courtship rituals?

Thinking Critically

1. What might be the adaptive value of sea turtle migration? Consider how this adaptive value has changed if, as a result of human activities, migration now puts sea turtles at greater risk than if they restricted their habitat to a single location. Discuss the possible evolutionary mechanisms by which the turtles' behavior may (or may not) adapt to these environmental pressures.
2. Using what you know about animal learning, propose experiments to test whether it might be possible to teach sea turtles "safer" reproductive behavior. As you do, discuss the following questions: What might be the broader, long-term ecological consequences of your proposals? Is it advisable for humans to attempt to alter the behavior of another species? Is this an ethical way of preserving a species that has become endangered due to human activities?
3. Behavioral ecologists have demonstrated that young tiger salamanders can become cannibals and devour other salamanders. Investigators observed that the cannibal salamanders preferred unrelated salamanders to their own cousins and preferred eating cousins to their siblings. Explain this behavior based on what you have learned in this chapter. (*Hint:* You may want to review the discussion of kin selection.)
4. Males and females have different reproductive strategies. What is meant by this statement? (Base your answer on evolutionary success.)
5. How many similarities between the transmission of information by symbolic language and by heredity can you think of? What differences?
6. Contrast the "language" of bees with human language.

Recommended Readings

Ellis, D.H., J.C. Bednarz, D.G. Smith, and S.P. Flemming, "Social Foraging Classes in Raptorial Birds," *BioScience*, Vol. 43, No. 1, January 1993, pp. 14–20. Many bird species have developed cooperative hunting.

Harbrecht, D., "Games Animals Play," *International Wildlife*, Vol. 23, No. 5, September–October 1993. An interesting discussion on why animals play.

Lohmann, K.J., "How Sea Turtles Navigate," *Scientific American*, Vol. 266, No. 1, January 1992, pp. 100–106. Research has begun to identify the biological compasses and maps that guide sea turtles as they migrate across hundreds of miles of ocean.

Page, R.E., G.E. Robinson, and M.K. Fondrk, "Genetic Specialists, Kin Recognition and Nepotism in Honey-Bee Colonies," *Nature*, Vol. 338, 1989.

Sherman, P.W., "Mate Guarding as Paternity Insurance in Idaho Ground Squirrels." *Nature* 338, 1989.

Sherman, P.W., J. Jarvis, and S.H. Braude, "Naked Mole Rats," *Scientific American*, Vol. 267, No. 2, August 1992, pp. 72–78. These animals have a social structure that resembles some of the insects.

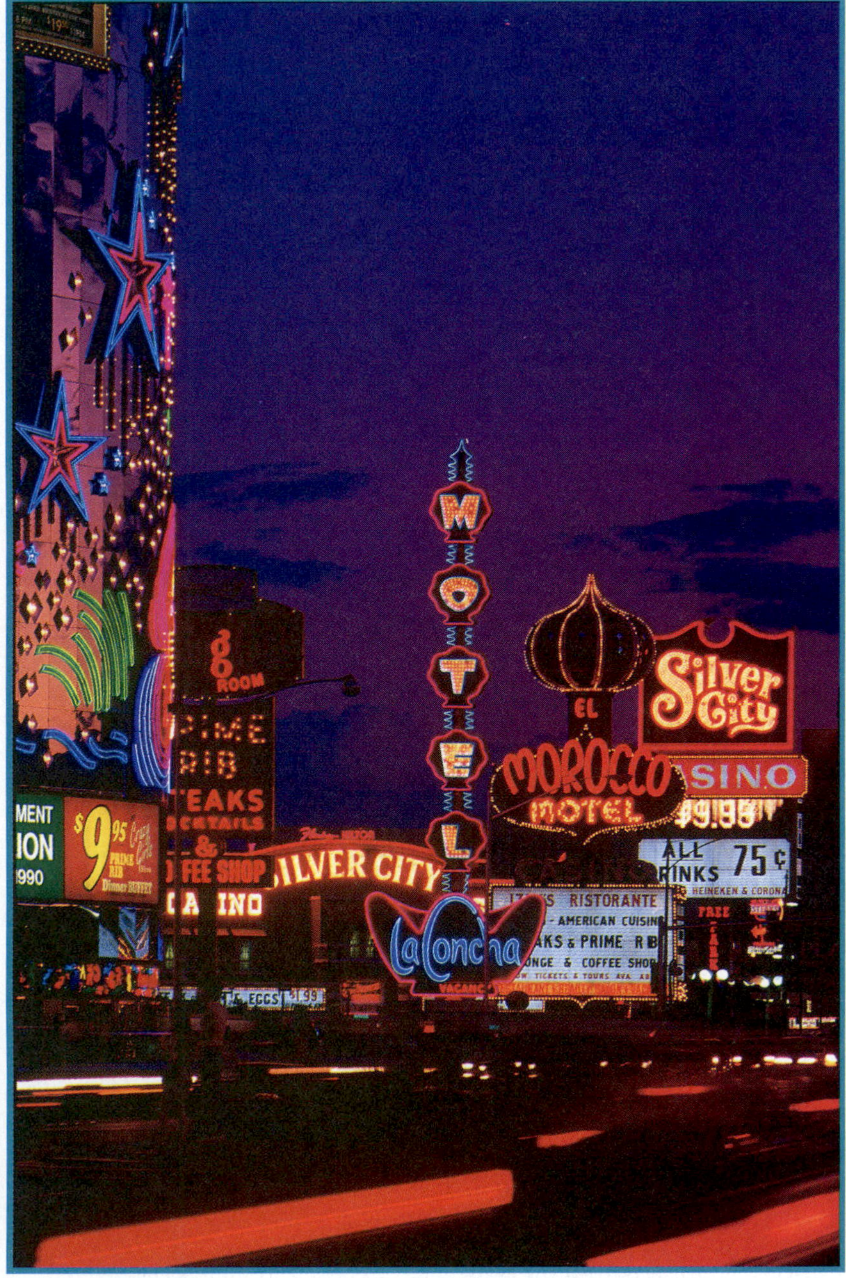

Las Vegas at night. *Electrical power and water use in this desert city are excessive.* *(Dembinsky Photo Associates)*

Too Many People Consuming Too Many Resources

In 1994, the population of the world exceeded 5.6 billion individuals—approximately 101 million more humans than the Earth supported in 1993.[1] The 1993 world population, in turn, was approximately 86 million more humans than in 1992.

As our numbers continue to grow during the next 100 years, environmental deterioration, hunger, persistent poverty, and health issues will become more critical. The need for food for increasing numbers of people living in arid regions around the world has already led to overuse of the land for grazing and crop production. As a result of such overuse, many of these formerly productive lands have been degraded into unproductive deserts.

Although the human population is increasing, not all countries have the same growth rate. **Highly developed countries,** such as the highly industrialized United States, Canada, Japan, and European countries, have low rates of population growth. **Less developed countries,** such as the relatively unindustrialized countries of Laos, Bangladesh, Niger, and Ethiopia, have high rates of population growth. **Moderately developed countries,** such as Mexico, Turkey, Thailand, and most of South America, fall in the middle with growth rates that are higher than in developed countries but lower than in less developed countries.

The relationships among population growth, natural resource utilization, and environmental degradation are complex, but two useful generalizations can be made. One, the resources essential to an individual's survival are small. However, a rapidly increasing *number* of people (found in many less developed and moderately developed countries) tends to overwhelm and deplete a country's soils, forests, and other natural resources.

Two, in highly developed countries, individual resource demands are large, far above the minimum requirements for survival. In order to satisfy their desires rather than their basic needs, people in more affluent countries exhaust natural resources and degrade the global environment through extravagant *consumption* and "throwaway" life-styles. A single child born in a country such as the United States, for example, causes a greater impact on the environment and on resource utilization than do a dozen or more children born in a country such as Niger.

Thus, the disproportionately large consumption of resources by people in highly developed countries affects natural resources and the environment as much as, or more than, the population explosion in the developing world. Discussion of population issues should therefore consider both concerns: too many people and too much consumption per person.

In this chapter, we focus on the dynamics of population growth characteristic of all living things, including humans. Then we describe the current human population.

[1] Unless otherwise noted, all population data in this chapter were obtained from the Population Reference Bureau, a private, nonprofit educational organization that disseminates demographic and population information.

Ecology of Populations

45

Learning Objectives

After you have studied this chapter you should be able to:

1. Distinguish among population, community, ecosystem, biosphere, and ecosphere.
2. Define population density and dispersion, and describe the main types of population dispersion.
3. Explain biotic potential.
4. List and explain the four factors that produce changes in population size.
5. Explain the differences between J-shaped and S-shaped growth curves.
6. Define environmental resistance and give its role in determining population growth and size.
7. Compare density-dependent factors and density-independent factors.
8. Distinguish between *K* strategies and *r* strategies, and give an example of each.
9. Summarize the history of human population growth.
10. Explain how developed and developing (less developed and moderately developed) countries differ in such population characteristics as infant mortality rate, total fertility rate, replacement-level fertility, and age structure.

Key Concepts

- ☐ Individual organisms are part of a larger organization—a group composed of members of the same species that live together in the same area at the same time. Such a group is called a population.
- ☐ Populations exhibit characteristics distinctive from those of the individuals of which they are composed. Some of the traits unique to populations are birth and death rates, growth rates, density, sex ratios, and age structure.
- ☐ A population responds in special ways to its environment, that is, to competition for resources, predation, and other environmental pressures. Population growth cannot increase indefinitely because of such environmental controls.
- ☐ The same parameters that affect other populations also affect the human population.

ECOLOGISTS STUDY THE HIGHEST LEVELS OF BIOLOGICAL ORGANIZATION

The concept of ecology was first developed in the 19th century by Ernst Haeckel, who also invented its name: *eco* from the Greek word for "household," and *logy* from the Greek word for "study." Thus, *ecology* literally means the study of one's house. It is known that living organisms and the physical environment interact in an immense and complicated web of relationships. **Ecology** is the study of the interactions among organisms and between organisms and their physical environment.

The levels of biological organization that interest ecologists are those above the level of the individual organism. Organisms are arranged into **populations** in which members of the same species live together in the same area at the same time (Fig. 45–1). **Population ecology** deals with the numbers of a particular organism that are found in an area and why those numbers change (or remain fixed) over time. Population ecologists try to determine the population processes that are common to all organisms.

Populations are organized into **communities.** A community consists of all of the populations of all of the different species that live and interact together within an area. A community ecologist might study how organisms interact with one another—including who eats whom—in a coral reef community or in an alpine meadow community. *Ecosystem* is a more inclusive term than is *community,* because an **ecosystem** is a community together with its physical environment. Thus, an ecosystem includes not only all of the interactions among the living organisms of a community but also all of the interactions between the organisms and their physical environment. An ecosystem ecologist, for example, might examine how temperature, light, precipitation, and soil factors affect the organisms living in a desert community or a coastal bay community (Fig. 45–2).

All of the communities of living things on Earth are organized into the **biosphere.** The organisms of the biosphere depend on one another and on other divisions of Earth's physical environment: the atmosphere, hydrosphere, and lithosphere. The **atmosphere** is the gaseous envelope surrounding Earth, the **hydrosphere** is Earth's supply of water (liquid and frozen, fresh and salty), and the **lithosphere** is the soil and rock of the Earth's crust.

(a)

(b)

FIGURE 45–1 A population is a group of organisms of the same species living in the same area at a given time. (***a***) Gooseneck barnacles live on a rocky shore in the subtidal zone. (***b***) A population of Mexican poppy blooms in the desert after the winter rains. (*a*, Dennis Drenner; *b*, Jim Brandenburg/Minden Pictures)

FIGURE 45–2 Many ecosystem studies require elaborate equipment. The floating rings extend downward into the water, enclosing part of the water column. (Courtesy of Dr. M.R. Reeve, National Science Foundation)

The term **ecosphere** encompasses the biosphere and its interactions with the atmosphere, hydrosphere, and lithosphere. Ecologists who study the biosphere or ecosphere examine the complex interrelationships among Earth's atmosphere, land, water, and living things.

DENSITY AND DISPERSION ARE IMPORTANT FEATURES OF POPULATIONS

By itself, a population number tells us relatively little. A thousand mice in a square kilometer (250 acres) is a very different matter from 1000 mice in a hectare (2.5 acres). Moreover, sometimes a population is too large to study in its entirety. Such a population is examined by sampling a part of it and then expressing the population in terms of density. For example, a large population expressed in terms of density might include the number of grass plants per square meter or the number of cabbage aphids per cabbage leaf. **Population density,** then, is the number of individuals of a species per unit of area at a given time.

Different environments vary in the population density of any species of organism that they can support. This density may also vary in a single habitat from season to season or year to year. As an example, consider red grouse populations in northwest Scotland, which were examined at two locations only 2.5 km (1.6 miles) apart. At one location, the population density remained stable during a three-year period, but at the other, it almost doubled in two years and then declined to its initial density once again. The reason was a change in the habitat. The area where the population density increased had been experimentally burned, and young plant growth produced after the burn was beneficial to the red grouse. So population density is determined in large part by external factors in the environment.

The geographic limit of a population's distribution is its **range.** The individuals within a population's range often exhibit characteristic patterns of **dispersion,** or spacing. Individuals may be dispersed uniformly (evenly spaced), randomly, or clumped into clusters.

Uniform dispersion occurs when individuals are more evenly spaced than would be expected from a random occupation of a given habitat (Fig. 45–3*a*). A nesting colony of seabirds, in which the birds place their nests at

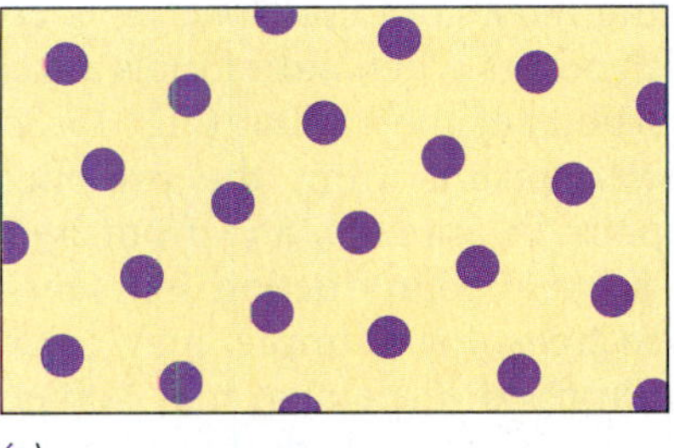

(*a*)

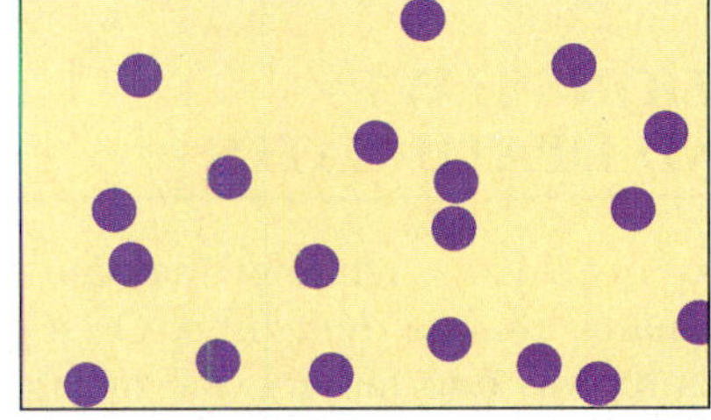

(*b*)

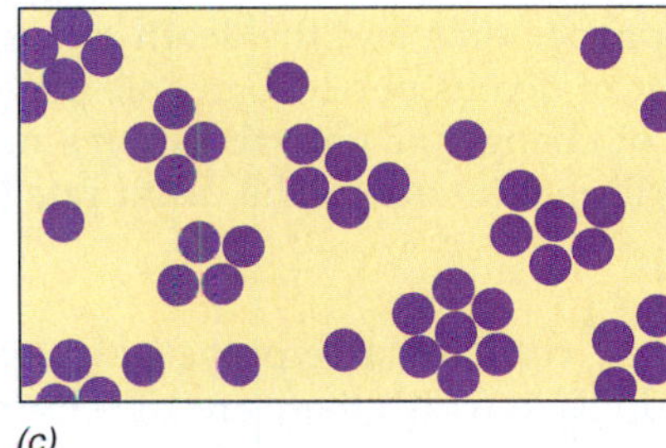

(*c*)

FIGURE 45–3 Individuals of a population may have different patterns of distribution in the habitat. (***a***) Uniform dispersion is usually caused by negative, or antagonistic, interactions among individuals of a population. (***b***) Random dispersion, which is not common in nature, occurs when individuals distribute themselves with no regard for the positions of other individuals in a population. (***c***) Clumped dispersion is usually the result of mutual attractions among individuals within a population.

a more or less equal distance from other birds, is an example of uniform dispersion. In this case, uniform dispersion occurs as a result of antagonistic interactions among the nesting birds. (Each bird places its nest just beyond the reach of surrounding nesting birds.) Uniform dispersion also occurs when competition among individuals is severe, when they wage chemical warfare on one another, and when they exhibit territorial behavior.

Random dispersion occurs when individuals in a population are spaced in an unpredictable or random way that is unrelated to the presence of others (Fig. 45–3*b*). Of the three major types of dispersion, random dispersion may seem the most likely to occur in nature, but it is not common or easy to observe. Random dispersion may occur infrequently because important environmental factors affecting dispersion usually do not occur at random. Flour beetle larvae in a container of flour are randomly dispersed, for example, but their environment is unusually homogeneous.

Perhaps the most common spacing is **clumped dispersion,** which occurs when individuals are concentrated in specific portions of the habitat (Fig. 45–3*c*). Clumped dispersion often results from the presence of family groups and pairs in animals, and from inefficient seed dispersal or asexual reproduction in plants. An entire grove of aspen trees, for example, may originate from a single seed. Clumped dispersion may sometimes be advantageous, as social animals derive many benefits from their association (Chapters 44 and 46).

POPULATIONS HAVE BIRTH AND DEATH RATES

Populations of organisms, whether they are sunflowers, eagles, or humans, change over time. On a *global* scale, this change is due to two factors: the number of births and the number of deaths in the population. For humans, the **birth rate** is often expressed as the number of births per 1000 people per year, and the **death rate** is expressed as the number of deaths per 1000 people per year.

The rate of change, or **growth rate** (r), of a population is the birth rate (b) minus the death rate (d).

$$r = b - d$$

As an example, consider a hypothetical population of 10,000 individuals in which there are 1000 births per year (or 100 births per 1000 people) and 500 deaths per year (or 50 deaths per 1000 people).

$$r = \underbrace{\frac{1000}{10{,}000}}_{b} - \underbrace{\frac{500}{10{,}000}}_{d}$$

$$= 0.1 - 0.05 = 0.05$$

A value of 0.05 for r means the population has an annual growth rate of 5%.

Another way to express the growth rate of a population is to determine the **doubling time,** which is the amount of time it would take for the population to double in size, assuming that its current growth rate does not change. Simplified, doubling time (t_d) is calculated as 0.7 divided by the growth rate.[2]

$$t_d = \frac{0.7}{r}$$

In our example ($r = 0.05$), the doubling time would be $0.7/0.05 = 14$ years.

In addition to the number of births and deaths, migration (movement from one region or country to another) must be considered when examining changes in populations on a *local* scale (Fig. 45–4). There are two types of migration: immigration and emigration. **Immigration** occurs when individuals enter a population and thus increase the size of the population. **Emigration** occurs when individuals leave a population and thus de-

[2] The actual formula involves calculus and is beyond the scope of this text.

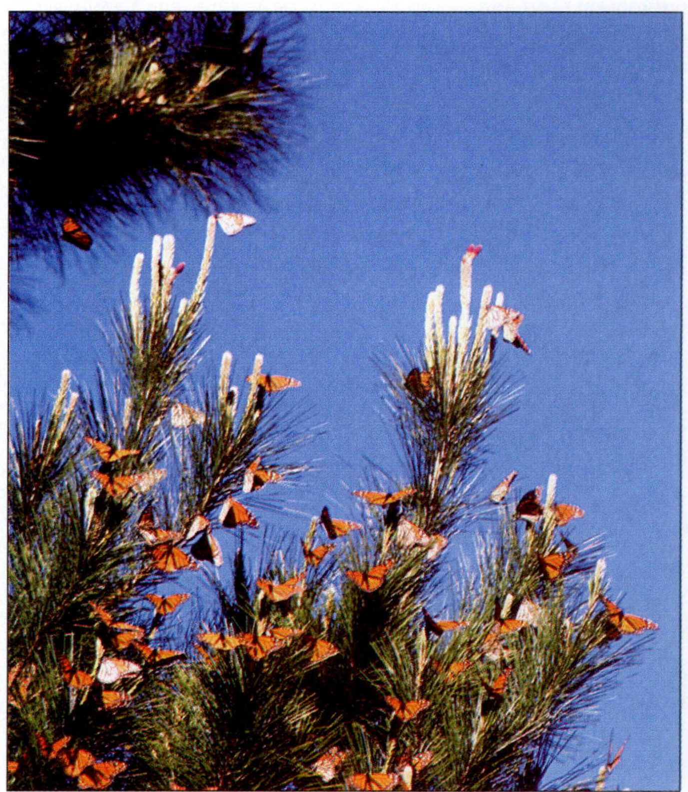

FIGURE 45–4 Migration is an important factor in the population size of migratory animals. Shown here is a gathering of monarch butterflies during the spring. Monarch butterflies migrate into the United States and Canada for the summer months and spend the winter in the mountains of central Mexico. A round trip often involves several generations. Somehow, individual monarchs know where to migrate without having seen the sites before. (Norbert Wu/Peter Arnold, Inc.)

Time	Number of bacteria
0	1
20 min	2
40 min	4
1 hour	8
1 hr, 20 min	16
1 hr, 40 min	32
2 hours	64
2 hr, 20 min	128
2 hr, 40 min	256
3 hours	512
3 hr, 20 min	1,024
3 hr, 40 min	2,048
4 hours	4,096
4 hr, 20 min	8,192
4 hr, 40 min	16,384
5 hours	32,768
5 hr, 20 min	65,536
5 hr, 40 min	131,072
6 hours	262,144
6 hr, 20 min	524,288
6 hr, 40 min	1,048,576
7 hours	2,097,152
7 hr, 20 min	4,194,304
7 hr, 40 min	8,388,608
8 hours	16,777,216
8 hr, 20 min	33,554,432
8 hr, 40 min	67,108,864
9 hours	134,217,728
9 hr, 20 min	268,435,456
9 hr, 40 min	536,870,912
10 hours	1,073,741,824

(a)

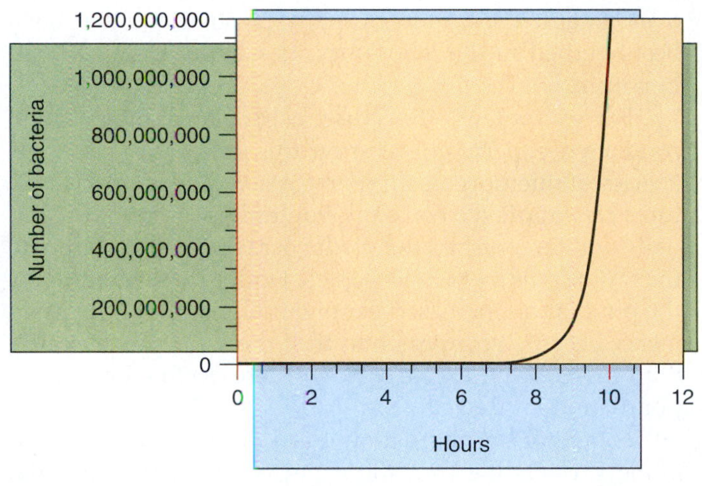

(b)

FIGURE 45–5 Exponential growth. (*a*) When bacteria divide every 20 minutes, their numbers increase exponentially. This set of figures assumes a zero death rate, but even if a certain percentage of each generation of bacteria died, exponential growth would still occur; it would just take longer to reach the very high numbers. (*b*) When these data are graphed, the curve of exponential growth has a characteristic "J" shape. The ideal conditions under which bacteria or other organisms reproduce exponentially rarely occur in nature, and when exponential growth does take place, it is of short duration.

crease its size. The growth rate of a local population of organisms must take into account birth rate (b), death rate (d), immigration (i), and emigration (e). The growth rate is equal to the value of the birth rate minus the death rate plus the value of immigration minus emigration:

$$r = (b - d) + (i - e)$$

For example, the growth rate of a population of 10,000 that has 1000 births, 500 deaths, 10 immigrants, and 100 emigrants in a given year is calculated as follows:

$$r = \left(\underbrace{\frac{1000}{10{,}000}}_{b} - \underbrace{\frac{500}{10{,}000}}_{d}\right) + \left(\underbrace{\frac{10}{10{,}000}}_{i} - \underbrace{\frac{100}{10{,}000}}_{e}\right)$$

$$= (0.1 - 0.05) + (0.001 - 0.01)$$

$$= 0.05 - 0.009 = 0.041$$

Note that, although emigration was greater than immigration in this example, the growth rate was still positive due to the very high birth rate.

Each Population Has a Characteristic Biotic Potential

The maximum rate at which a population could increase under ideal conditions is known as its **biotic potential.** Different species have different biotic potentials. A particular species' biotic potential is influenced by several factors, including the age at which reproduction begins, the percentage of the lifespan during which the organism is capable of reproducing, and the number of offspring produced during each period of reproduction. These factors determine whether a particular species has a large or small biotic potential.

Generally, larger organisms, such as blue whales and elephants, have smaller biotic potentials, whereas microorganisms have the greatest biotic potentials. Under ideal conditions, certain bacteria can reproduce by splitting in half every 20 minutes (Chapter 21). If all individuals survived, at this growth rate a single bacterium would increase to a population of more than one billion in just 10 hours!

If one were to plot this increase in number versus time, the graph would have a "J" shape that is characteristic of **exponential growth,** the constant growth rate that occurs under optimal conditions (Fig. 45–5). Regardless of which organism one is considering, whenever its biotic potential is plotted versus time, the shape of the curve is the same. The only variable is time. That is, it may take longer for an elephant population than for a bacterial population to reach a certain size, but its population will always increase exponentially under ideal conditions.

Nature Limits Population Growth

Certain populations may exhibit exponential growth for a short period of time. However, organisms cannot reproduce indefinitely at their biotic potentials because the environment sets limits, which are collectively called **environmental resistance.** Using the earlier example, bacteria would never be able to reproduce unchecked for an extended period of time because they would run out of food and living space, and poisonous metabolic wastes would accumulate in their vicinity. With crowding, they would also become more susceptible to parasites and predators. As their environment changed, their birth rate (*b*) would decline and their death rate (*d*) would increase due to shortages of food, increased predation, increased competition, and other environmental stresses. The environmental conditions might worsen to a point where *d* would exceed *b*, and the population would decrease. The number of organisms in a population, then, is controlled by the ability of the environment to support them.

Over longer periods of time, the growth rate for most organisms decreases to around zero. This leveling out occurs at or near the limits of the environment to support a population. The **carrying capacity (*K*)** represents the highest population that can be maintained for an indefinite period of time by a particular environment.

When a population is graphed over longer periods of time, the curve has a characteristic "S" shape that shows the population's initial exponential increase under ideal conditions (note the curve's "J" shape at the start), followed by a leveling out as the carrying capacity of the environment is approached (Fig. 45–6). Although the "S" curve is a simplification of actual population changes over time, it appears to fit many populations that have been studied in the laboratory, as well as a few studied in nature.

For example, G. F. Gause, a Russian ecologist who conducted experiments during the 1930s, grew a single species of microorganism (*Paramecium caudatum*) in a test tube. He supplied a constant, but limited, amount of food daily and replenished the media occasionally to eliminate the buildup of metabolic wastes. Under these conditions, the population increased exponentially at first. The paramecia became so numerous that the water was cloudy with them. But then their growth rate declined and their population leveled off.

Sometimes a population crash occurs when a population exceeds the carrying capacity, and environmental degradation results. In 1910, a small herd of 26 reindeer was introduced on one of the Pribilof Islands of Alaska. The herd's population increased exponentially for about 25 years until there were approximately 2000 reindeer, many more than the island could support. The reindeer overgrazed the vegetation until the plant life was almost wiped out. Then, in slightly over one decade as reindeer died from starvation, the number of reindeer plunged to 8, one-third the size of the original introduced population.

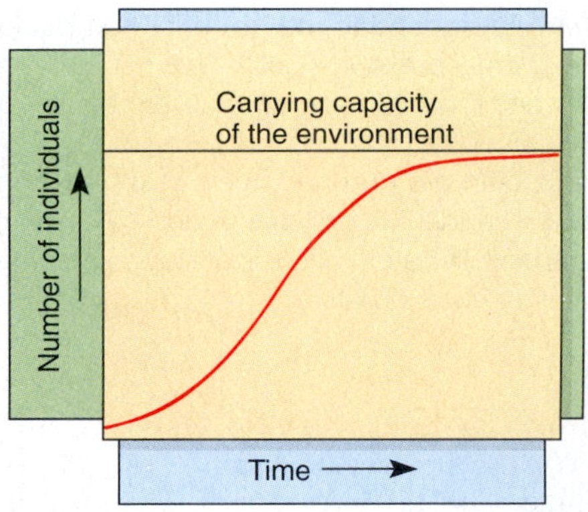

FIGURE 45–6 In many laboratory studies, exponential population growth slows as the carrying capacity of the environment is approached. This produces a curve with a characteristic "S" shape.

POPULATION GROWTH IS LIMITED BY THE ENVIRONMENT

Population growth is ultimately stopped by limitations of the environment to support organisms. Environmental factors that limit population growth fall into two categories: density-dependent and density-independent factors. These two factors vary in importance from organism to organism, and in many cases, interact to determine a population's size.

Density-Dependent Factors Influence Populations

If a change in population density affects the influence of an environmental factor on that population, then the environmental factor is said to be **density dependent.** As population density increases, density-dependent factors tend to slow population growth, by causing an increase in death rate and/or a decrease in birth rate. The effect of density-dependent factors on population growth increases as population density increases. Density dependence can also exert an effect on population growth when population density declines; a decrease in population density results in density-dependent factors enhancing population growth (by decreasing death rate and/or in-

(a)

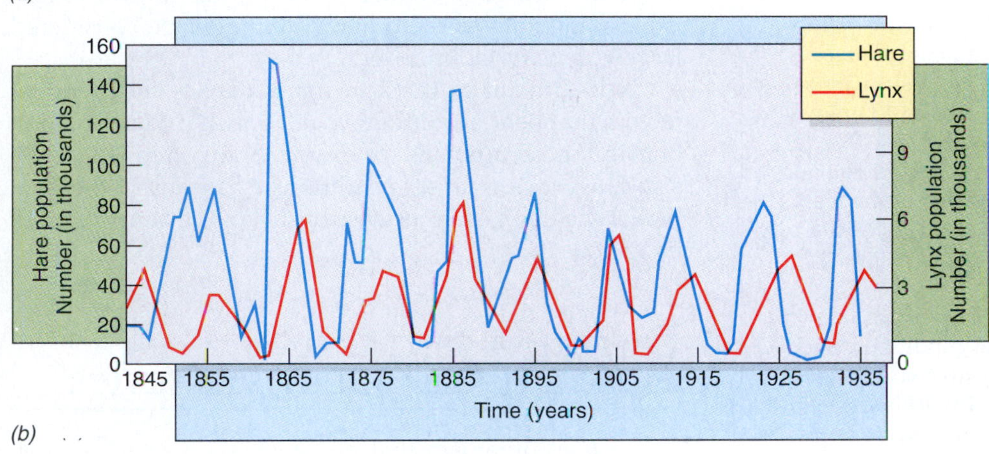

(b)

FIGURE 45–7 A classic case of cyclical oscillation in population density and of dependence of predator upon prey. (***a***) A Canada lynx pursues its primary prey, the snowshoe hare. (***b***) Changes in the relative abundance of the lynx and the snowshoe hare, as indicated by the number of pelts received by the Hudson Bay Company. (Ed Cesar/Photo Researchers, Inc.)

creasing birth rate). Density-dependent factors, then, tend to cause a population to maintain a relatively constant number near the carrying capacity of the environment. Predation, disease, and competition are examples of density-dependent factors.

A classic example of predation as a density-dependent factor is that of the snowshoe hare and Canada lynx (Fig. 45–7). The Hudson Bay Company has kept continuous records of the number of hare and lynx pelts bought from trappers since colonial times. These pelts provide an accurate measure of the populations of animals each year. When the population of hares is large, for example, trappers will catch large numbers of them.

Fluctuations in the number of hare skins are consistently followed, after a brief time lag, by similar fluctuations in the number of lynx skins. When the hare population increases, for example, there is a corresponding increase in the lynx population. It is hard to view the lynx as controlling the numbers of hares in this instance; rather, the predator population seems to be controlled by the size of the prey population. In this example, predation is an important density-dependent regulator for the predator population.

Predation is a density-dependent factor in most, but not all, predator–prey interactions. For example, there is no obvious density-dependent relationship between tawny owls and voles, their chief rodent prey, because the owl population remains relatively constant despite vole population fluctuations.

As the density of a population increases, the physical distance between organisms decreases, they meet one another more frequently, and the chance of their trans-

CONCEPT CONNECTIONS

Predator–Prey Dynamics *Population Genetics*

During the early 1900s, a small herd of moose wandered across the ice of frozen Lake Superior to an island, Isle Royale. In the ensuing years, they became successfully established on the island. In 1949, a few Canadian wolves wandered across the same frozen lake and discovered abundant moose prey. The wolves also remained and became established.

Moose has been the wolf's primary prey on Isle Royale ever since. Wolves hunt in packs by encircling a moose and trying to get it to run so it can be attacked from behind. (The moose is the wolf's largest, most dangerous prey. A standing moose is more dangerous than a running moose because, when standing, it can kick and slash its attackers with its hooves.)

Since 1958, wildlife biologists have studied the populations of moose and wolves on Isle Royale. The wolves primarily feed on the very old and very young in the moose population. Healthy moose in their peak reproductive years are not eaten.

Despite the fact that both populations appeared to be in a state of dynamic equilibrium, a new chapter recently began in the Isle Royale story. The wolf population has plunged, from 50 animals in 1980 to 13 animals in 1993. (The population rebounded slightly in 1994, to 17 wolves) Biologists think that the extreme genetic uniformity of the wolf population may be one of the reasons for its decline. Genetic studies of the Isle Royale wolves indicate that they are all descended from the same female.

Populations that lack genetic variability cannot adapt well to changing environmental conditions. Such populations often have a low reproductive success (recall the cheetah discussed in Chapter 18). In two recent years (1992 and 1993), only four wolf pups were born on the island, and all had the same mother.

It is not known if the wolf population on Isle Royale will recover or disappear altogether. Given their genetic uniformity, the outlook for their long-term survival is poor. Nevertheless, in nature, where populations are subjected to intense natural selection, the significance of genetic uniformity is not clear. Continued studies of the Isle Royale wolves should help address this important issue.

mitting infectious disease to one another increases. If the disease is fatal or markedly reduces the chances of its host's reproducing, the population will decrease to the level where disease transmission becomes less likely.

Competition for resources—living space, food, cover, water, minerals, and sunlight—occurs both *within* a population (**intraspecific competition**) and among populations of different species (**interspecific competition**). As population density increases, so does competition for resources. Eventually it may reach the point where many members of the population fail to obtain the minimum of whatever resource is in shortest supply. This raises the death rate and inhibits further population growth. For example, plants and sessile animals such as barnacles compete for space when their populations are large. In like manner, when the density of animals is high, they compete for food.

Density-Independent Factors Influence Populations

Any environmental factor that regulates the size of a population but is not influenced by changes in population density is called a **density-independent factor.** Random weather events that reduce an organism's population serve as density-independent factors. These often affect population density in unpredictable ways. For example, a very severe blizzard, hurricane, or fire may cause extreme and irregular reductions in populations of organisms vulnerable to them and thus might be considered largely density independent.

It is difficult to think of many density-independent factors that bear absolutely *no* relationship to population density. Social animals, for example, are often able to resist dangerous weather conditions by means of their behavior together, as in the case of sheep huddling together in a snowstorm.

DIFFERENT SPECIES HAVE DIFFERENT REPRODUCTIVE TACTICS

Imagine an organism that possesses the perfect life history strategy, ensuring that it can reproduce at the highest biotic potential possible. That is, our hypothetical organism produces the maximum number of offspring, and the majority of these offspring survive to reproductive maturity. Such an organism would have to mature soon after it was born so that it could begin reproducing at an early age. It would reproduce frequently throughout its life and produce large numbers of offspring each time. Further, it would have to be able to care for its young in order to assure their survival.

In nature, such an organism does not exist because if an organism puts all of its energy into being a perfect "reproductive machine," it would not expend any energy toward ensuring its own survival. For example, animals must use energy to hunt for food, and plants need energy to grow taller than surrounding plants in order to obtain adequate sunlight. Nature, then, requires organisms to

(a)

(b)

FIGURE 45–8 Two life history strategies, *r* selection and *k* selection. (***a***) Dandelions are *r* strategists: annuals that have an early maturity and produce many small seeds. Their population fluctuates from year to year but rarely approaches the carrying capacity of the environment. (***b***) Tawny owls are *K* strategists: they maintain a fairly constant population size at or near the carrying capacity. They mature slowly and have delayed reproduction and a larger body size. (*a*, © David Muench 1995; *b*, Stephen Dalton/Photo Researchers, Inc.)

compromise when expending energy. Living things, if they are to be successful, must do what is required to survive as individuals as well as to survive as a species.

Each species has its own life history strategy—its own reproductive characteristics, body size, habitat requirements, migration patterns, and so on—designed around this energy compromise. Although many different life histories exist, living organisms often fall into two main groups with respect to their life history strategies: *r*-selected species and *K*-selected species.

Populations described by the concept of ***r* selection** have a strategy in which evolution has selected traits that lead to a high growth rate. (Recall that *r* designates the growth rate. Because such organisms have a high *r*, they are known as ***r* strategists,** or ***r*-selected species.**) Small body size and one large brood during its lifetime are typical of many *r* strategists, which are usually opportunists found in variable, temporary, or unpredictable environments.

Some of the best examples of *r* strategists are common weeds, such as the dandelion (Fig. 45–8*a*). Dandelions are propagated by parachute-like fruits containing little seeds that are scattered widely by any strong breeze or wind. In temperate climates, any suitable habitat can be colonized by dandelions as soon as it becomes available.

The seeds of some varieties of dandelions are produced by apomixis, an asexual process for producing seeds (see Chapter 29). Upon germination, the new plant uses another form of asexual reproduction. The dandelion has a long taproot that splits longitudinally to form two roots, then four roots, and so on, so that natural clones of plants spread out from the original plant in genetically identical clumps.

Asexual reproduction has three key advantages for dandelions and other *r* strategists. First, asexual reproduction in plants does not require insect pollinators or even other plants, both of which may be in short supply in unpredictable habitats. Second, asexual reproduction is fast. Third, as long as the particular kind of habitat for which an *r* strategist is adapted is common, asexual reproduction preserves a genotype adapted to that habitat.

In populations described by the concept of *K* selection, evolution has selected traits that maximize *K* and reduce *r*. (Remember that *K* is the carrying capacity of the environment. The population of a species with *K* selection is maintained at or near the carrying capacity, and as a result, its individuals have little need to have a high growth rate.) These organisms, called ***K* strategists** or ***K*-selected species,** do not use environmental resources and energy to produce large numbers of offspring. They characteristically have a long lifespan with a slow development, late reproduction, large body size, and repeated reproductive cycles. Animals that are *K* strategists also invest in parental care of their young. *K* strategists are found in relatively constant or stable environments, where they have a high competitive ability.

Tawny owls are *K* strategists that pair-bond for life, with both members of a pair living and hunting on adjacent, well-defined territories (Fig. 45–8*b*). They regulate their reproduction in accordance with the resources, especially food, present in their territories. In an average year, 30% of the birds do not breed. If food supplies are more limited than initially indicated, many of those that breed fail to incubate their eggs. Rarely do the owls lay the maximum number of eggs, and often they delay breeding until late in the season when the rodent populations on which they depend have become large. Thus, tawny owls regulate their population behaviorally so that its number stays at or near the carrying capacity of the environment. Starvation, an indication that the tawny owl population has exceeded the carrying capacity, rarely occurs.

HUMAN POPULATION GROWTH FOLLOWS THE SAME PARAMETERS AS POPULATION GROWTH IN OTHER ORGANISMS

Now that we have examined some of the basic concepts of population ecology, we can apply those concepts to the human population. Examine Figure 45–9, which shows the world increase in the human population since the development of agriculture, approximately 10,000 years ago. Now look back at Figure 45–5 and observe how the human population is increasing exponentially. The characteristic J curve of exponential growth reflects the decreasing amount of time it has taken to add each additional billion people to our numbers. It took thousands of years for the human population to reach 1 billion, a milestone that took place in 1800. It took 130 years to reach a population of 2 billion (in 1930), 30 years to reach 3 billion (in 1960), 15 years to reach 4 billion (in 1975), and 12 years to reach 5 billion (in 1987). (The population is projected to reach 6 billion in 1999.)

Thomas Malthus (1766–1834), a British economist, was one of the first people to recognize that the human population cannot continue to increase indefinitely (Chapter 17). He pointed out that human population growth was not always desirable (a view contrary to the beliefs of his day) and that the human population was capable of increasing faster than was the food supply. He maintained that the inevitable consequences of population growth were famine, disease, and war.

The overall growth rate for the human population was 1.6% in 1994. This growth rate was not caused by an increase in the birth rate (*b*). In fact, the world birth rate has actually declined slightly during the past 200 years. The population growth rate was due instead to a large *decrease in the death rate* (*d*), which has occurred primarily because of greater food production, better medical care, and an increase in sanitation practices. For example, from about 1920 to 1994, the death rate in Mexico fell from approximately 40 to 6, whereas the birth rate dropped from approximately 40 to 28 (Fig. 45–10).

The human population has reached a turning point. Although our numbers continue to increase, the growth rate has declined over the past several years. Population experts at the United Nations and the World Bank have projected that the growth rate will continue to slowly decrease until zero population growth is attained. It is pro-

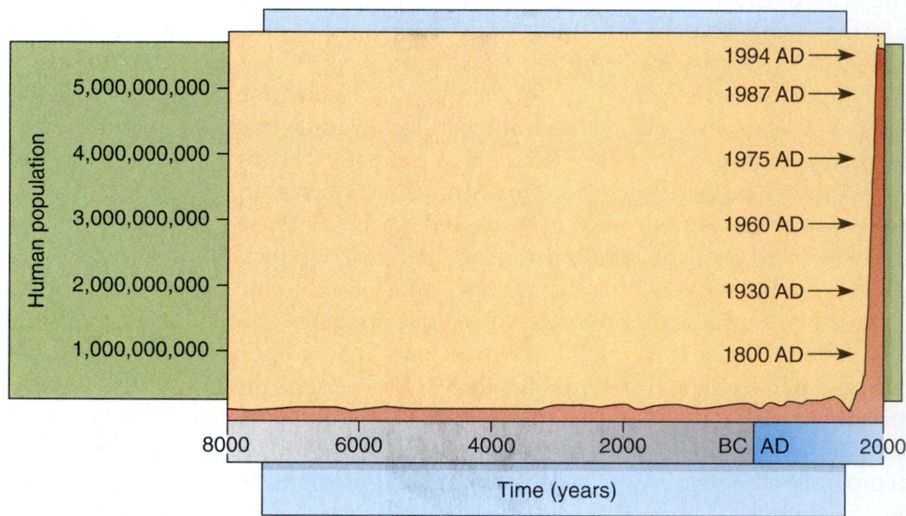

FIGURE 45–9 Human population growth. The human population has been increasing exponentially from the beginning of agriculture (10,000 years ago) to the present.

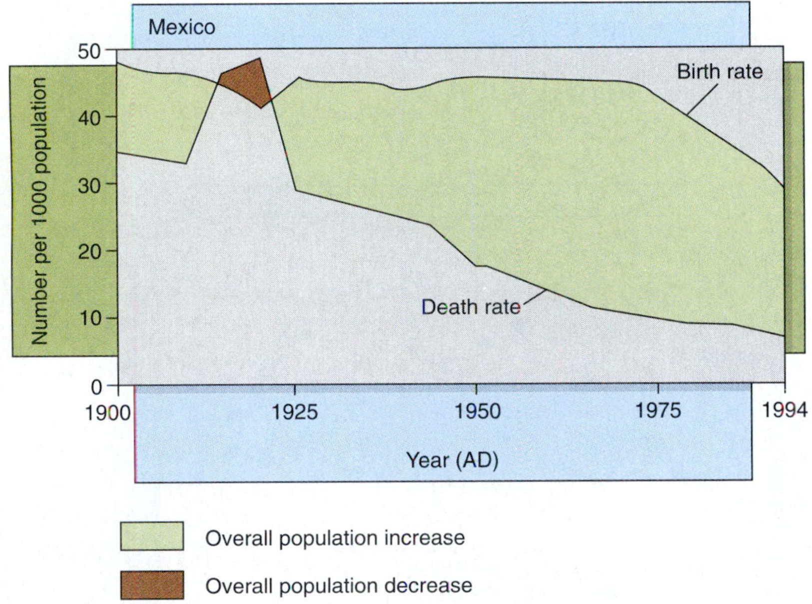

FIGURE 45–10 Birth rates, death rates, and population growth. In Mexico, the birth and death rates have generally declined in this century. Because the death rate has declined much more than the birth rate, however, Mexico has experienced a high growth rate.

jected that **zero population growth**—when the birth rate equals the death rate ($r = 0$)—will occur toward the end of the 21st century, when it is anticipated that the human population will level off at approximately 10.4 billion.[3] This number is almost twice what the population of the world was in 1994.

Population projections are "what if" exercises: given certain assumptions about future tendencies in birth rate, death rate, and migration, an area's population can be calculated for a given number of years into the future. Population projections indicate the changes that are upcoming but they must be interpreted with care because they vary depending on what assumptions have been made. For example, in projecting that the population will stabilize at 10.4 billion by the end of the 21st century, population experts assume that the average number of children born to each woman in all countries will have declined to just about 2.0 by 2040 A.D. (In 1994, the world average number of children born to each woman was 3.2.) If that decline does not occur by 2040 A.D., our population will not stabilize at 10.4 billion people by the end of the 21st century, but will stabilize later and at a higher number. For example, if the population were to continue to increase at its 1994 growth rate, there would be more than 27 billion humans toward the end of the 21st century. This number is almost five times larger than our current population.

The main unknown factor in this population growth scenario is the carrying capacity of our environment. No one knows how many humans can be supported by the Earth, and projections and estimates vary widely, depending on what assumptions are made. It is also not clear what will happen to the human population if or when the carrying capacity is approached. Optimists suggest that the human population will stabilize because of a decrease in the birth rate. Some experts take a more pessimistic view and predict that the widespread degradation of our environment caused by our ever-expanding numbers will result in a massive wave of human deaths.[4]

Not All Countries Have the Same Growth Rates

While world population numbers illustrate overall trends, they do not describe other important aspects of the human population story, such as population differences from country to country. **Demographics,** the branch of sociology that deals with population statistics, provides interesting information on the populations of various countries. As discussed in the Chapter Opener, countries can be classified into three groups—highly developed, moderately developed, and less developed—depending upon their growth rate, degree of industrialization, and relative prosperity (Table 45–1).

Developed countries (also called highly developed countries) have low growth rates, are highly industrialized, and have high per capita GNPs[5] relative to the rest

[3] This projection is based on many assumptions and will be affected by the actions that individuals and nations take between now and then.

[4] Some experts think that the human population has already exceeded Earth's carrying capacity.

[5] GNP stands for gross national product, the total value of a nation's annual output in goods and services.

FOCUS ON *The Environment*

The Collapse of the Newfoundland Cod Fishery

In 1497 John Cabot sailed from Bristol, England, on a voyage to find a rapid route to Asia. Instead, he discovered northern North America and with it, one of the richest fishing areas in the world, the Grand Banks, off the coast of Newfoundland. The key fish in these waters was the Atlantic cod, *Gadus morhua,* a slate-grey fish with a massive mouth, a small "goatee," and a typical weight of 3 to 4 kilograms (6.6 to 8.8 pounds). Cod were initially so abundant that fishermen simply dropped buckets over the sides of a ship and hauled in nature's bounty. Immediately a major European fishery developed to exploit this seemingly inexhaustible resource.

Over the last 300 years, a unique culture centered around fishing has developed in Newfoundland. By 1992, about 54,000 people obtained their livelihood either directly from catching cod and other fishes or indirectly from processing them. However, in that year the cod fishery collapsed. Most of the collapse was due to the disappearance of the northern cod off the northeast coast of Newfoundland and Labrador. In response, the Canadian government immediately announced a moratorium on fishing, which will last to the end of the 1990s, to give the fish stocks time to recover. With the moratorium was a compensation package totalling almost 2.5 billion dollars to tide the people over. How did this incredible disaster happen in the face of so much scientific knowledge, and what lessons can be learned to prevent similar events from occurring elsewhere?

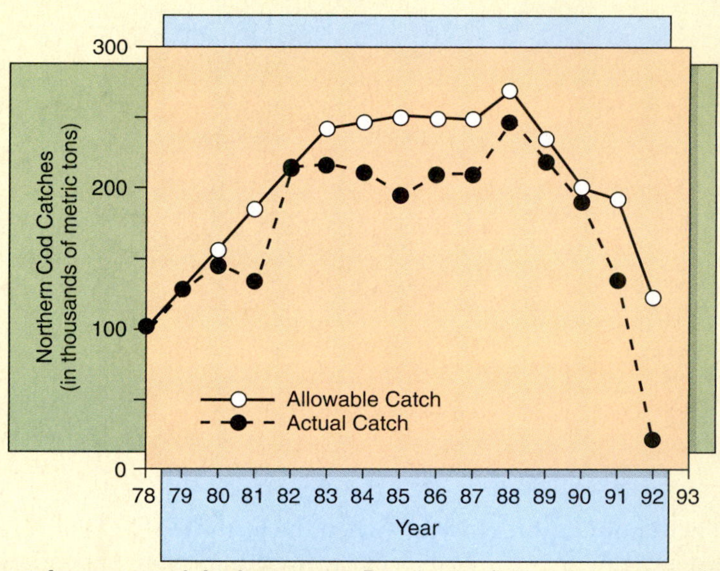

Warnings of an overexploited resource. Despite the fishing industry's technological efficiency and obvious economic motivation, they were consistently unable to catch as many fish as the government allowed. The reason eventually became apparent: cod stocks were far smaller than government estimates indicated. (Data from Department of Fisheries and Oceans, Ottawa, Canada)

Resources used by humans can be separated into two kinds: those that are renewable and those that aren't. **Renewable resources** are those which can last indefinitely if they are managed properly. They include such things as fish and wildlife species, forests, soil, and water. **Nonrenewable resources,** such as oil and natural gas reserves, are those that can only be used once, after which they are gone forever.

TABLE 45–1
Comparison of 1994 Population Data in Developed and Developing Countries

	Developed	*Developing*	
	(Highly Developed) United States	*(Moderately Developed)* Brazil	*(Less Developed)* Ethiopia
Fertility rate	2.1	3.0	6.9
Doubling time at current rate	98 years	40 years	22 years
Infant mortality rate	8.3 per 1000	66 per 1000	110 per 1000
Life expectancy at birth	76 years	67 years	52 years
Per capita GNP (U.S. $; 1992)	$23,120	$2770	$110
Women using modern contraception	69%	56%	3%

Common-property resources are a type of renewable resources. The heart of the overfishing of cod lies in the common-property nature of this resource. A society of those who fish is as distinctive as a society of those who farm, with the difference that those in the fishery have no secure claim to the underlying resource. Fishes only become private property when they are caught and removed from the water. This tends to result in a "free-for-all" race for the limited fish stocks. Each fisherman or boat owner therefore tries to maximize his or her share of the harvest within the limits dictated by nature or by government regulation. The American biologist Garrett Hardin stated this most eloquently with his term "tragedy of the commons." People ruthlessly competing for these common resources pursue their own interests without limit. They basically reason as follows: "If I don't use this resource and get as much as possible for myself, someone else will take it." The net result is the collapse or serious degradation of the resource. "Freedom of the commons brings ruin to all."

To comprehend how the cod fishery collapsed, we must appreciate some of the basic biology of the species and the major contributing factors in the fishing industry. Atlantic cod is a widespread species with 11 stocks occurring between northern Labrador and the Canada-U.S. border on the Georges Bank. Stocks are genetically distinct populations that occur in a certain area of the ocean. Females mature at about 7 years old and thereafter can spawn over one million eggs per year. Spawning occurs in large concentrations of adults during the cool months. On the Grand Banks, this occurs from April to June. On average only one egg in a million reaches maturity. Young cod eat plankton and bottom creatures, but the adults are voracious, feeding on other fishes such as herring and capelin.

Determining sensible government regulations for how many fishes can be harvested per year relies on knowing accurately how many are there in the first place. Fishes are very hard to count, especially on the high seas. Government scientists recommended that about 20% of the stock could be harvested per year, but the size of the stock was overestimated. As shown in the graph, the actual catch seldom equaled the allowable catch. The stock size was overestimated for 3 major reasons: (1) Intentional underreporting by fisherman of what they actually caught. Deep water trawlers catch everything in their path and maximum fishing effort occurred when cod were concentrated in spawning aggregations. Fishes unintentionally caught or that were too small were simply dumped overboard (already dead) and not reported. (2) Inaccurate estimates of survival of fishes from egg to maturity. (3) Highly efficient fishing technology with the development of sophisticated sonar tracking, which allowed the nets to be full until virtually the last fish was caught. This gave the appearance that there was still lots of fishes when in fact they were practically economically extinct. In addition, offshore fleets, particularly from Spain, Portugal, and Korea, fished outside Canada's 200 mile limit in two crucial areas of the Grand Banks, the Nose and the Tail. From 1986–1990, these fleets took almost 600,000 metric tons more fishes, including cod, than they should have. Lastly, the people in Atlantic Canada demanded too much from the fishery. It was the employer of last resort, with a historical tradition of the "right to fish." There were too many people fishing for too few cod, and governments assumed that the fishery could absorb the unemployed from other industries.

The collapse of the cod fishery is not unique. This century we have witnessed the collapse of many fisheries, including the Peruvian anchovy fishery in the early 1970s, the decline of the Alaskan salmon fishery in the 1960s and 1970s, and the collapse of the Antarctic whale stocks after World War II. Unless common good at both the national and international level can replace individual competition, fishermen will continue to argue amongst themselves, communities will fight one another for survival, and countries will continue to overexploit each other's stocks.

Essay contributed by Dr. Rudy Boonstra, Division of Life Sciences, The University of Toronto-Scarborough Campus, Scarborough, Ontario, Canada

of the world. Developed countries have the lowest birth rates. Indeed, some developed countries (such as Germany) have birth rates just below that needed to sustain the population and are thus declining slightly in numbers. Highly developed countries also have a very low **infant mortality rate** (the number of infant deaths per 1000 live births). The infant mortality rate of the United States was 8.3 in 1994, for example, compared with a 1994 world infant mortality rate of 63. Highly developed countries also have a longer life expectancy (76 years in the U.S. versus 65 years worldwide) and a higher average per capita GNP ($23,120 versus $4340 worldwide).

Developing countries fall into two subcategories: moderately developed and less developed. When compared with highly developed countries, the birth rates and infant mortality rates of moderately developed countries are higher. Moderately developed countries have a medium level of industrialization and their average per capita GNPs are lower than those of highly developed countries. Less developed countries have the highest birth rates, the highest infant mortality rates, the lowest life expectancies, and the lowest average per capita GNPs in the world.

One way to represent the population growth of a country is to determine the doubling time (the amount of time it would take for its population to double in size, assuming that its current growth rate does not change). A look at a country's doubling time can identify it as a highly, moderately, or less developed country: the shorter its doubling time, the less developed the country. At 1994 growth rates, the doubling time is 19 years for Togo, 22 years for Ethiopia, 31 years for Mexico, 98 years for the United States, and 330 years for Belgium.

It is also instructive to examine **replacement-level fertility,** which represents the number of children a couple must produce in order to "replace" themselves. Replacement-level fertility is usually given as 2.1 children in developed countries and 2.7 children in developing countries. The number is greater than 2.0 because some children die before they reach reproductive age. Thus, higher infant mortality rates are the main reason that the replacement levels in developing countries are greater than those in developed countries. Worldwide, the **total fertility rate,** the average number of children born to a woman during her lifetime, is 3.2, which is well above replacement levels in both developed and developing countries.

Demographers Recognize Four Demographic Stages

In the last 200 years, Europe shifted from relatively high to relatively low birth and death rates (Fig. 45–11). Based on observations of Europe as it became industrialized and urbanized, four demographic stages are known. Finland's population changes over the last 200 years are representative of the four demographic stages.

In the first stage, called the **preindustrial stage,** birth and death rates are high, and the population grows at a modest rate. Although women bear many children, the infant mortality rate is high. Intermittent famines, plagues, and wars also increase the death rate, so that the population grows slowly. Finland in the late 1700s is an example of the first demographic stage.

As a result of improved health care and more reliable food and water supplies that accompany the initiation of an industrial society, the second demographic stage, called the **transitional stage,** is characterized by a lowered death rate. However, because the birth rate is still high, the population grows rapidly. Finland in the mid-1800s was in stage 2, and today, many Latin American, Asian, and African countries are in the second demographic stage.

The third demographic stage, the **industrial stage,** takes place at some point during the industrialization process. It is characterized by a decline in the birth rate, which along with the relatively low death rate, slows population growth. For Finland, this occurred in the early 1900s.

The fourth demographic stage, sometimes called the **postindustrial stage,** is characterized by low birth and death rates. In countries that are heavily industrialized, people are better educated, more affluent, and tend to desire smaller families and take steps to limit family size. The population grows slowly or not at all in the fourth demographic stage—for example, in such developed countries as the United States, Canada, Australia, Japan, and most of western Europe, including Finland.

Once a country reaches the fourth demographic stage, is it correct to assume it will continue at low birth rates indefinitely? The answer is that we don't know. Low birth rates may be a permanent response to the socioeconomic factors that are a part of an industrialized, urbanized society. On the other hand, low birth rates may be a temporary response to socioeconomic factors such as the changing roles of women in developed countries. No one knows for sure.

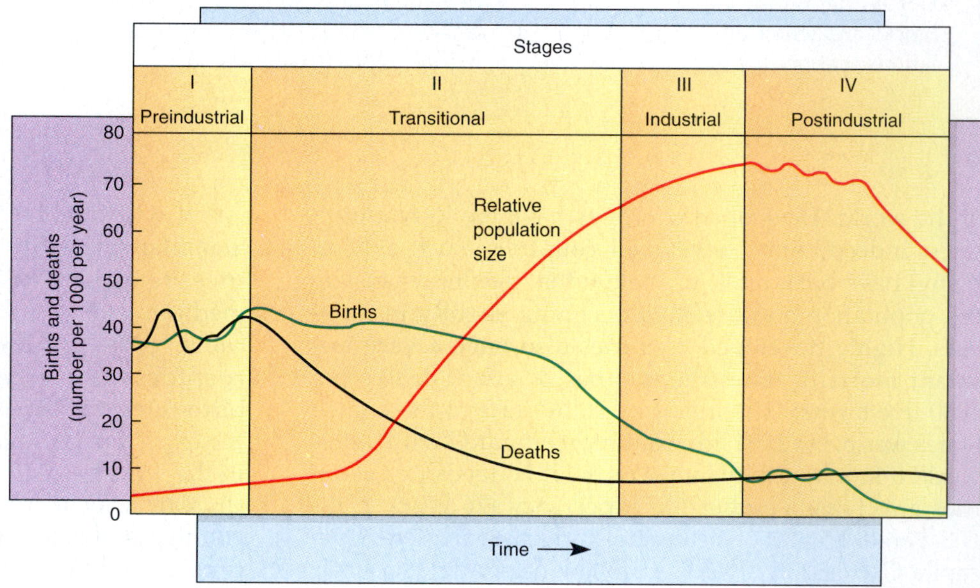

FIGURE 45–11 The demographic transition. The demographic transition consists of four demographic stages through which a population progresses as its society becomes industrialized. Note that the death rate declines first, followed by a decline in the birth rate.

TABLE 45–2
Fertility Changes in Selected Developing Countries

	*Total fertility rate**	
Country	*1960-65*	*1994*
Afghanistan	7.0	6.9
Bangladesh	6.7	4.9
Brazil	6.2	3.0
China	5.9	2.0
Egypt	7.1	3.9
Guatemala	6.9	5.4
India	5.8	3.6
Kenya	8.1	6.3
Mexico	6.8	3.2
Nepal	5.9	5.5
Nigeria	6.9	6.5
Thailand	6.4	2.2

* Total fertility rate = average number of children born to each woman during her lifetime.

All highly developed and moderately developed countries have followed this demographic transition to date. As a result, demographers generally assume that the same progress will occur in less developed countries as they become industrialized.

The population in many developing countries is beginning to approach stabilization. Fertility rates must decline in order for the population to stabilize (see Table 45–2 and note the general decline in total fertility rate from the 1960s to 1994 in selected developing countries). Worldwide, the total fertility rate in developing countries has decreased from an average of 6.1 children per woman in 1970 to 3.2 in 1994. Fertility rates have declined by at least 25% in the past decade in countries like Brazil, Indonesia, and Mexico.[6] The fertility rate continues to increase in only a few African countries—Burkina Faso, for example.

[6] Although the fertility rate has declined dramatically in these countries, remember that it is still greater than replacement-level fertility. Consequently, the population in these countries is still increasing.

The Age Structure of a Country Can Be Used to Predict Future Population Growth

In order to predict the future growth of a population, it is important to know its **age structure,** which is the percentage of the population at different ages. The number of males and females at each age, from birth to death, is represented in age structure diagrams (Fig. 45–12). Age structure diagrams are divided vertically in half to show sex ratios: one side represents the males in a population and the other side represents the females. The bottom one-third of the diagram represents prereproductive humans (from 0 to 14 years of age), the middle one-third corresponds to the reproductive ages (15 to 44 years), and the top one-third depicts postreproductive humans who are 45 years and older. The width of the individual diagrams at any given point is proportional to the population size; a broader width implies a larger population.

The overall shape of an age structure diagram indicates whether the population is increasing, stable, or shrinking. The age structure diagram of a country with a very high growth rate (for example, Nigeria or Venezuela) is shaped like a pyramid (Fig. 45–13*a*). The largest percentage of the population is in the prereproductive age group, providing the momentum for future population growth. When these children mature, they will become the parents of the next generation. Thus, *even if the fertility rate of such a country is at replacement levels, the population will continue to grow.* In contrast, the more tapered bases of the age structure diagrams of countries with stable or declining populations indicate a smaller ratio of children who will become the parents of the next generation.

The age structure diagram of a stable population, one that is neither growing nor shrinking, demonstrates that the number of people at the prereproductive age and at the reproductive age is approximately the same (Fig. 45–13*b*). Also, a larger percentage of the population is older (postreproductive) than in a rapidly increasing population. Many countries in Europe have stable populations.

In a population that is shrinking, the prereproductive age group is *smaller* than either the reproductive or postreproductive groups. Germany, Bulgaria, and Hungary are examples of countries with shrinking populations.

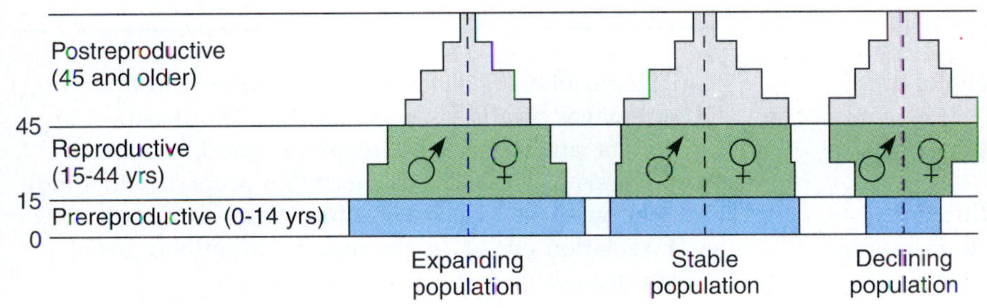

FIGURE 45–12 Generalized age structure diagrams for an expanding population, a stable population, and a population that is decreasing in size. In each, the left half of the diagram represents the males in the population, and the right half represents the females. Each diagram is divided horizontally into age groups, and the width of each segment represents the population size of that group.

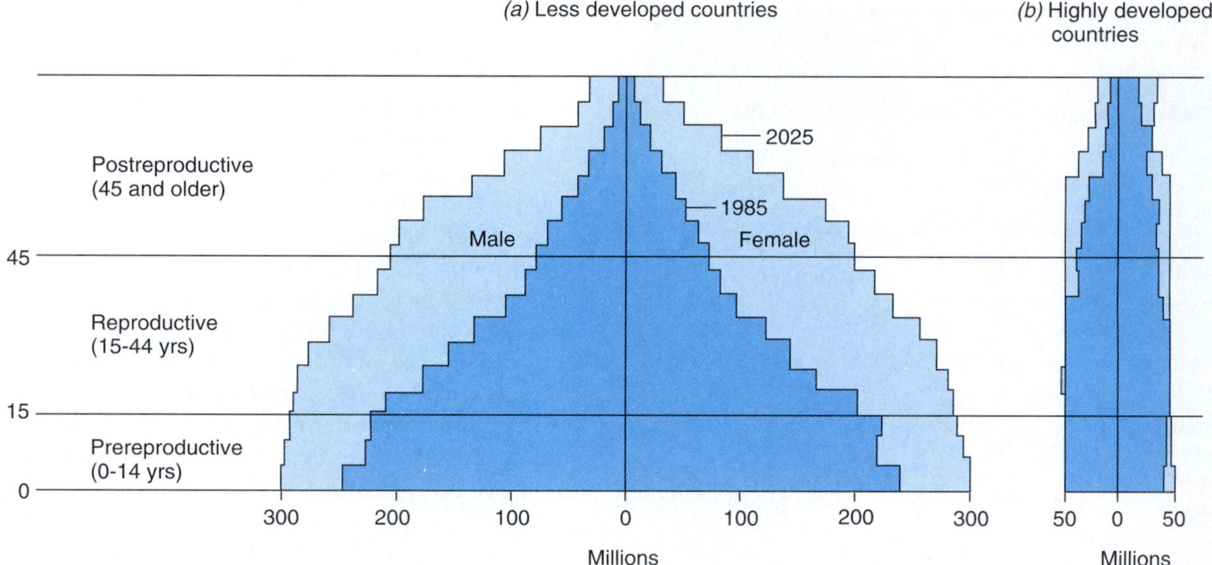

FIGURE 45–13 Age structure diagrams. A big difference exists between the age structure of less developed (***a***) and highly developed (***b***) countries. The dark blue region represents actual age distribution in 1985. The light blue region represents projected age distribution in 2025. These age structure diagrams indicate that less developed countries have a much higher percentage of young people than highly developed countries. As a result, less developed countries are projected to have greater population growth than highly developed countries.

Worldwide, one-third of the human population is under the age of 15. When these people enter their reproductive years, they have the potential to cause a large increase in the growth rate. Even if the birth rate does not increase, the growth rate will increase simply because there are more people reproducing.

Most of the world population increase that has occurred since 1950 has taken place in developing countries (as a result of the younger age structure and the higher-than-replacement level fertility rates of their populations). In 1950, 66.8% of the world's population was in the developing countries found in Africa, Asia (minus Japan), and Latin America; the remaining 33.2% of the population was in the developed countries found in Europe, the USSR, Japan, Australia, and North America. Between 1950 and 1994, the world's population had more than doubled in size, but most of that growth occurred in developing countries. As a reflection of this, in 1994 the number of people in developing countries had increased to 79% of the world's population. Most of the population increase that will occur during the next century will take place in developing countries. By 2020, it is estimated that about 85% of the people in the world will live in developing countries.

Chapter Summary

I. Ecology is the study of the relationships between organisms and their environment.
   A. A population is all of the members of the same species that live together.
   B. A community is all of the populations of different species living in the same area; an ecosystem is a community and its physical environment.
   C. The biosphere is all of the communities on Earth; the ecosphere comprises interactions among Earth's biosphere, atmosphere, hydrosphere, and lithosphere.

II. Populations of organisms have certain properties that individual organisms lack, such as birth rates and death rates.
   A. Population density is the number of individuals of a species per unit of area at a given time.

B. Population dispersion may be uniform, random, or clumped.

III. Population change results from differences among the number of births, deaths, immigration, and emigration.
   A. Biotic potential is the maximum rate at which a population can grow under ideal conditions.
   B. Environmental limitations prevent indefinite population growth. The collective total of all such limitations constitutes environmental resistance.
   C. The carrying capacity of the environment is the highest population that can be maintained for an indefinite period of time by a particular environment.

IV. Environmental factors limit population growth.
   A. Density-dependent factors are most effective at limiting population growth when the population density is high. Predation, disease, and competition are examples.
   B. Density-independent factors limit population growth but are not influenced by changes in population density. Hurricanes and fires are examples.

V. Each organism has its own life history strategy.
   A. An r strategy emphasizes a high growth rate. These organisms often have small body sizes, one major reproductive event during their lifetimes, short lifespans, and inhabit variable environments.
   B. A K strategy emphasizes maintenance of a population near the carrying capacity of the environment. These organisms often have large body sizes, repeated reproductive cycles, long lifespans, and inhabit stable environments.

VI. The principles of population ecology apply to humans as well as to other living organisms.
   A. Currently, the human population is increasing exponentially. The growth rate has declined slightly over the past several years, however, leading demographers to project that the world population will stabilize ($r = 0$) by the end of the 21st century.
   B. Highly developed countries have the lowest birth rates, the lowest infant mortality rates, and the longest life expectancies. They also have the highest per capita GNPs. Developing countries have the highest birth rates, the highest infant mortality rates, and the shortest life expectancies. They also have the lowest per capita GNPs.
   C. The age structure of a population greatly influences population dynamics. It is possible for a country to have replacement-level fertility and still experience population growth if the largest percentage of the population is in the prereproductive years.

Selected Key Terms

biosphere, p. 908
biotic potential, p. 911
birth rate, p. 910
carrying capacity, p. 912
clumped dispersion, p. 910
community, p. 908
death rate, p. 910
doubling time, p. 910
ecology, p. 908
ecosphere, p. 909
ecosystem, p. 908
emigration, p. 910
environmental resistance, p. 912
growth rate, p. 910
immigration, p. 910
infant mortality rate, p. 919
K selection, p. 915
population, p. 908
population density, p. 909
random dispersion, p. 910
replacement-level fertility, p. 920
r selection, p. 915
total fertility rate, p. 920
uniform dispersion, p. 909
zero population growth, p. 917

Post-Test

1. A/An _______________ is a group of organisms of the same species living in a particular area at a given time.
2. Population _______________ is the number of individuals of a species per unit of area at a given time.
3. _______________ _______________ is when individuals are grouped together in specific portions of the habitat.
4. The maximum rate at which a population could increase under ideal conditions is known as its _______________ _______________.
5. When graphing population number versus time, a J-shaped curve is characteristic of _______________ growth.
6. The biotic potential of organisms never occurs indefinitely because the environment sets limits, which are collectively called _______________ _______________.
7. The _______________ _______________ is the largest population that a particular environment can maintain for an indefinite period of time.
8. Predation, disease, and competition are examples of density-_______________ factors.
9. The effect of density-_______________ factors on population growth intensifies as population density increases.
10. Density-_______________ factors regulate the size of a population but are not influenced by changes in population density.

11. Organisms with ________________ selection have traits that allow them to have a high growth rate.
12. Organisms that are ________________ strategists do not use environmental resources and energy to produce large numbers of offspring.
13. Population experts project that ________________ ________________ ________________, when the human birth rate equals the death rate, will occur around 2089.
14. Developed countries have a low ________________ ________________ ________________, which is the number of infant deaths per 1000 live births.
15. The percentage of the population at different ages is known as its ________________ ________________.
16. Choose the sequence that reflects an increase in the level of biological organization. (a) community ⟶ ecosystem ⟶ population (b) population ⟶ individual organism ⟶ community (c) population ⟶ community ⟶ biosphere (d) ecosphere ⟶ biosphere ⟶ ecosystem
17. *Global* growth rate is a result of the interactions of: (a) birth rate and immigration (b) birth rate and death rate (c) birth rate and emigration (d) birth rate, death rate, immigration, and emigration.
18. According to Thomas Malthus, (a) the continual growth of the human population is good (b) the human population could increase faster than our ability to produce food (c) the human population will increase to 10.4 billion and then stabilize (d) the human population has been declining slowly since the beginning of the twentieth century.
19. Humans in developing countries have (a) the lowest birth rates in the world (b) the highest per capita GNPs in the world (c) the highest infant mortality rates in the world (d) longer life expectancies.
20. Which of the following age structure diagrams indicates a population that is growing? Explain your answer.

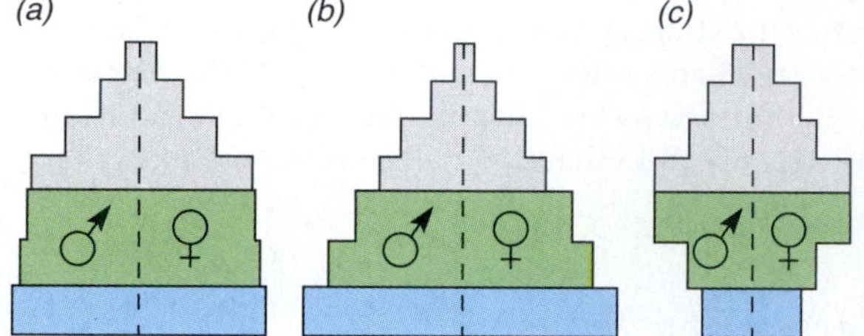

Review Questions

1. List and define the four components that comprise the ecosphere.
2. Define each of the following and explain its effect on a population: birth rate, death rate, immigration, and emigration.
3. Give several biological advantages for organisms having a clumped dispersion. What disadvantages are there?
4. What is the carrying capacity of an environment?
5. Explain the J-shaped and S-shaped population growth curves.
6. Give several examples of density-dependent and density-independent factors that limit population growth.
7. Explain how an infectious disease might be a density-dependent factor.
8. Explain how a population can have an increase in its growth rate when its birth rate is declining.
9. What is replacement-level fertility? Why is it higher in developing countries than in developed countries?

Thinking Critically

1. Suppose that you were asked to develop an official policy to reduce unnecessary resource consumption in the United States. What would you recommend?
2. As the world population continues to grow during your lifetime, what changes, if any, do you anticipate in your quality of life?
3. If a population of 100,000 has 2000 births per year and 1000 deaths per year, what is its growth rate? What is its doubling time?
4. Are humans r strategists or K strategists? Explain your answer.

5. If all of the women in the world suddenly started bearing children at replacement-level fertility rates, would the population stop increasing immediately? Why or why not?
6. How might humans decrease the carrying capacity of their environment? How might they increase it?
7. A reciprocal relationship exists between the number of people and the per capita resources consumed in estimating the human carrying capacity of a particular area. Explain the reciprocal relationship.

Recommended Readings

Daily, G.C., and P.R. Ehrlich, "Population, Sustainability, and Earth's Carrying Capacity," *BioScience,* Vol. 42, No. 10, November 1992. Explores the question of how many people (and what types of human lifestyles) the Earth can support indefinitely.

Durning, "Long on Things, Short on Time," *Sierra,* January/February 1993. Contains arguments for a low consumption, sustainable lifestyle.

Haub, C., "Understanding Population Projections," *Population Bulletin,* Vol. 42, No. 4, December 1987. A comprehensive evaluation of the challenges facing demographers as they calculate population projections.

Sadik, N., *The State of World Population 1990,* New York, United Nations Population Fund, 1990. Summarizes world population issues and our options during the 1990s.

Sherman, D., "Institutions: Zero Population Growth," *Environment,* Vol. 35, No. 9, November 1993. Examines ZPG, an organization founded to bring Earth's population into balance with its resources and the environment.

World Population Data Sheet of the Population Reference Bureau, Inc., 1994, Washington, D.C. An annual chart that provides current population data for all countries. Includes birth rates, death rates, infant mortality rates, total fertility rates, and life expectancies as well as other pertinent information.

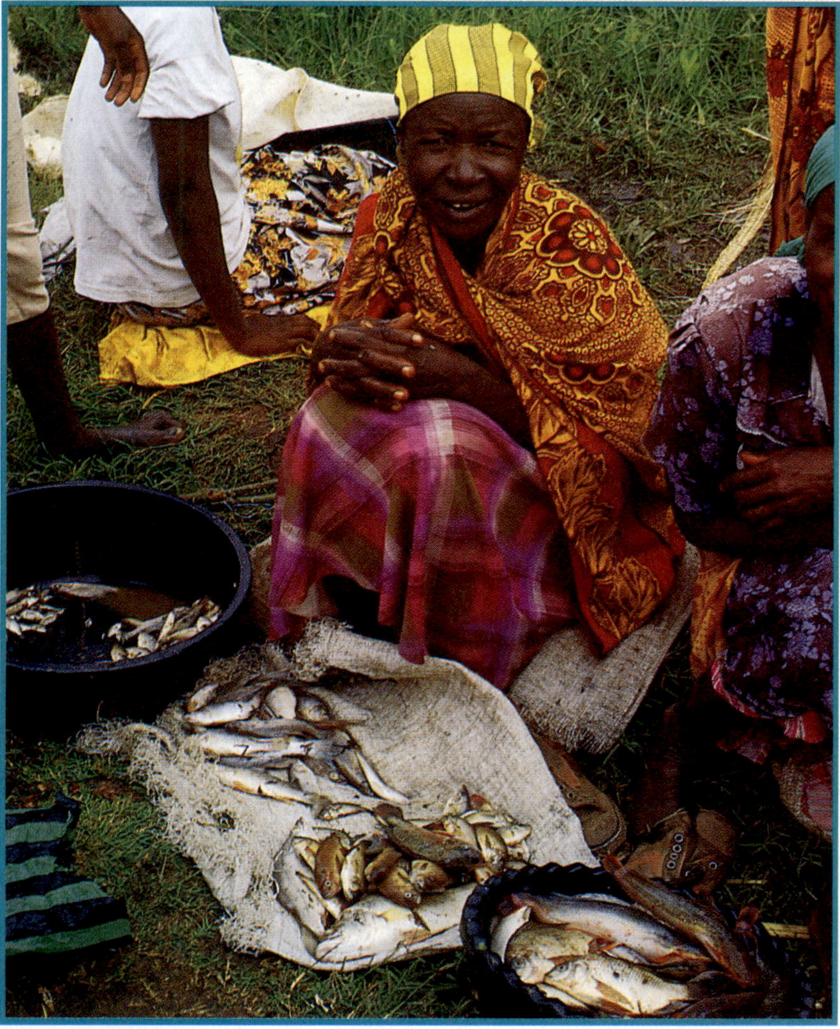

Woman with a mixed catch of nile perch (larger fishes) and cichlids (smaller fishes). (Mark Chandler, New England Aquarium)

THE DEATH OF A LAKE

The world's second largest freshwater lake, Lake Victoria in East Africa, is larger than the state of West Virginia. Until relatively recently, it was home to about 400 different species of small, colorful fishes known as cichlids (sik-lids). The different species of cichlids in Lake Victoria had remarkably different eating habits. Some grazed on algae; some consumed dead organic material at the bottom of the lake; and others were predatory and ate insects, shrimp, and other cichlid species. These fishes, which thrived throughout the lake ecosystem, provided much-needed protein to the diets of 30 million humans living in the lake's vicinity.

Today, the aquatic community is very different in Lake Victoria. More than half of the cichlids and other native fish species have disappeared. As a result of their mass extinction, the algal population has increased explosively (the algae-eating cichlids are almost all extinct). When the dense algal blooms die, their decomposition uses up the dissolved oxygen in the water. The bottom zone of the lake, once filled with cichlids, is empty because the deep water contains too little dissolved oxygen, and any fishes venturing into the anaerobic zone suffocate. Local fishermen, who once caught and ate hundreds of different types of fishes, now catch only three types.

The most important reason for the destruction of Lake Victoria's delicate ecological balance was the deliberate introduction of the Nile perch, a large and voracious predator, into the lake in 1960. It was thought by proponents of the introduction that the successful establishment of the Nile perch would stimulate the local economy and help the fishermen.

For about 20 years, the Nile perch did not appear to have an appreciable effect on the lake. But in 1980, fishermen noticed that they were harvesting increasing quantities of Nile perch and decreasing amounts of native fishes. By 1985, most of the annual catch was Nile perch, which was experiencing a population explosion fueled by an abundant food supply—the cichlids. More dramatic evidence of the lake's unhealthy state was revealed in 1987 when a study reported that the lake's deepest waters were depleted of oxygen. Many other changes in the lake's ecology, involving organisms from snails to aquatic insects, have also occurred.

The Lake Victoria story is far from finished. Once the Nile perch decimated the cichlids, it changed its feeding habits and now consumes freshwater shrimp and smaller Nile perch. A species that cannibalizes its own young for a major part of its food supply cannot persist indefinitely. Moreover, during the past decade or so, a large fishing industry, which processes Nile perch for export to countries such as Israel and the Netherlands, developed along the shores of Lake Victoria. Many are concerned that overfishing could collapse the Nile perch population, ultimately causing protein malnutrition in the humans living around the lake basin.

Biological communities often contain an astonishing assortment of organisms that interact with one another and are interdependent in a variety of ways. The introduction of a non-native species—in this case the Nile perch—usually generates havoc in an ecosystem, as it has in Lake Victoria. The cichlids helped to maintain a proper balance between producers, consumers, and decomposers. However, their decline put the entire community of organisms living in Lake Victoria in jeopardy. Human tampering of the Lake Victoria community, although well intentioned, upset the delicate balance between predators and prey, between producers and consumers. In this chapter, we examine some of the complex relationships among organisms that live together and interact dynamically with one another.

Communities of Organisms

46

Learning Objectives

After you have studied this chapter you should be able to:

1. Distinguish between a community and an ecosystem.
2. Characterize producers, consumers, and decomposers, and give the primary function of each in a community.
3. Define predation and describe the effects of natural selection on predator–prey relationships.
4. Explain symbiosis and distinguish among mutualism, commensalism, and parasitism.
5. Define ecological niche and distinguish between an organism's fundamental niche and its realized niche.
6. Summarize the concept of competitive exclusion.
7. Give several examples of limiting factors and discuss how they might affect an organism's ecological niche.
8. Summarize the main determinants of species diversity in a community.
9. Define ecological succession and distinguish between primary succession and secondary succession.
10. Describe the stages of secondary succession in an abandoned field.

Key Concepts

- ☐ A biological community consists of different populations of organisms that interact and live together. Each organism plays one of three main roles in community life: producer, consumer, or decomposer.
- ☐ The organisms that comprise a community interact with one another in many ways. Species compete with one another for food, water, living space, and other resources. Organisms kill and eat other organisms. Species form symbiotic associations with one another.
- ☐ Each species confronts the challenge of survival in its own unique way. The distinctive life style and role of an organism in a community is its ecological niche. An organism's ecological niche takes into account all aspects of its existence, that is, all of the physical, chemical, and biological factors that the organism needs to survive, remain healthy, and reproduce.
- ☐ Organisms are potentially able to exploit more resources and play a broader role in the life of their community than they actually do. Competition from other organisms constrains the actual ecological niche to narrower dimensions than it theoretically could assume.

Key Concepts, continued

☐ Succession is the orderly replacement of one community by another. Primary succession begins in an area that has not previously been inhabited, whereas secondary succession begins in an area where there was a preexisting community and a well-formed soil.

ORGANISMS LIVE TOGETHER IN COMMUNITIES AND ECOSYSTEMS

As you may recall from the previous chapter, the term *community* has a far broader sense in ecology than in everyday speech. For the biologist, a **community** is an association of organisms of different species living and interacting together. Thus you, your dog, and the fleas on your dog are all members of the same biological community! You could also add cockroaches, silverfish, dandelions, grasses, maple trees, and much more to the list.

Communities vary greatly in size, do not have precise boundaries, and are rarely completely isolated. They interact with and influence one another in countless ways—even when that interaction is not readily apparent. Furthermore, communities are nested within one another like Chinese boxes; that is, there are communities within communities. A forest is a community, but so is a rotting log in that forest. The log contains bacteria, fungi, slime molds, worms, insects, and perhaps even mice. The microorganisms living within the gut of a termite in the rotting log also form a community. On the other end of the scale, the entire living world can be considered a community.

Living things exist in a nonliving environment that is as essential to their lives as their interactions with other living organisms. Minerals, air, water, and sunlight are just as much a part of a honeybee's environment, for example, as the flowers that it pollinates and from which it takes nectar. A biological community and its nonliving environment together comprise an **ecosystem.** Although the living community is emphasized in this chapter, communities and their physical environments are inseparably linked together.

COMMUNITIES CONTAIN PRODUCERS, CONSUMERS, AND DECOMPOSERS

The organisms of a community can be divided into three categories based on how they get their nourishment: producers, consumers, and decomposers (Fig. 46–1). Most communities contain representatives of all three groups, which interact extensively with one another.

Sunlight is the source of energy that powers almost all life processes on Earth. **Producers,** also called **autotrophs** (from the Greek words *auto,* meaning "self," and *tropho,* meaning "nourishment") manufacture complex organic molecules from simple inorganic substances (generally carbon dioxide and water) usually using the energy of sunlight to do so. In other words, producers perform the process of photosynthesis. By incorporating the chemicals that they manufacture into their own biomass (living material), producers become potential food resources for other organisms. While plants are the most significant producers on land, algae and cyanobacteria are important producers in aquatic environments.

Animals, animal-like protists, and a very few predatory fungi and plants are **consumers,** that is, they use the biomass of other organisms as a source of food energy and body-building materials. Consumers are also called **heterotrophs** (from the Greek words *heter,* meaning "different," and *tropho,* meaning "nourishment"). Consumers that eat producers are called **primary consumers,** which usually means that they are exclusively **herbivores** (plant eaters). Grasshoppers and deer are examples of primary consumers. **Secondary consumers** eat primary consumers and include flesh-eating **carnivores,** which consume other animals. Lions and spiders are examples of carnivores. Other consumers, called **omnivores,** eat a variety of organisms, both plant and animal. Bears, pigs, and humans are examples of omnivores.

Some consumers, called **detritus feeders** (or **detritivores**), consume **detritus,** which is dead organic matter that includes animal carcasses, leaf litter, and feces. Detritus feeders are especially abundant in aquatic environments, where they burrow in the bottom muck and consume the organic matter that collects there. Earthworms are terrestrial detritus feeders, as are termites and maggots (the larvae of flies). Detritus feeders work together with decomposers to destroy dead organisms and waste products. An earthworm, for example, actually eats its way through the soil, digesting much of the organic matter the soil may contain. Earthworms also aerate the soil and redistribute its minerals and organic matter by their extensive tunneling.

Many consumers do not fit readily into a single category (herbivore, carnivore, omnivore, or detritivore).

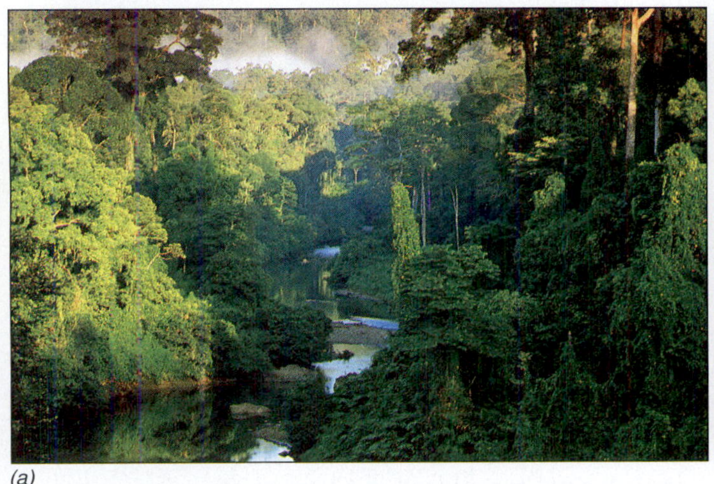
(a)

(b)

(c)

FIGURE 46–1 Producers, consumers, and decomposers. (***a***) The trees and other vegetation of the tropical rain forest are producers. (***b***) The three-toed sloth is a consumer in the tropical rainforest community. It eats mostly leaves; other consumers eat meat. (***c***) Decomposers such as this cup fungus reduce dead organisms to their mineral constituents, plus carbon dioxide and water. (*a*, Frans Lanting/Minden Pictures; *b*, Visuals Unlimited/A. Kerstitch; *c*, James L. Castner)

These organisms modify their food preferences to some degree as the need arises.

Decomposers (also called **saprophytes**) are heterotrophs that break down organic material and use the decomposition products to supply them with energy. They typically release simple inorganic molecules such as carbon dioxide and mineral salts that can then be reused by producers. Bacteria and fungi are important examples of decomposers. Dead wood, for example, is invaded first by sugar-metabolizing fungi that consume the wood's simple carbohydrates such as glucose or maltose. When these carbohydrates are exhausted, fungi, often aided by termites (with symbiotic bacteria in their guts), complete the digestion of the wood by breaking down cellulose, a complex carbohydrate that is the main component of wood.

Communities contain a balanced representation of all three ecological categories of organisms—producers, consumers, and decomposers. Producers and decomposers have indispensable roles in ecosystems. Producers provide both food and oxygen for all life. Decomposers are also necessary for the long-term survival of any community, because without them, dead organisms and waste products would accumulate indefinitely. Without decomposers, elements such as potassium, nitrogen, and phosphorus would remain permanently in dead organisms and therefore would be unavailable for use by new generations of living things. Consumers also play an important role in the balance of communities, as seen by the adverse changes caused by the decline of the cichlids in Lake Victoria (see Chapter Opener).

LIVING ORGANISMS INTERACT IN A VARIETY OF WAYS

No organism exists independent of other living things. The producers, consumers, and decomposers of a community interact with one another in a variety of complex ways, and each forms associations with other organisms. Three main types of interactions occur among species in

CONCEPT CONNECTIONS

Predation ⟷ *Animal Size* ⟷ *Evolution*

One evolutionary trend thought to be related to predation is a gradual increase in size of both predator and prey. Most predators are larger than their prey, and as a result, they are able to catch and kill their prey more easily. (Those few predators that are smaller than their prey overcome the size disadvantage by social behavior, that is, by hunting in packs.) In addition, because no species is completely free from the risk of predation, an increase in body size of the prey species offers some protection against predators. For example, adult whales are normally safe from attack by sharks and other predaceous fish. Thus, the selective pressures that accompany predation tend to favor an increase in size of both predator and prey. It is thought that the large size attained by both carnivorous dinosaurs (predators) and herbivorous dinosaurs (their prey) was the result of this evolutionary trend toward larger size.

Other advantages unrelated to predation exist for large body size. For example, larger animals can move more efficiently (in terms of energy consumption) than can smaller animals. Of course, a large animal has a greater energy requirement overall and must respire more than does a small animal. However, the amount of respiration per gram of body weight is less in large animals than it is in small animals.*

Despite the relative advantages of large size, animals occur in a wide range of sizes, and many small animals are extremely successful in their modes of life. In fact, most animal species that exist today have a small body size. Furthermore, the evolutionary trend toward *decreasing* size has been an important part of the evolution of certain animal groups, such as hummingbirds, shrews, and mites. Animal size as it relates to cell size and multicellularity was discussed in Chapter 31.

* A comparison with cars and buses may help to clarify this observation. A bus requires more gasoline to drive a given distance than does an automobile. However, people commuting to work in a bus will consume significantly less gasoline than the same number of people commuting in several compact cars equivalent to the weight of the bus.

a community: competition, predation, and symbiosis. Chapter 45 discussed competition (for resources within a population) and predation (as it relates to population ecology). We now examine how predation and symbiosis lead to diverse adaptations when species interact with one another. Competition between species will be discussed later in the chapter.

Natural Selection Shapes Both Prey and Predator

Predation is the consumption of one species, the **prey,** by another, the **predator.** It includes both animals eating other animals and animals eating plants. Predation has resulted in a coevolutionary "arms race," with the evolution of predator strategies—more efficient ways to catch prey—and prey strategies—better ways to escape the predator. A predator that is more efficient at catching prey exerts a strong selective force on its prey, which over time may evolve some sort of countermeasure. The countermeasure acquired by the prey in turn acts as a strong selective force on the predator. This interdependent evolution between two interacting species is known as **coevolution** (introduced in Chapter 29).

Pursuit and Ambush Are Two Predator Strategies

The brown pelican sights its prey—a fish—while in flight. Less than two seconds after diving into the water, it has its catch (Fig. 46–2). Killer whales, which hunt in packs, often herd salmon or tuna into a cove so that they are easier to catch. Any trait that increases hunting efficiency, such as the speed of a brown pelican or the cunning of killer whales, favors predators that pursue their prey. Because these carnivores must be able to process information quickly during the pursuit of prey, their brains are generally larger (relative to body size) than those of the prey they pursue.

FIGURE 46–2 Pursuit of prey requires speed. A brown pelican alights in the water having just caught a fish. (SharkSong/M. Kazmers/Dembinsky Photo Associates)

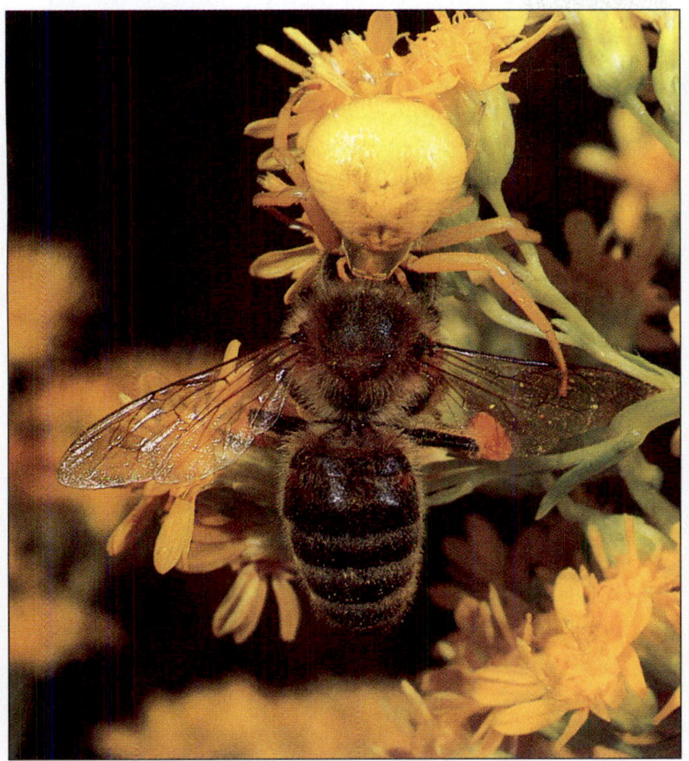

FIGURE 46–3 Ambush, an effective strategy for predators, relies on surprising the prey. A goldenrod spider attacks its insect prey. Its camouflage coloration enables the spider to blend into its surroundings (goldenrod flowers). (Patti Murray)

FIGURE 46–4 The common milkweed (*Asclepias syriaca*) is protected by its toxic chemicals. Its leaves are poisonous to most herbivores except monarch caterpillars (shown) and a few other insects. (Patti Murray)

Ambush is another effective way to catch prey. The goldenrod spider, for example, is the same color as the white or yellow flowers in which it hides (Fig. 46–3). This camouflage keeps unwary insects that visit the flower for nectar from noticing the spider until it is too late. Predators that are able to *attract* prey are particularly effective at ambushing. For example, a diverse group of deep sea fishes known as angler fishes, possess rodlike luminescent lures to attract food (see Fig. 48–18, page 984).

Chemical Protection Is an Effective Plant Defense Against Herbivores

Plants cannot escape predators by fleeing, but they possess a number of adaptations that protect them from being eaten. The presence of spines, thorns, tough leathery leaves, or even thick wax on leaves discourages foraging herbivores from grazing (see Fig. 27–*12a*, page 536).

Other plants produce an array of protective chemicals that are unpalatable or even toxic to herbivores. The active ingredients in such plants as marijuana, opium poppy, tobacco, and peyote cactus, for example, are thought to discourage the foraging of herbivores.

Milkweeds are an excellent example of the biochemical coevolution between plants and herbivores (Fig. 46–4). Milkweeds produce alkaloids and cardiac glycosides, chemicals that are poisonous to all animals except for a small group of insects. During the course of evolution, these insects acquired the ability to either tolerate or metabolize the milkweed toxins. As a result, they can eat milkweeds without being poisoned. These insects avoid competition from other herbivorous insects since few other insects are able to tolerate milkweed toxins. Predators also learn to avoid these insects, which accumulate the toxins in their tissues and are usually brightly colored to announce that fact. The black, white, and yellow banded caterpillar of the monarch butterfly (discussed later in this chapter) is an example of a milkweed feeder.

Animals Possess a Variety of Defensive Adaptations

Many animals, such as meadow voles and woodchucks, flee from predators by rapidly running home. Others have mechanical defenses (for example, the barbed quills

FIGURE 46–5 **Warning coloration.** The poison arrow frog (*Dendrobates tinctorius*) advertises its poisonous nature with its conspicuous coloring, warning away would-be predators. (Animals Animals © 1995 Michael Fogden)

of a porcupine and the shell of a pond turtle). Some animals live in groups—for example, a herd of antelope, colony of honeybees, school of anchovies, or flock of pigeons. Because the group has so many eyes, ears, and noses watching, listening, and smelling for predators, this social behavior decreases the likelihood of a predator catching one of them unawares.

Chemical defenses are also common among animal prey. The South American poison arrow frog (*Dendrobates*) has poison glands in its skin. Its bright yellow **warning coloration** prompts avoidance by experienced predators (Fig. 46–5). Snakes or other animals that have tried to eat a poisonous frog do not repeat their mistake! Other examples of warning coloration occur in the striped skunk, which sprays acrid chemicals from its anal glands, and the bombardier beetle, which sprays harsh chemicals at potential predators (see Fig. 7–11, page 142).

Some animals hide from predators by blending into their surroundings. Such camouflage is often enhanced by the animal's behavior. Many examples of camouflage come readily to mind (Fig. 46–6). Certain caterpillars resemble

(a)

(b)

(c)

FIGURE 46–6 **Examples of camouflage.** (***a***) Geometrid larvae are caterpillars that resemble twigs when resting on a plant or other object. Can you find the caterpillar? (***b***) Different katydid species have various forms of protective coloration to blend into their surroundings. The wings of this katydid even have venation patterns similar to leaves. Some katydids are mottled to match partly dead leaves; still others are perfectly camouflaged when resting on rain forest tree trunks. (***c***) The bay pipefish is closely related to the seahorse. Most species have thin, narrow bodies from 2.5–45 cm (1–18 in) in length. In addition to its protective coloration, its habit of holding its body in a position that resembles waving eel grass or algae aids in its camouflage. (*a*, *b*, James L. Castner; *c*, Doug Wechsler)

(a)

(b)

FIGURE 46–7 Batesian mimicry. In this example, a fly is the mimic and a yellowjacket wasp is the model. (***a***) Few would want to get close enough to this insect to discover that it is actually a fly. (***b***) A genuinely noxious insect, the yellowjacket wasp. (*a, b,* James L. Castner)

twigs so closely that you would never guess they are animals—until they move. Some katydids resemble leaves not only in color but in the pattern of veins in their wings. Pipefish have almost perfect camouflage coloration in green eel grass. Such camouflage has been preserved and accentuated by means of natural selection (Chapter 17).

Sometimes a harmless or edible species (called a *mimic*) is protected from predation by its resemblance to a species that is dangerous in some way (called a *model*). Such a strategy is called **batesian mimicry.** Numerous examples of batesian mimicry exist. For example, a harmless moth may look so much like a bee or wasp that predators avoid it (Fig. 46–7). Even a biologist would hesitate to pick it up!

In **mullerian mimicry,** different species, all of which are poisonous, harmful, or distasteful, resemble one another. Although their harmfulness protects them as individual species, their similar coloration works as an added advantage. Potential predators can more easily learn their common warning coloration than if each species had its own distinctive coloration. Viceroy and monarch butterflies are currently viewed as an example of mullerian mimicry (see Focus On the Process of Science: Batesian Butterflies Disproved).

Symbiosis Involves Close Associations in which One Species Usually Lives on or in the Other Species

Symbiosis is any intimate, long-term relationship or association between two or more species. The partners of a symbiotic relationship, called **symbionts,** may benefit from, be unaffected by, or be harmed by the relationship. The thousands, or even millions, of symbiotic associations in nature are all products of coevolution. There are three kinds of symbiosis: mutualism, commensalism, and parasitism.[1]

In Mutualism, Benefits Are Shared

Mutualism is a symbiotic relationship in which both partners benefit. The association between nitrogen-fixing bacteria of the genus *Rhizobium* and legumes (plants such as peas, beans, and clover) is an example of mutualism (Chapter 21). Nitrogen-fixing bacteria, which live in nodules in the roots of legumes, supply the plants with all of the nitrogen they need[2] (see Fig. 47–3, page 950). The legumes supply sugar to their bacterial symbionts.

Another example of mutualism is the association between reef-building coral animals and microscopic algae (see Chapter 24 Opener). These symbiotic algae, which are called zooxanthellae, live inside cells of the coral where they photosynthesize and provide the animal with

[1] Anton de Bary, an important 19th century biologist, coined the word symbiosis and defined it in a very broad way (a living together of different organisms). de Bary placed mutualism, commensalism, and parasitism under the concept of symbiosis. Although later workers often equated mutualism and symbiosis, most biologists today define symbiosis more broadly in recognition of de Bary's contributions.

[2] The soil, which is often deficient in nitrogen, is the source of nitrogen for plants without nitrogen-fixing bacteria.

(Text continues on page 935)

FOCUS ON *The Process of Science*

BATESIAN BUTTERFLIES DISPROVED

The monarch butterfly (*Danaus plexippus*) is an attractive insect found throughout much of North America (see figure). As a caterpillar, it feeds exclusively on milkweed leaves. The milky white liquid produced by the milkweed plant contains poisons that apparently do not harm the insect, but remain in its tissues for life. When a young bird encounters and eats its first monarch butterfly, it gags until the insect is regurgitated. Thereafter, the bird avoids eating the distinctively-marked insect.

Many people confuse the viceroy butterfly (*Limenitis archippus*) (see figure) with the monarch. The viceroy, which is found throughout most of North America, is approximately the same size, and the coloration and markings of its wings are almost identical to those of the monarch. As caterpillars, viceroys eat willow and poplar leaves, which do not contain poisonous substances.

During the past century, it was thought that the viceroy butterfly was a tasty food for birds, but that its close resemblance to monarchs gave it some protection against being eaten. In other words, birds that had learned to associate the distinctive markings and coloration of the monarch butterfly with its bad taste tended to avoid viceroys because they were similarly marked. The viceroy butterfly was therefore considered a classic example of batesian mimicry.

In 1991, two biologists—David Ritland and Lincoln Brower—at the University of Florida decided to test the long-held notion that birds like the taste of viceroys but avoid eating them because of their resemblance to monarchs. They pulled the wings off several different kinds of butterflies—including monarchs, viceroys, and other tasty species—and fed the seemingly identical wingless bodies to red-winged blackbirds. The results were surprising: monarchs and viceroys were *equally* distasteful to red-winged blackbirds.

(a)

(b)

Mullerian mimicry. Monarch (***a***) and viceroy (***b***) butterflies. (*a*, John Gerlach/Dembinsky Photo Associates; *b*, Sharon Cummings/Dembinsky Photo Associates)

As a result of this work, scientists are re-evaluating the evolutionary significance of different types of mimicry. Monarchs and viceroys appear to be an example of mullerian mimicry, in which two or more different species that are distasteful or poisonous have come to resemble one another during the course of evolution. This likeness provides an adaptive advantage because predators learn quickly to avoid all butterflies with the coloration and markings of monarchs and viceroys. As a result, fewer butterflies of either species dies, and more individuals survive to reproduce.

The butterfly study provides us with a useful reminder about the nature of science. Expansion of knowledge in science is an ongoing enterprise, and newly acquired evidence always takes precedence over older ideas (Chapter 2). Thus, scientific knowledge is not static, but continually changing.

carbon compounds as well as oxygen. Zooxanthellae have a stimulatory effect on the growth of corals, which deposit calcium carbonate skeletons around their bodies much faster when the algae are present. The coral in turn supplies its zooxanthellae with waste products such as ammonia, which the algae use to make nitrogen compounds for both partners.

Mycorrhizae are mutualistic associations between fungi and the roots of about 80% of all plants. The fungus absorbs essential minerals, especially phosphorus, from the soil and provides them to the plant. In return, the plant provides the fungus with sugars produced by photosynthesis. Plants grow more vigorously in the presence of mycorrhizae (see Fig. 22–8, page 411), and they are better able to tolerate environmental stresses such as drought and high soil temperatures. Often plants cannot maintain themselves if the fungi with which they normally form such critically important mutualistic associations are not present.

Commensalism Is Taking without Harming

Commensalism is a type of symbiosis in which one organism benefits and the other one is neither harmed nor helped. One example of commensalism is the relationship between two types of insects, silverfish and army ants. Certain types of silverfish move along in permanent association with the marching columns of army ants and share the food caught in their raids. The army ants gain no apparent benefit (or harm) from the silverfish. Another example of commensalism is the relationship between a tropical tree and its **epiphytes,** which are smaller plants that live attached to the bark of its branches (Fig. 46–8). The epiphyte anchors itself onto the tree but does not obtain nutrients or water directly from the tree. Its location on the tree enables an epiphyte to obtain adequate light, water (as rainfall dripping down the branches), and required minerals (washed out of the tree's leaves by rainfall). Thus, the epiphyte benefits from the association, whereas the tree remains largely unaffected.

Parasitism Is Taking at Another's Expense

Parasitism is a symbiotic relationship in which one member, the **parasite,** benefits and the other, the **host,** is adversely affected. The parasite obtains nourishment from its host, and although a parasite may weaken its host, it rarely kills it. (A parasite would have a difficult life if it kept killing off its hosts!) Some parasites, such as ticks, live outside the host's body; other parasites, such as tapeworms, live within the host.

When a parasite causes disease and sometimes the death of a host, it is called a **pathogen.** For example, humans sometimes get histoplasmosis, a serious (often fatal) disease caused by a fungus. Humans are infected when they breathe the spores of the fungus. The spores grow in the lungs, causing chronic coughing and fever. Eventually, the disease progresses to other organs of the body. The spores of the fungus are common in soils that have a high concentration of bird droppings, and the disease is more prevalent in warm tropical regions of the world.

Crown gall disease, which is caused by a bacterium, occurs in many different types of plants and results in millions of dollars of damage each year to ornamental and agricultural plants (Chapter 16). Crown gall bacteria, which live in the soil, enter plants through small wounds

FIGURE 46–8 Commensalism. Epiphytes are small plants that grow attached to the body of a tree. (Earth Scenes ©1995 Breck P. Kent)

such as those caused by insects. They cause galls (tumor-like growths), often at the crown (between the stem and the roots) of a plant. Although plants seldom die from crown gall disease, they are weakened, grow more slowly, and often succumb to other pathogens.

Many parasites do not cause disease. For example, humans can become infected with the pork tapeworm by eating poorly cooked pork that is infested with immature tapeworms. Once the tapeworm is inside the human digestive system, it attaches itself to the wall of the small intestine and grows rapidly by absorbing nutrients that pass through the small intestine. A single pork tapeworm that lives in the human digestive tract does not usually cause any noticeable symptoms. Some weight loss may be associated with a multiple infestation.

THE NICHE DESCRIBES AN ORGANISM'S ROLE IN THE COMMUNITY

We have seen that a diverse assortment of organisms inhabits each community and that these organisms obtain nourishment in a variety of ways. We have also examined some of the ways that living things interact to form interdependent relationships within the community. A biological description of an organism includes (1) whether it is a producer, consumer, or decomposer; (2) whether it is a predator and/or prey; and (3) the kinds of associations it forms. Other details are needed, however, to provide a complete picture.

Every species has its own distinctive role within the structure and function of an ecosystem. We call this role its ecological **niche** (Fig. 46–9). An ecological niche takes into account *all* aspects of the organism's existence—that is, all physical, chemical, and biological factors that the organism needs to survive, remain healthy, and reproduce. Among other things, the niche includes the physical surroundings in which an organism lives (its **habitat**). An organism's niche also encompasses what it consumes, what consumes it, what organisms it competes with, and how it interacts with and is influenced by the nonliving components of its environment (for example, light, temperature, and moisture). The niche, then, represents the totality of an organism's adaptations, its use of resources, and the life-style to which it is suited. Thus, a complete description of an organism's ecological niche involves numerous dimensions.

There are two aspects to an organism's niche in its community: one relates to the role an organism *could* play in the community, and the other relates to the role that it actually *fulfills.* The ecological niche of an organism may be far broader potentially than it is in actuality. Put differently, an organism is usually capable of utilizing much more of its environment's resources or of living in a wider assortment of habitats than it actually does. The potential ecological niche of an organism is its **fundamental niche,** but various factors such as competition with other species may exclude it from part of its fundamental niche. Thus, the life-style that an organism actually pursues and the resources that it actually utilizes comprise its **realized niche.**

FIGURE 46–9 Some songbird ecological niches. Each of these species of Dendroica, common warblers, spends most of its feeding time in different portions of the trees it frequents and also consumes somewhat different insect food. The colored regions indicate where each species spends at least half its foraging time. (After MacArthur.)

(a)

(b)

An example may help to make this distinction clear. The green anole (*Anolis carolinensis*), a lizard native to Florida and other southeastern states, perches on trees, shrubs, walls, or fences during the day waiting for insect and spider prey (Fig. 46–10*a*). In former years, these little lizards were widespread in Florida. Several years ago, however, a related species, the brown anole (*Anolis sagrei*), was introduced from Cuba into southern Florida and quickly became common (Fig. 46–10*b*). Suddenly, the green anoles became rare—apparently driven out of their habitat by the competition from the slightly larger brown anoles. Careful investigation disclosed, however, that green anoles were still around. They were now confined largely to the vegetation in wetlands and to the foliated crowns of trees, where they were less easily seen.

The habitat portion of the green anole's fundamental niche includes the trunks and crowns of trees, exterior house walls, and many other locations. Where they became established, Cuban anoles were able to drive green anoles out from all but the tree crowns, so that their realized niche became much smaller as a result of **interspecific competition** (competition between different species; Fig. 46–10*c,d*). Because all natural communities consist of numerous species, many of which compete to some extent, the complex interactions among them produce each one's realized niche.

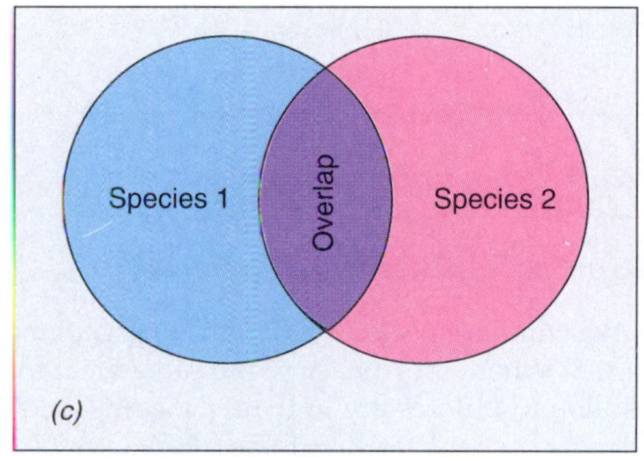

(c)

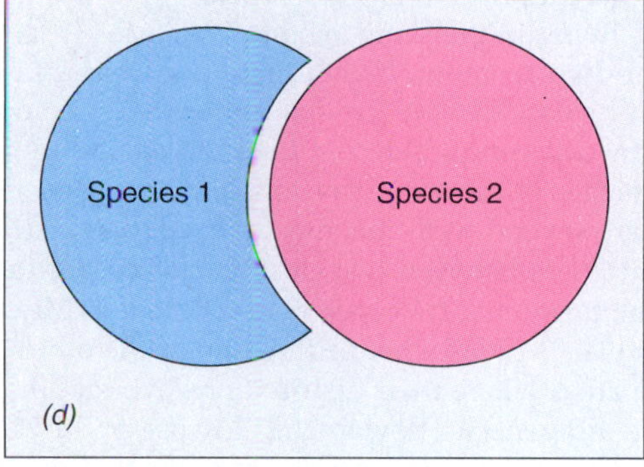

(d)

FIGURE 46–10 Competition can restrict an organism's realized niche. (***a***) The green anole is native to Florida. (***b***) The brown anole was introduced to Florida. (***c***) The fundamental niches of the two lizards overlap. Species 1 represents the green anole and species 2 represents the brown anole. (***d***) The brown anole was able to out-compete the green anole, restricting its niche. (*a*, Runk/Schoenberger, from Grant Heilman; *b*, Connie Toops)

Competition between Two Species with Identical or Similar Niches Leads to Competitive Exclusion

When two species are very similar, as are the green and Cuban anoles, their fundamental niches may overlap. However, no two species can indefinitely occupy the

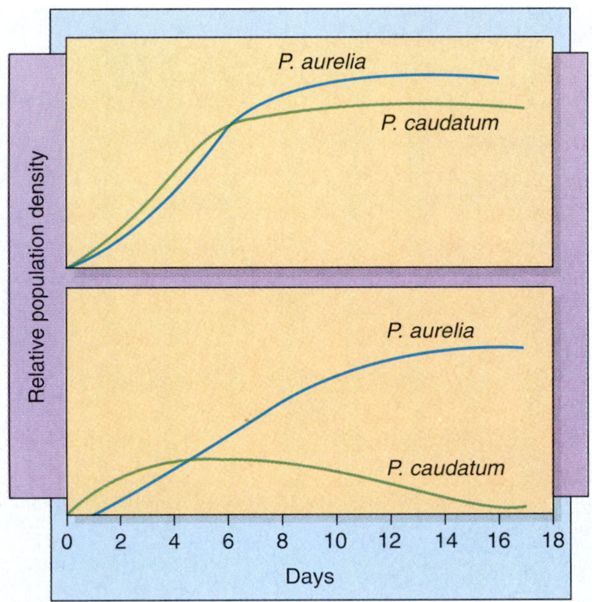

FIGURE 46–11 Competition among *Paramecium*. The top graph shows how a population of each species of *Paramecium* grows in a single-species environment; the bottom graph shows how they grow when in competition with each other. (After Gause)

same niche in the same community because **competitive exclusion** eventually occurs. In competitive exclusion, one species is excluded from a niche by another as a result of interspecific competition. Although it is possible for different species to compete for some necessary resource without being total competitors, two species with absolutely identical ecological niches cannot coexist. Coexistence *can* occur, however, if the overlap in the two species' niches is reduced. In the lizard example, direct competition between the two species was reduced as the Cuban anole excluded the green anole from most of its former physical habitat until the only place that remained open to it was the tree crowns.

The initial evidence that interspecific competition determines a species' realized niche came from a series of experiments conducted by the Russian biologist Gause in 1934 (Chapter 45). In one study, Gause grew two species of *Paramecium* (a type of protozoa)—*P. aurelia* and the larger *P. caudatum* (Fig. 46–11). When grown in separate test tubes, each species quickly increased its population to a high level, which it maintained for some time thereafter. When grown together, however, only *P. aurelia* thrived. *P. caudatum* dwindled and eventually died out. Under different sets of culture conditions, *P. caudatum*

CONCEPT CONNECTIONS

Competition ⟷ *Natural Selection*

Sometimes two similar species overlap in their geographical distribution. Where they occur in the same area, the two species tend to differ more in their structural, ecological, and behavioral characteristics than where each occurs in separate geographical areas. Such divergence in traits in two species living in the same geographical area is known as **character displacement.** It is thought that character displacement prevents the two groups from directly competing, since the differences between them give them different ecological niches in the same environment.

There are several well-documented examples of character displacement between two closely related species. For example, the flowers of two species of *Solanum* in Mexico are very similar in areas where either one or the other occurs, but in areas where their distributions overlap, there is a noticeable difference in flower size. Because of this difference, the flowers of the two species are pollinated by different kinds of bees. In other words, character displacement reduces interspecific competition, in this case for the same animal pollinator.

The body size and bill size of Darwin's finches (Chapter 17) provide another example of character displacement (see figure). On large islands in the Galapagos where *Geospiza fortis* and *G. fuliginosa* occur together, their bill depth is distinctive: *G. fuliginosa* has a smaller bill depth that enables it to crack small seeds, whereas *G. fortis* has a larger bill depth that enables it to crack medium-sized seeds. However, *G. fortis* and *G. fuliginosa* are also found separately on smaller islands. Where this occurs their bill depths tend to be the same intermediate size, perhaps because there is no competition from the other species.

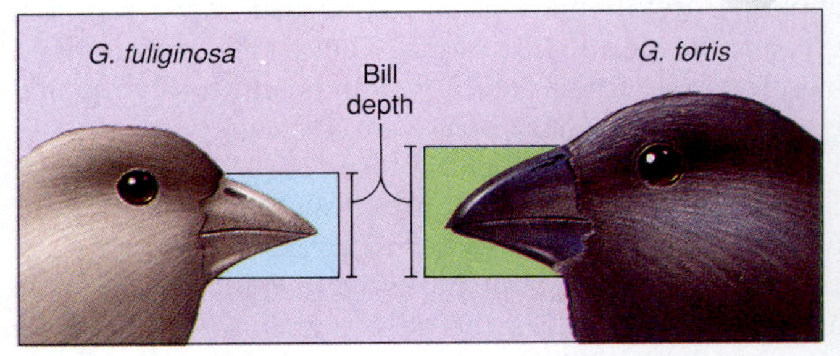

Character displacement in two species of finches from the Galapagos, *Geospiza fuliginosa* and *G. fortis*. Where the two species are found on the same island, *G. fuliginosa* has a smaller average bill depth than *G. fortis*.

prevailed over *P. aurelia*. Gause interpreted this to mean that, although one set of conditions favored one species, a different set favored the other. Nonetheless, because both species were similar, given time, one or the other would eventually triumph at the other's complete cost.

Sometimes there are apparent contradictions to the competitive exclusion principle. In Florida, for example, native and introduced fish seem to coexist in identical niches. Similarly, botanists have observed closely competitive plant species in the same location. Although such situations seem to contradict the concept of competitive exclusion, the realized niches of these organisms may differ significantly in some way that scientists do not yet understand.

Limiting Factors Restrict an Organism's Ecological Niche

The environmental factors that actually determine a species' ecological niche can be extremely difficult to identify. For this reason, the concept of the ecological niche is largely abstract, although some of its dimensions can be experimentally determined. Whatever environmental variable tends to restrict the ecological niche of an organism is called a **limiting factor.** (Recall from Chapter 45 that environmental factors can also limit the population size of a species.)

What factors actually determine the ecological niche of a species? A niche is basically determined by the total of a species' structural, physiological, and behavioral adaptations. Such adaptations determine, for example, a species' tolerance for environmental extremes. If any feature of its environment lies outside the bounds of its tolerance, then the organism cannot live there. Just as you would not expect to find a cactus living in a pond, you would not expect water lilies in a desert.

Most of the limiting factors that have been studied are simple variables such as the mineral content of soil, the extremes of temperature, and the amount of precipitation. Such investigations have disclosed that any factor that exceeds an organism's tolerance or that is present in quantities smaller than the minimum required limits the occurrence of that organism in a community. By their interaction, such factors help to define an organism's ecological niche.

Limiting factors that affect one part of an organism's life may apply throughout its life cycle. For instance, although adult blue crabs can live in almost fresh water, they cannot become permanently established in such areas because their larvae cannot tolerate fresh water. Similarly, the ring-necked pheasant, a popular game bird, has been widely introduced in North America but does not survive in the southern United States. The adult birds do well, but the eggs cannot develop properly in the high temperatures that are found there.

Communities Vary in Species Diversity

What determines the number of species in a community? There seems to be no single answer, but several factors appear significant. Species diversity is related to the abundance of potential ecological niches. An already complex community offers a greater variety of potential ecological niches than does a simple community. It may become even more complex if organisms potentially capable of filling those niches evolve or migrate into the community.

Species diversity is inversely related to geographical isolation of a community. Isolated island communities are usually much less diverse than are communities located on continental areas with similar environments. This is due partly to the difficulty encountered by many species in reaching and colonizing the island. Also, sometimes species become locally extinct as a result of random events. In isolated habitats such as islands or mountain tops, extinct species cannot be readily replaced. Isolated areas are likely to be small and to possess fewer potential ecological niches.

Generally, species diversity is inversely related to the environmental stress of a habitat. Only those species capable of tolerating extreme environmental conditions will be present in an environmentally stressed community. Thus, the species diversity of a polluted stream is low compared to that of a nearby pristine stream.

Similarly, the species diversity of high-latitude (further from the equator) communities exposed to harsh climates is less than that of lower-latitude (closer to the equator) communities with milder climates. Although the equatorial countries of Colombia, Ecuador, and Peru occupy only 2% of the Earth's land, they contain an astonishing 40,000 native plant species. The United States and Canada, with a significantly larger land area, possess a total of 700 native species of plants. Ecuador alone contains more than 1300 native species of birds—twice as many as the United States and Canada combined.

Species diversity is usually greater at the margins of distinctive communities than in their centers. This is because the **edges** (transitional zones where two or more communities meet) contain all or most of the ecological niches of the adjacent communities. The change in species composition produced at the edges is known as the **edge effect.**

Species diversity is reduced when any one species enjoys a decided position of dominance within a community so that it is able to appropriate a disproportionate share of available resources, thus crowding out, or outcompeting, many other species. Recall how the dominance of the Nile perch has reduced species diversity in Lake Victoria.

Species diversity is greatly affected by geological history. Tropical rain forests are very old, stable communities that have undergone few climatic changes in the entire history of the Earth. During this time, myriad species evolved in tropical rain forests, having experienced few or no abrupt climatic changes that might have led to their extinction. In contrast, glaciers have repeatedly altered temperate and arctic regions during Earth's history. An area recently vacated by glaciers will have a low species diversity because few species will as yet have had a chance to enter it and become established.

SUCCESSION IS COMMUNITY CHANGE OVER TIME

A community of organisms does not spring into existence full-blown but develops gradually through a series of stages. The process of community development over time, which involves species in one stage being replaced by different species, is called **succession.** An area is initially colonized by certain species that are replaced over time by others, which themselves may be replaced much later by still others.

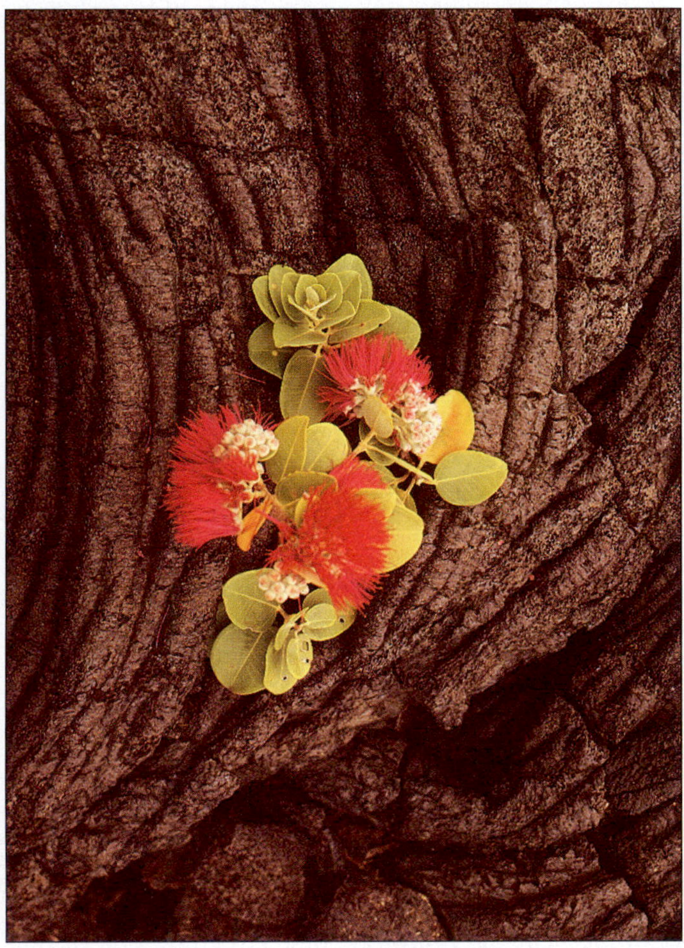

FIGURE 46–12 Primary succession. This view shows small plants growing on recently cooled volcanic lava in Hawaii. (© David Muench 1995)

Ecologists initially thought that succession inevitably led to a stable and persistent community, known as a **climax community.** But more recently, the traditional view has fallen out of favor. The apparent stability of a "climax" forest, for example, is probably the result of how long trees live relative to the human lifespan. It is now recognized that mature "climax" communities are not in a state of permanent equilibrium, but rather in a state of continual flux. A mature community changes in species composition and in the relative abundance of each species despite the fact that it retains a uniform appearance.

Succession is usually described in terms of the changes in the species composition of the vegetation (plants) of an area, although each successional stage also has its own characteristic kinds of animals and other organisms. The time involved in ecological succession is on the order of tens, hundreds, or thousands of years, not the millions of years involved in the evolutionary time scale.

Sometimes Communities Develop in a "Lifeless" Environment

Primary succession is the change in species composition over time in a habitat that has not previously been inhabited by organisms. No soil exists when primary succession begins. A bare rock surface such as recently formed volcanic lava (Fig. 46–12) and rock scraped clean by glaciers are examples of sites where primary succession might occur.

During Primary Succession on Bare Rock, the Rock Is Eventually Transformed into Soil

Although the details vary from one site to another, in primary succession on bare rock, one might first observe a community of lichens, which are dual organisms usually composed of a fungus and an alga or cyanobacterium (Chapter 22 and Fig. 46–13). Lichens are often the most important element in the **pioneer** community, which is the first community to appear in a succession. Lichens secrete acids that help to break the rock apart, which is how soil starts to form.

Over time, the lichen community may be replaced by mosses and drought-resistant ferns, followed in turn by tough grasses and herbs. Once sufficient soil has accumulated, grasses and herbs might be replaced by low shrubs, which in turn would be replaced by forest trees in several distinct stages. Primary succession on bare rock may take hundreds or thousands of years to proceed from a pioneer community to a forest community.

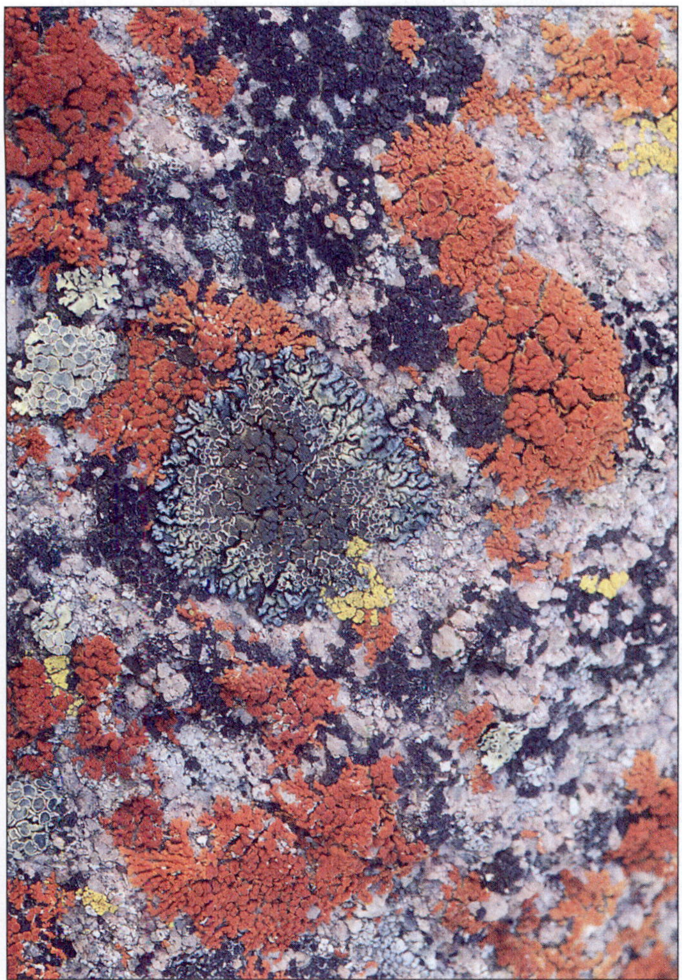

FIGURE 46–13 The role of lichens in primary succession. Some lichens grow as thin mats encrusted on rock. They are the first organisms to colonize bare rock. (Ed Reschke)

Primary Succession Also Occurs on Sand Dunes

Lake and ocean shores often have extensive sand dunes, which are deposited by wind and water. At first, these dunes are blown about by the wind. The sand dune environment is severe, with high temperatures during the day and low temperatures during the night. The sand may also be deficient in certain mineral nutrients needed by plants. As a result, few plants are able to tolerate the environmental conditions of a sand dune.

Grasses are common pioneer plants on sand dunes around the Great Lakes. As the grasses extend over the surface of the dune, their roots hold it in place, helping to stabilize the dune. At this point, mat-forming shrubs may invade, further stabilizing the dune. Much later, shrubs may be replaced by pines, which years later are replaced by oaks. Because the soil fertility remains low, oaks are rarely replaced by other forest trees.

Sometimes Communities Develop in an Environment where a Previous Community Existed

Secondary succession is the change in species composition over time in a habitat already substantially modified by a pre-existing community. Soil is already present at these sites. An open area caused by a forest fire (Fig. 46–14) and an abandoned field are common examples of sites in which secondary succession occurs.

Secondary succession on abandoned farmland has been studied extensively. Although it takes more than 100 years for secondary succession to occur at a single site, it is possible for a biologist to study old field succession in its entirety by observing different sites in the same area. By examining court records, the biologist can accurately determine when each field was abandoned.

Abandoned farmland in North Carolina, which has been studied extensively, is colonized by a predictable succession of communities. The first year after cultivation ceases, the field is dominated by crabgrass. During the second year, horseweed, a larger plant that outgrows the crabgrass, is the dominant species. Horseweed does not dominate more than one year, however, because decaying horseweed roots inhibit the growth of young horseweed seedlings. In addition, horseweed does not compete well with other plants that appear in the third year. During the third year after the last cultivation, other weeds—broomsedge, ragweed, and aster—become established. Typically, broomsedge out-competes aster because broomsedge is drought tolerant, whereas aster is not.

In years 5 to 15, the dominant plants in an abandoned field are young pines such as shortleaf pine and loblolly pine. Through the buildup of litter (pine needles and branches) on the soil, pines produce conditions that cause the earlier dominant plants to decline in importance. Over time, pines give up their dominance to hardwoods such as oaks. The replacement of pines by oaks depends primarily on the environmental changes produced by the pines. The pine litter causes soil changes, such as an increase in water-holding capacity, that are necessary in order for young hardwood seedlings to become established.

Animal Life Also Changes during Secondary Succession

As secondary succession proceeds, a progression of animal life follows the changes in vegetation. Although a

(a) (b)

FIGURE 46–14 Secondary succession after the Yellowstone fires of 1988. (***a***) Gray ash covers the forest floor after the fire. The trees, although dead, remain standing. (***b***) Less than one year later, in Spring 1989, young plants mark the beginning of secondary succession. Many of the dead trees have fallen over. The dominant plant at this stage is trout lily (*Erythronium*), which sprouts up rapidly after a fire because its underground parts are not killed by the fire. (*a*, Ted and Jean Reuther/Dembinsky Photo Associates; *b*, Stan Osolinski/Dembinsky Photo Associates)

few animals, for example, the short-tailed shrew, are found in all stages of abandoned farmland succession, most animals appear with certain stages and disappear with others. During the crabgrass and weed stages of secondary succession, the habitat is characterized by open fields that support grasshoppers, meadow voles, cottontail rabbits, and birds such as grasshopper sparrows and meadowlarks. As young pine seedlings become established, animals of open fields give way to animals commonly found in mixed herbaceous and shrubby habitats. Now white-tailed deer, white-footed mice, ruffed grouse, robins, and song sparrows are common, whereas grasshoppers, meadow mice, grasshopper sparrows, and meadowlarks disappear. As the pine seedlings grow into trees, animals of the forest replace those common in mixed herbaceous and shrubby habitats. Cottontail rabbits give way to red squirrels, and ruffed grouse, robins, and song sparrows are replaced by warblers and veeries. Thus, each stage of succession supports its own characteristic animal life.

Chapter Summary

I. A biological community consists of a group of organisms of different species that interact and live together. A community and its nonliving environment comprise an ecosystem.

II. The major roles of organisms in communities are those of producer, consumer, and decomposer.
   A. Producers are the photosynthetic organisms that are at the base of most food webs. They include plants and algae.
   B. Consumers are almost exclusively animals. They feed upon other organisms.
   C. Decomposers are bacteria and fungi that recycle the components of dead organisms and organic wastes by feeding on them.

III. Predation is the consumption of one species (the prey) by another (the predator).
   A. During coevolution between predator and prey, the predator evolves more efficient ways to catch prey, and the prey evolves better ways to escape the predator.
   B. Two effective predator strategies are pursuit and ambush.
   C. Plants possess a number of adaptations that protect them from being eaten, including spines, thorns, tough leathery leaves, and protective chemicals that are unpalatable or toxic to herbivores.
   D. Animals possess many strategies that help them avoid being killed and eaten.
      1. Many animals flee from predators, some have mechanical defenses, and some associate in groups.
      2. Animals that possess chemical defenses also exhibit warning coloration.
      3. Some animals hide from predators by blending into their surroundings.
      4. In batesian mimicry, a harmless or edible species resembles another species that is dangerous in some way. Predators avoid the mimic as well as the model.
      5. In mullerian mimicry, several different species—all of which are poisonous, harmful, or distasteful—resemble one another. Predators easily learn to avoid their common warning coloration.

IV. Symbiosis is any intimate association between two or more species.
   A. In mutualism, both partners benefit.
   B. In commensalism, one organism benefits and the other is unaffected.
   C. In parasitism, one organism (the parasite) benefits and the other (the host) is harmed.

V. The distinctive life-style and role of an organism in a community is its ecological niche. An organism's ecological niche takes into account all aspects of the organism's existence.
   A. Organisms are potentially able to exploit more resources and play a broader role in the life of their community than they actually do.
      1. The potential ecological niche for an organism is its fundamental niche, whereas the niche an organism actually occupies is its realized niche.
      2. Interspecific competition is one of the chief biological determinants of a species' ecological niche.
   B. It is thought that no two species can occupy the same niche in the same community for an indefinite period of time because competitive exclusion occurs. In competitive exclusion, one species is excluded by another as a result of competition for a resource in limited supply.
   C. An organism's limiting factors (such as the mineral content of soil, temperature extremes, and amount of precipitation) tend to restrict its ecological niche.

VI. Community complexity is related to a variety of factors.
   A. Community complexity is expressed in terms of species diversity.
   B. Species diversity is often great when there are many potential ecological niches, when a community is not isolated or severely stressed, at the edges of adjacent communities, and in communities with a long history.

VII. Succession is the orderly replacement of one community by another.
   A. Primary succession occurs in a habitat that has not previously been inhabited (for example, bare rock or shifting sand dunes).
   B. Secondary succession begins in an area where there was a pre-existing community and a well-formed soil (for example, abandoned farmland).

Selected Key Terms

batesian mimicry, p. 933
character displacement, p. 938
coevolution, p. 930
commensalism, p. 935
community, p. 928
competitive exclusion, p. 938
consumer, p. 928
decomposer, p. 929
detritus, p. 928
ecosystem, p. 928
edge effect, p. 939
fundamental niche, p. 936
habitat, p. 936
interspecific competition, p. 937
limiting factor, p. 939
mullerian mimicry, p. 933
mutualism, p. 933
niche, p. 936
parasitism, p. 935
pathogen, p. 935
predation, p. 930
primary consumer, p. 928
producer, p. 928
realized niche, p. 936
secondary consumer, p. 928
succession, p. 940
symbiosis, p. 933
warning coloration, p. 932

Post-Test

1. A/An ______________ is an association of different species living together in one area.
2. Ecologically speaking, mushrooms would be classified as ______________, and foxes would be classified as consumers.
3. ______________ eat meat, ______________ eat plants, and ______________ eat a variety of organisms.
4. Monarch and viceroy butterflies are an example of ______________ mimicry.
5. A symbiotic association in which organisms are beneficial to one another is known as ______________.
6. ______________, although adversely affecting their hosts, usually do not kill them.
7. The symbiotic association exemplified by silverfish and army ants that live together and share the food caught by the army ants is ______________.
8. An organism's ecological ______________ is the totality of its adaptations, use of resources, and the life-style to which it is fitted.
9. The ______________ niche is the ecological niche that an organism could potentially occupy.
10. Competition from other species helps to determine an organism's ______________ niche.
11. "Complete competitors cannot coexist" is a statement of the principle of ______________ ______________.
12. The ______________ ______________ signifies that species diversity is higher where two communities meet than at the center of either community.
13. Isolated communities such as islands are likely to be ______________ (more or less) diverse than communities that are adjacent to large areas of suitable habitat.
14. A succession of communities on bare rock is an example of ______________ succession.
15. Camouflage, toxic chemicals, and mimicry are examples of (a) symbiotic associations (b) adaptations of predators and prey (c) interspecific competition (d) parasitic adaptations.
16. A parasite obtains nourishment from a (a) pathogen (b) predator (c) host (d) producer (e) decomposer.
17. Mycorrhizae (a) are a symbiotic association between a fungus and the roots of a plant (b) are the pioneer stage in primary succession on bare rock (c) reside in corals where they photosynthesize and provide food for their hosts (d) cause root disease in many trees.
18. The change in communities over time is known as (a) evolution (b) succession (c) character displacement (d) commensalism (e) interspecific competition.
19. After how much time does pine become the dominant species during secondary succession on abandoned farmland? (a) 1 year (b) 2 years (c) 5–15 years (d) 100–250 years (e) 1000 years

Review Questions

1. How is a community different from an ecosystem?
2. How might one distinguish between a producer and a decomposer? A consumer and a decomposer? A producer and a consumer?
3. Give a specific example of each of the following: (a) camouflage, (b) chemical protection for a plant, (c) warning coloration, (d) camouflage of prey, and (e) batesian mimicry.
4. How is mutualism different from commensalism?
5. How do parasitism and commensalism differ?
6. Distinguish between an organism's habitat and its ecological niche.
7. Why is an organism's realized niche usually narrower, or more restricted, than its fundamental niche?
8. What is a limiting factor? Give at least two examples of limiting factors.
9. What important ecological concept did Gause demonstrate? Describe his experiment.
10. Define character displacement and give an example.

Thinking Critically

1. Some biologists have "rescued" a few cichlid species from Lake Victoria. These fishes are being maintained in aquaria throughout the world in the hope of someday reintroducing them into Lake Victoria. Do you think such reintroductions would be successful or unsuccessful? Why? What could biologists do to increase the likelihood of successful cichlid reintroductions?
2. How does predation differ from parasitism? How are they similar?
3. In what symbiotic relationships are humans involved?

4. Describe the ecological niche of humans. Do you think our realized niche has changed during the past 1000 years? Why or why not?
5. What kinds of environmental conditions trigger primary succession? Secondary succession?
6. What would happen if humans were to introduce polar bears to Antarctica? Should this be done? Why or why not?
7. The naturalist and early environmentalist John Muir once said, "When one tugs at a single thing in nature, he finds it attached to the rest of the world." What did he mean?

Recommended Readings

Baskin, Y., "Losing a Lake," *Discover,* March 1994. Focuses on the human link to Lake Victoria—the 30 million people who depend on the lake for food and a living.

DeVries, P.J., "Singing Caterpillars, Ants, and Symbiosis," *Scientific American,* October 1992. Fascinating symbiotic partnerships between ants and caterpillars protect the caterpillars from predators.

Gillis, A.M., "Sea Dwellers and Their Sidekicks," *Bioscience,* Vol. 43, No. 9, October 1993. Unusual mutualistic relationships between bioluminescent (light-producing) bacteria and squid or fish.

Grall, G., "Pillar of Life," *National Geographic,* Vol. 182, No. 1, July 1992. A complex community of organisms exists on wharf pilings in the Chesapeake Bay.

Kaufman, L., "Catastrophic Change in Species-Rich Freshwater Ecosystems," *Bioscience,* Vol. 42, No. 11, December 1992. The ecological demise of Lake Victoria.

Mohlenbrock, R.H., "Mount St. Helens, Washington," *Natural History,* June 1990. Secondary succession of areas devastated by the eruption of Mount St. Helens.

Tumlinson, J.H., W.J. Lewis, and L.E.M. Vet, "How Parasitic Wasps Find Their Hosts," *Scientific American,* March 1993. One way that wasps identify their caterpillar hosts is by recognizing chemicals produced by the plants on which caterpillars feed.

Wiley, J.P., Jr., "You Don't Have to Look Like a Fish to Succeed as One," *Smithsonian,* August 1992. Unusual and often bizarre adaptations of fish from South Australia.

A peregrine falcon. This bird of prey was almost extinct in the late 1960s as a result of pesticides concentrating in the food chain. (Dominique Braud/ Dembinsky Photo Associates)

Hard Times at the Top of the Food Chain

Pesticides, toxic chemicals that help control pests, have saved millions of human lives by killing disease-carrying insects and by increasing food crop yields. Modern agriculture depends on pesticides to produce blemish-free fruits and vegetables at a reasonable cost to consumers. However, pesticides also cause environmental and human health problems. And it appears that in many cases their harmful effects outweigh their benefits. For example, some pesticides (chlorinated hydrocarbons) persist in the environment and concentrate at higher levels in food chains.

Certain problems of chlorinated hydrocarbon pesticide use were first demonstrated by the effects of *d*ichloro-*d*iphenyl-*t*richloroethane (DDT) on many bird species, particularly birds of prey. Falcons, pelicans, bald eagles, ospreys, and a number of other birds are very sensitive to traces of DDT in their tissues. DDT causes these birds to lay eggs with extremely thin, fragile shells that usually break during incubation (causing the chick's death). After DDT was banned in the United States in 1972, the reproductive success of those birds improved.

This effect on birds is the result of two characteristics of DDT: its persistence in the environment and its biological magnification. Some pesticides, particularly chlorinated hydrocarbons, are extremely stable and may take many years to break down into less toxic forms. This **persistence** is caused by the novel chemical structures of these pesticides. Natural decomposers such as bacteria have not evolved ways to degrade them, so they accumulate in the environment and in the food chain.

Organisms higher on the food chain tend to carry greater concentrations of pesticides in their bodies than do those that are lower on the food chain. Pesticides, such as DDT, that cannot be metabolized or excreted simply get stored, usually in the organism's fatty tissues. The increasing concentration of pesticides in the tissues of organisms higher in the food chain is known as **biological magnification.**[1]

As an example of the concentrating effects of persistent pesticides, consider a hypothetical food chain: plant ⟶ insect ⟶ frog ⟶ hawk. When a pesticide is sprayed on a plant to control insects, it is extremely dilute; we will assign it an arbitrary concentration of "1" per leaf. Each insect grazing on the plant consumes 10 leaves, concentrating the pesticide in its tissues to a value of "10." (Assume that these insects are genetically resistant to the pesticide so that they stay alive.) A frog that eats ten insects laced with pesticide will end up with a pesticide level of "100." The top carnivore in this example, a hawk, will have a pesticide value of "1000" if it eats 10 contaminated frogs. While this example involves a bird at the top of the food chain, it is important to recognize that *all* top carnivores, from fish to humans, are at risk from biological magnification.

In this chapter, we examine several aspects of ecosystems, beginning with cycles of matter. We then examine how energy flows through ecosystems—by moving through food chains—and conclude the chapter with an overview of physical factors in the environment that profoundly influence living things.

[1] Other toxic substances besides pesticides may exhibit biological magnification, including radioactive isotopes, heavy metals like mercury, and industrial chemicals like PCBs.

Ecosystems

47

Learning Objectives

After you have studied this chapter you should be able to:

1. Compare how matter and energy operate in ecosystems.
2. Explain how carbon cycles between living organisms and the physical environment.
3. Outline the main steps in the nitrogen cycle.
4. Trace the path of phosphorus through an ecosystem.
5. Explain how water cycles between the atmosphere, land, and oceans.
6. Draw and explain typical pyramids of numbers, biomass, and energy.
7. Distinguish between gross primary productivity and net primary productivity.
8. Summarize the effects of the sun on Earth's climate.
9. Discuss the role of solar energy in the production of global air and water flow patterns.
10. Give three causes of regional differences in precipitation.

Key Concepts

- ☐ Matter, the material of which living things are composed, cycles from the living world to the abiotic (nonliving) physical environment and back again. All materials vital to life are continually recycled through ecosystems and so become available to new generations of organisms.
- ☐ Although matter is cyclic in ecosystems, energy flow is not. Energy moves through ecosystems in a linear, one-way direction. Once energy has been used to do biological work for a living organism, it is unavailable to other organisms. Energy cannot be recycled and reused.
- ☐ The abiotic environment, including climate, causes conditions that determine where species live as well as their distribution and range.

MATTER CYCLES THROUGH ECOSYSTEMS

Planet Earth has often been compared to a vast spaceship whose life support system consists of (1) the living things that inhabit it and (2) energy from the sun. Earth's living things produce oxygen, cleanse the air, adjust gases, transfer energy, and recycle waste products with great efficiency. Yet, none of these processes would be possible without the nonliving physical environment of our spaceship Earth.

The science of ecology deals with the abiotic environment as well as living organisms. Individual communities and their abiotic environments are **ecosystems.** Spaceship Earth, which encompasses the biosphere and its interactions with the hydrosphere, lithosphere, and atmosphere, is the **ecosphere** (Chapter 45).

Matter, the material of which living things are composed, moves in numerous cycles from the living world to the nonliving physical environment and back again. We call these cycles **biogeochemical cycles.** The Earth is essentially a closed system (a system from which matter cannot escape). The materials utilized by organisms cannot be "lost," although they can end up in locations that are outside the reach of organisms. Usually, however, materials are reused and are often recycled both within and among ecosystems.

Four different biogeochemical cycles of matter—carbon, nitrogen, phosphorus, and water—are representative of all biogeochemical cycles. These four cycles are particularly important to living things. Carbon, nitrogen, and water have gaseous components and so cycle over large distances with relative ease. Phosphorus, however, is an element that is completely nongaseous, and as a result, only local cycling of phosphorus occurs easily.

Carbon Dioxide Is the Pivotal Molecule of the Carbon Cycle

Carbon must be available to living things because proteins, carbohydrates, lipids, nucleic acids, and other molecules essential to life contain carbon. Carbon is present in the atmosphere as a gas, carbon dioxide (CO_2), which makes up approximately 0.03% of the atmosphere. It is also present in water as dissolved carbon dioxide, that is, as carbonate (CO_3^{2-}) and bicarbonate (HCO_3^-), and in rocks such as limestone. Carbon cycles between the nonliving environment, including the atmosphere, and living organisms (Fig. 47–1).

During photosynthesis, plants, algae, and cyanobacteria remove carbon dioxide from the air and **fix,** or incorporate, it into complex chemical compounds such as sugar (glucose; Chapter 9). The overall equation for photosynthesis is as follows:

$$\underset{\text{Carbon Dioxide}}{6CO_2} + \underset{\text{Water}}{12H_2O} \xrightarrow{\text{Light}} \underset{\text{Sugar}}{C_6H_{12}O_6} + \underset{\text{Oxygen}}{6O_2} + \underset{\text{Water}}{6H_2O}$$

Plants use glucose to make other compounds. Thus, photosynthesis incorporates carbon from the abiotic environment into the biological compounds of producers. These compounds are usually used as fuel for cellular

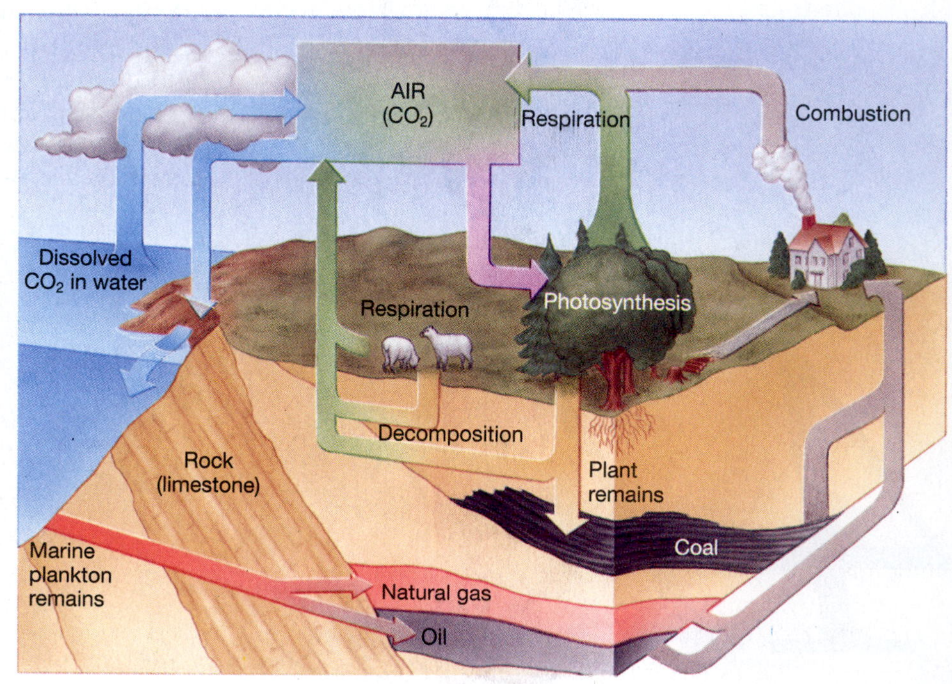

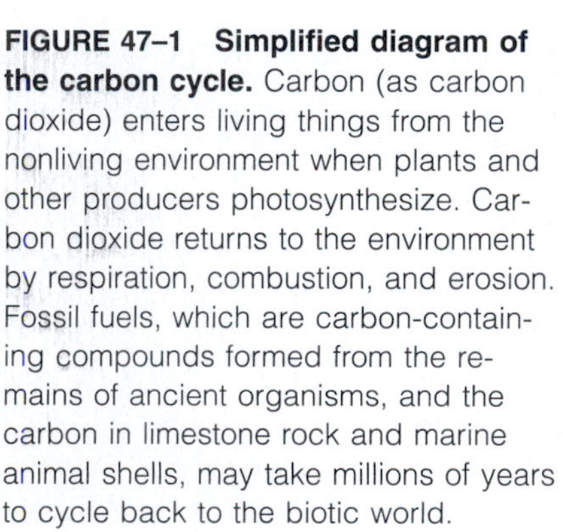

FIGURE 47–1 Simplified diagram of the carbon cycle. Carbon (as carbon dioxide) enters living things from the nonliving environment when plants and other producers photosynthesize. Carbon dioxide returns to the environment by respiration, combustion, and erosion. Fossil fuels, which are carbon-containing compounds formed from the remains of ancient organisms, and the carbon in limestone rock and marine animal shells, may take millions of years to cycle back to the biotic world.

respiration (Chapter 8) by the producer that made them, or by a consumer that eats the producer or by a decomposer that breaks down the remains of the producer or consumer. The overall equation for aerobic respiration is:

$$\underset{\text{Glucose}}{C_6H_{12}O_6} + \underset{\text{Oxygen}}{6O_2} + \underset{\text{Water}}{6H_2O} \longrightarrow$$

$$\underset{\text{Carbon Dioxide}}{6CO_2} + \underset{\text{Water}}{12H_2O} + \text{Energy for biological work}$$

Thus, carbon dioxide is returned to the atmosphere by the process of aerobic respiration. A similar carbon cycle occurs in aquatic ecosystems between aquatic organisms and dissolved carbon dioxide in the water.

Sometimes the carbon in biological molecules is not recycled back to the abiotic environment for some time. Much carbon is stored in the wood of trees where it may stay for several hundred years or even longer. In addition, millions of years ago, vast coal beds formed from the bodies of ancient trees that were buried under anaerobic conditions and did not decay fully (Chapter 23). Similarly, the oils of unicellular marine organisms probably gave rise to the underground deposits of oil and natural gas that accumulated in the geological past. Coal, oil, and natural gas, called **fossil fuels** because they formed from the remains of ancient organisms, are vast depositories of carbon compounds—the end products of photosynthesis that occurred millions of years ago.

The carbon in coal, oil, natural gas, and wood may be returned to the atmosphere by the process of burning, or **combustion.** In combustion, organic molecules are rapidly oxidized (combined with oxygen), converting them into carbon dioxide and water with an accompanying release of heat and light.

An even greater amount of carbon that leaves the carbon cycle for millions of years is incorporated into the shells of marine organisms. When these organisms die, their shells sink to the ocean floor and are covered by sediments. These shells form seabed deposits thousands of feet thick that are eventually cemented together to form sedimentary rock called limestone. The Earth's crust is dynamically active, and over millions of years, sedimentary rock on the bottom of the sea floor may lift to form land surfaces (for example, the summit of Mount Everest is composed of sedimentary rock). When limestone is exposed by the process of geological uplift, it slowly erodes, or wears away, by chemical and physical weathering processes. This returns the carbon to the water and atmosphere, where it is available to participate in the carbon cycle once again.

Thus, one process (photosynthesis) removes carbon from the abiotic environment and incorporates it into biological molecules, and three processes (cellular respiration, combustion, and erosion) return carbon to the water and atmosphere of the abiotic environment.

Human Activities Have Disturbed the Balance of the Carbon Cycle

Since the advent of the Industrial Revolution and continuing to the present, humans have burned increasing amounts of coal, oil, and natural gas. This increased rate of fossil fuel combustion, along with a greater combustion of wood as a fuel and the burning of large sections of tropical forest, has released carbon dioxide into the atmosphere at a rate greater than is removed by photosynthesis.

As a result, the level of atmospheric CO_2 is slowly and steadily increasing, and this increase may cause changes in climate called global warming. Global warming could cause a rise in sea level (as the polar ice caps melt), changes in precipitation patterns, death of forests, extinction of animals and plants, and problems for agriculture. It could result in the displacement of thousands or even millions of people, particularly from coastal areas. (Look at a map to get an idea of how many large cities are located on coasts.) A more complete discussion of rising levels of atmospheric CO_2 and global warming is found in Chapter 49.

Bacteria Are Essential to the Nitrogen Cycle

Nitrogen is crucial for all living things because it is an essential part of proteins and nucleic acids. At first glance it would appear that there is no possible shortage of nitrogen for living organisms: the Earth's atmosphere is about 80% nitrogen gas, N_2. However, molecular nitrogen is so stable that it does not readily combine with other elements. Therefore, living things cannot take nitrogen gas from the atmosphere and use it to manufacture proteins and nucleic acids. Molecular nitrogen must first be broken apart before the nitrogen can combine with other elements to form proteins and nucleic acids. The overall reaction that breaks up molecular nitrogen and combines it with such elements as oxygen or hydrogen requires a great deal of energy.

There are five steps in the nitrogen cycle: (1) nitrogen fixation; (2) nitrification; (3) assimilation; (4) ammonification; and (5) denitrification (Fig. 47–2). All of these steps except assimilation are performed by bacteria.

The first step in the nitrogen cycle is **nitrogen fixation,** which involves the conversion of gaseous nitrogen (N_2) to **ammonia (NH_3).** The process is called nitrogen fixation because nitrogen is *fixed* into a form that living things can use. Although considerable nitrogen is also fixed by combustion, volcanic action, and lightning discharges and by industrial means (all of these processes supply enough energy to break up molecular nitrogen),

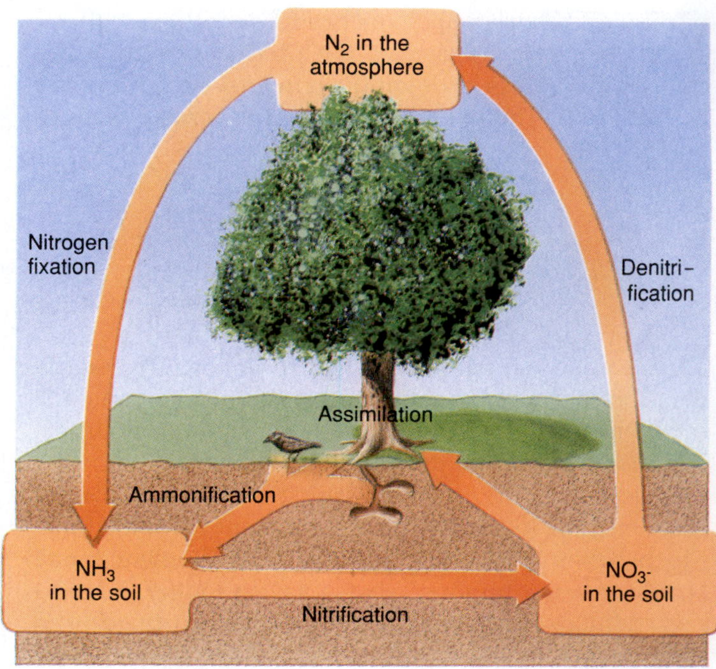

FIGURE 47–2 The nitrogen cycle has five steps. (*1*) Nitrogen-fixing bacteria, including cyanobacteria, convert atmospheric nitrogen (N_2) into ammonia (NH_3). (*2*) Ammonia is converted to nitrate (NO_3^-) by bacteria in the soil known as nitrifying bacteria. Nitrate is the main form of nitrogen absorbed by plants. (*3*) Plants assimilate nitrate when they produce proteins and nucleic acids; then animals eat plant proteins and produce animal proteins as a part of assimilation. (*4*) When plants and animals die, the nitrogen compounds in their remains are broken down by ammonifying bacteria. One of the products of this decomposition is ammonia. (*5*) Nitrogen is returned to the atmosphere by denitrifying bacteria, which convert nitrate to molecular nitrogen.

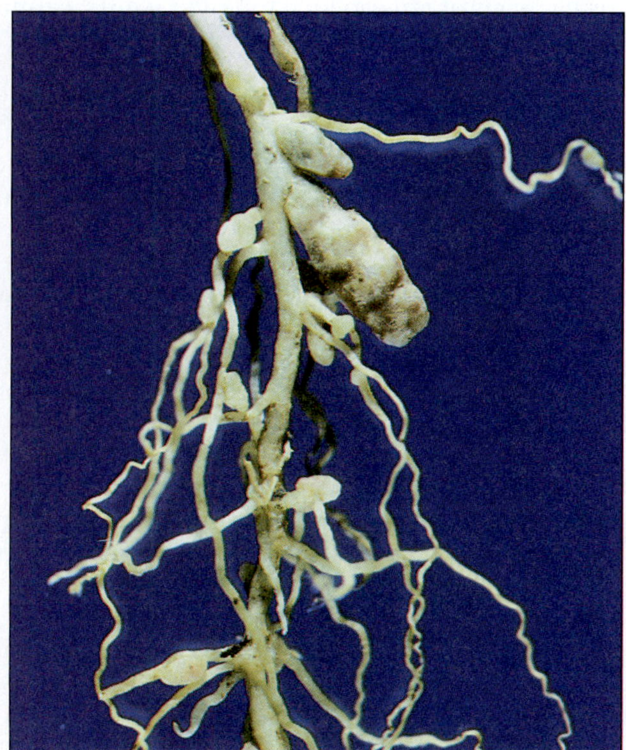

FIGURE 47–3 Root nodules of a clover plant (a legume). Mutualistic *Rhizobium* bacteria live in these nodules, using energy derived from sugars provided by their legume host. The bacteria fix nitrogen, some of which is used by the host plant. The ultimate death and decay of both partners enrich the soil with the fixed nitrogen. (Carolina Biological Supply Company)

most nitrogen fixation is biological. It is carried out by nitrogen-fixing bacteria, including cyanobacteria, in soil and aquatic environments. Nitrogen-fixing bacteria employ an enzyme called **nitrogenase** to break up molecular nitrogen and combine it with hydrogen.

Because nitrogenase functions only in the absence of oxygen, the bacteria that use nitrogenase must insulate the enzyme from oxygen by some means. Some nitrogen-fixing bacteria live beneath layers of oxygen-excluding slime on the roots of a number of plants. But the most important terrestrial nitrogen-fixing bacteria, *Rhizobium*, live in special swellings, or **nodules** (Fig. 47–3), on the roots of legumes such as beans or peas and some woody plants. The relationship between *Rhizobium* and their host plants is mutualistic: the bacteria receive carbohydrates from the plant, and the plant receives nitrogen in a form that it can use.

In aquatic habitats, most nitrogen fixation is performed by cyanobacteria. Filamentous cyanobacteria have special oxygen-excluding cells called **heterocysts** that function to fix nitrogen (Fig. 47–4). Some water ferns have cavities in which cyanobacteria live, in a manner comparable to how *Rhizobium* lives in root nodules of legumes. Other cyanobacteria fix nitrogen in symbiotic association with certain plants or as the photosynthetic partner of certain lichens.

The reduction of nitrogen gas to ammonia by nitrogenase is a remarkable accomplishment of living organisms that is achieved without the tremendous heat, pressure, and energy required to do the same thing during

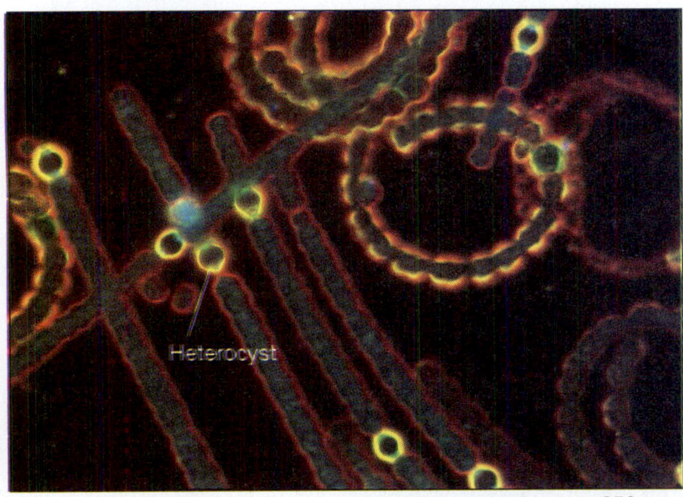

FIGURE 47–4 Many cyanobacteria fix nitrogen, often in association with plants. Shown is *Anabaena*, a cyanobacterium that has distinctive specialized cells, called heterocysts, where nitrogen-fixation occurs. (Dennis Drenner)

the manufacture of commercial fertilizers. Even so, nitrogen-fixing bacteria must consume the energy in 12 grams of glucose or the equivalent to fix a single gram of nitrogen biologically.

The second step is **nitrification,** which is the conversion of ammonia (NH_3) to **nitrate (NO_3^-).** Nitrification, a two-step process, is accomplished by soil bacteria. First, the soil bacteria *Nitrosomonas* and *Nitrococcus* convert ammonia to nitrite (NO_2^-). Then, the soil bacteria *Nitrobacter* oxidize nitrite to nitrate. The process of nitrification furnishes these bacteria, called nitrifying bacteria, with energy.

The third step is **assimilation,** in which plant roots absorb either nitrate (NO_3^-) and/or ammonia (NH_3) that were formed by nitrogen fixation and nitrification, and incorporate the nitrogen in these molecules into plant proteins and nucleic acids. When animals consume plant tissues, they assimilate nitrogen as well, by taking in plant nitrogen compounds and converting them to animal compounds.

The fourth step is **ammonification,** which is the conversion of biological nitrogen compounds into ammonia. Ammonification begins when living organisms produce nitrogen-containing waste products such as urea (in urine) and uric acid (in the wastes of birds). These substances, along with the nitrogen compounds in dead organisms, are decomposed, releasing the nitrogen into the abiotic environment as ammonia (NH_3). The bacteria that perform this process in both the soil and aquatic environments are called ammonifying bacteria. The ammonia produced by ammonification enters the nitrogen cycle and is available once again for the processes of nitrification or assimilation.

The fifth, and final, step of the nitrogen cycle is **denitrification,** which is the reduction of nitrate (NO_3^-) to gaseous nitrogen (N_2). Denitrifying bacteria reverse the action of nitrogen-fixing and nitrifying bacteria. That is, denitrifying bacteria return nitrogen to the atmosphere as nitrogen gas. Denitrifying bacteria are anaerobic, which means they prefer to live and grow where there is little or no free oxygen. For example, they are found deep in the soil near the water table, an environment that is nearly oxygen-free.

Humans Affect the Nitrogen Cycle by Producing Large Quantities of Nitrogen Fertilizer from Nitrogen Gas

While commercial fertilizer in itself is not bad, its overuse on the land can cause water pollution. Nitrate fertilizer is washed by rain into rivers and lakes where it stimulates an excessive growth of algae (known as an algal bloom). As these algae die, their decomposition robs the water of dissolved oxygen, which in turn causes other aquatic organisms, including fishes, to die of suffocation. Nitrates from fertilizers can also leach (filter) down through the soil and contaminate groundwater. Many people who live in rural areas and some cities drink groundwater, and groundwater contaminated by nitrates is dangerous, particularly for infants and small children.

The Phosphorus Cycle Lacks a Gaseous Component

Phosphorus, which does not exist in a gaseous state and therefore does not enter the atmosphere, cycles from the land to sediments in the oceans and back to the land (Fig. 47–5). As water runs over rocks containing phosphorus, it gradually wears away the surface and carries off inorganic **phosphate (PO_4^{-3})** molecules.

The eroding process of phosphorus rocks releases phosphate into the soil where it is taken up by plant roots, with the help of mycorrhizae (Chapter 22). Once in the plant's cells, phosphate is incorporated into a variety of biological molecules, including nucleic acids. Animals obtain most of their required phosphorus from the food they eat, although in some localities drinking water may contain a substantial amount of inorganic phosphate. Phosphate released by decomposers becomes part of the pool of inorganic phosphate in the soil that can be reused by plants.

Phosphorus cycles through aquatic communities in much the same way it does in terrestrial (land-dwelling) communities. Dissolved phosphate enters aquatic communities through absorption by algae and aquatic plants, which are in turn consumed by plankton and larger organisms. These are in turn eaten by a variety of fin and

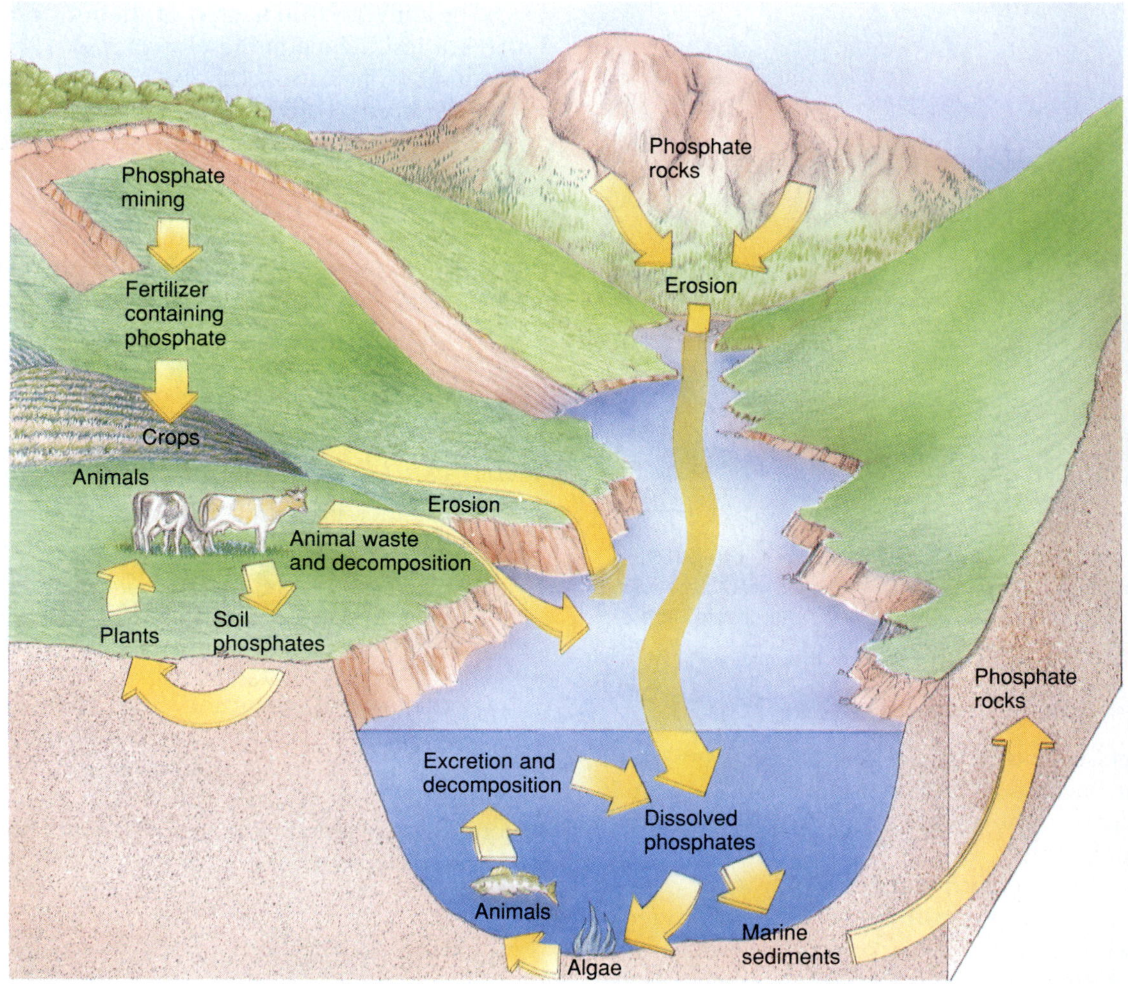

FIGURE 47–5 The phosphorus cycle in terrestrial and aquatic environments. Recycling of phosphorus (as phosphate, PO_4^{3-}) is slow because no biologically important form of phosphorus is gaseous. Phosphate that becomes part of marine sediments may take millions of years to solidify into rock, uplift as mountains, and erode to again become available to living things.

shell fish. Ultimately, decomposers that break down wastes and dead organisms release inorganic phosphate into the water, making it available once again for use by aquatic producers.

Phosphate can be lost from biological cycles. Some phosphate is carried from the land by streams and rivers to the ocean. It can be deposited on the sea floor, where it can remain for millions of years. The geological process of uplift may some day expose these sea floor sediments as new land surfaces, from which phosphate is once again eroded.

Some phosphate in the ocean finds its way back to the land. A small portion of the fishes and aquatic invertebrates are eaten by sea birds, which may defecate on the land where they roost. **Guano,** the manure of sea birds, contains large amounts of phosphate and nitrate. Once on land, these minerals may be absorbed by the roots of plants. The phosphate contained in guano may enter terrestrial food chains in this way, although the amounts involved are quite small.

Humans Affect the Natural Cycling of Phosphorus by Accelerating Its Long-Term Loss from the Land

Corn grown in Iowa may be used to fatten cattle in an Illinois feedlot. Beef from these cattle may be consumed by people living far away—for instance, in New York City. Thus, part of the phosphate absorbed from the Iowa soil by the roots of the corn plants ends up in feedlot wastes, which probably eventually wash into the Mississippi River, and from there into the ocean. More of this phosphate ends up in human wastes and is flushed down toilets into the New York City sewer system. Sewage treatment rarely removes phosphate, which causes water quality problems in rivers and lakes on its way to the ocean. To replace this steady loss of phosphate from their land, farmers must add phosphate fertilizer to their fields. More than likely, that fertilizer is produced in Florida from large deposits of mined phosphate rock. Thus, human activities speed up the slow movement of phosphate from rocks to soil to water.

In natural communities, very little phosphate is lost from the cycle, but few communities today are unaffected by human activities. Phosphate loss is accelerated from the soil by practices such as the clear-cutting of forest or by erosion from agricultural or residential land. For practical purposes, phosphate that washes from the land into the sea is permanently lost from the terrestrial phosphorus cycle, for it remains in the sea for millions of years.

Water Is Cycled in the Hydrologic Cycle

Water continuously circulates from the oceans to the atmosphere to the land and back to the oceans, providing us with a renewable supply of purified water on land. This complex cycle, known as the **hydrologic cycle,** results in a balance between water in the oceans, on the land, and in the atmosphere (Fig. 47–6). Water moves from the atmosphere to the land and oceans in the form of precipitation (rain, snow, sleet, or hail). When water evaporates from the ocean's surface, it forms clouds in the atmosphere. Water also evaporates from soil, streams, rivers, and lakes on land. In addition, **transpiration,** which is the loss of water vapor from land plants, adds water to the atmosphere (Chapter 27).

Water may evaporate from land and reenter the atmosphere directly. Alternatively, it may flow in rivers and streams to coastal **estuaries,** where fresh water meets the oceans. The movement of surface water from land to oceans is called **runoff,** and the area of land being drained by runoff is called a **watershed.** Water also seeps downward in the soil to become **groundwater,** where it is trapped and held for a time. Groundwater supplies water to the soil, to streams and rivers, and to plants.

Regardless of its physical form (solid, liquid, or vapor) or location, every molecule of water eventually moves through the hydrologic cycle. As in other cycles, water can be lost from the cycle for thousands of years (as glaciers and icecaps).

Tremendous quantities of water are cycled annually between the Earth and its atmosphere. (The amount of water entering the atmosphere each year is estimated at 95,000 cubic miles.) Approximately three-fourths of this water reenters the ocean directly as precipitation over water; the remaining amount falls on land.

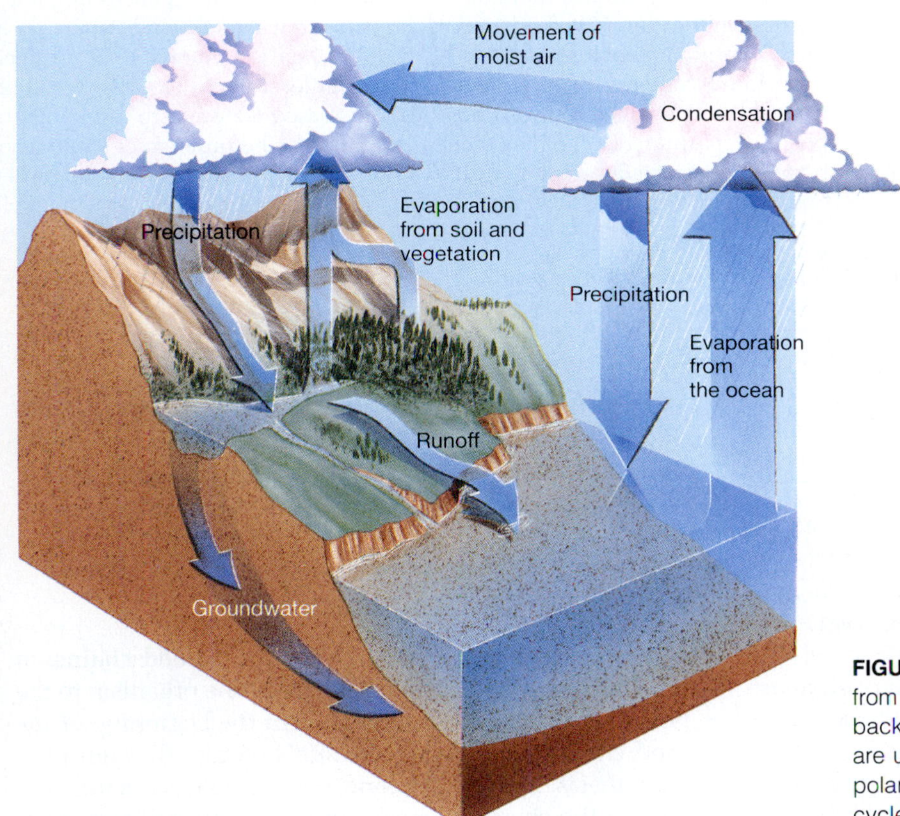

FIGURE 47–6 The hydrologic cycle. Water cycles from the oceans to the atmosphere to the land and back to the oceans. Although some water molecules are unavailable for thousands of years (locked up in polar ice, for example), all water molecules eventually cycle through the hydrologic cycle.

FOCUS ON *The Environment*

CHANGES IN THE ANTARCTIC FOOD WEB

Although the icy waters around Antarctica may seem to be a very inhospitable environment, a rich variety of life is found there. The base of the food web is microscopic algae, which are present in vast numbers. These algae are eaten by a huge population of tiny shrimplike animals called **krill,** which in turn support a variety of larger animals. One of the main consumers of krill is baleen whales, which filter krill out of the frigid water. Baleen whales include blue whales, humpback whales, and right whales. Krill are also consumed in great quantities by squid and fish. These in turn are eaten by other carnivores: toothed whales such as the sperm whale, elephant seals and leopard seals, king penguins and emperor penguins, and birds such as the albatross and petrel.

Humans have had an impact on the complex Antarctic food web as they have on most other ecosystems. Before the advent of whaling, baleen whales consumed huge quantities of krill. During the past 150 years—until a 1986 global ban on hunting all whales—whaling steadily reduced the numbers of large baleen whales in Antarctic waters. Many whale populations are so decimated that they are on the brink of extinction. As a result of fewer whales eating krill, more krill has been available for other krill-eating animals, whose populations have increased. Seals, penguins, and smaller baleen whales have replaced the large baleen whales as the main eaters of krill.

Now that commercial whaling is regulated, it is hoped that the number of large baleen whales will slowly increase, which appears to be the case for at least some species.* It is not known whether baleen whales will return to or be excluded from their former position of dominance (in terms of krill consumption) in the food web. Biologists will monitor changes in the Antarctic food web as the whale populations recover.

More recently, a human-related change has developed in the atmosphere over Antarctica that has the potential to cause far greater effects on the entire Antarctic food web. This change is the depletion of the ozone layer in the stratospheric region of the atmosphere. This ozone "hole" allows more of the sun's ultraviolet radiation to penetrate to the Earth's surface. Ultraviolet radiation contains more energy than does visible light. It is so energetic that it can break the chemical bonds of some biologically important molecules.

It is not known how increased ultraviolet radiation levels will affect organisms in the Antarctic food web, but biologists are most concerned about its effects on the lowest trophic levels of the food web. If algae and krill are harmed by higher levels of ultraviolet radiation, negative repercussions will extend throughout the entire Antarctic food web. A 1992 study confirmed that increased ultraviolet radiation is penetrating the surface waters around Antarctica and that algal productivity has declined by at least 6% to 12% as a result of increased exposure to ultraviolet radiation. The problem of stratospheric ozone depletion is discussed in detail in Chapter 49.

* As of 1994, the southern blue whale population did not appear to be growing in response to the moratorium on whaling. To compound the problem, it is extremely difficult to make accurate population estimates of blue whales.

THE FLOW OF ENERGY THROUGH ECOSYSTEMS IS LINEAR

The passage of energy in a one-way direction through an ecosystem is known as energy flow. Energy enters an ecosystem as radiant energy (sunlight), some of which is trapped by plants during photosynthesis. The energy, now in chemical form, is stored in bonds of organic molecules such as glucose. When these molecules are broken apart by cellular respiration, energy becomes available to do work such as repairing tissues, producing body heat, or reproducing. As the work is accomplished, energy escapes the living organism and dissipates into the environment as low-quality heat. Ultimately, this heat energy radiates into space. Thus, once energy has been used by living things, it becomes unavailable for reuse (Chapter 7).

Who Eats Whom in Ecosystems Describes the Path of Energy Flow

In an ecosystem, energy flow occurs in **food chains,** in which energy from food passes from one organism to the next in a sequence. Producers form the beginning of the food chain by capturing the sun's energy through photosynthesis. Herbivores (and omnivores) eat plants, obtaining the chemical energy of the producers' molecules

Adelie penguins are flightless birds adapted to swim and dive in the southern ocean around Antarctica. The main food of Adelie pengins is krill (*inset*). Vast numbers of these tiny animals are consumed by whales, squids, and fishes as well as penguins. (George Holton/Photo Researchers, Inc.; *inset*, Flip Nicklin/Minden Pictures)

as well as building materials from which they construct their own tissues. Herbivores are in turn consumed by carnivores (and omnivores), who reap the energy stored in the herbivores' molecules. At the end of food chains are decomposers, which respire organic molecules in the remains (carcasses and body wastes) of all other members of the food chain.

Simple food chains as just described rarely occur in nature, since few organisms eat just one kind of other organism. More typically, the flow of energy and materials through ecosystems take place in accordance with a range of choices of food on the part of each organism involved. In an ecosystem of average complexity, hundreds of alternative pathways are possible. Thus, a **food web,** which is a complex of interconnected food chains in an ecosystem, is a more realistic model of the flow of energy and materials through ecosystems (Fig. 47–7). (See Focus On The Environment: Changes in the Antarctic Food Web for an examination of how humans have affected the complex food web in Antarctic waters.)

The most important thing to remember about energy flow in ecosystems is that it is *linear,* or one-way. That is, energy can move along a food web from one organism to the next as long as it is not used. Once energy has been used by an organism, it becomes unavailable for any other living thing in the ecosystem.

FIGURE 47–7 A food web for a marsh. This food web is greatly simplified compared to what actually happens in nature. Many species are not included, and many links in the web are not shown.

Concept Connections

Food Webs ⟷ ***Predation*** ⟷ ***Mutualism***

Both negative and positive interactions occur in a food web. Because food webs are descriptions of "who eats whom," they indicate the negative effects that predators have on their prey. For example, consider the simple food chain: grass ⟶ field mouse ⟶ owl. The owl, which kills and eats mice, obviously exerts a negative effect on the mouse population; in like manner, field mice, which eat grass seeds, lower the grass population.

One trophic level in a food web also influences other trophic levels to which it is *not* directly linked. The producers and top carnivores do not exert direct effects on one another, yet each is affected by the other. In our example, the owl population indirectly helps the producer population by keeping the population of seed-eating mice under control. Likewise, the producers benefit owls by supporting a population of mice upon which the owl population feeds. These indirect interactions are mutualistic in nature (discussed in Chapter 46) and may be as important as direct, predator–prey interactions (discussed in Chapters 45 and 46) in food web dynamics.

Ecological Pyramids Help Us to Understand How Ecosystems Work

Each level in a food web is called a **trophic level** (Fig. 47–8). The first trophic level is formed by producers, the second trophic level by primary consumers (herbivores), the third trophic level by secondary consumers (carnivores), and so on.

Ecologists sometimes sum up the number of organisms at each trophic level, their biomass, or their relative energy content to construct **ecological pyramids.** There are three main types of pyramids: a pyramid of numbers, a pyramid of biomass, and a pyramid of energy.

A **pyramid of numbers** shows the number of organisms at each trophic level in a given ecosystem, with greater numbers illustrated by a wider pyramid (Fig. 47–9). In most pyramids of numbers, each successive trophic level is occupied by fewer organisms. Thus, the number of herbivores (such as zebras and wildebeests) is greater than the number of carnivores (such as lions) in a typical grassland.

A **pyramid of biomass** illustrates the total biomass at each successive trophic level. **Biomass** is a quantitative estimate of the total mass, or amount, of living material. Its units of measure vary: biomass may be represented as total volume, as dry weight, or as live weight. Typically, pyramids of biomass illustrate a progressive reduction of biomass in succeeding trophic levels (Fig. 47–10). On the assumption that there is, on the average, about a 90% reduction of biomass for each trophic level,[2] 10,000 kilograms of grass should be able to support 1000 kg of crickets, which in turn support 100 kg of frogs. By this logic, the biomass of frog-eaters (such as a heron) could only weigh, at the most, about 10 kg. From this brief exercise, you can see that although carnivores may eat no vegetation, a great deal of vegetation is still required to support them.

A **pyramid of energy** indicates the energy content (usually expressed in calories) in the biomass of each trophic level. These pyramids show that most energy dissipates into the environment when going from one trophic level to another (Fig. 47–11). Less energy reaches each successive trophic level from the level beneath it because some of the energy at the lower level is used by those organisms to perform work and some of it is lost (recall from Chapter 7 that no biological process is 100% efficient). Energy pyramids explain why there are few trophic levels: *Food webs are short because of the dramatic reduction in energy content that occurs at each trophic level.*

Ecosystems Vary in Productivity

The **gross primary productivity**[3] of an ecosystem is the *rate* at which energy is captured and stored in plant tissues (that is, in plant biomass) during photosynthesis. Thus, gross primary productivity is the total amount of photosynthesis in a given period of time. Of course, plants must respire to provide energy for their life processes, and cellular respiration acts as a drain on photosynthesis. Energy that remains in plant tissues after cellular respiration has occurred is called **net primary productivity.** That is, net primary productivity is the amount of biomass found in excess of that broken down by a plant's cellular respiration. Net primary productivity represents the *rate* at which this organic matter is actually incorporated into plant tissues so as to produce growth (Fig. 47–12).

$$\underbrace{\text{Net primary productivity}}_{\text{plant growth}} = \underbrace{\text{Gross primary productivity}}_{\text{total photosynthesis}} - \text{Plant respiration}$$

[2] The 90% reduction in biomass at each trophic level is an approximation; actual biomass reduction varies widely in nature.

[3] Gross and net primary productivity are referred to as *primary* because plants occupy the first position in food webs.

(Text continues on page 960)

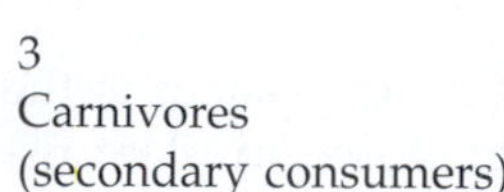

FIGURE 47–8 Trophic levels help us simplify food webs by grouping organisms based on their position in the food web. (***a***) Plants such as knapweed occupy the first trophic level in an ecosystem. The grasshopper eating the knapweed is a primary consumer and occupies the second trophic level. (***b***) The third trophic level is represented by the grasshopper mouse, which is eating a grasshopper. (***c***) The barn owl with a mouse in its talons is a top carnivore at the fourth trophic level. (*a*, Richard Kolar © 1995 Animals Animals; *b*, Tom McHugh/Photo Researchers, Inc.; *c*, John Mielcarek/Dembinsky Photo Associates)

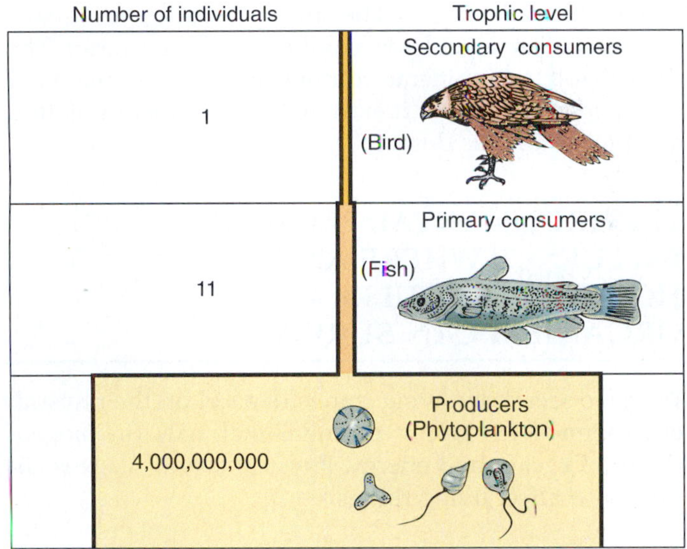

FIGURE 47–9 A pyramid of numbers is based on the number of organisms at each trophic level. Typically, there are more producers than primary consumers, more primary consumers than secondary consumers, and so on.

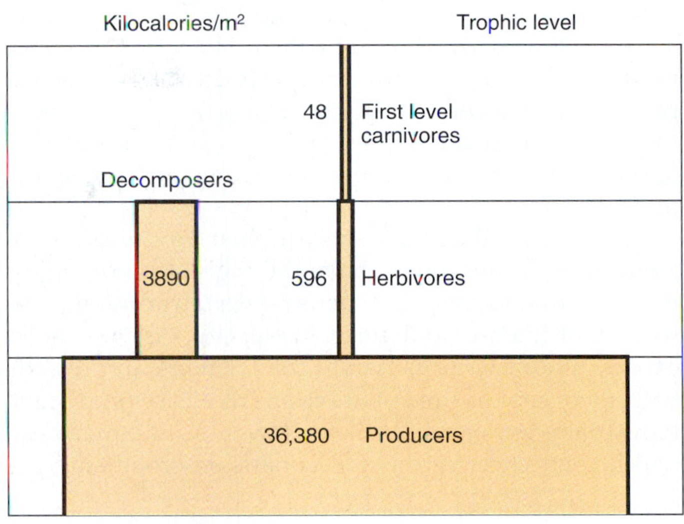

FIGURE 47–11 A pyramid of energy. The functional basis of ecosystem structure—energy flow—is represented by a pyramid of energy. Note the relatively large role played by decomposing bacteria. These decomposers actually operate on all trophic levels.

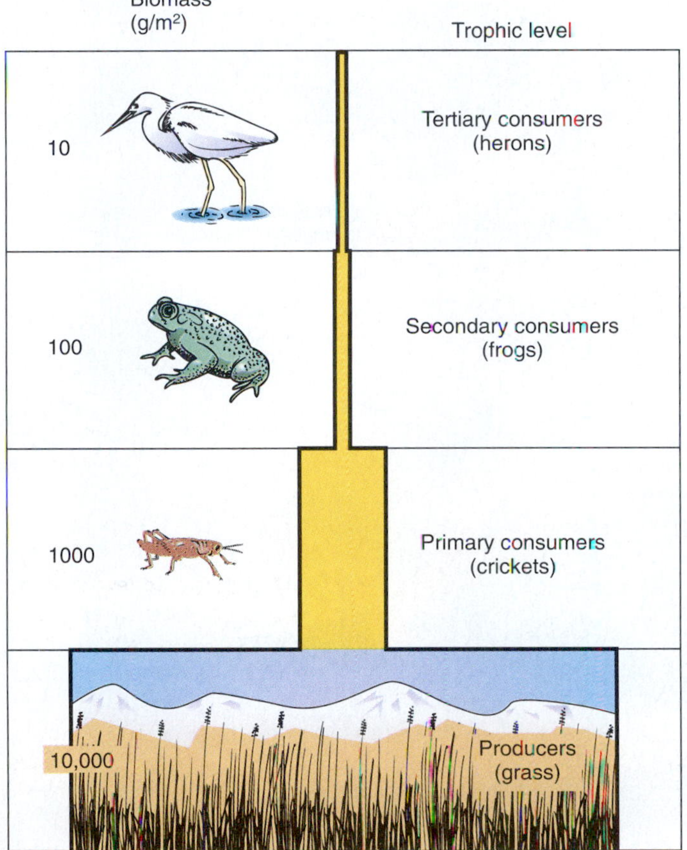

FIGURE 47–10 A pyramid of biomass for a hypothetical area of a temperate grassland. Pyramids of biomass are based on the biomass at each trophic level and typically resemble pyramids of numbers.

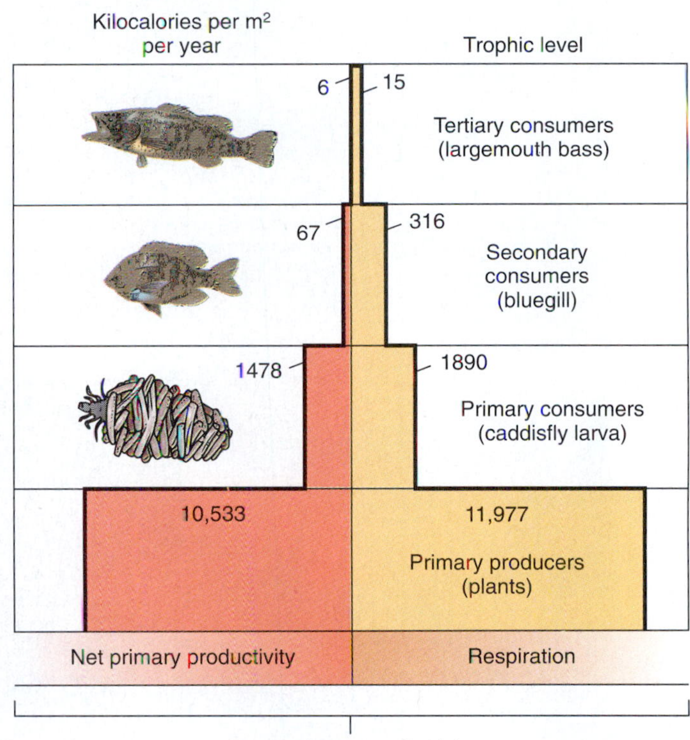

FIGURE 47–12 A pyramid of energy for a river ecosystem. Gross primary productivity and net primary productivity are illustrated in this pyramid. Measurements are in kilocalories per square meter per year.

Only the energy represented by net primary productivity is available for consumers, and of this energy only a portion is actually utilized by them. Both gross primary productivity and net primary productivity can be expressed in terms of kilocalories (of energy fixed by photosynthesis) per square meter per year or of dry weight (grams of carbon incorporated into tissue) per square meter per year.

Ecosystems differ strikingly in their productivity. On land, tropical rain forests have the highest productivity, probably due to their abundant rainfall, warm temperatures, and intense sunlight (Chapter 48). As you might expect, tundra with its harsh, cold winters and deserts with their lack of precipitation are the least productive terrestrial ecosystems. Wetlands (swamps and marshes), which connect terrestrial and aquatic environments, are extremely productive. The most productive aquatic ecosystems are algal beds, coral reefs, and estuaries. The unavailability of mineral nutrients in the sunlit region of the open ocean makes it extremely unproductive, equivalent to an aquatic desert.

ENVIRONMENTAL FACTORS INFLUENCE WHERE AND HOW SUCCESSFULLY AN ORGANISM CAN SURVIVE

We have seen how living things depend on the physical environment to supply essential materials (in biogeochemical cycles) and energy. Physical factors such as climate also affect living things.

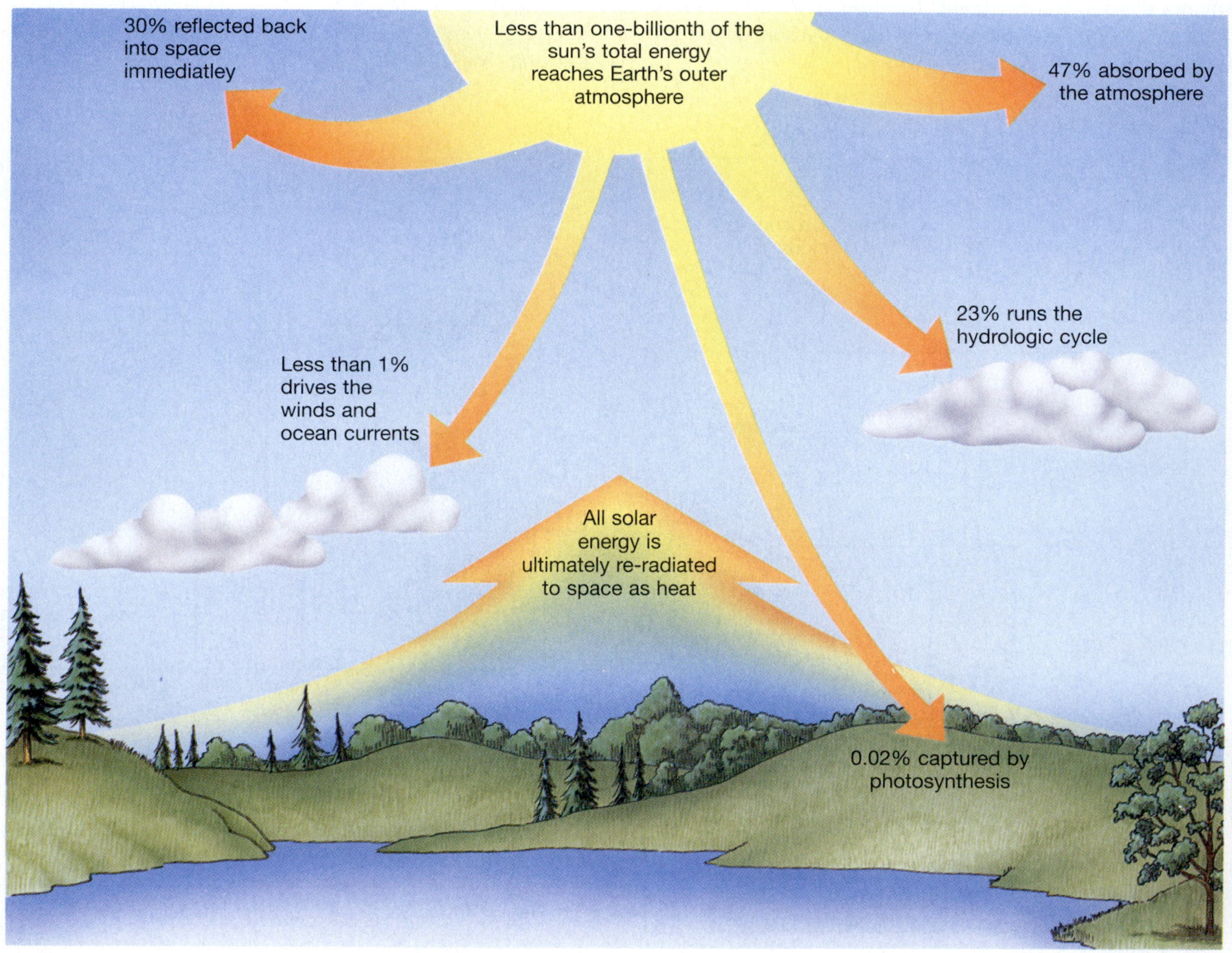

FIGURE 47–13 Sunlight intercepted by Earth. Solar energy travels out from the sun in all directions, and most of the energy produced by the sun never reaches the Earth. The solar energy that does warms the Earth's surface, powers the hydrologic cycle and other biogeochemical cycles, produces our climate, and powers almost all life through the process of photosynthesis, which converts solar energy into the chemical energy of organic molecules.

Climate refers to the average weather conditions that occur in a place over a period of years and includes temperature and precipitation. Day-to-day variations, day-to-night variations, and seasonal variations in temperature and precipitation are important aspects of climate. In addition, climate includes wind, humidity, fog, and cloud cover. We discuss climate in the remainder of this chapter.

The Sun Warms the Earth

The sun's energy, which is the product of a massive nuclear fusion reaction, is emitted into space in the form of electromagnetic radiation, especially light, infrared, and ultraviolet radiation. An infinitesimal portion of this energy—one-billionth of the sun's total production—strikes the Earth's atmosphere, and a minute part of this tiny trickle of energy operates the ecosphere.

Of the solar radiation that falls upon the Earth, 30% is immediately reflected away by clouds and surfaces, especially snow, ice, and oceans (Fig. 47–13). The remaining 70% is absorbed by the Earth and runs the water cycle, drives winds and ocean currents, powers photosynthesis, and warms the planet. Ultimately, however, all of this energy is lost by the continual radiation of long-wave infrared (heat) energy into space.

Solar Energy Is More Concentrated at the Equator and Less Concentrated at the Poles

The most significant local variation in the Earth's temperature is produced because the sun's energy does not uniformly reach all places on Earth. A combination of the Earth's roughly spherical shape and the tilted angle of its axis produces a great deal of variation in the exposure of the Earth's surface to the energy delivered by sunlight.

The principal difference this tilting makes is in the angles at which the sun's rays strike different areas of the Earth at any one time. On average, the sun's rays hit the Earth vertically near the Equator, making the sun more concentrated and producing higher temperatures. Near the poles the sun's rays hit more obliquely. As a result, they are spread over a larger surface area. Also, rays of light entering the atmosphere obliquely near the poles must pass through a deeper envelope of air than those entering near the Equator. This causes more of the sun's energy to be scattered and reflected back to space, which further lowers temperatures near the poles. Thus, the solar energy that reaches polar regions is less concentrated and produces lower temperatures.

Seasons are determined by the inclination of the Earth's axis (23.5°) as it rotates around the sun. During half of the year (March 21 to September 22), the Northern Hemisphere tilts *toward* the sun, concentrating the sunlight and making the days longer. Thus, the Northern Hemisphere is warm at this time. During the other half of the year (September 22 to March 21), the Northern Hemisphere tilts *away* from the sun (Fig. 47–14), giving it a lower concentration of sunlight and shorter days. The orientation of the Southern Hemisphere during this time of the year is *toward* the sun.

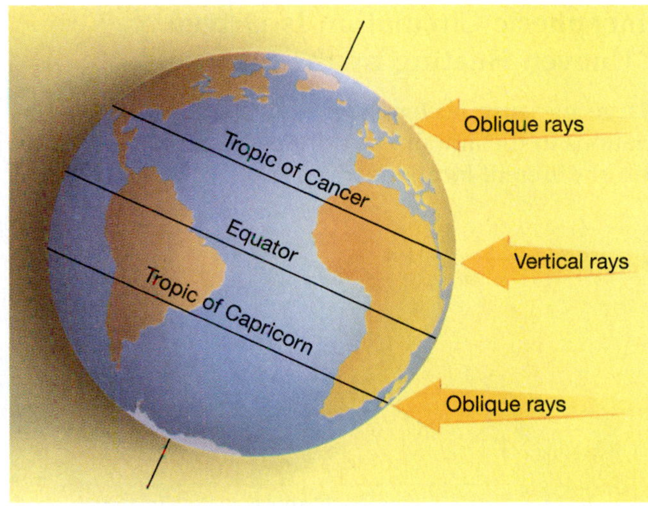

(*a*)

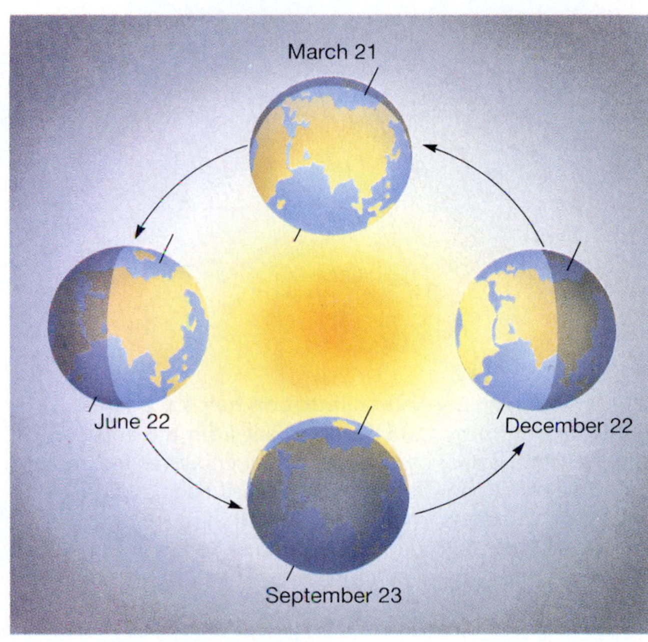

(*b*)

FIGURE 47–14 Variation in solar intensity on Earth. (***a***) The angle at which the sun's rays strike the Earth varies from one geographical location to another due to the spherical shape of the planet and its inclination on its axis. The month of June is represented here. (***b***) The inclination of Earth's axis remains the same as it travels around the sun. Thus, the sun's rays hit the Northern Hemisphere obliquely during the winter months and more directly during the summer. In the Southern Hemisphere, the sun's rays are oblique during winter months, which corresponds to summer in the Northern Hemisphere. At the equator, the sun's rays are approximately vertical at all times of the year.

Atmospheric Circulation Is Driven by Uneven Heating by the Sun

In large measure, differences in temperature caused by variations in the amount of solar energy that reaches the Earth at different locations drive the circulation of the atmosphere. The warm surface of the Earth near the Equator heats the air with which it comes into contact, causing this air to expand and rise. As the warm air rises, it flows away from the Equator, cools, and sinks again. Much of it recirculates back to the same areas it left, but the remainder flows toward the poles, where eventually it is chilled. Similar upward movements of warm air and its subsequent flow toward the poles occur at higher latitudes (further from the Equator) as well (Fig. 47–15). As air cools by contact with the polar ground and ocean, it sinks and flows toward the Equator, generally beneath the sheets of warm air flowing toward the pole at the same time. The constant motion of air transfers heat from the Equator toward the poles, and as the air returns, it cools the land over which it passes. This constant turnover does not equalize temperatures over the Earth's surface, but it does moderate them.

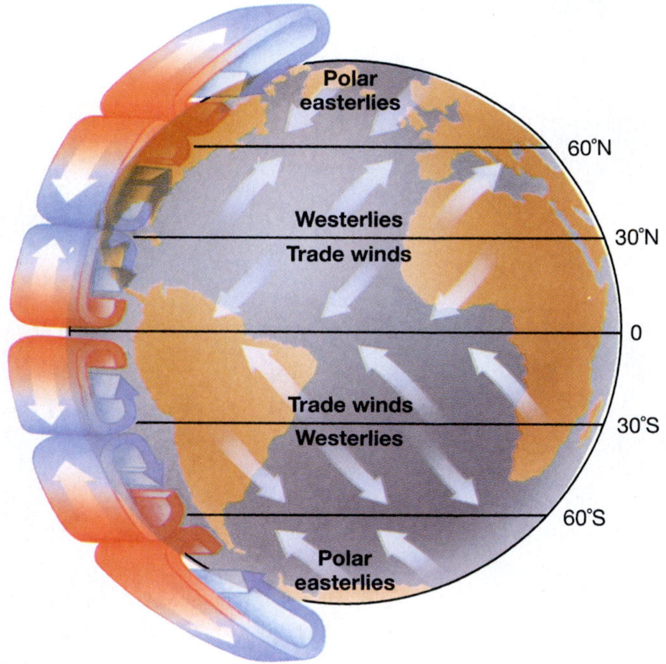

FIGURE 47–15 Atmospheric circulation helps moderate global temperature by transporting heat from the equator to the poles. The greatest solar energy input occurs at the equator, heating air most strongly in that area. The air rises and travels poleward, but is cooled in the process so that much of it descends again around 30 degrees latitude in both hemispheres. At higher latitudes the patterns of movement are more complex.

Surface Ocean Currents Are Driven by Winds and by the Earth's Rotation

Persistent prevailing winds blowing over the ocean produce mass movements of surface ocean water known as **currents** (Fig. 47–16). *Circular* ocean currents are called **gyres.** For example, in the North Atlantic, tropical trade winds tend to blow toward the west, whereas westerlies in the midlatitudes blow toward the east. This helps establish a clockwise gyre in the North Atlantic. Thus, surface ocean currents and winds tend to move in the same direction, although there are many variations to this general rule. Other factors that contribute to ocean currents include the Earth's rotation, varying densities of water, and the position of land masses.

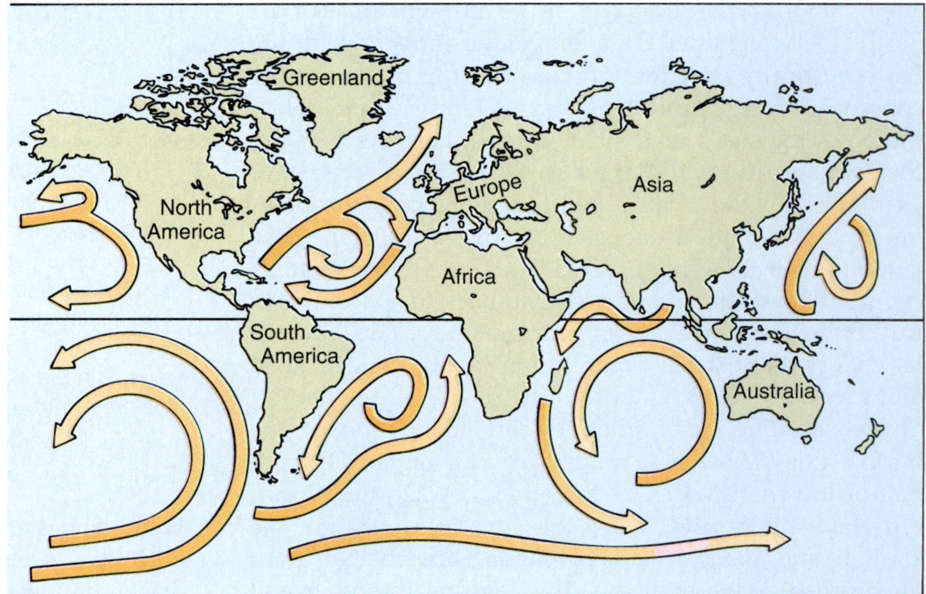

FIGURE 47–16 The basic pattern of surface ocean currents is largely caused by the action of winds. The main ocean current flow—clockwise in the Northern Hemisphere and counterclockwise in the Southern Hemisphere—results partly from the Earth's rotation.

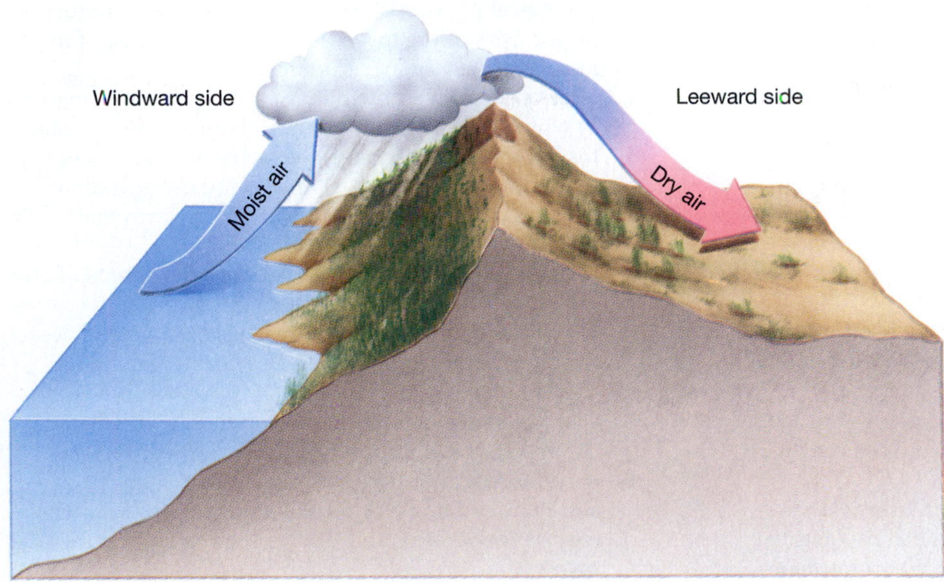

FIGURE 47–17 Rain shadow. When wind blows moist air over a mountain range, precipitation occurs on the windward side of the mountain, causing a dry rain shadow on the leeward side. Such a rain shadow occurs in Washington state east of the Cascade Mountains.

Precipitation Patterns Are Affected by Air and Water Movements and by Surface Features of the Land

The heavy rainfall areas of the tropics result mainly from the equatorial upwelling of moisture-laden air. High surface water temperatures cause evaporation of vast quantities of water from tropical oceans, and prevailing winds blow the resulting moist air over land masses. Heating of air by land surfaces warmed by the sun causes the moist air to rise. As it rises, the air cools and moisture condenses from water vapor and falls as precipitation. The air eventually returns to Earth on either side of the Equator between the Tropics of Cancer and Capricorn (between the latitudes of 23.5° North and 23.5° South). By then, most of its moisture has precipitated so that dry air returns to the Equator. This dry air makes little biological difference over the ocean, but lack of moisture in returning air produces some of the great tropical deserts, such as the Sahara Desert.

Air also dries from long journeys over land masses. Near the windward (side from which the wind blows) coasts of continents, rainfall may be heavy. However, continental interiors are usually dry, because they are far from oceans that replenish the water in the air that passes over them.

Moisture is also removed from air by mountains, which cause air masses to rise. If prevailing winds blow onto a mountain range, precipitation occurs primarily on the windward slopes of mountains. This situation occurs on the North American west coast, where the precipitation falls on western slopes of the mountains. Downwind, or leeward (in this case, east of the mountain range), a low-precipitation **rain shadow** (Fig. 47–17) develops, often generating deserts. Thus, some of the regional differences in precipitation result from the drying of air as it is returned to more equatorial areas, some from long travel over continents, and some from cooling produced by mountainous regions.

Chapter Summary

I. Biogeochemical cycles are the cycling of matter from the environment to living things and back to the environment.
   A. Carbon enters plants, algae, and cyanobacteria as CO_2, which is incorporated into organic molecules by photosynthesis. Cellular respiration, combustion, and erosion return CO_2 to the atmosphere, again making it available for producers.
   B. There are five steps in the nitrogen cycle.
      1. Nitrogen fixation is the conversion of nitrogen gas to ammonia.
      2. Nitrification is the conversion of ammonia to nitrate, one of the main forms of nitrogen used by plants.
      3. Assimilation is the biological conversion of nitrates or ammonia into proteins and other nitrogen-

containing compounds by plants. The conversion of plant proteins into animal proteins is also part of assimilation.

4. Ammonification is the conversion of organic nitrogen to ammonia.
5. Denitrification converts nitrate to nitrogen gas.

C. The phosphorus cycle has no biologically important gaseous compounds.
   1. Phosphorus erodes from rock as inorganic phosphate, which is absorbed from the soil by the roots of plants.
   2. Animals obtain the phosphorus they need from their diets. Decomposers release inorganic phosphate into the environment.
   3. Phosphorus can be lost from biological cycles for millions of years when it washes into the ocean and is deposited in sea beds.

D. The hydrologic cycle, which continually renews the supply of water that is so essential to life, involves an exchange of water between the land, the atmosphere, and living things.
   1. Water enters the atmosphere by evaporation and transpiration, and leaves the atmosphere as precipitation.
   2. On land, water filters through the ground or runs off to lakes, rivers, and oceans.

II. Energy flows through an ecosystem in a linear direction, from the sun to producer to consumer to decomposer. Much of this energy is converted to less useful heat as the energy moves from one organism to another.
   A. Trophic relationships may be expressed as food chains, or more realistically, as food webs, which show the multitude of alternative pathways that energy may take among the producers, consumers, and decomposers of an ecosystem.
   B. Ecological pyramids express the progressive reduction in numbers of organisms, biomass, and energy found in successively higher trophic levels.
   C. Gross primary productivity of an ecosystem is the rate at which energy accumulates as biomass during photosynthesis. Net primary productivity is the energy that remains (as biomass) after cellular respiration.

III. The unique planetary environment of the Earth makes life possible.
   A. Sunlight is the primary (almost the sole) source of energy available to the biosphere.
      1. Of the solar energy that reaches the Earth, 30% is immediately reflected away and the remaining 70% is absorbed.
      2. Ultimately, all absorbed solar energy is radiated into space as infrared (heat) radiation.
   B. A combination of the Earth's roughly spherical shape and the tilted angle of its axis concentrates solar energy at the equator and dilutes solar energy at the poles.
      1. The tropics are therefore hotter and less variable in climate than are temperate and polar areas.
      2. Seasons are determined by the inclination of the Earth's axis.
   C. Atmospheric heat transfer from the Equator to the poles produces both a movement of warm air toward the poles and of cool air toward the Equator, thus moderating the climate.
   D. Surface ocean currents result primarily from prevailing winds and the Earth's rotation.
   E. Precipitation is greatest where warm air passes over the ocean, absorbing moisture, and then is cooled, such as when humid air is forced upward by mountains. Deserts develop in the rain shadows of mountain ranges or in continental interiors.

Selected Key Terms

ammonia, p. 949
ammonification, p. 951
assimilation, p. 951
biogeochemical cycle, p. 948
biomass, p. 957
climate, p. 961
denitrification, p. 951
ecological pyramid, p. 957
food chain, p. 954
food web, p. 955
gross primary productivity, p. 957
net primary productivity, p. 957
nitrate, p. 951
nitrification, p. 951
nitrogen fixation, p. 949
phosphate, p. 951
rain shadow, p. 963
trophic level, p. 957

Post-Test

1. A community and its environment best defines a/an _______________.

2. Carbon dioxide enters living organisms by the biological process of _______________; carbon dioxide is returned to the atmosphere by the biological process of _______________ _______________.

3. Ammonia is oxidized to nitrate by _______________ bacteria.

4. The _______________ cycle does not have a gaseous component.

5. The global recycling of water is called the _______________ _______________.

6. The primary source of energy for almost all ecosystems is the _______________.

7. A/An _______________ _______________ is a complex of interconnected food chains in an ecosystem.

8. Each level in a food web is called a/an ______________ level.

9. In a food web, producers are eaten by ______________ ______________.

10. The three types of ecological pyramids are the pyramids of ______________, ______________, and ______________.

11. The quantitative estimate of the total mass, or amount, of living material in an area at a particular time is called ______________.

12. Net primary productivity equals gross primary productivity minus ______________ ______________.

13. The warmth and constancy of equatorial climates results mostly from the sun's rays hitting the surface ______________ (obliquely or vertically).

14. Mountain ranges may produce downwind arid ______________ ______________.

15. The movements of matter in ecosystems is ______________, whereas the movement of energy is ______________. (a) linear, cyclic (b) cyclic, linear (c) linear, linear (d) cyclic, cyclic

16. Biological nitrogen fixation is performed by microorganisms, such as nitrogen-fixing bacteria that form mutualistic associations with (a) fungi (b) mycorrhizae (c) roots of legumes (d) ammonifying bacteria.

17. Which of the following is a primary consumer? (a) elephant seal (b) krill (c) algae (d) king penguin (e) albatross

18. The amount of energy in any trophic level is always ______________ the amount of energy in the next level below. (a) greater than (b) equal to (c) less than

19. Where would you expect precipitation to be heavy? (a) continental interiors (b) the windward coasts of continents (c) the windward slopes of mountains (d) both a and c (e) both b and c

Review Questions

1. Why is the cycling of matter essential to the continuance of life on the Earth?
2. Diagram the carbon cycle, including the following biological processes: photosynthesis, cellular respiration, and combustion.
3. List and describe the five steps in the nitrogen cycle.
4. Why is the concept of a food web generally preferable to that of a food chain?
5. Define climate and list several features that climate includes.
6. What basic forces determine the circulation of the Earth's atmosphere? Describe the general directions of atmospheric circulation.
7. What basic forces produce the main ocean currents? Describe the general directions of main ocean currents.
8. What conditions produce regional differences in precipitation such as those that cause rain forests and deserts?
9. Draw a food web containing organisms found in the Antarctic.

Thinking Critically

1. Although DDT has been banned in the U.S. for over two decades, it still contaminates our food. Explain. Should we be concerned about our consumption of DDT? Why or why not?
2. Since photosynthesis is the conversion of CO_2 and H_2O to sugar and oxygen, and cellular respiration is the conversion of sugar and oxygen to CO_2 and H_2O, why doesn't each process cancel out the effect of the other in a plant that is both photosynthesizing and respiring?
3. Describe the simplest stable ecosystem that you can imagine.
4. How might humans disturb the temperature balance of the Earth?
5. Suggest a possible food chain that might have an inverted pyramid of numbers (that is, greater numbers of living organisms at higher trophic levels than at lower trophic levels).
6. Is it possible to have an inverted pyramid of energy? Why or why not?

Recommended Readings

Caraco, N.F., "Disturbance of the Phosphorus Cycle: A Case of Indirect Effects of Human Activity," *Tree*, Vol. 8, No. 2, February 1993. Humans have caused both direct and indirect effects on the phosphorus cycle.

Pimm, S.L., J.H. Lawton, and J.E. Cohen, "Food Web Patterns and Their Consequences," *Nature*, Vol. 340, No. 25, April 1991. A nice review article on current ecological knowledge of food webs.

Expanding deserts in Mauritania have engulfed a house, forcing the occupants to leave. *(Steve McCurry/ Magnum Photos, Inc.)*

EXPANDING DESERTS

Rangelands are semi-arid grasslands that are found in both temperate and tropical climates. They serve as important areas of food production for humans by providing fodder for domestic animals such as sheep, cattle, and goats. Grasses, the predominant vegetation of rangelands, have a fibrous root system, in which many roots form a diffuse network in the soil to anchor the plant. Plants with fibrous roots hold the soil in place quite well, thereby reducing soil erosion. If only the upper portion of the grass is eaten by animals, the roots can continue to develop, allowing the plant to recover and grow to its original size.

The **carrying capacity** of a rangeland is the maximum number of animals that rangeland plants can sustain (carrying capacity was defined in a broader sense in Chapter 45). When the carrying capacity of a rangeland is exceeded, grasses and other plants are **overgrazed,** that is, so much of the plant is consumed by the grazing animals that it cannot recover, and it dies. Overgrazing results in barren, exposed soil that is susceptible to erosion (see Chapter 28 Opener).

Most of the world's rangelands occur in areas that have extended natural droughts. Under normal conditions, native grasses survive severe drought: the aboveground portion of the plant dies back, but underground, the extensive root system remains alive and holds the soil in place. When the rains return, roots send forth new aboveground growth.

When overgrazing and extended drought occur together, however, once-fertile rangeland can be converted to desert. The lack of plant cover due to overgrazing allows wind to erode the soil. Even when the rains return, the land is so degraded that it cannot recover. Water erosion removes the little bit of remaining topsoil, and the sand that is left behind forms dunes. This process, which converts rangeland (or tropical dry forest) to desert, is called **desertification.** It ruins economically valuable land, forces out wildlife, and threatens endangered species.

Expansion of deserts is causing rapid deterioration of rangelands worldwide. It is calculated that 10% of the world's land surface has already been desertified (that is, its topsoil has been lost) and that an additional 25% is at risk.

Desertification is related to overpopulation (Chapter 45). In the 1970s, a devastating drought occurred in the African Sahel, an area south of the Sahara Desert, from Senegal to Sudan. Then, from 1980 to 1986, another disastrous drought struck the arid lands of East Africa, particularly those in Ethiopia, Sudan, and Mozambique. In both cases, people living in affected regions suffered greatly, with many children and adults starving after their livestock died and their crops failed. The semiarid lands of Africa have always had periodic droughts, but what made these particular droughts so devastating was the large number of people (and livestock) attempting to live on ecologically fragile land. They overwhelmed the land, degrading it by chopping down most of the trees for firewood and severely overgrazing the rangeland with their livestock. Had fewer people been living in such a marginal area, they might have been spared the horrors of starvation.

Rangelands are not the only areas to have experienced deleterious human impact. With our expanding population, humans have affected all major ecosystems in one way or another. This chapter describes Earth's major terrestrial (land) and aquatic ecosystems, each of which is characterized by distinctive environmental conditions (for example, climate and soil) and living organisms. Human impacts on each ecosystem are also considered.

Major Ecosystems of the World

48

Learning Objectives

After you have studied this chapter you should be able to:

1. Define biome.
2. Portray the eight major terrestrial biomes, giving attention to the climate, soil, and characteristic plants and animals of each.
3. Explain permafrost and describe where it is found and its effects on plant life.
4. Distinguish between temperate deciduous forests and temperate rain forests.
5. Identify the biomes that make the best agricultural lands and explain why they are superior.
6. Describe some of the water-conserving adaptations of organisms that live in deserts.
7. Relate at least one human effect on each of the biomes described in the chapter.
8. Explain the important environmental factors that affect aquatic ecosystems.
9. Distinguish among plankton, nekton, and benthos.
10. Briefly describe the various freshwater, estuarine, and marine ecosystems, giving attention to the environmental characteristics and representative organisms of each.

Key Concepts

- ☐ Climate, particularly temperature and precipitation, influences the distribution of living organisms.
- ☐ In each major kind of climate, a distinctive type of vegetation develops. For example, desert plants are associated with arid climates, grasses with semiarid climates, and forests with moist climates. Certain animals and other types of organisms are associated with each major type of vegetation.
- ☐ Major terrestrial ecosystems, called biomes, extend over large geographical areas. Each biome is characterized by a similar climate, soil, plants, animals, and other organisms, regardless of its geographical location. In like manner, aquatic organisms are characteristically assembled in each of the major aquatic ecosystems.

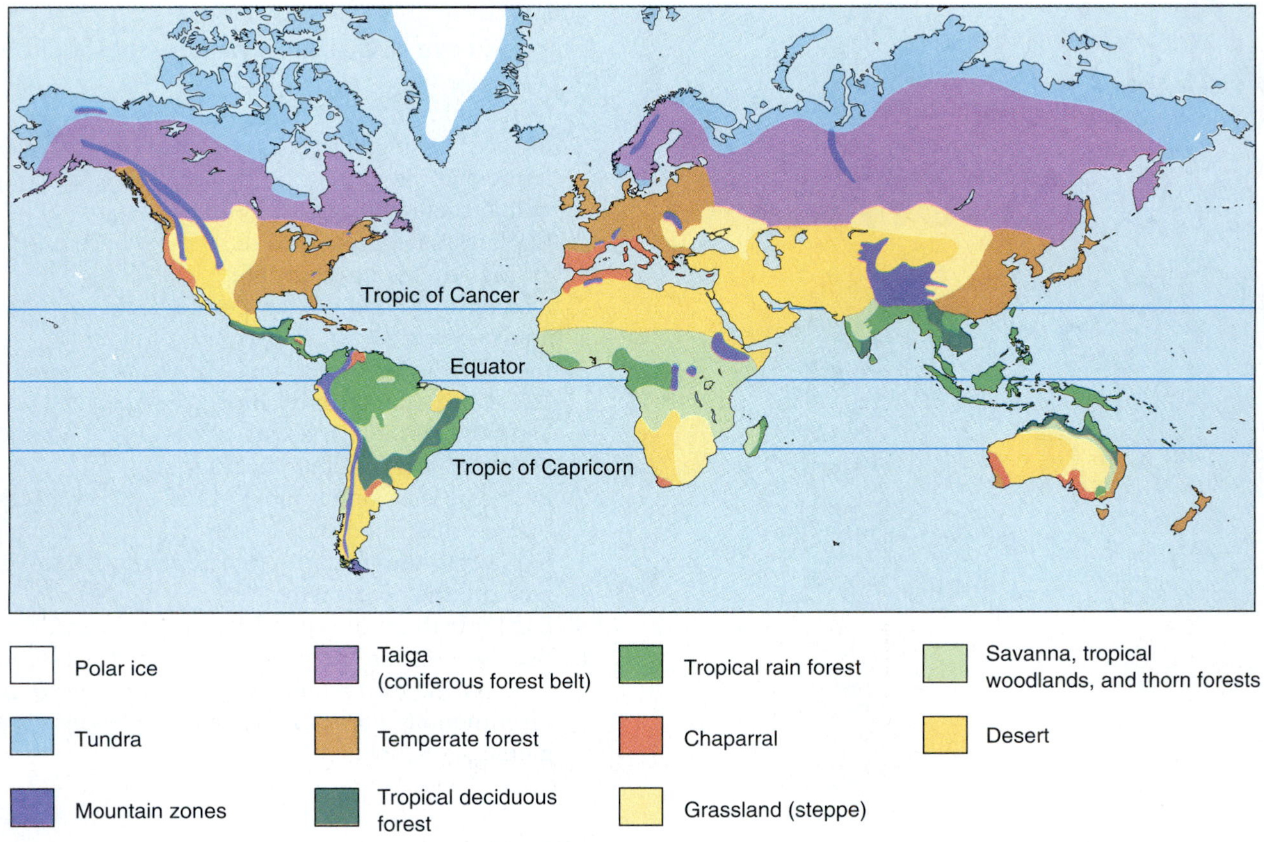

FIGURE 48–1 Earth's biomes (major terrestrial ecosystems). Biomes are distributed primarily in accordance with two factors, temperature and precipitation. At higher latitudes (further from the equator), temperature is the more important of the two. In temperate and tropical zones, precipitation is a significant determinant of community composition.

MAJOR TERRESTRIAL ECOSYSTEMS, CALLED BIOMES, ARE LARGELY DETERMINED BY CLIMATE

A **biome** is a large, relatively distinct ecosystem that is characterized by similar climate, soil, plants, and animals, regardless of where it occurs (Fig. 48–1). Examples of biomes include deserts, tropical rain forests, and tundra. A biome's boundaries are determined by climate more than any other factor. Tundra, the most northern biome, is colder and has shorter growing seasons than do other biomes. Few plants tolerate these conditions, and plant diversity there is less than in warmer biomes. In locations with the same latitude (and as a result, similar temperatures), precipitation becomes a more critical climatic factor than does temperature. Differences in precipitation produce the temperate communities of desert, grassland, and forest, in increasing order.

Tropical and subtropical biomes, located in lower latitudes near the Equator, lack pronounced temperature differences throughout the year. They are at least as varied as temperate biomes. Like temperate biomes, tropical and subtropical biomes are determined mainly by the amount and seasonality of precipitation they receive. Thus, there are not only tropical forests, but also tropical grasslands and tropical deserts. In the tropics, seasonal distribution of rainfall is especially important. Some tropical grasslands would be rain forests (in terms of the *amount* of precipitation they receive) except that almost all of their rainfall occurs during just two months of the year. Lush rain forest vegetation could scarcely persist for 10 months without water!

Tundra Consists of Cold, Boggy Plains of the Far North

The **tundra** exists in extreme northern latitudes wherever snow melts seasonally (Fig. 48–2). (The Southern Hemisphere has no equivalent of the Arctic tundra because it has no land in the proper latitudes.) Tundra has long,

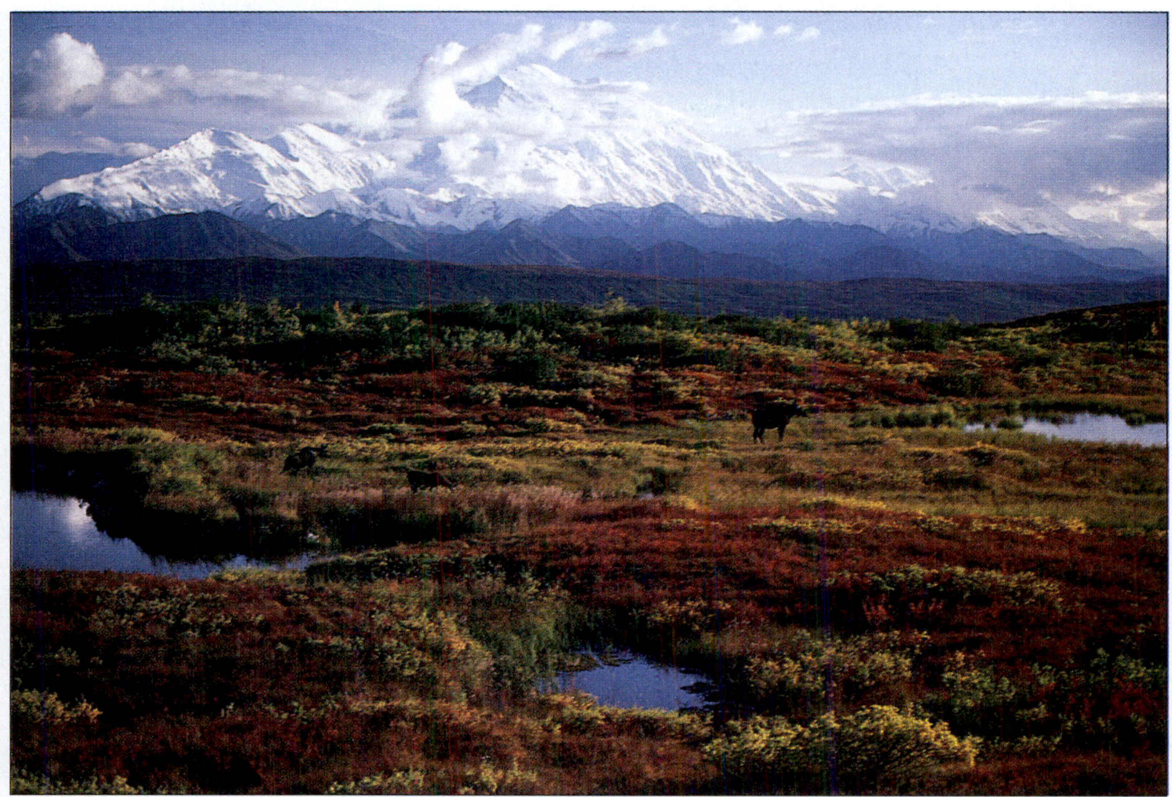

FIGURE 48–2 Tundra. During its short growing season, small, hardy plants grow in tundra, the northernmost biome that encircles the Arctic Ocean. (Michio Hoshino/Minden Pictures)

harsh winters and short summers. Although the growing season with its warmer temperatures is short (from 50 to 160 days depending on location), the days are long. In many places, the sun does not set at all for a considerable number of days in midsummer, although the amount of light at midnight is only one-tenth that at noon. There is little precipitation (10 to 25 centimeters, or 4 to 10 inches, per year) over much of the tundra, with most of it falling during summer months.

Tundra soils tend to be geologically young, since most of them were formed only after the last Ice Age.[1] These soils are usually nutrient-poor and have little organic litter. Although surface soil melts during summer, tundra has deeper layers of permanently frozen ground called **permafrost** (that varies in depth and thickness). Permafrost interferes with drainage (soil is usually waterlogged during the summer) and prevents the roots of larger plants from becoming established. Limited precipitation, in combination with low temperatures, flat topography (surface features), and permafrost produces a landscape of broad shallow lakes, sluggish streams, and bogs.

Few species are found in the tundra, but individual species present often exist in great numbers. Tundra is dominated by mosses, lichens (such as reindeer moss), grasses, and grasslike sedges. There are no readily recognizable trees or shrubs except in very sheltered localities, although dwarf willows and other dwarf trees only a few centimeters tall are common. Tundra plants seldom grow taller than 30 centimeters (12 inches) in open areas.

Year-round animal life of the tundra includes lemmings, voles, weasels, arctic foxes, snowshoe hares, ptarmigan, snowy owls, and musk-oxen. In the summer, caribou migrate north to the tundra to graze on sedges, grasses, and dwarf willow. Dozens of birds also migrate north in summer to nest and feed on abundant insects. (Mosquitos, blackflies, and deerflies survive the winter as eggs or pupae and occur in great numbers during summer weeks.) There are no reptiles or amphibians.

Tundra regenerates very slowly after it has been disturbed. Even casual use by hikers can cause damage. Long-lasting injury, likely to persist for hundreds of years, has occurred to large portions of the arctic tundra as a result of oil exploration and military use.

[1] Glacier ice, which occupied about 29% of the Earth's land during the last Ice Age, began retreating about 17,000 years ago. Today, glacier ice occupies about 10% of the land surface.

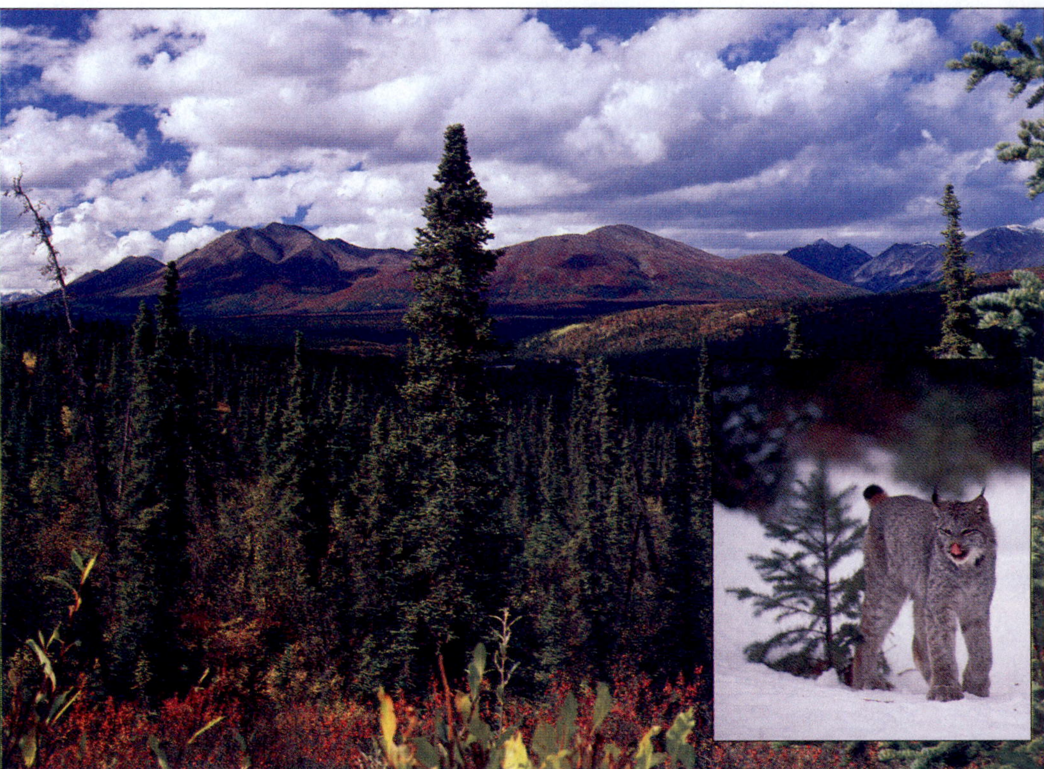

FIGURE 48–3 Taiga, or boreal forest. These coniferous forests occur in cold regions of the Northern hemisphere adjacent to the tundra. (***Inset***) A lynx steps into a clearing in the forest. Winters are harsh in the taiga, and unlike the lynx, many animals migrate or hibernate. (Charlie Ott/Photo Researchers, Inc.; (*Inset*) Carl R. Sams II/Dembinsky Photo Associates)

The Taiga, or Boreal Forest, Is an Evergreen Forest of the North

Just south of the tundra is the **taiga,** or **boreal forest,** which stretches across both North America and Eurasia, covering approximately 11% of the Earth's land (Fig. 48–3). A biome comparable to the taiga is not found in the Southern Hemisphere. Winters are extremely cold and severe, although not as harsh as in the tundra. The growing season of the boreal forest is somewhat longer than that of the tundra. Taiga receives little precipitation, perhaps 50 centimeters (20 inches) per year, and its soil is typically acidic, mineral poor, and characterized by a deep layer of partly decomposed pine and spruce needles at the surface. Permafrost is patchy and, where found, is often deep underneath the soil. Taiga contains numerous ponds and lakes in water-filled depressions that were dug in the ground by grinding ice sheets during the last Ice Age.

Deciduous trees such as aspen or birch, which shed their leaves in autumn, may form striking stands in the taiga, but overall, spruce, fir, and other conifers (cone-bearing evergreens) clearly dominate. Conifers have many drought-resistant adaptations such as needle-like leaves with a minimal surface area for water loss. Such an adaptation enables conifers to withstand the "drought" of northern winter months (roots cannot absorb water when the ground is frozen).

Animal life of the boreal forest consists of some larger species such as caribou (which migrate from the tundra to the taiga for winter), wolves, bears, and moose. However, most animal life is medium-sized to small, including rodents, rabbits, and fur-bearing predators like lynx, sable, and mink. Most species of birds abundant in the summer migrate to warmer climates for winter. Insects are abundant, but there are few amphibians and reptiles except in the southern taiga.

Most of the taiga is not well suited to agriculture because of its short growing season and mineral-poor soil. However, the boreal forest yields vast quantities of lumber and pulpwood (for making paper products), plus animal furs and other forest products.

Forests Occur in Temperate Areas where Precipitation Is Relatively High

In temperate latitudes, precipitation varies greatly with location. Continental interiors tend to be dry for a variety of reasons (Chapter 47). Permanent high-pressure areas, such as those over the Sahara Desert, may nudge moist air masses aside. Air passing over large land masses dries out since there is little opportunity to be recharged with fresh moisture.

Climate of the North American continent, especially the West, is dominated by rain shadows cast by mountain ranges (Chapter 47). As prevailing westerly winds push against the Cascade Mountains of the Pacific Northwest, masses of moist air from the Pacific Ocean are forced upward, where they cool and precipitate much of their moisture. Thus, the western slopes of mountains are so well watered that a temperate rain forest develops. Considerable precipitation falls in the upper reaches of eastern slopes also, but by the time air has sunk back to lower altitudes, most available moisture has fallen.

Temperate Rain Forest Is Characterized by Cool Weather, Dense Fog, and High Precipitation

A coniferous **temperate rain forest** occurs on the northwest coast of North America. Similar vegetation exists in southeastern Australia and in southern South America. Annual precipitation in this biome is high, from 200 to 380 centimeters (80 to 152 inches), and this is augmented by condensation of water from dense coastal fogs. The proximity of temperate rain forest to the coastline moderates the temperature, so that there is a narrow seasonal fluctuation in temperature; winters are mild and summers are cool. Temperate rain forest has relatively nutrient-poor soil, although its organic content may be high. (Needles and large fallen branches and trunks accumulate on the ground as litter that takes many years to decay and release nutrients back to the soil).

The dominant vegetation in the North American temperate rain forest is large evergreen trees such as western hemlock, Douglas fir, Sitka spruce, and western arborvitae. Temperate rain forest (like tropical rain forests) is rich in epiphytic vegetation, which include smaller plants that grow nonparasitically on larger trees (Fig. 48–4). Epiphytes in temperate rain forest are mainly mosses, club mosses, lichens, and ferns. Squirrels, deer, and numerous bird species are animals found in temperate rain forest.

Temperate rain forest is one of the richest wood producers in the world, supplying us with lumber and pulpwood. It is also one of the most complex ecosystems on Earth. Care must be taken to avoid over-harvesting original old-growth forest, however, because it takes hundreds of years to develop such an ecosystem. The logging industry typically clearcuts old-growth forest and replants with a monoculture of trees (of a single species) that it can harvest in 40- to 60-year cycles. Thus, the old-growth forest ecosystem, once harvested, never has a chance to redevelop.

FIGURE 48–4 Temperate rain forest. Note the epiphytes hanging from the branches of coniferous trees. This temperate biome is characterized by high amounts of precipitation. (***Inset***) Steller's jay eats pine seeds and acorns in the treetops. (Terry Donnelly/ Dembinsky Photo Associates; (*Inset*) Jim Roetzel/ Dembinsky Photo Associates)

Temperate Deciduous Forest Has a Dense Canopy of Broadleaf Trees that Overlie Saplings and Shrubs

Seasonality (hot summers and cold winters) are characteristic of **temperate deciduous forest,** which occurs in temperate areas where precipitation ranges from about 75 to 126 centimeters (30 to 50 inches) annually. Typically, soil of a temperate deciduous forest consists of a topsoil rich in organic material and a deep, clay-rich lower layer. As organic materials decay, mineral ions are released. If they are not immediately absorbed by the roots of living trees, these ions leach into the clay, where they may be retained.

Temperate deciduous forests of the northeastern and middle eastern United States are dominated by broad-leaved hardwood trees, such as oak, hickory, and beech, that lose their foliage annually (Fig. 48–5). In southern reaches, the number of broad-leaved evergreen trees, such as magnolia, increases.

Temperate deciduous forest originally contained a variety of large mammals, such as puma, wolves, bison, and other species now extinct, plus deer, bears, many small mammals, and birds. Both reptiles and amphibians abounded, together with a denser and more varied insect life than exists today.

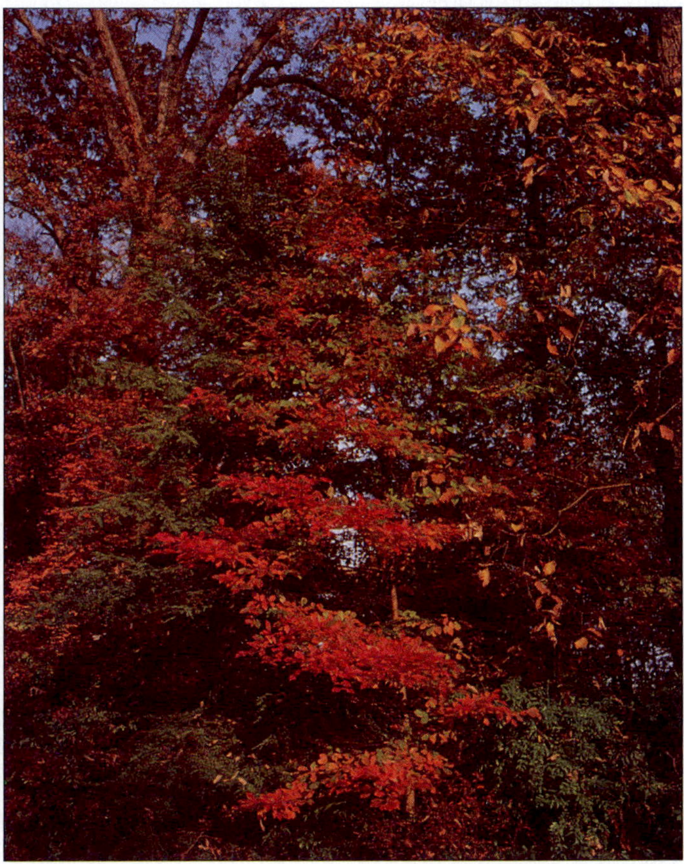

FIGURE 48–5 Temperate deciduous forest. The broad-leaf trees that dominate this biome are deciduous and will shed their leaves before winter. (Dennis Drenner)

Much of the original temperate deciduous forest was removed by logging and land clearing. Where it has been allowed to regenerate, temperate deciduous forest is often in a seminatural state, that is, highly modified by humans.

Worldwide, deciduous forests were among the first biomes to be converted to agricultural use. In Europe and Asia, for example, many soils that originally supported deciduous forests have been cultivated without a substantial loss in fertility for thousands of years by traditional agricultural methods. However, American farmers of the 18th and 19th centuries perceived land as a limitless resource. They often abandoned the wise soil conservation practices of their ancestors and allowed erosion and other forms of soil depletion to damage their land.

Grasslands Occur in Temperate Areas of Moderate Precipitation

Summers are hot, winters are cold, and rainfall is often uncertain in **temperate grasslands.** Annual precipitation averages 25 to 75 centimeters (10 to 30 inches). In grasslands with less precipitation, there is a tendency for minerals to accumulate in a marked layer just below the topsoil. (These minerals tend to wash out in areas with more precipitation.) Grassland soil contains considerable organic material because above-ground portions of many grasses die off each winter and contribute to the organic content of the soil, while roots and rhizomes (underground stems) survive underground. Also, many grasses are sod-formers, that is, their roots and rhizomes form a thick, continuous underground mat.

The North American Midwest is an excellent example of a moist temperate grassland. Here there are few trees, except for those that grow near rivers and streams, and grasses grow in great profusion in the thick, rich soil. Formerly, certain species of grasses grew as tall as a person on horseback, and the land was covered with herds of grazing animals, particularly bison. The principal predators were wolves, although in sparser, drier areas their place was taken by coyotes. Smaller animals included prairie dogs and their predators (foxes, black-footed ferrets, and various birds of prey), grouse, reptiles (such as snakes and lizards), and great numbers of insects.

Shortgrass prairies (Fig. 48–6), such as those found in South Dakota, are temperate grasslands with less precipitation than moist grasslands just described but with greater precipitation than deserts. They have less grass than moister grasslands, and some bare soil is occasionally exposed. Native grasses are drought-resistant. (Rangelands discussed in the Chapter Opener are shortgrass prairies.)

FIGURE 48–6 Temperate grassland. A western grassland of the North American plains is shown. Other temperate grasslands have more luxuriant vegetation because they receive more annual precipitation than this shortgrass prairie. (© David Muench 1994)

Development of the steel plow in postcolonial times, and later the tractor, spelled doom to the original North American grassland. This natural biome was so well suited to agriculture that little of it now remains. Almost nowhere can we seen an approximation of what our ancestors saw as they settled the Midwest. It is not surprising that the American Midwest, the Ukraine, and other moist temperate grasslands became the breadbaskets of the world. These habitats provide ideal growing conditions for crops such as corn and wheat, which also are grasses.

Chaparral Is a Thicket of Evergreen Shrubs and Small Trees

Some temperate habitats have mild winters with abundant rainfall combined with very dry summers. Such **Mediterranean climates,** as they are called, occur not only in the area around the Mediterranean Sea but also in California, southern and southwestern Australia, central Chile, and the Cape region of South Africa. In the North American Southwest, this environment is known as **chaparral.** Chaparral soil is thin and not very fertile. Frequent fires occur naturally in this habitat, particularly in late summer and autumn.

Chaparral vegetation looks strikingly similar in different areas of the world, even though the individual species are quite different. Chaparral is usually dominated by a dense growth of evergreen shrubs, but may contain drought-resistant pine or scrub oak trees (Fig. 48–7). During the rainy winter season, the habitat may be lush and green, but plants lie dormant during the hot, dry summer. Trees and shrubs often have hard, small, leathery leaves that resist water loss. Many plants are also specifically fire-adapted and grow best following a fire. Such growth is possible because fire releases minerals that were tied up in above-ground parts of plants that burned. Underground parts are not destroyed by fire, however, and with the new availability of essential minerals, plants sprout vigorously during winter rains. Mule deer, wood rats, chipmunks, lizards, and many different birds are common animals of the chaparral.

Fires that occur at irregular intervals in California chaparral vegetation are often quite costly because they consume expensive homes built on hilly chaparral landscape. Unfortunately, efforts to control naturally occurring

FIGURE 48–7 Chaparral. Hot, dry summers and mild, rainy winters characterize the Santa Monica Mountains, California. Chaparral vegetation consists primarily of drought-resistant evergreen shrubs. (Visuals Unlimited/John Cunningham)

fires sometimes backfire. Denser, thicker vegetation tends to accumulate when periodic fires are prevented; then, when a fire does occur, it is much more severe. Removing the chaparral vegetation, whose roots hold the soil in place, can also cause problems—witness the mud slides that sometimes occur during winter rains in these areas.

Deserts Are Arid Ecosystems

Deserts are dry areas found in both temperate and tropical regions. The atmosphere's low water content leads to temperature extremes of heat and cold, so that a major change in temperature occurs in a single, 24-hour period. Deserts vary greatly, depending upon the amount of precipitation they receive, which is generally less than 25 centimeters (10 inches) per year. A few deserts are so dry that virtually no plant life occurs in them. As a result of sparse vegetation, desert soil is low in organic material, but is often high in mineral content. In some regions, concentration of certain soil minerals reaches toxic levels.

Plant cover is spotty in deserts, and much desert soil is exposed. Both perennials (plants that live for more than two years) and annuals (plants that complete their life cycle in one growing season) are present. Plants in North American deserts include cacti, yuccas, Joshua trees, and widely scattered bunchgrass (Fig. 48–8). Desert plants tend to have leaves that are reduced or absent, an adaptation that conserves water. Other desert plants shed their

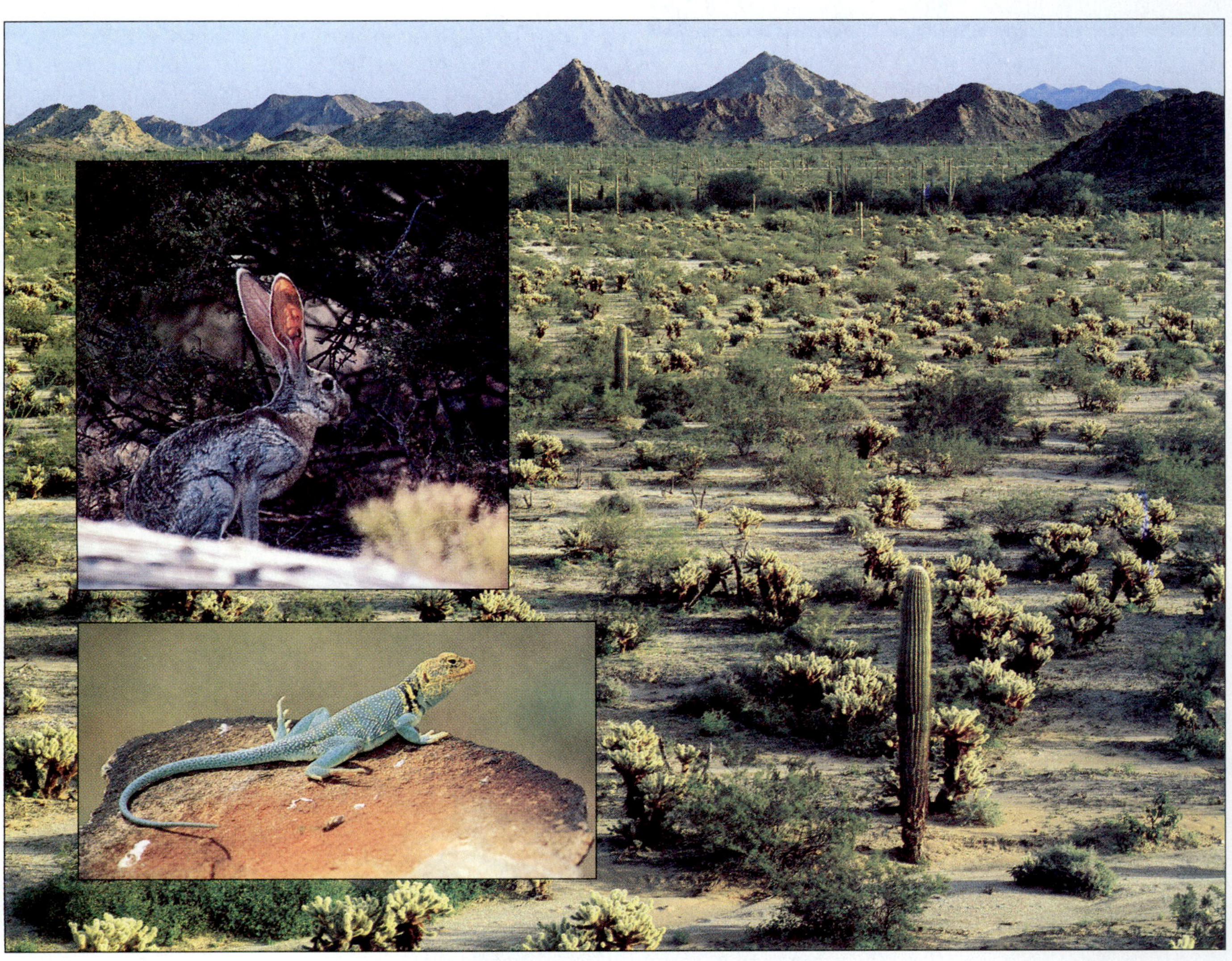

FIGURE 48–8 Desert. Inhabitants of deserts are remarkably adapted to the demands of their environment. The warmer deserts of North America that are characterized by summer rainfall frequently contain large cacti such as the giant saguaro, which grows 15 to 18 meters (50 to 60 feet) tall. (***Top inset***) An antelope jack-rabbit. This animal is so named because its huge ears resemble those of a jackass. Jack-rabbits never drink water, but obtain needed liquid from the plants on which they feed. (***Bottom inset***) A collared lizard suns on a rock and waits for insect prey. Note the bands across its neck (its "collar"). (Willard Clay/Dembinsky Photo Associates; (*Top inset*) Stan Osolinski/Dembinsky Photo Associates; (*Bottom inset*) Rod Planck/ Dembinsky Photo Associates)

FIGURE 48–9 Savanna. These grasslands with scattered acacia trees are found in eastern Africa. They formerly supported large herds of grazing animals and their predators, both of which are swiftly vanishing under pressure from pastoral and agricultural land use. (Frans Lanting/Minden Pictures)

leaves for most of the year, growing only during the brief moist season. Desert plants are noted for **allelopathy,** an adaptation in which toxic substances secreted by roots or shed leaves inhibit the establishment of competing plants nearby. (Allelopathy causes uniform dispersion; see Chapter 45.) Many desert plants are provided with defensive spines, thorns, or toxins to resist heavy grazing pressure often experienced in this food- and water-deficient environment.

Desert animals tend to be small. During the heat of the day, they remain under cover or return to shelter periodically. At night they come out to forage or hunt. In addition to desert-adapted insects, there are many specialized desert reptiles such as lizards, tortoises, and snakes. Mammals include such rodents as the American kangaroo rat, which does not have to drink water but can subsist solely on the water content of its food (primarily seeds and insects). In American deserts, there are also jack-rabbits (Fig. 48–8), and in Australian deserts, kangaroos. Carnivores like the African fennec fox and some birds of prey, especially owls, live on rodents and rabbits.

American deserts have been altered by humans in several ways. Off-road vehicles damage desert vegetation, which sometimes takes years to recover, and certain cacti and desert tortoises are rare as a result of poaching. Also, houses, factories, and farms built in desert areas require vast quantities of water, which must be imported from distant areas.

Savanna Is a Tropical Grassland with Scattered Trees

The **savanna** biome is a tropical grassland with widely scattered trees (Fig. 48–9). Savanna is found in areas of low rainfall or seasonal rainfall with prolonged dry periods. Temperatures in tropical savannas vary little throughout the year. Thus, seasons are regulated by precipitation rather than by temperature as in temperate grasslands. Annual precipitation is 85 to 150 centimeters (34 to 60 inches). Savanna soil is low in essential mineral nutrients, in part because the parent rock from which it is formed is infertile. Although the African savanna is best known, savanna also occurs in South America, western India, and northern Australia.

Savanna is characterized by wide expanses of grasses interrupted by occasional trees such as *Acacia,* which bristles with thorns that provide protection against herbivores. Both trees and grasses have fire-adapted features, such as extensive underground root systems, that enable them to survive periodic fires that sweep through savanna.

The world's greatest assemblage of hoofed mammals occurs in the African savanna. Here live great herds of herbivores, including wildebeest, antelope, giraffe, zebra, and elephants. Large predators, such as lions and hyenas, kill and scavenge the herds. In areas of seasonally varying rainfall, the herds and their predators migrate annually.

Savanna and other tropical grasslands are rapidly being converted to rangeland for cattle and other domesticated animals, which are replacing the big herds of wild animals. Severe overgrazing in places has converted marginal savanna into desert.

Tropical Rain Forests Are Lush Equatorial Forests

Tropical rain forests occur where temperatures are warm throughout the year and precipitation occurs almost daily. The annual precipitation of **tropical rain forests** is

FIGURE 48–10 Tropical rain forest. A broad view of tropical rainforest vegetation along a riverbank in Peru. (***Upper left***) The emerald tree boa lives in rain forests of northern South America where it feeds on birds and other small animals. (***Lower left***) A rainforest grasshopper in Peru. Insects, which represent the largest fraction of animal species known to science, are most diverse in tropical rain forests. (***Right***) The capuchin, an intelligent rainforest monkey that rarely leaves the treetops, is abundant in rain forests of Central and South America. (*Main photo, lower left inset*) James L. Castner; (*Upper left inset*) Joe McDonald © 1995 Animals Animals; (*Right inset*) Ken Cole © 1995 Animals Animals)

200 to 450 centimeters (80 to 180 inches). Much of this precipitation comes from locally recycled water that enters the atmosphere by transpiration (Chapter 27) of the forest's own trees.

Tropical rain forests are often located in areas with ancient, highly weathered, mineral-poor soil. Little organic matter accumulates in such soils. Since temperatures are high year round, decay organisms and detritus-feeding ants and termites decompose organic litter quite rapidly. Mineral nutrients from decomposing material are quickly absorbed by roots. Thus, minerals of tropical rain forests are tied up in the vegetation, rather than in the soil.

Tropical rain forests are very productive. High productivity is stimulated by the abundant solar energy and precipitation. Productivity is high here despite the scarcity of mineral nutrients in the soil.

Of all the biomes on land, the tropical rain forest is unexcelled in species diversity and variety. No one species dominates. Often one can travel for 0.4 kilometer (0.25 mile) or more without encountering two members of the same species of tree.

Trees of tropical rain forests are usually evergreen flowering plants (Fig. 48–10). Their roots are often shallow and form a mat almost 1 meter (about 3 feet) thick on the soil surface. Roots catch and absorb almost all mineral nutrients released from leaves and litter by decay processes. Swollen bases or braces called buttresses hold the trees upright and aid in the extensive distribution of shallow roots (Fig. 48–11).

A fully developed rain forest has at least three distinct stories of vegetation. The topmost story consists of crowns of occasionally very tall trees, some 50 meters (164

FIGURE 48–11 Buttress roots. Tropical rainforest trees typically possess elaborate systems of buttress roots that support them in the shallow, often wet soil.

feet) or more in height. They are exposed to direct sunlight. The middle story, which reaches a height of 30 to 40 meters (100 to 130 feet), forms a continuous canopy of leaves that lets in very little sunlight for the sparse understory. The understory itself consists of both smaller plants that are specialized for life in shade and seedlings of taller trees. Vegetation of tropical rain forests is not dense at ground level except near stream banks or where a fallen tree has opened the canopy. The continuous canopy of leaves overhead produces a dark, extremely moist habitat.

Tropical rain forest trees also support extensive epiphytic communities of smaller plants such as orchids and bromeliads. Although epiphytes grow in crotches of branches, on bark, or even on the leaves of their hosts, they only use their host trees for physical support, not for nourishment.

Since little light penetrates to the understory, many plants living there are adapted to climb upon already established host trees rather than to invest their meager photosynthetic resources in the dead cellulose tissues of their own trunks. Lianas (woody tropical vines), some as thick as a human thigh, twist up through the branches of tropical rain forest trees.

Rainforest animals include the most abundant and varied insect, reptile, and amphibian fauna on Earth (Fig. 48–10). Birds, too, are varied and often brilliantly colored. Most rainforest mammals, such as sloths and monkeys, live only in the trees, although a few large ground-dwelling mammals, including elephants, are also found in rain forests.

Unless strong conservation measures are initiated soon, human population growth and industrialization in tropical countries will spell the end of tropical rain forests *by the end of this century.* Scientists fear that many rain forest organisms will become extinct before they have even been scientifically described (see Chapter 20 Opener).

Concept Connections

Major Ecosystems ⟷ *Climate* ⟷ *Productivity*

In Chapter 47, we discussed the productivity of ecosystems. Recall that gross primary productivity is the rate at which energy is captured by producers of an ecosystem during photosynthesis. Net primary productivity (NPP), on the other hand, is the amount of tissue produced in excess of metabolic costs (that is, gross primary productivity minus cellular respiration).

Ecologists use different methods to measure primary productivity, depending on whether gross or net primary productivity is being assessed. Methods also vary from one type of ecosystem to another. On land, for example, ecologists might cut, dry, and weigh plants at the end of a growing season to measure NPP. However, this method would not be useful to measure NPP in the ocean, where microscopic algae are the main producers.

Although different methods are employed, it is possible to compare NPPs of diverse ecosystems, as in the following table:

Terrestrial Ecosystem	*NPP**
Tropical rain forest	1800
Temperate deciduous forest	1250
Boreal forest	800
Savanna	700
Temperate grassland	500
Tundra	140
Desert	70

*NPP is expressed as grams of tissue per square meter of land per year.

Certain general patterns are evident. As expected, NPP is greatest in the humid tropics and least in cold or arid ecosystems (tundra and desert). In the three forest ecosystems, water availability does not affect NPP as much as does temperature, which depends on light energy. (Water is still an important factor in forests, however. The warmer the climate, the greater the annual precipitation required to support forest vegetation.) In ecosystems with comparable annual temperatures—temperate deciduous forest, temperate grassland, and desert, for example—water availability (as annual precipitation) directly affects NPP. Availability of essential minerals such as nitrogen and phosphorus can also affect NPP.

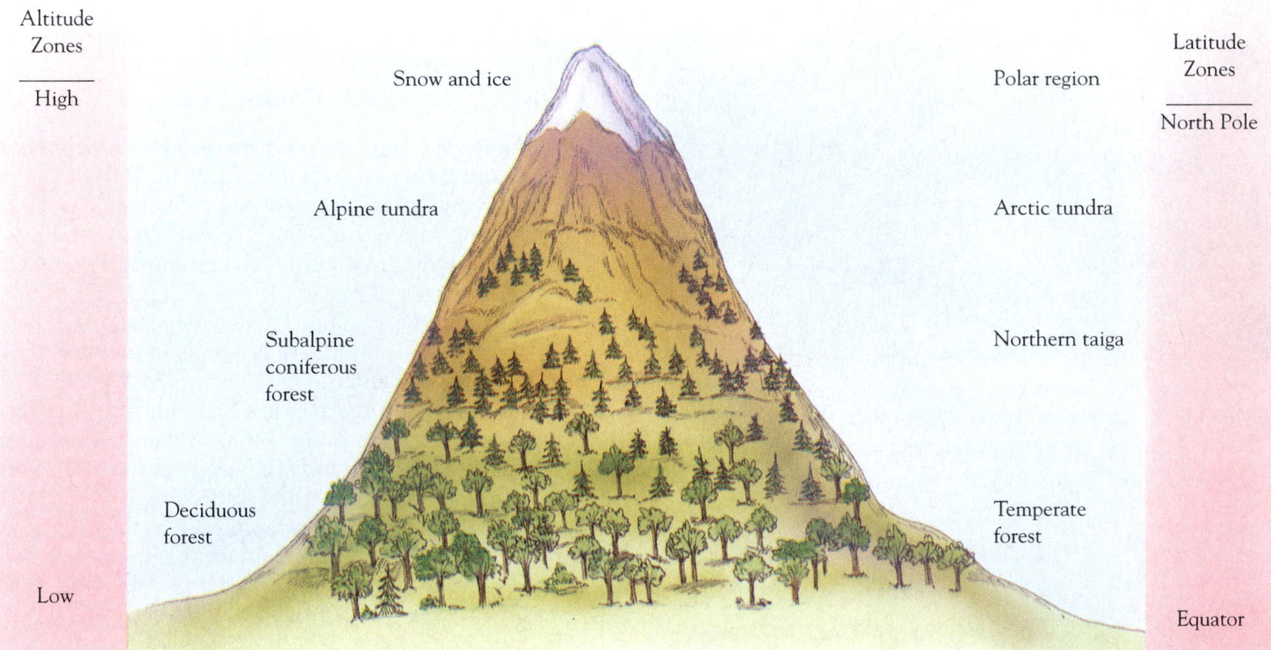

FIGURE 48–12 Patterns of vegetation at different altitudes. The cooler temperatures at higher elevations of a mountain produce a series of biomes similar to those encountered when going from the Equator toward the North Pole.

ALTITUDE ALSO AFFECTS ECOSYSTEMS

Hiking up a mountain is similar to traveling toward the North Pole with respect to the major patterns of vegetation encountered. This altitude–latitude similarity in vegetation occurs because as one climbs a mountain, it gets colder just like it does when one travels north. The types of plants growing on the mountain change as temperatures change (Fig. 48–12).

The base of a mountain in Colorado, for example, might be covered by deciduous trees, which shed their leaves every autumn. Above that altitude, where the climate is colder and more severe, a coniferous forest (called a subalpine forest) that resembles northern taiga grows. Higher still, where the climate is very cold, a kind of tundra occurs with vegetation composed of grasses, sedges, and small tufted plants. This tundra is called alpine tundra to distinguish it from arctic tundra. At the very top of the mountain, a permanent ice or snow cap might be found, similar to nearly lifeless polar land areas.

There are important environmental differences between high altitudes and high latitudes, however, that affect the types of organisms found. Alpine tundra typically lacks permafrost and receives more precipitation than does Arctic tundra. Also, high elevations of temperate mountains do not have the great extremes of day length that are associated with the changing seasons in biomes at high latitudes. Furthermore, the intensity of solar radiation is greater at high elevations than at high latitudes. At high elevations, the sun's rays pass through less atmosphere, which results in a greater exposure to ultraviolet radiation (less UV is filtered out by the atmosphere) than occurs at higher latitudes.

AQUATIC ECOSYSTEMS OCCUPY MOST OF THE EARTH'S SURFACE

Not surprisingly, aquatic ecosystems are different in almost all respects from terrestrial ecosystems. For example, in terrestrial ecosystems, temperature and precipitation are major determinants of plant and animal inhabitants, and light is relatively plentiful (except in certain environments such as the floor of the rain forest). Significant environmental factors in aquatic ecosystems are very different. Temperature is less important in aquatic habitats than it is on land, and water is obviously never an important limiting factor.

Salinity (the concentration of dissolved salts, such as sodium chloride, in a body of water) affects the types of organisms present in aquatic ecosystems, as does the amount of dissolved oxygen. Water also greatly interferes with the penetration of light, so floating aquatic organisms that photosynthesize must remain near the water's surface, and vegetation attached to the bottom can grow only in shallow water. In addition, low levels of essential mineral nutrients often limit the numbers and distribution of living things in certain aquatic environments.

Aquatic ecosystems contain three main categories of organisms: free-floating plankton, strongly swimming nekton, and bottom-dwelling benthos. **Plankton** are small or microscopic organisms that are relatively feeble swimmers. For the most part, they are carried about by currents and waves (Fig. 48–13). They are unable to swim far horizontally, but some species are capable of large daily vertical migrations and are found at different depths of water at different times of day or at different seasons.

Plankton are generally subdivided into two major categories: phytoplankton and zooplankton. **Phytoplankton** (photosynthetic cyanobacteria and free-floating algae of several types) are producers that are the base of most aquatic food webs. **Zooplankton** are nonphotosynthetic organisms that include protozoa (animal-like protists) and small animals, including larval stages of many animals that are large as adults.

Nekton are larger, stronger swimming organisms such as fish and turtles. **Benthos** are bottom-dwelling organisms that fix themselves to one spot (oysters and barnacles), burrow into the sand (many worms and echinoderms), or simply walk about on the bottom (lobsters and brittle stars).

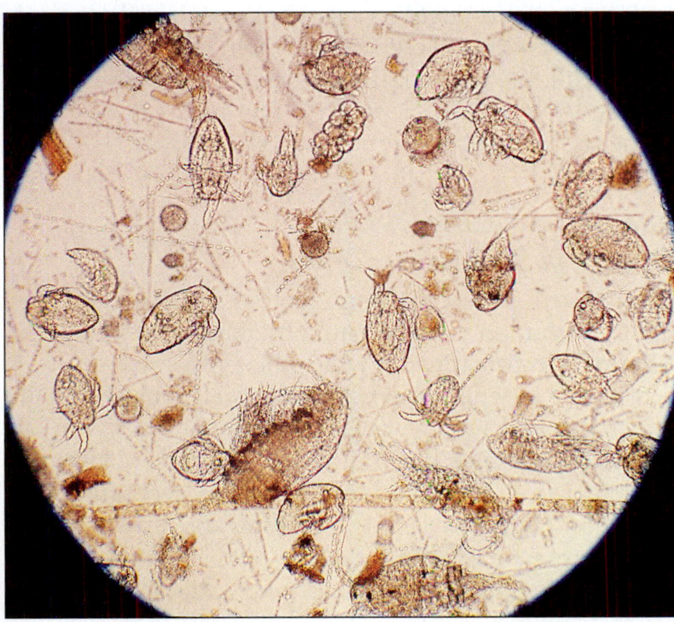

FIGURE 48–13 Marine plankton. Most of the organisms shown here are nauplius larvae, tiny immature crustaceans that hatch from eggs. Nauplius larvae eventually mature into several types of small crustaceans, including fairy shrimp, tadpole shrimp, and copepods, all of which graze on diatoms in the plankton and are in turn the food for larger aquatic organisms. (Runk/Schoenberger, from Grant Heilman)

The most fundamental division in aquatic ecology is probably between freshwater and saltwater habitats.

Freshwater Ecosystems Occupy a Relatively Small Portion of the Earth's Surface, Yet Are Ecologically Very Important

Freshwater ecosystems include rivers and streams (flowing water), lakes and ponds (standing water), and marshes and swamps (freshwater wetlands). Each type of freshwater ecosystem is distinguished by its own specific environmental conditions and characteristic living organisms.

Freshwater ecosystems have important ecological roles in the hydrologic cycle (Chapter 47). They assist in recycling precipitation that flows as surface runoff to the ocean. Large bodies of fresh water also help moderate daily and seasonal temperature fluctuations on nearby land.

Many Different Conditions Exist Along the Length of a River or Stream

The nature of a flowing-water ecosystem changes greatly from its source (where it begins) to its mouth (where it empties into another body of water). For example, headwater streams (small streams that are the sources of a river) are usually shallow, cold, swiftly-flowing, and therefore highly oxygenated. In contrast, rivers located downstream from headwaters are wider, deeper, not as cold, slower-flowing, and therefore less oxygenated.

Organisms found in flowing-water ecosystems vary greatly from one stream to another, depending primarily on the strength of the current. In streams with fast currents, inhabitants may have adaptations such as suckers to attach themselves to rocks so they are not swept away, or they may have flattened bodies to enable them to slip under or between rocks. Organisms in large, slow-moving streams and rivers do not need such adaptations, although they are typically streamlined (as are most aquatic organisms) to lessen resistance when moving through water. Where current is slow, plants and animals of the headwaters are replaced by those characteristic of ponds and lakes.

Unlike other freshwater ecosystems, streams and rivers depend on land for much of their energy. In headwater streams, for example, up to 99% of the energy input comes from detritus, dead organic material (such as leaves), carried from the land into streams and rivers by wind or surface drainage. Downstream, rivers contain more producers and therefore have a slightly lower dependence on detritus as a source of energy than in the headwaters.

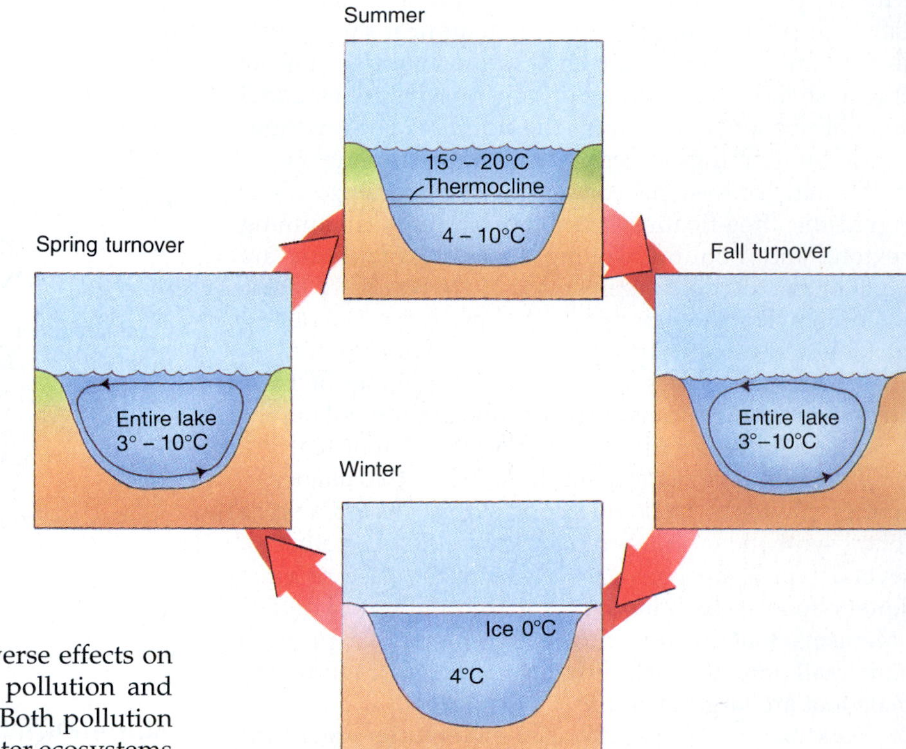

FIGURE 48–14 Turnover in a deep temperate lake. Fall and spring turnovers cause the mixing of the upper and lower layers of water. This circulation, which tends to equalize temperatures throughout the lake, brings oxygen to the oxygen-depleted depths and minerals to the mineral-deficient surface waters.

Human activities result in several adverse effects on rivers and streams. These include water pollution and dams, which are built to impound water. Both pollution and dams change the nature of flowing-water ecosystems downstream.

Standing Water Ecosystems Are Characterized by Zonation

A large lake has three basic layers or zones: the littoral, limnetic, and profundal zones. Smaller lakes and ponds typically lack a profundal zone. The **littoral zone** is a shallow water area along the shore of a lake or pond. It includes lake shore vegetation such as cattails and burreeds, plus several deeper-dwelling aquatic plants and algae. The littoral zone is the most highly productive zone of the lake (that is, photosynthesis is greatest here), in part because it receives nutrient inputs from surrounding land that stimulate the growth of plants and algae. Animals of the littoral zone include frogs and their tadpoles; turtles; worms; crayfish and other crustaceans; insect larvae; and many fish such as perch, carp, and bass. Here, too, at least in the quieter areas, one finds surface dwellers such as water striders and whirligig beetles.

The **limnetic zone** is the open water area away from the shore; it extends down as far as the sunlight penetrates. The main organisms of the limnetic zone are microscopic phytoplankton and zooplankton. Larger fish also spend some of their time in the limnetic zone, although they may visit the littoral zone to feed and reproduce. Owing to its depth, less vegetation grows here.

Beneath the limnetic zone of a large lake is the **profundal zone.** Because no light penetrates to this depth, producers do not live in the profundal zone. Food drifts into the profundal zone from the littoral and limnetic zones. Bacteria decompose dead plants and animals that reach the profundal zone, liberating minerals. These minerals are not effectively recycled because there are no producers to absorb and incorporate them into the food web. As a result, the profundal zone tends to be both mineral-rich and oxygen-deficient, with few forms of life occupying it other than anaerobic bacteria.

Thermal Stratification Occurs in Temperate Lakes

The marked layering of large temperate lakes caused by light penetration is accentuated by thermal stratification. **Thermal stratification,** in which the temperature changes sharply with depth, occurs because the summer sunlight penetrates and warms surface waters, making them less dense.[2] In summer, cool (and therefore more dense) water remains at the lake bottom, and is separated from warm (and therefore less dense) water above by a marked and abrupt temperature transition called the **thermocline.**

In temperate lakes, falling temperatures in autumn cause a mixing of lake waters called **fall turnover** (Fig. 48–14). (Since little seasonal temperature variation occurs in the tropics, such turnovers are not common there.) As surface water cools in fall, its density increases and eventually it displaces the less dense, warmer, mineral-rich

[2] The density of water is greatest at 4°C; both above and below this temperature, water is less dense.

water beneath. Warmer water then rises to the surface where it cools and sinks. This process of cooling and sinking continues until the lake reaches a uniform temperature throughout.

When winter comes, surface water cools below 4°C, its temperature of greatest density and, if it is cold enough, ice forms. Ice, which forms at 0°C, is less dense than is cold water; thus, ice forms on the surface, and the water on the lake bottom is warmer than the ice on the surface.

In spring, a **spring turnover** occurs as ice melts and surface water reaches 4°C. Surface water again sinks to the bottom, and bottom water returns to the surface. As summer arrives, thermal stratification occurs once again.

The mixing of deeper, nutrient-rich water with surface, nutrient-poor water during the fall and spring turnovers brings essential minerals to the surface. The sudden presence of large amounts of essential minerals in surface waters encourages the development of large algal populations, which form temporary **blooms** in the fall and spring.

Freshwater Wetlands Are Lands that Are Transitional between Aquatic and Terrestrial Ecosystems

Freshwater wetlands are usually covered by shallow water for at least part of the year and have characteristic soils and water-tolerant vegetation. They include marshes, in which grasslike plants dominate, and swamps, in which woody trees or shrubs dominate. Freshwater wetlands also include hardwood bottomland forests (lowlands along streams and rivers that are periodically flooded), prairie potholes (small, shallow ponds that first formed when glacial ice melted at the end of the last Ice Age), and peat moss bogs (peat-accumulating wetlands where mosses dominate).

At one time, wetlands were thought of as wastelands, areas that needed to be filled in or drained so that farms, housing developments, or industries could be built on them. Wetlands are also a breeding place for mosquitos and therefore were viewed as a menace to public health. Today, however, the crucial environmental services that wetlands provide are widely recognized, and wetlands are somewhat protected by law.

Wetland plants, which are highly productive, provide enough food to support a wide variety of organisms. Wetlands are valued as a wildlife habitat for migratory waterfowl and many other bird species, beaver, otters, muskrats, and game fish. Wetlands help to control flooding by acting as holding areas for excess water when rivers flood their banks. The floodwater stored in wetlands then drains slowly back into the rivers, providing a steady flow of water throughout the year. Wetlands also serve as groundwater recharging areas. One of their most important roles is to help cleanse and purify water by trapping and holding pollutants in the flooded soil.

Estuaries Occur where Fresh Water and Saltwater Meet

Where the sea meets the land there may be one of several kinds of ecosystems: a rocky shore, a sandy beach, an intertidal mud flat, or a tidal estuary. An **estuary** is a coastal body of water, partly surrounded by land, with access to the open sea and a large supply of fresh water from rivers. Estuaries in temperate climates usually contain **salt marshes,** areas dominated by grasses in which the salinity fluctuates between that of seawater and fresh water (Fig. 48–15). Many estuaries undergo marked variations in temperature, salinity, depth of light penetration, and other physical properties in the course of a year. To

FIGURE 48–15 A salt marsh at Assateague Island National Seashore in Maryland. The plants in this part of the Chesapeake Bay estuary are cordgrass (*Spartina*). (Connie Toops)

survive there, estuarine organisms must have a wide range of tolerance to such changes.

Estuaries are among the most fertile ecosystems in the world, often having a much greater productivity than do adjacent oceans or freshwater rivers. This high productivity is brought about by (1) the action of tides, which promote a rapid circulation of nutrients and help remove waste products; (2) the transport of nutrients from land into rivers and creeks that empty into the estuary; and (3) the presence of many plants, which provide an extensive photosynthetic carpet and whose roots and stems also mechanically trap much potential food material. As plants die, they decay, forming the basis of detritus food webs. The majority of commercially important fish and shellfish spend their larval stages in estuaries among protective tangles of decaying stems.

Salt marshes have often appeared to be worthless, empty stretches of land to uninformed people. As a result, salt marshes have been used as dumps and have become severely polluted or have been filled with dredged bottom material to form artificial land for residential and industrial development. A large part of the estuarine environment has been lost in this way, along with many of its benefits (the same benefits discussed in the section on freshwater wetlands): wildlife habitat, sediment and pollution trapping, flood control, and groundwater supply.

Mangrove forests, the tropical equivalent of salt marshes, cover perhaps 70% of tropical coastlines. Like salt marshes, mangrove forests provide valuable environmental services. They are breeding grounds and nurseries for several commercially important fishes, and their roots stabilize the soil, thereby preventing coastal erosion.

Marine Divisions Include the Intertidal Zone, Neritic Province, and Oceanic Province

Although oceans and lakes are comparable in many ways, there are many differences. Depths of even the deepest lakes, for example, do not approach those of oceanic abysses (Fig. 48–16), which are extremely deep areas that

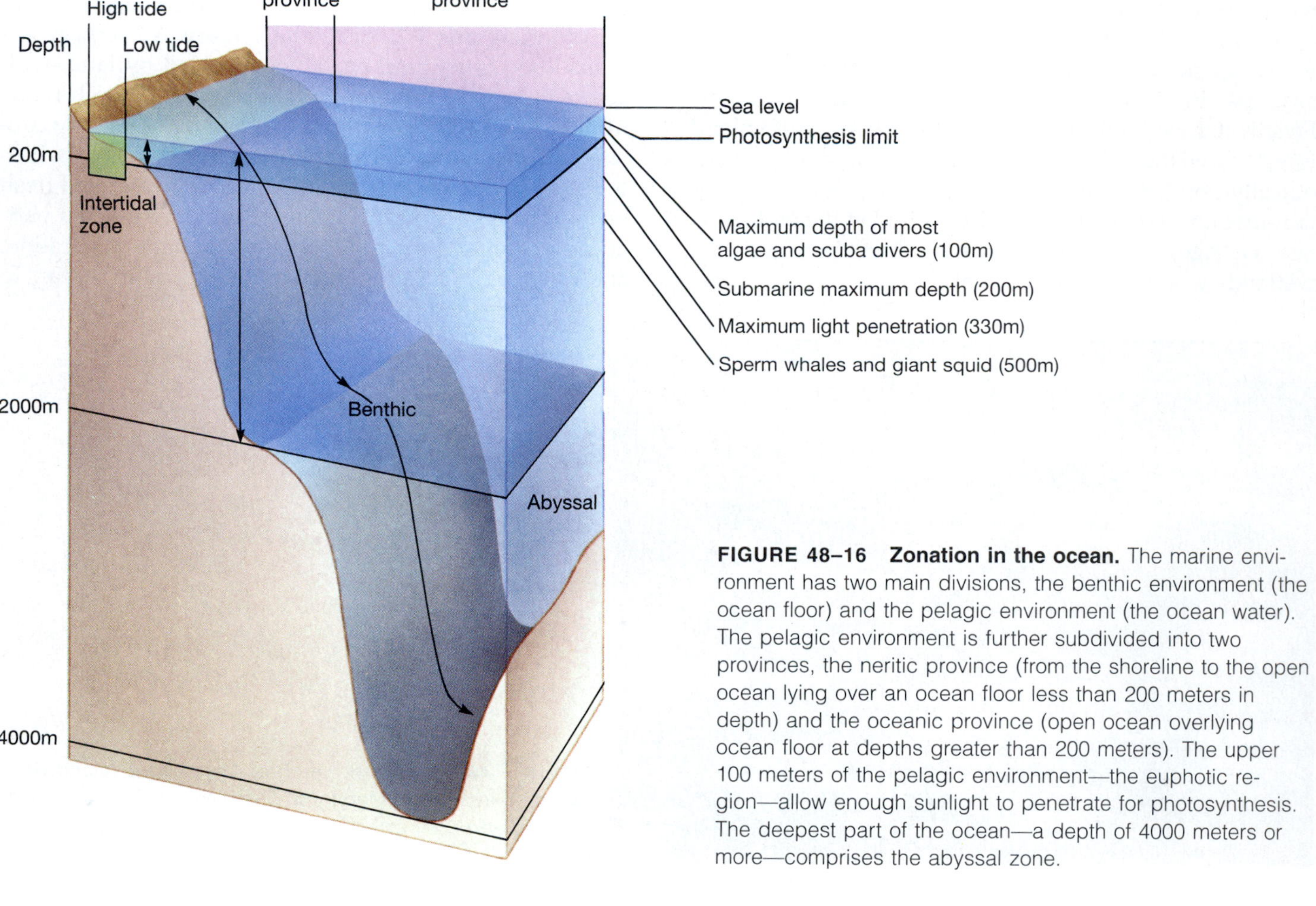

FIGURE 48–16 Zonation in the ocean. The marine environment has two main divisions, the benthic environment (the ocean floor) and the pelagic environment (the ocean water). The pelagic environment is further subdivided into two provinces, the neritic province (from the shoreline to the open ocean lying over an ocean floor less than 200 meters in depth) and the oceanic province (open ocean overlying ocean floor at depths greater than 200 meters). The upper 100 meters of the pelagic environment—the euphotic region—allow enough sunlight to penetrate for photosynthesis. The deepest part of the ocean—a depth of 4000 meters or more—comprises the abyssal zone.

extend more than 6 kilometers (3.6 miles) below the sunlit surface. Oceans are profoundly influenced by tides and currents. Gravitational pulls of both the sun and moon produce two tides a day throughout the oceans, but the height of those tides varies with the phases of the moon (a full moon causes the highest tides), the season, and the local topography.

The shoreline area between low and high tide is called the **intertidal zone.** Although high levels of light and nutrients, together with an abundance of oxygen, make the intertidal zone a biologically productive habitat, it is a very stressful one. If an intertidal beach is sandy, inhabitants must contend with a constantly shifting environment that threatens to engulf them and gives them scant protection against wave action. Consequently, most sand-dwelling organisms, such as mole crabs, are continuous and active burrowers. They are, however, able to follow tides up and down the beach, and so do not usually have any notable adaptations to survive drying or exposure.

A rocky shore provides a fine anchorage for seaweeds and animals. However, it is exposed to constant wave action and, when exposed to the air during low tides, to drying and temperature changes (Fig. 48–17). A typical rocky shore inhabitant will have some way of sealing in moisture, perhaps by closing its shell if it has one, plus a powerful means of anchoring itself to rocks. Mussels, for example, have horny, threadlike anchors, and barnacles have a special cement gland. Rocky shore intertidal algae (seaweeds) usually have thick, gummy polysaccharide coats, which dry out slowly when exposed, and flexible bodies not easily broken by wave action. Some rocky shore inhabitants hide in burrows or crevices at low tide, and some small semiterrestrial crabs run about the splash line, following it up and down the beach.

Offshore, the marine environment has two main divisions: the pelagic (ocean water) environment and the benthic (ocean floor) environment. Upper reaches of the pelagic environment comprise the **euphotic** region, which extends from the surface to a depth of approximately 100 meters (325 feet). Light penetrates the euphotic region in sufficient amounts to support photosynthesis. The pelagic environment is divided into two provinces: the neritic province and the oceanic province.

The **neritic province** is ocean water that extends from the shoreline to where the bottom reaches a depth of 200 meters (650 feet). The neritic province occurs over the **continental shelf,** the gently sloping sea bed around the margin of each continent. Nekton (such as shark, tuna, and porpoise) and larger benthos (such as coral, spiny lobster, and starfish) are mostly confined to shallower neritic waters (less than 60 meters, or 195 feet, deep) because that is where their food is found. Not only are there

(a)

(b)

FIGURE 48–17 Adaptations to life in a rocky intertidal zone. (***a***) The sea palm (*Postelsia*) is a sturdy seaweed that is common on the rocky Pacific coast from Vancouver Island to California. Its base is firmly attached to the rocky substrate, enabling it to withstand pounding wave action. (***b***) Limpets (*Acmaea digitalis*) adhere to rock in the intertidal zone of Point Lobos State Park, California. Limpets are mollusks whose shells help them resist pounding waves of their habitat. (*a*, François Gohier/Photo Researchers, Inc.; *b*, William E. Ferguson)

seaweeds on the bottom of shallower areas, but also large numbers of phytoplankton in the water itself.

The **oceanic province** is ocean water that overlies an ocean bottom deeper than 200 meters (650 feet). It comprises most of the ocean. In fact, about 88% of the ocean is more than 1.5 kilometers (0.9 mile) deep. Because light cannot penetrate these depths, the ocean province has few organisms. Most life that exists under the tremendous pressure and darkness of the abysses depends upon whatever food drifts down into its habitat from upper (lighted) regions.

Animals of the abyss are strikingly adapted to darkness and the scarcity of food (Fig. 48–18). Abyssal fish have huge jaws, for example, that enable them to swallow large food particles they might chance upon. (When an organism does not encounter food very often, it needs to eat as much as possible when food is present!) Many have light-producing organs, enabling them to contact one another for mating or food capture. Abyssal organisms are predators or scavengers (there is no other choice!) and live in dispersed populations.

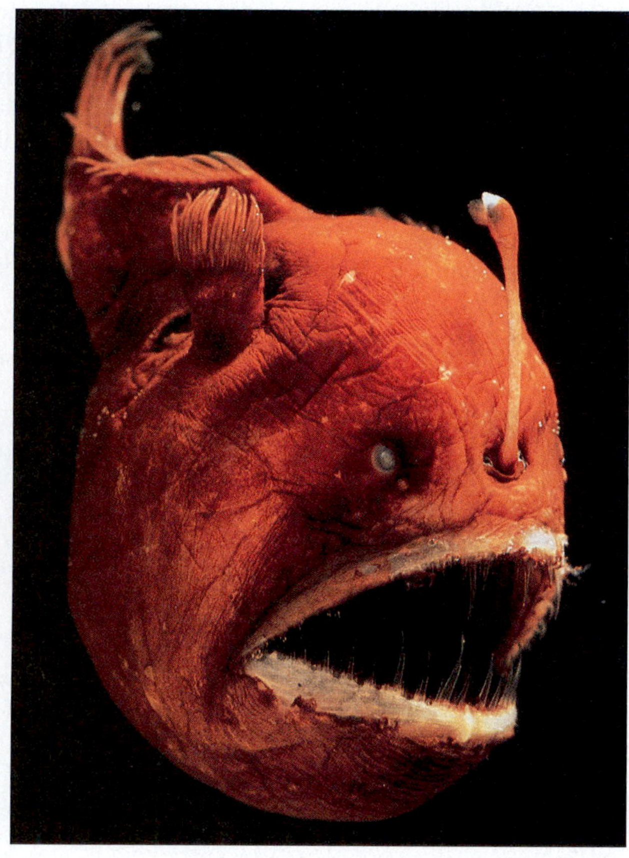

FIGURE 48–18 A female anglerfish discovered at a depth greater than 300 meters (1000 ft). Like many deep-sea fish, this one has weak or vestigial eyes and a luminous lure that may help it to locate a mate or attract prey. (© Norbert Wu 1991)

ALL TERRESTRIAL AND AQUATIC ECOSYSTEMS INTERACT WITH ONE ANOTHER

We have discussed terrestrial and aquatic ecosystems as clearly distinct and separate entities, but they intergrade at their boundaries. The transition zone where two ecosystems meet and intergrade is called an **ecotone.** For example, at the border between tundra and taiga, an extensive ecotone occurs that consists of tundra vegetation interspersed with small, scattered conifers. Such ecotones provide habitat diversity and are often populated by a greater variety and density of organisms than either adjacent ecosystem.

Ecosystems never exist in isolation. When parts of the Amazon rain forest flood annually, for example, fish leave stream beds and swim over the forest floor, where they play a role in dispersing the seeds of many species of plants. In the Antarctic, whose waters are much more productive than are its land areas, there are few terrestrial communities, but there are many seabirds and seals that form a connecting link between the two environments. These animals are supported exclusively by the ocean. Their waste products, such as cast-off feathers, when deposited on land support what lichens, mosses, insects, and bacteria may occur there.

Some inhabitants of terrestrial and aquatic ecosystems cover great distances— in the case of migratory fish and birds, even global distances. For example, many young albacore tuna migrate across the Pacific Ocean from California to Japan! Flycatchers spend their summers in North America and their winters in Central and South America. Like the flycatchers, many migratory birds commonly spend critical parts of their life cycles in entirely different countries, which can make their conservation difficult. It does little good, for instance, to protect a songbird in one country if inhabitants of the next put it in the cooking pot as soon as it lands in their neighborhood. This idea of large-scale interaction makes some ecological concepts difficult to grasp or apply.

Chapter Summary

I. A biome is a large, relatively distinct ecosystem characterized by a similar climate, soil, plants, and animals, regardless of where it occurs.
   A. Tundra, the northernmost biome, is characterized by a frozen layer of subsoil (permafrost) and low-growing vegetation adapted to extreme cold and a very short growing season.
   B. Taiga, or boreal forest, lies south of the tundra and is dominated by coniferous trees.
   C. Temperate forests occur where precipitation is relatively high.
      1. Temperate deciduous forests are dominated by broad-leaved trees that for the most part lose all their leaves seasonally.
      2. Temperate rain forests, such as occur on the northwest coast of North America, are dominated by conifers.
   D. Temperate grasslands typically possess a deep, mineral-rich soil and have moderate but uncertain precipitation.
   E. The chaparral biome is characterized by thickets of small-leaved shrubs and trees, and a climate of wet, mild winters and very dry summers.
   F. Deserts, produced by low rates of precipitation, possess communities whose organisms have specialized water-conserving adaptations. Deserts occur in both temperate and tropical areas.
   G. Tropical grasslands, called savannas, have widely scattered trees interspersed with grassy areas.
   H. Tropical rain forests are characterized by mineral-poor soil and very high rainfall that is evenly distributed throughout the year. Tropical rain forests have high species diversity with at least three stories of forest foliage and many epiphytes.

II. In aquatic ecosystems, important environmental factors include salinity, dissolved oxygen, and the availability of light.
   A. Aquatic life is ecologically divided into plankton (free-floating), nekton (strong-swimming), and benthos (bottom-dwelling).
   B. The microscopic phytoplankton are photosynthetic and are the base of food webs in most aquatic communities.

III. Freshwater ecosystems include flowing water (rivers and streams), standing water (lakes and ponds), and freshwater wetlands.
   A. Water flows in a current in flowing-water ecosystems.
      1. Flowing-water ecosystems depend on detritus from the land for much of their energy.
      2. The types of organisms in flowing-water ecosystems vary greatly, mostly due to the strength of currents, which are usually more swift in headwaters than downstream.
   B. Freshwater lakes are differentiated into zones on the basis of water depth and light penetration.
      1. The marginal littoral zone contains emergent vegetation and heavy growths of algae. The limnetic zone is open water away from shore. The deep, dark profundal zone holds little life other than decomposers.
      2. In summer, large temperate lakes exhibit thermal stratification in which the thermocline separates warmer water above from deep, cold water. Annual spring and fall turnovers remix these layers.
   C. Freshwater wetlands are transitional between freshwater and terrestrial ecosystems. They are covered by shallow water at least part of the year and have characteristic soils and vegetation.

IV. The main marine habitats include estuaries, the intertidal zone, the neritic province, and the oceanic province.
   A. Estuaries are very productive. They receive high nutrient inputs from adjacent lands and serve as important nursery areas for young stages of many aquatic organisms.
   B. Organisms of the intertidal zone possess adaptations that enable them to resist wave action and the extremes of being covered by water (high tide) and exposed to air (low tide).
   C. The neritic province is ocean water that extends from the shoreline to where the bottom reaches a depth of 200 meters (650 feet).
   D. The oceanic province is ocean water that overlies an ocean bottom deeper than 200 meters (650 feet).

V. Ecotones are transition zones where ecosystems meet and intergrade.

Selected Key Terms

benthos, p. 979
biome, p. 968
chaparral, p. 973
ecotone, p. 984
estuary, p. 981
euphotic region, p. 983
intertidal zone, p. 983
limnetic zone, p. 980
littoral zone, p. 980
nekton, p. 979
neritic province, p. 983
oceanic province, p. 984
permafrost, p. 969
plankton, p. 979
profundal zone, p. 980
savanna, p. 975
taiga, p. 970
temperate rain forest, p. 971
tropical rain forest, p. 975
tundra, p. 968

Post-Test

1. Tundra typically has little precipitation, a short growing season, and a permanently frozen underground layer of ______________.
2. ______________ could be described as the "spruce-moose" biome.
3. ______________ are the dominant vegetation of the taiga.
4. In the temperate zone, the deciding factor in whether forest or grassland is present is usually ______________ (temperature or precipitation).
5. In temperate deciduous forests, minerals leached from decomposing material accumulate in a subsurface layer of ______________.
6. The thickest, richest soil in the world occurs in temperate ______________.
7. ______________ is a thicket of evergreen shrubs and small trees found in areas of the North American Southwest that have dry summers and mild winters.
8. On land, species diversity is exceptionally high in tropical ______________ ______________.
9. The ______________ is a tropical biome in which widely spaced trees are interspersed with grassland.
10. Compared with terrestrial ecosystems, aquatic ecosystems are less variable in temperature and have less available ______________ and ______________.
11. Temperate zone lakes are thermally stratified, with warm and cold layers separated by a transitional ______________.
12. Emergent vegetation grows in the ______________ zone of freshwater lakes.
13. ______________ occur where freshwater and saltwater meet.
14. Organisms in aquatic environments fall into three categories: free-floating ______________, strong-swimming ______________, and bottom-dwelling ______________.
15. The ______________ ______________ is open ocean from the shoreline to where the seabed reaches a depth of 200 meters.
16. The dark ______________ province is almost entirely heterotrophic, living on an input of dead organisms from other marine environments.
17. Which of the following statements about biomes is *not* correct? (a) Much of the variation among biomes can be accounted for by variations in climate. (b) Biomes are completely separate entities that are not interconnected or interdependent. (c) Biomes are major terrestrial ecosystems. (d) Biomes extend over large geographical areas.
18. Mosses, lichens, and dwarf shrubs are the dominant vegetation of (a) tropical rain forest (b) temperate rain forest (c) tundra (d) chaparral (e) savanna.
19. Which of the following biomes has the shortest growing season? (a) tropical rain forest (b) savanna (c) temperate grassland (d) temperate deciduous forest (e) taiga
20. Which of the following statements about biomes is true? (a) Tundra has a mineral-rich soil with a high organic content. (b) Desert soils possess a high organic content but a low mineral content. (c) Chaparral soil is thin and not very fertile. (d) Temperate grassland soil is poor in both minerals and organic content.
21. Tropical rain forests (a) make excellent agricultural land when cleared (b) exhibit a low species diversity compared to other biomes (c) are dominated by the trees of evergreen flowering plants (d) are open woodlands with widely scattered trees.

Review Questions

1. What climate and soil factors produce the major terrestrial biomes?
2. List the main terrestrial biomes and give several representative plants and animals of each.
3. In which biome do you live? If your biome does not match the description given in this text, explain the discrepancy.
4. Which biomes are best suited for agriculture? Explain why each of the biomes you did not mention are unsuitable for agriculture.

5. Compare temperate and tropical rain forests.
6. What environmental factors are most important in determining the adaptations of the organisms found in aquatic habitats?
7. What are plankton? What is their role in aquatic ecosystems?
8. How do the inhabitants of a rocky beach differ from those of a sandy beach?
9. Explain the roles of phytoplankton, zooplankton, nekton, and benthos in aquatic food webs.
10. What are the horizontal zones of a lake? How does a temperate lake become thermally stratified?

Thinking Critically

1. If you were put in charge of reversing desertification in the African Sahel, what policies would you implement, and why?
2. In which biomes would migration be most common? Hibernation? Explain your answers.
3. Why do most animals of the tropical rain forest live in trees?
4. What would happen to the organisms in a river with a fast current if a dam were built? Explain your answer.
5. How does turnover affect lake productivity? Why?
6. What features of estuaries make them good breeding areas for marine animals?

Recommended Readings

Chadwick, D.H., "Roots of the Sky," *National Geographic,* Vol. 184, No. 4, October 1993. Wildlife flourishes in the patchy remnants of the North American prairie.

Goulding, M., "Flooded Forests of the Amazon," *Scientific American*, March 1993. During the rainy season, vast sections of the Amazonian rain forest are inundated, making them a uniquely aquatic ecosystem as well as terrestrial. Many of the organisms living here possess special adaptations that enable them to survive in such a changing environment.

"Life in the Deep," *Discover*, November 1993. A photo essay of several unusual organisms adapted to various parts of the marine environment.

Moffett, M., "These Plants Claw and Strangle Their Way to the Top," *Smithsonian*, September 1993. A striking assemblage of vines are found in tropical forests.

Swan, L.W., "The Aeolian Biome," *BioScience,* Vol. 42, No. 4, April 1992. Life in extreme environments such as barren rock and ice.

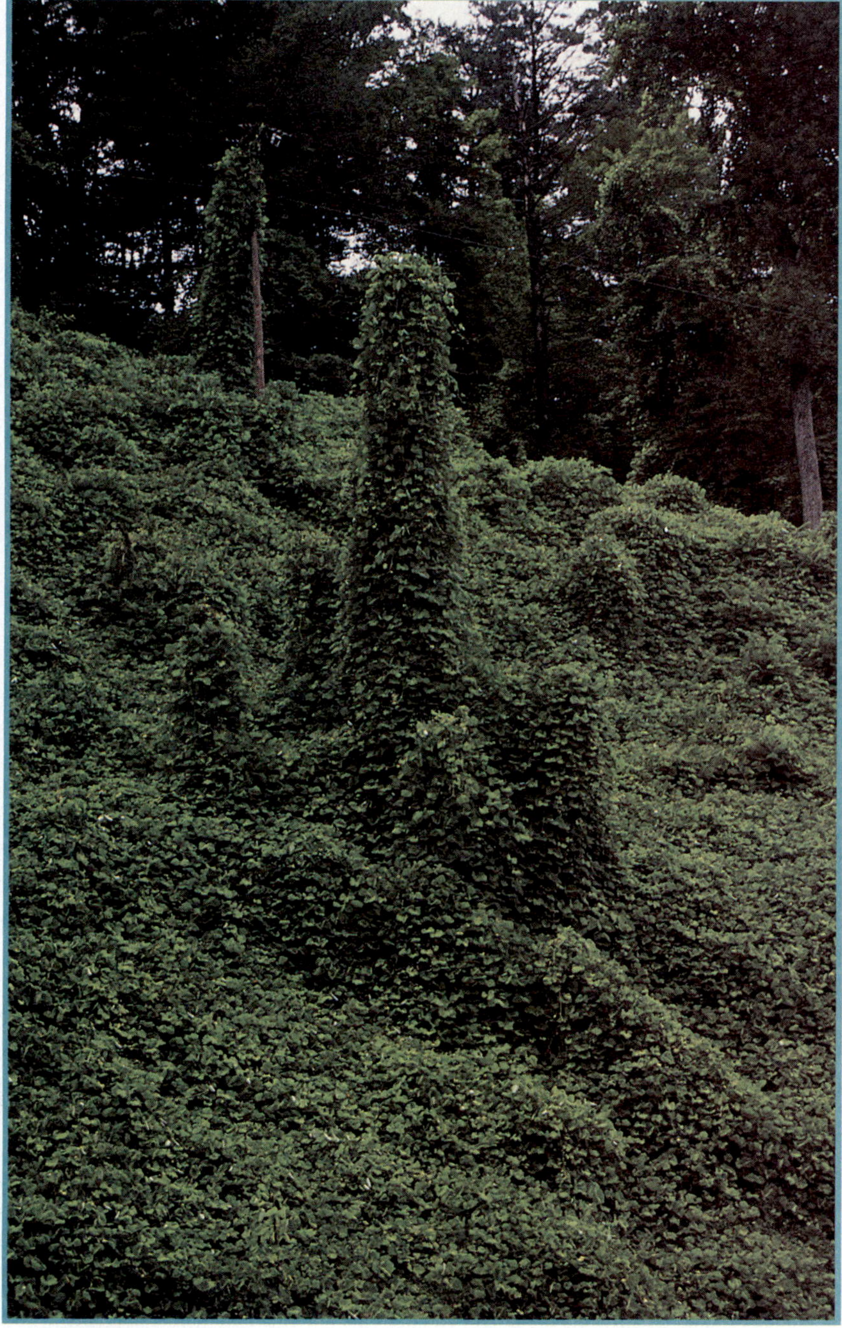

Kudzu growing over trees and other vegetation near Roanoke, Virginia. (*Stephen J. Shaluta, Jr./Dembinsky Photo Associates*)

WINNERS AND LOSERS

Certain human-produced atmospheric pollutants, such as carbon dioxide produced during the combustion of oil, coal, and natural gas, could cause air to retain heat, which in turn warms the Earth. The increasing presence of these pollutants in the atmosphere is raising concern that a major climate change may occur during the next century. Biologists are only beginning to assess the possible consequences of global warming, but it is clear that in many areas there could be drastic changes in both species diversity and species distribution. Two organisms—a mountain butterfly and an Asian vine widely introduced in the United States—provide us with a glimpse of what may be in store for many of the world's organisms.

A rare mountain butterfly (*Boloria acrocnema*) may become one of the first species in the United States to face extinction because of global warming. This little butterfly, a member of a group of butterflies called fritillaries, lives in an extremely restricted area: a few high-altitude, northeast-facing, snow-moistened slopes in the San Juan Mountains of western Colorado. Its habitat is surrounded by warmer, drier areas that the butterfly cannot tolerate. During the hot, dry summers of the late 1980s, one of two known populations of this fritillary species disappeared, and less than half of the other population survived the decade. Biologists concluded that the only reason for the butterfly's declining numbers was unusually warm, dry weather. In the summer of 1991, the fritillary was listed as an endangered species, but this federal protection will not keep it from becoming extinct if the climate continues to be as warm and dry throughout the 1990s as it was in the late 1980s.

Kudzu (*Pueraria lobata*) is a fast-growing, purple-flowered vine that has become a pest throughout much of the southern United States. It was imported to the United States from Japan and was planted widely throughout the South during the 1930s in an effort to halt soil erosion. Because it grows rapidly—as much as 0.3 meter (1 foot) per day under ideal growing conditions—kudzu has choked out many other plants, including trees, and has become a major nuisance from eastern Texas to Florida. Its current northernmost range extends into Maryland. Biologists who have studied kudzu conclude that low winter temperatures are the only reason the plant has not spread further north. Should global warming cause an average winter temperature to increase by 3°C, the vine may well spread north to Michigan, New York, and Massachusetts.

There are many environmental concerns today, in fact, too many to be considered in a single chapter. We therefore center our attention on four very serious environmental issues that are profoundly affecting humans and other living things: declining biological diversity, deforestation, global climate change, and ozone depletion in the stratosphere. Environmental issues that have been discussed in previous chapters include declining biological diversity (Chapters 1, 19, 20, 22, 24, 29, 44, and 46), acid rain and air pollution (Chapters 3, 27, and 38), chemical pollutants (Chapters 4, 21, 30, and 47), deforestation (Chapter 9), land degradation (Chapters 23, 28, and 48), and the most serious environmental problem—overpopulation (Chapters 39 and 45).

Environmental Problems

49

Learning Objectives

After you have studied this chapter you should be able to:

1. Distinguish among threatened species, endangered species, and extinct species.
2. Discuss at least four causes of declining biological diversity and identify the most important cause.
3. Compare in situ conservation and ex situ conservation.
4. Discuss the ecological benefits of forests.
5. State at least three reasons why forests are disappearing today.
6. Name at least three greenhouse gases and explain how greenhouse gases contribute to global warming.
7. Describe how global warming may affect sea level, precipitation patterns, living organisms (including humans), and food production.
8. Give examples of ways to prevent, mitigate, and adapt to global warming.
9. Distinguish between ozone in the lower atmosphere and ozone in the stratosphere.
10. Cite the potential effects of ozone destruction in the stratosphere.

Key Concepts

- ☐ Living organisms are an important natural resource that are not fully appreciated. Biological diversity contributes toward a sustainable environment—one that provides a life support system that enables humans as well as other species to survive. The current reduction in biological diversity caused by the extinction of many species results in lost opportunities and lost solutions to future problems.
- ☐ The world's forests provide many environmental services (habitat for wildlife, watershed protection, and prevention of soil erosion, to name a few) as well as commercially important timber. The greatest problem facing forests today is deforestation, the permanent removal of forest.
- ☐ Production of atmospheric pollutants that trap solar heat in the atmosphere will probably affect Earth's climate during the 21st century. Global warming could alter food production, destroy forests, cause a reduction in biological diversity, submerge coastal areas, and change precipitation patterns.
- ☐ The chemical destruction of the stratospheric ozone shield by gaseous air pollutants is permitting large amounts of solar ultraviolet radiation to penetrate to the Earth's surface. Living organisms, including humans, will be harmed by increased exposure to ultraviolet radiation.

SPECIES ARE DISAPPEARING FROM THE EARTH AT AN ALARMING RATE

Extinction, the death of a species, occurs when the last individual member of a species dies (Chapter 19). Extinction is an irreversible loss, because once a species is extinct it can never reappear. Biological extinction is the eventual fate of all species, much as death is the eventual fate of all life forms.

Although extinction is a natural biological process, it can be greatly accelerated by human activities. Burgeoning human populations have spread into almost all areas of the Earth, and whenever humans invade an area, the habitats of many plants and animals are disrupted or destroyed, which can lead to their extinction. For example, recall the dusky seaside sparrow, a small bird that became extinct in 1987 largely due to the destruction of its habitat (Chapter 19).

Earth's biological diversity is currently decreasing at an alarming rate, and it is likely that thousands of species will be eliminated within the next few decades (Table 49–1). As many as one-fourth of the higher plant fami-

TABLE 49–1
Wildlife at Risk

lies[1] may be extinct by the end of the 21st century, and countless animal species that depend upon those plants for food and habitat will probably also become extinct.

Some biologists fear that we are entering the greatest mass extinction episode in the Earth's history, but the current mass extinction differs from previous periods of mass extinction (Chapter 19) in several respects. First, its cause is directly attributable to human activities. Second, it is occurring in a tremendously compressed period of time (just a few decades as opposed to millions of years). Perhaps even more sobering, larger numbers of plant species are becoming extinct than in previous mass extinctions. Since plants are the bases of food webs, extinction of animals that depend on plants cannot be far behind.

A species is **endangered** when its numbers are so severely reduced that it is in danger of becoming extinct. When extinction is less imminent, but the population of a particular species is still quite low, the species is said to be **threatened.** Even though not extinct, endangered and threatened species represent a decline in biological diversity because their genetic variability is severely diminished. Endangered and threatened species are at greater risk of extinction than are species with greater genetic variability, because long-term survival and evolution depend upon genetic diversity.

A Number of Human Activities Contribute to Species Endangerment or Extinction

Species become endangered and extinct for a variety of reasons. These include destruction or modification of a habitat and environmental pollution. Humans also upset the delicate balance of living organisms in a given area by introducing new, exotic species or by controlling pests or predators. Too much hunting and commercial harvest are also factors (see Focus On The Environment: The Collapse of the Newfoundland Cod Fishery in Chapter 45).

Destruction of natural habitats is the primary cause of the current episode of species endangerment and extinction (Fig. 49–1). Building roads, parking lots, and buildings; clearing forest to grow crops or graze domestic animals; and logging forests for timber all take their toll on natural habitats. Draining marshes converts aquatic habitats to terrestrial ones, and building dams floods terrestrial habitats. Habitat destruction threatens the survival of species because most organisms require a particular type of environment in which they can live.

Even habitats left in their natural state are modified by human activities that produce acid precipitation and other forms of *pollution.* Acid precipitation (Chapter 3) is thought to have contributed to the decline of large stands of forest trees and the biological death of many freshwater lakes. Production of other types of pollutants also adversely affects wildlife (Fig. 49–2). Such pollutants include industrial and agricultural chemicals, organic pollutants from sewage, acid wastes seeping from mines, and thermal pollution from the heated waste water of power and industrial plants.

The *introduction of a foreign, or exotic, species* into an area where it is not native often upsets the balance among organisms living in that area. The foreign species may compete with native species for food or habitat, or may

FIGURE 49–1 Endangered species and habitat destruction. The California condor is a scavenger bird that lives off carrion and requires a large, undisturbed territory (hundreds of square kilometers) in order to find enough food. On the brink of extinction, this highly endangered species was no longer found in the wild from 1987 to 1992. Zoo-bred California condors were released into the wild beginning in 1992. (Tom McHugh/Photo Researchers, Inc.)

[1] Recall that a family is a level of organization in classification. A family consists of a number of related genera, each of which consists of a number of related species. When a family becomes extinct, all of the species of all of the genera comprising that family no longer exist.

FIGURE 49–2 Pollution, including mercury contamination, has been strongly implicated in the decline of the common loon. A common loon gives her chicks a ride. (Jean F. Stoick/ Dembinsky Photo Associates)

FIGURE 49–3 The introduction of exotic species often threatens native species. *Euglandia rosea,* an exotic species in Oahu, consumes a smaller native *Achatinella* species. (Robert J. Western)

prey on them. An introduced competitor or predator usually causes a greater negative effect on local organisms than do native competitors or predators. (Most introduced species lack natural agents—such as diseases, predators, and competitors—that would otherwise control them.) Although exotic species may be introduced into new areas by natural means, humans are usually responsible for such introductions, either knowingly or accidentally.

There are many examples of introduced exotic species causing local organisms to become endangered or extinct. In 1958, a carnivorous snail was introduced in Oahu, an island in Hawaii, as a way to control another snail species that had been introduced by humans and had become a pest (Fig. 49–3). The newly introduced species started consuming native snail species, already in jeopardy from human collectors and human-introduced rats, in large numbers. As a result, *Achatinella,* a Hawaiian snail genus with numerous species known for their rainbow-colored shells, has all but disappeared from Oahu.

Sometimes species become endangered or extinct as a result of deliberate *efforts to eradicate or control their numbers.* Many of these species prey on game animals or sometimes on livestock. Populations of large predators like the wolf, mountain lion, and grizzly bear have been decimated by ranchers, hunters, and government agents.

Predators are not the only animals vulnerable to human control efforts; some animals are killed because their life-styles cause problems for humans. The Carolina parakeet, a beautiful green, red, and yellow bird endemic to the Southeastern United States, was extinct by 1920, exterminated by farmers because it ate fruit from their fruit trees.

Prairie dogs and pocket gophers were poisoned and trapped because their burrows weaken the ground on which unwary cattle grazed. If the cattle stepped into the burrows, they may have been crippled. As a result of sharply decreased numbers of prairie dogs and pocket gophers, the black-footed ferret, a natural predator of these animals, became severely endangered. Indeed, none was found in the wild in the United States between 1986 and 1991. A small number had survived in zoos, and a successful captive breeding program enabled scientists to release 50 black-footed ferrets to the Wyoming prairie in September 1991.

In addition to hunting for predator and pest control, there are three other types of hunting: (1) *commercial hunting,* which is the killing of animals for profit, for example, by selling their fur; (2) *sport hunting,* which is the killing of animals for recreation; and (3) *subsistence hunting,* which is the killing of animals for food. Subsistence hunting has caused the extinction of certain species in the past, but it is not a major cause of extinction today, mainly because so few human groups still rely on subsistence hunting for their food supply. Sport hunting, also a major factor in the extinction of animals in the past, is now strictly controlled in most countries.

Commercial hunting, however, continues to endanger a number of large animals such as the tiger, cheetah, and snow leopard, whose beautiful furs are quite valuable. Rhinoceroses are slaughtered for their horns (used for dagger handles in the Middle East and as a medicine and an aphrodisiac in Asia), and bears are killed for their gall bladders (used in Asian medicine to treat ailments from indigestion to hemorrhoids). Although these animals are protected, demand for their products on the black market has caused them to be hunted illegally. When illegally obtained wildlife trophies are confiscated by police or park rangers, they are destroyed (Fig. 49–4).

In contrast to commercial hunting, in which target organisms are killed, *commercial harvesting* removes live organisms from the wild. Organisms that are commercially harvested end up in zoos, aquaria, research laboratories, and pet stores. Several million birds are commercially harvested each year for the pet trade, but unfortunately, many die in transit and many more die from improper treatment after they are in their owners' homes. At least nine bird species are now threatened or endangered because of commercial harvest. Although it is illegal to capture endangered animals from the wild, there is a thriving black market, mainly because collectors in the United States, Europe, and Japan will pay extremely large sums of money to obtain rare tropical birds. Imperial Amazon macaws, for example, fetch up to $30,000 each.

FIGURE 49–4 Burning of wildlife trophies. The Kenyan government destroys elephant tusks confiscated from poachers. (Steve Turner ©1995 Animals Animals)

Animals are not the only organisms threatened by commercial harvesting. A number of unique or rare plants have been collected from the wild to the point that they are classified as endangered. These include carnivorous plants, certain cacti, and orchids.

Two Types of Conservation Efforts to Save Wildlife Are In Situ Conservation and Ex Situ Conservation

In situ conservation, which includes the establishment of parks and reserves, concentrates on preserving biological diversity *in the wild.* A high priority of in situ conservation is the identification and protection of sites with a great deal of biological diversity. With increasing demands on land, however, in situ conservation cannot guarantee the preservation of all types of biological diversity. **Ex situ conservation** involves conserving biological diversity *in human-controlled settings.* Breeding captive species in zoos and seed storage of genetically diverse crops (Chapter 29) are examples of ex situ conservation.

Protecting Wildlife Habitats Helps Preserve Biological Diversity

Many nations are beginning to appreciate the need to protect their biological heritage and have set aside areas for wildlife habitats. Such natural areas offer the best strategy for protecting and preserving biological diversity. There are currently more than 3000 national parks, sanctuaries, refuges, forests, and other protected areas throughout the world. Some of these areas have been set aside to protect specific endangered species. The first such refuge, established in 1903 at Pelican Island, Florida, was set aside to protect the brown pelican. Today, the National Wildlife Refuge System of the United States has land set aside in over 400 refuges.

Many protected areas have multiple uses. National parks may serve recreational needs, for example, whereas national forests may be open for logging, grazing, and farming operations. The mineral rights to many refuges are privately owned, and some wildlife refuges have had oil, gas, and other mineral development. For example, the D'Arbonne Wildlife Refuge in Louisiana, which is a sanctuary for 145 species of birds, contains soil and water pollution from natural gas wells. Hunting is allowed in over one-half of the wildlife refuges in the United States, and military exercises are conducted in several. The Air Force, for example, conducts low-flying jet exercises and live fire exercises over portions of the Prieta Wildlife Refuge, an Arizona refuge established for bighorn sheep.

Certain parts of the world are critically short of protected areas. In addition to tropical rain forests, protected areas are needed in the tropical grasslands and savannas of Brazil and Australia and in the dry forests that are widely scattered around the world. The wildlife in tropical deserts is under-protected in northern Africa and Argentina, and the wildlife of many islands and lakes needs protection.

Zoos, Aquaria, and Botanical Gardens Often Make Attempts to Save Species on the Brink of Extinction

Human supervision under artificial conditions can maintain some species in danger of extinction. Eggs may be collected from the wild, for example, or the few remaining animals may be captured and bred in zoos or other research environments. Special techniques, such as artificial insemination and embryo transfer, are used to increase the number of offspring. In **artificial insemination,** sperm collected from a suitable male of a rare species is used to artificially impregnate a female (perhaps located in another zoo in a different city or even in another country). In **embryo transfer** (also called **host mothering**), a female of a rare species is treated with fertility drugs, which cause her to produce multiple eggs. Some of these eggs are collected, fertilized with sperm, and surgically implanted into a female of a related but less rare species, who later gives birth to offspring of the rare species. (Artificial insemination and host mothering in humans are discussed in the Chapter 43 Chapter Opener.)

There have been a few spectacular successes in captive breeding programs, in which large enough numbers of a species have been produced to reestablish small populations in the wild. Whooping cranes, which had reached the critically low population of 15 in 1941, have increased in number to over 100. Biologists are hoping to have the whooping crane removed from the endangered species list and classified as only threatened by the year 2000.

Attempting to save a species on the brink of extinction is expensive. Moreover, zoos, aquaria, and botanical gardens do not have the space to try to save *all* endangered species. For example, zoos have traditionally focused on large, charismatic animals, because the public is more interested in them. Such ex situ conservation efforts ignore the millions of less glamorous but ecologically important species (Chapter 22 Chapter Opener). However, they serve a useful purpose in that they educate the public about the value of wildlife conservation. Although ex situ conservation is a valuable strategy, it is clearly more effective to protect natural habitats so that different species do not become endangered in the first place.

Deforestation Is Occurring at an Unprecedented Rate

Permanent destruction of all tree cover in an area is **deforestation.** When forests are destroyed, soil fertility decreases and soil erosion increases. Increased sedimentation of waterways caused by soil erosion can harm aquatic ecosystems. In drier areas, deforestation can lead to the formation of deserts (Chapter 48 Chapter Opener). Deforestation also contributes to loss of biological diversity.

Forests, particularly on hillsides and mountains, provide nearby lowlands with some protection from floods by trapping and absorbing precipitation. When forest is cut down, the area cannot hold water nearly as well, and the total amount of surface runoff flowing into rivers and streams increases. This not only causes soil erosion, but puts lowland areas at extreme risk of flooding.

Regional and global climate changes are induced by deforestation. Transpiring trees release substantial amounts of moisture into the air (Chapter 27). This moisture falls back to Earth in the hydrologic cycle (Chapter 47). When forest is removed, rainfall declines and

Concept Connections

Soil — Nutrient Cycling

In a natural ecosystem, essential minerals cycle from the soil to living organisms, and back again to the soil when those organisms die and decay (Chapter 47). An agricultural system disrupts this pattern when crops are harvested. Much plant material, containing minerals, is removed from the cycle, sent to market, turned into waste and sewage, and deposited in another ecosystem. As a result, it fails to decay and release its nutrients back to the soil from which it came. Thus, over time, soil that is farmed inevitably loses its fertility.

When tropical rain forests are removed to produce agricultural land, mineral depletion is particularly severe. Recall that soils found in tropical rain forests are nutrient-poor because nutrients are stored in vegetation (Chapter 48). Any minerals released as dead organisms decay are promptly reabsorbed by plant roots and their mutualistic fungi (Chapter 22). If this did not occur, heavy rainfall would quickly wash the nutrients away. Nutrient reabsorption by vegetation is so effective that tropical rain forest soils can support luxuriant plant growth despite relative infertility of the soil, *as long as the forest remains intact.*

When a rain forest is cleared, its efficient nutrient cycling is disrupted. Removing vegetation that so effectively stores nutrients allows minerals to wash out of the system. Crops will grow in these soils for only a few years before the small mineral reserves in the soil are depleted.

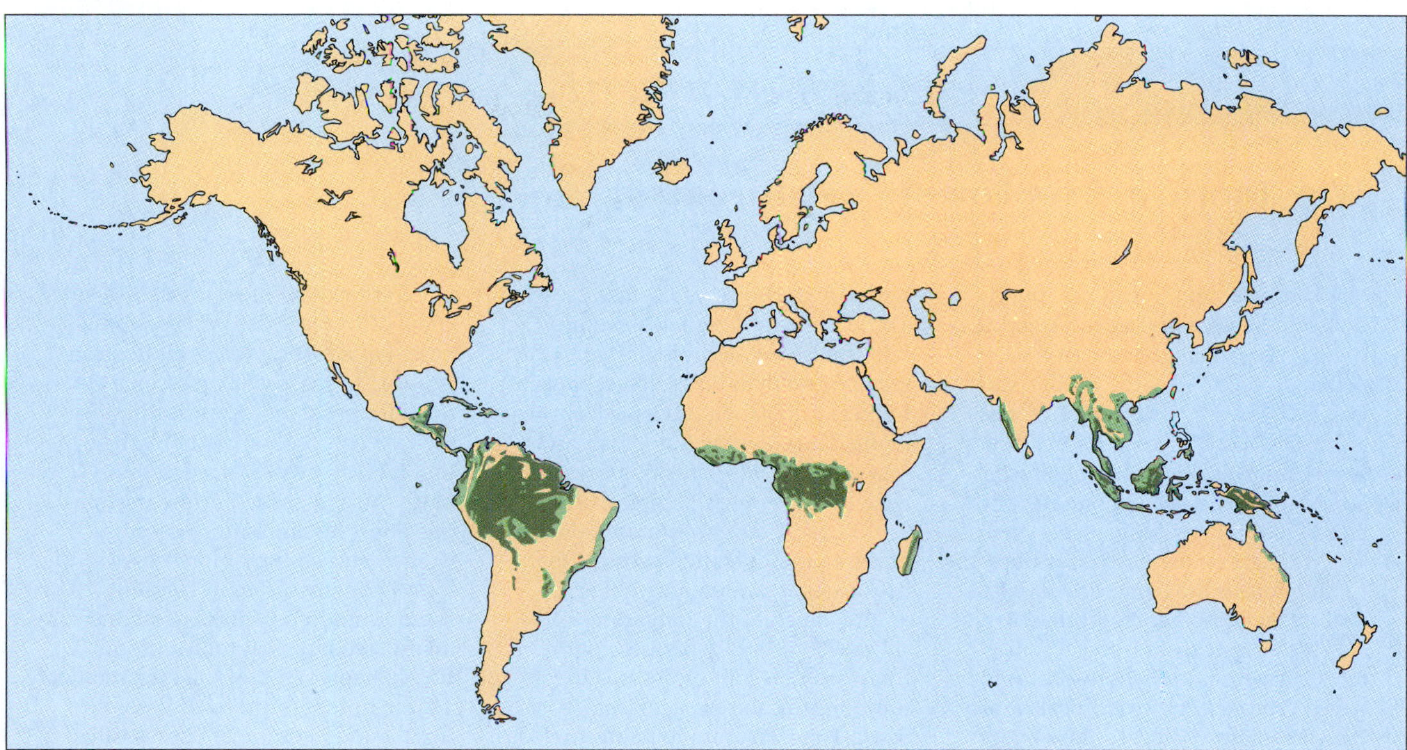

FIGURE 49–5 The distribution of tropical rain forests. Tropical rain forests are located in wet equatorial areas of South and Central America, Africa, and Asia. They originally covered the entire green area, but in recent years, human activities have destroyed forests in the light green area.

droughts become common in that region. Deforestation contributes to an increase in global temperature (discussed shortly) because deforestation results in a release of stored carbon into the atmosphere as carbon dioxide, which causes air to retain solar heat.

Where and Why Are Forests Disappearing?

In the past 1000 years, forests in temperate areas were largely cleared for housing and agriculture. Today, however, deforestation in the tropics is occurring much more rapidly and over a much larger area (Fig. 49–5). Most remaining undisturbed tropical rain forest, which occurs in the Amazon and Congo River basins of South America and Africa, is currently cleared and burned at a rate unprecedented in human history (see Focus On The Environment: Why Tropical Rain Forests Are Important). Tropical rain forests are also currently destroyed at an extremely rapid rate in southern Asia, Indonesia, Central America, and the Philippines. Ten countries account for 76% of tropical deforestation: Brazil, Indonesia, Zaire, Burma, Colombia, India, Malaysia, Mexico, Nigeria, and Thailand.

Three main causes of deforestation of tropical rain forests are subsistence agriculture, commercial logging, and cattle ranching.

Subsistence Agriculture Is the Most Important Cause of Deforestation

Subsistence agriculture, in which just enough food is produced to feed oneself and one's family, accounts for 60% of tropical deforestation. In many developing countries where tropical rain forests are located, the majority of people do not own land on which they live and work. Most subsistence farmers were displaced from traditional farmlands because of an inequitable destribution of land ownership. They have no place to go except into the forest, which they clear to grow food.

Subsistence farmers often follow access roads built by loggers. These poor rural people cut down the forest and allow it to dry, then they burn the area (Fig. 49–6) and plant crops immediately after burning. This is known as **slash-and-burn agriculture.** Yields from the first crop are often quite high because nutrients that had been in the trees are available in the soil after trees are burned. However, soil productivity declines at a rapid rate, so that subsequent crops are poor. In a very short time, people farming the land must move to new forests and repeat the process. Often, cattle ranchers then claim the land for

(Text continues on page 997)

FOCUS ON *The Environment*

Why Tropical Rain Forests Are Important

Tropical rain forests are found in South and Central America, Central Africa, and Southeast Asia. Although only 7% of the Earth's surface is covered by tropical rain forests, at least 50% of the Earth's species of plants, animals, and other organisms inhabit these forests. Human destruction of natural environments is occurring throughout the Earth, but tropical rain forests are being destroyed faster than any other environment.

Many species in tropical rain forests are found nowhere else in the world. The clearing of tropical rain forests therefore leads to their extinction. These organisms are important in their own right, but the mass extinction that is currently taking place in tropical rain forests has indirect ramifications as far away as North America. Birds that migrate from North America to Central America and the Caribbean have been declining in numbers. Not all migratory birds are declining at the same rate. Those birds that winter in tropical rain forests are declining at a much greater rate than birds that winter in tropical grasslands or tropical shrublands. Thus, destruction of tropical rain forests is also affecting organisms of the temperate region.

Tropical rain forests provide important environmental services that help to maintain the forest. For example, much of the rainfall in tropical rain forests is generated by the forest itself. If half of the existing rain forest in the Amazon region of South America were to be destroyed, precipitation in the remaining forest would decrease. As the land becomes drier, organisms adapted to moister conditions would be replaced by organisms able to tolerate the drier conditions. Because many of the original species would be unable to tolerate the drier conditions, they would become extinct.

Perhaps the most unsettling outcome of tropical deforestation is its potential effect on the evolutionary process. In the Earth's past, mass extinctions were followed during the next several million years by the evolution of many new species to replace those that died out. For example, after the dinosaurs became extinct, ancestral mammals evolved into the many different running, swimming, flying, and burrowing mammals that exist today. (Recall from Chapter 19 that the evolution of a large number of related species from an ancestral organism is called adaptive radiation.) In the past, tropical ecosystems such as tropical rain forests supplied the base of ancestral organisms from which adaptive radiations could occur. By destroying tropical rain forests, we may be reducing or eliminating nature's ability to restore its species through adaptive radiation.

FIGURE 49–6 Converting tropical rain forest in Brazil to agricultural land. Tropical forests are increasingly destroyed for slash-and-burn agriculture. (Dr. Nigel Smith ©1995 Earth Scenes)

grazing because land that is not rich enough to support crops can support livestock.

Slash-and-burn agriculture done on a small scale with long periods between cycles—20 to 100 years—is actually sustainable. But when *several hundred million people* try to obtain a living in this way, the land cannot lie uncultivated for an adequate recovery period.

Vast Tracts of Tropical Rain Forests Are Being Removed by Commercial Logging

Commercial logging, mostly for export abroad, accounts for 21% of tropical deforestation. Most tropical countries allow commercial logging to proceed at a much faster rate than is sustainable. For example, in parts of Malaysia, current logging practices remove the forest almost two times faster than is sustainable. If this continues, Malaysia will soon experience shortages of timber and will have to start importing logs. When that happens, Malaysia will have lost future revenues from its newly vanished forests, both from logging and from harvesting other forest products.

Cattle Ranching Also Causes Deforestation

Approximately 12% of tropical forest destruction occurs to provide open rangeland for cattle. After the forests are cleared, cattle are raised on the land for 6 to 10 years, after which time shrubby plants, known as **scrub savanna,** take over the range. Much of the beef raised on these ranches, which are often owned by foreign companies, is exported to fast-food chains.

Dry Tropical Forests Are Being Destroyed Primarily for Use as Fuel

Wood—perhaps one-half of the wood consumed worldwide—is used by much of the developing world for heating and cooking fuel (Fig. 49–7). Fuel wood consumption

(a)

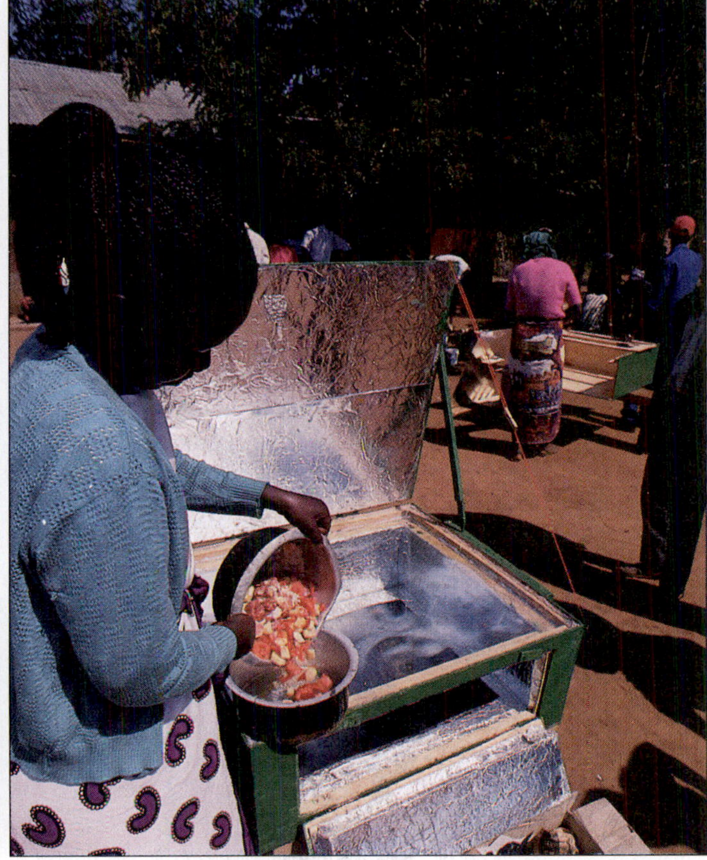

(b)

FIGURE 49–7 Using wood for domestic energy needs. (***a***) A woman in India lights a wood cooking fire. Half of the people in the world use wood fires to cook food. (***b***) A woman in Kenya prepares her first meal using a solar oven that she just built. In average sunlight, food will cook in two to four hours. Solar ovens, which can be built from inexpensive, easily available materials, reduce the number of trees destroyed for fuel. (*a,* Inga Spence/Tom Stack & Associates; *b,* ©1993 John Reader, Courtesy of Earthwatch Expeditions)

is a greater concern in dry tropical forests—tropical areas subjected to a wet season and a prolonged dry season—than in humid forests.

Often wood cut for fuel is converted to charcoal,[2] which is then used to power steel, brick, and cement factories. Charcoal production is extremely inefficient. When wood is converted to charcoal, about 70% of the energy in wood is lost.

PRODUCTION OF ATMOSPHERIC POLLUTANTS TRAPS SOLAR HEAT IN THE ATMOSPHERE AND MAY AFFECT THE EARTH'S CLIMATE

Imagine a world in which beautiful tropical islands, such as the Maldives in the Indian Ocean, disappear forever under the waves. Closer to home, consider the permanent flooding of Louisiana's bayous and the Florida Keys. Imagine forests in the southern United States dying from high temperatures and drought. Think of the effects of an annual hurricane season that is one month longer than it is now. Consider the consequences of tropical pests and diseases spreading northward into the United States and remaining through the winter.

Although none of these events has occurred, there is a chance that some or all of them will take place, possibly even during your lifetime. This is because Earth may become warmer during the next century than it has been for the past one million years. Unlike climate changes in the past, which occurred over thousands of years or longer, this change would take place in a matter of decades.

Although most scientists agree that the Earth will continue to warm, they disagree over how rapidly the warming will proceed, how severe it will be, whether or not it has started, and where warming will be most pronounced. As a result of these uncertainties, many people, including policymakers, are confused about what should be done.

Greenhouse Gases Cause Global Warming

Carbon dioxide (CO_2) and certain other trace gases, including methane (CH_4), ozone (O_3),[3] nitrous oxide (N_2O), and chlorofluorocarbons (CFCs), are accumulating in the atmosphere as a result of human activities. The concentration of atmospheric CO_2 has increased from about 280 parts per million (ppm) approximately 200 years ago (before the Industrial Revolution began) to 360 ppm today. And atmospheric CO_2 is still increasing, as are the levels of the other trace gases associated with global warming.

For example, every time you drive your car, combustion of gasoline in the car's engine releases CO_2 and N_2O, and triggers the production of O_3. Every day as tracts of rain forest are burned in the Amazon, CO_2 is released. One way that CFCs get into the atmosphere is from old, leaking refrigerators and air conditioners, whereas decomposition in landfills is a major source of CH_4.

Global warming occurs because these gases are able to retain heat in the atmosphere that normally would dissipate into space. Thus, the atmosphere warms (Fig. 49–8). Some heat from the atmosphere is transferred to oceans and raises their temperature as well. As atmosphere and oceans warm, the overall temperature of the Earth gets warmer. Because CO_2 and other gases trap the sun's radiation in much the same way that glass does in a greenhouse, global warming produced in this manner is known as the **greenhouse effect,** and gases that retain heat in the atmosphere are called **greenhouse gases.**

Global Warming Could Alter Food Production, Destroy Forests, Submerge Coastal Areas, and Displace Millions of People

Because global interactions among the atmosphere, oceans, and land are too complex and too large to study in the laboratory, climatologists develop models using computer simulations. A model, however, is only as good as the data and assumptions upon which it is based, and a number of uncertainties are built into models on global warming. For example, if global warming causes more low-lying clouds to form, they will block some sunlight and decrease warming. On the other hand, global warming may cause more high, thin cirrus clouds to form, which would actually increase the greenhouse effect. As new data about these uncertainties become available, they are used to make the models' predictions more precise.

Current models predict that a doubling of the concentration of CO_2 in the atmosphere will cause the average temperature of the Earth to increase by between 2° and 5°C before the end of the next century. Warming will not be uniform from region to region, however. At current rates of fossil fuel combustion and deforestation, scientists expect a doubling of CO_2 to occur within the next 50 years. However, the warming trend may be slower than the increasing CO_2 might indicate because oceans take longer than the atmosphere to absorb heat. The second half of the 21st century will probably experience greater warming than will the first half.

[2] Partially burning wood in a large kiln from which air is excluded converts wood into charcoal.

[3] Ozone in the lower atmosphere (the troposphere) is a greenhouse gas as well as a component of smog. Ozone in the upper atmosphere (the stratosphere) provides an important planetary service that will be discussed shortly.

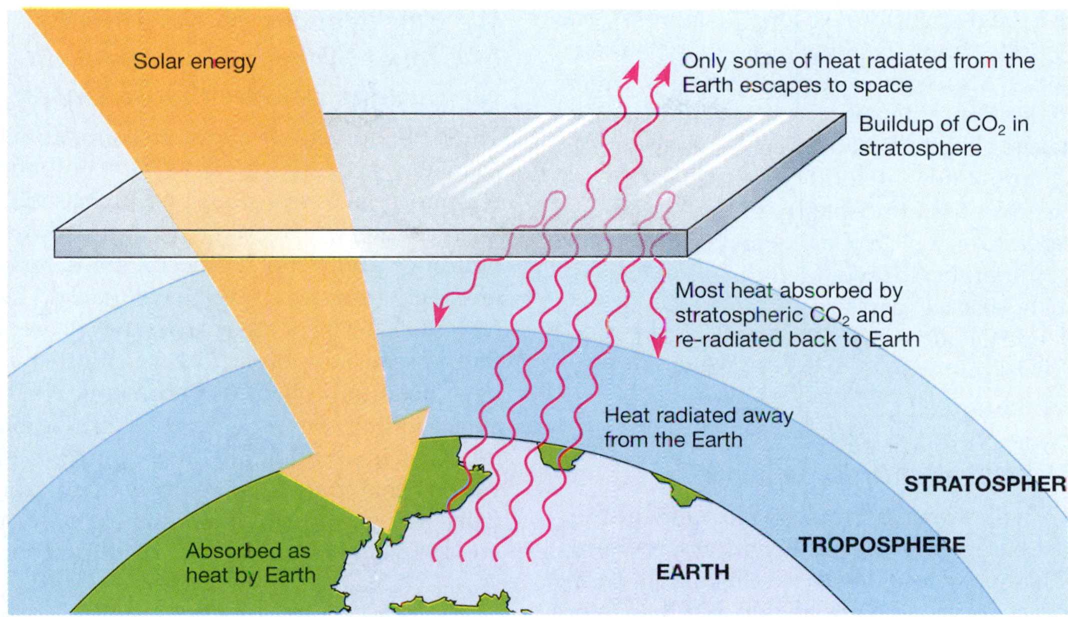

FIGURE 49–8 How carbon dioxide and the greenhouse effect promote global warming. Greenhouse gases in the atmosphere, which act like a glass roof on a greenhouse, trap heat near the surface of the Earth.

Sea Level Is Expected to Rise

If the overall temperature of the Earth increases by just a few degrees, there could be a major thawing of polar icecaps and glaciers. In addition, thermal expansion of oceans will probably occur because water, like other substances, expands when it is warmer. These two changes may cause the sea level to rise by as little as 0.2 meter (8 inches) or as much as 2.2 meters (7 feet) by 2050, flooding low-lying coastal areas. Coastal areas that are not inundated will be more likely to suffer damage from hurricanes and typhoons. These effects are certainly a cause for concern.

Countries most vulnerable to a rise in the sea level have dense populations that inhabit low-lying river deltas—countries such as Bangladesh, Egypt, Vietnam, and Mozambique. For example, a rising sea could cause Bangladesh to lose perhaps 25% of its land. Since 1970, at least 300,000 people in Bangladesh have been killed by tropical storms that caused increased flooding and created high waves that lashed the land. A rising sea level caused by global warming would put even more people at risk in this densely populated nation.

Precipitation Patterns Will Probably Change

Global warming is also expected to change precipitation patterns, causing some areas to have more frequent droughts. At the same time, other areas may be more likely to have flooding. The frequency and intensity of storms may increase also. All of these factors could affect the availability and quality of fresh water in many locations. It is projected that areas that are currently arid or semiarid will have the greatest water-shortage problems.

Living Organisms Will Probably Be Affected

Biologists have started to examine some of the potential consequences of global warming on wildlife. Each species will react to changes in temperature differently (recall the Chapter Opener). Some species will undoubtedly become extinct, particularly those with narrow temperature requirements, those confined to small reserves or parks, and those living in fragile ecosystems, whereas other species may survive in greatly reduced numbers and range. Ecosystems considered most vulnerable to the loss of species in the short-term are polar seas, coral reefs, mountain ecosystems, coastal wetlands, tundra, and boreal and temperate forests.

Some species may be able to migrate to new environments or adapt to the changing conditions in their present habitat. Also, some species may be unaffected by global warming, whereas others may come out of global warming as winners, with greatly expanded numbers and range. Those considered most likely to prosper include weeds, pests, and disease-carrying organisms that are already common and found in many different environments.

Biologists generally agree that global warming will have an unusually severe impact on plants because they are unable to move about when conditions change. Although seeds may be dispersed over long distances, seed dispersal has definite limits in terms of how fast a plant species can migrate. Moreover, soil characteristics, water availability, competition with other species, and human alterations of natural habitats all affect the rate at which plants can move into a new area.

Increasing CO_2 in the atmosphere and the resulting increase in temperature will probably not harm human health directly. Human health will be indirectly affected, however, as such disease carriers as malaria-infected mosquitos and encephalitis-infected flies expand their range into newly warm areas.

Agriculture Will Probably Be Affected

Global warming will increase problems for agriculture, which is already beset with the challenge of providing enough food for a hungry world without irreparably damaging the environment. Rising sea level will inundate river deltas, which are some of the world's best agricultural lands. Certain agricultural pests and disease-causing organisms will probably proliferate. Global warming will also increase the frequency and duration of droughts. To get an idea of how devastating droughts can be, consider the drought experienced in the American grain belt during the summer of 1988. Wheat yields were reduced by 40% and the United States, which is usually the world's largest exporter of grain, used more grain than it produced.

How Should We Deal with Global Warming?

Even if we were to immediately stop polluting the atmosphere with greenhouse gases, the Earth would still have some climate change because of the greenhouse gases that have accumulated during the past 100 years. The amount and severity of global warming depend upon how much additional greenhouse gases we add to the atmosphere.

Three different ways to approach the management of global warming exist: prevention, mitigation, and adaptation. We can *prevent* global warming by developing ways to prevent buildup of greenhouse gases in the atmosphere. Prevention is the ultimate and best solution to global warming because it is permanent. *Mitigation,* which involves ways to moderate or postpone global warming, gives us time to pursue other, more permanent solutions to global warming. Further, it gives us time to understand more fully how global warming operates so we can avoid some of its worst consequences. *Adaptation* is responding to changes brought about by global warming. Developing adaptive strategies to climate change implies an assumption that global warming is unavoidable.

We Can Respond to the Threat of Global Warming by Preventing the Buildup of Greenhouse Gases

The development of alternatives for fossil fuels (such as solar energy) offers a permanent solution to the global warming challenge caused by increased CO_2 emissions.[4] Solar energy, which can be used directly to heat water and buildings and generate electricity, is renewable and has fewer environmental problems than do fossil fuels. Solar and other forms of alternative energy are not practical for widespread adoption, however, until their technologies have been improved. Governments must increase available funding for research into renewable energy technologies. In addition, the invention of technological innovations that trap CO_2 being emitted from smokestacks (currently not possible) would help prevent global warming and yet allow us to use fossil fuels for energy.

We Can Respond to the Threat of Global Warming by Mitigating Its Effect

One of the most effective ways to mitigate global warming involves forests. As you know, atmospheric CO_2 is removed from the air by actively growing forests, which incorporate carbon into leaves, stems, and roots by photosynthesis. On the other hand, deforestation releases CO_2 into the atmosphere as trees are burned or decomposed. We can mitigate global warming by planting and maintaining new forests. In addition to planting new forests, we can help mitigate global warming by increasing the energy efficiency of automobiles and appliances, which would reduce the output of CO_2.

We Can Respond to the Threat of Global Warming by Adapting to Its Reality

Government planners and social scientists are developing a number of strategies to help adapt to global warming. For example, what should be done with people living in coastal areas? They can be moved inland away from the dangers of storm surges, although this solution has high societal and economic costs. An alternative plan, which is also extremely expensive, is to build dikes and levees to protect coastal land. The Dutch, who have been doing this sort of thing for several hundred years, are offering their technical expertise to several developing nations threatened by a rise in sea level.

[4] Alternatives to the other greenhouse gases will also have to be developed, but CO_2 is focused on here because it is produced in the greatest quantities and has the largest total effect of all of the greenhouse gases.

We also have to be able to adapt to shifting agricultural zones. Many temperate countries are in the process of evaluating semi-tropical crops to determine the best ones to substitute for traditional crops if/when the climate warms. Drought-resistant strains of lumber trees are being developed by large lumber companies now because the trees planted today will be harvested during the 21st century, when global warming may already be well advanced.

OZONE IS DISAPPEARING IN THE STRATOSPHERE

Ozone (O_3) is a form of oxygen that is a human-made pollutant in the lower atmosphere (it is a component of photochemical smog), but a naturally produced, essential part of the stratosphere. The **stratosphere**, which encircles our planet some 10 to 45 kilometers (6 to 28 miles) above the Earth's surface, contains a layer of ozone that shields the Earth from ultraviolet radiation coming from the sun (Fig. 49–9). Should ozone disappear from the stratosphere, the Earth would become unlivable for most forms of life. (Ozone in the lower atmosphere is converted back to oxygen in a few days and therefore does not replenish the ozone depleted from the stratosphere.)

The problem of stratospheric ozone depletion was demonstrated in 1984 with the discovery of a large hole, or thin spot, in the ozone layer over Antarctica (Fig. 49–10). Ozone levels decrease by as much as 67% here each year. There is a smaller thin spot, or hole, in the stratospheric ozone layer over the Arctic, but probably the most disquieting news is that *worldwide levels* of stratospheric ozone have been decreasing for several decades. The rate of ozone depletion during the 1980s was roughly three times the depletion rate of the 1970s.

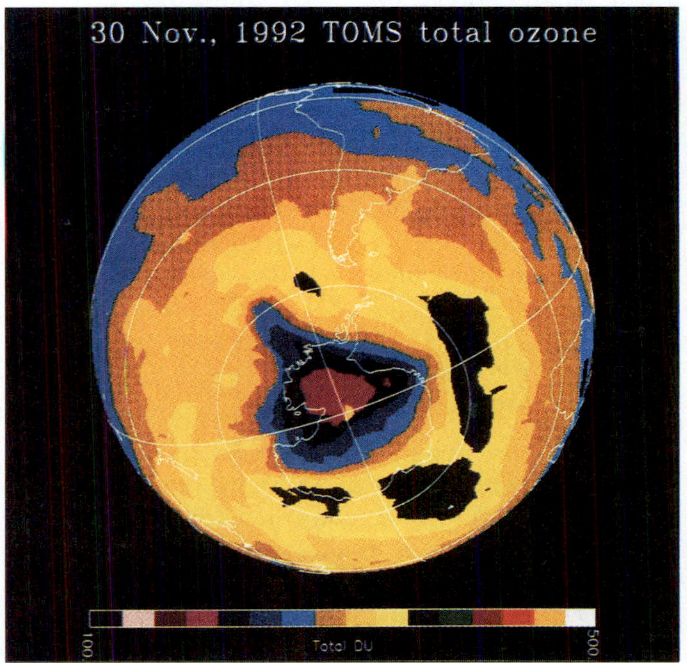

FIGURE 49–10 A computer-generated image of the Southern Hemisphere on November 30, 1992, reveals the ozone "hole" over Antarctica. The color scale at the bottom shows total ozone values. (NASA)

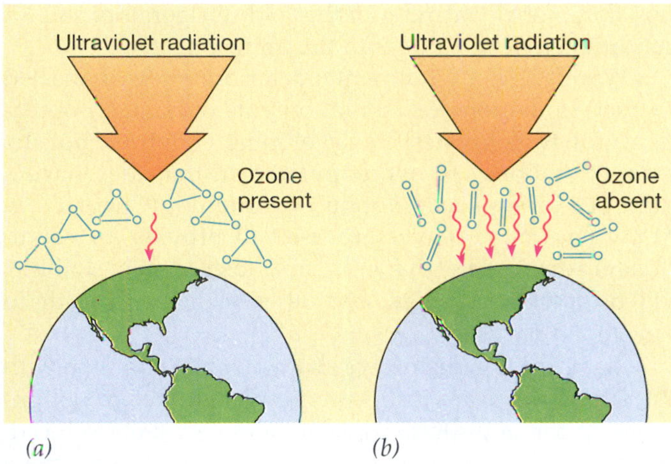

FIGURE 49–9 Ultraviolet radiation and the ozone layer. (***a***) Ozone absorbs ultraviolet radiation, effectively shielding the Earth. (***b***) When ozone is absent, more high-energy ultraviolet radiation penetrates the atmosphere to the Earth's surface, where its presence harms living things.

The Chemical Destruction of Ozone in the Stratosphere Is Caused by CFCs and Other Industrial Chemicals

The primary culprit responsible for ozone loss in the stratosphere is a group of commercially important compounds called chlorofluorocarbons, or CFCs. CFCs are used as propellants in some aerosol cans, as coolants in air conditioners and refrigerators (for example, Freon), as foam for insulation and packaging (for example, styrofoam), and as cleaners in the electronics industry. Additional compounds that also attack ozone include: halons (found in many fire extinguishers), methyl bromide (used as a pesticide in agriculture), methyl chloroform (used to degrease metals), and carbon tetrachloride (used in many industrial processes, including the manufacture of pesticides and dyes).

CFCs and similar compounds drift up to the stratosphere and become widely dispersed. In the stratosphere, ultraviolet radiation breaks down these chemicals into chlorine, flourine, and carbon. Under certain conditions found in the stratosphere, chlorine is capable of reacting with ozone, converting it into molecular oxygen. The chlorine is not altered by this process; as a result, a single chlorine atom can break down thousands and thousands of ozone molecules.

Ozone Depletion Adversely Affects Living Organisms

Ozone molecules in the stratosphere absorb incoming solar ultraviolet radiation. With depletion of the ozone layer, higher levels of ultraviolet radiation reach the Earth's surface. Excessive exposure to ultraviolet radiation is linked to a number of health problems in humans, including cataracts, skin cancer, and a weakened immune system. Much scientific evidence also documents crop damage from exposure to high levels of ultraviolet radiation.

Biologists are concerned that the ozone hole over Antarctica could possibly damage plankton that forms the base of the food web for the Southern Ocean (see Focus On The Environment: Changes in the Antarctic Food Web in Chapter 47). A 1992 study confirmed that increased ultraviolet radiation is penetrating surface waters around Antarctica and that productivity of Antarctic phytoplankton has declined by 6% to 12% as a result. If productivity of phytoplankton continues to decline, the food web of Antarctica may collapse.

The Global Community Is Cooperating to Prevent Significant Depletion of the Ozone Layer

In 1987, representatives from a number of countries met in Montreal to sign the Montreal Protocol, an agreement to significantly reduce CFC production by 50% by 1998. Then, in 1990, more than 90 countries signed an agreement to phase out all use of CFCs by 2000. Despite these agreements, the news about the ozone layer has continually worsened since 1990. As a result, some nations have taken even stricter measures to limit CFC production. At an international meeting in Copenhagen in 1992, representatives agreed to phase out CFCs, carbon tetrachloride, halons, methyl chloroform, and similar chemicals by early 1996. They also agreed to freeze production of other ozone-destroying chemicals.

The ozone depletion problem is far from over. Unfortunately, CFCs are extremely stable and will continue to deplete stratospheric ozone well into the 21st century. In the final analysis, however, ozone depletion may become an environmental success story. When faced with a serious environmental crisis, governments, scientists, industrialists, and citizens of many nations worked together to help solve the problem.

ENVIRONMENTAL PROBLEMS ARE INTERRELATED

Several connections have been pointed out between the four environmental problems discussed in this chapter. For example, deforestation, global warming, and ozone depletion will likely cause future extinctions. Similarly, any environmental problem you can think of—even if it was not discussed in this chapter—is related to other environmental concerns.

Consider solid waste, which threatens to bury us under mountains of garbage. Solid waste is placed in sanitary landfills, which are large areas of land that can no longer provide wildlife habitat except for a few species such as rats and gulls. Toxic substances in solid waste often escape into surrounding soil, water, and air, thereby harming wildlife. As organic material decays in landfills, methane, a potent greenhouse gas, is released. Solid waste further contributes to global warming because it contains materials that were discarded rather than reused or recycled. (It requires more energy to replace those materials that were discarded than to reuse or recycle them. Also, burning fossil fuels to supply that energy produces more CO_2.) Solid waste contributes to both ozone depletion and global warming in that old refrigerators and air conditioners leak CFCs into the atmosphere.

Worldwide environmental deterioration is connected to overpopulation. Recall that the rate of human population growth is greatest in developing countries, but developed nations are overpopulated in that they have a high per capita consumption of resources (Chapter 45). Many natural resources are used to provide air conditioners, disposable diapers, cars, fashionable clothing, video cassette recorders, and other "comforts" of life in developed nations.

As living organisms, we share much in common with the fate of other life forms on the planet. We are not immune to the environmental damage we have produced. We differ from other organisms, however, in our capacity to reflect on the consequences of our actions and alter our behavior accordingly. *Humans, both individually and collectively, are able to bring about change.* This talent is the key to any hope of ensuring the biosphere's survival.

Chapter Summary

I. A reduction in biological diversity is occurring worldwide.
  A. When the last individual member of a species dies, it becomes extinct.
    1. A species whose numbers are severely reduced so that it is in danger of extinction is called an endangered species.
    2. When extinction is less imminent but the population is quite low, a species is called a threatened species.
  B. Human activities that contribute to a reduction in biological diversity include habitat loss and disturbance, pollution, introduction of foreign species, pest and predator control, hunting, and commercial harvest. Of these, habitat loss is the most significant factor in declining biological diversity.
  C. Efforts to preserve biological diversity in the wild are known as in situ conservation. Ex situ conservation (for example, captive breeding) occurs in human-controlled settings.

II. Forests provide us with many ecological benefits, including watershed protection, soil erosion prevention, climate moderation, and wildlife habitat.
  A. The greatest problem facing the world forests today is tropical deforestation.
  B. In the tropics, forests are destroyed to (1) provide colonizers with temporary agricultural land, (2) produce timber, particularly for developed nations, (3) provide open rangeland for cattle, and (4) supply people with fuel wood.

III. Carbon dioxide and other greenhouse gases cause the air to retain heat, which warms the Earth.
  A. The increase in CO_2 and other greenhouse gases in the atmosphere is causing concerns about major climate changes that may occur during the next century.
    1. The combustion of fossil fuels produces pollutants, especially CO_2.
    2. Other greenhouse gases are methane, nitrogen oxide, CFCs, and ozone.
  B. Global warming may cause a rise in sea level, changes in precipitation patterns, the death of forests, the extinction of animals and plants, and problems for agriculture. It could result in the displacement of thousands or even millions of people.
  C. The challenge of global warming can be met by prevention (stop polluting the air with greenhouse gases), mitigation (slow down the rate of global warming), and/or adaptation (make adjustments to live with global warming).

IV. The ozone layer in the stratosphere helps shield Earth from damaging ultraviolet radiation.
  A. The total amount of ozone in the stratosphere is slowly declining, and large ozone holes develop over Antarctica and the Arctic each year.
  B. The attack on the ozone layer is caused by chlorofluorocarbons (CFCs) and similar chlorine-containing compounds.

Selected Key Terms

deforestation, p. 994
endangered species, p. 991
ex situ conservation, p. 993
extinction, p. 990
greenhouse effect, p. 998
greenhouse gases, p. 998
in situ conservation, p. 993
ozone, p. 1001
scrub savanna, p. 997
slash-and-burn agriculture, p. 995
stratosphere, p. 1001
subsistence agriculture, p. 995
threatened species, p. 991

Post-Test

1. ______________ is the permanent loss of a species from the Earth.

2. A/An ______________ species is severely reduced in number and is in danger of becoming extinct.

3. A/An ______________ species is not in imminent danger of extinction, but its numbers are low enough for concern.

4. ______________ hunting is done for profit, ______________ hunting is done for recreation, and ______________ hunting is done for food.

5. ______________ ______________ conservation concentrates on preserving biological diversity in the wild.

6. The permanent destruction of forests is known as ______________.

7. Poor rural people living in the tropics practice a type of agriculture known as ______________ and ______________.
8. The area of the world where deforestation is most acute is the ______________.
9. The five greenhouse gases are ______________, ______________, ______________, ______________, and ______________.
10. Driving a gasoline-powered automobile produces three greenhouse gases: ______________, ______________, and ______________.
11. Greenhouse gases retain ______________ in the atmosphere.
12. Biologists think that global warming will have the greatest impact on ______________ (plants, animals, or fungi).
13. ______________ is/are a human-made pollutant in the lower atmosphere but a natural (and beneficial) gas in the stratosphere.
14. ______________ is/are a human-made pollutant that causes global warming *and* ozone depletion in the stratosphere.
15. Stratospheric ozone is beneficial because it absorbs incoming ______________ radiation from the sun.
16. Although there are many reasons for declining biological diversity, ______________ is the most important. (a) sport hunting of animals (b) habitat loss or disturbance (c) pollution of lakes and other aquatic ecosystems (d) pest and predator control
17. The most important cause of deforestation of tropical rain forests is (a) commercial logging (b) subsistence agriculture (c) ranching (d) gathering firewood.
18. Current models predict that a doubling of atmospheric CO_2 will cause an increase in worldwide temperature of (a) 2°–5°C (b) 5°–10°C (c) 10°–20°C (d) 20°–25°C.
19. Which of the following activities would help mitigate global warming? (a) building coastal dikes (b) planting and maintaining forests (c) developing trees resistant to drought (d) growing semi-tropical crops in temperate areas
20. Worldwide, the amount of stratospheric ozone is (a) increasing (b) decreasing (c) remaining the same.

Review Questions

1. State one reason why we should protect biological diversity.
2. How does habitat destruction contribute to declining biological diversity?
3. State at least three environmental benefits that forests provide.
4. Explain the relationship between eating a hamburger at a fast food restaurant and tropical deforestation.
5. Discuss some of the environmental costs of deforestation.
6. Describe the greenhouse effect and list the five greenhouse gases.
7. It has been suggested that the wisest way to "use" fossil fuels would be to leave them in the ground. How would this affect global warming? Energy supplies?
8. Discuss and give an example of each of the three approaches—prevention, mitigation, and adaptation—that deal with global warming.
9. What environmental service does the stratospheric ozone layer provide?
10. Describe what is causing the chemical destruction of ozone in the stratosphere.

Thinking Critically

1. If the climate patterns change in the next century as predicted, why will so many species become extinct?
2. If most of the world's biological diversity were to disappear, how would it affect your life?
3. Would controlling the growth of the human population have any effect on biological diversity? Why or why not?
4. Why might captive breeding programs that reintroduce species into natural environments be doomed to failure? Base your answer on what you have learned about the genetics of populations that drop to a low number (such as the cheetah, discussed in Chapter 18).
5. If all of the world's tropical rain forests were destroyed, how would it affect your life?
6. If you were given the task of developing a federal policy for dealing with global climate change during the next 100 years, would you stress prevention, mitigation, or adapta-

tion? Explain. What do you think our current government is stressing?

7. If climate change occurs more rapidly than our forests can migrate naturally, humans can plant tree seeds in new locations (sort of an artificial migration). Based on what you have learned in this book, explain why such trees might not survive to form forest ecosystems.

8. How would a rise in sea level associated with global warming have a personal effect on you?

9. This statement was overheard in an elevator: "CFCs cannot cause the stratospheric ozone depletion over Antarctica because there are no refrigerators in Antarctica." Criticize the reasoning behind this statement.

Recommended Readings

Cohn, J.P., "The Flight of the California Condor," *Bioscience*, Vol. 43, No. 4, April 1993. Captive breeding of this endangered species has been so successful that it has been reintroduced into the wild.

Conniff, R., "From *Jaws* to laws—Now the Big, Bad Shark Needs Protection from Us," *Smithsonian*, May 1993. New conservation measures have been passed to help shark populations recovery from human predation.

Hansen, J. et al., "How Sensitive Is the World's Climate?" *Research & Exploration*, Vol. 9, No. 2, Spring 1993. Humans now have the capability to rival the natural forces of nature in determining climate. (This entire issue of *Research and Exploration*, which is a scholarly publication of the National Geographic Society, is devoted to global climate change.)

Line, L., "Silence of the Songbirds," *National Geographic*, Vol. 183, No. 6, June 1993. Populations of migratory songbirds that spend their summers in North America and their winters in Central and South America have declined dramatically during the past few years.

Reports to the Nation: Our Ozone Shield, University Corporation for Atmospheric Research, No. 2, Fall 1992. This booklet, which can be obtained by calling (303) 497–1982, summarizes the ozone depletion problem.

Ryan, J.C., "When Nature Loses Its Cool," *World-Watch*, Vol. 5, No. 5, September/October 1992. A sobering discussion of the effects that global warming will probably have on wildlife.

Tattersall, I., "Madagascar's Lemurs," *Scientific American*, January 1993. Habitat destruction in Madagascar has caused the extinction of hundreds of species of lemurs and threatened many others.

Vincent, J.R., "The Tropical Timber Trade and Sustainable Development," *Science*, Vol. 256, June 19, 1992. Tropical countries are exporting wood from their forests faster than can be sustained. This paper examines the root causes of this problem.

Appendix A

Post-Test Answers

Chapter 1

1. study of life 2. metabolism 3. homeostasis 4. DNA 5. adaptations 6. cells 7. organs 8. ecosystems 9. ecology 10. photosynthesis 11. water; energy (any order) 12. decomposers 13. mutations 14. selection 15. Protista 16. a 17. c 18. d 19. c 20. e

Chapter 2

1. biogenesis 2. deductive 3. Sampling 4. Inductive 5. hypothesis 6. experiment 7. theory 8. experiment 9. b 10. b 11. e 12. e 13. a

Chapter 3

1. atom 2. neutrons 3. orbitals 4. C; H; O 5. donors 6. covalent 7. reduction 8. acid 9. basic 10. buffer 11. capillary; adhesive 12. hydrogen 13. e 14. b 15. a 16. c 17. c 18. b 19. c 20. a

Chapter 4

1. b 2. c 3. d 4. c 5. e (a and b) 6. a 7. b 8. b 9. amino acids 10. primary 11. Cellulose 12. glycogen 13. macromolecule 14. Chitin 15. pentose

Chapter 5

1. resolving power 2. ribosomes 3. smooth ER 4. Golgi complex 5. mitochondria 6. chloroplasts 7. microtubules 8. Microfilaments 9. nucleus 10. envelope (membrane) 11. chromatin; chromosomes 12. genes 13. cell wall 14. vacuole 15. function(s) 16. e 17. c 18. b 19. a 20. b

Chapter 6

1. selectively 2. hydrophilic; hydrophobic 3. hydrophobic tails 4. Microvilli 5. plasmodesmata 6. Desmosomes 7. gap 8. diffusion 9. osmosis 10. hypertonic 11. isotonic 12. exocytosis 13. phagocytosis 14. fluid 15. active 16. contractile 17. c 18. c 19. c 20. d

Chapter 7

1. energy 2. potential 3. thermodynamics 4. free 5. Endergonic 6. equilibrium 7. 3 8. ATP 9. enzyme 10. activation energy 11. ase 12. coenzyme 13. competitive 14. c 15. d 16. c 17. b 18. a 19. d 20. a

Chapter 8

1. catabolism 2. oxygen 3. cytosol; mitochondria 4. NAD^+; NADH 5. acetyl CoA 6. citric acid 7. carbon dioxide 8. energy 9. ATP 10. anaerobes 11. fermentation 12. ethyl alcohol 13. lactic acid 14. 36–38 15. oxidized 16. a 17. c 18. b 19. a 20. b

Chapter 9

1. photosynthesis 2. chemical 3. glucose; oxygen (any order) 4. photons 5. chlorophyll 6. thylakoids 7. electrons 8. ADP; ATP 9. NADPH 10. protons; thylakoid 11. ATP 12. stroma 13. Calvin 14. CO_2 fixation 15. 6 16. c 17. a 18. d 19. d

Chapter 10

1. chromatin 2. diploid, 2n 3. haploid, n 4. homologous 5. cell cycle 6. mitosis 7. cytokinesis 8. interphase 9. S (Synthesis) 10. centromeres 11. meta-phase 12. anaphase 13. 2; 4 14. tetrads 15. crossing over 16. recombination 17. c 18. c 19. e 20. b 21. a 22. a

Chapter 11

1. Mendel 2. gene 3. alleles 4. phenotype; genotype 5. homozygous 6. dominant 7. gamete 8. chance 9. test 10. homologous 11. linked 12. codominant 13. epistasis 14. pleiotropy 15. environment 16. b 17. a 18. c 19. d 20. b

Answers to Genetics Problems, Chapter 11

1a. *ww*; all gametes would have *w*

1b. 1/4 purple

1c. none

2. 1/2

3. There are two possible answers: The father would be reddish or roan.

4. 4 possible blood types: A, B, AB, and O

5. Yes. Father must be type A or type AB (he must have at least one I^A allele).

6. 9:3:3:1 (9 yellow, tall: 3 green, tall: 3 yellow, short: 1 green, short)

7a. *PPSs* (purple, spiny), *PPss* (purple, smooth), *PpSs* (purple, spiny), and *Ppss* (purple, smooth)

7b. (5) *PPSs*

8a. All tall, round (all *TtRr*)

8b. 9:3:3:1

9. *Nn* × *nn*

10. *OoNn* × *ooNn*

Chapter 12

1. autosomes; sex chromosomes 2. Y 3. females; males 4. x-linked 5. hemizygous 6. mother 7. karyotype 8. nondisjunction 9. Klinefelter; Turner 10. PKU 11. hemoglobin 12. cystic fibrosis 13. amniocentesis 14. consanguineous 15. d 16. a 17. c 18. b 19. c

Answers to Genetics Problems, Chapter 12

1. 50% (or 1/2)

2. Father was colorblind. Parents were X^CX^c and X^cY.

3. 50% (or 1/2)

4. 25% (or 1/4)

5. 50% (or 1/2)

Chapter 13

1. nucleus 2. DNA 3. bacteriophages 4. DNA 5. double helix 6. nucleotides 7. base; phosphate; sugar (any order) 8. thymine; cytosine 9. GCCAGT 10. semiconservative 11. replication fork 12. DNA polymerase 13. 5′; 3′ 14. chromosomes 15. histone 16. d 17. a 18. c 19. d 20. a

Chapter 14

1. transcription 2. ribose; uracil 3. mRNA; rRNA; tRNA (any order) 4. mRNA 5. rRNA; proteins (any order) 6. tRNA 7. introns 8. exons 9. translation 10. codon; anticodon 11. genetic code 12. initiation; elongation; termination 13. nucleus; ribosome 14. ribosomes 15. mutations 16. a 17. b 18. b 19. a 20. e

Chapter 15

1. differentiation 2. totipotency 3. constitutive 4. transcription 5. operon 6. structural 7. promoter 8. operator 9. repressor 10. operator 11. inducible 12. repressible 13. enhancers 14. mRNA processing 15. feedback inhibition 16. off; on 17. presence; absence 18. a 19. c 20. d

Chapter 16

1. recombinant 2. restriction 3. vector 4. plasmids 5. restriction 6. ligase 7. recombinant 8. antibiotics 9. clone 10. genetic probe 11. hybridize 12. cDNA 13. polymerase chain reaction 14. Gene replacement 15. crown gall 16. b 17. a 18. c 19. d 20. d

Chapter 17

1. evolution 2. gene pool 3. Lamarck 4. populations 5. natural selection 6. survive 7. humans 8. neo-Darwinism 9. fossils 10. analogous 11. vestigial 12. biogeography 13. amino acids 14. genetic code 15. c 16. c 17. d 18. b 19. a

Chapter 18

1. population 2. allele frequencies 3. mutation 4. genetic drift 5. genetic bottleneck 6. migration 7. natural selection 8. stabilizing 9. directional 10. heterozygous advantage 11. species 12. temporal 13. mechanical 14. allopatric 15. allopolyploid 16. e 17. a 18. c 19. c 20. b

Chapter 19

1. macroevolution 2. development 3. adaptive radiation 4. extinct 5. Mass 6. punctuated equilibrium 7. gradualism 8. time 9. clay 10. H_2S 11. endosymbiont 12. Cenozoic 13. volcanoes 14. c 15. b 16. d 17. c 18. a

Chapter 20

1. taxonomy 2. genus; specific 3. *Penicillium* 4. family 5. division 6. Protista 7. Fungi 8. ancestor 9. homologous 10. derived 11. clocks 12. phenetic 13. cladistic 14. classical evolutionary 15. e 16. a 17. b

Chapter 21

1. DNA; RNA (any order) 2. capsid 3. bacteria 4. lytic 5. Prokaryotae 6. cell wall 7. saprobes 8. cocci; bacilli; spirilla 9. gram positive 10. plasmodium 11. cellular 12. red algae 13. brown 14. pseudopodia 15. sporozoa 16. b 17. c 18. d 19. c 20. a

Chapter 22

1. cell walls 2. spores 3. budding 4. septa 5. dikaryotic 6. zygospores 7. ascospores; asci 8. basidiocarp 9. basidiospore 10. gills 11. sexual 12. decomposers (or saprophytes) 13. parasites 14. roots 15. Morels; truffles (any order) 16. b 17. c 18. d 19. d 20. a 21. See Figure 22–5

Chapter 23

1. cuticle 2. green alga 3. stomata 4. alternation; generations 5. Water 6. gametophyte 7. sori 8. Horsetails 9. xylem; phloem 10. Seeds 11. fruit 12. cycads; ginkgoes; gnetophytes (any order) 13. flowers 14. monocots 15. endosperm 16. d 17. c 18. a 19. d 20. b

Chapter 24

1. invertebrates 2. bilaterally symmetrical 3. body cavity 4. echinoderms; chordates (any order) 5. spicules 6. eggs; sperm (any order) 7. cnidocytes 8. b 9. d 10. c 11. e 12. d 13. a 14. d 15. c 16. b 17. d 18. d 19. c 20. b 21. d 22. c 23. a 24. e

Chapter 25

1. notochord; nerve cord; gill; tail 2. Tunicates 3. vertebral column; cranium 4. ray-finned; tetrapods (amphibians)

5. amphibians 6. terrestrial; protects the embryo (and provides a water environment) 7. birds; mammals (any order) 8. lay eggs; pouches (marsupia) 9. e 10. d 11. e 12. c 13. d 14. e 15. *habilis*

Chapter 26

1. herbaceous 2. fibrous 3. bud scales 4. node 5. compound 6. opposite 7. vascular 8. collenchyma 9. tracheids 10. companion 11. cuticle 12. meristems 13. apical 14. leaf primordia; bud primordia (any order) 15. d 16. a 17. b 18. c 19. d 20. See Figure 26–1

Chapter 27

1. mesophyll 2. guard cells 3. xylem 4. lower 5. dicots 6. potassium 7. transpiration 8. circadian rhythms 9. abscission 10. Tendrils 11. storage 12. a 13. d 14. b 15. c 16. c 17. d

Chapter 28

1. primary; secondary 2. dicots 3. cortex 4. water; minerals 5. root cap 6. pericycle 7. xylem; pith 8. vascular cambium; cork cambium (any order) 9. periderm 10. translocation 11. root pressure 12. transpiration 13. pressure flow 14. macronutrients 15. b 16. c 17. d 18. d 19. d

Chapter 29

1. sexual 2. rhizome 3. tuber 4. asexual 5. carpel 6. insects 7. wind 8. double fertilization 9. suspensor 10. seed 11. fruit 12. Aggregate; multiple 13. accessory 14. wind 15. animals 16. d 17. c 18. c 19. b 20. d

Chapter 30

1. photoperiodism 2. darkness 3. red 4. Vernalization 5. pulvinus 6. Phototropism 7. thigmotropism 8. circadian rhythm 9. hormones 10. auxin 11. gibberellin 12. auxin 13. Cytokinins 14. ethylene 15. dormancy 16. a 17. b 18. d 19. c 20. e

Chapter 31

1. tissue 2. gland 3. squamous; cuboidal; columnar 4. intercalated disks 5. skeletal; cardiac (any order) 6. glial 7. endocrine 8. integumentary 9. negative 10. a 11. d 12. c 13. b 14. b 15. c 16. d 17. d 18. b 19. e

Chapter 32

1. epidermis; dermis 2. corneum 3. keratin 4. Hydrostatic 5. molt 6. endoskeletons 7. compact; spongy 8. lubricant; joints 9. myosin, actin 10. d, c, b, a, e, f, g 11. energy storage 12. glycogen; ATP 13. d 14. a 15. e 16. b 17. c 18. b

Chapter 33

1. reception 2. muscles; glands (any order) 3. glial; neurons 4. cell body 5. axon 6. cellular sheath 7. ganglion 8. sodium–potassium pumps 9. action potential (nerve impulse) 10. refractory period 11. synapse 12. neurotransmitter 13. reflex 14. afferent (sensory) 15. convergence 16. e 17. d 18. a 19. a 20. c

Chapter 34

1. nerve net 2. the central nervous system (CNS) 3. cerebral ganglia 4. brain; spinal cord 5. midbrain 6. olfactory; smell 7. association 8. protect the CNS 9. cerebrum 10. motor areas 11. beta 12. consciousness 13. c 14. d 15. a 16. b 17. e 18. sympathetic; parasympathetic 19. tolerance

Chapter 35

1. Sensory receptors 2. Proprioceptors 3. Photoreceptors; mechanoreceptors 4. energy; electrical; receptor potential 5. adaptation 6. gravity 7. spindles; Golgi tendon; joint 8. equilibrium 9. labyrinth; saccule, utricle; semicircular 10. otoliths; gravity 11. endolymph; ampulla 12. inner; sound 13. Corti 14. chemoreceptors 15. rhodopsins 16. c 17. d 18. e 19. a 20. a

Chapter 36

1. open 2. diffusion 3. plasma 4. anemia 5. thrombin 6. blood pressure 7. left atrium 8. liver 9. baroreceptors 10. mitral 11. vasoconstrictors 12. oxygen 13. arterial; blood 14. interstitial (tissue) 15. a 16. d 17. b 18. c 19. a 20. b 21. b

Chapter 37

1. immune response 2. antibodies (or immunoglobulins) 3. interferons 4. redness, heat, swelling (edema), pain 5. bone marrow; thymus 6. Suppressor 7. memory 8. plasma (differentiated B cells) 9. antibody 10. thymus 11. antigen-antibody 12. MHC 13. active 14. passive 15. immune response; transplanted 16. privileged 17. allergen; allergic 18. histamine; inflammation 19. c 20. b 21. c 22. d 23. e

Chapter 38

1. gas exchange (diffusion); moist; capillaries (blood vessels) 2. tracheal; spiracles 3. lungs 4. air sacs 5. trachea 6. alveoli (air sacs) 7. diaphragm 8. Decompression 9. inhaling dirty air 10. emphysema 11. cigarette smoking 12. a 13. a 14. e 15. e

Chapter 39

1. ingestion 2. Digestion 3. Elimination 4. one 5. two 6. amylase 7. dentin 8. d 9. e 10. a 11. e 12. c 13. a 14. c 15. a 16. c 17. e 18. b 19. c 20. a

Chapter 40

1. excretion 2. osmoregulation 3. uric acid 4. urea 5. protonephridia 6. metanephridia 7. Malpighian tubules 8. nephrons 9. capillaries; Bowman's capsule 10. afferent; efferent 11. filtrate 12. excreted in the urine 13. salt 14. collecting ducts; reabsorbed 15. a 16. c 17. e 18. d 19. e 20. c

Chapter 41

1. ducts; blood 2. hormones 3. cyclic AMP 4. hypothalamus 5. target 6. hypo- (too little); thyroid hormones 7. goiter 8. cortisol (glucocorticoids) 9. parathyroid 10. diabetes 11. a 12. d 13. a 14. e 15. b 16. e 17. b 18. e

Chapter 42

1. fragmentation 2. unfertilized 3. hermaphroditic 4. gamete; zygote 5. sterile 6. scrotum 7. therapeutic 8. female sterilization 9. 28; 14 10. c 11. e 12. a 13. c 14. a 15. b 16. a 17. e 18. c 19. c 20. d 21. c 22. b 23. e

Chapter 43

1. morphogenesis 2. cellular differentiation 3. cleavage 4. gastrulation 5. ectoderm; endoderm 6. neural plate 7. brain; spinal cord (any order) 8. amniotic fluid (amnion) 9. placenta 10. inner cell mass; embryo 11. implant; uterus (endometrium) 12. fetus 13. second; labor 14. a 15. b 16. a 17. a

Chapter 44

1. Behavior 2. Ethology 3. diurnal 4. crepuscular 5. sign stimulus 6. Innate; learned 7. Imprinting 8. stimulus 9. classical (Pavlovian) 10. social 11. society 12. Pheromones 13. dominance hierarchy 14. home range 15. conflict; population 16. pair bond 17. female 18. innate 19. culture 20. altruistic 21. Kin 22. making more genes 23. a 24. d 25. d 26. e

Chapter 45

1. population 2. density 3. Clumped dispersion 4. biotic potential 5. exponential 6. environmental resistance 7. carrying capacity 8. dependent 9. independent 10. independent 11. r 12. K 13. zero population growth 14. infant mortality rate 15. age structure 16. c 17. b 18. b 19. c 20. b

Chapter 46

1. community 2. decomposers 3. Carnivores; herbivores; omnivores 4. mullerian 5. mutualism 6. Parasites 7. commensalism 8. niche 9. fundamental 10. realized 11. competitive exclusion 12. edge effect 13. less 14. primary 15. b 16. c 17. a 18. b 19. c

Chapter 47

1. ecosystem 2. photosynthesis; cellular respiration 3. nitrifying 4. phosphorus 5. hydrologic cycle 6. sun 7. food web 8. trophic 9. primary consumers 10. numbers; biomass; energy (any order) 11. biomass 12. cellular respiration 13. vertically 14. rain shadows 15. b 16. c 17. c 18. c 19. e

Chapter 48

1. permafrost 2. Taiga 3. Lichens 4. precipitation 5. clay 6. grasslands 7. Chaparral 8. rain forests 9. savanna 10. light; oxygen (any order) 11. thermocline 12. littoral 13. Estuaries 14. plankton; nekton; benthos 15. neritic province 16. oceanic 17. b 18. c 19. e 20. c 21. c

Chapter 49

1. Extinction 2. endangered 3. threatened 4. Commercial; sport; subsistence 5. In situ 6. deforestation 7. slash; burn 8. tropics 9. CO_2; CH_4; O_3; N_2O; CFCs (any order) 10. CO_2; O_3; N_2O (any order) 11. heat 12. plants 13. Ozone (or O_3) 14. CFCs 15. ultraviolet 16. b 17. b 18. a 19. b 20. b

Appendix B Periodic Table of the Elements

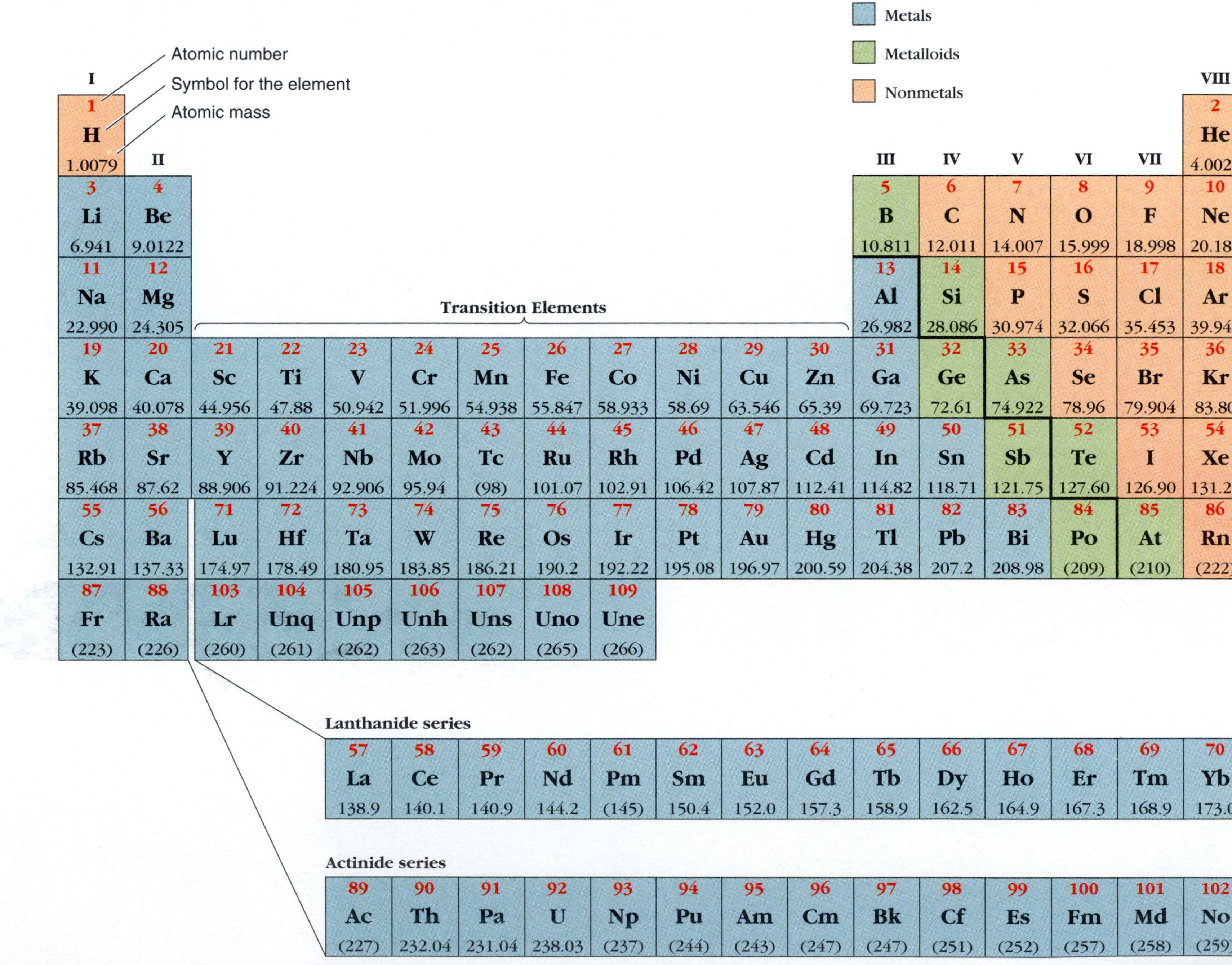

I	II											III	IV	V	VI	VII	VIII
1 **H** 1.0079																	2 **He** 4.0026
3 **Li** 6.941	4 **Be** 9.0122											5 **B** 10.811	6 **C** 12.011	7 **N** 14.007	8 **O** 15.999	9 **F** 18.998	10 **Ne** 20.180
11 **Na** 22.990	12 **Mg** 24.305											13 **Al** 26.982	14 **Si** 28.086	15 **P** 30.974	16 **S** 32.066	17 **Cl** 35.453	18 **Ar** 39.948
19 **K** 39.098	20 **Ca** 40.078	21 **Sc** 44.956	22 **Ti** 47.88	23 **V** 50.942	24 **Cr** 51.996	25 **Mn** 54.938	26 **Fe** 55.847	27 **Co** 58.933	28 **Ni** 58.69	29 **Cu** 63.546	30 **Zn** 65.39	31 **Ga** 69.723	32 **Ge** 72.61	33 **As** 74.922	34 **Se** 78.96	35 **Br** 79.904	36 **Kr** 83.80
37 **Rb** 85.468	38 **Sr** 87.62	39 **Y** 88.906	40 **Zr** 91.224	41 **Nb** 92.906	42 **Mo** 95.94	43 **Tc** (98)	44 **Ru** 101.07	45 **Rh** 102.91	46 **Pd** 106.42	47 **Ag** 107.87	48 **Cd** 112.41	49 **In** 114.82	50 **Sn** 118.71	51 **Sb** 121.75	52 **Te** 127.60	53 **I** 126.90	54 **Xe** 131.29
55 **Cs** 132.91	56 **Ba** 137.33	71 **Lu** 174.97	72 **Hf** 178.49	73 **Ta** 180.95	74 **W** 183.85	75 **Re** 186.21	76 **Os** 190.2	77 **Ir** 192.22	78 **Pt** 195.08	79 **Au** 196.97	80 **Hg** 200.59	81 **Tl** 204.38	82 **Pb** 207.2	83 **Bi** 208.98	84 **Po** (209)	85 **At** (210)	86 **Rn** (222)
87 **Fr** (223)	88 **Ra** (226)	103 **Lr** (260)	104 **Unq** (261)	105 **Unp** (262)	106 **Unh** (263)	107 **Uns** (262)	108 **Uno** (265)	109 **Une** (266)									

Lanthanide series

57 **La** 138.9	58 **Ce** 140.1	59 **Pr** 140.9	60 **Nd** 144.2	61 **Pm** (145)	62 **Sm** 150.4	63 **Eu** 152.0	64 **Gd** 157.3	65 **Tb** 158.9	66 **Dy** 162.5	67 **Ho** 164.9	68 **Er** 167.3	69 **Tm** 168.9	70 **Yb** 173.0

Actinide series

89 **Ac** (227)	90 **Th** 232.04	91 **Pa** 231.04	92 **U** 238.03	93 **Np** (237)	94 **Pu** (244)	95 **Am** (243)	96 **Cm** (247)	97 **Bk** (247)	98 **Cf** (251)	99 **Es** (252)	100 **Fm** (257)	101 **Md** (258)	102 **No** (259)

Appendix C
The Classification of Organisms

The system of cataloging organisms used here is described in Chapter 1 and in Part V. We have omitted many groups (especially extinct ones) in order to simplify the vast number of diverse categories of living organisms and their relationships to one another. Note that we have omitted the viruses from this survey, since they do not fit into any of the five kingdoms.

KINGDOM PROKARYOTAE: BACTERIA

Prokaryotic organisms that lack nuclear envelopes, mitochondria, and other membranous organelles. Typically unicellular, but some form colonies or filaments. The predominant mode of nutrition is heterotrophic, but some groups are photosynthetic or chemosynthetic. Reproduction is primarily asexual by fission. Bacteria are nonmotile or move by the beating of flagella. Flagella, when present, are solid (rather than the 9 + 2 type typical of eukaryotes). More than 10,000 species.

Subkingdom Archaebacteria

Anaerobic prokaryotic organisms with a number of features that set them apart from the rest of the bacteria: unusual cell wall composition, ribosomal RNA differences, lipid structure, and specific enzyme differences. Archaebacteria are found in extreme environments—hot springs, sea vents, dry and salty seashores, boiling muds, and near ash-ejecting volcanoes.

Phylum Methanocreatrices. *Methanogenic archaebacteria.* Anaerobes that produce methane from CO_2 and H_2. Found in sewage, swamps, and digestive tracts of humans and other animals.

Phylum 2. *Halophilic and thermoacidophilic bacteria.* This phylum, which has not yet been officially named, contains the halophiles, which live in extremely salty environments, and the thermoacidophiles, which normally grow in hot, acidic environments. None of these archaebacteria produce methane.

Subkingdom Eubacteria

Contains the vast majority of bacteria. This very large and very diverse group posseses three different cell wall conditions: no walls, gram-negative walls, and gram-positive walls. On the basis of their cell walls, the eubacteria are currently divided into three divisions. (Bacterial nomenclature and taxonomic practices do not currently correspond to those of other organisms.)

Division Tenericutes. *Bacteria that lack a rigid cell wall.* Contains the mycoplasmas, which are extremely small bacteria bounded by a plasma membrane.

Division Gracilicutes. *Gram-negative bacteria.* Possess a thin cell wall. Includes nitrogen-fixing aerobic bacteria, enterobacteria, spirochetes, cyanobacteria, rickettsias, chlamydias, and myxobacteria.

Division Firmicutes. *Gram-positive bacteria.* Possess a thick cell wall of peptidoglycan. All are nonphotosynthetic, and many produce spores. Includes the lactic acid bacteria, streptococci, staphylococci, clostridia, and actinomycetes.

KINGZDOM PROTISTA

Primarily unicellular or simple multicellular eukaryotic organisms that do not form tissues and that exhibit relatively little division of labor. Most modes of nutrition occur in this kingdom. Life cycles may include both sexually and asexually reproducing phases and may be extremely complex, especially in parasitic forms. Locomotion is by cilia, flagella, amoeboid movement, or by other means. Flagella and cilia have 9 + 2 structure.

Animal-like Protists

Phylum Zoomastigina. *Flagellates.* Single cells that move by means of flagella. Some free-living; many symbiotic; some pathogenic. Reproduction usually asexual by binary fission.

Phylum Rhizopoda. *Amoebas.* Naked single-celled protists whose movement is associated with pseudopods.

Phylum Ciliophora. *Ciliates.* Unicellular organisms that move by means of cilia. Reproduction is asexual by binary fission or sexual by conjugation. About 7200 species.

Phylum Apicomplexa. *Sporozoa.* Parasitic single-celled protists that reproduce by spores and have no means of locomotion. Reproduce by an unusual type of multiple fission. Some pathogenic. About 3900 species.

Phylum Foraminifera. *Foraminiferans.* Marine protists that produce shells (tests) with pores through which cytoplasmic projections are extended.

Plant-like Protists

Phylum Dinoflagellata. *Dinoflagellates.* Unicellular (some colonial), photosynthetic, biflagellate. Cell walls, composed of overlapping cell plates, contain cellulose. Contain chlorophylls *a* and *c* and carotenoids, including fucoxanthin. About 2100 species.

Phylum Bacillariophyta. *Diatoms.* Unicellular (some colonial), photosynthetic. Most nonmotile, but some move by gliding. Cell walls composed of silica rather than cellulose. Contain chlorophylls *a* and *c* and carotenoids, including fucoxanthin. About 5600 species.

Phylum Euglenophyta. *Euglenoids.* Unicellular, photosynthetic, two flagella (one of them very short). Flexible outer covering. Contain chlorophylls *a* and *b* and carotenoids. About 1000 species.

Phylum Chlorophyta. *Green algae.* Many aquatic. Unicellular, colonial, siphonous, and multicellular forms. Some motile and flagellated. Photosynthetic; contain chlorophylls *a* and *b* and carotenoids. About 7000 species.

Phylum Rhodophyta. *Red algae.* Most multicellular (some unicellular), many marine. Some (coralline algae) have bodies impregnated with calcium carbonate. No motile cells. Photosynthetic; contain chlorophyll *a*, carotenoids, phycocyanin, and phycoerythrin. About 4000 species.

Phylum Phaeophyta. *Brown algae.* Multicellular, often quite large (kelps). Photosynthetic; contain chlorophylls *a* and *c* and carotenoids, including fucoxanthin. Biflagellate reproductive cells. About 1500 species.

Fungal-like Protists

Phylum Myxomycota. *Plasmodial slime molds.* Spend part of life cycle as a thin, streaming, multinucleate plasmodium that creeps along on decaying leaves or wood. Flagellated or amoeboid reproductive cells; form spores in sporangia. About 500 species.

Phylum Acrasiomycota. *Cellular slime molds.* Vegetative (nonreproductive) form unicellular; move by pseudopods. Amoeba-like cells aggregate to form a multicellular pseudoplasmodium that eventually develops into a fruiting body that bears spores. About 70 species.

Phylum Oomycota. *Water molds.* Consist of branched, coenocytic mycelia. Cellulose and/or chitin in cell walls. Produce biflagellate asexual spores. Sexual stage involves production of oospores. Some parasitic. About 580 species.

KINGDOM FUNGI

All eukaryotic, mainly multicellular organisms that are heterotrophic with saprophytic or parasitic nutrition. Body form often a mycelium, and cell walls consist of chitin. No flagellated stages. Reproduce by means of spores, which may be produced sexually or asexually. Cells usually haploid or dikaryotic with brief diploid period following fertilization.

Division Zygomycota. *Zygomycetes.* Produce sexual resting spores called zygospores and asexual spores in a sporangium. Hyphae are coenocytic. Many are heterothallic (two mating types). About 765 species.

Division Ascomycota. *Ascomycetes.* Sexual reproduction involves formation of ascospores in little sacs called asci. Asexual reproduction involves production of spores called conidia, which pinch off from conidiophores. Hyphae usually have perforated septa. About 30,000 species.

Division Basidiomycota. *Basidiomycetes.* Sexual reproduction involves formation of basidiospores on a basidium. Asexual reproduction uncommon. Heterothallic. Hyphae usually have perforated septa. About 16,000 species.

Division Deuteromycota. *Imperfect fungi.* Sexual stage has not been observed. Most reproduce only by conidia. About 17,000 species.

KINGDOM PLANTAE

Multicellular eukaryotic organisms with differentiated tissues and organs. Cell walls contain cellulose. Cells frequently contain large vacuoles, photosynthetic pigments in plastids. Photosynthetic pigments are chlorophylls *a* and *b* and carotenoids. Nonmotile. Reproduce both asexually and sexually, with alternation of gametophyte and sporophyte generations.

Division Bryophyta. *Mosses.* Nonvascular plants that lack xylem and phloem. Marked alternation of generations with dominant gametophyte generation. Motile sperm. About 9000 species.

Division Pterophyta. *Ferns.* Vascular plants with a dominant sporophyte generation. Generally homosporous. Gametophyte is free-living and photosynthetic. Reproduce by spores. Motile sperm. About 11,000 species.

Division Psilotophyta. *Whisk ferns.* Vascular plants with a dominant sporophyte generation. Homosporous. Stem is distinctive because it branches dichotomously; plant lacks true roots and leaves. The gametophyte is subterranean and nonphotosynthetic and forms a mycorrhizal relationship with a fungus. Motile sperm. About 12 species.

Division Sphenophyta. *Horsetails.* Vascular plants with hollow, jointed stems and reduced scalelike leaves. Although modern representatives are small, some extinct species were treelike. Homosporous. Gametophyte a tiny photosynthetic plant. Motile sperm. About 15 species.

Division Lycophyta. *Club mosses.* Sporophyte plants are vascular with branching rhizomes and upright stems that bear microphylls. Although modern representatives are small, some extinct species were treelike. Some homosporous, others heterosporous. Motile sperm. About 1000 species.

Division Coniferophyta. *Conifers.* Heterosporous vascular plants with woody tissues (trees and shrubs) and needle-shaped leaves. Most are evergreen. Seeds are usually borne naked on the surface of cone scales. Nutritive tissue in the seed is haploid female gametophyte tissue. Nonmotile sperm. About 550 species.

Division Cycadophyta. *Cycads.* Heterosporous, vascular, dioecious plants that are small and shrubby or larger and palmlike. Produce naked seeds in conspicuous cones. Flagellated sperm. About 140 species.

Division Ginkgophyta. *Ginkgo.* Broad-leaved deciduous trees that bear naked seeds directly on branches. Dioecious. Contain vascular tissues. Flagellated sperms. The ginkgo tree is the only living representative. One species.

Division Gnetophyta. *Gnetophytes.* Woody shrubs, vines, or small trees that bear naked seeds in cones. Contain vascular tissues. Possess many features similar to flowering plants. About 70 species.

Division Anthophyta. *Flowering plants.* Largest, most successful group of plants. Heterosporous; dominant sporophytes with extremely reduced gametophytes. Contain vascular tissues. Bear flowers, fruits, and seeds (enclosed in a fruit; seeds contain endosperm as nutritive tissue). Double fertilization. About 235,000 species.

KINGDOM ANIMALIA: ANIMALS

Multicellular eukaryotic heterotrophs with differentiated cells. In most animals, cells are organized to form tissues, and tissues are organized to form organs; in complex forms specialized body systems carry on specific functions. Most animals have a well-developed nervous system and can respond rapidly to changes in their environment. Most are capable of locomotion during some time in their life cycle. Most animals reproduce sexually with large nonmotile eggs and flagellated sperm.

Subkingdom Parazoa

Tissue differentiation very limited; no organs. Adults sessile.

Phylum Porifera. *Sponges.* Mainly marine. Body perforated with many pores to admit water from which food is filtered by collar cells (choanocytes). Solitary or form colonies. Asexual reproduction by budding; external sexual reproduction in which sperm is released and swims to internal egg. Larva is motile. About 5000 species.

Subkingdom Eumetazoa

Tissues and organs; many at organ system level of organization.

Phylum Cnidaria. *Hydras, jellyfish, sea anemones, corals.* Tentacles surrounding mouth. Stinging cells (cnidocytes) that contain stinging structures called nematocysts. Polyp and medusa forms. Planula larva. Solitary or colonial. Marine, with a few freshwater forms. About 10,000 species.

Phylum Ctenophora. *Comb jellies.* Biradial symmetry. Free-swimming; marine. Two tentacles and eight longitudinal rows of cilia resembling combs; animal moves by means of these bands of cilia. About 100 species.

Protostomes

Spiral, determinate cleavage; mouth develops from blastopore.

Phylum Platyhelminthes. *Flatworms.* Acoelomate. Planarians are free-living; flukes and tapeworms are parasitic. Body dorsoventrally flattened; cephalization; three tissue layers. Simple nervous system with ganglia in head region. Excretory organs are protonephridia with flame cells. About 18,000 species.

Phylum Nemertea. *Ribbon worms.* Acoelomate. Long, dorsoventrally flattened body with complex proboscis armed with a hook for capturing prey. Simplest animal to have definite organ systems. Complete digestive tract. Circulatory system with blood. About 900 species.

Phylum Nematoda. *Roundworms.* Pseudocoelomate. *Ascaris,* hookworms, pinworms. Slender, elongated, cylindrical worms; covered with cuticle. Free-living and parasitic forms. About 12,000 species.

Phylum Rotifera. *Wheel animals.* Pseudocoelomate. Microscopic, wormlike animals. Anterior end has ciliated crown that looks like a wheel when the cilia beat. Posterior end tapers to a foot. Constant number of cells. About 2000 species.

Phylum Mollusca. *Snails, clams, squids, octopods.* Coelomate. Unsegmented, soft-bodied animals usually covered by a dorsal shell. Have a ventral, muscular foot. Most organs located above foot in visceral mass. A shell-secreting mantle covers the visceral mass and forms a mantle cavity which contains gills. Trochophore and/or veliger larva. About 60,000 species.

Phylum Annelida. *Segmented worms.* Coelomate. Polychaetes, earthworms, leeches. Both body wall and internal organs are segmented. Body segments separated by septa. Some have nonjointed appendages. Setae used in locomotion. Closed circulatory system; metanephridia; specialized regions of digestive tract. Trochophore larva. About 15,000 species.

Phylum Arthropoda. *Crabs, shrimp, insects, spiders, mites, ticks, centipedes, millipedes.* Coelomate. Segmented animals with paired, jointed appendages and a hard exoskeleton made of chitin. Open circulatory system with dorsal heart. Hemocoel occupies most of body cavity and coelom is reduced. About 1 million species.

Deuterostomes

Phylum Echinodermata. *Sea stars, sea urchins, sand dollars, sea cucumbers.* Coelomate. Marine animals that have pentaradial symmetry as adults, but bilateral symmetry as larvae. Endoskeleton of small, calcareous plates. Water vascular system; tube feet for locomotion. About 6000 species.

Phylum Hemichordata. *Acorn worms.* Coelomate. Marine animals with an anterior muscular proboscis, connected by a collar region to a long wormlike body. The larval form resembles an echinoderm larva. About 85 species.

Phylum Chordata. *Tunicates, lancelets, and vertebrates.* Coelomate. Notochord; pharyngeal gill groves; dorsal, tubular nerve cord; and postanal tail present at some time in life cycle. About 42,000 species.

Appendix D

Understanding Biological Terms

Your task of mastering new terms will be greatly simplified if you learn to dissect each new word. Many terms can be divided into a prefix, the part of the word that precedes the main root, the word root itself, and often a suffix, a word ending that may add to or modify the meaning of the root. As you progress in your study of biology, you will learn to recognize the more common prefixes, word roots, and suffixes. Such recognition will help you analyze new terms so that you can more readily determine their meaning and will also help you remember them.

Prefixes

a-, ab- from, away, apart (abduct, lead away, move away from the midline of the body)

a-, an-, un- less, lack, not (asymmetrical, not symmetrical)

ad- (also **af-, ag-, an-, ap-**) to, toward (adduct, move toward the midline of the body)

allo- different (allometric growth, different rates of growth for different parts of the body during development)

ambi- Both sides (ambidextrous, able to use either hand)

andro- male (androecium, the male portion of a flower)

anis- unequal (anisogamy, sexual reproduction in which the gametes are of unequal sizes)

ante- forward, before (anteflexion, bending forward)

anti- against (antibody, proteins that have the capacity to react against foreign substances in the body)

auto- self- (autotrophic, organisms that manufacture their own food)

bi- two (biennial, a plant that takes two years to complete its life cycle)

bio- life (biology, the study of life)

circum-, circ- around (circumcision, a cutting around)

co-, con- with, together (congenital, existing with or before birth)

contra- against (contraception, against conception)

cyt- cell (cytology, the study of cells)

di- two (disaccharide, a compound made of two sugar molecules chemically combined)

dis- apart (dissect, cut apart)

ecto- outside (ectoplasm, outer layer of cytoplasm)

end-, endo- within, inner (endoplasmic reticulum, a network of membranes found within the cytoplasm)

epi- on, upon (epidermis, upon the dermis)

ex-, e-, ef- out from, out of (extension, a straightening out)

extra- outside, beyond (extraembryonic membrane, a membrane that encircles and protects the embryo)

gravi- heavy (gravitropism, growth of a plant in response to gravity)

hemi- half (cerebral hemisphere, lateral half of the cerebrum)

hetero- other, different (heterozygous, unlike members of a gene pair)

homo-, hom- same (homologous, corresponding in structure; homozygous, identical members of a gene pair)

hyper- excessive, above normal (hypersecretion, excessive secretion)

hypo- under, below, deficient (hypotonic, a solution whose osmotic pressure is less than that of an isotonic solution)

in-, im- not (incomplete flower, a flower that does not have one or more of the four main parts)

inter- between, among (interstitial, situated between parts)

intra- within (intracellular, within the cell)

iso- equal, like (isotonic, equal osmotic concentration)

macro- large (macronucleus, a large, polyploid nucleus found in ciliates)

mal- bad, abnormal (malnutrition, poor nutrition)

mega- large, great (megakaryocyte, giant cell of bone marrow)

meso- middle (mesoderm, middle layer of the animal embryo)

meta- after, beyond (metaphase, the stage of mitosis after prophase)
micro- small (microscope, instrument for viewing small objects)
mono- one (monocot, a group of flowering plants with one cotyledon, or seed leaf, in the seed)
oligo- small, few, scant (oligotrophic lake, a lake deficient in nutrients and organisms)
oo- egg (oocyte, developing egg cell)
paedo- a child (paedomorphosis, the preservation of a juvenile characteristic in an adult)
para- near, beside, beyond (paracentral, near the center)
peri- around (pericardial membrane, membrane that surrounds the heart)
photo- light (phototropism, growth of a plant in response to the direction of light)
poly- many, much, multiple, complex (polysaccharide, a carbohydrate composed of many simple sugars)
post- after, behind (postnatal, after birth)
pre- before (prenatal, before birth)
pseudo- false (pseudopod, a temporary protrusion of a cell, i.e., "false foot")
retro- backward (retroperitoneal, located behind the peritoneum)
semi- half (semilunar, half-moon)
sub- under (subcutaneous tissue, tissue immediately under the skin)
super-, supra- above (suprarenal, above the kidney)
sym- with, together (sympatric speciation, evolution of a new species within the same geographical region as the parent species)
syn- with, together (syndrome, a group of symptoms which occur together and characterizes a disease)
trans- across, beyond (transport, carry across)

Suffixes

-able, -ible able (viable, able to live)
-ad used in anatomy to form adverbs of direction (cephalad, toward the head)
-asis, -asia, -esis condition or state of (hemostasis, stopping of bleeding)
-cide kill, destroy (biocide, substance that kills living things)
-emia condition of blood (anemia, a blood condition in which there is a lack of red blood cells)
-gen something produced or generated or something that produces or generates (pathogen, an organism that produces disease)
-gram record, write (electrocardiogram, a record of the electrical activity of the heart)
-graph record, write (electrocardiograph, an instrument for recording the electrical activity of the heart)
-ic adjective-forming suffix which means *of* or *pertaining to* (ophthalmic, of or pertaining to the eye)
-itis inflammation of (appendicitis, inflammation of the appendix)
-logy study of science or (cytology, study of cells)
-oid like, in the form of (thyroid, in the form of a shield)
-oma tumor (carcinoma, a malignant tumor)
-osis indicates disease (psychosis, a mental disease)
-pathy disease (dermopathy, disease of the skin)
-phyll leaf (mesophyll, the middle tissue of the leaf)
-scope instrument for viewing or observing (microscope, instrument for viewing small objects)

Some Common Word Roots

abscis cut off (abscission, the falling off of leaves or other plant parts)
angi, angio vessel (angiosperm, plants that produce seeds enclosed within a fruit or "vessel")
apic tip, apex (apical meristem, area of cell division located at the tips of plant stems and roots)
arthr joint (arthropods, invertebrate animals with jointed legs and segmented bodies)
aux grow, enlarge (auxin, a plant hormone involved in growth and development)
bi, bio life (biology, study of life)
blast a formative cell, germ layer (osteoblast, cell that gives rise to bone cells)
brachi arm (brachial artery, blood vessel that supplies the arm)
bry grow, swell (embryo, an organism in the early stages of development)
cardi heart (cardiac, pertaining to the heart)
carot carrot (carotene, a yellow, orange, or red pigment in plants)
cephal head (cephalad, toward the head)
cerebr brain (cerebral, pertaining to the brain)
cervic, cervix neck (cervical, pertaining to the neck)
chlor green (chlorophyll, a green pigment found in plants)
chondr cartilage (chondrocyte, a cartilage cell)
chrom color (chromosome, deeply staining body in nucleus)
cili small hair (cilium, a short, fine cytoplasmic hair projecting from the surface of a cell)
coleo a sheath (coleoptile, a protective sheath that encircles the stem in grass seeds)
conjug joined together (conjugation, a sexual phenomenon in certain protists)
cran skull (cranial, pertaining to the skull)
cyt cell (cytology, study of cells)
decid falling off (deciduous, a plant that sheds its leaves at the end of the growing season)
dehis split (dehiscent fruit, a fruit that splits open at maturity)
derm skin (dermatology, study of the skin)

ecol dwelling, house (ecology, the study of organisms in relation to their environment, i.e., "their house")
enter intestine (enterobacteria, bacteria that inhabit humans, particularly in their large intestine)
evol to unroll (evolution, descent with modification, or gradual directional changes)
fil a thread (filament, the thin stalk of the stamen in flowers)
gamet a wife or husband (gametangium, the part of a plant or protist that produces reproductive cells)
gastr stomach (gastrointestinal tract, the digestive tract)
glyc, glyco sweet, sugar (glycogen, storage form of glucose)
gon seed (gonad, an organ that produces gametes)
gutt a drop (guttation, loss of water as liquid "drops" from plants)
gymn naked (gymnosperm, a plant that produces seeds that are not enclosed within a fruit, i.e., "naked")
hem blood (hemoglobin, the pigment of red blood cells)
hepat liver (hepatic, of or pertaining to the liver)
hist tissue (histology, study of tissues)
hom, homeo same, unchanging, steady (homeostasis, reaching a steady state)
hydr water (hydrolysis, a breakdown reaction involving water)
leuk white (leukocyte, white blood cell)
menin membrane (meninges, the three membranes that envelop the brain and spinal cord)
morph form (morphogenesis, development of body form)
my, myo muscle (myocardium, muscle layer of the heart)
myc a fungus (mycelium, the vegetative body of a fungus)
nephr kidney (nephron, microscopic unit of the kidney)
neur, nerv nerve (neuromuscular, involving both the nerves and muscles)
occiput back part of the head (occipital, back region of the head)
ost bone (osteology, study of bones)
path disease (pathologist, one who studies disease processes)
ped, pod foot (bipedal, walking on two feet)
pell skin (pellicle, a flexible covering over the body of certain protists)
phag eat (phagocytosis, process by which certain cells ingest particles and foreign matter)
phil love (hydrophilic, a substance that attracts, i.e., "loves" water)
phloe bark of a tree (phloem, food-conducting tissue in plants which corresponds to bark in woody plants)
phyt plant (xerophyte, a plant adapted to xeric, or dry, conditions)
plankt wandering (plankton, microscopic aquatic protists that float or drift passively)
rhiz root (rhizome, a horizontal, underground stem that superficially resembles a root)
scler hard (sclerenchyma, cells that provide strength and support in the plant body)
sipho a tube (siphonous, a type of tubular body form found in certain algae)
som body (chromosome, deeply staining body in the nucleus)
sor heap (sorus, a cluster or "heap" of sporangia in the ferns)
spor seed (spore, a reproductive cell that gives rise to individual offspring in plants and protists)
stom a mouth (stoma, a small pore, i.e.., "mouth" in the epidermis of plants)
thigm a touch (thigmotropism, plant growth in response to touch)
thromb clot (thrombus, a clot within a blood vessel)
tropi turn (thigmotropism, growth of a plant in response to contact with a solid object, as when a tendril "turns" or wraps around a wire fence)
visc pertaining to an internal organ or body cavity (viscera, internal organs)
xanth yellow (xanthophyll, a yellowish pigment found in plants)
xyl wood (xylem, water-conducting tissue in plants, the "wood" of woody plants)
zoo an animal (zoology, the science of animals)

Glossary

abdomen (ab'doh-men) (1) In mammals, the region of the body between the diaphragm and the rim of the pelvis; (2) in arthropods, the posterior-most major division of the body.

abiotic Nonliving. Compare with biotic.

abortion Expulsion of an embryo or fetus before it is capable of surviving. May occur spontaneously (miscarriage) or may be induced.

abscisic acid (ab-sis'ik) A plant hormone involved in responses to stress and in dormancy.

abscission (ab-sizh'en) The normal (usually seasonal) falling off of leaves or other plant parts, such as fruits or flowers.

abscission zone The area at the base of the petiole where the leaf will break away from the stem.

absorption spectrum A measure of the amount of energy at specific wavelengths that has been absorbed as light passes through a substance. Each type of molecule has a characteristic absorption spectrum.

accessory fruit A fruit composed primarily of tissue other than ovary tissue. Apples and pears are accessory fruits.

acetylcholine (ah"see-til-koh'leen) A common neurotransmitter released by cholinergic neurons, including motor neurons.

acetyl CoA (ah-see'til) A key intermediate compound in metabolism; consists of an acetyl group covalently bonded to coenzyme A.

achene (a-keen') A simple, dry, indehiscent fruit with one seed. The fruit wall is separate from the seed coat. Sunflower fruits are achenes.

acid A substance that releases hydrogen ions (protons) in solution. Acids have a sour taste and unite with bases to form salts. Compare with base.

acid rain Atmospheric water that is acidic as a result of sulfur and/or nitrogen oxides forming acids when they react with water in the Earth's atmosphere; partially due to the combustion of coal. (Acid snow and acid fog are other forms of acid precipitation.)

acoelomate organisms (a-seel'oh-mate) Organisms that lack a body cavity (coelom). Flatworms are acoelomates.

actin (ak'tin) A contractile protein that, together with the protein myosin, is responsible for the ability of muscles to contract. Forms thin filaments in muscle fibers.

action potential The brief change in electrical activity developed across the plasma membrane of a muscle or nerve cell during activity; a neural impulse.

activation energy The energy required to initiate a chemical reaction.

active site Area of an enzyme that accepts a substrate and catalyzes its breakdown or its reaction with another substrate.

active transport Energy-requiring transport of a molecule across a membrane from a region of low concentration to a region of high concentration.

adaptation (1) Any feature of an organism that improves its chances of surviving and producing offspring. Adaptations are favored by natural selection. (2) A decrease in the response of a receptor that is subjected to repeated or prolonged stimulation.

adaptive radiation The evolution of a large number of related species from an unspecialized ancestral organism.

adenine (ad'eh-neen) A purine (nitrogenous base) that is a component of nucleic acids and of nucleotides important in energy transfer—for example, adenosine triphosphate (ATP). See nucleic acid.

adenosine triphosphate (ATP) (a-den'oh-seen) An organic compound containing adenine, ribose, and three phosphate groups; of prime importance for energy transfers in biological systems.

adhesion The attraction of the molecules of two different substances to one another. Compare with cohesion.

adrenal glands (ah-dree'nul) Paired endocrine glands, one located just superior to each kidney. Release hormones that help the body respond to stress and help regulate many aspects of metabolism.

adventitious root (ad"ven-tish'us) A root that arises in an unusual position on the plant.

aerobic (air-oh'bik) Growing or metabolizing only in the presence of molecular oxygen.

aerobic respiration The process by which cells utilize oxygen to break down organic molecules into waste products (such as carbon dioxide and water), with the release of energy that can be used for biological work.

afferent (af'fur-ent) Pertaining to a structure that leads toward another organ or structure, e.g., afferent neurons conduct impulses to the central nervous system.

age structure The percentage of a population, including the number of people of each sex, at different ages. Age structure diagrams represent the number of males and females at each age in a population.

aggregate fruit A fruit that develops from a single flower with many separate carpels, such as a raspberry.

agnathans (ag-na'thanz) Jawless fishes; class of vertebrates, including lampreys, hagfishes, and many extinct forms.

air pollution Various chemicals (gases, liquids, or solids) present in high enough levels in the atmosphere to harm humans, other animals, plants, or materials.

algae (al'gee) Single-celled or simple multicellular photosynthetic organisms; important producers in aquatic ecosystems.

allantois (a-lan'toe-iss) One of the extraembryonic membranes of reptiles, birds, and mammals.
allele frequency The relative number of times a particular allele occurs at its locus in all individuals of a population.
alleles (al-leels') Alternate forms of a gene that occupy corresponding positions on homologous chromosomes.
allelopathy An adaptation in which toxic substances secreted by roots or shed leaves inhibit the establishment of competing plants nearby.
allopatric speciation (al-oh-pa'trik) Speciation that occurs when one population becomes geographically separated from the rest of the species and subsequently evolves. Compare with sympatric speciation.
allopolyploid (al"oh-pol'ee-ploid) A polyploid formed by joining one or more sets of chromosomes from each of two different species.
allosteric site (al-oh-steer'ik) A site on an enzyme that enables a substance other than the normal substrate to bind to the enzyme and to change the shape and the activity of the enzyme.
allozyme (al'loh-zime) One of two or more slightly different versions of the same enzyme. The presence of allozymes may confer a selective advantage on the organism that possesses them.
alpine tundra A distinctive ecosystem located in the higher elevations of mountains, above the tree line.
alternation of generations A type of life cycle characteristic of plants. Plants spend part of their life in the haploid (gametophyte) stage and part in the diploid (sporophyte) stage.
altitude The height of an object above sea level. Compare with latitude.
alveolus (al-vee'o-lus) (1) An air sac of the lung through which gas exchange with the blood takes place; (2) a saclike unit of some glands.
ameboid motion The movement of a cell by means of pseudopodia (projections of the cell's surface by means of the slow oozing of its cellular contents).
amino acid (uh-mee'no) An organic compound containing an amino group ($—NH_2$) and a carboxyl group ($—COOH$). Amino acids may be linked together to form the polypeptide chains of protein molecules.
ammonification The conversion of nitrogen-containing organic compounds to ammonia (NH_3) by certain bacteria (ammonifying bacteria) in the soil; part of the nitrogen cycle.
amniocentesis (am"nee-oh-sen-tee'sis) Sampling of the amniotic fluid surrounding a fetus in order to obtain information about its development and genetic makeup.
amnion (am'nee-on) An extraembryonic membrane that forms a fluid-filled sac for the protection of the developing embryo.
amylase Starch-digesting enzyme, e.g., human salivary amylase or pancreatic amylase.
anabolism (an-ab'oh-lizm) The phase of metabolism in which simpler substances are combined to form more complex substances; requires energy. Compare with catabolism.
anaerobic (an"air-oh'bik) Growing or metabolizing in the absence of molecular oxygen.
analogous Similar in function or appearance but not in origin or development. Compare with homologous.
anaphase (an'uh-faze) The stage of mitosis between metaphase and telophase in which the chromatids separate and move toward opposite ends of the cell.
angiosperm (an'jee-oh-sperm") The traditional name for plants having flowers and seeds enclosed in fruits. Includes monocotyledons and dicotyledons.
anion (an'eye-on) A negatively charged ion such as Cl^-.
annelids (an'eh-lids) Segmented worms with true coeloms, such as earthworms.
annual A plant that completes its entire life cycle in one growing season. Compare with biennial and perennial.
annual ring A layer of wood (secondary xylem) formed in woody plants growing in temperate areas. One layer is usually formed per year.
anther (an'thur) The part of the stamen in flowers that produces microspores and, ultimately, pollen.
antheridium (an"thur-id'ee-im) The male gametangium in certain land plants. Antheridia produce sperm.
anthropoid A member of a suborder of primates that includes monkeys, apes, and humans.
antibiotic (an"ty-by-ot'ik) Substance produced by microorganisms that has the capacity, in dilute solutions, to inhibit the growth of or destroy bacteria and other microorganisms; used largely in the treatment of infectious diseases in humans and animals.
antibodies (an'-tih-bod"ees) Protein compounds (immunoglobulins) produced by plasma cells in response to specific antigens and having the capacity to react against the antigens.
anticodon (an'ty-koh"don) A sequence of three nucleotides in transfer RNA that is complementary to, and combines with, the three-nucleotide codon on messenger RNA, thus helping to specify the addition of a particular amino acid to the end of a growing peptide.
antigen (an'tih-jen) Any substance capable of stimulating an immune response; usually a protein or large carbohydrate that is foreign to the body.
anus (ay'nus) The distal end and opening of the digestive tract.
aorta (ay-or'tah) The largest and main systemic artery of the body; arises from the left ventricle and branches to distribute blood to all parts of the body; main artery leaving the heart in vertebrates.
apical dominance (ape'ih-kl) The inhibition of lateral buds by the apical meristem.
apical meristem (mehr'ih-stem) An area of cell division located at the tips of plant stems and roots. Apical meristems produce primary tissues.
apomixis (ap"uh-mix'us) A type of reproduction in which fruits and seeds are formed asexually, such as in dandelions.
aquatic Pertaining to the water. Compare with terrestrial.
arachnids (ah-rack'nids) Eight-legged arthropods such as scorpions and spiders.

archaebacteria (ar″kuh-bak-teer′ee-uh) Anaerobic prokaryotic organisms with a number of features that set them apart from the rest of the bacteria.

archegonium (ar″ke-go′nee-um) The female organ of certain land plants (such as mosses) in which an egg is produced.

arid land A fragile ecosystem in which lack of precipitation limits plant growth. Arid lands are found in both temperate and tropical regions; also called desert.

arteries Thick-walled blood vessels that carry blood away from the heart and toward the body organs.

arteriole (ar-teer′ee-ole) A very small artery. Vasoconstriction and vasodilation of arterioles help regulate blood pressure.

arthropod (ar′throh-pod) An invertebrate, such as an insect or crustacean, that has jointed appendages, a segmented body, and a hard exoskeleton.

artificial selection Selection by humans of traits that are desirable in plants or animals and breeding of only those individuals that possess the desired traits.

ascus A saclike spore case in some fungi that contains sexual spores called ascospores. Compare with basidium.

asexual reproduction Reproduction in which only one parent participates, there is no fusion of gametes, and in which the genetic makeup of parent and offspring is similar. Compare with sexual reproduction.

assimilation (of nitrogen) The conversion of inorganic nitrogen (nitrate NO_3^- or ammonia NH_3) to the organic molecules of living things; part of the nitrogen cycle.

atom The smallest quantity of an element that can retain the chemical properties of that element; composed of protons, neutrons, and electrons.

atomic mass A number that represents the sum of the number of protons and neutrons in the nucleus of an atom. The atomic mass represents the relative mass of an atom. Compare with atomic number.

atomic number A number that represents the number of protons in the nucleus of an atom. Each element has its own characteristic atomic number. Compare with atomic mass.

atrium (of the heart) (ay′tree-um) A chamber that receives blood from the veins.

autoimmune disease A disease in which the body produces antibodies against its own cells or tissues.

autonomic nervous system (aw-tuh-nom′ik) The portion of the peripheral nervous system that controls the visceral functions of the body, e.g., regulates smooth muscle, cardiac muscle, and glands, thereby helping to maintain homeostasis. See sympathetic and parasympathetic nervous systems.

autosome (aw′toh-sohm) A chromosome other than the sex (X and Y) chromosomes.

autotroph (aw′-toh-trof) An organism that obtains organic molecules by synthesizing them from inorganic materials. Also called producer.

auxin (awk-sin) A plant hormone involved in various aspects of plant growth and development.

axon (ax′on) The long extension of the neuron that transmits nerve impulses away from the cell body.

background extinction The continuous, low-level extinction of species that has been evident throughout much of the history of life. Compare with mass extinction.

bacteria (bak-teer′ee-uh) Unicellular, prokaryotic microorganisms. Most bacteria are decomposers, but some are parasites and some are autotrophs.

bacteriophage (bak-teer′ee-oh-fayj) Virus that can infect a bacterium (literally, "bacteria eater").

basal body (bay′sl) Structure that is similar to a centriole in its arrangement of microtubules and other components and that is involved in the organization and anchorage of a cilium or flagellum.

basal metabolic rate The amount of energy expended by the body just to keep alive, when no food is being digested and no voluntary muscular work is being done.

base A proton (H^+ ion) acceptor. Compare with acid.

basidium (ba-sid′ee-um) The clublike, spore-producing organ of certain fungi. Compare with ascus.

B cell (B lymphocyte) A type of white blood cell responsible for antibody-mediated immunity. When stimulated, B cells differentiate to become plasma cells that produce antibodies. Compare with T cell.

benthos (ben′thos) Bottom-dwelling sea organisms that fix themselves to one spot, burrow into the sand, or simply walk about on the ocean floor.

berry A simple, fleshy fruit in which the fruit wall is soft throughout. Tomatoes, bananas, and grapes are berries.

bicuspid (by-kus′pid) Having two points or cusps (as bicuspid teeth) or two flaps (as the mitral valve of the heart).

biennial (by-en′ee-ul) A plant that takes two years (i.e., two growing seasons) to complete its life cycle. Compare with annual and perennial.

bilateral symmetry A body shape with right and left halves that are approximately mirror images of one another. Compare with radial symmetry.

binary fission (by′nare-ee fish′un) Equal division of a cell or organism into two; usually a variety of asexual reproduction.

binomial nomenclature (by-nome′ee-ul) System of naming organisms by the combination of the names of genus and species.

biodiversity See biological diversity.

biogenesis The generalization that all living things arise (are propagated) only from pre-existing living things.

biogeochemical cycle Process by which matter cycles from the living world to the nonliving physical environment and back again. Examples of biogeochemical cycles include the carbon cycle, the nitrogen cycle, and the phosphorus cycle.

biogeography The study of the distribution of organisms.

biological clocks Means by which activities of plants or animals are adapted to regularly recurring changes.

biological diversity The number and variety of living organisms; includes diversity within a species (genetic diversity), among species (species diversity), and among ecosystems (ecological diversity); also called biodiversity.

biological magnification The increased concentration of toxic chemicals such as PCBs, heavy metals, and pesticides in the tissues of organisms higher in the food web.

biomass A quantitative estimate of the total mass, or amount, of living plant or animal material in a particular ecosystem.

biome (by'ohm) A large, relatively distinct ecosystem characterized by a similar climate, soil, plants, and animals, regardless of where it occurs on Earth.

biosphere All of Earth's living organisms. Compare with ecosphere.

biotechnology See genetic engineering.

biotic Living. Compare with abiotic.

biotic potential The maximum rate at which a population could increase when environmental conditions are optimal.

bipedal Walking on two feet.

birth rate The number of births per 1000 people per year.

blastocyst (blas'toh-sist) The blastula stage in the development of the mammalian embryo; a spherical mass consisting of a single layer of cells, the trophoblast, from which a small cluster, the inner cell mass, projects into a central cavity.

blastopore (blas'toh-pore) Primitive opening into the body cavity of an early embryo that may become the mouth or anus of the adult organism.

blastula (blas'tew-lah) In animal development, a hollow ball of cells produced by cleavage of a fertilized ovum.

B lymphocyte See B cell.

boreal forest See taiga.

bottleneck Genetic drift that may result from a sudden decrease in a population.

bronchiole (bronk'ee-ole) Tiny air duct in the lung that branches from a bronchus; divides to form air sacs (alveoli).

bronchus (bronk'us); pl. **bronchi** (bronk'eye) One of the branches of the trachea and its immediate branches within the lung.

bryophytes (bry'oh-fites) Members of the plant kingdom comprising mosses, liverworts, and hornworts.

bud An undeveloped shoot that can develop into flowers, stems, or leaves. Buds may be terminal (at the tip of the stem) or lateral (on the side of the stem).

budding Asexual reproduction in which a small part of the parent's body separates from the rest and develops into a new individual, eventually either taking up an independent existence or becoming a more or less independent member of the colony.

bud scale A modified leaf that covers and protects winter buds.

buffers Substances in a solution that tend to lessen the change in hydrogen ion concentration (pH) that otherwise would be produced by adding acids or bases.

bulb A globose, fleshy, underground bud. A bulb is a short stem with fleshy leaves, e.g., onion.

bulliform cell (bool-ee'form) A large, thin-walled cell found in the epidermis of some monocots; aids in the folding and unfolding of leaves associated with periods of drought.

bundle sheath A ring of cells surrounding the vascular bundle in dicot and monocot leaves.

callus (kal'us) Undifferentiated tissue in plant tissue culture.

calorie (kal'oh-ree) A unit of heat energy. The calorie used in the study of metabolism is the large Calorie or kilocalorie and is defined as the amount of heat required to raise one kilogram of water one degree Celsius.

Calvin cycle Cyclic series of reactions occurring in the light-independent phase of photosynthesis that fixes carbon dioxide and produces glucose.

calyx (kay'liks) The collective term for the sepals of a flower.

camouflage An organism's blending into its surroundings caused by its having a similar color, form, or behavior as its environment.

CAM plant A plant that photosynthesizes by crassulacean acid metabolism. CAM plants are better adapted to arid environments than non-CAM plants.

cancer A malignant tumor anywhere in the body. Cancers tend to spread throughout the body.

capillaries (kap'i-lare-eez) Microscopic blood vessels occurring in the tissues that permit exchange of materials between tissues and blood.

capsule (1) The portion of the moss sporophyte that contains spores; (2) a simple, dry, dehiscent fruit that opens along many seams or pores to release seeds, such as cotton fruits.

carbohydrate An organic compound containing carbon, hydrogen, and oxygen, in the approximate ratio of 1C:2H:1O, such as sugars, starches, and cellulose.

carbon cycle The worldwide circulation of carbon from the abiotic environment into living things and back into the abiotic environment.

carboxyl group (kar-bok'sil) A functional group characteristic of organic acids; group of atoms in a molecule arranged as $-C\begin{smallmatrix}{\nearrow}O\\ \searrow OH\end{smallmatrix}$.

carcinogen (kar'sin-oh-jen, kar-sin'oh-gen) A substance that causes cancer or accelerates its development.

cardiac (kar'dee-ak) Pertaining to the heart.

cardiac muscle Distinctive involuntary but striated type of muscle occurring in the vertebrate heart.

carnivore (kar'ni-vor) An animal that feeds on other animals; flesh-eater. Compare with herbivore and omnivore.

carotene (kare'oh-teen) Yellow to orange-red pigments found in carrots, sweet potatoes, leafy vegetables, etc.; can be converted in the animal body to vitamin A.

carpel (kar'pul) The female reproductive unit of a flower; carpels bear ovules.

carrying capacity The maximum population that a particular habitat can support and sustain (long-term).

cartilage Flexible skeletal tissue of vertebrates.

Casparian strip (kas-pare'ee-un) A band of waterproof material around the walls of the cells of the root endodermis.

catabolism The phase of metabolism in which complex organic molecules are broken down with the release of energy. Compare with anabolism.

catalyst (kat'al-list) A substance that regulates the speed at which a chemical reaction occurs without affecting the end point of the reaction and without being used up as a result of the reaction.

cation (kat'eye-un) An ion bearing a positive charge, e.g. Na^+.

cDNA A DNA molecule that was formed by reverse transcription of a mature mRNA transcript.

cell The basic structural and functional unit of life. Some simple organisms are composed of single cells, whereas complex organisms are composed of many cells.

cell cycle Cyclic series of events in the life of a dividing eukaryotic cell consisting of M (mitosis), cytokinesis, and the stages of interphase, which are the G_1 (first gap), S (DNA synthesis), and G_2 (second gap) phases.

cell differentiation See differentiation.

cell plate Forming cell wall that separates the two daughter cells produced by plant mitosis.

cellular respiration A cellular process in which the energy of organic molecules is released. Aerobic respiration is a type of cellular respiration.

cellulose (sel'yoo-lohs) A complex polysaccharide that composes the cell walls of plants.

cell wall A comparatively rigid supporting wall located exterior to the plasma membrane in plants, fungi, bacteria, and certain protists.

center of origin The particular place where a species originated or evolved.

central nervous system (CNS) In vertebrates, the brain and spinal cord.

centriole (sen'tree-ohl) One of a pair of small, dark-staining organelles lying near the nucleus in the cytoplasm of animal cells.

centromere (sen'tro-meer) Specialized constricted region of a chromatid that serves as the site of spindle fiber attachment during cell division; sister chromatids are joined in the vicinity of their centromeres.

cephalization The evolution of a head with sense organs and a concentration of nervous tissue.

cerebellum (ser-eh-bel'um) The deeply convoluted subdivision of the brain lying beneath the cerebrum that is concerned with the coordination of muscular movements; second largest part of the brain.

cerebral cortex (ser-ee'brul kor'tex) The outer layer of the cerebrum composed of grey matter and consisting mainly of nerve cell bodies.

cerebrum (ser-ee'brum) Largest subdivision of the brain; functions as the center for learning, voluntary movement, and interpretation of sensation.

CFCs See chlorofluorocarbons.

chaparral (shap"uh-ral') A biome with a Mediterranean climate (mild, moist winters and hot, dry summers). Chaparral vegetation is characterized by drought-resistant, small-leaved evergreen shrubs and small trees.

chemical evolution The origin of life from nonliving matter.

chemiosmosis During electron transport of aerobic respiration, the energy of a proton gradient established across the inner mitochondrial membrane is used to synthesize ATP.

chemoautotrophs (kee"moh-aw'toh-trofes) Autotrophic organisms that depend on the energy of pre-existing chemical compounds in their environment rather than on the energy of sunlight.

chemoreceptor (kee"moh-ree-sep'tor) A sensory receptor that responds to chemical stimuli.

chemotropism (kee"moh-tro'pizm) A growth response to a chemical stimulus.

chiasma (ky-az'muh); pl. **chiasmata** A site in a tetrad where homologous (nonsister) chromatids have undergone exchange by breakage and rejoining.

chitin (ky'tin) An insoluble, horny protein-polysaccharide that forms the exoskeleton of arthropods and the cell walls of many fungi.

chlorofluorocarbons (CFCs) Human-made organic compounds composed of carbon, chlorine, and fluorine that have a number of applications, including refrigerants and starting materials for certain plastics. In the atmosphere, chlorofluorocarbons break down stratospheric ozone.

chlorophyll (klor'oh-fil) Light-trapping green pigments found in most photosynthetic organisms.

chloroplast (klor'oh-plast) A chlorophyll-bearing intracellular organelle of plant cells; site of photosynthesis.

chordates (kor'dates) Phylum of animals that possess, at some time in their lives, a cartilaginous dorsal skeletal structure called a notochord; a dorsal, tubular nerve cord; pharynged gill slits; and a postanal tail.

chorion (kor'ee-on) An extraembryonic membrane in reptiles, birds, and mammals that forms an outer cover around the embryo, and in mammals contributes to the formation of the placenta.

chorionic villus sampling (CVS) A technique for diagnosing genetic abnormalities while the fetus is in the uterus. A sample of extra-embryonic cells that is genetically identical to the cells of the fetus is removed and studied.

chromatid (kroh'mah-tid) One of the two halves of a replicated chromosome.

chromatin (kroh'mah-tin) The complex of DNA and protein that makes up eukaryotic chromosomes.

chromosomes (kro'moh-soms) Rod-shaped bodies in the cell nucleus that contain the hereditary units, the genes.

cilium (sil'ee-um); pl. **cilia** One of many short hairlike structures that project from the surface of some cells and are used for locomotion or movement of materials across the cell surface; structurally like flagella, including a cylinder of 9 double microtubules and 2 central single microtubules, all covered by a plasma membrane.

circadian rhythm (sir-kay'dee-un) An internal rhythm that approximates the 24-hour day. Circadian rhythms are found in plants, animals, and other organisms.

citric acid cycle Aerobic series of chemical reactions in which fuel molecules are completely degraded to carbon dioxide and water with the release of metabolic energy; also known as the Krebs cycle.

cladistics A method of classification in which the criterion for classification is recency of common ancestry, rather than the degree of structural similarity.

cladogram A branching diagram that illustrates taxonomic relationships based on the principles of cladistics.

classical conditioning A type of learning in which an association is formed between some normal response to a stimulus and a new stimulus, after which the new stimulus elicits the response.

cleavage Series of cell divisions that converts the zygote to a multicellular blastula.

climax community The final, stable, and mature community in a successional series.

clone A population of cells produced by mitotic division from a single ancestral cell, or a population of genetically identical organisms asexually propagated from a single ancestor.

club moss Small vascular plants with branching rhizomes and upright stems that bear microphylls; club mosses reproduce by forming spores and are similar to ferns in certain respects. Extinct species were among the dominant plants of large swampy forests from which much of our coal formed.

cnidarians (nye-dare'ee-uns) Phylum of animals possessing stinging cells called cnidoblasts, a single body opening, two tissue layers, and radial symmetry, e.g., *Hydra.*

coastal wetland Marshes, bays, tidal flats, and swamps that are found along a coastline. See salt marsh and wetland.

cochlea The structure of the inner ear of mammals that contains the auditory receptors (organ of Corti).

codominance (koh"dom'in-ints) Condition in which both alleles of a locus are expressed in a heterozygote.

codon (koh'don) A trio of mRNA bases that specifies an amino acid or a signal to terminate the polypeptide.

coelom (see'lum) Body cavity that is completely lined by mesoderm.

coenzyme (koh-en'zime) An organic, nonprotein substance that serves as a cofactor; participates in the reaction by donating or accepting some reactant; loosely bound to enzyme. Most of the vitamins function as coenzymes.

coevolution The interdependent evolution of two or more species that occurs as a result of their interactions over a long period of time. Flowering plants and their animal pollinators are an example of coevolution because each has profoundly affected the other's characteristics.

cofactor A nonprotein substance needed by an enzyme for normal action; cofactors are metal ions or coenzymes.

cohesion The attraction of the molecules of a substance to one another, thereby holding it together. Compare with adhesion.

coleoptile (kol-ee-op'tile) A protective sheath that encloses the stem in grasses.

collagen (kol'ah-gen) Protein in connective tissue fibers.

collenchyma (kol-en'kih-mah) Living cells with moderately but unevenly thickened walls. Collenchyma cells help support the primary plant body.

combustion The process of burning by which organic molecules are rapidly oxidized, converting them into carbon dioxide and water with an accompanying release of heat and light.

commensalism A type of symbiosis in which one organism benefits and the other one is neither harmed nor helped. See symbiosis. Compare with mutualism and parasitism.

community An association of organisms of different species living together in a defined habitat with some degree of mutual interdependence. Compare with ecosystem.

companion cell A cell in the phloem of plants that is responsible for loading and unloading sugar into the sieve tube member for conduction.

competition The interaction among organisms that require the same resources in an ecosystem (such as food, living space, or other resources). See interspecific competition and intraspecific competition.

competitive exclusion The concept that no two species with identical living requirements can occupy the same ecological niche indefinitely. Eventually, one species will be excluded by the other as a result of interspecific (between species) competition for a resource in limited supply.

competitive inhibition Interference with enzyme action by an abnormal substrate that is not permanently bound to the active site and that competes with the normal substrate for the site.

complete flower A flower that possesses all four main parts: sepals, petals, stamens, and carpels.

compound eye An eye, such as that of an insect, composed of many light-sensitive units (called ommatidia).

condensation reaction A chemical reaction in which water is removed as two or more smaller molecules join together to form a larger molecule. Compare with hydrolysis.

cone (1) In botany, a reproductive structure in many gymnosperms that produces either microspores or megaspores. (2) In zoology, one of the conical photoreceptive cells of the retina that is particularly sensitive to bright light, and, by distinguishing light of various wavelengths, mediates color vision. Compare with rod.

conifer (kon'ih-fur) Any of a group of woody trees or shrubs (gymnosperms) that bear needle-like leaves and seeds in cones.

conjugation A sexual phenomenon in certain protists that involves exchange or fusion of a cell with another cell. The term is also used sometimes for DNA exchange in bacteria.

connective tissue Animal tissue consisting mostly of a matrix composed of cell products in which the cells are embedded, e.g., bone.

consumers Organisms that cannot synthesize their own food from inorganic materials and therefore must use the bodies of other organisms as a source of energy and body-building materials; also called heterotrophs. Compare with producers.

continental shelf The submerged, relatively flat ocean bottom that surrounds continents. The continental shelf extends out into the ocean to the point where the ocean floor begins a steep descent.

contraceptive A device or drug used to intentionally prevent pregnancy.

contractile vacuole (vak'yoo-ohl) A vacuole that expands, filling with water, and periodically contracts, ejecting the water from the cell.

control group In a scientific experiment, a group in which the experimental variable is kept constant. The control provides a standard of comparison in order to verify the results of an experiment.

convergent evolution The independent evolution of similar structures that carry on similar functions, in two or more organisms of widely different, unrelated ancestry.

copulation Sexual union; act of physical joining of two animals during which sperm cells are transferred from one animal to the other.

cork Cells produced by the cork cambium. Cork is dead at maturity and functions for protection.

cork cambium (kam′bee-um) Lateral meristem in plants that produces cork cells and cork parenchyma. Cork cambium and the tissues it produces make up the outer bark on a woody plant.

cork parenchyma (par-en′kih-mah) One or more layers of parenchyma cells produced by the cork cambium.

corm A short, thickened underground stem specialized for food storage and asexual reproduction, as exists in the gladiolus.

cornea Transparent anterior covering of the eye.

corolla (kor-ohl′ah) A collective term for the petals of a flower.

corpus luteum (loo-tee′um) The temporary endocrine tissue in the ovary that develops from the ruptured follicle after ovulation; secretes progesterone and estrogen.

cortex (kor′tex) (1) The outer layer of an organ. (2) In plants, the tissue beneath the epidermis in many nonwoody plants.

cotyledon (kot″i-lee′dun) The seed leaf of the embryo of a plant that may contain food stored for germination.

covalent bond Chemical bond involving one or more shared pairs of electrons.

cristae (kris′tee) Shelflike folds of the inner membrane of a mitochondrion.

crossing over The breaking and rejoining of homologous (nonsister) chromatids during early meiotic prophase I, resulting in an exchange of genetic material.

ctenophores (teen′oh-forz) Marine animals (comb jellies) whose outer surface is covered with eight rows of cilia resembling combs, by which the animal moves through the water.

cultural evolution The progressive addition of knowledge to the human experience.

cuticle (kew′tih-kl) (1) A waxy covering over the epidermis of the above-ground portion of land plants that reduces water loss from plant surfaces. (2) The outer covering of some animals.

cyanobacteria (sy-an′oh-bak-teer′ee-uh) Prokaryotic photosynthetic microorganisms that possess chlorophyll and produce oxygen by the photolysis of water. Formerly known as blue-green algae.

cycads (sih′kads) A group of gymnosperms that live mainly in tropical and semitropical regions and have stout stems and fernlike leaves.

cyclic AMP (cAMP) A form of adenosine monophosphate in which the phosphate is part of a ring-shaped structure; acts as a regulatory molecule and second messenger in organisms ranging from bacteria to humans.

cytochromes (sy′toh-kromz) The iron-containing heme proteins of the electron transport system that are alternately oxidized and reduced in biological oxidation.

cytokinesis (sy″toh-kih-nee′-sis) Stage of the cell cycle in which the cytoplasm is divided to form two daughter cells.

cytokinin (sy″toh-kih′nin) A plant hormone involved in various aspects of plant growth and development.

cytoplasm (sy′toh-plazm) General cellular contents exclusive of the nucleus.

cytosine See nucleic acid.

cytoskeleton Internal structure of microfilaments, intermediate filaments, and microtubules that gives shape and mechanical strength to cells.

cytosol Fluid component of the cytoplasm in which the organelles are suspended.

cytotoxic T cells T lymphocytes that destroy cancer cells and other pathogenic cells on contact; also called killer T cells.

deamination (dee-am-ih-nay′shun) Removal of an amino group (—NH_2) from an amino acid or other organic compound.

death rate The number of deaths per 1000 people per year.

deciduous (de-sid′yoo-us) Falling off at a certain time or stage of growth, such as some leaves, antlers, and the wings of some insects.

decomposer A heterotroph that breaks down organic material and uses the decomposition products to supply it with energy. Decomposers are microorganisms of decay. Also called saprophyte. Compare with detritivore.

deductive reasoning Reasoning that operates from generalities to specifics and can make relationships among data more apparent. Compare with inductive reasoning.

deforestation The removal of forest without adequate replanting.

dehiscent fruit (dee-his′sent) A simple, dry fruit that splits open along one or more seams at maturity. Compare with indehiscent fruit.

dehydrogenation (dee-hy″dro-jen-ay′shun) A form of oxidation in which hydrogen atoms are removed from a molecule.

demographics The branch of sociology that deals with population statistics (such as density and distribution).

demographic transition The process by which a country moves from high birth rates and high death rates to low birth rates and low death rates with an intervening spurt of population growth.

denature (dee-nay′ture) To alter the physical properties and three-dimensional structure of a protein, nucleic acid, or other macromolecule by mild treatment that does not break the primary structure but that does destroy its activity.

dendrite (den′drite) A short branch of a neuron that receives and conducts nerve impulses toward the cell body.

denitrification The conversion of nitrate (NO_3^-) to nitrogen gas (N_2) by certain bacteria (denitrifying bacteria) in the soil; part of the nitrogen cycle.

deoxyribose The 5-carbon sugar found in DNA.

depolarization (dee-pol″ar-ih-zay′shun) Change in electric charge of a plasma membrane that produces the action potential.

dermis (dur′mis) The layer of dense connective tissue beneath the epidermis in the skin of vertebrates.

desert See arid land.

desertification Degradation of once-fertile rangeland (or tropical dry forest) into nonproductive desert. Caused partly by soil erosion, deforestation, and overgrazing.

desmosomes (dez′moh-somz) Button-like plaques, present on the two opposing cell surfaces and separated by the intercellular space, that serve to hold the cells together.

detritivore An organism (such as an earthworm or crab) that consumes fragments of dead organisms; also called detritus feeder. Compare with decomposer.

detritus feeders (deh-try'tus) See detritivore.

deuterostome (doo'ter-oh-stome) A division of coelomate animals that includes the echinoderms and chordates; characterized by radial cleavage and development of the anus from the blastopore. Compare with protostome.

developed country A country industrialized; characterized by a low fertility rate, low infant mortality rate, and high per capita income. Developed countries include the United States, Canada, Japan, and European countries; also called highly developed country. Compare with developing country.

developing country A country not highly industrialized; characterized by a high fertility rate, high infant mortality rate, and low per capita income. Most developing countries are located in Africa, Asia, and Latin America. They fall into two subcategories—moderately developed and less developed. Compare with developed country.

development The orderly sequence of progressive changes in the life of an organism. Involves embryonic and other developmental processes.

dicotyledon (dy-kot-ih-lee'dun) A flowering plant with embryos having two seed leaves, or cotyledons; also known as a dicot. Compare with monocotyledon.

differentiation Development toward a more mature state; a process changing a relatively unspecialized cell to a more specialized cell. Also called cell differentiation.

diffusion The net movement of molecules from a region of high concentration to one of lower concentration, brought about by their kinetic energy.

dihybrid cross (dy-hy'brid) A genetic cross that takes into account the behavior of two distinct pairs of genes.

dikaryotic cells (dy-kare-ee-ot'ik) Cells having two nuclei.

dinoflagellates (dy"noh-flaj'eh-lates) Single-celled algae, surrounded by a shell of thick interlocking cellulose plates.

dioecious (dy-ee'shus) Having male and female reproductive structures on separate plants. Compare with monoecious.

diploid (dip'loid) or **2*n*** A chromosome number twice that found in gametes; containing two sets of chromosomes. Compare with haploid.

directional selection The gradual replacement of one phenotype with another due to environmental change which favors those phenotypes at one of the extremes of the normal distribution. Compare with disruptive selection and stabilizing selection.

disaccharide (dy-sak'ah-ride) A two-unit sugar, such as sucrose, that consists of two monosaccharide subunits.

disease A departure from the body's normal healthy state as a result of infectious organisms, environmental stresses, or some inherent weakness.

disruptive selection A special type of directional selection in which changes in the environment favor two or more variant phenotypes at the expense of the mean. Compare with directional selection and stabilizing selection.

distal Remote; farther from the main body axis.

diurnal (dy-ur'nl) Active during the daytime.

division A taxonomic category below that of kingdom; comparable to that of phylum.

DNA Deoxyribonucleic acid; present in a cell's chromosomes, DNA contains genetic information for all living organisms.

DNA ligase A repair enzyme that catalyzes the joining of broken parts of DNA. DNA ligase is used in recombinant DNA techniques.

DNA polymerase A group of enzymes that catalyzes DNA replication by adding nucleotides to a growing strand of DNA.

DNA replication The synthesis of DNA by using a complementary strand of DNA as a template.

dominant allele (al-leel') An allele that is always expressed when it is present, regardless of whether it is homozygous or heterozygous.

dormancy A resting condition, as when an organism is alive but is not growing and has a relatively inactive metabolism.

dorsal (dor'sl) Pertaining to the back of an animal.

double fertilization A process in the angiosperm reproductive cycle where there are two fertilizations: one results in the formation of a young plant; the second results in the formation of the endosperm.

doubling time The amount of time it takes for a population to double in size, assuming that its current rate of increase does not change.

Down syndrome An inherited defect in which individuals have abnormalities of the face, tongue, and other parts of the body, and are retarded in both their physical and mental development; usually results from a trisomy of chromosome 21.

drupe (droop) A simple, fleshy fruit in which the inner wall of the fruit is hard and stony. Peaches and cherries are drupes.

duodenum The first portion of the small intestine.

echinoderms (eh-kine'oh-derms) Spiny-skinned marine animals such as sea stars, sea urchins, and sea cucumbers.

ecological niche See niche.

ecological pyramid A graphic representation of the relative energy value at each trophic level. See pyramid of biomass, pyramid of energy, and pyramid of numbers.

ecology (ee-kol'uh-jee) A discipline of biology that studies the interrelationships among living things and their environments.

ecosphere The interactions among and between all of the Earth's living organisms and the air (atmosphere), land (lithosphere), and water (hydrosphere) that they occupy. Compare with biosphere.

ecosystem (ee'koh-sis-tem) The interacting system that encompasses a community and its nonliving, physical environment. Compare with community.

ecotone (ee'koh-tone) A fairly broad transition region between adjacent biomes; contains some organisms from each of the two biomes plus some that are characteristic of, and perhaps restricted to, the ecotone.

ectoderm (ek'toh-derm) The outer of the three embryonic germ layers that gives rise to the skin and nervous system. Compare with mesoderm and endoderm.

ectotherm An animal whose temperature fluctuates with that of the environment; may use behavioral adaptations to regulate temperature; cold-blooded. Compare with endotherm.

effector A muscle or gland that contracts or secretes in direct response to nerve impulses.

efferent (ef'fur-ent) Pertaining to a structure that leads away from another structure or organ, such as the efferent arteriole of the kidney nephron.

ejaculation A sudden expulsion, as in the ejection of semen from the penis.

electrochemical potential The potential energy possessed by a system in which there is a difference in electrical charge, as well as a concentration gradient of ions across a membrane.

electron A negatively charged subatomic particle located at some distance from the atomic nucleus.

electron transport chain A series of chemical reactions during which hydrogens or their electrons are passed along from one acceptor molecule to another, with the release of energy.

element Chemically, one of the 100 or so types of matter, natural or synthetic, composed of atoms with the same number of protons and electrons.

elimination The ejection of waste products, especially undigested food remnants, from the digestive tract (not to be confused with excretion).

embryo (em'bree-oh) A young organism before it emerges from the egg, seed, or body of its mother; the developing human organism until the end of the second month, after which it is referred to as a fetus.

embryo sac The female gametophyte generation in angiosperms.

emigration A type of migration in which individuals leave a population and thus decrease its size. Compare with immigration.

endangered species A species whose numbers are so severely reduced that it is in imminent danger of becoming extinct. Compare with threatened species.

endergonic (end"er-gon'ik) Nonspontaneous reaction requiring a net input of free energy. Compare with exergonic.

endocrine glands (en'doh-crin) Glands that secrete hormones into the blood or tissue fluid. Compare with exocrine glands.

endocytosis (en"doh-sy-toh'sis) The active transport of substances into a cell by the formation of invaginated regions of the plasma membrane that pinch off and become cytoplasmic vesicles. Compare with exocytosis.

endoderm (en'doh-derm) The inner germ layer of the early embryo; becomes the digestive tract and lining of its outgrowths—the liver, lungs, and pancreas. Compare with ectoderm and mesoderm.

endodermis (en"doh-der'mis) The innermost layer of the cortex in the plant root. Endodermis cells have a Casparian strip running around radial and transverse walls.

endogenous (en-doj'eh-nus) Produced within the body, or due to internal causes. Compare with exogenous.

endometrium (en"doh-mee'tree-um) Uterine lining.

endoplasmic reticulum (en"doh-plaz'mik reh-tik'yoo-lum) **(ER)** Organelle composed of numerous internal membranes within eukaryotic cells.

endorphins (en-dor'finz) Polypeptide neurotransmitters of certain brain and visceral neurons whose action is mimicked by opiate alkaloids.

endoskeleton (en"doh-skel'eh-ton) Bony and cartilaginous supporting structures *within* the body that provide support. Compare with exoskeleton.

endosperm (en'doh-sperm) The nutritive tissue that is found at some point in all angiosperm seeds.

endospore A highly resistant bacterial spore that develops within certain bacterial cells.

endosymbiont theory (en"doh-sim'bee-ont) That certain organelles such as mitochondria and chloroplasts originated as symbiotic prokaryotes that lived inside larger cells.

endotherm (en'doh-therm) An animal that uses metabolic energy to maintain a constant body temperature despite variations in environmental temperature; e.g., birds and mammals. Compare with ectotherm.

energy The capacity or ability to do work.

energy flow The passage of energy in a one-way direction through an ecosystem.

entropy (en'trop-ee) A quantitative measure of the randomness or disorder of a system.

environment All of the external conditions, both abiotic and biotic, that affect an organism or group of organisms.

environmental resistance Limits set by the environment that prevent organisms from reproducing indefinitely at their biotic potential.

enzyme (en'zime) An organic catalyst produced within a living organism that accelerates specific chemical reactions.

epicotyl (ep'ih-kot"il) The part of the axis of a plant embryo or seedling above the point of attachment of the cotyledons. Compare with hypocotyl.

epidermis (ep-ih-dur'mis) An outer layer of cells covering the body of plants and animals; functions primarily for protection.

epiphyte (ep'ih-fite) A small organism that grows on another organism but is not parasitic on it. Small plants that live attached to the bark of a tree are epiphytes.

epistasis (ep"ih-sta'sis) Condition in which certain alleles at one locus can alter the expression of alleles at a different locus.

epithelial tissue (ep-ih-theel'ee-al) The type of animal tissue that covers body surfaces, lines body cavities, and forms glands; also called epithelium.

equilibrium, dynamic A state in which there is no net change because two opposing processes occur at the same rate, e.g., when forward and reverse reactions of a reversible reaction occur at the same rate.

erythrocyte (er-eeth'roh-site) Vertebrate red blood cell.

esophagus (ee-sof'ah-gus) The muscular tube extending from the pharynx to the stomach.

estrus (es'trus) The recurrent period of sexual receptivity occurring around ovulation in female mammals having estrous cycles.

estuary A coastal body of water that connects to oceans, in which fresh water from the land mixes with salt water from the oceans.

ethology (ee-thol'oh-jee) The study of animal behavior under natural conditions from the point of view of adaptation.

ethylene (eth'ih-leen) A plant hormone involved in various aspects of plant senescence.

eubacteria (yoo'bak-teer"ee-ah) Prokaryotes other than the archaebacteria.

euglenoids (yoo-glee'noids) A group of unicellular protists that includes *Euglena.*

eukaryote (yoo-kare'ee-ote) An organism whose cells possess organelles surrounded by membranes. Compare with prokaryote.

evolution Cumulative genetic change in a population of organisms from generation to generation. Evolution leads to differences among populations and explains the origin of all of the organisms that exist today or have ever existed.

excretion (ek-skree'shun) The discharge from the body of a waste product of metabolism (not to be confused with the elimination of undigested food materials).

exergonic (ex"er-gon'ik) A reaction characterized by the release of energy. Compare with endergonic.

exocrine glands (ex'oh-crin) Glands that excrete their products through ducts in the epithelium onto the surface of the skin, such as sweat glands. Compare with endocrine glands.

exocytosis (ex"oh-sy-toh'sis) Export of materials from the cell by fusion of cytoplasmic vesicles with the plasma membrane. Compare with endocytosis.

exogenous (ek-sodj'eh-nus) Due to, or produced by, an external cause; not arising within the body. Compare with endogenous.

exon (1) A protein-coding region of a eukaryotic gene; (2) the RNA transcribed from such a region. Compare with intron.

exoskeleton (ex"oh-skel'eh-ton) An external skeleton such as the shell of mollusks or outer covering of arthropods; provides protection and sites of attachment for muscles. Compare with endoskeleton.

exponential growth Growth that occurs at a constant rate of increase over a period of time. When the increase in number versus time is plotted on a graph, exponential growth produces a characteristic J-shaped curve.

ex situ conservation Conservation efforts that involve conserving biological diversity in human-controlled settings. Compare with in situ conservation.

extinction The disappearance of a species. Extinction occurs when the last individual of a species dies. See background extinction and mass extinction.

F_1 generation (first filial) The first generation of filial offspring resulting from a genetic cross.

F_2 generation (second filial) The offspring of the F_1 generation.

facilitated diffusion The passage of ions and molecules, bound to specific carrier proteins, across a cell membrane down their concentration gradients. Facilitated diffusion does not require the cell to expend energy.

facultative parasite An organism that is normally saprophytic but, given the opportunity, becomes parasitic. Compare with obligate parasite.

fall turnover A mixing of the lake waters in temperate lakes, caused by falling temperatures in autumn. Compare with spring turnover.

fatty acid An organic acid composed of a long, unbranched chain of carbon and hydrogen atoms with a —COOH group at one end; fatty acids linked to glycerol form a fat.

feces Solidified waste products eliminated by the digestive tract of an animal.

feedback control System in which the accumulation or deficiency of the product of a reaction leads to a change in its rate of production. See negative feedback system and positive feedback system.

feedback system See negative feedback system.

fermentation Anaerobic respiration that utilizes organic compounds as both electron donors and acceptors.

fern An ancient group of vascular plants that reproduce by spores. The dominant sporophyte generation of ferns is typically a perennial with large, conspicuous leaves (fronds) and a horizontal, creeping rhizome.

fertilization Fusion of male and female gametes.

fetus The unborn human offspring from the third month of pregnancy to birth.

fiber (1) In plants, a type of sclerenchyma. Fibers are long, tapered cells with thick walls. (2) In animals, an elongated cell such as a muscle or nerve cell.

fibrous root system A root system in plants that has several main roots without a dominant root.

filament In flowering plants, the thin stalk of a stamen.

first law of thermodynamics Energy cannot be created or destroyed, although it can be transformed from one form to another. Compare with second law of thermodynamics.

fission (fish'un) Process of asexual reproduction in which an organism divides into two approximately equal parts.

fixed action pattern (FAP) An innate behavior triggered by a sign stimulus.

flagellum (flah-jel'um) Long, whiplike, movable cellular organelle that is used in locomotion. Eukaryote flagella are composed of two single microtubules surrounded by nine double microtubules.

flowering plant See angiosperm.

fluid mosaic model The modern picture of the plasma membrane (and other cell membranes) in which protein molecules float in a phospholipid bilayer.

follicle (fol'i-kl) (1) A simple, dry, dehiscent fruit that splits open along one seam to liberate the seeds; (2) a small sac of cells in the mammalian ovary that contains a maturing egg; (3) the pocket in the skin from which a hair grows.

food chain The successive series of organisms through which energy flows in an ecosystem. Each organism in the series eats or decomposes the preceding organism in the chain. See food web.

food web A complex interconnection of all of the food chains in an ecosystem.

foramen magnum The opening in the vertebrate skull through which the spinal cord passes.

forest decline A gradual deterioration (and often death) of many trees in a forest. The cause of forest decline is unclear at the present time, and it may involve a combination of factors.

fossil Parts of an ancient organism or traces left by previous life.

fossil fuel Combustible deposits in the Earth's crust. Fossil fuels are composed of the remnants of prehistoric organisms that existed millions of years ago. Examples include oil, natural gas, and coal.

founder effect Genetic drift that results from a small population colonizing a new area.

frond A leaf of a fern.

fruit In angiosperms, a mature, ripened ovary. Fruits contain seeds and usually serve for seed protection and dispersal.

fucoxanthin (few″koh-zan′thin) The brown pigment found in diatoms, brown algae, and dinoflagellates.

fundamental niche The potential ecological niche that an organism could have if there were no competition from other species. See niche. Compare with realized niche.

fungus; pl. **fungi** A heterotrophic eukaryote with cell walls and a body in the form of a mycelium of threadlike hyphae, or unicellular. Most fungi are decomposers; a few are parasitic.

gametangium (gam″uh-tan′gee-um) Reproductive structure of plants, protists, and fungi in which gametes are formed.

gamete (gam′eet) A cell that functions in sexual reproduction; an egg or sperm whose union, in sexual reproduction, initiates the development of a new individual.

gametophyte generation (gam-ee′toh-fite) The haploid or gamete-producing stage in the life cycle of a plant. Compare with sporophyte generation.

ganglion (gang′glee-on) A mass of cell bodies of neurons located outside the central nervous system.

gap junction Structure consisting of specialized regions of the plasma membranes of two adjacent cells containing numerous pores that allow passage of certain molecules and ions between them.

gastrula (gas′troo-lah) Early stage of embryonic development during which the three germ layers form.

gemma (jem′mah) Vegetative bud in bryophytes that develops asexually into a new plant.

gene A discrete unit of hereditary information that usually specifies a polypeptide. It consists of DNA and is part of the chromosomes.

gene amplification Process by which multiple copies of a gene are produced by selective replication, thus allowing for increased synthesis of the gene product. See polymerase chain reaction.

gene flow The movement of alleles between local populations, or demes, due to the migration of individuals. Gene flow can have significant evolutionary consequences.

gene pool All the genes present in a population.

gene replacement therapy Techniques that involve introducing normal copies of a gene into some of the cells of the body of a person afflicted with a genetic disorder.

genetic code Code consisting of groups of three bases in mRNA which specifies individual amino acids or translation start and stop signals.

genetic drift A random change in allele frequency in a small breeding population.

genetic engineering The ability to take a specific gene from one cell and place it into another cell where it is expressed; also called biotechnology.

genetic recombination See recombination, genetic.

genome (jee′nome) A complete set of hereditary factors contained in one haploid assortment of chromosomes.

genotype (jeen′oh-type) The complete genetic makeup of an organism.

genus (jee′nus) A rank in taxonomic classification above the species.

germ cells Cells within the body that give rise to gametes.

germination The resumption of growth in seeds or spores.

germ layer Primitive embryonic tissue layer; endoderm, mesoderm, or ectoderm.

gibberellin (jib″ur-el′lin) A plant hormone involved in various aspects of plant growth and development.

gill The respiratory organ of aquatic animals, usually a thin-walled projection from the body surface or from some part of the digestive tract.

ginkgoes An ancient gymnosperm group that consists of a single living representative (*Ginkgo*), which is a large deciduous tree with broad, fan-shaped leaves and naked fleshy seeds (on female trees).

gland Body structure specialized for secretion.

globulin One of a class of proteins in blood plasma, some of which (gamma globulins) function as antibodies.

glomerulus (glom-air′yoo-lus) In the nephron, a cluster of capillaries surrounded by Bowman's capsule; site of filtration in the vertebrate kidney.

glucagon (gloo′kah-gahn) A pancreatic hormone that stimulates the breakdown of glycogen, increasing the concentration of glucose in the blood. Compare with insulin.

glycogen (gly′koh-jen) A polysaccharide formed from glucose and stored primarily in liver and (to a lesser extent) muscle tissue; the major carbohydrate stored in animal cells.

glycolysis (gly-kol′ih-sis) The metabolic conversion of glucose into pyruvic acid with the production of ATP. Glycolysis is a metabolic pathway present in all living cells.

gnetophyte One of a small group of unusual gymnosperms that possess many features similar to flowering plants.

Golgi complex (goal′jee); also called **Golgi body** or **Golgi apparatus** Organelle composed of stacks of flattened membranous sacs; mainly responsible for modifying, packaging, and sorting proteins.

gonad (goh′nad) A gamete-producing gland; an ovary or testis.

gradualism The idea that evolutionary change of a species is due to a slow, steady accumulation of changes over time. Compare with punctuated equilibrium.

grain A simple, dry, indehiscent fruit in which the fruit wall is fused to the seed coat, making it impossible to separate the fruit from the seed. Corn kernels are grains.

granum (gran′um); pl. **grana** A stack of thylakoids within a chloroplast.

gravitropism (grav″ih-troh′pizm) Growth of a plant in response to gravity.

greenhouse effect The global warming of our atmosphere produced by the buildup of carbon dioxide and other greenhouse gases, which trap the sun's radiation in much the same way that glass does in a greenhouse. Greenhouse gases allow the sun's energy to penetrate to the Earth's surface but do not allow as much of it to escape as heat.

gross primary productivity The rate at which energy accumulates in an ecosystem (as biomass) during photosynthesis. Compare with net primary productivity.

growth rate The rate of change of a population; calculated by subtracting the death rate from the birth rate (in populations with little or no migration).

guanine (gwan′een) See nucleic acid.

guard cell A cell in the epidermis of plant stems and leaves. Two guard calls form a pore for gas exchange, collectively called a stoma.

guttation (gut-tay′shun) The appearance of water droplets on leaves, forced out through leaf pores by root pressure.

gymnosperms (jim′noh-sperms) Seed plants in which the seeds are not enclosed in an ovary; gymnosperms frequently bear their seeds in cones. Includes conifers, cycads, ginkgoes, and gnetophytes.

habitat The natural environment or place where an organism, population, or species lives.

habituation (hab-it″yoo-ay′shun) The process by which organisms become accustomed to a stimulus and cease to respond to it.

hair cell A mechanoreceptor found in a variety of sense organs, including the cochlea of the ear; detects motion or differences in pressure.

haploid (hap′loyd) or ***n*** The chromosome number characteristic of gametes or spores; half the diploid number. In plants, the chromosome number of body cells of the gametophyte generation. Compare with diploid.

Hardy-Weinberg law The principle that, regardless of dominance or recessiveness, allele frequencies do not change from generation to generation in a large population in the absence of evolution (natural selection, migration, genetic drift).

helper T cells T lymphocytes that facilitate immune responses, e.g., activate B lymphocytes so that they form an antibody-producing clone in response to an antigen.

hemizygous Possessing only one allele for a particular locus.

hemoglobin (hee′-moh-gloh″bin) The red, iron-containing protein pigment of erythrocytes that transports oxygen and carbon dioxide and aids in regulation of pH.

hemophilia (hee″moh-feel′ee-ah) "Bleeder's disease"; hereditary disease in which blood does not clot properly.

herbaceous (er-bay′shus) A nonwoody plant.

herbivore (erb′i-vore) An animal that feeds on plants or algae. Compare with carnivore and omnivore.

hermaphrodite (her-maf′roh-dite) An organism that produces both male and female gametes.

heterocysts (het′ur-oh-sists″) Oxygen-excluding cells of cyanobacteria that fix nitrogen.

heterospory (het″ur-os′pur-ee) Production of two types of spores in plants—microspores and megaspores. Compare with homospory.

heterothallic (het-ur-oh-thal′ik) Pertaining to an organism having two mating types; only by combining a plus and a minus strain can sexual reproduction occur.

heterotrophs (het′ur-oh-trofes) Organisms that cannot synthesize their own food from inorganic materials and therefore must live either at the expense of autotrophs, other heterotrophs, or upon decaying matter. Also called consumers.

heterozygous advantage A phenomenon in which the heterozygous condition confers some special advantage on an individual that either homozygous condition does not, i.e., *Aa* has a higher degree of fitness than does *AA* or *aa*.

heterozygous Possessing two different alleles of the same gene. Compare with homozygous.

hibernation The dormant state of decreased metabolism in which certain animals pass the winter.

highly developed country See developed country.

histones (his′tones) Small, positively charged (basic) proteins in the nucleus that bind to the negatively charged DNA.

holdfast The structure for attachment to solid surfaces found in multicellular algae.

homeobox (home′ee-oh-box) Specific DNA sequence found in many genes that are involved in controlling the development of the body plan.

homeostasis (home″ee-oh-stay′sis) The balanced internal environment of the body; the automatic tendency of an organism to maintain such a steady state.

homeotherm (home′ee-oh-therm) See endotherm.

homeotic gene (home″ee-ot′ik) A gene that controls the formation of specific structures during development. Such genes were originally identified through insect mutants in which one body part was substituted for another.

hominid (hah′min-id) Any of a group of extinct and living humans.

hominoids The apes and hominids.

homologous (hom-ol′ah-gus) Similar in basic structural plan and development, which is assumed to reflect a common genetic ancestry. Compare with analogous.

homologous chromosomes Chromosomes that are similar in morphology and genetic constitution. In humans, there are 23 pairs of homologous chromosomes, each containing one member from the mother and one member from the father.

homospory (hoh-mos′pur-ee) Production of one type of spore in plants. The spore gives rise to a bisexual gametophyte. Compare with heterospory.

homozygous (hoh″moh-zy′gus) Possessing a pair of identical alleles. Compare with heterozygous.

hormone One of many chemical messengers in multicellular organisms that usually travel in body fluids to target cells, where it combines with receptors and affects some aspect of metabolism, growth, or reproduction.

hornwort A type of bryophyte (a small, inconspicuous, nonvascular plant with a dominant gametophyte generation).

horsetails Vascular plants with hollow, jointed stems and reduced, scalelike leaves. Although modern representatives are small, some extinct species were treelike.

hybridization Interbreeding between members of different taxa.

hydrocarbons Organic compounds composed solely of hydrogen and carbon.

hydrogen bond A weak chemical bond that forms between a slightly positive hydrogen atom in one molecule and a slightly negative atom (usually oxygen) in another molecule. Hydrogen bonding is of primary importance in the structure of nucleic acids and proteins.

hydrologic cycle The water cycle, which includes evaporation, precipitation, and flow to the seas. The hydrologic cycle supplies terrestrial organisms with a continual supply of fresh water.

hydrolysis (hy-drol'ih-sis) The splitting of a molecule into smaller molecules by the addition of water between certain of its bonds, the hydroxyl group being incorporated into one fragment, and the hydrogen atom into the other. Compare with condensation reaction.

hydrophilic Attracted to water. Compare with hydrophobic.

hydrophobic Repelled by water. Compare with hydrophilic.

hydroponics Growing plants in an aerated solution of dissolved inorganic minerals, that is, without soil.

hydroxide ion A negatively charged particle consisting of oxygen and hydrogen, usually written OH^-.

hydroxyl (hy-drok'sil) Polar functional group, usually written —OH.

hypertension High blood pressure.

hypertonic (hyperosmotic) Referring to a solution with an osmotic pressure (or solute concentration) greater than that of the solution with which it is compared. Compare with isotonic and hypotonic.

hypha (hy'fah) One of the filaments composing the mycelium of a fungus.

hypocotyl (hy'poh-kah"tl) The part of the axis of a plant embryo or seedling below the point of attachment of the cotyledons. Compare with epicotyl.

hypothalamus (hy-poh-thal'uh-mus) Part of the brain that functions in regulating the pituitary gland, the autonomic system, emotional responses, body temperature, water balance, and appetite; located below the thalamus.

hypothesis An educated guess that might be true and is testable by observation and experimentation. Compare with theory.

hypotonic (hypo-osmotic) Referring to a solution with an osmotic pressure (or solute concentration) less than that of the solution with which it is compared. Compare with hypertonic and isotonic.

immigration A type of migration in which individuals enter a population and thus increase the population size. Compare with emigration.

immune response The production of antibodies or T cells in response to foreign antigens.

immunoglobulins (im-yoon"oh-glob'yoo-lins) See antibodies.

immunological tolerance (im-yoon"uh-loj'ih-kl) The ability of an organism to accept cells transplanted from a genetically distinct organism.

implantation The embedding of a developing embryo in the wall (endometrium) of the uterus.

imprinting A form of learning by which a young bird or mammal forms a strong social attachment to an individual (usually a parent) or object within a few hours after hatching or birth.

inbreeding Mating of genetically similar individuals. Homozygosity increases with each successive generation of inbreeding. Compare with outbreeding.

incomplete dominance Condition in which neither member of a pair of contrasting alleles is completely expressed when the other is present.

incomplete flower A flower lacking one or more of the four main parts: sepals, petals, stamens, and/or carpels.

indehiscent fruit (in"dee-his'ent) A simple, dry fruit that does not split open at maturity. Compare with dehiscent fruit.

independent assortment The mutually independent inheritance of genes that are located on separate chromosome pairs.

index fossil Certain key invertebrate fossils that are found in the same sedimentary layers in different geographical areas. Index fossils help geologists identify comparable layers in widely separate locations.

induced fit Hypothesis that the active site of certain enzymes becomes conformed to the shape of the substrate molecule.

inductive reasoning Reasoning that uses specific examples to draw a general conclusion or discover a general principle. Compare with deductive reasoning.

infant mortality rate The number of infant deaths per 1000 live births. (An infant is a child in its first year of life.)

infectious disease A disease caused by a microorganism (such as a bacterium or fungus) or infectious agent (such as a virus). Infectious diseases can be transmitted from one individual to another.

inflammation The response of body tissues to injury or infection, characterized clinically by heat, swelling, redness, and pain, and physiologically by increased vasodilation and capillary permeability.

innate behaviors Behaviors that are inherited and typical of the species.

inorganic fertilizer Plant nutrients (especially nitrates, phosphates, and potassium) that are manufactured commercially.

inorganic plant nutrient A nutrient such as phosphate or nitrate that stimulates plant or algal growth. Excessive amounts of inorganic plant nutrients, which may come from animal wastes and plant residues as well as fertilizer runoff, can cause both soil and water pollution.

insight learning Refers to innate (genetically programmed) behavior.

in situ conservation Conservation efforts that concentrate on preserving biological diversity in the wild. Compare with ex situ conservation.

instinct A genetically determined pattern of behavior or responses that is not based on the individual's previous experience.

insulin A hormone secreted by the islets of the pancreas that lowers blood glucose content. Compare with glucagon.

integumentary system (in-teg"yoo-men'tar-ee) The body's covering, including the skin and its nails, glands, hair, and other associated structures.

interferons (in"tur-feer'ons) Proteins produced by animal cells infected by a virus. Stimulates other cells to produce antiviral proteins.

intermediate filaments Cytoplasmic fibers that are part of the cytoskeletal network and intermediate in size between microtubules and microfilaments.

interneuron (in"tur-noor'on) An association neuron; carries impulses to other nerve cells and is not directly associated with either an effector or a sense receptor.

internode The section of a stem between two nodes.

interphase The period in the life cycle of a cell in which there is no visible mitotic division; period between mitotic divisions.

interspecific competition Competition between different species for a resource such as food. Compare with intraspecific competition.

intertidal zone The marine shoreline area between the low tide mark and the high tide mark.

intraspecific competition Competition within a species for a resource such as food. Compare with interspecific competition.

intron A nonprotein-coding region of a eukaryotic gene and also of the RNA transcribed from such a region. Introns do not appear in mature mRNA. Compare with exon.

invagination (in-vaj"ih-nay'shun) The infolding of one part within another, specifically a process of gastrulation in which one region folds in to form a double-layered cup.

inversion, chromosomal Turning a segment of a chromosome end for end and attaching it to the same chromosome.

invertebrates Collective term for animals that are not vertebrates and therefore do not possess backbones (sponges, jellyfish, worms, octopods, butterflies, sea stars, and so on).

in vitro Occurring outside a living organism (literally, "in a glass").

in vivo Occurring in a living organism.

involuntary muscle See smooth muscle.

ion An atom or a group of atoms bearing an electric charge, either positive (cation) or negative (anion).

ionic bond An electrostatic attraction between oppositely charged ions.

islets of Langerhans (eye'lets of lahng'er-hanz) The endocrine portion of the pancreas that secretes glucagon and insulin. These hormones regulate blood sugar level.

isomer (eye'soh-mur) One of two or more chemical compounds having the same chemical formula but a different structural formula, such as glucose and fructose.

isotonic (eye"soh-ton'ik) **(iso-osmotic)** Referring to a solution with identical concentrations of solute and solvent molecules, and hence the same osmotic pressure, as the solution with which it is compared. Compare with hypertonic and hypotonic.

isotope (eye'suh-tope) An alternate form of an element with a different number of neutrons but the same number of protons and electrons.

karyotype (kare'ee-oh-type) The chromosomal composition of an individual. Representations of the karyotype are generally prepared by photographing the chromosomes and arranging the homologous pairs according to size and centromere position.

keratin (kare'ah-tin) A horny, water-insoluble protein found in the epidermis of vertebrates and in nails, feathers, hair, and horns.

killer T cells See cytotoxic T cells.

kilocalorie See calorie.

kinetic energy The energy of a body that results from its motion. Compare with potential energy.

kinetochore (kin-eh'toh-kore) Portion of the chromosome centromere to which mitotic spindle fibers attach.

Krebs cycle See citric acid cycle.

lamella (lah-mel'ah) A thin leaf or plate, as of bone.

larva An immature, free-living form in the life history of some animals in which it may be unlike the parent.

larynx (lare'inks) The organ at the upper end of the trachea that contains the vocal cords.

lateral meristem An area of cell division located on the side of the plant. There are two lateral meristems, the vascular cambium and the cork cambium.

latitude The distance, measured in degrees north or south, from the equator. Compare with altitude.

learning A change in the behavior of an animal that results from experiences during its lifetime.

legume (leg'yoom) A simple, dry, dehiscent fruit that splits open along two seams to release its seeds.

less developed country A developing country with a low level of industrialization, a very high fertility rate, a very high infant mortality rate, and a very low per capita income (relative to highly developed countries). Compare with moderately developed and highly developed country.

leukocyte (loo'koh-site) White blood cell; defends the body against disease-causing organisms.

leukoplasts (loo'koh-plasts) Colorless plastids that act as centers for starch formation in certain kinds of plant cells.

lichens (ly'kenz) Compound organisms composed of symbiotic algae (or cyanobacteria) and fungi.

ligament (lig'uh-ment) A connective tissue cable or strap that connects bones to each other or holds other organs in place.

light-dependent reactions That portion of photosynthesis that requires the presence of light (light energy is converted to chemical energy of ATP and NADPH).

light-independent reactions That portion of photosynthesis that does not directly require light. See Calvin cycle.

lignin (lig′nin) The substance responsible for the hard, woody nature of plant stems and roots.

limiting factor Whatever environmental variable tends to restrict the growth, distribution, or abundance of a particular population.

limnetic zone The open water area away from the shore of a lake or pond that extends down as far as sunlight penetrates.

linkage The tendency for a group of genes located on the same chromosome to be inherited together.

lipase (lip′ase) Fat-digesting enzyme.

lipid Any of a group of organic compounds that are insoluble in water but soluble in fat solvents; lipids serve as a storage form of fuel and an important component of cell membranes.

littoral zone (lit′or-ul) The region of shallow water along the shore of a lake or pond.

liverwort A type of bryophyte (a small, inconspicuous, nonvascular plant with a dominant gametophyte generation).

locus The particular point on the chromosome at which the gene for a given trait occurs.

lumen (loo′men) The cavity or channel within a tube or tubular organ, such as a blood vessel or the digestive tract; the space enclosed by a membrane, such as the lumen of the endoplasmic reticulum.

lymph (limf) The colorless fluid within the lymphatic vessels that is derived from blood plasma and resembles it closely in composition; contains white cells; ultimately, returned to the blood.

lymph nodes A mass of lymph tissue surrounded by a connective tissue capsule; manufactures lymphocytes and filters lymph.

lymphocyte (limf′oh-site) White blood cell with nongranular cytoplasm that is responsible for immune responses.

lysis (ly′sis) The process of disintegration of a cell or some other structure.

lysosomes (ly′soh-somes) Intracellular organelles present in many animal cells; contain a variety of hydrolytic enzymes that act when the lysosome ruptures or fuses with another vesicle. Lysosomes function in development and in phagocytosis.

macroevolution (mak″roh-eh-voh-loo′shun) Large-scale evolutionary change; evolutionary change involving higher taxa, such as genera and orders, i.e., above the level of species.

macromolecule A very large molecule such as a protein or nucleic acid.

macrophage (mak′roh-faje) A large phagocytic cell capable of ingesting and digesting bacteria and cellular debris.

mandible (man′dih-bl) (1) The lower jaw of vertebrates; (2) an external mouthpart of certain arthropods, e.g., insects.

mantle Thin outside layer of mollusk body that is usually responsible for the production of the shell.

marsupials (mar-soo′pee-uls) A subclass of mammals, the Metatheria, characterized by the possession of an abdominal pouch in which the young are carried for some time after being born in a very undeveloped condition.

mass extinction The extinction of numerous species and higher taxa during a relatively short period of geological time. Compare with background extinction.

matrix (may′triks) (1) Nonliving material secreted by and surrounding connective tissue cells; (2) the interior of the compartment formed by the inner mitochondrial membrane.

medulla (meh-dul′uh) (1) The inner part of an organ, such as the medulla of the kidney; (2) the most posterior part of the brain, lying next to the spinal cord.

medusa A jellyfish; a stage in the life cycle of certain cnidarians in which the organism is free-swimming and umbrella-shaped. Compare with polyp.

megaphyll (meg′uh-fil) A leaf that contains multiple vascular strands. Megaphylls are found in ferns, gymnosperms, and angiosperms. Compare with microphyll.

megasporangium (meg″ah-spor-an′jee-um) A spore sac containing megaspores.

megaspore (meg′ah-spor) The haploid spore in heterosporous plants that gives rise to a female gametophyte.

meiosis (my-oh′sis) Division of the cell nucleus that produces haploid cells; produces gametes in animals and spores in plants.

melanin (mel′ah-nin) A dark brown to black pigment common in the integument of many animals and sometimes found in other organisms.

menopause The period (usually from 45 to 55 years of age) in women when the recurring menstrual cycle ceases.

menstruation (men-stroo-ay′shun) The monthly discharge of blood and degenerated uterine lining in the human female; marks the beginning of each menstrual cycle.

meristem (mer′ih-stem) A localized area of mitosis and growth in the plant body.

mesentery (mes′en-tare″ee) Internal membrane of vertebrates that holds the organs in place.

mesoderm (mez′oh-derm) The middle layer of the three basic tissue layers that develop in the early embryo; gives rise to connective tissue, muscle, bone, blood vessels, kidneys, and many other structures. Compare with ectoderm and endoderm.

mesophyll (mez′oh-fil) Photosynthetic cells in the interior of a leaf.

messenger RNA (mRNA) RNA that has been transcribed from DNA and that specifies the amino acid sequence of a protein in eukaryotes and prokaryotes. Compare with transfer RNA and ribosomal RNA.

metabolism The sum of all of the chemical processes that occur within a cell or organism; the transformations by which energy and matter are made available for use by the organism.

metamorphosis (met″ah-mor′fuh-sis) The process of development from one stage to another, such as from a larva to an adult.

metaphase (met'ah-faze) The stage of mitosis and meiosis during which the chromosomes line up along the equatorial plane.

metastasis (me-tas'tuh-sis) The spread of cancer cells from one organ or part of the body to another not directly connected to it.

metazoan A multicellular animal; any animal except for sponges.

microbodies Membrane-bound eukaryotic cellular structures containing enzymes.

microclimate Local variations in climate produced by differences in elevation, in the steepness and direction of slopes, and in exposure to prevailing winds.

microevolution Changes in allele frequencies that occur within a population over successive generations.

microfilaments Tiny rodlike structures with contractile properties that make up part of the internal skeletal framework of the cell. Microfilaments are composed of actin.

microphyll A leaf that contains one vascular strand; microphylls are found in horsetails and club mosses. Compare with megaphyll.

microsporangium (my"kroh-spor-an'jee-um) A spore sac containing microspores.

microspore The haploid spore in heterosporous plants that gives rise to a male gametophyte.

microtubules (my-kroh-too'bewls) Hollow, cytoplasmic cylinders, composed mainly of tubulin protein, that compose such organelles as flagella and centrioles and serve as a skeletal component of the cell.

microvilli (my-kroh-vil'ee) Minute projections of the plasma membrane that increase the surface area of the cell; found mainly in cells concerned with absorption or secretion, such as those lining the intestine or kidney tubules.

migration Movement of an organism (individual or population) from one place to another. See immigration and emigration.

mimicry (mim'ik-ree) An adaptation for survival in which an organism resembles some other living or nonliving object.

mitochondria (my"toh-kon'dree-ah) Spherical or elongated intracellular organelles that contain the electron transport system and enzymes for the citric acid cycle. Sometimes referred to as the power plants of the cell.

mitosis (my-toh'sis) Division of the cell nucleus resulting in the distribution of a complete set of chromosomes to each daughter cell; cytokinesis (division of the cytoplasm) usually occurs during the telophase stage of mitosis. Mitosis consists of four phases: prophase, metaphase, anaphase, and telophase.

mitotic spindle (my-tot'ik) See spindle.

moderately developed country A developing country with a medium level of industrialization, a high fertility rate, a high infant mortality rate, and a low per capita income (relative to highly developed countries). Compare with less developed country and highly developed country.

mold Multicellular fungi including the mildews, rusts, and mushrooms.

mole The amount of a chemical compound whose mass in grams is equivalent to its molecular weight, the sum of the atomic weights of its atoms.

molecule The smallest particle of a covalently bonded element or compound that has the composition and properties of a larger part of the substance.

molting The shedding and replacement of an outer covering such as hair, feathers, or exoskeleton.

monoacylglycerol (mon'o-as"il-glis'er-ol) A fat consisting of a glycerol chemically combined with a single fatty acid; also called monoglyceride.

monocotyledon (mon'o-kot-ih-lee'dun) A flowering plant with embryos having a single seed leaf, or cotyledon; also known as a monocot. Compare with dicotyledon.

monoecious (mon-ee'shus) Having separate male and female reproductive parts on the same plant. Compare with dioecious.

monohybrid cross A genetic cross that takes into account the behavior of alleles of a single locus.

monomer (mon'oh-mer) A simple molecule of a compound of relatively low molecular weight that can be linked with others to form a polymer.

monophyletic origin Describes a group that shares a common ancestor. Compare with polyphyletic origin.

monosaccharide (mon-oh-sak'ah-ride) A simple sugar; one that cannot be degraded by hydrolysis to a simpler sugar.

monosomy (mon'uh-soh"mee) Abnormal condition in which one member of a specific chromosome pair is absent.

monotremes (mon'oh-treems) Egg-laying mammals such as the duck-billed platypus of Australia.

morphogenesis (mor-foh-jen'eh-sis) The development of the form and structures of the body and its parts by precise movements of its cells.

moss A type of bryophyte (a small, inconspicuous, nonvascular plant with a dominant gametophyte generation).

multiple alleles (al-leels') Three or more alleles of a single locus (in a population), such as the alleles governing the ABO series of blood types.

multiple fruit A fruit that develops from many ovaries of many separate flowers to form one fruit. Pineapples are multiple fruits.

muscle An organ that produces movement by contraction.

mutagen (mew'tah-jen) Any agent that is capable of producing mutations.

mutant A gene that has undergone mutation (or an organism that bears such a gene). Compare with wild-type.

mutation A change in the DNA (a gene) of an organism. A mutation in reproductive cells may be passed on to the next generation, where it may result in birth defects or genetic disease.

mutualism A symbiotic relationship in which both partners benefit from the association. See symbiosis. Compare with parasitism and commensalism.

mycelium (my-seel'ee-um) The vegetative body of fungi and certain protists (water molds); consists of a branched network of hyphae.

mycorrhizae (my"kor-rye'zee) Mutualistic associations of fungi and plant roots that aid in the plant's absorption of essential minerals from the soil.

myelin (my'eh-lin) The white fatty material that forms a sheath around the axons of certain nerve cells, which are then called myelinated fibers.

myosin (my'oh-sin) A protein that, together with actin, is responsible for muscle contraction.

natural selection The tendency of organisms that possess favorable adaptations to their environment to survive and become the parents of the next generation; evolution occurs when natural selection results in changes in allele frequencies in a population; the mechanism of evolution first proposed by Charles Darwin.

negative feedback system A homeostatic mechanism in which a change in some condition triggers a response that counteracts, or reverses, the changed condition; examples include (1) the regulation of many biochemical processes and (2) how mammals maintain body temperature. Compare with positive feedback system.

nekton (nek'ton) Free-swimming aquatic organisms such as fish and turtles. Compare with plankton.

nematocyst (nem-at'oh-cyst) A minute stinging structure found within cnidocytes (stinging cells) in cnidarians; used for capturing prey.

neo-Darwinism See synthetic theory of evolution.

nephridium (neh-frid'ee-um) The excretory organ of the earthworm and other annelids which consists of a ciliated funnel, opening into the next anterior coelomic cavity and connected by a tube to the outside of the body.

nephron (nef'ron) The functional, microscopic unit of the vertebrate kidney.

neritic province (ner-ih'tik) Ocean water that extends from the shoreline to where the bottom reaches a depth of 200 meters. Compare with oceanic province.

nerve A large bundle of axons (or dendrites) wrapped in connective tissue that conveys impulses between the central nervous system and some other part of the body.

net primary productivity Energy that remains in an ecosystem (as biomass) after cell respiration has occurred. Compare with gross primary productivity.

neuroglia (noor-og'lee-ah) Supporting cells in the central nervous system.

neuron (noor'on) A nerve cell; a conducting cell of the nervous system that typically consists of a cell body, dendrites, and an axon.

neurotransmitter A chemical messenger used by neurons to transmit impulses across a synapse.

neutrons (noo'tronz) Electrically uncharged particles of matter existing along with protons in the atomic nucleus.

niche The totality of an organism's adaptations, its use of resources, and the life-style to which it is fitted. The niche describes how an organism utilizes materials in its environment as well as how it interacts with other organisms; also called ecological niche.

nitrification The conversion of ammonia (NH_3) to nitrate (NO_3^-) by certain bacteria (nitrifying bacteria) in the soil; part of the nitrogen cycle.

nitrogen cycle The worldwide circulation of nitrogen from the abiotic environment into living things and back into the abiotic environment.

nitrogen fixation The conversion of atmospheric nitrogen (N_2) to ammonia (NH_3) by certain bacteria; part of the nitrogen cycle.

node The area on a plant stem where the leaves attach.

noncompetitive inhibitor A substance that permanently destroys the ability of an enzyme to function.

nondisjunction Abnormal separation of homologous chromosomes or sister chromatids caused by their failure to disjoin (move apart) properly during cell division.

nonpolar covalent bond Chemical bond in which electrons are shared equally among the participating atoms and which does not produce any electrical charge within the molecule. Compare with polar covalent bond.

notochord (no'toe-kord) The flexible rod in the anteroposterior axis that serves as an internal skeleton in the embryos of all chordates.

nuclear envelope The double membrane system that surrounds the cell nucleus of eukaryotes.

nuclear pores Structures in the nuclear envelope that allow passage of certain molecules between the cytoplasm and the nucleus.

nucleic acid (noo-klay'ik) DNA or RNA. A polymer composed of nucleotides that contain the purine bases adenine and guanine and the pyrimidine bases cytosine and thymine or uracil.

nucleolus (new-klee'-oh-lus) Specialized structure in the nucleus formed from regions of several chromosomes; site of ribosome synthesis.

nucleosome (new'klee-oh-sohm) Repeating unit of chromatin structure consisting of a length of DNA wound around a complex of eight histone molecules (two of each of four different types) plus a DNA linker region associated with a fifth histone protein.

nucleotide (noo'klee-oh-tide) A molecule composed of a phosphate group, a five-carbon sugar (ribose or deoxyribose) and a nitrogenous base (purine or pyrimidine); one of the subunits of nucleic acids.

nucleus (new'klee-us) (1) That portion of an atom that contains the protons and neutrons, the core; (2) a cellular organelle containing DNA and serving as the control center of the cell; (3) a mass of nerve cell bodies in the central nervous system.

nymph A juvenile insect that often resembles the adult stage and that will become an adult without an intervening pupal stage.

obligate parasite An organism that can only exist as a parasite. Compare with facultative parasite.

oceanic province That part of the open ocean that overlies an ocean bottom deeper than 200 meters. Compare with neritic province.

omnivore (om'nih-vore) An animal that eats a variety of plant and animal material. Compare with herbivore and carnivore.

oncogene (on'koh-jeen) Any of a number of genes that usually play an essential role in cell growth or division and that cause the formation of a cancer cell when mutated; also known as cellular oncogenes.
oogenesis (oh"oh-jen'eh-sis) Production of female gametes (eggs). Compare with spermatogenesis.
operant (instrumental) conditioning A type of learning in which an animal is rewarded or punished for performing a behavior it discovers by chance.
operator The site on DNA to which repressor molecules are bound, thereby inhibiting the synthesis of mRNA by the genes in the adjacent operon; adjacent to the structural genes in the operon.
operon (op'er-on) In prokaryotes, a group of structural genes that are transcribed as a single message plus their associated regulatory elements. An operon is controlled by a single repressor.
optimal foraging The theory that animals feed in a manner that maximizes benefits and/or minimizes costs.
orbital Any one of the permissible patterns of motion of an electron in an atom or molecule.
organ A specialized structure made up of tissues and adapted to perform a specific function or group of functions, such as the heart or liver.
organelle One of the specialized structures within the cell, such as the mitochondria, Golgi complex, ribosomes, or contractile vacuole.
osmoregulation (oz"moh-reg-yoo-lay'shun) The active regulation of the osmotic pressure of body fluids so that they do not become excessively dilute or excessively concentrated.
osmosis (oz-moh'sis) Diffusion of water (the principal solvent in biological systems) through a selectively permeable membrane from a region of higher concentration of water to a region of lower concentration of water.
osmotic pressure A measure of the solute concentration of a solution.
osteocyte (os'tee-oh-site) A mature bone cell; an osteoblast that has become embedded within the bone matrix and occupies a lacuna.
osteon (os'tee-on) Spindle-shaped unit of bone composed of concentric layers of osteocytes; Haversian system of bone.
outbreeding The mating of individuals that are unrelated or from different populations. Compare with inbreeding.
ovary (oh'var-ee) (1) In animals, one of the paired female gonads; responsible for producing eggs and sex hormones; (2) in flowering plants, the base of the carpel that contains ovules; ovaries develop into fruits after fertilization.
overpopulation A situation in which a country or geographical area has more people than its resource base can support without damaging the environment.
oviduct (oh'vih-dukt) Tube that carries ova from the ovary to the uterus, cloaca, or body exterior. In humans, also called fallopian tube or uterine tube.
oviparous (oh-vip'ur-us) A type of development in which the young hatch from eggs laid outside the mother's body.
ovoviviparous (oh"voh-vih-vip'ur-us) A type of development in which the young hatch from eggs incubated within the mother's body.
ovule (ov'yool) The structure in the ovary that develops into a seed after fertilization.
ovum The female gamete, or egg.
oxidation The loss of electrons, or in organic chemistry, the loss of hydrogen from a compound. Compare with reduction.
oxidation-reduction reaction A chemical reaction in which one or more electrons are transferred from one atom or molecule to another.
oxidative phosphorylation (fos"for-ih-lay'shun) The production of ATP using energy derived from the transfer of electrons in the electron transport system of the mitochondria.
ozone A blue gas, O_3, that has a distinctive odor. Ozone is a humanmade pollutant in one part of the atmosphere (the troposphere) but a natural and essential component in another (the stratosphere).

pancreas (pan'kree-us) Large digestive gland located in the vertebrate abdominal cavity. The pancreas produces pancreatic juice containing digestive enzymes; also serves as an endocrine gland, secreting the hormones insulin and glucagon.
parasitism A symbiotic relationship in which one member (the parasite) benefits and the other (the host) is adversely affected. See symbiosis. Compare with mutualism and commensalism.
parasympathetic nervous system A division of the autonomic nervous system concerned primarily with the control of the internal organs; functions to conserve or restore energy. Compare with sympathetic nervous system.
parenchyma (par-en'kih-mah) Plant cells that are relatively unspecialized, are thin-walled, may contain chlorophyll, and are typically rather loosely packed; they function in photosynthesis and in the storage of nutrients.
parthenogenesis (par"theh-noh-jen'eh-sis) The development of an unfertilized egg into an adult organism; common among honey bees, wasps, and certain other arthropods.
passive immunity Temporary immunity derived from the immunoglobulins of another organism.
pathogen (path'oh-gen) An organism, usually a microorganism, capable of producing disease.
pellicle (pel'ih-kl) A flexible, proteinaceous covering over the body of certain protists.
penis The male organ of copulation in reptiles and mammals.
peptide (pep'tide) A compound consisting of two or more amino acids.
peptide bond A distinctive covalent carbon-to-nitrogen bond that links amino acids in peptides and proteins.
perennial (pur-en'ee-ul) A plant that grows year after year. Compare with annual and biennial.
pericycle (pehr'eh-sy'kl) A layer of meristematic cells in roots that gives rise to branch roots.
periderm (pehr'ih-durm) Layers of cells covering the surface of woody stems and roots (i.e., the outer bark). Anatomically, the periderm is composed of cork cells, cork cambium, and cork parenchyma, along with traces of primary tissues.

peripheral nervous system The nerves and receptors that lie outside the central nervous system.

peristalsis (pehr"ih-stal'sis) Powerful, rhythmic waves of muscular contraction and relaxation in the walls of hollow tubular organs, such as the ureter or parts of the digestive tract, that serve to move the contents through the tube.

peritoneum (pehr-ih-tuh-nee'um) The membrane lining the abdominal and pelvic cavities (the parietal peritoneum) and the membrane forming the outer layer of the stomach and intestine (the visceral peritoneum).

permafrost Permanently frozen subsoil characteristic of frigid areas such as the tundra.

persistence A characteristic of certain chemicals that are extremely stable and may take many years to be broken down into simpler forms by natural processes. Certain pesticides, for example, exhibit persistence and remain unaltered in the environment for years.

pest Any organism that interferes in some way with human welfare or activities.

pesticide Any toxic chemical used to kill pests.

petals The colored cluster of modified leaves that constitute the next-to-outermost portion of a flower.

petiole (pet'ee-ohl) The part of a leaf that attaches to a stem.

pH A number from 0 to 14 that indicates the degree of acidity or alkalinity of a substance. Mathematically, pH is the negative logarithm of the hydrogen ion concentration.

phagocytosis (fag"oh-sy-toh'sis) Literally, "cell-eating"; a type of endocytosis by which certain cells engulf food particles, microorganisms, foreign matter, or other cells.

pharynx (far'inks) Part of the digestive tract. In higher vertebrates it is bounded anteriorly by the mouth and nasal cavities and posteriorly by the esophagus and larynx; the throat region in humans.

phenotype (fee'noh-type) The physical or chemical expression of an organism's genes (see also genotype).

pheromone (feer'oh-mone) A substance secreted by one organism to the external environment that influences the development or behavior of other members of the same species.

phloem (floh'em) Vascular tissue that conducts food in plants.

phospholipids (fos"foh-lip'idz) Lipids similar to triglycerides in which a phosphorus-containing group occurs in place of one of the fatty acids. Phospholipids compose most of the plasma and internal membranes of cells.

phosphorylation (fos"for-ih-lay'shun) The introduction of a phosphate group into an organic molecule.

photon (foh'ton) A particle of electromagnetic radiation; one quantum of radiant energy.

photoperiodism (foh"toh-peer'ee-od-izm) The physiological response of animals and plants to variations of light and darkness.

photophosphorylation (foh"toh-fos-for-ih-lay'shun) The production of ATP in photosynthesis.

photoreceptor (foh'toh-ree-sep"tor) (1) A sense organ specialized to detect light; (2) a pigment that absorbs light before triggering a physiological response.

photorespiration (foh"toh-res-pur-ay'shun) The production of carbon dioxide and consumption of oxygen during photosynthesis at high light intensities by C-3 plants.

photosynthesis (foh"toh-sin'thuh-sis) The biological process that captures light energy and transforms it into the chemical energy of organic molecules (such as glucose), which are manufactured from carbon dioxide and water. Photosynthesis is practiced by plants, algae, and several kinds of bacteria.

photosystem A group of chlorophyll and other molecules located in the thylakoid membrane (in photoautotrophic eukaryotes) that emits electrons in response to light.

phototropism (foh"toh-troh'pizm) The growth response of an organism to the direction of light.

phylogeny (fy-loj'en-ee) The complete evolutionary history of a group of organisms.

phylum A taxonomic category below that of kingdom; comparable to that of division.

phytochrome (fy'toh-krome) A blue-green, proteinaceous pigment that is the photoreceptor for a wide variety of physiological responses, including initiation of flowering in certain plants.

phytoplankton (fy"toh-plank'tun) Smaller microscopic floating algae that are the base of most aquatic food webs. See plankton. Compare with zooplankton.

pinocytosis (pin"oh-sy-toh'sis) Cell-drinking; the engulfing and absorption of droplets of liquids by cells.

pioneer community The first organisms (such as lichens or mosses) to colonize an area and begin the first stage of ecological succession.

pith Large, thin-walled parenchyma cells found as the innermost tissue in many plants.

pituitary (pit-oo'ih-tehr"ee) Endocrine gland located below the hypothalamus; secretes a variety of hormones influencing a wide range of physiological processes.

placenta (plah-sen'tah) The partly fetal and partly maternal organ whereby materials are exchanged between fetus and mother in the uterus of placental mammals.

plankton Small or microscopic aquatic organisms that are relatively feeble swimmers and thus, for the most part, are carried about by currents and waves. Composed of phytoplankton and zooplankton. Compare with nekton.

plasma cells Cells that secrete antibodies; differentiated B lymphocytes.

plasma membrane A living, functional part of the cell through which all nutrients entering the cell and all waste products or secretions leaving it must pass; the surface membrane of the cell that acts as a selective barrier to passage of molecules and ions into the cell.

plasmids (plaz'midz) Small circular DNA molecules that carry genes separate from the main bacterial chromosome.

plasmodesmata Cytoplasmic channels connecting adjacent plant cells and allowing for the movement of small molecules and ions between cells.

plasmodium (plaz-moh'dee-um) (1) Multinucleate, ameboid mass of living matter that constitutes the vegetative phase of the life cycle of slime molds; (2) a single-celled organism that reproduces by spore formation and causes malaria.

plasmolysis (plaz-mol'ih-sis) The shrinkage of cytoplasm and the pulling away of the plasma membrane from the cell wall when a plant cell (or other walled cell) loses water, usually after being placed in a hypertonic environment.

plastids (plas'tidz) A family of membrane-bounded organelles occurring in photosynthetic eukaryote cells; examples are chloroplasts and leukoplasts.

platelets (playt'lets) Cell fragments in the blood that function in clotting; also called thrombocytes.

pleiotrophy (ply-aht'roh-pee) A gene that affects a number of different characteristics in a given individual.

ploidy (ploy'dee) Relating to the number of sets of chromosomes in a cell.

plumule (ploom'yool) The embryonic shoot of a seed plant.

polar covalent bond A chemical bond established by electron sharing that produces some difference in the charge of the ends of the molecule. Compare with nonpolar covalent bond.

polar transport The unidirectional movement of the plant hormone, auxin, from the stem tip to the roots.

pollen The immature male gametophytes of seed plants that produce sperm.

pollination In seed plants, the transfer of pollen from the male to the female part of the plant.

pollution An unwanted change in the atmosphere, water, or soil that can harm humans or other living organisms.

polygenes (pol'ee-jeens") Two or more pairs of genes that affect the same trait in an additive fashion.

polymer (pol'ih-mer) A molecule composed of repeating units of the same general type, such as a protein, nucleic acid, or polysaccharide.

polymerase chain reaction (PCR) A method to quickly and easily amplify small amounts of DNA into millions of copies. See gene amplification.

polymorphism (pol"ee-mor'fizm) (1) The existence of two or more phenotypically different individuals within the same species; (2) the presence of more than one allele for a given locus.

polypeptide A chain of many amino acids linked by peptide bonds.

polyphyletic origin Describes a group derived from two or more distinct ancestors. Compare with monophyletic origin.

polyploidy (pol'ee-ploy"dee) Possession of more than two sets of chromosomes per nucleus.

polyps (pol'ips) Hydra-like animals; the sessile stage of the life cycle of certain cnidarians. Compare with medusa.

polyribosomes A complex consisting of a number of ribosomes attached to an mRNA molecule during translation; also known as polysomes.

polysaccharide (pol-ee-sak'ah-ride) A carbohydrate consisting of many monosaccharide units; examples are starch, glycogen, and cellulose.

population A group of organisms of the same species that live in the same geographical area at the same time.

portal system A circulatory pathway in which blood flows from a vein draining one region to a second capillary bed in another organ, rather than directly to the heart; an example is the hepatic portal system.

positive feedback system A homeostatic mechanism in which a change in some condition triggers a response that intensifies the changing condition. Compare with negative feedback system.

post-translational processing The modification of newly formed polypeptide chains into functional proteins.

postzygotic isolating mechanism (post'zy-got'ik) A mechanism that restricts gene flow between species and ensures reproductive failure even though fertilization has taken place.

potassium (K^+) ion mechanism Mechanism by which plants open and close their stomata. The influx of potassium ions into the guard cells causes water to move in by osmosis, changing the shape of the guard cells and opening the pore.

potential energy Stored energy that is the result of the relative position of matter instead of its motion. Compare with kinetic energy.

predation Relationship in which a species kills and devours other organisms.

prehensile (pree"hen'sil) Adapted for grasping by wrapping around an object, as in a prehensile tail.

pre-mRNA RNA precursor to mRNA in eukaryotes; contains both introns and exons.

prezygotic isolating mechanism (pree"zy-got'ik) A mechanism that restricts gene flow between species by preventing mating from taking place.

primary consumer A consumer that eats producers; also called herbivore. Compare with secondary consumer.

primary growth An increase in the length of a plant. This growth occurs at the tips of the stems and roots due to the activity of apical meristems. Compare with secondary growth.

primary succession An ecological succession that occurs on land that has not previously been inhabited by plants; no soil is present initially. Compare with secondary succession.

producers Organisms, such as plants, that produce complex organic molecules from simple inorganic substances. In most ecosystems, producers are photosynthetic organisms. Also called autotrophs. Compare with consumers.

profundal zone The deepest zone of a large lake.

prokaryote (pro-kare'ee-ote) Organisms that lack membrane-bounded nuclei and other membrane-bounded organelles; the bacteria and cyanobacteria. Compare with eukaryote.

promoter A recognition signal encoded in DNA that functions to initiate transcription.

prophase The first stage in mitosis, during which the chromatin threads condense so that distinct chromosomes become evident, the nuclear envelope breaks down, and a spindle forms.

prostaglandins (pros"tah-glan'dinz) Derivatives of unsaturated fatty acids that produce a wide variety of hormone-like effects; synthesized by most cells of the body; sometimes called local hormones.

protective coloration The coloring of an organism so that it blends into its surroundings in such a way that it is difficult to see.

protein A large, complex organic compound composed of amino acid subunits; proteins are the principal structural components of cells.

protist (proh'tist) One of a vast kingdom of eukaryotic organisms, primarily single-celled or simple multicellular, mostly aquatic.

proton A basic physical particle present in the nuclei of all atoms that has a positive electrical charge and a mass of 1; a hydrogen ion consists of a single proton.

protostome (proh'toh-stome) Major division of the animal kingdom in which the blastopore develops into the mouth, and the anus forms secondarily; includes the annelids, arthropods, and mollusks. Compare with deuterostome.

protozoa (proh"toh-zoh'a) Single-celled, animal-like protists, including amebas, ciliates, flagellates, and sporozoans.

proximal Relatively near the body center.

pseudocoelom (soo"doh-see'lom) A body cavity between the mesoderm and endoderm; derived from the blastocoele.

pseudoplasmodium A stage in the life cycle of the cellular slime molds in which cells aggregate to form a sluglike structure.

pseudopod A temporary extension of an ameboid cell, which the cell uses for feeding and locomotion.

punctuated equilibrium The concept that evolution proceeds with periods of inactivity (i.e., periods of little or no change within a species) followed by very active phases, so that major adaptations or clusters of adaptations appear suddenly in the fossil record. Compare with gradualism.

pupa (pew'pah) A stage in the development of an insect, between the larva and the adult; a form that neither moves nor feeds and may be in a cocoon.

purines (pure'eenz) Nitrogenous bases with carbon and nitrogen atoms in two interlocking rings; components of nucleic acids, ATP, NAD^+, and other biologically active substances. Examples are adenine and guanine.

pyramid of biomass An ecological pyramid that illustrates the total biomass (for example, the total dry weight of all living organisms in a community) at each successive trophic level. See ecological pyramid. Compare with pyramid of energy and pyramid of numbers.

pyramid of energy An ecological pyramid that shows the energy flow through each trophic level of an ecosystem. See ecological pyramid. Compare with pyramid of biomass and pyramid of numbers.

pyramid of numbers An ecological pyramid that shows the number of organisms at each successive trophic level in a given ecosystem. See ecological pyramid. Compare with pyramid of biomass and pyramid of energy.

pyrimidines (pyr-im'ih-deenz) Nitrogenous bases composed of a single ring of carbon and nitrogen atoms; components of nucleic acids. Examples are thymine, cytosine, and uracil.

radial symmetry A body shape with equal parts radiating out like the spokes of a wheel around a central axis. Compare with bilateral symmetry.

radicle (rad'ih-kl) The embryonic root of a seed plant.

rain shadow An area on the downwind side of a mountain range with very little precipitation. Deserts often occur in rain shadows.

range The area of the Earth in which a particular species occurs.

realized niche The life-style that an organism actually pursues, including the resources that it actually utilizes. An organism's realized niche is narrower than its fundamental niche because of competition from other species. See niche. Compare with fundamental niche.

receptacle In botany, the end of a flower stalk where the floral parts are attached.

receptor (1) A specialized sensory neural structure that is excited by a specific type of stimulus; (2) a site on the cell surface specialized to combine with a specific substance such as a hormone or neurotransmitter.

recessive alleles Alleles not expressed in the heterozygous state.

recombinant DNA Any DNA molecule made by combining genes from different organisms.

recombination, genetic Formation in offspring of allele combinations that are not present in either parent; the result of crossing-over in meiosis, chromosome rearrangements, and mutation.

redox reactions (ree'dox) Chemical reactions in which one substance is oxidized and another reduced; involves the transfer of one or more electrons from one reactant to another.

red tide A population explosion, or bloom, of dinoflagellates.

reduction In chemistry, the gain of electrons by a substance or the chemical addition of hydrogen. Compare with oxidation.

reflex An automatic, involuntary response to a given stimulus that generally functions to restore homeostasis.

regulator genes Special genes that provide codes for the synthesis of repressor or activator proteins.

releaser A stimulus that triggers an unlearned behavior; a communication signal between members of a species.

repressor The protein substance produced by a regulator gene that represses protein synthesis of a specific gene.

reproductive isolation The reproductive barriers that prevent a species from interbreeding with another species. As a result, each species' gene pool is isolated from other species.

respiration (1) Cellular respiration is the process by which cells conserve the energy of food molecules in biologically useful forms, such as ATP; (2) organismic respiration is the act or function of gas exchange.

resting potential The membrane potential (difference in electrical charge) of an inactive neuron (about −70 millivolts).

restriction enzyme One of a class of enzymes that recognizes specific base sequences of DNA and cleaves the DNA molecule at that site. Bacteria produce these enzymes to cleave (and thereby inactivate) foreign DNA when it enters a cell. Used in recombinant DNA technology.

reticulum (reh-tik'yoo-lum) A general term referring to any network; for example, the endoplasmic reticulum is a network of membranes within the cell.

retina The innermost of the three layers of the eyeball, which is continuous with the optic nerve and contains the light-sensitive rod and cone cells.

retrovirus (ret'roh-vy"rus) An RNA virus that produces a DNA intermediate in its host cell.

reverse transcriptase Enzyme produced by retroviruses to enable the transcription of DNA from the viral RNA in the host cell.

reverse transcription The synthesis of DNA using RNA as a template; requires the enzyme reverse transcriptase, which is produced by retroviruses.

rhizoids (ry'zoids) Colorless, hairlike absorptive filaments analogous to roots that extend from the base of the stem of mosses, liverworts, and fern prothallia.

rhizome (ry'zome) A horizontal underground stem that gives rise to above-ground leaves.

ribosomal RNA (rRNA) A type of RNA that joins with various proteins to form a ribosome. Compare with transfer RNA and messenger RNA.

ribosomes (ry'boh-sohms) Organelles that are part of the protein synthesis machinery: consist of a larger and a smaller subunit, each composed of ribosomal RNA (rRNA) and ribosomal proteins.

ribozyme (ry'boh-zime) A molecule of RNA that has catalytic ability.

rickettsia (rih-ket'see-uh) A type of disease organism intermediate in size and complexity between a virus and a 0

RNA Ribonucleic acid. A single-stranded nucleic acid molecule necessary for protein synthesis; there are three forms—messenger RNA, transfer RNA, and ribosomal RNA.

RNA polymerase Family of enzymes that catalyze the synthesis of RNA molecules from DNA templates.

rod One of the rod-shaped, light-sensitive cells of the retina, which are particularly sensitive to dim light and mediate black and white vision. Compare with cone.

root cap A covering of cells over the root tip that protects the delicate meristematic tissue directly behind it.

root hair An extension of an epidermal cell in roots. Root hairs increase the absorptive capacity of roots.

runoff The movement of fresh water from precipitation and snowmelt to rivers, lakes, wetlands, and ultimately, the ocean.

salinity The concentration of dissolved salts (such as sodium chloride) in a body of water.

salt marsh A wetland dominated by grasses in which the salinity fluctuates between that of sea water and fresh water. Salt marshes are usually located in estuaries.

saprophytic (sap-roh-fit'ik) A type of heterotrophic nutrition in which organisms absorb their required nutrients from nonliving organic material.

sarcolemma (sar"koh-lem'mah) The muscle cell plasma membrane.

sarcomere (sar'koh-meer) A segment of a striated muscle cell located between adjacent Z-lines that serves as a unit of contraction.

sarcoplasmic reticulum System of vesicles in a muscle cell that surrounds the myofibrils and releases calcium in muscle contraction; a modified endoplasmic reticulum.

savanna A tropical grassland containing scattered trees; found in areas of low rainfall or seasonal rainfall with prolonged dry periods.

scientific method The steps a scientist uses to approach a problem (making observations, stating the problem, developing a hypothesis, making a prediction to be tested, performing an experiment, and using the results to support or refute the hypothesis, or to generate other hypotheses).

sclerenchyma (skler-en'kim-uh) Cells that provide strength and support in the plant body. Sclerenchyma cells are dead at maturity and have extremely thick walls.

sclerophyllous leaf A hard, small, leathery leaf that resists water loss; characteristic of perennial plants adapted to extremely dry habitats.

scrotum (skroh'tum) The external sac of skin found in most male mammals that contains the testes and their accessory organs.

secondary consumer An organism that consumes primary consumers; also called carnivore. Compare with primary consumer.

secondary growth An increase in the width of a plant due to the activity of lateral meristems (vascular cambium and cork cambium). Compare with primary growth.

secondary succession An ecological succession that takes place after some disturbance destroys the existing vegetation; soil is already present. Compare with primary succession.

second law of thermodynamics When energy is converted from one form to another, some of it is degraded into a lower-quality, less useful form. Thus, with each successive energy transformation, less energy is available to do work. Compare with first law of thermodynamics.

second messenger A substance within a cell that relays a message and (usually) triggers a response to a hormone located at the cell's surface; cyclic AMP and calcium are examples.

seed A plant reproductive body that is composed of a young, multicellular plant and nutritive tissue (food).

segregation Separation of homologous chromosomes during meiosis; results in the placement of a single allele of each pair into different gametes.

selectively permeable membrane A membrane that allows some substances to cross it more easily than others. Biological membranes are generally permeable to water but restrict the passage of many solutes.

semen Fluid composed of sperm suspended in various glandular secretions that is ejaculated from the penis during orgasm.

semi-arid land Land that receives more precipitation than a desert but is subject to frequent and prolonged droughts.

semicircular canals The passages in the vertebrate inner ear that contain structures that control the sense of equilibrium (balance).

semiconservative replication The type of replication characteristic of DNA in which each new double-stranded molecule of DNA consists of one strand from the original DNA molecule and one strand of newly synthesized DNA.

senescence The aging process.

sensitization An increased response by an animal to a stimulus that has been presented before.

sepals (see'puls) The outermost parts of a flower, usually leaflike in appearance, that protect the flower as a bud.

sessile (ses'sile) Permanently attached to one location. Coral animals, for example, are sessile.

sex-linked genes Genes located on a sex chromosome. In mammals, almost all sex-linked genes are located on the X-chromosome. See X-linked gene.

sexual dimorphism (dy-mor'fizm) Difference in body proportions, coloring, or other characteristics in the two sexes of a species.

sexual reproduction Type of reproduction in which two gametes (usually, but not necessarily, contributed by two different parents) fuse to form a zygote. Compare with asexual reproduction.

sieve tube member The cell that conducts food in the phloem of plants.

sign stimulus Any stimulus that elicits an innate response in an animal.

simple fruit A fruit that develops from a single ovary of a single flower.

skeletal muscle Voluntary (striated) muscle of vertebrates, so-called because it usually is directly or indirectly attached to some part of the skeleton.

slash-and-burn agriculture A type of agriculture in which the forest is cut down, allowed to dry, and burned; the crops that are planted immediately afterwards thrive because the ashes provide nutrients. In a few years, however, the soil is depleted and the land must be abandoned.

smooth muscle Vertebrate muscle tissue that lacks transverse striations; found mainly in sheets surrounding hollow organs, such as blood vessels; also called involuntary muscle because it is controlled by the autonomic nervous system.

sodium-potassium pump Cellular active transport mechanism that transports sodium out of, and potassium into, cells.

soil The uppermost layer of the Earth's crust that supports terrestrial plants, animals, and microorganisms. Soil is a complex mixture of inorganic minerals (from the parent rock), organic material, water, air, and living organisms.

soil erosion The wearing away or removal of soil from the land; caused by wind and flowing water. Although soil erosion occurs naturally from precipitation and runoff, human activities (such as clearing land) accelerate it.

solute (sol'yoot) The dissolved substance in a solution.

solvent A liquid substance, such as water, in which other materials may be dissolved.

somatic cell A cell of the body not involved in sexual reproduction.

somatic nervous system That part of the nervous system that keeps the body in adjustment with the external environment; includes the sensory receptors on the body surface and within the muscles, and the nerves that link them with the central nervous system.

speciation Evolution of a new species.

species A group of organisms with similar structural and functional characteristics that in nature breed only with one another and have a close common ancestry; a group of organisms with a common gene pool.

specific heat The amount of heat required to raise 1 gram of a substance 1°C.

sperm The motile, haploid male reproductive cell of animals and some plants and protists; spermatozoa.

spermatogenesis (spur"mah-toh-jen'eh-sis) The production of sperm by meiosis. Compare with oogenesis.

spermatozoa (spur-mah-toh-zoh'uh) Mature sperm cells.

sphincter (sfink'tur) A group of circularly arranged muscle fibers, the contractions of which close an opening, e.g., the pyloric sphincter at the end of the stomach.

spindle The intracellular apparatus, composed of microtubules, that separates chromosomes in cell division of eukaryotes.

spine A leaf that is modified for protection, such as a cactus spine; compare with thorn.

spiracle (speer'ih-kl) An opening for gas exchange, such as the opening on the body surface of a trachea in insects.

sporangium (spor-ran'jee-um) A spore case, found in plants and certain protists and fungi.

spore A reproductive cell that gives rise to individual offspring in plants, algae, fungi, and certain protozoa.

sporophyte generation (spor'oh-fite) The diploid portion of a plant life cycle that produces spores by meiosis. Compare with gametophyte generation.

spring turnover A mixing of the lake waters in temperate lakes that occurs in spring as ice melts and the surface water reaches 4°C, its temperature of greatest density. Compare with fall turnover.

stabilizing selection Natural selection that acts against extreme phenotypes and favors intermediate variants; associated with a population well-adapted to its environment. Compare with directional selection and disruptive selection.

stamen (stay'men) The male part of flowers that produces pollen.

standing crop The current plant biomass.

steroids (steer'oids) Complex molecules containing carbon atoms arranged in four interlocking rings, three of which contain six carbon atoms each and the fourth of which contains five; the male and female sex hormones and the adrenal cortical hormones of vertebrates are examples.

stigma That portion of the carpel where the pollen lands prior to fertilization.

stimulus A physical or chemical change in the internal or external environment of an organism potentially capable of provoking a response.

stolon An above-ground, horizontal stem with long internodes. Stolons often form buds that develop into separate plants.

stoma; pl. **stomata** Small pore flanked by specialized cells (i.e., guard cells) that are located in the epidermis of land plants; stomata allow for gas exchange necessary for photosynthesis.

stop codon See termination codon.

stratosphere The layer of the atmosphere between the troposphere and the mesosphere. It contains a thin ozone layer that protects life by filtering out much of the sun's ultraviolet radiation.

striated muscle See skeletal muscle.

strobilus (stroh'bil-us) A conelike structure that bears sporangia.

stroma The matrix of the chloroplast that surrounds the grana.

stromatolite (stroh-mat'oh-lite) A column-like rock that is composed of many minute layers of prokaryotic cells, usually cyanobacteria. Some stromatolites are over 3 billion years old.

style The neck connecting the stigma to the ovary of a carpel.

suberin (soo'ber-in) A waterproof material found in plants that occurs in the covering of leaf scars, in cork cells, and in the Casparian strip of endodermal cells.

subsistence agriculture The production of enough food to feed oneself and one's family with little left over to sell or reserve for bad times.

substrate A substance on which an enzyme acts; a reactant in an enzymatically catalyzed reaction.

succession The sequence of changes in a plant community over time. See primary succession and secondary succession.

supraorbital ridges (soop"rah-or'bit-ul) Prominent bony ridges above the eye sockets. Ape skulls have prominent supraorbital ridges.

surface area to volume ratio A mathematical relationship that shows that smaller cells have a larger surface area to volume ratio than larger cells. Constrains the size of cells because a large cell has more cytoplasm (i.e., volume) than its plasma membrane (i.e., surface area) can efficiently service.

suspensor In plant embryo development, a multicellular structure that anchors the embryo and aids in nutrient absorption from the endosperm.

symbionts The partners of a symbiotic relationship.

symbiosis (sim-bee-oh'sis) An intimate relationship between two or more organisms of different species. See commensalism, mutualism, and parasitism.

sympathetic nervous system A division of the autonomic nervous system; its general effect is to mobilize energy, especially during stress situations; prepares the body for fight-or-flight response. Compare with parasympathetic nervous system.

sympatric speciation (sim-pat'rik) The evolution of a new species within the same geographical region as the parent species. Compare with allopatric speciation.

synapse (sin'aps) The junction between two neurons or between a neuron and an effector (muscle or gland).

synapsis (sin-ap'sis) The pairing of homologous chromosomes during prophase I of meiosis.

syngamy (sin'gah-mee) Sexual reproduction; the union of the gametes in fertilization.

synthetic theory of evolution The synthesis of previous theories, especially of Mendelian genetics with Darwin's theory of evolution, to formulate a comprehensive explanation of evolution; also called neo-Darwinism.

systematics Study of the classification of organisms in an evolutionary context; sometimes used synonymously with taxonomy, but also interpreted more widely to include taxonomy and phylogenetic reconstructions.

taiga A region of coniferous forests (such as pine, spruce, and fir) in the northern hemisphere. The taiga is located just south of the tundra and stretches across both North America and Eurasia; also called boreal forest.

tap root A root system in plants that has one main root with smaller roots branching off it.

taxon A taxonomic group of any rank.

taxonomy (tax-on'ah-mee) The science of naming, describing, and classifying organisms.

T cell (T lymphocyte) Lymphocyte that is processed in the thymus. T cells have a wide variety of immune functions; responsible for cell-mediated immunity against invading organisms (e.g., viruses) inside host cells. Compare with B cell.

telophase (teel'oh-faze or tel'oh-faze) The last stage of mitosis when, having reached the poles, the chromosomes decondense and a nuclear envelope forms around each group.

temperate deciduous forest A forest biome that occurs in temperate areas where annual precipitation ranges from about 75 cm to 125 cm.

temperate grassland A grassland characterized by hot summers, cold winters, and less rainfall than is found in a temperate deciduous forest biome.

temperate rain forest A coniferous biome characterized by cool weather, dense fog, and high precipitation. Found on the north Pacific coast of North America.

tendon A connective tissue structure that joins a muscle to another muscle, or a muscle to a bone. Tendons transmit the force generated by a muscle.

tendril A leaf or stem that is modified for holding or attaching onto objects.

termination codon Any codon in mRNA that does not code for an amino acid (UAA, UAG, and UGA). This stops the translation of a polypeptide at that point; also called stop codon.

terrestrial Pertaining to the land. Compare with aquatic.

territoriality Behavior pattern in which one organism (usually a male) delineates a territory of its own and defends it against intrusion by other members of the same species and sex.

testis (tes'tis) The male gonad that produces sperm and the male hormone testosterone; in humans and certain other mammals, the testes are situated in the scrotal sac.

tetrad Association of a pair of homologous chromosomes during meiotic prophase I; a tetrad contains four chromatids.

thalamus (thal'uh-mus) The part of the brain that serves as a main relay center transmitting information between the spinal cord and the cerebrum.

theory A widely accepted idea supported by a large body of observations and experiments. Compare with hypothesis.

thermal stratification The marked layering (separation into warm and cold layers) of temperate lakes during the summer. See thermocline.

thermocline A marked and abrupt temperature transition in temperate lakes between warm surface water and cold deeper water. See thermal stratification.

thermodynamics (thurm″oh-dy-nam′iks) The branch of physics that deals with energy and its various forms and transformations.

thigmotropism (thig″moh-troh′pizm) Plant growth in response to contact with a solid object, such as plant tendrils.

thorax (1) The upper body of vertebrates. (2) The second major division of the arthropod body.

thorn A stem that is modified for protection; compare with spine.

threatened species A species in which the population is low enough for it to be at risk of becoming extinct, but not low enough that it is in imminent danger of extinction. Compare with endangered species.

threshold The potential that a neuron or other excitable cell must reach for an action potential to be initiated.

thrombocytes See platelets.

thylakoids (thy′lah-koids) Flat membranous sacs that occur in stacks inside the chloroplast; where light energy is converted into ATP and NADPH used in carbohydrate synthesis.

thymine (thy′meen) See nucleic acid.

thymus gland (thy′mus) An endocrine gland that functions as part of the lymphatic system; important in the development of immune responses.

thyroid gland An endocrine gland that lies anterior to the trachea and releases hormones that regulate the rate of metabolism.

tight junctions Specialized structures that form between some animal cells, producing a tight seal that prevents materials from passing through the spaces between the cells.

tissue A group of closely associated, similar cells that work together to carry out specific functions.

total fertility rate The average number of children born to a woman during her lifetime.

totipotency (toh-ti-poh′tun-cee) Ability of a cell (or nucleus) to provide information for the development of an entire organism.

trachea (tray′kee-uh) (1) Principal thoracic air duct of terrestrial vertebrates; (2) one of the microscopic air ducts branching throughout the body of insects.

tracheids (tray′kee-idz) A type of water-conducting cell in the xylem of plants.

transcription The synthesis of RNA from a DNA template.

transduction The transfer of a genetic fragment from one cell to another, e.g., from one bacterium to another, by a virus.

transfer RNA (tRNA) A form of RNA composed of about 70 nucleotides which serves in the synthesis of proteins. An amino acid is bound to a specific kind of tRNA and then arranged in order by the complementary pairing of the nucleotide triplet (codon) in mRNA and the triplet anticodon of tRNA. Compare with messenger RNA and ribosomal RNA.

transformation (1) The incorporation of genetic material by a cell that causes a change in its phenotype; (2) the conversion of a normal cell to a malignant cell.

translation Conversion of information provided by mRNA into a specific sequence of amino acids in the production of a polypeptide chain; the information in the mRNA is translated into a certain kind of protein.

translocation (1) The movement of materials (water, dissolved minerals, dissolved food) in the vascular tissues of a plant; (2) chromosome abnormality in which part of one chromosome has become attached to another.

transmission, neural Conduction of a neural impulse along a neuron or from one neuron to another.

transpiration Evaporation of water vapor from the leaves of a plant.

transposable element A segment of DNA that is able to move into and out of a chromosome; capable of turning genes on and off. Also called transposon.

transposon See transposable element.

triacylglycerol (try-as″il-glis′er-ol) The most common form of neutral fat consisting of three fatty acid chains chemically linked with a glycerol. Also called triglyceride.

trichome (trik′ome) A hair or other appendage growing out from the epidermis of plants.

trilobite (try′loh-bite) Marine arthropods of the Paleozoic era characterized by two dorsal longitudinal furrows that separated the body into three lobes.

triplet A sequence of three nucleotides in DNA that codes for a codon in messenger RNA.

trophic level Each level in a food web. All producers belong to the first trophic level, all herbivores belong to the second trophic level, and so on.

tropical dry forest A tropical forest where enough precipitation falls to support trees, but not enough to support the lush vegetation of a tropical rain forest. Many tropical dry forests occur in areas with pronounced rainy and dry seasons.

tropical rain forest A lush, species-rich forest biome that occurs in tropical areas where the climate is very moist throughout the year. Tropical rain forests are also characterized by old, infertile soils.

tropism (troh′pizm) A growth response in plants that is elicited by an external stimulus.

troposphere The atmosphere from the Earth's surface to the stratosphere. It is characterized by the presence of clouds, turbulent winds, and decreasing temperature with increasing altitude.

tubers Thickened underground stems that are adapted for food storage; found in plants such as the white potato.

tumor Mass of tissue that is growing in an uncontrolled manner; a neoplasm.

tundra The treeless biome in the far north that consists of boggy plains covered by lichens and small plants such as mosses. The tundra is characterized by harsh, very cold winters and extremely short summers.

turgor pressure (tur′gor) Hydrostatic pressure that develops within a walled cell, such as a plant cell, when the osmotic pressure of the cell's contents is greater than the osmotic pressure of the surrounding fluid.

uracil (yur'ah-sil) See nucleic acid.
urea (yur-ee'ah) The principal nitrogenous excretory product of mammals; one of the water-soluble end products of protein metabolism.
ureter (yoo-ree'tur) One of the paired tubular structures that conducts urine from the kidney to the bladder.
urethra (yoo-ree'thruh) The tube that conducts urine from the bladder to the outside of the body.
uric acid (yoor'ik) The principal nitrogenous excretory product of insects, birds, and reptiles; a relatively insoluble end product of protein metabolism; also occurs in mammals as an end product of purine metabolism.
uterus (yoo'tur-us) The womb; the hollow, muscular organ of the female reproductive tract in which the fetus undergoes development.

vaccine (vak-seen') The commercially produced antigen of a particular pathogen that stimulates the body to make antibodies and is not sufficiently strong to cause the disease's harmful effects.
vacuole (vak'yoo-ole) A membrane-bound sac found within the cytoplasm; may function in storage, digestion, transport, or water elimination.
vagina The elastic, muscular tube extending from the cervix to its orifice that receives the penis during sexual intercourse and serves as part of the birth canal.
valence The number of electrons that an atom can donate, accept, or share in the formation of chemical bonds.
vascular cambium A lateral meristem in plants that produces secondary xylem (wood) and secondary phloem (inner bark).
vector (1) Nucleic acid molecule such as a plasmid that transfers genetic information; (2) agent that transfers a parasite or a pathogen from one organism to another.
vein A blood vessel that carries blood from the tissues toward the heart.
ventral Referring to the belly aspect of an animal's body.
vernalization Promotion of flowering in certain plants by exposing them to a cold period.
vertebrates Chordates that possess a bony vertebral column; fish, amphibians, reptiles, birds, and mammals.
vesicle (ves'ih-kl) Any small sac, especially a small spherical membrane-bounded compartment within the cytoplasm.
vessel element A type of water-conducting cell in the xylem of plants.
vestigial (ves-tij'ee-ul) An evolutionary remnant of a formerly functional structure.
villus; pl. **villi** A minute, elongated projection from the surface of a membrane, e.g., villi of the mucosa of the small intestine.
viroids (vy'roids) Tiny, naked viruses consisting only of nucleic acid.
virus A tiny pathogen composed of a core of nucleic acid usually encased in protein and capable of infecting living cells. A virus is characterized by total dependence upon a living host.
vitamin A complex organic molecule required in very small amounts for the normal metabolic functioning of living cells.
viviparous (vih-vip'er-us) Bearing living young that develop within the body of the mother.

weathering process A chemical or physical process that helps form soil from rock; during weathering, the rock is gradually broken down into smaller and smaller particles.
wetland Land that is transitional between aquatic and terrestrial ecosystems and is covered with water for at least part of the year.
whisk fern Tropical or subtropical vascular plant with a dominant sporophyte generation that reproduces by forming spores and lacks true leaves and roots.
wild-type The phenotypically normal (naturally occurring) form of a gene or an organism. Compare with mutant.
woody A plant with secondary tissues (wood and bark).

X-linked gene Gene carried on an X chromosome. See sex-linked genes.
xylem (zy'lem) Vascular tissue that conducts water and dissolved minerals in certain plants.

yolk sac One of the extraembryonic membranes; a pouch-like outgrowth of the digestive tract of certain vertebrate embryos that grows around the yolk, digests it, and makes it available to the rest of the organism.

zero population growth When the birth rate equals the death rate. A population with zero population growth remains the same size.
zooplankton (zoh"oh-plank'tun) The nonphotosynthetic organisms present in plankton. See plankton. Compare with phytoplankton.
zoospore (zoh'oh-spore) A flagellated motile spore produced asexually.
zooxanthellae Algae that live inside coral animals and have a mutualistic relationship with them.
zygote (zy'gote) The diploid (2n) cell that results from the union of two haploid gametes; a fertilized egg.

Index

Italicized page numbers indicate a figure, *t* indicates a table, and *n* a footnote.

UNITS OF MEASURE

The Metric System

Length

Standard Unit = Meter

1 meter (m) = 39.37 in
1 centimeter (cm) = 0.39 in — 1 inch = 2.54 cm
1 kilometer (km) = 0.62 mi — 1 mile = 1.61 km

Prefixes and units of length

Prefix	Meaning	Unit
kilo	thousand	kilometer (km) = 1000 m
centi	one-hundredth	centimeter (cm) = 0.01 m = 10^{-2} m
milli	one-thousandth	millimeter (mm) = 10^{-3} m
micro	one-millionth	micrometer (μm) = 10^{-6} m
nano	one-billionth	nanometer (nm) = 10^{-9} m

These prefixes are also used in units of volume and mass

Volume

Standard Unit = Liter = 1000 cm^3

1 liter (l) = 1.06 qt — 1 qt = 0.94 l
1 milliliter (ml) = 0.03 fluid oz — 1 fluid oz = 30 ml
1 l = 0.26 gal — 1 gal = 3.79

Mass

Standard Unit = Kilogram = 1000 grams

·1 kilogram (kg) = 2.21 lb — 1 lb = 453.6 grams (g) = 0.45 kg